Electronic Configurations of the Elements

Element	K	L	M	4s	4p	4d	4f	5s	5p	5d	5f	5g	6s	6p	6d	6f	6g	6h	7
55. Cs	2	8	18	2	6	10		2	6				1						
56. Ba	2	8	18	2	6	10		2	6				2						
57. La	2	8	18	2	6	10		2	6	1			2						
58. Ce	2	8	18	2	6	10	2	2	6				2						
59. Pr	2	8	18	2	6	10	3	2	6				2						
60. Nd	2	8	18	2	6	10	4	2	6				2						
61. Pm	2	8	18	2	6	10	5	2	6				2						
62. Sm	2	8	18	2	6	10	6	2	6				2						
63. Eu	2	8	18	2	6	10	7	2	6				2						
64. Gd	2	8	18	2	6	10	7	2	6	1			2						
65. Tb	2	8	18	2	6	10	9	2	6				2						
66. Dy	2	8	18	2	6	10	10	2	6				2						
67. Ho	2	8	18	2	6	10	11	2	6				2						
68. Er	2	8	18	2	6	10	12	2	6				2						
69. Tm	2	8	18	2	6	10	13	2	6				2						
70. Yb	2	8	18	2	6	10	14	2	6				2						
71. Lu	2	8	18	2	6	10	14	2	6	1			2						
72. Hf	2	8	18	2	6	10	14	2	6	2			2						
73. Ta	2	8	18	2	6	10	14	2	6	3			2						
74. W	2	8	18	2	6	10	14	2	6	4			2						
75. Re	2	8	18	2	6	10	14	2	6	5			2						
76. Os	2	8	18	2	6	10	14	2	6	6			2						
77. Ir	2	8	18	2	6	10	14	2	6	7			2						
78. Pt	2	8	18	2	6	10	14	2	6	9			1						
79. Au	2	8	18	2	6	10	14	2	6	10			1						
80. Hg	2	8	18	2	6	10	14	2	6	10			2						
81. Tl	2	8	18	2	6	10	14	2	6	10			2	1					
82. Pb	2	8	18	2	6	10	14	2	6	10			2	2					
83. Bi	2	8	18	2	6	10	14	2	6	10			2	3					
84. Po	2	8	18	2	6	10	14	2	6	10			2	4					
85. At	2	8	18	2	6	10	14	2	6	10			2	5					
86. Rn	2	8	18	2	6	10	14	2	6	10			2	6					
87. Fr	2	8	18	2	6	10	14	2	6	10			2	6					1
88. Ra	2	8	18	2	6	10	14	2	6	10			2	6					2
89. Ac	2	8	18	2	6	10	14	2	6	10			2	6	1				2
90. Th	2	8	18	2	6	10	14	2	6	10			2	6	2				2
91. Pa	2	8	18	2	6	10	14	2	6	10	2		2	6	1				2
92. U	2	8	18	2	6	10	14	2	6	10	3		2	6	1				2
93. Np	2	8	18	2	6	10	14	2	6	10	5		2	6					2
94. Pu	2	8	18	2	6	10	14	2	6	10	6		2	6					2
95. Am	2	8	18	2	6	10	14	2	6	10	7		2	6					2
96. Cm	2	8	18	2	6	10	14	2	6	10	7		2	6	1				2
97. Bk	2	8	18	2	6	10	14	2	6	10	8		2	6	1				2
98. Cf	2	8	18	2	6	10	14	2	6	10	10		2	6					2
99. Es	2	8	18	2	6	10	14	2	6	10	11		2	6					2
100. Fm	2	8	18	2	6	10	14	2	6	10	12		2	6					2
101. Md	2	8	18	2	6	10	14	2	6	10	13		2	6					2
102. No	2	8	18	2	6	10	14	2	6	10	14		2	6					2
103. Lr	2	8	18	2	6	10	14	2	6	10	14		2	6	1				2

ADVANCED INORGANIC CHEMISTRY

A Comprehensive Text

FOURTH EDITION

ADVANCED INORGANIC CHEMISTRY

A Comprehensive Text

F. ALBERT COTTON

ROBERT A. WELCH DISTINGUISHED PROFESSOR OF CHEMISTRY
TEXAS A AND M UNIVERSITY
COLLEGE STATION, TEXAS, USA

and

GEOFFREY WILKINSON

SIR EDWARD FRANKLAND PROFESSOR OF INORGANIC CHEMISTRY
IMPERIAL COLLEGE OF SCIENCE AND TECHNOLOGY
UNIVERSITY OF LONDON, ENGLAND

Fourth Edition, completely revised from the original literature

A WILEY-INTERSCIENCE PUBLICATION
JOHN WILEY & SONS
NEW YORK • CHICHESTER • BRISBANE • TORONTO

Library of Congress Cataloging in Publication Data

Cotton, Frank Albert, 1930-
    Advanced inorganic chemistry.

    "A Wiley-Interscience publication."
    Includes index.
    1.   Chemistry, Inorganic. I. Wilkinson,
Geoffrey, 1921-   joint author.   II. Title.
QD151.2.C68 1980   546   79-22506
ISBN 0-471-02775-8

Printed in the United States of America

10 9 8 7 6 5 4 3 2

Preface to the Fourth Edition

It is remarkable how the subject of inorganic chemistry has not only grown but changed in form and emphasis since the early 1970s. In this fourth edition we have made major alterations in the arrangement of our material in an effort to reflect these changes.

The main purpose of our book has not changed. It is to provide the student, or other reader, with the knowledge necessary to read with comprehension the contemporary research literature in inorganic chemistry and certain areas of organometallic chemistry.

We have, as usual, also updated our coverage of all topics to include developments published as late as the first half of 1979. This, too, has entailed a significant increase in the amount of factual material. To keep the size of the book from getting out of hand we have omitted still more of the relatively elementary material included in earlier editions. There are now a number of more elementary textbooks, including our own *Basic Inorganic Chemistry,* where elementary topics are fully covered. For this reason we believe that the absence of this material here is pedagogically acceptable.

We again wish to thank all of those who have been kind enough to give us constructive suggestions and insight into fields where they are experts and we are not. We continue to be receptive to such contributions.

F. Albert Cotton
Geoffrey Wilkinson

College Station, Texas, USA
London, England
January 1980

Preface to the First Edition

It is now a truism that, in recent years, inorganic chemistry has experienced an impressive renaissance. Academic and industrial research in inorganic chemistry is flourishing, and the output of research papers and reviews is growing exponentially.

In spite of this interest, however, there has been no comprehensive textbook on inorganic chemistry at an advanced level incorporating the many new chemical developments, particularly the more recent theoretical advances in the interpretation of bonding and reactivity in inorganic compounds. It is the aim of this book, which is based on courses given by the authors over the past five to ten years, to fill this need. It is our hope that it will provide a sound basis in contemporary inorganic chemistry for the new generation of students and will stimulate their interest in a field in which trained personnel are still exceedingly scarce in both academic and industrial laboratories.

The content of this book, which encompasses the chemistry of all of the chemical elements and their compounds, including interpretative discussion in the light of the latest advances in structural chemistry, general valence theory, and, particularly, ligand field theory, provides a reasonable achievement for students at the B.Sc. honors level in British universities and at the senior year or first year graduate level in American universities. Our experience is that a course of about eighty lectures is desirable as a guide to the study of this material.

We are indebted to several of our colleagues, who have read sections of the manuscript, for their suggestions and criticism. It is, of course, the authors alone who are responsible for any errors or omissions in the final draft. We also thank the various authors and editors who have so kindly given us permission to reproduce diagrams from their papers: specific acknowledgements are made in the text. We sincerely appreciate the secretarial assistance of Miss C. M. Ross and Mrs. A. B. Blake in the preparation of the manuscript.

F. A. COTTON
Cambridge, Massachusetts

G. WILKINSON
London, England

Contents

PART ONE

Introductory Topics

1.	Nonmolecular Solids	3
2.	Symmetry and Structure	28
3.	Introduction to Ligands and Complexes	61
4.	Classification of Ligands by Donor Atoms	107
5.	Stereochemistry and Bonding in Main Group Compounds	195

PART TWO

Chemistry of the Main Group Elements

6.	Hydrogen	215
7.	The Group I Elements: Li, Na, K, Rb, Cs	253
8.	Beryllium and the Group II Elements: Mg, Ca, Sr, Ba, Ra	271
9.	Boron	289
10.	The Group III Elements: Al, Ga, In, Tl	326
11.	Carbon	352
12.	The Group IV Elements: Si, Ge, Sn, Pb	374
13.	Nitrogen	407
14.	The Group V Elements: P, As, Sb, Bi	438
15.	Oxygen	483
16.	The Group VI Elements: S, Se, Te, Po	502
17.	The Halogens: F, Cl, Br, I, At	542
18.	The Noble Gases	577
19.	Zinc, Cadmium, and Mercury	589

PART THREE

Chemistry of the Transition Elements

20. The Transition Elements and the Electronic Structures of Their Compounds   619
21. The Elements of the First Transition Series   689
    A. Titanium, 692
    B. Vanadium, 708
    C. Chromium, 719
    D. Manganese, 736
    E. Iron, 749
    F. Cobalt, 766
    G. Nickel, 783
    H. Copper, 798
22. The Elements of the Second and Third Transition Series   822
    A. Zirconium and Hafnium, 824
    B. Niobium and Tantalum, 831
    C. Molybdenum and Tungsten, 844
    D. Technetium and Rhenium, 883
    E. The Platinum Metals, 901
    F. Ruthenium and Osmium, 912
    G. Rhodium and Iridium, 934
    H. Palladium and Platinum, 950
    I. Silver and Gold, 966
23. The Lanthanides; also Scandium and Yttrium   981
24. The Actinide Elements   1005

PART FOUR

Special Topics

25. Metal Carbonyls and Other Complexes with π-Acceptor Ligands   1049
26. Metal-to-Metal Bonds and Metal Atom Clusters   1080
27. Transition Metal Compounds with Bonds to Hydrogen and Carbon   1113
28. Reaction Mechanisms and Molecular Rearrangements in Complexes   1183
29. Transition Metal to Carbon Bonds in Synthesis   1234
30. Transition Metal to Carbon Bonds in Catalysis   1265
31. Bioinorganic Chemistry   1310

Appendices

1. Units, Fundamental Constants, and Conversion Factors   1347

2. Ionization Enthalpies of the Atoms   1349
3. Enthalpies of Electron Attachment for Atoms   1351
4. Atomic Orbitals   1351
5. The Quantum States Derived from Electronic Configurations   1354
6. Magnetic Properties of Chemical Compounds   1359

Index   1367

Abbreviations in Common Use

1. Chemicals, Ligands, Radicals, etc.

Ac	acetyl, CH_3CO
acac	acetylacetonate anion
acacH	acetylacetone
AIBN	azoisobutyronitrile
am	ammonia (or occasionally an amine)
Ar	aryl or arene (ArH)
aq	aquated, H_2O
ATP	adenosine triphosphate
9-BBN	9-borabicyclo[3,3,1]nonane
bipy	2,2'-dipyridine, or bipyridine
Bu	butyl (superscript n, i, s or t, normal, iso, secondary or tertiary butyl)
Bz	benzyl
COD or cod	cycloocta-1,5-diene
COT or cot	cyclooctatetraene
Cp	cyclopentadienyl, C_5H_5
cy	cyclohexyl
depe	1,2-bis(diethylphosphino)ethane
depm	1,2-bis(diethylphosphino)methane
diars	o-phenylenebisdimethylarsine, o-$C_6H_4(AsMe_2)_2$
dien	diethylenetriamine, $H_2N(CH_2CH_2NH)_2H$
diglyme	diethyleneglycoldimethylether, $CH_3O(CH_2CH_2O)_2CH_3$
diop	{[2,2-dimethyl-1,3-dioxolan-4,5-diyl)bis(methylene)]bis(diphenylphosphine)}
diphos	any chelating diphosphine, but usually 1,2-bis(diphenylphosphino)ethane, dppe
DME	dimethoxyethane
DMF or dmf	N,N'-dimethylformamide, $HCONMe_2$
dmg	dimethylglyoximate anion
dmgH₂	dimethylglyoxime
dmpe	1,2-bis(dimethylphosphino)ethane
DMSO or dmso	dimethylsulfoxide, Me_2SO

dppe	1,2-bis(diphenylphosphino)ethane
DPPH	diphenylpicrylhydrazyl
dppm	bis(diphenylphosphino)methane
E	electrophile or element
$EDTAH_4$	ethylenediaminetetraacetic acid
$EDTAH_{4-n}^{n-}$	anions of $EDTAH_4$
en	ethylenediamine, $H_2NCH_2CH_2NH_2$
Et	ethyl
Fc	ferrocenyl
Fp	$Fe(CO)_2Cp$
glyme	ethyleneglycoldimethylether, $CH_3OCH_2CH_2OCH_3$
hfa	hexafluoroacetylacetonate anion
HMPA	hexamethylphosphoric triamide, $OP(NMe_2)_3$
L	ligand
M	central (usually metal) atom in compound
Me	methyl
Mes	mesityl
Me_6tren	tris-(2-dimethylaminoethyl)amine, $N(CH_2CH_2NMe_2)_3$
NBD or nbd	norbornadiene
NBS	N-bromosuccinimide
np^2	Bis-(2-diphenylphosphinoethyl)amine, $HN(CH_2CH_2PPh_2)_2$
np^3	Tris-(2-diphenylphosphinoethyl)amine, $N(CH_2CH_2PPh_2)_3$
$NTAH_3$	nitrilotriacetic acid, $N(CH_2COOH)_3$
OAc	acetate anion
ox	oxalate ion, $C_2O_4^{2-}$
Pc	phthalocyanine
Ph	phenyl, C_6H_5
phen	1,10-phenanthroline
pn	propylenediamine (1,2-diaminopropane)
PNP ($= np^2$)	Bis-(2-diphenylphosphinoethyl)amine, $HN(CH_2CH_2PPh_2)_2$
pp^3	Tris-(2-diphenylphosphinoethyl)phosphine, $P(CH_2CH_2PPh_2)_3$
PPN^+	$[(Ph_3P)_2N]^+$
Pr	propyl (superscripts, n or i)
py	pyridine
pz	pyrazolyl
QAS	Tris-(2-diphenylarsinophrnyl)arsine, $As(o\text{-}C_6H_4AsPh_2)_3$
QP	Tris-(2-diphenylphosphinophenyl)phosphine, $P(o\text{-}C_6H_4PPh_2)_3$
R	alkyl (preferably) or aryl group
R_F	Perfluoro alkyl group
S	solvent
sal	salicylaldehyde
sal_2en or salen	bis-salicylaldehydeethylenediimine

TAN	Tris-(2-diphenylarsinoethyl)amine, $N(CH_2CH_2AsPh_2)_3$
TAP	Tris-(3)dimethylarsinopropyl)phosphine, $P(CH_2CH_2CH_2AsMe_2)_3$
TAS	Bis-(3-dimethylarsinopropyl)methylarsine, $MeAs(CH_2CH_2CH_2AsMe_2)_2$
TCNE	tetracyanoethylene
TCNQ	7,7,8,8-tetracyanoquinodimethane
terpy	terpyridine
TFA	trifluoroacetic acid
THF or thf	tetrahydrofuran
TMED	N,N,N',N'-tetramethylethylenediamine
tn	1,3-diaminopropane(trimethylenediamine)
TPN (= np^3)	Tris-(2-diphenylphosphinoethyl)amine, $N(CH_2CH_2PPh_2)_3$
TPP	$meso$-tetraphenylporphyrin
tren	Tris-(2-aminoethyl)amine, $N(CH_2CH_2NH_2)_3$
trien	Triethylenetetraamine, $(CH_2NHCH_2CH_2NH_2)_2$
triphos	1,1,1-tris(diphenylphosphinomethyl)ethane
TSN	Tris-(2-methylthiomethyl)amine, $N(CH_2CH_2SMe)_3$
TSP	Tris-(2-methylthiophenyl)phosphine, $P(o\text{-}C_6H_4SMe)_3$
TSeP	Tris-(2-methylselenophenyl)phosphine, $P(o\text{-}C_6H_4SeMe)_3$
TTA	thenoyltrifluoroacetone, $C_4H_3SCPCH_2CPCF_3$
X	halogen or pseudohalogen

2. Miscellaneous

Å	Angstrom unit, 10^{-10} m
AOM	angular overlap model
asym	asymmetric or antisymmetric
bcc	body centered cubic
BM	Bohr magneton
b.p.	boiling point
ccp	cubic close packed
CFSE	crystal field stabilization energy
CFT	crystal field theory
CIDNP	chemically induced dynamic nuclear polarization
cm^{-1}	wave number
CT	charge transfer
d	decomposes
d-	dextorotatory
ESCA	electron spectroscopy for chemical analysis
esr or epr	electron spin (or paramagnetic) resonance
eV	electron volt
Ft	Fourier transform (for nmr)
g	g-values

(g)	gaseous state
h	Planck's constant
hcp	hexagonal close packed
HOMO	highest occupied molecular orbital
Hz	hertz, sec^{-1}
ICCC	International Coordination Chemistry Conference
ir	infrared
IUPAC	International Union of Pure and Applied Chemistry
l-	levorotatory
(l)	liquid state
LCAO	linear combination of atomic orbitals
LFSE	ligand field stabilization energy
LFT	ligand field theory
LUMO	lowest unoccupied molecular orbital
MO	molecular orbital
MOSE	molecular orbital stabilization energy
m.p.	melting point
nmr	nuclear magnetic resonance
P.E.	photoelectron (spectroscopy)
R	gas constant
(s)	solid state
SCE	saturated calomel electrode
SCF	self consistent field
SCF-Xα-SW	self-consistent field, Xα, scattered wave (form of MO theory)
sp or *spy*	square pyramid(al)
str	vibrational stretching mode
sub	sublimes
sym	symmetrical
tbp	trigonal bipyramid(al)
U	lattice energy
uv	ultraviolet
VB	valence bond
Z	atomic number
ϵ	molar extinction coefficient
ν	frequency (cm^{-1} or Hz)
μ	magnetic moment in Bohr magnetons
χ	magnetic susceptibility
θ	Weiss constant

ADVANCED INORGANIC CHEMISTRY

A Comprehensive Text

FOURTH EDITION

1

INTRODUCTORY TOPICS

CHAPTER ONE

Nonmolecular Solids

1-1. Introductory Remarks

Inorganic chemistry deals with substances having virtually every known type of physical and structural characteristic. It is not, therefore, easy to decide where to begin a book that will, in the end, have had something to say about many of them. There is no uniquely logical or convenient approach. No matter what order of topics is chosen, there are bound to be some carts before horses, since many facets of the subject interrelate in such a way that neither can be considered as simply prerequisite to the other.

We begin this edition with a discussion of substances that exist in the solid state as extended arrays rather than molecular units. Although there are doubtless some arguments for *not* starting a book this way, we have several reasons for doing so.

There are a great many important substances that are nonmolecular. Most of the elements themselves are nonmolecular. Thus more than half the elements are metals in which close-packed arrays of atoms are held together by delocalized electrons, while others, such as carbon, silicon, germanium, red and black phosphorus, and boron involve infinite networks of more localized bonds. There are also many compounds, such as SiO_2 and SiC, in which the array is held together by localized heteropolar bonds. The degree of polarity varies, of course, and this class of substances grades off toward the limiting case of the ionic arrays in which there are well-defined ions held together principally by the Coulombic forces between those of opposite charge.

There are also solids that consist neither of small, well-defined molecules nor of well-ordered infinite arrays of atoms; examples are the glasses and polymers, which, for reasons of space, are not explicitly discussed here. It is, of course, true that most molecular substances form a crystalline solid phase; but because of the relatively weak intermolecular interactions, crystallinity is usually of little chemical importance, though, of course, of enormous practical significance in that it facili-

tates the investigation of molecular structures, namely, by X-ray crystallography.

Finally, we note that the area of materials science, which is a marriage of applied
chemistry and applied physics, deals to a considerable extent with inorganic materials: semiconductors, superconductors, ceramics, refractory compounds, alloys,
and so on. Virtually all these are substances with nonmolecular structures.

1-2. Close Packing of Spheres

The packing of spherical atoms or ions in such a way that the greatest number occupy each unit of volume is one of the most fundamental structural patterns of
Nature. It is seen in its simplest form in the solid noble gases, where spherical atoms
are concerned, in a variety of ionic oxides and halides, where small cations can be
considered to occupy interstices in a close-packed array of the larger spherical
anions, and in metals where close-packed arrays of metal ions are permeated by
a cloud of delocalized electrons binding them together.

All close-packed arrangements are built by stacking of close-packed layers of
the type shown in Fig. 1-1a; it should be evident that this is the densest packing
arrangement in two dimensions. Two such layers may be brought together as shown

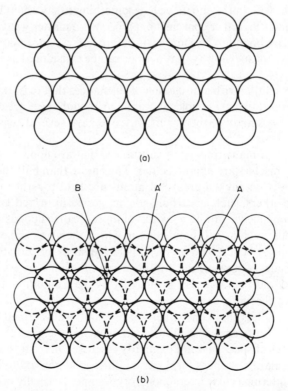

Fig. 1-1. (a) One close-packed layer of spheres. (b) Two close-packed layers showing how tetrahedral
(A, A') and octahedral (B) interstices are formed.

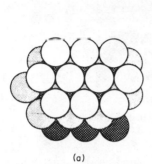 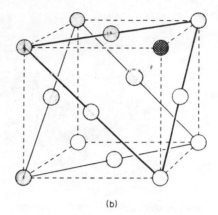

(a) (b)

Fig. 1-2. (a) The stacking of close-packed layers in the ABC pattern that on repetition, gives cubic closest packing *ccp*. (b) Another view of the *ccp* pattern, emphasizing its cubic symmetry.

in Fig. 1-1b, spheres of one layer resting in the declivities of the other; this is the densest packing arrangement of the two layers. It will be noted that between the two layers there exist interstices of two types: tetrahedral and octahedral.

When we come to add a third layer to the two already stacked, two possibilities arise. The third layer can be placed so that its atoms lie directly over those of the first layer, or with a displacement relative to the first layer, as in Fig. 1-2a. These two stacking arrangements may be denoted ABA and ABC, respectively. Each may be continued in an ordered fashion so as to obtain

> Hexagonal close packing (*hcp*): ABABAB ...
> Cubic close packing (*ccp*): ABCABCABC ...

It is immediately obvious that the *hcp* arrangement does indeed have hexagonal symmetry, but the cubic symmetry of what has been designated *ccp* may be less evident. Figure 1-2b provides another perspective, which emphasizes the cubic symmetry; it shows that the close-packed layers lie perpendicular to body diagonals of the cube. Moreover, the cubic unit cell is not primitive but face centered.

There are, of course, an infinite number of stacking sequences possible within the definition of close packing, all of them, naturally, having the same packing density. The *hcp* and *ccp* sequences are those of maximum simplicity and symmetry. Some of the more complex sequences are actually encountered in Nature, though far less often than the two just described.

In *any* close-packed arrangement, each atom has twelve nearest neighbors, six surrounding it in its own close-packed layer, three above and three below this layer. In the *hcp* structure each layer is a plane of symmetry and the set of nearest neighbors of each atom has D_{3h} symmetry. In *ccp* the set of nearest neighbors has D_{3d} symmetry. (See Section 2-4 for definition of these symbols.)

1-3. Metals

It is evident, even on the most casual inspection, that metals have many physical properties quite different from those of other solid substances. Although there are

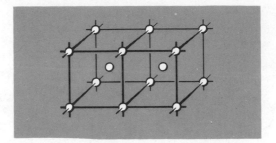

Fig. 1-3. A body-centered cubic (*bcc*) structure.

individual exceptions to each, the following may be cited as the characteristic properties of metals as a class: (1) high reflectivity, (2) high electrical conductance, decreasing with increasing temperature, (3) high thermal conductance, and (4) mechanical properties such as strength and ductility. An explanation for these properties, and for their variations from one metal to another, must be derived from the structural and electronic nature of the metal.

Metal Structures. Almost all metal phases have one of three basic structures, or some slight variation thereof, although there are a few exceptional structures that need not concern us here. The three basic structures are cubic and hexagonal close-packed, which have already been presented in Section 1-2, and body-centered cubic (*bcc*), illustrated in Fig. 1-3. In the *bcc* type of packing each atom has only eight instead of twelve nearest neighbors, although there are six next nearest neighbors that are only about 15% farther away. It is only 92% as dense an arrangement as the *hcp* and *ccp* structures. The distribution of these three structure types, *hcp, ccp,* and *bcc,* in the Periodic Table, is shown in Fig. 1-4. The majority of the metals deviate slightly from the ideal structures, especially those with *hcp* structures. For the *hcp* structure the ideal value of *c/a*, where *c* and *a* are the hexagonal unit-cell edges, is 1.633, whereas all metals having this structure have a smaller *c/a* ratio (usually 1.57–1.62); zinc and cadmium, for which *c/a* values are 1.86 and 1.89, respectively, are exceptions. Although such deviations cannot in general be predicted, their occurrence is not particularly surprising, since for a given atom its six in-plane neighbors are not symmetrically equivalent to the set of six lying above and below it, and there is consequently no reason for its bonding to those in the two nonequivalent sets to be precisely the same.

Metallic Bonding. The characteristic physical properties of metals as well as the high coordination numbers (either twelve or eight nearest neighbors plus six more that are not too remote) suggest that the bonding in metals is different from that in other substances. Clearly there is no ionic contribution, and it is also obviously impossible to have a fixed set of ordinary covalent bonds between all adjacent pairs of atoms, since there are neither sufficient electrons nor sufficient orbitals. Attempts have been made to treat the problem by invoking an elaborate resonance of electron-pair bonds among all the pairs of nearest neighbor atoms, and this approach has had a certain degree of success. However the main thrust of theoretical

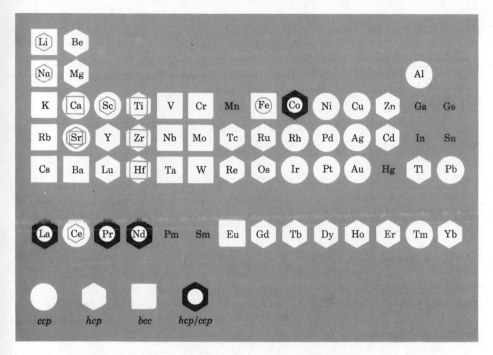

Fig. 1-4. The occurrence of hexagonal close-packed (*hcp*), cubic close-packed (*ccp*), and body-centered (*bcc*) structures among the elements. Where two or more symbols are used, the largest represents the stable form at 25°C. The symbol labeled *hcp/ccp* signifies a mixed ... ABCABABCAB ... type of close packing, with overall hexagonal symmetry. [Adapted, with permission, from H. Krebs, *Fundamentals of Inorganic Crystal Chemistry*, McGraw-Hill Book Co., 1968.]

work on metals is in terms of the *band theory,* which gives in a very natural way an explanation for the electrical conductance, luster, and other characteristically metallic properties. A detailed explanation of band theory would necessitate a level of mathematical sophistication beyond that appropriate for this book; however the qualitative features are sufficient.

Let us imagine a block of metal expanded, without change in the geometric relationships between the atoms, by a factor of, say, 10^6. The interatomic distances would then all be 10^2 greater, that is, about 300 to 500 Å. Each atom could then be described as a discrete atom with its own set of well-defined atomic orbitals. Now let us suppose the array contracts, so that the orbitals of neighboring atoms begin to overlap, hence to interact with each other. Since so many atoms are involved, this gives rise, at the actual internuclear distances in metals, to sets of states so close together as to form essentially continuous energy bands, as illustrated in Fig. 1-5. Spatially these bands are spread through the metal, and the electrons that occupy them are completely delocalized. In the case of sodium, shown in Fig. 1-5, the $3s$ and $3p$ bands overlap.

Another way to depict energy bands is that shown in Fig. 1-6. Here energy is plotted horizontally and the envelope indicates on the vertical the number of elec-

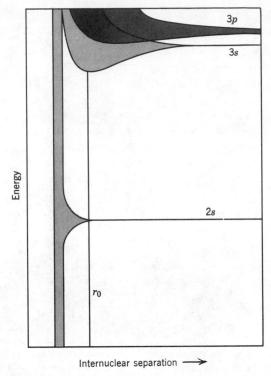

Fig. 1-5. Energy bands of sodium as a function of internuclear distance; r_0 represents the actual equilibrium distance. [Reproduced by permission from J. C. Slater, *Introduction to Chemical Physics*, McGraw-Hill Book Co., 1939.]

trons that can be accommodated at each value of the energy. Shading is used to indicate filling of the bands.

Completely filled or completely empty bands (Fig. 1-6a) do not permit net electron flow, and the substance is an insulator. Covalent solids can be discussed from this point of view (though it is unnecessary to do so) by saying that all electrons occupy low-lying bands (equivalent to the bonding orbitals), while the high-lying bands (equivalent to antibonding orbitals) are entirely empty. Metallic conductance occurs when there is a partially filled band, as in Fig. 1-6b; the transition metals, with their incomplete sets of *d* electrons, have partially filled *d* bands; and this accounts for their high conductances. The alkali metals have half-filled *s* bands formed from their *s* orbitals, as shown in Fig. 1-5; actually these *s* bands overlap the *p* bands, and it is because this overlap occurs also for the Ca group metals, where the atoms have filled valence shell *s* orbitals, that they nevertheless form metallic solids, as indicated in Fig. 1-6c.

Cohesive Energies of Metals. The strength of binding among the atoms in metals can conveniently be measured by the enthalpies of atomization. Figure 1-7 plots the energies of atomization of the metallic elements, lithium to bismuth, from their standard states. It is first notable that cohesive energy tends to maximize with el-

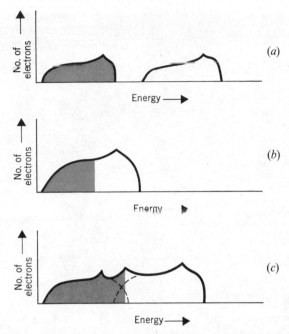

Fig. 1-6. Envelopes of energy bands; shading indicates filling.

ements having partially filled d shells (i.e., with the transition metals). However it is particularly with the elements near the middle of the second and third transition series, especially Nb–Ru and Hf–Ir, that the cohesive energies are largest, reaching 837 kJ mol^{-1} for tungsten. It is noteworthy that these large cohesive energies are principally due to the structural nature of the metals whereby high coordination numbers are achieved. For a *hcp* or *ccp* structure, there are six bonds per metal atom (since each of the twelve nearest neighbors has a half-share in each of the twelve bonds). Therefore each bond, even when cohesive energy is 800 kJ mol^{-1}, has an energy of only 133 kJ mol^{-1}, roughly half the C—C bond energy in diamond where each carbon atom has only four near neighbors.

1-4. Interstitial Compounds[1]

The term "interstitial compounds" refers primarily (though usage is sometimes more flexible) to combinations of the relatively large transition metal atoms with the small metalloid or nonmetal atoms such as hydrogen, boron, carbon, and nitrogen. The substances may be thought of as consisting of a metal host lattice with

[1]   L. E. Toth, *Transition Metal Carbides and Nitrides,* Academic Press, 1971; *Refractory Carbides,* G. V. Samsonov, Ed., Consultants Bureau, 1974; H. J. Goldschmidt, *Interstitial Alloys,* Plenum Press, 1967; E. K. Storms, *The Refractory Carbides,* Academic Press, 1967; B. Aronsson, T. Lundstrom, and S. Rundqvist, *Borides, Silicides, and Phosphides,* Wiley, 1965; *Transition Metal Hydrides,* E. L. Muetterties, Ed., Dekker, 1971; H. A. Johansen, *Surv. Progr. Chem.,* 1977, **8,** 57.

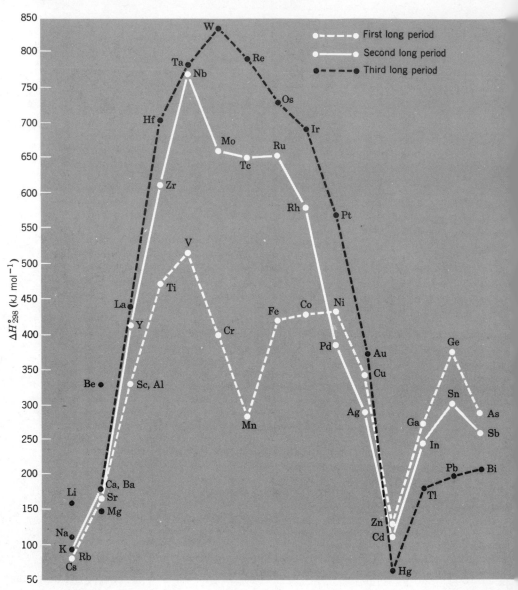

Fig. 1-7. Heats of atomization of metals ΔH°_{298} for M(s) → M(g). [Reproduced by permission from W. E. Dasent, *Inorganic Energetics,* Penguin Books, Ltd., 1970.]

the small nonmetal atoms occupying the interstices (octahedral or tetrahedral in, e.g., the *hcp* and *ccp* structures). However the array of metal atoms need not (and usually does not) correspond to any known packing arrangement of the pure metal. One common type of interstitial structure is that obtained by filling every octahedral interstice in a *ccp* structure with a small atom. This arrangement (Fig. 1-8a) is identical in geometric form to the well-known sodium chloride structure for ionic

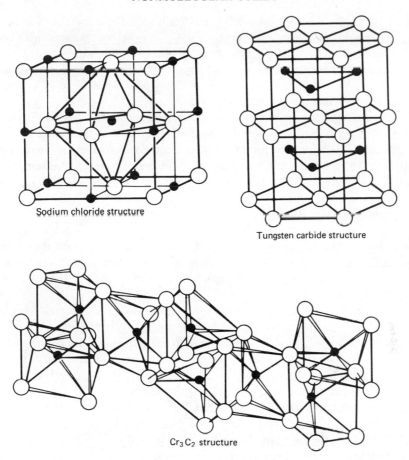

Sodium chloride structure

Tungsten carbide structure

Cr_3C_2 structure

Fig. 1-8. Three structures of interstital compounds; open circles are metal atoms; small filled circles are nonmetal atoms. (a) The NaCl-like structure obtained by filling all octahedral holes in a *ccp* metal array. (b) The WC structure. WN has the same structure, thus showing the unimportance of conventional valences. (c) The Cr_3C_2 structure.

materials that we shall discuss later. Figure 1-8b shows a structure in which the metal atoms themselves do not define a close-packed structure.

In general the complexity of the structures of interstitial compounds depends on the ratio, $R = r_x/r_M$, of the guest atom atomic radius r_x to that of the transition metal atom r_M. The atomic radii themselves are in most cases equal to one-half the internuclear distances for crystals of the pure elements, though there are exceptions, and small corrections are often made. It has been observed that when R is less than 0.59 the metal atoms form simple structures such as *hcp, ccp, bcc,* or the arrangement shown in Fig. 1-8b. Interstitial compounds with such structures are called Hägg compounds after the Swedish chemist who first noted the radius-ratio rule. When R exceeds 0.59, much more complicated structures, such as that of Cr_3C_2 (Fig. 1-8c), are adopted.

Not only are these substances rarely interstitial in the sense that the native metal structure is usually not retained, but they are generally not "compounds" either if that term is considered to imply exact stoichiometry. They are usually *phases* whose structure remains intact over a range of composition. For example, the "compound," VC actually varies in composition from $C/V = 37$ to 47%, and VC of exact 1:1 composition does not exist at all.

The interstitial compounds are characterized, typically, by great hardness (though also great brittleness) and high melting points, as well as retention of typically metallic characteristics such as luster and good conductivity for heat and electricity. At high temperatures many of them even acquire metallic-type mechanical properties such as malleability.

1-5. Ionic Crystal Structures

Ionic Radii. One of the major factors in determining the structures of the substances that can be thought of, at least approximately, as made up of cations and anions packed together, is ionic size, especially the ratio of radii of the two or more ions present.

It is obvious from the nature of wave functions that no ion or atom has a precisely defined radius. The only way radii can be assigned is to determine how closely the centers of two atoms or ions actually approach each other in solid substances and then to assume that such a distance is equal or closely related to the sum of the radii of the two atoms or ions. Even this procedure is potentially full of ambiguities, and further provisions and assumptions are required to get an empirically useful set of radii. The most ambitious attempt to handle this problem is that of Pauling, and some of his arguments and results are briefly summarized here.

We begin with the four salts NaF, KCl, RbBr, and CsI, in each of which the cation and anion are isoelectronic and the radius ratios ($r_{cation}/r_{anion} = r_+/r_-$) should be similar in all four cases. Two assumptions are then made:

1. The cation and anion are assumed to be in contact, so that the internuclear distance can be set equal to the sum of the radii.

2. For a given noble gas electron configuration, the radius is assumed to be inversely proportional to the effective nuclear charge felt by the outer electrons.

The implementation of these rules may be illustrated by using NaF, in which the internuclear distance is 2.31 Å. Hence

$$r_{Na^+} + r_{F^-} = 2.31 \text{ Å}$$

Next, using rules developed by Slater to estimate how much the various electrons in the $1s^2 2s^2 2p^6$ configuration shield the outer electrons from the nuclear charge, we obtain 4.15 for the shielding parameter. The effective nuclear charges, Z, felt by the outer electrons are then, for Na^+ with $Z = 11$,

$$11 - 4.15 = 6.85$$

and for F^-, with $Z = 9$:

$$9 - 4.15 = 4.85$$

According to rule 2, the radius ratio r_{Na^+}/r_{F^-} must be inversely proportional to these numbers; hence

$$r_{Na^+}/r_{F^-} = \frac{1}{6.85/4.85} = 0.71$$

Solving this and the previous equation for the sum of the radii simultaneously we obtain

$$r_{Na^+} = 0.95 \text{ Å}$$
$$r_{F^-} = 1.36 \text{ Å}$$

This method, with certain refinements, was used by Pauling to estimate individual ionic radii. Earlier, V. M. Goldschmidt, using a somewhat more empirical method, also estimated ionic radii. The radii for a number of important ions, obtained by the two procedures, are given in Table 1-1. A more recent set of most probable ionic radii is also given in Table 1-1. These are based on a combination of shortest interatomic distances and experimental electron density maps.

Important Ionic Crystal Structures. Figure 1-9 shows six of the most important structures found among essentially ionic substances. In an ionic structure each ion is surrounded by a certain number of ions of the opposite sign; this number is called the *coordination number* of the ion. In the first three structures shown, namely, the NaCl, CsCl, and CaF_2 types, the cations have the coordination numbers 6, 8, and 8, respectively.

We now ask why a particular compound crystallizes with one or another of these structures. To answer this, we first recognize that ignoring the possibility of metastability, which seldom arises, the compound will adopt the arrangement providing the greatest stability, that is, the lowest energy. The factors that contribute to the energy are the attractive force between oppositely charged ions, which will increase with increasing coordination number, and the forces of repulsion, which will increase very rapidly if ions of the same charge are "squeezed" together. Thus the optimum arrangement in any crystal should be the one allowing the greatest number of oppositely charged ions to "touch" without requiring any squeezing together of ions with the same charge. The ability of a given structure to meet these requirements will depend on the relative sizes of the ions.

Let us analyze the situation for the CsCl structure. We place eight negative ions of radius r^- around a positive ion with radius r^+ so that the M^+ to X^- distance is $r^+ + r^-$ and the adjacent X^- ions are just touching. Then the X^- to X^- distance, a, is given by

$$a = \frac{2}{\sqrt{3}}(r^+ + r^-) = 2r^-$$

or

$$\frac{r^-}{r^+} = 1.37$$

Now, if the ratio r^-/r^+ is greater than 1.37, the only way we can have all eight X^- ions touching the M^+ ion is to squeeze the X^- ions together. Alternatively, if r^-/r^+ is greater than 1.37, and we do not squeeze the X^- ions, they cannot touch the M^+ ion and a certain amount of electrostatic stabilization energy will be un-

TABLE 1-1
Goldschmidt (G),[a] Pauling (P)[a], and Ladd (L)[a,b] Ionic Radii (Å)

Ion	G	P	L	Ion	G	P	L
H^-	1.54	2.08	1.39	Pb^{2+}	1.17	1.21	—
F^-	1.33	1.36	1.19				
Cl^-	1.81	1.81	1.70	Mn^{2+}	0.91	0.80	0.93
Br^-	1.96	1.95	1.87	Fe^{2+}	0.83	0.76	0.90
I^-	2.20	2.16	2.12	Co^{2+}	0.82	0.74	0.88
				Ni^{2+}	0.68	0.69	—
O^{2-}	1.32	1.40	1.25	Cu^{2+}	0.72	—	—
S^{2-}	1.74	1.84	1.70				
Se^{2-}	1.91	1.98	1.81	Bi^{3+}	0.2	0.20	—
Te^{2-}	2.11	2.21	1.97	Al^{3+}	0.45	0.50	—
				Sc^{3+}	0.68	0.81	—
Li^+	0.78	0.60	0.86	Y^{3+}	0.90	0.93	—
Na^+	0.98	0.95	1.12	La^{3+}	1.04	1.15	—
K^+	1.33	1.33	1.44	Ga^{3+}	0.60	0.62	—
Rb^+	1.49	1.48	1.58	In^{3+}	0.81	0.81	—
Cs^+	1.65	1.69	1.84	Tl^{3+}	0.91	0.95	—
Cu^+	0.95	0.96	—				
Ag^+	1.13	1.26	1.27	Fe^{3+}	0.53	—	—
Au^+	—	1.37	—	Cr^{3+}	0.53	—	—
Tl^+	1.49	1.40	1.54				
NH_4^+	—	1.48	1.66	C^{4+}	0.15	0.15	—
				Si^{4+}	0.38	0.41	—
Be^{2+}	0.34	0.31	—	Ti^{4+}	0.60	0.68	—
Mg^{2+}	0.78	0.65	0.87	Zr^{4+}	0.77	0.80	—
Ca^{2+}	1.06	0.99	1.18	Ce^{4+}	0.87	1.01	—
Sr^{2+}	1.27	1.13	1.32	Ge^{4+}	0.54	0.53	—
Ba^{2+}	1.43	1.35	1.49	Sn^{4+}	0.71	0.71	—
Ra^{2+}	—	1.40	1.57	Pb^{4+}	0.81	0.84	—
Zn^{2+}	0.69	0.74	—				
Cd^{2+}	1.03	0.97	1.14				
Hg^{2+}	0.93	1.10	—				

[a] These radii are obtained by using the *rock-salt type of structure* as standard (i.e., six coordination); small corrections can be made for other coordination numbers; see A. P. Sinha, *Struct. Bonding,* 1976, **25**, 69. For effective ionic radii of M^{3+} in corundum-type oxides of Al, Cr, Ga, V, Fe, Rh, Ti, In, and Tl, see C. T. Prewitt *et al., Inorg. Chem.,* 1969, **9**, 1985, and for M^{4+} in rutile or closely related oxides of Si, Ge, Mn, Cr, V, Rh, Ti, Ru, Ir, Pt, Re, Os, Tc, Mo, W, Ta, Nb, Sn, and Pb, see D. B. Rogers *et al., Inorg. Chem.,* 1969, **8**, 841.

[b] M. F. C. Ladd, *Theor. Chim. Acta,* 1968, **12**, 333.

attainable. Thus when r^-/r^+ becomes equal to 1.37, the competition between attractive and repulsive Coulomb forces is balanced, and any increase in the ratio may make the CsCl structure unfavorable relative to a structure with a lower coordination number, such as the NaCl structure.

In the NaCl structure, in order to have all ions just touching but not squeezed, with radius r^- for X^- and r^+ for M^+ we have

$$2r^- = \sqrt{2}(r^+ + r^-)$$

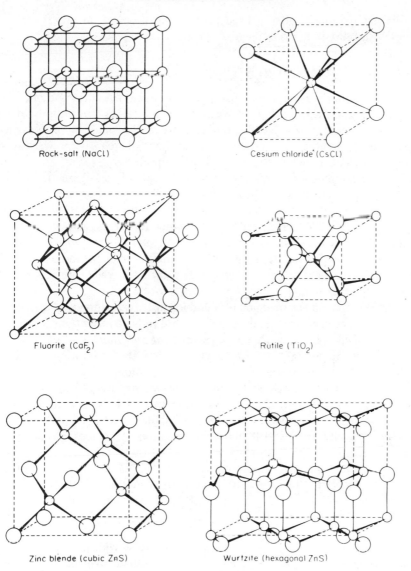

Rock-salt (NaCl)

Cesium chloride (CsCl)

Fluorite (CaF$_2$)

Rutile (TiO$_2$)

Zinc blende (cubic ZnS)

Wurtzite (hexagonal ZnS)

Fig. 1-9. Six important ionic structures. Small circles denote metal cations, large circles denote anions.

which gives for the critical radius ratio

$$\frac{r^-}{r^+} = 2.44$$

If the ratio r^-/r^+ exceeds 2.44, the NaCl structure becomes disfavored, and a structure with cation coordination number 4, for which the critical value of r^-/r^+ is 4.44, may be more favorable. To summarize, in this simple approximation, packing considerations lead us to expect the various structures to have the following ranges of stability in terms of the r^-/r^+ ratio:

CsCl and CaF_2 structures	$1 < r^-/r^+ < 1.37$
NaCl and rutile structures	$1.37 < r^-/r^+ < 2.44$
ZnS structures	$2.44 < r^-/r^+ < 4.44$

Obviously, similar reasoning may be applied to other structures and other types of ionic compound. Since the model we are using is a rather crude approximation, we must not expect these calculations to be more than a rough guide. We can certainly expect to find the CsCl structure in compounds where $r^- \approx r^+$, whereas when $r^- \gg r^+$ a structure such as that of ZnS will be preferred.

The more common ionic crystal structures shown in Fig. 1-9 are mentioned repeatedly throughout the text. The *rutile structure,* named after one mineralogical form of TiO_2, is very common among oxides and fluorides of the MF_2 and MO_2 types (e.g., FeF_2, NiF_2, ZrO_2, RuO_2), where the radius ratio favors coordination number 6 for the cation. Similarly, the *zinc blende* and *wurtzite structures,* named after two forms of zinc sulfide, are widely encountered when the radius ratio favors four-coordination, and the *fluorite structure* is common when eight-coordination of the cation is favored.

When a compound has stoichiometry and ion distribution opposite to that in one of the structures just mentioned, it may be said to have an *anti* structure. Thus compounds such as Li_2O, Na_2S, and K_2S, have the *antifluorite* structure in which the anions occupy the Ca^{2+} positions and the cations the F^- positions of the CaF_2 structure. The antirutile structure is sometimes encountered also.

Structures with Close Packing of Anions. Many structures of halides and oxides can be regarded as close-packed arrays of anions with cations in the octahedral and/or tetrahedral interstices. Even the NaCl structure can be thought of in this way (*ccp* array of Cl^- ions with all octahedral interstices filled), although this is not ordinarily useful. $CdCl_2$ also has *ccp* Cl^- ions with every other octahedral hole occupied by Cd^{2+}, and CdI_2 has *hcp* I^- ions with Cd^{2+} ions in half the octahedral holes. It is noteworthy that the $CdCl_2$ and CdI_2 structures (the latter appears in Fig. 1-10) are *layer structures.* The particular pattern in which cations occupy half the octahedral holes, is such as to leave alternate layers of direct anion-anion contact.

Corundum, the α form of Al_2O_3, has an *hcp* array of oxide ions with two-thirds of the octahedral interstices occupied by cations and is adopted by many other oxides (e.g., Ti_2O_3, V_2O_3, Cr_2O_3, Fe_2O_3, Ga_2O_3, and Rh_2O_3). The BiI_3 structure has an *hcp* array of anions with two-thirds of the octahedral holes in each alternate pair of layers occupied by cations, and it is adopted by $FeCl_3$, $CrBr_3$, $TiCl_3$, VCl_3, and many other AB_3 compounds. As indicated, all the *structures* just mentioned are adopted by numerous *substances.* The structures are usually named in reference to one of these substances. Thus we speak of the NaCl, $CdCl_2$, CdI_2, BI_3, and corundum (or α-Al_2O_3) structures.

Some Mixed Oxide Structures. There are a vast number of oxides (and also some stoichiometrically related halides) having two or more different kinds of cation. Most of them occur in one of a few basic structural types, the names of which are

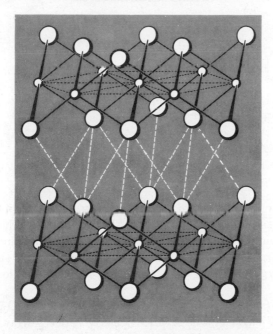

Fig. 1-10. A portion of the CdI_2 structure. Small spheres represent metal cations.

derived from the first or principal compound shown to have that type of structure. Three of the most important such structures are now described.

1. *The Spinel Structure.* The compound $MgAl_2O_4$, which occurs in Nature as the mineral spinel, has a structure based on a *ccp* array of oxide ions. One-eighth of the tetrahedral holes (of which there are two per anion) are occupied by Mg^{2+} ions and one-half of the octahedral holes (of which there is one per anion) are occupied by Al^{3+} ions. This structure, or a modification to be discussed below, is adopted by many other mixed metal oxides of the type $M^{II}M_2^{III}O_4$ (e.g., $FeCr_2O_4$, $ZnAl_2O_4$, and $Co^{II}Co_2^{III}O_4$), by some of the type $M^{IV}M_2^{II}O_4$ (e.g., $TiZn_2O_4$ and $SnCo_2O_4$), and by some of the type $M_2^IM^{VI}O_4$ (e.g., Na_2MoO_4 and Ag_2MoO_4). This structure is often symbolized as $A[B_2]O_4$, where brackets enclose the ions in the octahedral interstices. An important variant is the *inverse spinel structure*, $B[AB]O_4$, in which half the B ions are in tetrahedral interstices and the A ions are in octahedral ones along with the other half of the B ions. This often happens when the A ions have a stronger preference for octahedral coordination than do the B ions. As far as is known, all $M^{IV}M_2^{II}O_4$ spinels are inverse (e.g., $Zn[ZnTi]O_4$), as are many of the $M^{II}M_2^{III}O_4$ ones (e.g., $Fe^{III}[Co^{II}Fe^{III}]O_4$, $Fe^{III}[Fe^{II}Fe^{III}]O_4$, and $Fe[NiFe]O_4$), as well.

There are also many compounds with *disordered spinel structures* in which only a fraction of the A ions are in tetrahedral sites (and a corresponding fraction in octahedral ones). This occurs when the preferences of both A and B ions for octahedral and tetrahedral sites do not differ markedly.

2. *The Ilmenite Structure.* This is the structure of the mineral ilmenite,

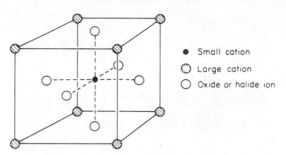

Fig. 1-11. The perovskite structure.

$Fe^{II}Ti^{IV}O_3$. It is closely related to the corundum structure except that the cations are of two different types. It is adopted by ABO_3 oxides when the two cations, A and B, are of about the same size, but they need not be of the same charge so long as their total charge is $+6$. Thus in ilmenite itself and in $MgTiO_3$ and $CoTiO_3$ the cations have charges $+2$ and $+4$, whereas in α-$NaSbO_3$ the cations have charges of $+1$ and $+5$.

   3. *The Perovskite Structure.* The mineral perovskite, $CaTiO_3$, has a structure in which the oxide ions and the large cation (Ca^{2+}) form a *ccp* array with the smaller cation (Ti^{4+}) occupying these octahedral holes formed exclusively by oxide ions, as shown in Fig. 1-11. This structure is often slightly distorted—in $CaTiO_3$ itself, for example. It is adopted by a great many ABO_3 oxides in which one cation is comparable in size to O^{2-} and the other much smaller, with the cation charges variable so long as their sum is $+6$. It is found in $Sr^{II}Ti^{IV}O_3$, $Ba^{II}Ti^{IV}O_3$, $La^{III}Ga^{III}O_3$, and $Na^{I}Nb^{V}O_3$, and $K^{I}Nb^{V}O_3$, and also in some mixed fluorides (e.g., $KZnF_3$ and $KNiF_3$).

1-6.   Energetics of Ionic Crystals

One may define a model in which so-called ionic crystals are treated as *purely ionic*—that is, only Coulombic forces plus some higher-order repulsive forces arising at close distances, because of the interpenetration of electron clouds, are assumed to be operating. It is then possible to correlate a great deal of thermodynamic data, but, as will be shown later, it is certain that this model is a considerable oversimplification and its success should not be taken as proof of its physical validity.

   Before developing this ionic model, a few points concerning the formulation of ions in the gas phase must be reviewed. To form cations, electrons must be detached from atoms; this process is characterized by a standard enthalpy ΔH°_{ion}. Similarly, to form an anion an electron must be attached to a neutral atom; this process also has a standard enthalpy ΔH°_{EA}, the enthalpy of electron attachment.* For use in subsequent calculations, these quantities are listed in Appendices 2 and 3.

---

* Much of the chemical literature uses I, the ionization potential, in electron volts, which is equal to ΔH°_{ion} if converted to kilojoules, and A, the so-called electron affinity, which is equal to $-\Delta H^\circ_{EA}$ if put in the same units. We shall not use these nonstandard (though commonplace) quantities.

TABLE 1-2
Madelung Constants for Several Structures

Structure type	M
NaCl	1.74756
CsCl	1.76267
CaF_2	5.03878
Zinc blende	1.63805
Wurtzite	1.64132

The enthalpy of vaporizing one mole of a crystalline, ionic compound to form an infinitely dilute gas consisting solely of the constituent ions has, of course, a positive sign. Traditionally this enthalpy has been designated the *lattice energy;* although this term lacks precision, we shall use it with the understanding that when precise definition is required it means the enthalpy of the process just described. The purely ionic description of "ionic" crystals, as stated above, allows an *ab initio* calculation of the lattice energy of such crystals, provided the structure of the crystal is known. Let us consider as an example sodium chloride, which has the structure shown in Fig. 1-9.

The energy, in joules, required to separate two opposite charges, $+e$ and $-e$, each in coulombs (C) initially at a distance r_0, in meters, to a distance of infinity is given by the equation

$$E = \frac{e^2}{4\pi\epsilon_0 r_0}$$

The factor $(4\pi\epsilon_0)^{-1}$ is required in the SI system of units; ϵ_0 is the permittivity (dielectric constant) of a vacuum that has the value $8.854 \times 10^{-12} \, C^2 \, m^{-1} \, J^{-1}$.

Let us call the shortest Na^+—Cl^- distance in NaCl r_0. Each Na^+ ion is surrounded by six Cl^- ions at the distance r_0, in meters, giving an energy term $6e^2/4\pi\epsilon_0 r_0$. The next closest neighbors to a given Na^+ ion are 12 Na^+ ions which, by simple trigonometry, lie $\sqrt{2}r_0$ away. Thus another energy term, with a minus sign because it is repulsive, is $-12e^2/\sqrt{2}r_0 4\pi\epsilon_0$. By repeating this sort of procedure, successive terms are found, leading to the expression

$$E = \frac{1}{4\pi\epsilon_0}\left(\frac{6e^2}{r_0} - \frac{12e^2}{\sqrt{2}r_0} + \frac{8e^2}{\sqrt{3}r_0} - \frac{6e^2}{2r_0} + \ldots\right)$$

$$= \frac{e^2}{4\pi\epsilon_0 r_0}\left(6 - \frac{12}{\sqrt{2}} + \frac{8}{\sqrt{3}} - \frac{6}{2} + \frac{24}{\sqrt{5}} - \ldots\right) \tag{1-1}$$

It is possible to derive a general formula for the infinite series and to find the numerical value to which it converges. That value is characteristic of the structure and independent of what particular ions are present. It is called the *Madelung constant,* M_{NaCl}, for the NaCl structure. It is actually an irrational number, whose value can be given to as high a degree of accuracy as needed, for example, 1.747 . . . , or 1.747558 . . . , or better. Madelung constants for many common ionic structures have been evaluated, and a few are given in Table 1-2 for illustrative purposes.

A unique Madelung constant is defined only for structures in which all ratios of interatomic vectors are fixed by symmetry. In the case of the rutile structure there are two crystal dimensions that can vary independently. There is a different Madelung constant for each ratio of the two independent dimensions.

When a mole (N ions of each kind, where N is Avogadro's number) of sodium chloride is formed from the gaseous ions, the total electrostatic energy released is given by

$$E_e = NM_{NaCl}\left(\frac{e^2}{4\pi\epsilon_0 r_0}\right) \tag{1-2}$$

This is true because the expression for the electrostatic energy of one Cl^- ion would be the same as that for an Na^+ ion. If we were to add the electrostatic energies for the two kinds of ion, the result would be twice the true electrostatic energy because each pairwise interaction would have been counted twice.

The electrostatic energy given by eq. 1-2 is not the actual energy released in the process

$$Na^+(g) + Cl^-(g) = NaCl(s) \tag{1-3}$$

Real ions are not rigid spheres. The equilibrium separation of Na^+ and Cl^- in NaCl is fixed when the attractive forces are exactly balanced by repulsive forces. The attractive forces are Coulombic and follow strictly a $1/r^2$ law. The repulsive forces are more subtle and follow an inverse r^n law, where n is >2 and varies with the nature of the particular ions. We can write, in a general way, that the total repulsive energy per mole at any value of r is

$$E_{rep} = \frac{NB}{r^n}$$

where B is a constant.

At the equilibrium distance, the net energy U for process 1-3 (where we now use algebraic signs in accord with convention), is given by

$$U = -NM_{NaCl}\left(\frac{e^2}{4\pi\epsilon_0 r_0}\right) + \frac{NB}{r_0^n} \tag{1-4}$$

Observe that the attractive forces produce an exothermic contribution and the repulsive ones an endothermic term.

The constant B can now be eliminated if we recognize that at equilibrium, when $r = r_0$ the energy U is a minimum by definition. The derivative of U with respect to r, evaluated at $r = r_0$ must equal zero. Differentiating eq. 1-4 we get

$$\left(\frac{dU}{dr}\right)_{r=r_0} = \frac{NM_{NaCl}e^2}{4\pi\epsilon_0 r_0^2} - \frac{nNB}{r_0^{n+1}} = 0$$

which can be rearranged and solved for B:

$$B = \frac{e^2 M_{NaCl}}{4\pi\epsilon_0 n} r_0^{n-1}$$

When this expression for B is substituted into eq. 1-4, we obtain

$$U = -\frac{NM_{NaCl}e^2}{4\pi\epsilon_0 r_0}\left(1 - \frac{1}{n}\right) \tag{1-5}$$

The value of n can be estimated as 9.1 from the measured compressibility* of NaCl.

In a form suitable for calculating numerical results, in kJ mol^{-1}, using r_0 in angstroms, eq. 1-5 becomes

$$U = -1389 \frac{M_{NaCl}}{r_0} \left(1 - \frac{1}{n}\right)$$

and inserting appropriate values of the parameters we obtain

$$U = -1389 \frac{1.747}{2.82} \left(1 - \frac{1}{9.1}\right)$$

$$U = -860 + 95 = -765 \text{ kJ mol}^{-1}$$

Notice that the repulsive energy equals about 11% of the Coulombic energy. The net result is not very sensitive to the exact value of n. If a value of $n = 10$ had been used, an error of only 9 kJ mol^{-1} or 1.2% would have been made.

Refinement of Lattice Energy Calculations. As noted, the Madelung constant is determined by the geometry of the structure only. Thus for a case, such as MgO, where the structure is the same but each ion has a charge of ± 2, the only modification required is to replace $-e^2$ by $(2e)(-2e) = -4e^2$ in the Coulombic energy term. In general, eq. 1-5 becomes

$$U = -\frac{NM_{NaCl}Z^2e^2}{4\pi\epsilon_0 r_0} \left(1 - \frac{1}{n}\right)$$

for any structure whose Madelung constant is M with ions of charges Z^+ and Z^-.

The value of n can be estimated for alkali halides by using the average of the following numbers:

He	5	Kr	10
Ne	7	Xe	12
Ar	9		

where the noble gas symbol denotes the noble-gas-like electron configuration of the ion. Thus for LiF an average of the He and Ne values $(5 + 7)/2 = 6$ would be used.

To make very accurate calculations of crystal energies, sometimes called lattice energies, certain refinements must be introduced. The main ones are as follows.

1. A more accurate, quantum expression for the repulsion energy.

2. A correction for van der Waals energy.

3. A correction for the "zero-point energy," the vibrational energy present even at 0° K.

The last two corrections are opposite in sign and often of similar magnitude. Thus for NaCl a refined calculation gives

Coulomb energy	−860
Repulsion energy	+99
van der Waals energy	−13
Zero-point energy	+8
	−766 kJ mol^{-1}

* Fractional change in volume per unit change in pressure, that is, $(\Delta V/V)/\Delta P$.

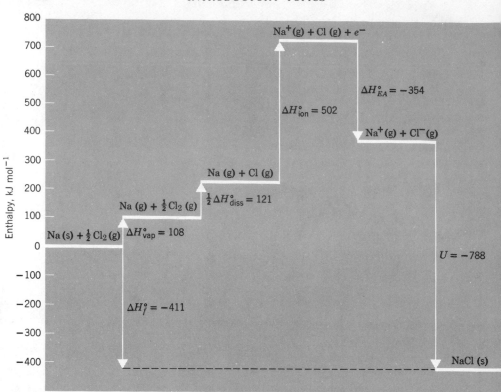

Fig. 1-12. The Born-Haber cycle for NaCl.

The Born-Haber Cycle. One test of whether an ionic model is a useful description of a substance such as sodium chloride is its ability to produce an accurate value of the enthalpy of formation. Note that we cannot make this test simply by measuring the enthalpy of reaction 1-3, or its reverse. The former is possible in principle but not experimentally feasible. The latter is not possible because sodium chloride does not vaporize cleanly to Na^+ and Cl^- but to NaCl, which then dissociates into atoms.

To handle the energy problem, we use a *thermodynamic cycle* called the Born-Haber cycle (Fig. 1-12). The basic idea is that the formation of NaCl(s) from the elements, $Na(s) + \frac{1}{2}Cl_2(g)$, whose enthalpy is by definition the enthalpy of formation of NaCl(s), can be broken down into a series of steps. If the enthalpies of these steps are added, algebraically, the result must be equal to ΔH_f° according to the law of conservation of energy, the first law of thermodynamics. We thus have the equation

$$\Delta H_f^\circ = \Delta H_{vap}^\circ + \frac{1}{2} \Delta H_{diss}^\circ + \Delta H_{EA}^\circ + \Delta H_{ion}^\circ + U \qquad (1\text{-}6)$$

where the enthalpy terms are for vaporization of sodium (ΔH_{vap}°), dissociation of $Cl_2(g)$ into gaseous atoms (ΔH_{diss}°), electron attachment to Cl(g) to give

$Cl^-(g)(\Delta H^\circ_{EA})$, ionization of $Na(g)$ to $Na^+(g) + e(\Delta H^\circ_{ion})$, and the formation of $NaCl(s)$ from gaseous ions (U).

More generally, any one of these energies can be calculated if all the others are known. For NaCl all the enthalpies except U in eq. 1-6 have been measured independently. The following summation can thus be made:

$$
\begin{array}{rcl}
\Delta H^\circ_f & & -411 \\
-\Delta H^\circ_{vap} & & -108 \\
-\tfrac{1}{2}\,\Delta H^\circ_{diss} & = & -121 \\
-\Delta H^\circ_{EA} & & 354 \\
-\Delta H^\circ_{ion} & & -502 \\
\hline
U & = & -788 \text{ kJ mol}^{-1}
\end{array}
$$

This result is very close (within 2.7%) of the value of U calculated by using the refined ionic model (-766 kJ mol^{-1}). Actually, even more precise calculations give a value that agrees to within less than 1%. This good agreement supports (but does not prove) the idea that the ionic model for NaCl is a useful one.

Often, the Born-Haber cycle, or a similar one, is used differently. If it is assumed that U calculated on the ionic model is correct, the cycle can be used to estimate some other energy term. For example, there is no convenient direct way to measure the enthalpy of formation of the gaseous CN^- ion. From a Born-Haber cycle for NaCN, where values of all the other enthalpies are available and U is calculated, it is found that ΔH_f for $CN^-(g)$ is 29 kJ mol^{-1}.

Caveat. It is important to realize that because the purely ionic model of some real compound such as NaCl affords a fairly accurate value for its lattice energy, this does not entitle us to conclude that the compound is literally ionic. Very refined X-ray diffraction studies indicate that the electron distribution does not correspond strictly to the requirements of Na^+ (10 electrons) and Cl^- (eighteen electrons), but shows somewhat greater density on Na and less on Cl. This is tantamount to the existence of some shared electron density, or, in other words, some covalence in the bonding. The purely ionic model is successful energetically mainly because the use of the experimental internuclear distances compensates for the assumption of exactly integral formal charges, ± 1, ± 2, and so on. Although we shall not go further into the matter here, the purely ionic model does not give structural predictions that are always correct; that is, the rules developed in Section 1-5 on the basis of fixed radius ratios and assuming purely electrostatic forces are *frequently* violated. Thus the purely ionic model is just that—a *model*. It is convenient but not literally true.

1-7. Covalent Solids

Elements. The elements that form extended covalent (as opposed to metallic) arrays are boron, all the Group IV elements except lead, also phosphorus, arsenic, selenium, and tellurium. All other elements form either only metallic phases or only molecular ones. Some of the elements above, of course, have allotropes of metallic or molecular type in addition to the phase or phases that are extended covalent arrays. For example, tin has a metallic allotrope (white tin) in addition to that with

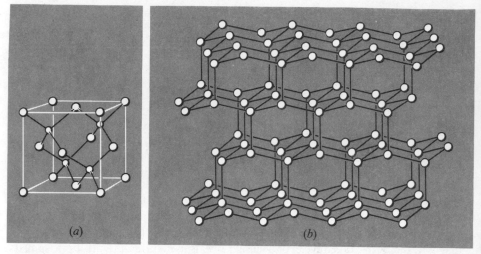

Fig. 1-13. The diamond structure seen from two points of view. (a) The conventional cubic unit cell. (b) A view showing how layers are stacked; these layers run perpendicular to the body diagonals of the cube.

the diamond structure (gray tin), and selenium forms two molecular allotropes containing Se_8 rings, isostructural with the rhombic form and the monoclinic form of sulfur. For tellurium we have a situation on the borderline of metallic behavior.

The structures of the principal allotropic forms of all the elements are discussed in detail as the chemistry of each element is treated. For illustrative purposes, we shall mention here only one such structure, the diamond structure, since this is adopted by several other elements and is a point of reference for various other structures. It is shown from two points of view in Fig. 1-13. The structure has a cubic unit cell with the full symmetry of the group T_d. However for some purposes it can be viewed as a stacking of puckered infinite layers. It will be noted that the zinc blende structure (Fig. 1-9) can be regarded as a diamond structure in which one-half the sites are occupied by Zn^{2+} (or other cation) and the other half are occupied by S^{2-} (or other anion) in an ordered way. In the diamond structure itself all atoms are equivalent, each being surrounded by a perfect tetrahedron of four others. The electronic structure can be simply and fairly accurately described by saying that each atom forms a localized two-electron bond to each of its neighbors.

Compounds. As soon as one changes from elements, where the adjacent atoms are identical and the bonds are necessarily nonpolar, to compounds, there enters the vexatious question of when to describe a substance as ionic and when to describe it as covalent. No attempt is made here to deal with this question in detail for the practical reason that, very largely, there is no need to have the answer—even granting, for the sake of argument only, that any such thing as "the answer" exists. Suffice it to say that bonds between unlike atoms all have some degree of polarity and (1) when the polarity is relatively small it is practical to describe the bonds as polar covalent ones and (2) when the polarity is very high it makes more sense to consider that the substance consists of an array of ions.

1-8. Defect Structures

All the foregoing discussion of crystalline solids has dealt with their perfect or ideal structures. Such perfect structures are seldom if ever found in real substances, and although low levels of imperfections have only small effects on their chemistry, the physical (i.e., electrical, magnetic, optical, and mechanical) properties of many substances are often crucially affected by their imperfections. It is, therefore, appropriate to devote a few paragraphs to describing the main types of imperfection, or defect, in real crystalline solids. We shall not, however, discuss the purely mechanical imperfections such as mosaic structure, stacking faults, and dislocations, all of which entail some sort of mismatch between lattice layers.

Stoichiometric Defects. There are some defects that leave the stoichiometry unaffected. One type is the *Schottky defect* in ionic crystals. A Schottky defect consists of vacant cation and anion sites in numbers proportional to the stoichiometry; thus there are equal numbers of Na^+ and Cl^- vacancies in NaCl and $2Cl^-$ vacancies per Ca^{2+} vacancy in $CaCl_2$. A Schottky defect in NaCl is illustrated in Fig. 1-14a.

When an ion occupies a normally vacant interstitial site, leaving its proper site vacant, the defect is termed a *Frenkel defect*. Frenkel defects are most common in crystals where the cation is much smaller than the anion, for instance, in AgBr, as illustrated schematically in Fig. 1-14b.

Nonstoichiometric Defects. These often occur in transition metal compounds, especially oxides and sulfides, because of the ability of the metal to exist in more than one oxidation state. A well-known case is "FeO," which consists of a *ccp* array of oxide ions with all octahedral holes filled by Fe^{2+} ions. In reality, however, some of these sites are vacant, whereas others—sufficient to maintain electroneutrality—contain Fe^{3+} ions. Thus the actual stoichiometry is commonly about $Fe_{0.95}O$. Another good example is "TiO," which can readily be obtained with compositions ranging from $Ti_{0.74}O$ to $Ti_{1.67}O$ depending on the pressure of oxygen gas used in preparing the material.

Nonstoichiometric defects also occur, however, even when the metal ion has but one oxidation state. Thus, for example, CdO is particularly liable to lose oxygen when heated, to give yellow to black solids of composition $Cd_{1+\gamma}O$. A comparable situation arises when NaCl is treated with sodium vapor, which it absorbs to give a blue solid of composition $Na_{1+\gamma}Cl$. An appropriate number ($N\gamma$) of anion sites are then devoid of anions but occupied by electrons. The electrons in these cavities behave roughly like the simple particle in a box, and there are excited states accessible at energies corresponding to the energy range of visible light. Hence these cavities containing electrons are color centers, commonly called *F* centers, from *Farbe* (German for color).

The existence of defects has a simple thermodynamic basis. The creation of a defect in a perfect structure has an unfavorable effect on the enthalpy; some

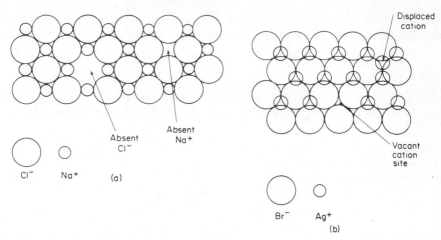

Fig. 1-14. (a) A Schottky defect in a layer of NaCl. (b) A Frenkel defect in a layer of AgBr. The defects need not be restricted to one layer, as shown.

Coulombic or bonding energy must be sacrificed to create it. However the introduction of some irregularities into an initially perfect array markedly increases the entropy; a $T\Delta S$ term large enough to cancel the unfavorable ΔH term will thus arise up to some limiting concentration of defects. It is possible to write an expression for the concentration of defects in equilibrium with the remainder of the structure just as though a normal chemical equilibrium were involved.

Finally, there are defects resulting from the presence of impurities. Some of these, when deliberately devised and controlled, constitute the basis for solid-state electronic technology. For example, a crystal of germanium (which has the diamond structure) can be "doped" with traces of gallium or arsenic. A gallium atom can replace an atom of germanium, but an electron vacancy is created. An electron can move into this hole, thus creating a hole elsewhere. In effect, the hole can wander through the crystal and under the influence of an electrical potential difference, it can travel through the crystal in a desired direction. The hole moves in the same direction as would a positive charge, and gallium-doped germanium is therefore called a p-type (for positive) semiconductor. Whenever an arsenic atom replaces a germanium atom, an electron is introduced into a normally unfilled energy band of germanium. These electrons can also migrate in an electric field, and the arsenic-doped material is called an n-type (for negative) semiconductor.

Doped silicon and germanium are technologically the most important types of semiconducting material. To ensure reproducible performance, the type and level of impurities must be strictly controlled; thus superpure silicon and germanium must first be prepared (cf. Chapter 12) and the desired doping then carried out. In addition to silicon or germanium as the basis for creating semiconductors, it is also possible to prepare certain isoelectronic III–V or II–VI compounds, such as GaAs or CdS. Holes or conduction electrons can then be introduced by variation of the stoichiometry or by addition of suitable impurities.

Semiconductor behavior is to be found in many other types of compound, for example, in $Fe_{1-\gamma}O$ and $Fe_{1-\gamma}S$, where electron transfer from Fe^{2+} to Fe^{3+} causes a virtual migration of Fe^{3+} ions, hence p-type conduction.

General References

Addison, W. E., *Allotropy of the Elements,* Oldbourne Press, 1966.

Dasent, W. E., *Inorganic Energetics,* Penguin Books, 1970. Excellent, readable coverage of basic principles.

Donahue, J., *The Structures of the Elements,* Wiley, 1974.

Galasso, F. S., *Structure and Properties of Inorganic Solids,* Pergamon Press, 1970.

Greenwood, N, N,, *Ionic Crystals, Lattice Defects and Non-Stoichiometry,* Butterworths, 1968. Excellent short introduction.

Hannay, N. B., *Solid-State Chemistry,* Prentice-Hall, 1967. Good general survey.

Johnson, D. A., *Some Thermodynamic Aspects of Inorganic Chemistry,* Cambridge University Press, 1968. A good outline of fundamentals.

Krebs, H., *Fundamentals of Inorganic Crystal Chemistry,* McGraw-Hill, 1968. Excellent discussion of structures and bonding.

Kröger, F. A., *Chemistry of Imperfect Crystals,* North Holland, 1964.

Ladd, M. F. C., *Structure and Bonding in Solid State Chemistry,* Horwood-Wiley, 1979.

McDowell, C. A., in *Physical Chemistry,* Vol. III, H. Eyring, D. Henderson, and W. Jost, Eds., Academic Press, 1969, p. 496. Evaluation of electron affinities.

Pearson, W. B., *The Crystal Chemistry and Physics of Metals and Alloys,* Wiley, 1972.

Vedeneyev, V. I., *et al., Bond Energies, Ionization Potentials, and Electron Affinities,* Arnold, 1965. Data tables.

Wells, A. F., *Structural Inorganic Chemistry,* 4th ed., Clarendon Press, 1975. A fascinating book with superb illustrations and a wealth of data; 1068 pages of text.

CHAPTER TWO

Symmetry and Structure

THE SYMMETRY GROUPS

Molecular symmetry and ways of specifying it with mathematical precision are important for several reasons. The most basic reason is that *all* molecular wave functions—those governing electron distribution as well as those for vibrations, nmr spectra, and so on—must conform, rigorously, to certain requirements based on the symmetry of the equilibrium nuclear framework of the molecule. When the symmetry is high these restrictions can be very severe. Thus from a knowledge of symmetry alone it is often possible to reach useful qualitative conclusions about molecular electronic structure and to draw inferences from spectra about molecular structures. The qualitative application of symmetry restrictions is most impressively illustrated by the crystal-field and ligand-field theories of the electronic structures of transition metal complexes, as described in Chapter 20, and by numerous examples of the use of infrared and Raman spectra to deduce molecular symmetry. Illustrations of the latter occur throughout the book, but particularly with respect to some metal carbonyl compounds in Chapter 25.

A more mundane use for the concept and notation of molecular symmetry is in the precise description of a structure. One symbol, such as D_{4h}, can convey precise, unequivocal structural information that would require long verbal description to duplicate. Thus if we say that the $Ni(CN)_4^{2-}$ ion has D_{4h} symmetry, we imply that (*a*) it is completely planar, (*b*) the Ni—C—N groups are all linear, (*c*) the C—Ni—C angles are all equal, at 90°, (*d*) the four CN groups are precisely equivalent to one another, and (*e*) the four Ni—C bonds are precisely equivalent to one another. The use of symmetry symbols has become increasingly common in the chemical literature, and it is now necessary to be familiar with the basic concepts and rules of notation to read many of the contemporary research papers in inorganic and, indeed, also organic chemistry with full comprehension. It thus seems appropriate to include a brief survey of molecular symmetry and the basic rules for specifying it.

28

2-1.   Symmetry Operations and Elements

When we say that a molecule has *symmetry,* we mean that *certain parts of it can be interchanged with others without altering either the identity or the orientation of the molecule.* The interchangeable parts are said to be equivalent to one another by symmetry. Consider, for example, a trigonal-bipyramidal molecule such as PF_5 (2-1). The three equatorial P—F bonds, to F_1, F_2, and F_3, are equivalent. They have

$$F_4$$

$$F_1 \cdots P \cdots F_3$$

$$F_2$$

$$F_5$$

(2-I)

the same length, the same strength, and the same type of spatial relation to the remainder of the molecule. Any permutation of these three bonds among themselves leads to a molecule indistinguishable from the original. Similarly, the axial P—F bonds, to F_4 and F_5, are equivalent. *But,* axial and equatorial bonds are different types (e.g., they have different lengths), and if one of each were to be interchanged, the molecule would be noticeably perturbed. These statements are probably self-evident, or at least readily acceptable, on an intuitive basis; but for systematic and detailed consideration of symmetry, certain formal tools are needed. The first set of tools is a set of *symmetry operations.*

Symmetry operations are geometrically defined ways of exchanging equivalent parts of a molecule. There are four kinds which are used conventionally and these are sufficient for all our purposes.

1. Simple rotation about an axis passing through the molecule by an angle $2\pi/n$. This operation is called a *proper rotation* and is symbolized C_n. If it is repeated

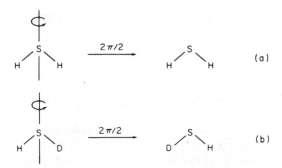

Fig. 2-1. The operation C_2 carries H_2S into an orientation indistinguishable from the original, but HSD goes into an observably different orientation.

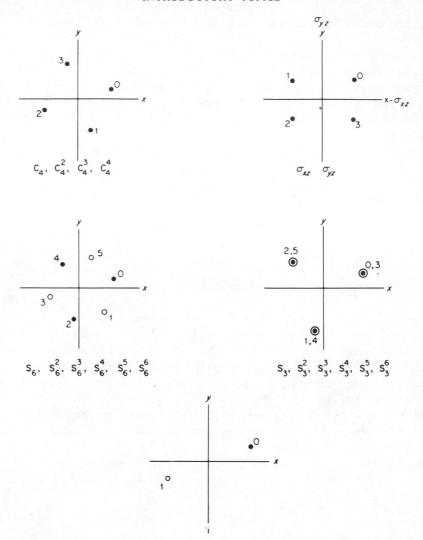

Fig. 2-2. The effects of symmetry operations on an arbitrary point, designated 0, thus generating sets of points.

n times, of course the molecule comes all the way back to the original orientation.

2. Reflection of all atoms through a plane that passes through the molecule. This operation is called *reflection* and is symbolized σ.

3. Reflection of all atoms through a point in the molecule. This operation is called *inversion* and is symbolized **i**.

4. The combination, in either order, of rotating the molecule about an axis passing through it by $2\pi/n$ and reflecting all atoms through a plane that is perpendicular to the axis of rotation is called *improper rotation* and is symbolized $\mathbf{S}_n$.

These operations are *symmetry operations if, and only if,* the appearance of the molecule is *exactly* the same after one of them is carried out as it was before. For instance, consider rotation of the molecule H_2S by $2\pi/2$ about an axis passing through S and bisecting the line between the H atoms. As shown in Fig. 2-1, this operation interchanges the H atoms and interchanges the S—H bonds. Since these atoms and bonds are equivalent, there is no physical (i.e., physically meaningful or detectable) difference after the operation. For HSD, however, the corresponding operation replaces the S—H bond by the S—D bond, and vice versa, and one can see that a change has occurred. Therefore, for H_2S, the operation C_2 is a symmetry operation; for HSD it is not.

These types of symmetry operation are graphically explained by the diagrams in Fig. 2-2, where it is shown how an arbitrary point (0) in space is affected in each case. Filled dots represent points above the xy plane and open dots represent points below it. Let us examine first the action of proper rotations, illustrated here by the C_4 rotations, that is, rotations by $2\pi/4 = 90°$. The operation C_4 is seen to take the point 0 to the point 1. The application of C_4 twice, designated C_4^2, generates point 2. Operation C_4^3 gives point 3 and, of course, C_4^4, which is a rotation by $4 \times 2\pi/4 = 2\pi$, regenerates the original point. The set of four points, 0, 1, 2, 3 are permutable, cyclically, by repeated C_4 proper rotations and are equivalent points. It will be obvious that in general repetition of a C_n operation will generate a set of n equivalent points from an arbitrary initial point, provided that point lies off the axis of rotation.

The effect of reflection through symmetry planes perpendicular to the xy plane, specifically, σ_{xz} and σ_{yz} is also illustrated in Fig. 2-2. The point 0 is related to point 1 by the σ_{yz} operation and to the point 3 by the σ_{xz} operation. By reflecting either point 1 or point 3 through the second plane, point 2 is obtained.

The set of points generated by the repeated application of an improper rotation will vary in appearance depending on whether the order of the operation, S_n, is even or odd, order being the number n. A crown of n points, alternately up and down, is produced for n even, as illustrated for S_6. For n odd there is generated a set of $2n$ points which form a right n-sided prism, as shown for S_3.

Finally, the operation **i** is seen to generate from point 0 a second point, 1, lying on the opposite side of the origin.

Let us now illustrate the symmetry operations for various familiar molecules as examples. As this is done it will be convenient to employ also the concept of *symmetry elements*. A symmetry element is an *axis* (line), *plane*, or *point* about which symmetry operations are performed. The existence of a certain symmetry operation implies the existence of a corresponding symmetry element, and conversely, the presence of a symmetry element means that a certain symmetry operation or set of operations is possible.

Consider the ammonia molecule (Fig. 2-3). The three equivalent hydrogen atoms may be exchanged among themselves in two ways: by proper rotations, and by reflections. The molecule has an axis of 3-fold proper rotation; this is called a C_3 axis. It passes through the N atom and through the center of the equilateral triangle defined by the H atoms. When the molecule is rotated by $2\pi/3$ in a clockwise direction H_1 replaces H_2, H_2 replaces H_3, and H_3 replaces H_1. Since the three H

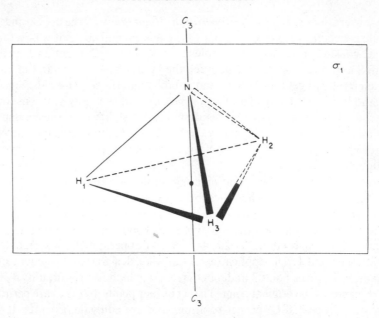

Fig. 2-3. The ammonia molecule, showing its 3-fold symmetry axis C_3, and one of its three planes of symmetry σ_1, which passes through H_1 and N and bisects the H_2—H_3 line.

atoms are physically indistinguishable, the numbering having no physical reality, the molecule after rotation is indistinguishable from the molecule before rotation. This rotation, called a C_3 or 3-fold proper rotation, is a symmetry operation. Rotation by $2 \times 2\pi/3$ also produces a configuration different, but physically indistinguishable, from the original and is likewise a symmetry operation; it is designated C_3^2. Finally, rotation by $3 \times 2\pi/3$ carries each H atom all the way around and returns it to its initial position. This operation, C_3^3, has the same net effect as performing no operation at all, but for mathematical reasons it must be considered as an operation generated by the C_3 axis. This, and other operations which have no net effect, are called *identity* operations and are symbolized by **E**. Thus, we may write $C_3^3 = E$.

The interchange of hydrogen atoms in NH_3 by reflections may be carried out in three ways; that is, there are three planes of symmetry. Each plane passes through the N atom and one of the H atoms, and bisects the line connecting the other two H atoms. Reflection through the symmetry plane containing N and H_1 interchanges H_2 and H_3; the other two reflections interchange H_1 with H_3, and H_1 with H_2.

Inspection of the NH_3 molecule shows that no other symmetry operations besides these six (three rotations, C_3, C_3^2, $C_3^3 \equiv E$, and three reflections, σ_1, σ_2, σ_3) are possible. Put another way, the only symmetry elements the molecule possesses are C_3 and the three planes that we may designate σ_1, σ_2, and σ_3. Specifically, it will be obvious that no sort of improper rotation is possible, nor is there a center of symmetry.

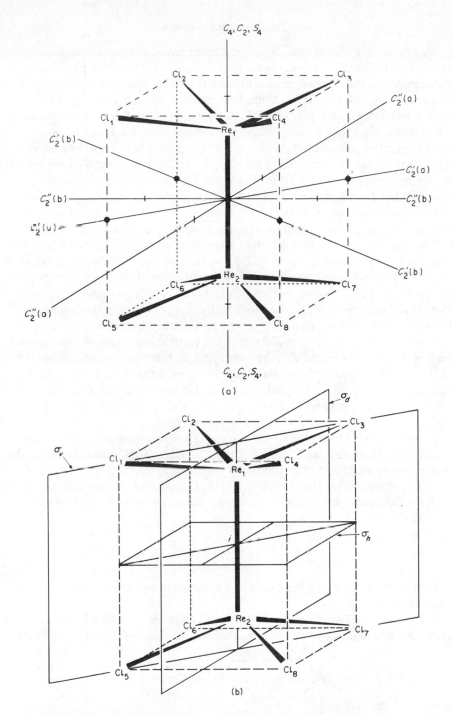

Fig. 2-4. The symmetry elements of the $Re_2Cl_8^{2-}$ ion. (a) The axes of symmetry. (b) One of each type of plane and the center of symmetry.

As a more complex example, in which all four types of symmetry operation and element are represented, let us take the $Re_2Cl_8^{2-}$ ion, which has the shape of a square parallepiped or right square prism (Fig. 2-4). This ion has altogether six axes of proper rotation, of four different kinds. First, the Re_1–Re_2 line is an axis of four-fold proper rotation, C_4, and four operations, $\mathbf{C}_4$, $\mathbf{C}_4^2$, $\mathbf{C}_4^3$, $\mathbf{C}_4^4 \equiv \mathbf{E}$, may be carried out. This same line is also a C_2 axis, generating the operation $\mathbf{C}_2$. It will be noted that the $\mathbf{C}_4^2$ operation means rotation by $2 \times 2\pi/4$, which is equivalent to rotation by $2\pi/2$, that is, to the $\mathbf{C}_2$ operation. Thus the C_2 axis and the $\mathbf{C}_2$ operation are implied by, not independent of, the C_4 axis. There are, however, two other types of C_2 axis that exist independently. There are two of the type that passes through the centers of opposite vertical edges of the prism, C_2' axes, and two more that pass through the centers of opposite vertical faces of the prism, C_2'' axes.

The $Re_2Cl_8^{2-}$ ion has three different kinds of symmetry plane (see Fig. 2-4b). There is a unique one that bisects the Re—Re bond and all the vertical edges of the prism. Since it is customary to define the direction of the highest proper axis of symmetry, C_4 in this case, as the vertical direction, this symmetry plane is horizontal and the subscript h is used to identify it, σ_h. There are then two types of vertical symmetry plane, namely, the two that contain opposite vertical edges, and two others that cut the centers of the opposite vertical faces. One of these two sets may be designated $\sigma_v^{(1)}$ and $\sigma_v^{(2)}$, the v implying that they are vertical. Since those of the second vertical set bisect the dihedral angles between those of the first set, they are then designated $\sigma_d^{(1)}$ and $\sigma_d^{(2)}$, the d standing for dihedral. Both pairs of planes are vertical and it is actually arbitrary which are labeled σ_v and which σ_d.

Continuing with $Re_2Cl_8^{2-}$, we see that an axis of improper rotation is present. This is coincident with the C_4 axis and is an S_4 axis. The $\mathbf{S}_4$ operation about this axis proceeds as follows. The rotational part, through an angle of $2\pi/4$, in the clockwise direction has the same effect as the $\mathbf{C}_4$ operation. When this is coupled with a reflection in the horizontal plane, σ_h, the following shifts of atoms occur:

$$
\begin{array}{llll}
Re_1 \rightarrow Re_2 & Cl_1 \rightarrow Cl_6 & Cl_5 \rightarrow Cl_2 \\
Re_2 \rightarrow Re_1 & Cl_2 \rightarrow Cl_7 & Cl_6 \rightarrow Cl_3 \\
                     & Cl_3 \rightarrow Cl_8 & Cl_7 \rightarrow Cl_4 \\
                     & Cl_4 \rightarrow Cl_5 & Cl_8 \rightarrow Cl_1
\end{array}
$$

Finally, the $Re_2Cl_8^{2-}$ ion has a center of symmetry i and the inversion operation $\mathbf{i}$ can be performed.

In the case of $Re_2Cl_8^{2-}$ the improper axis S_4 might be considered as merely the inevitable consequence of the existence of the C_4 axis and the σ_h, and, indeed,

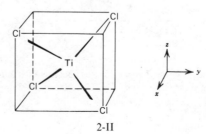

2-II

this is a perfectly correct way to look at it. However it is important to emphasize that there are cases in which an improper axis S_n exists without independent existence of either C_n or σ_h. Consider, for example, a tetrahedral molecule as depicted in 2-II, where the TiCl$_4$ molecule is shown inscribed in a cube and Cartesian axes, x, y, and z are indicated. Each of these axes is an S_4 axis. For example rotation by $2\pi/4$ about z followed by reflection in the xy plane shifts the Cl atoms as follows:

$$Cl_1 \rightarrow Cl_3 \qquad Cl_3 \rightarrow Cl_2$$
$$Cl_2 \rightarrow Cl_4 \qquad Cl_4 \rightarrow Cl_1$$

Note, however, that the Cartesian axes are not C_4 axes (though they are C_2 axes) and the principal planes (viz., xy, xz, yz) are not symmetry planes. Thus we have here an example of the existence of the S_n axis without C_n or σ_h having any independent existence. The ethane molecule in its staggered configuration has an S_6 axis and provides another example.

2-2. Symmetry Groups

The complete set of symmetry operations that can be performed on a molecule is called the *symmetry group* for that molecule. The word "group" is used here not as a mere synonym for "set" or "collection," but in a technical, mathematical sense, and this meaning must first be explained.

Introduction to Multiplying Symmetry Operations. We have already seen in passing that if a proper rotation C_n and a horizontal reflection σ_h can be performed, there is also an operation that results from the combination of the two which we call the improper rotation S_n. We may say that S_n is the product of C_n and σ_h. Noting also that the order in which we perform σ_h and C_n is immaterial,* we can write:

$$C_n \times \sigma_h = \sigma_h \times C_n = S_n$$

This is an algebraic way of expressing the fact that successive application of the two operations shown has the same effect as applying the third one. For obvious reasons, it is convenient to speak of the third operation as being the product obtained by multiplication of the other two.

The example above is not unusual. Quite generally, any two symmetry operations can be multiplied to give a third. For example, in Fig. 2-2 the effects of reflections in two mutually perpendicular symmetry planes are illustrated. It can be seen that one of the reflections carries point 0 to point 1. The other reflection carries point 1 to point 2. Point 0 can also be taken to point 2 by way of point 3 if the two reflection operations are performed in the opposite order. But a moment's thought will show that a direct transfer of point 0 to point 2 can be achieved by a C_2 operation about the axis defined by the line of intersection of the two planes. If we call the two reflections $\sigma(xz)$ and $\sigma(yz)$ and the rotation $C_2(z)$, we can write

$$\sigma(xz) \times \sigma(yz) = \sigma(yz) \times \sigma(xz) = C_2(z)$$

It is also evident that

* This is, however, a special case; in general, order of multiplication matters (see p. 38).

$$\sigma(yz) \times C_2(z) = C_2(z) \times \sigma(yz) = \sigma(xz)$$

and

$$\sigma(xz) \times C_2(z) = C_2(z) \times \sigma(xz) = \sigma(yz)$$

It is also worth noting that if any one of these three operations is applied twice in succession, we get no net result or, in other words, an identity operation, namely;

$$\sigma(xz) \times \sigma(xz) = E$$
$$\sigma(yz) \times \sigma(yz) = E$$
$$C_2(z) \times C_2(z) = E$$

Introduction to a Group. If we pause here and review what has just been done with the three operations $\sigma(xz)$, $\sigma(yz)$, and $C_2(z)$, we see that we have formed all the nine possible products. To summarize the results systematically, we can arrange them in the annexed tabular form. Note that we have added seven more multiplications, namely, all those in which the identity operation **E** is a factor. The results of these are trivial, since the product of any other, nontrivial operation with **E** must be just the nontrivial operation itself, as indicated.

	E	$C_2(z)$	$\sigma(xz)$	$\sigma(yz)$
E	E	$C_2(z)$	$\sigma(xz)$	$\sigma(yz)$
$C_2(z)$	$C_2(z)$	E	$\sigma(yz)$	$\sigma(xz)$
$\sigma(xz)$	$\sigma(xz)$	$\sigma(yz)$	E	$C_2(z)$
$\sigma(yz)$	$\sigma(yz)$	$\sigma(xz)$	$C_2(z)$	E

The set of operations **E**, $C_2(z)$, $\sigma(xz)$, and $\sigma(yz)$ evidently has the following four interesting properties:

1. There is one operation **E**, the identity, that is the trivial one of making no change. Its product with any other operation is simply the other operation.

2. There is a definition of how to multiply operations: we apply them successively. The product of any two is one of the remaining ones. In other words, this collection of operations is self-sufficient, all its possible products being already within itself. This is sometimes called the property of *closure*.

3. Each of the operations has an *inverse,* that is, an operation by which it may be multiplied to give **E** as the product. In this case, each operation is its own inverse, as shown by the occurrence of **E** in all diagonal positions of the table.

4. It can also be shown that if we form a triple product, this may be subdivided in any way we like without changing the result, thus

$$\sigma(xz) \times \sigma(yz) \times C_2(z)$$
$$= [\sigma(xz) \times \sigma(yz)] \times C_2(z) = C_2(z) \times C_2(z)$$
$$= \sigma(xz) \times [\sigma(yz) \times C_2(z)] = \sigma(xz) \times \sigma(xy)$$
$$= E$$

Products that have this property are said to obey the *associative law* of multiplication.

The four properties just enumerated are of fundamental importance. They are the properties—and the *only* properties—that any collection of symmetry operations must have to constitute a *mathematical group*. Groups consisting of symmetry operations are called *symmetry groups* or sometimes *point groups*. The latter term arises because all the operations leave the molecule fixed at a certain point in space. This is in contrast to other groups of symmetry operations, such as those that may be applied to crystal structures in which individual molecules move from one location to another.

The symmetry group we have just been examining is one of the simpler groups; but nonetheless, an important one. It is represented by the symbol C_{2v}, the origin of this and other symbols is discussed below. It is not an entirely representative group in that it has some properties that are *not* necessarily found in other groups. We have already called attention to one, namely, that each operation in this group is its own inverse; this is actually true of only three kinds of operation: reflections, twofold proper rotations, and inversion **i**. Another special property of the group C_{2v} is that all multiplications in it are *commutative*; that is, every multiplication is equal to the multiplication of the same two operations in the opposite order. It can be seen that the group multiplication table is symmetrical about its main diagonal, which is another way of saying that all possible multiplications commute. In general, multiplication of symmetry operations is *not* commutative, as subsequent discussion will illustrate.

For another simple, but more general, example of a symmetry group, let us recall our earlier examination of the ammonia molecule. We were able to discover six and only six symmetry operations that could be performed on this molecule. If this is indeed a complete list, they should constitute a group. The easiest way to see if they do is to attempt to write a multiplication table. This will contain 36 products, some of which we already know how to write. Thus we know the result of all multiplications involving **E**, and we know that

$$C_3 \times C_3 = C_3^2$$
$$C_3 \times C_3^2 = C_3^2 \times C_3 = E$$

It will be noted that the second of these statements means that C_3 is the inverse of C_3^2 and vice versa. We also know that **E** and each of the σ's is its own inverse. So all operations have inverses, thus satisfying requirement 3.

To continue, we may next consider the products when one σ_v is multiplied by another. A typical example is shown in Fig. 2-5a. When point 0 is reflected first through $\sigma^{(1)}$ and then through $\sigma^{(2)}$, it becomes point 2. But point 2 can obviously also be reached by a clockwise rotation through $2\pi/3$, that is, by the operation C_3. Thus we can write:

$$\sigma^{(1)} \times \sigma^{(2)} = C_3$$

If, however, we reflect first through $\sigma^{(2)}$ and then through $\sigma^{(1)}$, point 0 becomes

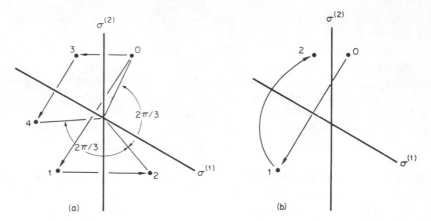

Fig. 2-5. The multiplication of symmetry operations. (a) Reflection times reflection. (b) Reflection followed by C_3.

point 4, which can be reached also by $C_3 \times C_3 = C_3^2$. Thus we write

$$\sigma^{(2)} \times \sigma^{(1)} = C_3^2$$

Clearly the reflections $\sigma^{(1)}$ and $\sigma^{(2)}$ do not commute. The reader should be able to make the obvious extension of the geometrical arguments just used to obtain the following additional products:

$$\sigma^{(1)} \times \sigma^{(3)} = C_3^2$$
$$\sigma^{(3)} \times \sigma^{(1)} = C_3$$
$$\sigma^{(2)} \times \sigma^{(3)} = C_3$$
$$\sigma^{(3)} \times \sigma^{(2)} = C_3^2$$

There remain, now, the products of C_3 and C_3^2 with $\sigma^{(1)}$, $\sigma^{(2)}$, and $\sigma^{(3)}$. Fig. 2-5b shows a type of geometric construction that yields these products. For example, we can see that the reflection $\sigma^{(1)}$ followed by the rotation C_3 carries point 0 to point 2, which could have been reached directly by the operation $\sigma^{(2)}$. By similar procedures all the remaining products can be easily determined. The complete multiplication table for this set of operations is given below.

	E	C_3	C_3^2	$\sigma^{(1)}$	$\sigma^{(2)}$	$\sigma^{(3)}$
E	E	C_3	C_3^2	$\sigma^{(1)}$	$\sigma^{(2)}$	$\sigma^{(3)}$
C_3	C_3	C_3^2	E	$\sigma^{(3)}$	$\sigma^{(1)}$	$\sigma^{(2)}$
C_3^2	C_3^2	E	C_3	$\sigma^{(2)}$	$\sigma^{(3)}$	$\sigma^{(1)}$
$\sigma^{(1)}$	$\sigma^{(1)}$	$\sigma^{(2)}$	$\sigma^{(3)}$	E	C_3	C_3^2
$\sigma^{(2)}$	$\sigma^{(2)}$	$\sigma^{(3)}$	$\sigma^{(1)}$	C_3^2	E	C_3
$\sigma^{(3)}$	$\sigma^{(3)}$	$\sigma^{(1)}$	$\sigma^{(2)}$	C_3	C_3^2	E

The successful construction of this table demonstrates that the set of six operations does indeed form a group. This group is represented by the symbol C_{3v}. The table shows that its characteristics are more general than those of the group C_{2v}. Thus it contains some operations that are not, as well as some which are, their own inverse. It also involves a number of multiplications that are not commutative.

2.3. Some General Rules for Multiplication of Symmetry Operations

In the preceding section several specific examples of multiplication of symmetry operations have been worked out. On the basis of this experience, the following general rules should not be difficult to accept:

1. The product of two proper rotations must be another proper rotation. Thus although rotations can be created by combining reflections [recall: $\sigma(xz) \times \sigma(yz) = C_2(z)$], the reverse is not possible.

2. The product of two reflections in planes meeting at an angle θ is a rotation by 2θ about the axis formed by the line of intersection of the planes [recall: $\sigma^{(1)} \times \sigma^{(2)} = C_3$ for the ammonia molecule].

3. When there is a rotation operation C_n and a reflection in a plane containing the axis, there must be altogether n such reflections in a set of n planes separated by angles of $2\pi/2n$, intersecting along the C_n axis [recall: $\sigma^{(1)} \times C_3 = \sigma^{(2)}$ for the ammonia molecule].

4. The product of two C_2 operations about axes that intersect at an angle θ is a rotation by 2θ about an axis perpendicular to the plane containing the two C_2 axes.

5. The following pairs of operations always commute:
   (a) Two rotations about the same axis.
   (b) Reflections through planes perpendicular to each other.
   (c) The inversion and any other operation.
   (d) Two C_2 operations about perpendicular axes.
   (e) C_n and σ_h, where the C_n axis is vertical.

2-4. A Systematic Listing of Symmetry Groups, with Examples

The symmetry groups to which real molecules may belong are very numerous. However they may be systematically classified by considering how to build them up using increasingly more elaborate combinations of symmetry operations. The outline that follows, though neither unique in its approach nor rigorous in its procedure, affords a practical scheme for use by most chemists.

The simplest nontrivial groups are those of order 2, that is, those containing but one operation in addition to E. The additional operation must be one that is its own inverse; thus the only groups of order 2 are:

$$C_s: E, \sigma$$
$$C_i: E, i$$
$$C_2: E, C_2$$

The symbols for these groups are rather arbitrary, except for C_2 which, we shall soon see, forms part of a pattern.

Molecules with C_s symmetry are fairly numerous. Examples are the thionyl halides and sulfoxides 2-III, and secondary amines 2-IV. Molecules having a center of symmetry as their *only* symmetry element are quite rare; two types are shown as 2-V and 2-VI. The reader should find it very challenging, though not impossible, to think of others. Molecules of C_2 symmetry are fairly common, two examples being 2-VII and 2-VIII.

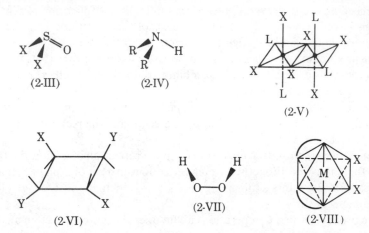

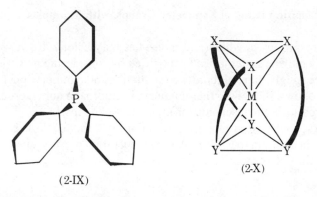

The Uniaxial or C_n Groups. These are the groups in which all operations are due to the presence of a proper axis as the sole symmetry element. The general symbol for such a group, and the operations in it, are

$$C_n: C_n, C_n^2, C_n^3, \ldots C_n^{n-1}, C_n^n \equiv E$$

A C_n group is thus of order n. We have already mentioned the group C_2. Molecules with pure axial symmetry other than C_2 are rare. Two examples of the group C_3 are shown in 2-IX and 2-X.

The C_{nv} Groups. If in addition to a proper axis of order n there is also a set of n vertical planes, we have a group of order $2n$, designated C_{nv}. This type of symmetry is found quite frequently and is illustrated in 2-XI to 2-XV, where the values of n are 2 to 6.

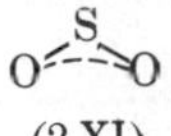

(2-XI)

(2-XII)

(2-XIII)

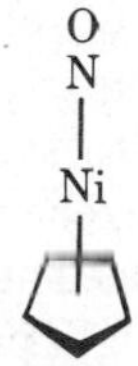

(2-XIV)

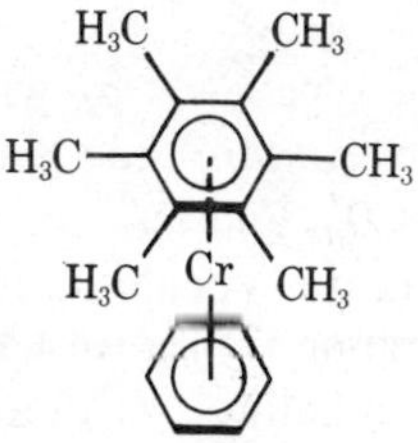

(2-XV)

The C_{nh} Groups. If in addition to a proper axis of order n there is also a horizontal plane of symmetry, we have a group of order $2n$, designated C_{nh}. The $2n$ operations include S_n^m operations that are products of C_n^m and σ_h for n odd, to make the total of $2n$. Thus for C_{3h} the operations are

$$C_3, C_3^2, C_3^3 \equiv E$$
$$\sigma_h$$
$$\sigma_h \times C_3 = C_3 \times \sigma_h = S_3$$
$$\sigma_h \times C_3^2 = C_3^2 \times \sigma_h = S_3^5$$

Molecules of C_{nh} symmetry with $n > 2$ are relatively rare; examples with $n = 2$, 3, and 4 are shown in 2-XVI to 2-XVIII.

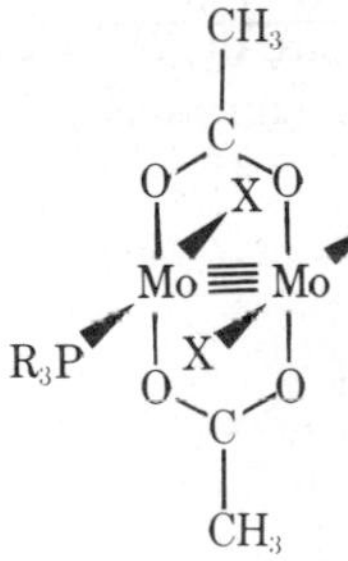

(2-XVI)

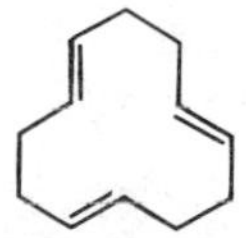

(2-XVII)

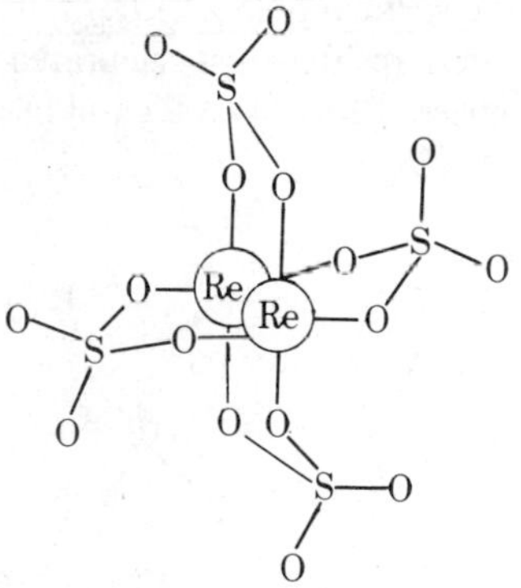

(2-XVIII)

The D_n Groups. When a vertical C_n axis is accompanied by a set of n C_2 axes perpendicular to it, the group is D_n. Molecules of D_n symmetry are, in general, rare, but there is one very important type, namely, the trischelates (2-XIX) of D_3 symmetry.

The D_{nh} Groups. If to the operations making up a D_n group we add reflection in a horizontal plane of symmetry, the group D_{nh} is obtained. It should be noted

(2-XIX)

that products of the type $C_2 \times \sigma_h$ will give rise to a set of reflections in vertical planes. These planes *contain* the C_2 axes; this point is important in regard to the distinction between D_{nh} and D_{nd}, mentioned next. The D_{nh} symmetry is found in a number of important molecules, a few of which are benzene (D_{6h}), ferrocene in an eclipsed configuration (D_{5h}), $Re_2Cl_8^{2-}$, which we examined above, (D_{4h}), $PtCl_4^{2-}$ (D_{4h}), and the boron halides (D_{3h}) and PF_5(D_{3h}). All right prisms with regular polygons for bases as illustrated in 2-XX and 2-XXI, and all bipyramids, as illustrated in 2-XXII and 2-XXIII, have D_{nh}-type symmetry.

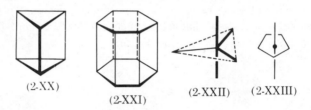

(2-XX) (2-XXI) (2-XXII) (2-XXIII)

The D_{nd} Groups. If to the operations making up a D_n group we add a set of vertical planes that bisect the angles between pairs of C_2 axes (note the distinction from the vertical planes in D_{nh}), we have a group called D_{nd}. The D_{nd} groups have no σ_h. Perhaps the most celebrated examples of D_{nd} symmetry are the D_{3d} and D_{5d} symmetries of $R_3W{\equiv}WR_3$ and ferrocene in their staggered configurations 2-XXIV and 2-XXV.

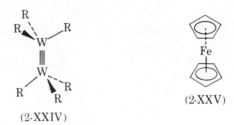

(2-XXIV) (2-XXV)

Two comments about the scheme so far outlined may be helpful. The reader may have wondered why we did not consider the result of adding to the operations of C_n *both* a set of $n\sigma_v$'s and a σ_h. The answer is that this is simply another way of getting to D_{nh}, since a set of C_2 axes is formed along the lines of intersection of the σ_h with each of the σ_v's. By convention, and in accord with the symbols used to designate the groups, it is preferable to proceed as we did. Second, in dealing with the D_{nh} type groups, if a horizontal plane is found, there must be only the n vertical planes *containing* the C_2 axis. If dihedral planes were also present, there would be, in all $2n$ planes and hence, as shown above, a principal axis of order $2n$, thus vitiating the assumption of a D_n type of group.

The S_n Groups. Our scheme has, so far, overlooked one possibility, namely, that a molecule might contain an S_n axis as its only symmetry element (except for others that are directly subservient to it). It can be shown that for n odd, the groups of operations arising would actually be those forming the group C_{nh}. For example, take the operations generated by an S_3 axis:

$$S_3$$
$$S_3^2 \equiv C_3^2$$
$$S_3^3 \equiv \sigma_h$$
$$S_3^4 = C_3$$
$$S_3^5$$
$$S_3^6 \equiv E$$

Comparison with the list of operations in the group C_{3h}, given on page 41, shows that the two lists are identical.

It is only when n is an even number that new groups can arise that are not already in the scheme. For instance, consider the set of operations generated by an S_4 axis:

$$S_4$$
$$S_4^2 \equiv C_2$$
$$S_4^3$$
$$S_4^4 \equiv E$$

This set of operations satisfies the four requirements for a group and is not a set that can be obtained by any procedure previously described. Thus S_4, S_6, etc., are new groups. They are distinguished by the fact that they contain no operation that is not an S_n^m operation, even though it may be written in another way, as with $S_4^2 \equiv C_2$ above.

Note that the group S_2 is not new. A little thought will show that the operation S_2 is identical with the operation $\mathbf{i}$. Hence the group that could be called S_2 is the one we have already called C_i.

An example of a molecule with S_4 symmetry is shown in 2-XXVI. Molecules with S_n symmetries are not very common.

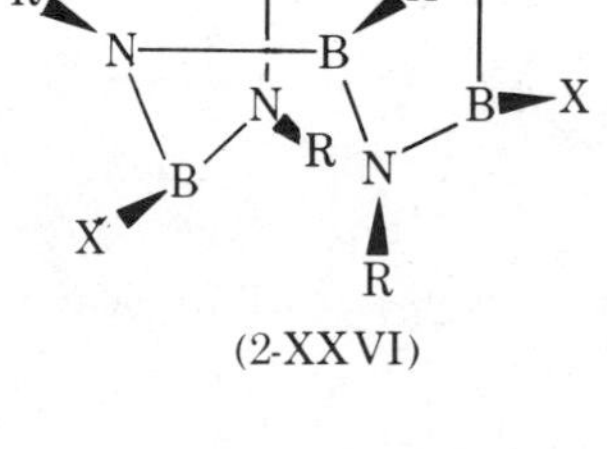

(2-XXVI)

Linear Molecules. There are only two kinds of symmetry for linear molecules. There are those represented by 2-XXVII that have identical ends. Thus in addition to an infinitefold rotation axis C_∞, coinciding with the molecular axis, and an infinite number of vertical symmetry planes, they have a horizontal plane of symmetry and an infinite number of C_2 axes perpendicular to C_∞. The group of these operations is $D_{\infty h}$. A linear molecule with different ends (2-XXVIII), has only C_∞ and the σ_v's as symmetry elements. The group of operations generated by these is called $C_{\infty v}$.

$$A—B—C—B—A \qquad A—B—C—D$$
$$\text{(2-XXVII)} \qquad \text{(2-XXVIII)}$$

2-5. The Groups of Very High Symmetry

The scheme followed in the preceding section has considered only cases in which there is a single axis of order equal to or greater than 3. It is possible to have symmetry groups in which there are several such axes. There are, in fact, seven such groups, and several of them are of paramount importance.

The Tetrahedron. We consider first a regular tetrahedron. Figure 2-6 shows some of the symmetry elements of the tetrahedron, including at least one of each kind. From this it can be seen that the tetrahedron has altogether 24 symmetry operations, which are as follows:

There are three S_4 axes, each of which gives rise to the operations S_4, $S_4^2 \equiv C_2$, S_4^3, and $S_4^4 \equiv E$. Neglecting the S_4^4's, this makes $3 \times 3 = 9$.

There are four C_3 axes, each giving rise to C_3, C_3^2 and $C_3^3 \equiv E$. Again omitting the identity operations, this makes $4 \times 2 = 8$.

There are six reflection planes, only one of which is shown in Fig. 2-6, giving rise to six σ_d operations.

Thus there are $9 + 8 + 6 +$ one identity operation $= 24$ operations. This group is called T_d. It is worth emphasizing that despite the considerable amount of symmetry, there is no inversion center in T_d symmetry. There are, of course, nu-

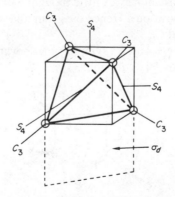

Fig. 2-6. The tetrahedron, showing some of its essential symmetry elements. All S_4 and C_3 axes are shown, but only one of the six dihedral planes σ_d.

merous molecules having full T_d symmetry, such as CH_4, SiF_4, ClO_4^-, $Ni(CO)_4$, and $Ir_4(CO)_{12}$, and many others where the symmetry is less but approximates to it.

If we remove from the T_d group the reflections, it turns out that the S_4 and S_4^3 operations are also lost. The remaining twelve operations (E, four C_3 operations, four C_3^2 operations and three C_2 operations) form a group, designated T. This group in itself has little importance, since it is very rarely, if ever, encountered in real molecules. However if we then add to the operations in the group T a different set of reflections in the three planes defined so that each one contains two of the C_2 axes, and work out all products of operations, we get a new group of 24 operations (E, four C_3, four C_3^2, three C_2, three σ_h, i, four S_6, four S_6^5) denoted T_h. This, too, is rare, but it occurs in some "octahedral" complexes in which the ligands are planar and arranged as in 2-XXIX. The important feature here is that each pair of ligands on each of the Cartesian axes is in a different one of the three mutually perpendicular planes, xy, xz, yz. Real cases are provided by $W(NMe_2)_6$ and several $M(NO_3)_6^{n-}$ ions in which the NO_3^- ions are bidentate.

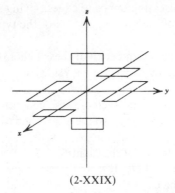

(2-XXIX)

The Octahedron and the Cube. These two bodies have the same elements, as shown in Fig. 2-7, where the octahedron is inscribed in a cube, and the centers of the six cube faces form the vertices of the octahedron. Conversely, the centers of the eight faces of the octahedron form the vertices of a cube. Figure 2-7 shows one

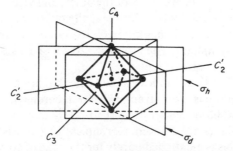

Fig. 2-7. The octahedron and the cube, showing one of each of their essential types of symmetry element.

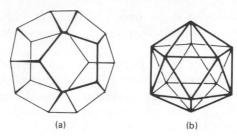

(a)   (b)

Fig. 2-8. The two regular polyhedra having I_h symmetry. (a) The pentagonal dodecahedron. (b) The icosahedron.

of each of the types of symmetry element that these two polyhedra possess. The list of symmetry operations is as follows:

There are three C_4 axes, each generating C_4, $C_4^2 \equiv C_2$, C_4^3, $C_4^4 \equiv E$. Thus there are $3 \times 3 = 9$ rotations, excluding C_4^4's.

There are four C_3 axes giving four C_3's and four C_3^2's.

There are six C_2' axes bisecting opposite edges, giving six C_2''s.

There are three planes of the type σ_h and six of the type σ_d, giving rise to nine reflection operations.

The C_4 axes are also S_4 axes and each of these generates the operations S_4, $S_4^2 \equiv C_2$ and S_4^3, the first and last of which are not yet listed, thus adding $3 \times 2 = 6$ more to the list.

The C_3 axes are also S_6 axes and each of these generates the new operations S_6, $S_6^3 \equiv i$, and S_6^5. The i counts only once, so there are then $(4 \times 2) + 1 = 9$ more new operations.

The entire group thus consists of the identity $+ 9 + 8 + 6 + 9 + 6 + 9 = 48$ operations. This group is denoted O_h. It is, of course, a very important type of symmetry since octahedral molecules (e.g., SF_6), octahedral complexes $[Co(NH_3)_6]^{3+}$, $IrCl_6^{3-}$], and octahedral interstices in solid arrays are very common. There is a group O, which consists of only the 24 proper rotations from O_h, but this, like T, is rarely if ever encountered in Nature.

The Pentagonal Dodecahedron and the Icosahedron.   These bodies (Fig. 2-8) are related to each other in the same way as are the octahedron and the cube, the vertices of one defining the face centers of the other, and vice versa. Both have the same symmetry operations, a total of 120! We shall not list them in detail but merely mention the basic symmetry elements: six C_5 axes; ten C_3 axes, fifteen C_2 axes, and fifteen planes of symmetry. The group of 120 operations is designated I_h and is often called the icosahedral group.

There is as yet no known example of a molecule that is a pentagonal dodecahedron, but the icosahedron is a key structural unit in boron chemistry, occurring in all forms of elemental boron as well as in the $B_{12}H_{12}^{2-}$ ion.

If the symmetry planes are omitted, a group called I consisting of only proper rotations remains. This is mentioned purely for the sake of completeness, since no example of its occurrence in Nature is known.

2-6. Molecular Dissymmetry and Optical Activity

Optical activity, that is, rotation of the plane of polarized light coupled with unequal absorption of the right- and left-circularly polarized components, is a property of a molecule (or an entire three-dimensional array of atoms or molecules) that is not superposable on its mirror image. When the number of molecules of one type exceeds the number of those that are their nonsuperposable mirror images, a net optical activity results. To predict when optical activity will be possible, it is necessary to have a criterion to determine when a molecule and its mirror image will not be identical, that is, superposable.

Molecules that are not superposable on their mirror images are called *dissymmetric*. This term is preferable to "asymmetric," which means "without symmetry," whereas dissymmetric molecules can and often do possess some symmetry, as will be seen.*

A compact statement of the relation between molecular symmetry properties and dissymmetric character is: *a molecule that has no axis of improper rotation is dissymmetric.*

This statement includes and extends the usual one to the effect that optical isomerism exists when a molecule has neither a plane nor a center of symmetry. It has already been noted that the inversion operation i is equivalent to the improper rotation S_2. Similarly, S_1 is a correct although unused way of representing σ, since it implies rotation by $2\pi/1$, equivalent to no net rotation, in conjunction with the reflection. Thus σ and i are simply special cases of improper rotations.

However even when σ and i are absent, a molecule may still be identical with its mirror image if it possesses an S_n axis of some higher order. A good example of this is provided by the $(—RNBX—)_4$ molecule shown in 2-XXVI. This molecule has neither a plane nor a center of symmetry, but inspection shows that it can be superposed on its mirror image. As we have noted, it belongs to the symmetry group S_4.

Dissymmetric molecules either have no symmetry at all, or they belong to one of the groups consisting only of proper rotation operations, that is, the C_n or D_n groups. (Groups T, O, and I are, in practice, not encountered, though molecules in these groups must also be dissymmetric.) Important examples are the bischelate and trischelate octahedral complexes 2-VIII, 2-X, and 2-XIX.

MOLECULAR SYMMETRY

2-7. Coordination Compounds

Historically it has been customary to treat *coordination compounds* as a special class separate from *molecular compounds*. On the basis of actual fact only (i.e.,

---

* Dissymmetry is sometimes called chirality, and dissymmetric chiral, from the Greek word $\chi\epsilon\iota\rho$ for hand, in view of the left-hand/right-hand relation of molecules that are mirror images.

neglecting the purely traditional reasons for such a distinction), there is very little, if indeed any, basis for continuing this dichotomy.

Coordination compounds are conventionally formulated as consisting of a *central atom* or ion surrounded by a set (usually 2 to 9) of other atoms, ions or small molecules, the latter being called *ligands*. The resulting conglomeration is often called a *complex* or, if it is charged, a *complex ion*. The set of ligands need not consist of several small, independent sets of atoms (or single atoms) but may involve fairly elaborate arrangements of atoms connecting those few that are directly bound to—or *coordinated* to—the central atom. However, there are many molecular compounds of which the same description may be given. Consider, for illustration, the following:

$$SiF_4 \qquad SiF_6^{2-} \qquad Cr(CO)_6 \qquad Co(NH_3)_6^{3+}$$
$$SF_6 \qquad PF_6^- \qquad Cr(NH_3)_6^{3+} \qquad CoCl_4^{2-}$$

Conventionally the first two are called molecules and five of the other six are called complexes; $Cr(CO)_6$ can be found referred to in either way depending on the context. Obviously, one basis for the different designations is the presence or absence of a net charge, only uncharged species being called molecules. Beyond this, which is really quite a superficial characteristic as compared with such basic ones as geometric and electronic structures, there is no logical reason for the division. The essential irrelevancy of the question of overall charge is well demonstrated by the fact that $Pt(NH_3)_2Cl_2$, $Cu(acac)_2$, $CoBr_2(Ph_3P)_2$, and scores of similar compounds are quite normally called complexes. The *molecules* are really only *complexes* that happen to have a charge of zero instead of $+n$ or $-m$.

Thus SiF_6^{2-}, PF_6^-, and SF_6 are isoelectronic and isostructural. Although the character of the bonds from the central atom to fluorine atoms doubtless varies from one to another, there is no basis for believing that SF_6 differs more from PF_6^- than the latter does from SiF_6^{2-}.

It might be argued that the terms "complex" or "coordination compound" should be applied only when the central atom, in some oxidation state, and the ligands, can be considered to exist independently, under reasonably normal chemical conditions. Thus Cr^{3+} and NH_3 would be said to so exist. However Cr^{3+} actually exists under normal chemical conditions not as such but as Cr^{3+} (aq) which is, in detail, $Cr(H_2O)_6^{3+}$, another species that would itself be called a complex. Again, in a similar vein, the argument that PF_6^- and SiF_6^{2-} can be considered to consist, respectively, of $PF_5 + F^-$ and $SiF_4 + 2F^-$, whereas there is no comparable breakdown of SF_6, is a poor one; once the set of six fluorine atoms is completed about the central atom, they become equivalent. The possibility of their having had different origins has no bearing on the nature of the final complex.

The terms "coordination compound" and "complex" may therefore be broadly defined to embrace all species, charged or uncharged, in which a central atom is surrounded by a set of outer or ligand atoms.

Having thus defined coordination compounds in a comprehensive way, we can proceed to discuss their structures in terms of only two properties: (1) *coordination number,* the number of outer, or ligand, atoms bonded to the central one, and (2) *coordination geometry,* the geometric arrangement of these ligand atoms and the consequent symmetry of the complex. We shall consider in detail coordination numbers 2 to 9, discussing under each the principal ligand arrangements. Higher coordination numbers will be discussed only briefly as they occur much less frequently.

Coordination Number 2. There are two geometric possibilities, linear and bent. If the two ligands are identical, the general types and their symmetries are: linear, L—M—L, $D_{\infty h}$; bent; L—M—L, C_{2v}. This coordination number is, of course, found in numerous molecular compounds of divalent elements, but is relatively uncommon otherwise. In many cases where stoichiometry might imply its occurrence, a higher coordination number actually occurs because some ligands form "bridges" between two central atoms. In terms of the more conventional types of coordination compound—those with a rather metallic element at the center—it is restricted mainly to some complexes of Cu^I, Ag^I, Au^I, and Hg^{II}. Such complexes have linear arrangements of the metal ion and the two ligand atoms, and typical ones are $[ClCuCl]^-$, $[H_3NAgNH_3]^+$, $[ClAuCl]^-$, and $[NCHgCN]$. The metal atoms in cations such as $[UO_2]^{2+}$, $[UO_2]^+$, and $[PuO_2]^{2+}$, which are linear, may also be said to have coordination number 2, but these oxo cations interact fairly strongly with additional ligands and their actual coordination numbers are much higher; it is true, however, that the central atoms have a specially strong affinity for the two oxygen atoms. Linear coordination also occurs in the several trihalide ions, such as I_3^- and $ClBrCl^-$.

Coordination Number 3.[1a] The two most symmetrical arrangements are planar (2-XXX) and pyramidal (2-XXXI), with D_{3h} and C_{3v} symmetry, respectively. Both these arrangements are found often among molecules formed by trivalent central elements. Among complexes of the metallic elements this is a rare coordination number; nearly all compounds or complexes of metal cations with stoichiometry MX_3 have structures in which sharing of ligands leads to a coordination for M that exceeds three. There are, however, a few exceptions, such as the planar HgI_3^- ion that occurs in $[(CH_3)_3S^+][HgI_3^-]$, the MN_3 groups that occur in $Cr(NR_2)_3$ and $Fe(NR_2)_3$, where R = $(CH_3)_3Si$, and the CuS_3 group found in $Cu[SC(NH_2)_2]_3Cl$ and $Cu(SPPh_3)_3ClO_4$.

In a few cases (e.g., ClF_3 and BrF_3), a T-shaped form (2-XXXII) of three-coordination (symmetry C_{2v}) is found.

(2-XXX)　　　　(2-XXXI)　　　　(2-XXXII)

Coordination Number 4. This is a highly important coordination number, occurring in hundreds of thousands of compounds, including, *inter alia,* most of those formed by the element carbon, essentially all those formed by silicon, germanium, and tin, and many compounds and complexes of other elements. There are three

[1a] P. G. Eller, D. C. Bradley, M. B. Hursthouse, and D. W. Meek, *Coord. Chem. Rev.,* 1977, **24,** 1 (a comprehensive review with 306 references).

principal geometries. By far the most prevalent is tetrahedral geometry (2-XXXIII), which has symmetry T_d when ideal. Tetrahedral complexes or molecules are almost the only kind of four-coordinate ones formed by non-transition elements; whenever the central atom has no electrons in its valence shell orbitals except the four pairs forming the σ bonds to ligands, these bonds are disposed in a tetrahedral fashion. With many transition metal complexes, square geometry (2-XXXIV) occurs be-

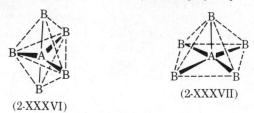

cause of the presence of additional valence shell electrons and orbitals (i.e., partially filled d orbitals), although there are also many tetrahedral complexes formed by the transition metals. In some cases (e.g., with Ni^{II}, Co^{II}, and Cu^{II} in particular), there may be only a small difference in stability between the tetrahedral and the square arrangement and rapid interconversions may occur.

Square complexes are also found with nontransitional central atoms when there are two electron pairs present beyond the four used in bonding; these two pairs lie above and below the plane of the molecule. Examples are XeF_4 and $(ICl_3)_2$. Similarly, when there is one "extra" electron pair, as in SF_4, the irregular arrangement, of symmetry C_{2v} (2-XXXV) is adopted. More detailed discussions of these non-transition element structures will be found in Chapter 5.

Coordination Number 5. Though less common than numbers 4 and 6, coordination number 5 is still very important.[1b] There are two principal geometries, and these may be conveniently designated by stating the polyhedra that are defined by the set of ligand atoms. In one case the ligand atoms lie at the vertices of a trigonal bipyramid (*tbp*) (2-XXXVI), and in the other at the vertices of a square pyramid (*sp*) (2-XXXVII). The *tbp* belongs to the symmetry group D_{3h}; the *sp* belongs to the group C_{4v}. It is interesting and highly important that these two

structures are similar enough to be interconverted without great difficulty. Moreover, a large fraction of the known five-coordinate complexes have structures that are intermediate between these two prototype structures. This ready deformability and interconvertibility gives rise to one of the most important types of stereochemical nonrigidity (cf. Section 28-13).

[1b] E. L. Muetterties and R. A. Schunn, *Q. Rev.*, 1966, **20**, 245; B. F. Hoskins and F. D. Whillan, *Coord. Chem. Rev.*, 1973, **11**, 343.

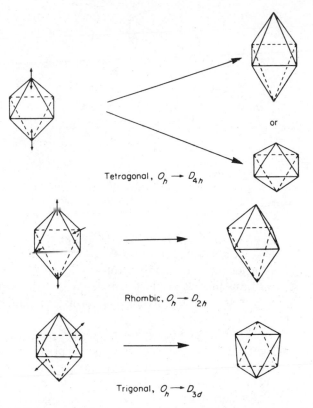

Tetragonal, $O_h \rightarrow D_{4h}$

or

Rhombic, $O_h \rightarrow D_{2h}$

Trigonal, $O_h \rightarrow D_{3d}$

Fig. 2-9. The three principal types of distortion found in real octahedral complexes.

On the whole the *tbp* seems to be somewhat more common than the *sp*, but there is no general predictive rule. For example the $[MCl_5]^{3-}$ ions (M = Cu, Cd, Hg) are *tbp*, but $[InCl_5]^{2-}$ and $[TlCl_5]^{2-}$ are *sp*,[2a] and there is one compound that contains both *tbp* and *sp* $[Ni(CN)_5]^{3-}$ ions in the same crystal.

There is one reported case of pentagonal planar coordination,[2b] in $[Te(S_2\text{-}COEt)_3]^-$ where two ligands are bidentate and one monodentate, but this unusual arrangement seems to be due to the presence of two stereochemically active lone pairs (see Chapter 5 for explanation).

Coordination Number 6. This is perhaps the most common coordination number, and the six ligands almost invariably lie at the vertices of an octahedron or a distorted octahedron. The very high symmetry, group O_h, of the regular octahedron, has been discussed in detail on page 45.

There are three principal forms of distortion of the octahedron. One is *tetragonal,* elongation or contraction along a single C_4 axis; the resultant symmetry is only D_{4h}. Another is *rhombic,* changes in the lengths of two of the C_4 axes so that no two are equal; the symmetry is then only D_{2h}. The third is a *trigonal* distortion, elongation

[2a] W. Clegg, D. A. Greenhalgh, and B. P. Straughan, *J. C. S. Dalton,* **1975,** 2591.
[2b] B. F. Hoskins and D. C. Pannan, *Aust. J. Chem.,* 1976, **29,** 2337.

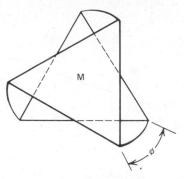

Fig. 2-10. A six-coordinate structure intermediate between the trigonal prism and antiprism projected down the 3-fold axis. The twist angle ϕ is measured in the plane of projection.

or contraction along one of the C_3 axes so that the symmetry is reduced to D_{3d}. These three distortions are illustrated in Fig. 2-9.

The tetragonal distortion most commonly involves an elongation of one C_4 axis and, in the limit, two trans ligands are lost completely, leaving a square, four-coordinate complex. The trigonal distortion transforms the octahedron into a trigonal antiprism.

Another type of six-coordinate geometry, much rarer but nonetheless important, is that in which the ligands lie at the vertices of a trigonal prism (2-XX); the ideal symmetry is D_{3h}. This arrangement has never been observed in a discrete ML_6 complex but only in complexes with chelating ligands and in a few metal sulfides, namely, MoS_2 and WS_2, where it was first seen many years ago, and more recently in MM'_3S_6 (M = Mn, Fe, Co, Ni; M' = Nb, Ta). The chelate complexes that best exemplify this type of coordination contain the 1,2-dithiolene or 1,2-diselenolene type ligands, RC(S)—C(S)R, RC(Se)—C(Se)R, discussed in Section 4-34, although a few other examples are known.[3]

It is important to recognize that there exists a continuous range of structures between the trigonal prism (2-XX) and the trigonal antiprism (Fig. 2-9). This can best be considered in terms of the twist angle ϕ, shown in Fig. 2-10, which is 0° for the prism and 60° for the antiprism. The drawing represents schematically a tris-chelate complex with an intermediate configuration. While there are some tris-(dithiolene)metal complexes with $\phi = 0$, there are others with angles between 0° and 60°. The tropolonato anion (Section 4-27) is another bidentate ligand prone to give intermediate angles. To explain the range of ϕ values observed, many factors have been considered.[4,5] With few if any exceptions, the metal ion itself will prefer $\phi = 60°$, but geometrical and electronic properties of the ligand can override this. If the ligand has a short "bite" (i.e., a short distance between the two atoms bonded to the metal), low ϕ is favored. It is also believed that direct interactions between the sulfur atoms in certain 1,2-dithiolene complexes favor low ϕ. There are a few

[3]   A. A. Diamantis, M. R. Snow, and J. A. Vanzo, *J. C. S. Chem. Comm.*, **1976**, 264.
[4]   A. Avdeef and J. P. Fackler, Jr., *Inorg. Chem.*, 1975, **14**, 2002.
[5]   M. Cowie and M. J. Bennett, *Inorg. Chem.*, 1976, **15**, 1596.

cases in which the ligand is all one hexadentate unit so designed as to impose a small or zero ϕ angle.[6]

Coordination Number 7.[7,8] There are three important geometrical arrangements: the pentagonal bipyramid (2-XXXVIII) of symmetry D_{5h}; the capped octahedron (2-XXXIX), symmetry C_{3v}, obtained by adding to an octahedral set a seventh li-

(2-XXXVIII)

(2-XXXIX)

(2-XL)

gand over the center of one face, which then becomes enlarged; and the capped trigonal prism (2-XL), symmetry C_{2v}, obtained by placing the seventh ligand over one of the rectangular faces of the trigonal prism. The available structural data suggest that Nature does not particularly favor any one of these except where a bias might be built into a polydentate ligand, and theory likewise implies that all three will, in general, have similar stability. Moreover, theory suggests that interconversions can occur without difficulty, so that seven-coordinate complexes should, in general, be stereochemically nonrigid.

Coordination Number 8.[9a,9b] It is conceptually convenient to begin with the most symmetrical polyhedron having eight vertices, namely, the cube, which has O_h symmetry. Cubic coordination occurs only rarely in discrete complexes,[9c] especially in those of uranium and some other actinides, although it occurs in various solid compounds where the anions form continuous arrays, as in the CsCl structure. Its occurrence is infrequent, presumably, because there are several ways in which the cube may be distorted so as to lessen repulsions between the X atoms while maintaining good M—X interactions.

The two principal ways in which the cube may become distorted are shown in Fig. 2-11. The first of these, rotation of one square face by 45° relative to the one opposite to it, lessens repulsions between nonbonded atoms while leaving M—X distances unaltered. The resulting polyhedron is the square antiprism (symmetry D_{4d}). It has square top and bottom and eight isosceles triangles for its vertical faces. The second distortion shown can be best comprehended by recognizing that the cube is composed of two interpenetrating tetrahedra. The distortion occurs when the vertices of one of these tetrahedra are displaced so as to decrease the two vertical angles, that is, to elongate the tetrahedron, while the vertices of the other one are displaced to produce a flattened tetrahedron. The resulting polyhedron is called a dodecahedron, or more specifically, to distinguish it from several other kinds of

6   P. B. Donaldson, P. A. Tasker, and N. W. Alcock, *J. C. S. Dalton,* **1977,** 1160.
7   R. Hoffmann, B. F. Beier, E. L. Muetterties, and A. R. Rossi, *Inorg. Chem.,* 1977, **16,** 511.
8   M. G. B. Drew, *Prog. Inorg. Chem.,* 1977, **23,** 67.
9a M. G. B. Drew, *Coord. Chem. Rev.,* 1977, **24,** 179 (covers coordination numbers 8 to 14, with 295 references).
9b J. Burdett, R. Hoffmann, and R. C. Fay, *Inorg. Chem.,* 1978, **17,** 2553.
9c A. R. Al-Karaghouli, R. O. Day, and J. S. Wood, *Inorg. Chem.,* 1978, **17,** 3702.

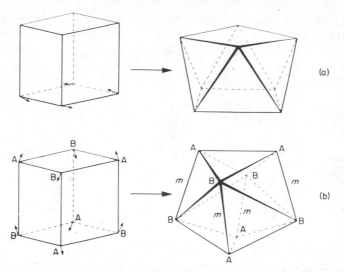

Fig. 2-11. The two most important ways of distorting the cube: (a) to produce a square antiprism; (b) to produce a dodecahedron.

dodecahedron, a triangulated dodecahedron. It has D_{2d} symmetry, and it is important to note that its vertices are not all equivalent but are divided into two bisphenoidal sets, those within each set being equivalent.

Detailed analysis of the energetics of M—X and X—X interactions suggests that there will in general be little difference between the energies of the square antiprism and the dodecahedral arrangement, unless other factors, such as the

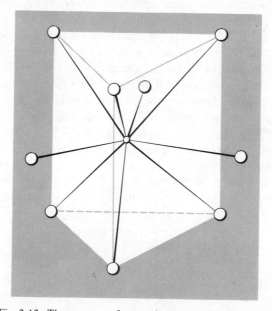

Fig. 2-12. The structure of many nine-coordinate complexes.

existence of chelate rings, energies of partially filled inner shells, exceptional opportunities for orbital hydridization, or the like, come into play. Both arrangements occur quite commonly, and in some cases [e.g., the $M(CN)_8^{n-}$ (M = Mo or W; n = 3 or 4) ions] the geometry varies from one kind to the other with changes in the counterion in crystalline salts and on changing from crystalline to solution phases.

A form of eight-coordination, which is a variant of the dodecahedral arrangement, is found in several compounds containing bidentate ligands in which the two coordinated atoms are very close together (ligands said to have a small "bite"), such as NO_3^- and O_2^{2-}. In these, the close pairs of ligand atoms lie on the m edges of the dodecahedron (see Fig. 2-11b); these edges are then very short. Examples of this are the $Cr(O_2)_4^{3-}$ and $Co(NO_3)_4^{2-}$ ions and the $Ti(NO_3)_4$ molecule.

Three other forms of octacoordination, which occur much less often and are essentially restricted to actinide and lanthanide compounds, are the hexagonal bipyramid (D_{6h}) (2-XLI), the bicapped trigonal prism (D_{3h}) (2-XLII) and the

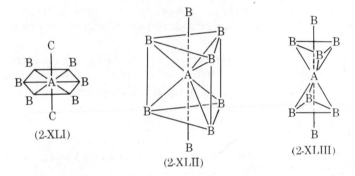

(2-XLI)   (2-XLII)   (2-XLIII)

bicapped trigonal antiprism (D_{3d}) (2-XLIII). The hexagonal bipyramid is restricted almost entirely to the oxo ions, where an OMO group defines the axis of the bipyramid, though it is occasionally found elsewhere.

Coordination Number 9.[9a,10] There are two regular structures, the tricapped trigonal prism (Fig. 2-12) and a capped square antiprism (Fig. 2-11a), where the ninth ligand lies over one square face. In both complexes and clusters (Section 2-8) the former is far more common; only three examples of the capped square antiprism have been reported, including one in the compound $RbTh_3F_{13}$ where both types of nine-coordinate polyhedra occur. Examples of coordination number 9 are restricted to lanthanide and actinide complexes and to $[ReH_9]^{2-}$, and the majority are polymeric, not discrete. Specific examples of the tricapped trigonal prism are $[ReH_9]^{2-}$ and the $[M(H_2O)_9]^{3+}$ ions with lanthanide ions. The rarer capped square antiprism occurs in $[ThF_8]^{4-}$ and the Cl-bridged $[LaCl(H_2O)_7]_2^{4+}$.

Higher Coordination Numbers.[9a] These are rare and found only for the largest metal ions; the geometry is usually not easily, if at all, classifiable in terms of simple prototypal shapes. Ten-coordinate species include $K_4Th(O_2CCO_2)_4 \cdot 4H_2O$, which has a bicapped square antiprism, and several others with capped dodecahedral geometry. Only two eleven-coordinate complexes have been clearly delineated. All

10   L. J. Guggenberger and E. L. Muetterties, *J. Am. Chem. Soc.*, 1976, **98**, 7221.

but one of the known cases of twelve-coordination comprise $M(LL)_6$ species with distorted icosahedral arrangements. Examples are $[Ce(NO_3)_6]^{2-}$ and other $[M(NO_3)_6]^{2-}$ ions and $[Pr(naph)_6]^{3+}$, where naph is 1,8-naphthyridine.

2-8. Cage and Cluster Structures[11]

The formation of polyhedral cages and clusters is now recognized as an important and widespread phenomenon, and examples may be found in nearly all parts of the Periodic Table. This section mentions each of the principal polyhedra and gives illustrations. Further details may be found under the chemistry of the particular elements and in the sections on metal atom clusters (Chap. 26) and polynuclear metal carbonyls (Section 25-1).

A cage or cluster is in a certain sense the antithesis of a complex; yet there are many similarities due to common symmetry properties. In each type of structure a set of atoms defines the vertices of a polyhedron, but in the one case—the complex—these atoms are each bound to one central atom and not to each other, whereas in the other—the cage or cluster—there is no central atom and the essential feature is a system of bonds connecting each atom directly to its neighbors in the polyhedron.

There are, however, some examples of clusters that also have a central atom, usually carbon. Thus they combine the features of both complexes and clusters, although the bonding between atoms in the polyhedron appears to be stronger and more important than bonds from outer atoms to the central atoms. Examples of these "filled clusters" are $Ru_6(CO)_{17}C$, $Co_6(CO)_{14}C^-$, and $Fe_5(CO)_{15}C$. The first two have an octahedral set of metal atoms with a C atom at the center, and the iron compound has a square pyramidal set of iron atoms with a C atom approximately in the basal plane (2-XLIV).

To a considerable extent the polyhedra found in cages and clusters are the same as those adopted by coordination compounds (e.g., the tetrahedron, trigonal bipyramid, octahedron), but there are also others (see especially the polyhedra with six vertices), and cages with more than six vertices are far more common than coordination numbers greater than 6. It should be noted that triangular clusters, as in $[Re_3Cl_{12}]^{3-}$ or $Os_3(CO)_{12}$, though not literally polyhedra, are not essentially different from polyhedral species such as $Mo_6Cl_8^{4+}$ or $Ir_4(CO)_{12}$, respectively.

Just as all ligand atoms in a set need not be identical, so the atoms making up a cage or cluster may be different; indeed, to exclude species made up of more than one type of atom would be to exclude the majority of cages and clusters, including some of the most interesting and important ones.

Four Vertices. Tetrahedral cages or clusters have long been known for the P_4,

[11]   R. B. King, in *Progress in Inorganic Chemistry,* Vol. 15, S. J. Lippard, Ed., Wiley-Interscience, 1972 (transition metal cluster compounds); P. Chini, G. Longoni, and V. G. Albano, in *Advances in Organometallic Chemistry,* Vol. 14, F. G. A. Stone and R. West, Eds., Academic Press, 1976 (review of metal clusters of the metal carbonyl type); J. D. Corbett, in *Progress in Inorganic Chemistry,* Vol. 21, S. J. Lippard, Ed., Wiley-Interscience, 1976 (cluster ions formed by main group metals); M. G. B. Drew, *Coord. Chem. Rev.,* 1977, **24,** 179 (structures of high coordination number complexes); K. Wade, *Adv. Inorg. Chem. Radiochem.,* 1976, **18,** 1 (an authoritative treatment of structural bonding patterns in cluster chemistry).

As$_4$, and Sb$_4$ molecules and in more recent years have been found in polynuclear metal carbonyls such as Co$_4$(CO)$_{12}$, Ir$_4$(CO)$_{12}$, [η^5-C$_5$H$_5$Fe(CO)]$_4$,* RSiCo$_3$-(CO)$_9$, Fe$_4$(CO)$_{13}^{2-}$, Re$_4$(CO)$_{12}$H$_4$, and a number of others; B$_4$Cl$_4$ is another well-known example and doubtless many more will be encountered.

Five Vertices. Polyhedra with five vertices are the trigonal bipyramid (*tbp*) and the square pyramid (*sp*). Both are found among the boranes and carbaboranes (e.g., the *tbp* in B$_3$C$_2$H$_5$ and the *sp* in B$_5$H$_9$), as well as among the transition elements. Examples of the latter are the *tbp* cluster, Pt$_3$Sn$_2$, in (C$_8$H$_{12}$)$_3$Pt$_3$(SnCl$_3$)$_2$ and the *sp* clusters in Fe$_5$(CO)$_{15}$C (2-XLIV) and the Fe$_3$(CO)$_9$E$_2$ (E = S or Se) species (2-XLV). The Os$_5$(CO)$_{16}$ molecule provides a recent example of a *tbp*.

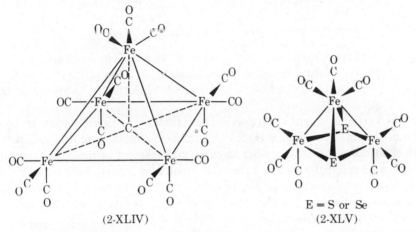

(2-XLIV)

E = S or Se
(2-XLV)

Six Vertices. Octahedral clusters and cages are rather numerous. Several are polynuclear metal carbonyls, such as Rh$_6$(CO)$_{16}$ and [Co$_6$(CO)$_{14}$]$^{4-}$.

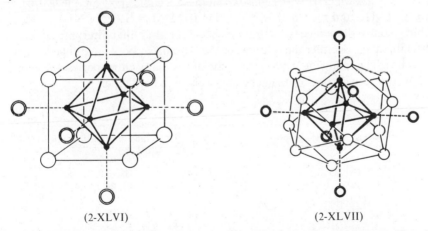

(2-XLVI)   (2-XLVII)

There is also an extensive series of metal atom cluster compounds formed by niobium, tantalum, molybdenum, and tungsten, all based on octahedral sets of metal atoms, the principal types being of M$_6$X$_8$ (2-XLVI) and M$_6$X$_{12}$ (2-XLVII) stoi-

* The η^5 is a symbol denoting the number of atoms (5 C) involved in π-bonding to the metal.

chiometry. The $B_6H_6^{2-}$ and $B_4C_2H_6$ species are also octahedral. The borane B_6H_{10}, however, has a pentagonal pyramid of boron atoms; B_6 octahedra also occur in the class of borides of general formula MB_6.

Still other, less regular geometries also occur—for example, a capped sp in $H_2Os_6(CO)_{18}$ (2-XLVIII) and a bicapped tetrahedron in $Os_6(CO)_{18}$ (2-XLIX).

• = Os(CO)$_3$
(2-XLVIII)

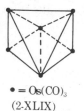

• = Os(CO)$_3$
(2-XLIX)

Seven Vertices. Such polyhedra are relatively rare. The isoelectronic $B_7H_7^{2-}$ and $B_5C_2H_7$ species have pentagonal bipyramidal (D_{5h}) structures. The $Os_7(CO)_{21}$ molecule has a capped octahedron of Os atoms with three CO groups on each Os.

Eight Vertices. Eight-atom polyhedral structures are very numerous. By far the most common polyhedron is the *cube*; this is in direct contrast to the situation with eightfold coordination, where a cubic arrangement of ligands is extremely rare because it is disfavored relative to the square antiprism and the triangulated dodecahedron in which ligand-ligand contacts are reduced. In the case of a cage compound, of course, it is the structure in which contacts between atoms are maximized that will tend to be favored (provided good bond angles can be maintained), since bonding rather than repulsive interactions exist between neighboring atoms.

The only known cases with eight *like* atoms in a cubic array are the hydrocarbon *cubane*, C_8H_8, and the $Cu_8(i\text{-MNT})_6^{4-}$ ion [i-MNT = $S_2CC(CN)_2^{2-}$]. The other cubic systems all involve two different species of atom that alternate as shown in (2-L). In all cases either the A atoms or the B atoms or both have appended atoms or groups. The following list collects some of the many cube species, the elements at the alternate vertices of the cube being given in bold type.

A (and appended groups)	B (and appended groups)
Mn(CO)$_3$	SEt
Os(CO)$_3$	O
PtMe$_3$ or **Pt**Et$_3$	Cl, Br, I, OH
CH$_3$**Zn**	OCH$_3$
Tl	OCH$_3$
η^5-C$_5$H$_5$**Fe**	S
Me$_3$As**Cu**	I
Ph**Al**	NPh
Co(CO)$_3$	Sb
FeSR	S

(2-L)

Although the polyhedron in cubane, or in any other $(AR)_6$ molecule, may have the full O_h symmetry of a cube, the A_4B_4-type structures can have at best tetrahedral, T_d, symmetry since they consist of two interpenetrating tetrahedra.

It must also be noted that only when the two interpenetrating tetrahedra happen to be exactly the same size will all the ABA and BAB angles be equal to 90°. Since the A and the B atoms differ, it is not in general to be expected that this will occur. In fact, there is, in principle, a whole range of bonding possibilities. At one extreme, represented by $[(\eta^5\text{-}C_5H_5)Fe(CO)]_4$, the members of one set of atoms (the Fe atoms) are so close together that they must be considered to be directly bonded, whereas the other set (the C atoms of the CO groups) are not at all bonded among themselves but only to those in the first set. In this extreme, it seems best to classify the system as having a tetrahedral cluster (of Fe atoms) supplemented by bridging CO groups.

At the other extreme are the A_4B_4 systems in which all A—A and B—B distances are too long to admit of significant A—A or B—B bonding; thus the system can be regarded as genuinely cubic (even if the angles differ somewhat from 90°). This is true of all the systems listed above. However the atoms in the smaller of the two tetrahedra tend to have some amount of direct interaction with one another, thus blurring the line of demarcation between the "cluster" and "cage" types.

Of the systems listed above those with RSFe and S groups are especially noteworthy because they occur in many of the biological electron carriers called ferredoxins (see Section 31-5 for details).

A relatively few species are known in which the polyhedron is, at least approximately, a triangulated dodecahedron (Fig. 2-11b). These are the boron species $B_8H_8^{2-}$, $B_6C_2H_8$, and B_8Cl_8.

Nine Vertices. Cages with nine vertices are rare. Representative ones are Bi_9^{5+} (in $B_{24}Cl_{26}$), $B_9H_9^{2-}$, and $B_7C_2H_9$, all of which have the tricapped trigonal prism structure (Fig. 2-12), and Sn_9^{4-}, which is a square antiprism capped on one square face[12].

Ten Vertices. Species with ten vertices are well known. In $B_{10}H_{10}^{2-}$ and $B_8C_2H_{10}$ the polyhedron 2-LI is a square antiprism capped on the square faces (symmetry D_{4d}). But there is a far commoner structure for 10-atom cages that is commonly called the *adamantane* structure after the hydrocarbon adamantane ($C_{10}H_{16}$), which has this structure; it is depicted in 2-LII and consists of two subsets of atoms: a set of four (A) that lie at the vertices of a tetrahedron and a set of six (B) that lie at the vertices of an octahedron. The entire assemblage has the T_d symmetry of the tetrahedron. From other points of view it may be regarded as a tetrahedron with

[12]   J. D. Corbett and P. A. Edwards, *J. Am. Chem. Soc.,* 1977, **99**, 3313.

(2-LI)

(2-LII)

a bridging atom over each edge or as an octahedron with a triply bridging atom over an alternating set of four of the eight triangular faces.

The adamantane structure is found in dozens of A_4B_6-type cage compounds formed mainly by the main group elements. The oldest recognized examples of this structure are probably the phosphorus(III) and phosphorus(V) oxides, in which we have P_4O_6 and $(OP)_4O_6$, respectively. Other representative examples include $P_4(NCH_3)_6$, $(OP)_4(NCH_3)_6$, $As_4(NCH_3)_6$, and $(MeSi)_4S_6$.
include $P_4(NCH_3)_6$, $(OP)_4(NCH_3)_6$, $As_4(NCH_3)_6$, and $(MeSi)_4S_6$.

Eleven Vertices.   Perhaps the only known eleven-atom cages are $B_{11}H_{11}^{2-}$ and $B_9C_2H_{11}$.

Twelve Vertices.   Twelve-atom cages are not widespread but play a dominant role in boron chemistry. The most highly symmetrical arrangement is the icosahedron (Fig. 2-8b), which has twelve equivalent vertices and I_h symmetry. Icosahedra of boron atoms occur in all forms of elemental boron, in $B_{12}H_{12}^{2-}$, and in the numerous carboranes of the $B_{10}C_2H_{12}$ type. A related polyhedron, the cuboctahedron (2-LIII) is found in several borides of stoichiometry MB_{12}.

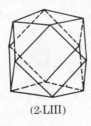

(2-LIII)

General References

Cotton, F. A., *Chemical Applications of Group Theory,* 2nd ed., Wiley-Interscience, 1971.

Wells, A. F., *Structural Inorganic Chemistry,* 4th ed., Clarendon Press, 1975.

Wyckoff, R. W. G., *Crystal Structures,* 2nd ed., Vols. 1–5, Wiley, 1963–1966. Encyclopedic and critical collection of structural data obtained by crystallography.

CHAPTER THREE

Introduction to Ligands and Complexes

GENERAL REMARKS

3.1 Introduction

It was known long before the Danish chemist S. M. Jørgensen (1837–1914) began his extensive studies on the synthesis of such "complex" compounds that the metal halides and other salts could give compounds with neutral molecules and that many of these compounds could easily be formed in aqueous solutions. The recognition of the true nature of "complexes" began with Alfred Werner (1866–1919) as set out in his classic work *Neuere Anschauungen auf dem Gebiete der anorganischen Chemie* (1906)[1]; he received the Nobel Prize for this work in 1913. Werner showed that neutral molecules were bound directly to the metal so that complex salts such as $CoCl_3 \cdot 6NH_3$ were correctly formulated $[Co(NH_3)_6]^{3+}Cl_3^-$. He also demonstrated that there were profound stereochemical consequences of the assumption that the molecules or ions (ligands) around the metal occupied positions at the corners of an octahedron or a square. The stereochemical studies of Werner were later followed by the ideas of G. N. Lewis and N. V. Sidgwick, who proposed that a chemical bond required the sharing of an electron pair. This led to the idea that a neutral molecule with an electron pair (Lewis base) can donate these electrons to a metal ion or other electron acceptor (Lewis acid). Well-known examples are the following:

---

[1]   Braunschweig, 1905; English translation of second edition as *New Ideas on Inorganic Chemistry*, by E. P. Hedley, London, 1911.

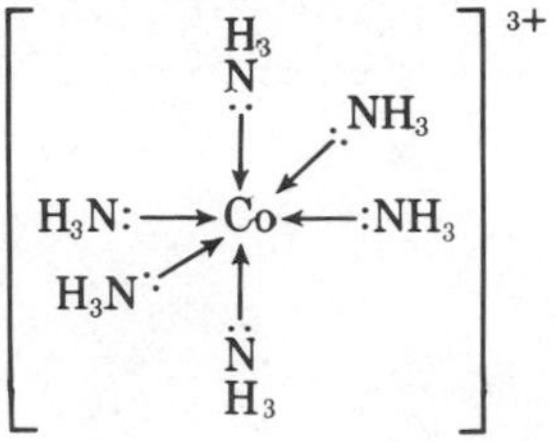

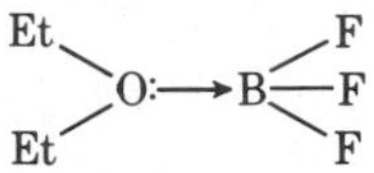

$$
\left[
\begin{array}{c}
\text{H}_3 \\
\text{N} \\
\vdots \\
\text{H}_3\text{N:} \rightarrow \text{Co} \leftarrow \text{:NH}_3 \\
\text{H}_3\text{N} \\
\vdots \\
\text{N} \\
\text{H}_3
\end{array}
\right]^{3+}
\qquad
\begin{array}{c}
\text{Et} \\
  \text{O:} \rightarrow \text{B} \\
\text{Et}
\end{array}
\begin{array}{c}
\text{F} \\
\text{F} \\
\text{F}
\end{array}
$$

We can now define a *ligand* as any molecule or ion that has at least one electron pair that can be so donated. Ligands may also be called Lewis bases; in the terms used in organic chemistry, they are nucleophiles. Metal ions or molecules such as BF_3 with incomplete valence electron shells are Lewis acids or electrophiles.

Although it is possible to regard even covalent compounds from the donor-acceptor point of view—for example, we could regard methane (CH_4) as composed of C^{4+} and four H^- ions—it is not a particularly profitable or realistic way of looking at such molecules. Nevertheless, in inorganic chemistry, ions such as H^-, F^-, Cl^-, NO_3^-, and SO_4^{2-}, and groups such as CH_3^- and $C_6H_5^-$, are commonly termed ligands even when they are bound in simple molecules by largely covalent bonds as in SF_6 or $W(CH_3)_6$. Although SiF_4 is normally called a molecule and SiF_6^{2-} a complex anion, the nature of the Si—F bonds in each species is essentially the same (See Section 2-7.).

There are innumerable ways of classifying ligands. One useful approach is based on the type of bonding interaction between the central atom and its surrounding neighbor atoms. The bonding details are dealt with later, but the distinction between two major types of ligand can be illustrated by posing two questions.

1. Why do molecules like water or ammonia give complexes with ions of *both* main group and transition metals—for example, $[Al(OH_2)_6]^{3+}$ or $[Co(NH_3)_6]^{3+}$, while other types of molecules such as PF_3 or CO give complexes only with transition metals?

2. Although PF_3 and CO give neutral complexes such as $Ni(PF_3)_4$ or $Cr(CO)_6$, why do NH_3, amines, oxygen compounds, and so on, not give complexes such as $Ni(NH_3)_4$?

There are two main classes of ligands:

(*a*) *Classical* or *simple donor ligands* act as electron-pair donors to acceptor ions or molecules, and form complexes with all types of Lewis acids, metal ions, or molecules.

(*b*) *Nonclassical ligands, π-bonding* or *π-acid ligands*, form compounds largely if not entirely with transition metal atoms. This interaction occurs because of the special properties of both metal and ligand. The metal has *d* orbitals that can be utilized in bonding; the ligand has not only *donor* capacity but also has *acceptor* orbitals. This latter distinction is perhaps best illustrated by comparison of an amine, $:NR_3$, with a tertiary phosphine, $:PR_3$. Both can act as bases toward H^+, but the P atom differs from N in that it has vacant $3d$ orbitals of low energy, whereas in

N the lowest energy d orbitals are at far too high an energy to use. Another example is that of CO, which has no measurable basicity to protons, yet readily reacts with metals like nickel that have high heats of atomization to give compounds like $Ni(CO)_4$.

Ligands may also be classified electronically, that is, according to the number of electrons that they contribute to a central atom when these ligands are regarded (sometimes artificially) as neutral species. Thus atoms or groups that can form a single covalent bond are regarded as *one-electron* donors—examples are F, SH, and CH_3. Any compound with an electron pair is a *two-electron donor* (e.g., $:NH_3$, $H_2O:$). Groups that can form a single bond and at the same time donate can be considered to be *three-electron donors*. For example, the acetate ion can be either a one- or a three-electron donor, namely,

A molecule with two electron pairs (e.g., $H_2NCH_2CH_2NH_2$) can be regarded as a *four-electron donor,* and so on. This classification method is useful in that it is an aid in electron counting, particularly for transition metal complexes whose valence shells contain eighteen electrons and whose stoichiometries correspond to what is called the *eighteen-electron rule* or noble gas formalism. This is merely a phenomenological way of expressing the tendency of a transition metal atom to use all its valence orbitals, namely, the five nd, the $(n + 1)s$, and the three $(n + 1)p$ orbitals as fully as possible in metal-ligand bonding. The sum of the number of valence electrons in the gaseous atom plus the number of electrons from neutral ligands may attain a maximum value of 18.

This is illustrated by the following examples:

$$Cr + 6CO = Cr(CO)_6$$
$$6 + (6 \times 2) = 18$$
$$Ru + H + CO_2CH_3 + 3PPh_3 = RuH(CO_2Me)(PPh_3)_3$$
$$8 + 1 + 3 + (3 \times 2) = 18$$

A third way of classifying ligands is *structurally,* that is, by the number of connections they make to the central atom. Where only one atom becomes closely connected (bonded) the ligand is said to be *unidentate* [e.g., the ligands in $Co(NH_3)_6^{3+}$, $AlCl_4^-$, $Fe(CN)_6^{3-}$]. When a ligand becomes attached by two or more atoms it is *bidentate, tridentate, tetradentate,* and so on, generally multidentate. Note that the Greek-derived corresponding prefixes (mono, ter, poly, etc.) are also used in the literature.

Bidentate ligands when bound entirely to one atom are termed *chelate,* as in 3-I, 3-II, and 3-III.

(3-1)   (3-II)   (3-III)

Another important role of ligands is as *bridging groups*. In many cases they serve as *unidentate* bridging ligands. This means that there is only *one* ligand atom that forms two (or even three) bonds to different metal atoms. For monoatomic ligands, such as the halide ions, and those containing only one possible donor atom, this *unidentate* form of bridging is, of course, the only possible one. A few examples are shown in 3-IV to 3-VI. Ligands having more than one atom that can be an electron donor often function as *bidentate* bridging ligands. Examples are shown in 3-VII and 3-VIII.

(3-IV)   (3-V)   (3-VI)   (3-VII)   (3-VIII)

A further classification is according to the nature of the donor atom of the ligand. Thus we may have carbon, nitrogen, phosphorus, oxygen, sulfur, and so on, donor atoms. Ligands classed this way are discussed in Chapter 4.

Under oxygen donors we can list not only H_2O but also Ph_3PO and SO_4^{2-}, and so on, whereas under carbon we can list CO, η-C_5H_5, $CH_3CH{=}CH_2$, and so on. For nitrogen we have all the aliphatic and aromatic amines plus a host of others, such as NO_2^- and NO.

Some unidentate ligands have two or more different donor sites so that the possibility of *linkage isomerism* arises. Some important ligands of this type, which are called *ambidentate ligands*,[2] are:

{M—NO₂	Nitro	{M—CN	Cyano
{M—ONO	Nitrito	{M—NC	Isocyano
{M—SCN	*S*-Thiocyanato	$R_2S \rightarrow M$, $R_2SO \rightarrow M$	S and O-Bonded
{M—NCS	*N*-Thiocyanato	‖ O	Dialkylsulfoxide

The next sections discuss primarily the chemistry of simple donor ligands. Much of this chemistry applies equally well to main group metal ions (e.g., Na^+, Ca^{2+}, Ga^{3+}, or Cd^{2+}) and to transition metal ions. It is a chemistry largely of aqua ions,

2   R. J. Balahura and N. A. Lewis, *Coord. Chem. Rev.*, 1976, **20**, 109; A. H. Norbury, *Adv. Inorg. Chem. Radiochem.*, 1975, **17**, 231 [NCO, NCS, and NCSe]; J. L. Burmeister, *Coord. Chem. Rev.*, 1966, **1**, 205; A. H. Norbury and A. P. Sinha, *Q. Rev.*, 1970, **24**, 69.

nitrogen donor ligands, such as ammonia or ethylenediamine, and halide ions, and it is chemistry of metal ions in positive oxidation states, usually 2+ and 3+.

Later sections consider complexes that have π-bonding ligands and also compounds that are called π *complexes,* which are those formed by unsaturated organic molecules. This is a chemistry largely of transition metals, often in formally low oxidation states such as -1, 0, and $+1$.

The borderline between π-bonding and non-π-bonding ligands is by no means clearly defined. Also the terms "nonclassical" versus "classical" are of limited validity, since Werner and his contemporaries studied complexes of cyanide ion and of tertiary phosphines; even pyridine, which they used extensively, is not solely a simple donor.

STABILITY OF COMPLEX IONS IN AQUEOUS SOLUTION[3]

3-2. Aqua Ions

In a fundamental sense metal ions simply dissolved in water are already complexed—they have formed aqua ions. The process of forming in aqueous solution what we more conventionally call complexes is really one of displacing one set of ligands, which happen to be water molecules, by another set. Thus the logical place to begin a discussion of the formation and stability of complex ions in aqueous solution is with the aqua ions themselves.

From thermodynamic cycles the enthalpies of plunging gaseous metal ions into water can be estimated and the results, 2×10^2 to 4×10^3 kJ mol^{-1} (see Table 3-1), show that these interactions are very strong indeed. It is of importance in understanding the behavior of metal ions in aqueous solution to know how many water molecules each of these ions binds by direct metal-oxygen bonds. To put it another way, if we regard the ion as being an aqua complex $[M(H_2O)_x]^{n+}$, which is then further and more loosely solvated, we wish to know the coordination number x and also the manner in which the x water molecules are arranged around the metal ion. Classical measurements of various types—for example, ion mobilities, apparent hydrated radii, entropies of hydration—fail to give such detailed information because they cannot make any explicit distinction between those water molecules directly bonded to the metal—the x water molecules in the inner coordination sphere—and additional molecules that are held less strongly by hydrogen bonds to the water molecules of the inner coordination sphere. There are, however, ways of answering the question in many instances, ways depending, for the most part, on modern physical and theoretical developments. A few illustrative examples will be considered here.

For the transition metal ions, the spectral and, to a lesser degree, magnetic

[3] (a) J. Burgess, *Metal Ions in Solution,* Horwood-Wiley, 1978 [covers most aspects of solution behavior (e.g., thermochemistry, kinetics, solvation, nonaqueous solutions]; (b) C. F. Baes and R. E. Mesmer, *Hydrolysis of Cations,* Wiley-Interscience, 1976.

TABLE 3-1

Enthalpies of Hydration[a] of Some Ions (kJ mol⁻¹)

H^+	−1091	Ca^{2+}	−1577	Cd^{2+}	−1807
Li^+	−519	Sr^{2+}	−1443	Hg^{2+}	−1824
Na^+	−406	Ba^{2+}	−1305	Sn^{2+}	−1552
K^+	−322	Cr^{2+}	−1904	Pb^{2+}	−1481
Rb^+	−293	Mn^{2+}	−1841	Al^{3+}	−4665
Cs^+	−264	Fe^{2+}	−1946	Fe^{3+}	−4430
Ag^+	−473	Co^{2+}	−1996	F^-	−515
Tl^+	−326	Ni^{2+}	−2105	Cl^-	−381
Be^{2+}	−2494	Cu^{2+}	−2100	Br^-	−347
Mg^{2+}	−1921	Zn^{2+}	−2046	I^-	−305

[a] Absolute values are based on the assignment of −1091 ± 10 kJ mol⁻¹ to H^+ (cf. H. F. Halliwell and S. C. Nyburg, *Trans. Faraday Soc.,* 1963, **59**, 1126). Each value probably has an uncertainty of at least $10n$ kJ mol⁻¹, where n is the charge of the ion.

properties depend on the constitution and symmetry of their surroundings. For example, the Co^{II} ion is known to form both octahedral and tetrahedral complexes. Thus we might suppose that the aqua ion could be either $[Co(H_2O)_6]^{2+}$ with octahedral symmetry, or $[Co(H_2O)_4]^{2+}$ with tetrahedral symmetry. It is found that the spectrum and the magnetism of Co^{II} in pink aqueous solutions of its salts with noncoordinating anions such as ClO_4^- or NO_3^- are very similar to the corresponding properties of octahedrally coordinated Co^{II} in general, and virtually identical with those of Co^{II} in such hydrated salts as $Co(ClO_4)_2 \cdot 6H_2O$ or $CoSO_4 \cdot 7H_2O$ where from X-ray studies octahedral $[Co(H_2O)_6]^{2+}$ ions are known definitely to exist. Complementing this, the spectral and magnetic properties of the many known tetrahedral Co^{II} complexes, such as $[CoCl_4]^{2-}$, $[CoBr_4]^{2-}$, $[Co(NCS)_4]^{2-}$, and $[py_2CoCl_2]$, which are intensely green, blue, or purple, are completely different from those of Co^{II} in aqueous solution. Thus there can scarely be any doubt that aqueous solutions of otherwise uncomplexed Co^{II} contain predominantly[4] well-defined, octahedral $[Co(H_2O)_6]^{2+}$ ions, further hydrated, of course. Evidence of similar character can be adduced for many of the other transition metal ions. For all the di- and tripositive ions of the first transition series, the aqua ions are octahedral $[M(H_2O)_6]^{2(or\ 3)+}$ species, although in those of Cr^{II}, Mn^{III}, and Cu^{II} there are definite distortions of the octahedra because of the Jahn-Teller effect (see Section 20-18). Information on aqua ions of the second and third transition series, of which there are only a few, however, is not so certain. It is probable that the coordination is octahedral in many, but higher coordination numbers may occur. For the lanthanide ions, $M^{3+}(aq)$, it is certain that the coordination number is higher.

For ions that do not have partly filled d shells, evidence of the kind mentioned is lacking, since such ions do not have spectral or magnetic properties related in a straightforward way to the nature of their coordination spheres. We are therefore not sure about the state of aquation of many such ions, although nmr and other

[4] However, there are also *small* quantities of tetrahedral $[Co(H_2O)_4]^{2+}$; see T. J. Swift, *Inorg. Chem.,* 1964, **3**, 526.

relaxation techniques have now supplied some such information. It should be noted that, even when the existence of a well-defined aqua ion is certain, there are vast differences in the average length of time that a water molecule spends in the coordination sphere, the so-called mean residence time. For Cr^{III} and Rh^{III} this time is so long that when a solution of $[Cr(H_2O)_6]^{3+}$ in ordinary water is mixed with water enriched in ^{18}O, many hours are required for complete equilibration of the enriched solvent water with the coordinated water. From a measurement of how many molecules of H_2O in the Cr^{III} and Rh^{III} solutions fail immediately to exchange with the enriched water added, the coordination numbers of these ions by water were shown to be 6. These cases are exceptional, however. Most other aqua ions are far more labile, and a similar equilibration would occur too rapidly to permit the same type of measurement. This particular rate problem is only one of several that are discussed more fully in Section 28-3.

Aqua ions are all more or less acidic[3]; that is, they dissociate in a manner represented by the equation

$$[M(H_2O)_x]^{n+} = [M(H_2O)_{x-1}(OH)]^{(n-1)+} + H^+ \qquad K_A = \frac{[H^+][M(H_2O)_{x-1}(OH)]}{[M(H_2O)_x]}$$

The acidities vary widely, as the following K_A values show:

M in $[M(H_2O)_6]^{n+}$	K_A
Al^{III}	1.12×10^{-5}
Cr^{III}	1.26×10^{-4}
Fe^{III}	6.3×10^{-3}

Coordinated water molecules in other complexes also dissociate in the same way, for example,

$$[Co(NH_3)_5(H_2O)]^{3+} = [Co(NH_3)_5(OH)]^{2+} + H^+ \qquad K \approx 10^{-5.7}$$
$$[Pt(NH_3)_4(H_2O)_2]^{4+} = [Pt(NH_3)_4(H_2O)(OH)]^{3+} + H^+ \qquad K \approx 10^{-2}$$

3-3. The "Stepwise" Formation of Complexes[3,5]

The thermodynamic stability of a species is a measure of the extent to which this species will form from, or be transformed into, other species under certain conditions

[5] L. D. Pettit and G. Brookes, *Essays Chem.*, 1977, **6**, 1; M. T. Beck, *Chemistry of Complex Equilibria,* Van Nostrand-Reinhold, 1970; S. Fronaeus, in *Techniques of Inorganic Chemistry,* Vol. 1, Wiley, 1963; F. L. C. Rossotti and H. Rossotti, *The Determination of Stability Constants,* McGraw-Hill, 1961; L. G. Sillén and A. E. Martell, Eds., *Stability Constants of Metal Ion Complexes,* Chemical Society, London (volumes collecting stability constant data); R. M. Smith and A. E. Martell, Eds., *Critical Stability Constants,* several volumes, Vol. I, 1974, Vol. II, 1976, Plenum Press; S. J. Ashcroft and C. J. Mortimer, *Thermochemistry of Transitional Metal Complexes,* Academic Press, 1970; J. J. Christensen, D. J. Eatough, and R. M. Izatt, *Handbook of Metal Ligand Heats and Related Thermodynamic Quantities,* 2nd ed., Dekker, 1975; J. Kragten, *Atlas of Metal Ligand Equilibria in Aqueous Solution,* Horwood-Wiley, 1977 (45 metals, 29 ligands); D. D. Perrin, *Stability Constants of Metal Ion Complexes, Part B, Organic Ligands,* Pergamon Press, 1979 (5000 references).

when the system has reached equilibrium. The kinetic stability of a species refers to the speed with which transformations leading to the attainment of equilibrium will occur. This section considers problems of thermodynamic stability, that is, the nature of equilibria once they are established.

If in a solution containing aquated metal ions M and unidentate ligands L, only soluble mononuclear complexes are formed, the system at equilibrium may be described by the following equations and equilibrium constants:

$$M + L = ML \qquad K_1 = \frac{[ML]}{[M][L]}$$

$$ML + L = ML_2 \qquad K_2 = \frac{[ML_2]}{[ML][L]}$$

$$ML_2 + L = ML_3 \qquad K_3 = \frac{[ML_3]}{[ML_2][L]}$$

$$\vdots \qquad \vdots \qquad\qquad \vdots \quad \vdots$$

$$ML_{N-1} + L = ML_N \qquad K_N = \frac{[ML_N]}{[ML_{N-1}][L]}$$

There will be N such equilibria, where N represents the maximum coordination number of the metal ion M for the ligand L, and N may vary from one ligand to another. For instance, Al^{3+} forms $AlCl_4^-$ and AlF_6^{3-} and Co^{2+} forms $CoCl_4^{2-}$ and $Co(NH_3)_6^{2+}$, as the highest complexes with the ligands indicated.

Another way of expressing the equilibrium relations is the following:

$$M + L = ML \qquad \beta_1 = \frac{[ML]}{[M][L]}$$

$$M + 2L = ML_2 \qquad \beta_2 = \frac{[ML_2]}{[M][L]^2}$$

$$M + 3L = ML_3 \qquad \beta_3 = \frac{[ML_3]}{[M][L]^3}$$

$$\vdots \qquad \vdots \qquad\qquad \vdots$$

$$M + NL = ML_N \qquad \beta_N = \frac{[ML_N]}{[M][L]^N}$$

Since there can be only N independent equilibria in such a system, it is clear that the K_i's and the β_i's must be related. The relationship is indeed rather obvious. Consider, for example, the expression for β_3. Let us multiply both numerator and denominator by $[ML][ML_2]$ and then rearrange slightly:

$$\beta_3 = \frac{[ML_3]}{[M][L]^3} \cdot \frac{[ML][ML_2]}{[ML][ML_2]}$$

$$= \frac{[ML]}{[M][L]} \cdot \frac{[ML_2]}{[ML][L]} \cdot \frac{[ML_3]}{[ML_2][L]}$$

It is not difficult to see that this kind of relationship is perfectly general, namely,

$$\beta_k = K_1 K_2 K_3 \ldots K_k = \prod_{i=1}^{i=k} K_i$$

The K_i's are called the *stepwise formation constants* (or stepwise stability constants), and the β_i's are called the *overall formation constants* (or overall stability constants); each type has its special convenience in certain cases.

In all the equilibria above we have written the metal ion without specifying charge or degree of solvation. The former omission is obviously of no importance, for the equilibria may be expressed as above whatever the charges. Omission of the water molecules is a convention that is usually convenient and harmless. It must be remembered when necessary. See, for example, the discussion of the chelate effect, in the next section.

With only a few exceptions, there is generally a slowly descending progression in the values of the K_i's in any particular system. This is illustrated by the data* for the Cd^{II}–NH_3 system where the ligands are uncharged and by the Cd^{II}–CN system where the ligands are charged.

$$Cd^{2+} + NH_3 = [Cd(NH_3)]^{2+} \qquad K = 10^{2.65}$$
$$[Cd(NH_3)]^{2+} + NH_3 = [Cd(NH_3)_2]^{2+} \qquad K = 10^{2.10}$$
$$[Cd(NH_3)_2]^{2+} + NH_3 = [Cd(NH_3)_3]^{2+} \qquad K = 10^{1.44}$$
$$[Cd(NH_3)_3]^{2+} + NH_3 = [Cd(NH_3)_4]^{2+} \qquad K = 10^{0.93} \; (\beta_4 = 10^{7.12})$$

$$Cd^{2+} + CN^- = [Cd(CN)]^+ \qquad K = 10^{5.48}$$
$$[Cd(CN)]^+ + CN^- = [Cd(CN)_2] \qquad K = 10^{5.12}$$
$$[Cd(CN)_2] + CN^- = [Cd(CN)_3]^- \qquad K = 10^{4.63}$$
$$[Cd(CN)_3]^- + CN^- = [Cd(CN)_4]^{2-} \qquad K = 10^{3.65} \; (\beta_4 = 10^{18.8})$$

Thus, typically, as ligand is added to the solution of metal ion, ML is first formed more rapidly than any other complex in the series. As addition of ligand is continued, the ML_2 concentration rises rapidly, while the ML concentration drops, then ML_3 becomes dominant, ML and ML_2 becoming unimportant, and so forth, until the highest complex ML_N is formed, to the nearly complete exclusion of all others at very high ligand concentrations. These relationships are conveniently displayed in diagrams such as those shown in Fig. 3-1.

A steady decrease in K_i values with increasing i is to be expected, provided there are only slight changes in the metal-ligand bond energies as a function of i, which is usually the case. For example, in the Ni^{2+}–NH_3 system to be discussed below, the enthalpies of the successive reactions $Ni(NH_3)_{i-1} + NH_3 = Ni(NH_3)_i$ are all within the range 16.7–18.0 kJ mol^{-1}.

There are several reasons for a steady decrease in K_i values as the number of ligands increases: (1) statistical factors, (2) increased steric hindrance as the number of ligands increases if they are bulkier than the H_2O molecules they replace, (3) Coulombic factors, mainly in complexes with charged ligands. The statistical factors may be treated in the following way. Suppose, as is almost certainly the case for Ni^{2+}, that the coordination number remains the same throughout the series $[M(H_2O)_N] \ldots [M(H_2O)_{N-n}L_n] \ldots [ML_N]$. The $[M(H_2O)_{N-n}L_n]$ species has n sites from which to lose a ligand, whereas the species $[M(H_2O)_{N-n+1}L_{n-1}]$ has

---

* Cd-NH_3 constants determined in $2M$ NH_4NO_3; Cd-CN^- constants determined in $3M$ $NaClO_4$.

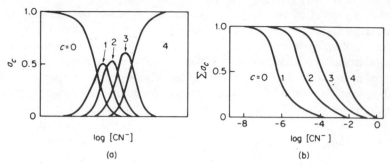

Fig. 3-1. Plots of the proportions of the various complexes $[Cd(CN)_c]^{(2-c)+}$ as a function of the ligand concentration:

$$\alpha_c = [Cd(CN)_c]/\text{total Cd} \qquad \Sigma\alpha_c = \sum_{c=0}^{4} [Cd(CN)_c]$$

[Reproduced by permission from F. J. C. Rossetti, in *Modern Coordination Chemistry*, J. Lewis and R. G. Wilkins, Eds., Interscience, 1960, p. 10.]

$(N - n + 1)$ sites at which to gain a ligand. Thus the relative probability of passing from $[M(H_2O)_{N-n+1}L_{n-1}]$ to $[M(H_2O)_{N-n}L_n]$ is proportional to $(N - n + 1)/n$. Similarly, the relative probability of passing from $[M(H_2O)_{N-n}L_n]$ to $[M(H_2O)_{N-n-1}L_{n+1}]$ is proportional to $(N - n)/(n + 1)$. Hence on the basis of these statistical considerations alone, we expect

$$K_{n+1}/K_n = \frac{N - n}{n + 1} \div \frac{N - n + 1}{n} = \frac{n(N - n)}{(n + 1)(N - n + 1)}$$

In the $Ni^{2+}-NH_3$ system ($N = 6$), we find the comparison between experimental ratios of successive constants and those calculated from the formula above to be as shown in Table 3-2. The experimental ratios are consistently smaller than the statistically expected ones, which is typical and shows that other factors are also of importance.

There are cases where the experimental ratios of the constants do not remain constant or change monotonically; instead, one of them is singularly large or small. There are several reasons for this: (1) an abrupt change in coordination number and hybridization at some stage of the sequence of complexes, (2) special steric effects that become operative only at a certain stage of coordination, and (3) an

TABLE 3-2
Comparison of Experimental and Statistical Formation Constants of $Ni^{2+}-NH_3$ Complexes

	Experimental	Statistical
K_2/K_1	0.28	0.417
K_3/K_2	0.31	0.533
K_4/K_3	0.29	0.562
K_5/K_4	0.36	0.533
K_6/K_5	0.2	0.417

abrupt change in electronic structure of the metal ion at a certain stage of complexation. Each of these is now illustrated.

Values of K_3/K_2 are anomalously low for the halogeno complexes of mercury(II); HgX_2 species are linear, whereas $[HgX_4]^{2-}$ species are tetrahedral. Presumably the change from sp to sp^3 hybridization occurs on going from HgX_2 to $[HgX_3]^-$. K_3/K_2 is anomalously small for the ethylenediamine complexes of Zn^{II}, and this is believed to be due to the change from sp^3 to sp^3d^2 hybridization if it is assumed that $[Znen_2]^{2+}$ is tetrahedral. For the Ag^+-NH_3 system $K_2 > K_1$, indicating that the linear, sp-hybridized structure is probably attained with $[Ag(NH_3)_2]^+$ but not with $[Ag(NH_3)(H_2O)_{3(or\ 5)}]^+$.

With 6,6'-dimethyl-2,2'-bipyridine (3-IX), many metal ions that form tris-2,2'-bipyridine complexes form only bis or mono complexes, or, in some cases, no isolable complexes at all, because of the steric hindrance between the methyl groups and other ligands attached to the ion.

(3-IX)

In the series of complexes of Fe^{II} with 1,10-phenanthroline (and also with 2,2'-bipyridine), K_3 is greater than K_2. This is because the tris complex is diamagnetic (i.e., the ferrous ion has the low-spin state t_{2g}^6—see Section 20-9 for the meaning of this symbol), whereas in the mono and bis complexes, as in the aqua ion, there are four unpaired electrons. This change from the $t_{2g}^4e_g^2$ to the t_{2g}^6 causes the enthalpy change for addition of the third ligand to be anomalously large because the e_g electrons are antibonding.

3-4. The Chelate Effect[6]

The term "chelate effect" refers to the enhanced stability of a complex system containing chelate rings as compared to the stability of a system that is as similar as possible but contains none or fewer rings. As an example, consider the following equilibrium constants:

$$Ni^{2+}(aq) + 6NH_3(aq) = [Ni(NH_3)_6]^{2+}(aq) \qquad \log\beta = 8.61$$
$$Ni^{2+}(aq) + 3en(aq) = [Ni\ en_3]^{2+}(aq) \qquad \log\beta = 18.28$$

The system $[Ni\ en_3]^{2+}$ in which three chelate rings are formed is nearly 10^{10} times as stable as that in which no such ring is formed. Although the effect is not always so pronounced, such a chelate effect is a very general one.

To understand this effect, we must invoke the thermodynamic relationships:

$$\Delta G° = -RT \ln\beta$$
$$\Delta G° = \Delta H° - T\Delta S°$$

[6] (a) R. T. Myers, *Inorg. Chem.*, 1978, **17**, 952; (b) D. Munro, *Chem. Br.*, 1977, **13**, 100; (c) C. F. Bell, *Principles and Applications of Metal Chelation*, Oxford University Press, 1977.

TABLE 3-3

Two Reactions Illustrating a Purely Entropy-Based Chelate Effect

$$Cd^{2+}(aq) + 4CH_3NH_2(aq) = [Cd(NH_2CH_3)_4]^{2+}(aq) \qquad \log \beta = 6.52$$
$$Cd^{2+}(aq) + 2H_2NCH_2CH_2NH_2(aq) = [Cd\ en_2]^{2+}(aq) \qquad \log \beta = 10.6$$

Ligands	ΔH° (kJ mol^{-1})	ΔS° (J mol^{-1} deg^{-1})	$-T\Delta S^\circ$(kJ mol^{-1})	ΔG° (kJ mol^{-1})
4CH$_3$NH$_2$	-57.3	-67.3	20.1	-37.2
2 en	-56.5	$+14.1$	-4.2	-60.7

Thus β increases as ΔG° becomes more negative. A more negative ΔG° can result from making ΔH° more negative or from making ΔS° more positive.

As a very simple case, consider the reactions, and the pertinent thermodynamic data for them, given in Table 3-3. In this case the enthalpy difference is well within experimental error; the chelate effect can thus be traced entirely to the entropy difference.

In the example first cited, the enthalpies make a slight favorable contribution, but the main source of the chelate effect is still to be found in the entropies. We may look at this case in terms of the following metathesis:

$$[Ni(NH_3)_6]^{2+}(aq) + 3\ en(aq) = [Ni\ en_3]^{2+}(aq) + 6NH_3(aq) \qquad \log \beta = 9.67$$

for which the enthalpy change is -12.1 kJ mol^{-1}, whereas $-T\Delta S^\circ = -55.1$ kJ mol^{-1}. The enthalpy change corresponds very closely to that expected from the increased CFSE* of [Ni en$_3$]$^{2+}$ which is estimated from spectral data to be -11.5 kJ mol^{-1} and can presumably be so explained.

As a final example, which illustrates the existence of a chelate effect despite an unfavorable enthalpy term, we may use the reaction

$$[Ni\ en_2(H_2O)_2]^{2+}(aq) + tren(aq) = [Ni\ tren(H_2O)_2]^{2+}(aq) + 2\ en(aq) \qquad \log \beta = 1.88$$
$$[tren = N(CH_2CH_2NH_2)_3]$$

For this reaction we have $\Delta H^\circ = +13.0$, $-T\Delta S^\circ = -23.7$, and $\Delta G^\circ = -10.7$ (all in kJ mol^{-1}). The positive enthalpy change can be attributed both to greater steric strain resulting from the presence of three fused chelate rings in Ni tren, and to the inherently weaker M—N bond when N is a tertiary rather than a primary nitrogen atom. Nevertheless, the greater number of chelate rings (3 vs. 2) leads to greater stability, owing to an entropy effect that is only partially canceled by the unfavorable enthalpy change.

Probably the main cause of the large entropy increase in each of the three cases we have been considering is the net increase in the number of unbound molecules—ligands *per se* or water molecules. Thus although 6 NH$_3$ displace 6 H$_2$O, making no net change in the number of independent molecules, it takes only 3 en molecules to displace 6 H$_2$O. Another more pictorial way to look at the problem is to visualize a chelate ligand with one end attached to the metal ion. The other end cannot then get very far away, and the probability of it, too, becoming attached

* For meaning of CFSE see Section 20-19.

TABLE 3-4

Factors Influencing Solution Stability of Complexes[a]

Enthalpy effects	Entropy effects
Variation of bond strength with electronegativities of metal ions and ligand donor atoms	Number of chelate rings Size of chelate ring
Ligand field effects	Changes of solvation on complex formation
Steric and electrostatic repulsion between ligands in complex	Arrangement of chelate rings
Enthalpy effects related to conformation of uncoordinated ligand	Entropy variations in uncoordinated ligands
Other Coulombic forces involving chelate ring formation	Effects resulting from differences in configurational entropies of the ligand in complex compound
Enthalpy of solution of ligands	Entropy of solution of ligands
Change in bond strength when ligand is charged (same donor and acceptor atom)	Entropy of solution of coordinated metal ions

[a] From R. T. Meyers, *Inorg. Chem.*, 1978, **17**, 952.

to the metal atom is greater than if this other end were instead another independent molecule, which would have access to a much larger volume of the solution.

The latter view provides an explanation for the decreasing magnitude of the chelate effect with increasing ring size, as illustrated by data such as those shown below for copper complexes of $H_2N(CH_2)_2NH_2$ and $H_2N(CH_2)_3NH_2(tn)$:

$$[Cu\ en_2]^{2+}(aq) + 2tn(aq) = [Cu\ tn_2]^{2+}(aq) + 2\ en(aq) \qquad \log \beta = -2.86$$

Of course, when the ring that must be formed becomes sufficiently large (seven-membered or more), it becomes more probable that the other end of the chelate molecule will contact another metal ion than that it will come around to the first one and complete the ring. Table 3-4 summarizes the factors influencing the stabilities of complexes.

3-5. The Macrocyclic Effect[7]

Just as a chelating *n*-dentate ligand gives a more stable complex (more negative $\Delta G°$ of formation) than *n* unidentate ligands of similar type, a phenomenon just discussed under the name *chelate effect,* so an *n*-dentate macrocyclic ligand (see Section 3-6) gives even more stable complexes than the most similar *n*-dentate open chain ligand. For example:

[7] F. P. Hinz and D. W. Margerum, *J. Am. Chem. Soc.*, 1974, **96**, 4993; *Inorg. Chem.*, 1974, **13**, 2841; G. F. Smith and D. W. Margerum, *J. C. S. Chem. Comm.*, **1975**, 807.

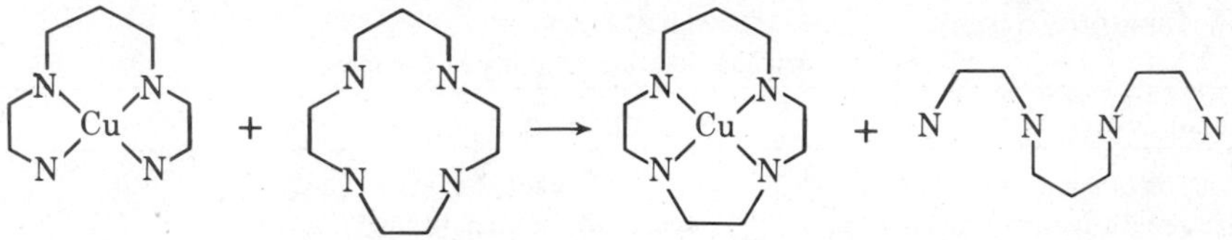

$$\log K = 5.2$$
$$\Delta G = -30 \text{ kJ mol}^{-1} \text{ at } 300°\text{K}$$

Just as in the case of the chelate effect, this so-called *macrocyclic effect* might result from either entropic or enthalpic contributions, or both, and similar controversy about their relative importance has arisen. The most recent results[8] indicate that entropy always favors the macrocycle, with a value of *ca.* 70 J mol^{-1} deg^{-1} for systems such as that shown above. The enthalpy contribution usually also favors the macrocycle, but by an amount that can vary a great deal from case to case. In the specific case shown, $\Delta H = -10$ kJ mol^{-1}. In general, the macrocyclic effect results from a favorable entropy change assisted, usually, by a favorable enthalpy change as well.

TYPES AND CLASSIFICATION OF LIGANDS

3-6.　Multi- or Polydentate Ligands

Regardless of whether π-bonding is involved, ligands can have various denticities, and we now illustrate some of the more important types.

Bidentate Ligands. These are very common and can be classified according to the size of the chelate ring formed as in the following examples:

Three-membered

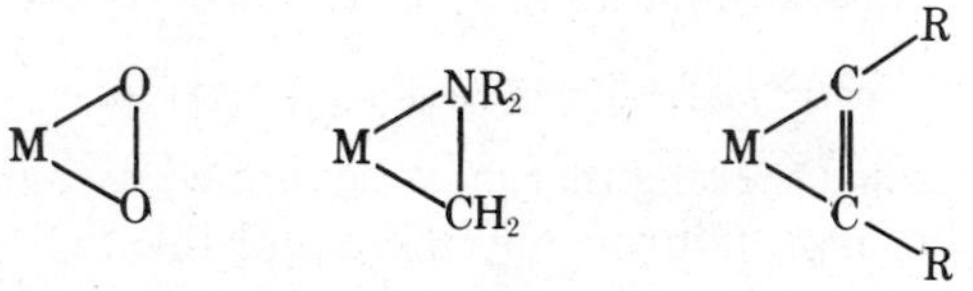

Four-membered

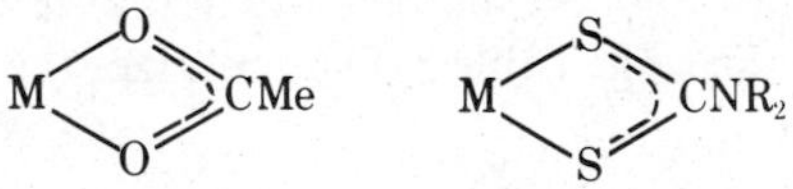

Five-membered

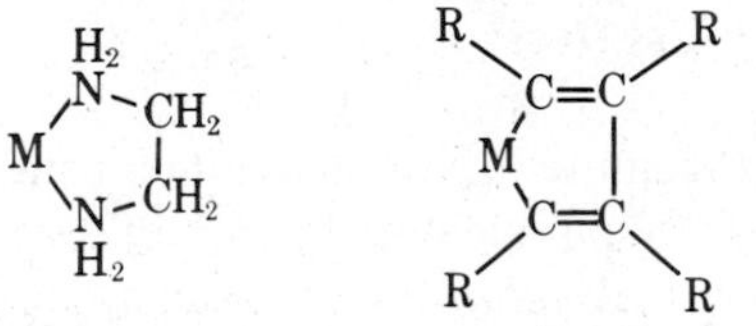

Six-membered

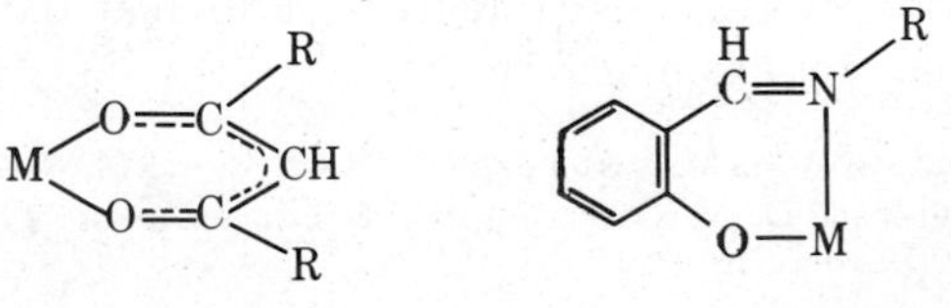

[8]　A. Anichini *et al., J. C. S. Dalton,* **1978,** 577.

Tridentate Ligands. Some are *obligate planar* such as

Terpyridine, terpy:

Acylhydrazones of salicylaldehyde:

and many similar ones where maintenance of the π conjugation markedly favors planarity. Such ligands must form complexes of the types 3-X or 3-XI.

(3-X)   (3-XI)

There are also many flexible tridentate ligands such as

Diethylenetriamine, dien:

$$HN \Big\langle \begin{matrix} CH_2CH_2NH_2 \\ CH_2CH_2NH_2 \end{matrix}$$

Bis(3-dimethylarsinylpropyl)methylarsine, triars:

$$CH_3As \Big\langle \begin{matrix} (CH_2)_3 \cdot As(CH_3)_2 \\ (CH_2)_3 \cdot As(CH_3)_2 \end{matrix}$$

which are about equally capable of meridional (3-X, 3-XI) and facial (3-XII) coordination.

(3-XII)

Quadridentate Ligands. There are four main types:
Open-Chain, Unbranched:

Triethylenetetramine, trien:

$$H_2N(CH_2)_2NH(CH_2)_2NH(CH_2)_2NH_2$$

Schiff bases derived from acetylacetone, for example,

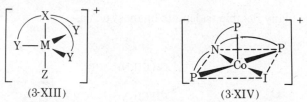

Tripod Ligands. These are of the type $X(\text{---}Y)_3$, where X is nitrogen, phosphorus, or arsenic, the Y groups are R_2N, R_2P, R_2As, RS, or RSe, and the connecting chains—are $(CH_2)_2$, $(CH_2)_3$, or *o*-phenylene. Some common ones are

$N(CH_2CH_2NH_2)_3$	tren
$N[CH_2CH_2N(CH_3)_2]_3$	Me_6tren
$N[CH_2CH_2P(C_6H_5)_2]_3$	np_3
$P[o\text{-}C_6H_4P(C_6H_5)_2]_3$	QP
$N(CH_2CH_2SCH_3)_3$	TSN
$As[o\text{-}C_6H_4As(C_6H_5)_2]_3$	QAS

The tripod ligands are used particularly to favor formation of trigonal-bipyramidal complexes of divalent metal ions, as shown schematically in 3-XIII, but they do not invariably give this result. For instance, whereas $Ni(np_3)I_2$ is trigonal-bipyramidal, $Co(np_3)I_2$ is square pyramidal (3-XIV).

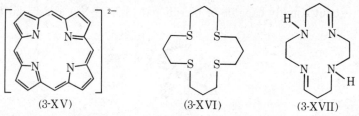

(3-XIII)   (3-XIV)

Macrocyclic. These may be (*a*) planar with unsaturated rings as in porphyrin (3-XV) and its derivatives, although, as discussed later in more detail (Section 4-14) the metal atom may be out of the plane of the nitrogen donor atoms, or (*b*) puckered with saturated rings as in the macrocycles 3-XVI and 3-XVII.

(3-XV)   (3-XVI)   (3-XVII)

Pentadentate and Higher-Dentate Ligands. Perhaps the best known hexadentate ligand is ethylenediaminetetraacetate ($EDTA^{4-}$), which can also be pentadentate as $EDTAH^{3-}$.

Other important multidentate ligands are the *crown ethers and cryptates* (Section 4-26). These are cyclic and polycylic ligands that form their most im-

$$\begin{array}{cc} {}^-OOCCH_2 & CH_2COO^- \\ {}^-OOCCH_2 & CH_2COO^- \end{array} NCH_2CH_2N$$

(EDTA)$^{4-}$

$$\begin{array}{cc} {}^-OOCCH_2 & CH_2COOH \\ {}^-OOCCH_2 & CH_2COO^- \end{array} NCH_2CH_2N$$

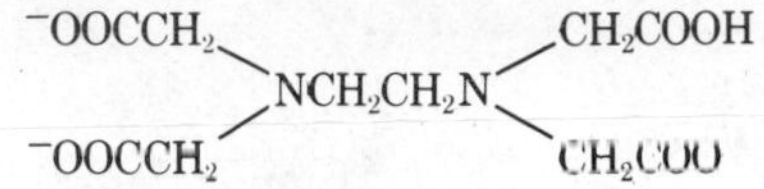

(EDTAH)$^{3-}$

portant complexes with alkali and alkaline earth ions. The macrocyclic polyethers, commonly called *crown ethers,* are typified by 3-XVIII and 3-XIX. Since the systematic names for such ligands are very unwieldy, a special nomenclature is used, in which 3-XVIII and 3-XIX are called, respectively, 15-crown-5 and dibenzo-18-crown-6. These examples should serve to show the rules for the simple nomenclature. Crown ethers with as many as ten oxygens are known and several (e.g., 3-XIX) are commercially available.

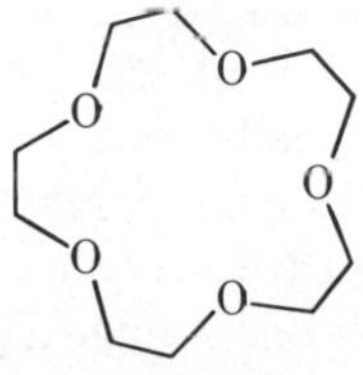

(3-XVIII) 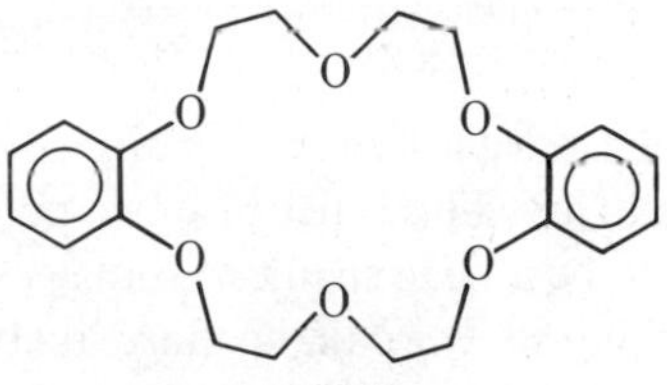   (3-XIX)

The cryptates are bicyclic species, most of which have the general formula 3-XX. Again a simplified code for naming them is a practical necessity. They are called "cryptate-*mmn*," where *m* and *n* are as defined in 3-XX. One of the commonest is cryptate-222.

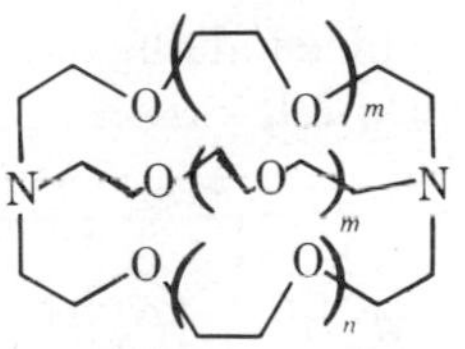

(3-XX)

These ligands have two characteristics that make them unusually interesting. Because they are chelating ligands of high denticity they give very high formation constants, and since the size of ion that will best fit the cavities can be predetermined by changing the ring size, these ligands can be designed to be selective.

In addition to the cryptates, which are synthesized apart from metal ions and then used to form complexes, there are other types of multicyclic ligand called *encapsulating ligands,* which are synthesized around the metal ion and cannot release it.[9] Two of these are 3-XXI and 3-XXII. An encapsulation complex allows studies to be carried out under extremely acidic or basic conditions since the metal ion, though it cannot be removed, can be oxidized or reduced. Such ligands also can enforce unusual coordination geometries; in the examples shown the coordination is much closer to trigonal prismatic than to octahedral.

[9] D. R. Boston and N. J. Rose, *J. Am. Chem. Soc.,* 1973, **95,** 4163; E. Larsen *et al., Inorg. Chem.,* 1972, **11,** 2652.

(3-XXI)   (3-XXII)

Ligands of Unusual Reach. Ordinarily bidentate ligands occupy cis positions around a metal ion. This is because two potential donor atoms separated by a chain long enough to be able to span two trans positions would have a very low probability of actually doing so. It would be more likely to form only one bond to a given metal atom, while using its second donor atom to coordinate to a different metal atom, or not at all. This is simply an inorganic example of the well-known problem in organic chemistry of synthesizing very large rings. By appropriately designing the connection between the donor atoms, however, ligands that span trans positions in a square complex or the two sites in a linear LML complex can be made.[10a] An example is 3-XXIII. Large chelate ring compounds[10b] with 12 to 72 membered rings can be made from flexible bidentate ligands [e.g., $Me_2N(CH_2)_nNMe_2$ or $R_2P(CH_2)_nPR_2$; see also Section 4-20].

H_2C CH_2

$(C_6H_5)_2P$ $P(C_6H_5)_2$

(3-XXIII)

3-7. Conformation of Chelate Rings[11]

Simple diagrams of chelate rings in which the ring conformation is ignored are adequate for many purposes. Indeed, in some cases, such as β-diketonate complexes,

[10]   (a) L. M. Venanzi *et al., Helv. Chim. Acta,* 1976, **59,** 2674, 2683, 2691; (b) B. L. Shaw *et al., J. C. S. Dalton,* **1979,** 496, 1109.
[11]   For a detailed discussion and references see C. J. Hawkins, *Absolute Configuration of Metal Complexes,* Wiley-Interscience, 1971, Chapter 3. See also J. K. Beattie, *Acc. Chem. Res.,* 1971, **4,** 253 (note that there are some differences in notation in these two articles); Y. Saito, *Topics in Stereochemistry,* Vol. 10, E. L. Eliel and N. L. Allinger, Eds., Wiley, 1978, p. 95.

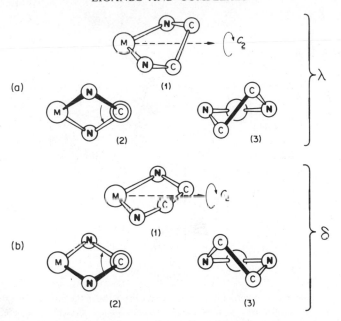

Fig. 3-2. Different ways of viewing the puckering of ethylenediamine chelate rings. The absolute configurations λ and δ are defined. [Reproduced by permission from C. J. Hawkins, Ref. 11.]

the rings are planar and no problem arises. However the relative stabilities and certain spectroscopic properties of many chelate complexes can be understood only by considering carefully the effects of the ring conformations, as in the important case of five-membered rings such as those formed by ethylenediamine.

Figure 3-2 shows three ways of viewing the puckered rings, and identifies the absolute configurations in the λ,δ notation. As indicated clearly in the figure, the chelate ring has as its only symmetry element a C_2 axis. It must therefore (see Section 2-6) be chiral, and the two forms of a given ring are enantiomorphs. When this source of enantiomorphism is combined with the two enantiomorphous ways, Λ and Δ, of orienting the chelate rings about the metal atom (Fig. 3-3), a number

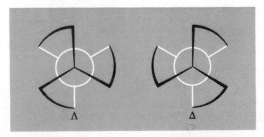

Fig. 3-3. Trischelate octahedral complexes (actual symmetry: D_3), showing how the absolute configurations Λ and Δ are defined according to the translation (twist) of the helices.

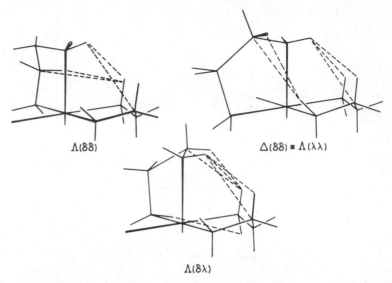

$\Lambda(\delta\delta)$ $\Delta(\delta\delta) \equiv \Lambda(\lambda\lambda)$

$\Lambda(\delta\lambda)$

Fig. 3-4. The different sets of repulsive interactions that exist between the three different pairs of ring conformations in octahedral ethylenediamine complexes; broken lines represent the significant repulsive interactions. [Reproduced by permission from C. J. Hawkins, Ref. 11.]

of diastereomeric molecules become possible, specifically, the following eight:

$\Lambda(\delta\delta\delta)$	$\Delta(\lambda\lambda\lambda)$
$\Lambda(\delta\delta\lambda)$	$\Delta(\lambda\lambda\delta)$
$\Lambda(\delta\lambda\lambda)$	$\Delta(\lambda\delta\delta)$
$\Lambda(\lambda\lambda\lambda)$	$\Delta(\delta\delta\delta)$

The two columns are here arranged so as to place an enantiomorphous pair on each line. In the following discussion we shall mention only members of the Λ series; analogous energy relationships must of course exist among corresponding members of the Δ series.

The relative stabilities of the four diastereomers have been extensively investigated. First, it can easily be shown that the diastereomers must, in principle, differ in stability because there are different nonbonded (repulsive) interactions between the rings in each case. Figure 3-4 shows these differences for any two rings in the complex. When any reasonable potential function is used to estimate the magnitudes of the repulsive energies, it is concluded that the order of decreasing stability is

$$\Lambda(\delta\delta\delta) > \Lambda(\delta\delta\lambda) > \Lambda(\delta\lambda\lambda) > \Lambda(\lambda\lambda\lambda)$$

However this is not the actual order because enthalpy differences between diastereomers are rather small ($2-3$ kJ mol^{-1}), and an entropy factor must also be considered. Entropy favors the $\delta\delta\lambda$ and $\delta\lambda\lambda$ species because they are three times as probable as the $\delta\delta\delta$ and $\lambda\lambda\lambda$ ones. Hence the best estimate of relative stabilities, which in fact agrees with all experimental data, becomes

$$\Lambda(\delta\delta\lambda) > \Lambda(\delta\delta\delta) \approx \Lambda(\delta\lambda\lambda) \gg \Lambda(\lambda\lambda\lambda)$$

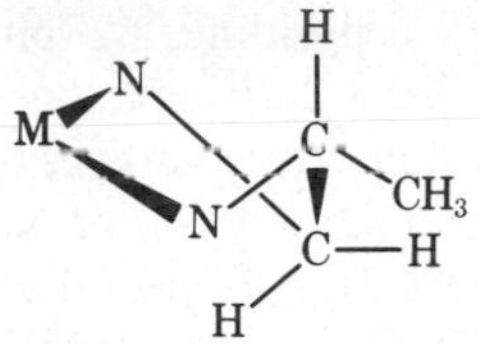

Fig. 3-5. The absolute configuration and expected conformation (i.e., with an equatorial CH_3 group) for an $M(l$-pn) chelate ring.

In crystalline compounds, the $\Delta(\delta\delta\delta)$ isomer (or its enantiomorph) has been found most often, but the other three have also been found. These crystallographic results probably prove nothing about the intrinsic relative stabilities, since hydrogen bonding and other intermolecular interactions can easily outweigh the small intrinsic energy differences.

Nmr studies of solutions of Ru^{II}, Pt^{IV}, Ni^{II}, Rh^{III}, Ir^{III}, and Co^{III} [M en$_3$]$^{n+}$ complexes have yielded the most useful data, and the general conclusions seem to be that the order of stability suggested above is correct and that ring inversions are very rapid. Both experiment and theory suggest that the barrier to ring inversion is only about 25 kJ mol^{-1}. Thus the four diastereomers of each overall form (Λ or Δ) are in labile equilibrium.

One of the interesting and important applications of the foregoing type of analysis is to the determination of absolute Λ or Δ configurations by using substituted ethylenediamine ligands of known absolute configuration. This is nicely illustrated by the [Co(l-pn)$_3$]$^{3+}$ isomers. The absolute configuration of l-pn [pn = 1,2-diaminopropane $NH_2CH(CH_3)—CH_2NH_2$] is known. It would also be expected from consideration of repulsions between rings in the tris complex (as indicated in Fig. 3-4) that pn chelate rings would always take a conformation that puts the CH_3 group in an equatorial position. Hence, an l-pn ring can be confidently expected to have the δ conformation shown in Fig. 3-5. Note that because of the extreme unfavorability of having axial CH_3 groups, only two tris complexes are expected to occur, namely, $\Lambda(\delta\delta\delta)$ and $\Delta(\delta\delta\delta)$. But by the arguments already advanced for en rings, the Λ isomer should be the more stable of these two, by 5 to 10 kJ mol^{-1}. Thus we predict that the most stable [Co(l-pn)$_3$]$^{3+}$ isomer must have the absolute configuration Λ about the metal.

In fact, the most stable [Co(l-pn)$_3$]$^{3+}$ isomer is the one with + rotation at the sodium-D line, and it has the same circular dichroism spectrum, hence the same absolute configuration as (+)-[Co en$_3$]$^{3+}$. The absolute configuration of the latter has been determined, and it is indeed Λ. Thus the argument based on conformational analysis is validated.

π-ACID OR π-BONDING LIGANDS: π COMPLEXES

The ligands for which π-bonding is important are carbon monoxide, isocyanides, substituted phosphines, arsines, stibines or sulfides, nitric oxide, various molecules with delocalized π-orbitals, such as pyridine, 2,2'-bipyridine, 1,10-phenanthroline, and with certain ligands containing 1,2-dithioketone or 1,2-dithiolene groups, such as the dithiomaleonitrile anion. Very diverse types of complex exist, ranging from binary molecular compounds such as $Cr(CO)_6$ or $Ni(PF_3)_4$ through mixed species such as $Co(CO)_3NO$ and $(C_6H_5)_3PFe(CO)_4$, to complex ions such as $[Fe(CN)_5CO]^{3-}$, $[Mo(CO)_5I]^-$, $[Mn(CNR)_6]^+$, $[Vphen_3]^+$, and $\{Ni[S_2C_2(CN)_2]_2\}^{2-}$.

In many of these complexes, the metal atoms are in low-positive, zero, or negative formal oxidation states. It is a characteristic of the ligands that they can stabilize low oxidation states. This property is associated with the fact that in addition to lone pairs, these ligands possess *vacant π orbitals*. These vacant orbitals accept electron density from filled metal orbitals to form a type of π bonding that supplements the σ bonding arising from lone-pair donation; high electron density on the metal atom—of necessity in low oxidation states—thus can be delocalized onto the ligands. The ability of ligands to accept electron density into low-lying empty π orbitals can be called π *acidity,* the word "acidity" being used in the Lewis sense.

There are many unsaturated organic molecules and ions that are also capable of forming complexes with transition metals in low oxidation states, and these are called π *complexes.* There is a qualitative difference from π-acid ligands. The latter form bonds to the metal involving σ orbitals and π orbitals whose nodal planes include the axis of the σ bond. For the π-complexing ligands such as alkenes, arenes, and allyl groups, *both* the donation and back-acceptance (see below) of electron density by the ligand are accomplished using ligand π orbitals. The metal is thus out of the molecular plane of the ligand, whereas with π-acid ligands the metal atom lies along the axes of the linear ligands or in the plane of planar ones.

In a third class of ligand that involves π bonding there are metal-oxygen or metal-nitrogen multiple bonds, as in $O{=}VCl_3$, MnO_4^-, and $N{\equiv}OsO_3^-$. Here the electron flow is in the opposite sense to that in the bonding of π-acid ligands (i.e., from p orbitals on O or N to the metal d orbitals).

We now consider bonding in the main classes of π bonding in a preliminary, qualitative way; a more sophisticated approach is given in Chapter 20.

π-BONDING LIGANDS

3-8.   Carbon Monoxide

Carbon monoxide is the most important π-bonding ligand. Thousands of compounds ranging from pure carbonyls like $Cr(CO)_6$ to mixed complexes like $RhH(CO)(PPh_3)_3$ or $[Ru(CO)Cl_5]^{3-}$ are known. The chemistry of carbonyls is discussed *in extenso* in Chapter 25.

The bonding of CO (and of other similar π acids) can be regarded as involving the following contributions:

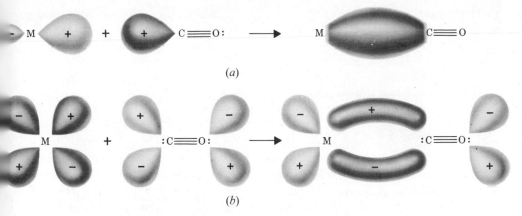

(a)

(b)

Fig. 3-6. (a) The formation of the metal ← carbon σ bond using an unshared pair on the C atom. (b) The formation of the metal → carbon π bond. The other orbitals on the CO are omitted for clarity.

1. Overlap of a filled carbon σ orbital with a σ-type orbital on the metal atom as in Fig. 3-6a. Electron flow C → M in such a dative overlap would lead to an unacceptable concentration of electron density on the metal atom when the latter is not a +2 or more highly charged ion. The metal therefore attempts to reduce this charge (Pauling's electroneutrality principle) by pushing electrons back to the ligand. This of course is possible *only* if the ligand has suitable acceptor orbitals.

2. A second dative overlap of a filled $d\pi$ or hybrid $dp\pi$ metal orbital with the empty, $p\pi$ orbital on carbon monoxide, which can act as a receptor of electron density (Fig. 3-6b).

This bonding mechanism is *synergic,* since the drift of metal electrons, referred to as "back-bonding", into CO orbitals, will tend to make the CO as a whole negative, hence to increase its basicity via the σ orbital of carbon; at the same time the drift of electrons to the metal in the σ bond tends to make the CO positive, thus enhancing the acceptor strength of the π orbitals. Thus up to a point the effects of σ-bond formation strengthen the π bonding, and vice versa. It may be noted here that dipole moment studies indicate that the moment of an M—C bond is only very low, about 0.5 D, suggesting a close approach to electroneutrality.

The main lines of physical evidence showing the multiple nature of the M—CO bonds are bond lengths and vibrational spectra. According to the preceding description of the bonding, as the extent of back-donation from M to CO increases, the M—C bond becomes stronger and the C≡O bond becomes weaker. Thus, the multiple bonding should be evidenced by shorter M—C and longer C—O bonds as compared to M—C single bonds and C≡O triple bonds, respectively. Actually very little information can be obtained from the CO bond lengths, because in the range of bond orders (2–3) concerned, CO bond length is relatively insensitive to bond order. The bond length in CO itself is 1.128 Å, while the bond lengths in metal carbonyl molecules are ~1.15 Å, a shift in the proper direction but of little quantitative significance owing to its small magnitude and the uncertainties (~0.02 Å) in the individual distances. For M—C distances, the sensitivity to bond order in

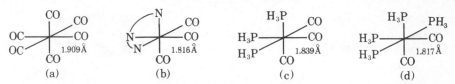

Fig. 3-7. Cr—C bond distances in (a) $Cr(CO)_6$, (b) fac-$[H_2N(CH_2)_2NH(CH_2)NH_2]Cr(CO)_3$, (c) fac-$(PH_3)_3Cr(CO)_3$, and (d) cis-$(PH_3)_4Cr(CO)_2$. [Data for (b) are from F. A. Cotton and D. C. Richardson, *Inorg. Chem.*, 1966, **5**, 1851. Data for (a) (c), and (d) are from G. Huttner and S. Schelle, *J. Crystallogr. Mol. Struct.*, 1971, **1**, 69.]

the range concerned (1–2) is relatively high, probably about 0.3 to 0.4 Å per unit of bond order, and good evidence for multiple bonding can therefore be expected from such data. However it is difficult to apply this criterion in that the length of an M—C single bond cannot be estimated easily because zerovalent metals do not form such bonds.

To estimate the extent to which the metal-carbon bonds are "shortened" we measure the lengths of M—CO bonds in the same molecule in which some other bond, M—X, exists, such that this bond must be single. Then, using the known covalent radius for X, estimating the single bond covalent radius of C to be 0.70 Å when an sp hybrid orbital is used (the greater s character makes this ~0.07 Å shorter than that for sp^3 carbon), the length for a single M—CO bond *in this molecule* can be estimated and compared with the observed value. Relatively few data suitable for this purpose are available.

Figure 3-7 shows some substitution products of $Cr(CO)_6$ in which three or four CO groups have been replaced by ligands such as aliphatic amine nitrogen, which has no capacity to compete with CO trans to it for π-bonding or PH_3, which has very little capacity to do so. We see that in such cases the remaining CO groups have even shorter Cr—C bonds because of even more extensive development of Cr—C π-back-bonding. The shortening is greater in (b) than in (c), since there is slight π bonding to the phosphorus atoms of PH_3. The shortening is also greater in (d) than in (c) because there are only two CO groups in (d) to compete for the available $d\pi$ electrons of the chromium atom.

A study of the electron density distribution in $Cr(CO)_6$ by a combination of X-ray and neutron diffraction[12] leads to the conclusion that the charge distribution is approximately $+0.15 \pm 0.12$ on Cr, $+0.09 \pm 0.05$ on C, and -0.12 ± 0.05 on O, and that the extent of σ charge transfer from C to Cr is almost exactly offset by the extent of π transfer from Cr to C, each of these transfers being 0.3 ± 0.1 electron.

From the vibrational spectra of metal carbonyls,[13] (Section 25-9) it is also possible to infer the existence and extent of M—C multiple bonding. This is most easily done by studying the CO stretching frequencies rather than the MC stretching frequencies, since the former give rise to strong sharp bands, well separated from all other vibrational modes of the molecules. MC stretching frequencies, on the other hand, are in the same range with other types of vibration (e.g., MCO bends);

---

[12]   B. Rees and A. Mitchler, *J. Am. Chem. Soc.*, 1976, **98**, 7918.

[13]   P. S. Braterman, *Struct. Bonding*, 1972, **10**, 57; 1974, **26**, 1. P. S. Braterman, *Metal Carbonyl Spectra*, Academic Press, 1975.

therefore assignments are not easy to make, nor are the so-called MC stretching modes actually pure MC stretching motions. The inferring of M—C bond orders from the behavior of C—O vibrations depends on the assumption that the valence of C is constant, so that a given increase in the M—C bond order must cause an equal decrease in the C—O bond order; this, in turn, will cause a drop in the CO vibrational frequency.

From the direct comparison of CO stretching frequencies in carbonyl molecules with the stretching frequency of CO itself, certain useful qualitative conclusions can be drawn. The CO molecule has a stretching frequency of 2143 cm^{-1}. Terminal CO groups in neutral metal carbonyl molecules are found in the range 2125–1850 cm^{-1}, showing the reduction in CO bond orders. Moreover, when changes are made that should increase the extent of M—C back-bonding, the CO frequencies are shifted to even lower values. Thus if some CO groups are replaced by ligands with low or negligible back-accepting ability, those CO groups that remain must accept $d\pi$ electrons from the metal to a greater extent to prevent the accumulation of negative charge on the metal atom. Thus the frequencies for $Cr(CO)_6$ are ~2100, ~2000, and ~1985 cm^{-1} (exact values vary with phase and solvent) whereas, when three CO's are replaced by amine groups that have essentially no ability to back-accept, as $Cr(dien)(CO)_3$ (Fig. 3-7b), there are two CO stretching modes with frequencies of ~1900 and ~1760 cm^{-1}. Similarly, when we go from $Cr(CO)_6$ to the isoelectronic $V(CO)_6^-$, when more negative charge must be taken from the metal atom, a band is found at ~1860 cm^{-1} corresponding to the one found at ~2000 cm^{-1} in $Cr(CO)_6$. A series of these isoelectronic species illustrating this trend, with their infrared-active CO stretching frequencies (cm^{-1}) is: $Ni(CO)_4$ (~2060); $Co(CO)_4^-$ (~1890); $Fe(CO)_4^{2-}$ (~1790). Conversely, a change that would tend to inhibit the shift of electrons from metal to CO π-orbitals, such as placing a positive charge on the metal, should cause the CO frequencies to *rise,* and this effect has been observed in several cases, the following being representative:

$Mn(CO)_6^+$, ~2090 $Mn(dien)(CO)_3^+$, ~2020, ~1900
$Cr(CO)_6$, ~2000 $Cr(dien)(CO)_3$, ~1900, ~1760
$V(CO)_6^-$, ~1860

To obtain semiquantitative estimates of M—C π bonding from vibrational frequencies, it is first necessary to carry out an approximate dynamical analysis of the CO stretching modes, thus to derive force constants for the CO groups. This procedure is discussed in Section 25-9.

As well as forming linear M—C—O groups, like most other ligands, CO can form bridges. Those of particular importance, shown as 3-XXIV and 3-XXV, are found associated with metal-metal bonds in polynuclear cluster carbonyls whose unusual properties are treated in detail in Chapters 25 and 26.

(3-XXIV)

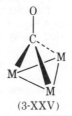

(3-XXV)

Although the carbon monosulfide molecule CS exists only in matrices or in the gas phase at high temperatures, we know of a number of *thiocarbonyl complexes*[14] both neutral and cationic, generally analogous to those of CO. Pure thiocarbonyls $M(CS)_n$ have not so far been made. An example is $W(CO)_5CS$. The CS compounds can be obtained from CS_2 or $CSCl_2$. The bonding of CS to a metal atom closely resembles that of CO but CS is somewhat more strongly bound. With CS the relative contributions of σ-donor versus π-acceptor properties appear to vary more than for CO, and there is a greater variation in CS bond lengths and stretching frequencies.[15]

The poorer C≡S multiple bonding in M—CS compounds means that the C atom in M—CS is markedly more electrophilic than the C atom in M—CO. This leads to greater reactivity toward electrophiles at sulfur and to nucleophiles at carbon. There are reactions[16] such as

$$\text{diphos}_2(CO)WCS + RSO_3F = \text{diphos}_2(CO)WCSR^+ + SO_3F^-$$
$$(CO)_5WCS + H_2NR = (CO)_5WC{=}NR + H_2S$$

$$\eta\text{-}C_5H_5Ru(CO)_2CS^+ + N_3^- = [\eta\text{-}C_5H_5Ru(CO)_2C(S)N_3] \xrightarrow{-N_2} \eta\text{-}C_5H_5Ru(CO)_2NCS$$
$$IrH(CS)(PPh_3)_3 + 2H_2 = IrH_2(SCH_3)(PPh_3)_3$$
$$IrCl_2(CS)(CO)(PPh_3)_2^+ + BH_4^- \rightarrow IrCl_2(SCH)(CO)(PPh_3)_2$$

Bridged CS complexes[17] such as 3-XXVI and 3-XXVII have also been made as well as a bridge type not known for CO, that is, $\text{diphos}_2(CO)WCSW(CO)\text{di-phos}_2$.

(3-XXVI)   (3-XXVII)

3-9. Dinitrogen[18]

The dinitrogen and carbon monoxide molecules are isoelectronic, but although metal carbonyls were discovered by Mond in 1890, the first dinitrogen complex

[14] I. M. Butler, *Acc. Chem. Res.*, 1977, **10**, 359; P. V. Yaneff *Coord. Chem. Rev.*, 1977, **23**, 183; G. Gattow and W. Behrend, *Carbon Sulfides and Their Inorganic and Complex Compounds*, Thieme, 1977.

[15] D. L. Lichtenberger and R. F. Fenske, *Inorg. Chem.*, 1976, **15**, 2015; M. A. Andrews, *Inorg. Chem.*, 1977, **16**, 496; I. S. Butler *et al.*, *Inorg. Chem.*, 1976, **15**, 2602.

[16] R. J. Angelici *et al.*, *Inorg. Chem.*, 1978, **17**, 1634; T. J. Collins and W. R. Roper, *J. Orgomet. Chem.*, 1978, **159**, 73; F. Faraone *et al.*, *J. C. S. Dalton*, **1979**, 931.

[17] R. J. Angelici *et al.*, *J. Organomet. Chem.*, 1978, **160**, 231; *Inorg. Chem.*, 1977, **16**, 1173; A. Efraty *et al.*, *Inorg. Chem.*, 1977, **16**, 3124; R. E. Wagner *et al.*, *J. Organomet. Chem.*, 1978, **148**, C35.

[18] D. Sellman, *Angew. Chem., Int. Ed.*, 1974, **13**, 639; A. D. Allen, *Chem. Rev.*, 1973, **73**, 11; J. Chatt and G. J. Leigh, *Chem. Soc. Rev.*, 1971, **1**, 121; W. G. Zumft, *Struct. Bonding*, 1976, **29**, 1; J. Chatt, J. R. Dilworth, and R. L. Richards, *Chem. Rev.*, 1978, **78**, 589.

was discovered only in 1965 by Allen and Senoff. Some chemistry of dinitrogen complexes is discussed in Section 29-17. The bonding in *linear M—N—N groups* is qualitatively similar to that in terminal M—CO groups; the same two basic components, $M \leftarrow N_2$ σ donation and $M \rightarrow N_2$ π acceptance, are involved. The major quantitative differences, which account for the lower stability of N_2 complexes, appear to arise from small differences in the energies of the MO's of CO and N_2. For CO the σ-donor orbital is weakly antibonding, whereas the corresponding orbital for N_2 is of bonding character. Thus N_2 is a significantly poorer σ donor than is CO. Now it is observed that in pairs of N_2 and CO complexes where the metal and other ligands are identical, the fractional lowerings of N_2 and CO frequencies are nearly identical. For the CO complexes, weakening of the CO bond, insofar as electronic factors are concerned, is due entirely to back-donation from metal $d\pi$-orbitals to CO π^* orbitals, with the σ donation slightly canceling some of this effect. For N_2 complexes, on the other hand, $N\equiv N$ bond weakening results from both σ donation and π back-acceptance. The very similar changes in stretching frequencies for these two ligands suggests then that N_2 is weaker than CO in both its σ-donor and π-acceptor functions. This in turn would account for the poor stability of N_2 complexes in general. Terminal dinitrogen compounds have N—N stretching frequencies in the region 1930–2230 cm^{-1} (N_2 has ν = 2331 cm^{-1}).

Bridging by the N_2 molecule is different from that of CO, and the following types[19] are known.

Symmetric linear	M—N—N—M
Asymmetric linear	M^1—N—N—M^2

Bent

$$M \diagdown \quad \diagup M$$
$$N—N$$

Perpendicular

$$M—\overset{N}{\underset{N}{\vert\vert\vert}}—M$$

For dinitrogen there is a further binding mode that does not occur with CO, save possibly in transition states in reactions and that is *sideways binding*. This bonding is discussed later when other molecules that bond sideways are treated (Section 3-14).

3-10.   Trivalent Phosphorus Compounds[20]

Compounds such as PF_3, PCl_3, $P(C_6H_5)_3$, and $P(OCH_3)_3$, as well as corresponding derivatives of arsenic and antimony are important π-bonding ligands. In particular,

[19]   M. Mercer, *J. C. S. Dalton*, **1974**, 1637; R. D. Sanner *et al.*, *J. Am. Chem. Soc.*, 1976, **98**, 8351, 8358; K. Jonas, *Angew. Chem., Int. Ed.*, 1973, **12**, 997; C. Kruger and Y.-H. Tsay, *Angew. Chem., Int. Ed.*, 1973, **12**, 998; K. Jonas *et al.*, *J. Am. Chem. Soc.*, 1976, **98**, 74.

[20]   (a) C. A. Tolman, *Chem. Rev.*, 1977, **77**, 313, Extensive review with ^{31}P nmr and other data; (b) J. Emsley and D. Hall, *The Chemistry of Phosphorus*, Wiley, 1976; (c) R. Mason and D. W. Meek, *Angew. Chem., Int. Ed.*, 1978, **17**, 183. (See also references in Section 4-20).

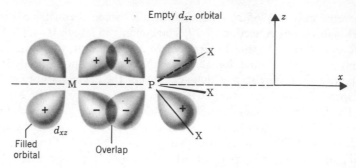

Fig. 3-8. The back-bonding from a filled metal d orbital to an empty phosphorus $3d$ orbital in the PX_3 ligand taking the internuclear axis as the z axis. An exactly similar overlap occurs in the yz plane using the d_{yz} orbitals.

PF_3 gives many compounds comparable to those given by CO. Tertiary phosphines and phosphites are much stronger Lewis bases than CO or PF_3 and will give complexes with non-transition-metal acceptors as well as being easily protonated. Tertiary phosphines and arsines also give many compounds with metals in high positive oxidation states [e.g., $ReCl_3(PPh_3)_3$ or $PtCl_4(PMe_2Ph)_2$]. In such compounds M—P bond lengths show no evidence for π-bonding, though the extent of π-bonding presumably increases as the oxidation state is lowered.

The π-bonding is shown in Fig. 3-8; it differs from that for CO in that the π-acceptor orbitals are the phosphorus $3d$ orbitals. Hence the bonding can be designated as $d\pi - d\pi$ whereas that of CO is $d\pi - p\pi$.

The extent of both donation from the lone pair on the P atom and back-donation depends on the nature of the groups attached to P. For PH_3 and $P(alkyl)_3$, π-acceptor ability is very low, but it becomes important with more electronegative groups. Analogous PX_3, AsX_3, and SbX_3 compounds differ very little, but the ligands having a nitrogen atom, which lacks π orbitals, cause significantly lower frequencies for the CO vibrations, as indicated by the CO stretching frequencies (cm^{-1}) in the following series of compounds:

$(PCl_3)_3Mo(CO)_3$	2040, 1991
$(AsCl_3)_3Mo(CO)_3$	2031, 1992
$(SbCl_3)_3Mo(CO)_3$	2045, 1991
dien $Mo(CO)_3$	1898, 1758

The pronounced effect of the electronegativity of the groups X is shown by the following CO stretching frequencies

$[(C_2H_5)_3P]_3Mo(CO)_3$	1937, 1841
$[(C_6H_5O)_3P]_3Mo(CO)_3$	1994, 1922
$[Cl_2(C_2H_5O)P]_3Mo(CO)_3$	2027, 1969
$(Cl_3P)_3Mo(CO)_3$	2040, 1991
$(F_3P)_3Mo(CO)_3$	2090, 2055

The most electronegative substituent, F in PF_3, will reduce very substantially the σ-donor character so that there will be less P $\rightarrow$ M electron transfer, and $Md\pi \rightarrow$

$Pd\pi$ transfer should be aided. The result is that PF_3 and CO are quite comparable in their π-bonding capacity.

Based on comparative spectroscopic data π ligands can be arranged in order of decreasing π acidity:

$$PF_3 > PCl_3 \sim AsCl_3 \sim SbCl_3 > PCl_2(OR) > PCl_2R > PCl(OR)_2 \sim P(OR)_3$$
$$> PR_3 \sim AsR_3 \sim SbR_3$$

X-Ray structure determination shows that in $(PhO)_3PCr(CO)_5$ the P—Cr bond is 0.11 Å shorter than in $(Ph_3P)Cr(CO)_5$, confirming the greater π acidity of phosphites. Spectroscopic data of other types have been used to confirm the extent of π bonding in PX_3 compounds.

Of at least as great importance to the chemistry of PX_3 compounds as the electronic factors are *steric factors*. Indeed these may be more important or even dominating in determining the stereochemistries and structures of compounds. Steric factors also affect rates and equilibria of dissociation reactions such as eq. 3-1 and the propensity of phosphine complexes to undergo oxidative-addition reactions (Chapter 29).

$$Pd(PR_3)_4 \underset{+PR_3}{\overset{-PR_3}{\rightleftarrows}} Pd(PR_3)_3 \underset{+PR_3}{\overset{-PR_3}{\rightleftarrows}} Pd(PR_3)_2 \tag{3-1}$$

The stereochemistry of phosphine ligands is the prime factor in many highly selective catalytic reactions of phosphine complexes, such as hydroformylation and asymmetric hydrogenation (Chapter 30).

The steric effects[20a] can be correlated with an easily measured parameter—the cone angle θ defined by a conical surface as in 3-XXVIII assuming a metal-phosphorus bond length of 2.28 Å (these are quite constant) that can just enclose the van der Waals surface of all ligand atoms over all rotational orientations about the M—P bond.* Triphenylphosphine has $\theta = 184 \pm 2°$ and $P(OCH_3)_3$ has $\theta = 107 \pm 2°$. It might have been expected that compounds with smaller cone angles would be better ligands, but since such compounds are stronger bases, it is not always easy to distinguish steric from electronic factors. However increasing the cone angle by having bulky groups tends to favor (*a*) lower coordination numbers, (*b*) the formation of less sterically crowded isomers, and (*c*) increased rates and equilibria

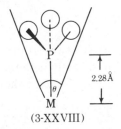

(3-XXVIII)

---

* A more sophisticated treatment calculates actual "ligand profiles," since the ligands are not regular cones. See G. Ferguson *et al., Inorg. Chem.*, 1978, **17**, 2965; J. D. Smith and J. D. Oliver, *Inorg. Chem.*, 1978, **17**, 2585. For an attempt to separate electronic and steric factors, see M. Zähres *et al., Angew. Chem., Int. Ed.*, 1979, **18**, 401.

TABLE 3-5
Dissociation Constants and θ Values for Phosphine Complexes of Nickel

L	K_d, M	$\theta(^\circ)$
$P(OEt)_3$	$< 10^{-10}$	109
PMe_3	$< 10^{-9}$	118
$P\left(O{-}\langle\bigcirc\rangle{-}Cl\right)_3$	2×10^{-10}	128
$P\left(O{-}\langle\bigcirc\rangle{-}Me\right)_3$	6×10^{-10}	128
$P(O\text{-}Pr^i)_3$	2.7×10^{-5}	130
PEt_3	1.2×10^{-2}	132
$P(o\text{-tol})_3$	4×10^{-2}	141
PMe_2Ph	5×10^{-2}	136
PPh_3	Fully dissociated	145

in dissociative reactions. Examples of dissociative equilibria are toward Co meso-porphyrin(IX)dimethyl ester in toluene at 25°C where the order is[21]

$$PBu_3^n \gg P(OMe)_3 > PPh_3 > AsPh_3$$

and for the reaction[20a]

$$NiL_4 \underset{}{\overset{K_d}{\rightleftharpoons}} NiL_3 + L$$

we have the values in Table 3-5. The dissociation of bulky phosphine ligands has profound consequences in catalytic reactions, since it can provide sites on a metal atom in which substrates can come together and react (Chap. 29).

Very bulky phosphines (e.g., with t-butyl groups) can also promote other effects such as (a) barriers to rotation about M—P bonds, (b) stabilization of unusual low coordination numbers and valencies, and (c) propensity to undergo cyclometallation reactions (Section 29-6).[22]

The importance of π acceptance in stabilizing bonds from poor σ donors to metal atoms is dramatically illustrated by the fact that $(CH_3)_3S^+$, which is isoelectronic with $(CH_3)_3P$, but certainly an exceedingly poor donor because of the positive charge, forms bonds to metal atoms and appears comparable to PCl_3 in its π acidity.[23]

3-11. Nitric Oxide[24]

The NO molecule is closely akin to CO except that it contains one more electron, which occupies a π^* orbital. Consistently with the general similarity of CO and NO, they form many comparable complexes, although, as a result of the presence

[21] T. Takayanagi, H. Yamamoto, and T. Kwen, *Bull. Soc. Chem. Jpn.*, 1975, **48**, 2618.

[22] For examples, see B. L. Shaw *et al., J. C. S. Dalton*, **1977**, 2285; **1978**, 257.

[23] R. D. Adams and D. F. Chodosh, *J. Am. Chem. Soc.*, 1978, **100**, 812.

[24] (a) F. Bottomley, *Acc. Chem. Res.*, 1978, **11**, 158; (b) R. Eisenberg and C. D. Meyer, *Acc. Chem. Res.*, 1975, **8**, 26; (c) J. H. Enemark and R. D. Feltham, *Coord. Chem. Rev.*, 1974, **13**, 339; (d) N. G. Connelly, *Inorg. Chim. Acta Rev.*, 1972, **6**, 48; (e) K. G. Caulton, *Coord. Chem. Rev.*, 1975, **14**, 317.

of the additional electron, NO also forms a class (bent MNO) with no carbonyl analogues.

Linear, Terminal MNO Groups. Just as the CO group reacts with a metal atom that presents an empty σ orbital and a pair of filled $d\pi$ orbitals, as illustrated in Fig. 3-6, to give a linear MCO grouping with a C $\rightarrow$ M σ-bond and a significant degree of M $\rightarrow$ C π-bonding, so the NO group engages in a structurally and electronically analogous reaction with a metal atom that may be considered, at least formally, to present an empty σ orbital and a pair of $d\pi$ orbitals containing only three electrons. The full set of four electrons for the $Md\pi \rightarrow \pi^*(NO)$ interactions is thus made up of three electrons from M and one from NO. In effect, NO contributes three electrons to the total bonding configuration under circumstances where CO contributes only two. Thus for purposes of formal electron "bookkeeping," the ligand NO can be regarded as a three-electron donor in the same sense as the ligand CO is considered a two-electron donor. This leads to the following very useful general rules concerning stoichiometry, which may be applied without specifically allocating the difference in the number of electrons to any particular (i.e., σ or π) orbitals:

1. Compounds isoelectronic with one containing an $M(CO)_n$ grouping are those containing $M'(CO)_{n-1}(NO)$, $M''(CO)_{n-2}(NO)_2$, and so on, where M', M'', and so on, have atomic numbers that are 1, 2, . . . , etc. less than M. Some examples are: $(\eta\text{-}C_5H_5)CuCO$, $(\eta\text{-}C_5H_5)NiNO$; $Ni(CO)_4$, $Co(CO)_3NO$, $Fe(CO)_2(NO)_2$, $Mn(CO)(NO)_3$; $Fe(CO)_5$, $Mn(CO)_4NO$.

2. Three CO groups can be replaced by two NO groups. Examples of pairs of compounds so related are

$$
\begin{array}{ll}
\text{Fe(CO)}_5, & \text{Fe(CO)}_2(\text{NO})_2 \\
\text{Mn(CO)}_4\,\text{NO}, & \text{Mn(CO)(NO)}_3 \\
\text{Cr(CO)}_6 & \text{Cr(NO)}_4
\end{array}
$$

It should be noted that the designation "linear MNO group" does not disallow a small amount of bending in cases where the group is not in an axially symmetric environment, just as with terminal MCO groups. Thus MNO angles of 161° to 175° may be found in "linear" MNO groups. Truly "bent MNO groups" have angles of 120 to 140° (see below).

In compounds containing both MCO and linear MNO groups, the M—C and M—N bond lengths differ by a fairly constant amount, ~ 0.07 Å, approximately equal to the expected difference in the C and N radii, and suggest that under comparable circumstances M—CO and M—NO bonds are typically about equally strong. In a chemical sense the M—N bonds appear to be stronger, since substitution reactions on mixed carbonyl nitrosyl compounds typically result in displacement of CO in preference to NO. For example, $Co(CO)_3NO$ reacts with a variety of R_3P, X_3P, amine, and RNC compounds, invariably to yield the Co-$(CO)_2(NO)L$ product.

The NO vibration frequencies for linear MNO groups substantiate the idea of extensive M—N π bonding, leading to appreciable population of NO π^* orbitals. Both the NO and O_2^+ species contain one π^* electron and their stretching frequencies are 1860 and 1876 cm$^-$, respectively. Thus the observed frequencies in the range 1800–1900 cm^{-1}, which are typical of linear MNO groups in molecules with small or zero charge, indicate the presence of approximately one electron pair shared between metal $d\pi$ and NO π^* orbitals.

Bridging NO Groups. These occur less commonly than bridging CO groups, but there are the same two types, doubly and triply bridging. A triply bridging NO group occurs in the compound $(\eta\text{-}C_5H_5)_3Mn_3(NO)_4$, which also contains three doubly bridging NO's.[25]

A symmetrical doubly bridging NO group occurs in

$$(\eta\text{-}C_5H_5)(NO)Cr(\mu\text{-}NO)(\mu\text{-}NH_2)Cr(NO)(\eta\text{-}C_5H_5)$$

and quite unsymmetrical doubly bridging NO groups occur in $\eta\text{-}C_5H_5(NO_2)Mn(\mu\text{-}NO)_2Mn(\eta\text{-}C_5H_5)NO$.

Just as with the corresponding types of bridging CO groups, the NO stretching frequencies decrease with the extent of the bridging. Thus in $(\eta\text{-}C_5H_5)_3Mn_3(NO)_4$ there are two bands due to the doubly bridging NO groups at 1543 and 1481 cm^{-1} and one from the triply bridging group at 1320 cm^{-1}. In $(\eta\text{-}C_5H_5)(NO)Cr(\mu\text{-}NO)(\mu\text{-}NH_2)Cr(NO)(\eta\text{-}C_5H_5)$ the terminal NO groups absorb at 1644 cm^{-1} and the bridging group has a frequency of 1505 cm^{-1}.

Bridging NO groups are also to be regarded as three-electron donors. The doubly bridging ones may be represented as

$$\overset{\cdot}{N} :: \overset{\cdot\cdot}{\underset{\cdot\cdot}{O}}$$

where the additional electron required to form two metal-to-nitrogen single bonds is supplied by one of the metal atoms. The situation is formally quite analogous to that for bridging halogen atoms.

Bent, Terminal MNO Groups. It has long been known that NO can form single bonds to univalent groups such as halogens and alkyl radicals, affording the bent species

$$\underset{X}{\overset{\cdot\cdot}{N}}{=}\overset{\cdot\cdot}{\underset{\cdot\cdot}{O}} \quad \text{and} \quad \underset{R}{\overset{\cdot\cdot}{N}}{=}\overset{\cdot\cdot}{\underset{\cdot\cdot}{O}}$$

Metal atoms with suitable electron configurations and partial coordination shells may bind NO in a similar way. This type of NO complex is formed when the incompletely coordinated metal ion L_nM would have a $t_{2g}^6 e_g$ configuration, thus being prepared to form one more single σ bond. Table 3-6 lists some compounds in which

[25] R. C. Elder, *Inorg. Chem.*, 1974, **13**, 1037.

TABLE 3-6
Some Compounds with Bent MNO Groups

	∠ MNO (deg)	ν_{NO} (cm^{-1})
[Co en$_2$Cl(NO)]$^+$	121	1611
[IrCl(CO)(PPh$_3$)$_2$NO]$^+$	124	1680
IrCl$_2$(NO)(PPh$_3$)$_2$	123	1560
IrI(CH$_3$)(NO)(PPh$_3$)$_2$	120	1525
[RuCl(NO)(PPh$_3$)$_2$NO]$^+$	136	1687[a]
Co[S$_2$CN(CH$_3$)$_2$]$_2$(NO)	139	1626

[a] The other NO group is of the linear MNO type and its stretching frequency is 1845 cm^{-1}.

this type of M—NO structure has been demonstrated by X-ray crystallography.

The NO stretching frequencies for the authenticated cases fall in the range 1525–1690 cm^{-1}, that is, generally lower than those for linear MNO systems, except perhaps when the latter occur in anionic complexes, such as [Cr(CN)$_5$(NO)]$^{4-}$ (ν_{NO} = 1515 cm^{-1}). Tentatively, at least, this may be used as a criterion of structure type. In organic nitroso compounds, RNO, ν_{NO} is generally found in the 1500–1600 cm^{-1} range.

Finally it may be noted that although many CS compounds are known, there are so far only a few with M—NS links.[26] These have been made by nucleophilic attacks of nitrido complexes on S$_8$, for example,

$$(Et_2NCS_2)_3Mo{\equiv}N + S = (Et_2NCS_2)_3MoNS$$

or by the reaction

$$Na[C_5H_5Cr(CO)_3] + \tfrac{1}{3}S_3N_3Cl_3 = C_5H_5Cr(CO)_2NS + CO + NaCl$$

In (η-C$_5$H$_5$)Cr(CO)$_2$NS the Cr—N—S group is essentially linear.

3-12. Isocyanides[27]

Isocyanide complexes can be obtained by direct substitution reactions of the metal carbonyls and in other ways. They include such crystalline, air-stable compounds as red Cr(CNPh)$_6$, white [Mn(CNCH$_3$)$_6$]I, and orange Co(CO)(NO)(CNC$_7$H$_7$)$_2$, all of which are soluble in benzene.

Isocyanides generally appear to be stronger σ-donors than CO, and various complexes such as [Ag(CNR)$_4$]$^+$, [Fe(CNR)$_6$]$^{2+}$, and [Mn(CNR)$_6$]$^{2+}$ are known where π bonding is of relatively little importance; derivatives of this type are not known for CO. However the isocyanides are capable of extensive back-acceptance of π electrons from metal atoms in low oxidation states. This is indicated qualitatively by their ability to form compounds such as Cr(CNR)$_6$ and Ni(CNR)$_4$, analogous to the carbonyls and more quantitatively by comparison of CO and CN stretching frequencies. As shown in Table 3-7, the extent to which CN stretching

26 P. Legzdins et al., J. C. S. Chem. Comm., **1978**, 1036; J. Am. Chem. Soc., 1978, **100**, 2247; J. Chatt et al., J. C. S. Dalton, **1979**, 1.

27 P. M. Treichel, Adv. Organomet. Chem., 1973, **11**, 21; F. Bonati and G. Minghetti, Inorg. Chim. Acta, 1974, **9**, 95.

TABLE 3-7

Lowering of CO and CN Frequencies in Analogous Compounds, Relative to Values for Free CO and CNAr[a]

Molecule[b]	Δv (cm^{-1}) for each fundamental mode		
$Cr(CO)_6$	43	123	160
$Cr(CNAr)_6$	68	140	185
$Ni(CO)_4$	15	106	
$Ni(CNAr)_4$	70	125	

[a] Ar represents C_6H_5 and p-$CH_3OC_6H_4$.

[b] Data for isonitriles from F. A. Cotton and F. Zingales, *J. Am. Chem. Soc.*, 1961, **83**, 351.

frequencies in $Cr(CNAr)_6$ and $Ni(CNAr)_4$ molecules are lowered relative to the frequencies of the free CNAr molecules exceeds that by which the CO modes of the corresponding carbonyls lie below the frequency of CO.

Structural evidence for the ability of isocyanides to form π bonds is provided by the Cr—C bond length[28] of 1.94 Å in $Cr(CNC_6H_5)_6$, which is very similar to the value of 1.91 Å in $Cr(CO)_6$. The same compound, however, also shows the ability, mentioned above, of isocyanides to function as good σ donors without extensive π-bonding. Although $Cr(CO)_6$ cannot be oxidized to $Cr(CO)_6^{n+}$ cations, the $Cr(CNPh)_6^{+,2+}$ ions can be isolated as PF_6^- or BPh_4^- salts that are thermally stable at room temperature.[29] There are several compounds containing bridging isocyanide groups[30a] of the types:

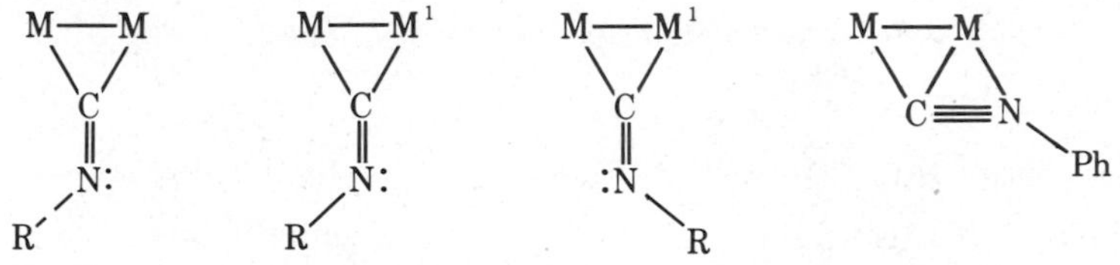

If the metals are different, there can be isomers, whereas for doubly bridging systems there will be syn and anti isomers as found in $(\eta$-$C_5H_5)_2Fe(CO)_2(CNMe)_2$[30b]; these may interconvert rapidly.

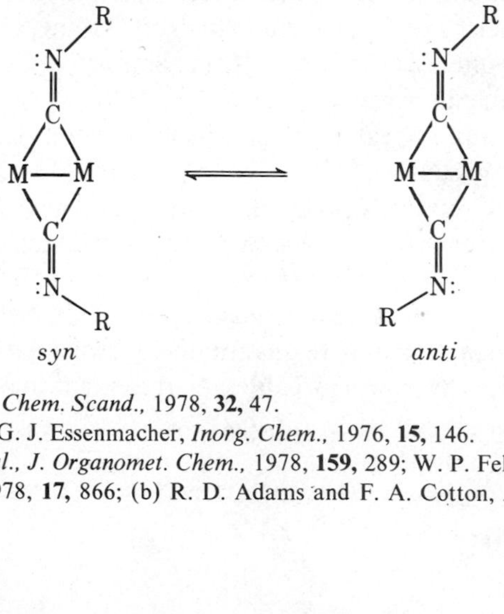

syn   anti

[28] E. Ljunstrom, *Acta Chem. Scand.*, 1978, **32**, 47.

[29] P. M. Treichel and G. J. Essenmacher, *Inorg. Chem.*, 1976, **15**, 146.

[30] (a) A. L. Balch *et al.*, *J. Organomet. Chem.*, 1978, **159**, 289; W. P. Fehlhammer *et al.*, *Angew. Chem., Int. Ed.*, 1978, **17**, 866; (b) R. D. Adams and F. A. Cotton, *Inorg. Chem.*, 1974, **13**, 249.

Finally the iso form of hydrogen cyanide can form complexes. These are usually obtained by protonation of cyanide complexes, for example,

$$Fe(phen)_2(CN)_2 + 2H^+ \rightleftharpoons [Fe(phen)_2(CNH)_2]^{2+}$$

3-13. Orders of π Bonding

We have dealt with the major π-bonding ligands, yet there are others in which π bonding plays a part. These ligands are (a) compounds of sulfur donors, such as thioethers, (b) ligands with extended π systems such as 2,2'-bipyridyl, 1,2-dithiolenes, and related sulfur ligands such as dithiocarbamates, and (c) compounds with multiply bonded groups such as M=O, M=N, and M=NR. All these ligands are considered under their particular donor atoms in Chapter 4.

It has been noted that from comparisons of CO stretching frequencies in substituted carbonyls it is possible to arrange a variety of ligands in order of π-bonding capacity, as done earlier, for phosphorus, arsenic, and antimony donors. Orders based on other sorts of data can be derived; for example, that from trans effect studies (Section 28-7) gives CO, CN^-, $C_2H_4 > PR_3$, $H^- > CH_3^-$, $SC(NH_2)_2 > C_6H_5^-$, NO_2^-, I^-, $SCN^- > Br^- > Cl^- > py$, NH_3, OH^-, OH_2, which also includes some non-π-bonding ligands. Such orders are semiquantitative at best, but there are evidently extremes of good and poor π-bonding ligands.

π COMPLEXES OF UNSATURATED ORGANIC MOLECULES

Molecules that have multiple bonds (C=C, C≡C, C=O, C=N, S=O, N=O, etc.) can form what are called π complexes with transition metals (Chapter 27).

3-14. Alkenes[31]

The most important π complexes are those of compounds with C=C bonds. The earliest known organotransition metal complex was discovered by Zeise in Copenhagen in about 1845, but the true constitution was not recognized until the 1950s. Zeise's salt, $K[Pt(C_2H_4)Cl_3]$, has ethylene bound to it as shown[32a] in Fig. 3-9a.

The key point is that the C=C axis of the coordinated alkene is perpendicular to one of the expected bond directions from the metal. The expected line of a bond orbital from the metal strikes the C=C bond at its midpoint (though for unsymmetrical alkenes, this need *not* be so).

Certain other molecules may be bound sideways, the most important of these being dinitrogen and dioxygen. The structures[32b] of the three square complexes, *trans*-$RhClL(PPr_3^i)_2$, L = C_2H_4, N_2, and O_2 are shown in Fig. 3-9b, c, and d.

[31] S. D. Ittel and J. A. Ibers, *Adv. Organomet. Chem.*, 1976, **14**, 33; D. P. Mingos, *Adv. Organomet. Chem.*, 1977, **15**, 1.

[32] (a) R. A. Love *et al.*, *Inorg. Chem.*, 1975, **14**, 2653; (b) C. Busetto *et al.*, *J. C. S. Dalton*, **1977**, 1828.

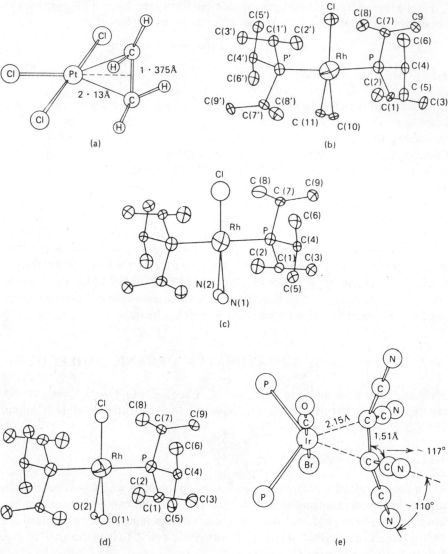

Fig. 3-9. (a) The structure of the ion in Zeise's salt. (b, c, d) The structures of ethylene, nitrogen, and oxygen complexes, *trans*-RhClL(PPr$^i_3$)$_2$. [Reproduced by permission from C. Busetto *et al.,* Ref. 32b.] (e) The structure of the tetracyanoethylene complex IrBr(CO)[(CN)$_2$C=C(CN)$_2$](PPh$_3$)$_2$. (PPh$_3$)$_2$.

The most generally useful description of the bonding was developed for copper-alkene complexes by M. J. S. Dewar and later extended to other transition metals. Fig. 3-10 illustrates the assumption that as with other π-bonding ligands like CO, there are *two* components to the total bonding: (*a*) overlap of the π-electron density of the olefin with a σ-type acceptor orbital on the metal atom and (*b*) a "back-bond" resulting from flow of electron density from filled metal d_{xz} or other

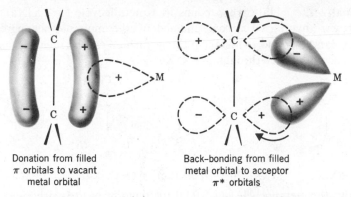

Donation from filled
π orbitals to vacant
metal orbital

Back–bonding from filled
metal orbital to acceptor
π* orbitals

Fig. 3 10. The molecular orbital views of olefin-metal bonding according to Dewar.

$d\pi$-$p\pi$ hybrid orbitals into *antibonding* orbitals on the carbon atoms. This view is thus similar to that discussed for the bonding of carbon monoxide and similar weakly basic ligands and implies the retention of appreciable "double-bond" character in the olefin. Of course, the donation of π-bonding electrons to the metal σ orbital and the introduction of electrons into the π-antibonding orbital both weaken the π-bonding in the olefin, and in every case except the anion of Zeise's salt there is significant lengthening of the olefin C—C bond.

The important qualitative idea about metal-olefin bonding is that the bonding has dual character. There is donation of those electrons initially forming the C—C π bond into a metal orbital of suitable symmetry, and there is donation of electrons from filled metal orbitals of suitable symmetry back into the π-antibonding orbitals of the olefin. As in the CO case, the two components are synergically related. As one component increases, it tends to promote an increase in the other. On both theoretical and experimental grounds, it appears that the metal-alkene bond is essentially electroneutral, with donation and back-acceptance approximately balanced.

Although as we have seen, in crystalline compounds the C—C axis of ethylene is perpendicular to the plane of the square coordination shell of the metal, in solution the nmr spectra show that there is rotation of the alkene about the metal-alkene axis. Extensive π-bonding would be expected to hinder the rotation as is the case for C=C bonds. However in some cases the barrier is relatively small; for example, in $Os(C_2H_4)(CO)(NO)(PPh_3)_2$ it is about 40 kJ mol^{-1}, whereas in other cases such as $(\eta$-$C_5H_5)W(CO)_2(C_2H_4)CH_3$ there appears to be more hindrance.[33]

In complexes containing tetracyanoethylene (Fig. 3-9e) or F_2C=CF_2[34] the C—C bond is about as long as a normal single bond and the angles within the $C_2(CN)_4$ or C_2F_4 ligand suggest that the carbon atoms bound to the metal approach tetrahedral hybridization. Indeed, it is possible to formulate the bonding as involving

[33] H. G. Alt, J. A. Schwärzle, and C. G. Kreiter, *J. Organomet. Chem.,* 1978, **153**, C7; W. Porzio and M. Zocchi, *J. Am. Chem. Soc.,* 1978, **100**, 2048; C. Bachmann, J. Demuynck, and A. Veillard, *J. Am. Chem. Soc.,* 1978, **100**, 2366.

[34] D. R. Russell and P. A. Tucker, *J. C. S. Dalton,* **1975**, 1752.

two normal $2e$-$2c$ metal-carbon bonds in a metallocycle (3-XXIXa) with approximately sp^3 hybridized carbon. A number of other molecules that have multiple bonds and can be bound to metals in the η^2 fashion can be regarded as forming metallocycles 3-XXIXb through 3-XXIXe.

| (3-XXIXa) | (3-XXIXb) | (3-XXIXc) | (3-XXIXd) | (3-XXIXe) |

Actually, the metallocycle view and the π-donor view are neither incompatible nor mutually exclusive but are complementary, with a smooth graduation of one description into the other. The one to be preferred in any given case depends on the extent to which the double bond of the ligand has been reduced to a single bond. From a formal point of view, however, the metallocycle view entails a problem with oxidation state. For example, a compound such as $Ni(C_2F_4)(CO)_3$ could be regarded as a nickel(II) rather than a nickel(0) complex. Clearly, in a compound such as $Pt(C_2H_4)_3$, it would be absurd to propose Pt^{VI}. It is best to regard molecules bound sideways as neutral ligands that do not alter the formal oxidation state.

Conjugated Alkenes. When two or more conjugated double bonds are engaged in bonding to a metal atom the interactions become more complex, though qualitatively the two types of basic, synergic components are involved. The case of the 1,3-butadiene unit is an important one and shows why it would be a drastic oversimplification to treat such cases as simply collections of separate monoolefin-metal interactions.

Two extreme formal representations of the bonding of 1,3-butadiene to a metal atom are possible (Fig. 3-11). The structure b would imply that bonds 1–2 and 3–4 should be longer than bond 2–3. In $C_4H_6Fe(CO)_3$ the bond lengths are approximately the same and ^{13}C–H coupling constants in the nmr spectra indicate that the hybridization at carbon still approximates to sp^2. However in some other compounds of conjugated cyclic alkenes, the pattern is of the long-short-long type, indicating some contribution from this extreme structure.

Alkynes. An alkyne $RC{\equiv}CR'$ can use only one pair of π electrons and bond to a metal atom in the same way as does an olefin. However complexes of this sort

(a)   (b)

Fig. 3-11. Two extreme formal representations of the bonding of a 1,3-butadiene group to a metal atom: (a) implies that there are two more or less independent monoolefin metal interactions; (b) depicts σ bonds to C-1 and C-4 coupled with a monoolefin metal interaction to C-2 and C-3.

are rare, and acetylenes are most commonly found[35] in a bridging posture, using both pairs of π electrons, as in 3-XXX. There may be at least one case[36] of an alkyne that may be thought of as a four-electron donor to one metal atom, but this is at best atypical.

(3-XXX)

3-15.   Aromatic Ring Systems[37]

Just as the π electrons of alkenes can interact with metal d orbitals, so can certain of the delocalized π-electron ring systems of aromatic molecules overlap with d_{xz} and d_{yz} metal orbitals.

The first example of this type of complex was the molecule $Fe(C_5H_5)_2$, now known as *ferrocene,* in which the 6π-electron system of the ion $C_5H_5^-$ is bound to the metal. Other aromatic systems with the "magic numbers" of 2, 6, and 10 for the aromatic electronic configuration are the carbocycles:

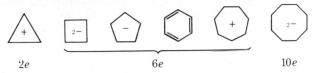

| 2e | 6e | 10e |

The C_5H_5, C_6H_6, and C_8H_8 rings are the most common in arene complexes, but the C_7H_7 and C_4H_4[38] systems also occur frequently. It should also be noted that for purposes of electron counting the ring system and the metal atom may be considered as neutral. For example, the total of eighteen electrons in ferrocene can be regarded as five per C_5H_5 ring plus eight from Fe.

Compounds are known that have only π-bonded rings such as ferrocene (3-XXXI), dibenzenechromium (3-XXXII), or $(C_8H_8)_2U$ (3-XXXIII), but there are many compounds with one ring and other ligands such as halogens, CO, RNC, and R_3P. Examples are η-$C_5H_5Mn(CO)_3$ and η-$C_5H_5Fe(CO)_2Cl$. The symbol η is used to signify that all carbon atoms of the ring are bonded to the metal atom. There are also molecules in which two different types of arene ring are present, such

35   W. I. Bailey, Jr., *et al., J. Am. Chem. Soc.,* 1978, **100,** 5764; V. W. Day *et al., J. Am. Chem. Soc.,* 1976, **98,** 8289.
36   L. Ricard *et al., J. Am. Chem. Soc.,* 1978, **100,** 1318.
37   (a) H. Werner, *Angew. Chem., Int. Ed.,* 1977, **16,** 1 (sandwich compounds); (b) W. E. Silverthorne, *Adv. Organomet. Chem.,* 1975, **13,** 48 (arene complexes); (c) R. E. Riley and R. E. Davis, *J. Am. Chem. Soc.,* 1976, **15,** 2735 (heterocyclic systems); (d) J. W. Lauher and R. Hoffmann, *J. Am. Chem. Soc.,* 1976, **98,** 1729 (bonding in C_5H_5M compounds); (e) K. R. Gordon and K. D. Warren, *Inorg. Chem.,* 1978, **17,** 987. Magnetic and spectroscopic data for $(C_5H_5)_2M$. See also Chapter 27.
38   A. Efraty, *Chem. Rev.,* 1977, **77,** 691; W. Stallings and J. Donohue, *J. Organomet. Chem.,* 1977, **139,** 143.

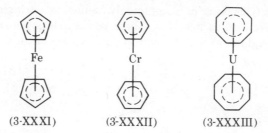

(3-XXXI)   (3-XXXII)   (3-XXXIII)

that the total number of π electrons they provide, plus those possessed by the metal atom itself, add to eighteen. For example, in 3-XXXIV,[39] there are five π electrons from C_5H_5, four from C_4R_4, and nine from Co. Similarly, we have $(\eta\text{-}C_5H_5)(\eta\text{-}C_6H_6)Mn$.

It is also possible for heterocyclic arene rings to form complexes,[40] examples being $(\eta\text{-}C_4H_4N)Mn(CO)_3$, $(\eta\text{-}C_4H_4S)Cr(CO)_3$, $(\eta\text{-}C_5H_5)(\eta\text{-}C_4H_4N)Fe$, $(\eta\text{-}C_5H_5)\text{-}(\eta\text{-}C_4H_4P)Fe$, and 3-XXXV.

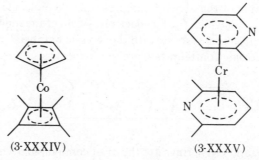

(3-XXXIV)   (3-XXXV)

The basic qualitative features of the bonding in ferrocene are well understood, and will serve to illustrate the basic principles for all $(\eta\text{-}C_nH_n)M$ bonding, although for $(C_8H_8)M$ systems there are a few additional points that are covered later.

The discussion of bonding does not depend critically on whether the preferred rotational orientation of the rings (see Fig. 3-12) in an $(\eta\text{-}C_5H_5)_2M$ compound is

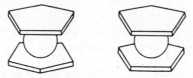

Fig. 3-12. Staggered and eclipsed configurations of an $(\eta\text{-}C_5H_5)_2M$ compound. In crystalline ferrocene there are molecules of different orientations randomly distributed throughout the crystal (P. Seiler and J. D. Dunitz, *Acta Cryst.*, 1979, **B35**, 1068). Also, the H-atoms of the rings are bent towards the metal (F. Takasagawa and J. F. Koetzle, *Acta Cryst.*, 1979, **B35**, 1074).

[39]   P. E. Riley and R. E. Davis, *J. Organomet. Chem.*, 1976, **113**, 157; A. Clearfield *et al.*, *J. Organomet. Chem.*, 1977, **135**, 229.
[40]   P. E. Riley and R. E. Davis, *Inorg. Chem.*, 1976, **15**, 2735; F. Mathey, A. Mitchler, and R. Weiss, *J. Am. Chem. Soc.*, 1977, **99**, 3537.

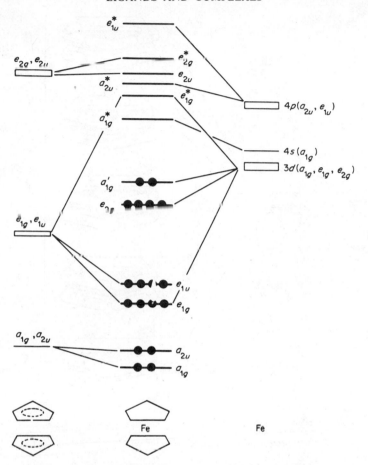

Fig. 3-13. An approximate MO diagram for ferrocene. Different workers often disagree about the exact order of the MOs; the order shown here, especially for the antibonding MOs, may be incorrect in detail, but the general pattern is widely accepted.

staggered (D_{5d}) or eclipsed (D_{5h}); nor is that question unequivocally settled. It is experimentally certain that in ferrocenes the barrier to rotation is only about 8 to 20 kJ mol^{-1}. The eclipsed configuration may be the more stable, but in condensed phases, especially crystals, where there are intermolecular energies of the same or greater magnitude than the barrier, either configuration may be found.

The bonding is best treated in the linear combination of atomic orbitals (LCAO-MO) approximation. A semiquantitative energy level diagram is given in Fig. 3-13. Each C_5H_5 ring, taken as a regular pentagon, has five π MOs, one strongly bonding (a), a degenerate pair that are weakly bonding (e_1), and a degenerate pair that are markedly antibonding (e_2), as shown in Fig. 3-14. The pair of rings taken together then has ten π orbitals and, if D_{5d} symmetry is assumed, so that there is a center of symmetry in the (η-C_5H_5)$_2$M molecule, there will be centrosymmetric (g) and antisymmetric (u) combinations. This is the origin of the

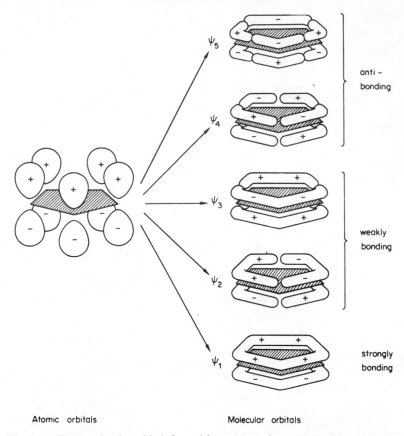

Atomic orbitals   Molecular orbitals

Fig. 3-14. The π molecular orbitals formed from the set of $p\pi$ orbitals of the C_5H_5 ring.

set of orbitals shown on the left of Fig. 3-13. On the right are the valence shell ($3d$, $4s$, $4p$) orbitals of the iron atom. In the center are the MOs formed when the ring π orbitals and the valence orbitals of the iron atom interact.

For $(\eta\text{-}C_5H_5)_2Fe$, there are eighteen valence electrons to be accommodated: five π-electrons from each C_5H_5 ring and eight valence shell electrons from the iron atom. It will be seen that the pattern of MOs is such that there are exactly nine bonding or nonbonding MOs and ten antibonding ones. Hence the eighteen electrons can just fill the bonding and nonbonding MOs, giving a closed configuration. Since the occupied orbitals are either of a type (which are each symmetric around the 5-fold molecular axis) or they are *pairs* of e_1 or e_2 type, which are also, *in pairs*, symmetrical about the axis, no intrinsic barrier to internal rotation is predicted. The very low barriers observed may be attributed to van der Waals forces directly between the rings.

Figure 3-13 indicates that among the principal bonding interactions is that giving rise to the strongly bonding e_{1g} and strongly antibonding e_{1g}^* orbitals. To give one concrete example of how ring and metal orbitals overlap, the nature of this particular important interaction is illustrated in Fig. 3-15. This particular interaction

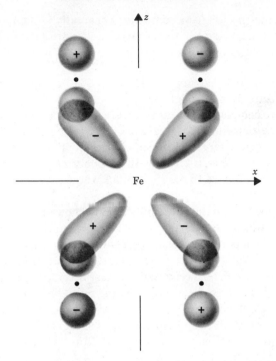

Fig. 3-15. Overlapping of one of the e_1-type d orbitals d_{xz} with an e_1-type π orbital to give a delocalized metal-ring bond: cross-sectional view taken in the xz plane.

is in general the most important single one because the directional properties of the e_1-type d orbitals (d_{xz}, d_{yz}) give an excellent overlap with the e_1-type ring π orbitals, as Fig. 3-15 shows.

Systems containing only one η-C_5H_5 ring include (η-C_5H_5)Mn(CO)$_3$, (η-C_5H_5)Co(CO)$_2$, (η-C_5H_5)NiNO, and (η-C_5H_5)CuPR$_3$. The ring-to-metal bonding in these cases can be accounted for by a conceptually simple modification of the picture given above for (η-C_5H_5)$_2$M systems. In each case a principal axis of symmetry can be chosen so as to pass through the metal atom and intersect the ring plane perpendicularly at the ring center; in other words, the C_5H_5M group is a pentagonal pyramid, symmetry C_{5v}. The single ring may then be considered to interact with the various metal orbitals in about the same way as do each of the rings in the sandwich system. The only difference is that opposite to this single ring is a different set of ligands which interact with the opposite lobes of, for example, the de_1 orbitals, to form their own appropriate bonds to the metal atom.

The MO picture of ring-to-metal bonding just given for ferrocene is *qualitatively* but *not quantitatively* applicable to other (η-C_5H_5)$_2$M compounds, and with obvious modifications, to other (η-arene)$_2$M molecules. Relative orbital energies can and do change as the metal is changed. Ferrocene itself has been treated by very elaborate MO calculations,[41] and at the other extreme a ligand field model has also

[41] P. S. Bagus, U. I. Walgren, and J. Almlof, *J. Chem. Phys.*, 1976, **64**, 2324.

been applied to the $(\eta\text{-}C_5H_5)_2M$ compounds generally.[42] Although it is clearly unrealistic to treat the bonding in these systems as electrostatic (i.e., M^{2+} and two negatively charged rings), this approach is of some utility in interpreting electronic absorption spectra.

It must also be noted that not all "$(\eta\text{-}C_5H_5)_2M$" compounds are as simple structurally or electronically as might naively be expected. Those involving titanium and niobium have complex binuclear structures, as described in detail under the chemistry of these elements. In the case of $(\eta\text{-}C_5H_5)_2Ni$ there are two electrons in excess of those required for an eighteen-electron configuration; nonetheless, the molecule appears to have a structure with parallel rings and the two additional electrons have been assigned to the e_{1g}^* orbitals where they are appreciably delocalized onto the rings.[43]

With manganese the situation is complicated but now understood: $(C_5H_5)_2Mn$ is a brown solid in which the Mn atoms have five unpaired electrons but show strong antiferromagnetic coupling. This coupling results from a chain structure[44] in the crystal (Fig. 3-16). At 159°C the crystal structure changes so that $(C_5H_5)_2Mn$ becomes isomorphous with ferrocene, and the color becomes light orange. However there are still five unpaired electrons; thus the $(\eta\text{-}C_5H_5)_2Mn$ molecule can be considered to be mainly ionic. It is also ionic, with five unpaired electrons, in the gas phase. However $(\eta\text{-}C_5H_4CH_3)_2Mn$ in the gas phase at about 25°C is a roughly 2–1 mixture of high-spin (ionic) molecules with Mn—C distances of 2.42 Å and low-spin (covalent) molecules with Mn—C = 2.14 Å.[45]

For a monocyclopentadienyl compound having two bulky substituents, $(R^1R^2C_5H_3)Co(PMe_3)_2$, rotational isomers have been observed.[46]

Compounds with two $\eta\text{-}C_5H_5$ rings that are *not parallel* are also numerous. They include a number of $(C_5H_5)_2MX_2$ compounds in which M = Ti, Zr, or Mo and X represents a univalent group such as a halogen, H or an R group, as well as others such as $(\eta\text{-}C_5H_5)_2Mo(CO)$ and $(\eta\text{-}C_5H_5)_2NbH(CO)$. The angle subtended at the metal atom by the centroids of the two rings is generally 130° to 135°.

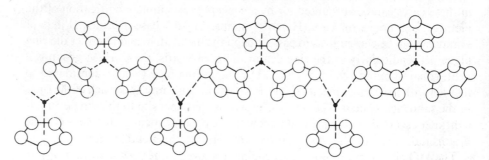

Fig. 3-16. The chain structure of $(C_5H_5)_2Mn$ in the low-temperature, antiferromagnetic crystal form.

[42] K. D. Warren, *Inorg. Chem.,* 1974, **13,** 1243.
[43] W. T. Scroggins, M. F. Rettig, and R. M. Wing, *Inorg. Chem.,* 1976, **15,** 1381.
[44] W. Bünder and E. Weiss, *Z. Naturforsch. B,* 1978, **33,** 1235.
[45] A. Almenningen, S. Samdal, and A. Haaland, *J. C. S. Chem. Comm.,* **1977,** 14.
[46] W. Hofmann, W. Büchner, and H. Werner, *Angew. Chem., Int. Ed.,* 1977, **16,** 795.

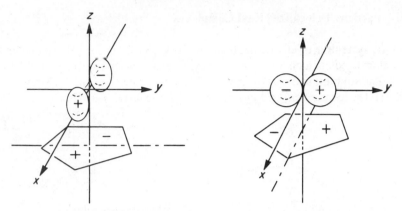

Fig. 3-17. Diagrams showing how p_x and p_y orbitals are symmetry-adapted to overlap the e_1 π orbitals of a C_5H_5 ring.

One of the features of $(\eta\text{-}C_5H_5)_2ReH$ and $(\eta\text{-}C_5H_5)_2MH_2$, (M = Mo, W) is that in addition to the orbitals involved in bonding to the C_5H_5 rings, there are three orbitals that may be used either for bonding to hydrogen or to hold lone pairs. Thus $(\eta\text{-}C_5H_5)_2WH_2$ has one lone pair and can accept a proton giving $(\eta\text{-}C_5H_5)_2WH_3^+$, whereas the molecule $(\eta\text{-}C_5H_5)_2TaH_3$ does not act as a base. The electronic structure of such bent sandwich compounds has been worked out in some detail.[47]

It is also possible to have covalent $(\eta\text{-}C_5H_5)M$ groups even when the metal atom has no valence shell d orbitals, provided it has p orbitals of suitable energy and size. As shown in Fig. 3-17, a pair of p_x and p_y orbitals can overlap with the e_1 π orbitals of C_5H_5 in much the same way as do d_{xz} and d_{yz} orbitals. The C_5H_5In and C_5H_5Tl molecules are the best documented cases of this type of bonding.

Bonding in $(\eta\text{-}C_8H_8)_2M$ Molecules.   Although there are a few cases—for example, $(\eta\text{-}C_8H_8)ZrCl_2(THF)$[48]—of one fully octagonal, planar C_8H_8 ring being symmetrically bonded to a d-block metal and it may be safely assumed that only metal d orbitals are used, the $(\eta\text{-}C_8H_8)_2M$ compounds are formed only by metal atoms like uranium and thorium where the participation of f orbitals might reasonably be expected. The highest half-occupied π orbital of a planar octagonal C_8H_8 ring is of type e_2 and can overlap much better with an e_2-type f orbital than with an e_2 type d orbital. Nonetheless there has been considerable dispute over the extent of such f-orbital participation; photoelectron spectra and MO calculations have been used to support arguments for both much[49] and little[50] f-orbital participation. It does seem likely that f-orbital participation is of some significance.

[47]   J. W. Lauer and R. Hoffmann, *J. Am. Chem. Soc.*, 1976, **98**, 1729; R. D. Wilson *et al.*, *J. Am. Chem. Soc.*, 1977, **99**, 1775; A. J. Schultz *et al.*, *Inorg. Chem.*, 1977, **16**, 3303.
[48]   D. J. Brauer and C. Kruger, *Inorg. Chem.*, 1975, **14**, 3053.
[49]   N. Edelstein *et al.*, *Inorg. Chem.*, 1976, **15**, 1397; J. P. Clark and J. C. Green, *J. C. S. Dalton*, 1977, 505; N. Rösch and A. Streitwieser, Jr., *J. Organomet. Chem.*, 1978, **145**, 195.
[50]   I. Fragola *et al.*, *J. Organomet. Chem.*, 1976, **122**, 357.

3-16. Partially Delocalized Enyl Complexes

Aromatic systems are fully delocalized, but other types of ligand are known that have delocalized open chain systems or cyclic systems with only partial delocalization. The simplest is the *allyl* group. This can be bound as in 3-XXXVI and 3-XXXVII, the former being designated η^3-allyl, the latter σ-allyl or η^1-allyl. Note

η^3 or π-Allyl
(3-XXXVI)

η^1 or σ-Allyl
(3-XXXVII)

that (1) the η^3-allyl group is a three-electron donor, and (2) the hydrogen atoms of the methylene groups are not equivalent. There are *syn*-(H_A) and *anti*-(H_B) protons that can be distinguished by nmr. Many allyls show nonrigid behavior in solution (Section 28-16).

Other more complicated, partly delocalized enyl systems are:

η-1-3-Cyclohexenyl   Three-Electron donor

η-1-5-Cyclohexadienyl   Five-Electron donor

Similar systems can be formed from seven-, eight-, and nine-membered ring hydrocarbons.

General References

Jensen, W. B., *Chem. Rev.,* 1978, **78**, 1. Lewis acids and base behavior, hard-strong acids and bases. Tables of donor strengths, and so on.

Hartley, F. R., *Chem. Soc. Revs.,* 1973, **2**, 163. Cis-trans effects of ligands.

Hoffmann, R., and M. M.-L. Chen, *Inorg. Chem.,* 1977, **16**, 503. Bonding of diatomic molecules to metals.

Malatesta, L., and C. Cenini, *Zerovalent Compounds of Metals,* Academic Press, 1974.

CHAPTER FOUR

Classification of Ligands
by Donor Atoms

Chapter 3 discussed the general features of ligands and the distinction between π-bonding and non-π-bonding types.

This chapter deals with ligands classified according to the atom of the ligand bound to the metal. Detailed chemistry of certain important classes of ligands is described separately, notably for metal carbonyls and related compounds (Chapter 25), cluster compounds, many of which are carbonyls (Chapter 26), and transition metal hydrido and organo compounds (Chapter 29).

HYDROGEN

The chemistry of binary hydrogen compounds except for boranes (Chapter 9), is discussed in Chapter 6, and the transition metal compounds with M—H bonds are dealt with in Chapter 27, and also in Chapter 29. Here we discuss only the ligand behavior of the simpler complex hydrido anions such as BH_4^-, $B_3H_8^-$ and AlH_4^-.

4-1. Complex Hydrido Anions

There are numerous complexes of *tetrahydridoborate*[1] (BH_4^-) with main group elements, *d*-group transition metals, lanthanides, and actinides. Some examples are $Al(BH_4)_3$, $[U(BH_4)_4]^-$, $(R_3P)_2CuBH_4$, $[Mo(CO)_4BH_4]^-$, and $Y(BH_4)_3(THF)_3$. They are usually obtained by interaction of the appropriate halide with $LiBH_4$ or $NaBH_4$ in a solvent like tetrahydrofuran or ether, or by interaction of B_2H_6 with metal alcoxides.

[1]   T. J. Marks and J. R. Kolb, *Chem. Rev.,* 1977, **77**, 263; B. G. Segal and S. J. Lippard, *Inorg. Chem.,* 1978, **17**, 844.

The ligand is bound *via hydrogen bridges:*

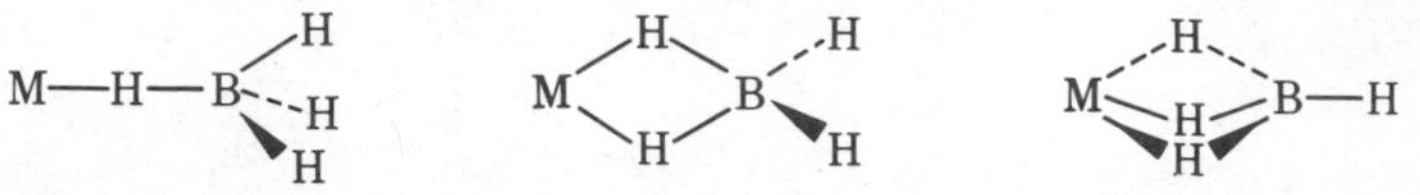

Few unidentate complexes are known, an example being $(Ph_2MeP)_3CuHBH_3$.[2] Bidentate complexes are most common. Many compounds have been structurally characterized, but infrared spectroscopy is a useful criterion; for example, bidentate compounds have an absorption at *ca.* 2500 cm^{-1}.

Nuclear magnetic resonance spectra show that the complexes are commonly nonrigid. In bi- and tridentate species the bridge and terminal hydrogen atoms are undergoing rapid intramolecular exchange and so appear equivalent on the nmr time scale.[3]

The octahydridotriborate ion $B_3H_8^-$ also forms complexes with M—H—B bonds,[4] for example, 4-I.

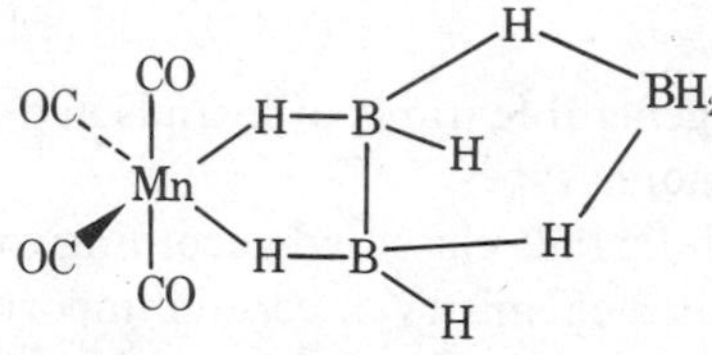

(4-I)

The more complicated borane anions also give complexes.[5] The tetrahydridoaluminate ion (AlH_4^-) forms few complexes, but a number of compounds that have groups of the following type are known.[6]

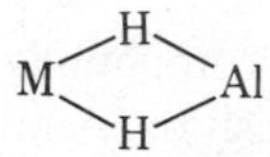

ATOMS OF GROUPS I TO III

Compounds of the elements in Groups I to III do not commonly act as ligands except in special cases, which are discussed separately.

[2] J. L. Atwood *et al., Inorg. Chem.,* 1978, **17**, 3558.

[3] P. L. Johnson *et al., J. Am. Chem. Soc.,* 1978, **100**, 2709; T. J. Marks, *et al., J. Am. Chem. Soc.,* 1977, **99**, 7539; S. W. Kirtley *et al., J. Am. Chem. Soc.,* 1977, **99**, 7154.

[4] S. J. Hildebrandt, D. F. Gaines, and J. C. Calabrese, *Inorg. Chem.,* 1978, **17**, 790; D. F. Gaines and S. J.Hildebrandt, *Inorg. Chem.,* 1978, **17**, 794.

[5] N. N. Greenwood *et al., J.C.S. Dalton,* **1978**, 237; *Pure Appl. Chem.,* 1977, **49**, 791; *Chem. Soc. Rev,* 1974, **3**, 231; T. P. Fehlner, *et al., J. Am. Chem. Soc.,* 1979, **101**, 4390.

[6] See, e.g., T. J. McNeese, S. S. Wreford, and B. M. Foxman, *J.C.S. Chem. Comm.,* **1978**, 500.

4-2. Group I and Group II

There are essentially no cases in which compounds of the electropositive elements in Groups I and II act as ligands in the acknowledged sense. The smallest of the alkali atoms, lithium, can be bound directly to other atoms in molecular compounds; also in polymeric organolithium compounds (Chapter 7) there is strong interaction between lithium and hydrogen atoms of an alkyl group leading to abnormally low C—H stretching frequencies. For the zinc-cadmium-mercury group these atoms, especially mercury, can be bound in compounds such as those of metal carbonyls like $Hg[Co(CO)_4]_2$.

4-3. Group III

The elements boron, aluminum, gallium, indium, and thallium form compounds of varying types. Those of indium and thallium are similar to those of zinc, cadmium, and mercury, noted above. There are a few compounds, mostly of transition metals, with bonds to aluminum, but the major ligand behavior is shown by boron compounds. These are rather special and are sufficiently important to be dealt with separately in Chapter 27; briefly, however, they comprise

1. Boranes acting as ligands but giving M—B as well as M—H—B bonds.
2. Carbaboranes (Chapter 9) acting as ligands comparable to η-C_5H_5 and giving sandwich-type π complexes.
3. η-Complexes of borazine ($B_3N_3H_6$) and related compounds analogous to those formed by arenes.

CARBON

4-4. Organometallic Compounds: General Survey of Types[7]

Organometallic compounds are those in which the *carbon* atoms of organic groups are bound to metal atoms. Thus we do not include in this category compounds in which carbon-containing components are bound to a metal through some other atom such as oxygen, nitrogen, or sulfur. For example, $(C_3H_7O)_4Ti$ is not considered to be an organometallic compound, whereas $C_6H_5Ti(OC_3H_7)_3$ is, because in the latter there is one direct linkage of the metal to carbon. Although organic

---

[7] G. Coates, M. L. H. Green, and K. Wade, *Organometallic Compounds,* 3rd ed., Vol. 1, *Main Group Elements,* 1967; Vol. 2, *Transition Elements,* 1968, Methuen; *Organometallic Chemistry,* Specialist Reports, Chemical Society, London; *Advances in Organometallic Chemistry, Academic Press; J. Organomet. Chem.* (annual reviews); E. Maslowsky, Jr, *Vibrational Spectra of Organometallic Compounds,* Wiley, 1977; J. D. Smith and D. R. M. Walton, *Adv. Organomet. Chem.,* 1975, **13**, 453 (guide to literature for main group elements); D. S. Matteson, *Organometallic Reaction Mechanisms,* Academic Press, 1974; *Comprehensive Organic Chemistry,* Parts 12–15, Vol. 3, Pergamon Press, 1979.

groups can be bound through carbon, in one way or another, to virtually all the elements in the Periodic Table, excluding the noble gases, the term organometallic is usually rather loosely defined and organo compounds of decidedly nonmetallic elements such as boron, phosphorus, and silicon are often included in the category. Specific compounds are discussed in the sections on the chemistry of the individual elements, since the organo derivatives are usually just as characteristic of any element as are, say, its halides or oxides. However, it is pertinent to make a few general comments here on the various types of compound.

1. *Ionic Compounds of Electropositive Metals.* The organometallic compounds of highly electropositive metals are usually ionic. Thus the alkali metal derivatives with the exception of those of lithium, which are fairly covalent, are insoluble in hydrocarbon solvents and are very reactive toward air, water, and so on. The alkaline earth metals calcium, strontium, and barium give poorly characterized compounds that are even more reactive and unstable than the alkali salts. The stability and reactivity of ionic compounds are determined in part by the stability of the carbanion. Compounds containing unstable anions (e.g., $C_nH_{2n+1}^-$) are generally highly reactive and often unstable and difficult to isolate; however where reasonably stable carbanions exist, the metal derivatives are more stable, though still quite reactive [e.g., $(C_6H_5)_3C^-$ Na^+ and $(C_5H_5^-)_2Ca^{2+}$].

2. *σ-Bonded Compounds.* Organo compounds in which the organic residue is bound to a metal by a normal two-electron covalent bond (albeit in some cases with appreciable ionic character) are formed by most metals of lower electropositivity and, of course, by nonmetallic elements. The normal valence rules apply in these cases, and partial substitution of halides, hydroxides, and so on, by organic groups is possible, as in $(CH_3)_3SnCl$ and CH_3SnCl_3, for example. In most of these compounds bonding is predominantly covalent and the chemistry is organic-like, although there are many differences in detail due to factors such as use of higher d-orbitals or donor behavior as in R_4Si, R_3P, R_2S, and so on, incomplete valence shells or coordinative unsaturation as in R_3B or R_2Zn, and effects of electronegativity differences between M—C and C—C bonds.

Although the existence of stable M—C bonds has long been regarded as a normal part of the chemistry of the non-transition metals and metalloids, compounds containing transition metal-to-carbon σ bonds have only in recent years been made in substantial numbers. The reasons for the relative rarity of such compounds are still a subject for investigation, but several points seem clear. First, an important pathway for decomposition of M—R bonds is by a shift of a β-hydrogen atom, followed by olefin elimination, namely;

$$M-CH_2-CHRR' \longrightarrow M \overset{\displaystyle H}{\underset{\displaystyle CCR'}{\overset{|}{\leftarrow}} \| \overset{CH_2}{}} \longrightarrow MH + CH_2{=}CRR'$$

This pathway for decomposition can be substantially inhibited either by (*a*) using groups such as CH_3, or sterically bulky alkyl groups like CH_2SiMe_3 or CH_2Ph that do not have β-hydrogen atoms, or (*b*) blocking the sites required for the transfer reaction by firmly held ligands as in the substitution-inert octahedral species $[RhC_2H_5(NH_3)_5]^{2+}$. Available bond energy data indicate that bonds between transition metals and carbon are comparable in strength to those with nontransition elements. Transition metal compounds are discussed in detail in Chapter 27.

3. *Nonclassically Bonded Compounds.* There are many compounds in which metal-to-carbon bonding cannot be explained as either ionic or covalent in the simple sense of a $2c$-$2e$ M—C bond or bonds. The largest and most important class of these "nonclassical" molecules comprises those formed primarily by the transition elements in which unsaturated groups are attached to metal atoms by interaction of the π electrons with metal orbitals. We have already discussed such ligands briefly in Chapter 3 and they are treated in more detail in Chapter 27.

Another, smaller class of nonclassical compounds is made up of those with *bridging alkyl groups.* The elements boron, aluminum, gallium, indium, and thallium all form fairly stable but reactive trialkyls and triaryls, those of boron, gallium, indium, and thallium being monomeric in the vapor and in solution. $(CH_3)_3In$ and $(CH_3)_3Tl$ form tetramers in their crystals, but the association is weak and uncertain and does not persist in other conditions. The aluminum compounds are unique in Group III in forming several reasonably stable dimers. Thus trimethylaluminum is a dimer in benzene solution and, partly, even in the vapor phase. $AlEt_3$ and $AlPr_3^n$ are also dimeric in benzene solution, but are almost completely dissociated in the vapor phase. $AlPr_3^i$ is a monomer in benzene. The molecules $Al_2(CH_3)_4(C_6H_5)_2$ and $Al_2(C_6H_5)_6$ are also known, as well as the polymeric $[Be(CH_3)_2]_x$. The structures of four of these substances are shown in Fig. 4-1. In all cases the bridging carbon atoms are equidistant from the metal atoms; in short, the $\overline{Al-C-Al-C}$ groups have D_{2h} symmetry.

The dimeric structure has also been established at low temperature in nondonor solvents by ^1H and ^{13}C nmr[8] spectroscopy, although at room temperature an exchange process occurs, leading to apparent equivalence of the terminal and bridging groups (see below).

The bridging is accomplished by means of Al—C—Al $3c$-$2e$ bonds where each Al atom supplies an s-p hybrid orbital and so also does the carbon atom. The situation then is as depicted in Fig. 4-2a. In the case of $Al_2(CH_3)_6$ this view of the bonding has been strongly supported by a low-temperature structure determination. In the case of the bridging phenyl groups, which lie perpendicular to the $\overline{AlCAlC}$ planes, the larger Al—C—Al angles and slight inequalities in the C—C distances about the rings have been taken to mean that the $p\pi$ orbital of the bridging carbon atom may also play some role in the bridge bonding, as indicated by the overlap depicted in Fig. 4-2b.

The fact that none of the alkyls of the Group III elements except those of aluminum are dimerized (except perhaps $GaEt_3$ and trivinylgallium, which appear to be dimers in solution) has not yet been satisfactorily explained. For the larger metals, the small M—C—M angles required to secure good overlap would introduce large repulsions between the bulky metal atoms, but this cannot explain why

[8]   G. A. Olah *et al., Proc. Nat. Acad. Sci. (U. S.),* 1977, **74,** 5217.

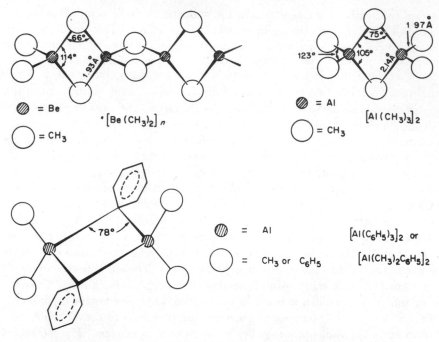

Fig. 4-1. The structures of $[Be(CH_3)_2]_n$ and several dimeric AlR_3 molecules.

$B(CH_3)_3$ does not dimerize, especially since hydrogen bridging is quite important in the boranes.

An important feature of coordinatively unsaturated alkyls, such as those just noted or those of magnesium, zinc, and so on, is the moderately rapid exchange of alkyl groups. The exchanges can be readily studied by nmr methods and it appears

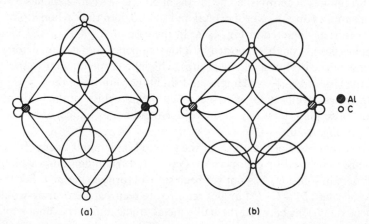

(a)   (b)

Fig. 4-2. Schematic indication of orbital overlaps in Al—C—Al bridge bonding: (a) for a methyl bridge; (b) possible π component in phenyl bridging.

that bridged transition states, or intermediates of the type 4-II and the like, provide

$$Me \diagdown \underset{Me}{\overset{Me}{M}} \diagup \underset{\underset{H_3}{C}}{\overset{\overset{H_3}{C}}{\cdots}} \diagdown \underset{Me}{\overset{Me}{M}} \diagup Me$$

(4-II)

the means for exchange. The rates of exchange reactions are usually slowed by the presence of donor ligands and, if the donor is strong enough to block the coordination sites, the exchange is stopped.

Bridging alkyl groups occur in lithium alkyls and are also known in transition metal compounds such as $\{Mn(CH_2SiMe_3)_2\}_n$ and $Re_3(CH_3)_9(PMe_3)_3$, as discussed in Chapter 27.

4-5. Cyanide[9]

Cyanide complexes are readily formed in aqueous solution by Zn^{2+}, Cd^{2+}, and Hg^{2+} Most transition metals of the d groups form complexes, a well-known example being ferrocyanide, $[Fe(CN)_6]^{4-}$. Some lanthanide complexes are known although other ligands are usually present.

As far as is known, it is always the carbon atom of unidentate CN that is bound to the metal. The π-acceptor behavior of CN^- does not seem to be nearly as high[10a] as for CO, NO^+, or RNC, which is, of course, reasonable in view of its negative charge. Since CN^- is a strong nucleophile, back-bonding need not be invoked to explain the stability of its complexes with metals in normal (i.e., II, III) oxidation states.

Types of Cyano Complex. The majority of cyano complexes have the general formula $[M^{n+}(CN)_x]^{(x-n)-}$ and are anionic, such as $[Fe(CN)_6]^{4-}$, $[Ni(CN)_4]^{2-}$, and $[Mo(CN)_8]^{3-}$. Mixed complexes, particularly of the type $[M(CN)_5X]^{n-}$, where X may be H_2O, NH_3, CO, NO, H, or a halogen, are also well known.

Cyanide can act as a *bridge* group usually of the end-on type M—C—N—M.[10b]

Linear bridges play an important part in the structures of many crystalline cyanides and cyano complexes. Thus AuCN, $Zn(CN)_2$, and $Cd(CN)_2$ are all polymeric with infinite chains.

The free anhydrous acids corresponding to many cyano anions can be isolated, examples being $H_3[Rh(CN)_6]$ and $H_4[Fe(CN)_6]$. These acids are thus different from those corresponding to many other complex ions, such as $[PtCl_6]^{2-}$ or $[BF_4]^-$,

[9] A. G. Sharpe, *The Chemistry of Cyano Complexes of Transition Metals,* Academic Press, 1976 (1800 references); W. P. Griffith, *Coord. Chem. Rev.,* 1975, **17**, 177 (Ti, Cr, V, Mn); L. H. Jones and B. I. Swanson, *Acc. Chem. Res.,* 1976, **9**, 128 (force constants).

[10a] M. S. Lazarus and T. S. Chow, *J. Chem. Phys.,* 1976, **64**, 3544.

[10b] D. N. Hendrickson *et al., Inorg. Chem.,* 1977, **16**, 924; *Inorg. Chem.,* 1974, **13**, 1911; P. L. Gaus and A. L. Crumbliss, *Inorg. Chem.,* 1976, **15**, 2080; D. Gaswick and A. Haim, *J. Inorg. Nucl. Chem.,* 1978, **40**, 437.

which cannot be isolated except as hydroxonium (H_3O^+) salts, and they are also different from metal carbonyl hydrides in that they contain no metal-hydrogen bonds. Instead, the hydrogen atoms are situated in hydrogen bonds between anions (i.e., MCN $\cdots$ H $\cdots$ NCM). Different sorts of structures arise depending on the stoichiometry. For example, in $H[Au(CN)_4]$ there are sheets. For octahedral anions there is a difference in structure depending on whether the number of protons equals half the number of cyanide ions. For $H_3[M(CN)_6]$ compounds an infinite, regular three-dimensional array is formed in which the hydrogen bonds are perhaps symmetrical, whereas in other cases the structures appear to be more complicated.

Metal-Cyanide Bonding. The cyanide ion occupies a very high position in the spectrochemical series (Section 20-12), gives rise to large nephelauxetic effects and produces a strong trans effect (Section 28-7). All these properties are accounted for most easily by postulating M—CN π bonding, and semiempirical MO calculations support this. From close analysis of the vibrational spectra of cyanide complexes, the existence of π bonding has been confirmed more directly, but it does not appear to be nearly as extensive as in carbonyls.

However the cyanide ion does have the ability to stabilize metal ions in low formal oxidation states, and it presumably does this by accepting electron density into its π^* orbitals. The fact that cyano complexes of zerovalent metals are generally much less stable (in a practical as opposed to a well-defined thermodynamic or chemical sense) than similar metal carbonyls has often been taken to show the poor π-acidity of CN^-, but it should be noted that the cyano compounds—for example, $[Ni(CN)_4]^{4-}$—are anionic and might tend to be more reactive for this reason alone.

In some instances cyano complexes are known in two or even three successive oxidation states, $[M(CN)_n]^{x-}$, $[M(CN)_n]^{(x+1)-}$, $[M(CN)_n]^{(x+2)-}$, and in this respect CN^- resembles ligands such as 2,2'-dipyridyl and 1,2-dithiolenes (Sections 4-10 and 4-34). Finally note that some cyano complexes may themselves act as ligands. One example[11] is that of $[Ni(CN)_4]^{2-}$, which gives with 2,2',2''- triaminotriethylamine (tren) the complex 4-III.

$$\left[\text{trenNi} \diagdown \begin{matrix} NC \\ NC \end{matrix} \diagup Ni \diagdown \begin{matrix} CN \\ CN \end{matrix} \diagup \text{Nitren} \right]^{2+}$$

(4-III)

4-6. Carbon Disulfide: Carbon Dioxide

Carbon disulfide complexes[12] were discovered at the same time as thiocarbonyl

[11] K. R. Mann, D. M. Duggan, and D. N. Hendrickson, *Inorg. Chem.,* 1975, **14**, 2577.

[12] P. V. Yaneff, *Coord. Chem. Rev.,* 1977, **23**, 183; G. Gattow and W. Behrendt, *Carbon Sulfides and Their Inorganic Complex Chemistry,* Thieme, 1977; H. Huber, G. A. Ozin, and W. J. Power, *Inorg. Chem.,* 1977, **16**, 2234; I. S. Butler, *Acc. Chem. Res.,* 1977, **10**, 359; M. Herberhold, M. Süss-Fink, and C. G. Kreiter, *Angew. Chem., Int. Ed.,* 1977, **16**, 193, 194; H. Le Bozec et al., *Inorg. Chem.,* 1978; **17**, 2568; U. Oemichen et al., *J. Organomet. Chem.,* 1978, **156**, C29.

complexes (page 86). Although rather weak end-on S-bonded η^1 complexes can be formed, most of the CS_2 complexes are of the three-membered ring η^2 type, examples being as follows:

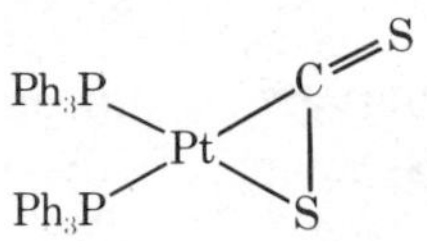

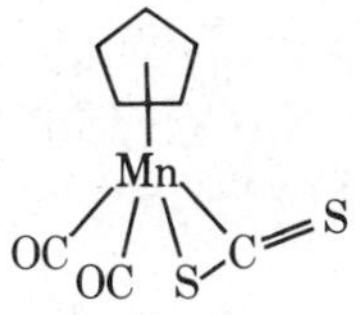

Bridging CS_2 complexes may be of these types:

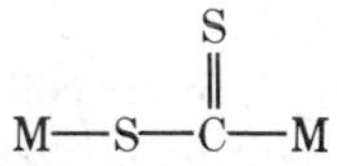

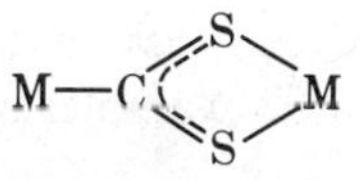

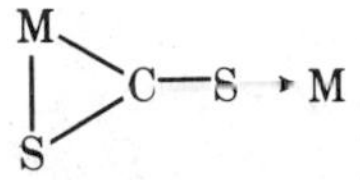

The η^2 complexes have strong CS bands *ca.* 1160 and 1140 cm^{-1}. The bound CS_2 is quite reactive and can be attacked by electrophiles[13a] and acetylenes[13b] sometimes to give carbene complexes (Chapter 27), for example,

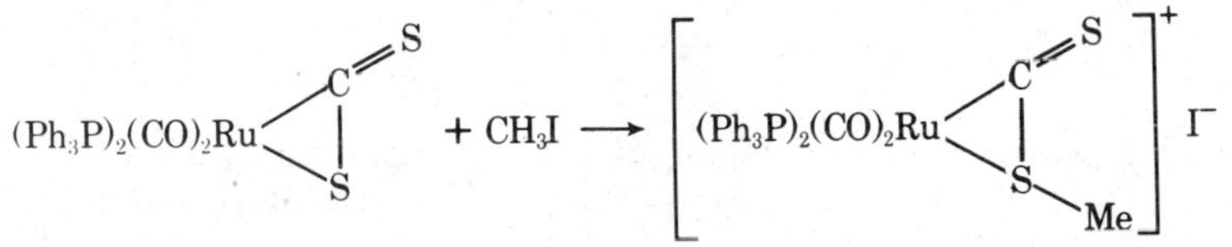

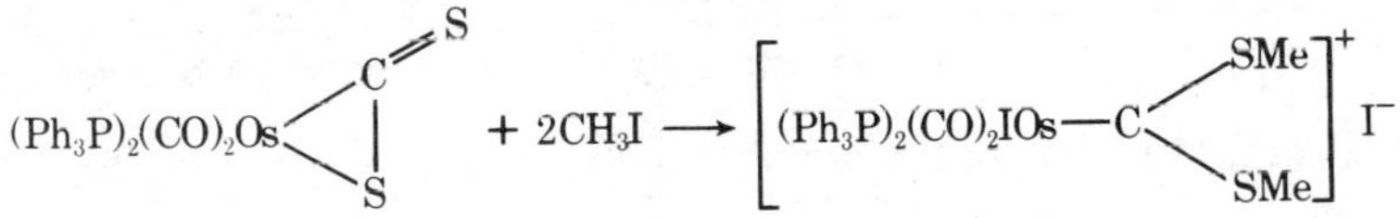

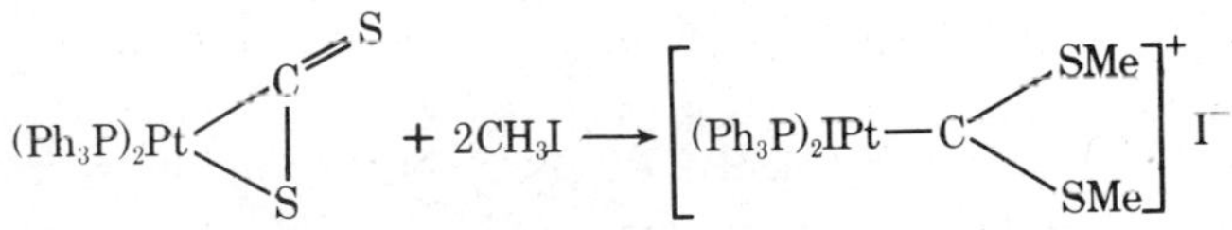

A number of *carbon dioxide* complexes[14] have been claimed, but some of these have turned out to be misnamed. Thus the action of CO_2 on $HRh(PR_3)_4$ gives a bridged *carbonato* complex

13a D. H. Farrar, R. O. Harris, and A. Walker, *J. Organomet. Chem.,* 1977, **124**, 125.

13b H. Le Bozec *et al., J. Am. Chem. Soc.,* 1978, **100**, 3946.

14 (a) S. Krogsrud *et al., Inorg. Chem.,* 1976, **15**, 2798; M. E. Volp'in and I. S. Kolomnikov, in *Organometallic Reactions,* E. I. Becker and M. Tsutsui, Eds., 1975, **5**, 313; G. Fachinetti *et al., J. Am. Chem. Soc.,* 1979, **101**, 1767; 1978, **100**, 7405; V. D. Bianco *et al., Inorg. Nucl. Chem. Lett.,* 1979, **15**, 187; (b) G. R. John *et al., J. Organomet. Chem.,* 1979, **169**, C23; (c) M. Aresta and C. F. Nobile, *Inorg. Chim. Acta,* 1977, **24**, L49.

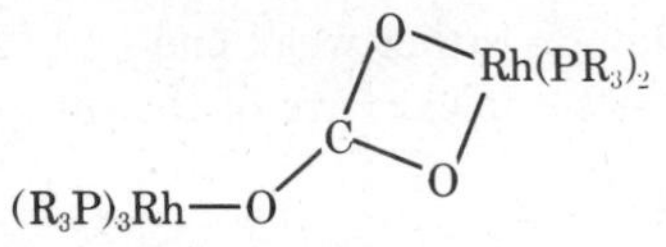

Osmium cluster anions with μ_2-CO_2 bridges are also known.[14b] However complexes such as $(Cy_3P)_2Ni(CO_2)$, and $RhCl(CO_2)(PR_3)_2$[14c] are authentic. The complexes appear to be of two types, one like the η^2-CS_2 complexes, the other, C-bonded:

The latter appear to have ir bands *ca.* 1550 and 1220 cm^{-1}, whereas the η^2-type bands are *ca.* 1660 and 1630 cm^{-1}; examples are respectively, probably, [Irdiars(CO_2)]$^+$ and $RhCl(CO_2)(Bu_3^nP)_2$.

SILICON, GERMANIUM, AND TIN

4-7.   Silicon, Germanium, and Tin(IV)[15]

The major types of complex containing silicon, germanium, or tin ligands are those with SiR_3, GeR_3, SnR_3 groups that are similar in stoichiometry to those of carbon compounds. They can be obtained by reactions such as

$$(\eta\text{-}C_5H_5)(CO)_3WNa + ClSiMe_3 = (\eta\text{-}C_5H_5)(CO)_3WSiMe_3 + NaCl$$

or, for silicon, by oxidative addition reactions (Section 29-3) of silanes, e.g.,

$$trans\text{-}Ir^ICl(CO)(PPh_3)_3 + Me_3SiH = Ir^{III}Cl(H)(SiMe_3)(CO)(PPh_3)_2$$

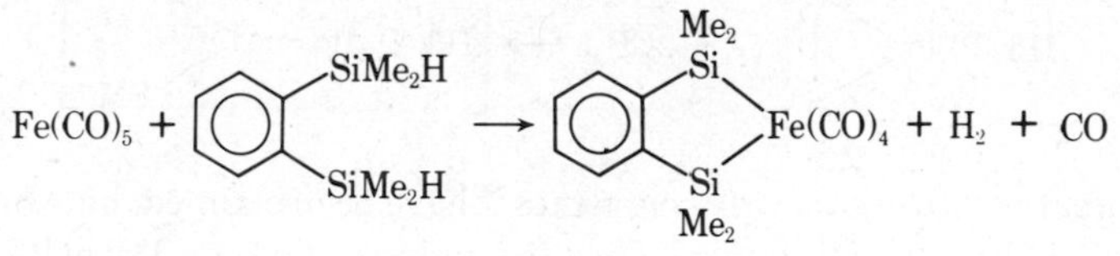

The SiR_3 groups have a very high trans effect. There are also compounds with bridging groups[16] such as the following:

---

[15] A. Bonny, *Coord. Chem. Rev.*, 1978, **25**, 229; E. H. Brooks and R. J. Cross, *Organomet. Chem. Rev., A,* 1970, **6**, 227; C. S. Cundy, B. M. Kingston and M. F. Lappert, *Adv. Organomet. Chem.,* 1973, **11**, 253; F. Höfler, *Topics Curr. Chem.,* 1974, **50**, 129.

[16] See, e.g., G. Schmidt and G. Etzvodt, *J. Organomet. Chem.,* 1977, **137**, 367.

4-8. Tin(II)[1]⁻

Divalent tin compounds have a lone electron pair (Chapter 12) and consequently can act as donors. There are numerous complexes of the trichlorotin(II) ion $SnCl_3^-$ [17b] and $Sn(\eta\text{-}C_5H_5)_2$; tin(II)carboxylates and β-diketonates can also act as ligands. Some germanium(II) and lead(II) compounds behave similarly.[18]

The $SnCl_3^-$ can generally displace Cl^-, CO, PF_3, and so on; for example,

$$PtCl_4^{2-} \xrightarrow{2SnCl_3^-} PtCl_2(SnCl_3)_2^{2-} \xrightarrow{3SnCl_3^-} Pt(SnCl_3)_5^{3-}$$

$$RhCl_3(aq) \xrightarrow{SnCl_3^-} [(Cl_3Sn)_2Rh^I(\mu\text{-}Cl)_2Rh^I(SnCl_3)_2]^{4-}$$

Platinum metal complexes with $SnCl_3^-$ dissolved in molten quaternary alkyl ammonium salts, $R_4N^+SnCl_3^-$ or $R_4N^+GeCl_3^-$, have been used as catalytic systems for hydrogenation of olefins and other reactions (cf. Chapter 30).[19]

Some metal-metal bonded complexes react with $SnCl_2$ in a type of "insertion reaction," (Chapter 29) where $SnCl_2$ could be considered to be showing carbenelike behavior, but the resulting compounds are clearly compounds of tin(IV):

$$(\eta\text{-}C_5H_5)_2Fe_2(CO)_4 \xrightarrow{SnCl_2} (\eta\text{-}C_5H_5)(CO)_2Fe\text{--}\overset{\overset{\displaystyle Cl}{|}}{\underset{\underset{\displaystyle Cl}{|}}{Sn}}\text{--}Fe(CO)_2(\eta\text{-}C_5H_5)$$

[17] (a) J. F. Young, *Adv. Inorg. Chem. Radiochem.,* 1968, **11,** 92; (b) T. Kruck *et al., Z. Naturfursch.,* 1978, **33b,** 129.

[18] P. G. Harrison *et al., J.C.S. Dalton,* **1975,** 2017; **1976,** 1054, 1608, 1612.

[19] G. W. Parshall, *J. Am. Chem. Soc.,* 1973, **94,** 8716.

NITROGEN

The types of organic nitrogen compound that can act as ligands are innumerable, and only the more important are discussed here. The bonding of N_2 and NO was discussed in Sections 3-9 and 3-11, respectively.

4-9. Ammonia and Amines[20]

Ammonia and substituted ammonias (including hydrazine and substituted hydrazines: see later) act as ligands toward both non-transition and transition metal ions, as well as giving adducts with Lewis acids.

Coordinated ammonia and amine ligands can undergo oxidation and condensation reactions.[21] Thus the trisethylenediamine complex $[Ruen_3]^{2+}$ may be oxidized, probably initially to a Ru^{III} complex, whereupon intramolecular oxidative dehydrogenation of the ligand occurs, to give an α-*diimine complex*:

$$
\left[en_2Ru \underset{NH_2\text{---}CH_2}{\overset{NH_2\text{---}CH_2}{}} \right]^{2+} \xrightarrow{-4e} \left[en_2Ru \underset{N\text{---}CH}{\overset{N\text{---}CH}{}} \right]^{2+} + 4H^+
$$

There is also a similar reaction for *o*-phenylenediamine complexes:

$$
\left[\underset{H_2}{\overset{H_2}{N}} Fe(CN)_4 \right]^{2-} \xrightarrow{-2e} \left[\underset{H}{\overset{H}{N}} Fe(CN)_4 \right]^{2+} + 2H^+
$$

o-Benzoquinonediimine complex

Diimine complexes can also be obtained in other ways—for example, by reduction of $FeCl_2(PMe_3)_2$ in CH_3CN by Na/Hg, the complex 4-IV is formed.[22]

$$
\begin{array}{c} MeC \!=\! N \\ \ \ | \qquad\ \ \diagdown \\ \ \ | \qquad\qquad Fe(PMe_3)_2 \\ MeC \!=\! N \diagup \end{array}
$$

(4-IV)

20 K. H. Schmidt and A. Muller, *Coord. Chem. Rev.,* 1976, **19,** 41 (vibrational spectra of pure amines); W. W. Wendlandt and T. P. Smith, *Thermal Properties of Transition Metal Amine Complexes,* Elsevier, 1967; D. A. House, *Coord. Chem. Rev.,* 1977, **23,** 223 (pentammines of Co^{III} and Cr^{III}, stereochemistry and reaction rates); S. T. Chow and C. A. McAuliffe, *Prog. Inorg. Chem.,* 1975, **19,** 51 (tridentate complexes of amino acids); D. A. Phipps, *J. Mol. Catal.,* 1979, **5,** 81 (amino acids); E. Uhlig, *Z. Chem.,* 1978; **18,** 440 (chelating pyridine complexes).

21 C. P. Guengerich and K. Schug, *J. Am. Chem. Soc.,* 1977, **99,** 3298; G. M. Brown *et al., Inorg. Chem.,* 1976, **15,** 190; G. G. Christoph and V. L. Goedken, *J. Am. Chem. Soc.,* 1973, **95,** 3869; D. F. Mahoney and J. K. Beattie, *Inorg. Chem.,* 1973, **12,** 2561; L. F. Lindsay and S. E. Livingstone, *Coord. Chem. Rev.,* 1967, **2,** 173; I. P. Evans, G. W. Everett, and A. M. Sargeson, *J. Am. Chem. Soc.,* 1976, **98,** 8041.

22 J. W. Rathke and E. L. Muetterties, *J. Am. Chem. Soc.,* 1975, **97,** 3272.

Finally, amine complexes may be oxidized to give nitrile complexes:

$$[(NH_3)_5RuNH_2CH_2R]^{3+} \rightarrow [(NH_3)_5RuN\equiv CR]^{2+}$$

Examples of condensation reactions are given later in this chapter when discussing template synthesis, but one example is the interaction of ammonia complexes with β-diketones in basic solution[23] to give nitrogen ligands that are comparable to β-diketonates (Section 4-27):

$$[Pt(NH_3)_6]^{4+} + CH_3COCH_2COCH_3 \xrightarrow{\text{base}} (NH_3)_4Pt \begin{bmatrix} \cdots \end{bmatrix}^{3+}$$

4-10. 2,2'-Bipyridine and Related Ligands[24]

The aromatic amines or rather polyimines, 2,2'-bipyridine (4-V), 1,10-phenanthroline (4-VI), and terpyridine (4-VII), differ considerably from aliphatic amines in that they can form complexes with metal atoms in a great range of oxidation

(4-V)   (4-VI)   (4-VII)

states. For example, there is the redox series[25]:

$$[Crdipy_3]^{3+} \rightleftarrows [Crdipy_3]^{2+} \rightleftarrows [Crdipy_3]^{+} \rightleftarrows [Crdipy_3]^{0} \rightleftarrows [Crdipy_3]^{-1}$$

For metal ions in "normal" oxidation states, the interaction of metal $d\pi$ orbitals with the ligand π^* orbitals is significant, but not exceptional. However, these ligands can stabilize metal atoms in very low formal oxidation states and in such complexes it is believed that there is extensive occupation of the ligand π^* orbitals, so that the compounds can often be best formulated as having radical anion ligands $L^{\cdot-}$. Most work has been carried out on bipy complexes, but it is apparent that phen and terpy afford very similar ones.

The methods of preparation are varied. Complexes involving transition metal ions in "normal" oxidation states can usually be obtained by conventional reactions

23 S. A. Brawner et al., Inorg. Chem., 1978, 17, 1304.
24 W. R. McWhinnie and J. D. Miller, Adv. Inorg. Chem., Radiochem., 1969, 12, 135; E. D. McKenzie, Coord. Chem. Rev., 1971, 6, 187; A. A. Schilt, Applications of 1,10-Phenanthroline and Related Compounds, Pergamon Press, 1969; W. A. McBryde, A Critical Review of Equilibrium Data for Proton and Metal Complexes of 1,10-Phenanthroline, 2,2'-Bipyridyl and Related Compounds, Pergamon Press, 1978.
25 M. C. Hughes and D. J. Macero, Inorg. Chem., 1976, 15, 2040.

and then reduced with a variety of reagents such as Na/Hg, Mg, or BH_4^-. The most general method employs Li_2bipy:

$$MX_y + yLi_2\text{bipy} + n(\text{bipy}) \xrightarrow{\text{THF}} M(\text{bipy})_n + yLiX + yLi(\text{bipy})$$

It is also noteworthy that many highly reactive organometallic compounds can be stabilized against hydrolysis, for example, by addition of these ligands. This is particularly true of R_2Zn, R_2Cd, and R_2Hg species.

The low-valent metal complexes are invariably colored, usually intensely so. For those containing transition metals, the bands responsible are believed to be mainly $d \rightarrow \pi^*$ charge-transfer bands. In other cases $\pi \rightarrow \pi^*$ ligand bands may also be active. For the ML_2 complexes of beryllium, magnesium, calcium, and strontium, electron spin resonance (esr) spectra show the presence of a ground, or low-lying excited, state that is a spin triplet. This can be best explained by postulating an M^{2+} cation and two radical anion ligands $L^{\doteq}$. Also for the *tris*-bipy complexes [Cr(bipy)$_3$]$^+$, [V(bipy)$_3$], and [Ti(bipy)$_3$]$^-$, esr data indicate that there is strong σ interaction with metal $4s$ orbitals, while the unpaired electrons are extensively delocalized on the ligands.

Aqueous solutions of bipy and phen complexes, such as [Fephen$_3$]$^{3+}$, [Rubipy$_2$py$_2$]$^{2+}$, or [Ptpy$_4$Cl$_2$]$^{2+}$ often show unexpected kinetic, equilibrium, and spectral behavior especially with OH$^-$, CN$^-$, and OR$^-$. This can be explained[26] by nucleophilic attack on the 2-position of the heterocyclic ring:

Although 2,2′-bipyridine and 1,10-phenanthroline usually give chelate complexes, unidentate complexes[27] can be formed as in [PtCl(PEt$_3$)$_2$(η^1-phen)]$^+$ and [Ir(bipy)$_2$[η^1-bipy)(H$_2$O)]$^+$; in solution they are fluxional and the platinum atom moves from one nitrogen atom to the other. This is so also for unidentate pyridazine and 1,8-naphthyridine complexes (see below).

26   R. D. Gillard *et al.*, *Trans. Met. Chem.*, 1977, **2**, 47; *J.C.S. Dalton*, **1979**, 190, 193; M. J. Blandamer, J. Burgess, and D. L. Roberts, *J.C.S. Dalton*, **1978**, 1086.
27   R. J. Watts, J. S. Harrington, and J. Van Houten, *J. Am. Chem. Soc.*, 1977, **99**, 2179; K. R. Dixon, *Inorg. Chem.*, 1977, **16**, 2618.

4-11. Other Nitrogen Heterocycles[28]

In addition to bipyridine and related heterocycles, there are numerous other N-heterocycles that give uni- or multidentate complexes. Some of the more important are:

Pyridazine	Pyrimidine	Purine

Pyrazine	1,8-Naphthyridine

Pyrazolate	Imidazolate

One of the most important areas of concern for metal binding with nucleotides, purines, and pyrimidines arises because of their presence in nucleic acids.[29] The action of certain metal complexes, notably cis-$PtCl_2(NH_3)_2$, as anticancer agents, is believed to arise through binding to nucleic acids. Other aspects of the binding of metals to nucleic acid include the attachment of lanthanide ions as shift reagents and fluorescent probes and the use of heavy metals to assist in X-ray structural determinations.

For unsubstituted *purines*, the most probable site for coordination is the imidazole nitrogen (N-9), which is protonated in the free neutral ligand. An example is the cobalt complex of adenine, $[Co(ad)_2(H_2O)_4]^+$ (4-VIII).

(4-VIII)

[28] M. Inoue and M. Kubo, *Coord. Chem. Rev.*, 1976, **21**, 1; B. C. Bunker *et al.*, *J. Am. Chem. Soc.*, 1978, **100**, 3805; W. E. Hatfield, *J.C.S. Dalton*, **1978**, 868 (pyridazine, pyrazine); J. G. Vos and W. L. Groenveld, *Inorg. Chim. Acta*, 1978, **27**, 173 (pyrazolate).

[29] B. E. Fischer and R. Bau, *Inorg. Chem.*, 1978, **17**, 27; D. J. Hodgson, *Prog. Inorg. Chem.*, 1977, **23**, 211 (stereochemistry of complexes); L. G. Marzilli, *Prog. Inorg. Chem.*, 1977, **23**, 256 (metal ion interactions); L. G. Marzilli and T. J. Kistenmacher, *Acc. Chem. Res.*, 1977, **10**, 146; G. Pneumatikakis *et al.*, *Inorg. Chem.*, 1978, **17**, 915 (Pd^{II}).

If the 9-position is blocked, the other imidazole nitrogen, N-7, is coordinated. Binding appears somewhat less likely through N-1 than through N-7; but of the three complexes established by X-ray crystallography, two also involve binding with both N-1 and N-7.

Imidazoles[30a] have been widely studied. Although the binding is usually through the N atom (4-IX), in some Ru^{II}, Ru^{III}, Fe^0, and Cr^0 complexes it is possible to have C-bonded groups[30b] (4-X).

(4-IX)   (4-X)   (4-XI)   (4-XII)

The C-bonded entity can be regarded as a carbene (4-XI) (see Chapter 27) or as a C-bound amidine[31] (4-XII). An example of a C-bonded species is the ruthenium(II) complex obtained as follows:

The N-bonded imidazoles commonly form bridges between two metal atoms[32] as in $[Cu_3(imH)_8(im)_2]^{4+}$, and in $\{Mn(im)(TPP)THF\}_n$, where TPP is tetraphenylporphyrin. Biimidazoles can act as mono or dianions,[33] for example, in rhodium(I) complexes:

30a   R. J. Sunderburg and R. B. Martin, *Chem. Rev.*, 1974, **74**, 471 (imidazole and histidine complexes).

30b   R. J. Sunderberg *et al.*, *J. Am. Chem. Soc.*, 1974, **96**, 381; *Inorg. Chem.*, 1977, **16**, 1470; S. S. Isied and H. Taube, *Inorg. Chem.*, 1976, **15**, 3070.

31   D. J. Doonan, J. E. Parks, and A. L. Balch, *J. Am. Chem. Soc.*, 1976, **98**, 2129.

32   G. Kolks *et al.*, *J. Am. Chem. Soc.*, 1976, **98**, 5720; J. T. Landrum *et al.*, *J. Am. Chem. Soc.*, 1978, **100**, 3232; M. S. Haddad *et al.*, *Inorg. Chem.*, 1979, **18**, 141.

33   S. W. Kaiser *et al.*, *Inorg. Chem.*, 1976, **15**, 2681.

4-12.   Ligands Derived by Deprotonation of Ammonia and Amines: Dialkylamido, Nitrene, and Nitrido Complexes

Ammonia can be deprotonated by alkali metals to give the anions NH_2^-, NH^{2-}, and N^{3-}, and all of these species can act as ligands.

There are numerous examples of the *amido* ligand NH_2 acting as a bridge, as in the ruthenium complex[34]

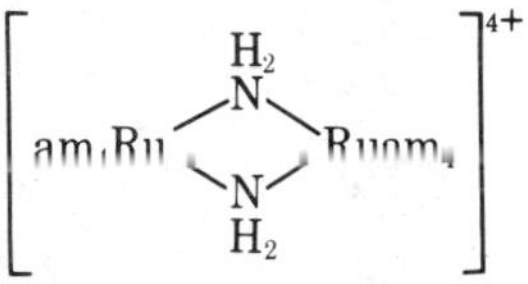

$$\left[am_4Ru \overset{\overset{\displaystyle H_2}{\underset{|}{N}}}{\underset{\underset{\displaystyle H_2}{\underset{|}{N}}}{}} Ruam_4 \right]^{4+}$$

The imido ion NH^{2-}, which is isoelectronic with O^{2-}, is not common as a ligand, although its alkyl and aryl derivatives NR are (see below). Some examples[35] of complexes, which have terminal or bent bridged NH groups, are the following species:

$$[(H_2O)_5Ru\overset{\overset{\displaystyle H}{|}}{\underset{\displaystyle \cdot\cdot}{N}}\,Cr(OH_2)_5]^{5+}$$

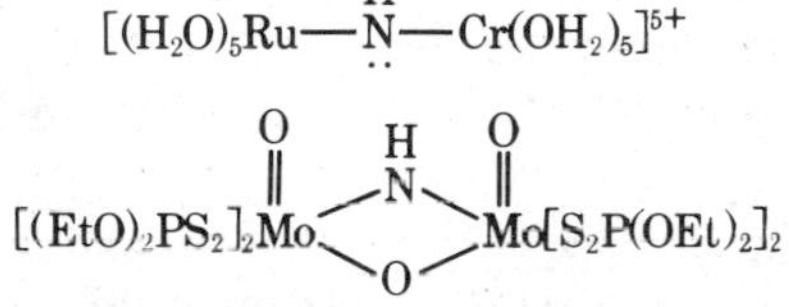

$$[(EtO)_2PS_2]_2Mo\overset{\overset{\displaystyle O}{\|}}{\underset{\displaystyle O}{}}\overset{\overset{\displaystyle H}{\underset{|}{N}}}{}\overset{\overset{\displaystyle O}{\|}}{}Mo[S_2P(OEt)_2]_2$$

$$(diphos)_2Cl_2Mo{=}NH$$

Nitrido Complexes have N^{3-} bound in the following ways:

1.   *Multiply Bonded Nitride* $M{\equiv}N$.[36]   Here the nitride ion is forming three covalent bonds to the metal; it is one of the strongest π donors known. The compounds are rather similar to those containing $M{=}O$ groups (Section 4-23). The complexes are largely those of molybdenum, tungsten, rhenium, ruthenium, and osmium, examples being $NReCl_2(PPh_3)_2$, $[NOsCl_5]^{2-}$, and $[NOsO_3]^-$. The $M{\equiv}N$ bonds are very short (*ca.* 1.16 Å) and the M—N stretching frequencies are in the region 950–1180 cm^{-1}.

2.   *N-Bridged Species.*   These are of the following types:

34   M. T. Flood *et al., Inorg. Chem.,* 1973, **12**, 2153.
35   R. P. Cheney and J. N. Armor, *Inorg. Chem.,* 1977, **16**, 3338; A. W. Edelblut, B. L. Haymore, and R. A. D. Wentworth, *J. Am. Chem. Soc.,* 1978, **100**, 2251.
36   W. P. Griffith, *Coord. Chem. Rev.,* 1972, **8**, 369. D. Pawson and W. P. Griffith, *J.C.S. Dalton,* **1975**, 417; C. D. Cowman *et al., Inorg. Chem.,* 1976, **15**, 1747.

$$L_nM-N-ML_n$$

Symmetric

$$L_nM-N-ML_n$$

Asymmetric

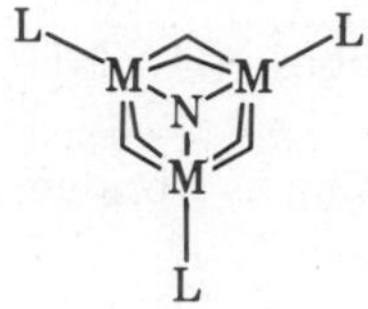

Triangular N-centered, where ⟨ = bidentate chelate anion
L = neutral ligand

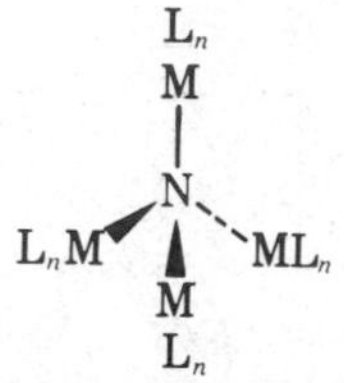

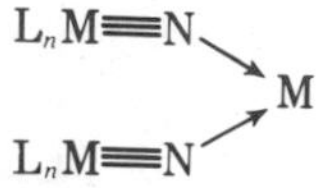

Only a few singly bridged species are known.[37] The bridges may be symmetrical ($M-N-M$) as in $[Ru_2NCl_8(H_2O)_2]^{3-}$ and $(TPPFe)_2N$ (where TPP = tetraphenylporphyrin), or asymmetric as in $[ReN(CN)_4]_n^{2-}$. Unlike oxo bridges (Section 4-23), there seem to be no bent MNM bridges. The linearity results from $M-N-M$ π bonding. The ruthenium complex is used for electroplating of ruthenium.

The triangular, N-centered complexes are much less common than O-centered triangles (Section 4-23) and no systematic way exists for making them. The metal atoms may be bridged by groups such as SO_4^{2-} or RCO_2^-, and the charge on the complex depends on the anion. Since NM_3^{III} has a 6+ charge, we can have complexes like $[Ir_3N(SO_4)_6(H_2O)_3]^{6-}$.

Tetra-bridged (μ^4) species are very uncommon, and the only well-defined example is the tetrahedral ion $[N(HgMe)_4]^+ClO_4^-$; this can be regarded as a type of quaternary ammonium ion.

The nitrogen atom in some of these compounds can be attacked. Thus tertiary phosphines react with $OsNCl_3(AsPh_3)_2$ to give phosphine imidate complexes $R_3PClOs-N=PR_3$ in which there is $d\pi$-$p\pi$-$d\pi$ bonding. The formation of a thionitrosyl complex with an $Mo-N=S$ group by attack of S_8[38] has already been mentioned (p. 93). An exceptional case of interaction is that with another metal atom as in the complex[39] 4-XIII, where the $Mo\equiv N$ distance (*ca.* 1.65 Å) is similar to that in $N\equiv Modtc_3$, whereas the $\equiv N \rightarrow Mo$ distance is 2.12 to 2.14 Å.

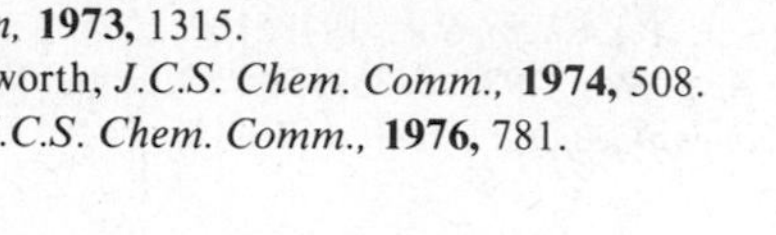

$$\left[\begin{array}{c} (R_2NCS_2)_3Mo\equiv N \\ \\ (R_2NCS_2)_3Mo\equiv N \end{array} \quad \diagdown \negthickspace Mo(S_2CNR_2)_3 \right]^{3+}$$

(4-XIII)

[37] D. A. Summerville and D. A. Cohen, *J. Am. Chem. Soc.,* 1976, **98**, 1747; D. Pawson and W. P. Griffith, *J.C.S. Dalton,* **1973,** 1315.

[38] J. Chatt and J. R. Dilworth, *J.C.S. Chem. Comm.,* **1974,** 508.

[39] M. W. Bishop *et al., J.C.S. Chem. Comm.,* **1976,** 781.

Dialkylamido ligands[40] are derived from secondary amines by deprotonation; for example,

$$Et_2NH + Bu^nLi = LiNEt_2 + C_4H_{10}$$

or by cleavage of a hydrazine

$$[(Me_3Si)_2N]_2 + 2Li = LiN(SiMe_3)_2$$

The complexes are obtained by reaction of lithium compounds with metal halides. Dialkylamides are closely related to alcoxides (Section 4-25) and alkyls, often having similar stoichiometries and structures [e.g., $Cr(NEt_2)_4$, $Cr(OBu^t)_4$, $Cr(CH_2SiMe_3)_4$]. The metal-nitrogen bond has some multiple bond character owing to flow to electrons from the lone pair into empty metal orbitals

This has the result that the M—NR_2 group is *planar* and that rotation of NR_2 about the M—N axis is considerably restricted. This can lead in certain complexes to the inequivalence of R groups in nmr spectra.

Dialkylamides can undergo insertion reactions (Chapter 29) as with CO_2 and CS_2 to give carbamates or dithiocarbamates; for example,

$$Me_3Ta(NMe_2)_2 + 2CS_2 = Me_3Ta(S_2CNMe)_2$$

Bulky dialkylamides are also very effective in preventing dimerization or polymerization reactions, so giving compounds with unusually low coordination numbers, such as two-coordination in $Be[N(SiMe_3)_2]_2$ or three-coordination in $Fe[N(SiMe_3)_2]_3$.

Alkylimido (Nitrene) Complexes.[41] By deprotonation of a primary amide we obtain the substituted imide RN^{2-}. The ligand can be bound to a metal by a double bond, with $p\pi$-$d\pi$ overlap. Although in certain compounds like

the M—N—C moiety is bent, in transition metal complexes like $O_3Os=NBu^t$ and $(Me_2N)_3Ta=NCMe_3$[42a] there is a linear M—N—C unit. The differences in chemical reactivity between the nucleophilic bent compounds and the electrophilic

40   M. H. Chisholm, M. Extine, W. Reichert, in *Inorganic Compounds with Unusual Properties,* American Chemical Society, 1976; D. C. Bradley and M. H. Chisholm, *Acc. Chem. Res.,* 1976, **9**, 273; M. F. Lappert, A. R. Sanger, R. C. Srivastava and P. P. Power, *Metal and Metalloidal Amides,* Horwood-Wiley, 1979; D. C. Bradley, *Adv. Inorg. Chem. Radiochem.,* 1972, **15**, 259.

41   S. Cenini and G. LaMonica, *Inorg. Chim. Acta,* 1976, **18**, 279; J. R. Dilworth *et al., J.C.S. Dalton,* 1979, 914, 921; B. L. Haymore *et al., J. Am. Chem. Soc.,* 1979, **101**, 2063; W. A. Nugent and R. L. Harlow, *J.C.S. Chem. Comm.,* 1979, 342.

42a W. A. Nugent and R. L. Harlow, *J.C.S. Chem. Comm.,* 1978, 579.

linear compounds reflects the increased π donation from the lone pair on nitrogen into low lying d orbitals of the metal, thus giving more triple-bond character to the M—N bond. Among the best known are the rhenium complexes, and these are made by condensation of oxorhenium species with aromatic primary amines such as aniline:

$$O{=}ReCl_3(PPH_3)_2 + ArNH_2 \longrightarrow ArN{=}ReCl_3(PPh_3)_2 + H_2O$$

The NH[42b] or NR ligand can be referred to as a *nitrene* just as CH_2 or CR_2 ligands can be considered as carbenes. There are a number of reactions in which nitrene complexes are believed to be intermediates. Isolation is usually impossible,[43] as for example, in the photolysis of $[Cr(NH_3)_5N_3]^{2+}$, which forms $Cr(NH_3)_5N^{2+}$, and in interaction of the complex $[(NH_3)_5IrNH_2Cl]^{3+}$ with hydroxide ion. However in the reduction of RNO_2 compounds by iron and other carbonyls (Chapter 30), intermediates with NR groups bound to metals have been isolated.[44] An example is the compound 4-XIV:

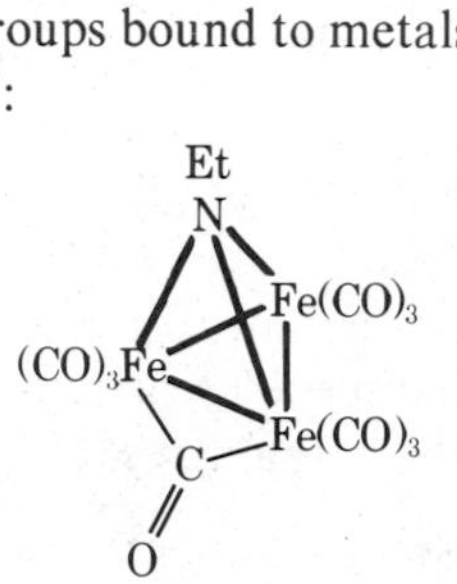

(4-XIV)

In some special cases, there may be other routes to nitrene complexes; for example,

$$IrCl(CO)(PMePh_2)_2 + \tfrac{1}{2}CF_3N{=}NCF_3 = Ir(NCF_3)Cl(CO)(PMePh_2)_2$$

The splitting of the N=N bond is doubtless favored by the electronegative CF_3 groups, since azo compounds generally retain the —N=N— bond on reaction with metal complexes (see later).

The reduction of $NbCl_4$ and $TaCl_4$ in CH_3CN by zinc gives complexes[45] with an unsaturated dinitrene ligand with linear M=C—N groups (4-XV) formed by dimerization of acetonitrile (cf. 4-IV).

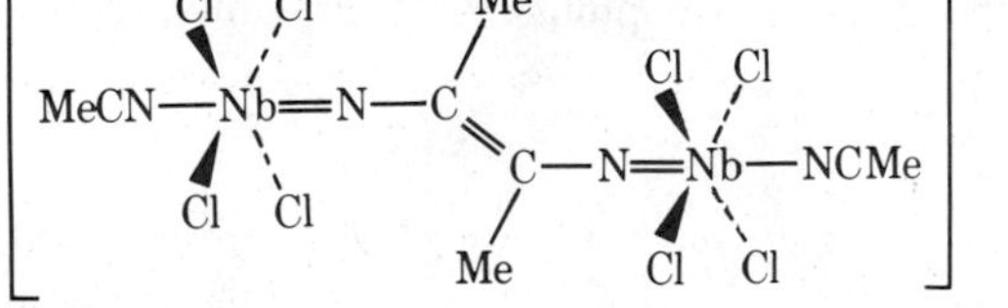

(4-XV)

42b J. Chatt *et al., Trans. Met. Chem.,* 1979; **4**, 59.
43   M. Katz and H. D. Gafney, *Inorg. Chem.,* 1978, **17**, 93; E. D. Johnson and F. Basolo, *Inorg. Chem.,* 1977, **16**, 554.
44   See, e.g., S. Aime *et al., J.C.S. Dalton,* **1978**, 534.
45   P. A. Finn *et al., J. Am. Chem. Soc.,* 1975, **97**, 220; F. A. Cotton and W. T. Hall, *Inorg. Chem.,* 1978, **17**, 3525.

Methyleneamido (amino) and Related Ligands. Alkylimido compounds of the type just discussed that have alkyl groups can undergo the reaction[46]

$$L_nM=N \overset{\cdots}{\underset{CH_2R}{}} \underset{H^+}{\overset{base}{\rightleftharpoons}} \left[L_nM-N\overset{\nwarrow}{=}C\overset{-H}{\underset{R}{\nearrow}} \right]^-$$

for example,

$$(PPh_2Me)_2Cl_3ReNCH_3 \underset{HCl}{\overset{py}{\rightleftharpoons}} (PPh_2Me)_2pyCl_2Re-N\overset{\diagdown}{\underset{CH_2}{}}$$

The ligand so formed, $-N=CH_2$, methyleneamido, or the substituted groups $-N=CR_2$, can be considered to resemble NO, since they can act not only as one-electron ligands but as three-electron ligands giving linear π-bonded groups. The bonding modes are thus:

| Bent | Linear | Bridged |

There are a number of compounds with symmetrical bridges.[47]

In addition to the deprotonation reactions mentioned above, there is a more general method of synthesis involving reaction of metal complex halides with $LiN=CR_2$,[48] which are in turn obtained from compounds such as $Ph_2C=NH$ or Bu_2^tC-NH by action of Bu^nLi.

4-13. Ligands with N—N Bonds[49]

There are a large number of complexes of ligands with N—N bonds. They have been much studied in recent years because of their relationship, real or imagined, to the problem of the conversion of dinitrogen to ammonia or hydrazine and the reactions of coordinated N_2 (Chapter 29).

The major ligand types and the ways in which they can be bound are given in Table 4-1.

Hydrazine usually acts as a reducing agent, but some compounds of N_2H_4 are known that have only one N coordinated, as in $Zn(N_2H_4)_2Cl_2$ or $(\eta\text{-}C_5H_5)Re(C-O)_2N_2H_4$.[50a] Substitution of alkyl or aryl groups as in NH_2NR_2 sterically inhibits

[46] J. Chatt et al., J.C.S. Dalton, **1976**, 2435.

[47] G. P. Khare and R. J. Doedens, Inorg. Chem., 1976, **15**, 86.

[48] M. Kilner et al., J.C.S. Dalton, **1974**, 639, 1620.

[49] J. R. Dilworth, Coord. Chem. Rev., 1976, **21**, 29 (an extensive review); D. Sutton, Chem. Soc. Rev., 1975, **4**, 443; A. Albini and H. Kisch, Topics Curr. Chem., 1976, **65**, 105 (diazene and diazo complexes); D. L. DuBois and R. Hoffmann, Nouv. J. Chim., 1977, **1**, 479 (MO picture of diazenido, 1,2-diazene, imido, and nitrido complexes).

[50a] D. Sellmann and E. Kleinschmidt, Z. Naturforsch., 1977, **32b**, 795.

TABLE 4-1
Ligands with N—N Bonds

Ligand	Structure
Diazenido (diazenato)[a]	
1,2-Diazene (azo)[b]	
1,2-Diazene (hydrazido,2-)[c]	
Hydrazino	
Hydrazido(1-)	M—NHNR$_2$ M—NHNHR
Triazenido	

[a] G. Butler, et al., J.C.S. Dalton, **1979**, 113; K. D. Schramm and J. A. Ibers, Inorg. Chem., 1977, **16**, 3287; J. A. Carrol et al., Inorg. Chem., 1977, **16**, 2462; D. Sutton, Chem. Soc. Rev., 1975, **4**, 443; J. Chatt et al., J. Chem. Soc., **1977**, 688; **1976**, 1520; W. A. Hermann et al., Angew. Chem., Int. Ed., **1976**, 164; M. Keubler et al., J.C.S. Dalton, **1975**, 1081; W. E. Carron, M. E. Deane, and F. J. Lalor, J.C.S. Dalton, **1974**, 1837; E. W. Abel and C. A. Burton, J. Organomet. Chem., 1979, **170**, 229.

[b] J. A. McCleverty, D. Seddon, and R. N. Whiteley, J.C.S. Dalton, **1975**, 839, D. Sellman, J. Organomet. Chem., 1973, **49**, C22.

[c] J. Chatt et al., J. Organomet. Chem., 1978, **160**, 165; J.C.S. Dalton, **1977**, 688; **1974**, 2074; M. Veith, Angew. Chem., Int. Ed., **1976**, 387; N. Wiberg, H. W. Häring, and O. Schneider, Angew. Chem., Int. Ed., **1976**, 386; M. Herberhold and K. Leonhard, Angew. Chem. Int. Ed., **1976**, 230.

the coordination of the substituted N atom, especially in the formation of octahedral complexes. However both nitrogen atoms of N_2H_4 and Me_2NNH_2 can bridge two metals as in the complex 4-XVI.[50b]

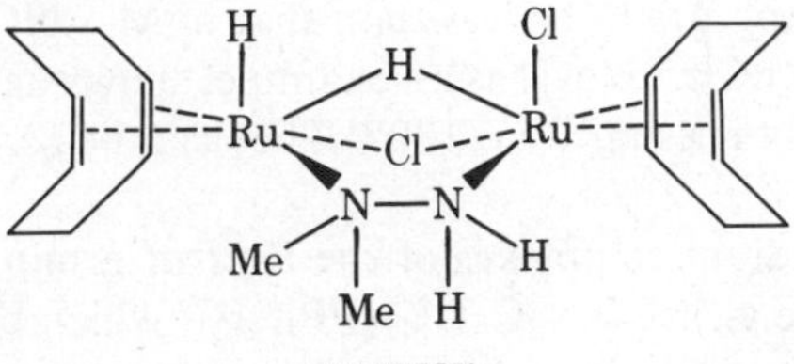

(4-XVI)

Diazenido. Compounds containing the ligand NNR may also be referred to in the literature as *arylazo, aryldiazo,* or *aryldiazenato.* The aryl compounds are the most common; they can be obtained from diazonium compounds, ArN_2^+, and from hydrazines such as ArCONNH. They are also formed by electrophilic or nucleophilic attacks on dinitrogen compounds (Section 30-4); for example,

$$(diphos)_2Mo(N_2) + RCOCl + HCl \xrightarrow{NEt_3} (diphos)_2ClMo(N_2COR)$$

$$\eta\text{-}C_5H_5(CO)_2MnN_2 \xrightarrow{LiPh} Li[\eta\text{-}C_5H_5(CO)_2Mn-N=N]$$
$$\underset{Ph}{|}$$

The ligand can be regarded as RN_2^+ and to be the analogue of NO^+, or as RN_2^-, the analogue of NO^-, or as neutral RN_2, the analogue of CO, and there has been the same type of discussion concerning the extent of metal-ligand π bonding. As with NO compounds, the N=N stretches vary widely from *ca.* 2095 cm^{-1} (indicating M—N π bonding) down to *ca.* 1440 cm^{-1} in bridging species.[51] The main distinction is between the following types[52] of complex:

(a)   (b)

but other types of bonding[53] of RN_2 groups are known namely:

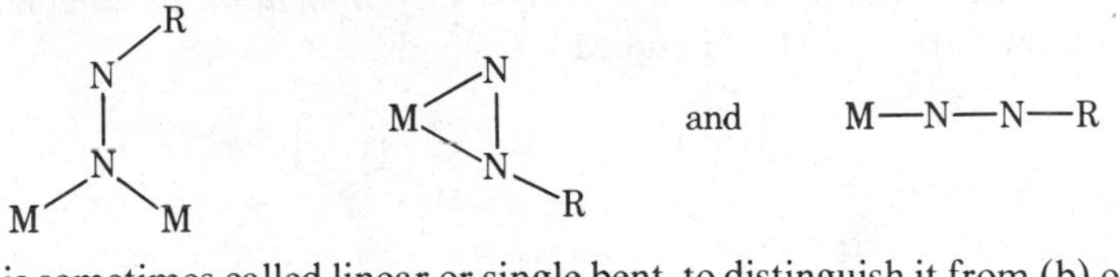

and   M—N—N—R

Type (a) is sometimes called linear or single bent, to distinguish it from (b) or double

---

[50b] T. V. Ashworth *et al., J.C.S. Dalton,* **1978**, 1036.
[51] K. D. Schramm and J. A. Ibers, *Inorg. Chem.,* 1977, **16**, 3287.
[52] D. T. Clark, *et al., Inorg. Chem.,* 1977, **16**, 1201; M. Cowie, B. L. Haymore, and J. A. Ibers, *J. Am. Chem. Soc.,* 1976, **98**, 7608.
[53] B. L. Haymore, *J. Organomet. Chem.,* 1977, **137**, C11.

bent. Type (a), of which $ArN_2RuCl_3(PPh_3)_2$ is an example, can be regarded as derived from RN_2^+. The M—N—N bonds are almost but not quite linear, the M—N—N angle usually being *ca.* 170°. There is considerable M—N π bonding, but the π character appears to be less than that in M—NO compounds.[53] Type (b), of which $[ArN_2RhCltriphos]^+$ is an example, show considerable variation in angles, but M—N—N is usually *ca.* 120°. They can be considered to be derived from RMN^-.

There are some iridium complexes of the *neutral* group $C_5Cl_4N_2$ acting as a two-electron donor [e.g., $IrCl(N_2C_5Cl_4)(PPh_3)_2$], which can be compared with $IrCl(CO)(PPh_3)_2$; these also have a type (a) bent structure.[54]

In the compound $[PhN=NMn(CO)_4]_2$, the double bridge is asymmetric, with Mn—N—N angles of 134° and 119°.[55] The bonding can be written as involving RNN as a neutral three-electron donor:

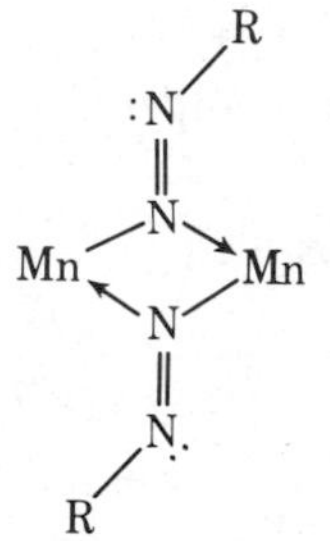

The unidentate ligand RN_2^+ can be protonated (see below), and it can be hydrogenated[56] to give a hydrazino complex; for example,

$$(Ph_3P)_2PtNNAr^+ \xrightarrow{H_2} (Ph_3P)_2Pt(H)(NH_2—NHAr)^+$$

Diazene Ligands. Diazenes or azo compounds RN=NR usually utilize the lone pairs on nitrogen as σ donors rather than use the π electrons of the double bond in ethylenelike bonding. Azo compounds of many types include azo dyestuffs. Both *cis* and *trans* azo compounds can give complexes normally unidentate but the *cis* azo compound pyridazine gives only bridged complexes.[57]

A feature of aromatic azo compounds is that the C—H of the ortho position on the aromatic ring is reactive and can undergo the *cyclometallation* reaction (Section 29-6), with formation of M—H and M—C bonds.

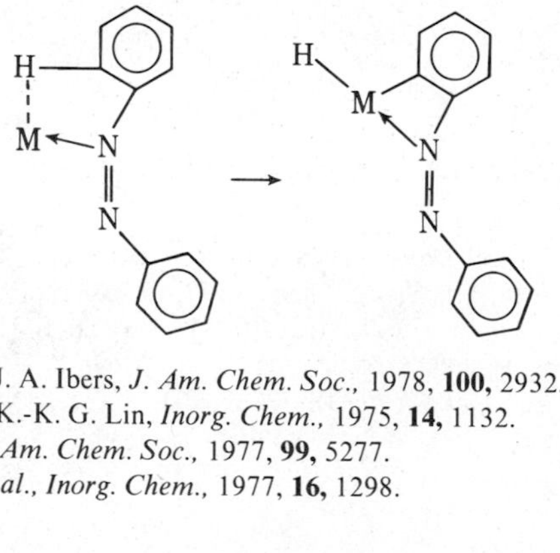

[54] K. D. Schramm and J. A. Ibers, *J. Am. Chem. Soc.*, 1978, **100**, 2932.
[55] M. R. Churchill and K.-K. G. Lin, *Inorg. Chem.*, 1975, **14**, 1132.
[56] S. Krogsrud *et al., J. Am. Chem. Soc.*, 1977, **99**, 5277.
[57] M. N. Ackermann *et al., Inorg. Chem.*, 1977, **16**, 1298.

Unsubstituted diazene complexes can be made,[58] for example, by oxidation of hydrazine complexes:

$$\eta\text{-}Cp(CO)_2MnNH_2\text{---}NH_2Cr(CO)_5 \xrightarrow{H_2O_2} \eta\text{-}Cp(CO)_2Mn\overset{H}{\underset{}{}}N{=}N\overset{Cr(CO)_5}{\underset{H}{}}$$

In the monosubstituted aryldiazene derivative 4-XVII, which is in the cis form, the metal-nitrogen bond is largely σ[59]:

$$\left[(Ph_3P)_2(CO)_2ClRu\text{---}N\overset{\ddot{N}-Ph}{\underset{H}{}} \right]^+$$

(4-XVII)

Such compounds are related to diazenato complexes by acid-base equilibria[60]:

$$M\text{---}N\overset{\ddot{N}-R}{\underset{H}{}} \underset{+H^+}{\overset{-H^+}{\rightleftharpoons}} M\text{---}N\overset{N-R}{\underset{\ddot{}}{}}$$

There are fewer examples of *N,N-disubstituted diazenes,* which are isomeric with RN=NR compounds. However the simple hydrogen compounds can be made, for example, by protonation of dinitrogen compounds

$$(diphos)_2Mo(N_2)_2 \xrightarrow{HBF_4} [(diphos)_2FMoNNH_2]^+$$

The M—N—NH$_2$ group is essentially linear.[61]

Diazenes with both nitrogen atoms bound to a metal in a three-membered ring are not common, but one example is $(R_3P)_2Ni(PhN{=}NPh)$.[62]

1,3-Triazenido Complexes.[63] The compounds ArNNHNAr are acids giving rise to a monoanion, ArNNNAr$^-$. These ligands can be unidentate, chelate, or μ_2-bridging:

$$M\text{---}N\overset{R}{\underset{\underset{\overset{\|}{:}N\diagdown_R}{N:}}{}} \qquad M\overset{\overset{N\diagup^R}{\diagdown}}{\underset{N\diagdown_R}{\diagup}}N \qquad \overset{R}{\diagdown}N{=}N{=}N\overset{R}{\diagup} \\ \underset{M\text{---}M}{}$$

58 D. Sellman and K. Jödden, *Angew. Chem. Int. Ed.,* 1977, **16,** 464.
59 B. L. Haymore and J. A. Ibers, *J. Am. Chem. Soc.,* 1975, **97,** 5369.
60 R. Mason *et al., J. Am. Chem. Soc.,* 1974, **96,** 261.
61 M. Hidai *et al., Inorg. Chem.,* 1976, **15,** 2694; G. A. Heath, R. Mason, and K. M. Thomas, *J. Am. Chem. Soc.,* 1974, **96,** 259.
62 S. D. Ittel and J. A. Ibers, *Inorg. Chem.,* 1975, **14,** 1183.
63 N. G. Connelly and Z. Demidowicz, *J.C.S. Dalton,* **1978,** 50 and references therein; P. I. van Vliet *et al., J. Organomet. Chem.,* 1976, **122,** 99; L. D. Brown and J. A. Ibers, *J. Am. Chem. Soc.,* 1976, **98,** 1597; *Inorg. Chem.,* 1976, **15,** 2788, 2794.

The structures of complexes closely resemble those of the considerably more important carboxylates (Section 4-28).

If the central N atom is replaced by CR' ($R' = H$, CH_3, etc.), we obtain N,N'-disubstituted *formamidino* or *alkyl*- or *arylamidino* ligands[64] (4-XVIII). If replaced by SR', we obtain the *sulfurdiimines*[65] (4-XIX).

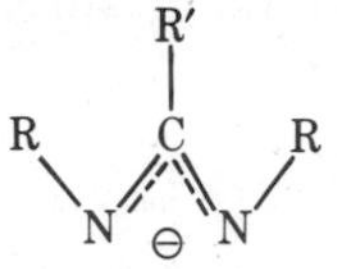

(4-XVIII) 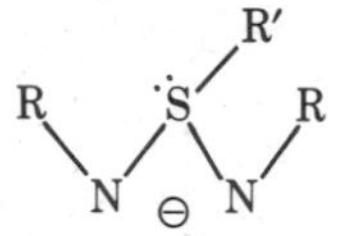   (4-XIX)

The neutral ligand $RN{=}S{=}NR$ also gives complexes of various types.[66]

Azide.[67]   The azide ion N_3^- can give unidentate complexes and also those that are bridged[68]:

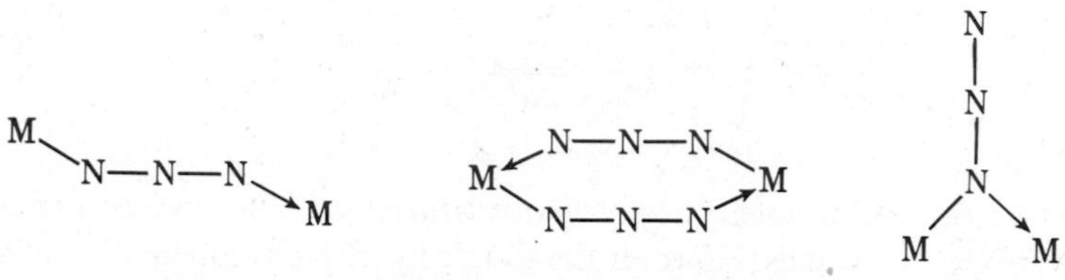

Examples of singly and doubly bridging azides are the nickel complexes $[Ni_2(N$-tetramethylcylam$)_2N_3]^+$ and $[Ni_2(tren)_2(N_3)_2]^+$, respectively; the symmetric bridging is found in $Cu(N_3)_2$.

4-14.   Macrocyclic Nitrogen Ligands[69]

The large ring compounds whose structures are such that several donor atoms can bind to a metal are most commonly nitrogen donors. However mixed nitrogen-oxygen, nitrogen-sulfur, and oxygen-sulfur donors are known, as well as macrocyclic oxygen and sulfur donors. Depending on the donor atoms, these can be designated N_4, N_2O_2, O_4, and so on.

[64]   W. H. de Roode *et al., J. Organomet. Chem.,* 1978, **154,** 273; 1978, **145,** 207; M. G. B. Drew and J. D. Wilkinson, *J.C.S. Dalton,* **1974,** 1973; F. A. Cotton *et al., Inorg. Chem.,* 1975, **14,** 2023, 2027.

[65]   K. Vrieze *et al., J. Organomet. Chem.,* 1977, 142, 337; 1978, **144,** 239.

[66]   R. Meij, *J.C.S. Chem. Comm.,* **1978,** 506.

[67]   Z. Dori and R. F. Ziolo, *Chem. Rev.,* 1973, **73,** 247.

[68]   F. Wagner *et al., J. Am. Chem. Soc.,* 1974, **96,** 2625; C. G. Pierpont *et al., Inorg. Chem.,* 1975, **14,** 604; D. M. Duggan and D. N. Hendrickson, *Inorg. Chem.,* 1973, **12,** 2422.

[69]   *Synthetic Multidentate Macrocyclic Compounds,* R. M. Izatt and J. J. Christensen, Eds., Academic Press, 1978 (this reference work discusses mainly oxygen macrocycles); D. H. Busch, *Acc. Chem. Res.,* 1978, **11,** 392; L. F. Lindoy, *Chem. Soc. Rev.,* 1975, **4,** 421 (transition metal complexes of synthetic macrocycle ligands); A. M. Tait, D. H. Busch, *et al., Inorg. Synth.,* 1978, **18,** Chapter 1; R. M. Izatt and J. J. Christensen, Eds., *Progress in Macrocyclic Chemistry,* Vol. 1, Wiley, 1979; G. A. Melson, Ed., *Coordination Chemistry of Macrocyclic Compounds,* Plenum, 1979.

The heterocyclic compounds can be broadly classed into those with conjugated π systems and those without. The former, especially macrocycles giving a set of four essentially coplanar N atoms, have been extensively studied in part because these types of systems are involved in chlorophyll, heme, vitamin B_{12} and other naturally occurring metal complexes (Chapter 31).

Macrocyclic complexes characteristically have the following properties.

1.   A marked kinetic inertness both to the formation of the complexes from the ligand and metal ion, and to the reverse, the extrusion of the metal ion from the ligand.

2.   They can stabilize high oxidation states that are not normally readily attainable, such as Cu^{III} or Ni^{III}.

3.   They have high thermodynamic stability—the formation constants for N_4 macrocycles may be orders of magnitude greater than the formation constants for nonmacrocyclic N_4 ligands.[70]

Thus for Ni^{2+} the formation constant for the macrocycle cyclam (4-XX) is about five orders of magnitude greater than that for the nonmacrocycle tetradentate 4-XXI:

(4-XX)   (4-XXI)

This "macrocyclic effect" has been discussed thermodynamically in Section 3-5.

Ligands with Conjugated π Systems

Phthalocyanines.[71] These were one of the earliest classes of synthetic N_4 macrocycles to be discovered. They are obtained by interaction of phthalonitrile with metal halides, in which the metal ion plays an essential role as a template. Complexes such as 4-XXII characteristically have exceptional thermal stabilities, subliming in vacuum around 500°C. They are also intensely colored and are an important class of commercial pigments. The solubility is usually very low, but sulfonated derivatives are soluble in polar solvents.

[70]   M. Kodama and E. Kimura, *J.C.S. Dalton,* **1978,** 1081; A. Anichini *et al., J.C.S. Dalton,* **1978,** 577; F. P. Hinz and D. W. Margerum, *J. Am. Chem. Soc.,* 1974, **96,** 4993.
[71]   A. B. P. Lever, *Adv. Inorg. Chem. Radiochem.,* 1965, **7,** 28.

(4-XXII)

Porphyrins.[72]   These ligands are derivatives of porphine (4-XXIII). They are especially important because many naturally occurring metal-containing natural products (chlorophylls, heme, cytochromes, etc.) contain related macrocyclic ligands.

(4-XXIII)   (4-XXIV)   (4-XXV)

Almost every metal in the Periodic Table can be coordinated to a porphyrin. The most widely used synthetic ligands apart from porphine itself are octaethylporphyrin (H_2OEP) (4-XXIV) and *meso*-tetraphenylporphyrin (H_2TPP) (4-XXV)[73]; in the latter the phenyl rings are free to rotate.

Although the most common form of coordination is that with a metal atom in the center of the plane and bound to four nitrogen atoms, nevertheless porphyrins can act as bi-, tri-, tetra-, or hexadentate ligands in which the metal atom lies out of the N_4 plane,[74] as discussed below. The metal atom is also 0.448 Å out of the plane in the iron(III)protoporphyrin IX dimethyl ester *p*-nitrobenzthiolate (4-XXVI),[75] where we use a diagrammatic representation of the N_4 prophyrin skeleton.

[72]   D. Dolphin, Ed., *The Porphyrins,* Academic Press, 1978. (an authoritative reference in seven volumes); J. F. Falk, *Porphyrins and Metalloporphyrins,* Elsevier, 1964; W. R. Scheidt, *Acc. Chem. Res.,* 1977, **10,** 339; K. M. Smith, Ed., *Porphyrins and Metalloporphyrins,* Elsevier, 1975; S. J. Chantrell *et al., Coord. Chem. Rev.,* 1975, **16,** 259 (MO calculations); J.-H. Fuhrhop, *Struct. Bonding,* 1975, **18,** 1; *Angew. Chem., Int. Ed.,* 1974, **13,** 321; 1976, **15,** 648; J.-H. Fuhrhop and K. M. Smith, *Laboratory Methods in Porphyrin and Metalloporphyrin Research,* Elsevier, 1975; P. Hambright, *Coord. Chem. Rev.,* 1971, **6,** 247; D. Dolphin and R. H. Felton, *Acc. Chem. Res.,* 1974, **7,** 26; J. W. Büchler, *Angew. Chem., Int. Ed.,* 1978, **17,** 407 (hemoglobin); F. R. Longo, Ed., *Porphyrin Chemistry Advances;* Wiley, 1979; J. W. Büchler *et al., Struct. Bonding,* 1978, **34,** 79.

[73]   S. S. Eaton and G. R. Eaton, *J. Am. Chem. Soc.,* 1977, **99,** 6594.

[74]   G. A. Taylor and M. Tsutsui, *J. Chem. Educ.,* 1975, **52,** 715; D. Ostfeld and M. Tsutsui, *Acc. Chem. Res.,* 1974, **7,** 52.

[75]   S. C. Tang *et al., J. Am. Chem. Soc.,* 1976, **98,** 2414.

(4-XXVI)

Complexes of synthetic porphyrins can be made by interaction of the ligand with a metal salt in a common solvent, usually dimethylformamide.[76] Although H_2OEP has high solubility in organic solvents, H_2TPP is less soluble, but sulfonated derivatives are soluble in both water and methanol. Complexes can also be obtained by interaction of H_2porph with metal carbonyls, acetylacetonates, alkyls, hydrides, and so on.

The radius of the central hole is, of course, fixed, although it can be altered to some extent by puckering of the rings, and it lies between 1.929 and 2.098 Å. This means that many metal atoms cannot fit *in* the hole and must form out-of-plane complexes. Some examples are the following:

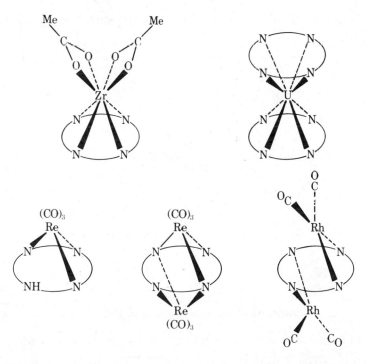

[76] W. Schneider, *Struct. Bonding*, 1975, **23**, 123 (kinetics and mechanism of metalloporphyrin formation).

Polymeric compounds with metal atoms between porphyrin groups are also known.

There have been recent attempts to synthesize porphyrin ligands that will act as models for hemoglobin (Section 31-3) in that absorption of oxygen is truly reversible, rather than as is usual with Fe^{II} porphyrin complexes, leading to irreversible oxidation to Fe^{III} porphyrin. These attempts involving "picket-fence" porphyrins are discussed in Section 4-24. Another type of modification is the porphyrin "crowned" by a cryptate (Section 4-26) ring that can accommodate two metal atoms[77] (4-XXVII). A cap can also be provided by pyridine groups as indi-

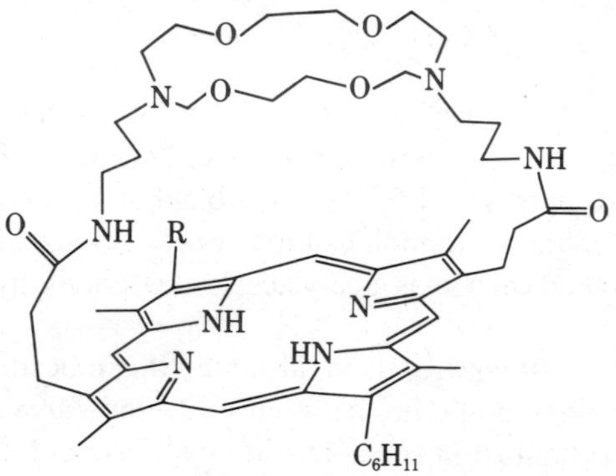

(4-XXVII)

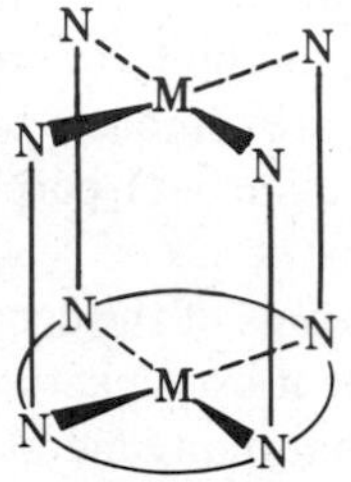

(4-XXVIII)

cated diagrammatically in 4-XXVIII. Yet another type is the *strati*-bisporphyrin[78] (4-XXIX).

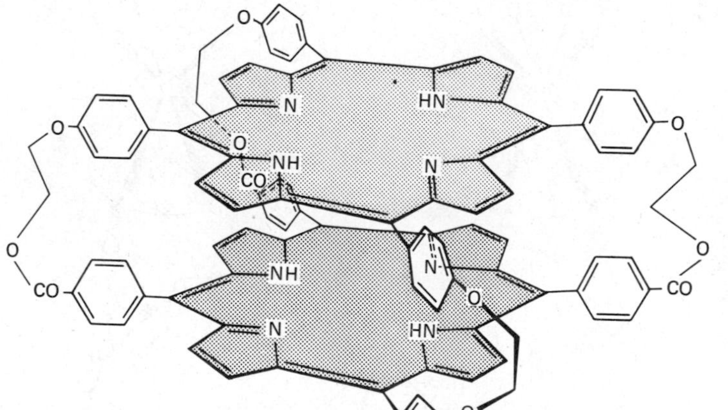

(4-XXIX)

Reproduced by permission from Ref. 78

[77]   C. K. Chang, *J. Am. Chem. Soc.,* 1979, **101,** 3413; 1977, **99,** 2819; D. A. Buckingham *et al., J. Am. Chem. Soc.,* 1978, **100,** 2899.

[78]   N. E. Kagan, D. Mauzerall, and R. B. Merryfield, *J. Am. Chem. Soc.,* 1977, **99,** 5484.

Other important conjugated macrocycles are the *corrins* (Section 31-9), which occur in vitamin B_{12}. Conjugated systems such as dibenzotetraaza-[14]-annulene (4-XXX)[79] have a smaller coordination radius than porphyrins and usually form out-of-plane complexes.

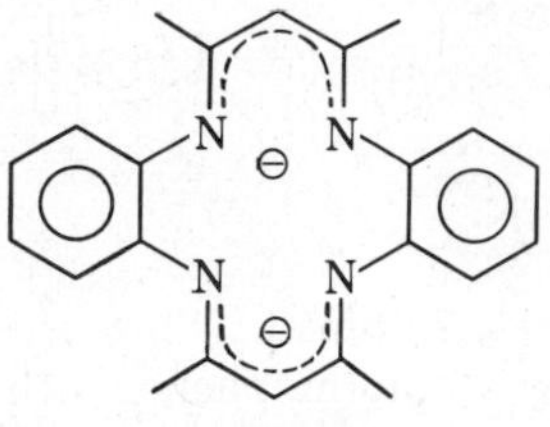

(4-XXX)

Template Syntheses.[80]   These are reactions in which the presence of the metal ion controls the synthesis. We have already noted this feature in the synthesis of phthalocyanins from phthalonitrile.

A great variety of nitrogen-containing macrocycles can be made by employing the Schiff base condensation reaction (eq. 4-1), often (but not necessarily) with

(4-1)

a metal ion as a template, and with subsequent hydrogenation to obtain a saturated system not subject to hydrolytic degradation by reversal of reaction 4-1. Some representative preparative reactions are eqs. 4-2, 4-3 and 4-4.

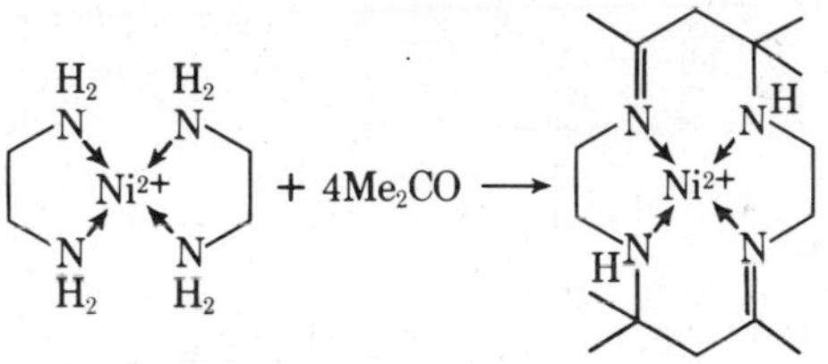

(4-2)

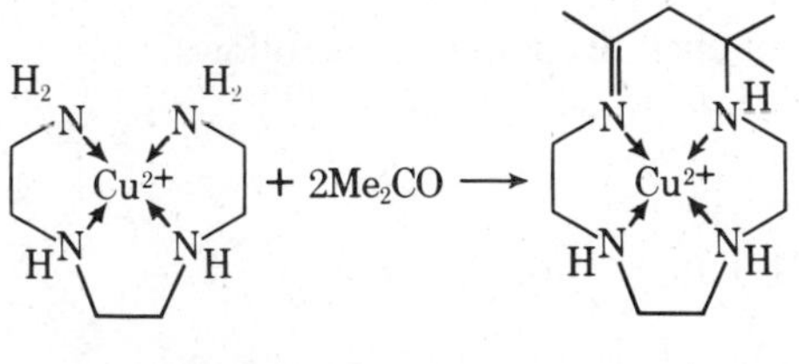

(4-3)

[79]   V. L. Goedken *et al., J. Am. Chem. Soc.,* 1976, **98,** 8391.
[80]   M. de S. Healy and A. J. Rest, *Adv. Inorg. Chem. Radiochem.,* 1978, **21,** 1; S.-M. Peng *et al., Inorg. Chem.,* 1978, **17,** 119, 820; N. F. Curtis, *Coord. Chem. Rev.,* 1968, **3,** 1; J. J. Christensen *et al., Chem. Rev.,* 1974, **74,** 651; D. St. C. Black and A. J. Hartshorn, *Coord. Chem. Rev.,* 1972–1973, **9,** 219; T. J. Marks and D. R. Stojakovic, *J. Am. Chem. Soc.,* 1978, **100,** 1695.

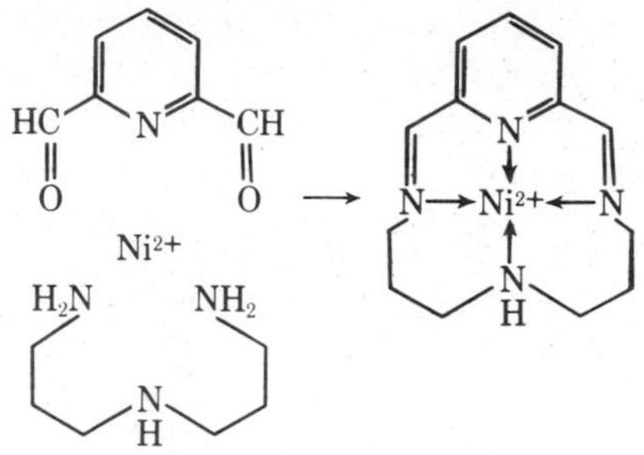

$$(4-4)$$

Other good examples[81] are as follows.

1.   The condensation of o-aminothiophenol with pyridine-1-carboxaldehyde gives benzthiaazoline in absence of metal ions:

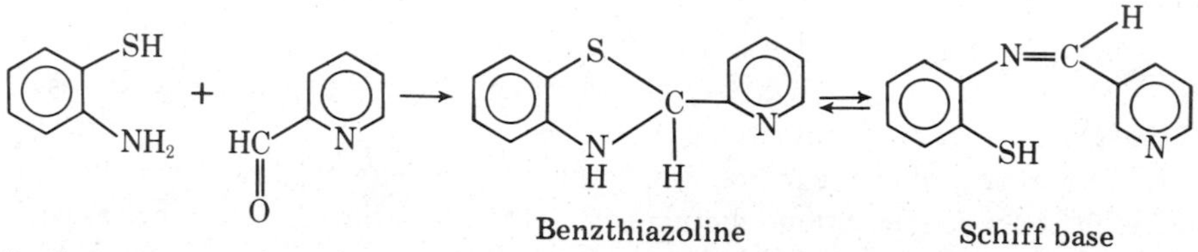

Benzthiazoline   Schiff base

In the presence of metal ions, any small amount of the Schiff base that may be in equilibrium with benthiaazoline will be removed to give a metal complex.

2.   The self-condensation of o-aminobenzaldehyde in presence of $BF_3 \cdot OEt_2$ in acetic acid gives a macrocycle[82] and when metal ion is present also, a complex, for example,

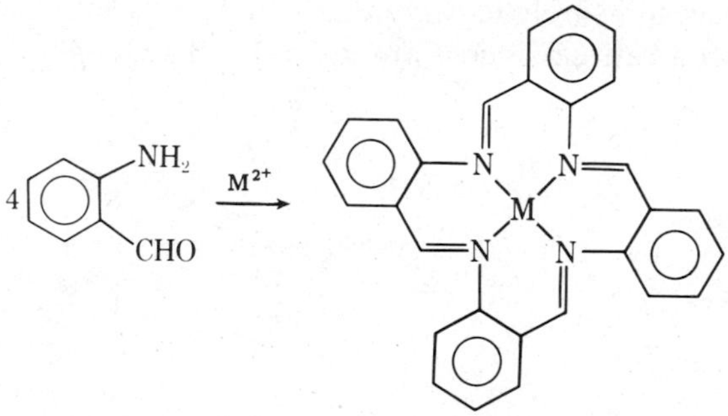

3.   The reaction of chelate complexes as follows:

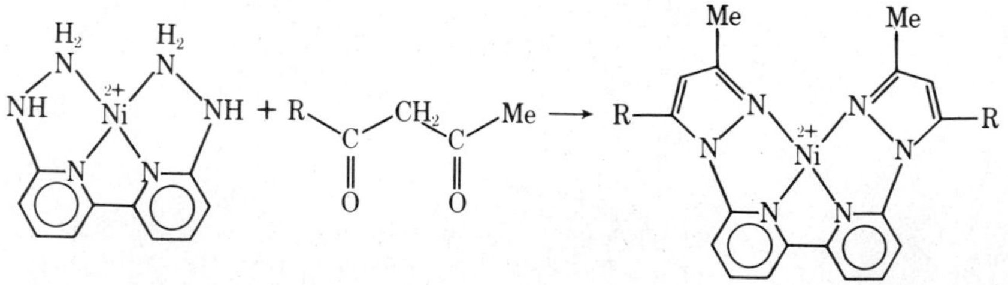

[81]   D. H. Busch et al., Inorg. Chem., 1977, **16**, 1716, 1721; J. Lewis and K. P. Wainwright, J.C.S. Dalton, **1978**, 440.

[82]   J. D. Goddard and T. Norris, Inorg. Nucl. Chem. Lett., 1978, **14**, 221.

Finally, note that there must be *some* control of the reaction by the size of the metal used. If the ion is too small or too large, no macrocyclic complex may be formed.[83]

Ligands Without Conjugated π Systems. There are a large number of macrocyclic N ligands that have double bonds in part of the ring only or are completely saturated.[84] There is often a very substantial difference between the conjugated and saturated macrocycles with regard to the rates of substitution reactions, which may be up to 10^{12} times greater for the conjugated systems. The enormous rate enhancement can be correlated with the lifetime of the leaving ligand in the inner coordination sphere of the metal complex. This profound difference presumably explains why biologically active metal systems invariably have highly unsaturated macrocyclic ligands.

Some examples of N_4 macrocycles have already been shown, others are the fourteen-membered ring compounds[85] 4-XXXI and 4-XXXII; there are similar tetraenes with fifteen- and sixteen-membered rings.[86]

Me$_4$[14]-1,3,8,10-tetraene-N$_4$
(4-XXXI)

Me$_2$[14]-1-ene-N$_4$
(4-XXXII)

4-15. Schiff Base Ligands[87]

Schiff base ligands are very diverse and usually contain both N and O donor atoms, although purely N donors are known. There are also N and S donors.

83 D. H. Cook *et al.*, *J.C.S. Dalton*, **1977**, 446; D. E. Fenton *et al.*, *J.C.S. Chem. Comm.*, **1977**, 623.
84 See, e.g., A. M. Tait, F. V. Lovecchio, and D. H. Busch, *Inorg. Chem.*, **1977**, **16**, 2206; R. F. Pasternack, M. A. Cobb, and N. Sutin, *Inorg. Chem.*, **1975**, **14**, 866; T. N. Margulis and L. J. Zompa, *J.C.S. Chem. Comm.*, **1979**, 430.
85 A. M. Tait and D. H. Busch, *Inorg. Chem.*, **1979**, **18**, 1555.
86 D. D. Riley, J. A. Stone, and D. H. Busch, *J. Am. Chem. Soc.*, **1976**, **98**, 1752.
87 R. H. Holm, G. W. Everett, Jr., and A. Chakravorty, *Prog. Inorg. Chem.*, **1966**, **7**, 83; M. Calligaris, G. Nardini, and L. Randaccio, *Coord. Chem. Rev.*, **1972**, **7**, 385; U. Casselato, M. Vidali, and P. A. Vigato, *Coord. Chem. Rev.*, **1977**, **23**, 31; D. G. Hodgson, *Prog. Inorg. Chem.*, **1975**, **19**, 173 (Schiff base dimeric complexes of first row transition metals); C. M. Harris and E. Sinn, *Coord. Chem. Rev.*, **1969**, **4**, 391 (Schiff base complexes as ligands); L. Sacconi, *Coord. Chem. Rev.*, **1966**, **1**, 192; S. Yamada, *Coord. Chem. Rev.*, **1966**, **1**, 415 (stereochemistry); C. A. McAuliffe *et al.*, *J.C.S. Dalton*, **1977**, 1762; D. E. Fenton and S. E. Gayda, *J.C.S. Dalton*, **1977**, 2095.

One of the best known Schiff base ligands is bis(salicyclaldehyde)ethylenedi-imine[88] (sal$_2$en):

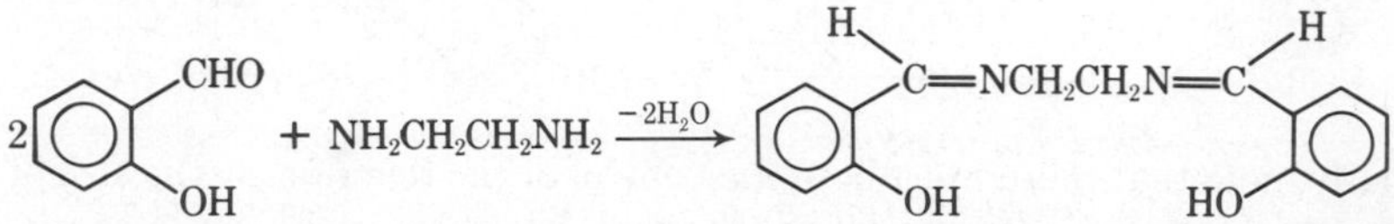

This is a bifunctional (two OH groups), tetradentate (2N, 2O) ligand. Other Schiff bases can be mono-, di-, or tetrafunctional and can have denticities of 6 or more with various donor atom combinations (e.g., for quinquedentate, N_3O_2; N_2O_3; N_2O_2P; N_2O_2S, etc.). Complexes of un-ionized or partly ionized Schiff bases are also known[89] (e.g., LaCl$_3$sal$_2$enH$_2$·aq).

Some representative types of complex that illustrate not only the formation of mononuclear but of binuclear and polymeric species are 4-XXXIII to 4-XXXVI.

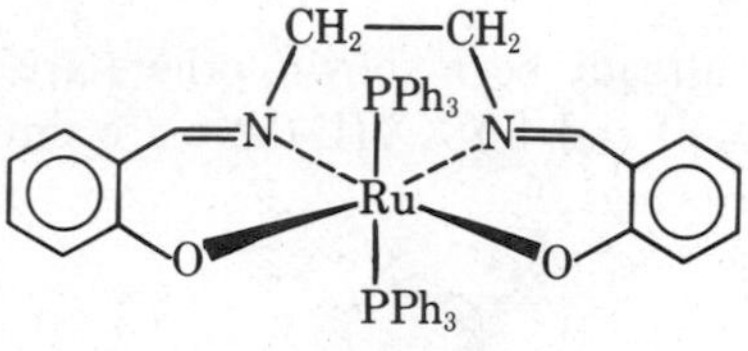

(4-XXXIII)

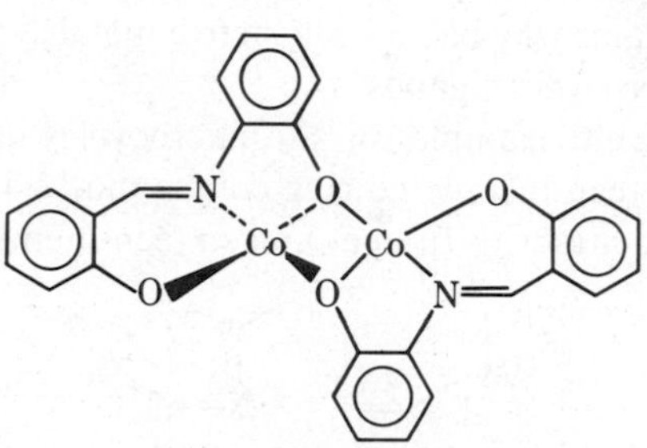

(4-XXXIV)

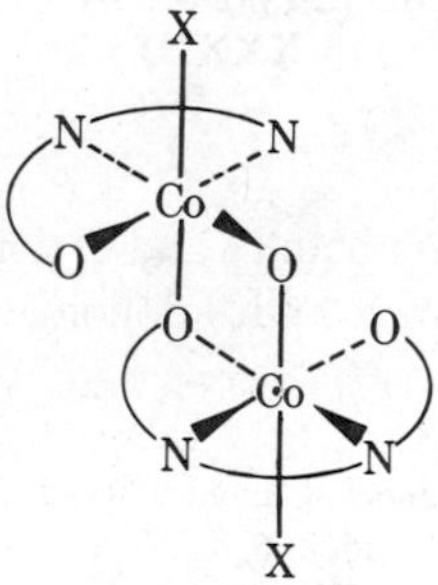

(4-XXXV)

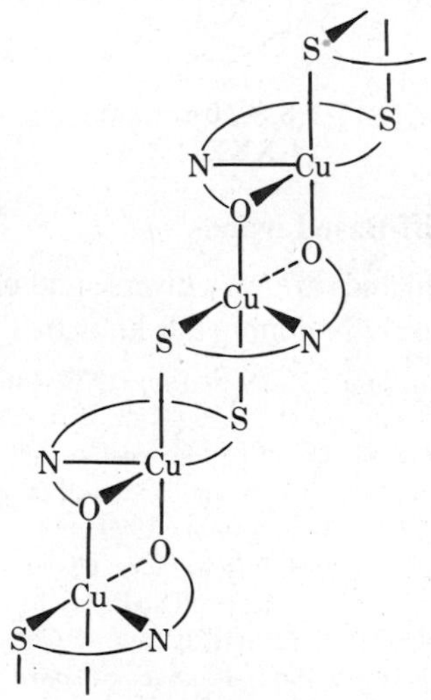

(4-XXXVI)

[88] M. D. Hobday and T. D. Smith, *Coord. Chem. Rev.,* 1972–1973, **9**, 311.
[89] J. I. Bullock and H.-A. Tajmir-Riahi, *J.C.S. Dalton,* **1978**, 36.

4-16.   Polypyrazolylborate Ligands[90]

The interaction of pyrazole (itself a ligand: see Section 4-11) and sodium borohydride or alkyl-substituted borohydrides leads to anionic ligands of the type 4-XXXVII and 4-XXXVIII designated $R_2Bpz_2^-$ and $RBpz_3^-$, respectively. The boron atom is tetrahedral, and bonding to metal is through only one of each of the pyrrole ring N atoms as shown,

(4-XXXVII)   (4-XXXVIII)

The bridge in $[Cu(HBpz_3)]_2$, shown diagramatically in 4-XXXIX, so far is unique.[91a]

(4-XXXIX)

The dipyrazolyl anions $R_2Bpz_2^-$ have a formal analogy to the β-diketonate ions (Section 4-27) and, like them, form complexes of the type $(R_2Bpz_2)_2M$. However because of the much greater steric requirements of the $R_2Bpz_2^-$ ligand, such compounds are always strictly monomeric. For steric reasons it appears to be difficult to make tris complexes, and only one example[91b] is known, namely, the anion $[V(H_2Bpz_2)_3]^-$.

The $RBpz_3^-$ ligands give a number of unusual complexes. These ligands themselves are unique in being the only trigonally tridentate, uninegative ligands. They form trigonally distorted octahedral complexes, $(RBpz_3)_2M^{0,+}$, with di- and trivalent metal ions, most of which are exceptionally stable. At least to a degree, an

90   A. Shaver, in *Organometallic Chemistry Reviews*, Vol. 3, Elsevier, 1977; S. Trofimenko, *Chem. Rev.*, 1972, **72**, 497.

91a   C. Mealli *et al.*, *J. Am. Chem. Soc.*, 1976, **98**, 711.

91b   P. Dapporto *et al.*, *Inorg. Chem.*, 1978, **17**, 1323.

analogy can be made between $RBpz_3^-$ and the cyclopentadienyl anion $C_5H_5^-$; both are six-electron, uninegative ligands. There are some mono-$RBpz_3^-$ complexes that bear considerable resemblance to half-sandwich complexes, $C_5H_5ML_x$ (see Chapter 27); thus $Mo(CO)_6$ reacts with $Na(RBpz_3)$ and NaC_5H_5 to give, respectively, $(RBpz_3)Mo(CO)_3^-$ and $(\eta\text{-}C_5H_5)Mo(CO)_3^-$.

4-17. Nitriles[92]

Acetonitrile and other organic cyano compounds are good donors and form complexes with most Lewis acids and metal ions. The bonding is usually end on through nitrogen (i.e., $RCN \rightarrow M$). Although "side-on" η_2 bonding (4-XL) has been pro-

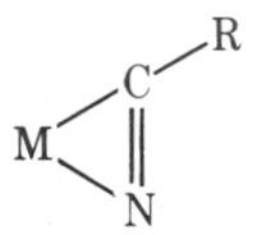

(4-XL)

posed on the basis of exceptionally low ir stretching frequencies, few cases have been definitely proved; that of $(R_3P)_2Pt(\eta_2\text{-}CF_3CN)$[93] is the most important. One compound, said to be $Ru^0(PPh_3)_4(\eta_2\text{-}MeCN)$ has been shown[94] to be the *ortho*metallated complex $RuH(C_6H_4PPh_2)(PPh_3)_2MeCN$.

Nitriles may reduce halides in higher oxidation states [e.g., interaction of $ReCl_5$ or WCl_6 gives $ReCl_4(MeCN)_2$ and $WCl_4(MeCN)_2$, respectively] together with chlorinated organic N compounds.

Nitrile complexes of copper are important in copper hydrometallurgy; nitrile complexes of ruthenium are involved in the catalytic dimerization of acrylonitrile, and nickel complexes are involved in the addition of HCN to olefins (Chapter 30).

Hydrolysis of a coordinated nitrile[95] appears to proceed by nucleophilic attack of OH^-:

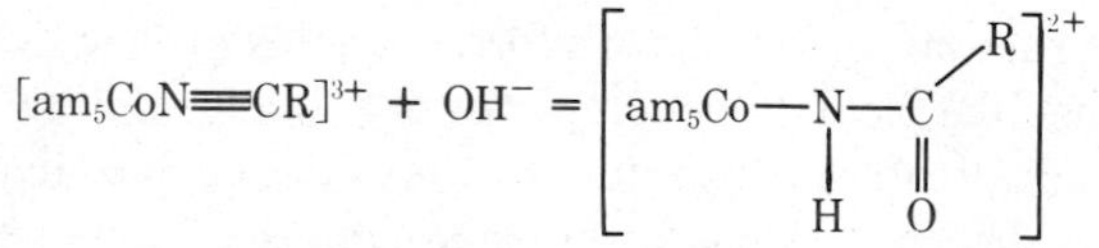

The catalytic hydration of nitriles to amides by rhodium and other transition metal complexes such as $Rh(OH)(CO)(PPh_3)_2$ probably proceeds by a cycle[96]

[92] B. N. Storhoff, *Coord. Chem. Rev.,* 1977, **23**, (transition metal complexes); R. A. Walton, *Q. Rev.,* 1965, **19**, 126.

[93] B. Storhoff and A. J. Infante, *Inorg. Chem.,* 1974, **13**, 3044; W. J. Bland, R. D. W. Kemmitt, and R. D. Moore, *J.C.S. Dalton,* **1973**, 1292; see also J. E. Sutton and J. I. Zink, *Inorg. Chem.,* 1976, **15**, 675.

[94] D. J. Cole-Hamilton and G. Wilkinson, *J.C.S. Dalton,* **1979**, 1283.

[95] R. J. Balahura *et al., J. Am. Chem. Soc.,* 1974, **96**, 2739.

[96] M. A. Bennett and T. Yoshida, *J. Am. Chem. Soc.,* 1973, **95**, 3030.

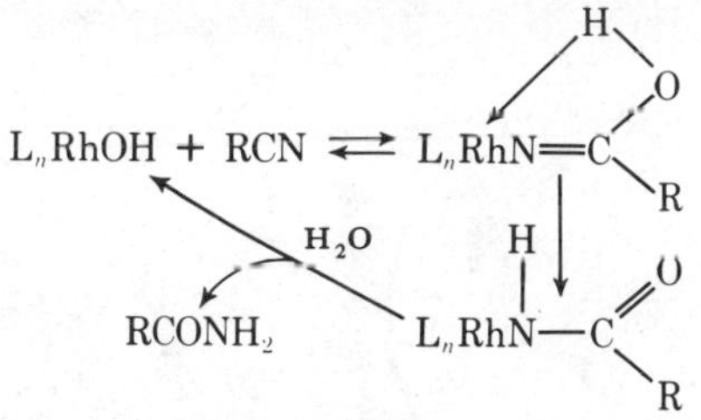

$$L_nRhOH + RCN \rightleftharpoons L_nRhN{=}C\begin{smallmatrix}H\\O\\R\end{smallmatrix}$$

RCONH$_2$ L$_n$RhN—C

Nucleophilic attacks on coordinated nitriles by aromatic amines and by alcohols give, respectively, complexes of amidines and imidate esters,[97] for example,

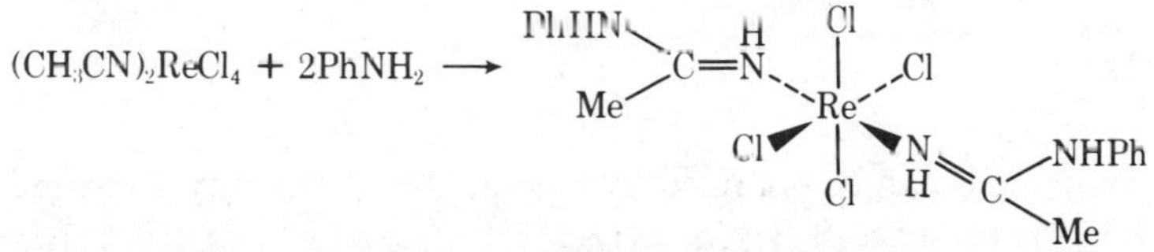

$$(CH_3CN)_2ReCl_4 + 2PhNH_2 \longrightarrow$$

We noted earlier in this chapter the reduction of nitrile ligands to give other types of complex (pp. 118, 126).

4-18. Oximes[98] and C-Nitroso Compounds

Oximes are derived by condensation of aldehydes and ketones with hydroxylamine. The best known are the *cis*-dioximes such as dimethylglyoxime (4-XLI) which, with Ni^{2+} in ammonia solution, gives the well-known red nickel complex 4-XLII.

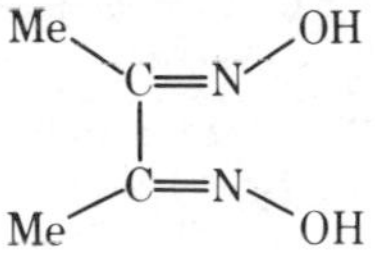

(4-XLI) 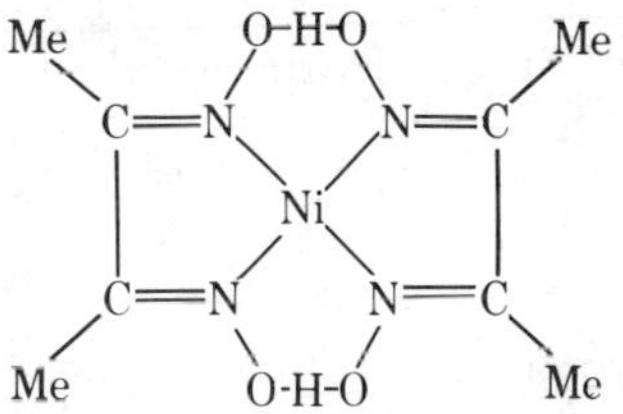   (4-XLII)

The important features here, apart from N$_4$ binding, are first the strong O—H—O hydrogen bonding, and second the stacking of the planar units parallel to each other in the crystal.

Oxime complexes have formed the basis for the construction of *clathrochelate or encapsulating* ligands in which three oxygen atoms of oxime ligands are "capped" by a group such as BF as shown in 4-XLIII (see also 3-XXI and 3-XXII).

[97] J. M. Castro and H. Hope, *Inorg. Chem.*, 1978, **17**, 1444; M. Wada and T. Shimohigashi, *Inorg. Chem.*, 1976; **15**, 954. See also A. M. Sargeson *et al., Acta Chem. Scand.*, 1978, **32A**, 789.

[98] R. C. Mehrotra *et al., Inorg. Chim. Acta*, 1975, **13**, 91; A. Chakravorty, *Coord. Chem. Rev.*, 1974, **13**, 1; B. Chatterjee, *Coord. Chem. Rev.*, 1978, **26**, 281 (hydroxamic acid complexes); A. Nakamura *et al., J.C.S. Dalton*, 1979, 488 (chiral oximates).

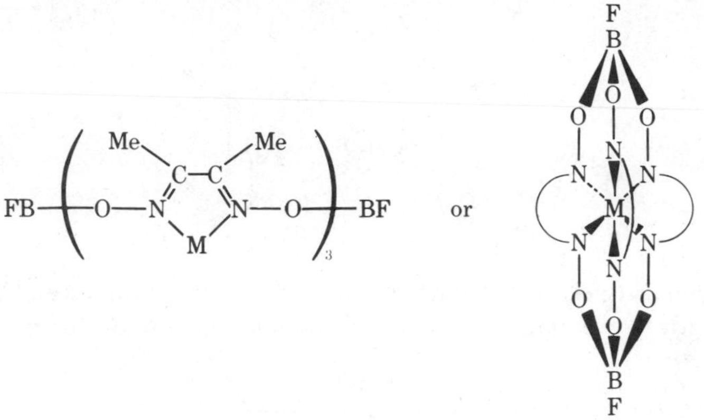

$$\left(\text{FB}\left[\begin{array}{c}\text{Me}\quad\text{Me}\\ \text{C—C}\\ \text{O—N}\quad\text{N—O}\\ \text{M}\end{array}\right]\text{BF}\right)_3 \quad\text{or}$$

(4-XLIII)

Such compounds can be obtained by interaction of dimethylglyoxime with a metal salt in the presence of BF_3 or $B(OH)_3$[99]; an example is $Fe(dmg)_3(BOH)_2$. Some of these clathrochelates, for example, $FB(O\ddot{N}CHC_5H_3\ddot{N})_3P^-$, can impose trigonal prismatic geometry. Instead of BF capping, clathrochelates with a *metal complex cap* [e.g., diethylenetriaminechromium(III)] can also be made.[100]

Monooximes with a different functional group[101] [e.g., 2(2-hydroxyethyl) imino-3-oximobutane] can give complexes such as 4-XLIV.

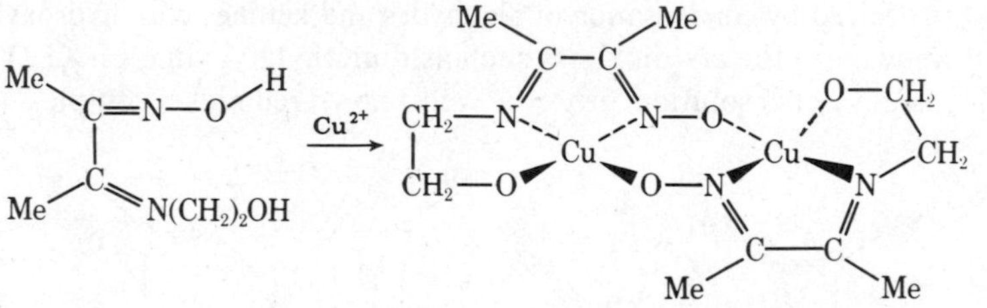

(4-XLIV)

Oxime complexes are used commercially for the extraction of metals by complexing and extraction into organic solvents.[102a] For example, copper is so removed from dilute copper sulfate solutions obtained by wet microbiological leaching of low-grade ores.

Oxime compounds that have the chelate function

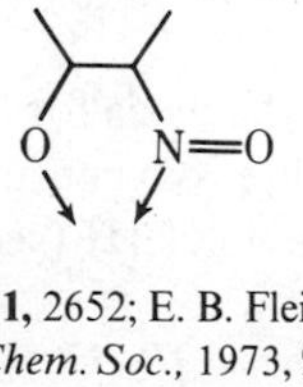

---

[99] E. Larsen et al., Inorg. Chem., 1972, **11**, 2652; E. B. Fleischer et al., Inorg. Chem., 1972, **11**, 2775; D. R. Boston and N. J. Rose, J. Am. Chem. Soc., 1973, **95**, 4163; M. R. Churchill and A. H. Reis, Jr., Inorg. Chem., 1972, **11**, 2239.

[100] R. S. Drago and J. H. Elias, J. Am. Chem. Soc., 1977, **99**, 6570.

[101] J. A. Bertrand, J. H. Smith, and P. G. Eller, Inorg. Chem., 1974, **13**, 1649.

[102a] A. W. Ashworth, Coord. Chem. Rev., 1975, **16**, 285.

occur in nature in the green iron(II) pigment ferroverdin.[102b] Such compounds are best regarded as derived from hydroxamic acids.

Monooximes of the type R_2C=NOH can be bound also in different ways (4-XLVa,b)[103]; in type a the ligand can be regarded as a three-electron donor

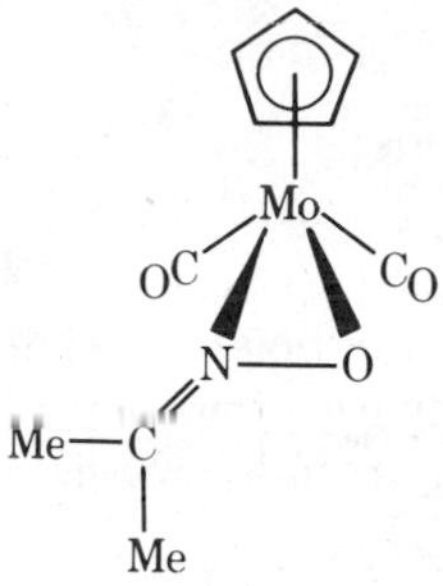

(4-XLVa) 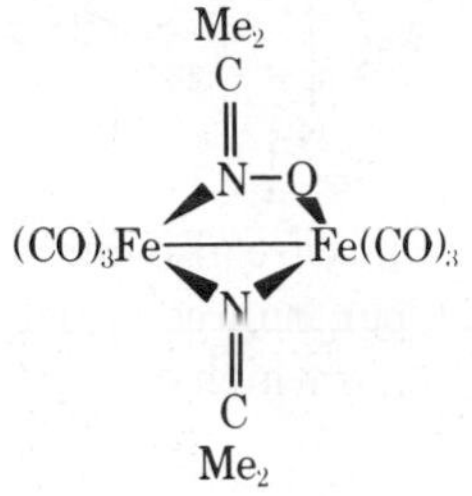   (4-XLVb)

Nitroso alkanes or arenes,[104] RNO, can be bound in complexes in several ways:

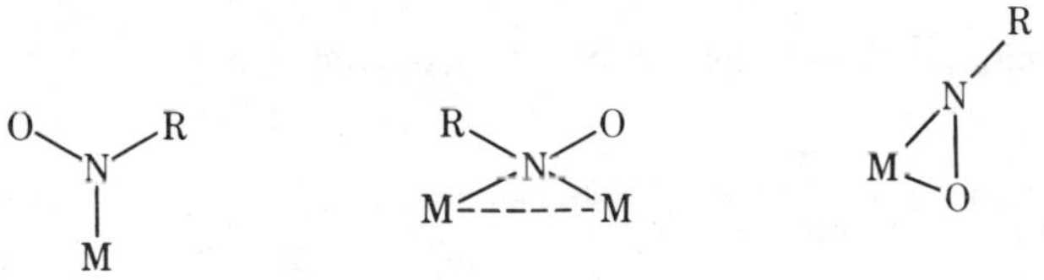

The complexes can be made directly from compounds like nitrosobenzene, PhNO, or from substituted hydroxylamines by reactions such as

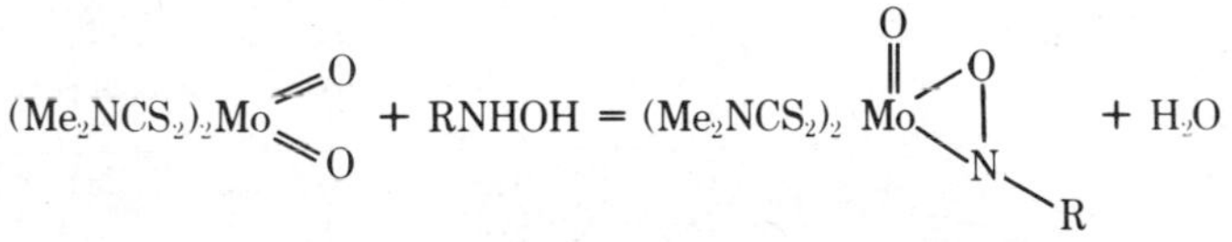

The three-membered rings have been termed[104a] *metallooxaziridines.*

Such compounds can undergo quite facile N—O bond cleavage[104a,b] and nitrene,

102b L. A. Epps *et al., Inorg. Chem.,* 1977, **16,** 2663.

103 (a) R. B. King and K. N. Chen, *Inorg. Chem.,* 1977, **16,** 1164; G. B. Khare and R. J. Doedens, *Inorg. Chem.,* 1977, **16,** 907. (b) S. Aime *et al.* Chem. Comm., **1976,** 370; G. P. Khare and R. J. Doedens, *Inorg. Chem.,* 1976, **15,** 86.

104 (a) K. B. Sharpless *et al., J. Am. Chem. Soc.,* 1978, **100,** 7061; F. Mares *et al., J. Am. Chem. Soc.,* 1978, **100,** 7063; D. B. Sams and R. L. Doedens, *Inorg. Chem.,* 1979, **18,** 151. (b) S. Otsuka *et al., Inorg. Chem.,* 1976, **15,** 657. (c) D. Mansuy *et al., J. Am. Chem. Soc.,* 1977, **99,** 6441; J. J. Watkins and A. L. Balch, *Inorg. Chem.,* 1975, **14,** 2720. (d) K. Wieghardt *et al., Angew. Chem. Int. Ed.,* 1979, **18,** 548, 549.

:NR, groups can be transferred to alkenes, cyclohexanone, or isocyanides, for example,

$$PhNO + Bu^tNC \xrightarrow{\ Ni(Bu^tNC)_4\ } PhN = CN^tBu$$

RNO compounds are also intermediates in the insertion of NO into metal alkyls (Section 29-11), but these can react further depending on whether the initial product is paramagnetic, in which case an N-methyl-N-nitroso-hydroxylaminato chelate is formed:

$$WMe_6 + NO \longrightarrow Me_5W\text{—}O\text{—}\overset{.}{N}\text{—}Me \xrightarrow{\ NO\ } Me_5W$$

or is diamagnetic, in which case MeN is transferred to NMe:

$$2ReOMe_4 + 2NO \longrightarrow 2ReO(ONMe)Me_3 \longrightarrow 2Me_3Re \overset{O}{\underset{O}{\diagup}} \overset{.}{\ } + MeN = NMe$$

A closely related complex made from hydroxylamine[104d] can be regarded either as protonated oxaziridine or merely as a η^2-hydroxylaminato complex:

$$+ 2NH_2OH \ = \qquad\qquad + H_2O$$

4-19. Other N Ligands

Biguanide [$H_2NC(NH)NHC(NH)NH_2$] has a large proton affinity in aqueous solution, its base strength being only slightly less than OH⁻ ($pK = 13.25$). It forms stable complexes[105a] as a bidentate ligand, has a strong preference for square coordination as in Ni(bgu)$^{2+}$, and finally tends to stabilize high oxidation states such as Ag^{III} in the complex 4-XLVI.

[105a] L. Fabbrizi *et al., Inorg. Chem.,* 1978, **17**, 494.

(4-XLVI)

Guanidines such as arginine and creatine can also give complexes.[105b] Thus a hydrogen atom of creatine can be replaced by PhHg to give the zwitterionic complex.

PHOSPHORUS, ARSENIC, ANTIMONY, AND BISMUTH[106]

In this section we deal only with phosphorus compounds as ligands. In general the behavior of analogous trivalent arsenic and antimony ligands resembles that of the phosphorus compounds. Relatively few compounds of antimony and even fewer of bismuth are known, and none are of great importance. Little further need be said concerning the complexes of compounds of arsenic, antimony, and bismuth. The σ-donor ability decreases in the order P > As > Sb > Bi, and bismuth is a very weak donor indeed. Steric effects due to the donor atom itself will increase as follows:

[105b] A. J. Canty *et al.*, *Inorg. Chem.*, 1978, **17**, 1467.

[106] For phosphines see *Organic Phosphorus Compounds*, G. M. Kosolapoff and L. Maier, Eds., Wiley; see also under phosphorus π bonding, Chapter 3, page 87; O. Stelzer, in *Topics in Phosphorus Chemistry*, Vol. 9, Wiley, 1977, p. 1 (extensive review); C. A. McAuliffe, Ed., *Transition Metal Complexes of Phosphorus, Arsenic and Antimony Ligands*, McMillan, 1973; C. A. McAuliffe and W. Levason, *Phosphine, Arsine and Stibine Complexes of Transition Metals*, Elsevier, 1978; W. A. Levason and C. A. McAuliffe, *Acc. Chem. Res.*, 1978, **11**, 363 (organostibine complexes); G. Booth, *Adv. Inorg. Chem. Radiochem.*, 1964, **6**, 1; W. Levason and C. A. McAuliffe, *Coord. Chem. Rev.*, 1976, **19**, 173 (P, As, and Sb complexes of main group elements); C. A. McAuliffe, *Adv. Inorg. Chem. Radiochem.*, 1975, **17**, 165 (complexes of open-chain tetradentate ligands); W. A. Levason and C. A. McAuliffe, *Adv. Inorg. Chem. Radiochem.*, 1972, **14**, 173 (complexes of bidentate P ligands); A. N. Hughes and K. Wright, *Coord. Chem. Rev.*, 1975, **15**, 239 (cyclic phosphine complexes); P. Rigo and A. Turco, *Coord. Chem. Rev.*, 1974, **13**, 133 (phosphorus-cyanide complexes); J. Verkade, *Coord. Chem. Rev.*, 1972–1973, **9**, 1, 106 (spectroscopic studies of phosphite complexes); D. G. Holah, A. N. Hughes, and K. Wright, *Coord. Chem. Rev.*, 1975, **15**, 239 (complexes of cyclic phosphines, phospholes, etc.); P. G. Eller *et al.*, *Coord. Chem. Rev.*, 1977, **24**, 1 (three-coordination review includes R_3P complexes of Group VIII metals); D. M. Roundhill *et al.*, *Coord. Chem. Rev.*, 1978, **26**, 263 (complexes of phosphinates and secondary phosphites).

$P < As < Sb < Bi$. Finally, steric effects of substituents will decrease in the order $P > As > Sb$.

We have discussed the bonding of phosphorus trihalides, notably PF_3, and tertiary phosphines, together with some other aspects of their ligand behavior in Chapter 3. We now consider additional types of P ligand.

4-20. Tertiary Phosphine Ligands

In addition to the unidentate R_3P ligands, there are a number of multidentate phosphines.

Diphosphine ligands (diphos) such as $Ph_2PCH_2CH_2PPh_2$ are almost invariably chelated, although there are examples [e.g., for $W(CO)_5$(diphos)] of unidentate and of bridging behavior.[107]

Bidentate phosphines with only one bridging group such as $Ph_2P—CH_2—PPh_2$[108] and $CH_3N(PF_2)_2$ tend to promote metal-metal interaction or bond formation because the two donor P atoms are so close together:

$$\begin{array}{c} CH_2 \\ R_2P \diagup \quad \diagdown PR_2 \\ | \qquad | \\ M----M \end{array}$$

The use of chelate phosphines with many bridging groups giving long flexible chains has quite a different effect. For example, the chelate phosphines $Bu_2^t P(CH_2)_{10}PBu_2^t$ or $Bu_2^t PC{\equiv}C(CH_2)_5C{\equiv}CPBu_2^t$ can give complexes[110] that have as many as 72 atoms in a ring of type 4-XLVII. Similar bidentates can span trans positions as shown diagrammatically in 4-XLVIII.[111]

$$\begin{array}{c} R_2P—CH_2—(CH_2)_8—CH_2—PR_2 \\ | \\ Cl—Pd—Cl \qquad\qquad Cl—Pd—Cl \\ | \\ R_2P—CH_2—(CH_2)_8—CH_2—PR_2 \end{array} \qquad \begin{array}{c} R_2P \\ \diagdown \\ M \\ \diagdown PR_2 \end{array}$$

(4-XLVII)   (4-XLVIII)

A rather special chelate[112] may form where triphenylphosphine is π-bonded via a phenyl ring in a compound like 4-XLIX. Substituted ferrocenes (Chapter 27) like 4-L can also be made.

[107] R. L. Keiter *et al.*, *J. Am. Chem. Soc.*, 1977, **99**, 5224; A. L. Balch, *J. Am. Chem. Soc.*, 1976, **98**, 8049.

[108] A. A. M. Ally *et al.*, *Angew. Chem. Int. Ed.*, 1978, **17**, 125.

[109] M. G. Newton *et al.*, *J.C.S. Chem. Comm.*, **1978**, 514.

[110] B. L. Shaw *et al.*, *J.C.S. Chem. Comm.*, **1977**, 311; *J.C.S. Dalton*, **1976**, 322; A. R. Sanger, *J.C.S. Dalton*, **1977**, 1971.

[111] L. M. Venanzi *et al.*, *Helv. Chim. Acta*, 1977, **60**, 2804, 2815, 2824 (see also Chap. 3, Ref. 10).

[112] C. Elschenbroich and F. Stohler, *Angew. Chem., Int. Ed.*, 1975, **14**, 174. For other references to PPh_3 acting as a π-arene see D. J. Cole-Hamilton and G. Wilkinson, *J.C.S. Dalton*, **1976**, 1995.

(4-XLIX)   (4-L)

Heterocyclic phosphorus compounds[113] such as phosphole (C_5H_5P) can bond either through P[114a] or be π-bonded as an arene.[114b] The compound 4-LI also acts[115] as a ligand in, for example, $Fe(CO)_4L^+$.

(4-LI)

It may be noted finally that tertiary phosphine ligands of the type R_2PH or R_2PCl can often undergo reactions whereby the H or Cl group is attacked.[116] Examples are

$$cis\text{-}Mo(CO)_4(PPh_2Cl)_2 \xrightarrow{RNH_2} cis\text{-}Mo(CO)_4(PPh_2NHR)_2$$

$$Ni(PCl_3)_4 \xrightarrow{MeOH} Ni[P(OMe)_3]_4$$

Macrocyclic Phosphine Complexes.[117] These may have only phosphorus or have mixed donor atoms of phosphorus, nitrogen, sulfur, and so on. Examples are 4-LII and 4-LIII.

(4-LII)   (4-LIII)

[113] R. E. Atkinson, *Rodd's Chemistry of Carbon Compounds,* Vol. IV, Part G, 2nd ed., Elsevier, 1978.

[114a] E. W. Abel and C. Towers, *J.C.S. Dalton,* **1979,** 814.

[114b] A. J. Ashe, III, and J. C. Colburn, *J. Am. Chem. Soc.,* 1977, **99,** 8099.

[115] R. W. Light and R. T. Paine, *J. Am. Chem. Soc.,* 1978, **100,** 2230; R. G. Montemayor, *J. Am. Chem. Soc.,* 1978, **100,** 2231.

[116] G. M. Gray and C. S. Kraihanzel, *J. Organomet. Chem.,* 1978, **146,** 23; J. von Seyerl *et al., Angew. Chem., Int. Ed.,* 1977, **16,** 858.

[117] R. E. Davis *et al., J. Am. Chem. Soc.,* 1978, **100,** 3642; T. A. Delbonno and W. Rosen, *J. Am. Chem. Soc.,* 1977, **99,** 8051, 8053, J. de O. Cabral *et al., Inorg. Chim. Acta,* 1977, **25,** L77; M. M. Taqui Khan and A. E. Martell, *Inorg. Chem.,* 1975, **14,** 676.

4-21. Phosphorus and Phosphorus Oxides

There are a few compounds in which P_2, P_3, or P_4 molecules act as ligands. Some rather unstable compounds of P_4 such as $RhCl(PPh_3)_2(P_4)$ and $[Fe(CO)_4]_3P_4$ are known, but the best characterized[118] are those of a tripod ligand $MeC(CH_2PPh_2)_3$ (L) that contain the cyclotriphosphorus (δ-P_3) group either as an end group or as a bridge; the P atoms can act as ligands and bind to $Cr(CO)_5$.

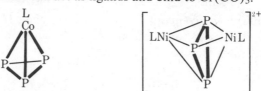

Some phosphido species containing only P atoms are noted later—these are considered to be derived from P^{3-}.

Since phosphorus *oxides* (Chapter 14) have lone pairs, these can also act as ligands, and compounds such as $Fe(CO)_4(P_4O_6)$, $W(CO)_5(P_4O_6)$ and $Cr(CO)_5(P_4O_7)$ are known.[119]

4-22. Phosphido-Bridged Species[120]

Phosphorus can act as a bridge between metal atoms in the form of P, PR and PR_2 (where R = H, F, Cl, CH_3, C_6H_5, etc.). The most common type of bridge is μ-PR_2. Phosphido-bridged species can be made directly by interaction of halide complexes with $LiPR_2$ or from sodium salts such as $NaCo(CO)_4$ with $PPhCl_2$ and in other ways. They are often obtained by reactions involving tertiary phosphine ligands in which a P—C bond of the ligand is cleaved. Some examples of such reactions are:

$$2\,Pt(PPh_3)_4 \longrightarrow \underset{Ph_3P}{\overset{Ph_3P}{>}}Pt\underset{Ph_2}{\overset{\overset{Ph_2}{P}}{\underset{P}{<}}}Pt\underset{PPh_3}{\overset{PPh_3}{<}} + Ph—Ph$$

$$2\,HIr(CO)(PPh_3)_3 \longrightarrow (Ph_3P)_2(CO)Ir\underset{\underset{Ph_2}{P}}{\overset{\overset{Ph_2}{P}}{<>}}Ir(CO)(PPh_3)_2 + 2\,C_6H_6$$

[118]   L. Sacconi *et al., Inorg. Chem.,* 1978, **17**, 3292; *Angew. Chem., Int. Ed.,* 1979, **18**, 72, 469; *J. Am. Chem. Soc.,* 1978, **100**, 2550; 1979, **101**, 1757.

[119]   M. L. Walker and J. L. Mills, *Inorg. Chem.,* 1977, **16**, 3033; *J. Organomet. Chem.,* 1976, **120**, 355.

[120]   C. M. Bartish and C. S. Kraihanzel, *Inorg. Chem.,* 1978, **17**, 735; J. C. Burt *et al., J.C.S. Dalton,* **1978**, 1385, 1387; E. Keller and H. Vahrenkamp, *Chem. Ber.,* 1977, **110**, 430; J. K. Burdett, *J.C.S. Dalton,* **1977**, 423; W. Clegg, *Inorg. Chem.,* 1976, **15**, 1609; W. Malisch *et al., Angew. Chem., Int. Ed.,* **1977**, 408; **1976**, 769; J. Reed *et al., Inorg. Chem.,* 1973, **12**, 2949; N. J. Taylor, *et al., J.C.S. Chem. Comm.,* **1975**, 448; E. A. V. Ebsworth, *et al., J.C.S. Dalton,* **1978**, 272; D. R. Fahey and J. E. Mahan, *J. Am. Chem. Soc.,* 1976, **98**, 4491; G. W. Bushnell *et al., J.C.S. Chem. Comm.,* **1977**, 709.

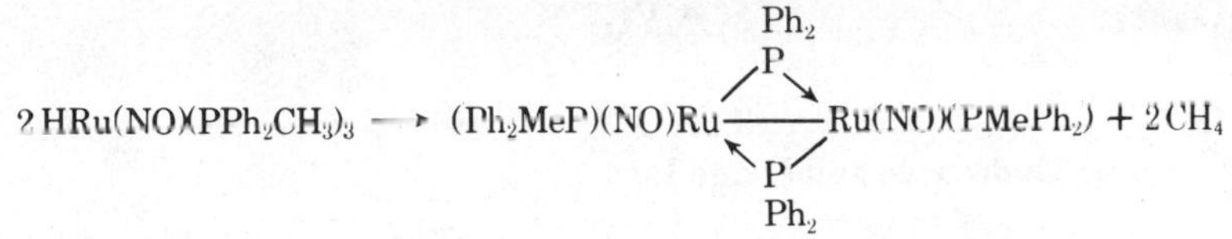

$$2\,HRu(NO)(PPh_2CH_3)_3 \longrightarrow (Ph_2MeP)(NO)Ru \overset{\overset{\textstyle Ph_2}{\underset{\textstyle |}{P}}}{\underset{\overset{\textstyle P}{\underset{\textstyle Ph_2}{}}}{}} Ru(NO)(PMePh_2) + 2\,CH_4$$

These elimination reactions are considered to involve intramolecular oxidative addition of the ligand (Section 29-4) followed by reductive elimination, namely,

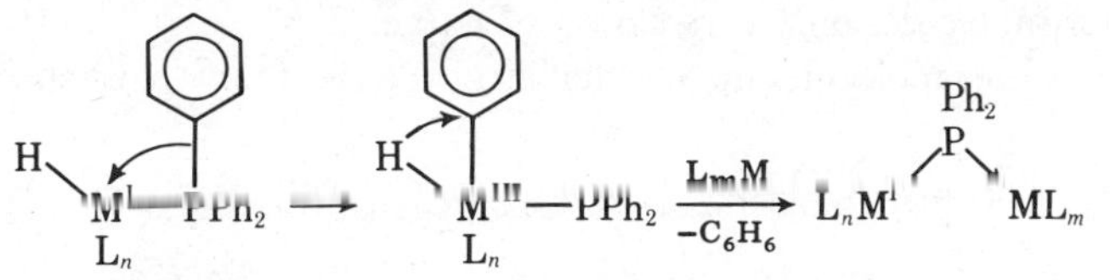

$$H\!-\!\underset{L_n}{M^I}\!-\!PPh_2 \longrightarrow H\!-\!\underset{L_n}{M^{III}}\!-\!PPh_2 \xrightarrow[-C_6H_6]{L_mM} L_nM^I \overset{\overset{\textstyle Ph_2}{\underset{\textstyle |}{P}}}{\diagdown} ML_m$$

or intermolecular transfers of the type

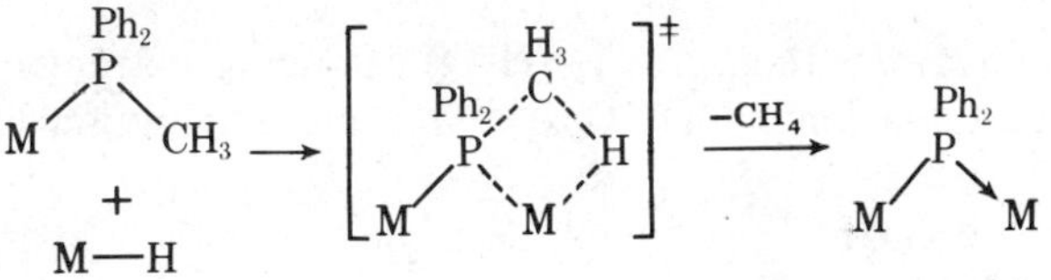

$$\underset{\underset{M-H}{+}}{M}\overset{\overset{\textstyle Ph_2}{\underset{\textstyle |}{P}}}{\diagdown}CH_3 \longrightarrow \left[\underset{M}{\overset{Ph_2}{}}\overset{\overset{\textstyle H_3}{\textstyle C}}{\underset{\textstyle P}{\diagup\diagdown}}\overset{\textstyle H}{\underset{\textstyle M}{}} \right]^{\ddagger} \xrightarrow{-CH_4} \underset{M}{\overset{Ph_2}{P}}\underset{M}{}$$

Although P—C bond cleavage also occurs in reactions of t-butyl phosphine with RhCl$_3$, for example, the product is a complex of Bu$_2^t$PH and is *not* a bridged species.[121]

An example of a μ_4-PPh bridged compound[122] is 4-LIV.

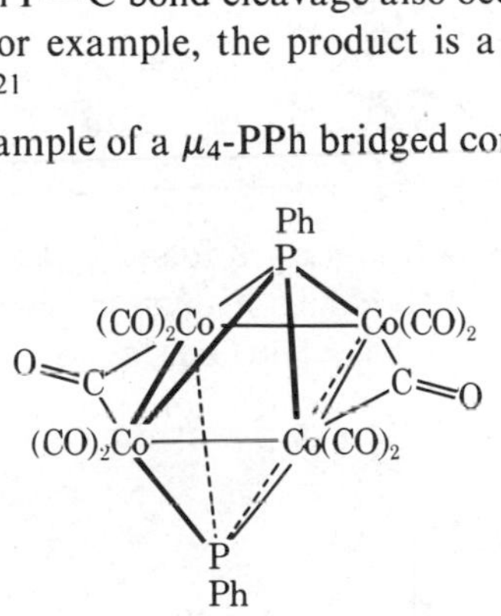

(4-LIV)

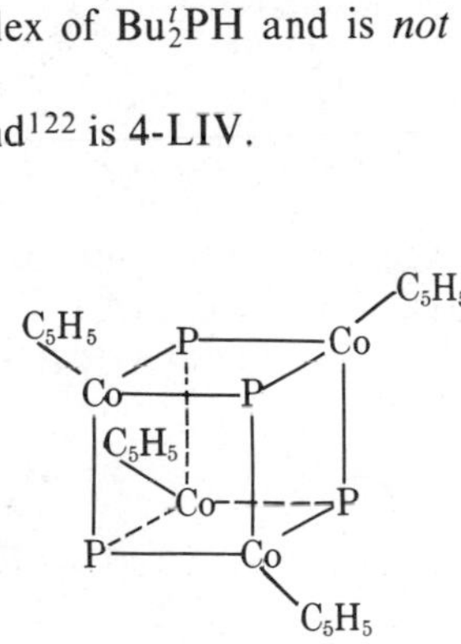

(4-LV)

There are finally a few compounds[123a] that have μ_3-P as in the cubane-type complex $[(\eta\text{-}C_5H_5)CoP]_4$ (4-LV), and encapsulated species such as $[Co_6(CO)_{14}(\mu\text{-}CO)_2P]^-$ and $[Rh_9(CO)_{21}P]^{2-}$ are known.[123b]

[121] R. G. Goel et al., J. Am. Chem. Soc., 1978, **100**, 3629.
[122] L. F. Dahl et al., J. Am. Chem. Soc., 1975, **97**, 6904; 1976, **98**, 5046. For μ_3-PH, see R. G. Austin and G. Urry, Inorg. Chem., 1977, **16**, 3359.
[123a] G. L. Simon and L. F. Dahl, J. Am. Chem. Soc., 1973, **95**, 7175.
[123b] P. Chini et al., J.C.S. Chem. Comm., **1979**, 188; J. L. Vidal et al., Inorg. Chem. 1979, **18**, 127.

OXYGEN

4-23. Water; Hydroxide and Oxide Ions

In aqueous solution metal ions are surrounded by water molecules; in some cases, such as the alkali ions, they are weakly bound, whereas in others, such as $[Cr(H_2O)_6]^{3+}$ or $[Rh(H_2O)_6]^{3+}$, they may be firmly bound and exchange with solvent water molecules only very slowly (Chapter 28).

The bound water molecules may be acidic, giving rise to *hydroxo* species,[124] such as

$$[M(H_2O)_x]^{n+} \rightleftharpoons [M(H_2O)_{x-1}(OH)]^{(n-1)+} + H^+$$

The acidities of aqua ions can vary by orders of magnitude, e.g.,

$$[Pt(NH_3)_4(H_2O)_2]^{4+} = [Pt(NH_3)_4(H_2O)(OH)]^{3+} + H^+ \qquad K \approx 10^{-2}$$
$$[Co(NH_3)_5(H_2O)]^{3+} = [Co(NH_3)_5(OH)]^{2+} + H^+ \qquad K \approx 10^{-5.7}$$

Some molten hydrates (e.g., $ZnCl_2 \cdot 4H_2O$) can act as extremely strong acids, even though the aqua ion $[Zn(H_2O)_4]^{2+}$ in aqueous solution is a very weak acid.[125]

A common feature of hydroxo complexes is the formation of hydroxo bridges[126] of the following types:

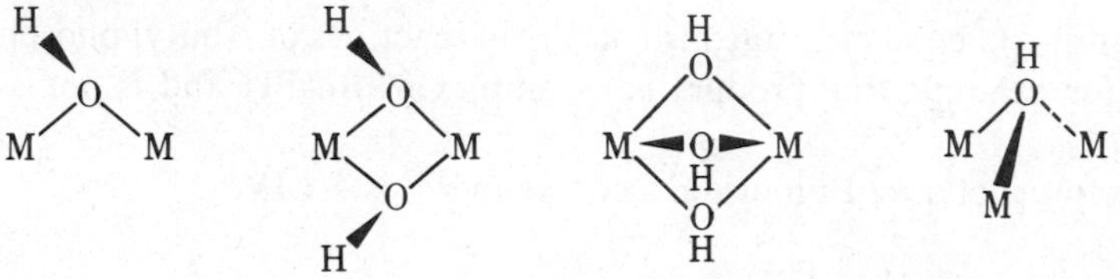

Double bridges are most common. Three μ_2-bridges are found in the π-arene complex $[ArRu(OH)_3RuAr]^+$, whereas a triply bridging, μ_3-hydroxo group occurs in the cubanelike ions $[(ArRuOH)_4]^{4+}$ (4-LVI), $[Pt(OH)Me_3]_4$, and some others.[127]

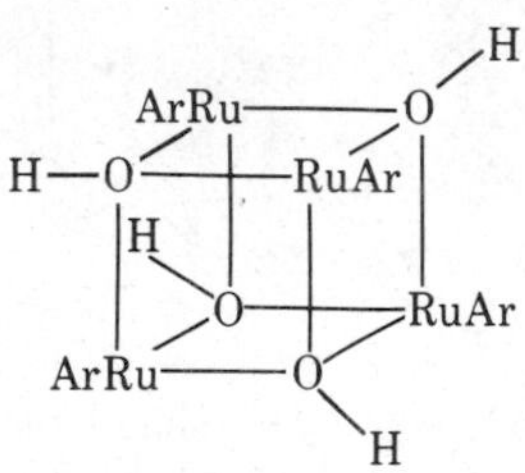

(4-LVI)

[124] C. F. Baes and R. E. Mesmer, *The Hydrolysis of Cations,* Wiley-Interscience, 1976; V. Baran, *Coord. Chem. Rev.,* 1971, **6,** 65.

[125] J. A. Duffy and M. D. Ingram, *Inorg. Chem.,* 1978, **17,** 2798.

[126] D. J. Hodgson, *Prog. Inorg. Chem.,* 1975, **19,** 173 (complexes of first-row transition elements).

[127] D. R. Robertson and T. A. Stephenson, *J. Organomet. Chem.,* 1976, **116,** C29; R. O. Gould *et al., J.C.S. Chem. Comm.,* **1977,** 222; S. Merlino *et al., Inorg. Chim. Acta,* 1978, **27,** 233.

The loss of a second proton from water can lead to the formation of *oxo compounds*,[128] which can be of several types:

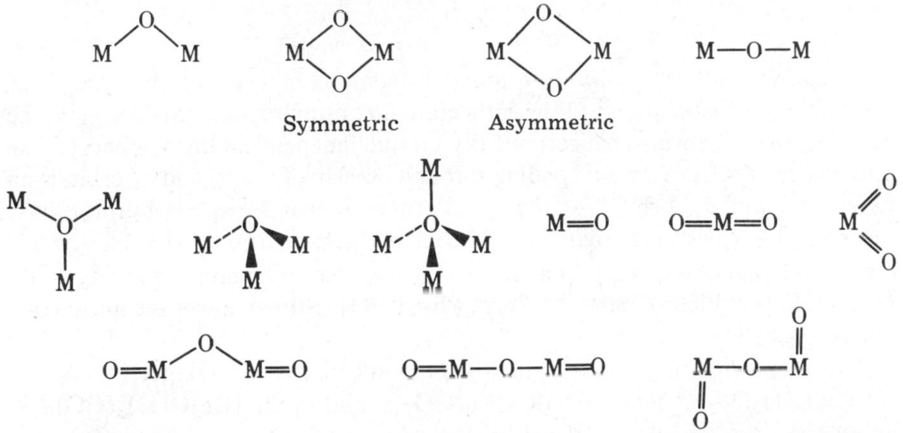

Symmetric	Asymmetric

The double bridged species are almost invariably symmetrical, but in an osmium compound there is an asymmetric bridge[129] with distances 1.78 and 2.22 Å.

The multiply bonded oxo group $M{=}O$, is found not only in oxo compounds and oxo anions of non-transition elements such as $O{=}SCL_2$, SO_4^{2-}, $O{=}PCl_3$, and PO_4^{3-}, but also in transition metal compounds such as vanadyl ($O{=}V^{2+}$), uranyl ($O{=}U{=}O^{2+}$), permanganate (MnO_4^-), and osmium tetraoxide (OsO_4). In all these cases the bond distances (*ca.* 1.59–1.66 Å) correspond to a double bond, and the $M{=}O$ infrared stretching frequencies usually lie in the 800 to 1000 cm^{-1} region for transition metal species. Protonation by acids will convert $M{=}O$ to $M{-}OH$. In transition metal compounds, the π component is best regarded as arising from $Op\pi \rightarrow Md\pi$ electron flow. Since this is the opposite of electron flow in π-bonding ligands of the CO type, it is not surprising that the latter are most stable in low oxidation states whereas $M{=}O$ bonds are most likely in high oxidation states.

The $M{=}O$ bonding is commonly affected by the nature of groups trans to oxygen—and oxygen has a strong trans effect (Section 28-7). Donors that increase electron density on the metal tend to reduce its acceptor properties, thus *lowering* the $M{-}O$ multiple bond character, hence the $M{-}O$ stretching frequency. Because of the strong trans effect, ligands trans to oxygen may be labile.

Dioxo compounds may be linear (trans) as in $O{=}U{=}O^{2+}$ or angular (cis) as in some molybdenum complexes and in ReO_2Me_3.

Singly bridged complexes may have either bent or linear bridges. The $M{-}O{-}M$ angle can vary from *ca.* 140° to 180° and to a large extent the angle seems to be determined by the steric requirements of the other ligands attached to the metal. There are few bent bridges except for those in $Cr_2O_7^{2-}$, $W_2O_{11}^{2-}$, and $P_2O_7^{2-}$. Examples are[130]:

128 W. P. Griffith, *Coord. Chem. Rev.*, 1970, **5**, 459 (M—O, M=O); K. S. Murray, *Coord. Chem. Rev.*, 1974, **12**, 1 (μ-oxo complexes of Fe^{III}).
129 B. A. Cartwright *et al.*, *J.C.S. Chem. Comm.*, **1978**, 853.
130 P. T. Cheng and S. C. Nyburg, *Inorg. Chem.*, 1975, **14**, 327; D. W. Phelps *et al.*, *Inorg. Chem.*, 1975, **14**, 2486.

$$(PPh_3)Cl(NO)Ir \overset{O}{\diagdown} Ir(NO)Cl(PPh_3)$$

$$[bipy_2(NO_2)Ru \overset{O}{\diagdown} Ru(NO_2)bipy]^{2+}$$

Linear M—O—M groups are found in some complexes of rhenium, iron, ruthenium, and osmium.[131] In the ruthenium and osmium ions, $[M_2OX_{10}]^{4-}$, the M—O—M unit forms an electronically unique independent chromophore. The linearity results from $d\pi$-$p\pi$ bonding through overlap of the p_x and p_y orbitals on O with d_{xz} and d_{yz} orbitals on the metal atoms. Linear M—O—M groups have infrared vibrations lower than those in bent bridges, *ca.* 260 cm^{-1} versus *ca* 570 cm^{-1}. In linear oxo species such as O≡Re—O—Re≡O again π bonding is important. Some silicon compounds (e.g., Ph_3Si—O—$SiPh_3$), also have linear Si—O—Si bonds.[132]

The *pyramidal M_3O* unit with μ_3-O occurs in the ion OHg_3^+, in $[W_3O_2$-$(O_2CR)_6(H_2O)_3]^{2+}$ ions,[133a] in $Os_4O_4(CO)_{12}$, and in the $[Re_3O(H)_3(CO)_9]^{2-}$ anion.[133b]

Oxo-centered complexes can have μ_4-tetrahedral oxygen in the center of a tetrahedron of divalent metal atoms as in $M_4O(O_2CMe)_4$.[134] The best-known complex is $Be_4O(O_2CMe)_4$, but Zn^{II} and Co^{IV} analogues are known. An iron(III) complex, $Fe_4^{III}O(O_2CMe)_{10}$, also has μ_4-O, and higher oxo iron polymers may have trigonal bipyramidal coordination, as in $[Fe_5O(O_2CMe)_{12}]^+$.[135]

Oxygen-centered triangles are found in the so-called basic carboxylates of trivalent metals such as Cr^{3+}, Mn^{3+}, and Fe^{3+}. They have the general formula $[M_3O(CO_2R)_6L_3]^+$, where L is a ligand such as H_2O or py. The structures 4-LVII

(4-LVII)

indicate that the M_3O group is planar on account of M—O π bonding. However in the pivalate $[Fe_3O(CO_2Me_3)_6(MeOH)_3]^+$ the O atom is 0.24 Å out of plane,[136]

[131] C. C. Ou *et al., J. Am. Chem. Soc.,* 1978, **100,** 4717; J. San Fillipo, Jr., *et al., Inorg. Chem.,* 1976, **15,** 269; 1977, **16,** 1016; R. J. H. Clark *et al., J. Am. Chem. Soc.,* 1977, **99,** 2473.

[132] C. Glidewell and D. C. Liles, *J.C.S. Chem. Comm.,* **1977,** 632.

[133a] A. Bino *et al., Inorg. Chem.,* 1978, **17,** 3245.

[133b] G. Ciani *et al., J.C.S. Dalton,* **1977,** 1667.

[134] J. Charalambous *et al., Inorg. Chim. Acta,* 1975, **14,** 53.

[135] J. Catterick *et al., J.C.S. Dalton,* **1977,** 1420.

[136] A. B. Blake and L. R. Frazer, *J.C.S. Dalton,* **1975,** 193.

and μ_3, out-of-plane oxygen atoms also occur in some molybdenum compounds that have a triangle of Mo atoms.[137]

For ruthenium, the O-centered complexes can be reduced[138]:

$$[Ru_3O(CO_2R)_6py_3]^+ \underset{-e}{\overset{+e}{\rightleftarrows}} [Ru_3O(CO_2R)_6py_3] \underset{O_2}{\overset{+2e}{\rightleftarrows}} Ru_3(CO_2R)_6py_3$$

In reduced species like $Ru_3O(CO_2R)_6py_3$ there are nonintegral oxidation states $(2III + II = 2\frac{2}{3})$ and the metal atoms are equivalent. Reduced forms of manganese and iron containing formally $M^{II,III,III}$ are known—for example, $Mn_3O(O_2C-Me)_6py_3$.[139]

Although most oxo-centered species have carboxylate bridges, some examples with SO_4^{2-} and NO_2^- bridges are known, as in the ion[140] $[Pt_3O(NO_2)_6]^{2-}$.

Finally amino acids can form oxo-centered species such as $[Fe_3O(ala-nine)_6(H_2O)_3]^{7+}$, and this type of structure is a possible candidate for the coordination of iron in the serum iron transport protein *ferritin*.[141]

4-24. Dioxygen, Superoxo, and Peroxo Ligands[142]

Molecular oxygen can be reduced by two one-electron processes without the O–O bond being broken

$$O_2 \underset{-e}{\overset{+e}{\rightleftarrows}} O_2^- \underset{-e}{\overset{+e}{\rightleftarrows}} O_2^{2-}$$

and each of these species can act as a ligand toward transition metals.

Molecular oxygen reacts reversibly with some metal complexes and such reactions play a key role in life processes, for example, in oxygenation of hemoglobin and myoglobin[143] (Chapter 31).

Compounds with O_2 groups can be obtained as follows:

1. From O_2 by reversible addition reactions with such coordinatively unsaturated complexes (Chapter 29) as

$$trans\text{-}Ir^ICl(CO)(PPh_3)_3 + O_2 \rightleftarrows Ir^{III}(O_2)Cl(CO)(PPh_3)_2$$
$$Co^{II}(acacen) + O_2 + Me_2NCHO \rightleftarrows Co^{III}(O_2)(acacen)(Me_2NCHO)$$

Note that in these two reactions the number of electrons formally transferred from

137 A. Bino *et al.*, *J. Am. Chem. Soc.*, 1978, **100**, 5252.
138 T. J. Meyer *et al.*, *Inorg. Chem.*, 1978, **17**, 3342; *J. Am. Chem. Soc.*, 1979, **101**, 2916.
139 A. R. E. Baikie *et al.*, *J.C.S. Chem. Comm.*, **1978**, 62.
140 A. E. Underhill and D. M. Watkins, *J.C.S. Dalton*, **1977**, 5.
141 E. M. Holt *et al.*, *J. Am. Chem. Soc.*, 1974, **96**, 2621.
142 A. P. B. Lever and H. B. Gray, *Acc. Chem. Res.*, 1978, **11**, 348; J. E. Lyons, in *Aspects of Homogeneous Catalysis*, Vol. 3, R. Ugo, Ed., D. Reidel, 1977; R. W. Erskine and B. O. Field, *Struct. Bonding*, 1976, **28**, 3; L. Vaska, *Acc. Chem. Res.*, 1976, **9**, 175; G. Henrici-Olivé and S. Olivé, *Angew. Chem., Int. Ed.*, 1974, **13**, 29; F. Basolo, J. A. Ibers, and B. M. Hoffman, *Chem. Rev.*, 1979, **79**, 139; *J. Chem. Educ.*, 1979, **56**, 157; *Acc. Chem. Res.*, 1976, **9**, 1384; G. McLendon and A. E. Martell, *Coord. Chem. Rev.*, 1976, **19**, 1; V. J. Choy and C. J. O'Connor, *Coord. Chem. Rev.*, 1972–1973, **9**, 145; J. Valentine, *Chem. Rev.*, 1973, **73**, 235; J. A. Connor and E. A. V. Ebsworth, *Adv. Inorg. Chem. Radiochem.*, 1964, **6**, 280 (peroxo).
143 O. Hayaishi, *Molecular Mechanisms of Oxygen Activation*, Academic Press, 1974 (oxygenases).

metal to ligand is 2 for the iridium complex and 1 for the cobalt complexes, corresponding to reduction of O_2 to O_2^{2-} and O_2^-, respectively.

2. From O_2 by *irreversible* reactions involving oxidation of the metal, notably cobalt complexes (Section 21-F-6), which form *bridged* complexes

$$Co_{aq}^{2+} + NH_3 + O_2 \rightarrow [am_5Co^{III}O_2Co^{III}am_5]^{5+}$$

Some of the amine complexes may have hydroxo bridges in addition to M—O—O—M bridges.[144] Note, of course that metal complexes are often irreversibly oxidized by O_2 *without* forming O_2 complexes—in some cases the metal is oxidized; for example, iron(II)tetrasulfophthalocyanin is oxidized to the iron(III) complex.[145] In others, the *ligand* may be oxidized and, for example, hydrogen atoms removed as in the conversion of ethylenediamine to imine complexes discussed earlier (p. 118). Destructive oxidation of organometallic and other complexes is a common feature. In most if not all these reactions peroxo and hydroperoxo intermediates are involved.

3. *From hydrogen peroxide* and metal aqua or other complex ions in aqueous solutions, e.g.,

$$HCrO_4^- + 2H_2O_2 + H^+ \rightarrow CrO(O_2)_2(H_2O) + 2H_2O$$

Peroxo species of titanium, niobium, tantalum, chromium, molybdenum, and tungsten have been well studied. Sometimes the same complex can be obtained either from O_2 or from H_2O_2, e.g.,

$$[Co^I diars_2]^+ \xrightarrow{O_2} [O_2Co^{III}diars_2]^+ \xleftarrow[\substack{-2H_2O \\ -2H^+}]{H_2O_2} cis\text{-}[Co^{III}(H_2O)_2 diars_2]^{3+}$$

Dioxygen complexes can broadly be classed in two groups:

1. Those containing *peroxo* (O_2^{2-}) *groups* that may be (*a*) part of a three-membered ring, (*b*) bridging staggered, or (*c*) bridging symmetrical

(a)   (b)   (c)

There is only one example of a symmetrical bridge, in the uranyl compound $[Cl_3O_2U\text{-}\mu(O_2)\text{-}UO_2Cl_3]^{4-}$ obtained by action of O_2 on uranyl sulfate (UO_2SO_4) in methanol.[146]

In peroxo compounds the O—O bond distances are *fairly constant* in the range 1.40–1.50 Å (O_2^{2-} = 1.49 Å) and *do not depend on the nature of the metal and its ligands.* The O—O stretching frequencies are in the 790 to 930 cm^{-1} region; for

[144] M. Zehnder and S. Fallab, *Helv. Chim. Acta,* 1975, **58**, 13; G. A. Lawrence *et al., Inorg. Chem.,* 1978, **17**, 3318.
[145] G. McLendon and A. E. Martell, *Inorg. Chem.,* 1977, **16**, 1812.
[146] R. Haegele and J. C. A. Boeyans, *J. C. S. Dalton,* **1977**, 648.

the triangular species these frequencies are around 850 cm^{-1}. It makes no difference whether compounds are made from O_2 or H_2O_2, and there is no correlation between reversible oxidation by O_2 and any bond parameters.

Although the bonding in the three-membered ring is most easily described by localized bonding, it can also be described by an MO treament[147] similar to that for the bonding of olefins or acetylenes (p. 95). Crudely, a σ bond is formed by filled $Op\pi \rightarrow Md\sigma$ bonding and back-bonding is due to $Md\pi \rightarrow O\pi^*$. Representative examples of three-membered ring compounds are oxygen adducts of planar d^8 metal complexes such as *trans*-IrCl(CO)(PPh$_3$)$_2$ (4-LVIII) and chromium peroxo complexes such as the dodecahedral $[Cr(O_2)_4]^{3-}$ ion 4-LIX.

(4-LVIII) (4-LIX)

2. The *superoxo* O_2^- *ion*[148] can be bound in the following ways:

The *bridged species* are mostly those of cobalt(III) or rhodium(III) formed by oxidation of peroxo complexes, e.g.,

The bridged superoxo complexes have O—O distances in the range 1.10–1.30 Å, e.g., 1.24 Å in $[(CN)_5Co(O_2)Co(CN)_5]^{5-}$,[149] which can be compared to that in O_2^- (1.33 Å). The O—O stretching frequencies lie in the 1075 to 1195 cm^{-1} region (O_2^-, 1145 cm^{-1}). The unpaired electron is delocalized over the metal atoms, according to electron paramagnetic resonance (epr) studies and lies in an MO of π symmetry relative to the planar MO$_2$M or MO$_2$ group.

147 S. Sakaki *et al., Inorg. Chem.,* 1978, **17**, 3183; Y. Ellinger *et al., Inorg. Chem.,* 1978, **17**, 2024; J. G. Norman, Jr., *Inorg. Chem.,* 1977, **16**, 1328.
148 A. M. Michelson, J. M. McCord, and I. Fridovich, Eds., *Superoxide and Superoxide Dismutases,* Academic Press, 1977.
149 G. McLendon *et al., Inorg. Chem.,* 1977, **16**, 1551.

End-on, unidentate, bent superoxo groups are found mainly in complexes of rhodium(III)[150] or cobalt(III),[151] such as $Co(O_2)$acacen (DMF) noted above, or $[Co(O_2)(CN)_5]^{3-}$.

The oxygen adducts of complexes of tetrapyrrole and other macrocyclic N ligands have been intensively studied because of the relation to natural oxygen transport molecules containing iron and copper. Considerable ingenuity has gone into trying to make truly reversible synthetic models for heme.[152] The problem is to prevent irreversible oxidation of the iron atom in the macrocycle from Fe^{II} to Fe^{III}.

One approach has been to construct what are termed "picket-fence" [153a] or basket handle[153b] porphyrins. Here, the way whereby the oxygen molecule can approach and leave the iron atom axially is sterically restricted by bulky groups (Fig. 4-3). What appears to be a much more realistic model for reversible binding of molecular oxygen has a heme group attached to a non-cross-linked polymer that confers water solubility (Fig. 4-4). Here the O_2 coordination site is shielded by both histidine and the polymer.[154] Heme systems are discussed further in Chapter 31.

4-25.   Alcohols, Alcoxides,[155] and Phenoxides

In solution in alcohols, particularly methanol, metal ions may be solvated just as in water, but the solvent molecules are usually readily displaced by stronger donor ligands such as water itself.

Just as coordinated water can lose a proton to give hydroxo complexes, so can alcohols:

$$M-O\underset{H}{\overset{R}{\diagdown}} \rightleftharpoons M-O^{\diagup R} + H^+$$

In the deprotonated form RO^-, *all* hydroxo compounds can act as ligands, for example, Schiff bases derived from hydroxo compounds, hydroxo acids, and so on.

Compounds derived from simple alcohols are called *alcoxides*. They are normally made by reactions of metal halides and alcohols in the presence of a hydrogen halide

---

[150]   R. D. Gillard *et al., J.C.S. Chem. Comm.,* **1977,** 58.

[151]   A. Dedieu *et al., J. Am. Chem. Soc.,* 1976, **98,** 5789; B.-K. Teo and W.-K. Li, *Inorg. Chem.,* 1976, **15,** 2005; M. Corrigan *et al., J.C.S. Dalton,* **1977,** 1478; G. B. Jameson *et al., J.C.S. Dalton,* **1978,** 191.

[152]   J. W. Buchler, *Angew. Chem., Int. Ed.,* 1978, **17,** 407; R. W. Erskine and B. O. Field, *Struct. Bonding,* 1976, **28,** 3; J.-H. Fuhrhop, *Angew. Chem., Int. Ed.,* 1976, **15,** 648; J. P. Collman, *Acc. Chem. Res.,* 1977, **10,** 265; F. Basolo, J. A. Ibers, and B. M. Hoffman, *Acc. Chem. Res.,* 1976, **9,** 384; W. M. Reiff *et al., Inorg. Chim. Acta,* 1977, **25,** 91.

[153a]   J. P. Collman *et al., J. Am. Chem. Soc.,* 1978, **100,** 2761; G. B. Jameson *et al., Inorg. Chem.,* 1978, **17,** 850, 858.

[153b]   M. Momenteau *et al., Nouv. J. Chim.,* 1979, **3,** 77.

[154]   E. Bayer and G. Holzbach, *Angew. Chim. Int. Ed.,* 1977, **16,** 117.

[155]   D. C. Bradley *et al., Metal Alkoxides,* Academic Press, 1978; *Adv. Inorg. Chem. Radiochem.,* 1972, **15,** 259; *Coord. Chem. Rev.,* 1967, **2,** 229; *Prog. Inorg. Chem.,* 1960, **2,** 203.

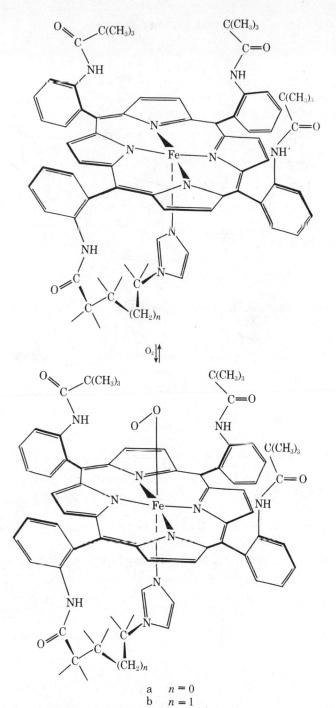

a $n = 0$
b $n = 1$

Fig. 4-3. Picket-fence porphyrins having an appended imidazole base. [Reproduced by permission from J. P. Collman, *et al.*, *Accounts Chem. Res.*, 1977, **10**, 265.]

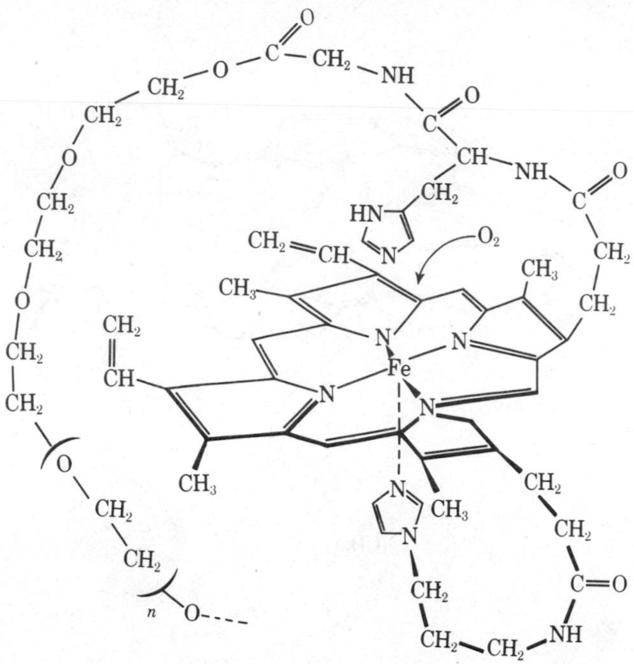

Fig. 4-4. Water soluble heme complex that binds O_2 reversibly. [Reproduced by permission from ref. 154.]

acceptor, e.g.,

$$TiCl_4 + 4C_2H_5OH + 4Et_3N = Ti(OC_2H_5)_4 + 4Et_3NH^+Cl$$

Although some alcoxides, particularly those with very bulky groups like *tert*-butyl can be monomeric, most alcoxides are polymeric with alcoxo bridge groups as in titanium tetraalcoxides (Fig. 4-5). Not only doubly bridging μ_2-RO groups are known, but also triply bridging μ_3- ones as in $Sn_6O_4(OMe)_4$, which has an adamantane Sn_6O_4 skeleton.[156] The polymerization is usually such as to attain maximum metal coordination. Alcoxides are readily hydrolyzed but are usually thermally stable, distillable liquids or volatile solids.

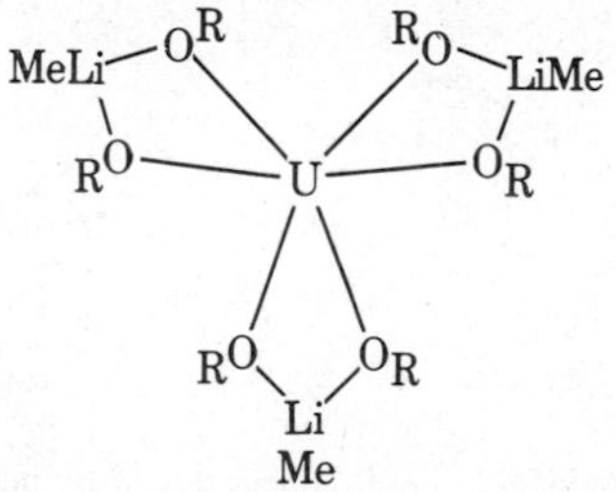

(4-LX) 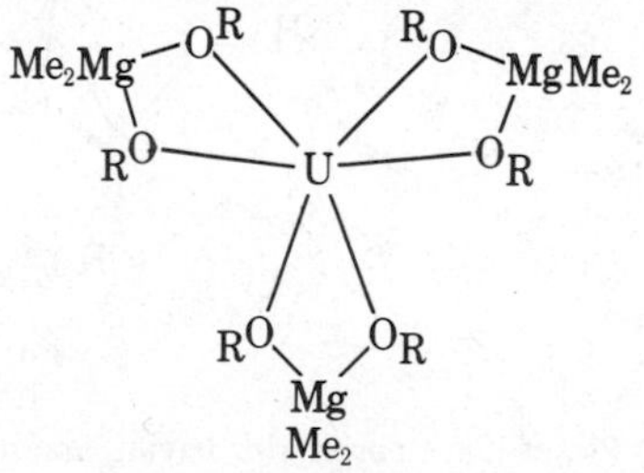   (4-LXI)

[156] P. G. Harrison *et al.*, *J.C.S. Chem. Comm.*, **1978**, 112.

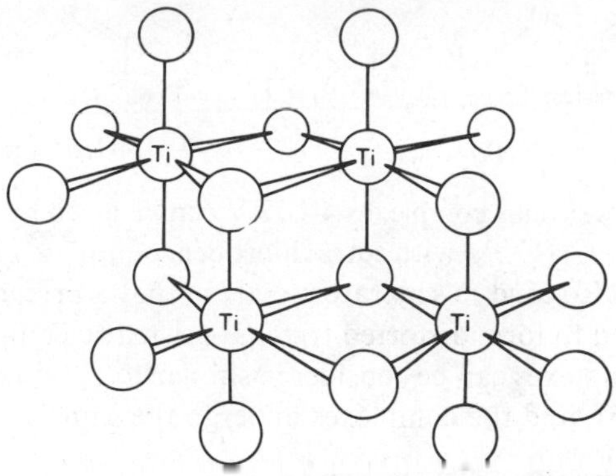

Fig. 4-5. The tetrameric structure of crystalline $[Ti(OC_2H_5)_4]_4$. Only Ti and O atoms are shown.

Not only can the alcoxo groups act as bridges to the same metal in the polymers, they may act as donors to other metal compounds. Thus $U(OPr^i)_6$ gives adducts with lithium, magnesium, and aluminum alkyls[157] as in 4-LX and 4-LXI.

Aromatic hydroxo compounds such as phenol readily form complexes that may have unidentate groups as in $W(OPh)_6$, or have phenoxo bridges as in (PhO)-Cl_2Ti-μ-$(OPh)_2TiCl_2(OPh)$.

A few cases are known where the phenoxide ion is π-bonded to the metal[158] as in 4-LXII. The C—O group becomes more like a keto group, and the bonding is delocalized and is similar to that in η-1-5-cyclohexadienyls (Chapter 27). Thus the compounds are best regarded as η-1-5-oxocyclohexadienyls rather than π arenes.

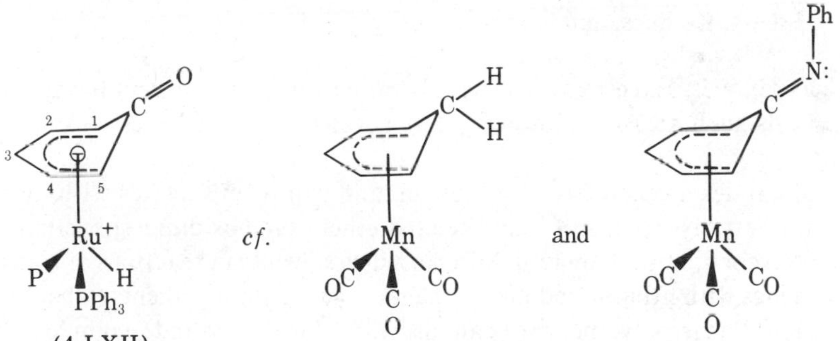

(4-LXII)   *cf.*   and

1,2-Dihydroxoarenes such as pyrocatechol can give chelates by bonding of the dianion (4-LXIII) as in $K_3[Cr(O_2C_6H_4)_3]$ or $(Et_4N)_2[Ti(O_2C_6H_4)_3]$.[159]

157   E. R. Sigurdson and G. Wilkinson, *J.C.S. Dalton,* **1977,** 812.
158   D. J. Cole-Hamilton, R. J. Young, and G. Wilkinson, *J.C.S. Dalton,* **1976,** 1995; W. S. Trahanovsky and R. A. Hall, *J. Am. Chem. Soc.,* 1977, **99,** 4850; C. White, S. J. Thompson, and P. M. Maitlis, *J. Organomet. Chem.,* 1977, **127,** 415.
159   K. N. Raymond *et al., J. Am. Chem. Soc.,* 1978, **100,** 7882; *Inorg. Chem.,* 1979, **18,** 1611. J. L. Martin and J. Takats, *Can. J. Chem.,* 1975, **53,** 572.

(4-LXIII)   (4-LXIV)

However formally similar complexes 4-LXIV can be given by *orthoquinones*.[160a] Thus interaction of $Cr(CO)_6$ with tetrachlorobenzoquinone (L) gives CrL_3; the oxygen atoms can also bind to separate metals so that a bridged quinone results. The tris species tend to form distorted trigonal prismatic complexes.

The quinone complexes can be considered similar to 1,2-dithiolenes and *o*-quinoneimines (p. 185), and the complexes undergo the same type of oxidation-reduction sequences[160b]:

Quinone   Semiquinone   Dihydroxo
                                      (pyrocatecholate)

In the paramagnetic semiquinone complex the electron is located on the ligand. The relative importance of the quinone versus dihydroxo species will depend on the basicity of the metal and the oxidizing ability of the quinone.

Paraquinones and their mono- and dianions also form complexes[161a]; the dianions can form *bridges* between two metals. The complexing of metals with a number of natural products that contain *o*-hydroxo aryl groups, quinol, or quinone moieties is probably important in nature.[161b]

4-26. Ethers, Ketones, and Esters

Ethers, ketones, and esters are generally rather weak bases except toward strong Lewis acids such as BF_3. However most covalent metal halides will form adducts.

Metal halides are commonly soluble in, and can form solvates with, tetrahydrofuran. Tetrahydrofuran[162a] and related ethers such as dimethoxyethane and ethyleneglycol dimethylether, are often used as solvents in reactions of transition metal halides with lithium, sodium, or magnesium alkylating agents (Chapter 27). Long-chain linear polyethers[162b] can also form "wrap-around" complexes with alkali ions similar to those described below.

[160a] C. G. Pierpont *et al., J. Am. Chem. Soc.,* 1978; **100,** 7894; *Inorg. Chem.,* 1979, **18,** 1616.
[160b] K. N. Raymond *et al., Inorg. Chem.,* 1979, **18,** 234; C. G. Pierpont *et al., Inorg. Chem.,* 1979, **18,** 1736.
[161a] S. L. Kessel and D. N. Hendrickson, *Inorg. Chem.,* 1978, **17,** 2630.
[161b] See, e.g., melanin complexes, C. C. Felix *et al., J. Am. Chem. Soc.,* 1978, **100,** 3922.
[162a] M. den Heijer and W. L. Dreissen, *Inorg. Chim. Acta,* 1979, **33,** 261.
[162b] G. Weber *et al., Angew. Chem., Int. Ed.,* 1979, **18,** 226; B. Tümmler *et al., J. Am. Chem. Soc.,* 1979, **101,** 2588.

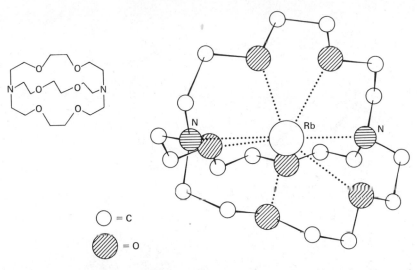

$= C$

$= O$

Fig. 4-6. The structure of the cation in the salt $[RbC_{18}H_{36}N_2O_6]SCN \cdot H_2O$. [Reproduced by permission from M. R. Truter, *Chem. Br.*, **1971**, 203.]

Crown Ethers and Cryptates.[163] The macrocyclic polyethers, termed "crown ethers" from their structural resemblance to crowns, were first synthesized by C. J. Pedersen in 1967 by reactions such as the following:

$$\text{OH} + (ClCH_2CH_2)_2O \xrightarrow[\text{2. H}^+]{\text{1, NaOH in BuOH}}$$

Dibenzo-18-crown-6

Ethers with from 3-20 oxygen atoms have been synthesized. The hydrogenated derivative of dibenzo-18-crown-6 is formally 2,5,8,15,18,21-hexaoxatricyclo[20.4.0.0⁹,¹⁴]-hexacosane but usually is called cyclohexyl-18-crown-6.

Related macropolycycles are the *cryptates*[164] or cryptands, which are N,O compounds such as $N[CH_2CH_2OCH_2CH_2OCH_2CH_2]_3N$, the structure of whose complex with Rb^+ is shown in Fig. 4-6. See also page 265 for further discussion.

Crown ethers have particularly large complexity constants for alkali metals—equilibrium constants for cyclohexyl-crown-6, for example, are in the order $K^+ > Rb^+ > Cs^+ > Na^+ > Li^+$. The cryptates also have high complexing ability espe-

[163] R. M. Izatt and J. J. Christensen, Eds., *Synthetic Multidentate Macrocyclic Compounds*, Academic Press, 1978 (a major reference, mainly on crown ethers and cryptates). C. J. Pedersen and K. K. Frensdorff, *Angew. Chem., Int. Ed.*, 1972, **11**, 16; D. J. Cram and J. M. Cram, *Acc. Chem. Res.*, 1978, **11**, 8; R. M. Izatt and J. J. Christensen, Eds., *Progress in Macrocyclic Chemistry*, Vol. 1, Wiley, 1979; G. W. Gokel and H. G. Hurst, *Synthesis*, **1976**, 168 (18-crown-6).
[164] J.-M. Lehn, *Acc. Chem. Res.*, 1978, **11**, 49; *Coord. Chem.*, **17**, IUPAC, Pergamon Press, 1977.

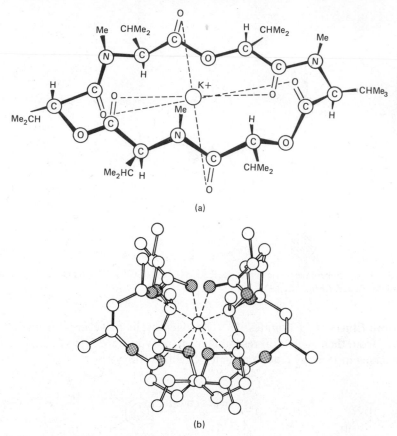

(a)

(b)

Fig. 4-7. Structures of K$^+$ salt of (a) [D-hydroxyisovaleric acid-N-methyl-L-valine]$_3$ or enniatin B and (b) nonactin. [Reproduced by permission from D. A. Fenton, *Chem. Soc. Rev.,* 1977, **6**, 325.]

cially for M^{2+} ions and will render even BaSO$_4$ soluble. They also have good complexing ability for transition metal ions (e.g., for lanthanides).[165]

Crown ethers find many uses. They will render salts such as KMnO$_4$ or KOH soluble in benzene or other aromatic hydrocarbons so increasing the facility for oxidation or base reactions. Species such as Sn$_9^{4-}$ or Pb$_5^{2-}$ can be also isolated as salts of crown-solvated alkali ions. The ethers are widely used as solvents in a variety of organic and organometallic reactions where solvation of alkali ions can effect improvements in rates.

A number of macrocyclic compounds of similar type exist in Nature[166] and are concerned with the complexing of Na$^+$ and K$^+$ to assist transport through the hydrophobic lipid bilayer of cell membranes. One such complex is valinomycin, which is discussed also in Section 7-7, another is enniatin-B (Fig. 4-7a gives the structure of its K$^+$ salt) and yet a third type isolated from *Actinomyces* species

[165]   O. A. Gansow *et al., J. Am. Chem. Soc.,* 1977, **99**, 7087.
[166]   D. E. Fenton, *Chem. Soc. Rev.,* 1977, **6**, 325; J. D. Dunitz and M. Dobler, in *Biological Aspects of Inorganic Chemistry,* A. W. Addison *et al.,* Eds., Wiley-Interscience, 1977.

are ether donors of the type:

The three-dimensional structure of the nonactin (R_1 to R_4 = CH_3) complex of K^+ is shown in Fig. 4-7b.

In addition to the O and O,N macrocycles there are other similar ligands, such as 4-LXV and 4-LXVI, which can bring two or more metal atoms into close proximity with each other.[167]

(4-LXV)   (4-LXVI)

Ketones are almost invariably bonded only through oxygen. However perfluoroacetone, $(CF_3)_2CO$, can be bound in a three-membered ring as in 4-LXVII, and the keto group of diphenylketene[168] (Ph_2C=C=O) can be similarly bound, but as a bridge to Cp_2Ti (4-LXVIII).

(4-LXVII)   (4-LXVIII)

[167]   R. Weiss *et al., J. Am. Chem. Soc.,* 1979, **101,** 3383.
[168]   G. Fachinetti *et al., J. Am. Chem. Soc.,* 1978, **100,** 1921.

4-27.   β-Ketoenolato[169] and Tropolonato Complexes[170]

β-Diketones have the property of forming stable anions as a result of enolization followed by ionization:

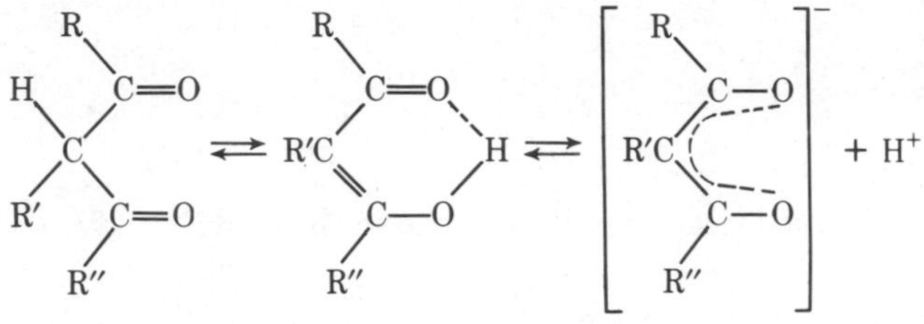

These β-ketoenolate ions form very stable chelate complexes with a great range of metal ions. The commonest such ligand is the acetylacetonate ion, acac⁻, in which $R = R'' = CH_3$ and $R' = H$. A general abbreviation for β-ketoenolate ions in general is dike.

Among the commonest types of diketo complex are those with the stoichiometries $M(dike)_3$ and $M(dike)_2$. The former all have structures based on an octahedral disposition of the six oxygen atoms. The trischelate molecules then actually have D_3 symmetry and exist as enantiomers. When there are unsymmetrical diketo ligands (i.e., those with $R \neq R''$), geometrical isomers also exist, as indicated in (4-LXIX). Such compounds have been of value in investigations of the mechanism of racemization of trischelate complexes, which are discussed in detail in Section 28-14.

cis   trans

(4-LXIX)

Tetradiketo complexes $M(β\text{-dike})_4$ are usually nonrigid.[171]

Substances of composition $M(dike)_2$ are almost invariably oligomeric, unless the R groups are very bulky ones, such as $(CH_3)_3C—$. Thus, for example, the acetylactonates of nickel(II), cobalt(II), and zinc(II) are trimeric, tetrameric, and trimeric, respectively, and the complexes containing the hindered β-diketonate with

[169] R. C. Mehrotra et al., Metal β-Diketonates and Allied Derivatives, Academic Press, 1978; J. P. Fackler, Prog. Inorg. Chem., 1966, 7, 361; D. W. Thompson, Struct. Bonding, 1971, 9, 27; J. J. Fortman and R. E. Sievers, Coord. Chem. Rev., 1971, 6, 331 (isomerism); D. Gibson, Coord. Chem. Rev., 1969, 4, 225 (C-bonded); D. P. Graddon, Coord. Chem. Rev., 1969, 4, (Lewis acid behavior); H. Musso et al., Angew. Chem., Int. Ed., 1971, 10, 225 (spectra); K. C. Joshi and V. N. Pathak, Coord. Chem. Rev., 1977, 22, 37 (fluorinated acacs).

[170] E. L. Muetterties and C. M. Wright, Q. Rev., 1967, 21, 109.

[171] R. C. Fay and J. K. Howie, J. Am. Chem. Soc., 1977, 99, 8111.

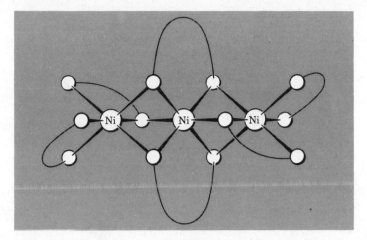

Fig. 4-8. The trimeric structure of nickel acetylacetonate. The unlabeled circles represent oxygen atoms, and the curved lines connecting them in pairs represent the remaining portions of the acetylacetonate rings. [Reproduced by permission from J. C. Bullen, R. Mason, and P. Pauling, *Inorg. Chem.*, 1965, **4**, 456.]

$R = R'' = (CH_3)_3C—$ are monomeric. These facts show that in $M(dike)_2$ molecules the M atoms are coordinately unsaturated; they prefer a coordination number of 6 (or at least 5) and attain such coordination numbers (usually 6) by sharing of oxygen atoms. The presence of bulky R groups sterically impedes such oligomerization. In the presence of good donors such as H_2O, ROH, or py, the metal atoms expand their coordination numbers from 4 to 5 or 6 by binding such donors. Thus complexes of the type $trans\text{-}M(dike)_2L_{1,2}$ are formed, instead of oligomers.

Oligomers have bridged β-diketonato groups, as shown for $Niacac_2$ in Fig. 4-8.

For neutral compounds, especially of acetylacetone, the methine CH group of the ring can undergo a wide range of substitution reactions similar to those of aromatic substances, even though the rings have little or no aromatic character. An unusual type of β-diketonato ligand is obtained by action of methyllithium on acyl metal carbonyls.[172]

$$(CO)_4Re\underset{\underset{CO}{|}}{\overset{\overset{Me}{\diagdown}{C=O}}{\diagup}} \xrightarrow{\text{LiMe}} (CO)_4Re\underset{\overset{|}{Me}}{\overset{\overset{Me}{\diagdown}{C=O}}{\overset{\diagdown}{C-O^-}}} \quad Li^+$$

These anions can form complexes with other metals in the normal way.

By use of 2 moles of methyllithium a tridentate ligand can be obtained and this too forms complexes:

[172] C. M. Lukehart, et al. *Inorg. Chem.*, 1979, **18**, 1297, 3150; *J. Organomet. Chem.*, 1978, **161**, 1; *J. Am. Soc.*, 1978, **100**, 2274; *Inorg. Chim. Acta.* 1978, **27**, 219.

$$CH_3CORe(CO)_5 + 2LiMe \longrightarrow \left[fac\text{-}(CO)_3Re \begin{array}{c} Me \\ C-O \\ \diagup \\ \\ C-O \\ | \\ Me \\ \diagdown \\ C-O \\ | \\ Me \end{array} \right]^{2-}$$

It is an analogue of the triacetylmethanide ion $(MeCO)_3C^-$ and also resembles the $RBpz_3^-$ ligand.

C-Bonded β-Diketonate Complexes.   These have the metal bound to the central carbon atom and are well known, notably for rhodium, iridium, palladium, and platinum.[173] One example is 4-LXX, which has both normal and C-bonded groups.

(4-LXX)   (4-LXXI)

Note that the C-bonded groups have two free $>C{=}O$ groups that can act as chelate donors to other metals,[174] as in 4-LXXI.

The intermediate formation of a C-bonded species is probably the reason for the existence, even at very low temperatures, of dynamic equilibria such as

Sometimes O-bonded complexes can be readily converted to C-bonded complexes,[173,175] as in the reactions

[173] T. Ito and A. Yamamoto, *J. Organomet. Chem.,* 1978, **161**, 61; M. A. Bennett and T. R. B. Mitchell, *Inorg. Chem.,* 1976, **15**, 2936.
[174] Y. Nakamura and K. Nakamoto, *Inorg. Chem.,* 1975, **14**, 63.
[175] S. Komiya and J. K. Kochi, *J. Am. Chem. Soc.,* 1977, **99**, 3695.

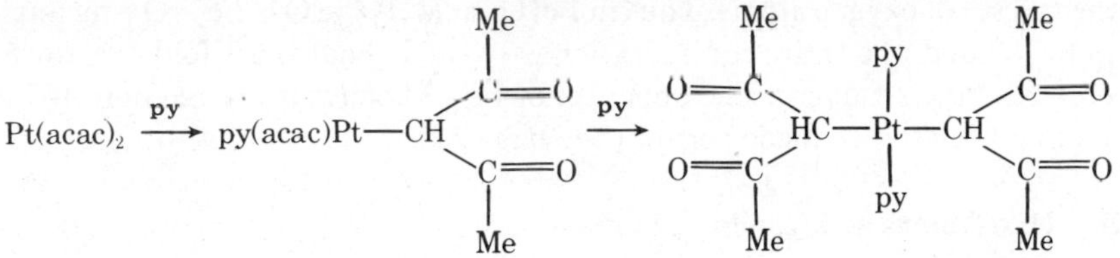

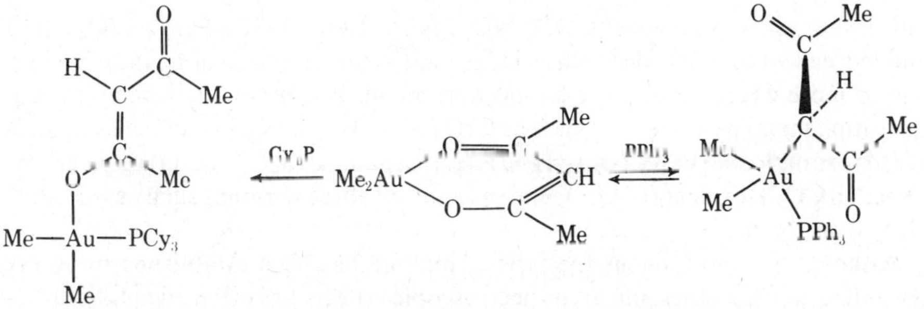

Finally note that β-diketones can occasionally act as *neutral* ligands,[176] being bound either through oxygen as in 4-LXXII or for certain metals that can form olefin complexes as shown in 4-LXXIII.

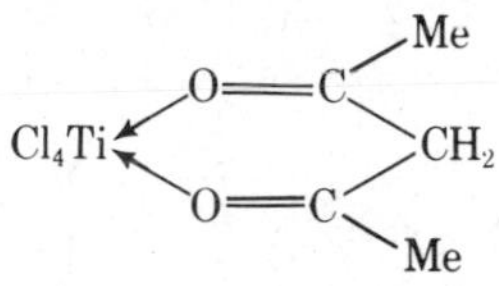

(4-LXXII) 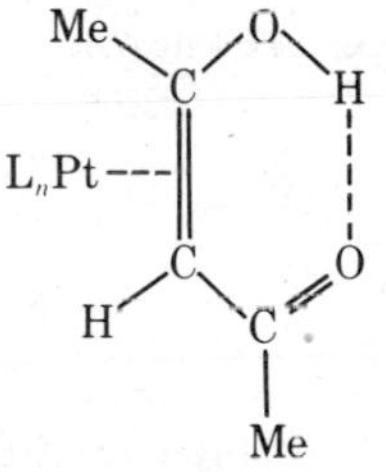   (4-LXXIII)

Tropolonates. A system similar to the β-diketonates is provided by *tropolone* and its anion 4-LXXIV. The tropolonato ion gives many complexes that are broadly similar to analogous β-diketonate complexes, although there are often very significant differences.

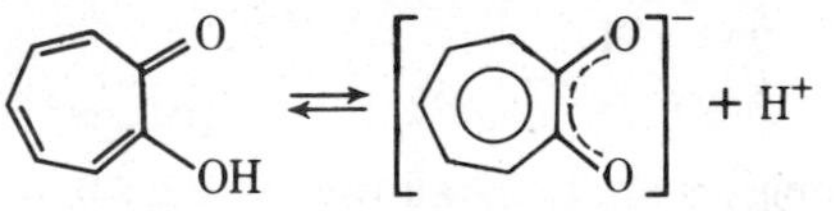

(4-LXXIV)

It should be noted that the tropolonato ion forms a five-membered chelate ring and that the "bite," that is, the O-to-O distance, is smaller than for β-diketonato ions. This leads in tristropolonato complexes to considerable distortion from an

[176] Y. Nakamura *et al., Inorg. Chem.,* 1972, **11**, 1573; *Chem. Lett.,* **1976**, 199.

octahedral set of oxygen atoms. Thus in $Fe(O_2C_7H_5)_3$ the O—Fe—O ring angles are only 78° and, the entire configuration is twisted about the 3-fold axis toward a more prismatic structure. The upper set of Fe—O bonds is twisted only 40° instead of 60° relative to the lower set (See page 52).

4-28.   Oxo Anions as Ligands

All the common oxo anions (CO_3^{2-}, NO_3^-, NO_2^-, SO_3^{2-}, SO_4^{2-}, PO_4^{3-}, ClO_4^-, etc.) and indeed less common ions such as iodate and tellurate can act as ligands in which one or more oxygen atoms are bound to a metal. It is less well recognized that transition metal oxo anions[177] such as CrO_4^-, MoO_4^{2-}, or WO_4^{2-} can act as ligands as, for example, in $[Fe_2(CrO_4)_4(H_2O)_2]^{2-}$. Organic oxo anions, notably carboxylates (RCO_2^-), oxalate ($C_2O_4^{2-}$), and organic-substituted anions such as $CH_3PO_3^-$ also act as ligands.

Although for most anions the type of binding has been established by X-ray crystallographic studies, infrared spectroscopic criteria are often adequate to distinguish different types of bonding.

Carbon Oxo Anions.   The commonest anions are carbonate and oxalate, but cyclic anions[178a] such as squarate ($C_4O_4^{2-}$) and croconate ($C_5O_5^{2-}$, Chapter 11) can act as ligands. *Carbonato* complexes[178b] are generally bidentate or unidentate, but some bridges are known.

Many *oxalato*[179] complexes are known and the types of linkage are as follows:

Typical chelate complexes are trisoxalato ions, e.g., $[Co\ ox_3]^{3-}$.

Carboxylate Ions.[180]   These ions form a particularly important class of ligands. They may be bound in several ways, as follows:

---

177   V. Debelle *et al.*, *Acta Crystallogr., B*, 1974, **30**, 2185.

178a   See A. J. Fatiadi, *J. Am. Chem. Soc.*, 1978, **100**, 2586.

178b   See A. R. Davis *et al.*, *J. Am. Chem. Soc.*, 1978, **100**, 6258.

179   K. L. Scott, K. Wieghardt, and A. G. Sykes, *Inorg. Chem.*, 1973, **12**, 655.

180   C. Oldham, *Prog. Inorg. Chem.*, 1968, **10**, 223; S. Herzog and W. Kalies, *Z. Chem.*, 1968, **8**, 81; C. D. Garner and B. Hughes, *Adv. Inorg. Chem., Radiochem.*, 1975, **17**, 2 (complexes of $CF_3CO_2^-$); J. Catterick and P. Thornton, *Adv. Inorg. Chem. Radiochem.*, 1977, **20**, 291 (polynuclear carboxylates).

$$
\underset{M-O}{\overset{R}{\underset{}{\big|}}}C{=}O
\qquad
M\langle\overset{O}{\underset{O}{}}\rangle C{-}R
\qquad
M\langle\overset{O}{\underset{O}{}}\rangle C{-}R
$$

$$
\underset{M-M}{\overset{R}{\underset{\big|\;\;\;\big|}{O^{\overset{C}{}}O}}}
\qquad
\underset{M\quad M}{\overset{R}{\underset{\big|\;\;\;\big|}{O^{\overset{C}{}}O}}}
\qquad
\underset{M\diagup O\;\;\;O\diagdown M}{\overset{R}{\overset{\big|}{C}}}
\qquad
\underset{M\diagup O\;\;\;O\diagdown M\atop M}{\overset{R}{\overset{\big|}{C}}}
\qquad
\underset{O\;\;\;O\atop M\;\;M}{\overset{R}{\overset{\big|}{C}}}
$$

Syn-syn   Syn-syn   Anti-anti   Anti-syn   Monatomic

The most important are unidentate, symmetrical chelate and symmetrical syn-syn bridging. The other forms are not common, but anti-anti single bridging occurs in $[MnsaI_2enCO_2Me]_n$, and anti-syn in $[(PhCH_2)_3SnCO_2Me]_n$.[181] Trifluoroacetates are known only in unidentate and bridging forms.

The main types of bonding can often be distinguished by ir and nmr spectra.[182a] The syn-syn bridging RCO_2^- ligand is extremely common and important in compounds with M—M quadruple bonds.[182b]

In "ionic" acetates or in aqueous solution, the "free" $CH_3CO_2^-$ ion has symmetric and antisymmetric C—O stretching modes at ~1415 and ~1570 cm^{-1}. These frequencies can vary by ±20 cm^{-1}. Since the symmetry of even the free ion is low and it gives two ir-active bands, evidence for the mode of coordination must be derived from the positions rather than the number of bands. When the carboxyl group is unidentate, one of the C—O bonds should have enhanced double-bond character and should give rise to a high-frequency band. Such bands are observed in the 1590 to 1650 cm^{-1} region and are considered to be diagnostic of unidentate coordination.

Symmetrical bidentate coordination, as in $Zn(CH_3CO_2)_2\cdot2H_2O$ and $Na[U\text{-}O_2(CH_3CO_2)_3]$, and symmetrical bridging, as in the $M_2(O_2CCH_3)_4L_2$ and $M_3O(O_2CCH_3)_6L_3$ types of molecules, leaves the C—O bonds still equivalent, and the effect on the frequencies is not easily predictable. In fact, no criteria for distinguishing these cases have been found. In general, multiple bands appear between 1400 and 1550 cm^{-1}, the multiplicity being attributable to coupling between CH_3CO_2 groups bonded to the same metal atom(s).

Straight-chain alkyl carboxylic acids derived from petroleum that have also a terminal cyclohexyl or cyclopentyl group are known as *naphthenic acids*.[183] They form complexes, presumably polymeric, with many transition metals, and these compounds are freely soluble in petroleum. Copper naphthenates are used as fungicides, aluminum naphthenate was used as a gelling agent in "napalm," and cobalt naphthenates are used in paints.

181   J. E. Davies, B. M. Gatehouse, and K. S. Murray, *J.C.S. Dalton,* **1973**, 2529.

182a   N. W. Alcock, V. M. Tracy, and T. C. Waddington, *J.C.S. Dalton,* **1976**, 2238.

182b   F. A. Cotton, *Chem. Soc. Rev.,* 1975, **4**, 27.

183   L. F. Hatch and S. Matar, *Hydrocarbon Process.,* **1977**, September, p. 165.

There are, of course, more complicated carboxylic acids such as ethylenediam-inetetraacetic acid that can function as multidentate ligands with both N and O bound to the metal. Also *hydroxo carboxylic acids* such as citrate readily form complexes in which both carboxylate and hydroxo groups are involved. Of such acids, probably the best studied are *tartrato complexes.*[184] A fairly common type of structure is one with bridges linking two metal atoms.

A particular example is the antimony complex "tartar emetic" (Fig. 4-9). Be-

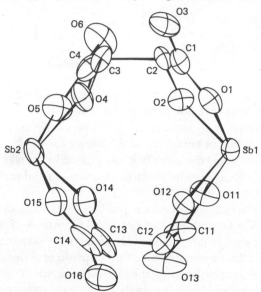

Fig. 4-9. Structure of the $Sb_2(d\text{-}C_4H_2O_6)_2^{2-}$ anion in tartar emetic. [Reproduced by permission from *Inorg. Chim. Acta*, 1974, **8**, 209.]

cause of the chirality and multiplicity of bonding possibilities, many isomers of tartrate complexes are possible, and the relative stabilities of these can be explained in terms of steric constraints of the binuclear structure and conformation of the tartrato groups, and depend strongly on the coordination geometry about the metal.

Carbamates.[185a]   Complexes of $R_2NCO_2^-$ are not as extensive or as useful as their sulfur analogues the dithiocarbamates. Carbamates are often obtained in insertion reactions of CO_2 (Section 29-4):

$$UCl_4 + 8R_2NH + 4CO_2 = U(O_2CNR_2)_4 + 4R_2NH_2^+Cl^-$$

$$Ti(NMe_2)_4 + 4CO_2 = Ti\left(\underset{O}{\overset{O}{\diagdown}} CNMe_2\right)_4$$

[184]   G. L. Robbins and R. E. Tapscott, *Inorg. Chem.*, 1976, **15**, 154; R. L. Belford and I. C. Paul, *Coord. Chem. Rev.*, 1969, **4**, 323; L. D. Pettit and J. L. M. Swash, *J.C.S. Dalton*, **1978**, 286.

[185a] F. Calderazzo *et al., Inorg. Chem.*, 1978, **17**, 471, 474; M. H. Chisholm and W. W. Reichert, *Inorg. Chem.*, 1978, **17**, 767.

Carbamates are usually chelate but may be bridging, as in the quadruply bonded $Cr_2(O_2CNEt_2)_4$ [185b].

The Nitrate Ion.[186] This ion has several structural roles, as follows:

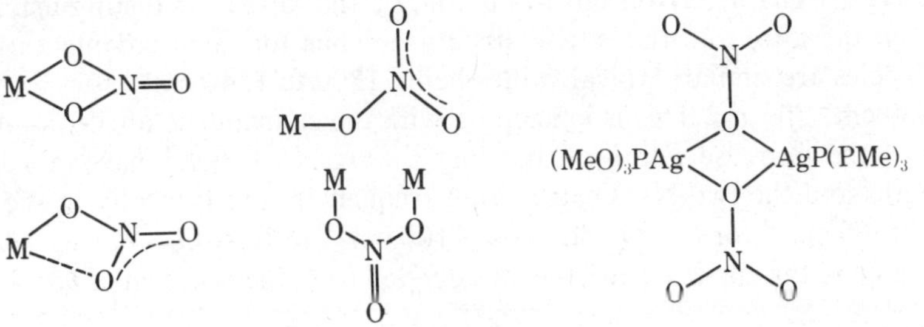

A great many nitrato complexes have been studied, and the symmetrical bidentate structure is preferred, with the unidentate form a less common alternative.

The free nitrate ion has relatively high symmetry (D_{3h}); thus its infrared spectrum is fairly simple. The totally symmetric N—O stretching mode is not ir-active, but the double degenerate N—O stretching mode gives rise to a strong bond at ~1390 cm^{-1}. There are also two ir-active deformation modes, one of which is doubly degenerate, at 830 and 720 cm^{-1}. When a nitrate ion becomes coordinated, its effective symmetry is reduced, causing the degeneracies to split and all modes (six) to be ir-active. Hence, it is possible to distinguish between ionic and coordinated nitrate groups.

Because the two commonest forms of coordinated nitrate ion have the same effective symmetry, hence the same number of ir-active vibrational modes, criteria for distinguishing between them must be based on the positions of the bands rather than their number. In practice, the situation is quite complex and there are no entirely straightforward criteria. This is because the array of frequencies depends on both the geometry and strength of coordination. Only one case of the single O-bridged complex is known.[187]

Nitrite Ion.[188] The NO_2^- ion can act as an N-ligand to form *nitro* compounds, and as O-bonded *nitrito:*

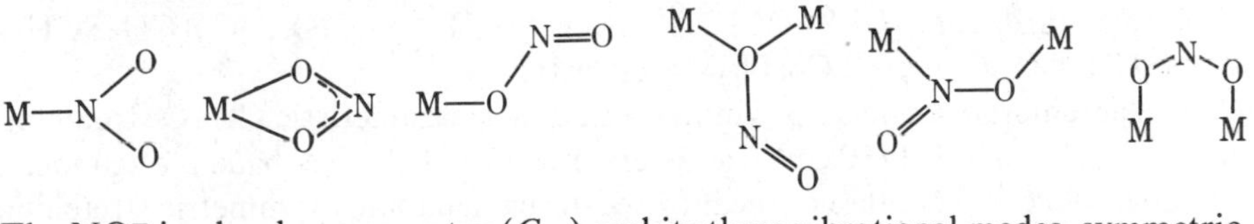

The NO_2^- ion has low symmetry (C_{2v}) and its three vibrational modes, symmetric N—O stretching v_s, antisymmetric N—O stretching v_{as}, and bending δ, are all

185b M. H. Chisholm *et al., Inorg. Chem.,* 1978, **17**, 3536.
186   C. C. Addison *et al., Q. Rev.,* 1971, **25**, 289; *Prog. Inorg. Chem.,* 1967, **8**, 195; *Adv. Inorg. Chem. Radiochem.,* 1964, **6**, 72; P. B. Critchlow and S. D. Robinson, *Coord. Chem. Rev.,* 1978, **25**, 69 (Pt metals).
187   J. H. Meiners *et al., Inorg. Chem.,* 1975, **14**, 632.
188   R. Birdy *et al., J.C.S. Dalton,* 1977, 1730; K. A. Klanderman *et al., Inorg. Chim. Acta,* 1977, **23**, 117.

ir-active to begin with. Thus the number of bands cannot change on coordination, and the use of ir spectra to infer structure must depend on interpretation of shifts in the frequencies. The δ vibration is rather insensitive to coordination geometry, but there are characteristic shifts of v_s and v_{as} that often can distinguish reliably between the nitro and the nitrito structure. Thus for nitro complexes the two frequencies are similar, typical values being 1300 to 1340 cm^{-1} for v_s and 1360 to 1430 cm^{-1} for v_{as}. This is in keeping with the equivalence of the N—O bond orders in the nitro case. For nitrito bonding, the two N—O bonds have very different strengths and the two N—O stretching frequencies are typically in the ranges 1400–1500 cm^{-1} for N=O and 1000–1100 cm^{-1} for N—O.

Related to the nitrite ion are the ions derived from the *aci* form of *nitroalkanes*, e.g.,

$$CH_3NO_2 \; \rightleftharpoons \; \underset{H}{\overset{H}{>}}C=N\underset{O}{\overset{O^{\ominus}}{<}} \; + \; H^+$$

In $Ni(O_2N=CHPh)_2$(tmed) the ligand is chelate through both oxygens as it is apparently also in $Ru(O_2N=CH_2)H(PPh_3)_3$ and some copper compounds such as $[CuphenO_2N=CH_2]^+$.[189] Some tin, lead, and mercury compounds[190] have unidentate groups

$$M-O\overset{\overset{\displaystyle O}{\|}}{N}CR'R''$$

However C-bonding, as $M-CH_2NO_2$, is also possible, as in the platinum compounds made by the reaction[191]:

$$PtCl_2(PEt_3)_2 + CH_3NO_2 \xrightarrow{Ag_2O} cis\text{-}PtCl(CH_2NO_2)(PEt_3)_2$$

This reaction implies initial oxidative addition, as H and CH_2NO_2, of nitromethane (Section 29-3), and indeed such additions are known.[192]

The Perchlorate Ion. This ion has relatively little tendency to serve as a ligand and is often used where an anion unlikely to coordinate is required. However there are cases, verified by X-ray crystallography, in which ClO_4^- is coordinated.[193] Examples are the $[Cl_3SnOClO_3]^{2-}$ ion, $(CH_3)_3SnClO_4$, $Co(CH_3SCH_2\text{-}CH_2SCH_3)_2(ClO_4)_2$, and $Co(Ph_2MeAsO)_4(ClO_4)_2$.

The perchlorate ion has T_d symmetry and the characteristic Cl—O stretching mode occurs at about 1110 cm^{-1}, as a very broad band. A weak band is often found at about 980 cm^{-1} because of the infrared-forbidden totally symmetric stretching

189   D.J. Cole-Hamilton and G. Wilkinson, *Nouv. J. Chim.,* 1977, **1**, 141 and references therein. J. Zagal *et al., J.C.S. Dalton,* **1974,** 85; N. Marsich and A. Camus, *J. Inorg. Nucl. Chem.,* 1977, **39**, 275.

190   J. Lorberth *et al., J. Organomet. Chem.,* 1973, **54**, 165, 177.

191   M. A. Cairns *et al., J. Organomet. Chem.,* 1977, **135**, C33.

192   W. Beck and K. Shorpp, *Chem. Ber.,* 1974, **107**, 1371, cf. also M. A. Bennett *et al., J. Am. Chem. Soc.,* 1973, **95**, 3028.

193   See R. C. Elder *et al., Inorg. Chem.,* 1978, **17**, 427.

frequency. In compounds that contain unidentate ClO_4^-, there are three bands, at about 1120, 1040, and 920 cm^{-1}, in accord with expectation for C_{3v} symmetry.

In compounds containing bidentate bridging perchlorate groups, such as $(CH_3)_3SnClO_4$, which has infinite —Sn—ClO$_4$—Sn—ClO$_4$— chains, there are four Cl—O stretching bands at ~1200, 1100, 1000, and 900 cm^{-1}. No bidentate chelate perchlorate complex has yet been reported. There is also some indication for complexing by perchlorate in solution,[194] and the reduction of ClO_4^- by reducing metal ions (e.g., Ru^{II}) doubtless proceeds via intial coordination.

Phosphorus Oxo Acids. The many types of phosphorus oxo acids, which may contain either P^{III} or P^V, are discussed in Chapter 14 along with organic-substituted ions such as dialkylphosphinates ($R_2PO_2^-$) and acids such as R_2POH and $(RO)_2POH$.[195]

Phosphato complexes may have PO_4^{3-}, HPO_4^{2-}, or $H_2PO_4^-$ coordinated. Few complexes have been studied crystallographically. In Co^{III}enPO$_4$ there is a bidentate chelate, whereas the pyrophosphate Co^{III}en$_2$(HP$_2$O$_7$) has a six-membered ring[196]:

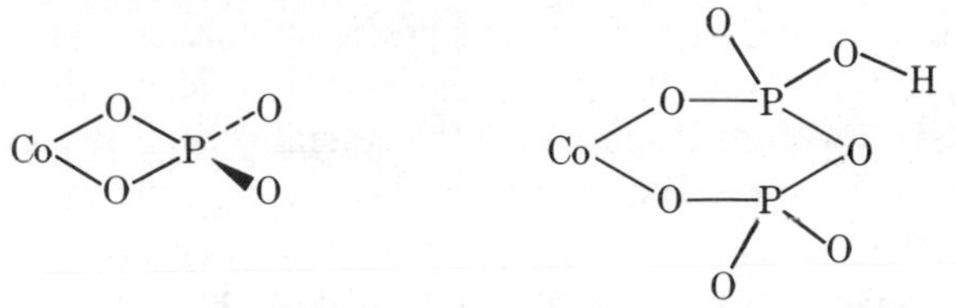

The various polyphosphoric anions (Chapter 14) also give complexes.

Sulfur Oxo Acids. The sulfate ions HSO_4^- and SO_4^{2-} form numerous complexes. *Sulfate* can be unidentate, bidentate chelate, or bidentate bridging:

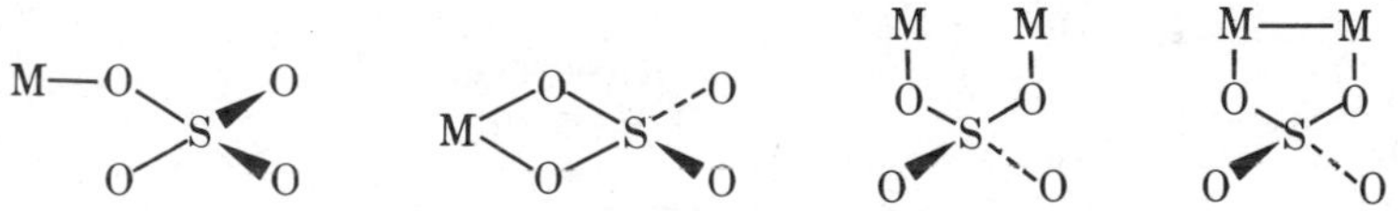

The free sulfate ion is tetrahedral (T_d), but when it functions as a unidentate ligand, the coordinated oxygen atom is no longer equivalent to the other three and the effective symmetry is lowered to C_{3v}. Since the M—O—S chain is normally bent, the actual symmetry is even lower, but this perturbation of C_{3v} symmetry does not measurably affect the infrared spectra. When two oxygen atoms become coordinated, either to the same metal ion or to different ones, the symmetry is lowered still further to C_{2v}.

The distinction between uncoordinated, unidentate and bidentate SO_4^{2-} by infrared spectra is very straightforward. Table 4-2 summarizes the selection rules

[194] L. Johansson, *Coord. Chem. Rev.*, 1974, **12**, 241.

[195] For references, see R. Cini *et al.*, *Inorg. Chem.*, 1977, **16**, 323; R. G. Sperline and D. M. Roundhill, *Inorg. Chem.*, 1977, **16**, 2612.

[196] B. Anderson *et al.*, *J. Am. Chem. Soc.*, 1977, **99**, 2652.

TABLE 4-2

Correlation of the Types and Activities of S—O Stretching Modes of SO_4^{2-}

State of SO_4^{2-}	Effective symmetry	Types and activities of modes[a] (R = Raman; I = ir)		
Uncoordinated	T_d	$\nu_1(A_1,R)$ ↓	$\nu_3(T_2,I,R)$	
Unidentate	C_{3v}	$\nu_1(A_1,I,R)$ ↓	$\nu_{3a}(A_1,I,R)\ \nu_{3b}(E,I,R)$	
Bidentate	C_{2v}	$\nu_1(A_1,I,R)$	$\nu_{3a}(A_1,I,R)\ \ \nu_{3b}(B_1,I,R)\ \ \nu_{3c}(B_2,I,R)$	

[a] ν_2 and ν_4 are not listed because they are O—S—O bending modes. Note also that the arrows drawn have only rough qualitative significance, since all modes of the same symmetry will be of mixed parentage in the higher symmetry.

for the S—O stretching modes in the three cases. It can be seen that uncoordinated SO_4^{2-} should have one, unidentate SO_4^{2-} three, and bidentate SO_4^{2-} four S—O stretching bands in the infrared. Note that the appearance of four bands for the bidentate ion is expected regardless of whether it is chelating or bridging.

Observed spectra are in accord with these predictions, except that ν_1 does appear weakly in the spectrum of the uncoordinated SO_4^{2-} ion. This is due to nonbonded interactions of SO_4^{2-} with its neighbors in the crystal, which perturb the T_d symmetry; the same environmental effects also cause the ν_3 band to be very broad.

Even though bridging and chelating sulfates cannot be distinguished on the basis of the number of bands they give, it appears that the former have bands at different frequencies than the latter, typical values being as follows:

Bridging	Chelating
1160–1200	1210–1240
~1110	1090–1176
~1130	995–1075
960–1000	930–1000

Sulfites (SO_3^{2-}) are always bound through oxygen, but *sulfinates*[197] (RSO_2^-) and *sulfenates* (RSO^-) may be bound either through O or through S.

The *p-toluene sulfonate ion* $p\text{-}MeC_6H_4SO_3^-$ can act as an O-ligand, both unidentate and bridging, in platinum complexes.[198] It may also be bound as a π complex through the aromatic ring as in $\eta\text{-}(p\text{-}MeC_6H_4SO_3)RuH(PPh_3)_2$.

[197] M. Lundeen *et al., Inorg. Chem.,* 1978. **17,** 701; D. L. Herting *et al., Inorg. Chem.,* 1978, **17,** 1649.
[198] J. Chatt and D. P. Mingos, *J. Chem. Soc., A,* **1969,** 1770.

4-29. Other Oxygen Ligands

Among the more common oxygen compounds not previously mentioned is the solvent *dimethylsulfoxide*.[199] This forms many complexes with transition metals that can be either O- or S-bonded:

These types can be distinguished by ir or by nmr spectra.[200]

Pyridine N-oxides and *phosphine oxides* (R_3PO) coordinate only through oxygen.[201]

Hydroxamates[202a] and their monothio[202b] and dithio analogues form complexes, usually as chelates. The NH proton can also show acidity as in the reaction

Hydroxamato ligands are present in microbial iron transport compounds called ferrichromes (see Section 31-2).

SULFUR

We deal only with *sulfur ligands,* and in general the analogous selenium or tellurium compounds behave similarly.[203]

4-30. Hydrogen Sulfide, Sulfide Ions, Thiols[204]

There are several complexes in which H_2S and sulfide ions act as ligands, although very commonly action of these gives merely insoluble sulfides, other ligands being displaced.

[199]   W. L. Reynolds, *Prog. Inorg. Chem.,* 1970, **12,** 1.
[200]   See, e.g., I. P. Evans *et al., J.C.S. Dalton,* **1973,** 204.
[201]   K. O. Joung *et al., J. Am. Chem. Soc.,* 1977, **99,** 7387; R. A. Jorge *et al., J.C.S. Dalton,* **1978,** 1102.
[202a] B. Chatterjee, *Coord. Chem. Rev.,* 1978, **26,** 281; K. N. Raymond *et al., J. Am. Chem. Soc.,* 1979, **101,** 2722; *Inorg. Chem.,* 1977, **16,** 807; K. S. Murray *et al., J. Am. Chem. Soc.,* 1978, **100,** 2251.
[202b] J. Leong and S. J. Bell, *Inorg. Chem.,* 1978, **17,** 1886.
[203]   S. E. Livingstone, *Q. Rev.,* 1965, **19,** 386; D. P. N. Satchell, *Chem. Soc. Rev.,* 1977, **6,** 345.
[204]   H. Vahrenkamp, *Angew. Chem. Int. Ed.,* 1975, **14,** 322.

Although water is a ubiquitous ligand, H_2S is not; it is easily deprotonated to SH^- or oxidized to sulfur, and comparatively few examples of H_2S as a ligand are known; examples are $W(CO)_5SH_2$ and the osmium and ruthenium clusters $M_3(CO)_9SH_2$.[205] The action of H_2S on aquapentammineruthenium, in presence of Eu^{2+} (to keep the system reduced) displaces water:

$$[am_5Ru(OH_2)]^{2+} + H_2S \rightleftharpoons [am_5RuSH_2]^{2+} \qquad K = 1.5 \times 10^{-3}\ M^{-1}$$

In absence of Eu^{2+}, $[am_5RuSH]^+$ is obtained.[206] A few other SH^- compounds are known,[207] e.g., $[FeLSH]^+$ and $[NiLSH]^{2+}$, where L is a tripod phosphine ligand, $[Cr(H_2O)_5SH]^{2+}$, $(Ph_3P)_2Pt(SH)_2$, and $(Et_3P)_2PtH(SH)$.

There are innumerable compounds of alkyl or aryl thiols M—SR.[208a] Both SH^- and SR^- commonly form μ_2-bridges[208b]:

Syn   Anti

Both syn- and anti-isomers can be obtained, and there may be *syn* → *anti* isomerization that proceeds via a bridge-opening mechanism.

There are many naturally occurring sulfur compounds, e.g., methionine $CH_3SCH_2CH(NH_3)CO_2$, that have —SH, SR, or —S—S— groups, and binding of heavy metal atoms to such sulfur groups commonly occurs. Examples of thiol complexes are given at various places in the text.

The *sulfide ion* S^{2-} forms an enormous number of compounds, but of course they are not usually regarded as complexes with ligands. However compounds with M=S groups similar to M=O exist, though they are rarer than oxo compounds.[209a] One example is S=WCl$_3$.

There are a great many compounds with bridging sulfur atoms; some like ferredoxins occur naturally and are of great importance (Chapter 31).

The known types of bridge[209b] are as follows:

[205]   M. Schmidt, et al., *Z. Naturforsch.*, 1978, **33b**, 1334; *Inorg. Chim. Acta*, 1979, **32**, L19; *Angew. Chem., Int. Ed.*, 1978, **17**, 598; B. F. G. Johnson *et al., J.C.S. Chem. Comm.*, **1978**, 551.
[206]   C. G. Kuehn and H. Taube, *J. Am. Chem. Soc.*, 1976, **98**, 689.
[207]   M. Di Vaira *et al., Inorg. Chem.*, 1978, **17**, 816; T. Ramasami and G. Sykes, *Inorg. Chem.*, 1976, **15**, 1016; I. M. Blacklaws *et al., J.C.S. Dalton*, **1978**, 753; H. Schmidtbauer and J. R. Mandl, *Angew. Chem., Int. Ed.*, 1977, **16**, 40.
[208a] See, e.g., D. Swenson *et al., J. Am. Chem. Soc.*, 1978, **100**, 1933 (for $[M(SPh)_4]^{2-}$ ions of Mn, Co, Ni, Zn, Cd).
[208b] See S. D. Killops and S. A. R. Knox, *J.C.S. Dalton*, **1978**, 1260; N. G. Connelly and G. A. Johnson, *J.C.S. Dalton*, **1978**, 1375.
[209a] D. A. Rice, *Coord. Chem. Rev.*, 1978, **25**, 199 (sulfide halides of transition metals).
[209b] For references see J. L. Vidal *et al., Inorg. Chem.*, 1978, **17**, 2574; A. Müller *et al., J.C.S. Chem. Comm.*, **1978**, 739; G. Christou, *et al., J.C.S. Chem. Comm.*, **1978**, 740.

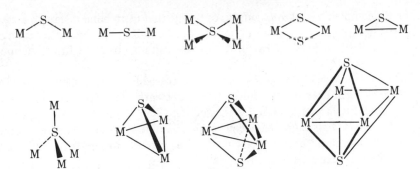

Of special importance in ferredoxins and model compounds for ferredoxins are M_4S_4 units, which have a cubane or distorted cubane structure[210] (4-LXXV).

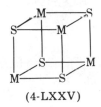

(4-LXXV)

Examples of such compounds are $[(\eta\text{-}C_5H_5)MoS]_4$, $[(NO)FeS]_4$, and $[(RS)_4Fe_4S_4]^{2-}$.

Disulfide ion (S_2^{2-}) can, like O_2^{2-}, coordinate either side on or as a bridge group:[211]

(a) (b) (c)

There are relatively few compounds with side-on S_2^{2-}, but examples are the compounds $Os(S_2)(CO)_2(PPh_3)_2$ and $[(C_2O_4)(O)Mo(S_2)_2]^{2-}$, and in the $Mo_3S_{13}^{2-}$ ion there are both side-on and b-type bridge groups. Type c is uncommon.

Bridged M—S—S—M complexes (type a) may be obtained by the reactions[212]:

$$[(NH_3)_5RuSO_2]^{2+} \xrightarrow{H_2S} [(NH_3)_5Ru\text{—}S\text{—}S\text{—}Ru(NH_3)_5]^{4+}$$

$$(H_2O)_5CrSH^{2+} \xrightarrow{I_2} [(H_2O)_5Cr\text{—}S\text{—}S\text{—}Cr(H_2O)_5]^{4+}$$

210' R. H. Holm, *Acc. Chem. Res.*, 1977, **10**, 427; see also for example, C. Mealli *et al., Inorg. Chem.*, 1978, **17**, 632; G. Bunzey and J. H. Enemark, *Inorg. Chem.*, 1978, **17**, 682; see also Chap. 32.
211 A. Müller *et al., Chem. Ber.*, 1979, **112**, 778; *Z. Naturforsch.*, 1979, **34**, 434; *Angew. Chem., Int. Ed.*, 1979, **18**, 168; B. Meunier and K. C. Prout, *Acta Crystallogr.*, B, 1979, **35**, 172.
212 (a) R. C. Elder and M. Trkula, *Inorg. Chem.*, 1977, **16**, 1048; (b) T. Ramasami, R. S. Taylor, and A. G. Sykes, *J.C.S. Chem. Comm.*, 1976, 385.

The Ru complex has a *trans*-sulfide S_2 bridge that from bond distances appears to be best described as an S_2^- group joining Ru^{II} and Ru^{III} atoms rather than S_2^{2-} joining two Ru^{III} atoms.[212a] Disulfide bridges can also be linked to three metal atoms as in $Mn_4S_4(CO)_{15}$.[213]

Polysulfide ions, notably S_4^{2-} and S_5^{2-}, can act as dinegative chelating ligands that thereby form heterocyclic sulfur rings. The earliest example, made in 1903 by boiling H_2PtCl_6 with NH_4S_x, is the red ion $[Pt^{IV}(S_5)_3]^{2-}$ in which there are three PtS_5 rings. This anion has been resolved and provides an unusual example of a purely inorganic chiral molecule[214a]; the barrier to ring inversion in $(\eta\text{-}C_5H_5)_2TiS_5$ has also been measured.[214b]

Other complexes can be made by reactions such as the following[215]:

$$(\eta\text{-}C_5H_5)_2WH_2 + 5S = (\eta\text{-}C_5H_5)_2W(S_4) + H_2S$$
$$(\eta\text{-}C_5H_5)_2TiCl_2 + Na_2S_5 = (\eta\text{-}C_5H_5)_2Ti(S_5) + 2NaCl$$

4-31. Thioethers[216]

In addition to simple thioethers (R_2S), we can have chelates [e.g., 2,5-dithiohexane, $MeS(CH_2)_2SMe$] and more complicated ligands, both open chain[216] and macrocyclic,[163,217] such as:

Thioethers have two lone pairs so that when one is involved in metal binding, we have potential for inversion as in NR_3 (Section 13-2) and also chirality:

Inversion can be studied by nmr and the barriers determined[218] in compounds such as $Cl_2Pt(SR_2)_2$; for sulfur these are in the range 51–56 kJ mol^{-1} and in corresponding R_2Se compounds, 60–66 kJ mol^{-1}.

[213]   V. Küllmer, E. Röttinger, and H. Vahrenkamp, *J.C.S. Chem. Comm.,* **1977**, 782.

[214a] R. D. Gillard and F. L. Wimmer, *J.C.S. Chem. Comm.,* **1978**, 936.

[214b] E. W. Abel *et al., J. Organomet. Chem.,* 1978, **160**, 75.

[215]   D. Coucouvanis *et al., J. Am. Chem. Soc.,* 1979, **101**, 3392.

[216]   R. C. Elder *et al., Inorg. Chem.,* 1978, **17**, 1296; M. Schmidt and G. G. Hoffman, *Phosphorus Sulfur,* 1978, **4**, 239, 249; C. A. McAuliffe, *Adv. Inorg. Chem. Radiochem.,* 1975, **17**, 165.

[217]   R. E. Simone and M. D. Glick, *J. Am. Chem. Soc.,* 1976, **98**, 762; N. W. Alcock *et al., J.C.S. Dalton,* **1978**, 394.

[218]   R. J. Cross *et al., J.C.S. Dalton,* **1976**, 1150; J. H. Eekhof *et al., J. Organomet. Chem.,* 1978, **161**, 183; E. W. Abel *et al., J. Organomet. Chem.,* 1978, **145**, C18; *J.C.S. Dalton,* **1977**, 42, 47.

Disulfides (RSSR) commonly undergo cleavage to give SR compounds,[203] but an intact, bridging, disulfide ligand is found in $(CO)_3Re(\mu\text{-}Br)_2(\mu\text{-}S_2Ph_2)\text{-}Re(CO)_3$.[219]

An N,S-tetradentate nickel complex can be made only by the reaction:

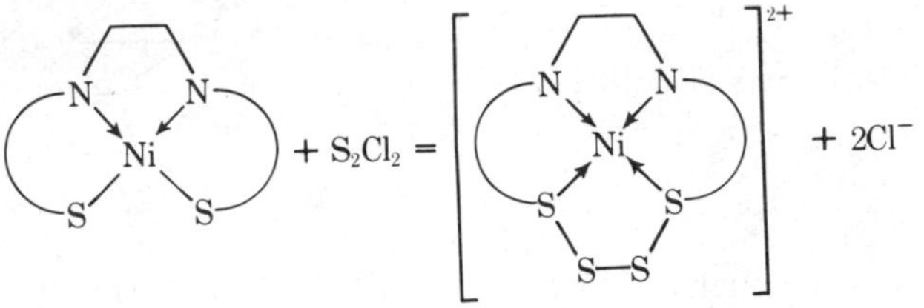

4-32. Sulfur Oxides

Despite the instability of lower oxides of sulfur, SO, $(SO)_2$, and S_2O (Chapter 16), all can be trapped as ligands.[220] Complexes containing the last two can be made by successive oxidations:

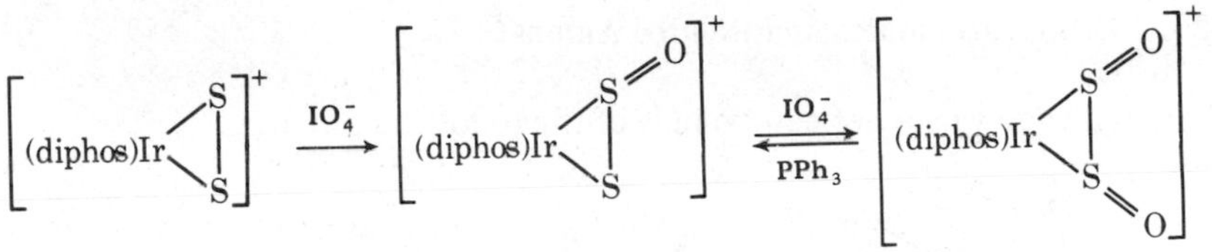

An SO bridge is found in $[(\eta\text{-}C_5H_5)Mn(CO)_2]_2SO$; it can be considered a derivative of thionylchloride $SOCl_2$.[221]

The most important ligand, however, is *sulfur dioxide*, which commonly forms complexes by direct interaction. The ligand may be bound in several different ways[222] as shown:

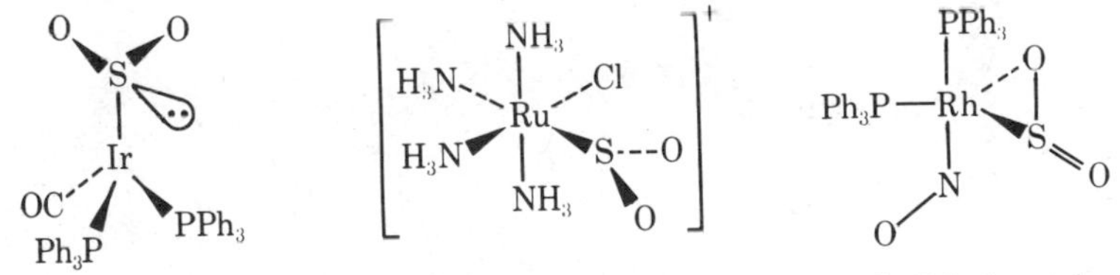

Pyramidal M—SO₂   Planar M—SO₂ η^2, Side-bound

219 I. Bernal *et al., Gazz. Chim. Ital.,* 1976, **106**, 971.
220 G. Schmidt and G. Ritter, *Angew. Chem., Int. Ed.,* 1975, **14**, 645.
221 M. Hofler and A. Baitz, *Chem. Ber.,* 1976, **109**, 314.
222 G. J. Kubas, *Inorg. Chem.,* 1979, **18**, 182; L. Sacconi *et al., Inorg. Chem.,* 1978, **17**, 3020; R. D. Wilson and J. A. Ibers, *J. Am. Chem. Soc.,* 1978, **100**, 2134; P. R. Blum and D. W. Meek, *Inorg. Chim. Acta,* 1977, **24**, L75; R. R. Ryan *et al., Inorg. Chem.,* 1979, **18**, 223, 227; M. Angoletta *et al., J.C.S. Dalton,* **1977**, 2131; M. Cowie *et al., Inorg. Chim. Acta,* 1978, **31**, L407.

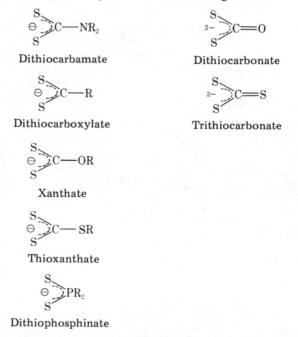

Bridge without M—M bond

Bridges with M—M bonds

In the compounds such as $[am_4ClRuSO_2]^+$ and $(\eta\text{-}C_5H_5)Rh(C_2H_4)SO_2$ that have planar $M\text{—}SO_2$ groups, there is evidently some M—S π-bonding with S acting as σ donor and π acceptor. The bonding of SO_2 has been investigated by MO methods.[223] The bridged molecules without M—M bonds can of course be regarded as derivatives of sulfuryl dichloride SO_2CL_2 rather than of SO_2. Similarly, what can be considered to be a bound sulfinate anion[224] is obtained by action of SO_2 on $\eta\text{-}C_5H_5Fe(CO)_2K$ to give the anion $\eta\text{-}C_5H_5Fe(CO)_2SO_2^-$.

4-33.   Dithiocarbamates and Related Anions[225]

There is a wide variety of compounds of the following anions:

Dithiocarbamate

Dithiocarbonate

Dithiocarboxylate

Trithiocarbonate

Xanthate

Thioxanthate

Dithiophosphinate

[223]   R. R. Ryan and P. G. Eller, *Inorg. Chem.*, 1976, **15**, 494;

[224]   C. R. Jablonski, *J. Organomet. Chem.*, 1977, **142**, C25.

[225]   (a) D. Coucouvanis, *Prog. Inorg. Chem.*, 1970, **11**, 233; (b) G. Gattow and W. Behrendt, *Carbon Sulfides and Their Inorganic and Complex Chemistry,* Thieme, 1977; J. Willemse *et al., Struct. Bonding,* 1976, **28**, 83.

Each of these ions may also have a corresponding monothio derivative that is bound by S and O, for example,

$$\underset{\text{Thiocarbamate}}{\ominus \underset{O}{\overset{S}{\diagdown}} C—NR_2} \qquad \underset{\text{Thiocarbonate}}{2- \underset{O}{\overset{S}{\diagdown}} C=O}$$

Dithiocarbamates and dithiophosphinates are probably the most important, and some complexes have commercial uses: dithiocarbamates as fungicides, dithiophosphinates as high-pressure lubricants.

The bonding of *dithiocarbamate* is variable, and the following forms are known:

$$\underset{\text{Unidentate}}{\overset{M—S}{\underset{S}{\diagdown}} C—NR_2} \qquad \underset{\text{Symmetrical chelate}}{M \overset{S}{\underset{S}{\diagdown}} C—NR_2} \qquad \underset{\text{Unsymmetrical chelate}}{M \overset{S}{\underset{S}{\diagdown}} C—NR_2}$$

The bridging $^-S_2CNR_2$ groups (shown diagrammatically) can have several forms[226]:

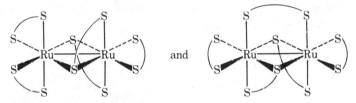

The compound $[Ru_2(S_2CNEt_2)_5]^+$, for example, has two isomers with chelate and different types of bridging dithiocarbamate:

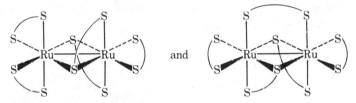

The unidentate and chelate types can be distinguished by ir and nmr spectra.[227] The nmr spectra of chelate dithiocarbamate (dtc) groups are commonly temperature dependent due to dynamic processes involving nonequivalence of the R groups, rotation about the C–N bond, and for trischelates (Mdtc₃), intramolecular metal-centered dynamic processes proceeding by a trigonal twist mechanism.[228]

A feature of the dithiocarbamate ligand is that uncommonly high oxidation states can be obtained as in $[Fe^{IV}(dtc)_3]^+$ and $[Ni^{IV}(dtc)_3]^+$.

The major resonance forms for dithiocarbamate are:

[226] A. R. Hendrickson, J. M. Hope and R. L. Martin, *J.C.S. Dalton,* **1976**, 2032; R. Hesse, *Arch. Kemi,* 1963, **20**, 481.

[227] See, e.g., J. R. Rowbottom and G. Wilkinson, *J.C.S. Dalton,* **1974**, 684.

[228] L. H. Pignolet *et al., Topics Curr. Chem.,* 1975, **56**, 91; *J. Am. Chem. Soc.,* 1973, **95**, 1125; *Inorg. Chem.,* 1974, **13**, 351, 2045, 2051.

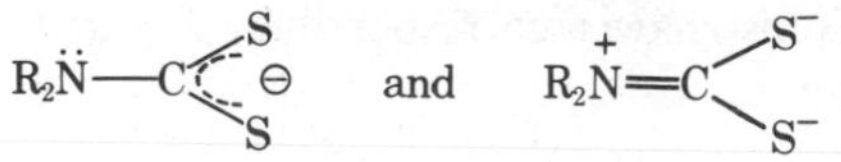

$$R_2\overset{..}{N}-C\overset{S}{\underset{S}{\diagup}}\ominus \quad \text{and} \quad R_2\overset{+}{N}=C\overset{S^-}{\underset{S^-}{\diagup}}$$

The NR_2 group has a strong electron-releasing effect, and when this is altered as in ligands such as

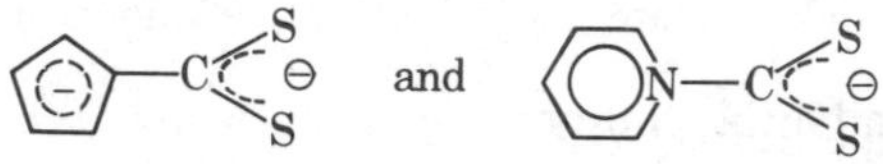

different behavior toward metals can be achieved.[229] Thus in the cyclopentadienyl compound, the driving force tending to make the ring aromatic leads to a dominant π-acceptor character at sulfur.

Thiocarbamates are similar,[230] with bonding modes such as the following:

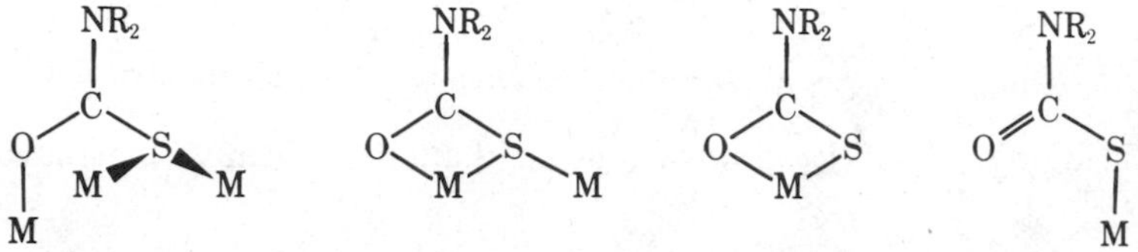

Little more need be said concerning the other ligands, which in general behave similarly. Thus dithiocarboxylates can act as chelate or as bridge groups.[231] Ions like $F_2PS_2^-$ also form complexes.[232a] In the complex $Mo^{IV}(O)(S_2CSR)_2$ one of the thioxanthate groups is η^3, being bound by two sulfur atoms and one carbon atom.[232b]

A rather special case of this type of ligand is comprised of the thioanions such as MoS_4^{2-}, $WO_2S_2^{2-}$, and WOS_3^{2-}, which can act as ligands just as oxo anions do.[233] Bonding is through sulfur as in these examples:

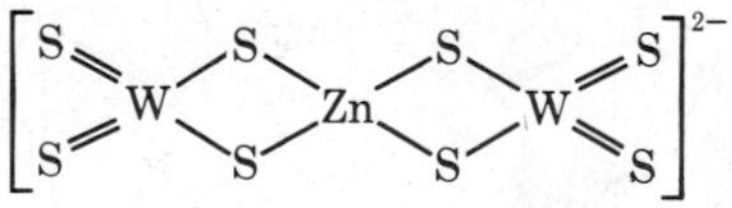

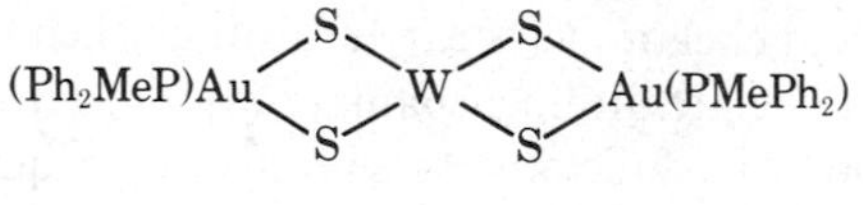

$$\left[\overset{S}{\underset{S}{\diagdown}}W\overset{S}{\underset{S}{\diagdown}}Zn\overset{S}{\underset{S}{\diagdown}}W\overset{S}{\underset{S}{\diagup}} \right]^{2-}$$

$(Ph_2MeP)Au\overset{S}{\underset{S}{\diagdown}}W\overset{S}{\underset{S}{\diagup}}Au(PMePh_2)$

[229]   R. D. Bereman and D. Nalewajek, *Inorg. Chem.,* 1977, **16,** 2687; A. G. El A'mma and R. S. Drago, *Inorg. Chem.,* 1977, **16,** 2975.

[230]   K. R. M. Springsteen, D. L. Green, and B. J. McCormick, *Inorg. Chim. Acta,* 1977, **23,** 13; J. G. M. Van der Linden *et al., Inorg. Chim. Acta,* 1977, **24,** 261.

[231]   M. Bonamico *et al., J.C.S. Dalton,* **1977,** 2315.

[232a]   M. V. Andreocci *et al., Inorg. Chem.,* 1978, **17,** 291.

[232b]   J. Hyde *et al., Inorg. Chem.,* 1978, **17,** 414.

[233]   I. Paulat-Böschen *et al., Inorg. Chem.,* 1978, **17,** 1440; A. Müller *et al., Angew. Chem. Int. Ed.,* 1978, **17,** 52; J. K. Stalick, et al., *J. Am. Chem. Soc.,* 1979, **101,** 2903.

In a particularly interesting case of a compound obtained from MoS_4^{2-}, the Fe_4S_4 clusters referred to earlier are bridged by MoS_4 groups in the ion $[Mo_2Fe_6S_6-(SEt)_8]^{3-}$. This may be a model for the molybdenum-iron complex in nitrogenase (Chapter 31).[234]

4-34.   1,2-Dithiolenes[235]

The 1,2-dithiolenes are a class of ligands that form a wide variety of compounds with metals in apparently many different oxidation states. This is more apparent than real because with ligands having extended π systems, delocalization of electrons onto the ligands occurs. It is very characteristic of dithiolene type ligands 4-LXXVI and related classes 4-LXXVII, 4-LXXVIII, and 4-LXXIX that re-

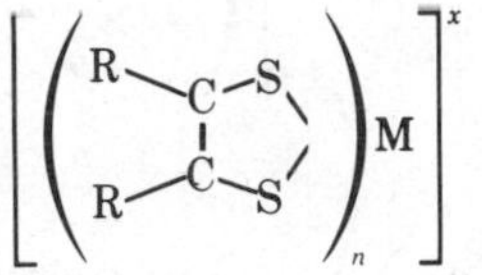

R = H, alkyl, C_6H_5, CF_3, CN
$n = 2; x = 0, -1, -2$
$n = 3; x = 0, -1, -2, -3$

(4-LXXVI)

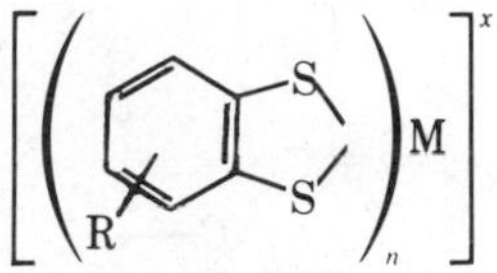

R = alkyl
$n = 2; x = 0, -1, -2$

(4-LXXVII)

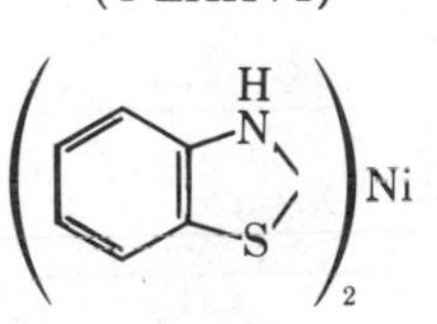

(4-LXXVIII)

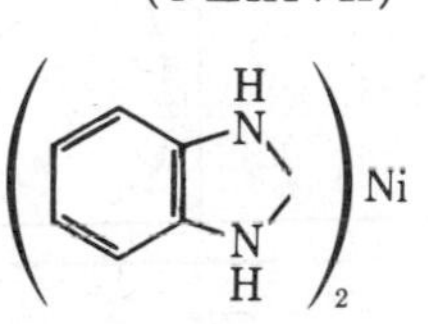

(4-LXXIX)

versible oxidation-reduction sequences between structurally similar molecules differing only in their electron populations can occur. Examples are:

$$\{Ni[S_2C_2(CN)_2]_2\} \underset{-e}{\overset{+e}{\rightleftharpoons}} \{Ni[S_2C_2(CN)_2]_2\}^- \underset{-e}{\overset{+e}{\rightleftharpoons}} \{Ni[S_2C_2(CN)_2]_2\}^{2-}$$

$$[CoL_2]_2^0 \underset{-e}{\overset{+e}{\rightleftharpoons}} [CoL_2]_2^{1-} \underset{-e}{\overset{+e}{\rightleftharpoons}} [CoL_2]_2^{2-} \underset{-2e}{\overset{+2e}{\rightleftharpoons}} 2[CoL_2]^{2-} \qquad [L = S_2C_2(CF_3)_2]$$

$$[CrL_3]^0 \underset{-e}{\overset{+e}{\rightleftharpoons}} [CrL_3]^{1-} \underset{-e}{\overset{+e}{\rightleftharpoons}} [CrL_3]^{2-} \underset{-e}{\overset{+e}{\rightleftharpoons}} [CrL_3]^{3-} \qquad [L = S_2C_2(CN)_2]$$

A few representative syntheses of dithiolene complexes are the following:

[234]   R. H. Holm et al., J. Am. Chem. Soc., 1979, **101**, 4140; 1978, **100**, 4630; D. Coucouvanis et al., J.C.S. Chem. Comm., **1979**, 361.

[235]   J. A. McCleverty, Prog. Inorg. Chem., 1968, **10**, 49; R. Eisenberg, Prog. Inorg. Chem., 1970, **12**, 295; D. Coucouvanis, Prog. Inorg. Chem., 1970, **11**, 233; R. J. H. Clark and P. C. Turtle, J.C.S. Dalton, **1977**, 2142; B.-K. Teo and P. A. Snyder-Robinson, Inorg. Chem., 1979, **18**, 1490.

$$NiCl_2 + Na_2^+[(NC)C(S)C(S)(CN)]^{2-} \xrightarrow{(C_2H_5)_4N^+}$$

$$\{[(C_2H_5)_4N]^+\}_2 \left[\left(\begin{array}{c} NC \\ \quad \\ NC \end{array} \begin{array}{c} S \\ C \\ C \\ \quad S \end{array} \right)_2 Ni \right]^{2-}$$

$$Ni(CO)_4 + 2(C_6H_5)_2C_2 + 4S \longrightarrow \left(\begin{array}{c} C_6H_5 \\ \quad \\ C_6H_5 \end{array} \begin{array}{c} S \\ C \\ C \\ \quad S \end{array} \right)_2 Ni \xrightarrow[(C_2H_5)_4N^+]{p\text{-}H_2NC_6H_4NH_2}$$

$$[(C_2H_5)_4N]^+ \left[\left(\begin{array}{c} C_6H_5 \\ \quad \\ C_6H_5 \end{array} \begin{array}{c} S \\ C \\ C \\ \quad S \end{array} \right)_2 Ni \right]^-$$

$$Ni(CO)_4 + 2 \begin{array}{c} F_3C \\ \quad C - S \\ \quad \| \\ \quad C - S \\ F_3C \end{array} \longrightarrow \left(\begin{array}{c} F_3C \\ \quad \\ F_3C \end{array} \begin{array}{c} S \\ C \\ C \\ \quad S \end{array} \right)_2 Ni \xrightarrow[\text{acetone} + (C_2H_5)_4N^+]{\text{spontaneously in}}$$

$$[(C_2H_5)_4N]^+ \left[\left(\begin{array}{c} F_3C \\ \quad \\ F_3C \end{array} \begin{array}{c} S \\ C \\ C \\ \quad S \end{array} \right)_2 Ni \right]^- \xrightarrow[+ (C_2H_5)_4N^+]{p\text{-}H_2NC_6H_4NH_2}$$

$$\{[(C_2H_5)_4N]^+\}_2 \left[\left(\begin{array}{c} F_3C \\ \quad \\ F_3C \end{array} \begin{array}{c} S \\ C \\ C \\ \quad S \end{array} \right)_2 Ni \right]^{2-}$$

For complexes containing only dithiolene ligands, four types of structure have been observed (Fig. 4-10). The planar D_{2h} structure is found for a majority of the structurally characterized bis complexes. The second structure type, observed in the remaining bis complexes, is dimeric, each metal atom being five-coordinate. The metal atoms are significantly displaced from the planes of the dithiolene ligands (by 0.2–0.4 Å), but the bridging linkages are relatively weak. The third type of structure is one having trigonal-prismatic D_{3h} coordination geometry. The inter-ligand S···S distances in this structure are rather short (3.0–3.1 Å), which suggests that there may be weak interactions directly between the sulfur atoms; this structure is found only in a few of the more highly oxidized or neutral tris complexes, one example being $Mo[Se_2C_2(CF_3)_2]_3$.

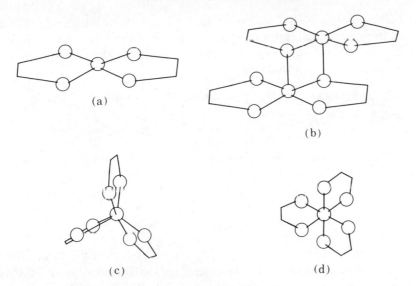

Fig. 4-10. The four basic structure types for "pure" dithiolene complexes: (a) square coordination (D_{2h} molecular symmetry); (b) five-coordinate dimer; (c) trigonal-prismatic coordination (D_{3h} symmetry); (d) octahedral coordination (D_3 symmetry).

In addition to "pure" dithiolene complexes, compounds are known with additional ligands such as CO, η-C_5H_5, and olefins.

The electronic structures of the 1,2-dithiolene complexes have provoked a great deal of controversy. The ring system involved can be written in two extreme forms:

The formal oxidation number of the metal differs by 2 in these two cases. In molecular-orbital terms the problem is one of the extent to which electrons are in metal d orbitals or delocalized over the ligand. Undoubtedly, in general, considerable delocalization occurs, which accounts for the ability of these complexes to exist with such a range of electron populations. The exact specification of orbital populations in any given case is a difficult and subtle question that we shall not discuss in detail here. The same problem arises for quinone (p. 162) and diimine (p. 118) complexes.

4-35. Other Sulfur Ligands

Sulfur Nitrides. Although NS cannot be isolated, a few complexes have N-bonded NS groups that are formally similar to NO (see p. 93).

A number of other N–S complexes[236] have been made by interaction of N_4S_4 (Section 16-9) with metal carbonyls or halides, but the constitution of these is often uncertain. Where hydroxylic solvents are used, hydrogen abstraction occurs, to give complexes such as the following:

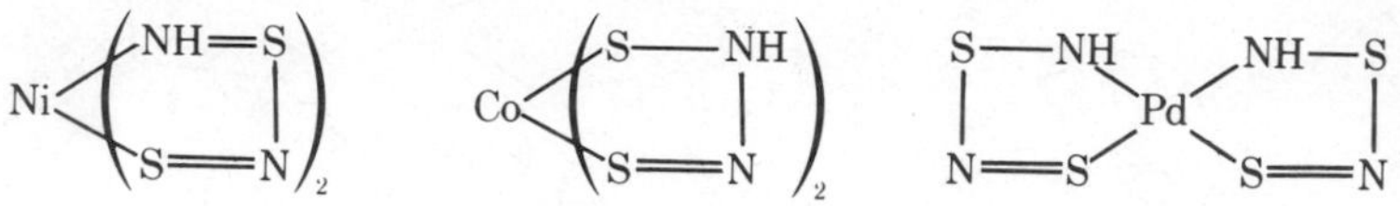

The *thiocyanate* ion is ambidentate, but toward heavier metals it is commonly S-bonded.[237] Bridged thiocyanates are well known, e.g.,

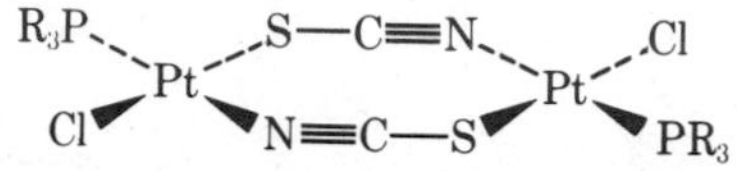

Sulfinate anions (RSO_2^-: see also p. 176) can be unidentate through sulfur or oxygen, or chelate through oxygen. An example of the S-bonded type is $Ph_3PAu-SO_2R$.[238] Sulfinyl halides can be added to $IrCl(CO)(PPh_3)_2$ oxidatively (Section 30-3); whereas CF_3SO_2Cl gives an O-bonded complex, $MeC_6H_4SO_2Cl$ gives an S-bonded, as well as a bidentate complex.[239] Although usually O-bonded, *sulfite* ion can be S-bonded[240] as in $[(NH_3)_5Co-SO_3]^+$.

Thiocarboxamido ligands[241] (R_2NCS) can be unidentate through sulfur, η^2-bidentate through carbon and sulfur (cf. CS_2, p. 114), or bridging.

Similar complexes of *dithio esters*[242] can be η^1 or η^2 as in

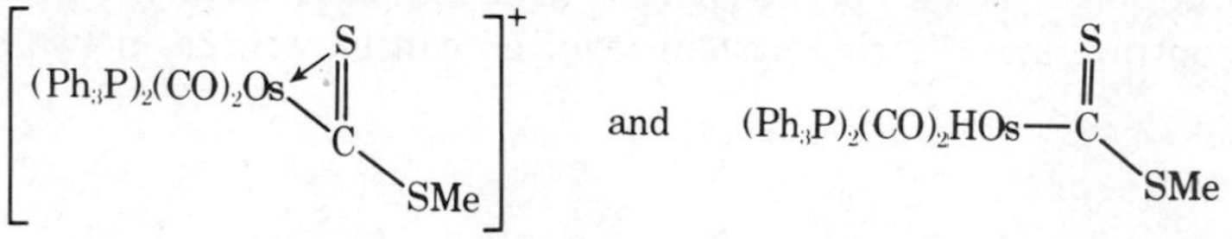

There are numerous compounds with S=C and S=P bonds, as well as others, that bond through sulfur. Most important perhaps are *thiourea* and substituted thioureas. These can be bound as follows:-

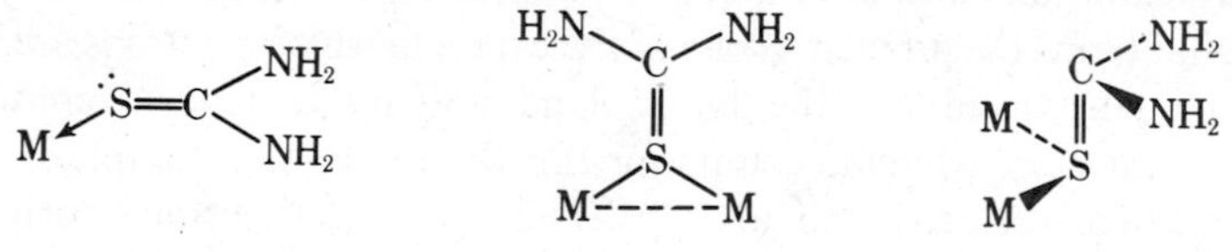

236   D. T. Haworth and G. Y. Lin, *J. Inorg. Nucl. Chem.*, 1977, **39**, 1838; K. J. Wynne and W. L. Jolly, *J. Inorg. Nucl. Chem.*, 1968, **30**, 2851; J. Weiss, *Fortschr. Chem. Forsch.*, 1966, **5**, 635.
237   A. H. Norbury, *Adv. Inorg. Chem. Radiochem.*, 1975, **17**, 232 (SCN, SeCN).
238   M. J. Mays and J. Bailey, *J.C.S. Dalton*, **1977**, 578.
239   D. M. Blake and Y. L. Chung, *J. Organomet. Chem.*, 1977, **134**, 327.
240   R. C. Elder and M. Trkula, *J. Am. Chem. Soc.*, 1974, **96**, 2635.
241   A. W. Gal *et al.*, *J. Organomet. Chem.*, 1978, **149**, 81.
242   J. M. Waters and J. A. Ibers, *Inorg. Chem.*, 1977, **16**, 3273.

Tertiary *phosphine sulfides* ($R_3P{=}S$, etc.), can be bound with sulfur unidentate and bridging or, in the case of $R_2P(S)CH_2CH_2P(S)R_2$, chelate. *1,3-Dithioketo-nates* are similar to the better known β-diketonates and give similar types of compounds.[243] Finally we note that thiosemicarbazones

$$H_2NC{-}N{-}N{=}CR_1R_2$$
$$\underset{S}{\|} \ \underset{H}{|}$$

and thiosemicarbazide

$$H_2NC \quad N \quad NH_2$$
$$\underset{S}{\|} \ \underset{H}{|}$$

usually act as chelates bonding through sulfur and the hydrazinic nitrogen atom.

HALIDE IONS[245]

4-36. General Remarks

All the halide ions have the ability to function as ligands and form, with various metal ions or covalent halides, complexes such as SiF_6^{2-}, $FeCl_4^-$, and HgI_4^{2-}, as well as mixed complexes along with other ligands, for example, $[Co(NH_3)_4Cl_2]^+$. We merely make some general remarks and cite some typical characteristics of such complexes, reserving detailed discussions for other places in connection with the chemistries of the complexed elements.

One of the important general questions that arises concerns the relative affinities of the several halide ions for a given metal ion. There is no simple answer to this, however. For crystalline materials it is obvious that lattice energies play an important role, and there are cases, such as BF_4^-, BCl_4^-, BBr_4^-, in which the last two are known only in the form of crystalline salts with large cations, where lattice

[243] W. L. Bowden *et al., Inorg. Chem.,* 1978, **17,** 256.

[244] C. Bellitto *et al., Inorg. Chim. Acta,* 1978, **27,** 269; M. J. M. Campbell, *Coord. Chem. Rev.,* 1975, **15,** 279.

[245] R. Colton and J. H. Canterford, *Halides of First Row Transition Elements,* 1969; *Halides of Second and Third Row Transition Elements,* 1968; D. Brown, *Halides of the Lanthanide and Actinides,* 1968, Wiley-Interscience. R. D. Peacock, *Advances in Fluorine Chemistry,* Vol. 7, Butterworths, 1973 (pentafluorides of transition metals); Y. Marcus, *Coord. Chem. Rev.,* 1967, **2,** 195, 257 (chloride complexes); T. A. O'Donnell, *Rev. Pure Appl. Sci.,* 1970, **20,** 159 (reactivity of higher fluorides); G. Hefter, *Coord. Chem. Rev.,* 1974, **12,** 221 (fluoride complexes in water); G. C. Allen and K. D. Warren, *Coord. Chem. Rev.,* 1975, **16,** 227 (electronic spectra MF_6^{n-}); R. A. Walton, *Coord. Chem. Rev.,* 1976, **21,** 63 (X-ray photoelectron spectra of halides and halide complexes).

energies are governing. In considering the stability of the complex ions in solution, it is important to recognize that (a) the stability of a complex involves, not only the absolute stability of the M—X bond, but also its stability relative to the stability of ion-solvent bonds, and (b) in general an entire series of complexes will exist, $M^{n+}(aq)$, $MX^{(n-1)+}(aq)$, $MX_2^{(n-2)+}(aq)$, $\cdots$, $MX_x^{(n-x)+}(aq)$, where x is the maximum coordination number of the metal ion. Of course these two points are of importance in all types of complex in solution.

A survey of all the available data on the stability of halide complexes shows that generally the stability decreases in the series $F > Cl > Br > I$, but with some metal ions the order is the opposite, namely, $F < Cl < Br < I$. No rigorous theoretical explanation for either sequence or for the existence of the two classes of acceptors relative to the halide ions has been given. It is likely that charge/radius ratio, polarizability, and the ability to use empty outer d orbitals for back-bonding are significant factors. From the available results it appears that for complexes where the replacement stability order is $Cl < Br < I$, the actual order of M—X bond strength is $Cl > Br > I$, so that ionic size and polarizability appear to be the critical factors.

The limiting factor in the formation of fluoro complexes for cations of small size and high charge is competitive hydrolysis; even at high concentrations many fluoro complexes are hydrolyzed and particularly so where the oxidation state is high. There is an empirical relation of wide applicability:

$$\log Q = -1.56 + \frac{0.48 Z_+^2}{r_+}$$

where Q is the formation constant for the reaction

$$M^{n+} + F^- = MF^{(n-1)+}$$

and Z_+ and r_+ are the charge and radius of the cation. It is to be emphasized that all complex fluoro "acids" such as HBF_4 and H_2SiF_6 are necessarily strong, since the proton can be bound only to a solvent molecule.

There are many references to the effect of steric factors in accounting for such facts as the existence of $FeCl_4^-(aq)$ as the highest ferric complex with Cl^-, whereas FeF_6^{3-} is rather stable, and similar cases such as $CoCl_4^{2-}$, SCl_4, $SiCl_4$ as the highest chloro species compared with the fluoro species CoF_6^{3-}, SF_6, SiF_6^{2-}. In many such cases thorough steric analysis, considering the probable bond lengths and van der Waals radii of the halide ions, shows that this steric factor alone cannot account for the differences in maximum coordination number, and this point requires further study.

Although iodo complexes are generally the least stable, and are dissociated or unstable in aqueous solution, a large number of complex anions can be made even where the metal M^{n+} is oxidizing toward I^-, provided nonaqueous media (e.g., nitriles, CH_3NO_2, or liquid HI) are used. The latter possibility arises because HCl has a free energy of formation, ca. 85 kJ mol^{-1} higher than that of HI, so that in the anhydrous reaction

$$MCl_6^{(6-n)-} + 6HI(l) = MI_6^{(6-n)-} + 6HCl$$

the equilibrium lies to the right and the greater volatility of HCl provides additional driving force.

Finally, it may be mentioned that in effecting the separation of metal ions, one can take advantage of halide complex formation equilibria in conjunction with anion-exchange resins. To take an extreme example, Co^{2+} and Ni^{2+}, which are not easily separated by classical methods, can be efficiently separated by passing a strong hydrochloric acid solution through an anion-exchange column. Co^{2+} forms the anionic complexes $CoCl_3^-$ and $CoCl_4^{2-}$ rather readily, whereas it does not seem that any anionic chloro complex of nickel is formed in aqueous solution even at the highest attainable activities of Cl^-; however tetrachloronickelates can be obtained in fused salt systems or in nonaqueous media. More commonly, effective separation depends on properly exploiting the difference in complex formation between two cations, both of which have some tendency to form anionic halide complexes.

4-37. Halide Bridges[246]

The formation of halide bridges is an important structural feature not only in complex compounds but in many simple molecular compounds. There are the following types:

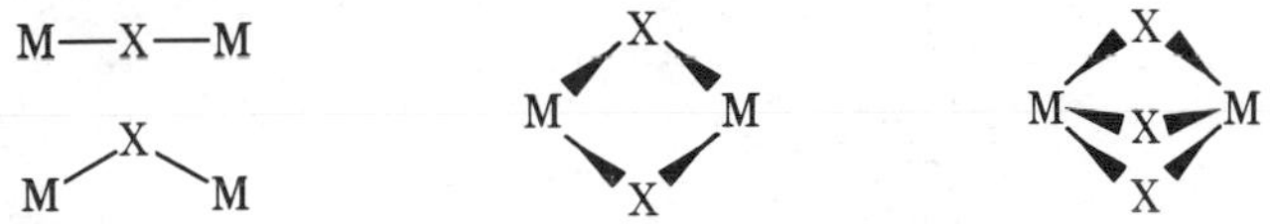

Bridges with two halogen atoms are most common, but there are compounds with single[247] and also *triple bridges,* as in $(R_3P)_2ClRuCl_3RuCl(PR_3)_2$.[248]

With Cl^- and Br^-, bridges are characteristically bent, whereas fluoride bridges may be either bent or linear. Thus in BeF_2 there are infinite chains, $\cdots BeF_2Be-F_2 \cdots$, with bent bridges, similar to the situation in $BeCl_2$. On the other hand, transition metal pentahalides afford a notable contrast. The pentachlorides dimerize (see 4-LXXX) with bent $M-Cl-M$ bridges, but the pentafluorides form cyclic tetramers (4-LXXXI) with linear $M-F-M$ bridges.

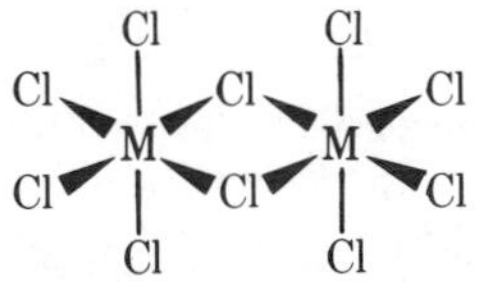

(4-LXXX) 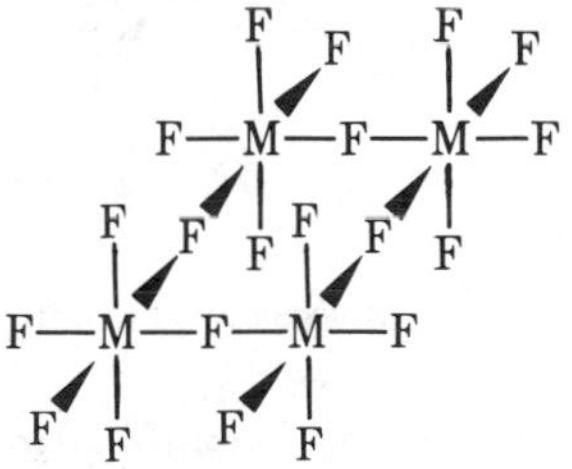   (4-LXXXI)

[246] D. J. Hodgson, *Prog. Inorg. Chem.,* 1975, **19,** 173; K. Seppelt, *Angew. Chem., Int. Ed.,* 1979, **18,** 186 (fluorides).

[247] See, e.g., L. Sacconi, P. Dapporto, and P. Stoppioni, *J. Am. Chem. Soc.,* 1977, **99,** 5595 (for Ni-I-Ni bridge); R. Mews, *Angew. Chem., Int. Ed.,* 1977, **16,** 56.

[248] R. A. Head and J. F. Nixon, *J.C.S. Dalton,* **1978,** 901.

The fluorides probably adopt the tetrameric structures with linear bridges, in part, because F is smaller than Cl, and this would introduce excessive M· · ·M repulsion in an

$$M\overset{\displaystyle F}{\underset{\displaystyle F}{<\ >}}M$$

system. Another notable example of a linear M—F—M bridge occurs in $K[(C_2H_5)_3Al—F—Al(C_2H_5)_3]$; the linearity of this Al—F—Al chain may be due in part to overlap of filled $2p\pi$ orbitals of fluorine with empty $3d\pi$ orbitals of the aluminum atoms, thus giving some π character to the Al—F bonds.

Polymeric gaseous halides and compounds such as $FeAuCl_6$ or $CrAl_3Cl_{12}$ all have halide bridges.[249]

The μ_2—X bonding can be interpreted by MO theory.[250] As indicated in Fig. 4-11a, each metal atom presents an empty σ orbital directed more or less toward the bridging halide X^-; these orbitals are ϕ_1 and ϕ_2. The X^- ion has four filled valence shell orbitals. One of these (which may be taken as an s orbital, a p orbital, or a hybrid), ϕ_3, will be directed down, as shown; the other, ϕ_4, is a pure p orbital. The metal orbitals may be combined (Fig. 4-11b) into a symmetric combination

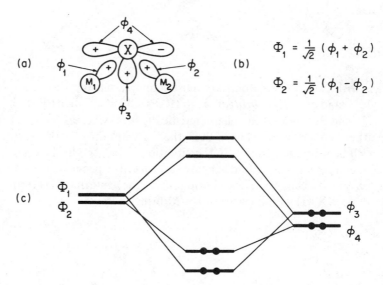

Fig. 4-11. Molecular orbital formation of a bridging halide system MXM.

ϕ_1, and an antisymmetric combination ϕ_2, which may interact with ϕ_3 and ϕ_4, respectively, to form bonding and antibonding orbitals, as in Fig. 4-11c. The four electrons then occupy the two bonding MOs, giving a mean M—X bond order of 1. It is also evident that the M—X—M angle is not sharply limited. In reality, angles

[249]   H. Schäfer, *Angew. Chem., Int. Ed.,* 1976, **15**, 713.
[250]   For detailed discussion see R. H. Sommerville and R. Hoffmann, *J. Am. Chem. Soc.,* 1976, **98**, 7240.

ranging from 60° to 180° are found, though the majority for Cl and Br are in the range 70–100°.

Many complexes with chloride bridges can be cleaved by Lewis bases:

$$[RhCl(CO)_2]_2 + 2PPh_3 \rightarrow 2RhCl(CO)_2(PPh_3)$$
$$(\pi\text{-allylPdCl})_2 + 2py \rightarrow \pi\text{-allylPdClpy}$$

The thermodynamics of some of these reactions have been studied.[251]

4-38. Halide Complex Ions as Ligands

A number of the fluorides mentioned at various places in the text have structures that can be considered to be composed of cations MF_n^+ and anions MF_n^- interacting strongly with each other. However the *fluoroborate ion* can act as a ligand, forming M—F—BF$_3$ links in complexes such as $[Nien_2(H_2O)BF_4]^+$ or $(Ph_3P)_3CuBF_4$.[252]

[251] M. P. Li et al., J. Am. Chem. Soc., 1977, **99**, 6900.
[252] A. A. G. Tomlinson et al., J.C.S. Dalton, **1972**, 1671; A. P. Gaughan, Jr., et al., Inorg. Chem., 1974, **13**, 1657.

Additional References

Blundell, T. L., and J. A. Jenkins, Chem. Soc. Rev., 1977, **6**, 139. Binding of heavy metals to proteins.

Davydova, S. L., and N. A. Plate, Coord. Chem. Rev., 1975, **16**, 195. High molecular weight, polymer ligands.

Dwyer, F. P., and D. P. Mellor, Eds., Chelating Agents and Metal Chelates, Academic Press, 1964.

Eller, P. G., et al., Coord. Chem. Revs, 1977, **24**, 1. Ligands forming three-coordinate complexes.

Smith, T. D., and J. R. Pilbow, Coord. Chem. Revs., 1974, **13**, 173. Dimeric complexes without metal-metal bonds. Bridge ligands.

Summerville, R. H., and R. Hoffmann, J. Am. Chem. Soc., 1979, **101**, 3821. Theory of $L_3M(\mu-X)_3ML_3$ bridged species; 50 references.

Tiglis, T., Inorg. Chim. Acta, 1973, **7**, 35. Pseudoallylic ligands A—B—C⁻.

Wentworth, R. A. D., Coord. Chem. Rev., 1972–1973, **9**, 171. Innocent ligands and octahedral versus trigonal prismatic structures.

O LIGANDS
Karayannis, N. M., et al., Coord. Chem. Rev., 1973, **11**, 93; 1976, **20**, 37. N-Oxides of aromatic amines, diimines, and diazenes.

Larpides, A., et al., Inorg. Chem., 1977, **16**, 3299. Malonates.

Vögtle and E. Weber, Angew. Chem. Int. Ed., 1979, **18**, 753. Multidentate acyclic neutral ligands and their complexation.

N LIGANDS
Langer, A. W., Ed., Polyamine Chelated Alkali Metal Compounds, ACS Advances in Chemistry Series Monograph No. 130, 1974.

N, S LIGANDS
Ali, M. A., and S. E. Livingstone, Coord. Chem. Rev., 1974, **13**, 101. Aminoethanethiol, o-aminothiophenol, thiosemicarbazides, etc.

O, S LIGANDS

Livingstone, S. E., *Coord. Chem. Rev.,* 1971, **7,** 59; M. E. Cox and J. Darken, *Coord. Chem. Rev.,* 1971, **7,** 29. Thio-β-diketonates.

S LIGANDS

Lindner, E., and G. Vitzhum, *Angew. Chem., Int. Ed.,* 1971, **10,** 315. Sulfinite complexes.

Lindoy, F., *Coord. Chem. Rev.,* 1969, **4,** 41. Reactions of S ligand complexes.

Wolterman, G. M., and H. J. Stoklosa, *Topics Curr. Chem.,* No. 35, 1973, Springer, Dithio- and diselenophosphate complexes.

CHAPTER FIVE

Stereochemistry and Bonding in Main Group Compounds

5-1. Introduction

In Chapter 2 we discussed molecular structures from the point of view of symmetry elements and geometrical description, without considering how these features have their origin in the bonding. We now address that problem, namely, finding the simplest and most useful relationships between the chemical bonds in a molecule and its geometry. The problem has two aspects. The one to which we devote most attention concerns the shapes of the molecules, that is, the angles between the bonds formed by a given atom. The second and shorter part of the discussion deals with certain aspects of multiple bonding, primarily with bond lengths. Since we are interested in simple, broadly used correlations, we neglect finer points, such as detailed electron density distributions in bonds. Single bonds are assumed to be symmetrical about the bond axis (i.e., to offer no inherent barrier to free rotation and not to be bent), although these idealizations are seldom exactly valid.

The difficulty in presenting this subject is that quite diverse viewpoints are taken by different workers in attempting to correlate electronic and molecular structures. Each approach has certain virtues, and yet they have very little common ground. It is not that their assumptions are particularly in conflict but rather that they seem unrelated, which is disconcerting because they all purport to be answering the same questions. Because of this situation, the discussion in this chapter necessarily has some of the same fragmented character. Each approach to the problem is presented along the lines generally used by its advocates. Following this, however, an attempt is made to compare, interrelate, and criticize the several models.

5-2.   The Valence Shell Electron Pair Repulsion (VSEPR) Model[1]

In the VSEPR model the arrangement of bonds around a central atom is considered to depend on how many valence shell electron pairs, each occupying a localized one- or two-center orbital, are present, and on the relative sizes and shapes of these orbitals. The first rule is as follows: (1) *the pairs of electrons in a valence shell adopt that arrangement which maximizes their distance apart; that is, the electron pairs behave as if they repel each other.*

Each electron pair is assumed to occupy a reasonably well-defined region of space, and other electrons are effectively excluded from this space. For electron pairs in the same valence shell, the arrangements that maximize their distance apart are listed in Table 5-1.

TABLE 5-1
Predicted Arrangements of Electron Pairs in One Valence Shell

Number of pairs	Polyhedron defined
2	Linear
3	Equilateral triangle
4	Tetrahedron
5	Trigonal bipyramid
6	Octahedron
7	Monocapped octahedron
8	Square antiprism
9	Tricapped trigonal prism

To apply rule 1 for the qualitative prediction of molecular shapes where only single bonds and unshared pairs are concerned, we compute the total number of electron pairs, bonding and nonbonding, select the appropriate arrangement in Table 5-1, and assign the electron pairs to it. In the cases of two, three, or four pairs, the results are immediately obvious as shown in Fig. 5-1. For example, an AB_4 molecule must be tetrahedral, an AB_3E molecule (E represents an unshared pair) must be pyramidal, and an AB_2E_2 molecule must be bent. There are no known exceptions to these predictions. Table 5-2 lists a few molecules and ions of the types AB_2E_2 and AB_3E, giving the angles.

TABLE 5-2
Angles (deg) in Some AB_2E_2 and AB_3E Molecules

AB_2E_2		AB_3E					
Molecule	Angle	Molecule	Angle	Molecule	Angle	Molecule	Angle
H_2O	104.5	NH_3	107.3	NF_3	102.1	OCl_2	111
H_2S	92.2	PH_3	93.3	PF_3	97.8	PCl_3	100.3
H_2Se	91	AsH_3	91.8	AsF_3	96.2	$AsCl_3$	98.7
H_2Te	89.5	SbH_3	91.3	SbF_3	88	$SbCl_3$	99.5
OF_2	103.2						

[1]   R. J. Gillespie, *Molecular Geometry,* van Nostrand-Reinhold, 1972 (this book should be consulted for more detailed discussion and references to earlier literature).

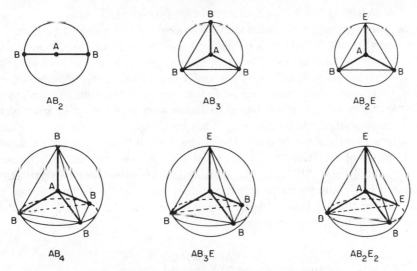

Fig. 5-1. Prediction of the shapes of molecules when the valence shell of the central atom contains two, three, or four electron pairs.

Molecules with five electron pairs pose additional problems. For the AB_5 molecules themselves, there are two plausible arangements: trigonal bipyramid (*tbp*) and square pyramid (*sp*) as discussed on page 50. The two differ little in stability, and the ease of interconvertibility gives rise to one of the more important forms of stereochemical nonrigidity (see Section 28-13). The VSEPR model does, however, favor the *thp* and for nearly every AB_5 molecule known, where A is a nontransition element, the *tbp* structure is found. The $InCl_5^{2-}$ and $TlCl_5^{2-}$ ions have *sp* structures, but since they occupy sites of 4-fold symmetry in the crystal,[2] it may well be that this structure is imposed by crystal packing. Similarly, it has also been argued that the closer approach of $SbPh_5$ to *sp* than to *tbp* is a crystal packing effect.[3] All other AB_5 species are *tbp*'s, and a few representative cases are listed in Table 5-3.

TABLE 5-3
Some Trigonal Bipyramidal Molecules and Their Bond Lengths

Molecule	Axial		Equatorial	
PF_5	2F	1.577	3F	1.534
$PF_3(CH_3)_2$	2F	1.643	F	1.553
			$2CH_3$	1.798
AsF_5	2F	1.711	3F	1.656
PCl_5(gas)	2Cl	2.19	3Cl	2.04
$SbPh_4OH$	Ph	2.22	3Ph	2.11–2.14
	OH	2.05		

[2]   D. S. Brown, F. W. B. Einstein, and D. G. Tuck, *Inorg. Chem.*, 1969, **8**, 14.
[3]   C. P. Brock and J. A. Ibers, *Acta Crystallogr., A*, 1976, **31**, 38.

For the $AB_4E, \ldots, AB_2E_3$ cases there is the further problem of whether non-bonding pairs will be in axial or equatorial positions. To deal with this question (and some others) an additional rule is needed: (2) *a nonbonding pair of electrons takes up more room on the surface of an atom than a bonding pair.* This is a plausible idea inasmuch as a nonbonding pair is under the influence of only one nucleus, whereas a bonding pair is constrained by two nuclei. Application of rule 2 to the molecules under discussion requires us to decide whether nonbonding pairs prefer axial or equatorial positions of the *tbp.* It has been argued that there is more room for the larger lone pairs in equatorial positions, though this may not be as obvious as proponents of the theory assert. However with this assumption, the structures of AB_4E, AB_3E_2, and AB_2E_3 molecules as well as AB_5 molecules are consistently accounted for. Some pertinent structural data are given in Table 5-4, and the structures for the AB_4E and AB_3E_2 types are illustrated in Fig. 5-2. All AB_2E_3 species (XeF_2, ICl_2^-, etc.) are linear.

TABLE 5-4
Structures of Some AB_4E and AB_3E_2 Molecules[a]

Molecule	B and A—B distance (Å)	BAB angle (°)	B′ and A—B′ distance (Å)	B′AB′ angle (°)
SF_4	F, 1.545	102	F, 1.646	173
$Se(CH_3C_6H_4)_2Cl_2$	C, 1.93	107	Cl, 2.38	178
$Se(CH_3C_6H_4)_2Br_2$	C, 1.95	108	Br, 2.55	177
$Te(CH_3)_2Cl_2$	C, 2.09	98	Cl, 2.48	172
$S(CH_2)_4TeI_2$	C, 2.16	100	I, 2.85–2.99	176
ClF_3	F, 1.598	—	F, 1.698	175
BrF_3	F, 1.721	—	F, 1.810	172

[a] See Fig. 5-2 for sketches defining B and B′ ligands, and the B′AB′ angles.

With species involving six pairs of valence shell electrons, rules 1 and 2 lead unambiguously to correct structures. Table 5-5 lists some AB_5E and AB_4E_2 molecules; the characteristic structures are shown in Fig. 5-3. Rule 2 is invoked to account for the square structures of AB_4E_2 molecules. It is clear that a lone pair encounters less repulsion from a *cis* bonding pair than from a *cis* lone pair; hence the lone pairs are *trans* to each other.

For the higher numbers of valence shell pairs, 7, 8, and 9, there are only a few non-transition-element complexes known. In the case of IF_7 the structure appears to be a pentagonal bipyramid, contrary to the entry in Table 5-1. However with these higher numbers, the predictions of preferred arrangements necessarily become

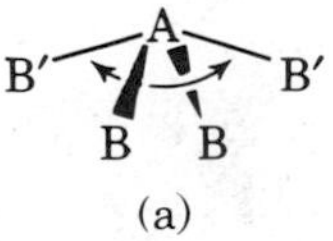

(a) 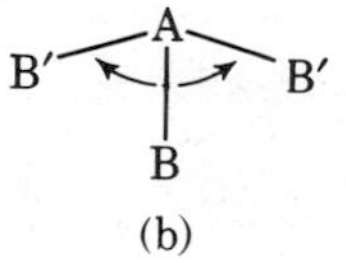   (b)

Fig. 5-2. The shapes of molecules of types (a) AB_4E and (b) AB_3E_2, showing how the B′AB′ angles are defined.

TABLE 5-5
Structures of Some AB₅E and AB₄E₂ Molecules[a]

Molecule	B and A—B distance (Å)	B′ and A—B′ distance (Å)	BAB′ angle (°)
ClF₅	4F, 1.72	F, 1.62	—[b]
BrF₅	4F, 1.78	F, 1.68	85
XeOF₄	4F, 1.95	O, 1.70	91 ± 2
XeF₄	4F, 1.953		
ICl₄⁻	4Cl, 2.46		
BrF₄⁻	4F, 1.88		

[a] See Fig. 5-3a for sketch defining B and B′ atoms in AB₅E.
[b] Not determined.

less certain because the repulsive energy of the set of electron pairs does not have a pronounced minimum for any one configuration and atom-atom interactions assume greater importance. In fact, for IF_7, XeF_6 (which is an AB_6E case), and species of coordination > 6 in general, the lack of any uniquely favored geometric arrangement results in stereochemical nonrigidity, a phenomenon discussed in detail in Section 28-13.

In addition to predicting, or at least correlating, gross geometrical features, as just outlined, the VSEPR model can give a consistent account of certain finer details of the structures. Thus according to rule 2 the greater size of lone pairs might be expected to result in the angles between their axes and bond axes being greater than the ideal values for the polyhedron concerned, with the observable results that the bond angles would be smaller. Tables 5-2 to 5-5 indicate that this is precisely the case for AB_3E, AB_2E_2, AB_4E, AB_3E_2, and AB_5E molecules.

To account for other details two more rules, both natural extensions of the scheme, are required:

Rule 3: *The size of a bonding electron pair decreases with increasing electronegativity of the ligand.*

Rule 4: *The two electron pairs of a double bond (or the three electron pairs of a triple bond) take up more room than does the one electron pair of a single bond.*

Using rule 3 one can rationalize some of the trends in Table 5-2. For instance, the angles in NF_3 and F_2O are less than those in NH_3 and H_2O. Similarly, in a set of halo molecules AB_2E_2 or AB_3E, the BAB angles increase in the order F < Cl < Br ≈ I. There are, however, often exceptions when hydrides are considered, since

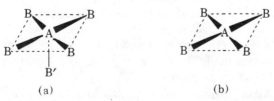

(a)   (b)

Fig. 5-3. The shapes of molecules of types (a) AB₅E and (b) AB₄E₂.

the PH_3, AsH_3, and SH_2 angles are less than those in any trihalide of the same element.

Rule 4 accounts for the fact that angles in which multiple bonds are involved are generally larger than those involving only single bonds. A few representative examples are shown in Table 5-6. It should be noted that when the double bond

TABLE 5-6
Bond Angles in Some Molecules Containing a Double Bond

Molecule	Angles (°)		
	XCX	XCO	XCC
F_2CO	108	126	
Cl_2CO	111	124	
$(NH_2)_2CO$	118	121	
F_2SO	93	107	
Br_2SO	96	108	
$H_2C{=}CF_2$	110		125
$H_2C{=}CCl_2$	114		123
OPF_3	103		
$OPCl_3$	104		

is to an atom less electronegative than those to which the single bonds are directed, the operation of rule 3 reinforces the effect of rule 4.

Predictions concerning relative lengths of bonds are also possible by using these rules. Thus for AB_5, AB_4E, AB_3E_2, and AB_5E molecules the A—B bonds and A—B' bonds differ in length by about 0.1 Å. In the first three cases, in which there is a trigonal-bipyramidal distribution of electron pairs, the axial bonds are longer. In AB_5E it is the four basal bonds that are longer. In the cases of five electron pairs with a *tbp* arrangement, the axial pairs have three neighboring pairs on lines at only 90° away, while the equatorial ones have only two such closely neighboring pairs; equilibrium is thus attained when the axial pairs move to a somewhat greater distance from the central atom, lengthening the axial bonds relative to the equatorial ones. In the case of AB_5E, the greater size of the lone pair will act more strongly on the bonding pairs *cis* to it, thus lengthening the set of basal bonds.

5-3. The Hybridization or Directed Valence Theory

According to the hybridization theory, bond directions are determined by a set of hybrid orbitals on the central atom which are used to form bonds to the ligand atoms and to hold unshared pairs. Thus AB_2 molecules are linear owing to the use of linear sp hybrid orbitals. AB_3 and AB_2E molecules should be equilateral triangular and angular, respectively, owing to use of trigonal sp^2 hybrids. AB_4, AB_3E, and AB_2E_2 molecules should be tetrahedral, pyramidal, and angular, respectively, because sp^3 hybrid orbitals are used. These cases are, of course, very familiar and involve no more than an octet of electrons.

For the AB_5, AB_4E, AB_3E_2, and AB_2E_3 molecules, the hybrids must now include d-orbitals in their formation. The hybrid orbitals used must obviously be of the sp^3d

type, but an ambiguity arises because there are two such sets, viz., $sp^3d_{z^2}$ leading to tbp geometry and $sp^3d_{x^2-y^2}$ leading to sp geometry. There is no way to predict with certainty which set is preferred, and doubtless the difference between them cannot be great. Since we know experimentally that AB_5 molecules nearly all have tbp structures, the same arrangement is assumed for the AB_4E cases, etc. Even this *ad hoc* assumption does not solve all difficulties, since the position preferred by lone pairs must be decided and there is no simple physical model here (as there was in the VSEPR approach) to guide us. A preference by lone pairs for equatorial positions has to be assumed. With these assumptions, a consistent correlation of all the structures in this five-electron-pair class is possible.

For AB_6 molecules octahedral sp^3d^2 hybrids are used. AB_5E molecules must, naturally, be sp. For AB_4E_2 molecules there is nothing in the directed valence theory itself to show whether the lone pairs should be *cis* or *trans*. The assumption that they must be *trans* leads to consistent results.

It may be seen that if we assume formation of hybrid orbitals to determine the basic geometry, it is possible then to borrow a portion of the VSEPR dogma, namely, rule 2, that a lone pair takes up more room than a bonding pair to rationalize the finer features of certain structures (e.g., the bond angles in AB_4E, AB_3E_2, AB_5E, and AB_4E_2 molecules). In short, one may reject the view that electron-pair repulsion is the primary factor in stereochemistry and assume instead that directed hybrid orbitals have a basic role but still allow that electron-pair repulsions enter into the problem at a secondary level.

5-4. The Three-Center Bond Model

The three-center bond model approach is predicated on two main ideas: (1) that the use of outer d orbitals of the central atom is so slight that they may be neglected altogether, and (2) that the persistent recurrence of bond angles close to 90° and 180° in AB_n molecules suggests that orbitals perpendicular to one another, namely, p orbitals, are being used.

Two types of chemical bond are considered. First there is the ordinary two-center, two-electron ($2c$-$2e$) bond, formed by the overlap of a p orbital of the central atom with a σ orbital of an outer atom. Second, there is the linear three-center, four-electron ($3c$-$4e$) bond formed from a p orbital of the central atom and the σ orbitals of two outer atoms.

For molecules with an octet, or less, of electrons in the valence shell of the central atom, the hybridization theory, employing sp, sp^2, and sp^3 orbitals remains valid. This model was proposed for molecules in which five or more electron pairs on the central atom must be accounted for. Since it is, as we shall note below, undoubtedly too great a simplification, it will be applied only to a few illustrative cases. It is worth mentioning, however, since it has a certain heuristic value.

In molecules of the AB_4E type, the central atom is considered to use p_x and p_y orbitals to bind the B atoms (cf. Fig. 5-2a) while the B' atoms are bound using the p_z orbital to form a $3c$-$4e$ bond. Again some supplementary assumption, such as rule 2 of the VSEPR model, must then be invoked to explain the nonlinearity of

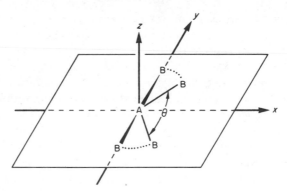

Fig. 5-4. The coordinate axes used for linear and bent AB_2 molecules.

the B'AB' set. The remaining electron pair is postulated to be occupying a pure s orbital and to have no stereochemical role. In an analogous way AB_3E_2 molecules are postulated to involve one $2c$-$2e$ bond, formed from a single p orbital of A to the B atom (Fig. 5-2b) and a $3c$-$4e$ bonding system for the B'AB' set, with the two other electron pairs in the s and remaining p orbital. For an octahedral molecule, three mutually perpendicular $3c$-$4e$ bonding systems are postulated, while for the AB_5E case, two such $3c$-$4e$ systems and one $2c$-$2e$ bond are employed.

It is clear that this model does, at least qualitatively, accord with the variation in bond lengths in these molecules. It allots a full electron pair to all the shorter A—B bonds and makes the longer ones members of a $3c$-$4e$ bond system.

5-5. The Correlation Diagram Approach

Correlation diagrams were first applied to relatively simple cases many years ago by Walsh; the approach led to certain generalizations, called Walsh's rules, relating the shapes of triatomic molecules to their electronic structures. The basic approach is to calculate, or estimate, the energies of molecular orbitals for two limiting structures, say, linear and bent (to 90°) for an AB_2 molecule, and draw a diagram showing how the orbitals of one configuration correlate with those of the other. Then, depending on which orbitals are occupied, one or the other structure can be seen to be preferred. By means of approximate MO theory, implemented by digital computers, this approach has been extended and generalized in recent years.[4,5]

Triatomic Molecules.   This case is elaborated in some detail to expound the method. Other cases then are treated more summarily. The coordinate system for the AB_2 molecule is shown in Fig. 5-4.

The AB_2 molecule has C_{2v} symmetry when it is bent and, when linear, $D_{\infty h}$ symmetry. To simplify notation, however, the linear configuration is considered

[4a] R. M. Gavin, Jr., *J. Chem. Educ.,* 1969, **46,** 413.

[4b] B. M. Gimarc, *J. Am. Chem. Soc.,* 1970, **92,** 266; 1971, **93,** 593, 815.

[5]   This approach is also widely appplicable to transition metal compounds as well as to those of main group elements. The extended Hückel theory (EHT) provides a convenient method of calculating the energies, as shown by R. Hoffmann and collaborators in many recent papers: see e.g., M. Elian and R. Hoffmann, *Inorg. Chem.,* 1975, **14,** 1058.

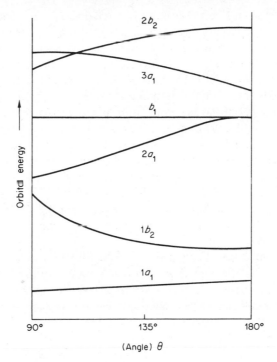

Fig. 5-5. Orbital correlation diagram for AB_2 triatomic molecules, where A uses only s and p orbitals.

to be simply an extremum of the C_{2v} symmetry; therefore the labels given to the orbitals through the range $90° \leqslant \theta < 180°$ are retained even when $\theta = 180°$. The symbols used to label the orbitals are derived from the orbital symmetry properties in a systematic way, but a detailed explanation is not given here.[6] For present purposes, these designations may be treated simply as labels.

The A atom of a AB_2 molecule will be assumed to have only s, p_x, p_y, and p_z orbitals in its valence shell, whereas each of the B atoms is allowed only a single orbital oriented to form a σ bond to A. In the linear configuration p_x^A and p_z^A are equivalent nonbonding orbitals labeled $2a_1$ and b_1, respectively. The orbitals s^A and p_y^A interact with σ_1^B and σ_2^B, the σ orbitals on the B atoms, to form one very strongly bonding orbital, $1a_1$, one less strongly bonding orbital, $1b_2$, and two antibonding σ orbitals, $3a_1$ and $2b_2$. The ordering of these orbitals and, in more detail, the approximate values of their energies can be estimated by an MO calculation. Similarly, for the bent molecule the MO energies may be estimated. Here only p_z^A is nonbonding; spacings and even the order of the other orbitals is a function of the angle of bending θ. The complete pattern of orbital energies, over a range of θ, as obtained with typical input parameters is shown in Fig. 5-5. Calculations in the Hückel approximation are simple to perform and give the correct general features

---

[6] For an explanation of the symbols, consult any introductory book on chemical or physical applications of group theory, e.g., F. A. Cotton, *Chemical Applications of Group Theory,* 2nd ed., Wiley-Interscience, 1971, p. 86, or H. H. Jaffé and M. Orchin, *Symmetry in Chemistry,* Wiley, 1965, p. 61.

of the diagram[4a] but for certain cases (e.g., AB_2E_2, as noted below) very exact computations are needed for an unambiguous prediction of structure.

From the approximate diagram it is seen that an AB_2 molecule (one with no lone pairs) is more stable when linear than when bent. The $1b_2$ orbital drops steadily in energy from $\theta = 90°$ to $\theta = 180°$, while the energy of the $1a_1$ orbital is fairly insensitive to angle. From an AB_2E molecule the results are ambiguous, because the trend in the energy of the $2a_1$ orbital approximately offsets that of the $1b_2$ orbital. The correct result may be a sensitive function of the nature of A and B. Important examples of AB_2E molecules are carbenes (Chap. 11) CR_2, whose structures vary with changes in R. For AB_2E_2 molecules, the result should be the same as for AB_2E, since the energy of the b_1 orbital is independent (in this rough approximation) of the angle. Thus it is not clear in this approach that AB_2E_2 molecules should *necessarily* be bent, but all known ones are. For AB_2E_3 molecules the behavior of the $3a_1$ orbital would clearly favor a linear structure, and this is in accord with all known facts.

Tetraatomic Molecules. For AB_3E molecules with a C_3 axis of symmetry, very accurate calculations of energy as a function of angle have been made because of the relevance of this to the problem of barriers to inversion in pyramidal molecules. In the best of these calculations it is found that the relative energies of the pyramidal and the planar configurations are such that the pyramidal configuration is always the more stable, but the energy difference varies considerably (from $\sim$25 kJ mol^{-1} for NH_3 to more than 150 kJ mol^{-1} for some AsX_3 and SbX_3 molecules). Also, interelectronic and internuclear repulsions as well as bond energies play an important role in determining the configurations.[7]

For molecules with more than an octet of valence shell electrons on the central atom, the employment of correlation diagrams has been less systematic. Instead, some individual cases have been treated to see what distortions from assumed idealized geometries might be expected. For example, the T-shaped ClF_3 molecule has been treated as indicated in Fig. 5-6, where the results suggest that the angle ϕ should be $\sim$10° less than 90°, in semiquantitative agreement with observation.

5-6.   Some Qualitative Failures of the Simple Theories

The Alkaline Earth Halides.   For these XMX molecules all the foregoing simple schemes make incorrect predictions. They are two-electron-pair cases in the VSEPR model, or, alternatively, they use sp hybridization; in either case linearity is expected. The same prediction is made by the correlation diagram in Fig. 5-5. If only p orbitals are used by the central atom, severely bent molecules are expected. If we consider the additional possibility that these molecules are essentially ionic, $X^-M^{2+}X^-$, linearity is again expected. In fact,[8] some are linear and some are not, as shown in Table 5-7.

[7]   Cf. A. Rauk, L. C. Allen, and K. Mislow, *Angew. Chem., Int. Ed.,* 1970, **9**, 400, for a summary and references.

[8]   (a) D. White *et al., J. Chem. Phys.,* 1973, **59**, 6645; (b) M. L. Leseicki and J. W. Nibler, *J. Chem. Phys.,* 1976, **64**, 871; (c) M. Guido and G. Gigli, *J. Chem. Phys.,* 1976, **65**, 1397.

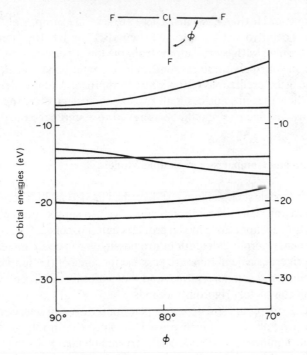

Fig. 5-6. Orbital energies (eV) as a function of angle for ClF_3 as calculated in the extended Hückel approximation. [Adapted from R. M. Gavin, Jr., Ref. 4a.]

The failure of the simple theories can be remedied in several ways, but it is worthwhile to stress that we have here an excellent reminder that the simple theories can never be trusted uncritically. The pattern of deviations from linearity suggests that either, or both, of two factors need to be invoked. If a basically ionic model is used, polarization of the central cation by the anions must be considered. Such polarization, which might account for bending, would be greatest with the largest cations and the smallest anions, which is consistent with the observed pattern of bending. Alternatively, it has been suggested that increasing participation of metal d orbitals as the metal atoms increase in atomic number could be added to a covalent picture of the bonding. It is by no means obvious, however, that this would need to occur in such a way as to cause bending.

TABLE 5-7
Structures of Alkaline Earth Molecules[a]

	F	Cl	Br	I
Be	1	1	1	1
Mg	? (~160°?)	1	1	1
Ca	b (~140°)	1	1	1
Sr	b (~108°)	b (~120°)	1	1
Ba	b (~100°)	b (~100°)	b	b

[a] 1 = linear; b = bent, with estimated angle, where available, in parentheses.

Halocomplexes of Heavier Elements. There are several examples of AB_6E species that are regular octahedra. Examples are the $SeBr_6^{2-}$ and $TeBr_6^{2-}$ ions, and a few other hexahalo anions with central atoms from the lower periods. These octahedral structures are, of course, *not* failures for the three-center bond model, but they are contrary to the usual predictions of the VSEPR approach and the directed valence theory. They show that in some cases all valence shell electrons cannot be treated equally, but instead one pair should be assigned to a stereochemically inactive *s* orbital.

5-7.   Criticism and Comparison of Simple Models

Why are simple models such as those just described used at all? Historically because truly rigorous calculations were impracticable, even on simple molecules like H_2O, and because a broadly applicable model provides unifying concepts that are valuable even if, and when, rigorous calculations are possible on specific molecules, one at a time. Today rigorous calculations are possible for some of the simpler molecules, but they are neither cheap nor routine, and molecules like PCl_5 and ClF_3 remain inaccessible by completely rigorous methods.

Simple models, therefore, still deserve a place in the art and science of chemistry, provided they are not misused or overused. Each of the simple models we have discussed has its limitations, and none can be trusted blindly. Each of them produces *mostly* correct predictions about structure; therefore they are useful for correlating information even if their rather extreme premises, hence their implications with regard to the actual electron distributions, are not to be taken literally. Let us now note a few of the main weaknesses of each.

The VSEPR Model.   This is in one sense the least sophisticated approach, making little explicit use of quantum mechanics. Its emphasis on interelectronic repulsions instead of on forming bonds is in sharp contrast to the other models in which repulsions are mentioned only a secondary stage to refine details of structure prediction. Two recent efforts to explore in detail the validity (as opposed to the empirical usefulness) of the VSEPR model, using H_2O and H_2S, have led to the conclusions[9,10] that "the theoretical basis of the VSEPR model is on weak ground," that "some of the assumptions of VSEPR theory are borne out by our calculations, but many are not," and that "a simple explanation of the equilibrium bond angles in H_2O and H_2S . . . remains to be given."

Although the detailed analyses are complicated and cannot be fully summarized here, two main considerations leading to these adverse judgments may be mentioned. First, we note that the repulsion between electron pairs that forms the basis for the VSEPR model consists of both the obvious electrostatic repulsion and the Pauli repulsion. The latter is a manifestation of the Pauli exclusion principle, which forbids electrons with the same spin to occupy the same spatial orbital. Pauli repulsion does not have the simple behavior of electrostatic repulsion. Thus the cal-

[9]   M. B. Hall, *Inorg. Chem.,* 1978, **17**, 2261.
[10]   W. E. Palke and B. Kirtman, *J. Am. Chem. Soc.,* 1978, **100**, 5717.

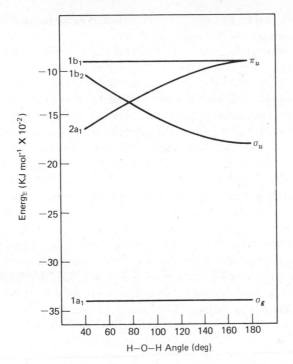

Fig. 5-7. A correlation diagram calculated by the extended Hückel method for H_2O. [Adapted from M. B. Hall, Ref. 9.]

culations for H_2O indicate that the Pauli repulsion is greatest between the bond pairs, rather than between the lone pairs as presumed in the VSEPR model.

Second, an inherent equivalence of the four electron pairs that is taken for granted in the VSEPR model is often unjustified. These electron pairs occupy the s and three p orbitals, as such or in some combinations called hybrids, and the presumed equivalence is more or less invalidated by any difference in the energies of these orbitals. In all many-electron atoms there must be a difference: the ns orbital is more stable than the np orbitals, and this difference increases rapidly across each period from lower to higher atomic number. For the oxygen atom, the $2s$ orbital is about 2300 kJ mol^{-1} more stable than the $2p$ orbitals. This means that there is a strong tendency for one electron pair to be mainly in the $2s$ orbital, hence *not* equivalent to the other three pairs.

The Walsh diagram (Fig. 5-5) pertains to a situation in which there is only a small energy difference between the ns and the np orbitals of the central atom, and, as stated in discussing that general-purpose diagram, it is not clear whether an AB_2E_2 molecule ought necessarily to be bent. In the diagram calculated expressly for H_2O (Fig. 5-7), the lowest level is practically pure $2s$ and its energy is essentially constant for all angles. It can be determined from this diagram that the energy is minimized at an angle of 106°, essentially in accord with the experimental value of 104.5°. Crudely speaking, the H_2O molecule has its characteristic angle *not* because it tends to have a tetrahedral angle (109.5°) that is then slightly reduced by lone-pair to lone-pair and lone-pair to bond-pair repulsions, but because the O—H bonds are formed mainly by oxygen $2p$ orbitals. The angle of 90°, which might be expected as a first approximation, is enlarged by Pauli repulsion between the bonding pairs, and perhaps by slight mixing of $2s$ character into the bonding orbitals.

The Hybridization or Directed Valence Model.   This model also is subject to certain basic criticisms despite its considerable success. For all cases in which there are more than four electron pairs in the valence shell of the central atom it is necessary to postulate that at least one d orbital becomes fully involved in the bonding. There are both experimental and theoretical reasons for believing that this is too drastic an assumption. Some recent MO calculations and other theoretical considerations suggest that although the valence shell d orbitals make a significant contribution to the bonding in many cases, they never play as full a part as do the valence shell p orbitals. Fairly direct experimental evidence in the form of nuclear quadrupole resonance studies of the ICl_2^- and ICl_4^- ions shows that in these species, d-orbital participation is very small; this participation is probably greater in species with more electronegative ligand atoms, such as PF_5, SF_6, and $Te(OH)_6$, but not of equal importance with the contributions of the s and p orbitals.

The Three-Center Bond Model.   The *total* neglect of both the ns orbital and all d orbitals on the central atom, which is the essential feature here, is obviously a "ruthless approximation." It is moderately successful because the ns orbital is, in many cases, so much more stable than the np orbitals that it plays only a slight role (cf. H_2O, just discussed) and because outer d orbitals are seldom more than partly involved in bond formation. However this model has only qualitative value. For example, in dealing with SF_4, it would imply that two of the S—F bonds are of the $2c$-$2e$ type and the other two are of the $3c$-$2e$ type. This corresponds to bond orders of 1.0 and 0.5; yet the bond lengths differ by only about 0.1 Å, albeit in the implied sense. Clearly, with such a small difference in bond lengths the bond orders cannot differ as greatly as this model would require. A related observation is that the "axial" fluorine atoms of SF_4 appear to be more "ionic" than the "equatorial" ones in the sense of having lower inner shell ionization potentials.[11] This, too, is qualitatively in accord with expectation for $3c$-$2e$ vs. $2c$-$2e$ bonds but quantitatively short of expectation.

5-8. $d\pi$-$p\pi$ Bonds

Although the role of central atom d orbitals in the formation of σ bonds to outer atoms has been a controversial subject, there is another role for d orbitals where their actual participation has been more generally accepted for some time, although here, too, the exact *extent* of that participation is subject to some differences of opinion.

While the heavier non-transition elements show little tendency to engage their p orbitals in π-bond formation, they do form at least partial π bonds to lighter el-

[11]   R. W. Shaw, Jr., T. X. Carroll, and T. D. Thomas, *J. Am. Chem. Soc.,* 1973, **95,** 5870.

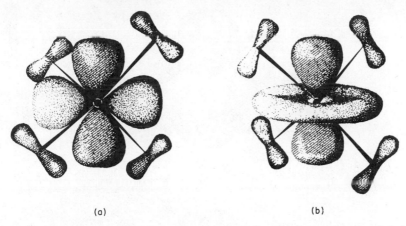

(a)   (b)

Fig. 5-8. Quasi-perspective view of the overlap of $p\pi$ orbitals of the B atoms in a tetrahedral AB_4 molecule with (a) the $d_{x^2-y^2}$ and (b) the d_{z^2} orbitals of the A atom.

ements, especially to oxygen and nitrogen, by using their outer d orbitals. The experimental indications of this are chiefly the high bond-stretching force constants and the shortness of bonds compared to the force constants and bond strengths to be expected for single bonds. Photoelectron·spectroscopy has provided evidence for such bonding even in H_3SiCl.

Tetrahedral Molecules. We first consider what possibilities exist in principle, that is, on the basis of compatibility of orbitals, for forming $d\pi$-$p\pi$ bonds. For a tetrahedral AB_4 molecule such as SiF_4 or PO_4^{3-}, each of the B atoms has two filled $p\pi$ orbitals perpendicular to the A—B bond axis and perpendicular to each other. The central atom A is assumed to use its s and p orbitals for σ bonding. A detailed examination of the suitability of the d orbitals of A for overlapping with the $p\pi$ orbitals on the B atoms shows that they all are able to do so, but two namely, d_{z^2} and $d_{x^2-y^2}$, are particularly well suited for this. Each of these two would be expected to have about $\sqrt{3}$ times as much overlap with the $p\pi$ orbitals of the four B atoms as one of the other three d orbitals. The principal $d\pi$-$p\pi$ overlap possibilities appear schematically in Fig. 5-8.

Bond length data in the series of ions SiO_4^{4-}, PO_4^{3-}, SO_4^{2-}, ClO_4^- indicate that such $p\pi$-$d\pi$ bonding actually does occur. As Table 5-8 indicates, the X—O bonds are all short relative to values reasonably expected for single bonds, and, moreover,

TABLE 5-8
Bond Lengths and $d\pi$-$p\pi$ Overlaps in XO_4^{n-} Ions

Ion	Observed X—O distance (Å)	Estimated X—O single-bond distance (Å)	Shortening	$p\pi$-$d\pi$ Overlap
SiO_4^{4-}	1.63	1.76	0.13	0.33
PO_4^{3-}	1.54	1.71	0.17	0.46
SO_4^{2-}	1.49	1.69	0.20	0.52
ClO_4^-	1.46	1.68	0.22	0.57

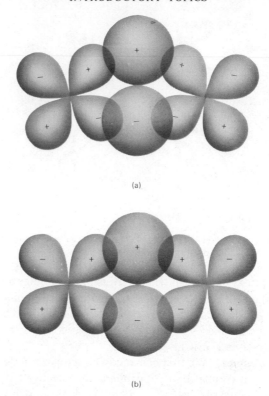

(a)

(b)

Fig. 5-9. Overlaps of a central $p\pi$ orbital with $d\pi$ orbitals on outer atoms for (a) bent and (b) linear configurations.

the trend in shortening is closely parallel to the calculated trends in $d\pi$-$p\pi$ overlaps.

Similarly, in SiF$_4$, even after due allowance for the effect of ionic-covalent resonance in strengthening, hence shortening the Si—F bonds, they appear to be around 0.13 Å shorter than the length expected for single bonds, implying that π bonding is present as well.

Other Molecules. In less symmetrical molecules the detailed analysis of π bonding is more difficult because the d orbitals of the central atom can interact with different types of outer atoms to different degrees. However by utilizing the idea that bond order and bond length are inversely related, it is possible to deduce approximate, relative degrees of π bonding in various compounds containing SiO, PO, SO, and ClO groups, as well as in other cases, to a more limited extent. For PO and SO bonds the data are most extensive. The various types of P—O bond vary in length from $\sim$1.68 to $\sim$1.40 Å, and it has been suggested that the last value, about the shortest observed distance for any SO, PO, SN, or PN bond, corresponds to about a double bond (i.e., to a π-bond order of about 1). However it is possible that this is an underestimate, since in SF$_3$N, where the S—N bond must be more nearly triple than double, the length is 1.42 Å.

For the molecules and ions SF_2O_2, PF_3O, ClO_3F, ClO_2^-, ClO_3^-, and ClO_4^-, *ab initio* MO calculations indicate significant *d*-orbital participation in both σ and π bonds.

In certain instances the existence of $d\pi$-$p\pi$ bonding is indicated by the overall molecular geometry. Thus the Si_3N and Ge_3N skeletons are planar in $(SiH_3)_3N$ and $(GeH_3)_3N$, and in $(Me_3Si)_2NBeN(SiMe_3)_2$ the $Si_2NBeNSi_2$ group has an allenelike configuration (Si_2NBe groups in two perpendicular planes). In each case the Si—N or Ge—N bonds are 0.05 to 0.15 Å shorter than expected for single bonds. These structural features can be accounted for by assuming that in the planar configuration about the N atom there is enough $N(2p_z) \rightarrow Si(3d)$ or $Ge(3d)$ π bonding to stabilize this configuration relative to the pyramidal one found in most other R_3N molecules.

Molecules containing the Si—O—Si unit[12] provide a particularly good illustration of the structural effects of $p\pi \rightarrow d\pi$ bonding, showing that it occurs to a significant extent without the configuration going completely linear. Thus in nine cases the Si—O distances are in the range 1.54–1.64 Å (all but two being 1.62–1.64 Å) and the Si—O—Si angles, when not constrained by belonging to a ring, are 145 $\pm$ 10°. These distances are all shorter by 0.20 to 0.25 Å than the expected Si—O single bond distance, but the $p\pi \rightarrow d\pi$ bonding that causes this shortening occurs without close approach to linearity. Figure 5-9 shows that the pertinent orbital overlap, though best at an angle of 180°, can be quite significant even at smaller angles.

[12] B. Csakvari et al., *J. Organomet. Chem.*, 1976, **107**, 287.

General References

Bartell, L. S., *J. Chem. Educ.*, 1968, **45**, 754. Contrasts different models.

Coulson, C. A., *J. Chem. Soc.*, **1964**, 1442. Specifically concerned with noble gas compounds, but the bonding theory pertains equally to many other systems, including all halogen compounds with more than four pairs of valence shell electrons.

Cruickshank, D. W. J., *J. Chem. Soc.*, **1961**, 5486. Broad discussion of $p\pi \rightarrow d\pi$ bonding; early but still valuable.

Fergusen, J. E., *Stereochemistry and Bonding in Inorganic Chemistry*, Prentice-Hall, 1974.

Malm, J. G., H. Selig, J. Jortner, and S. Rice, *Chem. Rev.*, 1965, **65**, 199. Theory of bonding as discussed specifically for Xe compounds has wide applicability.

Mitchell, K. A. R., *Chem. Rev.*, 1969, **69**, 157. Use of outer *d* orbitals in bonding.

Pettit, L. D., *Q. Rev.*, 1971, **25**, 1. Multiple bonding in inorganic compounds.

2

CHEMISTRY OF THE MAIN GROUP ELEMENTS

CHAPTER SIX

Hydrogen

GENERAL REMARKS

6-1. Introduction

Three isotopes of hydrogen are known: ^1H, ^2H (deuterium or D), and ^3H (tritium or T). Although isotope effects are greatest for hydrogen, justifying the use of distinctive names for the two heavier isotopes, the chemical properties of H, D, and T are essentially identical except in matters such as rates and equilibrium constants of reactions; in addition, diverse methods of isotope separation are known.[1] The normal form of the element is the diatomic molecule; the various possibilities are H_2, D_2, T_2, HD, HT, DT.*

Naturally occurring hydrogen contains 0.0156% deuterium, whereas tritium occurs naturally in only minute amounts, believed to be of the order of 1 in 10^{17}.

Tritium[2] is formed continuously in the upper atomsphere in nuclear reactions induced by cosmic rays. For example, fast neutrons arising from cosmic ray reactions can produce tritium by the reaction $^{14}N(n, {}^3H)^{12}C$. Tritium is radioactive (β^-, 12.4 years) and is believed to be the main source of the minute traces of ^3He found in the atmosphere. It can be made artificially in nuclear reactors, for example, by the thermal neutron reaction $^6Li(n,\alpha)^3H$, and is available for use as a tracer in studies of reaction mechanisms.

Deuterium[3] as D_2O is separated from water by fractional distillation or elec-

---

1 H. K. Rae, Ed., *Separation of Hydrogen Isotopes,* ACS Symposium Series, No. 68, 1978.

2 E. A. Evans, *Tritium and Its Compounds,* 2nd ed., Halstead-Wiley, 1974; E. Buncel and C. C. Lee, *Tritium in Organic Chemistry,* Elsevier, 1978.

3 G. Vasaru, D. Ursu, A. Mihaila, and P. Szentgyorgi, *Deuterium and Heavy Water, A Selected Bibliography, 1932–1974,* Elsevier, 1975.

* Molecular H_2 (and D_2) have *ortho* and *para* forms in which the nuclear spins are aligned or opposed, respectively. This leads to very slight differences in bulk physical properties, and the forms can be separated by gas chromatography.

trolysis and by utilization of very small differences in the free energies of the H and D forms of different compounds, the H_2O–H_2S system being particularly favorable in large-scale use:

$$HOH(l) + HSD(g) = HOD(l) + HSH(g) \qquad K \approx 1.01$$

Deuterium oxide is available in ton quantities and is used as a moderator in nuclear reactors, both because it is effective in reducing the energies of fast fission neutrons to thermal energies and because deuterium has a much lower capture cross section for neutrons than has hydrogen, hence does not reduce appreciably the neutron flux. Deuterium is widely used in the study of reaction mechanisms and in spectroscopic studies.

Although the abundance on earth of molecular hydrogen is trivial, hydrogen in its compounds has one of the highest of abundances. Hydrogen compounds of all the elements other than the noble gases are known, and many of these are of transcendental importance. Water is the most important hydrogen compound; others of great significance are hydrocarbons, carbohydrates and other organic compounds, ammonia and its derivatives, sulfuric acid, and sodium hydroxide. Hydrogen forms more compounds than any other element.

Molecular hydrogen[4] is a colorless, odorless gas (f.p. 20.28°K), virtually insoluble in water. It is most easily prepared on a small scale by the action of dilute acids on metals such as zinc or iron and by electrolysis of water.

Industrially hydrogen is used mainly as a feed stock in synthesis of ammonia and other bulk chemicals and is produced for this purpose by the following reactions (where CH_4 is used to represent natural gas, not necessarily pure methane):

$$CH_4 + H_2O \rightarrow CO + 3H_2$$
$$CO + H_2O \rightarrow CO_2 + H_2$$

The second reaction is called the water gas shift reaction, and catalysts are used to accelerate it. Essentially pure H_2 may be obtained by scrubbing out the CO_2. (See Chapter 30 for discussion of CO/H_2 mixtures, called synthesis gas.) Heavier hydrocarbons are also used to some extent. Although little used at present, reaction of coal with H_2O seems likely to become more important in the future:

$$C + H_2O \rightarrow CO + H_2$$
$$CO + H_2O \rightarrow CO_2 + H_2$$

Direct commercial production of hydrogen from water, "water splitting," which is possible only by electrolysis, is currently uneconomic because the electricity is itself generated by burning fossil fuels. Considerable research is going into the possibility of water splitting using thermal energy[5] or light,[6] but no such process is yet practical.

4   K. E. Cox and K. D. Williamson, Eds., *Hydrogen: Its Technology and Implications*, CRC Press, 1976.

5   F. Schreiner and E. H. Appelman, *Surv. Prog. Chem.*, 1977, **8**, 171; C. E. Bamberger, J. Braunstein, and D. M. Richardson, *J. Chem. Educ.*, 1978, **55**, 561.

6   S. M. Kuznicki and E. M. Eyring, *J. Am. Chem. Soc.*, 1978, **100**, 6790; K. Kalyanasundaram, *Nouv. J. Chem.*, 1979, **3**, 511.

Hydrogen is not exceptionally reactive. It burns in air to form water and will react with oxygen and the halogens explosively under certain conditions. At high temperatures the gas will reduce many oxides either to lower oxides or to the metal. In the presence of suitable catalysts and above room temperature it reacts with N_2 to form NH_3. With electropositive metals and most nonmetals it forms hydrides.

A number of reactions of hydrogen and organic molecules are catalyzed by transition metals; these processes are covered in Chapter 30. For example, in the presence of suitable catalysts, usually Group VIII metals or their compounds, a great variety of both inorganic and organic substances can be reduced. Heterogeneous hydrogenation may be carried out in gas phase or in solution, and a number of transition metal ions and complexes can react with hydrogen, transferring it to a substrate homogeneously in solution.

The dissociation of hydrogen is highly endothermic, and this accounts in part for its rather low reactivity at low temperatures:

$$H_2 = 2H \qquad \Delta H_0^0 = 434.1 \text{ kJ mol}^{-1}$$

In its low-temperature reactions with transition metal species heterolytic splitting may occur to give H^-, bound to the metal, and H^+; the energy involved is much lower, probably ~ 125 kJ mol^{-1}. At high temperature, in arcs at high current density, in discharge tubes at low hydrogen pressure, or by ultraviolet irradiation of hydrogen, atomic hydrogen can be produced. It has a short half-life (~ 0.3 sec). The heat of recombination is sufficient to produce exceedingly high temperatures, and atomic hydrogen has been used for welding metals. Atomic hydrogen is exceedingly reactive chemically, being a strong reducing agent.

6-2. The Bonding of Hydrogen

The chemistry of hydrogen depends mainly on three electronic processes:

1. *Loss of the Valence Electron.* The 1s valence electron may be lost to give the hydrogen ion H^+, which is merely the proton. Its small size ($r \sim 1.5 \times 10^{-13}$ cm) relative to atomic sizes ($r \sim 10^{-8}$ cm) and its small charge result in a unique ability to distort the electron cloud surrounding other atoms; the proton accordingly never exists, as such, except in gaseous ion beams; in condensed phases it is invariably associated with other atoms or molecules.

2. *Acquisition of an Electron.* The hydrogen atom can acquire an electron, attaining the $1s^2$ structure of He, to form the hydride ion H^-. This ion exists as such essentially only in the saline hydrides formed by the most electropositive metals (Section 6-14).

3. *Formation of an Electron-Pair Bond.* The majority of hydrogen compounds contain an electron-pair bond. The number of carbon compounds of hydrogen is legion, and most of the less metallic elements form numerous hydrogen derivatives. Many of these are gases or liquids.

The chemistry of many of these compounds is highly dependent on the nature of the element (or the element plus its other ligands) to which hydrogen is bound.

Particularly dependent is the degree to which compounds undergo dissociation in polar solvents and act as acids:

$$HX \rightleftharpoons H^+ + X^-$$

Also important for chemical behavior is the electronic structure and coordination number of the molecule as a whole. This is readily appreciated by considering the covalent hydrides BH_3, CH_4, NH_3, OH_2, and FH. The first not only dimerizes (see below) but is a Lewis acid, methane is chemically unreactive and neutral, ammonia has a lone pair and is a base, water can act as a base or as a very weak acid, and FH is appreciably acidic in water.

Except in H_2 itself, where the bond is homopolar, all other H—X bonds possess polar character to some extent. The dipole may be oriented either way, and important chemical differences arise accordingly. Although the term "hydride" might be considered appropriate only for compounds with H negative, many compounds that act as acids in polar solvents are properly termed covalent hydrides. Thus although HCl and $HCo(CO)_4$ behave as strong acids in aqueous solution, they are gases at room temperature and are undissociated in nonpolar solvents.

4. *Unique Bonding Features.* The nature of the proton and the complete absence of any shielding of the nuclear charge by electron shells allow other forms of chemical activity that are either unique to hydrogen or particularly characteristic of it. Some of these are the following, which are discussed in some detail subsequently:

(*a*)   The formation of numerous compounds, often nonstoichiometric, with metallic elements. They are generally called *hydrides* but cannot be regarded as simple saline hydrides (Section 6-16).

(*b*)   *Formation of hydrogen bridge bonds* in electron-deficient compounds such as in 6-I or transition metal complexes as in 6-II.

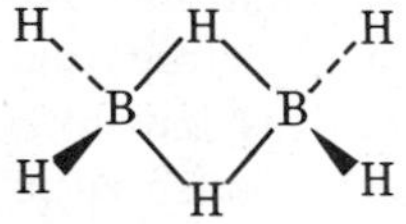

(6-I) 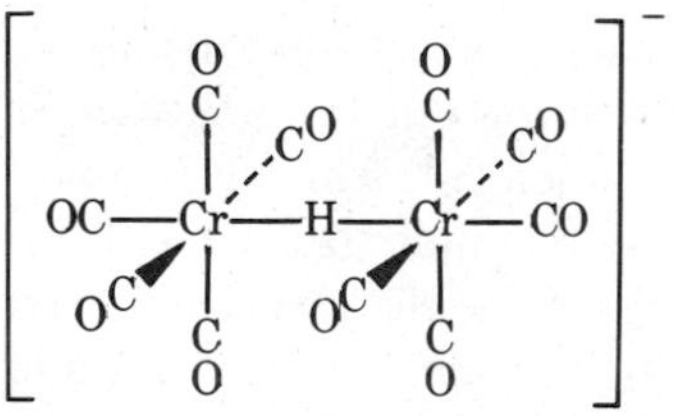   (6-II)

The bridge bonds in boranes and related compounds are discussed in Chapter 9; the transition metal compounds are described in detail in Section 25-6 and Chapter 27.

(*c*)   *The Hydrogen Bond.* This bond is important not only because it is essential to an understanding of much other hydrogen chemistry but also because it is one of the most intensively studied examples of intermolecular attraction. Hydrogen bonds dominate the chemistry of water, aqueous solutions, hydroxylic solvents, and OH-containing species generally, and they are of crucial importance in biological systems, being responsible *inter alia* for the linking of polypeptide chains in proteins and the base pairs in nucleic acids.

THE HYDROGEN BOND,
HYDRATES, HYDROGEN ION, AND ACIDS

6-3. The Hydrogen Bond[7]

"Hydrogen bond" is the term given to the relatively weak secondary interaction between a hydrogen atom bound to an electronegative atom and another atom that is also generally electronegative and has one or more lone pairs enabling it to act as a base. We can thus refer to proton donors XH and proton acceptors Y and can give the following generalized representation of a hydrogen bond.

$$\overset{\delta-}{X}-\overset{\delta+}{H}\cdots\cdots\overset{}{Y}$$

Such interaction is strongest when both X and Y are first-row elements; the main proton donors are N—H, O—H and F—H, and the most commonly encountered hydrogen bonds are O—H$\cdots$O and N—H$\cdots$O. The groups P—H, S—H, Cl—H, and Br—H can also act as proton donors, and so even can C—H, provided the C—H bond is relatively polar as it is when the carbon is bound to electronegative groups as in $CHCl_3$ or the carbon atom is in an sp-hybridized state as in HCN or RC$\equiv$CH. The acceptor atoms can be N, O, F, Cl, Br, I, S, or P, but carbon never acts as an acceptor other than in certain π systems noted below.

Much of the earlier experimental evidence for hydrogen bonding came from comparisons of the physical properties of hydrogen compounds. Classic examples are the apparently abnormally high boiling points of NH_3, H_2O, and HF (Fig. 6-1), which imply association of these molecules in the liquid phase. Other properties such as heats of vaporization provided further evidence for association. Physical properties reflecting association are still useful in detecting hydrogen bonding, but the most satisfactory evidence comes from X-ray and neutron diffraction crystallographic studies and from infrared and nuclear magnetic resonance spectra.

[7] (a) W. C. Hamilton and J. A. Ibers, *Hydrogen Bonding in Solids,* Benjamin, 1968; (b) G. C. Pimentel and A. L. McClellan, *The Hydrogen Bond,* Freeman, 1960; *Ann. Rev. Phys. Chem.,* 1971, **22,** 347; (c) A. K. Covington and P. Jones, Eds., *Hydrogen-Bonded Solvent Systems,* Taylor and Francis, 1968 (symposium report on various topics); (d) S. N. Vinogradov and R. H. Linnel, *Hydrogen Bonding,* Van Nostrand-Reinhold, 1971; (e) J. C. Speakman, *Struct. Bonding,* 1972, **12,** 141 (very short hydrogen bonds); (f) A. Novak, *Struct. Bonding,* 1974, **18,** 177 (crystal and spectral data); (g) P. A. Kollman and L. C. Allen, *Chem. Rev.,* 1972, **72,** 283 (theory of the hydrogen bond); (h) G. E. Bacon, *Neutron Scattering in Chemistry,* Butterworths, 1977 (Chapter 4 deals with structural studies including hydrogen bonds and oxonium ions); (i) R. P. Bell, *The Proton in Chemistry,* 2nd ed., Chapman & Hall, 1973; (j) M. D. Joesten and L. J. Schaad, *Hydrogen Bonding,* Dekker, 1974; (k) P. Shuster, P. G. Zundel, and C. Sandorfy, Eds., *The Hydrogen Bond: Recent Developments in Theory and Experiments,* 3 vols. North Holland, 1976; (l) R. D. Green, *Hydrogen Bonding by C-H Groups,* Macmillan, London, Wiley, New York, 1974.

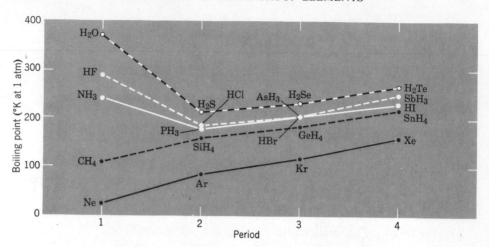

Fig. 6-1. Boiling points of some molecular hydrides.

Although H atoms are often observable in X-ray studies, their positions seldom can be ascertained with any accuracy. However neutron diffraction data can usually give quite precise locations, because the scattering of neutrons of thermal energies is roughly similar for all nuclei, regardless of atomic number, whereas the scattering of X-rays depends on electron density and is lowest for hydrogen. Even if accurate location of hydrogen atoms is not possible, the overall X—Y distance is significant. If X—Y distances are significantly shorter than normal van der Waals contacts for nonbonded atoms by, say, 0.2 Å, we can be fairly certain of the presence of a hydrogen bond. For O—H···O, distances below approximately 3 Å indicate hydrogen bonding. Thus in crystalline $NaHCO_3$ there are four kinds of O—O distance (between O's of different HCO_3^- ions) having values of 2.61, 3.12, 3.15, and 3.19 Å. The last three correspond to van der Waals contacts, but the first, 2.61 Å, corresponds to the H-bonded pair O—H···O.

The $NaHCO_3$ structure also serves to illustrate the results of a typical neutron diffraction study. As shown in 6-III, the HCO_3^- ions form infinite chains and the neutron results reveal the location of the proton. This example has two features that are entirely characteristic of most hydrogen bonds: (1) the X—H···Y chain is nearly (occasionally exactly) linear, and (2) the bond is unsymmetrical. The X—H distance is only a little (ca. 0.1 Å) longer than for the isolated X—H bond, but the H···Y distance is far greater than the normal H—Y bond distance.

(6-III)

TABLE 6-1
Some Parameters of Hydrogen Bonds

Bond	Compound	Bond energy (kJ mol^{-1})	Depression of stretching frequency (cm^{-1})	Bond length^a (Å)	X—H distance (Å)
F—H—F	KHF$_2{}^b$	~150	~2450	2.292(4)	1.145
F—H···F	HF(g)	~28.6	700	2.55	
O—H···O	(HCOOH)$_2$	29.8	~460	2.67	
O—H···O	H$_2$O(s)	~21	~430	2.76	0.97
O—H···O	B(OH)$_3$			2.74	1.03
N—H···N	Melamine	~25	~120	3.00	
N—H···Cl	N$_2$H$_5$Cl		~460	3.12	
C—H···N	(HCN)$_n$		180	3.2	1.0

^{a} The distance between the hydrogen-bonded atoms X and Y.
^{b} H. L. Carell and J. Donohue *Isr. J. Chem.*, 1972, **10**, 195.

X-Ray and neutron diffraction are, of course, very powerful methods, but they are applicable only to crystalline solids. Spectroscopic methods are therefore valuable. Infrared and Raman spectra can detect hydrogen-bond formation because when an X—H bond participates in H-bond formation three main changes occur: (1) ν_{X-H} decreases, (2) the ν_{X-H} band becomes broader and its intensity increases, (3) the X—H wagging mode(s) increase in frequency. Proton nmr spectra give evidence of H-bond formation by showing a shift (usually to lower fields) of the proton involved.

The strengths of hydrogen bonds may be defined as the enthalpy of the process:

$$X—H \cdots Y \rightarrow X—H + Y$$

They vary from approximately zero when the overall X-to-Y distance is so long that it is about equal to the van der Waals contact distance for X—H with Y, to rather high values (>100 kJ mol^{-1}) for a few very short bonds. In general, such energies are rarely measurable and estimates of the relative strengths of H-bonds are based on the magnitude of infrared shifts and the shortness of the X-to-Y distances. Table 6-1 lists a few examples of hydrogen bonds and illustrates the ranges of certain properties.

The O—H···O bonds, which are the most common and have been most extensively studied, range from about 3.0 to 2.31 Å in length. Those from 3.0 to 2.8 Å are considered long and are relatively weak. The majority are in the range 2.8–2.6 Å and probably have energies in the range 15–40 kJ mol^{-1}. These are all unsymmetrical, like that shown in 6-III.

As hydrogen bonds become extremely short and strong (<2.6 Å for O—H···O bonds), they become more *likely* to be symmetrical, but recent neutron work has

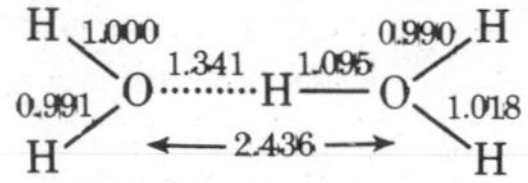

$$H \underset{0.991}{\overset{1.000}{\diagdown}} O \underset{}{\overset{1.341}{\cdots\cdots\cdots}} H \underset{}{\overset{1.095}{\text{———}}} O \underset{1.018}{\overset{0.990}{\diagup}} H$$

$$H \diagup \quad \longleftarrow 2.436 \longrightarrow \quad \diagdown H$$

Fig. 6-2. The $H_5O_2^+$ ion in sulfosalicylic acid trihydrate, based on results cited in detail in Bacon.[7h] Distances in Angstroms.

refuted the idea that they necessarily become symmetrical below some critical $X \cdots Y$ separation. It is now virtually certain that there is no unique correlation between the $X \cdots Y$ and $X—H$ distances in any $X—H \cdots Y$ system; the complete and detailed environment must be taken into consideration. For example, in sulfosalicylic acid trihydrate, which has a bond that is very short, the bond is also very unsymmetrical, as shown in Fig. 6-2. However in $HBr \cdot 2H_2O$ there is a similar $H_2O \cdots H \cdots OH_2^+$ unit with a very similar $O \cdots O$ distances (2.40 Å), but the proton is very nearly centered, the two $O \cdots H$ distances differing by only 0.05 Å.[8] There are a few cases in which very short $O \cdots H \cdots O$ bonds are exactly symmetrical within experimental uncertainty.

It has long been known conclusively that the $F \cdots H \cdots F$ bond in KHF_2, which is extremely short (2.26 Å), is symmetrical. This led to the view that any bond as short and strong as this one would surely have to be symmetrical. However in $p\text{-}CH_3C_6H_4NH_3^+ HF_2^-$ it is found[9] that the bond is decidedly unsymmetrical ($H \cdots F$ distances of 1.025 and 1.235 Å), even though it is just as short as that in KHF_2. In the p-toluidine case the FHF^- ion is in a very unsymmetrical environment of hydrogen bonding near neighbors.

The shortest $O \cdots H \cdots O$ distances ever reported (2.31–2.33 Å)[10] occur in compound 6-IV.

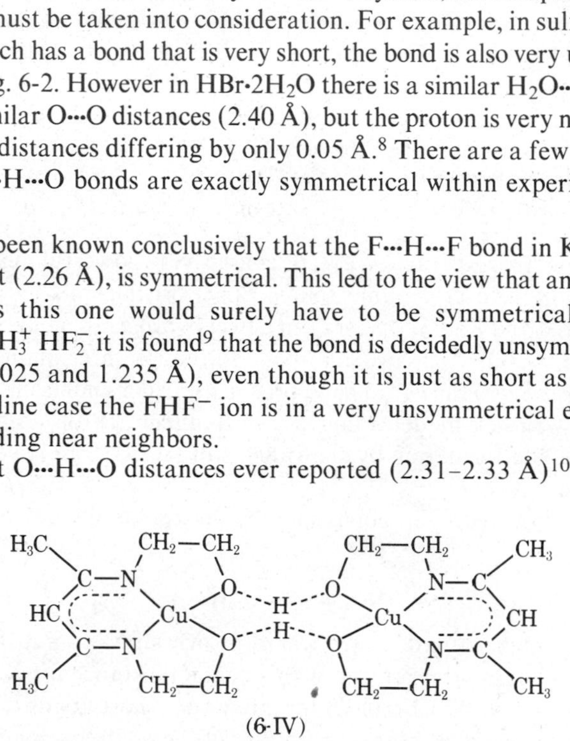

(6-IV)

The locations of the protons have not yet been fixed accurately by neutron diffraction. X-Ray work suggests that they are about 1.1 Å from one O atom and 1.3 Å from the other and somewhat off the $O \cdots O$ line, with $O—H—O$ angles of about 150°. Another extraordinary feature of this compound is that it shows an extremely large antiferromagnetic coupling between the Cu^{2+} ions (-94 cm^{-1}) even over a distance of $ca.$ 5 Å. This coupling must be effected by a superexchange mechanism through the hydrogen bonds.

8   R. Attig and J. M. Williams, $Angew. Chem., Int. Ed.,$ 1976, **15**, 491.
9   J. M. Williams and L. F. Schneemeyer, $J. Am. Chem. Soc.,$ 1973, **95**, 5780.
10   J. A. Bertrand $et al., Inorg. Chem.,$ 1976, **15**, 2965.

Finally, we note that though there are several authentic cases of bifurcated H bonds (6-V), this phenomenon is extremely rare, presumably because it is not

$$X-H \begin{matrix} \diagup Y \\ \diagdown Y' \end{matrix}$$

(6-V)

possible for both Y and Y' to simultaneously make good contact with the same X—H proton. In earlier X-ray work the hydrogen dinitrate ion $NO_3^- \cdots H^+ \cdots NO_3^-$ had been observed, and it was tentatively suggested that the proton might be at the center of a roughly tetrahedral arrangement of oxygen atoms. A definitive neutron study[11] shows only a conventional $O \cdots H \cdots O$ system, however, with overall distance of 2.47 Å and the proton centered, though slightly off line.

Theory of H-Bonds.[12]   For the majority of H-bonds that are unsymmetrical and have medium to long overall lengths, the bonding force is mainly electrostatic. In short the proton in a polar $X^{\delta-}$—$H^{\delta+}$ bond is attracted to the negative and/or polarizable atom Y. Given that the bond is essentially electrostatic, another question arises. If unshared electron pairs are concentrated along the direction of hybrid orbitals, will the proton approach the atom Y preferentially along these directions? In other words, does the proton see the atom Y as a structureless concentration of negative charge or as an atomic dipole? The answer to this question is not entirely clear-cut because in most cases where the angle θ in 6-VI is in accord with the latter

$$\begin{matrix} Y\text{-}\llcorner\text{-}H\text{---}X \\ \diagup \theta \\ Z \end{matrix}$$

(6-VI)

idea, it is possible to attribute this to steric requirements, as in carboxylic acid dimers or o-nitrophenol, or it can be equally well explained on the simpler theory as in the case of HCN polymers that are linear. However the case of the $(HF)_n$ polymer (Fig. 6-3) and a few others seem to lend strong support to the hypothesis of preferred directions, since there appears to be no other reason for the structure not to be linear.

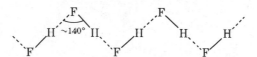

Fig. 6-3. The structure of crystalline hydrogen fluoride.

For the shortest H bonds covalent forces doubtless come into play. Since it is not possible for the hydrogen atom to form two electron-pair bonds simultaneously,

[11]   J. Roziere, M. Roziere-Boris, and J. M. Williams, *Inorg. Chem.,* 1976, **15,** 2490.
[12]   L. C. Allen, *J. Am. Chem. Soc.,* 1975, **97,** 6921.

this must be formulated in terms of a three-center, two-electron bond picture.

6-4. Ice and Water[13]

The structural natures of ice and, *a fortiori,* of water are very complex matters that can be treated but briefly here.

There are nine known modifications of ice, each stable over a certain range of temperature and pressure. Ordinary ice, ice I, which forms from liquid water at 0°C and 1 atm, has a rather open structure built of puckered, six-membered rings (Fig. 6-4). Each H_2O is tetrahedrally surrounded by the oxygen atoms of four neighboring molecules, and the whole array is linked by unsymmetrical hydrogen bonds. The O—H···O distance is 2.75 Å (at 100°K), and the H atoms lie 1.01 Å from one oxygen and 1.74 Å from the other. Each oxygen atom has two near and two far hydrogen atoms, but there are six distinct arrangements, two being illustrated in Fig. 6-5, all equally probable. However the existing arrangement at any one oxygen eliminates certain of these at its neighbors. A rigorous analysis of the probability of any given arrangement in an entire crystal leads to the conclusion that at the absolute zero ice I should have a disordered structure with a zero-point entropy of 3.4 J mol^{-1} deg^{-1}, in excellent agreement with experiment. This result in itself constitutes a good proof that the hydrogen bonds are unsymmetrical; if they

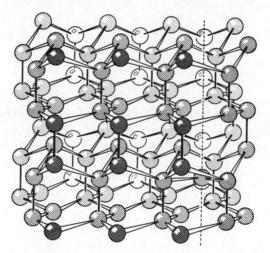

Fig. 6-4. The structure of ice I. Only the oxygen atoms are shown.

[13]   R. A. Horne, *Survey of Progress in Chemistry,* Vol. 4, Academic Press, 1968 (a good review); D. Eisenberg and W. Kautzman, *The Structure and Properties of Water,* Oxford University Press, 1968; A. H. Narten and H. A. Levy, *Science,* 1969; **165,** 447; N. H. Fletcher, *The Chemical Physics of Ice,* Cambridge University Press, 1970; F. Franks, Ed., *Water: A Comprehensive Treatise,* Plenum Press, 1972, *et seq.;* R. A. Horne, Ed., *Water and Aqueous Solutions: Structure Thermodynamics and Transport Processes,* Wiley, 1972; P. Krindel and I. Eliezer, *Coord. Chem. Rev.,* 1971, **3,** 217 (water structure models).

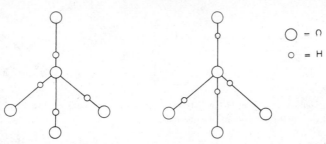

Fig. 6-5. Two possible configurations about an oxygen atom in ice.

were symmetrical, there would be a unique, ordered structure, hence no zero-point entropy. Entirely similar considerations confirm the presence of a network of unsymmetrical H bonds in KH_2PO_4 and $Ag_2H_3IO_6$, whereas the absence of zero-point entropy in $K[FHF]$ accords with the symmetrical structure of the FHF^- ion.

The structure of liquid water is still controversial: it is not random, as in liquids consisting of more or less spherical nonpolar molecules. Instead, water is highly structured owing to the persistence of hydrogen bonds; even at 90°C only a few percent of the water molecules appear not to be hydrogen-bonded. Still, there is considerable disorder, or randomness, as befits a liquid.

In an attractive, though not universally accepted, model of liquid water the liquid consists at any instant of an imperfect network, very similar to the network of ice I, but differing in that (a) some interstices contain water molecules that do not belong to the network but instead disturb it, (b) the network is patchy and does not extend over long distances without breaks, (c) the short-range ordered regions are constantly disintegrating and re-forming (they are "flickering clusters"), and (d) the network is slightly expanded compared to ice I. The fact that water has a slightly higher density than ice I may be attributed to the presence of enough interstitial water molecules to more than offset the expansion and disordering of the ice I network. This model of water receives strong support from X-ray scattering studies.

It may be noted that alcohols, which are similar to water in many respects, cannot form three-dimensional arrays, hence only linear or cyclic polymers exist.

6-5. Hydrates[7a,14]

Crystalline hydrates, especially of metal ions, are of enormous importance in inorganic chemistry, but hydrates of organic substances, especially those with N—H and O—H bonds are also common. For metal ions, the negative end of the water dipole (i.e. oxygen) is always bound to the metal, and the lone pairs on it can be directed toward the metal and involved in the bonding; they can, however, also act as acceptors for H bonds, and H atoms of coordinated water are normally hydrogen bonded. Thus there is an extreme flexibility in hydrogen bonding, and this allows stabilization in lattices of many different types of hydrate structure.

[14] H. Veillard, *J. Am. Chem. Soc.*, 1977, **99**, 7194.

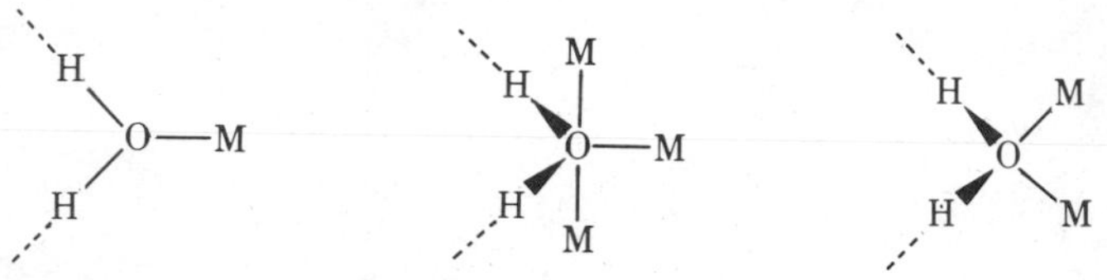

Fig. 6-6. Idealized structures for H-bonded coordinated water in hydrates. The metal M can also be replaced by HY in hydrates of organic compounds.

Some of the main structures are given in Fig. 6-6. It must be particularly noted that *gross distortions* from the angles of the ideal models shown are the *rule* and also that large deviations from linearity of the H bonds also occur.

6-6. Clathrates, Gas Hydrates, and Other Enclosure Compounds[15]

There are certain substances formed by combining one stable compound with another or with an atomic or molecular element without the existence of any chemical bonds between the two components. This occurs when one of the compounds can crystallize in a very open structure containing cavities, holes, or channels in which atoms or molecules of the other can be trapped. Those compounds in which the *host* lattice contains cavities like cages are the most important type; they are called *clathrate* compounds, from the Latin *clathratus,* meaning "enclosed or protected by crossbars or grating." Most of these, and the important ones certainly, involve hydrogen bonding.

One of the first clathrate systems to be investigated in detail, and still one of the best understood, comprises the substances in which the host is β-quinol. Crystallization of solutions of quinol[p-dihydroxybenzene, p-$C_6H_4(OH)_2$] in water or alcohol under presence of 10 to 40 atm of, say, krypton, produces crystals, often up to 1 cm long, which are readily distinguishable from the crystals of ordinary quinol (α-quinol), even visually. These crystals contain the noble gas trapped in the lattice of β-quinol. When the crystals are dissolved in water, or heated, the gas is released. The crystals are stable at room temperature and can be kept for years. Quinol clathrates have also been prepared enclosing O_2, N_2, CO, NO, CH_4, SO_2, HCl, HBr, Ar, Kr, Xe, HCOOH, HCN, H_2S, CH_3OH, or CH_3CN.

X-Ray analysis indicates that in β-quinol, three quinol molecules form an approximately spherical cage of free diameter $\sim$4 Å, with the quinol molecules bound together by hydrogen bonds. The free volumes are in the form of isolated cavities, and the apertures leading from one cage to another through the crystal are very small in diameter. Molecules trapped within these cavities during formation of the crystal are unable to escape. As a molecule approaches the cage walls, it experiences repulsive forces. Since three quinol molecules are required to form each cavity, the limiting ratio of quinol to trapped atom or molecule for the composition of clathrates is 3:1. This ratio is reached for acetonitrile, but for the noble gases various com-

---

[15]   J. E. D. Davies, *J. Chem. Educ.,* **1977,** 536; M. Hagen, *Clathrate Inclusion Compounds,* Reinhold, 1962.

position ranges may be obtained depending on conditions—for example, $C_6H_4(OH)_2/Kr$, 3:0.74; $C_6H_4(OH)_2/Xe$, 3:0.88—and normally the cages are incompletely filled.

Since the free diameter of the quinol cage in a clathrate compound is ~4 Å, only molecules of appropriate size may be expected to be trapped. Thus although CH_3OH forms quinol clathrates, C_2H_5OH is too large and does not. On the other hand, not all small molecules may form clathrates. Helium does not, the explanation being that the He atom is too small and can escape between the atoms of the quinol molecules that form the cage. Similarly, neon has not been obtained in a quinol clathrate yet. Water, although of a suitable size, also does not form a clathrate; in this case the explanation cannot be a size factor but may lie in the ability of water molecules to form hydrogen bonds, which enables them to approach the cage walls and escape through gaps in the walls.

A second important class of clathrates are the *gas hydrates*. When water is solidified in the presence of certain atomic or small molecular gases, as well as some substances such as $CHCl_3$ that are volatile liquids at room temperature, it forms one of several types of very open structure in which there are cages occupied by the gas or other guest molecules. These structures are far less dense than the normal form of ice and are unstable with respect to the latter in the absence of the guest molecules. There are two common gas hydrate structures, both cubic. In one the unit cell contains 46 molecules of H_2O connected to form six medium-size and two small cages. This structure is adopted when atoms (Ar, Kr, Xe) or relatively small molecules (e.g., Cl_2, SO_2, CH_3Cl) are used, generally at pressures greater than 1 atm for the gases. Complete filling of only the medium cages by atoms or molecules X would give a composition $X \cdot 7.67H_2O$, whereas complete filling of all eight cages would lead to $X \cdot 5.76H_2O$. In practice, complete filling of all cages of one or both types is seldom attained, and these formulas therefore represent limiting rather than observed compositions; for instance, the usual formula for chlorine hydrate (see Section 17-3) is $Cl_2 \cdot 7.3H_2O$. The second structure, often formed in the presence of larger molecules of liquid substances (thus sometimes called the liquid hydrate structure) such as chloroform and ethyl chloride, has a unit cell containing 136 water molecules with eight large cages and sixteen smaller ones. The anesthetic effect of substances such as chloroform is due to the formation of liquid hydrate crystals in brain tissue.

A third notable class of clathrate compounds, salt hydrates, is formed when tetraalkylammonium or sulfonium salts crystallize from aqueous solution with high water content, for example, $[(C_4H_9)_4N]C_6H_5CO_2 \cdot 39.5H_2O$ or $[(C_4H_9)_3S]F \cdot 20H_2O$. The structures of these substances are very similar to the gas and liquid hydrate structures in a general way, though they differ in detail. These structures consist of frameworks constructed mainly of hydrogen-bonded water molecules but apparently including also the anions (e.g., F^-) or parts of the anions (e.g., the O atoms of the benzoate ion). The cations and parts of the anions (e.g., the C_6H_5C part of the benzoate ion) occupy cavities in an incomplete and random way.

An additional relationship between the gas hydrate and the salt hydrate structures has been revealed by the discovery that bromine hydrate ($Br_2 \cdot {\sim}8.5H_2O$), crystallizes in neither of the cubic gas hydrate structures but rather is nearly iso-structural with the tetragonal tetrabutylammonium salt hydrates. Its ideal limiting composition would be $Br_2 \cdot 8.6H_2O$.

Although not classifiable as clathrate compounds, many other crystalline substances have holes, channels, or honeycomb structures that allow inclusion of foreign molecules, and many studies have been made in this field. Urea is an example of an organic compound that in the crystal has parallel continuous uniform capillaries; it may be utilized to separate straight-chain hydrocarbons from branched-chain ones, the latter being unable to fit into the capillaries.

Among inorganic lattices that can trap molecules, the best known are the so-called molecular sieves, which are discussed in Section 12-7.

6-7.   The Hydrogen Ion

For the reaction

$$H(g) = H^+(g) + e$$

the ionization potential 13.59 eV (ΔH = 569 kJ mol^{-1}), is higher than the first ionization potential of xenon and is high by comparison with lithium or cesium and indeed many other elements. Hence with the possible exception of HF, bonds from hydrogen to other elements must be mainly covalent. For HF the bond energy is 5.9 eV. For a purely ionic bond the energy can be estimated as the sum of (1) 13.6 eV to ionize H, (2) −3.5 eV to place the electron on F, and (3) −15.6 eV as an upper limit on the electrostatic energy of the ion pair H^+F^- at the observed internuclear distance in HF. The sum of these terms is −5.5 eV as an upper limit, which is not too far below the actual bond energy. For HCl, on the other hand, the experimental bond energy is 4.5 eV, whereas for a purely ionic situation we would have the sum +13.6 − 3.6 −11.3 = −1.5 eV as an upper limit. Thus purely electrostatic bonding cannot nearly explain the stability of HCl.

Hydrogen can form the hydrogen ion *only* when its compounds are dissolved in media that *solvate* protons. The solvation process thus provides the energy required for bond rupture; a necessary corollary of this process is that the proton H^+ never exists in condensed phases, but occurs always as solvates (H_3O^+, R_2OH^+, etc.). The order of magnitude of these solvation energies can be appreciated by considering the solvation reaction in water (estimated from thermodynamic cycles):

$$H^+(g) + xH_2O = H^+(aq) \qquad \Delta H = -1091 \text{ kJ mol}^{-1}$$

Compounds that furnish solvated hydrogen ions in suitable polar solvents, such as water, are *protonic acids*.

The nature of the hydrogen ion in water, which should more correctly be called the hydroxonium ion H_3O^+, is discussed below. The hydrogen ion in water is customarily referred to as "the hydrogen ion," implying H_3O^+. The use of other terms, such as hydroxonium, is somewhat pedantic except in special cases. We shall usually write H^+ for the hydrogen ion and assume it to be understood that the ion is aquated, since in a similar manner many other cations (Na^+, Fe^{2+}, Zn^{2+}, etc.) are customarily written as such, although there also it is understood that the actual species present in water are aquated species, for example, $[Fe(H_2O)_6]^{2+}$.

Water itself is weakly ionized:

$$2H_2O = H_3O^+ + OH^- \quad \text{or} \quad H_2O = H^+ + OH^-$$

Other cases of such *self-ionization* of a compound, where one molecule solvates a proton originating from another, are known; for example, in pure sulfuric acid

$$2H_2SO_4 = H_3SO_4^+ + HSO_4^-$$

and in liquid ammonia

$$2NH_3 = NH_4^+ + NH_2^-$$

In aqueous solutions, the hydrogen ion concentration is often given in terms of pH, defined as $-\log_{10}[H^+]$, where $[H^+]$ is the hydrogen ion activity, which may be considered to approximate to the molar concentration of H^+ ions in very dilute solutions.

At 25° the ionic product of water is

$$K_w = [H^+][OH^-] = 1 \times 10^{-14} \ M^2$$

This value is significantly temperature dependent. When $[H^+] = [OH^-]$, the solution is neutral and $[H^+] = 1 \times 10^{-7} \ M$; that is, pH = 7.0. Solutions of lower pH are acidic; those of higher pH are alkaline.

The standard hydrogen electrode provides the reference for all other oxidation-reduction systems.[16] The hydrogen half-cell or hydrogen electrode is

$$H^+(aq) + e = \frac{1}{2} H_2(g)$$

By definition, the potential of this system is zero ($E^0 = 0.000$ V) at all temperatures when an inert metallic electrode dips into a solution of hydrogen ions of unit activity (i.e., pH = 0) in equilibrium with H_2 gas at 1 atm pressure. The potentials of all other electrodes are then referred to this defined zero. However the absolute potentials of other electrodes may be either greater or smaller; thus some must have positive and others negative potentials relative to the standard hydrogen electrode. This subject is not properly an aspect of the chemistry of hydrogen, but it is discussed briefly as a matter of convenience.

The difficulties that are sometimes caused by the so-called electrochemical sign conventions have arisen largely because the term "electrode potential" has been used to mean two distinct things:

1. *The Potential of an Actual Electrode.* For example, a zinc rod in an aqueous solution of zinc ions at unit activity ($a = 1$) at 25° has a potential of -0.7627 V relative to the standard hydrogen electrode. There is no ambiguity about the sign because if this electrode and a hydrogen electrode were connected with a salt bridge,

---

[16]   G. Charlot, A. Collumeau, and M. J. C. Marchon, *Oxidation-Reduction Potentials of Inorganic Substances in Aqueous Solution.* Butterworths, 1971; G. Milazzo, S. Caroli, and V. K. Sharma, Eds., *Tables of Standard Electrode Potentials,* Wiley, 1978.

it would be necessary to connect the zinc rod to the negative terminal of a poten-tiometer and the hydrogen electrode to the positive terminal to measure the potential between them. Physically, the zinc electrode is richer in electrons than the hydrogen electrode.

2. *The Potential of a Half-Reaction.* Using the same chemical systems as an example, and remembering also that the Gibbs free energy of the standard hydrogen electrode is also defined as zero, we can write:

$$Zn + 2H^+(a = 1) \rightarrow Zn^{2+}(a = 1) + H_2(g) \quad \Delta G^0 = -147.5 \text{ kJ}$$
$$Zn \qquad\qquad \rightarrow Zn^{2+}(a = 1) + 2e^- \quad E^0 = -\Delta G^0/nF = +0.7627V$$
$$Zn^{2+}(a = 1) + H_2(g) \rightarrow Zn + 2H^+(a = 1) \quad \Delta G^0 = +147.5 \text{ kJ}$$
$$Zn^{2+}(a = 1) + 2e^- \rightarrow Zn \qquad\qquad E^0 = -\Delta G^0/nF = -0.7627V$$

Since metallic zinc does actually dissolve in acid solutions under conditions specified in the definition of a standard electrode, the standard change in Gibbs free energy must be negative for the first pair of reactions and positive for the second pair. The potential of the zinc couple, defined by $\Delta G^0 = -nFE^0$ (n = number of electrons = 2, F = the Faraday), has to change sign accordingly. The half-reaction

$$Zn \rightarrow Zn^{2+} + 2e^-$$

involves oxidation, and its potential is an *oxidation potential* whose sign is that of the so-called American sign convention. The half-reaction

$$Zn^{2+} + 2e^- \rightarrow Zn$$

involves reduction, and its potential is a *reduction potential* associated with the European sign convention. There is no doubt about which potential is relevant, provided the half-reaction to which it refers is written out in full.

Inspection shows that the reduction potential has the same sign as the potential of the actual electrode. For this reason we adopt the IUPAC recommendation that *only reduction potentials be called electrode potentials.* Every half-reaction is therefore written in the form

$$ox + ne^- \rightleftharpoons red$$

and the Nernst equation for the electrode potential E is

$$E = E^0 + \frac{2.3026 \, RT}{nF} \log_{10} \frac{\text{activity of oxidant}}{\text{activity of reductant}} \qquad (6-1)$$

where E^0 is the standard electrode potential, R the gas constant, and T the absolute temperature. Alternatively, we may sometimes speak of the electrode potential of a couple e.g., Fe^{3+}/Fe^{2+}, giving it the sign appropriate to the half-reaction written as a reduction.

For pure water, in which the H^+ activity is only 10^{-7} mol 1^{-1} at 25° the electrode potential, according to eq. 6-1 is more negative than the standard potential, that is, hydrogen becomes a better reductant:

$$H^+(aq)(10^{-7} M) + e = \frac{1}{2} H_2 \qquad E_{298} = -0.414 \text{ V}$$

In a basic solution, where the OH^- activity is $1M$, the potential is -0.83 V. In the absence of overvoltage (a certain lack of reversibility at certain metal surfaces), hydrogen is liberated from pure water by reagents whose electrode potentials are more negative than -0.414 V. Similarly, certain ions (e.g., the U^{3+} ion, for which the U^{4+}/U^{3+} standard potential is -0.61 V) will be oxidized by water, liberating hydrogen.

Many electropositive metals or ions, even if they do not liberate hydrogen from water, will be oxidized by a greater concentration of hydrogen ions—thus the reactions of zinc or iron are normally used to prepare hydrogen from dilute acids.

Finally a word on rates of acid-base reactions. All the protons in water are undergoing rapid migration from one oxygen atom to another, and the life time of an individual H_3O^+ ion in water is only approximately 10^{-13} sec. The rate of reaction of H_3O^+ with a base such as OH^- in water is very fast but also is diffusion controlled. Reaction occurs when the solvated ions diffuse to within a critical separation, whereupon the proton is transferred by concerted shifts across one or more solvent molecules hydrogen-bonded to the base.

6-8. Oxonium Ions

The H_3O^+ ion mentioned above in connection with aqueous solutions of acids is known to occur as such in some crystalline acid hydrates[17a], and higher oxonium ions, particularly $H_5O_2^+$, have also been shown to exist. In acids generally either the proton must be present as some kind of oxonium ion or it must be H-bonded to some suitable atom in the acid molecule. For some acids, such as $H_2PtCl_6 \cdot 2H_2O$, only the hydrates can be prepared, since there would be no suitable place for the protons in an anhydrous form.

On the other hand, both hydrated and anhydrous forms of other acids such as $H_4[Fe(CN)_6]$ are known, and in the anhydrous form there are H-bonds M—CN—H—NC—M. There are also cases where no oxonium ion is present, such as hydrated oxalic acid, $(COOH)_2 \cdot 2H_2O$, which has a three-dimensional H-bonded structure, and phosphoric acid hemihydrate, which is $2PO(OH)_3 \cdot H_2O$. It is worth noting that adducts of acids are sometimes not all that they might seem; thus $CH_3CN \cdot 2HCl$ could have contained the HCl_2^- ion, but it is actually $[CH_3(Cl)C=NH_2]^+Cl^-$. The nitric acid adducts of certain metal complex nitrate salts also do not contain oxonium ions, but rather the ions $[H(NO_3)_2]^-$ and $[H(NO_3)_4]^{3-}$, which have $O—H \cdots O$ bonds; other H-bonded anions HX_2^- or HXY^- are known where X may be F, Cl, CO_3, RCOO, and so on.

The structural role of H_3O^+ in a crystal often closely resembles that of NH_4^+; thus $H_3O^+ ClO_4^-$ and $NH_4^+ ClO_4^-$ are isomorphous. The important difference is that compounds of H_3O^+ and other oxonium ions generally have much lower melting points than have NH_4^+ salts. The structure of the H_3O^+ ion is that of a

[17a] G. D. Mateescu and G. M. Benedikt, *J. Am. Chem. Soc.*, 1979, **101**, 3959.

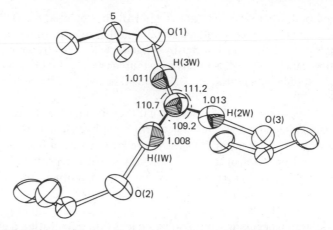

Fig. 6-7. The H_3O^+ ion in p-toluenesulfonic acid monohydrate. The immediate surroundings to which it is hydrogen bonded are also shown, and the angles (deg) and O—H distances (Å) are indicated. [Reproduced by permission from Lundgren and Williams, Ref. 17b.]

rather flat pyramid, as shown in Fig. 6-7 for the one in $[p$-$CH_3C_6H_4SO_3^-][H_3O^+]$, for which an accurate neutron structure is available.[17b]

The $H_5O_2^+$ ion consists essentially of two water molecules united by a hydrogen bond. In four representative examples where neutron data are available, the H bonds are all short but they vary in their degree of symmetry. Figure 6-2 shows one that is highly unsymmetrical, whereas that in $HBr\cdot2H_2O$ is nearly but not entirely centered. In o-$C_6H_4(COOH)SO_3H\cdot3H_2O$ there is one with a distance of 2.414 Å that is even more nearly centered,[18] and in $[trans$-$Coen_2Cl_2]Cl\cdot HCl\cdot2H_2O$ there is an H bond 2.431 Å long that appears to be exactly centered.[19a]

There is good evidence that oxonium ions other than H_3O^+ also can exist in solution. Thus the presence of species such as $H_9O_4^+$ has been used to explain many properties of aqueous acid solutions such as the extraction of metal ions into organic solvents. An example is the extraction of the ion $AuCl_4^-$ from hydrochloric acid solutions into benzene containing tributyl phosphate.

The extent and structural nature of solvation of the OH^- ion has remained much less known than for the hydrogen cation. Recently, however, the nature of primary and secondary solvation of OH^- has been observed in a complex crystal structure.[19b] Primary solvation gives an $[H$—O—H—O—$H]^-$ unit where the O$\cdots$O separation of 2.29Å is the shortest ever reported. This central $H_3O_2^-$ ion is then surrounded by four additional water molecules each forming a hydrogen bond to a lone pair on one of the $H_3O_2^-$ oxygen atoms.

17b J. Lundgren and J. M. Williams, *J. Chem. Phys.*, 1973, **58**, 788.
18   R. Attig and J. M. Williams, *Inorg. Chem.*, 1976, **15**, 3057.
19a J. Roziere and J. M. Williams, *Inorg. Chem.*, 1976, **15**, 1174.
19b K. N. Raymond *et al.*, *J. Am. Chem. Soc.*, 1979, **101**, 3688.

STRENGTHS OF PROTONIC ACIDS[20]

One of the most important characteristics of hydrogen compounds (HX) is the extent to which they ionize in water or other solvents, that is, the extent to which they act as acids. The strength of an acid depends not only on the nature of the acid itself but very much on the medium in which it is dissolved. Thus CF_3COOH and $HClO_4$ are strong acids in water, whereas in 100% H_2SO_4 the former is nonacidic and the latter only a very weak acid. Similarly, H_3PO_4 is a base in 100% H_2SO_4. Although acidity can be measured in a wide variety of solvents, the most important is water, for which the pH scale was discussed above.

6-9. Binary Acids

Although the intrinsic-strength of H—X bonds is one factor, other factors are involved, as the following consideration of the appropriate thermodynamic cycles for a solvent system indicates. The intrinsic strength of H—X bonds and the thermal stability of covalent hydrides seem to depend on the electronegativities and size of the element X. The variation in bond strength in some binary hydrides is shown in Fig. 6-8. There is a fairly smooth *decrease* in bond strength with *increasing Z* in a periodic group and a general *increase across* any period.

For HX dissolved in water we may normally (but not always—see below) assume that dissociation occurs according to the equation:

$$HX(aq) = H^+(aq) + X^-(aq)$$

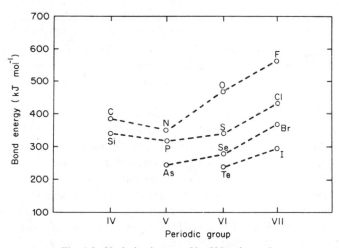

Fig. 6-8. Variation in mean H—X bond energies.

[20] D. D. Perrin, *Dissociation of Inorganic Acids and Bases in Aqueous Solution,* Butterworths, 1970; E. J. King, *Acid-Base Equilibria,* Vol. 4, Topic 15, *International Encyclopedia of Physics and Chemistry,* Pergamon Press, 1965; R. P. Bell, *Acids and Bases; Their Quantitative Behavior,* 2nd ed., Methuen, 1969.

TABLE 6-2

Free Energy Changes (kJ mol^{-1}) for Dissociation of HX Molecules in Water at 298°

Process	HF	HCl	HBr	HI
HX(aq) = HX(g)	23.9	−4.2	−4.2	−4.2
HX(g) = H(g) + X(g)	535.1	404.5	339.1	272.2
H(g) = H$^+$(g) + e	1,320.2	1,320.2	1,320.2	1,320.2
X(g) + e = X$^-$(g)	−347.5	−366.8	−345.4	−315.3
H$^+$(g) + X$^-$(g) = H$^+$(aq) + X$^-$(aq)	−1,513.6	−1,393.4	−1,363.7	−1,330.2
HX(aq) = H$^+$(aq) + X$^-$(aq)	18.1	−39.7	−54.0	−57.3
pK_a(= $\Delta G°/5.71$)	3.2	−7.0	−9.5	−10

The dissociation constant K is related to the change in Gibbs free energy by the relation

$$\Delta G^0 = -RT \ln K \qquad (6\text{-}2)$$

and the free energy change is in turn related to the changes in enthalpy and entropy via the relation

$$\Delta G = \Delta H - T\Delta S \qquad (6\text{-}3)$$

in which R is the gas constant and T is the absolute temperature, which we shall take to be 298° in the following discussion. The dissociation process may be considered to be the sum of several other reactions (i.e., as one step in a thermodynamic cycle). Table 6-2 summarizes the Gibbs free energy changes for these several steps.

HF in aqueous solution is a very weak acid; the pK values for the halogen acids change only slightly (3 units overall) from HI to HBr to HCl, but there is a precipitous change, by 10.2 units, on going to HF. The classic thermodynamic explanation for this is that the HF bond is so much stronger than those of the other HX molecules that even an unusually high free energy of solvation of F$^-$ cannot compensate for it; thus most dissolved HF molecules remain undissociated. However it has been argued recently that this is not correct.[21] From infrared studies of the four aqueous HX acids it was inferred that the H$_3$O$^+$ ion in aqueous HF is indeed present in large quantity but with a perturbed spectrum indicative of strong H bonding to F$^-$. In other words HF *is* dissociated, but tight ion pairs, F$^-$···H$^+$—OH$_2$, unique to F$^-$, which is a far better participant in H bonding than Cl$^-$, Br$^-$, or I$^-$, reduce the thermodynamic activity coefficient of H$_3$O$^+$. The structural distinction between F$^-$···H$^+$—OH$_2$ and F—H···OH$_2$ is a delicate one, however, and in any case the important points are that HF is a weak acid and that in some sense or other this is because the F—H bond, whether in the former (as now proposed) or the latter of the foregoing species, is exceptionally strong.

[21]   P. A. Giguére, *Chem. Phys. Lett.,* 1976, **41**, 598; P. A. Giguére and S. Turrell, *Can. J. Chem.,* 1976, **54**, 3477.

6-10. Oxo Acids

The second main class of acid behavior is shown by compounds with X—OH groups; these are called oxo acids, and they generally have a formula of the type H_nXO_m, for example, H_3PO_4.

For oxo acids certain useful generalizations may be made concerning (a) the magnitude of K_1 and (b) the ratios of successive constants, K_1/K_2, K_2/K_3, etc. The value of K_1 seems to depend on the charge on the central atom. Qualitatively it is very reasonable to suppose that the greater the positive charge, the more will the process of proton loss be favored on electrostatic grounds. It has been found that if this positive charge is taken to be the so-called formal charge, semiquantitative correlations are possible. The formal charge in an oxo acid H_nXO_m is computed in the following way, assuming the structure of the acid to be $O_{m-n}X(OH)_n$. Each X—(OH) bond is formed by sharing one X electron and one OH electron and is thus *formally* nonpolar. Each X—O bond is formed by using two X electrons and thus represents a net loss of one electron by X. Therefore the formal positive charge on X is equal to the number of X—O bonds, hence equal to $(m-n)$. It may be seen from the data in Table 6-3 that with the exception of the acids listed in brackets, which are special cases to be discussed presently, the following relations between $m-n$ (or formal positive charge on X) and the values of K_1 hold:

TABLE 6-3
Strengths of Oxo Acids, H_nXO_m, in Water

$(m-n)$	Examples	$-\log K_1$ (pK_1)	$-\log K_2$ (pK_2)	$-\log K_3$ (pK_3)
0	HClO	7.50	—	—
	HBrO	8.68	—	—
	H_3AsO_3	9.22	?	?
	H_4GeO_4	8.59	13	?
	H_6TeO_6	8.80	?	?
	[H_3PO_3]	1.8	6.15	—
	H_3BO_3	9.22	?	?
1	H_3PO_4	2.12	7.2	12
	H_3AsO_4	3.5	7.2	12.5
	H_5IO_6	3.29	6.7	~15
	H_2SO_3	1.90	7.25	—
	H_2SeO_3	2.57	6.60	—
	$HClO_2$	1.94	—	—
	HNO_2	3.3	—	—
	[H_2CO_3]	6.38 (3.58)	10.32	—
2	HNO_3	Large neg. value	—	—
	H_2SO_4	Large neg. value	1.92	—
	H_2SeO_4	Large neg. value	2.05	—
3	$HClO_4$	Very large neg. value	—	—
	$HMnO_4$	Very large neg. value	—	—
−1(?)	[H_3PO_2]	2	?	?

For $m-n = 0$: $pK_1 \sim 8.5 \pm 1.0$ $(K \sim 10^{-8}$ to $10^{-9})$
For $m-n = 1$: $pK_1 \sim 2.8 \pm 0.9$ $(K \sim 10^{-2}$ to $10^{-4})$
For $m-n \geq 2$: $pK_1 \ll 0$ (the acid is very strong)

It will also be noted that with very few exceptions, the difference between successive pK's is 4 to 5.

H_3PO_3 obviously is out of line with the other acids having $m-n = 0$ and seems to fit fairly well in the group with $m-n = 1$. This is, in fact, where it belongs, since there is independent evidence (Section 14-14) that its structure is $OPH*(OH)_2$ with H* bonded directly to P. Similarly, H_3PO_2 has a pK_1 that would class it with the $m-n = 1$ acids where it, too, belongs, since its structure is $OP(H*)_2(OH)$, with the two H* hydrogen atoms directly bound to P.

Carbonic acid is exceptional in that the directly measured pK_1, 6.38, does not refer to the process

$$H_2CO_3 = H^+ + HCO_3^-$$

since carbon dioxide in solution is only partly in the form of H_2CO_3, but largely present as more loosely hydrated species $CO_2(aq)$. When a correction is made for the equilibrium

$$CO_2(aq) + H_2O = H_2CO_3(aq)$$

the pK_1 value of 3.58 is obtained which falls in the range for other $m-n = 1$ acids (see also Section 11-5).

We may note finally that many metal ions whose solutions are acidic can be regarded as oxo acids. Thus although the hydrolysis of metal ions is often written as shown here for Fe^{3+}:

$$Fe^{3+} + H_2O = Fe(OH)^{2+} + H^+$$

it is just as valid thermodynamically and much nearer to physical reality to recognize that the ferric ion is coordinated by water molecules and to write:

$$[Fe(H_2O)_6]^{3+} = [Fe(H_2O)_5(OH)]^{2+} + H^+ \qquad K_{Fe^{3+}} \approx 10^{-3}$$

From this formulation it becomes clear why the ferrous ion, with a lower positive charge, is less acidic or, in alternative terms, less hydrolyzed than the ferric ion:

$$[Fe(H_2O)_6]^{2+} = [Fe(H_2O)_5(OH)]^+ + H^+ \qquad K_{Fe^{2+}} \ll K_{Fe^{3+}}$$

It should be noted that one cannot necessarily compare the acidity of the bivalent ion of one metal with that of the trivalent ion of *another* metal in this way, however. There appears to be no good general rule concerning the acidities of hydrated metal ions at the present time, although some attempts have been made at correlations.

6-11. General Theory of Ratios of Successive Constants

It was shown many years ago by Niels Bjerrum that the ratios of successive acid dissociation constants could be accounted for in a nearly quantitative way by

TABLE 6-4
K_1/K_2 Ratio for Dicarboxylic Acids,
$HOOC(CH_2)_nCOOH$

n	1	2	3	4	5	6	7	8
K_1/K_2	1120	29.5	17.4	12.3	11.2	10.0	9.5	9.3

electrostatic considerations. Consider any bifunctional acid HXH:

$$HXH = HX^- + H^+ \qquad K_1$$
$$HX^- = X^{2-} + H^+ \qquad K_2$$

There is a purely statistical effect that can be considered in the following way. For the first process, dissociation can occur in two ways (i.e., there are two protons, either of which may dissociate), but recombination in only one; whereas in the second process dissociation can occur in only one way, but recombination in two (i.e., the proton has two sites to which it may return, hence twice the probability of recombining). Thus on purely statistical grounds one would expect $K_1 = 4K_2$. Bjerrum observed that for the dicarboxylic acids $HOOC(CH_2)_nCOOH$, the ratio K_1/K_2 was always greater than 4 but decreased rapidly as n increased (see Table 6-4). He suggested the following explanation. When the two points of attachment of protons are close together in the molecule, the negative charge left at one site when the first proton leaves strongly restrains the second one from leaving by electrostatic attraction. As the separation between the sites increases, this interaction should diminish.

By making calculations using Coulomb's law,* Bjerrum was able to obtain rough agreement with experimental data. The principal difficulty in obtaining quantitative agreement lies in a choice of dielectric constant, since some of the lines of electrostatic force run through the molecule ($D \sim 1-10$), others through neighboring water molecules (D uncertain), and still others through water having the dielectric constant (~ 82) of pure bulk water. Nearly quantitative agreement is obtained using more elaborate models that take into account the variability of the dielectric constant. The important point here for our purposes is to recognize the physical principles involved without necessarily trying to obtain quantitative results.

Thus the large separation in successive pK's for the oxo acids is attributable to the electrostatic effects of the negative charge left by the dissociation of one proton on the remaining ones. In bifunctional binary acids, where the negative charge due to the removal of one proton is concentrated on the very atom to which the second proton is bound, the separation of the constants is extraordinarily great: K_1 and K_2 for H_2S are $\sim 10^{-7}$ and $\sim 10^{-14}$, respectively, whereas for water we have

$$H_2O = H^+ + OH^- \qquad K_1 = 10^{-14}$$
$$OH^- = H^+ + O^{2-} \qquad K_2 < 10^{-36} \text{ (est)}$$

* $F \propto q_1q_2/Dr$, where F is the force; q_1 and q_2 the charges separated by r; and D the dielectric constant of the medium between them.

6-12. Pure Acids and Relative Acidities[22]

The concepts of hydrogen ion concentration and pH discussed above are meaningful only for dilute aqueous solutions of acids. A widely used means of gauging acidity in other media and at high concentrations is the Hammett acidity function H_0, which is defined in terms of the behavior of one or more indicator bases B, for which there is the protonation equilibrium

$$B + H^+ = BH^+$$

The acidity function is defined as

$$H_0 = pK_{BH^+} - \log \frac{[BH^+]}{[B]}$$

In very dilute solutions

$$K_{BH^+} = \frac{[B][H^+]}{[BH^+]} .$$

so that in water, H_0 becomes synonymous with pH. By using suitable organic bases (e.g., p-nitroaniline) and suitable indicators over various ranges of concentration and acidities or by nmr methods, it is possible to interrelate values of H_0 for strong acids extending from dilute solutions to the pure acid.

For a number of strong acids in aqueous solution up to concentrations about $8M$, the values of H_0 are very similar. This suggests that the acidity is independent of the anion. The rise in acidity with increasing concentration can be fairly well predicted by assuming that the hydrogen ion is present as $H_9O_4^+$, so that protonation can be represented as

$$H_9O_4^+ + B \rightleftharpoons BH^+ + 4H_2O$$

Values of H_0 for some pure liquid acids are given in Table 6-5. It is to be noted particularly that for HF the acidity can be very substantially increased by the addition of a Lewis acid or fluoride ion acceptor, for example:

$$2HF + SbF_5 \rightleftharpoons H_2F^+ + SbF_6^-$$

but its acidity is decreased by addition of NaF owing to formation of the HF_2^- ion.

TABLE 6-5
The Hammett Acidity Function H_0 for Several Acids

Acid	$-H_0$	Acid	$-H_0$
$HSO_3F + SbF_5$ (25 mol %)	21.5	HF	11
$HF + SbF_5$ (0.6 mol %)	21.1	$HF + NaF$ (1M)	8.4
HSO_3F	15	H_3PO_4	5.0
$H_2S_2O_7$	15	H_2SO_4 (63% in H_2O)	4.9
H_2SO_4	12	HCOOH	2.2

[22]   C. H. Rochester, *Acidity Functions,* Academic Press, 1970.

Antimony pentafluoride is commonly used as the Lewis acid, since it is comparatively easy to handle, being a liquid, and is commercially available. However other
fluorides such as BF_3, NbF_5, and TaF_5 behave in a similar way. Acid media with
H_0 values above about 6 are often referred to as *superacids,* since they are upward
of 10^6 times as strong as a $1 M$ aqueous solution of a strong acid.[23a] The addition
of SbF_5 to HSO_3F dramatically raises the $-H_0$ value from 15 for 0% SbF_5 to *ca.*
17 at 0.4 mol % SbF_5 and finally to 21.5 at 25 mol %, the latter being the highest
$-H_0$ value known. The SbF_5—FSO_3H system, which is very complicated, has been
thoroughly investigated by nuclear magnetic resonance and Raman spectra. The
acidity is due to the formation of the $H_2SO_3F^+$ ion. The equilibria depend on the
ratios of the components; with low ratios of SbF_5 to FSO_3H the main ones are the
following:

$$SbF_5 + HSO_3F \rightleftharpoons$$

$$H(F_5SbSO_3F) + HSO_3F \rightleftharpoons \quad + H_2SO_3F^+$$

$$2H(F_5SbSO_3F) \rightleftharpoons \quad + H_2SO_3F^+$$

At higher ratios the solutions appear to contain also the ions SbF_6^- and $[F_5Sb—$
$F—SbF_5]^-$, which occur in solutions of SbF_5 in liquid HF, together with HS_2O_6F
and HS_3O_9F, which occur in SO_3—FSO_3H solutions. These species are generated
by the additional reactions

$$HSO_3F \rightleftharpoons HF + SO_3$$
$$3SbF_5 + 2HF \rightleftharpoons HSbF_6 + HSb_2F_{11}$$
$$2HSO_3F + 3SO_3 \rightleftharpoons HS_2O_6F + HS_3O_9F$$

[23a] R. J. Gillespie, *Endeavour,* 1973, **32,** 541; *J. Am. Chem. Soc.,* 1973, **95,** 5173; G. A. Olah, *Angew.*
Chem., Int. Ed., 1973, **12,** 173; R. J. Gillespie and T. E. Peel, *Adv. Phys. Org. Chem.,* 1971, **9,** 1;
J. Sommer *et al., J. Am. Chem. Soc.,* 1976, **98,** 2671, 1978, **100,** 2576.

The SbF_5—HSO_3F solutions are very viscous and are normally diluted with liquid sulfur dioxide so that better resolution of nmr spectra is obtained. Although the equilibria appear not to be appreciably altered for molecular ratios SbF_5:HSO_3F < 0.4, in more concentrated SbF_5 solutions the additional equilibria noted above are shifted to the left by removal of SbF_5 as the stable complex $SbF_5 \cdot SO_2$, which can be obtained crystalline (see Section 16-12). This results in a lowering of the acidity of the system; on the contrary, if SO_3 is added the acidity is increased by raising the concentration of $H_2SO_3F^+$, and the strongest known acid is SbF_5—$HSO_3F \cdot nSO_3$ ($n \geq 3$).

There has been extensive study of very strong acids especially FSO_3H—SbF_5—SO_2, HF—SbF_5, and HCl—Al_2Cl_6 for the protonation of weak bases. Virtually all organic compounds can be protonated and the resultant species characterized by nuclear magnetic resonance. Thus formic acid at $-60°$ gives equal amounts of the protonated species 6-VII and 6-VIII and protonated formaldehyde 6-IX; fluorobenzene gives the ion 6-X.

(6-VII)   (6-VIII)   (6-IX)   (6-X)

The superacid media can induce hydride abstraction, H–D exchange, and other reactions even with saturated hydrocarbons.[23b] Carbonium ions are formed, some of which, notably the trimethylcarbonium ion, are quite stable:

$$Me_3CH \rightarrow Me_3C^+ \leftarrow CH_3CH_2CH_2CH_3$$

It is postulated that the attack by H^+ occurs on the electron density of the C—H and C—C single bonds and not on the C and H atoms themselves. The order of reactivity, qualitatively, is: tertiary CH > C—C > secondary CH ≫ primary CH. Even molecular H_2 may be protonated, since H_2–D_2 exchange is observed in the superacids, probably through a planar H_3^+ transition state. The reactions possibly involve "pentacoordinate" carbonium ions, e.g., where the two hydrogen atoms are bound to carbon by closed three-center bonds.

$$R_3CH + H^+ \rightleftharpoons \left[R_3C \overset{H}{\underset{H}{\cdots}} \right]^+ \rightleftharpoons CR_3^+ + H_2$$

The carbonium ions can undergo complex reactions. Thus methane can give carbonium ions with C—C bonds by condensation reactions of the type:

$$CH_4 \underset{}{\overset{H^+}{\rightleftharpoons}} H_2 + CH_3^+ \overset{CH_4}{\longrightarrow} C_2H_7^+ \rightleftharpoons H_2 + C_2H_5^+, \text{ etc.}$$

23b G. A. Olah et al., Angew. Chem., Int. Ed., 1978, 17, 909.

CH_3^+ ions have been detected by trapping[23c] with CO and subsequent hydrolysis of the acylium ion with water to give acetic acid, e.g.,

$$CH_3^+ + CO \rightarrow CH_3CO^+ \xrightarrow{H_2O} CH_3COOH + H^+$$

The use of $HCl-Al_2Cl_6$ or $HF-SbF_5$ to isomerize straight-chain to branched-chain alkanes, or vice versa, has potential industrial value, and indeed such acidic media are already important in many organic reactions of hydrocarbons such as isomerization, acetylation, and alkylation.

Considerably more study has been given to protonation of organic compounds, but a number of inorganic compounds have also been studied. Thus many metal carbonyl and organometallic complexes may be protonated on the metal or on the ligand (Sections 27-9, 29-1), e.g.:

$$Fe(CO)_5 + H^+ \rightleftharpoons HFe(CO)_5^+$$
$$(\eta^5\text{-}C_5H_5)_2Fe + H^+ \rightleftharpoons (\eta^5\text{-}C_5H_5)_2FeH^+$$
$$C_8H_8Fe(CO)_3 + H^+ \rightleftharpoons C_8H_9Fe(CO)_3^+$$

Even protonated carbonic acid, or more properly, the trihydroxycarbonium ion $C(OH)_3^+$, has been observed in solutions of carbonates or bicarbonates in $FSO_3H-SbF_5-SO_2$ solutions at $-78°$; the ion is stable to $0°$ in absence of SO_2. It was suggested that $C(OH)_3^+$ might be involved even in biological systems at very acid sites in enzymes such as carbonic anhydrase.

6-13. Properties of Some Common Strong Acids[24]

In Table 6-6 are collected some properties of the more common and useful strong acids in their pure states.

Hydrogen Fluoride.[25] The acid HF is made by the action of concentrated H_2SO_4 on CaF_2 and is the principal source of fluorine compounds (Chapter 17). It is commercially available in steel cylinders, with purity approximately 99.5%; it can be purified further by distillation. Although liquid HF attacks glass rapidly, it can be handled conveniently in apparatus constructed either of copper or Monel metal or of materials such as polytetrafluoroethylene (Teflon or PTFE) and Kel-F (a chlorofluoro polymer).

The high dielectric constant is characteristic of hydrogen-bonded liquids. Since HF forms only a two-dimensional polymer, it is less viscous than water. In the vapor, HF is monomeric above $80°$, but at lower temperatures the physical properties are best accounted for by an equilibrium between HF and a hexamer, $(HF)_6$, which has a puckered ring structure. Crystalline $(HF)_n$ has zigzag chains (Fig. 6-3).

[23c] H. Hogeveen et al., Rec. Trav. Chim. Pays-Bas, 1969, **88**, 703, 719; J.C.S. Chem. Commn., **1969**, 921.

[24] E. F. Caldin and V. Gold, Eds., Proton Transfer Reactions, Chapman & Hall, 1975 (chapter by R. J. Gillespie on acids).

[25] (a) M. Kilpatrick and J. G. Jones, in The Chemistry of Non-Aqueous Solvents, Vol. 2, J. J. Lagowski, Ed., Academic Press, 1967; (b) H. H. Hyman and J. J. Katz, in Non-Aqueous Solvent Systems, T. C. Waddington, Ed., Academic Press, 1965.

TABLE 6-6
Properties of Some Strong Acids in the Pure State

Acid	M.p. (°C)	B.p. (°C)	κ^a	ϵ^b
HF	−83.36	19.74	1.6×10^{-6} (0°)	84 (0°)
HCl	−114.25	−85.09	3.5×10^{-9} (−85°)	14.3 (−114°)
HBr	−86.92	−66.78	1.4×10^{-10} (−84°)	7.33 (−86°)
HI	−50.85	−35.41	8.5×10^{-10} (−45°)	3.57 (−45°)
HNO_3	−41.59	82.6	3.72×10^{-2} (25°)	
$HClO_4$	−112	(109° extrap.)		
HSO_3F	−88.98	162.7	1.085×10^{-4} (25°)	~120 (25°)
H_2SO_4	10.3771	~270 d^c	1.044×10^{-2} (25°)	110 (20°)

^{a} Specific conductance in ohm^{-1} cm^{-1}. Values are often very sensitive to impurities.
^{b} Dielectric constant divided by that of a vacuum.
^{c} Constant-boiling mixture (338° C) contains 98.33% of H_2SO_4; d = with decomposition.

After water, liquid HF is one of the most generally useful of solvents. Indeed in some respects it surpasses water as a solvent for both inorganic and organic compounds, which often give conducting solutions as noted above; it can also be used for cryoscopic measurements.

The self-ionization equilibria in liquid HF are:

$$2HF \rightleftharpoons H_2F^+ + F^- \qquad K \sim 10^{-10}$$

$$F^- + HF \rightleftharpoons HF_2^- \overset{HF}{\rightleftharpoons} H_2F_3^-, \text{ etc.}$$

The formation of the stable hydrogen-bonded anions accounts in part for the extreme acidity. In the liquid acid the fluoride ion is the conjugate base, and ionic fluorides behave as bases. Fluorides of M^+ and M^{2+} are often appreciably soluble in HF, and some such as TlF are very soluble.

The only substances that function as "acids" in liquid HF are those such as SbF_5 noted above, which increase the concentration of H_2F^+. The latter ion appears to have an abnormally high mobility in such solutions.

Reactions in liquid HF are known that illustrate also amphoteric behavior, solvolysis, or complex formation. Although HF is waterlike, it is not easy, because of the reactivity, to establish an emf series, but a partial one is known.

In addition to its utility as a solvent system, HF as either liquid or gas is a useful fluorinating agent, converting many oxides and other halides into fluorides.

In *aqueous solution,* HF differs from the other halogen acids in that it is a weak acid. In 5 to 15M aqueous solution the acidity increases owing to ionization to H_3O^+, HF_2^-, and more complex $(H_nF_{n+1})^-$ species.

Hydrogen Chloride, Bromide, and Iodide.[26a] These three hydrogen halides are very similar to each other and differ notably from hydrogen fluoride. They are

[26a] F. Klanberg, in *The Chemistry of Non-Aqueous Solvents,* Vol. 2, J. J. Lagowski, Ed., Academic Press, 1967; M. E. Peach and T. C. Waddington, in *Non-Aqueous Solvent Systems,* T. C. Waddington, Ed., Academic Press, 1965; and papers by T. C. Waddington, mainly in *J. Chem. Soc.*

normally pungent gases; in the solid state they have hydrogen-bonded zigzig chains and there is probably some hydrogen bonding in the liquid. Hydrogen chloride is made by the action of concentrated H_2SO_4 on concentrated aqueous HCl or NaCl; HBr and HI may be made by catalytic reaction of $H_2 + X_2$ over platinized silica gel or, for HI, by interaction of iodine and boiling tetrahydronaphthalene. The gases are soluble in a variety of solvents, especially polar ones. The solubility in water is not exceptional[26b]; in moles of HX per mole of solvent at 0° and 1 atm the solubilities in water, 1-octanol, and benzene, respectively, are : HCl, 0.409, 0.48, 0.39; HBr, 1.00, 1.30, 1.39; HI, 0.065, 0.173, 0.42.

The self-ionization is very small:

$$3HX \rightleftharpoons H_2X^+ + HX_2^-$$

Liquid HCl has been fairly extensively studied as a solvent, and many organic and some inorganic compounds dissolve giving conducting solutions:

$$B + 2HCl \rightleftharpoons BH^+ + HCl_2^-$$

The low temperatures required and the short liquid range are limitations, but conductimetric titrations are readily made.

Salts of the ion H_2Cl^+ have not been isolated, but salts of the HCl_2^- and HBr_2^- ions, which have X—H—X distances of 3.14 and 3.35 Å, respectively, are not uncommon. These distances, like that in HF_2^- (2.26 Å), are ~0.5 Å shorter than the sum of van der Waals radii and suggest that there are strong hydrogen bonds.[27]

Nitric Acid.[28] Nitric acid is made industrially by oxidation of ammonia with air over platinum catalysts. The resulting nitric oxide (Section 13-5) is absorbed in water in the presence of air to form NO_2, which is then hydrated. The normal concentrated aqueous acid (ca. 70% by weight) is colorless but often becomes yellow as a result of photochemical decomposition, which gives NO_2:

$$2HNO_3 \xrightarrow{h\nu} 2NO_2 + H_2O + \tfrac{1}{2}O_2$$

The so-called fuming nitric acid contains dissolved NO_2 in excess of the amount that can be hydrated to $HNO_3 + NO$.

Pure nitric acid can be obtained by treating KNO_3 with 100% H_2SO_4 at 0° and removing the HNO_3 by vacuum distillation. The pure acid is a colorless liquid or white crystalline solid; the latter decomposes above its melting point according to the equation given above for the photochemical decomposition, hence must be stored below 0°.

The pure acid has the highest self-ionization of the pure liquid acids. The initial protolysis

$$2HNO_3 \rightleftharpoons H_2NO_3^+ + NO_3^-$$

[26b] W. Gerrard, *Chem. Ind. (London)*, **1969**, 295.

[27] D. G. Tuck, *Prog. Inorg. Chem.*, 1968, **9**, 161.

[28] W. H. Lee, in *The Chemistry of Non-Aqueous Solvents*, Vol. 2, J. J. Lagowski, Ed., Academic Press, 1967; S. A. Stern, J. T. Mullhaupt, and W. B. Kay, *Chem. Rev.*, 1960, **60**, 195 (an exhaustive review of the physical properties).

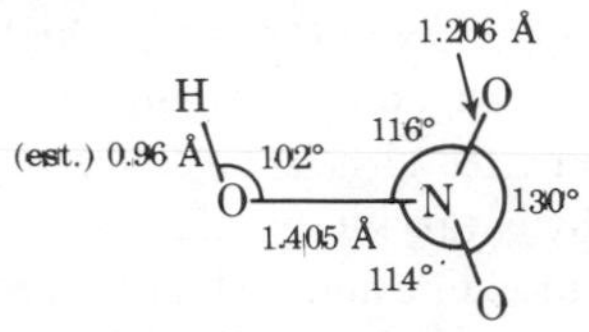

Fig. 6-9. The structure of nitric acid in the vapor.

is followed by rapid loss of water:

$$H_2NO_3^+ = H_2O + NO_2^+$$

so that the overall self-dissociation is

$$2HNO_3 \rightleftharpoons NO_2^+ + NO_3^- + H_2O$$

Pure nitric acid is a good ionizing solvent for electrolytes, but unless they produce the NO_2^+ or NO_3^- ions (Section 13-7), salts are sparingly soluble.

In dilute aqueous solution, nitric acid is approximately 93% dissociated at $0.1 M$ concentration. Nitric acid of concentration below $2M$ has little oxidizing power. The concentrated acid is a powerful oxidizing agent and, of the metals, only gold, platinum, rhenium, and iridium are unattacked, although a few others such as aluminum, iron, and copper are rendered "passive," probably owing to formation of an oxide film; magnesium alone can liberate hydrogen and then only initially from dilute acid. The attack on metals generally involves reduction of nitrate. Aqua regia (ca. 3 vol. of conc. HCl + 1 vol. of conc. HNO_3) contains free chlorine and ClNO, and it attacks gold and platinum metals, its action being more effective than that of HNO_3 mainly because of the complexing function of chloride ion. Similarly, some metals, notably tantalum, are quite resistant to HNO_3 but dissolve with extreme vigor if HF is added, to give TaF_6^- or similar ions. Nonmetals are usually oxidized by HNO_3 to oxo acids or oxides. The ability of nitric acid, especially in the presence of concentrated sulfuric acid, to nitrate many organic compounds is attributable to the formation of the nitronium ion, NO_2^+ (see discussion in Section 13-7).

Gaseous nitric acid has a planar structure (Fig. 6-9), although hindered rotation of OH relative to NO_2 probably occurs.

Perchloric Acid.[29] Perchloric acid ($HClO_4$) is commercially available in concentrations 70 to 72% by weight. The water azeotrope with 72.5% of $HClO_4$ boils at 203°, and although some chlorine is produced, which can be swept out by air, there is no hazard involved. The anhydrous acid is best prepared by vacuum distillation of the concentrated acid in presence of the dehydrating agent $Mg(ClO_4)_2$; it reacts explosively with organic material. The pure acid is stable at room temperature for only 3 to 4 days, decomposing to give $HClO_4 \cdot H_2O$ (84.6% acid) and Cl_2O_7.

The most important applications of aqueous perchloric acid involve its use as an oxidant. However at concentrations below 50% and temperatures not exceeding

29   G. S. Pearson, *Adv. Inorg. Chem. Radiochem.*, 1966, **8**, 177 (an exhaustive review).

50 to 60°, there is no release of oxygen. The hot concentrated acid oxidizes organic materials vigorously or even explosively; it is a useful reagent for the destruction of organic matter, especially after pretreatment with, or in the presence of, sulfuric or nitric acid. The addition of concentrated $HClO_4$ to organic solvents such as ethanol should be avoided where possible, even if the solutions are chilled.

Sulfuric Acid.[30] Sulfuric acid is prepared on an enormous scale by the lead chamber and contact processes.[31] In the former, SO_2 oxidation is catalyzed by oxides of nitrogen (by intermediate formation of nitrosylsulfuric acid, $HOSO_2O-NO$); in the latter, heterogeneous catalysts such as platinum are used for the oxidation. Pure sulfuric acid (H_2SO_4) is a colorless liquid that is obtained from the commercial 98% acid by addition first of sulfur trioxide or oleum and then titration with water until the correct specific conductance or melting point is achieved.

The phase diagram of the H_2SO_4–H_2O system is complicated, and eutectic hydrates such as $H_2SO_4 \cdot H_2O$ (m.p. 8.5°) and $H_2SO_4 \cdot 2H_2O$ (m.p. −38°) occur.

In pure crystalline H_2SO_4 there are SO_4 tetrahedra with S—O distances 1.42, 1.43, 1.52, and 1.55 Å, linked by strong hydrogen bonds. There is also extensive hydrogen bonding in the concentrated acid.

Pure H_2SO_4 shows extensive self-ionization resulting in high conductivity. The equilibrium

$$2H_2SO_4 \rightleftharpoons H_3SO_4^+ + HSO_4^- \qquad K_{10°} = 1.7 \times 10^{-4} \ mol^2 \ kg^{-2}$$

is only one factor, since there are additional equilibria due to dehydration:

$$2H_2SO_4 \rightleftharpoons H_3O^+ + HS_2O_7^- \qquad K_{10°} = 3.5 \times 10^{-5} \ mol^2 \ kg^{-2}$$
$$H_2O + H_2SO_4 \rightleftharpoons H_3O^+ + HSO_4^- \qquad K_{10°} = 1 \ mol \ kg^{-1}$$
$$H_2S_2O_7 + H_2SO_4 \rightleftharpoons H_3SO_4^+ + HS_2O_7^- \qquad K_{10°} = 7 \times 10^{-2} \ mol \ kg^{-1}$$

Estimates of the concentrations in 100% H_2SO_4 of the other species present, namely, H_3O^+, HSO_4^-, $H_3SO_4^+$, $HS_2O_7^-$, and $H_2S_2O_7$ can be made; for example, at 25°, HSO_4^- is 0.023 molar.

Pure H_2SO_4 and dilute oleums have been much studied as solvent systems,[32a] but interpretation of the cryoscopic and other data is often complicated. Sulfuric acid is not a very strong oxidizing agent, although the 98% acid has some oxidizing ability when hot. The concentrated acid reacts with many organic materials, removing the elements of water and sometimes causing charring, for example, of carbohydrates. Many substances dissolve in the 100% acid, often undergoing protonation. Alkali metal sulfates and water also act as bases. Organic compounds may also undergo further dehydration reactions, for example:

$$C_2H_5OH \overset{H_2SO_4}{\rightleftharpoons} C_2H_5OH_2^+ + HSO_4^- \overset{H_2SO_4}{\longrightarrow} C_2H_5HSO_4 + H_3O^+ + HSO_4^-$$

[30] R. J. Gillespie and E. A. Robinson, in *Non-Aqueous Solvents*, T. C. Waddington, Ed., Academic Press, 1965; W. M. Lee, in *The Chemistry of Non-Aqueous Solvents*, Vol. 2, J. J. Lagowski, Ed., Academic Press, 1967.

[31] A. Phillips, *Chem. Br.*, 1977, **13**, 471.

[32a] R. J. Gillespie, in *Inorganic Sulphur Chemistry*, G. Nickless, Ed., Elsevier, 1968; A. Vincent and R. F. M. White, *J. Chem. Soc. A*, **1970**, 2179; M. Liler, *Reaction Mechanisms in Sulfuric Acid and Other Strong Acid Solutions*, Academic Press, 1971.

Because of the strength of H_2SO_4, salts of other acids may undergo solvolysis, for example:

$$NH_4ClO_4 + H_2SO_4 \rightleftharpoons NH_4^+ + HSO_4^- + HClO_4$$

There are also examples of acid behavior. Thus H_3BO_3, which behaves initially as a base, gives quite a strong acid:

$$H_3BO_3 + 6H_2SO_4 \rightleftharpoons B(HSO_4)_3 + 3H_3O^+ + 3HSO_4^-$$
$$B(HSO_4)_3 + HSO_4^- \rightleftharpoons B(HSO_4)_4^-$$

The addition of SO_3 to H_2SO_4 gives what is known as *oleum* or fuming sulfuric acid $(SO_3)_n \cdot H_2O$; the constitution of concentrated oleums is controversial, but with equimolar ratios the major constituent is pyrosulfuric (disulfuric) acid $(H_2S_2O_7)$. At higher concentrations of SO_3, Raman spectra indicate the formation of $H_2S_3O_{10}$ and $H_2S_4O_{13}$. Pyrosulfuric acid[32b] has higher acidity than H_2SO_4 and ionizes thus:

$$2H_2S_2O_7 \rightleftharpoons H_2S_3O_{10} + H_2SO_4 \rightleftharpoons H_3SO_4^+ + HS_3O_{10}^-$$

The acid protonates many materials; $HClO_4$ behaves as a weak base, and CF_3COOH is a nonelectrolyte in oleum.

Fluorosulfuric Acid.[33] Fluorosulfuric acid is made by the reaction:

$$SO_3 + HF = FSO_3H$$

or by treating KHF_2 or CaF_2 with oleum at ~250°. When freed from HF by sweeping with an inert gas, it can be distilled in glass apparatus. Unlike $ClSO_3H$, which is explosively hydrolyzed by water, FSO_3H is relatively slowly hydrolyzed.

Fluorosulfuric is one of the strongest of pure liquid acids. It is commonly used in presence of SbF_5 as a protonating system, as noted above (p. 249). An advantage over other acids is its ease of removal by distillation in vacuum. The self-ionization

$$2FSO_3H \rightleftharpoons FSO_3H_2^+ + FSO_3^-$$

is much lower than for H_2SO_4 and consequently interpretation of cryoscopic and conductometric measurements is fairly straightforward.

In addition to its solvent properties, FSO_3H is a convenient laboratory fluorinating agent. It reacts readily with oxides and salts of oxo acids at room temperature. For example, K_2CrO_4 and $KClO_4$ give CrO_2F_2 and ClO_3F, respectively.

Trifluoromethylsulfonic Acid.[34] This very strong $(-H_0 = 15.1)$, useful acid $(CF_3SO_3H$, b.p. 162°) is often given the trivial name "triflic" acid and its salts called "triflates." It is very hygroscopic and forms the monohydrate $CF_3SO_3H \cdot H_2O$

[32b] R. J. Gillespie and K. C. Malhotra, *J. Chem. Soc., A,* **1968,** 1933.

[33] A. W. Jacke, *Adv. Inorg. Chem., Radiochem.,* 1974, **16,** 177.

[34] R. D. Howells and J. D. McCown, *Chem. Rev.,* 1977, **77,** 69 (a review with 302 references); R. J. Phillips and T. E. Peel, *J. Am. Chem. Soc.,* 1973, **95,** 5173.

(m.p. 34°). Its salts are similar to perchlorates but are nonexplosive. In crystal structures the $CF_3SO_3^-$ ion is less likely to be disordered than is perchlorate.

BINARY METALLIC HYDRIDES[35-38]

Figure 6-10 represents a rough attempt to classify the various hydrides.

```
H                                                                                    He
Li   Be                                                              B   C   N   O   F  Ne
Na   Mg                                                             Al   Si  P   S   Cl Ar
K    Ca   Sc   Ti*  V*   Cr*  Mn*  Fe*  Co*  Ni*   Cu    Zn  Ga  Ge  As  Se  Br  Kr
Rb   Sr   Y    Zr*  Nb*  Mo*  Tc*  Ru*  Rh*  Pd*   Ag    Cd  Zn  Sn  Sb  Te  I   Xe
Cs   Ba  La-Lu Hf*  Ta*  W*   Re*  Os*  Ir*  Pt*   Au    Hg  Tl  Pb  Bi  Po  At  Rn
Fr   Ra   Ac              U,Pu
Saline            Transition metal hydrides            Borderline    Covalent hydrides
hydrides                                                hydrides
```

Fig. 6-10. A classification of the hydrides. The starred elements are the transition elements for which complex molecules or ions containing M—H bonds are known.

6-14.   The Hydride Ion H⁻; Saline Hydrides

The formation of the unipositive ion H^+ (or H_3O^+, etc.) suggests that hydrogen should be classed with the alkali metals in the Periodic Table. On the other hand, the formation of the hydride ion might suggest an analogy with the halogens. Such attempts at classification of hydrogen with other elements can be misleading. The tendency of the hydrogen atom to form the negative ion is much lower than for the more electronegative halogen elements. This may be seen by comparing the energetics of the formation reactions:

$\frac{1}{2}H_2(g) \rightarrow H(g)$ $\Delta H = 218$ kJ mol⁻¹ $\frac{1}{2}Br_2(g) \rightarrow Br(g)$ $\Delta H = 113$ kJ mol⁻¹

$H(g) + e \rightarrow H^-(g)$ $\Delta H = -67$ kJ mol⁻¹ $Br(g) + e \rightarrow Br^-(g)$ $\Delta H = -345$ kJ mol⁻¹

$\frac{1}{2}H_2(g) + e \rightarrow H^-(g)$ $\Delta H = +151$ kJ mol⁻¹ $\frac{1}{2}Br_2(g) + e \rightarrow Br^-(g)$ $\Delta H = -232$ kJ mol⁻¹

Thus, owing to the endothermic character of the H⁻ ion, only the most electropositive metals—the alkalis and the alkaline earths—form saline or saltlike hydrides, such as NaH and CaH_2. The ionic nature of the compounds is shown by their high conductivities just below or at the melting point and by the fact that on electrolysis of solutions in molten alkali halides hydrogen is liberated at the *anode.*

[35]   W. M. Mueller, J. P. Blackledge, and G. G. Libowtiz, *Metal Hydrides,* Academic Press, 1968.
[36]   G. Alefield and J. Völki, Eds., *Hydrogen in Metals,* Springer, 1977, 1978.
[37]   D. G. Westlake, C. B. Satterthwaite, and J. H. Weaver, *Phys. Today,* November 1978, p. 32.
[38]   R. Bau, Ed., *Transition Metal Hydrides,* ACS Advances in Chemistry Series No. 167 (most of this book deals with molecular hydrides, but there are several chapters on solid state systems).

TABLE 6-7
The Saline Hydrides and Some of Their Properties

Salt	Structure	Heat of formation $-\Delta H_f(298\ K)(kJ\ mol^{-1})$	M—H distance (Å)	Apparent radius of H⁻ (Å)[a]
LiH	NaCl type	91.0	2.04	1.36
NaH	NaCl type	56.6	2.44	1.47
KH	NaCl type	57.9	2.85	1.52
RbH	NaCl type	47.4	3.02	1.54
CsH	NaCl type	49.9	3.19	1.52
CaH₂	Slightly distorted *hcp*	174.5	2.33[b]	1.35
SrH₂	Slightly distorted *hcp*	177.5	2.50	1.36
BaH₂	Slightly distorted *hcp*	171.5	2.67	1.34
MgH₂	Rutile type	74.5	—	1.30

[a] See text.
[b] Although half the H⁻ ions are surrounded by four Ca²⁺ and half by three Ca²⁺, the Ca—H distances are the same.

X-Ray and neutron diffraction studies show that in these hydrides the H⁻ ion has a crystallographic radius between those of F⁻ and Cl⁻. Thus the electrostatic lattice energies of the hydride and the fluoride and chloride of a given metal will be similar. These facts and a consideration of the Born-Haber cycles lead us to conclude that *only* the most electropositive metals *can* form ionic hydrides, since in these cases relatively little energy is required to form the metal ion.

The known saline hydrides and some of their physical properties are given in Table 6-7. The heats of formation of the saline hydrides, compared with those of the alkali halides, which are about 420 kJ mol⁻¹, reflect the inherently small stability of the hydride ion.

For the relatively simple two-electron system in the H⁻ ion, it is possible to calculate an effective radius for the free ion, the value 2.08 Å having been obtained. It is of interest to compare this with some other values, specifically, 0.93 Å for the He atom, ~0.5 Å for the H atom, 1.81 Å for the crystallographic radius of Cl⁻, and 0.30 Å for the covalent radius of hydrogen, as well as with the values of the "apparent" crystallographic radius of H⁻ given in Table 6-7. The values in the table are obtained by subtracting the Goldschmidt radii of the metal ions from the experimental M—H distances. The value 2.08 Å for the radius of free H⁻ is at first sight surprisingly large, being more than twice that for He. This results because the H⁻ nuclear charge is only half that in He and the electrons repel each other and screen each other (~30%) from the pull of the nucleus. Table 6-7 indicates that the apparent radius of H⁻ in the alkali hydrides never attains the value 2.08 Å and also that it decreases markedly with decreasing electropositive character of the metal. The generally small size is probably attributable in part to the easy compressibility of the rather diffuse H⁻ ion and partly to a certain degree of covalence in the bonds.

Preparation and Chemical Properties. The saline hydrides are prepared by direct interaction at 300 to 700°. For complete reaction of lithium, the temperature must be approximately 725°. Sodium normally reacts with H_2 only above 200° and the reaction is slow because a coating of an inert hydride forms. However studies on continuously clean surfaces show that the reaction obeys first-order kinetics with an activation energy of *ca.* 70 kJ mol^{-1}. Dispersion of sodium in mineral oil increases the reactivity, but NaH in a very reactive form can be prepared at room temperature and pressure by interaction of H_2 with sodium and naphthalene ($Na^+Ca_{10}H_8^-$) in tetrahydrofuran, with titanium isopropoxide as catalyst.

The saline hydrides are crystalline solids, white when pure but usually gray owing to traces of metal. They can be dissolved in molten alkali halides and on electrolysis of such a solution, for example, CaH_2 in $LiCl + KCl$ at 360°, hydrogen is released at the anode. They react instantly and completely with any substance affording even the minutest traces of H^+, such as water, according to the reaction:

$$MH + H^+ = M^+ + H_2$$

The standard potential of the H_2/H^- couple has been estimated to be -2.25 V, making H^- one of the most powerful reducing agents known.

The hydrides are reactive toward air and water, and those of rubidium, cesium, and barium may ignite spontaneously in moist air. Thermal decomposition at high temperatures gives the metal and hydrogen; LiH alone can be melted (m.p. 688°) and it is unaffected by oxygen below red heat or by chlorine or dry HCl.

LiH is seldom used except for the preparation of the more useful complex hydride $LiAlH_4$ discussed in Sect. 10-9. However NaH and CaH_2 are used. NaH is available as a dispersion in mineral oil; although the solid reacts violently with water, the reaction of the dispersion is less violent. It is used extensively in organic synthesis[39] and for the preparation of $NaBH_4$.

CaH_2 reacts smoothly with water and is a useful source of hydrogen (one liter per gram); it is a convenient drying agent for organic solvents and for gases.[39]

6-15. Hydrides of More Covalent Nature

The hydrides BeH_2 and MgH_2 may be obtained by thermal decomposition of the alkyls $[Me_3C]_2Be$ and Et_2Mg, respectively, but MgH_2 can be made by direct interaction of magnesium under pressure or from the alloy Mg_2Cu at 1 atm and 300°, or in a reactive form, by interaction of NaH and $MgBr_2$ in diethyl ether. ZnH_2 has recently been made[40] and is fairly stable.

BeH_2 is difficult to obtain pure, but it is believed to have a polymeric structure with bridging hydrogen atoms as in the boranes (Chapter 9). MgH_2 has a rutile-type structure (Table 6-7) like MgF_2; it also has a low heat of formation and is less stable thermally than the true saline hydrides.

The chemistry of aluminum is surprisingly complex. There is evidence for AlH_3 and Al_2H_6 in the gas phase and AlH_3 is known to form at least six different solid

[39] See L. F. Fieser and M. Fieser, *Reagents for Organic Synthesis,* Wiley, 1967; J. Plesek and S. Hermanck, *Sodium Hydride: Its Use in the Laboratory and Technology,* Butterworths, 1968.
[40] J. J. Watkins and E. C. Ashby, *Inorg. Chem.,* 1974, **13,** 2350.

phases,[41] as well as solid etherates. α-AlH$_3$, which appears to be the most stable solid form, has Al atoms octahedrally surrounded by H atoms in a structure very similar to that of AlF$_3$. AlH$_3$ is a useful reducing agent in organic chemistry, and the products formed are significantly different from those of reduction by LiAlH$_4$. Nitriles can be reduced to amines via complex intermediates, and the reduction of alkyl halides is slower than with LiAlH$_4$ so that carboxyl or ester groups in compounds RCOOH and RCOOR' can be reduced preferentially in presence of R"X.

There are a number of complex hydrides of aluminum; they are discussed in Section 10-9. where also the Lewis acid behavior of AlH$_3$ is considered.

There is no evidence for a comparable gallium hydride, but an unstable viscous oil that shows bands in the infrared spectrum due to Ga—H has been obtained by the reaction:

$$Me_3N\cdot GaH_3(s) + BF_3(g) \xrightarrow{-15°} GaH_3(I) + Me_3NBF_3(g)$$

There is also little evidence for hydrides of indium and thallium; the hydrides of the metallic elements of Groups IV and V are covalent volatile compounds and are considered under their respective elements in later chapters.

6-16.   Transition Metal Hydrides[42]

Hydrogen reacts with many transition metals or their alloys on heating, to give compounds commonly called hydrides even though in some cases they clearly do not contain hydride ions. Many of these metal hydride systems are exceedingly complicated, showing the existence of more than one phase, often with wide divergences from stoichiometry. The most extensive studies have been made on the most electropositive elements, the lanthanides and actinides, and the titanium and vanadium groups of the d-block elements.

Satisfactory theoretical understanding of these substances has been slow to develop. Simple models emphasizing hydridic character, or protonic character or, again, covalent character for the hydrogen, have all been discussed, but it has been convincingly shown in recent years that only by elaborate band structure calculations can a true appreciation of their electronic structures and properties be obtained.[43]

Lanthanide Hydrides. The metals such as lanthanum or neodymium react with H$_2$ at 1 atm and at or slightly above room temperature, to give black solids of graphite-like appearance. These products are pyrophoric in air and react vigorously with water. There are the phases MH$_2$ and MH$_3$ which are nonstoichiometric (e.g.,

[41]   F. M. Brower et al., J. Am. Chem. Soc., 1976, **98**, 2450.

[42]   T. R. P. Gibb, Prog. Inorg. Chem., 1962, **3**, 315; V. I. Mikheeva, Hydrides of the Transition Elements, U.S. Atomic Energy Commission. A.E.C.-tr-5224, 1962. Office of Technical Service, Department of Commerce, Washington, D.C. (a review with 678 references).

[43]   A. C. Switendick, in Ref. 38, Chapter 19.

$LaH_{2.87}$). Normally europium and ytterbium give only the dihydride phase, but higher ratios (e.g., $YbH_{2.55}$) can be obtained at 350° under pressure.

The hydrides appear to be predominantly ionic and to contain M^{3+} ions even in the MH_2 phase where the odd valence electron is probably located in a metallic conduction band as in the so-called dihalides such as LaI_2 (Chapter 23); in $YbH_{2.55}$ there is some evidence for both Yb^{2+} and Yb^{3+}.

Actinide Hydrides. Thorium and other actinides form complex systems with nonstoichiometric and stoichiometric phases. Uranium hydride is of some importance chemically because it is often more suitable for the preparation of uranium compounds than is the massive metal. Uranium reacts rapidly and exothermically with hydrogen at 250 to 300° to give a pyrophoric black powder. The reaction is reversible:

$$U + \tfrac{3}{2}H_2 - UH_3 \qquad \Delta H_f^0 = -129 \text{ kJ mol}^{-1}$$

The hydride decomposes at somewhat higher temperatures to give extremely reactive, finely divided metal. A study of the isostructural deuteride by X-ray and neutron diffraction shows that the deuterium atoms lie in a distorted tetrahedron equidistant from four uranium atoms; no U—U bonds appear to be present, and the U—D distance is 2.32 Å. The stoichiometric hydride UH_3 can be obtained, but the stability of the product with a slight deficiency of hydrogen is greater.

Some typical useful reactions are the following:

$$UH_3 \begin{cases} \xrightarrow{\text{H}_2\text{O, 350°}} UO_2 \\[6pt] \xrightarrow{\text{Cl}_2\text{, 200°}} UCl_4 \\[6pt] \xrightarrow{\text{H}_2\text{S, 450°}} US_2 \\[6pt] \xrightarrow{\text{HF, 400°}} UF_4 \\[6pt] \xrightarrow{\text{HCl, 250–300°}} UCl_3 \end{cases}$$

Hydrides of _d_-Block Transition Metals. Titanium, zirconium, and hafnium absorb hydrogen exothermically to give nonstoichiometric materials such as $TiH_{1.7}$ and $ZrH_{1.9}$. These and the similar hydrides of vanadium, niobium, and tantalum are grayish-black solids similar in appearance and reactivity to the finely divided metal. They are fairly stable in air but react when heated with air or acid reagents. The titanium and zirconium hydrides are used as reducing agents in metallurgical and other processes.

The affinity of many of the other _d_-block elements for hydrogen is small or zero with the exception of the following two special cases.

Palladium.[44] One of the unique characteristics of metallic palladium and Pd–Ag

---

[44]   F. A. Lewis, _The Palladium-Hydrogen System,_ Academic Press, 1967 (a very detailed review). See also T. B. Flanagan and W. A. Oates, in Ref. 38, Chapter 30.

or Pd–Au alloys is the high rate of diffusion of hydrogen gas through a metal membrane compared to the rates for other metals such as nickel or iridium. There is no doubt that pressure-temperature-composition curves indicate the presence of palladium hydride phases.

Copper. There has been much discussion on whether a true hydride exists. It appears that an insoluble CuH with a wurtzite structure (p. 15) can be obtained by reduction of Cu^{2+} solutions by hypophosphorous acid. An amorphous hydride soluble in organic solvents such as pyridine or alkylphosphines can be obtained from the reaction of CuI and $LiAlH_4$ in pyridine.

General References

Alefield, G., and J. Volki, Eds., *Hydrogen in Metals,* Vols. I, II, Springer, 1978.

Augustine, R. L., *Catalytic Hydrogenation,* Arnold-Dekker, 1965.

Bau, R., Ed., *Transition Metal Hydrides,* ACS Advances in Chemistry Series No. 167, 1978.

Emmett, P. H., Ed., *Catalysis,* Vols. 3–5, Reinhold, 1955–1957. These volumes cover various aspects of hydrogenation and other reactions involving hydrogen.

Frankenberg, W. G., V. I. Komarewsky, and E. K. Rideal, Eds., *Advances in Catalysis,* Academic Press, annually from 1948. These volumes discuss various aspects of hydrogenation and other reactions involving hydrogen.

Freifelder, M., *Practical Catalytic Hydrogenation; Technique and Applications,* Wiley, 1971.

Hajós, A., *Complex Hydrides and Related Reducing Agents in Organic Synthesis,* Elsevier, 1979.

Pourbaix, M., *Atlas of Electrochemical Equilibria in Aqueous Solution,* Pergamon Press, 1966.

Rylander, P. N., *Catalytic Hydrogenation over Platinum Metals,* Academic Press, 1967.

Serjeant, E. P., and B. Dempsey, *Ionization Constants of Organic Acids in Aqueous Solutions,* Pergamon Press, 1979.

Siegel, B., *J. Chem. Educ.,* 1961, **38,** 484. A review of the reactions of atomic hydrogen.

Sokolskii, D. V., *Hydrogenation in Solutions,* Oldbourne Press, 1965. Comprehensive Russian source book.

Tanabe, K., *Solid Acids and Bases; Their Catalytic Properties,* Academic Press, 1971. Properties of oxides, sulfates, clay minerals, etc.

Wiberg, E., and E. Amberger, *Hydrides of Elements of Main Groups I–IV,* Elsevier, 1971.

CHAPTER SEVEN

The Group I Elements: Li, Na, K, Rb, Cs

GENERAL REMARKS

7-1. Introduction

The closely related elements lithium, sodium, potassium, rubidium, and cesium, often termed the alkali metals, have a single s electron outside a noble gas core. Some relevant data are listed in Table 7-1.

TABLE 7-1
Some Properties of Group I Metals

Element	Electronic configuration	Metal radius (Å)	Ionization enthalpies (kJ mol^{-1}) 1st	2nd × 10^{-3}	M.p. (°C)	B.p. (°C)	$E°$ [a] (V)	E_{diss} [b] (kJ mol^{-1})
Li	[He]2s	1.52	520.1	7.296	180.5	1326	−3.02	108.0
Na	[Ne]3s	1.86	495.7	4.563	97.8	883	−2.71	73.3
K	[Ar]4s	2.27	418.7	3.069	63.7	756	−2.92	49.9
Rb	[Kr]5s	2.48	402.9	2.640	38.98	688	−2.99	47.3
Cs	[Xe]6s	2.65	375.6	2.26	28.59	690	−3.02	43.6
Fr	[Rn]7s							

[a] For $M^+(aq) + e = M(s)$.
[b] Energy of dissociation of the diatomic molecule M_2.

253

As a result of the low ionization enthalpies for the outer electrons and the sphericity and low polarizability of the resulting M^+ ions the chemistry of these elements is essentially that of their $+1$ ions. No other cations are known, or, in view of the values of the second ionization enthalpies, expected. It is interesting that the Na^- ion probably exists in the stable (at $-10°$) compound $[Na\ (2,2,2\text{-crypt})]^+Na^-$, where 2,2,2-crypt represents a cryptate ligand (cf. p. 163 for its structure). This substance forms[1] on cooling a saturated solution of sodium in ethylamine in the presence of 2,2,2-crypt, and its crystal structure is very similar to that of $[Na(2,2,2\text{-crypt})]^+I^-$. Although the reaction $2Na(g) \rightarrow Na^+(g) + Na^-(g)$ is endothermic by 438 kJ mol^{-1}, the lattice energy and complexing of the cation together overcome this.

Nuclear magnetic resonance has proved useful in studying some aspects of the chemistry of lithium (^7Li, 92.6%, $S = 3/2$), sodium (^{23}Na, 100%, $S = 3/2$), and cesium (^{133}Cs, 100%, $S = 7/2$).

Although the chemistry of the elements is predominantly cationic, some degree of covalent bonding occurs in certain cases. The gaseous diatomic molecules (Na_2, Cs_2, etc.) are covalently bonded, and the bonds to oxygen, nitrogen, and carbon in various chelate and organometallic compounds doubtless have some slight covalent character. In these cases, especially the organometallic compounds, only lithium seems to exhibit significant covalence. There are, however, some curious suboxides, known only for rubidium and cesium, in which there are covalent M—M bonds. These compounds are highly colored, often metallic in appearance; the well-characterized ones[2] include Rb_9O_2, $Rb_{12}O_2$, $Cs_{11}O_3$, and $Cs_{21}O_3$. The first of these consists of confacial bioctahedra of Rb atoms with one oxygen atom in the center of each; the second contains these Rb_9O_2 units separated by an intervening "mortar" of Rb atoms. $Cs_{11}O_3$ contains units consisting of three metal octahedra fused together (7-I), each with an oxygen atom in its center, and $Cs_{21}O_3$ also contains these units separated by a "mortar" of additional Cs atoms.

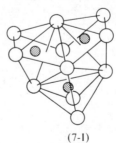

(7-I)

The element *francium* is formed in the natural radioactive decay series and in nuclear reactions. All its isotopes are radioactive with short half-lives. The ion behaves as would be expected from its position in the group.

Of all the groups in the Periodic Table, the Group I metals show most clearly and with least complication the effect of increasing size and mass on chemical and

[1]   J. L. Dye, *J. Chem. Educ.*, 1977, **54**, 322.
[2]   A. Simon, *Struct. & Bonding*, 1979, **36**, 81.

physical properties. Thus all of the following *decrease* through the series: (*a*) melting points and heats of sublimation of the metals; (*b*) lattice energies of all salts except those with the very smallest anions (because of irregular radius ratio effects); (*c*) effective hydrated radii and the hydration energies (see Table 7-2); (*d*) ease of thermal decomposition of nitrates and carbonates; (*e*) strength of the covalent bonds in the M_2 molecules; (*f*) heats of formation of fluorides, hydrides, oxides, and carbides (because of higher lattice energies with the smaller cations). Other trends also can readily be found.

TABLE 7-2

Data on Hydration of Aqueous Group I Ions

	Li^+	Na^+	K^+	Rb^+	Cs^+
Crystal radii[a] (Å)	0.86	1.12	1.44	1.58	1.84
Approximate hydrated radii (Å)	3.40	2.76	2.32	2.28	2.28
Approximate hydration numbers[b]	25.3	16.6	10.5	—	9.9
Hydration energies (kJ mol^{-1})	519	406	322	293	264
Ionic mobilities (at ∞ dil., 18°)	33.5	43.5	64.6	67.5	68

[a] Ladd radii; for Pauling radii see Table 1-1.
[b] From transference data.

The reactivity of the Group I metals toward all chemical reagents except nitrogen increases with increasing electropositive nature (Li → Cs). Usually the least reactive, lithium is only rather slowly attacked by water at 25°, whereas sodium reacts vigorously, potassium inflames, and rubidium and cesium react explosively. With liquid Br_2, Li, and Na barely react, whereas the others do so violently. Lithium does not replace the weakly acidic hydrogen in $C_6H_5C{\equiv}CH$, whereas the other alkali metals do so, yielding hydrogen gas. However with N_2, Li is uniquely reactive to give a ruby-red crystalline nitride Li_3N (Mg also reacts to give Mg_3N_2); at 25° this reaction is slow, but it is quite rapid at 400° and has been studied in detail. Both Li and Mg can be used to remove nitrogen from other gases. When heated with carbon, both Li and Na react to form the *acetylides* Li_2C_2 and Na_2C_2. The heavier alkali metals also react with carbon, but give nonstoichiometric interstitial compounds where the metal atoms enter between the planes of carbon atoms in the lamellar graphite structure. This difference may be attributed to size requirements for the metal, both in the ionic acetylides ($M_2^I C_2^2$) and in the penetration of the graphite.

A particularly fundamental chemical difference between lithium and its congeners, attributable to cation size, is the reaction with oxygen. When the metals are burnt in air or oxygen at 1 atm, lithium forms the oxide Li_2O, with only a trace of Li_2O_2, whereas the other alkali oxides (M_2O) react further, giving as principal products the peroxides M_2O_2 and (from K, Rb, and Cs) the superoxides MO_2.

Although the molten metals Na to Cs are miscible in all proportions, Li is immiscible with K, Rb, or Cs and miscible with Na only above 380°.

The Li^+ ion is exceptionally small and has, therefore, an exceptionally high charge-radius ratio, comparable to that of Mg^{2+}. The properties of a number of *lithium compounds* are therefore anomalous (in relation to the other Group I elements) but resemble those of magnesium compounds. Many of the anomalous properties arise because the salts of Li^+ with small anions are exceptionally stable owing to their very high lattice energies, whereas salts with large anions are relatively unstable owing to poor packing of very large with very small ions. LiH is stable to approximately 900°, but NaH decomposes at 350°. Li_3N is stable, whereas Na_3N does not exist at 25°. Lithium hydroxide decomposes at red heat to Li_2O, whereas the other hydroxides MOH sublime unchanged; LiOH is also considerably less soluble than the other hydroxides. The carbonate Li_2CO_3 is thermally much less stable relative to Li_2O and CO_2 than are other alkali metal carbonates M_2CO_3. The solubilities of Li^+ salts resemble those of Mg^{2+}. Thus LiF is sparingly soluble (0.27 g/100 g H_2O at 18°) and can be precipitated from ammoniacal NH_4F solutions; LiCl, LiBr, LiI and, especially, $LiClO_4$ are soluble in solvents such as ethanol, acetone, and ethyl acetate, and LiCl is soluble in pyridine.

$NaClO_4$ is less soluble than $LiClO_4$ in various solvents by factors of 3 to 12, whereas $KClO_4$, $RbClO_4$, and $CsClO_4$ have solubilities only 10^{-3} of that of $LiClO_4$. Since the spherical ClO_4^- ion is virtually nonpolarizable and the alkali metal perchlorates form ionic crystals, the high solubility of $LiClO_4$ is mainly attributable to strong solvation of the Li^+ ion. LiBr in hot concentrated solution has the unusual property of dissolving cellulose. Lithium sulfate, in contrast to the other M_2SO_4 salts, does not form alums; it is also not isomorphous with the other sulfates.

The elements copper, silver, and gold, the so-called coinage metals, also have a +1 oxidation state but their single s electron is outside a relatively reactive, polarizable d^{10} shell and their overall chemical resemblance to the alkalis is very slight. They are best considered as close relatives of the transition metals, which they resemble in much of their chemistry, such as formation of complexes and variable oxidation state.

It is pertinent that there are other ions that have chemical behavior closely resembling that of the Group I ions:

1. The most important of these are the ammonium ions NH_4^+, RNH_3^+, ..., R_4N^+. Salts of NH_4^+ generally resemble those of potassium quite closely in their solubilities and crystal structures.

2. The thallium(I) ion Tl^+ behaves in certain respects as an alkali metal ion (although in others more like Ag^+). Its ionic radius (1.54 Å) is comparable to that of Rb^+, although it is more polarizable. Thus thallous hydroxide is a water-soluble, strong base, which absorbs carbon dioxide from the air to form the carbonate. The sulfate and some other salts are isomorphous with the alkali metal salts.

3. A variety of other types of monopositive, essentially spherical cation often behave like alkali metal ions of comparable size. For example, the very stable di-(η^5-cyclopentadienyl)cobalt(III) ion and its analogues with similar "sandwich" structures (Chapter 27) have precipitation reactions similar to those of Cs^+, and $[(\eta^5\text{-}C_5H_5)_2Co]OH$ is a strong base that absorbs carbon dioxide from the air and forms insoluble salts with large anions.

THE ELEMENTS

7-2. Preparation, Properties, Uses

Sodium and potassium are in high abundance (2.6 and 2.4%) in the lithosphere and occur in large deposits of sodium chloride and carnallite ($KCl \cdot MgCl_2 \cdot 6H_2O$). Lithium, rubidium, and cesium have much lower abundances and occur mainly in a few silicate minerals.

Lithium and sodium are obtained by electrolysis of fused salts or of low-melting eutectics such as $CaCl_2 + NaCl$. Potassium cannot readily be made by electrolysis, owing to its low melting point and ready vaporization, and it is made by treating molten KCl with sodium vapor in a countercurrent fractionating tower. Rubidium and cesium are made similarly. All the metals are best purified by distillation. Because there is only one valence electron per metal atom, the binding energies in the close-packed metal lattices are relatively weak and the metals are consequently very soft and have low melting points. Liquid alloys of the metals are known, the most important being the Na–K alloys. The eutectic mixture in this system contains 77.2% of K and melts at $-12.3°$. This alloy, which has a wide liquid range and high specific heat, has been considered as a coolant for nuclear reactors, but sodium is used for this purpose. Lithium is relatively light (density 0.53 g/cm^3) and has the highest melting and boiling points and also the longest liquid range of all the alkali metals; it has also an extraordinarily high specific heat. These properties should make it an excellent coolant in heat exchangers, but it is also very corrosive—more so than other liquid metals—which is a great practical disadvantage; it is used to deoxidize, desulfurize, and generally degas copper and copper alloys.

Sodium metal may be dispersed by melting on various supporting solids (sodium carbonate, kieselguhr, etc.) or by high-speed stirring of a suspension of the metal in various hydrocarbon solvents held just above the melting point of the metal. Dispersions of the latter type are commercially available; they may be poured in air, and they react with water only with effervescence. They are often used synthetically where sodium shot or lumps would react too slowly. Sodium and potassium, when dispersed on supports such as carbon or K_2CO_3, are used as catalysts for various reactions of alkenes, notably the dimerization of propene to 4-methyl-1-pentene (cf. the use of Li alkyls discussed below).

The high electrode potentials of the alkali metals make it natural to consider them as possible anode materials for batteries, but only recently has progress been made to overcome serious practical difficulties.[3] The most promising efforts to date employ a LiAl alloy anode, an FeS/FeS_2 cathode, and a molten eutectic LiCl–KCl mixture. The operating temperature is *ca.* 475° with voltages in the range 1.2–1.8

---

[3]   E. J. Cairns and R. K. Steunenberg, *Prog. High Temp. Phys. Chem.,* 1973, **5,** 63.

V. Significant progress has also been made with a battery having a liquid sodium anode, a sulfur cathode, and a β-alumina ($Na_2O\cdot11Al_2O_3$) "electrolyte," and operating at about 300° to give about 2 V.

Another potential use for large quantities of lithium will be to make tritium, by reaction 7-1, for use in the thermonuclear reaction 7-2, which is currently considered the only practical basis for potential power generation by nuclear fusion.

$$Li + n \rightarrow {}^3H + \alpha \tag{7-1}$$

$${}^3H + {}^2H \rightarrow {}^4He + n + 17.6 \text{ MeV} \tag{7-2}$$

Studies of the spectra of Group I metal vapors at about the boiling points of the metals show the presence of ~1% of diatomic molecules whose dissociation energies decrease with increasing atomic number (Table 7-1). These molecules provide the most unambiguous cases of covalent bonding of the alkalis; some $s-p$ hybridization is considered to be involved.

All the metals are highly electropositive and react directly with most other elements. The reactions of substances in liquid sodium have been studied in some detail in view of the use of the metal as a reactor coolant. As noted above, the reactivities toward air and water increase down the group. In air Li, Na, and K tarnish rapidly and the other metals must be handled in an inert atmosphere, as must Na–K alloys. Although Li, Na, K, and Rb are silvery in appearance, Cs has a distinct golden-yellow cast. The metals dissolve in mercury with considerable vigor, to give amalgams. Sodium amalgam (Na/Hg) is a liquid when containing little sodium but a solid when rich in sodium; it is a useful reducing agent and can be employed with aqueous solutions.

The metals also dissolve with reaction in alcohols, to give the alcoxides. Sodium or potassium in ethanol or *tert*-butyl alcohol is commonly used in organic chemistry as a reducing agent and also provides a source of the nucleophilic alcoxide ions. The similar dialkylamides ($LiNR_2$), which are also widely useful synthetically, are best prepared indirectly by reaction of butyllithium (p. 266) with R_2NH.

7-3.   Solutions of Metals in Liquid Ammonia and Other Solvents

The Group I metals, and to a lesser extent calcium, strontium, barium, europium, and ytterbium, are soluble in liquid ammonia and certain other solvents, giving solutions that are blue when dilute. These solutions conduct electricity *electrolytically* and measurements of transport numbers suggest that the main current carrier, which has an extraordinarily high mobility, is the solvated electron. Solvated electrons are now known to be formed in aqueous or other polar media by photolysis, radiolysis with ionizing radiations such as X-rays, electrolysis, and probably some chemical reactions. The high reactivity of the electron and its short lifetime (in 0.75 M $HClO_4$, 6×10^{-11} sec; in neutral water, $t_{1/2}$ *ca.* 10^{-4} sec) make detection of such low concentrations difficult. Electrons can also be trapped in ionic lattices or in frozen water or alcohol when irradiated and again blue colors are observed.

In very pure liquid ammonia the lifetime of the solvated electron may be quite long (1% decomposition per day), but under ordinary conditions initial rapid decomposition occurs with water present, and with glass this is followed by a slower decomposition.

Solutions in ammonia and other solvents have been extensively studied and it is agreed that in *dilute solutions* the metal is dissociated into solvated metal ions M^+ and electrons. The broad absorption around 15,000 Å accounts for the common blue color; since the metal ions are colorless, this absorption must be associated with the solvated electrons. Magnetic and electron spin resonance studies show the presence of "free" electrons, but the decrease in paramagnetism with increasing concentration suggests that the ammoniated electrons can associate to form dia-magnetic species containing electron pairs. Although there may be other equilibria, the data can be accommodated by equilibria such as the following:

$$Na(s) \text{ (dispersed)} \rightleftharpoons Na \text{ (in solution)} \rightleftharpoons Na^+ + e$$
$$2e \rightarrow e_2$$

Just how the electrons are associated with the ammonia molecules or the solvated metal ions is still a matter of discussion.[4] However the most satisfactory models assume that the electron is not localized but is "smeared out" over a large volume so that the surrounding solvent molecules experience electronic and orientational polarization. The electron is trapped in the resultant polarization field, and repulsion between the electron and the electrons of the solvent molecules leads to the for-mation of a cavity within which the electron has the highest probability of being found. In ammonia this is estimated to be approximately 3 to 3.4 Å in diameter; this cavity concept is based on the fact that solutions are of much lower density than the pure solvent; that is, they occupy far greater volume than that expected from the sum of the volumes of metal and solvent.

There is evidence for formation of metal ion clusters as the concentration of metal increases. At concentrations $3M$ or above, the solutions are copper colored and have a metallic luster, and in various physical properties, such as their exceedingly high electrical conductivities, they resemble liquid metals. When 20% solutions of Li in liquid ammonia are cooled, a golden-yellow conducting solid $Li(NH_3)_4$ is ob-tained. Similar behavior is also shown by the Group II elements, which also give solids, usually nonstoichiometric but approximating to $M(NH_3)_6$.

The metals are also soluble to varying degrees in other amines, and sodium and potassium are soluble in hexamethylphosphoramide, $P(NMe_2)_3$. Fairly stable so-lutions of potassium, rubidium, and cesium have been obtained in tetrahydrofuran, ethylene glycol dimethyl ether, and even in diethyl ether containing cyclic polyethers that form complexes with the alkali metal ions. The general properties of these solutions, insofar as they have been determined in view of the attack on the solvents, appear to be similar to those of the amine and liquid ammonia solutions; the alkali metal concentrations in saturated ether solutions are, however, only $\sim 10^{-4}$ mol l^{-1}.

To make possible the formation of such metal solutions, the dielectric constant of the solvent is important in the same way as in the solution of an ionic solid, namely, to diminish the forces of attraction between the oppositely charged par-

4   M. R. C. Symons, *Chem. Soc. Rev.*, 1976, **5**, 337.

ticles—in this case, M^+ ions and electrons. Furthermore, if the solvent molecules immediately surrounding these particles interact strongly with them, the energy of the system is further lowered. The details of the interaction of the electrons with the surrounding solvent molecules are still debatable, but evidently the metal ions are solvated in the same way as they would be in a solution of a metal salt in the same solvent (see discussion below).

The ammonia and amine solutions of alkali metals are widely used in preparing both organic and inorganic compounds. Thus lithium in methylamine shows great selectivity in its reducing properties, but both this reagent and lithium in ethylenediamine are quite powerful and can reduce aromatic rings to cyclic monoolefins. Sodium in liquid ammonia is probably the most widely used system for preparative purposes. The ammonia solution is moderately stable, but the decomposition reaction

$$Na + NH_3(l) = NaNH_2 + \tfrac{1}{2}H_2$$

can occur photochemically and is catalyzed by transition metal salts. Sodium amide can be conveniently prepared by treatment of sodium with liquid ammonia in the presence of a trace of ferric chloride. Amines react similarly.

For the amides of K, Rb, and Cs it has been shown that the reaction

$$e + NH_3 \rightleftharpoons NH_2^- + \tfrac{1}{2}H_2 \qquad K = 5 \times 10^4$$

is reversible, but for $LiNH_2$ and $NaNH_2$ which are insoluble in liquid ammonia, we have, for example

$$Na^+(am) + e^-(am) + NH_3(l) = NaNH_2(s) + \tfrac{1}{2}H_2 \qquad K = 3 \times 10^9$$

COMPOUNDS OF THE GROUP I ELEMENTS

7-4. Binary Compounds

The Group I metals react directly with most nonmetals to give one or more binary compounds; they also form numerous alloys and compounds with other metals such as lead and tin.

The most important are the *oxides,* obtained by combustion as noted in Section 7-1. Although sodium normally gives Na_2O_2, it will take up further oxygen at elevated pressures and temperatures to form NaO_2. The per- and superoxides of the heavier alkalis can also be prepared by passing stoichiometric amounts of oxygen into their solutions in liquid ammonia, and ozonides (MO_3) are also known. The structures of the ions O_2^{2-}, O_2^-, and O_3^- and of their alkali salts are discussed in Section 15-7. The increasing stability of the per- and superoxides as the size of the alkali ions increases is noteworthy and is a typical example of the stabilization of larger anions by larger cations through lattice energy effects.

Owing to the highly electropositive character of the metals, the various oxides (and also sulfides and similar compounds) are readily hydrolyzed by water according to the following equations:

$$M_2O + H_2O = 2M^+ + 2OH^-$$
$$M_2O_2 + 2H_2O = 2M^+ + 2OH^- + H_2O_2$$
$$2MO_2 + 2H_2O = O_2 + 2M^+ + 2OH^- + H_2O_2$$

The oxide Cs_2O has the *anti*-$CdCl_2$ structure and is the only known oxide with this type of lattice. An abnormally long Cs—Cs distance and a short Cs—O distance imply considerable polarization of the Cs^+ ion.

Both rubidium and cesium form nonstoichiometric metallic suboxides.

The *hydroxides* MOH are white crystalline solids soluble in water and in alcohols. They can be sublimed unchanged at 350 to 400°, and the vapors consist mainly of dimers $(MOH)_2$. KOH at ordinary temperatures is monoclinic with each K surrounded by a distorted octahedron of O atoms while the OH groups form a zigzag hydrogen-bonded chain with O to O distances of 3.35 Å, which rules out any significant hydrogen bonding. There is also a cubic high-temperature form.

Measurements of the proton affinities of MOH in the gas phase show that the base strength increases from lithium to cesium, but this order need not be observed in aqueous or alcoholic solutions where the base strength of the hydroxide is reduced by solvent effects and hydrogen bonding. In suspension in nonhydroxylic solvents such as 1,2-dimethoxyethane, the hydroxides are exceedingly strong bases and can coveniently be used to deprotonate a wide variety of weak bases such as PH_3 ($pK \approx 27$) or C_5H_6 ($pK \approx 16$). The driving force for the reaction is provided by the formation of the stable hydrate:

$$2KOH(s) + HA = K^+A^- + KOH \cdot H_2O(s)$$

The alkali metals form a multitude of compounds with the elements of Groups IIIB to VIB, only a few of which can be thought of in *simple* ionic terms (i.e., in terms of M^+ ions and anions with complete octets, such as Na_2S, K_3P). The vast majority are far richer in the metalloidal element (e.g., NaP_7, $SrSi_2$, $LiGe$) and contain complex polynuclear anionic structures. These materials are structurally and electronically transitional between ionic compounds and alloys. Most of them can be made either by direct reaction of the elements or by reaction of liquid ammonia solution of the alkali metals with compounds of the metalloidal components. Such substances are often called *Zintl* compounds or phases, after the chemist who first extensively studied them.[5] Many of them are mentioned elsewhere in connection with the metal or metalloid element.

7-5. Ionic Salts

Salts of the bases MOH with virtually all acids are known. For the most part they are colorless, crystalline, ionic solids. Those that are colored owe this property to the anions, except in special cases. The colors of metal ions are due to absorption

5   H. Schäfer, B. Eisenmann, and W. Müller, *Angew. Chem., Int. Ed.,* 1973, **12**, 694.

of light of proper energy to excite electrons to higher energy levels; for the alkali metal ions with their very stable noble gas configurations, the energies required to excite electrons to the lowest available empty orbitals could be supplied only by quanta far out in the vacuum ultraviolet (the transition $5p^6 \rightarrow 5p^56s$ in Cs^+ occurs at about 1000 Å). However colored crystals of compounds such as NaCl are sometimes encountered. This is due to the presence in the lattice of holes and free electrons, called color centers, and such chromophoric disturbances can be produced by irradiation of the crystals with X-rays and nuclear radiation. The color results from transitions of the electrons between energy levels in the holes in which they are trapped. These electrons behave in principle similarly to those in solvent cages in the liquid ammonia solutions, but the energy levels are differently spaced and consequently the colors are different and variable. Small excesses of metal atoms produce similar effects, since these atoms form M^+ ions and electrons that occupy holes where anions would be in a perfect crystal.

The structures and stabilities of the ionic salts are determined in part by the lattice energies and by radius ratio effects, which have been discussed in Chapter 1. Thus the Li^+ ion is usually tetrahedrally surrounded by water molecules or negative ions, although $Li(H_2O)_6^+$ has also been found.[6a] On the other hand, the large Cs^+ ion can accommodate eight near-neighbor Cl^- ions, and its structure is different from that of NaCl, where the smaller cation Na^+ can accommodate only six near neighbors. The Na^+ ion appears to be six-coordinate in some nonaqueous solvents.[6b]

The salts are generally characterized by high melting points, by electrical conductivity of the melts, and by ready solubility in water. They are seldom hydrated when the anions are small, as in the halides,[6c] because the hydration energies of the ions are insufficient to compensate for the energy required to expand the lattice. Owing to its small size, the Li^+ ion has a large hydration energy, and it is often hydrated in its solid salts when the same salts of other alkalis are unhydrated (viz., $LiClO_4 \cdot 3H_2O$). For salts of *strong* acids, the lithium salt is usually the *most* soluble in water of the alkali metal salts, whereas for *weak* acids the lithium salts are usually *less* soluble than those of the other alkalis.

The large size of the Cs^+ and Rb^+ ions frequently allows them to form ionic salts with rather unstable anions, such as various polyhalide anions (Section 17-15) and the superoxides already mentioned.

Since there are few salts that are not appreciably water soluble, there are few important *precipitation reactions* of the aqueous ions. Generally the larger the M^+ ion, the more numerous are its insoluble salts. Thus sodium has very few insoluble salts; the mixed sodium zinc and sodium magnesium uranyl acetates [e.g., $NaZn(UO_2)_3(CH_3COO)_9 \cdot 6H_2O$], which may be precipitated almost quantitatively under carefully controlled conditions from dilute acetic acid solutions, are useful for analysis. The perchlorates and hexachloroplatinates of K, Rb, and Cs are rather

[6a] A. Sequeira et al., *Acta Crystallogr., B*, 1975, **31**, 1735.

[6b] N. Ahmeed and M. C. Day, *J. Inorg. Nucl. Chem.*, 1978, **40**, 1383.

[6c] For a detailed discussion of the factors involved in the solubilities of alkali halides in water, see J. Elson, *J. Chem. Educ.*, 1969, **46**, 86.

insoluble in water and virtually insoluble in 90% ethanol. These heavier ions may also be precipitated by cobaltinitrite ion $[Co(NO_2)_6]^{3-}$ and various other large anions. Sodium tetraphenylborate $NaB(C_6H_5)_4$, which is moderately soluble in water, is a useful reagent for precipitating the tetraphenylborates of K, Rb, and Cs from neutral or faintly acid aqueous solutions, and quantitative gravimetric determinations of these ions may be made.

Sodium chloride and some other alkali metal salts (MX) can be precipitated from aqueous solution[7a] with the ligand 7-II as, for example, NaL_3Cl, in which each Na^+ has distorted octahedral coordination by NH_2 groups from six different ligands.[7b]

(7-II)

7-6. The M^+ Ions in Solution

The first coordination sphere, or primary hydration shell, for Li^+ in aqueous solution is doubtless tetrahedral. Only tetrahydrated salts are formed except when there is also hydration of the anions. X-ray scattering studies show that the primary hydration number of K^+ is 4, and since Na^+ is known to form the stable $Na(NH_3)_4^+$ ion in liquid ammonia (see below), it too presumably has a first coordination sphere of four water molecules. There is no direct evidence regarding Rb^+ or Cs^+, but a higher number, probably 6, seems likely.

In all cases electrostatic forces operate beyond the first shell, and additional water molecules are bound in layers of decreasing firmness. The larger the cation itself, the less it binds outer layers. Thus although the crystallographic radii *increase* down the group the hydrated radii *decrease* (Table 7-2). The hydration number of Li^+ is very large and Li^+ salt solutions generally deviate markedly from ideal solution behavior, showing abnormal colligative properties such as very low vapor pressures and freezing points. Also, hydration energies of the gaseous ions decrease. The decrease in size of the hydrated ions is manifested in various ways. The mobility of the ions in electrolytic conduction increases, and so generally does the strength of binding to ion-exchange resins.

In a cation-exchange resin two cations compete for attachment at anionic sites in the resin, as in the following equilibrium:

$$A^+(aq) + [B^+R^-](s) = B^+(aq) + [A^+R^-](s)$$

where R represents the resin and A^+ and B^+ the cations. Such equilibria have been measured quite accurately, and the order of preference of the alkali cations is usually $Li^+ < Na^+ < K^+ < Rb^+ < Cs^+$, although irregular behavior does occur in some

[7a] N. P. Marullo *et al., Inorg. Chem.,* 1974, **13,** 115.
[7b] L. A. Duvall and D. P. Miller, *Inorg. Chem.,* 1974, **13,** 120.

cases. The usual order may be explained if we assume that the binding force is essentially electrostatic and that under ordinary conditions the ions within the water-logged resin are hydrated approximately as they are outside it. Then the ion with the smallest hydrated radius (which is the one with the largest "naked" radius) will be able to approach most closely to the negative site of attachment and will hence be held most strongly according to Coulomb's law. The efficiency of separating alkali metal ions on cation exchangers can be increased significantly by adding chelating agents like EDTA to the eluting solution. These agents bind more strongly to the ions that are less strongly held to the resin, thus enhancing the separation factors.

Of importance in connection with the solubility of the metals in liquid ammonia are ammonia solvates such as the $[Na(NH_3)_4]^+$ ion, which is formed on treatment of NaI with liquid ammonia. $[Na(NH_3)_4]I$ is a liquid of fair thermal stability. It freezes at 3° and at 25° has an equilibrium pressure of NH_3 of 420 torr; thus it must be kept in an atmosphere of ammonia with at least this pressure at 25°. The infrared and Raman spectra indicate the complex ion $[Na(NH_3)_4]^+$ to be tetrahedral with Na—N bonds about as strong as the Zn—N bonds in $[Zn(NH_3)_4]^{2+}$ or the Pb—C bonds in $Pb(CH_3)_4$. Bending and rocking frequencies, however, are quite low, suggesting that the Na—N bonding is mainly due to ion-dipole forces. Thus it may be assumed that Na^+ and other metal ions in the dilute liquid ammonia, amine, and ether solutions are strongly solvated in the same way.

The effectiveness of tetrahydrofuran and the dimethyl ethers of ethylene and diethylene glycols ("glyme" and "diglyme," respectively) as media for reactions involving sodium may be due in part to the slight solubility of the metal, but the solvation of ions by ether molecules undoubtedly provides the most important contribution. Indeed there are numerous very stable and often crystalline solvates for the M^+ (and also M^{2+}) ions with polydentate cyclic and polycyclic ligands and they are discussed in detail in Section 7-7.

7-7.   Complexes

The alkali metal ions are only weakly complexed by simple anions and hardly at all by monodentate neutral ligands. Chelation is a necessary condition for significant complexation, and complexes are formed with β-diketones, nitrophenols, 1-nitroso-2-naphthol, etc.[8] Some of these, such as the hexafluoroacetylacetone, are sublimable at 200°, though the metal-ligand bonds are doubtless quite polar. The anhydrous β-diketonates are usually insoluble in organic solvents, indicating an ionic nature, but in presence of additional coordinating ligands, including water, they may become soluble even in hydrocarbons; for example, sodium benzoylacetylacetonate dihydrate is soluble in toluene, whereas tetramethylethylenediaminelithium hexafluoroacetylacetonate is monomeric in benzene.

This behavior has allowed the development of solvent-extraction procedures for alkali metal ions. Thus not only can the trioctylphosphine oxide adduct Li-$(PhCOCHCOPh)[OP(octyl)_3]_2$ be extracted from aqueous solutions into p-xylene, but also this process can be used to separate lithium from other alkali metal ions.

8   A. J. Layton et al., J. Chem. Soc., A, **1970,** 1894.

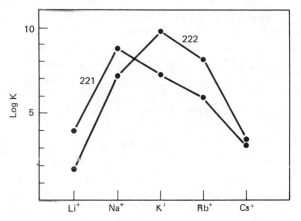

Fig. 7-1.   Stability constants for cryptate-221 and -222 versus alkali ion. [Adapted from J. M. Lehn and J. P. Sauvage, *J. Am. Chem. Soc.*, 1975, **97**, 6700.]

Furthermore Cs^+ can be extracted from aqueous solutions into hydrocarbons by 1,1,1-trifluoro-3-(2'-thenoyl)acetone (TTA) in presence of nitromethane.

The most stable and important complexes of the alkali metal ions are those with macrocyclic polyethers and cryptates (defined on p. 77). Such ligands bind alkali metal ions much more strongly than any other ligands and with great selectivity as well.[9,10] The affinity of such a ligand for an ion is strongly dependent on how well the ion fits into the cavity that the ligand can provide for it. At the same time the strength of complexation and to some extent the selectivity also depend on the solvent. The size effect is the simplest to visualize and has been examined in detail. Illustrations of selectivity, provided by cryptate-221 and cryptate-222 in 95–5 methanol/water solvent, are shown in Fig. 7-1. For cryptate-222, for example, K^+ fits the cavity very well, but Li^+ and Na^+ are too small to make good contacts with the oxygen atoms and Rb^+ and Cs^+ are too large to enter without appreciable steric strain. An example of an alkali cation, Rb^+, occupying the cavity in cryptate-222, and coordinated by the six oxygen and two nitrogen atoms, is presented in Fig. 4-6, which gives the results of an X-ray study of a crystalline compound.

Macrocyclic compounds with ion-binding properties occur naturally and play a role in transporting alkali and alkaline earth ions across membranes in living systems. Many of these are small cyclic polypeptides, of which valinomycin (7-III), is perhaps the best-known example. Several other important naturally-occurring complexes have already been discussed (p. 164).

$$\left(\left(\begin{array}{c} \underset{\displaystyle\overset{\displaystyle CH(CH_3)_2}{|}}{O-\underset{H}{C}-\underset{\|}{C}-N-\underset{H}{\overset{CH(CH_3)_2}{C}}-\underset{H}{\overset{}{C}}-O-\underset{H}{\overset{CH_3}{C}}-\underset{\|}{C}-N-\underset{H}{\overset{CH(CH_3)_2}{C}}-\underset{\|}{C}} \end{array}\right)_3\right) \qquad (7\text{-}III)$$

[9]   W. E. Morf and W. Simon, *Helv. Chim. Acta*, 1971, **54**, 2683.
[10]   J. M. Lehn and J. P. Sauvage, *J. Am. Chem. Soc.*, 1975, **97**, 6700.

It may be noted finally that $^7Li(92.7\%)$ gives nuclear magnetic resonance signals comparable to those given by ^1H, so that complex formation in aqueous solutions can be studied; in this way it was shown that nitrilotriacetic acid $[N(CH_2COOH)_3, H_3NTA]$ probably forms a complex ion $[Li(NTA)_2]^{5-}$ in solution.

7-8.　Organometallic Compounds[11]

One of the most important areas of the chemistry of Group I elements is that of their organic compounds. The most important are those of lithium; organosodium compounds, and to a lesser extent organopotassium ones, are of limited use.

Lithium Alkyls and Aryls.　One of the largest uses of metallic lithium, industrially and in the laboratory, is for the preparation of organolithium compounds. These are of great importance and utility; in their reactions they generally resemble Grignard reagents, although they are usually more reactive. Their preparation is best accomplished by using an alkyl or aryl chloride (eq. 7-3) in benzene or petroleum; ether solutions can be used, but these solvents are attacked slowly by the lithium compounds. Metal-hydrogen exchange (eq. 7-4), metal-halogen exchange (eq. 7-5) and metal-metal exchange (eq. 7-6) may also be used.

$$C_2H_5Cl + 2Li = C_2H_5Li + LiCl \tag{7-3}$$

$$n\text{-}C_4H_9Li + Fe = Fe + n\text{-}C_4H_{10} \tag{7-4}$$

$$n\text{-}C_4H_9Li + \quad = \quad + n\text{-}C_4H_9Br \tag{7-5}$$

$$2Li + R_2Hg = 2RLi + Hg \tag{7-6}$$

n-Butyllithium in hexane, benzene, or ethers is commonly used for such reactions. Methyllithium is also prepared by exchange through the interaction of $n\text{-}C_4H_9Li$ and CH_3I in hexane at low temperatures, whence it precipitates as insoluble white crystals.

Organolithium compounds all react rapidly with oxygen, being usually spontaneously flammable in air, with liquid water, and with water vapor.

Organolithium compounds are among the very few alkali metal compounds that have properties—solubility in hydrocarbons or other nonpolar liquids and high volatility—typical of covalent substances. They are generally liquids or low-melting solids, and molecular association is an important structural feature.

[11]　G. E. Coates and K. Wade, *Organometallic Compounds,* 3rd ed., Vol. 1, Methuen, 1967.

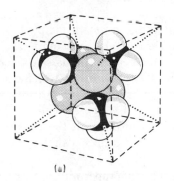
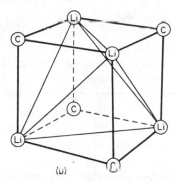

(a)   (b)

Fig. 7-2.   The structure of $(CH_3Li)_4$. (a) Showing the tetrahedral Li_4 unit with the CH_3 groups located symmetrically above each face of the tetrahedron. [Adapted from E. Weiss and E. A. C. Lucken, *J. Organomet. Chem.*, 1964, **2**, 197.] The structure can also be regarded as derived from a cube (b).

In the crystals of methyl- (Fig. 7-2) and ethyllithium (EtLi, m.p. 90°) the lithium atoms are at the corners of a tetrahedron with the alkyl groups centered over the facial planes. Although the CH_3 group is symmetrically bound to three Li atoms, the α carbon of the C_2H_5 group is closer to one Li atom than the other two. In ether solution methyllithium forms complexes with LiBr and LiI, believed to have formulas Li_4Me_3Br, $Li_4Me_2Br_2$, and Li_4Me_3I, presumably by Br or I replacing CH_3 in the tetrahedral structure of Li_4Me_4.[12]

The alkyl bridge bonding is of the electron-deficient multicenter type found in beryllium and aluminum alkyls and in boranes. Aggregate formation is due to Li—C—Li rather than to Li—Li bonding interactions, although there may be some direct Li . . . Li contribution of a secondary nature. Varied estimates of the importance of this have been given, and it remains a controversial point.

Lithium alkyls and aryls are associated in solutions, but the nature of the species depends on the nature of the solvent, the steric nature of the organic radical, and the temperature. Cryoscopy and nmr study with ^{13}C, ^7Li, and ^1H resonances show that in hydrocarbon solvents MeLi, EtLi, Pr^nLi, and some others are hexamers, but *tert*-butyllithium, which presumably is too bulky, is only tetrameric. On addition of ethers or amines, or in these as solvent, solvated tetramers are formed. The formation of dimers, or aggregates less than tetramers, seems not to occur.

However when chelating ditertiary amines, notably tetramethylethylenediamine (TMED, $Me_2NCH_2CH_2NMe_2$) are used, comparatively stable monomeric alkyllithium complexes are obtained. The alkyls and aryls also form complexes with other metal alkyls such as those of magnesium, cadmium, and zinc, e.g.,

$$2LiC_6H_5 + Mg(C_6H_5)_2 = Li[Mg(C_6H_5)_4]$$

This type of complexing, as well as the rates and mechanisms of alkyl-exchange reactions in solution, have been studied by nmr methods.

It is not surprising that there are wide variations in the comparative reactivities of Li alkyls depending on the differences in aggregation and ion-pair interactions.

12   D. P. Novak and T. L. Brown, *J. Am. Chem. Soc.*, 1972, **94**, 3793.

An example is benzyllithium, which is monomeric in tetrahydrofuran and reacts with a given substrate more than 10^4 times as fast as the tetrameric methyllithium. The monomeric TMED complexes noted above are also very much more reactive than the corresponding aggregated alkyls. Alkyllithiums can polylithiate acetylenes, acetonitrile, and other compounds; thus $CH_3C\equiv CH$ gives Li_4C_3, which can be regarded as a deviative of C_3^{4-}. Polylithio compounds, especially of polyacetylenes, have reactivity patterns that are not entirely understood.[13] The difficulties center around the extent to which the π-electronic structures resemble those expected for simple anions, an approximation that is obviously at variance with the presence of covalence in the Li—C bonds.

Perlithio compounds can be made by reaction of lithium vapor with either carbon vapor[14] or CCl_4 or C_2Cl_6.[15] The Li_4C_3 so obtained is evidently the same as that already known; the other substances have not yet been fully characterized. The structures of these perlithio compounds are unknown. For Li_4C_3 *ab initio* MO calculations suggest a curious structure with an $LiCC(Li)_2CLi$ arrangement, not one of the hydrocarbonlike Li_3C—$C\equiv CLi$ or Li_2C=C=CLi_2 structures.[16]

Reactions of lithium alkyls are generally considered to be carbanionic, but in reactions with alkyl halides free radicals have been detected by esr. Lithium alkyls are widely employed as stereospecific catalysts for the polymerization of alkenes, notably isoprene, which gives up to 90% of 1,4-*cis*-polyisoprene; numerous other reactions with alkenes have been studied. The TMED complexes again are especially active: not only do they polymerize ethylene, but they even metallate benzene and aromatic compounds, as well as reacting with hydrogen at 1 atm to give LiH and alkane.

7-9. Organosodium and -potassium Compounds[17]

The organosodium and -potassium compounds are all essentially ionic and are not soluble to any appreciable extent in hydrocarbons; they are exceedingly reactive, being sensitive to air and hydrolyzed vigorously by water. Although alkyl- and particularly arylsodium derivatives can be prepared for use as reaction intermediates *in situ,* they are seldom isolated. However methylpotassium, which is a highly pyrophoric substance, has been obtained by the reaction:

$$(CH_3)_2Hg + 2K\text{-}Na \rightarrow 2CH_3K + Na\text{-}Hg$$

KCH_3 as well as $RbCH_3$, have the NiAs-type structure (Fig. 16-4), and the isolated methyl groups are presumably in the lattice as rotating and/or rotationally disordered pyramidal CH_3^- ions.[18]

13   W. Priester and R. West, *J. Am. Chem. Soc.,* 1976, **98,** 8426 and earlier references cited therein.

14   L. A. Shimp and R. J. Lagow, *J. Am. Chem. Soc.,* 1976, **95,** 1343.

15   C. Chung and R. J. Lagow, *J. C. S. Chem. Comm.,* **1972,** 1078.

16   E. D. Jemmis *et al., J. Am. Chem. Soc.,* 1977, **99,** 5796.

17   M. Schlosser, *Angew. Chem. Int. Ed.,* 1964, **3,** 287, 362.

18   E. Weiss and H. Köster, *Chem. Ber.,* 1977, **110,** 717.

Sodium and potassium alkyls can be used for metallation reactions as, for example, in eq. 7-4. They can also be prepared from sodium or potassium dispersed on an inert support material, and such solids act as carbanionic catalysts for the cyclization, isomerization, or polymerization of alkenes. The so-called alfin catalysts for copolymerization of butadiene with styrene or isoprene to give rubbers consist of sodium alkyl (usually allyl) and alcoxide (usually isopropoxide) and NaCl, which are made simultaneously in hydrocarbons.

More important are the compounds formed by acidic hydrocarbons such as cyclopentadiene, indene, and acetylenes. These are obtained by reaction with sodium in liquid ammonia or, more conveniently, sodium dispersed in tetrahydrofuran, glyme, diglyme, or dimethylformamide.

$$3C_5H_6 + 2Na \rightarrow 2C_5H_5^-Na^+ + C_5H_8$$
$$RC{\equiv}CH + Na \rightarrow RC{\equiv}C^-Na^+ + {}^1/_2H_2$$

Many aromatic hydrocarbons, as well as aromatic ketones, triphenylphosphine oxide, triphenylarsine, azobenzene, etc., can form highly colored *radical anions*[19] when treated at low temperatures with sodium or potassium in solvents such as tetrahydrofuran. For the formation of such anions it must be possible to delocalize the negative charge over an aromatic system. Species such as the benzenide $(C_6H_6^-)$, naphthalenide, or anthracenide ions can be detected and characterized spectroscopically and by electron spin resonance.[20] The sodium-naphthalene system $Na^+[C_{10}H_8]^-$ in an ether is widely used as a powerful reducing agent (e.g., in nitrogen-fixing systems employing titanium catalysts and for the production of complexes in low oxidation states. The blue solution of sodium and benzophenone in tetrahydrofuran, which contains the "ketyl" or radical ion, is a useful and rapid reagent for the removal of traces of oxygen from nitrogen.

[19]   E. de Boer, *Adv. Organomet. Chem.,* 1965, **2**, 115; E. T. Kaiser and J. L. Kevan, *Radical Ions,* Wiley, 1968.

[20]   N. Hirota, *J. Amer. Chem. Soc.,* 1968, **90**, 3603; K. Höfelmann, J. Jagur-Grodzinski, and M. Szwarc, *J. Am. Chem. Soc.,* 1969, **91**, 4645.

Further Reading

The Alkali Metals, Special Publication No. 22, Chemical Society, London, 1967. Symposium reports on wide variety of topics varying from binary compounds to reactions in the liquid metals.

Advances in Chemistry Series, No. 19, "*Handling and Uses of Alkali Metals,*" American Chemical Society, Washington, D.C., 1957. Recovery, handling, and manufacture of metals, hydrides and oxides of Li, Na, and K.

Advances in Chemistry Series, No. 130, *Polyamine Chelated Alkali Metal Compounds,* American Chemical Society, Washington, D.C., 1974.

Bulletins of the Foote Mineral Company, Route 2, Exton, Pa. 19341. A variety of publications giving physical and chemical data and describing uses of lithium and its compounds.

Fatt, I., and M. Tashima, *Alkali Dispersions,* Van Nostrand, 1962. Extensive review with preparative details.

Jackson, C. B., Ed., *Liquid Metals Handbook,* 3rd ed. (Sodium, NaK Suppl), Atomic Energy Commission and Bureau of Ships, U.S. Department of the Navy, 1955.

Jolly, W. L., *Metal-Ammonia Solutions,* Dowden, Hutchinson and Ross, Stroudsburg, Pa., 1972.

Kaiser, E. M., "Organometallic Chemistry of the Heavier Alkali Metals, 1974," *J. Organomet. Chem.,* 1975, **103,** 1–38.

Kapoor, P. N. and R. C. Mehrotra, *Coord. Chem. Rev.,* 1974, **14,** 1. Coordination compounds with covalent characteristics.

Kaufmann, D. W., *Sodium Chloride* (American Chemical Society Monograph No. 145), Reinhold, 1960. An Encyclopedic account of salt.

Lagowski, J. J., and M. J. Sienko, Eds., *Metal-Ammonia Solutions,* Butterworths, 1970.

Mellor's Comprehensive Treatise on Inorganic and Theoretical Chemistry, Vol. II, Suppl. 2, Li, Na (1961); Supplement 3, K, Rb, Cs, Fr (1963), Longmans Green.

Midgley, D., "Alkali-Metal Complexes in Aqueous Solution," *Chem. Soc. Rev.,* 1975, **4,** 549.

Perel'man, F. M., *Rubidium and Caesium,* Pergamon Press, 1965. Comprehensive reference book.

Structure and Bonding, Vol. 16, Springer-Verlag, 1973. This entire volume deals with alkali metal complexes of organic ligands, especially the crown ether and cryptate types.

Wakefield, B. J., *The Chemistry of Organolithium Compounds,* Pergamon Press, 1974.

CHAPTER EIGHT

Beryllium and the Group II Elements: Mg, Ca, Sr, Ba, Ra

GENERAL REMARKS

8-1. Group Relationships

Some pertinent data for Group II elements are given in Table 8-1. Briefly, beryllium has unique chemical behavior with a predominantly covalent chemistry, although forming a cation $[Be(H_2O)_4]^{2+}$. Magnesium, the second-row element, has a chemistry intermediate between that of Be and the heavier elements, but it does not stand in as close relationship with the predominantly ionic heavier members as might have been expected from the similarity of Na, K, Rb, and Cs. It has con-

TABLE 8-1
Some Physical Parameters for the Group II Elements

Element	Electronic configuration	M.D. (°C)	Ionization enthalpies (kJ mol^{-1}) 1st	Ionization enthalpies (kJ mol^{-1}) 2nd	E^0 for M^{2+} (aq) + 2e = M(s) (V)	Ionic radii (Å)[a]	Charge radius
Be	[He]$2s^2$	1278	899	1757	-1.70[b]	0.31	6.5
Mg	[Ne]$3s^2$	651	737	1450	-2.37	0.78	3.1
Ca	[Ar]$4s^2$	843	590	1146	-2.87	1.06	2.0
Sr	[Kr]$5s^2$	769	549	1064	-2.89	1.27	1.8
Ba	[Xe]$6s^2$	725	503	965	-2.90	1.43	1.5
Ra	[Rn]$7s^2$	700	509	979	-2.92	1.57	1.3

[a] Ladd radii.
[b] Estimated.

siderable tendency to covalent bond formation, consistent with the high charge/ radius ratio. For instance, like beryllium, its hydroxide can be precipitated from aqueous solutions, whereas hydroxides of the other elements are all moderately soluble, and it readily forms bonds to carbon.

The metal atomic radii are smaller than those of the Group I metals owing to the increased nuclear charge; the number of bonding electrons in the metals is twice as great, so that the metals have higher melting and boiling points and greater densities.

All are highly electropositive metals, however, as is shown by their high chemical reactivities, ionization enthalpies, standard electrode potentials and, for the heavier ones, the ionic nature of their compounds. Although the energies required to vaporize and ionize these atoms to the M^{2+} ions are considerably greater than those required to produce the M^+ ions of the Group I elements, the high lattice energies in the solid salts and the high hydration energies of the $M^{2+}(aq)$ ions compensate for this, with the result that the standard potentials are similar to those of the Li-Cs group.

The potential E^0 of beryllium is considerably lower than those of the other elements, indicating a greater divergence in compensation by the hydration energy, the high heat of sublimation and the ionization enthalpy. As in Group I, the smallest ion crystallographically (i.e. Be^{2+}) has the largest hydrated ionic radius.

All the M^{2+} ions are smaller and considerably less polarizable than the isoelectronic M^+ ions. Thus deviations from complete ionicity in their salts due to polarization of the cations are even less important. However for Mg^{2+} and, to an exceptional degree for Be^{2+}, polarization of anions by the cations does produce a degree of covalence for compounds of Mg and makes covalence characteristic for Be. Accordingly only an estimated ionic radius can be given for Be; the charge/ radius ratio is greater than for any other cation except H^+ and B^{3+}, which again do not occur as such in crystals. The closest ratio is that for Al^{3+} and some similarities between the chemistries of Be and Al exist. Examples are the resistance of the metal to attack by acids owing to formation of an impervious oxide film on the surface, the amphoteric nature of the oxide and hydroxide, and Lewis acid behavior of the chlorides. However Be shows just as many similarities to zinc, especially in the structures of its binary compounds (see Sections 19-6 to 19-8) and in the chemistry of its organic compounds. Thus BeS (zinc blende structure) is insoluble in water, although Al_2S_3, CaS, etc., are rapidly hydrolyzed.

Calcium, strontium, barium, and radium form a closely allied series in which the chemical and physical properties of the elements and their compounds vary systematically with increasing size in much the same manner as in Group I, the ionic and electropositive nature being greatest for radium. Again the larger ions can stabilize certain large anions: the peroxide and superoxide ions, polyhalide ions, etc. Some examples of systematic group trends in the series Ca–Ra are: (a) hydration tendencies of the crystalline salts increase; (b) solubilities of sulfates, nitrates, chlorides, etc. (fluorides are an exception) decrease; (c) solubilities of halides in ethanol decrease; (d) thermal stabilities of carbonates, nitrates, and peroxides increase; (e) rates of reaction of the metals with hydrogen increase. Other similar trends can be found.

All isotopes of *radium* are radioactive, the longest-lived isotope being ^{226}Ra (α; ~1600 years). This isotope is formed in the natural decay series of ^{238}U and was first isolated by Pierre and Marie Curie from pitchblende. Once widely used in radiotherapy, it has largely been supplanted by radioisotopes made in nuclear reactors.

The elements zinc, cadmium, and mercury, which have two electrons outside filled penultimate d shells, are also classed in Group II. Although the difference between the calcium and zinc subgroups is marked, zinc, and to a lesser extent cadmium, show some resemblance to beryllium or magnesium in their chemistry. We shall discuss these elements separately (Chapter 19), but it may be noted here that zinc, which has the lowest second ionization enthalpy in the Zn, Cd, Hg group, still has a value (1726 kJ mol^{-1}) similar to that of beryllium (1757 kJ mol^{-1}), and its standard potential (-0.76 V) is considerably less negative than that of magnesium.

There are a few ions with ionic radii and chemical properties similar to those of Sr^{2+} or Ba^{2+}, notably those of the +2 lanthanides (Section 23-14) and especially the europous ion Eu^{2+} and its more readily oxidized analogues Sm^{2+} and Yb^{2+}. Because of this fortuitous chemical similarity, europium is frequently found in Nature in Group II minerals, and this is a good example of the geochemical importance of such chemical similarity.

8-2. Lower Oxidation States

Although the differences between the first and second ionization enthalpies especially for beryllium might suggest the possibility of a stable +1 state, there is no evidence to support this. Calculations using Born-Haber cycles show that owing to the much greater lattice energies of MX_2 compounds, MX compounds would be unstable and disproportionate:

$$2MX \rightarrow M + MX_2$$

Thorough studies of the calcium, strontium, and barium halide systems show that contrary to some early reports, it is not possible to isolate any monohalides, even metastable ones.[1] The alleged CaCl, whose structure has been described,[2] is actually CaHCl.[1]

There is some evidence for Be^{I} in fused chloride melts, for example:

$$Be + Be^{II} = 2Be^{I}$$

but no Be^{I} compound has been isolated. Some studies of the dissolution of Be from anodes suggested Be$^+$ as an intermediate, but subsequent work showed that disintegration of the metal occurs during dissolution, so that the apparent effect is one of the metal going into solution in the +1 state—too much metal is lost for the amount of current passed. The anode sludge, a mixture of Be and Be(OH)$_2$, had been considered to be due to disproportionation of Be$^+$, but photomicrography

[1]   H. H. Emons *et al.*, *Z. Anorg. Allg. Chem.*, 1963, **323**, 114.
[2]   Cf. R. W. G. Wyckoff, *Crystal Structures*, 2nd ed., Vol. 1, Interscience, 1963, pp. 155–156.

indicates that the beryllium in the sludge is due merely to spallation of the anode.

On the other hand, similar studies of anodic dissolution of Mg in pyridine and aqueous salt solutions do provide some evidence for transitory Mg^+ ions, which would account for evolution of H_2 at or near the anode. Electrically generated Mg^+ ions have been used to reduce organic compounds.

BERYLLIUM[3]

8-3.   Covalency and Stereochemistry

As a result of the small size, high ionization enthalpy, and high sublimation energy of beryllium, its lattice and hydration energies are insufficient to provide complete charge separation and the formation of simple Be^{2+} ions. In fact, in all compounds whose structures have been determined, even those of the most electronegative elements (i.e. BeO and BeF_2), there appears to be substantial covalent character in the bonding. On the other hand, to allow the formation of two covalent bonds —Be—, it is clear that unpairing of the two $2s$ electrons is required. Where free BeX_2 molecules occur, the Be atom is promoted to a state in which the two valence electrons occupy two equivalent sp hybrid orbitals and the X—Be—X system is linear. However in such a linear molecule the Be atom has a coordination number of only 2 and there is a strong tendency for Be to achieve maximum (fourfold) coordination, or at least threefold coordination. Maximum coordination is achieved in several ways:

1.   Polymerization may occur through bridging, as in solid $BeCl_2$ (Fig. 8-1). The coordination of Be is not exactly tetrahedral, since the ClBeCl angles are only 98°, which means that the $BeCl_2Be$ units are somewhat elongated in the direction of the chain axis. In such a situation the exact sizes of the angles are determined by competing factors. If the ClBeCl angles were opened to 109°, the BeClBe angles would decrease to 71°, which would probably weaken the Be—Cl bonds; Be··· Be repulsions would also be increased. The observed value is a compromise giving the lowest free energy.

The chloride, like all the beryllium halides, consists of linear triatomic molecules

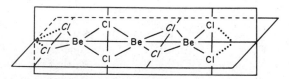

Fig. 8-1.   The structure of polymeric $BeCl_2$ in the crystal. The structure of $Be(CH_3)_2$ is similar.

3   D. A. Everest, *The Chemistry of Beryllium,* Elsevier, 1964; H. E. Stockinger, *Beryllium, Its Industrial Hygiene Aspects,* Academic Press, 1966; L. B. Tepper, H. L. Hardy, and R. I. Chamberlain, *Toxicity of Beryllium Compounds,* Elsevier, 1961.

in the gas phase at high temperatures, but at lower temperatures there are appreciable amounts ($\sim$20% at 560°) of a dimer in which Be is, presumably, three-coordinate. In the compounds of the type $M^IBe_2Cl_5$ (M^I = K, Rb, Tl, NO, NH_4) the anion 8-I resembles a portion of the $(BeCl_2)_n$ chain.[4]

$$(8\text{-}I)$$

2. Many alcoxides $[Be(OR)_2]_n$ are known, association being generally observed, leading to three- or four-coordinate Be. For example, $[Be(OMe)_2]_n$ is a high polymer, insoluble in hydrocarbon solvents. Even $[Be(OBu^t)_2]_3$, which is soluble, is trimeric (possibly with structure 8-II). $\{Be[OC(CF_3)_3]_2\}_2$ is a highly volatile dimer.[5] The only monomer known is 8-III.

$$(8\text{-}II)$$

$$(8\text{-}III)$$

3. By functioning as Lewis acids, many beryllium compounds attain maximum coordination of the metal atom. Thus the chloride forms etherates $[Cl_2Be(OR_2)_2]$, and complex ions such as BeF_4^{2-} and $[Be(H_2O)_4]^{2+}$ exist. In chelate compounds such as the acetylacetonate $[Be(acac)_2]$, four approximately tetrahedral bonds are formed, with the C—O and Be—O bond lengths equivalent.

In addition to the tetrahedral BeF_4^{2-} ion, some substances of composition $M^IBe_2F_5$ contain infinite sheet anions (i.e., not the chains present in the chloro analogue above) in which there are hexagonal rings of BeF_4 tetrahedra sharing corners, structurally analogous to the sheet silicate anions of empirical formula $Si_2O_5^{2-}$ (cf. Section 12-6).[6]

The packing in crystals is almost invariably such as to give beryllium a coordination number of 4, with a tetrahedral configuration. In binary compounds, the structures are often those of the corresponding zinc compounds. Thus the low-temperature form of BeO has the wurtzite structure (Fig. 1-9); the most stable $Be(OH)_2$ polymorph has the $Zn(OH)_2$ structure; and BeS has the zinc blende

[4]   J. MacCordick, *Ber.*, 1974, **107**, 1066.
[5]   R. A. Anderson and G. E. Coates, *J.C.S., Dalton*, **1975**, 1244.
[6]   Y. LeFur and S. Aleonard, *Acta Crystallogr.*, B, 1972, **28**, 2115, and earlier papers cited therein.

structure (Fig. 1-9). Be_2SiO_4 is exceptional among the orthosilicates of the alkaline earths, the rest of which have structures giving the metal ion octahedral coordination, in having the Be atoms tetrahedrally surrounded by oxygen atoms. It may be noted that Be with F gives compounds often isomorphous with oxygen compounds of silicon; thus $BaBeF_4$ is isomorphous with $BaSO_4$ and $NaBeF_3$ with $CaSiO_3$, and there are five different corresponding forms of Na_2BeF_4 and Ca_2SiO_4.

Three-coordinate Be occurs in some cases, for example, the gaseous dimers Be_2Cl_4 and Be_2Br_4 and 8-II. In Be phthalocyanine the metal is perforce surrounded by four nitrogen atoms in a plane. This compound constitutes an example of a *forced configuration,* since the Be atom is held strongly in a rigid environment.

There are very few examples of Be compounds existing at room temperature in which Be is two-coordinate with *sp* linear bonds. One example is 8-III, and others are the monomeric compounds di-*tert*-butylberyllium [$Be(CMe_3)_2$] and the silazane,[7] $Be[N(SiMe_3)_2]_2$.

It is to be noted especially that beryllium compounds are exceedingly poisonous, particularly if inhaled, and great precautions must be taken in handling them.

8-4. Elemental Beryllium

The most important mineral is *beryl* [$Be_3Al_2(SiO_3)_6$], which often occurs as large hexagonal prisms. The extraction from ores is complicated.[8] The metal is obtained by electrolysis of $BeCl_2$, but since the melt has very low electrical conductivity (about 10^{-3} that of NaCl), sodium chloride is also added.

The gray metal is rather light (1.86 g/cm³) and quite hard and brittle. Since the absorption of electromagnetic radiation depends on the electron density in matter, beryllium has the lowest stopping power per unit mass thickness of all suitable construction materials. It is used for "windows" in X-ray apparatus and has other special applications in nuclear technology. Like aluminum, metallic beryllium is rather resistant to acids unless finely divided or amalgamated, owing to the formation of an inert and impervious oxide film on the surface. Thus although the standard potential (-1.85 V) would indicate rapid reaction with dilute acids (and even H_2O), the rate of attack depends greatly on the source and fabrication of the metal. For very pure metal the relative dissolution rates are HF $>$ H_2SO_4 $\sim$ HCl $>$ HNO_3. The metal dissolves rapidly in $3M$ H_2SO_4 and in $5M$ NH_4F, but very slowly in HNO_3. Like aluminum, it dissolves also in strong bases, forming what is called the beryllate ion.

8-5. Binary Compounds

The white crystalline *oxide* BeO is obtained on ignition of beryllium or its compounds in air. It resembles Al_2O_3 in being highly refractory (m.p. 2570°) and in having polymorphs; the high-temperature form ($>800°$) is exceedingly inert and

[7]   A. H. Clark and A. Haaland, *Acta Chem. Scand.*, 1970, **24**, 3024.
[8]   *Chem. Eng. News,* **1965,** April 19, p. 70.

dissolves readily only in a hot syrup of concentrated H_2SO_4 and $(NH_4)_2SO_4$. The more reactive forms dissolve in hot alkali hydroxide solutions or fused $KHSO_4$.

Addition of OH^- ion to $BeCl_2$ or other beryllium solutions gives the *hydroxide*. This is amphoteric, and in alkali solution the "beryllate" ion, probably $[Be(OH)_4]^{2-}$, is obtained. When these solutions are boiled, the most stable of several polymorphs of the hydroxide can be crystallized.

Beryllium halides, all four of which are known, are deliquescent and cannot be obtained from their hydrates by heating, since HX is lost as well as H_2O. The fluoride is obtained as a glassy hygroscopic mass by heating $(NH_4)_2BeF_4$. The glassy form has randomly oriented chains of $\cdots F_2BeF_2Be\cdots$ similar to those in $BeCl_2$ and $BeBr_2$ but disordered. Two crystalline modifications are known, which appear to be structurally analogous to the quartz and cristobalite modifications of SiO_2 (Section 12-6). BeF_2 melts ($555°$) to a viscous liquid that has low electrical conductivity. The polymerization in the liquid may be lowered by addition of LiF, which forms the BeF_4^{2-} ion.

Beryllium chloride is prepared by passing CCl_4 over BeO at 800°C. On a small scale the chloride and bromide are best prepared pure by direct interaction in a hot tube. The white crystalline chloride (m.p. 405°) dissolves exothermically in water; from HCl solutions the salt $[Be(H_2O)_4]Cl_2$ can be obtained. $BeCl_2$ is readily soluble in oxygenated solvents such as ethers. In melts with alkali halides, chloroberyllate ions $[BeCl_4]^{2-}$ may be formed, but this ion does not exist in aqueous solution.

On interaction of Be with NH_3 or N_2 at 900–1000° the *nitride* Be_3N_2 is obtained as colorless crystals, readily hydrolyzed by water. The metal reacts with ethylene at 450° to give BeC_2.

8-6. Complex Chemistry

Oxygen Ligands. In strongly acid solutions the *aqua ion* $[Be(H_2O)_4]^{2+}$ occurs, and crystalline salts with various anions can be readily obtained. The water in such salts is more firmly retained than is usual for aquates, indicating strong binding. Thus the sulfate is dehydrated to $BeSO_4$ only on strong heating, and $[Be(H_2O)_4]Cl_2$ loses no water over P_2O_5. Solutions of beryllium salts are acidic; this may be ascribed to the acidity of the aqua ion, the initial dissociation being

$$[Be(H_2O)_4]^{2+} \rightleftharpoons [Be(H_2O)_3(OH)]^+ + H^+$$

The addition of soluble carbonates to beryllium salt solutions gives only basic carbonates. Beryllium salt solutions also have the property of dissolving additional amounts of the oxide or hydroxide. This behavior is attributable to the formation of complex species with Be—OH—Be or Be—O—Be bridges. The rapidly established equilibria[9] involved in the hydrolysis of the $[Be(H_2O)_4]^{2+}$ ion are very complicated and depend on the anion, the concentration, the temperature, and the pH. The main species, which will achieve four-coordination by additional water molecules, are considered to be $Be_2(OH)^{3+}$, $Be_3(OH)_3^{3+}$ (probably cyclic), and

⁹ R. E. Mesmer and C. F. Baes, Jr., *Inorg. Chem.,* 1967, **6**, 1951; G. Schwartzenbach and H. Wenger, *Helv. Chim. Acta,* 1969, **52**, 644.

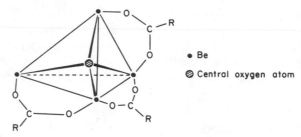

Fig. 8-2. The structure of the basic carboxylate complexes $Be_4O(OOCR)_6$. Only three RCOO groups are shown.

possibly $Be_5(OH)_7^{3+}$. The $[Be_3(OH)_3]^{3+}$ ion has been shown to predominate at pH 5.5 in perchlorate solution.[10] Various crystalline hydroxo complexes have been isolated. In concentrated alkaline solution the main species is $[Be(OH)_4]^{2-}$.

Other complexes of oxygen ligands are mainly adducts of beryllium halides or alkyls with ethers, ketones, etc. [e.g., $BeCl_2(OEt_2)_2$]. There are also neutral complexes of β-diketones and similar compounds, of which the acetylacetonate is the simplest, and solvated cationic species such as $[Be(DMF)_4]^{2+}$. The most unusual complexes have the formula $Be_4O(OOCR)_6$ and are formed by refluxing the hydroxide with carboxylic acids. These white crystalline compounds are soluble in organic solvents, even alkanes, but are insoluble in water and lower alcohols; they are inert to water but are hydrolyzed by dilute acids; in solution they are un-ionized and monomeric; X-ray study has shown that they have the structures illustrated in Fig. 8-2. The central oxygen atom is tetrahedrally surrounded by the four beryllium atoms (this being one of the few cases, excepting solid oxides, in which oxygen is four-coordinate), and each beryllium atom is tetrahedrally surrounded by four oxygen atoms. Zinc also forms such complexes, as does the ZrO^{2+} ion, with benzoic acid. The zinc complexes are rapidly hydrolyzed by water, in contrast to those of beryllium. The acetate complex has been utilized as a means of purifying beryllium by solvent extraction from an aqueous solution into an organic layer. When $BeCl_2$ is dissolved in N_2O_4 in ethyl acetate, crystalline $Be(NO_3)_2 \cdot 2N_2O_4$ is obtained. When heated at 50° this gives $Be(NO_3)_2$, which at 125° decomposes to N_2O_4 and volatile $Be_4O(NO_3)_6$. The structure of the latter appears to be similar to that of the acetate but with bridging nitrate groups. The basic nitrate is insoluble in nonpolar solvents.

The only halogeno complexes are the *tetrafluoroberyllates*, which are obtained by dissolving BeO or $Be(OH)_2$ in concentrated solutions or melts of acid fluorides such as NH_4HF_2. The tetrahedral ion has a crystal chemistry similar to that of SO_4^{2-}, and corresponding salts (e.g., $PbBeF_4$ and $PbSO_4$) usually have similar structures and solubility properties. BeF_2 readily dissolves in water to give mainly $BeF_2(H_2O)_2$ according to ^9Be nmr spectra. In $1M$ solutions of $(NH_4)_2BeF_4$ the ion BeF_3^- occurs to 15 to 20%.

The interaction between Cl^- and $[Be(H_2O)_4]^{2+}$ is very small and may be outer sphere in nature.

10   M. K. Cooper, D. E. J. Garman, and D. W. Yaniuk, *J.C.S. Dalton,* **1974**, 1281.

Other Complexes. The stability of complexes with ligands containing nitrogen or other atoms is lower than those of oxygen ligands. Thus $[Be(NH_3)_4]Cl_2$ is thermally stable but is rapidly hydrolyzed in water. When $BeCl_2$ is treated with the Li salt of 2,2′-bipyridine, a green paramagnetic complex is formed which is best regarded as a complex of Be^{2+} with the bipyridinyl radical anion.

Most of the other nitrogen complexes are derived from the hydride (p. 249) or organoberylliums, although compounds are known such as $[Be(NMe_2)_2]_3$, which have a central four-coordinate Be and terminal three-coordinate Be atoms with both bridge and terminal NMe_2 groups.

8-7.   Organoberyllium Compounds

Although beryllium alkyls can be obtained by the interaction of $BeCl_2$ with lithium alkyls or Grignard reagents, they are best made in a pure state by heating the metal and a mercury dialkyl, for example:

$$HgMe_2 + Be \xrightarrow{110°} BeMe_2 + Hg$$

The alkyl can be collected by sublimation or distillation in a vacuum. On the other hand, the aryls are made by reaction of a lithium aryl in a hydrocarbon with $BeCl_2$ in diethyl ether in which the LiCl formed is insoluble, for example:

$$2LiC_6H_5 + BeCl_2 \rightarrow 2LiCl\downarrow + Be(C_6H_5)_2$$

The beryllium alkyls are liquids or solids of high reactivity, being spontaneously flammable in air and violently hydrolyzed by water. Dimethylberyllium is a chain polymer (cf. $BeCl_2$, Fig. 8-1) with bridging CH_3 groups; for the bonding, see page 112. In the vapor, $BeMe_2$ is monomeric and linear (sp); it is also monomeric in ether, presumably as the complex $Me_2Be(OEt_2)_2$. The alkyls readily undergo exchange reactions in solution and, as in Grignard reagents, the equilibrium

$$BeR_2 + BeX_2 \rightleftharpoons 2RBeX$$

lies to the right.

An excellent method of preparing beryllium aryls, $Be(aryl)_2$, is by the reaction[11]:

$$3BeEt_2 + 2B(aryl)_3 \rightarrow 3Be(aryl)_2 + 2BEt_3$$

The o- and m-tolyl compounds are dimers, presumed to have structure 8-IV:

(8·IV)

[11]   G. E. Coates and R. C. Srivastava, *J.C.S. Dalton*, **1972**, 1541.

The structure of the sparingly soluble diphenylberyllium is not known.

The higher alkyls are progressively less highly polymerized; diethyl- and diiso-propylberyllium are dimeric in benzene, but the *tert*-butyl compound is monomeric; the same feature is found in aluminum alkyls.

As with several other elements, notably magnesium and aluminum, there are close similarities between the alkyls and hydrides, especially in the complexes with donor ligands. For the polymeric alkyls, especially $BeMe_2$, strong donors such as Et_2O, Me_3N, or Me_2S are required to break down the polymeric structure. Mixed hydrido alkyls are known; thus pyrolysis of diisopropylberyllium gives a colorless, nonvolatile polymer:

$$x(iso\text{-}C_3H_7)_2Be \xrightarrow{200°} [(iso\text{-}C_3H_7)BeH]_x + xC_3H_6$$

However, above 100° the *tert*-butyl analogue gives pure BeH_2 (p. 249). With tertiary amines, reactions of the following types may occur:

$$BeMe_2 + Me_3N \rightarrow Me_3N\cdot BeMe_2$$
$$2BeH_2 + 2R_3N \rightarrow [R_3NBeH_2]_2$$

The trimethylamine hydrido complex appears to have structure 8-V.

(8-V)

Beryllium alkyls give colored complexes with 2,2′-bipyridine [e.g., bipy-Be$(C_2H_5)_2$, which is bright red]; the colors of these and similar complexes with aromatic amines given by beryllium, zinc, cadmium, aluminum, and gallium alkyls are believed to be due to electron transfer from the M—C bond to the lowest un-occupied orbital of the amine.

Beryllium forms cyclopentadienyl compounds, some of which have unorthodox structures because of the small size of the beryllium atom. In $(C_5H_5)BeX$ molecules, where X is an ordinary univalent atom or group (e.g., Cl, Br, —CH_3 or $HC\equiv C—$), the ring is attached to the metal in a symmetrical, pentahapto fashion giving the entire molecule C_{5v} symmetry.[12,13] For the very air-sensitive $(C_5H_5)_2Be$, the structure has been more difficult to ascertain with certainty.[14] A perfect D_{5d} or D_{5h} structure like that found for the transition metal compounds definitely does not occur, and in view of the size of the Be atom, could not be expected. One ring does appear to be symmetrically bonded to Be, and the other is variously described as σ-bonded, "slipped," or ionically bonded. The beryllium atoms appear to be disordered in the crystals at both 25° and −120°, making definitive interpretation impossible.

[12]   D. A. Drew and A. Haaland, *J.C.S. Chem. Comm.*, **1971**, 1551.
[13]   A. Haaland and D. P. Novak, *Acta Chem. Scand.*, A, 1974, **28**, 153.
[14]   C. Wong *et al.*, *Acta Crystallogr.*, B, 1972, **28**, 1662; *Inorg. Nucl. Chem. Lett.*, 1973, **9**, 667.

MAGNESIUM, CALCIUM, STRONTIUM, BARIUM, AND RADIUM

8-8.   Occurrence; The Elements

Except for radium, the Group II elements are widely distributed in minerals and in the sea. They occur in substantial deposits such as *dolomite* ($CaCO_3 \cdot MgCO_3$), *carnallite* ($MgCl_2 \cdot KCl \cdot 6H_2O$), and *barytes* ($BaSO_4$). Calcium is the third most abundant metal terrestrially.

Magnesium is produced in several ways. An important source is dolomite from which, after calcination, the calcium is removed by ion exchange using seawater. The equilibrium is favorable because the solubility of $Mg(OH)_2$ is lower than that of $Ca(OH)_2$:

$$Ca(OH)_2 \cdot Mg(OH)_2 + Mg^{2+} \rightarrow 2Mg(OH)_2 + Ca^{2+}$$

The most important processes for preparation of magnesium are (*a*) the electrolysis of fused halide mixtures (e.g., $MgCl_2 + CaCl_2 + NaCl$) from which the least electropositive metal Mg is deposited, and (*b*) the reduction of MgO or of calcined dolomite ($MgO \cdot CaO$). The latter is heated with ferrosilicon:

$$CaO \cdot MgO + FeSi = Mg + \text{silicates of Ca and Fe}$$

and the magnesium is distilled out. MgO can be heated with coke at 2000° and the metal deposited by rapid quenching of the high-temperature equilibrium, which lies well to the right:

$$MgO + C \rightleftharpoons Mg + CO$$

Magnesium, which currently sells for about twice the price of aluminum, may in the long run replace it in many applications because the supply available in seawater is virtually unlimited.[15]

Calcium and the other metals are made only on a relatively small scale, by electrolysis of fused salts or reduction of the halides with sodium.

Radium is isolated in the processing of uranium ores; after coprecipitation with barium sulfate, it can be obtained by fractional crystallization of a soluble salt.

Magnesium is a grayish-white metal with a surface oxide film that protects it to some extent chemically—thus it is not attacked by water, despite the favorable potential, unless amalgamated. It is readily soluble in dilute acids and is attacked by most alkyl and aryl halides in ether solution to give Grignard reagents.

Magnesium metal in the form of an exceptionally reactive black powder suspended in ethers can be obtained by reducing magnesium salts with molten sodium or potassium.[16] In this form magnesium reacts with many otherwise inert alkyl halides to give previously unavailable Grignard reagents.

---

[15]   *Trends in the Usage of Magnesium,* National Technical Information Service, Springfield, Va. 22161, Report No. PB254, 1976.
[16]   R. D. Rieke and S. E. Bales, *J. Am. Chem. Soc.,* 1974, **96,** 1775.

Calcium and the other metals are soft and silvery, resembling sodium in their chemical reactivities, although somewhat less reactive. These metals are also soluble, though less readily and to a lesser extent than sodium, in liquid ammonia, giving blue solutions similar to those of the Group I metals (p. 258). These blue solutions are also susceptible to decomposition (with the formation of the amides) and have other chemical reactions similar to those of the Group I metal solutions. They differ, however, in that moderately stable metal ammines such as $Ca(NH_3)_6$ can be isolated on removal of solvent at the boiling point.

8-9. Binary Compounds

Oxides. The oxides MO are obtained most readily by calcination of the carbonates. They are white crystalline solids with ionic, NaCl-type lattices. Magnesium oxide is relatively inert, especially after ignition at high temperatures, but the other oxides react with water, evolving heat, to form the hydroxides. They also absorb carbon dioxide from the air. Magnesium hydroxide is insoluble in water ($\sim 1 \times 10^{-4}$ g/l at 20°) and can be precipitated from Mg^{2+} solutions; it is a much weaker base than the Ca–Ra hydroxides, although it has no acidic properties and unlike $Be(OH)_2$ is insoluble in excess of hydroxide. The Ca–Ra hydroxides are all soluble in water, increasingly so with increasing atomic number [$Ca(OH)_2$, ~ 2 g/l; $Ba(OH)_2$, ~ 60 g/l at $\sim 20°$], and all are strong bases.

There is no optical transition in the electronic spectra of the M^{2+} ions, and they are all colorless. Colors of salts are thus due only to colors of the anions or to lattice defects. The oxides may also be obtained with defects, and BaO crystals with $\sim 0.1\%$ excess of metal in the lattice are deep red.

Halides. The anhydrous halides can be made by dehydration (Section 17-8) of the hydrated salts. For rigorous studies, however, magnesium halides are best made* by the reaction

$$Mg + HgX_2 \xrightarrow{\text{boiling ether}} MgX_2(\text{solv}) + Hg$$

Magnesium and calcium halides readily absorb water. The tendency to form hydrates, as well as the solubilities in water, decrease with increasing size, and strontium, barium, and radium halides are normally anhydrous. This is attributed to the fact that the hydration energies decrease more rapidly than the lattice energies with increasing size of M^{2+}.

The fluorides vary in solubility in the reverse order (i.e., Mg < Ca < Sr < Ba) because of the small size of the F^- relative to the M^{2+} ion. The lattice energies decrease unusually rapidly because the large cations make contact with one another without at the same time making contact with the F^- ions.

The alkaline earth halides are all typically ionic solids, but can be vaporized as molecules, the structures of which are not all linear (cf. p. 205). On account of its dispersion and transparency properties, CaF_2 is used for prisms in spectrometers and for cell windows (especially for aqueous solutions). It is also used to provide a stabilizing lattice for trapping lanthanide +2 ions (cf. Chapter 23).

---

* According to E. C. Ashby and R. C. Arnott, *J. Organomet. Chem.,* 1968, **14**, 1.

Carbides. All the metals in the Ca–Ba series or their oxides react directly with carbon in an electric furnace to give the carbides MC_2. These are ionic acetylides whose general properties [hydrolysis to $M(OH)_2$ and C_2H_2, structures, etc.] are discussed in Chapter 11. Magnesium at $\sim500°$ gives MgC_2 but, at 500 to 700° with an excess of carbon, Mg_2C_3 is formed, which on hydrolysis gives $Mg(OH)_2$ and propyne and is presumably ionic, that is, $(Mg^{2+})_2(C_3^{4-})$.

Other Compounds. Direct reaction of the metals with other elements can lead to binary compounds such as borides, silicides, arsenides, and sulfides. Many of these are ionic and are rapidly hydrolyzed by water or dilute acids. At $\sim300°$, magnesium reacts with nitrogen to give colorless, crystalline Mg_3N_2 (resembling Li and Be in this respect). The other metals also react normally to form M_3N_2, but other stoichiometries are known. An interesting compound is Ca_2N, which has an *anti*-$CdCl_2$ type of layer structure[17], as does Cs_2O. In Ca_2N, however, there is one "excess" electron per formula unit. These excess electrons evidently occupy delocalized energy bands within metal atom layers, causing a lustrous, graphitic appearance.

The *hydrides* are discussed on page 247; a complex salt $KMgH_3$ has been prepared.[18]

8-10.   Oxo Salts, Ions, and Complexes

All the elements of Group II form *oxo salts,* those of magnesium and calcium often being hydrated. The carbonates are all rather insoluble in water, and the solubility products decrease with increasing size of M^{2+}. The same applies to the sulfates; magnesium sulfate is readily soluble in water, and calcium sulfate has a hemihydrate $2CaSO_4 \cdot H_2O$ (plaster of Paris) that readily absorbs more water to form the very sparingly soluble $CaSO_4 \cdot 2H_2O$ (gypsum); Sr, Ba, and Ra sulfates are insoluble and anhydrous. The nitrates of Sr, Ba, and Ra are also anhydrous, and the last two can be precipitated from cold aqueous solution by addition of fuming nitric acid. Magnesium perchlorate is used as a drying agent.

For water, acetone, and methanol solutions, nuclear magnetic resonance studies have shown that the coordination number of Mg^{2+} is 6, although in liquid ammonia it appears to be 5. The $[Mg(H_2O)_6]^{2+}$ ion is not acidic and in contrast to $[Be(H_2O)_4]^{2+}$ can be dehydrated fairly readily; it occurs in a number of crystalline salts.

Complexes. Only magnesium and calcium show any appreciable tendency to form complexes and in solution, with a few exceptions, these are with oxygen ligands. $MgBr_2$, MgI_2, and $CaCl_2$ are soluble in alcohols and some other organic solvents, as is $Mg(ClO_4)_2$; cationic solvated ions (see above) may be formed in these solvents. Adducts of ethers are known [e.g., $MgBr_2(OEt_2)_2$ and $MgBr_2(THF)_4$].

The only example of a halomagnesium anion $MgCl_4^{2-}$, shown by vibrational spectra to be tetrahedral, is afforded by $[NMe_4]_2MgCl_4$.[19]

[17]   E. T. Keve and A. C. Skapski, *Inorg. Chem.,* 1968, **7**, 1757.
[18]   E. C. Ashby, R. Kovar, and R. Arnott, *J. Am. Chem. Soc.,* 1970, **92**, 2182.
[19]   J. E. D. Davies, *J. Inorg. Nucl. Chem.,* 1974, **36**, 1711.

Both Mg^{2+} and Ca^{2+} form some stable five-coordinate complexes, of which $[M(OAsMe_3)_5](ClO_4)_2$ is an example.[20]

Oxygen chelate compounds, among the most important being those of the ethylenediaminetetraacetate (EDTA) type, readily form complexes in alkaline aqueous solution, e.g.:

$$Ca^{2+} + EDTA^{4-} = [Ca(EDTA)]^{2-}$$

The complexing of calcium by EDTA and also by polyphosphates is of some importance, not only for removal of calcium ions from water, but also for the volumetric estimation of calcium. Nitrogen ligands generally form weak complexes that exist in the solid state and dissociate in aqueous solution. Both Mg and Ca halides absorb NH_3 or amines to give, e.g., $[Mg(NH_3)_6]Cl_2$. Calcium, strontium, and barium perchlorates give nine-coordinate ions $[M\ dien_3]\ (ClO_4)_2$, but again these exist only in the solid state.

An important exception to this rule is provided by the magnesium complexes of tetrapyrrole systems, the parent compound of which is porphine (8-VI). These conjugated heterocycles provide a rigid planar environment for Mg^{2+} (and similar) ions. The most important of such derivatives are the *chlorophylls* and related compounds, which are of transcendental importance in photosynthesis in plants. The structure of chlorophyll-*a*, one of the many chlorophylls, is 8-VII.

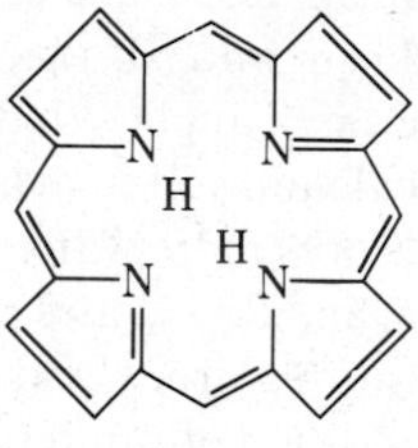

(8-VI)

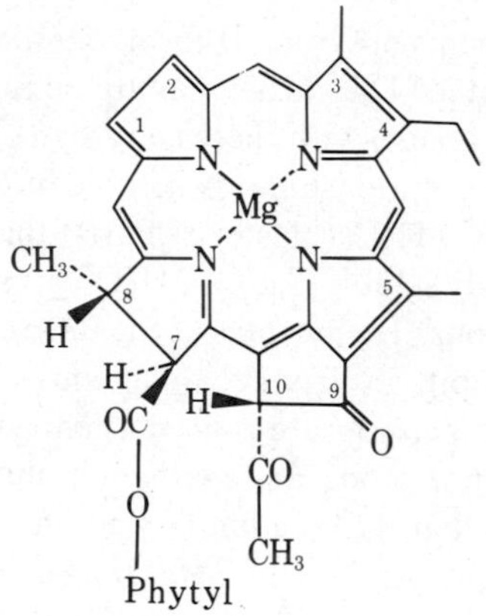

(8-VII)

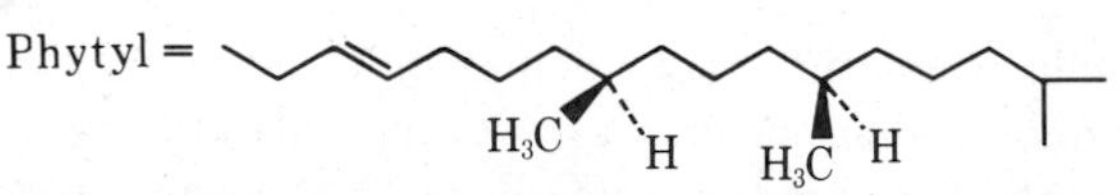

In such porphine compounds the Mg atom is formally four-coordinate but further interaction with either water or other solvent molecules is a common, if not universal, occurrence; furthermore, in chlorophyll, interaction with the keto group at position 9 in *another* molecule is also established. It also appears that five-coordination is preferred over six-coordination as in the structure of magnesium tetraphenylporphyrin hydrate, where the Mg atom is out of the plane of the N atoms

20   Y. S. Ng, G. A. Rodley, and W. T. Robinson, *Inorg. Chem.*, 1976, **15**, 303.

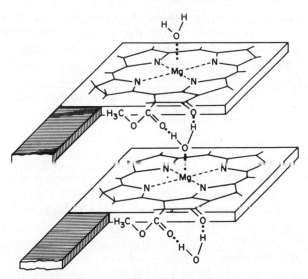

Fig. 8-3. Structure illustrating the chlorophyll-*a*–water–chlorophyll-*a* interaction. The dimensions of the ring and the phytyl chain are not to scale. [Reproduced by permisssion from K. Ballschmiter and J. J. Katz, *J. Am. Chem. Soc.,* 1969, **91**, 2661.]

and is approximately square pyramidal. Although Mg and other metalloporphyrins can undergo oxidation by one-electron changes, for Mg it is the macrocycle, not the metal, that is involved.

In chlorophylls hydrogen-bonding interactions lead to polymerization (Fig. 8-3); the hydrates may be monomeric or dimeric in benzene, but ordered aggregates of colloidal dimensions are formed in dodecane. Where a polar solvent is not present, association via coordination of the keto group at position 9 occurs as in solutions of anhydrous chlorophyll in alkanes.

The role of chlorophyll in the photosynthetic reduction of CO_2 by water in plants is to provide a source of electrons that may continue to be supplied for a time in the dark. Electron spin resonance studies of light-irradiated chlorophyll show that radicals are formed. These are probably of the type 8-VIII. The electrons are transmitted through chlorophyll micelles to other intermediates involved in the reduction of CO_2.

$$(8\text{-VIII})$$

The Ca^{2+}, Sr^{2+}, and Ba^{2+} ions can be complexed by the crown ethers[21] and the

[21] D. G. Parsons and J. N. Wingfield, *Inorg. Chem. Acta,* 1976, **18**, 263.

cryptate[22] ligands (*cf.* p. 163). As with the alkali ions, Na^+ and K^+, there are also complex substances, both natural and synthetic, that have special affinity for Ca^{2+}, thus exercise a controlling effect on the many biological processes in which this ion is involved. An example is the antibiotic called A23187, a monocarboxylic acid that binds and transports Ca^{2+} ions across natural and artificial membranes. It is a tridentate (N, O, O) ligand and has been shown to form a seven-coordinate calcium complex $Ca(O—O—N)_2(H_2O)$.[23]

8-11. Organomagnesium Compounds

The organic compounds of calcium, strontium, and barium are relatively obscure and of little utility, although some, such as the RMI and Cp_2M types can be prepared in good yields.[24] Magnesium compounds, of which the Grignard reagents are the best known, are probably the most widely used of all organometallic compounds. They are employed for the synthesis of alkyl and aryl compounds of other elements as well as for a host of organic syntheses.

Magnesium compounds are of the types RMgX—the Grignard reagents—and MgR_2. The former are made by direct interaction of the metal with an organic halide RX in a suitable solvent, usually an ether such as diethyl ether or tetrahydrofuran. The reaction is normally most rapid with iodides RI, and iodine may be used as an initiator. For most purposes RMgX reagents are used *in situ*. The species MgR_2 are best made by the dry reaction

$$HgR_2 + Mg(excess) \rightarrow Hg + MgR_2$$

The dialkyl or diaryl is then extracted with an organic solvent. Both RMgX, as solvates, and R_2Mg, are reactive, being sensitive to oxidation by air and to hydrolysis by water.

For diethylmagnesium $Mg(C_2H_5)_2$ the structure is that of a chain polymer, similar to that of $Be(CH_3)_2$, with bridging alkyl groups but again tetrahedral Mg. A special case is that of magnesium cyclopentadienide $Mg(C_5H_5)_2$, which has a "sandwich" structure similar to that of ferrocene, but with $C_5H_5^-$ and Mg^{2+}. This compound is readily made by direct action of cyclopentadiene vapor on hot Mg, or by thermal decomposition of C_5H_5MgBr, which in turn is made by action of cyclopentadiene (C_5H_6) on C_2H_5MgBr in solution.

There has been prolonged controversy concerning the nature of Grignard reagents. Discordant results have often been obtained because of failure to eliminate impurities, such as traces of water or oxygen, which can aid or inhibit the attainment of equilibrium, and the occurrence of exchange reactions. Although recent work has given a reasonable understanding, the following discussion probably applies only to Grignard reagents prepared under strict conditions, not to those normally prepared without special precautions in the laboratory.

X-Ray diffraction studies on certain crystalline Grignard reagents have been made. In the structures of $C_6H_5MgBr\cdot2(Et_2O)$ and $C_2H_5MgBr\cdot2(Et_2O)$ the Mg

[22] B. Metz, D. Moras, and R. Weiss, *Acta Crystallogr., B,* 1973, **29,** 1377, 1382, 1388.
[23] G. D. Smith and W. L. Duax, *J. Am. Chem. Soc.,* 1976, **98,** 1578.
[24] B. G. Gowenlock, W. E. Lindsell, and B. Singh, *J.C.S. Dalton,* **1978,** 657.

atom is, essentially, tetrahedrally surrounded by C, Br, and two oxygen atoms of
the ether as in 8-IX. For less sterically demanding ethers such as tetrahydrofuran,
higher coordination numbers may occur, as in $CH_3MgBr\cdot3THF$ which is *tbp* (cf.
also $MgBr_2\cdot2Et_2O$ but $MgBr_2\cdot4THF$). Thus it is now clear that in crystals the basic
Grignard structure is $RMgX\cdot n$(solvent).

$$
\begin{array}{c}
Br \\
\big\uparrow \\
R\blacktriangleright Mg\blacktriangleleft OEt_2 \\
\big\downarrow \\
OEt_2
\end{array}
\qquad (8\text{-}IX)
$$

The nature of Grignard reagents *in solution* is complex and depends critically
on the alkyl and halide groups and on the solvent, concentration, and temperature.
Quite generally, the equilibria involved are of the type:

$$
RMg \underset{X}{\overset{X}{\diagdown\diagup}} MgR \;\rightleftarrows\; 2RMgX \;\rightleftarrows\; R_2Mg + MgX_2 \;\rightleftarrows\; \underset{R}{\overset{R}{\diagdown}} Mg \underset{X}{\overset{X}{\diagup\diagdown}} Mg
$$

Solvation (not shown) occurs and association is predominantly by halide rather
than by carbon bridges, except for methyl compounds, where bridging by CH_3
groups may occur.

In dilute solutions and in more strongly donor solvents the monomeric species
normally predominate; but in diethyl ether at concentrations exceeding $0.1 M$ association occurs, and linear or cyclic polymers may be present. The behavior of
several compounds is shown in Fig. 8-4, which includes the halides $MgBr_2$ and
MgI_2.

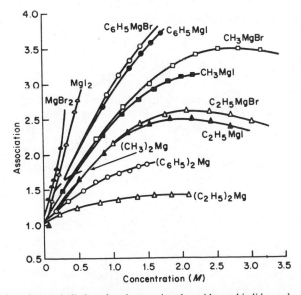

Fig. 8-4. Association of several alkyl- and arylmagnesium bromides and iodides and related magnesium
compounds in diethyl ether. [Reproduced by permission from F. W. Walker and E. C. Ashby, *J. Am.
Chem. Soc.,* 1969, **91**, 3845.]

Nmr spectra normally do not distinguish between RMgX and R_2Mg in solution because of rapid exchange of alkyl or aryl groups via a transition state such as 8-X. However for C_6F_5MgBr and $(C_6F_5)_2Mg$ the distinction can be made at room temperature, although for normal aryls and alkyls lower temperatures are required. At temperatures below $-70°$ the exchange rates are slow and nmr spectra of $(CH_3)_2Mg$ and CH_3MgBr can be resolved; the distinction can also be made in presence of hexamethylphosphoramide at $25°$.

$$R-Mg\overset{\displaystyle R}{\underset{\displaystyle X}{\diamond}}Mg-R$$

(8-X)

More complicated magnesium alkyls can be prepared. Thus $MgMe_2$ dissolves in Al_2Me_6 to give species with methyl bridges. Stable complexes with certain amines can also be obtained, for example, with N,N,N',N'-tetramethylethylenediamine (TMED), which gives $C_6H_5MgBr(TMED)$ and $R_2Mg(TMED)$. With other amines, elimination of alkane may occur to give complexes similar to those of Be, with bridging alkylamido groups and three-coordinate Mg.

$$Me_2Mg + (Me_3Al)_2 \rightleftarrows Me_2Al\overset{\displaystyle Me}{\underset{\displaystyle Me}{\diamond}}Mg\overset{\displaystyle Me}{\underset{\displaystyle Me}{\diamond}}AlMe_2$$

Further Reading

Bell, N. A., "Beryllium Halides and Pseudohalides," in *Adv. Inorg. Chem. Radiochem.*, 1972, **14,** 255.

Bellamy, R. G., and N. A. Hall, *Extraction and Metallurgy of Uranium, Thorium and Beryllium,* Pergamon Press, 1965.

Darwin, F. E., and J. H. Buddery, *Beryllium* (No. 7 of *Metallurgy of the Rarer Metals*), Butterworths, 1960.

Hausner, H. H., *Beryllium, Its Metallurgy and Properties,* University of California Press, 1965.

Kapoor, P. N. and R. C. Mehrotra, *Coord. Chem. Revs.,* 1974, **14,** 1. Coordination compounds of alkaline earth metals with covalent characteristics.

Kharasch, M. S., and O. Reinmuth, *Grignard Reactions of Non-Metallic Substances,* Constable and Co., and Prentice-Hall, 1954.

Mantell, C. L., and C. Hardy, *Calcium Metallurgy and Technology,* Reinhold, 1945.

Pannell, E. V., *Magnesium: Its Production and Use,* Pitman, 1948.

Pinkus, A. G., *Coord. Chem. Revs.,* 1978, **25,** 173. Three-coordinate Mg compounds.

Roberts, C. S., *Magnesium and Its Alloys,* Wiley, 1960.

Williams, R. J. P., *Q. Rev.,* 1970, **24,** 331. Biochemistry of Ca, Mg, Na, and K.

CHAPTER NINE

Boron

GENERAL REMARKS

9-1. Electronic Structure and Bonding

The first ionization potential of boron, 8.296 eV, is rather high, and the next two are much higher. Thus the total energy required to produce B^{3+} ions is far more than would be compensated by lattice energies of ionic compounds or by hydration of such ions in solution. Consequently, simple electron loss to form a cation plays no part in boron chemistry. Instead, covalent bond formation is of major importance, and boron compounds usually resemble those of other nonmetals, notably silicon, in their properties and reactions.

Despite the $2s^2 2p$ electronic structure, boron is always trivalent and never monovalent. This is because the total energy released in formation of three bonds in a BX_3 compound exceeds the energy of formation of one bond in a BX compound by more than enough to provide for promotion of boron to a hybridized valence state of the sp^2 type, wherein the three sp^2 hybrid orbitals lie in one plane at angles of 120°. It would therefore be expected, and is indeed found, without exception, that all monomeric, three-covalent boron compounds (trihalides, trialkyls, etc.) are planar with X—B—X bond angles of 120°. The covalent radius for trigonally hybridized boron is not well defined but probably lies between 0.85 and 0.90 Å. There are apparently substantial shortenings of many B—X bonds, and this has occasioned much discussion. For example, the estimated B—F, B—Cl and B—Br distances would be ~1.52, ~1.87, and ~1.99 Å, whereas the actual distances in the respective trihalides are 1.310, 1.75, and 1.87 Å.

Three factors appear to be responsible for the shortness of bonds to boron:

1. Formation of $p\pi$-$p\pi$ bonds using filled $p\pi$ orbitals of the halogens and the vacant $p\pi$ orbital of boron. This is probably most important in BF_3, but of some significance in BCl_3 and BBr_3 as well.

2. Strengthening, hence shortening, of the B—X bonds by ionic-covalent reso-

nance, especially for B—F and B—O bonds because of the large electronegativity differences. Evidence that this is important, in addition to the dative $p\pi$-$p\pi$ bonding, is afforded by the fact that even in BF_3 complexes such as $(CH_3)_3\overset{+}{N}\overset{-}{B}F_3$ and BF_4^-, where the $p\pi$-$p\pi$ bond must be largely or totally absent, the B—F bonds are still apparently shortened.

3. Because of the incomplete octet in boron, repulsions between nonbonding electrons may be somewhat less than normal, permitting closer approach of the bonded atoms.

Elemental boron has properties that place it on the borderline between metals and nonmetals. It is a semiconductor, not a metallic conductor, and chemically it must be classed as a nonmetal. In general, boron chemistry resembles that of silicon more closely than that of aluminum, gallium, indium, and thallium. The main resemblances to Si and differences from Al are the following:

1. The similarity and complexity of the boric and silicic acids is notable. Boric acid, $B(OH)_3$, is weakly but definitely acidic, and not amphoteric, whereas $Al(OH)_3$ is mainly basic with some amphoteric behavior.

2. The hydrides of B and Si are volatile, spontaneously flammable, and readily hydrolyzed, whereas the only binary hydride of Al is a solid, polymeric material. However, structurally the boron hydrides are unique, having unusual stoichiometries and configurations and unusual bonding because of their *electron-deficient* nature.

3. The boron halides (not BF_3), like the silicon halides, are readily hydrolyzed, whereas the aluminum halides are only partially hydrolyzed in water.

4. B_2O_3 and SiO_2 are similar in their acidic nature, as shown by the ease with which they dissolve metallic oxides on fusion to form borates and silicates, and both readily form glasses that are difficult to crystallize. Certain oxo compounds of B and Si are structurally similar, specifically the linear $(BO_2)_x$ and $(SiO_3)_x$ ions in metaborates and pyroxene silicates, respectively.

5. However, despite dimerization of the halides of Al and Ga and of the alkyls of Al, they behave as acceptors and form adducts similar to those given by boron halides and alkyls, for example, $Cl_3\overset{-}{Al}\overset{+}{N}(CH_3)_3$. Aluminum, like boron, also forms volatile alcoxides such as $Al(OC_2H_5)_3$, which are similar to borate esters $B(OR)_3$.

9-2. Acceptor Behavior

In BX_3 compounds the boron octet is incomplete; boron has a low-lying orbital that it does not use in bonding owing to a shortage of electrons, although partial use is made of it in the boron halides through B—X multiple bonding. The alkyls and halides of aluminum make up this insufficiency of electrons by forming dimers with alkyl or halogen bridges, but the boron compounds do not. The reason or reasons for this difference are not known with certainty. The size factor may be important for BCl_3 and BBr_3, since the small boron atom may be unable to coordinate strongly to four atoms as large as Cl and Br. The stability of BCl_4^- and BBr_4^- ions only in crystalline salts of large cations such as Cs^+ or $(CH_3)_4N^+$ might suggest this. The necessity of sacrificing a certain amount of B—X $p\pi$-$p\pi$ bond energy would also detract from the stability of dimers relative to monomers. The size factor cannot be controlling for BF_3, however, since BF_4^- is quite stable. Here the donor power of the fluorine already bonded to another boron atom may be so low that the energy of the bridge bonds would not be sufficient to counterbalance the energy required to break the $B—F\pi$ bonding in the monomer. Such phenomena are often difficult to explain with certainty.

There is one interesting case in which a tricoordinate boron unit does polymerize to provide the boron atom with an octet, viz., H_2BCN. Cyclic polymers, in which there are four to nine units, are formed via $—CN \rightarrow B$ bonds. The structure of the hexamer[1] shows essentially tetrahedral boron atoms. Evidently one CN group is insufficient to provide enough electron density to boron by π donation, yet at the same time the donation of the nitrogen lone pair takes precedence over hydrogen bridge formation.

The incomplete octet in BX_3 compounds causes them to behave as acceptors (Lewis acids), in which boron achieves its maximum coordination with approximately sp^3 hybridization, toward many Lewis bases, such as amines, phosphines, ethers, and sulfides. Examples are $(CH_3)_3NBCl_3$, $(CH_3)_3PBH_3$, and $(C_2H_5)_2OBF_3$.

The relative strengths of the boron halides as Lewis acids are in the order $BBr_3 > BCl_3 \geqslant BF_3$, which is the opposite to that expected either on steric grounds or from electronegativity. It can be explained, at least partially, in terms of boron-halogen π bonding. In an addition compound this π bonding is largely or completely lost, so that addition compounds of the trihalide with the strongest π bonding will be the most destabilized by loss of the energy of π bonding. Calculations indicate that the π-bonding energies of the trihalides are in the order $BF_3 \geqslant BCl_3 > BBr_3$. However certain properties of the BX_3 adducts with donor molecules suggest that the donor-to-boron bonds may themselves increase in strength in the order $BF_3 < BCl_3 < BBr_3$. No satisfactory explanation has been given for this.

The inability of a weak donor such as CO to form a genuine complex with BF_3 has recently been demonstrated in a positive way by structural characterization of BF_3CO as simply a "van der Waals molecule" with the BF_3 group still planar and $B—C = 2.89$ Å. BF_3N_2 and BF_3Ar are similar.[2a]

Boron also completes its octet by forming both anionic and cationic complexes. The former include such species as BF_4^-, BH_4^-, $B(C_6H_5)_4^-$, and $BH(OR)_3^-$, as well as chelates such as $[B(o\text{-}C_6H_4O_2)_2]^-$ and the salicylato complex 9-I, which has been partially resolved by fractional crystallization of its strychnine salt.

(9-I)

(9-II)

[1] A. T. McPhail and D. L. McFadden, *J.C.S. Dalton*, **1975**, 1784.
[2a] K. J. Janda *et al., J. Am. Chem. Soc.*, 1978, **100**, 8074.

The cationic species[2b] are of three main types, some containing B–H bonds (see later) such as 9-II, but other types are known, e.g.,

$$(C_6H_5)_2BCl + dipy + AgClO_4 \xrightarrow{MeNO_2} [Ph_2Bdipy]^+ + ClO_4^- + AgCl$$

These cations have considerable hydrolytic stability, though they are attacked by base; 9-II has been resolved and is optically stable in acid at 25°.

THE ELEMENT[3]

9-3.   Occurrence, Isolation, and Properties

The most abundant boron mineral is *tourmaline,* a complex aluminosilicate containing about 10% of boron. The principal boron ores are borates, such as borax, $Na_2B_4O_5(OH)_4 \cdot 8H_2O$, which occur in large beds in arid parts of California and elsewhere.

Natural boron consists of two isotopes, ^{10}B (19.6%) and ^{11}B (80.4%). Isotopically enriched boron compounds can be made and are useful in spectroscopic and reaction mechanism studies. The boron nuclear spins (^{10}B, $S = 3$; ^{11}B, $S = 3/2$) are also useful in structure elucidation. For an example, see page 309.

It is exceedingly difficult to prepare elemental boron in a state of high purity because of its high melting point and the corrosiveness of the liquid. It can be prepared in quantity but low purity (95–98%) in an amorphous form by reduction of B_2O_3 with magnesium, followed by vigorous washing of the material so obtained with alkali, hydrochloric acid, and hydrofluoric acid. This amorphous boron is a dark powder that may contain some microcrystalline boron but also contains oxides and borides.

The preparation of pure boron in crystalline form is a matter of considerable complexity and difficulty even when only small research-scale quantities are required. There are several allotropic forms.

α-Rhombohedral boron has been obtained by pyrolysis of BI_3 on tantalum, tungsten, and boron nitride surfaces at 800 to 1000°, by pyrolysis of boron hydrides, and by crystallization from boron-platinum melts at 800 to 1200°. It is the most dense allotrope, and its structure consists of B_{12} icosahedra (cf. Fig. 2–8b) packed together in a manner similar to cubic closest packing of spheres; there are bonds between the icosahedra that are, however, weaker than those within the icosahedra.

A tetragonal form of boron, which can be obtained by reduction of BBr_3 with H_2 on a tantalum or tungsten filament at 1200 to 1400°, consists of layers of B_{12} icosahedra connected by single boron atoms.

[2b] G. E. Ryschtewitsch, in *Boron Hydride Chemistry,* E. L. Muetterties, Ed., Academic Press, 1975, p. 223.

[3]   V. I. Matkovich, Ed., *Boron and Refractory Borides,* Springer, 1977.

β-Rhombohedral boron is invariably obtained by crystallization of fused boron. It is built of B_{12} icosahedra packed together, with B—B bonds between them, in a more complicated way than in α-rhombohedral boron. The β-rhombohedral form is the thermodynamically stable one over a considerable range of temperature, though sluggish attainment of equilibria makes this a difficult point to establish with certainty. The melting point is $2250 \pm 50°C$.

Crystalline boron is extremely inert chemically. It is unaffected by boiling HCl or HF, only slowly oxidized by hot, concentrated nitric acid when finely powdered, and either not attacked or only very slowly attacked by many other hot concentrated oxidizing agents.

BORON COMPOUNDS[4]

9-4. Borides[5]

Compounds of boron with elements less electronegative than itself (i.e., metals) are called borides. Often compounds of boron with rather less metallic or metalloidal elements (e.g., P, As) are also termed borides. Borides of most but not all elements are known. They are generally hard, refractory substances and fairly inert chemically, and they often possess very unusual physical and chemical properties. For example, the electrical and thermal conductivities of ZrB_2 and TiB_2 are about 10 times greater than those of the metals themselves, and the melting points are more than 1000° higher. Some of the lanthanide hexaborides are among the best thermionic emitters known. The monoborides of phosphorus and arsenic are promising high-temperature semiconductors, and higher borides of some metalloids (e.g., AsB_6) are remarkably inert to chemical attack.

Industrially, borides are prepared in various ways, including reduction of metal oxides by mixtures of carbon and boron carbide, electrolysis in fused salts, and direct combination of the elements.

The reduction by sodium borohydride of aqueous solutions of transition metal ions such as Ni^{2+}, Co^{2+}, and Rh^{3+}, gives black materials that usually contain not only boron but hydrogen. These substances have been used for catalytic hydrogenations of CO and olefins.[6]

The borides do not conform to the ordinary concepts of valence either in stoichiometry or in structure. With only a few exceptions, borides are of one of the following main types:

1. *Borides with Isolated Boron Atoms.* These include most of those with low

4   E. L. Muetterties, Ed., *The Chemistry of Boron and Its Compounds,* Wiley, 1967.
5   B. Aronsson, T. Lundstrom, and S. Rundqvist, *Borides, Silicides, and Phosphides,* Methuen, 1965.
6   J. M. Pratt and G. Swinden, *J.C.S. Chem. Commun.,* **1969,** 1321; P. C. Maybury, R. W. Mitchell, and M. F. Hawthorne, *J.C.S. Chem. Comm.,* *1974,* 534; R. Wade *et al., Catal. Rev. Sci. Eng.,* 1976, **14,** 211.

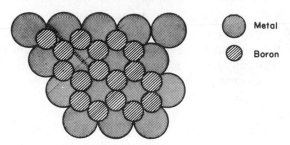

Fig. 9-1. Parallel layers of metal atoms and boron atoms in MB_2 compounds.

B-to-M ratios such as M_4B, M_3B, M_2B, M_5B_2, and M_7B_3. In the M_4B and M_2B structures, boron atoms lie in triangular-prismatic or square-antiprismatic holes between multiple layers of metal atoms. In the others, the metal atoms are arranged in approximately close-packed arrays, with the boron atoms in triangular-prismatic interstices.

2. *Borides with Single and Double Chains of Boron Atoms.* As the proportion of boron atoms increases, so do the possibilities for boron–boron linkages. In V_3B_2 there are pairs of boron atoms. In one modification of Ni_4B_3, two-thirds of the boron atoms form infinite, zigzag chains, while one-third are isolated from other boron atoms; in another modification all the boron atoms are members of chains. MB compounds all have structures with single chains, while in many M_3B_4 compounds there are double chains.

3. *Borides with Two-Dimensional Nets.* These are represented by MB_2 and M_2B_5 compounds and include some of the best electrically conducting, hardest, and highest melting of all borides. The crystal structures of the MB_2 compounds are unusually simple, consisting of alternating layers of close-packed metal atoms and "chicken wire" sheets of boron atoms, as shown in Fig. 9-1.

4. *Borides with Three-Dimensional Boron Networks.* The major types have formulas MB_4, MB_6, and MB_{12}. MB_4 compounds may be of several types insofar as structural details are concerned. ThB_4 and CeB_4 contain rather open networks of boron atoms interpenetrating a network of metal atoms. Perhaps as many as 20 other MB_4 compounds have the same structure. The MB_6 structure is fairly easy to visualize with the help of Fig. 9-2: it can be thought of as a CsCl structure, with B_6 octahedra in place of the Cl^- ions; however the B_6 octahedra are closely linked along the cube edges, so that the boron atoms constitute an infinite three-dimensional network. The MB_{12} compounds also have cubic structures consisting of M atoms and B_{12} cubooctahedra (9-III) packed in the manner of NaCl; again, the B_{12} polyhedra are closely linked to one another. Heating NaB_6 at 1000° in argon

(9-III)

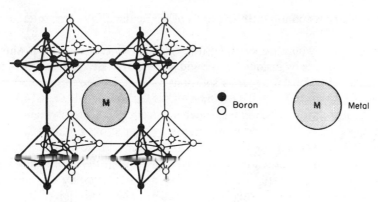

Fig. 9-2. Atomic arrangement in many MB_6 compounds.

converts it into NaB_{15}, in which there are icosahedra linked both directly and through B_3 chains.

Boron nitride, which can be obtained by interaction of boron with ammonia at white heat, is a slippery white solid with a layer structure very similar to that of graphite (p. 357). The units, instead of being hexagonal carbon rings, have alternate B and N atoms 1.45 Å apart with angles of 120° (sp^2 at B). The distance between the sheets is 3.34 Å. The analogy of C—C and B—N further discussed below is heightened by the conversion of graphitelike BN under high temperature and pressure into a cubic form with a diamondlike structure and by the formation of alkali metal intercalation compounds. The cubic form is extremely hard and will scratch diamond. The nitride is stable in air but slowly hydrolyzed by water.

Boron sulfide (B_2S_3) has a layer structure[7] with both B_3S_3 and B_2S_2 rings linked by sulfur bridges; there is no resemblance to B_2O_3.

9-5. Oxygen Compounds of Boron

The oxygen-containing substances are among the most important compounds of boron, comprising nearly all the naturally occurring forms of the element. The structures of such compounds consist mainly of trigonal BO_3 units with the occasional occurrence of tetrahedral BO_4 units. B—O bond energies are 560 to 790 kJ, rivaled only by the B—F bond in BF_3 (640 kJ) in strength.

Boron Oxides.[8a]   The principal oxide, B_2O_3, is obtained by fusing boric acid. It usually forms a glass and can be crystallized only with the greatest difficulty. This glass is believed to consist of randomly oriented B_3O_3 rings connected by bridging oxygen atoms.[8b]

B_2O_3 is acidic, reacting with water to give boric acid, $B(OH)_3$, and, when fused, dissolves many metal oxides to give borate glasses. Both the glassy and the crys-

[7]   M. Diercks and B. Krebs, *Angew. Chem., Int. Ed.,* 1977, **16**, 313.
[8a] R. D. Srivastava and M. Farber, *Chem. Rev.,* 1978, **78**, 627.
[8b] G. E. Jellison Jr., *et al., J. Chem. Phys.,* 1977, **66**, 802.

talline substances contain infinite chains of triangular BO_3 units, interconnected by weaker B—O bonds.

Borates.   Many borates occur naturally, usually in hydrated form. Anhydrous borates can be made by fusion of boric acid and metal oxides, and hydrated borates can be crystallized from aqueous solutions. The stoichiometry of borates—for example, $KB_5O_8 \cdot 4H_2O$, $Na_2B_4O_7 \cdot 10H_2O$, CaB_2O_4, and $Mg_3B_7O_{13}Cl$—gives little idea of the structures of the anions, which are cyclic or linear polymers formed by linking together of BO_3 and/or BO_4 units by shared oxygen atoms. The main principles for determining these structures are similar to those for silicates, to which the borates are structurally and often physically similar in forming glasses.

In contrast to the borates, the carbonate ion of superficially similar structure forms no polymeric species; this is attributable to the formation of strong C—O π bonds.

Examples of complex anhydrous borate anions are the ring anion 9-IV in $K_3B_3O_6$ and the infinite chain anion 9-V in CaB_2O_4.

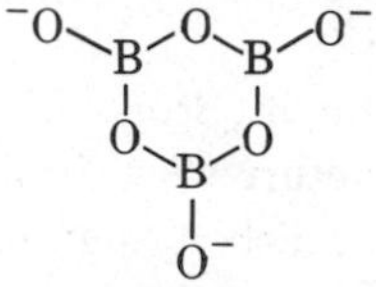

(9-IV) 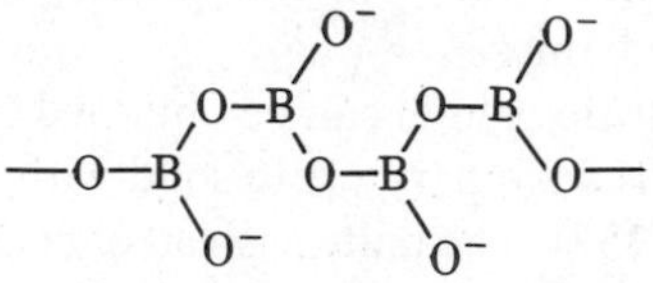   (9-V)

The diborate ion $O_2BOBO_2^{4-}$ is known, and a linear ion O—B—O$^-$ occurs in an apatite.[9]

Hydrated borates also contain polyanions in the crystal, but not all the known polyanions exist as such in solution (see below); only those containing one or more BO_4 groups appear to be stable. Important features of the structures are the following:

1. Both trigonal BO_3 and tetrahedral BO_4 groups are present, the ratio of BO_4 to total B being equivalent to the ratio of the charge on the anion to total boron atoms. Thus $KB_5O_8 \cdot 4H_2O$ has one BO_4 and four BO_3, whereas $Ca_2B_6O_{11} \cdot 7H_2O$ has four BO_4 and two BO_3 groups.

2. The basic structure is a six-atom boroxine ring whose stability depends on the presence of one or two BO_4 groups. Anions that do not have BO_4 groups, such as metaborate, $B_3O_6^{3-}$ or metaboric acid, $B_3O_3(OH)_3$, hydrate rapidly and lose their original structures. Certain complex borates can be precipitated or crystallized from solution, but this does not constitute evidence for the existence of such anions in solution, since other less complex anions can readily recombine during the crystallization process.

3. Other discrete and chain-polymer anions can be formed by linking of two or more rings by shared tetrahedral boron atoms, in some cases with dehydration (cf. metaborate, below).

Boric Acid and Borate Ions in Solution.   The hydrolysis of boron halides, hydrides, and so on, gives boric acid, $B(OH)_3$, or its salts. The acid is usually obtained

---

9   C. Calvo and R. Fagiani, *J.C.S. Chem. Comm.*, **1974**, 714.

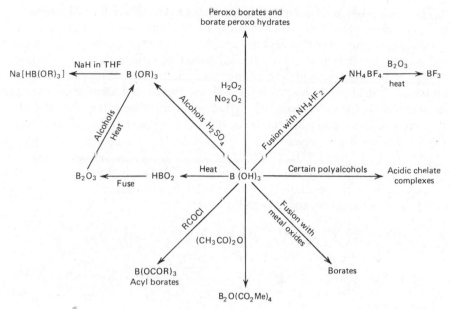

Fig. 9-3. Some reactions of boric acid.

from borax and it forms white, needlelike crystals in which $B(OH)_3$ units are linked together by hydrogen bonds to form infinite layers of nearly hexagonal symmetry; the layers are 3.18 Å apart, which accounts for the pronounced basal cleavage. Some reactions of boric acid are given in Fig. 9-3.

Since boric acid is moderately soluble in water, with a large negative heat of solution, the solubility increases markedly with temperature. It is a very weak and exclusively monobasic acid that acts not as a proton donor, but as a Lewis acid, accepting OH^-:

$$B(OH)_3 + H_2O \rightleftharpoons B(OH)_4^- + H^+ \qquad pK = 9.00$$

The $B(OH)_4^-$ ion occurs in several minerals. At concentrations $\leqslant 0.025\ M$, essentially only mononuclear species $B(OH)_3$ and $B(OH)_4^-$ are present. Boric acid–borate buffer mixtures serve as pH standards, and they occur in natural aqueous systems and in detergents.

In concentrated solutions polymeric ions[10] are also present, e.g.,

$$2B(OH)_3 + B(OH)_4^- \rightleftharpoons B_3O_3(OH)_4^- + 3H_2O$$

Equilibrium between the various ions is rapidly established as shown by rapid exchange between $B(OH)_3$ labeled with ^{18}O and borates. The determination of the species in solution has been largely based on Raman spectra and comparison with the structures known for hydrated borates from X-ray diffraction.[11]

[10] R. E. Mesmer, C. F. Baes, Jr., and F. H. Sweeton, *Inorg. Chem.,* 1972, **11,** 537; H. D.Smith, Jr., and R. W. Wiersema, *Inorg. Chem.,* 1972, **11,** 1152.

[11] L. Maya, *Inorg. Chem.,* 1976, **15,** 2179.

The species $B_5O_6(OH)_4^-$, $B_3O_3(OH)_4^-$, and $B_4O_5(OH)_4^{2-}$ are formed successively with increasing pH. These ions, which can be designated (5.1), (3.1), (4.2), etc., to indicate the number of B atoms and the charge (which corresponds to the number of four-coordinate B atoms) have the boroxine ring structures 9-VI, 9-VII, and 9-VIII, which have been confirmed in crystalline borates. With increasing pH, attack on neutral trigonal boron is favored, but there is a discontinuity when the number of BO_4 groups exceeds 50% so that tetraborate goes directly to $B(OH)_4^-$. In dilute solutions depolymerization rapidly occurs; thus the mononuclear species is formed when borax, which contains the ion 9-VIII, is dissolved. Although the hydrated diborate ion $(HO)_3BOB(OH)_3^{2-}$ occurs in crystals, it does not occur in solution. However a molecular compound with a B—O—B group does exist; this is the so-called boron acetate (Fig. 9-3), which has two acetato bridges, an oxo bridge and unidentate acetates as end groups.[12]

(9-VI)   (9-VII)   (9-VIII)

Borate Esters.   Boric acid is readily converted to alkyl or aryl orthoborates, $B(OR)_3$, by action of alcohols and sulfuric acid. These compounds are usually colorless liquids. A common test for boron involves treating compounds with methanol and acid and observing the green color imparted to a flame by $B(OMe)_3$. With polyhydroxy alcohols or carboxylic acids that have cis-hydroxy groups, boric acid forms 1:1 or 1:2 chelate complexes,[13] stepwise; these may be very stable (e.g., 9-IX, 9-X). In the 1:1 complexes the acidity of the OH groups exceeds that in $B(OH)_3$, so that if glycerol is added to a boric acid solution, this can be titrated using aqueous NaOH.

(9-IX)   (9-X)

Steric considerations are very critical in the formation of these complexes. Thus 1,2- and 1,3-diols in the cis form only, such as cis-1,2-cyclopentanediol, are active, and only o-quinols react. Indeed, the ability of a diol to affect the acidity of boric

---

[12]   D. Dal Negro, L. Ungareti, and A. Perrotti, *J.C.S. Dalton,* **1972,** 1639.
[13]   R. Pitzer and L. Babcock, *Inorg. Chem.,* 1977, **16,** 1677; R. P. Oertal, *Inorg. Chem.,* 1972, **11,** 544.

acid is a useful criterion of the configuration where *cis-trans* isomers are possible.

Peroxoborates. Treatment of borates with hydrogen peroxide or of boric acid with sodium peroxide leads to products variously formulated as $NaBO_3 \cdot 4H_2O$ or $NaBO_2 \cdot H_2O_2 \cdot 3H_2O$, which are extensively used in washing powders because they afford H_2O_2 in solution. The crystal structure has been found to contain $[B_2(O_2)_2(OH)_4]^{2-}$ units with two peroxo groups bridging the tetrahedral boron atoms. When this salt is heated, paramagnetic solids containing O_2^-, O_3^- and a peroxoborate radical are formed.

Metaborates. When heated, boric acid loses water stepwise:

$$B(OH)_3 \underset{H_2O}{\overset{heat}{\rightleftarrows}} HBO_2 \underset{H_2O}{\overset{heat}{\rightleftarrows}} B_2O_3$$

The intermediate substance, metaboric acid (HBO_2), exists in three modifications. If the $B(OH)_3$ is heated below 130°, HBO_2-III is formed. This has a layer structure in which B_3O_3 six-rings are joined by hydrogen bonding between OH groups on the boron atoms. On continued heating of HBO_2-III at 130 to 150°, HBO_2-II is formed; this has a more complex structure containing both BO_4 tetrahedra and B_2O_5 groups in chains linked by hydrogen bonds. Finally, on heating of HBO_2-II above 150°, cubic HBO_2-I is formed in which all boron atoms are four-coordinate.

9-6. Trihalides of Boron[14]

Compounds of the type BX_3 exist for all the halogens. The mixing of two halides leads to essentially a statistical equilibrium, e.g.,

$$BF_3 + BCl_3 \rightleftarrows BFCl_2 + BF_2Cl.$$

Although not isolatable, photoelectron, microwave, and nmr spectra of the individual halides can be identified.[15] These redistribution reactions presumably involve transitory formation of dimers such as $F_2BFClBCl_2$, which may dissociate to BF_2Cl + BCl_2F

Some reactions of the halides are summarized in Fig. 9-4.

Boron Trifluoride. This pungent, colorless gas (b.p. −101°) is prepared by heating B_2O_3 with NH_4BF_4 or with CaF_2 and concentrated H_2SO_4. It reacts with water to form two "hydrates," which may be written $BF_3 \cdot H_2O$ and $BF_3 \cdot 2H_2O$. These hydrates melt at 10.18° and 6.36°, respectively, and they are un-ionized in the solid state. Both hydrates partially dissociate into ions in their liquid phases, presumably as follows:

$$2(BF_3 \cdot H_2O) \rightleftarrows [H_3O{-}BF_3]^+ + [BF_3OH]^-$$

$$BF_3 \cdot 2H_2O \rightleftarrows H_3O^+ + [BF_3OH]^-$$

[14]   J. S. Hartman and J. M. Miller, *Adv. Inorg. Chem. Radiochem.*, 1978, **21**, 147 (adducts of mixed trihalides); A. G. Massey, *Adv. Inorg. Chem. Radiochem.*, 1967, **10**, 1; G. A. Olah, Ed., *Friedel-Crafts and Related Reactions*, Vols. 1, II, Wiley-Interscience, 1963 (contains much chemistry of boron halides).

[15]   H. W. Kroto *et al.*, *J.C.S. Chem. Comm.*, **1975**, 810.

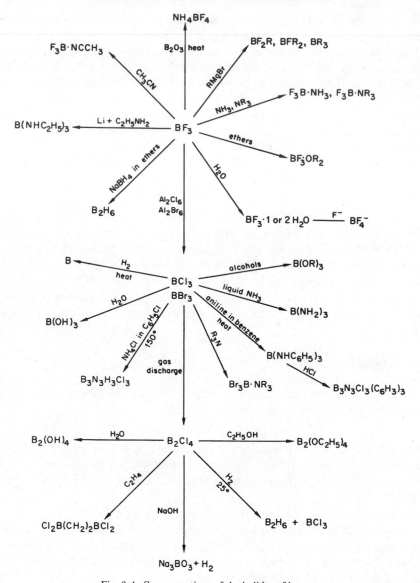

Fig. 9-4. Some reactions of the halides of boron.

Above about 20° they decompose extensively, giving off BF_3. When relatively small amounts of BF_3 are passed into water, a solution of fluoroboric acid (see below) is obtained:

$$4BF_3 + 6H_2O \rightarrow 3H_3O^+ + 3BF_4^- + B(OH)_3$$

Boron trifluoride is a Lewis acid and readily unites with water, ethers, alcohols, amines, phosphines, etc., to form adducts. BF_3 is commonly available as its diethyl etherate $(C_2H_5)_2\ddot{O}BF_3$. Because of its potency as a Lewis acid and its greater resistance to hydrolysis compared with BCl_3 and BBr_3, BF_3 is widely used to promote various organic reactions, such as (a) ethers or alcohols + acids → esters + H_2O or ROH; (b) alcohols + benzene → alkylbenzenes + H_2O; (c) polymerization of olefins and olefin oxides; and (d) Friedel-Crafts-like acylations and alkylations. In the first two cases, the effectiveness of BF_3 must depend on its ability to form an adduct with one or both of the reactants, thus lowering the activation energy of the rate-determining step in which H_2O or ROH is eliminated by breaking of C—O bonds. However the exact mechanisms of these reactions are not known, nor are those of the olefin and olefin oxide polymerizations.

In the case of the Friedel-Crafts-like reactions, isolation of certain intermediates at low temperatures has provided a fairly definite idea of the function of the BF_3. Thus the ethylation of benzene by ethyl fluoride proceeds as in eq. 9-1. With benzene, HF, and BF_3, compound 9-XI can be isolated at low temperatures. It will be seen that the BF_3 is not really "catalytic" but must be present in the stoichiometric amount, since it is consumed in the process of tying up the HF as HBF_4.

$$C_2H_5F + BF_3 \longrightarrow [C_2H_5^{\delta+} \cdots F \cdots ^{\delta-}BF_3] \xrightarrow{C_6H_6}$$

$$\left[\text{\includegraphics} \begin{array}{c} H \\ C_2H_5 \end{array} \right]^+ + BF_4^- \longrightarrow \text{\includegraphics} C_2H_5 + HBF_4 \quad (9\text{-}1)$$

$$\left[\text{\includegraphics} H_2 \right]^+ BF_4^-$$

(9-XI)

Fluoroborate Anions. Fluoroboric acid does not, of course, exist in the pure state. Solid salts, such as those of NH_4^+ or Na^+, are commercially available. Ammonium tetrafluoroborate is made by fusing NH_4HF_2 and B_2O_3.

The reaction of boric acid with aqueous HF to form the BF_4^- ion is slow except in strongly acid solution, and the major species is BF_3OH^-.

$$B(OH)_3 + 3F^- + 2H^+ = BF_3OH^- + 2H_2O$$

In aqueous solutions, $BF_2(OH)_2^-$ and $BF(OH)_3^-$ also occur, and in concentrated solutions above pH 6.5, $B_3O_3F_6^{3-}$ is present.[11] The BF_4^- ion is tetrahedral, and fluoroborates resemble the corresponding perchlorates in their solubilities and crystal structures. Although BF_4^- is normally a noncoordinating anion, in certain cases (see Section 4-38) coordination does occur.

Other Trihalides. *Boron trichloride* is a liquid at room temperature under slight pressure (b.p. 12.5°), and the *bromide* boils at 90°. Both fume in moist air and are completely hydrolyzed by water, e.g.,

$$BCl_3 + 3H_2O \rightarrow B(OH)_3 + 3HCl$$

The compounds are prepared by direct interaction of the elements at elevated temperatures.

The rapid hydrolysis by water could indicate that these halides are stronger Lewis acids than BF_3, and indeed, various studies have shown that the order[16] of Lewis acidity is generally $BBr_3 > BCl_3 \approx BH_3 > BF_3 > BMe_3$.

The triiodide is prepared by the action of iodine on $NaBH_4$ or of HI on BCl_3 at red heat. It is an unstable white solid that polymerizes on standing, is explosively hydrolyzed by water, and acts as a Lewis acid.[17]

Tetrachloroborates are obtained by addition of BCl_3 to alkali chlorides at high pressures, by cold milling at room temperatures, or by the reaction

$$[(C_2H_5)_4N]^+Cl^- + BCl_3 \xrightarrow{CHCl_3} [(C_2H_5)_4N]^+BCl_4^-$$

The stability of these salts and the corresponding tetrabromoborates and tetraiodoborates is greatest with the largest cations. With a given cation, the stability order is $MBCl_4 > MBBr_4 > MBI_4$, tetraiodoborates occurring only with the largest cations. Mixed ions such as BF_3Cl^- also exist.

9-7. Boron Halides with B—B bonds and Related Compounds

Diboron Tetrahalides[18] (B_2X_4). The fluoride and the chloride are gaseous and liquid, respectively, at 25°. The fluoride can be made by interaction of BF_3 and B at 1850°, and the BF so obtained on condensation at −196° is allowed to react with BF_3.

The chloride can be made on a 10 g scale by cocondensation of BCl_3 and Cu vapor at −196°

$$2BCl_3 + 2Cu = B_2Cl_4 + 2CuCl$$

or by electric discharge in BCl_3 vapor.[19]

The X_2B—BX_2 molecules are planar in the crystal for both the fluoride and the chloride, but in the gas F_2BBF_2 is planar (D_{2h}), whereas Cl_2BBCl_2 is staggered (D_{2d}).[20] It has been estimated that for the latter the planar form is only slightly (*ca.* 7.5 kJ mol^{-1}) less stable than the staggered; the staggered form is favored by Cl $\cdots$ Cl repulsions, but in the crystal these are dominated by intermolecular packing forces. For B_2F_4, conjugation of B=F bonds favors the planar form by *ca.* 2 kJ mol^{-1}:

B_2Cl_4 is very reactive, and among its most important reactions are protolytic ones and addition to olefins:

16   See, e.g., D. C. Mente, J. L. Mills, and R. E. Mitchell, *Inorg. Chem.*, 1975, **14**, 123.
17   J. R. Blackborow and J. C. Lockhart, *J.C.S. Dalton,* **1973**, 1303.
18   P. L. Timms, *Acc. Chem. Res.*, 1973, **12**, 118.
19   J. P. Brennan, *Inorg. Chem.*, 1974, **13**, 490.
20   D. D. Danielson, J. V. Patton, and K. Hedberg, *J. Am. Chem. Soc.*, 1977, **99**, 6484.

$$B_2Cl_4 + 4ROH \rightarrow B_2(OR)_4 + 4HCl \ (R = alkyl, aryl \ or \ H)$$

$$B_2Cl_4 + RCH=CHR \rightarrow Cl_2BCHR—CHRBCl_2$$

Both B_2F_4 and B_2Cl_4 react avidly with oxygen, the latter burning with a beautiful green flame:

$$6B_2Cl_4 + 3O_2 \rightarrow 2B_2O_3 + 8BCl_3$$

Other halides. Tetraboron tetrachloride (B_4Cl_4) is a pale yellow, volatile solid (m.p. 95°), obtained by mercury arc discharge on B_2Cl_4 vapor. The molecule has a B_4 tetrahedron with Cl bound to each B (T_d symmetry).

Thermal decomposition or electric discharges in B_2Cl_4 or B_2Br_4 or mixtures thereof[21] give a variety of polynuclear molecular compounds such as B_8Cl_8, B_9Cl_9, B_9Br_9, and B_8ClBr_7. B_8Cl_8 has a distorted trigonal dodecahedron (D_{2d}) of B atoms each bearing one Cl.

The halide B_9Cl_9 can undergo reversible oxidation-reduction to give the ion $B_9Cl_9^{2-}$.[22]

Some polynuclear fluorides are known, but they decompose readily to intractable polymers.

Oxide. Interaction of B and B_2O_3 at 1300° gives B_2O_2 on condensation at −196°. An oxide can also be obtained from the alcoxide $B_2(OR)_4$ by hydrolysis at pH 7 to $B_2(OH)_4$ and dehydration of this at 250° to $(BO)_n$.

THE BORANES (BORON HYDRIDES) AND RELATED COMPOUNDS.[23]

In a remarkable series of papers from 1912 to 1936, Alfred Stock and his co-workers prepared and chemically characterized the following hydrides of boron (boranes): B_2H_6, B_4H_{10}, B_5H_9, B_5H_{11}, B_6H_{10}, and $B_{10}H_{14}$. With the exception of diborane (B_2H_6), which was prepared by thermal decomposition of higher boranes, Stock prepared these hydrides by the action of acid on magnesium boride (MgB_2), obtaining in this way a mixture of volatile, reactive, and air-sensitive (some spontaneously flammable) compounds. To handle compounds with these properties, Stock developed the glass vacuum line and techniques for using it.

Preparative methods for the boranes are numerous and highly varied; Stock's original method is now used only for B_6H_{10}. Most preparations begin with B_2H_6 and involve a pyrolysis under a variety of conditions and often in the presence of H_2 or other reagents. The very important $B_{10}H_{14}$, for example, is obtained by py-

[21] M. S. Reason and A. G. Massey, *J. Inorg. Nucl. Chem.,* 1975, **37,** 1593.
[22] R. M. Kaffani and E. H. Wong, *J.C.S. Chem. Comm.,* **1978,** 462.
[23] (a) E. L. Muetterties, Ed., *Boron Hydride Chemistry,* Academic Press, 1975; A. B. Burg, *Chem. Tech.,* **1977,** 50; (b) A. Stock, *The Hydrides of Silicon and Boron,* Cornell University Press, 1933; (c) R. L. Hughes, I. C. Smith, and E. W. Lawless, *Production of Boranes and Related Research,* Academic Press, 1967; (d) A. Pelter and K. Smith, in *Comprehensive Organic Chemistry,* Part 14-2, Vol. 3, Pergamon Press, 1979.

rolysis of B_2H_6 at about 100°, and B_5H_9 is formed on pyrolysis of B_2H_6 in presence of hydrogen at 250°. An electric discharge through B_2H_6 or other boranes is sometimes used, but this method gives only small yields. The best method for any given borane may be peculiar to it, e.g., reaction of B_5H_{11} with the surface of crystalline hexamethylene-tetramine to give B_9H_{15}, and the carefully controlled hydrolysis of the hydroxonium ion salt of $B_{20}H_{18}^{2-}$ to give a mixture of the isomers of $B_{18}H_{22}$.

Table 9-1 lists the better characterized boranes, most of whose structures are now known.

The number of boron atoms is indicated by a latin prefix and the number of H atoms is given with an arabic number in parentheses: for example, pentaborane(9) is B_5H_9; pentaborane(11) is B_5H_{11}. If there is only one compound of B_n, the number can be omitted. Additional nomenclature is discussed later.

9-8.   Structure and Bonding in Boranes and Their Derivatives[23a,24]

The stoichiometries of the boranes, from the simplest B_2H_6 to the most complex, together with the number of electrons available, do not permit of structures or bonding schemes like those for hydrocarbons or other "normal" compounds of the lighter nonmetals. X-Ray crystallography and other studies show that the structures of the boranes are quite unlike hydrocarbon structures. Figure 9-5 shows a few of them. Not only are the structures unique, but in all of the boranes there is the problem of *electron deficiency,* that is, there are not enough electrons to permit the formation of conventional two-electron bonds ($2c$–$2e$ bonds) between all adjacent pairs of atoms. To rationalize the structures in terms of acceptable bonding prescriptions, *multicenter bonding* of various sorts must be widely employed.

For diborane itself $3c$–$2e$ bonds are required to explain the B—H—B bridges. The terminal B—H bonds may be regarded as conventional $2c$–$2e$ bonds. Thus each boron atom uses two electrons and two roughly sp^3 orbitals to form $2c$–$2e$ bonds to two hydrogen atoms. The boron atom in each BH_2 group still has one electron and two hybrid orbitals for further bonding. The plane of the two remaining orbitals is perpendicular to the BH_2 plane. When two such BH_2 groups approach each other as in Fig. 9-6, with hydrogen atoms also lying, as shown, in the plane of the four empty orbitals, two B—H—B $3c$–$2e$ bonds are formed. The total of four electrons required for these bonds is provided by the one electron carried by each H atom and by each BH_2 group.

We have just seen that two structure-bonding elements are used in B_2H_6, viz., $2c$–$2e$ BH groups and $3c$–$2e$ BHB groups. To account for the structures and bonding of the higher boranes, these elements as well as three others are required. The three others are: $2c$–$2e$ BB groups; $3c$–$2e$ open BBB groups; and $3c$–$2e$ closed BBB groups. These five structure-bonding elements may be conveniently represented in the following way:

[24]   R. W. Rudolph, *Acc. Chem. Res.,* 1976, **9**, 446.

TABLE 9-1

The More Important Properties of Boranes[a]

Formula	Name	Melting point (°C)	Boiling point (°C)	Reaction with air, at 25°	Thermal stability	Reaction with water
B_2H_6	Diborane(6)	−164.85	−92.59	Spontaneously flammable	Fairly stable at 25°	Instant hydrolysis
B_4H_{10}	Tetraborane(10)	−120	18	Not spontaneously flammable if pure	Decomposes fairly rapidly at 25°	Hydrolysis in 24 hr
B_5H_9	Pentaborane(9)	−46.8	60	Spontaneously flammable	Stable at 25°; slow decomposition 150°	Hydrolyzed only on heating
B_5H_{11}	Pentaborane(11)	−122	65	Spontaneously flammable	Decomposes very rapidly at 25°	Rapid hydrolysis
B_6H_{10}	Hexaborane(10)	−62.3	108	Stable	Slow decomposition at 25°	Hydrolyzed only on heating
B_6H_{12}	Hexaborane(12)	−82.3	80–90	—	Liquid stable few hours at 25°	Quantitative, to give B_4H_{10}, $B(OH)_3$, H_2
$B_{10}H_{14}$	Decaborane(14)	99.5	213 (extrap.)	Very stable	Stable at 150°	Slow hydrolysis
$B_{14}H_{18}$[b]	Tetradecaborane(18)	liquid	—	Stable	Decomposes ~100°	—
$B_{14}H_{20}$[c]	Tetradecaborane(20)	solid	—	Stable	—	—
$B_{20}H_{16}$	Icosaborane(16)	196–199	—	Stable	—	Irreversibly gives $B_{20}H_{16}(OH)_2^{2-}$ and $2H^+$

[a] Other hydrides: B_8H_{12}, B_8H_{14}, B_8H_{16}, B_8H_{18}, n-B_9H_{15}, i-B_9H_{15}, $B_{10}H_{16}$, $B_{13}H_{19}$, $B_{15}H_{23}$, $B_{16}H_{20}$, n-$B_{18}H_{22}$, i-$B_{18}H_{22}$, $B_{20}H_{26}$ (8 isomers, see G. M. Brown et al., Inorg. Chem., 1979, 18, 1951. Transient species detected in gas phase: BH_3, B_2H_4, B_3H_7, B_4H_8, B_8H_{12}, B_9H_{13}.

[b] S. Hermanek et al., Inorg. Chem., 1975, 14, 2250.

[c] J. C. Hoffman, D. C. Moody, and R. Schaeffer, J. Am. Chem. Soc., 1975, 97, 1621.

Terminal 2c-2e boron—hydrogen bond   B—H

3c-2e Hydrogen bridge bond   B $\overset{H}{\diagup\diagdown}$ B

2c-2e Boron—boron bond   B—B

Open 3c-2e boron bridge bond   B $\overset{B}{\diagup\diagdown}$ B

Closed 3c-2e boron bond   B $\overset{\overset{\displaystyle B}{|}}{\diagup\diagdown}$ B

By using these five elements, "semitopological" descriptions of the structures and bonding in all of the boranes may be given, as shown by Lipscomb. The scheme is capable of elaboration into a comprehensive, semipredictive tool for correlating the structural data not only of the stable compounds but of transient species such as B_2H_4, B_3H_9, or B_4H_{12}.[25] A few examples are illustrated in Fig. 9-7.

The semitopological scheme does not always provide the best description of bonding in the boranes and related species such as the polyhedral borane anions and carbaboranes to be discussed below. Where there is symmetry of a high order it is often more convenient to use a highly delocalized molecular-orbital description of the bonding. For instance, in B_5H_9 where the four basal boron atoms are equivalently related to the apical boron atom, it is *possible* to depict a resonance hybrid involving the localized B $\overset{B}{\diagup\diagdown}$ B and B—B elements, viz.:

but it is ultimately neater and simpler to formulate a set of seven five-center molecular orbitals with the lowest three occupied by electron pairs. When one approaches the hypersymmetrical species such as $B_{12}H_{12}^{2-}$, use of the full molecular symmetry in an MO treatment becomes the only practical course.

Full MO treatments can, of course, be carried out even for the boranes in which the localized structure-bonding elements used in the semitopological theory appear to be well-defined. In such cases the completely general MOs are equivalent to, and can be readily transformed into, more localized components corresponding to the structure-bonding elements just mentioned.

In addition to the desire to develop a consistent set of principles to explain the structures of boranes and to predict new ones, one of the main motivations for theoretical study of the electronic structures of these molecules is to understand their chemical reactivity. One of the most important types of reaction that the boranes (and also the borane anions and carbaboranes) undergo is electrophilic substitution. Although there is no *a priori* basis for presuming it to be so, it is an empirical fact that those boron atoms to which bonding theory assigns the greatest negative charge are those preferentially attacked in electrophilic substitution. For example, in $B_{10}H_{14}$ charge distributions calculated from MO treatments assign considerable ($\sim 0.25\ e$) excess negative charge to boron atoms 2 and 4 (see Fig. 9-5

[25]   W. N. Lipscomb et al., Inorg. Chem., 1978, 17, 3443.

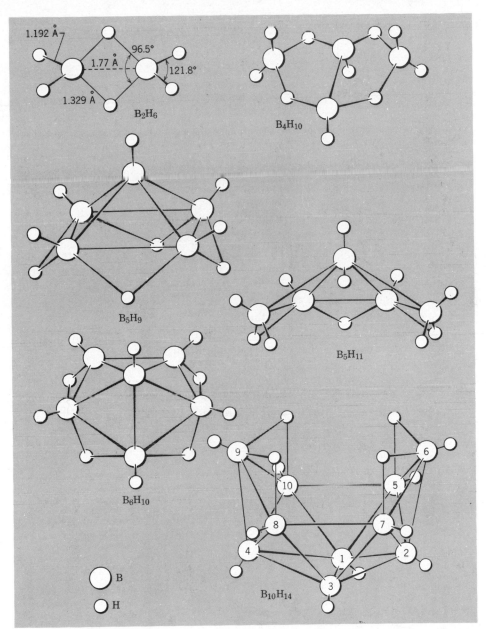

Fig. 9-5. Structures of some boranes.

for numbering), approximate neutrality to boron atoms 1 and 3, and positive charge to all others. Experiments show consistently that only positions 1, 2, 3, and 4 can be substituted electrophilically and that positions 2 and 4 are perhaps slightly preferred. Similar agreement between experimental results and calculated charge

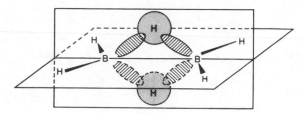

Fig. 9-6. The approach of two properly oriented BH_2 radicals and two H atoms leads to the formation of two 3c-2e B—H—B bonds.

distributions has been obtained for $B_{10}C_2H_{12}$. Thus it is believed that, at least in a qualitative sense, computations of electronic structures are worth making to gain clues as to preferred positions of reactivity.

Structural Study by Nmr.[26] Though X-ray crystallography is the most precise source of structural information on boranes, nuclear magnetic resonance is an important adjunct, especially in elucidating the course of substitution reactions. This is particularly true with respect to the polyhedral borane anions and carbaboranes discussed below. Primarily, it is the abundant [11]B isotope that is studied; proton-resonance spectra are rarely useful because the signals are broad, complex multiplets as a result of splitting by the [11]B and [10]B nuclei.

Figure 9-8 gives some general features of [11]B and ^{1}H nmr spectra, and Fig. 9-9 shows a specific example, that of $B_{10}H_{14}$, together with the assignment in terms of the conventional numbering scheme (Fig. 9-5). Each type of B atom is represented by a signal with intensity proportional to the number of such nuclei; the signals are all doublets because of splitting by the proton attached by a 2c-2e bond to each boron atom. Splitting by bridging protons is not resolved. The insert in Fig.

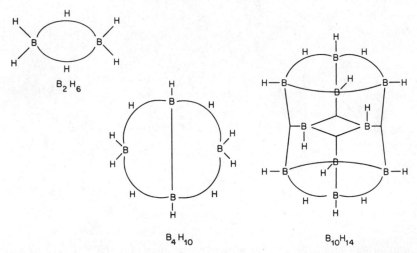

Fig. 9-7. Valence descriptions of boron hydrides in terms of Lipscomb's "semitopological" scheme.

[26] G. R. Eaton and W. N. Lipscomb, *Nmr Studies of Boron Hydrides and Related Compounds*, Benjamin, 1969.

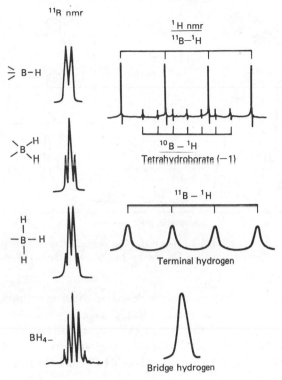

Fig. 9-8. Some general features of ^{11}B and ^1H nmr spectra. [Reproduced by permission from S. G. Shore, in *Boron Hydride Chemistry*, E. L. Muetterties, Ed., Academic Press, 1975.]

9-9 shows the principal difference (minor changes occur elsewhere in the spectrum) observed in the spectrum of 2-iododecaborane. It is quite clear that this molecule has the iodine substituent in the 2 (or 4)-position. Note the lack of splitting when no H is bonded to the boron atom.

Finally it must be noted that many boron hydride compounds are fluxional in the sense that some or all hydrogen atoms are migrating. Consequently nmr spectra in solution may suggest higher symmetry than is consistent with known or presumed molecular structures.[27]

Further Nomenclature. It is now appropriate to outline briefly further points of nomenclature concerning the polyhedral boranes and of their derivatives such as carbaboranes. The boron or boron + carbon, boron + phosphorus, etc., skeletons of the various polyhedra are designated by the Greek terms *closo* (closed), *nido* (nestlike), *arachno* (weblike), and *hypho* (netlike); the order indicates increasing openness.

Figure 9-10 presents idealized polyhedral structures for *closo-, nido-,* and *arachno*-boranes and heteroboranes showing their relationship to each other.

[27] H. Beall and C. H. Bushweller, *Chem. Rev.*, 1973, **73**, 465 (dynamic processes in boranes, carboranes, etc.).

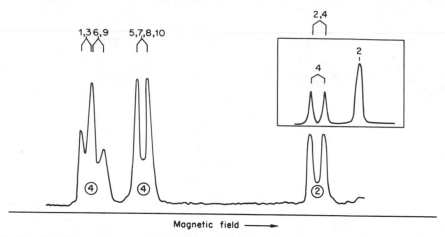

Magnetic field ⟶

Fig. 9-9. The ^{11}Bnmr. spectrum of $B_{10}H_{14}$ at 64 mHz. The assignment is shown at the top and circled numbers give intensities of the multiplets. The insert is part of the spectrum of $2\text{-}B_{10}H_{13}I$.

1.   *Closo* molecules have a complete closed polyhedron with triangular faces. The best known *closo* molecules are the $B_nH_n^{2-}$ ions ($n = 6$ to 12) and the dicarbaboranes $B_{n-2}C_2H_n$ ($n = 5$ to 12). For *closo* compounds the boron atoms are numbered in sequential planes perpendicular to the long axis in a clockwise direction as in 9-XII. The only neutral *closo*-borane is $B_{20}H_{16}$, which has two B_{10} frameworks fused together.

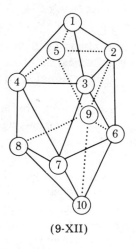

Boron skeleton of decahydro-
closo-decaborate(2–)
closo-$B_{10}H_{10}^{2-}$. There is an
H atom on each boron.

(9-XII)

2.   *Nido* molecules have nonclosed structures, e.g., decaborane(14) in Fig. 9-5. They can be considered to arise by removing the highest connected vertex in a *closo* structure, as indicated in Fig. 9-10 by the diagonal upward arrows.

    Nido molecules may be obtained when *closo* molecules are reduced, e.g.,

$$closo\text{-}B_9C_2H_{11} + 2e = nido\text{-}B_9C_2H_{11}^{2-}$$

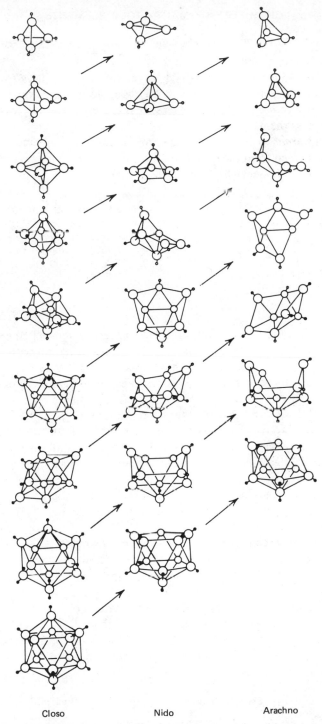

Closo   Nido   Arachno

Fig. 9-10. Idealized polyhedral boron frameworks for *closo-*, *nido-*, and *arachno-*boranes and heteroboranes. Bridge hydrogens and BH_2 groups are not shown, but where appropriate they lie around the open face of the framework. The lines linking boron atoms merely illustrate cluster geometry. [Reproduced by permission from R. W. Rudolph, *Acc. Chem. Res.*, 1976, **9**, 446.]

or are opened up by action of donors, e.g.,

$$closo\text{-}B_4C_2H_6 + NR_3 = nido\text{-}B_4C_2H_6NR_3$$

3.   *Arachno* molecules are obtained by removing from the *nido* cluster the highest connected atom of the open face. *Hypho* molecules are even more opened out than *arachno* or *nido* molecules. There are relatively few examples known, and we do not discuss them.

In *nido* and *arachno* compounds the boron atoms are numbered from a plane projection as viewed from the open side. Interior atoms are numbered first, then peripheral atoms, all clockwise from 12:00 as in 9-XIII.

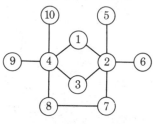

Boron skeleton of *nido*-decaborane $B_{10}H_{14}$. There is an H on each boron with hydride bridges between 5–6, 6–7, 8–9, 9–10.

(9-XIII)

Structure-Bonding Correlation.   It has been recognized that there are rule-of-thumb correlations of the *closo, nido,* and *arachno* structures with the number of electrons assigned to bonding with in the central cage. Though not entirely rigorous or reliable, these correlations,[28] sometimes called Wade's rules, are useful and have an approximate basis in MO theory. They apply most easily to B_nH_n, $B_{n-x}C_xH_n$, B_nH_{n+x}, and certain other heteroboranes (e.g., $B_{11}SH_{11}$). A B—H unit contributes two electrons to cage bonding. Added electrons that make the species an anion are assumed to belong to the cage. Each additional H atom adds an electron to the cage. CH is assumed to contribute 3 electrons while S and P retain one localized lone pair while contributing their remaining electrons—4 and 3, respectively—to the cage. The rules then say that for n-membered *closo, nido,* and *arachno* cages, the pertinent numbers of electrons are $2n + 2$, $2n + 4$, and $2n + 6$, respectively.

To illustrate, let us consider some *closo* systems. In any $B_nH_n^{2-}$ ion there are $2n + 2$ electrons; hence all such anions are *closo.* Similarly, any $B_{n-2}C_2H_n$ or $B_{n-1}CH_n^-$ type carbaborane system is *closo.* Other examples are $B_{11}SH_{11}$ and $B_{10}CPH_{11}$.

The two reactions above, where reduction or addition of a (two-electron) donor to a *closo* molecule gave *nido* molecules, are understandable on Wade's rules, since $2n + 2$ species are changed to $2n + 4$ species. Other *nido* species that satisfy the $2n + 4$ rule are $B_3C_2H_7$ and B_8SH_{10}. The borane B_5H_9 is *nido*, but B_5H_{11} corresponds to $2n + 6$ and is *arachno.*

[28]   K. Wade, *Adv. Inorg. Chem. Radiochem.,* 1976, **18,** 1; R. E. Williams, *Adv. Inorg. Chem. Radiochem.,* 1976, **18,** 67.

9-9.   Chemistry of Some Boranes

Diborane (B_2H_6).[29]   Diborane is the starting material for the synthesis of other boranes and their derivatives and is a useful reagent in organic chemistry.[30]

It can be prepared by action of 85% phosphoric acid on $NaBH_4$ (p. 316), but a convenient method is to drop boron trifluoride etherate into a solution of $NaBH_4$ in ethyleneglycoldimethylether (diglyme).

$$3NaBH_4 + 4BF_3 = 3NaBF_4 + 2B_2H_6$$

Diborane is available in cylinders and as various BH_3 solvates. The gas is highly toxic. When pure it does not react with oxygen at 25°, but nevertheless it must be handled with care. It burns evolving much heat:

$$B_2H_6 + 3O_2 = B_2O_3 + 3H_2O \qquad \Delta H = -2137.7 \ kJ \ mol^{-1}$$

This reaction shows why boranes were once seriously considered as potential rocket fuels. The thermal decomposition of B_2H_6 is extremely complicated and the products depend on the conditions.[31]

Borane (BH_3) can be obtained in the gas phase by thermal decomposition of $H_3B\cdot PF_3$ but cannot be isolated.

Some reactions of diborane are shown in Fig. 9–11. Although rapidly hydrolyzed by water, B_2H_6 reacts with alcohols in two stages, the last slow so that the intermediate compounds $(RO)_2BH$ can be isolated as adducts with NMe_3. With glycol, a dioxoborolane is formed:

$$2 \ \begin{matrix} CH_2OH \\ | \\ CH_2OH \end{matrix} \ + \ B_2H_6 \ \xrightarrow{\ Et_2O\ } \ 2 \ \begin{matrix} O \\ \diagdown \\ O \end{matrix} BH \ + \ 2H_2$$

Many types of reaction of B_2H_6 initially involve addition to Lewis bases to give unstable adducts (B_2H_6L). These can then undergo either symmetric cleavage to given borane adducts or unsymmetrical cleavage.

Many *borane adducts* ($H_3B\cdot L$) are known, and when the relative stabilities of these compounds are considered it is clear that BH_3 must be regarded as a stronger Lewis acid than BF_3.[32] Some adducts such as those of phosphorus compounds of the type $PX(CF_3)(OMe)$ are very stable.[33]

With some Lewis bases (e.g., NH_3) the reaction can be complicated, but asymmetric cleavage of B_2H_6 occurs to give borohydride cations of the type

[29]   L. H. Long, *Prog. Inorg. Chem.*, 1972, **15**, 1; *Adv. Inorg. Chem. Radiochem.*, 1974, **16**, 201; T. P. Fehlner and D. J. Pasto, in *Boron Hydride Chemistry*, E. L. Muetterties, Ed., Academic Press, 1975.

[30]   C. F. Lane, *Chem. Rev.*, 1976, **76**, 773 (an extensive review); C. F. Lane, in *Synthetic Reagents*, Vol. 3, Wiley, 1977; H. C. Brown, *Organic Syntheses via Boranes*, Wiley, 1975.

[31]   H. Fernandez, J. Grotewold, and C. M. Previtali, *J.C.S. Dalton* **1973**, 2090.

[32]   See, e.g., M. Durand *et al., J.C.S. Dalton,* **1977**, 57.

[33]   A. B. Burg, *Inorg. Chem.*, 1977, **16**, 379.

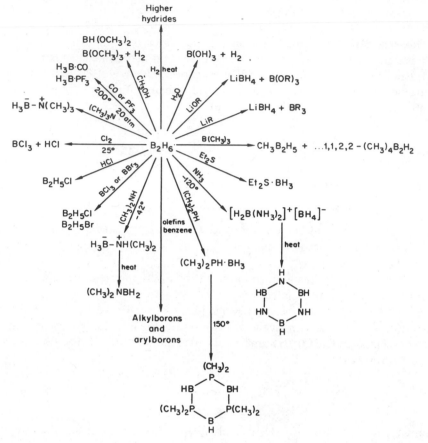

Fig. 9-11. Some reactions of diborane.

$[L_2BH_2]^+BH_4^-$. Some adducts with amines and phosphines undergo further changes on heating to give cyclic compounds (Fig. 9–11; see also p. 323). The addition of anions to B_2H_6 leads to borohydride anions (e.g., BH_4^- as discussed on p. 315).

Some of the most important uses of diborane are in the reduction of organic substances and in the hydroboration reaction discussed later (p. 321). B_2H_6 differs from BH_4^- as a reductant for organic substances in that it attacks at electron-rich positions, whereas BH_4^- reacts by nucleophilic attack on an electrophilic center. Thus B_2H_6 will reduce both RCOOH and RCOOR; whereas $LiBH_4$ reduces only RCOOR'. B_2H_6 reduces azo compounds, aldehydes, and nitriles.

Polyhedral Boranes.[34] We cannot discuss in any detail the extremely diverse and extensive chemistry of the polyhedral boranes. Some properties are listed in Table 9–1. Broad classes of chemical reactions are (a) acidity and the formation of borane anions, (b) substitution of H by halogens, organic groups, etc., (c) adduct

[34] R. L. Middaugh, H. Beall, and S. G. Shore, in *Boron Hydride Chemistry*, E. L. Muetterties, Ed., Academic Press, 1975.

formation, (*d*) electron addition to give anions, (*e*) skeletal additions and formation of compounds that have hetero atoms (C, Si, P, etc.) in the skeleton.

Two of the best studied are the *nido* boranes B_5H_9 and $B_{10}H_{14}$.

Pentaborane(9)[35] is a toxic liquid that can detonate in air. It can be halogenated exclusively in the apical position (see Fig. 9-5) by halogens in presence of aluminum halide catalysts. Interaction with alkenes at *ca.* 150° gives alkyl derivatives.

Decaborane(14) has been made on a multiton scale as a rocket fuel by pyrolysis of B_2H_6 at 150° in presence of dimethylether. It is readily halogenated, and iodine gives 1- and 2-iododecaborane(14) and also 1,2-diiodo- and 2,4-diiododecaborane(14). Alkylation occurs with $AlCl_3$ and alkyl halides to give various substituted derivatives consistent with electrophilic attack at B atoms of highest electron density (i.e., 1, 3, 2, 4). Adducts are formed with neutral donors [e.g., $B_{10}H_{14}$ 6,9-$(CH_3CN)_2$] or with anions [e.g., $B_{10}H_{14}CN^-$, which readily loses H_2 to give $B_{10}H_{12}CN^-$].

Many of the lower boranes function as monoprotic acids,[36] and it is a *bridging H atom* that acts as the source of H^+. The conjugate base, a borane anion of the type discussed below then has a B—B bond. This B—B bond is susceptible to insertion of electrophilic reagents; thus attack by BH_3 acting as an electrophile leads to *polyhedral expansion* of the boron framework. Some examples are

$$B_4H_{10} \xrightarrow{KH} B_4H_9^- \xrightarrow{BH_3} B_5H_{12}^- \xrightarrow{H^+} B_5H_{11}$$

$$B_5H_9 \xrightarrow{KH} B_5H_8^- \xrightarrow{BH_3} B_6H_{11}^- \xrightarrow{H^+} B_6H_{12}$$

Within a given class of boranes, the acidity increases with the size of the B framework. Thus for the *nido* series $B_5H_9 < B_6H_{10} < B_{10}H_{14} < B_{16}H_{20} < n\text{-}B_{18}H_{22}$, and for the *arachno* series $B_4H_{10} < B_5H_{11}$.

9-10. Borane Anions

We have noted that the BH_4^- ion can be considered as arising from Lewis acid behavior of BH_3 toward hydride ion. The polyhedral borane anions are *nido* and *closo* types. Coordination of the tetrahydroborate ion has been discussed in Chapter 4, page 107. Metalloborane complexes are discussed in Section 27-18.

Tetrahydridoborate. This tetrahedral ion, BH_4^-, commonly called borohydride, and some of its substituted derivatives such as $[BH_3CN]^-$ or $[BH(OMe)_3]^-$ are widely used as reducing agents and sources of H^- ion in both inorganic and organic chemistry (Section 6-14), and borohydrides of a great many metallic elements have been made. Typical preparative reactions are:

$$4NaH + B(OCH_3)_3 \xrightarrow{\sim 250°} NaBH_4 + 3NaOCH_3$$

$$NaH + B(OCH_3)_3 \xrightarrow{THF} NaBH(OCH_3)_3$$

[35]   D. F. Gaines, *Acc. Chem. Res.*, 1973, **6**, 416.
[36]   See, e.g., R. J. Remmel *et al.*, *J. Am. Chem. Soc.*, 1975, **97**, 5395.

$$2\text{LiH} + \text{B}_2\text{H}_6 \xrightarrow{\text{ether}} 2\text{LiBH}_4$$

$$\text{AlCl}_3 + 3\text{NaBH}_4 \xrightarrow{\text{heat}} \text{Al(BH}_4)_3 + 3\text{NaCl}$$

$$\text{UF}_4 + 2\text{Al(BH}_4)_3 \longrightarrow \text{U(BH}_4)_4 + 2\text{AlF}_2\text{BH}_4$$

NaBH_4 is representative of the alkali borohydrides, and it is the most common one. It is a white crystalline substance, stable in dry air and nonvolatile. Though insoluble in diethyl ether, it dissolves in water, tetrahydrofuran; glymes (ethylene glycol ethers), and pyridine. Lithium borohydride is soluble in diethyl ether, but there is strong lithium-hydrogen interaction of the type[37]

$$(\text{Et}_2\text{O})_2\text{Li} \underset{\text{H}}{\overset{\text{H}}{\diagdown\!\!\!\diagup}} \text{BH}_2.$$

Aqueous solutions of BH_4^- slowly hydrolyze, though they are more stable when alkaline. Studies of the hydrolysis of BH_4^- in D_2O and of BD_4^- in H_2O provide good evidence that transient BH_5 (cf., CH_5^+) is a five-coordinate, fluxional intermediate of lifetime $ca.$ 10^{-10} sec.[38] Thus much of the gas evolved from BH_4^- in D_2O is HD and this and other observations can be accommodated by the scheme

Sodium borohydride reacts rapidly with methanol but rather slowly with ethanol, in which it is quite soluble. As a reducing agent for organic compounds,[39] NaBH_4 has the limitation that it is insoluble in nonpolar solvents. However phase transfer catalytic reductions can be made using cationic surfactants[40]; alternatively long-chain quaternary ammonium borohydrides can be used.

The cyanoborohydride BH_3CN^- has the advantage that it can be used in moderately acid solutions because it is more stable toward hydrolysis. There are also adducts of trialkylborons with LiH (e.g., LiBHMe_3).[41] The compound LiBHEt_3 is an exceptionally powerful S_N2 nucleophile toward alkyl halides and the replacement of H by C_2H_5 increases the nucleophilicity of LiBH_4 by a factor of 10^5. This can be attributed to the greater ease of H transfer from the weaker Lewis acid BEt_3 than from BH_3.[42]

[37] A. E. Shirk and D. F. Shriver, *J. Am. Chem. Soc.,* 1973, **95**, 5901; E. C. Ashby, F. R. Dobbs, and H. P. Hopkins, Jr., *J. Am. Chem. Soc.,* 1973, **95**, 2823.

[38] R. Willem, *J.C.S. Dalton,* **1979**, 33, and references quoted. See also B. S. Meeks, Jr. and M. M. Kreevoy, *Inorg. Chem.,* 1979, **18**, 2185.

[39] E. R. H. Walker, *Chem. Soc. Rev.,* 1976, **5**, 23.

[40] J. P. Masse and E. R. Parayne, *J.C.S. Chem. Comm.,* **1976**, 438.

[41] H. C. Brown *et al., J. Am. Chem. Soc.,* 1977, **99**, 6237; *Inorg. Chem.,* 1977, **16**, 2229.

[42] G. B. Dunks and K. P. Ordonez, *Inorg. Chem.,* 1978, **17**, 1514.

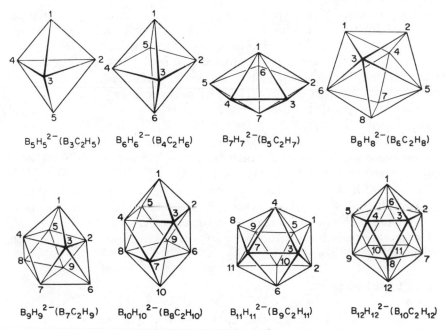

$B_5H_5^{2-}(B_3C_2H_5)$ $B_6H_6^{2-}(B_4C_2H_6)$ $B_7H_7^{2-}(B_5C_2H_7)$ $B_8H_8^{2-}(B_6C_2H_8)$

$B_9H_9^{2-}(B_7C_2H_9)$ $B_{10}H_{10}^{2-}(B_8C_2H_{10})$ $B_{11}H_{11}^{2-}(B_9C_2H_{11})$ $B_{12}H_{12}^{2-}(B_{10}C_2H_{12})$

Fig. 9-12. The triangulated-polyhedral structures of $B_nH_n^{2-}$ and $B_{n-2}C_2H_n$ species; conventional numbering schemes are indicated.

Polyhedral Borane Anions. There are many other borane anions containing from 3 to 20 boron atoms, and those of formula $B_nH_n^{2-}$ ($n = 6$ to 12) are well characterized. Some are obtained as above by acid dissociation of boranes, but others are made in different ways.

Interaction of $NaBH_4$ and BF_3OEt_2 in diglyme or similar ether gives the ion $B_3H_8^-$

$$5BH_4^- + 4BF_3OEt_2 \rightarrow 2B_3H_8^- + 3BF_4^- + 2H_2 + 4Et_2O$$

whereas the same reaction but using excess BF_3 gives $B_{11}H_{14}^-$

$$17BH_4^- + 20BF_3OEt_2 \xrightarrow{105°} 2B_{11}H_{14}^- + 15BF_4^- + 20H_2 + 20Et_2O$$

Pyrolysis of NaB_3H_8 at 200 to 230° in vacuum gives salts of BH_4^-, $B_{10}H_{10}^{2-}$, and $B_{12}H_{12}^{2-}$, and air oxidation of $B_9H_9^{2-}$ in tetrahydrofuran gives $B_8H_8^{2-}$ and $B_7H_7^{2-}$. The borane anions have full triangulated-polyhedral *closo* structures (Fig. 9-12).

The ions $B_{10}H_{10}^{2-}$ and $B_{12}H_{12}^{2-}$ are hydrolytically stable and have been most thoroughly studied. The most useful syntheses are:

$$B_{10}H_{14} + 2R_3N \xrightarrow[\text{xylene}]{\text{boiling}} 2(R_3NH)^+ + B_{10}H_{10}^{2-} + H_2$$

$$5B_2H_6 + 2NaBH_4 \xrightarrow[\text{diglyme}]{180°} 2Na^+ + B_{12}H_{12}^{2-} + 13H_2$$

For these ions there is an enormous body of substitution chemistry, reminiscent of aromatic hydrocarbon chemistry. Attack by electrophilic reagents is the most important general reaction type. Representative attacking species are RCO^+, CO^+, $C_6H_5N_2^+$, and Br^+. These reactions proceed most readily in strongly acid media, and, in general, $B_{10}H_{10}^{2-}$ is much more susceptible to substitution than is $B_{12}H_{12}^{2-}$. Less reactive nucleophiles such as $C_6H_5N_2^+$ fail to attack $B_{12}H_{12}^{2-}$ and attack $B_{10}H_{10}^{2-}$ selectively at the apical (1, 10)-positions. Both $B_{10}H_{10}^{2-}$ and $B_{12}H_{12}^{2-}$ can be partially and completely halogenated; the rate of reaction of $B_{10}H_{10}^{2-}$ in aqueous or ethanolic solution with halogens decreases in the order $Cl_2 > Br_2 > I_2$ and also diminishes with increasing substitution.[43] The perhalogeno ions have extremely high thermal stabilities and are also highly resistant to hydrolysis.

Oxidation of $B_{10}H_{10}^{2-}$ in aqueous solution by Fe^{3+} gives an ion $B_{20}H_{18}^{2-}$ that has two polyhedra linked by two three-center B—B—B bonds between the 6, 10, and 6′,10′ positions. This in turn is reduced by sodium in liquid ammonia to $B_{20}H_{18}^{4-}$, which has the polyhedra linked only by a B—B bond between the 10 positions. Isomers of $B_{20}H_{18}^{4-}$, can be made.

9-11. Carbaboranes[44]

Carbaboranes, or carboranes as they are sometimes called, are compounds that contain carbon atoms incorporated into a polyhedral borane. The CH group is isoelectronic with, thus may replace, the BH^- group. Thus polyhedral carbaboranes may be considered as *formally* derived from $B_nH_n^{2-}$ ions (e.g., two replacements lead to molecules of general formula $B_{n-2}C_2H_n$). Some structures are shown in Fig. 9–12.

There are (*a*) neutral, two-carbon carbaboranes $B_nC_2H_{n+2}$ ($n = 3$ to 10) (*b*) two-carbon carbaborane anions derived from them, and (*c*) one-carbon carbaborane anions ($B_nCH_{n+1}^-$). Isomeric forms of carbaboranes have been isolated. Carbaboranes are mainly *closo* and *nido,* but there are four *arachno* carbaboranes and one *hypho*.[45] All such compounds may act as ligands toward transition metals, and this chemistry is discussed in Section 27-17.

Carbaboranes are obtained by interaction of boranes or borane adducts with acetylenes. Some examples are the following

$$nido\text{-}B_5H_9 + C_2H_2 \rightarrow nido\text{-}4,5\text{-}B_4C_2H_8 + \tfrac{1}{2}B_2H_6$$

$$nido\text{-}4,5\text{-}B_4C_2H_8 \xrightarrow[\text{1–3 sec}]{450°} closo\text{-}B_3C_2H_5 \quad 40\%$$

$$closo\text{-}B_5C_2H_7 \quad 40\%$$
$$closo\text{-}B_4C_2H_6 \quad 20\%$$

$$nido\text{-}B_{10}H_{14} + 2Et_2S \xrightarrow{\text{reflux}} B_{10}H_{12}(SEt_2)_2 + H_2$$

[43] K.-G. Bührens and W. Preetz, *Angew. Chem., Int. Ed.,* 1977, **16**, 173.

[44] H. Beall and T. Onak, in *Boron Hydride Chemistry,* E. L. Muetterties, Ed., Academic Press, 1975, R. N. Grimes, *Carboranes,* Academic Press, 1971; G. B. Dunks and M. F. Hawthorne, *Acc. Chem Res.,* 1973, **6**, 124 (nonicosahedral carbaboranes).

[45] J. Duben, S. Hermonek, and B. Stibr, *J.C.S. Chem. Comm.,* **1978**, 287.

$$B_{10}H_{12}(SEt_2)_2 + RC \equiv CR' \longrightarrow 1,2\text{-}closo\text{-}B_{10}H_{10}C_2RR' + 2Et_2S + H_2$$

$$1,2\text{-}closo\text{-}B_{10}H_{10}C_2RR' \xrightarrow{450°} 1,7\text{-}closo\text{-}B_{10}H_{10}C_2RR'$$

Carbon-substituted derivatives can be obtained using substituted acetylenes or by reaction sequences such as the following for $1,2\text{-}closo\text{-}B_{10}H_{10}C_2H_2$ where a conventional and self-explanatory abbreviation is used for the $B_{10}C_2H_{10}$ group.

The chemistry of C-substituted carbaboranes is quite similar to that of conventional carbon systems, and indeed organic-type reactions have led to the synthesis of thousands of carbaborane derivatives. A simple example is chlorination, where 1,2- and $1,7\text{-}B_{10}H_{10}C_2H_2$ react at the most negative boron atoms (cf. $B_{10}H_{14}$) to give $B_{10}C_2H_{12-x}Cl_x$ molecules.

Carbaborane Anions. Both 1,2- and $1,7\text{-}closo\text{-}B_{10}C_2H_{12}$ can be attacked by ethoxide ion or other strong bases to give isomeric *nido* carborane anions $B_9C_2H_{12}^-$, viz.,

$$closo\text{-}B_{10}C_2H_{12} + EtO^- + 2EtOH = nido\text{-}B_9C_2H_{12}^- + B(OEt)_3 + H_2$$

This removal of a BH^{2+} unit from the parent carbaborane can be regarded as a nucleophilic attack at the most electron-deficient B atom.

MO calculations show that the carbon atoms in the $B_{10}C_2H_{12}$ molecules (and in carbaboranes in general) have considerable electron-withdrawing power; thus the most electron-deficient boron atoms are those *adjacent* to C, namely, those at positions 3 and 6 in $1,2\text{-}B_{10}C_2H_{12}$ and those at positions 2 and 3 in $1,7\text{-}B_{10}C_2H_{12}$. Therefore, the direct products of the reactions of the isomeric $B_{10}C_2H_{12}$ molecules with strong base should be the isomeric $B_9C_2H_{11}^{2-}$ ions shown in Fig. 9–13. These ions themselves are strong bases and acquire protons to form the $B_9C_2H_{12}^-$ ions. It is not known precisely how the twelfth hydrogen atom is bound in these species, but it presumably resides somewhere on the open face of the now incomplete icosahedron and is probably very labile.

Both the isomeric $nido\text{-}B_9C_2H_{12}^-$ ions can be protonated by anhydrous acids to the neutral *nido*-carbaboranes $B_9C_2H_{13}$, which are themselves strong acids

$$B_9C_2H_{12}^- \underset{-H^+}{\overset{H^+}{\rightleftharpoons}} nido\text{-}B_9C_2H_{13}$$

and at 150° these in turn can be converted to the *closo*-carbaboranes $B_9C_2H_{11}$

$$nido\text{-}B_9C_2H_{13} \xrightarrow{150°} closo\text{-}B_9C_2H_{11} + H_2$$

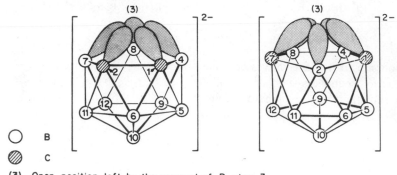

(3) Open position left by the removal of B atom 3

Fig. 9-13. Structures of the isomeric $B_9C_2H_{11}^{2-}$ ions.

Several of the lower *closo*-carbaboranes can then be obtained by the reaction sequence:

$$B_9C_2H_{11} \xrightarrow{H_2CrO_4} B_7C_2H_{13} \xrightarrow{200°} \begin{cases} 1,7\text{-}B_6C_2H_8, \\ 1,6\text{-}B_8C_2H_{10} \end{cases} \text{good yield} \\ B_7C_2H_9 \quad \text{low yield}$$

The $1,6\text{-}B_8C_2H_{10}$ rearranges quantitatively to the 1,10-isomer at 350°.

Boranes with Hetero Atoms Other than Carbon. Closely related to the carbaboranes are compounds with other atoms present, notably phosphorus, arsenic, and sulfur. The formalism for relating isoelectronic species is $P \equiv BH^-$ and $S \equiv BH^{2-}$.

The *thiaborane anion* $B_{10}SH_{10}^{2-}$ is isoelectronic with $B_9C_2H_{11}^{2-}$ and is presumably isostructural with it. An example of a carbaphosphaborane is $B_{10}CPH_{11}$. The 1,2-isomer is obtained in the reaction

$$Na_3B_{10}H_{10}CH + PCl_3 \rightarrow 1,2\text{-}B_{10}CPH_{11} + 3NaCl$$

Like $1,2\text{-}B_{10}C_2H_{12}$, $1,2\text{-}B_{10}CPH_{11}$ can be thermally converted into its 1,7- and 1,12-isomers.

The heteroboranes, like boranes and carbaboranes, form transition metal complexes (Section 27-17).

9-12. Organoboron Compounds[46]

There are thousands of boron compounds with B—C, B—O—C, B—S—C, B—N—C, etc., bonds, whose chemistry is essentially organic. True organoboron compounds contain at least one B—C bond, however.

[46] T. Onak, *Organoboron Chemistry,* Academic Press, 1975; H. Steinberg and R. J. Brotherton, *Organoboron Chemistry,* Vols. I, II, Wiley, 1964, 1967; K. Smith, *Chem. Soc. Rev.,* 1974, **3**, 443 (preparation of organoboranes: reagents for organic synthesis); H. C. Brown, *Organic Syntheses via Boranes,* Wiley, 1975; H. C. Brown, *Boranes in Organic Chemistry,* Cornell University Press, 1972; G. M. L. Cragg, *Organoboranes in Organic Synthesis,* Dekker, 1973; J. Weill-Raynal, *Synthesis,* **1976,** 633 (C—C bond formation); G. M. L. Cragg and K. R. Koch, *Chem. Soc. Rev.,* 1977, **6**, 393; A. Pelter and K. Smith, *Comprehensive Organic Chemistry,* Part 14, Vol. 3, Pergamon Press, 1979.

The alkyl and aryl borons can be made from the halides by conventional methods using lithium or Grignard reagents. The lower alkyls are reactive substances inflaming in air, but the aryls are stable. Compounds can be of the types R_3B, R_2BX, and RBX_2. Mixed compounds can be made by reactions such as[47]

$$R_3B + MeSH \xrightarrow{-MeR} R_2BSH \xrightarrow{Br_2} R_2BBr + RSSR$$

Boron can also be present in heterocyclic compounds of various types; for example, pyrolysis of BMe_3 gives a boraadamantane[48] that is, of course, a carbaborane, $Me_6B_6C_4H_4$. Of the many types of compound, we mention only a few.

Boronic and *boronous* acids, $RB(OH)_2$ and $R_2B(OH)$, and their esters and anhydrides, can be made by reactions such as

$$BF_3OEt_2 + C_6H_5MgBr \rightarrow C_6H_5BF_2 \xrightarrow{H_2O} C_6H_5B(OH)_2$$

These stable and water-soluble compounds have acidities that depend on the R group. Dehydration of $RB(OH)_2$ by heating gives a boroxine.

All ring compounds with planar alternant B and O atoms are called *boroxines*.

Hydroboration.[30,46] This reaction was discovered and has been developed by H. C. Brown and his students. It involves interaction of alkenes or alkynes with boranetetrahydrofuran, H_3BSMe_2, or B_2H_6 prepared *in situ,* to give, initially, alkylborons RBH_2, R_2BH, or R_3B. In effect, H and BH_2 groups are added *anti*-Markownikoff and cis, viz.,

With simple unhindered alkenes the reaction proceeds to trialkylboron, but with trisubstituted and some disubstituted ones (e.g., cyclohexene) the reaction may be stopped at the R_2BH stage; tetrasubstituted alkenes react readily only to give RBH_2.

From the alkylborons so prepared, hydrocarbons, alcohols, ketones, and other compounds may be obtained by reaction with other reagents. The importance lies in the fact that the resulting hydration or hydrogenation products are *anti*-Markownikoff and cis

47 A. Pelter *et al., J.C.S. Dalton,* **1976,** 2087.
48 M. P. Brown *et al., J.C.S. Dalton,* **1977,** 1862.

With RBH_2 and R_2BH the reaction with carbon monoxide followed by H_2O_2 leads to aldehydes and ketones respectively. Polyenes may also undergo hydroboration, e.g.,

A unique hydroboration reagent[49] is 9-borobicyclo[3.3.1]nonane or 9-BBN (9-XIV). This liquid (b.p. 195°, 12 torr) is made by interaction of diborane and cyclooocta-1,5-diene in tetrahydrofuran. It is quite stable in absence of air and is commercially available. In the solid and in solution it is dimeric with hydrogen bridges.

(9-XIV)

9-BBN shows remarkable regioselectivity toward acyclic olefins; for example, hydroboration followed by oxidation converts hex-1-ene into hexan-1-ol in 99.9% yield. The organoboranes obtained using 9-BBN are also far more resistant to isomerization than most organoboranes. These undergo thermal isomerization in presence of H^- ion at 75 to 160°, (where boron evidently migrates along a carbon chain) to give a mixture of organoboranes approaching the thermodynamic equilibrium distribution. Again this leads to high selectivity in reactions.

Tetraalkyl and -aryl Borates. When boron halides are treated with four equivalents of alkylating agent, the trialkyl or triaryl reacts further to form an anion of the type BR_4^-. The most important such compound is *sodium tetraphenylborate,* $Na[B(C_6H_5)_4]$; this is soluble in water and is stable in weakly acid solution; it gives insoluble precipitates with larger cations such as K^+, Rb^+, or Me_4N^+, that are suitable for gravimetric analysis. The ion can also act as a ligand wherein one phenyl ring is bound in an arene complex (p. 99).

In lithium tetramethylborate,[50] as in lithium alkyls themselves (p. 266) polymerization occurs via strong interaction of lithium with the hydrogen atoms of the

49   H. C. Brown *et al., J. Am. Chem. Soc.,* 1974, **96,** 7765; 1976, **98,** 7107; 1977, **99,** 3427.
50   W. E. Rhine, G. Stucky, and S. W. Peterson, *J. Am. Chem. Soc.,* 1975. **97,** 6401.

methyl groups. There are planar sheets of lithium atoms bridged by BMe₄ groups through bridges of the types 9-XV and 9-XVI.

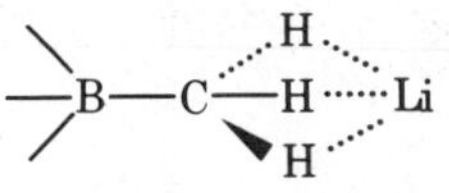

(9-XV)

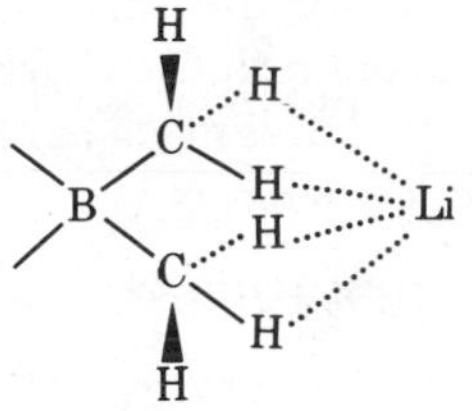

(0 XVI)

9-13.   Boron-Nitrogen Compounds: Borazines.[51]

The group —NR'—BR— is similar to —CR'=CR— and can replace it in many compounds. The analogy may be justified by assuming that the electron distribution in the N—B bond can be described by the resonance 9-XVIIa,b whereby appreciable π bonding is introduced.

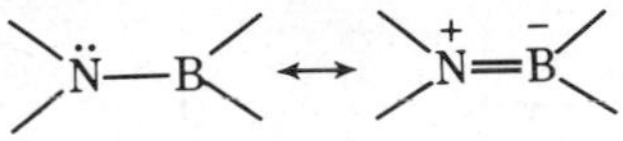

(9-XVIIa)   (9-XVIIb)

There is evidence that such N–B bonds have appreciable π character, while at the same time they lack the polarity that would be engendered by 9-XVIIb. This apparent paradox is explained by the existence of considerable polarity in the σ bond, in a direction opposite to that in the π bond. Thus the net, or actual, polarity is only the difference between the two.

There are a number of carbon heterocycles such as 9-XVIII and 9-XIX containing one or two—BR'—NR—units, but perhaps the most interesting compounds are borazine (9-XX) and its derivatives.

(9-XVIII) 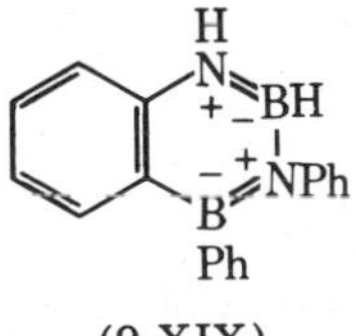   (9-XIX) 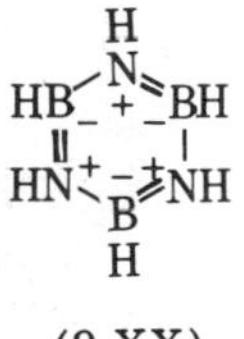   (9-XX)

Borazine has an obvious formal resemblance to benzene, and the physical properties of the two compounds are very similar.

Both MO calculations and experimental results for the transmission of substituent effects through the B_3N_3 ring indicate that the π electrons are partially delocalized.[52] Complete delocalization is not to be expected, since the Nπ orbitals should be of appreciably lower energy than the Bπ orbitals. The planarity of the

[51]   D. F. Gaines and J. Borlin, in *Boron Hydride Chemistry*, E. L. Muetterties, Ed., Academic Press, 1975.
[52]   See, e.g., B. Wrackmeyer and H. Nöth, *Chem. Ber.,* 1976, **109**, 348 (¹¹B, ¹⁴N nmr data).

borazine molecule is shown by MO calculations to be stabilized by the π bonding, and the calculations also suggest that the N-to-B π-electron drift is actually outweighed by B to N σ drift, so that the nitrogen atoms are relatively negative.

The retention of negative charge by nitrogen means that chemically, borazine is not like benzene. It is much more reactive and, unlike benzene, readily undergoes addition reactions such as

$$B_3N_3H_6 + 3HX \rightarrow (\text{—}H_2N\text{—}BHX\text{—})_3 \qquad (X = Cl, OH, OR, \text{etc.})$$

Borazine decomposes slowly on storage and is hydrolyzed at elevated temperatures to NH_3 and $B(OH)_3$. It is of interest that borazine resembles benzene in forming arene-metal complexes (Section 27-15); thus the hexamethylborazine complex, $B_3N_3(CH_3)_6Cr(CO)_3$ resembles $C_6(CH_3)_6Cr(CO)_3$ but is thermally less stable.

Reaction sequences by which borazine and substituted borazines may be synthesized are the following:

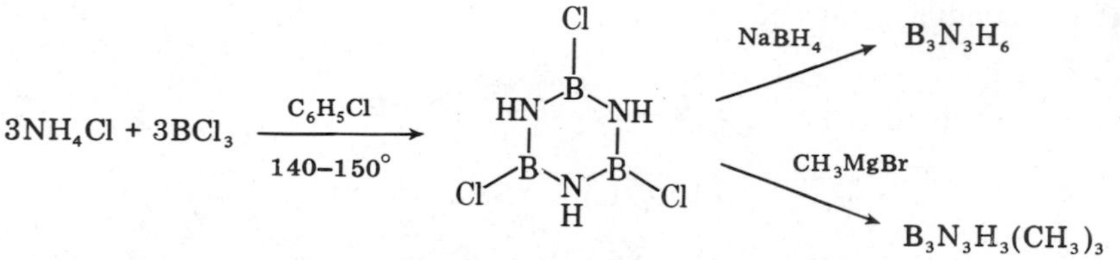

$$CH_3NH_2 + BCl_3 \xrightarrow[C_6H_5Cl]{\text{Boiling}} Cl_3B \cdot NH_2CH_3 \ (\text{m.p. } 126\text{–}128°)$$

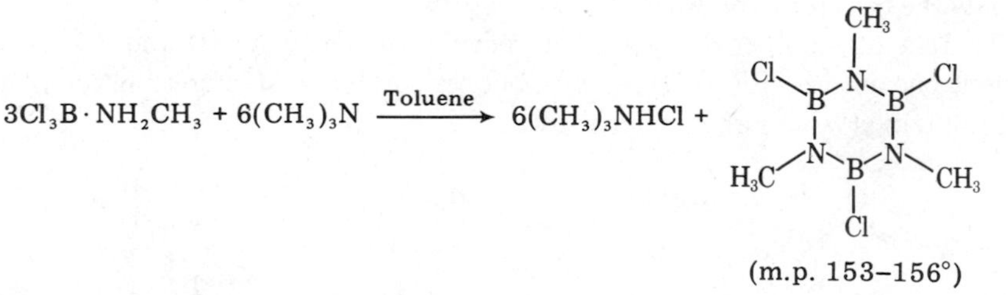

There is also an extensive chemistry of compounds similar to B—N compounds with B—P, B—As, B—S, etc., bonds. There are simple adducts (e.g., H_3PBH_3) and also cyclic compounds such as $(Me_2PBH_2)_3$ with alternant B and P atoms. This compound and its arsenic analogue are extraordinarily stable and inert, a fact that has been attributed to a drift of electron density from the BH_2 groups into the *d* orbitals of P or As. This serves to reduce the hydridic nature of the hydrogen atoms, making them less susceptible to reaction with protonic reagents, and also to offset the $\bar{B}$—$\overset{+}{P}$, $\bar{B}$—$\overset{+}{As}$ polar character, which the σ bonding alone tends to produce. Another example is tribromoborthiin ($B_3S_3Br_3$), which has a six-membered BS ring with bromine atoms on B.

General References

Gmelin, *Handbook of Inorganic Chemistry,* Springer-Verlag. New Supplement Series, *Boron Compounds,* Parts 1 to 17. Covers most of the aspects including oxides, boric acid, and borates (Vols. 28, 44, 48), boron halogen compounds (Vol. 34), tetrahydroborate (Vol. 33), carboranes (Vols. 27, 42, 43), and borazine (Vol. 51).

Muetterties, E. L., *The Chemistry of Boron and Its Compounds,* Wiley, 1967.

Nöth, H, and B. Wrackmeyer, *Nmr Spectroscopy of Boron Compounds,* Vol. 14 of *Nmr Basic Principles and Progress,* Springer, 1978.

CHAPTER TEN

The Group III Elements: Al, Ga, In, Tl

GENERAL REMARKS

10-1. Electronic Structures and Valences

The electronic structures and some other important fundamental properties of the Group III elements are listed in Table 10.1.

TABLE 10-1
Some Properties of the Group III Elements

Property	Al	Ga	In	Tl
Electronic configuration	$[Ne]3s^23p$	$[Ar]3d^{10}4s^24p$	$[Kr]4d^{10}5s^25p$	$[Xe]4f^{14}5d^{10}6s^26p$
Ionization enthalpies (kJ mol^{-1})				
1st	576.4	578.3	558.1	589.0
2nd	1814.1	1969.3	1811.2	1958.7
3rd	2741.4	2950.0	2689.3	2862.8
4th	11563.0	6149.7	5571.4	4867.7
M.p. (°C)	660	29.8	157	303
B.p. (°C)	2327	ca. 2250	2070	1553
$E°$ for $M^{3+} + 3e = M(V)$.^{a}	−1.66	−0.35	−0.34	0.72
Radii M^{3+} (Pauling)(Å)	0.50	0.62	0.81	0.95
Radii M^+ (Å)		1.13	1.32	1.40

^{a} $In^+ + e = In$, $E° = -0.25$ V; $Tl^+ + e = Tl$, $E° = -0.3363$ V.

Although the first and smallest member of Group III, namely, boron, has no cationic chemistry, the remaining four elements form cationic complexes such as $[Ga(H_2O)_6]^{3+}$; they also form many molecular compounds that are covalent.

While the main oxidation state is III, there are certain compounds in which the oxidation state is formally II. These, however, may have metal-metal bonds as in $[Cl_3Ga—GaCl_3]^{2-}$. The univalent state becomes progressively more stable as the group is descended and for thallium the Tl^I-Tl^{III} relationship is a dominant feature of the chemistry. This occurrence of an oxidation state two below the group valence is sometimes called the *inert pair* effect, which first makes itself evident here, although it is adumbrated in the low reactivity of mercury in Group II, and it is much more pronounced in Groups IV and V. The term refers to the resistance of a pair of s electrons to be lost or to participate in covalent bond formation. Thus mercury is difficult to oxidize, allegedly because it contains only an inert pair ($6s^2$), Tl readily forms Tl^I rather than Tl^{III} because of the inert pair in its valence shell ($6s^2 6p$), etc. The concept of the inert pair does not actually tell us anything about the ultimate reasons for the stability of certain lower valence states, but it is useful as a label and is often encountered in the literature. The true cause of the phenomenon is not intrinsic inertness, that is, unusually high ionization potential of the pair of s electrons, but rather the decreasing strengths of bonds as a group is descended. Thus, for example, the sum of the second and third ionization enthalpies (kJ mol^{-1}) is lower for In, 4501, than for, Ga 4916, with Tl, 4820, intermediate. There is, however, a steady decrease in the mean thermochemical bond energies, for example, among the trichlorides: Ga, 242; In, 206; Tl, 153 kJ mol^{-1}. The relative stabilities of oxidation states differing in the presence or absence of the inert pair are further discussed in connection with the Group IV elements. Recent theoretical work shows that relativistic effects make an important contribution to the inert pair effect.[1]

In the trihalide, trialkyl, and trihydride compounds there are some resemblances to the corresponding boron chemistry. Thus MX_3 compounds behave as Lewis acids and can accept either neutral donor molecules or anions to give tetrahedral species; the acceptor ability generally decreases in the order Al > Ga > In, with the position of Tl uncertain. There are, however, notable distinctions from boron. These are in part due to the reduced ability to form multiple bonds and to the ability of the heavier elements to have coordination numbers exceeding four. Thus although boron gives $Me_2\bar{B}=\overset{+}{N}Me_2$, Al, Ga, and In give dimeric species, e.g., $[Me_2AlNMe_2]_2$, in which there is an NMe_2 bridging group and both the metal and nitrogen atoms are four-coordinate. Similarly, the boron halides are all monomeric, whereas those of Al, Ga, and In are all dimeric. The polymerization of trivalent Al, Ga, In, and Tl compounds to achieve coordination saturation is general, and four-membered rings appear to be a common way despite the valence angle stain implied. Second, compounds such as $(Me_3N)_2AlH_3$ have trigonal-bypyramidal structures, which of course are impossible for boron adducts.

The stereochemistries and oxidation states of Group III are summarized in Table 10-2.

[1] K. S. Pitzer, *Acc. Chem. Res.*, 1979, **12**, 271; P. Pyykko and J.-P. Desclaux, *ibid.*, 1979, **12**, 276.

TABLE 10-2

Oxidation States and Stereochemistries of Group III Elements

Oxidation state	Coordination number	Geometry	Examples
+1	6	Distorted octahedral	TlF
+2	4	Tetrahedral	$[Ga_2Cl_6]^{2-}$, $[In_2Cl_6]^{2-}$
+3	3	Planar	$In[Co(CO)_4]_3$[a]
	4	Tetrahedral	$[AlCl_4]^-$, $[GaH_4]^-$, $Al_2(CH_3)_6$, Ga_2Cl_6
	5	Tbp	$AlCl_3(NMe_3)_2$, $[In(NCS)_5]^{2-}$, $Me_2GaClphen$
		Sp	$[InCl_5]^{2-}$
	6	Octahedral	$[Al(H_2O)_6]^{3+}$, $Ga(acac)_3$, $[AlF_6]^{3-}$, $[GaCl_2phen_2]^+$

[a] W. R. Robinson and D. P. Schüssler, *Inorg. Chem.,* 1973, **12**, 848.

THE ELEMENTS

10-2. Occurrence, Isolation, and Properties

Aluminum, the commonest metallic element in the earth's crust (8.8 mass %) occurs widely in Nature in silicates such as micas and feldspars, as the hydroxo oxide (*bauxite*), and as *cryolite* (Na_3AlF_6). The other three elements are found only in trace quantities. Gallium and indium occur in aluminum and zinc ores, but the richest sources contain less than 1% of gallium and still less indium. Thallium is widely distributed and is usually recovered from flue dusts from the roasting of certain sulfide ores, mainly pyrites.

Aluminum is prepared on a vast scale from bauxite. This is purified by dissolution in sodium hydroxide and reprecipitation using carbon dioxide. It is then dissolved in cryolite at 800 to 1000° and the melt is electrolyzed. Aluminum is a hard, strong, white metal. Although highly electropositive, it is nevertheless resistant to corrosion because a hard, tough film of oxide is formed on the surface. Thick oxide films, some with the proper porosity when fresh to trap particles of pigment, are often electrolytically applied to aluminum. Aluminum is soluble in dilute mineral acids, but is passivated by concentrated nitric acid. If the protective effect of the oxide film is overcome, for example by scratching or by amalgamation, rapid attack even by water can occur. The metal is attacked under ordinary conditions by hot alkali hydroxides, halogens, and various nonmetals. Highly purified aluminum is quite resistant to acids and is best attacked by hydrochloric acid containing a little cupric chloride or in contact with platinum, some H_2O_2 also being added during the dissolution.

Gallium, indium, and *thallium* are usually obtained by electrolysis of aqueous solutions of their salts; for Ga and In this possibility arises because of large overvoltages for hydrogen evolution on these metals. They are soft, white, comparatively reactive metals, dissolving readily in acids; however thallium dissolves only slowly in sulfuric or hydrochloric acid, since the Tl^1 salts formed are only sparingly soluble.

Gallium, like aluminum, is soluble in sodium hydroxide. The elements react rapidly at room temperature, or on warming, with the halogens and with nonmetals such as sulfur.

The exceptionally low melting point of gallium has no simple explanation. Since its boiling point (2070°) is not abnormal, gallium has the longest liquid range of any known substance and finds use as a thermometer liquid.

CHEMISTRY OF THE TRIVALENT STATE

BINARY COMPOUNDS

10-3. Oxygen Compounds[2a]

Stoichiometrically there is only one oxide of aluminum, namely, *alumina* (Al_2O_3). However this simplicity is compensated by the occurrence of various polymorphs, hydrated species, and so on, the formation of which depends on the conditions of preparation. There are two forms of anhydrous Al_2O_3, namely, α-Al_2O_3 and γ-Al_2O_3. Other trivalent metals (Ga, Fe) form oxides that crystallize the same two structures. In α-Al_2O_3 the oxide ions form a hexagonal close-packed array and the aluminum ions are distributed symmetrically among the octahedral interstices. The γ-Al_2O_3 structure is sometimes regarded as a "defect" spinel structure, that is, as having the structure of spinel with a deficit of cations (see below).

α-Al_2O_3 is stable at high temperatures and also indefinitely metastable at low temperatures. It occurs in Nature as the mineral *corundum* and may be prepared by heating γ-Al_2O_3 or any hydrous oxide above 1000°. γ-Al_2O_3 is obtained by dehydration of hydrous oxides at low temperatures ($\sim$450°). α-Al_2O_3 is very hard and resistant to hydration and attack by acids, whereas γ-Al_2O_3 readily takes up water and dissolves in acids. The Al_2O_3 that is formed on the surface of the metal has still another structure, namely, a defect rock salt structure; there is an arrangement of Al and O ions in the rock salt ordering with every third Al ion missing.

There are several important hydrated forms of alumina corresponding to the stoichiometries AlO·OH and Al(OH)₃. Addition of ammonia to a boiling solution of an aluminum salt produces a form of AlO·OH known as *boehmite*, which may be prepared in other ways also. A second form of AlO·OH occurs in Nature as the mineral *diaspore*. The true hydroxide Al(OH)₃ is obtained as a crystalline white precipitate when carbon dioxide is passed into alkaline "aluminate" solutions. It occurs in Nature as *gibbsite*. What is known as β-alumina normally contains alkali metal ions and has the ideal position $M_2^{\mathrm{I}}O \cdot Al_2O_3$.[2b]

[2a] R. D. Srivastava and M. Farber, *Chem. Rev.*, 1978, **78**. 627 (Thermodynamic prospects of Group III oxides).

[2b] W. A. England *et al.*, *J.C.S. Chem. Comm.*, **1976**, 895; R. Collongues *et al.*, in *Solid Electrolytes*, P. Hagenmuller and W. Van Goel; Eds., Academic Press, 1978.

Aluminas used in chromatography or as catalyst supports[2c] are prepared by heating hydrated oxide to various temperatures so that the surfaces may be partially or wholly dehydrated. The activity of the alumina depends critically on the treatment, subsequent exposure to moist air, and other factors.

The *gallium oxide* system is a similar system, affording a high-temperature α- and a low-temperature γ-Ga_2O_3. The trioxide is formed by heating the nitrate, the sulfate, or the hydrous oxides that are precipitated from Ga^{III} solutions by the action of ammonia. β-Ga_2O_3 contains both tetrahedrally and octahedrally coordinated gallium with Ga—O distances of 1.83 and 2.00 Å, respectively. The hydrous oxides GaOOH and $Ga(OH)_3$ are similar to their Al analogues.

Indium gives yellow In_2O_3, which is known in only one form, and a hydrated oxide $In(OH)_3$. *Thallium* has only the brown-black Tl_2O_3, which begins to lose oxygen at about 100° to give Tl_2O. The action of NaOH on Tl^{III} salts gives what appears to be the oxide, whereas with Al, Ga, and In the initial products are basic salts.

Aluminum, gallium, and thallium form mixed oxides with other metals. There are, first, aluminum oxides containing only traces of other metal ions. These include ruby (Cr^{3+}) and blue sapphire (Fe^{2+}, Fe^{3+}, and Ti^{4+}). Synthetic ruby, blue sapphire, and white sapphire (gem-quality corundum) are now produced synthetically in large quantities. Second are mixed oxides containing macroscopic proportions of other elements, such as the minerals *spinel* ($MgAl_2O_4$) and *crysoberyl* ($BeAl_2O_4$). The spinel structure has been described and its importance as a prototype for many other $M^{II}M_2^{III}O_4$ compounds noted (p. 000). Alkali metal compounds such as $NaAlO_2$, which can be made by heating Al_2O_3 with sodium oxalate at 1000° are also ionic mixed oxides.

10-4. Halides[3]

All four halides of each element in Group III are known, with one exception. The compound TlI_3, obtained by adding iodine to thallous iodide, is not thallium(III) iodide, but rather thallium(I) triiodide [$Tl^I(I_3)$]. This situation may be compared with the nonexistence of iodides of other oxidizing cations such as Cu^{2+} and Fe^{3+}, except that here a lower-valent compound fortuitously has the same stoichiometry as the higher-valent one.

The monomers MX_3 are formed at high temperatures and are believed to be planar, like BX_3. Proof of planarity has been obtained from several matrix-isolated species,[4a] though for $AlCl_3$ it was found that with N_2 present in the matrix there is a pyramidal species postulated to be an N_2 complex $N_2 \cdot AlCl_3$.[4b]

[2c] C. S. John and M. S. Scurrell, *Catalysis,* Vol. 1, 136. Specialist Report, Chemical Society, London, 1977; G. H. Posner, *Angew. Chem., Int. Ed.,* 1978, **17**, 487 (organic reactions on Al_2O_3 surfaces).

[3] (a) *Friedel-Crafts and Related Reactions,* Vol. 1, G. A. Olah, ed., Wiley, 1963 (much information on Al and Ga halides and their complexes)

(b) J. Carty, *Coord. Chem. Rev.,* 1969, **4**, 29 (halides and complexes of Ga, In and Tl).

(c) R. A. Walton, *Coord. Chem. Rev.,* 1971, **6**, 1 (halides and complexes of thallium)

[4a] I. R. Beattie *et al., J.C.S. Dalton,* **1976**, 666.

[4b] I. R. Beattie *et al., J. Chem. Phys.,* 1976, **64**, 1909; J. S. Shirk and A. E. Shirk, 1976, *J. Chem. Phys.,* 1976, **64**, 910.

The coordination numbers found in the crystalline halides are shown in Table 10-3. The fluorides of Al, Ga, and In are all high melting [1290°, 950° (subl), 1170°, respectively], wheras the chlorides, bromides, and iodides have lower melting points. There is, in general, good correlation between melting points and coordination number. Thus the three chlorides have the following melting points: $AlCl_3$, 193° (at 1700 mm); $GaCl_3$, 78°; $InCl_3$, 586°.

TABLE 10-3
Coordination Numbers of Metal Atoms in Group III Halides

	F	Cl	Br	I
Al	6	6	4	4
Ga	6	4	4	4
In	6	6	6	4
Tl	6	6	4	

The halides with coordination numbers 4 can be considered to consist of discrete dinuclear molecules (Fig. 10-1), and since there are no strong lattice forces, the

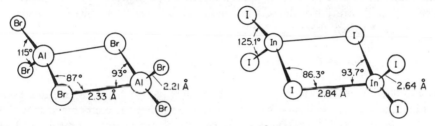

Fig. 10-1. The structures of Al_2Br_6 and In_2I_6.

melting points are low. In the vapor, aluminum chloride is also dimeric, so that there is a radical change of coordination number on vaporization, and these covalent structures persist in the vapor phase at temperatures not too far above the boiling points.

The halides dissolve readily in many nonpolar solvents such as benzene, in which they are dimeric. The enthalpies of dissociation, $Al_2X_6(g) = 2AlX_3(g)$ have been measured and are 46 to 63 kJ mol^{-1}. As Fig. 10-1 shows, the configuration of halogen atoms about each metal atom is roughly, though far from exactly, tetrahedral. The formation of such dimers is attributable to the tendency of the metal atoms to complete their octets. The dimers may be split by reaction with donor molecules, giving complexes such as R_3NAlCl_3. The halides dissolve in water, giving acidic solutions from which hydrates may be obtained. In acetonitrile dissociation occurs with the formation of $[Al(MeCN)_5Cl]^+$ and $AlCl_4^-$ ions.[4c]

Aluminum chloride–sodium chloride has been much used as a molten salt medium (mp. 173°) for electrolytic and other reactions (e.g., with S_8 and with metal halides). The equilibria involved are:

[4c] I. R. Beattie, et al., J.C.S. Dalton, 1979, 528.

$$2AlCl_3(l) = Al_2Cl_6 \qquad K = 2.86 \times 10^7$$
$$AlCl_4^- + AlCl_3 = Al_2Cl_7^- \qquad K = 2.4 \times 10^4$$
$$2AlCl_4^- = Al_2Cl_7^- + Cl^- \qquad K = 1.06 \times 10^{-7}$$

The interaction of $AlCl_3$ with some other metal halides (e.g., $CaCl_2$, $CrCl_3$, UCl_5) leads to mixed halides with halogen bridges.[5] These species may be quite volatile—for example, the vapor pressure of $NdCl_3$ is increased by 10^{13} at 600°. Melts with $GaCl_3$ and $InCl_3$ have similar properties. A typical example is

$$2FeCl_3 + Al_2Cl_6 \rightleftharpoons 2FeAlCl_6$$

Thallium(III) Halides. The chloride, which is most commonly used, can be prepared by the sequence:

$$Tl, \text{ or } TlCl, \text{ or } Tl_2CO_3 \xrightarrow{\text{ClNO}} TlCl_3 \cdot NOCl \xrightarrow{\text{heat}} TlCl_3$$

Solutions of $TlCl_3$ and $TlBr_3$ in CH_3CN, which are useful for preparative work, are conveniently obtained by treating solutions of the monohalides with Cl_2 or Br_2. Solid $TlCl_3$ loses chlorine at about 40°and above, to give the monochloride, and the tribromide loses bromine at even lower temperatures to give first "$TlBr_2$," which is actually $Tl^I[Tl^{III}Br_4]$. The fluoride is stable to about 500°. These facts provide a very good illustration of the way in which the stability of the lower valence state dominates thallium chemistry.

10-5. Other Binary Compounds

The Group III elements form various compounds such as carbides, nitrides, phosphides, and sulfides, commonly by direct interaction of the elements.

Aluminum carbide (Al_4C_3) is formed from the elements at temperatures of 1000 to 2000°. It reacts instantly with water to produce methane, and since there are discrete carbon atoms (C—C = 3.16 Å) in the structure, it may be considered as a "methanide" with a C^{4-} ion, but this is doubtless an oversimplification.

The *nitrides* AlN, GaN, and InN are known. Only aluminum reacts directly with nitrogen. GaN is obtained on reaction of Ga or Ga_2O_3 at 600 to 1000° with NH_3 and InN by pyrolysis of $(NH_4)_3InF_6$. All have a wurtzite structure (Fig. 1-9). They are fairly hard and stable, as might be expected from their close structural relationship to diamond and the diamond-like BN.

Aluminum and especially gallium and indium form 1:1 compounds with Group V elements, the so-called III-V compounds, such as GaAs.[6] These compounds have semiconductor properties similar to those of elemental silicon and germanium, to which they are electronically and structurally similar. They can be obtained by

5   C. W. Schläpfer and C. Rohrbasser, *Inorg. Chem.*, 1978, **17**, 1623; G. N. Papatheodorou *et al.*, *Inorg. Nucl. Chem. Lett.*, 1979, **15**, 51; *Inorg. Chem.*, 1979, **18**, 385; F. P. Emmeneger *et al.*, *Inorg. Chem.*, 1977, **16**, 343, 2957; *Z. Anorg. Allg. Chem.*, 1977, **436**, 127; D. R. Taylor and E. M. Larson, *J. Inorg. Nucl. Chem.*, 1979, **41**, 481.

6   *Gallium Arsenide and Related Compounds,* Institute of Physics, Bristol, 1977; R. K. Willardson and H. L. Goering, Eds., Reinhold, Vol. 1, 1962; R. K. Willardson and A. C. Beer, *Semiconductors and Semi-Metals,* Academic Press, 1966 *et seq.*

direct interaction or in other ways. Thus GaP can be obtained as pale orange single crystals by the reaction of phosphorus and Ga_2O vapor at 900 to 1000°.

COMPLEX COMPOUNDS[7]

10-6.   The Aqua Ions; Oxo Salts, Aqueous Chemistry

The Group III elements form a wide variety of salts including hydrated chlorides, nitrates, sulfates, and perchlorates, as well as sparingly soluble phosphates.

In aqueous solution the octahedral $[M(H_2O)_6]^{3+}$ ions are quite acidic. The Tl^{III} aquo ion has two *trans* water molecules that are more strongly bound that the others (cf. stability of $TlCl_2^+$ below). For the reaction

$$[M(H_2O)_6]^{3+} = [M(H_2O)_5(OH)]^{2+} + H^+$$

the following constants have been determined: $K_a(Al)$, 1.12×10^{-5}; $K_a(Ga)$, 2.5×10^{-3}; and $K_a(In)$, 2×10^{-4}; $K_a(Tl)$, $\sim 7 \times 10^{-2}$. Although little emphasis can be placed on the exact numbers, the orders of magnitude are important, for they show that aqueous solutions of the M^{III} salts are subject to extensive hydrolysis. Indeed, salts of weak acids (sulfides, carbonates, cyanides, acetates, etc.) cannot exist in contact with water.

The hydrolytic reaction above is the simplest of several.[8] The hydroxo species can form a dimer:

$$2[Al(H_2O)_6]^{3+} \rightleftharpoons [(H_2O)_4Al \underset{\underset{H}{O}}{\overset{\overset{H}{O}}{<>}} Al(H_2O)_4]^{4+} + 2H^+ \qquad K = 10^{-6 \cdot 95} \ (30°)$$

This dimeric ion is also found in some crystalline salts. At pH 5 as much as 90% of the Al is in the form of polymers, and at high degrees of hydrolysis the main ion appears to be $[Al_{13}O_4(OH)_{24}(H_2O)_{12}]^{7+}$. This occurs in the salt $Na_3[Al_{13}O_4(OH)_{24}(H_2O)_{12}(SO_4)_4]$.

The replacement of H_2O in the aqua ion by other ligands such as Me_2SO, THF, and SO_4^{2-} can be studied by nmr methods using ^{27}Al resonances.[9] Ligand exchange rates and determination of coordination numbers in solution can be studied using ^{17}O nmr.

For Ga^{3+} in perchlorate solution the main ion is $[Ga(H_2O)_6]^{3+}$. At 25° hydrolysis is very slow, giving GaOOH. On addition of hydrochloric acid to $[Ga(H_2O)_6]^{3+}$

[7]   (a) A. J. Carty and D. G. Tuck, *Prog. Inorg. Chem.*, 1975, **19**, 243 (an exhaustive review of indium complexes); (b) R. A. Walton, *Coord. Chem. Rev.*, 1971, **6**, 1 (thallium halides and complexes).

[8]   R. C. Turner, *Can. J. Chem.*, 1976, **54**, 1910; 1975, **53**, 2811; D. N. Water and M. S. Henty, *J.C.S. Dalton*, **1977**, 243; D. R. Gilden et al., *Inorg. Chem.*, 1977, **16**, 1257.

[9]   J. J. Delpuech, *J. Am. Chem. Soc.*, 1975, **97**, 3375.

the tetrachlorogallate ion is formed via successive equilibria,[10] and so at some value of n there must be a change in coordination number:

$$[Ge(H_2O)_6]^{3+} \underset{HCl}{\overset{HCl}{\rightleftharpoons}} [GaCl_n(H_2O)_m]^{(3-n)} + \overset{HCl}{\rightleftharpoons} GaCl_4^-$$

The $GaCl_4^-$ ion can be extracted from aqueous solutions by ether (see below).

Aluminates and gallates. The hydroxides are amphoteric:

$$Al(OH)_3(s) = Al^3 + 3OH^- \qquad K \approx 5 \times 10^{-33}$$
$$Al(OH)_3(s) = AlO_2^- + H^+ + H_2O \qquad K \approx 4 \times 10^{-13}$$
$$Ga(OH)_3(s) = Ga^{3+} + 3OH^- \qquad K \approx 5 \times 10^{-37}$$
$$Ga(OH)_3(s) = GaO_2^- + H^+ + H_2O \qquad K \approx 10^{-15}$$

and not only the hydroxides and oxides, but also the metals, dissolve in alkali bases as well as in acids. The oxides and hydroxides of In and Tl are, by contrast, purely basic; hydrated Tl_2O_3 is precipitated from solution even at pH 1 to 2.5. The nature of the so-called aluminate and gallate solutions has been a problem. For the aluminum system from pH 8 to 12, according to Raman spectra, the main species appears to be a polymer with octahedral Al and OH bridges, but, at pH > 13 and concentrations below *ca.* 1.5 M, ^{27}Al nmr, infrared, and Raman spectra, as well as ion-exchange studies, indicate a tetrahedral $Al(OH)_4^-$ ion. Above 1.5M there is condensation to give the ion $[(HO)_3AlOAl(OH)_3]^{2-}$, which occurs in the crystalline salt $K_2[Al_2O(OH)_6]$ and has an angular Al—O—Al bridge.

The indium and thallium aqua ions are known in ClO_4^- solution, but in presence of halide and other complexing anions complex species such as $InSO_4^+(aq)$ or the very stable linear $TlCl_2^+(aq)$ are formed.

A particularly important class of aluminum salts, the *alums* are structural prototypes and give their name to a large number of analogous salts formed by other elements. They have the general formula $MAl(SO_4)_2 \cdot 12H_2O$ in which M is almost any common univalent, monatomic cation except for Li^+, which is too small to be accommodated without loss of stability of the structure. The crystals are made up of $[M(H_2O)_6]^+$, $[Al(H_2O)_6]^{3+}$, and two SO_4^{2-} ions. There are actually three structures, all cubic, consisting of the ions above, but differing slightly in details depending on the size of the univalent ion. Salts of the same type, $M^IM^{III}(SO_4)_2 \cdot 12H_2O$, having the same structures are formed by many other trivalent metal ions, including those of titanium, vanadium, chromium, manganese, iron, cobalt, gallium, indium, rhenium, and iridium, and all such compounds are referred to as alums. The term is used so generally that alums containing aluminum are designated, in a seeming redundancy, as aluminum alums.

10-7. Halide Complexes and Adducts

Fluorides. The hydrated fluorides $AlF_3 \cdot nH_2O$ (n = 3 or 9) can be obtained by dissolving Al in aqueous HF. The nonahydrate is very soluble in water, and ^{19}F nmr spectra show the presence of $(H_2O)_3AlF_3$ as well as the ions AlF_4^-, $AlF_2(H_2O)_4^+$, and $AlF(H_2O)_5^{2+}$. At high fluoride concentrations and in crystalline solids the AlF_6^{3-} ion is also formed. The gallium system is similar.

10   S. F. Lincoln *et al., J.C.S. Dalton,* **1975,** 669.

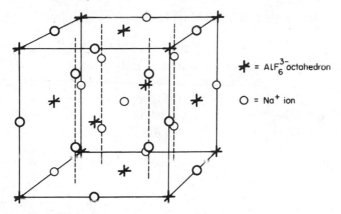

Fig. 10-2. The cubic structure of cryolite (Na_3AlF_6).

The most important fluoro salt is *cryolite,* whose structure (Fig. 10-2) is adopted by many other salts containing small cations and large octahedral anions and, in its *anti* form, by many salts of the same type as $[Co(NH_3)_6]I_3$. It is closely related to the structures adopted by many compounds of the types $M_2^+[AB_6]^{2-}$ and $[XY_6]^{2+}Z_2^-$. The last two structures are essentially the fluorite (or antifluorite) structures (see Fig. 1-9), except that the anions (or cations) are octahedra whose axes are oriented parallel to the cube edges. The unit cell contains four formula units.

Chlorides. Aluminum and gallium form only the tetrahedral ions MCl_4^-. Crystalline salts can be obtained, for example, by action of Et_4NCl on M_2Cl_6 in an organic solvent.

The tetrahaloaluminates are hydrolyzed by water, but gallium can be extracted from 8 M HCl solutions into ethers, where the ether phase contains $GaCl_4^-$ ions. Oxonium salts of the type $[(Et_2O)_nH]^+MCl_4^-$ of Al, Ga, In, and also Fe, have been isolated as viscous oils by reaction of the chloride in ether with hydrogen chloride. The formation of $AlCl_4^-$ and $AlBr_4^-$ ions is essential to the functioning of Al_2Cl_6 and Al_2Br_6 as Friedel-Crafts catalysts, since in this way the necessary carbonium ions are simultaneously formed. The spectra and structures of intermediates in the Friedel-Crafts reaction have been studied.[11] There are two types, one molecular (10-I), the other ionic (10-II), depending on the solvent. In CH_2Cl_2 only 10-I is formed.

$$\begin{array}{c} R \\ \diagdown \\ C{=}O \longrightarrow AlX_n \\ \diagup \\ X \end{array} \qquad\qquad [R{-}C{-}O]^+[AlX_{n+1}]^-$$

$$(10\text{-}I) \qquad\qquad\qquad\qquad (10\text{-}II)$$

[11] B. Chevrier and R. Weiss, *Angew Chem., Int. Ed.,* 1974, **13**, 1; J. Wilinski and R. J. Kurland, *J. Am Chem. Soc.,* 1978, **100**, 2233.

The oxocarbenium ion can then react further, for example,

$$RCO^+ + C_6H_6 \rightarrow [RCOC_6H_5]^+ \rightarrow RCOC_6H_5 + H^+$$

Although indium and thallium also form salts of MCl_4^- ions, they differ in forming complexes of higher coordination number (e.g., $InCl_5^-$, InF_6^{3-}, [In-$Br_5(H_2O)]^{2-}$, $[TlCl_5(H_2O)]^{2-}$). The $InCl_4^-$ ion is stabilized in organic solvents and like $GaCl_4^-$ can be extracted from HCl solutions. Among five-coordinate species both sp (e.g., $InCl_5^{2-}$ in the Et_4N^+ salt) and tbp [e.g., $InCl_3(PPh_3)_2$] structures have been found.

In aqueous solution the equilibria are complex, but a simplified scheme is

$$
\begin{array}{l}
[InCl_4(H_2O)]^- \rightleftharpoons [InCl_4(H_2O)_2]^- \\
Cl^-\downarrow\uparrow \\
[InCl_5(H_2O)]^- \rightleftharpoons [InX_5]^{2-} + H_2O \\
Cl^-\downarrow\uparrow \\
InCl_6^{3-}
\end{array}
$$

Pseudohalide complexes, similar to those of halides [e.g., $In(NCS)_5^{2-}$ and $In(NCS)_6^{3-}$] may also be obtained, depending on the size of the cation.[12]

For TlI_4^-, the stability of iodide in contact with Tl^{III} is a result of the stability of the ion, since TlI_3 is itself unstable relative to $Tl^I(I_3)$. Thallium alone forms the ion $Tl_2Cl_9^{3-}$, which has the confacial bioctahedron structure. (cf. 22-C-I).

Cationic Complexes. Apart from the aqua ions and partially substituted species such as $[GaCl(H_2O)_5]^{2+}$ noted above, complexes with pyridine, bipyridine, or phenanthroline are known, e.g., for gallium,[13] $[GaCl_2phen_2]Cl$, and $[Gaphen_3]Br_3$. For thallium(III) complexes the best route appears to be oxidation of thallium(I) halide in acetonitrile solution by halogen, followed by addition of ligands to get complex ions.[14]

Neutral Adducts. The trihalides (except the fluorides), and other R_3M compounds such as the trialkyls, triaryls, mixed R_2MX compounds and AlH_3, all function as Lewis acids, forming 1:1 adducts with a great variety of Lewis bases. This is one of the most important aspects of the chemistry of the Group III elements. The Lewis acidity of the AlX_3 groups (where $X = Cl, CH_3$, etc.) has been extensively studied thermodynamically, and basicity sequences for a variety of donors have been established.

As already indicated, the MX_3 molecules (X = halide) react with themselves to form dimeric molecules in which each metal atom has distorted tetrahedral coordination. Even in mixed organo halo compounds such as $(CH_3)_2AlCl$, this type of dimerization with *halogen atom bridges* occurs.

There are both four- and five-coordinate species. In the *tbp* complexes $MX_3(NMe_3)_2$, the nitrogen ligands are in axial positions; in $GaMe_2Clphen$ the phen bridges are axial and equatorial positions.[13]

Certain of the adducts, especially with bipyridine, will ionize in polar media to

12   J. J. Habeeb and D. G. Tuck, *J.C.S. Dalton,* **1973,** 96.
13   A. T. McPhail *et al., J.C.S. Dalton,* **1976,** 1657.
14   R. A. Walton, *Inorg. Nucl. Chem. Lett.,* **1976,** 767.

form cations such as $[GaCl_2bipy_2]^+$ or anions like $[InCl_4bipy]^-$. Although the 1:1 tetrahydrofuran adduct of $AlCl_3$ is molecular, the 1:2 adduct is $[AlCl_2(TH-F)_4]^+AlCl_4^-$.[15]

10-8. Chelate and Other Complexes

The most important octahedral complexes of the Group III elements are those containing chelate rings. Typical are those of β-diketones, pyrocatechol (10-III), dicarboxylic acids (10-IV), and 8-quinolinol (10-V). The neutral complexes dissolve readily in organic solvents, but are insoluble in water. The acetylacetonates have low melting points ($<200°$) and vaporize without decomposition. The anionic complexes are isolated as the salts of large univalent cations. The 8-quinolinolates are used for analytical purposes. Tropone (T) gives an eight-coordinate anion of indium in $Na[InT_4]$.

(10-III)   (10-IV)   (10-V)

Aluminum β-diketonates have been much studied by nmr methods because of their stereochemical nonrigidity (Chapter 28).[16] The *carboxylates* of indium[17] and thallium are obtained by dissolving the oxides in acid. Thallium acetate and trifluoroacetate are most important, since they are used extensively as reagents in organic synthesis.[18] Certain other thallium compounds have been used also. The trifluoroacetate will directly "thallate" aromatic compounds to give aryl thallium di(trifluoroacetate), e.g., $C_6H_4Tl(CO_2CF_3)_2$ (cf. aromatic mercuration, Section 19-9.).

Dithiocarbamates. Sulfur complexes of Al, Ga, and In are rather uncommon, but the trisdithiocarbamate $M(dtc)_3$ can be made by interaction of MCl_3 with Na dtc. Gallium and In compounds differ structurally from most other trisdithiocarbamates in their close approach to a prismatic arrangement of S atoms.[19]

Alcoxides. These are all polymeric even in solution in inert solvents. Only those of aluminum, particularly the isopropoxide, which is widely used in organic chemistry as a reducing agent for aldehydes and ketones, are of importance, They can be made by the reactions

15 J. Derouault, *Inorg. Chem.,* 1977, **16,** 3207, 3214.
16 M. Pickering *et al., J. Am. Chem. Soc.,* 1976, **98,** 4503.
17 J. J. Habeeb and D. J. Turk, *J.C.S. Dalton,* **1973,** 243.
18 A. McKillop and E. C. Taylor, *Adv. Organomet. Chem.,* 1973, **11,** 147; *Acc. Chem. Res.,* 1970, **10,** 338; M. Fieser and L. F. Fieser, *Reagents for Organic Synthesis,* Wiley; R. J. Ouellette, *Oxid. Org. Chem.,* 1973, **2,** 135.
19 K. Dymock *et al., J.C.S. Dalton,* **1976,** 28; H. Abrahamson *et al., Inorg. Chem.,* 1975, **14,** 2070; A. H. White *et al., Aust. J. Chem.,* 1978, **31,** 1927.

$$Al + 3ROH \xrightarrow[\text{catalyst, warm}]{\text{1\% HgCl}_2 \text{ as}} (RO)_3Al + \tfrac{3}{2} H_2$$

$$AlCl_3 + 3RONa \rightarrow (RO)_3Al + 3NaCl$$

The alcoxides hydrolyze vigorously in water. The *tert*-butoxide is a cyclic dimer (10-VI) in solvents, whereas the isopropoxide is tetrameric (10-VII) at ordinary temperature according to ^{1}H, ^{13}C, and ^{27}Al nmr spectra,[20] but trimeric at elevated temperatures. Terminal and bridging alcoxo groups can be distinguished by nmr spectra. Other alcoxides can exist also as dimers and trimers. Mixed alcoxides, e.g., $(MeOAlH_2)_n$ and $[MeOAlMe_2]_2$, are also known.

(10-VI)   (10-VII)

Nitrato Complexes. Gallium and thallium form *nitrato* complexes by the reaction of N_2O_5 with $NO_2^+[GaCl_4]^-$ or $TlNO_3$, respectively. The gallium ion in $NO_2^+[Ga(NO_3)_4]^-$ appears to have unidentate NO_3 groups, by contrast with $[Fe(NO_3)_4]^-$, which has bidentate groups and is eight-coordinate, although the radius of Ga^{III} (0.62 Å) is only slightly smaller than that of Fe^{III} (0.64 Å).

10-9. Hydrides and Complex Hydrides

Aluminum hydride[21] is obtained by interaction of $LiAlH_4$ with 100% H_2SO_4 in tetrahydrofuran:

$$2LiAlH_4 + H_2SO_4 = 2AlH_3 + 2H_2 + Li_2SO_4$$

In the interaction of $LiAlH_4$ with $AlCl_3$, intermediates like $AlHCl_2$ are formed.

The white hydride is thermally unstable. It has a three-dimensional lattice isostructural with AlF_3. It is a useful reducing agent in organic chemistry, and the products formed are significantly different from those of reduction by $LiAlH_4$. Nitriles can be reduced to amines via complex intermediates, and the reduction of alkyl halides is slower than with $LiAlH_4$ so that carboxyl or ester groups in

20   J. W. Akitt and R. H. Duncan. J. *Magn.* Resonance, 1974, **15**, 162.
21   For references, see P. R. Oddis and M. G. H. Wallbridge, *J.C.S. Dalton*, **1978**, 572.

compounds RCOOH and RCOOR' can be reduced preferentially in presence of R"X.

There is no evidence for a comparable gallium hydride, but an unstable viscous oil that shows bands in the infrared spectrum due to Ga—H has been obtained by the reaction:

$$\text{Me}_3\text{NGaH}_3(s) + \text{BF}_3(g) \xrightarrow{-15°} \text{GaH}_3(l) + \text{Me}_3\text{NBF}_3(g)$$

There is also little evidence for hydrides of In and Tl.

There is an extensive range of complex hydrides that can be regarded as arising from Lewis acid behavior of MH_3. Thus adducts can be readily formed with donor molecules such as NR_3 and PR_3, or with H^- and other anions.

Hydride Anions. Both Al and Ga hydride anions are obtained by the reaction

$$4\text{LiH} + \text{MCl}_3 \xrightarrow{\text{Et}_2\text{O}} \text{LiMH}_4 + 3\text{LiCl}$$

However for AlH_4^- the sodium salt can be obtained by direct interaction:

$$\text{Na} + \text{Al} + 2\text{H}_2 \xrightarrow[150°/2000 \text{ p.s.i.}/24h]{\text{THF}} \text{NaAlH}_4$$

The salt is obtained by precipitation with toluene and can be efficiently converted into the lithium salt:

$$\text{NaAlH}_4 + \text{LiCl} \xrightarrow{\text{Et}_2\text{O}} \text{NaCl}(s) + \text{LiAlH}_4$$

The most important compound is *lithium aluminum hydride,*[22] a nonvolatile crystalline solid, stable below 120°, that is explosively hydrolyzed by water. In the crystal there are tetrahedral AlH_4^- ions with an average Al—H distance of 1.55 Å. The Li^+ ions each have four near hydrogen neighbors (1.88–2.00 Å) and a fifth that is more remote (2.16 Å).

Lithium aluminum hydride is soluble in diethyl and other ethers. In ethers, the Li^+, Na^+, and R_4N^+ salts of AlH_4^- and GaH_4^- tend to form three types of species depending on the concentration and on the solvent, namely, either loosely or tightly bound aggregates or ion pairs.[23] Thus LiAlH_4 is extensively associated in diethylether, but at low concentrations in tetrahydrofuran there are ion pairs. Nmr studies suggest that not only for AlH_4^- but also for ClO_4^- and I^- there is a solvated ion Li(THF)_4^+ since all three salts have an infrared band at 420 cm^{-1} associated with Li^+ in the solvent cage. NaAlH_4 is insoluble in diethyl ether.

Some reactions of LiAlH_4 are given in Fig. 10-3, LiAlH_4 accomplishes many otherwise tedious and difficult reductions in organic chemistry[22,24] as well as being a useful inorganic reductant. There may be differences in rates of reaction of different salts, (e.g., Li^+ and Na^+) due not only to differences in the species in solution

[22]   J. S. Pizey, *Lithium Aluminium Hydride,* Horwood-Wiley, 1977.
[23]   A. E. Shirk and D. F. Shriver, *J. Am. Chem. Soc.,* 1973, **95,** 5904; E. C. Ashby *et al., J. Am. Chem. Soc.,* 1973, **95,** 2823.
[24]   E. R. H. Walker, *Chem. Soc. Rev.,* 1976, **5,** 23 (functional group selectivity).

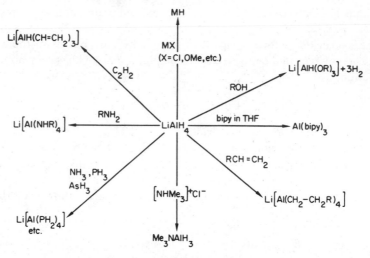

Fig. 10-3. Some reactions of lithium aluminum hydride.

but to involvement of Li$^+$ in the reaction. Thus in the reduction of ketones[25] in tetrahydrofuran (S), Li$^+$ appears to have a stronger tendency to coordination with the >C=O group, and the pathway probably involves some such sequence as

In addition to AlH$_4^-$, the octahedral *hexahydrido ion* AlH$_6^{3-}$ is known. Although salts were first made by the reactions

$$2NaH + NaAlH_4 \xrightarrow{\text{heat in heptane}} Na_3AlH_6$$

$$2LiH + LiAlH_4 \xrightarrow{\text{Et}_3\text{Al as catalyst}} Li_3AlH_6$$

they are best prepared by direct interaction of the metals and hydrogen under pressure. The commercial benzene-soluble reducing agent "Red-Al" is Na[Al-H(OMe)$_2$OEt].

25   E. C. Ashby and J. R. Boone, *J. Am. Chem. Soc.*, 1976, **98**, 5524.

Lithium gallium hydride decomposes slowly even at 25° to give LiH, Ga, and H_2. Comparing the MH_4^- ions of B, Al, and Ga, the thermal and chemical stabilities vary according to the ability of MH_3 to act as an acceptor for H^-. The order, and also the M—H force constants of MH_4^-, are in the order B > Al ≫ Ga.

Donor Adducts. These are similar to borane adducts, the stability order being B > Al > Ga and also similar to adducts of the halides and alkyls, where the stability order is halides > alkyls > hydrides. The most studied adducts are the trialkylamine alanes (alane = AlH_3). Trimethylamine gives both 1:1 and 2:1 adducts, but the latter are stable only in the presence of an excess of amine:

$$Me_3NAlCl_3 + 3LiH \xrightarrow{Et_2O} Me_3NAlH_3 + 3LiCl$$
$$Me_3HN^+Cl^- + LiAlH_4 \xrightarrow[-60°]{Et_2O} Me_3NAlH_3 + LiCl + H_2$$

$$3LiAlH_4 + AlCl_3 + 4NMe_3 \rightarrow 4Me_3NAlH_3 + 3LiCl$$
$$Me_3NAlH_3 + Me_3N \rightleftharpoons (Me_3N)_2AlH_3$$

The monoamine is a white, volatile, crystalline solid (m.p. 75°), readily hydrolyzed by water, which slowly decomposes to $(AlH_3)_n$. It is monomeric and tetrahedral. The bisamine is trigonal-bipyramidal with axial N atoms. Tetrahydrofuran also gives 1:1 and 2:1 adducts, but diethyl ether, presumably for steric reasons, gives only the 1:1 compound, though a mixed THF-Et_2O adduct exists.

There are similar monoamine gallanes where the Ga—H stretch at ~1850 cm^{-1}, compared to ~1770 cm^{-1} for the alane, suggests a stronger M—H bond, and indeed the gallanes are less sensitive to hydrolysis. The $(Me_3N)_2GaH_3$ compound is unstable above -60°. Preparation of the gallanes illustrates a useful principle regarding the use of a weak donor as solvent:

$$\underset{\text{strong—weak}}{Me_3NGaH_3} + \underset{\text{strong}}{BF_3} \xrightarrow[\text{weak}]{Me_2S} \underset{\text{weak—weak}}{Me_2SGaH_3} + \underset{\text{strong—strong}}{Me_3NBF_3}$$

Because the weak-weak, strong-strong combination is favored over two weak-strong adducts, the net effect is to displace the strong donor Me_3N by the weaker one Me_2S. Compounds of the type $[HAl(NR_2)_2]_n$ and others with Al—N—Al bridges are discussed later.

Finally, we note aluminum borohydride $[Al(BH_4)_3$, b.p. 44°], and mixed compounds like $Me_2Al(BH_4)$; all these have $Al(\mu\text{-}H)_2BH_2$ bridges.[26] (See Chapter 4, p. 108).

10-10. Organometallic Compounds[27]

The aluminum organometallic compounds are by far the most important and best known. They may be prepared by the classical reaction of aluminum with the ap-

[26]   P. R. Oddy and M. L. M. Wallbridge, *J.C.S. Dalton,* **1978**, 572, **1976**, 869.

[27]   (a) T. Mole and E. A. Jeffery, *Organoaluminium Compounds,* Elsevier, 1972 (an exhaustive reference text comprehensive to 1971). (b) A. McKillop and E. C. Taylor, *Adv. Organomet. Chem.,* 1973, **11**, 147 (organothallium compounds). (c) A. G. Lee, *Organomet. React.,* 1975, **5**, 1 (reactions of organothallium compounds). (d) Y. Yamamoto and H. Nozaki, *Angew. Chem., Int. Ed.,* 1978, **17**, 169 (selective organic reactions with organoaluminums). G. Zweifel in *Comprehensive Organic Chemistry,* Vol. 3, Part 15.3 Pergamon Press, 1979.

propriate organomercury compound:

$$2Al + 3R_2Hg \rightarrow 2R_3Al(\text{or } [R_3Al]_2) + 3Hg$$

or by reaction of Grignard reagents with $AlCl_3$:

$$RMgCl + AlCl_3 \rightarrow RAlCl_2, R_2AlCl, R_3Al$$

More direct methods suitable for large-scale use are now available. These proce-dures stemmed from studies by K. Ziegler that showed that aluminum hydride or $LiAlH_4$ reacts with olefins to give alkyls or alkyl anions—a reaction specific for B and Al hydrides:

$$AlH_3 + 3C_nH_{2n} \rightarrow Al(C_nH_{2n+1})_3$$
$$LiAlH_4 + 4C_nH_{2n} \rightarrow Li[Al(C_nH_{2n+1})_4]$$

The mechanism of the hydroalumination reaction of alkenes is not well known, but data[28] for the interaction of the trimer $(Bu_2^iAlH)_3$ with 4-octyne is consistent with formation of a π complex between the π-electron density of the acetylene and the Al—H bond:

$$(R_2AlH)_3 \rightleftharpoons 3R_2AlH$$

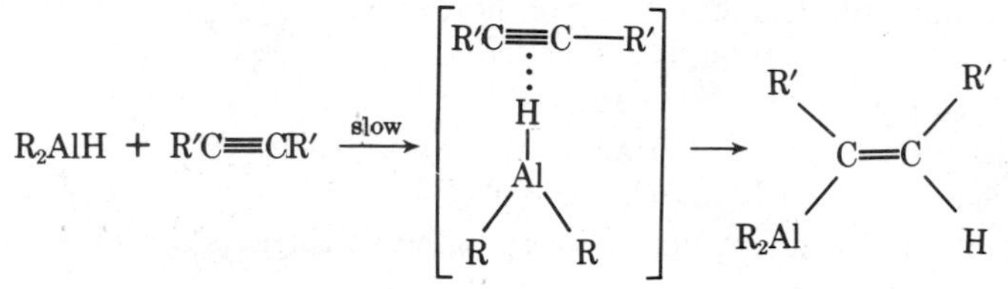

Although $(AlH_3)_n$ cannot be made by direct interaction of Al and H_2, nevertheless in the presence of aluminum alkyl the following reaction to give the dialkyl hydride can occur:

$$Al + \tfrac{3}{2}H_2 + 2AlR_3 \rightarrow 3AlR_2H$$

This hydride will then react with olefins:

$$AlR_2H + C_nH_{2n} \rightarrow AlR_2(C_nH_{2n+1})$$

The direct interaction of Al, H_2, and olefin is used to give either the dialkyl hy-drides of the trialkyls.

It may be noted in connection with the direct synthesis of $NaAlH_4$ mentioned above that, if ethylene (or other olefin) is present and $AlEt_3$ is used as catalyst, the direct interaction gives $Na[AlH_{4-n}Et_n]$.

Other technically important compounds are the "sesquichlorides" such as $Me_3Al_2Cl_3$ or $Et_3Al_2Cl_3$. These compounds can be made by direct interaction of Al or Mg-Al alloy with the alkyl chloride. This reaction fails for propyl and higher alkyls since the alkyl halides decompose in presence of the alkyl aluminum halides to give HCl, alkenes, and so on.

The lower aluminum alkyls are reactive liquids, inflaming in air and explosively

28 J. J. Eisch et al., J. Am. Chem. Soc., 1974, **96**, 7276; J. Organomet. Chem., 1974, **64**, 41.

sensitive to water. All other derivatives are similarly sensitive to air and moisture though not all are spontaneously flammable.

Not only are the $R_n AlCl_{3-n}$-type compounds dimerized through Cl bridges, as might be expected, but so also are the lower binary alkyls and the aryls. In these, there are bridging carbon atoms that participate in forming $3c$-$2e$ bonds supplemented in the aryls by additional interactions (see pages 111 and 346). The structures of four Al_2R_6 molecules, with R = CH_3, C_6H_5, cyclo-C_3H_5,[29] and vinyl[29] are well established. Even in the case of $[AlPh_2(C{\equiv}CPh)]_2$ where $C{\equiv}C$ π electrons are available, a single carbon atom forms the bridge[30] (10-VIII).

$$
\begin{array}{c}
\text{Al} \\
\text{PhCC} \diamondsuit \text{CCPh} \\
\text{Al}
\end{array}
$$

(10-VIII)

For trimethylaluminum at $-75°$ the proton nuclear resonance spectrum exhibits separate resonances for the terminal and bridging methyl groups, but on warming, these begin to coalesce, and at room temperature only one sharp peak is observed. This indicates that the bridging and terminal methyl groups can exchange places across a relatively low energy barrier. The exchange process[31] in Me_2Al_6 involves dissociation to monomer and reformation to the dimer:

$$Me_6Al_2 \rightleftarrows 2Me_3Al$$

The extent of the dissociation for Al_2Me_6 is very small, 0.0047% at $20°$. Exchange in aryl aluminums may involve only a partial dissociation, e.g.,

$$
(Me_2ArAl)_2 \rightleftarrows
\begin{array}{c}
\text{Me} \quad\quad \text{Ar} \quad\quad \text{Me} \\
\diagdown \diagup \quad \diagdown \diagup \\
\text{Al} \quad\quad \text{Al} \\
+ \quad\quad - \\
\diagup \quad\quad \diagdown \\
\text{Me} \quad\quad\quad \text{Me} \\
\text{Ar}
\end{array}
$$

The alkyls are Lewis acids, combining with donors such as amines, phosphines, ethers, and thioethers to give tetrahedral, four-coordinate species. Thus Me_3N-$AlMe_3$ in the gas phase has C_{3v} symmetry with staggered methyl groups. With tetramethylhydrazine and $(CH_3)_2NCH_2N(CH_3)_2$, five-coordinate species that appear to be of the kind shown in 10-IX are obtained, although at room temperature exchange processes cause all methyl groups and all ethyl groups to appear equivalent in the proton nuclear resonance spectrum. With $(CH_3)_2NCH_2CH_2N(CH_3)_2$ a complex is formed that has an AlR_3 group bound to each nitrogen atom. Aluminum alkyls also combine with lithium alkyls:

$$(C_2H_5)_3Al + LiC_2H_5 \xrightarrow{\text{in benzene}} LiAl(C_2H_5)_4$$

[29] J. P. Oliver et al., J. Am. Chem. Soc., 1971, 93, 1035; 1976, 98, 3995.
[30] G. D. Stucky et al., J. Am. Chem. Soc., 1974, 96, 1941.
[31] T. B. Stanford, Jr., and K. L. Henold, Inorg. Chem., 1975, 14, 2426; M. B. Smith; J. Organomet. Chem., 1972, 46, 31, 211.

$$
\begin{array}{c}
\diagdown \mathrm{N} \diagup \\
\mathrm{R} \diagdown \quad | \quad \diagdown \\
\quad \mathrm{Al} - \mathrm{N} - \\
\mathrm{R} \diagup \quad | \quad \diagdown \\
\quad \mathrm{R}
\end{array}
$$

(10-IX)

X-Ray study has shown that $LiAl(C_2H_5)_4$ is built up of chains of alternating tetrahedral $Al(C_2H_5)_4^-$ and Li^+ in such a way that each lithium atom is tetrahedrally surrounded by four α-carbon atoms, close enough to indicate weak Li—C bonds; vibrational spectra suggest that $LiAlMe_4$ is similar.

Triethylaluminum, the sesquichloride $(C_2H_5)_3Al_2Cl_3$, and alkyl hydrides are used together with transition metal halides or alcoxides or organometallic complexes as catalysts (e.g., Ziegler catalysts) for the polymerization of ethylene, propene, and a variety of other unsaturated compounds, as discussed in Chapter 30. They are also used widely as reducing and alkylating agents for transition metal complexes.

In alkylation of metal halides usually only *one* alkyl group is transferred, since the dialkyl aluminum halides are much less powerful alkylating agents than the trialkyls, e.g.,

$$WCl_6 + 6AlMe_3 = WMe_6 + 6AlClMe_2$$

The alkylaluminums react with virtually all compounds that have acidic hydrogens, giving alkane. In some cases, alkylation occurs as, e.g.,

$$R_3COH \xrightarrow[\substack{\text{toluene} \\ 100-120°}]{Me_3Al} R_3CMe$$

but in others, complex alkylaluminum compounds are obtained (e.g., with pyrazole, β-diketonates, phosphoric acid).

Compounds with Al—N Bonds.   The interaction of alkylaluminums or mixed alkyls such as $AlClEt_2$ with primary or secondary amines gives adducts initially, as we have seen, but on heating these can lose alkane to give species that have Al—N—Al bonds, each Al becoming four-coordinate by interaction with a nitrogen atom lone pair:

$$2Et_2AlBr \cdot NH_2{}^iBu \xrightarrow{120°} (EtBrAlNH{}^iBu)_2 + 2C_2H_6$$

$$Me_3Al + H_2NMe \rightarrow [MeAlNMe]_n + 2CH_4$$

The compounds may be dimers, trimers, tetramers, etc.[32] The dimers have a four-membered Al_2N_2 ring (10-X), which may have isomers if the groups are different. The tetramers have an Al_4N_4 cubic cage (10-XI) comparable to the C_8 cage in the hydrocarbon cubane. A more complex cage has been formed in the heptamer $(MeAlNMe)_7$ cf..10-XV below.

[32]   J. D. Smith *et al.*, *J.C.S. Dalton*, **1979**, 1206; **1976**, 1433; G. Del Piero *et al.*, *J. Organomet. Chem.*, 1977, **129**, 288, **137**, 265; S. Cucinella *et al.*, *J. Organomet. Chem.*, 1976, **121**, 137.

(10-X)   (10-XI)

Chelate amines can give either monomers or dimers 10-XII and 10-XIII, where steric factors of the alkyl groups appear to be most important in determining the nature of the product.

(10-XII)   (10-XIII)

There are similar *poly(N-alkylimino alanes)*[33] that are made by reactions such as

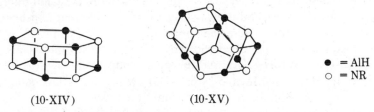

where n ranges from 2 to at least 35, but quite commonly takes values of 4, 6, or 8, giving compounds with the cage structures 10-XI, 10-XIV, and 10-XV.

(10-XIV)   (10-XV)

$\bullet$ = AlH
$\circ$ = NR

The compounds have been used as reducing agents, for hydrogenation,[34] and as polymerization catalysts. Polymers with Al—N backbones are obtained by interaction of R_3NAlH_3 with, e.g., ethylenediamine or acetonitrile.

The trialkyls of Ga,[35] In, and Tl resemble those of aluminum, but they have been less extensively investigated and are increasingly less stable. One signal point of difference is the lack of dimerization of the alkyls of B, Ga, In, and Tl at ordinary temperatures, with the exception of the unusual polymerization of crystalline Me_3In and Me_3Tl.

33   S. Cucinella *et al., J. Organomet. Chem.,* 1979, **173**, 263.
34   S. Cucinella *et al., J. Organomet. Chem.,* 1979, **164**, 1.
35   R. A. Kovar *et al., Inorg. Chem.,* 1975, **14**, 2089.

The In and Tl phenyls also exhibit weak intermolecular interaction, probably involving donation of electrons from a phenyl ring to a vacant p orbital of the metal in another molecule.

A number of dialkyl compounds of Ga, In, and Tl are well characterized and are stable even in aqueous solutions. Thus interaction of Me_2GaCl and ammonia gives $[Me_2Ga(NH_3)_2]^+Cl^-$, whereas partial hydrolysis of $Me_3Ga \cdot Et_2O$ gives crystals of the OH-bridged tetramer $[Me_2GaOH]_4$, which is soluble in acids, to give the cation $[Me_2Ga(H_2O)_2]^+$ and in bases to give $[Me_2Ga(OH)_2]^-$. Thallium gives very stable ionic derivatives of the type R_2TlX (X = halogens, SO_4^{2-}, CN^-, NO_3^-, etc.), which resemble compounds R_2Hg in being unaffected by air and water. The ion $(CH_3)_2Tl^+$ in aqueous solution and in salts is linear, like R_2Hg and the ions Me_2Pb^{2+} and Me_2Sn^{2+}; the reason for the difference in structure of Me_2Ga^+ and Me_2Tl^+ is not obvious.

In crystals of thallium salts, as in dimethyltin salts (Section 12-12), the anions may act as bridges. Thus the sulfinate $(Me_2TlSO_2Me)_2$ is dimeric with sulfinate bridges. Additional coordination can certainly occur, and a complex $[Me_2Tl\ py]ClO_4$ has been isolated. The $[Me_2Tl\ py]^+$ ion appears to be T-shaped on the basis of spectroscopic measurements. Bis(pentafluorophenyl)thallium halides give what appear to be five-coordinate adducts such as $(C_6F_5)_2TlCl(bipy)$.

10-11.   Transition Metal Complexes[36]

The interaction of sodium salts of carbonylate ions, e.g., $[Co(CO)_4]^-$, (see Chapter 25) with the metal chlorides, commonly gives compounds that are readily soluble in organic solvents. Examples are $AlCo_3(CO)_9$, $In[Mn(CO)_5]_3$, and $Tl[Co(CO)_4]_3$. These compounds have metal-metal bonds and are similar to those given, e.g., by mercury(II). These are relatively few Al compounds of this type, but interaction of aluminum alkyls with metal carbonyls can give adducts that appear to have M—C—O—AlR_3 bonds. Where the elimination of alkane can occur, as with transition metal hydrides, more complicated compounds are formed. Thus the reaction of $(\eta\text{-}C_5H_5)_2MoH_2$ with $AlMe_3$ gives compounds of stoichiometry $[MoH(C_5H_5)(C_5H_4)]_2Al_3Me_5$ and $[Mo(C_5H_4)_2]Al_4Me_6$ that have complicated structures[37] involving aluminum bridge bonds of the types

LOWER VALENT COMPOUNDS

Since the elements in Group III have the outer electron configurations ns^2np, it is natural to consider whether monovalent ions might exist. It may be recalled that there is no evidence for B^I under chemically important conditions.

[36] A. T. T. Hsieh, *Inorg. Chim. Acta*, 1975, **14**, 87 (an extensive review).
[37] M. L. H. Green et al., *J.C.S. Dalton*, **1976**, 1993.

10-12. Aluminum and Gallium

There is no evidence that compounds containing Al^I exist at ordinary temperatures. Anodic oxidation of aluminum at high current densities evidently produces lower-valent aluminum ions, either Al^I or Al^{II}, or both, but they are ephemeral. There is no doubt that *gaseous* Al^I halide molecules exist at high temperatures, and their spectroscopic properties are well known. In the chloride system the equilibrium

$$AlCl_3(g) + 2Al(s) \rightleftharpoons 3AlCl(g)$$

has been thoroughly studied, and its use in purifying aluminum has been proposed. The reaction proceeds to the right at high temperatures, but reverses readily at low temperatures. Similarly, it has been shown that gaseous Al_2O and AlO molecules exist above 1000°, but no solid oxide containing lower-valent aluminum has been shown definitely to exist under ordinary conditions, although Al_2O and other M_2O species can be trapped in inert gas matrices at low temperatures.

A zerovalent complex $Al(bipy)_3$ is formed by reduction of $AlCl_3$ with Li bipyridyl in THF; it is exceedingly air-sensitive, green, and paramagnetic ($\mu = 2.32$ Bohr magnetons). See, however, Section 4-10.

Gallium(I) *compounds* have been prepared in the gas phase at high temperatures by reactions such as

$$Ga_2O_3(s) + 4Ga(l) \xrightarrow{700°} 3Ga_2O(g)$$
$$Ga(l) + SiO_2(s) \rightleftharpoons Si(\text{in Ga}) + 2Ga_2O(g)$$
$$GaCl_3(g) \underset{1100°}{\rightleftharpoons} GaCl(g) + Cl_2$$

Although GaCl has not been isolated pure, Ga_2O and Ga_2S can be, but the latter solid is nonstoichiometric. The best-known compounds are the "dihalides" GaX_2; these are known to have the saltlike structure $Ga^I[Ga^{III}X_4]$; the Ga^I ion can also be obtained in other salts such as $Ga[AlCl_4]$. Fused $GaCl_2$ is a typical conducting molten salt. These halides are prepared by the reaction[38]

$$2GaX_3 + 4Ga \xrightarrow[60°]{\text{benzene}} 3Ga_2X_2$$

With X = I, a stoichiometric amount of Ga must be used, otherwise Ga_4I_6 is obtained; this is thought to be $(Ga^+)_2(Ga_2I_6^{2-})$. Salts of the type $[GaL_4][GaCl_4]$ have been obtained with sulfur, selenium, and arsenic donors, but a dioxan complex $Ga_2Cl_4(C_4H_8O_2)_2$ has a Ga—Ga bond (2.406 Å).

Gallium(II) Compounds. The anodic dissolution of Ga in $6M$ HCl or HBr at 0° followed by addition of Me_4NX precipitates white crystalline salts of the ion $[Ga_2X_6]^{2-}$ that are stable and diamagnetic. In the D_{3d} ions there is a Ga—Ga bond of 2.39 Å for X = Cl, and 2.41 Å when X = Br, so that formally, gallium is in the II oxidation state.[39a] The treatment of $GaCl_2$ with dioxane gives a crystalline solvate

[38]   J. C. Beamish *et al.*, *Inorg. Chem.*, 1979, **18**, 220.
[39a] C. A. Evans *et al.*, *J.C.S. Dalton*, **1973**, 983; K. L. Brown and D. Hall, *J.C.S. Dalton*, **1973**, 1843; H. J. Cumming *et al.*, *Crystallogr. Struct. Comm.*, 1974, **3**, 107.

(diox)Cl_2Ga-GaCl$_2$(diox), in which each Ga atom has a distorted tetrahedral configuration and the Ga—Ga distance is 2.406(1) Å.[39b] The chalconides GaS, GaSe, and GaTe, obtained by direct interaction, also have Ga—Ga units in the structure.

10-13. Indium(I, II)

Indium(I) can be obtained in low concentration in aqueous solution by using an indium metal anode in 0.01M perchloric acid. It is rapidly oxidized by both H^+ ion and by air, and is also unstable to disproportionation:

$$In^{3+} + 3e = In \qquad E° = -0.343 \text{ V}$$
$$In^{3+} + 2e = In^+ \qquad E° = -0.426 \text{ V}$$
$$In^+ + e = In \qquad E° = -0.178 \text{ V}$$

The halides InX (X = Cl, Br, I) are all isostructural with the low-temperature form of TlI. These compounds as well as InO, In_2S, and In_2Se are obtained by solid state reactions: they are unstable to water. In^I is more stable in acetonitrile, and it is possible to isolate ClO_4^-, BF_4^- and PF_6^- salts of In^+ by reaction of In/Hg with silver salts in acetonitrile.[40]

The "dihalides," i.e., $In^I[InX_4]$, are most easily made[41] by refluxing indium with halogen in xylene:

$$2In + 3X_2 = 2InX_3$$
$$2InX_3 + In = 3InX_2$$

There is a definite In^I compound C_5H_5In (cf. Tl^+ analogue, later) that can be used as a source of other In^I compounds by treating it with acids such as HX or even 8-hydroxyquinoline[42]:

$$C_5H_5In + HX = InX + C_5H_6$$

The extremely reactive In^{2+}(aq) ion has been produced in reduction of In^{3+} with hydrated electrons. The dihalides are believed to be similar to those of gallium, i.e., $In^+(InX_4^-)$, but unambiguous In^{II} compounds are made by the reaction[43]:

$$2R_4NX + 2InX_2 \xrightarrow{\text{xylene}} [R_4N]_2[In_2X_6]$$

10-14. Thallium[44]

The unipositive state is quite stable; in aqueous solution it is distinctly more stable than Tl^{III}:

$$Tl^{3+} + 2e = Tl^+ \qquad E° = +1.25 \text{ V } [E_f = +0.77, 1M \text{ HCl}; +1.26, 1M \text{ HClO}_4]$$

[39b] J. C. Beamish, R. W. H. Small and I. J. Worrall, *Inorg. Chem.*, 1979, **18**, 220.

[40] M. Ashraf *et al., J.S.C. Dalton*, **1977**, 170.

[41] B. H. Greeland and D. G. Tuck, *Inorg. Chem.*, 1976, **15**, 475.

[42] J. J. Habeeb and D. G. Tuck, *J.C.S. Dalton*, **1975**, 1815; **1976**, 866.

[43] B. H. Freeland *et al., Inorg. Chem.*, 1976, **15**, 2144.

[44] A. G. Lee, *Coord. Chem. Rev.*, 1972, **8**, 289 (complexes of Tl^I).

The thallous ion is not very sensitive to pH, although the thallic ion is extensively hydrolyzed to $TlOH^{2+}$ and the colloidal oxide even at pH 1 to 2.5; the redox potential is hence very dependent on pH as well as on presence of complexing anions. Thus as indicated by the potentials above, the presence of Cl^- stabilizes Tl^{3+} more (by formation of complexes) than Tl^+ and the potential is thereby lowered.

The colorless thallous ion has a radius of 1.54 Å, which can be compared with those of K^+, Rb^+, and Ag^+ (1.44, 1.58, and 1.27 Å). In its chemistry this ion resembles either the alkali or silver(I) ions.

The thallous ion has been proposed as a probe for the behavior of K^+ in biological systems. The two isotopes ^{203}Tl and ^{205}Tl (70.48%) both have nuclear spin, and nmr signals are readily detected both in solutions and in solids; also the Tl^I (and Tl^{III}) resonances are very sensitive to the environment and have huge solvent-dependent shifts. For Tl^+ it is possible to correlate shifts with solvating ability, for example[45]; hence the utility as a probe in biological systems.

In crystalline salts, the Tl^+ ion is usually six- or eight-coordinate. The yellow hydroxide is thermally unstable, giving the black oxide Tl_2O at about 100°. The latter and the hydroxide are readily soluble in water to give strongly basic solutlions that absorb carbon dioxide from the air: TlOH is a weaker base than KOH, however. Many thallous salts have solubilities somewhat lower than those of the corresponding alkali salts, but otherwise are similar to and quite often isomorphous with them. Examples of such salts are the cyanide, nitrate, carbonate, sulfate, phosphates, perchlorate, and alums. Thallous solutions are exceedingly poisonous and in traces cause loss of hair.

Thallous sulfate, nitrate, and acetate are moderately soluble in water, but—except for the very soluble TlF—the halides are sparingly soluble. The chromate and the black sulfide Tl_2S, which can be precipitated by hydrogen sulfide from weakly acid solutions, are also insoluble. Thallous chloride also resembles silver chloride in being photosensitive: it darkens on exposure to light. Incorporation of Tl^I halides into alkali halides gives rise to new absorption and emission bands because complexes of the type that exist also in solutions, most notable TlX_2^- and TlX_4^{3-}, are formed; such thallium-activated alkali halide crystals are used as phosphors (e.g., for scintillation radiation dectors). Thallous chloride is insoluble in ammonia, unlike AgCl.

Thallous fluoride is essentially ionic, with layers held together by Tl $\cdots$ F interactions. The ion is octahedral, with two cis fluorines at longer distances. It is also found that the Tl^+ ion is far from being spherical.[46]

With the exception of those with halide, oxygen, and sulfur ligands, Tl^I gives rather few complexes. The dithiocarbamates $Tl(S_2CNR_2)$, made from aqueous Tl_2SO_4 and the sodium dithiocarbamates, are useful reagents for the synthesis of other metal dithiocarbamates from the chlorides in organic solvents, since the Tl dithiocarbamate is soluble and TlCl is precipitated on reaction. The structure of

[45] J. J. Dechter and J. I. Zink, *J. Am. Chem. Soc.*, 1976, **98**, 845; *Inorg. Chem.*, 1976, **15**, 1690; R. W. Briggs and J. F. Hinton, *J. Solution Chem.*, 1978, **7**, 1.
[46] N. W. Alcock and H. D. B. Jenkins, *J.C.S. Dalton*, **1974**, 1907.

the n-propyl complex shows that it is polymeric with $[TlS_2CNPr_2]_2$ dimeric units linked by Tl—S bonds.

Electron-exchange reactions in the Tl^I—Tl^{III} system have been intensively studied and appear to be two-electron transfer processes; various Tl^{III} complexes participate under appropriate conditions.

The only known Tl^I organo compound is the polymeric TlC_5H_5[47] precipitated on addition of aqueous TlOH to cyclopentadiene; InC_5H_5 is similar. In the gas phase these compounds consist of discrete molecules having 5-fold symmetry. The metal atoms lie over the centers of the rings and are apparently bound by forces that are mainly covalent. TlC_5H_5 is a very useful reagent for the synthesis of other metal cyclopentadienyl compounds (Chapter 27).

When oxygen, nitrogen, and sulfur bound to organic groups are also bound to Tl^I, the Tl—X bond appears to be more covalent than the bond to alkali metal ions in similar compounds. Thallium compounds tend to be polymeric rather than ionic. Thus the acetylacetonate is a linear polymer with four-coordinate Tl. The alcoxides, which are obtained by the reaction

$$4Tl + 4C_2H_5OH \rightarrow (TlOC_2H_5)_4 + 2H_2$$

are liquids with the exception of the crystalline methoxide. All are tetramers and the methoxide has a distorted cube structure, with the Tl and O atoms at corners of regular tetrahedra of different size, so that oxygen is four-coordinate. Vibrational studies indicate that there are weak direct $Tl\cdots Tl$ interactions in these molecules.

Tl^I and Tl^{III} form compounds with transition metals as in $TlCo(CO)_4$. These compounds are mainly "salts" of carbonylate anions and tend to be ionic. Metal-metal bonds may also be cleaved, e.g.,

$$[(\eta\text{-}C_5H_5)(CO)_3Mo]_2 + 2Tl = 2TlMo(CO)_3(\eta\text{-}C_5H_5)$$

10-15. Thallium(II)

So far, no stable compounds of Tl^{II} exist, but the existence of the Tl^{2+} ion has been inferred from rates of reactions involving Tl^+ or Tl^{3+} with one-electron oxidants or reductants, respectively.

Transient Tl^{2+} ions that decay with a half-life of 0.5 msec can be formed by flash photolysis of Tl^{3+} solutions,

$$H_2O + Tl^{3+} \overset{h\nu}{\rightarrow} Tl^{2+} + OH^\bullet + H^+$$

The hydroxyl radicals can be scavenged by Tl^+:

$$Tl^+ + OH^\bullet \rightarrow TlOH^+$$
$$TlOH^+ + H^+ \rightarrow Tl^{2+}$$

Some redox reactions and the disproportionation reaction

$$2Tl^{2+} \rightarrow Tl^+ + Tl^{3+}$$

[47]   C. K. Anderson et al., J.C.S. Chem. Comm., 1978, 709.

can be studied and estimates made for the potentials

$$Tl^{3+} + e = Tl^{2+} \qquad E = +0.33 \text{ V}$$
$$Tl^{2+} + e = Tl^{+} \qquad E = 2.22 \text{ V}$$

The latter shows that Tl^{2+} is a more powerful oxidant than Co^{3+} (Section 22-F-3).

General References

Lee, A. G., *The Chemistry of Thallium*, Elsevier, 1971.
Sheka, I. A., I. S. Chaus, and T. T. Mityuveva, *The Chemistry of Gallium*, Elsevier, 1966.

CHAPTER ELEVEN

Carbon

GENERAL REMARKS

There are more compounds of carbon than of any other element except hydrogen, and most of them are best regarded as organic chemicals.

The electronic structure of the carbon atom in its ground state is $1s^2 2s^2 2p^2$, with the two $2p$ electrons unpaired, following Hund's rule. To account for the normal four-covalence of carbon, we must consider that it is promoted to a valence state based on the configuration $2s 2p_x 2p_y 2p_z$. The ion C^{4+} does not arise in any normal chemical process: something approximating to the C^{4-} ion may possibly exist in some carbides. In general, however, carbon forms covalent bonds.

Some cations, anions, and radicals of moderate stability can occur, and there is abundant evidence from the study of organic reaction mechanisms for transient species of these types.

Carbonium ions,[1] also called *carbocations,*[2] are of the type $R^1R^2R^3C^+$. The triphenylmethyl cation, one of the earliest known, owes its stability primarily to the fact that the positive charge is highly delocalized, as indicated by canonical structures of the type 11-Ia–d. It behaves in some respects like other large univalent cations (Cs^+, R_4N^+, R_4As^+, etc.) and forms insoluble salts with large anions such as BF_4^- and $GaCl_4^-$. There is good evidence that the cation has a propellerlike arrangement for the phenyl groups, which are bound to the central atom by coplanar sp^2 trigonal bonds.

| (11-Ia) | (11-Ib) | (11-Ic) | (11-Id) |

[1] G. A. Olah and P. von R. Schleyer, Eds., *Carbonium Ions,* Vols. 1 and 2, Wiley, 1968, 1970; G. A. Olah, *Angew. Chem., Int. Ed.,* 1973, **12**, 173.

[2] L. A. Telkowski and M. Saunders, in *Dynamic Nuclear Magnetic Resonance Spectroscopy,* L. M. Jackman and F. A. Cotton, Eds., Academic Press, 1975.

Simpler alkyl cations, not stabilized by resonance like Ph_3C^+ or the tropylium ion $C_7H_7^+$, are not only important intermediates in many reactions but can be produced and studied by nmr in highly acidic or superacidic (cf. Section 6-12) media.[2] One of the most important empirical facts about such carbocations is that their relative stabilities vary $R_3C^+ \gg R_2CH^+ \gg RCH_2^+$. *Carbanions*[3] are of the type $R^1R^2R^3C^-$ and generally have no permanent existence, except in cases where the negative charge can be effectively delocalized. The triphenylmethyl carbanion (11-II) is a good example, as is also the cyclopentadienyl anion (11-III). In fact, since the negative charge in the latter case is equally delocalized on all the carbon atoms, the anion is a regular planar pentagon and the π-electron density distribution can be well represented by 11-IV.

| (11-IIa) | (11-IIb) | (11-IIc) | (11-IId) |

| (11-IIIa) | (11-IIIb) | (11-IIIc) | (11-IV) |

Some stable carbanions, isolable in crystalline salts, are 11-Va, b, c; these all have a planar set of bonds to the central carbon atom.

(a) R = —CN
(b) R = —C(CN)$_2$
(c) R = —NO$_2$

(11-V)

There are also a number of *radicals* that are fairly long-lived, such as the triphenylmethyl radical. Here again the stability is due mainly to delocalization—in this case of the odd electron—in a set of structures like those of $(C_6H_5)_3C^-$ with the odd electron in place of the electron pair.

The three prototypal species CH_3^+, CH_3, and CH_3^- have been subjected to much experimental and theoretical study. The cation and the radical are planar, whereas the CH_3^- ion is, as expected, pyramidal. Many stable carbanions are planar, of course, because they owe their stability to delocalization of the electron pair into π molecular orbitals.[4]

[3]   D. J. Cram, *Fundamentals of Carbanion Chemistry,* Academic Press, 1965; M. Szwarc, *Carbanions, Living Polymers and Electron Transfer Processes,* Wiley-Interscience, 1968; U. Schöllkopf, *Angew. Chem. Int. Ed.,* 1970, **9,** 763.

[4]   G. D. Stucky et al., *J. Am. Chem. Soc.,* 1972, **94,** 7333, 7339, 7346; B. Klewe et al., *Acta Chem. Scand.,* 1972, **26A,** 1049, 1058, 1874, 1921.

The MeS(O)·CMePh anion may be pyramidal, but its rapid racemization shows that a planar transition state for inversion is readily accessible.[5]

Divalent Carbon Compounds.   There are a number of :CRR′ species, generally called *carbenes,* which play a role in many reactions even though they are short-lived.[6a] A general means of generating carbenes is by photolysis of diazoalkanes; this is done in the presence of the substrate with which the carbene is intended to react, such as an olefin:

$$R_2CN_2 \xrightarrow{h\nu} N_2 + R_2C: \xrightarrow{>C=C<} \overset{\displaystyle R_2}{\underset{\displaystyle >C\text{---}C<}{C}}$$

Carbenes so generated are, however, very energetic and their reactions are often indiscriminate. Other methods of generating carbenes are therefore used for practical synthetic purposes but, in many if not all of these, truly free carbenes may never be formed. Thus carbenes, especially halocarbenes, may be conveniently generated from organomercury compounds, thermally or by action of sodium iodide, e.g.,

$$PhHgCF_3 + NaI + >C=C< \xrightarrow[\text{benzene}]{\text{reflux in}} \overset{\displaystyle C}{\underset{\displaystyle C}{|}}\!CF_2 + PhHgI + NaF$$

The structures and ground states of carbenes are difficult to establish with certainty. Experimental data and calculations indicate[6b] that CH_2 has a bent triplet ground state with the lowest singlet about 34 kJ mol^{-1} higher. It is likely that most other carbenes have structures with two unpaired electrons, with the exceptions of dihalocarbenes and those with O, N, or S attached to the divalent carbon. These exceptional compounds probably have no unpaired electrons.

It may be noted that SiF_2 and other Group IV MX_2 compounds (Chapter 12) can be considered to have carbenelike behavior.

Carbene complexes of transition metals are discussed in Chapter 27.

Catenation.   A key feature of carbon chemistry is the formation of chains or rings of C atoms, not only with single but also with multiple bonds (11-VI, 11-VII, 11-VIII). Clearly an element must have a valence of at least 2 and must form strong bonds with itself to do this. Sulfur and silicon are the elements next most inclined to catenation but are far inferior to carbon in this respect.

$$R\text{---}(C=C\text{---})_n R \qquad\qquad R\text{---}(C\equiv C\text{---})_n R$$

(11-VI)   (11-VII)

(11-VIII)

[5]   M. B. D'Amore and J. I. Brauman, *J.C.S. Chem. Comm.,* **1973**, 398.

[6a]   M. Jones, Jr., and R. A. Moss, Eds., *Carbenes,* Wiley, 1973; T. L. Gilchrist and C. W. Rees, *Carbenes, Nitrenes and Arynes,* Nelson, 1969; W. Kirmse, *Carbene Chemistry,* 2nd ed., Academic Press, 1971.

[6b]   J. F. Harrison, *Acc. Chem. Res.,* 1974, **7**, 378; R. K. Lengel and R. N. Zare, *J. Am. Chem. Soc.,* 1978, **100**, 7495.

TABLE 11-1
Some Bond Energies Involving Carbon, Silicon and Sulfur

Bond	Energy (kJ mol^{-1})	Bond	Energy (kJ mol^{-1})
C—C	356	C—O	336
Si—Si	226	Si—O	368
S—S	226	S—O	~330

The unusual stability of catenated carbon compounds, compared with those of silicon and sulfur, can be appreciated by considering the bond-energy data shown in Table 11-1. Thus the simple *thermal* stability of ···C—C—C··· chains is high because of the intrinsic strength of C—C bonds. The relative stabilities toward oxidation follow from the fact that C—C and C—O bonds are of comparable stability, whereas for Si, and probably also for S, the bond to oxygen is considerably stronger. Thus given the necessary activation energy, compounds with a number of Si—Si links are converted very exothermically into compounds with Si—O bonds.

Coordination Numbers. In virtually all its stable compounds carbon forms four bonds and has coordination numbers of 2 (≡C— or =C=), 3 (=C<) or 4, with linear, triangular (planar), and tetrahedral geometries, respectively. In interstitial carbides (p. 9) and certain metal cluster compounds (Section 26-4) carbon atoms are found with coordination numbers of 5 or 6, and CO is a genuine example of coordination number 1 in a stable substance. Compounds with formulas (cation)-CBr$_5$ have recently been made,[7a] but X-ray study[7b] shows CBr$_4$ tetrahedra linked into bands by weak interactions with Br$^-$ ions.

THE ELEMENT

Naturally occurring carbon has the isotopic composition ^{12}C 98.89%, ^{13}C 1.11%. Only ^{13}C has nuclear spin (S = $\frac{1}{2}$), which provides a useful means of probing the structure and bonding in carbon compounds. The nmr measurements are more difficult than are those for ^{1}H, partly because ^{13}C generally relaxes slowly, so that only low power levels may be used and partly because the abundance is low unless ^{13}C-enriched samples are used; with Fourier transform spectroscopy measurements can be made using natural abundance.

The radioisotope ^{14}C (β^-, 5570 years), which is widely used as a tracer, is made by thermal neutron irradiation of lithium or aluminum nitride ^{14}N$(n, p)^{14}$C. It is available not only as CO$_2$ or carbonates but also in numerous labeled organic compounds. Its formation in the atmosphere and absorption of CO$_2$ by living organisms provide the basis of radiocarbon dating.

[7a] F. Effenberger *et al.*, *Chem. Ber.*, 1976, **109**, 306.
[7b] H. J. Linder, H. J. Lindner and B. K.-v. Gross, *Chem. Ber.*, 1976, **109**, 314.

11-1.   Allotropy of Carbon: Diamond; Graphite[8]

The two best-known forms of carbon, diamond and graphite, differ in their physical and chemical properties because of differences in the arrangement and bonding of the atoms. Diamond is denser than graphite (diamond, 3.51 g cm^{-3}; graphite, 2.22 g cm^{-3}), but graphite is the more stable, by 2.9 kJ mol^{-1} at 300°K and 1 atm pressure. From the densities it follows that to transform graphite into diamond, pressure must be applied, and from the known thermodynamic properties of the two allotropes it can be estimated that they would be in equilibrium at 300°K under a pressure of ~15,000 atm. Of course, equilibrium is attained extremely slowly at this temperature, and this property allows the diamond structure to persist under ordinary conditions.

The energy required to vaporize graphite to a monoatomic gas is an important quantity, since it enters into the estimation of the energies of all bonds involving carbon. It is not easy to measure directly because even at very high temperatures, the vapor contains appreciable fractions of C_2, C_3, etc. Spectroscopic studies established that the value had to be either ~520, ~574, or 716.9 kJ mol^{-1}, depending on the process measured spectroscopically. The composition of vapors has been determined mass spectrographically with sufficient accuracy to show that the low values are unacceptable; hence it is now certain that the exact value is 716.9 kJ mol^{-1} at 300°K. In using older tables of bond energies, attention should be paid to the value that was used for the heat of sublimation of graphite.

Diamond.[9a]   This form of carbon is almost invariably found with the cubic structure (see Fig. 1-13). There is also a hexagonal form (lonsdaleite) found in certain meteorites and also available synthetically, in which the puckered layers are stacked in an ABAB··· pattern instead of the ABCABC··· pattern. The hexagonal form is probably unstable toward the cubic, since unlike the cubic, it contains some eclipsed bonds.

Diamond is one of the hardest solids known, and it has high density and index of refraction. It is produced from graphite only under high pressure, and an appreciable rate of conversion requires high temperatures as well. Naturally occurring diamonds must have been formed when such conditions were provided by geological processes. Since at least 1880 recognition of these requirements has led many workers to attempt the production of synthetic diamonds. Until 1955 all such attempts ended in failure, inadequately proved claims, and even in a bogus report of success. Modern knowledge of the thermodynamics of the process indicates that none of the conditions of temperature and pressure reported could have been sufficient for success.

The phase diagram for carbon is summarized in Fig. 11-1. Although graphite can be directly converted into diamond at temperatures of *ca.* 3000°K and pressures above 125 kbar, in order to obtain useful rates of conversion, a transition-metal catalyst such as chromium, iron, or platinum is used. It appears that a thin film

---

[8]   Cf. J. Donohue, *The Structures of the Elements,* Wiley, 1974, pp. 250–262, for details.
[9a] S. Tolansky, *History and Uses of Diamond,* Methuen, 1962; J. E. Fields, Ed., *Properties of Diamond,* Academic Press, 1979.

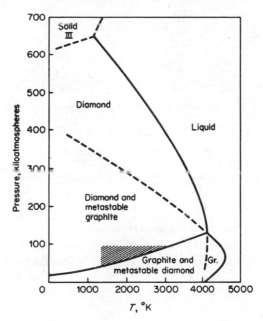

Fig. 11-1. Carbon phase diagram. Shaded area is the most favorable for catalyzed graphite–diamond conversion. [Adapted from F. P. Bundy, *J. Chem. Phys.,* 1963, **38**, 618, 631.]

of molten metal forms on the graphite, dissolving some and reprecipitating it as diamond, which is less soluble. Diamonds up to 0.1 carat of high industrial quality can be routinely produced at competitive prices, and about 40% of the world's supply of industrial-quality diamonds are now synthetic ones. Even gem-quality stones up to about 5 mm in diameter can be obtained, although these are not economically competitive with natural diamonds. By accidental or deliberate introduction of trace impurities, colored stones can be produced—nitrogen causes a yellow coloration, and boron gives blue.[9b]

The chemical reactivity of diamond is much lower than that of carbon in the form of macrocrystalline graphite or the various amorphous forms. Diamond can be made to burn in air by heating it to 600 to 800°C.

Graphite. Graphite has a layer structure as indicated in Fig. 11-2. The separation of the layers is 3.35 Å, which is about equal to the sum of van der Waals radii and indicates that the forces between layers should be relatively slight. Thus the observed softness and particularly the lubricity of graphite can be attributed to the easy slippage of these layers over one another. It will be noted that within each layer each carbon atom is surrounded by only three others. After forming one σ bond with each neighbor, each carbon atom would still have one electron and these are paired up into a system of π bonds (11-IX). Resonance with other structures having different but equivalent arrangements of the double bonds makes all C—C distances equal at 1.415 Å. This is a little longer than the C—C distance in benzene, where

[9b] R. H. Wentorf, Jr., *Adv. High Pressure Res.,* 1974, **4**, 1.

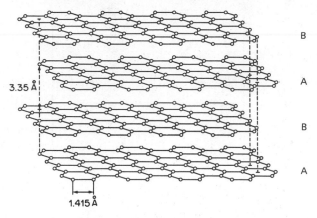

Fig. 11-2. The normal structure of graphite.

the bond order is 1.5, and agrees with the assumption that the bond order in graphite is ~1.33.

(11-IX)

Actually two modifications of graphite exist, differing in the ordering of the layers. In no case do all the carbon atoms of one layer lie directly over those in the next layer, but, in the structure shown in Fig. 11-2, carbon atoms in every other layer are superposed. This type of stacking, which may be designated (ABAB . . .), is apparently the most stable and exists in the commonly occurring hexagonal form of graphite. There is also a rhombohedral form, frequently present in naturally occurring graphite, in which the stacking order is (ABCABC...); that is, every third layer is superposed. It seems that local areas of rhombic structure can be formed by mechanical deformation of hexagonal crystals and can be removed by heat treatment.

The many forms of so-called amorphous carbon, such as charcoals, soot, and lampblack, are all actually microcrystalline forms of graphite, In some soots the microcrystals are so small that they contain only a few unit cells of the graphite structure. The physical properties of such materials are mainly determined by the nature and magnitude of their surface areas. The finely divided forms, which present relatively vast surfaces with only partially saturated attractive forces, readily absorb large amounts of gases and solutes from solution.[10] Active carbons impregnated

[10]   M. Smisek and S. Cerny, *Active Carbon*, Elsevier, 1970; J. S. Mattson and M. B. Mark, Jr., *Activated Carbon*, Dekker, 1971.

with salts of palladium, platinum, or other metals are widely used as industrial catalysts.

An important aspect of graphite technology is the production of very strong fibers by pyrolysis, at 1500°C or above, of oriented organic polymer fibers (e.g., those of polyacrylonitrile, polyacrylate esters, or cellulose). When incorporated into plastics the reinforced materials are light and very strong. Other forms of graphite such as foams, foils, or whiskers can also be made.

Other Forms. At least four other forms—"carbon III" shown in Fig. 11-1, a very rare mineral called chaoite, and other cubic and hexagonal forms,[11] all rare and poorly understood—are also known

11-2. Intercalation Compounds of Graphite[12]

The very loose, layered structure of graphite makes it possible for many molecules and ions to penetrate between the layers, forming interstitial or lamellar compounds. There are two basic types: those in which the graphite, which has good electrical conductivity, becomes nonconducting and those in which high electrical conductivity remains and is enhanced. Only two substances of the first type are known, namely, graphite oxide and graphite fluoride. Graphite oxide is obtained by treating graphite with strong aqueous oxidizing agents such as fuming nitric acid or potassium permanganate. Its composition is not entirely fixed and reproducible, but approximates to C_2O with a little hydrogen always present; the layer separation increases to 6 to 7 Å and it is believed that the oxygen atoms are present in C—O—C bridges across *meta* positions, and in keto and enol groups, the latter being fairly acidic; the graphite layers thus lose their unsaturated character and buckle.

Graphite fluoride, also called poly(carbon monofluoride), is obtained by direct fluorination of graphite at ~600°. When lower temperatures are used, gray or black materials deficient in fluorine are obtained, but under proper conditions white, stoichiometric $(CF)_x$ can be reproducibly obtained.[13] Actually, with the usual small particles of graphite (10^2–10^3 Å), the formation of CF_2 groups at the edges of layers leads to a stoichiometry around $CF_{1.12}$. The layer spacing is ~8 Å and the layers are most likely buckled. $(CF)_x$ has lubricating properties like those of graphite, but it is superior in resisting oxidation by air up to at least 700°. The structure that is currently considered most probable for $(CF)_x$ is shown in Fig. 11-3.

Electrically conducting intercalation compounds, also called lamellar compounds, are formed by insertion of various atoms, molecules, or ions between the layers of graphite. For any given guest species there is generally a whole series of stoichiometric compositions obtainable, each corresponding to a "stage." This term refers to the frequency with which the layers of the graphite host are invaded, as shown in Fig. 11-4. For a stage n compound every nth layer contains guest species, so that the highest concentration of guest occurs in the stage 1 compound. This

[11] A. G. Whittaker and G. M. Wolten, *Science,* 1972, **178,** 54.
[12] L. B. Ebert, *Ann. Rev. Mater. Sci.,* 1976, **6,** 181; J. E. Fischer and T. E. Thompson, *Phys. Today,* **1978,** 36.
[13] R. J. Lagow *et al, J. Am. Chem. Soc.,* 1974, **96,** 2628; *Inorg. Chem.,* 1972, **11,** 2568.

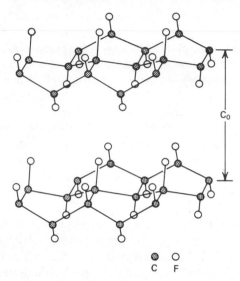

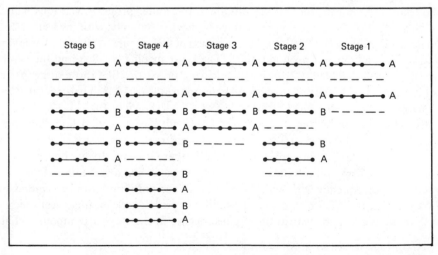

Fig. 11-3. Model for poly(carbon monofluoride) consisting of an infinite array of cis- and trans-linked cyclohexane boats.

tendency of graphite to fill only some layers fully is unique among lamellar hosts. In contrast, others such as the transition metal sulfide hosts (e.g., TiS_2) almost always give stage 1 compounds only, but with continuously variable filling of each layer.

Among the earliest lamellar compounds to be discovered were the graphite–alkali metal ones, which are easily made by direct thermal reaction. The stage 1 com-

Fig. 11-4. The various stages of graphite intercalation compounds: the letters A and B refer to the stacking pattern of carbon layers, and the two carbon layers flanking a guest layer are always equivalent (i.e., have their carbon atoms superposed).

pounds have the composition C_8M for cesium, rubidium, and potassium, but C_6Li for lithium. Sodium does not form a compound more concentrated than $C_{64}Na$. All the alkali metal compounds ignite in air and react explosively with water. Their electrical conductivities are higher than that of graphite and some stage 1 compounds are superconductors.

Among the many other guests in lamellar compounds of graphite are Cl_2, Br_2, a great variety of halides, oxides, and sulfides of metals (e.g., $FeCl_3$, UCl_4, FeS_2, and MoO_3). All of the latter form the lamellar compounds spontaneously—simply on contact. Still others are formed by electrolysis of the reactant using a graphite anode, for example, with sulfuric acid. In many cases the guest is present as anions and the graphite host is cationic. With very strong oxidizing agents,[14] compounds of exceptionally high conductivity can be obtained. Some graphite/AsF_5 materials have in-plane conductivity exceeding that of copper.[15]

NONMOLECULAR COMPOUNDS

11-3.   Carbides[16]

Compounds in which carbon is combined with elements of similar or lower electronegativity, especially metals, are called carbides. Thus compounds with oxygen, sulfur, nitrogen, phosphorus, halogens, etc., are not called carbides, nor are compounds with hydrogen so designated.

The general preparative methods for carbides of all types include: (*a*) direct union of the elements at high temperature (2200° and above); (*b*) heating a compound of the metal, particularly the oxide, with carbon; and (*c*) heating the metal in the vapor of a suitable hydrocarbon. Carbides of copper, silver, gold, zinc, and cadmium, also commonly called acetylides, are prepared by passing acetylene into solutions of the metal salts; with Cu, Ag and Au, ammoniacal solutions of salts of the unipositive ions are used to obtain Cu_2C_2, Ag_2C_2, and Au_2C_2 (uncertain), whereas for Zn and Cd the acetylides ZnC_2 and CdC_2 are obtained by passing acetylene into petroleum solutions of dialkyl compounds. The Cu and Ag acetylides are explosive, being sensitive to both heat and mechanical shock.

1. *Saltlike Carbides.* The most electropositive metals form carbides having physical and chemical properties indicating that they are essentially ionic. The colorless crystals are hydrolyzed by water or dilute acids at ordinary temperatures, and hydrocarbons corresponding to the anions C^{4-} (CH_4), C_2^{2-} (C_2H_2), and C_3^{4-} (C_3H_4) are formed. For example, Be_2C and Al_4C_3 are methides, the former having an antifluorite structure. Al_4C_3 hydrolyzes according to

$$Al_4C_3 + 12H_2O \rightarrow 4Al(OH)_3 + 3CH_4$$

[14]   N. Bartlett *et al., J.C.S., Chem. Comm.,* **1978,** 200.
[15]   E. R. Falardeau, *J.C.S. Chem. Comm.,* **1977,** 389.
[16]   T. Ya. Kosolapova, *Carbides,* Plenum Press, 1971; H. A. Johnson, *Surv. Prog. Chem.,* 1977, **8,** 57.

There are a great many carbides that contain C_2^{2-} ions, or anions that can be so written to a first approximation. For the M_2^I compounds, where M^I may be one of the alkali metals or one of the coinage metals, and for the $M^{II}C_2$ compounds where M^{II} may be an alkaline earth metal, zinc, or cadmium, and for $M_2^{III}(C_2)_3$ compounds in which M^{III} is aluminum, lanthanum, praseodymium, or terbium, this description is probably a very good approximation. In these cases, the postulation of C_2^{2-} ions requires that the metal ions be in their normal oxidation states. In those instances where accurate structural parameters are known, the C—C distances lie in the range 1.19–1.24 Å. The compounds react with water and the C_2^{2-} ions are hydrolyzed to give acetylene only, e.g.,

$$Ca^{2+}C_2^{2-} + 2H_2O \rightarrow HCCH + Ca(OH)_2$$

There are, however, a number of carbides that have the same structures as those discussed above, meaning that the carbon atoms occur in discrete pairs, but they cannot be satisfactorily described as C_2^{2-} compounds. These include YC_2, TbC_2, YbC_2, LuC_2, UC_2, Ce_2C_3, Pr_2C_3, and Tb_2C_3. For all the MC_2 compounds in this list, neutron-scattering experiments show that (a) the metal atoms are essentially trivalent and (b) the C—C distances are 1.28 to 1.30 Å for the lanthanide compounds and 1.34 Å for UC_2. These facts and other details of the structures are consistent with the view that the metal atoms lose not only the electrons necessary to produce C_2^{2-} ions (which would make them M^{2+} ions) but also a third electron, mainly to the antibonding orbitals of the C_2^{2-} groups, thus lengthening the C—C bonds (cf. C—C = 1.19 Å in CaC_2). There are actually other, more delocalized, interactions among the cations and anions in these compounds, since they have metallic properties. The M_2C_3 compounds have the metals in their trivalent states, C—C distances of 1.24 to 1.28 Å and also direct metal-metal interactions. These carbides, which cannot be represented simply as aggregates of C_2^{2-} ions and metal atoms in their normal oxidation states, are hydrolyzed by water to give only 50 to 70% of HCCH, while C_2H_4, CH_4 and H_2 are also produced. There is no detailed understanding of these hydrolytic processes.

Most of the MC_2 acetylides have the CaC_2 structure which is derived from the NaCl structure with the $[C—C]^{2-}$ ions lying lengthwise in the same direction along the cell axes, thus causing a distortion from cubic symmetry to tetragonal symmetry with one axis longer than the other two. In thorium carbide the C_2^{2-} ions are lying flat in parallel planes in such a way that two axes are equally lengthened with respect to the third. These structures are shown in Fig. 11-5. Li_2C_2 has a structure similar to that of CaC_2.

2. *Interstitial Carbides.* These have already been discussed from a structural point of view in Section 1-4. They characteristically have very high melting points, great hardness, and metallic conductivity.

3. *Covalent Carbides.* Although other carbides (e.g., Be_2C) are at least partially covalent, the two elements that approach carbon closely in size and electronegativity, namely, silicon and boron, give completely covalent compounds. Silicon carbide (SiC), technically known as carborundum, is an extremely hard, infusible, and chemically stable material made by reducing SiO_2 with carbon in an electric

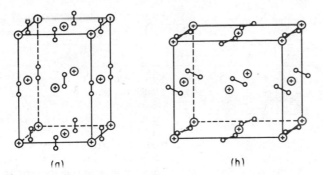

Fig. 11-5. The structures of (a) CaC$_2$ and (b) ThC$_2$ (the latter is somewhat simplified).

furnace. It occurs in three structural modifications, in all of which there are infinite three-dimensional arrays of Si and C atoms, each tetrahedrally surrounded by four of the other kind. Interestingly, no evidence has ever been obtained for a germanium carbide of this or any other type.

Boron carbide (B$_4$C) is also an extremely hard, infusible, and inert substance, made by reduction of B$_2$O$_3$ with carbon in an electric furnace, and has a very unusual structure. The carbon atoms occur in linear chains of three, and the boron atoms in icosahedral groups of twelve (as in crystalline boron itself). These two units are then packed together in a sodium chloride-like array. There are, of course, covalent bonds between carbon atoms and boron atoms as well as between boron atoms in different icosahedra.

SIMPLE MOLECULAR COMPOUNDS

Some of the more important inorganic carbon compounds and their properties are listed in Table 11-2.

11-4. Carbon Halides

Carbon tetrafluoride is an extraordinarily stable compound. It is the end product in the fluorination of any carbon-containing compound. A useful laboratory preparation, for example, involves the fluorination of silicon carbide. The SiF$_4$ also formed is removed easily by passing the mixture through 20% NaOH solution. The CF$_4$ is unaffected, whereas the SiF$_4$ is immediately hydrolyzed; the difference exists because in CF$_4$ carbon is coordinately saturated, whereas silicon in SiF$_4$ has 3d orbitals available for coordination of OH$^-$ ions in the first step of the hydrolysis reaction.

Carbon tetrachloride is a common solvent; it is fairly readily photochemically decomposed and also quite often readily transfers chlorine to various substrates, CCl$_3$ radicals often being formed simultaneously at high temperatures (300–500°). It is often used to convert oxides into chlorides. Although it is thermodynamically

TABLE 11-2
Some Simple Compounds of Carbon

Compound	M.p. (°C)	B.p. (°C)	Remarks
CF_4	-185	-128	Very stable
CCl_4	-23	76	Moderately stable
CBr_4	93	190	Decomposes slightly on boiling
CI_4	171	—	Decomposes before boiling; can be sublimed under low pressure
COF_2	-114	-83	Easily decomposed by H_2O
$COCl_2$	-118	8	"Phosgene"; highly toxic
$COBr_2$	—	65	Fumes in air; $COBr_2 + H_2O \rightarrow CO_2 + 2HBr$
$CO(NH_2)_2$	132	—	Isomerized by heat to $NH_4^+NCO^-$
CO	-205	-190	Odorless and toxic
CO_2	-57 (5.2 atm)	-79	
C_3O_2	—	6.8	Evil-smelling gas
COS	-138	-50	Flammable; slowly decomposed by H_2O
CS_2	-109	46	Flammable and toxic
$(CN)_2$	-28	-21	Very toxic; colorless; water-soluble
HCN	-13.4	25.6	Very toxic; high dielectric constant (116 at 20°) for the associated liquid

unstable with respect to hydrolysis, the absence of acceptor orbitals on carbon makes attack very difficult.

Carbon tetrabromide is a pale yellow solid at room temperature. It is insoluble in water and other polar solvents but soluble in some nonpolar solvents such as benzene.

Carbon tetraiodide is a bright red, crystalline material possessing an odor like that of iodine. Both heat and light cause decomposition to iodine and tetraiodoethylene. The tetraiodide is insoluble in water and alcohol, though attacked by both at elevated temperatures, and soluble in benzene. It may be prepared by the reaction:

$$CCl_4 + 4C_2H_5I \xrightarrow{AlCl_3} CI_4 + 4C_2H_5Cl$$

The increasing instability, both thermal and photochemical, of the carbon tetra-halides with increasing weight of the halogen correlates with a steady decrease in the C—X bond energies:

C—F 485;   C—Cl 327;   C—Br 285;   C—I 213 kJ mol⁻¹

The *carbonyl halides* (X_2CO: X = F, Cl, Br), or mixed (e.g., ClBrCO) are hydrolytically unstable compounds. Urea [$(NH_2)_2CO$] is related but more stable. They all have longer CO bonds than do simple ketones because of partial double bonding of X to C, which weakens the C-to-O π bonding.

11-5. Carbon Oxides

There are four stable oxides of carbon: CO, CO_2, C_3O_2, and $C_{12}O_9$. The last is the anhydride of mellitic acid (11-X) and not discussed, nor are unstable oxides such as C_2O, C_2O_3, and CO_3.

(11-X)

Carbon monoxide is formed when carbon is burned with a deficiency of oxygen. The following equilibrium exists at all temperatures, but is not rapidly attained at ordinary temperatures:

$$2CO(g) \rightleftharpoons C(s) + CO_2(g)$$

Carbon monoxide is made industrially on a huge scale, usually mixed with hydrogen as "synthesis gas," and is used for a variety of large-scale organic syntheses. This chemistry is discussed in Chapter 30 (Sections 30-7 to 30-10). A convenient laboratory preparation of CO is by the dehydrating action of concentrated sulfuric acid on formic acid:

$$HCOOH \xrightarrow{-H_2O} CO$$

Although CO is an exceedingly weak Lewis base, one of its most important properties is the ability to act as a donor ligand toward transition metals (p. 82). For example, nickel metal reacts with CO to form $Ni(CO)_4$, and iron reacts under more forcing conditions to give $Fe(CO)_5$; many other carbonyl complexes are also known. Metal carbonyls are discussed in detail in Chapter 25.

Carbon monoxide is very toxic, rapidly giving a bright red carbonyl complex with the hemoglobin of blood. CO reacts with alkali metals in liquid ammonia to give the alkali metal "carbonyls"; these white solids contain the $[OCCO]^{2-}$ ion.

Carbon dioxide is obtained by combustion of carbon in the presence of an excess of oxygen or by treating carbonates with dilute acids. Its important properties should already be familiar. It undergoes a number of "insertion" reactions (Section 29-7) similar to those of CS_2 noted below, and a few complexes with transition metals are known (Section 4-6).

Carbon suboxide[17] (C_3O_2), an evil-smelling gas, is formed by dehydrating malonic acid with P_2O_5 in vacuum at 140 to 150°, or, better, by thermolysis of diacetyltartaric anhydride. The C_3O_2 molecule is linear and can be represented by the structural formula $O{=}C{=}C{=}C{=}O$. It is stable at $-78°$, but at 25° it polymerizes, forming yellow to violet products. Photolysis of C_3O_2 gives C_2O, which will react with olefins:

$$C_2O + CH_2{=}CH_2 \rightarrow CH_2CCH_2 + CO$$

It reacts but slowly with water to give malonic acid, of which it is the anhydride, but more rapidly with stronger nucleophiles:

$$C_3O_2 + 2H_2O \rightarrow HO_2CCH_2CO_2H$$
$$C_3O_2 + 2NHR_2 \rightarrow R_2NCOCH_2CONR_2$$

[17]   T. Kappe and E. Ziegler, *Angew. Chem., Int. Ed.*, 1974, **13**, 491.

A higher oxide C_5O_2 has been claimed, but its existence has not been shown conclusively.

Carbonic Acids. Though CO is formally the anhydride of formic acid, its solubility in water and bases is slight. It will give formates when heated with alkalis, however. As just noted, C_3O_2 gives malonic acid. However, CO_2 is by far the most important carbonic acid anhydride, combining with water to give "carbonic acid," for which the following equilibrium constants are conventionally written:

$$\frac{[H^+][HCO_3^-]}{[H_2CO_3]} = 4.16 \times 10^{-7}$$

$$\frac{[H^+][CO_3^{2-}]}{[HCO_3^-]} = 4.84 \times 10^{-11}$$

The equilibrium quotient in the first equation is not really correct. It assumes that all CO_2 dissolved and undissociated is present as H_2CO_3, which is not true. In fact, the greater part of the dissolved CO_2 is only loosely hydrated, so that the correct first dissociation constant, using the "true" activity of H_2CO_3, has a value of about 2×10^{-4}, which is more nearly in agreement with expectation for an acid with the structure $(HO)_2CO$.

The rate at which CO_2 comes into equilibrium with H_2CO_3 and its dissociation products when passed into water is measurably slow, and this indeed is what has made possible an analytical distinction between H_2CO_3 and the loosely hydrated $CO_2(aq)$. This slowness is of great importance physiologically and in biological, analytical, and industrial chemistry.

The slow reaction can easily be demonstrated by addition of a saturated aqueous solution of CO_2 on the one hand and of dilute acetic acid on the other to solutions of dilute NaOH containing phenolphthalein indicator. The acetic acid neutralization is instantaneous whereas with the CO_2 neutralization it takes several seconds for the color to fade.

The neutralization of CO_2 occurs by two paths. For pH < 8 the principal mechanism is direct hydration of CO_2:

$$\begin{array}{lll} CO_2 + H_2O = H_2CO_3 & \text{(slow)} & \\ H_2CO_3 + OH^- = HCO_3^- + H_2O & \text{(instantaneous)} & \end{array} \qquad (11\text{-}1)$$

The rate law is pseudo-first order,

$$-d(CO_2)/dt = k_{CO_2}(CO_2); \qquad k_{CO_2} = 0.03 \text{ sec}^{-1}$$

At pH > 10 the predominant reaction is direct reaction of CO_2 and OH^-:

$$\begin{array}{lll} CO_2 + OH^- = HCO_3^- & \text{(slow)} & \\ HCO_3^- + OH^- = CO_3^{2-} + H_2O & \text{(instantaneous)} & \end{array} \qquad (11\text{-}2)$$

where the rate law is

$$-d(CO_2)/dt = k_{OH}(OH^-)(CO_2); \quad k_{OH} = 8500 \text{ sec}^{-1} (\text{mol}/\text{l})^{-1}$$

This can be interpreted, of course, merely as the base catalysis of (11-1). In the pH range 8–10 both mechanisms are important. For each hydration reaction (11-1, 11-2) there is a corresponding dehydration reaction:

$$H_2CO_3 \rightarrow H_2O + CO_2 \qquad k_{H_2CO_3} = k_{CO_2} \times K = 20 \; sec^{-1}$$
$$HCO_3^- \rightarrow CO_2 + OH^- \qquad k_{HCO_3^-} = k_{OH} \times KK_w/K_a = 2 \times 10^{-4} \; sec^{-1}$$

Hence for the equilibrium

$$H_2CO_3 \rightleftharpoons CO_2 + H_2O$$
$$K = (CO_2)/(H_2CO_3) = k_{H_2CO_3}/k_{CO_2} = ca. \; 600 \qquad (11\text{-}3)$$

It follows from eq. 11-3 that K_a, the true ionization constant of H_2CO_3, is greater than the apparent constant, as noted above.

An etherate of H_2CO_3 is obtained by interaction of HCl with Na_2CO_3 at low temperatures in dimethyl ether. The resultant white crystalline solid (m.p. 179), which decomposes at about 5°, is probably $OC(OH)_2 \cdot O(CH_3)_2$.

There are also planar, cyclic anions with aromatic character,[18] such as $C_4O_4^{2-}$ (squarate), $C_5O_5^{2-}$ (croconate), and $C_6O_6^{2-}$.

Carbamic acid $[O{=}C(OH)NH_2]$ can be regarded as derived from carbonic acid by substitution of $-NH_2$ for $-OH$. This is only one example of the existence of compounds that are related in this way; $-NH_2$ and $-OH$ are isoelectronic and virtually isosteric and frequently give rise to isostructural compounds. If the second OH in carbonic acid is replaced by NH_2, we have urea. Carbamic acid is not known in the free state, but many salts are known, all of which are unstable to water, however, because of hydrolysis:

$$H_2NCO_2^- + H_2O \rightarrow NH_4^+ + CO_3^{2-}$$

11-6. Compounds with C—N Bonds; Cyanides and Related Compounds

An important area of "inorganic" carbon chemistry is that of compounds with C—N bonds. The most important species are the cyanide, cyanate, and thiocyanate ions and their derivatives. We can regard many of these compounds as being pseudohalogens or pseudohalides, but the analogies, although reasonably apt for cyanogen, $(CN)_2$, are not especially valid in other cases.

·1. *Cyanogen*.[19] This flammable gas (Table 11-2) is stable even though it is unusually endothermic ($\Delta Hf_{298}^\circ = 297 \; kJ \; mol^{-1}$). It can be prepared by oxidation of HCN using (a) O_2 with a silver catalyst, (b) Cl_2 over activated carbon or silica, or (c) NO_2 over calcium oxide-glass; the last reaction allows the NO produced to be recycled:

$$2HCN + NO_2 \rightarrow (CN)_2 + NO + H_2O$$

Cyanogen can also be obtained from the cyanide ion by aqueous oxidation using Cu^{2+} (cf. the $Cu^{2+} - I^-$ reaction):

$$Cu^{2+} + 2CN^- \rightarrow CuCN + \tfrac{1}{2}(CN)_2$$

[18]   For references, see A. J. Fatiadi, *J. Am. Chem. Soc.*, 1978, **100**, 2586.
[19]   T. K. Brotherton and J. W. Lynn, *Chem. Rev.*, 1959, **59**, 841; H. E. Williams, *Cyanogen Compounds*, 2nd ed., Arnold, 1948 (describes most CN compounds including cyanides); G. J. Jantz, in *Cyanogen and Cyanogen-like Compounds as Dienophiles in 1,4-Cycloaddition Reactions*, J. Hamer, Ed., Academic Press, 1967.

or acidified peroxodisulfate. A better procedure for dry $(CN)_2$ employs the reaction:

$$Hg(CN)_2 + HgCl_2 \rightarrow Hg_2Cl_2 + (CN)_2.$$

This reaction also gives some paracyanogen, $(CN)_n$. Although pure $(CN)_2$ is stable, the impure gas may polymerize at 300 to 500°. The solid polymer reverts to $(CN)_2$ at 800 to 850° but decomposes above this temperature. The structure of $(CN)_n$ has been inferred from infrared spectroscopy to be 11-XI. The cyanogen molecule NCCN is linear, and the bonding is reasonably well represented by N≡C—C≡N.

(11-XI)

It dissociates into CN radicals, and, like RX and X_2 compounds, it can oxidatively add to lower-valent metal atoms giving dicyano complexes, e.g.,

$$(Ph_3P)_4Pd + (CN)_2 \rightarrow (Ph_3P)_2Pd(CN)_2 + 2Ph_3P$$

A further resemblance to the halogens is the disproportionation in basic solution:

$$(CN)_2 + 2OH^- \rightarrow CN^- + OCN^- + H_2O$$

Thermodynamically this reaction can occur in acid solution, but it is rapid only in base. Cyanogen has a large number of reactions, some of which are shown in Fig. 11-6. A stoichiometric mixture of O_2 and $(CN)_2$ burns, producing one of the hottest flames (ca. 5050°K) known from a chemical reaction.

2. *Hydrogen Cyanide.* Like the hydrogen halides, HCN is a covalent, molecular substance, but capable of dissociation in aqueous solution. It is an extremely poisonous (though less so than H_2S), colorless gas and is evolved when cyanides are treated with acids. It is made commercially on a large scale by reaction[20] 11-4.

$$CH_4 + NH_3 \xrightarrow[Pt]{1200°} HCN + 3H_2 \qquad \Delta H = -247 \text{ kJ mol}^{-1} \qquad (11\text{-}4)$$

HCN condenses at 25.6° to a liquid with a very high dielectric constant (107 at 25°). Here, as in similar cases, such as water, the high dielectric constant is due to association of intrinsically very polar molecules by hydrogen bonding. Liquid HCN is unstable and can polymerize violently in the absence of stabilizers: in aqueous solutions polymerization is induced by ultraviolet light.

Hydrogen cyanide is thought to have been one of the small molecules in the earth's primeval atmosphere and to have been an important source or intermediate in the formation of biologically important chemicals.[21] Among the many poly-

[20] E. Koberstein, *Ind. Eng. Chem. Prod. Res. Div.*, 1973, **12**, 444.
[21] See S. Yuasa and M. Ishigami, *Geochem. J.*, 1977, **11**, 247; J. P. Ferris and E. H. Edelson, *J. Org. Chem.*, 1978 **43**, 3989.

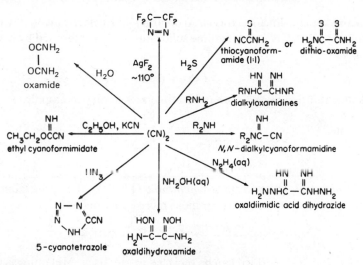

Fig. 11-6. Some reactions of cyanogen. Other products may also be obtained by fluorination (e.g., CF$_3$N=NCF$_3$).

merized products of HCN are the trimer aminomalononitrile, HC(NH$_2$)(CN)$_2$, the tetramer diaminomalononitrile, and polymers of high molecular weight. Furthermore, under pressure with traces of water and ammonia, HCN pentamerizes to adenine, and HCN can also act as a condensing agent for amino acids to give polypeptides.

In aqueous solution HCN is a very weak acid (pK$_{25°}$ = 9.21), and solutions of soluble cyanides are extensively hydrolyzed.

Important industrial uses of HCN are for synthesis of methyl methacrylate and to form adiponitrile (for adipic acid and nylon) by addition to 1,3-butadiene in presence of nickel(0) phosphite complexes (Section 30-3). Waste HCN is also oxidatively hydrolyzed to give oxamide for use as fertilizer (Section 30-15).

3. *Cyanides.* Sodium cyanide is now made by absorbing gaseous HCN in NaOH or Na$_2$CO$_3$ solution. It used to be made by the reaction of molten sodium with ammonia first to give NaNH$_2$, which reacts with carbon to give sodium cyanamide NaNCN and finally NaCN according to the stoichiometry

$$NaNH_2 + C = NaCN + H_2$$

In crystalline alkali cyanides at normal temperatures the CN$^-$ ion is rotationally disordered and is thus effectively spherical with a radius of 1.92 Å. Hence NaCN has the NaCl structure.

The main use of NaCN is in the extraction of gold and silver from their ores by the formation of cyano complexes. The cyanide ion as a ligand has been discussed in Section 4-5. The ions Ag$^+$, Hg$_2^{2+}$, and Pb^{2+} give insoluble cyanides.

Calcium cyanamide (CaNCN) is made in an impure form, largely for fertilizer use, by the reaction

$$CaC_2 + N_2 \xrightarrow{ca.\ 1100°} CaNCN + C \qquad \Delta H = -297 \text{ kJ mol}^{-1}$$

The cyanamide ion is linear and is isostructural and isoelectronic with CO_2.

Cyanamide (H_2NCN), a crystalline solid (mp. 45°), is prepared by hydrolysis of CaNCN:

$$CaNCN + H_2O + CO_2 = CaCO_3 + H_2NCN$$

In alkaline solution at 80° cyanamide dimerizes to dicyandiamide

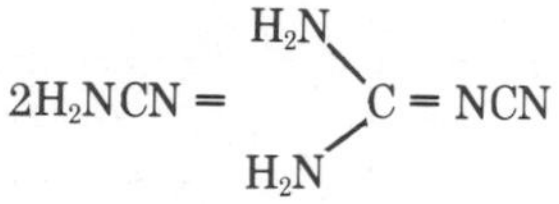

$$2H_2NCN = \begin{array}{c} H_2N \\ \diagdown \\ \diagup \\ H_2N \end{array} C = NCN$$

and this in turn may be converted to *melamine* (11-XII), the cyclic trimer of cyanamide, by heating in NH_3. Melamine is more easily made from urea,

$$6NH_2CONH_2 \xrightarrow{\text{100 atm, 300°}} C_3N_3(NH_2)_3 + 6NH_3 + 3CO_2$$

and the CO_2 and NH_3 formed can be recycled to give urea. Melamine is used for polymers and plastics.

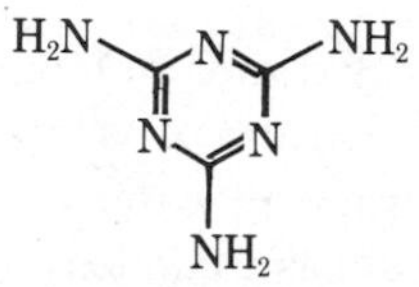

(11-XII)

4. *Cyanogen Halides.* The most important compound is *cyanogen chloride* (b.p. 13°), which is obtained by action of Cl_2 on HCN, by electrolysis of aqueous solutions of HCN and NH_4Cl, and in other ways. It may be polymerized thermally to *cyanuric chloride,* which has the cyclic triazine structure 11-XIII, similar to that of melamine. The chlorine atoms in $C_3N_3Cl_3$ are labile and there is an extensive organic chemistry of triazines, since these compounds are widely used in herbicides and dye stuffs.

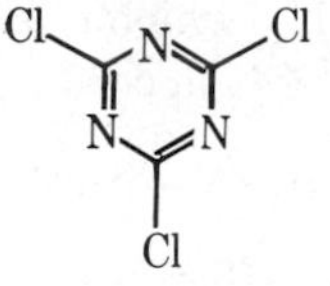

(11-XIII)

Fluorination of $C_3N_3Cl_3$ gives $C_3N_3F_3$, which can be cracked to give FCN. Although this is stable as a gas (b.p. −46°), it polymerizes at 25°. BrCN is similar to ClCN. ICN is made by treating $Hg(CN)_2$ with I_2.

The cyanogen halides generally behave like other halogenoids.

Compounds between CN and other halogenoid radicals are known, such as NCN_3 formed by the reaction

$$BrCN + NaN_3 = NaBr + NCN_3$$

5. *Cyanate*[22] *and its Analogous S, Se, and Te Ions.* The linear cyanate ion OCN^- is obtained by mild oxidation of aqueous CN^-, e.g.:

$$PbO(s) + KCN(aq) \rightarrow Pb(s) + KOCN(aq)$$

The free acid, $K = 1.2 \times 10^{-4}$, decomposes in solution to NH_3, H_2O, and CO_2. There is little evidence for $(OCN)_2$, but covalent compounds such as $P(NCO)_3$ and some metal complexes are known. The compounds are usually prepared from halides by interaction with $AgNCO$ in benzene or NH_4OCN in acetonitrile or liquid SO_2. In such compounds or complexes, either the O or N atoms of OCN can be bound to other atoms and this possibility exists also for SCN. In general most nonmetallic elements seem to be N-bonded.

Thiocyanates[23a] are obtained by fusing alkali cyanides with sulfur; the reaction of S with KCN is rapid and quantitative, and S in benzene or acetone can be titrated with KCN in 2-propanol with bromothymol blue as indicator. Thiocyanogen is obtained by oxidation of aqueous SCN^- with MnO_2:

$$(SCN)_2 + 2e = 2SCN^- \qquad E^\circ = +0.77 \text{ V}$$

but since it is rapidly decomposed by water it is best made by action of Br_2 on AgSCN in an inert solvent. In the free state $(SCN)_2$ rapidly and irreversibly polymerizes to brick red polythiocyanogen, but it is most stable in CCl_4 or CH_3COOH solution, where it exists as NCSSCN.

The SCN^- ion is a good ligand and the numerous thiocyanate complexes, which may be either S- or N-bonded, are usually stoichiometrically analogous to halide complexes. Similar complexes of $SeCN^-$ are known. Nonmetallic thiocyanates are usually S-bonded. The $TeCN^-$ ion in its $(Ph_3P)_2N^+$ salt is linear.[23b]

11-7. Compounds with C—S Bonds[24]

All the substances discussed in this section behave as ligands and were discussed as such in Chapter 4.

Carbon Disulfide. This compound is prepared on a large scale by direct interaction of C and S at high temperatures. A similar yellow liquid CSe_2 is made by action of CH_2Cl_2 on molten selenium; it has a worse smell than CS_2 but, unlike it, is nonflammable. The selenide slowly polymerizes spontaneously, but CS_2 does so only under high pressures, to give a black solid having structure 11-XIV.

(11-XIV)

[22] S. Patai, Ed., *The Chemistry of Cyanates and Their Thio Derivatives,* Wiley, 1977.

[23a] A. A. Newman, Ed., *The Chemistry and Biochemistry of Thiocyanic Acid and Its Derivatives,* Academic Press, 1975.

[23b] A. S. Foust, *J.C.S. Chem. Comm.,* **1979,** 414.

[24] G. Gattow and W. Behrend, *Carbon Sulfides and Their Inorganic and Complex Chemistry,* Thieme, 1977 (an extensive monograph).

In addition to its high flammability in air, CS_2 is a very reactive molecule and has an extensive chemistry, much of it organic. It is used to prepare carbon tetrachloride industrially:

$$CS_2 + 3Cl_2 \rightarrow CCl_4 + S_2Cl_2$$

Carbon disulfide is one of the small molecules that readily undergo the "insertion reaction" (Section 29-7) where the

$$\begin{matrix} —S—C— \\ \| \\ S \end{matrix}$$

group is inserted between Sn—N, Co—Co, or other bonds. Thus with metal dialkylamides, dithiocarbamates are obtained:

$$M(NR_2)_4 + 4CS_2 \rightarrow M(S_2CNR_2)_4$$

Important reactions of CS_2 involve nucleophilic attacks on carbon by the ions SH^- and OR^- and by primary or secondary amines, which lead respectively to thiocarbonates, xanthates, and dithiocarbamates, e.g.:

$$SCS + :SH^- \longrightarrow S_2CSH^- \xrightarrow{OH^-} CS_3^{2-}$$

$$SCS + :OCH_3^- \longrightarrow CH_3OCS_2^-$$

$$SCS + :NHR_2 \xrightarrow{OH^-} R_2NCS_2^-$$

Thiocarbonates. Thiocarbonates are readily formed by the action of SH^- on CS_2 in alkaline solution (cf. reaction above), and numerous yellow salts containing the planar ion are known. Heating CS_3^{2-} with S affords orange tetrathiocarbonates, which have the structure $[S_3C—S—S]^{2-}$. The free acids can be obtained from both these ions as red oils, stable at low temperatures.

Dithiocarbamates; Thiuram Disulfides.[25] Dithiocarbamates are normally prepared as alkali metal salts by action of primary or secondary amines on CS_2 in presence of, say, NaOH. The zinc, manganese, and iron dithiocarbamates are extensively used as agricultural fungicides, and zinc salts as accelerators in the vulcanization of rubber. Alkali metal dithiocarbamates are usually hydrated and are dissociated in aqueous solution. When anhydrous, they are soluble in organic solvents in which they are associated.

On oxidation of aqueous solutions by H_2O_2, Cl_2, or $S_2O_8^{2-}$, thiuram disulfides, of which the tetramethyl is the commonest, are obtained:

$$I_2 + 2Me_2NCS_2^- \longrightarrow \begin{matrix} Me_2NC—S—S—CNMe_2 + 2I^- \\ \| \qquad\qquad \| \\ S \qquad\qquad S \end{matrix}$$

Thiuram disulfides, which are strong oxidants, are also used as polymerization initiators (for, when heated, they give radicals) and as vulcanization accelerators.

[25] G. D. Thorn and R. A. Ludwig, *The Dithiocarbamates and Related Compounds*, Elsevier, 1962.

Also tetraethylthiuram disulfide is "Antabuse," an agent for rendering the body allergic to ethanol.

General References

Blackman, L. C. F., Ed., *Modern Aspects of Graphite Technology,* Academic Press, 1970.

Carbon, Pergamon Press. A review journal on physical properties.

Davidson, H. W., *et al., Manufactured Carbon,* Pergamon Press, 1968.

Kühle, F. *Angew. Chem. Int. Ed.* 1973, 12, 630. Reactive monomeric derivatives of carbonic acid: ClCN, CS_3^{2-}, COS, $COCl_2$, carbodiimides, etc.

Reynolds, W. N., *Physical Properties of Graphite,* Elsevier, 1968.

Walker, P. L., Ed., *Chemistry and Physics of Carbon,* Vols. I and II, Dekker, 1966.

CHAPTER TWELVE

The Group IV Elements:
Si, Ge, Sn, Pb

GENERAL REMARKS

12-1. Group Trends

There is no more striking example of an enormous discontinuity in general properties between the first- and the second-row elements followed by a relatively smooth change toward more metallic character thereafter than in Group IV. Little of the chemistry of silicon can be inferred from that of carbon. Carbon is strictly nonmetallic; silicon is essentially nonmetallic; germanium is metalloid; tin and especially lead are metallic. Some properties of the elements in Group IV are given in Table 12-1.

Catenation. Though not as extensive as in carbon chemistry, catenation is an important feature of Group IV chemistry in certain types of compound. Extensive chains occur in silicon and germanium hydrides (up to Si_6H_{14} and Ge_9H_{20}), in Si halides (only Ge_2Cl_6 is known), and in certain organo compounds. For tin and lead, catenation occurs only in organo compounds. Lead and tin, along with certain other metallic main group elements, (e.g., bismuth), react with alkali metals in liquid ammonia to form compounds such as Na_4Pb_9 and Na_2Sn_5 that contain polyhedral clusters of metal atoms. Silicon also forms a few cluster anions such as the Si_4 ion in $BaSi_2$.

However there is a general if not entirely smooth decrease in the tendency to catenation in the order $C \gg Si > Ge \approx Sn \gg Pb$, which may be ascribed partly to diminishing strength of the C—C, Si—Si, Ge—Ge, Sn—Sn, and Pb—Pb bonds (Table 12-2).

Bond Strengths. The strengths of single covalent bonds between Group IV atoms and other atoms (Table 12-2) generally decrease from Si to Pb. In some cases there is an initial rise from C to Si followed by a decrease. These energies do not, of course,

374

TABLE 12-1
Some Properties of the Group IV Elements[a]

Element	Electronic structure	M.p. (°C)	B.p. (°C)	Ionization enthalpies (kJ mol^{-1}) 1st	2nd	3rd	4th	Electro- nega- tivity[b]	Covalent radius[c] (Å)
C	[He]$2s^22p^2$	>3550^d	4827	1086	2353	4618	6512	2.5–2.6	0.77
Si	[Ne]$3s^23p^2$	1410	2355	786.3	1577	3228	4355	1.8–1.9	1.17
Ge	[Ar]$3d^{10}4s^24p^2$	937	2830	760	1537	3301	4410	1.8–1.9	1.22
Sn	[Kr]$4d^{10}5s^25p^2$	231.9	2260	708.2	1411	2942	3928	1.8–1.9	1.40^e
Pb	[Xe]$4f^{14}5d^{10}6s^26p^2$	327.5	1744	715.3	1450	3080	4082	1.8	1.44^f

^{a} For further detail see E. A. V. Ebsworth.[1]
^{b} H. O. Pritchard and H. A. Skinner, *Chem. Rev.,* 1955, **55**, 745; see also A. J. Smith, W. Adcock, and W. Kitching, *J. Am. Chem. Soc.,* 1970, **92**, 6140.
^{c} Tetrahedral (i.e., sp^3 radii).
^{d} Diamond.
^{e} Covalent radius of Sn^{II}, 1.63 Å.
^{f} Ionic radius of Pb^{2+}, 1.21; of Pb^{4+}, 0.775 Å.

reflect the ease of heterolysis of bonds, which is the usual way in chemical reactions; thus, for example, in spite of the high Si—Cl or Si—F bond energies, compounds containing these bonds are highly reactive. Since the charge separation in a bond is a critical factor, the bond ionicities must also be considered when interpreting the reactivities toward nucleophilic reagents. Thus Si—Cl bonds are much more reactive than Si—C bonds because, though stronger, they are more polar, Si$^{\delta+}$—Cl$^{\delta-}$ rendering the silicon more susceptible to attack by a nucleophile such as OH$^-$.

Two other points may be noted; (a) there is a steady decrease in M—C and M—H bond energies; (b) M—H bonds are stronger than M—C bonds.

Electronegativities. The electronegativities of the Group IV elements have been a contentious matter. Although C is generally agreed to be the most electronegative element, certain evidence, some of it suspect, has been interpreted as indicating that Ge is more electronegative than Si or Sn. It is to be remembered that electro-

TABLE 12-2
Approximate Average Bond Energies

Group IV[a] element	Energy of bond (kJ mol^{-1}) with: Self	H	C	F	Cl	Br	I	O
C	356	416		485	327	285	213	336
Si	210–250	323	250–335	582	391	310	234	368
Ge	190–210	290	255	·465	356	276	213	
Sn	105–145	252	193		344	272	187	

^{a} Data derived mainly from MX$_4$-type compounds that are unstable or nonexistent when M = Pb. Pb—C in PbEt$_4$ = 128.8 kJ mol^{-1}; see C. F. Shaw and A. L. Allred, *Organomet. Chem. Rev., A,* 1970, **5**, 96.

[1]   E. A. V. Ebsworth, in *The Organometallic Compounds of Group IV Elements,* Vol. 1, Part 1, A. G. MacDiarmid, Ed., Dekker, 1968.

negativity is a very qualitative matter, and it seems most reasonable to accept a slight progressive decrease Si → Pb.

It can be noted that Zn in HCl reduces only germanium halides to the hydrides, which suggests a higher electronegativity for Ge than for Si or Sn. Also, dilute aqueous NaOH does not affect GeH_4 or SnH_4, but SiH_4 is rapidly hydrolyzed by water containing a trace of OH^-. This is consistent with, though not necessarily indicative of, the Ge—H or Sn—H bonds either being nonpolar or having the positive charge on hydrogen. Finally, germanium halides are hydrolyzed in water only slowly and reversibly.

The Divalent State. The term "lower valence" indicates the use of fewer than four electrons in bonding. Thus although the *oxidation state* of carbon in CO is usually formally taken to be 2, this is only a formalism, and carbon uses more than two valence electrons in bonding. True divalence is found in carbenes (p. 354) and in a few SiX_2 compounds discussed below; the high reactivity of carbenes may result from the greater accessibility of the sp^2 hybridized lone pair in the smaller carbon atom. The stable divalent compounds of the other elements can be regarded as carbenelike in the sense that they are bent with a lone pair and undergo the general type of carbene reactions to give two new bonds to the element, that is,

However the divalent state becomes increasingly stable down the group and is dominant for lead.

Inspection of Table 12-1 clearly shows that this trend cannot be explained exclusively in terms of ionization enthalpies, since these are essentially the same for all the elements Si—Pb. The "inert pair" concept is not particularly instructive either, especially since the nonbonding electrons are known not to be inert in a stereochemical sense (see below).

Other factors that undoubtedly govern the relative stabilities of the oxidation states are promotion energies, bond strengths for covalent compounds, and lattice energies for ionic compounds. Taking first the promotion energies, it is rather easy to see why the divalent state becomes stable if we remember that the M—X bond energies generally decrease in the order Si—X, Ge—X, Sn—X, Pb—X(?). (For methane the factor that stabilizes CH_4 relative to $CH_2 + H_2$, despite the much higher promotional energy required in forming CH_4, is the great strength of the C—H bonds and the fact that two more of these are formed in CH_4 than in CH_2.) Thus if we have a series of reactions $MX_2 + X_2 = MX_4$ in which the M—X bond energies are decreasing, it is obviously possible that this energy may eventually become too small to compensate for the $M^{II} \rightarrow M^{IV}$ promotion energy and the MX_2 compound becomes the more stable. The progression is illustrated by ease of addition of chlorine to the dichlorides:

$$GeCl_2 + Cl_2 \rightarrow GeCl_4 \text{ (very rapid at } 25°)$$
$$SnCl_2 + Cl_2 \rightarrow SnCl_4 \text{ (slow at } 25°)$$
$$PbCl_2 + Cl_2 \rightarrow PbCl_4 \text{ (only under forcing conditions)}$$

Note that even $PbCl_4$ decomposes except at low temperatures, while $PbBr_4$ and PbI_4 do not exist, probably owing to the reducing power of Br^- and I^-. For ionic compounds matters are not so simple but, since the sizes of the (real or hypothetical) ions, M^{2+} and M^{4+}, increase down the group, it is possible that lattice energy differences no longer favor the M^{4+} compound relative to the M^{2+} compound in view of the considerable energy expenditure required for the process

$$M^{2+} \rightarrow M^{4+} + 2e$$

Of course, there are few compounds of the types MX_2 or MX_4 that are entirely covalent or ionic (almost certainly no ionic MX_4 compounds), so that the arguments above are oversimplifications, but they indicate roughly the factors involved. For solutions no simple argument can be given, since Sn^{4+} and Pb^{4+} probably have no real existence.

It should be noted that compounds appearing to contain group IV elements in formal oxidation states between II and IV are generally not built up of individual atoms in these oxidation states. For example, Ge_5F_{12} consists of four Ge^{II} and one Ge^{IV} suitably coordinated and linked by fluorine atoms.[2a] Similarly Sn_3F_8 is built of octahedral $Sn^{IV}F_6$ and pyramidally coordinated $Sn^{II}F_3$ in 1:2 ratio, linked by shared F atoms.[2b]

The only authentic stable compounds of group IV elements with oxidation number III are of the types 12-I and 12-II.

$(Me_3Si)_2CH$ and $CH(SiMe_3)_2$ on M, with $CH(SiMe_3)_2$ below.

(12-I) a. M = Si
       b. M = Ge
       c. M = Sn

$(Me_3Si)_2N$ and $N(SiMe_3)_2$ on M, with $N(SiMe_3)_2$ below.

(12-II) a. M = Ge
        b. M = Sn

Type 12-Ia has a half life of only *ca.* 10 min, but the others have lives of 3 to 12 months.[3] The stability of these radicals is presumably due to the great steric hindrance to attack resulting from the bulk of the ligands.

Tin, like iron, is an element for which Mössbauer spectroscopy can be broadly useful. Using the Sn^{119m} nucleus, isomer shift (IS) measurements allow one to determine oxidation numbers and to some extent to estimate structural and bonding features. The IS values are usually in the order $Sn^{II} > Sn^0 > Sn^{IV}$, though this cannot be trusted absolutely.[4]

Multiple Bonding. Silicon, germanium, tin, and lead do not form from multiple bonds using $p\pi$ orbitals. Consequently numerous types of carbon compound, such as alkenes, alkynes, ketones, and nitriles, have no analogues. Although stoichio-

---

[2a] J. C. Taylor and P. W. Wilson, *J. Am. Chem. Soc.*, 1973, **95**, 1834.

[2b] M. F. A. Dove, R. King and T. J. King, *J. C. S. Chem. Comm.*, **1973**, 944.

[3] M. J. S. Gynane *et al.*, *J. C. S. Dalton*, **1977**, 2004; A. Hudson, M. F. Lappert, and P. W. Lednor, *J. C. S. Dalton*, **1976**, 2360.

[4] P. G. Harrison and J. J. Zuckerman, *Inorg. Chim. Acta*, 1977, **21**, L3.

metric similarities may occur [e.g., CO_2, SiO_2, $(CH_3)_2CO$, $(CH_3)_2SiO$], there is no structural or chemical relationship, and reactions that might have been expected to yield a carbonlike product do not; for example, dehydration of silanols $R_2Si(OH)_2$ produces $R_2Si(OH)$—O—$SiR_2(OH)$ and $(R_2SiO)_n$.

However for silicon, transient intermediates[5] with Si—C $p\pi$-$p\pi$ bonding can be obtained by photolysis or thermolysis of suitable precursors, e.g.,

$$H_2Si\diamond \xrightarrow[N_2]{560°} [H_2Si{=}CH_2] + CH_2{=}CH_2$$

The compound $Me_2Si{=}C(SiMe_3)_2$ has been detected in the gas phase by mass spectroscopy, and there is a similar compound $Me_2Si{=}NSiMe_3$.[6]

There is also an iron complex with a π-sila allyl group, Me_2Si—CH—CH_2, where C—Si multiple bonding is presumably involved.[7]

Multiple bonding of the $d\pi$-$p\pi$ type for silicon is quite well established,[8] especially in bonds to O and N. It is important to note, however, that this does not necessarily lead to conjugation in the sense usual for carbon multiple bond systems. Thus nmr contact shifts for the tetrahedral tropone iminates of Si, Ge, and Sn (12-III) show negligible conjugation with the π system of the ring. Observations of the following types provide evidence for $d\pi$-$p\pi$ bonding.

1. Trisilylamine $[(H_3Si)_3N]$ differs from $(H_3C)_3N$ in being planar rather than pyramidal and in being a very weak Lewis base. Other compounds such as $(H_3Si)_2NH$ and 12-IV are also planar; H_3SiNCO has a linear Si—N—C—O chain

(12-III)   (12-IV)

in the vapor phase, although there is some bending with $\angle$Si—N—C = 158° and $\angle$N—C—O = 176° in the crystalline solid.[9] These observations can be explained by supposing that nitrogen forms dative π bonds to the silicon atoms. In the planar state of $N(SiH_3)_3$, the nonbonding electrons of nitrogen would occupy the $2p_z$ orbital, if we assume that the N—Si bonds are formed using sp_xp_y trigonal hybrid orbitals of nitrogen. Silicon has empty $3d$ orbitals, which are of low enough energy to be able to interact appreciably with the nitrogen $2p_z$ orbital. Thus the N—Si π bonding is due to the kind of overlap indicated in Fig. 12-1. It is the additional

5   A. G. Brook and J. W. Harris, *J. Am. Chem. Soc.,* 1976, **98,** 3381; R. L. E. Gusel'nikov, N. S. Nametkin, and V. M. Udovin, *Acc. Chem. Res.,* 1975, **8,** 18; C. M. Golino, R. D. Bush, and L. H. Sommer, *J. Am. Chem. Soc.,* 1975, **97,** 7371.

6   N. Wiberg and G. Preiner, *Angew. Chem., Int. Ed.,* 1978, **17,** 362.

7   H. Sakurai *et al., J. Am. Chem. Soc.,* 1976, **98,** 7453.

8   K. Hafner *et al., d Orbitals in the Chemistry of Silicon, Phosphorus and Sulfur,* Springer, 1977.

9   M. J. Barrow *et al., J. C. S. Chem. Commun.,* **1977,** 744.

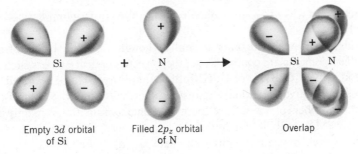

Empty $3d$ orbital   Filled $2p_z$ orbital   Overlap
of Si   of N

Fig. 12-1. Formation of $d\pi$–$p\pi$ bond between Si and N in trisilylamine.

bond strength to be gained by this $p\pi$–$d\pi$ bonding that causes the NSi$_3$ skeleton to take up a planar configuration, whereas with N(CH$_3$)$_3$, where the carbon has no low-energy d orbitals, the σ bonding alone determines the configuration, which is pyramidal as expected.

(H$_3$Si)$_3$P is pyramidal, however, indicating that the second-row atom P is less able to contribute a p orbital to $d\pi$–$p\pi$ bonding than is nitrogen.

Similarly, H$_3$GeNCO is *not* linear in gas phase.[10] Evidently effective π bonding in the linear structure occurs only for Si—N and not Ge—N bonds.

2.   The disilyl ethers (R$_3$Si)$_2$O all have large angles at oxygen (140–180°), and both electronic and steric explanations have been suggested. Electronically, overlap between filled oxygen $p\pi$ orbitals and silicon $d\pi$ orbitals would improve with increasing angle and in the limit might favor linearity. When R = C$_6$H$_5$ the angle at oxygen is 180°, but there may also be a strong steric factor here because of the very large R groups.[11]

3.   Silanols, such as (CH$_3$)$_3$SiOH, are stronger acids than the carbon analogues and form stronger hydrogen bonds; for Ph$_3$MOH the acidities are in the order C $\approx$ Si $\gg$ Sn. The hydrogen bonding can be ascribed to Si—O π bonding involving one of the two unshared pairs of the silanol oxygen and the $3d$ orbital of Si to give a situation somewhat similar electronically to that of the nitrogen atom in an imine R$_2$C=NH. One unshared pair still remains on the oxygen, which is consistent with the failure of the *base* character of the silanol to be much lowered, in spite of its stronger acidity, compared with the analogous alcohol; the base order is C $\approx$ Si < Ge < Sn.

A similar situation arises with the acid strength of R$_3$MCOOH, where the order is Si $\geq$ Ge > C; in this case the $d\pi$–$p\pi$ bonding probably acts to stabilize the anion. The order of π bonding, C > Si > Ge $\geqslant$ Sn > Pb, is obtained from hydrogen bonding and nmr studies on amines. Thus N[Si(CH$_3$)$_3$]$_3$ is virtually nonbasic, the germanium compound is about as basic as a tertiary amine, and the tin compound is more basic than any organic amine. The same order[12] is found in RMX$_3$ when X = alkyl, but when X = halogen it is C < Si < Ge < Sn.

10   J. D. Murdock and D. W. H. Rankin, *J. C. S. Chem. Commun.,* **1972,** 748.
11   C. Glidewell and D. C. Liles, *J. C. S. Chem. Commun.,* **1979,** 93.
12   G. M. Whitesides et al., *J. Organomet. Chem.,* 1970, **22,** 365.

Stereochemistry.[13] The stereochemistries of Group IV compounds are given in Table 12-3.

IV-Oxidation State. Silicon is normally, though not exclusively, tetrahedral, and the other elements are commonly so; the expected optical isomers such as those of $SiMePhEt(C_6H_4COOH)$ or $GeHMePh(\alpha\text{-naphthyl})$ can be resolved. In view of the possibility of valence shell expansion by utilization of the outer d orbitals, coordination numbers of 5 and 6 are common. Pentacoordination is mainly confined to (a) the ions MX_5^- stabilized in lattices by large cations; (b) compounds, especially of Si, with various oxygen and nitrogen chelates; (c) adducts with organic bases such as MX_4L and R_3MXL; and (d) for tin, polymeric compounds of the type R_3SnX where X acts as a bridge group. Octahedral coordination is well known for all four elements. Only one case, that of the pyrocatecholate ester, $Si(O_2C_6H_4)_2$, of planar silicon is known (see Table 12-3).

For the ions and adducts, it cannot be predicted whether a complex will involve five- or six-coordination, since that depends on delicate energy balances.

Silicon forms silicides containing a variety of Si_n^{x-} anions, including "Si^{4-}" in Ba_2Si,[14] which has an $anti\text{-}PbCl_2$ structure with octahedrally coordinated Si^{4-}.

II-Oxidation State. For Sn^{II}, and to a lesser extent for Ge^{II} and Pb^{II}, it has been shown that the pair of electrons that is unused in bonding has important effects on the stereochemistry.[15a]

Thus in the blue-black form of SnO, each tin atom is surrounded by five oxygen atoms at approximately the vertices of an octahedron, the sixth vertex being presumably occupied by the lone pair. This is called a ψ-octahedral arrangement. In $SnCl_2$, SnS, SnSe (orthorhombic form), $SnCl_2 \cdot 2H_2O$, $K_2SnCl_4 \cdot H_2O$, and $SnSO_4$, there are ψ-tetrahedral groupings, that is, atoms at three corners of a tetrahedron and a lone pair of electrons at the fourth. Thus $SnCl_2 \cdot 2H_2O$ has a pyramidal $SnCl_2OH_2$ molecule, the second H_2O not being coordinated (it is readily lost at 80°), while $K_2SnCl_4 \cdot H_2O$ consists of ψ-tetrahedral $SnCl_3^-$ ions and Cl^- ions. The ψ-tetrahedral SnF_3^- ion is also known, and the $Sn_2F_5^-$ ion consists of two SnF_3^- ions sharing a fluorine atom. Other Sn^{II} compounds such as $SnCl_2$ or SnS similarly involve three-coordination but with a bridge group between the metal atoms.

An important consequence is that solvated $SnCl_2$, the $SnCl_3^-$ ion, and also other Sn^{II} and Ge^{II} compounds with lone pairs, such as $Sn(acac)_2$, can act as a donor ligand toward transition metals.

Stereochemically active lone pairs are also recognized for Ge^{II} and Pb^{II}. Thus the Ge^{II} atoms in Ge_5F_{12}[2a] and Pb^{II} in PbO have square pyramidal coordination. In $Pb(O_2PPh_2)_2$ the lead atom has ψ-trigonal bipyramidal coordination, with the lone pair in an equatorial position.[15b]

---

[13] B. J. Aylett, *Prog. Stereochem.*, 1969, **4**, 213; see also *Organometallic Compounds of Group IV Elements*, Vol. 1, Part 1, A. G. McDiarmid, Ed., Dekker, 1969; J. A. Zubieta and J. J. Zuckerman, *Prog. Inorg. Chem.*, 1978, **24**, 251 (an exhaustive survey of tin stereochemistry).

[14] A. Widera and H. Schäfer, *Z. Naturforsch.*, 1976, **31b**, 1434.

[15a] P. Colamarino et al., *Inorg. Chem.*, 1976, **15**, 800.

[15b] E. Hough and D. G. Nicholson, *J. C. S. Dalton*, **1976**, 1782.

TABLE 12-3

Valence and Stereochemistry of Group IV Elements

Valence	Coordination number	Geometry[a]	Examples
Si^0	6	Octahedral	$Si(bipy)_3$
Si^{II}, Ge^{II}, Sn^{II}, Pb^{II} [b]	2	ψ-Trigonal (angular)	$SiF_2(g)$, $SnCl_2(g)$, $Pb(C_5H_5)_2$, $GeF_2(g)$
	3	Pyramidal	$SnCl_2\cdot 2H_2O$, $SnCl_3^-$, $SnC_2(s)$, $Sn_2F_5^-$
	4	ψ-tbp	Pb^{II} in Pb_3O_4, $Sn(S_2CNR_2)_2$ $Sn(\beta$-like$)_2$
	5	ψ-Octahedral	SnO (blue-black form)
	6	Octahedral	PbS(NaCl type), GeI_2(CdI$_2$ type)
	7	Complex	$[SC(NH_2)_2]_2PbCl$
	6, 7	ψ-Pentagonal bipyramid + complex ψ-8-coord.[c]	$Sn^{II}[Sn(EDTA)H_2O]\cdot H_2O$
Si^{III}, Ge^{III}, Sn^{III}	3	Nonplanar	See page 377
Si^{IV}, Ge^{IV}, Sn^{IV}, Pb^{IV}	4	Tetrahedral	SiO_4 (silicates, SiO_2), SiS_2, $SiCl_4$, $PbMe_4$, GeH_4
	4	Planar	$Si(O_2C_6H_4)_2^{[d]}$
	5	Tbp	$Me_3SnClpy$, $SnCl_5^-$, SiF_5^-, $RSiF_4^-$, Me_3SnF (copolymer)
	5	?	$[SiPh_3bipy]^+$
	6	Octahedral	SiF_6^{2-}, $[Si(acac)_3]^+$, $SnCl_6^{2-}$, GeO_2, SnO_2 (rutile str.), $PbCl_6^{2-}$, cis-$SnCl_4(OPCl_3)_2$, SnF_4, $trans$-$GeCl_4py_2$, Pb^{IV} in Pb_3O_4, $Sn(S_2CNEt_2)_4$ $[M(C_2O_4)_3]^{2-}$, M = Si, Ge or Sn
	7	Distorted	$[Sn(O_2CMe)_5]^{-}$ [e]
	8	Dodecahedral	$Sn(NO_3)_4$; $Pb(O_2CMe)_4$, $Sn(O_2CMe)_4^{[e]}$

[a] ψ indicates that a coordination position is occupied by a lone pair.
[b] P. G. Harrison, *Coord. Chem. Rev.*, 1976, **20**, 1.
[c] F. P. van Remoortere et al., *Inorg. Chem.*, 1971, **10**, 1511.
[d] H. Meyer and G. Nagorsen, *Angew. Chem. Int. Ed.*, 1979, **18**, 551; E.-U. Würthwein and P. von R. Schleyer, *ibid.*, 553.
[e] N. W. Alcock and V. L. Tracy, *Acta Cryst.*, B, 1979, **35**, 80.

THE ELEMENTS

12-2. Occurrence, Isolation, and Properties

Silicon is second only to oxygen in weight percentage of the earth's crust ($\sim$28%) and is found in an enormous diversity of silicate minerals. Germanium, tin, and lead are relatively rare elements ($\sim$10^{-3} wt %), but they are well known because of their technical importance and the relative ease with which tin and lead are obtained from natural sources.

Silicon is obtained in the ordinary commercial form by reduction of SiO_2 with carbon or CaC_2 in an electric furnace. Similarly, germanium is prepared by reduction of the dioxide with carbon or hydrogen. Silicon and germanium are used as semiconductors, especially in transistors. For this purpose exceedingly high purity is essential, and special methods are required to obtain usable materials. Methods for silicon vary in detail, but the following general procedure is followed.

1.   Ordinary, "chemically" pure silicon is converted, by direct reaction, into a silicon halide or into $SiHCl_3$. This is then purified (of B, As, etc.) by fractional distillation in quartz vessels.

2.   The SiX_4 or $SiHCl_3$ is then reconverted into elemental silicon by reduction with hydrogen in a hot tube, or on a hot wire when X is Cl or Br,

$$SiX_4 + 2H_2 \rightarrow Si + 4HX$$

or by direct thermal decomposition on a hot wire when X is I. Very pure Si can also be obtained by thermal decomposition of silane, SiH_4.

3.   Pure silicon is then made "superpure" (impurities $<10^{-9}$ at. %) by zone refining. In this process a rod of metal is heated near one end so that a cross-sectional wafer of molten silicon is produced. Since impurities are more soluble in the melt than they are in the solid, they concentrate in the melt, and the melted zone is then caused to move slowly along the rod by moving the heat source. This carries impurities to the end. The process may be repeated. The impure end is then removed.

Superpure germanium is made in a similar way. Germanium chloride is fractionally distilled and then hydrolyzed to GeO_2, which is then reduced with hydrogen. The resulting metal is zone melted.

Tin and lead are obtained from the ores in various ways, commonly by reduction of their oxides with carbon. Further purification is usually effected by dissolving the metals in acid and depositing the pure metals electrolytically.

Silicon is ordinarily rather unreactive. It is attacked by halogens giving tetrahalides, and by alkalis giving solutions of silicates. It is not attacked by acids except hydrofluoric; presumably the stability of SiF_6^{2-} provides the driving force here. A highly reactive form of silicon has been prepared by the reaction

$$3CaSi_2 + 2SbCl_3 \rightarrow 6Si + 2Sb + 3CaCl_2$$

Silicon so prepared reacts with water to give SiO_2 and hydrogen. It has been suggested that it is a graphitelike allotrope, but proof is as yet lacking, and its reactivity may be due to a state of extreme subdivision, as in certain reactive forms of amorphous carbon.

Germanium is somewhat more reactive than silicon and dissolves in concentrated sulfuric and nitric acids. Tin and lead dissolve in several acids and are rapidly attacked by halogens. They are attacked slowly by cold alkali, rapidly by hot, to form stannates and plumbites. Lead often appears to be much more noble and unreactive than would be indicated by its standard potential of -0.13 V. This low reactivity can be attributed to a high overvoltage for hydrogen and also in some cases to insoluble surface coatings. Thus lead is not dissolved by dilute sulfuric and concentrated hydrochloric acids.

12-3. Allotropic Forms

Silicon and germanium are normally isostructural with diamond. By use of very high pressures, denser forms with distorted tetrahedra have been produced. The graphite structure is peculiar to carbon, which is understandable because such a structure requires the formation of $p\pi-p\pi$ bonds.

Tin has two crystallite modifications, with the equilibria

$$\alpha\text{-Sn} \underset{\text{``grey''}}{\overset{18°}{\rightleftharpoons}} \beta\text{-Sn} \underset{\text{``white''}}{\overset{232°}{\rightleftharpoons}} \text{Sn(1)}$$

α-Tin, or gray tin (density at $20° = 5.75$), has the diamond structure. The metallic form, β or white Sn (density at $20° = 7.31$), has a distorted close-packed lattice. The approach to ideal close packing accounts for the considerably greater density of the β-metal compared with the diamond form.

The most metallic of the Group IV elements, lead, exists only in a *ccp*, metallic form. This is a reflection both of its preference for divalence rather than tetravalence and of the relatively low stability of the Pb—Pb bond.

COMPOUNDS OF GROUP IV ELEMENTS

12-4. Hydrides

The Group IV hydrides are listed in Table 12-4; all are colorless.

Silanes. Monosilane (SiH_4) is best prepared on a small scale by heating SiO_2 and $LiAlH_4$ at 150 to 170°. On a larger scale the reduction of SiO_2 or alkali silicates is possible by means of an $NaCl—AlCl_3$ eutectic (m.p. 120°) containing Al metal, or with hydrogen at 400 atm and 175°. The original Stock procedure (cf. boranes, p. 303) of acid hydrolysis of magnesium silicide (prepared by direct interaction of Mg and Si or SiO_2) gives a mixture of silanes. Chlorosilanes may also be reduced by $LiAlH_4$.

Only SiH_4 and Si_2H_6 are indefinitely stable at 25°; the higher silanes decompose giving hydrogen and mono- and disilane, possibly indicating SiH_2 as an intermediate.

The hydridic reactivity of the Si—H bond in silanes and substituted silanes is

TABLE 12-4
Hydrides and Halides of Group IV Elements

Hydrides		Fluorides and chlorides[a]		
MH_4	Other	MF_4	MCl_4	Other
SiH_4 b.p.—112°	$Si_2H_6 \rightarrow Si_6H_{14}$[b,c] b.p.—145°	SiF_4 b.p.—86°	$SiCl_4$ b.p.—57.6°	$Si_2Cl_6 \rightarrow Si_6Cl_{14}$ b.p.—145° $Si_2F_6 \rightarrow Si_{16}F_{34}$ b.p.—18.5°
GeH_4 b.p.—88°	$Ge_2H_6 \rightarrow Ge_9H_{20}$[b] b.p.—29°	GeF_4 m.p.—37°	$GeCl_4$ b.p.83°	Ge_2Cl_6 m.p. 40°
SnH_4 b.p.—52.5°	Sn_2H_6	SnF_4 subl. 704°	$SnCl_4$ b.p. 114.1°	
PbH_4		PbF_4	$PbCl_4$ d. 105°	

[a] All MX_4 compounds except $PbBr_4$ and PbI_4 are known, as well as mixed halides of Si (e.g., SiF_3I, $SiFCl_2Br$) and even $SiFClBrI$; see F. Höfler and W. Veigl, *Angew. Chem., Int. Ed.,* 1971, **10**, 919. Mixed Si-Ge hydrides are also known.

[b] Species and their isomers are separable by gas-liquid chromatography.

[c] Cyclosilanes are also known; see E. Hengge and D. Kovar, *Angew. Chem., Int. Ed.,* 1977, **16**, 403.

similar and may be attributed to charge separation $Si^{\delta+}$—$H^{\delta-}$ that results from the greater electronegativity of H than of Si. Silanes are spontaneously flammable in air, e.g.,

$$Si_4H_{10} + {}^{13}\!/_2\, O_2 \rightarrow 4SiO_2 + 5H_2O$$

Although silanes are stable to water and dilute mineral acids, rapid hydrolysis occurs with bases:

$$Si_2H_6 + (4 + 2n)H_2O \rightarrow 2SiO_2{\cdot}nH_2O + 7H_2$$

The silanes are strong reducing agents. With halogens they react explosively at 25°, but controlled replacement of H by Cl or Br may be effected in presence of AlX_3 to give halogenosilanes such as SiH_3Cl.

Monogermane together with Ge_2H_6 and Ge_3H_8 can be made by heating GeO_2 and $LiAlH_4$ or by addition of $NaBH_4$ to GeO_2 in acid solution. Higher germanes are made by electric discharge in GeH_4. Germanes are less flammable than silanes, although still rapidly oxidized in air, and the higher germanes increasingly so. The germanes are resistant to hydrolysis, and GeH_4 is unaffected by even 30% NaOH.

Stannane (SnH_4) is best obtained by interaction of $SnCl_4$ and $LiAlH_4$ in ether at −30°. It decomposes rapidly when heated and even at 0° where it gives β-tin. Although it is stable to dilute acids and bases, 2.5 M NaOH causes decomposition to Sn and a little stannate. SnH_4 is easily oxidized and can be used to reduce organic compounds (e.g., C_6H_5CHO to $C_6H_5CH_2OH$, and $C_6H_5NO_2$ to $C_6H_5NH_2$). With concentrated acids at low temperatures, the solvated stannonium ion is formed by the reaction

$$SnH_4 + H^+ \rightarrow SnH_3^+ + H_2$$

Plumbane (PbH_4) is said to be formed in traces when magnesium-lead alloys are hydrolyzed by acid or when lead salts are reduced cathodically, but doubt has been expressed about its actual existence.[16]

All the elements form organohydrides R_nMH_{4-n}, and even lead derivatives are stable; they are readily made by reduction of the corresponding chlorides with $LiAlH_4$. There are also a number of compounds of transition metals with silyl groups [e.g., $H_3SiCo(CO)_4$].

Perhaps the most important reaction of compounds with an Si—H bond, such as Cl_3SiH or Me_3SiH, and one that is of commercial importance, is the Speier, or hydrosilation, reaction of alkenes, e.g.:

$$RCH=CH_2 + HSiCl_3 \rightarrow RCH_2CH_2SiCl_3$$

Normally chloroplatinic acid is used as a catalyst. The mechanism of reaction is discussed in Section 30-2.

SiH_4 and GeH_4 react with the alkali metals to form H_3SiM and H_3GeM compounds, which have saltlike structures. For example, with M = K and Rb they have the NaCl structure with the H_3Si and H_3Ge ions rotating or rotationally disordered on their sites.[17]

The unusual reducing properties of $SiHCl_3$ are discussed below.

12-5. Halides

The more important halides are given in Table 12-4.

Fluorides. These compounds, which are of limited utility, are obtained by fluorination of the other halides or by direct interaction; GeF_4 is best made by heating $BaGeF_6$. Tetrafluorides of Si and Ge are hydrolyzed by an excess of water to the hydrous oxides; the main product from SiF_4 and H_2O in the gas phase is $F_3SiOSiF_3$. In an excess of aqueous HF, the hexafluoro anions MF_6^{2-} are formed. SnF_4 is polymeric, with Sn octahedrally coordinated by four bridging and two nonbridging F atoms. PbF_4 is made by action of F_2 on PbF_2; a supposed preparation by the action of HF on $Pb(O_2CMe)_4$ in $CHCl_3$ at 0° gives the much more reactive $Pb(O_2CMe)_2F_2$.

Silicon Chlorides. $SiCl_4$ is made by chlorination of Si at red heat. Si_2Cl_6 can be obtained by interaction of $SiCl_4$ and Si at high temperatures or, along with $SiCl_4$ and higher chlorides, by chlorination of a silicide such as that of calcium. The higher members, which have highly branched structures, can also be obtained by amine-catalyzed reactions such as

$$5Si_2Cl_6 \rightarrow Si_6Cl_{14} + 4SiCl_4$$
$$3Si_3Cl_8 \rightarrow Si_5Cl_{12} + 2Si_2Cl_6$$

and by photolysis of $SiHCl_3$.[18] The products are separated by fractional distillation.

All the chlorides are immediately and completely hydrolyzed by water, but careful hydrolysis of $SiCl_4$ gives $Cl_3SiOSiCl_3$ and $(Cl_3SiO)_2SiCl_2$.

[16]   C. J. Porritt, *Chem. Ind. (London)*, **1975**, 258.
[17]   G. Thirase et al., *Z. Anorg. Allg. Chem.*, 1975, **417**, 221.
[18]   K. G. Sharp et al., *J. Am. Chem. Soc.*, 1975, **97**, 5610.

Hexachlorodisilane (Si_2Cl_6), is a useful reducing agent for compounds with oxygen bound to S, N, or P; under mild conditions, at 25° in $CHCl_3$ chlorooxosilanes are produced. It is particularly useful for converting optically active phosphine oxides $R^1R^2R^3PO$ into the corresponding phosphine. Since the reduction is accompanied by configurational inversion, the intermediacy of a highly nucleophilic $SiCl_3^-$ ion (cf. PCl_3) has been proposed:

$$Si_2Cl_6 + O{=}P\cdots \longrightarrow Cl_3SiOP^+\cdots + SiCl_3^-$$

$$\cdots P + Cl_3SiOSiCl_3 \longleftarrow OSiCl_3^- + \cdots P^+SiCl_3$$

The postulation of $SiCl_3^-$ can also accommodate the equally useful, clean, selective reductions by trichlorosilane (b.p. 33°) and also the formation of $C{=}C$ and $Si{-}C$ bonds by reaction of $SiHCl_3$ with CCl_4, RX, RCOCl, and other halogen compounds in presence of amines. In these cases the hypothetical $SiCl_3^-$ could be generated by the reaction

$$HSiCl_3 + R_3N \rightleftarrows R_3NH^+ + SiCl_3^-$$

followed by

$$SiCl_3^- + Cl_3C{-}CCl_3 \rightarrow SiCl_4 + Cl^- + Cl_2C{=}CCl_2$$
$$SiCl_3^- + RX \rightarrow [R^- + XSiCl_3] \rightarrow RSiCl_3 + X^-$$

There is some precedent for the postulation of the $SiCl_3^-$ ion, since trisubstituted organosilanes (R_3SiH), react with bases to give silyl ions (R_3Si^-).

Chloride Oxides.   A variety of chlorooxosilanes, both linear and cyclic, is known. Thus controlled hydrolysis of $SiCl_4$ with moist ether, or interaction of Cl_2 and O_2 on hot silicon, gives $Cl_3SiO(SiOCl_2)_nSiCl_3$, where $n = 1$ to 4.

Germanium Tetrachloride.   This differs from $SiCl_4$ in that only partial hydrolysis occurs in aqueous 6 to 9 M HCl and there are equilibria involving species of the type $[Ge(OH)_nCl_{6-n}]^{2-}$; the tetrachloride can be distilled and separated from concentrated HCl solutions of GeO_2.

Tin and Lead Tetrachlorides.   These are also hydrolyzed completely only in water and in presence of an excess of acid form chloroanions, as discussed below.

12-6.   Oxygen Compounds of Silicon[19a]

Silica.[19b]   Pure SiO_2 occurs in only two forms, *quartz* and *cristobalite*. The silicon is always tetrahedrally bound to four oxygen atoms, but the bonds have

[19a] W. Eitel, Ed., *Silicate Science*, Vols. 1–6, Academic Press; W. A. Deer, R. A. Howie, and J. Zussman, *Rock Forming Minerals*, Vols. I–V, Longmans; N. V. Belov, *Crystal Chemistry of Large Cation Silicates*, Consultants Bureau, 1963; A. A. Hodgson, *Fibrous Silicates*, Royal Institute of Chemistry (London), Lecture Series No. 4, 1964; M. D. Britten, *Angew. Chem., Int. Ed.*, 1976, **15**, 346 (glass).

[19b] R. K. Iler, *The Chemistry of Silica: Solubility, Polymerisation, Colloid and Surface Properties and Biochemistry*, Wiley, 1979; K. K. Unger, *Porous Silica*, Elsevier, 1979.

considerable ionic character. In cristobalite the Si atoms are placed as are the C atoms in diamond, with the O atoms midway between each pair. In quartz there are helices, so that enantiomorphic crystals occur, and these may be easily recognized and separated mechanically.

The interconversion of quartz and cristobalite on heating requires breaking and re-forming of bonds, and the activation energy is high. However the rates of conversion are profoundly affected by the presence of impurities, or by the introduction of alkali metal oxides or other "mineralizers." Studies of the system have shown that what was believed to be another form of quartz, tridymite, is a solid solution of mineralizer and silica.

Slow cooling of molten silica or heating any form of solid silica to the softening temperature gives an amorphous material that is glassy in appearance and is indeed a glass in the general sense, that is, a material with no long-range order but rather a disordered array of polymeric chains, sheets or three-dimensional units.

Dense forms of SiO_2, called coesite and stishovite, were first made under drastic conditions (250–1300° at 35–120 k atm), but they were subsequently identified in meteor craters where the impact conditions were presumably similar; stishovite has the rutile structure.[20] Both are chemically more inert than normal SiO_2 to which they revert on heating.

Silica is relatively unreactive towards Cl_2, H_2, acids, and most metals at ordinary or slightly elevated temperatures, but it is attacked by fluorine, aqueous HF, alkali hydroxides, fused carbonates, etc.

Silicates.   When alkali metal carbonates are fused with silica ($\sim$1300°), CO_2 is driven off and a complex mixture of alkali silicates is obtained. When the mixtures are rich in alkali, the products are soluble in water, but with low alkali contents they become quite insoluble. Presumably the latter contain very large, polymeric anions. Aqueous sodium silicate solutions appear to contain the ion $[SiO_2(OH)_2]^{2-}$ according to Raman spectra but, depending on the pH and concentration, tetrameric and other polymerized species are also present.

Most of our understanding of silicate structures comes from studies of the many naturally occurring (and some synthetic) silicates of heavier metals. The basic unit of structure is the SiO_4 tetrahedron. These tetrahedra occur singly or, by sharing oxygen atoms, in small groups, in small cyclic groups, in infinite chains or in infinite sheets.

Simple Orthosilicates.   A few silicates are known in which there are simple, discrete orthosilicate, (SiO_4^{4-}) anions. In such compounds the associated cations are coordinated by the oxygen atoms, and various structures are found depending on the coordination number of the cation. In phenacite (Be_2SiO_4) and willemite (Zn_2SiO_4) the cations are surrounded by a tetrahedrally arranged set of four oxygen atoms. There are a number of compounds of the type M_2SiO_4, where M^{2+} is Mg^{2+}, Fe^{2+}, Mn^{2+}, or some other cation with a preferred coordination number of 6, in which the SiO_4^{4-} anions are so arranged as to provide interstices with six oxygen atoms at the apices of an octahedron in which the cations are found. In zircon

20   W. H. Baur and A. A. Kahn, *Acta Crystallogr. B*, 1971, **27**, 2133.

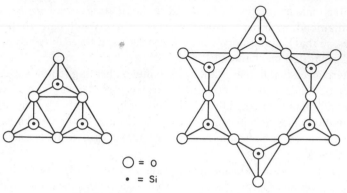

$\bigcirc$ = O
$\bullet$ = Si

Fig. 12-2. Examples of cyclic silicate anions.

($ZrSiO_4$) the Zr^{4+} ion is eight-coordinate, although not all Zr—O distances are equal. It may be noted that, although the M—O bonds are probably more ionic than the Si—O bonds, there is doubtless some covalent character to them, and these substances should not be regarded as literally ionic in the sense $[M^{2+}]_2[SiO_4^{4-}]$ but rather as somewhere between this extreme and the opposite one of giant molecules. There are also other silicates containing discrete SiO_4 tetrahedra.

Other Discrete, Noncyclic Silicate Anions. The simplest of the condensed silicate anions—that is, those formed by combining two or more SiO_4 tetrahedra by sharing of oxygen atoms—is the pyrosilicate ion $Si_2O_7^{6-}$. This ion occurs in thortveitite ($Sc_2Si_2O_7$), hemimorphite [$Zn_4(OH)_2Si_2O_7$], and in at least three other minerals. It is interesting that the Si—O—Si angle varies from 131° to 180° in these substances. In the rare earth silicates $M_2O_3 \cdot 2SiO_2$, there are $O_3SiO-Si(O)_2OSiO_3^{6-}$ ions.[21a]

Cyclic Silicate Anions. The structures of two such cyclic ions, $Si_3O_9^{6-}$ and $Si_6O_{18}^{12-}$, are shown schematically in Fig. 12-2. It should be clear that the general formula for any such ion must be $Si_nO_{3n}^{2n-}$. The ion $Si_3O_9^{6-}$ occurs in benitoite ($BaTiSi_3O_9$) and probably in $Ca_2BaSi_3O_9$. The ion $Si_6O_{18}^{12-}$ occurs in beryl ($Be_3-Al_2Si_6O_{18}$).

Infinite Chain Anions. These are of two main types: the *pyroxenes*, which contain single-strand chains of composition $(SiO_3^{2-})_n$ (Fig. 12-3) and the *amphiboles*, which contain double-strand, cross-linked chains or bands of composition $(Si_4O_{11}^{6-})_n$. Note that the general formula of the anion in a pyroxene is the same as in a silicate with a cyclic anion. Silicates with this general stoichiometry are often called "metasilicates," especially in older literature. There is actually neither metasilicic acid nor any discrete metasilicate anion. With the exception of the few "metasilicates" with cyclic anions, such compounds contain infinite chain anions.

Examples of pyroxenes are enstatite ($MgSiO_3$), diopside [$CaMg(SiO_3)_2$], and spodumene [$LiAl(SiO_3)_2$], the last being an important lithium ore. In the lithium compound there is one unipositive and one tripositive cation instead of two dipositive

[21a] J. Felsche, *Naturwissenschaften*, 1972, **59**, 35.

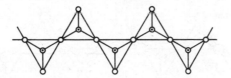

Fig. 12-3. A linear chain silicate anion.

cations. Indeed, the three compounds cited illustrate very well the important principle that within rather wide limits, *the specific cations or even their charges are unimportant as long as the total positive charge is sufficient to produce electroneutrality*. This may be easily understood in terms of the structure of the pyroxenes in which the $(SiO_3)_n$ chains lie parallel and are held together by the cations that lie between them. Obviously the exact identity of the individual cations is of minor importance in such a structure.

A typical amphibole is tremolite, $Ca_2Mg_5(Si_4O_{11})_2(OH)_2$. Although it would not seem to be absolutely necessary, amphiboles apparently always contain some hydroxyl groups attached to the cations. Aside from this, however, they are structurally similar to the pyroxenes, in that the $(Si_4O_{11}^{6-})_n$ bands lie parallel and are held together by the metal ions lying between them. Like the pyroxenes and for the same reason, they are subject to some variability in the particular cations incorporated.

Because of the strength of the $(SiO_3)_n$ and $(Si_4O_{11})_n$ chains in the pyroxenes and amphiboles, and also because of the relative weakness and lack of strong directional properties in the essentially electrostatic forces between them via the metal ions, we might expect such substances to cleave most readily in directions parallel to the chains. This is in fact the case, dramatically so in the various asbestos minerals,[21b] which are all amphiboles.

Infinite Sheet Anions. When SiO_4 tetrahedra are linked into infinite two-dimensional networks as shown in Fig. 12-4, the empirical formula for the anion is $(Si_2O_5^{2-})_n$. Many silicates have such sheet structures with the sheets bound together by the cations that lie between them. Such substances might thus be expected to cleave readily into thin sheets, and this expectation is confirmed in the micas, which are silicates of this type.

Framework Minerals. The next logical extension in the progression above from simple SiO_4^{4-} ions to larger and more complex structures would be to three-dimensional structures in which every oxygen is shared between two tetrahedra. The empirical formula for such a substance would be simply $(SiO_2)_n$; that is, we should have silica. However, if some silicon atoms in such a three-dimensional framework structure are replaced by aluminum the framework must be negatively charged and there must be other cations uniformly distributed through it. Aluminosilicates of this type are the feldspars, zeolites, and ultramarines, which (except for the last) are among the most widespread, diverse, and useful silicate minerals in Nature. Moreover, many synthetic zeolites have been made in the laboratory, and have

[21b] L. Michaels and S. S. Chissick, Eds., *Asbestos*, Vol. 1, *Properties, Applications, and Hazards,* Wiley, 1979.

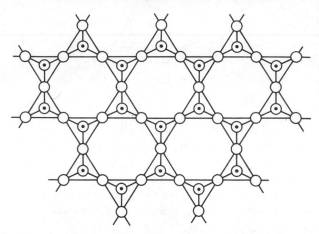

Fig. 12-4. Sheet silicate anion structure idealized. [For a real example, see A. K. Pant, *Acta Crystallogr., B,* 1968, **24,** 1077.]

important uses (see below). The feldspars are the major constituents of igneous rocks and include such minerals as orthoclase ($KAlSi_3O_8$), which may be written $K[(AlO_2)(SiO_2)_3]$ to indicate that one-fourth of the oxygen tetrahedra are occupied by Al atoms, and anorthite ($CaAl_2Si_2O_8$ or $Ca[(AlO_2)_2(SiO_2)_2]$), in which half the tetrahedra are AlO_4 and half SiO_4.

A semiprecious deep blue gem called lapis lazuli has been known from ancient times and is available in synthetic forms under the name "ultramarine." These are aluminosilicates of the sodalite type that contain sulfur in the form of the radical anions S_3^- and S_2^-. The former, always present, causes a deep blue color, and when S_2^- is also present a green hue is produced.[22]

12-7. Zeolites[23]

The zeolites are the most important framework silicates. A zeolite may be defined as an aluminosilicate with a framework structure enclosing cavities occupied by large ions and water molecules, both of which have considerable freedom of movement, permitting ion exchange and reversible dehydration. The framework consists of an open $(Al,SiO_2)_\infty$ arrangement of corner-sharing tetrahedra plus enough cations to give electroneutrality and many water molecules to occupy the cavities. Some typical cavities occurring in zeolites are shown in Figs. 12-5 and 12-6. Typical formulas for naturally occurring zeolites are $Ca_6Al_{12}Si_{24}O_{72}\cdot40H_2O$ for

22 F. A. Cotton, J. B. Harmon, and R. M. Hedges, *J. Am. Chem. Soc.,* 1976, **98,** 1417; R. J. H. Clark and D. J. Cobbold, *Inorg. Chem.,* 1978, **11,** 3169.
23 D. W. Breck, *Zeolite Molecular Sieves,* Wiley, 1974; W. M. Meier and J. B. Uytterhoeven, Eds., *Molecular Sieves,* Advances in Chemistry Series, ACS Monograph No. 121, 1973; J. A. Rabo, Ed., *Zeolite Chemistry and Catalysis,* ACS Monograph No. 171, 1976; P. A. Jacobs, *Carboniogenic Activity of Zeolites,* Elsevier, 1977; J. R. Kratzer, Ed., *Molecular Sieves,* ACS Symposium Series, No. 40, 1977; R. M. Barrer, *Zeolites and Clay Minerals,* Academic Press, 1978; L. B. Sand and F. A. Mumpton, *Natural Zeolites, Occurrence, Properties, and Uses,* Pergamon Press, 1978.

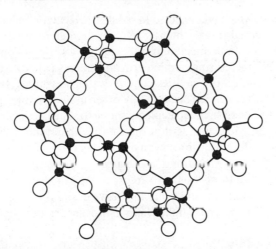

Fig. 12-5. The arrangement of AlO_4 and SiO_4 tetrahedra that gives the cubooctahedral cavity in some zeolites and felspathoids; dot represents Si or Al.

the mineral chabazite and $Na_{13}Ca_{11}Mg_9K_2Al_{55}Si_{137}O_{384} \cdot 235H_2O$ for faujasite. The latter gives one an idea of the extreme variability possible in the cation composition. Many of the natural zeolites and a number of synthetic ones can be made by hydrothermal synthesis. The three main uses that zeolites have, or have had, are as (1) ion exchangers, (2) molecular sieves, (3a) catalysts, (3b) catalyst supports.

Ion Exchange. Zeolites are now seldom used in this way, having been largely replaced by synthetic ion-exchange resins. However there are still a few specialized uses such as removal of ammonia from wastewater and in treatment of radioactive wastes.

Molecular Sieves. To obtain a molecular sieve, the water of hydration is driven out by heating to about 350° in vacuum. In a typical case, $Na_{12}[Al_{12}Si_{12}O_{48}] \cdot 27H_2O$, a synthetic called A zeolite, one is left with the anhydrous cubic microcrystals, in which the AlO_4 and SiO_4 tetrahedra are linked together to form a ring of eight oxygen atoms on each face of the unit cube and an irregular ring of six oxygen atoms across each corner. In the center of the unit cell is a large cavity about

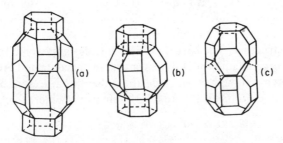

Fig. 12-6. Cavities of different dimensions in (a) chabazite $(Ca_6Al_{12}Si_{24}O_{72} \cdot H_2O)$, (b) gmelinite $[Na_2Ca)_4Al_8Si_{16}O_{48} \cdot 24H_2O]$, and (c) erionite, $(Ca_{4.5}Al_9Si_{27}O_{72} \cdot 27H_2O]$.

11.4 Å in diameter, which is connected to six identical cavities in adjacent unit cells by the eight-membered rings, which have inner diameters of about 4.2 Å. In addition, the large cavity is connected to eight smaller cavities, about 6.6 Å in diameter, by the six-membered rings, which provide openings about 2.0 Å in diameter. In the hydrated form all the cavities contain water molecules. In the anhydrous state the same cavities may be occupied by other molecules brought into contact with the zeolite, provided such molecules are able to squeeze through the apertures connecting cavities. Molecules within the cavities then tend to be held there by attractive forces of electrostatic and van der Waals types. Thus the zeolite will be able to absorb and strongly retain molecules just small enough to enter the cavities. It will not absorb at all those too big to enter, and it will absorb weakly very small molecules or atoms that can enter but also leave easily. For example, the zeolite under discussion will absorb straight-chain hydrocarbons but not branched-chain or aromatic ones.

Catalysis.   Dehydrated zeolites possess a wide spectrum of potent catalytic activity, and some of their capabilities may be enhanced and modified by deliberate introduction and variation of cations other than the alkali and alkaline earth cations initially present.

Catalytic cracking, hydroisomerization, and re-forming of petroleum to produce high-octane fuel are the principal catalytic uses of zeolites.[24] In 1969 90% of the refiners in the United States were using zeolite-based cracking catalysts and in the U.S.S.R. zeolite catalysts have replaced all amorphous silica-alumina catalysts. The practical zeolite catalysts vary considerably in structure and composition. Both faujasite and mordenite structures, with Ca^{2+}, tripositive lanthanide, or H^+ ions are used for cracking, and others containing highly dispersed noble metals are useful for isomerization and re-forming. High acidity, either protonic or Lewis, is a key factor in creating carbonium ions that can rearrange to produce the lower molecular weight, branched hydrocarbons having the required combustion properties.

New zeolite catalysts useful in many commercially significant reactions, such as conversion of methanol to gasoline, are still being developed.[25]

12-8.   Oxygen Compounds of Germanium, Tin, and Lead

Oxides and Hydroxides.   The oxides GeO_2, SnO_2, and PbO_2 are all well-characterized compounds. GeO_2 is known in the quartz, stishovite, and crystobalite[26] structures analogous to those of SiO_2. SnO_2 exists in three different modifications of which the rutile form (in the mineral cassiterite) is most common; PbO_2 shows only the rutile structure. The basicity of the dioxides appears to increase from Si to Pb. SiO_2 is purely acidic, GeO_2 is less so and in concentrated HCl gives $GeCl_4$, SnO_2 is amphoteric, though when made at high temperatures, or by dissolving Sn in hot concentrated HNO_3 it is, like PbO_2, remarkably inert to chemical attack.

[24]   *Hydrocarbon Processing,* Refining Handbook Issue, September 1978.
[25]   G. T. Kokotailo *et al., Nature (London),* 1978, **272,** 437, **275,** 119; see also Section 30-7, Ref. 58.
[26]   E. Hauser *et al., Monatsh.* 1970, **101,** 715; 1971, **102,** 1006.

There is little evidence that there are true hydroxides, $M(OH)_4$, and the products obtained by hydrolysis of hydrides, halides, alcoxides, and so on, are best regarded as hydrous oxides. Thus the addition of OH^- to Sn^{IV} solutions gives a white gelatinous precipitate that when heated is dehydrated through various intermediates and gives SnO_2 at $600°$.

Oxo Anions. The known chemistry of germanates is much less extensive than that of silicates. Although there are many similarities, there are some structural differences because Ge more readily accepts higher coordination numbers. In $K_2Ge_8O_{17}$, for example, six of the Ge atoms are in GeO_4 tetrahedra but two are five-coordinate,[27] as silicon never is under normal conditions. Among compounds homologous to their silicon analogues there are metagermanates and orthogermanates that have been obtained in crystalline form with structures analogous to the corresponding meta- and orthosilicates. Thus $SrGeO_3$ contains a cyclic $Ge_3O_9^{6-}$ ion. Germanates containing the $Ge(OH)_6^{2-}$ ion are also known. In dilute aqueous solutions the major germanate ions appear to be $[GeO(OH)_3]^-$, $[GeO_2(OH)_2]^{2-}$, and $\{[Ge(OH)_4]_8(OH)_3\}^{3-}$. Fusion of SnO_2 or PbO_2 with K_2O gives K_2MO_3, which has chains of edge-shared MO_5 square pyramids. Crystalline alkali metal stannates and plumbates can be obtained as trihydrates, for instance, $K_2SnO_3 \cdot 3H_2O$. Such materials contain the octahedral anions $Sn(OH)_6^{2-}$ and $Pb(OH)_6^{2-}$.

12-9. Complexes of Group IV Elements

Most of the complexes of germanium, tin, and lead in the IV oxidation state contain halide ions or donor ligands that are oxygen, nitrogen, sulfur, or phosphorus compounds.

Anionic Species. Silicon forms only fluoro anions, normally SiF_6^{2-}, whose high stability accounts for the incomplete hydrolysis of SiF_4 in water:

$$2SiF_4 + 2H_2O \rightarrow SiO_2 + SiF_6^{2-} + 2H^+ + 2HF$$

The ion is usually made by attack of HF on hydrous silica and is stable even in basic solution. Although the salts that crystallize are normally those of the SiF_6^{2-} ion, the pentafluorosilicate ion is found in compounds such as $[Ph_4As][SiF_5]$. Typical reactions producing SiF_5^- salts are

$$SiO_2 + HF(aq) + R_4N^+Cl \xrightarrow{CH_3OH} [R_4N]SiF_5$$

$$SiF_4 + [R_4N]F \xrightarrow{CH_3OH} [R_4N]SiF_5$$

Nmr data for the ion and also for similar species $RSiF_4^-$ and $R_2SiF_3^-$ indicate *tbp* structures, but above $-60°$ exchange processes are occurring.

Germanium, tin, and lead also form hexafluoro anions; for example, dissolution of GeO_2 in aqueous HF followed by the addition of KF at $0°$ gives crystals of K_2GeF_6. The Ge and Sn anions are hydrolyzed by bases, but the Pb salts are hy-

[27] E. Fay, H. Vollenke and A. Wittmann, *Z. Kristallogr.,* 1974, **138,** 439.

drolyzed even by water. A great variety of tin species $SnF_{6-n}X_n^{2-}$ have been studied by nmr spectroscopy, and the equilibrium constants have been estimated, e.g.,

$$SnF_6^{2-} + H_2O \rightleftharpoons SnF_5(OH_2)^- + F^- \qquad K = 2.3 \times 10^{-6}$$
$$SnF_6^{2-} + OH^- \rightleftharpoons SnF_5(OH)^{2-} + F^- \qquad K = 7.7 \times 10^6$$

Anhydrous hexafluorostannates can be made by dry fluorination of the stannates, $M_2^ISnO_3 \cdot 3H_2O$.

The hexachloro ions of germanium and tin are normally made by the action of HCl or M^ICl on MCl_4. The thermally unstable yellow salts of $PbCl_6^{2-}$ are obtained by action of HCl and Cl_2 on $PbCl_2$. Under certain conditions, pentachloro complexes of Ge and Sn may be stabilized, for example, by the use of $(C_6H_5)_3C^+$ as the cation or by the interaction of MCl_4 and $(C_4H_9)_4N^+Cl^-$ in $SOCl_2$ solution.

Other anionic species include the ions $[Sn(NO_3)_6]^{2-}$ and also thiostannate, written SnS_3^{2-} but of uncertain structure, obtained by dissolving SnS_2 in alkali or ammonium sulfide solution. The most extensive series are the oxalates $[Mox_3]^{2-}$ (where M = Si, Ge, or Sn) and other carboxylates.

Cationic Species. There are comparatively few cationic complexes, the most important being the octahedral β-diketonates and tropolonates (T) of Si and Ge such as $[Ge\ acac_3]^+$ and SiT_3^+. "Siliconium" ions can also be formed by reactions such as

$$Ph_3SiX + bipy \xrightarrow{CH_2Cl_2} \left[\begin{array}{c} Ph \\ \diagdown \\ Ph \end{array} \begin{array}{c} N \diagdown \\ Si - N \\ \diagup \\ Ph \end{array} \right]^+ + X^-$$

or by oxidation[28a] of $SiCl_2(bipy)_2$ to cis-$[SiCl_2(bipy)_2]^{2+}$.[28b]

Neutral Species; Adducts. These are numerous and quite varied in type. The majority are six-coordinate, examples being $trans$-$SnCl_2(\beta$-dike$)_2$,[29] $SnCl_2(S_2CNEt_2)_2$, and $Sn[(OC_2H_4)_2N(C_2H_4OH)]_2$.[30] Both lower and higher coordination numbers also occur, examples being 5 in $PhSi(o$-$C_6H_4O_2)_2$ and presumably 7 or 8 in $Sn(S_2CNEt_2)_4$.

The tetrahalides are prone to add additional ligands to form adducts that are usually six-coordinate. Typical examples are $trans$-SiF_4py_2, cis-SiF_4bipy, $SiCl_4L_2$ (L = py, PMe_3), and numerous cis-$SnX_2(L$—$L)$ and $trans$-SnX_4L_2 compounds,[31,32] of which $trans$-$SnCl_4(PEt_3)_2$ is characterized by X-ray crystallography.[33] The tin halides add neutral Hacac to give $SnCl_4(Hacac)$, which base converts to $SnCl_4(acac)^-$.[34] $SnCl_4$ is a considerably stronger Lewis acid than $SnBr_4$ or SnI_4. The germanium halides form complexes similar to those of silicon and tin.

[28a] D. Kummer and T. Seshadri, *Angew. Chem. Int. Ed.*, 1975, **14**, 699.
[28b] G. Sawitzky *et al.*, *Chem. Ber.*, 1978, **111**, 3705.
[29] G. A. Miller and E. O. Schlemper, *Inorg. Chem.*, 1973, **12**, 667.
[30] H. Follner, *Monatsh.*, 1972, **103**, 1438.
[31] N. Ohkaku and K. Nakamoto, *Inorg. Chem.*, 1973, **12**, 2440, 2446.
[32] P. G. Harrison, B. C. Lane, and J. J. Zuckerman, *Inorg. Chem.*, 1972, **11**, 1537.
[33] G. G. Mather, G. M. McLaughlin, and A. Pidcock, *J. C. S. Dalton*, 1973, 1823.
[34] D. W. Thompson, J. F. Lefelhocz, and K. S. Wong, *Inorg. Chem.*, 1972, **11**, 1139.

12-10.   Other Compounds

Alcoxides, Carboxylates, and Oxo Salts.   All four elements of Group IV form alcoxides, but those of silicon, e.g., $Si(OC_2H_5)_4$, are the most important; the surface of glass or silica can also be alcoxylated. Alcoxides are normally obtained by the standard method:

$$MCl_4 + 4ROH + 4\ amine \rightarrow M(OR)_4 + 4\ amine \cdot HCl$$

Silicon alcoxides are hydrolyzed by water, eventually to hydrous silica, but polymeric hydroxo alcoxo intermediates occur.

Of the carboxylates, lead tetraacetate is the most important because it is used in organic chemistry[35] as a strong but selective oxidizing agent. It is made by dissolving Pb_3O_4 in hot glacial acetic acid or by electrolytic oxidation of Pb^{II} in acetic acid. In oxidations the attacking species is generally considered to be $Pb(OOCMe)_3^+$, which is isoelectronic with the similar oxidant $Tl(OOCMe)_3$, but this is not always so, and some oxidations are known to be free radical in nature. The trifluoroacetate is a white solid, which will oxidize even heptane to give CF_3CO_2R species, from which the alcohol ROH is obtained by hydrolysis; benzene similarly gives phenol.

The tetraacetates of Si, Ge, Sn, and Pb also form complex anions such as $[Pb(O_2CMe)_6]^{2-}$ or $[Sn(O_2CMe)_5]^-$. For $M(O_2CMe)_4$, Si and Ge are four-coordinate with unidentate acetate; Pb has only bidentate acetates, whereas the smaller Sn has a very distorted dodecahedron.[36]

Oxo Salts.   These are few. Tin(IV) sulfate, $Sn(SO_4)_2 \cdot 2H_2O$, can be crystallized from solutions obtained by oxidation of Sn^{II} sulfate; it is extensively hydrolyzed in water.

Tin(IV) nitrate is obtained as a colorless volatile solid by interaction of N_2O_5 and $SnCl_4$; it contains bidentate NO_3^- groups giving dodecahedral coordination. The compound reacts with organic matter.

Sulfides.   Lead disulfide is not known, but for the other elements direct interaction of the elements gives MS_2. The silicon and germanium compounds are colorless crystals hydrolyzed by water. The structures of SiS_2 and GeS_2 are chains of tetrahedral MS_4 linked by the sulfur atoms. SnS_2 has a CaI_2 lattice, each Sn atom having six sulfur neighbors.

Silicon—Nitrogen Compounds.   There is a very extensive chemistry of compounds with nitrogen bound to silicon. The amide $Si(NH_2)_4$ is made by action of NH_3 on $SiCl_4$; on being heated, it gives an imide and finally the nitride Si_3N_4.

12-11.   The Divalent State[37a]

Silicon.[37b]   Divalent silicon species are thermodynamically unstable under normal conditions. However SiX_2 species have been identified in high-temperature

[35]   R. N. Butler, in *Synthetic Reagents,* Vol. 3, J. S. Pizey, Ed., Wiley, 1977.

[36]   N. W. Alcock and V. L. Tracy, *Acta Crystallogr., B,* 1979, **35,** 80.

[37a]   P. G. Harrison, *Coord. Chem. Rev.,* 1976, **20,** 1 (structural chemistry of divalent Ge, Sn, and Pb).

[37b]   H. Bürger and R. Eugen, *Topics Curr. Chem.,* 1974, No. 50, p. 1.

reactions and have been trapped by rapid chilling to liquid nitrogen temperature. The best studied compound[38] is SiF_2, but SiO, SiS, SiH_2, $SiCl_2$, and some other species are known.

At *ca.* 1150° and low pressures SiF_4 and Si react to give SiF_2 in *ca.* 50% yield:

$$SiF_4 + Si \rightleftharpoons 2SiF_2$$

The compound is stable for a few minutes at 10^{-4} cm pressure, whereas CF_2 has $t_{1/2} \approx 1$ sec and GeF_2 is a stable solid at room temperature. It is diamagnetic and the molecule is angular, with a bond angle of 101° both in the vapor and in the condensed phase. The reddish-brown solid gives an esr spectrum and presumably contains also $\cdot SiF_2(SiF_2)_n SiF_2 \cdot$ radicals. When warmed it becomes white, cracking to give fluorosilanes up to $Si_{16}F_{34}$.

In the gas phase, SiF_2 reacts with oxygen but is otherwise not very reactive, but allowing the solid to warm in presence of various compounds (e.g., CF_3I, H_2S, GeH_4, or H_2O) gives insertion products such as $SiF_2H(OH)$ and H_3GeSiF_2H. The corresponding chloride exists for only milliseconds at 10^{-4} cm pressure, since it readily reacts with an excess of $SiCl_4$. By reaction with other halides, (e.g., BCl_3), mixed compounds such as Cl_3SiBCl_2 can be prepared.

There are a number of ways to generate transient $(CH_3)_2Si$, dimethylsilylene, and the reactions of this compound have been studied.[39]

Germanium.   The germanium dihalides are quite stable. GeF_2, a white crystalline solid (m.p. 111°), is formed by action of anhydrous HF on Ge in a bomb at 200° or by reaction of Ge and GeF_4 above 100°. It is a fluorine-bridged polymer, the Ge atom having a distorted trigonal-bipyramid arrangement of four atoms and an equatorial lone pair. The compound reacts exothermically with solutions of alkali metal fluorides to give the hydrolytically stable ion GeF_3^-; in fluoride solutions the ion is oxidized by air, and in strong acid solutions by H^+, to give GeF_6^{2-}. GeF_2 vapors contain oligomers $(GeF_2)_n$ where $n = 1$ to 3.

The other dihalides are less stable than GeF_2 and similar to each other. They can be prepared[40] by the reactions

$$Ge + GeX_4 \rightarrow 2GeX_2$$

In the gas phase or isolated in noble gas matrices,[41] they are bent with angles of 90 to 100°. The solids react to complete their octets (e.g., with donors,[42,43] to produce pyramidal $LGeX_2$ molecules), or with butadiene:

$$GeI_2 + R_3P \longrightarrow R_3PGeI_2$$

---

[38]   D. L. Perry *et al., Inorg. Chem.,* 1978, **17**, 1364, and references therein.
[39]   D. Seyferth and D. C. Annarelli, *J. Am. Chem. Soc.,* 1975, **97**, 7162.
[40]   M. D. Curtis and P. Wolber, *Inorg. Chem.,* 1972, **11**, 431.
[41]   W. A. Guillory *et al., J. Chem. Phys.,* 1972, **56**, 1423; **57**, 1116.
[42]   J. Escudie *et al., J. Organomet. Chem.,* 1977, **124**, C45.
[43]   P. Jützi *et al., Angew. Chem., Int. Ed.,* 1973, **12**, 1002.

Other divalent germanium compounds include salts of $GeCl_3^-$, the sulfide GeS, and a white to yellow hydroxide of no definite stoichiometry that is converted by NaOH to a brown material that has Ge—H bonds.[44]

Germanium[45] and tin[46] form surprisingly stable, diamagnetic β-diketonato complexes such as $M(acac)_2$ and $M(acac)X$, where M is Ge, Sn, and X is Cl or I. The $M(dike)_2$ types can be distilled or sublimed; they are soluble and monomeric in benzene and other hydrocarbon solvents.

Germanium, like tin and lead, forms an $(\eta\text{-}C_5H_5)_2M$ molecule, first reported[47a] in 1973. It also forms GeR_2 compounds with very bulky R groups and analogous to those of tin that are discussed below; there are also the $M[N(SiMe_3)_2]_2$ (M = Ge, Sn, Pb) compounds.[47b]

Tin. The fluoride and chloride are obtained by reaction of tin with gaseous HF or HCl. $SnBr_2$ is obtained by dissolving tin in aqueous HBr, distilling off constant boiling HBr/H_2O, and cooling.[48] The tin atoms are nine- and eight-coordinate in $SnCl_2$ and $SnBr_2$, respectively. SnF_2 has a unique structure with an eight-membered ring of alternating Sn and F atoms and one terminal F on each trigonal pyramidal Sn atom.[49] Water hydrolyzes the halides, but they dissolve in solutions containing excess halide ion to give SnX_3^- ions. With the fluoride, SnF^+ and $Sn_2F_5^-$ can also be detected, and the latter is known in crystalline salts. SnF_2 forms $[SnF]^+[SbF_6]^-$, and similar compounds with F^- acceptors. Tin(II) fluoride is used in toothpastes, presumably as a source of fluoride to harden dental enamel.

The halides readily dissolve in donor solvents such as acetone, pyridine, or DMSO, and pyramidal adducts SnX_2L are formed. The lone pair in $SnCl_2L$ or $SnCl_3^-$ can be utilized, and numerous transition metal complexes with tin(II) chloride as ligand are known (Section 4-8); Mössbauer studies suggest that such species are best regarded as Sn^{IV} complexes, however. The oxidation on complex formation is more clearly illustrated by the "carbenelike" reactions of $SnCl_2$ with metal-metal bonds, where an "insertion reaction" (Section 30-7) occurs, e.g.,

$$[\eta^5\text{-}C_5H_5(CO)_2Fe]_2 + Sn^{II}Cl_2 \rightarrow \eta^5\text{-}C_5H_5(CO)_2Fe\text{—}Sn^{IV}Cl_2\text{—}Fe(CO)_2\eta^5\text{-}C_5H_5$$

The very air-sensitive tin(II) ion Sn^{2+} occurs in acid perchlorate solutions, which may be obtained by the reaction

$$Cu(ClO_4)_2 + Sn/Hg = Cu + Sn^{2+} + 2ClO_4^-$$

Hydrolysis gives the following main reactions[50] at pH 2.7 to 3.7 in $3M$ $NaClO_4$:

$$Sn^{2+} + H_2O = Sn(OH)^+ + H^+ \qquad \log K = -3.70 \pm 0.02$$
$$3Sn^{2+} + 4H_2O = Sn_3(OH)_4^{2+} + 4H^+ \qquad \log K = -6.8 \pm 0.03$$

[44]   D. J. Yang, W. L. Jolly, and A. O'Keefe, *Inorg. Chem.*, 1977, **16**, 2980.

[45]   A. Rogers and S. R. Stobart, *J. C. S. Chem. Comm.*, 1976, 52.

[46]   P. F. R. Ewings, P. G. Harrison, and D. E. Fenton, *J. C. S. Dalton*, **1975**, 821.

[47a]   J. V. Scibelli and M. D. Curtis, *J. Am. Chem. Soc.*, 1973, **95**, 924.

[47b]   M. F. Lappert *et al., J. C. S. Chem. Comm.*, **1979**, 369.

[48]   J. Anderson, *Acta Chem. Scand.*, 1975, **A29**, 956.

[49]   R. C. McDonald, H. H. Hau, and K. Eriks, *Inorg. Chem.*, 1976, **15**, 762.

[50]   S. Gobom, *Acta. Chem. Scand.*, 1976, **A30**, 745.

The trimeric, probably cyclic, ion appears to provide the nucleus of several basic tin(II) salts obtained from aqueous solutions at fairly low pH. Thus the nitrate appears to be $Sn_3(OH)_4(NO_3)_2$ and the sulfate $Sn_3(OH)_2OSO_4$, which contains the cyclic $[Sn_3O(OH)_2]^{2+}$ ion.[51] All Sn^{II} solutions are readily oxidized by oxygen and, unless stringently protected from air, normally contain some Sn^{IV}. The chloride solutions are often used as mild reducing agents:

$$SnCl_6^{2-} + 2e = SnCl_3^- + 3Cl^-   \qquad E^0 = ca.\ 0.0\ V\ (1M\ HCl,\ 4\ M\ Cl^-)$$

The addition of aqueous ammonia to Sn^{II} solutions gives the white hydrous oxide that is dehydrated to black SnO when heated in suspension at 60 to 70° in 2 M NH_4OH; when heated at 90 to 100° in presence of hypophosphite, it is dehydrated to a red modification. The hydrous oxide obtained as just described and in other ways from aqueous solution is not a true hydroxide. It has the composition $3SnO \cdot H_2O$, and the structure contains $Sn_6O_4(OH)_4$ units with all eight oxygen atoms joined by H bonds in a regular cube superimposed on an octahedron of tin atoms. This structure is derived from two of the cyclic $Sn_3(OH)_4^{2+}$ cations by condensation and loss of $4H^+$. What appears to be a true tin(II) hydroxide has been made by the anhydrous reaction

$$2R_3SnOH + SnCl_2 \rightarrow Sn(OH)_2 + 2R_3SnCl$$

as a white, amorphous solid.[52] The hydrous oxide is amphoteric and dissolves in alkali hydroxide to give solutions of stannites, which may contain the ion $[Sn(OH)_6]^{4-}$. These solutions are quite strong reducing agents; on storage they deposit SnO, and at 70 to 100° they disproportionate slowly to β-tin and Sn^{IV}. The only characterized oxostannite is the deep yellow $K_2Sn_2O_3$ made by direct interaction at 550°; it has a perovskite structure with half the anions missing.[53]

A large number of other Sn^{II} compounds are known, including carboxylates[54] and carboxylato anion complexes, a perchlorate, thiocyanate, and phosphite. The methoxide $Sn(OMe)_2$ is a useful synthetic reagent in tin(II) chemistry.[55] The dithiocarbamate $Sn(S_2CNEt_2)_2$ has chelate groups and a lone pair.[56]

The bulky dialkylamides have been noted above, but $Sn(NEt_2)_2$, a white crystalline solid, is dimeric in solution,[57a] like $[Sn(OBu^t)_2]_2$, probably with the structure 12-V. Associated or chelate compounds are white; monomeric dialkylamides are red.[57b]

(12-V)

[51]   C. G. Davies et al., J. C. S. Dalton, 1975, 2241.
[52]   W. D. Honnick and J. J. Zuckerman, Inorg. Chem., 1976, 15, 3034.
[53]   R. M. Braun and R. Hoppe, Angew. Chem., Int. Ed., 1978, 17, 449.
[54]   P. G. Harrison and E. W. Thornton, J. C. S. Dalton, 1978, 1274.
[55]   W. D. Honnick and J. J. Zuckerman, Inorg. Chem., 1978, 17, 501.
[56]   P. F. R. Ewings, P. G. Harrison, and T. J. King, J. C. S. Dalton, 1976, 1399.
[57a] W. Petz, J. Organomet. Chem., 1979, 165, 199.
[57b] P. J. Corvan and J. J. Zuckerman, Inorg. Chim. Acta, 1979, 34, L255.

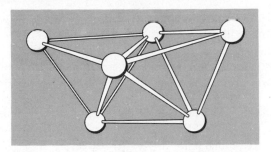

Fig. 12-7. The three face-sharing tetrahedra of Pb atoms in the $Pb_6O(OH)_6^{4+}$ cluster

There are a number of Sn^{II} (and also Pb^{II}) chalcogenide, halide, and mixed compounds (e.g., $CsSnBr_3$, Pb_4Cl_6S) that are intensely colored, even black, the color becoming more intense as the atomic number of the chalcogen or halogen increases. These appear to have unusual structures and bonding. For example, $CsSnBr_3$, has an ideal perovskite structure (no stereochemically active lone pair) and is a semiconductor. It is proposed[58] that the halide and chalcogenide ions use outer d orbitals to form energy bands that are partially populated by the $5s$ or $6s$ electrons of Sn^{II} or Pb^{II}.

Lead. Lead has a well-defined cationic chemistry. There are several crystalline salts, but except for $Pb(NO_3)_2$ and $Pb(OCOCH_3)_2 \cdot 2H_2O$ (which ionizes incompletely in water), most lead salts are sparingly soluble (PbF_2, $PbCl_2$) or insoluble ($PbSO_4$, $PbCrO_4$, etc.) in water. The halides, unlike the tin(II) halides, are always anhydrous and have complex crystal structures with distorted close-packed halogen lattices. In water, species such as PbX^+ are formed, and on addition of an excess of halogen acid, $PbX_n^{(n-2)-}$; with the fluoride, only PbF^+ occurs, but with Cl^- several complex ions are formed.

The oxide has two forms: litharge, red with a layer structure, and massicott, yellow with a chain structure.[59] The oxide Pb_3O_4 (red lead), which is made by heating PbO or PbO_2 in air, behaves chemically as a mixture of PbO and PbO_2, but the crystal contains $Pb^{IV}O_6$ octahedra linked in chains by sharing of opposite edges, the chains being linked by Pb^{II} atoms each bound to three oxygen atoms.

The lead(II) ion is partially hydrolyzed in water. In perchlorate solutions the first equilibrium appears to be

$$Pb^{2+} + H_2O = PbOH^+ + H^+ \qquad \log K \approx -7.9$$

but the main aqueous species is $Pb_4(OH)_4^{4+}$. In chloride solution basic complexes with OH groups are formed.[60] Crystalline salts containing polymeric anions can be obtained by dissolving PbO in perchloric acid and adding an appropriate quantity of base. X-Ray investigation[61] of the structure of the Pb_6 species $[Pb_6O(OH)_6]^{4+}(ClO_4^-)_4 \cdot H_2O$ reveals three tetrahedra of Pb atoms that share faces. The middle tetrahedron has an O atom near the center (Fig. 12-7); OH groups lie

58   J. D. Donaldson *et al.*, *J. C. S. Dalton*, **1975**, 1500; **1977**, 996.
59   D. M. Adams and D. C. Stevens, *J. C. S. Dalton*, **1977**, 1096.
60   P. Tsai and R. P. Cooney, *J. C. S. Dalton*, **1976**, 1631.
61   A. Olin and R. Soderqvist, *Acta Chem. Scand.*, 1972, **26A**, 3505.

on the faces of the two end tetrahedra. Addition of more base gives the hydrous oxide, which dissolves in an excess of base to give the plumbate ion. If aqueous ammonia is added to a lead acetate solution, a white precipitate of a basic acetate is formed that on suspension in warm aqueous ammonia and subsequent drying gives very pure lead oxide, the most stable form of which is the tetragonal red form.

Lead(II) also forms numerous complexes that are mostly octahedral, although a phosphorodithioate, $Pb(S_2PPr_2^i)_2$, is polymeric, with six Pb—S bonds and a stereochemically active lone pair.

12-12. Organo Compounds of Group IV Elements[62]

There is an exceedingly extensive chemistry of the Group IV elements bound to carbon, and some of the compounds, notably silicon-oxygen polymers and alkyl-tin and -lead compounds, are of commercial importance; germanium compounds appear to offer little utility.

Divalent Compounds. The alkyls of the type MR_2 (M = Ge, Sn) that have long been known are not in fact true compounds of Ge^{II} or Sn^{II} but are either cyclic oligomers with M—M bonds, long polymers with unknown end groups, or telomers $[R(MR_2)R]$ with only approximately correct analyses. Recently, however, some genuine MR_2 compounds of Ge, Sn, and also Pb have been made with the bulky R group $(SiMe_3)_2CH$—.[63a] These MR_2 compounds are highly colored monomers in solution and the gas phase, but form less intensely colored crystals (which become colorless, or nearly so, at 77°K) containing M—M bonded dimers. The chemistry of SnR_2 has been investigated in some detail; it readily undergoes oxidative additions (e.g., with CH_3I, HCl, Br_2), serves as a donor (e.g., to form $R_2SnCr(CO)_5$), and has weak acceptor properties.

The $(R_2Sn)_2$ dimers have the structure shown in Fig. 12-8a and are diamagnetic. The proposed explanation for these properties is the type of overlap shown in Fig. 12-8b, in which double bent bonding is implied. Some related compounds have $(Me_3Si)_2N$ and $(Me_3C)_2P$ ligands, and these have four-membered Sn_2P_2 rings.[63b]

Other divalent compounds are the cyclopentadienyls $(\eta\text{-}C_5H_5)_2M$

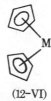

(12–VI)

[62] G. E. Coates, M. L. H. Green, and K. Wade, *Organometallic Compounds,* Vol. I, Methuen, 1967 (an excellent account of Ge and Pb chemistry); A. G. MacDiarmid, Ed., *Organometallic Compounds of Group IV Elements,* Vol. 1, Part I (nature of bond to C; Si—C); Part II (Ge, Sn, Pb to C bonds) (additional volumes are in preparation), Dekker; R. Weiss, *Organometallic Compounds,* Vol. 2, 1972, Part I (Si—halogen bonds), Part II (Ge-, Sn-, Pb-halogen bonds); Springer; supplement, 1973.

[63a] M. F. Lappert *et al., J. C. S. Dalton,* **1976,** 2268, 2275, 2286.

[63b] W. W. du Mont and H.-J. Kroth, *Angew. Chem., Int. Ed.,* 1977, **16,** 792.

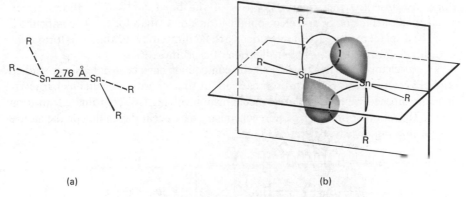

(a)   (b)

Fig. 12-8. (a) The structure of $Sn_2[CH(SiMe_3)_2]_4$. (b) Proposed bonding.

(M = Ge, Sn, Pb). In the crystals the molecules lose their individuality and long, zigzag chains with bridging C_5H_5 rings are formed. However there are bent molecules (12-VI), in the gas phase and in solution, and the lone pair is available for donation to Lewis acids, to form adducts such as $(\eta\text{-}C_5H_5)_2SnBF_3$ or $(\eta\text{-}C_5H_5)_2SnAlCl_3$.[64]

Tetravalent Compounds.   The general formula is $R_{4-n}MX_n$ (n = 0 to 3), where R is alkyl or aryl and X is any of a wide variety of atoms or groups (H, halogen, OR', NR$_2$', SR', Mn(CO)$_5$, etc.). The elements may also form part of a heterocyclic ring, e.g., $(R_2MO)_3$.

For a given class of compound, members with C—Si and C—Ge bonds have higher thermal stability and lower reactivity than those with bonds to Sn and Pb. In catenated compounds similarly, Si—Si and Ge—Ge bonds are more stable and less reactive than Sn—Sn and Pb—Pb bonds; for example, Si_2Me_6 is very stable, but Pb_2Me_6 blackens in air and decomposes rapidly in CCl_4, although it is fairly stable in benzene.

The bonds to carbon are usually made via interaction of lithium, mercury, or aluminum alkyls or RMgX and the Group IV halide, but there are many special synthetic methods, some of which are noted below.

Silicon and Germanium.[65]   The organo compounds of Si and Ge are very similar in their properties, although Ge compounds have been less extensively studied. We discuss only Si compounds.

Silicon-carbon bond dissociation energies are less than those of C—C bonds but are still quite high, in the region 250–335 kJ mol^{-1}. The tetra-alkyls and -aryls are hence thermally quite stable; $Si(C_6H_5)_4$, for example, boils unchanged at 530°.

The chemical reactivity of Si—C bonds is generally greater than that of C—C

64   P. G. Harrison and J. A. Richards, *J. Organomet. Chem.,* 1976, **108**, 35; A. K. Holliday, P. H. Makin, and R. H. Puddephat, *J. C. S. Dalton,* **1976**, 435.
65   V. Bazant, V. Chvalovsky, and J. Rathowsky, *Organosilicon Compounds,* Academic Press, 1965 (comprehensive reference volumes); M. Lesbre, P. Mazerolles, and J. Satgé, *The Organic Compounds of Germanium,* Wiley, 1971; E. W. Colvin, *Chem. Soc. Rev.,* 1978, **7**, 15 (organosilicon compounds in organic synthesis); *Comprehensive Organic Chemistry,* Vol. 3, Part 13, I. Fleming, (Si), Part 15-4, R. C. Poller, (Ge-Pb), Pergamon Press, 1979.

bonds because (a) the greater polarity of the bond $Si^{\delta+}$—$C^{\delta-}$ allows easier nucleophilic attack on Si and electrophilic attack on C than for C—C compounds, and (b) displacement reactions at silicon are facilitated by its ability to form five-coordinate transition states by utilization of d orbitals.

The mechanisms of reactions of silicon compounds have been extensively studied and are complicated. Substitution reactions at four-coordinate silicon character-istically proceed via an associative mechanism involving five-coordinate transition states. Retention or inversion of stereochemistry may occur depending on the nature of the entering or leaving groups,[66] viz.,

$$R^2\text{---}\underset{\underset{R^3}{|}}{\overset{\overset{R^1}{|}}{Si}}\text{—}X \xrightarrow{+Y} \left[\underset{Y}{\overset{X}{\underset{|}{\triangle}}} \begin{array}{cc} R^1 & \\ R^3 & R^2 \end{array} \right] \xrightarrow{-X} R^2\text{---}\underset{\underset{R^1}{|}}{\overset{\overset{R^3}{|}}{Si}}\text{—}Y$$

A very important and characteristic feature of organosilicon (and -germanium) chemistry, setting it strikingly apart from carbon chemistry, is the great ease with which R_3Si (and R_3Ge) groups migrate; a factor of up to 10^{12} as compared to analogous carbon compounds is typical. Among the best studied migration reactions are anionic 1,2 shifts, represented generally by the equation:

$$\underset{R_3Si}{\overset{\diagdown}{\underset{\diagup}{X}}}\text{—}\overset{\bar{\;}}{Y}\overset{\diagup}{\diagdown} \longrightarrow \left[\underset{R_3}{\overset{X\text{—}Y}{\underset{Si}{\diagdown\diagup}}} \right]^{-} \longrightarrow \overset{\diagdown}{\underset{\diagup}{X}}\text{—}\underset{SiR_3}{\overset{\diagup}{\underset{\diagdown}{Y}}}$$

where X—Y may be N—N, O—N, or S—C. As indicated, a transition state in-volving five-coordinate Si (possible by use of a d orbital) is postulated; since carbon has no valence shell d orbitals, it cannot form such a transition state easily, and such 1,2-shifts are "forbidden" in the Woodward-Hoffmann sense.

Radicals are less important in silicon than in carbon chemistry.[67] However silicon radicals have been detected in solution by esr and have been isolated in matrices. They are made by hydrogen abstraction with t-butoxy and other radicals generated photochemically, e.g.,

$$R_3SiH + Me_3CO\cdot \rightarrow R_3Si\cdot + Me_3COH$$

The stable $R_3Si\cdot$ radicals with extremely bulky R groups have been noted earlier (p. 377).

A comparison of the rates of reactions such as

$$p\text{-}XC_6H_4\,MR_3 + H_2O \rightarrow R_3MOH + C_6H_5X$$

in aqueous-methanolic $HClO_4$ gives the order $Si(1) < Ge(36) \ll Sn(3 \times 10^5) \ll \ll Pb(2 \times 10^8)$, which suggests that with increasing size, there is increased avail-

[66] See R. J. P. Corriu and G. F. Lanneau, *J. Organomet. Chem.*, 1974, **67**, 243, and references therein.

[67] I. M. T. Davidson and A. V. Howard, *J. C. S. Faraday, I*, 1975, **71**, 69.

ability of outer orbitals, which allows more rapid initial solvent coordination to give the five-coordinated transition state.

Alkyl- and Arylsilicon Halides. These compounds are of special importance because of their hydrolytic reactions. They may be obtained by normal Grignard procedures from $SiCl_4$, or, in the case of the methyl derivatives, by the Rochow process, in which methyl chloride is passed over a heated, copper-activated silicon:

$$CH_3Cl + Si(Cu) \rightarrow (CH_3)_nSiCl_{4-n}$$

The halides are liquids that are readily hydrolyzed by water, usually in an inert solvent. In certain cases, the silanol intermediates R_3SiOH, $R_2Si(OH)_2$, and $RSi(OH)_3$ can be isolated, but the diols and triols usually condense under the hydrolysis conditions to siloxanes that have Si—O—Si bonds.[68] The exact nature of the products depends on the hydrolysis conditions and linear, cyclic, and complex cross-linked polymers of varying molecular weights can be obtained. They are often referred to as silicones; the commercial polymers usually have $R = CH_3$, but other groups may be incorporated for special purposes.

Controlled hydrolysis of the alkyl halides in suitable ratios can give products of particular physical characteristics. The polymers may be liquids, rubbers, or solids, which have in general high thermal stability, high dielectric strength, and resistance to oxidation and chemical attack.

Examples of simple siloxanes are $Ph_3SiOSiPh_3$ and the cyclic trimer or tetramer $(Et_2SiO)_{3(or\ 4)}$; linear polymers contain $-SiR_2-O-SiR_2-O-$ chains, whereas the cross-linked sheets have the basic unit

$$-O-\overset{\displaystyle R}{\underset{\displaystyle O}{\overset{\displaystyle |}{\underset{\displaystyle |}{Si}}}}-O-$$

Tin.[69] Where the compounds of tin differ from those of Si and Ge they do so mainly because of a greater tendency of Sn^{IV} to show coordination numbers higher than 4 and because of ionization to give cationic species.

Trialkyltin compounds of the type R_3SnX, of which the best studied are the CH_3 compounds ($X = ClO_4$, F, NO_3, etc.), are of interest in that they are always associated in the solid by anion bridging (12-VII and 12-VIII); the coordination of the tin atom is close to *tbp* with planar $SnMe_3$ groups. When X is RCOO the com-

[68]   R. J. H. Voorhoeve, *Organohalosilanes: Precursors to Silicones,* Elsevier, 1967; W. Noll et al., *Chemistry and Technology of Silicones,* Academic Press, 1968; H. A. Liebhafsky, *Silicones under the Monogram,* Wiley, 1978 (historical account of G. E. research).

[69]   G. J. M. van der Kerk, *Chem. Tech.,* **1978,** 356; J. J. Zuckerman, Ed. *Organotin Compounds* ACS Advances in Chemistry Ser. No. 157, 1976; R. C. Poller, *Organotin Chemistry,* Academic Press, 1970; W. P. Neuman, *The Organic Chemistry of Tin* (transl. R. Moser), Wiley, 1970; R. Okawara and M. Wada, *Adv. Organomet. Chem.,* 1967, **5,** 137 (structural aspects of organotin compounds); B. Y. K. Ho and J. J. Zuckerman, *J. Organomet. Chem.,* 1973, **49,** 1 (structural aspects).

pounds may in addition be monomeric with unidentate or bidentate carboxylate groups.

$$
\begin{array}{cc}
\text{(12-VII)} & \text{(12-VIII)}
\end{array}
$$

The R_3SnX (and also R_3PbX) compounds also form 1:1 and 1:2 adducts with Lewis bases, and these also generally appear to contain five-coordinate Sn, with the alkyl groups in axial positions. In water the perchlorate and some other compounds ionize to give cationic species, e.g., $[Me_3Sn(H_2O)_2]^+$.

Dialkyltin compounds (R_2SnX_2) have behavior similar to that of the trialkyl compounds. Thus the fluoride Me_2SnF_2 is again polymeric, with bridging F atoms, but Sn is octahedral and the Me—Sn—Me group is linear. However the chloride and bromide have low melting points (90° and 74°) and are essentially molecular compounds, only weakly linked by halogen bridges.[70] The nitrate $Me_2Sn(NO_3)_2$ is strictly molecular with bidentate nitrate groups.[71]

The halides also give conducting solutions in water and the aqua ion has the linear C—Sn—C group characteristic of the dialkyl species (cf. the linear species Me_2Hg, Me_2Tl^+, Me_2Cd, Me_2Pb^{2+}), probably with four water molecules completing octahedral coordination. The linearity in these species appears to result from maximizing of s character in the bonding orbitals of the metal atoms. The ions Me_2SnCl^+ and Me_2SnOH^+ also exist, and in alkaline solution $trans$-$[Me_2Sn(OH)_4]^{2-}$.

Catenated linear and cyclic organotin compounds are relatively numerous and stable. For example, the reaction of sodium in liquid ammonia with $Sn(CH_3)_2Cl_2$ gives "$[Sn(CH_3)_2]_n$," which consists mainly of linear molecules with chain lengths of 12 to 20 (and perhaps more), as well as at least one cyclic compound, $[Sn(CH_3)_2]_6$. There is no evidence for branching of chains. Similar results have been obtained with other alkyl and aryl groups; for example, the cyclic hexamer and nonamer of Et_2Sn, the cyclic pentamer and hexamer of Ph_2Sn, and the cyclic tetramer of Bu_2^iSn have been isolated, as well as linear species. In some cases the terminal groups of the linear species are SnR_2H. The structure of $[Ph_2Sn]_6$ has an Sn_6 ring in a chair configuration, with the Sn—Sn bonds of about the same length as those in gray tin.

Finally, organotin hydrides (R_3SnH), which can be made by $LiAlH_4$ reduction of the halide or in other ways, are useful reducing agents in organic chemistry; some of the reactions are known to proceed by free-radical pathways. The hydrides undergo additional reactions with alkenes or alkynes similar to the hydrosilation reaction, which provides a useful synthetic method for organotin compounds containing functional groups. In contrast to the addition of Si—H (Section 30-2) the

[70]   See N. W. Alcock and J. F. Sawyer, *J. C. S. Dalton*, **1977**, 1090.
[71]   J. Hilton, E. K. Nunn, and S. C. Wallwork, *J. C. S. Dalton*, **1973**, 173.

hydrostannation reaction is free radical for nonactivated C—C bonds, but for activated bonds an ionic mechanism may additionally be involved.

Some examples of the use of tin hydrides are:

$$R_3SnH + CH_2{=}CHCN \longrightarrow R_3SnCH_2CH_2CN$$

$$Ti(NMe_2)_4 + 4Ph_3SnH \longrightarrow (Ph_3Sn)_4Ti + 4NHMe_2$$

Organotin compounds are probably used in more ways than any other metal organics: as stabilizers for polyvinylchloride (R_2SnX_2), as wood preservatives (tributyltin oxide), and as antifouling compounds, catalysts, fungicides, and so on.

Lead.[72] There is an extensive organolead chemistry, but the most important compounds are $PbMe_4$ and $PbEt_4$, which are made in vast quantities for use as antiknock agents in gasoline. Although lead alkyls and aryls can be made by alkylation of Pb^{II} compounds, the two tetraalkyls are made otherwise.

The major commercial synthesis is by the interaction of a sodium-lead alloy with CH_3Cl or C_2H_5Cl in an autoclave at 80 to 100°, without solvent for C_2H_5Cl but in toluene at a higher temperature for CH_3Cl. The reaction is complicated and not fully understood, and only a quarter of the lead appears in the desired product:

$$4NaPb + 4RCl \rightarrow R_4Pb + 3Pb + 4NaCl$$

The required recycling of the lead is disadvantageous, and electrolytic procedures have been developed. One process involves electrolysis of $NaAlEt_4$ with a lead anode and mercury cathode; the sodium formed can be converted into NaH and the electrolyte regenerated:

$$4NaAlEt_4 + Pb \rightarrow 4Na + PbEt_4 + 4AlEt_3$$
$$4Na + 2H_2 \rightarrow 4NaH$$
$$4NaH + 4AlEt_3 + 4C_2H_4 \rightarrow 4NaAlEt_4$$

Another process involves electrolysis of solutions of Grignard reagents in ethers at lead anodes; $RMgCl$ is regenerated by adding RCl:

$$4RMgCl \rightleftharpoons 4R^- + 4MgCl^+$$

$$4R^- + Pb \xrightarrow{-4e} PbR_4$$

$$4MgCl^+ \xrightarrow{+4e} 2Mg + 2MgCl_2$$

The alkyls are nonpolar, highly toxic liquids[73]; the methyl member begins to decompose around 200° and the ethyl member around 110°, by free-radical mechanisms. Lead antiknock compounds are responsible for a major part of lead

[72] H. Shapiro and F. W. Frey, *The Organic Compounds of Lead*, Wiley, 1968.
[73] D. Bryce-Smith, *Chem. Br.*, 1970, **54** (toxicology of lead); M. E. Milburn, *Chem. Soc. Rev.*, 1979, **8**, 63 (Environmental Lead in Perspective).

contamination in the environment. Their use is unnecessary, since lead-free gasolines can be easily made, though at somewhat higher cost than unleaded ones.

General References

Abrikosov, N. Kh., *et. al., Semiconducting II-VI, IV-VI and V-VI Compounds,* Plenum Press, 1969.

Bazant, V., J. Joklik, and J. Rathousky, *Angew. Chem., Int. Ed.,* 1968, **7,** 112. Direct synthesis of $R_n SiX_{4-n}$.

Borisov, S. N., M. G. Voronkov, and E. Ya Lukevits, *Organosilicon Heteropolymers and Heterocompounds,* Heyden Press, 1969.

Cundy, C. S., B. M. Kingston, and M. F. Lappert, *Adv. Organomet. Chem.,* 1973, **11,** 253. Organometallic complexes, silicon–transition metal, or silicon–carbon–transition metal bonds.

Davidov, V. I., *Germanium,* Gordon and Breach, 1966.

Ebsworth, E. A. V., *Volatile Silicon Compounds,* Pergamon Press, 1962.

Glockling, F., *The Chemistry of Germanium,* Academic Press, 1962.

Ho, T.-L., *Synthesis,* **1979,** 1. Reduction of organic compounds by low-valent Group IVb species.

Kuhn, A. T., *Electrochemistry of Lead,* Academic Press, 1979.

Petrosyan, V. S., N. S. Yashina, and O. A. Reutov, *Adv. Organomet. Chem.,* 1976, **14,** 63. Methyltin halides and their molecular complexes.

Unger, K. K., *Porous Silica: Its Properties and Uses as Support in Column Ligand Chromatography,* Elsevier, 1979.

West, R., *Adv. Organomet. Chem.,* 1977, **16,** 1. 1,2-Anionic rearrangements of organosilicon and -germanium compounds.

Zubieta, J. A., and J. J. Zuckerman, *Prog. Inorg. Chem.,* 1978, **24,** 251. Structural tin chemistry.

CHAPTER THIRTEEN

Nitrogen

GENERAL REMARKS

13-1. Introduction

The electronic configuration of the nitrogen atom in its ground state (^4S) is $1s^2 2s^2 2p^3$, with the three $2p$ electrons distributed among the p_x, p_y, and p_z orbitals with spins parallel. Nitrogen forms an exceedingly large number of compounds, most of which are to be considered organic rather than inorganic. It is one of the most electronegative elements, only oxygen and fluorine exceeding it in this respect.

The nitrogen atom may complete its octet in several ways:

1. *Electron Gain to Form the Nitride Ion* N^{3-}. This ion occurs only in the saltlike nitrides of the most electropositive elements, for example, Li_3N. Many nonionic nitrides exist and are discussed later in this chapter.

2. *Formation of Electron-Pair Bonds.* The octet can be completed either by the formation of three single bonds, as in NH_3 or NF_3, or by multiple-bond formation, as in nitrogen itself, $:N{\equiv}N:$, azo compounds, $-\ddot{N}{=}\ddot{N}-$, nitro compounds, RNO_2, and so on.

3. *Formation of Electron-Pair Bonds with Electron Gain.* The completed octet is achieved in this way in ions such as the amide ion $NH_2{}^-$ and the imide ion NH^{2-}.

4. *Formation of Electron-Pair Bonds with Electron Loss.* Nitrogen can form four bonds, provided an electron is lost, to give positively charged ions R_4N^+ such as NH_4^+, $N_2H_5^+$, and $(C_2H_5)_4N^+$. The ions may sometimes be regarded as being formed by protonation of the lone pair:

$$H_3N: + H^+ \rightarrow [NH_4]^+$$

or generally

$$R_3N: + RX \rightarrow R_4N^+ + X^-$$

407

Failure to Complete the Octet. There are a few relatively stable species in which, *formally,* the octet of nitrogen is incomplete. The classic examples are NO and NO_2 together with nitroxides R_2NO and the ion $(O_3S)_2NO^{2-}$; all have one unpaired electron. Nitroxide radicals are used as "spin labels," since they can be attached to proteins, membranes, etc., and inferences about their environment drawn from the characteristics of the observed esr signals.

Expansion of Octet. Normally, no expansion of the octet is permissible, but there are reactions that may indicate pentacoordinate nitrogen intermediates, or more likely transition states.

Thus when molten ammonium trifluoroacetate (m.p. 130°) is treated with LiH, deuteration studies show isotopic scrambling.[1] Although it could be that D^- impinges only on the hydrogen of NH_4^+, another possibility is

$$D^- + \overset{+}{N}H_4 \rightleftharpoons \left[\begin{array}{c} H \\ | \\ D-N-H \\ \diagup \ \diagdown \\ H \quad H \end{array} \right] \rightleftharpoons NH_3D + H^-$$

Also the synthesis of NF_4^+ from NF_3, F_2 and AsF_5 could involve an NF_5 species, though this seems unlikely (see Section 13-8).

Formal Oxidation Numbers. Classically, formal oxidation numbers ranging from -3 (e.g., in NH_3) to $+5$ (e.g., in HNO_3) have been assigned to nitrogen; though useful in balancing redox equations, they have no physical significance.

13-2. Types of Covalence in Nitrogen; Stereochemistry

In common with other first-row elements, nitrogen has only four orbitals available for bond formation, and a maximum of four $2c$-$2e$ bonds may be formed. However since formation of three electron-pair bonds completes the octet $:N(:R)_3$, and the nitrogen atom then possesses a lone pair of electrons, four $2c$-$2e$ bonds can only be formed either (*a*) by coordination, as in donor-acceptor complexes, e.g., $F_3\bar{B}-\overset{+}{N}(CH_3)_3$, or in amine oxides, e.g., $(CH_3)_3\overset{+}{N}-\bar{O}$, or (*b*) by loss of an electron, as in ammonium ions NH_4^+, NR_4^+. This loss of an electron gives a valence state configuration for nitrogen (as N^+) with four unpaired electrons in sp^3 hybrid orbitals analogous to that of neutral carbon, while, as noted above, gain of an electron (as in NH_2^-) leaves only two electrons for bond formation. In this case the nitrogen atom (as N^-) is isoelectronic with the neutral oxygen atom, and angular bonds are formed. We can thus compare, sterically, the following isoelectronic species:

Tetrahedral	*Angular*
NH_4^+ and CH_4	$:\overset{..}{O}:$ and $:\overset{..}{N}:^-$
(also BH_4^-)	

It may be noted that the ions NH_2^-, OH^-, and F^- are isoelectronic and have comparable sizes. The amide, imide, and nitride ions, which can be considered to be

[1]   G. A. Olah *et al., J. Am. Chem. Soc.,* 1975, **97**, 3559.

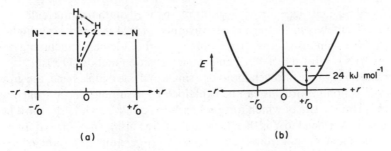

Fig, 13-1, Diagrams illustrating the inversion of NH_3 (see text).

members of the isoelectronic series NH_4^+, NH_3, NH_2^-, NH^{2-}, N^{3-}, occur as discrete ions only in salts of highly electropositive elements.

In all nitrogen compounds where the atom forms two or three bonds, there remain, respectively, two pairs or one pair of nonbonding or lone-pair electrons. The lone pairs have a profound effect on stereochemistry and are also responsible for the donor properties of the atom possessing them. To illustrate the important chemical consequences of nonbonding electron pairs, we consider one of the most important types of molecule, namely, NR_3, as exemplified by NH_3 and amines.

Three-Covalent Nitrogen. Molecules of this type are invariably pyramidal, except in special cases such as the planar N-centered triangular iridium complexes (p. 124), e.g., $[Ir_3N(SO_4)_6(H_2O)_3]^{4-}$, the trisilylamine compound, or when multiple bonding is involved. The bond angles vary according to the groups attached to the nitrogen atom. Pyramidal molecules of the kind NRR'R" should be chiral. No optical isomers have ever been isolated, however, because molecules of this type execute a motion known as inversion, in which the nitrogen atom oscillates through the plane of the three R groups much as an umbrella can turn inside out. As the nitrogen atom crosses from one side of the plane to the other (from one equilibrium position, say $+ r_0$, to the other, $-r_0$, in Fig. 13-1a), the molecule goes through a state of higher potential energy, as shown in the potential energy curve (Fig. 13-1b). However this "potential energy barrier" to inversion is only 23.4 kJ mol^{-1}, and the frequency of the oscillation is 2.387013×10^{10} Hz in NH_3. In simple alkylamines generally, barriers are in the range 16–40 kJ mol^{-1}; thus isolation of optical isomers can never be expected. However heteroatom substitution and incorporation of N into strained rings raises the barrier, and in some cases invertomers have been separated. For the heavier elements in Group V, the inversion barriers of XH_3 are much higher; a qualitative MO explanation has been provided.[2]

Multiple Bonding in Nitrogen and Its Compounds. Like its neighbors carbon and oxygen, nitrogen readily forms multiple bonds, differing in this respect from its heavier congeners phosphorus, arsenic, antimony, and bismuth. Nitrogen thus forms many compounds for which there are no analogues among the heavier elements. Thus whereas phosphorus, arsenic, and antimony form tetrahedral molecules

[2] C. C. Levin, *J. Am. Chem. Soc.*, 1975, **97**, 5649; W. Cherry and N. Epiotis, *J. Am. Chem. Soc.*, 1976, **98**, 1135.

P_4, As_4, and Sb_4, nitrogen forms the multiple-bonded diatomic molecule $:N\equiv N:$, with an extremely short internuclear distance (1.094 Å) and very high bond strength. Nitrogen also forms triple bonds to other elements including carbon (CH_3—$C\equiv N$), sulfur ($F_3S\equiv N$), and some transition metals (O_3OsN^-).

In compounds where nitrogen forms one single and one double bond, the grouping X—$\ddot{N}$=Y is nonlinear. This can be explained by assuming that nitrogen uses a set of sp^2 orbitals, two of which form σ bonds to X and Y, while the third houses the lone pair. A π bond to Y is then formed using the nitrogen p_z orbital. In certain cases stereoisomers result from the nonlinearity, for example, *cis*- and *trans*-azobenzenes (13-Ia and 13-Ib) and the oximes (13-IIa and 13-IIb). These are interconverted more easily than are *cis*- and *trans*-olefins, but not readily.

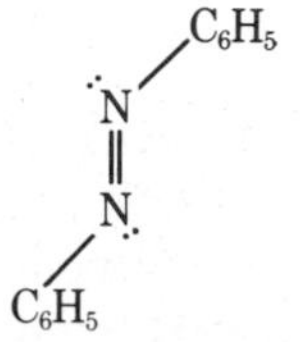

(13-Ia) 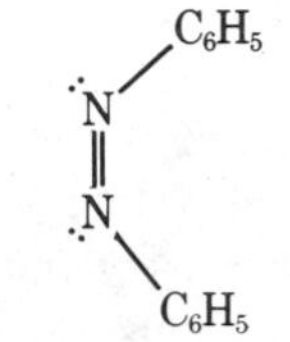   (13-Ib) 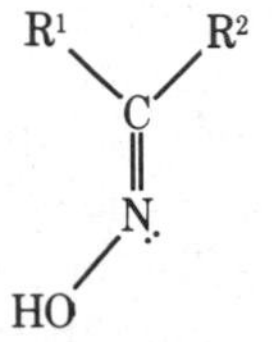   (13-IIa) 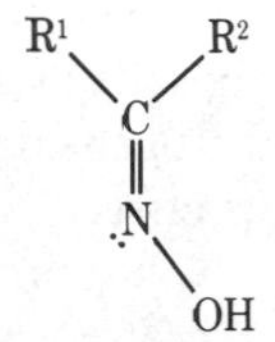   (13-IIb)

Multiple bonding occurs also in oxo compounds. For example, NO_2^- (13-III) and NO_3^- (13-IV) can be regarded as resonance hybrids in the valence bond approach. From the MO viewpoint, one considers the existence of a π MO extending symmetrically over the entire ion and containing the two π electrons.

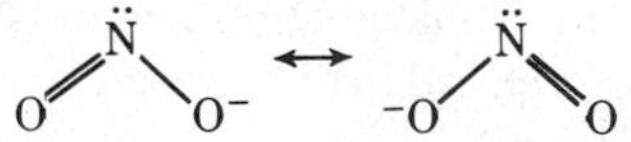

(13-IIIa)   (13-IIIb)

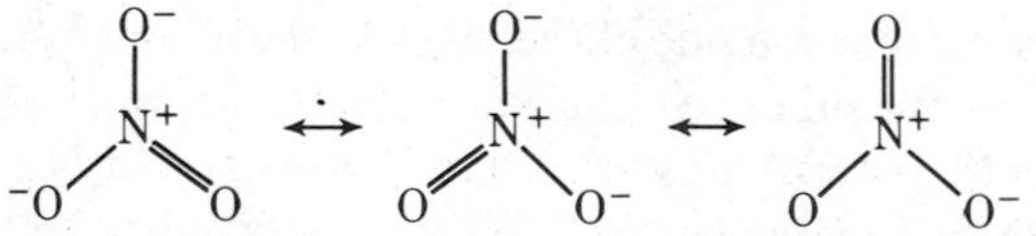

(13-IVa)   (13-IVb)   (13-IVc)

An unusual but significant case of multiple bonding occurs in trisilylamine [$N(SiH_3)_3$] and other compounds with Si—N bonds, as discussed on page 378, where $d\pi$–$p\pi$ bonds are involved. Another example is tetrasilylhydrazine, which appears to have the structure shown in Fig. 13-2, in which the $NNSi_2$ groups are each planar. This should be compared with the structure of hydrazine (N_2H_4, p. 418). In F_2PNH_2 the configuration at N is also planar.[3] This might again be attributed to P—N π bonding, though this explanation has been disputed.[4]

Another unusual case with trigonal planar nitrogen is compound 13-V, but the planarity seems to result from steric rather than electronic factors.[5]

3   A. H. Brittain *et al., J. Am. Chem. Soc.,* 1971, **93**, 6772.
4   I. G. Csizmadia, *J. C. S. Chem. Comm.,* **1974**, 432.
5   J. G. Verkade, *et al., J. Am. Chem. Soc.,* 1977, **99**, 631, 6607.

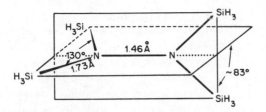

Fig. 13-2. The molecular structure of tetrasilylhydrazine as determined by electron diffraction. The dihedral angle could be 90°.

$$
\begin{array}{c}
\underset{|}{\overset{\text{CH}_2\text{CH}_2\text{O}}{\text{N}}}\!\!\!\diagdown \\
\text{CH}_2\text{CH}_2\text{O} \diagup \text{P}\!=\!\text{S} \\
\\
\text{CH}_2\text{CH}_2\text{O} \diagup
\end{array}
$$

(13-V)

Donor Properties of Three-Covalent Nitrogen; Four-Covalent Nitrogen. As noted above, the formation of approximately tetrahedral bonds to nitrogen occurs principally in ammonium cations (R_4N^+), amine oxides ($R_3N^+\!\!-\!O^-$), and in Lewis acid-Lewis base adducts (e.g., $R_3\overset{+}{N}\!\!-\!\bar{B}X_3$). In the amine oxides and these adducts the bonds must have considerable polarity; in the amine oxides, for instance, $N \rightarrow O$ donation cannot be effectively counterbalanced by any back-donation to N. In accord with this, the stability of amine oxides decreases as the R_3N basicity decreases, since the ability of N to denote to O is the major factor. Similarly, $R_3N \rightarrow BX_3$ complexes have stabilities that roughly parallel the R_3N basicity for given BX_3. When R is fluorine, basicity is minimal and $F_3N \rightarrow BX_3$ compounds are unknown. It is therefore curious that F_3NO is an isolable compound (see p. 436). Evidently the extreme electronegativity of fluorine coupled with the availability of $p\pi$ electrons on oxygen allows structures 13-VI to contribute to stability.

$$
\begin{array}{ccccc}
\text{F}\diagdown & & \text{F}\diagdown & & \text{F}^- \\
\text{F}\!-\!\text{N}^+\!\!=\!\!\text{O} & \leftrightarrow & \text{F}^-\diagup\text{N}^+\!\!=\!\!\text{O} & \leftrightarrow & \text{F}\!-\!\text{N}^+\!\!=\!\!\text{O} \\
\text{F}^-\diagup & & \text{F}\diagup & & \text{F}\diagup
\end{array}
$$

(13-VI)

Catenation and N—N Single-Bond Energies. Unlike carbon and a few other elements, nitrogen has little tendency for catenation, primarily owing to the weakness of the N—N single bond. If we compare the approximate single-bond energies in $H_3C\!-\!CH_3$, $H_2N\!-\!NH_2$, $H\!-\!O\!-\!O\!-\!H$, and $F\!-\!F$ (ca. 350, 160, 140, and 150 kJ mol^{-1}, respectively), it is clear that there is a profound drop between C and N. This difference is most probably attributable to the effects of repulsion between nonbonding lone-pair electrons. The strength of the N—N bond, and also of the O—O bond, decreases with increasing electronegativity of the attached groups; increasing electronegativity would perhaps have been expected to reduce repulsion between lone pairs, but it obviously will also weaken any homonuclear σ bond.

There are a few types of compound containing chains of three or more nitrogen atoms with some multiple bonds such as $R_2N—N=NR_2$, $R_2N—N=N—NR_2$, $RN=N—NR—NR_2$, $RN=N—NR—N=NR$, and $RN=N—NR—N=N—NR—N=NR$, where R represents an organic radical (some R's may be H, but known compounds contain only a few H's). There are also cyclic compounds containing rings with up to five consecutive nitrogen atoms. Many of these compounds are not particularly stable, and all are traditionally in the realm of organic chemistry.

Hydrogen Bonding. Since it is one of the most electronegative elements, nitrogen, along with oxygen, fluorine and, to a lesser extent, chlorine, enters extensively into hydrogen-bond formation in its hydrogen compounds, both as a proton donor, $N—H \cdots X$, and as a proton receptor, $\geq N \cdots H—X$.

THE ELEMENT

13-3.   Occurrence and Properties

Nitrogen occurs in Nature mainly as dinitrogen (N_2), an inert diatomic gas (m.p. 63.1°K, b.p. 77.3°K) that comprises 78% by volume of the earth's atmosphere. Naturally occurring nitrogen consists of ^{14}N and ^{15}N with an absolute ratio $^{14}N/^{15}N = 272.0$. The isotope ^{15}N is often useful as a tracer, and it has been found possible to prepare nitric acid containing up to 99.8% ^{15}N by efficient fractionation of the system.[6a]

$$^{15}NO(g) + H^{14}NO_3(aq) = {}^{14}NO(g) + H^{15}NO_3(aq) \qquad K = 1.055$$

The $H^{15}NO_3$ produced can be used to prepare any desired ^{15}N-labeled nitrogen compound.

The heat of dissociation of dinitrogen is extremely large:

$$N_2(g) = 2N(g) \qquad \Delta H = 944.7 \text{ kJ mol}^{-1} \qquad K_{25°} = 10^{-120}$$

Because the reaction is endothermic, the equilibrium constant increases with increasing temperature, but still, even at 3000° and ordinary pressures, there is no appreciable dissociation. The great strength of the $N\equiv N$ bond is principally responsible for the chemical inertness of N_2 and for the endothermicity of most simple nitrogen compounds, even though they may contain strong bonds. Thus $E(N\equiv N) \approx 6E(N—N)$, whereas $E(C\equiv C) \approx 2.5E(C—C)$. Dinitrogen is notably unreactive in comparison with isoelectronic, triply bonded systems such as $X—C\equiv C—X$, $:C\equiv O:$, $X—C\equiv N:$, and $X—N\equiv C:$. Both $—C\equiv C—$ and $—C\equiv N$ groups are known to serve as donors by using their π electrons. The inability of N_2 to form stable linkages in this way may be attributed to its electron configuration, which

[6a] Cf. T. Taylor and W. Spindel, in *Proceedings of the International Symposium on Isotope Separation,* P. Kistemacher, J. Biegeleisen, and O. Cier, Eds., North Holland, 1958.

is $\cdots(\pi)^4(\sigma_2)^2$; that is, the π-bonding electrons are even more tightly bound than the σ-bonding electrons, and the latter are themselves tightly bound ($\Delta H_{ion} = 1496$ kJ mol^{-1}). In acetylene, on the other hand, the electron configuration is $\cdots$ $(\sigma_g)^2(\pi_u)^4$, and the ΔH_{ion} of the π_u electrons is only 1100 kJ mol^{-1}. The N_2 molecule can, however, form complexes similar to those formed by CO, though to a much more limited extent, in which there are M—N≡N and M—C≡O configurations. The very different abilities of the two molecules to function in this way are attributable to several quantitative differences in their qualitatively similar electronic structures.

Dinitrogen is obtained commercially by liquefaction and fractionation of air; it usually contains some argon and, depending on the quality, upwards of ~30 ppm of oxygen. The oxygen may be removed by admixture with a little hydrogen and treatment with a platinum catalyst, by passing the gas over hot copper or other metal, or by bubbling it through aqueous solutions of Cr^{2+} or V^{2+} ions. Spectroscopically pure nitrogen is conveniently prepared by thermal decomposition of sodium azide or barium azide, for example:

$$2NaN_3 \rightarrow 2Na + 3N_2$$

The only reactions of N_2 at room temperature are with metallic lithium to give Li_3N, with certain transition metal complexes, and with nitrogen-fixing bacteria, either free-living or symbiotic on root nodules of clover, peas, beans, etc.

At elevated temperatures dinitrogen becomes more reactive, especially when catalyzed, typical reactions being:

$$N_2(g) + 3H_2(g) = 2NH_3(g) \qquad K_{25°} = 10^3 \text{ atm}^{-2}$$
$$N_2(g) + O_2(g) = 2NO(g) \qquad K_{25°} = 5 \times 10^{-31}$$
$$N_2(g) + 3Mg(s) = Mg_3N_2(s)$$
$$N_2(g) + CaC_2(s) = C(s) + CaNCN(s)$$

Active Nitrogen.[6b] When gaseous molecular nitrogen is subjected to an electrical discharge, under suitable conditions, a very reactive form of nitrogen is generated, accompanied by a yellow afterglow that may persist for several seconds after the discharge has been terminated. The high reactivity is largely due to the presence of ground state (^4S) nitrogen atoms. These have a relatively long lifetime in a vessel suitably "poisoned" to minimize wall recombination. The ternary collision process

$$2N(^4S) + X \rightarrow N_2 + X$$

where X is molecular or atomic nitrogen, is at least partially responsible for populating the excited molecular states, which in turn lead to the afterglow. The afterglow is due mainly to emission of the first positive band system, $N_2(B^3\pi_g) \rightarrow N_2(A^3\Sigma_u^+)$ of the molecular nitrogen spectrum, though some other band systems also contribute.

6b A. N. Wright and C. A. Winkler, *Active Nitrogen,* Academic Press, 1968; R. Brown and C. A. Winkler, *Angew. Chem., Int. Ed.,* 1970, **9,** 181.

NITROGEN COMPOUNDS

13-4. Nitrides

As with carbides, there are three general classes. Ionic nitrides are formed by magnesium, calcium, barium, strontium, zinc, cadmium, lithium, and thorium. Their formulas correspond to what would result from combination of the normal metal ions with N^{3-} ions. They are all essentially ionic compounds and are properly written as $(Ca^{2+})_3(N^{3-})_2$, $(Li^+)_3N^{3-}$, etc. Nitrides of the M_3N_2 type are often anti-isomorphous with oxides of M_2O_3 type. This does not in itself mean that, like the oxides, they are ionic. However their ready hydrolysis to ammonia and the metal hydroxides makes this seem likely. The ionic nitrides are prepared by direct union of the elements or by loss of ammonia from amides on heating, for example:

$$3Ba(NH_2)_2 \rightarrow Ba_3N_2 + 4NH_3$$

There are various covalent "nitrides" (BN, S_4N_4, P_3N_5, etc.), and their properties vary greatly depending on the element with which nitrogen is combined. Such substances are therefore discussed under the appropriate element.

The transition metals form nitrides[7a] that are analogous to the transition metal borides and carbides in their constitution and properties. The nitrogen atoms often occupy interstices in the close-packed metal lattices. These nitrides are often not exactly stoichiometric (being nitrogen deficient), and they are metallic in appearance, hardness, and electrical conductivity, since the electronic band structure of the metal persists. Like the borides and carbides, they are chemically very inert, extremely hard, and have very high melting points. They are usually prepared by heating the metal in ammonia at 1100 to 1200°. A representative compound VN melts at 2570° and has a hardness between 9 and 10.

13-5. Nitrogen Hydrides

Ammonia. NH_3 may be generated in the laboratory by treatment of an ammonium salt with a base:

$$NH_4X + OH^- \rightarrow NH_3 + H_2O + X^-$$

Hydrolysis of an ionic nitride is a convenient way of preparing ND_3 (or NH_3):

$$Mg_3N_2 + 6D_2O \rightarrow 3Mg(OD)_2 + 2ND_3$$

Industrially[7b] ammonia is obtained by the Haber process, in which the reaction

$$N_2(g) + 3H_2(g) = 2NH_3(g) \qquad \Delta H = -46 \text{ kJ mol}^{-1}; \qquad K_{25°} = 10^3 \text{ atm}^{-2}$$

is carried out in the presence of a catalyst at pressures of 10^2 to 10^3 atm and temperatures of 400 to 550°. Although the equilibrium is most favorable at low tem-

---

[7a] R. Juza, *Adv. Inorg. Chem. Radiochem.*, 1967, **9**, 81 (1st transition series).

[7b] F. Bottomley and R. C. Burns, Eds., *Treatise on Dinitrogen Fixation*, Wiley, 1979.

TABLE 13-1
The Ammonia System and the Water System

Ammonia system		Water system	
Class of compound	Example	Class of compound	Example
Acids	$NH_4^+X^-$	$NH_4^+Cl^-$	$H_3O^+ X^-$	$H_3O^+Cl^-$
Bases	Amides	$Na^+ NH_2^-$	Hydroxides	Na^+OH^-
	Imides	$(Li^+)_2 NH^{2-}$	Oxides	$(Li^+)_2O^{2-}, Mg^{2+}O^{2-}$
	Nitrides	$(Mg^{2+})_3(N^{3-})_2$		

perature, even with the best available catalysts elevated temperatures are required to obtain a satisfactory rate of conversion. The best catalyst is α-iron containing some oxide to widen the lattice and enlarge the active interface.

Ammonia is a colorless pungent gas with a normal boiling point of $-33.35°C$ and a freezing point of $-77.7°C$. The liquid has a large heat of evaporation (1.37 kJ g^{-1} at the boiling point) and is therefore fairly easily handled in ordinary laboratory equipment. Liquid ammonia resembles water in its physical behavior, being highly associated because of the polarity of the molecules and strong hydrogen bonding.[8] Its dielectric constant ($\sim$22 at $-34°$; cf. 81 for H_2O at $25°$) is sufficiently high to make it a fair ionizing solvent.[9] A system of nitrogen chemistry with many analogies to the oxygen system based on water has been built up. Thus we have the comparable self-ionization equilibria:

$$2NH_3 = NH_4^+ + NH_2^- \qquad K_{-50°} = [NH_4^+][NH_2^-] = \sim 10^{-30}$$
$$2H_2O = H_3O^+ + OH^- \qquad K_{25°} = [H_3O^+][OH^-] = 10^{-14}$$

Table 13-1 presents a comparison of the ammonia and the water systems.

Liquid ammonia has lower reactivity than H_2O toward electropositive metals, such metals reacting immediately with water to evolve hydrogen. Liquid ammonia, on the other hand, dissolves many electropositive metals to give blue solutions containing metal ions and solvated electrons (see also p. 258).

Because $NH_3(l)$ has a much lower dielectric constant than water, it is a better solvent for organic compounds but generally a poorer one for ionic inorganic compounds. Exceptions occur when complexing by NH_3 is superior to that by water. Thus AgI is exceedingly insoluble in water but $NH_3(l)$ at $25°$ dissolves 207 g/100 ml. Primary solvation numbers of cations in $NH_3(l)$ appear similar to those in H_2O (e.g., 5.0 ± 0.2 and 6.0 ± 0.5 for Mg^{2+} and Al^{3+}, respectively), but there may be some exceptions.[10] Thus Ag^+ appears to be primarily linearly two-coordinate in

[8] For Raman spectra see: D. J. Gardiner et al., J. Raman Spectrosc., 1973, 1, 87; A. T. Lemley et al., J. Phys. Chem., 1973, 77, 2185.
[9] J. J. Lagowski, Ed., The Chemistry of Non-Aqueous Solvents. Academic Press, 1967; G. Jander, M. Spandau, and C. C. Addison, Eds., Chemistry in Anhydrous Ammonia, Vol. 1, Part I, Inorganic and General Chemistry, 1966; Part II, Organic Reactions, 1963, Wiley-Interscience; W. L. Jolly and C. J. Hallada, in Non-Aqueous Solvent Systems, T. C. Waddington, Ed., Academic Press, 1965; G. W. A. Fowles, in Developments in Inorganic Nitrogen Chemistry, Vol. 1, C. B. Colburn, Ed., Elsevier, 1966; J. J. Lagowski, J. Chem. Educ., 1978, 55, 752; D. Nicholls, Inorganic Chemistry in Liquid Ammonia, Elsevier, 1979.
[10] P. Gans and J. B. Gill, J.C.S. Dalton, 1976, 779.

H_2O but tetrahedrally coordinated as $[Ag(NH_3)_4]^+$ in $NH_3(l)$. It has also been suggested that $[Zn(NH_3)_4]^{2+}$ may be the principal species in $NH_3(l)$ as compared to $[Zn(H_2O)_6]^{2+}$ in H_2O.

Reactions of Ammonia. Ammonia reacts with both oxygen and water. Normal combustion in air follows reaction 13-1. However ammonia can be

$$4NH_3(g) + 3O_2(g) = 2N_2(g) + 6H_2O(g) \qquad K_{25°} = 10^{228} \qquad (13\text{-}1)$$

made to react with oxygen as shown in eq. 13-2, even though the process of eq. 13-1 is thermodynamically much more favorable,

$$4NH_3 + 5O_2 = 4NO + 6H_2O \qquad K_{25°} = 10^{168} \qquad (13\text{-}2)$$

by carrying out the reaction at 750 to 900° in the presence of a platinum or platinum-rhodium catalyst.[11] This can easily be demonstrated in the laboratory by introducing a piece of glowing platinum foil into a jar containing gaseous NH_3 and O_2; the foil will continue to glow because of the heat of reaction 13-2, which occurs only on the surface of the metal, and brown fumes will appear owing to the reaction of NO with the excess of oxygen to produce NO_2. Industrially the mixed oxides of nitrogen are then absorbed in water to form nitric acid:

$$2NO + O_2 \rightarrow 2NO_2$$
$$3NO_2 + H_2O \rightarrow 2HNO_3 + NO, \text{ etc.}$$

Thus the sequence in industrial utilization of atmospheric nitrogen is as follows:

$$N_2 \xrightarrow[\text{Haber process}]{H_2} NH_3 \xrightarrow[\text{Ostwald process}]{O_2} NO \xrightarrow{O_2 + H_2O} HNO_3(aq)$$

Ammonia is extremely soluble in water. Two stable crystalline hydrates are formed at low temperatures, $NH_3 \cdot H_2O$ (m.p. 194.15°K) and $2NH_3 \cdot H_2O$ (m.p. 194.32°K), in which the NH_3 and H_2O molecules are linked by hydrogen bonds. The substances contain neither NH_4^+ and OH^- ions nor discrete NH_4OH molecules. Thus $NH_3 \cdot H_2O$ has chains of H_2O molecules linked by hydrogen bonds (2.76 Å). These chains are cross-linked by NH_3 into a three-dimensional lattice by O—H$\cdots$N (2.78 Å) and O$\cdots$H—N bonds (3.21–3.29 Å). In aqueous solution ammonia is probably hydrated in a similar manner. Although aqueous solutions are commonly referred to as solutions of the weak base NH_4OH, called "ammonium hydroxide," this is to be discouraged, since there is no evidence that undissociated NH_4OH exists and there is reason to believe that it probably does not. Solutions of ammonia are best described as $NH_3(aq)$, with the equilibrium written as

$$NH_3(aq) + H_2O = NH_4^+ + OH^- \qquad K_{25°} = \frac{[NH_4^+][OH^-]}{[NH_3]} = 1.81 \times 10^{-5} \ (pK_b = 4.75)$$

In an odd sense NH_4OH might be considered a *strong* base, since it is completely dissociated in water. A $1M$ solution of NH_3 is only $0.0042M$ in NH_4^+ and OH^-.

Nuclear magnetic resonance measurements show that the hydrogen atoms of

[11]   F. Sperner and W. Hofmann, *Platinum Met. Rev.*, 1976, **20**, 12.

NH_3 rapidly exchange with those of water by the process

$$H_2O + NH_3 = OH^- + NH_4^+$$

but there is only slow exchange between NH_3 molecules in the vapor phase or in the liquid if water is completely removed.

Ammonium Salts. There are many rather stable crystalline salts of the tetrahedral NH_4^+ ion; most of them are water soluble, like alkali metal salts. Salts of strong acids are fully ionized, and the solutions are slightly acidic:

$$NH_4Cl = NH_4^+ + Cl^- \qquad K \approx \infty$$
$$NH_4^+ + H_2O = NH_3 + H_3O^+ \qquad K_{25°} = 5.5 \times 10^{-10}$$

Thus a $1 M$ solution will have a pH of $\sim$4.7. The constant for the second reaction is sometimes called the hydrolysis constant; however it may equally well be considered to be the acidity constant of the cationic acid NH_4^+, and the system regarded as an acid-base system in the following sense:

$$NH_4^+ + H_2O = H_3O^+ + NH_3(aq)$$
$$\text{Acid} \quad \text{Base} \quad \text{Acid} \quad \text{Base}$$

Ammonium salts generally resemble those of potassium and rubidium in solubility and, except where hydrogen bonding effects are important, in structure, since the three ions are of comparable (Pauling) radii: $NH_4^+ = 1.48$ Å, $K^+ = 1.33$ Å, $Rb^+ = 1.48$ Å.

Many ammonium salts volatilize with dissociation around $300°$, for example:

$$NH_4Cl(s) = NH_3(g) + HCl(g) \qquad \Delta H = 177 \text{ kJ mol}^{-1}; K_{25°} = 10^{-16}$$
$$NH_4NO_3(s) = NH_3(g) + HNO_3(g) \qquad \Delta H = 171 \text{ kJ mol}^{-1}$$

Some salts that contain oxidizing anions decompose when heated, with oxidation of the ammonia to N_2O or N_2 or both. For example:

$$(NH_4)_2Cr_2O_7(s) = N_2(g) + 4H_2O(g) + Cr_2O_3(s) \qquad \Delta H = -315 \text{ kJ mol}^{-1}$$
$$NH_4NO_3(l) = N_2O(g) + 2H_2O(g) \qquad \Delta H = -23 \text{ kJ mol}^{-1}$$

Ammonium nitrate volatilizes reversibly at moderate temperatures; at higher temperatures, irreversible decomposition occurs exothermically, giving mainly N_2O. This is the reaction by which N_2O is prepared commercially. At still higher temperatures, the N_2O itself decomposes into nitrogen and oxygen. Ammonium nitrate can be caused to detonate when initiated by another high explosive, and mixtures of ammonium nitrate with TNT or other high explosives are used for bombs. The decomposition of liquid ammonium nitrate can also become explosively rapid, particularly when catalyzed by traces of acid and chloride; there are a number of instances of disastrous explosions of ammonium nitrate in bulk following after fires. Moreover, ammonium perchlorate is important as an oxidizer in solid propellants for rocket fuels, and its thermal decomposition has been studied in detail.

Tetraalkylammonium ions (R_4N^+), prepared generally by the reaction

$$R_3N + RI = R_4N^+I^-$$

are often of use in inorganic chemistry when large univalent cations are required. Various R_4N radicals in the form of apparently crystalline amalgams ($\sim$12

Hg/R$_4$N) can be obtained either electrolytically or by reduction of R$_4$NX with Hg/Na in media where the resulting NaX is insoluble.

Hydrazine. Hydrazine (N$_2$H$_4$) may be thought of as derived from ammonia by replacement of a hydrogen atom by the —NH$_2$ group. It might therefore be expected to be a base, but somewhat weaker than NH$_3$, which is the case. It is a bifunctional base:

$$N_2H_4(aq) + H_2O = N_2H_5^+ + OH^- \qquad K_{25°} = 8.5 \times 10^{-7}$$
$$N_2H_5^+(aq) + H_2O = N_2H_6^{2+} + OH^- \qquad K_{25°} = 8.9 \times 10^{-16}$$

and two series of hydrazinium salts are obtainable. Those of N$_2$H$_5^+$ are stable in water, and those of N$_2$H$_6^{2+}$ are, as expected from the foregoing equilibrium constant, extensively hydrolyzed. Salts of N$_2$H$_6^{2+}$ can be obtained by crystallization from aqueous solution containing a large excess of the acid, since they are usually less soluble than the monoacid salts.

As another consequence of its basicity, hydrazine, like NH$_3$, can form coordination complexes with both Lewis acids and metal ions (Section 4-13). Just as with respect to the proton, electrostatic considerations (and, in these cases, also steric considerations) militate against bifunctional behavior.

Anhydrous N$_2$H$_4$ (m.p. 2°, b.p. 114°), a fuming colorless liquid with a high dielectric constant (ϵ = 52 at 25°), is surprisingly stable in view of its endothermic nature (ΔH_f^0 = 50 kJ mol^{-1}). It will burn in air, however, with considerable evolution of heat, which accounts for interest in it and certain of its alkylated derivatives as potential rocket fuels.

$$N_2H_4(l) + O_2(g) = N_2(g) + 2H_2O(l) \qquad \Delta H° = 622 \text{ kJ mol}^{-1}$$

At 25°C N$_2$H$_4$ is 100% in the *gauche* form 13-VII (cf. N$_2$F$_4$, below).

(13-VII)

Aqueous hydrazine is a powerful reducing agent in basic solution; in many of such reactions, diimine (see below) is an intermediate. One reaction, which is quantitative with some oxidants (e.g., I$_2$), is

$$N_2 + 4H_2O + 4e = 4OH^- + N_2H_4(aq) \qquad E^0 = -1.16 \text{ V}$$

However NH$_3$ and HN$_3$ are also obtained under various conditions. Air and oxygen, especially when catalyzed by multivalent metal ions in basic solution, produce hydrogen peroxide:

$$2O_2 + N_2H_4(aq) \rightarrow 2H_2O_2(aq) + N_2$$

but further reaction occurs in presence of metal ions:

$$N_2H_4 + 2H_2O_2 \rightarrow N_2 + 4H_2O$$

In acid solution, hydrazine can reduce halogens:

$$N_2H_4(aq) + 2X_2 \rightarrow 4HX + N_2$$

The preparation of hydrazine has been the subject of much study. Many reactions produce it in small amounts under certain conditions, for example:

$$N_2 + 2H_2 \rightarrow N_2H_4$$
$$N_2O + 2NH_3 \rightarrow N_2H_4 + H_2O + N_2$$
$$2NH_3(g) + \tfrac{1}{2}O_2 \rightarrow N_2H_4 + H_2O$$
$$N_2O + 3H_2 \rightarrow N_2H_4 + H_2O$$

However none of these has ever been developed into a practical method because there are competing, and thermodynamically more favorable, reactions, such as

$$2NH_3 + \tfrac{3}{2}O_2 = N_2 + 3H_2O$$
$$3N_2O + 2NH_3 = 4N_2 + 3H_2O$$
$$N_2O + H_2 = N_2 + H_2O$$

The last three reactions are good illustrations of the effect of the great stability of N_2 on nitrogen chemistry.

The only practical methods for preparing hydrazine in quantity are the Raschig synthesis, discovered in the first decade of this century, and a variant thereof. The overall reaction, carried out in aqueous solution, is

$$2NH_3 + NaOCl \rightarrow N_2H_4 + NaCl + H_2O$$

The reaction proceeds in two steps:

$$NH_3 + NaOCl \rightarrow NaOH + NH_2Cl \quad \text{(fast)}$$
$$NH_3 + NH_2Cl + NaOH \rightarrow N_2H_4 + NaCl + H_2O$$

However there is a competing and parasitic reaction that is rather fast once some hydrazine has been formed:

$$2NH_2Cl + N_2H_4 \rightarrow 2NH_4Cl + N_2$$

To obtain appreciable yields, it is necessary to add some gelatinous material, which serves two essential purposes. First, it sequesters heavy metal ions that catalyze the parasitic reaction: even the part per million or so of Cu^{2+} in ordinary water will almost completely prevent the formation of hydrazine if no catalyst is used. Since simple sequestering agents such as EDTA are not as beneficial as gelatin, the latter is assumed to have a positive catalytic effect as well. Yields of 60 to 70% are obtained under optimum conditions. Anhydrous hydrazine may be obtained by distillation over NaOH or by precipitating $N_2H_6SO_4$, which is then treated with liquid NH_3 to precipitate $(NH_4)_2SO_4$. A more recent variant of the Raschig process involves the use of a ketone to catalyze the reaction of Cl_2 with NH_3.

A recent, potentially viable process[12] avoids the use of chlorine compounds that require a lot of energy and provide disposal problems. This involves the sequence:

[12]   *Chem. Eng. News,* 1974, September 16, p. 19.

in which catalytic quantities of phosphate are used to serve as a carrier via a peroxoanion for oxidation of the intermediate imine. The azine is insoluble in the aqueous medium and is hydrolyzed separately.

Diazine, N_2H_2, and Other Nitrogen Hydrides.[13] Diazine can be obtained[14] mixed with NH_3, by microwave discharge in gaseous hydrazine, and in the pure state by the reaction (Tos = p-toluenesulfonate)

The diamagnetic yellow compound is unstable above $-180°$, decomposing mainly to N_2, H_2, and N_2H_4. Thermolysis of the potassium salt gives the *cis* isomer, and the cesium salt gives $N=NH_2$, isodiazene; both are very unstable. *Trans*-$CH_3N=NH$ and both *cis*- and *trans*-$CH_3N=NCH_3$ are known.[15]

Diazine also has a transient existence in solution during oxidations of hydrazine by two-electron oxidants (molecular oxygen, peroxides, chloramine-T, etc.):

$$N_2H_4 \xrightarrow{-2H} N_2H_2$$

It is also formed in alkaline cleavage of chloramine:

$$H_2NCl \xrightarrow{OH^-} HNCl^-$$

$$HNCl^- + H_2NCl \xrightarrow{-Cl^-} HN-NH_2 \xrightarrow{-HCl} HN=NH$$
$$\quad\quad\quad\quad\quad\quad\quad\quad\quad\quad | $$
$$\quad\quad\quad\quad\quad\quad\quad\quad\quad\quad Cl$$

The existence of N_2H_2 has been shown by, *inter alia,* the stereospecific *cis*-hydrogenation of C=C bonds by hydrazine and an oxidant.

Some other unstable hydrides (e.g., N_4H_2 and N_4H_4) are known,[13] and interaction of NH_3 on a silver zeolite produces Ag^+ complexes of triazane, N_3H_5, and cyclotriazane (N_3H_3).[16]

[13] N. Wiberg, *Chimia,* 1976, **30**, 426; *Angew. Chem. Int. Ed.,* 1975, **14**, 177, 178.
[14] C. Willis, *et al., Can. J. Chem.,* 1973, **51**, 3605; 1974, **52**, 1006; N. Wiberg, G. Fischer, and H. Backhuber, *Angew. Chem., Int. Ed.,* 1977, **16**, 780.
[15] M. N. Ackerman *et al., J. Am. Chem. Soc.,* 1977, **99**, 1661.
[16] Y. Kim, J. W. Gilje, and K. Seff, *J. Am. Chem. Soc.,* 1977, **99**, 7057.

Hydrazoic Acid and Azides.[17a] Although hydrazoic acid (HN_3), is a hydride of nitrogen in a formal sense, it has no essential relationship to NH_3 and N_2H_4. The sodium salt is prepared by the reactions

$$3NaNH_2 + NaNO_3 \xrightarrow{175°} NaN_3 + 3NaOH + NH_3$$

$$2NaNH_2 + N_2O(g) \xrightarrow{190°} NaN_3 + NaOH + NH_3$$

and the free acid can be obtained in solution by the reaction

$$N_2H_5^+ + HNO_2 \xrightarrow{\text{aq. soln.}} HN_3 + H^+ + 2H_2O$$

Many other oxidizing agents attack hydrazine to form small amounts of HN_3 or azides. Hydrazoic acid ($pK_a^{25} = 4.75$), obtainable pure by distillation from aqueous solutions, is a colorless liquid (b.p. 37°) and dangerously explosive. Azides of many metals are known: those of heavy metals are generally explosive; lead, mercury, and barium azides explode on being struck sharply and are used in detonation caps.

Azides of electropositive metals are not explosive and, in fact, decompose smoothly and quantitatively when heated to 300° or higher, for example,

$$2NaN_3(s) = 2Na(l) + 3N_2(g)$$

Azide ion also functions as a ligand in complexes of transition metals.[17b] In general, N_3^- behaves rather like a halide ion and is commonly considered to be a pseudohalide, although the corresponding pseudohalogen ($N_3)_2$ is not known.

The azide ion itself is symmetrical and linear (N—N, 1.16 Å), and its electronic structure may be represented in valence bond theory as

$$:\ddot{N}\!=\!\overset{+}{N}\!=\!\ddot{N}: \longleftrightarrow :N\!\equiv\!\overset{+}{N}\!:\!\overset{2-}{N}: \longleftrightarrow :\overset{2-}{\ddot{N}}\!:\!\overset{+}{N}\!\equiv\!N:$$

In covalent azides, on the other hand, the symmetry is lost, as is evident in HN_3 and CH_3N_3 (Fig. 13-3). In such covalent azides the electronic structure is a resonance hybrid:

$$R\!:\!\ddot{N}\!=\!\overset{+}{N}\!=\!\ddot{N}: \longleftrightarrow R\!:\!\overset{..}{\ddot{N}}\!:\!\overset{+}{N}\!\equiv\!N:$$

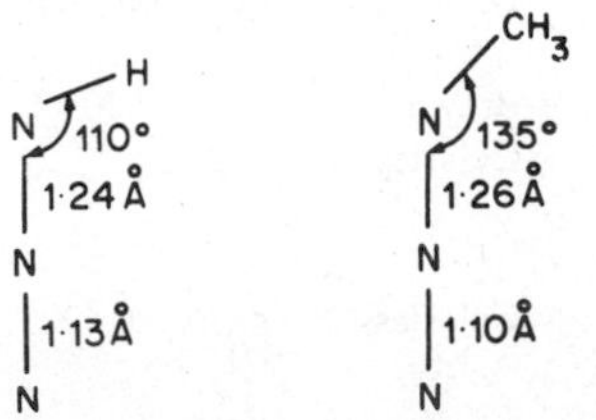

Fig. 13-3. Structures of HN_3 and CH_3N_3.

[17a] H. D. Fair and R. F. Walker, Eds., *Energetic Materials,* Vols. 1, 2, Plenum Press, 1977.
[17b] Z. Dori and R. F. Ziolo, *Chem. Rev.,* 1973, **73**, 247.

Hydroxylamine. Since hydrazine may be thought of as derived from ammonia by replacement of one hydrogen by NH_2, so hydroxylamine (NH_2OH) is obtained by replacement of H by OH. Like hydrazine, hydroxylamine is a weaker base than NH_3:

$$NH_2OH(aq) + H_2O = NH_3OH^+ + OH^- \qquad K_{25°} = 6.6 \times 10^{-9}$$

Hydroxylamine is prepared by reduction of nitrates or nitrites either electrolytically or with SO_2, under very closely controlled conditions. It is also made, in 70% yield, by H_2 reduction of NO_2 in HCl solution with platinized active charcoal as catalyst. Free hydroxylamine is a white solid (m.p. 33°) that must be kept at 0°C to avoid decomposition. It is normally encountered as an aqueous solution and as salts, e.g., $[NH_3OH]Cl$, $[NH_3OH]NO_3$ and $[NH_3OH]_2SO_4$, which are stable, water-soluble, white solids. Although hydroxylamine can serve as either an oxidizing or a reducing agent, it is usually used as the latter.

The reaction

$$4Fe^{3+} + 2NH_3OH^+ = 4Fe^{2+} + N_2O + 6H^+ + H_2O$$

is quantitative but mechanistically complex.[18]

13-6. Oxides of Nitrogen

The known oxides of nitrogen are listed in Table 13-2, and their structures are shown in Fig. 13-4.

Nitrous Oxide. Nitrous oxide (N_2O) is obtained by thermal decomposition of ammonium nitrate in the melt at 250 to 260°:

$$NH_4NO_3 \rightarrow N_2O + 2H_2O$$

The contaminants are NO, which can be removed by passage through iron (II) sulfate solution and 1 to 2% of nitrogen. The NH_4NO_3 must be free from Cl^-, since

TABLE 13-2
Oxides of Nitrogen

Formula	Name	Color	Temperatures (°C)	Remarks
N_2O	Nitrous oxide	Colorless	m. −90.8; b. −88.5	Rather unreactive
NO	Nitric oxide	Colorless	m. −163.6; b. −151.8	Moderately reactive
N_2O_3	Dinitrogen trioxide	Dark blue	f.p. −100.6; d. 3.5	Extensively dissociated as gas
NO_2	Nitrogen dioxide	Brown	m. −11.20	Rather reactive
N_2O_4	Dinitrogen tetroxide	Colorless	b. 21.2	Extensively dissociated to NO_2 as gas and partly as liquid
N_2O_5	Dinitrogen pentoxide	Colorless	m. 30; d. 47	Unstable as gas; ionic solid
NO_3; N_2O_6	—	—	—	Not well characterized and quite unstable

[18] G. Bengtsson, *Acta Chem. Scand.,* 1973, **27,** 1717.
[19] R. Kummel and F. Pieschel, *Z. Anorg. Allg. Chem.,* 1973, **396,** 90.

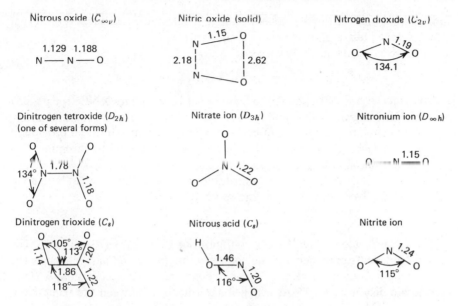

Fig. 13-4. The structures and point group symmetries of some nitrogen oxides and anions (angles in degrees; bond lengths in Å).

this catalytically causes decomposition to N_2.[19] However heating HNO_3 or H_2SO_4 solutions of NH_4NO_3 with small amounts of Cl^- gives almost pure N_2O. The gas is also produced in the reduction of nitrites and nitrates under certain conditions and by decomposition of hyponitrites.

Nitrous oxide is relatively unreactive, being inert to the halogens, alkali metals, and ozone at room temperature. It will oxidize some low-valent transition metal complexes and forms the complex $[Ru(NH_3)_5N_2O]^{2+}$ (Section 22-F-4). At elevated temperatures it decomposes to nitrogen and oxygen, reacts with alkali metals and many organic compounds, and supports combustion. Apart from its anesthetic role, its chief commercial use is as an aerosol propellant.

Nitric Oxide. Nitric oxide (NO) is formed in many reactions involving reduction of nitric acid and solutions of nitrates and nitrites. For example, with $8M$ nitric acid:

$$8HNO_3 + 3Cu \rightarrow 3Cu(NO_3)_2 + 4H_2O + 2NO$$

Reasonably pure NO is obtained by the following aqueous reactions:

$$2NaNO_2 + 2NaI + 4H_2SO_4 \rightarrow I_2 + 4NaHSO_4 + 2H_2O + 2NO$$
$$2NaNO_2 + 2FeSO_4 + 3H_2SO_4 \rightarrow Fe_2(SO_4)_3 + 2NaHSO_4 + 2H_2O + 2NO$$

or dry:

$$3KNO_2(l) + KNO_3(l) + Cr_2O_3(s) \rightarrow 2K_2CrO_4(s, l) + 4NO$$

Commercially it is obtained by catalytic oxidation of ammonia as already noted. Direct combination of the elements occurs only at very high temperatures, and to isolate the small amounts so formed (a few volume percent at 3000°) the equilibrium mixture must be rapidly chilled. Though much studied, this reaction has not been developed into a practical commercial synthesis.

Nitric oxide reacts instantly with O_2:

$$2NO + O_2 \rightarrow 2NO_2$$

It also reacts with F_2, Cl_2, and Br_2 to form the nitrosyl halides XNO (see p. 435) and with CF_3I to give CF_3NO and I_2. It is oxidized to nitric acid by several strong oxidizing agents; the reaction with permanganate is quantitative and provides a method of analysis. It is reduced to N_2O by SO_2 and to NH_2OH by chromium (II) ion, in acid solution in both cases.

Nitric oxide is thermodynamically unstable at 25° and 1 atm and at high pressures it readily decomposes[20] in the range 30–50°:

$$3NO \rightarrow N_2O + NO_2$$

The NO molecule has the electron configuration $(\sigma_1)^2(\sigma_1{*})^2(\sigma_2, \pi)^6(\pi{*})$. The unpaired $\pi{*}$ electron renders the molecule paramagnetic and partly cancels the effect of the π-bonding electrons. Thus the bond order is 2.5, consistent with an interatomic distance of 1.15 Å, which is intermediate between the triple-bond distance in NO^+ (see below) of 1.06 Å and representative double-bond distances of ~1.20 Å.

Nitric oxide dimerizes in the solid state and dimers also persist in the vapor at the boiling point.[21a] The binding energy of the dimer is less than 10 kJ mol^{-1}, consistent with the long N—N bond. The dimer has no unpaired spins but feeble intrinsic temperature-independent paramagnetism. Unstable forms can be isolated in matrices.

The electron in the $\pi{*}$ orbital is relatively easily lost ($\Delta H_{ion} = 891$ kJ mol^{-1}), to give the *nitrosonium ion* NO^+, which has an extensive and important chemistry. Because the electron removed comes out of an antibonding orbital, the bond is stronger in NO^+ than in NO: the bond length decreases by 0.09 Å and the vibration frequency rises from 1840 cm^{-1} in NO to 2150–2400 cm^{-1} (depending on environment) in NO^+. Numerous ionic compounds of NO^+ are known.

When N_2O_3 or N_2O_4 is dissolved in concentrated sulfuric acid, the ion is formed:

$$N_2O_3 + 3H_2SO_4 = 2NO^+ + 3HSO_4^- + H_3O^+$$
$$N_2O_4 + 3H_2SO_4 = NO^+ + NO_2^+ + 3HSO_4^- + H_3O^+$$

The isolable compound $NO^+HSO_4^-$, nitrosonium hydrogen sulfate, is an important intermediate in the lead-chamber process for manufacture of sulfuric acid. Its saltlike constitution has been shown by electrolysis, conductivity studies, and cryoscopic measurements. The compounds $NO^+ClO_4^-$ and $NO^+BF_4^-$, both isostructural with the corresponding ammonium and H_3O^+ compounds, are known;

[20] T. P. Melia, *J. Inorg. Nucl. Chem.*, 1965, **27**, 95.
[21a] S. Skaarup, P. N. Skanke, and J. E. Boggs, *J. Am. Chem. Soc.*, 1976, **98**, 6106.

many others such as $(NO)_2PtCl_6$, $NOFeCl_4$, $NOAsF_6$, $NOSbF_6$, and $NOSbCl_6$ may be made in the following general ways:

$$NO + MoF_6 \rightarrow NO^+MoF_6^-$$
$$ClNO + SbCl_5 \rightarrow NO^+SbCl_6^-$$

All such salts are readily hydrolyzed:

$$NO^+ + H_2O \rightarrow H^+ + HNO_2$$

and they must be prepared and handled under anhydrous conditions.

In alkaline solution at $0°$, SO_3^{2-} reacts with NO to give a white crystalline solid, potassium N-nitrosohydroxylamine-N-sulfonate ($K_2SO_3N_2O_2$)[21b]:

$$O\dot{N} + :SO_3^{2-} = [O\dot{N} \leftarrow SO_3]^{2-}$$
$$[O\dot{N}SO_3]^{2-} + \dot{N}O = [O=N \diagdown_{N}^{\diagup} SO_3]^{2-}$$
$$\underset{O}{\overset{|}{}}$$

Other species with N_2O_2 groups are obtained by interaction of amines with NO; alcohol in base also gives $[O_2N_2CH_2N_2O_2]^{2-}$.

The NO^+ ion is isoelectronic with CO, and, like CO, will form bonds to metals. Thus, for example, analogous to nickel carbonyl, $Ni(CO)_4$, there is the isoelectronic $Co(CO)_3NO$. These transition metal nitrosyl complexes are discussed in Sections 3-10 and 25-2), but we note here that the compound responsible for the brown ring in the test for nitrates is a nitrosyl complex of iron with the formula $[Fe(H_2O)_5NO]^{2+}$.

Dinitrogen Trioxide. This oxide is best obtained by interaction of stoichiometric quantities of NO and O_2 or NO and N_2O_4 as an intensely blue liquid and a pale blue solid. It is formally the anhydride of nitrous acid, and dissolution of an equimolar mixture of NO and NO_2 in alkalis gives virtually pure nitrite; in the gas phase nitrous acid is formed.[22]

The dissociation of N_2O_3:

$$N_2O_3 = NO + NO_2$$

begins to be significant above $-30°$. There appears to be some self-ionization[23] in the liquid;

$$N_2O_3 \rightleftharpoons NO^+ + NO_2^-$$

[21b] R. Longhi, R. O. Ragsdale, and R. S. Drago, *Inorg. Chem.*, 1962, **1**, 768; T. L. Nunes and R. E. Powell, *Inorg. Chem.*, 1970, **9**, 1916.
[22] A. J. Vosper, *J.C.S. Dalton*, **1976**, 135.
[23] A. W. Shaw, A. J. Vosper, and M. Pritchard, *J.C.S. Dalton*, **1974**, 2172.

The molecule is polar in the sense $ON^{\delta+}$—$NO_2^{\delta-}$. The stable structure in the solid has a long N—N bond, and this structure persists in the liquid and gas (Fig. 13-4).

Nitrogen Dioxide and Dinitrogen Tetroxide. These two oxides, NO_2 and N_2O_4, exist in a strongly temperature-dependent equilibrium

$$2NO_2 \rightleftharpoons N_2O_4$$

Brown,   Colorless,
paramagnetic   diamagnetic

both in solution and in the gas phase, where ΔH°_{298} for dissociation is 57 kJ mol^{-1}. In the solid state the oxide is wholly N_2O_4. Partial dissociation occurs in the liquid; it is pale yellow at the freezing point and contains 0.01% of NO_2, which increases to 0.1% in the deep red-brown liquid at the boiling point, 21.15°. In the vapor at 100° the composition is NO_2 90%, N_2O_4 10%, and dissociation is complete above 140°. Molecular beam mass spectrometric studies indicate that trimers and tetramers occur,[24] but there is no evidence for such species under normal conditions.

The monomer NO_2 has an unpaired electron and its properties, red-brown color and ready dimerization to colorless and diamagnetic N_2O_4, are not unexpected for such a radical. NO_2 can also lose its odd electron fairly readily (ΔH_{ion} = 928 kJ mol^{-1}) to give NO_2^+, the *nitronium ion,* discussed below.

Although other forms can exist in inert matrixes, the most stable form of N_2O_4 is that shown in Fig. 13-4. This molecule has unusual features[25]: namely, the planarity and the long N—N bond, 1.78 Å compared to 1.47 in H_2N—NH_2. Molecular orbital calculations suggest that although the long N—N bond is of σ type, it is long because of delocalization of the electron pair over the whole molecule, with large repulsion between doubly occupied MOs of NO_2. The coplanarity results from a delicate balance of forces favoring the skew and planar forms. The barrier to rotation about the N—N bond is estimated to be about 9.6 kJ mol^{-1}.

The mixed oxides are obtained by heating metal nitrates, by oxidation of nitric oxide in air, and by reduction of nitric acid and nitrates by metals and other reducing agents. The gases are highly toxic and attack metals rapidly. They react with water:

$$2NO_2 + H_2O = HNO_3 + HNO_2$$

the nitrous acid decomposing, particularly when warmed:

$$3HNO_2 = HNO_3 + 2NO + H_2O$$

The thermal decomposition

$$2NO_2 \rightleftharpoons 2NO + O_2$$

begins at 150° and is complete at 600°.

The oxides are fairly strong oxidizing agents in aqueous solution, comparable in strength to bromine:

$$N_2O_4 + 2H^+ + 2e = 2HNO_2 \qquad E^0 = +1.07 \text{ V}$$

[24]   S. E. Novick, B. J. Howard, and W. Klemperer, *J. Chem. Phys.,* 1972, **57**, 5619.
[25]   For references, see: D. C. Frost, C. A. McDowell, and N. P. C. Westwood, *J. Electron Spectrosc. Relat. Phenom.,* 1977, **10**, 293.

The mixed oxides, "nitrous fumes," are used in organic chemistry as selective oxidizing agents; the first step is hydrogen abstraction:

$$RH + NO_2 = R\cdot + HONO$$

and the strength of the C—H bond generally determines the nature of the reaction.

Nitrogen dioxide and nitric oxide, commonly referred to as NO_x, are both of concern in atmospheric pollution being produced in combustion.

Dinitrogen tetroxide has been extensively studied as a nonaqueous solvent. The electrical conductivity of the liquid is quite low, and self-ionization is very endothermic.[26]

$$N_2O_4 \rightleftharpoons NO^+ + NO_3^- \qquad \Delta H = 49.8 \text{ kJ mol}^{-1}$$

It forms molecular addition compounds with a great variety of nitrogen, oxygen, and aromatic donor compounds. Systems involving liquid N_2O_4 mixed with an organic solvent are often very reactive; for example, they dissolve relatively noble metals to form nitrates, often solvated with N_2O_4. Thus copper reacts vigorously with N_2O_4 in ethyl acetate to give crystalline $Cu(NO_3)_2 \cdot N_2O_4$, from which anhydrous, volatile (at 150–200°) cupric nitrate is obtained (Section 21-H-3). Some of the compounds obtained in this way may be formulated as nitrosonium salts, for example, $Zn(NO_3)_2 \cdot 2N_2O_4$ as $(NO^+)_2[Zn(NO_3)_4]^{2-}$.

X-ray diffraction studies of $Cu(NO_3)_2 \cdot N_2O_4$[27] confirm that no molecular N_2O_4 is present; it consists of NO^+ ions and polymeric nitrato anions. The complex $Fe(NO_3)_3(1.5N_2O_4)$ also appears to have a nitrato anion $[Fe(NO_3)_4]^-$, but with a cation $N_4O_6^{2+}$ that can be regarded as NO_3^- bound to three NO^+ groups.[28]

In anhydrous acids N_2O_4 dissociates ionically, as in H_2SO_4 above, and in anhydrous HNO_3 dissociation is almost complete:

$$N_2O_4 = NO^+ + NO_3^-$$

The dissociation in H_2SO_4 is complete in dilute solution; at higher concentrations undissociated N_2O_4 is present, and at very high concentrations nitric acid is formed:

$$N_2O_4 + 3H_2SO_4 = NO^+HSO_4^- + HNO_3 + HSO_4^- + SO_3 + H_3O^+$$

The $NOHSO_4$ actually crystallizes out. The detailed mechanism and intermediates are undoubtedly complex.

Dinitrogen Pentoxide. The oxide N_2O_5 is usually obtained by dehydration of nitric acid with P_2O_5: it is not too stable (sometimes exploding) and is distilled in a current of ozonized oxygen.

$$2HNO_3 + P_2O_5 \rightarrow 2HPO_3 + N_2O_5$$

It is, conversely, the anhydride of nitric acid:

$$N_2O_5 + H_2O = 2HNO_3$$

It is deliquescent, readily producing nitric acid by the reaction above.

[26]   R. Andinos, *J. Chim. Phys.,* 1972, **69**, 1263.
[27]   L. J. Blackwell, T. J. King, and A. Morris, *J.C.S. Chem. Comm.,* **1973**, 644.
[28]   C. C. Addison *et al., J.C.S. Chem. Comm.,* **1973**, 347.

The gaseous compound appears to have a structure of type 13-VIII

(13-VIII)

with a bent N—O—N group, although this angle may be near 180°. Solid N_2O_5 in its stable form is nitronium nitrate ($NO_2^+NO_3^-$), but when the gas is condensed on a surface at ~90°K, the molecular form is obtained and persists for several hours. On warming to ~200°K, however, the latter rapidly changes to $NO_2^+NO_3^-$. The structures of the two ions are shown in Fig. 13-4.

As with N_2O_4, ionic dissociation occurs in anhydrous H_2SO_4, HNO_3, or H_3PO_4 to produce NO_2^+, for instance,

$$N_2O_5 + 3H_2SO_4 \rightleftharpoons 2NO_2^+ + 3HSO_4^- + H_3O^+$$

Many gas phase reactions of N_2O_5 depend on dissociation to NO_2 and NO_3, with the latter then reacting further as an oxidizing agent. These reactions are among the better understood complex inorganic reactions.

In the N_2O_5-catalyzed decomposition of ozone, the steady state concentration of NO_3 can be high enough to allow its absorption spectrum to be recorded.

Nitrosylazide, Nitrylazide. Although they have the stoichiometry of nitrogen oxides (i.e., N_4O, N_4O_2), these unstable compounds[29] are azides of NO^+ and NO_2^+, respectively. They are obtained in solution by action of NO^+ and NO^{2+} salts (e.g., of BF_4^-) with sodium azide in an organic solvent but have not been isolated. N_4O decomposes at −50° and N_4O_2 above −10°.

13-7.   Oxo Acids of Nitrogen

Hyponitrous Acid.[30]   Sodium hyponitrite is best obtained by reduction of nitric oxide with sodium in presence of benzophenone in dimethoxyethane-toluene.[31] The salt can be extracted with water and crystallized by addition of ethanol. The silver salt, which is insoluble in water, is used in syntheses of alkyl hyponitrites. The free acid, $pK = 7$, can be obtained by action of HCl on the silver salt; it is moderately stable in solution. Hyponitrites of the alkali metals react with CO_2 to give N_2O.

The hyponitrite ion has the trans configuration (13-IX).

(13-IX)   (13-X)

Hyponitrites undergo various oxidation-reduction reactions in acid and alkaline

[29]   M. P. Doyle, J. J. Maciejko, and S. C. Busman, *J. Am. Chem. Soc.,* 1973, **95**, 952.
[30]   M. N. Hughes, *Q. Rev.,* 1968, **22**, 1.
[31]   G. D. Mendenhall, *J. Am. Chem. Soc.,* 1974, **96**, 5000.

solutions, depending on conditions; they usually behave as reducing agents, however.

There is a compound, *nitramide*, which is also a weak acid ($K_{25°} = 2.6 \times 10^{-7}$) and is an isomer of hyponitrous acid. Its structure has been shown to be 13-X.

The Oxohyponitrite or trioxodinitrate(II) Ion ($N_2O_3^{2-}$). The interaction of hydroxylamine and an alkyl nitrate in methanol containing sodium methoxide at 0° gives the salt $Na_2N_2O_3$.

Solutions are readily oxidized by air and decompose to give N_2O and NO_2^-, possibly by way of HNO or NOH as an intermediate, since at pH 4 to 8 the decomposition is first order in $HN_2O_3^-$,[32] namely,

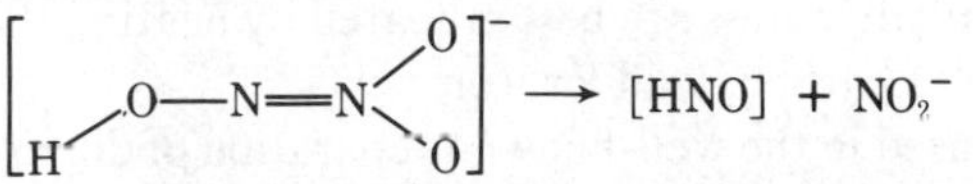

$$\left[H{-}O{-}N{=}N{\overset{\displaystyle O}{\underset{\displaystyle O}{\diagup}}} \right]^- \longrightarrow [HNO] + NO_2^-$$

$$2[HNO] \longrightarrow N_2O + H_2O$$

The species HNO, which is isoelectronic with O_2, and its dimer HON=NOH, have been detected[33] in the gas phase in the reaction

$$\cdot H + NO = HNO$$

The structure[34a] of the $N_2O_3^{2-}$ ion (13-XI) in $Na_2N_2O_5{\cdot}H_2O$ indicates that the N=N bond is similar to that in azobenzenes and that bonding in the ion can be represented approximately by 13-XII.

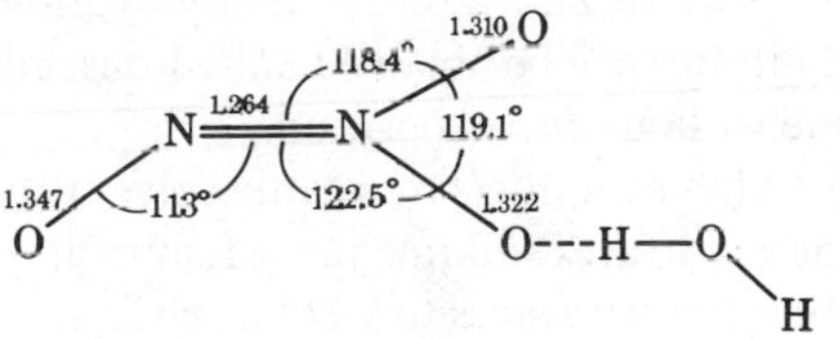

(13-XI)

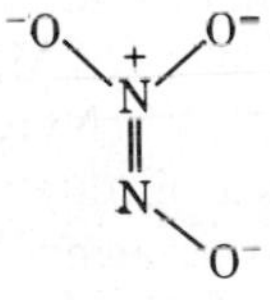

(13-XII)

Nitrous Acid. Solutions of the weak acid HNO_2($pK_a^{25} = 5.22$) are easily made by acidifying solutions of nitrites. The aqueous solution can be obtained free of salts by the reaction

$$Ba(NO_2)_2 + H_2SO_4 \rightarrow 2HNO_2 + BaSO_4(s)$$

The acid is unknown in the liquid state, but it can be obtained in the vapor phase; the *trans* form (Fig. 13-4) has been shown to be more stable than the *cis* form by about 2.1 kJ mol^{-1}. In the gas phase the following equilibrium is rapidly established:

$$NO + NO_2 + H_2O = 2HNO_2 \qquad K_{20°} = 1.56 \; atm^{-1}$$

Aqueous solutions of nitrous acid are unstable and decompose rapidly when

[32] F. T. Bonner and B. Ravid, *Inorg. Chem.*, 1975, **14**, 558; M. N. Hughes and P. E. Wimbledon, *J. C. S. Dalton*, **1976**, 703.

[33] See G. A. Gallup, *Inorg. Chem.*, 1975, **14**, 563.

[34a] H. Hope and M. R. Sequeira, *Inorg. Chem.*, *1973*, **12**, 286.

heated, according to the equation

$$3HNO_2 = HNO_3 + H_2O + 2NO$$

This reaction is reversible. Nitrous acid can behave both as an oxidant, e.g., toward I^-, Fe^{2+}, or $C_2O_4^{2-}$:

$$HNO_2 + H^+ + e = NO + H_2O \qquad E^0 = 1.0 \text{ V}$$

and as a reducing agent[34b]:

$$NO_3^- + 3H^+ + 2e = HNO_2 + H_2O \qquad E^0 = 0.94 \text{ V}$$

Nitrites of the alkali metals are best prepared by heating the nitrates with a reducing agent such as carbon, lead, or iron.

Nitrous acid is used in the well-known preparation of diazonium compounds in organic chemistry. Numerous organic derivatives of the NO_2 group are known. They are of two types: nitrites (R—ONO) and nitro compounds (R—NO$_2$). Similar tautomerism occurs in inorganic complexes, in which either oxygen or nitrogen is the actual donor atom when NO_2^- is a ligand. (Section 4-28).

The nitrite ion is bent (Fig. 13-4), as expected.

Peroxonitrous acid (HOONO), is formed as an intermediate in the oxidation of HNO_2 to HNO_3 by H_2O_2. Although the acid is unstable, the anion is stable in alkaline solution, imparting a yellow color.

Nitric Acid and Nitrates. The acid has already been discussed (p. 243). Nitrates of almost all metallic elements are known. They are frequently hydrated and most are soluble in water. Many metal nitrates can be obtained anhydrous, and a number of these, e.g., $Cu(NO_3)_2$, sublime without decomposition.

Anhydrous nitrates often have high solubility in organic solvents and are best regarded as covalent.[35] Alkali metal nitrates sublime in a vacuum at 350 to 500°, but decomposition occurs at higher temperatures to yield nitrites or, at very high temperatures, oxides or peroxides. NH_4NO_3 gives N_2O and H_2O (cf. above). In neutral solution, nitrates can be reduced only with difficulty, and the mechanism of the reduction is still obscure. Aluminum or zinc in alkaline solution produces NH_3.

Nitrate, when covalently bound in anhydrous nitrates, or in nitrato complexes (see Section 4-28) can be unidentate ($MONO_2$), bidentate (MO_2NO), bridging bidentate [$MON(O)OM$], or bridging terdentate [$MON(OM)OM$]. The interaction of $NaNO_3$ and Na_2O at 300° gives a product believed to contain the orthonitrate ion NO_4^{3-}; it is very readily hydrolyzed, regenerating NO_3^-.[36]

The Nitronium Ion. This ion (NO_2^+) is directly involved, not only in the dissociation of nitric acid itself, but also in nitration reactions and in solutions of nitrogen oxides in nitric and other strong acids.

[34b] P. L. Asquith and B. J. Tyler, *J.C.S. Chem. Comm.*, **1970**, 744.
[35]   K. F. Chew *et al.*, *J.C.S. Dalton*, **1975**, 1315.
[36]   M. Jansen, *Angew. Chem., Int. Ed.*, **1977**, **16**, 534.

Detailed kinetic studies on the nitration of aromatic compounds[37] first led to the idea that the attacking species was the NO_2^+ ion generated by ionizations of the following types:

$$2HNO_3 = NO_2^+ + NO_3^- + H_2O$$
$$HNO_3 + H_2SO_4 = NO_2^+ + HSO_4^- + H_2O$$

The importance of the first type is reflected in the fact that addition of ionized nitrate salts to the reaction mixture will retard the reaction. The actual nitration process can then be formulated as

The dissociation of nitric acid in various media has been confirmed by cryoscopic studies, and nitrogen oxides have also been found to dissociate to produce nitronium ions as noted above. Spectroscopic studies have confirmed the presence of the various ions in such solutions. For example, the NO_2^+ ion can be identified by a Raman line at about 1400 cm^{-1}.

Crystalline *nitronium salts* can be made by reactions such as:

$$N_2O_5 + HClO_4 \rightarrow [NO_2^+ClO_4^-] + HNO_3$$
$$N_2O_5 + FSO_3H \rightarrow [NO_2^+FSO_3^-] + HNO_3$$
$$HNO_3 + 2SO_3 \rightarrow [NO_2^+HS_2O_7^-]$$

The first two of these reactions are really just metatheses, since N_2O_5 in the solid and in anhydrous acid solution is $NO_2^+NO_3^-$. The other reaction is one between an acid anhydride, SO_3, and a base(!), $NO_2^+OH^-$.

Nitronium salts are thermodynamically stable, but very reactive chemically. They are rapidly hydrolyzed by moisture; in addition $NO_2^+ClO_4^-$, for example, reacts violently with organic matter, but it can actually be used to carry out nitrations in nitrobenzene solution.

The Hydroxylamine-N,N-Disulfonate and Nitroso Disulfonate Ions. The interaction of nitrite and bisulfite ions gives the hydroxylamine-N,N-disulfonate ion $HON(SO_3)_2^{2-}$ or, in base, $ON(SO_3)_2^{3-}$.

$$2HSO_3^- + HNO_2 \rightarrow HON(SO_3)_2^{2-} \xrightarrow{OH^-} ON(SO_3)_2^{3-}$$

In the salt $Na_3[ON(SO_3)_2]\cdot H_2O$ the N atom is pyramidal and is bound to oxygen and to two S atoms of the sulfonate.[38] The colorless ion can be oxidized electrolytically in a one-electron step or by $KMnO_4$ in ammonia solution to give the *nitrosodisulfonate ion* $ON(SO_3)_2^{2-}$. This ion is violet and paramagnetic in solution. In salts of large cations it is also paramagnetic. However the potassium salt (Fremy's

[37]   R. B. Moodie and K. Schofield, *Acc. Chem. Res.*, 1976, **9**, 287; L. F. Albright and C. Hanson, Eds., *Industrial and Laboratory Nitrations*, ACS Symposium Series No. 22, 1975. G. A. Olah *et al.*, *Proc. Nat. Acad. Sci. (U.S.)*, 1978, **75**, 1045; G. Davis and N. Cook, *Chem. Tech.*, **1977**, 626.
[38]   J. S. Rutherford and B. E. Robertson, *Inorg. Chem.*, 1975, **14**, 2537.

salt)[39] is dimorphic with yellow and orange-brown forms. The yellow monoclinic version, though nearly diamagnetic, has a thermally accessible triplet state. The triclinic form is paramagnetic with magnetic interaction between neighboring ions, which are arranged:

$$
\begin{array}{c}
O_3\bar{S} \quad\quad SO_3^- \\
\diagdown\quad\diagup \\
N\text{-----}O \\
| \quad\quad\quad | \;\; 1.28 \\
O\underset{2.86}{\text{-----}}N \\
\diagup\quad\diagdown \\
\bar{S}O_3 \quad\quad SO_3^-
\end{array}
$$

13-8. Halogen Compounds of Nitrogen

Binary Halides. In addition to NF_3, NF_2Cl, $NFCl_2$, and NCl_3, we have N_2F_2, N_2F_4, and the halogen azides XN_3 (X = F, Cl, Br, or I). With the exception of NF_3 the halides are reactive, potentially hazardous substances, some of them like $NFCl_2$ explosive, others not. Only the fluorides are important.

Nitrogen Trifluoride. NF_3, plus small amounts of dinitrogen difluoride (N_2F_2) is obtained by electrolysis of NH_4F in anhydrous HF, whereas electrolysis of molten NH_4F constitutes a preferred preparative method for N_2F_2. The following reactions have all been proposed as good synthetic ones for the several nitrogen fluorides:

$$
2NF_2H \xrightarrow[\text{pH 1 - 2}]{\text{FeCl}_3(\text{aq})} N_2F_4 \;(\sim100\%)
$$

$$
NH_3 + F_2 \;(\text{diluted by } N_2) \xrightarrow[\text{reactor}]{\text{copper-packed}}
\begin{cases}
NF_3 \\
N_2F_4 \\
N_2F_2 \\
NHF_2
\end{cases}
$$

(Predominant product depends on conditions, esp. F_2/NH_3 ratio.)

$$
2NF_2H + 2KF \rightarrow 2KHF_2 + N_2F_2
$$

N_2F_2 may also be prepared by photolysis of N_2F_4 in presence of Br_2.

Nitrogen trifluoride (b.p. $-129°$) is a very stable gas that normally is reactive only at about 250 to 300° but reacts readily with $AlCl_3$ at 70°:

$$
2NF_3 + 2AlCl_3 \rightarrow N_2 + 3Cl_2 + 2AlF_3
$$

It is not affected by water or most other reagents at room temperature and does not decompose when heated in the absence of reducing metals; when heated in presence of fluorine acceptors such as copper, the metal is fluorinated and N_2F_4 is obtained. The NF_3 molecule has a pyramidal structure but a very low dipole moment, and it appears to be totally devoid of donor properties.

---

[39] B. J. Wilson, J. M. Hayes, and J. A. Durbin, *Inorg. Chem.*, 1976, **15**, 1702; B. D. Perlson and D. B. Russell, *Inorg. Chem.*, 1975, **14**, 2097.

The Tetrafluoroammonium Ion.[40] The interaction of NF_3, F_2, and a strong Lewis acid such as BF_3, AsF_5, or SbF_5 under pressure, ultraviolet irradiation at low temperatures, or glow discharge in a sapphire apparatus, gives salts of the ion NF_4^+, e.g.,

$$NF_3 + F_2 + BF_3 \xrightarrow{h\nu} NF_4^+ BF_4^-$$

$$NF_3 + F_2 + AsF_5 \rightarrow NF_4^+ AsF_6^-$$

In quartz, oxygen abstraction leads also to the formation of O_2^+ salts. The radical ion $\cdot NF_3^+$ has also been characterized, e.g., in γ-radiation of NF_4^+ salts or uv irradiation of NF_3, F_2, and AsF_5 mixtures.

NF_4^+ salts are quantitatively hydrolyzed:

$$NF_4^+ + (1 + x)H_2O = NF_3 + H_2F^+ + xH_2O_2 + \tfrac{1}{2}(1 - x)O_2$$

The unstable difluoroammonium ion[41] can be made from the explosive NHF_2 by the reaction

$$NHF_2 + HF + AsF_5 \xrightarrow{-78°} NH_2F_2^+ AsF_6^-$$

Tetrafluorohydrazine. N_2F_4, also a gas (b.p. $-73°$), is best prepared by the reaction of NF_3 with copper mentioned above. Its structure is similar to that of hydrazine, but differs in consisting of comparable fractions of *gauche* and *trans* forms, the latter being slightly more stable, by *ca.* 2 kJ mol^{-1}. It is interesting that N_2F_4 dissociates readily in the gas and the liquid phase according to the equation

$$N_2F_4 = 2NF_2 \qquad \Delta H_{298°} = 84 \text{ kJ mol}^{-1}$$

which accounts for its high reactivity. The esr and electronic spectra of the difluoroamino radical $\cdot NF_2$ indicate that it is bent (cf. OF_2, O_3^-, SO_2^-, ClO_2) with the odd electron in a relatively pure π molecular orbital.

Since N_2F_4 dissociates so readily, it shows reactions typical of free radicals; thus it abstracts hydrogen from thiols:

$$2NF_2 + 2RSH \rightarrow 2HNF_2 + RSSR$$

and undergoes other reactions such as:

$$N_2F_4 + Cl_2 \xrightarrow{h\nu} 2NF_2Cl \qquad K_{25°} = 1 \times 10^{-3}$$

$$RI + NF_2 \xrightarrow{uv} RNF_2 + \tfrac{1}{2}I_2$$

$$RCHO + N_2F_4 \rightarrow RCONF_2 + NHF_2$$

$$R_FSF_5 + N_2F_4 \rightarrow R_FNF_2$$

It reacts explosively with H_2 in a radical chain reaction. It also reacts at 300° with NO and rapid chilling in liquid nitrogen gives the unstable purple nitrosodi-

[40] K. O. Christie *et al., Inorg. Chem.,* 1978, **17**, 759, 3189; 1977, **16**, 353, 849, 937, 2238; *J. Fluorine Chem.,* 1976, **8**, 541.

[41] K. O. Christie, *Inorg. Chem.,* 1975, **14**, 2821.

fluoroamine $ONNF_2$. N_2F_4 is hydrolyzed by water, but only after an inhibition period. In HF(l) it reacts with SbF_5 to give $N_2F_3^+SbF_6^-$.[42]

Difluorodiazene (Dinitrogen Difluoride). N_2F_2 is a gas consisting of two isomers 13-XIII and 13-XIV. The *cis* isomer predominates ($\sim$90%) at 25° and is the

b.p. – 105.7°   b.p. – 111.4°
(13-XIII)   (13-XIV)

more reactive. Isomerization to the equilibrium mixture is catalyzed by stainless steel. The pure *trans* form can be obtained in about 45% yield by the reaction

$$2N_2F_4 + 2AlCl_3 \rightarrow N_2F_2 + 3Cl_2 + 2AlF_3 + N_2$$

Nitrogen Trichloride, Tribromide, and Triiodide.[43]   The chloride NCl_3 is formed in the chlorination of slightly acid solutions of NH_4Cl and may be continuously extracted into CCl_4. When pure, it is a pale yellow oil (b.p. $\sim$71°). It is endothermic ($\Delta H_f^0 = 232$ kJ mol^{-1}), explosive, photosensitive, and, generally, very reactive. The vapor has been employed in bleaching flour. The molecule is pyramidal, with $N—Cl = 1.753$ Å and $Cl—N—Cl = 107°\ 47'$.[44]

Nitrogen tribromide is similar to NCl_3 and is hydrolyzed in base.[45]

$$2NBr_3 + 3OH^- = N_2 + 3Br^- + 3HOBr$$

Concentrated aqueous ammonia reacts with I_2 at 23° to give black, explosive crystals of $(NI_3 \cdot NH_3)_n$ that contains zigzag chains of NI_4 tetrahedra sharing corners, with NH_3 molecules lying between the chains and linking them together; $NI_3 \cdot 3NH_3$ is similar.

Haloamines.   These are compounds of the type H_2NX and HNX_2, where also H may be replaced by an alkyl radical. Only H_2NCl (chloramine), HNF_2, and H_2NF have been isolated; $HNCl_2$, H_2NBr, and $HNBr_2$ probably exist but are quite unstable. It is believed that on chlorination of aqueous ammonia, NH_2Cl is formed at pH > 8.5, $NHCl_2$ at pH 4.5 to 5.0, and NCl_3 at pH < 4.4. Difluoroamine, a colorless, explosive liquid (b.p. 23.6°), can be obtained as above or by H_2SO_4 acidification of fluorinated aqueous solutions of urea; the first product, H_2NCONF_2, gives HNF_2 on hydrolysis. It can be converted into chlorodifluoroamine, $ClNF_2$, by action of Cl_2 and KF.[45]

NF_3 is devoid of donor properties, but NHF_2 is a weak donor:

$$NHF_2(g) + BF_3(g) = HF_2NBF_3(s) \qquad \Delta H = -88 \text{ kJ mol}^{-1}$$

Oxo Halides.[46]   The known compounds and some of their properties are listed

42   K. O. Christie and C. J. Schack, *Inorg. Chem.,* 1978, **17**, 2749.
43   J. Jander, *Adv. Inorg. Chem. Radiochem.,* 1976, **19**, 2.
44   G. Cazzoli, P. G. Favero, and A. D. Borgo, *J. Mol. Spectra,* 1974, **50**, 82.
45   G. W. Inman, Jr., T. F. LaPointe, and J. D. Johnson, *Inorg. Chem.,* 1976, **15**, 3037.
46   R. Schmutzler, *Angew. Chem., Int. Ed.,* 1968, **7**, 440 (oxofluorides).

TABLE 13-3
Physical Properties of Nitrosyl Halides, HalNO, and Nitryl Halides, HalNO$_2$

Property	FNO[a]	ClNO	BrNO	FNO$_2$[a]	ClNO$_2$
Color of gas	Colorless	Orange-yellow	Red	Colorless	Colorless
Melting point (°C)	−133	−62	−56	−166	−145
Boiling point (°C)	−60	−6	∼0	−72	−15
Structure	Bent	Bent	Bent	Planar[b]	Planar[b]
X—N distance (Å)	1.52	1.95 ± 0.01	2.14 ± 0.02	1.35	1.840 ± 0.002
N—O distance (Å)	1.13	1.14 ± 0.02	1.15 ± 0.04	1.23	1.202 ± 0.001
X—N—O angle (deg)	110	116 ± 2	114		
O—N—O angle (deg)				125 (assumed)	130.6 ± 0.2

[a] Uncertainties in structure parameters not known.
[b] Molecular symmetry, C_{2v}.

in Table 13-3; the energetics and MO description of these molecules, isomers such as FON, which can be isolated in most matrices, and other possible NOF compounds have been discussed.[47]

The nitrosyl halides can all be obtained by direct union of the halogens with nitric oxide and also in other ways. They are increasingly unstable in the series FNO, ClNO, BrNO. ClNO is always slightly impure, decomposing (to Cl$_2$ and NO) to the extent of about 0.5% at room temperature, and BrNO is decomposed to ∼7% at room temperature and 1 atm.

All three are reactive and are powerful oxidizing agents, able to attack many metals. All decompose on treatment with water producing HNO$_3$, HNO$_2$, NO, and HX.

The only known nitryl halides, which may be regarded as derivatives of nitric acid where a halogen atom replaces OH, are FNO$_2$ and ClNO$_2$. The former is conveniently prepared by the reaction:

$$N_2O_4 + 2CoF_3(s) \xrightarrow{300°} 2FNO_2 + 2CoF_2(s)$$

ClNO$_2$ is not obtainable by direct reaction of NO$_2$ and Cl$_2$, but is easily made in excellent yield by the reaction

$$ClSO_3H + HNO_3 \text{ (anhydrous)} \xrightarrow{0°} ClNO_2 + H_2SO_4$$

Both compounds are quite reactive; both are decomposed by water:

$$XNO_2 + H_2O = HNO_3 + HX$$

Reaction of NO$_2$ and F$_2$ at −30° gives *nitrosyl hypofluorite* (ONOF),[48] which has been studied also spectroscopically by matrix isolation.[47]

[47] R. R. Smardzewski and W. B. Fox, *J. Chem. Phys.*, 1974, **60**, 2104, 2980; *J. Am. Chem. Soc.*, 1974, **96**, 304; P. S. Ganguli and H. A. McGee, Jr., *Inorg. Chem.*, 1972, **11**, 3071.
[48] J. E. Sicre and H. J. Schumaker, *Z. Anorg. Allg. Chem.*, 1971, **385**, 131.

Interaction of NCl_3, $SOCl_2$, and $SbCl_5$ in CCl_4 gives the yellow salt $[ONCl_2]^+SbCl_6^-$, which is stable to 145°.[49]

Halogen Nitrates. Two of these highly reactive substances are known: $ClONO_2$ (b.p. 22.3°) and $FONO_2$ (b.p. −46°). $ClONO_2$ is not known to be intrinsically explosive, but it reacts explosively with organic matter; $FONO_2$ is liable to explode. The best preparative reactions appear to be[50]

$$ClF + HNO_3 \rightarrow ClONO_2 + HF$$
$$F_2 + HNO_3 \rightarrow FONO_2 + HF$$

Trifluoramine Oxide. This stable, toxic, and oxidizing gas (b.p. −87.6°), which is resistant to hydrolysis, can be prepared[51] by the reactions

$$2NF_3 + O_2 \xrightarrow[\text{discharge}]{\text{electric}} 2NF_3O$$

$$3NOF + 2IrF_6 \rightarrow 2NOIrF_6 + NF_3O$$

or by fluorination of NO, provided the gas mixture is quenched. Spectroscopic data indicate that it is a "tetrahedral" molecule (C_{3v} symmetry). Since NF_3 is inactive as a simple donor and lacks additional orbitals for a multiple interaction with an additional atom or group, the stability of NF_3O is somewhat surprising. Its electronic structure is perhaps best represented by the resonance forms shown in 13-VI. The only chemical reactions so far reported are with strong F^- acceptors (e.g., AsF_5, SbF_5), giving salts of NF_2O^+.

[49]   K. Dehnicke et al., Angew. Chem., Int. Ed., 1977, **16**, 545.

[50]   C. J. Schack, Inorg. Chem., 1967, **6**, 1938.

[51]   W. B. Fox et al., J. Am. Chem. Soc., 1970, **92**, 9240; P. J. Bassett and D. R. Lloyd, J. Chem. Soc., A, **1971**, 3377.

General References

Addison, C. C., et al., Q. Rev., 1971, **25**, 289. Structural aspects of coordinated nitrate.

Advanced Propellant Chemistry, ACS Advances in Chemistry Series, No. 54. Chemistry of N_2O_5, N-F compounds, NO_2^+, and $N_2H_5^+$ perchlorates.

Colburn, C. B., Ed., Developments in Inorganic Nitrogen Chemistry, Elsevier, Vol. 1, 1966: Bonding, azides, S-N compounds, N ligands, P-N compounds, N compounds of B, Al, Ga, In, and Tl; reactions in liquid NH_3; Vol. 2, 1973: NO, N_2O_5, C-Cl compounds.

Fair, D. H., and R. F. W. Wallace, Eds., The Technology of Inorganic Azides, Plenum Press, 1976.

Forrester, A. R., and F. A. Neugebauer, Organic N-Centered Radicals and Nitroxide Radicals, Springer, 1979.

Hardy, R. W. F., and W. S. Silver, Eds., Treatise on Dinitrogen Fixation, Wiley-Interscience, 1979.

Honti, G. D., Ed., The Nitrogen Industry, Akademia Kiado.

Klimisch, R. L., and J. G. Larson, Eds., The Catalytic Chemistry of Nitrogen Oxides, 1975, Plenum Press.

Kosower, E. M., Acc. Chem. Res., 1971, **4**, 193. Reactions of monosubstituted diazenes including metal complexes.

Lwowski, W., Nitrenes, Wiley, 1970.

Mellor's Comprehensive Treatise on Inorganic and Theoretical Chemistry, Vol. VIII, Suppls. I and II, Longmans Green, 1967. Encyclopedic coverage of the inorganic chemistry of nitrogen.

National Academy of Sciences, *Nitrogen Oxides,* Washington, D.C., 1977. Medical and biologic effects of atmospheric pollutants.

Postgate, J. R., *et al.,* Eds., *The Chemistry and Biochemistry of Nitrogen Fixation,* Plenum Press, 1971; *Recent Developments in Nitrogen Fixation,* Academic, Press, 1977.

Rozantsev, E. G., *Free Nitroxyl Radicals,* Plenum Press, New York, 1970.

Smith, P. A. S., *The Chemistry of Open-Chain Organic Nitrogen Compounds,* Vols. 1 and II, Benjamin, 1966. Contains much of inorganic interest.

Thompson, J. C., *Electrons in Liquid Ammonia,* Clarendon Press, 1976.

CHAPTER FOURTEEN

The Group V Elements:
P, As, Sb, Bi

GENERAL REMARKS

14-1. Group Trends and Stereochemistry

The electronic structures and some other properties of the elements in Group V are listed in Table 14-1. The valence shells have a structure formally similar to that of nitrogen, but beyond the stoichiometries of some of the simpler compounds—NH_3, PH_3, NCl_3, $BiCl_3$, for example—there is little resemblance between the characteristics of these elements and those of nitrogen.

The elements phosphorus, arsenic, antimony, and bismuth show a considerable range in chemical behavior. There are fairly continuous variations in certain properties and characteristics, although in several instances there is no regular trend, for example, in the ability of the pentoxides to act as oxidizing agents. Phosphorus, like nitrogen, is essentially covalent in all its chemistry, whereas arsenic, antimony, and bismuth show increasing tendencies to cationic behavior. Although the electronic structure of the next noble gas could be achieved by electron gain, considerable energies are involved (e.g., $\sim$1450 kJ mol^{-1} to form P^{3-} from P); thus significantly ionic compounds such as Na_3P are few. The loss of valence electrons is similarly difficult to achieve because of the high ionization enthalpies. The 5+ ions do not exist, but for trivalent antimony and bismuth cationic behavior does occur. BiF_3 seems predominantly ionic, and salts such as $Sb_2(SO_4)_3$ and $Bi(NO_3)_3 \cdot 5H_2O$, as well as salts of the oxo ions SbO^+ and BiO^+, exist.

Some of the more important trends are shown by the oxides, which change from acidic for phosphorus to basic for bismuth, and by the halides, which have increasingly ionic character: PCl_3 is instantly hydrolyzed by water to $HPO(OH)_2$, and the other trihalides give initially clear solutions that hydrolyze to As_2O_3, $SbOCl$,

TABLE 14-1
Some Properties of P, As, Sb, and Bi

Property	P	As	Sb	Bi
Electronic structure	$[Ne]3s^23p^3$	$[Ar]3d^{10}4s^24p^3$	$[Kr]4d^{10}5s^25p^3$	$[Xe]4f^{14}5d^{10}6s^26p^3$
Sum of 1st three ionization enthalpies $[(kJ\ mol^{-1})/10^3]$	5.83	5.60	5.05	5.02
Electronegativity[a]	2.06	2.20	1.82	1.67
Radii (Å)				
Ionic	$2.12(P^{3-})$		$0.92(Sb^{3+})$	$1.08(Bi^{3+})$
Covalent[b]	1.10	1.21	1.41	1.52
Melting point (°C)	44.1 (α-form)	814 (36 atm)	603.5	271.3

[a] Allred-Rochow type.
[b] For trivalent state.

and BiOCl, respectively. There is also an increase in the stability of the lower oxidation state with increasing atomic number; thus Bi_2O_5 is the most difficult to prepare and the least stable pentoxide.

Although oxidation states or oxidation numbers can be, and often are, assigned to these elements in their compounds, they are of rather limited utility except in the formalities of balancing equations. The important valence features concern the number of covalent bonds formed and the stereochemistries. The general types of compound and stereochemical possibilities are given in Table 14-2.

The differences between the chemistries of N and P, which are due to the same factors that are responsible for the C—Si and O   S differences, can be summarized as follows:

Nitrogen	Phosphorus
(a) Very strong $p\pi$-$p\pi$ bonds	Unstable $p\pi$-$p\pi$ bonds
(b) $p\pi$-$d\pi$ bonding is rare	Weak to moderate but important $d\pi$-$p\pi$ and $d\pi$-$d\pi$ bonding
(c) No valence expansion	Valence expansion

Point a leads to facts such as the existence of $P(OR)_3$ but not of $N(OR)_3$, nitrogen giving instead $O=N(OR)$, and the structural differences between nitrogen oxides and oxo acids and oxides such as P_4O_6 or P_4O_{10} and the polyphosphates. A few unstable compounds with short half-lives that can be considered as having $p\pi$ $p\pi$ bonding have been detected in the gas phase by microwave or mass spectrometry; examples are $CH_2=PCl$, $CF_2=PH$, and $HC\equiv P$.[1]

Point b is associated with rearrangements such as

$$\text{\Large $\diagdown$}_{\diagup}P-OH \rightleftharpoons H-\text{\Large $\diagup$}_{\diagdown}P=O$$

and with the existence of phosphonitrilic compounds, $(PNCl_2)_n$. Furthermore, although PX_3, AsX_3, and SbX_3 (X = halogen, alkyl or aryl), like NR_3 compounds, behave as donors owing to the presence of lone pairs, there is one major difference:

[1]   M. J. Hopkinson et al., J.C.S., Chem. Comm., **1976**, 513.

TABLE 14-2

Group V Compound Types and Stereochemical Possibilities

Formal valence	Number of bonds to other atoms	Geometry	Examples
1	3	Trigonal plane	$PhP[Mn(CO)_2C_5H_5]_2$ (see text)
3	2	Angular	PH_2^-, $[P(NMe_2)_2]^+AlCl_4^-$ [a]
	3	Pyramidal	PH_3, $AsCl_3$, $SbPh_3$
	4	Tetrahedral	PH_4^+, PPh_4^+, PCl_4^+,
			$Me_2\overset{+}{P}{\large\langle}{}^{CH_2-}_{CH_2-}$
		ψ-tbp	KSb_2F_7, $SbCl_3(PhNH_2)$, $K_2[Sb_2(tart)_2]\cdot 3H_2O$, $SbOCl$
	5	ψ-Octahedral	SbF_5^{2-}, $[Sb_4F_{16}]^{4-}$, Sb_2S_3, $SbCl_3(PhNH_2)_2$
	6	Octahedral	$[Bi_6O_6(OH)_3]^{3+}$, $[As(DMF)_6]^{3+}$
5	3	See text	$P[N(SiMe_3)_2](NSiMe_3)_2$
	5	tbp	PF_5, AsF_5, $SbCl_5$, $AsPh_5$, Ph_3XSb—O—$SbXPh_3$ [b]
		sp [c]	$SbPh_5$
	6	Octahedral	PF_6^-, $Sb(OH)_6^-$, $SbBr_6^-$

[a] M. G. Thomas et al., J. Am. Chem. Soc., 1974, **96**, 2641.

[b] F. C. March and G. Ferguson, J.C.S. Dalton 1975, 1291.

[c] This may be only a crystal packing effect, since Sb(p-tolyl)$_5$ is tbp; see C. P. Brock and J. A. Ibers, Acta Crystallogr., A, 1976, **31**, 38.

the nitrogen atom can have no function other than simple donation, because no other orbital is accessible; but P, As, and Sb have empty d orbitals of fairly low energy.[2] Thus when the atom to which the P, As, or Sb donates has electrons in orbitals of the same symmetry as the empty d orbitals, back-donation resulting in overall multiple-bond character may result. This factor is especially important for the stability of complexes with transition elements where $d\pi$-$d\pi$ bonding contributes substantially to the bonding (see Fig. 3-8). The consequences of vacant d orbitals are also evident on comparing the amine oxides R_3NO on the one hand, with R_3PO or R_3AsO on the other. In the N-oxide the electronic structure can be represented by the single canonical structure $R_3\overset{+}{N}$—$\bar{O}$, whereas for the others the bonds to oxygen have multiple character and are represented as resonance hybrids:

$$R_3\overset{+}{P}—\bar{O} \leftrightarrow R_3P{=}O \leftrightarrow R_3\overset{+}{P}{\equiv}\bar{O}$$

These views are substantiated by the shortness of the P—O bonds (~ 1.45 as compared with ~ 1.6 Å for the sum of the single-bond radii) and by the normal bond lengths and high polarities of N—O bonds. The amine oxides are also more chemically reactive, the P—O bonds being very stable indeed, as would be expected from their strength, ~ 500 kJ mol^{-1}.

[2]   H. Kwart and K. King, d-Orbitals in the Chemistry of Silicon, Phosphorus and Sulfur, Springer-Verlag, 1977.

Point c is responsible for phenomena such as the Wittig reaction (p. 467) and for the existence of compounds such as $P(C_6H_5)_5$, $P(OR)_5$, $[P(OR)_6]^-$, and $[PR_4]^+[PR_6]^-$ in which the coordination number is 5 or 6. The extent to which hybridization employing $3d$ orbitals is involved is somewhat uncertain, since the d levels are rather high for full utilization and the higher states may be stabilized in part by electrostatic forces; it is significant that the higher coordination numbers for P^V are most readily obtained with more electronegative groups such as halogens, OR, or phenyl.

The stereochemistry of compounds of P^V, As^V, and Sb^V gains complexity because the *tbp* and *sp* configurations differ little in energy; the *tbp* structure is inherently more stable, for example, in PF_5, by 12 to 16 kJ mol^{-1}. Most of the molecular species MX_5 have *tbp* configurations, but there are a few [SbPh$_5$, SbF$_5$, and perhaps Sb(C$_3$H$_5$)$_5$] for which the *sp* structure is more stable; the *sp* structure can be stabilized by the presence of unsaturated ring systems with oxygen bound to phosphorus.[3a] Detailed MO studies have been made of substitutional effects in PX_5 systems; for the most stable *tbp* structure it appears that atoms or groups that act as *π acceptors prefer axial sites*, and *π donors the equatorial ones*.[3b] The similar energies of the two configurations provide a pathway for stereochemical nonrigidity (Section 28-13), which is an essential factor in the chemistry of the pentavalent compounds.

The only naturally occurring isotope of phosphorus, ^{31}P, has a nuclear spin of $\frac{1}{2}$ and a large magnetic moment. Nuclear magnetic resonance spectroscopy has accordingly played an extremely important role in the study of phosphorus compounds. For antimony, the isotope ^{121}Sb is suitable for Mössbauer spectroscopy.

Both phosphorus and arsenic show a significant tendency to catenation, forming a series of cyclic compounds $(RP)_n$ and $(RAs)_n$ where $n = 3$ to 6, as well as some R_2PPR_2 and R_2AsAsR_2 compounds.

Stereochemically Active Lone Pairs in Five- and Six-Coordinated As, Sb, and Bi. Table 14-2 indicates that the five-coordinate compounds have different stereochemistries depending on the formal valence state of the element. This is in accord with the principles discussed in Chapter 5. In the pentavalent compounds the central atom has only the five bonding pairs in its valence orbitals and the usual trigonal-bipyramidal arrangement is adopted. In the trivalent species (e.g., SbF_5^{2-}) there are six electron pairs; this anion is isoelectronic with BrF_5 and has the same ψ-octahedral structure (see p. 199), where the five bonding and one nonbonding pair of electrons are approximately at the vertices of an octahedron and the larger charge cloud of the nonbinding pair makes the F—Sb—F angles somewhat less than 90°.

Another good example is $[Sb^{III}ox_3]^{3-}$ which is ψ-seven coordinate. There are now many other cases of structures whose distortions can be attributed to the presence of stereochemically active lone pairs.[4] Examples are the dithiocarbamates

[3a] R. R. Holmes, *J. Am. Chem. Soc.*, 1975, **97**, 5379.
[3b] R. Hoffmann, J. M. Howell, and E. L. Muetterties, *J. Am. Chem. Soc.*, 1972, **94**, 3047.
[4] S. L. Lawton *et al.*, *Inorg. Chem.*, 1974, **13**, 135.

of arsenic, antimony, and bismuth [M(S$_2$CNEt$_2$)$_3$], which have three long and three short M—S bonds.[5]

Planar P, As, and Sb Compounds. There are unusual compounds in which the elements, formally in the I oxidation state, form trigonal planar compounds of type 14-I; $d\pi$-$p\pi$ interaction is doubtless involved in the bonding.[6] In the second[7] (14-II), phosphorus is formally pentavalent.

$$
\begin{array}{c}
\text{Ph} \\
| \\
110° \diagup \text{Sb} \\
\text{Cp(CO)}_2\text{Mn} \diagdown_{140°} \diagdown \text{Mn(CO)}_2\text{Cp}
\end{array}
\qquad\qquad
\begin{array}{c}
\text{Me}_3\text{SiN} \\
\diagdown \\
\text{P}\!-\!\text{N(SiMe}_3)_2 \\
\diagup \\
\text{Me}_3\text{SiN}
\end{array}
$$

(14-I)   (14-II)

Phosphorus Radicals. Action of light in presence of an activated alkene as halogen acceptor leads to R$_2$P radicals, namely,

$$[(Me_3Si)_2CH]_2PCl \xrightarrow{h\nu} [(Me_3Si)_2CH]_2P$$

The stability of the radical is connected with the bulk of the bis(trimethylsilyl)-methyl group.[8]

14-2. The Elements

Occurrence. Phosphorus occurs in various orthophosphate minerals, notably *fluorapatite* [3Ca$_3$(PO$_4$)$_2$·Ca(F, Cl)$_2$]. Arsenic and antimony occur more widely, though in lower total abundance, and are often associated with sulfide minerals, particularly those of copper, lead, and silver. Bismuth ores are uncommon, the sulfide being the most important; bismuth also occurs in other sulfide minerals.

Phosphorus. The element is obtained by reduction of phosphate rock with coke and silica in an electric furnace. Phosphorus volatilizes as P$_4$ molecules (partly dissociated above 800° into P$_2$) and is condensed under water as white phosphorus, a white, soft, waxy solid (m.p. 44.1°, b.p. 280°):

$$2Ca_3(PO_4)_2 + 6SiO_2 + 10C = P_4 + 6CaSiO_3 + 10CO$$

There are three main forms of phosphorus—white, black, and red—but there are numerous allotropes, some of dubious validity.[9,10]

White phosphorus, in the liquid and solid forms, consists of tetrahedral P$_4$

5   C. L. Raston and A. H. White, *J.C.S. Dalton,* **1976,** 791.
6   G. Hüttner *et al., Angew. Chem., Int. Ed.,* 1978, **17,** 843, 844.
7   S. Pohl, E. Niecke, and B. Krebs, *Angew. Chem., Int. Ed.,* 1975, **14,** 261.
8   M. J. S. Gynane *et al., J.C.S., Chem. Comm.,* **1976,** 623.
9   J. Donohue, *The Structures of the Elements,* Wiley, 1974.
10 D. E. C. Corbridge, *The Structural Chemistry of Phosphorus,* Elsevier, 1974.

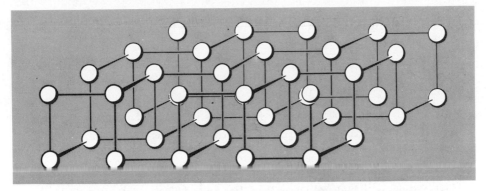

Fig. 14-1. The arrangement of atoms in the double layers found in crystalline black phosphorus.

molecules; in the vapor below 800°, where measurable dissociation to P_2 occurs, the element is also as P_4 molecules. The P—P distances are 2.21 Å; the P—P—P angles, of course, are 60°. The low angle indicates considerable strain, and the strain energy has been estimated to be about 96 kJ mol^{-1}. This means that the total energy of the six P—P bonds in the molecule is that much smaller than would be the total energy of six P—P bonds of the same length formed by phosphorus atoms with normal bond angles. Thus the structure of the molecule is consistent with its high reactivity. It is most likely that pure $3p$ orbitals are involved, even though the bonds are bent, since hybridization such as pd^2, which would give 60° angles, would require a rather large promotion energy. MO calculations also indicate that d-orbital participation is negligible.

   Black phosphorus is flaky, with a metallic, graphitelike appearance. It is obtained by heating white P either under very high pressure or at 220 to 370° for 8 days in the presence of mercury as a catalyst and with a seed of black P. The structure consists of double layers, each P atom being bound to three neighbors (Fig. 14-1). The closest P—P distances within each double layer are 2.23 Å (i.e., normal single-bond distances). The entire structure consists of a stacking of these double layers with the shortest P—P distance between layers at 3.59 Å.

   Red phosphorus is made by heating P_4 at 400° for several hours. The color and other physical properties depend on the method of preparation. Commercial red phosphorus is amorphous with pyramidal phosphorus linked in a random network.

   The main forms of phosphorus show considerable difference in chemical reactivity; white is by far the most reactive and black the least. White P is stored under water because it inflames in air, whereas the red and black are stable in air; indeed, black P can be ignited only with difficulty. White P is soluble in organic solvents such as CS_2 and benzene. Some reactions of both red and white P are shown in Fig. 14-2.

   In the P_4 molecules there should be a lone pair directed outward from each P atom; thus the molecule might be expected to have donor ability. Somewhat unstable complexes, e.g., $RhCl(PR_3)_2(P_4)$ have been reported (see p. 150).

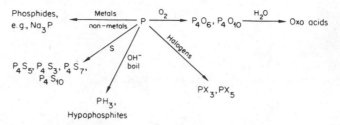

Fig. 14-2. Some typical and important reactions of red and white phosphorus.

Arsenic, Antimony, and Bismuth. These elements are obtained by reduction of their oxides with hydrogen or carbon. For As and Sb unstable yellow allotropes, presumably containing tetrahedral As_4 and Sb_4 molecules, can be obtained by rapid condensation of vapors. They are easily transformed into the normal forms having a bright and metallic appearance.

They have structures with puckered sheets of covalently bound atoms stacked in layers, rather similar to black phosphorus. There are other allotropes.[9] When heated, the metals burn in air to form the oxides. They react readily with halogens and some other nonmetals and form alloys with various other metals. They are unaffected by dilute nonoxidizing acids, but with nitric acid, As gives arsenic acid, Sb gives the trioxide, and Bi dissolves to form the nitrate.

BINARY COMPOUNDS

14-3. Phosphides, Arsenides, Antimonides, and Bismuthides

Electropositive metals, such as the alkalis and alkaline earths, react directly with P, As, and Sb to give a range of compounds. A few (e.g., Na_3P, Sr_3P_2) correspond to simple ionic compositions (though they should not be considered literally as ionic), but many are richer in P, As, or Sb and contain polyhedral anions. Examples are Sr_3P_{14}, Ba_3As_{14}, and $[Na(2,2,2-crypt]_3Sb_7$ all of which[11] contain X_7 cluster anions with structure 14-III.

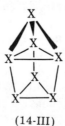

(14-III)

[11]   H. G. von Schnering et al., Angew. Chem., Int. Ed., 1977, **16**, 857; D. G. Adolphson, J. D. Corbett, and D. J. Merryman, J. Am. Chem. Soc., 1976, **98**, 7234.

Compositions such as LiAs, MgP_4, and MP_5 (M = lanthanide) contain rings, chains, or sheets of P or As atoms.[12] Although there are numerous cluster cations of bismuth (Section 14-5), there are few anions[13]; but $[K \text{ crypt}]_2^+ Bi_4^{2-}$, which is black, has a square ion like Te_4^{2+}.

With some less electropositive elements compounds like CdP_2, which contains infinite spiral chains of P atoms with alternating P—P distances of 2.05 and 2.39 Å, are formed. There are also a few cases (e.g., PtP_2), in which discrete P_2 units occur.

Many of the transition metals form metal-rich phosphides (e.g., Fe_2P), which are commonly hard, brittle, gray-black substances, metallic in appearance and properties, and very stable both thermally and chemically. They are reminiscent of the interstitial carbides and nitrides of the transition metals and have P atoms surrounded by up to ten metal atoms.

14-4. Hydrides

The gases MH_3 can be obtained by treating phosphides or arsenides of electropositive metals with acids or by reduction of sulfuric acid solutions of arsenic, antimony, or bismuth with an electropositive metal or electrolytically. The stability falls rapidly down the group, so that SbH_3 and BiH_3 are very unstable thermally, the latter having been obtained only in traces. The average bond energies are in accord with this trend in stabilities: $E_{N—H}$, 391; $E_{P—H}$, 322; $E_{As—H}$, 247; and $E_{Sb—H}$, 255 kJ mol^{-1}.

Phosphine (PH_3)[14] is readily obtained by action of dilute acid on calcium or aluminum phosphide, by pyrolysis of H_3PO_3 or, in a purer state, by action of KOH in PH_4I. On a large scale PH_3 is made by action of NaOH on white phosphorus, which also forms sodium hypophosphite (p. 473).

The molecule is pyramidal with an HPH angle of 93.7°. Phosphine, when pure, is not spontaneously flammable, but often inflames owing to traces of P_2H_4 or P_4 vapor. It is readily oxidized by air when ignited, and explosive mixtures may be formed. It is also exceedingly poisonous. Unlike NH_3, it is not associated in the liquid state and it is only sparingly soluble in water; pH measurements show that the solutions are neither basic nor acidic—the acid constant is $\sim10^{-29}$ and the base constant $\sim10^{-26}$. However it does react with some acids to give phosphonium salts (p. 460), and in basic solution we have

$$\tfrac{1}{4}P_4 + 3H_2O + 3e = PH_3 + 3OH^- \qquad E^0 = -0.89 \text{ V}$$

The proton affinities of PH_3 and NH_3 (eq. 14-1) differ considerably.

$$EH_3(g) + H^+(g) = EH_4^+(g) \qquad (14\text{-}1)$$
$$\Delta H^0 = 770 \text{ kJ mol}^{-1} \text{ for E = P}$$
$$\Delta H^0 = 866 \text{ kJ mol}^{-1} \text{ for E = N}$$

[12]   H. G. von Schnering et al., Z. Anorg. Allg. Chem., 1976, **422**, 219, 226.
[13]   A. Cisar and J. D. Corbett, Inorg. Chem., 1977, **16**, 2482.
[14]   E. Fluck, The Chemistry of Phosphine, Vol. 35, Topics of Current Chemistry, Springer, 1973.

Also, the barrier to inversion for PH_3 is 155 kJ mol^{-1} as compared with only 24 kJ mol^{-1} for NH_3. Quite generally the barriers for R_3P and R_3N compounds differ by about this much. Like other PX_3 compounds, PH_3 (and also AsH_3) forms complexes with transition metals[15] [e.g., cis-$Cr(CO)_3(PH_3)_3$].

Arsine (AsH_3) is extremely poisonous. Its ready thermal decomposition to arsenic, which is deposited on hot surfaces as a mirror, is utilized in tests for arsenic, for example, the well-known Marsh test, where arsenic compounds are reduced by zinc in HCl solution.

Stibine is very similar to arsine but even less stable.

All these hydrides are strong reducing agents and react with solutions of many metal ions, such as Ag^I and Cu^{II}, to give the phosphides, arsenides, or stibnides, or a mixture of these with the metals.

Phosphorus alone forms other hydrides. *Diphosphine* (P_2H_4) is obtained along with phosphine by hydrolysis of calcium phosphide and can be condensed as a yellow liquid. It is spontaneously flammable and decomposes on storage to form polymeric, amorphous yellow solids, insoluble in common solvents and of stoichiometry approximating to, but varying around, P_2H. Unlike N_2H_4 diphosphine has no basic properties. On photolysis it gives P_3H_5. It exists mainly in the *gauche* form (cf. hydrazine, p. 418).[16]

14-5. Halides

The binary halides are of two main types, MX_3 and MX_5 (Table 14-3). All trihalides except PF_3 are best obtained by direct halogenation, keeping the element in excess, whereas all the pentahalides may be prepared by treating the elements with an excess of the appropriate halogen.

Mixed trihalides can be detected in mixtures, but it is not certain, despite some older claims, that any can be isolated in pure form since the equilibria, e.g.,

$$PCl_3 + PBr_3 \rightleftharpoons PCl_2Br + PClBr_2$$

are labile.

All the trihalides are rapidly hydrolyzed by water and are rather volatile. The gaseous molecules have pyramidal structures, and some form molecular lattices. The iodides AsI_3, SbI_3, and BiI_3 crystallize in layer lattices with no discrete molecules. BiF_3 has an ionic lattice, and SbF_3 has an intermediate structure in which SbF_3 molecules (Sb—F = 1.92 Å) are linked through F bridges (Sb$\cdots$F, 2.61 Å) to give each Sb^{III} a very distorted octahedral environment.

Phosphorus Trifluoride. This is a colorless gas, best made by fluorination of PCl_3. It forms complexes with transition metals similar to those formed by carbon monoxide (p. 88). Like CO, it is highly poisonous because of the formation of a hemoglobin complex. Unlike the other trihalides, PF_3 is hydrolyzed only slowly by water, but it is attacked by alkalis.

[15] See, e.g., R. A. Schunn, *Inorg. Chem.*, 1973, **12**, 1573.
[16] S. Elbel et al., *Inorg. Chem.*, 1976, **15**, 1235.

TABLE 14-3
Group V Binary Halides[a,b]

Halides and temperatures (°C)											
Fluorides			Chlorides			Bromides			Iodides		
PF_3	b	−101.8	PCl_3	b	76.1	PBr_3	b	173.2	PI_3	m	61.2
AsF_3	b	62.8	$AsCl_3$	b	103.2	$AsBr_3$	m	31.2	AsI_3	m	140.0
							b	221.0			
SbF_3	m	292.0	$SbCl_3$	m	73.17	$SbBr_3$	m	97.0	SbI_3	m	171.0
BiF_3	m	725.0	$BiCl_3$	m	233.5	$BiBr_3$	m	219.0	BiI_3	m	408.6
PF_5	b	−84.5	PCl_5	subl	160	PBr_5	d	106.0	—		
AsF_5	b	−52.8	$AsCl_5{}^c$	m	−50	—			—		
$(SbF_5)_4$	m	8.3	$SbCl_5$	m	479	—			—		
	b	150.0									
BiF_5	m	151.4	—			—			—		
	b	230.0									

[a] Many mixed halides are known (e.g., PF_2Cl, $SbBr_2I$, PF_3Cl_2) and can be made by redistribution reactions: $PCl_3 + PBr_3 \rightleftarrows PCl_2Br + PClBr_2$

Though they have not been isolated they can be identified spectroscopically.

[b] Mixed fluoropentahalides (e.g., PF_4Cl, PF_4H, SbF_3Cl_2) exist.

[c] Phase studies show that $AsCl_5$ does not exist in stable equilibria at room temperature. It appears to decompose to $AsCl_3 + Cl_2$ at its melting point of *ca.* −50° (cf. K. Seppelt, *Angew. Chem., Int. Ed.,* 1976, **15**, 377). Raman spectra show it to have a *tbp* structure.

Although PF_3 has no Lewis acid behavior, it does have rather weak basic properties and in FSO_3H SbF_5 SO_2 solution it (and indeed other PX_3 halides) is protonated on phosphorus as shown by ^{31}P nmr spectra; thus HPF_3^+ has a quartet of doublets.[17] It also forms a weak complex with $AlCl_3$.[18]

Phosphorus trichloride is violently hydrolyzed by water to give phosphorous acid or, under special conditions, other acids of lower-valent phosphorus. It also reacts readily with oxygen to give $OPCl_3$. The hydrolysis of PCl_3 may be contrasted with that of NCl_3, which gives $HOCl$ and NH_3. Figure 14-3 illustrates some of the important reactions of PCl_3. Many of these reactions are typical of other MX_3 compounds and also, with obvious changes in formulas, of $OPCl_3$ and other oxo halides.

Arsenic, Antimony, and Bismuth Trihalides. Arsenic trihalides are similar to those of phosphorus in both physical and chemical properties. However they have appreciable electrical conductances and chemical evidence suggests that this may be due to autoionization. Thus, the addition of KF or SbF_5 to liquid AsF_3 increases the conductance, and the compounds $KAsF_4$ and $SbF_5 \cdot AsF_3 (AsF_2^+ SbF_6^-)$ can be isolated.

SbF_3, a white, readily hydrolyzed solid, finds considerable use as a moderately active fluorinating agent. Both AsF_3 and SbF_3 can function as F^- acceptors, though the product is seldom the simple MF_4^- ion.

[17] L. J. Vande Griend, and J. G. Verkade, *J. Am. Chem. Soc.,* 1975, **97**, 5958.
[18] E. R. Elton, R. G. Montemayor, and R. W. Parry, *Inorg. Chem.,* 1974, **13**, 2267.

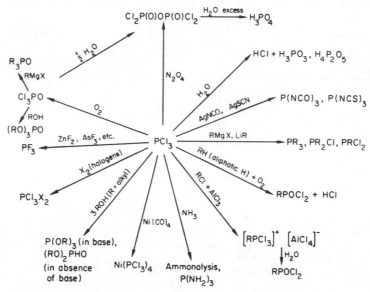

Fig. 14-3. Some important reactions of PCl₃. Many of these are typical for other MX₃ compounds as well as for MOX₃ compounds.

Antimony trichloride differs from its P and As analogues in that it dissolves in a limited amount of water to give a clear solution that, on dilution, gives insoluble oxo chlorides such as $SbOCl$ and $Sb_4O_5Cl_2$. There is, however, no evidence to suggest that any simple Sb^{3+} ion exists in the solutions.

The chloride has a high dielectric constant, and many chlorides dissolve in the melt[19] to give conducting solutions:

$$MCl + SbCl_3 \rightleftharpoons M^+ + SbCl_4^-$$

Bismuth trichloride, a white, crystalline solid, is hydrolyzed by water to BiOCl but BiOCl redissolves in concentrated HCl, and $BiCl_3$ is obtained on evaporating such solutions.

Pentahalides. Fluorides. PF_5 is prepared by the interaction of PCl_5 with CaF_2 at 300 to 400°. It is a very strong Lewis acid and forms complexes with amines,[20] ethers, and other bases as well as with F^- in which phosphorus becomes six-coordinate. However these organic complexes are less stable than those of BF_3 and are rapidly decomposed by water and alcohols. Like BF_3, PF_5 is a good catalyst, especially for ionic polymerization.

The only arsenic pentahalide stable at room temperature, AsF_5, is similar to PF_5. The action of chlorine on AsF_3 at 0° gives a compound whose conductivity in an excess of AsF_3 suggests that it may be $[AsCl_4]^+[AsF_6]^-$.

Antimony pentafluoride is a viscous liquid that is associated even in the vapor

[19]   E. C. Baughan, in *The Chemistry of Non-Aqueous Solvents,* J. J. Lagowski, Ed., Academic Press, 1976 (SbCl₃ as solvent).

[20]   See, e.g., W. S. Sheldrick, *J.C.S. Dalton,* **1974,** 1402 for the structure of PyPF₅.

state at the boiling point. Nmr studies of the liquid suggest that each antimony atom is surrounded octahedrally by six fluorine atoms, two *cis* fluorines being shared with adjacent octahedra. It is not certain whether SbF_5 is dimeric or whether there are higher polymers but polymers are favored. In the crystal it is tetrameric with a structure similar to that of $(RhF_4)_4$, Fig. 22-E-1, but with alternating fluorine bridge angles of 170° and 141°. $SbCl_4F$ and $SbCl_3F_2$ are also tetramers with *cis*-F bridges.[21]

SbF_5 vapor from 140° to 350° consists of polymers, apparently similar to the cis polymers in the liquid plus a small amount of monomer that increases with temperature. The monomer appears to have a *tbp* structure.[22] PF_5 and AsF_5 molecules are well known to have the *tbp* structure.

Bismuth pentafluoride, made by direct fluorination of liquid bismuth at 600° with fluorine at low pressure, is a white crystalline solid and an extremely powerful fluorinating agent.

AsF_5 and SbF_5, and, to a lesser extent PF_5, are potent fluoride ion acceptors, forming MF_6^- ions or more complex species. The PF_6^- ion is a common and convenient "noncomplexing" anion, which has even less coordinating ability than ClO_4^- or BF_4^-.

In liquid HF, PF_5 is a nonelectrolyte, but AsF_5 and SbF_5 give conducting solutions, presumably because of the reactions

$$2HF + 2AsF_5 \rightarrow H_2F^+ + As_2F_{11}^- \text{ (can be isolated as } Et_4N^+ \text{ salt)}$$
$$2HF + SbF_5 \rightarrow H_2F^+ + SbF_6^-$$

We have discussed the use of SbF_5 to enhance the acidities of HF and HSO_3F (p. 239), and the complex F_5SbOSO is mentioned elsewhere (p. 528).

Chlorides. Phosphorus(V) chloride has a *tbp* molecular structure in the gaseous state, but the solid is $[PCl_4]^+[PCl_6]^-$. The tetrahedral PCl_4^+ ion (P—Cl = 1.91 Å) can be considered to arise here by transfer of Cl^- to the Cl^- acceptor PCl_5. It is not therefore surprising that many salts of the PCl_4^+ ion are obtained when PCl_5 reacts with other Cl^- acceptors, e.g.,

$$PCl_5 + TiCl_4 \rightarrow [PCl_4]_2^+[Ti_2Cl_{10}]^{2-} \text{ and } [PCl_4]^+[Ti_2Cl_9]^-$$
$$PCl_5 + NbCl_5 \rightarrow [PCl_4]^+[NbCl_6]^-$$

Pyridine complexes $[PCl_4L]^+$ and $[PCl_4L_2]^+$ are also known.[23a]

In polar solvents like nitrobenzene or acetonitrile, there appears to be dissociation:

$$2PCl_5 \rightleftharpoons PCl_4^+ + PCl_6^-$$
$$PCl_5 \rightleftharpoons PCl_4^+ + Cl^-$$

but in nonpolar solvents, PCl_5 is molecular; in CCl_4, it may be dimeric, but in benzene it is monomeric.[23b]

[21] J. G. Ballard, F. Birchall, and D. R. Slim, *J.C.S. Chem. Comm.,* **1976,** 653.
[22] I. R. Beattie et al., *J.C.S.Dalton,* **1976,** 1381.
[23a] K. B. Dillon, R. N. Reeve, and T. C. Waddington, *J.C.S. Dalton,* **1977,** 2382.
[23b] R. W. Suter et al., *J. Am. Chem. Soc.,* 1973, **95,** 1474.

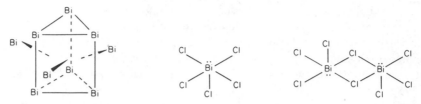

Fig. 14-4. The structures of the species present in "BiCl," which is, in fact, $Bi_{24}Cl_{28}$.

Solid phosphorus(v) bromide is also ionic, $[PBr_4]^+Br^-$. There is even an $AsCl_4^+$ ion, which occurs with such large anions as PCl_6^-, $AlCl_4^-$, and $AuCl_4^-$, even though $AsCl_5$ does not exist.

Arsenic pentachloride is unstable above $-50°$; it is made by photochlorination of $AsCl_3$ at $-105°$.[24]

Antimony pentachloride is a fuming liquid, normally yellow, but colorless when pure. In the solid, it is a dimer $Cl_4Sb(\mu\text{-}Cl_2)SbCl_4$. $SbCl_5$ is a powerful chlorinating agent.

The mixed pentavalent halides and also organic substituted halides appear to have *tbp* structures and to follow the general rule that the equatorial positions are preferred by the less electronegative substituents. For example, AsF_2Ph_3 has axial F atoms.[25] The mixed hydrofluorides PHF_4 and PH_2F_3 are also known.

The relative stabilities, $PCl_5 \gg AsCl_5 \ll SbCl_5$, and the nonexistence of $BiCl_5$ have been attributed[24] to differences in the ionization energies of the trichlorides. The As—Cl bonds in $AsCl_5$ seem to be about as strong as the P—Cl and Sb—Cl bonds. If we consider the oxidative addition to proceed via the sequence

$$AsCl_3 \xrightarrow{-e} AsCl_3^+ \xrightarrow{+Cl} AsCl_4^+ \xrightarrow{+Cl^-} AsCl_5$$

and the overall enthalpy, -88.4 kJ mol^{-1} for PCl_5 (g) and 68 kJ mol^{-1} for $SbCl_5$ (g), then the difference, assuming similar bond energies, is due to the ionization energy: PCl_3, 1012.3 kJ mol^{-1}, and $AsCl_3$ 1127.9 kJ mol^{-1}. The higher value for $AsCl_3$ means that the promotional energy into the excited state is accordingly raised; this may be attributed to a $3d$ contraction comparable to the "lanthanide" contraction.

Lower Halides.   Phosphorus and arsenic form the so-called tetrahalides, P_2Cl_4, P_2I_4, and As_2I_4, which decompose on storage to the trihalide and nonvolatile yellow solids and are readily decomposed by air and water. P_2I_4 has a I_2P—PI_2 structure with a *trans* rotomeric orientation in the solid but probably staggered (like N_2H_4, p. 418) in CS_2 solution.

It has long been known that when metallic bismuth is dissolved in molten $BiCl_3$ a black solid of approximate composition BiCl can be obtained. This solid is $Bi_{24}Cl_{28}$, and it has an elaborate constitution, consisting of four $BiCl_5^{2-}$, one $Bi_2Cl_8^{2-}$, and two Bi_9^{5+} ions, the structures of which are depicted in Fig. 14-4. The electronic

24   K. Seppelt, *Z. Anorg. Allg. Chem.*, 1977, **434**, 5.
25   A. Augustine, G. Ferguson, and F. C. March, *Can. J. Chem.*, 1975, **53**, 1647.

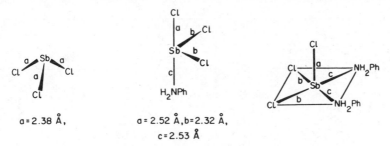

a = 2.38 Å,   a = 2.52 Å, b = 2.32 Å,
c = 2.53 Å

Fig. 14-5. The structures of SbCl$_3$ and two of its aniline complexes. [Reproduced by permission from R. Hulme, D. Mullen, and J. C. Scrutton, *Acta Crystallogr., A*, 1969, 25, S. 171.]

structure of the Bi$_9^{5+}$ ion, a metal atom cluster, has been treated in terms of delocalized molecular orbitals.

Other low-valent species[26] present in various molten salt solutions are Bi$^+$, Bi^{3+}, Bi$_5^{3+}$, and Bi$_8^{2+}$.

14-6. Complexes of the Halides

Phosphorus and arsenic trihalides have only slight Lewis acidity but the PBr$_4^-$ ion has the expected SF$_4$-like structure with a lone pair.

Antimony. SbCl$_3$ forms complexes with neutral donors; those with PhNH$_2$ have structures showing that the N—Sb bonds are relatively weak and that the lone pair is stereochemically active, as illustrated in Fig. 14-5. The variations in Sb—Cl distances are in very good accord with the three-center bond model discussed in Section 5-4.

SbCl$_3$ also forms 2:1 and 1:1 complexes with aromatic hydrocarbons such as naphthalene or *p*-xylene,[27] where weak interaction between SbCl$_3$ and the π cloud occurs. An unusual complex [(η-C$_5$H$_5$)Fe(CO)$_2$Cl]$_4$(SbCl$_3$)$_4$ has an Sb$_4$Cl$_4$ cubanelike nucleus with η-C$_5$H$_5$(CO)$_2$Fe units bound to the bridge Cl atoms.[28]

Complex anions of antimony(III) are particularly well known (Table 14-4, Figs. 14-6 and 14-7); the stereochemistries are consistent with the presence of lone pairs.

Bismuth. Bismuth(III) forms numerous halo complexes,[29a] often similar in formula to those of Sb^{III}, but structurally quite different. In the case of Bi^{III} it appears that the lone pair is only slightly if at all stereochemically active[29b] (cf. TeX$_6^{2-}$ ions, which are regular octahedra). Thus in [(CH$_3$)$_2$NH$_2$]$_3$BiBr$_6$ there are BiBr$_6^{3-}$ ions that are only slightly distorted octahedra and in "BiBr$_5^{2-}$," "BiBr$_4^-$" salts there are chains of halogen-bridged octahedra that are but slightly distorted.

26   J. D. Corbett, *Progr. Inorg. Chem.*, 1976, **21**, 140.
27   R. Hulme and D. J. E. Mullen, *J.C.S. Dalton*, **1976**, 802.
28   V. Trinh-Toan and L. F. Dahl, *Inorg. Chem.*, 1976, **15**, 2953.
29a B. Ya. Spivakov *et al.*, *J. Inorg. Nucl. Chem*, 1979, **14**, 453.
29b cf. E. A. Myers *et al.*, *J. Phys. Chem.*, 1967, **71**, 3531; 1968, **72**, 532, 3117.

TABLE 14-4

Complex Anions of Antimony(III)[a]

Halide	Complex anion	Structure
SbF$_3$[b]	SbF$_5^{2-}$	sp
	Sb$_2$F$_7^-$	Chain with SbF$_3$ and tbp SbF$_4^-$
	Sb$_4$F$_{16}^{4-}$	See Fig. 14-7
SbCl$_3$	SbCl$_4^-$	Chain (Fig. 14-6a)
	SbCl$_5^{2-}$	sp in NH$_4^+$ and K$^+$ salts[c] (Fig. 14-6b)
	SbCl$_6^{3-}$	Octahedral in [Co(NH$_3$)$_6$]$^{3+}$ salt and in solution
SbBr$_3$	Sb$_2$Br$_9^{3-}$	Br$_3$Sb(μ-Br)$_3$SbBr$_3$
	Sb$_2$Br$_{11}^{3-}$	Has Sb$_2$Br$_9^{3-}$ + Br$_2$ molecule acting as bridge to give three-dimensional network[d]

[a] Note that in salts the structure may be distorted with particular cations.

[b] R. Fourcade and G. Mascherpa, *Rev. Chim. Min.,* 1978, **15**, 295; J. G. Ballard *et al., J.C.S. Dalton,* **1976**, 2409.

[c] R. K. Wismer and R. A. Jacobson, *Inorg. Chem.,* 1974, **13**, 1678.

[d] C. B. Hubb and R. A. Jacobson, *Inorg. Chem.,* 1972, **11**, 2247.

Pentahalides. These and their related complexes have rather limited and straightforward coordination chemistry, consisting primarily of the acquisition by PF$_5$, AsF$_5$, and SbF$_5$ of F$^-$ ions to form, usually, the MF$_6^-$ ions, though occasionally more complex species such as As$_2$F$_{11}^-$, F$_5$SbFSbF$_5^-$, and Sb$_2$F$_{11}^-$.[30] A crystalline fluoride of stoichiometry Sb$_{11}$F$_{45}$ made by action of F$_2$ on Sb appears to have [Sb$_6$F$_{13}$]$^{5+}$ polymeric chains and distorted SbF$_6^-$ anions.[31]

The use of SbF$_5$ for the generation of superacids has already been discussed (p. 239). The combination of methyl fluoride and SbF$_5$ in SO$_2$ or SO$_2$ClF acts as a very strong methylating agent for organic substances. In SO$_2$ the solvent is in fact

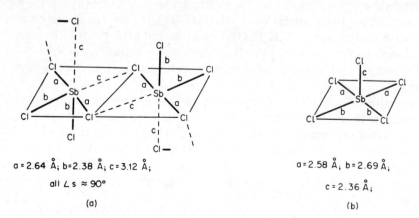

$a = 2.64$ Å; $b = 2.38$ Å; $c = 3.12$ Å;

all $\angle$ s ≈ 90°

(a)

$a = 2.58$ Å; $b = 2.69$ Å;

$c = 2.36$ Å;

(b)

Fig. 14-6. (a) A portion of the anion chain in (pyH)SbCl$_4$. (b) The SbCl$_5^{2-}$ ion found in (NH$_4$)$_2$SbCl$_5$.

[30]   A. J. Edwards and R. J. C. Sills, *J.C.S. Dalton,* **1974**, 1726.

[31]   A. J. Edwards and D. R. Slim, *J.C.S. Chem. Comm.,* **1974**, 178.

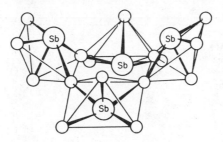

Fig. 14-7. The structure of the $Sb_4F_{16}^{4-}$ ion in $KSbF_4$.

methylated to give $MeOSO^+$, but in SO_2F_2 as solvent the donor complex CH_3F $\rightarrow SbF_5$ can be characterized by ^{19}F nmr.[32]

Phosphorus(V) fluorides show decreasing acceptor strength in the order PF_5 > $ArPF_4$ > $alkylPF_4$ > R_2PF_3.[33] Phosphorus pentachloride forms 1:1 adducts with pyridine, but 1,10-phenanthroline gives $[PCl_4phen]PCl_6$.[34] The PCl_6^- ion also occurs as noted above in $PCl_5(s)$. Antimony pentachloride also forms the ion $SbCl_6^-$.

14-7. Oxides

The following oxides of the Group V elements are known, and most are structurally characterized by X-ray or electron diffraction (bold type) or spectroscopically (italics):

P_4O_6	**As_4O_6**	**Sb_4O_6**	Bi_2O_3
P_4O_7			
P_4O_8	As_2O_4?	**Sb_2O_4**	
P_4O_9			
P_4O_{10}	As_2O_5	Sb_2O_5	Bi_2O_5?

Phosphorus Oxides. The pentoxide P_4O_{10}, so-called for historical reasons, is usually the main product of burning phosphorus; conditions can be optimized to make it the sole product. It is a white, crystalline solid that sublimes at 360° (1 atm) and this affords a good method of purifying it from its commonest impurities, which are nonvolatile products of incipient hydrolysis. The hexagonal crystalline form (H form) obtained on sublimation contains P_4O_{10} molecules (Fig. 14-8a) in which a P_4 tetrahedron has six edge-bridging oxygen atoms and four terminal ones. These molecules, which persist in the gas phase, have full tetrahedral (T_d) symmetry, and the P—O(br) and P—O(t) distances are 1.604 and 1.429 Å, respectively, by gas phase electron diffraction. Other crystalline and glassy forms, obtained by heating the H form, have infinite sheets (Fig. 14-8b); note that the local environment of each phosphorus atom is the same as that in the P_4O_{10} molecule.

The most important chemical property of P_4O_{10} is its avidity for water. It is one of the most effective drying agents known at temperatures below 100°. It reacts

32 J.-Y. Calves and R. J. Gillespie, *J. Am. Chem. Soc.*, 1977, **99**, 1788.
33 A. H. Cowley, P. J. Wisiam, and M. Sanchez, *Inorg. Chem.*, 1977, **16**, 1451.
34 K. B. Dillon, R. N. Reeve, and T. C. Waddington, *J.C.S. Dalton*, **1977**, 1410.

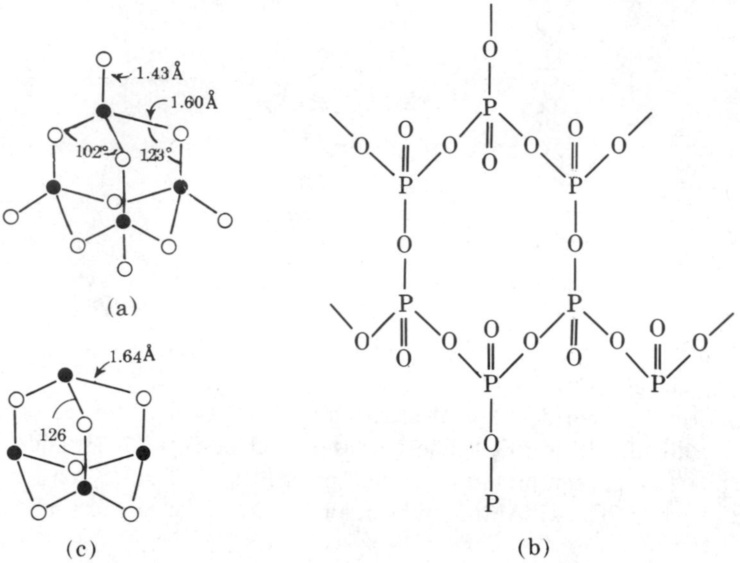

(a)

(b)

(c)

14-8\ The structures of (a) P_4O_{10} molecules (T_d); (b) P_4O_{10} sheets; (c) P_4O_6 molecules.

with water to form a mixture of phosphoric acids (see below) whose composition depends on the quantity of water and other conditions. It will even extract the elements of water from many other substances themselves considered good dehydrating agents; for example, it converts pure HNO_3 into N_2O_5 and H_2SO_4 into SO_3. It also dehydrates many organic compounds (e.g., converting amides into nitriles). With alcohols it gives esters of simple and polymeric phosphoric acids depending on reaction conditions.

The breakdown of P_4O_{10} with various reagents (alcohols, water, phenols, ethers, alkyl phosphates, etc.) is a very general one and is illustrative also of the general reaction schemes for the breakdown of P_4S_{10} and for the reaction of P_4 with alkali to give PH_3, hypophosphite, and so on. The reaction initially involves breaking a P—O—P bridge. Thus an alcohol reacts with P_4O_{10} as in eq. 14-2, followed by further reaction at the next most anhydridelike linkage, until eventually products containing only one P atom are produced (eq. 14-3).

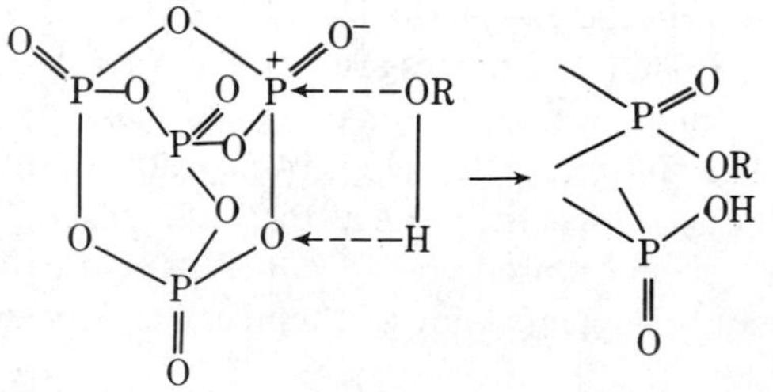

$$(14\text{-}2)$$

$$P_4O_{10} + 6\,ROH \longrightarrow 2\,(RO)_2PO\cdot OH + 2\,RO\cdot PO(OH)_2 \qquad (14\text{-}3)$$

The fusion of P_4O_{10} with basic oxides gives solid phosphates of various types, their nature depending on experimental conditions.

P_4O_6. The trioxide, again so named for historical reasons, has a structure of tetrahedral symmetry similar to that of P_4O_{10} except that the four terminal oxygen atoms are not present (Fig. 14-8c), and the P—O(br) distances, 1.638 Å, are slightly longer. It is a colorless, volatile (m. 23.8°, b. 175°) compound formed in about 50% yield when white phosphorus is burned in an oxygen-deficient atmosphere. It is difficult to separate by distillation from traces of unchanged phosphorus, but irradiation with ultraviolet light changes the white phosphorus into red, from which the P_4O_6 can be separated by dissolution in organic solvents. The chemistry of P_4O_6 is not well known, but it appears to be complex.

When heated above 210°, P_4O_6 decomposes into red P and other oxides PO_x. It reacts vigorously with chlorine and bromine to give the oxo halides and with iodine in a sealed tube to give P_2I_4. It is stable to oxygen at room temperature. When it is shaken vigorously with an excess of cold water, it is hydrated exclusively to phosphorous acid, H_3PO_3, of which it is formally the anhydride; P_4O_6 apparently cannot be obtained by dehydration of phosphorous acid. The reaction of phosphorus trioxide with *hot* water is very complicated, producing among other products PH_3, phosphoric acid, and elemental P; it may be noted in partial explanation that phosphorous acid itself, and all trivalent phosphorus acids generally, are thermally unstable, for example,

$$4H_3PO_3 \rightarrow 3H_3PO_4 + PH_3$$

P_4O_7, P_4O_8, P_4O_9. These intermediate oxides have structures in which one, two, and three terminal oxygen atoms are added to the P_4O_6 structure. White P_4O_7 is obtained in 50 to 55% yield by thermal decomposition of P_4O_6 in diglyme at 140° for 9 hr:[35]

$$P_4O_6 \rightarrow P_4O_7 + \text{yellow-orange } PO_x \text{ polymer}$$

The polymer has not been characterized. P_4O_8 and P_4O_9 are obtained by controlled partial oxidation of P_4O_6.

Arsenic Oxides. *Arsenic trioxide* (As_4O_6), formed on burning the metal in air, has in the ordinary form the same structure as P_4O_6. In other crystalline forms there are AsO_3 pyramids joined through oxygen atoms to form layers. The ordinary form is soluble in various organic solvents as As_4O_6 molecules and in water to give solutions of "arsenious acid." *Arsenic pentoxide* cannot be obtained by direct reaction of arsenic with oxygen. It is prepared by oxidation of As with nitric acid followed by dehydration of the arsenic acid hydrates so obtained. It readily loses oxygen when heated to give the trioxide. It is very soluble in water, giving solutions of arsenic acid. It has a structure similar to that of β-Ga_2O_3 with cations in adjacent tetrahedral and octahedral oxygen environments.[36]

---

[35]   M. L. Walker and J. L. Mills, *Synth. React. Inorg. Met.-Org. Chem.,* 1975, **5,** 29.
[36]   M. Jansen, *Z. Anorg. Allg. Chem.,* 1978, **441,** 5.

Antimony Oxides.[37a] Antimony *trioxide* is obtained by direct reaction of the element with oxygen. In the vapor and in the solid below 570° it consists of P_4O_6-type molecules; the high-temperature solid form is polymeric. It is insoluble in water or dilute nitric and sulfuric acids, but soluble in hydrochloric and certain organic acids. It dissolves in bases to give solutions of antimonates(III). The yellow *pentoxide* is made mainly by action of oxygen at high pressure and temperature on Sb_2O_3; it has a structure similar to that of Nb_2O_5 with octahedral SbO_6 groups.[37b]

When either the tri- or the pentoxide is heated in air at about 900°, a white insoluble powder of composition SbO_2 is obtained. Both α- and β-forms are recognized, and the former has been shown[37c] to consist of Sb^V in octahedral interstices and $Sb^{III}O_4$ pyramidal units.

Bismuth Oxides. The only well-established oxide of bismuth is Bi_2O_3, a yellow powder soluble in acids to give bismuth salts; being insoluble in alkalis, however, it has no acidic character. From solutions of bismuth salts, alkali, or ammonium hydroxide precipitates a *hydroxide,* $Bi(OH)_3$. Like the oxide, this compound is completely basic. *Bismuth(V) oxide* is extremely unstable and has never been obtained in pure form. The action of extremely powerful oxidizing agents on Bi_2O_3 gives a red-brown powder that rapidly loses oxygen at 100°.

The oxides of the Group V elements clearly exemplify two important trends that are manifest to some extent in all main groups of the Periodic Table: (1) the stability of the higher oxidation state decreases with increasing atomic number, and (2) in a given oxidation state the metallic character of the elements, therefore the basicity of the oxides, increase, with increasing atomic number. Thus P^{III} and As^{III} oxides are acidic, Sb^{III} oxide is amphoteric, and Bi^{III} oxide is strictly basic.

14-8. Sulfides and Other Chalcogenides

Like the oxides, sulfides and other chalcogenides are mostly based structurally on a tetrahedral array of P or As atoms with bridging S or Se atoms and, less commonly, terminal S or Se, but the details differ a good deal. We shall discuss in detail only the phosphorus sulfides. The known structures are shown in Fig. 14-9. In many cases these structures are known quantitatively from X-ray crystallography, but for some only the qualitative structure is available from ^{31}P nmr.

The molecular sulfides P_4S_3, P_4S_7, P_4S_9, and P_4S_{10} are prepared by heating together red phosphorus and sulfur in the formula ratios. Melts of composition intermediate between P/S = 4:3 and P/S = 4:7 contain at least five other molecular phosphorus sulfides. These cannot be efficiently separated, but may be prepared separately in other ways.

α- and β-P_4S_4.[38] Iodine adds across a P—P bond in P_4S_3 to give β-$P_4S_3I_2$,[39]

[37]   (a) C. A. Cody *et al., Inorg. Chem.,* 1979, **18,** 1572. (b) M. Janson, *Angew. Chem., Int. Ed.,* 1978, **17,** 137; E. Schwarzmann, H. Rumel, and W. Berndt, *Z. Naturforsch.,* 1977, **32b,** 617. (c) P. S. Gopalakrishnan and H. Manohar, *Crystallogr. Struct. Comm.,* 1975, **4,** 203.

[38]   A. M. Griffin, P. C. Minshall, and G. M. Sheldrick, *J.C.S., Chem. Commun.,* **1976,** 809.

[39]   G. J. Penney and G. M. Sheldrick, *J. Chem. Soc., A,* **1971,** 1100.

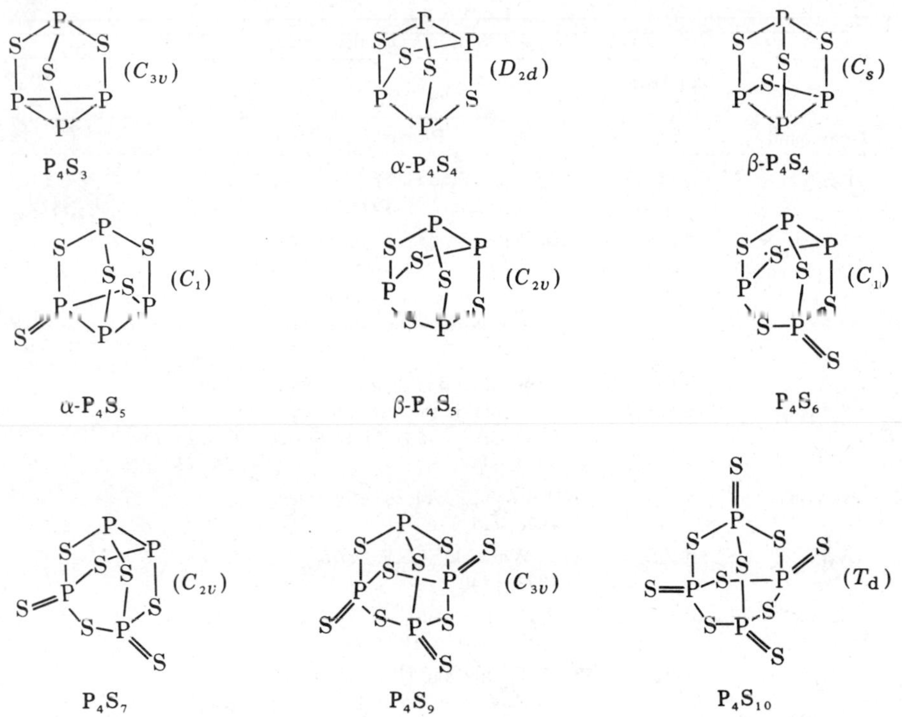

P_4S_3 (C_{3v}) $\alpha\text{-}P_4S_4$ (D_{2d}) $\beta\text{-}P_4S_4$ (C_s)

$\alpha\text{-}P_4S_5$ (C_1) $\beta\text{-}P_4S_5$ (C_{2v}) P_4S_6 (C_1)

P_4S_7 (C_{2v}) P_4S_9 (C_{3v}) P_4S_{10} (T_d)

14-9 Structures of P_4S_n molecules (see refs. 38, 40, 41).

which slowly isomerizes to $\alpha\text{-}P_4S_3I_2$. The latter may also be obtained directly by heating the elements together. The scheme below shows the quantitative conversion of α- and $\beta\text{-}P_4S_3I_2$ to α- and $\beta\text{-}P_4S_4$ by bis(trimethytin)sulfide in CS_2:

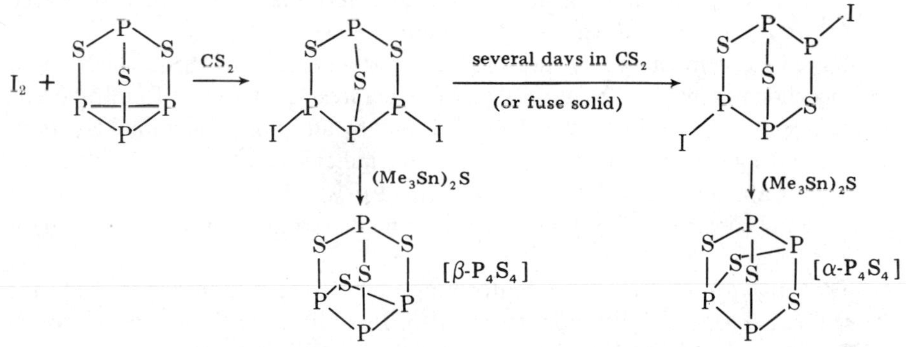

α- and $\beta\text{-}P_4S_5{}^{40}$ are obtained by the following reactions:

$$5P_4S_3 + 12Br_2 \rightarrow 3\alpha\text{-}P_4S_5 + 8PBr_3$$
$$P_4S_7 + 2Ph_3P \rightarrow \beta\text{-}P_4S_5 + 2Ph_3PS$$

[40]   A. M. Griffin and G. M. Sheldrick, *Acta Crystallogr., B,* 1975, **31**, 2738.

TABLE 14-5
Other Group V Molecular Chalcogenides

Compound	Molecular structure type	Remarks or reference
P_4Se_5	α-P_4S_5	C. J. Penney and G. M. Sheldrick, *J. Chem. Soc., A,* **1971,** 245.
$P_4Se_3I_2$	β-$P_4S_3I_2$	G. J. Penney and G. M. Sheldrick, *Acta Crystallogr., B,* 1970, **26,** 2092.
As_4S_3	P_4S_3	This molecule is found in both α- and β-dimorphite. H. J. Whitfield, *J.C.S. Dalton,* **1973,** 1737.
α-As_4S_4 β-As_4S_4	α-P_4S_4} α-P_4S_4}	{ α is the mineral realgar; β is synthetic. Both contain the same molecular unit. E. J. Porter and G. M. Sheldrick, *J.C.S. Dalton,* **1972,** 1347.
γ-As_4S_4	β-P_4S_4	A. Kutoglu, *Z. Anorg. Allg. Chem.,* 1976, **419,** 176.
As_4S_5	β-P_4S_5	H. J. Whitfield, *J.C.S. Dalton,* **1973,** 1740.
As_4Se_3	P_4S_3	T. J. Barstow and H. J. Whitfield, *J.C.S. Dalton,* **1977,** 959.
As_4Se_4	α-P_4S_4	T. J. Barstow and H. J. Whitfield, *J.C.S. Dalton,* **1973,** 1739; E. J. Smail and G. M. Sheldrick, *Acta Crystallogr., B,* 1973, **29,** 2014.

In the second reaction P_4S_6 is an intermediate, but has not been isolated pure as it forms mixed crystals with P_4S_7. Its structure is inferred from those of P_4S_7 and β-P_4S_5. Ph_3P is a general reagent for removal of terminal S and can also be used to prepare P_4S_9 from P_4S_{10} and β-P_4S_4 from α-P_4S_5.

All the P_4S_n compounds are stable in CS_2 solution except for P_4S_4, which slowly disproportionates into α-P_4S_5 and P_4S_3. All structures derive from a P_4 tetrahedron by replacement of P—P units by P—S—P and by addition of terminal S atoms. It is interesting that the mass spectrum of P_4S_{10} indicates that the $P_2S_5^+$ ion is three times more abundant than the most abundant $(P_4S_{10}^+)$ P_4S_n ion.[41] It appears that a tetrahedral core $P_4(\mu$-$S)_6$ entails much accumulated strain at the S—P—S and/or P—S–P angles.

Other Group V chalcogenides conforming to the same structural pattern as the phosphorus sulfides are listed in Table 14-5. In addition to these, arsenic forms As_2S_3 and As_2Se_3, which have layer structures, and As_2S_5 (of unknown structure). As_2S_3 and As_2S_5 can be precipitated from aqueous solutions of As^{III} and As^V by H_2S. They are insoluble in water but acidic enough to dissolve in alkali sulfide solutions to form thio anions.

Antimony forms Sb_2S_3 either by direct combination of the elements or by pre-

---

[41] G. J. Penney and G. M. Sheldrick, *J. Chem. Soc., A,* **1971,** 243.

cipitation with H_2S from Sb^{III} solutions; it dissolves in an excess of sulfide to give anionic thio complexes, probably mainly SbS_3^{3-}. Sb_2S_3, as well as Sb_2Se_3 and Bi_2S_3, have a ribbonlike polymeric structure in which each Sb atom and each S atom is bound to three atoms of the opposite kind, forming interlocking SbS_3 and SSb_3 pyramids (see p. 515). So-called antimony(V) sulfide (Sb_2S_5) is not a stoichiometric substance and according to Mössbauer spectroscopy contains only Sb^{III}.

Bismuth gives dark brown Bi_2S_3 on precipitation of Bi^{III} solutions by H_2S: it is not acidic. A sulfide BiS_2 is obtained as gray needles by direct interaction at 1250° and 50 kbar; its structure is unknown but may be $Bi^{3+}[BiS_4]^{3-}$.

OTHER COMPOUNDS

14-9.   Oxo Halides

The most important oxo halides are the *phosphoryl halides* (X_3PO), in which X may be F, Cl, or Br. The commonest, Cl_3PO, is obtained by the reactions

$$2PCl_3 + O_2 \rightarrow 2Cl_3PO$$
$$P_4O_{10} + 6PCl_5 \rightarrow 10Cl_3PO$$

The reactions of Cl_3PO are much like those of PCl_3 (Fig. 14-3). The halogens can be replaced by alkyl or aryl groups by means of Grignard reagents, and by alcoxo groups by means of alcohols; hydrolysis by water yields phosphoric acid. Cl_3PO also has donor properties toward metal ions, and many complexes are known. Distillation of the Cl_3PO complexes of $ZrCl_4$ and $HfCl_4$ can be used to separate zirconium and hafnium, and the very strong $Cl_3PO\cdot Al_2Cl_6$ complex has been utilized to remove Al_2Cl_6 from adducts with Friedel-Crafts reaction products.

All X_3PO molecules have a pyramidal PX_3 group, with the oxygen atom occupying the fourth position to complete a distorted tetrahedron. Corresponding compounds X_3PS and X_3PSe exist.

More complex oxo halides containing P—O—P bonds may have linear or ring structures. The compound $Cl_2(O)P$—O—$P(O)Cl_2$ is obtained either by oxidation of PCl_3 with N_2O_4 or by partial hydrolysis of Cl_3PO; the fluorine analogues exist.

Antimony and *bismuth* form the important oxo halides SbOCl and BiOCl, which are insoluble in water. They are precipitated when solutions of Sb^{III} and Bi^{III} in concentrated HCl are diluted. They have quite different but complicated, layer structures.

Fluorination of a 1:1 mixture of $AsCl_3$ and As_2O_3 has been reported to give F_3AsO (b. 26°) but this has been questioned. Cl_3AsO has been made by action of ozone on $AsCl_3$ in an inert solvent[42] or noble gas matrix.[43] It decomposes slowly

[42]   K. Seppelt, *Angew. Chem., Int. Ed.,* 1976, **15,** 766.
[43]   F. W. S. Benfield *et al., J.C.S. Chem. Comm.,* **1976,** 856.

at $-25°$ and rapidly at $0°$:

$$3Cl_3AsO \rightarrow AsCl_3 + Cl_2 + As_2O_3Cl_4$$

The latter is a polymer, stable to $150°$.

14-10. Phosphonium Compounds

Although organic derivatives of the type $[MR_4]^+X^-$ are well known for M = P, As, and Sb (see p. 467), only phosphorus gives the prototype PH_4^+, and this does not form any very stable compounds. As noted in Section 14-4, the proton affinity of PH_3 is substantially less than that of NH_3. The PH_4^+ ion is tetrahedral, with P—H = 1.414 Å as compared to P—H = 1.44 Å in PH_3. The best-known phosphonium salt is the iodide, which is formed as colorless crystals on mixing of gaseous HI and PH_3. The chloride and bromide are even less stable; the dissociation pressure of PH_4Cl into PH_3 and HCl reaches 1 atm below $0°$. The estimated basicity constant of PH_3 in water is about 10^{-26}, and phosphonium salts are completely hydrolyzed by water, releasing the rather insoluble gas PH_3:

$$PH_4I(s) + H_2O \rightarrow H_3O^+ + I^- + PH_3(g)$$

PH_3 dissolves in very strong acids such as $BF_3·H_2O$ and $BF_3·CH_3OH$ where it is protonated to PH_4^+.

The most readily produced compound is tetra(hydroxymethyl)phosphonium chloride obtained by the interaction of phosphine with formaldehyde in hydrochloric acid solution:

$$PH_3 + 4CH_2O + HCl \rightarrow [P(CH_2OH)_4]^+Cl^-$$

It is a white crystalline solid, soluble in water and it is available commercially. On addition of base it forms $P(CH_2OH)_3$.

14-11. Phosphorus-Nitrogen Compounds[44]

There is a very extensive and important chemistry of compounds with P—N and P=N bonds. R_2N—P bonds are particularly stable and occur widely in combination with bonds to other univalent groups such as alkyl, aryl, and halogen. The most important class of compounds are the *phosphazenes*.

Phosphazenes. These are cyclic or chain compounds that contain alternating phosphorus and nitrogen atoms with two substituents on each phosphorus atom. The three main structural types are the cyclic trimer 14-IV, the cyclic tetramer, 14-V, and the oligomer or high polymer 14-VI. A few ten-, twelve- and even six-

---

[44] H. R. Allcock, *Phosphorus-Nitrogen Compounds,* Academic Press, 1972; *Chem. Rev.,* 1972, **72,** 315; R. A. Shaw, *Pure Appl. Chem.,* 1975, **44,** 317; M. Berman, *Adv. Inorg. Chem. Radiochem.,* 1972, **14,** 1 (phosphazotrihalides, NPX$_3$); G. M. Kosalopoff and L. Maier, *Organic. Phosphorus Compounds,* Vol. 6, Wiley-Interscience, 1973; S. S. Krishnamurthy, A. C. Sau, and M. Woods, *Adv. Inorg. Chem. Radiochem.,* 1978, **21,** 41 (cyclophosphazenes).

teen-membered rings are also known.[45] The alternating sets of single and double bonds in 14-IV to 14-VI are written for convenience but (see below) should not be taken literally

(14-IV)   (14-V)   (14-VI)

Hexachlorocyclotriphosphazene, $(NPCl_2)_3$, is a key intermediate in the synthesis of many other phosphazenes and is readily prepared as follows:

$$nPCl_5 + nNH_4Cl \xrightarrow{\text{in } C_2H_2Cl_4 \text{ or } C_6H_5Cl} (NPCl_2)_n + 4nHCl$$

This reaction produces a mixture of $[NPCl_2]_n$ species, but under selected conditions high yields of the trimer and tetramer can be obtained. These can be readily separated. The trimer, a white crystalline solid (m. 113°) that sublimes readily in vacuum at 50° is commercially available; it is the source for synthesis of linear polymers discussed below.

The majority of phosphazene reactions involve replacement of halogen atoms by other groups to give partially or fully substituted derivatives, e.g.,

$$(NPCl_2)_3 + 6NaF \xrightarrow{MeCN} (NPF_2)_3 + 6NaCl$$

$$(NPCl_2)_3 + 6\,NaOR \rightarrow [NP(OR)_2]_3 + 6NaCl$$
$$[NP(OR)_2]_3 + xRNH^- \rightarrow N_3P_3(OR)_{6-x}(NHR)_x + xOR^-$$

The mechanism of these reactions, especially with organometallic reagents, is not fully understood, but they appear to proceed by S_N2 attack on P by the nucleophile.[46]

In partially substituted molecules, many isomers are of course possible. These isomers can usually be separated and characterized (e.g., by nmr spectroscopy).[47]

The rings in $(NPF_2)_x$ where $x = 3$ or 4 are planar but larger rings are not planar.[48a] For other $(NPX_2)_n$ compounds the six-rings are planar or nearly so, but larger rings are generally nonplanar with NPN angles of $\sim$120° and PNP angles of $\sim$132°. Figure 14-10 shows the structures of $(NPCl_2)_3$ and $(NPClPh)_4$. The P–N distances, which are generally equal or very nearly so in these ring systems, lie in

[45]   M. W. Dougill and N. L. Paddock, *J.C.S. Dalton,* **1974,** 1022.
[46]   H. R. Allcock and L. A. Smeltz, *J. Am. Chem. Soc.,* 1976, **98,** 4143.
[47]   For example, see J. L. Schumitz and H. R. Allcock, *J. Am. Chem. Soc.,* 1975, **97,** 2433.
[48a] J. G. Hartsuicker and A. J. Wagner, *J.C.S. Dalton,* **1978,** 1425.

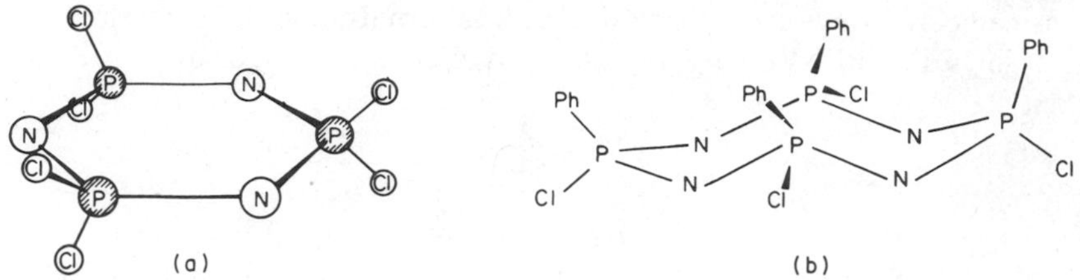

Fig. 14-10. The structures of two representative cyclic phosphazenes: (a) $(NPCl_2)_3$; (b) All-*cis*-$(NPClPh)_4$.

the range 1.55–1.61 Å; they are thus shorter than the expected single-bond length of ~1.75–1.80 Å. Considerable attention has been paid to the nature of the P—N π bonding, which the P—N distances indicate is appreciable, but the matter is still subject to controversy. The main question concerns the extent of delocalization, that is, whether there is complete delocalization all around the rings to give them a kind of aromatic character, or whether there are more localized "islands" within the NPN segments. Of course there may be considerable differences between the essentially planar rings and those that are puckered. The problem is a complicated one owing to the large number of orbitals potentially involved and to the general lack of ring planarity, which means that rigorous assignment of σ and π character to individual orbitals is impossible.

These rings are conformationally flexible, and π bonding is only one of many factors that influence the conformations, which may be understood on the basis of the same approach as that used for organic systems, namely, by using an empirical force field in which potential functions for bond twisting, bending and stretching, and nonbonded repulsions are included.[48b]

Compared to the linear phosphazenes, the cyclophosphazenes have few uses, but spiro compounds such as 14-VII crystallize with channels and can act as tunnel clathrates for aromatic molecules, olefins, etc.[49]

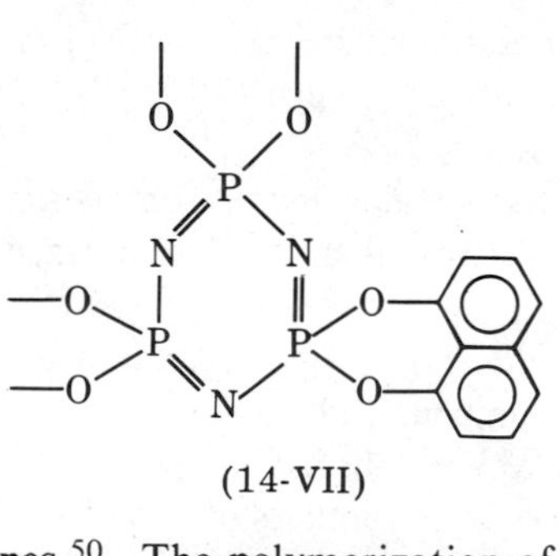

(14-VII)

Linear Polyphosphazenes.[50]   The polymerization of molten $(NPCl_2)_3$ above 230° to high molecular weight materials was first reported by Stokes in 1897; above

[48b] R. H. Boyd and L. Kesner, *J. Am. Chem. Soc.*, 1977, **99**, 4248.
[49]   H. R. Allcock, *Angew. Chem., Int. Ed.*, 1978, **17**, 81.
[50]   H. R. Allcock, *Chem. Tech.*, **1975**, 552; *Angew. Chem., Int. Ed.*, 1977, **16**, 147.

350° the polymers degrade to give mixtures of cyclic species. $(NPF_2)_3$ also polymerizes to an amber, rubbery elastomer. Similar polymers with hydrophobic substituents should be, and indeed may be, very stable, but it has proved impossible to polymerize the corresponding organic substituted trimers $(NPR_2)_3$ because of thermodynamic factors associated with the bulky R groups. However a major breakthrough by H. R. Allcock and co-workers was the demonstration that carefully prepared $(NPCl_2)_n$ undergoes the same type of substitutions as, say, $(NPCl_2)_3$. The important point is that $(NPCl_2)_n$ soluble in organic solvents is required. Unless $(NPCl_2)_3$ is rigorously purified from traces of PCl_5 left from the synthesis, cross-linking usually occurs to give insoluble materials. However, with care, soluble polymers that have molecular weights up to 3 to 4×10^6 (i.e., a degree of polymerization exceeding 15,000) can be obtained. Such polymers are readily substituted:

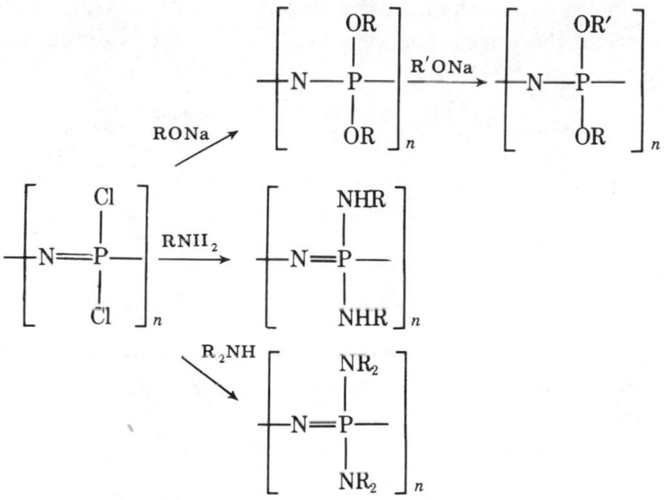

The trimer $(NPF_2)_3$ can also be polymerized at 350° and the organic soluble polymer reacted with magnesium or lithium alkyls or aryls to give partly substituted materials.[51] Clearly by having mixed substituents, a uniquely wide range of polymers can be made, and this distinguishes the phosphazenes from all other types of polymeric material, organic or inorganic.

The polymers can be glasses, tough or flexible solids, or rubbers, but they do not normally form crystals. The materials can be fabricated into fibers, woven fabrics, flexible films, or tubes, or used to impregnate other fabrics as waterproofing or flame-retarding agents. Some organophosphazene polymers remain flexible down to −90°, and since they are hydrocarbon resistant, they can be used for fuel pipes, O-rings, etc. Others, with substituents such as OCH_2CF_3 are water repellent and do not react with living tissue; thus they can be used for heart valves and other replacement parts for the body. Biodegradable polymers derived from amino acids,

51   H. R. Allcock, D. P. Patterson, and T. L. Evans, *J. Am. Chem. Soc.*, 1977, **99**, 6095.

such as

$$
\left[-N{=}P- \begin{array}{c} NHCH_2CO_2Et \\ | \\ | \\ NHCH_2CO_2Et \end{array} \right]_n
$$

will hydrolyze slowly to harmless products (viz., amino acid, phosphate, and NH_3) and may be used for sutures. The phosphazenes can also be used as ligands (donor N). A water-soluble polybis(dimethylaminophosphazene) can be attached also to hemin to provide a good model for hemoglobin (Chapter 31), since in concentrated solution reversible binding of oxygen is observed.

Other P-N Compounds. There are other compounds, two of which have been mentioned, such as $R_2NPR'_2$, R_2NPX_2, and $R_2N{-}P{=}NR$. The latter type of compound, aminoiminophosphanes, are phosphorus nitrogen ylids. They may be either monomeric if there are no bulky substituents on the N atoms, but may be dimers with bulky ligands:

$$
(Me_3Si)_2N{-}P{=}N(SiMe_3) \xrightarrow{25°}
$$

The N_2P_2 ring compounds, aminocyclodiphosphazanes, or diazadiphosphetidines may have planar rings and cis-trans isomers, depending on the nature of the groups attached to P and N.[52]

Other types of ring with single bonds are known. Thus the reaction

$$
EtPCl_2 + (Me_3Si)_2NMe \longrightarrow 2\,Me_3SiCl +
$$

gives cyclotetra (λ^3)-phosphazanes where the seven-membered ring has a crown structure with the nitrogen atoms almost planar.[53]

There are also heterocycles, one example being synthesized as follows[54]:

$$
MeN(PF_2)_2 + B_2H_4(PF_3)_2 \xrightarrow{-23°}
$$

52   R. Keat *et al., J.C.S. Dalton,* **1979**, 1224.
53   W. Zeiss, W. Schwarz, and H. Hess, *Angew. Chem., Int. Ed.,* 1977, **16**, 407.
54   R. J. Paine, *J. Am. Chem. Soc.,* 1977, **99**, 3884.

The material known as *phospham*, a highly cross-linked polymer of stoichiometry $(PN_2H)_n$ can be made by the reaction:

$$4NH_3(g) + 2P(red) \underset{}{\overset{500°}{\rightleftharpoons}} 2PN_2H(s) + 5H_2$$

Its structure is presumably:

$$\left[\begin{array}{c} HN- \\ | \\ -P=N- \\ | \end{array}\right]_n$$

It is possible that phosphonitrilic amides, which have potential as fertilizers, can be obtained by action of ammonia on phospham at high pressures.[55]

Last there are two-coordinate phosphorus cations[56] made by reactions such as:

$$(Me_2N)_2PCl + AlCl_3 \rightarrow [(Me_2N)_2P]^+AlCl_4^-$$
$$(Me_2N)_2PF + PF_5 \rightarrow [(Me_2N)_2P]^+PF_6^-$$

These have $N \rightarrow P \, \pi$ bonding with the NPN group planar:

$$\left[\begin{array}{c} Me\text{—}N \overset{\diagup Me}{\underset{\diagdown}{\diagdown}} \\ \qquad\qquad P: \\ Me\text{—}N \underset{\diagdown Me}{\diagup} \end{array}\right]^+$$

14-12. Organic Compounds

There is a vast chemistry of organophosphorus compounds,[57] and even for arsenic, antimony, and bismuth, the literature[58] is voluminous. Consequently only a few topics can be discussed here. It must also be noted that we discuss only the compounds that have P—C bonds. Many compounds sometimes referred to as organophosphorus compounds that are widely used as insecticides, nerve poisons, etc., as a result of their anticholinesterase activity, do *not*, in general, contain P—C bonds. They are usually organic esters of phosphates or thiophosphates; examples

[55] J. M. Sullivan, *Inorg. Chem.*, 1976, **15**, 1055.

[56] M. G. Thomas, C. W. Schultz, and R. W. Parry, *Inorg. Chem.*, 1977, **16**, 994; K. Dimroth, *Topics Curr. Chem.*, 1973, **38**, 1.

[57] *Organophosphorus Compounds*, Specialist Periodical Reports, Chemical Society, London; A. J. Kirby and S. G. Warren, *The Organic Chemistry of Phosphorus*, Elsevier, 1967 (reaction mechanisms); G. M. Kosolapoff and L. Maier, Eds., *Organic Phosphorus Compounds*, Vol. 1, Wiley-Interscience, 1972 (includes metal compounds), Vol. 7, 1976; W. E. McEwan and K. D. Berlin, *Organophosphorus Stereochemistry*, Parts I and II, Wiley-Halsted, 1975; H. Harnisch, *Angew. Chem., Int. Ed.*, 1976, **15**, 468 (compounds with P—C bonds).

[58] Houben-Weyl, *Methoden der Organische Chemie*, Band XIII-8, Thieme, 1978; G. O. Doak and L. D. Freedman, *Organometallic Compounds of Arsenic, Antimony and Bismuth*, Wiley-Interscience, 1970; P. G. Harrison, *Organomet. Chem. Rev.*, 1970, **5**, 183 (Bi); M. Dub, *Organometallic Compounds*, Vol. 3, 2nd ed., Springer (As, Sb, Bi); R. C. Poller, *Comprehensive Organic Chemistry*, Vol. 3, Part 15.5, Pergamon Press, 1979 (Sb, Bi).

are the well-known malathion, and parathion, which is $(EtO)_2P^V(S)$-$(OC_6H_4NO_2)$.[59] Compounds with P—C bonds are almost entirely synthetic, though a few rare examples occur in Nature.

With a few exceptions, mentioned at the end of this section, the organo derivatives are compounds with only three or four bonds to the central atom. They may be prepared in a great variety of ways, the simplest being by treatment of halides or oxo halides with Grignard reagents:

$$(O)MX_3 + 3RMgX \rightarrow (O)MR_3 + 3MgX_2$$

Trimethylphosphine is spontaneously flammable in air, but the higher trialkyls are oxidized more slowly. The phosphine oxides R_3MO, which may be obtained from the oxo halides as shown above or by oxidation of the corresponding R_3M compounds by H_2O_2 or air, are all very stable.

The P—O bonds are very short (e.g., 1.483 Å in Ph_3PO[60]), suggesting a bond order greater than 2. In R_3PS compounds the P—S values are close to the double-bond values.[61]

There are good methods[62] for preparing optically pure dissymmetric phosphine oxides, abcPO, e.g., $(CH_3)(C_3H_7)(C_6H_5)PO$. It is then possible to reduce these to optically pure phosphines with either retention or inversion. The reductant $HSiCl_3$ (p. 386) accomplishes this with either retention or inversion, depending on the base used in conjunction with it. Hexachlorodisilane reduces with inversion, and to account for this the following mechanism has been proposed:

$$abcPO + Si_2Cl_6 \rightarrow abcP^+OSiCl_3 + SiCl_3^-$$
$$SiCl_3^- + abcP^+OSiCl_3 \rightarrow Cl_3SiP^+abc + SiCl_3O^- \quad \text{(inversion)}$$
$$Cl_3SiP^+abc + SiCl_3O^- \rightarrow Cl_3SiOSiCl_3 + Pabc \quad \text{(attack of } Cl_3SiO^- \text{ on } SiCl_3)$$

Interestingly, the same reagent removes S from abcPS with retention; it is presumed that the first step is similar, but that $SiCl_3^-$ then attacks sulfur, rather than phosphorus:

$$abcPS + Si_2Cl_6 \rightarrow abcP^+SSiCl_3 + SiCl_3^-$$
$$abcPSSiCl_3 + SiCl_3^- \rightarrow abcP + Cl_3SiSSiCl_3$$

Trialkyl- and triarylphosphines, -arsines, and -stibines, and chelating di- and triphosphines and -arsines are widely used as π-acid ligands (p. 87, 147).

Toward trivalent boron compounds, gas phase calorimetric studies of the reactions (M = P, As, Sb)

$$Me_3M_{(g)} + BX_{3(g)} = Me_3MBX_{3(s)}$$

show[63] that the order of base strength is P > As > Sb, and the acid strength of BX_3 toward $(CH_3)_3P$ is $BBr_3 > BCl_3 \approx BH_3 > BF_3$ as found by other studies. The *oxides* R_3MO also form many complexes, but they function simply as donors. Trialkyl-

[59] M. Eto, *Organophosphorus Pesticides: Chemistry and Biochemistry,* CRC Press, 1974.
[60] G. Ruban and V. Zabel, *Crystallogr. Struct. Comm.,* 1976, **5**, 671.
[61] C. J. Williams *et al., J. Am. Chem. Soc.,* 1975, **97**, 6352.
[62] K. Mislow *et al., J. Am. Chem. Soc.,* 1968, **90**, 4842; 1969, **91**, 7023.
[63] D. C. Mente, J. L. Mills, and R. E. Mitchell, *Inorg. Chem.,* 1975, **14**, 123.

and triarylphosphines, -arsines, and -stibines generally react with alkyl and aryl halides to form *quaternary salts:*

$$R_3M + R'X \rightarrow [R_3R'M]^+X^-$$

The stibonium compounds are the most difficult to prepare and are the least common. These quaternary salts, except the hydroxides, which are obtained as syrupy masses, are white crystalline compounds. The tetraphenylphosphonium and -arsonium ions are useful for precipitating large anions such as ReO_4^- and ClO_4^-, and complex anions of metals.

Triphenylphosphine, a white crystalline solid (m 80°), is a particularly important ligand for transition metal complexes and is used industrially in the rhodium-catalyzed hydroformylation process (Section 30-8). It is also widely used in the *Wittig reaction* for olefin synthesis. This reaction involves the formation of alkylidene-triphenylphosphoranes from the action of butyllithium or other base on the quaternary halide, for example,

$$[(C_6H_5)_3PCH_3]^+Br^- \xrightarrow{\text{n-butyllithium}} (C_6H_5)_3P{=}CH_2$$

This intermediate reacts very rapidly with aldehydes and ketones to give zwitterionic compounds (14-VIII), which eliminate triphenylphosphine oxide under mild conditions to give olefins (14-IX):

$$(C_6H_5)_3P{=}CH_2 \xrightarrow{\text{cyclohexanone}} \text{(14-VIII)} \rightarrow \text{(14-IX)} + (C_6H_5)_3PO$$

(14-VIII)   (14-IX)

Alkylidenephosphoranes.[64a] These are compounds such as $Me_3P{=}CH_2$, $Et_3P{=}CH_2$, $Me_2EtP{=}CH_2$, and $Et_3P{=}CHMe$, all colorless liquids, stable for long periods in an inert atmosphere.

In these ylids the P—C distances range from 1.66 Å in Ph_3PCH_2, clearly indicative of double-bond character, to 1.74 Å. A single P—C bond would be 1.80 to 1.85 Å long. An authentic, localized P=C bond is found in the compound R—P=CPh_2, where R = mesityl.[64b]

Trimethylphosphinemethylene gives rise to a wide variety of transition metal complexes with M—C bonds (Sections 27-4 and 27-5). It can be made by the reaction[65]:

$$Me_4PBr + NaNH_2 \xrightarrow{\text{reflux}} Me_3P{=}CH_2 + NaBr + NH_3$$

although at 25° this reaction gives $Me_3P{=}N{-}P(Me_2)CH_2$, which can be protonated to give $[Me_3P{=}NPMe_3]^+$.

Finally, we note that there are *phosphacumulene* ylids.[66] Although cumulenes

64a K. Dimroth, *P-C Double Bonds,* Vol. 38, *Topics in Current Chemistry,* Springer, 1973.
64b T. C. Klebach, R. Lourens, and F. Bickelhaupt, *J. Am. Chem. Soc.,* 1978, **100**, 4887.
65   H. Schmidbauer and H.-J. Fuller, *Angew. Chem., Int. Ed.,* 1976, **15**, 501.
66   H. J. Bestman, *Angew. Chem., Int. Ed.,* 1977, **16**, 349.

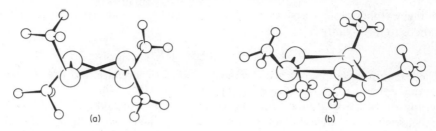

Fig. 14-11. The structures of (a) $(CF_3P)_4$ and (b) $(CF_3P)_5$. Large, medium, and small circles represent P, C, and F atoms, respectively.

proper must be linear $(R_2C{=}C{=}CR_2)$, there is d-orbital participation in the phosphorus compounds, which are bent[67]:

$$Ph_3P{\Large =}\overset{\displaystyle C}{}{=}PPh_3$$

Cyclopolyphosphines and -arsines.[68] These are compounds of general formula $(RP)_n$ and $(RAs)_n$, $n = 3$ to 6. $(C_2F_5P)_3$ is best known for $n = 3$ and only $(PhP)_6$, $(PhAs)_6$, and the three isomeric $[(MeC_6H_4)As]_6$ for $n = 6$. The four- and five-membered rings predominate, with four-membered ones being favored by bulky substituents. The puckered structures adopted by these rings are illustrated by representative compounds in Fig. 14-11.

These compounds in general are thermally stable, though often reactive. Typical preparative reactions are:

$$nRPH_2 + nRPCl_2 \rightarrow 2(RP)_n + 2nHCl$$
$$RAsO_3Na_2 + H_3PO_2 \rightarrow (RAs)_n$$

The preferred conformation[69] for R_3E_3 is 14-X, but as might be expected, an all-cis arrangement is enforced in 14-XI.[70]

(14-X)   (14-XI)

[67] P. J. Carrol and D. D. Titus, *J.C.S. Dalton*, **1977**, 824.
[68] L. R. Smith and J. R. Mills, *J. Organomet. Chem.*, 1975, **84**, 1; *J. Am. Chem. Soc.*, 1976, **98**, 3852.
[69] M. Baudler *et al.*, *Z. Naturforsch.*, 1976, **31b**, 1305, 1311.
[70] J. Ellermann and H. Schossner, *Angew. Chem., Int. Ed.*, 1974, **13**, 601.

The phenyl compound $(PhP)_5$ reacts with alkali metals in tetrahydrofuran to give red solutions:

$$(PhP)_5 + \frac{10}{n} Na = \frac{5}{n} Na_2(PPh)_n$$

The anionic species depend on the nature of M and on the stoichiometry, but the potassium species appear to involve equilibria such as

where the metal ions are doubtless solvated by the ether.[71]

There are a few linear triphosphines that all contain CF_3 and they are obtained by the reactions

$$2(CF_3)_2PI + CF_3PH_2 + 2(CH_3)_3N \rightarrow (CF_3)_2P—P(CF_3)—P(CF_3)_2 + 2(CH_3)_3NHI$$
$$2(CF_3)_2PCl + CH_3PH_2 + 2(CH_3)_3N \rightarrow CH_3P[P(CF_3)_2]_2 + 2(CH_3)_3NHCl$$

The extensive series of dimethylarsenic compounds, often called "cacodyl" compounds (e.g., Me_2AsCl, cacodyl chloride) is worthy of mention. There are also diarsenic tetraalkyls such as dicacodyl, $Me_2AsAsMe_2$, and $Et_2AsAsEt_2$.

A phosphorus-to-carbon triple bond evidently occurs in $HC{\equiv}P$, which is a pyrophoric substance that polymerizes slowly even at $-130°$. The $C{\equiv}P$ bond length is only 1.54 Å.

Pentavalent Compounds (R_5M). These tend to be more stable with the heavier elements, whereas the corresponding R_3M compounds become less stable. The pentaphenyl compounds Ph_5M are perhaps the best characterized pure organic derivatives, but a host of mixed compounds, especially of antimony, of the types Ph_4SbX and Ph_3SbX_2 ($X = OH$, OR, or halogen) are known. In all but two cases these molecules have *tbp* structures, with the more electronegative ligands at axial positions. A configuration closer to *sp* has been found for crystalline Ph_5Sb, and an *sp* configuration has been postulated on spectroscopic evidence for penta(cyclopropyl)antimony. However, since $(p\text{-tolyl})_5Sb$ has a *tbp* structure in the crystal, though it is nonrigid in solution except at $-130°$, it seems likely that the *sp* structure in crystalline Ph_5Sb is occasioned by packing interactions and is not inherently preferred.[72]

Although alkyl compounds are well characterized for antimony, phosphorus and arsenic alkyls are unstable, and attempts to make pentamethylphosphorane by the same method as succeeds for Ph_5P gives the ylid, viz.,

$$Ph_4PBr + PhLi \rightarrow Ph_5P + LiBr$$
but
$$Me_4PI + MeLi \rightarrow Me_3P{=}CH_2 + CH_4 + LiI$$
and
$$Ph_4PBr + MeLi \rightarrow Ph_3P{=}CH_2 + C_6H_6 + LiBr.$$

[71]   P. R. Hoffman and K. G. Caulton, *J. Am. Chem. Soc.*, 1975, **97**, 6370.
[72]   For references, see G. L. Kuykendall and J. L. Mills, *J. Organomet. Chem.*, 1976, **118**, 123.

However the formation of ylids can be prevented by incorporation of P into heterocycles as in 14-XII to 14-XIV.[73] The spirophosphoranes and phosphorus(V) esters of type 14-XV are approximately square pyramidal.[74]

(14-XII)   (14-XIII)

(14-XIV)   (14-XV)

The antimony compounds $Me_3Sb(NO_3)_2$ and $Me_3Sb(ClO_4)_2$, which appear to be molecular with *tbp* structures in the solid, dissolve in water and ionize, apparently to give the planar cation $(CH_3)_3Sb^{2+}$.

Aromatic Heterocycles. It may also be noted that phosphorus, arsenic, antimony, and bismuth all form analogues to pyridine, and these have significant aromatic character as a result of π interactions.[75] Also, there are P and As analogues of pyrrole that also have significant participation of the P and As lone pairs in forming an aromatic sextet.[76]

14-13.   Aqueous Cationic Chemistry

Apart from the quaternary salts mentioned in the preceding section, there is no cationic chemistry of P and As. Although the reactions

$$H_2O + OH^- + AsO^+ \leftarrow As(OH)_3 \rightarrow As^{3+} + 3OH^-$$

may occur to some slight extent, there is little direct evidence for the existence of significant concentrations of either of the cations even in strong acid solutions.

Antimony has some definite cationic chemistry, but only in the trivalent state, the basic character of Sb_2O_5 being negligible. Cationic compounds of Sb^{III} are mostly of the "antimonyl" ion SbO^+, although some of the "Sb^{3+}" ion, such as $Sb_2(SO_4)_3$, are known. Antimony salts readily form complexes with various acids in which the antimony forms the nucleus of an anion.

In sulfuric acid from 0.5 to $12M$, Sb^V appears to be present as the $[Sb_3O_9]^{3-}$ ion. For Sb^{III} in sulfuric acid, the species present vary markedly with the acid concentration, namely,

[73]   H. Schmidbauer, P. Holl, and F. H. Köhler, *Angew. Chem., Int. Ed.,* 1977, **16,** 722.
[74]   R. R. Holmes *et al., Inorg. Chem.,* 1978, **17** 3265; 1979, **18,** 1653.
[75]   A. J. Ashe, R. R. Sharp, and J. W. Tolan, *J. Am. Chem. Soc.,* 1976, **98,** 5451.
[76]   N. D. Epiotes and W. Cherry, *J. Am. Chem. Soc.,* 1976, **98,** 4365.

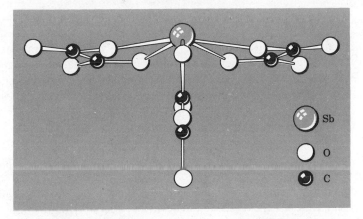

Fig. 14-12. The $Sb(C_2O_4)_3^{3-}$ ion projected on a plane approximately perpendicular to the basal plane of the pentagonal pyramid.

$$SbO^+ \text{ and/or } Sb(OH)_2^+ \qquad <1.5M \text{ } H_2SO_4$$
$$SbOSO_4^-, Sb(SO_4)_2^- \qquad \sim 1.0\text{--}18 \text{ } M \text{ } H_2SO_4$$

In salts the $Sb(C_2O_4)_3^{3-}$ ion has a ψ-pentagonal-bipyramid structure (Fig. 14-12) with a lone pair at one axial position.

The tartrate complex of antimony(III), *tartar emetic*, $K_2[Sb_2(d\text{-}C_4O_6H_2)_2]\cdot 3H_2O$, has been known in medicine for over 300 years and is used for treatment of schistosomiasis and leishmaniasis; the toxic side effects can be mediated by penicillamine.[77] In the salts the ion has a binuclear structure[78a] (Fig. 4-9, p. 172) and the Sb atom has ψ-*tbp* geometry. This coordination is also found in the complex $K[As(C_6H_4O_2)_2]$ derived from catechol (*o*-dihydroxybenzene).

Only for bismuth can it be said that there is an extensive true cationic chemistry. Aqueous solutions contain well-defined hydrated cations, but there is no evidence for the simple aqua ion $[Bi(H_2O)_n]^{3+}$. In neutral perchlorate solutions the main species is $[Bi_6O_6]^{6+}$ or its hydrated form, $[Bi_6(OH)_{12}]^{6+}$, and at higher pH $[Bi_6O_6(OH)_3]^{3+}$ is formed. The $[Bi_6(OH)_{12}]^{6+}$ species contains an octahedron of Bi^{3+} ions with an OH^- bridging each edge. Vibrational analysis suggests some weak bonding directly between the Bi atoms.

There is considerable evidence for the association of Bi^{3+} ion with nitrate ions in aqueous solution. The nitrate ions appear to be mainly bidentate, and all members of the set $Bi(NO_3)(H_2O)_n^{2+}\cdots Bi(NO_3)_4^-$ appear to occur. From acid solution various hydrated crystalline salts such as $Bi(NO_3)_3\cdot 5H_2O$, $Bi_2(SO_4)_3$, and double nitrates of the type $M_3^{II}[Bi(NO_3)_6]_2\cdot 24H_2O$ can be obtained. Treatment of Bi_2O_3 with HNO_3 gives basic salts such as $BiO(NO_3)$ and $Bi_2O_2(OH)(NO_3)$. Similar salts, generally insoluble in water, are precipitated on dilution of strongly acid solutions of Bi compounds. The nitrate $Bi_6O_4(OH)_4(NO_3)_6H_2O$ has a Bi_6 octahedron with face-bridging μ_3-oxo groups.[78b]

[77] S. O. Wandiga, *J.C.S. Dalton*, **1975**, 1894.
[78a] M. E. Gress and R. A. Jacobson, *Inorg. Chim. Acta*, 1974, **8**, 209.
[78b] B. Sundvall, *Acta Chem. Scand.*, 1979, **33A**, 219.

THE OXO ANIONS

The oxo anions in both lower and higher states are a very important part of the chemistry of phosphorus and arsenic and comprise the only real aqueous chemistry of these elements. For the more metallic antimony and bismuth, oxo anion formation is less pronounced, and for bismuth only ill-defined "bismuthates" exist.

14-14. Oxo Acids and Anions of Phosphorus

All phosphorus oxo acids have POH groups in which the hydrogen atom is ionizable; hydrogen atoms in P—H groups are not ionized. There is a vast number of oxo acids or ions, some of them of great technical importance; but with the exception of the simpler species, they have not been well understood structurally until quite recently. We can attempt to deal only with some structural principles, and some of the more important individual compounds. The *oxo anions* are of main importance, since in many cases the free acid cannot be isolated, though its salts are stable. Both lower (P^{III}) and higher (P^V) acids are known.

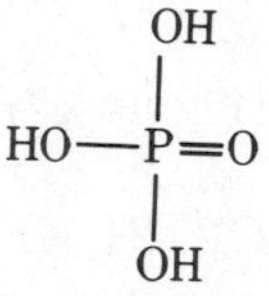

(14-XVIa) 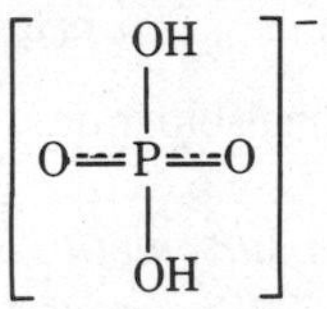   (14-XVIb) 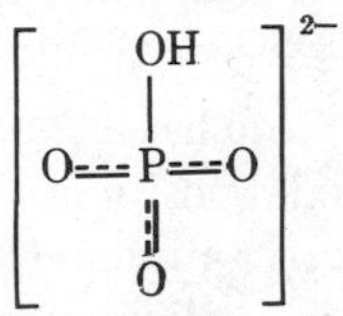   (14-XVIc) 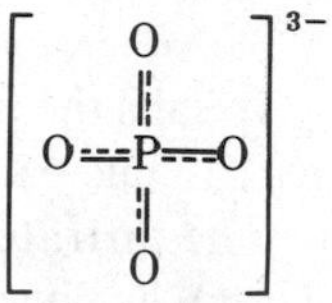   (14-XVId)

The principal higher acid is orthophosphoric acid, 14-XVIa, and its various anions 14-XVIb to 14-XVId, all of which are tetrahedral. The phosphorus(III) acid, which might naively have been considered to be $P(OH)_3$, has in fact the four-connected, tetrahedral structure 14-XVIIa; it is only difunctional, and its anions are 14-XVIIb and 14-XVIIc. Only in the triesters of phosphorous acid $[P(OR)_3]$ do we encounter three-connected phosphorus, and even these, as will be seen below, have a tendency to rearrange to four-connected species.

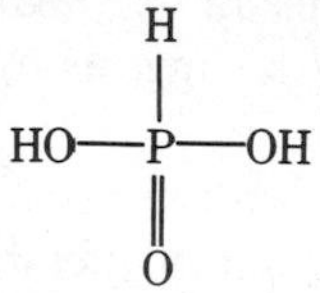

(14-XVIIa) 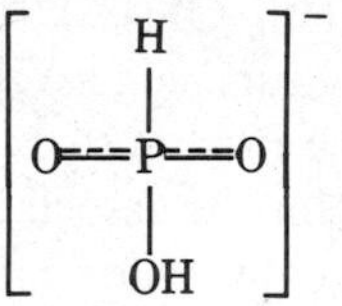   (14-XVIIb) 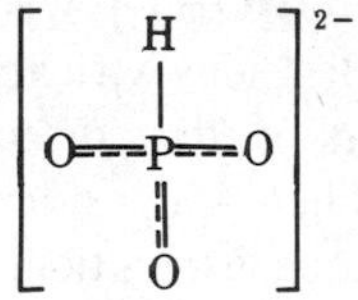   (14-XVIIc)

Similarly, the acid of formula H_3PO_2, hypophosphorous acid, also has a four-connected structure (14-XVIIIa), as does its anion 14-XVIIIb.

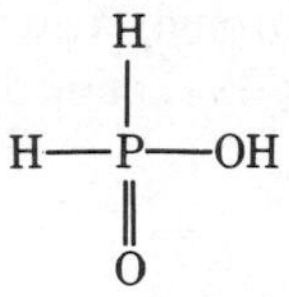

(14-XVIIIa) 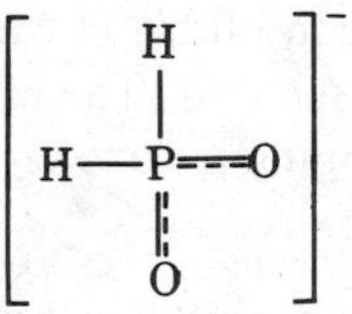   (14-XVIIIb)

Lower Acids. *Hypophosphorous Acid,* $H[H_2PO_2]$. The salts are usually prepared by boiling white phosphorus with alkali or alkaline earth hydroxide. The main reactions appear to be

$$P_4 + 4OH^- + 4H_2O \rightarrow 4H_2PO_2^- + 2H_2$$
$$P_4 + 4OH^- + 2H_2O \rightarrow 2HPO_3^{2-} + 2PH_3$$

The calcium salt is soluble in water, unlike that of phosphite or phosphate; the free acid can be made from it or obtained by oxidation of phosphine with iodine in water. Both the acid and its salts are powerful reducing agents, being oxidized to orthophosphate. The pure white crystalline solid is a monobasic acid ($pK = 1.2$); other physical studies, such as nmr, confirm the presence of a PH_2 group, and the anion has been characterized crystallographically. Either or both of the hydrogen atoms can be replaced, by indirect methods, with alkyl groups to give mono- or dialkyl *phosphonous* compounds.

Phosphorous Acid, $H_2[HPO_3]$. As noted above, this acid and its mono and diesters have a P—H bond. The free acid is obtained by treating PCl_3 or P_4O_6 with water; when pure, it is a deliquescent colorless solid (m. 70.1°, $pK = 1.8$). The presence of the P—H bond has been demonstrated by a variety of structural studies as well as by the formation of only mono and di series of salts. It is oxidized to orthophosphate by halogen, sulfur dioxide, and other agents, but the reactions are slow and complex. The mono-, di-, and triesters can be obtained from reactions of alcohols or phenols with PCl_3 alone or in the presence of an organic base as hydrogen chloride acceptor. They can also be obtained directly from white phosphorus by the reaction[79]

$$P_4 + 6OR^- + 6CCl_4 + 6ROH = 4P(OR)_3 + 6ClHCl_3 + 6Cl^-$$

RPO_3^{2-} ions are called *phosphonate* ions.

The *phosphite triesters* $P(OR)_3$ (cf. above) are notable for forming donor complexes with transition metals and other acceptors. They are readily oxidized to the respective phosphates:

$$2(RO)_3P + O_2 \rightarrow 2(RO)_3PO$$

They also undergo the Michaelis-Arbusov reaction with alkyl halides, forming dialkyl phosphonates:

$$P(OR)_3 + R'X \longrightarrow [(RO)_3PR'X] \longrightarrow RO-\overset{\overset{\displaystyle O}{\|}}{\underset{\underset{\displaystyle OR}{|}}{P}}-R' + RX$$

Phosphonium intermediate

The methyl ester easily undergoes spontaneous isomerization to the dimethyl ester of methylphosphonic acid:

$$P(OCH_3)_3 \longrightarrow CH_3PO(OCH_3)_2$$

[79]   C. Brown *et al., J.C.S. Chem. Comm.,* **1978**, 7.

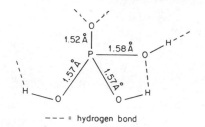

--- = hydrogen bond

Fig. 14-13. Structure of anhydrous orthophosphoric acid.

Higher Acids. *Orthophosphoric acid* (H_3PO_4), commonly called phosphoric acid, is one of the oldest known and most important phosphorus compounds. It is made in vast quantities, usually as 85% syrupy acid, by the direct reaction of ground phosphate rock with sulfuric acid and also by the direct burning of phosphorus and subsequent hydration of the oxide P_4O_{10}. The pure acid is a colorless crystalline solid (m.p. 42.35°). It is very stable and has essentially no oxidizing properties below 350 to 400°. At elevated temperatures it is fairly reactive toward metals and is reduced; it will also then attack quartz. Fresh molten H_3PO_4 has appreciable ionic conductivity suggesting autoprotolysis:

$$2H_3PO_4 \rightleftharpoons H_4PO_4^+ + H_2PO_4^-$$

Pyrophosphoric acid is also produced:

$$2H_3PO_4 \rightarrow H_2O + H_4P_2O_7$$

but this conversion is temperature dependent and is slow at room temperature.

The acid is tribasic: at 25°, $pK_1 = 2.15$, $pK_2 = 7.1$, $pK_3 \approx 12.4$. The pure acid and its crystalline hydrates have tetrahedral PO_4 groups connected by hydrogen bonds (Fig. 14-13). These persist in the concentrated solutions and are responsible for the syrupy nature. For solutions of concentration less than ~50%, the phosphate anions are hydrogen bonded to the liquid water rather than to other phosphate anions.

Phosphates of most metal ions and other cations are known. Some of these are of enormous commercial and practical importance, for example, ammonium phosphate fertilizers, and alkali phosphate buffers. Natural phosphate minerals are *all* orthophosphates, the major one being fluorapatite; hydroxoapatites, partly carbonated, make up the mineral part of teeth. The role of traces of F^- in strengthening dental enamel is presumably connected with these structural relationships, but a detailed explanation of the phenomenon is still lacking.[80a]

Orthophosphoric acid and phosphates form complexes with many transition metal ions. The precipitation of insoluble phosphates from fairly concentrated acid solution (3-6M HNO_3) is characteristic of 4+ cations such as those of Ce, Th, Zr, U, Pu, etc. Phosphates of B, Al, Zr, etc., are used industrially as catalysts for a variety of reactions.[80b]

Despite the well-established formation of phosphato complexes, few crystallo-

[80a] F. Freund and R. M. Knobel, *J.C.S. Dalton,* **1977,** 1136.
[80b] J. B. Moffatt, *Catal. Rev. Sci. Eng.,* 1978, **18,** 199.

graphic determinations have been made on simple complexes, but bidentate phosphate ligands have been confirmed in the following cases[80c,80d]:

Most biological phosphate chemistry (see later) is probably metal catalyzed.

Large numbers of *phosphate esters* are known, normally of the type $(RO)_3PO$, where R is an alkyl or aryl group. Some of them, such as tri-*n*-butyl phosphate and di-*t*-butylphosphinic acid $[(Me_3C)_2PO(OH)^{81}]$, are used for solvent extraction of metal ions, such as actinides and lanthanides, from aqueous solutions. There is also a variety of sulfur compounds such as $(RO)_3P{=}S$ and $(RS)_3PO$.

There are several phosphate esters that have five or six oxygen atoms surrounding phosphorus. An example is the octahedral tris(*o*-phenylenedioxo)phosphate ion, which is made from a catechol phosphazene (14-VII) compound[82]:

The interaction of $P(OMe)_5$ and KOMe in crown-18-ether gives the octahedral ion $[P(OMe)_6]^-$.

Condensed Phosphates. Condensed phosphates are those containing more than one P atom and having P—O—P bonds. We may note that the *lower* acids can also give condensed species, although we shall deal here only with a few examples of phosphates.

There are three main building units in condensed phosphates: the end unit (14-XIX), middle unit (14-XX), and branching unit (14-XXI). These units can be distinguished not only chemically—for example, the branching points are rapidly attacked by water—but also by ^{31}P nmr spectra. The units can be incorporated into

80c B. Anderson *et al.*, *J. Am. Chem. Soc.*, 1977, **99**, 2652.
80d A. Bino and F. A. Cotton, *Angew. Chem., Int. Ed.*, 1979, **18**, 462.
81   M. E. Druyan *et al.*, *J. Am. Chem. Soc.*, 1976, **98**, 4801.
82   M. Gallagher *et al.*, *J.C.S. Chem. Comm.*, **1976**, 320; H. R. Allcock and E. C. Bissel, *J. Am. Chem. Soc.*, 1973, **95**, 3154; J. Gloede and H. Gross, *Tetrahedron Lett.*, **1976**, 917.

$$\text{(14-XIX)} \qquad \text{(14-XX)} \qquad \text{(14-XXI)}$$
$$PO_{3.5}^{2-} \qquad\qquad PO_3^- \qquad\qquad PO_{2.5}$$

either (a) chain or *polyphosphates,* containing two to ten P atoms, (b) simple ring *metaphosphates,* containing three to ten or more P atoms, (c) infinite chain *metaphosphates,* (d) *ultraphosphates* that have branching units; of these, P_4O_{10} is the extreme example, having *only* units of type 14-XXI.

Linear polyphosphates are salts of anions of general formula $[P_nO_{3n+1}]^{(n+2)-}$ Examples are $M_4^I P_2O_7$ (14-XXII), a pyrophosphate or dipolyphosphate, and $M_5^I P_3O_{10}$ (14-XXIII), a tripolyphosphate.

$$\text{(14-XXII)} \qquad\qquad \text{(14-XXIII)} \qquad\qquad \text{(14-XXIV)}$$

Cyclic polyphosphates are salts of anions of general formula $[P_nO_{3n}]^{n-}$. Examples are $M_3P_3O_9$, a trimetaphosphate (14-XXIV), and $M_4P_4O_{12}$, a tetrametaphosphate. The eight-membered ring of the $P_4O_{12}^{4-}$ ion is puckered with equal P—O bond lengths.

An example of an infinite chain metaphosphate is provided by $Li_2(NH_4)$-(P_3O_9).[83]

Condensed phosphates are usually prepared by dehydration of orthophosphates under various conditions of temperature (300–1200°) and also by appropriate hydration of dehydrated species, as, for example,

$$(n-2)NaH_2PO_4 + 2Na_2HPO_4 \xrightarrow{\text{heat}} Na_{n+2}P_nO_{3n+1} + (n-1)H_2O$$
$$\text{Polyphosphate}$$

$$nNaH_2PO_4 \xrightarrow{\text{heat}} (NaPO_3)_n + nH_2O$$
$$\text{Metaphosphate}$$

They can also be prepared by controlled addition of water or other reagents to P_4O_{10}, by treating chlorophosphates with silver phosphates, etc. The complex mixtures of anions that can be obtained are separated by using ion-exchange or chromatographic procedures as illustrated in Fig. 14-14.

Polyphosphoric acid, which is obtained by self-condensation of H_3PO_4 on heating at > 400°, is a viscous liquid that becomes mobile at about 100°. It is also a source of polyphosphate ions on dissolution in water, but in addition it has many uses as such in organic chemistry (for condensations, cyclizations, etc.).[84]

[83]   M. T. Averbach-Pouchet, A. Durif, and J. C. Gurtel, *Acta Crystallogr. B,* 1976, **33**, 2440.
[84]   See M. Fieser and L. F. Fieser, *Reagents for Organic Syntheses,* Wiley, New York.

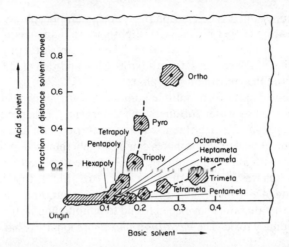

Fig. 14-14. Chromatographic separation of complex phosphate mixtures. Corner of a two-dimensional paper chromatogram, showing the positions of the pentameta- through octametaphosphate rings in relation to the positions of the well-known ring and chain phosphates. The basic solvent traveled 23 cm in 24 h whereas the acid solvent traveled 11.5 cm in 5.5 hr.

Condensed phosphates may also be obtained by chemical dehydration using carbodiimides[85] in solvents such as dimethyl sulfoxide or tetramethylurea. Of particular interest is that H_3PO_4, chain, and ring acids may be condensed to an oxo-bridged acid $H_2P_4O_{11}$. This acid, 1,5-μ-oxotetrametaphosphoric acid, is the acid that would be formed by scission of a single P—O—P bond in P_4O_{10} (cf. eq. 14-2). For example, we have the ring tetrametaphosphoric acid condensing with one equivalent of cyclohexylcarbodiimide to give the μ-oxo species; this may be further dehydrated to give a compound that is essentially the P_4O_{10} molecule solvated by a carbodiimide molecule.

Tetrametaphosphoric   1,5-μ-Oxotetrametaphosphoric

85   T. Glonek et al., Inorg. Chem., 1974, 13, 2337; T. Glonek, T. C. Myers, and J. R. van Wazer, J. Am. Chem. Soc., 1975, 97, 206.

The most important cyclic phosphate is tetrametaphosphate, which can be prepared by heating copper nitrate with slightly more than an equimolar amount of phosphoric acid (75%) slowly to 400°. The sodium salt can be obtained by treating a solution of the copper salt with Na_2S. Slow addition of P_4O_{10} to ice water gives $\sim75\%$ of the P as tetrametaphosphate.

Condensed phosphates form soluble complexes with many metals, and chain phosphates are used as water softeners and detergents. This practice has led to trouble because phosphates in effluents, acting as nutrients for algae, have resulted in eutropification of lakes.[86] Phosphates can be largely removed in sewerage plants by precipitation with aluminum and iron(III) sulfate at pH 5 to 6.

Halogenophosphates.[87] As with many other oxo anions, fluorine can replace OH in phosphate to give mono- and difluorophosphate salts and esters. The dialkyl monofluorophosphate esters have been found to inhibit cholinesterase in the body and to be exceedingly toxic. The *hexafluorophosphate* ion, PF_6^-, has been discussed (p. 449).

Phosphate Esters in Biology. Many of the most essential chemicals in life processes are phosphate esters. These include the genetic substances DNA and RNA (representative fragments of the chains appear as 14-XXV and 14-XXVI, re-

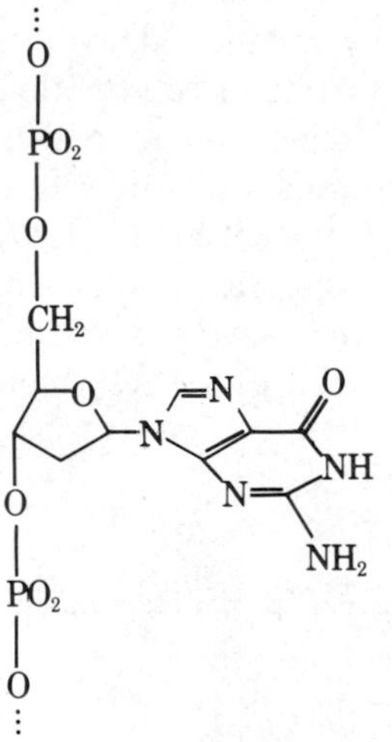

(14-XXV)

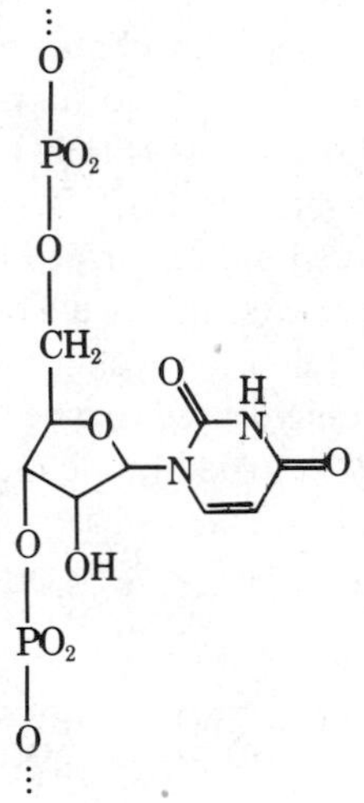

(14-XXVI)

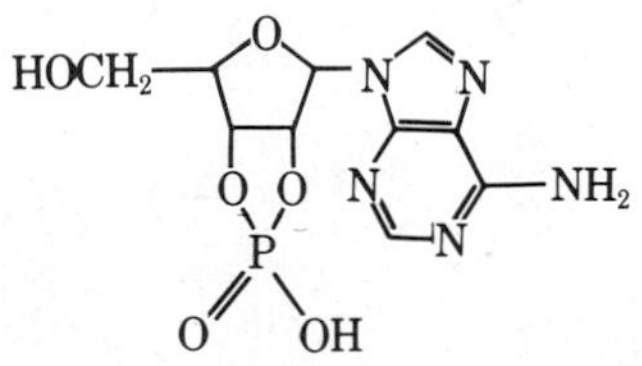

(14-XXVII)

[86] D. Gleisberg *et al.*, *Angew. Chem., Int. Ed.*, 1976, **15**, 354.
[87] K. Dehnicke and A. F. Shidada, *Struct. Bonding*, 1976, **28**, 51.

spectively), as well as cyclic AMP (adenosine monophosphate), 14-XXVII. In addition, the transfer of phosphate groups between ATP and ADP (eq. 14-4), is of fundamental importance in the energetics of biological systems. All the biological

$$\text{Adenosine}-O-\overset{\overset{\displaystyle H}{\underset{\displaystyle O}{\|}}}{P}-O-\overset{\overset{\displaystyle H}{\underset{\displaystyle O}{\|}}}{P}-O-\overset{\overset{\displaystyle O^-}{\underset{\displaystyle O}{\|}}}{P}-OH$$

ATP

$$\underset{-H_2O}{\overset{+H_2O}{\rightleftharpoons}} \quad \text{Adenosine}-O-\overset{\overset{\displaystyle H}{\underset{\displaystyle O}{\|}}}{P}-O-\overset{\overset{\displaystyle H}{\underset{\displaystyle O}{\|}}}{P}-OH + H_2PO_4^- \quad (14\text{-}4)$$

ADP

reactions involving formation and hydrolysis of these and other phosphate esters, and polyphosphates are effected by enzyme catalysts, many of which contain metal ions as parts of their structure, or require them as coenzymes.

In no small measure because of the importance of such substances and processes as those just mentioned, the hydrolysis of phosphate esters has received much fundamental study. Triesters are attacked by OH$^-$ at phosphorus and by H$_2$O at

$$OP(OR)_3 \underset{H_2{}^{18}O}{\overset{{}^{18}OH^-}{<}} \begin{array}{l} OP(OR)_2({}^{18}OH) + RO^- \\[1em] OP(OR)_2(OH) + R^{18}OH \end{array} \quad (14\text{-}5)$$

carbon, as shown in eq. 14-5. Diesters, which are strongly acidic (eq. 14-6), are completely in the anionic form at normal (and physiological) pHs:

$$RO-\overset{\overset{\displaystyle O}{\|}}{\underset{\underset{\displaystyle OH}{|}}{P}}-OR \rightleftharpoons R'OPO_2OR^- + H^+ \qquad K \approx 10^{-1.5} \qquad (14\text{-}6)$$

They are thus relatively resistant to nucleophilic attack by either OH$^-$ or H$_2$O, and this is why enzymic catalysis is indispensable to achieve useful rates of reaction.

Relatively little has been firmly established concerning the mechanisms of most phosphate ester hydrolyses, especially the many enzymic ones. However, the following three basic mechanistic types can be envisioned for cases where nucleophilic attack on phosphorus is the rate-limiting step:

1. One-step nucleophilic displacement (S_N2) with inversion:

$$H_2O(\text{or } OH^-) \quad \overset{O}{\underset{{}^-O}{\diagdown}}\overset{}{\underset{OR'}{P}}-OR \longrightarrow HO-\overset{O}{\underset{{}^-O}{\diagup}}\overset{}{\underset{OR'}{P}} \quad + HOR$$

2. Release of a short-lived "metaphosphate" group, which rapidly recovers the four-connected orthophosphate structure:

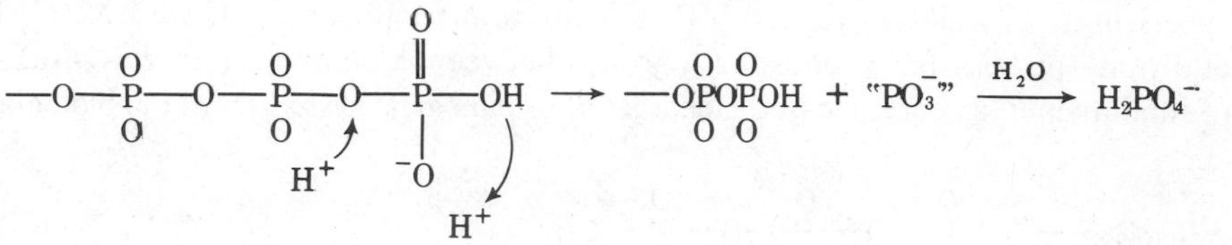

Although trapping experiments show that PO_3^- has a finite lifetime,[88] it is severely electron deficient at P and is rapidly attacked by nucleophiles:

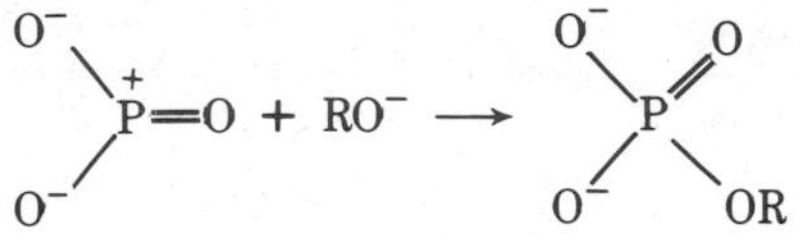

This difference from nitrate (NO_3^-) has been attributed to the presence of low-energy, unoccupied, and in-plane π-type orbitals, available for attack by nucleophiles.[89]

3. Nucleophilic attack in which a cyclic five-coordinate intermediate is formed, which then pseudorotates (cf. Section 28-13):

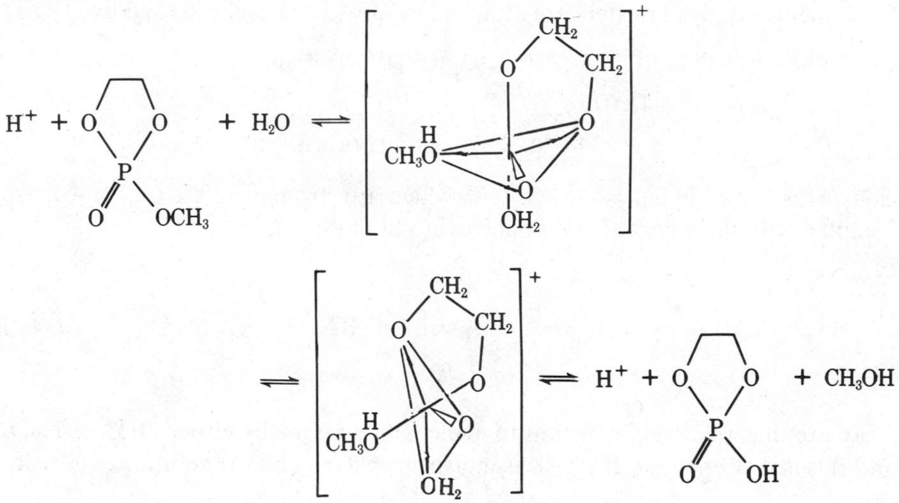

14-15. Oxo Acids and Anions of Arsenic, Antimony, and Bismuth

Arsenic. Raman spectra show that in acid solutions of As_4O_6 the only detectable species is the pyramidal $As(OH)_3$. In basic solutions ($[OH^-]/[As^{III}]$ ratios of 3.5–15) the four pyramidal species $As(OH)_3$, $As(OH)_2O^-$, $As(OH)O_2^{2-}$, and AsO_3^{3-} appear to be present. In solid salts the arsenite ion is known in the AsO_3^{3-} form, as well as in more complex ones. Arsenite ($AsO(OH)_2^-$) catalyzes the hydration of CO_2.[90]

88 F. H. Westheimer *et al., J. Am. Chem. Soc.,* 1975, **97**, 6873.
89 L. M. Loew, *J. Am. Chem. Soc.,* 1976, **98**, 1630.
90 W. C. Copenhafer and P. H. Rieger, *J. Am. Chem. Soc.,* 1978, **100**, 3776.

Arsenic acid (H_3AsO_4) is obtained by treating arsenic with concentrated nitric acid to give white crystals, $H_3AsO_4 \cdot \frac{1}{2}H_2O$. Unlike phosphoric acid, it is a moderately strong oxidizing agent in acid solution, the potentials being

$$H_3AsO_4 + 2H^+ + 2e = HAsO_2 + 2H_2O \qquad E° = 0.559 \text{ V}$$
$$H_3PO_4 + 2H^+ + 2e = H_3PO_3 + H_2O \qquad E° = -0.276 \text{ V}$$

Arsenic acid is tribasic but somewhat weaker ($pK_1 = 2.3$) than phosphoric acid. The arsenates generally resemble orthophosphates and are often isomorphous with them.

Condensed arsenic anions are much less stable than the condensed phosphates and, owing to rapid hydrolysis, do not exist in aqueous solution. Dehydration of KH_2AsO_4 gives three forms, stable at different temperatures, of metaarsenate; one form is known to contain an infinite chain polyanion, like that in one form of KPO_3.

There are also fluoroarsenates, such as the $M_2^I[As_2F_8O_2]$ compounds, which contain arsenic atoms octahedrally coordinated by four fluoride ions and two bridging oxygen atoms.

Antimony. No lower acid is known, but only the hydrated oxide, $Sb_2O_3(aq)$; the antimonites are well-defined salts, however.

Addition of dilute alkali hydroxide to $SbCl_5$ gives solutions containing the $[Sb^V(OH)_6]^-$ ion, which can be obtained in crystalline salts; the sodium salt is unusual in being the least soluble of the alkali salts.[91]

There do not appear to be finite SbO_4^{3-} ions under any circumstances. Some "antimonates" obtained by heating oxides, for example, $M^I SbO_3$, $M^{III}SbO_4$, and $M_2^{II}Sb_2O_7$, contain SbO_6 octahedra and differ only in the manner of linking in the lattice. They are best regarded as mixed oxides.

Bismuth. When $Bi(OH)_3$ in strongly alkaline solution is treated with chlorine or other strong oxidizing agents, "bismuthates" are obtained, but never in a state of high purity. They can be made, for example, by heating Na_2O_2 and Bi_2O_3 which gives $NaBi^VO_3$. This yellow-brown solid dissolves in $0.5M$ $HClO_4$ to give a solution that is stable in absence of light for several days.[92] The Bi^V/Bi^{III} potential appears to be $+2.03$ V, which suggests that Bi^V is one of the most powerful oxidants in aqueous solution, being comparable to peroxodisulfate ($S_2O_8^{2-}/2SO_4^{2-}$, $E° = 2.01$ V) or ozone (O_3, $2H^+/O_2$, H_2O, $E° = 2.07$ V). The precise nature of the bismuth (V) species is unknown, but it could be $\lfloor Bi(OH)_6 \rfloor^-$.

91 M. J. Blandamer, J. Burgess, and R. D. Peacock, *J.C.S. Dalton,* **1974,** 1084.
92 M. H. Ford-Smith and J. J. Habeeb, *J.C.S. Dalton,* **1973,** 461.

General References

Mellor's Comprehensive Treatise on Inorganic and Theoretical Chemistry, Suppl. III to Vol. VIII, *Phosphorus,* Longmans Green, 1971.
Berlin, K. D., and G. B. Butler, *Chem. Rev.,* 1960, **60,** 243. Preparation and properties of phosphine oxides.

Corbridge, D. E. C., *The Structural Chemistry of Phosphorus*, Elsevier, 1974. An exhaustive compedium of structure and bond distances for P and its compounds. *Phosphorus: An outline of Its Chemistry, Biochemistry and Technology*, Elsevier, 1978.

Emsley, J., *Chem. Br.*, **1977,** 459. Phosphate cycles in Nature.

Emsley, J., and Hall, D., *The Chemistry of Phosphorus*, Harper & Row, 1976. An authoritative reference text.

Granzow, A., *Angew. Chem., Int. Ed.*, 1978, **17,** 177. Phosphorus flame retardants.

Grayson, M., and E. J. Griffiths, Eds., *Topics in Phosphorus Chemistry*, Vol. 1, Wiley, 1964. A series of detailed reviews.

Grayson, M., and L. Horner, Eds., *Phosphorus*, Vol. 1, Gordon and Breach, 1971. Journal devoted to P and Group V.

Griffith, E. J., *et al.*, *Environmental Phosphorus Handbook*, Wiley, 1973.

Holmes, R. R., *Acc. Chem. Res.*, 1979, **12,** 257. Cyclic pentacoordinated molecules.

Hudson, R. F., *Structure and Mechanism in Organophosphorus Chemistry*, Academic Press, 1965.

Kramich, L. K., Ed., *Compounds Containing As-N Bonds*, Wiley, 1976.

Luckenback, R., *Dynamic Stereochemistry of Five Coordinate Phosphorus and Related Elements*, Thieme, 1973.

Mann, F. G., *The Heterocyclic Derivatives of Phosphorus, Arsenic, Antimony, and Bismuth*, 2nd ed., Wiley-Interscience, 1970. A comprehensive treatise.

Payne, D. S., in *Non-Aqueous Solvent Systems*, Academic Press, 1965. Group V halides and oxo halides as solvents.

Sheldrick, W. S., *Topics in Current Chemistry*, Vol. 73, Springer. Stereochemistry of five- and six-coordinate phosphorus derivatives.

Toy, A. D. F., *The Chemistry of Phosphorus*, Pergamon Press, 1975.

Van Wazer, J. R., *Phosphorus and Its Compounds*, Vol. I, Interscience, 1958. A comprehensive account of all phases of phosphorus chemistry; Vol. II, 1961, technology, biological functions, applications.

CHAPTER FIFTEEN

Oxygen

GENERAL REMARKS

15-1. Types of Oxides

The oxygen atom has the electronic structure $1s^2 2s^2 2p^4$. Oxygen forms compounds with all the elements except helium, neon, and possibly argon, and it combines directly with all the other elements except the halogens, a few noble metals, and the noble gases, either at room or at elevated temperatures. The earth's crust contains about 50% by weight of oxygen. Most inorganic chemistry is concerned with its compounds, if only in the sense that so much chemistry involves the most important oxygen compound—water.

As a first-row element, oxygen follows the octet rule, and the closed-shell configuration can be achieved in ways that are similar to those for nitrogen, namely, by (a) electron gain to form O^{2-}, (b) formation of two single covalent bonds (e.g., R—O—R) or a double bond (e.g., O=C=O), (c) gain of one electron and formation of one single bond (e.g., in OH^-), and (d) formation of a three- or four-covalent bonds (e.g., R_2OH^+, etc.).

There is a variety of disparate binary oxygen compounds. The change of physical properties is attributable to the range of bond types from essentially ionic to essentially covalent.

The formation of the oxide ion (O^{2-}) from molecular oxygen requires the expenditure of a considerable energy, ~ 1000 kJ mol^{-1}:

$$\tfrac{1}{2} O_2(g) = O(g) \qquad \Delta H = 248 \text{ kJ mol}^{-1}$$

$$O(g) + 2e = O^{2-}(g) \qquad \Delta H = 752 \text{ kJ mol}^{-1}$$

Moreover, in the formation of an ionic oxide, energy must be expended in vaporizing and ionizing the metal atoms. Nevertheless, many essentially ionic oxides exist (e.g., CaO) and are very stable because the energies of lattices containing the relatively small (1.40 Å), doubly charged oxide ion are quite high. In fact, the lattice energies are often sufficiently high to allow the ionization of metal atoms to unusually high oxidation states. Many metals form oxides in oxidation states not encountered in their other compounds, except perhaps in fluorides or some complexes. Examples of such higher oxides are MnO_2, AgO, and PrO_2. Many of these higher ionic oxides are nonstoichiometric.

In some cases the lattice energy is still insufficient to permit complete ionization, and oxides having substantial covalent character, such as BeO or B_2O_3, are formed. Finally, at the other extreme there are numerous oxides, such as CO_2, the nitrogen and phosphorus oxides, SO_2, and SO_3, that are essentially covalent molecular compounds. Such compounds are gases or volatile solids or liquids. Even in "covalent" oxides, unusually high *formal* oxidation states are often found, as in OsO_4, CrO_3, SO_3, etc.

In some oxides containing transition metals in very low oxidation states, metal "*d* electrons" enter delocalized conduction bands and the materials have metallic properties. An example is NbO.

In terms of chemical behavior, it is convenient to classify oxides according to their acid or base character in the aqueous system.

Basic Oxides.　Although X-ray studies show the existence of discrete oxide ions O^{2-} (and also peroxide, O_2^{2-}, and superoxide, O_2^-, ions, discussed below), these ions cannot exist in any appreciable concentration in aqueous solution owing to the hydrolytic reaction

$$O^{2-}(s) + H_2O = 2OH^-(aq) \qquad K > 10^{22}$$

We have also for the per- and superoxide ions:

$$O_2^{2-} + H_2O = HO_2^- + OH^-$$
$$2O_2^- + H_2O = O_2 + HO_2^- + OH^-$$

Thus only those ionic oxides that are insoluble in water are inert to it. Ionic oxides function, therefore, as *basic anhydrides*. When insoluble in water, they usually dissolve in dilute acids, for example

$$MgO(s) + 2H^+(aq) \rightarrow Mg^{2+}(aq) + H_2O$$

although in some cases, MgO being one, high-temperature ignition produces a very inert material, quite resistant to acid attack.

Acidic Oxides.　The covalent oxides of the nonmetals are usually acidic, dissolving in water to produce solutions of acids. They are termed *acid anhydrides*. Insoluble oxides of some less electropositive metals of this class generally dissolve in bases. Thus:

$$N_2O_5(s) + H_2O \rightarrow 2H^+(aq) + 2NO_3^-(aq)$$
$$Sb_2O_5(s) + 2OH^- + 5H_2O \rightarrow 2Sb(OH)_6^-$$

Basic and acidic oxides will often combine directly to produce salts, such as:

$$Na_2O + SiO_2 \xrightarrow{\text{fusion}} Na_2SiO_3$$

Amphoteric Oxides. These oxides behave acidically toward strong bases and as bases toward strong acids:

$$ZnO + 2H^+(aq) \rightarrow Zn^{2+} + H_2O$$
$$ZnO + 2OH^- + H_2O \rightarrow Zn(OH)_4^{2-}$$

Other Oxides. There are various other oxides, some of which are relatively inert, dissolving in neither acids nor bases, for instance, N_2O, CO, and MnO_2; when MnO_2 (or PbO_2) does react with acids (e.g., concentrated HCl), it is a redox, not an acid-base, reaction.

There are also many oxides that are nonstoichiometric. These commonly consist of arrays of close-packed oxide ions with some of the interstices filled by metal ions. However if there is variability in the oxidation state of the metal, nonstoichiometric materials result. Thus iron(II) oxide generally has a composition in the range $FeO_{0.90}$–$Fe_{0.95}$, depending on the manner of preparation. There is an extensive chemistry of mixed metal oxides (see also p. 16).

It may be noted further that when a given element forms several oxides, the oxide with the element in the highest formal oxidation state (usually meaning more covalent) is more acidic. Thus for chromium we have: CrO, basic; Cr_2O_3, amphoteric; and CrO_3, fully acidic.

The Hydroxide Ion.[1] Discrete hydroxide ions (OH^-) exist only in the hydroxides of the more electropositive elements such as the alkali metals and alkaline earths. For such an ionic material, dissolution in water results in formation of aquated metal ions and aquated hydroxide ions:

$$M^+OH^-(s) + nH_2O \rightarrow M^+(aq) + OH^-(aq)$$

and the substance is a strong base. In the limit of an extremely covalent M—O bond, dissociation will occur to varying degrees as follows:

$$MOH + nH_2O \rightleftharpoons MO^-(aq) + H_3O^+(aq)$$

and the substance must be considered an acid. Amphoteric hydroxides are those in which there is the possibility of either kind of dissociation, the one being favored by the presence of a strong acid:

$$M\text{—}O\text{—}H + H^+ = M^+ + H_2O$$

the other by strong base:

$$M\text{—}O\text{—}H + OH^- = MO^- + H_2O$$

because the formation of water is so highly favored, that is,

$$H^+ + OH^- = H_2O \qquad K_{25°} = 10^{14}$$

Hydrolytic reactions of metal ions can be written

$$M(H_2O)_x^{n+} = [M(H_2O)_{x-1}(OH)]^{(n-1)+} + H^+$$

[1] (a) R. F. W. Bader, in *The Chemistry of the Hydroxyl Group,* S. Patai, Ed., Wiley-Interscience, 1971 (theoretical and physical properties of the hydroxyl ion and group). (b) J. R. Jones, *Chem. Br.,* **1971,** 336 (highly basic media and applications).

Thus we may consider that the more covalent the M—O bond tends to be, the more acidic are the hydrogen atoms in the aquated ion, but at present there is no extensive correlation of the acidities of aqua ions with properties of the metal.

The behavior of OH^- and O^{2-} as ligands has been outlined in Section 4-23.

The formation of hydroxo bridges occurs at an early stage in the precipitation of hydroxides or, in some cases more accurately, hydrous oxides. In the case of Fe^{3+}, precipitation of $Fe_2O_3 \cdot xH_2O$ [commonly, but incorrectly, written $Fe(OH)_3$] proceeds through the stages

$$[Fe(H_2O)_6]^{3+} \xrightarrow{-H^+} [Fe(H_2O)_5OH]^{2+} \rightarrow [(H_2O)_4Fe(OH)_2Fe(H_2O)_4]^{4+} \xrightarrow{-xH^+}$$

$$\text{pH} < 0 \qquad\qquad 0 < \text{pH} < 2 \qquad\qquad \sim2 < \text{pH} < \sim3$$

$$\text{colloidal } Fe_2O_3 \cdot xH_2O \xrightarrow{-yH^+} F_2O_3 \cdot zH_2O \text{ ppt}$$

$$\sim3 < \text{pH} < \sim5 \qquad\qquad \text{pH} \sim 5$$

The $H_3O_2^-$ ion has been noted on page 232.

15-2. Covalent Compounds; Stereochemistry of Oxygen

Two-Coordinate Oxygen. The majority of oxygen compounds contain two-coordinate oxygen, in which the oxygen atom forms two single bonds to other atoms and has two unshared pairs of electrons in its valence shell. Such compounds include water, alcohols, ethers, and a variety of other covalent oxides. In the simple two-coordinate situations, where the σ bonds are not supplemented by π bonds to any significant extent, the X—O—X group is always bent, typical angles are 104.5° in H_2O and 111° in $(CH_3)_2O$.

In many cases, where the X atoms of the X—O—X group have orbitals (usually d orbitals) capable of interacting with the lone-pair orbitals of the oxygen atom, the X—O bonds acquire some π character. Such interaction causes shortening of the X—O bonds and generally widens the X—O—X angle. The former effect is not easy to document, since an unambiguous standard of reference for the pure single bond is generally lacking. However the increases in angle are self-evident, for example, $(C_6H_5)_2O$ (124°) and Si—O—Si angle in quartz (142°). In the case of H_3Si—O—SiH_3 the angle is apparently $>150°$.

The limiting case of π interaction in X—O—X systems occurs when the σ bonds are formed by two diagonal sp hybrid orbitals on oxygen, thus leaving two pairs of π electrons in pure p orbitals; these can then interact with empty $d\pi$ orbitals on the X atoms so as to stabilize the linear arrangement. Many examples of this are known, e.g., $[Cl_5Ru$—O—$RuCl_5]^{4-}$.

Three-Coordinate Oxygen. Both pyramidal and planar geometries are found. The pyramidal type are represented mainly by *oxonium ions* (e.g., H_3O^+, R_2OH^+, ROH_2^+, R_3O^+) and by donor-acceptor complexes such as $(C_2H_5)_2OBF_3$, although there are examples of both geometric types among trinuclear μ_3-oxo complexes (cf. Section 4-23). The formation of oxonium ions is analogous to the formation of ammonium ions such as NH_4^+, RNH_3^+ ⋯ R_4N^+, except that oxygen is less basic and the oxonium ions are therefore less stable. Water, alcohol, and ether molecules serving as ligands to metal ions presumably also have pyramidal structures, at least for the most part. Like NR_3 compounds (p. 409), OR_3^+ species undergo rapid inversion.

Four-Coordinate Oxygen. Attainment of this coordination number is not common, but there are a number of well-documented examples. Certain ionic or partly ionic oxides (e.g., PbO) have this coordination number, and it sometimes forms the center of polynuclear complexes. Examples are $Mg_4OBr_6 \cdot 4C_4H_{10}O$, $Cu_4OCl_6(Ph_3PO)_4$ and the long-known $M_4O(OOCR)_6$ compounds, where M = Be or Zn.

Unicoordinate, Multiply Bonded Oxygen. There are, of course, innumerable examples of XO groups, where the order of the XO bond may vary from essentially unity, as in amine oxides, $\rightarrow \overset{+}{N}:\overset{..}{O}:^-$, through varying degrees of π bonding up to a total bond order of 2, or a little more. The simplest π-bonding situation occurs in ketones where one well-defined π bond occurs perpendicular to the molecular plane. In most inorganic situations, such as R_3PO or R_3AsO compounds, tetrahedral oxo ions such as PO_4^{3-}, ClO_4^-, MnO_4^-, OsO_4, or species such as $OsCl_4O_2^{2-}$, the opportunity exists for two π interactions between X and O, in mutually perpendicular planes that intersect along the X—O line. Indeed, the symmetry of the molecule or ion is such that the two π interactions *must* be of equal extent. Thus, in principle the extreme limiting structure 15-Ib must be considered to be mixed with 15-Ia. In general, available evidence suggests that a partially polar bond

$$\overset{+}{X}:\overset{\bar{\;}}{\underset{..}{O}}: \leftrightarrow \overset{\bar{\;}}{X}:::\overset{+}{O}:$$

(15-Ia)   (15-Ib)

of order approaching 2 results; it is important, however, to note the distinction from the situation in a ketone, since a π bond order of 1 in this context does not mean one full π interaction but rather two mutually perpendicular π interactions of order 0.5.

Catenation. As with nitrogen, catenation occurs only to a very limited extent. In peroxides and superoxides there are two consecutive oxygen atoms. Only in O_3, O_3^-, and the few $R_FO_3R_F$ molecules are there well-established chains of three oxygen atoms. There is no confirmed report of longer chains.

The weakness of O—O single bonds in H_2O_2 and O_2 and O_2^{2-} is doubtless due to repulsive effects[2] of the electron pairs in these small atoms (cf. also F_2, Chapter 17).

THE ELEMENT

15-3. Occurrence, Properties, and Allotropes

Oxygen occurs in Nature in three isotopic species: ^{16}O (99.759%), ^{17}O (0.0374%), and ^{18}O (0.2039%)[3]. The rare isotopes, particularly ^{18}O, can be concentrated by

---

[2] P. Politzer, *Inorg. Chem.*, 1977, **16**, 3350.
[3] D. Staschewski, *Angew Chem. Int. Ed.*, 1974, **13**, 357 (stable isotopes of oxygen and their uses).

fractional distillation of water, and concentrates containing up to 97 at. % ^{18}O or up to 4 at. % ^{17}O and other labeled compounds are commercially available. ^{18}O has been widely used as a tracer in studying reaction mechanisms of oxygen compounds. ^{17}O has a nuclear spin 5/2, but nuclear magnetic resonance has not been widely used because of the low abundance of this isotope and appreciable quadrupole moment, making observation of signals difficult. However enriched materials and Fourier transform nmr techniques are leading to wider use.[4] Inorganic uses have included study of exchange[4b] between H_2O and oxo anions, and aqua complex ions such as $[Co(NH_3)_5H_2O]^{3+}$, and identification of different types of O atoms in polymolybdates such as $Mo_6O_{19}^{2-}$.

Dioxygen occurs in two allotropic forms; the common, stable O_2, and ozone (O_3). O_2 is paramagnetic in the gaseous, liquid, and solid states and has the rather high dissociation energy of 496 kJ mol^{-1}. The valence bond theory in its usual form would predict the electronic structure :$\ddot{O}{=}\ddot{O}$: which, though accounting for the strong bond, fails to account for the paramagnetism. However, the MO approach, even in first approximation, correctly accounts for the triplet ground state ($^3\Sigma_g^-$) having a double bond. There are several low-lying singlet states that are important in photochemical oxidations; these are discussed shortly. Like NO, which has one unpaired electron in an antibonding ($\pi*$) MO, oxygen molecules associate only weakly, and true electron pairing to form a symmetrical O_4 species apparently does not occur even in the solid. Both liquid and solid O_2 are pale blue.

Ozone.[5]   Ozone is usually prepared by the action of a silent electric discharge on O_2; concentrations up to 10% of O_3 can be obtained in this way. Ozone gas is perceptibly blue and is diamagnetic. Pure ozone can be obtained by fractional liquefaction of O_2–O_3 mixtures. There is a two-phase liquid system; one with 25% of ozone is stable, but a deep purple phase with 70% of ozone is explosive, as is the deep blue pure liquid (b.p. $-112°$). The solid (m.p. $-193°$) is black-violet. Small quantities of ozone are formed in electrolysis of dilute sulfuric acid, in some chemical reactions producing elemental oxygen, and by the action of ultraviolet light on O_2.

Ozone is very endothermic

$$O_3 = \tfrac{3}{2}O_2 \qquad \Delta H = -142 \text{ kJ mol}^{-1}$$

but it decomposes only slowly at 250° in absence of catalysts and ultraviolet light.

Ozone is an important natural constituent of the atmosphere, being principally concentrated (up to *ca.* 27% by weight) between altitudes of 15 and 25 km. Its formation is caused by solar ultraviolet radiation in the range 240–300 nm via the reactions

4   (a) W. G. Klemperer, *Angew. Chem. Int. Ed.,* 1978, **17,** 246 (an extensive review); see also *Inorg. Chem.,* 1979, **18,** 93, for examples; (b) W. C. Copenhaver and P. H. Rieger, *J. Am. Chem. Soc.,* 1978, **100,** 3776; R. K. Murmann and K. C. Giese, *Inorg. Chem.,* 1978, **17,** 1160.

5   J. S. Murphy and J. R. Orr, *Ozone Chemistry and Technology,* Franklin Institute Press, 1975 (a review of literature, 1961–1974); *Ozone Chemistry and Technology,* ACS Advances in Chemistry Series No. 21, 1959.

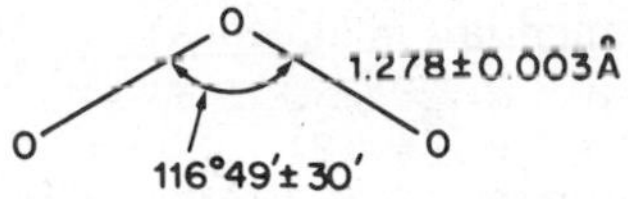

Fig. 15-1. The structure of ozone (O_3).

$$O_2 \xrightarrow{h\nu} 2O \tag{15-1}$$

$$O + O_2 \longrightarrow O_3 \tag{15-2}$$

Ozone itself absorbs uv radiation from 200 to 360 nm. This leads partly to a reversal of reaction 15-2 and thus a steady state concentration is established. The net result of all these processes is absorption and conversion to heat of considerable solar uv radiation that would otherwise strike the earth's surface. Destruction of any significant percentage of this ozone could have serious effects (e.g., increased surface temperature, high incidence of skin cancer), and some of man's activities are capable of destroying stratospheric ozone.[6] Supersonic aircraft, which fly in the ozone layer, discharge NO and NO_2, and these can catalyze the decomposition of ozone via the following reactions:

$$O_3 + NO \rightarrow O_2 + NO_2$$
$$NO_2 + O \rightarrow O_2 + NO$$
$$\text{or} \quad NO_2 + O_3 \rightarrow O_2 + NO_3$$
$$NO_3 \xrightarrow{h\nu} O_2 + NO$$

However the situation is somewhat more complicated, and the net effect of supersonic aircraft varies with the type of engine and mode of operation. The Concorde-type plane, as currently operated, appears to have no significant effect. Chlorofluorocarbons[6] such as $CFCl_3$ and CF_2Cl_2, widely used as foam-blowing agents, aerosol propellants, and refrigerants, are photochemically decomposed to give Cl atoms, and these catalyze ozone decomposition via the mechanism

$$O_3 + Cl \rightarrow O_2 + ClO$$
$$ClO + O \rightarrow O_2 + Cl$$

The rate of chlorofluorocarbon release into the atmosphere prevailing in 1977 could reduce the ozone layer significantly by the year 2000, according to present kinetic analysis, but there may be additional factors not yet included in the analysis that could alter the conclusion—for better or worse.

The structure of O_3 is shown in Fig. 15-1. Since the O—O bond distances are 1.49 Å in HOOH (single bond) and 1.21 Å in O_2 (~ double bond), it is apparent that the O—O bonds in O_3 must have considerable double-bond character.

Chemical Properties of Dioxygen. Although oxygen combines directly with almost all other elements, it does so usually only at elevated temperatures. However O_2 will react with certain transition metal complexes, sometimes reversibly, and the ligand behavior of O_2 and the related ions O_2^- and O_2^{2-} is discussed in Section 4-24.

[6] L. Dotto and H. Schiff, *The Ozone War*, Doubleday, 1978.

We have the following potentials in aqueous solution:

$$O_2 + 4H^+ + 4e = 2H_2O \qquad E^0 = +1.229 \text{ V}$$
$$O_2 + 2H_2O + 4e = 4OH^- \qquad E^0 = +0.401 \text{ V}$$
$$O_2 + 4H^+(10^{-7}M) + 4e = 2H_2O \qquad E^0 = +0.815 \text{ V}$$

It can be seen that neutral water saturated with O_2 is a fairly good oxidizing agent. For example, although Cr^{2+} is just stable toward oxidation in pure water, in air-saturated water it is rapidly oxidized; Fe^{2+} is oxidized (only slowly in acid, but rapidly in base) to Fe^{3+} in presence of air, although in air-free water Fe^{2+} is quite stable:

$$Fe^{3+} + e = Fe^{2+} \qquad E^0 = +0.77 \text{ V}$$

The rate of oxidation of various substances (e.g., ascorbic acid) may be vastly increased by catalytic amounts of transition metal ions, especially Cu^{2+}, where a Cu^I–Cu^{II} redox cycle is involved.

Oxygen is readily soluble in organic solvents, and merely pouring such liquids in air serves to saturate them with oxygen. This should be kept in mind when determining the reactivity of air-sensitive materials in solution in organic solvents.[7] Note that many organic substances such as ethers readily form peroxides or hydroperoxides in air.

Measurements of electronic spectra of alcohols, ethers, benzene, and even saturated hydrocarbons show that there is interaction of the charge-transfer type with the oxygen molecule. However there is no true complex formation, since the heats of formation are negligible and the spectral changes are due to contact between the molecules at van der Waals distances. The classic example is that of N,N-dimethylaniline, which becomes yellow in air or oxygen but colorless again when the oxygen is removed by nitrogen. Such weak charge-transfer complexes make certain electronic transitions in molecules more intense; they are also a plausible first stage in photooxidations.

The precise mechanism of the reduction of dioxygen has been much studied.[8a] There is no evidence for four- or two-electron reduction steps, as would be suggested by the overall reactions noted above, or the following:

$$O_2 + 2H^+ + 2e = H_2O_2 \qquad E^0 = +0.682 \text{ V}$$
$$O_2 + H_2O + 2e = OH^- + HO_2^- \qquad E^0 = -0.076 \text{ V}$$

It is clear[8b] that the first step is a one-electron reduction to the *superoxide radical ion* O_2^-. The potential for the reduction

$$O_2 + e = O_2^-$$

ranges from *ca.* -0.2 to -0.5 V depending on the medium. The O_2^- ion is a moderate reducing agent, comparable to dithionite, and a *very* weak oxidizing agent. Thus most of the oxidation by O_2 is due to *peroxide* ions HO_2^- and O_2^{2-}, formed by re-

[7]   D. F. Shriver, *The Manipulation of Air-Sensitive Compounds,* McGraw-Hill, 1969.

[8a]   J. Wilshire and D. T. Sawyer, *Acc. Chem. Res.,* 1979, **12,** 105; D. T. Sawyer et al., *Inorg. Chem.,* 1979, **18,** 1971.

[8b]   D. T. Sawyer et al., *J. Am. Chem. Soc.,* 1978, **100,** 627; *Inorg. Chem.,* 1977, **16,** 3379; R. K. Sen et al., *Inorg. Chem.,* 1977, **16,** 3379.

actions such as

$$H^+ + O_2^- + e = HO_2^-$$
$$2O_2^- + H_2O = O_2 + HO_2^- + OH^-$$

The superoxide radical ion is known to have a significant lifetime in aprotic solvents,[9] and even in *alkaline* aqueous solutions. It is present in all organisms that use O_2 and is destroyed by the metalloenzyme superoxide dismutase.[10] The ion will add irreversibly to copper complexes.[11]

The coordinated O_2 molecule in complexes (Section 4-24) is more reactive than the free molecule, and various substances not directly oxidized under mild conditions can be oxidized by metal complexes (Section 29-17), mostly by radical reactions.

Singlet O_2 and Photochemical Oxidations.[12] The lowest-energy electron configuration of the O_2 molecule, which contains two electrons in π^* orbitals, gives rise to three states, as tabulated below. Oxygen molecules in excited singlet states, especially the $^1\Delta_g$ state, which has a much longer lifetime than the $^1\Sigma_g^+$ state, react with a variety of unsaturated organic substrates to cause limited, specific oxidations, a very typical reaction being a Diels-Alder-like 1,4-addition to a 1,3-diene:

State	π_a^*	π_b^*	Energy
$^1\Sigma_g^+$	↑	↑	155 kJ ($\sim$13,000 cm^{-1})
$^1\Delta_g$	↑↓	—	92 kJ ($\sim$8,000 cm^{-1})
$^3\Sigma_g^-$	↑	↑	0 (ground state)

There are three ways of generating the singlet oxygen molecules: (1) photochemically by irradiation in presence of a sensitizer, (2) chemically, (3) in an electrodeless discharge. The last is inefficient and impractical. The photochemical route is believed to proceed as follows, where "sens" represents the photosensitizer (typically a fluorescein derivative, methylene blue, certain porphyrins or certain polycyclic aromatic hydrocarbons):

$$^1sens \xrightarrow{h\nu} {}^1sens^*$$
$$^1sens^* \rightarrow {}^3sens^*$$
$$^3sens^* + {}^3O_2 \rightarrow {}^1sens + {}^1O_2$$
$$^1O_2 + substrate \rightarrow products$$

9   L. Lee-Ruff, *Chem. Soc. Rev.,* 1977, **6,** 195 (organic chemistry of O_2^-).
10 I. Fridovich, *Acc. Chem. Res.,* 1972, **5,** 321; A. M. Michaelson, J. M. McCord, and I. Fridovich, Eds., *Superoxide and Superoxide Dismutases,* Academic Press, 1977.
11 M. G. Simić and M. Z. Hoffman, *J. Am. Chem. Soc.,* 1977, **99,** 2370.
12 B. Rånby and J. F. Rabek, Eds., *Singlet Oxygen: Reaction with Organic Compounds and Polymers,* Wiley, 1978; A. P. Schaap, Ed., *Singlet Molecular Oxygen,* Wiley, 1976; J. Bland, *J. Chem. Educ.,* **1976,** 274 (singlet O_2 in biochemistry); H. H. Wasserman and R. W. Murray, Eds., *Singlet Oxygen,* Academic Press, 1979.

Energy transfer from triplet excited sensitizer ^{3}sens*, to $^3O_2$ to give $^1O_2$ is a spin-allowed process.

The photochemical generation of singlet oxygen is widely used in the fine chemicals industry for selective oxidations—many tons a year are generated—since it reacts electrophilically rather than in a free-radical fashion. It is also probably important in biological oxidation.

Singlet oxygen can be generated in a number of chemical reactions. In the reactions:

$$H_2O_2 + Cl_2 = 2Cl^- + 2H^+ + O_2$$
$$H_2O_2 + ClO^- = Cl^- + H_2O + O_2$$

the accompanying red chemiluminescent glow is due to the excited oxygen molecule trapped in the bubbles.

The base-catalyzed disproportionation[13] of H_2O_2 gives a ratio of singlet to triplet oxygen of about 1:3:

$$H_2O_2 + HO^- = HOO^- + H_2O$$
$$H_2O_2 + HOO^- = H_2O + HO^- + O_2$$

Both these reactions of H_2O_2 may be concerted two-electron transfers, e.g.,

Singlet O_2 is deactivated by various transition metal complexes.[14]

Chemical Properties of Ozone. The O_3 molecule is a much more powerful oxidant than is O_2 and reacts with most substances at 25°. It is often used in organic chemistry.[15] The oxidation mechanisms doubtless involve free-radical chains as well as peroxo intermediates.

Some compounds form ozone adducts, e.g.,

$$(PhO)_3P + O_3 \xrightarrow{-78°} (PhO)_3P(O_3) \xrightarrow{-15°} (PhO)_3PO + O_2(^1\Delta_g)$$

Complexes are also formed with aromatic compounds and other π systems.[16]

The reaction

$$O_3 + 2KI + H_2O = I_2 + 2KOH + O_2$$

is quantitative and can be used to determine O_3. The overall potentials in aqueous solution are:

$$O_3 + 2H^+ + 2e = O_2 + H_2O \qquad E^0 = +2.07 \text{ V}$$
$$O_3 + H_2O + 2e = O_2 + 2OH^- \qquad E^0 = +1.24 \text{ V}$$
$$O_3 + 2H^+ (10^{-7}M) + 2e = O_2 + H_2O \qquad E^0 = +1.65 \text{ V}$$

[13] L. L. Smith and M. L. Kulig, *J. Am. Chem. Soc.*, 1976, **98** 1027.

[14] H. Furne and K. E. Russell, *Can. J. Chem.*, 1978, **56**, 1595.

[15] P. S. Bailey, *Ozonation in Organic Chemistry*, Vol. 1, Academic Press, 1978; L. F. Fieser and M. Fieser, *Reagents for Organic Synthesis*, Wiley; R. Criegee, *Angew. Chem., Int. Ed.*, 1975, **14**, 745.

[16] P. S. Bailey et al., *J. Am. Chem. Soc.*, 1978, **100**, 894, 899.

In acid solution O_3 is exceeded in oxidizing power only by fluorine, the perxenate ion, atomic oxygen, OH radicals, and a few other such species. The rate of decomposition of ozone drops sharply in alkaline solutions, the half-life being *ca.* 2 min in $1M$ NaOH at $25°$, 40 min at $5M$, and 83 h at $20M$; the ozonide ion (see below) is also more stable in alkaline solution.

OXYGEN COMPOUNDS

Most oxygen compounds are described in this book during treatment of the chemistry of other elements. Water and the hydroxonium ion have already been discussed (Chapter 6). A few important compounds and classes of compounds are mentioned here.

15-4. Oxygen Fluorides

Since fluorine is more electronegative than oxygen, it is logical to call its binary compounds with fluorine oxygen fluorides rather than fluorine oxides, although the latter names are sometimes seen. Oxygen fluorides have been intensively studied as potential rocket fuel oxidizers.

Oxygen Difluoride, OF_2. This is prepared by passing fluorine rapidly through 2% sodium hydroxide solution, by electrolysis of aqueous HF–KF solutions, or by action of F_2 on moist KF. It is a pale yellow poisonous gas (b.p. $145°$). It is relatively unreactive and can be mixed with H_2, CH_4, or CO without reaction, although sparking causes violent explosion. Mixtures of OF_2 with Cl_2, Br_2, or I_2 explode at room temperature. It is fairly readily hydrolyzed by base:

$$OF_2 + 2OH^- \rightarrow O_2 + 2F^- + H_2O$$

It reacts more slowly with water, but explodes with steam:

$$OF_2 + H_2O \rightarrow O_2 + 2HF$$

and it liberates other halogens from their acids or salts:

$$OF_2 + 4HX(aq) \rightarrow 2X_2 + 2HF + H_2O$$

Metals and nonmetals are oxidized and/or fluorinated; in an electric discharge even xenon reacts to give a mixture of fluoride and oxide fluoride.

Dioxygen Difluoride, O_2F_2. The difluoride is a yellow-orange solid (m.p. $109.7°K$), obtained by high-voltage electric discharges on mixtures of O_2 and F_2 at 10 to 20 mm pressure and temperatures of 77 to $90°K$. It decomposes into O_2 and F_2 in the gas at $-50°$ with a half-life of about 3 hr. It is an extremely potent fluorinating and oxidizing agent, and under controlled conditions OOF groups may be transferred to a substrate. Many substances explode on exposure to dioxygen difluoride at low temperatures, and even C_2F_4 is converted into COF_2, CF_4,

CF_3OOCF_3, etc. In presence of F^- acceptors it forms dioxygenyl salts (see below):

$$O_2F_2 + BF_3 \rightarrow O_2{}^+BF_4^- + \tfrac{1}{2} F_2$$

O_2F_2 has been used for oxidizing primary aliphatic amines to the corresponding nitroso compounds.

The structure of O_2F_2 (15-II) is notable for the shortness of the O—O bond (1.217 Å, cf. 1.48 Å in H_2O_2 and 1.49 in O_2^{2-}) and the relatively long O—F bonds (1.575 Å) compared with those in OF_2 (15-III) (1.409 Å). A plausible but distinctly *ad hoc* explanation for this is that each singly occupied π^* orbital of the O_2 molecule interacts with a singly occupied fluorine σ orbital to form two $3c$-$2e$ OOF bonds in roughly perpendicular planes, as illustrated in (15-IV). Thus a strong, essentially double O—O bond is retained while relatively weak O—F bonds are formed.

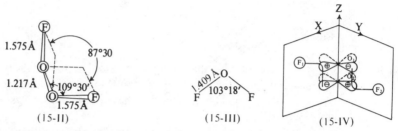

Other, very unstable oxygen fluorides have been reported. The only certain one is O_4F_2, which decomposes even at $-183°$ slowly. Some oxygen fluoride chloride compounds are known also.

15-5.   The Dioxygenyl Cation

The O_2^+ ion was first obtained by the interaction of oxygen with PtF_6, which gives the orange solid O_2PtF_6, isomorphous with $KPtF_6$. Other salts can be made by reactions[17] such as

$$O_2 + BF_3 + \tfrac{1}{2} F_2 \xrightarrow[-78°]{h\nu} O_2^+BF_4^-$$

$$2O_2 + 2GeF_4 + F_2 \xrightarrow{h\nu} 2O_2^+GeF_5^-$$

$$O_2 + Pt + 3F_2 \xrightarrow{280°} O_2^+PtF_6$$

Clearly, large, inoxidizable anions are required to stabilize O_2^+. Some of the salts are quite volatile (e.g., O_2RhF_6 will sublime at room temperature (but are readily hydrolyzed by water. The O_2^+ ion is found, as expected, to be paramagnetic and the O—O stretching frequency is *ca.* 1900 cm^{-1}. Spectroscopic study of gaseous O_2^+ gives an O—O distance of 1.12 Å (cf. 1.09 Å in isoelectronic NO).

[17]   A. J. Edwards, *et al., J.C.S. Dalton,* **1974**, 1129. K. O. Christie *et al., Inorg. Chem.,* 1976, **15**, 127.

15-6.   Hydrogen Peroxide[18]

The main process for the synthesis of H_2O_2 is the autooxidation in an organic solvent such as alkylbenzenes of an alkyl anthraquinol; in the original I.G. Farbenindustrie process 2-ethylanthraquinol was used:

OH

Et

O_2 →

O

Et

$+ H_2O_2$

The H_2O_2 is extracted with water and the 20 to 40% solution so obtained purified by solvent extraction. The anthraquinone solution also has to be purified by removal of degradation products before reduction back to the quinol on a supported platinum or nickel catalyst:

O

Et

H_2, Pd →

OH

Et

OH

A second process (Shell) involves oxidation of isopropanol to acetone and H_2O_2 in either vapor or liquid phases at 15 to 20 atm and *ca.* 100°C.

$Me_2C \overset{H}{\underset{OH}{\diagdown}}$ $\overset{O_2}{\longrightarrow}$ $Me_2C \overset{OOH}{\underset{OH}{\diagdown}}$ $\longrightarrow$ $Me_2C=O + H_2O_2$

Some secondary reactions also occur giving peroxides of aldehydes and acids. The water-peroxide-acetone-isopropanol mixture is fractionated by distillation.

The older electrolytic process, based on the conversion of sulfate to peroxodisulfate (Section 16-17), and hydrolysis of the latter to H_2O_2 is now hardly used.

Dilute solutions of H_2O_2 are concentrated by vacuum distillation; higher concentrations have added stabilizers.

Pure H_2O_2 is a colorless liquid (b. 150.2°, m. −0.43°) that resembles water in many of its physical properties, although it is denser (1.44 at 25°C). The pure liquid has a dielectric constant at 25° of 93, and a 65% solution in water has a dielectric constant of 120. Thus both the pure liquid and its aqueous solutions are potentially excellent ionizing solvents, but its utility in this respect is limited by its strongly oxidizing nature, and its ready decomposition in the presence of even traces of many

[18]   (a) C. A. Crampton *et al.,* in *The Modern Inorganic Chemicals Industry,* R. Thompson, Ed., Special Publication No. 31, Chemical Society, London, 1977, and references therein; (b) P. A. Giguère, *Peroxyde d'Hydrogène et Polyoxydes d'Hydrogène,* Vol. 4, Compléments au Nouveau Traité de Chimie Minérale, Manon, 1975; (c) W. C. Schumb, C. N. Satterfield, and R. L. Wentworth, *Hydrogen Peroxide,* ACS Monograph No. 128, Reinhold, 1955.

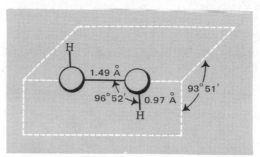

Fig. 15-2. The structure of hydrogen peroxide.

heavy metal ions according to the equation:

$$2H_2O_2 = 2H_2O + O_2 \qquad \Delta H = -99 \text{ kJ mol}^{-1} \tag{15-3}$$

In dilute aqueous solution it is more acidic than water:

$$H_2O_2 = H^+ + HO_2^- \qquad K_{20^\circ} = 1.5 \times 10^{-12}$$

The molecule H_2O_2 has a skew, chain structure (Fig. 15-2). There is only a low barrier to internal rotation about the O—O bond. In the liquid state H_2O_2 is even more highly associated via hydrogen bonding than is H_2O.

Its oxidation-reduction chemistry in aqueous solution[8a] is summarized by the following potentials:

$$
\begin{array}{ll}
H_2O_2 + 2H^+ + 2e = 2H_2O & E^0 = 1.77 \text{ V} \\
O_2 + 2H^+ + 2e = H_2O_2 & E^0 = 0.68 \text{ V} \\
HO_2^- + H_2O + 2e = 3OH^- & E^0 = 0.87 \text{ V}
\end{array}
$$

from which it can be seen that hydrogen peroxide is a strong oxidizing agent in either acid or basic solution; only toward very strong oxidizing agents such as MnO_4^- will it behave as a reducing agent.

Dilute or 30% hydrogen peroxide solutions are widely used as oxidants. In acid solution oxidations with hydrogen peroxide are most often slow, whereas in basic solution they are usually fast. Decomposition of hydrogen peroxide according to reaction 15-3, which may be considered a self-oxidation, occurs most rapidly in basic solution; hence an excess of H_2O_2 may best be destroyed by heating in basic solution.

The oxidation of H_2O_2 in aqueous solution by Cl_2, MnO_4^-, Ce^{4+}, etc., and the catalytic decomposition caused by Fe^{3+}, I_2, MnO_2, etc., have been studied. In both cases, by using labeled H_2O_2 it has been shown that the oxygen produced is derived entirely from the peroxide, not from water. This suggests that oxidizing agents do not break the O—O bond but simply remove electrons. In the case of oxidation by chlorine, a mechanism of the following kind is consistent with the lack of exchange of ^{18}O between H_2O_2 and H_2O:

$$Cl_2 + H_2{}^{18}O_2 \rightarrow H^+ + Cl^- + H^{18}O^{18}OCl$$
$$H^{18}O^{18}OCl \rightarrow H^+ + Cl^- + {}^{18}O_2$$

It is important to recognize, however, that very many reactions involving H_2O_2 (and also O_2) in solutions are free-radical ones. Metal-ion-catalyzed decomposition of H_2O_2 and other reactions can give rise to radicals of which HO_2 and OH are most important. HO_2 has been detected in ice irradiated at low temperature and also in aqueous solutions where H_2O_2 interacts with Ti^{3+}, Fe^{2+}, or Ce^{IV} ions.

There is ordinarily no exchange of oxygen isotopes between H_2O_2 and H_2O in the liquid phase, even in presence of strong acids. However HSO_3F catalyzes the exchange, possibly via the intermediate $H_3O_2^+$.[19]

Hydrogen peroxide has been estimated to be more than 10^6 times less *basic* than H_2O. However on addition of concentrated H_2O_2 to tetrafluoroboric acid in tetrahydrothiophene 1,1-dioxide (sulfolane) the conjugate cation $H_3O_2^+$ can be obtained. The solutions are very powerful, but unselective, oxidants for benzene, cyclohexane, and other organic materials.

15-7. Peroxides, Superoxides, and Ozonides[20]

Ionic Peroxides. Peroxides that contain O_2^{2-} ions are known for the alkali metals, calcium, strontium, and barium. Sodium peroxide is made commercially, by air oxidation of sodium, first to Na_2O, then to Na_2O_2; it is a yellowish powder, very hygroscopic though thermally stable to 500°, which contains also, according to esr studies, about 10% of the superoxide. Barium peroxide, which was originally used for making dilute solutions of hydrogen peroxide by treatment with dilute sulfuric acid, is made by action of air or O_2 on BaO; the reaction is slow below 500° and BaO_2 decomposes above 600°.

The ionic peroxides with water or dilute acids give H_2O_2, and all are powerful oxidizing agents. They convert all organic materials into carbonate even at moderate temperatures. Na_2O_2 also vigorously oxidizes some metals; e.g., Fe violently gives FeO_4^{2-}, and Na_2O_2 can be generally employed for oxidizing fusions. The alkali peroxides also react with CO_2:

$$2CO_2(g) + 2M_2O_2 \rightarrow 2M_2CO_3 + O_2$$

Peroxides can also serve as reducing agents for such strongly oxidizing substances as permanganate.

A number of other electropositive metals such as magnesium, the lanthanides, or uranyl ion also give peroxides; these are intermediate in character between the ionic ones and the essentially covalent peroxides of metals such as zinc, cadmium, and mercury. The addition of H_2O_2 to solutions of, e.g., Zn^{2+} or UO_2^{2+}, gives impure peroxides.

A characteristic feature of the ionic peroxides is the formation of well-crystallized hydrates and H_2O_2 adducts. Thus $Na_2O_2 \cdot 8H_2O$ can be obtained by adding ethanol

[19] S.-K. Chung and P. Decapite, *J. Org. Chem.*, 1978, **43**, 2935.
[20] I. I. Vol'nov, *Peroxides, Superoxides, and Ozonides of Alkali and Alkaline Earth Metals*, Consultants Bureau–Plenum Press, 1966; N.-G. Vannerberg, *Prog. Inorg. Chem.*, 1962, **4**, 125; A. W. Petrocelli *et al., J. Chem. Educ.*, 1962, **39**, 557; 1963; **40**, 146.

to 30% H_2O_2 in concentrated NaOH at 15°, or by rapid crystallization of Na_2O_2 from iced water. The alkaline earths all form the octahydrates ($M^{II}O_2\cdot 8H_2O$). They are isostructural, containing discrete peroxide ions to which the water molecules are hydrogen-bonded, giving chains of the type $\cdots O_2^{2-}\cdots(H_2O)_8\cdots O_2^{2-}\cdots(H_2O)_8\cdots$.

Superoxides. The action of oxygen at pressures near atmospheric on potassium, rubidium, or cesium gives yellow to orange crystalline solids of formula MO_2. NaO_2 can be obtained only by reaction of Na_2O_2 with O_2 at 300 atm and 500°. LiO_2 cannot be isolated, and the only evidence for it is the similarity in the absorption spectra of the pale yellow solutions of lithium, sodium, and potassium on rapid oxidation of the metals in liquid ammonia at −78° by oxygen. Alkaline earth, magnesium, zinc, and cadmium superoxides occur only in small concentrations as solid solutions in the peroxides. Tetramethylammonium superoxide has been obtained as a yellow solid (m.p. 97°), which dissolves in water with evolution of O_2; it is soluble in aprotic solvents such as CH_3CN giving solutions containing the O_2^- ion (see p. 491).

There is clearly a direct correlation between superoxide stability and electropositivity of the metal concerned.

The paramagnetism of the compounds corresponds to one unpaired electron per two oxygen atoms, consistent with the existence of O_2^- ions, as first suggested for these oxides by Pauling. Crystal structure determinations show the existence of such discrete O_2^- ions. The compounds KO_2, RbO_2, and CsO_2 crystallize in the CaC_2 structure (Fig. 11-5), which is a distorted NaCl structure. NaO_2 is cubic owing to the disorder in the orientation of the O_2^- ions. The superoxides are very powerful oxidizing agents. They react vigorously with water:

$$2O_2^- + H_2O = O_2 + HO_2^- + OH^-$$
$$2HO_2^- = 2OH^- + O_2 \quad \text{(slow)}$$

The reaction with CO_2, which involves peroxocarbonate intermediates, is of some technical use for removal of CO_2 and regeneration of O_2 in closed systems. The overall reaction is:

$$4MO_2(s) + 2CO_2(g) = 2M_2CO_3(s) + 3O_2(g)$$

Ozonides.[21] Reaction of O_3 with the hydroxides of alkali metals (M) gives the ozonides:

$$3MOH(s) + 2O_3(g) = 2MO_3(s) + MOH\cdot H_2O(s) + \tfrac{1}{2}O_2(g)$$

Their thermal stabilities decrease in the order $CsO_3 > KO_3 > NaO_3 > LiO_3$; decomposition leads to $MO_2 + O_2$. NH_4O_3 has also been reported. There is some evidence for the formation of O_3^- in the decomposition of alkaline H_2O_2 and in radiolytic reactions.

The ozonide ion is paramagnetic with one unpaired electron. It has C_{2v} symmetry with an angle of 111°. All ozonides are yellow, orange, or red owing to an absorption band in the 400–600 nm region.

[21] L. Andrews, *J. Chem. Phys.*, 1975, **63**, 4465.

TABLE 15-1
Various Bond Values for Oxygen Species[a]

Species	O—O Distance (Å)	Number of π^* electrons	$\nu_{OO}(cm^{-1})$
O_2^+	1.12	1	1905
O_2	1.21	2	1580
O_2^-	1.33	3	1097
O_2^{2-}	1.49	4	802

[a] L. Vaska, *Acc. Chem. Res.,* 1976, **9**, 175.

Materials of composition approximating to M_2O_3 (M = alkali metal) are almost certainly mixtures of peroxide and superoxide and there is no evidence for the existence of an O_3^{2-} ion.

The various $O_2^{n\pm}$ species, from O_2^+ to O_2^{2-}, provide an interesting illustration of the effect of varying the number of antibonding electrons on the length and stretching frequency of a bond, as the data in Table 15-1 show.

15-8.   Other Peroxo Compounds

A large number of *organic peroxides* and *hydroperoxides* are known. Peroxo carboxylic acids (e.g., peroxoacetic acid, $CH_3CO\cdot OOH$) can be obtained by action of H_2O_2 on acid anhydrides. Peroxoacetic acid is commercially made as 10 to 55% aqueous solutions containing some acetic acid by interaction of 50% H_2O_2 and acetic acid, with H_2SO_4 as catalyst at 45 to 60°; the dilute acid is distilled under reduced pressure. It is also made by air oxidation of acetaldehyde. The peroxo acids are useful oxidants and sources of free radicals [e.g., by treatment with $Fe^{2+}(aq)$]. Benzoyl peroxide and cumyl hydroperoxide are moderately stable and widely used as polymerization initiators and for other purposes where free-radical initiation is required.

Organic peroxo compounds are also obtained by *autooxidation* of ethers, unsaturated hydrocarbons, and other organic materials on exposure to air. The autooxidation is a free-radical chain reaction that is initiated almost certainly by radicals generated by interaction of oxygen and traces of metals such as copper, cobalt, or iron.[22] The attack on specific reactive C—H bonds by a radical X· gives first R·, then hydroperoxides, which can react further:

$$RH + X\cdot \rightarrow R\cdot + HX$$
$$R\cdot + O_2 \rightarrow RO_2\cdot$$
$$RO_2\cdot + RH \rightarrow ROOH + R\cdot$$

Peroxide formation can lead to explosions if oxidized solvents are distilled. Peroxides are best removed by washing with acidified $FeSO_4$ solution or, for ethers and hydrocarbons, by passage through a column of activated alumina. Peroxides are absent when the $Fe^{2+} + SCN^-$ reagent gives no red color.

22   J. F. Black, *J. Am. Chem. Soc.,* 1978, **100**, 527.

There is a great variety of inorganic peroxo compounds where O is replaced by —O—O— groups. Typical are peroxo anions such as peroxosulfates 15-V and 15-VI. All peroxo acids yield H_2O_2 on hydrolysis. Peroxodisulfate, as the ammonium salt, is commonly used as a strong oxidizing agent in acid solution, for example, to convert C into CO_2, Mn^{2+} into MnO_4^-, or Ce^{3+} into Ce^{4+}. The last two reactions are slow and normally incomplete in the absence of silver ion as a catalyst.

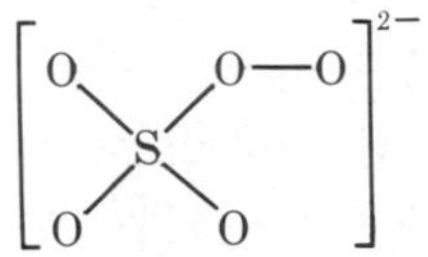

(15-V)
Peroxomonosulfate

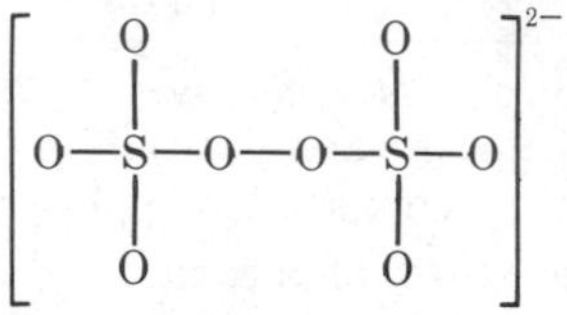

(15-VI)
Peroxodisulfate

It is important to make the distinction between true peroxo compounds, which contain —O—O— groups, and compounds such as $2Na_2CO_3 \cdot 3H_2O_2$[23] or $Na_4P_2O_7 \cdot nH_2O_2$, which contain H_2O_2 of crystallization hydrogen-bonded to the anion. Many detergents contain sodium peroxoborate $Na[(HO)_2B(\mu\text{-}O_2)_2B(OH)_2] \cdot 6H_2O$ or sodium carbonate peroxohydrate.

Fluorinated Peroxides.[24] There are a number of peroxo compounds that contain fluorine groups, and many of these are reasonably stable. Some examples are

$(FSO_2)OO(SO_2F)$ Peroxodisulfonyldifluoride
SF_5OOSF_5 Bispentafluorosulfurperoxide
CF_3OOCF_3 Bisperfluoromethylperoxide

The compounds are usually prepared by fluorination of oxygen compounds, e.g.,

$$SO_3 + F_2 \xrightarrow[160°]{AgF_2} (FSO_2)OO(SO_2F)$$

$$COF_2 + ClF_3 \xrightarrow[250°]{KF} CF_3OOCF_3$$

23   M. A. A. F. de C. T. Carrondo *et al., J.C.S. Dalton,* **1977,** 2323.
24   R. A. De Marco and J. M. Shreeve, *Adv. Inorg. Chem. Radiochem.,* 1974, **16,** 109.

General References

Brilkina, T. G., and V. A. Shuschunov, *Reactions of Organometallic Compounds with Oxygen and Peroxides,* Iliffe, 1969.

Connor, J. A., and E. A. V. Ebsworth, *Adv. Inorg. Chem. Radiochem.,* 1964, **6,** 280. Peroxo compounds of transition metals.

Franks, F., Ed., *Water: A Comprehensive Treatise,* 5 vols., Plenum Press, 1972–1975.

Hayaishi, O., *Molecular Oxygen in Biology,* North Holland—Elsevier, 1974.

Hitchmann, M. I., *Measurement of Dissolved Oxygen,* Wiley, 1978.

Hoare, P. J., *The Electrochemistry of Oxygen,* Interscience, 1968.

Metzner, H., Ed., *Photosynthetic Oxygen Evolution,* Academic Press, 1978.

Ochiai, E.-I., *J. Inorg. Nucl. Chem.,* 1975, **37,** 1503. Bioinorganic chemistry of O_2.

Perst, H., *Oxonium Ions in Organic Chemistry,* Academic Press, 1971.

Pitts, J. N., Jr., and B. J. Finlayson, *Angew. Chem. Int. Ed.,* 1975, **14,** 1. Mechanism of photochemical air pollution.

Severn, D., *Organic Peroxides,* Vols., I–III, Wiley Interscience, 1972.

Vol'novi, I. I., *Russ. Chem. Rev.,* **1972,** 314. Inorganic peroxo compounds.

Vaska, L., *Acc. Chem. Res.,* 1976, **9,** 175. Dioxygen-metal complexes.

Wells, A. F., *Structural Inorganic Chemistry,* 4th ed., Clarendon Press, 1975. Contains much information on oxide and hydroxide structures.

Yost, D. M., and H. Russell, *Systematic Inorganic Chemistry (of the 5th and 6th Group Elements),* Prentice-Hall, 1946. Excellent on selected aspects.

CHAPTER SIXTEEN

The Group VI Elements: S, Se, Te, Po

GENERAL REMARKS

16-1. Electronic Structures, Valences, and Stereochemistries

Some properties of the elements in Group VI are given in Table 16-1.

The atoms are two electrons short of the configuration of the next noble gas, and the elements show essentially nonmetallic covalent chemistry except for polonium and to a very slight extent tellurium. They may complete the noble gas configuration by forming (a) the *chalconide* ions S^{2-}, Se^{2-}, and Te^{2-}, although these ions exist only in the salts of the most electropositive elements, (b) two electron-pair bonds [e.g., $(CH_3)_2S$, H_2S, SCl_2, etc.], (c) ionic species with one bond and one negative charge (e.g., HS^-, RS^-), or (d) three bonds and one positive charge (e.g., R_3-S^+).

In addition to such divalent species, the elements form compounds in *formal* oxidation states IV and VI with four, five, or six bonds; tellurium may give an

TABLE 16-1
Some Properties of the Group VI Elements

Element	Electronic structure	Melting point (°C)	Boiling point (°C)	Radius X^{2-} (Å)	Covalent radius —X— (Å)	Electro- negativity
S	$[Ne]3s^23p^4$	119[a]	444.6	1.90	1.03	2.44
Se	$[Ar]3d^{10}4s^24p^4$	217	684.8	2.02	1.17	2.48
Te	$[Kr]4d^{10}5s^25p^4$	450	990	2.22	1.37	2.01
Po	$[Xe]4f^{14}5d^{10}6s^26p^4$	254	962	2.30		1.76

[a] For monoclinic S, see text.

TABLE 16-2
Compounds of Group VI Elements and Their Stereochemistries

Valence	Number of bonds	Geometry	Examples
II	2	Angular	Me_2S, H_2Te, S_n
	3	Pyramidal	Me_3S^+
	4	Square	$Te[SC(NH_2)_2]_2Cl_2$
IV	2	Angular	SO_2
	3	Pyramidal	SF_3^+, OSF_2, SO_3^{2-}, Me_3TeBPh_4[b]
		Trigonal-planar	$(SeO_2)_n$
	4	ψ-Trigonal-bipyramidal[a]	SF_4, RSF_3, Me_2TeCl_2
		Tetrahedral	Me_3SO^+
	5	ψ-Octahedral (square-pyramidal)	$SeOCl_2py_2$, SF_5^-, TeF_5^-
	6	Octahedral	$SeBr_6^{2-}$, PoI_6^{2-}, $TeBr_6^{2-}$
VI	3	Trigonal-planar	$SO_3(g)$, $S(NCR)_3$[c]
	4	Tetrahedral	SeO_4^{2-}, $SO_3(s)$, SeO_2Cl_2
	5	Trigonal-bipyramidal	SOF_4
	6	Octahedral	RSF_5, SeF_6, $Te(OH)_6$
	8(?)	?	TeF_8^{2-} (?)

[a] Lone pairs are equatorial.
[b] R. F. Ziolo and J. M. Troup, *Inorg. Chem.*, 1979, **18**, 2271.
[c] O. Glemser et al., *Angew. Chem., Int. Ed.*, 1977, **16**, 789; has S=N multiple bonds by $p\pi$–$p\pi$ overlap.

eight-coordinate ion TeF_8^{2-}. Some examples of compounds of Group VI elements and their stereochemistries are listed in Table 16-2.

16-2.   Group Trends

There are great differences between the chemistry of oxygen and that of sulfur, with more gradual variations through the sequence S, Se, Te, Po. Differences from oxygen are attributable, among other things, to the following:

1.   The lower electronegativities of the S–Po elements lessens the ionic character of those of their compounds that are formally analogous to those of oxygen, alters the relative stabilities of various kinds of bonds, and drastically lessens the importance of hydrogen bonding, although weak S⋯H—S bonds do indeed exist.

2.   The maximum coordination number is not limted to 4, nor is the valence limited to 2, as in the case of oxygen, since d orbitals may be utilized in bonding. Thus sulfur forms several hexacoordinate compounds (e.g., SF_6), and for tellurium 6 is the characteristic coordination number.

3.   Sulfur has a strong tendency to catenation, so that it forms compounds having no oxygen, selenium, or tellurium analogues, for example: polysulfide ions S_n^{2-}, sulfanes XS_nX (where X may be H, halogen, —CN, or —NR$_2$), and the polysulfuric acids $HO_3SS_nSO_3H$ and their salts.

Sulfur has a great propensity to form S_n rings not only in the element, but also in compounds such as S_8O and sometimes rings with heteroatoms (p. 519). Sulfur-sulfur bonds are also very important in organic systems, and in Nature in compounds like cysteine, certain proteins, and enzymes. The S—S bonds[1] are very variable and flexible; the lengths vary from 1.8 to 3.0 Å, the angles from 90° to 100°, and the dihedral angles from 0 to 180°. Also the bond energies may be as high as 430 kJ mol^{-1}. These variations lead to considerable differences in the properties and reactivities of compounds with S—S bonds.

Although selenium and tellurium have a smaller tendency to catenation, they form rings (Se only) and long chains in their elemental forms. None of these chains is branched, because the valence of the element is only 2.

Gradual changes of properties are evident with increasing size, decreasing electronegativity, and so on, such as:

1. Decreasing thermal stability of the H_2X compounds. Thus H_2Te is considerably endothermic.

2. Increasing metallic character of the elements.

3. Increasing tendency to form anionic complexes such as $SeBr_6^{2-}$, $TeBr_6^{2-}$, and PoI_6^{2-}.

4. Decreasing stability of compounds in high formal positive oxidation states.

5. Emergence of cationic properties for Po and, very marginally, for Te. Thus TeO_2 and PoO_2 appear to have ionic lattices and they react with hydrohalic acids to give Te^{IV} and Po^{IV} halides, and PoO_2 forms a hydroxide $Po(OH)_4$. There are also some ill-defined "salts" of Te and Po, such as $Po(SO_4)_2$ and $TeO_2 \cdot SO_3$.

Use of *d* Orbitals.[2] In addition to the ability of the S—Po elements to utilize *d* orbitals in hybridization with *s* and *p* orbitals to form more than four σ bonds to other atoms, sulfur particularly and also selenium, appear to make frequent use of $d\pi$ orbitals to form multiple bonds. Thus, for example, in the sulfate ion, where the *s* and *p* orbitals are used in σ bonding, the shortness of the S—O bonds suggests that there must be considerable multiple-bond character. The only likely explanation for this is that empty $d\pi$ orbitals of sulfur accept electrons from filled $p\pi$ orbitals of oxygen (see p. 209). Similar $d\pi$–$p\pi$ bonding occurs in some phosphorus compounds, but it seems to be more prominent with sulfur.

Secondary Bonding. Both Se and Te have a strong tendency in crystalline compounds to form secondary bonds.[3] In certain compounds, in addition to the normal single bonds, there are additional bonding interactions resulting in distances considerably longer than single bonds, but shorter than van der Waals distances. The situation is not dissimilar to H bonding (p. 219). An example is the interaction between F^- and TeF_3^+ in TeF_4, where there are chains $F—TeF_3—F—TeF_3—$ with the "normal" Te—F distance 1.89 Å and the asymmetric bridges at 2.08 and 2.26 Å. $TeCl_4$ (p. 524) is similar, though the structure is different. Where secondary

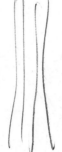

[1] R. Steudel, *Angew. Chem. Int. Ed.*, 1975, **14**, 655.
[2] H. Kwart and K. King, *d-Orbitals in the Chemistry of Silicon Phosphorus and Sulfur*, Springer, 1977.
[3] N. W. Alcock, *Adv. Inorg. Chem. Radiochem.*, 1972, **15**, 1.

bonding cannot occur, as in Me_3TeBPh_4, the cation is pyramidal (see Table 16-2). A different type is found in divalent complexes such as $Te[S_2P(OMe)_2]_2$ (16-I).

$$\begin{array}{c} \text{MeO} \\ \text{MeO} \end{array} \!\! P \begin{array}{c} \text{S} \quad 2.44 \text{ Å} \quad \text{S} \\ \text{Te} \\ \text{S} \quad 3.31 \text{ Å} \quad \text{S} \end{array} P \begin{array}{c} \text{OMe} \\ \text{OMe} \end{array}$$

(16-I)

This type of secondary interaction is found in other compounds of heavy nonmetallic elements that are coordinatively unsaturated.

Both selenium and tellurium compounds are generally toxic. Tellurium has no biological role, but traces of selenium are essential, and the enzyme glutathione peroxidase is selenium dependent.[4] Selenium may also have anticancer properties.[5]

THE ELEMENTS

16-3. Occurrence

Sulfur occurs widely in Nature as the element, as H_2S and SO_2, in numerous sulfide minerals and in sulfates such as *anhydrite* ($CaSO_4$), etc. It occurs in crude oils and in coal and as H_2S in natural gas, from which it is recovered in large quantities via the reaction

$$2H_2S + SO_2 = 3S + 2H_2O$$

The main use of sulfur is for manufacture of sulfuric acid, but it is used in vulcanization of rubber and to make CS_2 (for CCl_4) and sulfides such as P_2S_5.

Selenium and tellurium are much less abundant than sulfur and frequently occur as selenide and telluride impurities in metal sulfide ores. They are recovered from flue dusts of combustion chambers for sulfur ores, particularly those of silver and gold, and from lead chambers in sulfuric acid manufacture.

Polonium occurs in U and Th minerals as a product of radioactive decay series. It was first isolated from pitchblende, which contains less than 0.1 mg of Po per ton. The most accessible isotope is ^{210}Po (α, 138.4d) obtained in gram quantities by irradiation of bismuth in nuclear reactors:

$$^{209}\text{Bi}(n,\gamma)^{210}\text{Bi} \rightarrow {}^{210}\text{Po} + \beta^-$$

Polonium is separated from Bi by sublimation or in a variety of chemical ways. The study of polonium chemistry is difficult owing to the intense α-radiation, which causes damage to solutions and solids, evolves much heat, and makes necessary special handling techniques for protection of the chemist.

[4]   T. C. Stadtman, *Science,* 1974, **183,** 915; H. E. Ganther in *Selenium,* R. A. Zingaro and W. C. Cooper, Eds., Van Nostrand, 1974; L. Klayman and W. H. H. Gunther, Eds., *Organic Selenium Compounds,* Wiley, 1973.

[5]   G. N. Schrauzer *et al., Bioinorg. Chem.,* 1978, **8,** 387.

16-4. Elemental Sulfur[6]

The structural relationships of sulfur in all three phases are exceedingly complex, and there has been considerable confusion concerning new allotropes, which subsequently turned out to be mixtures or to be impure, and also concerning nomenclature. We deal only with the main, well-established species.

Solid Sulfur. All modifications of crystalline sulfur contain either (*a*) sulfur rings, which may have from 6 to 20 sulfur atoms and are referred to as cyclohexa-, cycloocta-, etc., sulfur, or (*b*) chains of sulfur atoms, referred to as catenasulfur (S_∞).

1. *Cyclooctasulfur* (S_8). This is the most common form and has three main allotropes (crystal forms): S_α, S_β, and S_γ.

Orthorhombic sulfur (S_α) is thermodynamically the most stable form and occurs in large yellow crystals in volcanic areas. It can be grown from solutions, although the crystals then usually contain solvent. Its structure is shown in Fig. 16-1. At 368.46°K (95.5°C) S_α transforms to the high-temperature form *monoclinic sulfur* (S_β). The enthalpy of the transition is small (0.4 kJ g-atom^{-1} at 95.5°C) and the process is slow, so that it is possible by rapid heating of S_α to attain the melting point of S_α (112.8°); S_β melts at 119°. Monoclinic S_β crystallizes from sulfur melts, and although slow conversion to S_α occurs, the crystals can be preserved for weeks. Its structure contains S_8 rings as in S_α, but differently packed.

Monoclinic sulfur (S_γ, m.p. 106.8°C) is obtained by decomposition of copper(I) ethyl xanthate in pyridine. It transforms slowly into S_β and/or S_α but is stable in the 95–115° region. It contains crown S_8 rings.[7]

2. *Cyclohexasulfur* (S_6). This is rhombohedral sulfur (S_ρ) and is obtained by the following reaction in ether:

$$S_2Cl_2 + H_2S_4 = S_6 + 2HCl$$

or by addition of concentrated HCl to a solution of $Na_2S_2O_3$ at $-10°$. The polythionate chains initially produced undergo ring closure to give S_6. Extraction of the precipitate with benzene and crystallization gives orange crystals, but extremely complicated procedures are required to obtain high purity. S_6 can also be made by the reaction.

$$H_2S_2 + S_4Cl_2 \rightarrow S_6 + 2HCl$$

It decomposes quite rapidly and is chemically very much more reactive than S_8 because the ring is considerably more strained; the reactions may be profoundly affected by impurities and light.

3. *Other Cyclosulfurs.*[8] By controlled reactions of sulfur chlorides with sulfanes

6   S. W. Benson, *Chem. Rev.* 1978, **78**, 23; J. Kao and N. L. Allinger, *Inorg. Chem.*, 1977, **16**, 35; B. Meyer, *Chem. Rev.*, 1976, **76**, 367; *Adv. Inorg. Chem. Radiochem.*, 1976, **18**, 287; M. Schmidt, *Angew. Chem. Int. Ed.*, 1973, **12**, 445; T. Chivers and I. Drummond, *Chem. Soc. Rev.*, 1973, **2**, 233.
7   L. K. Templeton, D. H. Templeton, and A. Zalkin, *Inorg. Chem.*, 1976, **15**, 1999.
8   R. Reinhardt *et al.*, *Angew Chem. Int. Ed.*, 1978, **17**, 57; M. Schmidt *et al.*, *Z. Anorg. Allg. Chem.*, 1974, **405**, 153.

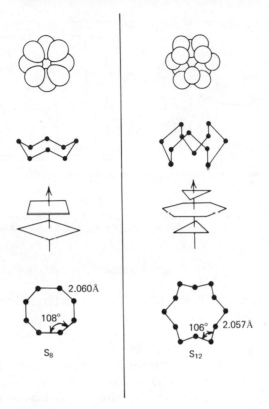

Fig. 16-1. The structures of S_8 and S_{12}. [From B. Meyer, *Chem. Rev.*, 1976, **76**, 367, by permission.]

or with the compound $(\eta^5\text{-}C_5H_5)_2TiS_5$, which contains a five-sulfur chain, thermodynamically unstable allotropes containing 7-, 8-, 9-, 10-, 11-, 12-, 18-, and 20-membered puckered rings can be obtained, e.g.,

$$2H_2S_4 + 2S_2Cl_2 \longrightarrow S_{12} + 4HCl$$

$$(\eta^5\text{-}C_5H_5)_2TiS_5 + S_nCl_2 \xrightarrow{CS_2} (\eta^5\text{-}C_5H_5)_2TiCl_2 + S_{5+n}$$

The cyclosulfurs S_{18} (d. 126°) and S_{20} (d. 121°) are quite stable, but other sulfurs are thermally unstable and sensitive to light and air, although they can be preserved at low temperatures. The seven-membered ring in S_7 has an unusually long S—S bond (2.18 Å vs. normal, *ca.* 2.06 Å).

4. *Catenasulfur.* When molten sulfur is poured into ice water, the so-called plastic sulfur is obtained; although normally this has S_8 inclusions, it can be obtained as long fibers by heating S_α in nitrogen at 300° for 5 min and quenching a thin stream in ice water. These fibers can be stretched under water and appear to contain helical chains of sulfur atoms with about 3.5 atoms per turn. Unlike the other sulfur allotropes, catenasulfur is insoluble in CS_2; it transforms slowly to S_α.

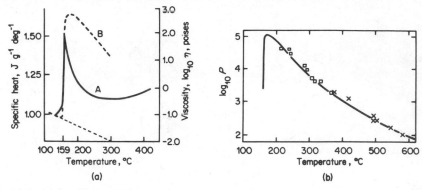

Fig. 16-2. (a) Specific heat (A) and viscosity (B) of liquid sulfur. (b) Chain length (P) as a function of temperature, X from magnetic susceptibility measurements and □ from esr measurements. [Reproduced by permission from B. Meyer, Ed., *Elemental Sulfur–Chemistry and Physics,* Interscience, 1965.]

5. Decomposition of N_4S_4 at high temperature and pressure gives a gray-black form of sulfur insoluble in CS_2.[9]

Liquid Sulfur. Precisely what happens when S_8 melts is still not fully understood; doubtless much depends on the level of impurities.

On melting, S_8 first gives a yellow, transparent, mobile liquid, which becomes brown and increasingly viscous above about 160°. The viscosity reaches a maximum at about 200° and thereafter falls until at the boiling point (444.60°) the sulfur is again a rather mobile, dark red liquid. Figure 16-2a shows the viscosity and specific heat as a function of temperature. Although S_8 rings persist in the liquid up to *ca.* 193° the changes in viscosity are due to ring cleavage and the formation of chains, as well as other sulfur ring species with $n = 6, 7, 12, 18, 20,$ and >20 in equilibrium.[10] The average degree of polymerization is shown in Fig. 16-2b. The sulfur chains must have radical ends; they reach their greatest average length, 5–8 $\times 10^5$ atoms, at about 200°, where the viscosity is highest. The quantitative behavior of the system is sensitive to certain impurities, such as iodine, which can stabilize chain ends by formation of S—I bonds. In the formation of polymers, almost every broken S—S bond of an S_8 ring is replaced by an S—S bond in a linear polymer, and the overall heat of the polymerization is thus expected to be close to zero. An enthalpy of 13.4 kJ mol^{-1} of S_8 converted into polymer has been found at the critical polymerization temperature (159°).

The color changes on melting are due to an increase in the intensity and a shift of an absorption band to the red. This is associated with the formation above *ca.* 250° of the red species S_3 and S_4, which comprise 1 to 3% of sulfur at its boiling point.

The other cyclosulfurs undergo similar decompositions and form polymers (e.g., S_{10} polymerizes at 60°, S_7 at 45°).

Sulfur Vapor. In addition to S_8, sulfur vapor contains $S_3, S_4, S_5, S_7,$ and probably other S_n species in a temperature-dependent equilibrium. S_2 molecules predominate

9   G. C. Vezzoli and J. Abel, *Science,* 1978, **200,** 765.
10   R. Steudel and H.-J. Maüsle, *Angew. Chem., Int. Ed.,* 1979, **18,** 152; 1978, **17,** 56.

at higher temperatures; above 2200° and at pressures below 10^{-7} cm sulfur atoms predominate. The S_2 species can be rapidly quenched in liquid nitrogen to give a highly colored solid, unstable above $-80°$, which contains S_2 molecules. These have two unpaired electrons (cf. O_2). The electronic absorption bands of S_2 in the visible region account for the deep blue color of hot sulfur vapor.

16-5.   The Structures of Elemental Selenium, Tellurium, and Polonium[11]

Three thermodynamically unstable, red monoclinic allotropes of selenium that contain cyclooctaselenium (Se_8) are known.[12] However the stable form consists of gray trigonal, metal-like crystals that may be grown from hot solutions of Se in aniline or from melts. The structure, which has no sulfur analogue, contains infinite spiral chains of selenium atoms. Although there are fairly strong single bonds between adjacent atoms in each chain, there is evidently weak metallic interaction between the neighboring atoms of different chains. Selenium is not comparable with most true metals in its electrical conductivity in the dark, but it is markedly photoconductive and is widely used in photoelectric devices.

The one form of *tellurium* is silvery-white, semimetallic, and isomorphous with gray Se. Like the latter it is virtually insoluble in all liquids except those with which it reacts.

In vapors the concentration of paramagnetic Se_2 and Te_2 molecules and Se and Te atoms is evidently much higher under comparable conditions of temperature and pressure than for sulfur, indicating decreased tendency toward catenation.

The trend toward greater metallic character in the elements is complete at polonium. Whereas sulfur is a true insulator (specific resistivity, in $\mu\Omega$-cm $= 2 \times 10^{23}$), selenium (2×10^{11}) and tellurium (2×10^{5}) are intermediate in their electrical conductivities, and the temperature coefficient of resistivity in all three cases is negative, which is usually considered characteristic of nonmetals. Polonium in each of its two allotropes has a resistivity typical of true metals ($\sim 43 \mu\Omega$-cm) and a positive temperature coefficient.

16-6.   Reactions of the Elements

The allotropes of S and Se containing cyclo species are soluble in CS_2 and other nonpolar solvents such as benzene and cyclohexane. The solutions are light sensitive, becoming cloudy on exposure, and may also be reactive toward air; unless very special precautions are taken sulfur contains traces of H_2S and other impurities that can have substantial effects on rates of reactions. The nature of the sulfur produced on photolysis is not well established, but such material reverts to S_8 slowly in the dark or rapidly in presence of triethylamine. From the solvent CHI_3, sulfur crystallizes as a charge-transfer compound $CHI_3 \cdot 3S_8$, with I⋯S bonds; isomorphous compounds with PI_3, AsI_3, and SbI_3 are also known. It is probable that similar charge-transfer complexes wherein the S_8 ring is retained are first formed in re-

[11]   D. M. Chizhikov and P. Schastivy, *Selenium and Selenides,* Collet, 1968; *Tellurium and Tellurides,* Collet, 1970.

[12]   O. Foss and V. Janickis, *J.C.S. Chem. Comm.,* **1977,** 834.

actions of sulfur with e.g., bromine. When heated, sulfur, selenium, and tellurium burn in air to give the dioxides MO_2, and the elements react when heated with halogens, most metals, and nonmetals. They are not affected by nonoxidizing acids, but the more metallic polonium will dissolve in concentrated HCl as well as in H_2SO_4 and HNO_3.

It has long been known that sulfur, selenium, and tellurium will dissolve in oleums to give blue, green, and red solutions, respectively, which are unstable and change in color when kept or warmed. The nature of the colored species has been controversial, but it is now known, largely from the work of R. J. Gillespie and his coworkers, that they are cyclic polycations[13] in which the element is formally in a fractional oxidation state. Although S_8 can be heated at 75° in 100% H_2SO_4 without attack, oxidation occurs with time and increasing SO_3 concentration, forming species such as S_4^{2+}, S_8^{2+}, and S_{16}^{2+}, together with SO_2. The oleum solutions of sulfur give esr spectra because of the presence of one or more radicals of the type S_n^+ in low concentration. The S_5^+ radical has been identified with certainty in 65% oleum.[14]

It is difficult to isolate solids from oleum solutions, but crystalline salts have been obtained by selective oxidations of the elements with SbF_5 or AsF_5 in liquid HF, or with $S_2O_6F_2$ in HSO_3F. Interaction of the element and its halide in molten $AlCl_3$ or anodic oxidation in $NaCl–AlCl_3$ melts also give salts.[15] Some representative reactions are:

$$S_8 + 3SbF_5 = S_8^{2+} + 2SbF_6^- + SbF_3$$
$$2S_8 + S_2O_6F_2 = S_{16}^{2+} + 2SO_3F^-$$
$$Se_8 + 3AsF_5 = Se_8^{2+} + 2AsF_6^- + AsF_3$$
$$7Te + TeCl_4 + 4AlCl_3 = 2Te_4^{2+} + 4AlCl_4^-$$

Structural characterization of many of the cations has been accomplished. The yellow S_4^{2+} and Se_4^{2+}, and the red Te_4^{2+} ions are square and are believed to have a six π-electron quasi-aromatic system (16-II). The E_8^{2+} (E = S, Se, Te) ions have a cyclic structure (16-III) with one transannular distance short enough to be considered a bond, though a weaker one. There are a number of E_6^{2+} ions including Te_6^{2+} and the mixed ones,[16] $Te_3S_3^{2+}$ and $Te_2Se_4^{2+}$, which have structures of the type 16-IV. Using liquid SO_2 as a solvent, mixed Te/Se ions of composition E_{10}^{2+} with structures 16-V are obtained.[17]

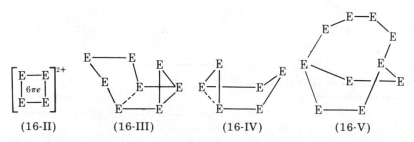

(16-II)　　　　(16-III)　　　　(16-IV)　　　　(16-V)

13　R. J. Gillespie and J. Passmore, *Adv. Inorg. Chem. Radiochem.*, 1975, **17**, 49.
14　H. S. Low and R. A. Beaudet, *J. Am. Chem. Soc.*, 1976, **98**, 3849.
15　R. Fehrmann et al., *Inorg. Chem.*, 1978, **17**, 1195; 1977, **16**, 2259, 2089.
16　R. J. Gillespie et al., *Inorg. Chem.*, 1977, **16**, 892.
17　P. Boldrini et al., *Inorg. Chem.*, 1976, **15**, 765.

The Te_8^{2+} ion[18] in $Te_6(As F_6)_4 \cdot 2As F_3$ is a unique trigonal prism cluster. The only other species of this type is prismane C_6H_6; the Bi_9^{5+} ion has the prism tricapped.

The difference between oxygen, which forms O_2 and O_2^+ only, and the heavier elements is due to the reluctance of the latter to form π bonds.

Sulfur is also soluble, with reaction, in organic amines such as piperidine, to give colored solutions containing N,N'-polythiobisamines in which there are free radicals (about 1 per 10^4 S atoms):

$$2RR'NH + S_n \rightarrow (RR'N)_2S_{n-1} + H_2S$$

Many sulfur reactions are catalyzed by amines, and such S—S bond-breaking reactions to give free radicals may be involved.

Sulfur and selenium react with many organic molecules. For example, saturated hydrocarbons are dehydrogenated. The reaction of sulfur with alkenes and other unsaturated hydrocarbons is of enormous technical importance: hot sulfurization results in the vulcanization (formation of S bridges between carbon chains) of natural and synthetic rubbers.

It is clear that all reactions of S_8 or other cyclo species require that the initial attack open the ring to give sulfur chains or chain compounds. Many common reactions can be rationalized by considering a nucleophilic attack on S—S bonds. Some typical reactions are

$$S_8 + 8CN^- \rightarrow 8SCN^-$$
$$S_8 + 8Na_2SO_3 \rightarrow 8Na_2S_2O_3$$
$$S_8 + 8Ph_3P \rightarrow 8Ph_3PS$$

Such reactions cannot possibly proceed by what, according to the stoichiometry, would be ninth-order reactions. It appears that the rate-determining step is the initial attack on the S_8 ring and that subsequent steps proceed very rapidly, so that the reactions can be assumed to proceed as follows:

$$S_8 + CN^- \rightarrow SSSSSSSSCN^-$$
$$S_6 - S - SCN^- + CN^- \rightarrow S_6SCN^- + SCN^-, \text{ etc.}$$

BINARY COMPOUNDS

16-7. Hydrides

The dihydrides, H_2S, H_2Se, and H_2Te are extremely poisonous gases with revolting odors; the toxicity of H_2S far exceeds that of HCN. They are readily obtained by the action of acids on metal chalconides. H_2Po has been prepared only in trace quantities, by dissolving magnesium foil plated with Po in 0.2 M HCl. The thermal stability and bond strengths decrease from H_2S to H_2Po. Although pure H_2Se is thermally stable to 280°, H_2Te and H_2Po appear to be thermodynamically unstable with respect to their constituent elements. All behave as very weak acids in aqueous solution, and the general reactivity and also the dissociation constants increase with

[18]   R. J. Gillespie et al., Inorg. Chem., 1979, **18**, 3086.

increasing atomic number. *Hydrogen sulfide* dissolves in water to give a solution about 0.1 M under 1 atm pressure. The dissociation constants are

$$H_2S + H_2O = H_3O^+ + HS^- \qquad K = 1 \times 10^{-7}$$
$$HS^- + H_2O = H_3O^+ + S^{2-} \qquad K = \sim 10^{-17}$$

In acid solution, H_2S is also a mild reducing agent.

Sulfanes. The compounds H_2S_2 through H_2S_6 have been isolated in pure states whereas higher members are so far known only in mixtures. All are reactive yellow liquids whose viscosities increase with chain length. They may be prepared in large quantities by reactions such as

$$Na_2S_n(aq) + 2HCl(aq) \rightarrow 2NaCl(aq) + H_2S_n(l)(n = 4 - 6)$$
$$S_nCl_2(l) + 2H_2S(l) \rightarrow 2HCl(g) + H_2S_{n+2}(l)$$
$$S_nCl_2(l) + 2H_2S_2(l) \rightarrow 2HCl(g) + H_2S_{n+4}(l)$$

The oils from the first reaction can be cracked and fractionated to give pure H_2S_2 through H_2S_5, whereas the higher sulfanes are obtained from the other reactions. Although the sulfanes are all thermodynamically unstable with respect to the reaction

$$H_2S_n(l) = H_2S(g) + (n - 1)S(s)$$

these reactions, which are believed to be free-radical in nature, are sufficiently slow for the compounds to be stable for considerable periods.

16-8. Metal Chalconides

Most metallic elements react directly with S, Se, Te and, so far as is known, Po. Often they react very readily, mercury and sulfur, for example, at room temperature. Binary compounds of great variety and complexity of structure can be obtained. The nature of the products usually also depends on the ratios of reactants, the temperature of reaction, and other conditions. Many elements form several compounds and sometimes long series of compounds with a given chalconide. We give here only the briefest account of the more important sulfur compounds. Generally, selenides and tellurides are similar.

Ionic Sulfides; Sulfide Ions. Only the alkalis and alkaline earths form sulfides that appear to be mainly ionic. They are the only sulfides that dissolve in water and they crystallize in simple ionic lattices, for example, an antifluorite lattice for the alkali sulfides and a rock salt lattice for the alkaline earth sulfides. Essentially only SH^- ions are present in aqueous solution, owing to the low second dissociation constant of H_2S. Although S^{2-} is present in concentrated alkali solutions, it cannot be detected below about $8M$ NaOH owing to the reaction

$$S^{2-} + H_2O = SH^- + OH^- \qquad K = \sim 1$$

The alkali and alkaline earth hydrosulfides can be made by action of H_2S on the metal in liquid ammonia.[19]

When aqueous sulfide solutions are heated with sulfur, solutions containing largely S_4^{2-} and S_3^{2-} are obtained.[20] These polysulfide ions are the only ones stable

[19]   J. A. Kaeser *et al.*, *Inorg. Chem.*, 1973, **12**, 3019.

[20]   W. F. Giggenbach, *Inorg. Chem.* 1974, **13**, 1724, 1730.

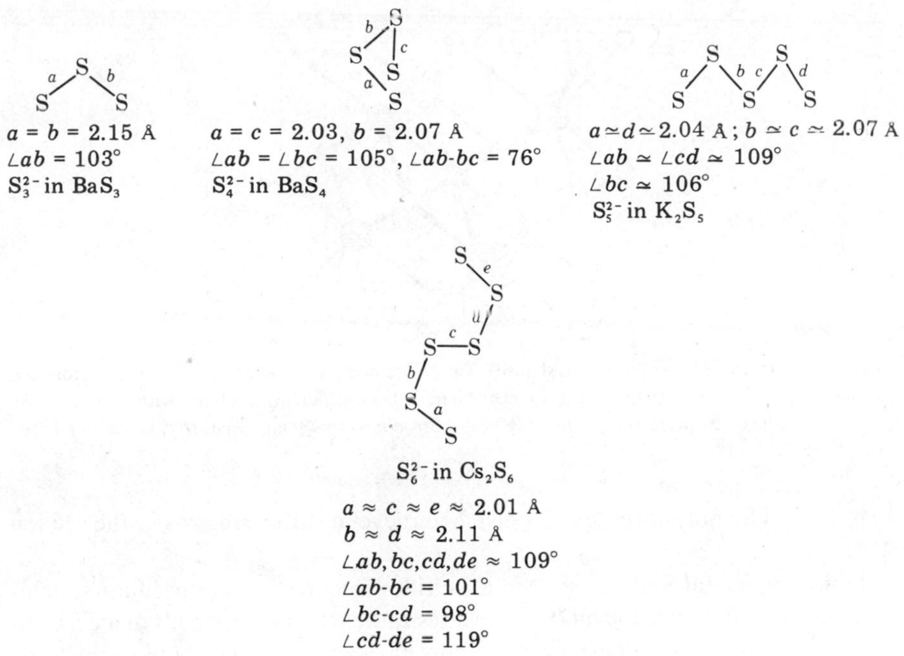

$a = b = 2.15$ Å
$\angle ab = 103°$
S_3^{2-} in BaS_3

$a = c = 2.03, b = 2.07$ Å
$\angle ab = \angle bc = 105°, \angle ab\text{-}bc = 76°$
S_4^{2-} in BaS_4

$a \simeq d \simeq 2.04$ Å $; b \simeq c \simeq 2.07$ Å
$\angle ab \simeq \angle cd \simeq 109°$
$\angle bc \simeq 106°$
S_5^{2-} in K_2S_5

S_6^{2-} in Cs_2S_6
$a \approx c \approx e \approx 2.01$ Å
$b \approx d \approx 2.11$ Å
$\angle ab, bc, cd, de \approx 109°$
$\angle ab\text{-}bc = 101°$
$\angle bc\text{-}cd = 98°$
$\angle cd\text{-}de = 119°$

Fig. 16-3. Structure of representative sulfide ions.

in aqueous solution, but other salts can be prepared in dry ways.[21] All have S_n chains (Fig. 16-3).

High-density power sources can be obtained from lithium and sodium-sulfur batteries. The sulfides present in these systems are M_2S, M_2S_2, M_2S_4, and M_2S_5.[22]

Although polyselenide and polytelluride ions are less common, the Se_3^{2-} and Te_3^{2-} ions are known.[23]

When alkali polysulfides are dissolved in polar solvents such as acetone, dimethylformamide, or dimethylsulfoxide, deep blue solutions are formed. Certain sulfur-containing minerals, notably lapis lazuli and ultramarine, are blue. There has been much discussion concerning the species responsible for the color, but it now seems[24] in all cases that the absorption band (λ_{max} 610 nm) is associated with the radical ion S_3^-. In ultramarine some S_2^- (λ_{max} 400 nm) is also present.

The behavior of S^{2-}, SH^-, S_2^{2-}, and S^{2-} as ligands has been discussed in Chapter

[21] G. Weddigen et al., J. Chem. Res. (S), 1978, 96; B. Kelly and P. Woodward, J. C. S. Dalton, 1976, 1314; G. J. Jantz et al., Inorg. Chem., 1976, 15, 1751, 1755; R. Rahman et al., Q. Rev., 1970, 24, 208.

[22] J. R. Birk and R. K. Stennenberg, in New Uses of Sulfur, J. R. West Ed., ACS Advances in Chemistry Series No. 140, 1975; D.-G. Oei, Inorg. Chem., 1973, 12, 438, 435.

[23] A. Cisar and J. D. Corbett, Inorg. Chem., 1977, 16, 632.

[24] R. J. M. Clark and D. G. Cobbold, Inorg. Chem., 1978, 17, 3169: F. A. Cotton, J. B. Harmon and R. M. Hedges, J. Am. Chem. Soc., 1976, 98, 1417.

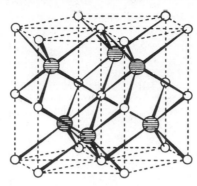

Fig. 16-4. Structure of NiAs (As atoms shaded). The Ni atom in the center of the diagram is surrounded octahedrally by six atoms and has also two near Ni neighbors, which are coplanar with four of the As atoms. [Reproduced by permission from A. F. Wells, *Structural Inorganic Chemistry*, Clarendon Press, 1945, p. 387.]

4 (p. 177). The polysulfur species give heterocyclic sulfur rings as in the red ion $[Pt(S_5)_3]^{2-}$.

Transition Metal Sulfides.[25]   Metal sulfides frequently have peculiar stoichiometries, are often nonstoichiometric phases rather than compounds in a classical sense, and are often polymorphic, and many of them are alloylike or semimetallic in behavior. Sulfides tend to be much more covalent than the corresponding oxides, with the result that quite often there is only limited and occasionally no stoichiometric analogy between the oxides and the sulfides of a given metal. Very often, indeed possibly in most cases, when there is a sulfide and an oxide of identical empirical formula they have different structures. A few examples are considered.

Several transition metal sulfides (e.g., FeS, CoS, and NiS) adopt a structure called the *nickel arsenide structure,* illustrated in Fig. 16-4. In this structure each metal atom is surrounded octahedrally by six sulfur atoms, but also is approached fairly closely by two other metal atoms. These metal-metal distances are 2.60 to 2.68 Å in FeS, CoS, and NiS, and at such distances there must be a considerable amount of metal-metal bonding, thus accounting for their alloylike or semimetallic character. Note that such a structure is not in the least likely for a predominantly ionic salt, requiring as it would the close approach of dipositive ions.

Another class of sulfides of considerable importance is the *disulfides,* represented by FeS_2, CoS_2, and others. All these contain discrete S_2 units with an S—S distance almost exactly equal to that to be expected for an S—S single bond. These assume one of two closely related structures. First there is the *pyrite structure* named after the polymorph of FeS_2 that exhibits it. This structure may be visualized as a distorted NaCl structure. The Fe atoms occupy Na positions and the S_2 groups are placed with their centers at the Cl positions but turned in such a way that they are not parallel to any of the cube axes. The *marcasite structure* is very similar but somewhat less regular.

[25]   D. J. Vaughan and J. R. Craig, *Mineral Chemistry of Metallic Sulphides,* Cambridge University Press, 1978.

FeS is a good example of a well-characterized nonstoichiometric sulfide. It has long been known that a sample with an Fe/S ratio precisely unity is rarely encountered, and in the older literature such formulas as Fe_6S_7 and $Fe_{11}S_{12}$ have been assigned to it. The iron-sulfur system assumes the nickel arsenide structure over the composition range 50–55.5 at. % of sulfur and, when the S/Fe ratio exceeds unity, some of the iron positions in the lattice are vacant in a random way. Thus the very attempt to assign stoichiometric formulas such as Fe_6S_7 is meaningless. We are dealing not with *one* compound, in the classical sense, but with a *phase* that may be perfect, that is, FeS, or may be deficient in iron. The particular specimen that happens to have the composition Fe_6S_7 is better described as $Fe_{0.858}S$.

An even more extreme example of nonstoichiometry is provided by the Co–Te (and the analogous Ni–Te) system. Here, a phase with the nickel arsenide structure is stable over the entire composition range CoTe to $CoTe_2$. It is possible to pass continuously from the former to the latter by progressive loss of Co atoms from alternate planes (see Fig. 16-4) until, at $CoTe_2$, every other plane of Co atoms present in CoTe has completely vanished.

Typical of a system in which many different phases occur (each with a small range of existence so that each may be encountered in nonstoichiometric form) is the Cr–S system, where six phases occur in the composition range $CrS_{0.95}$ to $CrS_{1.5}$.

Although there are differences, the chemistry of selenides and tellurides is generally similar to that of sulfides.

Nonmetallic Binary Sulfides. Most nonmetallic (or metalloid) elements form sulfides that if not molecular, have polymeric structures involving sulfide bridges. Thus silicon disulfide (16-VI) consists of infinite chains of SiS_4 tetrahedra sharing edges, whereas Sb_2S_3 and Bi_2S_3 are isomorphous (16-VII), forming infinite bands that are then held in parallel strips in the crystal by weak secondary bonds.

(16-VI)   (16-VII)

16-9. Sulfur-Nitrogen Compounds[26]

Unlike NO, the radical NS is unstable and detectable only in the gas phase at low pressures. However, it can be trapped as a ligand in transition metal complexes (p. 93).

There is a very extensive chemistry of compounds with S—N bonds. Among the most important and best studied are S_4N_4 and a range of compounds[27] that have

[26]   A. J. Banister and J. A. Durrant, *J. Chem. Res.* (S), **1978,** 152 (bond distances and angles); H. W. Roesky, *Angew. Chem., Int. Ed.,* 1979, **18,** 91 (cyclic SN compounds); *Chem. Ztg.,* 1974, **98,** 121; H. G. Heal, *Adv. Inorg. Chem. Radiochem.,* 1972, **15,** 375.

[27]   T. Chivers *et al., Inorg. Chem.,* 1978, **17,** 318; *J.C.S. Chem. Comm.,* **1978,** 212, 391, 642; R. Bartetzko and R. Gleiter, *Inorg. Chem.,* 1978, **17,** 998.

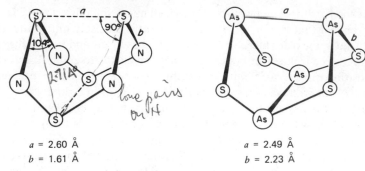

$a = 2.60$ Å $a = 2.49$ Å
$b = 1.61$ Å $b = 2.23$ Å

Fig. 16-5. Structures of N_4S_4 and As_4S_4. Both have D_{2d} symmetry.

planar S—N rings and follow a $(4n + 2)$ π-electron rule (e.g., S_2N_2, $S_4N_3^+$, $S_3N_3^-$, $S_4N_4^{2+}$, $S_5N_5^+$). The ion $S_3N_2^+$ is anomalous.

Tetrasulfurtetranitride.   This is the source material for synthesis of many S–N compounds. It is best made by action of ammonia on disulfurdichloride in CCl_4. The reaction is complicated, but two intermediates have been isolated. One is the volatile unstable gas NSCl, the other the yellow crystalline salt $S_4N_3^+$ Cl$^-$. The following known reactions constitute a plausible but speculative mechanism:

$$S_2Cl_2 + NH_3 \longrightarrow NSCl$$
$$NSCl + S_2Cl_2 = [S_3N_2Cl]^+Cl^- + SCl_2$$

$$[S_3N_2Cl]^+Cl^- \xrightarrow{\text{NH}_3, \ \text{S}_2\text{Cl}_2} [S_4N_3]^+Cl^-$$

$$[S_4N_3]^+Cl^- \xrightarrow{\text{NH}_3} S_4N_4$$

S_4N_4 forms thermochromic crystals (m.p. 185°) that are orange-yellow at 25°, red above 100°, and almost colorless at $-190°$. The compound must be handled with care, since grinding, percussion, friction, or rapid heating can cause it to explode.

The structure of S_4N_4 is a cage with a square set of N atoms and a bisphenoid of S atoms (Fig. 16-5), which is in interesting contrast to the structure of As_4S_4 (realgar) also shown in Fig. 16-5. The S···S distance (2.60 Å) is longer than the normal S—S single-bond distance ($\sim$2.08 Å) but short enough to indicate significant interaction; even the S-to-S (linked by N) distance (2.71 Å) is indicative of direct S···S interaction. A low-temperature X-ray study indicates electron density along the shorter S···S line. The N—S distances and the angles in N_4S_4 and also the $S_4N_5^-$ below, suggest the presence of lone pairs on the N and S atoms.

A satisfactory account of both the geometric and the electronic structures is provided by MO theory.[28] There are long single S—S bonds, lone pairs on the N atoms, and some electrons delocalized over the cyclic $(NS)_4$ framework.

Some reactions of S_4N_4 are given in Fig. 16-6. In addition to its conversion to S_2N_2 and (SN), discussed below, it undergoes two main types of reaction: (*a*) reactions in which the S—N ring is preserved as in adducts of BF_3 or $SbCl_5$ [29] or in the reduction to $S_4N_4H_2$; and (*b*) reactions in which ring cleavage occurs with

28   R. D. Harcourt, *J. Inorg. Nucl. Chem.*, 1977, **39**, 237.
29   U. Thewalt, *Angew. Chem., Int. Ed.*, 1976, **15**, 765.

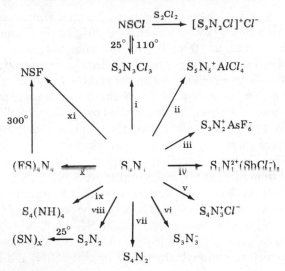

Fig. 16-6. Some reactions of S_4N_4. Note that S_4N_4 forms complexes with transition metals. See G. G. Alenge and A. J. Banister, *J. Inorg. Nuclear Chem.*, 1978, **40**, 203; I. S. Butler and T. Sawai, *Can. J. Chem.*, 1977, **55**, 3838.

reorganization to form other S—N rings. These products may be cationic[30] as when S_4N_4 reacts with Lewis acids like SbF_5; examples are the ions $S_3N_2^+$ and $S_4N_4^{2+}$. They may also be anionic, as in the reaction of S_4N_4 and N_3^- (see below).

The Disulfur Nitride Ion; Disulfur Dinitride. Although the NO_2^+ ion has long been known, only recently has the NS_2^+ ion been obtained[31] by the reaction in liquid SO_2:

$$S_7NH \xrightarrow{SbCl_5} [S_2N]^+ SbCl_6^-$$

The colorless crystalline S_2N_2 is made by pumping S_4N_4 vapor through silver wool.[32] It has a nearly square structure (16-VIII). Like S_4N_4 it can explode and is also thermally polymerized to $(SN)_x$ probably via a radical intermediate S—N—S—N

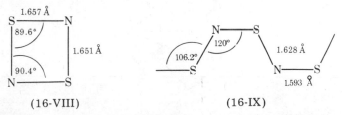

(16-VIII)   (16-IX)

Polythiazyl.[33] A polymer $(SN)_x$ has been known since 1910, but only recently,

30 R. J. Gillespie *et al.*, *J.C.S. Chem. Comm.*, **1977**, 253; *Can. J. Chem.*, 1975, **53**, 3147.
31 R. J. Gillespie *et al.*, *Inorg. Chem.*, 1978, **17**, 2875.
32 R. D. Smith, *J.C.S. Dalton*, **1979**, 478.
33 A. G. McDiarmid *et al.*, *J. Am. Chem. Soc.*, 1976, **98**, 3844; R. H. Baughman *et al.*, *J.C.S. Chem. Comm.*, **1977**, 49; R. J. Nowak *et al.*, *J.C.S. Chem. Comm.*, **1977**, 9; M. M. Labes *et al.*, *Chem. Rev.*, 1979, **79**, 1.

when it was first obtained in a pure state, has it become of importance. Very pure $(SN)_x$ can be made directly by heating S_4N_4 at 70° and condensing the vapor on glass or polymer surfaces at 10 to 30°. It can be obtained as epitaxial oriented crystalline films and shows strong optical and electrical anisotropy. The lustrous gold diamagnetic solid is a metallic conductor and is superconducting at 0.26°K. The conduction is along the axis of the S—N chains, which are parallel and zigzig (16-IX). Calculations[34] support the idea implied by the near equality of the S—N distance, namely, that delocalization along the chains is essentially complete.

A copper-colored polymer $(SNBr_{0.25})_x$ is also known.[35]

Cyclothiazenium Cations. There are several cations with SN rings.[31,36] Some can be obtained from S_4N_4 as above. Examples are $S_3N_2^+$, $S_4N_3^+$, and $S_5N_5^+$. The nomenclature is trithiadiazenium, tetrathiatriazenium, pentathiapentazenium, etc. The cyclic ions as well as S_2N_2 and S_4N_2 can be regarded as quasi-aromatic systems. Thus $S_5N_5^+$ has fourteen π electrons, $S_4N_3^+$, ten, S_2N_2, six, etc.

With $AlCl_3$ in $SOCl_2$, N_4S_4 gives $[S_5N_5]^+AlCl_4^-$, and oxidation of S_4N_4 with AsF_5 or trifluoromethanesulfonylanhydride[37] gives the $S_3N_2^+$ radical ion:

$$S_4N_4 \;+\; (CF_3SO_2)_2O \;\longrightarrow\; S_3N_2^+SO_3CF_3^- \;+\; \underset{\text{(ring structure with NSO}_2\text{CF}_3)}{\boxed{}}$$

The action of Cl_2 or SO_2Cl_2 on S_4N_4 or $S_3N_2Cl^+$ Cl^- gives the cyclic trimer $(SNCl)_3$, trichlorocyclotrithiazene.[38] This compound can be depolymerized by heat to SNCl, a colorless gas that repolymerizes at 25°. Finally the very stable tetrathiazyltriazene ion $S_4N_3^+$ is made by action of dry HCl on S_4N_4 in CCl_4; it has a planar structure.

Cyclothiazenium Anions. The $S_4N_5^-$ ion is made by the unusual reaction[39]:

$$Me_3Si—N{=}S{=}N—SiMe_3 \xrightarrow{\text{MeOH}} [NH_4]^+[S_4N_5]^-$$

It has the S_4N_4 structure but with an additional S—N—S bridge (16-X).

(16-X)

[34] M. Whangbo, R. Hoffmann, and R. B. Woodward, *Proc. Roy. Soc. Ser. A.*, 1979, **366**, 23.
[35] M. Akta *et al.*, *J.C.S. Chem. Comm.*, **1977**, 846.
[36] A. J. Banister *et al.*, *J.C.S. Dalton*, **1976**, 928.
[37] H. W. Roesky and A. Hamaza, *Angew. Chem., Int. Ed.*, 1976, **15**, 226.
[38] G. G. Alange *et al.*, *Inorg. Nucl. Chem. Lett.*, 1979, **15**, 175.
[39] W. Flues *et al.*, *Angew Chem., Int. Ed.*, 1976, **15**, 379.

Interaction of S_4N_4 with N_3^- gives either $S_4N_5^-$ or $S_3N_3^-$, depending on whether the Li and Na, or the tetralkylammonium azides, respectively, are used.[40] $K^+S_3N_3^-$ is the major product of the interaction of $S_4N_4H_4$ and KH or the reduction of S_4N_4 by K in 1,2-dimethyoxyethane.

Other S–N Compounds. In addition to S_4N_4, the reaction of S_2Cl_2 with NH_3 at 0° gives the *imide* S_7NH. However it is best made by action of sodium azide on S_8 in dimethylformamide, followed by hydrolysis of the resulting blue solution.

S_7NH is an acid, pK ~5 (i.e., 10^{34} difference from NH_3, pK = 39!). In hexamethylphosphoramide S_7NH reacts to give S_7N^-, which is in equilibrium with NS_4^-.

$$S_7N^- \rightleftharpoons NS_4^- + \tfrac{3}{8} S_8 \qquad K = 0.13$$

The NS_4^- ion has an infrared spectrum similar to that of CS_4^{2-} and is probably[41a] $S{=}N(S)(S_2)^-$.

On heating the mercury salt of S_7NH, the compound S_4N_2 is obtained[41b]:

$$Hg(S_7N)_2 = HgS + S_4N_2 + \tfrac{9}{8} S_8$$

There are other imides. $S_4(NH)_4$ can be made by reduction of S_4N_4 with $SnCl_2$ in ethanol-benzene. It is a white solid and gives complexes with various metal ions.[41c] Reduction of S_4N_4 with hydrazine gives a mixture S_{8-n} $(NH)_n$ and some $S_{11}NH$.

The entire series of compounds S_7NR to $S_4(NR)_4$, where R can be H or CH_3, is known. All have puckered eight-membered rings like S_8. There may be some small amounts of N → S π bonding, since S—N distances are somewhat ($\sim$0.06 Å) shorter than S—N single bonds and the configuration at N is nearly planar (mean angle *ca.* 119.0°).[42]

There is an extensive chemistry of compounds with S or N bonds to carbon and other elements. Thus interaction of S_4N_4 or S_4N_2 with olefins gives ring compounds with S–C bonds.[43] *Sulfaminic chloride* ($S_3N_3O_3Cl_3$) has a slightly puckered NS ring with each S bound to O and to an axial chlorine. The compound $S_8N_4(SO_2CF_3)_4$ has a twelve-membered ring.[44]

16-10. Halides

The halides are listed in Table 16-3.

Sulfur Fluorides. Some reactions of S–F compounds are shown in Fig. 16-7. Direct fluorination of S_8 yields mainly SF_6 and traces of SF_4 and S_2F_{10}.

Lower Fluorides.[45] The compounds SF_2 and S_2F_4 (believed to be $F_3S{\cdot}SF$) are

40   T. Chivers et al., Inorg. Chem., 1978, **17**, 3668, 318; J. Am. Chem. Soc., 1979, **101**, 4517.
41a T. Chivers and I. Drummond, Inorg. Chem., 1974, **13**, 1222.
41b H. G. Heal and R. J. Ramsey, J. Inorg. Nucl. Chem., 1975, **37**, 286.
41c S. N. Nabi, J. C. S. Dalton, **1977**, 1152.
42   See A. L. MacDonald and J. Trotter, Can. J. Chem., 1973, **51**, 2504.
43   R. R. Adkins and A. G. Turner, Inorg. Chim. Acta., 1977, **25**, 233.
44   B. Krebs et al., Angew. Chem., Int. Ed., 1978, **17**, 778.
45   F. Steel, Adv. Inorg. Chem. Radiochem., 1974, **16**, 297.

TABLE 16-3

The Group VI Binary Halides

(m = m.p.; b = b.p.; d = decomposes; subl = sublimes; all in °C)

Fluorides	Chlorides	Bromides	Iodides
Sulfur			
S_2F_2,[a] m − 165, b − 10.6	S_2Cl_2,[b] m − 80, b 138	S_2Br_2,[b] m − 46, d 90	S_2I_2, d − 30[c]
$[SF_2]$[d]	SCl_2, m − 78, b 59		
S_2F_4			
SF_4, m − 121, b − 40	SCl_4, d − 31		
SF_6, subl − 65, m − 51			
S_2F_{10}, m − 53, b 29			
Selenium			
SeF_4, m − 10, b 106	Se_2Cl_2	Se_2Br_2, d in vapor	
SeF_6, sub − 47, m − 35	$SeCl_2$, d in vapor	$SeBr_2$, d in vapor	
	$SeCl_4$, subl 191	$SeBr_4$	
Tellurium[e]			
TeF_4, m 130	Te_4Cl_{16} m 223, b 390	Te_4Br_{16}, m 388, b 414 d in vapor	Te_4I_{16}, m 280 d 100
TeF_6, subl − 39, m − 38			

[a] Isomeric mixture of FSSF, m − 133°, and F₂SS, b − 10.6°.

[b] Also the dichlorosulfanes S_nCl_2, 2 < n < 100(?) (see, e.g., F. Fetier and M. Kulus, *Z. Anorg. Allg. Chem.*, 1969, **364**, 241) and dibromosulfanes S_nBr_2, n > 2.

[c] D. K. Padma, *J. Indian Chem. Soc.*, 1974, **12**, 417.

[d] Detected by microwave spectroscopy among gaseous products of radiofrequency discharge in SF_6. Molecular parameters (S—F = 1.589 Å, ∠FSF = 98° 16′; μ = 1.05 D), but not bulk properties, are known (D. R. Johnson and F. X. Powell, *Science*, 1969, **164**, 950).

[e] Subhalides such as Te_2Br and TeI can be obtained by interaction of TeX_4 with SnX_2 (R. Kniep and D. Katryniok, *J.C.S. Dalton*, **1977**, 2048).

obtained by action of SCl_2 vapor at low pressure with KF or HgF_2 at 150°. Action of S_8 or SCl_2 on AgF, HgF_2, or KF at 120 to 160° gives FSSF, SSF_2, and SF_4.

SF_2 is stable only in highly dilute gas streams, but S_2F_4 is stable to −75°. Of the two isomers of S_2F_2, $F_2S{=}S$ is stable to 250°, but traces of HF or BF_3 catalyze its conversion to SF_4; FSSF can be handled only in the gas phase at low temperatures.

Sulfur Tetrafluoride. SF_4 is best made by reaction of SCl_2 with NaF in acetonitrile at 70 to 80°. SF_4 is an extremely reactive substance, instantly hydrolyzed by water to SO_2 and HF, but its fluorinating action is remarkably selective. It will convert C=O and P=O groups smoothly into CF_2 and PF_2, and COOH and P(O)OH groups into CF_3 and PF_3 groups, without attack on most other functional or reactive groups that may be present. Compounds of the type $ROSF_3$, which may be intermediates in the reaction with keto groups, have been prepared. SF_4 is also quite useful for converting metal oxides into fluorides, which are (usually) in the same oxidation state.

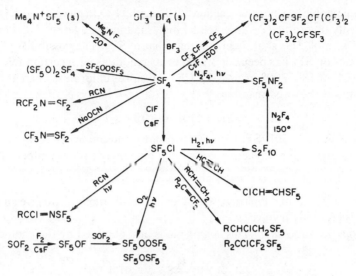

Fig. 16-7. Some reactions of sulfur-fluorine compounds.

Aryl-substituted fluorides can be readily obtained by the reaction

$$(C_6H_5)_2S_2 + 6AgF_2 \rightarrow 2C_6H_5SF_3 + 6AgF$$

which is carried out in trichloro- or trifluoromethane. The arylsulfur trifluorides are more convenient laboratory fluorinating agents than SF_4 in that they do not require pressure above atmospheric. The structure of SF_4 and of substituted derivatives RSF_3 is that of a trigonal bipyramid with an equatorial position occupied by the lone pair.

Sulfur Hexafluoride. The hexafluoride is normally very resistant to attack, and extreme conditions are often required. Thus SF_6 resists molten KOH and steam at 500°. It reacts with O_2 when a platinum wire is exploded electrically and with some red-hot metals.

SF_6 may be reduced by sodium in liquid ammonia or slowly by $LiAlH_4$ in either.[46] It also reacts with $AlCl_3$ at 200° to give sulfur chloride, with SO_3 at 250° to give SO_2F_2, and with C or CS_2 at 500° and 4000 atm.[47] Because of its inertness and high dielectric strength and molecular weight, it is used as a gaseous insulator in high-voltage generators and other electrical equipment.

SF_6 and its substituted derivatives (see below) have, or may be presumed to have, octahedrally bonded sulfur. In SF_6 the S—F bonds are about 0.2 Å shorter than expected for S–F single bonds. The low reactivity, particularly toward hydrolysis, which contrasts with the very high reactivity of SF_4, is presumably due to a combination of factors including high S—F bond strength and the coordinate saturation and steric hindrance of sulfur, augmented in the case of SF_6 by the lack of polarity of the molecule. The low reactivity is mainly due to *kinetic factors,* not to ther-

[46] D. K. Padma *et al., J. Fluorine Chem.,* 1972–1973, **2**, 113.
[47] A. P. Hagen and B. W. Calloway, *Inorg. Chem.,* 1975, **14**, 2825.

modynamic stability, since the reaction of SF_6 with H_2O to give SO_3 and HF would be decidedly favorable ($\Delta F = -460$ kJ mol^{-1}), and the average bond energy in SF_4 (326 kJ mol^{-1}) is slightly higher than that of SF_6. The possibility of electrophilic attack on SF_6 has been confirmed by its reactions with certain Lewis acids. Thus Al_2Cl_6 at 180–200° gives AlF_3, Cl_2, and sulfur chlorides, whereas the thermodynamically allowed reaction

$$SF_6 + 2SO_3 \rightarrow 3SO_2F_2$$

proceeds slowly at 250°. SF_6 also reacts rapidly and quantitatively with sodium in ethylene glycol dimethyl ether containing biphenyl at room temperature:

$$8Na + SF_6 \rightarrow Na_2S + 6NaF$$

Electron transfer from a biphenyl radical ion to an SF_6 molecule to give an unstable SF_6^- ion is probably involved.

Disulfur Decafluoride. This compound is best obtained by the photochemical reaction

$$2SF_5Cl + H_2 \overset{h\nu}{\rightleftharpoons} S_2F_{10} + 2HCl$$

It is extremely poisonous (the reason for which is not clear), being similar to phosgene in its physiological action. It is not dissolved or hydrolyzed by water or alkalis and is not very reactive. In S_2F_{10} each S atom is octahedral and the S–S bond is unusually long (2.21 Å vs. *ca.* 2.08 A expected for a single bond), whereas the S—F bonds are, as in SF_6, about 0.2 Å shorter than expected for an S—F single bond. At room temperature it shows scarcely any chemical reactivity, though it oxidizes the iodide in an acetone solution of KI. At elevated temperatures, however, it is a powerful oxidizing agent, generally causing destructive oxidation and fluorination, presumably owing to initial breakdown to free radicals:

$$S_2F_{10} \rightarrow 2SF_5\cdot$$

$$SF_5\cdot \rightarrow SF_4 + F\cdot$$

Substituted Sulfur Fluorides. There is an extensive chemistry of substituted sulfur fluorides of the types RSF_3 and RSF_5; examples of the former were mentioned above. The SF_5 derivatives bear considerable resemblance to CF_3 derivatives, with the principal difference that in reactions with organometallic compounds the SF_5 group is fairly readily reduced, whereas the CF_3 group is not.

The mixed halide SF_5Cl is an important intermediate (Fig. 16-7). Although it can be made by interaction of S_2F_{10} with Cl_2 at 200–250° it is best made by the CsF-catalyzed reaction:

$$SF_4 + ClF \xrightarrow{25°,\ 1\ hr,\ CsF} SF_5Cl$$

A probable intermediate is the salt $CsSF_5$, which dissociates significantly above 150°:

$$CsF + SF_4 \underset{150°}{\overset{100°}{\rightleftharpoons}} CsSF_5$$

SF_5Cl is a colorless gas (b.p. $-15.1°$, m.p. $-64°$), which is more reactive than SF_6, being readily attacked by OH^- and other nucleophiles,[48] though it is inert to acids. Its hydrolysis and its powerful oxidizing action toward many organic substances are consistent with the charge distribution $F_5S^{\delta-} - Cl^{\delta+}$. Its radical reactions with olefins and fluoroolefins resemble those of CF_3I.

The very reactive yellow *pentafluorosulfur hypofluorite* is one of the few known hypofluorites; it is obtained by the catalytic reaction

$$SOF_2 + 2F_2 \xrightarrow{\text{CsF, 25°}} SF_5OF$$

In the absence of CsF, SOF_4 is obtained and this reacts separately with CsF to give SF_5OF.

The hydroxide $SF_5OH^{[49]}$ and the hydroperoxide, $SF_5OOH^{[50]}$ are also known; the former gives the anion SF_5O^-.

Finally the carbon compounds F_5SCH_3 and $F_4S{=}CH_2$ can be made by the sequence

$$SF_5Cl \xrightarrow{CH_2=C=O} SF_5CH_2COCl \xrightarrow{H_2O} SF_5CH_2CO_2H$$

$$SF_5CH_2CO_2Ag \xrightarrow[-CO_2]{Br_2} SF_5CH_2Br \xrightarrow{Zn/HCl} SF_5CH_3$$

$$\Big\downarrow BuLi$$

$$SF_5CH_2Li \xrightarrow[-LiF]{-70°} F_4S{=}CH_2$$

Both are quite stable gases. F_4SCH_2 is *not* an ylid; there is no free rotation about the S—C bond, and this and the low reactivity are consistent with an S=C bond.[51]

Sulfur Chlorides. The chlorination of molten sulfur gives S_2Cl_2, an orange liquid of revolting smell. By using an excess of chlorine and traces of $FeCl_3$, $SnCl_4$, I_2, etc., as catalyst at room temperature, an equilibrium mixture containing *ca.* 85% of SCl_2 is obtained. The dichloride readily dissociates within a few hours:

$$2SCl_2 \rightleftharpoons S_2Cl_2 + Cl_2$$

but it can be obtained pure as a dark red liquid by fractional distillation in presence of some PCl_5, small amounts of which will stabilize SCl_2 for some weeks.

Sulfur chlorides are used as a solvent for sulfur (giving dichlorosulfanes up to about $S_{100}Cl_2$), in the vulcanization of rubber, as chlorinating agents, and as intermediates. Specific higher chlorosulfanes can be obtained by reactions such as

$$2SCl_2 + H_2S_4 \xrightarrow{-80°} S_6Cl_2 + 2HCl$$

The sulfur chlorides are readily hydrolyzed by water. In the vapor S_2Cl_2 [52a] has a Cl—S—S—Cl chain with C_2 symmetry (S—S = 1.95 Å, S—Cl = 2.05 Å),

48   T. Kitazume and J. M. Shreeve, *J. Am. Chem. Soc.,* 1977, **99**, 3690.
49   D. D. DesMarteau, *J. Am. Chem. Soc.,* 1972, **94**, 8933; K. Seppelt, *Z. Anorg. Allg. Chem.,* 1977, **428**, 35.
50   K. Seppelt, *Angew. Chem., Int. Ed.,* 1976, **15**, 44.
51   G. Kleeman and K. Seppelt, *Angew. Chem., Int. Ed.,* 1978, **17**, 516.
52a C. J. Marsden *et al., J.C.S. Chem. Comm.,* **1979**, 399.

whereas SCl_2 has a C_{2v} structure with S—Cl = 2.014 (5) Å and /Cl—S—Cl = 102.8(2)°.[52b]

Action of Cl_2 on sulfur chlorides at −80° gives SCl_4 as yellow crystals. It dissociates above −31°; it may be $SCl_3^+ Cl^-$.

Sulfur Iodides. No stable compounds with S—I bonds are known, except for S_7I^+, which is obtained by interaction of S_8 with I_2 and SbF_5 as the SbF_6^- salt. The ring has a chair conformation similar to that in S_7.[53]

Selenium and Tellurium Halides

Tetrafluorides. SeF_4 is similar to SF_4 and, being a liquid, (m.p. −9.5°; b.p. 106°) and easier to handle than SF_4, has some advantage as a fluorinating agent.[54] It and TeF_4 are best made by the reaction

$$EO_2 + 2SF_4 \rightarrow EF_4 + 2SOF_2 \qquad E = Se, Te$$

TeF_4, a colorless crystalline solid (m.p. 130°), has a chain structure with distorted square pyramidal TeF_4 units linked by *cis*-Te—F—Te single bridges; it can be regarded as $TeF_3^+F^-$ with strong interaction between the ions (cf. $TeCl_4$).

Hexafluorides. These appear to be somewhat more reactive than SF_6. SeF_6, which is very toxic and is hydrolyzed in the respiratory tract, appears to be unreactive to water at 25°, but TeF_6 reacts slowly,[55] giving $Te(OH)_6$. Hydroxofluorides, $Te(OH)_{6-n}F_n$, are intermediates:

$$Te(OH)_6 + nHF \rightleftharpoons Te(OH)_{6-n}F_n + nH_2O$$

Both Se and Te give chlorides (MF_5Cl) and hydroxides (MF_5OH); the latter are acidic, forming salts of the ions MF_5O^-.[56]

Chlorides and Bromides. These are similar to the bromides of sulfur but are thermally more stable, though readily hydrolyzed. Only $TeCl_4$ is stable in the vapor, $SeCl_4$ dissociating to $SeCl_2$ and Cl_2; $SeBr_4$ dissociates even in solution. In the vapor[57] $TeCl_4$ is ψ-*tbp* (C_{2v}), but in the crystal it has a cubanelike structure Te_4Cl_{16} (16-XI).

(16-XI)

[52b] J. T. Murray *et al.*, *J. Chem. Phys.*, 1976, **65**, 985.
[53] J. Passmore *et al.*, *J.C.S. Chem. Comm.*, **1976**, 689.
[54] G. A. Olah *et al.*, *J. Am. Chem. Soc.*, 1974, **96**, 925.
[55] U. Elgad and H. Selig, *Inorg. Chem.*, 1975, **14**, 140; G. W. Fraser and G. D. Meikle, *J.C.S. Chem. Comm.*, **1974**, 624.
[56] E. Mayer and F. Sladky, *Inorg. Chem.*, 1975, **14**, 589.
[57] I. R. Beattie *et al.*, *J.C.S. Dalton*, **1974**, 1747.

The halides Se_4Cl_{16} and Te_4Br_{16} are similar, but the iodide Te_4I_{16} differs in having a structure like that of titanium alcoxides (p. 161) and containing TeI_6 octahedra.[58]

Tellurium Subhalides.[59] The halides $TeCl_2$, Te_2Br, and β-TeI have structures with chains of Te atoms, some of which bear halogens; α-TeI is a molecule, Te_4I_4.

Polonium Halides. Polonium halides are similar to those of tellurium, being volatile above 150° and soluble in organic solvents. They are readily hydrolyzed and form complexes, for example, $Na_2[PoX_6]$, isomorphous with those of tellurium. There is tracer evidence for the existence of a volatile polonium fluoride. The metal is also soluble in hydrofluoric acid, and complex fluorides exist.

16-11. Halide and Other Complexes of Sulfur, Selenium, and Tellurium

1. *Hexavalent Complexes.* These are mainly salts[60] of TeF_7^- and TeF_8^{2-}, the adduct $TeF_6(NMe_3)_2$ and various alcoxides formed by interaction of TeF_6 with alcohols.[61]

2. *Tetravalent Complexes.* Since the compounds MX_4 have a lone pair, they could be expected to act as donors, and since they are only four-coordinate, also to act as Lewis acids.

Fluorides. These are weak acids and give only the ions MF_5^-. Those of Te are readily obtained by dissolving TeO_2 in 40% HF.

With strong Lewis acids like BF_3 or SbF_5, usually 1:1 complexes are formed. In the crystal these adducts may be described as containing MF_3^+ ions (e.g., $SF_3^+BF_4^-$ [62]), but there is clearly secondary bonding to the anions, and for SeF_4 there appear to be cations[63] such as

$$[F_3Se-F-\overset{\displaystyle F}{\underset{\displaystyle F}{\overset{|}{\underset{|}{B}}}}-F-SeF_3]^+$$

In presence of strong bases such as tertiary amines and in polar solvents, solvated ions $[TeF_3L_2]^+$ are formed.

Chlorides and Bromides. These act as Lewis acids normally forming octahedral anions (MX_6^{2-}), which despite the presence of a putative lone pair have O_h symmetry (see discussion, p. 206).[64] K_2SeCl_6 is easily isolated by saturating a KCl solution of H_2SeO_3 with HCl gas.

58   V. Paulat and B. Krebs, *Angew. Chem., Int. Ed.,* 1976, **15**, 39.
59   M. Takeda and N. N. Greenwood, *J.C.S. Dalton,* **1976**, 631; P. Klaeboe *et al., Acta Chem. Scand.,* 1978, **32A**, 565.
60   H. Selig, S. Sarig, and S. Abramowitz, *Inorg. Chem.,* 1974, **13**, 1500.
61   G. W. Fraser and G. D. Meikle, *J.C.S. Dalton,* **1975**, 1033.
62   D. D. Gibler *et al., Inorg. Chem.,* 1972, **11**, 2325.
63   M. Brownstein and R. J. Gillespie, *J.C.S. Dalton,* **1973**, 67.
64   R. J. H. Clark and M. L. Duarte, *J.C.S. Dalton,* **1976**, 2081.

Tellurium forms an unusual anion $Te_3Cl_{13}^-$ in which one $TeCl_3^+$ corner group from Te_4Cl_{16} (16-XI) is removed by $Ph_3C^+Cl^-$.[65a]

Interaction with Lewis acids again gives cations[65b] as in $MCl_3^+AlCl_4^-$, which may be solvated in polar media to give, e.g., $[MCl_3(MeCN)_2]^+$. Similar cations may be formed from substituted halides like Me_3TeCl.

A few neutral complexes like $trans$-py_2TeCl_4 are also known for both Se and Te.

Other Tetravalent and Divalent Complexes.   There are rather few but some are known, an example being the catecholate[66] (16-XII).

(16-XII)

The three elements form a variety of complexes, especially with sulfur ligands such as dithiocarbamate or dithiophosphinate.[67] Although usually planar, most of these compounds are quite distorted, with short and long bonds.

Tellurium forms a number of complex halides in the II as well as the IV state, some of the best known being those with thiourea (tu) or substituted thioureas as ligands. The red Te^{IV} compounds are made by treating TeO_2 in concentrated hydrochloric acid solution with, e.g., tetramethylthiourea (Me_4tu):

$$TeO_2 + 4HCl + 2Me_4tu \rightarrow trans\text{-}Te(Me_4tu)_2Cl_4 + 2H_2O$$

The structures of these compounds are octahedral with trans sulfur atoms showing no evidence (cf. TeX_6^{2-}) of stereochemical influence of the lone pair.

This ligand can further act as a reducing agent in methanolic $4M$ HCl.

$$Te(Me_4tu)_2Cl_4 \xrightarrow{heat} Te^{II}(Me_4tu)Cl_2 + (Me_4tu)^{2+} + 2Cl^-$$

$$Te^{II}(Me_4tu)Cl_2 + Me_4tu \xrightarrow[\text{heat}]{MeOH} Te^{II}(Me_4tu)_2Cl_2$$
$$\text{Red} \qquad\qquad\qquad \text{Yellow}$$

In $TeCl_2(ethylenethiourea)_4 \cdot 2H_2O$ there is a $[TeS_4]^{2+}$ square,[68] but for $TeCl_2(entu)_2$ also square, there can also be cis and trans isomers.

Finally in the ion $[Te^{II}(OSCOEt)_3]^-$ there are two chelate groups and one unidentate xanthate group.[69]

16-12.   Oxides

The principal oxides are given in Table 16-4.

Sulfur Monoxide; Disulfur Monoxide.   The extremely reactive biradical SO can

[65a] B. Krebs and V. Paulet, *Angew. Chem., Int. Ed.*, 1976, **15**, 666.

[65b] A. J. Edwards, *J.C.S. Dalton*, **1978**, 1723.

[66]   O. Lindqvist, *Acta Chem. Scand.*, 1967, **21**, 1473.

[67]   N. J. Brøndmo et al., *Acta Chem. Scand.*, 1975, **A29**, 93; R. O. Gould et al., *J.C.S. Dalton*, **1976**, 908.

[68]   R. C. Elder et al., *Inorg. Chem.*, 1977, **16**, 2700.

[69]   B. F. Hoskins and C. D. Pannan, *Aust. J. Chem.*, 1976, **29**, 2337.

TABLE 16-1
Oxides of S, Se, Te,[a] and Po

$\begin{bmatrix} S_2O \\ SO \end{bmatrix}$			
SO_2	SeO_2	TeO_2	$PoO_2[PoO(OH)_2]$
b.p. $-10.07°$	subl $315°$	m.p. $733°$	
m.p. $-75.5°$			
SO_3	SeO_3	TeO_3	
m.p. $16.8°$ (γ)	m.p. $120°$	d $400°$	
b.p. $44.8°$		Te_2O_5	
		d $>400°$	
S_nO $n = 5-8$			
S_7O_2			

[a] See M. Takeda and N. N. Greenwood, *J.C.S. Dalton*, **1975**, 2207 and W. A. Dutton and W. C. Cooper, *Chem. Revs.*, 1966, **66**, 637. Te also gives mixed Te^{IV}, Te^{VI} oxides and mixed oxides like Na_2TeO_4 (E. Gutierrez-Rios *et al.*, *J.C.S. Dalton*, **1975**, 915, 918).

be obtained by heating thiran oxides.[70] This and the dimer $(SO)_2$ have only millisecond lifetimes.

Disulfur monoxide[71] is a colorless gas that can be obtained by heating CuO and S_8 in vacuum or by passing SO_2Cl_2 over warm Ag_2S, etc. It is more stable than SO and can be kept at low pressures for periods up to days, but it polymerizes readily even at low temperatures. It has an angular structure, SSO.

Dioxides. The dioxides are obtained by burning the elements in air, though small amounts of SO_3 also form in the burning of sulfur. Sulfur dioxide is also produced when many sulfides are roasted in air and when sulfur-containing fuels such as oils and coals are burned. It presents a major pollution and ecological problem.[72] For example, in Norway, the pH of lakes is decreasing because of the sulfuric acid formed from SO_2 pollution originating in the British Isles and southern Europe. SO_2 can be removed from flue gases by solid slurries of, say, calcium hydroxide, but there is then still an enormous sludge problem.

Selenium and tellurium dioxides are also obtained by treating the metals with hot nitric acid to form H_2SeO_3 and $2TeO_2 \cdot HNO_3$, respectively, and then heating these to drive off water or nitric acid.

The dioxides differ considerably in structure. SO_2 is a gas, SeO_2 is a white volatile solid, and TeO_2 is a nonvolatile white solid. Gaseous SO_2 and SeO_2 are bent symmetrical molecules; in each case the short S—O and Se—O bond distances imply that there is considerable multiple bonding. There may be $p\pi-p\pi$ bonding as well as $p\pi-d\pi$ bonding due to the overlap of filled $p\pi$ orbitals of oxygen with vacant

[70] B. F. Bonini *et al.*, *J.C.S. Chem. Comm.*, **1976**, 431.
[71] S.-Y. Tang and C. W. Brown, *Inorg. Chem.*, 1975, **14**, 2856.
[72] G. E. Likens, *Chem. Eng. News*, **1976**, Nov. 22, 29; C. F. Cullis, *Chem. Br.* **1978**, 384; R. B. Engdahl and H. S. Rosenberg, *Chem. Tech.*, **1978**, 118.

$d\pi$ orbitals of sulfur. SO_2 solidifies to form a molecular lattice as far as is known.

Selenium dioxide (16-XIII) forms chains in which each Se forms three primary bonds (ca. 1.78 Å) to oxygen in a flat pyramid but with three secondary bonds (2.72 Å) to oxygen atoms in another chain. Tellurium dioxide (and PoO_2) exists in two forms both related to a distorted TiO_2 structure (p. 15). The α form has four short bonds with Te at the apex of a distorted square pyramid, but there are also two longer, secondary cis bonds to oxygen (16-XIV).

(16-XIII)

(16-XIV)

Sulfur Dioxide. This is a weak reducing agent in acid solution, but a stronger one in basic solution, where the sulfite ion is formed.

Liquid SO_2 is a useful nonaqueous solvent.[73] Although it is a relatively poor ionizing medium ($\epsilon = 15.1$ at 0°), it dissolves many organic and inorganic substances and is often used as a solvent for nmr studies as well as in preparative reactions. There is little evidence for self-ionization of liquid SO_2, and the conductivity (3×10^{-8} to 2×10^{-7} $\Omega\text{-cm}^{-1}$) is mainly a reflection of the purity.

Sulfur dioxide has lone pairs and can act as a Lewis base; it can also act as a Lewis acid. With certain amines,[74] crystalline 1:1 charge-transfer complexes are formed in which electrons from nitrogen are presumably transferred to antibonding acceptor orbitals localized on sulfur. One of the most stable is $Me_3N{\cdot}SO_2$ (16-XV); here the dimensions of the SO_2 molecule appear to be unchanged by complex formation. SO_2 also forms weak complexes with halide ions in both aqueous and

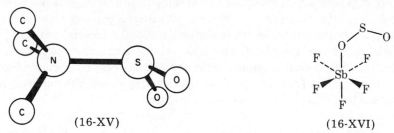

(16-XV)

(16-XVI)

nonaqueous solutions. The interaction is in the order F > Cl > Br > I.[75] SO_2 can also bond to halide ion complexes [e.g., $(R_3P)_2MePtI{\cdot}SO_2$ [76a]] and to S of thiol in

[73] N. H. Lichtin, *Prog. Phys. Org. Chem.,* 1963, **1**, 75; T. C. Waddington, in *Non-Aqueous Solvent Systems,* T. C. Waddington, Ed., Academic Press, 1965; C. C. Addison, W. Karcher, and H. Hecht, *Chemistry in Liquid Dinitrogen Tetroxide and Sulphur Dioxide,* Vol. III, *Chemistry in Non-aqueous Ionizing Solvents,* Pergamon Press, 1968; D. F. Burrow, in *The Chemistry of Non-Aqueous Solvents,* Vol. 3, J. J. Lagowski, Ed., Academic Press, 1970.

[74] R. R. Lucchese et al., *J. Am. Chem. Soc.,* 1976, **98**, 7617.

[75] P. G. Eller and G. J. Kubas, *Inorg. Chem.,* 1978, **17**, 894.

[76a] M. R. Snow and J. A. Ibers, *Inorg. Chem.,* 1973, **12**, 1221.

$(Ph_2MeP)_3CuSPh\cdot SO_2$.[76b] The ion SO_2F^- (p. 540) can be regarded as an especially stable solvate.

Although crystals are formed with quinol and some other hydrogen-bonding compounds, these are clathrates or, in the case of $SO_2\cdot 7H_2O$, a clathrate hydrate (p. 227).

Sulfur dioxide also forms complexes with a number of transition metal species (Section 4-32). In the crystalline compound $SbF_5\cdot SO_2$,[77] which is of interest because of the use of SO_2 as a solvent for superacid systems (p. 240), the SO_2 is bound as in 16-XVI.

It may be noted that in SO_2 solutions of $CH_3F + SbF_5$ there is in fact a reaction in which SO_2 is attacked[78]:

$$SO_2 + CH_3F + 2SbF_5 \rightarrow [CH_3{-}O{=}S{=}O]^+Sb_2F_{11}^-$$
$$CH_3OSO\ Sb_2F_{11} \rightleftharpoons CH_3OSOF + 2SbF_5$$
$$CH_3OSOF + CH_3F + SbF_5 + SO_2 \rightarrow (CH_3OSO)_2F^+SbF_6^-$$

Sulfur dioxide also undergoes insertion reactions with compounds with metal-carbon bonds (Section 29-9), e.g.,

$$(CH_3)_4Sn + SO_2 = (CH_3)_3SnSO_2CH_3$$

Cyclosulfur Oxides. The oxidation of cyclosulfurs with trifluoroperoxo acetic acid gives[79] oxides:

$$S_n + CF_2CO_3H \rightarrow S_nO + CF_3CO_2H \qquad n = 5\text{--}8$$
$$S_8 + 4CF_3CO_3H \rightarrow S_7O_2 + 4CF_3CO_2H + SO_2$$

Although S_5O exists only in solution, S_6O, S_7O, and S_8O are solids with S–S distances between 2.0 and 2.2 Å.

Sulfur trioxide. The only important trioxide in this group, SO_3, is obtained by reaction of sulfur dioxide with molecular oxygen, a reaction that is thermodynamically very favorable but extremely slow in absence of a catalyst. Platinum sponge, V_2O_5, and NO serve as catalysts under various conditions. SO_3 reacts vigorously with water to form sulfuric acid. Commercially, for practical reasons, SO_3 is absorbed in concentrated sulfuric acid, to give oleum (p. 245), which is then diluted. SO_3 is also used as such for preparing sulfonated oils and alkyl arenesulfonate detergents. It is also a powerful but generally indiscriminate oxidizing agent; however, it will selectively oxidize pentachlorotoluene and similar compounds to the alcohol.

The free molecule, in the gas phase, has a planar, triangular structure that may be considered to be a resonance hybrid involving $p\pi$-$p\pi$ S—O bonding, as in 16-XVII, with additional π bonding via overlap of filled oxygen $p\pi$ orbitals with empty sulfur $d\pi$ orbitals, to account for the very short S—O distance of 1.41 Å.

[76b] P. G. Eller and G. J. Kubas, *J. Am. Chem. Soc.*, 1977, **99**, 4346.

[77] D. M. Byler and D. F. Shriver, *Inorg. Chem.*, 1976, **15**, 32.

[78] P. E. Petersen et al., *J. Am. Chem. Soc.*, 1976, **98**, 2660; G. A. Olah et al., *J. Am. Chem. Soc.*, 1976, **98**, 2661; J.-Y. Calves and R. J. Gillespie, *J.C.S. Chem. Comm.*, **1976**, 506.

[79] R. Steudel et al., *Angew Chem., Int. Ed.*, 1978, **17**, 611, 134; 1977, **16**, 716; *Chem. Ber.*, 1977, **110**, 423.

(16-XVII)

In view of this affinity of S in SO_3 for electrons, it is not surprising that SO_3 functions as a fairly strong Lewis acid toward the bases that it does not preferentially oxidize. Thus the trioxide gives crystalline complexes with pyridine, trimethylamine, or dioxane, which can be used, like SO_3 itself, as sulfonating agents for organic compounds.

The structure of solid SO_3 is complex. At least three well-defined phases are known. There is first γ-SO_3, formed by condensation of vapors at $-80°$ or below. This icelike solid contains cyclic trimers with structure 16-XVIII.

A more stable, asbestos-like phase, β-SO_3, has infinite helical chains of linked SO_4 tetrahedra (16-XIX), and the most stable form α-SO_3, which also has an asbestos-like appearance, presumably has similar chains cross-linked into layers.

(16-XVIII)   (16-XIX)

Liquid γ-SO_3, which is a monomer-trimer mixture, can be stabilized by the addition of boric acid. In the pure state it is readily polymerized by traces of water.

Selenium Trioxide is made by dehydration of H_2SeO_4 by P_2O_5 at 150 to 160°; it is a strong oxidant and is rapidly rehydrated by water. SeO_3 dissolves in liquid HF to give *fluoroselenic acid* ($FSeO_3H$; cf. FSO_3H, p. 246), a viscous fuming liquid.

Tellurium trioxide is made by dehydration of $Te(OH)_6$. This orange compound reacts only slowly with water but dissolves rapidly in bases to give tellurates.

OXO ACIDS

16-12.   General Remarks

The oxo acids of sulfur are by far the most important and the most numerous. Some are not known as such, but like phosphorus oxo acids occur only as anions and salts. In Table 16-5 the various oxo acids of sulfur are grouped according to structural

TABLE 16-5
Principal Oxo Acids of Sulfur

Name	Formula	Structure[a]
Acids Containing One Sulfur Atom		
Sulfurous	H_2SO_3[b]	SO_3^{2-} (in sulfites)
Sulfuric	H_2SO_4	$O-\overset{\overset{\displaystyle O}{\|}}{\underset{\underset{\displaystyle OH}{\|}}{S}}-OH$
Acids Containing Two Sulfur Atoms		
Thiosulfuric	$H_2S_2O_3$	$HO-\overset{\overset{\displaystyle OH}{\|}}{\underset{\underset{\displaystyle O}{\|}}{S}}-S$
Dithionous	$H_2S_2O_4$[b]	$HO-\overset{\overset{\displaystyle O}{\|}}{S}-\overset{\overset{\displaystyle O}{\|}}{S}-OH$
Disulfurous	$H_2S_2O_5$[b]	$HO-\overset{\overset{\displaystyle O}{\|}}{S}-\overset{\overset{\displaystyle O}{\|}}{\underset{\underset{\displaystyle O}{\|}}{S}}-OH$
Dithionic	$H_2S_2O_6$	$HO-\overset{\overset{\displaystyle O}{\|}}{\underset{\underset{\displaystyle O}{\|}}{S}}-\overset{\overset{\displaystyle O}{\|}}{\underset{\underset{\displaystyle O}{\|}}{S}}-OH$
Disulfuric	$H_2S_2O_7$	$HO-\overset{\overset{\displaystyle O}{\|}}{\underset{\underset{\displaystyle O}{\|}}{S}}-O-\overset{\overset{\displaystyle O}{\|}}{\underset{\underset{\displaystyle O}{\|}}{S}}-OH$
Acids Containing Three or More Sulfur Atoms		
Polythionic	$H_2S_{n+2}O_6$	$HO-\overset{\overset{\displaystyle O}{\|}}{\underset{\underset{\displaystyle O}{\|}}{S}}-S_n-\overset{\overset{\displaystyle O}{\|}}{\underset{\underset{\displaystyle O}{\|}}{S}}-OH$
Peroxo Acids		
Peroxomonosulfuric	H_2SO_5	$HOO-\overset{\overset{\displaystyle O}{\|}}{\underset{\underset{\displaystyle O}{\|}}{S}}-OH$
Peroxodisulfuric	$H_2S_2O_8$	$HO-\overset{\overset{\displaystyle O}{\|}}{\underset{\underset{\displaystyle O}{\|}}{S}}-O-O-\overset{\overset{\displaystyle O}{\|}}{\underset{\underset{\displaystyle O}{\|}}{S}}-OH$

[a] In most cases the structure given is inferred from the structure of anions in salts of the acid.
[b] Free acid unknown.

type. This classification is to some extent arbitrary, but it corresponds with the order in which we discuss these acids. None of the oxo acids in which there are S—S bonds has any known Se or Te analogue.

16-14. Sulfurous Acid[80]

SO_2 is quite soluble in water; such solutions, which possess acidic properties, have long been referred to as solutions of sulfurous acid (H_2SO_3). However H_2SO_3 either is not present or is present only in infinitesimal quantities in such solutions. The so-called hydrate $H_2SO_3 \cdot \sim 6H_2O$ is the gas hydrate $SO_2 \cdot \sim 7H_2O$. The equilibria in aqueous solutions of SO_2 are best represented as

$$SO_2 + xH_2O = SO_2 \cdot xH_2O \text{ (hydrated } SO_2)$$
$$[SO_2 \cdot xH_2O = H_2SO_3 \qquad K \lll 1]$$
$$SO_2 \cdot xH_2O = HSO_3^- \text{ (aq)} + H_3O^+ + (x - 2)H_2O$$

and the first acid dissociation constant for "sulfurous acid" is properly defined as follows:

$$K_1 = \frac{[HSO_3^-][H^+]}{[\text{total dissolved } SO_2] - [HSO_3^-] - [SO_3^{2-}]} = 1.3 \times 10^{-2}$$

Although sulfurous acid does not exist, there are two series of salts, the *hydrogen sulfites* (HSO_3^-) and *sulfites* (SO_3^{2-}).

The lighter alkali ions give sulfites or disulfites (see below), and it appears[81] that larger ions such as Rb^+, Cs^+, or R_4N^+ are necessary to stabilize HSO_3^-. Both the $SO_2(OH)^-$ and SO_3^{2-} ions are pyramidal.

In aqueous solutions it has been suggested that the tautomers 16-XXa and 16-XXb are possible, but detailed study of the system[82a] provides no evidence for them up to 0.2M.

(16-XXa)   (16-XXb)

Heating solid bisulfites or passing SO_2 into their aqueous solutions affords *disulfites*:

$$2MHSO_3 \overset{\text{heat}}{\rightleftharpoons} M_2S_2O_5 + H_2O$$

$$HSO_3^-(aq) + SO_2 = HS_2O_5^- \text{ (aq)}$$

Whereas diacids [e.g., disulfuric ($H_2S_2O_7$), discussed in Chapter 6, p. 246], usually have oxygen bridges, the disulfite ion has an S—S bonds, hence an unsymmetrical structure, $O_2S—SO_3(C_s)$. Some important reactions of sulfites are shown in Fig. 16-8.

Aqueous solutions of SO_2 and of sulfites are commonly used as reducing agents:

$$SO_4^{2-} + 4H^+ + (x - 2)H_2O + 2e = SO_2 \cdot xH_2O \qquad E^0 = 0.17 \text{ V}$$

$$SO_4^{2-} + H_2O + 2e = SO_3^{2-} + 2OH^- \qquad E^0 = -0.93 \text{ V}$$

[80]   For thermodynamic functions, see J. W. Cobble *et al.*, *Inorg. Chem.*, 1972, **11**, 1669.
[81]   R. Maylor, J. B. Gill, and D. C. Goodall, *J.C.S. Dalton*, **1972**, 2001.
[82a] E. Mayor, A. Treinin, and J. Wilf, *J. Am. Chem. Soc.*, 1972, **94**, 47.

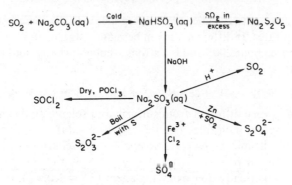

Fig. 16-8. Some reactions of sulfites.

There are tautomeric forms of diesters of sulfurous acids, *dialkyl sulfites* [OS(OR)$_2$], and *alkanesulfonic esters* [RSO$_2$(OR)], comparable to the tautomers (16-XXa) and (16-XXb) discussed earlier (p. 532) for the hydrogen sulfite ion itself.

16-15. Selenous and Tellurous Acids

SeO$_2$ dissolves in water to give solutions that contain selenous acid, OSe(OH)$_2$. Raman spectra show that it is negligibly dissociated in aqueous solution, whereas in half- and fully neutralized solutions the pyramidal ions SeO$_2$(OH)$^-$ and SeO$_3^{2-}$ are formed, salts of which can be isolated. Above *ca.* 0.1M, diselenate ions are formed:

$$2SeO(OH)_2^- = Se_2O_5^{2-} + H_2O$$

There are also ions H(SeO$_3$)$_2^{3-}$, H$_2$(SeO$_3$)$_2^{2-}$, and H$_3$(SeO$_3$)$_2^-$, as well as H$_4$(SeO$_3$)$_2$, and these probably dimerize by way of hydrogen bonded bridges Se—O—H—O—Se.[82b] Crystalline H$_2$SeO$_3$, though efflorescent, can be isolated; it has layers of pyramidal SeO$_3$ groups connected by hydrogen bonds. The acid and its salts are moderately strong oxidizing agents:

$$H_2SeO_3 + 4H^+ + 4e = Se + 3H_2O \qquad E^0 = 0.74 \text{ V}$$

and they oxidize SO$_2$, HI, H$_2$S, etc.

TeO$_2$ is virtually insoluble in water but dissolves in strong bases to give solutions from which crystalline tellurites (e.g., Na$_2$TeO$_3$) may be isolated. Acidification precipitates H$_2$TeO$_3$. In alkaline solution the main ions are probably[83] TeO$_2$(OH)$^-$ and TeO$_3^{2-}$.

16-16. Selenic and Telluric Acids (for Sulfuric Acid see p. 245)

Vigorous oxidation of selenites or fusion of selenium with potassium nitrate gives *selenic acid* (or its salts). The free acid forms colorless crystals (m.p. 57°). It is very similar to sulfuric acid in its formation of hydrates, in acid strength, and in the

[82b] A. D. Fowless and D. R. Stranks, *Inorg. Chem.*, 1977, **16**, 1271, 1282.
[83] M. R. Masson, *J. Inorg. Nucl. Chem.*, 1976, **38**, 545.

properties of its salts, most of which are isomorphous with the corresponding sulfates and hydrogen sulfates. It differs mainly in being less stable, and it evolves oxygen when heated above about 200° and is a strong, though usually not kinetically fast, oxidizing agent:

$$SeO_4^{2-} + 4H^+ + 2e = H_2SeO_3 + H_2O \qquad E^0 = 1.15 \text{ V}$$

Telluric acid is very different from sulfuric and selenic acids and has hydrogen-bonded octahedral molecules $Te(OH)_6$ in the crystal.

The acid or its salts may be prepared by oxidation of tellurium or TeO_2 by H_2O_2, Na_2O_2, CrO_3, or other powerful oxidizing agents. It is a moderately strong, but, like selenic acid, kinetically slow oxidizing agent ($E^0 = 1.02$ V). It is a very weak dibasic acid with $K_1 \approx 10^{-7}$. Tellurates of various stoichiometries are known and most, if not all, contain TeO_6 octahedra. Examples are $K[TeO(OH)_5]\cdot H_2O$, $Ag_2[TeO_2(OH)_4]$ and Hg_3TeO_6. Tellurates such as $BaTeO_4$ that can be made by heating TeO_2 and metal oxides are *not* isostructural with sulfates. $MgTeO_4$ is isostructural with $MgWO_4$ and contains again TeO_6 octahedra.[84]

16-16. Peroxo Acids

No peroxo acid containing Se or Te is known.

Peroxodisulfuric acid can be obtained from its NH_4^+ or Na^+ salts, which can be crystallized from solutions after electrolysis of the corresponding sulfates at low temperatures and high current densities. The $S_2O_8^{2-}$ ion has the structure O_3S—O—O—SO_3, with approximately tetrahedral angles about each S atom.

The peroxodisulfate ion is one of the most powerful and useful of oxidizing agents:

$$S_2O_8{}^{2-} + 2e = 2SO_4^{2-} \qquad E^0 = 2.01 \text{ V}$$

However, the reactions may be complicated mechanistically and in many of them there is good evidence for the formation of the radical ion SO_4^- by one-electron reduction:

$$S_2O_8^{2-} + e = SO_4^{2-} + SO_4^-$$

The oxidations by $S_2O_8^{2-}$ often proceed slowly but become more rapid in presence of catalysts, the silver ion being commonly used for this purpose. The precise mechanism is not quite clear, but it appears that a weak 1:1 complex is first formed between Ag^+ and $S_2O_8^{2-}$, the rapidly reacting oxidizing species being Ag^{II}.

Peroxomonosulfuric acid (Caro's acid) is obtained by hydrolysis of peroxodisulfuric acid:

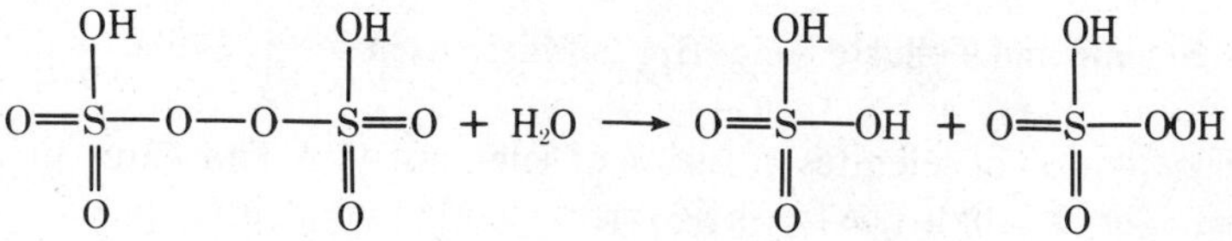

84   A. W. Sleight *et al., Inorg. Chem.,* 1972, **11**, 1157; E. Gutierrez-Rios *et al., J.C.S. Dalton,* **1978**, 915, 918.

and also by the action of concentrated hydrogen peroxide on sulfuric acid or chlorosulfuric acid:

$$H_2O_2 + H_2SO_4 \rightarrow HOOSO_2OH + H_2O$$
$$H_2O_2 + HSO_3Cl \rightarrow HOOSO_2OH + HCl$$

The salts such as $KHSO_5$ can be obtained only impure, admixed with K_2SO_4 and $KHSO_4$; aqueous solutions decompose to give mainly O_2 and SO_4^{2-} with small amounts of H_2O_2 and $S_2O_8^{2-}$.

16-17. Thiosulfuric Acid

Thiosulfates are readily obtained by reaction of H_2S with aqueous solutions of sulfites. The mechanism appears to involve the following two principal steps:

$$
\begin{aligned}
&1. \ 2HS^- + HSO_3^- \rightarrow 3S + 3OH^- \\
&2. \ 3S + 3HSO_3^- \rightarrow 3S_2O_3^{2-} + 3H^+ \\
\hline
&3. \ 2HS^- + 4HSO_3^- \rightarrow 3S_2O_3^{2-} + 3H_2O
\end{aligned}
$$

Step 2 is a reversible equilibrium, since $S_2O_3^{2-}$ decomposes in acid solution to give elemental sulfur and HSO_3^-. In addition to kinetics, evidence supporting this mechanism is provided[85] by labeling the sulfur in HS^- with ^{35}S. When the $S_2O_3^{2-}$ is hydrolyzed by acid, the elemental sulfur produced contains $\frac{2}{3}$ the percentage of ^{35}S as did the HS^- used. The ion has an SSO_3 structure, with S—S and S—O distances 2.013(3) and 1.468(4) Å that imply some S—S π bonding and considerable S—O π bonding.

16-18. Dithionous Acid

The reduction of sulfites in aqueous solutions containing an excess of SO_2, usually by zinc dust, or of SO_3 by formate in aqueous methanol, gives the dithionite ion $S_2O_4^{2-}$.[86a] Solutions of this ion are not very stable and decompose in a complex way according to the stoichiometry

$$2S_2O_4^{2-} + H_2O \rightarrow S_2O_3^{2-} + 2HSO_3^-$$

Oxygen-free dithionite solutions show a strong esr signal because of the radical ion SO_2^-, and there is the reversible dissociation

$$S_2O_4^{2-} \rightleftharpoons 2SO_2^-$$

The solutions are very rapidly oxidized by molecular O_2; initially H_2O_2 is formed, but this is slowly destroyed.[86b]

Acidification causes rapid decomposition, producing some elemental sulfur. The Zn^{2+} and Na^+ salts are commonly used as powerful and rapid reducing agents in alkaline solution:

$$2SO_3^{2-} + 2H_2O + 2e = 4OH^- + S_2O_4^{2-} \qquad E^0 = -1.12 \text{ V}$$

[85]   G. W. Heunisch, *Inorg. Chem.*, 1977, **16**, 1411.
[86a] G. Ertl *et al.*, *Angew. Chem., Int. Ed.*, 1979, **18**, 313.
[86b] C. Creutz and N. Sutin, *Inorg. Chem.*, 1974, **13**, 2041.

Fig. 16-9. Structure of the dithionite ion $S_4O_4^{2-}$ in $Na_2S_2O_4$.

Since the important reducing species is the SO_2^- radical ion, this potential is not very significant.

In the presence of 2-anthraquinonesulfonate as catalyst (Fieser's solution) aqueous $Na_2S_2O_4$ efficiently removes oxygen from inert gases.

The structure of the dithionite ion (Fig. 16-9) has several remarkable features. The oxygen atoms, which must bear considerable negative charge, are closely juxtaposed by the eclipsed (C_{2v}) configuration and by the small value of the angle α, which would be 35° for sp^3 tetrahedral hybridization at the sulfur atom. Second, the S—S distance is much longer than S—S bonds in disulfides, polysulfides, etc., which are in the range $ca.$ 2.0–2.15 Å. The long bond is believed to be due to weakening by repulsion of lone pairs on sulfur resulting from dp hybridized bonding. The weak bonding is consistent with dissociation to SO_2^-.

15-20. Dithionic Acid

Although $H_2S_2O_6$ might at first sight appear to be the simplest homologue of the polythionates ($S_nO_6^{2-}$), discussed in the next section, dithionic acid and its salts do not behave like the polythionates. Furthermore, from a structural point of view, dithionic acid is not a member of the polythionate series, since dithionates contain no sulfur atom bound only to other sulfur atoms as do $H_2S_3O_6$ and all higher homologues $H_2S_nO_6$. The dithionate ion has a D_{3d} O_3SSO_3 structure with approximately tetrahedral bond angles about each sulfur, and the S—O bond length[87] (1.45 Å; cf. 1.44 Å in SO_4^{2-}) again suggests considerable double-bond character.

Dithionate is usually obtained by oxidation of sulfite or SO_2 solutions with manganese(IV) oxide:

$$MnO_2 + 2SO_3^{2-} + 4H^+ = Mn^{2+} + S_2O_6^{2-} + 2H_2O$$

Other oxo acids of sulfur that are formed as by-products are precipitated with barium hydroxide, and $BaS_2O_6 \cdot 2H_2O$ is then crystallized. Treatment of aqueous solutions of this with sulfuric acid gives solutions of the free acid, which may be used to prepare other salts by neutralization of the appropriate bases. Dithionic acid is a moderately stable strong acid that decomposes slowly in concentrated solutions and when warmed. The ion itself is stable, and solutions of its salts may be boiled without decomposition. Although it contains sulfur in an intermediate oxidation state, it resists most oxidizing and reducing agents, presumably for kinetic reasons.

[87]   S. J. Cline et al., J.C.S. Dalton, 1977, 1662.

15-20. Polythionates[88a]

The polythionate anions have the general formula $[O_2SS_nSO_3]^{2-}$. The free acids are not stable, decomposing rapidly into S, SO_2, and sometimes SO_4^{2-}. Also, no acid salt is known. The well-established polythionate anions are those with $n = 1$ to 4. They are named according to the total number of sulfur atoms viz.,: trithionate, $S_3O_6^{2-}$; tetrathionate, $S_4O_6^{2-}$, etc. In all these anions there are sulfur chains, whose conformations are very similar to those of S_8, whereas the configurations about the end sulfur atoms —S—SO_3 are approximately tetrahedral.[88b] They can be regarded as derivatives of sulfanes, hence the name *sulfanedisulfonic acids;* e.g., tetrathionates can be called disulfanedisulfonates. In addition to these structurally well-characterized polythionates, others containing up to 20 sulfur atoms have been prepared.

Polythionates are obtained by reduction of thiosulfate solutions with SO_2 in presence of As_2O_3 and by the reaction of H_2S with an aqueous solution of SO_2, which produces a solution called Wackenroder's liquid. A general reaction said to produce polythionates up to very great chain lengths is

$$6S_2O_3^{2-} + (2n - 9)H_2S + (n - 3)SO_2 \rightarrow 2S_nO_6^{2-} + (2n - 12)H_2O + 6OH^-$$

Many polythionates are best made by selective preparations, e.g., the action of H_2O_2 on cold saturated sodium thiosulfate:

$$2S_2O_3^{2-} + 4H_2O_2 \rightarrow S_3O_6^{2-} + SO_4^{2-} + 4H_2O$$

Tetrathionates are obtained by treatment of thiosulfates with iodine in the reaction used in the volumetric determination of iodine:

$$2S_2O_3^{2-} + I_2 \rightarrow 2I^- + S_4O_6^{2-}$$

Various species containing Se and Te are also known such as $Se_nS_2O_6^{2-}$ ($2 \leqslant n \leqslant 6$), $O_3S_2SeS_2SO_3^{2-}$, and $O_3S_2TeS_2O_3^{2-}$.

16-12. Oxohalides

Compounds like SF_5OF and SF_5OOF and the ions F_5SO^- were noted earlier. We now discuss compounds with EO, EO_2, μ-O, and μ-O_2 groups. The *thionyl* and *selenyl* halides are:

SOF_4	SOF_2	$SOCl_2$	$SOBr_2$	$SOFCl$
$SeOF_4$	$SeOF_2$	$SeOCl_2$	$SeOBr_2$	

With the exception of SOF_2, which reacts only slowly with water, the compounds are rapidly, sometimes violently hydrolyzed.

The most common compound is *thionyl dichloride,* which is made by the reaction

$$SO_2 + PCl_5 = SOCl_2 + POCl_3$$

[88a] J. Janickis, *Acc. Chem. Res.,* 1969, **2**, 316.
[88b] K. Marvy, *Acta. Chem., Scand.,* 1971, **25A**, 2580; 1973, **27A**, 1705; M. B. Ferrari *et al., J.C.S. Chem. Comm.,* **1977**, 8.

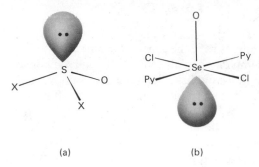

(a)   (b)

Fig. 16-10. Structures of (a) thionyl halides (SOX_2) and (b) $SeOCl_2py_2$.

It is used to prepare anhydrous metal halides from hydrated chlorides or hydroxides, since the only products of hydrolysis are gases:

$$SOCl_2 + H_2O = 2HCl + SO_2$$

Other representative syntheses are:

$$SOCl_2 + 2HF(l) = SOF_2 + 2HCl$$

$$SOCl_2 + 2HBr \overset{0°}{=} SOBr_2 + 2HCl$$

$$SeO_2 + SeCl_4 \xrightarrow{CCl_4} 2SeOCl_2$$

The dihalides are stable at ordinary temperatures but decompose on heating. They have pyramidal structures (Fig. 16-10a) with a lone pair, and the S—O bond has an order of about 2 as indicated by the bond distance of *ca.* 1.45 Å (S—O single bond = 1.7 Å). The bond order decreases $OSF_2 > OSCl_2 > OSBr_2$; that is, the most electronegative halogen encourages the greatest oxygen $\rightarrow$ sulfur $p\pi$–$d\pi$ bonding.

Thionyl and selenyl halides can act both as weak Lewis *bases* using lone pairs on oxygen, and as weak Lewis *acids* using vacant *d* orbitals. The structure of $SeOCl_2py_2$ is shown in Fig. 16-10b.

Sulfuryl halides are SO_2F_2, SO_2Cl_2, SO_2FCl, and SO_2FBr. *Sulfuryl dichloride* is formed by action of Cl_2 on SO_2 in presence of a little $FeCl_3$ catalyst. It is stable to almost 300°. It fumes in moist air and is hydrolyzed reasonably rapidly with liquid water. It can be used as a chlorinating agent. Sulfuryl difluoride, a rather inert gas, is made by fluorination of SO_2Cl_2 or by heating barium fluorosulfate

$$Ba(SO_3F)_2 \xrightarrow{500°} SO_2F_2 + BaSO_4$$

SO_2F_2 is soluble in water and can be removed from it by pumping.[89] Hydrolysis is very slow in water but rapid in base solutions:

[89]   T. M. Churchill and M. Lustig, *J. Inorg. Nucl. Chem.,* 1974, **36,** 1426.

TABLE 16-6
Structures of Some Oxofluorides

Compound	Structure	Compound	Structure
SO_3F_2	F—S—OF (with O above and O below S)	$Se_2O_2F_8{}^a$	F_4Se / O \ SeF_4 (bridged O structure)
$S_2O_5F_2$	F—S—O—S—F (each S with two O)	$S_3O_8F_2$	F—S—O—S—O—S—F (each S with two O)
S_2OF_{10}	F_5S —O— $SF_5{}^b$	$Te_5O_4F_{22}{}^c$	$F_2Te(OTeF_5)_4$

[a] Planar M_2O_2 rings in this and Te analogue. See H. Oberhammer and K. Seppelt, *Inorg. Chem.*, 1979, **18**, 2226; I. R. Beattie *et al.*, *J.C.S. Dalton*, **1976**, 1380.

[b] Se and Te analogues are also known (H. Oberhammer and K. Seppelt, *Angew. Chem., Int. Ed.*, 1978, **17**, 69).

[c] Octahedral Te (H. Pritzkow and K. Seppelt, *Inorg. Chem.*, 1977, **16**, 2685).

$$SO_2F_2 + OH^- \longrightarrow \left[HOS \begin{smallmatrix} O \\ \cdots F \\ \cdots F \\ O \end{smallmatrix} \right]^- \longrightarrow HOSF + F^-$$

It also reacts with other nucleophiles in aqueous solution, probably again by nucleophilic attack on S and displacement of F^-. Thus NH_3 gives $SO_2(NH_2)_2$, and $C_6H_5O^-$ gives $C_6H_5OSO_2F$.

The halides MO_2X_2 have distorted tetrahedral structures with the S—O multiple-bond order decreasing $SO_2F_2 > SOF_2 > SO_2Cl_2 > SOCl_2$.

Other oxohalides, which are liquids or gases, are listed in Table 16-6, along with their structures.

Peroxodisulfuryl difluoride (b.p. 67°, m.p. −55.4°) is obtained by fluorination of SO_3:

$$2SO_3 + F_2 \xrightarrow{170°, \, 3 \, hr} FS\text{—}O\text{—}O\text{—}SF$$
(each S bearing two O)

At 120° the O—O bond breaks and $S_2O_6F_2$ dissociates, forming an intense brown radical:

$$S_2O_6F_2 = 2FSO_3 \quad \Delta H_{diss} = 94 \text{ kJ mol}^{-1}$$

The compound is a strong oxidant and is a versatile reagent for the preparation of other sulfur oxo fluorides; thus F_2CO gives CF_3OOSO_2F, and Br^- and I^- give $X(SO_3F)_4^-$.

16-22.   Halooxo Acids

Fluorosulfurous acid exists only as salts which are formed by the action of SO_2 on alkali fluorides, e.g.,

$$KF + SO_2 \rightleftharpoons KSO_2F$$

The salts have a measurable dissociation pressure at normal temperatures but are useful and convenient mild fluorinating agents, e.g.,

$$(PNCl_2)_3 + 6KSO_2F \rightarrow (PNF_2)_3 + 6KCl + 6SO_2$$
$$C_6H_5COCl + KSO_2F \rightarrow C_6H_5COF + KCl + SO_2$$

As noted above (p. 528) the stability of other anions SO_2X^- is rather low; they can be regarded as weak charge-transfer complexes.[90]

Halogenosulfuric acids (FSO_3H, p. 246, $ClSO_3H$, and $BrSO_3H$) can be regarded as derived from SO_2X_2 by replacement of one halogen by OH.

Chlorosulfuric acid, a colorless fuming liquid, explosively hydrolyzed by water, forms no salts. It is made by treating SO_3 with dry HCl. Its main use is for the sulfonation of organic compounds.

Bromosulfuric acid, prepared from HBr and SO_3 in liquid SO_2 at $-35°$, decomposes at its melting point ($8°$) into Br_2, SO_2, and H_2SO_4.

[90]   M. A. Khan *et al., J.C.S. Dalton,* **1978,** 915, 918.

General References

JOURNALS AND PERIODICALS

Phosphorus, Sulfur and Related Elements, (formerly *Int. J. Sulfur Chem*) Gordon and Breach.
Organic Compounds of Sulphur, Selenium and Tellurium, Specialist Report, Chemical Society, London, Vol. 1, 1970.
Topics in Sulfur Chemistry, A. Senning, Ed., G. Thieme, Vol. 1, 1976.

TEXTS, REFERENCE BOOKS, ETC

Abrahams, S. C., *Q. Rev.,* 1956, **10,** 407. An excellent review of the stereochemistry of S, Se, Te, Po and O.
Bagnall, K. W., *The Chemistry of Se, Te and Po,* Elsevier, 1966. General inorganic chemistry.
Banks, R. E., and R. B. Haszeldine, *Adv. Inorg. Chem. Radiochem.,* 1961, **3,** 408. Polyfluoroalkyl derivatives of S and Se.
Benson, S. W., *Chem. Rev.,* 1978, **78,** 23. Thermodynamics and kinetics of sulfur-containing molecules and radicals.
Breslow, D. S., and H. Skolnik, *Multisulfur and Sulfur-Oxygen, Five- and Six-Membered Heterocycles,* Wiley, 1967.
Cady, G. H., *Adv. Inorg. Chem. Radiochem.,* 1960, **2,** 105. Fluorine-containing sulfur compounds.
Cilento, G., *Chem. Rev.,* 1960, **60,** 147. Comprehensive survey of effects of utilization of *d* orbitals in organic sulfur chemistry.
Clive, D. L. J., *Modern Organoselenium Chemistry,* Pergamon Press, 1978.
Cooper, W., Ed., *The Physics of Selenium and Tellurium,* Pergamon Press, 1969.
Cooper, W., Ed., *Tellurium,* Van Nostrand, 1971. Extraction, chemistry, and application.

Cruickshank, D. W. J., *J. Chem. Soc.*, **1961**, 5486. A detailed discussion of $d\pi-p\pi$ bonding in S—O bonds.

Foss, O., *Adv. Inorg. Chem. Radiochem.*, 1960, **2**, 237. A comprehensive account of compounds with S—S bonds.

George, J. W., *Prog. Inorg. Chem.*, 1960, **2**, 33. Halides and oxohalides of the elements of Groups Vb and VIb.

Husar, R. B., J. P. Lodge, and D. J. Moore, *Sulfur in the Atmosphere*, Pergamon Press, 1978.

Janssen, M., Ed., *Organosulfur Chemistry*, Wiley, 1967.

Karchner, J. H., *Analytical Chemistry of Sulfur and Its Compounds*, Part 1, Wiley-Interscience 1970.

Kharasch, N., and C. Y. Meyers, Eds., *The Chemistry of Organosulfur Compounds*, Pergamon Press. Several volumes (includes some inorganic topics, e.g., acids).

Liler, M., *Reaction Mechanisms in Sulfuric Acid and Other Strong Acid Solutions*, Academic Press, 1971.

Meyer, B., *Sulfur, Energy and Environment*, Elsevier, 1977.

Milligan, B., and J. M. Swan, *Rev. Pure Appl. Chem.*, 1962, **12**, 73. Salts of S-aryl and alkyl thiosulfuric esters (Bunte salts).

Nickless, G., Ed., *Inorganic Sulfur Chemistry*, Elsevier, 1968. An extensive treatise.

Oae, S., *The Organic Chemistry of Sulfur*, Plenum Press, 1977.

Parker, A. J., and N. Kharasch, *Chem. Rev.*, 1959, **59**, 583. Scission of S—S bonds.

Price, C. C., and S. Oae, *Sulfur Bonding*, Ronald Press, 1962. Chemical properties and bonding in sulfides, sulfoxides, sulfones, etc.

Quarterly Reports on Sulfur Chemistry, and *Annual Reports on Mechanism and Reactions of Sulfur Compounds*, Intra Science Research Foundation, Santa Monica, Calif.

Rao, S. R., *Xanthates and Related Compounds*, Dekker, 1971.

Reid, E. E., *Organic Chemistry of Bivalent Sulfur*, Chemical Publishing Co., 1963. Vol. V covers CS_2, thiourea, etc.

Ring, N. H., Ed., *Inorganic Polymers*, Academic Press, 1978.

Roy, A. B., and P. A. Trudinger, *The Biochemistry of Inorganic Compounds of Sulphur*, Cambridge University Press, 1970.

Schmidt, M., *Inorg. Macromol. Rev.*, Vol. 1, 1969. Sulfur-containing polymers.

Steudel, R., *Angew Chem., Int. Ed.*, 1975, **14**, 655. Properties of S—S bonds.

Sulphur Manual, Texas Gulf Sulphur Co., New York, 1959.

Taller, W. N., Ed., *Sulphur Data Book*, McGraw-Hill, 1954.

Thorn, G. D., and R. A. Ludwig, *The Dithiocarbamates and Related Compounds*, Elsevier, 1962.

Tobolsky, A. V., Ed., *The Chemistry of Sulfides*, Wiley, 1968. Symposium on various topics, mostly polymers and biological systems, but includes N—S chemistry.

van der Heijde, H. B., in *Organic Sulfur Compounds*, N. Kharasch, Ed., Pergamon Press, 1961. Inorganic sulfur acids.

West, J. R., *New Uses of Sulfur*, ACS Advances in Chemistry Series, No. 740, 1975.

Yost, D. M., and H. Russell, *Systematic Inorganic Chemistry* (*of the 5th and 6th Group Elements*), Prentice-Hall, 1946. Selected aspects of chemistry of S, Se, and Te.

CHAPTER SEVENTEEN

The Group VII Elements:
F, Cl, Br, I, At

GENERAL REMARKS

17-1. Electronic Structures and Valences

Some important properties of the Group VII elements (halogens) are given in Table 17-1. Since the atoms are only one electron short of the noble gas configuration, the elements readily form the anion X^- or a single covalent bond. Their chemistries are essentially completely nonmetallic, and in general, the properties of the elements and their compounds change progressively with increasing size. As in other groups there is a much greater change between the first-row element fluorine and the second-row element chlorine than between other pairs, but with the exception of the Li–Cs group there are closer similarities within the group than in any other in the Periodic Table.

The high reactivity of fluorine results from a combination of the low F—F bond energy and the high strength of bonds from fluorine to other atoms. The small size and high electronegativity of the F atom account for many of the other differences between fluorine and the other halogens.

The coordination number of the halogens in the -1 state is normally only 1, but exceptions are found in HXH^+ cations, in polyhalide ions such as $FClF^-$ and I_3^-, and when X^- ions occur as bridging ligands (see Section 4-37), where the coordination number is 2. There are also triply bridging X^- ions in some metal atom cluster compounds (see Section 26-5).

Positive *formal* oxidation numbers and higher coordination numbers may be assigned to the central halogen atoms in several classes of compounds such as the halogen fluorides (e.g., ClF_3, ClF_5, BrF_5, IF_7), oxo compounds (e.g., Cl_2O_7 or I_2O_5), oxofluorides (e.g., F_3BrO, FIO_3), and a few other cases such as CF_3IF_2. It should

TABLE 17-1
Some Properties of the Halogen Atoms and Molecules

Element	Atomic ground state configuration	Ionization enthalpy of atom (kJ mol^{-1})	Electron attachment enthalpy of atom (kJ mol^{-1})	Dissociation enthalpy of molecule (kJ mol^{-1})	B.p. of X$_2$ (°C)	M.p. of X$_2$ (°C)	Crystal radius X$^-$ (Å)	Covalent radius X (Å)
F	$1s^22s^22p^5$	1681	328.0	157^a	−118	−233	1.19	0.71
Cl	[Ne]$3s^23p^5$	1250	348.8	242	−34.6	−103	1.70	0.99
Br	[Ar]$3d^{10}4s^24p^5$	1139	324.6	193	58.76	−7.2	1.87	1.14
I	[Kr]$4d^{10}5s^25p^5$	1007	295.3	150	184.35	113.5	2.12	1.33
At	[Xe]$4f^{14}5d^{10}6s^26p^5$		270(±20)					—

^{a} Cf. J. Berkowitz and A. C. Wahl, *Adv. Fluorine Chem.*, 1973, **6**, 147.

be remembered that such formal oxidation numbers, though pragmatically useful in certain ways, bear no relation to actual charges.

For fluorine there is little evidence for positive behavior even in the formal sense. In oxygen fluorides the F atom is probably somewhat negative with respect to oxygen; whereas in ClF evidence from chlorine nuclear quadrupole coupling shows that the actual charge distribution involves partial positive charge on Cl.

Bond polarities in other halogen compounds indicate the importance of forms such as I^+Cl^- in ICl or I^+CN^- in ICN. In general, when a halogen atom forms a bond to another atom or group more electronegative than itself, the bond will be polar with a partial positive charge on the halogen.[1] Examples are the Cl atoms in SF_5OCl, CF_3OCl, FSO_2OCl, and O_3ClOCl, the bromine atoms in O_3ClOBr and FSO_2OBr, and the iodine atom in $I(NO_3)_3$.

Even for the heaviest member of the group, astatine, there is little evidence for any extensive "metallic" behavior.

THE ELEMENTS

None of the halogens occur in the elemental state in Nature; this is because of their high reactivity. All exist as diatomic molecules, which, being homonuclear, are without permanent electrical polarity. In the condensed phases, only weak van der Waals forces operate; hence the trend in melting and boiling points of the halogens parallels that in the noble gases. In both cases the dominating factor is the increasing magnitude of the van der Waals forces as size and polarizability of the atoms or molecules increase. The increase in color of the elements and of their covalent compounds with increasing size is due in general to a progressive shift of electronic absorption bands to longer wavelengths in the absorption spectrum.

17-2.   Fluorine

Fluorine occurs widely in Nature, notably as *fluorspar* (CaF_2), *cryolite* (Na_3AlF_6), and *fluorapatite* [$3Ca_3(PO_4)_2Ca(F,Cl)_2$]. It is more abundant (0.065%) than chlorine (0.055%) in the earth's crust.

The estimated standard potential of fluorine ($E^0 = +2.85$ V) clearly indicates why early attempts to prepare the element by electrolytic methods in aqueous solution suitable for chlorine ($E^0 = +1.36$ V) failed. The element was first isolated in 1886 by Moissan, who pioneered the chemistry of fluorine and its compounds. The colorless gas is obtained by electrolysis of HF. Although anhydrous HF is a poor conductor, the addition of anhydrous KF gives conducting solutions. The most commonly used electrolytes are KF·2–3HF, which is molten at 70 to 100°, and KF–HF, which is molten at 150 to 270°. When the melting point begins to be too high because of HF consumption, the electrolyte can be regenerated by resaturation with HF from a storage tank. There have been many designs for fluorine cells; these

[1]   J. Schack and K. O. Christe, *Isr. J. Chem.*, 1978, **17**, 20.

are constructed of steel, copper, or Monel metal, which become coated with an unreactive layer of fluoride.

Steel or copper cathodes with ungraphitized carbon anodes are used. Although fluorine is often handled in metal apparatus, it can be handled in the laboratory in glass apparatus provided traces of HF, which attack glass rapidly, are removed by passing the gas through sodium or potassium fluoride with which HF forms the bifluorides MHF_2.

Fluorine is the chemically most reactive of all the elements and combines directly at ordinary or elevated temperatures with all the elements other than nitrogen, oxygen, and the lighter noble gases, often with extreme vigor. It also attacks many other compounds, particularly organic compounds, breaking them down to fluorides; organic materials often inflame and burn in the gas.

The low F—F bond energy, which is so important in the high reactivity of F_2 (in both the kinetic and thermodynamic senses), is best explained by the small size[2] and high nuclear charge of the fluorine atom, which causes decreased overlap of the bonding orbitals and increased repulsion between the nonbonding orbitals on the two fluorine atoms. It should be noted that the O—O bond in peroxides and the N—N bond in hydrazines are also relatively weak for similar reasons.

17-3. Chlorine[3a]

Chlorine occurs in Nature mainly as sodium chloride in seawater and in various inland salt lakes, and as solid deposits originating presumably from the prehistoric evaporation of salt lakes. Chlorine is prepared industrially almost entirely by electrolysis of brine:

$$Na^+ + Cl^- + H_2O \rightarrow Na^+ OH^- + \tfrac{1}{2}Cl_2 + \tfrac{1}{2}H_2$$

A disadvantage of brine electrolysis is the use of mercury as the electrode; mercury is inevitably lost, constituting a major pollution hazard. A process producing chlorine independently would be advantageous. The old Deacon process

$$2HCl + \tfrac{1}{2}O_2 \rightleftharpoons Cl_2 + H_2O$$

though it has an unfavorable equilibrium, may become economic when nitrogen oxides are used as catalyst and the equilibrium is shifted by removal of water by sulfuric acid. It can utilize the HCl by-product generated in the production of chlorinated hydrocarbons from methane or ethane and chlorine. The equilibrium of the Deacon process may also be shifted by removal of Cl_2 as dichloroethane or chlorinated methanes in the following oxychlorination reaction.

$$2HCl + \tfrac{1}{2}O_2 + C_2H_4 \rightarrow C_2H_4Cl_2 + H_2O$$

The dichloroethane is pyrolyzed to vinyl chloride, which in turn is used for polymerization to polyvinyl chloride (PVC).

---

[2] P. Politzer, *Inorg. Chem.*, 1977, **16**, 3350.
[3a] G. C. White, *The Handbook of Chlorination*, Van Nostrand-Reinhold, 1972 (manufacture and uses).

Chlorine and hydrogen can also be recovered by electrolysis of warm 22% hydrochloric acid, which is obtained as a by-product in chlorination processes.

Chlorine is a greenish gas. It is moderately soluble in water with which it reacts (see p. 556). When chlorine is passed into dilute solutions of $CaCl_2$ at $0°$, feathery crystals of "chlorine hydrate," $Cl_2 \cdot 7.3H_2O$, are formed. This substance is a clathrate of the gas-hydrate type (see p. 227), having all medium holes and $\sim 20\%$ of the small holes in the structure filled with chlorine molecules.

17-4. Bromine

Bromine occurs principally as bromide salts of the alkalis and alkaline earths in much smaller amounts than, but along with, chlorides. Bromine is obtained from brines and seawater by chlorination at a pH of ~ 3.5 and is swept out in a current of air.

Bromine is a dense, mobile, dark red liquid at room temperature. It is moderately soluble in water (33.6 g l^{-1} at $25°$) and miscible with nonpolar solvents such as CS_2 and CCl_4. Like Cl_2 it gives a crystalline hydrate, which is, however, structurally different from that of chlorine.

17-5. Iodine

Iodine occurs as iodide in brines and in the form of sodium and calcium iodates. Also, various forms of marine life concentrate iodine. Production of iodine involves either oxidizing I^- or reducing iodates to I^- followed by oxidation to the elemental state. Exact methods vary considerably depending on the raw materials. A commonly used oxidation reaction, and one suited to laboratory use when necessary, is oxidation of I^- in acid solution with MnO_2 (also used for preparation of Cl_2 and Br_2 from X^-).

Iodine is a black solid with a slight metallic luster. At atmospheric pressure it sublimes (violet vapor) without melting. Its solubility in water is slight (0.33 g l^{-1} at $25°$). It is readily soluble in nonpolar solvents such as CS_2 and CCl_4 to give violet solutions; spectroscopic studies indicate that "dimerization" occurs in solutions to some extent:

$$2I_2 \rightleftharpoons I_4$$

Iodine solutions are brown in solvents such as unsaturated hydrocarbons, liquid SO_2, alcohols, and ketones, and pinkish brown in benzene (see below).

Iodine forms the well-known blue complex with the amylose form of starch. From resonance Raman and ^{129}I Mossbauer spectroscopy it has been shown that the color is caused by a linear array of I_5^- ($I_2I^-I_2$) repeating units held inside the amylose helix.[3b]

[3b] T. J. Marks et al., J. Am. Chem. Soc., 1978, **100**, 3215.

17-6. Astatine (At), Element 85[3c]

Isotopes of element 85 have been identified as short-lived products in the natural decay series of uranium and thorium. The element was first obtained in quantities sufficient to afford a knowledge of some of its properties by the cyclotron reaction: $^{209}Bi(\alpha,2n)^{211}At$. The element was named astatine from the Greek for "unstable." About 20 isotopes are known, the longest lived being ^{210}At with a half-life of only 8.3 hr; consequently macroscopic quantities cannot be accumulated, although some compounds, HAt, CH_3At, AtI, $AtBr$, and $AtCl$, have been detected mass spectroscopically. Our knowledge of the chemistry of At is based mainly on tracer studies, which show that it behaves about as one might expect by extrapolation from the other halogens. The element is rather volatile. It is somewhat soluble in water, from which it may, like iodine, be extracted into benzene or carbon tetrachloride. Unlike iodine, however, it cannot be extracted from basic solutions.

The At^- ion is produced by reduction with SO_2 or zinc but not iron(II) ion (which gives some indication of the oxidation potential of At^-). This ion is carried down in AgI or TlI precipitates. Bromine and, to some extent, iron(III) ion oxidize it to what appears to be AtO^- or $HOAt$. $HOCl$ or hot $S_2O_8^{2-}$ oxidizes it to an anion carried by IO_3^-, therefore probably AtO_3^-. Astatine is also carried when $[Ipy_2]^+$ salts are isolated, indicating that $[Atpy_2]^+$ can exist. In $0.1 M$ acid, the potentials appear to be:

$$AtO_3^- \overset{1.5}{\text{——}} HOAt(?) \overset{1.0}{\text{——}} At \overset{0.3}{\text{——}} At^-$$

There is some evidence for the ion At^+.

17-7. Charge-Transfer Compounds of Halogens[4]

As noted above, iodine gives solutions in organic solvents whose color depends on the solvent. It has been shown that for the brown solutions in donor solvents, solvation and 1:1 complex formation $I_2 \cdots S$ occur; such interactions are said to be of the "charge-transfer" type and the complexes are called charge-transfer complexes. This name derives from the nature of the interaction, in which the bonding energy is attributable to a partial transfer of charge. The ground state of the system can be described as a resonance hybrid of 17-I and 17-II, with the former predominating. An electronic transition to an excited state, which is also a resonance hybrid

[3c] E. H. Appelman, *MTP Int. Rev. Sci., Inorg. Chem., Ser. 1*, 1972, **3**, 181; W. A. Chalkin *et al., Chem. Zeit.*, 1977, **101**, 470.

[4] L. J. Andrews and R. M. Keefer, *Molecular Complexes in Organic Chemistry*, Holden-Day, 1964; C. K. Prout and J. D. Wright, *Angew. Chem., Int. Ed.*, 1968, **7**, 659 (structures of solid donor-acceptor complexes); H. A. Bent, *Chem. Rev.*, 1968, **68**, 587 (general donor-acceptor interactions); R. S. Mulliken and W. B. Person, *Molecular Complexes*, Wiley-Interscience, 1969; and in *Physical Chemistry*, Vol. 3, D. Henderson, Ed., Academic Press, 1969; M. J. Blandamer and M. F. Fox, *Chem. Rev.*, 1970, **70**, 59; R. Foster, *Organic Charge Transfer Complexes*, Academic Press, 1969; for a good example, I_2-C_2H_5OH, see L. M. Julien, W. E. Bennett, and W. B. Person, *J. Am. Chem. Soc.*, 1969, **91**, 6195.

of 17-I and 17 II, with the latter predominating, is characteristic of these complexes. It usually occurs near or in the visible region and is the cause of their typically intense colors. This transition is called a charge-transfer transition.

$$X_2 \cdots S \leftrightarrow X_2^- S^+$$

$$(17\text{-I}) \qquad (17\text{-II})$$

Although iodine has been most extensively studied, chlorine and bromine show similar behavior. For a given group of donors, the frequency of the intense charge-transfer absorption band in the ultraviolet is dependent on the ionization potential of the donor solvent molecule, and electronic charge can be transferred either from a π-electron system as in benzene or from long pairs as in ethers or amines. Charge-transfer spectra and complexes are of importance elsewhere in chemistry.

For halogens, and interhalogens such as ICl, charge-transfer compounds can in fact be isolated in the crystalline state, though low temperatures are often required. Thus dioxane with Br_2 gives a compound with a chain structure (17-III) where the Br—Br distance (2.31 Å) is only slightly longer than in Br_2 itself (2.28 Å). In the benzene compound 17-IV the halogen molecules lie along the axis perpendicular to the center of the ring. Other crystals such as $(CH_3)_3NI_2$ and $(CH_3)_2COI_2$ all have one halogen linked to the donor atom and the second pointing away, as in N⋯X—X. In many of the compounds, especially with O and N donors, there is considerable similarity to hydrogen-bonding interactions.

(17-III)

(17-IV)

In certain cases, interaction of the halogen and the donor may become sufficiently strong for the X—X bond to be broken. Thus for adducts of the type $R_2Y \cdot X_2$, where Y is O, S, Se, or Te, the components R_2Y and X_2 retain their identities when Y is more electronegative than X. Otherwise there is an oxidative addition (Section 29-3) of X_2 to Y, and the resulting complex [e.g., $(p\text{-}ClC_6H_4)_2SCl_2$ or Me_2TeCl_2] has a trigonal ψ-bipyramidal structure with axial halide atoms. For pyridine the final crystalline product is [py—I—py]I; of course the latter could be regarded as a coordination complex of I^+.

In certain cases charge-transfer complexes may be intermediates in halogen reactions. Thus I_2 first reacts with KCNS in water to give a yellow solution, after which there is a slow reaction in neutral or basic solution:

$$I_2 + SCN^- \xrightarrow{\text{rapid}} I_2NCS^- \xrightarrow{\text{slow}} ICN + SO_4^{2-}$$

Complexes of Br_2 with dioxane and crown ethers can be used as selective reagents for bromination of olefins.[5]

HALIDES

Except for helium, neon, and argon, all the elements in the Periodic Table form halides, often in several oxidation states, and halides generally are among the most important and common compounds. The ionic and covalent radii of the halogens are shown in Table 17-1.

There are almost as many ways of classifying halides as there are types of halide—and this is many. There are not only binary halides that can range from simple molecules with molecular lattices to complicated polymers and ionic arrays, but also oxohalides, hydroxo halides, and other complex halides of various structural types.

17-8. Preparation of Anhydrous Halides

Although preparations of individual halogen compounds are mentioned throughout the text, anhydrous halides are of such great importance in chemistry that a few of the more important general methods of preparation can be noted.

1. *Halogenation of the Elements.* This method is perhaps the most important preparative method for all halides. Higher fluorides, such as AgF_2 or CrF_4, usually require the use of elemental fluorine, with the metal or a lower fluoride or other salt. For chlorides, bromides, and iodides of transition metals, elevated temperatures are usually necessary in dry reactions. Reaction with Cl_2 or Br_2 is often more rapid when tetrahydrofuran or some other ether is used as the medium, the halide being obtained as a solvate. Where different oxidation states are possible, F_2 and Cl_2 usually give a higher state than bromine or iodine does. Nonmetals, such as phosphorus, usually react readily without heating, and their reaction with fluorine may be explosive. Elemental Mn, Fe, Co, Ni, Cu, and Zn can be oxidized electrochemically in presence of Cl_2, Br_2, or I_2 to afford the anhydrous metal(II) halides in a form very useful for synthesis.[6a]

2. *Halogen Exchange.* This method is especially important for fluorides, many of which are normally obtained from the chlorides by action of various metal fluorides such as SbF_3, SbF_3Cl_2, and some alkali metal fluorides; this type of replacement is much used for organic fluorine compounds (p. 572).

Many Cl, Br, and I compounds undergo rapid halogen exchange with either the

[5] K. H. Parnell and A. Mayr, *J.C.S. Chem. Comm.,* **1979,** 132.

[6a] J. Habeeb, L. Neilson, and D. G. Tuck, *Inorg. Chem.,* 1978, **17,** 306.

elements X_2 or the acids HX. An excess of reagent is usually required, since equilibrium mixtures are normally formed.

3. *Halogenation by Halogen Compounds.* This is an important method, particularly for metal fluorides and chlorides. The reactions involve treatment of anhydrous compounds, often oxides, with halogen compounds such as BrF_3, CCl_4, hexachlorobutadiene, and hexachloropropene at elevated temperatures:

$$NiO + ClF_3 \longrightarrow NiF_2$$

$$UO_3 + CCl_2{=}CCl{-}CCl{=}CCl_2 \xrightarrow{reflux} UCl_4$$

$$Pr_2O_3 + 6NH_4Cl(s) \xrightarrow{300°} 2PrCl_3 + 3H_2O + 6NH_3$$

$$Sc_2O_3 + CCl_4 \xrightarrow{600°} ScCl_3$$

4. *Dehydration of Hydrated Halides.* Hydrated halides are usually obtained easily from aqueous solutions. They can sometimes be dehydrated by heating them in a vacuum, but often this leads to oxo halides or impure products. Various reagents can be used to effect dehydration. For example, $SOCl_2$ is often useful for chlorides. Another fairly general reagent is 2,2-dimethoxypropane.

$$[Cr(H_2O)_6]Cl_3 + 6SOCl_2 \xrightarrow{reflux} CrCl_3 + 12 HCl + 6SO_2$$

$$MX_n \cdot mH_2O + mCH_3C(OCH_3)_2CH_3 \longrightarrow MX_n + m(CH_3)_2CO + 2mCH_3OH$$

In many cases the acetone and/or methanol becomes coordinated to the metal, but gentle heating or pumping usually gives the solvate-free halide.

17-9. Binary Ionic Halides[6b]

Most metal halides are substances of predominantly ionic character, although partial covalence is important in some. Actually, of course, there is a uniform gradation from halides that are for all practical purposes purely ionic, through those of intermediate character, to those that are essentially covalent. As a rough guide, we can consider to be basically ionic the halides in which the lattice consists of discrete ions rather than definite molecular units, although there may still be considerable covalence in the metal-halogen interaction, and the description "ionic" should never be taken entirely literally. As a borderline case *par excellence,* which clearly indicates the danger of taking such rough classifications as "ionic" and "covalent" or even "ionic" and "molecular" too seriously, we have $AlCl_3$ (p. 331); this has an extended structure in which aluminum atoms occupy octahedral interstices in a close-packed array of chlorine atoms; this kind of nonmolecular structure could accommodate an appreciably ionic substance. Yet $AlCl_3$ melts at a low temperature (193°) to a molecular liquid containing Al_2Cl_6, these molecules being much like the Al_2Br_6 molecules that occur in both solid and liquid states of

6b   J. Portier, *Angew. Chem., Int. Ed.,* 1976, **15,** 475 (ionic fluorides).

the bromide. Thus although $AlCl_3$ cannot be called simply a molecular halide, it is an oversimplification to call it an ionic one.

The relatively small radius of F^- (1.19 Å) is almost identical with that of the oxide O^{2-} ion (1.25 Å); consequently many fluorides of monovalent metals and oxides of bivalent metals are ionic with similar formulas and crystal structures—for example, CaO and NaF. The compounds of the other halogens with the same formula usually form quite different lattices due to the change in the anion/cation radii ratios and may even give molecular lattices. Thus chlorides and other halides often resemble sulfides, just as the fluorides often resemble oxides. In several cases the fluorides are completely ionic, whereas the other halides are covalent; for example, CdF_2 and SrF_2 have the CaF_2 lattice (nearly all difluorides have the fluorite or rutile structure), but $CdCl_2$ and $MgCl_2$ have layer lattices with the metal atoms octahedrally surrounded by chlorine atoms.

Many metals show their highest oxidation state in the fluorides. Let us consider the Born-Haber cycle in eq. 17-1:

$$M(s) \xrightarrow{S} M(g) \xrightarrow{\Delta H_{ion}^{(4)}} M^{4+}(g) \searrow$$
$$\qquad\qquad\qquad\qquad\qquad\qquad MX_4(s) \qquad\qquad (17\text{-}1)$$
$$2X_2(g) \xrightarrow{2D} 4X(g) \xrightarrow{4A} 4X^-(g) \nearrow$$

The value of $(A - D/2)$, the energy change in forming 1 g-ion of X^- from $\frac{1}{2}$ mole of X_2, is ~250 kJ for all the halogens, and S is small compared to $\Delta H_{ion}^{(4)}$ in all cases. Although the structure of MX_4, hence the lattice energy, may not be known to allow us to say whether $4(A - D/2)$ plus the lattice energy will compensate for $(\Delta H_{ion}^{(4)} + S)$, we can say that the lattice energy, hence the potential for forming an ionic halide in a high oxidation state, will be greatest for fluoride, since generally, for a given cation size, the greatest lattice energy will be available for the smallest anion, that is, F^-.

However for very high oxidation states, which are formed notably with transition metals, for example, WF_6 or OsF_6, the energy available is quite insufficient to allow ionic crystals with, say, W^{6+} or Os^{6+} ions; consequently such fluorides are gases, volatile liquids, or solids resembling closely the covalent fluorides of the nonmetals. It cannot be reliably predicted whether a metal fluoride will be ionic or molecular, and the distinction between the types is not always sharp.

In addition to the tendency of high cation charge to militate against ionicity, as just noted, coordination number plays an important role in determining the character of a halide. For a halide of formula MX_n where M is relatively large and has a high coordination number and n is small, the coordination number of M can be satisfied only by having a packing arrangement whereby each X atom is shared so that more than n of them may surround each M atom. Usually such structures are in fact essentially ionic, but the nonmolecular, hence nonvolatile, character of the substance is a consequence of the packing regardless of the degree of ionicity of the bonds. At the other extreme, if n is large and M has a coordination number of n, the halide will be molecular and probably volatile. The sequence KCl, $CaCl_2$,

$ScCl_3$, $TiCl_4$ shows these effects, since the first three have nonmolecular (and essentially ionic) structures, whereas $TiCl_4$ is a volatile molecular substance. Similarly, for a given metal with various oxidation numbers, the lower halide(s) tend to be nonvolatile (and ionic), and the higher one(s) are more covalent and molecular. This is illustrated by $PbCl_2$ vs. $PbCl_4$ and UF_4 (a solid of low volatility) vs. UF_6, which is a gas at $25°$.

Most ionic halides dissolve in water to give hydrated metal ions and halide ions. However the lanthanide and actinide elements in the $+3$ and $+4$ oxidation states form fluorides insoluble in water. Fluorides of Li, Ca, Sr, and Ba also are sparingly soluble, the lithium compound being precipitated by ammonium fluoride. Lead gives a sparingly soluble salt PbClF, which can be used for gravimetric determination of F^-. The chlorides, bromides, and iodides of Ag^I, Cu^I, Hg^I, and Pb^{II} are also quite insoluble. The solubility through a series of mainly ionic halides of a given element, $MF_n \rightarrow MI_n$, may vary in either order. When all four halides are essentially ionic, the solubility order is iodide $>$ bromide $>$ chloride $>$ fluoride, since the governing factor is the lattice energies, which increase as the ionic radii decrease. This order is found among the alkali, alkaline earth, and lanthanide halides. On the other hand, if covalence is fairly important, it can reverse the trend, making the fluoride most and the iodide least soluble, as in the familiar cases of silver and mercury(I) halides.

17-10.   Molecular Halides[6c]

Molecular halides are usually volatile, though this will not be so if they are polymeric, as, for example, Teflon $(-CF_2-)_n$. There are also many cases (e.g., $AlCl_3$ and SnF_4) of substances that can exist as molecules in the gas phase, but because of the tendency of the metal atom to have a higher coordination number (4 and 6, respectively, in the cases above) the solids have extended array structures. Most of the electronegative elements and the metals in high oxidation states (V, VI) form molecular halides. A unique but very important group of molecular halides are the hydrogen halides (see p. 242). These form molecular crystals and are volatile, although they readily and extensively dissociate in polar media such as H_2O. The H—X bond energies and the thermal stabilities decrease markedly in the order HF $>$ HCl $>$ HBr $>$ HI, that is, with increasing atomic number of the halogen. The same trend is found, in varying degrees, among the halides of all elements giving a set of molecular halides, such as those of C, B, Si, and P. Interhalogen compounds are discussed below.

Molecular Fluorides.   Many molecular fluorides exist, but it is clear that because of the high electronegativity of fluorine, the bonds in such compounds tend to be very polar. Because of the low dissociation energy of F_2 and the relatively high energy of many bonds to F (e.g., C—F, 486; N—F, 272; P—F, 490 kJ mol^{-1}), molecular fluorides are often formed very exothermically; this is just the opposite of the situation with nitrogen, where the great strength of the bond in N_2 makes nitrogen compounds mostly endothermic. Interestingly, in what might be considered

   [6c] K. Seppelt, *Angew. Chem., Int. Ed.,* 1979, **18,** 186 (fluorides and oxofluorides of nonmetals).

a direct confrontation between these two effects, the tendency of fluorine to form exothermic compounds wins. Thus for NF_3 we have

$$\tfrac{1}{2}N_2(g) = N(g) \qquad \Delta H \approx 475 \text{ kJ mol}^{-1}$$
$$\tfrac{3}{2}F_2(g) = 3F(g) \qquad \Delta H \approx 232 \text{ kJ mol}^{-1}$$
$$N(g) + 3F(g) = NF_3(g) \qquad -3E_{N-F} \approx -3(272) = -816 \text{ kJ mol}^{-1}$$
$$\text{Therefore } \tfrac{1}{2}N_2(g) + \tfrac{3}{2}F_2(g) = NF_3(g) \qquad \Delta H \approx -109 \text{ kJ mol}^{-1}$$

The high electronegativity of fluorine often has a profound effect on the properties of molecules in which several F atoms occur. Representative are facts such as (*a*) CF_3COOH is a strong acid, (*b*) $(CF_3)_3N$ and NF_3 have no basicity, and (c) CF_3 derivatives in general are attacked much less readily by electrophilic reagents in anionic substitutions than are CH_3 compounds. The CF_3 group may be considered as a kind of large pseudohalogen with an electronegativity about comparable to that of Cl.

Reactivity. The detailed properties of a given molecular halide depend on the particular elements involved, and these are discussed where appropriate in other chapters. However a fairly general property of molecular halides is their easy hydrolysis to produce the hydrohalic acid and an acid of the other element. Typical examples are:

$$BCl_3 + 3H_2O \rightarrow B(OH)_3 + 3H^+ + 3Cl^-$$
$$PBr_3 + 3H_2O \rightarrow HPO(OH)_2 + 3H^+ + 3Br^-$$
$$SiCl_4 + 4H_2O \rightarrow Si(OH)_4 + 4H^+ + 4Cl^-$$

When the central atom of a molecular halide has its maximum stable coordination number, as in CCl_4 or SF_6, the substance usually is quite unreactive toward water or even OH^-. However this does not mean that reaction is thermodynamically unfavorable, but only that it is kinetically inhibited, since there is no room for nucleophilic attack. Thus for CF_4 the equilibrium constant for the reaction

$$CF_4(g) + 2H_2O(l) = CO_2(g) + 4HF(g)$$

is *ca.* 10^{23} The necessity for a means of attack is well illustrated by the failure of SF_6 to be hydrolyzed, whereas SeF_6 and TeF_6 are hydrolyzed at 25° through expansion of the coordination sphere, which is possible only for selenium and tellurium.

OXIDES, OXO ACIDS, AND THEIR SALTS

17-11. Oxides

The oxides of chlorine, bromine, and iodine are listed in Table 17-2. Oxides of *fluorine* were discussed in Section 15-4; although these were called oxygen fluorides because of the greater electronegativity of fluorine, those of the remaining halogens are conventionally and properly called *halogen oxides,* since oxygen is the more electronegative element, though not by a very great margin relative to chlorine.

TABLE 17-2
Oxides of the Halogens

Fluorine[a]	B.p. (°C)	M.p. (°C)	Chlorine	B.p. (°C)	M.p. (°C)	Bromine[b]	B.p. (°C)	Iodine[b]
F_2O	−145	−224	Cl_2O	~4	−116	Br_2O ~0°C −18		I_2O_4
			Cl_2O_3					
F_2O_2	−57	−163	ClO_2	~10	−5.9	Br_3O_8 or ~80°C		I_4O_9
						BrO_3		
			Cl_2O_4	44.5	−117			
			Cl_2O_6		3.5	BrO_2 ~40°C		I_2O_5 300°C
			Cl_2O_7	82	−91.5			

(handwritten brace linking I_2O_4 and I_4O_9 labeled 100°C)

[a] See page 493.
[b] Decompose on heating.

All the oxides may be formally considered to be anhydrides or mixed anhydrides of the appropriate oxo acids, but this aspect of their chemistry is of little practical consequence. Several of the oxides, ClO_2, Cl_2O, Cl_2O_3, and Cl_2O_7, are prone to explode; they appear to be shock sensitive rather than thermally sensitive. The oxides are uncommon and not useful, except for ClO_2 and Cl_2O, which find wide use as commercial bleaching agents (paper pulp, flour, etc.).

Chlorine Oxides. Probably the best characterized is *dichlorine monoxide* (Cl_2O).[7] It is a yellowish-red gas at room temperature. It explodes rather easily to Cl_2 and O_2 when heated or sparked. It dissolves in water, forming an orange-yellow solution that contains some HOCl, of which it is formally the anhydride; the hypochlorite anion OCl^- is formed in alkaline solutions. The Cl_2O molecule is angular (111°) and symmetrical, with Cl—O = 1.71 Å. It is prepared by treating freshly prepared yellow mercuric oxide with chlorine gas or with a solution of chlorine in carbon tetrachloride:

$$2Cl_2 + 2HgO \rightarrow HgCl_2 \cdot HgO + Cl_2O$$

Chlorine dioxide[8] is also highly reactive and is liable to explode very violently; apparently mixtures with air containing less than ~50 mm partial pressure of ClO_2 are safe. ClO_2 is made on a fairly large scale, but it is always produced where and as required. The best preparation is the reduction of $KClO_3$ by moist oxalic acid at 90°, since the CO_2 liberated also serves as a diluent for the ClO_2. Commercially the gas is made by the exothermic reaction of sodium chlorate in 4 to 4.5M sulfuric acid containing 0.05 to 0.25M chloride ion with sulfur dioxide:

$$2NaClO_3 + SO_2 + H_2SO_4 \rightarrow 2ClO_2 + 2NaHSO_4$$

ClO_2 is a yellowish gas at room temperature. The molecule is angular (118°) with Cl—O = 1.47 Å. Although ClO_2 is an odd molecule, it has no marked tendency to dimerize, perhaps because the electron is more effectively delocalized than in

[7] J. J. Renard and H. I. Bolker, *Chem. Rev.*, 1976, **76**, 487.
[8] G. Gordon, R. G. Kieffer, and D. H. Rosenblatt, *Prog. Inorg. Chem.*, 1972, **15**, 201; W. J. Masschelein, *Chlorine Dioxide,* Ann Arbor Science–Wiley, 1979.

other odd molecules such as NO_2. It is soluble in water, and solutions with up to 8 g l^{-1} are stable in the dark, but in light decompose slowly to HCl and $HClO_3$. In alkaline solution a mixture of chlorite and chlorate ions is formed fairly rapidly. Acid solutions are much more stable, but reduction to $HClO_2$ occurs first, followed by decomposition to HCl + $HClO_3$. The photolysis of ClO_2 at low temperature affords what is believed to be Cl_2O_3 as a dark brown solid that explodes readily.

Dichlorine tetraoxide, which is structurally $ClOClO_3$ and commonly is called chlorine perchlorate, is stable for only short periods of time at 25°. $FOClO_3$ is somewhat more stable, but $BrOClO_3$ is even less stable. The halogen perchlorates are useful intermediates for the synthesis of fluorocarbon perchlorates and anhydrous metal perchlorates.[9]

Dichlorine hexaoxide is an unstable red oil. In the solid state it has the ionic structure[10] $ClO_2^+ClO_4^-$, suggesting that the molecule may have an unsymmetrical oxygen-bridged structure rather than a symmetrical one with a Cl—Cl bond.

Dichlorine heptooxide is the most stable chlorine oxide. It is a colorless liquid formed by dehydration of perchloric acid with P_2O_5 at −10°, followed by vacuum distillation with precautions against explosions. It reacts with water and OH$^-$ to generate ClO_4^-. Electron diffraction shows the structure $O_3ClOClO_3$, with a ClOCl angle of 118.6°. The reaction of Cl_2O_7 with alcohols yields alkyl perchlorates ($ROClO_3$), which find use as intermediates in synthesis.[11] It reacts similarly with amines to yield R_2NClO_3 or $RHNClO_3$.[12]

Bromine Oxides. The bromine oxides are all of very low thermal stability. Br_2O, a dark brown liquid, decomposes at an appreciable rate above −50°. Br_3O_8 (also claimed to be BrO_3) is a white solid unstable above −80° except in an atmosphere of ozone. BrO_2, a yellow solid unstable above about −40°, is believed to be a dimer with a Br—Br bond.[13] On thermal decomposition it gives Br_2O_3, whose structure is uncertain, both OBrOBrO and BrOBrO$_2$ being consistent with its vibrational spectrum.[14]

Iodine Oxides. Of these, white crystalline iodine pentoxide is the most important and is made by the reaction

$$2HIO_3 \underset{}{\overset{240°}{\rightleftharpoons}} I_2O_5 + H_2O$$

It has IO_3 pyramids sharing one oxygen to give O_2IOIO_2 units, but quite strong intermolecular I···O interactions lead to a three-dimensional network. This compound is stable up to about 300°, where it melts with decomposition to iodine and oxygen. It is the anhydride of iodic acid and reacts immediately with water. It reacts as an oxidizing agent with various substances such as H_2S, HCl, and CO. One of its important uses is as a reagent for the determination of CO, the iodine that is

9   C. J. Schack, D. Pilipovich, and K. O. Christe, *J. Inorg. Nucl. Chem.,* Suppl. **1976,** 207.
10 A. C. Pavia, J. L. Pascal, and A. Potier, *Compt. Rend., C,* 1971, **272,** 1425.
11 K. Baum and C. D. Beard, *J. Am. Chem. Soc.,* 1974, **96,** 3233.
12 K. Baum and C. D. Beard, *J. Am. Chem. Soc.,* 1974, **96,** 3237; D. Baumgarten *et al., Z. Anorg. Allg. Chem.,* 1974, **405,** 77.
13 J.-L. Pascal and J. Potier, *J.C.S. Chem. Commun.,* **1973,** 446.
14 J.-L. Pascal *et al., Compt. Rend., C,* 1974, **279,** 43.

produced quantitatively being then determined by standard iodometric procedures:

$$5CO + I_2O_5 \rightarrow I_2 + 5CO_2$$

The other oxides of iodine I_2O_4 and I_4O_9 are of less certain nature. They decompose when heated at $\sim100°$ to I_2O_5 and iodine, or to iodine and oxygen. The yellow solid I_2O_4, which is obtained by partial hydrolysis of $(IO)_2SO_4$ (discussed below), appears to have a network built up of polymeric I—O chains that are cross-linked by IO_3 groups. I_4O_9, which can be made by treating I_2 with ozonized oxygen, can be regarded as $I(IO_3)_3$, similarly cross-linked.

17-12.   Oxo Acids and Anions[15]

The known oxo acids of the halogens are listed in Table 17-3. The chemistry of these acids and their salts is very complicated. Solutions of all the acids and of several of the anions can be obtained by reaction of the free halogens with water or aqueous bases. We discuss these reactions first; the term "halogen" refers to chlorine, bromine, and iodine, only.

Reaction of Halogens with H_2O and OH^-.   A considerable degree of order can be found in this area if full and proper use is made of thermodynamic data in the form of oxidation potentials and equilibrium constants and if the relative rates of competing reactions are also considered. The basic thermodynamic data are given in Table 17-4. From these, all necessary potentials and equilibrium constants can be derived by use of elementary thermodynamic relationships.

The halogens are all to some extent soluble in water. However in all such solutions there are species other than solvated halogen molecules, since a disproportionation reaction occurs *rapidly*. Two equilibria serve to define the nature of the solution:

$$X_2(g,l,s) = X_2(aq) \qquad K_1$$
$$X_2(aq) = H^+ + X^- + HOX \qquad K_2$$

The values of K_1 for the various halogens are: Cl_2, 0.062; Br_2, 0.21; I_2, 0.0013. The values of K_2 can be computed from the potentials in Table 17-4 to be 4.2×10^{-4} for Cl_2, 7.2×10^{-9} for Br_2 and 2.0×10^{-13} for I_2. We can also estimate

TABLE 17-3
Oxo Acids of the Halogens

Fluorine	Chlorine	Bromine	Iodine
HOF	HOCl[a]	HOBr[a]	HOI[a]
	HOClO[a]	HOBrO[a]	—
	HOClO_2[a]	HOBrO_2[a]	HOIO_2
	HOClO_3	HOBrO_3[a]	HOIO_3, (HO)_5IO, H_4I_2O_9

[a] Stable only in solution

[15]   F. Solymosi, *Structure and Stability of Salts of Halogen Oxyacids in the Solid Phase*, Wiley, 1977.

TABLE 17-4

Standard Potentials for Reactions of the Halogens (V)

Reaction	Cl	Br	I
$H^+ + HOX + e = \frac{1}{2}X_2(g,l,s) + H_2O$	1.63	1.59	1.45
$3H^+ + HXO_2 + 3e = \frac{1}{2}X_2(g,l,s) + 2H_2O$	1.64	—	—
$6H^+ + XO_3^- + 5e = \frac{1}{2}X_2(g,l,s) + 3H_2O$	1.47	1.52	1.20
$8H^+ + XO_4^- + 7e = \frac{1}{2}X_2(g,l,s) + 4H_2O$	1.42	1.59^a	1.34
$\frac{1}{2}X_2(g,l,s) + e = X^-$	1.36	1.07	0.54^b
$XO^- + H_2O + 2e = X^- + 2OH^-$	0.89	0.76	0.49
$XO_2^- + 2H_2O + 4e = X^- + 4OH^-$	0.78	—	—
$XO_3^- + 3H_2O + 6e = X^- + 6OH^-$	0.63	0.61	0.26
$XO_4^- + 4H_2O + 8e = X^- + 8OH^-$	0.56	0.69^a	0.39

[a] Calculated from data of G. K. Johnson et al., Inorg Chem., 1970, **9**, 119.

[b] Indicates that I^- can be oxidized by oxygen in aqueous solution.

from

$$\frac{1}{2}X_2 + e = X^-$$

and

$$O_2 + 4H^+ + 4e = 2H_2O \qquad E^0 = 1.23 \text{ V}$$

that the potentials for the reactions

$$2H^+ + 2X^- + \frac{1}{2}O_2 = X_2 + H_2O$$

are -1.62 V for fluorine, -0.13 V for chlorine, 0.16 V for bromine, and 0.69 V for iodine.

Thus for saturated solutions of the halogens in water at 25° we have the results shown in Table 17-5. There is an appreciable concentration of hypochlorous acid in a saturated aqueous solution of chlorine, a smaller concentration of HOBr in a saturated solution of Br_2, but only a negligible concentration of HOI in a saturated solution of iodine.

Hypohalous Acids. The compound HOF,[16] a colorless solid, melts at $-117°$ to a pale yellow liquid. It is a gas at ambient temperature, highly reactive towards water, and has a half-life for spontaneous decomposition to HF and O_2 of ca. 30 min at 25°. It is prepared, with difficulty, by reaction of F_2 with H_2O at low temperature. The molecule has the smallest angle (97°) known at an unconstrained oxygen atom; for HOCl (see below) the angle is 103°. A number of compounds

TABLE 17-5

Equilibrium Concentrations in Aqueous Solutions of the Halogens at 25° (mol l^{-1})

	Cl_2	Br_2	I_2
Total solubility	0.091	0.21	0.0013
Concentration X_2(aq)	0.061	0.21	0.0013
$[H^+] = [X^-] = [HOX]$	0.030	1.15×10^{-3}	6.4×10^{-6}

[16] E. H. Appelman, Acc. Chem. Res., 1973, **6**, 113.

containing covalently bound OF groups, and called hypofluorites, are known,[17] examples being CF_3OF, SF_5OF, O_3ClOF, and FSO_2OF.

The other HOX compounds are also unstable. In water their dissociation constants are: HOCl, 3.4×10^{-8}; HOBr, 2×10^{-9}, HOI, 1×10^{-11}. As can be readily seen, reaction of halogens with water does not constitute a suitable method for preparing aqueous solutions of the hypohalous acids owing to the unfavorable equilibria. A useful general method is interaction of the halogen and a well-agitated suspension of mercuric oxide:

$$2X_2 + 2HgO + H_2O \rightarrow HgO \cdot HgX_2 + 2HOX$$

In the vapor phase, HOCl is formed in the equilibrium

$$H_2O(g) + Cl_2O(g) \rightleftharpoons 2HOCl(g)$$

The reaction of HgO and I_2 can be used to form organic hypoiodites from alcohols; these can be used as oxidants.[18a] The hypohalous acids are good oxidizing agents, especially in acid solution (see Table 17-4). HOCl is a chlorinating agent for aromatic compounds; the reactive intermediates[18b] seem to be $ClO^\bullet$ and/or H_2OCl^+, depending on conditions, but never (as often suggested) Cl^+.

The hypohalite ions can all be produced in principle by dissolving the halogens in base according to the general reaction

$$X_2 + 2OH^- \rightarrow X^- + XO^- + H_2O$$

and for these reactions the equilibrium constants are all favorable—7.5×10^{15} for Cl_2, 2×10^8 for Br_2, and 30 for I_2—and the reactions are rapid.

However the situation is complicated by the tendency of the hypohalite ions to disproportionate further in basic solution to produce the halate ions:

$$3XO^- = 2X^- + XO_3^-$$

For this reaction the equilibrium constant is in each case very favorable, that is, 10^{27} for ClO^-, 10^{15} for BrO^-, and 10^{20} for IO^-. Thus the actual products obtained on dissolving the halogens in base depend on the rates at which the hypohalite ions initially produced undergo disproportionation, and these rates vary from one to the other and with temperature.

The disproportionation of ClO^- is slow at and below room temperature. Thus when chlorine reacts with base "in the cold," reasonably pure solutions of Cl^- and ClO^- are obtained. In hot solutions ($\sim 75°$) the rate of disproportionation is fairly rapid and under proper conditions, good yields of ClO_3^- can be secured.

The disproportionation of BrO^- is moderately fast even at room temperature. Consequently solutions of BrO^- can only be made and/or kept at around $0°$. At temperatures of 50 to $80°$ quantitative yields of BrO_3^- are obtained:

$$3Br_2 + 6OH^- \rightarrow 5Br^- + BrO_3^- + 3H_2O$$

The rate of disproportionation of IO^- is very fast at all temperatures, so that

[17] M. Lustig and J. M. Shreeve, *Adv. Fluorine Chem.*, 1973, **7**, 175.
[18a] A. Godsen and H. A. H. Lane, *J. Chem. Soc., B*, **1969**, 995.
[18b] C. G. Swain and D. R. Crist. *J. Am. Chem. Soc.*, 1972, **94**, 3195.

it is unknown in solution. Reaction of iodine with base gives IO_3^- quantitatively according to an equation analogous to that for Br_2.

It remains now to consider the equilibria of the oxo anions not yet mentioned and their kinetic relations to those we have discussed. *Halite ions* and *halous acids* do not arise in the hydrolysis of the halogens. HOIO apparently does not exist, HOBrO is doubtful, and HOClO is not formed by disproportionation of HOCl if for no other reason than that the equilibrium constant is quite unfavorable:

$$2HOCl = Cl^- + H^+ + HOClO \qquad K \sim 10^{-5}$$

The reaction

$$2ClO^- = Cl^- + ClO_2^- \qquad K \sim 10^7$$

is favorable, but the disproportionation of ClO^- to ClO_3^- and Cl^- (see above) is so much more favorable that the first reaction is not observed.

Finally, we must consider the possibility of production of *perhalate ions* by disproportionation of the halate ions. Since the acids $HOXO_2$ and $HOXO_3$ are all strong, these equilibria are independent of pH. The reaction

$$4ClO_3^- = Cl^- + 3ClO_4^-$$

has an equilibrium constant of 10^{29}, but it takes place only very slowly in solution even near 100°; hence perchlorates are not readily produced. Neither perbromate nor periodate can be obtained in comparable disproportionation reactions because the equilibrium constants are 10^{-33} and 10^{-53}, respectively.

The only definitely known halous acid, *chlorous acid,* is obtained in aqueous solution by treating a suspension of barium chlorite with sulfuric acid and filtering off the precipitate of barium sulfate. It is a relatively weak acid ($K_a \approx 10^{-2}$) and cannot be isolated in the free state. *Chlorites* ($MClO_2$) themselves are obtained by reaction of ClO_2 with solutions of bases:

$$2ClO_2 + 2OH^- \rightarrow ClO_2^- + ClO_3^- + H_2O$$

Chlorites are used as bleaching agents. In alkaline solution the ion is stable to prolonged boiling and up to a year at 25° in absence of light. In acid solutions, however, the decomposition is rapid and is catalyzed by Cl^-:

$$5HOClO \rightarrow 4ClO_2 + Cl^- + H^+ + 2H_2O$$

but the reaction sequence is complicated.

Halic Acids.   Only *iodic acid* is known in the free state. This is a stable white solid obtained by oxidizing iodine with concentrated nitric acid, hydrogen peroxide, ozone, or various other strong oxidizing agents. It can be dehydrated to its anhydride I_2O_5, as already noted. Salts such as KHI_2O_6 exist in the solid state, probably owing to favorable lattice energies and low solubility. In aqueous solutions the predominant species is IO_3^-. *Chloric* and *bromic acids* are best obtained in solution by treating the barium halates with sulfuric acid.

All the halic acids are strong acids and are powerful oxidizing agents; the mechanisms of reduction (e.g., by I^-) are very complicated. The halate ions (XO_3^-)

are all pyramidal,[19] as is to be expected from the presence of an octet, with one unshared pair, in the halogen valence shell.

Iodates of certain +4 metal ions—notably those of Ce, Zr, Hf, and Th—can be precipitated from $6M$ nitric acid to provide a useful means of separation.

Perchlorates. Although disproportionation of ClO_3^- to ClO_4^- and Cl^- is thermodynamically very favorable, the reaction occurs only very slowly in solution and does not constitute a useful preparative procedure. Perchlorates are commonly prepared by electrolytic oxidation of chlorates. The properties of perchloric acid were discussed on page 244.

Perchlorates of almost all electropositive metals are known. Except for a few with large cations of low charge, such as $CsClO_4$, $RbClO_4$, and $KClO_4$, they are readily soluble in water. Solid perchlorates containing the tetrahedral ClO_4^- ion are often isomorphous with salts of other tetrahedral anions (e.g., MnO_4^-, SO_4^{2-}, BF_4^-). A particularly important property of the perchlorate ion is its slight tendency to serve as a ligand in complexes. Thus perchlorates are widely used in studies of complex ion formation, the *assumption* being made that no appreciable correction for the concentration of perchlorate complexes need be considered. This may often be true for aqueous solutions, but it is well known that when no other donor is present to compete, perchlorate ion exercises a donor capacity and can be monodentate, bridging bidentate; or chelating bidentate.[20] This is illustrated by structures of compounds such as $(CH_3)_3SnClO_4$, $Co(MeSC_2H_4SMe)_2(ClO_4)_2$, and $(Ph_3BiOBiPh_3)(ClO_4)_2$.[21a]

The use of perchlorate as an ion for the isolation of crystalline salts of organometallic ions such as $(\eta^5\text{-}C_5H_5)_2Fe^+$ is to be avoided, since such salts are often dangerously explosive: the use of the trifluoromethanesulfonate ion, $CF_3SO_3^-$, PF_6^-, or BF_4^-, which behave very much like ClO_4^-, is preferable, but even BF_4^- and PF_6^- may act as ligands. Although ClO_4^- is potentially a good oxidant

$$ClO_4^- + 2H^+ + 2e = ClO_3^- + H_2O \qquad E^0 = 1.23 \text{ V}$$

in aqueous solution it is reduced only by Ru^{II}, V^{II}, V^{III}, and Ti^{III}. Despite the more favorable potential for reduction by Eu^{2+} or Cr^{2+}, no reaction occurs, for reasons that are not entirely clear.

Perbromic Acid and Perbromates. Perbromates were prepared only in 1968; previously there were many papers justifying theoretically their nonexistence. This provides an excellent example of the folly of concluding the nonexistence of certain compounds until all conceivable preparative methods have been exhausted.[21b]

The potential

$$BrO_4^- + 2H^+ + 2e = BrO_3^- + H_2O \qquad E^0 = +1.76 \text{ V}$$

shows that only the strongest oxidants can form perbromate. Probably kinetic reasons are responsible for the failure of ozone ($E = +2.07$ V) and $S_2O_8^{2-}$ ($E =$

[19]   G. Kemper, A. Vos, and H. M. Rietveld, *Can. J. Chem.,* 1972, **50,** 1134; L. Y. Y. Chan and F. W. B. Einstein, *Can. J. Chem.,* 1971, **49,** 468.

[20]   K. O. Christe and C. J. Schack, *Inorg. Chem.,* 1974, **13,** 1452.

[21a]   F. C. March and G. Ferguson, *J.C.S. Dalton,* **1975,** 1291.

[21b]   E. H. Appelman et al., *J. Am. Chem. Soc.,* 1979, **101,** 929.

⊢ 2.01 V) to cause oxidation. Perbromate is a stronger oxidant than ClO_4^- (1.23 V) or IO_4^- (1.64 V); there is no convincing explanation of this anomaly, and similar instability of the high oxidation state occurs in other first long-period elements (e.g., for Se in the S, Se, Te group).

Small amounts of perbromic acid or perbromates can be obtained by oxidation of BrO_3^- electrolytically or by the action of XeF_2. The best preparation involves oxidation of BrO_3^- by fluorine in $5M$ NaOH solution; by a rather complicated procedure, pure solutions can be obtained:

$$BrO_3^- + F_2 + 2OH^- \rightarrow BrO_4^- + 2F^- + H_2O$$

Solutions of $HBrO_4$ can be concentrated up to $6M$ (55%) without decomposition and are stable indefinitely even at 100°. More concentrated solutions, up to 83%, can be obtained but these are unstable; the hydrate $HBrO_4 \cdot 2H_2O$ can be crystallized. The ion is tetrahedral, with Br—O = 1.61 Å (cf. Cl—O in ClO_4^-, 1.45 Å; and I—O in IO_4^-, 1.79 Å).

In dilute solution perbromate is a sluggish oxidant at 25° and is slowly reduced by I^- or Br^- but not by Cl^-. However the $3M$ acid readily oxidizes stainless steel, and the $12M$ acid rapidly oxidizes Cl^- and will explode in contact with tissue paper. Above $6M$ the solutions are erratically, but not explosively, unstable.

Pure potassium perbromate is stable up to 275°, where it decomposes to $KBrO_3$, and even NH_4BrO_4 is stable to around 170°.

Periodic Acid and Periodates. Periodic acid exists in solution as the tetrahedral ion IO_4^-, as well as in several hydrated forms. The complexity of the periodates is similar to that found for the oxo acids of Sb and Te, and periodates often resemble tellurates in their stoichiometries.

The main equilibria in acid solutions are:

$$H_5IO_6 = H^+ + H_4IO_6^- \qquad K = 1 \times 10^{-3}$$
$$H_4IO_6^- = IO_4^- + 2H_2O \qquad K = 29$$
$$H_4IO_6^- = H^+ + H_3IO_6^{2-} \qquad K = 2 \times 10^{-7}$$

In aqueous solutions at 25° the periodate ion IO_4^- predominates; the hydrated species are termed orthoperiodates. The various pH-dependent equilibria are established rapidly.

Salts of periodic acid are of several types. The commonest are the acid salts such as $NaH_4IO_6 \cdot H_2O$, $Na_2H_3IO_6$, and $Na_3H_2IO_6$; on addition of CsOH to H_5IO_6 solutions, however, the periodate $CsIO_4$ is precipitated. The free acid H_5IO_6 can be dehydrated to $H_4I_2O_9$ at 80° and to HIO_4 at 100°.

In alkaline solution periodate dimerizes:

$$2IO_4^- + 2OH^- \rightarrow H_2I_2O_{10}^{4-}$$

The dimeric ion in solution is presumably the same as that found as a discrete ion $[O_3(OH)IO_2IO_3(OH)]^{4-}$ in the crystalline salt $K_4H_2I_2O_{10} \cdot 8H_2O$; the ion has two IO_6 octahedra sharing one edge. The Cs^+ salt, which can be made by heating $CsIO_4$ with an excess of CsOH, is dehydrated to $Cs_4I_2O_9$ at 60°.

Most periodates, other than those with the tetrahedral IO_4^- ion, contain an IO_6 octahedron of some sort, but K_3IO_5 is known to contain square-pyramidal IO_5^{3-}

ions.[22] H_5IO_6 is protonated in strong acids to give $I(OH)_6^+$, which can be isolated in solid salts.[23]

The chief characteristic of periodic acids is that they are powerful oxidizing agents that usually react smoothly and rapidly. They are thus useful for analytical purposes, for example, to oxidize manganous ion to permanganate. Ozone (derived from O atoms) may be liberated in the reactions, but not hydrogen peroxide.

Periodic acid or its salts are also commonly used in organic chemistry.

INTERHALOGEN COMPOUNDS AND IONS

17-13. General Survey

The halogens form compounds that are binary and ternary combinations of themselves. Except for BrCl, ICl, ICl_3, and IBr, the compounds are all halogen fluorides (Table 17-6, below) such as ClF, BrF_3, IF_5, and IF_7. Ternary compounds occur only as polyhalide ions; the principal types are listed in Tables 17-7 and 17-8. All stable interhalogen molecules are of the type XX'_n where n is an odd number and there are no unpaired electrons, and for $n \geq 3$, X' is the lighter halogen.[24] No ternary interhalogen compounds are known, although attempts have been made to prepare them. This is probably because any ternary molecules formed can readily redistribute to form a mixture of the (presumably) more stable binary compounds and/or elemental halogens. Another general observation is that stability of the compounds with higher n increases as X becomes larger and X' smaller.

The structures and other physical properties[25] of the interhalogen compounds are all known, to varying degrees of accuracy. These structures and the reasons for them in terms of the electronic configuration of the molecules are discussed in Chapter 5.

Chemically, the interhalogens are all rather reactive. They are corrosive oxidizing substances and attack most other elements, producing mixtures of the halides. They are all more or less readily hydrolyzed (some, e.g., BrF_3, being dangerously explosive in this respect), in some cases according to the equation[26a]

$$XX' + H_2O = H^+ + X'^- + HOX$$

The diatomic compounds often add to ethylenic double bonds and may react with the heavier alkali and alkaline earth metals to give polyhalide salts.

The *diatomic compounds* are ClF, BrF, BrCl, IBr, and ICl. In their physical properties they are usually intermediate between the constituent halogens. They

[22] H. Dölling and M. Trömel, *Naturwissenschaften,* 1973, **60**, 153.
[23] H. Siebert and U. Woerner, *Z. Anorg. Allg. Chem.,* 1973, **398**, 193.
[24] However at very low temperatures radical species, such as ClF_2, ClF_4, and ClF_6, can be generated and survive long enough for epr study. Cf. K. Nishikida *et al., J. Am. Chem. Soc.,* 1975, **97**, 3526.
[25] H. C. Heung and A. Anderson, *Can. J. Chem.,* 1974, **52**, 1081.
[26a] K. O. Christe, *Inorg. Chem.,* 1972, **11**, 1220.

are of course polar, whereas the halogen molecules are not. ClF is colorless; BrF, BrCl, ICl, and IBr are red or reddish brown. IF is unknown except in minute amounts observed spectroscopically: it is apparently too unstable with respect to disproportionation to IF_5 and I_2 to permit its isolation. The other isolable diatomic compounds have varying degrees of stability with respect to disproportionation and fall in the following stability order, where the numbers in parentheses represent the disproportionation constants for the gaseous compounds and the elements in their standard states at 25°: ClF (2.9×10^{-11}) > ICl (1.8×10^{-3}) > BrF (8×10^{-3}) > IBr (5×10^{-2}) > BrCl (0.34). BrF also disproportionates according to:

$$3BrF = BrF_3 + Br_2$$

ClF may be prepared by direct interaction at 220 to 250° and it is readily freed from ClF_3 by distillation, but it is best prepared by interaction of Cl_2 and ClF_3 at 250 to 350°. BrF also results on direct reaction of Br_2 with F_2, but it has never been obtained in high purity because of its ready disproportionation. Iodine monochloride is obtained as brownish-red tablets (β form) by treating liquid chlorine with solid iodine in stoichiometric amount, and cooling to solidify the liquid product. It readily transforms to the α-form, ruby red needles. BrCl is unstable[26b]

$$2BrCl \rightleftharpoons Br_2 + Cl_2 \qquad K = 0.145 \ (25° \text{ in } CCl_4)$$

and has only recently been obtained pure.[27a] IBr, a solid resulting from direct combination of the elements, is endothermic and extensively dissociated in the vapor. It is used instead of Br_2 in some industrial processes.[27b] Despite the general instability of the BrX compounds, the fluorosulfate $BrOSO_2F$, obtained by treating Br_2 with $S_2O_6F_2$, is stable to 150°.

Iodine trichloride (ICl_3) is also formed (like ICl) by treatment of liquid chlorine with the stoichiometric quantity of iodine, or with a deficiency of iodine followed by evaporation of the excess of chlorine. It is a fluffy orange powder, unstable much above room temperature.

The most important and most intensively studied compounds are the halogen fluorides.

17-14.   The Halogen Fluorides[28]

These compounds and some of their important physical properties are listed in Table 17-6.

The preparations of ClF and BrF have already been mentioned. ClF_3 may be prepared by direct combination of the elements at 200 to 300° and is available commercially. It is purified by converting it into $KClF_4$ by the action of KF and

[26b] T. Surles and I. Popov, *Inorg. Chem.*, 1969, **8**, 2049.

[27a] M. Schmeisser and K. H. Tytko, *Z. Anorg. Allg. Chem.*, 1974, **403**, 231.

[27b] J. F. Mills and J. A. Schneider, *Ind. Eng. Chem. Prod. Res. Dev.*, 1973, **12**, 160.

[28] K. O. Christe, in *Proceedings of the 24th International Conference of Pure and Applied Chemistry*, Vol. 4, Butterworths, 1974.

TABLE 17-6
Some Physical Properties of Halogen Fluorides

	M.p. (°C)	B.p. (°C)	Specific conductivity[a] at 25°C (Ω^{-1} cm^{-1})	Structure
ClF	−156.6	−100.1	—	
ClF$_3$	−76.3	11.75	3.9×10^{-9}	Planar; distorted "T"
ClF$_5$	−103	−14	—	Square pyramidal
BrF	−33	20	—	
BrF$_3$	9	126	$>8.0 \times 10^{-3}$	Planar; distorted "T"
BrF$_5$	−60	41	9.1×10^{-8}	Square pyramidal
IF$_3$	—	—	—	
IF$_5$	10	101	5.4×10^{-6}	Square pyramidal[b]
IF$_7$	6.45[c]	—	—	Pentagonal bipyramidal[d]

[a] Values in the literature may be very inaccurate in view of the possibility of hydrolysis by traces of water.
[b] R. D. Burbank and G. R. Jones, *Inorg. Chem.*, 1974, **13**, 1071.
[c] Triple point; sublimes 4.77° at 1 atm.
[d] However there are small but significant dynamic distortions from D_{5h} symmetry according to H. H. Eysel and K. Seppelt, *J. Chem. Phys.*, 1972, **56**, 5081.

thermally decomposing the salt at 130 to 150°. ClF$_5$ can be made by interaction of F$_2$ and ClF$_3$ above 200° but is best prepared by the reaction:

$$KCl + 3F_2 \xrightarrow[\text{bomb}]{200°} KF + ClF_5$$

ClF$_5$ is a colorless gas; it is less stable thermally but also less reactive than ClF$_3$, and above 165° there is the equilibrium

$$ClF_5 \rightleftharpoons ClF_3 + F_2$$

The other halides are best prepared by the reactions:

$$Br_2 + 3F_2 \xrightarrow{200°} 2BrF_3$$
$$BrF_3 + F_2 \longrightarrow BrF_5$$
$$I_2 + 5F_2 \xrightarrow{25°} 2IF_5$$
$$KI + 4F_2 \longrightarrow KF + IF_7$$

BrF$_3$ and IF$_5$ are formed by the reaction of X$_2$ with AgF in HF to give AgX plus XF, followed by disproportionation of XF, but the products are not easily purified.[29] IF$_3$ is a yellow powder obtained by fluorination of I$_2$ in Freon at −78°, it decomposes to I$_2$ and IF$_5$ above −35°.

The halogen fluorides are very reactive, and with water or organic substances they react vigorously or explosively. They are powerful fluorinating agents for inorganic compounds or, when diluted with nitrogen, for organic compounds. The most useful compounds are ClF, ClF$_3$, and BrF$_3$. Although only qualitative data

[29] J. L. Russell and A. W. Jache, *J. Inorg. Nucl. Chem., Suppl.*, **1976**, 81.

TABLE 17-7
Principal Fluorohalogen Cations and Anions

Parent	Cation	Anion	Parent	Cation	Anion
ClF	Cl_2F^+	ClF_2^-	IF		IF_2^-
ClF_3	ClF_2^+	ClF_4^-	IF_3	IF_2^+	IF_4^-
ClF_5	ClF_4^+	—			IF_6^{3-}
	ClF_6^+		IF_5	IF_4^+	IF_6^-
BrF		BrF_2^-	IF_7	IF_6^+	IF_8^-
BrF_3	BrF_2^+	BrF_4^-			
BrF_5	BrF_4^+	BrF_6^-			
	BrF_6^+				

are usually available, the order of reactivity is approximately $ClF_3 > BrF_3 > BrF_5 > IF_7 > ClF > IF_5 > BrF$.

Certain compounds, notably ClF, BrF_3, and IF_5, have high entropies of vaporization, and BrF_3 has appreciable electrical conductance. To account for these observations, association by fluorine bridging, in addition to self-dissociation,

$$2BrF_3 \rightleftharpoons BrF_2^+ + BrF_4^-$$

have been postulated. An analogy with other solvent systems can be made. In liquid BrF_3, for example, the "acid" would be BrF_2^+ and the "base" BrF_4^-. Indeed, suitable compounds such as $BrF_3 \cdot SbF_5$ (or $BrF_2^+ \cdot SbF_6^-$) and $KBrF_4$ dissolve in BrF_3 to give highly conducting solutions.

A characteristic property of most halogen fluorides is their amphoteric character; that is, with strong bases, such as alkali metal fluorides, they can form anions, and with strong Lewis acids such as SbF_5 they can form cations:

$$XF_{n+1}^- \xleftarrow{+F^-} XF_n \xrightarrow{-F^-} XF_{n-1}^+$$

A few long-known reactions are given below, and Table 17-7 lists the principal established cations and anions.

$$2ClF + AsF_5 \rightarrow FCl_2^+ AsF_6^-$$
$$ClF + CsF \rightarrow Cs^+ClF_2^-$$
$$ClF_3 + CsF \rightarrow Cs + ClF_4^-$$
$$ClF_5 + SbF_5 \rightarrow ClF_4^+ SbF_6^-$$
$$ClF_3 + AsF_5 \rightarrow ClF_2^+ AsF_6^-$$
$$IF_5 + CsF \rightarrow Cs^+IF_6^-$$

Among recent results in this area are the following. As expected, the IF_4^- ion has been shown[30] to be planar, and an entire series of $[IF_{2n}(O_2CCF_3)_{6-2n}]^{3-}$ from IF_6^{3-} to $[I(O_2CCF_3)_6]^{3-}$ has been established.[31] The pentafluorides ClF_5, BrF_5, and IF_5 all form adducts with AsF_5 and SbF_5 that appear to contain XF_4^+ ions, e.g., $[BrF_4^+][Sb_2F_{11}^-]$.[32,33] The XF_6^+ cations (X = Cl, Br, I) have all been obtained in

30 K. O. Christe and D. Naumann, *Inorg. Chem.*, 1973, **12**, 59.
31 K. O. Christe and D. Naumann, *Spectrochim. Acta*, 1973, **29A**, 2017; D. Nauman, H. Dolhaine, and W. Stopschinski, *Z. Anorg. Allg. Chem.*, 1972, **394**, 133.
32 K. O. Christe and W. Sawodny, *Inorg. Chem.*, 1973, **12**, 2879.
33 M. D. Lind and K. O. Christe, *Inorg. Chem.*, 1972, **11**, 608; T. Surles *et al.*, *J. Inorg. Nucl. Chem.*, 1972, **34**, 3561.

solution and characterized by ^{19}F nmr, and a number of solid salts can be prepared.[34] ClF_6^+ and BrF_6^+ deserve special interest, because ClF_7 and BrF_7 do not exist. Therefore the preparation of these cations by simple F^- abstraction from the parent molecules is not possible, and powerful fluorinating oxidizers such as PtF_6 and $KrF^+SbF_6^-$ are required for their syntheses.

IF_7 is only a weak F^- acceptor but does react with NOF and CsF to form $[IF_8]^-$ salts, which have not been structurally characterized.[35]

Another route to at least some XF_n^- ions is by fluorination of simple halides, e.g.,

$$CsCl + 2F_2 \rightarrow CsClF_4$$

The question of how ionic these varous XF_n^+ and XF_n^- compounds are is not always easily answered, even when X-ray results are available. For example, in the SbF_6^- salts of ClF_2^+ and BrF_2^+ the Cl and Br atoms have two close fluorine neighbors but also two more distant ones (belonging mainly to the SbF_6^- ions), so that to describe them literally as XF_2^+ salts is rather an oversimplification. Generally, the importance of fluorine bridging increases with increasing size of the central atom (i.e., from chlorine fluorides, to bromine fluorides, to iodine fluorides).

The IF_6^- ion, as its Cs^+ salt, appears to have a symmetry lower than octahedral (cf. XeF_6 and discussion, p. 582).

Note that although BrF_6^- is well defined, there is no evidence for ClF_6^-, suggesting that ClF_5 is coordinatively saturated with one localized lone pair and five F atoms; in agreement with this, ClF_5 appears not to associate.

Substituted Halogen Fluorides. Finally it may be noted that in addition to the mixed $[IF_{2n}(O_2CCF_3)_{6-2n}]^{3-}$ anions mentioned above, there are neutral molecules derived formally from IF_3 or IF_5 in which some F atoms are replaced by R,[36] OR,[37] or R_F[38] (a perfluoro alkyl or aryl). These arise by the following reactions:

$$CH_3I + XeF_2 \rightarrow CH_3IF_2 \quad \text{(stable for several hours)}$$
$$IF_5 + (CH_3)_3Si(OCH_3) \rightarrow IF_{5-n}(OMe)_n \quad n = 1-4$$
$$4ClF_3 + 3R_FI \rightarrow 3R_FIF_4 + 2Cl_2$$
$$IF_5 + nSO_3 \rightarrow IF_{5-n}(OSO_2F)_n$$

17-15. Other Interhalogen Ions

In addition to the fluoro ions discussed above, several other types are known; examples are listed in Table 17-8.

Anionic Species. One of the earliest to be recognized is I_3^- whose formation

[34] K. O. Christe, *Inorg. Chem.,* 1973, **12,** 1580; F. Q. Roberto, *Inorg. Nucl. Chem. Lett.,* 1972, **8,** 737; K. O. Christe and R. D. Wilson, *Inorg. Chem.,* 1975, **14,** 694; R. J. Gillespie and G. J. Schrobilgen, *Inorg. Chem.,* 1974, **13,** 1230; K. O. Christe, J. F. Hon, and D. Pilipovich, *Inorg. Chem.,* 1973, **12,** 84; M. Brownstein and H. Selig, *Inorg. Chem.,* 1972, **11,** 658; F. A. Hohorst, L. Stein, and E. Gebert, *Inorg. Chem.,* 1975, **14,** 2233.

[35] C. J. Adams, *Inorg. Nucl. Chem. Lett.,* 1974, **10,** 831.

[36] J. A. Gibson and A. F. Janzen, *J.C.S. Chem. Commun.,* **1973,** 739.

[37] G. Oates, J. M. Winfield, and O. R. Chambers, *J.C.S. Dalton,* **1974,** 1381.

[38] J. M. Winfield *et al., J.C.S. Dalton,* **1974,** 119, 509.

TABLE 17-8
Principal Types of Polyhalide Ion[a] (Other than Fluoro)

Cations	Anions[b]		Cations	Anions[b]	
X_n^+	X_3^-	X_5^-	X_n^+	X_3^-	X_5^-
Br_2^+	Cl_3^-		I_5^+	I_2Cl^-	I_5^-
I_2^+	Br_3^-	ICl_4^-	ICl_2^+	IBr_2^-	
Cl_3^+	I_3^-	$IBrCl_3^-$	$IBrCl^+$	ICl_2^-	
Br_3^+	Br_2Cl^-	$I_2Br_2Cl^-$	IBr_2^+	$IBrCl^-$	
I_3^+	$BrCl_2^-$	I_4Cl^-	I_2Cl^+	$IBrF^-$	
			I_2Br^+		

[a] Also I_7^-, I_6Br^-, Br_6Cl^-, I_9^-; for fluoro species, see Table 17-7.
[b] Usually as salts of large univalent cations (e.g., Cs^+, Et_4N^+).

accounts for the increased solubility of I_2 in water on addition of KI. Few of the other ions are stable in aqueous solution, the most stable being I_3^-. In nonaqueous media such as CH_3OH or CH_3CN, stabilities are substantially higher. Many of the ions, especially the larger ones, exist only in crystalline salts with large cations such as Cs^+ or R_4N^+. The ions or salts are usually prepared by direct interaction between X^- and X_2, e.g.,

$$KI + Cl_2 \rightarrow KICl_2 \leftarrow KCl + ICl$$

Of the simple X_3^- ions, Cl_3^- is the least stable but is formed when concentrated solutions of Cl^- are saturated with chlorine:

$$Cl^-(aq) + Cl_2 \rightleftharpoons Cl_3^-(aq) \qquad K \approx 0.2$$

The F_3^- ion (nor any other one with F at the center) does not exist under normal chemical circumstances, but it is formed in an argon matrix at 15°K by reaction of MF with F_2. It appears on spectroscopic evidence to be linear and symmetrical.[39]

There is slight evidence in the bromine system for Br_5^-, but the series of polyions I_5^-, I_7^-, and I_9^- is well established for iodine.

The X_3^- species are linear but not necessarily symmetrical.[40] In solution I_3^- appears to be symmetrical, as it is in $Ph_4As^+I_3^-$; but in CsI_3 there are two I—I distances (2.83 and 3.03 Å), and long polymeric chains are present in benzamide hydrogen triiodide. The anions in $CsBr_3$, CsI_2Br, and $CsIBr_2$ are also unsymmetrical.

In the Cl_3^-, $BrCl_2^-$, and ICl_2^- ions, nuclear quadrupole resonance (nqr) studies show that the negative charge is increasingly concentrated on the outer Cl atoms in the order above.[41a]

The I_5^- ion (Fig. 17-1) can best be described as I^- with two I_2 molecules fairly weakly coordinated to it; and I_7^- appears to be similar but with even weaker bonds between I_3^- and two I_2 molecules—whether such an arrangement really constitutes a discrete "ion" is questionable.

[39] B. S. Ault and L. Andrews, *Inorg. Chem.*, 1977, **16**, 2024.
[40] T. Surles *et al.*, *J. Inorg. Nucl. Chem.*, 1973, **35**, 668; *Inorg. Nucl. Chem. Lett.*, 1973, **9**, 437, 1131.
[41a] E. F. Riedel and R. D. Willet, *J. Am. Chem. Soc.*, 1975, **97**, 701.

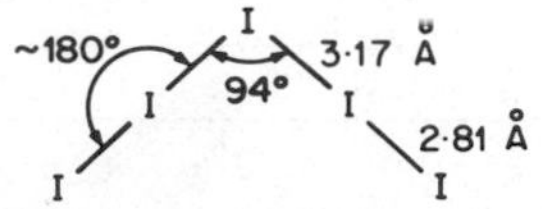

Fig. 17-1. Structure of the pentaiodide ion I_5^- in $[Me_4N]I_5$.

The largest polyiodide anion, I_{16}^{4-} in a theobromine salt, is nearly planar and is best described[41b] as $I_3^- \cdots I_2 \cdots I_3^- \cdots I_3^- \cdots I_2 \cdots I_3^-$.

Cationic Species. There is no evidence for any simple X^+ salts, but complexed forms of Br^+ and I^+ are known (see below) and X_2^+ and X_3^+ ions exist under a variety of circumstances. Reactions of Br_2 and I_2 with the oxidant $S_2O_6F_2$ afford SO_3F^- salts of these cations,[42] but chlorine gives only the covalent molecule $ClOSO_2F$ under similar circumstances, though there is some epr evidence for an ion that may be $OClF^+$ or O_2ClF^+. The Cl_3^+ and Br_3^+ ions are formed in the following reactions:

$$ClF + Cl_2 + AsF_5 \rightarrow Cl_3^+ \, AsF_6^-$$
$$O_2^+ AsF_6^- + \tfrac{3}{2}Br_2 \rightarrow Br_3^+ \, AsF_6^- + O_2$$

Mixed cations, such as ICl_2^+, IBr_2^+, I_2Br^+, I_2Cl^+, and $BrICl^+$, can be obtained either as SO_3F^- salts[43] or $SbCl_6^-$ salts.[44]

Iodine-containing cations are the most numerous and generally the best characterized, and there is even an I_5^+ ion,[45] but even they exist only in media with low nucleophilic properties, since in more strongly donor solvents the ions are either attacked or are complexed to give species such as $[py_2I]^+$.

Solutions of I_2^+ and I_3^+ can be obtained by oxidation in oleum:

$$2I_2 + 6H_2S_2O_7 \rightarrow 2I_2^+ + 2HS_3O_{10}^- + 5H_2SO_4 + SO_2$$
$$3I_2 + 6H_2S_2O_7 \rightarrow 2I_3^+ + 2HS_3O_{10}^- + 5H_2SO_4 + SO_2$$

while ICl_3 in oleum gives ICl_2^+.

Molten iodine is electrically conducting, and this property can be ascribed to self-ionization:

$$3I_2 \rightleftharpoons I_3^+ + I_3^-$$

Complexes of Cationic Halogens. Although the I^+ ion appears not to exist, compounds of the types $[Ipy_2]^+X^-$ ($X = NO_3^-$, RCO_2^-, or ClO_4^-), or $IpyX$, where the anion is also coordinated as in $IpyONO_2$ or $IpyOCOR$, are known.

Such compounds are generally prepared by treatment of a silver salt with the stoichiometric amount of iodine and an excess of pyridine in $CHCl_3$, e.g.,

$$AgNO_3 + I_2 + 2py = [I \, py_2]NO_3 + AgI\downarrow$$

After removal of AgI the complexes can be isolated from the solution. Electrolysis

[41b] F. H. Herbstein and M. Kapon, *J.C.S. Chem. Comm.*, **1975**, 677.

[43] W. W. Wilson and F. Aubke, *Inorg. Chem.*, 1974, **13**, 326.

[44] J. Shamir and M. Lustig, *Inorg. Chem.*, 1973, **12**, 1108; D. J. Merryman and J. D. Corbett, *Inorg. Chem.*, 1974, **13**, 1258.

[45] D. J. Merryman et al., *J.C.S. Chem. Comm.*, **1972**, 779.

of $[Ipy_2]^+NO_3^-$ in chloroform yields iodine at the cathode; however compounds of the type pyIOCOR give only feebly conducting solutions in acetone. Direct interaction of I_2 and pyridine gives a compound that X-ray studies show to contain the planar ion $[pyIpy]^+$ along with I_3^- and I_2 molecules. The only other ion of this type, $[(NH_2)_2CS]_2I^+\cdot I^-$, is obtained by grinding iodine and thiourea together.[46a]

Similar but less stable complexes of chlorine and bromine are known, including $[Br(quinoline)_2^+][ClO_4^-]$, which has been shown by X-ray study to contain a discrete cation with a linear N—Br—N chain.[46b]

OTHER COMPOUNDS

17-16. Halogen Oxofluorides[47a]

There are a number of halogen oxofluorides, and their chemistry is broadly similar to that of the halogen fluorides, viz., (1) the central halogen (Cl, Br, I) may be in various formal oxidation states, (2) they are generally very reactive and serve as oxidants and fluorinating agents, (3) they react with F^- donors and acceptors to give mixed oxofluoro anions and cations. The known compounds are listed in Table 17-9.

$FClO_2$ (b.p. $-6°$) is formed by partial hydrolysis of chlorine fluorides or by the reaction

$$6NaClO_3 + 4ClF_3 \xrightarrow{25°} 6NaF + 2Cl_2 + 3O_2 + 6FClO_2$$

It reacts with F^- acceptors to give ClO_2^+ salts. $FBrO_2$[47b] is prepared by the reaction

$$2KBrO_3 + BrF_5 \xrightarrow{HF} 2K[BrF_2O_2] + BrO_2F$$

Some $K[BrF_4O]$, which forms as an intermediate, may also be isolated. The $BrF_2O_2^-$ ion has an SF_4-like structure with the O atoms equatorial, whereas BrF_4O^- is sp,

TABLE 17-9.
Halogen Oxofluorides

$FClO_2$	$FBrO_2$	FIO_2
F_3ClO		F_3IO
$FClO_3$	$FBrO_3$	FIO_3
F_3ClO_2		F_3IO_2
		F_5IO

46a H. Hope and G. H. Y. Lin, *J.C.S. Chem. Comm.*, **1970**, 179.
46b N. W. Alcock and G. B. Robertson, *J.C.S. Dalton*, **1975**, 2483.
47a K. O. Christe *et al.*, *Inorg. Nucl. Chem. Lett.*, **1975**, 161.
47b R. J. Gillespie and P. H. Spekkens, *Isr. J. Chem.*, 1978, **17**, 11.

like XeF_4O. FIO_2 is a white solid, readily hydrolyzed but not generally well characterized. It is obtained by fluorination of I_2O_5.

ClF_3O^{48} can be prepared in various ways; it is a colorless gas or liquid (b.p. 29°) or white solid (m.p. −42°), and has an SF_4-like structure with O in one of the equatorial positions. It reacts as an oxygenating and/or fluorinating agent and it also forms adducts with both Lewis acids and Lewis bases as illustrated by the following two reactions

$$ClF_3O + AsF_5 \rightarrow [ClF_2O^+][AsF_6^-]$$
$$ClF_3O + CsF \rightarrow [Cs^+][ClF_4O^-]$$

The ClF_2O^+ ion has a pyramidal structure like that of SF_2O and ClF_4O^- has a *sp* structure with O at the apex, like that of XeF_4O.

F_3IO, obtained by the reaction

$$I_2O_5 + 3IF_5 \rightarrow 5IOF_3$$

is a white solid that readily disproportionates:

$$2IOF_3 \rightarrow FIO_2 + IF_5$$

Perchloryl fluoride ($FClO_3$) is perhaps the most important compound of this class. It can be prepared by the action of F_2 or FSO_3H on $KClO_4$, but is best made by the solvolytic reaction of $KClO_4$ with a superacid (e.g., a mixture of HF and SbF_5):

$$KClO_4 + 2HF + SbF_5 \xrightarrow{40-50°} FClO_3 + KSbF_6 + H_2O$$

The toxic gas (m.p. −147.8°, b.p. −46.7°) is thermally stable to 500° and resists hydrolysis. At elevated temperatures it is a powerful oxidizing agent and has selective fluorinating properties, especially for replacement of H by F in CH_2 groups. It can also be used to introduce ClO_3 groups into organic compounds (e.g., C_6H_5Li gives $C_6H_5ClO_3$). In these reactions it appears that the nucleophile attacks the chlorine atom, e.g.,

$$RO^- + FClO_3 \longrightarrow \quad\quad \longrightarrow ROClO_3 + F^-$$

Perbromyl fluoride[49] (m.p. −110°) is made similarly, but it is more reactive than ClO_3F and is hydrolyzed by base:

$$BrO_3F + 2OH^- = BrO_4^- + H_2O + F^-$$

presumably by initial associative attack of OH^- on Br.

F_3ClO_2, a colorless gas or liquid (b.p. −21.6°) or white solid (m.p. −81.2°) has a *tbp* structure with the oxygen atoms in equatorial positions. It forms the $F_2ClO_2^+$

48   K. O. Christe *et al., Inorg. Chem.,* 1972, **11,** 2189, 2192, 2196, 2201, 2205, 2209.
49   G. K. Johnson, P. A. G. O'Hare, and E. H. Appelman, *Inorg. Chem.,* 1972, **11,** 800.

ions with Lewis bases. The iodine analogue F_3IO_2 is structurally more complex. The crystal[50] contains centrosymmetric dimers (17-V), which appear to occur in the vapor also, but solutions appear to contain polymers. IO_2F_3 reacts with AsF_5, SbF_5, NbF_5, and some other similar molecules to form polymeric adducts in which oxo bridges connect both IF_4 and MF_4 units with some pairs of oxygen atoms cis and others trans. With F^- donors it also gives the cis and trans octahedral ions $[IO_2F_4]^-$, which can also be made by reaction of IO_4^- with HF or by partial hydrolysis of IF_7 in HF.[51a]

(17-V)

The fluoroxosulfate ion, O_3SOF^-, can be isolated by action of F_2 on $CsSO_4$ solutions. It is a powerful but unstable oxidant.[51b]

17-17. Compounds of Iodine(III)

Many compounds of iodine(III) are known, including some with organic groups. The only bromo analogues are $Br(OSO_2F)_3$ and $K[Br(OSO_2F)_4]$. Among the compounds in which I^{III} is combined with oxoanions are $I(OSO_2F)_3$,[52] $I(NO_3)_3$,[53] $I(OCOCH_3)_3$, IPO_4, and $I(OClO_3)_3$.[54a] These compounds contain essentially covalent I—O bonds. Preparative methods include the following:

$$I_2 + AgClO_4 \xrightarrow[-85°]{\text{ether}} I(OClO_3)_3 + AgI$$

$$I_2 + 6ClOClO_3 \longrightarrow 2I(OClO_3)_3 + 3Cl_2$$

$$CsI + 4ClOClO_3 \longrightarrow Cs[I(OClO_3)_4] + 2Cl_2$$

$$I_2 + HNO_3 \text{ (fuming)} + (CH_3CO)_2O \longrightarrow I(OCOCH_3)_3$$

$$I_2 + HNO_3 \text{ (conc)} + H_3PO_4 \longrightarrow IPO_4$$

The compounds are sensitive to moisture and are not stable much above room temperature. They are hydrolyzed with disproportionation of the I^{III}, as illustrated for IPO_4 thus:

$$5IPO_4 + 9H_2O \rightarrow I_2 + 3HIO_3 + 5H_3PO_4$$

Covalent I^{III} is known also in the compound triphenyliodine $(C_6H_5)_3I$ and a large number of diaryliodonium salts, such as $(C_6H_5)_2I^+X^-$, where X may be one of a number of common anions. Aryl compounds[54b] such as $C_6H_5ICl_2$ are also well

50   L. E. Smart, *J.C.S. Chem. Comm.*, **1977**, 519.
51a   H. Selig and U. Elgad, *J. Inorg. Nucl. Chem., Suppl.*, **1976**, 91.
51b   E. H. Appelman, *et al.*, *J. Am. Chem. Soc.*, 1979, **101**, 3384.
52   F. Aubke and D. D. Desmarteau, *Fluorine Chem. Rev.*, 1977, **8**, 73.
53   K. O. Christe, C. J. Schack, and R. D. Wilson, *Inorg. Chem.*, 1974, **13**, 2378.
54a   K. O. Christe and C. J. Schack, *Inorg. Chem.*, 1972, **11**, 1682.
54b   N. W. Alcock *et al.*, *J.C.S. Dalton*, **1979**, 854.

known and can be prepared by direct interactions of C_6H_5I with X_2. They can be regarded as *tbp*, with axial chlorine atoms and equatorial phenyl group and lone pairs. The compounds can be used for chlorination of alkenes.

Oxo Compounds.   The so-called iodosyl sulfate $(IO)_2SO_4$, which is a yellow solid obtained by the action of H_2SO_4 on I_2O_5 and I_2, has polymeric $(I—O)_n$ chains cross-linked by the anion. Similarly, HIO_3 in H_2SO_4 gives monomeric IO_2HSO_4 in dilute solution, but at high concentrations polymerization occurs and a white solid, $I_2O_5SO_3$, can be obtained.

17-18.   Organic Fluorine Chemistry[55]

Organic molecules in which most or all of the hydrogen atoms are replaced by fluorine atoms have special, and often valuable, properties, and they are often made using inorganic reagents. Hence they merit brief discussion here.

The main preparative methods are the following:

1.   *Replacement of Other Halogens by Means of Metal Fluorides.*   The driving force for a reaction

$$R—Cl + MF = R—F + MCl$$

depends in part on the free-energy difference of MF and MCl, which is approximately equal to the difference in lattice energies. The larger the cation M, the more favorable tends to be the free energy for the reaction above. Thus among the alkalis, LiF is the poorest and CsF the most effective. For AgF the difference in lattice energies is small owing to contributions of nonionic bonding in AgCl, so that AgF is a very powerful fluorinating agent.

Other fluorinating agents, each having particular advantages under given conditions, are SbF_3 ($+SbCl_5$ catalyst), HgF_2, KHF_2, ZnF_2, and AsF_3. Examples of some fluorinations are:

$$C_6H_5PCl_2 + AsF_3 \xrightarrow{25°} C_6H_5PF_2 + AsCl_3$$

$$C_6H_5CCl_3 + SbF_3 \longrightarrow C_6H_5CF_3 + SbCl_3$$

2.   *Replacement of Other Halogens Using Hydrogen Fluoride.*   Anhydrous HF together with catalysts such as $SbCl_5$ or CrF_4 at temperatures of 50 to 150° and pressures of 50 to 500 psi will effect processes such as

$$2CCl_4 + 3HF \longrightarrow CCl_2F_2 + CCl_3F + 3HCl$$
$$\text{b.p. } -29.8° \quad \text{b.p. } 23.7°$$

$$CHCl_3 + 2HF \longrightarrow CHClF_2 + 2HCl$$

$$CCl_3COCCl_3 \xrightarrow[\substack{Cr \\ catalyst}]{HF} CF_3COCF_3$$

[55]   W. A. Sheppard and C. M. Sharts, *Organic Fluorine Chemistry,* Benjamin, 1970; R. D. Chalmers, *Fluorine in Organic Chemistry,* Wiley, 1973 (comprehensive monographs); R. E. Banks, *Organofluorine Chemicals and Their Industrial Applications,* Wiley, 1979.

3. *Electrolytic Replacement of Hydrogen by Fluorine.* Electrolysis of organic compounds in liquid HF at voltages ~4.5–6 below that required for the liberation of fluorine, in steel cells with Ni anodes and steel cathodes, causes fluorination at the anode. Examples of such fluorinations are

$$(C_2H_5)_2O \rightarrow (C_2F_5)_2O$$
$$C_8H_{18} \rightarrow C_8F_{18}$$
$$(CH_3)_2S \rightarrow CF_3SF_5 + (CF_3)_2SF_4$$
$$(C_4H_9)_3N \rightarrow (C_4F_9)_3N$$

4. *Direct Replacement of Hydrogen by Fluorine.* Although most organic compounds inflame or explode when mixed with fluorine in normal circumstances, direct fluorination of many compounds is possible under appropriate conditions. There are two main techniques.

Catalytic fluorination involves mixing the reacting compound and fluorine diluted with nitrogen in presence of the catalyst. The catalyst may be copper gauze, silver-coated copper gauze, or cesium fluoride. An example is

$$C_6H_6 + 9F_2 \xrightarrow{\text{Cu},265°} C_6F_{12} + 6HF$$

A recently developed procedure[56] giving high yields involves the reaction of the substrate in the *solid* state (liquids or gases are frozen) with fluorine diluted with helium over a rather long period (12–36 hr) at a low temperature in presence of a heat sink in the form of the reactor and containers. The purpose is to allow the heat generated in the exothermic reaction (overall for replacement of H by F, *ca.* 420 kJ mol^{-1}), which could lead to C—C bond breaking, to be slowly dissipated. The replacement reaction appears to proceed by several steps, each less exothermic than the C—C average bond strength, so that provided the reaction time allows completion of individual reactions, fluorination without degradation is possible. Examples of materials that can be fluorinated in this way are polystyrene, anthracene and other polynuclear hydrocarbons, and compounds containing atoms other than carbon such as carbaboranes, β-trichloroborazines, phthalocyanine, and many organometallic compounds.

Inorganic fluorides such as cobalt(III) fluoride have also commonly been used for the vapor phase fluorination of organic compounds, e.g.,

$$(CH_3)_3N \xrightarrow{\text{CoF}_3} (CF_3)_3N + (CF_3)_2NF + CF_3NF_2 + NF_3$$

5. *Replacement of Oxygen by Fluorine.* A particularly useful and selective fluorinating agent for oxygen compounds is SF_4 (p. 520); e.g., ketones RR'CO may be converted to RR'CF$_2$. The CsF-catalyzed interaction of CO_2 and F_2 gives difluorobis(fluorooxy)methane, $CF_2(OF)_2$.

The high stabilities of many of fluoroorganic compounds owe something to the very high C—F bond energy (486 kJ mol^{-1}; cf. C—H 415 and C—Cl 332 kJ mol^{-1}), but organic fluorides are not necessarily particulary stable thermodynamically; rather, the low reactivities of fluorine derivatives must be attributed

56   R. J. Lagow and J. L. Margrave, *Prog. Inorg. Chem.*, 1979, 26, 69.

to the impossibility of expanding the octet of fluorine and carbon and the inability of, say, water to coordinate to fluorine or carbon as the first step in hydrolysis. Because of the small size of the F atom, H atoms can be replaced by F atoms with the least introduction of strain or distortion, as compared with replacement by other halogen atoms. The F atoms also effectively shield the carbon atoms from attack. Finally, since C bonded to F can be considered to be effectively oxidized (whereas in C—H it is reduced), there is no reaction with gaseous oxygen. Fluorocarbons are attacked only by hot metals (e.g., molten Na). When pyrolyzed, fluorocarbons tend to split at C—C rather than C—F bonds.

The replacement of H by F leads to increased density, but not to the same extent as by other halogens. Completely fluorinated derivatives C_nF_{2n+2} have very low boiling points for their molecular weights and low intermolecular forces; the weakness of these forces is also shown by the very low coefficient of friction for polytetrafluoroethylene $(-CF_2-CF_2-)_n$.

Among the commercially important organic fluorine compounds are the chlorofluorocarbons such as CCl_2F_2 and $CClF_3$, which were used as aerosol propellants and are working fluids in refrigerators, $CF_3CHBrCl$ and some similar compounds, used as anesthetics, and $CHClF_2$, which is used for making tetrafluoroethylene:

$$2CHClF_2 \xrightarrow{500-1000°} CF_2{=}CF_2 + 2HCl$$

Tetrafluoroethylene (b.p. $-76.6°$) can be polymerized thermally or in aqueous emulsion by use of oxygen, peroxides, etc., as free-radical initiators. A convenient laboratory source of C_2F_4 is thermal cracking of the polymer at 500 to 600°.

A rather bizarre use of certain perfluoroorganics is as "synthetic blood." The blood of small mammals can be entirely replaced by the fluorocarbon, and the animal remains alive while slowly restoring its real blood.[57] The fluorocarbon, which is essentially nontoxic, has the capacity to dissolve, transport, and release oxygen by simple solubility.

The fluorinated carboxylic acids are notable for their strongly acid nature—for example, for CF_3COOH, $K_a = 5.9 \times 10^{-1}$, whereas CH_3COOH has $K_a = 1.8 \times 10^{-5}$. Many standard reactions of carboxylic acids can be carried out leaving the fluoroalkyl group intact:

$$C_3F_7COOH \xrightarrow[C_2H_5OH]{H_2SO_4} C_3F_7COOC_2H_5 \xrightarrow{NH_3} C_3F_7CONH_2 \begin{matrix} \xrightarrow{P_2O_5} C_3F_7CN \\ \\ \xrightarrow{LiAlH_4} C_3F_7CH_2NH_2 \end{matrix}$$

Perfluoroalkyl halides $C_nF_{2n+1}X$ (R_FX) readily undergo free-radical reactions on heating or irradiation, (e.g., $CF_3I \rightarrow CF_3^{\bullet} + I^{\bullet}$, $\Delta H \approx 115$ kJ mol^{-1}) but do not readily form Grignard reagents. CF_3I is an important reagent for preparing numerous compounds containing CF_3 groups via radical pathways. A newer method is to generate CF_3 radicals from C_2F_6 in a glow discharge (plasma reactor). For

[57]   J. G. Reiss and M. LeBlanc, *Angew. Chem., Int. Ed.,* 1978, **17**, 621; T. Vanleer, *Omni,* 1979, **1**, 14.

example, this provides a nearly quantitative synthesis[58] of pure $(CF_3)_2Hg$, which is then a useful reagent for synthesis of CF_3 compounds. Other CF_3 derivatives worthy of note are CF_3NO, which can be copolymerized with $CF_2{=}CF_2$ to form very stable elastomers, and CF_3OF, which is thermally stable to 450° but readily gives oxidative addition products with simple substrates, e.g.:

$$CF_3OF + SO_2 \rightarrow CF_3OSO_2F$$
$$CF_3OF + SO_3 \rightarrow CF_3OOSO_2F$$
$$CF_3OF + C_2F_4 \rightarrow CF_3OCF_2CF_3$$
$$CF_3OF + SF_4 \rightarrow CF_3OSF_5$$

CF_3OF is also highly useful for selective fluorination of aromatic compounds.[59]

[58]   R. J. Lagow et al., J. Am. Chem. Soc., 1975, **97**, 518.
[59]   S. Rozen and O. Lerman, J. Am. Chem. Soc., 1979, **101**, 2782.

General References

Brown, D., *Halides of Lanthanides and Actinides*, Wiley-Interscience, 1968.

Canterford, J. H., and R. Colton, *Halides of First Row Transition Metals*, 1969; *Halides of Second and Third Row Transition Metals*, Wiley-Interscience, 1968.

Downs, A. J., and Adams, C. J., *The Chemistry of Chlorine, Bromine, Iodine and Astatine*, Pergamon Press, 1975.

Eagers, R. Y., *Toxic Properties of Inorganic Fluorine Compounds*, Elsevier, 1969.

Emeléus, H. J., *The Chemistry of Fluorine and Its Compounds*, Academic Press, 1969.

Emeléus, H. J., and J. C. Tatlow, Eds., *Journal of Fluorine Chemistry*, Vol. 1, 1971.

Gillespie, R. J., and M. J. Morton, *Q. Rev.*, 1971, **25**, 553. Halogen and interhalogen cations.

Gutmann, V., Ed., *Halogen Chemistry*, Academic Press, 1967. Three volumes on various aspects of chemistry.

Jolles, Z. E., Ed., *Bromine and Its Compounds*, Benn, London, and Academic Press, New York, 1966. Reference book.

Kubo, M., and D.Nakamura, *Adv. Inorg. Chem. Radiochem.*, 1966, **8**, 257. Nuclear quadrupole resonance studies, especially of halogen compounds.

Lawless, E. W., and I. C. Smith, *Inorganic High Energy Oxidizers*, Dekker, 1969. Properties of F_2, fluorohalogen, and fluoronitrogen compounds.

Mellor, J. W., *Comprehensive Treatise on Inorganic Chemistry*, Suppl. II, Pt. I, Longmans Green, 1956. Contains all the Group VII elements.

Neumark, H. R., et al., *The Chemistry and Chemical Technology of Fluorine*, Wiley-Interscience, 1967.

O'Donnell, T. A., *The Chemistry of Fluorine*, Pergamon Press, 1975.

Pennman, R. A., *Inorg. Chem.*, 1969, **8**, 1379. Use of molar refractivity for ascertaining composition of transition metal fluoro complexes.

Scherer, O., *Technische organische Fluor Chemie, Topics in Current Chemistry*, Vol. 2, Part 13/14, Springer-Verlag, 1970.

Schmeisser, M., and K. Brändle, *Adv. Inorg. Chem. Radiochem.*, 1963, **5**, 41. Oxides and oxofluorides of halogens; a comprehensive review.

Schumaker, J. C., Ed., *Perchlorates*, ACS Monograph No. 146, Reinhold, 1960. An extensive review of perchloric acid and perchlorates.

Sconce, J. S., Ed., *Chlorine: Its Manufacture, Properties, and Uses*, ACS Monograph No. 154, Reinhold, 1962.

Simons, J. H., Ed., *Fluorine Chemistry,* Vols. I–V, Academic Press, 1954. Comprehensive reference books on special topics in fluorine chemistry.

Tarrant P., Ed., *Fluorine Chemistry Review,* Vol. 1, Dekker, 1967. Various topics.

Tatlow, J. C., *et al,* Eds., *Advances in Fluorine Chemistry.* Review volumes on all aspects of fluorine chemistry.

Treichel, P. M., and F. G. A. Stone, *Adv. Organomet. Chem.,* 1964, **1,** 143. Fluorocarbon derivatives of metals.

Zinov'ev, A. A., *Russ. Chem. Rev.,* **1963,** 268. Perchloric acid and Cl_2O_7.

CHAPTER EIGHTEEN

The Noble Gases

THE ELEMENTS

18-1. Group Trends

The closed-shell electronic structures of the noble gas atoms are extremely stable, as shown by the high ionization enthalpies, especially of the lighter members (Table 18-1). The elements are all low-boiling gases whose physical properties vary sys-

TABLE 18-1
Some Properties of the Noble Gases

	Outer shell configuration	Atomic number	First ionization enthalpy kJ mol^{-1}	Normal b.p. (°K)	ΔH_{vap} (kJ mol^{-1})	% by volume in the atmosphere	Promotion energy (kJ mol^{-1}), $ns^2np^6 \rightarrow$ $ns^2np^5(n+1)s$
He	$1s^2$	2	2372	4.18	0.09	5.24×10^{-4}	—
Ne	$2s^22p^6$	10	2080	27.13	1.8	1.82×10^{-3}	1601
Ar	$3s^23p^6$	18	1520	87.29	6.3	0.934	1110
Kr	$4s^24p^6$	36	1351	120.26	9.7	1.14×10^{-3}	955
Xe	$5s^25p^6$	54	1169	166.06	13.7	8.7×10^{-6}	801
Rn	$6s^26p^6$	86	1037	208.16	18.0		656

tematically with atomic number. The boiling point of helium is the lowest of any known substance. The boiling points and heats of vaporization increase monotonically with increasing atomic number.

The heats of vaporization are measures of the work that must be done to overcome interatomic attractive forces. Since there are no ordinary electron-pair interactions between noble gas atoms, these weak forces (of the van der Waals or London type) are proportional to the polarizability and inversely proportional to the ionization enthalpies of the atoms; they increase therefore as the size and diffuseness of the electron clouds increase.

The ability of the noble gases to enter into chemical combination with other atoms is very limited, only krypton, xenon, and radon having so far been induced to do so, and only bonds to F, Cl, O, and N are stable. This ability would be expected to increase with decreasing ionization enthalpy and decreasing energy of promotion to states with unpaired electrons. The data in Table 18-1 for ionization enthalpies and for the lowest-energy promotion process show that chemical activity should increase down the group. Apparently the threshold of actual chemical activity is reached only at Kr. The chemical activity of Xe is markedly greater. That of Rn is presumably still greater, but it is difficult to assess because the half-life of the longest-lived isotope, ^{222}Rn, is only 3.825 days, so that only tracer studies can be made.

18-2.   Occurrence, Isolation, and Applications

The noble gases occur as minor constituents of the atmosphere (Table 18-1). Helium is also found as a component (up to ~7%) in certain natural hydrocarbon gases in the United States. This helium undoubtedly originated from decay of radioactive elements in rocks, and certain radioactive minerals contain occluded helium that can be released on heating. All isotopes of radon are radioactive and are occasionally given specific names (e.g., actinon, thoron) derived from their source in the radioactive decay series; ^{222}Rn is normally obtained by pumping off the gas from radium chloride solutions. Ne, Ar, Kr, and Xe are obtainable as products of fractionation of liquid air.

The main uses of the gases are in welding (argon provides an inert atmosphere), in gas-filled electric light bulbs and radio tubes (argon), and in discharge tubes (neon); radon has been used therapeutically as an α-particle source in the treatment of cancer. Helium as liquid is extensively used in cryoscopy and as an inert protective gas in chemical reactions.

The amounts of He and Ar formed by radioactive decay in minerals can be used to determine the age of the specimen. For example, in the course of the decay of ^{238}U, eight α-particles are produced; these acquire electrons to form He atoms by oxidizing other elements present. If the rock is sufficiently impermeable, the total He remains trapped therein. If the amounts of trapped helium and remaining ^{238}U are measured, the age of the specimen can be calculated, for one-eighth of the atoms of He represent the number of ^{238}U atoms that have decayed. A correction must be applied for thorium, which also decays by α emission and generally occurs in small amounts with uranium. Argon arises in potassium-containing minerals by electron capture of ^{40}K; a complication arises here, since ^{40}K also decays by β emission to ^{40}Ca, and the accuracy of the age determination in this case depends on accurate determination of the branching ratio of ^{40}K.

18-3. Special Properties of Helium[1a]

Naturally occurring helium is essentially all ^{4}He, although ^{3}He occurs to the extent of $\sim 10^{-7}$ at. %. ^{3}He can be made in greater quantities by nuclear reactions and by β^- decay of tritium.

The pressure-temperature diagram near absolute zero for helium is shown in Fig. 18-1. Its most remarkable feature is that helium has no triple point; that is,

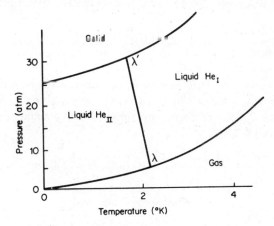

Fig. 18-1. Helium phase diagram near absolute zero.

there is *no* combination of temperature and pressure at which solid, liquid, and gas coexist in equilibrium. Helium is the only substance known to lack a triple point. It is also the only one that cannot be solidified at atmospheric pressure. These departures of helium from the universal pattern are due to a quantum effect. Its zero-point energy is so high that it outweighs the weak interatomic forces which, without the application of external pressure, are not strong enough to bind the helium atoms into the crystalline state. Still more remarkable, however, is the transition that takes place across the line λ–λ' $\sim 2.2°$K from He$_I$ to He$_{II}$. This is marked by a huge specific heat anomaly, He$_{II}$ having much lower entropy. Nevertheless, He$_{II}$ has the random ordering of a true liquid. It is, however, a unique liquid in exhibiting the phenomenon of superconductivity. It has immeasurably low viscosity and readily forms films only a few hundred atoms thick that flow apparently without friction, even up over the edges of a vessel. No fully satisfactory explanation of these properties has yet been devised.

[1a] J. Wilks, *The Properties of Liquid and Solid Helium*, Oxford University Press, 1967; R. H. Kropschot, B. W. Birmingham, and D. B. Mann, Eds., *The Technology of Liquid Helium*, U.S. National Bureau of Standards Monograph No. 111, Government Printing Office, 1968; S. J. Putterman and I. Rudnick, *Phys. Today*, 1971, **24**, 39.

THE CHEMISTRY OF THE NOBLE GASES

After his observation that O_2 reacts with PtF_6 to give the compound $[O_2^+][PtF_6^-]$ (p. 494), N. Bartlett in 1962 recognized that since the ionization enthalpy of xenon is almost identical with that of the oxygen molecule, an analogous reaction should occur with xenon. He then confirmed this prediction by obtaining a red crystalline solid, originally believed to be "$XePtF_6$" (see below), by direct interaction of Xe with PtF_6. This discovery led to rapid and extensive developments in xenon chemistry.

18-4. The Chemistry of Xenon

Xenon reacts directly only with fluorine, but compounds in oxidation states from II to VIII are known, some of which are exceedingly stable and can be obtained in large quantities. The more important compounds and some of their properties are given in Table 18-2.

TABLE 18-2
Principal Xenon Compounds

Oxidation state	Compound	Form	M.p. (°C)	Structure	Remarks
II	XeF_2	Colorless crystals	129	Linear	Hydrolyzed to Xe + O_2; v. soluble in HF(l)
IV	XeF_4	Colorless crystals	117	Square	Stable, $\Delta H_f^{298°} = -284$ kJ mol^{-1}
VI	XeF_6	Colorless crystals	49.6	Complex	Stable, $\Delta H_f^{298°} = -402$ kJ mol^{-1}
	$CsXeF_7$	Colorless solid		See text	Dec. >50°
	Cs_2XeF_8	Yellow solid		Archim. antiprism[a]	Stable to 400°
	$XeOF_4$	Colorless liquid	−46	ψ-Octahedral[b]	Stable
	XeO_2F_2	Colorless crystals	31	ψ-Tbp[b]	Metastable
	XeO_3	Colorless crystals		ψ-Tetrahedral[b]	Explosive, $\Delta H_f^{298°} = +402$ kJ mol^{-1}; hygroscopic; stable in solution
	$nK^+[XeO_3F^-]_n$	Colorless crystals		Sp (F bridges)	Very stable
VIII	XeO_4	Colorless gas		Tetrahedral	Explosive
	XeO_6^{4-}	Colorless salts		Octahedral	Anions $HXeO_6^{3-}$, $H_2XeO_6^{2-}$, $H_3XeO_6^-$ also exist

[a] In the salt $(NO^+)_2[XeF_8]^{2-}$ from XeF_6 and NOF.
[b] Lone pair present.

Fluorides.   The equilibrium constants for the reactions

$$Xe + F_2 \rightarrow XeF_2$$
$$XeF_2 + F_2 \rightarrow XeF_4$$
$$XeF_4 + F_2 \rightarrow XeF_6$$

have been either measured or calculated for the range 25–500°. The studies show unequivocally that only these three binary fluorides exist. The equilibria are established rapidly only above 250°, and this is the lower limit for thermal preparative methods. All three fluorides are volatile, readily subliming at room temperature. They can be stored indefinitely in nickel or Monel metal containers, but XeF_4 and XeF_6 are particularly susceptible to hydrolysis and traces of water must be rigorously excluded.

Xenon Difluoride.   This is best obtained by interaction of F_2 and an excess of Xe at high pressure, but there are other methods such as the interaction of Xe and O_2F_2 at −118° and procedures in which XeF_2 is trapped from mixtures of F_2 and Xe at low pressures. It is soluble in water, giving solutions $0.15M$ at 0°, that evidently contain XeF_2 molecules. The hydrolysis is slow in dilute acid but rapid in basic solution:

$$XeF_2 + 2OH^- \rightarrow Xe + \tfrac{1}{2}O_2 + 2F^- + H_2O$$

The solutions, which have a pungent odor due to XeF_2, are powerful oxidizing agents (e.g., HCl gives Cl_2, Ce^{III} gives Ce^{IV}), and the estimated potential is

$$XeF_2(aq) + 2H^+ + 2e = Xe + 2HF(aq) \qquad E^0 = +2.64V$$

XeF_2 also acts as a mild fluorinating agent for organic compounds[1b]; for example, in solution or in the vapor phase benzene is converted into C_6H_5F.

Xenon Tetrafluoride. XeF_4 is the easiest fluoride to prepare, and essentially quantitative conversion is obtained when a 1:5 mixture of Xe and F_2 is heated in a nickel vessel at 400° and *ca.* 6 atm pressure for a few hours. Its properties are similar to those of XeF_2 except regarding its hydrolysis, as discussed below. With hydrogen, XeF_4 rapidly gives Xe and HF, and with fluorine under pressure it forms XeF_6. XeF_4 specifically fluorinates the ring in substituted arenes such as toluene.

Xenon Hexafluoride.   This preparation requires more severe conditions, but at high pressure (>50 atm) and temperature (>250°) quantitative conversion may be obtained. The solid is colorless but becomes yellow when heated and gives a yellow liquid and vapor. It reacts rapidly with quartz:

$$2XeF_6 + SiO_2 \rightarrow 2XeOF_4 + SiF_4$$

and it is extremely readily hydrolyzed.

The XeF_6 molecule would not be expected to be a stereochemically rigid, octahedral molecule because there are seven electron pairs in the valence shell of the

---

[1b]   R. Filler, *Isr. J. Chem.,* 1978, **17**, 71 (review).

Xe atom. A variety of experimental data confirm this, and theoretical work[2] harmonizes all the data in terms of a substantially distorted but stereochemically nonrigid (see Section 28-13) structure. The structure of the solid[3a] is extremely complex. There are at least four crystalline forms, three of which consist of tetramers and the fourth of both tetramers and hexamers. As shown in Fig. 18-2, these oligomers are built of square pyramidal XeF_5^+ units bridged by F^- ions.

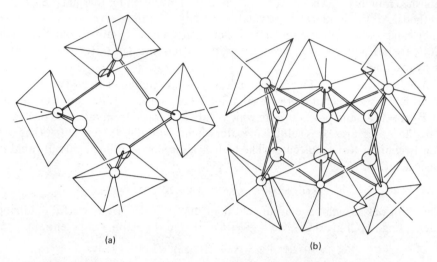

(a)

(b)

Fig. 18-2. (a) Tetrameric and (b) hexameric units in one of the crystal forms of XeF_6.

XeF_6 gives a yellow solution in $(F_5S)_2O$ at 25°, indicative of XeF_6 molecules; but as the temperature is lowered the solution becomes colorless, and nmr results at $-118°C$ show that tetramers are present.[3b]

Xenon Oxofluorides. There are two, $XeOF_4$ and XeO_2F_2, which are stable and well characterized. Suitable preparative reactions are:

$$XeF_6 + H_2O = XeOF_4 + 2HF$$
$$XeO_3 + XeOF_4 = 2XeO_2F_2$$
$$2XeO_3 + XeF_6 = 3XeO_2F_2$$
$$XeO_3 + 2XeF_6 = 3XeOF_4$$

In the first of these reactions the quantity of H_2O must be limited to the stoichiometric quantity to avoid forming the explosive XeO_3 (see below). $XeOF_4$ is square pyramidal (C_{4v}), whereas XeO_2F_2 is like SF_4, with F atoms in axial positions (C_{2v}).[4]

The compounds $XeOF_2$ and XeO_3F_2 are unstable and of little importance. XeO_2F_4 has been observed in the mass spectrometer.[5]

2   K. S. Pitzer and L. S. Bernstein, *J. Chem. Phys.*, 1975, **63**, 3849.
3a R. D. Burbank and G. R. Jones, *J. Am. Chem. Soc.*, 1974, **96**, 43.
3b H. Rupp and K. Seppelt, *Angew. Chem., Int. Ed.*, 1974, **13**, 613.
4   S. W. Petersen, R. D. Willett, and J. L. Houston, *J. Chem. Phys.*, 1973, **59**, 453.
5   J. L. Huston, *J. Am. Chem. Soc.*, 1971, **93**, 5255.

Fluorocations. One of the characteristic reactions of the binary fluorides is the transfer of F^- to strong fluoride acceptors to form compounds containing cations of the type $X_nF_m^+$. Actually these compounds are not fully ionic, since the fluoroanions form fluoro bridges to the cations. Compounds in this class formed by XeF_2 are of 2:1, 1:1, and 1:2 stoichiometries, and examples are $2XeF_2 \cdot SbF_5$, $XeF_2 \cdot AsF_5$, and $XeF_2 \cdot 2RuF_5$. Among other fluoride ion acceptors forming such compounds are TaF_5, NbF_5, PtF_5, IrF_5, and OsF_5. These compounds have been studied extensively by X-ray crystallography[6] and Raman spectroscopy.[7] There are two types of cation so formed, XeF^+ and $Xe_2F_3^+$. The way in which XeF^+ typically interacts with an F atom in the accompanying anion is shown for $[XeF^+][RuF_6^-]$ in Fig. 18-3, and the structure of $Xe_2F_3^+$ appears in 18-I.

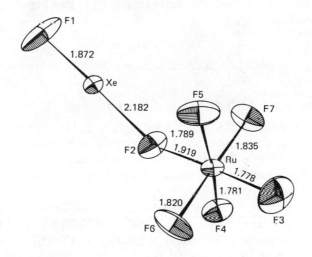

Fig. 18-3. The $[XeF^+][RuF_6^-]$ structural unit. [Reproduced by permission from Ref. 6.]

$$\text{F} \underset{F}{\overset{\text{Xe}}{\diagup}} \underset{151°}{\overbrace{\qquad}} \underset{F}{\overset{\text{Xe} \ 1.90 \text{ Å}}{\diagdown}} \quad \text{F } 2.14 \text{ Å}$$

(18-I)

Based on present knowledge of XeF_2 chemistry, it is believed that Bartlett's original reaction is best written

$$Xe + 2PtF_6 \xrightarrow{25°} [XeF^+][PtF_6^-] + PtF_5 \xrightarrow{60°} [XeF^+][Pt_2F_{11}^-]$$

The product obtained at 25° appears to contain both $[XeF^+][PtF_6^-]$ and $[PtF_5^-]$.

6   N. Bartlett *et al., Inorg. Chem.,* 1973, **12,** 1717.
7   R. J. Gillespie and B. Landa, *Inorg. Chem.,* 1973, **12,** 1383; B. Frlec and J. H. Holloway, *J.C.S. Dalton,* **1975,** 535.

The other fluorides have comparable though less extensive cation chemistry. For example, XeF_4 reacts with BiF_5 to give $[XeF_3^+][BiF_6^-]$[8], and XeF_6 forms the compounds $[XeF_5^+][RuF_6^-]$[6] and $[XeF_5^+]_2[PdF_6^{2-}]$[9] and others.[10] The presence of XeF_5^+ cations in such compounds is consistent with their incipient formation in solid XeF_6 (cf. Fig. 18-2). In the reaction of XeF_6 with AuF_3 and F_2 the compound $[Xe_2F_{11}^+][AuF_6^-]$ is obtained, in which the cation consists of two XeF_5^+ units bridged by F^- to give the overall composition $Xe_2F_{11}^+$.[11]

When dissolved in PF_5, AsF_5, and SbF_5; the oxofluorides $XeOF_4$ and XeO_2F_2 also form cations believed to be $XeOF_3^+$ and XeO_2F^+ that have been observed by Raman spectroscopy.[12a]

Other Xenon Fluorine Compounds. XeF_6 acts as a fluoride ion acceptor, since it reacts with alkali metal fluorides (other than LiF) to give heptafluoro- or octafluoroxenates(VI):

$$XeF_6 + RbF \rightarrow RbXeF_7$$

The Rb and Cs salts are well characterized; above 20° and 50° respectively, they decompose:

$$2MXeF_7 \rightarrow XeF_6 + M_2XeF_8$$

This formation of XeF_7^- and XeF_8^{2-} salts from XeF_6 resembles the behavior of UF_6. The Rb and Cs octafluoroxenates are the most stable xenon compounds yet made and decompose only above 400°; they hydrolyze in the atmosphere to give xenon-containing oxidizing products. The sodium fluoride adduct of XeF_6 decomposes below 100° and can be used to purify XeF_6.

Oxofluoro anions can also be obtained, but not directly from $XeOF_4$. The very stable salts $MXeO_3F$ are obtained when a solution of XeO_3 is treated with KF or CsF; X-ray study shows that the anion is polymeric, i.e., $(XeO_3F^-)_n$ with chains of XeO_3 groups linked by an angular Xe—F—Xe bridge,[12b] giving distorted square-pyramidal coordination of Xe. An apparent ten valence electron shell in XeO_3F^- is really, because of polymerization, a twelve electron shell (cf. BrF_3 and I_2Cl_6).

XeF_2 reacts[13-15] with strong oxo acids (e.g., HSO_3F) or their anhydrides (e.g., POF_2—O—POF_2) to give products in which one or both F atoms are replaced. Examples are:

$$XeF_2 + HSO_3F \rightarrow FXeOSO_2F, \ Xe(OSO_2F)_2$$
$$XeF_2 + P_2O_3F_4 \rightarrow FXe(OPOF_2), \ Xe(OPOF_2)_2$$

[8] R. J. Gillespie *et al.*, *J.C.S. Dalton*, **1977**, 2234.
[9] K. Leary *et al.*, *Inorg. Chem.*, 1973, **12**, 1726.
[10] R. J. Gillespie and G. J. Schrobilgen, *Inorg. Chem.*, 1974, **13**, 765.
[11] K. Leary, A. Zalkin, and N. Bartlett, *Inorg. Chem.*, 1974, **13**, 775.
[12a] R. J. Gillespie, B. Landa, and G. J. Schrobilgen, *Inorg. Chem.*, 1976, **15**, 1256.
[12b] D. J. Hodgson and J. A. Ibers, *Inorg. Chem.*, 1969, **8**, 326.
[13] N. Bartlett *et al.*, *Inorg. Chem.*, 1972, **11**, 1124.
[14] M. Eisenberg and D. D. Desmarteau, *Inorg. Chem.*, 1972, **11**, 1901.
[15] M. Wecksberg *et al.*, *Inorg. Chem.*, 1972, **11**, 3063.

The products are subject to slow thermal decomposition at or not far above room temperature, and several are known to detonate.

XeF_4 and XeF_6 react with HSO_3F as follows[16]:

$$XeF_4 + HSO_3F \rightarrow FXe(OSO_2F), \ Xe(OSO_2F)_2 + S_2O_6F_2$$
$$XeF_6 + HSO_3F \rightarrow F_5Xe(OSO_2F) + HF$$

$F_5Xe(OSO_2F)$ is a white solid, thermally stable at $22°$ but decomposing above $73°$.

Xenon-Oxygen Compounds. Both XeF_4 and XeF_6 are violently hydrolyzed by water to give Xe^{VI}, evidently in the form of undissociated XeO_3:

$$3XeF_4 + 6H_2O \rightarrow XeO_3 + 2Xe + \tfrac{3}{2}O_2 + 12HF$$
$$XeF_6 + 3H_2O \rightarrow XeO_3 + 6HF$$

Aqueous solutions, colorless, odorless, and stable, as concentrated as $11M$ in Xe^{VI}, have been obtained. They are nonconducting. On evaporation, XeO_3 is obtained.

Xenon trioxide is a white deliquescent solid and is dangerously explosive; its formation is one of the main reasons that great care must be taken to avoid water in studies of XeF_6. The molecule is pyramidal (C_{3v}). XeO_3 can be quantitatively reduced by iodide:

$$XeO_3 + 6H^+ + 9I^- \rightarrow Xe + 3H_2O + 3I_3^-$$

There is some evidence that xenate esters are formed in the violent reactions with alcohols.

XeO_3 reacts with $RbCl$ and $CsCl$ to give compounds with the composition $M_9(XeO_3Cl_2)_4Cl$, which consist of M^+ cations, Cl^- anions, and infinite chain anions $-XeO_3Cl-Cl-XeO_3Cl-Cl-$, in which each Xe is surrounded by a very distorted octahedron of three O atoms ($Xe-O \approx 1.77$ Å) and three Cl atoms ($Xe-Cl \approx 2.96$ Å). These are the only examples of compounds stable at room temperature that contain $Xe-Cl$ bonds.[17a]

In water XeO_3 appears to be present as XeO_3 molecules, but in base solutions we have

$$XeO_3 + OH^- \rightleftharpoons HXeO_4^- \quad K = 1.5 \times 10^{-3}$$

The main species in solutions, $HXeO_4^-$, slowly disproportionates to produce Xe^{VIII} and Xe:

$$2HXeO_4^- + 2OH^- \rightarrow XeO_6^{4-} + Xe + O_2 + 2H_2O$$

Aqueous Xe^{VIII} arises, not only in the disproportionation above, but also when ozone is passed through a dilute solution of Xe^{VI} in base. These yellow *perxenate* solutions are powerful and rapid oxidizing agents.

Stable, insoluble perxenate salts can be precipitated from Xe^{VIII} solutions (e.g., $Na_4XeO_6 \cdot 8H_2O$, $Na_4XeO_6 \cdot 6H_2O$, $Ba_2XeO_6 \cdot 1.5H_2O$); the first two contain XeO_6

[16] D. D. Desmarteau and M. Eisenberg, *Inorg. Chem.*, 1972, **11**, 2641.
[17a] R. D. Willett, S. W. Peterson, and B. A. Cole, *J. Am. Chem. Soc.*, 1977, **99**, 8203.

octahedra. The solutions of sodium perxenate are alkaline owing to hydrolysis, and the following equilibrium constants have been estimated:

$$HXeO_6^{3-} + OH^- = XeO_6^{4-} + H_2O \quad K < 3$$
$$H_2XeO_6^{2-} + OH^- = HXeO_6^{3-} + H_2O \quad K \sim 4 \times 10^3$$

so that even at pH 11 to 13 the main species is $HXeO_6^{3-}$.[17b] From the equilibria it follows that for H_4XeO_6, pK_3 and pK_4 are $\sim 4 \times 10^{-11}$ and $< 10^{-14}$, respectively. Hence by comparison with H_6TeO_6 and H_5IO_6, H_4XeO_6 appears to be an anomalously weak acid.

Solutions of perxenates are reduced by water at pH 11.5 at a rate of about 1% per hour, but in acid solutions almost instantaneously:

$$H_2XeO_6^{2-} + H^+ \rightarrow HXeO_4^- + \tfrac{1}{2}O_2 + H_2O$$

This reduction appears to proceed almost entirely through the formation of OH radicals according to the scheme

$$Xe^{VIII} + H_2O \rightleftharpoons Xe^{VII} + OH^{\cdot}$$
$$Xe^{VII} + H_2O \rightleftharpoons Xe^{VI} + OH^{\cdot}$$
$$2OH^{\cdot} \rightarrow H_2O_2$$
$$Xe^{VIII} + H_2O_2 \rightarrow Xe^{VI} + O_2$$

Xenon tetroxide is a highly unstable and explosive gas formed by action of concentrated H_2SO_4 on barium perxenate. It is tetrahedral (Xe—O = 1.736 Å),[17c] according to electron diffraction studies.

The aqueous chemistry of xenon is briefly summarized by the potentials:

$$\text{Acid solution:} \quad H_4XeO_6 \xrightarrow{\text{2.36 V}} XeO_3 \xrightarrow{\text{2.12 V}} Xe \;.$$

$$XeF_2 \xrightarrow{\text{2.64 V}} Xe$$

$$\text{Alkaline solution:} \quad HXeO_6^{3-} \xrightarrow{\text{0.94 V}} HXeO_4^- \xrightarrow{\text{1.26 V}} Xe$$

The only compound in which Xe^{IV} is bound exclusively to oxygen is $Xe(OTeF_5)_4$, made by the reaction[17d]

$$3XeF_4 + 4B(OTeF_5)_3 \xrightarrow{0°C} 3Xe(OTeF_5)_4 + 4BF_3$$

Xe—N Bonds. XeF_2 reacts with $(SO_2F)_2NH$ to form HF and $FXeN(SO_2F)_2$,[18] a white solid, readily hydrolyzed and decomposing at 70°. This compound reacts with AsF_5 to form $[(FO_2S)_2NXe^+][AsF_6^-]$, an unstable adduct, thought to be

17b G. Downey, H. Claassen, and E. H. Appelman, *Inorg. Chem.,* 1971, **10**, 1817.
17c G. Gunderson, K. Hedberg, and J. L. Huston, *Acta Crystallogr., A,* 1969, **25**, 5124.
17d D. Lentz and K. Seppelt, *Angew. Chem. Int. Ed.,* 1978, **17**, 356.
18   D. D. DesMarteau, *J. Am. Chem. Soc.,* 1978, **100**, 6270.

ionic, which loses AsF_5 on pumping to give $\{[(FO_2S)_2NXe]_2F^+\}[AsF_6^-]$. The $(FO_2S)_2N-$ group may be construed as a fluorinelike pseudohalogen, and the cations may be compared with the XeF^+ and $Xe_2F_3^+$ cations.

18-5. Other Noble Gas Chemistry

Xenon dichloride can be formed by microwave discharge through, or photolysis of, $Xe-Cl_2$ mixtures and trapped in a xenon matrix, where it is identified by Raman spectroscopy.[19] There is evidence for the CH_3Xe^+ ion in the gas phase from ion cyclotron resonance spectroscopy. It has also been shown[20] that XeF_6 forms a graphite intercalation compound of composition $C_{19.1\pm0.2}XeF_6$.

Krypton difluoride is obtained when an electric discharge is passed through Kr and F_2 at $-183°$, or when the gases are irradiated with high-energy electrons or protons. It is a volatile white solid that decomposes slowly at room temperature. It is a highly reactive fluorinating agent. The KrF_2 molecule has been shown to be linear.[21]

The KrF_2 molecule is thermodynamically unstable, whereas XeF_2 is stable, as the following enthalpies show:

$$KrF_2(g) = Kr(g) + F_2(g) \qquad \Delta H° = -63 \text{ kJ mol}^{-1}$$
$$XeF_2(g) = Xe(g) + F_2(g) \qquad \Delta H° = 105 \text{ kJ mol}^{-1}$$

These energetic relationships are understandable on the basis of rigorous quantum mechanical calculations,[22] which justify the view that in both difluorides there is considerable ionic character. The bonding can be simply represented by the following resonance picture:

$$F-Xe^+F^- \leftrightarrow F^-Xe^+-F$$

Since the ionization enthalpies (Table 18-1) of Kr and Xe differ by about 182 kJ mol^{-1}, the experimental difference in $\Delta H_f°$ values, viz., $105 - (-63) = 168$ kJ mol^{-1} is well explained by this picture.

No other molecular fluoride of Kr has been isolated, but the cationic species KrF^+ and $Kr_2F_3^+$ are known.[23] These are formed in reactions of KrF_2 with strong fluoride acceptors such as AsF_5 and SbF_5, and compounds have formulas such as $KrF^+Sb_2F_{11}^-$, $KrF^+SbF_6^-$, and $Kr_2F_3^+AsF_6^-$.

The study of radon chemistry is difficult because of the radioactivity of all isotopes. It appears that a radon fluoride, of unknown composition, forms readily but decomposes under attempts to vaporize it.[24] Theoretical arguments[25] suggest that it may be ionic.

[19]   I. R. Beattie *et al., J.C.S. Dalton,* **1975,** 1659.
[20]   H. Selig *et al., J. Am. Chem. Soc.,* 1976, **98,** 1601.
[21]   R. D. Burbank, W. E. Falconer, and W. A. Sunder, *Science,* 1972, **178,** 1285.
[22]   P. S. Bagus *et al., J. Am. Chem. Soc.,* 1975, **97,** 7216.
[23]   R. J. Gillespie and C. J. Schrobilgen, *Inorg. Chem.,* 1976, **15,** 22.
[24]   L. Stein, *Science,* 1970, **168,** 362.
[25]   K. S. Pitzer, *J.C.S. Chem. Comm.,* **1975,** 1760.

The only compound with a bond to carbon is $Xe(CF_3)_2$ made by the action of CF_3 radicals on XeF_2. It has a half life of only *ca* 30 min at room temperature.[26]

General References

Bartlett, N., *The Chemistry of the Noble Gases,* Elsevier, 1971.

Cook, G. A., Ed., *Argon, Helium, and the Rare Gases,* 2 vols., Wiley-Interscience, 1961. Comprehensive, exclusive of chemical behavior.

Hawkins, D. T., W. E. Falconer, and N. Bartlett, *Noble Gas Compounds,* A Bibliography, Plenum Press, 1978. References; 1962–1976.

Holloway, J. H., *Noble Gas Chemistry,* Methuen, 1968.

Klein, M. L., and Venables, J. A., *Rare Gas Solids,* Vol. 1, 1976; Vol. 2, 1977, Academic Press.

Moody, G. J., *J. Chem. Educ.,* 1974, **51,** 628.

Sladky, F., *Noble Gases,* MTP International Review of Science, Inorganic Chemistry, Series 1, Vol. 3, Butterworths, 1972.

[26]   R. J. Lagon *et al., J. Am. Chem. Soc.,* 1979, **101,** 5833.

CHAPTER NINETEEN

Zinc, Cadmium, and Mercury

GENERAL REMARKS

19-1. Position in the Periodic Table: Group Trends

Zinc, cadmium, and mercury follow copper, silver, and gold and have two s electrons outside filled d shells. Some of their properties are given in Table 19-1. Whereas in Cu, Ag, and Au the filled d shells may fairly readily lose one or two d electrons to give ions or complexes in the II and III oxidation states, this does not occur for the Group II elements.

Divergence from +2 valence occurs in the following cases:

1. The univalent metal-metal bonded ions M_2^{2+}, of which only Hg_2^{2+} is ordinarily stable.

2. Mercury ions in low partial oxidation states, notably $Hg^{0.33+}$.

3. A shortlived (*ca* 5 sec) Hg^{III} complex. Since the third ionization enthalpies are extremely high, compensating energy from solvation or lattice formation is normally impossible.

TABLE 19-1
Some Properties of the Group IIb Elements

Property	Zn	Cd	Hg
Outer configuration	$3d^{10}4s^2$	$4d^{10}5s^2$	$5d^{10}6s^2$
Ionization enthalpies (kJ mol^{-1})			
1st	906	867	1006
2nd	1726	1625	1799
3rd	3859	3666	3309
Melting point (°C)	419	321	−38.87
Boiling point (°C)	907	767	357
Heat of vaporization (kJ mol^{-1})	130.8	112	61.3
E^0 for $M^{2+} + 2e = M$ (V)	−0.762	−0.402	0.854
Radii of divalent ions (Å)	0.69	0.92	0.93

Since these elements form no compound in which the d shell is other than full, they are regarded as non-transition elements, whereas by the same criteria Cu, Ag, and Au are considered as transition elements. Also the metals are softer and lower melting, and Zn and Cd are considerably more electropositive, than their neighbors in the transition groups. However there is some resemblance to the d-group elements in their ability to form complexes, particularly with ammonia, amines, halide ions, and cyanide. For complexes, even with CN^-, it must be borne in mind that the possibility of $d\pi$ bonding between the metal and the ligand is very much lowered compared to the d-transition elements, owing to the electronic structure, and no carbonyl, nitrosyl, olefin complex, etc., of the type given by transition metals is known.

The chemistries of Zn and Cd are very similar, but that of Hg differs considerably and cannot be regarded as homologous. As examples we quote the following. The hydroxide $Cd(OH)_2$ is more basic than $Zn(OH)_2$, which is amphoteric, but $Hg(OH)_2$ is an extremely weak base. The chlorides of Zn and Cd are essentially ionic, whereas $HgCl_2$ gives a molecular crystal. Zinc and cadmium are electropositive metals, but Hg has a high positive standard potential; furthermore, the Zn^{2+} and Cd^{2+} ions somewhat resemble Mg^{2+} (see Section 19-9), but Hg^{2+} does not. Though all the M^{2+} ions readily form complexes, those of Hg^{2+} have formation constants greater by orders of magnitude than those for Zn^{2+} or Cd^{2+}.

All three elements form a variety of covalently bound compounds, and the polarizing ability of the M^{2+} ions is larger than would be predicted by comparing the radii with those of the Mg—Ra group, a fact that can be associated with the greater ease of distortion of the filled d shell. Compared to the noble gaslike ions of the latter elements the promotional energies $ns^2 \rightarrow nsnp$ (433, 408, and 524 kJ mol^{-1} for Zn, Cd, and Hg, respectively) involved in the formation of two covalent bonds are also high, and this has the consequence, particularly for Hg, that further ligands can be added only with difficulty. This is probably the main reason that two-coordination is the commonest for Hg.

19-2. Stereochemistry

The stereochemistry of the elements in the II state is summarized in Table 19-2. Since there is no ligand field stabilization effect in Zn^{2+} and Cd^{2+} ions because of their completed d shells, their stereochemistry is determined solely by considerations of size, electrostatic forces, and covalent bonding forces. The effect of size is to make Cd^{2+} more likely than Zn^{2+} to assume a coordination number of 6. For example, ZnO crystallizes in lattices where the Zn^{2+} ion is in tetrahedral holes surrounded by four oxide ions, whereas CdO has the rock salt structure. Similarly $ZnCl_2$ has Zn^{2+} in tetrahedral holes with Cl^- in hexagonally packed layers,[1] whereas Cd^{2+} in $CdCl_2$ is octahedrally coordinated.

In their complexes, Zn, Cd, and Hg commonly have coordination numbers 4, 5, and 6, with 5 especially common for zinc. The coordination number for all three

[1]   J. Brynestad, *Inorg. Chem.*, 1978, **17**, 1376.

TABLE 19-2
Stereochemistry of Divalent Zinc, Cadmium, and Mercury

Coordination number	Geometry	Examples
2	Linear[a]	$Zn(CH_3)_2$, HgO, $Hg(CN)_2$
3	Planar	$[Me_3S]^+HgX_3^-$, $[MeHg\ bipy]^+$, $Hg(SiMe_3)_3^-$
4	Tetrahedral	$[Zn(CN)_4]^{2-}$, $ZnCl_2(s)$, ZnO, $[Cd(NH_3)_4]^{2+}$, $HgCl_2(OAsPh_3)_2$
	Planar	Bis(glycinyl)Zn
5	Tbp	Terpy $ZnCl_2$, $[Zn(SCN)\ tren]^+$, $[Co(NH_3)_6][CdCl_5]$
	Sp	$Zn(acac)_2 \cdot H_2O$
6	Octahedral	$[Zn(NH_3)_6]^{2+}$ (solid only), CdO, $CdCl_2$, $[Hg\ en_3]^{2+}$, $[Hg(C_5H_4NO)_6]^{2+}$
7	Pentagonal bipyramid[b]	$[Zn(H_2dapp)(H_2O)_2]^{2+}$
	Dist. pentagonal bipyramid[c]	$Cd(quin)(H_2O)(NO_3)_2$
8	Dist. square antiprism	$[Hg(NO_2)_4]^{2-\ d}$
	Dodecahedral	$(Ph_4As)_2Zn(NO_3)_4^e$

[a] For references, see N. B. Behrens et al., Inorg. Chim. Acta, 1978, 31, L471; J. P. Oliver, J. Am. Chem. Soc., 1978, 100, 7761.

[b] H_2dapp = 2,6-diacetylpyridine(2′-pyridylhydrazone), see D. Wester and G. J. Palenik, Inorg. Chem., 1976, 15, 755.

[c] Chelate nitrate groups (A. F. Cameron et al., J.C.S. Dalton, 1973, 2130).

[d] Bidentate NO_2^- (L. F. Power et al., J.C.S. Dalton, 1976, 93).

[e] The bidentate nitrate is asymmetric (Zn—O, 2.06, 2.58 Å), reflecting a tendency for Zn to be four-coordinate, unlike the bonding in $M(NO_3)_4$ species (M = Mn^{II}, Sn^{IV}, Ti^{IV}, and Fe^{II}), where the nitrate is symmetrical (C. Bellito et al., J.C.S. Dalton, 1976, 989.)

elements in organo compounds is usually 2, but only Hg commonly forms linear bonds in other cases (e.g., in HgO). Indeed, linear two-coordination is more characteristic for Hg^{II} than for any other metal species.

THE ELEMENTS[2]

19-3. Occurrence, Isolation, and Properties

The elements have relatively low abundance in Nature (of the order of 10^{-6} of the earth's crust for Zn and Cd), but have long been known because they are easily obtained from their ores.

Zinc occurs widely in a number of minerals, but the main source is *sphalerite* [(ZnFe)S], which commonly occurs with galena (PbS); cadmium minerals are

[2] L. G. Hepler and G. Olufsson, Chem. Rev., 1975, 75, 585 (thermodynamic properties, chemical equilibria, and standard potentials for mercury and its compounds; an authoritative critical compendium).

scarce, but as a result of its chemical similarity to Zn, Cd occurs by isomorphous replacement in almost all zinc ores. There are numerous methods of isolation, initially involving flotation and roasting; Zn and Pb are commonly recovered simultaneously by a blast furnace method. Cadmium is invariably a by-product and is usually separated from zinc by distillation or by precipitation from sulfate solutions by zinc dust:

$$Zn + Cd^{2+} = Zn^{2+} + Cd \qquad E = +0.36 \text{ V}$$

The only important ore of mercury is *cinnabar* (HgS); this is roasted to the oxide, which decomposes at *ca.* 500°, the mercury vaporizing.

Properties. Some properties of the elements are listed in Table 19-1. Zinc and cadmium are white, lustrous, but tarnishable metals. Like Be and Mg, with which they are isostructural, their structures deviate from perfect hexagonal close packing by elongation along the 6-fold axis. Mercury is a shiny liquid at ordinary temperatures.[3] All are remarkably volatile for heavy metals, mercury, of course, uniquely so. Mercury gives a monoatomic vapor and has an appreciable vapor pressure (1.3 × 10^{-3} mm) at 20°. It is also surprisingly soluble in both polar and nonpolar liquids. The solubility[4] in air-free water at 25° is 6.39×10^{-7} g l^{-1}

Because of its high volatility and toxicity, mercury should always be kept in stoppered containers and handled in well-ventilated areas. Mercury is readily lost from dilute aqueous solutions and even from solutions of mercuric salts owing to reduction of these by traces of reducing materials and by disproportionation of Hg_2^{2+}.

Both Zn and Cd react readily with nonoxidizing acids, releasing hydrogen and giving the divalent ions, whereas Hg is inert to nonoxidizing acids. Zinc also dissolves in strong bases because of its ability to form zincate ions (see below), commonly written ZnO_2^{2-}:

$$Zn + 2OH^- \rightarrow ZnO_2^{2-} + H_2$$

Although cadmiate ions are known, Cd does not dissolve in bases.

Zinc and cadmium react readily when heated with oxygen, to give the oxides. Although mercury and oxygen are unstable with respect to HgO at room temperature, their rate of combination is exceedingly slow; the reaction proceeds at a useful rate at 300 to 350°, but, around 400° and above, the stability relation reverses and HgO decomposes rapidly into the elements:

$$HgO(s) = Hg(s) + \tfrac{1}{2}O_2 \qquad \Delta H_{diss} = 160 \text{ kJ mol}^{-1}$$

This ability of mercury to absorb oxygen from air and regenerate it again in pure form was of considerable importance in the earliest studies of oxygen by Lavoisier and Priestley.

All three elements react directly with halogens and with nonmetals such as sulfur, selenium, and lead.

Zinc and cadmium form many alloys, some, such as brass, being of technical

[3] M. C. Wilkinson, *Chem. Rev.*, 1972, **82**, 575 (surface properties of Hg).
[4] I. Sanemasa, *Bull. Chem. Soc. Jap.*, **1975**, 1795.

importance. Mercury combines with many other metals, sometimes with difficulty but sometimes, as with sodium or potassium, very vigorously, giving *amalgams*. Some amalgams have definite compositions; that is, they are compounds, such as Hg_2Na. Some of the transition metals do not form amalgams, and iron is commonly used for containers of mercury. Sodium amalgams and amalgamated zinc are frequently used as reducing agents for aqueous solutions.

THE UNIVALENT STATE

The univalent state is of importance only for mercury, but unstable species of Zn^I and Cd^I exist. Although the last two have the formula M_2^{2+}, there is evidence that highly unstable, strongly reducing Zn^+ and Cd^+ ions can be obtained when aqueous solutions of Zn^{2+} and Cd^{2+} are irradiated.

When zinc is added to molten $ZnCl_2$ at 500 to 700°, a yellow, diamagnetic glass is obtained on cooling. According to Raman and other spectra, this glass contains Zn_2^{2+}. It is soluble in warm saturated $ZnCl_2$ solution, and the resulting greenish-yellow solution is stable for some days, but on dissolution in CH_3OH or acetone, decomposition and precipitation of Zn occur in less than a minute.

Similarly, when cadmium is dissolved in molten cadmium halides, very dark red melts are obtained. The high color may be due to the existence of both Cd^I and Cd^{II} joined by halide bridges, since mixed valence states in complexes are known to give intense colors in many other cases. If, for example, aluminum chloride is added to the $Cd–CdCl_2$ melt, only a greenish-yellow melt is obtained, and phase studies here and for the bromide have shown the presence of Cd^I. Yellow solids can be isolated; they are diamagnetic and can be formulated $(Cd_2)^{2+}(AlCl_4^-)_2$.[5] When such solids are added to donor solvents or to water, cadmium metal is at once formed, together with Cd^{2+}, so that it is not surprising that there is no evidence for Cd^I in aqueous solution. The stabilization by the tetrahaloaluminate ions is presumably due to lowering of the difference between the lattice energies of the two oxidation states and lessening of the tendency to disproportionate. A similar case of this type of stabilization has been noted for Ga^I, as in $Ga^+AlCl_4^-$.

The force constants for the M—M bonds, obtained from Raman spectra, clearly show the stability order: Zn_2^{2+}, 0.6; Cd_2^{2+}, 1.1; Hg_2^{2+}, 2.5 mdyne Å^{-1}, which may be compared with values of 0.98 for K_2 and 1.7 for Na_2 and I_2.

19-4. The Mercurous Ion, Mercurous-Mercuric Equilibria, and Mercurous Compounds

The mercurous ion Hg_2^{2+} is readily obtained from mercuric salts by reduction in aqueous solution and is readily reoxidized to $[Hg(H_2O)_2]^{2+}$.

The main lines of evidence showing the binuclearity of Hg_2^{2+} are:

1. Mercurous compounds are diamagnetic both as solids and in solution, whereas Hg^+ would have an unpaired electron.

5 J. D. Corbett, *Prog. Inorg. Chem.*, 1976, **21**, 135.

TABLE 19-3
Mercury-Mercury Bond Lengths in Mercurous Compounds

Salt	Hg—Hg (Å)	Salt	Hg—Hg (Å)
Hg_2F_2	2.51	$Hg_2(NO_3)_2 \cdot 2H_2O$	2.54
Hg_2Cl_2	2.53	$Hg_2(BrO_3)_2$	2.51
Hg_2Br_2	2.58	$Hg_2SO_4{}^a$	2.50
Hg_2I_2	2.69		

$$\begin{array}{c} O \\ \| \\ \end{array}$$

[a] Contains chains, $-O-Hg-Hg-O-S-O-$ (E. Dorm, *Acta Chem. Scand.*, 1969, **23**, 1607).

$$\begin{array}{c} \| \\ O \end{array}$$

2. X-Ray determination of the structures of several mercurous salts shows the existence of individual Hg_2^{2+} ions. The Hg—Hg distances are far from constant (Table 19-3) and show some correlation with the electronegativity of the anion.

3. The Raman spectrum of an aqueous solution of mercurous nitrate contains a strong line that can only be attributed to an Hg—Hg stretching vibration.

4. There are various kinds of equilibria for which constant equilibrium quotients can be obtained only by considering the mercurous ion to be Hg_2^{2+}. For example, suppose we add an excess of mercury to a solution initially X molar mercuric nitrate. An equilibrium between Hg, Hg^{2+}, and mercurous ion will be reached (see below); depending on the assumed nature of mercurous ion, the following equilibrium quotients can be written:

$$Hg(l) + Hg^{2+} = Hg_2^{2+} \qquad K = \frac{[Hg_2^{2+}]}{[Hg^{2+}]} = \frac{f}{1-f}$$

$$Hg(l) + Hg^{2+} = 2Hg^+ \qquad K' = \frac{[Hg^+]^2}{[Hg^{2+}]} = \frac{(2fX)^2}{(1-f)X} = \frac{4f^2X}{1-f}$$

where f represents the fraction of the initial Hg^{2+} determined by analysis or otherwise to have disappeared when equilibrium is reached. It is found that when values of K and K' are calculated from experimental data at different values of X, the former are substantially constant but the latter are not.

5. The electrical conductances of solutions of mercurous salts resemble closely, in magnitude and variation with concentration, the conductances of uni-divalent rather than uni-univalent electrolytes.

The greater strength of the Hg—Hg bond in Hg_2^{2+} compared with Cd—Cd in Cd_2^{2+} is reflected also on comparison of the bond energies $HgH^+ > CdH^+$ in the spectroscopic ions, and qualitatively the stability of Hg_2^{2+} is probably related to the large electron attachment enthalpy of Hg^+. The electron attachment enthalpy of M^+ (equal to the first ionization enthalpy of the metal) is 135 kJ mol^{-1} greater for Hg^+ than for Cd^+ because the $4f$ shell in Hg shields the $6s$ electrons relatively poorly. The high ionization enthalpy of Hg also accounts for the so-called inert pair phenomenon, namely, the exceptionally noble character of mercury and its low energy of vaporization. ·

Hg^I–Hg^{II} Equilibria. An understanding of the thermodynamics of these equilibria is essential to an understanding of the chemistry of the mercurous state. The important thermodynamic values are the potentials

$$Hg_2^{2+} + 2e = 2Hg(l) \qquad E^0 = 0.7960 \text{ V} \qquad (19\text{-}1)$$
$$2Hg^{2+} + 2e = Hg_2^{2+} \qquad E^0 = 0.9110 \text{ V} \qquad (19\text{-}2)$$
$$Hg^{2+} + 2e = Hg(l) \qquad E^0 = 0.8535 \text{ V} \qquad (19\text{-}3)$$

For the disproportionation reaction, which is rapid and reversible,

$$Hg_2^{2+} = Hg(l) + Hg^{2+} \qquad E^0 = -0.131 \text{ V} \qquad (19\text{-}4)$$

and we then have

$$K = \frac{[Hg^{2+}]}{[Hg_2^{2+}]} = 6.0 \times 10^{-3}$$

The implication of the standard potentials is clearly that only oxidizing agents with potentials in the range -0.79 to -0.85 V can oxidize mercury to Hg^I but not to Hg^{II}. Since no common oxidizing agent meets this requirement, it is found that when mercury is treated with an excess of oxidizing agent it is entirely converted into Hg^{II}. However when mercury is in at least 50% excess, only Hg^I is obtained, since according to eq. 19-4 Hg(l) readily reduces Hg^{2+} to Hg_2^{2+}.

The kinetics of oxidation of Hg_2^{2+} in perchlorate solution by one- and two-electron oxidants suggest that the pathways are different.[6] For one-electron oxidants, breaking of the Hg—Hg bond is involved, viz.,

$$(Hg^I)_2 + e \rightarrow Hg^I + Hg^{II}$$
$$Hg^I + e \rightarrow Hg^{II} \text{ fast}$$

whereas for two-electron oxidants,

$$(Hg^I)_2 \rightarrow Hg^0 + Hg^{II} \text{ fast}$$
$$Hg^0 + 2e = Hg^{II}$$

The equilibrium constant for reaction 19-4 shows that although Hg_2^{2+} is stable with respect to disproportionation, it is stable only by a small margin. Thus any reagents that reduce the activity (by precipitation or complexation) of Hg^{2+} to a significantly greater extent than they lower the activity of Hg_2^{2+} will cause disproportionation of Hg_2^{2+}. Since there are many such reagents, such as NH_3, amines, OH^-, CN^-, SCN^-, S^{2-}, and acacH, the number of stable Hg^I compounds is rather restricted.

Thus when OH^- is added to a solution of Hg_2^{2+}, a dark precipitate consisting of Hg and HgO is formed; evidently mercurous hydroxide, if it could be isolated, would be a stronger base than HgO. Similarly, addition of sulfide ions to a solution of Hg_2^{2+} gives a mixture of Hg and the extremely insoluble HgS. Mercurous cyanide does not exist because $Hg(CN)_2$ is so slightly dissociated though soluble. The reactions in these cases are

$$Hg_2^{2+} + 2OH^- \rightarrow Hg(l) + HgO(s) + H_2O$$
$$Hg_2^{2+} + S^2 \rightarrow Hg(l) + HgS$$
$$Hg_2^{2+} + 2CN^- \rightarrow Hg(l) + Hg(CN)_2(aq)$$

[6]   D. Davies et al., J. Am. Chem. Soc., 1973, 95, 7250.

The reactions with OH^- and CN^- have been studied kinetically.[7] The rate-determining step appears to be the cleavage of the Hg—Hg bond, viz.,

$$Hg_2^{2+} + OH^- \rightleftharpoons Hg_2OH^+ \text{ (rapid equilibrium)}$$
$$Hg_2OH^+ \rightarrow Hg + HgOH^+$$
$$Hg_2OH^+ + OH^- \rightarrow Hg + Hg(OH)_2$$

Mercury(I) Compounds. The best known are the *halides*. The fluoride is unstable toward water, being hydrolyzed to HF and unisolable mercurous hydroxide (which disproportionates as above). The other halides are highly insoluble, which precludes the possibilities of hydrolysis or disproportionation to give Hg^{II} halide complexes. *Mercurous nitrate* is known only as the dihydrate $Hg_2(NO_3)_2 \cdot 2H_2O$, which contains the ion $[H_2O—Hg—Hg—OH_2]^{2+}$; a *perchlorate* $Hg_2(ClO_4)_2 \cdot 4H_2O$ is also known. Both are very soluble in water, and the halides and other relatively insoluble salts of Hg_2^{2+} may conveniently be prepared by adding the appropriate anions to their solutions. Other known mercurous salts are the sparingly soluble sulfate, chlorate, bromate, iodate, and acetate.

Mercurous ion forms few *complexes*; this may in part be due to a low tendency for Hg_2^{2+} to form coordinate bonds, but it is probably mainly because mercuric ion will form even more stable complexes with most ligands, for example, CN^-, I^-, amines, and alkyl sulfides, so that the Hg_2^{2+} disproportionates. Nitrogen ligands of low basicity tend to favor Hg_2^{2+}, and there are relatively stable complexes with aniline, $[Hg_2(PhNH_2)]^{2+}$, and with 1,10-phenanthroline.

Complexes can readily be obtained in solution with oxygen donor ligands that form essentially ionic metal-ligand bonds, hence no strong complexes with mercury(II). Such ligands are oxalate, succinate, pyrophosphate, and tripolyphosphate. Pyrophosphate gives the species $[Hg_2(P_2O_7)_2]^{6-}$ (pH range 6.5–9) and $[Hg_2(P_2O_7)OH]^{3-}$ for which stability constants have been measured.

Some oxygen donor complexes, e.g., $[Hg_2(OPPh_3)_6]^{2+}$, can be isolated.[8]

19-5. Mercury in Oxidation State 0.33

Like S, Se, and Te, (p. 510) mercury can be oxidized by AsF_5 in liquid SO_2,[9] the nature of the products depending on the mole ratio:

$2Hg + 3AsF_5 = Hg_2(AsF_6)_2 + AsF_3$	Colorless solution
$3Hg + 3AsF_5 = Hg_3(AsF_6)_2 + AsF_3$	Yellow solution
$4Hg + 3AsF_5 = Hg_4(AsF_6)_2 + AsF_3$	Red solution

The intermediate species are in equilibrium with Hg and each other, and the final overall reaction is

$$6Hg + 3AsF_5 = 2Hg_3AsF_6 + AsF_3$$

Neutron diffraction study[10] of the insoluble silvery-golden product Hg_3AsF_6, which behaves as an isotropic semiconductor, shows that there are two orthogonal,

7　I. Sanemasa, *Inorg. Chem.*, 1977, **16**, 2786; 1976, **15**, 1973.
8　D. L. Kepert *et al.*, *J.C.S. Dalton*, **1973**, 392, 1657.
9　N. D. Miro *et al.*, *J. Inorg. Nucl Chem.*, 1978, **40**, 1351.
10　J. M. Williams *et al.*, *Inorg. Chem.*, 1978, **17**, 646.

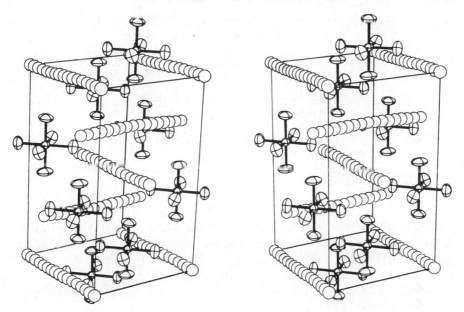

Fig. 19-1. A stereoscopic view of the unit cell of Hg_3AsF_6 showing disordered model for the Hg chains. Each Hg position is only partially occupied. [Reproduced by permission from Reference 10.]

nonintersecting linear chains of $Hg^{0.33+}$ ions passing through a lattice of octahedral AsF_6 ions (Fig. 19-1). The Hg chains are not quite linear, and there is also a defect of one AsF_6^- in 20, so that there is an apparent nonstoichiometry $Hg_{2.82}(AsF_6)_{0.94}$. Mercury will also dissolve in FSO_3H to give a yellow solution and will give a yellow melt in NaCl:

$$Hg + Hg_2Cl_2 + 2AlCl_3 \rightarrow Hg_3(AlCl_4)_2$$

DIVALENT ZINC AND CADMIUM COMPOUNDS

19-6. Oxides and Hydroxides

The *oxides* ZnO and CdO are formed on burning the metals in air or by pyrolysis of the carbonates or nitrates; oxide smokes can be obtained by combustion of the alkyls, those of cadmium being exceedingly toxic. Zinc oxide is normally white but turns yellow on heating; cadmium oxide varies in color from greenish yellow through brown to nearly black, depending on its thermal history. These colors are the result of various kinds of lattice defect. Both oxides sublime without decomposition at very high temperatures.

The *hydroxides* are precipitated from solutions of salts by addition of bases. The solubility products of $Zn(OH)_2$ and $Cd(OH)_2$ are about 10^{-11} and 10^{-14}, respectively, but $Zn(OH)_2$ is more soluble than would be expected from this constant

TABLE 19-4

Structures[a] of Zn and Cd Oxides and Chalconides[b]

Metal	O	S	Se	Te
Zn	W, Z	Z, W	Z	Z, W
Cd	NaCl	W, Z	W, Z	Z

[a]   W = wurtzite structure; Z = zinc blende structure; NaCl = rock salt structure.

[b]   Where two polymorphs occur, the one stable at lower temperatures is listed first.

owing to the equilibrium

$$Zn(OH)_2(s) = Zn(OH)_2(aq) \qquad K = 10^{-6}$$

Zinc hydroxide readily dissolves in aqueous NaOH, but $Cd(OH)_2$ dissolves only in concentrated base. From such solutions crystalline zincates or cadmiates,[11] e.g., $Na_2[Cd(OH)_4]$, may be obtained. In aqueous solution the principal zincate ion appears to be $[Zn(OH)_3(H_2O)]^-$ according to Raman spectra.[11]

Both Zn and Cd hydroxide readily dissolve in an excess of strong ammonia to form the ammine complexes, for example, $[Zn(NH_3)_4]^{2+}$.

19-7.   Sulfides, Selenides, and Tellurides

The sulfides, selenides, and tellurides are all crystalline substances, insoluble in water. Three structures are represented among the eight compounds as shown in Table 19-4. All of these structures have been described on pages 13 to 16. In the NaCl structure, the cation is octahedrally surrounded by six anions, whereas in the other two the cation is tetrahedrally surrounded by anions. It will be seen from Table 19-4 that zinc and cadmium prefer tetrahedral coordination in their chalconides.

19-8.   Halides

All four halides of both zinc and cadmium are known. Some of their relevant properties are given in Table 19-5.

Both ZnF_2 and CdF_2 show distinct evidence of being considerably more ionic than the other halides of the same element. Thus they have higher melting and boiling points, and they are considerably less soluble in water. The latter property is attributable not only to the high lattice energies of the fluorides, but also to the fact that the formation of halo complexes in solution, which enhances the solubility of the other halides, does not occur for the fluorides (see below).

The structures of the chlorides, bromides, and iodides may be viewed as close-packed arrays of halide ions, but there is a characteristic difference in that zinc

[11]   F. Ichikawa and T. Sato, *J. Inorg. Nucl. Chem.,* 1973, **35**, 2594; S. K. Sharma and M. D. Reed, *J. Inorg. Nucl. Chem.,* 1976, **38**, 1971.

[12]   D. H. McDaniel, *Inorg. Chem.,* 1979, **18**, 1412.

TABLE 19-5
Some Properties of the Zinc and Cadmium Halides

Halide	Solubility in water (mol 1^{-1})	M.p. (°C)	B.p. (°C)	Structure
ZnF_2	1.57 (20°)	872	1502	Rutile
$ZnCl_2$[a]	31.8 (25°)	275	756	⎫
$ZnBr_2$	20.9 (25°)	394	697	⎬ *ccp* anions with Zn in
ZnI_2	13 (25°)	446	(Sublimes) ⎭ tetrahedral interstices
CdF_2	0.29 (25°)	1110	1747	Fluorite
$CdCl_2$	7.7 (20°)	868	980	⎫ Close-packed anions with Cd in
$CdBr_2$	4.2 (20°)	560	1136	⎬ octahedral interstices
CdI_2	2.3 (20°)	387	(Sublimes) ⎭

[a] Very pure $ZnCl_2$ has but one form; some reported polymorphs contain OH groups (see Ref. 1).

ions occupy tetrahedral interstices whereas cadmium ions occupy octahedral ones. Raman spectra show that depending on the concentration, the species present in aqueous solutions of $ZnCl_2$ are $[Zn(H_2O)_6]^{2+}$, $ZnCl^+(aq)$, $ZnCl_2(aq)$, and $[ZnCl_4(H_2O)_2]^{2-}$, but no indication was found of $[ZnCl_3]^-$ or $[ZnCl_4]^{2-}$ (see below).

Zinc chloride is so soluble in water that mole ratios H_2O to $ZnCl_2$ can easily be less than 2:1. Both zinc and cadmium halides are quite soluble in alcohol, acetone, and similar donor solvents, and in some cases adducts can be obtained.

Aqueous solutions of cadmium halides appear, superficially, to be incompletely dissociated, that is, to be weak electrolytes. Although there are significant amounts of the undissociated halides (CdX_2) and polymeric species present in moderately concentrated solutions, there are other species also present (Table 19-6). Thus the solutions are best regarded as systems containing all possible species in equilibrium rather than simply as solutions of a weak electrolyte.

19-9. Oxo Salts and Aqua Ions

Salts of oxo acids such as the nitrate, sulfate, sulfite, perchlorate, and acetate are soluble in water. The Zn^{2+} and Cd^{2+} ions are rather similar to Mg^{2+}, and many of their salts are isomorphous with magnesium salts, for example, $Zn(Mg)SO_4 \cdot 7H_2O$ and $M_2^ISO_4 \cdot Hg(Mg)SO_4 \cdot 6H_2O$. The aqua ions are quite strong acids, and aqueous solutions of salts are hydrolyzed. In perchlorate solution the only species

TABLE 19-6
Approximate Concentrations of Dissociated and Undissociated Species in 0.5M $CdBr_2$ Solution at 25°

Species	Concentration (M)	Species	Concentration (M)
Cd^{2+}	0.013	Br^-	0.200
$CdBr^+$	0.259	$CdBr_2$	0.164
$CdBr_3^-$	0.043	$CdBr_4^{2-}$	0.021

for Zn, Cd (and Hg) below 0.1 M are the MOH^+ ions, e.g.,

$$Cd^{2+}(aq) + H_2O \rightleftharpoons CdOH^+(aq) + H^+ \qquad \log\beta = 6.08$$

For more concentrated cadmium solutions, the principal species is Cd_2OH^{3+}:

$$2Cd^{2+}(aq) + H_2O \rightleftharpoons Cd_2OH^{3+}(aq) + H^+$$

In presence of complexing anions (e.g., halide), species such as $Cd(OH)Cl$ or $CdNO_3^+$ may be obtained.

The hydrate $ZnCl_2 \cdot 4H_2O$ is a liquid at ambient temperature and acts as a strong protic acid.[12]

On distillation of the normal acetate in a vacuum, zinc forms a basic acetate $Zn_4O(OCOCH_3)_6$, isomorphous with that of Be. It is a crystalline solid rapidly hydrolyzed by water, unlike the beryllium compound, the difference being due to the possibility of coordination numbers exceeding 4 for zinc.

Carbonates and hydroxo carbonates are also known.

19-10. Complexes of Zinc and Cadmium

Complex anions with halides are formed by both metals in aqueous solution, but constants differ widely; they are also many orders of magnitude smaller than those of Hg^{2+}, as is clear from Tables 19-7 and 19-8. The exact values are not important, since ionic strength effects are rather large, but certain qualitative features can be discerned. The formation of fluoro complexes is restricted, and none has been isolated as a solid. There is evidence for the attainment of all four stages of complexation by both zinc and cadmium with Cl^-, Br^-, and I^-, the cadmium complexes being moderately stable, whereas those of zinc are of rather low stability. The ZnX_4^{2-} complexes can be isolated as salts of large cations, but cadmium can also form $CdCl_3^-$. Zn^{2+} tends to form stronger bonds to F and O, whereas Cd^{2+} is more strongly bound to Cl, S, and P ligands.

Complex cations with ammonia and amine ligands are well defined and can be obtained as crystalline salts.

Zinc complexes of *dithiocarbamates* and of other sulfur compounds are important accelerators in the vulcanization of rubber by sulfur. The isostructural Zn and Cd compounds $[M(S_2CNEt_2)_2]_2$ achieve five-coordination by dimerizing as

TABLE 19-7

Equilibrium Constants for Some Typical Complexes of Zn,
Cd, and Hg ($M^{2+} + 4X = [MX_4]$; $K = [MX_4]/[M^{2+}][X]^4$)

X	\multicolumn{3}{c}{K}		
	Zn^{2+}	Cd^{2+}	Hg^{2+}
Cl^-	1	10^3	1.3×10^{15}
Br^-	10^{-1}	10^4	9.2×10^{20}
I^-	10^{-2}	10^6	5.6×10^{29}
NH_3	10^9	10^7	2.0×10^{19}
CN^-	10^{21}	10^{19}	1.9×10^{41}

TABLE 19-8
Some Formation Constants of Zinc and Cadmium Halide Complexes (at 25°)

	Halogen	Log K_1	Log K_2	Log K_3	Log K_4	Medium
Zn	F	0.75	Not obs.	Not obs.	Not obs.	0.5–1.0 M NaClO$_4$
	Cl	−1.0 to +1.0	−1.0 to +1.0	−1.0 to +1.0	−1.0 to +1.0	Variable
	Br	−0.60	−0.37	−0.73	0.44	Ionic str. = 4.5
	I	−2.93	1.25	−0.07	−0.59	Ionic str. = 4.5
Cd	F	0.46	0.07	Not obs.	Not obs.	1.0 M NaClO$_4$
	Cl	1.77	1.45	−0.25	−0.05	2.1 M KNO$_3$
	Br	1.97	1.25	0.24	0.15	1 M KNO$_3$
	I	2.96	1.33	1.07	1.00	1.6 M KNO$_3$

shown for cadmium in 19-I; one metal-sulfur bond is considerably longer that the others. The dimers may be split by amines to form five-coordinate 1 to 1 adducts There are also compounds of the type ZnX$_2$(dtc)$_2$ and [Zndtc$_3$]$^-$ in which there are one chelate and two formally unidentate S$_2$CNR$_2$ groups. The Zn atom is in an essentially distorted tetrahedral environment[13] (19-II).

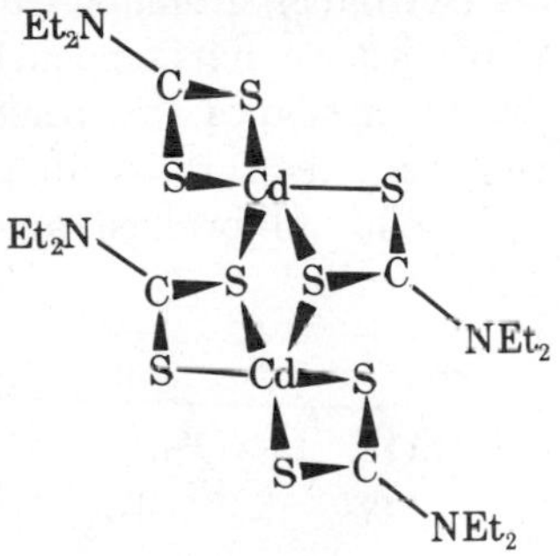

(19-I)

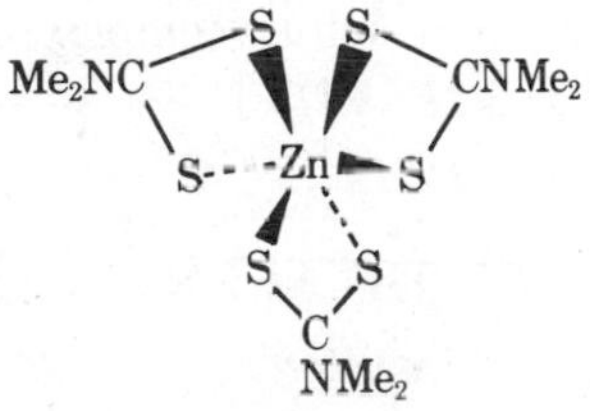

(19-II)

The complex ion derived from mercaptobenzthiazol (mbtH, 19-III), which also acts as a vulcanization accelerator, is similar.

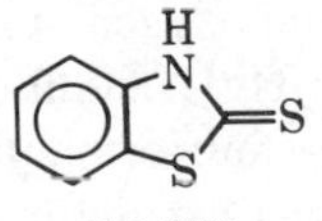

(19-III)

It is possible that the quasi-unidentate groups are susceptible to attack by S$_8$ to give a perthiocarbamate or mercaptobenzthiolate group:

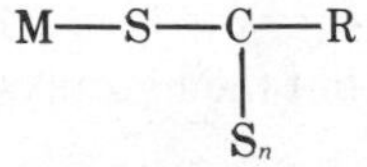

$$M—S—C—R$$
$$\overset{|}{S_n}$$

Such polysulfide ions could react with rubber, forming intermediates that degrade to give the sulfur cross-linked vulcanizate.[14] By contrast with the dithiocarbamates,

13   J. A. McCleverty and N. J. Morrison, *J.C.S. Dalton,* **1976**, 2169.
14   C. C. Ashworth *et al., J.C.S. Chem. Comm.,* **1976**, 743.

the compound $[C_6H_5CS_2]_2Zn$ has two bidentate groups, with zinc being coordinated by four sulfur atoms in a distorted tetrahedron.[15]

Zinc β-diketonates readily form five-coordinate adducts with water, alcohols, and nitrogen bases.

19-11. The Biological Role of Zinc and Cadmium

Zinc appears to be one of the most important metals biologically.[16] It is probably second only to iron among the heavy metals (i.e., excluding Na^+, Ca^{2+}, and Mg^{2+}). More than 25 zinc-containing proteins have been identified, most of them enzymes. These are discussed in detail in Chapter 31.

The biological role of *cadmium* is unclear, but low concentrations may be required, since metallothionein contains cadmium—the only biological material known to do so. There is a mechanism in the body that controls the level of zinc, but not the level of cadmium, which is one of the five most toxic metals. Cadmium accumulates in the kidneys, liver, and blood vessels and has a biological half-life in humans of 10 to 30 years, much longer than that of mercury.[17] The action may involve bonding to cysteine residues as for Hg (see later), although Cd has less affinity for sulfur than has mercury and it approaches Zn in its tendency to bind to carboxylate groups. Various model systems have been studied; thus penicillamine ($SCMe_2CH\overset{+}{N}H_3CO_2^-$) bonds through both sulfur and the carboxylate groups.[18] Other cadmium sulfur complexes include the unusual bridged aggregate[19] $[Cd_{10}(SCH_2CH_2OH)_{16}]^{4+}$.

DIVALENT MERCURY COMPOUNDS

19-12. Mercuric Oxide and Sulfide

Red mercuric oxide is formed on gentle pyrolysis of mercurous or mercuric nitrate, by direct interaction of mercury and oxygen at 300 to 350°, or as red crystals by heating of an alkaline solution of K_2HgI_4. Addition of OH^- to aqueous Hg^{2+} gives a yellow precipitate of HgO; the yellow form differs from the red only in particle size. The usual form of the oxide has a structure with zigzag chains —Hg—O— Hg— (Hg—O = 2.03 Å, ∠HgOHg = 109°, ∠OHgO = 179°); there is only weak bonding between the chains, the shortest Hg—O distance here being 2.82 Å. In the solid state, mixed oxides such as $Hg_2Nb_2O_7$ are known: all these contain linear O—Hg—O groups.

No hydroxide can be isolated, but the hydrolysis of mercuric ion in perchlorate

[15]   B. Bonamico *et al., J.C.S. Dalton,* **1972,** 2515.
[16]   M. F. Dunn, *Struct. Bonding,* 1975, **23,** 61 (zinc catalysis in enzymes and small molecules).
[17]   R. L. Singal *et al., Fed. Proc.,* 1976, **35,** 75.
[18]   A. J. Carty and N. J. Taylor, *Inorg. Chem.,* 1977, **16,** 177.
[19]   R. A. Haberkorn, *et al., Inorg. Chem.,* 1976, **15,** 2408.

solutions can be interpreted in terms of the equilibria[20]

$$Hg^{2+} + H_2O = Hg(OH)^+ + H^+ \qquad K = 2.6 \times 10^{-4}$$
$$Hg(OH)^+ + H_2O = Hg(OH)_2 + H^+ \qquad K = 2.6 \times 10^{-3}$$

No polynuclear species appear to be formed. The mercuric ion appears to be solvated by six water molecules.

Mercuric sulfide (HgS) is precipitated from aqueous solutions as a black, highly insoluble compound. The solubility product is 10^{-54}, but the sulfide is somewhat more soluble than this figure would imply because of some hydrolysis of Hg^{2+} and S^{2-} ions. The black sulfide is unstable with respect to a red form identical to the mineral cinnabar and changes into it when heated or digested with alkali polysulfides or mercurous chloride. The red form has a distorted sodium chloride lattice with Hg—S chains similar to those in HgO. Another form, occurring as the mineral metacinnabarite, has a zinc blende structure, as have the selenide and telluride.

19-13. Mercuric Halides[21a]

The *fluoride* of mercury is essentially ionic and crystallizes in the fluorite structure; it is almost completely decomposed, even by cold water, as would be expected for an ionic compound that is the salt of a weak acid and an extremely weak base. Not only does mercury(II) show no tendency to form covalent Hg—F bonds, but no fluoro complex is known.

In sharp contrast to the fluoride, the other halides show marked covalent character. *Mercuric chloride* crystallizes in an essentially molecular lattice, the two short Hg—Cl distances being about the same length as the Hg—Cl bonds in gaseous $HgCl_2$, whereas the next shortest distances are much longer (see Table 19-9).

In the larger lattice of $HgBr_2$ each Hg atom is surrounded by six Br atoms, but two are so much closer than the other four that it can be considered that perturbed $HgBr_2$ molecules are present. The normal red form of HgI_2 has a layer structure with HgI_4 tetrahedra linked at some of the vertices. However at 126° it is converted into a yellow molecular form.

In the vapor all three halides are distinctly molecular, as they are also in solutions. Relative to ionic HgF_2, the other halides have very low melting and boiling points

TABLE 19-9
Hg—X Distances in Mercuric Halides (Å)

Compound	Solid			Vapor
	Two at	Two at	Two at	
HgF_2		Eight at 2.40		—
$HgCl_2$	2.25	3.34	3.63	2.28 ± 0.04
$HgBr_2$	2.48	3.23	3.23	2.40 ± 0.04
HgI_2 (red)		Four at 2.78		2.57 ± 0.04

[20] See R. D. Hancock and F. Marsicano, *J.C.S. Dalton*, **1976**, 1832.
[21a] P. A. W. Dean, *Prog. Inorg. Chem.*, 1978, **24**, 109 (coordination chemistry of HgX₂).

TABLE 19-10
Some Properties of Mercuric Halides

Halide	M.p. (°C)	B.p. (°C)	Solubility (moles/100 moles at 25°)			
			H_2O	C_2H_5OH	$C_2H_5OCOCH_3$	C_6H_6
HgF_2	645 d.	—	Hydrolyzes	Insol.	Insol.	Insol.
$HgCl_2$	280	303	0.48	8.14	9.42	0.152
$HgBr_2$	238	318	0.031	3.83	—	—
HgI_2	257	351	0.00023	0.396	0.566	0.067

(Table 19-10). They also show marked solubility in many organic solvents. In aqueous solution they exist almost exclusively (∼99%) as HgX_2 molecules, but some hydrolysis occurs, the principal equilibrium being, e.g.,

$$HgCl_2 + H_2O \rightleftharpoons Hg(OH)Cl + H^+ + Cl^-$$

In organic solvents like benzene, there appear to be dimers,[21b] probably $XHg(\mu\text{-}X)_2HgX$.

19-14. Mercuric Oxo Salts

Among the mercuric salts that are totally ionic, hence highly dissociated in aqueous solution, are the nitrate, sulfate, and perchlorate. Because of the great weakness of mercuric hydroxide, aqueous solutions of these salts tend to hydrolyze extensively and must be acidified to be stable.

In aqueous solutions of $Hg(NO_3)_2$ the main species are $Hg(NO_3)_2$, $HgNO_3^+$, and Hg^{2+}, but at high concentrations of NO_3^- the complex anion $[Hg(NO_3)_4]^{2-}$ is formed.

Mercuric carboxylates, especially the acetate and the trifluoroacetate, are of considerable importance because of their utility in attacking unsaturated hydrocarbons, as discussed below. They are made by dissolving HgO in the hot acid and crystallizing. The trifluoroacetate is also soluble in benzene, acetone, and tetrahydrofuran, which increases its utility, while the acetate is soluble in water and alcohols.

Other salts such as the oxalate and phosphates are sparingly soluble in water.

Mercuric ions catalyze a number of reactions of complex compounds such as the aquation of $[Cr(NH_3)_5X]^{2+}$. Bridged transition states, e.g.,

$$[(H_2O)_5CrCl]^{2+} + Hg^{2+} = [(H_2O)_5Cr—Cl—Hg]^{4+}$$

are believed to be involved.

19-15. Mercuric Complexes

A number of the mercuric complexes have been mentioned above. The Hg^{2+} ion has indeed a strong tendency to complex formation, and the characteristic coor-

21b I. Eliezer and G. Algavish, *Inorg. Chim. Acta,* 1974, **9,** 257.

dination numbers and stereochemical arrangements are two-coordinate, linear and four-coordinate, tetrahedral. Octahedral coordination is less common; a few three- and five-coordinate complexes are also known. There appears to be considerable covalent character in the mercury-ligand bonds, especially in the two-coordinate compounds. The most stable complexes are those with C, N, P, and S as ligand atoms.

Halide and Pseudohalide Complexes. In the halide systems, depending on the concentration of halide ions, there are equilibria such as

$$HgX^+ \rightleftharpoons HgX_2 \rightleftharpoons HgX_3^- \rightleftharpoons HgX_4^{2-}$$

At 1 M Cl$^-$ the main species is $[HgCl_4]^{2-}$, but in tributyl phosphate as solvent the most stable ion is $[HgCl_3]^-$; at 10^{-1} M Cl$^-$ the concentrations of $HgCl_2$, $HgCl_3^-$, and $HgCl_4^{2-}$ are about equal. In $Me_3S^+HgCl_3^-$ the ion is polymeric with mercury *tbp* with equatorial terminal Cl atoms and axial, bridged Cl atoms.[22a] In aqueous solutions[22b] $HgCl_3^-$ appears to be planar and solvated by axial water molecules in a *tbp* structure, whereas HgI_3^- is solvated by only one H_2O, giving a tetrahedral species.

Mercuric cyanide $[Hg(CN)_2]$, which contains discrete molecules with linear C—Hg—C bonds, is soluble in CN$^-$ to give $Hg(CN)_3^-$ and $Hg(CN)_4^{2-}$ only. SCN$^-$ is similar to CN$^-$ in its complexing behavior.

Oxo Ion Complexes. Several of these exist, e.g., $[Hg(SO_3)_2]^{2-}$ and $[Hg\ ox_2]^{2-}$. The yellow crystals formed by adding KNO_2 to $Hg(NO_3)_2$ solution contain $K_2[Hg(NO_2)_4]\cdot KNO_3$, where the nitrite ion is bidentate, giving an eight-coordinate, very distorted square antiprism. The acetate and the tropolonate have O—Hg—O bonds but with some much weaker interaction to the other oxygen atom of the ligand.[23]

Nitrogen, Phosphorus, and Sulfur Ligands. Many complexes of these ligands are known, including macrocycle complexes (see below). For sulfur species, Hg—S bonding always occurs, and the name "mercaptan" for RSH compounds stems from the affinity for mercury.

Since the toxic effects of Hg^{2+} are quite different from those of $MeHg^+$ described below, the former being nephrotic, the types of complex formed by the two ions could be expected to differ, and indeed they do. With amino acids containing SH groups, the binding to Hg^{2+} and $MeHg^+$ differs strikingly. The latter form monomeric nonpolar complexes MeHgSR, whereas Hg^{2+} complexes are polymeric and polar.[24]

The dithiocarbamate $Hg(S_2CNEt_2)_2$ has two forms, one isomorphous with the Zn and Cd analogues (19-I), the other a helical polymer.[25]

Tertiary phosphines (R_3As, R_2S, py, bipy, etc.) generally give mercury(II) halide

22a   P. Biscarini *et al., J.C.S. Dalton,* **1977,** 664.
22b   T. R. Griffiths and R. A. Anderson, *J.C.S. Chem. Comm.,* **1979,** 61.
23   K. Dietrich and H. Musso, *Angew. Chem., Int. Ed.,* 1975, **14,** 358.
24   A. J. Carty and N. J. Taylor, *J.C.S. Chem. Comm.,* **1976,** 214.
25   P. C. Healy and A. H. White, *J.C.S. Dalton,* **1973,** 284; H. Iwasaki, *Acta Crystallogr., B,* 1973, **29,** 2115.

complexes that are either monomeric (HgX_2L_2) or dimeric as in 19-IV, but more highly bridged structures such as 19-V may also be formed.[26]

(19-IV)   (19-V)

Although there is a tendency to form ammonobasic compounds, a variety of *amines* form complexes with Hg^{II} and the affinity of Hg^{II} for nitrogen ligands in aqueous solution exceeds that of the transition metals. In addition to the ammonia and amine complexes of the type $Hg(NH_3)_2X_2$, tetraammines such as $[Hg(NH_3)_4](NO_3)_2$ can be prepared in saturated aqueous ammonium nitrate. The ion $[Hg\ en_3]^{2+}$ has octahedral Hg^{II}, as have complexes of the type $[HgL_6](ClO_4)_2$ obtained when suitable oxygen donors are added to Hg^{II} perchlorate in ethanol.

In the chelates $[Hg\ en_2]^{2+}$ and $[Hg\ en(SCN)_2]$, mercury is in a distorted tetrahedral environment.[27]

19-16. Novel Compounds of Mercury(II) with Nitrogen[28]

It has been known since the days of alchemy that when Hg_2Cl_2 is treated with aqueous ammonia, a black residue is formed, and this reaction is still used in qualitative analysis to identify Hg_2Cl_2. These residues contain nitrogen compounds of Hg^{II} plus metallic mercury, and the Hg^{II} compounds can be obtained directly from Hg^{II} salts.

There are three known products of the reaction of $HgCl_2$ with ammonia, the proportion of any one of them depending on the conditions. The products $Hg(NH_3)_2Cl_2$, $HgNH_2Cl$, and $Hg_2NCl\cdot H_2O$ are formed according to the following equations:

$$HgCl_2 + 2NH_3 \rightleftharpoons Hg(NH_3)_2Cl_2(s)$$
$$HgCl_2 + 2NH_3 \rightleftharpoons HgNH_2Cl(s) + NH_4^+ + Cl^-$$
$$2HgCl_2 + 4NH_3 + H_2O \rightleftharpoons Hg_2NCl\cdot H_2O + 3NH_4^+ + 3Cl^-$$

The equilibria seem to be labile, so that the product obtained can be controlled by varying the concentrations of NH_3 and NH_4^+. In concentrated NH_4Cl solution, the diammine $Hg(NH_3)_2Cl_2$ is precipitated, whereas with dilute ammonia and no excess of NH_4^+, the amide $HgNH_2Cl$ is formed. The compound $Hg_2NCl\cdot H_2O$ is probably not produced in a pure state by the reaction above, but it can be obtained by treating the compound $Hg_2NOH\cdot 2H_2O$ (*Millon's base*) with hydrochloric acid. Millon's base itself is made by the action of aqueous ammonia on yellow mercuric oxide.

The diammine has been shown to consist of discrete tetrahedral molecules. The

[26] See, e.g., S. Akyüz *et al., J.C.S. Dalton,* **1976,** 1746; N. A. Bell *et al., J.C.S. Chem. Commun.* **1976,** 1039.
[27] T. Duplaničic *et al., J.C.S. Dalton,* **1976,** 887.
[28] D. Breitinger and K. Broderson, *Angew. Chem., Int. Ed.,* 1970, **9,** 357.

amide has infinite chains —Hg—NH$_2$—Hg—NH$_2$—, where N—Hg—N segments are linear and the bonds about nitrogen are tetrahedral; the chloride ions lie between the chains. The analogous bromide has the same structure.

Millon's base has a three-dimensional framework of composition Hg$_2$N with the OH$^-$ ions and water molecules occupying rather spacious cavities and channels. Many salts of Millon's base are known, for example, Hg$_2$NX·nH$_2$O (X = NO$_3^-$, ClO$_4^-$, Cl$^-$, Br$^-$, or I$^-$; n = 0–2). In these the framework appears to remain essentially unaltered; thus it is an ion exchanger similar to a zeolite.

Returning to the dark residues given by mercurous chloride, one or both of Hg$_2$NH$_2$Cl and Hg$_2$NCl·H$_2$O are present together with free metal. The insolubility of these compounds causes the disproportionation of the Hg$_2^{2+}$, for example,

$$Hg_2Cl_2(s) + 2NH_3 = Hg(l) + HgNH_2Cl(s) + NH_4^+ + Cl^-$$

The action of aqueous ammonia on organomercury compounds such as C$_6$H$_5$HgCl also gives ionic amido derivatives (C$_6$H$_5$Hg)$_2$NH$_2^+$.

There is no evidence for any intermediate ammonobasic or ammine compound of mercury(I).

19-17. Compounds with Metal-to-Mercury Bonds

A wide variety of compounds are known in which the Hg atom is bound to other metals, very commonly transition metal atoms, in complexes. These are all best formulated as Hg^{II} compounds, and usually the bonds are linear, M—Hg—M or M—HgX as in [en$_2$Rh-Hg-Rhen$_2$]$^{4+}$ or η^5 C$_5$H$_5$(CO)$_3$MoHgCl.[29a] However, in [η^5-C$_5$H$_5$(CO)$_3$MoHgMo]$_4$ there is a Mo$_4$Hg$_4$ cube with each Hg bound to three molybdenum atoms in the cube and to the Mo of the organometallic group.[29b]

Zinc and cadmium commonly form similar compounds, and some representative examples are (CO)$_4$CoMCo(CO)$_4$ (M = Zn, Cd, or Hg), Fe(CO)$_4$(HgCl)$_2$, and η^5-C$_5$H$_5$Fe(CO)$_2$-Hg-Co(CO)$_4$. The general nature of compounds with metal-metal bonds is outlined in Chapter 26.

Some of these compounds are made by action of HgCl$_2$ on hydrido complexes or on carbonylate anions (Section 25-6), e.g.,

$$2C_5H_5Mo(CO)_3^- Na^+ + HgCl_2 \rightarrow [C_5H_5Mo(CO)_3]_2Hg + 2NaCl$$
$$(Ph_2MeAs)_3RhHCl_2 + HgCl_2 \rightarrow (Ph_2MeAs)_3Rh(HgCl)Cl_2 + HCl$$

Mercury(II) compounds HgX$_2$ or RHgX may also undergo the oxidative-addition reaction (Section 29-3) in which metal-mercury bonds are formed,[30] e.g.,

$$trans\text{-}IrCl(CO)(PPh_3)_2 + HgCl_2 = IrCl_2(HgCl)(CO)(PPh_3)_2$$

The HgCl$_2$ *molecule* may also add as such. Thus organic sulfides R$_2$S give compounds such as (R$_2$S)HgCl$_2$. The adduct (η^5-C$_5$H$_5$)(CO)$_2$Co·HgCl$_2$ has the

29a M. J. Albright *et al.*, *J. Organomet. Chem.*, 1978, **161**, 221; J. Gulens and F. A. Anson, *Inorg. Chem.*, 1973, **12**, 2568.
29b J. Deutscher *et al.*, *Angew Chem., Int. Ed.*, 1977, **16**, 704.
30   J. Kuyper, *Inorg. Chem.*, 1978, **17**, 1458; O. A. Reutov *et al.*, *J. Organomet. Chem.*, 1978, **160**, 7; P. D. Brotherton *et al.*, *J.C.S. Dalton*, 1976, 1799.

structure (19-VI) in which the axial positions of the trigonal bipyramid are occupied by weakly coordinated chlorine atoms of adjacent molecules, whereas in [Ar-Mo(CO)$_3$·2HgCl$_2$]$_2$ there is a grouping as shown in 19-VII.[31]

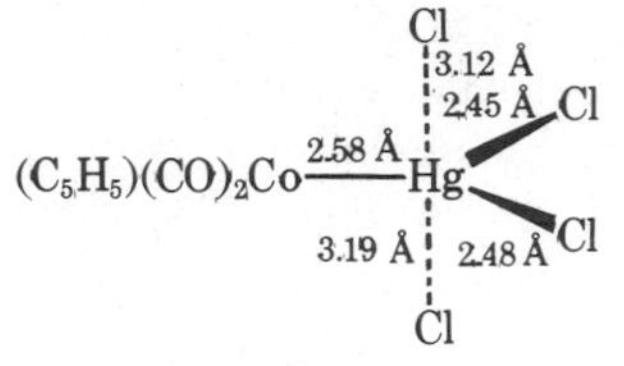

(19-VI)

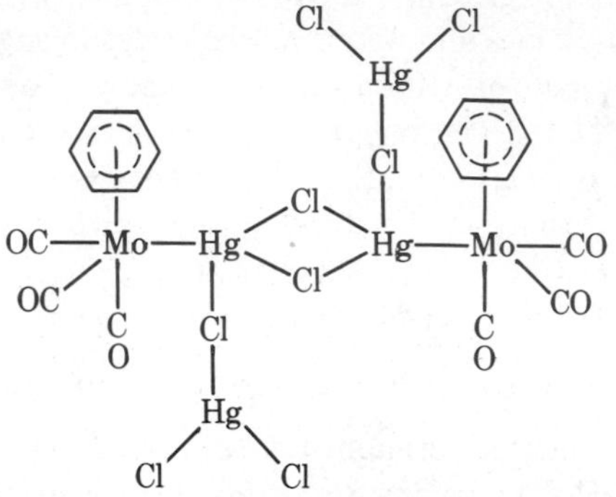

(19-VII)

TRIVALENT MERCURY

No compounds of Hg^{III} have yet been isolated, but Born-Haber cycles suggest that Hg^{III} could be accessible, and it would provide a unique case of the removal of a d electron from a Group IIB element.

The electrochemical oxidation[32a] of acetonitrile solutions at −78° of the Hg^{II} complex [Hg cyclam](BF$_4$)$_2$ provides evidence for Hg^{III}. Cyclam, or 1,4,8,11-tetraazacyclotetradecane, is not readily oxidized. A species of half-life about 5 sec but characterizable by cyclic voltammetry, epr, and electronic absorption spectroscopy is consistent with a paramagnetic Hg^{III} complex.

The estimated potential is

$$[Hg\ cyclam]^{3+} + e = [Hg\ cyclam]^{2+} \qquad E = 1.6\ to\ 1.8\ V$$

ORGANOMETALLIC COMPOUNDS

19-18. Organozinc and Organocadmium Compounds[32b]

Organozinc compounds are historically important because they were the first organometallic compounds to be prepared; their discovery by Sir Edward Frankland in 1849 played a decisive part in the development of modern ideas of chemical bonding. The cadmium compounds are also of interest, since their mild reactivities

[31] M. R. Snow et al., J.C.S. Dalton, **1976**, 35.
[32a] R. L. Deming et al., J. Am. Chem. Soc., 1976, **98**, 4132.
[32b] I. Sheverdina and K. A. Kocheskov, The Organic Compounds of Zinc and Cadmium, North Holland, 1967.

toward certain organic functional groups give them unique synthetic potentiali-
ties.[33]

Organozinc compounds of the types R_2Zn and $RZnX$ are known; $EtZnI$ is a
polymer with iodide bridges; each iodine atom forms three bonds to zinc, two long
and one normal.[34] The chloride and bromide may have cubane-type structures.
Except for Bu^nCdCl, only R_2Cd compounds have been isolated.

The constitution of RMX in solution has presented a problem similar to that for
Grignard reagents. Monomeric RMX species predominate in ethers, e.g.,

$$Me_2Cd + CdI_2 \xrightleftharpoons{THF} 2MeCdI \qquad K \geq 100$$

For the perfluorophenyls, exchange between C_6H_5MX and $(C_6H_5)_2M$ is sufficiently slow that the Schlenk equilibria can be studied by ^{19}F nmr.[35]

There is self-exchange in Me_2Cd and with Zn, Ga, and In alkyls *via* alkyl-bridged
species.

The zinc alkyls can be obtained by thermal decomposition of RZnI, which is
prepared by the reaction of alkyl iodides with a zinc-copper couple:

$$C_2H_5I + Zn(Cu) \rightarrow C_2H_5ZnI \xrightarrow{heat} \tfrac{1}{2}(C_5H_5)_2Zn + \tfrac{1}{2}ZnI_2$$

The alkyls may also be prepared, and the diaryls most conveniently obtained, by
the reaction of zinc metal with an organomercury compound:

$$R_2Hg + Zn \rightarrow R_2Zn + Hg$$

or by reaction of zinc chloride with organolithium, -aluminum or -magnesium.

The best preparation of R_2Cd compounds is by treatment of the anhydrous
cadmium halide with RLi or RMgX. The reaction of cadmium metal with alkyl
iodides in dimethylformamide or $(CH_3)_2SO$ gives RCdI in solution.

The R_2Zn and R_2Cd compounds are nonpolar liquids or low-melting solids,
soluble in most organic liquids. The lower alkyl zinc compounds are spontaneously
flammable, and all react vigorously with oxygen and with water. The cadmium
compounds are less sensitive to oxygen but are less stable thermally.

Both zinc and cadmium compounds react readily with compounds containing
active hydrogen, such as alcohols:

$$R_2M + R'OH \rightarrow RMOR' + RH$$

and are generally similar to RLi or RMgX, although their lower reactivity allows
selective alkylations not possible with the more standard reagents. An important
example is the use of the cadmium compounds in the synthesis of ketones from acyl
chlorides:

$$2RCOCl + R'_2Cd \rightarrow 2RCOR' + CdCl_2$$

[33]   P. R. Jones and P. J. Desio, *Chem. Rev.*, 1978, **78**, 491 (less familiar reactions of organocad-
miums).

[34]   P. T. Moseley and H. M. M. Shearer, *J.C.S. Dalton*, **1973**, 64.

[35]   D. F. Evans and R. F. Phillips, *J.C.S. Dalton*, **1973**, 878.

With lithium alkyls and aryls, complexes such as $Li[ZnPh_3]$ and $Li_2[CdMe_4]$ may be formed; hydrido complexes such as $Li[(Et_2Zn)_2H]$ appear to have Zn—H—Zn bridges.

19-19.   Organomercury Compounds[36,37a]

A vast number of organomercury compounds are known, some of which have useful physiological properties. They are of the types RHgX and R_2Hg. They are commonly made by the interaction of $HgCl_2$ and RMgX, but Hg—C bonds can also be made in other ways discussed below.

The *RHgX compounds* are crystalline solids whose properties depend on the nature of X. When X is an atom or group that can form covalent bonds to mercury (e.g., Cl, Br, I, CN, SCN, or OH), the compound is a covalent nonpolar substance more soluble in organic liquids than in water. When X is SO_4^{2-} or NO_3^-, the substance is saltlike and presumably quite ionic, for instance, $[RHg]^+NO_3^-$. Acetates behave as weak electrolytes. For iodides or thiocyanates, complex anions (e.g., $RHgI_2^-$ and $RHgI_3^{2-}$) may be formed. A particularly important species due to its environmental significance is the ion $CH_3Hg(OH_2)^+$ discussed later.

The *dialkyls and diaryls* are nonpolar, volatile, toxic, colorless liquids or low-melting solids. Unlike the Zn and Cd alkyls, they are much less affected by air or water, presumably because of the low polarity of the Hg—C bond and the low affinity of mercury for oxygen. However they are photochemically and thermally unstable, as would be expected from the low bond strengths, which are of the order 50 to 200 kJ mol^{-1}. In the dark, mercury compounds can be easily kept for months. The decomposition generally proceeds by homolysis of the Hg—C bond and free-radical reactions.

All RHgX and R_2Hg compounds have linear bonds, but deviation from linearity has been claimed in a few cases, and in solution in particular, solvation effects may contribute to nonlinearity. The cyclopentadienyl $[(C_5H_5)_2Hg, C_5H_5HgX]$ and indenyl $[(C_9H_7)_2Hg]$ compounds are fluxional (see Chapter 28).

The principal utility of dialkyl- and diarylmercury compounds, and a very valuable one, is in the preparation of other organo compounds by interchange, e.g.,

$$\frac{n}{2} R_2Hg + M \rightarrow R_nM + \frac{n}{2} Hg$$

This reaction proceeds essentially to completion with the Li and Ca groups, and with Zn, Al, Ga, Sn, Pb, Sb, Bi, Se, and Te, but with In, Tl, and Cd reversible equilibria are established. Partial alkylation of reactive halides can be achieved, e.g.,

$$AsCl_3 + Et_2Hg \rightarrow EtHgCl + EtAsCl_2$$

[36]   R. C. Larock, *Angew. Chem., Int. Ed.*, 1978, **17,** 26 (organomercury compounds in organic synthesis).

[37a] L. G. Makarova and A. N. Nesmeyanov, *The Organic Chemistry of Mercury,* North Holland, 1967.

The special class of *perhalogenoalkylmercury* compounds (e.g., $PhHgCX_3$ or $PhHgCX_2Y$) may be used as sources of halogenocarbenes for transfer reactions to organic substrates. A convenient preparation of such a compound is

$$PhHgCl + CHX_3 + t\text{-}C_4H_9OK \xrightarrow{\text{benzene}} PhHgCX_3 + KCl + t\text{-}C_4H_9OH$$

There is much known concerning mechanisms of reaction of organomercury compounds, but only brief mention can be made here.[37b] Exchange reactions of the type:

$$RHgBr + {}^*HgBr_2 \rightleftharpoons R{}^*HgBr + HgBr_2$$

have been studied using tracer mercury. It was shown that the electrophilic substitution, S_E2, proceeds with full retention of configuration when an optically active group, *sec*-butyl, is present. The reaction, which is also catalyzed by anions, is believed to proceed through a cyclic transition state such as 19-VIII. Reactions such as

$$R_2Hg + HgX_2 = 2RHgX$$

have equilibrium constants of 10^5 to 10^{11} and proceed at rates that are slow and solvent dependent. Nmr studies have also shown that in solutions of RHgI compounds, there is relatively fast exchange of R groups and, again, a mechanism involving a cyclic intermediate or transition state (19-IX) has been postulated

(19-VIII)   (19-IX)

Organomercury Compounds in the Environment.[38] The deleterious effect of mercury compounds in the environment was first noted in Sweden, where the effluent from paper mills contained mercury. Later it was shown that human disasters such as those at Minimata, Japan, and in Iraq, had resulted from mercury poisoning. The effect is ascribed to the methylmercury ion, which induces irreversible complex disturbances of the central nervous system. The mercury released to the environ-

[37b] F. R. Jensen and R. Rickborn, *Electrophilic Substitution of Organomercurials,* McGraw-Hill, 1968; O. A. Reutov and I. P. Beletskaya, *Reactions of Organometallic Compounds,* North Holland, 1968 (despite the title, this is mainly concerned with mercury compounds); D. E. Matteson, *Organomet. Chem. Rev., A,* 1969, **4,** 263 (a good critical review of electrophilic displacements by mercury compounds); R. E. Dessy and W. Kitching, *Adv. Organomet. Chem.,* 1966, **4,** 268 (reaction mechanisms involving Hg—C bonds, including mercuration and oxomercuration); W. Kitching, *Organomet. Chem. Rev.,* 1968, **3,** 35, 61 (mercuration and oxomercuration).

[38] *Assessment of Mercury in the Environment, U.S. National Academy of Sciences,* 1978; J. M. Wood, in *Advances in Environmental Science,* Vol. 2, J. N. Pitts and R. L. Metcalf, Eds., Wiley-Interscience, 1971; L. T. Friberg and J. J. Vostal, *Mercury in the Environment,* CRC Press, 1972; D. L. Rabenstein, *J. Chem. Educ.,* 1978, **55,** 292 (toxicology of MeHg+).

ment as metal (e.g., by losses from electrolytic cells used for NaOH and Cl_2 production or in compounds such as mercury seed dressings or fungicides) is converted to CH_3Hg^+ by a biological methylation. It is known that vitamin B_{12} and model compounds such as methylcobaloximes or methylpentacyanocobaltate (Section 21-F-9) that have Co—CH_3 bonds will transfer the CH_3 to Hg^{2+}. There are a number of microorganisms that can perform the same function, doubtless by similar routes.

In aqueous solutions[39] the CH_3Hg^+ ion is hydrated, and there are pH-dependent reactions giving mercury-substituted oxonium ions:

$$CH_3Hg(OH_2)^+ + OH^- \rightleftharpoons CH_3HgOH + H_2O$$
$$CH_3Hg(OH_2)^+ + CH_3HgOH \rightleftharpoons (CH_3Hg)_2OH^+ + H_2O$$
$$CH_3HgOH + (CH_3Hg)_2OH^+ \rightleftharpoons (CH_3Hg)_3O^+ + H_2O$$

The ion, which can have coordination numbers up to 4, binds to S and Se very strongly. Since such reactions with proteins, peptides, etc., are presumably involved in toxic behavior (binding to pyrimidines, nucleotides, and nucleosides can also occur[40]), there has been much study of complexes with cysteine, methionine, other amino acids, etc.[41]

The formation constants for such sulfur compounds are ca. 10^8 times greater than the formation constants by the NH_2 group. It appears that cysteine can neutralize to some extent the toxic effects of MeHg$^+$ in sea urchin cells. Metallothionein, an SH-rich protein, and also selenium compounds, can act the same way. The cysteinate has been shown[42] to have the structure $H_3CHgSCH_2CH$-$(\overset{+}{N}H_3)(CO_2^-)$ with a linear C—Hg—S group. Other complexes such as [MeHg-bipy]$^+$, where mercury is planar and also three-coordinate, are known.[43]

Some Special Cases. Mercury(II) complexes of β-diketones are abnormal and have metal to γ-carbon bonds like platinum(II) compounds. Certain of these compounds undergo keto-enol tautomerism,[44] viz.,

39   D. L. Rabenstein, *Acc. Chem. Res.*, 1978, **11**, 100.
40   S. Mansy et al., *J. Am. Chem. Soc.*, 1974, **96**, 1762.
41   A. J. Canty et al., *J. Am. Chem. Soc.*, 1977, **99**, 643; *J.C.S. Dalton*, **1977**, 1157, 1801. K. Stanley et al., *Inorg. Chim. Acta*, 1978, **27**, L111.
42   N. J. Taylor et al., *J.C.S. Dalton*, **1975**, 439.
43   A. J. Canty et al., *Inorg. Chem.*, 1976, **15**, 425; *J.C.S. Dalton*, **1976**, 2018.
44   R. H. Fish, *J. Am. Chem. Soc.*, 1974, **96**, 6664.

Mercurous perchlorate and nitrate react with acetone to give complex species that have acetone bound to mercury as the enolate ion. However ethanol reacts in basic solution with HgO to give a polymeric material. This is probably derived by condensation of the unknown molecule $C(HgOH)_4$ to give a polymeric oxonium ion with C—Hg—OH—Hg—C bridges. This is suggested because the polymer dissolves in carboxylic acids to give compounds $C(HgCO_2R)_4$, which have tetrahedral central carbon and linear C—Hg—O bonds.[45] These carboxylates can also be obtained by quite another route:

$$B(OMe)_3 + Li + CCl_4 \rightarrow C[B(OMe)_2]_4 \xrightarrow{Hg(O_2CMe)_2} C[Hg(CO_2Me)]_4$$

Crystalline water-soluble salts of the type $(RHg)_3O^+$ and the compound $C(HgI)_4$ are also known.

Mercuration and Oxomercuration. An important reaction for the formation of Hg—C bonds, and one that can be adapted to the synthesis of a wide variety of organic compounds, is the addition of mercuric salts, notably the acetate, trifluoroacetate, or nitrate to unsaturated compounds.[37b,46]

The simplest reaction, the mercuration reaction of aromatic compounds, is commonly achieved by the action of mercuric acetate in methanol, e.g.,

Even aromatic organometallic compounds such as tricarbonylcyclobutadienyliron, $C_4H_4Fe(CO)_3$ (Section 27-11), can be mercurated.

By use of aryl compounds prepared in this way, the arylation of alkenes and a variety of other organic unsaturated compounds can be achieved. A palladium compound, usually Li_2PdCl_4, is used as a transfer agent, and organopalladium species are believed to be unstable intermediates; by use of air and a Cu^{II} salt the reaction can be made catalytic, since the Pd metal formed in the reaction is dissolved by Cu^{II} and the Cu^I so produced is oxidized by air (cf. the Wacker process, Section 30-14). A typical reaction is

Although Hg-arene complexes have been postulated as intermediates in aromatic mercuration, only recently have such complexes been isolated[47] by reactions such as

$$Hg(SbF_6)_2 + 2C_6H_6 \xrightarrow{SO_2(l)} (C_6H_6)_2 \cdot Hg(SbF_6)_2$$

[45] D. Grdenić et al., J.C.S. Chem. Comm., **1974**, 646; D. S. Matteson et al., J. Am. Chem. Soc., 1970, **92**, 231.

[46] See L. F. Fieser and M. Fieser, Reagents for Organic Synthesis, Wiley, 1967, et seq.

[47] L. C. Damude and P. A. W. Dean, J.C.S. Chem. Comm., **1978**, 1083.

and for benzene, we have the equilibrium

$$Hg^{2+} + C_6H_6 \rightleftharpoons Hg(C_6H_6)^{2+} \qquad K_1 = 0.48 \ l \ mol^{-1} \ (308°K)$$

The interaction of mercuric salts is not confined to arenes, but with alkenes there appears to be a general reversible reaction

$$\begin{array}{c}\diagdown \\ / \end{array}C=C\begin{array}{c}\diagup \\ \diagdown \end{array} + HgX_2 \rightleftharpoons \overset{X}{\underset{HgX}{\begin{array}{c}\diagdown \\ - \end{array}C-C\begin{array}{c}\diagup \\ \diagdown \end{array}}} \tag{19-5}$$

In most cases, the reactions have to be carried out in an alcohol or other protic medium, so that further reaction with the solvent is normally complete and the reaction is called *oxomercuration*, e.g.,

$$\begin{array}{c}\diagdown \\ / \end{array}C=C\begin{array}{c}\diagup \\ \diagdown \end{array} + Hg(OCOCH_3)_2 + C_2H_5OH \longrightarrow \overset{C_2H_5O}{\underset{HgOCOCH_3}{\begin{array}{c}\diagdown \\ - \end{array}C-C\begin{array}{c}\diagup \\ \diagdown \end{array}}} + CH_3COOH$$

The evidence that HgX_2 adds across the bond is usually indirect, often, by observing the products on hydrolysis, e.g.,

$$CH_2{=}CH_2 + Hg(NO_3)_2 \xrightarrow{OH^-} HOCH_2CH_2Hg^+ + 2NO_3^-$$

or by removal of Hg as $HgCl_2$ by action of HCl, which reverses the addition:

$$\begin{array}{c}\diagdown \\ / \end{array}C=C\begin{array}{c}\diagup \\ \diagdown \end{array} + Hg(O_2CMe)_2 \longrightarrow \overset{MeCO_2}{\underset{HgO_2CMe}{\begin{array}{c}\diagdown \\ - \end{array}C-C\begin{array}{c}\diagup \\ \diagdown \end{array}}} \xrightarrow{HCl} \begin{array}{c}\diagdown \\ / \end{array}C=C\begin{array}{c}\diagup \\ \diagdown \end{array} + HgCl_2 + 2MeCO_2H$$

The reversibility of the initial addition is readily established by using $Hg(O-COCF_3)_2$, and since the latter is soluble in nonpolar solvents, the equilibrium constants for the addition reaction 19-5, where $X = OCOCF_3$, can readily be measured and the dependence on the nature of the alkene and the solvent studied.

$$\overset{Hg^{2+}}{\underset{(19\text{-}X)}{\diagup\diagdown}}\qquad\qquad \overset{\overset{X}{\mid}\ Hg}{\underset{(19\text{-}XI)}{\diagup\diagdown}}$$

In reactions such as the above, mercurinium ions of the type 19-X and 19-XI are believed to be intermediates. Indeed in $FSO_3H\text{-}SbF_5\text{-}SO_2$ at $-70°$ long-lived ions have been obtained[48] by reactions such as

$$CH_3OCH_2CH_2HgCl \xrightarrow{H^+} CH_3OH_2^+ + CH_2CH_2Hg^{2+} + HCl$$

$$\bigcirc + Hg(OCOCF_3)_2 \xrightarrow{H^+} \bigcirc\!\!:Hg^{2+}$$

48 G. A. Olah and P. R. Clifford, *J. Am. Chem. Soc.*, 1973, **95**, 6067.

This type of addition has been used for the simple preparation, from alkenes and other unsaturated substances, of alcohols, ethers, and amines; the additions of HgX_2 are carried out in water, alcohols, or acetonitrile, respectively. The mercury is removed from the intermediate by reduction with sodium borohydride. Such additions are also useful in that the products are those in the Markownikoff direction. Two examples are the following:

$$\text{RCH}=\text{CH}_2 + \text{CH}_3\text{CN} + \text{Hg(NO}_3)_2 = \underset{\underset{\underset{\text{ONO}_2}{|}}{\overset{|}{\underset{}{\text{N}=\text{CCH}_3}}}}{\text{RCHCH}_2\text{HgNO}_3} \xrightarrow[\text{NaOH}]{\text{NaBH}_4} \underset{\underset{\underset{\text{O}}{\|}}{\overset{|}{\text{NHCCH}_3}}}{\text{RCHCH}_3}$$

The catalytic activity of mercuric salts in sulfuric acid solutions for hydration of acetylenes doubtless proceeds by routes similar to the above; the overall reaction is

$$\text{RC}\equiv\text{CR}' + \text{H}_2\text{O} \xrightarrow{\text{H}^+} \underset{\underset{\text{OH H}}{| \ \ |}}{\text{R}-\text{C}=\text{C}-\text{R}'} \longrightarrow \underset{\underset{\text{O}}{\|}}{\text{R}-\text{C}-\text{CH}_2\text{R}'}$$

Acetylene itself gives acetaldehyde.

Finally it is of interest that methanolic solutions of Hg^{II} acetate readily absorb carbon monoxide at atmospheric pressure, and the resulting compound can be converted by halide salts into compounds of the type $XHgC(O)OCH_3$. It has been shown that carbon monoxide is, in effect, inserted between the Hg and O of a solvolyzed mercuric ion, though the mechanism is not established in detail:

$$Hg(OCOCH_3)_2 + CH_3OH \rightleftharpoons CH_3COOHgOCH_3 + CH_3COOH$$
$$CH_3COOHg-OCH_3 + CO \rightleftharpoons CH_3COOHgC(O)OCH_3$$

The CO can be regenerated from the compounds by heating or by action of concentrated hydrochloric acid. Under pressures of 25 atm, reactions such as

$$RHgNO_3 + CO + CH_3OH \rightarrow RCOOCH_3 + Hg + HNO_3$$

can be carried out.

General References

Specialist Periodical Reports, Chemical Society, Vol. 4 of *The Inorganic Chemistry of the Transition Elements,* 1976. Contains the first specialist report on Zn, Cd, and Hg, including bioinorganic chemistry.

Specialist Periodical Reports, Chemical Society, *Organometallic Chemistry.* Contain Zn, Cd, and Hg references.

Chizhikov, D. M., *Cadmium,* Pergamon Press, 1966. Mainly the technology of production.

Fleischer, A., and J. J. Lander, *Zinc–Silver Oxide Batteries,* Wiley, 1971.

McAuliffe, C. A., Ed., *The Chemistry of Mercury,* Macmillan, 1977. Coordination chemistry and organic and biochemistry of mercury.

Miller, M. W., and T. W. Clarkson, Eds., *Mercury, Mercurials and Mercaptans,* Thomas Books, 1973. Includes toxicology.

Roberts, H. L., *Adv. Inorg. Chem. Radiochem.,* 1968, **11,** 309. An excellent account of mercury chemistry, containing useful thermodynamic and other data.

Taylor, M. J., *Metal to Metal Bonded States of Main Group Elements,* Academic Press, 1975.

3

CHEMISTRY OF THE TRANSITION ELEMENTS

CHAPTER TWENTY

The Transition Elements and the Electronic Structures of Their Compounds

20-1. Definition and General Characteristics of Transition Elements

The transition elements may be strictly defined as those that *as elements,* have partly filled d or f shells. Here we shall adopt a slightly broader definition and include also elements that have partly filled d or f shells in any of their commonly occurring oxidation states. This means that we treat the coinage metals copper, silver, and gold as transition metals, since Cu^{II} has a $3d^9$ configuration, Ag^{II} a $4d^9$ configuration, and Au^{III} a $5d^8$ configuration. From a purely chemical point of view it is also appropriate to consider these elements as transition elements because their chemical behavior is, on the whole, quite similar to that of other transition elements.

With our broad definition in mind, we find that there are now some 56 transition elements, counting the heaviest elements through the one of atomic number 104. Clearly the majority of all known elements are transition elements. All these transition elements have certain general properties in common:

1. They are all metals.
2. They are almost all hard, strong, high-melting, high-boiling metals that conduct heat and electricity well. In short, they are "typical" metals of the sort we meet in ordinary circumstances.
3. They form alloys with one another and with other metallic elements.
4. Many of them are sufficiently electropositive to dissolve in mineral acids, although a few are "noble"—that is, they have such low electrode potentials that they are unaffected by simple acids.
5. With very few exceptions they exhibit variable valence, and their ions and compounds are colored in one if not all oxidation states.

6.   Because of partially filled shells, they form at least some paramagnetic compounds.

This large number of transition elements is subdivided into three main groups: (*a*) the main transition elements or *d*-block elements, (*b*) the lanthanide elements, and (*c*) the actinide elements.

The main transition group or *d* block includes the elements that have partially filled *d* shells only. Thus the element scandium, with the outer electron configuration $4s^2 3d$, is the lightest member. The eight succeeding elements, Ti, V, Cr, Mn, Fe, Co, Ni, and Cu, all have partly filled $3d$ shells either in the ground state of the free atom (all except Cu) or in one or more of their chemically important ions (all except Sc). This group of elements is called the *first transition series*. At zinc the configuration is $3d^{10}4s^2$, and this element forms no compound in which the $3d$ shell is ionized, nor does this ionization occur in any of the next nine elements. It is not until we come to yttrium, with ground state outer electron configuration $5s^2 4d$, that we meet the next transition element. The following eight elements, Zr, Nb, Mo, Tc, Ru, Rh, Pd, and Ag, all have partially filled $4d$ shells either in the free element (all but Ag) or in one or more of the chemically important ions (all but Y). This group of nine elements constitutes the *second transition series*.

Again there follows a sequence of elements in which there are never *d*-shell vacancies under chemically significant conditions until we reach the element lanthanum, with an outer electron configuration in the ground state of $6s^2 5d$. Now, if the pattern we have observed twice before were to be repeated, there would follow eight elements with enlarged but not complete sets of $5d$ electrons. This does not happen, however. The $4f$ shell now becomes slightly more stable than the $5d$ shell, and through the next fourteen elements, electrons enter the $4f$ shell until at lutetium it becomes filled. Lutetium thus has the outer electron configuration $4f^{14}5d6s^2$. Since both La and Lu have partially filled *d* shells and no other partially filled shells, it might be argued that both these should be considered as *d*-block elements. However for chemical reasons, it would be unwise to classify them in this way, since all the fifteen elements La ($Z = 57$) through Lu ($Z = 71$) have very similar chemical and physical properties, those of lanthanum being in a sense prototypal; hence these elements are called the *lanthanides,* and their chemistry is considered separately in Chapter 23. Since the properties of Y are extremely similar to, and those of Sc mainly like, those of the lanthanide elements proper, and quite different from those of the regular *d*-block elements, we treat them also in Chapter 23.

For practical purposes, then, the *third transition series* begins with hafnium, having the ground state outer electron configuration $6s^2 5d^2$, and embraces the elements Ta, W, Re, Os, Ir, Pt, and Au, all of which have partially filled $5d$ shells in one or more chemically important oxidation states as well as (except Au) in the neutral atom.

Continuing on from mercury, which follows gold, we come via the noble gas radon and the radioelements Fr and Ra to actinium, with the outer electron configuration $7s^2 6d$. Here we might expect, by analogy to what happened at lanthanum, that in the following elements electrons would enter the $5f$ orbitals, producing a lanthanidelike series of fifteen elements. What actually occurs is, unfortunately, not so simple. Although immediately following lanthanum the 4f orbitals become decisively more favorable than the 5d orbitals for the electrons entering in the succeeding elements, there is apparently not so great a difference between the 5f and 6d orbitals until later. Thus for the elements immediately following Ac, and their ions, there may be electrons in the 5f or 6d orbitals or both. Since it appears that later on, after four or five more electrons have been added to the Ac configuration, the 5f orbitals do become definitely the more stable, and since the elements from about americium on do show moderately homologous chemical behavior, it has become accepted practice to call the fifteen elements beginning with Ac the *actinide elements*.

There is an important distinction, based on electronic structures, between the three classes of transition elements. For the d-block elements the partially filled shells are d shells, 3d, 4d, or 5d. These d orbitals project well out to the periphery of the atoms and ions so that the electrons occupying them are strongly influenced by the surroundings of the ion and, in turn, are able to influence the environments very significantly. Thus many of the properties of an ion with a partly filled d shell are quite sensitive to the number and arrangement of the d electrons present. In marked contrast to this, the 4f orbitals in the lanthanide elements are rather deeply buried in the atoms and ions. The electrons that occupy them are largely screened from the surroundings by the overlying shells (5s, 5p) of electrons; therefore reciprocal interactions of the 4f electrons and the surroundings of the atom or the ion are of relatively little chemical significance. This is why the chemistry of all the lanthanides is so homologous, whereas there are seemingly erratic and irregular variations in chemical properties as one passes through a series of d-block elements. The behavior of the actinide element lies between those of the two types described above because the 5f orbitals are not so well shielded as are the 4f orbitals, although not so exposed as are the d orbitals in the d-block elements.

20-2. Position in the Periodic Table

Figure 20-1 shows in a qualitative way the relative variations in the energies of the atomic orbitals as a function of atomic number in neutral atoms. It is well to realize that in a multielectron atom—one with, say, 20 or more electrons—the energies of all the levels are more or less dependent on the populations of all the other levels. Hence the diagram is rather complicated.

In hydrogen all the subshells of each principal shell are equienergic, but in more complex atoms the s, p, d, f, g, etc., subshells split apart and drop to lower energies. This descent in energy occurs because the degree to which an electron in a particular orbital is shielded from the nuclear charge by all the other electrons in the atom is insufficient to prevent a steady increase in the *effective nuclear charge* felt by that electron with increasing atomic number. In other words, each electron is imperfectly shielded from the nuclear charge by the other electrons. The energy of an electron in an atom is given by

$$E = -\frac{2\pi^2\mu e^4(Z^*)^2}{n^2h^2}$$

(20-1)

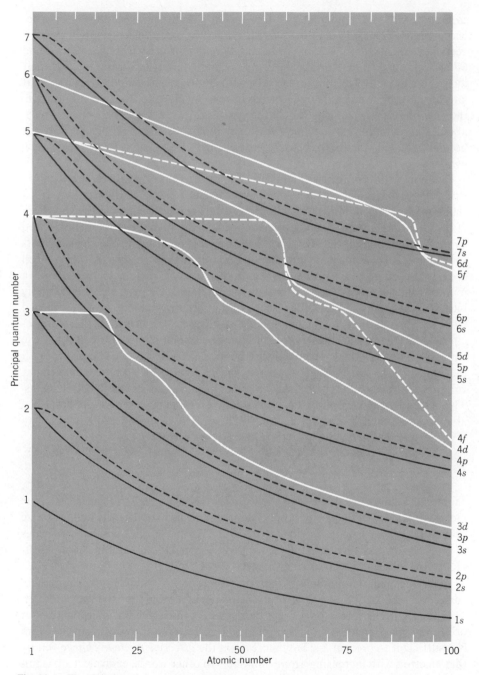

Fig. 20-1. The variation of the energies of atomic orbitals with increasing atomic number in neutral atoms (energies not strictly to scale).

where $Z*$ is the effective nuclear charge, and the energy of the electron falls as $Z*$ increases. The diagram is complicated, however, because all subshells do not drop in parallel fashion, since the several subshells of the same principal shell are shielded to different degrees by the core of electrons beneath.

From Fig. 20-1 we see that the $1s$, $2s$, $2p$, $3s$, and $3p$ levels occur in that sequence in all known atoms. Thus through the atoms (H to Ar) in which this sequence of orbitals is being filled, they are filled in that order. While the filling of this set of orbitals is taking place, the energies of the higher and as yet unfilled orbitals are being variously affected by the screening power of these first eighteen electrons. In particular, the $3d$ levels, which penetrate the argon core rather little, have scarcely dropped in energy when we reach argon ($Z = 18$), whereas the $4s$ and $4p$ levels, especially the former, which penetrate the argon core quite a bit, have dropped rather steeply. Thus when two more electrons are added to the argon configuration to give the potassium and calcium atoms, they enter the $4s$ orbital, which has fallen below the $3d$ orbitals. As these two electrons are added, the nuclear charge is also increased by two units. Since the $3d$ orbitals penetrate the electron density in the $4s$ orbitals very considerably, the net result is that the effective nuclear charge for the $3d$ orbitals increases rather abruptly, and they now drop well below the $4p$ orbitals to about the level of the $4s$ orbital. The next electron therefore enters the $3d$ shell, and scandium has the configuration $[Ar]4s^23d$. This $3d$ electron screens the $4p$ levels more effectively than it screens the remaining $3d$ orbitals, so the latter remain the lowest available orbitals and the next electron is also added to the $3d$ shell to give Ti, with the configuration $[Ar]4s^23d^2$. This process continues in a similar way until the entire $3d$ shell is filled. Thus at Zn we have the configuration $[Ar]4s^23d^{10}$, and the $4p$ orbitals, now the lowest available ones, become filled in the six succeeding elements.

The same sequence of events is repeated again in the elements following krypton, which has the electron configuration $[Ar]3d^{10}4s^24p^6$. Because of the way in which the shielding varies, the $4d$ levels, which in a one-electron atom would be next in order of stability, are higher in energy than the $5s$ and $5p$ orbitals, so that the next two electrons added go into the $5s$ orbitals, giving the alkali and alkaline earth elements Rb and Sr. But the shielding of the $4d$ orbitals by these $5s$ electrons is very poor, so that the $4d$ orbitals feel strongly the increase of two units of nuclear charge and take a sharp drop, becoming appreciably more stable than the $5p$ orbitals, and the next electron added becomes a $4d$ electron. Thus the next element, Y, is the first member of the second transition series. This series is completed at Ag, configuration $[Kr]4d^{10}5s^2$; then six $5p$ electrons are added to make Xe, the next noble gas.

At Xe ($Z = 54$) the next available orbitals are the $6s$ and $6p$ orbitals. The $4f$ orbitals are so slightly penetrating with respect to the Xe core that they have scarcely gained any stability, but the more penetrating $6s$ and $6p$ levels have gained a good deal. Hence the next two electrons added are $6s$ electrons, giving again an alkali and an alkaline earth element Cs and Ba, respectively. However the $6s$ shell scarcely shields the $4f$ orbitals, so the latter abruptly feel an increase in effective nuclear charge and suffer a steep drop in energy. At the same time, however, the energy of the $5d$ levels also drops abruptly, just as did that of $(n-1)d$ levels previously as electrons are added to the ns level, and the final situation is one in which, at Ba, the $6s$, $5d$, and $4f$ levels are all of about the same energy. The next entering electron, in the element lanthanum, enters a $5d$ orbital, but the following element, cerium, has the configuration $6s^24f^2$. Through the next twelve elements electrons continue to enter the $4f$ orbitals, and it is likely that even at cerium they are intrinsically more stable than the $5d$'s. Certainly they are so by the time we reach ytterbium, with the configuration $6s^24f^{14}$. Now, with the $6s$ and $4f$ shells full, the next lowest levels are unequivocally the $5d$'s, and from lutetium, with the configuration $6s^24f^{14}5d$, through mercury, with the configuration $[Xe]6s^24f^{14}5d^{10}$, the ten $5d$ electrons are added. Chemically, lanthanum and lutetium, each of which has a single $5d$ electron, are very similar to each other, and all the elements in between, with configurations $[Xe]4f^n6s^2$, have chemical properties intermediate between those of lanthanum and lutetium. Consequently these fifteen elements are all considered to be members of one class, the lanthanides. Hafnium, $[Xe]4f^{14}5d^26s^2$, through gold are the eight elements that we regard as the members of the third transition series.

Following mercury there are six elements in which electrons enter the $6p$ orbitals until the next noble gas, radon, is reached. Its configuration is $[Xe]4f^{14}5d^{10}6s^26p^6$. Because of their relatively nonpenetrating character, the $5f$ orbitals have dropped so much more slowly than have the $7s$ and $7p$ orbitals that the next two electrons beyond the radon core are added to the $7s$ level, and again an alkali and an alkaline earth element are formed, namely, Fr, $[Rn]7s$, and Ra, $[Rn]7s^2$. But, again in analogy to the situation one row up in the Periodic Table, both the $5f$ and the $6d$ orbitals penetrate the $7s$ orbitals very considerably; they are thus abruptly stabilized relative to the $7p$ orbitals, and the next electrons added enter them. It appears that as we proceed through actinium and the following elements, the energies of the $6d$ and $5f$ orbitals remain for a while so similar that the exact configuration is determined by interelectronic forces of the sort discussed in Section 20-3. In the case of protactinium it is not certain whether the ground state is $[Rn]7s^26d^3$, $[Rn]$-$7s^26d^25f$, $[Rn]7s^26d5f^2$, or $[Rn]7s^25f^3$. These four configurations doubtless differ very little in energy, and for chemical purposes the question of which is actually the lowest is not of great importance. The next element, uranium, appears definitely to have the configuration $[Rn]7s^25f^36d$, and the elements thereafter are all believed to have the configurations $[Rn]7s^25f^n6d$. The important point is that around actinium, the $6d$ and $5f$ levels are of almost the same energy, with the $5f$'s probably becoming slowly more stable later on.

20-3.   Electron Configurations of the Atoms and Ions

The discussion in the preceding section is incomplete because it takes account only of the shielding of a given electron from the nuclear charge by other electrons in the atom. One electron may help to determine the orbital occupied by another electron not only in this indirect way but also because of direct interactions between the electrons. These direct interactions cause the differences in energy between different states derived from the same configuration, as explained in Appendix 5.

In cases where the energies of two orbitals differ by an amount comparable to or less than the energies arising from electron-electron interactions, it is not possible to predict electron configurations solely by consideration of the order of orbital energies.

The dominance of interelectronic interactions over orbital energy differences is well illustrated by the "special stability" of half-filled shells. Examples are found in the first transition series and in the lanthanides, notably the boxed columns in the annexed series.

	Sc	Ti	V	Cr	Mn	Fe	Co	Ni	Cu	Zn
$4s$	2	2	2	1	2	2	2	2	1	2
$3d$	1	2	3	5	5	6	7	8	10	10

	Sm	Eu	Gd	Tb
$6s$	2	2	2	2
$5d$	0	0	1	0
$4f$	6	7	7	9

Half-filled shells have an amount of exchange energy considerably greater than would be interpolated from the energies of the configurations to either side of them. Hence there is a driving force either to take an electron "out of turn," as with Cr and Cu, or to shunt an excess electron to another shell of similar energy, to achieve or maintain the half-filled arrangement. All spins are parallel, giving maximum spin multiplicity in these half-filled shells.

In the second transition series the irregularities become more complex, as shown in the annexed series Y to Cd. No simple analysis is possible here;

	Y	Zr	Nb	Mo	Tc	Ru	Rh	Pd	Ag	Cd
$5s$	2	2	1	1	1	1	1	0	1	?
$4d$	1	2	4	5	6	7	8	10	10	10

both nuclear-electron and electron-electron forces play their roles in determining these configurations. Although a preference for the filled $4d$ shell is evident at the end of the series and the elements Nb and Mo seem to prefer the half-filled shell, the configuration of Tc shows that this preference is not controlling throughout this series.

It is well to point out that the interelectronic forces and variations in total nuclear charge play a large part in determining the configurations of ions. We cannot say that because $4s$ orbitals become occupied before $3d$ orbitals they are always more stable. If this were so, we should expect the elements of the first transition series to ionize by loss of $3d$ electrons, whereas, in fact, they ionize by loss of $4s$ electrons first. Thus it is the net effect of all the forces—nuclear-electronic attraction, shielding of one electron by others, interelectronic repulsions and the exchange forces—that determines the stability of an electron configuration; and unfortunately there are many cases in which the interplay of these forces and their sensitivity to changes in nuclear charge and the number of electrons present cannot be simply described.

20-4. Magnetism in Transition Metal Chemistry

Since transition metal atoms and ions characteristically have incomplete d or f shells, their magnetic properties are a source of valuable information. Even when a transition metal compound is diamagnetic, this is informative to a degree that it is not for a compound of a main group element, because the pairing up of all d or f electrons in an incomplete set implies certain limits on the relative energies and degeneracies of the occupied orbitals. This section covers a few key points concerning the relationship of observed magnetic moments for paramagnetic compounds to their electronic structures. Basic definitions (e.g., of diamagnetism, paramagnetism), and a discussion of how magnetic moments are related to magnetic susceptibilities can be found in Appendix 6. In this book we give magnetic moments in Bohr magnetons (BM).

Electrons determine the magnetic properties of matter in two ways. First, eacl electron is, in effect, a magnet in itself. From a pre-wave-mechanical viewpoint, the electron may be regarded as a small sphere of negative charge spinning on its axis. Then, from completely classical considerations, the spinning of charge produces a magnetic moment. Second, an electron traveling in a closed path around a nucleus, again according to the pre-wave-mechanical picture of an atom, will also produce a magnetic moment, just as does an electric current traveling in a loop of wire. The magnetic properties of any individual atom or ion will result from some combination of these two properties, that is, the inherent *spin moment* of the electron and the *orbital moment* resulting from the motion of the electron around the nucleus. These physical images should not, of course, be taken too literally, for they have no place in wave mechanics, nor do they provide a basis for quantitatively correct predictions. They are qualitatively useful conceptual aids, however.

As mentioned above, the magnetic moments of atoms, ions, and molecules are expressed here in *Bohr magnetons;* this unit is defined in terms of fundamental constants as

$$1 \text{ B.M.} = \frac{e\,h}{4\pi mc} \tag{20-2}$$

where e is the electronic charge, h is Planck's constant, m is the electron mass, and c is the speed of light. This is *not,* however, the moment of a single electron. Because of certain features of quantum theory, the relationship is a little more complicated.

The magnetic moment μ_s of a single electron is given, according to wave mechanics, by the equation

$$\mu_s \text{ (in BM)} = g\sqrt{s(s+1)} \tag{20-3}$$

in which s is simply the absolute value of the spin quantum number and g is the gyromagnetic ratio, more familiarly known as the "g factor." The quantity $\sqrt{s(s+1)}$ is the value of the angular momentum of the electron; thus g is the ratio of the magnetic moment to the angular momentum, as its name is intended to suggest. For the free electron, g has the value 2.00023, which may be taken as 2.00 for most purposes. From eq. 20-3 we can calculate the spin magnetic moment of the electron as

$$\mu_s = 2\sqrt{\tfrac{1}{2}(\tfrac{1}{2}+1)} = \sqrt{3} = 1.73 \text{ BM}$$

TABLE 20-1
"Spin-Only" Magnetic Moments for Various Numbers of Unpaired Electrons

Number of unpaired electrons	S	μ_s(BM)
1	$\frac{1}{2}$	1.73
2	1	2.83
3	$\frac{3}{2}$	3.87
4	2	4.90
5	$\frac{5}{2}$	5.92
6	3	6.93
7	$\frac{7}{2}$	7.94

Thus any atom, ion, or molecule having one unpaired electron (e.g., H, Cu^{2+}, ClO_2) should have a magnetic moment of 1.73 BM from the electron spin alone. This may be augmented or diminished by an orbital contribution (see below).

There are transition metal ions having one, two, three, . . . up to seven unpaired electrons. As indicated in Appendix 5, the spin quantum number for the ion as a whole S is the sum of the spin quantum number $s = \frac{1}{2}$ for the individual electrons. For example, in the manganese(II) ion with five unpaired electrons, $S = 5(\frac{1}{2}) = \frac{5}{2}$; and in the gadolinium(III) ion with seven unpaired electrons, $S = 7(\frac{1}{2}) = \frac{7}{2}$. Thus we can use eq. 20-3, substituting S for s, to calculate the magnetic moment due to the electron spins alone, the so-called spin-only moment, for any atom or ion, provided we know the total spin quantum number S. The results are summarized in Table 20-1 for all possible real cases.

In the two examples chosen above, namely, Mn^{II} and Gd^{III}, the observed values of their magnetic moments agree very well with the spin-only values in Table 20-1. Generally, however, experimental values differ from the spin-only ones, usually being somewhat greater. This is because the orbital motion of the electrons also makes a contribution to the moment. The theory by which the exact magnitudes of these orbital contributions may be calculated is by no means simple, and we give here only a superficial and pragmatic account of the subject. More detailed discussion of a few specific cases can be found in several places later in the text.

For Mn^{II}, Fe^{III}, Gd^{III}, and other ions whose ground states are S states, there is no orbital angular momentum even in the free ion. Hence there cannot be any orbital contribution to the magnetic moment, and the spin-only formula applies exactly.* In general, however, the transition metal ions in their ground states, D or F being most common, do possess orbital angular momentum. Wave mechanics shows that for such ions, if the orbital motion makes its full contribution to the magnetic moments, they will be given by

$$\mu_{S+L} = \sqrt{4S(S + 1) + L(L + 1)} \qquad (20\text{-}4)$$

* Because of certain high-order effects and also, in part, because of covalence in metal-ligand bonds, slight departures (i.e., a few tenths of a BM) from the spin-only moments are sometimes observed.

TABLE 20-2

Theoretical and Experimental Magnetic Moments (BM) for Various Transition Metal Ions

Ion	Ground state quantum numbers S	L	Spectroscopic symbol	μ_S	μ_{S+L}	Observed moments
V^{4+}	$\frac{1}{2}$	2	^2D	1.73	3.00	1.7–1.8
Cu^{2+}	$\frac{1}{2}$	2	^2D	1.73	3.00	1.7–2.2
V^{3+}	1	3	^3F	2.83	4.47	2.6–2.8
Ni^{2+}	1	3	^3F	2.83	4.47	2.8–4.0
Cr^{3+}	$\frac{3}{2}$	3	^4F	3.87	5.20	~3.8
Co^{2+}	$\frac{3}{2}$	3	^4F	3.87	5.20	4.1–5.2
Fe^{2+}	2	2	^5D	4.90	5.48	5.1–5.5
Co^{3+}	2	2	^5D	4.90	5.48	~5.4
Mn^{2+}	$\frac{5}{2}$	0	^6S	5.92	5.92	~5.9
Fe^{3+}	$\frac{3}{2}$	0	^6S	5.92	5.92	~5.9

in which L represents the orbital angular momentum quantum number for the ion.

Table 20-2 lists magnetic moments actually observed for the common ions of the first transition series together with the calculated values of μ_S and μ_{S+L}; observed values of μ frequently exceed μ_S, but seldom are as high as μ_{S+L}. This is because the electric fields of other atoms, ions, and molecules surrounding the metal ion in its compounds restrict the orbital motion of the electrons so that the orbital angular momentum, hence the orbital moments, are wholly or partially "*quenched*." In some cases (e.g., d^3 and d^8 ions in octahedral environments and d^7 ions in tetrahedral ones) the quenching of orbital angular momentum in the ground state is expected to be complete according to the simplest arguments, and yet such systems deviate from spin-only behavior. However when the effect of spin-orbit coupling is considered, it is found that orbital angular momentum is mixed into the ground state from the first excited state of the system. This phenomenon is discussed quantitatively for the d^7 ion Co^{II} in Section 21-F-3. In the case of a d^3 ion in an octahedral environment, the orbital contribution is introduced in *opposition* to the spin contribution and moments slightly below the spin-only value are therefore observed, as for Cr^{III}.

20-5. Chemical Bonding for *d*-Block Elements

From the point of view of the bonding in their compounds, the *d*-block transition elements and the main group elements have a basic difference: in the former *d* orbitals play the most important role, whereas the role of *s* and *p* orbitals, especially the latter, is secondary; for the main group elements, *s* and *p* orbitals, especially *p* orbitals, are of key importance and *d* orbitals play a secondary—often entirely negligible—role. Because of this, an entire repertoire of approximations and formalisms has been developed—and widely used—for the transition elements that have no relevance to main group chemistry. Before we expound the chemistry of the transition elements, these concepts are presented.

We shall begin with the most thorough and rigorous approach to the electronic structure of any compound, namely, a molecular orbital treatment. Once this picture is understood, the validity and short-comings of the various approximations and formalisms (crystal field theory, ligand field theory, the angular overlap model, the simple valence bond theory, etc.) can be more readily recognized and evaluated. This is a counterhistorical approach, since the models and approximations were developed first, before the availability of the high-speed digital computers required to carry out the rigorous MO calculations. However the logical and pedagogical advantages of a presentation in the reverse order are indubitable.

It should be emphasized that in this chapter we are concerned only with compounds lacking metal-metal (M—M) bonds, that is, with Werner complexes in the broadest sense of that term. The bonding in non-Wernerian compounds, namely, those containing M—M bonds, both discrete two-center bonds of various orders as well as the delocalized multicenter bonding in many metal atom cluster compounds, is covered in Chapter 26.

THE MOLECULAR ORBITAL THEORY

20-6. Qualitative Introduction to the MO Theory

The first task in working out the MO treatment for a particular type of complex is to find out which orbital overlaps are or are not possible because of the inherent symmetry requirements of the problem. This can be done quite elegantly and systematically by using some principles of group theory, but such an approach is outside the scope of this discussion. Instead we shall simply present the results that are obtained for octahedral complexes, illustrating them pictorially. It may be noted that ultimately, for the experimental inorganic chemist, this pictorial representation is much more important than mathematical details, for it provides a basis for visualizing the bonding and for thinking concretely about it.

The molecular orbitals we shall use here will be of the linear combination of atomic orbitals (LCAO) type. Our method for constructing them, which we shall apply specifically to octahedral complexes, takes the following steps:

1.   We note that there are nine valence shell orbitals of the metal ion to be considered. Six of these—d_{z^2}, $d_{x^2-y^2}$, s, p_x, p_y, and p_z—have lobes lying along the metal-ligand bond directions (i.e., are suitable for σ bonding), whereas three, namely, d_{xy}, d_{yz}, d_{zx}, are so oriented as to be suitable only for π bonding.

2.   We assume initially that each of the six ligands possesses one σ orbital. These individual σ orbitals must then be combined into six "symmetry" orbitals, each constructed to overlap effectively with a particular one of the six metal ion orbitals that are suitable for σ bonding. Each of the metal orbitals must then be combined with its matching symmetry orbital of the ligand system to give a bonding and an antibonding molecular orbital.

3.   If the ligands also possess π orbitals, these too must be combined into

"symmetry" orbitals constructed to overlap effectively with the metal ion π orbitals, and the bonding and antibonding MOs then formed by overlap.

Complexes with No π Bonding. The six σ symmetry orbitals are indicated in Fig. 20-2, in which they are illustrated pictorially, expressed algebraically as normalized linear combinations of the individual ligand σ orbitals, and juxtaposed with the metal ion orbitals with which they are matched by symmetry. On the left side of Fig. 20-2 are the symmetry symbols A_{1g}, E_g, and T_{1u}, for these orbitals. These symbols are of group theoretical origin, and they stand for the symmetry orbital class to which belong the metal orbital, the matching symmetry orbital of the ligand system, and the molecular orbitals that will result from the overlap of these two. They are very commonly used simply as convenient labels, but they also carry information. The symbol A_{1g} always represents a single orbital that has the full symmetry of the molecular system. The symbol E_g represents a pair of orbitals that are equivalent except for their orientations in space, whereas T_{1u} represents a set of three orbitals that are equivalent except for their orientations in space. The subscripts g and u are used to indicate whether the orbital(s) is centrosymmetric (g from the German *gerade* meaning even) or anticentrosymmetric (u from the German *ungerade* meaning uneven).

The final step now needed to obtain the molecular orbitals themselves is to allow each metal orbital to overlap with its matching symmetry orbital of the ligand system. As usual two combinations are to be considered: one in which the matched orbitals unite with maximum positive overlap, thus giving a bonding MO, and the other in which they unite with maximum negative overlap to give the corresponding antibonding MO. This process is illustrated for the pair p_z and Σ_z in Fig. 20-3. From the energy point of view, these results may be expressed in the usual type of MO energy-level diagram (Fig. 20-3, right). Note there that the p_z and Σ_z orbitals are not assumed to have the same energies, for in general they do not. To a first approximation, the energies of the bonding and antibonding MOs lie equal distances below and above, respectively, the mean of the energies of the combining orbitals.

In just the same way, the other metal ion orbitals combine with the matching symmetry orbitals of the ligand system to form bonding and antibonding MOs. The MOs of the same symmetry class—which are equivalent except for their spacial orientations—have the same energies, but orbitals of different symmetry classes do not in general have the same energies, since they are not equivalent. The energy level diagram that results when all the σ interactions are considered is shown in Fig. 20-4. Here we name the orbitals only by their symmetry designations, using the asterisk to signify that a molecular orbital is antibonding. It should be noted in Fig. 20-4 that the three metal ion d orbitals that are suitable for forming π bonds but not σ bonds are given their appropriate symmetry label T_{2g} and shown as remaining unchanged in energy, since we are considering ligands having no π orbitals with which they might interact.

Certain implications of this energy level diagram deserve special attention. In general, in MO diagrams of this type, it may be assumed that if a molecular orbital is much nearer in energy to one of the atomic orbitals used to construct it than to

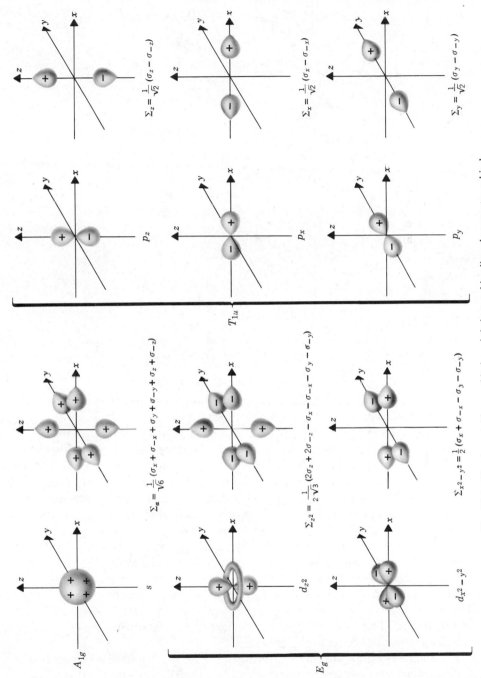

Fig. 20-2. The six metal ion σ orbitals and their matching ligand symmetry orbitals.

A_{1g}

s

$\Sigma_a = \frac{1}{\sqrt{6}}(\sigma_x + \sigma_{-x} + \sigma_y + \sigma_{-y} + \sigma_z + \sigma_{-z})$

T_{1u}

p_z

$\Sigma_z = \frac{1}{\sqrt{2}}(\sigma_z - \sigma_{-z})$

p_x

$\Sigma_x = \frac{1}{\sqrt{2}}(\sigma_x - \sigma_{-x})$

p_y

$\Sigma_y = \frac{1}{\sqrt{2}}(\sigma_y - \sigma_{-y})$

E_g

d_{z^2}

$\Sigma_{z^2} = \frac{1}{2\sqrt{3}}(2\sigma_z + 2\sigma_{-z} - \sigma_x - \sigma_{-x} - \sigma_y - \sigma_{-y})$

$d_{x^2-y^2}$

$\Sigma_{x^2-y^2} = \frac{1}{2}(\sigma_x + \sigma_{-x} - \sigma_y - \sigma_{-y})$

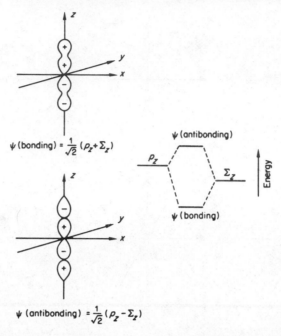

Fig. 20-3. *Left:* orbital pictures showing the bonding and antibonding molecular orbitals that are the z components of the T_{1u} sets; *right:* energy level diagram showing how the energies of the various orbitals are related.

the other one, it has much more the character of the first one than of the second. On this basis then, Fig. 20-4 implies that the six bonding σ MOs, three T_{1u}'s, the A_{1g}, and the two E_g's, have more the character of ligand orbitals than they do of metal orbitals. It can then be said that electrons occupying these orbitals will be mainly "ligand electrons" rather than "metal electrons," though they will partake of metal ion character to some significant extent. Conversely, electrons occupying any of the antibonding MOs are to be considered as predominantly metal electrons. Any electrons in the T_{2g} orbitals will be *purely* metal electrons when there are no ligand π orbitals, as in the case being considered.

The two MOs in the center of the diagram, T_{2g} and E_g^*, should be especially noted. It is into one (T_{2g}) or both of these that the metal d electrons will go when the complex is formed. Since the T_{2g} orbitals have exclusively metal d character when there is no π bonding and the E_g^* orbitals have largely metal d character, the electrons that occupy these orbitals in the complex can still be thought of as "metal d electrons." The significance of this will be apparent later when the crystal field and ligand field theories are discussed.

Complexes with π Bonding. If the ligands have π orbitals, filled or unfilled, it is necessary to consider their interactions with the T_{2g} d orbitals, that is, the d_{xy}, d_{yz}, and d_{zx} orbitals. In the simplest case each ligand has a pair of π orbitals mutually perpendicular, making $6 \times 2 = 12$ altogether. From group theory it is found that these may be combined into four triply degenerate sets belonging to the sym-

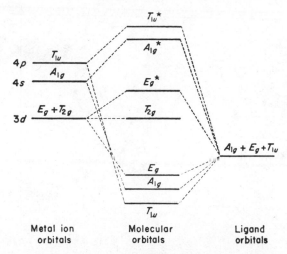

Fig. 20-4. The qualitative MO energy level diagram for an octahedral complex between a metal ion of the first transition series and six ligands that do not possess π orbitals.

metry classes T_{1g}, T_{2g}, T_{1u}, and T_{2u}. Those in the classes T_{1g} and T_{2u} will remain rigorously nonbonding. (We use the terms "bonding," "nonbonding," and "antibonding" with reference to the metal-ligand interactions regardless of the character of the orbitals in respect to bonding between atoms within polyatomic ligands.) This is because the metal ion does not possess any orbitals of these symmetries with which they might interact. The T_{1u} set can interact with the metal ion p orbitals, which are themselves a set with T_{1u} symmetry, and in a quantitative discussion it would be necessary to make allowance for this. However in a qualitative treatment we may assume that since the p orbitals are already required for the σ bonding, we need not consider π bonding by means of T_{1u} orbitals, which are thus nonbonding. This then leaves us with only the T_{2g} set of symmetry orbitals to overlap with the metal ion T_{2g} d orbitals.

The ligand π orbitals may be simple $p\pi$ orbitals as in the Cl^- ion, simple $d\pi$ orbitals as in phosphines or arsines, or molecular orbitals of a polyatomic ligand as in CO, CN^-, or pyridine. When they are simple $p\pi$ or $d\pi$ orbitals, it is quite easy to visualize how they combine to form the proper symmetry orbitals for overlapping with the metal ion orbitals. This is illustrated for $p\pi$ orbitals in Fig. 20-5.

The effects of π bonding via molecular orbitals of the T_{2g} type on the energy levels must now be considered. These effects vary depending on the energy of the ligand π orbitals relative to the energy of the metal T_{2g} orbitals and on whether the ligand π orbitals are filled or empty. Let us consider first the case involving empty π orbitals of higher energy than the metal T_{2g} orbitals. This situation is found in complexes where the ligands are phosphines or arsines, for example. As shown in Fig. 20-6a, the net result of the π interaction is to stabilize the metal T_{2g} orbitals (which, of course, also acquire some ligand orbital character in the process) relative to the metal E_g^* orbitals. In effect, the π interaction causes the Δ value for the complex to be greater than it would be if there were only σ interactions.

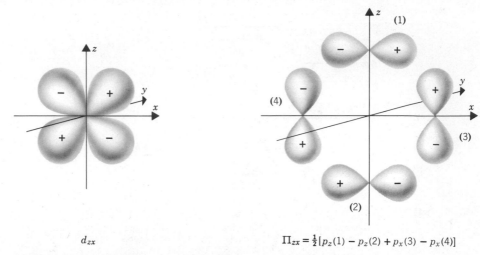

d_{zx} $\Pi_{zx} = \frac{1}{2}[p_z(1) - p_z(2) + p_x(3) - p_x(4)]$

Fig. 20-5. *Right:* the symmetry orbital, made up of ligand *p* orbitals, that has the proper symmetry to give optimum interaction with the metal ion d_{xz} orbital (*left*). There are analogous symmetry orbitals, π_{xy} and π_{yz}, that are similarly related to the metal ion d_{xy} and d_{yz} orbitals.

In a second important case the ligands possess only filled π orbitals of lower energy than the metal T_{2g} orbitals. As shown in Fig. 20-6b, the interaction here destabilizes the T_{2g} orbitals relative to the E_g^* orbitals, thus diminishes the value of Δ. This is probably the situation in complexes of metal ions in their normal oxidation states, especially the lower ones, with the ligand atoms oxygen and fluorine.

There are also important cases in which the ligands have both empty and filled π orbitals. In some, such as the Cl^-, Br^-, and I^- ions, these two types are not directly interrelated, the former being outer *d* orbitals and the latter valence shell *p* orbitals. In others, such as CO, CN^-, and pyridine, the empty and filled π orbitals are the antibonding and bonding $p\pi$ orbitals. In such cases the net effect is the result of competition between the interaction of the two types of ligand π orbitals with the metal T_{2g} orbitals, and simple predictions are not easily made. Figure 20-7 is

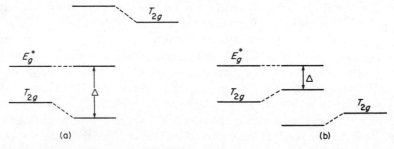

Fig. 20-6. Energy level diagrams showing how π interactions can affect the value of Δ. (a) Ligands have π orbitals of higher energy than the metal T_{2g} orbitals; (b) ligands have π orbitals of lower energy than the metal T_{2y} orbitals.

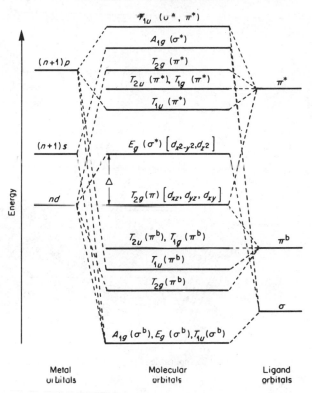

Fig. 20-7. A qualitative MO diagram for an $M(CO)_6$ or $[M(CN)_6]^{n-}$ compound. [Adapted from H. B. Gray and N. A. Beach, *J. Am. Chem. Soc.,* 1963, **85**, 2922.]

a diagram for octahedral metal-cyano complexes and carbonyls that is based on some rough calculations of overlaps and comparisons with absorption spectra. Though it has only qualitative value, and even in this sense may not be entirely correct, most workers would probably consider it to be essentially correct for the Group VI hexacarbonyls and cyano complexes of Fe^{II}, Ru^{II}, and Os^{II}. However the order of some of the MOs may well be changed by large changes in the metal orbital energies, as, for example, on going from a normal oxidation state of the metal to an unusually low or high one.

Of course all other types of complex besides octahedral ones can be treated, both qualitatively and quantitatively, by MO theory. To illustrate this qualitatively, we mention explicitly tetrahedral and square ones. Tetrahedral species can be divided roughly into two broad classes:

1. Oxo species in which the formal oxidation number of the metal is high (≥ 6) and there must be very extensive π bonding. Examples, including some that have been subject to prolonged controversy, are MnO_4^-, MnO_4^{2-}, CrO_4^{2-}, and MoO_4^{2-}.

2. Complexes in which the metals are in lower oxidation states, such as $+2$ or $+3$, and the ligands are halide ions, amine-N atoms or RO^- ions. A diagram that would be qualitatively applicable to most complexes of the second type is given in Fig. 20-8.

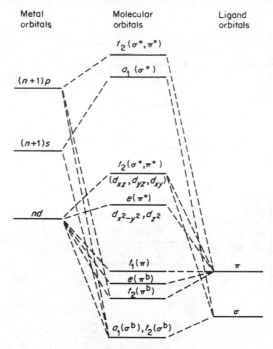

Fig. 20-8. An MO diagram that would be applicable to most tetrahedral complexes of transition metal ions in lower oxidation states.

For most square complexes, the general form of the MO diagram is as shown in Fig. 20-9.

Numerical MO Calculations. Even with the current availability of fast digital computers, very few efforts have been made to carry out complete, nonempirical calculations on transition metal complexes,[1] and the results of these are not necessarily more useful to the chemist just because they are more elaborate. Nevertheless, it is likely that work of this nature will in future become more common, and in certain cases it will be of great importance because of giving results free of doubts regarding the validity of empirical assumptions. Such *ab initio* calculations are particularly important in cases such as organometallic compounds and metal carbonyl type compounds (Cf. Section 20-7) and metal-metal bonded compounds (cf. Chapter 26) where the common simple methods to be discussed presently, such as crystal field theory or the angular overlap model, are of doubtful value.

To illustrate the results of a "complete" calculation, Fig. 20-10 shows the results in diagrammatic form for the square $[CuCl_4]^{2-}$ ion. It will be seen that in key qualitative respects, this diagram is similar to that in Fig. 20-9. Its advantages are that the numerical results from which it was drawn give a detailed breakdown on the percentage of ligand and metal contributions to each molecular orbital and the orbital energies should have at least approximate quantitative significance.

[1]   A. Veillard and J. Demuynck, in *Applications of Electronic Structure Theory,* H. F. Schaeffer, III, Ed., Plenum Press, 1977, pp. 187–219.

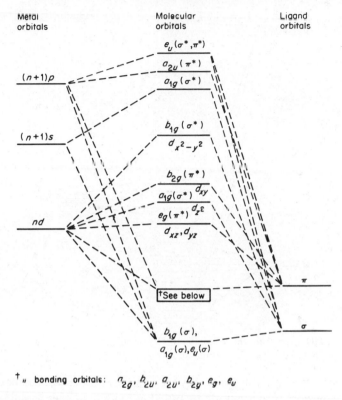

Fig. 20-9. A qualitative MO diagram for a square ML_4 complex. This arrangement should be essentially correct for $PtCl_4^{2-}$.

There is, of course, an entire range of MO methods lying between the two extremes of *ab initio* calculations on the one hand and the essentially qualitative approach based mainly on symmetry properties on the other. In increasing order of sophistication, a few of these methods are:

1. *The Extended Hückel (Wolfsberg-Helmholz) Method.* MO energies are assumed to be proportional to overlap integrals, and overlaps between all but nearest neighbor atoms usually are neglected. A great variety of problems have been treated by this method, whose virtue lies mainly in providing backup for essentially qualitative arguments,[2] not in providing accurate MO energies.

2. *The Fenske-Hall Method.* This is an approximate Hartree-Fock method involving simplified procedures for estimating the more difficult integrals.[3] It is more quantitative than the extended Hückel method and does not employ empirical values for parameters.

3. *The Xα Methods.*[4] These vary in a number of particulars and are of special

---

[2] See numerous papers, mainly in *J. Am. Chem. Soc.* and *Inorg. Chem.* by R. Hoffmann and many collaborators for skillful examples of this type of application.

[3] M. B. Hall and R. F. Fenske, *Inorg. Chem.,* 1972, **11**, 768.

[4] K. H. Johnson, *Ann. Rev. Phys. Chem.,* 1975, **26**, 39.

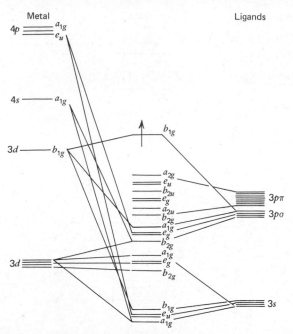

Fig. 20-10. An MO diagram for the square $[CuCl_4]^{2-}$ ion obtained by an *ab initio* (Hartree-Fock) calculation. [Reproduced by permission from Ref. 1.]

interest to inorganic chemists, since they are semiquantitative and can be as readily applied to molecules with heavy atoms as to those with light ones. Results obtained by this method are cited frequently in Chapter 26 in connection with M—M bonds.

20-7. MO Theory for Some Special Cases

Metal Carbonyls. We have already discussed the key, qualitative features of an individual M—CO bond (Section 3-7), these being the presence of considerable dative π bonding from M to CO and consequent weakening of the C—O π bonding. A number of calculations have been reported[5] on entire metal carbonyl molecules, such as $Cr(CO)_6$,[5] $Fe(CO)_5$,[1] and $Ni(CO)_4$.[6] Covalent bonding, both σ and π, is so strong in such molecules that only MO theory can give a quantitative account of the overall molecular electronic structure.

It is interesting that there has been some controversy among theoreticians using different methods as to the magnitude of the M—C π bonding. However it now appears[7] that all the calculations support the view that this is a key feature. The best calculations on the $M(CO)_n$ molecules all show that there is enough π back-donation from metal to CO to slightly more than offset the σ donation, with the

[5] I. H. Hillier and V. R. Saunders, *Mol. Phys.*, 1971, **22**, 1025.
[6] J. Demuynk and A. Veillard, *Theor. Chim. Acta*, 1973, **28**, 241.
[7] S. Larsson and M. Braga, *Int. J. Quantum Chem.*, 1979, **15**, 1.

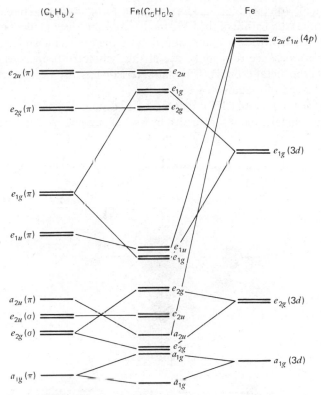

$(C_5H_5)_2$ $Fe(C_5H_5)_2$ Fe

Fig. 20-11. An MO diagram for ferrocene obtained from an *ab initio* (Hartree-Fock) calculation. [Reproduced by permission from Ref. 1.]

result that the metal atoms have a slightly positive charge (*ca.* $+0.5e$) and the each oxygen atom has a slight negative charge (*ca.* $-0.1e$).

Metal-Arene Complexes. Several of these have been treated, but the most attention has been given to the prototypal ferrocene $(\eta^5\text{-}C_5H_5)_2Fe$. The results of an *ab initio* calculation[8] are shown in Fig. 20-11. It may be seen that these results incorporate the key point emphasized in Section 3-15, namely, that the e_{1g}-type $3d$ orbitals (d_{xz}, d_{yz}) overlap with appropriate ring orbitals (cf. Fig. 3-15) to provide the main source of metal-ring bonding. However a comparison of the qualitative diagram (Fig. 3-13) and the one in Fig. 20-10 reveals many differences in detail.

20-8. The Crystal Field and Ligand Field Theories

The crystal field theory (CFT), first expounded by H. Bethe in 1929, treats a metal ion in a complex (or in a crystal environment) as though it were subject to a purely electrostatic perturbation by its ligands or nearest neighbors, and these are treated simply as point charges or point dipoles. The ligand field theory (LFT) introduces

8   M.-M. Coutière, J. Demuynk, and A. Veillard, *Theor. Chim. Acta,* 1972, **27,** 281.

some empirically dictated, ad hoc modifications that compensate partly for the physically unrealistic nature of the CFT model. As Van Vleck pointed out long ago, the CFT and LFT provide useful results despite their extreme simplicity and unphysical nature because they make use of the symmetry properties of both the structure and the metal orbitals. This characteristic has also been described as the formalism of "operator equivalence." [9]

The key idea of CFT and LFT is that the 5-fold degeneracy of the metal d orbitals is lifted when a metal atom is surrounded by a set of negative charges or the negative

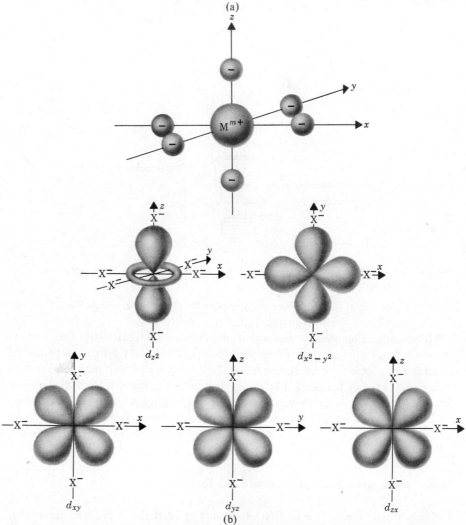

Fig. 20-12. (a) Six negative charges arranged octahedrally around a central M^{m+} ion with a set of Cartesian axes for reference. (b) The distribution of electron density in the five d orbitals with respect to a set of six octahedrally arranged negative charges.

[9]　J. S. Griffith, *J. Chem. Phys.*, 1964, **41**, 516 (see this reference for citations of earlier landmark papers).

ends of dipoles. The pattern of energies produced depends on the symmetry of the arrangement of ligands. Let us consider first an octahedral complex MX_6, in which the ligands X are anions that are treated as charges e^- arranged as in Fig. 20-12a. Figure 20-12b shows the spatial distributions of electronic charge in the d orbitals of the metal atom M.

It is immediately evident that the d_{xy}, d_{yz}, and d_{zx} orbitals are equivalent and place electron density between the M—X axes, whereas for the d_{z^2} and $d_{x^2-y^2}$ orbitals electron density is directed toward the negative ligands. It can be shown, though it is not obvious from Fig. 20-12, that the latter two orbitals are equivalent. The effect of the six negatively charged ligands is to split the d orbitals into two sets, one triply degenerate and the other doubly degenerate. The former are more stable, and the energy difference (the "splitting") may be denoted Δ_0.

For a tetrahedral set of negative charges a similar pictorial argument (as well, of course, as an exact numerical calculation) shows that an electron in the d_{xy}, d_{yz}, and d_{zx} orbitals is less repelled by the charges than one in the d_{z^2} or $d_{x^2-y^2}$ orbital. A splitting of magnitude Δ_t is produced. The effects of both octahedral and tetrahedral "crystal fields" are shown in Fig. 20-13. Each subset of orbitals is denoted

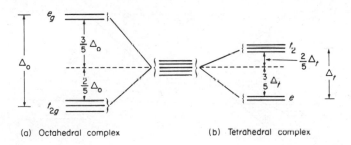

(a) Octahedral complex   (b) Tetrahedral complex

Fig. 20-13. Energy level diagrams showing the splitting of a set of d orbitals by octahedral and tetrahedral electrostatic crystal fields.

by a symbol that indicates its symmetry. We cannot explain the code in full, but the following definitions are worth remembering: t is used for triply degenerate orbitals, e for doubly degenerate ones, and a or b for singly degenerate ones.

The arrangement in Fig. 20-13 follows a convention whereby the mean energy of the split orbitals is equated to that of the original unsplit set. Physically this corresponds to supposing that the metal atom is first placed at the center of a sphere bearing a uniform charge of $6e^-$ (or $4e^-$). This of course would drastically change the energy of a set of ten electrons occupying the d orbitals but would not split the orbitals. If this charge were then caused to collect, in equal portions, at the vertices of the octahedron or tetrahedron, the total energy of the set of electrons would be unchanged, but some would increase in energy and others would decrease because of the orbital splitting. To keep the total energy of the ten electrons (or five orbitals) unchanged, the upward and downward movements must be inversely proportional to the degeneracies, as shown. Exact algebraic analysis also shows that for the purely electrostatic, point charge model $\Delta_t = -\frac{4}{9}\,\Delta_0$.

Electrostatic splitting diagrams are relatively easy to deduce for the octahedral and tetrahedral cases because of the high symmetry. This is also true for a cubic arrangement; since a cube is made up of two complementary tetrahedra, the pattern is the same as for one tetrahedron but the splitting is twice as large, i.e., $\Delta_c = 2\Delta_t = -\frac{8}{9}\Delta_0$.

For other arrangements two or more splitting parameters Δ_i are required and the diagrams are more complex. For a symmetrical linear arrangement with the charges along $+z$ and $-z$, it is easy to see qualitatively that the orbital energies must increase in the order $d_{x^2-y^2}, d_{xy} < d_{xz}, < d_{z^2}$; the ratio of the two splitting energies is not fixed by symmetry alone and can be obtained only by calculation. Its value depends on the distance from M to the negative charges and on the radial wave function for the orbitals.

For the two regular five-coordinate geometries sp and tbp, there are several splitting parameters whose ratios vary with the M—X distances and, in the sp, with the X—M—X angles that are not fixed by symmetry. Hence no unique diagram exists in either case, but those given in Fig. 20-14 are typical and show the important qualitative features.

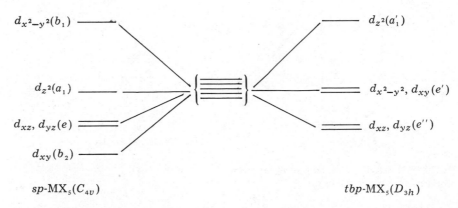

Fig. 20-14. Orbital splitting diagrams for sp and tbp complexes. The sp pattern is not unambiguously derivable from qualitative arguments but is supported by experimental data [B. Bosnich, W. G. Jackson and S. T. D. Lo, *Inorg. Chem.*, 1975, 14, 2998].

Finally we consider the splitting of the d orbitals in tetragonally distorted octahedral complexes and in planar complexes. We begin with an octahedral complex MX_6, from which we slowly withdraw two trans ligands. Let these be the two on the z axis. As soon as the distance from M^{m+} to these two ligands becomes greater than the distance to the other four, new energy differences among the d orbitals arise. First of all, the degeneracy of the e_g orbitals is lifted, the z^2 orbital becoming more stable than the $(x^2 - y^2)$ orbital. This happens because the ligands on the z axis exert a much more direct repulsive effect on a d_{z^2} electron than upon a $d_{x^2-y^2}$ electron. At the same time the 3-fold degeneracy of the t_{2g} orbitals is also lifted. As the ligands on the z axis move away, the yz and zx orbitals remain equivalent to each other, but they become more stable than the xy orbital because their spatial distribution makes them more sensitive to the charges along the z axis than is the xy orbital. Thus for a small tetragonal distortion of the type considered, we may draw the energy level diagram shown in Fig. 20-15. It should be obvious that for the opposite type of tetragonal distortion, that is, one in which two *trans* ligands lie closer to the metal ion than do the other four, the relative energies of the split components will be inverted.

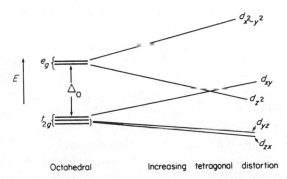

Octahedral   Increasing tetragonal distortion

Fig. 20-15. Energy level diagram showing the further splitting of the d orbitals as an octahedral array of ligands becomes progressively distorted by the withdrawal of two *trans* ligands, specifically those lying on the z axis.

As Fig. 20-15 shows, it is in general *possible* for the tetragonal distortion to become so large that the z^2 orbital eventually drops below the xy orbital. Whether this will *actually happen* for any particular case, even when the two *trans* ligands are completely removed so that we have the limiting case of a square, four-coordinated complex, depends on quantitative properties of the metal ion and the ligand concerned. Semiquantitative calculations with parameters appropriate for square complexes of Co^{II}, Ni^{II}, and Cu^{II} lead to the energy level diagram shown in Fig. 20-16, in which the z^2 orbital has dropped so far below the xy orbital that it is nearly

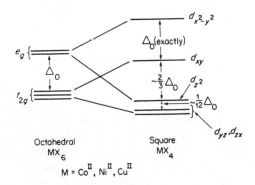

Octahedral   Square
MX$_6$ MX$_4$

M = Co^{II}, Ni^{II}, Cu^{II}

Fig. 20-16. Approximate energy level diagram for corresponding octahedral and square complexes of some metal ions in the first transition series.

as stable as the (yz, zx) pair. As Fig. 20-15 indicates, the d_{z^2} level might even drop below the (d_{xz}, d_{yz}) levels, and in fact experimental results suggest that in some cases (e.g., $PtCl_4^{2-}$) it does.

20-9. Orbital Splitting and Magnetic Properties

As already noted in Section 20-4, the unpaired electrons in many compounds of transition elements provide an important tool in studying their chemistry. Experimental measurements can indicate the number of unpaired electrons, and we require a simple way of relating these facts to structure and bonding. It is now shown that the number of unpaired electrons is usually related simply to the pattern and magnitude of the splitting of the d orbitals. The quantitative accuracy of any calculated splitting pattern is less important than the qualitative correctness of the pattern; thus the following discussion, though usually treated as an application of crystal field theory, is more generally valid.

It is a general rule that if a group of n or fewer electrons occupies a set of n degenerate orbitals, they will spread themselves among the orbitals and give n unpaired spins. This is Hund's first rule, or the rule of maximum multiplicity. It means that pairing of electrons is an unfavorable process; energy must be expended to make it occur. If two electrons are not only to have their spins paired but also to be placed in the same orbital, there is a further unfavorable energy contribution because of the increased electrostatic repulsion between electrons that are compelled to occupy the same regions of space. Let us suppose now that in some hypothetical molecule we have two orbitals separated by an energy ΔE and that two electrons are to occupy these orbitals. Referring to Fig. 20-17, we see that when we place one electron

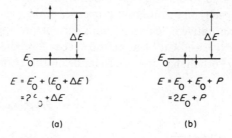

(a)   (b)

Fig. 20-17. A hypothetical two-orbital system showing two possible distributions of two electrons and the resulting total energies.

in each orbital, their spins will remain uncoupled and their combined energy will be $2E_0 + \Delta E$. If we place them both in the lower orbital, their spins will have to be coupled to satisfy the exclusion principle, and the total energy will be $2E_0 + P$, where P stands for the energy required to cause pairing of two electrons in the same orbital. Thus whether this system will have distribution a or b for its ground state depends on whether ΔE is greater or less than P. If $\Delta E < P$, the triplet state a will be the more stable; if $\Delta E > P$, the singlet state b will be the more stable.

Octahedral Complexes. We first apply an argument of the type outlined above to octahedral complexes, using the d orbital-splitting diagram previously deduced from CFT. As indicated in Fig. 20-18, we may place one, two, and three electrons in the d orbitals without any possible uncertainty about how they will occupy the orbitals. They will naturally enter the more stable t_{2g} orbitals with their spins all parallel, and this will be true irrespective of the strength of the crystal field as measured by the magnitude of Δ. Furthermore, for ions with eight, nine, and ten d electrons, there is only one possible way in which the orbitals may be occupied to give the lowest energy (see Fig. 20-18). For each of the remaining configurations,

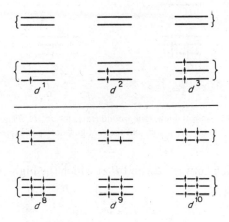

Fig. 20-18. Sketches of the unique ground state occupancy schemes for d orbitals in octahedral complexes with d configurations d^1, d^2, d^3, d^8, d^9, d^{10}.

d^4, d^5, d^6, and d^7, two possibilities exist, and the question of which represents the ground state can be answered only by comparing the values of Δ_0 and P, an average pairing energy. The two configurations for each case, together with simple expressions for their energies, are set out in Fig. 20-19. The configurations with the maximum possible number of unpaired electrons are called the *high-spin* configurations, and those with the minimum number of unpaired spins are called the *low-spin* or *spin-paired* configurations. These configurations can be written out in a notation similar to that used for electron configurations of free atoms, whereby we list each occupied orbital or set of orbitals, using a right superscript to show the number of electrons present. For example, the ground state for a d^3 ion in an octahedral field is t_{2g}^3; the two possible states for a d^5 ions in an octahedral field are t_{2g}^5 and $t_{2g}^3 e_g^2$. This notation is further illustrated in Fig. 20-19. The energies are referred to the energy of the unsplit configuration (the energy of the ion in a spherical shell of the same total charge) and are simply the sums of $-\frac{2}{5}\Delta_0$ for each t_{2g} electron, $+\frac{3}{5}\Delta_0$ for each e_g electron, and P for every pair of electrons occupying the same orbital.

For each of the four cases where high- and low-spin states are possible, we may obtain from the equations for the energies given in Fig. 20-19 the following ex-

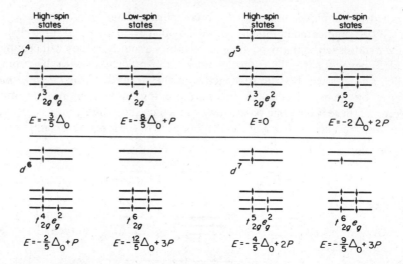

Fig. 20-19. The possible high-spin and low-spin ground states for d^4, d^5, d^6, and d^7 ions in octahedral crystal fields, including the notation for writing out the configurations and expressions for their energies, derived as explained in the text.

pression for the relation between Δ_0 and P at which the high- and low-spin states have equal energies:

$$\Delta_0 = P$$

The relationship is the same in all cases and means that the spin state of any ion in an octahedral electrostatic field depends simply on whether the magnitude of the field as measured by the splitting energy Δ_0 is greater or less than the mean pairing energy P for the particular ion. For a particular ion of the d^4, d^5, d^6, or d^7 type, the stronger the crystal field, the more likely it is that the electrons will crowd

TABLE 20-3

Crystal Field Splittings Δ_0 and Mean Electron Pairing Energies P for Several Transition Metal Ions

Config-uration	Ion	P^a (cm^{-1})	Ligands	Δ (cm^{-1})	Spin state Predicted	Observed
d^4	Cr^{2+}	23,500	$6H_2O$	13,900	High	High
	Mn^{3+}	28,000	$6H_2O$	21,000	High	High
d^5	Mn^{2+}	25,500	$6H_2O$	7,800	High	High
	Fe^{3+}	30,000	$6H_2O$	13,700	High	High
d^6	Fe^{2+}	17,600	$6H_2O$	10,400	High	High
			$6CN^-$	33,000	Low	Low
	Co^{3+}	21,000	$6F^-$	13,000	High	High
			$6NH_3$	23,000	Low	Low
d^7	Co^{2+}	22,500	$6H_2O$	9,300	High	High

[a] It can be shown that these energies in the complexes should probably be $\sim$20% lower than the free-ion values given. It can be seen, however, that even if they are decreased by this amount the correct spin states are still predicted.

as much as possible into the more stable t_{2g} orbitals, whereas in the weaker crystal fields, where $P > \Delta_0$, the electrons will remain spread out over the entire set of d orbitals as they do in the free ion. For ions of the other types, d^1, d^2, d^3, d^8, d^9, and d^{10}, the number of unpaired electrons is fixed at the same number as in the free ion irrespective of how strong the crystal field may become.

Approximate theoretical estimates of the mean pairing energies for the relevant ions of the first transition series have been made from spectroscopic data. Table 20-3 lists these energies, along with Δ_0 values for some complexes (derived by methods to be described in the next section). It will be seen that the theory developed above affords correct predictions in all cases and that the mean pairing energies vary irregularly from one metal ion to another, as do the values of Δ_0 for a given set of ligands. Thus as Table 20-3 shows, the d^5 systems should be exceptionally stable in their high-spin states, whereas the d^6 systems should be exceptionally stable in their low-spin states. These expectations are in excellent agreement with the experimental facts.

Tetrahedral Complexes. It is found that for the d^1, d^2, d^7, d^8, and d^9 cases only high-spin states are possible, whereas for d^3, d^4, d^5 and d^6 configurations both high-spin and low-spin states are in principle possible. The existence of low-spin states would require that $\Delta_t > P$. Since Δ_t values are only about half as great as Δ_0 values, it is to be expected that low-spin tetrahedral complexes of ions of the first transition series with d^3, d^4, d^5, and d^6 configurations would be scarce or unknown. None has so far been shown to exist, and there seems little chance that any will be found.

Square and Tetragonally Distorted Octahedral Complexes. These two cases may be considered together because, as noted earlier, they merge into one another.

Let us consider as an example the d^8 system in an octahedral environment that is then subjected to a tetragonal distortion. We have already seen (Fig. 20-15) how a decrease in the electrostatic field along the z axis splits apart the $(x^2 - y^2)$ and z^2 orbitals. We have also seen that if the tetragonal distortion, that is, the disparity between the contributions to the electrostatic potential of the two z-axis ligands and the other four becomes sufficiently great, the z^2 orbital may fall below the xy orbital. In either case the two least stable d orbitals are now no longer degenerate but are separated by some energy Q. Now the question whether the tetragonally distorted d^8 complex will have high or low spin depends on whether the pairing energy P is greater or less than the energy Q. Figure 20-20(a) shows the case of a "weak" tetragonal distortion, that is, one in which the second highest d orbital is still d_{z^2}.

Figure 20-20(b) shows a possible arrangement of levels for a strongly tetragonally distorted octahedron, or for the extreme case of a square, four-coordinate complex (cf. Fig. 20-16), and the low-spin form of occupancy of these levels for a d^8 ion. In this case, owing to the large separation between the highest and the second highest orbitals, the high-spin configuration is impossible of attainment with the pairing energies of the real d^8 ions (e.g., Ni^{II}, Pd^{II}, Pt^{II}, Rh^{I}, Ir^{I} and Au^{III}), which normally occur, and all square complexes of these species are diamagnetic (unless the ligands have unpaired electrons). Similarly, for a d^7 ion in a square complex, as exemplified

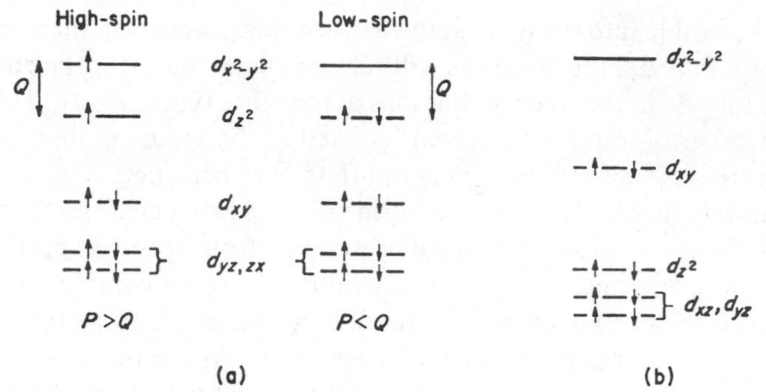

Fig. 20-20. Energy level diagrams showing the possible high-spin and low-spin ground states for a d^8 system (e.g., Ni^{2+}) in a tetragonally distorted octahedral field. (a) Weak tetragonal distortion; (b) strong distortion or square field.

by certain Co^{II} complexes, only the low-spin state with one unpaired electron should occur, and this is in accord with observation.

Five-Coordinate Complexes. We shall consider only the two regular geometries: *sp* and *tbp*. It should be stressed, however, that in practice distortions of these limiting geometries are more the rule than the exception and may have a significant effect on the magnetic properties. By far the majority of cases studied are for $Co^{II}(d^7)$ and $Ni^{II}(d^8)$.

Sp. Employing a diagram similar to that in Fig. 20-14, we first conclude that only the two larger splittings can ever be large enough to lead to spin pairing. Thus d^2 and d^3 configurations will always be high spin; $d^4 - d^8$ configurations can either be high or low spin depending on the magnitude of these larger splittings. Experimental data are relatively scarce, since most work on five-coordination has employed polydentate ligands predisposed to favor a *tbp* configuration. For d^6, d^7, and d^8 configurations in complexes of the type $[M(OER_3)_4(ClO_4)]^+$ (E = P or As), the spin states are high, but the spin states are low for essentially all other ligand sets having, at least roughly, *sp* geometry.

Tbp. From the orbital diagram, Fig. 20-14, where the $e'-e''$ energy difference (δ_1) is small, comparable to Δ_t values, while the a'_1-e' energy difference (δ_2) is larger, comparable to Δ_0 values, we can make the following inferences:

1. d^2 complexes must be high spin.

2. d^3 and d^4 complexes are likely to be high spin, since δ_1 is unlikely to exceed spin-pairing energies.

3. d^5 to d^8 complexes may have either high- or low-spin configurations depending on whether δ_2 is smaller or greater than the mean spin-pairing energy.

High-Spin, Low-Spin Crossovers. One may envisage the occurrence of critical orbital splittings of a magnitude comparable to P so that high- and low-spin states will have about the same energy.

A number of such cases have now been found, and a few salient points are presented here. For convenience and also to conform to the literature, we use a notation that describes the entire electronic state of the ion. For example, the high-spin configuration $t_{2g}^3 e_g^2$ of a d^5 ion in an octahedral environment gives rise to a state denoted $^6A_{1g}$, where the superscript 6 is the spin multiplicity. This is equal to the number of unpaired electrons plus one. Table 20-4 lists the main systems in which spin-crossover phenomena have been studied. All those listed are found in six-coordinate, distorted octahedral structures, but there are important cases in other geometries, especially five-coordinate, that are not discussed explicitly here.

TABLE 20-4
Some "Octahedral" Spin-Crossover Cases

d^n	Low spin configuration and state	High-spin configuration and state	Example
d^5	$^2T_{2g}(t_{2g}^5)$	$^6A_{1g}(t_{2g}^3 e_g^2)$	Fe^{III}
d^6	$^1A_{1g}(t_{2g}^6)$	$^5T_{2g}(t_{2g}^4 e_g^2)$	Fe^{II}, Co^{III}
d^7	$^2E_g(t_{2g}^6 e)$	$^4T_{1g}(t_{2g}^5 e_g^2)$	Co^{II}

The iron(III) d^5 case is well represented by a variety of tris(dithiocarbamato) complexes Fe(S$_2$CNR$_2$)$_3$, which have trigonally distorted "octahedral" configuration of six sulfur atoms. With most R groups the $^2T_{2g}$ state lies several hundred reciprocal centimeters below the $^6A_{1g}$ state. Thus at low temperatures the effective magnetic moment μ_{eff} tends toward a value of ~2.1 BM, which is characteristic of a t_{2g}^5 configuration. As the temperature increases, molecules begin to populate the high-spin state and the average effective magnetic moment rises, following a sigmoidal curve that appears to be approaching an asymptotic limit. This limiting value must, of course, be less than the μ_{eff} for a pure high-spin complex, since it will never be possible to excite thermally all the molecules into the high-spin state. This behavior is shown in Fig. 20-21, for a typical case where ΔE, the energy difference between the low- and high-spin states is between 50 and 250 cm^{-1}; this is the crucial range, since thermal energies vary from ~70 cm^{-1} at 100°K to ~200 cm^{-1} at room temperature. If ΔE becomes greater than ~29 kJ mol^{-1} (>10 times RT at the highest temperature of measurement) thermal population of the high-spin

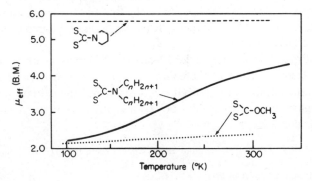

Fig. 20-21. Variation of μ_{eff} with temperature for some FeL$_3$ complexes, where L is a dithiocarbamate or xanthate.

excited state is negligible—or nearly so—and simple, temperature-independent, low-spin behavior is observed. This is the case when the R,R' groups are cyclohexyl, or when NRR' is replaced by OR (to give xanthates instead of dithiocarbamates). Conversely, the replacement of NRR' by pyrrolidino causes the high-spin state to become more stable by an energy much in excess of RT and we have simple, temperature-independent high-spin behavior.

It should be noted that although the spin-crossover behavior just discussed is relatively simple for solutions (it is formally no different from any ordinary chemical equilibrium, A $\rightleftarrows$ B, between isomers), it becomes much more complex in the solid state. The experimental results can be fitted only approximately by considering a simple Boltzmann distribution of molecules between two states. Because of the redistribution of electrons from nonbonding to antibonding orbitals, the metal-ligand bond lengths and perhaps other structural features change, the packing of the molecules in the crystal may change, vibrational frequencies and energies change, and so forth. The physical techniques employed in efforts to follow these changes in detail include study of the pressure dependence of metal-ligand vibrations[10] and X-ray crystallography. For $Fe(S_2CNEt_2)_3$ which has $\mu_{eff} = 2.2$ BM at 79°K and 4.3 BM at 297°K, it has been found that the crystal packing and molecular dimensions are significantly different at these two temperatures.[11] The Fe—S distances change from 2.306 Å at 79°K, which is consistent with the low-spin $^2T_{2g}$ state, to 2.357 Å, indicative of at least partial conversion to the $^6A_{1g}$ state.

Spin-crossover phenomena are quite common in d^6 systems, especially complexes of iron(II),[12] and in some Co^{III} compounds. Several of these iron(II) cases have shown very clearly the inadequacy of treating solids as though a true (and simple) Boltzmann equilibrium occurs.[13]

Various Co^{II} complexes afford examples of the d^7 spin-crossover situation. About fifteen different systems of the type $[CoL_2]X_2$, where L is a tridentate ligand such as terpyridyl,[14] have been shown to give lower ^2E_g states with $^4T_{1g}$ states lying only a few hundred wave numbers-higher. These compounds are quite sensitive to the identity of the counterions X and to the crystal packing. Thus at 25°C $[Co(terpy)_2](SCN)_2$ has $\mu_{eff} \approx 4.0$ BM and the bromide has $\mu_{eff} \approx 2.9$. Moreover, there are real differences in the Co—N distances in the two cases, and as the bromide shifts from the low- to high-spin state, the Br^- ions move to different positions in the crystal. These results, and others, clearly show that spin-crossover phenomena in crystals are quite complicated.

Magnetic Properties of the Heavier Elements. Whereas a simple interpretation of magnetic susceptibilities of the compounds of first transition series elements usually gives the number of unpaired electrons, hence the oxidation state and d-orbital configuration, more complex behavior is often encountered in compounds of the heavier elements.

[10]   R. J. Butcher, J. R. Ferraro, and E. Sinn, *Inorg. Chem.,* 1976, **15**, 2077.
[11]   J. G. Liepolt and P. Coppens, *Inorg. Chem.,* 1973, **12**, 2269.
[12]   H. A. Goodwin, *Coord. Chem. Rev.,* 1976, **18**, 293.
[13]   J. R. Sams and T. B. Tsin, *Inorg. Chem.,* 1976, **15**, 1544.
[14]   C. L. Raston and A. H. White, *J.C.S. Dalton,* **1974**, 1803; **1976**, 7.

One important characteristic of the heavier elements is that they tend to give *low-spin* compounds, which means that in oxidation states characterized by an odd number of d electrons there is frequently only one unpaired electron, and ions with an even number of d electrons are very often diamagnetic. There are two main reasons for this intrinsically greater tendency to spin pairing. First, the $4d$ and $5d$ orbitals are spatially larger than $3d$ orbitals so that double occupation of an orbital produces significantly less interelectronic repulsion. Second, a given set of ligand atoms produces larger splittings of $5d$ than of $4d$ orbitals and in both cases larger splittings than for $3d$ orbitals (see p. 663).

When there are unpaired electrons, the susceptibility data are often less easily interpreted. For instance, low-spin octahedral Mn^{III} and Cr^{II} complexes have t_{2g}^4 configurations, hence two unpaired electrons. They have magnetic moments in the neighborhood of 3.6 BM that can be correlated with the presence of the two un-paired spins, these alone being responsible for a moment of 2.83 BM, plus a con-tribution from unquenched orbital angular momentum. Now, Os^{IV} also forms octahedral complexes with t_{2g}^4 configurations, but these commonly have moments of the order of 1.2 BM; such a moment, taken at face value, has little meaning and certainly does not give any simple indication of the presence of two unpaired elec-trons. Indeed, in older literature it was naïvely taken to imply that there was only one unpaired electron, from which the erroneous conclusion was drawn that the osmium ion was in an odd oxidation state instead of the IV state.

Similar difficulties arise in other cases, and their cause lies in the *high spin-orbit coupling constants* of the heavier ions. Figure 20-22 shows how the effective magnetic moment of a t_{2g}^4 configuration depends on the ratio of the thermal energy kT to the spin-orbit coupling constant λ. For Mn^{III} and Cr^{II}, λ is sufficiently small that at room temperature ($kT \approx 200$ cm^{-1}) both these ions fall on the plateau of the curve, where their behavior is of the familiar sort. Os^{IV}, however, has a spin-orbit coupling constant that is an order of magnitude higher, and at room temperature kT/λ is still quite small. Thus at ordinary temperatures octahedral Os^{IV} compounds should (and do) have low, strongly temperature-dependent magnetic moments. Obviously if measurements on Os^{IV} compounds could be made at sufficiently high temperatures—which is usually impossible—they would have "normal" moments, and, conversely, at very low temperatures Mn^{III} and Cr^{II} compounds would show "abnormally" low moments.

The curve shown in Fig. 20-22 for the t_{2g}^4 case arises because of the following effects of spin-orbit coupling. First the spin-orbit coupling splits the lowest triplet state in such a way that in the component of lowest energy the spin and orbital moments cancel one another completely. When λ, hence this splitting, are large compared with the available thermal energy, the Boltzmann distribution of systems among the several spin-orbit split components is such that most of the systems are in the lowest one that makes no contribution at all to the average magnetic moment. At 0°K, of course, all systems would be in this nonmagnetic state and the substance would become entirely diamagnetic. Second however, the spin-orbit coupling causes an interaction of this lowest nonmagnetic state with certain high-lying excited states so that the lowest level is not actually entirely nonmagnetic at all temperatures,

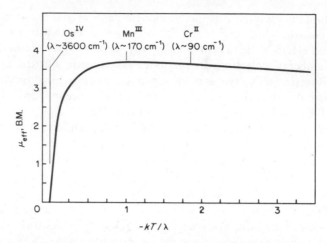

Fig. 20-22. Curve showing the dependence on temperature and on the spin-orbit coupling constant λ of the effective magnetic moment of a d^4 ion in octahedral coordination.

and in the temperature range where kT/λ is much less than unity the effective magnetic moment varies with the square root of the temperature.

Similar difficulties arise for d^1 ions in octahedral fields, when the spin-orbit coupling constant is large. For example, if $\lambda = 500$ (as for Zr^{III}) the nonmagnetic ground state, which splits off from the $^2T_{2g}$ term under the influence of spin-orbit coupling, will be so low that a temperature-independent susceptibility corresponding to an effective moment of only $\sim$0.8 BM at room temperature will be observed. Again, this moment as such has no unique interpretation in terms of the number of unpaired electrons for the ion.

Those systems for which fairly complicated behavior is expected (only octahedral coordination being considered here) are d^1, d^2, d^7, d^8, and d^9. The d^6 systems have no paramagnetism (unless there is some of the temperature-independent type), since they have t_{2g}^6 configurations with no unpaired electrons. The d^3 systems have magnetic moments that are rigorously temperature independent regardless of the magnitude of λ. The d^5 systems have moments that vary with temperature only for very low values of kT/λ, and even then the temperature dependence is not severe; nevertheless, these systems can show complicated behavior because of intermolecular magnetic interactions in compounds that are not magnetically dilute.

20-10. The Angular Overlap Model

Many important properties of transition metal compounds and complexes depend to a good approximation only on the arrangement of the metal d orbitals (strictly speaking, on the molecular orbitals that are made up primarily of metal d orbital contributions) and the relative magnitudes of the differences in their energies. That is why the crystal field and ligand field theories have been useful, despite their incompleteness relative to a full MO treatment and despite certain physically un-realistic features and ad hoc assumptions.

However since the CFT and LFT deal with ligands as if they are point charges or dipoles that affect the metal d electrons only by imposing an electrostatic field on them, these treatments cannot sort out the interplay between σ- and π-bonding interactions. This deficiency and its seriousness were recognized a long time ago, but only in the last decade has a feasible way been developed to overcome the problem while still preserving the essential simplicity of focusing attention on the "d orbitals": the *angular overlap model (AOM)*, which is explained in this sec-tion.

It should be kept in mind that although the AOM is more detailed than, hence superior to, CFT in its modeling of the bonding in complexes, it is still far from a complete MO treatment. It is still, as its name says, only a *model*. It is, however, an effective one for many purposes and its use has increased rapidly in recent years. It has, indeed, effectively replaced CFT or LFT for many purposes. Moreover the way in which the AOM has been implemented gives it an additional practical ad-vantage, namely, that it can readily be used for complexes that have little or no symmetry.

The AOM, like CFT, provides a parameterized set of relationships between the energies of the d orbitals, in which the absolute values of the parameters are not available from the theory itself but are obtained by fitting the parameterized re-lationships to experimental data. The idea is, once again, that the number of pa-rameters needing to be evaluated will be smaller than the total number of experi-mental data potentially available, so that from the results of a few experiments, via the parametric equations, other data can be predicted, or rationalized.

We explain here the conceptual basis of the AOM and give some simple illus-trations. However for "operating instructions" complete with all necessary algebraic details and tables, the reader must consult one or more of a number of detailed accounts.[15] Additional discussion of applications of the AOM to specific problems may be found in various reviews and research papers.[16]

The AOM is based on the premise that the energetic effects of an interaction between any given ligand orbital ϕ_L and any given metal d orbital Φ_d may be formulated in the simple way shown in Fig. 20-23. The energies of ψ_b and ψ_a, the resultant bonding and antibonding orbitals, respectively, are shown from simplified MO theory to be given by the following expression:

$$E(\psi_a) - E(\Phi_d) = E(\phi_L) - E(\psi_b) = \frac{[E(\Phi_d) + E(\phi_L)]^2}{E(\Phi_d) - E(\phi_L)} S^2 \qquad (20\text{-}5)$$

[15] C. E. Schäffer, *Struct. Bonding,* 1973, **14**, 69; *Proc. Roy. Soc. London, A,* 1967, **297**, 96; C. E. Schäffer and C. K. Jørgensen, *Mol. Phys.* 1965, **9**, 401; C. E. Schäffer, *Pure Appl. Chem.,* 1970, **24**, 361; M. Gerloch and R. C. Slade, *Ligand Field Parameters,* Cambridge University Press, 1973; S. F. A. Kettle, *J. Chem. Soc., A,* **1966**, 420; E. Larsen and G. N. LaMar, *J. Chem. Educ.,* 1974, **51**, 633.

[16] J. K. Burdett, *Struct. Bonding,* 1976, **31**, 67; D. W. Smith, *Struct. Bonding,* 1978, **35**, 87; P. E. Hoggard and H. H. Schmidtke, *Inorg. Chim. Acta,* 1979, **34**, 77; J. K. Burdett, *Adv. Inorg. Chem. Radiochem.,* 1978, **21**, 113.

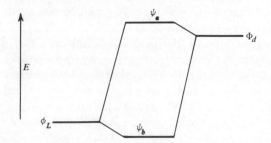

Fig. 20-23. How overlap of a ligand orbital ϕ_L with a metal d orbital Φ_d will lead to the formation of an antibonding orbital ψ_a, mainly Φ_d in character, and a bonding orbital ψ_b, mainly φ_L in character.

The symbol S represents the overlap integral between Φ_d and ϕ_L. This expression is only a rough approximation for several reasons. However if it is used for every one of the interactions between a d orbital and a ligand σ or π orbital, there will be a good deal of cancelation of errors insofar as the *relative energies* of the d orbitals are concerned.

It should be noted that for a given type of ϕ_L [say, the σ orbitals on the six NH_3 molecules in an $M(NH_3)_6$ complex] the fraction containing only $E(\Phi_d) \pm E(\phi_L)$ terms is a constant.

Moreover, the magnitude of the overlap S depends on both the radial and angular parts of the d-orbital wave functions, but for a given central metal atom the radial distribution is fixed, and S varies only with the angle made by the M—L bond axis with the coordinates used to specify the metal d orbitals. Hence we may break the total overlap down into a product of a constant radial factor S_R and an angular factor S_A and write

$$S^2 = S_R^2 S_A^2$$

If we now represent by e, which is an energy, all the factors that are constant for any one complex, we may rewrite the expression (20-5) as (20-6).

$$E(\psi_a) - E(\Phi_d) = E(\phi_L) - E(\psi_b) = eS_A^2 \tag{20-6}$$

Thus the shifts in energy for either the "d orbital" or the "ligand orbital" are proportional to the square of an "angular overlap" integral. This explains the origin of the name "angular overlap model."

Although eq. 20-6 deals with shifts in both the d orbitals (to give ψ_a) and the ligand orbital (to give ψ_b), our attention henceforth is confined to the quantity $E(\Phi_d) - E(\phi_a)$, which is the shift in the energy of a d orbital from its initial position. When we have these quantities for each of the d orbitals, we have the complete d-orbital splitting pattern.

Let us now look in more detail at two specific ligand-metal interactions when we have a ligand placed on the z axis of the coordinate system used to define the d orbitals, as in Fig. 20-24. Part (a) shows a ligand σ orbital ϕ_σ overlapping with the d_{z^2} orbital and (b) shows the overlap of a ligand π orbital ϕ_π with the d_{yz} orbital. It is clear that the overlaps have their maximum possible values for the orientations

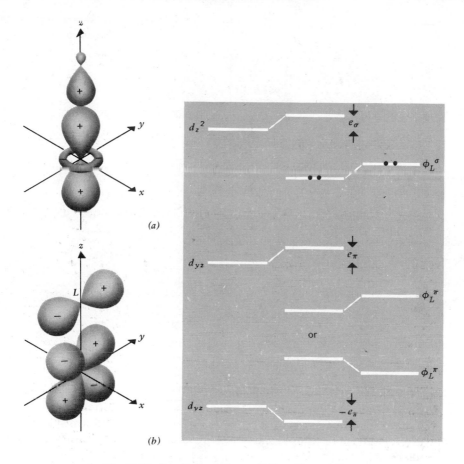

(a)

(b)

Fig. 20-24. Standard σ and π metal-ligand interactions.

shown and that any shift of the ligand atom off the z axis will diminish the overlap.

For the σ overlap in Fig. 20-24 it is clear that if the energy of a filled ϕ_σ is initially lower than that of the d_{z^2} orbital, the interaction raises the energy of the latter. For the orientation shown we can designate that rise in energy as e_σ. Now, if the ligand is moved off the z axis (but always keeping the same internuclear distance) so that the M—L line makes an angle χ with the z axis, as in Fig. 20-25, the overlap integral will be lower than when $\chi = 0$. The energy will be $(S^\chi/S^0)^2 e_\sigma$. The magnitude of the ratio S^χ/S^0 is easy to calculate as a function of S^χ by considering the algebraic expression for the angular factor of the d_{z^2} wave function, which is $3 \cos \chi - 1$.

As χ increases from 0 the amplitude of the wave function, at constant r, decreases until at $\chi = (109.5°)/2$ it equals zero. It then increases, through negative values to $-\frac{1}{2}$ times its value at $\chi = 0$, when $\chi = 90°$. The overlap integrals will follow the

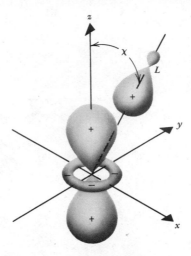

Fig. 20-25. Approach of a ligand σ orbital to a metal d_{z^2} orbital at an angle χ from the z axis.

same course, since the amplitude of the ligand σ orbital is constant and r is constant. Hence from a knowledge of the shape of the d orbital we can tell what relative effect a ligand σ orbital at any point on a sphere of constant radius will have on the energy of the d_{z^2} orbital.

In an analogous way, the relative effect of a ligand σ or π orbital at any point on the sphere on the energy of each of the d orbitals may be computed. To get the total effect of an entire set of ligands on each d orbital, the effects are calculated for each ligand and added.

Returning to Fig. 20-24, it must be noted that there are two broad possibilities for π interactions. Either the ligand π orbital is a stable filled one, in which case it raises the energy of the d_{yz} orbital by e_π, or it is an empty higher-energy orbital (such as an outer d orbital of PX_3 or an antibonding π orbital of CO) in which case it makes the d_{xz} orbital more stable by e_π; algebraically, the change in the energy of d_{xz} is $-e_\pi$ in the latter case.

The actual application of the AOM to deriving the pattern of d orbitals for a given complex is quite simply accomplished by using a table of "angular overlap factors," one for each d orbital. These factors state the contribution of a σ or a π orbital of a ligand to the energy of each d orbital as a function of its polar coordinates. Tables of angular overlap factors are presented in several of the articles cited earlier.[15]

When applied to octahedral and tetrahedral complexes, the AOM predicts exactly the same splitting patterns as does the CFT, although these energies are given as functions of the e_σ and e_π parameters, namely, $\Delta_0 = 3e_\sigma - 4e_\pi$ and $\Delta_t = -\frac{4}{9}\Delta_0$. In the following discussions of spectra we shall use the symbols Δ_0 and Δ_t, since we are referring to the experimental splitting, not to the parametric treatment by which it is obtained.

ELECTRONIC ABSORPTION SPECTRA*

20-11. Octahedral and Tetrahedral Complexes

d^1 and d^9 Systems. Let us first consider the simplest possible case, an ion with a d^1 configuration, lying at the center of an octahedral field, for example, the Ti^{III} ion in $[Ti(H_2O)_6]^{3+}$. The d electron will occupy a t_{2g} orbital. On irradiation with light of frequency v, equal to Δ_0/h, where h is Planck's constant and Δ_0 is the energy difference between the t_{2g} and the e_g orbitals, it should be possible for such an ion to capture a quantum of radiation and convert that energy into energy of excitation of the electron from the t_{2g} to the e_g orbital. The absorption band that results from this process is found in the visible spectrum of the hexaaquatitanium(III) ion (Fig. 20-26) and is responsible for its violet color. Three features of this absorption band are of importance: its position, its intensity, and its breadth.

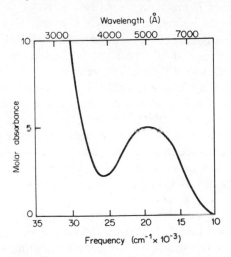

Fig. 20-26. The visible absorption spectrum of $[Ti(H_2O)_6]^{3+}$.

In discussing the positions of absorption bands in relation to the splittings of the d orbitals, it is convenient and common practice to use the same unit, the reciprocal centimeter or wave number, abbreviated cm^{-1}, for both the unit of frequency in the spectra and the unit of energy for the orbitals. With this convention, we see that the spectrum of Fig. 20-26 tells us that Δ_0 in $[Ti(H_2O)_6]^{3+}$ is 20,000 cm^{-1}. Since there are 83.7 cm^{-1} per kilojoule, this means that the splitting energy is ~240 kJ mol^{-1}, which is comparable with the usual values of chemical bond energies.

Turning now to the intensity of this absorption band in the $[Ti(H_2O)_6]^{3+}$ ion, we note that it is extremely weak by comparison with absorption bands found in many other systems.

---

* The d-orbital splitting patterns and the splittings of free-ion states are discussed mainly in terms of the crystal field theory, since this is the most convenient way to convey the essential points.

Allowed electronic transitions typically have a molar absorbances of 10^3 to 10^5, whereas the band shown in Fig. 20-26 has a peak absorbance of only ~ 5. This is because the transition involved is from one centrosymmetric orbital to another, and according to quantum theory such an "even-even" transition should be entirely forbidden (i.e., have no intensity at all). This will be true of all the spectral transitions ("d–d" transitions) in which electrons move from one orbital to another in the set of "d orbitals" that have been split apart by interactions with the ligands. These transitions in octahedral complexes do indeed give observable absorption bands, albeit weak ones, primarily because certain vibrations of the octahedral complex cause deviations from the centrosymmetric character of a rigid octahedron. For tetrahedral complexes the "d–d" transitions are even stronger, by about a factor of 10,* because here the entire complex is lacking a center of symmetry. Nonetheless, the orbitals involved are largely metal d orbitals into which relatively small amounts of ligand orbitals have been mixed, and this means that the transitions are still strongly "even-even," hence not as strong as the genuinely allowed "even-odd" or "odd-even" transitions normally are.

It will also be noted that the band in Fig. 20-26 is several thousand wave numbers broad, rather than a sharp line at a frequency precisely equivalent to Δ_0. This also is a result of the vibrations of the complex and is discussed in detail later.

A qualitative interpretation of the d–d spectrum of a d^9 ion can be given as easily as for a d^1 ion by virtue of the *hole formalism,* according to which a d^{10-n} configuration can be treated as n holes in the d shell. These holes are in many ways equivalent to n positrons, and in the framework of the CFT we may then say that in effect the energy ordering of orbitals for n electrons can be inverted to get the orbital pattern for n holes. We can thus look upon a Cu^{II} ion in an octahedral environment as a one-positron ion in an octahedral field and deduce that in the ground state the positron will occupy an e_g orbital from which it may be excited by radiation providing energy Δ_0 to a t_{2g} orbital.

Experimentally it is found, however, that the absorption band of the Cu^{II} ion in aqueous solution is not a simple, symmetrical band but instead appears to consist of several nearly superposed bands. The observant reader may have noticed that the absorption band of the $[Ti(H_2O)_6]^{3+}$ ion is not quite a simple, symmetrical band either. In each case these complications are traceable to distortions of the octahedral environment that are required by the Jahn-Teller theorem (see p. 678).

d^2–d^8 Ions; Energy Level Diagrams.

To interpret the spectra of complexes in which the metal ions have more than one but less than nine d electrons, we must employ an energy level diagram based upon the Russell-Saunders states of the relevant d^n configuration in the free (uncomplexed) ion. Readers unfamiliar with the way in which these states arise in multielectron configurations, and with the notation used for them, will find an explanation in Appendix 5.

It can be shown that just as the set of five d orbitals is split apart by the electrostatic field of surrounding ligands to give two or more sets of lower degeneracy,

* See Fig. 20-29 for an illustration.

TABLE 20 5
Splitting of Russell-Saunders States in Octahedral and Tetrahedral Electrostatic Fields

State of free ion	States in the crystal field
S	A_1
P	T_1
D	$E + T_2$
F	$A_2 + T_1 + T_2$
G	$A_1 + E + T_1 + T_2$
H	$E + 2T_1 + T_2$

so also are the various Russell-Saunders states of a d^n configuration. The number and types of the components into which an octahedral or tetrahedral field will split a state of given L is the same regardless of the d^n configuration from which it arises, and these facts are summarized in Table 20-5. The designations of the states of the ion in the crystal field are the *Mulliken* symbols; their origin is in group theory, but they may be regarded simply as labels. The main letters in these symbols have the following meanings: A and B designate singly degenerate states, E designates doubly degenerate states, and T indicates triple degeneracy. Left superscripts can be applied just as they are to the letter symbols for free ion states (i.e., S, P, D, etc.) to indicate spin multiplicity.

Although the states into which a given free ion state is split are the same in number and type in both octahedral and tetrahedral fields, the pattern of energies is inverted in one case relative to the other. This is quite analogous to the results in the d^1 case, as we have already seen (Fig. 20-13).

A discussion of the way in which the energies of the crystal field states are calculated would be beyond the scope of this book. The inorganic chemist can use the energy level diagrams without knowing how they are obtained; it is, of course, necessary to know how to interpret them properly. At this point we examine several of them in some detail, to explain their interpretation. Others are introduced subsequently in discussing the chemistry of particular transition elements.

We look first at the energy level diagram for a d^2 system in an octahedral field, (Fig. 20-27). The ordinate is in energy units (usually cm^{-1}), and the abscissa is in units of crystal field splitting energy as measured by Δ_0, the splitting of the one-electron orbitals. At the extreme left are the Russell-Saunders states of the free ion. It may be seen that each of these splits in the crystal field into the components specified in Table 20-5. Three features of this energy level diagram particularly to be noted because they will be found in all such diagrams are:

1. States with identical designations never cross.

2. States of the complex have the same spin multiplicity as the free ion states from which they originate.

3. States that are the only ones of their type have energies that depend linearly on the crystal field strength, whereas when there are two or more states of identical designation, their lines will in general show curvature. This is because such states interact with one another.

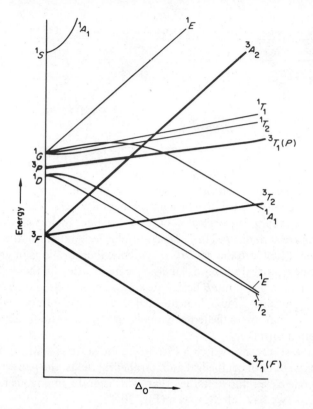

Fig. 20-27. The complete energy level diagram for the d^2 configuration in an octahedral crystal field. The heavier lines are those for the triplet states.

It is interesting to note in Fig. 20-27 that a triplet state lies lowest at all field strengths shown, and since its slope is as steep as the slope of any other state, it will continue to be the lowest state, no matter how intense the crystal field may become. This is in complete agreement with our previous conclusion, based on the simple splitting diagram for the d orbitals, which showed that the two d electrons would have their spins parallel in an octahedral field, irrespective of how strong the field might be.

To use this energy level diagram to predict or interpret the spectra of octahedral complexes of d^2 ions, for example, the spectrum of the $[V(H_2O)_6]^{3+}$ ion, we first note that there is a quantum mechanical selection rule that forbids transitions between states of different spin multiplicity. This means that in the present case only three transitions, those from the $^3T_1$ ground state to the three triplet excited states $^3T_2$, $^3A_2$, and $^3T_1(P)$, will occur. Actually, spin-forbidden transitions, that is, those between levels of different spin multiplicity, do occur very weakly because of weak spin-orbit interactions, but they are several orders of magnitude weaker than the spin-allowed ones and are ordinarily not observed.

Experimental study of the $[V(H_2O)_6]^{3+}$ ion reveals just three absorption bands with energies of about 17,000, 25,000, and 38,000 cm^{-1}. Using an energy level

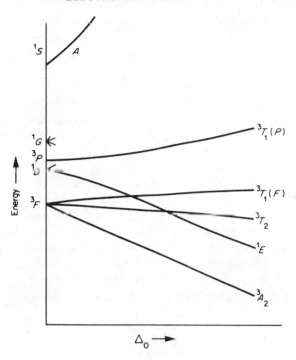

Fig. 20-28. Partial energy level diagram for a d^8 ion in an octahedral field, showing the triplet states and only the lowest singlet state.

diagram like that in Fig. 20-27, in which the separations of the free-ion states are adjusted to match exactly those appropriate for V^{III}, one finds that when Δ_0 is 21,500 cm^{-1}, the three transitions are expected at 17,300, 25,500, and 38,600 cm^{-1}, in excellent agreement with observation.

We look next at the energy level diagram for a d^8 ion in an octahedral field (Fig. 20-28), restricting our attention to the triplet states except for the lowest energy singlet state that comes from the ^1D state of the free ion. This has been included to show that for this system the ground state will always be a spin triplet no matter how strong the crystal field, a result in agreement with the conclusion previously drawn from consideration of the distribution of eight electrons among a set of five d orbitals. On comparing the arrangements of the three components derived from the ^3F ground state for the d^2 and the d^8 cases, we note that one pattern is the inverse of the other. This is a manifestation of the hole formalism for the d^2–d^{10-2} configurations that is fundamentally analogous to the d^1–d^{10-1} example we previously examined.

Similar energy level diagrams may be drawn for d^n systems in tetrahedral crystal fields. There is an interesting relationship between these and the ones for certain systems in octahedral fields. We have already seen that the splitting pattern for the d orbitals in a tetrahedral field is just the inverse of that for the d orbitals in an octahedral field. A similar inverse relationship exists between the energy level diagrams of d^n systems in tetrahedral and octahedral fields. The components into which each Russell-Saunders state is split are reversed in their energy order in the tetrahedral compared to the octahedral cases. Furthermore, the splittings in the tetrahedral case will be $4/9$ those in the octahedral case, according to either a CFT or an AOM approach.

Finally, there are rather extensive qualitative similarities between the energy level diagrams of groups of the d^n systems because of the combined effects of reversals in the splitting patterns by changing from an octahedral to a tetrahedral field and by changing from a d^n to a d^{10-n} configuration. When we go from d^n in an octahedral field to d^n in a tetrahedral field, all splittings of the Russell-Saunders states are inverted. But the same inversions occur on changing from the d^u configuration in an octahedral (tetrahedral) field to the d^{10-n} configuration in an octahedral (tetrahedral) field. These relations, combined with the fact that for the free ions, the Russell-Saunders states of the pairs of d^n and d^{10-n} systems are identical in number, type, and relative (though certainly not absolute) energies, mean that various pairs of configuration-environment combinations have qualitatively identical energy level diagrams in crystal fields, and that these differ from others only in the reversal of the splittings of the individual free ion states. These relations are set out in Table 20-6.

TABLE 20-6

Relations Between Energy Level Diagrams for Various d^n Configurations in Octahedral and Tetrahedral Crystal Fields

Octahedral d^1 and tetrahedral d^9	reverse[a] of	Octahedral d^9 and tetrahedral d^1
Octahedral d^2 and tetrahedral d^8	reverse of	Octahedral d^8 and tetrahedral d^2
Octahedral d^3 and tetrahedral d^7	reverse of	Octahedral d^7 and tetrahedral d^3
Octahedral d^4 and tetrahedral d^6	reverse of	Octahedral d^6 and tetrahedral d^4
Octahedral d^5	identical with	Tetrahedral d^5

[a] "Reverse" means that the order of levels coming from each free-ion state is reversed; it does *not* mean that the diagram as a whole is reversed.

The foregoing description and illustrations of energy level diagrams indicate that they may be used to determine from observed spectral bands the magnitudes of Δ_0 and Δ_t in complexes. It may be noted in the diagrams for the d^2 and d^8 systems that there are three spin-allowed absorption bands whose positions are all determined by the one* parameter, Δ_0 or Δ_t. Thus in these cases the internal consistency of the theory may be checked.

Certain generalizations may be made about the dependence of the magnitudes of Δ values on the valence and atomic number of the metal ion, the symmetry of the coordination shell, and the nature of the ligands. For octahedral complexes containing high-spin metal ions, it may be inferred from the accumulated data for a large number of systems that:

1. Δ_0 values for complexes of the first transition series are 7500 to 12,500 cm^{-1} for divalent ions and 14,000 to 25,000 cm^{-1} for trivalent ions.

* This is only approximately true because the separation between the ^3F and ^3P states is not the same in the complexed ion as it is in the free ion; therefore this separation becomes a second parameter, in addition to Δ, to be determined from experiment.

2. Δ_0 values for corresponding complexes of metal ions in the same group and with the same valence increase by 25 to 50% on going from the first transition series to the second and by about this amount again from the second to the third. This is well illustrated by the Δ_0 values for the complexes $[Co(NH_3)_6]^{3+}$, $[Rh(NH_3)_6]^{3+}$, and $[Ir(NH_3)_6]^{3+}$, which are, respectively, 23,000, 34,000, and 41,000 cm^{-1}.

3. Δ_t values are about 40 to 50% of Δ_0 values for complexes differing as little as possible except in the geometry of the coordination shell, in agreement with theoretical expectation.

4.   The dependence of Δ values on the identity of the ligands follows a regular order known as the spectrochemical series, which is now explained.

20-12.   The Spectrochemical Series

It has been found by experimental study of the spectra of a large number of complexes containing various metal ions and various ligands, that ligands may be arranged in a series according to their capacity to cause d-orbital splittings. This series, for the more common ligands, is: $I^- < Br^- < Cl^- < F^- < OH^- < C_2O_4^{2-} \sim H_2O < -NCS^- < py \sim NH_3 < en < bipy < o\text{-phen} < NO_2^- < CN^-$. The idea of this series is that the d-orbital splittings, hence the relative frequencies of visible absorption bands for two complexes containing the same metal ion but different ligands, can be predicted from the series above whatever the particular metal ion may be. Naturally, one cannot expect such a simple and useful rule to be universally applicable. The following qualifications must be remembered in applying it:

1.   The series is based on data for metal ions in common oxidation states. Because the nature of the metal-ligand interaction in an unusually high or unusually low oxidation state of the metal may differ qualitatively in certain respects from that for the metal in a normal oxidation state, striking violations of the order shown may occur for complexes in unusual oxidation states.

2.   Even for metal ions in their normal oxidation states, inversions of the order of adjacent or nearly adjacent members of the series are sometimes found.

20-13.   Further Remarks on Intensities and Line Widths, Illustrated by Manganese(II)

Figure 20-29 shows the spectra of octahedrally coordinated Mn^{2+}, in $[Mn(H_2O)_6]^{2+}$, and tetrahedrally coordinated Mn^{2+}, in the $[MnBr_4]^{2-}$ ion. Let us look first at the $[Mn(H_2O)_6]^{2+}$ spectrum. The spectra of other octahedrally coordinated Mn^{2+} complexes are quite similar. The most striking features are (a) the weakness of the bands, (b) the large number of bands, and (c) the great variation in the widths of the bands, with one extremely narrow indeed.

The bands in $[Mn(H_2O)_6]^{2+}$ are more than 100 times weaker than those normally found for d–d crystal field transitions, which is why the ion has such a pale color; the finely ground solid appears to be white. The ground state of the d^5 system in a weak octahedral field has one electron in each d orbital, and their spins are parallel, making it a spin sextuplet. This corresponds to the ^6S ground state of the free ion, which is not split by the ligand field. This, however, is the only sextuplet

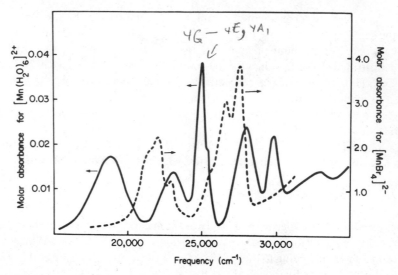

Fig. 20-29. The d–d absorption spectra of the $[Mn(H_2O)_6]^{2+}$ ion (solid curve) and the $[MnBr_4]^{2-}$ ion (dashed curve).

state possible, for every conceivable alteration of the electron distribution $t_{2g}^3 e_g^2$ results in the pairing of two or four spins, thus making quartet or doublet states. Hence all excited states of the d^5 system have different spin multiplicity from the ground state, and transitions to them are spin forbidden. Because of weak spin-orbit interactions, such transitions are not totally absent, but they are very weak. As a rough rule, spin-forbidden transitions give absorption bands ∼100 times weaker than those for similar but spin-allowed transitions.

To understand the number and widths of the spin-forbidden bands, we refer to an energy level diagram (Fig. 20-30) in which all spin doublet states are omitted. Most of these are of very high energy, and transitions to them from the sextuplet ground state are doubly spin forbidden, hence never observed. It is seen that there are four Russell-Saunders states of the free ion that are quartets, and their splittings as a function of ligand field strength are shown. The observed bands of $[Mn(H_2O)_6]^{2+}$ can be fitted by taking Δ equal to about 8600 cm^{-1}, as indicated by the vertical dashed line. The diagram shows that to the approximation used to calculate it, the ^4E and $^4A_1$ states arising from the ^4G term are degenerate. This is very nearly but not exactly so, as the slight shoulder on the sharp band at ∼25,000 cm^{-1} shows.

It will also be noted that there are three states, the $^4A_2$ state from ^4F, the ^4E state from ^4D, and the $(^4E_1, {}^4A_1)$ state from ^4G, whose energies are independent of the strength of the ligand field. Such a situation, which never occurs for upper states of the same spin multiplicity as the ground state, makes it unusually easy to measure accurately the decrease in the interelectronic repulsion parameters.

Theoretical considerations show that the widths of spectral bands due to d–d transitions should be proportional to the slope of the upper state relative to that of the ground state. In the present case (the ground state energy is independent of

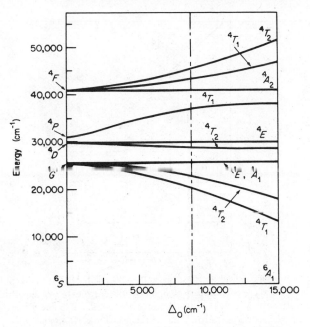

Fig. 20-30. Partial energy level diagram for the Mn^{II} ion, showing only the ^6S state and the quartet states. The separations of the Russell-Saunders states at $\Delta = 0$ are those appropriate for the $[Mn(H_2O)_6]^{2+}$ ion (*not* the actual free Mn^{2+} ion), and the broken vertical line is at the Δ value (8600 cm^{-1}) for this species.

the ligand field strength), this means that the bandwidths should be proportional to the slopes of the lines for the respective upper states as they are seen in Fig. 20-30. Comparison of the spectrum of $[Mn(H_2O)_6]^{2+}$ with the energy level diagram shows that this expectation is well fulfilled. Thus the narrowest bands are those at ~25,000 and ~29,500 cm^{-1}, which correspond to the transitions to upper states with zero slope. The widths of the other lines are also seen to be greater in proportion to the slopes of the upper state energy lines.

It is easy to grasp qualitatively why the bandwidths are proportional to the slopes. As the ligand atoms vibrate back and forth, the strength of the ligand field Δ also oscillates back and forth about a mean value corresponding to the mean position of the ligands. Now, if the separation between the ground and excited states is a sensitive function of Δ, the energy difference will vary considerably over the range in Δ that corresponds to the range of metal-ligand distances covered in the course of the vibrational motion. If, on the other hand, the energy separation of the two states is rather insensitive to Δ, only a narrow range of energy will be encompassed over the range of the vibration. This argument is illustrated in Fig. 20-31.

Tetrahedral complexes of Mn^{II} are yellow-green, and the color is more intense than that of the octahedral complexes. A typical spectrum is shown in Fig. 20-29. First it will be noted that the molar absorbance values are in the range 1.0–4.0, whereas for octahedral Mn^{II} complexes (see Fig. 20-29), they are in the range 0.01–0.04. This increase by a factor of about 100 in the intensities of tetrahedral

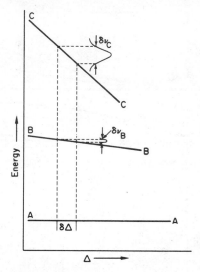

Fig. 20-31. The relation of bandwidth to the slope of the upper state relative to that of the ground state; $A–A$ gives the energy of the ground state, $B–B$ and $C–C$ the energies of two upper states as functions of the ligand field strength Δ, $\delta\Delta$ represents the range of variation of Δ due to the ligand vibrations, and $\delta\nu_B$ and $\delta\nu_C$ are the widths of the bands due to the transitions $A \rightarrow B$ and $A \rightarrow C$.

complexes over octahedral ones is entirely typical. The reasons for it are not known with complete certainty, but it is thought to be due in part to mixing of metal p and d orbitals in the tetrahedral environment, which is facilitated by overlap of metal d orbitals with ligand orbitals in the tetrahedral complexes.

It may also be seen in Fig. 20-29 that there are six absorption bands in two groups of three, just as there are in $[Mn(H_2O)_6]^{2+}$, but they are here much closer together. This is to be expected, since the Δ value for the tetrahedral complex $[MnBr_4]^{2-}$ should be less than that for $[Mn(H_2O)_6]^{2+}$. The energy level diagram (Fig. 20-30) indicates that the uppermost band in the group at lower energy should be due to the transition to the field-strength-independent $(^4E_1, {}^4A_1)$ level, and this band does seem to be quite narrow, although it is partly overlapped by the other two in the group.

20-14. Charge-Transfer Spectra

We have so far considered only those electronic transitions in which electrons move between orbitals that have predominantly metal d-orbital character. Thus the charge distribution in the complex is about the same in the ground and the excited state. There is another important class of transitions in which the electron moves from a molecular orbital centered mainly on the ligands to one centered mainly on the metal atom, or *vice versa*. In these, the charge distribution is considerably different in ground and the excited states (at least we naively assume it is), and so they are called *charge-transfer transitions*.

As just implied, there are two broad classes: ligand-to-metal $(L \rightarrow M)$ and metal-to-ligand $(M \rightarrow L)$. In general, the former are better understood. In most

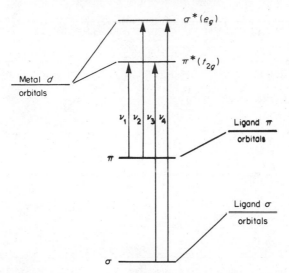

Fig. 20-32. Partial MO diagram for an octahedral MX_6 complex, showing the four main classes of $L \rightarrow M$ charge-transfer transitions.

cases charge-transfer (CT) processes are of higher energy than d–d transitions; thus they usually lie at the extreme blue end of the visible spectrum, or in the ultraviolet region. Also, nearly all observed CT transitions are fully allowed, hence the CT bands are strong. Extinction coefficients are typically 10^3 to 10^4, or more. There are, of course, many forbidden CT transitions that give rise to weak bands, these are seldom observed because they are covered up by the strong CT bands.

$L \rightarrow M$ Transitions in Octahedral Complexes. The first CT spectra to be systematically studied, and still among the best understood, are the $L \rightarrow M$ transitions in hexahalo complexes. Fig. 20-32 shows a partial MO diagram for such complexes, and the four types of transition that would be expected are indicated. Actually, each of the transitions shown is a group of transitions, since the excited orbital configuration (and in many cases also the ground configuration) gives rise to several different states of similar but not identical energies.

Transitions of the ν_1 type will obviously be of lowest energy. Second, since the π and π^* orbitals involved are both approximately nonbonding, they will not vary steeply with M—L distance as the ligands vibrate. Thus, according to the type of argument applied in Section 20-13, the bands for these transitions should be relatively narrow. A third factor that should assist in identifying the ν_1 set of bands is that they will be missing whenever the $\pi^*(t_{2g})$ orbitals are filled (i.e. in d^6 complexes). Table 20-7 indicates how observed transitions in a variety of complexes have been assigned to the ν_1 class. Two trends in these bands support the assigned $L \rightarrow M$ nature of these transitions. For a given metal ion (e.g., Os^{4+}), their energies decrease in the sequence MCl_6, MBr_6, MI_6, which is the order of decreasing ionization potentials (i.e., easier oxidizability) of the halogen atoms. As the oxidation state of the metal increases (e.g., $RuCl_6^{3-}$, $RuCl_6^{2-}$) its orbitals should be deeper, thus the transition should go to lower energy.

TABLE 20-7

L → M CT Transitions (cm^{-1} × 10^{-3}) in Some Hexahalo Complexes[a,b]

d^n	Complex	ν_1 Set[c]	ν_2 Set	ν_4 Set
$4d^4$	$RuCl_6^{2-}$	17.0–24.5 (0.65–3.0)	36.0–41.0 (12–18)	—
$4d^5$	$RuCl_6^{3-}$	25.5–32.5 (0.60–2.1)	43.6(16)	—
$5d^4$	$OsCl_6^{2-}$	24.0–30.0 (1.0–8.0)	47.0(20)	—
	$OsBr_6^{2-}$	17.0–25.0 (1.6–7.5)	35.0–41.0 (10–15)	—
	OsI_6^{2-}	11.5–18.5 (2.5–6.0)	27.0–35.5 (8.0–9.2)	44.6(41)
$5d^6$	$PtBr_6^{2-}$	—	27.0–33.0 (7.0–18)	44.2(70)
	PtI_6^{2-}	—	20.0—30.0 (8.0–13)	40.0–43.5 (40–60)

[a] Data and assignments from C. K. Jørgensen, *Mol. Phys.*, 1959, **2**, 309; *Adv. Chem. Phys.*, 1963, **5**, 33.

[b] Numbers in parentheses are molar extinction coefficients (×10^{-3}).

[c] Band half-widths, 400–1000 cm^{-1}.

Transitions of the ν_2 type should give the lowest-energy CT bands in t_{2g}^6 complexes (e.g., those of the PtX_6^{2-} type). Since the transition is from a mainly non-bonding level to a distinctly antibonding one, the bands should be fairly broad. The transitions assigned to the ν_2 sets all have half-widths of 2000 to 4000 cm^{-1}. The shifts in energy of these bands with change of halogen and change of metal oxidation state are again as expected for L → M transitions.

Transitions of the ν_3 set are all expected to be broad and weak and are not observed. The ν_4 transitions have been observed in a few cases, but in many cases they must lie beyond the range of observation.

L → M Transitions in Tetrahedral Complexes. For the tetrahalo complexes (e.g., the NiX_4^{2-}, CoX_4^{2-} and MnX_4^{2-} species) strong L → M CT spectra can be observed and assigned in much the same way as for the octahedral MX_6^{n-} complexes.

Of course CT spectra are not restricted, like d–d spectra, to transition metal complexes. For example, the series of complexes $HgCl_4^{2-}$, $HgBr_4^{2-}$, HgI_4^{2-} have absorptions at 43,700, 40,000, and 31,000 cm^{-1} that may be assigned as L → M CT transitions.

M → L Transitions. Transitions of this type can be expected only when the ligands possess low-lying empty orbitals and the metal ion has filled orbitals lying higher than the highest filled ligand orbitals. The best examples are provided by complexes containing CO, CN, or aromatic amines (e.g. pyridine, bipyridine, or phenanthroline) as ligands.

In the case of the octahedral metal carbonyls $Cr(CO)_6$ and $Mo(CO)_6$, pairs of intense bands at 35,800 and 44,500 cm^{-1} for the former and 35,000 and 43,000 cm^{-1} for the latter have been plausibly assigned to transitions from the bonding (mainly metal) to the antibonding (mainly ligand) components of the metal-ligand π-bonding interactions.

For $Ni(CN)_4^{2-}$ three medium-to-strong bands at 32,000, 35,200, and 37,600 cm^{-1} have been assigned as transitions from the three types of filled metal d orbitals, d_{xy}, d_{z^2}, and (d_{xz}, d_{yz}) to the lowest-energy orbital formed from the $\pi*$ orbitals of the set of CN groups.

OPTICAL ACTIVITY

The most important optically active inorganic compounds, or at least those most intensively studied in recent years, are complexes of transition metals containing two or three chelate rings. The special importance of transition metal complexes in the study of optical activity is due to two things. First, there are several transition metal ions, especially Co^{III}, Cr^{III}, Rh^{III}, Ir^{III}, and Pt^{IV}, that give such kinetically inert complexes that resolution of optical isomers is possible and the rates of racemization are slow enough for spectroscopic studies to be carried out conveniently. Second, detailed studies of optical activity require as a starting point a reasonably clear understanding of the energy levels and electronic spectra of the compounds to be studied, and the studies are greatly facilitated if the compounds have easily observable absorption bands in the visible region, both of which requirements are well met by transition metal compounds.

20-15. Basic Principles and Definitions

Optical activity in a molecule can be expected when and only when the molecule is so structured that it cannot be superposed on its mirror image. Such a molecule is said to be *dissymmetric,* or *chiral.* Chiral, from the Greek word for hand, is useful in emphasizing the left-hand to right-hand type of relationship between nonsuperposable mirror images or enantiomorphs.

The general conditions for the existence of chirality have been specified earlier (p. 47). In short, a dissymmetric or chiral molecule must have either no element or symmetry or, at most, only proper axes of symmetry.

The six-coordinate chelate complexes of the types M(bidentate ligand)$_3$ and *cis*-M(bidentate ligand)$_2$X$_2$, which have symmetries D$_3$ and C$_2$, respectively, fulfill this condition and are the commonest cases in which the "center of dissymmetry" is the metal ion itself.

The simplest way in which optical activity may be observed is doubtless already familiar. It consists in the observation that the plane of polarization of plane-polarized monochromatic light is rotated on passing through a solution containing one or the other—or an excess of one or the other—of two enantiomorphic molecules.

To appreciate more fully this phenomenon and some others closely related to it, the nature of plane-polarized light must be considered in more detail. When observed along the direction of propagation, a beam of plane-polarized light appears to have its electric vector, which oscillates as a sine wave with the frequency of the light, confined to one plane. There is also an oscillating magnetic vector confined to a perpendicular plane, but we shall not be specifically interested in this.

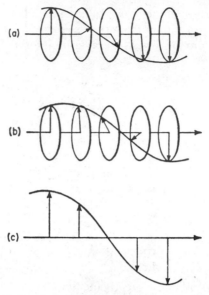

Fig. 20-33. (a) Right-circularly polarized light. (b) Left-circularly polarized light. (c) The plane-polarized resultant of a and b. Horizontal arrow gives direction of propagation, and arrows perpendicular to this direction denote the instantaneous spatial direction of the electric vector. [Reproduced by permission from S. F. Mason, *Chem. Br.*, **1965**, 245.]

It is useful to think of a beam of plane-polarized light as the resultant of two coterminous beams of right- and left-circularly polarized light that have equal amplitudes and are in phase. A circularly polarized beam is one in which the electric vector rotates uniformly about the direction of propagation by 2π during each cycle. Figure 20-33 shows how two such beams, circularly polarized in opposite senses, give a plane-polarized resultant. The most important property of the two circularly polarized components to be noted here is that they are enantiomorphous to each other, that is, one is the nonsuperposable mirror image of the other.

Now, just as there are molecules AB consisting of two separately dissymmetric halves, say A(+),A(−), and B(+),B(−), which are different substances [e.g., the diastereoisomers A(+)B(+) and A(−)B(+)] having numerically different physical properties, so the physical interactions of the two circularly polarized beams with a given enantiomorph of a dissymmetric molecule will be quantitatively different. The two important differences are (1) the refractive indices for left- and right-circularly polarized light n_l and n_r, respectively, will be different, and (2) the molar absorbances ϵ_l and ϵ_r, will be different.

If only the refractive index difference existed, the rotation of the plane of polarization would be explained as shown in Fig. 20-34, since the retarding of one circularly polarized component relative to the other can be seen to have this net effect.

Actually, the simultaneous existence of a difference between ϵ_l and ϵ_r means that the rotated "plane" is no longer strictly a plane. This can be seen in Fig. 20-35; since one rotating electric vector is not exactly equal in length to the other after the two

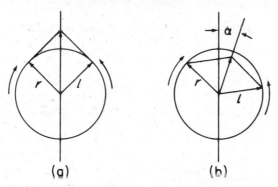

Fig. 20-34. (a) An instantaneous view along the direction of propagation of the two vectors **r** and **l** of the circularly polarized beams and their resultant, which lies in a vertical plane. (b) If $n_l > n_r$, the beam **l** is retarded relative to beam **r**, thus causing the plane of the resultant to be tilted by the angle α.

components have traversed the optically active medium, their resultant describes an ellipse, whose principal axis defines the "plane" of the rotated beam and whose minor axis is equal to the absolute difference $|\epsilon_l - \epsilon_r|$. This difference is usually very small, so that it is a very good approximation to speak of rotating "the plane"; however it can be measured and constitutes the *circular dichroism*. It is most important that both the optical rotation and the circular dichroism are dependent on wavelength, especially in the region of an electronic absorption band of the atom or ion lying at the "center of dissymmetry." Moreover, at a given wavelength the values of $n_l - n_r$ and $\epsilon_l - \epsilon_r$ for one enantiomorph are equal and opposite to those for the other enantiomorph. The variations of $n_l - n_r$ and $\epsilon_l - \epsilon_r$ with wavelength for a pair of enantiomorphs in the region of an absorption band with a maximum at λ_0 are illustrated schematically in Fig. 20-36. The variation of the angle of rotation with wavelength is called *optical rotatory dispersion* (ORD). This, together with the circular dichroism (CD) and the attendant introduction of ellipticity into

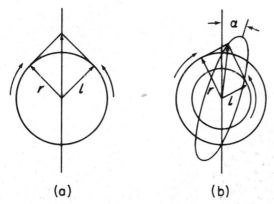

Fig. 20-35.   (a) Same as Fig. 20-34a, (b) If $n_l > n_r$, and also $\epsilon_l > \epsilon_r$, the vectors **r** and **l** will be affected qualitatively as shown, and they will then give rise to a resultant that traces out the indicated ellipse. To make the diagram clear, the quantity $\epsilon_l - \epsilon_r$ is vastly exaggerated compared to real cases.

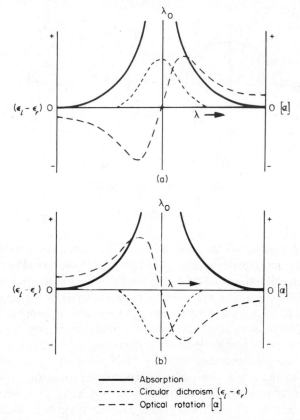

(a)

(b)

———— Absorption

------- Circular dichroism $(\epsilon_l - \epsilon_r)$

– – – – Optical rotation $[\alpha]$

Fig. 20-36. The Cotton effect, as manifested in circular dichroism $\epsilon_l - \epsilon_r$ and optical rotatory dispersion $[\alpha]$, as would be given by a dissymmetric compound with an absorption band centered at λ_0, on the assumption that there are no other absorption bands close by. (a) A positive Cotton effect. (b) A negative Cotton effect.

the rotated beam, are, all together, called the *Cotton effect,* in honor of the French physicist Aimé Cotton, who made pioneering studies of the wavelength-dependent aspects of these phenomena in 1895.

20-16. Applications

Cotton effects are studied today by inorganic chemists for two principal purposes. First, they can be used in a fairly empirical way to correlate the configurations of related dissymmetric molecules, thus to follow the steric course of certain reactions. Second, both theoretical and experimental work is in progress to establish generally reliable criteria for determining spectroscopically the absolute configurations of molecules.

   Before summarizing the results in each of these areas, we mention the types of optically active complex that have been most studied (Fig. 20-37).

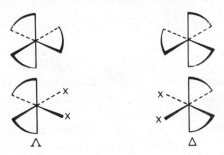

Fig. 20-37. The principal types of optically active chelate molecule and the IUPAC nomenclature. [IUPAC Bulletin No. 33, 1968, p. 68.]

The use of ORD and CD data in making empirical correlations of configurations is increasingly important. The basic idea is simply that similar electronic transitions in similar molecules should have the same signs for CD and ORD effects when the molecules have the same absolute chirality. The chief uncertainty in the conclusions arises from failure to satisfy adequately the criteria of "similarity" in the nature of the electronic transitions and in the structures of the molecules themselves. As an illustration, we use a case where the required similarities are obviously present. Figure 20-38 shows the CD curves for $(+)$ [Co en$_3$]$^{3+}$ and $(+)$[Co(l-pn)$_3$]$^{3+}$, where the $(+)$ signs indicate that these are the enantiomers having positive values of $[\alpha]$ at the sodium D line. Clearly these two complex ions must have the same absolute

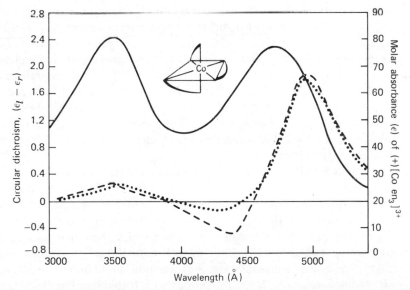

Fig. 20-38. (a) The visible absorption spectrum of $(+)$ [Co en$_3$]$^{3+}$ (solid curve). (b) The circular dichroism of $(+)$[Co en$_3$]$^{3+}$ (dotted curve). (c) The circular dichroism of $(+)$[Co(l-pn)$_3$]$^{3+}$ (dashed curve). The absorption spectrum of $(+)$[Co(l-pn)$_3$]$^{3+}$ is almost identical with that of $(+)$[Co en$_3$]$^{3+}$. The small sketch shows the absolute configuration Λ of $(+)$[Co en$_3$]$^{3+}$ as determined by anomalous scattering of X-rays.

configuration. That of the (+) [Co en$_3$]$^{3+}$ ion has been determined by means of anomalous X-ray scattering as Λ.

It must not be thought that the identical rotational directions at the sodium D line (or at any other single wavelength) could itself have been taken as a criterion of identical absolute configuration. Such relationships fail so frequently as to make the "criterion" useless: for example, (+)[Co|en$_2$(NH$_3$)Cl]$^{2+}$| and |(−)[Co en$_2$-(NCS)Cl]$^+$ have the same configuration.

When the ligands themselves have electronic transitions in the ultraviolet region that persist in perturbed but identifiable form in the complexes, there is a reliable method of assigning absolute configurations directly from observed ORD or, better, CD data without employing any reference compound of known configuration. This is possible because when two or three such ligands are in close proximity in the complex, their individual electric dipole transition moments can couple to produce exciton splittings. Detailed analysis shows that each component of the split band will have a different sign for its circular dichroism and that this sign may be predicted from the absolute configuration by an argument not dependent on numerical accuracy.

The types of ligand that lend themselves to this treatment are bipyridine and phenanthroline, which have strong near-uv transitions. In the case of [Fe(phen)$_3$]$^{2+}$ the correctness of the method has been confirmed by an X-ray crystallographic determination of absolute configuration.

Magnetic Circular Dichroism (MCD).[17,18]   All substances, whether chiral or not, rotate the plane of polarization of light and exhibit ORD and CD effects when placed in a magnetic field that has a component in the direction of propagation of the polarized radiation. Phenomenologically these effects, collectively called the *Faraday effect,* are analogous to ordinary optical activity and the *Cotton effect,* but their interpretation and application to chemical problems are more complicated and have only rather recently received detailed study. Because all substances (including solvents, cell windows, etc.) exhibit the Faraday effect, it is, in general, possible to get interpretable results only by measuring MCD through the electronic absorption bands of the complex of interest. This can be done with apparatus normally used for conventional CD measurements by surrounding the sample cell with a small superconducting solenoid.

The principles of MCD pertinent to metal complexes may be illustrated by considering electronic transitions in an atom. Any orbitally degenerate state (e.g., a P, D, F, etc. state) of an atom consists of components having different angular momenta. The energies of these components differ little if at all in the absence of a magnetic field, but when a magnetic field is applied, they diverge in energy (Zeeman splitting) as shown in Fig. 20-39 for a ^1P state. If this ^1P state is an excited state of a system with a ^1S ground state, the transition is split into $^1S \rightarrow {}^1P_{+1}$ and $^1S \rightarrow {}^1P_{-1}$ components; the $^1S \rightarrow {}^1P_0$ component is forbidden. The crucial point is that the two allowed components are sensitive to the polarization of the light, with the former allowed for right-circularly polarized (rcp) light and the latter for

[17]   P. N. Shatz and A. J. McCaffery, *Q. Rev.,* 1969, **23**, 552.
[18]   P. J. Stevens, *Ann. Rev. Phys. Chem.,* 1974, **25**, 201.

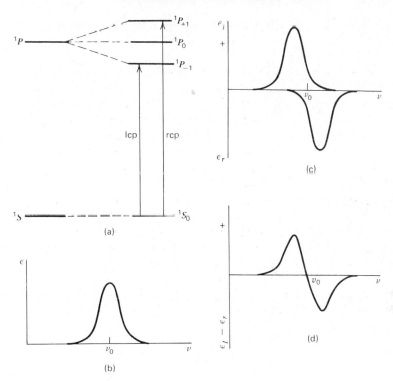

Fig. 20-39. The Faraday effect for an atomic $^1S \rightarrow {}^1P$ transition, showing the origin of a positive Faraday A term in the MCD spectrum. [Adapted from Ref. 17.]

left-circularly polarized (lcp) light. This means that for the $^1S \rightarrow {}^1P$ transition in a magnetic field, if we were to measure the absorptions of rcp and lcp light, we would see one absorption for each component, separated by the Zeeman splitting as shown in Fig. 20-39c. A measurement of the CD spectrum ($\epsilon_l - \epsilon_d$ vs. frequency) would then give a curve of the form shown in 20-39d. This type of curve is known as a Faraday A term and its appearance constitutes proof that the transition has a degenerate upper state. Note that if the nature of the degenerate state is such as to give a Zeeman splitting pattern opposite to that shown, the shape of the CD curve will be reversed· such negative A terms have been observed, and the positive or negative character of an A term provides further information on the assignment of the band.

In the event that the ground state itself is degenerate, hence undergoes Zeeman splitting, we get the results sketched in Fig. 20-40. The resultant CD spectrum, called a Faraday C term, is again a powerful form of evidence in making assignments in appropriate cases. The following example provides an illustration.

For the $Fe(CN)_6^{3-}$ ion there are three charge-transfer bands in visible and near-uv spectra as shown in Fig. 20-41. Theory suggested that all must involve transitions from orbitals of t_{1u} and t_{2u} symmetry on the ligands to the vacancy in the metal d orbitals of t_{2g} symmetry but could not directly show which transition was which.

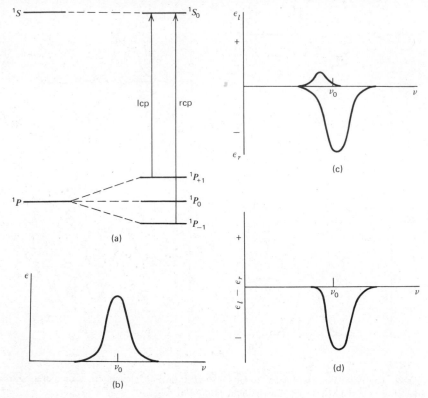

Fig. 20-40. The Faraday effect for atomic $^1P \rightarrow {}^1S$ transition, showing the origin of a Faraday C term in the MCD spectrum. [Adapted from Ref. 17.]

However it was possible to show theoretically that a $t_{1u} \rightarrow t_{2g}$ transition should have a positive MCD effect and a $t_{2u} \rightarrow t_{2g}$ transition should have a negative one. From the observed MCD spectrum (Fig. 20-41), it is then clear that the $t_{2u} \rightarrow t_{2g}$ transition lies between the two $t_{1u} \rightarrow t_{2g}$ transitions. In favorable circumstances the shape and/or magnitude of the MCD effect can show that a transition involves an excited state that is degenerate.

STRUCTURAL AND THERMODYNAMIC CONSEQUENCES OF PARTLY FILLED d SHELLS

As noted at the beginning of this chapter, the chemistry of the transition elements differs strongly from that of the main group elements because of the participation of d and f orbitals in the bonding. One particular aspect of this has to do with the consequences of these shells, particularly the d shells, being partially filled, and also split into subshells that vary in their bonding (or antibonding) character. The remaining sections of this chapter examine the way in which this combination of splitting and partial filling causes remarkable variations in the structural and thermodynamic properties of the d-block elements.

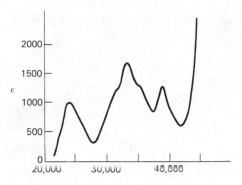

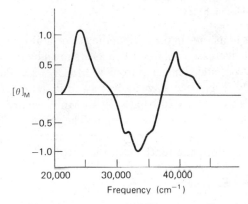

Fig. 20-41. The absorption spectrum (upper) and MCD spectrum (lower) of the $[Fe(CN)_6]^{3-}$ ion showing the three charge-transfer transitions and the signs of their Faraday C terms. [Adapted from Ref. 17.]

20-17. Ionic Radii

The seemingly irregular variations in ionic radii of the dipositive transition metal ions of the $3d$ series are plotted in Fig. 20-42. The points for Cr^{2+} and Cu^{2+} are indicated with open circles because the Jahn-Teller effect, discussed below, makes it impossible to obtain these ions in truly octahedral environments, thus rendering the assessment of their "octahedral" radii somewhat uncertain. A smooth curve has also been drawn through the points for Ca^{2+}, Mn^{2+}, and Zn^{2+} ions, which have the electron configurations $t_{2g}^0 e_g^0$, $t_{2g}^3 e_g^2$, and $t_{2g}^6 e_g^4$, respectively. This curve may be considered to indicate the expected radii for all the ions, if they had their electrons equally distributed over all the d orbitals. In fact, the plot of the actual radii shows two "festoons" as a consequence of the electrons preferring t_{2g} orbitals to e_g orbitals.

The t_{2g} orbitals are approximately nonbonding, whereas the e_g orbitals are distinctly antibonding. Thus in going from Ca^{2+} to Ti^{2+} to V^{2+}, electrons are being added to the nonbonding t_{2g} orbitals only, causing essentially no interference with the metal-ligand bonds. If, however, each of the two and three d electrons were

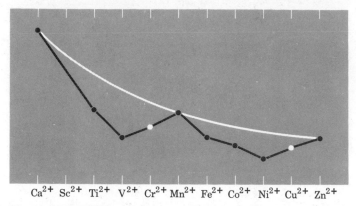

Ca^{2+} Sc^{2+} Ti^{2+} V^{2+} Cr^{2+} Mn^{2+} Fe^{2+} Co^{2+} Ni^{2+} Cu^{2+} Zn^{2+}

Fig. 20-42. The relative ionic radii of divalent ions of the first transition series. The broken line is a theoretical curve explained in the text.

spending two-fifths of its time in the antibonding e_g orbitals, the metal-ligand bond lengths would have decreased only as much as is shown by the smooth curve. The two additional electrons added in going from V^{2+} to Cr^{2+} and then Mn^{2+}, however, enter the e_g orbitals, where they have a severely antibonding effect and cause the metal-ligand bonds to increase in length. With Mn^{2+} we are back at a uniform distribution of d electrons, one in each orbital. The second festoon can obviously be explained in the same way—the sixth, seventh, and eighth d electrons entering t_{2g} orbitals while the ninth and tenth enter the e_g orbitals.

20-18. Jahn-Teller Effects[19]

In 1937 Jahn and Teller proved the theorem that any nonlinear molecular system in a degenerate electronic state will be unstable and will undergo some kind of distortion that will lower its symmetry and split the degenerate state. This theorem has great practical importance in understanding the structural chemistry of certain transition metal ions. To illustrate, we begin with the Cu^{2+} ion. Suppose this ion finds itself in the center of an octahedron of ligands. This ion may be thought of as possessing one hole in the e_g orbitals; hence the electronic state of the ion is a degenerate E_g state. According to the Jahn-Teller theorem then, the octahedron cannot remain perfect at equilibrium but must become distorted in some way.

The dynamic reason for the distortion can be appreciated by simple physical reasoning. Let us suppose that of the two e_g orbitals, it is the $(x^2 - y^2)$ orbital that is doubly occupied, and the z^2 orbital is only singly occupied. In simple CFT terms this means that the four negative charges or the negative ends of dipoles in the xy plane will be more screened from the electrostatic attraction of the Cu^{2+} ion than will the two charges on the z axis. Naturally, then, the latter two ligands will be drawn in somewhat more closely than the other four. If, conversely, the z^2 orbital is doubly occupied and the (x^2-y^2) orbital only singly occupied, the four ligands

[19]   R. Engelman, *The Jahn-Teller Effect in Molecules and Crystals,* Wiley-Interscience, 1972; I. B. Bersuker, *Coord. Chem. Rev.,* 1975, **14,** 357.

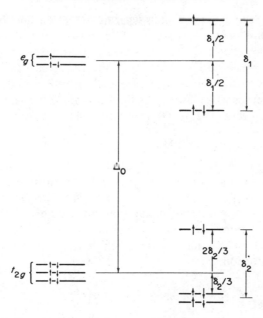

Fig. 20-43. Schematic diagram of the splittings caused by an elongation of an octahedron along one axis. The various splittings are not to the same scale: δ_1 and δ_2 are smaller relative to Δ_0 than is indicated.

in the xy plane will be drawn more closely to the cation than will the other two on the z axis. It is also possible that the unpaired electron could be in an orbital that is some linear combination of (x^2-y^2) and z^2, in which case the resulting distortion would be some related combination of the simple ones considered above. These simple considerations call attention to several important facts relating to the operation of the Jahn-Teller theorem:

1. The theorem predicts only that for degenerate states a distortion must occur; it gives no explicit prediction of the geometrical nature of the distortion or how great it will be. A prediction of the nature and magnitude of the distortion requires detailed calculations of the energy of the entire complex as a function of all possible types and degrees of distortion. Such calculations, however, are extremely laborious and few have been attempted.

2. There is one general restriction on the distortions, namely, if the undistorted configuration has a center of symmetry, so must the distorted equilibrium configuration.

More insight into the energy problem may be had by considering what happens to the d-orbital energies when there occurs a small distortion of the type in which the octahedron becomes stretched along its z axis. The effects are shown in Fig. 20-43. In the interest of clarity, the various splittings are not drawn to scale: both the splittings due to the distortion are much smaller than Δ_0 and, as noted below, δ_2 is much smaller than δ_1. It should also be noted that each of the splittings obeys a center-of-gravity rule. The two e_g orbitals separate so that one goes up as much as the other goes down; the t_{2g} orbitals separate so that the doubly degenerate pair goes down only half as far as the single orbital goes up. It can be seen that for the d^9 case, there is no net energy change for the t_{2g} electrons, since four are stabilized by $\delta_2/3$ and two are destabilized by $2\delta_2/3$. For the e_g electrons, however, a net stabilization occurs, since the energy of one electron is raised by $\delta_1/2$, but two electrons have their energies lowered by this same amount; the net lowering of the electronic energy is thus $\delta_1/2$. It is this stabilization that provides the driving force for the distortion.

It is easy to see from Fig. 20-43 that, for both the configurations $t_{2g}^6 e_g$ and $t_{2g}^6 e_g^3$, distortion of the octahedron will cause stabilization; thus we predict, as could also be done directly from the Jahn-Teller theorem, that distortions are to be expected in the octahedral complexes of ions with these configurations, but not for ions having t_{2g}^6, $t_{2g}^6 e_g^2$, or $t_{2g}^6 e_g^4$ configurations. In addition, it should also be obvious from the foregoing considerations that a high-spin d^4 ion, having the configuration $t_{2g}^3 e_g$, will also be subject to distortion. Some real ions having those configurations which are subject to distortion are:

$$t_{2g}^3 e_g:\ \text{High-spin Cr}^{\text{II}} \text{ and Mn}^{\text{III}}$$
$$t_{2g}^6 e_g:\ \text{Low-spin Co}^{\text{II}} \text{ and Ni}^{\text{III}}$$
$$t_{2g}^6 e_g^3:\ \text{Cu}^{\text{II}}$$

There are ample data to show that such distortions do occur and that they take the form of elongation of the octahedron along one axis. Indeed, in a number of Cu^{II} compounds the distortions of the octahedra around the cupric ion are so extreme that the coordination is best regarded as virtually square and, of course, Cu^{II} forms many square complexes. Specific illustrations of distortions in the compounds of these several ions are mentioned when the chemistry of the elements is described in Chapter 21.

The Jahn-Teller theorem also applies to excited states, although there the effect is a complicated dynamic one because the short life of an electronically excited state does not permit the attainment of a stable equilibrium configuration of the complex. We may consider the $[Ti(H_2O)_6]^{3+}$, $[Fe(H_2O)_6]^{2+}$, and $[CoF_6]^{3-}$ ions. The first of these has an excited state configuration e_g. The presence of the single e_g electron causes the excited state to be split, and it is this that accounts for the broad, flat contour of the absorption band of $[Ti(H_2O)_6]^{3+}$ as seen in Fig. 20-26. In both $[Fe(H_2O)_6]^{2+}$ and $[CoF_6]^{3-}$ the ground state has the configuration $t_{2g}^4 e_g^2$, and the excited state with the same number of unpaired electrons has the configuration $t_{2g}^3 e_g^3$. Thus the excited states of these ions are subject to Jahn-Teller splitting into two components, and this shows up very markedly in their absorption spectra, as Fig. 20-44 shows for $[CoF_6]^{3-}$.

Jahn-Teller distortions can also be caused by the presence of one, two, four, or five electrons in the t_{2g} orbitals of an octahedrally coordinated ion, as can easily be seen in the lower part of Fig. 20-43. If one t_{2g} electron is present, distortion by elongation on one axis will cause stabilization by $\delta_2/3$. Distortion by flattening along one axis would produce a splitting of the t_{2g} orbitals which is just the reverse of that shown in Fig. 20-43 for elongation, and thus would cause stabilization by twice as much, namely, $2\delta_2/3$. The same predictions can obviously be made for the t_{2g}^4 case.

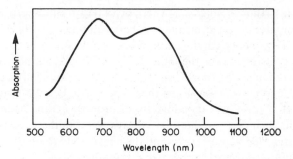

Fig. 20-44. The absorption spectrum of the $[CoF_6]^{3-}$ ion in $K_2Na[CoF_6]$, showing the splitting due to a Jahn-Teller distortion of the excited state with the configuration $t_{2g}^3e^3$.

For a t_{2g}^2 configuration (if we assume—with good reason, since δ_2 will be much less than the electron-pairing energy—that pairing of electrons will not occur) the elongation distortion would be favored, since it will provide a total stabilization of

$$2 \times \frac{\delta_2}{3} = \frac{2\delta_2}{3}$$

whereas the flattening would give a net stabilization energy of only

$$\frac{2\delta_2}{3} - \frac{\delta_2}{3} = \frac{\delta_2}{3}$$

For the t_{2g}^5 case elongation is again predicted to cause the greater stabilization. There is, however, little experimental confirmation of these predictions of Jahn-Teller effects for partially filled t_{2g} shells, since the effects are expected, theoretically, to be much smaller than those for partially filled e_g orbitals. Theory shows that for a given amount of distortion $\delta_2 \ll \delta_1$. Thus the stabilization energies, which are the driving forces for the distortions, are evidently not great enough to cause well-defined, clearly observable distortions in cases of partially occupied t_{2g} orbitals, which concentrate their electrons between the metal-ligand bonds.

In terms of MO theory, the e_g orbitals are antibonding, and a change in the population of these orbitals should strongly affect the metal-ligand bond strength, whereas t_{2g} orbitals are nonbonding in respect to metal-ligand σ interaction, though they may have antibonding or bonding π character. Since σ bonding is usually far more important than π bonding, changes in the t_{2g} population have much less influence on the metal-ligand bond strengths.

Lest the impression be left that the operation of the Jahn-Teller effect is always clear and recognizable structurally, it may be noted that there are a number of cases involving Cu^{II} complexes in which simple tetragonal distortions are *not* observed. For example, in a series of compounds $M_2^I M^{II}[Cu(NO_2)_6]$, the structures are cubic at high temperature but become orthorhombic at lower temperatures.[20] For the analogous Ni compounds the structure is always cubic. The different behavior is evidently due to the operation of Jahn-Teller effects for Cu^{2+}, but the details are

[20]   M. D. Joesten, S. Takagi, and P. G. Lenhert, *Inorg. Chem.,* 1977, **16**, 2680.

subtle. In some orthorhombic cases the $Cu(NO_2)_6$ group approximates to a tetragonally elongated octahedron and in others to a tetragonally flattened octahedron. The X-ray evidence combined with epr results suggests that an elongation is the preferred distortion for the individual $Cu(NO_2)_6$ octahedron and that for such a system the long and short Cu—N distances are typically 2.31 and 2.04 Å, with $g_{\parallel} \approx 2.25$ and $g_{\perp} \approx 2.06$. When there appears to be a flattened octahedron, it is believed that there is a dynamically disordered arrangement of elongated $Cu(NO_2)_6$ units such that one short axis always points in the same crystal direction and the other short O_2NCuNO_2 axis and the long O_2NCuNO_2 axis of the complex are rapidly scrambled over the other two crystal directions. The observed g values are in good agreement with those based on this model, namely $g \approx 2.06$ and $g \approx \frac{1}{2}(2.06 + 2.25) \approx 2.16$, as are the Cu—N distances. Similar observations have been made on some compounds containing the $[Cu\ en_3]^{2+}$ ion[21] and the $[Cu(pyO)_6]^{2+}$ ion.[22]

20-19. Thermodynamic Effects of d-Orbital Splitting

Just as in the case of metal-ligand bond lengths discussed in Section 20-17, the fact that electrons are entering d orbitals of different energies as we progress through a transition series has the effect of causing plots of various measures of thermodynamic stability (e.g., $\Delta H°$ for formation of complexes) to be nonlinear. Examples of this phenomenon are provided by enthalpies of hydration (eq. 20-7) and lattice energies (eq. 20-8).

$$M^{2+}(g) + \infty H_2O(l) = [M(H_2O)_n]^{2+}(aq) \tag{20-7}$$

$$M^{2+}(g) + 2X^-(g) = MX_2(s) \tag{20-8}$$

Data for these two processes in the first transition series are shown graphically in Fig. 20-45. In both cases the metal ions are surrounded octahedrally by the ligands.

An analysis of this behavior may be given in either crystal field or MO terms. According to CFT the d orbitals are split into the t_{2g} set, which are stabilized by $\frac{2}{5} \Delta_0$, and the e_g set, which are destabilized by $\frac{3}{5} \Delta_0$. Thus we can calculate for a given high-spin d^n ion a crystal field stabilization energy (CFSE) in units of Δ_0, by algebraically summing the quantities $+\frac{2}{5}$ for each t_{2g} electron and $-\frac{3}{5}$ for each e_g electron. For example, for a high-spin d^7 ion (Co^{2+}) we have $5(\frac{2}{5}) + 2(-\frac{3}{5}) = \frac{4}{5}$. Table 20-8 shows these numbers for each d^n ion.

For each M^{2+} aqua ion or each MCl_2 compound the value of Δ_0 can be obtained from the observed "d–d" absorption spectrum. The actual CFSEs (in kJ mol^{-1}) can thus be obtained. It is found that when these are subtracted from the experimental values of ΔH_{hyd} and ΔH_{latt}, points are obtained that fall on the smooth curves of Fig. 20-45 to within experimental error. The general rise in energy of the smooth curves reflects the steady increase in the strength of the metal-ligand bonds

[21]   I. Bertini, D. Gatteschi, and A. Scozzafava, *Inorg. Chem.*, 1977, **16**, 1973.
[22]   J. S. Wood, E. de Boer, and C. P. Keijzers, *Inorg. Chem.*, 1979, **18**, 904.

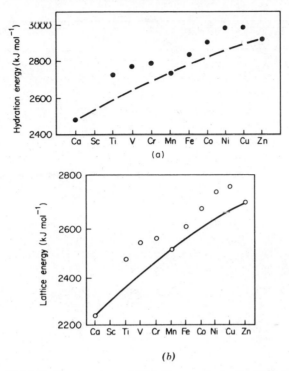

Fig. 20-45. Plots of (a) hydration energies and (b) lattice energies for the dichlorides of the divalent elements of the first transition series (kJ mol^{-1}). Also shown are smooth curves passing through the d^0, d^5, and d^{10} cases.

as the metal ions become smaller and the overlap of $4s$ and $4p$ orbitals, as well as $3d$ orbitals, with the ligand donor orbitals increases.

An alternative analysis of the data in these "double-humped" curves can be given using a more direct molecular orbital analysis.[23] For a qualitative discussion the

TABLE 20-8
Stabilization Energies of d^n Configurations in Octahedral Complexes

n in d^n	CFSE/Δ_0	MOSE/$e_\sigma S_\sigma^2$
0	0	12
1	$2/5$	12
2	$4/5$	12
3	$6/5$	12
4	$3/5$	8
5	0	6
6	$2/5$	6
7	$4/5$	6
8	$6/5$	6
9	$3/5$	3
10	0	0

[23] J. K. Burdett, *J.C.S. Dalton*, **1976**, 1725.

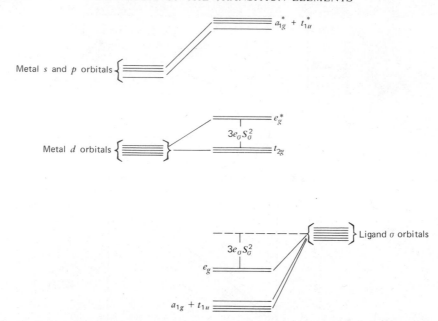

Fig. 20-46. A schematic MO diagram showing the stabilization and destablization energies for the e_g and e_g^* MOs, respectively, and also showing qualitatively the formation of σ bonds by means of the metal s and p orbitals.

AOM may be used, but with attention being paid to the stabilization of the ligand donor orbitals as well as to the destabilization of the metal d orbitals. The key features, for the type of complex in which only σ bonding is important, are shown in Fig. 20-46, which is obtained straightforwardly from the arguments outlined in Sections 20-6 and 20-10. The stabilization of the e_g orbitals and the destabilization of the e_g^* orbitals by $3e_\sigma S_\sigma^2$ is a result of the AOM analysis. Figure 20-46 also indicates that interactions of the metal s and p orbitals with the ligand σ orbitals gives rise to additional bonding MOs (a_{1g}, t_{1u}) and antibonding MOs (a_{1g}^*, t_{1u}^*). Initially we shall be concerned only with the e_g and e_g^* orbitals.

The MO analysis involves calculating for each d^n case the MO analogue of the CFSE, which is the molecular orbital stabilization energy (MOSE). For the d^0 case there are four e_g electrons, each stabilized by $3e_\sigma S_\sigma^2$. In units of $e_\sigma S_\sigma^2$, the stabilization energy is 12. In the d^1 to d^3 cases electrons are added to the non-bonding t_{2g} orbitals, but this does not change the MOSE, which remains 12. At d^4 one electron now enters the e_g^* orbital and the MOSE is diminished by 3 to a value of 9. By obvious extension of this argument all the MOSE values listed in Table 20-8 are obtained.

Figure 20-47 shows again the experimental values of ΔH_{hyd} and indicates how the "double-humped" curve can be explained in MO terms. We see that the experimental points are built on two ascending lines of σ-bonding stabilization, separated by $6e_\sigma S_\sigma^2$ and again by $6e_\sigma S_\sigma^2$ as calculated from the MO diagram in Fig. 20-46. The main reason the "base lines" themselves rise is that there are steadily

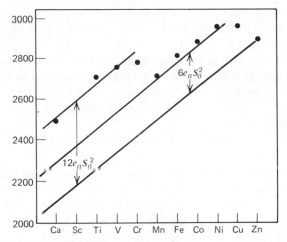

Fig. 20-47. The enthalpies of hydration ΔH_{hyd} of the M^{2+} ions from Ca^{2+} to Zn^{2+} with lines showing how their variations are explained in MO theory.

increasing contributions to the total metal-ligand bonding from the metal s and p orbitals, which get progressively lower in energy, hence progressively more useful for bonding.

Formation Constants of Complexes. It is a fairly general observation that the equilibrium constants for the formation of analogous complexes of the divalent metal ions of Mn through Zn with ligands that contain nitrogen as the donor atom fall in the following order of the metal ions: $Mn^{2+} < Fe^{2+} < Co^{2+} < Ni^{2+} < Cu^{2+} > Zn^{2+}$. There are occasional exceptions to this order, sometimes called the Irving-Williams order, which may be attributed to the occurrence of spin pairing in strong crystal fields. Spin pairing, naturally, affects the relative energies in a different way. The generality of the foregoing order of stability constants receives a natural explanation in terms of CFSEs, or MOSEs. Since the magnitudes of stability constants are proportional to the antilogarithms of standard free energy changes, the order above is also that of the $-\Delta G°$ values for the formation reactions. The standard free energies of formation are related to the enthalpies by the relation

$$-\Delta G° - -\Delta H° + T\Delta S°$$

and there are good reasons for believing that entropies of complex formation are substantially constant in this series of ions. We thus finally conclude that the order of formation constants above is also the order of $-\Delta H°$ values for the formation reactions. Indeed, in a few cases direct measurements of the $\Delta H°$ values have shown this to be true.

The formation of a complex in aqueous solution involves the displacement of water molecules by ligands. If the metal ion concerned is subject to crystal field stabilization, as is, for example, Fe^{2+}, this stabilization will be greater in the complex than in the aqua ion, since the nitrogen-containing ligand will be further along in the spectrochemical series than is H_2O (see p. 663). For Mn^{2+}, however, there is no CFSE in the hexaaqua ion or in the complex, so that complexation cannot cause any increased stabilization. Thus the Fe^{2+} ion has more to gain by combining with the ligands than does Mn^{2+}, and it accordingly shows a greater affinity for them. Similarly, of two ions, both of which experience crystal field stabilization, the one that experiences the greater amount from both the ligand and from H_2O will also experience the larger increase on replacement of H_2O by the ligand. Thus, in general, the order of ions in the stability series follows their order in regard to crystal field stabilization energies.

Octahedral Versus Tetrahedral Coordination. This section considers a phenomenon that is structural but depends directly on d-orbital-splitting energies, namely, variation in the relative stability of octahedral and tetredral complexes for various metal ions. It should be clearly understood that the $\Delta H°$ of transformation of a tetrahedral complex of a given metal ion to an octahedral complex of the same ion, as represented, for instance, by the equation

$$[MCl_4]^{2-} + 6H_2O = [M(H_2O)_6]^{2+} + 4Cl^- \tag{20-9}$$

is a quantity to which the difference in CFSEs of the octahedral and tetrahedral species makes only a small contribution. The metal-ligand bond energies, polarization energies of the ligands, hydration energies, and other contributions all play larger roles, and the calculation of $\Delta H°$ for a particular metal M is a difficult and, as yet, insuperable problem. However if a reaction of this type occurs (actually or hypothetically as a step in a thermodynamic cycle) for a series of metal ions increasing regularly in atomic number, say, the ions Mn^{2+}, Fe^{2+}, $\cdots$, Cu^{2+}, Zn^{2+}, it is reasonable to suppose that the various factors contributing to $\Delta H°$ will all change uniformly *except* the differences in the CFSEs. The latter therefore might be expected to play a decisive role, despite being inherently small parts of the entire $\Delta H°$ in each individual case, in determining an irregular variation in the equilibrium constants for such reactions from one metal ion to another. There are two cases in which experimental data corroborate this expectation.

For reaction 20-9 carried out hypothetically in the vapor phase, the enthalpies have been estimated from thermodynamic data for the metals M in the series Mn^{2+}, Fe^{2+}, $\cdots$, Cu^{2+}, Zn^{2+}. At the same time, from the spectra of the $[M(H_2O)_6]^{2+}$ and $[MCl_4]^{2-}$ ions the values of Δ_0 and Δ_t have been evaluated and the differences between the two CFSEs calculated. Figure 20-48 compares these two sets of quantities. It is evident that the qualitative relationship is very close even though some quantitative discrepancies exist. The latter may well be due to inaccuracies in the ΔH values since these are obtained as net algebraic sums of the independently measured enthalpies of several processes. The qualitatively close agreement between the variation in the enthalpies and the CFSE difference justifies the conclusion that it is the variations in CFSE's that account for gross qualitative stability relations such as the fact that tetrahedral complexes of Co^{II} are relatively stable while those of Ni^{II} are not.

A second illustration of the importance of CFSEs in determining stereochemistry is provided by the *site preference* problem in the mixed metal oxides that have the spinel or inverse spinel structures. These structures have been described on page

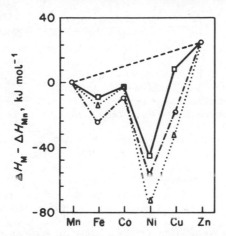

Fig. 20-48. Enthalpies of reaction 20-9 expressed as differences between ΔH for a particular metal and ΔH for the Mn^{2+} compounds. Squares and circles are "experimental" values derived in different ways from thermodynamic data; triangles are values calculated from CFSE differences. In each case the ΔH values are plotted relative to the interpolated value between Mn^{2+} and Zn^{2+} as indicated by the straight line between the points for these two ions. [Reproduced with permission from A. B. Blake and F. A. Cotton, *Inorg. Chem.*, 1964, **3**, 9.]

17, but the reason for the occurrence of inverse spinels was not given. In every case of inversion, an explanation in terms of CFSEs can be given. For example, $NiAl_2O_4$ is inverted, that is, Ni^{2+} ions occupy octahedral interstices and half the aluminum ions occupy tetrahedral ones. It could *not* be predicted that this would necessarily occur simply because the CFSE of Ni^{2+} is much greater in octahedral than in tetrahedral environments, because there are other energy differences that oppose exchanging the sites of Ni^{2+} ions and Al^{3+} ions. However it can be said that Ni^{2+} is the ion most likely to participate in such an inversion and if inversion is to occur at all, it must occur with $NiAl_2O_4$. For $FeAl_2O_4$, in contrast, CFSE differences would again dictate a qualitative preference by Fe^{2+} for the octahedral site, but inversion does not occur.

Another example of the role of CFSEs in determining site preferences is provided by the series of oxides Fe_3O_4, Mn_3O_4, and Co_3O_4, of which only the first is inverted. All energy changes connected with inversion should be similar in the three compounds except the differences in CFSEs, and these are just such as to favor inversion in Fe_3O_4 but not in the others. Thus transfer of the d^5 Fe^{3+} ion involves no change in CFSE, but transfer of the high-spin d^6 Fe^{2+} ion from a tetrahedral to an octahedral hole produces a net gain in CFSE. For Mn_3O_4, transfer of the d^5 Mn^{2+} ion makes no change in CFSE, but transfer of the d^4 Mn^{3+} ion from an octahedral to a tetrahedral hole would decrease the CFSE, so that the process of inverting Mn_3O_4 is disfavored. For Co_3O_4 the transfer of the Co^{2+} ions to octahedral holes would be only slightly favored by the CFSEs, whereas the transfer of a low-spin d^6 Co^{3+} ion from an octahedral hole to a tetrahedral one where it would presumably become high-spin would cause an enormous net decrease in CFSE, so that here, even more than in the case of Mn_3O_4, we do not expect inversion.

General References

Ballhausen, C. J., *Introduction to Ligand Field Theory*, McGraw-Hill, 1962. An excellent book for the inorganic chemist to learn the quantitative aspects of LFT; also reviews experimental data.

Ballhausen, C. J., *Molecular Electronic Structures of Transition Metal Complexes*, McGraw-Hill, 1979. Covers all basic theory for transition metal complexes.

Burdett, J. K., *Adv. Inorg. Chem. Radiochem.*, 1978, **21**, 113–146.

Cotton, F. A., *Chemical Applications of Group Theory*, 2nd ed., Wiley, 1971. Chapter 9 discusses the symmetry basis of LFT.

Dieke, G. H., *Spectra and Energy Levels of Rare Earth Ions in Crystals*, Wiley-Interscience, 1968. Application of CFT to the complex spectra of the lanthanide ions.

Dunn, T. M., in *Modern Coordination Chemistry*, J. Lewis and R. G. Wilkins, Eds., Interscience, 1960, p. 229. Survey of LFT.

Fackler, J. P., Jr., *Symmetry in Chemical Theory*, Dowden, Hutchinson and Ross, 1973. A collection of literature articles, including many classic papers on CFT and MO theory of transition metal complexes.

Ferguson, J., *Prog. Inorg. Chem.*, 1970, **12**, 159. A review of experimental data and assignments of d–d spectra for complexes of the first transition series.

Figgis, B. N., *Introduction to Ligand Fields*, Wiley, 1966. Good introduction to mathematical LFT.

George, P., and D. S. McClure, *Prog. Inorg. Chem.*, 1959, **1**, 38. Effects of crystal field splittings on thermodynamic properties of compounds.

Gerloch, M., and R. C. Slade, *Ligand Field Parameters*, Cambridge University Press, 1973.

Griffith, J. S., *The Theory of Transition Metal Ions*, Cambridge University Press, 1961. A comprehensive mathematical treatise.

Jørgensen, C. K., *Modern Aspects of Ligand Field Theory*, North Holland, 1971.

Jørgensen, C. K., *Prog. Inorg. Chem.*, 1970, **12**, 101. A good survey of the author's very elaborate methods of analyzing and correlating CT spectra.

König, E. and S. Kremer, *Ligand Field Energy Level Diagrams*, Plenum Press, 1977.

Lever, A. B. P., *Inorganic Electronic Spectroscopy*, Elsevier, 1968. Semiquantitative introduction.

Mabbs, F. E. and D. J. Machin, *Magnetism and Transition Metal Complexes*, Chapman Hall, 1973.

Orgel, L. E., *An Introduction to Transition Metal Chemistry, Ligand Field Theory*, 2nd ed., Methuen and John Wiley, 1966. A comprehensive, nonmathematical introduction to LFT.

Owen, J., and J. H. M. Thornley, *Rep. Prog. Phy.*, 1966, **29**, 675. A literature review on crystal field and MO theory of transition metal complexes at a rigorous level.

Schläfer, H. L., and G. Gliemann, *Basic Principles of Ligand Field Theory*, Wiley-Interscience, 1969. Quantitative introduction; references only to about 1965, however.

Theissing, H. H., and P. J. Caplan, *Spectroscopic Calculations for a Multielectron Ion*, Wiley-Interscience, 1966. An exhaustive discussion of the energy levels and spectra of the Cr^{3+} ion.

Williams, A. F., *A Theoretical Approach to Inorganic Chemistry*, Springer, 1979.

CHAPTER TWENTY–ONE

The Elements of the
First Transition Series

GENERAL REMARKS

We discuss in this chapter the elements of the first transition series, titanium through copper. There are two main reasons for considering these elements apart from their heavier congeners of the second and third transition series: (1) In each group (e.g., V, Nb, Ta) the first-series element always differs appreciably from the heavier elements, and comparisons are of limited use; and (2) the aqueous chemistry of the first-series elements is much simpler, and the use of ligand field theory in explaining both the spectra and magnetic properties of compounds has been far more extensive. The ionization enthalpies for the first-series atoms are listed in Appendix 2. The separate Sections 21-A through 21-H summarize the oxidation states and stereochemistries for each element; we do not specify, except in cases of special interest, distortions from perfect geometries that can be expected in octahedral d^1 and d^2 (slight), high-spin octahedral d^4 (two long coaxial bonds), low-spin octahedral d^4 (slight), high-spin octahedral d^6, d^7 (slight), or low-spin octahedral d^7, d^8 molecules (two long coaxial bonds). A few other general features of the elements can be mentioned here.

The energics of the $3d$ and $4s$ orbitals in the neutral atoms are quite similar, and their configurations are $3d^n4s^2$ except for Cr, $3d^54s^1$, and Cu, $3d^{10}4s^1$, which are attributable to the stabilities of the half-filled and the filled d shells, respectively. Since the d orbitals become stabilized relative to the s orbital when the atoms are charged, the predominant oxidation states in ionic compounds and complexes of non-π-bonding ligands are II or greater. Owing to its electronic structure, copper has a higher second ionization enthalpy than the other elements and the Cu^I state is important. The high values of third ionization enthalpies also indicate why it is difficult to obtain oxidation states for nickel and copper greater than II. Although ionization enthalpies give some guidance concerning the relative stabilities of oxidation states, this problem is a very complex one and not amenable to ready generalization. Indeed it is often futile to discuss relative stabilities of oxidation states because some oxidation states may be perfectly stable under certain conditions (e.g., in solid compounds, in fused melts, in the vapor at high temperatures, in absence of air) but nonexistent in aqueous solutions or in air. Thus there is no aqueous chemistry of Ti^{2+}, yet crystalline $TiCl_2$ is stable up to about 400° in absence of air; also, in fused potassium chloride, titanium and titanium trichloride give Ti^{II} as the main species and Ti^{IV} is in vanishingly small concentrations; on the other hand, in aqueous solutions in air only Ti^{IV} species are stable.

However it is sometimes profitable to compare the relative stabilities of ions differing by unit charge when surrounded by similar ligands with similar stereochemistry, as in the case of the $Fe^{3+}-Fe^{2+}$ potentials (p. 755), or with different anions. In these cases, as elsewhere, many factors are usually involved; some of these have already been discussed, but they include (a) ionization enthalpies of the metal atoms, (b) ionic radii of the metal ions, (c) electronic structure of the metal ions, (d) the nature of the anions or ligands involved with respect to their polarizability, donor $p\pi$- or acceptor $d\pi$-bonding capacities, (e) the stereochemistry either in a complex ion or a crystalline lattice, and (f) nature of solvents or other media. In spite of the complexities there are a few trends to be found, namely:

1. From Ti to Mn the highest valence, which is usually found only in oxo compounds or fluorides or chlorides, corresponds to the total number of d and s electrons in the atom. The stability of the highest state decreases from Ti^{IV} to Mn^{VII}. After Mn (i.e., for Fe, Co, and Ni) the higher oxidation states are difficult to obtain.

2. In the characteristic oxo anions of the valence states IV to VII, the metal atom is tetrahedrally surrounded by oxygen atoms, whereas in the oxides of valences up to IV the atoms are usually octahedrally coordinated.

3. The oxides of a given element become more acidic with increasing oxidation state and the halides more covalent and susceptible to hydrolysis by water.

4. In the II and III states, complexes in aqueous solution or in crystals are usually either in four- or six-coordination and, across the first series, generally similar in respect to stoichiometry and chemical properties.

5. The oxidation states less than II, except for Cu^{I}, are found only with π-acid-type ligands or in organometallic compounds.

Finally we reemphasize that the occurrence of a given oxidation state as well as its stereochemistry depend very much on the experimental conditions and that species that cannot have independent existence under ordinary conditions of temperature and pressure in air may be the dominant species under others. In this connection we may note that transition metal ions may be obtained in a particular configuration difficult to produce by other means through incorporation by isomorphous substitution in a crystalline host lattice, for example, tetrahedral Co^{3+} in other oxides, tetrahedral V^{3+} in the $NaAlCl_4$ lattice, as well as by using ligands of fixed geometry such as phthalocyanins.

Although some discussion of the relationships between the first, second, and third transition series is useful, we defer this until the next chapter.

In the discussion of individual elements we have kept to the traditional order; that is, elemental chemistries are considered separately, with reference to their

TABLE 21-1
Standard Potentials (Acid Solution) for +2 and +3 States[a] (V)

	Ti	V	Cr	Mn	Fe	Co	Ni	Cu[b]
$M^{2+} + 2e = M$	-1.6	-1.18	-0.91	-1.18	-0.44	-0.28	-0.24	$+0.34$
$M^{3+} + e = M^{2+}$	-0.37	-0.25	-0.41	$+1.54$	$+0.77$	$+1.84$	—	—

[a] Some potentials depend on acidity and complexing anions, e.g., for Fe^{3+}–Fe^{2+} in $1M$ acids: HCl, $+0.70$; $HClO_4$, $+0.75$; H_3PO_4, $+0.44$; $0.5M$ H_2SO_4, $+0.68$ V.

[b] $Cu^{2+} + e = Cu^+$, $E^0 = +0.15$ V; $Cu^+ + e = Cu$, $E^0 = +0.52$ V.

oxidation state. However it is possible to organize the subject matter from the standpoint of the d^n electronic configuration of the metal. This can bring out useful similarities in spectra and magnetic properties in certain cases and has a basis in theory (Chapter 20); nevertheless the differences in chemical properties of d^n species due to differences in the nature of the metal, its energy levels, and especially the charge on the ion, often exceed the similarities. Nonetheless, such cross-considerations (e.g., in the d^6 series V^{-1}, Cr^0, Mn^I, Fe^{II}, Co^{III}, Ni^{IV}) can provide a useful exercise for students.

Before we consider the chemistries, a few general remarks on the various oxidation states of the first-row elements are pertinent.

The II State. All the elements Ti–Cu inclusive form well-defined binary compounds in the divalent state, such as oxides and halides, which are essentially ionic. Except for Ti, they form well-defined aqua ions $[M(H_2O)_6]^{2+}$; the potentials are summarized in Table 21-1.

In addition, all the elements form a wide range of complex compounds, which may be cationic, neutral or anionic depending on the nature of the ligands.

The III State. All the elements form at least some compounds in this state, which is the highest known for copper, and then only in certain complex compounds. The fluorides and oxides are again generally ionic, although the chlorides may have considerable covalent character, as in $FeCl_3$.

The elements Ti–Co form aqua ions, although the Co^{III} and Mn^{III} ones are readily reduced. In aqueous solution certain anions readily form complex species and for Fe^{3+}, for example, one can be sure of obtaining the $Fe(H_2O)_6^{3+}$ ion only at high acidity (to prevent hydrolysis) and when noncomplexing anions such as ClO_4^- or $CF_3SO_3^-$ are present. There is an especially extensive aqueous complex chemistry of the substitution-inert octahedral complexes of Cr^{III} and Co^{III}.

The trivalent halides, and indeed also the halides of other oxidation states, generally act readily as Lewis acids and form neutral compounds with donor ligands [e.g., $TiCl_3(NMe_3)_2$] and anionic species with corresponding halide ions (e.g., VCl_4^-, FeF_6^{3-}).

The IV State. This is the most important oxidation state of Ti, and the main chemistry is that of TiO_2 and $TiCl_4$ and its derivatives. This is also an important state for vanadium, which forms the vanadyl ion VO^{2+} and many derivatives, cationic, anionic, and neutral, containing the VO group. For the remaining elements Cr–Ni, the IV state is found mainly in fluorides, fluoro complex anions, and cation complexes; however an important class of compounds are the salts of the oxo ions and other oxo species.

The V, VI, and VII States. These occur only as $Cr^{V,VI}$, $Mn^{V,VI,VII}$, and $Fe^{V,VI}$, and apart from the fluorides CrF_5, CrF_6, and oxofluorides MnO_3F, the main chemistry is that of the oxo anions $M^nO_4^{(8-n)-}$. All the compounds in these oxidation states are powerful oxidizing agents.

The Lower Oxidation States I, 0, −I. All the elements form some compounds, at least, in these states but only with ligands of π-acid type. There are few such compounds for Ti, and they are confined essentially to bipyridine complexes [e.g., Ti(bipy)$_3$]. An exception, of course, is the Cu^I state for copper, where some insoluble binary Cu^I compounds such as CuCl are known, as well as complex compounds. In absence of complexing ligands the Cu$^+$ ion has only a transitory existence in water, although it is quite stable in CH_3CN. Finally, it is also possible to classify the chemistry according to the ligands present. An extensive survey of ligand types has been given in Chapters 3 and 4, and Chapters 25 to 27 deal with entire classes of compounds, encompassing all transition elements with specific classes of ligands (e.g., CO and other π acids) and organic moieties.

21-A. TITANIUM[1a]

Titanium is the first member of the d-block transition elements and has four valence electrons $3d^24s^2$. Titanium(IV) is the most stable and common oxidation state; compounds in lower oxidation states, −I, 0, II, and III, are quite readily oxidized to Ti^{IV} by air, water, or other reagents. The energy for removal of four electrons is high, so that the Ti^{4+} ion does not have a real existence and Ti^{IV} compounds are generally covalent. In this IV state there are some resemblances to the elements Si, Ge, Sn, and Pb, especially Sn. The estimated ionic radii (Sn^{4+} = 0.71, Ti^{4+} = 0.68 Å) and the octahedral covalent radii (Sn^{IV} = 1.45, Ti^{IV} = 1.36 Å) are similar; thus TiO$_2$ (rutile) is isomorphous with SnO$_2$ (cassiterite) and is similarly yellow when hot. Titanium tetrachloride, like SnCl$_4$, is a distillable liquid readily hydrolyzed by water, behaving as a Lewis acid, and giving adducts with donor molecules; SiCl$_4$ and GeCl$_4$ do not give stable, solid, molecular addition compounds with ethers, although TiCl$_4$ and SnCl$_4$ do so—a difference that may be attributed to the ability of the halogen atoms to fill the coordination sphere of the smaller Si and Ge atoms. There are also similar halogeno anions such as TiF$_6^{2-}$, GeF$_6^{2-}$, TiCl$_6^{2-}$, SnCl$_6^{2-}$, and PbCl$_6^{2-}$, some of whose salts are isomorphous; the Sn and Ti nitrates M(NO$_3$)$_4$ are also isomorphous. There are other similarities such as the behavior of the tetrachlorides on ammonolysis to give amido species. It is a characteristic of Ti^{IV} compounds that they undergo hydrolysis to species with Ti—O bonds, in many of which

[1a] R. J. H. Clark, *The Chemistry of Titanium and Vanadium,* Elsevier, 1968 (mainly chemistry of compounds); J. Barksdale, *Titanium: Its Occurrence, Chemistry and Technology,* 2nd ed., Ronald Press; R. I. Jaffee and N. E. Promisel, Eds., *The Science, Technology and Applications of Titanium,* Pergamon Press, 1970.

TABLE 21-A-1
Oxidation States and Stereochemistry of Titanium

Oxidation state	Coordination number	Geometry	Examples
Ti^{-1}	6	Octahedral	Ti bipy$_3^-$
Ti^0	6	Octahedral	Ti bipy$_3$, Ti(CO)$_6{}^a$
Ti^{II}, d^2	4	Dist. tetrahedral	$(\eta$-C$_5$H$_5)_2$Ti(CO)$_2$
	6	Octahedral	TiCl$_2$
Ti^{III}, d^1	3	Planar	Ti[N(SiMe$_3)_2]_3$
	5	Tbp	TiBr$_3$(NMe$_3)_2$
	6^b	Octahedral	TiF$_6^{3-}$, Ti(H$_2$O)$_6^{3+}$, TiCl$_3$·3THF
Ti^{IV}, d^0	4^b	Tetrahedral	TiCl$_4$
		Dist. tetrahedral	$(\eta^5$-C$_5$H$_5)_2$TiCl$_2$
	5	Dist. tbp,	K$_2$Ti$_2$O$_5$
		sp	TiO(porphyrin)
	6^b	Octahedral	TiF$_6^{2-}$, Ti(acac)$_2$Cl$_2$
			TiO$_2$, [Cl$_3$POTiCl$_4]_2$
	7	ZrF$_7^{3-}$ type	[Ti(O$_2$)F$_5]^{3-}$
		Pentagonal bipyramidal	Ti$_2$(ox)$_3$·10H$_2$O, TiCp(S$_2$CNEt$_7)_3{}^d$
	8	Dist. dodecahedral	TiCl$_4$diars$_2{}^e$
			Ti(S$_2$CNEt$_2)_4$

[a] In inert gas matrix, *ca.* 100K; cf. R. Busby *et al., Inorg. Chem.,* 1977, **16**, 822.
[b] Most common state.
[c] Distortions occur in some forms of TiO$_2$ and in BaTiO$_3$.
[d] R. C. Fay *et al., Inorg. Chem.,* 1978, **17**, 3498.
[e] See 21-A-II.

there is octahedral coordination by oxygen; Ti—O—C bonds are well known, and compounds with Ti—O—Si and Ti—O—Sn bonds are known.

The stereochemistry of titanium compounds is summarized in Table 21-A-1.

21-A-1. The Element

Titanium is relatively abundant in the earth's crust (0.6%). The main ores are *ilnenite* (FeTiO$_3$) and *rutile,* one of the several crystalline varieties of TiO$_2$. It is not possible to obtain the metal by the common method of reduction with carbon because a very stable carbide is produced; moreover, the metal is rather reactive toward oxygen and nitrogen at elevated temperatures. Because the metal appears to have certain uniquely useful properties, however, several expensive methods for its purification are used. In addition to a proprietary electrolytic method, there is the older Kroll process in which ilmenite or rutile is treated at red heat with carbon and chlorine to give TiCl$_4$, which is fractionated to free it from impurities such as FeCl$_3$. The TiCl$_4$ is then reduced with molten magnesium at ~800° in an atmosphere of argon. This gives metallic titanium as a spongy mass from which the excess of Mg and MgCl$_2$ is removed by volatilization at ~1000°. The sponge may then be fused in an atmosphere of argon or helium in an electric arc and cast into ingots.

Extremely pure titanium can be made on the laboratory scale by the van Arkel-de Boer method (also used for other metals) in which TiI_4[1b] that has been carefully purified is vaporized and decomposed on a hot wire in a vacuum.

The metal has a hexagonal close-packed lattice and resembles other transition metals such as iron and nickel in being hard, refractory (m.p. 1680° ± 10°, b.p. 3260°), and a good conductor of heat and electricity. It is, however, quite light in comparison to other metals of similar mechanical and thermal properties and unusually resistant to certain kinds of corrosion; therefore it has come into demand for special applications in turbine engines and industrial chemical, aircraft, and marine equipment.

Although rather unreactive at ordinary temperatures, titanium combines directly with most nonmetals, for example, hydrogen, the halogens, oxygen, nitrogen, carbon, boron, silicon, and sulfur, at elevated temperatures. The resulting nitride (TiN), carbide (TiC), and borides (TiB and TiB_2) are interstitial compounds that are very stable, hard, and refractory.

The metal is not attacked by mineral acids at room temperature or even by hot aqueous alkali. It dissolves in hot HCl, giving Ti^{III} species, whereas hot nitric acid converts it into a hydrous oxide that is rather insoluble in acid or base. The best solvents are HF or acids to which fluoride ions have been added. Such media dissolve titanium and hold it in solution as fluoro complexes.

TITANIUM COMPOUNDS

21-A-2. The Chemistry of Titanium(IV), d^0

The most important oxidation state for titanium is IV, and we consider its chemistry first.

Binary Compounds.

Halides.   The tetrachloride $TiCl_4$ is one of the most important titanium compounds, since it is the usual starting point for the preparation of most other Ti compounds. It is a colorless liquid (m.p. −23°, b.p. 136°) with a pungent odor. It fumes strongly in moist air and is vigorously, though not violently, hydrolyzed by water:

$$TiCl_4 + 2\,H_2O \rightarrow TiO_2 + 4HCl$$

With some HCl present or a deficit of H_2O, partial hydrolysis occurs, giving oxo chlorides; Raman spectra of the yellow solution of $TiCl_4$ in aqueous HCl indicate that the species present is $[TiO_2Cl_4]^{4-}$ or $[TiOCl_5]^{3-}$ but not $[TiCl_6]^{2-}$, since the oxo chloride $TiOCl_2$ in HCl gives the same spectrum.

$TiBr_4$ and a metastable form[1b] of TiI_4 are crystalline at room temperature and are isomorphous with SiI_4, GeI_4, and SnI_4, having molecular lattices. The fluoride is obtained as a white powder by action of F_2 on Ti at 200°; it sublimes readily and

[1b] E. G. M. Tornqvist and W. F. Libby, *Inorg. Chem.*, 1979, **18**, 1792.

is hygroscopic; its structure is not known. All the halides behave as Lewis acids; with neutral donors such as ethers they give adducts, and with halide ions give the respective halogeno complex anions (see below).

Titanium Oxide; Complex Oxides; Sulfide. The dioxide TiO_2 has three crystal modifications, *rutile, anatase,* and *brookite,* all of which occur in Nature. In rutile, the commonest, the titanium is octahedrally coordinated, and this structure has been discussed on page 16, since it is a common one for MX_2 compounds. In anatase and brookite there are very distorted octahedra of oxygen atoms about each titanium, two being relatively close. Although rutile has been assumed to be the most stable form because of its common occurrence, thermochemical data indicate that anatase is 8 to 12 kJ mol^{-1} more stable than rutile.

The dioxide is used as a white pigment. Naturally occurring forms are usually colored, sometimes even black, owing to the presence of impurities such as iron. Pigment-grade material is generally made by hydrolysis of titanium(IV) sulfate solution[2a] or vapor-phase oxidation of $TiCl_4$ with oxygen. The solubility of TiO_2 depends considerably on its chemical and thermal history. Strongly roasted specimens are chemically inert.

The precipitates obtained on adding base to Ti^{IV} solutions are best regarded as hydrous TiO_2. This substance dissolves in concentrated alkali hydroxide, to give solutions from which may be obtained hydrated "titanates" having formulas such as $M_2^ITiO_3 \cdot nH_2O$ and $M_2^ITi_2O_5 \cdot nH_2O$ but of unknown structure.

There are also a number of Ti_nO_{2n-1} phases, $n = 3$ to 10, in which there are oxygen vacancies and Ti^{III} ions; these have complex magnetic properties.[2b]

The photolysis of chemisorbed water on titanium dioxide yields both H_2 and O_2 and in presence of N_2, the hydrogen evolved forms NH_3.[3]

A considerable number of materials called "titanates"[4a] are known, and some are of technical importance. Nearly all have one of the three major mixed metal oxide structures (p. 16), and indeed the names of two of the structures are those of the titanium compounds that were the first found to possess them, namely, *ilmenite* ($FeTiO_3$), and *perovskite* ($CaTiO_3$). Other titanites with the ilmenite structure are $MgTiO_3$, $MnTiO_3$, $CoTiO_3$, and $NiTiO_3$, and others with the perovskite structure are $SrTiO_3$ and $BaTiO_3$. There are also titanates with the spinel structure such as Mg_2TiO_4, Zn_2TiO_4, and Co_2TiO_4.

BaO and TiO_2 react to form an extensive series of phases[4b] from simple ones such as $BaTiO_3$ (commonly called barium titanate) and Ba_2TiO_4 to $Ba_4Ti_{13}O_{30}$ and $Ba_6Ti_{17}O_{40}$, the general formula being $Ba_xTi_yO_{x+2y}$. All are of technical interest because of their ferroelectric properties, which may be qualitatively understood as follows. The Ba^{2+} ion is so large relative to the small ion Ti^{4+} that the latter can literally "rattle around" in its octahedral hole. When an electric field is applied

[2a] J. F. Duncan and R. G. Richards, *N. Z. J. Sci.,* 1976, **19,** 179, 185.

[2b] J. F. Houlihan and L. N. Mulay, *Inorg. Chem.,* 1974, **13,** 745, and references therein.

[3]   G. N. Schrauzer and T. D. Guth, *J. Am. Chem. Soc.,* 1978, **100,** 7189

[4a] See F. L. Bowden, *Inorganic Chemistry, Transition Elements,* Vol. 6, Chemical Society Specialist Report, 1978, p. 10.

[4b] T. Negas *et al., J. Solid State Chem.,* 1974, **9,** 297.

to a crystal of this material, it can be highly polarized because each of the Ti^{4+} ions is drawn over to one side of its octahedron thus causing an enormous electrical polarization of the crystal as a whole.

The compound Ba_2TiO_4 has discrete, somewhat distorted TiO_4 tetrahedra and the structure is related to that of β-K_2SO_4 or β-Ca_2SiO_4.

TiS_2, like the disulfides of Zr, Hf, V, Nb, and Ta, has a layer structure; two adjacent close-packed layers of S atoms have Ti atoms in octahedral interstices. These "sandwiches" are then stacked so that there are adjacent layers of S atoms. In 1969 it was first reported that Lewis bases such as aliphatic amines can be intercalated between these adjacent sulfur layers, and it has since been shown that qualitatively similar intercalation compounds can be made with all the MS_2 and MSe_2 compounds for M = Ti, Zr, Hf, V, Nb, and Ta.[5a] Many of these have potentially useful electrical properties, including superconductivity, and may be compared with the intercalation compounds of graphite (p. 359)

Even earlier,[5b] alkali metal intercalates M^IMY_2 (where M^I is any alkali metal, M = Ti, Zr, Hf, V, Nb, Ta, and Y = S, Se, Te) were known. The lithium intercalates can be made very smoothly by treating MY_2 with butyllithium.

Titanium(IV) Complexes.

Aqueous Chemistry; Oxo Salts.   There is no simple aquated Ti^{4+} ion because of the high charge-to-radius ratio, and in aqueous solutions hydrolyzed species occur and basic oxo salts or hydrated oxides may be precipitated. Although there have been claims for a titanyl ion TiO^{2+}, its existence is unproved except in TiO(porphyrin), where the Ti—O bond is very short: 1.619 Å.[6] In $TiOSO_4 \cdot H_2O$ there are infinite zigzag—Ti—O—Ti—O—chains with SO_4^{2-} ions and H_2O coordinated to complete octahedra about the Ti atoms. $TiO(acac)_2$ is a dimer with a $Ti(\mu$-$O)_2Ti$ ring,[7] and $(NH_4)_2TiO(C_2O_4)_2 \cdot H_2O$ contains cyclic tetrameric anions with a central eight-ring $(-Ti-O-)_4$. The bridging O atoms occupy two cis positions of an octahedron about Ti, with the oxalato ions at the other four.[8]

For aqueous solutions of Ti^{IV} in $2M$ $HClO_4$, ion-exchange studies are consistent with the presence of monomeric 2+ ions but not higher polymers.[9] Whether the 2+ species is TiO^{2+} or $Ti(OH)_2^{2+}$ and the extent to which there is further hydrolysis are unresolved questions:

$$TiO^{2+} + H_2O = TiO(OH)^+ + H^+$$
$$Ti(OH)_2^{2+} + H_2O = Ti(OH)_3^+ + H^+$$

In sulfuric acid these and other species such as $Ti(OH)_3HSO_4$ and $Ti(OH)_2HSO_4^+$ have been invoked. On increase in the pH, polymerization and further hydrolysis eventually give colloidal or precipitated hydrous TiO_2.

[5a] See R. R. Chianelli *et al., Inorg. Chem.,* 1975, **14,** 1691.

[5b] See D. W. Murphy *et al., Inorg. Chem.,* 1976, **15,** 17 for references.

[6]   P. N. Dwyer *et al., Inorg. Chem.,* 1975, **14,** 1782.

[7]   G. D. Smith, C. N. Caughlan, and J. A. Campbell, *Inorg. Chem.,* 1972, **11,** 2989.

[8]   G. M. H. van de Velde, S. Harkema, and P. J. Gellings, *Inorg. Nucl. Chem. Lett.,* 1973, **9,** 1169.

[9]   J. D. Ellis and A. G. Sykes, *J. C. S. Dalton,* **1973,** 537.

The halides are excellent starting materials for the preparation of complexes and other compounds because they are very labile. When $TiCl_4$ and $TiBr_4$ are mixed, nmr[10] and Raman[11] studies show that all five $TiCl_nBr_{4-n}$ species promptly appear in essentially statistical proportions and exchange halogens at rates of $>10^4$ sec^{-1}. Similarly when $TiCl_4$ and TiF_4 are mixed in $CH_3OCH_2CH_2OCH_3$ (glyme), all possible mixed octahedral molecules cis-$Ti(glyme)Cl_nF_{4-n}$ are quickly formed.[12]

Titanium tetrachloride accelerates a number of organic reactions,[13] acting either as a Lewis acid or as a powerful dehydrating agent, e.g.,

$$HCO_2H + C_2NH_2 \rightarrow HCONHC_2 + H_2O$$

Anionic Complexes. The solutions obtained by dissolving Ti, TiF_4, or hydrous oxides in aqueous HF contain various fluoro complex ions but predominantly the very stable TiF_6^{2-} ion, which can be isolated as crystalline salts, in which it is distorted octahedral.[14] In moderately polar solvents (e.g., SO_2, CH_3CN) a variety of mono- and polynuclear and, six-coordinate fluoro complexes, such as $[TiF_5(solv)]^-$ and $Ti_2F_9^-$ are formed.[15a]

Genuine five-coordinate anions $TiCl_5^-$ and $TiBr_5^-$ appear to be formed in CH_2Cl_2 solution on mixing equimolar amounts of R_4NX and TiX_4. Large cations (Bu_4N^+, Ph_4As^+) precipitate these TiX_5^- ions, but with smaller cations (Pr_4N^+, Et_4N^+) the dinuclear $Ti_2X_{10}^{2-}$ ions are formed.[15b] The reaction of $TiCl_4$ with PCl_5 in $POCl_3$ or $SOCl_2$ gives PCl_4^+ salts of the edge-bridged $Ti_2Cl_{10}^{2-}$ ion and the face-bridged bioctahedral $Ti_2Cl_9^-$ ion.[15c]

In aqueous HCl, $TiCl_4$ gives oxo chloro complexes, but from solutions saturated with gaseous HCl salts of the $[TiCl_6]^{2-}$ ion may be obtained. These are better made by interaction of $TiCl_4$ with KCl or of $TiCl_4$ and $(Me_4N)Cl$ in $SOCl_2$ solution. The green or yellow salts are hydrolyzed in water to oxo species.

Simple Adducts of TiX_4. All the halides form adducts of the types of TiX_4L and TiX_4L_2, many of which are crystalline solids, soluble in organic solvents. They have been extensively studied by spectroscopic techniques. Most such adducts contain octahedrally coordinated titanium, the monoadducts being dimerized through halogen bridges, e.g., $[TiCl_4(OPCl_3)]_2$ and $[TiCl_4(MeCO_2Et)]_2$, whereas the diadducts such as $TiCl_4(OPCl_3)_2$ have cis configurations. With certain chelating ligands such as diars both six-coordinate 21-A-I and eight-coordinate 21-A-II complexes are obtained.

10 R. G. Kidd, R. W. Mathews, and H. G. Spinney, *J. Am. Chem. Soc.,* 1972, **94,** 6686.
11 R. J. H. Clark and C. J. Willis, *Inorg. Chem.,* 1971, **10,** 1118.
12 R. S. Borden, P. A. Loeffler, and D. S. Dyer, *Inorg. Chem.,* 1972, **11,** 2481.
13 T. Mukaiyama, in *New Synthetic Methods,* Vol. 6, Verlag Chemie, 1979; *Angew. Chem., Int. Ed.,* 1977, **16,** 817 (use of $TiCl_4$ in organic synthesis).
14 I. W. Forrest and A. P Lane, *Inorg. Chem.,* 1976, **15,** 265.
15a P. A. W. Dean and B. J. Ferguson, *Can. J. Chem.,* 1974, **52,** 667; H. G. Lee, D. S. Dyer, and R. O. Ragsdale, *J. C. S. Dalton,* **1976,** 1325.
15b C. S. Creaser and J. A. Creighton, *J. Inorg. Nucl. Chem.,* 1979, **41,** 469.
15c T. J. Kistenmacher and G. D. Stucky, *Inorg. Chem.,* 1971, **10,** 122.

(21-A-I)   (21-A-II)

β-Diketonates and Other Chelates. TiCl$_4$ and other TiX$_4$ species react with a great variety chelating ligands, of which Hacac and other β-ketoenols are the most common. Compounds of the type *cis*-Ti(dike)$_2$X$_2$ (X = halogen, pseudo-halogen, OR) are among the most common.[16,17] They are characteristically fluxional and/or labile toward ligand-exchange reactions.[16]

Complexes of the type Ti(dike)Cl$_3$ are also easily prepared. In the solid, Ti(acac)Cl$_3$ is a Cl-bridged dimer giving each metal atom octahedral coordination.[18] Although monomers are present in certain solvents such a nitrobenzene, it is questionable whether these are five-coordinate as suggested[18] or six-coordinate due to solvent coordination. There is indirect evidence that authentic five-coordinate species may exist in some Ti(OR)$_3$(MeNCH$_2$CH$_2$NMe$_2$)-type compounds.[19]

Schiff bases (e.g., sal$_2$enH$_2$) react directly with TiCl$_4$ to give complexes such as 21-A-III. These are also obtained by oxidation of the initial products of reaction of Schiff bases with TiCl$_3$(THF)$_3$, but other more complicated reactions in which Ti^{III} oxidation is accompanied by ring hydrogenation also occur to give Ti^{IV} complexes of partly saturated ligands.[20]

(21-A-III)   (21-A-IV)

Reaction of TiCl$_4$ with K[BH(pz)$_3$] gives presumably octahedral HB(pz)$_3$TiCl$_3$.[21] With NaS$_2$CNR$_2$ the products are TiCl$_{4-n}$(S$_2$CNR$_2$)$_n$, where *n* depends on how much NaS$_2$CNR$_2$ is used; the products are six-; seven-, and eight-coordinate for *n* = 2, 3, and 4, respectively. The six-coordinate ones are *cis*-octahedral, the seven-coordinate ones pentagonal bipyramidal, and the Ti(S$_2$CNR$_2$)$_4$ molecules dodecahedral.[22] The similar Ti(OSCNR$_2$)$_4$ molecule is interesting because it has a dodecahedral structure (21-A-IV) in which the sulfur atoms occupy adjacent positions.[23]

[16]   P. H. Bird, A. R. Fraser, and C. F. Lau, *Inorg. Chem.*, 1973, **12**, 1322; J. F. Harrod and K. R. Taylor, *Inorg. Chem.*, 1975, **14**, 1541.
[17]   R. C. Fay *et al.*, *Inorg. Chem.*, 1974, **13**, 1309; 1975, **14**, 282.
[18]   N. Serpone *et al.*, *Inorg. Chem.*, 1977, **16**, 2381.
[19]   E. C. Alyea and P. H. Merrel, *Inorg. Nuclear Chem. Lett.*, 1973, **9**, 69.
[20]   F. L. Bowden and D. Ferguson, *J. C. S. Dalton*, **1974**, 460.
[21]   J. K. Kouba and S. S. Wreford, *Inorg. Chem.*, 1976, **15**, 2313.
[22]   A. H. Bhat *et al.*, *Inorg. Chem.*, 1974, **13**, 886.
[23]   W. L. Steffen and R. C. Fay, *Inorg. Chem.*, 1978, **17**, 2114, 2120.

The orthophenylenedithiolate complex $[Ti(S_2C_6H_3CH_3)_3]^{2-}$ is octahedral,[24] although many similar dithiolene complexes are trigonal prismatic,[25] which is consistent with the view that the higher-lying d orbitals of Ti do not interact effectively with the lone pairs on the sulfur atoms.

Peroxo Complexes. One of the most characteristic reactions of aqueous Ti^{IV} solutions is the development of an intense orange color on addition of hydrogen peroxide, and this reaction can be used for the colorimetric determination of either Ti or of H_2O_2. Detailed studies of the system show that below pH 1, the main peroxo species is probably $[Ti(O_2)(OH)aq]^+$; in less acid solutions rather complex polymerization processes lead eventually to a precipitate of a peroxohydrate. Various crystalline salts can be isolated, e.g., of the ions $[Ti(O_2)F_5]^{3-}$, $[Ti(O_2)(SO_4)_2]^{2-}$, and $[Ti_2O(O_2)_2(dipic)_2]^{2-}$ (dipic = 2,6-pyridinedicarboxylate). The X-ray structure[26] of the related complexes (21-A-V) shows clearly the correctness of the peroxo formula, since the O—O distances are 1.46 Å.

$$X = H_2O, F^-$$

(21-A-V)

Solvolytic Reactions of $TiCl_4$; Alcoxides and Related Compounds. Titanium tetrachloride reacts with a wide variety of compounds containing active hydrogen atoms, such as those in OH groups, with removal of HCl. The replacement of chloride is usually incomplete in absence of an HCl acceptor such as an amine or the ethoxide ion.

Alcoxides. These have been much studied and are generally typical of other transition metal alcoxides, (Section 4–25). The compounds can be obtained by reactions such as

$$TiCl_4 + 4ROH + 4NH_3 \rightarrow Ti(OR)_4 + 4 NH_4Cl$$
$$TiCl_4 + 3EtOH \rightarrow 2HCl + TiCl_2(OEt)_2 \cdot EtOH$$

The titanium alcoxides are liquids or solids that can be distilled or sublimed and are soluble in organic solvents such as benzene. They are exceedingly readily hydrolyzed by even traces of water, the ease decreasing with increasing chain length of the alkyl group; such reactions give polymeric species with OH or O bridges.

It is characteristic of alcoxides that unless the OR groups are extremely bulky or the compounds are in extremely dilute solutions, they exist as polymers with bridging OR groups and the structure of $[Ti(OC_2H_5)_4]_4$ shown in Fig. 4–5 is representative. In solutions, there may be various degrees and types of polymerization depending on solvent and OR group. In benzene it appears that for primary OR groups trimers are formed, whereas with secondary and tertiary OR, only monomers

24 J. L. Martin and J. Takats, *Inorg. Chem.*, 1975, **14**, 73.
25 M. Cowie and M. J. Bennett, *Inorg. Chem.*, 1976, **15**, 1584, 1589, 1595.
26 D. Schwarzenbach, *Helv. Chim. Acta,* 1972, **55**, 2990.

are present in significant amount. For $Ti(OR)_4$ with primary OR in $CDCl_3$ solution ^{13}C nmr indicates strong association to trimers and further aggregation of the trimers; for secondary OR groups trimer formation is incomplete.[27a] The alcoxides are often referred to commercially as "alkyl titanates" and used in heat-resisting paints where eventual hydrolysis to TiO_2 occurs. The compound $[TiCl_2(OPh)_2]_2$ is a dimer with bridging phenoxide groups.[27b]

Nitrogen Compounds. Nitrogen compounds with N—H bonds appear to react with titanium halides to give initially an adduct, from which hydrogen halide is eliminated by base catalysis. Thus the action of diluted gaseous ammonia on $TiCl_4$ gives the addition product, but with an excess of ammonia ammonolysis occurs and up to three Ti—Cl bonds are converted into Ti—NH_2 bonds. With increasing replacement, the remaining Ti—Cl bonds become more ionic and even liquid ammonia ammonolyzes only three bonds. Primary and secondary amines react in a similar way to give orange or red solids such as $TiCl_2(NHR)_2$ and $TiCl_3NR_2$, which can be further solvated by the amine.

The action of lithium alkylamides $LiNR_2$ on $TiCl_4$ leads to liquid or solid compounds of the type $Ti[N(C_2H_5)_2]_4$, which, like the alcoxides, are readily hydrolyzed by water with liberation of amine. Similar dialkylamides are known also for both Ti^{III} and Ti^{II}. It is characteristic of $Ti(NR_2)_4$ compounds to undergo insertion reactions; with CS_2, for example, the dithiocarbamates $Ti(S_2CNR_2)_4$ are formed.

In some cases NR_2 groups function as bridges in cyclic oligomers; thus TiF_4 reacts with $Ti(NMe_2)_4$ to give $[TiF_2(NMe_2)_2]_4$ in which distorted TiF_3N_3 octahedra are linked by μ-F and μ-NMe_2 groups.[28] The compound $(TiCl_2NSiMe_3)_x$ is polymerized in two ways: first, pairs of $TiCl_2NSiMe_3$ are joined by μ-NSiMe to form four-membered rings, and these are then joined into infinite chains by μ-Cl.[29a] It has been claimed that $(TiCl_2NSiMe)_x$ forms a monomeric dipyridine adduct containing a Ti=N bond, but this has not been substantiated by X-ray work.

Other Titanium(IV) Compounds. The anhydrous nitrate is a volatile compound (m.p. 58°), made by the action of N_2O_5 on hydrated Ti^{IV} nitrate, and having the structure shown in Fig. 21-A-1.[29b] It resembles other anhydrous nitrates (such as those of Sn^{IV}, Co^{III}, and Cu^{II}) in being an extremely efficient nitrating agent for organic compounds. For aromatic compounds the pattern of substitution is similar to that obtained with HNO_3/H_2SO_4 (Section 13–7) where the NO_2^+ ion is the attacking species. With support from an *ab initio* calculation of the electronic structure of $Ti(NO_3)_4$, a mechanism whereby NO_2^+ is generated has been postulated.[30] The sulfate, of unknown structure, is made by the reaction

$$TiCl_4 + 6SO_3 \rightarrow Ti(SO_4)_2 + 2S_2O_5Cl_2$$

[27a] C. E. Holloway, *J. C. S. Dalton,* **1976,** 1050.
[27b] G. Wilkinson *et al., J. C. S. Dalton,* **1978,** 454.
[28]   W. S. Sheldrick, *J. Fluorine Chem.,* 1974, **4,** 415.
[29a] N. W. Alcock, M. Pierce-Butler, and G. R. Willey, *J. C. S. Dalton,* **1976,** 707.
[29b] A. A. McDowell *et al., J. C. S. Chem. Comm.,* **1979,** 427.
[30]   C. D. Garner, I. H. Hillier, and M. F. Guest, *J. C. S. Dalton,* **1975,** 1934.

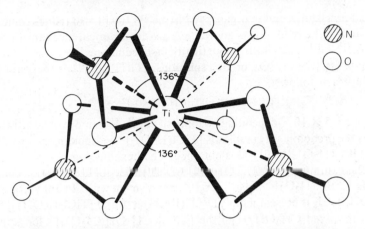

Fig. 21-A-1. The structure of titanium(IV) nitrate. Each NO_3 group is bidentate, and the groups are disposed so that the N atoms form a slightly distorted tetrahedron with D_{2d} symmetry. [Reproduced by permission from C. D. Garner and F. C. Wallwork, *J. Chem. Soc., A*, **1966**, 1496.]

21-A-3. The Chemistry of Titanium(III), d^1

There is an extensive chemistry of solid compounds and Ti^{III} species in solution.

Binary Compounds. The most important compound is the chloride $TiCl_3$, which has several crystalline forms. It can be made by H_2 reduction of $TiCl_4$ vapor at 500 to 1200°; this and other high-temperature methods give the violet α-form. The reduction of $TiCl_4$ by aluminum alkyls in inert solvents gives a brown β-form, which is converted into the α-form at 250 to 300°. Two other forms are known, but these and the α-form have layer lattices containing $TiCl_6$ groups, whereas β-$TiCl_3$ is fibrous with single chains of $TiCl_6$ octahedra sharing edges. The latter is of particular importance because the stereospecific polymerization of propene (Section 30-5) depends critically on the structure of the β form.

The trichloride is oxidized by air and reacts with donor molecules to give adducts of general formula $TiCl_3 \cdot nL (n = 1-6)$. When heated above 500°, $TiCl_3$ disproportionates (see below). The fluoride, bromide, and iodide of Ti^{III} are also known but have little importance.

The *oxide* Ti_2O_3 (corundum structure) is obtained by reducing TiO_2 at 1000° in a stream of H_2. It is rather inert and is attacked only by oxidizing acids. The addition of OH^- ions to aqueous Ti^{III} solutions gives a purple precipitate of the hydrous oxide.

Solution Chemistry and Complexes of Titanium(III). Aqueous solutions of the $[Ti(H_2O)_6]^{3+}$ ion can be readily obtained by reducing aqueous Ti^{IV} either electrolytically or with zinc. The violet solutions reduce oxygen and hence must be handled in a nitrogen or hydrogen atmosphere:

$$\text{"}TiO^{2+}\text{" (aq)} + 2H^+ + e = Ti^{3+} + H_2O \qquad E^0 = ca.\ 0.1\ V$$

The Ti^{III} ion is extensively used as a reducing agent in both inorganic and organic analytical procedures and in organic syntheses.[31a] It will also reduce some oxygen ligands and in the reduction of pyridine N-oxide, a radical mechanism has been proposed.[31b] Ti^{III} is generally a kinetically fast reductant.[32]

In dilute $HClO_4$, H_2SO_4, or HCl solutions, $[Ti(H_2O)_6]^{3+}$ is the main species. It hydrolyzes as follows:

$$[Ti(H_2O)_6]^{3+} = [Ti(OH)(H_2O)_5]^{2+} + H^+$$

with $K = ca.$ 5×10^{-3} depending on ionic strength.[33] In more concentrated HCl solutions the predominant complex is $[TiCl(H_2O)_5]^{2+}$, although on crystallization $trans$-$[TiCl_2(H_2O)_4]Cl \cdot 2H_2O$ is obtained. The hexaqua ion also occurs in other salts such as alums, e.g., $CsTi(SO_4)_2 \cdot 12H_2O$. A variety of bromo/aqua complexes, including $[TiBr_5(H_2O)]^{2-}$ and $[TiBr_6]^{3-}$, are also known.[34] In anhydrous MeOH and EtOH, $TiCl_3$ is present mainly as $[Ti(ROH)_6]^{3+}$ and $[TiCl_2(ROH)_2]^+$ ions,[35] whereas in mixed CH_3OH/H_2O the $[TiCl_2(CH_3OH/H_2O)_4]^+$ ion appears to predominate.[36] EDTA and other ligands of the polyaminocarboxylic acid type react with aqueous Ti^{III} to form both mono- and polynuclear complexes.[37] In pyridine $TiCl_3$ is present mainly as $py_2Cl_2Ti(\mu\text{-}Cl)_2TiCl_2py_2$ according to esr,[38] and it reacts with oxygen to form $Cl_3TiOTiCl_3 \cdot 4py$.[39]

In its complexes Ti^{III} appears to prefer octahedral coordination. Titanium(III) halides readily react with simple donors such as CH_3CN, CH_3CO_2H, or THF to form complexes[40a] of the types $TiCl_3L_3$ and $[TiCl_4L_2]^-$. In $TiCl_3py_3 \cdot py$ the configuration is mer because a fac arrangement is disfavored by interligand repulsion; however the diglyme complex $TiCl_3L$ does have a fac structure.[40b] There are also some $TiCl_3L_2$ species formed with phosphines that may be (but are again structurally uncharacterized) five-coordinate.[41] $TiCl_3(THF)_3$ reacts with Hdike to form $TiCl_2(THF)_2(dike)$.[42a] Although, as noted earlier, Ti^{III} can reduce the ligands, Schiff base complexes can be made, e.g., by Zn reduction of $TiCl_2(sal_2en)$.[42b]

Other Ti^{III} compounds of significance include $Ti[N(SiMe_3)_2]_3$, which has planar, three-coordinate Ti^{III} (as do the V^{III}, Cr^{III}, and Fe^{III} analogues); the bonding in these molecules and the reasons for their stability as three-coordinate species are not well understood.[43]

[31a]   J. E. McMurry, *Acc. Chem. Res.*, 1974, **7**, 281.
[31b]   R. O. Ragsdale *et al.*, *J. C. S. Dalton*, **1976**, 2449.
[32]   T. P. Logan and J. P. Birk, *Inorg. Chem.*, 1973, **12**, 580, 2464.
[33]   P. Chaudhuri and H. Diebler, *J. C. S. Dalton*, **1977**, 596.
[34]   S. E. Adnitt *et al.*, *J. C. S. Dalton*, **1974**, 644.
[35]   B. Pittel and W. H. E. Schwarz, *Z. Anorg. Allg. Chem.*, 1973, **396**, 152.
[36]   I. B. Goldberg and W. F. Goeppinger, *Inorg. Chem.*, 1972, **11**, 3129.
[37]   D. J. Cookson, T. D. Smith, and J. R. Pilbrow, *J. C. S. Dalton*, **1977**, 1396.
[38]   S. G. Carr and T. D. Smith, *J. C. S. Dalton*, **1972**, 1887.
[39]   C. D. Schmulbach, *Inorg. Chem.*, 1974, **13**, 2026.
[40a]   G. R. Hoff and C. H. Brubaker, Jr., *Inorg. Chem.*, 1971, **10**, 2063; L. P. Podmore, P. W. Smith, and R. Stoessiger, *J. C. S. Dalton*, **1973**, 209.
[40b]   M. G. B. Drew and J. A. Hutton, *J. C. S. Dalton*, **1978**, 1176.
[41]   C. D. Schmulbach, C. H. Kolich, and C. C. Hinckley, *Inorg. Chem.*, 1972, **11**, 2841.
[42a]   L. E. Manzer, *Inorg. Chem.*, 1978, **17**, 1552.
[42b]   M. Pasquali *et al.*, *J. C. S. Dalton*, **1978**, 545.
[43]   M. F. Lappert *et al.*, *J. C. S. Dalton*, **1976**, 1737.

TiCl$_4$ reacts with (Ph$_3$P)$_3$Pt to give products that contain [Ph$_3$P)$_3$PtCl]$^+$ ions and polynuclear anions, thought to be Ti$_3$Cl$_{11}^-$ and Ti$_5$Cl$_{19}^-$ that contain both Ti^{III} and Ti^{IV}, but definitive structural data are not available.[44]

Finally, Ti^{III} in a zeolite is said to allow photochemical cleavage of water to yield H$_2$ (cf. TiO$_2$ above).[45]

Electronic Structure. The Ti^{III} ion is a d^1 system, and in an octahedral ligand field the configuration must be t_{2g}. One absorption band is expected ($t_{2g} \rightarrow e_g$ transition), and has been observed in several compounds. The spectrum of the [Ti(H$_2$O)$_6$]$^{3+}$ ion is discussed on page 657. The violet color of the hexaqua ion is attributable to the band that is placed to permit some blue and most red light to be transmitted.

Although a d^1 ion in an electrostatic field of perfect O_h symmetry should show a highly temperature-dependent magnetic moment as a result of spin orbit coupling, with μ_{eff} becoming 0 at 0°K, the combined effects of distortion and covalence (which causes delocalization of the electron) cause a leveling out of μ_{eff}, which has in general been found to vary from not less than about 1.5 BM at 80°K to ~1.8 BM at about 300°K. Room temperature values of μ_{eff} are generally close to 1.7 BM.

For a number of (η^5-C$_5$H$_5$)$_2$Ti^{III}(μ-L)$_2$Ti^{III} (η^5-C$_5$H$_5$)$_2$ systems, and other polynuclear ones, there is a weak magnetic coupling between the Ti^{III} atoms.[46]

21-A-4. The Chemistry of Titanium(II), d^2

Compounds of divalent titanium are few, and Ti^{II} has no aqueous chemistry because of its oxidation by water, although it has been reported that ice-cold solutions of TiO in dilute HCl contain Ti^{II} ions which persist for some time. The well-defined compounds are TiCl$_2$, TiBr$_2$, TiI$_2$, and TiO. The halides are best obtained by reduction of the tetrahalides with titanium:

$$\text{TiX}_4 + \text{Ti} \rightarrow 2\text{TiX}_2$$

or by disproportionation of the trihalides:

$$2\text{TiX}_3 \rightarrow \text{TiX}_2 + \text{TiX}_4$$

(the volatile tetrahalides are readily removed). The equilibria and thermodynamics of these interactions have been studied in detail. In presence of AlCl$_3$, Ti^{II} and Ti^{III} become more volatile, presumably because of the formation of species such as TiAl$_2$Cl$_8$, TiAlCl$_6$, and TiAl$_3$Cl$_{11}$[47a]; a halide Ti$_7$Cl$_{16}$ that contains Ti—Ti bonds has also been obtained.[47b]

The oxide, which is made by heating Ti and TiO$_2$, has the NaCl structure but is normally nonstoichiometric.

44   S. Wongnawa and E. P. Schram, *Inorg. Chem.*, 1977, **16**, 1001.
45   S. M. Kuznicki and E. M. Eyring, *J. Am. Chem. Soc.*, 1978, **100**, 6790.
46   B. F. Fieselmann, D. N. Hendrickson, and G. D. Stucky, *Inorg. Chem.*, 1978, **17**, 2078, and earlier references therein.
47a M. Soslie and H. A. Oye, *Inorg. Chem.*, 1978, **17**, 2473.
47b H. Schäfer et al., *Angew. Chem., Int. Ed.*, 1979, **18**, 325.

There are a few complexes, such as the halides $[TiCl_5]^{3-}$, $[TiCl_4]^{2-}$, and adducts with CH_3CN, pyridine, or other ligands.

The Ti^{2+} ion isolated in place of Na^+ in an NaCl crystal shows the expected d–d transitions of (Section 20–11), viz., $^3T_{1g} \rightarrow {}^3T_{2g}$ and $^3T_{1g} \rightarrow {}^3T_{1g}(P)$, from which $\Delta_0 = 8520$ cm^{-1} and $B = 572$ cm^{-1} are calculated.[48]

21-A-5.   Organometallic Compounds[49]

Organotitanium compounds have been intensely studied, initially mainly because of the discovery by Ziegler and Natta that ethylene and propylene can be polymerized by $TiCl_3$–aluminum alkyl mixtures in hydrocarbons at 25° and 1 atm pressure (Section 30-5). More recently, organic compounds have been found to react with molecular nitrogen and to act as catalysts in a number of other reactions.

Molecular alkyls of both Ti^{IV} and Ti^{III} can be made using bulky, elimination stabilized groups (Section 27-3). Examples are $Ti(CH_2Ph)_4$, $Ti(CH_2SiMe_3)_4$, and $Ti[CH(SiMe_3)_2]_3$.[50] Although CH_3TiCl_3 is stable at 25°, the yellow $TiMe_4$ is unstable above ca. $-40°$. However both compounds form thermally stable adducts with donor ligands, although even these are sensitive to air and water. Similarly, the vinyl compound CH_2=$CHTiCl_3$., though it decomposes rapidly above $-30°$, forms the adducts $C_2H_3TiCl_3(THF)_2$ and $C_2H_3TiCl_3(glyme)$, which decompose only slowly at 25°.[51] The adducts of $MeTiCl_3$ with bidentate ligands such as glyme are six-coordinate with a preferred meridional arrangement of the Cl atoms.[52]

The η^5-C_5H_5 (hereafter Cp) compounds of titanium are among the most important of its organo derivatives. The red, crystalline Cp_2TiCl_2, (m.p. 230°) is the principal starting material for much of the chemistry. It is readily made by the reaction of NaC_5H_5 with $TiCl_4$, and has a quasi-tetrahedral structure (21-A-VI).

(21-A-VI)   (21-A-VII)

It behaves as a homogeneous catalyst for alkene polymerization in presence of aluminum alkyls and has an extremely varied chemistry, involving reduction to Ti^{III} and Ti^{II} species, loss of one ring to give $CpTiX_3$ compounds, and replacement of halogen by other unidentate ligands. Some reactions are shown in Fig. 21-A-2.

[48]   W. E. Smith, *J. C. S. Chem. Comm.*, **1972**, 1121.
[49]   Houben-Weyl, *Methods in Organic Chemistry*, Vol. 13, Pt. 7, Thieme, 1976; P. C. Wailes, R. S. P. Coutts, and H. Weigold, *Organometallic Chemistry of Titanium, Zirconium and Hafnium*, Academic Press, 1974.
[50]   M. F. Lappert *et al.*, *J. C. S. Dalton*, **1978**, 734.
[51]   B. J. Hewitt, A. K. Holiday, and R. J. Puddephat, *J. C. S. Dalton*, **1973**, 801.
[52]   R. J. H. Clark and A. J. McAlles, *J. C. S. Dalton*, **1972**, 604.

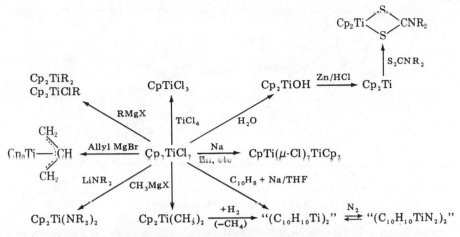

Fig. 21-A-2. Some reactions of dicyclopentadienyl compounds of titanium(II), -(III), and -(IV) (Cp = η^5-C_5H_5).

The alkyl and aryl derivatives e.g., (Cp$_2$TiClR] may be quite stable. The tetracyclopentadienyl (C_5H_5)$_4$Ti, which has the alkyl-like structure (21-A-VII), is fluxional in two senses: (*a*) as in other cases (Section 28-16) the σ-bonded rings undergo rapid shifts, and (*b*) the two types of ring interchange their roles rapidly, so that at 25° all twenty protons give only a single broad line.

The reduction of (C_5H_5)$_2$TiCl$_2$ to Ti^{III} species can be carried out in solutions or in solid state reactions. Cp$_2$Ti^{III}Cl can also be obtained by reaction of TlC$_5$H$_5$ with TiCl$_3$ and converted to Cp$_2$TiBH$_4$ and to alkyls Cp$_2$TiR.

Chiral tetrahedral titanium compounds[53] have been resolved for the first time using species of the type 21-A-VIII.

$$R = CHMePh; R' = Ph$$
$$R = CMe_2Ph; R' = OPh$$

(21-A-VIII)

The red dicarbonyl Cp$_2$Ti(CO)$_2$, which is best made by reduction of Cp$_2$TiCl$_2$ with Al in THF under CO,[54] has a carbenelike chemistry and readily undergoes oxidative-addition reactions[55] (Section 29-3) with loss of CO (Fig. 21-A-3). The acyls have been shown to be η^2 like those of other earlier transition elements (Section 27-4). With hydrogen, there is a remarkable reaction in which CO is converted to CH$_4$ and the blue cluster Cp$_6$Ti$_6$O$_8$ is formed.[56] The latter has an octahedron of

53   J. Tirouflet *et al., J. Am. Chem. Soc.,* 1975, **97**, 6272; *J. Organomet. Chem.* 1975, **101**, 71.
54   B. Demerseman *et al., J. Organomet. Chem.,* 1975, **101**, C24.
55   G. Fachinetti *et al., J. C. S. Dalton,* **1978**, 1398; **1977**, 2297, **1974**, 2433.
56   K. G. Caulton *et al., J. Am. Chem. Soc.,* 1977, **99**, 5829.

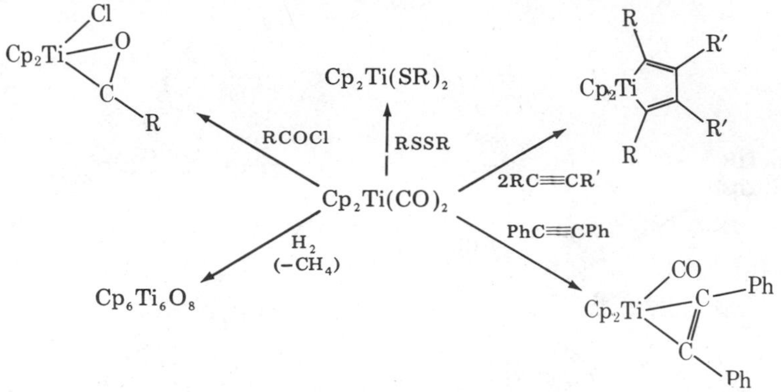

Fig. 21-A-3. Some reactions of $Cp_2Ti(CO)_2$.

Ti atoms, μ_3-O bridges on each face, and a Cp group on each Ti. The chemistry of methyl compounds includes not only Cp_2TiMe_2 and the bridged methyl compound $Cp_2Ti(\mu\text{-}Me_2)AlMe_2$, but also a remarkable series of methylene bridged compounds (Section 27-4) obtained[57] by reactions such as

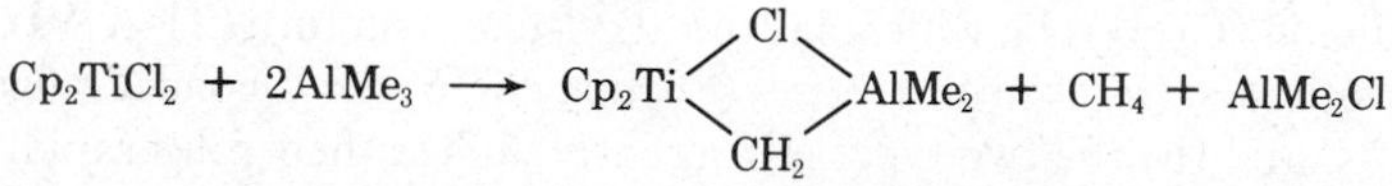

$$Cp_2TiCl_2 + 2AlMe_3 \longrightarrow Cp_2Ti\underset{CH_2}{\overset{Cl}{\big<}}AlMe_2 + CH_4 + AlMe_2Cl$$

These species will act as CH_2 transfer agents, homologating olefins and converting, e.g., R_2CO into $R_2C{=}CH_2$. The dimethyl in presence of Al alkyls also polymerizes olefins.[58] Finally we can note a number of other species such as Cp_2TiS_5 and monocyclopentadienyls like $CpTiCl_3$, $CpTi(S_2CNR_2)_3$, $CpTi(\eta^7\text{-}C_7H_7)$, and the remarkable heteropoly anion[59] $[CpTi(PW_{11}O_{39})]^{4-}$.

The interaction of $TiCl_2$ with NaC_5H_5 first gave a green compound originally thought to be the ferrocene analogue Cp_2Ti. However the chemistry of "titanocene" has turned out to be extraordinarily complicated and for the most part, unique. This green compound has been obtained in several other ways, including hydrogenolysis of Cp_2TiMe_2 and action of various reducing agents on Cp_2TiCl_2.[60]

Although the structure has been much debated, nmr data favor the structure 21-A-IX (X = H), in which there is a bridging fulvalene moiety. Although X-ray

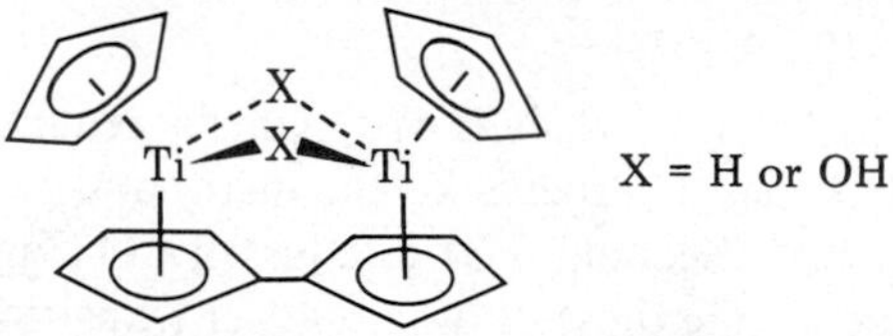

X = H or OH

(21-A-IX)

57   F. N. Tebbe *et al., J. Am. Chem. Soc.*, 1978, **100**, 3611.
58   A. Andreson *et al., Angew. Chem., Int. Ed.*, 1976, **15**, 630.
59   R. K. C. Ho and W. G. Klemperer, *J. Am. Chem. Soc.*, 1978, **100**, 6773.
60   G. P. Pez, *J. C. S. Chem. Comm.*, **1977**, 560; *J. Am. Chem. Soc.*, 1976, **98**, 8072 and references quoted; G. P. Pez and S. C. Kwan, *J. Am. Chem. Soc.*, 1976, **98**, 8079.

$$(\eta^5\text{-}C_5Me_5)_2Ti(CH_3)_2$$

$$\downarrow \begin{array}{c} \text{toluene} \\ 100^\circ \end{array} \Big| -CH_4$$

$$(\eta^5\text{-}C_5Me_5)(\eta^6\text{-}C_5Me_4CH_2)TiCH_3$$

$$\Big\downarrow +H_2, -CH_4$$

$$[(\eta^5\text{-}C_5Me_5)_2Ti]_2 \;\rightleftharpoons\; (\eta^5\text{-}C_5Me_5)_2Ti \xrightarrow{\;H_2\;} (\eta^5\text{-}C_5Me_5)_2TiH_2$$

$$\Big| \searrow CO$$

$$N_? \qquad (\eta^5\text{-}C_5Me_5)_2Ti(CO)_2$$

$$\Big\downarrow$$

$$[(\eta^5\text{-}C_5Me_5)_2Ti]_2N_2 \xrightarrow{\;N_2\;} (\eta^5\ C_5Me_5)_2Ti(N_2)_2$$

Fig. 21-A-4. Some important reactions involving $(\eta^5\text{-}C_5Me_5)_2Ti$.

evidence is still lacking, the corresponding hydroxide 21-A-IX (X = OH) formed on hydrolysis has been structurally characterized.[61]

A more reactive form of "titanocene," which reacts with H_2, N_2, NH_3 etc., is obtained by reduction of Cp_2TiCl_2 with potassium naphthalenide in THF.[60,62] X-Ray study of its THF adduct shows that it is $\mu(\eta^1:\eta^5\text{-cyclopentadien-yl})$tris($\eta$-cyclopentadienyl) dititanium (21-A-X).

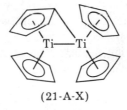

(21-A-X)

The reactivity is probably due to the "open" side of the Ti—Ti bond that can be readily attacked.

Because of the reactivity of the C_5H_5 compounds of titanium due to ring-to-metal hydrogen transfer, the chemistry of analogous C_5Me_5 compounds has been explored.[63] It has proved to be extensive and interesting in its own right. Some of the key reactions are shown in Fig. 21-A-4. The N_2 bridged compound has structure 21-A-XI. In contrast to the chemistry of the C_5H_5 analogue, the pentamethyl compound dissociates into a monomer $(\eta^5\text{-}C_5Me_5)_2Ti$ having two unpaired electrons and reacts with hydrogen to form a simple dihydride, which has an unusually low Ti—H stretching frequency and reversibly loses H_2.

61 L. J. Guggenberger and F. N. Tebbe, *J. Am. Chem. Soc.*, 1976, **98**, 4137.
62 J. N. Armor, *Inorg. Chem.*, 1978, **17**, 203, 213.
63 J. E. Bercaw et al., *J. Am. Chem. Soc.*, 1978, **100**, 3078, and references therein; see also J. D. Zeinstra et al., *J. Organomet. Chem.*, 1979, **170**, 39, for a linear TiN$_2$Ti group.

(21-A-XI)

In contrast to the structural complexity of "titanocene," there are several (η-arene)$_2$Ti species that appear to be simple monomers with "sandwich" structures, although these compounds, with the arenes being benzene, toluene, or mesitylene, have been obtained only by reaction of titanium vapor with the cold arene.[64]

21-B.   VANADIUM

The maximum oxidation state of vanadium is V, but for this there is little similarity, other than in some of the stoichiometry, to the chemistry of elements of the P group. The chemistry of V^{IV} is dominated by the formation of oxo species, and a wide range of compounds with VO^{2+} groups is known. There are four well-defined cationic species, $[V^{II}(H_2O)_6]^{2+}$, $[V^{III}(H_2O)_6]^{3+}$, $V^{IV}O^{2+}$ aq, and $V^V O_2^+$ aq, and none of these disproportionates because the ions become better oxidants as the oxidation state increases; both V^{II} and V^{III} ions are oxidized by air. As with Ti, and in common with other transition elements, the vanadium halides and oxohalides behave as Lewis acids, forming adducts with neutral ligands and halogeno complex ions with halide ions.

The oxidation states and stereochemistries for vanadium are summarized in Table 21-B-1.

21-B-1.   The Element

Vanadium has an abundance in Nature of about 0.02%. It is widely spread, but there are few concentrated deposits. Important minerals are *patronite* (a complex sulfide), *vanadinite* [$Pb_5(VO_4)_3Cl$], and *carnotite* [$K(UO_2)VO_4 \cdot \tfrac{3}{2}H_2O$]. The last of these is more important as a uranium ore, but the vanadium is usually recovered as well. Vanadium also occurs widely in certain petroleums, notably those from Venezuela, and it can be isolated from them as oxovanadium(IV) porphyrins. V_2O_5 is recovered from flue dusts after combustion.

Vanadium occurs as V^{III} and V^{IV} in a few living systems, notably ascidians such

[64]   M. T. Anthony, M. L. H. Green, and E. Young, *J. C. S. Dalton*, **1975**, 1419.

TABLE 21-B-1
Oxidation States and Stereochemistry of Vanadium

Oxidation state	Coordination number	Geometry	Examples
V^{-1}, d^6	6	Octahedral	$V(CO)_6^-$, $Li[V(bipy)_3]\cdot 4C_4H_8O$
V^0, d^5	6	Octahedral	$V(CO)_6$, $V(bipy)_3$, $V[C_2H_4(PMe_2)_2]_3$
V^I, d^4	6	Octahedral	$[V(bipy)_3]^+$
		Tetragonal pyramidal	$\eta^5\text{-}C_5H_5V(CO)_4$
V^{II}, d^3	6	Octahedral	$[V(H_2O)_6]^{2+}$, $[V(CN)_6]^{4-}$
V^{III}, d^2	3	Planar	$V[N(SiMe_3)_2]_3$, $V[CH(SiMe_3)_2]_3$ [a]
	4	Tetrahedral	$[VCl_4]^-$
	5	Tbp	$trans\text{-}VCl_3(SMe_2)_2$, $VCl_3(NMe_3)_2$
	6 [b]	Octahedral	$[V(NH_3)_6]^{3+}$, $[V(C_2O_4)_3]^{3-}$, VF_3
	7	Pentagonal bipyramidal	$K_4[V(CN)_7]\cdot 2H_2O$
V^{IV}, d^1	4	Tetrahedral	VCl_4, $V(NEt_2)_4$, $V(CH_2SiMe_3)_4$
	5	Tetragonal pyramidal	$VO(acac)_2$, $PCl_4^+VCl_5^-$
		?	$[VO(SCN)_4]^{2-}$
		Tbp	$VOCl_2$ $trans\text{-}(NMe_3)_2$
	6 [b]	Octahedral	VO_2(rutile), K_2VCl_6, $VO(acac)_2py$, Vacac_2Cl_2
	8	Dodecahedral	$VCl_4(diars)_2$, $V(S_2CMe)_4$
V^V, d^0	4	Tetrahedral (C_{3v})	$VOCl_3$
	5	Tbp	$VF_5(g)$
		Sp	$CsVOF_4$
	6 [b]	Octahedral	$VF_5(s)$, VF_6^-, V_2O_5 (very distorted, almost tbp with one distant O); $[VO_2ox_2]^{3-}$, V_2S_5 [c]
	7	Pentagonal bipyramidal	$VO(NO_3)_3\cdot CH_3CN$, $VO(Et_2NCS_2)_3$

[a] G. A. Barker et al., J.C.S. Dalton, **1978**, 734.
[b] Most important states
[c] E. Diemann and A. Müller, Z. Anorg. Allg. Chem., 1978, **444**, 181.

as sea squirts and tunicates.[1] A vanadyl complex of a hydroxyiminopropionate ligand has been isolated from the mushroom *Amanita muscaria*.[2]

Very pure vanadium is rare because, like titanium, it is quite reactive toward oxygen, nitrogen, and carbon at the elevated temperatures used in conventional thermometallurgical processes. Since its chief commercial use is in alloy steels and cast iron, to which it leads ductility and shock resistance, commercial production is mainly as an iron alloy, *ferrovanadium*. The very pure metal can be prepared by the de Boer-van Arkel process (p. 694). It is reported to melt at ~1700°, but

[1] W. R. Biggs and J. H. Swinehart, in *Metal Ions in Biological Systems*, H. Sigel, Ed., Dekker, 1976; R. Good and D. T. Sawyer, *Inorg. Chem.*, 1976, **15**, 819; N. M. Senozan, *J. Chem. Educ.*, 1974, **51**, 503.

[2] H. Kneifel and E. Bayer, *Angew Chem., Int. Ed.*, 1974, **13**, 508.

addition of carbon (interstitially) raises the melting point markedly: vanadium containing 10% of carbon melts at ~2700°. The pure, or nearly pure, metal resembles titanium in being corrosion resistant, hard and steel gray. In the massive state it is not attacked by air, water, alkalis, or nonoxidizing acids other than HF at room temperature. It dissolves in nitric acid and aqua regia.

At elevated temperatures it combines with most nonmetals. With oxygen it gives V_2O_5 contaminated with lower oxides, and with nitrogen the interstitial nitride VN. Arsenides, silicides, carbides, and other such compounds, many of which are definitely interstitial and nonstoichiometric, are also obtained by direct reaction of the elements.

VANADIUM COMPOUNDS[3]

21-B-2. Vanadium Halides[4]

The halides of vanadium are listed in Table 21-B-2 together with some of their reactions.

The *tetrachloride* is obtained not only from $V + Cl_2$ but also by the action of CCl_4 on red-hot V_2O_5 and by chlorination of ferrovanadium (followed by distillation to separate VCl_4 from Fe_2Cl_6). It is an oil that is violently hydrolyzed by water to give solutions of oxovanadium(IV) chloride; its magnetic and spectral properties

TABLE 21-B-2
The Halides of Vanadium

$VF_5{}^a$	$\xrightarrow{PCl_3}$	VF_4	$\xrightarrow{\sim 150°{}^b}$	VF_3	$\xrightarrow[115°]{H_2+HF}$	VF_2
Colorless, m.p. 19.5° b.p. 48°		Lime-green, subl > 150°		Yellow-green		Blue
		$25°\Big\uparrow\begin{smallmatrix}HF\\ in\\ CClF_3\end{smallmatrix}$		$600°\Big\uparrow HF(g)$		$600°\Big\uparrow HF(g)$
		$VCl_4{}^a$	$\underset{Cl_2}{\overset{reflux}{\rightleftharpoons}}$	VCl_3	$\xrightarrow{>450°{}^b}$	VCl_2
		Red-brown, b.p. 154°		Violet		Pale green
		$[VBr_4{}^c]$	$\underset{Br_2}{\overset{>-23°}{\rightleftharpoons}}$	$VBr_3{}^a$	$\xrightarrow{>280°}$	VBr_2
		Magenta		Black		Red-brown
		$[VI_4(g)]$	$\longleftarrow$	$VI_3{}^d$	$\xrightarrow{>280°}$	VI_2
				Brown		Dark violet

[a] Made by direct interaction at elevated temperatures, F_2, 300°; Cl_2, 500°; Br_2, 150°.
[b] Disproportionation reaction, (e.g., $2VCl_3 = VCl_2 + VCl_4$).
[b] Isolated from vapor at ~550° by rapid cooling; decomposes above −23°.
[d] Made in a temperature gradient with V at >400°, I_2 at 250 to 300°.

3   R. J. H. Clark, *The Chemistry of Titanium and Vanadium*, Elsevier, 1968; D. Nicholls, *Coord. Chem. Rev.,* 1966, **1**, 379 (extensive review on the coordination chemistry of vanadium compounds); J. O. Hill, I. G. Worsley, and L. G. Hepler, *Chem. Rev.,* 1971, **71**, 127 (thermodynamic properties and oxidation potentials).
4   R. A. Walton, *Prog. Inorg. Chem.,* 1972, **16**, 1 (also Ti, Cr, Re).

confirm its nonassociated tetrahedral nature. It has a high dissociation pressure and loses chlorine slowly when kept, but rapidly on boiling, leaving VCl_3. The latter may be decomposed to VCl_2, which is then stable (m.p. 1350°):

$$2VCl_3(s) \rightarrow VCl_2(s) + VCl_4(g)$$
$$VCl_3(s) \rightarrow VCl_2(s) + \tfrac{1}{2}Cl_2(g)$$

The bromide system is similar but there is only indirect evidence for VI_4 in the vapor phase. The trihalides have the BI_3 structure in which each metal atom is at the center of a nearly perfect octahedron of halogen atoms.

In the crystal the *pentafluoride,* VF_5, and CrF_5 with which it is isostructural, have a cis-bridged infinite polymer structure (21-B-I).

(21-B-I)

This structure persists in the liquid, explaining the high viscosity (cf. SbF_5), although for VF_5 (not CrF_5) there is evidence from vibrational spectra for the presence of VF_5 monomer.[5]

All the halides act as Lewis acids. The oxohalides are discussed below.

21-B-3.   The Chemistry of Vanadium(V)

Vanadium(V) Oxide.   Vanadium(V) oxide is obtained on burning the finely divided metal in an excess of oxygen, although some quantities of lower oxides are also formed. The usual method of preparation is by heating so-called ammonium metavanadate:

$$2NH_4VO_3 \rightarrow V_2O_5 + 2NH_3 + H_2O$$

It is thus obtained as an orange powder that melts at about 650° and solidifies on cooling to orange, rhombic needle crystals. Addition of dilute H_2SO_4 to solutions of NH_4VO_3 gives a brick red precipitate of V_2O_5. This has slight solubility in water ($\sim$0.007 g l^{-1}) to give pale yellow acidic solutions. Although mainly acidic, hence readily soluble in bases, V_2O_5 also dissolves in acids. That the V^V species so formed are moderately strong oxidizing agents is indicated by the evolution of chlorine when V_2O_5 is dissolved in hydrochloric acid; V^{IV} is produced. This oxide is also reduced by warm sulfuric acid. The following standard potential has been estimated:

$$VO_2^+ + 2H^+ + e = VO^{2+} + H_2O \qquad E^0 = 1.0 \text{ V}$$

Vanadates.[6]   Vanadium pentoxide dissolves in sodium hydroxide to give colorless

5   S. D. Brown et al., J. Chem. Phys., 1976, **64,** 260.
6   F. Corighano and S. Di Pasquale, J.C.S. Dalton, **1978,** 1329; S. E. O'Donnell and M. T. Pope, J.C.S. Dalton, **1976,** 2290; D. M. Druskovich and D. L. Kepert, J.C.S. Dalton, **1975,** 947; A. Gaghani et al., Inorg. Chem., 1974, **13,** 1715; J. B. Goddard and A. M. Gonas, Inorg. Chem., 1973, **12,** 575; B. W. Clare et al., J.C.S. Dalton, **1973,** 2476, 2479, 2481.

solutions and in the highly alkaline region, pH > 13, the main ion is VO_4^{3-}. As the basicity is reduced, a series of complicated reactions occurs. A protonated species is first formed:

$$VO_4^{3-} + H_2O \rightleftharpoons VO_3(OH)^{2-} + OH^- \qquad pK = 1.0$$

and this then aggregates into binuclear and subsequently more complex species depending on the concentration and pH.

In the pH range 2–6 the main species is the orange *decavanadate ion*, which can exist in several protonated forms.

$$V_{10}O_{28}^{6-} + H^+ \rightleftharpoons V_{10}O_{27}(OH)^{5-}$$
$$V_{10}O_{27}(OH)^{5-} + H^+ \rightleftharpoons V_{10}O_{26}(OH)_2^{4-}$$
$$V_{10}O_{26}(OH)_2^{4-} + H^+ \rightleftharpoons V_{10}O_{25}(OH)_3^{3-}$$

$$V_{10}O_{25}(OH)_3^{3-} + H^+ \rightleftharpoons V_{10}O_{25}(OH)_4^{2-} \underset{OH^-}{\overset{H^+}{\rightleftharpoons}} VO_2^+$$

The $V_{10}O_{25}(OH)_4^{2-}$ ion is very unstable and with further acid rapidly gives the dioxovanadium(V) ion VO_2^+. In alkaline solution breakup of the $V_{10}O_{28}$ unit is much slower.

The decavanadate ion occurs in salts such as $K_2Zn_2V_{10}O_{28}\cdot16H_2O$ and $Ca_3V_{10}O_{28}\cdot18H_2O$, and its structure is shown in Fig. 21-B-1a; it has ten VO_6 octahedra linked together.[7a]

The structure of the ion in solution is evidently the same as that in crystals according to Raman spectra and to ^{18}O and ^{51}V nmr studies.[7b]

When solutions of the orange dodecavanadate are heated, less soluble vanadates such as $K_3V_5O_{14}$ or KV_3O_8 are precipitated. Other crystalline salts such as KVO_3 and $Na_4V_2O_7\cdot18H_2O$ can be obtained. The structure of the former is shown in Fig. 21-B-1b and has chains of VO_4 tetrahedra sharing corners; $KVO_3\cdot H_2O$ (Fig. 21-B-1c) has chains of linked VO_5 polyhedra. The ion $HV_4O_{12}^{3-}$, obtained by action of V_2O_5 with Bu^tOH in EtOH, has a cyclic anion with four VO_4 units formed by linking corners (cf. tetrametaphosphate).[8]

Vanadium(V) Oxo Halides. These are $VOX_3 (X = F, Cl, or Br)$, VO_2F, and VO_2Cl. VOF_3, made by the action of F_2 on V_2O_5 at 450°, has a sheet structure with both VFV and VF_2V bridges. The oxotrichloride, is made by the action of Cl_2 on $V_2O_5 + C$ at *ca.* 300° and is a yellow, readily hydrolyzed liquid; it is monomeric, with C_{3v} symmetry.[9] Interaction of $VOCl_3$ with alcohols gives alcoxides such as $VO(OBu^t)_3$ that can be regarded as vanadate esters; they can be converted to polymeric carboxylates.[10] The oxohalides give adducts such as $VOCl_3(NEt_3)_2$, $VOCl_3(MeCN)_2$, and the ions VOX_4^-.

The Dioxovanadium(V) Ion; Vanadium(V) Complexes. When vanadates are strongly acidified the ion VO_2^+, probably *cis*-$[VO_2(H_2O)_4]^+$, is formed. It is

[7a] H. T. Evans, Jr., *Perspect. Struct. Chem.*, 1971, **4**, 1.

[7b] O. W. Howarth and M. Jarrold, *J.C.S. Dalton*, **1978**, 503; R. K. Murmann *et al., Inorg. Chem.*, 1978, **17**, 1160; 1977, **16**, 146; W. G. Klemperer and W. Shum, *J. Am. Chem. Soc.*, 1977, **99**, 3544; S. E. O'Donnell and M. T. Pope, *J.C.S. Dalton*, **1976**, 2290.

[8] J. Fuchs *et al., Angew. Chem., Int. Ed.*, 1976, **15**, 374.

[9] R. J. H. Clark, and P. D. Mitchell, *J.C.S. Dalton*, **1972**, 2429.

[10] F. Preuss *et al., J. Inorg. Nucl. Chem.*, 1973, **35**, 3723.

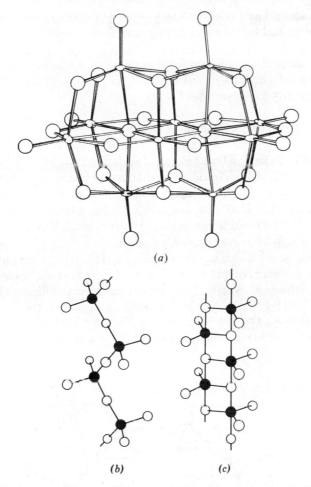

(a)

(b)   (c)

Fig. 21-B-1. The structures in the crystalline state of (a) the decavanadate ion $V_{10}O_{28}^{6-}$, (b) the anion in KVO_3, (c) the anion in $KVO_3 \cdot H_2O$.

complexed when anions other than ClO_4^-, $CF_3SO_3^-$, etc., are present; for example, in HCl the anion is *cis*-$[VO_2Cl_4]^{3-}$.

There are other complex anions such as *cis*-$[VO_2EDTA]^{3-}$ and *cis*-$[VO_2ox_2]^{3-}$. All such species have *cis*-dioxo groups, and they give bands in infrared and Raman spectra characteristic for M=O groups. The cis arrangement for dioxo compounds of metals with no d electrons is preferred over the trans arrangement found in some other metal dioxo systems (e.g., RuO_2^{2+}) because the strongly π-donating O ligands then have exclusive share of one $d\pi$ orbital each (d_{xz}, d_{yz}) and share a third one (d_{xy}), whereas in the trans configuration they would have to share two $d\pi$ orbitals and leave one unused.

The dissolution of V_2O_5 in 30% H_2O_2, or the addition of H_2O_2 to acidic V^V solutions, gives red peroxo complexes in which oxygen atoms in the vanadate are replaced by one or more O_2^{2-} groups. Several complex peroxovanadates have been isolated and some, like $[VO(O_2)_2NH_3]^-$, and $[V(O)(O_2)_2ox]^{3-}$, which have a distorted pentagonal bipyramidal configuration, have been structurally characterized.[11,12]

There are some other non-oxovanadium(V) complexes, mostly those derived via Lewis acid behavior of VF_5 and VOX_3. Thus addition of KF to VF_5 in liquid HF gives the easily hydrolyzed KVF_6.

21-B-4. The Chemistry of Vanadium(IV), d^1

Vanadium(IV) Oxide and Oxo Anions. The dark blue oxide VO_2 is obtained by mild reduction of V_2O_5, a classic method being by fusion with oxalic acid; it is amphoteric, being about equally readily soluble in both noncomplexing acids, to give the blue ion $[VO(H_2O)_5]^{2+}$, and in base. It has a distorted rutile structure; one bond (the $V=O$) is much shorter that the others in the VO_6 unit (note that the Ti—O distances in TiO_2 are essentially equal).

When strong base is added to a solution of $[VO(H_2O)_5]^{2+}$, a gray hydrous oxide $VO_2 \cdot nH_2O$, is formed at *ca.* pH 4. This dissolves to give brown solutions from which brown-black salts (e.g., $Na_{12}V_{18}O_{42} \cdot 24H_2O$) can be crystallized. These contain the ion $V_{18}O_{42}^{2-}$ (Fig. 21-B-2), which is believed to occupy the position in V^{IV} chemistry held by $V_{10}O_{28}^{6-}$ in V^V chemistry.[13] The ion is unstable in dilute solutions giving an uncharacterized monomeric species. Some heteropolymolybdates con-

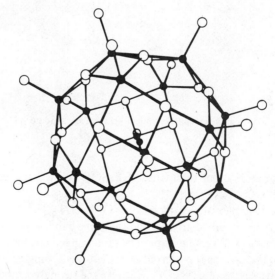

Fig. 21-B-2. The structure of the vanadium(IV) isopolyanion $V_{18}O_{42}^{12-}$.

[11] F. W. B. Einstein *et al., Inorg. Chem.,* 1975, **14**, 1785; 1973, **12**, 829.
[12] N. Vuletić and C. Djordjević *J.C.S. Dalton,* **1973,** 1137.
[13] G. K. Johnson and E. O. Schlemper, *J. Am. Chem. Soc.,* 1978, **100**, 3645.

taining V^{IV} are also known.[14] Fusion of VO_2 with alkaline earth oxides yields solids described as $M^{II}VO_3$ and $M_2^{II}VO_4$.

Oxovanadium(IV) Ion; Complexes.[15] The oxovanadium(IV) ion is obtained by mild reduction of the VO_2^+ ion or by oxidation by air of V^{3+} solutions:

$$VO^{2+} + 2H^+ + e = V^{3+} + H_2O \qquad E^0 = +0.34\ V$$
$$VO_2^+ + 2H^+ + e = VO^{2+} + H_2O \qquad E^0 = +1.0\ V$$

The interaction of V_2O_5 with ethanolic HCl gives a solution containing $VOCl_5^{3-}$ that can conveniently be used as a source of V^{IV} oxo complexes. Also $VOCl_3$ can be reduced by hydrogen to the deliquescent solid $VOCl_2$.

VO^{2+} ion forms some solid compounds with oxo anions in which the VO unit is preserved but coordinated by oxygen atoms from the anions.[16] Examples are α- and $\beta\ VO(SO)_4$, $VO(MoO_4)$, $VO(Se_2O_5)$, and $VO(SiP_2O_8)$. In all these the short V–O distance is in the range 1.59–1.68 Å. The VO units are arranged in linear chains, V=O···V=O···V=O···, with long O···V distances of about 2.5 Å and four equatorial V–O bonds (2.0 Å) that complete the distorted octahedra.

There is an extremely wide range of complexes that, depending on the nature of the ligands, are cationic, neutral, or anionic and either five-coordinate, where the stereochemistry is that of the square pyramid, or six-coordinate, containing a distorted octahedron. Examples are $[VO\ bipy_2Cl]^+$, $VOacac_2$, $[VO(NCS)_4]^{2-}$, and $VO(S_2CNEt_2)_2$.[17]

Although most oxovanadium(IV) complexes are blue, certain Schiff base complexes may vary from yellow to maroon. Earlier suggestions that such colors, together with reductions in the V O stretching frequencies, indicated polymerization or V–O···V–O···V–O interaction are erroneous. Oxovanadium complexes range from distorted trigonal bipyramidal as in the Schiff base complex 21-B-II to square pyramidal as in 21-B-III.[18a] The V–O stretching frequencies and also V–O distances can vary considerably from *ca.* 1.56 Å in 22-B-II to 1.615 Å.[18b]

(22-B-II) (22-B-III)

14 J. J. Altenan *et al., Inorg. Chem.,* 1975, **14**, 417.
15 A. Syamal, *Coord. Chem. Rev.,* 1975, **16**, 309; J. Selbin, *Coord. Chem. Rev.,* 1966, **1**, 293; *Chem. Rev.,* 1965, **65**, 153; *Angew. Chem., Int. Ed.,* 1966, **5**, 712.
16 C. E. Rice *et al., Inorg. Chem.,* 1976, **15**, 345.
17 K. Henrick *et al., J.C.S. Dalton,* **1976**, 26; T. L. Riechel *et al., Inorg. Chem.,* 1976, **15**, 1900.
18a M. Pasquali *et al., J.C.S. Dalton,* **1977**, 139.
18b R. D. Bereman and D. Nalewajek, *J. Inorg. Nucl. Chem.,* 1978, **40**, 1313.

Several five-coordinate species, though not the Schiff base ones,[19] commonly will add ligands, becoming octahedral. Thus the hydrate $VOSO_4 \cdot 5H_2O$ has an octahedral structure with one H_2O *trans* to O at 2.22 Å, three other H_2O molecules *cis* at 2.04 Å, and the oxygen of the monodentate OSO_3^{2-} at 1.98 Å in the fourth *cis* position. The V=O stretching frequency is quite sensitive to the nature of the *trans* ligand, and donors that increase the electron density on the metal thereby reduce its acceptor properties toward O, hence lower the V—O multiple-bond character and the stretching frequency. The epr and electronic spectra of VO species are sensitive also to solvents for the same reason, although H-bonding solvents appear to form hydrogen bonds to the V=O group rather than occupy the sixth site.

Because of the strong VO π bonding in oxovanadium(IV) compounds, interpretation of the electronic spectra is not as simple as it would be for an ordinary octahedral complex. A further result of this π bonding is the strong trans effect labilizing groups trans to oxygen, e.g., in $[VO(H_2O)_5]^{2+}$.

The oxovanadium(IV) moiety can also be bound to pentaammine ruthenium(II) in a manner similar to N_2, CH_3CN, etc., (Section 22-F-4); $[Ru(NH_3)_5H_2O]^{2+}$ reacts[20] to give a very stable cation $[(NH_3)_5Ru^{II}-O-V^{IV}(H_2O)_n]^{4+}$. Similar species but with $V^{II}-O-V^{IV}$ bonds have been invoked as intermediates in electron transfer reactions[21] between, e.g., $[V(H_2O)_6]^{2+}$ and $[VO(H_2O)_5]^{2+}$ or $[Vedta]^{2-}$ and $[VOedta]^{2-}$.

Non-oxo Vanadium(IV) Compounds.   There are numerous adducts of the halides VX_4, including neutral ones such as $VCl_4(PMe_3)_2$ and anions such as VCl_5^- obtained by reaction of VCl_4 with Me_4NCl in $SOCl_2$.

Tetrahedral compounds such the $V(CH_2SiMe_3)_4$, $V(NR_2)_4$, or $[V(OR)_4]_2$ can be made from VCl_4. The dialkylamides undergo insertion reactions and with CS_2 give the dodecahedral *dithiocarbamates* [e.g., $V(S_2CNEt_2)_4$]; the latter can be made in other ways.[22]

There is then a variety of Schiff base complexes such as $VCl_2(sal_2en)$ and the catecholate $V(O_2C_6H_4)_2$.[23]

21-B-5.   The Chemistry of Vanadium(III), d^2

Vanadium(III) Oxide.   This black, refractory substance is made by reduction of V_2O_5 with hydrogen or carbon monoxide. It has the corundum structure but is difficult to obtain pure, since it has a marked tendency to become oxygen deficient without change in structure. Compositions as low in oxygen as $VO_{1.35}$ are reported to retain the corundum structure.

[19]   C. E. Mannix and A. P. Zipp, *J. Inorg. Nucl. Chem.*, 1979, **41**, 59.
[20]   H. De Smedt *et al.*, *Inorg. Chem.*, 1974, **13**, 91.
[21]   F. J. Kristine *et al.*, *J.C.S. Chem. Comm.*, **1976**, 994.
[22]   O. Piovesana and G. Cappuccilli, *Inorg. Chem.*, 1972, **11**, 1543; M. Bonamico *et al.*, *J.C.S. Dalton*, **1974**, 1298; D. C. Bradley *et al.*, *J.C.S. Dalton*, **1973**, 2228.
[23]   J. P. Wilshire and D. T. Sawyer, *J. Am. Chem. Soc.*, 1978, **100**, 3972; M. Pasquali *et al.*, *J.C.S. Chem. Comm.*, **1975**, 534; A. A. Diamantis *et al.*, *J.C.S. Chem. Comm.*, **1976**, 264.

V_2O_3 is entirely basic and dissolves in acids to give solutions of the V^{III} aqua ion or its complexes. From these solutions addition of OH^- gives the hydrous oxide, which is very easily oxidized in air.

The Aqua Ion and Complexes. The blue aqua ion $[V(H_2O)_6]^{3+}$ can be obtained as above or by electrolytic or chemical reduction of V^{IV} or V^V solutions. Such solutions, and also others, of V^{III} are subject to aerial oxidation in view of the potential

$$VO^{2+} + 2H^+ + e = V^{3+} + H_2O \qquad E^0 = 0.34 \text{ V}$$

The ion hydrolyzes partially to VO^+ and $V(OH)^{2+}$.

When solutions of V^{2+} and VO^{2+} are mixed, V^{3+} is formed, but VOV^{4+}, a brown intermediate species that has an oxo bridge, occurs; this is similar to a chromium(III) species $CrOCr^{4+}$ obtained when Cr^{2+} is oxidized under conditions where a Cr^{IV} complex might be expected—by the two electron oxidant Tl^{3+}.

Vanadium(III) forms a number of complex ions, mostly octahedral e.g., $[V(H_2O)_6]^{3+}$, $[VCl_2(MeOH)_4]^+$,[24] $[Vox_3]^{3-}$, $[V(NCS)_6]^{3-}$, and $V[S_2P(OEt)_2]_3$, but a seven-coordinate cyanide complex $K_4[V(CN)_7]\cdot2H_2O$, can be obtained as red crystals by action of KCN on VCl_3 in dilute HCl.[25]

The $[V(H_2O)_6]^{3+}$ ion occurs in *alums* $M^IV(SO_4)_2\cdot12H_2O$. The ammonium alum is obtained as air-stable, blue-violet crystals by electrolytic reduction of NH_4VO_3 in H_2SO_4. The hydrated halides $VX_3\cdot6H_2O$, have the structure *trans*-$[VCl_2(H_2O)_4]Cl\cdot2H_2O$ as found in similar hydrates of Fe^{III} and Cr^{III}.[26] The bromide and some bromo complexes can be made by heating V_2O_5 with ethanolic HBr (as noted above, HCl gives VO^{2+} species).

The electronic structure of an octahedrally coordinated d^2 ion, of which V^{III} is the example *par excellence,* is discussed on page 660. It need be added here only that data for a number of V^{III} octahedral complexes, such as $V(H_2O)_6^{3+}$, VF_6^{3-}, $V(C_3H_2O_4)_3^{3-}$, and V^{3+} substituted into α-Al_2O_3, have been interpreted satisfactorily in terms of the ligand field model, although in general it has been found necessary to take account of the effects of trigonal distortion (to D_{3d}) of the basically octahedral field.

Other Vanadium(III) Complexes. As with other halides, there are adducts such as $VX_3(NMe_3)_3$ and VX_3py_3, and anionic species (e.g., VCl_4^- and $V_2Cl_9^{3-}$).

21-B-6. The Chemistry of Vanadium(II)

The dipositive state of vanadium is the least known of its oxidation states. The black oxide VO has a rock salt type lattice, but it shows a marked tendency to nonstoichiometry, being obtainable with anywhere from ~45 to ~55 at. % oxygen. It has a metallic luster and rather good electrical conductivity of a metallic nature. There is probably considerable V—V bonding. The oxide is basic and dissolves in mineral acids, giving V^{II} solutions.

24 Y. Doi and M. Tsutsui, *J. Am. Chem. Soc.,* 1978, **100**, 324.
25 R. A. Levenson, *et al., Inorg. Chem.,* 1974, **13**, 2761.
26 F. Donovan *et al., J.C.S. Dalton,* **1976**, 1741; **1975**, 894.

Aqueous Solutions, Salts, and Complexes. Electrolytic or zinc reduction of acidic solutions of V^V, V^{IV} or V^{III} produces violet air-sensitive solutions containing the $[V(H_2O)_6]^{2+}$ ion. These are strongly reducing (Table 21-1) and are oxidized by water with evolution of hydrogen even though the standard potential V^{3+}/V^{2+} would indicate otherwise. The oxidation of V^{2+} by air is complicated and appears to proceed in part by direct oxidation to VO^{2+} and in part by way of an intermediate species of type VOV^{4+}, noted earlier.

Several crystalline salts contain the $[V(H_2O)_6]^{2+}$ ion, although the hydrate $VCl_2 \cdot 4H_2O$ is actually $trans$-$VCl_2(H_2O)_4$. The most important are the sulfate $VSO_4 \cdot 6H_2O$, which is formed as violet crystals on addition of ethanol to reduced sulfate solutions, and the double sulfates (Tutton salts) $M_2[V(H_2O)_6](SO_4)_2$, where $M = NH_4^+$, K^+, Rb^+, or Cs^+. From the V^{2+} solutions amine complexes, e.g., $[Ven_3]Cl_2 \cdot H_2O$, can be isolated.

The electronic absorption spectra are consistent with octahedral aqua ions both in crystals and in solution, and the energy level diagram is analogous to that for Cr^{III} (p. 730). The magnetic moments of the sulfates lie close to the spin-only value (3.87 BM).

In aqueous solution $[V(H_2O)_6]^{2+}$ is kinetically inert because of its d^3 configuration (cf. Cr^{3+}), and substitution reactions are relatively slow. Although F^- and SCN^- form weak complexes, there is little evidence for complexing with Cl^-, Br^-, I^-, or SO_4^{2-}. Reorganization of the precursor complex is believed to be the rate-determining step in many reductions by V^{2+}, since the rates of redox reactions are similar to those of substitution, e.g.,

$$V(H_2O)_6^{2+} + SCN^- \rightarrow (H_2O)_5VNCS^+ + H_2O$$

and reduction reactions appear to proceed by a substitution-controlled inner-sphere mechanism (Section 28-10). However in the reduction of $IrCl_6^{2-}$ by V^{2+}, the rate constant is higher by *ca.* 10^5, so that here an outer-sphere mechanism is involved.

Vanadium(II) complex fluorides can be made by fusing VCl_2 and MF^{27}; $K_4[V(NCS)_6] \cdot EtOH$ is obtained[28] by action of KSCN on $VSO_4 \cdot 6H_2O$.

The complex VCl_2py_4 can be used[29] for coupling of activated halides such as benzyl bromide:

$$PhCH_2Br \rightarrow (PhCH_2)_2$$

21-B-7. Carbonyl and Organovanadium Compounds

Carbonyl and organovanadium compounds are known in the -1 to $+5$ oxidation states, but the most extensive chemistry is that of η^5-cyclopentadienyl, arene and carbonyl complexes.

[27]   R. F. Williamson and W. O. J. Boo, *Inorg. Chem.,* 1977, **16**, 646, 649.

[28]   G. Trageser and H. H. Eysel, *Inorg. Chem.,* 1977, **16**, 713.

[29]   T. A. Cooper, *J. Am. Chem. Soc.,* 1973, **95**, 4158.

Unlike titanium, vanadium forms a simple, octahedral carbonyl:

$$VCl_3 + 4Na + 6CO \xrightarrow[\substack{160° \\ 200\ atm}]{diglyme} [Na\ diglyme_2][V(CO)_6] + 3NaCl$$

$$\downarrow \substack{HCl; \\ Et_2O}$$

$$V(CO)_6$$

This green-black compound is unusual in that it is paramagnetic; it undergoes substitution reactions typical of other metal carbonyls (Chapter 25). In the salt $[(Ph_3P)_2N]^+[V(CO)_6]^-$ the anion is octahedral,[30] but in $[V^{II}(thf)_4][V(CO)_6]_2$ there is an unusual linear grouping $(CO)_5VCO—V^{II}—OCV(CO)_5$, with tetrahydrofuran coordinated to the central V^{II}.[31] Substituted anions such as $[V(CO)_4diphos]^-$ and $[V(CO)_5PPh_3]^-$ are known.

The chemistry of σ-alkyls and aryls is less well developed than for some other elements, but $V[CH(SiMe_3)_2]_3$, $V(CH_2SiMe_3)_4$, and $VO(CH_2SiMe_3)_3$ are all isolable. Unstable alkyls are present in the solutions of V halides and Al alkyls, which are used in the Ziegler-Natta type of reaction for the copolymerization of styrene, butadiene, and dicyclopentadiene to give synthetic rubbers. The dicyclopentadienyl derivatives of V^{II}, V^{III}, and V^{IV} are well established, and unlike the unusual Ti^{II} compound (p. 706), the V^{II} complex is the simple paramagnetic $(\eta^5\text{-}C_5H_5)_2V$. This undergoes oxidative additions with many compounds to give V^{III} or V^{IV} compounds (e.g., Cp_2VCS_2, Cp_2VCl_2).

Mixed complexes such as $\eta^5\text{-}C_5H_5V(CO)_4$, $\eta^5\text{-}C_5H_5V(CO)_3H^-$,[32] and $\eta^5\text{-}C_5H_5V(\eta^7\text{-}C_7H_7)$ and the diarenes [e.g., $V(C_6H_6)_2$] are known, as well as some alkene complexes.

By matrix isolation methods the unstable species $V_2(CO)_{10}$ and also $V(N_2)_6$ have been characterized.[33]

21-C. CHROMIUM

For chromium, as for Ti and V, the highest oxidation state is that corresponding to the total number of $3d$ and $4s$ electrons. Although Ti^{IV} is the most stable state for titanium and V^V is only mildly oxidizing, chromium(VI), which exists only in oxo species such as CrO_3, CrO_4^{2-}, and CrO_2F_2, is strongly oxidizing. Apart from stoichiometric similarities, chromium resembles the Group VI elements of the sulfur group only in the acidity of the trioxide and the covalent nature and ready hydrolysis of CrO_2Cl_2.

The intermediate states Cr^V and Cr^{IV} have very restricted chemistry. The very low formal oxidation states are found largely in carbonyl- and organometallic-type compounds, which are discussed in Chapters 25 and 27. With ligands such as aryl

[30] R. D. Wilson and R. Bau, *J. Am. Chem. Soc.*, 1974, **96**, 7601.
[31] M. Schneider and E. Weiss, *J. Organomet. Chem.*, 1976, **121**, 365.
[32] R. G. Bergman *et al.*, *J. Am. Chem. Soc.*, 1978, **100**, 7902.
[33] H. Huber *et al.*, *J. Am. Chem. Soc.*, 1976, **98**, 3176.

TABLE 21-C-1
Oxidation States and Stereochemistry of Chromium

Oxidation state	Coordination number	Geometry	Examples
Cr^{-II}		?	$Na_2[Cr(CO)_5]$
Cr^{-I}		Octahedral	$Na_2[Cr_2(CO)_{10}]$
Cr^0	6	Octahedral	$Cr(CO)_6$, $[Cr(CO)_5I]^-$, $Cr(bipy)_3$
Cr^I, d^5	6	Octahedral	$[Cr(bipy)_3]^+$, $[Cr(CNR)_6]^+$
Cr^{II}, d^4	4	Dist. tetrahedral	$CrCl_2(MeCN)_2$, $CrI_2(OPPh_3)_2$
	5	Tbp	$[Cr(Me_6 \, tren)Br]^+$
	6	Dist.^{a} octahedral	CrF_2, $CrCl_2$, CrS
	5 or 6	Cr—Cr quaduple bond	$Cr_2(O_2CR)_4L_2$, $Cr_2[(CH_2)_2P(CH_3)_2]_4$
	7	?	$[Cr(CO)_2(diars)_2X]X$
Cr^{III}, d^3	3	Planar	$Cr(NPr_2)_3$
	4	Dist. tetrahedral	$[PCl_4]^+[CrCl_4]^-$, $[Cr(CH_2SiMe_3)_4]^-$
	5	Tbp	$CrCl_3(NMe_3)_2$
	6^b	Octahedral	$[Cr(NH_3)_6]^{3+}$, $Cr(acac)_3$, $K_3[Cr(CN)_6]$
Cr^{IV}, d^2	4	Tetrahedral	$Cr(OC_4H_9)_4$, Ba_2CrO_4, $Cr(CH_2SiMe_3)_4$
	6	Octahedral	K_2CrF_6, $[Cr(O_2)_2en]\cdot H_2O$
Cr^V, d^1	4	Tetrahedral	CrO_4^{3-}
	5	?	CrF_5
	5	Sp	$CrOCl_4^-$
	6	Octahedral	$K_2[CrOCl_5]$
	8	Quasi-dodecahedral	K_3CrO_8 (see text)
Cr^{VI}, d^0	4	Tetrahedral	CrO_4^{2-}, CrO_2Cl_2, CrO_3

^{a} Four short and two long bonds.
^{b} Most stable state.

isocyanides, bipy, terpy, and phen, which are better donors and somewhat less obligatory back-acceptors than CO, it is possible to generate stable Cr^I compounds. With the isocyanides electrochemical or Ag^+ oxidation may be used to obtain from $Cr(CNR)_6$ the +1 and +2 ions,[1] with one and two unpaired electrons, respectively. With the amine ligands the entire series of $Cr(LL)_3^n$ with $n = -1, 0, +1, +2$, and +3 is obtained and the compounds are electrochemically interconvertible.[2] In the more electron-rich ones electrons enter orbitals with appreciable (for −1, predominant) ligand character.

The most stable and generally important states are Cr^{II} and Cr^{III}. This dominance of the II and III states that begins here persists through the following transition elements. We shall discuss these states first.

The oxidation states and stereochemistry are summarized in Table 21-C-1.

21-C-1. The Element

The chief ore is *chromite* ($FeCr_2O_4$), which is a spinel with Cr^{III} on octahedral sites and Fe^{II} on the tetrahedral ones. If pure chromium is not required—as for use in ferrous alloys—the chromite is reduced with carbon in a furnace, affording the

[1] P. M. Treichel and G. J. Essenmacher, *Inorg. Chem.*, 1976, **15**, 146; 1977, **16**, 800.
[2] M. C. Hughes and D. J. Macero, *Inorg. Chem.*, 1976, **15**, 2040.

carbon-containing alloy ferrochromium:

$$FeCr_2O_4 + 4C \rightarrow Fe + 2Cr + 4CO$$

When pure chromium is required, the chromite is first treated with molten alkali and oxygen to convert the Cr^{III} to chromate(VI), which is dissolved in water and eventually precipitated as sodium dichromate. This is then reduced with carbon to Cr^{III} oxide:

$$Na_2Cr_2O_7 + 2C \rightarrow Cr_2O_3 + Na_2CO_3 + CO$$

This oxide is then reduced with aluminum:

$$Cr_2O_3 + 2Al \rightarrow Al_2O_3 + 2Cr$$

Chromium is a white, hard, lustrous, and brittle metal (m.p. 1903° ± 10°). It is extremely resistant to ordinary corrosive agents, which accounts for its extensive use as an electroplated protective coating. The metal dissolves fairly readily in nonoxidizing mineral acids, for example, hydrochloric and sulfuric acids, but not in cold aqua regia or nitric acid, either concentrated or dilute. The last two reagents passivate the metal in a manner which is not well understood. The electrode potentials of the metal are:

$$Cr^{2+} + 2e = Cr \qquad E^0 = -0.91 \text{ V}$$
$$Cr^{3+} + 3e = Cr \qquad E^0 = -0.74 \text{ V}$$

Thus it is rather active when not passivated, and it readily displaces copper, tin, and nickel from aqueous solutions of their salts.

At elevated temperatures chromium unites directly with the halogens, sulfur, silicon, boron, nitrogen, carbon, and oxygen.

CHROMIUM COMPOUNDS

21-C-2. Binary Compounds

Halides. These are listed in Table 21-C-2. The anhydrous Cr^{II} halides are obtained by action of HF, HCl, HBr, or I_2 on the metal at 600 to 700° or by reduction of the trihalides with H_2 at 500 to 600°. $CrCl_2$ is the most common and most important of these halides, dissolving in water to give a blue solution of Cr^{2+} ion.

Of the Cr^{III} halides the red-violet chloride, which can be prepared in a variety of ways (e.g., by the action of $SOCl_2$ on the hydrated chloride) is singularly important. It can be sublimed in a stream of chlorine at about 600°, but if heated to such a temperature in the absence of chlorine it decomposes to Cr^{II} chloride and chlorine. The flaky or leaflet form of $CrCl_3$ is a consequence of its crystal structure, which is of an unusual type. It consists of a cubic close-packed array of chlorine atoms in which two-thirds of the octahedral holes between *every other* pair of Cl planes are occupied by metal atoms. The alternate layers of chlorine atoms with no metal atoms between them are held together only by van der Waals' forces; thus the crystal has pronounced cleavage parallel to the layers. $CrCl_3$ is the only substance known to have this exact structure, but $CrBr_3$, as well as $FeCl_3$ and triiodides

TABLE 21-C-2

Halides of Chromium

Halogen	Cr^{II}	Cr^{III}	Higher and mixed oxidation states		
F	CrF_2	$CrF_3{}^a$	$CrF_4{}^b$	CrF_5	$CrF_6{}^c$
		Green, m.p. 1404°	Green, subl 100°	Red, m.p. 30°	Yellow
			$Cr_2F_5{}^d$		
Cl	$CrCl_2$	$CrCl_3$	$CrCl_4{}^e$		
		Violet, m.p. 1150°			
Br	$CrBr_2$	$CrBr_3$	$CrBr_4{}^e$		
		Black, subl			
I	CrI_2	CrI_3			
		Black, dec			

[a] Melts only in a closed system; in an open system disproportionates above 600° to give CrF_5.

[b] Becomes brown on slightest contact with moisture.

[c] Unstable above −100°.

[d] Often nonstoichiometric; contains regular $Cr^{III}F_6$ and highly distorted $Cr^{II}F_6$ octahedra sharing corners and edges.

[e] Not known as solids; appear to exist in vapors formed when the trihalides are heated in an excess of the halogen.

of As, Sb, and Bi, have a structure that differs only in that the halogen atoms are in hexagonal rather than cubic close packing.

Chromic chloride does not dissolve at a significant rate in pure water, but it dissolves readily in presence of Cr^{II} ion or reducing agents such as $SnCl_2$ that can generate some Cr^{II} from the $CrCl_3$. This is because the process of solution can then take place by electron transfer from Cr^{II} in solution *via* a Cl bridge to the Cr^{III} in the crystal. This Cr^{II} can then leave the crystal and act on a Cr^{III} ion elsewhere on the crystal surface, or perhaps it can act without moving. At any rate, the "solubilizing" effect of reducing agents must be related in this or some similar way to the mechanism by which chromous ions cause decomposition of otherwise inert Cr^{III} complexes in solution (Section 28-10).

Chromic chloride forms adducts with a variety of donor ligands. The tetrahydrofuranate $CrCl_3 \cdot 3THF$, which is obtained as violet crystals by action of a little zinc on $CrCl_3$ in THF, is a particularly useful material for the preparation of other chromium compounds such as carbonyls or organo compounds, as it is soluble in organic solvents.

Chromium(IV) fluoride is made by fluorination of the metal at ca. 350°; at 350 to 500° CrF_5 is obtained. The hexafluoride is formed in low yield with fluorine at 200 atm pressure and 400° in a bomb. CrF_5 is a powerful fluorinating agent, converting Xe to XeF_2, forming CrF_6^- salts with Cs^+ and NO^+, and reacting with strong F^- acceptors to give CrF_4^+ as in $CrF_4Sb_2F_{11}$.[3] In the liquid state CrF_5 is a polymer with cis bridges.

Oxides. Only Cr_2O_3, CrO_2, and CrO_3 are of importance. The green oxide α-Cr_2O_3, which has the corundum structure (p. 16), is formed on burning the metal

3　　S. D. Brown *et al.*, *J. Chem. Phys.*, 1976, **64**, 260; *J. Fluorine Chem.*, 1976, **7**, 19.

in oxygen, on thermal decomposition of Cr^{VI} oxide or ammonium dichromate, or on roasting the hydrous oxide $Cr_2O_3 \cdot nH_2O$. The latter, normally obtained by adding hydroxide to aqueous Cr^{III} at room temperature, has variable water content. It is often called chromic hydroxide, but there is in fact a true, crystalline hydroxide,[4] $Cr(OH)_3(H_2O)_3$ that can be prepared by slow addition of base to a cold solution of $[Cr(H_2O)_6]^{3+}$. The crystalline material quickly becomes amorphous at higher temperatures.

If ignited too strongly Cr_2O_3 becomes inert toward both acid and base, but otherwise it and its hydrous form are amphoteric, dissolving readily in acid to give aqua ions $[Cr(H_2O)_6]^{3+}$, and in concentrated alkali to form "chromites."

Chromium oxide and chromium supported on other oxides such as Al_2O_3 are important catalysts for a wide variety of reactions.

Chromium(IV) oxide (CrO_2) is normally synthesized by hydrothermal reduction of CrO_3. It has an undistorted rutile structure (i.e., no M—M bonds as in MoO_2). It is ferromagnetic and has metallic conductance, presumably because of delocalization of electrons into energy bands formed by overlap of metal d and oxygen $p\pi$ orbitals.

Chromium(VI) oxide (CrO_3) can be obtained as an orange-red precipitate on adding sulfuric acid to solutions of Na or K dichromate. The red solid is thermally unstable above its melting point 197°, losing oxygen to give Cr_2O_3 after various intermediate stages. It is readily soluble in water and is highly poisonous.

The crystal structure of CrO_3 consists of infinite chains of CrO_4 tetrahedra sharing corners.

Interaction of CrO_3 and organic substances is vigorous and may be explosive. However CrO_3 is widely used in organic chemistry as an oxidant, commonly in acetic acid as solvent. The mechanism has been much studied and is believed to proceed initially by the formation of chromate esters (when pure, they are highly explosive) that undergo C—H bond cleavage as the rate-determining step to give Cr^{IV} as the first product; the general scheme appears to be

$$H_2A + Cr^{VI} \rightleftharpoons Cr^{IV} + A \text{ (slow)}$$
$$Cr^{IV} + Cr^{VI} \rightleftharpoons 2Cr^{V}$$
$$Cr^{V} + H_2A \rightleftharpoons Cr^{III} + A$$

The oxides give rise to various mixed metal oxides; those containing the higher oxidation states are discussed later. Cr_2O_3 can be fused with a number of M^{II} oxides to give crystalline $M^{II}O \cdot Cr_2O_3$ compounds having the spinel structure (p. 17) with Cr^{III} ions in the octahedral holes. Sodium metal reacts[5] with each of the oxides Cr_2O_3, CrO_2, CrO_3, as well as with Na_2CrO_4, to give the "chromite" $NaCrO_2$, in which both cations have octahedral coordination.

Other Binary Compounds.   The chromium sulfide system is very complex, with two forms of Cr_2S_3 and several intermediate phases between these and CrS. Rhombohedral Cr_2S_3 has complex electrical and magnetic properties.

[4]   R. Giovanali, W. Stadelmann, and W. Feitknecht, *Helv. Chim. Acta,* 1973, **56**, 839; V. von Mevenburg, O. Siroky, and G. Schwarzenbach, *Helv. Chim. Acta,* 1973, **56**, 1009.
[5]   M. G. Barker and A. J. Hooper, *J. C. S. Dalton,* **1976**, 1093.

21-C-3.   The Chemistry of Chromium(II)

Mononuclear Compounds; The Aqua Ion.   This ion is bright blue and is best obtained in solution by dissolving the very pure metal in deoxygenated, dilute mineral acids or by reducing Cr^{III} solutions electrolytically or with Zn/Hg. The ion is readily oxidized:

$$Cr^{3+} + e = Cr^{2+} \qquad E^0 = -0.41 \text{ V}$$

and the solutions must be protected from air—even then, they decompose at rates varying with the acidity and the anions present, by reducing water with liberation of hydrogen.

The aqua ion has been extensively used as a reductant in mechanistic studies, since the resulting Cr^{III} species are substitution inert and can provide evidence as to the participation of bridging groups in the electron transfer step. This aspect is treated in detail in Section 28-10. It has been proposed that oxidation of the aqua ion by O_2 gives first the CrO_2Cr group, which undergoes protonation to $[(H_2O)_4Cr(\mu\text{-}OH)_2Cr(H_2O)_4]^{2+}$; this in turn splits to give $[Cr(H_2O)_6]^{3+}$, which according to labeling studies, contains all atoms originally in the O_2.

An important reaction of the aqua ion is to give $CrR^{2+}(aq)$ species; there are many variants, all of which appear to involve radical mechanisms. $CHCl_3$, for example, gives $[Cr(CHCl_2)(H_2O)_5]^{2+}$, and many other alkyl halides react similarly. $R\cdot$ radicals generated by pulse radiolysis in aqueous media react readily with the aqua ion.[6] The rates and mechanism of Cr—C bond cleavage by H_3O^+,[6] by Br_2, or in other ways[7] have been much studied. Electrophilic attack on the Cr—C bond appears to be the usual mechanism.

Chromium(II) compounds have been extensively used to reduce organic compounds.[8] Thus alkyl halides can be reduced to alkanes, particularly if aqueous DMF is used as solvent and ethylenediamine is present; the intermediate alkyl, e.g., $[Cr^{III}R\,en_2(H_2O)]^{2+}$, is hydrolytically and protolytically unstable, giving rise to RH.

The aqua ion reacts with NO to give $CrNO^{2+}(aq)$, which is presumably a NO^- complex of Cr^{3+}.[9]

Mononuclear Complexes.   The Cr^{II} halides form numerous complexes. A series of Cr^{II} ammine complexes can be obtained by bubbling NH_3 through ethanolic solutions of the halides[10]. The hexammines $Cr(NH_3)_6X_2$ are formed initially but readily lose NH_3 to form $Cr(NH_3)_5X_2$ and, on heating $Cr(NH_3)_2X_2$. All have four unpaired electrons. $[Cren_3]^{2+}$ and $[Cren_2X_2]$ type complexes are formed from $CrCl_2$ and ethylenediamine in ethanol, but water displaces the diamine. As a d^4 ion Cr^{2+} may form either high-spin ($4e^-$, $\mu \approx 4.9$ BM) or low-spin ($2e^-$, $\mu \approx 3$ BM) octahedral complexes. Most, like the ammines just mentioned, are high spin, but with strong field ligands low-spin complexes such as $[Cr(CN)_6]^{4-}$, $[Cr(bi$-

[6]   H. Cohen and D. Meyerstein, *Inorg. Chem.*, 1974, **13**, 2434.

[7]   J. H. Espenson *et al.*, *J. Am. Chem. Soc.*, 1974, **96**, 1008; *Inorg. Chem.*, 1976, **15**, 1886.

[8]   J. R. Hanson, *Synthesis*, **1974**, 1.

[9]   J. N. Armor and M. Buchbinder, *Inorg. Chem.*, 1973, **12**, 1086.

[10]   L. F. Larkworthy and J. M. Tabatabai, *J. C. S. Dalton*, **1976**, 814.

[11]   A. Earnshaw *et al.*, *J. C. S. Dalton*, **1977**, 2209.

py)$_3$]$^{2+}$, [Cr(phen)$_3$]$^{2+}$, cis-Cr(phen)$_2$(NCS)$_2$[11], and Cr(diars)$_2$X$_2$ (X = Cl, Br, I)[12] are obtained.

It is likely that all "octahedral" high-spin complexes suffer Jahn-Teller distortion of the type that causes tetragonal elongation, and in the limit this leads to four-coordinate square complexes, of which several examples are now well documented: Cr(acac)$_2$,[13] Cr(H$_2$Bpz$_2$)$_2$,[14] and Cr[N(SiMe$_3$)$_2$]$_2$(THF)$_2$.[15] The Cr^{II} tetraphenyl porphyrin is another example[16]; this takes up two py ligands and changes to a low-spin, $trans$-Cr(porphyrin)(py)$_2$ complex (μ = 2.9 BM).[17]

A few five-coordinate complexes are known, of which the best characterized are the tbp type with tripod ligands, viz., [Cr(Me$_6$tren)Br]$^+$, [Cr(PN$_2$)I]$^+$ and [Cr(pn^{3}Br]$^+${pn^3 = (Et$_2$NCH$_2$CH$_2$)$_2$N(CH$_2$CH$_2$PPh$_2$)}. These are high-spin complexes with the two expected d–d transitions (p. 642) $e'' \rightarrow a_1$ and $e' \rightarrow a_1'$.

Chromium(II) occurs in a number of compounds of the double-salt type such as M$_2^I$CrX$_4$ (M^I = K, Rb, Cs; X = Cl, Br), KCrCl$_3$, and Na$_2$Cr(SO$_4$)$_2$·6H$_2$O. None of these appear to contain discrete complexes. Those of the first type, for example, have structures of the K$_2$NiF$_4$ type in which each transition metal ion is surrounded by a planar set of four anions, each of which forms a 180° bridge to a neighbor, plus two unshared anions above and below, thus completing a distorted octahedron. These have high-spin Cr^{II} but are exceptional in displaying ferromagnetic coupling.[18] KCrCl$_3$ and CsCrCl$_3$ also have high spin Cr^{II} ions in octahedra consisting of some shared and some unshared Cl$^-$ ions, but here the coupling is antiferromagnetic.[19]

CrCl$_2$ dissolves in CH$_2$Cl$_2$ containing R$_4$NCl to give an anion of apparent composition Cr$_3$Cl$_{10}^{4-}$ but unknown structure[20]; it reacts with (R$_4$N)$_2$MoCl$_6$ to give [MoCrCl$_9$]$^{3-}$, a mixed analogue of the [M$_2$Cl$_9$]$^{3-}$ (M = Cr, Mo, W) ions.

Binuclear Compounds; Quadruple Bonds. One of the earliest Cr^{II} compounds discovered (1844) was the acetate hydrate Cr$_2$(O$_2$CCH$_3$)$_4$(H$_2$O)$_2$. It was long recognized as anomalous because it is red and almost diamagnetic (the weak paramagnetism is due to Cr^{III} impurities), whereas the mononuclear Cr^{II} compounds are blue or violet and strongly paramagnetic. In recent times the acetate and numerous other carboxylato compounds have been shown to have the binuclear structure[21] shown in Fig. 21-C-1. In every case the Cr$_2$(O$_2$CR)$_4$ unit has ligands L lying at moderate distances (2.22–2.44 Å) along the Cr–Cr axis. In those without additional ligands the Cr$_2$(O$_2$CR)$_4$ molecules serve as axial ligands to one another. In all these cases the Cr—Cr bond lengths are in the range 2.28–2.54 Å.

Within the past few years a number of new dinuclear Cr^{II} compounds have been

12   F. Mani, P. Stoppioni, and L. Sacconi, *J. C. S. Dalton,* **1975,** 461.
13   F. A. Cotton, C. E. Rice, and G. W. Rice, *Inorg. Chim. Acta,* 1977, **24,** 231.
14   P. Dapporto, F. Mani, and C. Mealli, *Inorg. Chem.,* 1978, **17,** 1323.
15   D. C. Bradley *et al., J. C. S. Chem. Comm.,* **1972,** 567.
16   W. R. Scheidt and C. A. Reed, *Inorg. Chem.,* 1978, **17,** 710.
17   C. A. Reed *et al., Inorg. Chem.,* 1978, **17,** 2666.
18   P. Day, *Acc. Chem. Res.,* 1979, **12,** 236.
19   D. H. Leech and D. J. Machin, *J. C. S. Dalton,* **1975,** 1609. See also L. F. Larkworthy, J. K. Trigg, and A. Yavari, *J. C. S. Dalton,* **1975,** 1879.
20   M. S. Matson and R. A. D. Wentworth, *Inorg. Chem.,* 1976, **15,** 2139.
21   F. A. Cotton and G. W. Rice, *Inorg. Chem.,* 1978, **17,** 2004.

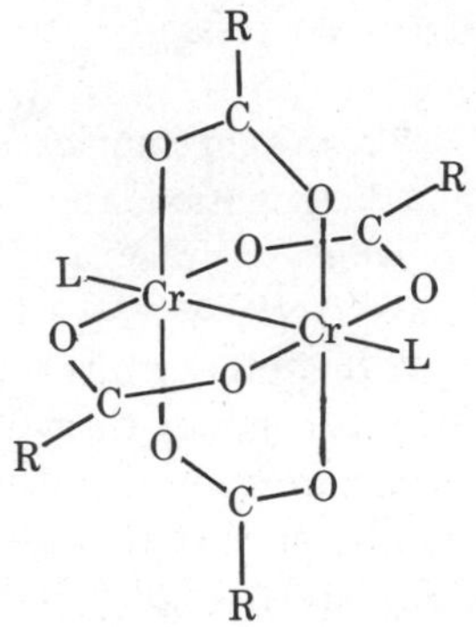

Fig. 21-C-1. The general structure of Cr[II] carboxylate compounds.

discovered, in which the Cr—Cr distances are extraordinarily short.[22] Some of these are shown in Table 21-C-3, with their Cr—Cr bond lengths. It can be seen that some of the bridging ligands in these compounds are sterically and electronically very similar to the carboxyl group except that all have groups such as RN or H_2C in place of O that for steric and/or other reasons prevent axial ligands from binding to the

TABLE 21-C-3
Some Representative Compounds with Very Short Cr—Cr Quadruple Bonds

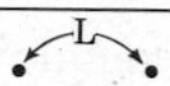	d_{Cr-Cr} (Å)
	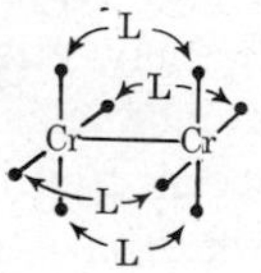 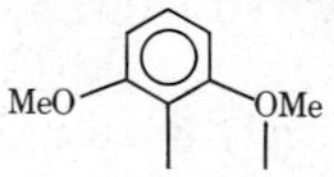 1.85
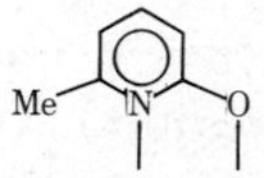	1.85
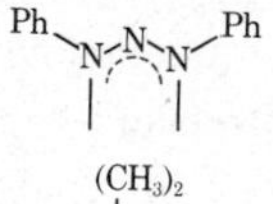	1.86
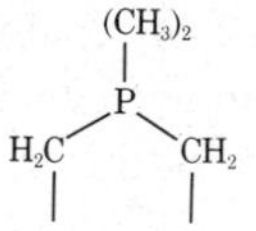	1.89
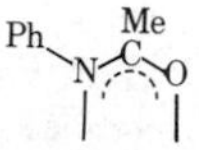	1.86
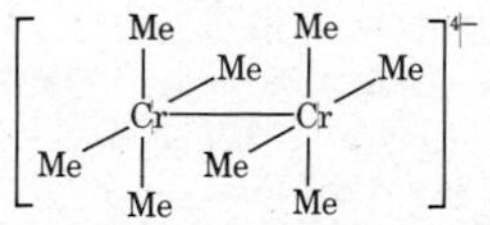	1.98

22   A. Bino, F. A. Cotton, and W. Kaim, J. Am. Chem. Soc., 1979, 101, 2506.

chromium atoms. In the case of $[Cr_2Me_8]^{4-}$ the absence of axial ligands is presumably due to the high negative charge already present. It thus appears that Cr—Cr quadruple bonds are extremely strong and short when unperturbed but that the formation of Cr—L axial bonds can compensate for the loss of some of the Cr—Cr bonding and cause the Cr—Cr bonds to become considerably longer (see Section 26-6).

The quadruply bonded Cr_2^{4+} compounds vary greatly in their chemical properties as well as in their bond lengths. The second compound shown in Table 21-C-3 is probably the most generally stable Cr^{II} compound presently known. It appears to be stable in the normal atmosphere for many months, whereas all $Cr_2(O_2CR)_4L_2$ species are attacked by air in a matter of minutes.

21-C-4. The Chemistry of Chromium(III), d^3

Chromium(III) Complexes. There are literally thousands of chromium(III) complexes that with a few exceptions, are all hexacoordinate. The principal characteristic of these complexes in aqueous solutions is their relative kinetic inertness.

Ligand displacement reactions of Cr^{III} complexes are only about 10 times faster than those of Co^{III}, with half-times in the range of several hours. It is largely because of this kinetic inertness that so many complex species can be isolated as solids and that they persist for relatively long periods of time in solution, even under conditions of marked thermodynamic instability.

The hexaaqua ion $[Cr(H_2O)_6]^{3+}$, which is regular octahedral, occurs in aqueous solution and in numerous salts such as the violet hydrate $[Cr(H_2O)_6]Cl_3$ and in an extensive series of alums $M^ICr(SO_4)_2\cdot12H_2O$. The chloride has three isomers, the others being the dark green *trans*-$[CrCl_2(H_2O)_4]Cl\cdot2H_2O$, which is the normal commercially available salt, and pale green $[CrCl(H_2O)_5]Cl_2\cdot H_2O$. The aqua ion is acidic ($pK = 4$), and the hydroxo ion condenses to give a dimeric hydroxo bridged species:

$$[Cr(H_2O)_6]^{3+} \underset{H^+}{\overset{-H^+}{\rightleftharpoons}} [Cr(H_2O)_5OH]^{2+} \rightleftharpoons [(H_2O)_5Cr\overset{OH}{\underset{OH}{\diamondsuit}}Cr(H_2O)_5]^{4+}$$

On addition of further base, soluble polymeric species of high molecular weight and eventually dark green gels are formed.

The ammonia and ammine complexes are the most numerous chromium derivatives and the most extensively studied. They include the pure ammines $[CrAm_6]^{3+}$, the mixed ammine-aqua types, that is, $[CrAm_{6-n}(H_2O)_n]^{3+}$ ($n = 0$-$4, 6$), the mixed ammine-acido types, that is, $[CrAm_{6-n}R_n]^{(3-n)+}$ ($n = 1$-$4, 6$), and mixed ammine-aqua-acido types, for example, $[CrAm_{6-n-m}(H_2O)_n R_m]^{(3-m)+}$. In these general formulas Am represents the monodentate ligand NH_3 or half of a polydentate amine such as ethylenediamine, and R represents an acido ligand such as a halide, nitro, or sulfate ion. These ammine complexes provide examples of virtually all the kinds of isomerism possible in octahedral complexes.

The preparation of polyammine complexes sometimes presents difficulties, partly because in neutral or basic solution hydroxo or oxobridged polynuclear complexes are often formed. Such polyammines are often conveniently prepared from the Cr^{IV} peroxo species, noted below; thus the action of HCl on $[Cr^{IV}en(H_2O)(O_2)_2]\cdot H_2O$ forms the blue salt $[Cr^{III}en(H_2O)_2Cl_2]Cl$.

(21-C-I)   (21-C-II)

The majority of polynuclear complexes are of one of the types 21-C-I or 21-C-II. In the former there is a single bridging group, which is usually O or OH. Some representative reactions involving such compounds are shown below.

$$[(NH_3)_5Cr(OH)Cr(NH_3)_5]^{5+} \underset{H^+}{\overset{OH^-}{\rightleftharpoons}} [(NH_3)CrOCr(NH_3)_5]^{4+}$$

$\downarrow H_2O$ (one day, 100°) $\downarrow OH^-$ (several days, 25°)

$$[(NH_3)_5Cr(OH)Cr(NH_3)_4(H_2O)]^{5+} \overset{H^+}{\longleftarrow} [(NH_3)_5Cr(OH)Cr(NH_3)_4(OH)]^{4+}$$

The oxo-bridged complex has a linear Cr—O—Cr group, indicating $d\pi$-$p\pi$ bonding as in other cases of M—O—M groups. Even in the "acid rhodo" complex $[(NH_3)_5Cr(OH)Cr(NH_3)_5]^{5+}$ the bridge is nearly linear ($\angle Cr$—O—$Cr = 166°$), and there is considerable magnetic coupling.[23] A large number of the 21-C-II type of complex are known, especially with X = Y = OH. Many in which the non-bridging ligands are bidentate chelates such as en, phen, and $H_2C(CO_2)_2^{2-}$, have been carefully studied both structurally and magnetically.[24] All show a repulsion between the Cr^{III} atoms (distances >3.0 Å) and weak but significant magnetic interactions that are nearly always antiferromagnetic, although when the chelating ligands are $CH_2(CO_2)_2^{2-}$ it is ferromagnetic. Other bridged dichromium(III) complexes have also been studied; Cr^{III}—Cr^{III} systems are almost as popular as Cu^{II}—Cu^{II} ones for studies of such magnetic interactions.

Anionic complexes are also common and are of the type $[CrX_6]^{3-}$, where X may be F^-, Cl^-, NCS^-, CN^-, but they may also have lower charges if neutral ligands are present as in the ion $[Cr(NCS)_4(NH_3)_2]^-$.[25] Complexes of bi- or polydentate anions are also known, one example being $[Cr\ ox_3]^{3-}$.

A different type of anionic complex is represented by the $Cr_2X_9^{3-}$ ions,[26] which have a confacial bioctahedron structure similar to $W_2Cl_9^{3-}$ (p. 864) except that the Cr^{3+} ions repel each other from the centers of their octahedra and the magnetic moments are normal, indicating that there is no Cr—Cr bond.

As expected, Cr^{III} can also form complexes of other types, including neutral

[23] J. T. Veal *et al., Inorg. Chem.,* 1973, **12,** 2928.
[24] W. E. Hatfield, D. J. Hodgson, *et al., Inorg. Chem.,* 1975, **14,** 1127; 1976, **15,** 1605; *J. C. S. Dalton,* **1977,** 1662; J. Josephsen and E. Pedersen, *Inorg. Chem.,* 1977, **16,** 2534.
[25] For other examples, see M. A. Bennett, R. J. H. Clark, and A. D. J. Goodwin, *Inorg. Chem.,* 1967, **6,** 1621.
[26] M. S. Matson and R. A. D. Wentworth, *Inorg. Chem.,* 1976, **15,** 2139 and references therein.

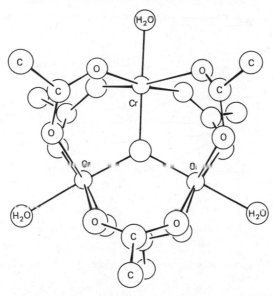

Fig. 21-C-2. The structure of $[Cr_3O(O_2CMe)_6(H_2O)_3]^+$

complexes with β-diketonates and similar ligands [e.g., Cr acac$_3$ and Cr(O-COCF$_3$)$_3$]. It also forms oxo-centered trinuclear carboxylates (p. 154) such as $[Cr_3O(O_2CMe)_6(H_2O)_3]^+$, which has the structure shown in Fig. 21-C-2. The central oxygen atom lies in the plane of the M$_3$ triangle, and its presence requires the M atoms to be so far apart (>3.2 Å) that there is no M—M bonding and these species are not to be considered metal atom *cluster* compounds (Chapter 26) in the proper sense of that term. There are weak magnetic interactions between the metal atoms, and these have been studied in some detail[27] in the Cr^{III} case.

The coordination number 3 occurs in dialkylamides [e.g., Cr(NPr$_2^i$)$_3$]. A combination of steric factors and multiple bonding has been proposed to explain the stability of such monomers. Another of the rare nonoctahedral Cr^{III} complexes is that shown in 21-C-III, where there is a very distorted pentagonal bipyramid.[28]

(21-C-III)

27   L. Dubicki and P. Day, *Inorg. Chem.*, 1972, **11**, 1868.
28   G. J. Palenik, *Inorg. Chem.*, 1976, **15**, 1814.

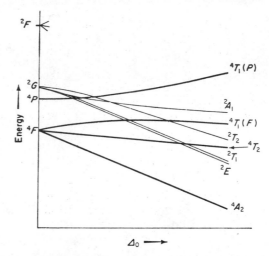

Fig. 21-C-3. Partial energy level diagram for a d^3 ion in an octahedral field (also for a d^7 ion in a tetrahedral field). The quartet states are drawn with heavier lines.

Compounds of formula $M^I CrX_4$, such as $KCrF_4$[29] and $[PCl_4][CrCl_4]$,[30] contain CrX_6 octahedra with some sharing of the X atoms.

Finally, as noted above, $CrCl_3$ forms numerous adducts with ethers, nitriles, amines, and phosphines, which have formulas $CrCl_3 \cdot 2L$ or $CrCl_3 \cdot 3L$. The adduct $CrCl_3 \cdot 2NMe_3$ provides one of the very few examples of an authenticated nonoctahedral Cr^{III} complex: here X-ray studies confirm the trigonal-bipyramidal structure with axial amine groups. Halide-bridged complexes such as $[CrCl_3(PR_3)_2]_2$ have been prepared by direct interaction.

Electronic Structures of Chromium(III) Complexes. The magnetic properties of the octahedral Cr^{III} complexes are uncomplicated. All such complexes must have three unpaired electrons irrespective of the strength of the ligand field, and this has been confirmed for all known mononuclear complexes. More sophisticated theory further predicts that the magnetic moments should be very close to, but slightly below, the spin-only value of 3.88 BM; this, too, is observed experimentally.

The spectra of Cr^{III} complexes are also well understood in their main features. A partial energy level diagram (Fig. 21-C-3) indicates that three spin-allowed transitions are expected, and these have been observed in a considerable number of complexes. Indeed, the spectrochemical series was originally established by Tsuchida using data for Cr^{III} and Co^{III} complexes. In the aqua ion the bands are found at 17,400, 24,700, and 37,000 cm^{-1}.

Ruby, natural or synthetic, is $\alpha\text{-}Al_2O_3$ containing occasional Cr^{III} ions in place of Al^{III} ions. The environment of the Cr^{III} in ruby is thus a slightly distorted (D_{3d}) octahedron of oxide ions. The frequencies of the spin-allowed bands of Cr^{III} in ruby

[29] J. C. Dewan and A. J. Edwards, *J. C. S. Chem. Comm.*, **1977**, 533.
[30] K. R. Seddon and V. H. Thomas, *Inorg. Chem.*, 1978, **17**, 749.

indicate that the Cr^{III} ions are under considerable compression, since the value of Δ_0 calculated is significantly higher than in the $[Cr(H_2O)_6]^{3+}$ ion or in other oxide lattices and glasses. Also, in ruby, spin-forbidden transitions from the $^4A_2$ ground state to the doublet states arising from the ^2G state of the free ion are observed. The transitions to the ^2E and $^2T_1$ states give rise to extremely sharp lines because the slopes of the energy lines for these states are the same as that for the ground state (except in extremely weak fields).

The same doublet states play a key role in the operation of the ruby laser. In this device a large single crystal of ruby is irradiated with light of the proper frequency to cause excitation to the $^4T_2(F)$ state. The exact magnitudes of certain energy differences and relaxation times are such, in the ruby, that the system rapidly makes a radiationless transition (i.e., by loss of energy to the crystal lattice in the form of vibrations) to the ^2E and $^2T_1$ states, instead of decaying directly back to the ground state. The systems then return from these doublet states to the ground state by stimulated emission of very sharp lines that are in phase with the stimulating radiation. Thus bursts of extremely intense, monochromatic, and coherent (all emitters in phase) radiation are obtained, which are of use in communication and as sources of energy.

Organochromium(III) Complexes.[31] Mention has been made of complexes with stable Cr—C bonds, but in addition there are a number of other alkyls and aryls of formula CrR_3L_n, where L is usually an ether molecule such as tetrahydrofuran. These are obtained by action of lithium alkyls or Grignard reagents on $CrCl_3 \cdot 3THF$. The alkyls are rather unstable and their modes of decomposition have been much studied, but the aryls such as $Cr(C_6H_5)_3 \cdot 3THF$ are considerably more stable. Some anionic methyl complex anions, e.g., $Li_3[Cr(CH_3)_6]$, are also reasonably stable, as is the six-coordinate chelate complex $[Cr\{(CH_2)_2PMe_2\}_3]$ (see Chapter 27).

21-C-5. The Chemistry of Chromium(IV), d^2

Chromium(IV) compounds are neither numerous nor important. The oxide CrO_2 is well defined, and there are some mixed oxides such as the deep green Na_4CrO_4[32] and the blue-black $M_2^{II}CrO_4$ (M^{II} = Sr, Ba). The latter are stable in air and contain discrete CrO_4^{4-} groups with magnetic moments of $\sim$2.8 BM. Preparative reactions are:

$$NaCrO_2 + 2Na_2O \xrightarrow{410°} Na_4CrO_4 + Na \uparrow$$

$$SrCrO_4 + Cr_2O_3 + 5Sr(OH)_2 \xrightarrow{1000°} 3Sr_2CrO_4 + 5H_2O \uparrow$$

$M_3^{II}CrO_5$ and $M_4^{II}CrO_6$ compounds have also been reported. There are also some peroxo species, mentioned in Section 21-C-8.

A number of $M_2^I CrF_6$ and $M^{II}CrF_6$ compounds[33] may be obtained by fluorination of stoichiometric mixtures of $CrCl_3$ and M^ICl or $M^{II}Cl_2$.

[31] R. P. A. Sneeden, *Organochromium Chemistry,* Academic Press, 1975.
[32] M. G. Barker and A. J. Hooper, *J. C. S. Dalton,* **1975,** 2487.
[33] G. Siebert and R. Hoppe, *Z. Anorg. Allg. Chem.,* 1972, **391,** 113, 126.

The most unusual Cr^{IV} compounds are the species with Cr—C, Cr—N and Cr—O bonds, exemplified by $Cr(CH_2SiMe_3)_4$, $Cr(NEt_2)_4$, and $Cr(OBu)_4$. These are surprisingly stable, blue, volatile, monomeric, paramagnetic substances. The alkyl is obtained by oxidation of the Cr^{III} species obtained by action of $Me_3Si\text{-}CH_2MgCl$ on $CrCl_3\cdot3THF$, and the dialkylamides are similarly obtained by using $LiNR_2$; the alcoxides can be made from the dialkylamides by action of alcohols. The magnetic and electronic absorption properties are consistent with a distorted tetrahedral structure.

21-C-6. The Chemistry of Chromium(V), d^1

There are relatively few stable compounds containing Cr^V, but esr spectra suggest that it may be present in various chromium-containing oxide lattices that have been suitably oxidized or reduced. The Cr^{5+} ions in such oxides are believed to be the active sites in chromium-containing alumina catalysts for polymerizing ethylene.

Alkali and alkaline earth chromates(V) [e.g., Li_3CrO_4, Na_3CrO_4, $Ca_3(CrO_4)_2$] that appear to contain discrete, tetrahedral CrO_4^{3-} ions are known.[32,34] They are dark green hygroscopic solids that hydrolyze with disproportionation to Cr^{III} and Cr^{VI} and have magnetic susceptibilities consistent with one unpaired electron per anion. There is kinetic evidence that one-electron reduction of $HCrO_4^-$ gives H_3CrO_4 in acid solution.[35] When K_3CrO_4 is decomposed by water, or thermally, the O_2 generated is at least partly in the $^1\Delta$ state as inferred by reaction products with olefins.[36]

There is a series of compounds $M^ICr_3O_8$, where M^I is any of the alkali metals. However structural studies[37a] show that these do *not* contain Cr^V. They are built of $Cr^{VI}O_4$ tetrahedra and $Cr^{III}O_6$ octahedra.

The oxofluoride $CrOF_3$ has been prepared in impure form by the action of ClF_3, BrF_3 on CrO_3 or $K_2Cr_2O_7$. $CrOCl_3$ is made by reaction of CrO_3 with $SOCl_2$. Some halo and oxohalo complexes are also known.[37b] The moisture-sensitive fluorooxochromates(V) $KCrOF_4$ and $AgCrOF_4$ can be obtained by treating CrO_3 mixed with KCl or $AgCl$ with BrF_3.

Complexes of the oxochromium(V) ion CrO^{3+} are among the most stable Cr^V compounds. The preparation of pure, authentic salts of the $[CrCl_4O]^-$ and $[CrCl_5O]^{2-}$ ions requires care, and much of the earlier literature describing these and analogous $[CrBr_5O]^{2-}$ salts is in error.[38] Both series are best obtained by dissolving CrO_3 in glacial acetic acid saturated with HCl and adding the appropriate cation (e.g., pyH^+, Ph_4As^+, Cs^+). CrO_2Cl_2 reacts with PCl_5 to give $[PCl_4]\text{-}[CrCl_4O]$.[39] The $[CrCl_4O]^-$ ion has a structure with C_{4v} symmetry; the Cr—O

34   R. Olazcuaga, J.-M. Reau, and G. LeFlem, *Compt. Rend., C,* 1972, **275,** 135.
35   J. H. Espenson, *Acc. Chem. Res.,* 1970, **3,** 347.
36   J. W. Peters et al., *J. Am. Chem. Soc.,* 1972, **94,** 4348.
37a   J. J. Foster and A. N. Hambly, *Aust. J. Chem.,* 1976, **29,** 2137.
37b   E. A. Sneddon et al., *Trans. Met. Chem.,* 1978, **3,** 318.
38   K. R. Seddon and V. H. Thomas, *J. C. S. Dalton,* **1977,** 2195.
39   B. Gahan et al., *J. C. S. Dalton,* **1977,** 1726.

distance is quite short (1.52 Å) and the O—Cr—Cl angle is 104.5°.[39] Its electronic structure has been much studied.[40] For $[CrCl_4O]^-$ ν_{Cr-O} is 1020 cm^{-1} and for $[CrCl_5O]^{2-}$ it is 930 cm^{-1}; each has a magnetic moment in the range 1.62–1.84 BM, depending on the compound. A related anion (21-C-IV) is said to be the most stable of all Cr^V compounds, since it can be made in a hot EtOH/H$_2$O mixture.[41]

$$(21\text{-}C\text{-}IV)$$

Besides H$_3$CrO$_4$ mentioned above, other short-lived Cr^V species occur in solution. Dissolution of chromates(VI) in 65% oleum evolves O$_2$ to give blue solutions having magnetism consistent with the presence of a Cr^V species. On the basis of transient esr signals, Cr^V species have been postulated in the reduction of Cr^{VI} by oxalic acid or isopropanol. Somewhat more stable Cr^V intermediates are observed on reduction of CrO$_3$ by citric acid or (CH$_3$CH$_2$)(CH$_3$)(OH)CCOOH.[42]

21-C-7. The Chemistry of Chromium(VI), d^0

Chromate and Dichromate Ions. In basic solutions above pH 6, CrO$_3$ forms the tetrahedral yellow *chromate* ion CrO$_4^{2-}$; between pH 2 and pH 6, HCrO$_4^-$ and the orange-red *dichromate* ion Cr$_2$O$_7^{2-}$ are in equilibrium; and at pH values below 1 the main species is H$_2$CrO$_4$. The equilibria are the following:

$$\begin{array}{ll} HCrO_4^- \rightleftharpoons CrO_4^{2-} + H^+ & K = 10^{-5.9} \\ H_2CrO_4 \rightleftharpoons HCrO_4^- + H^+ & K = 4.1 \\ Cr_2O_7^{2-} + H_2O \rightleftharpoons 2HCrO_4^- & K = 10^{-2.2} \end{array}$$

In addition there are the base-hydrolysis equilibria:

$$\begin{array}{l} Cr_2O_7^{2-} + OH^- \rightleftharpoons HCrO_4^- + CrO_4^{2-} \\ HCrO_4^- + OH^- \rightleftharpoons CrO_4^{2-} + H_2O \end{array}$$

which have been studied kinetically for a variety of bases.

The pH-dependent equilibria are quite labile, and on addition of cations that form insoluble chromates (e.g., Ba^{2+}, Pb^{2+}, Ag$^+$) the chromates and not the dichromates are precipitated. Furthermore, the species present depend on the acid used, and only for HNO$_3$ and HClO$_4$ are the equilibria as given above. When hydrochloric acid is used, there is essentially quantitative conversion into the chlorochromate ion; with sulfuric acid a sulfato complex results:

$$\begin{array}{l} CrO_3(OH)^- + H^+ + Cl^- \rightarrow CrO_3Cl^- + H_2O \\ CrO_3(OH)^- + HSO_4^- \rightarrow CrO_3(OSO_3)^{2-} + H_2O \end{array}$$

[40] C. D. Garner *et al., Inorg. Chem.,* 1976, **15**, 1287; *J. C. S. Dalton,* **1976**, 2258.
[41] C. J. Willis *et al., J. C. S. Dalton,* **1975**, 45.
[42] M. Krumpolc and J. Roček, *J. Am. Chem. Soc.,* 1979, **101**, 3206.

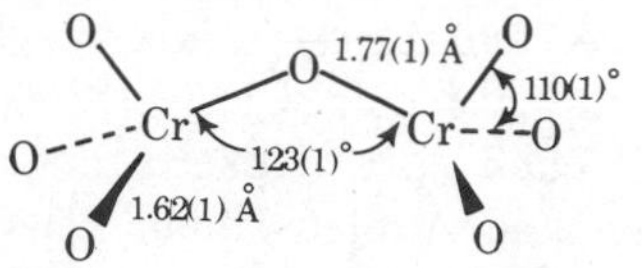

Fig. 21-C-4. The structure of the dichromate ion as found in $Rb_2Cr_2O_7$. [P. Löfgren and K. Waltersson, *Acta Chem. Scand.*, 1971, **25**, 35].

Orange potassium chlorochromate can be prepared simply by dissolving $K_2Cr_2O_7$ in hot $6M$ HCl and crystallizing. It can be recrystallized from HCl but is hydrolyzed by water:

$$CrO_3Cl^- + H_2O \rightarrow CrO_3(OH)^- + H^+ + Cl^-$$

The potassium salts of CrO_3F^-, CrO_3Br^-, and CrO_3I^- are obtained similarly. They owe their existence to the fact that dichromate, though a powerful oxidizing agent, is kinetically slow in its oxidizing action toward halide ions.

Acid solutions of dichromate are strong oxidants:

$$Cr_2O_7^{2-} + 14H^+ + 6e = 2Cr^{3+} + 7H_2O \qquad E^0 = 1.33 \text{ V}$$

The mechanism of oxidation of Fe^{2+} and other common ions by Cr^{VI} has been studied in detail; with one- and two-electron reductants, respectively, Cr^V and Cr^{IV} are initially formed. The reaction with H_2O_2 in acid solution has a very complex and imperfectly understood mechanism.[43]

The chromate ion in basic solution, however, is much less oxidizing:

$$CrO_4^{2-} + 4H_2O + 3e = Cr(OH)_3(s) + 5OH^- \qquad E^0 = -0.13 \text{ V}$$

Chromium(VI) does not give rise to the extensive and complex series of polyacids and polyanions characteristic of the somewhat less acidic oxides of V^V, Mo^{VI}, and W^{VI}. The reason for this is perhaps the greater extent of multiple bonding (Cr=O) for the smaller chromium ion. However compounds containing the $Cr_3O_{10}^{2-}$ and $Cr_4O_{13}^{3-}$ ions are known and the structures have been determined.[44] They continue the pattern set by the $Cr_2O_7^{2-}$ ion (Fig. 21-C-4) in having chains of CrO_4 tetrahedra sharing corners. In the limit of course we have CrO_3, which consists of infinite chains of corner-sharing tetrahedra —O—CrO_2—O—.

Oxohalides. The most important oxohalide is chromyl chloride (CrO_2Cl_2), a deep red liquid (b.p. 117°). It is formed by the action of hydrogen chloride on chromium(VI) oxide:

$$CrO_3 + 2HCl \rightarrow CrO_2Cl_2 + H_2O$$

by warming dichromate with an alkali metal chloride in concentrated sulfuric acid:

$$K_2Cr_2O_7 + 4KCl + 3H_2SO_4 \rightarrow 2CrO_2Cl_2 + 3K_2SO_4 + 3H_2O$$

and in other ways. It is photosensitive but otherwise rather stable, although it vigorously oxidizes organic matter, sometimes selectively. It is hydrolyzed by water to chromate ion and hydrochloric acid.

[43]   S. Funahashi, F. Uchida, and M. Tanaka, *Inorg. Chem.*, 1978, **17**, 2784.
[44]   P. Löfgren, *Chem. Scripta*, 1974, **5**, 91.

Chromyl fluoride (CrO_2F_2), made by fluorination of CrO_2Cl_2 or in many other ways[45] is a very stable red-brown gas. When CrO_3 is fluorinated the product may be either CrO_2F_2 or $CrOF_4$[46] (m.p. 55°, b.p. 95°); at the highest temperatures CrF_5 and CrF_4 are formed. The related $CrO_2(O_2CCF_3)_2$ and $CrO_2(NO_3)_2$, a red solid, are also known.[47]

21-C-8. Peroxo Complexes of Chromium(IV), (V), and (VI)

Like other transition metals, notably Ti, V, Nb, Ta, Mo, and W, chromium forms peroxo compounds (p. 155) in the higher oxidation states. They are all more or less unstable, and in the solid state some of them are dangerously explosive or flammable in air.

When acid dichromate solutions are treated with hydrogen peroxide, a deep blue color rapidly appears but does not persist long. The overall reaction is

$$2HCrO_4^- + 3H_2O_2 + 8H^+ \rightarrow 2Cr^{3+} + 3O_2 + 8H_2O$$

but depending on the conditions, the intermediate species may be characterized. At temperatures below 0°, green cationic species are formed:

$$2HCrO_4^- + 4H_2O_2 + 6H^+ \rightarrow Cr_2(O_2)^{4+} + 3O_2 + 8H_2O$$
$$6HCrO_4^- + 13H_2O_2 + 16H^+ \rightarrow 2[Cr_3(O_2)_2]^{5+} + 9O_2 + 24H_2O$$

The blue species, which is one of the products at room temperature,

$$HCrO_4^- + 2H_2O_2 + H^+ \rightarrow CrO(O_2)_2 + 3H_2O$$

decomposes fairly readily, giving Cr^{3+}, but it may be extracted into ether where it is more stable and, on addition of pyridine to the ether solution, the compound $pyCrO_5$, a monomer in benzene and essentially diamagnetic, is obtained. These facts lead to the formulation of the blue species as aqua, ether, or pyridine adducts of the molecule $CrO(O_2)_2$ containing Cr^{VI}.

The structures of $pyCrO_5$ and $bipyCrO_5$ are shown in Fig. 21-C-5.

The action of H_2O_2 on neutral or slightly acidic solutions of K, NH_4, or Tl dichromate leads to diamagnetic, blue-violet, violently explosive salts, believed to contain the ion $[Cr^{VI}O(O_2)_2OH]^-$.

On treatment of alkaline chromate solutions with 30% hydrogen peroxide—and after further manipulations—the red-brown peroxochromates $M_3^I CrO_8$, can be isolated. They are paramagnetic with one unpaired electron per formula unit, and K_3CrO_8 forms mixed crystals with K_3NbO_8 and K_3TaO_8; the heavy metals are pentavalent in both. The $[Cr(O_2)_4]^{3-}$ ion has the dodecahedral structure shown in Fig. 21-C-5. This ion can be converted to the previously mentioned $[Cr(O_2)_2O(OH)]^-$ ion, and there is evidence[48] that the following equilibrium is attained:

$$[Cr(O_2)_4]^{3-} + 2H^+ + H_2O = [Cr(O_2)_2(O)(OH)]^- + \tfrac{3}{2} H_2O_2$$

Red brown   Violet

45   P. J. Green and G. L. Gard, *Inorg. Chem.*, 1977, **16**, 1243.
46   A. J. Edwards, W. E. Falconer, and W. A. Sunder, *J. C. S. Dalton*, **1974**, 541.
47   S. D. Brown and G. L. Gard, *Inorg. Chem.*, 1973, **12**, 483.
48   B. L. Bartlett and D. Quane, *Inorg. Chem.*, 1973, **12**, 1925.

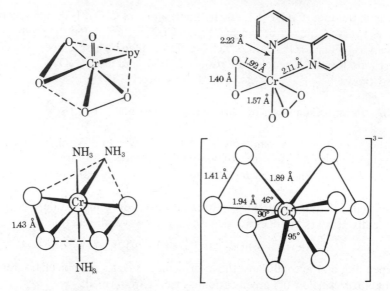

Fig. 21-C-5. The structures of some peroxo chromium species mentioned in the text.

When the reaction mixture used in preparing $(NH_4)_4CrO_8$ is heated to 50° and then cooled to 0°, brown crystals of $(NH_3)_3CrO_4$ are obtained. The structure of this molecule is also shown in Fig. 21-C-5.

The action of H_2O_2 on aqueous solutions of CrO_3 containing ethylenediamine or other amines gives chromium(IV) complexes such as the olive green [Cr en(H_2O)$(O_2)_2$]·H_2O. As noted above, these are useful sources for the preparation of Cr^{III} ammine complexes.

21-D. MANGANESE

As with Ti, V, and Cr, the highest oxidation state of manganese corresponds to the total number of $3d$ and $4s$ electrons. This VII state occurs only in the oxo compounds MnO_4^-, Mn_2O_7, and MnO_3F, and these compounds show some similarity to corresponding compounds of the halogens, for example, in the instability of the oxide. Manganese(VII) is powerfully oxidizing, usually being reduced to Mn^{II}. The intermediate oxidation states are known, but only a few compounds of Mn^V have been characterized; nevertheless, Mn^V species are frequently postulated as intermediates in the reduction of permanganates. Although Mn^{II} is the most stable state, it is quite readily oxidized in alkaline solution. The oxidation states and stereochemistry of manganese are summarized in Table 21-D-1.

21-D-1. The Element

Manganese is relatively abundant, constituting about 0.085% of the earth's crust. Among the heavy metals, only iron is more abundant. Although widely distributed,

TABLE 21-D-1
Oxidation States and Stereochemistry of Manganese

Oxidation state	Coordination number	Geometry	Examples
Mn^{-III}	4	Tetrahedral	$Mn(NO)_3CO$
Mn^{-II}	4 or 6	Square	$[Mn(phthalocyanine)]^{2-}$
Mn^{-I}	5	*Tbp*	$Mn(CO)_5^-$, $[Mn(CO)_4PR_3]^-$
	4 or 6	Square	$[Mn(phthalocyanine)]^-$
Mn^0	6	Octahedral	$Mn_2(CO)_{10}$
Mn^I, d^6	6	Octahedral	$Mn(CO)_5Cl$, $K_5[Mn(CN)_6]$, $[Mn(CNR)_6]^+$
Mn^{II}, d^5	4	Tetrahedral	$MnCl_4^{2-}$, $MnBr_2(OPR_3)_2$, $\{Mn(CH_2SiMe_3)_2\}_n$
	4	Square	$[Mn(H_2O)_4]SO_4 \cdot H_2O$, $Mn(S_2CNEt_2)_2$
	6^a	Octahedral	$[Mn(H_2O)_6]^{2+}$, $[Mn(SCN)_6]^{4-}$
	5	?	$[Mn(dienMe)X_2]$
		Tbp	$[Mn(trenMe_6)Br]Br$
	7	NbF_7^{2-} structure	$[Mn(EDTA)H_2O]^{2-}$
		Pentagonal bipyramidal	$MnX_2(N_5$ macrocycle$)^b$
	8	Dodecahedral	$(Ph_4As)_2Mn(NO_3)_4$
Mn^{III}, d^4	5	*Sp*	$[bipyH_2][MnCl_5]$, $MnXsal_2en$, $MnX(Ph_4porph)^c$
	6^a	Octahedral	$Mn(acac)_3$, $[Mn\ ox_3]^{3-}$, $MnF_3(dist.)$, $Mn(S_2CNR_2)_3$
	7	?	$[Mn(EDTA)H_2O]^-$
Mn^{IV}, d^3	4	Tetrahedral	$Mn(1-norbornyl)_4$
	6	Octahedral	MnO_2, $Mn(SO_4)_2$, $MnCl_6^{2-}$, $Mn(S_2CNR_2)_3^+$
Mn^V, d^2	4	Tetrahedral	MnO_4^{3-}
Mn^{VI}, d^1	4	Tetrahedral	MnO_4^{2-}
Mn^{VII}, d^0	3	Planar	MnO_3^+
	4^a	Tetrahedral	MnO_4^-, MnO_3F

[a] Most common states.
[b] M. G. B. Drew *et al., J.C.S. Dalton,* **1977**, 438.
[c] A. Tulinsky and B. M. L. Chen, *J. Am. Chem. Soc.,* 1977, **99**, 3647.

it occurs in a number of substantial deposits, mainly oxides, hydrous oxides, or carbonate. It also occurs in nodules on the Pacific seabed together with Ni, Cu, and Co.[1]

The metal is obtained from the oxides by reduction with Al. A large use of Mn is in ferromanganese for steels.

Manganese is roughly similar to iron in its physical and chemical properties, the chief difference being that it is harder and more brittle but less refractory (m.p. 1247°). It is quite electropositive and readily dissolves in dilute, nonoxidizing acids. It is not particularly reactive toward nonmetals at room temperatures, but at elevated temperatures it reacts vigorously with many. Thus it burns in chlorine to give $MnCl_2$, reacts with fluorine to give MnF_2 and MnF_3, burns in nitrogen above 1200° to give Mn_3N_2, and combines with oxygen, giving Mn_3O_4, at high temperatures.

[1]   W. E. Krumbein, Ed., *Environmental Biogeochemistry and Geomicrobiology,* Vol. 3, Ann Arbor Pulishers, 1978; *Manganese Recovery Technology* (NMAB-323) National Technical Information Service, Springfield, Va.

It also combines directly with boron, carbon, sulfur, silicon, and phosphorus, but not with hydrogen.

MANGANESE COMPOUNDS[2]

21-D-2. The Chemistry of Divalent Manganese, d^5

The divalent state is the most important and generally the most stable oxidation state for the element. In neutral or acid aqueous solution it exists as the very pale pink hexaaqua ion $[Mn(H_2O)_6]^{2+}$, which is quite resistant to oxidation as shown by the potentials:

$$MnO_4^- \quad \underset{\overline{}}{Mn^{3+}\xrightarrow{\ 1.6\ V\ } Mn^{2+} \xrightarrow{\ -1.18\ V\ } Mn}$$

$$1.5\ V$$

In basic media, however, the hydroxide $Mn(OH)_2$ is formed, and this is very easily oxidized even by air, as shown by the potentials:

$$MnO_2 \cdot yH_2O \xrightarrow{\ -0.1\ V\ } Mn_2O_3 \cdot xH_2O \xrightarrow{\ -0.2\ V\ } Mn(OH)_2$$

Binary Compounds. Manganese(II) oxide is a gray-green to dark green powder made by roasting the carbonate in hydrogen or nitrogen or by action of steam on $MnCl_2$ at 600°.[3] It has the rock salt structure and is insoluble in water. Manganese(II) hydroxide is precipitated from Mn^{2+} solutions by alkali metal hydroxides as a gelatinous white solid that rapidly darkens because of oxidation by atmospheric oxygen. $Mn(OH)_2$ is a well-defined compound—not an indefinite hydrous oxide—having the same crystal structure as magnesium hydroxide. It is only very slightly amphoteric:

$$Mn(OH)_2 + OH^- = Mn(OH)_3^- \qquad K \approx 10^{-5}$$

Manganese(II) sulfide is a salmon-colored substance precipitated by alkaline sulfide solutions. It has a relatively high K_{sp} (10^{-14}) and redissolves easily in dilute acids. It is a hydrous form of MnS and becomes brown when left in air owing to oxidation. If air is excluded, the salmon-colored material changes on long storage, or more rapidly on boiling, into green, crystalline, anhydrous MnS.

MnS, MnSe, and MnTe have the rock salt structure. They are all strongly antiferromagnetic, as are also the anhydrous halides. A superexchange mechanism is believed responsible for their antiferromagnetism.

Manganese(II) Salts. Manganese(II) forms an extensive series of salts with all common anions. Most are soluble in water, although the phosphate and carbonate are only slightly so. Most of the salts crystallize from water as hydrates. With noncoordinating anions the salts contain $[Mn(H_2O)_6]^{2+}$, but $MnCl_2 \cdot 4H_2O$

---

[2]   T. A. Zordan and L. G. Hepler, *Chem. Rev.*, 1968, **68**, 737 (thermodynamic properties and reduction potentials).

[3]   C. E. Bamberger and D. M. Richardson, *J. Inorg. Nucl. Chem.*, 1977, **39**, 151.

contains cis-$MnCl_2(H_2O)_4$ units, and $MnCl_2 \cdot 2H_2O$ has polymeric chains with $trans$-$Mn(H_2O)_2Cl_4$ octahedra sharing edges.[4] The sulfate $MnSO_4$ is obtained on fuming down sulfuric acid solutions. It is quite stable and may be used for manganese analysis, provided no other cations giving nonvolatile sulfates are present.

Manganese(II) Complexes. Manganese(II) forms many complexes, but the equilibrium constants for their formation in aqueous solution are not high compared to those for the divalent cations of succeeding elements (Fe^{II}–Cu^{II}), as noted on page 685, because the Mn^{II} ion is the largest of these and it has no ligand field stabilization energy in its complexes (except in the few of low spin). Many hydrated salts ($Mn(ClO_4)_2 \cdot 6H_2O$, $MnSO_4 \cdot 7H_2O$, etc.) contain the $[Mn(H_2O)_6]^{2+}$ ion, and the direct action of ammonia on anhydrous salts leads to the formation of ammoniates, some of which have been shown to contain the $[Mn(NH_3)_6]^{2+}$ ion. Chelating ligands such as ethylenediamine and oxalate ions form complexes isolable from aqueous solution. Some EDTA complexes, e.g., $[Mn(OH_2)EDTA]^{2-}$, are seven-coordinate, and certain tridentate amines give five-coordinate species.

In aqueous solution the formation constants for halogeno complexes are very low, e.g.,

$$Mn^{2+}_{aq} + Cl^- \rightleftharpoons MnCl^+_{aq} \qquad K \approx 3.85$$

but when ethanol or acetic acid is used as solvent salts of complex anions of varying types may be isolated, such as

MnX_3^-:	Octahedral with perovskite structure
MnX_4^{2-}:	Tetrahedral (green-yellow) or polymeric octahedral with halide bridges (pink)
$MnCl_6^{4-}$:	Only Na and K salts known; octahedral
$Mn_2Cl_7^{3-}$:	With Me_3NH^+, linear chain of face-sharing $MnCl_6$ octahedra; also discrete $MnCl_4^{2-}$, tetrahedral[5]

The precise nature of the product obtained depends on the cation used and also on the halide and the solvent, but MnI_2 gives only MnI_4^{2-}. By contrast, the thiocyanates $M_4^I[Mn(NCS)_6]$ can be crystallized as hydrates from aqueous solution. Salts of ions such as $trans$-$[MnCl_4(H_2O)_2]^{2-}$ and $trans$-$[Mn_2Cl_6(H_2O)_4]^{2-}$ are also known.

Manganese(II) acetylacetonate is probably an oligomer, but there is no definite structural information. It reacts readily with water and other donors, forming octahedral species [e.g., $Mn(acac)_2(H_2O)_2$]. The diethyldithiocarbamate is an infinite polymer[6] as shown in 21-D-I.

(21-D-I)

[4] J. N. McElearney et al., Inorg. Chem., 1973, **12**, 906.
[5] R. E. Caputo et al., Inorg. Chem., 1976, **15**, 820; J. N. McElearney, Inorg. Chem., 1976, **15**, 823.
[6] M. Ciampolini et al., J.C.S. Dalton, **1975**, 2051.

Although the normal coordination number of Mn^{II} is 6, the MnX_4^{2-} ions are tetrahedral. Also Mn^{2+} ions are known to occupy tetrahedral holes in certain glasses and to substitute for Zn^{II} in ZnO. Tetrahedral Mn^{II} has a green-yellow color, far more intense than the pink of the octahedrally coordinated ion, and it very often exhibits intense yellow-green fluorescence. Many commercial phosphors are manganese-activated zinc compounds, wherein Mn^{II} ions are substituted for some of the Zn^{II} ions in tetrahedral surroundings, as for example in Zn_2SiO_4.

A wide variety of adducts of anhydrous manganese(II) halides with ligands such as py, MeCN, R_3P, and R_3AsO, mostly of stoichiometry MnX_2L_2 are known. Most are octahedral or tetrahedral but a few, e.g., $[Mn(trenMe_6)Br]Br$ and $MnCl_2(2\text{-methylimidazole})_3$, are five-coordinate.[7]

Manganese(II) porphyrins[8] and phthalocyanines are known. The tetraphenyl-porphyrin Mn(TPP), which is made by reduction of $Mn^{III}Cl(TPP)$ with $Cr(acac)_2$ in toluene, has the Mn atom out of the N_4 plane. The TPP and related porphyrin complexes readily add other ligands to become five-coordinate. These adducts react reversibly with oxygen at $-78°$ in toluene,[9] e.g.,

$$Mn(por)py + O_2 \rightleftharpoons Mn(por)O_2 + py$$

The oxygen is believed to be bound as O_2^{2-} in a three-membered ring with Mn^{IV}, thus different from Co and Fe porphyrin adducts, which have end-on O_2^-.

At higher temperatures irreversible oxidation to complexes with Mn^{III}–O–Mn^{III} bonds occurs, as also happens when Mn^{II} phthalocyanins are oxidized.

A reversible system for absorption of O_2 at 1 atm is said to be formed on refluxing $MnBr_2$ and tertiary phosphines in THF; the colorless solution turns purple with dry O_2. The nature of the system is not understood.[10]

Electronic Spectra of Manganese(II) Compounds. As noted in Chapter 20, the high-spin d^5 configuration has certain unique properties; manganese(II), as the most prominent example of this configuration, has been discussed (p. 663); only a few remarks need be made here.

The majority of Mn^{II} complexes are high spin. In octahedral fields this configuration gives spin-forbidden as well as parity-forbidden transitions, thus accounting for the extremely pale color of such compounds. In tetrahedral environments, the transitions are still spin forbidden but no longer parity forbidden; these transitions are therefore about 100 times stronger and the compounds have a noticeable pale yellow-green color. The high spin d^5 configuration gives an essentially spin-only, temperature-independent magnetic moment of ~5.9 BM.

At sufficiently high values of Δ_0, a t_{2g}^5 configuration gives rise to a doublet ground state; for Mn^{II} the pairing energy is high and only a few of the strongest ligand sets, e.g., those in $[Mn(CN)_6]^{4-}$, $[Mn(CN)_5NO]^{3-}$, and $[Mn(CNR)_6]^{2+}$, can accomplish this.

[7]   F. L. Phillips et al., Acta Crystallogr., 1976, B**32**, 687.
[8]   L. J. Boucher, Coord. Chem. Rev., 1972, **7**, 289 (porphyrin complexes of Mn^{II}, Mn^{III}, and Mn^{IV}).
[9]   F. Basolo et al., J. Am. Chem. Soc., 1978, **100**, 7253, 4416.
[10]   C. A. McAuliffe et al., J.C.S. Chem. Comm., **1979**, 736.

In the square environment provided by phthalocyanine, Mn^{II} has a $^4A_{1g}$ ground state.

21-D-3.   The Chemistry of Manganese(III), d^4.[11]

Binary Compounds.   The oxides are the most important.[12] The final product of oxidation of Mn or MnO at 470 to 600° is Mn_2O_3. A 1000° this decomposes giving black Mn_3O_4 (*hausmannite*), which is a spinel, $Mn^{II}Mn_2^{III}O_4$. The higher or lower oxides can be interconverted by proper choice of temperature and oxygen partial pressure.[13] A brown hydrous oxide of stoichiometry $MnO(OH)$ is formed when $Mn(OH)_2$ is oxidized by air. Although the mineral *manganite* [γ-$MnO(OH)$] formally contains Mn^{III}, magnetic measurements suggest that it contains both Mn^{II} and Mn^{IV}. Manganese(III) occurs in other mixed oxide systems including the alkali ones $LiMnO_2$, $NaMnO_2$, and $K_6Mn_2O_6$, the latter containing discrete $Mn_2O_6^{6-}$ ions with the Al_2Cl_6-type structure.[14]

Manganese(III) fluoride is obtained on fluorination of $MnCl_2$ or other compounds and is a red-purple solid instantaneously hydrolyzed by water. It has been used as a fluorinating agent. The black trichloride can be made by action of HCl on Mn^{III} acetate or by chlorination of $MnO(OH)$ in CCl_4 at low temperatures. Although it decomposes about $-40°$, its purple solutions in diethylether are reasonably stable at $-10°$ and can be used to make various adducts such as $(R_4N)_2 MnCl_5$ or $MnCl_3(PR_3)_3$.[15]

The Manganese(III) Ion.   The aqua ion[16] can be obtained by electrolytic or peroxosulfate oxidation of Mn^{2+} solutions, or by reduction of MnO_4^-. The ion plays a central role in the complex redox reactions of the higher oxidation states of manganese in aqueous solutions.

It is most stable in acid solutions, since it is very readily hydrolyzed:

$$Mn^{3+} + H_2O = MnOH^{2+} + H^+ \qquad \log K = 0.4$$

Under proper conditions the Mn^{3+}–Mn^{2+} couple is reversible and $E^0 = 1.559$ V in $3M$ $LiClO_4$.

The Mn^{III} ion slowly oxidizes water, liberating oxygen:

$$2Mn^{3+} + H_2O = 2Mn^{2+} + 2H^+ + \tfrac{1}{2}O_2$$

Manganese(III) Complexes.[17]   The Mn^{III} state can be stabilized in aqueous solution by complexing anions such as $C_2O_4^{2-}$, SO_4^{2-}, and $EDTA^{4-}$, but even the most stable species $[MnEDTA(H_2O)]^-$ undergoes decomposition because of slow oxidation of the ligand.

---

[11]   W. Levason and C. A. McAuliffe, *Coord. Chem. Rev.*, 1972, **7**, 353 [review of chemistry of oxidation states (III)–(VII)].

[12]   See M. B. Robinson and P. Day, *Adv. Inorg. Chem. Radiochem.*, 1967, **10**, 288.

[13]   R. Pompe, *Acta Chem. Scand.*, 1976, **30A**, 370.

[14]   G. Brachtel and R. Hoppe, *Naturwissenschaften*, 1976, **63**, 339.

[15]   W. Levason and C. A. McAuliffe, *J. Inorg. Nucl. Chem.*, 1975, **37**, 340.

[16]   G. Biedermann and R. Palombari, *Acta Chem. Scand.*, 1978, **32A**, 381.

[17]   G. Davies, *Coord. Chem. Rev.*, 1969, **21**, 199.

One of the best known compounds is the so-called manganic acetate. This dark red substance is obtained as a hydrate by action of $KMnO_4$ on a hot solution of Mn^{II} acetate in glacial acetic acid; the "anhydrous" compound can be obtained by crystallization from acetic acid containing acetic anhydride. It has the stoichiometry $[Mn_3O(CO_2Me)_6]^+CO_2Me^-$, and the oxo-centered cation has the same type of structure as other M^{III} oxocarboxylates (p. 154). The adducts will oxidize alkene (to lactones[18]), and aromatic hydrocarbons like toluene. It also catalyzes the decarboxylation of carboxylic acids and the reaction[19]

$$C_6H_6 + CH_3NO_2 \xrightarrow{CH_3CO_2H} C_6H_5CH_2NO_2$$

All these reactions involve free radicals.

The oxo-centered acetate can be reduced by a one-electron step to a neutral species $Mn_3O(CO_2Me)_6py_3$, which formally has two Mn^{III} and one Mn^{II} atoms, but doubtless all are equivalent as in similar ruthenium compounds and have nonintegral oxidation states.

The acetate has often been used as a starting material to make other Mn^{III} species; since it is still commonly assumed to be a simple acetate $Mn(CO_2Me)_3$, some of the products may well be misformulated. A recent example so obtained is $Mn(3,5$-di-t-butylcatecholate$)_3$.[20]

The dark brown crystalline *acetylacetonate* $[Mn(acac)_3]$ is readily obtained by oxidation of basic solutions of Mn^{2+} by air or chlorine in the presence of acetylacetone. Like the acetate, it acts as an oxidant, coupling phenols and, in presence of donors such as Me_2SO, initiating the free-radical polymerization of acrylonitrile and styrene.

Halogeno Complexes.   Dissolution of $MnO(OH)$ in HF followed by addition of CsF gives $Cs[MnF_4(OH_2)_2]$, which is octahedral with *trans*-H_2O molecules.[21] Anhydrous salts of MnF_4^-, MnF_5^{2-}, and MnF_6^{3-} are also known, as are chloro anions such as $MnCl_5^{2-}$ and neutral complexes like $MnCl_3(dioxan)_2$.[22]

The biological role of manganese, (Section 31-11) has stimulated much study of complexes of manganese in the II and IV states especially. Manganese is an essential metal in several biological systems that are involved in electron transfer reactions. The most important systems are those of photosynthesis in plants and the bacterial enzyme superoxide dismutase[23], which catalyzes the decomposition of O_2^-. The precise nature and role of the manganese species is not clear. Nevertheless searches for model systems have led to study of polydentate[24] and macro-

[18]   E. I. Heiba *et al.*, *J. Am. Chem. Soc.*, 1974, **96**, 7977.

[19]   M. E. Kurz and R. T. Y. Chen, *J.C.S. Chem. Comm.*, **1976**, 968.

[20]   D. T. Sawyer *et al.*, *J. Am. Chem. Soc.*, 1978, **100**, 989.

[21]   P. Bukovec and V. Kaučič, *J.C.S. Dalton*, **1977**, 945.

[22]   R. Uson *et al.*, *Trans. Met. Chem.*, 1976, **1**, 122.

[23]   A. M. Michaelson, J. M. McCord, and I. Fridovitch, Eds., *Superoxide and Superoxide Dismutases,* Academic Press, 1977; G. D. Lawrence and D. T. Sawyer, *Coord. Chem. Revs.*, 1978, **27**, 173.

[24]   D. T. Sawyer *et al.*, *J. Am. Chem. Soc.*, 1976, **98**, 6698; *Inorg. Chem.*, 1979, **18**, 706; D. C. Dabrowiak *et al.*, *Inorg. Chem.*, 1977, **16**, 540; W. M. Coleman, and L. T. Taylor, *Inorg. Chem.*, 1977, **16**, 1114.

cyclic[25] ligands such as polyamino carboxylates, 8-quinolinolates, polyhydroxo compounds, and porphyrins. As an example we quote the complexes[26] of the D-gluconate ion (21-D-II) that is resistant to oxidation. Stable complexes in all three

$$
\begin{array}{cccc}
\text{OH} & \text{OH} & \text{H} & \text{OH} \\
| & | & | & | \\
\text{HOH}_2\text{C}-\text{C}-\text{C}-\text{C}-\text{C}-\text{CO}_2^- \\
| & | & | & | \\
\text{H} & \text{H} & \text{OH} & \text{H}
\end{array}
$$

(21-D-II)

oxidation states are formed by the dianion GH_3^{2-}, viz.,

$$[Mn^{II}(GH_3)_2]^{2-}, \ [Mn^{III}(GH_3)_2OH]^{2-}, \ [Mn^{IV}(GH_3)_2(OH)_3]^{3-}$$

Although the oxidation-reduction potentials are rather similar to those found for Mn in photosynthetic systems and some O_2 evolution can be observed from solutions of the Mn^{IV} complex under some conditions, there is little connection evident with the biosystem. Schiff base complexes[27] have also been much studied. Typical is the square pyramidal $MnXsal_2en$, where X can be halide or organic groups like Me and Ph. Oxidation of the $Mn^{II}sal_2en$ complexes by O_2 gives probably $Mn(OH)sal_2en$.[28] As noted above, a number of Mn^{III} or Mn^{IV} compounds formed by oxidation with O_2 have hydroxo or oxo bridges. A simple complex that has a linear Mn—O—Mn bridge is $K_6[(CN)_5MnOMn(CN)_5]$, which is obtained by reaction of $KMnO_4$ with KCN in aqueous solution.[29]

Electronic Structure of Mn^{III} Compounds. The $^5E_g(t_{2g}^3e_g)$ state for octahedral Mn^{III} is subject to a Jahn-Teller distortion. Because of the odd number of e_g electrons, this distortion should be appreciable (p. 680), and resemble the distortion in Cr^{II} and Cu^{II} compounds. Indeed, a considerable elongation of two trans bonds with little difference in the lengths of the other four has been observed in many Mn^{III} compounds. For example, MnF_3 has the same basic structure as VF_3 where each V^{3+} ion is surrounded by a regular octahedron of F^- ions, except that two Mn—F distances are 1.79 Å, two more are 1.91 Å, and the remaining two are 2.09 Å. There is also distortion of the spinel structure of Mn_3O_4 where Mn^{2+} ions are in tetrahedral interstices and Mn^{3+} ions in octahedral interstices: each of the latter tends to distort its own octahedron, and the cumulative effect is that the entire lattice is distorted from cubic to elongated tetragonal.

$Mn(acac)_3$ was long believed not to show Jahn-Teller distortion. However it is now known that there are two forms,[30] one of which shows a substantial tetragonal

25 L. J. Boucher, *Coord. Chem. Rev.*, 1972, **7**, 289; A. Tulinsky and B. M. L. Chen, *J. Am. Chem. Soc.*, 1977, **99**, 3647.
26 D. T. Sawyer *et al., Inorg. Chem.*, 1979, **18**, 706.
27 L. J. Boucher *et al., Inorg. Chem.*, 1977, **16**, 1360; 1975, **14**, 1289; K. Dey and R. L. De, *J. Inorg. Nucl. Chem.*, 1977, **39**, 153.
28 C. J. Boreham and B. Chiswell, *Inorg. Chim. Acta*, 1977, **24**, 77.
29 R. F. Ziolo *et al., J. Am. Chem. Soc.*, 1977, **96**, 7911.
30 J. P. Fackler, Jr., and A. Avdeef, *Inorg. Chem.*, 1974, **13**, 1864; V. W. Day *et al., J. Am. Chem. Soc.*, 1979, **101**, 1853.

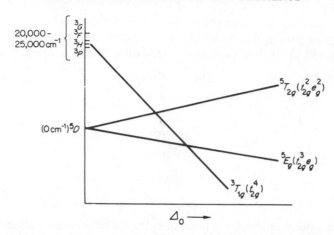

Fig. 21-D-1. Simplified energy level diagram for the d^4 system Mn^{III} in octahedral surroundings.

elongation (two Mn—O = 2.12 Å; four Mn—O = 1.93 Å) as do other high-spin complexes such as porphyrin adducts; the other form shows a moderate tetragonal compression (two Mn—O = 1.95 Å; four Mn—O = 2.00 Å).

The very strong acidity of the aqua ion has been attributed[16] to a strong ligand field stabilizing a distorted ion, possibly $[Mn(OH)(H_2O)_5]^{2+}$.

A simplified energy level diagram for d^4 systems is shown in Fig. 21-D-1. It is consistent with the existence of both high-spin and low-spin octahedral complexes. Because the next quintet state (^5F, derived from the d^3s configuration) lying $\sim$110,000 cm^{-1} above the ^5D ground state of the free ion is of such high energy, only one spin-allowed absorption band ($^5E_g \rightarrow {}^5T_{2g}$) is to be expected in the visible region. For $[Mn(H_2O)_6]^{3+}$ and tris(oxalato)- and tris(acetylacetonato)manganese(III) a rather broad band appears around 20,000 cm^{-1} and the red or red-brown colors of high-spin Mn^{III} compounds may be attributed to such absorption bands. However the spectra of some six-coordinate Mn^{III} complexes are not so simple, and they are difficult to interpret in all their details, presumably because both static and dynamic Jahn-Teller effects perturb the simple picture based on O_h symmetry.

The only low-spin manganese(III) compounds are salts of the $[Mn(CN)_6]^{3-}$ ion. Manganese(II) in the presence of an excess of CN$^-$ is readily oxidized, even by a current of air, with the production of this ion that is first isolated from the solution as the Mn^{II} salt $Mn_3^{II}[Mn(CN)_6]_2$, from which other salts are obtained. For $[Mn(CN)_6]^{3-}$ there appears to be no transition likely below a frequency where it would be obscured by strong ultraviolet bands, and none has been observed.

21-D-4. The Chemistry of Manganese(IV), d^3

Binary Compounds. The most important compound is *manganese dioxide*, which is a gray to black solid occurring in ores such as *pyrolusite*, where it is usually nonstoichiometric. When made by the action of oxygen on manganese at a high temperature, it has the rutile structure found for many other oxides MO_2 (e.g., those of Ru, Mo, W, Re, Os, Ir, and Rh). However as normally made by heating $Mn(NO_3)_2 \cdot 6H_2O$ in air ($\sim$530°), it is nonstoichiometric. A hydrated form is obtained by reduction of aqueous $KMnO_4$ in basic solution. Manganese(IV) occurs in a number of mixed oxides.

Manganese dioxide is inert to most acids except when heated, but it does not dissolve to give Mn^{IV} in solution: instead it functions as an oxidizing agent, the exact manner of this depending on the acid. With HCl, chlorine is evolved:

$$MnO_2 + 4HCl \rightarrow MnCl_2 + Cl_2 + 2H_2O$$

and this reaction is often used for small-scale generation of the gas in the laboratory. With sulfuric acid at 110°, oxygen is evolved and an Mn^{III} acid sulfate is formed. Hydrated manganese dioxide is used in organic chemistry for the oxidation of alcohols and other compounds.[31]

The *tetrafluoride* MnF_4, obtained by direct interaction, is an unstable blue solid decomposing slowly to MnF_3 and F_2.

Complexes. The only cationic species are the dithiocarbamates $[Mn(S_2CNR_2)_3]^+$, obtained as the dark purple BF_4^- salts by air oxidation[32] of the Mn^{III} dithiocarbamates in CH_2Cl_2 in presence of BF_3 (cf. Fe^{IV} and Co^{IV} analogues), or by electrochemical oxidation.[33] The salt K_2MnF_6, obtained by reaction of $KMnO_4$ in 40% HF, is stable,[34] but the $[MnCl_6]^{2-}$ ion is not, although it can be studied in a K_2SnCl_6 matrix where it is obtained by reduction of $KMnO_4$ by strong HCl in presence of K_2SnCl_6.[35] There are 2,2′-bipyridyl and 1,10-phenanthroline complexes that have a two oxo bridges.[36] They can be reduced stepwise to III–IV and III–III complexes, e.g.,

$$[phen_2Mn^{IV}\underset{O}{\overset{O}{<>}}Mn^{IV}phen_2]^{4+} \underset{-e}{\overset{+e}{\rightleftharpoons}} [phen_2MnO_2Mnphen_2]^{3+} \underset{-e}{\overset{+e}{\rightleftharpoons}} [phen_2Mn^{III}O_2Mn^{III}phen_2]^{2+}$$

Red-brown

The Mn^{IV} complex can be used for one-electron oxidations in aqueous media.

There are also a number of similar complexes of Schiff bases, polyhydroxo, and macrocyclic ligands that contain Mn^{IV}. One example is $[Mn^{IV}sal_2enO]_n$, which is a polymer with Mn—O—Mn bridges. Another example,[37a] formed by a tetradenate Schiff base, L, is

$$LMn^{IV}\underset{O}{\overset{O}{<>}}Mn^{IV}L$$

31 A. J. Fatiadi, *Synthesis*, **1976**, 65, 133 (review with 697 references).
32 R. Y. Saleh and D. K. Strand, *Inorg. Chem.*, 1974, **13**, 3017.
33 A. R. Hendrickson *et al.*, *Inorg. Chem.*, 1974, **13**, 1933.
34 A. M. Black and C. D. Flint, *J.C.S. Dalton*, **1974**, 977.
35 P. J. McCarthy and R. D. Bereman, *Inorg. Chem.*, 1973, **12**, 1909.
36 M. M. Morrison, and D. J. Sawyer, *J. Am. Chem. Soc.*, 1977, **99**, 257; S. R. Cooper *et al.*, *J. Am. Chem. Soc.*, 1978, **100**, 7248.
37a L. J. Boucher and C. G. Coe, *Inorg. Chem.*, 1975, **14**, 1289.

The high-spin sorbitolate[37b] complex, $[Mn(C_6H_{12}O_6)_3]^{2-}$, appears to be octahedral.

Manganese(IV) has also been obtained in heteropolyniobate and heteropolyvanadate ions, where it is known or assumed to have an octahedral environment.

Manganese (IV) alkyls are noted later.

21-D-5. The Chemistry of Manganese(V), d^2

There are few authenticated examples of compounds in the Mn^V state. The *oxohalide* MnOCl$_3$,[38] which decomposes above 0° to give MnCl$_3$ and is readily hydrolyzed, is formed by reducing KMnO$_4$ dissolved in HSO$_3$Cl with sucrose.

The so-called *hypomanganates* (MnO_4^{3-}) can be obtained as bright blue salts by reduction of MnO_4^- with an excess of SO_3^{2-}:

$$MnO_4^- + e = MnO_4^{2-} \qquad E = +0.56 \text{ V}$$
$$MnO_4^{2-} + e = MnO_4^{3-} \qquad E = ca. +0.3 \text{ V}$$

Alkali-metal salts are deliquescent and easily hydrolyzed. The MnO_4^{3-} ion is also formed by disproportionation when MnO$_2$ dissolves in concentrated KOH, and its spectra have been obtained when the ion is trapped in a lattice of Ca$_2$ClPO$_4$.

Hypomanganate cyclic esters (21-D-III) as intermediates have been postulated as in the oxidation of olefins by KMnO$_4$, but it seems[39] that at least the detectable intermediates are Mn^{IV} species.

(21-D-III)

21-D-6. The Chemistry of Manganese(VI), d^1, and -(VII), d^0

Manganese(VI) is known in only one environment, namely, as the deep green *manganate* ion MnO_4^{2-}. This ion is formed on oxidizing MnO$_2$ in fused KOH with potassium nitrate, air, or other oxidizing agent, or by evaporating KMnO$_4$ and KOH solutions. Only two salts, K$_2$MnO$_4$, and several hydrated forms of Na$_2$MnO$_4$, have been isolated pure. Both are very dark green.

The manganate ion is stable only in very basic solutions. In acid, neutral, or only slightly basic solutions it readily disproportionates in a kinetically complex way,[40] according to the equation:

$$3MnO_4^{2-} + 4H^+ = 2MnO_4^- + MnO_2(s) + 2H_2O \qquad K \sim 10^{58}$$

[37b] D. T. Richens and D. T. Sawyer, *J. Am. Chem. Soc.*, 1979, **101**, 3681.
[38] J. P. Jasinsku and S. L. Holt, *Inorg. Chem.*, 1975, **14**, 1267.
[39] E. M. Karchefski *et al.*, *Inorg. Chim. Acta*, 1978, **31**, L457.
[40] J. H. Sulter *et al.*, *Inorg. Chem.*, 1974, **13**, 1444.

Manganese(VII). The best known compounds are salts of the *permanganate* ion MnO_4^-. Sodium and potassium permanganates are made on a large scale by electrolytic oxidation of basic solutions of MnO_4^{2-}.

Solutions of MnO_4^- are intrinsically unstable, decomposing slowly but observably in acid solution:

$$4MnO_4^- + 4H^+ = 3O_2(g) + 2H_2O + 4MnO_2(s)$$

In neutral or slightly alkaline solutions in the dark, decomposition is immeasurably slow. It is, however, catalyzed by light; thus standard permanganate solutions should be stored in dark bottles.

In basic solution permanganate functions as a powerful oxidizing agent:

$$MnO_4^- + 2H_2O + 3e = MnO_2(s) + 4OH^- \qquad E^0 = +1.23 \text{ V}$$

In very strong base and with an excess of MnO_4^-, however, manganate ion is produced:

$$MnO_4^- + e = MnO_4^{2-} \qquad E^0 = +0.56V$$

In acid solution permanganate is reduced to Mn^{2+} by an excess of reducing agent:

$$MnO_4^- + 8H^+ + 5e = Mn^{2+} + 4H_2O \qquad E^0 = +1.51 \text{ V}$$

but because MnO_4^- oxidizes Mn^{2+}:

$$2MnO_4^- + 3Mn^{2+} + 2H_2O = 5MnO_2(s) + 4H^+ \qquad E^0 = +0.46 \text{ V}$$

the product in presence of an excess of permanganate is MnO_2.

The mechanisms of oxidations by permanganate are very complicated and involve several bimolecular steps. One example is the oxidation of chloride ion where MnO_4^- is first reduced to Mn^{3+}:

$$MnO_4^- + 4Cl^- + 8H^+ \rightarrow Mn^{3+} + 2Cl_2 + 4H_2O$$

The Mn^{3+} is only very slowly reduced further to Mn^{2+} unless a catalyst such as Cu^{2+} or Ag^+ is present. The rate law suggests that the rate-determining step involves the interaction of a chloro oxo species (cf. below) and Cl^-:

$$HMnO_3Cl^+ + Cl^- \rightarrow Cl_2 + HMn^VO_3$$

The acid $HMnO_4$ does not exist, but on concentration of aqueous solutions a violet, unstable material is obtained that appears to contain a type of heteropoly anion[41]

$$[H_3O]_2[Mn^{IV}(Mn^{VII}O_4)_6] \cdot 11H_2O$$

Potassium permanganate is widely used as an oxidant. By addition of crown ethers, $KMnO_4$ can be made to dissolve (up to $0.06M$) in benzene. Quaternary ammonium salts, e.g., $[Bu_4^iN][MnO_4]$, m.p., $120°$, are also soluble in organic solvents[42] and can oxidize hydrocarbons and other materials.

41   R. Krebs and K.-D. Hasse, *Angew. Chem., Int. Ed.,* 1974, **13**, 603.
42   H. J. Schmidt and H. J. Schäfer, *Angew. Chem., Int. Ed.,* 1979, **18**, 68; T. Sala and M. V. Sargent, *J.C.S. Chem. Commun.,* **1978**, 253.

The permanganate ion has no unpaired electron, but it does have a small temperature-independent paramagnetism (Appendix 6). A fully satisfactory theoretical treatment of the electronic structures and spectra of the MnO_4^{n-} ions has proved difficult to obtain.[43a]

Manganese(VII) Oxide and Oxo Halides. The addition of small amounts of $KMnO_4$ to concentrated H_2SO_4 gives a clear green solution where the ionization

$$KMnO_4 + 3H_2SO_4 \rightleftharpoons K^+ + MnO_3^+ + H_3O^+ + 3HSO_4^-$$

appears to occur. The electronic absorption spectra are consistent with a planar trigonal ion MnO_3^+

With larger amounts of $KMnO_4$, the explosive oil Mn_2O_7 separates. This can be extracted into CCl_4 or chlorofluorocarbons in which it is reasonably stable and safe. By interaction of Mn_2O_7 and $ClSO_3H$, the green, volatile, explosive liquids MnO_3Cl and $MnOCl_3$ and the highly unstable brown MnO_2Cl_2 can be obtained.[43b] Interaction of $KMnO_4$ and FSO_3H gives the green liquid MnO_3F, which also explodes at room temperature.

21-D-7. Organometallic Compounds

Alkyls. Stable manganese dialkyls have not been obtained until recently, and these have bulky alkyl groups (Section 29-3)[44] that are polymers or oligomers. Thus $[Mn(CH_2SiMe_3)_2]_n$ is a linear polymer, but the neophyl CH_2CMe_2Ph compound is a dimer with the phenyl group blocking a manganese coordination position (Fig. 21-D-2).

Lithium tetraalkylmanganates, such as $[Li\ tmed]_2[Mn(CH_3)_4]$ (tmed = tetramethylethylenediamine) are also known. Although these Mn^{II} alkyls can be oxidized to green Mn^{IV} alkyls, these are thermally unstable and photosensitive. The most stable is $Mn(1\text{-norbornyl})_4$.

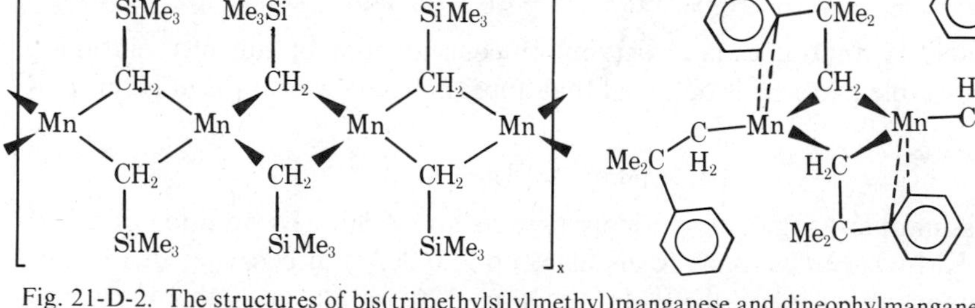

Fig. 21-D-2. The structures of bis(trimethylsilylmethyl)manganese and dineophylmanganese (from ref. 44, by permission).

[43a] J. T. Wrobleski and G. J. Long, *J. Chem. Educ.*, **1977**, 75.
[43b] T. S. Briggs, *J. Inorg. Nucl. Chem.*, 1968, **30**, 2866.
[44] R. A. Andersen *et al.*, *J.C.S. Dalton*, **1976**, 2204.

Cyclopentadienyl and Related Compounds. This type of compound is discussed in Sections 3-15 and 29-12. The compound $(\eta\text{-}C_5H_5)_2Mn$ is brown and antiferromagnetic below $180°$,[45,46] and between this temperature and its m.p. $193°$ is pink and high spin. The brown form is polymeric with zigzag chains and C_5H_5 bridges as shown in Fig. 3-16, which evidently allow antiferromagnetic interaction between the Mn^{2+} ions. The bonding is essentially ionic. The pink form appears to have the ferrocene structure, which is found in the gas phase.[47]

Chemically, $(\eta\text{-}C_5H_5)_2Mn$ behaves as an ionic cyclopentadienide. With CO it is converted to $\eta\text{-}C_5H_5Mn(CO)_3$, a stable yellow compound that has been given the name cymantrene, since it has an extensive organic chemistry—like that of ferrocene. The methyl compound $\eta\text{-}MeC_5H_4Mn(CO)_3$ is a liquid and has been used as an antiknock agent in fuels.

Low Oxidation States. Manganese carbonyl $Mn_2(CO)_{10}$ and the manganese(I) compounds $Mn(CO)_5X$ (X = H, Cl, Br, I, CH_3 etc.) are discussed in Chapter 25.

Other Mn^I derivatives are isocyanides[48] $[Mn(RNC)_6]^+$ and the cyanide $[Mn(CN)_6]^{5-}$, which is obtained by reduction of $K_4[Mn(CN)_6]$ by K/Hg.[49] In the reduction[50] of the Mn^{II} sulfonated phthalocyanin complex, there is also evidence for Mn^I and Mn^0.

21-E. IRON

With iron the trends already noted in the relative stabilities of oxidation states continue, except that there is now no compound or chemically important circumstance in which the oxidation state is equal to the total number of valence shell electrons, which in this case is eight. The highest oxidation state known is VI, and it is rare. The only oxidation states of importance in the ordinary aqueous and related chemistry of iron are II and III.

The Mössbauer effect, though it has been observed for about one-third of the elements, is of major importance only for iron and to a lesser extent tin. For iron the effect depends on the fact that the nuclide ^{57}Fe, which is formed in the decay of ^{57}Co, has an excited state ($t_{1/2} \sim 10^{-7}$ sec) at 14.4 keV above the ground state; this can lead to a very sharp resonance absorption peak. Thus if γ-radiation from the ^{57}Co source falls on an absorber where the iron nuclei are in an environment identical with that of the source atoms, resonant absorption of γ-rays occurs. However if the Fe nuclei are in a different environment, no absorption occurs and the radiation is transmitted and can be measured. To obtain resonant absorption,

45   J. C. Smart and J. L. Robbins, *J. Am. Chem. Soc.*, 1979, **101**, 892, 3853.
46   A. Haaland, *Inorg. Nucl. Chem. Letts.*, 1979, **15**, 267.
47   W. Bünder and E. Weiss, *Z. Naturforsch.*, 1978, **33b**, 1235; A. Haaland, *Inorg. Nucl. Chem. Lett.*, 1979, **15**, 267.
48   P. M. Treichel and H. J. Mueh, *Inorg. Chem.*, 1977, **16**, 167.
49   J. Hauck, *Inorg. Nucl. Chem. Lett.*, 1976, **12**, 893.
50   D. J. Cookson *et al.*, *J.C.S. Dalton*, **1976**, 211.

it is then necessary to impart a velocity to the absorber, relative to the source. This motion changes the energy of the incident quanta (Döppler effect), so that at a certain velocity there is correspondence with the excitation energy of the nuclei in the absorber. The shifts in the absorption position relative to stainless steel as arbitrary zero are customarily expressed in velocities (mm sec^{-1}) rather than in energies. The shift in the resonance absorption depends both on the chemical environment and on temperature.

The chemical or isomer shift δ is a linear function of electron density (due to electrons occupying s orbitals) at the nucleus. This, in turn, is influenced by many factors (oxidation state, spin state, s character in the σ bonds, $d\pi$ back-bonding, ionicity), and correlations are not simple. For low-spin complexes δ is rather independent of oxidation state from -2 to $+2$. Even $Fe(CN)_6^{4-}$ and $Fe(CN)_6^{3-}$ have almost identical δ values. For high-spin compounds, however, δ varies markedly with formal oxidation state and indeed provides an excellent means of establishing it. A few examples of how Mössbauer spectra can be employed in studying the chemistry of iron are given now; others are mentioned as appropriate later.

In general, the Mössbauer spectra of compounds containing two or more iron atoms can give evidence for the occurrence of structurally nonequivalent iron atoms, but the converse procedure is highly dangerous, namely, concluding that all iron atoms must be equivalent when no resolution into separate peaks is observed. In certain cases when the environment of the iron atom is unsymmetrical the consequent electric field gradient interacts strongly enough with the nuclear quadrupole to produce substantial splitting of the resonance, and this observation can often be informative about structure.

As implied though not explicitly stated in the opening description, Mössbauer spectra can be recorded only on nuclei bound in a rigid solid environment. Therefore they are primarily used to study crystalline substances, though solutions that have been frozen to glasses are also suitable.

Prussian blue, formed from Fe^{3+} and $Fe(CN)_6^{4-}$, and Turnbull's blue, formed from Fe^{2+} and $Fe(CN)_6^{3-}$, have been shown to be identical, both having the composition $Fe_4[Fe(CN)_6]_3$ (i.e., ferric ferrocyanide). High-spin Fe^{III} and low-spin Fe^{II} were each identified, indicating that individual iron atoms have distinct, well-defined electron configurations with lifetimes of at least 10^{-7} sec. For further discussion of Prussian blue see page 762.

For a series of ions $[Fe(CN)_5L]^{n-}$ information about the π-acid character of the ligands L has been obtained, since π donation from Fe to L deshields the Fe nucleus. The strongest π acid has the highest s-electron density at the nucleus, hence the smallest δ. The order of δ values was found to be $NO < CO < CN^- < SO_3^{2-} < Ph_3P < NO_2^- < NH_3$.

The oxidation states and stereochemistry of iron are summarized in Table 21-E-1.

21-E-1.   The Element

Iron is the second most abundant metal, after aluminum, and the fourth most

TABLE 21-E-1
Oxidation States and Stereochemistry of Iron

Oxidation state	Coordination number	Geometry	Examples
Fe^{-II}	4	Tetrahedral	$Fe(CO)_4^{2-}$, $Fe(CO)_2(NO)_2$
Fe^0	5	Tbp	$Fe(CO)_5$, $(Ph_3P)_2Fe(CO)_3$, $Fe(PF_3)_5$
	6	Octahedral(?)	$Fe(CO)_5H^+$, $Fe(CO)_4PPh_3H^+$
Fe^I, d^7	6	Octahedral	$[Fe(H_2O)_5NO]^{2+}$
Fe^{II}, d^6	4	Tetrahedral	$FeCl_4^{2-}$, $FeCl_2(PPh_3)_2$
	5	Tbp	$[FeBr(Me_6tren)]Br$
	5	Sp	$[Fe(ClO_4)(OAsMe_3)_4]ClO_4$
	6^a	Octahedral	$[Fe(H_2O)_6]^{2+}$, $[Fe(CN)_6]^{4-}$
	8	Dodecahedral (D_{2h})	$[Fe(1,8\text{-naphthyridine})_4](ClO_4)_2$
Fe^{III}, d^5	3	Trigonal	$Fe[N(SiMe_3)_2]_3$
	4	Tetrahedral	$FeCl_4^-$, Fe^{III} in Fe_3O_4
	5	Sp	$FeCl(dtc)_2,^b$ $Fe(acac)_2Cl$
	5	Tbp	$Fe(N_3)_5^{2-}$, $FeCl_5^{2-}$
	6^a	Octahedral	Fe_2O_3, $[Fe(C_2O_4)_3]^{3-}$, $Fe(acac)_3$, $FeCl_6^{3-}$
	7	Approx. pentagonal bipyramidal	$[FeEDTA(H_2O)]^-$
	8	Dodecahedral	$[Fe(NO_3)_4]^-$
Fe^{IV}, d^4	4	Tetrahedral(?)	$Fe(1\text{-norbornyl})_4^c$
	6	Octahedral	$[Fe(diars)_2Cl_2]^{2+}$
Fe^{VI}, d^2	4	Tetrahedral	FeO_4^{2-}

[a] Most common states.
[b] dtc = dithiocarbamate.
[c] B. K. Bower and H. G. Tennent, *J. Am. Chem. Soc.*, 1972, **94**, 2512.

abundant element in the earth's crust. The core of the earth is believed to consist mainly of iron and nickel, and the occurrence of many iron meteorites suggests that it is abundant throughout the solar system. The major iron ores are *hematite* (Fe_2O_3), *magnetite* (Fe_3O_4), *limonite* [$FeO(OH)$], and *siderite* ($FeCO_3$).

The technical production and the metallurgy of iron are not discussed here. Chemically pure iron can be prepared by reduction of pure iron oxide (which is obtained by thermal decomposition of ferrous oxalate, carbonate, or nitrate) with hydrogen, by electrodeposition from aqueous solutions of iron salts, or by thermal decomposition of iron carbonyl.

Pure iron is a white, lustrous metal (m.p. 1528°). It is not particularly hard, and it is quite reactive. In moist air it is rather rapidly oxidized to give a hydrous oxide that affords no protection because it flakes off, exposing fresh metal surfaces. In a very finely divided state, metallic iron is pyrophoric. It combines vigorously with chlorine on mild heating and also with a variety of other nonmetals including the other halogens, sulfur, phosphorus, boron, carbon, and silicon. The carbide and silicide phases play a major role in the technical metallurgy of iron.

The metal dissolves readily in dilute mineral acids. With nonoxidizing acids and in absence of air, Fe^{II} is obtained. With air present or when warm dilute nitric acid is used, some of the iron goes to Fe^{III}. Very strongly oxidizing media such as concentrated nitric acid or acids containing dichromate passivate iron. Air-free water and dilute air-free hydroxides have little effect on the metal, but hot concentrated sodium hydroxide attacks it. In presence of air and water iron rusts to give a hydrated ferric oxide.

At temperatures up to 906° the metal has a body-centered lattice. From 906° to 1401°, it is cubic close packed, but at the latter temperature it again becomes body centered. It is ferromagnetic up to its Curie temperature of 768°, where it becomes simply paramagnetic.

Because of its high abundance, iron is often found as an impurity in other metal compounds. For example, corundum (γ-Al$_2$O$_3$) of gem quality is sapphire, and its colors are caused by small amounts of iron. Fe^{3+}—O—Fe^{3+} units cause a yellow color; Fe^{3+}—O—Fe^{2+} and/or Fe^{3+}—O—Ti^{4+} cause blue and green coloration.[1]

IRON COMPOUNDS

21-E-2. The Oxides of Iron

Because of the fundamental structural relationships between them, we discuss the oxides of iron together, rather than separately under the different oxidation states. Three such compounds are known. They all tend to be nonstoichiometric, but the ideal compositions of the phases are FeO, Fe$_2$O$_3$, and Fe$_3$O$_4$.

Iron(II) oxide is obtained by thermal decomposition of iron(II) oxalate in a vacuum as a pyrophoric black powder that becomes less reactive if heated to higher temperatures. The crystalline substance can be obtained only by establishing equilibrium conditions at high temperature, then rapidly quenching the system, since at lower temperatures FeO is unstable with respect to Fe and Fe$_3$O$_4$; slow cooling allows disproportionation. FeO has the rock salt structure. The FeO referred to thus far is iron defective (see below), having a typical composition of Fe$_{0.95}$O. Essentially stoichiometric FeO has been prepared from Fe$_{0.95}$O and Fe at 1050°K and 50 katm; it is about 0.4% less dense.

The brown hydrous ferric oxide FeO(OH) exists in several forms depending on the method of preparation, e.g., by hydrolysis of iron(III) chloride solutions at elevated temperatures or by oxidation of iron(II) hydroxide. When heated at 200° the final product is the red-brown α-Fe$_2$O$_3$, which occurs in Nature as the mineral hematite. It has the corundum structure where the oxide ions form a *hexagonally* close-packed array with Fe^{III} ions occupying octahedral interstices. By careful oxidation of Fe$_3$O$_4$ or by heating one of the modifications of FeO(OH) (lepidocrocite) one obtains γ-Fe$_2$O$_3$, which may be regarded as a *cubic* close-packed array of oxide ions with the Fe^{III} ions distributed randomly over both the octahedral and the tetrahedral interstices. There is also a third, rare, form designated β-Fe$_2$O$_3$.[2]

[1] J. Ferguson and P. E. Fielding, *Aust. J. Chem.*, 1972, **25**, 1371.
[2] H. Braun and K. J. Gallagher, *Nature (London)*, 1972, **240**, 13.

Finally, there is Fe_3O_4, a mixed $Fe^{II}-Fe^{III}$ oxide that occurs in Nature in the form of black, octahedral crystals of the mineral magnetite. It can be made by ignition of Fe_2O_3 above 1400°. It has the inverse spinel structure (p. 687). Thus the Fe^{II} ions are all in octahedral interstices, whereas the Fe^{III} ions are half in tetrahedral and half in octahedral interstices of a cubic close-packed array of oxide ions. The electrical conductivity, which is 10^6 times that of Fe_2O_3 is probably due to rapid valence oscillation between the Fe sites.

21-E-3. Other Binary Compounds

Halides. Only iron(II) (ferrous) and iron(III) (ferric) halides are known. The anhydrous ones are:

FeF_3	$FeCl_3$	$FeBr_3$	—
FeF_2	$FeCl_2$	$FeBr_2$	FeI_2

The three ferric halides can be obtained by direct halogenation of the metal. The iodide does not exist in the pure state, though some may be formed in equilibrium with FeI_2 and iodine. In effect, iron(III) is too strong an oxidizing agent to coexist with such a good reducing agent as I^-. In aqueous solution Fe^{3+} and I^- react quantitatively:

$$Fe^{3+} + I^- \rightarrow Fe^{2+} + \frac{1}{2}I_2$$

The fluoride is white, having only spin-forbidden electronic transitions in the visible spectrum (cf. Mn^{II}) and no low-energy charge-transfer band. $FeCl_3$ and $FeBr_3$ are red-brown because of charge-transfer transitions. All three halides have nonmolecular crystal structures with Fe^{III} ions occupying two-thirds of the octahedral holes in alternate layers. In the gas phase there are dimeric molecules, presumably consisting of tetrahedra sharing an edge, and monomers. All three will decompose to the Fe^{II} halides on strong heating in a vacuum.

Only iron(III) chloride is commonly encountered, as the yellow hexahydrate obtained by evaporation of aqueous solutions on a steam bath. Hydrates of the fluoride and bromide are also known.

The iron(II) halides are all known in both anhydrous and hydrated forms. The iodide and bromide can be prepared by reaction of the elements, though iron must be present in excess in the case of $FeBr_2$. For FeF_2 and $FeCl_2$ it is necessary to use HF or HCl to avoid forming the trihalides. $FeCl_3$ may be reduced to $FeCl_2$ by heating it in hydrogen, by treatment of a tetrahydrofuran solution with an excess of iron filings, or by refluxing it in chlorobenzene.

Iron dissolves in the aqueous hydrohalic acids and from these solutions the hydrated halides $FeF_2 \cdot 8H_2O$ (colorless), $FeCl_2 \cdot 6H_2O$ (pale green), and $FeBr_2 \cdot 4H_2O$ (pale green), may be crystallized. $FeCl_2 \cdot 6H_2O$ contains $trans$-$[FeCl_2(H_2O)_4]$ units.

Iron forms many binary compounds with the Group V and Group VI elements. Many are nonstoichiometric and/or interstitial. The sulfides are the most common. Iron(II) sulfide (FeS) and FeS_2 have been discussed (p. 514). Iron(III) sulfide is unstable and can be prepared and stored only with difficulty.[3] By treatment of

[3] A. H. Stiller *et al.*, *J. Am. Chem. Soc.*, 1978, **100**, 2553.

aqueous Fe^{III} with the stoichiometric quantity of aqueous Na_2S at 0° or below, it is obtained as a black, air-sensitive solid that slowly decomposes at 20°C.

21-E-4.   Aqueous and Coordination Chemistry of Iron(II), d^6

Iron(II) forms salts with virtually every stable anion, generally as green, hydrated, crystalline substances isolated by evaporation of aqueous solutions. The sulfate and perchlorate contain octahedral $[Fe(H_2O)_6]^{2+}$ ions. An important double salt is Mohr's salt $(NH_4)_2SO_4 \cdot FeSO_4 \cdot 6H_2O$, which is fairly stable toward both air oxidation and loss of water. It is commonly used in volumetric analysis to prepare standard solutions of iron(II) and as a calibration substance in magnetic measurements. Many other ferrous compounds are more or less susceptible to superficial oxidation by air and/or loss of water of crystallization, thus making them unsuitable as primary standards. This behavior is particularly marked for $FeSO_4 \cdot 7H_2O$, which slowly effloresces and becomes yellow-brown when kept.

Iron(II) carbonate, hydroxide, and sulfide may be precipitated from aqueous solutions of ferrous salts. Both the carbonate and the hydroxide are white, but in presence of air they quickly darken owing to oxidation. The sulfide also undergoes slow oxidation.

$Fe(OH)_2$ is somewhat amphoteric. It readily redissolves in acids, but also in concentrated sodium hydroxide. If 50% NaOH is boiled with finely divided iron and then cooled, fine blue-green crystals of $Na_4[Fe(OH)_6]$ are obtained. The strontium and barium salts may be precipitated similarly.

By reaction of Na_2O with Fe^{II} oxide, the compound Na_4FeO_3 is obtained. This contains the remarkable planar, carbonatelike $[FeO_3]^{4-}$ ion,[4] with Fe—O = 1.88 Å.

Iron(II) hydroxide can be obtained as a definite crystalline compound having the brucite $[Mg(OH)_2]$ structure.

Aqueous Chemistry. Aqueous solutions of iron(II), not containing other complexing agents, contain the pale blue-green hexaaqua iron(II) ion $[Fe(H_2O)_6]^{2+}$. The potential of the Fe^{3+}–Fe^{2+} couple, 0.771 V, is such that molecular oxygen can convert ferrous into ferric ion in acid solution:

$$2Fe^{2+} + \tfrac{1}{2}O_2 + 2H^+ = 2Fe^{3+} + H_2O \qquad E^0 = 0.46 \text{ V}$$

In basic solution, the oxidation process is still more favorable:

$$\tfrac{1}{2}Fe_2O_3 \cdot 3H_2O + e = Fe(OH)_2(s) + OH^- \qquad E^0 = -0.56 \text{ V}$$

Thus ferrous hydroxide almost immediately becomes dark when precipitated in presence of air and is eventually converted into $Fe_2O_3 \cdot nH_2O$.

Neutral and acid solutions of ferrous ion oxidize *less* rapidly with increasing acidity (even though the potential of the oxidation reaction becomes more positive). This is because Fe^{III} is actually present in the form of hydroxo complexes, except in extremely acid solutions, and there may also be kinetic reasons.

The oxidation of Fe^{II} to Fe^{III} in neutral solutions by molecular oxygen has been

4   H. Rieck and R. Hoppe, *Naturwissenschaften*, 1974, **61**, 126.

the subject of much speculation and may involve a reaction between $FeOH^+$ and O_2OH^-. The related problem of oxidation of Fe^{2+} by H_2O_2 (Fenton's reagent) is complicated and involves radicals generated by the reaction[5]

$$Fe^{II} + H_2O_2 \rightarrow Fe^{III}(OH) + OH$$

Complexes. Iron(II) forms a number of complexes, most of them octahedral. Ferrous complexes can normally be oxidized to ferric complexes and the Fe^{II}–Fe^{III} aqueous system provides a good example of the effect of complexing ligands on the relative stabilities of oxidation states:

$$[Fe(CN)_6]^{3-} + e = [Fe(CN)_6]^{4-} \qquad E^0 = 0.36 \text{ V}$$
$$[Fe(H_2O)_6]^{3+} + e = [Fe(H_2O)_6]^{2+} \qquad E^0 = 0.77 \text{ V}$$
$$[Fe(phen)_3]^{3+} + e = [Fe(phen)_3]^{2+} \qquad E^0 = 1.12 \text{ V}$$

The ferrous halides combine with gaseous ammonia forming several ammoniates, of which the highest are the hexaammoniates that contain the ion $[Fe(NH_3)_6]^{2+}$. Other anhydrous ferrous compounds also absorb ammonia. The ammine complexes are not stable in water, however, except in saturated aqueous ammonia. With chelating amine ligands, many complexes stable in aqueous solution are known. For example, ethylenediamine forms the entire series:

$$[Fe(H_2O)_6]^{2+} + en = [Fe(en)(H_2O)_4]^{2+} + 2H_2O \qquad K = 10^{4.3}$$
$$[Fe(en)(H_2O)_4]^{2+} + en = [Fe(en)_2(H_2O)_2]^{2+} + 2H_2O \qquad K = 10^{3.3}$$
$$[Fe(en)_2(H_2O)_2]^{2+} + en = [Fe(en)_3]^{2+} + 2H_2O \qquad K = 10^2$$

With ligands like phen, dipy, and others[6] supplying imine nitrogen donor atoms, stable low-spin (i.e., diamagnetic), octahedral, or distorted octahedral complexes are formed. β-diketones form $Fe(dike)_2$ complexes. $Fe(acac)_2$ is known to be a polymer in solution, and in the solid it has an extraordinary tetrameric structure[7] in which each iron atom is six-coordinate as a result of both oxygen bridging and weak Fe—C bonds (Fe—C = 2.79 Å).

The brown-ring test for nitrates and nitrites depends on the fact that under the conditions of the test, nitric oxide is generated. This combines with ferrous ion to produce a brown complex[8a] $[Fe(H_2O)_5NO]^{2+}$.

The hexacyanoferrate(II) ion, commonly called ferrocyanide, is a very stable and well-known complex of iron(II). The free acid $H_4[Fe(CN)_6]$ can be precipitated as an ether addition compound (probably containing oxonium ions R_2OH^+) by adding ether to a solution of the ion in strongly acidic solution; the ether can then be removed to leave the acid as a white powder. It is a strong tetrabasic acid when dissolved in water; in the solid the protons are bound to the nitrogen atoms of the CN groups with intermolecular hydrogen bonding. $H_4Fe(CN)_6$ dissolves without decomposition in liquid HF and is protonated[8b] to give $[Fe(CNH)_6]^{2+}$.

5   For references, see C. Walling *et al., Inorg. Chem.* 1970, **9**, 931; *J. Am. Chem. Soc.,* 1971, **93**, 4275.

6   K. B. Mertes, P. W. R. Corfield, and D. H. Busch, *Inorg. Chem.,* 1977, **16**, 3227.

7   F. A. Cotton and G. W. Rice, *Nouv. J. Chim.,* 1977, **1**, 301.

8a L. Burlamacchi, G. Martini, and E. Tiezzi, *Inorg. Chem.,* 1969, **8**, 2021.

8b R. J. Gillespie and R. Hulme, *J.C.S. Dalton,* **1973**, 1261.

Several diphosphine complexes trans-[FeCl$_2$(diphos)$_2$] can be prepared, and by treatment of these with LiAlH$_4$ in THF the hydrido complexes trans-[FeHCl(diphos)$_2$] and trans-[FeH$_2$(diphos)$_2$] can be obtained. These hydrido complexes are easily oxidized by air but have good thermal stability.

Only a few tetrahedral Fe^{II} complexes are known. These include the FeCl$_4^{2-}$ salts with large cations, of which FeCl$_4^{2-}$ is best characterized,[9] a few FeL$_4^{2+}$ complexes where L is, for example, Ph$_3$PO or (Me$_2$N)$_3$PO, and two cases where the donor atoms are sulfur: Fe[SP(Me$_2$)NP(Me$_2$)S]$_2$ and 21-E-I.[10] These, especially the latter, are interesting as models for the iron in rubredoxin (Section 31-5).

(21-E-I)   (21-E-II)

Iron(II), like Co^{II} and Ni^{II}, forms *five-coordinate complexes* with the tripod ligands (Section 3-6), typical ones being Fe(np$_3$)X$^+$ and Fe(pp$_3$)X$^+$, where X is a halogen.[11] Their structures may all be regarded as essentially *tbp*, but appreciable distortions occur in some cases. The number of unpaired electrons varies from 4 to 0 depending on the nucleophilicity and electronegativity of the donor atoms, and spin-state crossover equilibria occur in some cases.

In [Fe(napy)$_4$](ClO$_4$)$_2$, where napy is 21-E-II, as in the analogous complexes of Mn^{II}, Co^{II}, Ni^{II}, Cu^{II}, and Zn^{II}, the naphthyridine ligands give eight-coordination of the dodecahedral type.[12]

The goal of understanding the biochemistry of heme proteins such as hemoglobin, myoglobin, and the cytochromes (Section 31-4) has stimulated much research on iron complexes Fe(LLLL)X$_2$ with tetradentate macrocyclic rings LLLL having nitrogen donor atoms.[13] Ring size, extent of unsaturation, and other factors have been varied to see what factors are important in controlling structure, spin-state, and the tendency toward association. As ring size decreases, *d*-orbital splitting increases dramatically, as might be expected. However there are a number of complications, such as the formation of cis as well as trans complexes and we shall not try to summarize the details here.

Electronic Structures of Iron(II) Complexes. The ground state ^5D of a d^6 configuration is split by octahedral and tetrahedral ligand fields into $^5T_2$ and ^5E states; there are no other quintet states; hence only one spin-allowed d–d transition occurs if one of these is the ground state. All tetrahedral complexes are high spin, and the $^5E \rightarrow {}^5T_2$ band typically occurs at ~4000 cm^{-1}. The magnetic moments are normally 5.0 to 5.2 BM, owing to the spins of the four unpaired electrons and a small,

9   J. W. Lauher and J. A. Ibers, *Inorg. Chem.*, 1975, **14**, 348.
10 R. W. Lane *et al., J. Am. Chem. Soc.*, 1977, **99**, 84.
11 L. Sacconi and M. DiVaira, *Inorg. Chem.*, 1978, **17**, 810, and earlier references therein.
12 R. L. Bodner and D. G. Hendricker, *Inorg. Chem.*, 1973, **12**, 33.
13 D. H. Busch *et al., Inorg. Chem.*, 1976, **15**, 387, and numerous earlier papers.

second-order orbital contribution. For high-spin octahedral complexes, e.g., $Fe(H_2O)_6^{2+}$, the $^5T_{2g} \rightarrow {}^5E_g$ transition occurs in the visible or near-infrared region ($\sim$10,000 cm^{-1} for the aqua ion) and is broad or even resolvably split owing to a Jahn-Teller effect in the excited state, which derives from a $t_{2g}^3 e_g^3$ configuration. Magnetic moments are around 5.2 BM in magnetically dilute compounds.

For Fe^{II} quite strong ligand fields are required to cause spin pairing, but a number of low-spin complexes, such as $Fe(CN)_6^{4-}$, $Fe(CNR)_6^{2+}$, and $[Fe(phen)_3]^{2+}$ are known. The best characterized exceptions are those with 2-chloro-phen and 2-methyl-phen as ligands.[14] The former has four unpaired electrons at all temperatures from 4 to 300°K; the latter is a temperature-dependent mixture of quintet and singlet states. Comparison with the known structure[15a] of $Fe(phen)_3^{3+}$, even allowing for some difference between the Fe^{II} and Fe^{III} radii, makes it clear that ligands with Cl or CH$_3$ at the 2-position will be unable to approach the iron atom as closely as phen itself. Fe phen$_2$(CN)$_2$ is also diamagnetic, though most Fe phen$_2$X$_2$ complexes are high spin. When X = SCN or SeCN a spin state crossover situation occurs (cf. p. 648), and the magnetic moment is temperature-dependent, ranging from $\sim$5.1 BM at 300°K to $\sim$1.5 BM at $\leqslant$150°K.[15b] Some other crossover cases that have been well studied are Fe(HBpz$_3$)$_2$, and several complexes with 2-pyridylmethylamine and some similar ligands.

Although it can be shown that for strict octahedral symmetry no d^6 ion can have a ground state with two unpaired electrons (only 4 or 0), this might be possible in six-coordinate complexes in which there are significant departures from O_h symmetry in the ligand field. Perhaps the best documented examples are complexes of the type [Fe(LL)$_2$ox] and [Fe(LL)$_2$mal], where LL represents a bidentate diamine ligand such as o-phen or bipy, and ox, mal represent oxalato and maleato ions. These complexes have magnetic susceptibilities that follow the Curie-Weiss law over a broad temperature range, with $\mu \approx 3.90$ BM (part of which is due to a temperature-independent paramagnetism).

For a strictly square, four-coordinate complex such as phthalocyanin iron(II), the extreme tetragonality of the ligand field apparently places one d orbital ($d_{x^2-y^2}$) at high energy, and the six electrons adopt a high-spin distribution among the remaining four, thus giving a triplet ground state, independent of temperature.

21-E-5. Aqueous and Coordination Chemistry of Iron(III)

Iron(III) occurs in salts with most anions, except those that are incompatible with it because of their character as reducing agents. Examples obtained as pale pink to nearly white hydrates from aqueous solutions are Fe(ClO$_4$)$_3$·10H$_2$O, Fe(NO$_3$)$_3$·9(or 6)H$_2$O, and Fe$_2$(SO$_4$)$_3$·10H$_2$O.

Aqueous Chemistry. One of the most conspicuous features of ferric iron in aqueous solution is its tendency to hydrolyze and/or to form complexes. It has been

---

[14] W. M. Rieff and G. J. Long, *Inorg. Chem.*, 1974, **13**, 2150.

[15a] A. Zalkin, D. H. Templeton, and T. Ueki, *Inorg. Chem.*, 1973, **12**, 1641.

[15b] E. Koenig and K. Madeja, *Inorg. Chem.*, 1968, **7**, 1848.

establisned that the hydrolysis (equivalent in the first stage to acid dissociation of the aqua ion) is governed in its initial stages by the following equilibrium constants:

$$[Fe(H_2O)_6]^{3+} = [Fe(H_2O)_5(OH)]^{2+} + H^+ \qquad K = 10^{-3.05}$$
$$[Fe(H_2O)_5(OH)]^{2+} = [Fe(H_2O)_4(OH)_2]^+ + H^+ \qquad K = 10^{-6.31}$$
$$2[Fe(H_2O)_6]^{3+} = [Fe(H_2O)_4(OH)_2Fe(H_2O)_4]^{4+} + 2H^+ \qquad K = 10^{-2.91}$$

The formulation of the dimer above, with two μ-OH groups, was for a long time accepted, but recent work[16] makes a linear Fe—O—Fe bridge seem far more likely. Even at pH's of 2 to 3, the extent of hydrolysis is very great, and to have solutions containing Fe^{III} mainly (say ~99%) in the form of the pale purple hexaaqua ion the pH must be around zero. As the pH is raised above 2 to 3, more highly condensed species than the dinuclear one noted above are formed, attainment of equilibrium becomes sluggish, and soon colloidal gels are formed; ultimately, hydrous ferric oxide is precipitated as a red-brown gelatinous mass. In basic solutions equilibria are reached *very* slowly (4–8 weeks) and the precipitates are often basic salts such as $Fe(OH)_{2.7}Cl_{0.3}$ or $Fe(OH)_2NO_3$.[17] There is no evidence for a stoichiometric hydroxide $Fe(OH)_3$, and in the absence of any anions other than OH^-, the precipitates are best written $Fe_2O_3 \cdot nH_2O$, though $FeO(OH)$ may sometimes be a component.

The various hydroxo species, such as $[Fe(OH)(H_2O)_5]^{2+}$, are yellow because of charge-transfer bands in the ultraviolet region, which have tails coming into the visible region. Thus aqueous solutions of ferric salts even with noncomplexing anions are yellow unless strongly acid.

Hydrous iron(III) oxide is readily soluble in acids but also to a slight extent also in strong bases. When concentrated solutions of strontium or barium hydroxide are boiled with ferric perchlorate, the hexahydroxoferrates(III) $M_3^{II}[Fe(OH)_6]_2$ are obtained as white crystalline powders. With alkali metal hydroxides, substances of composition M^IFeO_2 can be obtained; these can also be made by fusion of Fe_2O_3 with the alkali metal hydroxide or carbonate in the proper stoichiometric proportion. Moderate concentrations of what is presumably the $[Fe(OH)_6]^{3-}$ ion can be maintained in strongly basic solutions.

Ferric iron in aqueous solution is rather readily reduced by many reducing agents, such as I^-, as noted above. It also oxidizes sulfide ion, so that ferric sulfide precipitated on addition of H_2S or a sulfide to an Fe^{III} solution rapidly changes to a mixture of iron(II) sulfide and colloidal sulfur. Adding carbonate or hydrogen carbonate to an iron(III) solution precipitates the hydrous oxide.

Iron(III) Complexes.[18] The majority of iron complexes are octahedral, but tetrahedral and square pyramidal ones are also important.

The hexaaqua ion exists in very strongly acid solutions of ferric salts, in the several ferric alums $M^IFe(SO_4)_2 \cdot 12H_2O$, and presumably also in the highly hydrated crystalline salts.

16 J. M. Knudsen *et al.*, *Acta Chem. Scand., A*, 1975, **29**, 833.
17 P. R. Danesi *et al.*, *Inorg. Chem.*, 1973, **12**, 2089.
18 S. A. Cotton, *Coord. Chem. Rev.*, 1972, **8**, 185.

The affinity of iron(III) for amine ligands is very low. No simple ammine complex exists in aqueous solution; addition of aqueous ammonia only precipitates the hydrous oxide. Chelating amines, for example, EDTA, do form some definite complexes, among which is the seven-coordinate $[Fe(EDTA)H_2O]^-$ ion. Also, those amines such as 2,2'-bipyridine and 1,10-phenanthroline that produce ligand fields strong enough to cause spin pairing form fairly stable complexes, isolable in crystalline form with large anions such as perchlorate.

Iron(III) has its greatest affinity for ligands that coordinate by oxygen, viz., phosphate ions, polyphosphates, and polyols such as glycerol and sugars. With oxalate the trisoxalato complex $[Fe(C_2O_4)_3]^{3-}$ is formed, and with β-diketones the neutral $[Fe(dike)_3]$ complexes are formed. Formation of complexes with β-diketones is the cause of the intense colors that develop when they are added to solutions of ferric ion, and this serves as a useful diagnostic test for them.

Ferric ion forms complexes with halide ions and SCN^-. Its affinity for F^- is quite high, as shown by the equilibrium constants

$$Fe^{3+} + F^- = FeF^{2+} \qquad K_1 \approx 10^5$$
$$FeF^{2+} + F^- = FeF_2^+ \qquad K_2 \approx 10^5$$
$$FeF_2^+ + F^- = FeF_3 \qquad K_3 \approx 10^3$$

The corresponding constants for chloro complexes are only ~ 30, ~ 5, ~ 0.1, respectively. In very concentrated HCl the tetrahedral $FeCl_4^-$ ion[19] is formed and its salts with large cations may be isolated. The $FeBr_4^-$ ion is less common but also well characterized.[20] $FeCl_3$ also forms simple tetrahedral adducts [e.g., $FeCl_3(THF)$].[21] Tetrahedral FeF_4^- does not appear to exist, $CsFeF_4$, for example, having FeF_6 octahedral involving shared F^- ions.[22] The complexes with SCN^- are an intense red, and this serves as a sensitive qualitative and quantitative test for ferric ion; $Fe(SCN)_3$ and/or $Fe(SCN)_4^-$ may be extracted into ether. Fluoride ion, however, will discharge this color. In the solid state, FeF_6^{3-} ions are known, but in solutions only species with fewer F atoms occur. Other halo complexes that occur in crystalline compounds are $FeCl_5^{2-}$,[23a] $FeCl_6^{3-}$, and $Fe_2Cl_9^{3-}$. The $FeCl_6^{3-}$ ion trapped at low concentration in a $Co(NH_3)_6InCl_6$ host has been thoroughly studied by epr.[23b] Fe^{III} forms various six-coordinate complexes with sulfur ligands such as $[Fe(SCN)_6]^{3-}$, the $Fe(S_2CNR_2)_3$ compounds where the short "bite" of the dithiocarbamato ligand gives three small S—Fe—S angles and reduces the symmetry to D_3,[24] and the interesting $[Fe(S_2C_2O_2)_3]^{3-}$ where the mode of coordination of the thiooxalate ions can be controlled by varying the counterions. Sulfur atoms are coordinated when the countercations are K^+ and the complex is low spin. When

19   F. A. Cotton and C. A. Murillo, *Inorg. Chem.*, 1975, **14**, 2467.
20   G. D. Sproul and G. D. Stucky, *Inorg. Chem.*, 1972, **11**, 1647; M. Vala, P. Mongan, and P. J. McCarthy, *J.C.S. Dalton*, **1972**, 1870; C. D. Flink and P. Greenough, *J. Chem. Phys.*, 1972, **56**, 5771.
21   L. S. Brenner and C. A. Root, *Inorg. Chem.*, 1972, **11**, 652.
22   D. Babel, F. Wall, and G. Heger, *Z. Naturforsch.*, 1974, **29b**, 139.
23a C. S. Creaser and J. A. Creighton, *J. Inorg. Nucl. Chem.*, 1979, **41**, 469.
23b E. W. Stout, Jr., and B. B. Garrett, *Inorg. Chem.*, 1973, **12**, 2565.
24   P. E. Healy and E. Sinn, *Inorg. Chem.*, 1975, **14**, 109.

K^+ is replaced by $(Ph_3P)_2Ag^+$, the preference of Ag for S reverses the dithiooxalate ions and the Fe^{III} becomes high spin.[25]

With the cyanide ion, salts of the hexacyanoferrate ion and the free acid $H_3[Fe(CN)_6]$ are well known. In contrast to $[Fe(CN)_6]^{4-}$, the $[Fe(CN)_6]^{3-}$ ion is quite poisonous; for kinetic reasons the latter dissociates and reacts rapidly, whereas the former is not labile. There are a variety of substituted ions $[Fe(CN)_5X]$, (X = H_2O, NO_2, etc.), of which the best known is the nitroprusside ion $[Fe(CN)_5NO]^{2-}$; this is attacked by OH^- to give $[Fe(CN)_5NO_2]^{2-}$.

Mixed metal oxides containing Fe^{III} are almost invariably close-packed arrays of oxygen atoms with Fe^{III} in the interstices. However the compound $K_6Fe_2O_6$ contains discrete $[Fe_2O_6]^{6-}$ anions consisting of FeO_4 tetrahedra sharing an edge.[26]

Oxo-Bridged Complexes. This is an aspect of iron(III) chemistry that has recently received considerable attention because of its possible relevance to understanding certain aspects of the biochemistry of iron (Section 31-6), although there does not, at present, appear to be any specific biochemical problem that is directly modeled by these substances. Nonetheless they are of considerable fundamental interest. The binuclear systems are of two types: doubly bridged ones that can be described as two octahedra sharing an edge, and those with a linear, or nearly linear, Fe—O—Fe bridge.

Some representative doubly bridged complexes[27,28] are shown schematically as 21-E-III, 21-E-IV, and 21-E-V.

(21-E-III)   (21-E-IV)

(21-E-V)

25   F. J. Hollander and D. Coucouvanis, *Inorg. Chem.,* 1974, **13,** 2381.
26   R. Rieck and R. Hoppe, *Z. Anorg. Allg. Chem.,* 1974, **408,** 151.

In all cases there is antiferromagnetic coupling between the iron atoms, but it is relatively weak,[27-29] and the magnetic properties can be understood in terms of two high-spin d^5 ions interacting via a magnetic exchange parameter J (see Appendix 5) with a value of 0 to -15 cm^{-1}.

The μ-O species show considerably stronger coupling. Their magnetic moments also approach zero toward 0°K and they tend to a high temperature limit of only *ca.* 1.9 BM per Fe. They have linear or nearly linear Fe—O—Fe units (angles $\geq 165°$) with rather short, strong Fe—O bonds. All these facts seem to imply that there is strong π bonding across the bridge leading to the situation of each Fe^{III} atom having one unpaired electron and these then being coupled antiferromagnetically. However very extensive studies[30] of several representative complexes of this type have led to the conclusion that the electronic structure is best described by strong antiferromagnetic coupling ($-J \approx 100$ cm^{-1}) of two high-spin d^5 Fe^{III} ions. Representative complexes of this type include [Fe$_2$O(HEDTA)$_2$]$^{2-}$ and [Fe$_2$O(EDTA)$_2$]$^{4-}$, where the iron atoms are six-coordinate, and [Fe$_2$(tetraphenylporphine)$_2$O] and [Fe$_2$(sal$_2$en)$_2$O], where they are five-coordinate.

Iron(III) also forms basic carboxylates based on the[Fe$_3$O(O$_2$CR)$_6$(H$_2$O)$_3$]$^+$ ion,[31a] which has the characteristic O-centered structure (page 154). In addition to the usual alkyl R groups, amino acids (i.e., R = CR'HNH$_3^+$) may also be incorporated.[31b] In all cases there are medium antiferromagnetic interactions ($-J \approx 30$ cm^{-1}), but there are some puzzling features suggesting that all three Fe–Fe interactions may not be equivalent,[32] which is difficult to reconcile with the symmetrical structures.

Electronic Structures of Iron(III) Compounds. Iron(III) is isoelectronic with manganese(II), but much less is known of the details of Fe^{III} spectra because of the much greater tendency of the trivalent ion to have charge-transfer bands in the near-ultraviolet region with strong low-energy wings in the visible that obscure the very weak, spin-forbidden d–d bands. Insofar as they are known, however, the spectral features of iron(III) ions in octahedral surroundings are in accord with theoretical expectations.

Magnetically iron(III), like manganese(II), is high spin in nearly all its complexes, except those with the strongest ligands, exemplified by [Fe(CN)$_6$]$^{3-}$, [Fe(bipy)$_3$]$^{3+}$, [Fe(phen)$_3$]$^{3+}$, and other tris complexes with imine nitrogen atoms as donors.[33] In the high-spin complexes the magnetic moments are always very close to the spin-only value of 5.9 BM because the ground state (derived from the ^6S state of the free ion) has no orbital angular momentum and there is no effective mechanism for introducing any by coupling with excited states. The low-spin complexes,

27 J. A. Thich *et al., J. Am. Chem. Soc.,* 1976, **98**, 1425.
28 J. A. Bertrand *et al., Inorg. Chem.,* 1974, **13**, 125, 927.
29 R. G. Wollmann and D. N. Hendrickson, *Inorg. Chem.,* 1978, **17**, 926.
30 H. B. Gray *et al., J. Am. Chem. Soc.,* 1972, **94**, 2683; ACS Advances in Chemistry Series Monograph No. 100, 1971, p. 365.
31a J. Catterick, P. Thornton, and B. W. Fitzsimmons, *J.C.S. Dalton,* **1977**, 1420.
31b R. Thundathil and S. L. Holt, *Inorg. Chem.,* 1976, **15**, 745.
32 G. L. Long *et al., J.C.S. Dalton,* **1973**, 573.
33 L. F. Warner, *Inorg. Chem.,* 1977, **16**, 2814; D. R. Eaton, W. R. McClellan, and J. F. Weiher, *Inorg. Chem.,* 1968, **7**, 2040.

with t_{2g}^5 configurations, usually have considerable orbital contributions to their moments at about room temperature, values of ~2.3 BM being obtained. The moments are, however, intrinsically temperature dependent, and at liquid nitrogen temperature (77°K) they decrease to ~1.9 BM. There is evidence[34] of very high covalence and electron delocalization in low-spin complexes such as $[Fe(phen)_3]^{3+}$ and $[Fe(bipy)_3]^{3+}$.

Five-coordinate complexes may be high or low spin depending on the ligands; some of the important high-spin ones have already been encountered in oxo-bridged dinuclear complexes. The $Fe(S_2CNR_2)_2X$ (X = Cl, Br, I) complexes that form readily on treating $Fe(S_2CNR_2)_3$ with halogens have three unpaired electrons. These molecules have very distorted sp configuration with X axial (actual symmetry C_{2v}), and with coordinate axes as defined in 21-E-VI the electron configuration is $d^2_{x^2-y^2}$, d^1_{xz}, d^1_{yz}, $d^1_{z^2}$, d^0_{xy}, according to magnetic anisotropy measurements.[35]

(21-E-VI)

Quite a few iron(III) complexes, especially some $Fe(S_2CNR_2)_3$ types,[36] provide good examples of spin-crossover as discussed earlier (p. 648).

21-E-6. Compounds with Iron in Mixed Oxidation States

Prussian Blue and Related Compounds. It has long been known that treating a solution of Fe^{III} with hexacyanoferrate(II) yields a blue precipitate called *Prussian blue* and that treating a solution of Fe^{II} with hexacyanoferrate(III) yields a blue precipitate called *Turnbull's blue*. It has been clear for some time that these substances are identical, both being iron(III) hexacyanoferrate(II). It was also shown many years ago that the structure is based on a cubic array of iron atoms with CN ions on the edges of the cubes, but precise structural characterization was not achieved because single crystals could not be obtained. There would be equal numbers of Fe^{III} and Fe^{II} atoms in the type of cubic array envisioned, and to reconcile this with composition, two possibilities were generally considered:

1. The correct formula is $Fe_4[Fe(CN)_6]_3 \cdot xH_2O$ and one-fourth of the Fe^{III} occupy the centers of cubes, as do water molecules, rather than the corners.

2.   The correct formula is $M^I Fe^{III}[Fe^{II}(CN)_6] \cdot yH_2O$, and the centers of cubes are occupied by M^I ions and water molecules, where M^I may be Na, K, or Rb.

[34]   P. B. Merrithew, C.-C. Lo, and A. J. Modestino, *Inorg. Chem.,* 1973, **12**, 1927.
[35]   P. Ganguli, V. R. Marathe, and S. Mitra, *Inorg. Chem.,* 1975, **14**, 970.
[36]   P. Ganguli and V. R. Marathe, *Inorg. Chem.,* 1978, **17**, 543.

The most recent work,[37] based on single-crystal X-ray data, suggests a different interpretation in which all Fe^{III} and Fe^{II} are at cube corners but with one-fourth of the $[Fe^{II}(CN)_6]^{4-}$ sites occupied by water molecules. In addition, water molecules occupy positions about Fe^{III} atoms where CN ions are missing, and there is one H_2O in the center of each cube. This structure corresponds to $Fe_4Fe(CN)_6]_3 \cdot 16H_2O$, which agrees well with the fourteen to sixteen water molecules typically found by analysis and gives a satisfactory calculated density. There is a so-called soluble form of Prussian blue that does contain potassium or other alkali metal, but not in stoichiometric quantity as implied by the formula above.

A fundamentally similar structure appears to exist in compounds such as iron(III) hexacyanoferrate(III), $Fe^{III}[Fe^{III}(CN)_6]$, white, insoluble $K_2Fe^{II}[Fe^{II}(CN)_6]$, and $Cu_2^{II}[Fe^{II}(CN)_6]$.

Although Turnbull's blue is not the expected ferrous ferricyanide, that compound can be obtained when $Fe_4[Fe(CN)_6]_3 \cdot xH_2O$ is pyrolyzed at 400° in vacuum, causing loss of all H_2O.[38]

Iron-Sulfur Clusters. This is another area of iron chemistry that has been enormously stimulated by the desire to understand biochemical systems, specifically, the behavior of nonheme iron-sulfur proteins (Section 31-5). A number of bi- and tetranuclear iron complexes containing bridging sulfur atoms have been synthesized,[39] and some of them appear to be identical in their inner regions to the iron-sulfur clusters in the biological systems.[39,40] They all exist in several overall states of oxidation, some of which correspond formally to mixtures of Fe^{II} and Fe^{III}; but there do not appear in any case to be distinct integral oxidation numbers for individual metal atoms.

The two most important types of iron-sulfur cluster are those represented schematically by 21-E-VII and 21-E-VIII.

(21-E-VII)   (21-E-VIII)

In most cases the group X is an RS— or half of an —S——S— ligand, but compounds in which X = Cl, Br, or I are also known.[41]

The redox chemistry of these systems is extensive. The 2Fe–2S clusters are known

37   H. J. Buser et al., Inorg. Chem., 1977, **16**, 2704.
38   J. G. Cosgrove, R. L. Collins, and D. S. Murty, J. Am. Chem. Soc., 1973, **95**, 1083.
39   S. J. Lippard, Acc. Chem. Res., 1973, **6**, 282; R. H. Holm, Acc. Chem. Res., 1977, **10**, 427.
40   J. Cambray et al., Inorg. Chem., 1977, **16**, 2565.
41   G. B. Wong, M. A. Bobrik, and R. H. Holm, Inorg. Chem., 1978, **17**, 578.

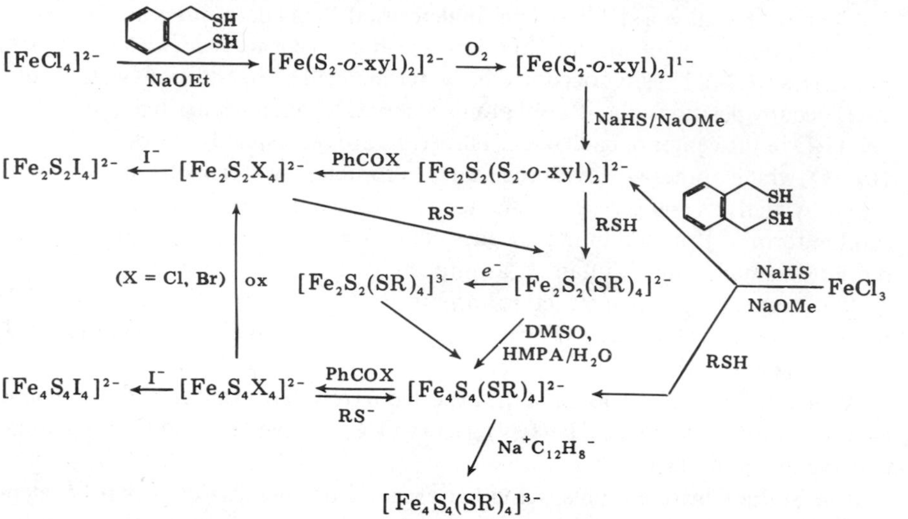

Fig. 21-E-1. Some important reactions of iron-sulfur cluster anions (See R. H. Holm, *Accts. Chem. Res.*, 1977, **10**, 427.)

in the series $[Fe_2S_2X_4]^{2-,3-,4-}$ and the $[Fe_4S_4X_4]$ ones may have charges of -1 through -4. The strong interactions between the iron atoms that allow delocalization rather than trapping of the II and III valence for iron occur at least in part, and perhaps entirely, via super exchange across the sulfur bridges.[42]

The compounds can be prepared relatively easily by reactions shown in Fig. 21-E-1. Note that an important role is played by mononuclear-binuclear and binuclear-tetranuclear interconversions.

Another class of polynuclear iron-sulfur compounds, some of which have been known for a very long time, are those containing also NO. These include the red Roussin ester $(NO)_2Fe(\mu\text{-SEt})_2Fe(NO)_2$, the black Roussin salts 21-E-IX, and the cubanelike $Fe_4(\mu_3\text{-S})_4(NO)_4$.[43]

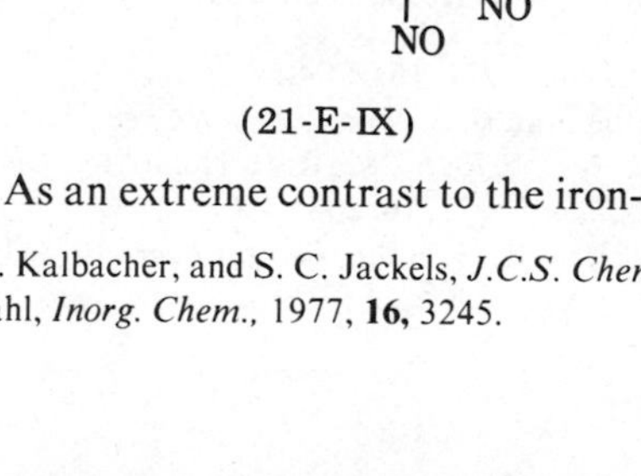

(21-E-IX)

Other Compounds. As an extreme contrast to the iron-sulfur clusters, there are

[42] J. G. Norman, Jr., B. J. Kalbacher, and S. C. Jackels, *J.C.S. Chem. Comm.*, **1978**, 1027.
[43] C. T. Chu and L. F. Dahl, *Inorg. Chem.*, 1977, **16**, 3245.

mixed valence compounds in which there is essentially complete trapping of valences at different sites. Among these are the $[(CN)_5Fe-LL-Fe(CN)_5]^n$ ions where LL is pyrazine or 4,4'-bipyridyl, for example, and all three combinations of oxidation states (+2, +2), (+2, +3), and (+3, +3) for iron atoms can be obtained.[44] There is an almost statistical equilibrium among them in solution, showing that Fe–Fe interactions are negligible. Spectroscopic study shows that there is only slight electron delocalization across the bridge.

21-E-7. Higher Oxidation States

The states IV, V, and VI are known, though V is known only in $M_3^IFeO_4$ compounds, which are rarely if ever obtained pure and are not well characterized.[45] The Fe^V appears to have three unpaired electrons.

Iron(IV). This state is known in only a few environments. The most easily prepared and stable type of Fe^{IV} compound comprises the trisdithiocarbamate cations[46] $Fe(S_2CNR_2)_3^+$, which can be obtained by electrochemical or air oxidation of $Fe(S_2CNR_2)_3$ and isolated as salts of BF_4^- or ClO_4^-. The cations, with structures very similar to those of the $Fe(S_2CNR_2)_3$ molecules,[47] have two unpaired electrons.

The mixed oxides Sr_2FeO_4 and Ba_2FeO_4 have long been known. They are made by the reaction

$$M_3^{II}[Fe(OH)_6]_2 + M^{II}(OH)_2 + \tfrac{1}{2}O_2 \xrightarrow{800-900°} 2M_2^{II}FeO_4 + 7H_2O$$

These contain no discrete FeO_4^{4-} ion but are mixed metal oxides, the barium one having the spinel structure.

The cationic species $[Fe(diars)_2X_2]^{2+}$ (X = Cl or Br) are obtained by oxidation of the $[Fe(diars)_2X_2]^+$ ions with $15M$ nitric acid. Their magnetic properties (μ_{eff} = 2.98 BM at 293°K) are consistent with there being two unpaired electrons in a tetragonally distorted set of t_{2g} orbitals.

There is finally the remarkable purple alkyl, Fe(1-norbornyl)$_4$, whose stability is doubtless due to the steric bulk of the ligand.[48]

Iron(VI). The Na, K, Rb, Cs, and Ba salts of the FeO_4^{2-} ion can be obtained a number of ways, but pure materials can be prepared only by certain methods.[49] These deep blue compounds contain discrete tetrahedral FeO_4^{2-} ions that have two unpaired electrons. The alkali metal salts are isomorphous with β-K_2SO_4. The FeO_4^{2-} ion is moderately stable in very basic aqueous solution but decomposes in neutral or acid solution according to the equation:

$$2FeO_4^{2-} + 10H^+ = 2Fe^{3+} + \tfrac{1}{2}O_2 + 5H_2O$$

[44] F. Felix and A. Ludi, *Inorg. Chem.*, 1978, **17**, 1782.
[45] I. G. Kokarovtseva, I. N. Belyaev, and L. V. Semenyakova, *Russ. Chem. Rev.*, 1972, **41**, 929; W. A. Levason and C. A. McAuliffe, *Coord. Chem. Rev.*, 1974, **12**, 151.
[46] E. A. Pasek and D. K. Straub, *Inorg. Chem.*, 1972, **11**, 259; R. M. Golding et al., *Aust. J. Chem.*, 1972, **25**, 2567; G. Cauquis and D. Laehend, *Inorg. Nucl. Chem. Lett.* 1973, **9**, 1095; D. Petridis et al., *Inorg. Chem.*, 1979, **18**, 505.
[47] R. L. Martin et al., *J. Am. Chem. Soc.*, 1974, **96**, 3647.
[48] B. K. Bower and H. G. Tennant, *J. Am. Chem. Soc.*, 1972, **94**, 2512.
[49] R. J. Audette and J. W. Quail, *Inorg. Chem.*, 1972, **11**, 1904.

It is an even stronger oxidizing agent than permanganate and can oxidize NH_3 to N_2, Cr^{II} to CrO_4^{2-}, and arsenite to arsenate. It is a powerful oxidizing agent in organic chemistry.[50]

21-F. COBALT

The trends toward decreased stability of the very high oxidation states and the increased stability of the II state relative to the III state, which have been noted through the series Ti, V, Cr, Mn, and Fe, persist with cobalt. Indeed, the first trend culminates in the complete absence of oxidation states higher than IV under chemically significant conditions. The oxidation state IV is represented by only a few compounds such as $Co(1\text{-norbornyl})_4$, Cs_2CoF_6, CoO_2, Ba_2CoO_4, and a heteropolymolybdate. The III state is relatively unstable in simple compounds, but the low-spin complexes are exceedingly numerous and stable, especially where the donor atoms (usually N) make strong contributions to the ligand field. There are also some important complexes of Co^I; this oxidation state is better known for cobalt than for any other element of the first transition series except copper.

The oxidation states and stereochemistry are summarized in Table 21-F-1.

21-F-1.   The Element

Cobalt always occurs in Nature in association with nickel and usually also with arsenic. The most important cobalt minerals are *smaltite* ($CoAs_2$) and *cobaltite* ($CoAsS$), but the chief technical sources of cobalt are residues called "speisses," which are obtained in the smelting of arsenical ores of nickel, copper, and lead. The separation of the pure metal is somewhat complicated and of no special relevance here.

Cobalt is a hard, bluish-white metal (m.p. 1493°, b.p. 3100°). It is ferromagnetic with a Curie temperature of 1121°. It dissolves slowly in dilute mineral acids, the Co^{2+}/Co potential being -0.277 V, but it is relatively unreactive. It does not combine directly with hydrogen or nitrogen; in fact, no hydride or nitride appears to exist. The metal will combine with carbon, phosphorus, and sulfur on heating. It also is attacked by atmospheric oxygen and by water vapor at elevated temperatures, giving CoO.

COBALT COMPOUNDS

21-F-2.   Binary Cobalt Compounds: Simple Salts

Oxides.[1]   On heating the metal in oxygen at 1100° or on heating cobalt carbonate, nitrate, etc., cobalt(II) oxide is obtained as an olive green substance. It

50   D. H. Williams and J. T. Riley, *Inorg. Chim. Acta,* 1974, **8,** 177.

1   (a) J. S. Choi and C. H. Yo, *Inorg. Chem.,* 1974, **13,** 1720; (b) W. Burrow and R. Hoppe, *Angew. Chem., Int. Ed.,* 1979, **18,** 61.

TABLE 21-F-1
Oxidation States and Stereochemistry of Cobalt

Oxidation state	Coordination number	Geometry	Examples
Co^{-1}, d^{10}	4	Tetrahedral	$Co(CO)_4^-$, $Co(CO)_3NO$
Co^0, d^9	4	Tetrahedral	$K_4[Co(CN)_4]$, $Co(PMe_3)_4$
Co^1, d^8	4	Tetrahedral	$CoBr(PR_3)_3$
	5[a]	Tbp	$[Co(CO)_3(PR_3)_2]^+$
			$HCo(PF_3)_4$,[b] $[Co(NCMe)_5]^+$
	5	Sp	$[Co(NCPh)_5]ClO_4$[c]
	6	Octahedral	$[Co(bipy)_3]^+$
Co^{II}, d^7	2	Linear	$Co[N(SiMe_3)_2]_2$
	4[a]	Tetrahedral	$[CoCl_4]^{2-}$,
			$CoBr_2(PR_3)_2$, Co^{II} in Co_3O_4
	5	Tbp	$[Co(Me_6tren)Br]^+$,
			$CoH(BH_4)(PCy_3)_2$[d]
	5	Sp	$[Co(ClO_4)(MePh_2AsO)_4]^+$,
			$[Co(CN)_5]^{3-}$,
			$[Co(CNPh)_5]^{2+}$
	6[a]	Octahedral	$CoCl_2$, $[Co(NH_3)_6]^{2+}$
	8	Dodecahedral	$(Ph_4As)_2[Co(NO_3)_4]$
Co^{III}, d^6	4	Tetrahedral	In a 12-heteropolytungstate; in garnets
	5	Sp	$Co(corrole)PPh_3$[e]
	6[a]	Octahedral	$[Co en_2Cl_2]^+$, $[Cr(CN)_6]^{3-}$,
			$ZnCo_2O_4$, CoF_3, $[CoF_6]^{3-}$
Co^{IV}, d^5	4	Tetrahedral	$Co(1-norbornyl)_4$
	6	Octahedral	$[CoF_6]^{2-}$
Co^V, d^4	4	Tetrahedral?	K_3CoO_4

[a] Most common states.
[b] CoP_4 forms an almost regular tetrahedron.
[c] C. A. L. Becker, *J. Inorg. Nucl. Chem.,* 1975, **37**, 703.
[d] M. Nakajima *et al., J.C.S. Dalton,* **1977**, 385.
[e] P. B. Hitchcock and G. M. McLaughlin, *J.C.S. Dalton,* **1976**, 1927.

normally has a slight excess of oxygen and is a *p*-type semiconductor. It has the NaCl structure.

On heating at 400 to 500° in air the oxide Co_3O_4 is obtained. This is a normal spinel with Co^{2+} ions in tetrahedral interstices and Co^{3+} ions in octahedral interstices. Other oxides Co_2O_3 and CoO_2[1a] and a red oxocobaltate(II) $Na_{10}[Co_4O_9]$,[1b] are known.

Halides. The anhydrous halides CoX_2 may be made by dehydration of hydrated halides and for CoF_2 by action of HF on $CoCl_2$. The chloride is bright blue.

The action of fluorine or other fluorinating agents on cobalt halides at 300 to 400° gives *cobalt(III) fluoride,* a dark brown substance commonly used as a fluorinating agent (p. 573). It is reduced by water.

Sulfide. From Co^{2+} solutions, a black solid CoS is precipitated by action of H_2S.

Salts. Cobalt(II) forms an extensive group of simple and hydrated salts. All hydrated salts are red or pink and contain the $[Co(H_2O)_6]^{2+}$ ion or other octahedrally coordinated ions. Addition of hydroxide ion to Co^{2+} solutions gives *cobalt(II) hydroxide,* which may be pink or blue depending on the conditions; only the pink form is stable. It is amphoteric, dissolving in concentrated hydroxide to give a deep blue solution containing $[Co(OH)_4]^{2-}$ ions, from which crystalline salts can be obtained.

Cobalt(III) forms few simple salts, but the green hydrated fluoride $CoF_3 \cdot 3.5H_2O$ and the blue hydrated sulfate $Co_2(SO_4)_3 \cdot 18H_2O$ separate on electrolytic oxidation of Co^{2+} in 40% HF and $8M$ H_2SO_4, respectively. Dark blue alums $M^ICo(SO_4)_2 \cdot 12H_2O$ are also known; they are reduced by water.

21-F-3. Complexes[2] of Cobalt(II), d^7

The aqua ion $[Co(H_2O)_6]^{2+}$ is the simplest complex of cobalt(II).

In aqueous solutions containing no complexing agents, the oxidation to Co^{III} is very unfavorable:

$$[Co(H_2O)_6]^{3+} + e = [Co(H_2O)_6]^{2+} \qquad E^0 = 1.84 \text{ V}$$

However electrolytic or O_3 oxidation of cold acidic perchlorate solutions of Co^{2+} gives $[Co(H_2O)_6]^{3+}$, which is in equilibrium with $[Co(OH)(H_2O)_5]^{2+}$. At $0°$, the half-life of these diamagnetic aqua ions is about a month.[3] In presence of complexing agents such as NH_3, which form stable complexes with Co^{III}, the stability of trivalent cobalt is greatly improved:

$$[Co(NH_3)_6]^{3+} + e = [Co(NH_3)_6]^{2+} \qquad E^0 = 0.1 \text{ V}$$

and in basic media:

$$CoO(OH)(s) + H_2O + e = Co(OH)_2(s) + OH^- \qquad E^0 = 0.17 \text{ V}$$

Water rapidly reduces Co^{3+} at room temperature and this relative instability of uncomplexed Co^{III} is evidenced by the rarity of simple salts and binary compounds, whereas Co^{II} forms such compounds in abundance.

Cobalt(II) forms numerous complexes, mostly either octahedral or tetrahedral but five-coordinate and square species are also known.

There are more *tetrahedral complexes* of cobalt(II) than for other transition metal ions. This is in accord with the fact that for a d^7 ion, ligand field stabilization energies disfavor the tetrahedral configuration relative to the octahedral one to a smaller extent than for any other d^n ($1 \leqslant n \leqslant 9$) configuration, although it should be carefully noted that this argument is valid only in comparing the behavior of one metal ion to another, not for assessing the absolute stabilities of the configu-

---

[2]   A. V. Ablov and N. M. Samus, *Coord. Chem. Rev.,* 1975, **17,** 253 (dioxime complexes); L. Sacconi *et al., Coord. Chem. Rev.,* 1973, **11,** 343; 1972, **8,** 351 (five-coordinate species).

[3]   G. Davies and B. Warnquist, *J.C.S. Dalton,* **1973,** 900.

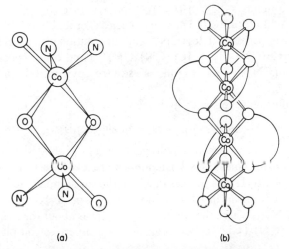

(a)   (b)

Fig. 21-F-1. Schematic representations of the structures of (a) the dimer of bis(N-methylsalicylaldi-minato)cobalt(II) and (b) the tetramer of bis(acetylacetonato)cobalt(II).

rations for any particular ion (See page 686). Co^{2+} is the only d^7 ion of common occurrence.

Because of the small stability difference between octahedral and tetrahedral Co^{II} complexes, there are several cases in which the two types with the same ligand are both known and may be in equilibrium. There is always some $[Co(H_2O)_4]^{2+}$ in equilibrium with $[Co(H_2O)_6]^{2+}$.

Tetrahedral complexes $[CoX_4]^{2-}$ are generally formed with monodentate anionic ligands such as Cl^-, Br^-, I^-, SCN^-, N_3^-, and OH^-; with a combination of two such ligands and two neutral ones, tetrahedral complexes of the type CoL_2X_2 are formed. With ligands that are bidentate monoanions, tetrahedral complexes are formed in some cases (e.g., with N-alkylsalicylaldiminato and bulky β-diketonate anions). With the less hindered ligands of this type, association to give a higher coordination number often occurs. Thus in bis(N-methylsalicylaldiminato)cobalt(II) a dimer with five-coordinate Co atoms is formed (Fig. 21-F-1a), whereas $Co(acac)_2$ is a tetramer in which each Co atom is six-coordinate (Fig. 21-F-1b).

Planar complexes are formed with several bidentate monoanions such as dimethylglyoximate, aminooxalate, o-aminophenoxide, dithioacetylacetonate, and the disulfur ligands discussed in Section 4-34. Several neutral bidentate ligands also give planar complexes, although it is either known or reasonable to presume that the accompanying anions are coordinated *to some degree,* so that these complexes could also be considered as very distorted octahedral ones. Examples are $[Co\ en_2](AgI_2)_2$ and $[Co(CH_3SC_2H_4SCH_3)_2](ClO_4)_2$. With the tetradentate ligands bis(salicylaldehydeethylenediiminato) ion and porphyrins, planar complexes are also obtained. The dimethylglyoximate complex is discussed further in Section 21-F-9.

The addition of KCN to aqueous cobalt(II) solutions gives a green solution and a purple solid. X-ray diffraction study of the purple salt $Ba_3[Co_2(CN)_{10}]\cdot 13H_2O$

shows that the anion has the $Mn_2(CO)_{10}$ type structure (p. 1052) with a Co—Co bond of 2.794 Å.[4] In solution the primary species is probably $[Co(CN)_5(H_2O)]^{3-}$, but there are also species with less CN^-, and the solution is unstable, reacting with water to give $[Co(CN)_5H]^{3-}$, $[Co(CN)_5OH]^{3-}$, and other species. In the salt $[NEt_2Pr_2^i]_3[Co(CN)_5]$ the anion is square pyramidal, as Co^{II} is in $[Co(CNPh)_5]^{2+}$.[5]

The green solution reacts with hydrogen

$$2[Co(CN)_5]^{3-} + H_2 = 2[Co^{III}(CN)_5H]^{3-}$$

and is a catalyst for homogeneous hydrogenation of conjugated alkenes, especially under phase transfer conditions.[6] The ion also reacts with C_2F_4, C_2H_2, SO_2, or $SnCl_2$ to give cobalt(III) complexes in which the small molecule is "inserted" between two cobalt atoms as in $[(NC)_5Co-CF_2CF_2-Co(CN)_5]^{6-}$ and $K_6[(CN)_5Co\,CH{=}CHCo(CN)_5]$, where the configuration is trans about the double bond.[7] The red-brown $[Co(CN)_5O_2]^{3-}$ is best regarded as a superoxo complex of Co^{III} (see later).[8]

The benzoate $Co_2(CO_2Ph)_4L_2$ (L = quinoline) has a carboxylato-bridged structure but the Co—Co distance is very long and there is no metal–metal bond.[9]

Cobalt(II) forms other types of five-coordinate species, mainly with polydentate ligands such as the quadridentate tripod ligands and certain tridentate ligands. The geometry varies, some approaching the trigonal-bipyramidal and others the square-pyramidal limiting cases; many have an intermediate (C_{2v}) arrangement. Interest in these complexes has centered mainly on correlating their electronic structures with molecular symmetry and the atoms constituting the ligand set; these points are mentioned below in connection with electronic structures.

Electronic Structures of Cobalt(II) Compounds. As already noted, cobalt(II) occurs in a great variety of structural environments; because of this the electronic structures, hence the spectral and magnetic properties of the ion, are extremely varied. We shall try to mention here each of the principal situations, giving its chief spectral and magnetic characteristics and citing representative cases.

High-Spin Octahedral and Tetrahedral Complexes. For qualitative purposes the partial energy level diagram in Fig. 21-F-2 is useful. In each case there is a quartet ground state and three spin-allowed electronic transitions to the excited quartet states. Quantitatively the two cases differ considerably, as might be inferred from the simple observation that octahedral complexes are typically pale red or purple, whereas many common tetrahedral ones are an intense blue. In each case

[4]   L. D. Brown, K. N. Raymond, and S. Z. Goldberg, *J. Am. Chem. Soc.,* 1972, **94**, 7664; G. L. Simon, A. W. Adamson, and L. F. Dahl, *J. Am. Chem. Soc.,* 1972, **94**, 7654.

[5]   L. D. Brown and K. N. Raymond, *Inorg. Chem.,* 1975, **14**, 2590; F. A. Furnak, D. R. Greig, and K. N. Raymond, *Inorg. Chem.,* 1975, **14**, 2585.

[6]   D. L. Reger *et al., J. Mol. Catal.,* 1978, **4**, 315; *Tetrahedron Lett., 1979,* 115; T. Funabiki *et al., J.C.S. Chem. Comm., 1978,* 63; *J.C.S. Dalton,* 1973, 1812; H. J. Clase *et al., J.C.S. Dalton,* **1973**, 2546.

[7]   O. Sala *et al., Inorg. Chim. Acta,* 1977, **22**, 155.

[8]   L. D. Brown and K. N. Raymond, *Inorg. Chem.,* 1975, **14**, 2595.

[9]   J. Catterick *et al., J.C.S. Dalton,* **1977**, 223.

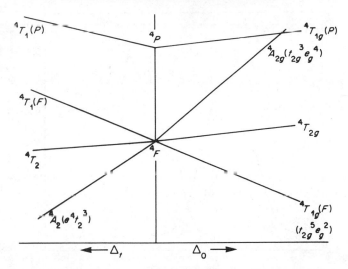

Fig. 21-F-2. Schematic energy level diagram for quartet states of a d^7 ion in tetrahedral and octahedral ligand fields.

the visible spectrum is dominated by the highest energy transition, $^4A_2 \rightarrow {^4T_1}(P)$ for tetrahedral and $^4T_{1g}(F) \rightarrow {^4T_{1g}}(P)$ for octahedral complexes; but in the octahedral systems the $^4A_{2g}$ level is usually close to the $^4T_{1g}(P)$ level and the transitions to these two levels are close together. Since the $^4A_{2g}$ state is derived from a $t_{2g}^3 e_g^4$ electron configuration, and the $^4T_{1g}(F)$ ground state is derived mainly from a $t_{2g}^5 e_g^2$ configuration, the $^4T_{1g}(F) \rightarrow {^4A_{2g}}$ transition is essentially a two-electron process; thus it is weaker by about a factor of 10^{-2} than the other transitions. In the tetrahedral systems, as illustrated in Fig. 21-F-3, the visible transition is generally about an order of magnitude more intense and displaced to lower energies, in accord with the observed colors mentioned above. For octahedral complexes, there is one more spin-allowed transition ($^4T_{1g}(F) \rightarrow {^4T_{2g}}$) which generally occurs in the near-infrared region. For tetrahedral complexes there is also a transition in the near-infrared region [$^4A_2 \rightarrow {^4T_1}(F)$], as well as one of quite low energy ($^4A_2 \rightarrow {^4T_2}$), which is seldom observed because it is in an inconvenient region of the spectrum (1000–2000 nm) and it is orbitally forbidden. The visible transitions in both cases, but particularly in the tetrahedral case, generally have complex envelopes because a number of transitions to doublet excited states occur in the same region, and these acquire some intensity by means of spin orbit coupling.

The octahedral and tetrahedral complexes also differ in their magnetic properties. Because of the intrinsic orbital angular momentum in the octahedral ground state, there is consistently a considerable orbital contribution, and effective magnetic moments for such compounds around room temperature are between 4.7 and 5.2 BM. For tetrahedral complexes the ground state acquires orbital angular momentum only indirectly through mixing in the $^4T_2$ state by a spin-orbit coupling perturbation. First-order perturbation theory leads to the expression

$$\mu = 3.89 - \frac{15.59 \, \lambda'}{\Delta_t}$$

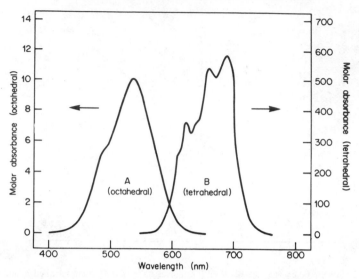

Fig. 21-F-3. The visible spectra of $[Co(H_2O)_6]^{2+}$ (curve A) and $[CoCl_4]^{2-}$ (curve B).

where 3.89 is the spin-only moment for three unpaired electrons and λ' is the effective value of the spin-orbit coupling constant (which is inherently negative). Since λ' varies little from one complex to another, orbital contributions vary inversely with the strength of the ligand field. For example among the tetrahalo complexes we have: $CoCl_4^{2-}$, 4.59 BM; $CoBr_4^{2-}$, 4.69 BM; CoI_4^{2-}, 4.77 BM.

Low-Spin Octahedral Complexes. A sufficiently strong ligand field ($\Delta_0 \geq$ 15,000 cm^{-1}) can cause a ^2E state originating in the ^2G state of the free ion to become the ground state. The electron configuration here is mainly $t_{2g}^6 e_g$; thus a Jahn-Teller distortion would be expected. Consequently perfectly octahedral low-spin Co(II) complexes must be rare or nonexistent. Very few six-coordinate low-spin Co(II) complexes have actually been reported and none has been structurally characterized. As noted above, the $Co(CNC_6H_5)_6^{2+}$ ion in solution is axially distorted. It appears that ligands tending to give a strong enough field to cause spin pairing give five- rather than six-coordinate complexes, or dinuclear ones such as $Co_2(CNCH_3)_{10}^{4+}$.

Square Complexes. All these are low spin with magnetic moments of 2.2 to 2.7 BM at 300°. Their spectra are complex, and neither magnetic nor spectral properties of such compounds have been treated in detail. There are some data to suggest that the unpaired electron occupies the d_{z^2} orbital, as might be expected.

Five-Coordinate Complexes. Both high-spin (three unpaired electrons) and low-spin (one unpaired electron) configurations are found for both trigonal-bipyramidal and square-pyramidal as well as for intermediate configurations. The patterns of orbital energies in the two limiting geometries are discussed on page 642, and from these we may write the following configurations for the four spin-structure combinations:

Spin	D_{3h}	C_{4v}
High	$(e'')^4(e'^2(a_1'))$	$b_2^2 e^3 a_1 b_1$
Low	$(e'')^4(e')^3$	$b_2^2 e^4 a_1$

It appears that the relationships between spin state, geometry, and nature of the ligand atoms are closely interlocked, so that no simple relationship between any two of these factors has been found. However it does appear that spin state and the nature of the donor atom set are roughly correlated independently of geometry in such a way that the more heavy atom (e.g., P, As, Br, or S) donors (as compared with O and N) are present, the greater is the tendency to spin pairing; this is hardly surprising. For high-spin complexes with fairly regular geometry, e.g., [Co(Me$_6$-tren)Br]$^+$ (C_{3v}) and [Co(Ph$_2$MeAsO)$_4$(ClO$_4$)]ClO$_4$ (C_{4v}), detailed and reasonably convincing spectral assignments have been made. For irregular geometries and for low-spin complexes there are more uncertainties.

21-F-4. Complexes[10] of Cobalt(III), d^6

The complexes of cobalt(III) are exceedingly numerous. Because they generally undergo ligand-exchange reactions realtively slowly, they have, from the days of Werner and Jørgensen, been extensively studied and a large fraction of our knowledge of the isomerism, modes of reaction, and general properties of octahedral complexes as a class is based on studies of Co^{III} complexes. Almost all discrete Co^{III} complexes are octahedral, though tetrahedral and square-antiprismatic Co^{III} occur in a few solid state situations, and a square-pyramidal, paramagnetic complex is formed by the quadridentate ligand[11] 21-F-I.

(21-F-I)

Co^{III} shows a particular affinity for nitrogen donors, and the majority of its complexes contain ammonia, amines such as ethylenediamine, nitro groups, or nitrogen-bonded SCN groups, as well as halide ions and water molecules. In general, these complexes are synthesized in several steps beginning with one in which the aqua Co^{II} ion is oxidized in solution, typically by molecular oxygen (see later discussion) or hydrogen peroxide and often a surface-active catalyst such as activated charcoal, in the presence of the ligands. For example, when a vigorous stream of air is drawn for several hours through a solution of a cobalt(II) salt, CoX$_2$ (X = Cl, Br, or NO$_3$), containing ammonia, the corresponding ammonium salt and some

[10]   W. Levason and C. A. McAuliffe, *Coord. Chem. Rev.*, 1974, **12**, 151 [Co^{III}, Co^{IV}, Co^V compounds]; L. I. Katzin and I. Eliezer, *Coord. Chem. Rev.*, 1972, **7**, 331 (circular dichroism).
[11]   N. A. Bailey *et al.*, *J.C.S. Dalton*, **1977**, 763.

activated charcoal, good yields of hexammine salts are obtained:

$$4CoX_2 + 4NH_4X + 20NH_3 + O_2 \rightarrow 4[Co(NH_3)_6]X_3 + 2H_2O$$

In the absence of charcoal, replacement usually occurs to give, for example, $[Co(NH_3)_5Cl]^{2+}$ and $[Co(NH_3)_4(CO_3)]^+$. Similarly, on air oxidation of a solution of $CoCl_2$, ethylenediamine, and an equivalent quantity of its hydrochloride salt, tris(ethylenediamine)cobalt(III) chloride is obtained.

$$4CoCl_2 + 8en + 4en \cdot HCl + O_2 = 4[Co(en)_3]Cl_3 + 2H_2O$$

However a similar reaction in acid solution with the hydrochloride gives the green *trans*-dichlorobis(ethylenediamine)cobalt(III) ion as the salt *trans*-$[Co\ en_2Cl_2]$-$[H_5O_2]Cl_2$, which loses HCl on heating. This *trans* isomer may be isomerized to the red racemic *cis* isomer on evaporation of a neutral aqueous solution at 90 to 100°. Both the *cis* and the *trans* isomers are aquated when heated in water:

$$[Co\ en_2Cl_2]^+ + H_2O \rightarrow [Co\ en_2Cl(H_2O)]^{2+} + Cl^-$$
$$[Co\ en_2Cl(H_2O)]^{2+} + H_2O \rightarrow [Co(en)_2(H_2O)_2]^{3+} + Cl^-$$

and on treatment with solutions of other anions are converted into other $[Co\ en_2X_2]^+$ species, for example,

$$[Co\ en_2Cl_2]^+ + 2NCS^- \rightarrow [Co\ en_2(NCS)_2]^+ + 2Cl^-$$

These few reactions are illustrative of the very extensive chemistry of Co^{III} complexes with nitrogen-coordinating ligands.

In addition to the numerous mononuclear ammine complexes of Co^{III}, there are a number of polynuclear ammine complexes in which hydroxo (OH^-), peroxo (O_2^{2-}), amido (NH_2^-), and imido (NH^{2-}) groups function as bridges. Some typical complexes of this class are

$$[(NH_3)_5Co—O—O—Co(NH_3)_5]^{4+}, \quad \{(NH_3)_3Co(OH)_3Co(OH)_3Co(NH_3)_3\}^{3+}$$
$$\text{and } [(NH_3)_4Co(OH)(NH_2)Co(NH_3)_4]^{4+}$$

Some other Co^{III} complexes of significance are the hexacyano complex $[Co(CN)_6]^{3-}$, the oxygen-coordinated complexes such as carbonates,[12] e.g., the green ion $[Co(CO_3)_3]^{3-}$ and *cis*-$[Co(CO_3)_2py_2]^-$, cobalt(III) acetylacetonate, and salts of the trisoxalatocobalt(III) anion.

Cobalt(III) acetate is made by ozonation of Co^{II} acetate in acetic acid: it usually forms a green oil, but crystals of $Co_3O(CO_2Me)_6(MeCO_2H)_3$ have been said[13] to have the same oxo-centered structure of many trivalent metal acetates (p. 154); however in a pyridine adduct both chelate and bridge acetate groups are present in the oxo-centered structure.[14] The acetate oxidizes various saturated hydrocarbons and alkyl side chains in aromatic hydrocarbons,[15a] and a cobalt-catalyzed process

[12]   G. Davies and Y.-W. Hung, *Inorg. Chem.*, 1976, **15**, 704, 1358.
[13]   J. J. Ziolkowski *et al., Inorg. Chim. Acta*, 1973, **7**, 473.
[14]   S. Uemura *et al., J.C.S. Dalton*, **1973**, 2565.
[15a]   S. R. Jones and J. M. Mellor, *J.C.S. Perkin II*, **1977**, 511; L. Verstraelen *et al., J.C.S. Perkin II*, **1976**, 1285; O. Onopohenko and J. G. D. Schulz, *J. Org. Chem.*, 1975, **40**, 3338.

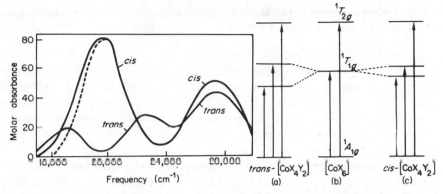

Fig. 21-F-4. *Left:* the visible spectra of *cis-* and *trans-*[Co en₂F₂]⁺. The broken line shows where the low-frequency side of the $A_{1g} \rightarrow T_{1g}$ band of the *cis* isomer would be if the band were completely symmetrical. The asymmetry is caused by slight splitting of the $^1T_{1g}$ state. *Right:* diagrammatic representation (not to scale) of the energy levels involved in the transitions responsible for the observed bands of octahedral Co^{III} complexes: (b) the levels for a regular octahedral complex [CoX₆]; (a) and (c) the splittings caused by the replacement of two ligands X by two ligands Y.

is used commercially for the oxidation of toluene to benzoic acid.[15b] Cobalt(III) in aqueous acid solution commonly oxidizes organic compounds.

The trifluoromethylsulfonate,[16] whose structure is not clear from the stoichiometry $Co(CF_3SO_3)_3$, is also a powerful oxidant for organic substances.

Electronic Structures of Cobalt(III) Complexes. The free Co^{III} ion d^6 has qualitatively the same energy level diagram as does Fe^{II}. However with Co^{III} the $^1A_{1g}$ state originating in one of the high-energy singlet states of the free ion drops very rapidly and crosses the $^5T_{2g}$ state at a very low value of Δ. Thus all known octahedral Co^{III} complexes, including even $[Co(H_2O)_6]^{3+}$ and $[Co(NH_3)_6]^{3+}$, have diamagnetic ground states, except for $[Co(H_2O)_3F_3]$ and $[CoF_6]^{3-}$, which are paramagnetic with four unpaired electrons.

The visible absorption spectra of Co^{III} complexes thus may be expected to consist of transitions from the $^1A_{1g}$ ground state to other singlet states. The two absorption bands found in the visible spectra of regular octahedral Co^{III} complexes represent transitions to the upper states $^1T_{1g}$ and $^1T_{2g}$. In complexes of the type CoA₄B₂, which can exist in both *cis* and *trans* configurations, certain spectral features are diagnostic of the *cis* or *trans* configuration (Fig. 21-F-4).

The origin of these features lies in the splitting of the $^1T_{1g}$ state by the environments of lower than O_h symmetry, as also shown diagrammatically in Fig. 21-F-4. Theory shows that splitting of the $^1T_{2g}$ state will always be slight, whereas the $^1T_{1g}$ state will be split markedly in the *trans*-isomer whenever there is a substantial difference in the positions of the ligands, A and B, in the spectrochemical series. Moreover, because the cis isomer lacks a center of symmetry it may be expected to have a somewhat more intense spectrum than the *trans*-isomer. These predictions are nicely borne out by the spectra of *cis-* and *trans-*[Co en₂F₂]⁺.

15b *Hydrocarbon Processing, Petrochemical Handbook,* 1977, **56**, 134.
16   R. Tang and J. K. Kochi, *J. Inorg. Nucl. Chem.,* 1973, **35**, 3845.

21-F-5.   The Oxidation of Cobalt(II) Complexes by Molecular Oxygen; Peroxo and Superoxo Species; Oxygen Carriers[17]

The interaction of cobalt complexes in solution with molecular oxygen has been the subject of intensive study. Under certain conditions (p. 774) oxidation ultimately to cobalt(III) complexes occurs. However in the absence of charcoal or other catalysts, or by the choice of suitable ligands, the intermediate peroxo and superoxo species (Section 4-24) can be isolated. Some of these complexes behave as reversible carriers of O_2 and have been studied intensively because of their potential utility and as models for natural oxygen transport systems (Chapter 31).

We first consider the situation for ligands like NH_3 or CN^-. The first in the sequence of steps may involve oxidative addition of O_2 to give a transient Co^{IV} species, which then reacts with another Co^{II} species to give a binuclear peroxo-bridged species such as $[(NH_3)_5CoOOCo(NH_3)_5]^{4+}$ or $[(NC)_5CoOO-Co(CN)_5]^{6-}$. These species are isolable as moderately stable solid salts but decompose fairly easily in water or acids. The open-chain species $[(NH_3)_5Co-O_2Co(NH_3)_5]^{4+}$ can be cyclized in presence of base to

$$[(NH_3)_4Co\underset{NH_2}{\overset{O_2}{\diagup\diagdown}}Co(NH_3)_4]^{3+}$$

It seems safe to assume that all such species, open-chain or cyclic, contain low-spin Co^{III} and bridging peroxide (O_2^{2-}) ions; in $[(NH_3)_5CoOOCo(NH_3)_5]^{4+}$ the O—O distance (1.47 Å) is the same as in H_2O_2.

These O_2-bridged binuclear complexes can often be oxidized in a one-electron step to species such as $[(NH_3)_5CoO_2Co(NH_3)_5]^{5+}$ and

$$[(NH_3)_4Co\underset{NH_2}{\overset{O_2}{\diagup\diagdown}}Co(NH_3)_4]^{4+}$$

These ions were first prepared by Werner, who formulated them as peroxo-bridged complexes of Co^{III} and Co^{IV}. Esr data have shown, however, that the single unpaired electron is distributed equally over both cobalt ions, thus ruling out that description. The problem of how best to formulate these complexes has been settled by X-ray structural data as in Fig. 21-F-5. The O—O distances may be compared with that (1.28 Å) characteristic of superoxide ion, O_2^-. The unpaired electron formally belonging to O_2^- resides in a molecular orbital of π symmetry relative to the planar Co—O—O—Co groupings and is delocalized over these four atoms. The cobalt atoms are formally described as Co^{III} ions.

In the case of cyanide[18] we have the system

[17]   G. A. Lawrence *et al., Inorg. Chem.,* 1978, **17,** 3317; W. R. Harris *et al., Inorg. Chem.,* 1978, **17,** 889; G. McLendon and A. E. Martell, *Coord. Chem. Rev.,* 1976, **19,** 1; F. Basolo, B. M. Hoffman, and J. A. Ibers, *Acc. Chem. Res.,* 1975, **8,** 384; J. S. Valentine, *Chem. Rev.,* **1973,** 235; E.-I Ochia, *J. Inorg. Nucl. Chem.,* 1973, **35,** 3375; G. Henrici-Olivé and S. Olivé, *Angew. Chem., Int. Ed.,* 1974, **13,** 29; R. Lancashire *et al., J.C.S. Dalton,* **1979,** 66.

[18]   V. M. Miskowski *et al., Inorg. Chem.,* 1975, **14,** 2318; F. R. Fronczek *et al., Inorg. Chem.,* 1975, **14,** 611; T. C. Strekas and T. G. Spiro, *Inorg. Chem.,* 1975, **14,** 1421.

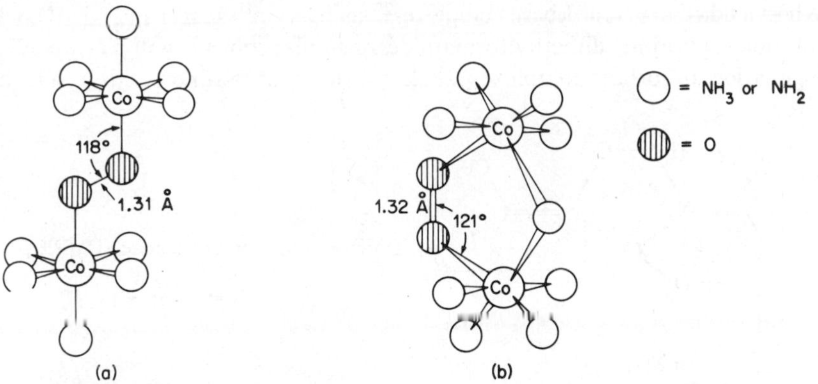

Fig. 21-F-5. The structures of (a) $[(NH_3)_5CoO_2Co(NH_3)_5]^{5+}$ and (b) $[(NH_3)_4Co\langle^{O_2}_{NH_2}\rangle Co(NH_3)_4]^{4+}$ showing the octahedral coordination about each cobalt ion and the angles and distances at the bridging superoxo groups. The five-membered ring in (b) is essentially planar. Both peroxo and superoxo complexes have these two types of bridged structure.

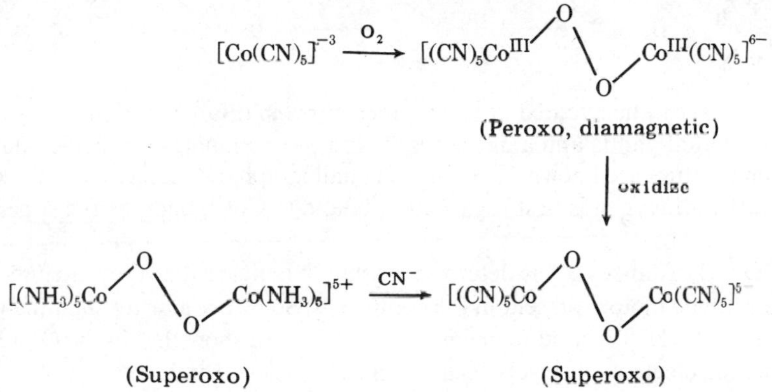

With polyfunctional ligands different behavior may be obtained, and we may have (a) oxidation of Co^{II} to Co^{III} as before, (b) formation of mononuclear oxygen adducts in addition to the possibility of bridged species as above, or (c) oxidation of the ligand as in the case[19] of the macrocycle:

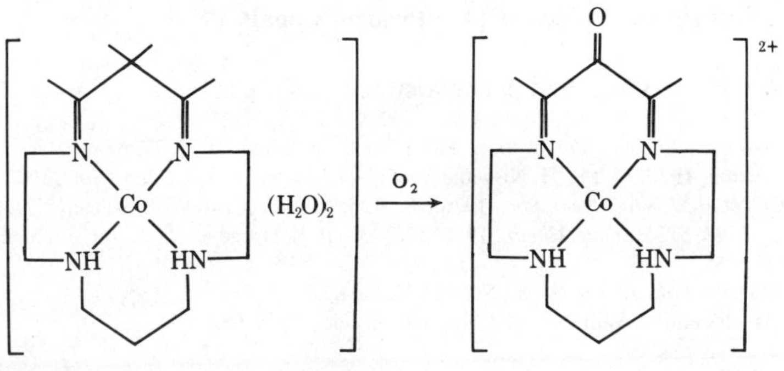

[19] B. Durham et al., Inorg. Chem., 1977, 16, 271.

The best studied are Schiff base complexes[20] such as $Co^{II}(acacen)$ (21-F-II), which in solution in pyridine, dimethylformamide, or similar solvents will pick up oxygen. Oxygenation of other complexes such as those of amino acids[21] has been studied.

$$+ \ DMF \ + \ O_2 \ \longrightarrow \ Co(acacen)(DMF)O_2$$

$$K = 21 \text{ at } -10°$$

(21-F-II)

The oxygenation reactions are usually reversible only at low temperatures, since the complex is either irreversibly oxidized or bridged peroxo or superoxo species are formed:

$$B—Co^{II} + O_2 \ \rightleftarrows \ B—Co^{III}—O^{\diagdown O} \ \longrightarrow \ B—Co—O^{\diagup O—Co—B}$$

Dimerization can be avoided at low temperatures in dilute solutions or by choice of appropriate ligands and axial bases B, and some monomeric species stable at room temperature are known. The X-ray crystallographic structures that are known confirm that dioxygen is best regarded as bound to Co^{III}, bent, as the superoxide ion O_2^-.

The O—O distances so far determined seem to indicate that they are dependent on all electronic factors present in the complexes. Some are shorter than the O—O distance in O_2^- (1.28 Å) and some longer. ESR spectra show that in the O_2 adducts the unpaired electron is largely located on the oxygen atoms.

Although the oxidation of substrates by O_2 in presence of metal complexes generally appears to be free radical, it has been claimed that the catalytic oxidation of Bu_3P to Bu_3PO by $Co(acac)_2$ proceeds by a nonradical path involving intermediates with Co—O—O—Co and Co—O bonds.[22]

21-F-6. Tetravalent Cobalt, d^5; Pentavalent Cobalt, d^4

Compounds in the tetra- and pentavalent classes are few and not well character-

[20] F. Basolo *et al., J.C.S. Chem. Comm.,* **1979,** 5; *J.C.S. Dalton,* **1975,** 97, 674; E. Cesarotti *et al., J.C.S. Dalton,* **1977,** 757; R. H. Niswander and L. T. Taylor, *J. Am. Chem. Soc.,* 1977, **99,** 5935; R. S. Gall *et al., J. Am. Chem. Soc.,* 1976, **98,** 5135; A. Avdeef and W. P. Schaefer, *J. Am. Chem. Soc.,* 1976, **98,** 5135; *Inorg. Chem.,* 1976, **15,** 1432; R. S. Drago *et al., J. Am. Chem. Soc.,* 1976, **98,** 5144.

[21] A. E. Martell *et al., J. Am. Chem. Soc.,* 1976, **98,** 8378.

[22] R. P. Hanzlik and D. Williamson, *J. Am. Chem. Soc.,* 1976, **98,** 6571.

ized.[10] Fluorination of Cs_2CoCl_4 gives Cs_2CoF_6, with a crystal structure isomorphous with that of Cs_2SiF_6 and a magnetic moment rising from 2.46 BM at 90°K to 2.97 BM at 294°K. The reflectance spectrum has been assigned to an octahedrally coordinated t_{2g}^5 ion. The high magnetic moments could be due to a large orbital contribution in the $^2T_{2g}$ ground state or to partial population of a $^6A_{1g}(t_{2g}^3 e_g^2)$ state. The action of oxidizing agents (e.g., Cl_2, O_2, or O_3) on strongly alkaline Co^{II} solutions produces a black material believed to be hydrous CoO_2, at least in part, but it is ill characterized. Ba_2CoO_4, a red-brown substance obtained by oxidation of $2Ba(OH)_2$ and $2Co(OH)_2$ at 1050°, and a heteropolymolybdate of Co^{IV}, namely, $3K_2O \cdot CoO_2 \cdot 9\ MoO_3 \cdot 6H_2O$, have been reported.

The best characterized Co^{IV} compound is the remarkable alkyl tetra(1-norbornyl)cobalt,[23] which is made from $CoCl_2$ and Li norbornyl; how oxidation occurs is not known. The brown compound is paramagnetic and reasonably stable to air and heat. It should be possible to make other similar Co^{IV} compounds.

Sodium peroxide and Co_3O_4 on heating give the moisture- and CO_2-sensitive Na_4CoO_4, which has tetrahedral $Co^{IV}O_4$ units with Na^+ ions coordinated to oxygen atoms.[24]

21-F-7. Complexes of Cobalt(I), d^8

Most cobalt(I) complexes contain π-acid ligands such as CO, RNC, or R_3P; however there are a few with N ligands, e.g., the reduced species of vitamin B_{12} and cobalt oximes discussed later. Similar macrocyclic complexes of cobalt(II) with the ligands such as 21-F-III can be reduced by hydrated electrons from pulse radiolysis, but these Co^I species have only a short lifetime and are powerful reducing agents as well as nucleophiles.[25]

(21-F-III)

[23] B. K. Bower and H. G. Tennant, *J. Am. Chem. Soc.*, 1972, **94**, 2512.
[24] M. Jansen, *Z. Anorg. Allg. Chem.*, 1975, **417**, 35.
[25] A. M. Tate *et al.*, *J. Am. Chem. Soc.*, 1976, **98**, 86.

21-F-8.   Complexes of Cobalt(0), (I), (II), and (III) with π-Acceptor Ligands, Phosphorus Donors[26]

There is an extensive chemistry particularly of cobalt(I) in which phosphines or phosphites are bound to cobalt either alone or with other ligands like MeCN or CO. Some representative examples are the following:

$Co(PMe_3)_4$ $CoCl(PR_3)_3$ $CoCl_2(PR_3)_2$ $CoH_3(PPh_3)_3$
$Co_2[P(OMe)_3]_8$ $CoCl(CO)_2(PR_3)_2$ $CoH[P(OPh)_3]_4^+$ $CoMe_3(PMe_3)_3$
                       $CoH(N_2)(PPh_3)_3$ $CoH_2[P(OR)_3]_4^+$
                       $CoH(PF_3)_4$
                       $CoH(CO)(PPh_3)_3$
                       $Co(CO)_2(PR_3)_3^+$
                       $Co[P(OEt)_3]_5^+$

The hydrido species such as $HCo[P(OPh)_3]_3(MeCN)$ may be catalysts for homogeneous hydrogenation of unsaturated substances.

Phosphine derivatives obtained from $Co_2(CO)_8$ (Chapter 25) invariably have CO groups present also, but these are either dimeric Co^0 species, or Co^I and Co^{-I} species formed by disproportionation:

$$Co_2(CO)_8 + 2PPh_3 \xrightarrow{0°, \text{benzene}} Co_2(CO)_6(PPh_3)_2 \xrightarrow[\substack{\text{polar} \\ \text{solvent} \\ -CO}]{\text{heat}} [Co^I(CO)_3(PPh_3)_2][Co^{-I}(CO)_4]$$

However in one case a dark green, air-stable paramagnetic species is formed[27]:

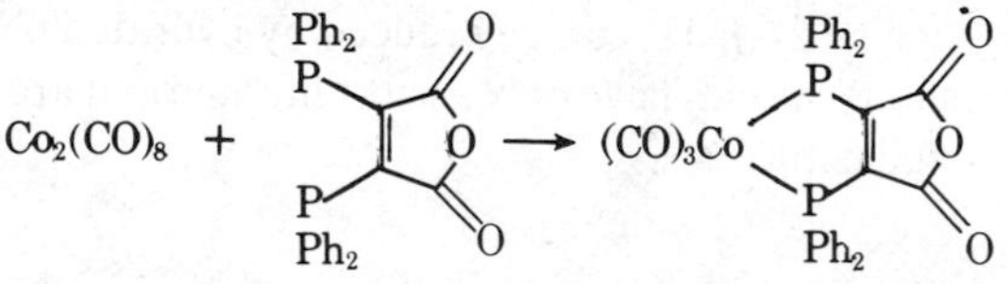

The usual method of making cobalt(I) species is reduction of the corresponding cobalt(II) complexes, or of $CoCl_2$ in presence of the ligand, by metals such as zinc or sodium.

The precise nature of the product depends very much on the nature of PR_3, and on the solvent, temperature, and metal used. Some representative reactions are shown in Fig. 21-F-6.

Using trimethylphosphine, reduction by Mg in tetrahydrofuran gives the brown, tetrahedral, paramagnetic $Co(PMe_3)_4$.[28] Cobalt(0) species are also known for phosphites. Although $Co[P(OPr^i)_3]_4$ is monomeric, for $P(OMe)_3$ and $P(OEt)_3$ there are both paramagnetic monomers and diamagnetic dimers, presumably because the steric hindrance is less than in the isopropyl.[28] The isopropylphosphite

[26]   E. L. Muetterties and P. L. Watson, *J. Am. Chem. Soc.*, 1978, **100**, 6978; G. Pilloni *et al.*, *J. Organomet. Chem.*, 1977, **134**, 305; L. W. Gosser, *Inorg. Chem.*, 1977, **16**, 427, 430, 1348; J. M. Whitfield *et al.*, *J.C.S. Dalton*, **1977**, 407; E. Bordignon *et al.*, *Inorg. Chem.*, 1974, **13**, 935; L. Sacconi *et al.*, *Inorg. Chem.*, 1975, **14**, 1380, 1790.
[27]   D. Fenske, *Angew Chem., Int. Ed.*, 1976, **15**, 381.
[28]   M. C. Rakowski and E. L. Muetterties, *J. Am. Chem. Soc.*, 1977, **99**, 739.

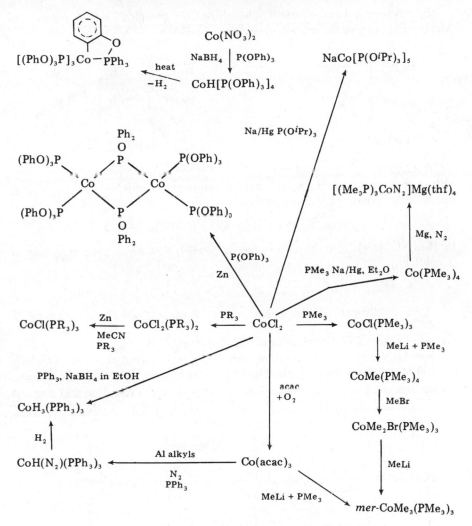

Fig. 21-F-6. Reactions of Cobalt Compounds with PR₃ ligands.

also gives the compound Na{Co[P(OPr^i)₃]₅}, which may have the oxygen atoms of the phosphite coordinated to the sodium ion, since the compound is very soluble in pentane.

Several cobalt(III) complexes are known with H and CH₃ ligands. The former can be made by oxidative addition reactions:

$$CoH(N_2)(PPh_3)_3 + H_2 = CoH_3(PPh_3)_3 + N_2$$
$$CoH[P(OR)_3]_4 + HX = CoH_2[P(OR)_3]_4^+ + X^-$$

The methyls[29] are made as in Fig. 21-F-6.

29   H.-F. Klein and H. H. Karsch, *Chem. Ber.,* 1975, **108,** 944, 956.

The *cis*-dimethyl-*mer*-trimethylphosphine Co^{III} complex 21-F-IV reacts with Me$_3$P=CH$_2$ to give the metallocyclic alkyl 21-F-V.

(21-F-IV)   (21-F-V)

Other Ligands. Isocyanides may be formed either by reduction of CoCl$_2$(CNR)$_4$ with an active metal, N$_2$H$_4$, S$_2$O$_4^{2-}$, etc., or by interaction of RNC with Co$_2$(CO)$_8$, which leads again to disproportionation:

$$Co_2(CO)_8 + 5RNC = [Co(CNR)_5]^+[Co(CO)_4]^- + 4CO$$

In [Co(NCMe)$_5$]ClO$_4$ the cation has trigonal-bipyramidal geometry, but in [Co(NCPh)$_5$]ClO$_4$ it is square pyramidal.[30] The reaction

$$CoX_2 + 2CH_2{=}CHCOOEt + Mn \xrightarrow{\text{MeCN}} MnX_2 + Co(MeCN)_x(\overset{\overset{\textstyle CH_2}{\|}}{CHCO_2Et})_2$$

gives a tetrahedral olefin complex of ethylfumarate.[31]

There is an extensive chemistry of monocyclopentadienyl compounds, of which (η-C$_5$H$_5$)Co(CO)$_2$ is one example. The CO may be replaced by phosphines, alkenes, isocyanides, etc. The dicarbonyl itself can be reduced[32] to a radical ion that contains formally cobalt (0.5)—or Co^0 + Co^I, and can be isolated as its (Ph$_3$P)$_2$N$^+$ salt.

21-F-9. Cobaloximes and Related Compounds[33]

Since the recognition of Vitamin B$_{12}$ and the existence in Nature of Co—C bonds (Chapter 31), there has been much study of "model" systems. These have been mainly complexes of dimethyl and other glyoximes (the so-called cobaloximes), and Schiff base and macrocyclic ligand complexes. The main interest has centered

[30]   C. A. L. Becker, *J. Inorg. Nucl. Chem.*, 1975, **37**, 703.
[31]   G. Agnes *et al.*, *J. Organomet. Chem.*, 1977, **129**, 401.
[32]   N. E. Schore *et al.*, *J. Am. Chem. Soc.*, 1977, **99**, 1781.
[33]   G. N. Schrauzer, *Angew. Chem., Int. Ed.*, 1977, **16**, 233; 1976, **15**, 417.

on the reduction to Co^I species and on the formation and properties of Co—C bonds.[34]

Reduction of cobaloximes, like the reduction of B_{12}, gives blue or green Co^I species that are very powerful reducing agents and nucleophiles. Thus they may react with water to give hydridocobalt(III) species, especially if tertiary phosphine ligands are also present.

In the Schiff base complexes such as 21-F-VI

(21-F-VI)

the strength of the bonding of the trans ligand is determined mainly by the inductive effects of the R group, although there is some evidence that the nature of the cis ligand atoms is important.[35] The reactivity of the Co—C bond and its ease of photolysis also are powerfully affected by the nature of the trans ligand.[36]

Finally we note that such alkylcobalt(III) species can also undergo a one-electron oxidation to give organocobalt(IV) radical ions, stable at low temperatures ($< -50°$). The C atom bound to the metal is very susceptible to nucleophilic attack.[37]

21-G.   NICKEL

The trend toward decreased stability of higher oxidation states continues with nickel, so that only Ni^{II} occurs in the ordinary chemistry of the element. However there is a complex array of stereochemistries associated with this species. The oxidation numbers 0 and +1 are found largely under conditions where such numbers have little physical meaning. The higher oxidation states, Ni^{III} and Ni^{IV} occur in very few compounds, and in many of these it is not clear whether it is really the metal atom rather than the ligand that is oxidized.

Nickel has an enormous, and important, organometallic chemistry,[1a] and many of its aspects are noted in Chapters 27, 29, and 30.

34   See, e.g., D. Dodd and M. D. Johnson, *J. Organomet. Chem.*, 1973, **52**, 1; M. W. Witman and J. H. Weber, *Inorg. Chim. Acta*, 1977, **23**, 263; D. A. Stotter and J. Trotter, *J.C.S. Dalton*, **1977**, 868.

35   W. D. Hemphill and D. G. Brown, *Inorg. Chem.*, 1978, **17**, 766.

36   P. A. Milton and T. L. Brown, *J. Am. Chem. Soc.*, 1977, **99**, 1390.

37   J. Topich and J. Halpern, *Inorg. Chem.*, 1979, **18**, 1339.

1a   P. W. Jolly and G. Wilke, *The Organic Chemistry of Nickel*, Vol. I., *Organonickel Complexes*, Academic Press, 1974.

TABLE 21-G-1

Oxidation States and Stereochemistry of Nickel

Oxidation state	Coordination number	Geometry	Examples
Ni^{-1}	4?	?	$[Ni_2(CO)_6]^{2-}$
Ni^0	3	?	$Ni[P(OC_6H_4\text{-}o\text{-}Me)_3]_3$
	4	Tetrahedral	$Ni(PF_3)_4$, $[Ni(CN)_4]^{4-}$, $Ni(CO)_4$
	5	?	$NiH[P(OEt)_3]_4^+$
Ni^I, d^9	4	Tetrahedral	$Ni(PPh_3)_3Br$
Ni^{II}, d^8 [a]	4 [b]	Square	$NiBr_2(PEt_3)_2$, $[Ni(CN)_4]^{2-}$, $Ni(DMGH)_2$,[c]
	4 [b]	Tetrahedral	$[NiCl_4]^{2-}$, $NiCl_2(PPh_3)_2$
	5	Sp	$[Ni(CN)_5]^{3-}$, BaNiS, $[Ni_2Cl_8]^{4-}$
	5	Tbp	$[NiX(QAS)]^+$, $[Ni(CN)_5]^{3-}$,
			$[NiP\{CH_2CH_2CH_2AsMe_2\}_3CN]^+$
	6 [b]	Octahedral	NiO, $[Ni(NCS)_6]^{4-}$, $KNiF_3$, $[Ni(NH_3)_6]^{2+}$,
			$[Ni(bipy)_3]^{2+}$
	6	Trigonal prism	NiAs
Ni^{III}, d^7	5	Tbp	$NiBr_3(PR_3)_2$
	6	Octahedral (dist.)	$[Ni(diars)_2Cl_2]^+$, $[NiF_6]^{3-}$
Ni^{IV}, d^6	6	Octahedral (dist.)	K_2NiF_6, $[Ni(Bu_2dtc)_3]^+$ [d] $\{Ni[Se_2C_2(CN)_2]_3\}^{2-}$

[a] Three-coordinate $NiCl_3^-$ may be present in molten $CsAlCl_4$ at 500°.
[b] Most common state.
[c] Square set of nitrogen atoms about Ni with long Ni—Ni bonds.
[d] $Bu_2dtc = N,N'$-dibutyldithiocarbamate.

The oxidation states and stereochemistry of nickel are summarized in Table 21-G-1.

21-G-1. The Element

Nickel occurs in Nature mainly in combination with arsenic, antimony, and sulfur, for example, as *millerite,* (NiS), as a red nickel ore that is mainly NiAs, and in deposits consisting chiefly of NiSb, $NiAs_2$, NiAsS, or NiSbS. The most important deposits commercially are *garnierite,* a magnesium-nickel silicate of variable composition, and certain varieties of the iron mineral *pyrrhotite* (Fe_nS_{n+1}) which contain 3 to 5% Ni. Elemental nickel is also found alloyed with iron in many meteors, and the central regions of the earth are believed to contain considerable quantities. The metallurgy of nickel is complicated in its details, many of which vary a good deal with the particular ore being processed. In general, the ore is transformed to Ni_2S_3, which is roasted in air to give NiO, and this is reduced with carbon to give the metal.[1b] Some high-purity nickel is made by the *carbonyl process:* carbon monoxide reacts with impure nickel at 50° and ordinary pressure or with nickel-copper matte under more strenuous conditions, giving volatile $Ni(CO)_4$, from which metal of 99.90 to 99.99% purity is obtained on thermal decomposition at 200°.

[1b] J. R. Boldt, Jr., *The Winning of Nickel,* Methuen, 1967.

Nickel is silver-white, with high electrical and thermal conductivities (both $\sim 15\%$ of those of silver) and m.p. 1452°, and it can be drawn, rolled, forged, and polished. It is quite resistant to attack by air or water at ordinary temperatures when compact and is therefore often electroplated as a protective coating. Because nickel reacts but slowly with fluorine, the metal and certain alloys (Monel) are used to handle F_2 and other corrosive fluorides. It is also ferromagnetic, but not as much as iron. The finely divided metal is reactive to air, and it may be pyrophoric under some conditions.

The metal is moderately electropositive:

$$Ni^{2+} + 2e = Ni \qquad F° = -0.24 \text{ V}$$

and dissolves readily in dilute mineral acids. Like iron, it does not dissolve in concentrated nitric acid because it is rendered passive by this reagent.

NICKEL COMPOUNDS

21-G-2. The Chemistry of Divalent Nickel, d^8

Binary Compounds. *Nickel(II) oxide,* a green solid with the rock salt structure, is formed when the hydroxide, carbonate, oxalate, or nitrate of nickel(II) is heated. It is insoluble in water but dissolves readily in acids.

The *hydroxide* $Ni(OH)_2$ may be precipitated from aqueous solutions of Ni^{II} salts on addition of alkali metal hydroxides, forming a voluminous green gel that crystallizes [$Mg(OH)_2$ structure] on prolonged storage. It is readily soluble in acid ($K_{sp} = 2 \times 10^{-16}$), and also in aqueous ammonia owing to the formation of ammine complexes. When a concentrated solution of NaOH is added to a considerable molar excess of dilute $Ni(ClO_4)_2$ solution, a soluble hydroxo species is formed that is believed, on the basis of equilibrium and kinetic studies, to be $[Ni(OH)]_4^{4+}$ with a cubic structure consisting of interpenetrating Ni_4 and $(OH)_4$ tetrahedra. $Ni(OH)_2$ is, however, not amphoteric.

Addition of sulfide ions to aqueous solutions of nickel(II) ions precipitates black NiS. This is initially freely soluble in acid, but like CoS, on exposure to air it soon becomes insoluble owing to oxidation to Ni(OH)S. Fusion of Ni, S, and BaS gives $BaNiS_2$, which forms black plates; this product is metallic and has Ni in square-pyramidal coordination.

All four nickel *halides* are known in the anhydrous state. Except for the fluoride, which is best made indirectly, they can be prepared by direct reaction of the elements. All the halides are soluble in water (the fluoride only moderately so), and from aqueous solutions they can be crystallized as the hexahydrates, except for the fluoride, which gives $NiF_2 \cdot 3H_2O$. Lower hydrates are obtained from these on storage or heating.

On addition of CN^- ions to aqueous Ni^{II} the *cyanide* is precipitated in a green hydrated form. When heated at 180 to 200° the hydrate is converted into the yellow-brown, anhydrous $Ni(CN)_2$. The green precipitate readily redissolves in an excess of cyanide to form the yellow $[Ni(CN)_4]^{2-}$ ion, which is both thermodynamically very stable (log $\beta_4 \approx 30.5$) and kinetically slow to release CN^- ion.[2] Many hydrated salts of this ion, for example, $Na_2[Ni(CN)_4]\cdot 3H_2O$, may be crystallized from such solutions. In strong cyanide solutions a further CN^- is taken up to give the red $[Ni(CN)_5]^{3-}$ ion, the structure of which is discussed below. Nickel(II) thiocyanate is also known, as a yellow-brown hydrated solid that reacts with an excess of SCN^- to form complex ions $[Ni(NCS)_4]^{2-}$ and $[Ni(NCS)_6]^{4-}$.

Other Binary Nickel(II) Compounds. A number of binary nickel compounds, probably all containing Ni^{II} but not all stoichiometric, may be obtained by direct reaction of nickel with various nonmetals such as P, As, Sb, S, Se, Te, C, and B. Nickel appears to form a nitride Ni_3N. The existence of a hydride is doubtful, although the finely divided metal absorbs hydrogen in considerable amounts.

Salts of Oxo Acids. A large number of these are known. They occur most commonly as hydrates, for example, $Ni(NO_3)_2\cdot 6H_2O$, $NiSO_4\cdot 7H_2O$, and most of them are soluble in water. Exceptions are the carbonate $NiCO_3\cdot 6H_2O$, which is precipitated on addition of alkali hydrogen carbonates to solutions of Ni^{II}, and the phosphate $Ni_3(PO_4)_2\cdot 7(?)H_2O$.

Aqueous solutions of Ni^{II} not containing strong complexing agents contain the green hexaaquanickel(II) ion, $[Ni(H_2O)_6]^{2+}$, which also occurs in a number of hydrated nickel(II) salts [e.g., $Ni(NO_3)_2\cdot 6H_2O$, $NiSO_4\cdot 6H_2O$, $NiSO_4\cdot 7H_2O$, $Ni(ClO_4)_2\cdot 6H_2O$].

21-G-3.   Stereochemistry and Electronic Structures of Nickel(II) Complexes

Nickel(II) forms a large number of complexes encompassing coordination numbers 4, 5, and 6, and all the main structural types (viz., octahedral, trigonal-bipyramidal, square-pyramidal, tetrahedral, and square). Moreover, it is characteristic of Ni^{II} complexes that complicated equilibria, which are generally temperature dependent and sometimes concentration dependent, often exist between these structural types. This section describes the characteristics of the individual structural types separately, using as examples mainly the compounds that exist completely or almost completely in one form or another. The next section discusses the configurational equilibria.

Octahedral Complexes. The maximum coordination number of nickel(II) is 6. A considerable number of neutral ligands, especially amines, displace some or all of the water molecules in the octahedral $[Ni(H_2O)_6]^{2+}$ ion to form complexes such as trans-$[Ni(H_2O)_2(NH_3)_4](NO_3)_2$, $[Ni(NH_3)_6](ClO_4)_2$, and $[Ni\ en_3]SO_4$. Such ammine complexes are characteristically blue or purple in contrast to the bright green of the hexaaquanickel ion. This is because of shifts in the absorption bands when H_2O ligands are replaced by others lying toward the stronger end of the spectrochemical series. This can be seen in Fig. 21-G-1, where the spectra of

2   W. C. Crouse and D. W. Margerum, *Inorg. Chem.,* 1974, **13**, 1437.

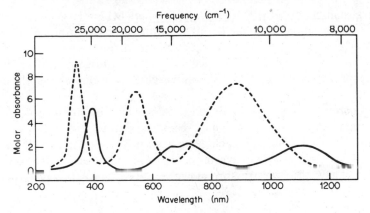

Fig. 21-G-1. Absorption spectra of $[Ni(H_2O)_6]^{2+}$ (solid curve) and $[Ni\ en_3]^{2+}$ (dashed curve).

$[Ni(H_2O)_6]^{2+}$ and $[Ni\ en_3]^{2+}$ are shown. These spectra can readily be interpreted by referring to the energy level diagram for d^8 ions (p. 661). Three spin-allowed transitions are expected, and the three observed bands in each spectrum may thus be assigned as shown in Table 21-G-2. It is a characteristic feature of the spectra of octahedral nickel(II) complexes, exemplified by those of $[Ni(H_2O)_6]^{2+}$ and $[Ni\ en_3]^{2+}$, that molar absorbances of the bands are at the low end of the range (1–100) for octahedral complexes of the first transition series in general, namely, between 1 and 10. The splitting of the middle band in the $[Ni(H_2O)_6]^{2+}$ spectrum is due to spin-orbit coupling that mixes the $^3T_{1g}(F)$ and ^1F_g states, which are very close in energy at the Δ_0 value given by $6H_2O$, whereas in the stronger field of the 3en they are so far apart that no significant mixing occurs. An analysis of the spectrum at the level of the discussion above is adequate for almost all chemical purposes, but it is worth noting that at higher resolution the spectra show much greater complexity and require much more sophisticated analysis.[3]

Magnetically, octahedral nickel(II) complexes have relatively simple behavior. From both a simple d-orbital splitting diagram (p. 645) and the energy level diagram (p. 661), it follows that they all should have two unpaired electrons, and this is found always to be the case, the magnetic moments ranging from 2.9 to 3.4 BM, depending on the magnitude of the orbital contribution.

$CsNiCl_3$ has a type of structure that is often considered to be the generic one

TABLE 21-G-2
Spectra of Octahedral Nickel(II) Complexes

Transition	Approximate band positions (cm^{-1})	
	$[Ni(H_2O)_6]^{2+}$	$[Ni\ en_3]^{2+}$
$^3A_{2g} \to {}^3T_{2g}$	9,000	11,000
$^3A_{2g} \to {}^3T_{1g}(F)$	14,000	18,500
$^3A_{2g} \to {}^3T_{1g}(P)$	25,000	30,000

[3]   A. F. Schreiner and D. J. Hamm, *Inorg. Chem.*, 1973, **12**, 2037.

for a large number of $M^IM^{II}X_3$ compounds, where M^I is a large univalent cation and M^{II} is a transition metal ion. In this structure there are infinite parallel chains of NiX_6 octahedra sharing opposite faces, and the M^I ions lie between these chains in an ordered pattern. The M^{II} ions generally show significant antiferromagnetic coupling that is believed to occur by a superexchange mechanism via the bridging X ions.[4]

Five-Coordinate Nickel(II) Complexes.   A considerable number of both trigonal-bipyramidal and square-pyramidal complexes occur and high- ($S = 1$) and low-spin ($S = 0$) examples of each geometry are known. Many of the trigonal-bipyramidal complexes contain one of the tripod ligands (Section 3–6) such as pp_3 and np_3 and have the type structure shown in 21-G-I. The fifth ligand is typically

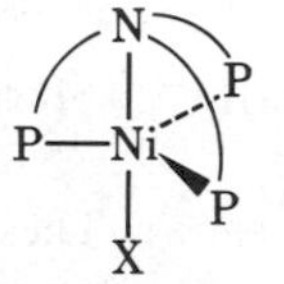

(21-G-I)

a halide ion, though H, SR, RSO_3, and R also occur[5]; thus the complex is a +1 cation. The symmetry cannot, of course, ever be the full D_{3h} symmetry of a true *tbp*, but such cases are usually called *tbp* complexes for simplicity. With few exceptions they are low spin (i.e., diamagnetic). $[Ni(Me_6tren)Br]^+$ is high spin, but replacement of ligand N atoms by P, As, S, Se, etc. causes larger ligand field splittings and stabilizes the low-spin configuration. There are other low-spin, five-coordinate Ni^{II} complexes with $[NiL_5]^{2+}$ and $[NiL_3X_2]$ compositions, as well as $[NiL_4X]^+$ types, and in many cases the ligands L are separate,[6] as in $[Ni(SbMe_3)_5]^{2+}$, $[Ni(PMe_3)_4Br]^+$, and $Ni(SbMe_3)_2X_2$.

The $[Ni(CN)_5]^{3-}$ ion is usually found with *sp* geometry,[7] but in $[Cr(en)_3]$-$[Ni(CN)_5]\cdot1\cdot5H_2O$ there are two crystallographically independent $[Ni(CN)_5]^{3-}$ ions, one with *sp* and the other with *tbp* geometry. However when this compound is dehydrated or subjected to pressure[8] the crystal structure changes and the *tbp* one becomes *sp*.

There are relatively few *sp* complexes, most of which, like $[Ni(Ph_2MeAsO)_4(ClO_3)]^+$ and $[Ni(Me_3AsO)_5](ClO_4)_2$,[9] are high spin. The complexes $[Ni(p_3)SO_4]$ and $[Ni(p_3)SeO_4]$, where $p_3 = CH_3C(CH_2PPh_2)_3$ and the oxo anions are bidentate, are *sp* and paramagnetic in the crystal but become mainly diamagnetic in solution.[10]

Tetrahedral Complexes.   These are mainly of the following stoichiometric types:

4   G. L. McPherson *et al., Inorg. Chem.,* 1973, **12,** 1196; 1974, **13,** 2230.
5   L. Sacconi *et al., Inorg. Chem.,* 1977, **16,** 1518, 1669, 2377.
6   M. Dartiguenave *et al., Inorg. Chem.,* 1978, **17,** 3503; 1977, **16,** 440.
7   F. A. Jurnak and K. N. Raymond, *Inorg. Chem.,* 1974, **13,** 2387.
8   L. J. Basile *et al., Inorg. Chem.,* 1974, **13,** 496.
9   Y. S. Ng, G. A. Rodley, and W. T. Robinson, *Inorg. Chem.,* 1976, **15,** 303.
10 C. Benelli *et al., Inorg. Chem.,* 1977, **16,** 182.

NiX_4^{2-}, NiX_3L^-, NiL_2X_2, and $Ni(L—L)_2$, where X represents a halogen or SPh,[11] L a neutral ligand such as a phosphine, phosphine oxide, or arsine, and L—L is one of several types of bidentate ligand (e.g., 21-G-II to 21-G-IV). These three bidentate

| (21-G-II) | (21-G-III) | (21-G-IV) |

ligands all contain sufficiently bulky substituents on, or adjacent to, the nitrogen atoms to render planarity of the $Ni(L—L)_2$ molecule sterically impossible. When small substituents are present, planar or nearly planar complexes are formed. It must be stressed that except for the NiX_4^{2-} species, a rigorously tetrahedral configuration cannot be expected. However in some cases there are marked distortions even from the highest symmetry possible, given the inherent shapes of the ligands. Thus in $Ni(L—L)_2$ molecules the most symmetrical configuration possible would have the planes of the two L—L ligands perpendicular. Most often, however, this dihedral angle differs considerably from 90°; for example, when L—L is 21-G-III the angle is 82°, and when L—L is 21-G-II it is only 76°. Thus the term "tetrahedral" is sometimes used very loosely (i.e., does not imply a regular tetrahedron); since all the so-called tetrahedral species are paramagnetic with two unpaired electrons, it would perhaps be better to simply call them paramagnetic rather than tetrahedral. Indeed, the most meaningful way to distinguish between "tetrahedral" and "planar" four-coordinate nickel(II) complexes is to consider that for a given ligand set ABCD, there is a critical value of the dihedral angle between two planes, such as A—Ni—B and C—Ni—D. When the angle exceeds this value the molecule will be paramagnetic; it may be called "tetrahedral" even though the dihedral angle is appreciably less than 90°. Conversely, when the angle is below the critical value the complex will be diamagnetic; it may be called "planar" even if the limit of strict planarity is not actually attained.

For regular or nearly regular tetrahedral complexes there are characteristic spectral and magnetic properties. Naturally the more irregular the geometry of a paramagnetic nickel(II) complex the less likely it is to conform to these specifications. In T_d symmetry the d^8 configuration gives rise to a $^3T_1(F)$ ground state. The transition from this to the $^3T_1(P)$ state occurs in the visible region ($\sim 15,000$ cm^{-1}) and is relatively strong ($\epsilon \approx 10^2$) compared to the corresponding $^3A_{2g} \rightarrow$ $^3T_{1g}$ transition in octahedral complexes. Thus tetrahedral complexes are generally strongly colored and tend to be blue or green unless the ligands also have absorption bands in the visible region. Because the ground state $^3T_1(F)$ has much inherent orbital angular momentum, the magnetic moment of truly tetrahedral Ni^{II} should be about 4.2 BM at room temperature. However even slight distortions reduce this

[11]   D. G. Holah and D. Coucouvanis, *J. Am. Chem. Soc.*, 1975, **97**, 6917.

markedly (by splitting the orbital degeneracy). Thus fairly regular tetrahedral complexes have moments of 3.5 to 4.0 BM; for the more distorted ones the moments are 3.0 to 3.5 BM (i.e., in the same range as for six-coordinate complexes).

The $NiCl_4^{2-}$ ion, a representative tetrahedral complex, has been studied in great detail spectroscopically[12] at 2.2°K. The observations can all be accounted for in detail by a parameterized crystal field model, with $\Delta_t \approx 3500$ cm^{-1}.

Planar Complexes. For the vast majority of four coordinate nickel(II) complexes, planar geometry is preferred. This is a natural consequence of the d^8 configuration, since the planar ligand set causes one of the d orbitals ($d_{x^2-y^2}$) to be uniquely high in energy and the eight electrons can occupy the other four d orbitals but leave this strongly antibonding one vacant. In tetrahedral coordination, on the other hand, occupation of antibonding orbitals is unavoidable. With the congeneric d^8 systems Pd^{II} and Pt^{II} this factor becomes so important that no tetrahedral complex is formed.

Planar complexes of Ni^{II} are thus invariably diamagnetic. They are frequently red, yellow, or brown owing to the presence of an absorption band of medium intensity ($\epsilon \approx 60$) in the range 450–600 nm, but other colors do occur when additional absorption bands are present.

As important examples of square complexes, we may mention yellow $Ni(CN)_4^{2-}$, red bis(dimethylglyoximato)nickel(II) (21-G-V), the red β-keto-enolate complex (21-G-VI), the yellow to brown $Ni(PR_3)_2X_2$ compounds in which R is alkyl, and

(21-G-V)   (21-G-VI)

complexes containing homologues of the ligands 21-G-III and 21-G-IV in which the substituents on nitrogen are small. A complex with entirely sulfur donors (21-G-VII) is known.[13]

(21-G-VII)

Ni—Ni Bonded Binuclear Compounds. Nickel forms a moderately large and very heterogeneous group of compounds in oxidation states of 2, 1.5, and 1, in which

[12]   V. J. Koester and T. M. Dunn, *Inorg. Chem.,* 1975, **14**, 1811.
[13]   P. H. Davis, L. K. White, and R. L. Belford, *Inorg. Chem.,* 1975, **14**, 1753.

there are Ni—Ni bonds varying in length from 2.32 to 2.79 Å. Unlike the M—M bonds formed by some of the early transition elements such as Cr, Mo, W, Re, and Ru, many of these Ni—Ni bonds have received little theoretical discussion and are not well understood. A few examples are shown in 21-G-VIII to 21-G-XI.

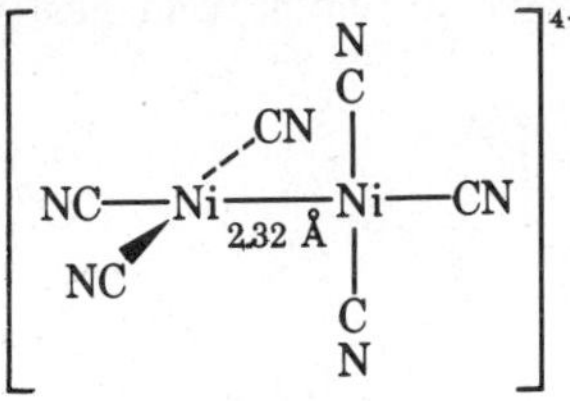

(21-G-VIII)[14] 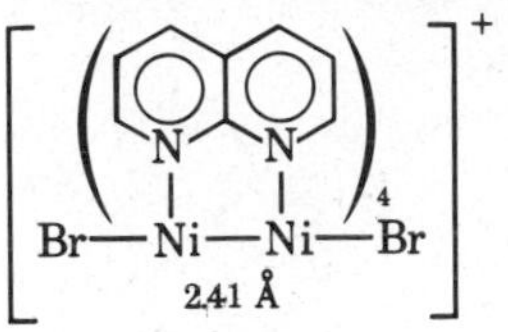 (21-G-IX)[15] 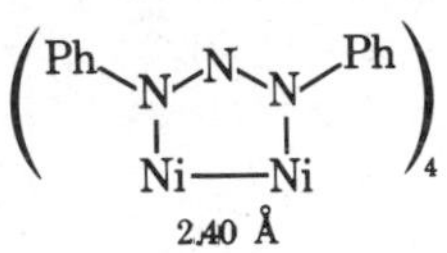 (21-G-X)[16]

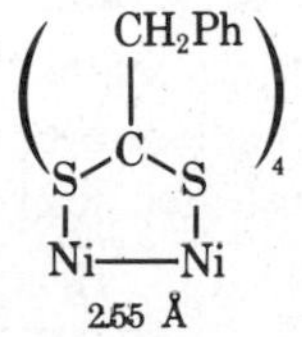

(21-G-XI)[17]

21-G-4. "Anomalous" Properties of Nickel(II) Complexes; Conformational Changes

A considerable number of nickel(II) complexes do not behave consistently in accord with expectation for any discrete structural types, and they have in the past been termed "anomalous." All the "anomalies" can be satisfactorily explained in terms of several types of conformational or other structural change and, ironically, so many examples are now known that the term "anomalous" is no longer appropriate. The three main structural and conformational changes that nickel(II) complexes undergo are described and illustrated below.

1. *Formation of Five- and Six-Coordinate Complexes by Addition of Ligands to Square Ones.* For any square complex NiL_4, the following equilibria with additional ligands L' must in principle exist:

$$ML_4 + L' = ML_4L'$$
$$ML_4 + 2L' = ML_4L_2'$$

In the case where L = L' = CN, only the five-coordinate species is formed, but in most cases equilibria strongly favor the six-coordinate species that have trans structures and two unpaired electrons. The complex 21-G-VI, for example, is normally prepared in water or alcohol and first isolated as a green, paramagnetic dihydrate or dialcoholate. Heating then drives off the H_2O or C_2H_5OH to leave the red, diamagnetic square complex.

[14] O. Jarchow, *Z. Kristallogr.* 1972, **136**, 122.
[15] L. Sacconi, C. Mealli, and D. Gatteschi, *Inorg. Chem.,* 1974, **13**, 1985.
[16] M. Corbett *et al., Aust. J. Chem.,* 1975, **28**, 2377.
[17] M. Bonamico, G. Dessy, and V. Fares, *J.C.S. Dalton,* **1977**, 2315.

An interesting series of compounds is provided by the series[18] based on $[NiL]^{2+}$, shown as 21-G-XII. In the perchlorate $[NiL](ClO_4)_2$ the nickel atom is four-coordinate and the compound is red and diamagnetic. The compounds $[NiLX]X$ (X = Cl, Br, I) are blue or green and have two unpaired electrons; the cationic complex is five-coordinate *sp*. $[NiL(NCS)_2]$ is octahedral, violet, and has two unpaired electrons.

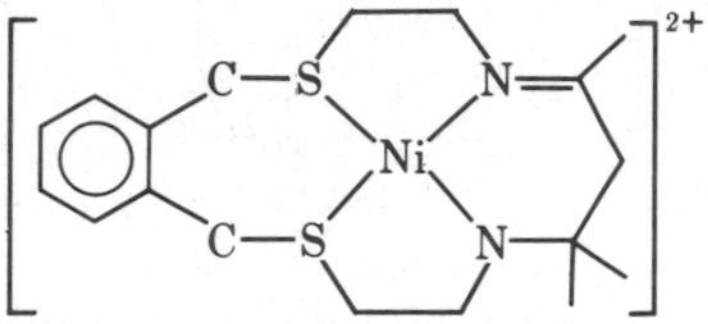

(21-G-XII)

Well-known examples of the square-octahedral ambivalence are the Lifschitz salts, complexes of nickel(II) with substituted ethylenediamines, especially the stilbenediamines, one of which is illustrated in 21-G-XIII. Many years ago Lifschitz

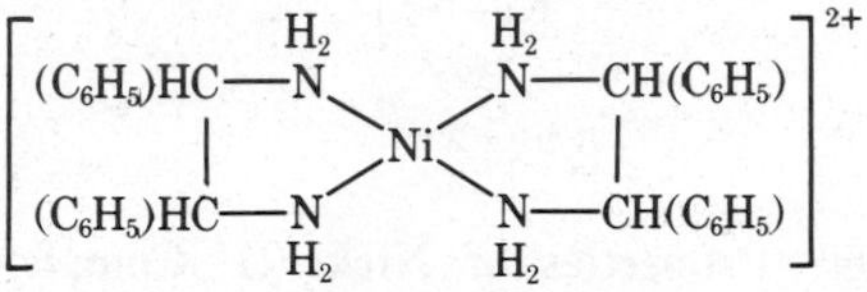

(21-G-XIII)

and others observed that such complexes were sometimes blue and paramagnetic and at other times yellow and diamagnetic, depending on many factors such as temperature, identity of the anions present, the solvent in which they are dissolved or from which they were crystallized, exposure to atmospheric water vapor, and the particular diamine involved. The bare experimental facts bewildered chemists for several decades, and many hypotheses were promulgated in an effort to explain some or all of them. It is now recognized that the yellow species are square complexes, as typified by 21-G-XIII, and the blue ones are octahedral complexes, derived from the square ones by coordination of two additional ligands—solvent molecules, water molecules, or anions—above and below the plane of the square complex.

The complex 21-G-XIV is interesting in that it contains, side by side, low-spin, square-coordinated and high-spin, octahedrally coordinated nickel atoms.[19a]

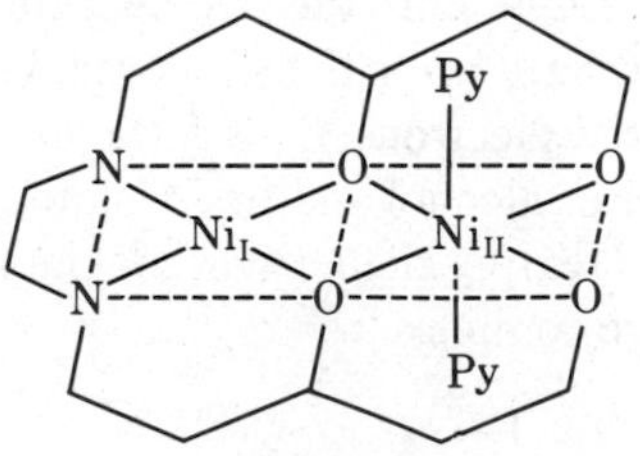

(21-G-XIV)

[18] F. L. Urbach and D. H. Busch, *Inorg. Chem.*, 1973, **12**, 408.
[19a] M. D. Glick *et al.*, *Inorg. Chem.*, 1976, **15**, 2259.

2. *Monomer-Polymer Equilibria.* In many cases four-coordinate complexes associate or polymerize, to give species in which the nickel ions become five- or six-coordinate. In some cases the association is very strong and the four-coordinate monomers are observed only at high temperatures; in others the position of the equilibrium is such that both red, diamagnetic monomers and green or blue, paramagnetic polymers are present in a temperature- and concentration-dependent equilibrium around room temperature. A clear example of this situation is provided by various β-ketoenolate complexes. When the β-ketoenolate is the acetylacetonate ion, the trimeric structure shown in Fig. 4-8 is adopted. As a result of the sharing of some oxygen atoms, each nickel atom achieves octahedral coordination; the situation is comparable to, but different in detail from, that found for $[Co(acac)_2]_4$. This trimer is very stable and only at temperatures around 200° (in a noncoordinating solvent) do detectable quantities of monomer appear. It is, however, readily cleaved by donors such as H_2O or pyridine, to give six-coordinate monomers. When the methyl groups of the acetylacetonate ligand are replaced by the very bulky $C(CH_3)_3$ group, trimerization is completely prevented and the planar monomer 21-G-VI results. When groups sterically intermediate between CH_3 and $C(CH_3)_3$ are used, temperature- and concentration-dependent purple monomer–green trimer equilibria are observed in non-coordinating solvents.[19b]

Partial dimerization, presumably to give five-coordinate, high-spin nickel(II), is known to be the cause of anomalous behavior in some instances. Thus although N-(n-alkyl)salicylaldiminato complexes of Ni[II] are, in general, planar, diamagnetic monomers in chloroform or benzene, when the alkyl group is CH_3 there is an equilibrium between the diamagnetic monomer and a paramagnetic dimer, presumably as shown in 21-G-XV and 21-G-XVI.

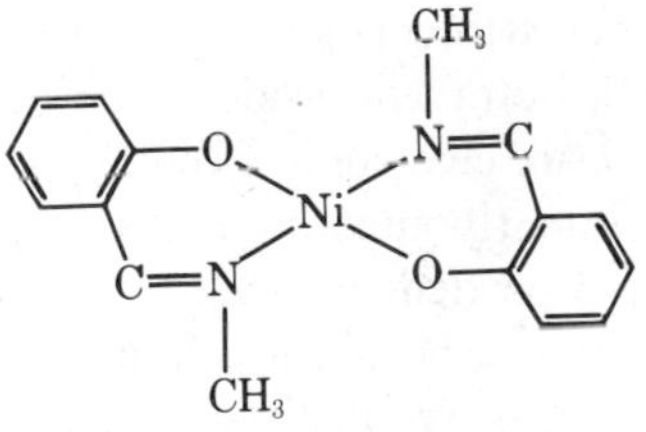

(21-G-XV)

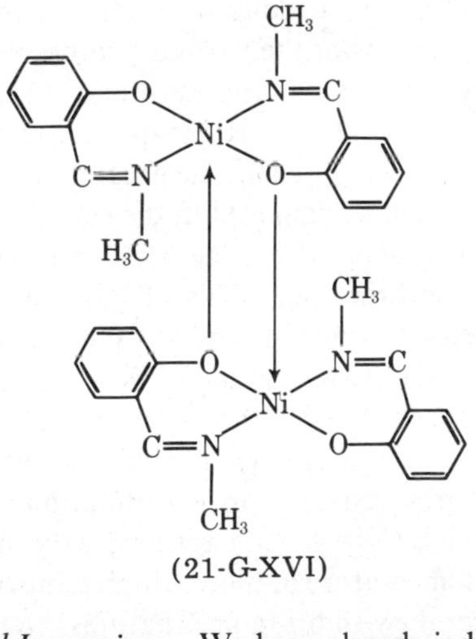

(21-G-XVI)

3. *Square-Tetrahedral Equilibria and Isomerism.* We have already indicated that nickel(II) complexes of certain stoichiometric types, namely, the bishalobis-phosphino and bissalicylaldiminato types may have either square or tetrahedral

[19b] C. S. Chamberlain and R. S. Drago, *Inorg. Chim. Acta,* 1979, **32**, 75.

structures, depending on the identity of the ligands. For example, in the NiL_2X_2 cases, when L is triphenylphosphine, tetrahedral structures are found, whereas the complexes with trialkylphosphines generally give square complexes. Perhaps it is then not very surprising that a number of NiL_2X_2 complexes in which L represents a mixed alkylarylphosphine exist in solution in an equilibrium distribution between the tetrahedral and square forms. Careful studies have shown that the influence of the varying R groups in the phosphines is almost entirely electronic rather than steric.[20] For example, in CH_2Cl_2 solution, for PPh_2Bu and $P(cyclohexyl)_3$, which do not differ greatly in size, the mole fractions of the tetrahedral form are 1.00 and 0.00, respectively. At 25° the rate constants for conversion of tetrahedral into planar isomers are in the range 10^5–10^6 sec^{-1} with enthalpies of activation of around 45 kJ mol^{-1}.

In some cases it is possible to isolate two crystalline forms of the compound, one yellow to red and diamagnetic, the other green or blue with two unpaired electrons. There is even $Ni[(C_6H_5CH_2)(C_6H_5)_2P]_2Br_2$, in which both tetrahedral and square complexes are found together in the same crystalline substance.

Planar-tetrahedral equilibria in compounds of type 21-G-XVII have been well

(21-G-XVII)

studied by nmr. In the tetrahedral forms unpaired electron spin density from the nickel atoms is introduced into the ligand π system, which results in large shifts in the positions of the various proton nuclear magnetic resonances. To a certain extent the position of the equilibrium is a function of steric factors, that is, of the repulsion between the R groups on the nitrogen atoms of one ligand and various parts of the other ligand, the greater degree of repulsion encountered in the square configuration tending to shift the equilibrium to the tetrahedral side. However some ring substituents affect the equilibrium by means of electronic effects as well.

Thermochromism.   This phenomenon is frequently encountered among Ni^{II} complexes. It comes in general from temperature-dependent variability of structure, which causes variation in the d–d absorption bands. Specific mechanisms differ from case to case. In some cases it appears that there are only relatively small relocations of ligands within a qualitatively fixed type of symmetry,[21] whereas in others gross changes in coordination geometry occur. In $(NR_xH_{4-x})_2NiCl_4$ compounds (X = 1, 2, 3), reversible thermochromism, from yellow-brown or green at low temperature to blue at high temperature, appears to be due to changes from octahedral (with bridging Cl atoms) to tetrahedral coordination.[22]

20   L. Que, Jr., and L. H. Pignolet, *Inorg. Chem.,* 1973, **12,** 156.
21   L. Fabrizzi, M. Micheloni, and P. Paoletti, *Inorg. Chem.,* 1974, **13,** 3019.
22   J. R. Ferraro and A. T. Sherren, *Inorg. Chem.,* 1978, **17,** 2498.

21-G-5. Higher Oxidation States[23]

Oxides, Hydroxides, and Halides. For Ni^{IV} there are no well-documented simple compounds, but there are complexes (see below).

Nickel(III) fluoride[24a] has been prepared as an impure black, noncrystalline solid, marginally stable at 25°. There is no good evidence for Ni_2O_3, but there are two proved crystalline forms of black $NiO(OH)$. The more common β-$NiO(OH)$ is obtained by the oxidation of nickel(II) nitrate solutions with bromine in aqueous potassium hydroxide below 25°. It is readily soluble in acids; on aging, or by oxidation in hot solutions, a Ni^{II}—Ni^{III} hydroxide of stoichiometry $Ni_3O_2(OH)_4$ is obtained. The oxidation of alkaline nickel sulfate solutions by $NaOCl$ gives a black "peroxide" "$NiO_2 \cdot nH_2O$," This is unstable, being readily reduced by water, but it is a useful oxidizing agent for organic compounds.[24b]

The compound $NaNiO_2$ and several related ones also seem to be genuine. They can be made by bubbling oxygen through molten alkali metal hydroxides contained in nickel vessels at about 800°. Other oxides and oxide phases can be made by heating NiO with alkali or alkaline earth oxides in oxygen. These mixed oxides evolve oxygen on treatment with water or acid.

The Edison or nickel-iron battery, which uses KOH as the electrolyte, is based on the reaction:

$$Fe + 2NiO(OH) + 2H_2O \underset{charge}{\overset{discharge}{\rightleftharpoons}} Fe(OH)_2 + 2Ni(OH)_2 \ (\sim 1.3 \ V)$$

but the mechanism and the true nature of the oxidized nickel species are not fully understood.

Complexes of Nickel(IV). The most firmly established and simplest example is the $[NiF_6]^{2-}$ ion, which occurs as red K, Rb, Cs, and other salts,[25] of which the potassium one is most often made. Other compounds in which Ni^{IV} is coordinated by electronegative atoms are a heteropolymolybdate anion of composition $[NiMo_9O_{32}]^{6-}$, the heteropolyniobate complex $[NiNb_{12}O_{38}]^{12-}$, and the periodates $Na(K)NiIO_6 \cdot nH_2O$. X-ray photoelectron spectra (for Ni $2p$ electrons) confirmed the oxidation number IV in K_2NiF_6 but raised some questions of interpretation in other cases.[26] Thus several dicarbollide complexes, as well as the 1,2-dithiete or 1,2-dithiol complexes of the type $[Ni(S_2C_2R_2)_2]^{2-}$, are most realistically regarded as Ni^{II} species, although from a purely formal point of view they might be considered to contain Ni^{IV}

Original formulations of sulfur-containing Ni^{IV} complexes with Ni—S—Ni bridges are incorrect and divalent Ni^{II} complexes of the type 21-G-XVIII are present.

23   W. Levason and C. A. McAuliffe, *Coord. Chem. Rev.*, 1974, **12**, 151.
24a T. L. Court and M. F. A. Dove, *J.C.S. Dalton*, **1973**, 1995.
24b M. V. George and K. S. Balachandran, *Chem. Rev.*, 1975, **75**, 491.
25   K. O. Christe, *Inorg. Chem.*, 1977, **16**, 2238.
26   C. A. Tolman *et al.*, *Inorg. Chem.*, 1973, **12**, 2770; L. O. Pont *et al.*, *Inorg. Chem.* 1974, **13**, 483.

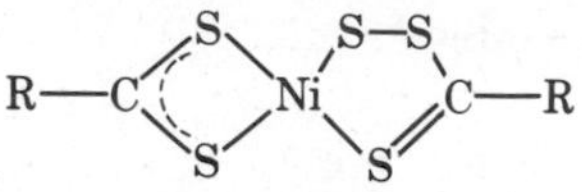

(21-G-XVIII)

Bromine oxidizes $Ni(Bu_2dtc)_2$ (Bu_2dtc = N,N'-dibutyldithiocarbamate) to $Ni(Bu_2dtc)_3Br$, and X-ray study shows this to contain a $[Ni(Bu_2dtc)_3]^+$ cation with a trigonally distorted (D_3) octahedral structure. There is also a selenium analogue.[27] These might seem rather obviously to contain Ni^{IV} coordinated by Bu_2dtc^-, or the analogous selenium ligands, but in view of the expected oxidizing nature of Ni^{IV} and the known ease of oxidation of dtc^- ions, this is probably only a naive formalism.

There is a recently discovered group of apparently genuine Ni^{IV} complexes with oxime type ligands (though certain older oxime complexes have been shown to be spurious). The NiL_2 complex where L is 21-G-XIX, has been structurally characterized.[28] It has octahedral coordination with Ni—N distances *ca.* 0.16Å shorter than those in Ni^{II} complexes with NiN_6 octahedra. The $[NiL_2]^{2+}$ complex in which L is 21-G-XX and a complex with the related hexadentate ligand 21-G-XXI, which are dark violet and diamagnetic, can be isolated as perchlorates.[29] They are relatively easy to prepare, stable under normal laboratory conditions, and have reversible redox chemistry connecting them with protonated (on oxygen) Ni^{II} complexes.

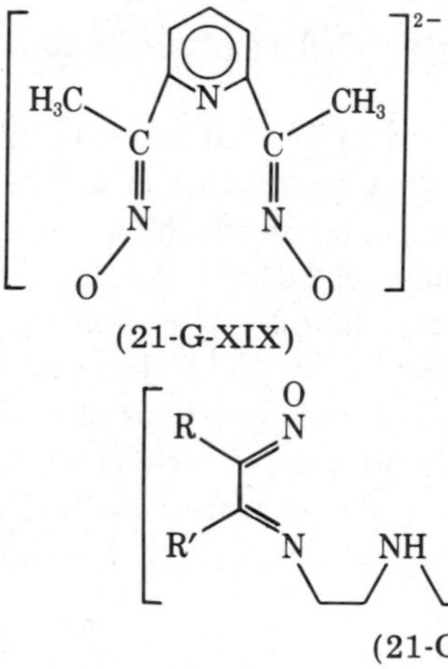

(21-G-XIX)

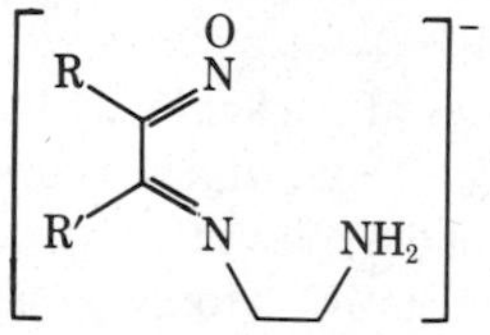

(21-G-XX)

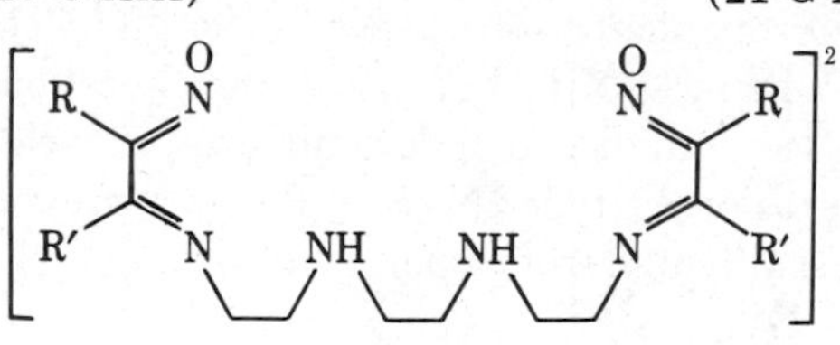

(21-G-XXI)

Complexes of Nickel(III). Violet K_3NiF_6 is a genuine Ni^{III} complex (contrary to earlier suggestions that it might be a mixture of Ni^{II} and Ni^{V}), and the NiF_6^{3-} octahedra are elongated by the Jahn-Teller effect expected for the $t_{2g}^6 e_g^1$ configu-

[27] P. T. Beurskens and J. A. Cras, *J. Cryst. Mol. Struct.*, **1971**, 63. See, however, W. Dietzsch, *et al., Inorg. Chem.*, 1978, **17**, 1665.

[28] G. Sproul and G. D. Stucky, *Inorg. Chem.*, 1973, **12**, 2898; J. H. Takemoto, *Inorg. Chem.* 1973, **12**, 949.

[29] A. Chakravorty *et al., Inorg. Chem.*, 1975, **14**, 2178; 1976, **15**, 2916; 1977, **16**, 2597.

ration.[30] The cation in *trans*-[Ni(diars)$_2$Cl$_2$]Cl also shows distortion because of preferred occupation of the d_{z^2} orbital by the one e_g electron[31]; the Ni—Cl distances are *ca.* 0.17 Å longer than in the analogous Co^{III} complex.

The five-coordinate NiX$_3$(PR$_3$)$_2$ complexes have long been known, and Ni-Br$_3$(PMe$_2$Ph)$_2$ has been shown crystallographically to be trigonal bipyramidal with the phosphine ligands axial.

The cyclam complex 21-G-XXII is a stable solid,[32] though rapidly attacked by moisture, and a number of related Ni^{III} species exist in solution.[32] A moderately stable EDTA complex can be obtained in solution, using OH radicals generated by pulse radiolysis to effect oxidation of the Ni^{II} complex.[33] Much less stable complexes of en and glycine have been generated in a similar way.[34] With various triply deprotonated polypeptide amides [e.g., triglycylamide, H(NHCH$_2$CO)$_3$NH$_2$] Ni^{III} complexes are formed in solution.[35]

(21-G-XXII)

The ligand 21-G-XXIII forms stable tris complexes[36] of Ni^{III}, and 21-G-XXIV, *bi,* gives compounds[37] like KNi(*bi*)$_2$. Finally there is the compound[38] Ni-(S$_2$CNBu$_2$)$_2$I.

(21-G-XXIII)   (21-G-XXIV)

21-G-6. Lower Oxidation States, −1, 0, +I

Compounds of Ni^{-1}, which are very few, and those of zerovalent nickel,[39] are formed mainly with ligands having strong π-acceptor properties; the oxidation

30   E. Alter and R. Hoppe, *Z. Anorg. Allg. Chem.,* 1974, **405**, 167; D. Reinen, C. Freibel, and V. Propach, *Z. Anorg. Allg. Chem.,* 1974, **408**, 187.
31   P. K. Bernstein, *et al., Inorg. Chem.,* 1972, **11**, 3040.
32   E. S. Gore and D. H. Busch, *Inorg. Chem.,* 1973, **12**, 1; F. V. Lovecchio, E. S. Gore, and D. H. Busch, *J. Am. Chem. Soc.,* 1974, **96**, 3109.
33   J. Lati, J. Koresh and D. Meyerstein, *Chem. Phys. Lett.,* 1972, **33**, 286.
34   J. Lati and D. Meyerstein, *Inorg. Chem.,* 1972, **11**, 2397.
35   D. W. Margerum *et al., J. Am. Chem. Soc.,* 1979, **101**, 1631.
36   R. S. Drago and E. I. Baucom, *Inorg. Chem.,* 1972, **11**, 2064.
37   J. J. Bour, P. J. Birker, and J. J. Steggerda, *Inorg. Chem.,* 1971, **10**, 1202.
38   J. Willemse, P. H. F. M. Reuwette, and J. A. Cras, *Inorg. Nucl. Chem. Lett.,* 1972, **8**, 389.
39   L. Malatesta and S. Cenini, *Zerovalent Compounds of Metals,* Academic Press, 1974.

numbers have no physical significance. Many such compounds are discussed in Chapters 25, 27, 29, and 30, but a few are mentioned here because of their close relationship certain to Ni^I complexes.

The majority of nickel(I) complexes contain phosphine ligands, or closely related ones, and have tetrahedral or *tbp* structures. They are paramagnetic as expected for d^9 configurations. The tetrahedral compounds $Ni(PPh_3)_3X$ (X = Cl, Br, I) were among the first of this sort to be isolated; they decompose only slowly in air and are stable for long periods of time in nitrogen. The compound $Ni\{CH_3C-(CH_2PPh_2)_3\}I$ also has a tetrahedral structure.[40] With the tripod ligand np₃ an entire series of compounds with *tbp* structures, $Ni(np_3)X$ (X = Cl, Br, I, CN, CO, H) can be prepared by reaction of nickel halides with the np_3 ligand in presence of borohydride, followed, if necessary, by methathesis with other X groups.[41]

Nickel(II) complexes of certain tetranitrogen-donating macrocycles can be reduced by solvated electrons (from pulse radiolysis of water) to Ni^I species. These, however, react rapidly as reducing agents and have not been isolated as solid compounds.[42]

The $Ni(np_3)$ system, already mentioned, has remarkably complex behavior.[43] It appears that the $Ni(np_3)$ unit retains its integrity, but the metal varies from Ni^0 to Ni^I to Ni^{II}, and under certain conditions products isolated from reactions are mixtures of at least two oxidation states, thereby giving variable and intermediate spectra, magnetic moments, and hydrogen content. The ligand H may be introduced deliberately (but not quantitatively) using borohydride, or it may originate by abstraction from solvent.

An interesting Ni^0 complex that has been isolated and structurally characterized[44] is $Ni(np_3)SO_2$. It has a structure of type 21-G-I (X = SO_2). This can be considered as a case of $Ni(np_3)$ serving as a donor to a π^* orbital of SO_2, much like NMe_3 in Me_3NSO_2 (p. 528).

With phosphine PR_3 and phosphite $P(OR)_3$ ligands, Ni^0 forms either three- or four-coordinate complexes,[45] the relative stabilities depending largely on the steric requirements of the ligands, through the rates of the equilibrium reactions

$$NiL_4 = NiL_3 + L$$

are strongly influenced by electronic factors. The four-coordinate species undergo many oxidative addition reactions[46] to form Ni^{II} complexes.

21-H. COPPER

Copper has a single s electron outside the filled $4d$ shell but cannot be classed in Group I, since it has little in common with the alkalis except formal stoichiometries

40   P. Dapporto, G. Fallani, and L. Sacconi, *Inorg. Chem.*, 1974, **13**, 2847.
41   L. Sacconi *et al., Inorg. Chem.*, 1974, **13**, 2850; 1975, **14**, 1380; R. Barbucci, A. Bencini, and D. Gatteschi, *Inorg. Chem.*, 1977, **16**, 2117.
42   A. M. Tait, M. Z. Hoffman, and E. Hayon, *Inorg. Chem.*, 1976, **15**, 934.
43   L. Sacconi, A. Orlandini, and S. Midollini, *Inorg. Chem.*, 1974, **13**, 2850.
44   C. Mealli *et al., Inorg. Chem.*, 1978, **17**, 3020.
45   C. A. Tolman, W. C. Seidel, and L. W. Gosser, *J. Am. Chem. Soc.*, 1974. **96**, 53.
46   C. A. Tolman and E. J. Lukosius, *Inorg. Chem.*, 1977, **16**, 940.

TABLE 21-H-1
Oxidation States and Stereochemistry of Copper

Oxidation state	Coordination number	Geometry	Examples
Cu^I, d^{10}	2	Linear	Cu_2O, $KCuO$, $CuCl_2^-$
	3	Planar	$K[Cu(CN)_2]$, $[Cu(SPMe_3)_3]ClO_4$
	4^a	Tetrahedral	CuI, $[Cu(CN)_4]^{3-}$, $[Cu(MeCN)_4]^+$
	4	Dist. planar	CuL^b
	5	Sp	$[CuLCO]^b$
Cu^{II}, d^9	5	Tbp	$[Cu(bipy)_2I]^+$, $[CuCl_5]^{2-}$, $[Cu_2Cl_8]^{4-}$
	5	Sp	$[Cu(DMGH)_2]_2(s)$
	$4^{a,c}$	Tetrahedral (dist.)	$(N$-isopropylsalicylaldiminato$)_2Cu$ $Cs_2[CuCl_4]$
	$4^{a,c}$	Square	CuO, $[Cupy_4]^{2+}$, $(NH_4)_2[CuCl_4]$
	$6^{a,c}$	Dist. octahedral	K_2CuF_4, $K_2[CuEDTA]$, $CuCl_2$
	6	Octahedral	$K_2Pb[Cu(NO_2)_6]$
	7	Pentagonal bipyramidal	$[Cu(H_2O)_2\ dps]^{2+}$ d
	8	Dist. dodecahedron	$Ca[Cu(CO_2Me)_4]\cdot 6H_2O$
Cu^{III}, d^8	4	Square	$KCuO_2$, $CuBr_2(S_2CNBu_2)$
	6	Octahedral	K_3CuF_6
Cu^{IV}, d^7	6	?	Cs_2CuF_6

[a] Most common states.

[b] R. R. Gagne et al., Inorg. Chem., 1978, **17**, 3563:

$$L =$$

[c] These three cases are often not sharply distinguished; see text.

[d] dps = 2,6-diacetylpyridinebissemicarbazone (D. Wester and G. J. Palenik, J. Am. Chem. Soc., 1974, **96**, 7565).

in the +1 oxidation state. The filled d shell is much less effective than is a noble gas shell in shielding the s electron from the nuclear charge, so that the first ionization enthalpy of Cu is higher than those of the alkalis. Since the electrons of the d shell are also involved in metallic bonding, the heat of sublimation and the melting point of copper are also much higher than those of the alkalis. These factors are responsible for the more noble character of copper, and the effect is to make the compounds more covalent and to give them higher lattice energies, which are not offset by the somewhat smaller radius of the unipositive ion compared to the alkali ions in the same period—Cu^+, 0.93; Na^+, 0.95; and K^+, 1.33 Å.

The second and third ionization enthalpies of Cu are very much lower than those of the alkalis and account in part for the transition metal character shown by the existence of colored paramagnetic ions and complexes in the II, III, and IV oxidation states. Even in the I oxidation state numerous transition-metal-like complexes are formed (e.g., those with olefins).

There is only moderate similarity between copper and the heavier elements Ag and Au, but some points are noted in the later discussions of these elements (Chapter 22).

The oxidation states and stereochemistry of copper are summarized in Table 21-H-1. Stable copper(0) compounds are not confirmed, but reactive intermediates appear to occur in some reactions.

21-H-1. The Element

Copper is widely distributed in Nature as metal, in sulfides, arsenides, chlorides, carbonates, etc. It is extracted[1] from ores usually by wet processes, e.g., by leaching with sulfuric acid, from which it is recovered by reduction with scrap steel, solvent extraction with oximes as chelates, etc. Copper is refined by electrolysis.

Copper is a tough, soft, and ductile reddish metal, second only to silver in its high thermal and electrical conductivities. It is used in many alloys such as brasses and is completely miscible with gold. It is only superficially oxidized in air, sometimes acquiring a green coating of hydroxo carbonate and hydroxo sulfate.

Copper reacts at red heat with oxygen to give CuO and, at higher temperatures, Cu_2O; with sulfur it gives Cu_2S or a nonstoichiometric form of this compound. It is attacked by halogens but is unaffected by nonoxidizing or noncomplexing dilute acids in absence of air. Copper readily dissolves in nitric acid and sulfuric acid in presence of oxygen. It is also soluble in ammonia, ammonium carbonate or potassium cyanide solutions in presence of oxygen, as indicated by the potentials.

$$Cu + 2NH_3 \xrightarrow{-0.12 \text{ V}} [Cu(NH_3)_2]^+ \xrightarrow{-0.01 \text{ V}} [Cu(NH_3)_4]^{2+}$$

It is also soluble in acid solutions containing thiourea, which stabilizes Cu^I as a complex; acid thiourea solutions are also used to dissolve copper deposits in boilers.[2]

COPPER COMPOUNDS

21-H-2. The Copper(I) State, d^{10}

Copper(I) compounds are diamagnetic and, except where color results from the anion or charge-transfer bands, colorless.

The relative stabilities of the copper(I) and copper(II) states are indicated by

[1]   D. S. Fett, *Acc. Chem. Res.*, 1977, **10**, 99; J. C. Yannopoulous and J. C. Agarwal, Eds., *Extractive Metallurgy of Copper*, Metallurgy Society, AIME, 1976.
[2]   J. G. Frost *et al.*, *Inorg. Chem.*, 1976, **15**, 940.

the following potential data:

$$Cu^+ + e = Cu \qquad E^0 = 0.52 \text{ V}$$
$$Cu^{2+} + e = Cu^+ \qquad E^0 = 0.153 \text{ V}$$

whence

$$Cu + Cu^{2+} = 2Cu^+ \qquad E^0 = -0.37 \text{ V}; \qquad K = [Cu^{2+}]/[Cu^+]^2 = \sim 10^6$$

The relative stabilities of Cu^I and Cu^{II} in aqueous solution depend very strongly on the nature of anions or other ligands present and vary considerably with solvent or the nature of neighboring atoms in a crystal.

In aqueous solution only low equilibrium concentrations of Cu^+ ($<10^{-2}M$) can exist (see below) and the only simple copper(I) compounds that are stable to water are the highly insoluble ones such as $CuCl$ or $CuCN$. This instability toward water is due partly to the greater lattice and solvation energies and higher formation constants for complexes of the copper(II) ion, so that ionic Cu^I derivatives are unstable. Of course numerous copper(I) cationic or anionic complexes are stable in aqueous solution.

The equilibrium $2Cu^I \rightleftharpoons Cu + Cu^{II}$ can readily be displaced in either direction. Thus with CN^-, I^-, and Me_2S, Cu^{II} reacts to give the Cu^I compound; with anions that cannot give covalent bonds or bridging groups (e.g., ClO_4^- and SO_4^{2-}) or with complexing agents that have their greater affinity for Cu^{II}, the Cu^{II} state is favored—thus ethylenediamine reacts with $CuCl$ in aqueous potassium chloride solution:

$$2CuCl + 2en = [Cu\ en_2]^{2+} + 2Cl^- + Cu^0$$

That the latter reaction also depends on the geometry of the ligand, that is, on its chelate nature, is shown by differences in the $[Cu^{2+}]/[Cu^+]^2$ equilibrium with chelating and nonchelating amines. Thus for ethylenediamine, K is $\sim 10^5$, for pentamethylenediamine (which does not chelate) 3×10^{-2}, and for ammonia 2×10^{-2}. Hence in the last case the reaction is

$$[Cu(NH_3)_4]^{2+} + Cu^0 = 2[Cu(NH_3)_2]^+$$

The lifetime of the Cu^+ aqua ion in water depends strongly on conditions. Usually disproportionation is very fast (<1 sec), but ca. $0.01M$ solutions prepared in $0.1M$ $HClO_4$ at $0°$ by the reaction

or by reduction of Cu^{2+} with V^{2+} or Cr^{2+}, may last for several hours if air is excluded.

An excellent illustration of how the stability of Cu^I relative to Cu^{II} may be affected by solvent is the case of acetonitrile.[3] The Cu^+ ion is very effectively solvated by CH_3CN, and the copper(I) halides have relatively high solubilities (e.g., CuI, 35 g per kg CH_3CN) vs. negligible solubilities in H_2O. Cu^I is more stable than Cu^{II}

[3] A. J. Parker, *Search*, 1973, **4**, 426. See also T. Ogura *et al.*, *Trans. Metal Chem.*, 1978, **3**, 342.

in CH_3CN and the latter is, in fact, a comparatively powerful oxidizing agent. The tetrahedral ion $[Cu(MeCN)_4]^+$ can be isolated[4] in salts with large anions (e.g., ClO_4^-, PF_6^-).

Copper(I) Binary Compounds.   The *oxide* and the *sulfide* are more stable than the corresponding Cu^{II} compounds at high temperatures. Cu_2O is made as a yellow powder by controlled reduction of an alkaline solution of a Cu^{2+} salt with hydrazine or, as red crystals, by thermal decomposition of CuO. A yellow "hydroxide" is precipitated from the metastable Cu^+ solution mentioned above. Cu_2S is a black crystalline solid prepared by heating copper and sulfur in absence of air.

Nearly colorless KCuO, prepared by heating together K_2O and Cu_2O, contains square $Cu_4O_4^{4-}$ rings with O atoms at the corners; there are similar Ag and Au compounds.

Copper(I) chloride and bromide are made by boiling an acidic solution of the copper(II) salt with an excess of copper; on dilution, white CuCl or pale yellow CuBr is precipitated. Addition of I^- to a solution of Cu^{2+} forms a precipitate that rapidly and quantitatively decomposes to CuI and iodine. CuF is unknown. The halides have the zinc blende structure (tetrahedrally coordinated Cu^+). CuCl and CuBr are polymeric in the vapor state, and for CuCl the principal species appears to be a six-ring of alternating Cu and Cl atoms with Cu—Cl $\sim$ 2.16 Å. White CuCl becomes deep blue at 178° and melts to a deep green liquid.

The halides are highly insoluble in water, log K_{sp} values at 25° being -4.49, -8.23, and -11.96 for CuCl, CuBr, and CuI, respectively. Solubility is enhanced by an excess of halide ions (owing to formation of, e.g., $CuCl_2^-$, $CuCl_3^{2-}$, and $CuCl_4^{3-}$) and by other complexing species such as CN^-, NH_3, and $S_2O_3^{2-}$.

Other relatively common Cu^I compounds are the *cyanide,* conveniently prepared by the reaction:

$$2Cu^{2+}(aq) + 4CN^-(aq) \rightarrow 2CuCN(s) + C_2N_2$$

and soluble in an excess of cyanide to give the ions $Cu(CN)_2^-$, $Cu(CN)_3^{2-}$, and $Cu(CN)_4^{3-}$. Although compounds such as $KCu(CN)_2$ and $NaCu(CN)_2 \cdot 2H_2O$ are known, they do not contain a simple ion analogous to $[Ag(CN)_2]^-$, but instead have infinite chains (see p. 805). However, in $Na_2[Cu(CN)_3] \cdot 3H_2O$ there is a discrete three-coordinate ion[5].

Copper(I) Carboxylates, Alcoxides, and Sulfate.   Copper(I) forms several carboxylates that have widely different structures.

Copper(I) acetate forms white air- and moisture-sensitive crystals by reduction of anhydrous Cu^{II} acetate by Cu in pyridine or MeCN. Its structure 21-H-I is one with an infinite planar chain of binuclear and dimeric units bridged through Cu and O atoms so that each Cu is bound to three O and one Cu in a distorted square with a short Cu—Cu distance (2.556 Å).[6] By contrast, like the trifluoroacetate

[4]   I. Czoregh *et al., Acta Crystallogr., B,* 1975, **31**, 314.

[5]   C. Kappenstein and R. P. Hugel, *Inorg. Chem.,* 1978, **17**, 1945; 1977, **16**, 250.

[6]   R. D. Mountz *et al., Inorg. Chem.,* 1974, **13**, 803; D. A. Edwards and R. Richards, *J.C.S. Dalton,* **1973**, 2463.

(21-H-I)   (21-H-II)

in $[CuO_2CCF_3]_4 \cdot 2C_6H_6$, the benzoate is a tetramer $[CuO_2CPh]_4$, in which the Cu atoms are in a parallelogram (internal angles 114.5°, 65.5°, 108.0°, and 71.2°: 21-H-II) with bridging carboxylate groups.[7] The compound 1,3-dimethyltriazeno copper(I) has a similar structure, with Cu—Cu distances of *ca.* 2.6 Å. This is only one type of Cu_4 polynuclear structure (see later). The shortest known Cu—Cu distance, 2.386 Å, occurs in a Cu^I—Cu^{II} naphthyridine complex.[8]

Copper(I)trifluoromethanesulfonate (triflate)[9] can be isolated as a white crystalline, but air-sensitive complex $[Cu(O_3SCF_3)]_2 \cdot C_6H_6$ by interaction of Cu_2O and trifluoromethanesulfonic anhydride in benzene. The benzene is readily displaced by a variety of olefins to give cationic olefin complexes (see below). The complex also catalyzes the cyclopropanation of olefins by use of diazoalkanes $RCHN_2$.

Copper(I) alcoxides (CuOR) are interesting yellow substances[10] that can be made, e.g., by the reactions:

$$CuCl + LiOR = LiCl + CuOR$$
$$(CuMe)_n + nC_6H_5OH = nCuOC_6H_5 + nCH_4$$

The methoxide is insoluble, but others are sublimable and soluble in ethers. The *t*-butoxide is a tetramer $[CuOCMe_3]_4$, with alcoxo bridges (see p. 806). Alcoxides react with organic halides R′I to give ethers ROR′. The butoxide will metallate acidic hydrocarbons such as C_5H_6 or $PhC\equiv CH$ and reacts with CO_2 and NHR_2 to give the carbamate.[11]

The *sulfate,* a grayish water-sensitive solid, is made by the reaction

7   M. G. B. Drew *et al., J.C.S. Dalton,* **1977,** 299.
8   D. Gatteschi, C. Mealli and L. Sacconi, *Inorg. Chem.,* 1976, **15,** 2774.
9   R. G. Salomon and J. K. Kochi, *J. Am. Chem. Soc.,* 1973, **95,** 3300.
10 T. Tsuda *et al., J. Am. Chem. Soc.,* 1972, **94,** 658; G. M. Whitesides *et al., J. Am. Chem. Soc.,* 1974, **96,** 2829.
11 T. Tsuda *et al., J.C.S. Chem. Commun.,* **1978,** 815.

$$Cu_2O + (CH_3)_2SO_4 \xrightarrow{100°} Cu_2SO_4 + (CH_3)_2O$$

Copper(I) Complexes.[12] Copper(I) complexes are usually obtained by (*a*) direct interaction of ligands with copper(I) halides or the triflate, (*b*) reduction of corresponding copper(II) compounds, or (*c*) reduction of Cu^{2+} in presence of, or by, the ligand.

The stoichiometries of the compounds give little clue to their structures, which can be very complicated, being mononuclear, binuclear with halide bridges, polynuclear and the copper atom two-, three-, or four-coordinate, or infinite chains.[12,13]

Mononuclear species can be of the structural types 21-H-III to 21-H-VII. For

| CuL_4^+ | $CuXL_3$ | $CuXL_2$ | $CuXL_2$ | CuX_2^- |
| (21-H-III) | (21-H-IV) | (21-H-V) | (21-H-VI) | (21-H-VII) |

$CuXL_2$ species where X is NO_3^- or BH_4^-, distorted tetrahedra may be formed[13] as in 21-H-VI. Some examples of mononuclear species[14] are $[Cu(PPh_3)_3]BF_4$, $[Cu(SPMe_3)_3]ClO_4$, and the previously mentioned $[Cu(MeCN)_4]ClO_4$. However compounds of a particular stoichiometry (e.g., CuXL or $CuXL_2$) may have more than one structure, depending on the nature of X and L, as we shall see. Halide complexes[12,15] may have four types of structure in solid complexes, as follows;

1. Discrete ions $CuCl_2^-$ (21-H-VII).
2. Infinite chains of $CuCl_4$ tetrahedral units sharing edges $[Cu(NH_3)_4]$-Cu_2Cl_4.
3. Infinite chains of $CuCl_4$ tetrahedral units sharing corners K_2CuCl_3.
4. Infinite double chains of $CuCl_4$ tetrahedra sharing corners $CsCu_2Cl_3$.

Binuclear Species. These are of formulas $Cu_2X_2L_4$ and $Cu_2X_2L_3$ and structures 21-H-VIII and 21-H-IX. An example of the latter, which has both three and four-coordinate copper(I), is $Cu_2Cl_2(PPh_3)_3$.

(21-H-VIII)   (21-H-IX)

[12] F. J. Jardine, *Adv. Inorg. Chem. Radiochem.*, 1975, **17**, 116 (an extensive review).
[13] For references, see I. F. Taylor *et al.*, *Inorg. Chem.*, 1974, **13**, 2835; J. T. Gill *et al.*, *Inorg. Chem.*, 1976, **15**, 1155.
[14] M. B. Dines, *J. Inorg. Nucl. Chem.*, 1976, **38**, 1380; A. P. Gaughan *et al.*, *Inorg. Chem.*, 1974, **13**, 1657; J. A. Tiethof *et al.*, *Inorg. Chem.*, 1973, **12**, 1171.
[15] G. A. Bowmaker *et al.*, *J.C.S. Dalton*, **1976**, 2329; D. D. Axtell, *J. Am. Chem. Soc.*, 1973, **95**, 4555.

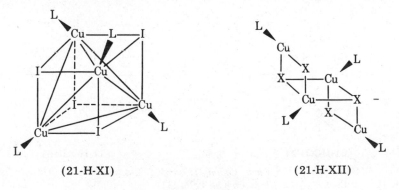

(21-H-X)

Chain Structures. As noted above, CuX_3^{2-} and $Cu_2X_3^-$ have single and double chains, respectively of linked CuX_4 tetrahedra. Another type of chain is that in $NaCu(CN)_2 \cdot 2H_2O$ where there is a spiral of almost planar trigonal Cu^I atom linked by CN bridges (21-H-X).

Tetrameric Structures. Copper benzoate and other compounds discussed above have four Cu atoms that although part of a ring, in themselves form a parallelogram. Tetrameric Cu_4^I complexes may have the structures in which the four Cu atoms are in (*a*) a parallelogram, rectangle, or a square; (*b*) most commonly, at the vertices of a tetrahedron, regular or slightly distorted, and (*c*) in a halogen-bridged step structure.

We consider first the important complexes $(CuXL)_4$ where X is halogen and L is usually a tertiary phosphine[16] but may be pyridine.[17] For these complexes there are two limiting structures, each of which may have distortion. The first is the cubane structure 21-H-XI in which there is a Cu_4 tetrahedron with a triply bridging halide and a ligand on each Cu atom, which is four-coordinate. The second has the step form 21-H-XII with double and triple halide bridges and two four-coordinate, tetrahedral and two three-coordinate, trigonal copper atoms. The silver analogues (Section 22-I-2) are similar. Which structure a complex has depends on the sizes of the metal and halide atoms and on the steric bulk of the ligand.

(21-H-XI)   (21-H-XII)

16   M. R. Churchill *et al.*, *Inorg. Chem.*, 1979, **18**, 1660.
17   C. L. Raston and A. H. White, *J.C.S. Dalton,* **1976**, 2153.

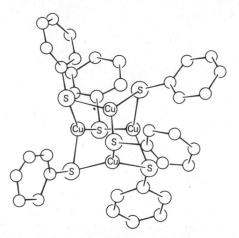

Fig. 21-H-1. The structure of the $[Cu_4(\mu_2\text{-}SPh)_6]^{2-}$ aggregate [Reproduced by permission from I. G. Dance and J. C. Calabrese, *Inorg. Chim. Acta,* 1976, **19**, L41.]

Thus for the smaller metal, Cu, cubanes are formed with iodide and small phosphines and with chloride and a large phosphine, e.g., $(CuIPEt_3)_4$ and $(CuClPPh_3)_4$, whereas step structures are found for $(CuBrPPh_3)_4$ and Cu_4I_4-$(Ph_2PCH_2CH_2PPh_2)_2$. The Cu—Cu distances in these structures range from 2.7 to 3.2 Å and indicate weak to moderate Cu—Cu bonding.

Other compounds that have a tetrahedral or distorted tetrahedral Cu_4 core are a number of sulfur complexes such as $Cu_4(S_2PPr^i_2)_4$, $[Cu_4(\mu_2\text{-}SPh)_6]^{2-}$, and $\{Cu_4[SC(NH_2)_2]_6\}^{4+}$. The last two[18] are representatives of what appears to be a characteristic structure for such sulfur complexes. There is a Cu_4 tetrahedron, but the atoms are bridged by sulfur to give a $\underline{Cu_4S_6}$ core that has an adamantane-type structure with linked six-membered $\underline{CuSCuSCuS}$ rings. The structure of the thiophenol complex is shown in Fig. 21-H-1. Finally, other compounds with planar Cu_4 cores include two compounds, $[Cu(CH_2SiMe_3)]_4$[19a] and $[CuOBu^t]_4$[19b] that have planar, eight-membered rings with linear, two-coordinate Cu^I (21-H-XIIIa, XIIIb).

(21-H-XIIIa)

(21-H-XIIIb)

[18]   (a) I. G. Dance *et al., J.C.S. Chem. Comm.,* **1976**, 68, 103; *Aust. J. Chem.,* 1978, **31**, 2195; (b) E. H. Griffith *et al., J.C.S. Chem. Comm.,* **1976**, 432.

[19a]   J. A. J. Jarvis *et al., J.C.S. Dalton,* **1977**, 999.

[19b]   T. Greiser and E. Weiss, *Chem. Ber.,* 1976, **109**, 3142.

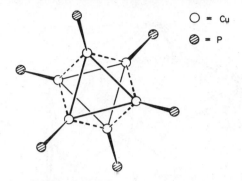

O = Cu

Ø = P

Fig. 21-H-2. The Cu_6P_6 group in $[CuH(PPh_3)]_6$. The Cu—Cu distances shown as solid lines have an average value of 2.65 Å; those shown by broken lines average 2.54 Å.

Pentanuclear complexes are so far uncommon, but one example is $[Cu_5(\mu_2\text{-}SBu^t)_6]^-$, which has trigonal bipyramidal Cu^I with bridging SR groups. The ion $[Cu_5(\mu_2\text{-}SPh)_7]^{2-}$ has four Cu atoms with trigonal planar coordination and one that is two-coordinate and linear.[18a]

Hexanuclear complexes are confined to the hydride $[CuHPPh_3]_6 \cdot Me_2NCHO$, which is obtained by treating $(CuClPPh_3)_4$ with $Na[HB(OMe)_3]$ in dimethylformamide.[20] It has the structure shown in Fig. 21-H-2; the H atoms were not located but probably lie as bridges along the six longer Cu—Cu edges.

Octa- and decanuclear complexes are mainly of sulfur ligands and have Cu_8S_{12} and $Cu_{10}S_{16}$ cores. Examples are $\{Cu_8[S_2CC(CN)_2]_{12}\}^{2-}$ and $[Cu_8(dts)_6]^{4-}$, where dts is the dithiosquarate ion $C_4O_2S_2^{2-}$ (cf. p. 367).[21]

The Cu_8S_{12} core has a cube of Cu atoms inside an icosahedral array of S atoms (Fig. 21-H-3).

Cu—Cu Interaction in Cu^I Aggregate Compounds. The short Cu···Cu distances in the various Cu aggregates, ranging from 2.38 Å (which is nearly 0.2 Å shorter than the Cu—Cu distance in metallic copper) to about 2.8 Å, raise the question of whether significant direct Cu—Cu bonding occurs. The short distances might be considered to imply that there is such bonding, but it has also been argued that the close approach of the metal atoms might result from ligand stereochemical requirements, since in *every* case there are ligand bridges. It is also pertinent that the oxidation state of copper in these compounds is I with a preferred d^{10} configuration that would not participate in Cu—Cu bonding. In view of these conflicting considerations, MO calculations were made for several representative systems.[22] The results are that direct Cu—Cu bonding is at best weak, and possibly negligible. The compounds are thus best referred to as aggregates, not clusters, whose defining property is the existence of metal-metal bonds (Chapter 26).

20   M. R. Churchill *et al., Inorg. Chem.,* 1972, **11,** 1818.
21   D. Coucouvanis *et al., J. Am. Chem. Soc.,* 1977, **99,** 6268, 8057.
22   A. Avdeef and J. P. Fackler, Jr., *Inorg. Chem.,* 1978, **17,** 2182; P. K. Mehrotra and R. Hoffmann, *Inorg. Chem.,* 1978, **17,** 2187.

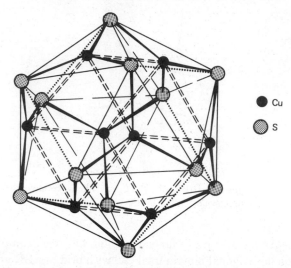

Fig. 21-H-3. The idealized Cu_8S_{12} aggregate. The cube of Cu atoms and its enclosing icosahedron of S atoms are viewed along a direction close to an S_6 symmetry axis. The six links between S atoms are denoted by dots. [Reproduced by permission from I. G. Dance and J. C. Calabrese, *Inorg. Chim. Acta*, 1976, **19**, L41.]

Copper(I) Organo Compounds.[23] Copper(I), but not copper(II), forms a variety of compounds with Cu—C bonds.

The *alkyls* and *aryls* may be obtained by interaction of copper(I) halides with lithium or Grignard results. The *alkyls* usually decompose readily but methyl copper, a bright yellow polymer insoluble in organic solvents, is reasonably stable; it can be used in certain organic syntheses,[24] but the use of lithium alkyl cuprates (see below) is more common. Alkyls can be stabilized by phosphine ligands as in e.g., $Bu^nPCu(CH_2)_3CH_3$.

The *aryls*[25] (and the trimethylsilylmethyl discussed earlier) are aggregates of Cu_4, Cu_6, or Cu_8 polyhedra with the organic group bridging the edges. The fluoro aryls, e.g., $[CuC_6F_5]_4$, and perfluoroalkyls are more stable than hydrocarbon analogues; the aryl $C_6H_4(2\text{-}NMe_2)$ has been much used, since the N atom can also occupy a coordination site and stabilize the molecules.

Mixed aryl (R) compounds such as $R_6Cu_6(C\equiv CR)_2$ and $R_4Cu_4Ag_2(CF_3SO_3)_2$ are also known.

The decarboxylation of copper perfluorobenzoate in quinoline gives $(CuC_6F_5)_n$. The catalytic action of copper or copper salts on decarboxylation reactions of carboxylic acids presumably involves organo intermediates.

[23]   (a) A. Camus *et al.*, *Inorg. Chim. Acta*, 1977, **23**, 131; A. E. Jukes, *Adv. Organomet. Chem.*, 1974, **12**, 215; (b) J. F. Normant, *Synthesis*, **1972**, 63; G. H. Posner, in *Organic Reactions*, Vol. 19, Wiley, 1972 (organocopper compounds in organic synthesis).

[24]   Y. Yamamoto *et al.*, *J.C.S. Chem. Comm.*, **1976**, 452.

[25]   J. C. Noltes *et al.*, *J.C.S. Dalton*, **1978**, 1800; *J. Organomet. Chem.*, 1978, **159**, 441; 1977, **133**, 113; *J. Org. Chem.*, 1977, **42**, 2047; *Inorg. Chem.*, 1977, **16**, 1728; *J.C.S. Chem. Comm.*, **1977**, 203.

Lithium Alkyl Cuprates.[26] These important species are made by interaction of Cu^I with LiR. They are not usually isolated but are used *in situ* in ether or similar solvent for a wide variety of organic syntheses. They are especially useful for C—C bond formation by interaction with organic halides:

$$LiCuR_2 + R'X \rightarrow R—R' + LiX + CuR$$

The species present in solutions depend on the solvent and the ratio of LiR to Cu.[27] Thus in tetrahydrofuran and Me_2O there can be $LiCu_2Me_3$, $LiCuMe_2$, and Li_2CuMe_3, whereas in diethyl ether there are $Li_2Cu_3Me_5$, $LiCuMe_2$, and Li_2-$CuMe_3$. Some of the species may well be dimers (e.g., $Li_2Cu_2Me_4$). The sort of structures proposed are 21-H-XIVa and 21-H-XIVb. A lithium aryl cuprate

(21-H-XIVa)   (21-H-XIVb)

$Li_2Cu(C_6H_4Me)_3 \cdot 2H_2O$ is also known.[28] The coupling reaction with organic halides may proceed via an oxidative addition reaction (Section 29-3) such as

$$Li_2Cu_2Me_4 + MeI = Li_2Cu^ICu^{II}Me_5^+ + I^-$$
$$Li_2Cu^ICu^{II}Me_5^+ = C_2H_6 + Li_2Cu_2^IMe_3^+$$
$$Li_2Cu_2Me_3^+ + LiMe = Li_2Cu_2Me_4 + Li^+$$

The interaction of lithium methyl cuprates with $LiAlH_4$ in ether leads to *complex hydrides*[29a] such as $Li_2Cu_3H_5$ and Li_5CuH_6, e.g.,

$$LiCuMe_2 + LiAlH_4 = LiCuH_2\downarrow + LiAlH_2Me_2$$

Alkene Complexes. Copper(I) forms complexes with olefins quite readily. They may be made by direct interaction of CuCl or $CuCF_3SO_3$ with olefins or by reduction of copper(II) salts by trialkylphosphites in ethanol in presence of alkene. The crystalline compounds obtained when using chelating alkenes such as norbornadiene or cyclic polyenes usually have polymeric structures such as that for the cyclooctatetraene complex 21-H-XV. Usually only one double bond is coordinated, and dissociation is relatively easy.

Cationic complexes[9,29b] are readily obtained by displacement of benzene from $(CuSO_3CF_3)_2 \cdot C_6H_6$ (p. 803). These are thermally stable and often soluble in organic solvents. They are of the types LCu^+ and L_2Cu^+, where L is a chelating diolefin such as cycloocta-1,5-diene (21-H-XVI). Cationic complexes not only of

[26] H. O. House, *Acc. Chem. Res.,* 1976, **9**, 59.
[27] J. San Fillipo, Jr., *Inorg. Chem.,* 1978, **17**, 275; E. C. Ashby *et al., J. Am. Chem. Soc.,* 1977, **99**, 5312.
[28] G. Van Koten *et al., J. Organomet. Chem.,* 1977, **140**, C23.
[29a] E. C. Ashby *et al., Inorg. Chem.,* 1977, **16**, 1437, 2043; *Tetrahedron Lett.,* **1977**, 3695.
[29b] R. G. Salomon and J. K. Kochi, *J. Am. Chem. Soc.,* 1973, **95**, 1889.

(21-H-XV)   (21-H-XVI)

olefins but of other ligands L and diethylenetriamine(dien) can be made by the general reaction[30]

$$Cu I + dien + L \xrightarrow{MeOH} [Cu(dien)L]^+ + I^-$$

An ethylene complex is made[31] by the action of Cu on $Cu(ClO_4)_2$ in aqueous solution under an ethylene atmosphere:

$$Cu + Cu^{2+} + 2C_2H_4 \rightleftharpoons 2Cu(C_2H_4)^+$$

Carbonyls.[32]   A carbonyl $Cu(CO)^+$ can be obtained similarly using CO:

$$Cu + Cu^{2+} + 2CO \rightleftharpoons 2Cu(CO)^+$$

but in 98% H_2SO_4 under CO pressure, $Cu(CO)_3^+$ and $Cu(CO)_4^+$ can be obtained.[33a]

Aqueous solutions of chlorocuprates(I) absorb CO and the species formed is probably $[Cu(CO)Cl_2]^-$.[33b] Aqueous ammonia solutions of CuCl also absorb CO quantitatively and the CO is regenerated on acidification.

A variety of other carbonyls are known, e.g., $\eta^5\text{-}C_5H_5Cu(CO)$ and $[Cuen_2\text{-}(CO)]^+$. The compounds $Cu(CO)[HB(pz)_3]$ and $[Cu(dien)CO]BPh_4$ have been studied crystallographically.[34]

Acetylene Compounds.   Copper(I) chloride in concentrated hydrochloric acid absorbs acetylene to give colorless species such as $CuCl(C_2H_2)$ and $[CuCl_2(C_2H_2)]^-$. These halide solutions can also catalyze the conversion of acetylene into vinylacetylene (in concentrated alkali chloride solution) or to vinyl chloride (at high HCl concentration), and the reaction of acetylene with hydrogen cyanide to give acrylonitrile is also catalyzed.

Copper(I) ammine solutions react with acetylenes containing the $HC\equiv C$ group to give yellow or red precipitates, which are believed to have the structure 21-H-

30   M. Pasquali *et al., J. Am. Chem. Soc.,* 1978, **100**, 4918.
31   T. Ogura, *Inorg. Chem.,* 1976, **15**, 2301.
32   M. I. Bruce, *J. Organomet. Chem.,* 1972, **44**, 209 (review).
33a H. Souma *et al., Inorg. Chem.,* 1976, **15**, 968.
33b M. A. Busch and T. C. Franklin, *Inorg. Chem.,* 1979, **18**, 521.
34   M. Pasquali *et al., J.C.S. Chem. Comm.,* **1978**, 921; *Inorg. Chem.,* 1978, **17**, 1684; R. R. Gagne *et al., Inorg. Chem.,* 1979, **18**, 771.

XVII. Propynylcopper dissolves in triethylphosphine in toluene to give a cyclic polymer of similar type $[Et_3PCuC{\equiv}CMe]_3$.

$$
\begin{array}{ccc}
| & & R \\
Cu & & | \\
\uparrow & & C \\
R{-}C{\equiv}C{-}Cu & \leftarrow & \| \\
& & C \\
& & | \\
& & Cu \\
& & \uparrow \\
R{-}C{\equiv}C{-}Cu & \leftarrow &
\end{array}
$$

(21-H-XVII)

Copper(I) acetylides provide a useful route to the synthesis of a variety of organic acetylenic compounds and heterocycles, by reaction with aryl and other halides. A particularly important indirect use, where acetylides are probable intermediates, is the oxidative dimerization of acetylenes. A common procedure is to use the N,N,N',N'-tetramethylethylenediamine complex of CuCl in a solvent, or CuCl in pyridine-methanol, and oxygen as oxidant.[35]

$$2RC{\equiv}CH + 2Cu^{2+} + 2py \rightarrow RC{\equiv}C{-}C{\equiv}CR + 2Cu^+ + 2pyH^+$$

Finally an unusual cluster compound $Cu_4Ir_2(PPh_3)_2(C_2Ph)_8$ formally containing Cu^0 and Ir^{IV} has been made by the reaction of *trans*-$IrCl(CO)(PPh_3)_2$ and $(PhC{\equiv}CCu)_n$. It has an octahedron of metal atoms with *trans*-Ir atoms. Each Ir is bound to one PPh_3 and four acetylide groups.[36]

21-H-3. The Copper(II) State, d^9

The dipositive state is the most important one for copper. Most Cu^I compounds are fairly readily oxidized to Cu^{II} compounds, but further oxidation to Cu^{III} is more difficult. There is a well-defined aqueous chemistry of Cu^{2+}, and a large number of salts of various anions, many of which are water soluble, exist in addition to a wealth of complexes.

Stereochemistry.[37] The d^9 configuration makes Cu^{II} subject to Jahn-Teller distortion if placed in an environment of cubic (i.e., regular octahedral or tetrahedral) symmetry, and this has a profound effect on all its stereochemistry. With only one possible exception, mentioned below, it is never observed in these regular environments. When six-coordinate, the "octahedron" is severely distorted, as indicated by the data in Table 21-H-2. The typical distortion is an elongation along one 4-fold axis, so that there is a planar array of four short Cu—L bonds and two *trans* long ones. In the limit, of course, the elongation leads to a situation indistinguishable from square coordination as found in CuO and many discrete com-

[35] See I. Bodek and G. Davies, *Inorg. Chem.*, 1978, **17**, 1814.
[36] M. R. Churchill and S. A. Bezman, *Inorg. Chem.*, 1974, **13**, 1416.

TABLE 21-H-2
Interatomic Distances in Some Copper(II) Coordination Polyhedra

Compound	Distances (Å)
$CuCl_2$	4Cl at 2.30, 2Cl at 2.95
$CsCuCl_3$	4Cl at 2.30, 2Cl at 2.65
$CuCl_2 \cdot 2H_2O$	2O at 2.01, 2Cl at 2.31, 2Cl at 2.98
$CuBr_2$	4Br at 2.40, 2Br at 3.18
CuF_2	4F at 1.93, 2F at 2.27
$[Cu(H_2O)_2(NH_3)_4]$ in $CuSO_4 \cdot 4NH_3 \cdot H_2O$	4N at 2.05, 1O at 2.59, 1O at 3.37
K_2CuF_4	2F at 1.95, 4F at 2.08

plexes of Cu^{II}. Thus the cases of tetragonally distorted "octahedral" coordination and square coordination cannot be sharply differentiated.

Chloro Complexes.[38]   These are varied with complex structures. Salts of stoichiometry M^ICuCl_3 usually contain $[Cu_2Cl_6]^{2-}$ ions and large cations (e.g., Ph_4P^+) to keep the anions well separated; the $[Cu_2Cl_6]^{2-}$ ions are formed by two tetrahedra sharing an edge. For smaller cations, the dimers become linked by long Cu—Cl bonds giving infinite chains with five or six-coordinate Cu^{II}. $CsCuCl_3$ is unique in forming infinite chains of distorted octahedra sharing opposite triangular faces.

For $[CuCl_4]^{2-}$ the best *ab initio* MO calculations predict a flattened tetrahedral (D_{2d}) geometry. This is found in $M_2^I[CuCl_4]$ when the cation is very large, thus isolating the anions (Fig. 21-H-4). With smaller cations, linking of $CuCl_4$ units occurs as for $[Cu_2Cl_6]^{2-}$ units above and a linked two-dimensional layer structure results in which Cu^{II} is in a tetragonally elongated octahedron. $(NH_4)_2CuCl_4$ contains square $[CuCl_4]^{2-}$ ions.

Some compounds with formulas corresponding to $[CuCl_5]^{3-}$ do, in fact, contain such discrete *tbp* anions if the cations are large and tripositive, e.g., $[M(NH_3)_6]^{3+}$ However with smaller cations (e.g., $M_3^ICuCl_5$ of the alkalis) there are discrete

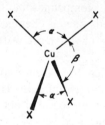

Fig. 21-H-4. Squashed tetrahedral structures of $[CuX_4]^{2-}$ ions in Cs_2CuX_4 salts; $\alpha > \beta$.

[37]   J. Gazo *et al.*, *Coord. Chem. Rev.*, 1976, **18**, 253; B. J. Hathaway and D. E. Billig, *Coord. Chem. Rev.*, 1970, **5**, 143 (electronic properties and stereochemistry of Cu^{II}).

[38]   D. W. Smith, *Inorg. Chim. Acta*, 1977, **22**, 107; *Coord. Chem. Rev.*, 1976, **21**, 93; H. M. Helis *et al.*, *Inorg. Chem.*, 1977, **16**, 2412; J. R. Wasson *et al.*, *Inorg. Chem.*, 1977, **16**, 458; L. P. Battaglia *et al.*, *Inorg. Chem.*, 1979, **18**, 148.

$[CuCl_4]^{2-}$ ions of D_{2d} structure, together with isolated Cl^- ions. In other cases, tetragonally distorted $[CuCl_6]$ units may be formed by sharing of Cl^- ions.

Other Complexes. In $K_2Pb[Cu(NO_2)_6]$ there is a *regular* octahedron of nitrogen atoms about the Cu^{2+} ion in the room-temperature cubic form. The subtle low-temperature behavior of these systems has been discussed earlier (p. 681).

Distorted tetrahedral coordination occurs in some bis(salicylaldiminato) complexes 21-H-XVIII, where R is bulky. When R = $(CH_3)_2CH$ and $(CH_3)_3C$ the

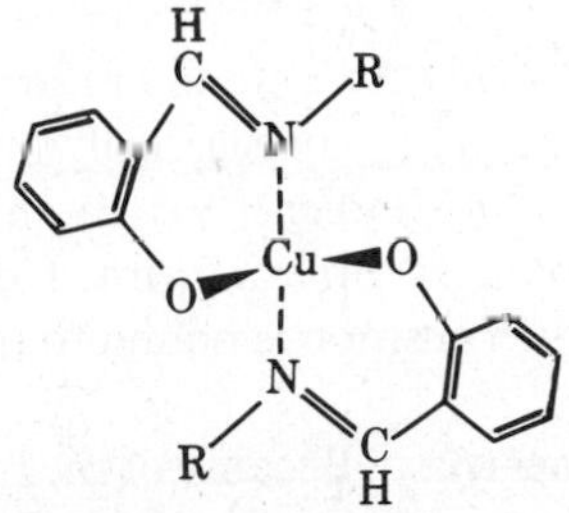

(21-H-XVIII)

angles between the two chelate ring planes in the crystalline materials are 60° and 54°, respectively, and there are reasons to believe these distortions persist in solution. However, in most cases, with small R groups, the coordination is planar. Again, in the dipyrromethene complex 21-H-XIX, steric interference of methyl groups renders a planar configuration impossible. The angle between the mean planes of the two ligands is 66°.

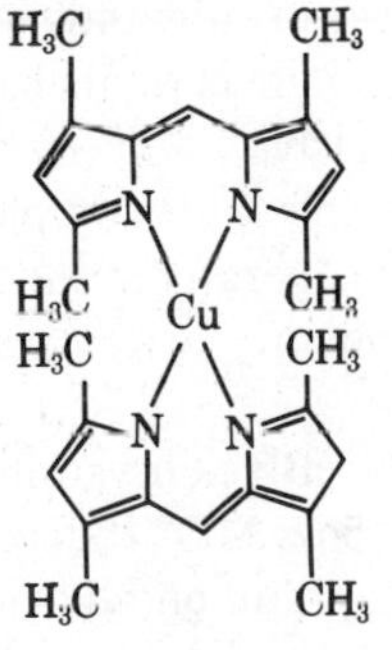

(21-H-XIX)

Numerous *planar complexes* are known. With a few exceptions such as the salicylaldiminato and dipyrromethene complexes just mentioned, *neutral four-coordinate complexes containing chelating ligands have planar coordination.* Variants on this include some cases where additional ligands complete a very elongated octahedron and many where there is dimerization of the type shown schematically in 21-H-XX for the β-form of bis(8-quinolinolato)copper(II), in which each metal atom becomes five-coordinate.

(21-H-XX)

Trigonal-bipyramidal coordination is found in several cases. In $Cu(terpy)Cl_2$ and $[Cu(dipy)_2I]I$ the *tbp* symmetry is necessarily imperfect.

Other *tbp* compounds are $[Cu(Me_4dien)(N_3)]Br$ and $[Cu(NH_3)_5][Ag(SCN)_3]$; copper(II) dimethylglyoximate is dimeric with a distorted square-pyramidal configuration in which the fifth position is bound to the oxygen of one of the NO groups.

Spectral and Magnetic Properties. Because of the relatively low symmetry (i.e., less than cubic) of the environments in which the Cu^{2+} ion is characteristically found, detailed interpretations of the spectra and magnetic properties are somewhat complicated, even though one is dealing with the equivalent of a one-electron case. Virtually all complexes and compounds are blue or green. Exceptions are generally caused by strong ultraviolet bands—charge-transfer bands—tailing off into the blue end of the visible spectrum, thus causing the substances to appear red or brown. The blue or green colors are due to the presence of an absorption band in the 600–900 nm region of the spectrum. The envelopes of these bands are generally unsymmetrical, seeming to encompass several overlapping transitions, but definitive resolution into the proper number of subbands with correct locations is difficult. Only when polarized spectra of single crystals have been measured has this resolution been achieved unambiguously.

The magnetic moments of simple Cu^{II} complexes (those lacking Cu—Cu interactions, described below) are generally in the range 1.75–2.20 BM, regardless of stereochemistry and independently of temperature except at extremely low temperatures ($<5°K$).

Binary Copper(II) Compounds. Black crystalline CuO is obtained by pyrolysis of the nitrate or other oxo salts; above 800° it decomposes to Cu_2O. The hydroxide is obtained as a blue bulky precipitate on addition of alkali hydroxide to cupric solutions; warming an aqueous slurry dehydrates this to the oxide. The hydroxide is readily soluble in strong acids and also in concentrated alkali hydroxides, to give deep blue anions, probably of the type $[Cu_n(OH)_{2n+2}]^{2-}$. In ammoniacal solutions the deep blue tetraammine complex is formed.

Copper(II), and also copper(III), form oxometallate ions of various structural types, but mainly containing linked CuO_4 planar units. Examples are Ba_2CuO_3, Sr_2CuO_3, and $BaCuO_2$.[39]

The *halides* are the colorless CuF_2, with a distorted rutile structure, the yellow

[39]   H. Müller-Buschbaum, *Angew. Chem., Int. Ed.,* 1977, **16,** 674; *Z. Anorg. Allg. Chem.,* 1977, **428,** 120.

chloride and the almost black bromide, the last two having structures with infinite parallel bands of square CuX_4 units sharing edges. The bands are arranged so that a tetragonally elongated octahedron is completed about each copper atom by bromine atoms of neighboring chains. $CuCl_2$ and $CuBr_2$ are readily soluble in water, from which hydrates may be crystallized, and also in donor solvents such as acetone, alcohol and pyridine.

Salts of Oxo Acids. The most familiar compound is the blue hydrated sulfate $CuSO_4 \cdot 5H_2O$, which contains four water molecules in the plane with O atoms of SO_4 groups occupying the axial positions, and the fifth water molecule H-bonded in the lattice. It may be dehydrated to the virtually white anhydrous substance. The hydrated nitrate cannot be fully dehydrated without decomposition. The anhydrous nitrate is prepared by dissolving the metal in a solution of N_2O_4 in ethyl acetate and crystallizing the salt $Cu(NO_3)_2 \cdot N_2O_4$, which probably has the constitution $[NO^+][Cu(NO_3)_3^-]$. When heated at 90° this solvate gives the blue $Cu(NO_3)_2$ which can be sublimed without decomposition in a vacuum at 150–200°. There are two forms of the solid, both possessing complex structures in which Cu^{II} ions are linked together by nitrate ions in an infinite array. However discrete molecules with the kind of structure shown in 21-H-XXI occur in the vapor phase.

$$O-N \underset{\sim 120°}{\overset{O}{\diagup}} Cu \underset{\sim 70°}{\overset{O}{\diagup}} N-O \qquad Cu-O \sim 2.0\,\text{Å}$$

(21-H-XXI)

Aqueous Chemistry and Complexes of Copper(II).[40] Most Cu^{II} salts dissolve readily in water and give the aqua ion, which may be written $[Cu(H_2O)_6]^{2+}$, but it must be kept in mind that two of the water molecules are farther from the metal atom than the other four. Addition of ligands to such aqueous solutions leads to the formation of complexes by successive displacement of water molecules. With NH_3, for example, the species $[Cu(NH_3)(H_2O)_5]^{2+} \dots [Cu(NH_3)_4(H_2O)_2]^{2+}$ are formed in the normal way, but addition of the fifth and sixth molecules of NH_3 is difficult. In fact, the sixth cannot be added to any significant extent in aqueous media but only in liquid ammonia. The reason for this unusual behavior is connected with the Jahn-Teller effect. Because of it, the Cu^{II} ion does not bind the fifth and sixth ligands strongly (even the H_2O). When this intrinsic weak binding of the fifth and sixth ligands is added to the normally expected decrease in the stepwise formation constants (p. 69), the formation constants K_5 and K_6 are very small indeed. Similarly, it is found with ethylenediamine that $[Cu\ en(H_2O)_4]^{2+}$ and $[Cu\ en_2(H_2O)_2]^{2+}$ form readily, but $[Cu\ en_3]^{2+}$ is formed only at extremely high concentrations of en. Many other amine complexes of Cu^{II} are known, and all are much more intensely blue than the aqua ion. This is because the amines produce

[40]   B. J. Hathaway and A. A. G. Tomlinson, *Coord. Chem. Rev.,* 1970, **5**, 1 (ammonia complexes); B. J. Hathaway and D. E. Billig, *Coord. Chem. Rev.,* 1970, **5**, 143 (mononuclear complexes); H. Sigel, *Angew. Chem., Int. Ed.,* 1974, **13**, 394 (complexes with mixed ligands); H. S. Maslen and T. N. Waters, *Coord. Chem. Rev.,* 1975, **17**, 137 (Schiff base complexes).

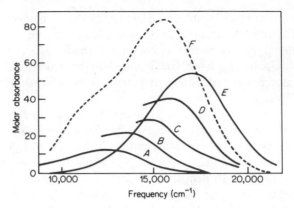

Fig. 21-H-5. Absorption spectra of $[Cu(H_2O)_6]^{2+}$ (*A*) and of the ammines in $2M$ ammonium nitrate at $25°$: $[Cu(NH_3)(H_2O)_5]^{2+}$(*B*), $[Cu(NH_3)_2(H_2O)_4]^{2+}$ (*C*), $[Cu(NH_3)_3(H_2O)_3]^{2+}$ (*D*), $[Cu(NH_3)_4(H_2O)_2]^{2+}$(*E*), and $[Cu(NH_3)_5H_2O]^{2+}$ (*F*).

a stronger ligand field, which causes the absorption band to move from the far red to the middle of the red region of the spectrum. For example, in the aqua ion the absorption maximum is at $\sim$800 nm, whereas in $[Cu(NH_3)_4(H_2O)_2]^{2+}$ it is at $\sim$600 nm, as shown in Fig. 21-H-5. The reversal of the shifts with increasing takeup of ammonia for the fifth ammonia is to be noted, indicating again the weaker bonding of the fifth ammonia molecule.

In halide solutions the equilibrium concentrations of the various possible species depend on the conditions; although $CuCl_5^{3-}$ has only a low formation constant, it is precipitated from solutions by large cations of similar charge (see above).

Many other Cu^{II} complexes may be isolated by treating aqueous solutions with ligands. When the ligands are such as to form neutral, water-insoluble complexes, as in the following equation, the complexes are precipitated and can be purified by recrystallization from organic solvents. The bis(acetylacetonato)copper(II) complex is another example of this type.

$$Cu^{2+} (aq) + 2 \ \ \text{(benzaldehyde phenolate structure)} \longrightarrow Cu \left(\text{(chelate structure)} \right)_2$$

Although as noted above (p. 801), addition of CN^- normally leads to reduction to CuCN, in presence of nitrogen donors like 1,10-phenanthroline reduction is inhibited and five-coordinate complexes like $[Cu \ phen_2(CN)]^+$ and $[Cu-phen_2(CN)_2]$ are obtained.[41]

Multidentate ligands that coordinate through oxygen or nitrogen, such as amino acids, form cupric complexes, often of considerable complexity.

The well-known blue solutions formed by addition of tartrate to Cu^{2+} solutions (known as Fehling's solution when basic and when *meso*-tartrate is used) may

[41]   M. Wicholas and T. Wolford, *Inorg. Chem.,* 1974, **13,** 316.

contain monomeric, dimeric, or polymeric species at different pH values. One of the dimers, $Na_2[Cu\{(\pm)C_4O_6H_2\}]\cdot5H_2O$, has square coordinate Cu^{II}, two tartrate bridges, and a Cu—Cu distance of 2.99 Å.

Halide complexes have been noted earlier. The main characteristic is the large number of salts of differing stoichiometries and cation-dependent structures that may be crystallized from solutions.

Polynuclear Copper(II) Complexes. These are less common than for Cu^I. Oxidation of $(CuClPEt_3)_4$ and similar complexes by oxygen leads to Cu^{II} complexes of the type $Cu_4OCl_6L_4$ (L = $OPEt_3$, py, NH_3) or, in $[Cu_4OCl_{10}]^{4-}$, the chloride ion. The structures 21-H-XXII have a μ_4-oxygen atom at the center of a Cu_4 tet-

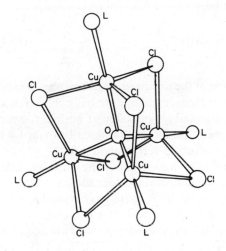

21–H–XXII

rahedron [cf. $Be_4O(CO_2R)_6$, p. 278]; each Cu is bound to a ligand L, and three Cl atoms act as bridges, so that Cu^{II} is approximately tbp[42] (see also p. 812).

A different type of compound, $Cu_4Cl_4(OR)_4$, where the alkoxide is 2-diethyl-aminoethanolate, has a cubane-type structure with a Cu_4O_4 cube and tetrahedral Cu^{II}.[43]

A purple complex of D-penicillamine of stoichiometry $\{Cu_8^ICu_6^{II}-[SCMe_2CH(NH_2)CO_2]_{12}Cl\}^{5-}$ contains both Cu^I and Cu^{II} and has resemblances to the copper(I) thiols discussed above. It may be related to copper enzymes (Chapter 31), but we can note that it has a Cl atom at the center of a Cu_8^I cluster.

Copper(II) Carboxylates.[44] These are readily made by action of the acid on the oxide or carbonate. They are binuclear with four carboxylate bridges (21-H-XXIII). A similar compound is the 1,3-triazinate 21-H-XXIV. In these compounds

[42] R. C. Dickenson *et al.*, *Inorg. Chem.*, 1977, **16,** 1530; M. R. Churchill *et al.*, *Inorg. Chem.*, 1975, **14,** 2496.
[43] J. W. Hall *et al.*, *Inorg. Chem.*, 1977, **16,** 1572.
[44] R. J. Doedens, *Prog. Inorg. Chem.*, 1976, **21,** 209.

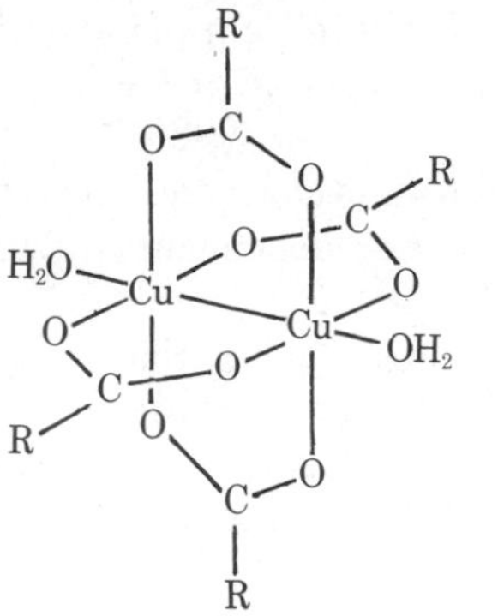

(21-H-XXIII) 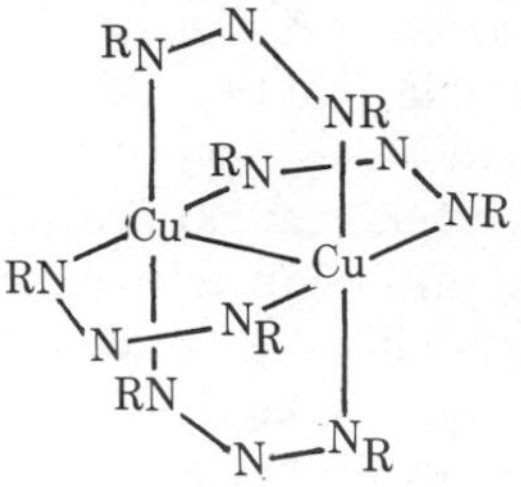   (21-H-XXIV)

there is weak coupling of the unpaired electrons, one on each Cu^{II} ion, giving rise to a singlet ground state with a triplet state lying only a few kJ mol^{-1} above it; the latter state is thus appreciably populated at normal temperatures and the compounds are paramagnetic. At 25° μ_{eff} is typically about 1.4 BM per Cu atom and the temperature dependence is very pronounced.

Phenomenologically the interaction in the dinuclear acetate and many other compounds with similar temperature dependences of the magnetic moment can be described as antiferromagnetic couplings of the unpaired spins on the adjacent Cu^{II} atoms, without invoking any Cu—Cu bonding.

Attempts to specify this interaction in detail have been plagued by controversy, and there are still differences of opinion. The interaction appears basically to be one between orbitals of δ symmetry, but whether this is primarily a direct interaction between $d_{x^2-y^2}$ orbitals of the two metal atoms or one that is substantially transmitted through the π orbitals of the bridging RCO_2^- groups is not clear.

21-H-4.   Copper(III) and Copper(IV) Compounds, d^8 and d^7

Oxides. Some alkali and alkaline earth compounds (e.g., $NaCuO_2$) obtained by heating the oxides in oxygen contain Cu^{III} and $BaCuO_{2.63}$ may contain Cu^{IV}.[45]

Complexes. Until recently only a few Cu^{III} complexes were known, but it now appears that Cu^{III} has an important biological role (Chapter 31), and a number of Cu^{III} complexes of deprotonated peptides and other ligands have been made.

With the exception of K_3CuF_6 all are diamagnetic with square or five-coordination of Cu^{III}. There are some relatively simple Cu^{III} compounds:

1. K_3CuF_6 is obtained as a pale green solid by fluorination of KCl + CuCl. This is the only high-spin octahedral Cu^{III} complex.

2.   Oxidation of $Cu(S_2CNBu_2^i)$ by Br_2 in CS_2 gives violet needles $CuBr_2(S_2CNBu_2^i)$.

45   M. Arjomand and D. J. Machin, *J.C.S. Dalton,* **1975**, 1061.

3. Peroxosulfate oxidation in alkaline solution of the biuret complex of Cu^{II}, K_2Cubi_2 ($biH_2 = H_2NCONHCONH_2$) gives $KCubi_2$ in which Cu^{III} has square coordination[46].

4. Oxidation by alkaline ClO^- of Cu^{2+} in presence of iodate and tellurate gives[47] $[Cu(IO_4OH)_2]^{5-}$ and $\{Cu[TeO_4(OH)_2]_2\}^{5-}$.

5. Oxidation of Cu^{2+} in presence of oxalodihydrazide and acetaldehyde gives a very stable violet anion that has five-coordinate Cu^{III} (21-H-XXV).[48]

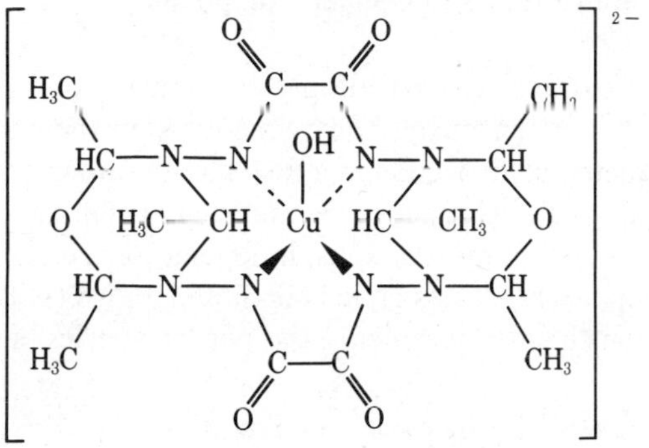

(21-H-XXV)

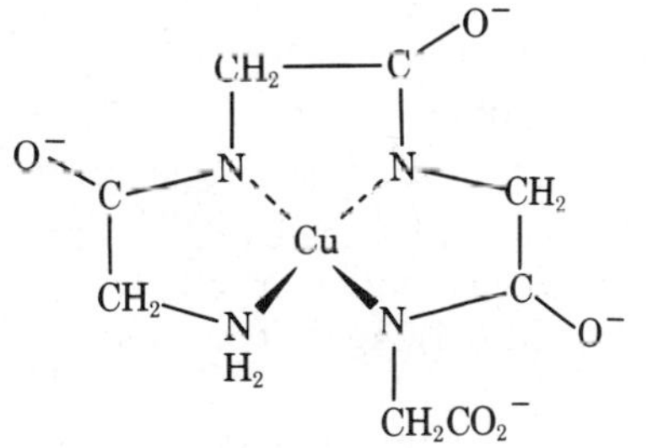

(21-H-XXVI)

In presence of a variety of peptides, deprotonated peptide complexes of Cu^{III} are quite reasonably stable in alkaline solution.[49] The Cu^{III}–Cu^{II} potentials are very sensitive to the nature of the ligand and vary from 0.45 to 1.02 V, e.g., for the tetraglycine (GH_4) complex 21-H-XXVI:

$$[Cu^{III}H_{-3}G]^- + e = [Cu^{II}H_{-3}G]^{2-} \qquad E^0 = +0.631 \text{ V}$$

A cationic Cu^{III} complex that is stable in acid solution has been made by electrolysis; it has deprotonated diglycylethylenediamine $NH_2CH_2CONHCH_2CH_2NHCOCH_2NH_2$ as ligand.[50]

46 P. J. M. W. L. Birker, *J.C.S. Chem. Comm.*, **1977**, 444; *Inorg. Chem.*, **1977**, **16**, 2478.

47 A. Balikungeri *et al.*, *Inorg. Chim. Acta*, 1977, **22**, 7.

48 W. E. Keyes *et al.*, *J. Am. Chem. Soc.*, 1977, **99**, 4527.

49 D. W. Margerum *et al.*, *Inorg. Chem.*, 1979, **18**, 444, 966; *J. Am. Chem. Soc.*, 1979, **101**, 1636; J. R. Kincaid *et al.*, *J. Am. Chem. Soc.*, 1978, **100**, 335.

50 P. Stevens *et al.*, *J. Am. Chem. Soc.*, 1978, **100**, 3632.

Copper(III) appears to have a role in biological redox systems such as galactose oxidase,[51] where Cu^{III} is reduced to Cu^I, which is then reoxidized by O_2 via a Cu^{II}—O_2 species.

Copper(IV) Compounds. These are few, and the only well-established one is the orange-red paramagnetic salt Cs_2CuF_6, obtained by fluorination of $CsCuCl_3$ at high pressure and temperature.[52]

21-H-5. Catalytic Properties of Copper Compounds[53]

Copper compounds catalyze an exceedingly varied array of reactions, often those also involving O_2, heterogeneously, homogeneously, in the gas phase, in organic solvents, and in aqueous solution. Copper also has an important biological role as a catalyst (see Chapter 31). Copper compounds have many uses in organic chemistry[54] for oxidations, coupling reactions, halogenations, etc., and some of these, involving organocopper(I) species have been noted above (p. 808).

Although reactions involving radicals and peroxo species of the type

$$Cu^+ + O_2 = CuO_2^+$$
$$CuO_2^+ + H^+ = Cu^{2+} + HO_2$$
$$Cu^+ + HO_2 = Cu^{2+} + HO_2^-$$
$$H^+ + HO_2^- = H_2O_2$$

have been invoked for oxidations in aqueous solution, it seems most likely that the actual oxidants are Cu^{II} or Cu^{III} species and that the function of O_2 is to reoxidize the Cu^I so produced. Although Cu^{II} species are usually taken to be the oxidants, the oxidation of ascorbic acid by O_2 probably involves Cu^{III},[55] and also as noted above, Cu^{III} appears to be involved in some biological oxidations.

In the Wacker process for oxidation of ethylene to acetaldehyde by Pd-Cu chloride systems, (Section 30-14), again the purpose of O_2 seems to be only to reoxidize Cu^I.

Since the enzymes that oxidize phenols and cleave aromatic C—C bonds contain copper (Chapter 31), simpler systems that do these reactions catalytically have been much studied, notably copper(I) chloride in pyridine and other donor solvents such as N-methylpyrrolidin-2-one.

In some cases oxygen appears to be directly involved, but in others oxidation by Cu^{II} species in the absence of oxygen has been confirmed,[56] where the catechol and quinone are of course complexed to copper:

[51] G. A. Hamilton et al., J. Am. Chem. Soc., 1978, **100**, 1899; 1976, **98**, 626.
[52] W. Harnischmacher and R. Hoppe, Angew. Chem., Int. Ed., 1973, **12**, 582.
[53] O. A. Chaltykyan, Copper Catalytic Reactions, Consultants Bureau, 1966.
[54] See M. Fieser and L. F. Fieser, Reagents for Organic Synthesis, several volumes, Wiley.
[55] R. F. Jameson and N. J. Blackburn, J.C.S. Dalton, **1976**, 534, 1596.
[56] M. M. Rogič and T. R. Demmin, J. Am. Chem. Soc., 1978, **100**, 5472; D. C. Brown et al., Tetrahedron Lett., 1977, **16**, 1363.

From the brown solutions formed in N-methylpyrrolidin-2-one (L) according to the equation

$$4CuCl + O_2 \xrightarrow{L} L_3Cu_4Cl_4O_2$$

a derivative of the actual tetrameric catalyst species has been crystallized and found to have a structure similar to the μ_4-oxo-centered Cu^{II} species discussed earlier (see 21-H-XXII),[57] and this is shown in 21-H-XXVII.

(21-H-XXVII)

It is believed that for oxidations using these systems for reactions such as

$$2\,C_6H_5OH + O_2 = O=\!\!\!\!\langle\quad\rangle\!\!-\!\!\langle\quad\rangle\!\!=\!\!O + 2\,H_2O$$

to be catalytic, the copper aggregate must have a basic terminal Cu—O bond, since the aggregate $Cu_4OCl_6L_4$ with all four terminal sites occupied by ligands is inactive. It seems unlikely that peroxo species are involved, and since at least in some cases[56] two-electron transfers appear to be involved, this is consistent with reduction of a Cu_4^{II} aggregate.

[57] G. Davies, M. R. Churchill, *et al., J.C.S. Chem. Comm.*, **1978**, 1045; see also Ref. 35.

CHAPTER TWENTY–TWO

The Elements of the Second and Third Transition Series

GENERAL COMPARISONS WITH THE FIRST TRANSITION SERIES

In general, the second and the third transition-series elements of a given Group have similar chemical properties but both show pronounced differences from their light congeners. A few examples will illustrate this generalization. Although Co^{II} forms a considerable number of tetrahedral and octahedral complexes and is the characteristic state in ordinary aqueous chemistry, Rh^{II} occurs only in a few complexes and Ir^{II} is virtually unknown. Similarly, the Mn^{2+} ion is very stable, but for Tc and Re the II oxidation state is known only in a few complexes. Cr^{III} forms an enormous number of cationic amine complexes, whereas Mo^{III} and W^{III} form only a few such complexes, none of which is especially stable. Again, Cr^{VI} species are powerful oxidizing agents, whereas Mo^{VI} and W^{VI} are quite stable and give rise to an extensive series of polynuclear oxo anions.

This is not to say that there is no valid analogy between the chemistry of the three series of transition elements. For example, the chemistry of Rh^{III} complexes is in general similar to that of Co^{III} complexes, and here, as elsewhere, the ligand field bands in the spectra of complexes in corresponding oxidation states are similar. On the whole, however, there are certain consistent differences of which the above-mentioned comparisons are only a few among many obvious manifestations.

Some important features of the elements and comparison of these with the corresponding features of the first series are the following:

1. *Radii.* The radii of the heavier transition atoms and ions are known only in a few cases. An important feature is that the filling of the $4f$ orbitals through the lanthanide elements causes a steady contraction, called the *lanthanide contraction* (Section 23-1), in atomic and ionic sizes. Thus the expected size increases

of elements of the third transition series relative to those of the second transition series, due to an increased number of electrons and the higher principal quantum numbers of the outer ones, are almost exactly offset, and there is in general much less difference in atomic and ionic sizes between the two heavy atoms of a group whereas the corresponding atom and ions of the first transition series are significantly smaller. Recent studies have shown that relativistic effects in the third transition series also have a significant effect on radii, ionization energies and other properties.[1]

2. *Oxidation states.* For the heavier transition elements, higher oxidation states are in general much more stable than for the elements of the first series. Thus the elements Mo, W, Tc, and Re form oxo anions in high valence states which are not especially easily reduced, whereas the analogous compounds of the first transition series elements, when they exist, are strong oxidizing agents. Indeed, the heavier elements form many compounds such as RuO_4, WCl_6, and PtF_6 that have no analogues among the lighter ones. At the same time, the chemistry of complexes and aquo ions of the lower valence states, especially II and III, which plays such a large part for the lighter elements, is of relatively little importance for most of the heavier ones.

3. *Aqueous chemistry.* Aqua ions of low and medium valence states are not in general well defined or important for any of the heavier transition elements, and some, such as Zr, Hf, and Re, do not seem to form any simple cationic complexes. For most of them anionic oxo and halo complexes play a major role in their aqueous chemistry although some, such as Ru, Rh, Pd, and Pt, do form important cationic complexes as well.

4. *Metal-metal bonding.* In general, although not invariably, the heavier transition elements are more prone to form strong M—M bonds than are their congeners in the first transition series. The main exceptions to this are the polynuclear metal carbonyl compounds and some related ones, where analogous or similar structures are found for all three elements of a given family. Aside from these, however, it is common to find that the first-series metal will form few or no M—M bonded species whereas the heavier congeners form an extensive series. Examples are the $M_6X_{12}^{n+}$ species formed by Nb and Ta, with no V analogs at all, the Mo_3^{IV} and W_3^{IV} oxo clusters for which no chromium analogues exist, and the $Tc_2Cl_8^{3-}$ and $Re_2Cl_8^{2-}$ ions, which have no manganese analogues.

5. *Magnetic properties.* In general, the heavier elements have magnetic properties that are less useful to the chemist than was the case in the first transition series. For one thing there is a much greater tendency to form low-spin complexes, which means that those with an even number of electrons are usually diamagnetic and therefore lack informative magnetic characteristics. Moreover, as previously explained (p. 650) the paramagnetic complexes usually have complicated behavior in which magnetic moments differ considerably from spin-only values and often vary markedly with temperature.

[1] K. S. Pitzer, *Acc. Chem. Res.,* 1979, **12**, 271; P. Pyykko and J.-P. Desclaux, *ibid.,* 1979, **12**, 276.

22-A. ZIRCONIUM AND HAFNIUM[1]

The chemistries of zirconium and hafnium are more·nearly identical than for any other two congeneric elements. This is due in considerable measure to the effect of the lanthanide contraction having made both the atomic and ionic radii (1.45 and 0.74 Å for Zr and Zr^{4+}; 1.44 and 0.75 Å for Hf and Hf^{4+}) essentially identical.

The oxidation states and stereochemistries are summarized in Table 22-A-1.

TABLE 22-A-1
Oxidation States and Stereochemistry of Zirconium and Hafnium

Oxidation state	Coordination number	Geometry	Examples
Zr^0	6	Octahedral(?)	$[Zr(bipy)_3]$?
Zr^I, Hf^I, d^3			
Zr^{II}, d^2		Complex sheet and cluster structures; see text	
Zr^{III}, Hf^{III}, d^1	6	Octahedral	$ZrCl_3, ZrBr_3, ZrI_3, HfI_3$
Zr^{IV}, Hf^{IV}, d^0	4	Tetrahedral	$ZrCl_4(g), Zr(CH_2C_6H_5)_4$
	6	Octahedral	$Li_2ZrF_6, Zr(acac)_2Cl_2, ZrCl_6^{2-}, ZrCl_4(s)$
	7	Pentagonal bipyramidal	$Na_3ZrF_7, Na_3HfF_7, K_2CuZr_2F_{12}\cdot 6H_2O$
		Capped trigonal prism	$(NH_4)_3ZrF_7$
		See text, Fig. 22-A-2	ZrO_2, HfO_2(monoclinic)
	8	Square antiprism	$Zr(acac)_4, Zr(SO_4)_2\cdot 4H_2O$
		Dodecahedron	$[Zr(C_2O_4)_4]^{4-}, [ZrX_4(diars)_2],$ $[Zr_4(OH)_8(H_2O)_{16}]^{8+}$

These elements, because of the larger atoms and ions, differ from titanium in having more basic oxides, having somewhat more extensive aqueous chemistry, and more commonly attaining higher coordination numbers, 7 and 8. They have a more limited chemistry of the III oxidation state.

22-A-1. The Elements

Zirconium occurs widely over the earth's crust but not in very concentrated deposits. The major minerals are *baddeleyite,* a form of ZrO_2, and *zircon* ($ZrSiO_4$). The chemical similarity of zirconium and hafnium is well exemplified in their geochemistry, for hafnium is found in Nature in all zirconium minerals in the range of fractions of a percent of the zirconium content. Separation of the two elements is extremely difficult, even more so than for adjacent lanthanides, but it can now be accomplished satisfactorily by ion-exchange or solvent-extraction fractionation methods.

[1]   E. M. Larson, *Adv. Inorg. Chem. Radiochem.,* 1970, **13**, 1 (an extensive review); D. L. Kepert, *The Early Transition Metals,* Academic Press, 1972, pp. 62–140.

Zirconium metal (m.p. $1855° \pm 15°$), like titanium, is hard and corrosion resistant, resembling stainless steel in appearance. It is made by the Kroll process
(p. 693). Hafnium metal (m.p. $2222° \pm 30°$) is similar. Like titanium, these metals
are fairly resistant to acids, and they are best dissolved in HF where the formation
of anionic fluoro complexes is important in the stabilization of the solutions. Zirconium will burn in air at high temperatures, reacting more rapidly with nitrogen
than with oxygen, to give a mixture of nitride, oxide, and oxide nitride
(Zr_2ON_2).

22-A-2. Compounds of Zirconium(IV) and Hafnium(IV)

Halides. The tetrahalides MCl_4, MBr_4, and MI_4 are all tetrahedral monomers
in the gas phase,[2] but the solids are polymers with halide bridging. $ZrCl_4$, a white
solid subliming at 331°, has the structure shown in Fig. 22-A-1 with zigzag chains

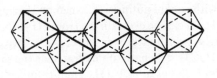

Fig. 22-A-1. The zigzag $ZrCl_6$ chains in $ZrCl_4$.

of $ZrCl_6$ octahedra; $ZrBr_4$, $HfCl_4$, and $HfBr_4$ are known to be isotypic. $ZrCl_4$ resembles $TiCl_4$ in its chemical properties. It may be prepared by chlorination of
heated zirconium, zirconium carbide, or a mixture of ZrO_2 and charcoal; it fumes
in moist air, and it is hydrolyzed vigorously by water. Hydrolysis proceeds only part
way at room temperature, affording the stable oxide chloride

$$ZrCl_4 + 9H_2O \rightarrow ZrOCl_2 \cdot 8H_2O + 2HCl$$

$ZrCl_4$ and the other halides combine with many neutral donors such as ethers,
esters, amines, $POCl_3$, and CH_3CN to form adducts that are generally six-coordinate unless there is steric hindrance.[3]

Hexachlorozirconates can be obtained by adding CsCl or RbCl to solutions of
$ZrCl_4$ in concentrated HCl. The $ZrCl_6^{2-}$ ion is unstable in solutions and even $15M$
HCl solutions contain some cationic hydroxo species and the oxide chloride can
be crystallized from such solutions.

$ZrCl_4$ also combines with two moles of certain diarsines (as do $TiCl_4$, $HfCl_4$,
and several other tetrahalides of these Group IV elements) to form $ZrCl_4(diars)_2$,
which has the dodecahedral type of eight-coordinate structure; 1:1 complexes are
also formed, but these are probably dimeric with octahedral Zr. With
$CH_3SCH_2CH_2SCH_3$ (dth), the compound $ZrCl_4(dth)_2$ is formed, and this appears
to be isostructural with $ZrCl_4(diars)_2$. $ZrCl_4$ reacts with carboxylic acids above
100° to give $Zr(RCO_2)_4$ compounds that appear to contain eight-coordinate

[2]   R. J. H. Clark, B. K. Hunter, and D. M. Rippon, *Inorg. Chem.*, 1972, **11**, 56.
[3]   E. M. Larsen and T. E. Henzler, *Inorg. Chem.*, 1974, **13**, 581.

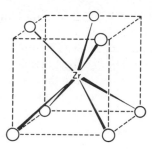

Fig. 22-A-2. Coordination geometry in the baddeleyite form of ZrO_2.

molecules. The eight-coordinate, distorted dodecahedral $Zr(NCS)_4(bipy)_2$ is prepared by reaction of bipy with $K_2Zr(NCS)_6$.[4]

$ZrBr_4$ and ZrI_4 are similar to $ZrCl_4$. ZrF_4 is a white crystalline solid subliming at 903°; unlike the other halides, it is insoluble in donor solvents; it has an eight-coordinate structure with square antiprisms joined by sharing fluorines. Hydrated fluorides $ZrF_4 \cdot 1$ or $3H_2O$, can be crystallized from $HF-HNO_3$ solutions. The trihydrate has an eight-coordinate structure with two bridging fluorides, $(H_2O)_3F_3ZrF_2ZrF_3(H_2O)_3$. The hafnium hydrate has the same stoichiometry but a different structure with chains of $HfF_4(H_2O)$ units linked through four bridging fluorine atoms.

Zirconium Oxide and Mixed Oxides. Addition of hydroxide to zirconium(IV) solutions causes the precipitation of white gelatinous $ZrO_2 \cdot nH_2O$, where the water content is variable; no true hydroxide exists. On strong heating, this hydrous oxide gives hard, white, insoluble ZrO_2. This has an extremely high melting point (2700°), exceptional resistance to attack by both acids and alkalis, and good mechanical properties; it is used for crucibles and furnace cores. ZrO_2 in its monoclinic (baddeleyite) form and one form of HfO_2 are isomorphous and have a structure in which the metal atoms are seven-coordinate, as shown in Fig. 22-A-2. Three other forms of ZrO_2 have been described, but none has the rutile structure so often found among MO_2 compounds.

A number of compounds called "zirconates" may be made by combining oxides, hydroxides, nitrates, etc., of other metals with similar zirconium compounds and firing the mixtures at 1000 to 2500°. These, like their titanium analogue, are mixed metal oxides; there are no discrete zirconate ions known. $CaZrO_3$ is isomorphous with perovskite. By dissolving ZrO_2 in molten KOH and evaporating off the excess of solvent at 1050°C, the crystalline compounds $K_2Zr_2O_5$ and K_2ZrO_3 may be obtained. The former contains ZrO_6 octahedra sharing faces to form chains that in turn, share edges and corners with other chains. The latter contains infinite chains of ZrO_5 square pyramids (22-A-I).

Stereochemistry.[5] The Zr^{4+} ion is relatively large, highly charged, and spherical, with no partly filled shell to give it stereochemical preferences. Thus it is not sur-

4   E. J. Peterson, R. B. von Dreele, and T. M. Brown, *Inorg. Chem.,* 1976, **15**, 309.
5   T. E. MacDermott, *Coord. Chem. Rev.,* 1973, **11**, 1.

(22-A-I)   (22-A-II)

prising that zirconium(IV) compounds exhibit high coordination numbers and a great variety of coordination polyhedra. This is well illustrated by the fluoro complexes. Li_2ZrF_6 and $CuZrF_6\cdot4H_2O$ contain the octahedral ZrF_6^{2-} ion, and ZrF_7^{3-} ions are known with two different structures; the pentagonal bipyramid in Na_3ZrF_7 and a capped trigonal prism (22-A-II) in $(NH_4)_3ZrF_7$. Pentagonal bipyramids sharing an edge form the $Zr_2F_{12}^{4-}$ ion in $K_2CuZr_2F_{12}\cdot6H_2O$. $N_2H_6ZrF_6$ also contains bicapped trigonal prisms sharing F atoms. ZrF_8 units of several types are found: $Cu_2ZrF_8\cdot12H_2O$ has discrete square antiprisms; $Cu_3Zr_2F_{14}\cdot16H_2O$ contains $Zr_2F_{14}^{6-}$ ions formed by two square antiprisms sharing an edge.

The chalcogenides ZrY_2 and HfY_2 (Y = S, Se, Te) have layered structures and are intrinsic semiconductors.[6] They form intercalation compounds,[7] mostly quite similar to those already discussed for titanium (p. 696).

α-Zirconium phosphate $Zr(HPO_4)_2\cdot H_2O$ has a unique crystal structure in which sheets of Zr^{4+} ions are sandwiched between sheets of HPO_4^{2-} ions so that each Zr^{4+} is in an almost perfect octahedral set of oxygen atoms. Between these sandwich layers are sheets of P—OH groups and water molecules.[8] This material serves as a cation exchanger, since the P—OH hydrogen ions are replaceable, not only in aqueous media[9a] but in molten salts,[9b] and catalytic properties are also present in materials containing transition metal cations replacing H.

With oxygen ligands high coordination numbers and varied stereochemistry are also prevalent. Thus $M_6^I[Zr_2(OH)_2(CO_3)_6]\cdot nH_2O$ (M^I = K or NH_4) contain OH-bridged binuclear units in which each Zr atom is in dodecahedral eight-coordination, and ZrO_7 pentagonal bipyramids are found in $ZrMo_2O_7$-$(OH)_2(H_2O)_2$.

Aqueous Chemistry and Complexes. ZrO_2 is more basic than TiO_2 and is virtually insoluble in an excess of base. There is a more extensive aqueous chemistry of zirconium because of a lower tendency toward complete hydrolysis. Nevertheless hydrolysis does occur, and the state of Zr^{IV} in aqueous solution is under no circumstances simple. There is no positive indication of Zr^{4+} or ZrO^{2+} under any conditions in solution. Some polymeric form of ZrO^{2+} (the "zirconyl ion") seems dominant. This idea receives some support from nmr measurements of hydration

[6]   L. Brattas and A. Kjekshus, *Acta Chem. Scand.*, 1973, **27**, 1290.
[7]   R. P. Clement *et al.*, *Inorg. Chem.*, 1978, **17**, 2754.
[8]   J. M. Troup and A. Clearfield, *Inorg. Chem.*, 1977, **16**, 3311.
[9a] S. Allulli *et al.*, *J.C.S. Dalton*, **1976**, 2115.
[9b] S. Allulli *et al.*, *J.C.S. Dalton*, **1976**, 1816.

numbers.[10] In very acid solutions (pH = 1 or less) $Zr(OH)^{3+}$ and $Hf(OH)^{3+}$ ions appear to exist.[11]

The most important "zirconyl" salt is $ZrOCl_2 \cdot 8H_2O$, which crystallizes from dilute hydrochloric acid solutions and contains the ion $[Zr_4(OH)_8(H_2O)_{16}]^{8+}$. Here the Zr atoms lie in a distorted square, linked by pairs of hydroxo bridges and also bound to four water molecules, so that the Zr atom is coordinated by eight oxygen atoms in a distorted dodecahedral arrangement. $ZrOBr_2 \cdot 8H_2O$ has a similar structure. Consistent with this structure, only half the water can be lost without major structural change.[12]

Both zirconium and hafnium form many other basic salts, such as sulfates[13,14] and chromates,[14] in which there are infinite chains of composition $[M(\mu\text{-}OH)_2]_n^{2n+}$. In addition to the bridging OH groups, the metal ions are coordinated by oxygen atoms of the anions and achieve coordination numbers of 7 or 8, with geometries of pentagonal bipyramid and square antiprism, respectively.

In acid solutions, except concentrated HF, where ZrF_6^{2-} and HfF_6^{2-} ions only (MF_7^{3-} species cannot be detected) are present, acid solutions of Zr^{IV} contain polymeric, partially hydrolyzed species. In 1 to $2M$ perchloric acid $[Zr_3(OH)_4]^{8+}$ and $[Zr_4(OH)_8]^{8+}$ are thought to be the major species. In $2.8M$ HCl the main species appears to be trinuclear, perhaps $[Zr_3(OH)_6Cl_3]^{3+}$, and as noted above, the stable phase that crystallizes from HCl solutions, $ZrOCl_2 \cdot 8H_2O$, contains tetramers. Studies of solutions are complicated by slowness in the attainment of equilibrium. Chelating agents such as EDTA and NTA form complexes with Zr^{IV}; the $\{Zr[N(CH_2COO)_3]_2\}^{2-}$ ion has been shown to be dodecahedral.

Acetylacetonates of the types $M(acac)_2X_2$ and $M(acac)_3X$ are known. The former have *cis* octahedral configurations but appear to dissociate, whereas the latter are seven-coordinate and except for X = I, do not dissociate.

Zirconium forms mono-[15] and dithiocarbamate complexes, $Zr(SOCNR_2)_4$, and $Zr(S_2CNR_2)_4$, which have dodecahedral structures and appear to be highly fluxional.[16] Both Zr and Hf form dithiolene (*dith*) complex anions of composition $[M(dith)_3]^{2-}$, about which there are the usual ambiguities as to oxidation number.[17]

Some seemingly simple zirconium salts are best regarded as essentially covalent molecules or as complexes; examples are the carboxylates $Zr(OCOR)_4$, the tetrakis(acetylacetonate), the oxalate, and the nitrate. Like its Ti analogue, the last of these is made by heating the initial solid adduct of N_2O_5 and N_2O_4 obtained in the reaction

$$ZrCl_4 + (4 + x + y)\, N_2O_5 \xrightarrow{30°} Zr(NO_3)_4 \cdot xN_2O_5 \cdot yN_2O_4 + 4NO_2Cl.$$

[10]   A. Fratiello, G. A. Vidulich, and F. Mako, *Inorg. Chem.,* 1973, **12,** 470.
[11]   B. Norén, *Acta. Chem. Scand.,* 1973, **27,** 1369.
[12]   D. A. Powers and H. B. Gray, *Inorg. Chem.,* 1973, **12,** 2721.
[13]   I. J. Bear and W. G. Mumme, *Rev. Pure Appl. Chem.,* 1971, **21,** 189.
[14]   M. Hansson, *Acta Chem. Scand. A,* 1973, **27,** 2455, 2614, 3467.
[15]   R. C. Fay *et al., Inorg. Chem.,* 1978, **17,** 2114, 2120.
[16]   E. L. Muetterties, *Inorg. Chem.,* 1974, **13,** 1011.
[17]   J. L. Martin and J. Takats, *Inorg. Chem.,* 1975, **14,** 73.

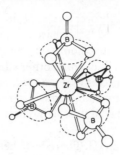

Fig. 22-A-3. The structure of $Zr(BH_4)_4$. [Reproduced by permission from P. H. Bird and M. R. Churchill, *J. C. S. Chem. Comm.*, 1907, 403; cf. also V. Plato and K. Hedberg, *Inorg. Chem.*, 1971, **10**, 590.]

It forms colorless sublimable crystals, and ir and Raman spectra suggest that the molecule is isostructural with $Ti(NO_3)_4$ and $Sn(NO_3)_4$. It is soluble in water but insoluble in toluene, whose ring it nitrates. Hafnium gives only $Hf(NO_3)_4 \cdot N_2O_5$. Nitrato complexes $M(NO_3)_6^{2-}$ are also known but not structurally character-ized.

Both elements form the borohydrides $M(BH_4)_4$, which according to both X-ray crystallography (Fig. 22-A-3) and vibrational spectra[18] have a highly symmetrical (T_d) structure in which each BH_4 ion is connected by three bridging H atoms to the central metal atom.

22-A-3. Lower Oxidation States

Zirconium(III) and Hafnium(III). There is no aqueous or other conventional so-lution chemistry yet known. The best known compounds are the trihalides $ZrCl_3$, $ZrBr_3$, and ZrI_3, which have a structure consisting of close-packed halide layers between which equally spaced zirconium (or hafnium) atoms occupy one-third of the octahedral interstices, to form infinite linear chains of metal atoms, perpen-dicular to the layers. In terms of coordination polyhedra the structure may be de-scribed as infinite parallel chains of MX_6 octahedra sharing opposite faces.[19] There are various synthetic methods, such as reduction of ZrX_4 with Al or Zr in molten AlX_3.[20] In this process the HfX_4 compounds are reduced more slowly (ten times for the chloride), thus allowing separation of Zr from contaminating Hf. However the behavior of $ZrCl_3$ in molten $AlCl_3$ is complicated.[21]

The trihalides can also be synthesized by reactions of ZrX_4 with ZrX in sealed systems at temperatures of 435 to 600°. It is found that they are nonstoichiometric[22] phases rather than distinct compounds. Their ranges of phase stability (i.e., values of x in ZrX_x) are: for Cl, 2.94 to 3.03; for Br, 2.87 to 3.23; for I, 2.83 to 3.43. The

18 T. A. Keiderling, *et al., Inorg. Chem.*, 1975, **14**, 576; N. Davies *et al., J.C.S. Dalton*, **1973**, 1962.
19 J. Kleppinger, J. C. Calabrese, and E. M. Larsen, *Inorg. Chem.*, 1975, **14**, 3128.
20 E. M. Larsen *et al., Inorg. Chem.*, 1974, **13**, 574.
21 B. Gilbert, G. Mamantov, and K. W. Fung, *Inorg. Chem.*, 1975, **14**, 1802.
22 R. L. Daake and J. D. Corbett, *Inorg. Chem.*, 1978, **17**, 1192.

color of each phase varies markedly over its composition range (e.g., from olive green to bluish black for $ZrBr_x$). For the HfI_3 phase a composition as high as $HfI_{3.5}$ can be obtained. Whether these "off-stoichiometric" compositions involve replacements of the type $4M^{III} \rightarrow 3M^{IV} + [\quad]$ or some other variation is not known.

Aside from some Cp_2Zr^{III} species that we shall not discuss here, complexes of Zr^{III} and Hf^{III} are essentially limited to adducts of the trihalides with simple Lewis bases.[23] Many of these have puzzling stoichiometry,[3] and the structures are all unknown.

Oxidation Numbers Less Than III. There have been fragmentary data showing the probable existence of lower halides ZrX_x, $1 \leq x < 3$, but only recently have thorough and rigorous studies begun to appear.[24] The monohalides of Zr and Hf are obtained by reactions of MX_4 with M at 800 to 850°. The structures consist of stacked, hexagonally packed layers of either all metal atoms or all X atoms, with a stacking sequence $\cdots XMMX \cdots XMMX \cdots$. There are two slightly different stacking patterns. The $X \cdots X$ interlayer distances (e.g., *ca.* 3.60 Å for chlorides) are normal van der Waals contacts, and the M—X distances (e.g., 2.63 Å in ZrCl) are appropriate for single bonds. Within the adjacent metal atom layers there are two sets of distances. Within one layer Zr—Zr distances are 3.42 Å, and between layers they are 3.09 Å; these may be compared to an average distance of 3.20 Å in α-Zr. The monohalides have great thermal stability (m.p. > 1100°) and metallic reflectivity, and they cleave like graphite. They are metallic in character. A compound of composition Hf_2S also appears to have a structure with double layers of metal atoms.

ZrCl and ZrBr react at 25° with H_2 to form discrete, though slightly nonstoichiometric phases, $ZrXH_{0.5}$ and ZrXH. It appears that the H atoms are inserted between the double metal layers of the initial halides[25].

An oxide ZrO, said to have the NaCl structure, has been mentioned but not further characterized. By vapor transport reactions at high temperatures (800–1000°) in sealed tantalum containers, the compounds ZrI_2 and $ZrCl_{2.5}$ which contain octahedral Zr_6 clusters can be obtained.[24]

There are a few compounds of zerovalent zirconium, but they are not well characterized. The reduction of $ZrCl_4$ with lithium in the presence of bipyridine in THF gives the violet $Zr(dipy)_4$ where, doubtless there is considerable delocalization of electrons over the ligands. The reaction of KCN or RbCN with $ZrCl_3$ in liquid ammonia is reported to give $M_5^I Zr(CN)_5$.[26]

22-A-4. Organometallic Compounds[27]

The organozirconium and -hafnium chemistry is similar to that of titanium (p. 704). Zirconium and hafnium form elimination-stabilized alkyls MR_4, such as benzyl

23   D. A. Miller and R. D. Bereman, *Coord. Chem. Rev.,* 1973, **9,** 107.
24   J. D. Corbett *et al., Inorg. Chem.,* 1976, **15,** 1820; 1977, **16,** 2029; *J. Am. Chem. Soc.,* 1978, **100,** 652.
25   A. W. Struss and J. D. Corbett, *Inorg. Chem.,* 1977, **16,** 360.
26   D. Nichols and T. A. Ryan, *Inorg. Chim. Acta,* 1977, **21,** L17–L18.
27   P. C. Wailes, R. S. P. Coutts, and H. Weigold, *Organometallic Chemistry of Titanium, Zirconium and Hafnium,* Academic Press 1974.

or allyl, and these form adducts with Lewis bases.[28] The most important chemistry is that with $(\eta\text{-}C_5H_5)_2M$ units (Chapter 27) such as Cp_2ZrHCl. This compound is used in the hydrozirconation reaction of olefins (Section 29-10).

The permethyl compound $(Me_5C_5)_2ZrH_2$ causes noncatalytic reduction of CO by H_2, and a dinitrogen complex $(Me_5C_5)_2(N_2)Zr(\mu\text{-}N_2)Zr(N_2)(Me_5C_5)_2$ has been characterized.[29] The compound Cp_4Zr has three η^5 rings and one η^1 ring in the solid state.[30]

Finally zirconium compounds are active in Ziegler-Natta type catalysis (Section 30-5) but usually give oligomers that are alk-1-enes rather than polymers.

22-B. NIOBIUM AND TANTALUM[1]

Niobium and tantalum, though metallic in many respects, have chemistries in the V oxidation state that are very similar to those of typical nonmetals. They have little cationic chemistry but form numerous anionic species. Their halides and oxide halides, which are their most important simple compounds, are mostly volatile and are readily hydrolyzed. In their lower oxidation states they form an extraordinarily large number of metal atom cluster compounds. Only niobium forms lower states in aqueous solution. The oxidation states and stereochemistries (excluding those in the cluster compounds) are summarized in Table 22-B-1.

22-B-1. The Elements

Niobium is 10 to 12 times more abundant in the earth's crust than tantalum. The main commercial sources of both are the *columbite-tantalite* series of minerals, which have the general composition $(Fe/Mn)(Nb/Ta)_2O_6$, with the ratios Fe/Mn and Nb/Ta continuously variable. Niobium is also obtained from *pyrochlore,* a mixed calcium-sodium niobate. Separation and production of the metals is complex. Both metals are bright, high-melting (Nb, 2468°; Ta, 2996°), and very resistant to acids. They can be dissolved with vigor in an HNO_3–HF mixture, and very slowly in fused alkalis.

NIOBIUM AND TANTALUM COMPOUNDS[1]

22-B-2. Niobium(V) and Tantalum(V), d^0

Oxygen Compounds. The oxides Nb_2O_5 (which has numerous modifications[2]) and Ta_2O_5 are dense white powders that are relatively inert chemically. They are

28 J. J. Felten and W. P. Anderson, *Inorg. Chem.,* 1973, **12,** 2334.
29 J. Bercaw *et al., J. Am. Chem. Soc.,* 1978, **100,** 2716, 3078.
30 J. L. Atwood *et al., J. Am. Chem. Soc.,* 1978, **100,** 5238.
1   F. Fairbrother, *The Chemistry of Niobium and Tantalum,* Elsevier, Amsterdam, 1967; J. O. Hill, I. G. Worsley, and L. G. Hepler, *Chem. Rev.,* 1971, **71,** 127 (oxidation potentials and thermodynamic data).
2   K. M. Nimmo and J. S. Anderson, *J.C.S. Dalton,* **1972,** 2328.

TABLE 22-B-1
Oxidation States and Stereochemistries of Niobium and Tantalum

Oxidation state	Coordination number	Geometry	Examples
Nb^{-I}, Ta^{-I}	6	Octahedral (?)	$[M(CO_6)]^-$
Nb^I, Ta^I, d^4	7	π Complex	$(C_5H_5)M(CO)_4$
	7	Dist. capped octahedron (nonrigid)	$TaH(CO)_2(diphos)_2$ [a]
Nb^{II}, Ta^{II}, d^3	6	Octahedral	NbO
Nb^{III}, Ta^{III}, d^2	6	Trigonal prism	$LiNbO_2$ [b]
		Octahedral	$Nb_2Cl_9^{3-}$
	7	Complex	$TaCl_3(CO)(PMe_2Ph)_3 \cdot EtOH$ [c]
	8	?	$K_5[Nb(CN)_8]$
Nb^{IV}, Ta^{IV}, d^1	6	Octahedral	$(NbCl_4)$, $TaCl_4py$, MCl_6^{2-}
	7	Dist. pentagonal bipyramidal	K_3NbF_7
	8	Nonrigid in solution	$TaH_4(diphos)_2$
		Square antiprism	$Nb(\beta\text{-dike})_4$ [d]
		Dodecahedral	$Nb(SCN)_4dipy_2$, [e] $K_4Nb(CN)_8 \cdot 2H_2O$
		π Complex	$(\eta\text{-}C_5H_5)_2NbMe_2^{\cdot}$
Nb^V, Ta^V, d^0	4	Tetrahedral	$ScNbO_4$
	5	*Tbp*	$MCl_5(vapor)$, $TaMe_5$, $Nb(NR_2)_5$
		Dist. tetragonal pyramid	$Nb(NMe_2)_5$
	6	Octahedral	$NaMO_3$ (perovskite), $NbCl_5 \cdot OPCl_3$, $TaCl_5 \cdot S(CH_3)_2$, TaF_6^-, $NbOCl_3$, M_2Cl_{10}, MCl_6^-
	6	Trigonal prism	$[M(S_2C_6H_4)_3]^-$ [f]
	7	Dist. pentagonal bipyramidal	$NbO(S_2CNEt_2)_3$
		Pentagonal bipyramidal, fluxional	$S{=}Ta(S_2CNEt_2)_3$, [h] $Ta(NMe_2)\text{-}(S_2CNMe_2)_3$, [i] $(S_2CNR_2)_2TaMe_3$
	8	Bicapped trigonal prism	$[Nb(trop)_4]^+$ [j]
		Square antiprism	Na_3TaF_8
		Dodecahedral	$Ta(S_2CNMe_2)_4^+$ [k]
	9	π Complex	$(\eta^5\text{-}C_5H_5)_2TaH_3$

[a] P. Meakin et al., *Inorg. Chem.*, 1974, **13**, 1023.
[b] G. Meyer and R. Hoppe, *Angew. Chem. Int. Ed.*, 1974, **13**, 744.
[c] G. Bandoli et al., *J. C. S. Dalton*, **1978**, 373.
[d] T. J. Pinnavaia et al., *J. Am. Chem. Soc.*, 1975, **97**, 2712.
[e] E. J. Peterson et al., *Inorg. Chem.*, 1976, **15**, 309.
[f] M. Cowie and M. J. Bennett, *Inorg. Chem.*, 1976, **15**, 1589.
[g] J. C. Dewan et al., *J. C. S. Dalton*, **1973**, 2082.
[h] E. J. Peterson, P. B. Von Dreele, and T. M. Brown, *Inorg. Chem.*, 1978, **17**, 1410.
[i] M. H. Chisholm, F. A. Cotton and M. W. Extine, *Inorg. Chem.*, 1978, **17**, 2000.
[j] A. R. Davis and F. W. B. Einstein, *Inorg. Chem.*, 1975, **14**, 3030.
[k] D. F. Lewis and R. C. Fay, *Inorg. Chem.*, 1976, **15**, 2219.

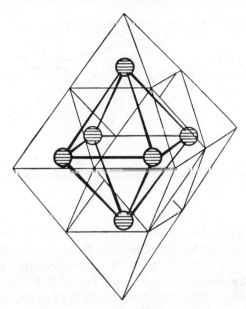

Fig. 22-B-1. Structure of the $M_6O_{19}^{8-}$ ions (M = Nb or Ta). The central oxygen atom is shared by all six octahedra. [Reproduced by permission from W. M. Nelson and R. S. Tobias, *Inorg. Chem.*, 1963, **2,** 985.]

scarcely if at all attacked by acids except concentrated hydrofluoric, they can also be dissolved by fusion with an alkali hydrogen sulfate or an alkali carbonate or hydroxide. They are obtained by dehydrating the hydrous oxides ("niobic" and "tantalic" acids) or by roasting certain other compounds in an excess of oxygen. The hydrous oxides, of variable water content, are gelatinous white precipitates obtained on neutralizing acid solutions of Nb^V and Ta^V halides. The Ta_2O_5 phase can exist with an excess of Ta atoms as interstitials over the range $TaO_{2.0}$–$TaO_{2.5}$; it then shows metallic conductance.

Niobate and tantalate isopolyanions can be obtained by fusing the oxides in an excess of alkali hydroxide or carbonate and dissolving the melts in water. The solutions are stable only at higher pH: precipitation occurs below pH $\sim$7 for niobates and $\sim$10 for tantalates. The only species that appear to be present in solution are the $[H_xM_6O_{19}]^{(8-x)-}$ ions (x = 0, 1, or 2), despite frequent claims for others. The structure of the $M_6O_{19}^{8-}$ ions, found in crystals and believed to persist in solutions, is shown in Fig. 22-B-1.

Heteropolyniobates and -tantalates are not well known, but a few of the former have been prepared and characterized.

With the exception of a few insoluble lanthanide niobates and tantalates (e.g., $ScNbO_4$) that contain discrete, tetrahedral MO_4^{3-} ions, the coordination number of Nb^V and Ta^V with oxygen is essentially always 6. The various "niobates" and "tantalates" are really mixed metal oxides.[3] Thus, for example, the M^IXO_3 compounds are perovskites (p. 18).

[3] See, e.g., J. C. Dewan *et al.*, *J. C. S. Dalton*, **1978,** 968.

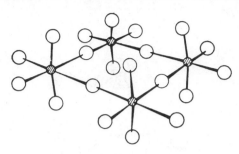

Fig. 22-B-2. The tetrameric structures of NbF_5 and TaF_5 (also MoF_5 and, with slight distortion, RuF_5 and OsF_5): Nb—F bond lengths: 2.06 Å (bridging), 1.77 Å (nonbridging). [Adapted by permission from A. J. Edwards, *J. Chem. Soc.*, **1964**, 3714.]

Fluorides and Fluoride Complexes. The pentafluorides are made by direct fluorination of the metals or the pentachlorides. Both are volatile white solids (Nb: m.p. 80°, b.p. 235°; Ta: m.p. 95°, b.p. 229°), giving colorless liquids and vapors. They have the tetrameric structure shown in Fig. 22-B-2; they appear to be polymeric also when molten. The oxide fluorides MOF_3 and MO_2F are known.

The metals and the pentoxides dissolve in aqueous HF to give fluoro complexes, whose composition depends markedly on the conditions. Addition of CsF to niobium in 50% HF precipitates $CsNbF_6$, while in weakly acidic solutions hydrolyzed species $[NbO_xF_y\cdot 3H_2O]^{5-2x-y}$, occur and components such as $K_2NbOF_5\cdot H_2O$ may be isolated. Raman and ^{19}F nmr spectra show that $[NbOF_5]^{2-}$ is present in aqueous solution up to about 35% in HF, and $[NbF_6]^-$ becomes detectable beginning at about 25% HF. $[NbF_6]^-$ is normally the highest fluoro complex of Nb^V formed in solution, although in 95–100% HF NbF_7^{2-} may possibly be present. Salts containing the NbF_7^{2-} ion can be crystallized from solutions with very high F^- concentrations. From solutions of low acidity and high F^- concentration, salts of the $[NbOF_6]^{3-}$ ion can be isolated.

There are crystalline tantalum fluoro compounds such as $KTaF_6$, K_2TaF_7, and K_3TaF_8. The TaF_8^{3-} ion, like NbF_7^{2-}, may be stabilized by crystal forces, since in aqueous HF or NH_4F solutions Raman spectra show the presence of TaF_6^- and TaF_7^{2-} ions only. In anhydrous HF solutions of $KTaF_6$ and K_2TaF_7 the only species identified by ^{19}F nmr spectra is TaF_6^-. In HF solutions tantalum can be separated from niobium by selective extraction into isobutyl methyl ketone.

The hexafluoro anions can also be made by the dry reaction:

$$M_2O_5 + 2KCl \xrightarrow{BrF_3} 2KMF_6$$

and the $[MOF_6]^{3-}$ salts can be prepared by bromination of the metals in methanol followed by addition of NH_4F or KF.

The ions $M_2F_{11}^-$, obtained by the reaction

$$MF_5 + Bu_4^nNBF_4 \xrightarrow{CH_2Cl_2} (Bu_4^nN)[MF_6]$$

$$(Bu_4^nN)[MF_6] + MF_5 \rightarrow (Bu_4^nN)[M_2F_{11}]$$

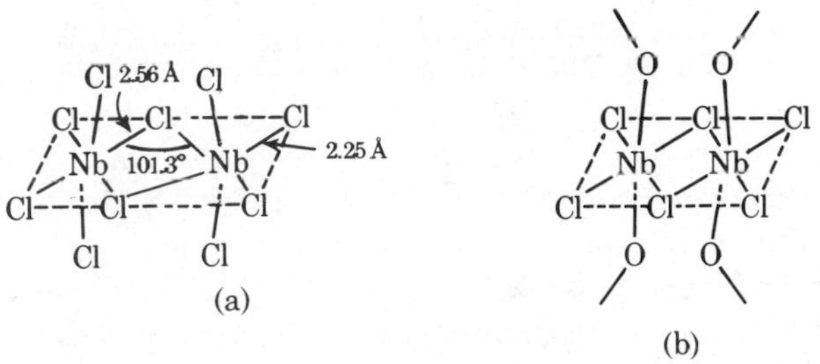

(a)

(b)

Fig. 22-B-3(a). The dinuclear structure of $NbCl_5$ in the solid; the octahedra are distorted as shown. The Nb-Nb distance, 3,951 Å indicates no metal-metal bonding. (b) The structure of $NbOCl_3$ in the crystal. The oxygen atoms form bridges between infinite chains of the planar Nb_2Cl_6 groups.

have two octahedra linked by an M—F—M bridge.[4] There are also some peroxo fluorides[5] like $[Ta(O_2)F_4(2\text{-MepyO})]^-$.

The pentafluorides also react as Lewis acids, giving adducts such as $NbF_5{\cdot}OEt_2$ and $TaF_5{\cdot}SEt_2$; the 2:1 adducts are of the type $[NbF_4L_4]^+ [NbF_6]^-$. The clear solutions formed on dissolving the fluorides in water, which are stable if not boiled, may well contain $[MF_4(H_2O)_4]^+$ and MF_6^-.

Other Halides of Niobium(V) and Tantalum(V). All six of these are yellow to brown or purple-red solids best prepared by direct reaction of the metals with excess of the halogen. The halides are soluble in various organic liquids such as ethers and CCl_4. They are quickly hydrolyzed by water to the hydrous pentoxides and the hydrohalic acid. The chlorides give clear solutions in concentrated hydrochloric acid, forming oxochloro complexes. All the pentahalides, which melt and boil between 200 and 300°, can be sublimed without decomposition in an atmosphere of the appropriate halogen; in the vapor they are monomeric and probably trigonal bipyramidal. Crystalline $NbCl_5$ has the dimeric structure shown in Fig. 22-B-3a; $NbBr_5$, $TaCl_5$, and $TaBr_5$ are isostructural. In CCl_4 and $MeNO_2$, both $NbCl_5$ and $TaCl_5$ are dimeric but in coordinating solvents adducts are formed. It appears probable that NbI_5 has a hexagonal close-packed array of iodine atoms with niobium atoms in octahedral interstices; TaI_5 is not isomorphous, and its structure is unknown.

Oxide Halides. The following are known.

$$NbOCl_3 \quad NbOBr_3 \quad NbOI_3 \quad NbOI_2{}^6 \quad NbO_2I$$
$$TaOCl_3 \quad TaOBr_3$$

The chlorides are white, and the bromides yellow, volatile solids; they are, however, less volatile than the corresponding pentahalides, and small amounts of the oxide halides that often arise in the preparation of the pentahalides in systems not scrupulously free from oxygen can be rather easily separated by fractional sublimation.

The MOX_3 compounds are monomeric in the vapor, but the lower volatility

[4]   S. Braunstein, *Inorg. Chem.,* 1973, **12,** 584.
[5]   J. C. Dewan et al., *J. C. S. Dalton,* **1977,** 978, 981.
[6]   J. Rijnsdorp and F. Jellinek, *J. Less Common Met.,* 1978, **61,** 79.

compared to that of MX_5 is understandable in view of the polymeric structures (see Fig. 22-B-3b for $NbOCl_3$). The niobium iodides also have polymeric structures.

The compounds can be obtained by controlled reaction between MX_5 and O_2 or, for $TaOCl_3$, by pyrolysis of $TaCl_5OEt_2$. They are all readily hydrolyzed; but from solutions in concentrated HX and M^+X^- or $R_4N^+X^-$, salts of ions such as $[NbOCl_5]^{2-}$, $[NbOF_5]^{2-}$, or $[NbOCl_4]^-$ can be obtained.

Some neutral adducts are known, e.g., $NbOCl_3(MeCN)_2$.[7]

Lewis Acid Behavior of the Pentahalides. The pentahalides form octahedral anions MCl_6^- in fused mixtures with alkali halides or with $R_4N^+Cl^-$ in CH_2Cl_2. With most O, N, S, and P donors, the halides form adducts that may be either six-(usually) or seven-coordinate.[8] For $NbCl_5$, however, the interaction with amines often leads to reduction to Nb^{IV} complex, (e.g., pyridine gives $NbCl_4py_2$).

The chlorides and bromides can also abstract oxygen from certain donor solvents (cf. VCl_4, $MoCl_5$) to give the oxohalides:

$$NbCl_5 + 3Me_2SO \rightarrow NbOCl_3 \cdot 2Me_2SO + Me_2SCl_2$$
$$Me_2SCl_2 \rightarrow ClCH_2SMe + HCl$$

They also act as Friedel-Crafts catalysts and will cause polymerization of acetylenes to arenes.

Complexes of Niobium(V) and Tantalum(V).[9] In addition to the anionic or neutral complexes formed by Lewis acid behavior of MX_5 (or MOX_3), there is an extensive chemistry of compounds in which halogen is replaced by alcoxide (OR), dialkylamide (NR_2), and alkyl (CR_3) groups. These compounds may be coordinately unsaturated, and they form anionic complexes like $[Ta(C_6H_5)_6]^-$ or neutral adducts.

Some reactions of $TaCl_5$ that are also, in general, typical for $NbCl_5$ are given in Fig. 22-B-4.

Oxygen Ligands. The dimeric *alcoxides* $M_2(OR)_{10}$, obtained by action of alcohols and an amine on the pentachlorides, have two alcoxo bridges. These may be cleaved by donors to give monomeric compounds like $Nb(OMe)_5py$. Partially substituted compounds can be made with alcohols like 2-methoxyethanol, which gives the six-coordinate $MCl_3(OCH_2CH_2OMe)_2$.[10]

An unusual compound with three oxygen bridges is the porphyrin complex $(por)Nb(\mu\text{-}O)_3Nbpor$, which has seven-coordinated niobium.[11] β-diketonato and tropolonato complexes are eight-coordinate[12] and of the type $[M(\beta\text{-dike})_4]^+$.

Nitrogen Ligands. The *dialkylamides* are monomeric; they can undergo in-

[7]   L. G. Hubert-Pfalzgraf and A. A. Pinkerton, *Inorg. Chem.,* 1977, **16**, 1895.

[8]   A. Thompson *et al., J. Less Common Met.,* 1978, **61**, 1, 31; J. D. Wilkins, *J. Inorg. Nucl. Chem.,* 1975, **37**, 2095; J. C. Dewan *et al., J. C. S. Dalton,* **1975**, 2031.

[9]   R. C. Mehrotra *et al., Inorg. Chim. Acta,* 1976, **16**, 237 (review of organic derivatives of Nb and Ta including halide adducts, alcoxides, and organometallic compounds).

[10]   L. G. Hubert-Pfalzgraf and J. G. Riess, *Chimia,* 1976, **30**, 481; *J. C. S. Dalton,* **1975**, 2854; L. G. Hubert-Pfalzgraf, *Inorg. Chim. Acta,* 1975, **12**, 229; A. A. Pinkerton *et al., Inorg. Chem.,* 1976, **15**, 1196; A. A. Jones and J. D. Wilkins, *J. Inorg. Nucl. Chem.,* 1975, **37**, 95.

[11]   J. F. Johnson and W. R. Scheidt, *Inorg. Chem.,* 1978, **17**, 1280.

[12]   A. R. Davis and F. W. B. Einstein, *Inorg. Chem.,* 1975, **14**, 3030.

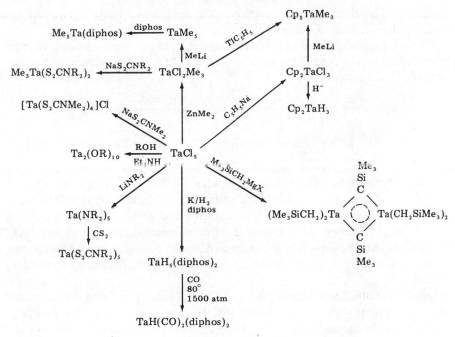

Fig. 22-B-4. Some reactions of tantalum pentachloride.

sertion reactions with CS_2 and CO_2:

$$Ta(NMe_2)_5 + 5CS_2 = Ta(S_2CNMe_2)_5$$
$$Nb(NMe_2)_5 + 5CO_2 = Nb(O_2CNMe_2)_5$$

The resulting dithiocarbamates and carbamates contain two unidentate and three bidentate ligands and are eight-coordinate.

The carbamates will exchange with labeled CO_2, probably by way of partly decarbonylated intermediates[13]:

$$(Me_2NCO_2)_4Nb\!-\!O\!-\!\overset{\displaystyle |}{\underset{\displaystyle \overset{\|}{O}}{C}}\!-\!NMe_2 \underset{+CO_2}{\overset{-CO_2}{\rightleftharpoons}} (Me_2NCO_2)_4NbNMe_2$$

Sulfur Ligands. The dithiocarbamates are made as above or by direct interaction,[14] in which case cationic species may be obtained, viz.,

$$TaCl_5 + NaS_2CNMe_2 \xrightarrow[\text{reflux}]{CH_2Cl_2} [Ta(S_2CNMe_2)_4]Cl\cdot CH_2Cl_2$$

With NaS_2CNEt_2 in acetonitrile an unusual complex of stoichiometry $TaS(S_2C-NEt_2)_3$ was obtained, and this is seven-coordinate with a Ta=S group.[15]

Compounds with an MS_6 grouping are those trigonal prismatic ions

13 M. H. Chisholm and M. Extine, *J. Am. Chem. Soc.,* 1975, **97,** 1623.
14 R. C. Fay *et al., J. Am. Chem. Soc.,* 1975, **97,** 7179.
15 E. J. Peterson *et al., Inorg. Chem.,* 1978, **17,** 1410.

$[M(S_2C_6H_4)_3]^-$ obtained by interaction of the diethylamides $M(NEt_2)_5$ with the sodium salt of toluene-3,4-dithiol.[16]

Hydride Ligands. Like Mo, W, Re, Os, and Ir, the elements Nb and Ta form polyhydrides of the type $MH_5(dmpe)_2$ (dmpe = $Me_2P(CH_2)_2PMe_2$). They can be made by the reactions[17]:

$$NbCl_5 \xrightarrow{Me_2Zn} Me_3NbCl_2 \xrightarrow[\text{diphos}]{MeLi} Me_5Nb(diphos)_4 \xrightarrow{H_2/500 \text{ atm}} NbH_5(diphos)_2$$

$$TaCl_5 \xrightarrow[\text{dmpe}]{K/H_2} TaH_5(diphos)_2 \xleftarrow[\text{dmpe}]{H_2} [TaPh_6]^- \xleftarrow[\text{ether}]{PhLi} TaCl_5$$

Other Compounds. These include nitrides, sulfides, silicides, selenides, and phosphides, as well as many alloys. Definite hydride phases also appear to exist.

There are no simple salts such as sulfates and nitrates. Sulfates such as $Nb_2O_2(SO_4)_3$ probably have oxo bridges and coordinated sulfato groups. In HNO_3, H_2SO_4, or HCl solutions, Nb^V can exist as cationic, neutral, and anionic species, hydrolyzed, polymeric, and colloidal forms in equilibrium, depending on the conditions.

Organometallic Chemistry of Nb^V and Ta^V. There is an extensive chemistry of these elements with M—C σ bonds, η-C_5H_5, and cyclooctatetraene groups.

Niobium and tantalum methyls, (Me_3MCl_2), were among the first known stable transition metal methyls. Other partially substituted alkyls,[18] as well as the pentamethyl,[19] are now known. These compounds are made by the reactions:

$$2TaCl_5 + 3ZnMe_2 \xrightarrow{\text{ether}} 2TaCl_2Me_3 + 3ZnCl_2$$

$$TaCl_2Me_3 + 2LiMe \xrightarrow{\text{ether}} TaMe_5 + 2LiCl$$

The niobium compounds are less stable than the tantalum ones. The formation of adducts such as $TaMe_5(diphos)$ increases the stability however. All these compounds (*a*) act as Lewis bases, adding neutral ligands or forming anions, and (*b*) can undergo insertion reactions into the M—CH_3 group (e.g., with NO^{20} or CS_2^{21}). Furthermore, the halide atoms can be replaced by other groups such as acac, Schiff bases, or $NR_2,$[21] to give compounds such as $Me_3Ta(acac)_2$.

Tantalum is one of the few elements for which compounds with Ta=CH_2, Ta=CR_2, and Ta≡CR bonds are well established. This chemistry is considered in Section 27-7. Of the dicyclopentadienyl compounds, Cp_2MX_3, the most inter-

[16] J. L. Martin and J. Takats, *Inorg. Chem.*, 1975, **14**, 73, 1359; M. Cowie and M. J. Bennett, *Inorg. Chem.*, 1976, **15**, 1589.

[17] F. N. Tebbe, *J. Am. Chem. Soc.*, 1973, **95**, 5823; R. R. Schrock, *J. Organomet. Chem.*, 1976, **121**, 373.

[18] G. W. A. Fowles *et al.*, *J. C. S. Dalton*, **1973**, 961; **1974**, 1080; C. Santini-Scampucci and J. G. Riess, *J.C.S. Dalton*, **1973**, 2436.

[19] R. R. Schrock, *J. Organomet. Chem.*, 1976, **122**, 209.

[20] M. G. B. Drew and J. D. Wilkins, *J. C. S. Dalton*, **1974**, 198, 1973.

[21] C. Santini-Scampucci and G. Wilkinson, *J. C. S. Dalton*, **1976**, 807.

esting are the trihydrides.[22] The niobium compound reacts with ethylene to give the niobium(III) compounds $Cp_2Nb(C_2H_4)H$ and $Cp_2Nb(C_2H_4)C_2H_5$, both of which are unusual examples of hydrido-ethylene and ethyl-ethylene complexes.[23]

The compounds Cp_2MH_3 also catalyze the exchange of D_2 with aromatic compounds at 1 to 2 atm and 80 to 100°.[24] The pentahydride $TaH_5(diphos)_2$, will also do this. The mechanism involves dissociation to form the monohydrido species and oxidative addition of benzene, viz.,

$$Cp_2MH_3 \xrightarrow{-H_2} Cp_2MH \xrightarrow{+C_6H_6} Cp_2M\underset{H}{\overset{C_6H_5}{\diagdown}}$$

The cyclooctatetraene complexes, e.g., $[Nb(C_8H_8)_3]^-$ and $CH_3Ta(C_8H_8)_2$, have η^4, butadienelike, and η^3-allyl-like bonding of the rings to the metal; they are nonrigid.[25]

22-B-3. Niobium(IV) and Tantalum(IV), d^1

Oxides. The only certain lower oxides are NbO and NbO_2. TaO_x compositions from $x = 2$ to 2.5 comprise a Ta_2O_5 phase with interstitial Ta atoms, not discrete phases or compounds. NbO_2 has a rutile-type structure with pairs of fairly close (2.80 Å) Nb atoms, presumably singly bonded to each other. NbO, which has only a narrow range of homogeneity, has metallic luster and excellent electrical conductivity of the metallic type.

Sulfides. The disulfides MS_2 give intercalation compounds.[26] In $NbS_2py_{0.5}$ there are Nb—S sheets, with the pyridine molecules perpendicular to the sheets. The TaS_2-ammonia intercalate has ionic layers, $TaS_2^{x-}(NH_4^+)_x(NH_3)_{1-x}$.

Halides. All the tetrahalides except TaF_4 are known. *Niobium(IV) fluoride* is a black, involatile, paramagnetic solid in which each Nb atom lies at the center of an octahedron.

The other six halides differ from NbF_4 and resemble each other in their structures and in being diamagnetic.

The *tetrachlorides* and *tetrabromides* are all brown-black or black isomorphous solids, obtained by reduction of the pentahalides with H_2, Al, Nb, or Ta at elevated temperatures. Crystalline $NbCl_4$ has linear chains with chlorine bridges (22-B-I) and a short Nb—Nb bond that accounts for the diamagnetism[27] as well as a longer Nb—Nb bond. The chlorine atoms are also bent away from the short Nb—Nb bond.

[22] R. D. Wilson et al., J. Am. Chem. Soc., 1977, **99**, 1775; J. A. Labinger and J. Schwarz, J. Am. Chem. Soc., 1975, **97**, 1596; I. H. Elson et al., J. Am. Chem. Soc., 1974, **96**, 7374; F. N. Tebbe, J. Am. Chem. Soc., 1973, **95**, 5413.

[23] L. J. Guggenberger et al., J. Am. Chem. Soc., 1974, **96**, 5420; Inorg. Chem., 1973, **12**, 294.

[24] U. Klabunde and G. W. Parshall, J. Am. Chem. Soc., 1972, **94**, 9081.

[25] L. J. Guggenberger, and R. R. Schrock, J. Am. Chem. Soc., 1975, **97**, 6693.

[26] R. Schöllhorn et al., Angew. Chem., Int. Ed., 1977, **16**, 199; J.C.S. Chem. Comm., **1976**, 863.

[27] R. D. Taylor et al., Inorg. Chem., 1977, **16**, 721.

3.794 Å   3.029Å

(22-B-I)

Niobium(IV) iodide is obtained on heating NbI_5 to 300°. It is diamagnetic and trimorphic. One form contains infinite chains of octahedra with the Nb atoms off center so as to form pairs with Nb—Nb distances of 3.31 Å. TaI_4 appears to be similar. The latter can be made most easily by allowing TaI_5 to react with an excess of pyridine to give TaI_4py_2 which, on heating, loses pyridine to give TaI_4.

Complexes.[28]   *Halide Complexes.* The salt K_3NbF_7 is made by the reaction[29]

$$4NbF_5 + Nb + 15KF \rightarrow 5K_3NbF_7$$

It is isostructural with $(NH_4)_3Zr^{IV}F_7$ and $K_3Nb^VOF_6$.

The *adducts* are paramagnetic and mostly of the type MX_4L_2,[30] e.g., $NbBr_4py_2$, $NbCl_4(thiophene)_2$, and $NbCl_4 (THF)_2$.

They are believed to have *cis* octahedral configurations. In some cases there are two forms (e.g., red and green $NbBr_4py_2$), which may be *cis*- and *trans*-isomers.

There are a few 1:1 adducts; these are only slightly paramagnetic, and halogen-bridged bioctahedral structures with M—M bonds have been postulated. With the chelating ligands eight-coordinate species such as $NbCl_4(diphos)_2$ may be obtained.

Other Complexes. In addition to pseudohalide analogues of NbX_4L_2 complexes, such as the red $Nb(SCN)_4dipy_2$, there are the compounds $Nb(NCS)_4$, $K_2[Nb(SCN)_6]$[31] and $K_4[Nb(CN)_8]$[32]; the latter is made by electrolytic reduction of $NbCl_5$ and KCN in methanol followed by air oxidation of the Nb^{III} complex $K_5[Nb(CN)_8]$ first formed.

The *β-diketonates* [e.g., $Nb(acac)_4$] have an unpaired electron and are not isostructural with the Zr and Hf analogues. Tantalum(IV) chloride reacts by abstracting oxygen, and analogous complexes are not obtained.

There are several Nb^{IV} alcoxides[33] such as $Nb(OEt)_4$, $[NbCl(OEt)_3py]_2$, $NbCl_3(OR)bipy$, and salts of the ion $[NbCl_5OEt]^{2-}$. There are also cyclopentadienyl compounds of stoichiometry Cp_2NbX_2 (X = CH_3, C_6H_5,[34] H, Cl, Br, I).

[28]   D. A. Miller and R. D. Bereman, *Coord. Chem. Revs.*, 1972–1973, **9**, 107 [an extensive review of complexes of Nb^{IV} and Ta^{IV} (and Zr^{III} and Hf^{III})].

[29]   L. O. Gilpatrick and L. M. Toth, *Inorg. Chem.*, 1974, **13**, 2242.

[30]   L. E. Manzer, *Inorg. Chem.*, 1977, **16**, 525; E. Samuel *et al., Nouv. J. Chim.*, **1977**, 93.

[31]   T. M. Brown *et al., Inorg. Chem.*, 1976, **15**, 309; 1972, **11**, 2697; P. M. Kiernan, and W. P. Griffith, *J. C. S. Dalton*, **1975**, 2489.

[32]   P. M. Kiernan *et al., Inorg. Chim. Acta.* 1979, **33**, L119.

[33]   N. Vuletić and D. Djordjević, *J. C. S. Dalton*, **1973**, 550.

[34]   I. H. Elson and J. K. Kochi, *J. Am. Chem. Soc.*, 1975, **97**, 1263; 1974, **96**, 7374.

The paramagnetic hydride and also the tetrahydride $TaH_4(diphos)_2$ are obtained from the M^V hydrides by H-abstraction using di-*tert*-butylperoxide,[35] viz.,

$$\left.\begin{array}{l} Cp_2NbH_3 \\ (diphos)_2TaH_5 \end{array}\right\} \xrightarrow{\text{Bu}^t\text{OOBu}^t} \left\{\begin{array}{l} Cp_2NbH_2 \\ (diphos)_2TaH_4 \end{array}\right.$$

Although niobium(V) in aqueous chloride or sulfate solutions can be reduced with zinc, the resulting blue species are poorly characterized, and it is by no means certain that they contain Nb^{IV}.

22-B-4. Niobium(III) and Tantalum(III) Compounds

The reduction of the thiophene complexes $NbX_4(SC_4H_8)_2$ with Na/Hg gives compounds[36] such as $Nb_2Br_6(SC_4H_8)_3$. This has a structure (22-B-II) with essentially two octahedral Nb atoms sharing a face, two bridging Br atoms, and a bridging S atom with a Nb=Nb double bond.

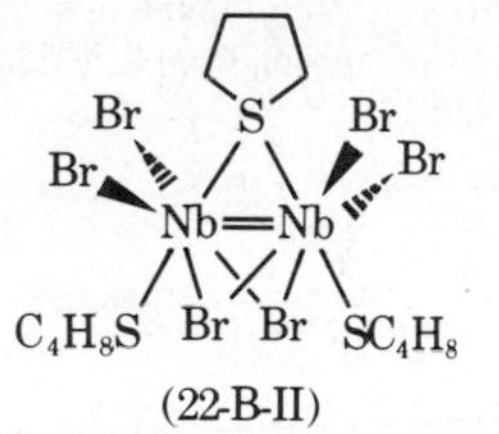

(22-B-II)

Ions of the type $M_3^INb_2X_9$ have three halide bridges.

The cyano complex $[Nb(CN)_8]^{5-}$ is made by electrolytic reduction of $NbCl_5$ + KCN in methanol.

Either reduction of $NbCl_4$ or $TaCl_4$ by zinc in CH_3CN or reaction of CH_3CN with the $M_2X_6(SC_4H_8)_3$ compounds of type 22-B-II mentioned above allows preparation of M^V compounds containing the central dimerized acetonitrile ligand shown in 22-B-III.[37]

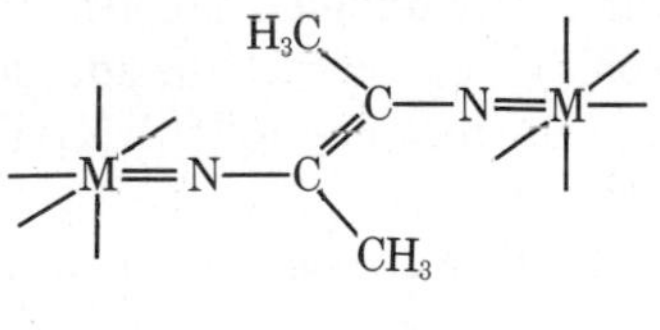

22-B-III

22-B-5. Niobium and Tantalum Cluster Compounds

The reduction of the pentahalides by the same metal alone, or better, in presence of sodium chloride gives cluster compounds that contain the basic M_6X_{12} unit.[38]

[35] L. E. Manzer, *Inorg. Chem.*, 1977, **16**, 525.

[36] J. L. Templeton et al., *Inorg. Chem.*, 1978, **17**, 1263.

[37] P. A. Finn et al., *J. Am. Chem. Soc.*, 1975, **97**, 220; F. A. Cotton and W. T. Hall, *ibid.*, 1979, **101**, 5094.

[38] F. W. Koknat et al., *Inorg. Chem.*, 1974, **13**, 1699.

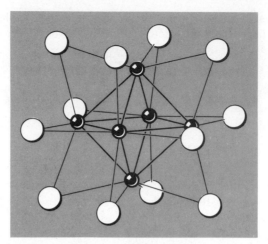

Fig. 22-B-5. The structure of the $[M_6X_{12}]^{n+}$ unit found in many halogen compounds of lower-valent niobium and tantalum. [Reproduced by permission from L. Pauling, *The Nature of the Chemical Bond*, 3rd ed., Cornell University Press, 1960.]

Thus for niobium we have

$$14NbCl_5 + 16Nb + 20NaCl \xrightarrow{12\ hr,\ 850°} 5Na_4Nb_6Cl_{18}$$

The extraction of the melt with very dilute HCl gives a green solution from which, by addition of concentrated HCl, the black solid $Nb_6Cl_{14}\cdot8H_2O$ is obtained. Corresponding tantalum compounds are obtained similarly, and mixed metal species, e.g., $[(Ta_5MoCl_{12})Cl_6]^{2-}$, can be obtained by the reduction of $TaCl_5 + MoCl_5$ in $NaAlCl_4$–$AlCl_3$ melts by Al.[39]

The M_6Cl_{14} compounds are formulated $[M_6Cl_{12}]^{2+}$ with the additional halide ions, water, or other ligand molecules filling the vacant outer coordination sites on the metal cluster, whose structure is given in Fig. 22-B-5. Here there is an octahedron of metal atoms with twelve bridging halogen atoms along the twelve edges of the octahedron. The anionic species, where the additional halide ions are bound to the six metal atoms, are thus formulated $\{[M_6X_{12}]X_6\}^{4-}$. There are also the hydroxides $M_6X_{12}(OH)_2$.[40]

These clusters can be oxidized electrochemically or chemically (e.g., by Cl_2 or I_2),[41] e.g.,

$$[(M_6X_{12})X_6]^{4-} \underset{}{\overset{-e}{\rightleftarrows}} [(M_6X_{12})X_6]^{3-} \underset{}{\overset{-e}{\rightleftarrows}} [M_6X_{12}X_6]^{2-}$$

Diamagnetic   Paramagnetic   Diamagnetic

The one unpaired electron on the paramagnetic species is delocalized over all six equivalent M atoms. X-Ray study[42] of the diamagnetic ion $[Nb_6Cl_{12}Cl_6]^{2-}$ shows the same basic features that appear in Fig. 22-B-5.

[39]   J. L. Meyer and R. E. McCarley, *Inorg. Chem.*, 1978, **17**, 1867.
[40]   N. Brnicevic and H. Schäfer, *Z. Anorg. Allg. Chem.*, 1978, **441**, 215, 230.
[41]   F. W. Koknat and R. E. McCarley, *Inorg. Chem.*, 1974, **13**, 295.
[42]   R. A. Field *et al.*, *J. C. S. Dalton*, **1973**, 1858.

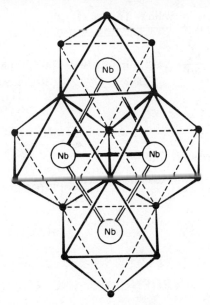

Fig. 22-B-6. The key structural unit in $M^1Nb_4X_{11}$ compounds. Small solid circles are halogen atoms and double lines represent M—M bonds. Many of the peripheral halogen atoms are shared with other units.

A variety of other compounds such as $Nb_6Cl_{15} \cdot 7H_2O$, $Ta_6Br_{15} \cdot 8H_2O$, $K_4Nb_6Cl_{18}$, and $(Et_4N)_4Nb_6Cl_{18}$ have been isolated.

There are also clusters of stoichiometry $M_4X_{11}^-$ (Fig. 22-B-6) as well as a considerable range of anhydrous halides with formal oxidation numbers ranging from 2 to 3.1. A cluster containing a π-bonded hexamethyl benzene is $[Nb_3(\mu\text{-}Cl)_6(C_6Me_6)_3]^{n+}$, which is diamagnetic for $n = 1$ and paramagnetic for $n = 2$.[43]

22-B-6. Niobium and Tantalum Compounds in Low Oxidation States

There are a number of Nb and Ta compounds in low oxidation states, most of them containing carbon monoxide.

The neutral carbonyls have not been isolated, but reduction of MCl_5 with Na in diglyme at 120° under 200 to 350 atm of CO gives the $[Nadiglyme_2]^+$ salts of the anion $[M(CO)_6]^-$. Some phosphine-substituted anions such as $[M(CO)_5PPh_3]^-$ and $[M(CO)_4diphos]^-$ can also be made.[44]

Under very high pressure of CO the pentahydride reacts:

$$TaH_5(diphos)_2 \xrightarrow{CO} TaH(CO)_2(diphos)_2$$

43   S. Z. Goldberg, et al., J. Am. Chem. Soc., 1977, **99**, 110.
44   J. E. Ellis and R. A. Faltynek, Inorg. Chem., 1976, **15**, 3168.

Another Ta^I species is yellow $TaCl(CO)_2(diphos)_2$, obtained by the sequence[45]

$$TaCl_5 + diphos \xrightarrow[THF]{Na} \underset{Blue}{TaCl_4(diphos)} \xrightarrow{Na} \underset{Red\text{-}brown}{TaCl_2(diphos)_2}$$

$$\Big\downarrow \begin{matrix} CO \\ +Na\ naphthalenide \end{matrix}$$

$$TaCl(CO)_2(diphos)_2$$

This species can then react further:

$$TaCl(CO)_2(diphos)_2 \xrightarrow{Na\ naphthalenide} [Ta(CO)_2(diphos)_2]^-$$

$$\Big\downarrow CH_3I$$

$$CH_3Ta(CO)_2(diphos)_2$$

22-C.   MOLYBDENUM[1,2] AND TUNGSTEN[2]

Molybdenum and tungsten are similar chemically, although there are differences between them in various types of compound that are not easy to explain. Thus some compounds of the same type differ noticeably in their reactivities toward various reagents: for example, $Mo(CO)_6$ but not $W(CO)_6$ reacts with acetic acid to give the quadruply bonded dimetal tetraacetate.

Except for compounds with π-acid ligands, there is not a great deal of similarity to chromium. The divalent state, well defined for Cr, is not well known for Mo and W except in strongly M—M bonded compounds; and the high stability of Cr^{III} in its complexes has no counterpart in Mo or W chemistry. For the heavier elements, the higher oxidation states are more common and more stable against reduction.

Both Mo and W have a wide variety of stereochemistries in addition to the variety of oxidation states, and their chemistry is among the most complex of the transition elements. Uranium has sometimes been classed with Mo and W in Group VI, and indeed there are some valid, though often rather superficial, similarities; the three elements form volatile hexafluorides, oxide halides, and oxo anions that are similar in certain respects. There is little resemblance to the sulfur group except in regard to stoichiometric similarities (e.g., SeF_6, WF_6, SO_4^{2-}, MoO_4^{2-}), and such comparisons are not profitable.

The oxidation states and stereochemistry are summarized in Table 22-C-1.

[45]   S. Datta and S. S. Wreford, *Inorg. Chem.*, 1977, **16**, 1134.

[1]   E. I. Stiefel, *Prog. Inorg. Chem.*, 1977, **22**, 1.

[2]   D. L. Kepert, *The Early Transition Metals,* Academic Press 1972, Chapter 4; I. Dellieu, F. M. Hall, and L. G. Hepler, *Chem. Rev.*, 1976, **76**, 283 (thermodynamics equilibria and potentials for Cr, Mo, and W compounds).

TABLE 22-C-1

Oxidation states and stereochemistry of Molybdenum and Tungsten

Oxidation state	Coordination number	Geometry	Examples
Mo^{-II}, W^{-II}	5	?	$[Mo(CO)_5]^{2-}$
Mo^0, W^0, d^6	6	Octahedral	$W(CO)_6$, $py_3Mo(CO)_3$, $[Mo(CO)_5I]^-$, $[Mo(CN)_5NO]^{4-}$, $Mo(N_2)_2(diphos)_2$
Mo^I, W^I, d^5	6^a	π-Complex	$(C_6H_6)_2Mo^+$, η^5-$C_5H_5MoC_6H_6$,
	7^a		$[\eta^5$-$C_5H_5Mo(CO)_3]_2$
	6	?	$MoCl(N_2)(diphos)_2$
Mo^{II}, W^{II}, d^4		π Complex	η^5-$C_5H_5W(CO)_3Cl$
	5	M—M quadruple bond	$Mo_2(O_2CR)_4$, $[Mo_2Cl_8]^{4-}$, $[W_2(CH_3)_8]^{4-}$
	6	Octahedral	$Mo(diars)_2X_2$, $trans$-$Me_2W(PMe_3)_4$
	7	Capped trigonal prism	$[Mo(CNR)_7]^{2+}$
	9	Cluster compounds	Mo_6Cl_{12}, W_6Cl_{12}
Mo^{III}, W^{III}, d^3	4	M—M triple bond	$Mo_2(OR)_6$, $W_2(NR_2)_6$
	6	Octahedral	$[Mo(NCS)_6]^{3-}$, $[MoCl_5]^{3-}$, $[W_2Cl_9]^{3-}$
	7	?	$[W(diars)(CO)_3Br_2]^+$
	8	Dodecahedral (?)	$[Mo(CN)_7(H_2O)]^{4-}$
Mo^{IV}, W^{IV}, d^2	8^a	π Complex	$(\eta^5$-$C_5H_5)_2WH_2$, $(\eta^5$-$C_5H_5)_2MoCl_2$
	9^a	π Complex	$(\eta^5$-$C_5H_5)_2WH_3$
	4	Tetrahedral	$Mo(NMe_2)_4$
	6	Octahedral	$[Mo(NCS)_6]^{2-}$, $[Mo(diars)_2Br_2]^{2+}$, $WBr_4(MeCN)_2$, $MoOCl_2(PR_3)_3$
	6	Trigonal prism	MoS_2
	8	Dodecahedral or square antiprism	$[Mo(CN)_8]^{4-}$, $[W(CN)_8]^{4-}$, $Mo(S_2CNMe_2)_4$, $M(picolinate)_4$
Mo^V, W^V, d^1	5	Tbp	$MoCl_5(g)$
	6	Octahedral	$Mo_2Cl_{10}(s)$, $[MoOCl_5]^{2-}$, WF_6^-
	8	Dodecahedral or square antiprism	$[Mo(CN)_8]^{3-}$, $[W(CN)_8]^{3-}$
Mo^{VI}, W^{VI}, d^0	4	Tetrahedral	MoO_4^{2-}, MoO_2Cl_2, WO_4^{2-}, WO_2Cl_2
	5?	?	$WOCl_4$, $MoOF_4$
	6	Octahedral	MoO_6, WO_6 in polyacids, WCl_6, $WOCl_4(s)$, MoF_6, $[MoO_2F_4]^{2-}$, MoO_3(dist.), WO_3(dist.)
	7	Dist. pentagonal bipyramid	$WOCl_4(diars)$, $K_2[MoO(O_2)ox]$
	8	?	MoF_8^{2-}, WF_8^{2-}, $[WMe_8]^{2-}$
	9	?	$WH_6(Me_2PhP)_3$

^{a} If C_6H_6 and η^5-C_5H_5 occupy three coordination sites.

Molybdenum is one of the biologically active transition elements.[3] It is intimately involved in the functioning of enzymes called nitrogenases, which cause atmospheric N_2 to be reduced to NH_3 or its derivatives, in enzymes concerned with reduction

[3] K. B. Swedo and J. H. Enemark, *J. Chem. Educ.*, 1979, **56**, 70; R. C. Bray, *Enzymes*, 1975, **12**, 299; R. C. Bray and J. C. Swan, *Struct. Bonding*, 1974, **11**, 107.

of nitrate, and in still other biological processes. Several aspects of molybdenum chemistry have been vigorously studied in the past decade, principally because of their possible relation to the biological processes. We refer to this point at appropriate places in this chapter, but the main discussion of molybdenum biochemistry and related matters appears in Chapter 31.

22-C-1.　The Elements

In respect to occurrence (abundance $\sim 10^{-4}\%$), metallurgy, and properties of the metals, molybdenum and tungsten are remarkably similar.

Molybdenum occurs chiefly as *molybdenite* (MoS_2) but also as molybdates such as $PbMoO_4$ (*wulfenite*) and $MgMoO_4$. Tungsten is found almost exclusively in the form of tungstates, the chief ones being *wolframite* (a solid solution and/or mixture of the isomorphous substances $FeWO_4$ and $MnWO_4$), *scheelite* ($CaWO_4$), and *stolzite* ($PbWO_4$).

The small amounts of MoS_2 in ores are concentrated by the foam flotation process; the concentrate is then converted into MoO_3 which, after purification, is reduced with hydrogen to the metal. Reduction with carbon must be avoided because this yields carbides rather than the metal.

Tungsten ores are concentrated by mechanical and magnetic processes and the concentrates attacked by fusion with NaOH. The cooled melts are leached with water, giving solutions of sodium tungstate from which hydrous WO_3 is precipitated on acidification. The hydrous oxide is dried and reduced to metal by hydrogen.

In the powder form in which they are first obtained both metals are dull gray, but when converted into the massive state by fusion are lustrous, silver-white substances of typically metallic appearance and properties. They have electrical conductances approximately 30% that of silver. They are extremely refractory; Mo melts at 2610° and W at 3410°.

Neither metal is readily attacked by acids. Concentrated nitric acid initially attacks molybdenum, but the metal surface is soon passivated. Both metals can be dissolved–tungsten only slowly, however—by a mixture of concentrated nitric and hydrofluoric acids. Oxidizing alkaline melts such as fused KNO_3–NaOH or Na_2O_2 attack them rapidly, but aqueous alkalis are without effect.

Both metals are inert to oxygen at ordinary temperatures, but at red heat they combine with it readily to give the trioxides. They both combine with chlorine when heated, but they are attacked by fluorine, yielding the hexafluorides, at room temperature.

The chief uses of both metals are in the production of alloy steels; even small amounts cause tremendous increases in hardness and strength. "High-speed" steels, which are used to make cutting tools that remain hard even at red heat, contain W and Cr. Tungsten is also extensively used for lamp filaments. The elements give hard, refractory, and chemically inert interstitial compounds with B, C, N, or Si on direct reaction at high temperatures. Tungsten carbide is used for tipping cutting tools, etc.

22-C-2. Oxides, Sulfides,[4a] and Oxo Acids

Oxides. Many molybdenum and tungsten oxides are known. The simple ones are MoO_3, WO_3, MoO_2, and WO_2. Other, nonstoichiometric oxides have been characterized and have complicated structures.

The ultimate products of heating the metals or other compounds such as the sulfides in oxygen are the *trioxides*. They are not attacked by acids but dissolve in bases to form molybdate and tungstate solutions, which are discussed below.

MoO_3 is a white solid at room temperature but becomes yellow when hot and melts at 795° to a deep yellow liquid. It is the anhydride of molybdic acid, but it does not form hydrates directly, although these are known (see below). MoO_3 has a rare type of layer structure in which each molybdenum atom is surrounded by a distorted octahedron of oxygen atoms.

WO_3 is a lemon yellow solid (m.p. 1200°); it has a slightly distorted form of the cubic rhenium trioxide structure (p. 888).

Molybdenum(IV) oxide (MoO_2), is obtained by reducing MoO_3 with hydrogen or NH_3 below 470° (above this temperature reduction proceeds to the metal) and by reaction of molybdenum with steam at 800°. It is a brown-violet solid with a coppery luster, insoluble in nonoxidizing mineral acids but soluble in concentrated nitric acid with oxidation of the molybdenum to Mo^{VI}. The structure is similar to that of rutile but so distorted that strong Mo—Mo bonds are formed.[4b] WO_2 is similar. Mo—Mo and W—W distances are 2.51 and 2.49 Å.

Although older literature describes simple oxides in intermediate oxidation states [e.g., Mo_2O_5 and $MoO(OH)_3$], these are apparently not genuine. There are, however, a large number of oxides of composition MO_x ($2 < x < 3$) obtainable by simply heating MoO_3 with Mo at 700° or WO_3 with W at 1000° or by heating the trioxides in a vacuum. These are all intensely colored, usually blue or purple, and have three types of structure:

1. An example of a *shear structure* is shown in Fig. 22-C-1. A structure in which *all* octahedra share each corner (but never an edge) with six neighboring octahedra would have the composition MO_3. To the extent that some oxygen atoms are now shared by three metal atoms, the O/M ratio drops below 3. In the example shown, the composition is Mo_8O_{23} (i.e., of the general type M_nO_{3n-1}). Other shear structures in which there are more such triply shared corners have the general composition M_nO_{3n-2}, of which $W_{20}O_{58}$ is an actual example. Along the shear plane the structure is locally similar to that throughout the MO_2 structure, but M—M distances remain long (*ca.* 3.25 Å), and no M—M bonding occurs.

2. There are a number of compounds that have seven-coordinate as well as six-coordinate metal atoms [e.g., Mo_5O_{14} ($MoO_{2.80}$), $Mo_{17}O_{47}$($MoO_{2.77}$), and $W_{18}O_{49}$($WO_{2.72}$)]. Figure 22-C-2 shows the structure of Mo_5O_{14}; the occurrence of some Mo atoms in pentagonal bipyramids can clearly be seen.

[4a] G. A. Tsigdinos and G. W. Moh, in *Aspects of Molybdenum and Related Chemistry,* Vol. 76, *Topics in Current Chemistry,* Springer, 1978.

[4b] J. Ghose et al., *Solid State Chem.,* 1976, **19**, 365.

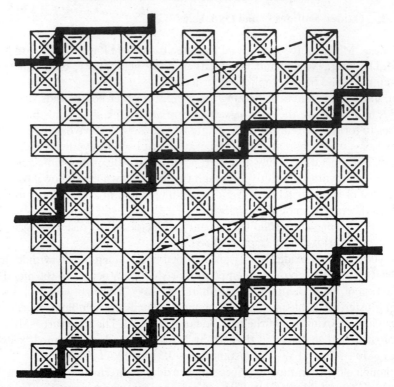

Fig. 22-C-1. One plane of the idealized Mo_8O_{23} structure. The shear planes that disturb the otherwise perfect checkerboard arrangement of MoO_6 octahedra sharing only corners are shown by heavy lines. The dashed lines outline the true unit cell.

3. In a few cases, also, there are some tetrahedrally coordinated metal atoms (e.g., Mo_4O_{11}).

Mixed Oxides. On fusion of MoO_3 or WO_3 with alkali or alkaline earth oxides, mixed oxide systems are obtained that are not related to the molybdates or tungstates made in aqueous solutions (see below). These usually have chain structures with linked MoO_6 polyhedra, but the stability of a particular type of structure depends on the cation size. Tungstates may differ from molybdates; thus $K_2Mo_4O_{13}$ has a chain, but $K_2W_4O_{13}$ has WO_6 octahedra linked by corners to give six-membered rings with the K^+ ions in the tunnels so formed.[4c]

The Blue Oxides. Also called *molybdenum blue and tungsten blue,* these are obtained by mild reductions (e.g., by Sn^{II}, SO_2, N_2H_4, H_2S, etc.) of acidified solutions of molybdates and tungstates or of suspensions of MoO_3 and WO_3 in water. Moist tungsten(VI) oxide will acquire a blue tint merely on exposure to ultraviolet light.

The "blue oxides" of Mo contain both oxide and hydroxide. There appears to be an entire series of "genotypic" compounds (i.e., having the same basic structure

4c Cf. K. Viswanathan, *J.C.S. Dalton,* **1974**, 2170, for references.

Fig. 22-C-2. The structure of Mo_5O_{14} showing linking of MoO_6 octahedra and MoO_7 pentagonal bipyramids. Each layer shares polyhedral corners with identical layers above and below the plane of the page.

but differing in the charges on cations and anions), with (olive green) $MoO(OH)_2$ as one limit and MoO_3 as the other. The compounds in which the mean oxidation state of Mo is between 5 and 6 are the blue ones [e.g., $MoO_{2.0}(OH)$ and $MoO_{2.5}(OH)_{0.5}$]. A detailed electronic explanation for the blue color has not been found, although the general idea that Mo_3 metal atom clusters might be responsible has been suggested. In the case of the "blue oxides" of W, similar general results have been obtained. These compounds are comparable to the heteropoly blues (p. 861).

Tungsten Bronzes.[5]   The reduction of sodium tungstate with hydrogen at red heat gives a chemically inert substance with a bronzelike appearance. Similar materials are obtained by vapor phase reduction of WO_3 with alkali metals.

The tungsten bronzes are nonstoichiometric substances of general formula $M_n^I WO_3$ ($0 < n \leqslant 1$). The colors vary greatly with composition from golden yellow for $n \approx 0.9$ to blue-violet for $n \approx 0.3$. Tungsten bronzes with $n > 0.3$ are extremely inert and have semimetallic properties, especially metallic luster, and good electrical

[5]   D. J. M. Bevan and P. Hagenmuller, *Nonstoichiometric Compounds: Tungsten Bronzes, Vanadium Bronzes, and Related Compounds,* Pergamon Press, 1975; F. R. Gamble and T. H. Geballe, "Inclusion Compounds," *Treatise on Solid State Chemistry,* Vol. 3, *Crystalline and Noncrystalline Solids,* N. B. Hannay, Ed., Plenum Press, 1976, Chapter 2.

conductivity in which the charge carriers are electrons. Those with $n < 0.3$ are semiconductors. They are insoluble in water and resistant to all acids except hydrofluoric, and they can be oxidized by oxygen in presence of base to give tungstates(VI):

$$4NaWO_3 + 4NaOH + O_2 \rightarrow 4Na_2WO_4 + 2H_2O$$

Structurally the sodium tungsten bronzes may be regarded as defective $M^I WO_3$ phases having the perovskite structure. In the defective phase $M_n^I WO_3$ there are $(1 - n)$ W^{VI} atoms, and $(1 - n)$ of the Na sites of the pure $NaWO_3$ phase are unoccupied. It appears that completely pure $NaWO_3$ has not been prepared, although phases with sodium enrichment up to perhaps $n \sim 0.95$ are known. The cubic structure collapses to rhombic and then triclinic for $n < \sim 0.3$. In the limit of $n = 0$ we have, of course, WO_3, which is known to have a triclinically distorted ReO_3 structure (p. 888). The cubic ReO_3 structure is the same as the perovskite structure with all the large cations removed. Thus the actual range of composition of the tungsten bronzes is approximately $Na_{0.3}WO_3$–$Na_{0.95}WO_3$.

The semimetallic properties of the tungsten bronzes are associated with the fact that no distinction can be made between W^V and W^{VI} atoms in the lattice, all W atoms appearing equivalent. Thus the n "extra" electrons per mole (over the number for WO_3) are distributed throughout the lattice, delocalized in energy bands somewhat similar to those of metals.

Sulfides.[4a]   Of the known sulfides Mo_2S_3, MoS_4, Mo_2S_5, MoS_3, and MoS_2, only the last three are important. The tungsten sulfides WS_2 and WS_3, are similar to their molybdenum analogues.

MoS_2 can be prepared by direct combination of the elements, by heating molybdenum(VI) oxide in hydrogen sulfide, or by fusing molybdenum(VI) oxide with a mixture of sulfur and potassium carbonate. It is the most stable sulfide at higher temperatures, and the others that are richer in sulfur revert to it when heated in a vacuum. It dissolves only in strongly oxidizing acids such as aqua regia and boiling concentrated sulfuric acid. Chlorine and oxygen attack it at elevated temperatures giving $MoCl_5$ and MoO_3, respectively.

MoS_2 has a structure built of close-packed layers of sulfur atoms stacked to create trigonal prismatic interstices that are occupied by Mo atoms. The stacking is such as to permit easy slippage of alternate layers; thus MoS_2 has mechanical properties (lubricity) similar to those of graphite.

Brown hydrous MoS_3, obtained on passing H_2S into slightly acidified solutions of molybdates, dissolves on digestion with alkali sulfide solution to give brown-red thiomolybdates (see below). Hydrated Mo_2S_5 is precipitated from Mo^V solutions. Both MoS_3 and Mo_2S_5 can be dehydrated.

Simple Molybdates and Tungstates.   The trioxides of molybdenum and tungsten dissolve in aqueous alkali metal hydroxides, and from these solutions the simple or normal molybdates and tungstates can be crystallized. They have the general formulas $M_2^I MoO_4$ and $M_2^I WO_4$ and contain the discrete tetrahedral ions MoO_4^{2-} and WO_4^{2-}. These are regular in alkali metal and a few other salts but may be distorted in salts of other cations. It is now certain that the MoO_4^{2-} and WO_4^{2-} ions are also tetrahedral in aqueous solution. Although both molybdates and tungstates can be reduced in solution (see below), they lack the powerful oxidizing property so characteristic of chromates(VI). The normal tungstates and molybdates of many other metals can be prepared by metathetical reactions. The alkali metal, ammonium, magnesium, and thallous salts are soluble in water, whereas those of other metals are nearly all insoluble.

When solutions of molybdates and tungstates are made weakly acid, polymeric anions are formed, but from more strongly acid solutions substances often called molybdic or tungstic acid are obtained. At room temperature the yellow $MoO_3 \cdot 2H_2O$ and the isomorphous $WO_3 \cdot 2H_2O$ crystallize, the former very slowly. From hot solutions, monohydrates are obtained rapidly. These compounds are oxide hydrates. $MoO_3 \cdot 2H_2O$ contains sheets of MoO_6 octahedra sharing corners and is best formulated as $[MoO_{4/2}O(H_2O)] \cdot H_2O$ with one H_2O bound to Mo, the other hydrogen bonded in the lattice.[6a]

No simple dimolybdate ion can be prepared from aqueous solutions of Mo^{VI} (see below), but the discrete $[Mo_2O_7]^{2-}$ can be obtained[6b] by addition of $[Bu_4^nN]OH$ to a CH_3CN solution of $[Bu_4^nN]_4Mo_8O_{26}$. The ion consists of two MoO_4 tetrahedra sharing a vertex, and it retains its structure in organic solvents, even on addition of water; but in presence of small counterions such as Na^+ or even Me_4N^+ it is converted to $[Mo_7O_{24}]^{6-}$. It should be noted that the many long-known compounds with formulas $M_2^I M_2O_7$ all have one of several structures containing infinite chain anions in which there are both MO_6 octahedra and MO_4 tetrahedra.[7]

The MoO_4^{2-} ion can serve as a ligand.[8] It also reacts with $o\text{-}C_6H_4(NH_2)(SH)$ to give $Mo(C_6H_4SNH)_3$, which has a trigonal pyramidal structure[9]; this complex is evidently similar to the better known dithiolene complexes (p. 185). A number of complexes with hydroxo compounds such as glycerol, tartrate ion, and sugars are known but have not been defined structurally. With diethylenetriamine the unique octahedral complex $MoO_3(dien)$ is obtained.

The simple thiomolybdate and -tungstate ions MS_4^{2-} are well known (and also $MoSe_4^{2-}$) in the form of alkali metal salts; MoS_4^{2-} is unstable to acid, but H_2WS_4 is known as an unstable red solid. Both MoS_4^{2-} and WS_4^{2-} can act as bidentate ligands.[10] There are also more complex thiomolybdate anions[11] such as $[(S_2)_2Mo(S_2)_2Mo(S_2)_2]^{2-}$ and $[Mo_3S_{13}]^{2-}$ that have S_2^{2-} ligands.

6a B. Krebs, *Acta Crystallogr., B,* 1972, **28**, 2222.

6b V. W. Day, W. G. Klemperer, *et al., J. Am. Chem. Soc.,* 1977, **99**, 6146.

7   A. W. Armour, M. G. B. Drew, and P. C. H. Mitchell, *J.C.S. Dalton,* **1975**, 1493; B. M. Gatehouse and P. Leverett, *J.C.S. Dalton,* **1976**, 1316.

8   R. S. Taylor, *Inorg. Chem.,* 1977, **16**, 116.

9   E. I. Stiefel *et al., Inorg. Chem.,* 1978, **17**, 897; K. Yamanouchi and J. H. Enemark, *Inorg. Chem.,* 1978, **17**, 2911.

10- A. Müller *et al., Chem. Ber.,* 1979, **112**, 778; *Angew. Chem., Int. Ed.,* 1976, **15**, 663; 1978, **17**, 52.

11 A. Müller *et al., Z. Anorg. Allg. Chem.,* 1978, **444**, 178.

22-C-3.   Isopoly and Heteropoly Acids and Their Salts[12a]

A prominent feature of the chemistry of molybdenum and tungsten is the formation of numerous polymolybdate(VI) and polytungstate(VI) acids and their salts. V^V, Nb^V, Ta^V, and U^{VI} show comparable behavior, but to a more limited extent.

The poly acids of molybdenum and tungsten are of two types: (*a*) the *isopoly acids* and their related anions, which contain only molybdenum or tungsten along with oxygen and hydrogen, and (*b*) the *heteropoly acids* and anions, which contain one or two atoms of another element in addition to molybdenum or tungsten, oxygen, and hydrogen. The polyanions consist primarily of octahedral MoO_6 or WO_6 groups, so that the conversion of MoO_4^{2-} or WO_4^{2-} into polyanions requires an increase in coordination number. It is still not clear why only certain metal oxo ions can polymerize, why for these metals only certain species (e.g., $Mo_7O_{24}^{6-}$, $HW_6O_{21}^{5-}$ or $Ta_6O_{18}^{6-}$) predominate under a given set of conditions, or why, for chromate, polymerization stops at $Cr_2O_7^{2-}$. The ability of the metal and oxygen orbitals to overlap to give substantial π bonding, $M{=}O$, must surely be involved (see p. 153), as must the base strength of the oxygen atoms and the ability of the initial protonated species $MO_3(OH)^-$ to expand its coordination sphere by coordination of water molecules. The size of the metal ion is clearly important.

X-Ray studies of crystalline compounds of a number of salts of isopoly and heteropoly anions have been made and are discussed below. By using these structures as a guide, considerable headway has been made in the interpretation of solution studies, and, especially in recent work, reliable, internally consistent data have been obtained. It is to be noted, however, that the X-ray studies do not show positions of hydrogen atoms, and although the basic units determined crystallographically often persist in solutions, hydration and protonation in solution depend on the conditions. Also, if a salt with a particular structure crystallizes from solution under certain conditions, this does not necessarily mean that the same anion is the major species—or that it even exists!—in solution. Recently it has been shown that the use of ^{17}O nmr allows direct observation of solution structures.[12b] There is a correlation of the ^{17}O chemical shifts with the structural role (terminal, $\mu_2{-}O$, $\mu_3{-}O$) of the oxygen atoms. It turns out that many of the structures found in crystals do appear to persist in solution, but this point should never be taken for granted.

Isopolymolybdates.[13]   The kinetics and equilibria for processes occurring as basic solutions of MoO_4^{2-} are acidified are very complex, and we cannot give more than an outline of a few key points.[14]

In strongly basic solution Mo(VI) is present only as MoO_4^{2-}. On addition of acid, the following protonation equilibria are established:

[12a] H. T. Evans, Jr., *Perspectives in Structural Chemistry*, Vol. IV, Wiley, 1971.
[12b] W. G. Klemperer *et al.*, *J. Am. Chem. Soc.*, 1976, **98**, 2345; *Angew. Chem., Int. Ed.*, 1978, **17**, 246.
[13]   K.-H. Tytko and O. Glemser, *Adv. Inorg. Chem. Radiochem.*, 1976, **19**, 239.
[14]   J. J. Cruywagen and E. F. C. H. Rohwer, *Inorg. Chem.*, 1975, **14**, 3137.

	K	ΔH (kJ mol^{-1})	ΔS (J mol^{-1} K^{-1})	
$MoO_4^{2-} + H^+ \overset{K_1}{=} HMoO_4^-$	$10^{3.7}$	20	140	(22-C-1)
$HMoO_4^- + H^+ + 2H_2O \overset{K_2}{=}$ $Mo(OH)_6$	$10^{3.7}$	-49	-92	(22-C-2)

It might be considered surprising that K_2 is as large as K_1, especially since the incorporation of two water molecules should cause a very unfavorable entropy change, as indeed it does. The ΔH and ΔS values for the first protonation are perfectly normal for a reaction of its type. The high value of K_2 is due to the large negative enthalpy change, and this may be attributed to the formation of two new Mo—O bonds while the number of O—H bonds is maintained (remember that H^+ belongs to H_3O^+).

On the basis of the equilibria above, it is possible to understand that the reaction

$$2HMoO_4^- = Mo_2O_7^{2-} + H_2O \qquad (22\text{-}C\text{-}3)$$

is not observed, even though the analogous reaction is very important for Cr^{VI} (cf. p. 733). In fact, as has long been known, no polynuclear Mo^{VI} species containing fewer than seven molybdenum atoms is observed in solution. In other words, in addition to the two equilibria above, the system is described by

$$7MoO_4^{2-} + 8H^+ = Mo_7O_{24}^{6-} + 4H_2O \qquad (22\text{-}C\text{-}4)$$

together with, at more acid pH's,

$$Mo_7O_{24}^{6-} + 3H^+ + HMoO_4^- = Mo_8O_{26}^{4-} + 2H_2O \qquad (22\text{-}C\text{-}5)$$

The absence of any detectable amounts of polynuclear species between the mononuclear ones and $Mo_7O_{24}^{6-}$ may be understood as follows. In view of the large value for K_2, $Mo(OH)_6$ is present in concentrations less than that of $HMoO_4^-$ by only a factor of $K_2[H^+]$. Thus reactions 22-C-6

$$Mo(OH)_6 + HMoO_4^- = (HO)_5MoOMoO_3^- + H_2O \qquad (22\text{-}C\text{-}6)$$
$$(HO)_5MoOMoO_3^- + HMoO_4^- = (HO)_4Mo(OMoO_3)_2^{2-} + H_2O$$
$$\vdots \qquad \vdots \qquad \vdots \qquad \vdots$$
$$(HO)Mo(OMoO_3)_5^{5-} + HMoO_4 = Mo(OMoO_3)_6^{6-} \quad + \quad H_2O$$
$$Mo(OMoO_3)_6^{6-} = Mo_7O_{24}^{6-}$$

can compete effectively with reaction 22-C-3. The species $Mo(OMoO_3)_6^{6-}$, with six MoO_4 tetrahedra, each attached through an oxygen atom to the central Mo atom, can rearrange internally to afford the final $Mo_7O_{24}^{6-}$ structure, which consists entirely of octahedra. Of course such a rearrangement may take place partially at earlier stages, and the extent of protonation of any species can vary, so the foregoing should not be taken literally.

Figure 22-C-3 shows some of the structures of polymolybdates, in a conventional schematic way that suggests, incorrectly, that the MoO_6 octahedra are regular.

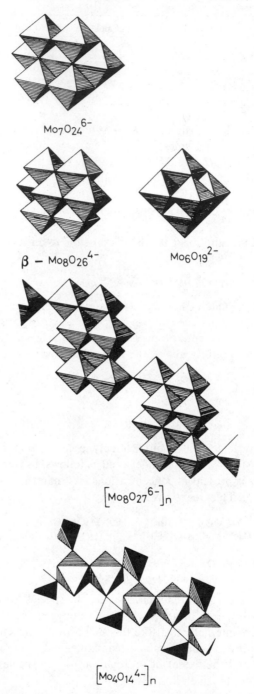

$Mo_7O_{24}{}^{6-}$

$\beta - Mo_8O_{26}{}^{4-}$

$Mo_6O_{19}{}^{2-}$

$[Mo_8O_{27}{}^{6-}]_n$

$[Mo_4O_{14}{}^{4-}]_n$

Fig. 22-C-3. Some of the known polymolybdate structures. [Reproduced by permission from Ref. 13.]

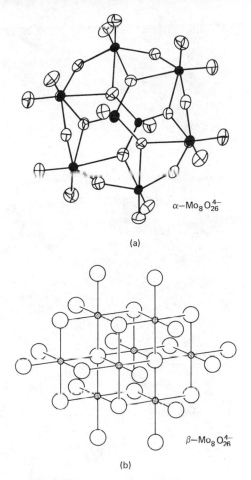

(a)

(b)

Fig. 22-C-4. The structures of the isomers of the $[Mo_8O_{28}]^{4-}$ ion: (a) α isomer, and (b) β isomer

Actually in all cases (including tungstates and the heteropoly acids, too) the metal atoms are displaced toward the outer (terminal) oxygen atoms. Nonetheless, structural representations of this kind are widely used because they convey a good overall picture.

The $[Mo_8O_{26}]^{4-}$ ion exists in two isomeric forms, (Fig. 22-C-4). In α-$[Mo_8O_{26}]^{4-}$ two of the metal atoms are in distorted tetrahedra, above and below a crown of six octahedra formed by edge sharing. It is quite similar to certain heteropoly anions, e.g., $[As_2Mo_6O_{26}]^{6-}$, discussed later. The α and β structures can coexist in solution; the equilibria and the products precipitated depend strongly on the cations,[15] and crystalline compounds containing pure α or pure β isomers can be obtained. Moreover, the interconversion of the two structures appears to

15   W. G. Klemperer and W. Shum, *J. Am. Chem. Soc.*, 1976, **98**, 8291; V. W. Day, W. G. Klemperer, *et al.*, *J. Am. Chem. Soc.*, 1977, **99**, 952.

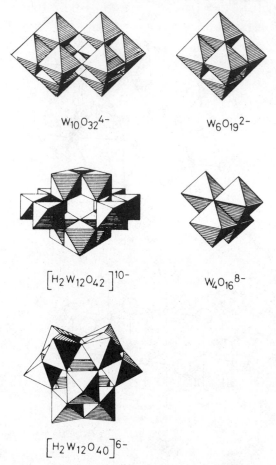

$W_{10}O_{32}^{4-}$

$W_6O_{19}^{2-}$

$[H_2W_{12}O_{42}]^{10-}$

$W_4O_{16}^{8-}$

$[H_2W_{12}O_{40}]^{6-}$

Fig. 22-C-5. Some isopolytungstate ions of known structure. [Reproduced by permission from Ref. 13.]

occur rapidly by an internal rearrangement for which a detailed series of steps can be envisioned.

The $[Mo_6O_{19}]^{2-}$ ion is found in solids but does not appear to exist (in significant concentration) in solution. The two polymeric species shown in Fig. 22-C-3 are, of course, also known only in crystals.

Isopolytungstates. Although the processes by which acidified tungstate solutions give rise to isopoly anions must be broadly similar to those occurring in molybdate solutions, there are such differences in detail that virtually no polyion in either system has a counterpart in the other. Figure 22-C-5 shows the main isopoly-tungstate ions of known structure. Only the $[W_6O_{19}]^{2-}$ ion[16] has a molybdenum analogue.

[16]   J. Fuchs, W. Freiwald, and H. Hartl, *Acta Crystallogr., B,* 1978, **34,** 1764.

TABLE 22-C-2
Some Heteropoly Anions with Buried Heteroatoms

Formula type	Central group	M = Mo	M = W
$X^{+n}M_{12}O_{40}^{(8-n)-}$	XO_4	Si^{IV}, Ge^{IV}, P^V, As^V, Ti^{IV}, Zr^{IV}	B^{III}, Si^{IV}, Ge^{IV}, P^V, As^V, Al^{III}, Fe^{III}, Co^{II}, Co^{III}, Cu^I, Cu^{II}, Zn^{II}, Cr^{III}, Mn^{IV}, $Te^{IV a}$ $Ga^{III b}$
$X_2^{+n}M_{18}O_{62}^{(16-2n)-}$	XO_4	P^V, As^V	P^V, As^V
$X_2^{+n}Z_4^{+m}M_{18}O_{70}H_4^{(28-2n-4m)-}$	XO_4	—	$X = P^V$, As^V $Z = Mn^{II}$, Co^{II}, Ni^{II} Cu^{II}, Zn^{II}
$X^{+n}M_9O_{32}^{(10-n)-}$	XO_6	Mn^{IV}, Ni^{IV}	—
$X^{+n}M_6O_{24}^{(12-n)-}$	XO_6	Te^{VI}, I^{VII}	Ni^{IV}, Te^{VI}, I^{VII}
$X^{In}M_6O_{24}H_6^{(6-n)-}$	XO_6	Al^{III}, Cr^{III}, Co^{III}, Fe^{III}, Ga^{III}, Rh^{III}, Mn^{II}, Co^{II}, Ni^{II}, Cu^{II}, Zn^{II}	Ni^{II}
$X_2^{+n}M_{10}O_{38}H_4^{(12-2n)-}$	XO_6	Co^{III}	—
$X^{+n}M_{12}O_{42}^{(12-n)-}$	XO_{12}	Ce^{IV}, Th^{IV}, U^{IV}	—

[a] Existence of anion, or membership of series, requires confirmation.
[b] Closely related 11-tungstate.

Heteropoly Anions.[17] These can be formed either by acidification of solutions containing the requisite simple anions, or by introduction of the heteroelement after first acidifying the molybdate or tungstate:

$$HPO_4^{2-} + MoO_4^{2-} \xrightarrow{\text{H}^+, 25°} [PMo_{12}O_{40}]^{3-}$$

$$WO_4^{2-} \xrightarrow{\text{H}^+ \text{ to pH } \sim 6 \quad Co^{2+}, 100°C} [Co_2W_{11}O_{40}H_2]^{8-}$$

Interconversions are also possible, viz.,

$$[P_2W_{18}O_{62}]^{6-} \xrightarrow{\text{HCO}_3^-, 25°} [P_2W_{17}O_{61}]^{10-}$$

Many of the heteropoly anions are quite robust toward excess acid and may be protonated to give the *heteropoly acids* both in solution and as crystalline hydrates.

This is a truly enormous class of compounds, and we note here only the main types. The largest and best known group is composed of those with the hetero atom(s) enshrouded by a cage of MO_6 octahedra. Table 22-C-2 lists most of those that have known structures. Six of the important structures that predominate in this group are shown schematically in Fig. 22-C-6.

The $[X^{+n}M_{12}O_{40}]^{(8-n)-}$ structure is often called the *Keggin structure* after its discoverer. It has full tetrahedral (T_d) symmetry and although very compact, it

[17] T. J. R. Weakley, *Structure and Bonding,* Vol. 18, Springer, 1974, p. 131. G. A. Tsigdinos, in *Aspects of Molybdenum and Related Chemistry,* Vol. 76, *Topics in Current Chemistry,* Springer, 1978.

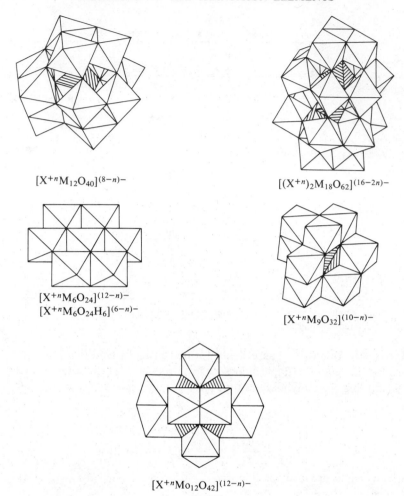

$[X^{+n}M_{12}O_{40}]^{(8-n)-}$

$[(X^{+n})_2M_{18}O_{62}]^{(16-2n)-}$

$[X^{+n}M_6O_{24}]^{(12-n)-}$
$[X^{+n}M_6O_{24}H_6]^{(6-n)-}$

$[X^{+n}M_9O_{32}]^{(10-n)-}$

$[X^{+n}Mo_{12}O_{42}]^{(12-n)-}$

Fig. 22-C-6. Structures of six important types of heteropolymolybdate and -tungstate ions with "enshrouded" heteroatoms. The general formula of each is shown below it.

accommodates a variety of heterocations that differ considerably in size, as Table 22-C-2 shows. There are many heteropolyanions with structures that can be regarded as modifications of the Keggin structure. One such modification is obtained by twisting one of the four sets of three octahedra in the Keggin structure by 60° and reattaching all the same corners. This spoils the T_d symmetry, leaving only one 3-fold axis, that around which the 60° rotation was made. This "isomeric" Keggin structure differs little in stability from the true Keggin structure, and the anions $[XW_{12}O_{40}]^{4-}$ with X = Si, Ge, for example, can be isolated in either form by varying the conditions of acidification in the preparation.

The structure of the $[(X^{+n})_2M_{18}O_{62}]^{(16-2n)-}$ ion (sometimes called the Dawson structure) is closely related to that of the Keggin ion. If three adjacent corner-linked MO_6 octahedra are removed from the Keggin structure, to leave a fragment with a set of three octahedra over a ring of six, we have one-half of this M_{18} anion. These two halves are then linked by corner sharing as shown in Fig. 22-C-6. Actually there are two such ways to link the halves. The way shown gives a structure of D_{3h} symmetry, but by rotating one half 60° relative to the other an isomer with D_{3d} symmetry is obtained; both occur.

The $[X^{+n}M_6O_{24}]^{(12-n)-}$ ion, its protonated form $[X^{+n}M_6O_{24}H_6]^{(6-n)-}$, and the $[X^{+n}M_9O_{32}]^{(10-n)-}$ ion are among those in which the heteroatom finds itself in an octahedron of oxygen atoms. The latter is known for only two cases, viz., those in which $X^{+n} = Mn^{+4}$ or Ni^{+4} and $M = Mo$, but these are of interest because the hetero ions are examples of unusual oxidation states stabilized by the unusual "ligand."

The $[X^{+n}M_{12}O_{42}]^{(12-n)-}$ ion provides an icosahedral set of twelve oxygen ligands for the hetero cation. In addition to Ce^{4+}, Th^{4+}, and U^{4+}, which have already been found in such a structure, a number of other M^{4+} ions from the lanthanides and actinides can probably be expected to form such anions.

Just as the heteropoly (and isopoly) ions are built up by acidification of solutions of the mononuclear oxo ions, the action of strong base on the heteropoly ions (and on the isopoly ions) will eventually degrade them entirely to mononuclear anions. There are well-defined steps in both processes, and intermediates can be observed[18] and even isolated in crystalline compounds.[19] Examples of species that have been well established are fragments of Keggin anions, especially those in which formal loss of "MO" has occurred, e.g., $[PW_{11}O_{39}]^{7-}$ or $[SiMo_{11}O_{39}]^{8-}$. There are also the "enneatungsto" compounds, such as $Na_{10}SiW_9O_{34}\cdot18H_2O$, that are thought to contain half-Dawson structures,[19] but this lacks proof. The $[P_2Mo_5O_{23}]^{6-}$ ion is an example of one of the intermediates that has been characterized structurally.[20] It consists of a cyclic assemblage of MoO_6 octahedra with four edge junctions and one corner junction, and there is a PO_4 tetrahedron fused over the hole on each side. Another example of this type of heteropoly anion in which the heteroatoms are exposed rather than buried is the $[As_2Mo_6O_{26}]^{6-}$ ion, whose structure has been determined by ^{17}O nmr.[21] This structure is much like the $[X^{+n}M_6O_{24}]^{(12-n)-}$ structure except that the central, octahedrally coordinated X^{+n} ion is replaced by two AsO^{+3} groups, one above and the other below the central cavity in the six-membered ring of MoO_6 octahedra.

Organoheteropoly Anions. A very new development in the poly ion field that is of interest because of its potential relationship to catalysis by metal oxide surfaces, is the preparation of anions having organic radicals bound to the surface of the anion. Among the simpler examples are those shown in Fig. 22-C-7. The $[(Me_2As)Mo_4O_{14}OH]^{2-}$ ion is easily prepared by acidification of a stoichiometric

18   R. Massart et al., Inorg. Chem., 1977, **16**, 2916.
19   G. Hervé and A. Tézé, Inorg. Chem., 1977, **16**, 2115.
20   J. Fischer, L. Ricard, and P. Toledano, J.C.S. Dalton, **1974**, 941.
21   M. Filowitz and W. G. Klemperer, J.C.S. Chem. Comm., **1976**, 233.

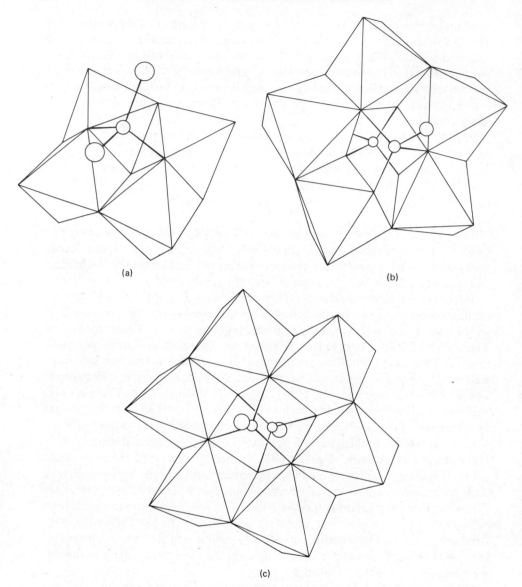

Fig. 22-C-7. Structures of some organoheteropoly anions. (a) The $[(Me_2As)Mo_4O_{14}OH]^{2-}$ ion. The OH is thought to be at the juncture of all four MoO_6 octahedra. (b) The $[(MeP)_2Mo_5O_{21}]^{4-}$ ion. (c) The $[(MeAs)_2Mo_6O_{24}]^{4-}$ ion. Large circles represent methyl groups, small circles P or As atoms.

mixture of MoO_4^{2-} and $(CH_3)_2AsO_2^-$ ions.[22] Somewhat larger ions of the same type are typified by the $[MeP)_2Mo_5O_{21}]^{4-}$ and $[(MeAs)_2Mo_6O_{24}]^{4-}$ ions.[23]

Steric factors are very important in these systems, as indicated by the fact that

[22] M. T. Pope, C. O. Quicksall, et al., J. Am. Chem. Soc., 1975, 97, 4146.
[23] M. T. Pope, et al., Inorg. Chem., 1976, 15, 2778; J. Am. Chem. Soc., 1979, 101, 2731.

instead of exact phosphorus and arsenic analogues we have the different formulas and structures just mentioned. Apparently the smaller, buckled Mo_5O_{21} ring, formed by a combination of both edge and corner sharing, gives smaller sets of three O atoms, better suited to the formation of the RPO_3 group, whereas the Mo_6O_{24} unit provides larger triangles of oxygen atoms that better fit the larger As atom.

More complex systems can be made by reaction of organometal and organometalloidal halides such as $RSnCl_3$ or $RAsCl_2$, with fragments of the Keggin ions, the latter being, for example, $[PW_{11}O_{39}]^{7-}$ or $[SiMo_{11}O_{39}]^{8-}$. The entering RSn or RAs unit binds to three oxygen atoms replacing the missing MoO or WO groups.[24] If an $(\eta\text{-}C_5H_5)TiCl_3$ group is used[24,25a] the $(\eta\text{-}C_5H_5)Ti$ unit is introduced. Finally, some formylated species with Mo—O—CHO and Mo—OCH$_2$O— groups have been made.[25b]

Heteropoly Blues. Anions of the Keggin and Dawson types can undergo reduction to blue mixed-valence species without loss of structure. These reduction products are formed rapidly and reversibly at least to the stage of six-electron reduction, but further reduction may involve isomerizations that are irreversible. The question of whether the added electrons enter delocalized orbitals or tend to be localized on individual Mo or W atoms has received considerable attention. The role of the heteroatoms is in general insignificant. Even when these are transition metal ions, as in $[Co^{II}W_{12}O_{40}]^{6-}$ and $[Fe^{III}W_{12}O_{40}]^{5-}$, the epr spectra of doubly reduced blues obtained from them continue to show the signals characteristic of the Co^{II} and Fe^{III} species present prior to reduction. The weight of evidence at present suggests that added electrons are localized on Mo or W atoms but that hopping occurs (with the mean residence time at any one site being $ca.$ 10^{-8} sec at $77°K$) and that very strong spin coupling occurs via a superexchange mechanism. Neither true delocalization nor M—M bonding appears to occur.

22-C-4. Halides and Halo Complexes

The more important halides are listed in Table 22-C-3. Those containing metal atom clusters or presumed to contain Mo—Mo quadruple bonds are discussed in Sections 22-C-9 and 22-C-7, respectively. Those with metal oxidation states III to VI, are discussed here.

Hexahalides. The MF_6 compounds are volatile, colorless liquids, readily hydrolyzed. MoF_6 is more reactive, less stable, and a considerably stronger oxidizing agent.[26] The existence of $MoCl_6$ is very doubtful, but WCl_6 and WBr_6 are both obtained by direct halogenation of the metal. WCl_6 can be volatilized to a monomeric vapor and is soluble in liquids such as CS_2, CCl_4, EtOH, and Et_2O, whereas WBr_6, a dark blue solid, gives WBr_5 on moderate heating. Both are hydrolyzed to tungstic acid.

Pentahalides. Treatment of molybdenum carbonyl with fluorine diluted in ni-

[24]   W. H. Knoth, *J. Am. Chem. Soc.*, 1979, **101**, 759.
[25a] R. K. C. Ho and W. G. Klemperer, *J. Am. Chem. Soc.*, 1978, **100**, 6772.
[25b] W. G. Klemperer *et al.*, *J.C.S. Chem. Comm.*, **1979**, 256; *J. Am. Chem. Soc.*, 1979, **101**, 491.
[26]   A. M. Bond, I. Irvine, and T. A. O'Connell, *Inorg. Chem.*, 1977, **16**, 841.

TABLE 22-C-3
The Fluorides and Chlorides of Molybdenum and Tungsten

III	IV	V	VI[a]
MoF_3 Yellow-brown, nonvolatile	MoF_4 Tan, nonvolatile	$(MoF_5)_4$ Yellow, m.p. 67°, b.p. 213°	MoF_6 Colorless, m.p. 17.5°, b.p. 35.0°
	WF_4 Red-brown, nonvolatile	$(WF_5)_4$ Yellow, disprop. 25°	WF_6 Colorless, m.p. 2.3°, b.p. 17.0°
$MoCl_3$ Dark red	$MoCl_4$ Dark red	$(MoCl_5)_2$ Green-black, m.p. 194°, b.p. 268°	
	WCl_4 Black	$(WCl_5)_2$ Green-black	WCl_6 Blue-black, m.p. 275°, b.p. 346°

[a] Also WBr_6, WF_5Cl, WCl_5F, WCl_4F_2.

trogen at −75° gives a product of composition Mo_2F_9. The nature of this substance has not been investigated, but when it is heated to 150° it yields the nonvolatile MoF_4 as a residue and the volatile MoF_5 condenses in cooler regions of the apparatus. MoF_5 is also obtained by the reactions:

$$5MoF_6 + Mo(CO)_6 \xrightarrow{25°} 6MoF_5 + 6CO$$

$$Mo + 5MoF_6 \longrightarrow 6MoF_5$$

$$Mo + F_2(\text{dilute}) \xrightarrow{400°} MoF_5$$

WF_5 is obtained by quenching the products of reaction of W with WF_6 at 800 to 1000°K. It disproportionates above 320°K into WF_4 and WF_6. Crystalline MoF_5 and WF_5 (and WOF_4, p. 866) have the tetrameric structure common to many pentafluorides.

Mo_2Cl_{10}, which is formed on direct chlorination of the metal, is moderately volatile and monomeric in the vapor, probably having a trigonal-bipyramidal structure. In the crystal, however, chlorine-bridged dimers are formed so that each molybdenum atom is hexacoordinate. Mo_2Cl_{10} is paramagnetic, the magnetic moment ($\mu_{\text{eff}} = 1.64$ BM at 293°K), indicating only negligible coupling of electron spins of the two molybdenum atoms (Mo—Mo = 3.84 Å). Mo_2Cl_{10} is soluble in benzene and also in more polar organic solvents. It is monomeric in solution and is presumably solvated. It readily abstracts oxygen from oxygenated solvents to give oxo species, and it is also reduced by amines to give amido complexes. It is rapidly hydrolyzed by water. Some of its reactions are shown in Figs. 22-C-8 and 22-C-9, which set out the preparative methods for lower chlorides and oxide chlorides.

Green WCl_5 and black WBr_5 are prepared by direct halogenation, the conditions being critical, especially the temperature. The chloride is isostructural with Mo_2Cl_{10} (W—W = 3.81 Å).[27]

[27] F. A. Cotton and C. E. Rice, *Acta Crystallogr., B,* 1978, **34,** 2833.

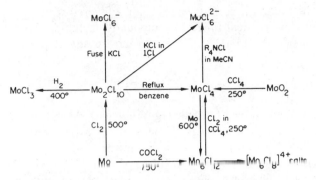

Fig. 22-C-8. Preparation of molybdenum chlorides and chloro complexes.

Tetrahalides. These include MoF_4 and WF_4, the former arising on disproportionation of Mo_2F_9 as noted above, and both by reduction of the hexahalides with hydrocarbons (e.g., benzene at $\sim110°$). Both are nonvolatile. $MoCl_4$, which is very sensitive to oxidation and hydrolysis, exists in three forms. By the reaction

$$MoCl_5 + C_2Cl_4 \rightarrow \alpha\text{-}MoCl_4$$

a form isomorphous with $NbCl_4$ (p. 839) is obtained. The same form, contaminated with carbon, is also obtained when $MoCl_5$ is reduced with hydrocarbons. On being heated to 250° in the presence of $MoCl_5$, the α form changes into the high temperature β form. α-$MoCl_4$ has partial spin pairing through Mo—Mo interactions, whereas the β form has an *hcp* array of Cl atoms, with Mo atoms so distributed in octahedral interstices that there is no Mo—Mo bonding. The third form, obtained by treatment of MoO_2 with carbon in an N_2-borne stream of CCl_4 vapor, has a magnetic moment of *ca.* 1.9 BM and its structure is unknown.[28]

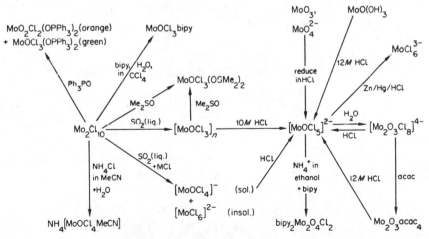

Fig. 22-C-9. Some preparations and reactions of molybdenum pentachloride and of oxomolybdenum compounds.

28   A. D. Westland and V. Uzelac, *Inorg. Chim. Acta*, 1977, **23**, L37.

WCl_4, obscure until recently, has become a favored starting material for preparing many lower-valent tungsten compounds, including those with triple and quadruple bonds (Section 22-C-7). It can most easily be prepared by the reaction:

$$2WCl_6 + W(CO)_6 \xrightarrow[\text{at reflux}]{C_6H_5Cl} 3WCl_4$$

It can also be obtained by reducing WCl_6 with Al in a thermal gradient; it disproportionates at 500° to $WCl_2 + 2WCl_5$ and is isotypic with α-$MoCl_4$. $MoBr_4$, WBr_4, and WI_4 all exist but are not well known.

Trihalides.   These include MoF_3, $MoCl_3$, $MoBr_3$, MoI_3, WCl_3, WBr_3, and WI_3. MoF_3, a nonvolatile brown solid in which Mo atoms are found in octahedra of F atoms, is obtained by reaction of Mo with MoF_6 at ~400°. $MoCl_3$ has two polymorphs, with *hcp* and *ccp* arrays of Cl atoms and, in each case, Mo atoms in pairs of adjacent octahedral holes at a distance of 2.76 Å across a common edge. The magnetic properties confirm the expected M—M interaction. WCl_3 has an especially interesting cluster structure (p. 883).

Halogeno Complexes.   Molybdenum(III), but not tungsten(III), forms complexes of the type $[MX_6]^{3-}$. Prolonged electrolytic reduction of a solution of MoO_3 in concentrated hydrochloric acid gives a solution of Mo^{III} in the form of chloro complexes, of which $[MoCl_6]^{3-}$ and $[MoCl_5(H_2O)]^{2-}$ can be precipitated with the larger alkali metal cations. The salts are red and fairly stable in dry air. In dilute solution the $MoCl_6^{3-}$ ion is readily aquated. Thus dissolution of K_3MoCl_6 in aqueous CF_3SO_3H followed by ion-exchange purification gives the yellow air-sensitive aqua ion $[Mo(H_2O)_6]^{3+}$. K_3MoCl_6 reacts with molten KHF_2 to produce brown cubic K_3MoF_6. The magnetic behavior of MoX_6^{3-} salts is usually straightforward, with essentially temperature-independent effective magnetic moments of about 3.8 BM, corresponding to the expected t_{2g}^3 configuration.

The $[MoCl_6]^{3-}$ or $[MoCl_5(H_2O)]^{2-}$ ions (or the bromo analogues) react with pyridine or substituted pyridines to give neutral complexes of the type MoX_3py_3, one of which has been shown to be the *mer* isomer.[29] A less direct but useful way of getting six-coordinate Mo^{III} compounds is the following route[30]:

$$MoCl_4(EtCN)_2 \xrightarrow{THF} MoCl_4(THF)_2 \xrightarrow[\text{in THF/CH}_2\text{Cl}_2]{Zn}$$

$$MoCl_3(THF)_3 \xrightarrow{L} MoCl_3(THF)_2L, \; MoCl_3(THF)L_2, \; MoCl_3L_3$$

W^{III} forms no $[WX_6]^{3-}$ ions, but both elements form the $[M_2X_9]^{3-}$ ions, which have the confacial bioctahedral structure 22-C-I:

(22-C-I)

29   J. V. Brencic and I. Leban, *Z. Anorg. Allg. Chem.*, 1978, **445**, 251.
30   M. W. Anker *et al., J.C.S. Dalton*, **1975**, 2639.

which is also possessed by $Cr_2Cl_9^{3-}$. The strength of the interaction between the two d^3 ions increases markedly in the series Cr, Mo, W. Thus in $Cr_2X_9^{3-}$ there is no M—M bond, and the Cr atoms actually repel each other. The magnetic and spectral properties are those of essentially unperturbed d^3 ions. In $W_2X_9^{3-}$ at the opposite extreme, the interaction is very strong, causing a marked distortion of the structure (W—W, 2.41 Å) and resulting in the absence of unpaired electrons. There must be one σ and two π bonds between the W atoms. In $Mo_2X_9^{3-}$ the situation is intermediate. The three M—M interactions, which are very strong in $W_2Cl_9^{3-}$, are of lower strength, giving Mo—Mo distances of 2.67 and 2.78 Å for $Cs_3Mo_2Cl_9$ and $Cs_3Mo_2Br_9$, respectively. In the former, as in both $W_2X_9^{3-}$ species, there is a small temperature-independent paramagnetism. In $Cs_3Mo_2Br_9$, however, the magnetism is temperature dependent, suggesting perhaps that the Mo—Mo interaction is weak enough to allow some unpairing of electrons at 300°K.

$K_3W_2Cl_9$ reacts with pyridine (and several substituted pyridines) as well as with alcohols to afford compounds in which one W—Cl—W bridge is opened, giving structures with two octahedra sharing an edge; the formulas are $W_2Cl_6py_4$ and $W_2Cl_4(OR)_2(ROH)_4$.

The $Mo_2Cl_9^{2-}$, $W_2Cl_9^{2-}$, and $W_2Br_9^{2-}$ ions can be obtained by oxidation of the 3– species ($W_2Cl_9^{2-}$ only) or indirectly[31]:

$$MoCl_5 + Cl^- + Mo(CO)_4Cl_3^- \rightarrow Mo_2Cl_9^{2-} + 4CO$$
$$W_2Cl_9^{3-} + \tfrac{1}{2}Cl_2 \rightarrow W_2Cl_9^{2-} + Cl^-$$
$$4W(CO)_5Br^- + 7BrCH_2CH_2Br \rightarrow 2W_2Br_9^{2-} + 20CO + 7C_2H_4$$

Halo complexes are also formed by M^{IV} and M^V. The pentachlorides and WBr_5 give rise to MX_6 complexes (e.g., by reaction of $MoCl_5$ with R_4NCl in CH_2Cl_2). These M^V complexes can sometimes be decomposed thermally to the M^{IV} complex, viz.,

$$2M^IWCl_6 \xrightarrow{280\text{-}300°} M_2WCl_6 + WCl_6$$
$$\text{Green} \qquad\qquad\qquad \text{Red}$$

Other ways of making M^{IV} complexes, in which the course of the reaction is often not understood, include direct reductive reaction of $MoCl_5$ with ligands:

$$MoCl_5 + \begin{Bmatrix} py \\ dipy \\ RCN \end{Bmatrix} \rightarrow \begin{Bmatrix} MoCl_4py_2 \\ MoCl_4dipy \\ MoCl_4(RCN)_2 \end{Bmatrix}$$

The ion $[Cl_5WOWCl_5]^{4-}$ is obtained on incomplete reduction of WO_4^{2-} by tin in concentrated HCl, and $K_4[W_2Cl_{10}O]$ can be isolated as an analytically pure solid. The WOW bridge is linear, and the tungsten atoms are equivalent W^{IV} (d^2) atoms strongly antiferromagnetically coupled through the bridge.[32a]

Compounds such as K_2WCl_6, and also the orange compound WCl_4py_2 obtained from it by action of pyridine or by treating WCl_4 with pyridine, have magnetic

[31]   W. H. Delphin and R. A. D. Wentworth, *J. Am. Chem. Soc.*, 1973, **95**, 7921; J. L. Templeton, R. A. Jacobson, and R. A. McCarley, *Inorg. Chem.*, 1977, **16**, 3320.

[32a] J. San Filippo, Jr., P. J. Fagan, and F. J. DiSalvo, *Inorg. Chem.*, 1977, **16**, 1016, and references therein.

moments that are much below the spin-only value for two unpaired electrons. Some compounds are certainly antiferromagnetic, and since WCl_6^{2-} salts have crystal structures similar to $IrCl_6^{2-}$ salts (i.e., antiferromagnetic interaction occurs through neighboring chlorine atoms), this explanation is probably general.

Hexafluoromolybdates(IV)—for example, the dark brown Na_2MoF_6—can be obtained by reduction of MoF_6 with an excess of NaI; the hexafluoromolybdates(IV) are much more stable with respect to hydrolysis than are the Mo^V species.

The hexafluoromolybdate(V) and hexafluorotungstate(V) anions can be obtained as Na, K, Rb or Cs salts by the reaction:

$$W(Mo)(CO)_6 + M^II + IF_5 \rightarrow M^I[W(Mo)F_6] + 6CO + \text{unidentified products}$$

Here, IF_5 serves as the fluorinating agent and the solvent. It is reported that the product of this reaction may also be K_3MoF_8, K_3WF_8, or K_2WF_8, depending on exact conditions. M^IWF_7 compounds have also been reported. None of these hepta- or octacoordinate compounds are structurally characterized.

The pentachlorides react with ROH and RO^- to yield various types of alcoxo complex. Tungsten forms an extensive series of complexes that includes the paramagnetic $[M(OR)Cl_5]^-$ and $[M(OR)_2Cl_4]^-$ ions as well as the diamagnetic $W_2Cl_2(OR)_8$ and $W_2Cl_4(OR)_6$ molecules, which presumably contain octahedrally coordinated metal atoms, bridging Cl atoms, and W—W bonds. $MoCl_5$ reacts with alcohols and amines to give products of the types $MoCl_3(OR)_2$ and $MoCl_3(NRR')_2$, which appear generally to be dinuclear with bridging chlorine atoms. The chief products of reaction with phenols are of the type $[MoCl_2(OAr)_3]_2$. Reactions of halides with $LiNR_2$ are mentioned later in connection with M—M triple bonds.

22-C-5. Oxide Halides

Oxidation State VI. These are of two stoichiometric types: MOX_4 and MO_2X_2. The molybdenum compounds are less stable than those of tungsten and are all fairly rapidly hydrolyzed by water. They are obtained as by-products in the halogenation of the metals unless the metal is first scrupulously reduced and the reaction system vigorously purged of oxygen.

$MoOF_4$ and WOF_4 can both be prepared by the same types of reaction, namely,

$$\left.\begin{array}{l} M + O_2 + F_2 \\ MO_3 + F_2 \\ MOCl_4 + HF \end{array}\right\} \rightarrow MOF_4 \quad (M = \text{Mo or W})$$

They are both colorless, volatile solids, not as reactive as the hexafluorides. $MoOF_4$ has octahedral units linked by bridging F's into infinite chains, but WOF_4 has the tetrameric NbF_5 structure (p. 834).

WO_2F_2 has been reported, but there is some doubt of its actual existence.

MoO_2F_2 can be obtained by the action of HF on MoO_2Cl_2; it is a white solid subliming at 270° at 1 atm.

$MoOCl_4$ (green crystals, m.p. 101–103°), is best made by interaction of $MoCl_5$ and O_2 or by refluxing MoO_3 with $SOCl_2$; it decomposes to $MoOCl_3$ and Cl_2 even at 25° and is readily reduced by organic solvents to Mo^V species.

MoO_2Cl_2 is best made by the action of chlorine on heated, dry MoO_2. It is fairly volatile and dissolves with hydrolysis in water, although in strong HCl an oxide halide species, possibly $Cl_2(O)MoO_2MoOCl_2$, exists.

When MoO_3 is treated with dry hydrogen chloride at 150 to 200° a pale yellow, very volatile compound soluble in various polar organic solvents is obtained; it has stoichiometry $MoO_2Cl_2 \cdot H_2O$ but is possibly a dimer.

The two tungsten oxide chlorides are formed together when WO_3 is heated in CCl_4, phosgene, or PCl_5 vapor. They are easily separated, since $WOCl_4$ is much more volatile than WO_2Cl_2. On strong heating above 200° the following reaction occurs:

$$2WO_2Cl_2 \rightarrow WO_3 + WOCl_4$$

$WOCl_4$ forms scarlet crystals in which $WOCl_4$ units are linked in chains with W—O—W—O—W bonds[32b] and a red monomeric vapor and is in general highly reactive. It is violently hydrolyzed by water. WO_2Cl_2 occurs as yellow crystals and is not nearly so reactive as $WOCl_4$; it is hydrolyzed only slowly by cold water.

Oxidation State V. The four principal compounds are black $MoOCl_3$ and $MoOBr_3$, olive $WOCl_3$, and brown to black $WOBr_3$. Methods of preparation include:

$$WOX_4 + \tfrac{1}{3}Al \rightarrow WOX_3 + \tfrac{1}{3}AlX_3$$
$$2W + WO_3 + \tfrac{9}{2}Br_2 \rightarrow 3WOBr_3$$
$$MoOCl_4 + C_6Cl_6 \rightarrow MoOCl_3$$

All four oxide halides occur in a crystalline form isotypic with $NbOCl_3$, but $MoOCl_3$ has a second (monoclinic) form that has the structure shown in Fig. 22-C-10.

Thio analogues of the oxide halides are also known for Mo^V, W^V, and W^{VI}.

22-C-6. Aqua and Oxo Complexes*

Aqua Ions. These are known for oxidation numbers II to V. The first has been shown to be the dinuclear, presumably quadruply bonded $Mo_2^{4+}(aq)$ ion and is formed when a solution of $K_4[Mo_2(SO_4)_4]$ is treated with $Ba(CF_3SO_3)_2$ in dilute CF_3SO_3H solution. It is a powerful reducing agent but thermally stable at 25°. The Mo^{3+} (aq) ion is also extremely reactive toward oxygen, and great care is required to prepare a pure solution. This is best done by dissolving an $MoCl_6^{3-}$ salt in CF_3SO_3H and separating the Mo^{3+} (aq) ion from Cl^- on a cation-exchange column. Its magnetic moment (3.69 BM) and absorption spectrum support its formulation as octahedral $[Mo(H_2O)_6]^{3+}$.

[32b] K. Ijima and S. Shibata, *Bull. Chem. Soc. Jap.,* 1974, **47,** 1393.

* Reference 1 is a thorough review of these topics to the end of 1974 for molybdenum. Only later work is referenced here.

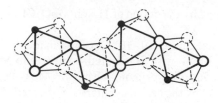

Fig. 22-C-10. A portion of an infinite chain of $MoCl_5O$ octahedra that occurs in monoclinic $MoOCl_3$. Solid and broken circles represent, respectively, upper and lower chlorine atoms; dots are oxygen atoms.

The "aqua ion" of Mo^{4+} is a partly hydrolyzed, dinuclear species. According to ion-exchange behavior, the dark red cation formed on heating equimolar amounts of Mo^{III} and Mo^V in $1M$ aqueous p-$CH_3C_6H_4SO_3H$ has a charge of $+4$, and titration with permanganate shows that it contains Mo^{IV}. It is completely stable in this noncomplexing acid medium and also in CF_3COOH and CF_3SO_3H, but it slowly attacks $HClO_4$. The most reasonable formula for it is $[Mo_2O_2]^{2+}$ (aq) which, in more detail, is probably 22-C-II.[33a]

$$\left[\begin{array}{c} \overset{\displaystyle H_2O \qquad OH_2}{\underset{\displaystyle OH_2 \quad OH_2}{H_2O \diagdown \overset{|}{\underset{Mo}{}} \diagdown \overset{O}{\diagup} \diagdown \overset{|}{\underset{Mo}{}} \diagdown OH_2}} \\ H_2O \diagup \overset{|}{\underset{OH_2}{Mo}} \overset{O}{\diagup} \overset{|}{\underset{OH_2}{Mo}} \diagdown OH_2 \end{array} \right]^{4+}$$

(22-C-II)

The aqua ion of Mo^V is binuclear with a charge of $+2$, and a plausible structure in agreement with these data is 22-C-III. Kinetic studies[33b] show that oxidation to Mo^{VI} generally requires bridge cleavage as a first step.

$$\left[\begin{array}{c} \overset{\displaystyle O \qquad\quad O}{H_2O \diagdown \overset{\|}{\underset{Mo}{}} \diagdown \overset{O}{\diagup} \diagdown \overset{\|}{\underset{Mo}{}} \diagdown OH_2} \\ H_2O \diagup \overset{|}{\underset{OH_2}{Mo}} \overset{O}{\diagup} \overset{|}{\underset{OH_2}{Mo}} \diagdown OH_2 \end{array} \right]^{2+}$$

(22-C-III)

No aqua ions of tungsten in any oxidation state have ever been reported.

The interrelationships in the chemistry of Mo^{III}, Mo^{IV}, and Mo^V in essentially noncomplexing aqueous solution are quite complicated. Electrochemical and spectroscopic studies have given the results[34] shown in Fig. 22-C-11.

M^{IV} Oxo Complexes. Although a few of these with Mo^{IV} were apparently made many years ago, they were not recognized as such. Very recently has it become clear that there are two main series, both trinuclear. By addition of oxalate ions to a so-

33a M. Ardon, A. Bino, and G. Yahav, *J. Am. Chem. Soc.*, 1976, **98**, 2338.
33b A. G. Sykes *et al., Inorg. Chem.*, 1977, **16**, 1377.
34   P. Chalilpoyil and F. C. Anson, *Inorg. Chem.*, 1978, **17**, 2418.

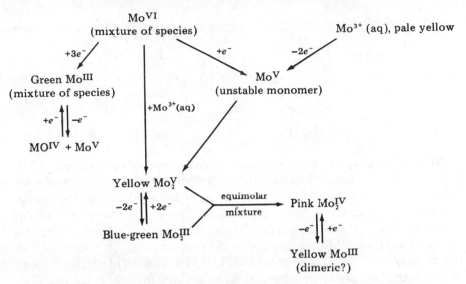

Fig. 22-C-11. Relationship of the uncomplexed forms of Mo^{III}, Mo^{IV}, Mo^{V} and Mo^{VI} in aqueous CF_3SO_3H. Adapted from Ref. 34.

lution of the dinuclear aqua ion $[Mo_2O_2]^{4+}$, one obtains the trinuclear species[35] shown in Fig. 22-C-12a. It is not clear how or why the conversion from the di- to trinuclear structure occurs. A tungsten complex of very similar structure, $[W_3O_4F_9]^{5-}$, in which the positions occupied by the oxalate and H_2O oxygen atoms are taken by F atoms, is also known.[36] The $[Mo_3S(S_2)_6]^{2-}$ ion[37] is basically similar, as shown in Fig. 22-C-12b, and all these structures are reminiscent of the Mo_3O_{13} groupings to be found in a series of mixed metal oxides[38] of the type $M_2^{II}Mo_3O_8$ (Fig. 22-C-12c). The M—M distances in all cases imply the existence of a set of three M—M single bonds in the M_3 triangles.

The second structural type,[39] in which there is again a set of M—M single bonds, is obtained by treating $M(CO)_6$ with a mixture of a carboxylic acid and its anhydride $RCOOH/(RCO)_2O$, pouring the reaction mixture on a cation-exchange column and eluting with an acid such as CF_3SO_3H or HBF_4. The products contain trinuclear cations of the general type $[M_3O_2(RCO_2)_6(H_2O)_3]^{2+}$, one of which is illustrated in Fig. 22-C-12d. The central arrangement here consists of an M_3O_2 trigonal bipyramid with axial O atoms.

In addition to these trinuclear M^{IV} oxo species, there are also some mononuclear octahedral ones such as 22-C-IV and 22-C-V.[40]

35 A. Bino, F. A. Cotton, and Z. Dori, *J. Am. Chem. Soc.*, 1978, **100**, 5252.
36 K. Mennemann and R. Mattes, *Z. Anorg. Allg. Chem.*, 1977, **437**, 175.
37 A. Müller *et al.*, *Angew. Chem., Int. Ed.*, 1978, **17**, 535.
38 W. H. McCarroll, *Inorg. Chem.*, 1977, **16**, 3351.
39 A. Bino, F. A. Cotton, Z. Dori, *et al.*, *Inorg. Chem.*, 1978, **17**, 3245.
40 R. E. Desimone and M. D. Glick, *Inorg. Chem.*, 1978, **17**, 3574; M. R. Churchill and F. J. Rotella, *Inorg. Chem.*, 1978, **17**, 668.

(22-C-IV)   (22-C-V)

M^V Oxo Species.   An extensive range of Mo^V compounds can be obtained by reduction of molybdates or MoO_3 in acid solution either chemically (e.g., by shaking with mercury) or electrolytically. The nature of the resultant species depends critically on the anions present and on conditions of pH and concentration. Probably the most important species, and one that is often used as a source material for preparation of other Mo^V compounds, is the emerald green ion $[MoOCl_5]^{2-}$ or the closely related $[MoOCl_4]^-$ and $[MoOCl_4(H_2O)]^-$ ions, all of which are readily interconverted by varying (or removing) the ligand trans to the Mo=O bond. There are also bromo and iodo analogues[41a]. These ions can be obtained by reduction of Mo^{VI} in aqueous HX, by oxidation of $Mo_2(O_2CCH_3)_4$ in aqueous HX, or by dissolving $MoCl_5$ in aqueous acid. $MoCl_5$ will also react with many organic compounds or solvents (e.g., Me_2SO or Ph_3PO) to abstract oxygen and form oxo-molybdenum(V) complexes such as $MoOCl_3(OSMe_2)_2$.

When WO_4^{2-} is reduced in $12M$ HCl, the blue $[WOCl_5]^{2-}$ ion is obtained. There are also various neutral complexes of the types $WOCl_3L_2$ and $WOCl_3(LL)$.[41b]

Although the mononuclear complexes just mentioned are important, the oxo chemistry of Mo^V is dominated by dinuclear complexes. Curiously, similar dinuclear oxo complexes of W^V are not in general known. These dinuclear oxo complexes of Mo^V are of two main types: singly bridged ones that exist in cis and trans rotamers 22-C-VIa and 22-C-VIb, and doubly bridged ones 22-C-VIc, which are cis. In most cases some or all of the ligands (not shown in these sketches) are chelating. Examples of type 22-C-VIa are $Mo_2O_3(S_2COEt)_4$ and $Mo_2O_3(S_2CNPr_2)_4$; complexes of type 22-C-VIb are $Mo_2O_3[S_2P(OEt)_2]_4$ and $Mo_2O_3(LL)_4$ in which LL represents o-thiopyridine.[42] The doubly bridged structure 22-C-VIc is found in $[Mo_2O_4(C_2O_4)_2(H_2O)_2]^{2-}$, for example, where the water molecules are only weakly bound trans to the Mo=O bonds. In all mono- and dinuclear oxo complexes of Mo^V ligands trans to Mo=O bonds are rather weakly bonded and are often entirely absent.

(22-C-VIa)   (22-C-VIb)   (22-C-VIc)

[41a]   A. Bino and F. A. Cotton, *Inorg. Chem.*, 1979, **18**, 2710.
[41b]   W. Levason, C. A. McAuliffe, and F. P. McCullough, Jr., *Inorg. Chem.*, 1977, **16**, 2911.
[42]   F. A. Cotton, P. E. Fanwick, and J. W. Fitch, III, *Inorg. Chem.*, 1978, **17**, 3254.

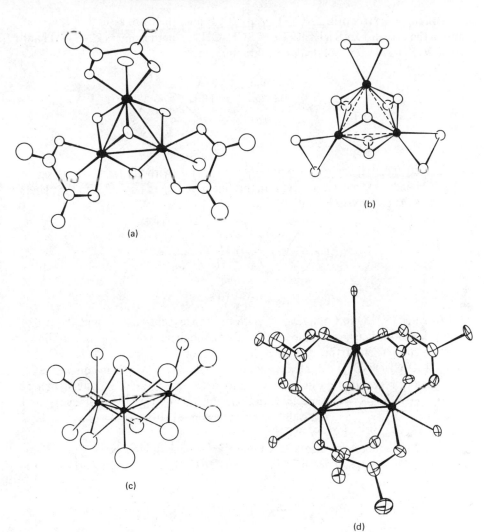

Fig. 22-C-12. Trinuclear M_3^{IV} complexes. (a) The $[Mo_3O_4(C_2O_4)_3(H_2O)_3]^{2-}$ ion. (b) The $[Mo_3S(S_2)_6]^{2-}$ ion. (c) The Mo_3O_{13} fragment in $Zn_2Mo_3O_8$ and similar mixed metal oxides. (d) $[W_3O_2(O_2CCH_3)_6(H_2O)_3]^{2+}$. Solid circles represent metal atoms.

There are many thio analogues to the doubly bridged oxo complexes. For example, the entire series[43] $Mo_2X_2(\mu\text{-}X)_2(LL)_2$, where LL represents S_2CNEt_2, is known, viz., $Mo_2O_2(\mu\text{-}O)_2(LL)_2$, $Mo_2O_2(\mu\text{-}O)(\mu\text{-}S)(LL)_2$, ..., $Mo_2OS(\mu\text{-}S)_2(LL)_2$, $MoS_2(\mu\text{-}S)_2(LL)_2$. The sulfur atoms first replace bridging oxygen atoms and then terminal ones, and there are no ligands trans to Mo=O or Mo=S bonds in any of these molecules.

43   F. A. Schultz et al., *Inorg. Chem.*, 1978, **17**, 1758; J. T. Huncke and J. H. Enemark, *Inorg. Chem.*, 1978, **17**, 3698.

Although cis structure 22-C-VIc is usually found, the trans structure can exist, and in the case of $Mo_2S_4(LL)_2$ ($LL = SCH_2CH_2S$), both isomers 22-C-VIIa and 22-C-VIIb have been isolated and characterized.[44]

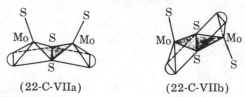

(22-C-VIIa)   (22-C-VIIb)

The tetraphenylporphyrin complex[45] 22-C-VIII, with its trans, trans set of Mo—O bonds, is structurally atypical; this linear arrangement is due to the planar nature of the porphyrin ligand.

$$O= \overset{N\frown N}{\underset{N\_N}{\Huge(}} Mo \underset{}{\Huge)} -O- \overset{N\frown N}{\underset{N\_N}{\Huge(}} Mo \underset{}{\Huge)} =O$$

(21-C-VIII)

Mo^{VI} and W^{VI} Oxo Complexes. With but few exceptions, such as the trioxo complex $MoO_3(dien)$, mentioned earlier, the mononuclear ones are six-coordinate MoO_2^{2+} species, with the oxygen atoms cis. Examples are the $[MoO_2Cl_4]^{2-}$ and $[MoO_2Cl_2(H_2O)_2]$ complexes obtained by dissolving MoO_3 in aqueous HCl, the former predominating in $12M$ and the latter in $6M$ acid. Adducts such as $MoO_2Cl_2(OPPh_3)_2$[46a] and species with chelating ligands, e.g., $[MoO_2(acac)_2]$ and $[MoO_2(S_2CNEt_2)_2]$, are also well known. Tungsten(VI) forms a few oxo complexes such as WOF_5^-, $WO_2F_4^{2-}$, $WO_2Cl_2^{2-}$, and $WO_3F_3^{3-}$.

With Mo^{VI} there are also a few binuclear species having both Mo=O and Mo—O—Mo groups. The anion in $K_2[Mo_2O_5(C_2O_4)_2(H_2O)_2]$ has the centro-

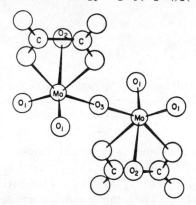

Fig. 22-C-13. The structure of the anion in $K_2[Mo_2O_5(C_2O_4)_2(H_2O)_2]$.

44   G. Bunzey and J. H. Enemark, *Inorg. Chem.*, 1978, **17**, 682.
45   J. F. Johnson and W. R. Scheidt, *Inorg. Chem.*, 1978, **17**, 1280.
46a R. J. Butcher *et al.*, *J.C.S. Dalton*, **1979**, 668.

symmetric structure in Fig. 22-C-13; the bridge group Mo—O—Mo is linear and symmetrical.

There is an interesting tetranuclear W^V–W^{VI} complex $[W_4O_8Cl_8(H_2O)_4]^{2-}$, with the symmetrical structure shown in 21-C-IX; the two "d" electrons are fully delocalized over all four tungsten atoms.[46b]

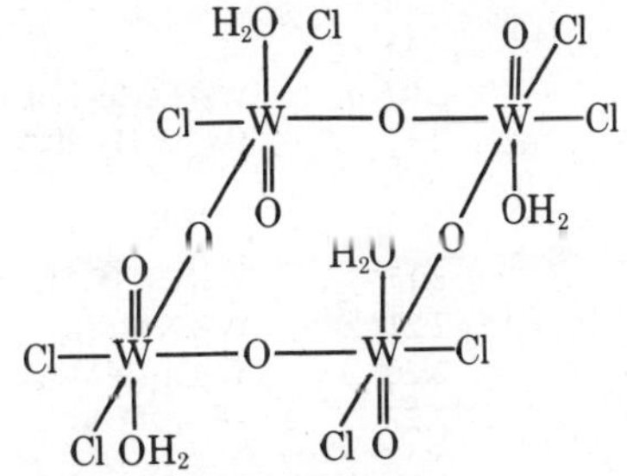

(22-C-IX)

There is evidence that ligand-exchange reactions are faster for Mo^{VI} than for W^{VI}, and it has been suggested that this may contribute to the preference for Mo in living systems (Chapter 31).[47]

Both molybdenum and tungsten form peroxo complexes, some of which are specific stoichiometric reagents for epoxidation of olefins (Section 30-15). Structures of two Mo complexes of this type,[48] containing Mo^{VI}, are 22-C-X and 22-C-XI. Both have pentagonal bipyramidal coordination. The O—O distances in the peroxo ligands are 1.45 to 1.47 Å.

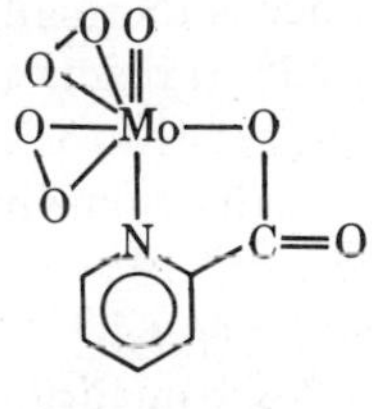

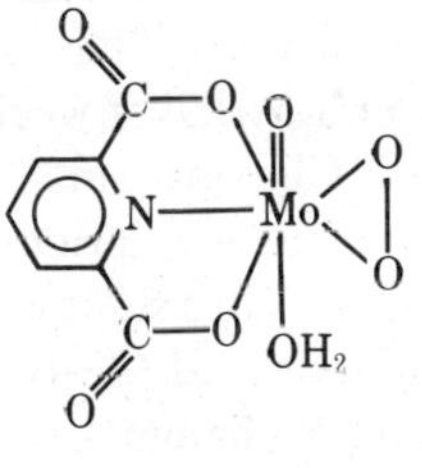

(22-C-X)   (22-C-XI)

22-C-7. Compounds with M—M Triple and Quadruple Bonds

The existence of compounds with triple and quadruple metal-metal bonds was recognized only recently (*ca.* 1964), but already an extensive chemistry has been developed. A general treatment of compounds of all elements containing metal-metal bonds, including discussion of the electronic structures, is given in Chapter 26. On present knowledge molybdenum must be considered the most prolific former

[46b] Y. Jeannin *et al.*, *Inorg. Chem.*, 1978, **17**, 374.
[47]   K. F. Miller and R. A. D. Wentworth, *Inorg. Chem.*, 1978, **17**, 2769.
[48]   S. E. Jacobson, R. Tang, and F. Mares, *Inorg. Chem.*, 1978, **17**, 3055.

TABLE 22-C-3

Some Examples of Triple and Quadruple Mo—Mo and W—W Bonds and their M—M Distances (Å)

Quadruple bonds			
$Mo_2(O_2CCH_3)_4$	2.09		
$[Mo_2Cl_8]^{4-}$	2.14		
$Mo_2(S_2CCH_3)_4$	2.14		
$Mo_2Cl_2(PEt_3)_2(CH_3OH)_2$	2.14		
$Mo_2(6\text{-Me-2-oxopyridine})_4$	2.07	$W_2(6\text{-Me-2-oxopyridine})_4$	2.16
$[Mo_2(CH_3)_8]^{4-}$	2.15	$[W_2(CH_3)_8]^{4-}$	2.26
Triple bonds			
$Mo_2(CH_2SiMe_3)_6$	2.17	$W_2(CH_2SiMe_3)_6$	2.26
$Mo_2(NMe_2)_6$	2.21	$W_2(NMe_2)_6$	2.29
$Mo_2Cl_2(NMe_2)_4$	2.20	$W_2Cl_2(NMe_2)_4$	2.29
$Mo_2(OCH_2CMe_3)_6$	2.22		
$[Mo_2(HPO_4)_4]^{2-}$	2.23		
		$W_2(O_2CNMe_2)_6$	2.28

of multiple M—M bonds. Table 22-C-3 lists some representative species and their M—M bond lengths.

Quadruple Bonds.[49] These occur in a larger and more diverse group of compounds than do the triple bonds. Molybdenum is known to form several hundred such compounds, many of which are stable in the atmosphere at room temperature, whereas tungsten has so far yielded only a few. The reason for this great difference remains obscure. A convenient entry into the Mo_2^{4+} compounds is by reaction of $Mo(CO)_6$ with acetic acid in diglyme, which produces $Mo_2(O_2CCH_3)_4$ (22-C-XII) in better than 80% yield. This yellow substance is thermally stable but is partly decomposed over a period of weeks by air at 25°. It reacts at *ca.* 0° with concentrated aqueous HCl to give, in >80% yield, the red $[Mo_2Cl_8]^{4-}$ ion (22-C-XIII), which can be isolated as a variety of air-stable salts. Some important reactions of these two key species are summarized in Fig. 22-C-14.

Among the reactions of $Mo_2(O_2CCH_3)_4$ and $[Mo_2Cl_8]^{4-}$ are many simple (i.e., nonredox) ligand-exchange reactions in which the quadruply bonded Mo_2^{4+} unit remains intact. There are also reactions in which the products are mononuclear, such as $[Mo(CNR)_7]^{2+}$ where there is no oxidation, and $[MoOX_4(H_2O)]^-$ where oxidation occurs. The reaction of $[Mo_2Cl_8]^{4-}$ with sulfuric acid in presence of O_2 gives $[Mo_2(SO_4)_4]^{3-}$ in which the loss of one electron reduces the Mo—Mo bond order to 3.5, and reaction with phosphoric acid in O_2 gives $[Mo_2(HPO_4)_4]^{2-}$, which contains only a triple bond.[50a] The decrease in bond order from 4.0 to 3.5 to 3.0 in the series $[Mo_2(SO_4)_4]^{4-}$, $[Mo_2(SO_4)_4]^{3-}$, $[Mo_2(HPO_4)_4]^{2-}$ is accompanied by a steady increase in bond lengths (2.11, 2.16, 2.23 Å, respectively).

The reaction of $Mo_2(O_2CCH_3)_4$ with gaseous HCl, HBr, or HI at *ca.* 300° gives dihalides, β-MoX_2, which are different from the long known "$MoCl_2$," which (see

[49] F. A. Cotton, *Chem. Soc. Rev.*, 1975, **4**, 27; *Acc. Chem. Res.*, 1978, **11**, 225.
[50a] A. Bino and F. A. Cotton, *Angew. Chem. Int. Ed.*, 1979, **18**, 462.

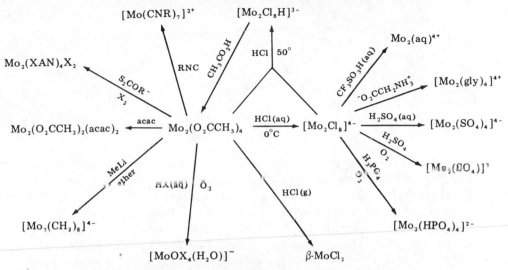

Fig. 22-C-14. Some important reactions involving Mo-Mo quadruple bonds.

Section 22-C-9) is a Mo_6 cluster compound. Though these powders have not been characterized by crystallography, their chemistry indicates that the quadruply bonded Mo_2 unit is present.[50b] The reaction of $Mo_2(O_2CCH_3)_4$ or $Mo_2Cl_8^{4-}$ with aqueous HX at 50° gives the hydrido bridged $[Mo_2X_8H]^{3-}$ ions,[51] 22-C-XIV, which have but one unpaired electron, retain a strong Mo—Mo bond, and can be reconverted to $Mo_2(O_2CCH_3)_4$ by CH_3CO_2H. These reactions are examples of oxidative addition and reductive elimination reactions involving dinuclear rather than the usual mononuclear complexes.

(22-C-XII)

(22-C-XIII)

(22-C-XIV)

One of the few stable compounds with a W—W quadruple bond is 22-C-XV, which has Mo_2 and Cr_2 analogues. All three are easily made by reaction of the ligand with $M(CO)_6$.[52]

50b H. D. Glicksman and R. A. Walton, *Inorg. Chem.*, 1978, **17**, 200.
51   A. Bino and F. A. Cotton, *J. Am. Chem. Soc.*, 1979, **101**, 4150.
52   F. A. Cotton *et al.*, *J. Am. Chem. Soc.*, 1978, **100**, 4725.

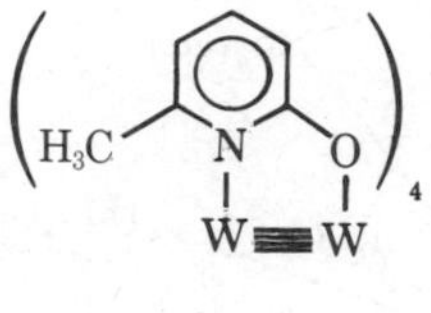

(22-C-XV) 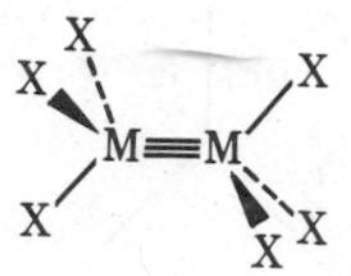   (22-C-XVI)

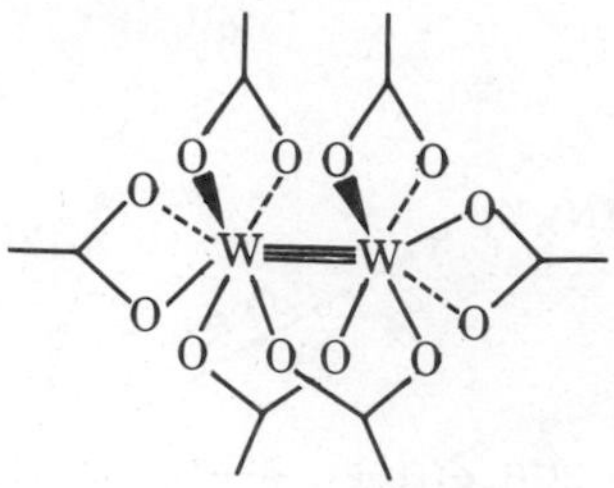

(22-C-XVII)

Triple Bonds.[53]   These are found mainly in compounds of stoichiometry M_2X_6, with staggered structures 22-C-XVI, where X represents one of the univalent groups R, NR_2, OR. It is significant that tungsten forms these triply bonded compounds as easily as does molybdenum. Some examples and their bond lengths are mentioned in Table 22-C-3. The chemistry of the $M_2(NR_2)_6$ compounds has been most thoroughly studied. They are obtained by reactions of $MoCl_5$, WCl_4, or WCl_6 with $LiNR_2$ reagents. The use of $LiNMe_2$ and WCl_6 gives a mixture of $W(NMe_2)_6$ and $W_2(NMe_2)_6$, which cocrystallize and are difficult to separate; but using an alcohol with a very bulky R group, the following process effects a separation of the mono- and dinuclear species.

$$W(NMe_2)_6 \cdot W_2(NMe_2)_6 \xrightarrow{6Bu^tOH} W_2(OBu^t)_6 + 6HNMe_2 + W(NMe_2)_6$$

Some reactions of the $M_2(NR_2)_6$ compounds are shown in Fig. 22-C-15. The conversion to $M_2Cl_2(NR_2)_4$ is important because from the chloride a number of other substituted compounds, including the bromide, iodide, and various alkyls can be obtained. The carboxylation reactions proceed differently with the Mo and W compounds, only the $W_2(NR_2)_6$ giving complete carboxylation. The product $W_2(O_2CNR_2)_6$ has a very complex structure in which the triple bond is retained, but each W atom is bonded to six oxygen atoms, as shown in 22-C-XVII.

A combined carboxylation and thermolysis reaction,[54] shown below, leads to smooth conversion of a triple bond to a quadruple bond:

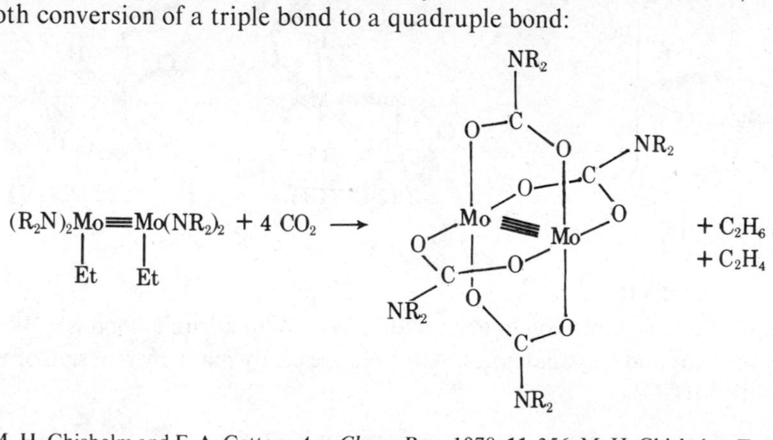

$$(R_2N)_2Mo{\equiv}Mo(NR_2)_2 + 4\ CO_2 \longrightarrow$$

53   M. H. Chisholm and F. A. Cotton, *Acc. Chem. Res.*, 1978, **11**, 356; M. H. Chisholm, *Trans. Met. Chem.*, 1978, **3**, 321.
54   M. H. Chisholm, D. A. Haitko, and C. A. Murillo, *J. Am. Chem. Soc.*, 1978, **100**, 6261.

W$_2$(O$_2$CNR$_2$)$_6$

$\uparrow$ CO$_2$

M$_2$Cl$_2$(NR$_2$)$_4$ $\xleftarrow{\text{Me}_3\text{SiCl}}$ M$_2$(NR$_2$)$_6$ $\xrightarrow{\text{ROH}}$ M$_2$(OR)$_6$ $\xrightarrow{\text{CO}}$ Mo$_2$(OR)$_6$(CO)

$\downarrow$ R'Li $\downarrow$ CS$_2$ $\downarrow$ CO$_2$ $\downarrow$ excess CO

M$_2$R'$_2$(NR$_2$)$_4$ W(S$_2$CNEt$_2$)$_3$ W$_2$(OBu^t)$_4$(O$_2$COBu^t)$_2$ Mo(CO)$_6$

Fig. 22-C-15. Some reactions of M$_2$(NR$_2$)$_6$ compounds.

22-C-8. Other Non-Oxo Complexes

In addition to the triply and quadruply bonded substances (and metal clusters to be discussed in the next section), there are numerous important complexes containing Mo and W in lower oxidation states. Most of these contain CO, phosphines, or other π-acceptor-type ligands.

Carbonyl, Tertiary Phosphine, and Related Species. The main chemistry of the carbonyls is discussed in Chapter 25. Mo(CO)$_6$ reacts with liquid chlorine at $-78°$ to give yellow, diamagnetic, apparently dinuclear, seven-coordinate [Mo-(CO)$_4$Cl$_2$]$_x$. This reacts readily with Ph$_3$P and Ph$_3$As to give, for example, Mo(Ph$_3$P)$_2$(CO)$_3$Cl$_2$. The Mo(CO)$_4$X$_2$ (X = Cl or Br) species react with an excess of an isocyanide to give diamagnetic, seven-coordinate Mo(CNR)$_5$X$_2$ species.

Mo(CO)$_6$ readily reacts with N, P, and As donors with displacement of one to four CO groups. Further reactions with Cl$_2$, Br$_2$, or I$_2$ afford a variety of six- and seven-coordinate Mo^I, Mo^{II}, and Mo^{III} complexes.

There is a wide variety of both Mo and W complexes of tertiary phosphines and other donor ligands that do not contain CO. These are made from the halides or complexes such as MoCl$_4$(EtCN)$_2$.

For the carbonyl compounds, the scheme in Fig. 22-C-16 is representative.

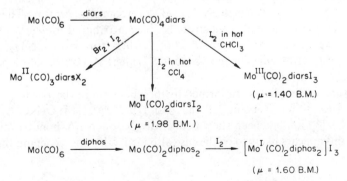

Fig. 22-C-16. Preparation of some carbonyl-containing molybdenum arsine and phosphine complexes.

The magnetic properties of most such complexes are complicated. The seven-coordinate Mo^{II} species have temperature-independent magnetic moments of 0.4 to 1.1 BM, but evidently have no unpaired electrons. The Mo^{I} and Mo^{III} species are believed to have single unpaired electrons; the Mo^{II} species presumably have two, although their magnetic moments are all in the range 1.4–2.0 BM.

In addition to the $Mo(CNR)_5X_2$ molecules noted above, $[Mo(CNR)_6I]^+$ and $[Mo(CNR)_7]^{2+}$ ions can be made reaction of $Mo(CO)_6$ with RNC and I_2 or by alkylation and reduction of $[Mo(CN)_8]$.[55a] Both types have capped trigonal prismatic geometry.[55a]

Cyano Complexes. The best known and most thoroughly studied cyano complexes are the octacyano ions $M(CN)_8^{3-}$ and $M(CN)_8^{4-}$ (M = Mo or W). The interest here has centered on their structures, which appear to vary with environment. The similar energies of dodecahedral (D_{2d}) and square-antiprismatic (D_{4d}) structures, and the attendant fluxional character for the former, have been mentioned (p. 53). In solid compounds the structures found for the $M(CN)_8^{n-}$ ions are as follows (where M indicates that both the Mo and W compounds have the stated structure):

$$\left. \begin{array}{l} Na_3M(CN)_8 \cdot 4H_2O \qquad\quad D_{4d} \\ [(n\text{-}C_4H_9)_4N]_3[Mo(CN)_8] \quad D_{2d} \end{array} \right\} M^V \text{ species}$$

$$\left. \begin{array}{l} K_4M(CN)_8 \cdot 2H_2O \qquad\quad D_{2d} \\ K_4M(CN)_8 \cdot 6H_2O \qquad\quad D_{4d} \end{array} \right\} M^{IV} \text{ species}$$

Thus the surroundings play a decisive role in stabilizing one or the other geometry. Most Raman and infrared studies of solutions have been inconclusive, though Raman spectra unequivocally favor the D_{2d} structure for $Mo(CN)_8^{4-}$ in aqueous solution. Esr studies were at first believed to favor D_{4d} symmetry or a fluxional D_{2d} structure rearranging through a D_{4d} intermediate.

The $M(CN)_8^{4-}$ ions in aqueous solution are photochemically converted, through several intermediates, into isolable species long believed to be $[M(CN)_4(OH)_4]^{4-}$ but more recently shown to be six-coordinate, *trans*-dioxo complexes $[MO_2(CN)_4]^{4-}$.

A cyano complex reported for Mo^{III} is $K_4Mo(CN)_7 \cdot 2H_2O$, which readily oxidizes to $K_4Mo(CN)_8 \cdot 2H_2O$. It has $\mu_{eff} = 1.75$ BM at room temperature. The apparent presence of only one unpaired electron has been attributed to d-orbital splitting in the necessarily low symmetry of either an $[Mo(CN)_7]^{4-}$ or an $[Mo(CN)_7(H_2O)]^{4-}$ ion. The molybdenum (II) ion, $Mo(CN)_7^{5-}$ has a pentagonal bipyramidal structure.[55b]

Thiocyanate complexes are formed by molybdenum in the III, IV, and V oxidation states, the last being of the oxo type, e.g., $[MoO(NCS)_5]^{2-}$. The $[Mo(NCS)_6]^{3-}$ ion has been shown conclusively to have *N*-bonded thiocyanate ions, and this appears likely to be the case also in all other molybdenum thiocyanato species.

[55a] S. J. Lippard *et al., Inorg. Chem.,* 1978, **17**, 2127.
[55b] M. G. B. Drew *et al., J.C.S. Dalton,* **1979,** 1213.

Dinitrogen Complexes. The chemistry of molybdenum and tungsten complexes containing N_2 as a ligand has been studied extensively (see also Section 29-17) to understand the chemistry of natural nitrogen-fixing systems, and perhaps to suggest new processes that could be developed in a practical way. The bis(dinitrogen) complexes are generally prepared by reactions of higher-valent halo complexes already containing the phosphine ligands with strong reducing agents (e.g., Na/Hg) in presence of N_2 gas. For example,[56]

$$WCl_4(PMe_2Ph)_2 + N_2 \xrightarrow[\text{Na/Hg}]{\text{excess } PMe_2Ph} W(N_2)_2(PMe_2Ph)_4$$

$$MoCl_5 + 4R_3P + N_2 \xrightarrow[\text{Na/Hg}]{\text{THF}} \text{trans } Mo(N_2)_2(R_3P)_4$$

Acylation occurs thermally and alkylation photochemically according to the following scheme,[57] where M represents Mo or W:

$$M(N_2)_2(dppe)_2 \xrightarrow{-N_2} M(N_2)(dppe)_2 \xrightarrow{+RCOCl}$$

$$ClM(dppe)_2(NNCR) \xrightarrow{+HCl} Cl_2M(dppe) \left(N\begin{smallmatrix} H \\ \\ NCR \end{smallmatrix} \right)$$

On treatment with protonic acids,[58] the bis-N_2 complexes react differently depending on the phosphine present and the metal. For both Mo and W:

$$M(N_2)_2(dppe)_2 \xrightarrow{H_2SO_4} [M(dppe)_2(HSO_4)(NNH_2)]^+ + N_2$$

For the $M(N_2)_2(PMePh_2)_4$ species ammonia is formed (*ca.* 1.9 NH_3 per W atom; *ca.* 0.7 NH_3 per Mo atom) from one N_2 ligand; as before, the other one is released as N_2. Schematically, the ammonia is thought to arise by a succession of steps such as:

$$M-N\equiv N \xrightarrow{H^+} M-N=NH \xrightarrow{H^+} M=N-NH_2 \xrightarrow{H^+} M-NHNH_2$$

$$\xrightarrow{H^+} M=NH + NH_3 \xrightarrow{H^+} M-NH_2 \xrightarrow{H^+} M^{VI} + NH_3$$

There is evidence for some of the postulated intermediates in this scheme.

Miscellaneous. There is an extensive chemistry of the $(\eta^5\text{-}C_5H_5)M$ group, in which it combines with a collection of (usually four) ligands to form complexes such as $(\eta^5\text{-}C_5H_5)W(CO)_3Cl$ and $(\eta^5\text{-}C_5H_5)Mo(CO)_2(allyl)$.

There are a number of interesting nitrosyl complexes. $Mo(NO)_2Cl_2$ is a dark green polymer obtained by reaction of ClNO with $Mo(CO)_6$; it reacts with additional donors L to form $Mo(NO)_2Cl_2L_2$ compounds. NO reacts with Mo_2Cl_{10} in benzene to give a dark red solid, thought (but not proved) to be $MoCl_5NO$, which

[56] B. Bell, J. Chatt, and G. J. Leigh, *J.C.S. Dalton*, **1972**, 2492; T. A. George and M. E. Noble, *Inorg. Chem.*, 1978, **17**, 1678.
[57] J. Chatt *et al.*, *J.C.S. Dalton*, **1977**, 688.
[58] J. Chatt, A. J. Pearman, and R. L. Richards, *J.C.S. Dalton*, **1977**, 16, 1852, 2139.

then reacts with many other ligands to give species such as $[MoCl_4LNO]^-$, $[Mo-Cl_3(NO)L_2]$, $[MoCl_3(NO)L_2]^-$, and $[MoCl(NO)L_4]$, all having $\nu(NO)$ bands in the range 1540–1715 cm^{-1}. This chemistry is complex and not yet well understood.[59] Reactions of $M_2(OR)_6$ compounds also give nitrosyl complexes[60] such as 22-C-XVIII. These are unusual among nitrosyl complexes in having only fourteen-electron configurations, very short M—N distances, and very low (1560–1640 cm^{-1}) N—O stretching frequencies.

(22-C-XVIIIa)   (22-C-XVIIIb)

The trisacetylacetonate $Mo(acac)_3$ is a purple-brown, air-sensitive compound obtained by heating $Mo(CO)_6$ or K_3MoCl_6 with Hacac.

Molybdenum and tungsten form a series of polyhydrides such as $MoH_4(PMePh_2)_4$ and $WH_6(PMe_2Ph)_3$. In most cases the nmr spectra show equivalence of the hydrogen atoms and the molecules are fluxional; they undergo protonation with fluoroboric acid to give the ions $MoH_3(PR_3)_3^+$ and $WH_5(PR_3)_4^+$.[61a]

A few Mo^{IV} and W^{IV} complexes of note are the eight-coordinate picolinates[61b] $M(pic)_4$, obtained by reaction of $M(CO)_6$ with Hpic, and the volatile $Mo(NMe_2)_4$, which contains nearly regular tetrahedral molecules.[62] Mononuclear $Mo(NMe_2)_4$ reacts with isopropanol, however, according to eq. 22-C-7 to give the dinuclear product shown as 22-C-XIX. This diamagnetic molecule, with an Mo—Mo distance of only 2.52 Å, is believed to contain a double bond between the metal atoms.[63]

$$2Mo(NMe_2)_4 + 8Pr^iOH \longrightarrow Mo_2(OPr^i)_8 + 8HNMe_2 \qquad (22\text{-}C\text{-}7)$$

(22-C-XIX)

[59]   F. King and G. J. Leigh, *J.C.S. Dalton*, **1977**, 429.
[60]   M. H. Chisholm *et al.*, *Inorg. Chem.*, 1979, **18**, 116.
[61a] E. Carmona-Guzman and G. Wilkinson, *J.C.S. Dalton*, **1977**, 1716.
[61b] C. J. Donahue and R. D. Archer, *Inorg. Chem.*, 1977, **16**, 2903; 1978, **17**, 1677.
[62]   M. H. Chisholm, F. A. Cotton, and M. W. Extine, *Inorg. Chem.*, 1978, **17**, 1329.
[63]   M. H. Chisholm *et al.*, *Inorg. Chem.*, 1978, **17**, 2944.

Tungsten has recently provided two similar complexes that illustrate the formation of single and double metal-metal bonds in circumstances where the assignment of such bond orders is relatively unambiguous.[64] Compounds 22-C-XX and 22-C-XXI are both diamagnetic, and each contains the same planar $Et_2NCS_2W(\mu\text{-}S)_2W(S_2CNEt_2)$ group but differs in the other four ligands. In 22-C-XX the W^V atoms are separated by 2.79 Å and a W—W single bond is assigned, whereas in 22-C-XXI the W^{IV} atoms are separated by only 2.53 Å and are united by a double bond.

(22-C-XX)

(22-C-XXI)

22-C-9.   Metal Atom Cluster Compounds

A number of compounds of low-valent (mainly oxidation state II) molybdenum and tungsten contain metal atom clusters. The key structural unit is that shown in Fig. 22-C-17, consisting of an octahedron of metal atoms with a bridging atom on each triangular face; the entire unit has full O_h symmetry. Detailed dimensions are available for the $(Mo_6Cl_8)^{4+}$ and $(Mo_6Br_8)^{4+}$ units.

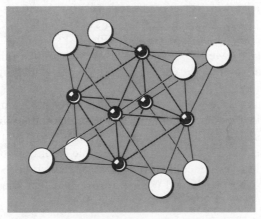

Fig. 22-C-17. The key structural unit $M_6X_8^{4+}$ found in metal atom cluster compounds of Mo^{II} and W^{II}.

[64]   A. Bino et al., Inorg. Chem., 1978, **17**, 2946.

The $(M_6X_8)^{4+}$ units have the capacity to coordinate six electron-pair donors, one to each metal atom along a 4-fold axis of the octahedron, and as far as is known, they always do so. Thus in molybdenum dichloride α-$MoCl_2$, $(Mo_6Cl_8)^{4+}$ units are connected by bridging chlorine atoms (four per unit) and there are nonbridging Cl atoms in the remaining two coordination positions. In $(Mo_6Br_8)Br_4(H_2O)_2$ the six outer positions are occupied by the four Br atoms and two water molecules. The Mo—Mo distances are 2.62 to 2.64 Å consistent with the view that the metal-metal bonds are single.

The usual synthetic routes to these cluster compounds begin with preparation of the anhydrous halides. The preparation of Mo_6Cl_{12}, a yellow, nonvolatile solid is shown in Fig. 22-C-8. The tungsten halides $W_6X_{12}(X = Cl$ or $Br)$ are obtained on disproportionation of WX_4 at 450 to 500° or by reduction of WCl_5 or WBr_5 with aluminum in an appropriate temperature gradient. W_6I_{12} is obtained by fusing W_6Cl_{12} with a tenfold excess of a KI-LiI mixture at 540°.

The mechanisms of formation of these clusters from simple (probably mononuclear[65]) starting materials are little understood, but the recent characterization of several octahedral fragments,[65,66] such as $\{[Mo_5Cl_8]Cl_5\}^{2-}$ and $\{[Mo_4I_7]I_4\}^{2-}$, has given some hope of clarifying this. By heating together $MoCl_5$, Mo, and S, Se, or Te in a sealed tube, compounds of composition $Mo_6Cl_{10}S$ (or the Se or Te analogue) are obtained. These contain $[Mo_6Cl_7S]^{4+}$ unit connected by bridging Cl atoms.[67] The μ_3-S so closely resembles the μ_3-Cl atoms that the cluster unit is randomly disordered. There are also mixed compounds having formulas $Mo_6S_6Br_2$ and $Mo_6S_6I_2$; like the mixed metal sulfides mentioned later, these are superconductors below $ca.$ 14°K.[68a]

The bridging groups in the $(M_6X_8)^{4+}$ units undergo replacement reactions only slowly, whereas the six outer ligands are labile; replacement reactions for the latter apparently occur mainly by a dissociative mechanism. It is thus possible to obtain compounds in which a wide variety of ligands occupy the outer positions, such as mixed halides, i.e., $(M_6X_8)Y_4$, a number of salts of the $[(M_6X_8)Y_6]^{2-}$ anions, and many complexes such as $[(Mo_6Cl_8)Cl_3(Ph_3P)_3]^+$, $[(Mo_6Cl_8)Cl_4(Ph_3P)_2]$, $[(Mo_6Cl_8)L_6]^{4+}$ (L = Me_2SO or Me_2NCHO), and comparable tungsten compounds. In aqueous solution the $(M_6X_8)^{4+}$ units are unstable to strongly nucleophilic groups such as OH^-, CN^-, and SH^-.

The molybdenum species show little tendency to act as reducing agents, despite the low formal oxidation number, but the tungsten compounds are fairly reactive reductants in aqueous media. It may be recalled that the $(Nb_6X_{12})^{2+}$ and $(Ta_6X_{12})^{2+}$ species (p. 841) can sustain reversible oxidation to the 3+ and 4+ species. There is little evidence that oxidation (or reduction) is possible for the $(Mo_6X_8)^{4+}$ species, but W_6Cl_{12} and W_6Br_{12} are oxidized by free halogen at elevated

[65]   K. Jödden and H. Schäfer, *Z. Anorg. Allg. Chem.,* 1977, **430**, 5.
[66]   R. E. McCarley *et al., J. Am. Chem. Soc.,* 1978, **100**, 6257.
[67]   C. Perrin *et al., J. Solid State Chem.,* 1978, **25**, 197.
[68a]   M. Sergent *et al., J. Solid State Chem.,* 1977, **22**, 87.

temperatures. In the case of W_6Br_{12} the products[68b] are W_6Br_{14}, W_6Br_{16}, and W_6Br_{18} if the temperature is kept below $150°$ (above which WBr_6 is obtained). In all these there has been a two-electron oxidation of the W_6Br_8 group. W_6Br_{14} may be formulated as $(W_6Br_8)Br_6$; the others contain bridging Br_4^{2-} units and are formulated as $(W_6Br_8)Br_4(Br_4)_{2/2}$ and $(W_6Br_8)Br_2(Br_4)_{4/2}$.

The reaction of Cl_2 with W_6Cl_{12} at $100°$ results in an intriguing structural change. The product is of stoichiometry WCl_3 and has been shown to contain the $(W_6Cl_{12})^{6+}$ unit, isostructural with the $(M_6X_{12})^{n+}$ units found characteristically in the cluster compounds of Nb and Ta (p. 842); the complete formulation of WCl_3 is $(W_6Cl_{12})Cl_6$.

There is an interesting series of mixed sulfides (which are commonly nonstoichiometric) with formulas $M_xMo_3S_4$ whose structures[69] contain Mo_6S_8 units having distorted forms of the $[Mo_6Cl_8]^{4+}$ structure shown in Fig. 22-C-17. The distortions in these clusters arise because the number of electrons available for Mo—Mo bonds is between 20 and 24 instead of the full complement of 24; this results in partial occupancy of a triply degenerate orbital of the octahedral structure which, in turn, causes a Jahn-Teller distortion. These compounds are important because they are superconductors (below about $15°K$) that have exceptional resistance to the quenching effect of magnetic fields on the superconductivity.

22-D. TECHNETIUM AND RHENIUM[1]

Technetium and rhenium are very similar chemically and differ considerably from manganese, despite similarities in the stoichiometries of a few compounds (e.g., the series MnO_4^-, TcO_4^-, ReO_4^-, and the metal carbonyls). The most stable and characteristic oxidation state for manganese is the II state, for which the chemistry is mainly that of the high-spin Mn^{2+} cation. Technetium and rhenium have little cationic chemistry, form few compounds in the II oxidation state, and have extensive chemistry in the IV and, especially, the V states. The TcO_4^- and ReO_4^- ions are much less oxidizing than MnO_4^-. A characteristic feature of Re^{III} in its halides is the formation of metal-metal bonds. Technetium also does this to some extent, whereas manganese forms no such compounds at all. Indeed rhenium shows a marked tendency to form M—M bonds in oxidation states up to at least IV. The ion $Re_2X_9^-$ has Re—Re = 2.71 Å, and in $La_4Re_6O_{19}$ where the mean oxidation number is +4.33 there are $Re(O)_2Re$ groups with the Re—Re distance 2.42 Å.

The oxidation states and stereochemistries of compounds of the elements are summarized in Table 22-D-1.

[68b] H. Schäfer and R. Siepmann, Z. Anorg. Allg. Chem., 1968, 357, 273; R. Siepmann and H.-G. von Schnering, Z. Anorg. Allg. Chem., 1968, 357, 289.

[69] J. Guillevic, O. Bars, and D. Grandjean, Acta Crystallogr., B, 1976, 32, 1338, 1342.

[1] R. Colton, The Chemistry of Rhenium and Technetium, Wiley-Interscience, 1966; R. D. Peacock, The Chemistry of Technetium and Rhenium, Elsevier, 1966; K. Schwochau, Chem. Ztg. 1978, 100, 329 (Tc review).

TABLE 22-D-1

Oxidation States and Stereochemistry of Technetium and Rhenium[a]

Oxidation state	Coordination number	Geometry	Examples
Tc^{-1}, Re^{-1}, d^8	5	?	$[Re(CO)_5]^-$
Tc^0, Re^0, d^7	6	Octahedral	$Tc_2(CO)_{10}, Re_2(CO)_{10}$
Tc^1, Re^1, d^6	6	π Complex	$\eta^5\text{-}C_5H_5Re(CO)_2C_5H_8, \eta^5\text{-}C_5H_5Re(CO)_3$
		Octahedral	$Re(CO)_5Cl, K_5[Re(CN)_6], Re(CO)_3py_2Cl, [(CH_3C_6H_4NC)_6Re]^+$
			$ReCl(N_2)(PR_3)_4$
Tc^{II}, Re^{II}, d^5	6	Octahedral	$ReCl(N_2)(PR_3)_3^+, ReH_2(NO)(PPh_3), ReCl_2(NO)(PPh_3)_2L$
			$Re(diars)_2Cl_2, Te(diars)_2Cl_2$
		Dinuclear (3° bond)	$Re_2Cl_4(RP_3)_4$
Tc^{III}, Re^{III}, d^4	5	π-Complex	$(\eta^5\text{-}C_5H_5)_2ReH, (\eta^5\text{-}C_5H_5)_2ReH_2^+$
		Tbp	$Ph_3Re(PhEt_2)_2$
	6	Octahedral	$[Te(diars)_2Cl_2]^+, ReCl_2acac(PPh_3)_2, trans\text{-}TcCl(acac)_2PPh_3,$
			$mer\text{-}ReCl_3(PR_3)_3$
		Trigonal prism	$Re(S_2C_2Ph_2)_3$
	7	Pentagonal bipyramidal	$ReH_3(diphos)_2, K_4[Re(CN)_7]\cdot 2H_2O$
		Dinuclear (4° bond)	$Re_2X_8^{2-}$
		Metal atom cluster	$Re_3X_9L_3, Re_3Cl_3(CH_2SiMe_3)_6, Re_3Me_9$
			$(Me_3SiCH_2)_4Re(N_2)Re(CH_2SiMe_4)_4$
Tc^{IV}, Re^{IV}, d^3	5	?	$K_2TcI_6, K_2ReCl_6, ReI_4py_2, TcCl_4, ReCl_4, [Re_2OCl_{10}]^{4-},$
	6^a	Octahedral	$ReCl_4diars$

	C.N.	Geometry	Examples
	7	?	$[\text{ReCOdiars}_2\text{I}_2](\text{ClO}_4)_2$
	5	Metal atom cluster	$\text{Re}_3(\text{CH}_2\text{SiMe}_3)_{12}$
$\text{Tc}^{V}, \text{Re}^{V}, d^2$	5	Tbp?	$\text{ReCl}_5(g), \text{ReF}_5, \text{NReCl}_2(\text{PPh}_3)_2$
	6^a	Sp	$[\text{ReOX}_4]^-$
		Octahedral	$\text{ReOCl}_3(\text{PPh}_3)_2, [\text{ReOCl}_5]^{2-}, \text{Re}_2\text{Cl}_{10}, \text{Tc(NCS)}_6^-$
	7	?	ReOCl_3TAS
	8	Dodecahedral(?)	$[\text{Re(diars)}_2\text{Cl}_4]^+$
		Dist. (C_s)	$\text{ReH}_5(\text{PEtPh}_2)_3$
$\text{Tc}^{VI}, \text{Re}^{VI}, d^1$	5	Sp	$\text{ReOMe}_4, \text{ReOCl}_4$
	6	Octahedral	$\text{ReO}_2, \text{ReF}_6, \text{ReMe}_6$
	7	?	ReOCl_6^{2-}
	8	Square antiprism	ReF_8^{2-}
		Dodecahedral	$[\text{ReMe}_8]^{2-}$
$\text{Tc}^{VII}, \text{Re}^{VII}, d^0$	4	Tetrahedral	$\text{ReO}_4^-, \text{TcO}_4^-, \text{ReO}_3\text{Cl}$
	5	Tbp	$cis\text{-ReO}_2\text{Me}_3$
	6	Octahedral	$\text{ReO}_3\text{Cl}_3^{2-}$
	7	Pentagonal bipyramidal	ReF_7
	9	Tricapped trigonal prism	ReH_9^{2-}

^{a} Most common states.

^{b} Some compounds in nonintegral oxidation states are also known (see text).

22-D-1. The Elements

Although its existence was predicted much earlier from the Periodic Table, rhenium was first detected, by its X-ray spectrum, only in 1925; later Noddack, Berg, and Tacke isolated about a gram of rhenium from molybdenite. Rhenium is now recovered on a fairly substantial scale from the flue dusts in the roasting of molybdenum sulfide ores and from residues in the smelting of some copper ores. The element is usually left in oxidizing solution as perrhenate ion ReO_4^-. After concentration, the perrhenate is precipitated by addition of potassium chloride as the sparingly soluble salt $KReO_4$.

All isotopes of technetium are unstable toward β decay or electron capture, and traces exist in Nature only as fragments from the spontaneous fission of uranium. The element was named technetium by the discoverers of the first radioisotope— Perrier and Segré. Three isotopes have half-lives greater than 10^5 years, but the only one that has been obtained on a macro scale is ^{99}Tc (β^-, 2.12×10^5 years). Technetium is recovered from waste fission product solutions after removal of plutonium and uranium. It is an interesting irony that the supply of technetium, which does not exist in Nature, might easily be made to exceed that of Re, which does, because of the increasing number of reactors and the very low ($\sim 10^{-9}\%$) abundance of Re in the earth's crust.

The metals resemble platinum in appearance but are usually obtained as gray powders; Re has a higher melting point ($3180°$) than any metal except W ($3400°$). Re and Tc are both obtained by thermal decomposition of NH_4MO_4 or $(NH_4)_2MCl_6$ in H_2. Technetium can also be made by electrolysis of NH_4TcO_4 in $2M$ H_2SO_4, with continuous addition of H_2O_2 to reoxidize a brown solid also produced. Rhenium can be electrodeposited from H_2SO_4 solutions, although special conditions are required to obtain coherent deposits. Both metals crystallize in an *hcp* arrangement. They burn in oxygen above $400°$ to give the oxides M_2O_7 that sublime; in moist air the metals are slowly oxidized to the oxo acids. The latter are also obtained by dissolution of the metals in concentrated nitric acid or hot concentrated sulfuric acid. The metals are insoluble in hydrofluoric or hydrochloric acid but are conveniently dissolved by warm bromine water. Rhenium, but not technetium, is soluble in hydrogen peroxide.

Technetium now has few uses. However, the TcO_4^- ion is said to be an excellent corrosion inhibitor for steels. The short-lived nuclide ^{99m}Tc, $t_{1/2} = 6$ hr, obtained by neutron capture in ^{98}Mo, is now used extensively in body-scanning diagnostic techniques. The Tc is introduced in the form of water-soluble Tc^{IV} complexes made by reduction of TcO_4^- by Sn^{II} in presence of a complexing agent such as diethylenetriaminepentaacetic acid or hydroxyethylidene phosphonic acid.[2] The biggest use for rhenium is alloyed with Pt on an alumina support as a catalyst for petroleum re-forming; the Pt-Re alloy has a longer life than Pt alone.[3]

---

[2] M. Molter, *Chem Ztg.*, 1979, **103**, 41; E. Deutch *et al.*, *J. Am. Chem. Soc.*, 1979, **101**, 4581.
[3] R. Burch, *Platinum Met. Rev.*, 1978, **22**, 57; B. D. Nicol, *J. Catal.*, 1977, **46**, 438.

TECHNETIUM AND RHENIUM COMPOUNDS[1,4]

22-D-2. Oxides and Sulfides

The known *oxides* of Re and Tc are shown in Table 22-D-2. The *heptaoxides*, obtained by burning the metals, are volatile. If acid solutions containing TcO_4^- are evaporated, the oxide is driven off, a fact that can be utilized to isolate and separate technetium; rhenium is not lost from acid solutions on evaporation (i.e., at 100°), but can be distilled from hot concentrated H_2SO_4. The heptaoxides readily dissolve in water, giving acidic solutions, and Re_2O_7 is deliquescent. The oxides differ structurally and in various physical properties. The structure of Re_2O_7 consists of an infinite array of alternating ReO_4 tetrahedra and ReO_6 octahedra sharing corners, whereas Tc_2O_7 consists of molecules in which TcO_4 tetrahedra share an oxygen atom and the Tc—O—Tc chain is linear. On evaporation of aqueous solutions of Re_2O_7 over P_2O_5 slightly yellow crystals of so-called perrhenic acid are obtained. These are actually $Re_2O_7(H_2O)_2$, which is binuclear with both tetrahedral and octahedral rhenium atoms, as in Re_2O_7 itself [i.e., O_3Re—O—$ReO_3(H_2O)_2$]. The Re—O—Re bond is essentially linear.

The lower oxides can be obtained either by thermal decomposition of NH_4MO_4 or by heating M_2O_7 + M, at 200 to 300°. The hydrated dioxides, $MO_2\cdot2H_2O$, can be obtained by addition of base to M^{IV} solutions, for instance, of $ReCl_6^{2-}$, by electrolytic reduction of ReO_4^- at Pt electrodes,[5] or for Tc, by reduction of TcO_4^- by Zn in HCl. Sodium borohydride reduction of TcO_4^- gives metal, but reduction of ReO_4^- gives mixed hydrous oxides ReO_2 and Re_2O_3.[6] The oxide Re_2O_5 has been

TABLE 22-D-2
Oxides of Rhenium and Technetium[a]

Rhenium		Technetium	
Oxide	Color	Oxide	Color
$Re_2O_3\cdot xH_2O$	Black		
ReO_2	Brown	TcO_2	Black
ReO_3	Red	$TcO_3(?)$	
Re_2O_5	Blue		
Re_2O_7	Yellow (m.p. 220°)	Tc_2O_7	Yellow (m.p. 119.5°)

[a] Lower hydrated oxides formulated as $ReO\cdot H_2O$ and $Re_2O\cdot2H_2O$, are obtained by Zn reduction of weakly acid ReO_4^- solutions; they are not fully investigated.

[4] (a) *Tc*: K. V. Kotegov et al., *Adv. Inorg. Nucl. Chem.*, 1969, **11**, 1; (b) *Re*: G. Rouschias, *Chem. Rev.*, 1974, **74**, 531; J. E. Fergusson, *Coord. Chem. Rev.*, 1966, **1**, 459; M. A. Ryashentseva and Kh. M. Minachev, *Russ. Chem. Rev.*, 1969, **38**, 944; W. H. Davenport et al., *Ind. Eng. Chem.*, 1968, **60**, 11 (catalytic properties of Re and its compounds).
[5] G. A. Mazzochini et al., *Inorg. Chim. Acta*, 1975, **13**, 209.
[6] R. A. Pacer, *J. Inorg. Nucl. Chem.*, 1976, **38**, 817.

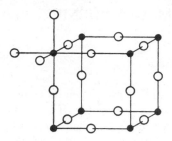

Fig. 22-D-1. The ReO_3 structure. Each metal atom lies at the center of an octahedron of oxygen atoms. This structure is closely related to the perovskite structure, since the latter is obtained from this one by insertion of a large cation into the center of the cube.

made by electrolytic reduction of perrhenate in sulfuric acid solution; it decomposes above 200°.

Both TcO_2 and ReO_2 have distorted rutile structures isotypic to that of MoO_2, so Re—Re interactions presumably exist.

Rhenium(VI) oxide (ReO_3), has a structure (Fig. 22-D-1) also found for CrO_3, WO_3, etc.,: it is usually termed the ReO_3 structure and is closely related to the perovskite structure (p. 18).

The pairs of *sulfides* TcS_2, ReS_2 and Tc_2S_7, Re_2S_7 are isomorphous. The black heptasulfides are obtained by saturation of 2 to 6M hydrochloric acid solutions of TcO_4^- or ReO_4^- with hydrogen sulfide. The precipitation is sensitive to conditions and is often incomplete. Treatment of neutral solutions of the oxo anions with thioacetamide or sodium thiosulfate followed by acidification gives a better yield. An excess of sulfur in the precipitates may be extracted with CS_2. ReS_3 is obtained by reduction of Re_2S_7 with hydrogen.

The disulfides are obtained by heating the heptasulfides with sulfur in a vacuum; they are commonly nonstoichiometric.

Rhenium sulfides are effective catalysts for hydrogenation of organic substances and they have the advantage over heterogeneous platinum metal catalysts in that they are not poisoned by sulfur compounds. An inorganic reduction that they catalyze is that of NO to N_2O at 100°.

22-D-3. Halides of Technetium and Rhenium

The halides of Tc and Re are listed in Table 22-D-3.

Technetium. Fluorination of Tc at 400° gives TcF_6 that is readily hydrolyzed to a black hydrous oxide.

The tetrachloride is obtained as paramagnetic red crystals by the action of carbon tetrachloride on Tc_2O_7 in a bomb, and is the major product on direct chlorination of the metal. $TcCl_4$ has a structure very similar to that of $ZrCl_4$ (Fig. 22-A-1) in which there are linked $TcCl_6$ octahedra. Neither the magnetic behavior nor its structure indicate metal-metal bonding, in marked contrast to $ReCl_4$ (see below); the Tc—Tc distance is 3.62 Å.

Rhenium. The heptafluoride resembles the only other heptahalide IF_7; it is

TABLE 22-D-3
The Halides of Technetium and Rhenium

	$TcCl_4$ Red-brown, subl >300		$TcCl_6$ M.p. 25°	
	ReF_4 Blue subl >300°	ReF_5 Greenish-yellow, m.p. 48°	TcF_6 Golden-yellow, m.p. 33°	ReF_7 Pale yellow
Re_3Cl_9 Dark red	$ReCl_4$ Black	$ReCl_5$ Dark red-brown, m.p. 261°	ReF_6 Pale yellow, m.p. 18.7°	
Re_3Br_9 Red-brown	$ReBr_4$ Dark red	$ReBr_5$ Dark brown	— [a]	
ReI_2 Black	Re_3I_9 Black	ReI_4 Black		

[a] Rhenium hexachloride is said to be green-black (m.p. 20°), but this is not unequivocally established; see J. Burgess et al., *J.C.S. Dalton*, **1973**, 501.

obtained by fluorination of Re at 400° under pressure. At 120° and 1 atm, ReF_6 is obtained. This is octahedral, and strong spin-orbit coupling of the unpaired electron is reflected in a very low magnetic moment and in the electronic spectrum. On partial hydrolysis[7] it gives blue $ReOF_4$ and on complete hydrolysis, hydrated ReO_2, $HReO_4$, and HF. Interaction of ReF_6 with H_2 in liquid HF at 25° gives ReF_5[8]; reduction with Re at 500° gives ReF_4.

The most important halide and a common starting material in rhenium syntheses is *rhenium pentachloride* (Re_2Cl_{10}), which is obtained on chlorination of Re at *ca.* 600° as a dark red-brown vapor that condenses to a dark red solid. It is rapidly hydrolyzed by water or base[9]:

$$3ReCl_5 + 16OH^- = 2ReO_2 \cdot 2H_2O + ReO_4^- + 15\ Cl^- + 4H_2O$$

Re_2Cl_{10} is reduced by many ligands and solvents such as Et_2O and MeCN, often to Re^{IV} complexes (see later). Some of its reactions are given in Fig. 22-D-3. In the crystal, $ReCl_6$ octahedra share an edge; the magnetic properties imply substantial exchange even though the distance (3.74 Å) is too large for Re—Re bonding. Re_2Cl_{10} thermally decomposes to Re_3Cl_9.

Rhenium tetrachloride can be prepared in several ways, e.g.,

$$2ReCl_5 + SbCl_3 \rightarrow 2ReCl_4 + SbCl_5.$$
$$3ReCl_5 + Re_3Cl_9 \rightarrow 6ReCl_4$$
$$2ReCl_5 + CCl_2{=}CCl_2 \rightarrow 2ReCl_4 + C_2Cl_6$$

Rhenium(IV) chloride has a structure consisting of zigzag chains of Re_2Cl_9 confacial octahedra where an end Cl atom is shared between two bioctahedra. The Re—Re distance (2.73 Å) indicates bonding. Although in both $ReCl_4$ and $TcCl_4$

[7] R. J. Paine, *Inorg. Chem.*, 1973, **12**, 1457.
[8] R. J. Paine and L. B. Asprey, *Inorg. Chem.*, 1975, **14**, 1111.
[9] J. Burgess et al., *J.C.S. Dalton*, **1973**, 501.

there are distorted MCl_6 octahedra, as noted above, there is no metal-metal bonding in the latter.

Bromination of rhenium at 600° gives the pentabromide, which decomposes readily to Re_3Br_9 when heated. The *tetrabromide* and *tetraiodide* can be made by careful evaporation of solutions of $HReO_4$ in an excess of HBr or HI. The tetraiodide is unstable and when heated at 350° in a sealed tube gives ReI_3. At 110° in nitrogen, ReI_2 is obtained; this is diamagnetic and is believed to be polymeric with Re-Re bonds.

22-D-4. Complex Halide Ions of the Oxidation States IV–VII

As usual, the halides act at Lewis acids. Thus ReF_6 reacts with HF(aq) to give $[ReOF_5]^-$,[10] and with KI in liquid SO_2 it is reduced to give KRe^VF_6. The rhenium(IV) fluoroanion ReF_6^{2-} is stable in aqueous solution, but the others are hydrolyzed.

Although $ReCl_6^-$ is formed as $[PCl_4][ReCl_6]$ when rhenium metal reacts with PCl_5 at 500°, the most important of all the halogeno ions are the yellow *hexachlororhenate(IV)* and yellow-green *hexachlorotechnate* ions MCl_6^{2-}. These are obtained by reduction of the oxo ions, MO_4^- in hydrochloric acid solution by KI. The salts have solubilities similar to those of K_2PtCl_6 (Section 22-H-3), those with large cations being insoluble. K_2ReCl_6 hydrolyzes in water to hydrous ReO_2 but it is stable in HCl solution.

22-D-5. Trirhenium Nonachloride and Its Derivatives.[11a]

Rhenium(III)chloride is a trimer (Re_3Cl_9). It is obtained by thermal decomposition of Re_2Cl_{10} as a nonvolatile, dark mauve, crystalline solid. Although not isomorphous with Re_3Cl_9, the bromide and iodide[11b] also have Re_3X_9 units.

The basic structure of an isolated Re_3X_9 unit is shown in Fig. 22-D-2a; In Re_3Cl_9 itself (Fig. 22-D-2b) these Re_3Cl_9 units are linked together by halogen bridges using the terminal halogens marked X′ in Fig. 22-D-2a to distinguish them from the bridging Cl atoms of the planar Re_3Cl_3 unit and the vacant site marked L. As discussed separately (Chapter 26), the Re—Re bonds in the triangle have bond order 2 (*ca.* 2.48 Å). The Re_3 units are so stable that they can persist in vapors at 600° and can be detected mass spectrometrically.

Trirhenium nonachloride is important because, like $ReCl_5$, it is a useful starting material for synthesis of other rhenium(III) complexes. Some of its reactions are shown in Fig. 22-D-3. The reactions are of the following types.

1. Addition of ligands to the rhenium atoms is possible for H_2O, tetrahydrofuran, and other ligands, giving neutral adducts $Re_3X_9L_3$. Also attachment of additional

[10]   J. H. Holloway, and J. B. Raynor, *J.C.S. Dalton,* **1975,** 737.
[11a] R. A. Walton, *Prog. Inorg. Chem.,* 1976, **21,** 105.
[11b] H. D. Glicksman and R. A. Walton, *Inorg. Chem.,* 1978, **17,** 200.

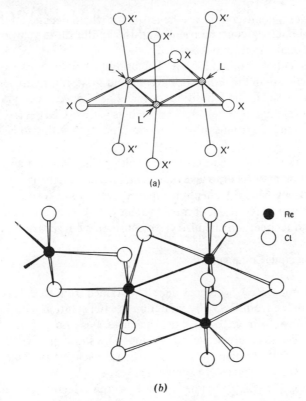

(a)

(b)

Fig. 22-D-2. (a) Sketch of the isolated Re_3X_9 unit, which has D_{3h} symmetry. The bridging halogens of the planar Re_3X_3 unit are marked X, the terminal halogen atoms X'. The positions on each Re where addition ligands may be coordinated are marked by ←L. When L is a halide ion, we obtain the anions $[(Re_3X_3)X'_{6+n}]^{n-}$ ($n = 1, 2,$ or 3 depending on the nature of the cation). (b) The structure of Re_3Cl_9 showing the linking of Re_3 units by chloride bridges.

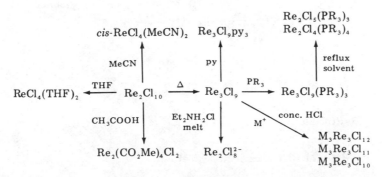

Fig. 22-D-3 Some reactions of Re_2Cl_{10} and Re_3Cl_9. Note that Re_3Cl_9 obtained from Re_2Cl_{10} is usually unreactive and is activated by exposure to air, giving $Re_3Cl_9(H_2O)_n$ or conversion to $Re_3Cl_9(THF)_3$ or $Re_3Cl_9(PPh_3)_3$.

891

halide ions gives anionic complexes the nature of which depend on the size of the cation, crystal-packing considerations, and the equilibria in solution. From one to three additional anions may be added.

2. Replacement of terminal Re—X. The bridging halides of the Re_3X_3 unit are much less reactive than the terminal X atoms, and these in turn are less labile than halogen atoms added in the L positions of Fig. 22-D-2a. Accordingly it is possible to obtain via exchange reactions species such as $Cs[Re_3Cl_3Br_7(H_2O)_2]$ with the planar Re_3Cl_3 core, Br atoms in X', and one L position with two H_2O molecules in L positions.

3. With certain ligands, notably tertiary phosphines, under more forcing conditions, reduction may occur to give reduced trimeric clusters or more commonly, cleavage reactions whereby rhenium dimeric species are formed. These dimer species, which can be in III, II, or mixed valency states, are an important feature of rhenium and technetium chemistry; we consider them separately.

22-D-6. Dirhenium Compounds[12]

The dinuclear anion, $Re_2Cl_8^{2-}$ is conveniently obtained from Re_3Cl_9 (Fig. 22-D-3); it can also be made from ReO_4^- by reduction in HCl solution with H_2 or H_3PO_2. This ion, like Re_2Cl_{10} and Re_3Cl_9, holds a central position in rhenium chemistry (Fig. 22-D-4). The structure of the $Re_2X_8^{2-}$ ions[13a] is shown in Fig. 22-D-5; the ions have an eclipsed conformation and a Re—Re quadruple bond (Chapter 26). Similar $Re_2Br_8^{2-}$, and $Re_2I_8^{2-}$,[13b] ions are known.

Curiously, the $[Tc_2Cl_8]^{2-}$ ion has never been reproducibly characterized, but

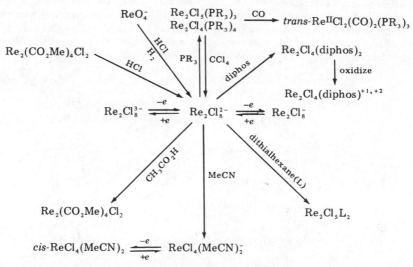

Fig. 22-D-4. Some reactions of dirhenium species in II, III and mixed oxidation states.

[12] For references, see R. A. Walton *et al.*, *Inorg. Chem.*, 1978, **17**, 3203, 3197, 2674, 2383; 1976, **15**, 1630; 1975, **14**, 1987.
[13a] F. A. Cotton and W. T. Hall, *Inorg. Chem.*, 1977, **16**, 1867.
[13b] W. Preetz and L. Rudzik; *Angew. Chem., Int. Ed.*, 1979, **18**, 150.

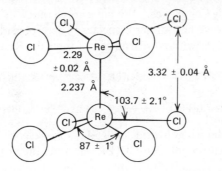

Fig. 22-D-5. The structure of the $Re_2Cl_8^{2-}$ ion in $CsRe_2Cl_8 \cdot H_2O$.

the $[Tc_2Cl_8]^{3-}$ ion, which has been thoroughly studied is known. The only compound presently known that contains a quadruple bond between two Tc atoms is
$Tc_2(O_2CCMe_3)_4Cl_2.$[14]

As shown for $Re_2Cl_8^{2-}$ in Fig. 22-D-4, the $Re_2X_8^{2-}$ ions can be reduced without
cleavage of the Re—Re bond either electrolytically or chemically.

There are a number of other rhenium(III) species that have metal-metal bonds;
examples are given in Fig. 22-D-6.

Fig. 22-D-6. Complexes of rhenium(III) with quadruple Re-Re bonds.

14   F. A. Cotton and L. D. Gage, *Nouv. J. Chim.*, 1977, **1**, 441.

Tertiary Phosphine Complexes.   We have noted that $Re_2Cl_8{}^{2-}$ reacts with tertiary phosphines to give new dimeric species. These phosphine species also can undergo redox reactions without disruption of the Re—Re bond,[15] e.g.,

$$Re_2^{II}X_4(PR_3)_4 \underset{}{\overset{-e}{\rightleftharpoons}} [Re_2X_4(PR_3)_4]^+ \underset{}{\overset{-e}{\rightleftharpoons}} [Re_2^{III}X_4(PR_3)_4]^{2+}$$

$$Re_2Cl_4(PR_3)_4 \xrightarrow[CCl_4]{O_2/HCl} Re_2Cl_5(PR_3)_3, Re_2Cl_6(PR_3)_2$$

The interaction of sodium borohydride on $Re_2Cl_8^{2-}$ in presence of tertiary phosphines can also give mixtures of $Re_2Cl_4(PR_3)_4$ and the rhenium(IV) hydride $Re_2H_8(PR_3)_4$.[16]

Several alkyl compounds with Re—Re bonds are also known (see below) and also the Re^{III} thiocyanate $(Bu_4N)_2[Re_2(NCS)_8]$[17]; the latter is cleaved by diphosphines to give $[Re^{III}(NCS)_4(diphos)]^-$.

OXO COMPOUNDS

For rhenium particularly, oxo compounds are predominant and of key importance in the higher oxidation states, especially V and VII. It is convenient to discuss all the oxo compounds and complexes together rather than under the separate oxidation numbers.

22-D-7. Simple Oxo Anions

The MO_4^- Ions.   The pertechnetates and perrhenates are among the most important compounds formed by these elements. The aqueous acids or their salts are formed on oxidation of all technetium or rhenium compounds by nitric acid, hydrogen peroxides, or other strong oxidizing agents. Pure perrhenic acid has not been isolated, but a red crystalline product, claimed to be $HTcO_4$, has been obtained; both acids are strong acids in aqueous solution. The solubilities of alkali perrhenates[18] generally resemble those of the perchlorates, but pertechnetates are more soluble in water than either (cf. $KReO_4$ 9.8 g l^{-1}, $KTcO_4$ 126 g l^{-1} at 20°). Highly insoluble precipitates, suitable for gravimetric determination, are given by tetraphenylarsonium chloride and nitron with both anions.

The tetrahedral TcO_4^- and ReO_4^- ions are quite stable in alkaline solution, unlike MnO_4^-. They are also much weaker oxidizing agents than MnO_4^-, but they are reduced by HCl, HBr, or HI. In acid solutions the ions can be extracted into various organic solvents such as tributyl phosphate, and cyclic amines extract them from basic solution. Such extraction methods of purification suffer from difficulties because of reduction of the ions by organic material. The anions can be readily

[15] R. A. Walton et al., J. Am. Chem. Soc., 1978, **100**, 4424, 991.
[16] P. Brant and R. A. Walton, Inorg. Chem., 1978, **100**, 2674.
[17] T. Nimry and R. A. Walton, Inorg. Chem., 1977, **16**, 2829.
[18] For AgReO4, see A. A. Woolf, J. Less Common Met., 1978, **61**, 151.

absorbed by anion-exchange resins, from which they can be eluted by perchloric acid. The ReO_4^- ion can function as a ligand that coordinates more strongly than ClO_4^- or BF_4^- but less strongly than Cl^- or Br^-.

When $Ba(ReO_4)_2$ and $BaCO_3$ are heated in stoichiometric proportions and for optimal temperatures and times, $Ba_3(ReO_5)_2$ and $Ba_5(ReO_6)_2$, the so-called meso- and orthoperrhenates, respectively, are obtained. Their structures are uncertain though spectroscopic data suggest discrete ReO_6^{5-} octahedra in the latter.

Mixed oxides such as $MgReO_4$ and Cr_2ReO_6 can be obtained at high temperature and pressure.[19]

Reduction of the TcO_4^- in aqueous solution containing various anions polarographically shows a number of reduction steps, but the nature of the species present is obscure. Potential data[20] for Re and Tc are also of very limited value. Reduction in presence of ligands gives oxo Re^V or Re^{IV} complexes (see later) and reduction in HCl solution can give $ReCl_6^{2-}$ or $Re_2Cl_8^{2-}$ ions as discussed earlier. Cathodic reduction of Me_4N^+ salts in acetonitrile gives the violet TcO_4^{2-} and olive ReO_4^{2-} ions that are paramagnetic (d^1) and very air-sensitive.[21a] In $0.1M$ aqueous $NaOH$[21b] the TcO_4^{2-} ion has a lifetime of about 10 msec; the TcO_4^{2-}/TcO_4^- couple has an estimated potential of -0.61 V.

22-D-8. Oxohalides and Their Complexes

Oxohalides. These are listed in Table 22-D-4.

The *pertechnetyl or perrhenyl fluorides* (MO_3F) can be made by action of liquid HF on NH_4TcO_4 or $KReO_4$,[22] The chloride ReO_3Cl is best made by the reaction[23]

$$ReCl_5 + 3Cl_2O \underset{250°}{\overset{CCl_4}{\rightleftharpoons}} ReO_3Cl + 5Cl_2$$

Tetrachlorooxorhenium ($ReOCl_4$) is made by action of SO_2Cl_2 on Re at 350° or by interaction of O_2 with $ReCl_5$ at *ca.* 300°; it is square pyramidal but weakly associated by long intermolecular Re⋯Cl bonds.[24]

Oxohalogeno Anions. The most common of these are MOX_4^- and MOX_5^{2-}, but others such as $ReO_3Cl_3^{2-}$ are known.

Salts of $ReOCl_5^{2-}$ are best prepared by addition of large cations to $ReCl_5$ in concentrated HCl. In such solutions the main species[25] is probably $[ReOCl_4(H_2O)]^-$.

Addition of CsCl to $ReOCl_4$ in $SOCl_2$ gives $Cs[ReOCl_5]$.

Rhenium alone forms the oxo-bridged species $[Cl_5Re—O—ReCl_5]^{4-}$, in which

19 A. W. Sleight, *Inorg. Chem.,* 1975, **14,** 597; S. Kemmler-Sack *et al., Z. Anorg. Allg. Chem.,* 1978, **444,** 190.
20 R. H. Busey *et al., J. Phys. Chem.,* 1969, **73,** 1039.
21a L. Astheimer and K. Schwochau, *J. Inorg. Nucl. Chem.,* 1976, **38,** 1131.
21b E. Deutsch *et al., J.C.S. Chem. Comm.,* **1978,** 1038.
22 J. Binenboyin *et al., Inorg. Chem.,* 1974, **13,** 319.
23 K. Dehnicke and W. Liese, *Chem. Ber.,* 1977, **110,** 3959.
24 A. J. Edwards, *J.C.S. Dalton,* **1972,** 582.
25 M. Pavlova *et al., J. Inorg. Nucl. Chem.,* 1974, **36,** 1623, 3845.

TABLE 22-D-4

Oxohalides of Technetium and Rhenium[a]

V	VI	VII	
	$(TcOF_4)_n$[b]	TcO_3F	
	Blue	Yellow, m.p. 18.3°	
$TcOCl_3$	$TcOCl_4$	TcO_3Cl	
Brown, subl ∼500°	Purple, m.p. ∼35°	Colorless liquid, b.p. ∼25°	
$TcOBr_3$			
Brown			
$ReOF_3$	$(ReOF_4)_n$[b]	ReO_3F	$ReOF_5$[c]
Black, nonvolatile	Blue, m.p. 107.8°	Yellow, m.p. 147°	Cream, m.p. 34.5°
$ReOCl_3(?)$[a]	$ReOCl_4$,	ReO_3Cl	ReO_2F_3[d]
	Green-brown, m.p. 30°	Colorless liquid, b.p. 131°	Pale yellow, m.p. 90°
$ReOBr_3(?)$	$ReOBr_4$	ReO_3Br	
	Blue, dec. >80°	Colorless, m.p. 39.5°	

[a] Other, more complex compounds are known (e.g., $Re_2O_4Cl_5$, an adduct of ReO_3Cl, and $ReOCl_4$, with one oxygen of the former giving an Re—O—Re bridge) (A. J. Edwards, *J.C.S. Dalton*, **1976**, 2419).

[b] In gas phase MOF_4 is C_{3v} but in crystal, there are octahedra linked by *cis*-fluoride bridges (R. T. Paine and R. S. McDowell, *Inorg. Chem.*, 1974, **13**, 2366).

[c] J. Burgess *et al.*, *J.C.S. Dalton*, **1977**, 2149.

[d] *Tbp* with equatorial O (I. R. Beattie *et al.*, *J.C.S. Dalton*, **1977**, 1481).

there is a linear ReORe group. The diamagnetism of this anion is attributable to π interactions through the bridging oxygen atom.

Oxohalide Complexes. The oxohalides may form adducts with various ligands directly (e.g., $ReOCl_4OPCl_3$[26]), but the most important class of compounds are those of oxorhenium(V) with phosphine ligands of stoichiometry $ReOX_3(PR_3)_2$ (X = Cl or Br). These are readily obtained by interaction of ReO_4^- with PR_3 in ethanolic HCl. There is a striking difference between technetium and rhenium, since under similar conditions, the complexes *trans*-$TcCl_4(PR_3)_2$ and *mer*-$TcCl_3(PR_3)_3$ are obtained from TcO_4^-, whereas a very long reaction time is required to convert $ReOCl_3(PPh_3)_2$ to $ReCl_3(PPh_3)_3$.[26,27]

Some of the more important reactions of $ReOCl_3(PPh_3)_2$ are shown in Fig. 22-D-7. The halide ion (or other ligand) opposite to the Re=O bond is labile; in ethanol, for example, it is rapidly replaced, giving the compound $ReOX_2(OEt)$-$PR_3)_2$. Two of the three possible isomers of $ReOX_3(PR_3)_2$ are known, the trans phosphine isomer being green, the *fac* one blue.[28]

Technetium(V) forms oxo complexes such as $[TcOCl_4]^-$, and related ones with sulfur ligands[29] are important in the use of ^{99m}Tc in radiopharmaceutical applications as noted on page 886. An example is 22-D-I.

[26]   A. H. Al Mowli and A. L. Porte, *J.C.S. Dalton*, **1976**, 50.

[27]   U. Mazzi *et al.*, *J. Inorg. Nucl. Chem.*, 1976, **38**, 721.

[28]   J. E. Fergusson and P. F. Heveldt, *J. Inorg. Nucl. Chem.*, 1976, **38**, 2231.

[29]   A. Davison *et al.*, *J. Am. Chem. Soc.*, 1978, **100**, 5570; J. E. Smith, F. A. Cotton, *et al.*, *ibid.*, 1978, **100**, 5571; J. E. Smith, *et al.*, *Inorg. Chem.*, 1979, **18**, 1832.

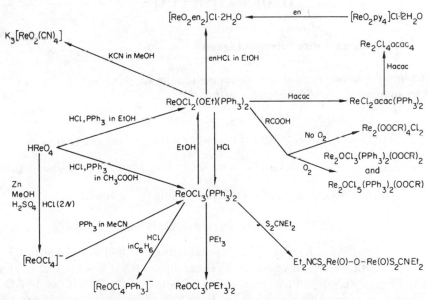

(22-D-I)

Complexes with ReO₂ and Re₂O₃ Groups. Rhenium forms numerous complexes that have a *trans*-dioxo unit.[30] Some of the syntheses are shown in Fig. 22-D-7. The compounds $ReO(OEt)I_2L_2$ (L = py or PPh_3) can also be converted to dioxo species by the reactions

$$ReO(OEt)I_2L_2 \xrightarrow{H_2O} ReO_2I_2L_2 \xrightarrow{CNS^-} ReO_2(SCN)_2L_2$$

and with moist pyridine, $ReOCl_3(PPh_3)_2$, $ReO(OEt)Cl_2(PPh_3)_2$, or $ReOCl_4^-$ react to give the orange salt *trans*-$[ReO_2py_4]Cl \cdot 2H_2O$, probably via the sequence[31] shown in Fig. 22-D-8, where the intermediate dimer has been isolated.

The O=Re-O-Re=O complexes may be *linear* as in $Re_2O_3(S_2NCEt_2)_4$ or *bent* with either *cis-* or *trans*-Re=O groups as in $Re_2O_3(CH_2SiMe_3)_6$[32]; the criteria for the formation of linear vs. bent bridges are not established.

Fig. 22-D-7. Some reactions of $ReOCl_3(PPh_3)_3$. Related compounds (i.e., with various PR_3, AsR_3, or SbR_3 groups in place of PPh_3) are known and have similar though not always identical reactions.

30 See M. C. Chakravorti *et al., Inorg. Chim. Acta,* 1976, **19,** 249; *J. Inorg. Nucl. Chem.,* 1975, **37,** 1991; D. L. Toppen and R. K. Murmann, *Inorg. Chem.,* 1973, **12,** 1615.

31 C. J. L. Lock and G. Turner, *Can J. Chem.,* 1978, **56,** 179; 1977, **55,** 333.

32 K. Mertis *et al., J.C.S. Dalton,* **1975,** 607.

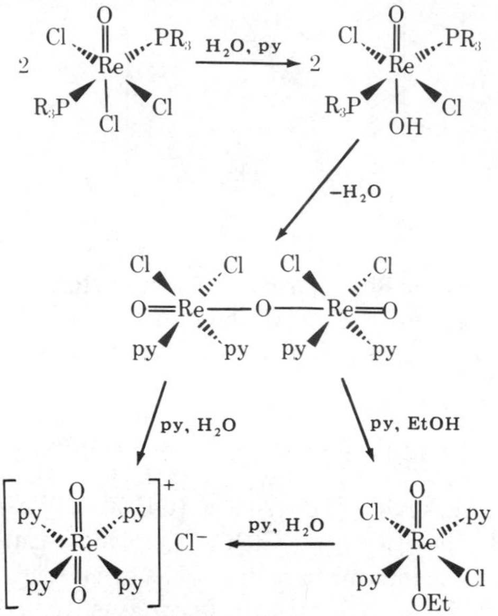

Fig. 22-D-8. Reactions of ReOCl$_3$(PPh$_3$)$_2$ with pyridine.

OTHER COMPLEXES

22-D-9. Complexes with Nitrogen Ligands

Rhenium forms a variety of complexes containing Re=NR and Re≡N bonds, representative preparative reactions being

$$ReOCl_3(PPh_3)_2 + PhNH_2 \rightarrow Re(NPh)Cl_3(PPh_3)_2 + H_2O$$

$$ReO(OEt)Cl_2(PPh_3)_2 + N_2H_4 \cdot 2HCl + H_2O \rightarrow ReNCl_2(PPh_3)_2 + PPh_3O + EtOH + NH_4Cl$$

$$ReOCl_3(PPh_3)_2 + RNH{-}NHR \cdot 2HCl \xrightarrow{PPh_3} Re(NR)Cl_3(PPh_3)_2 + RNH_3Cl + PPh_3O + HCl$$

The alkylimino complexes react with Cl$_2$ according to the equation

$$Re(NR)Cl_3(PPh_3)_2 \xrightarrow[CCl_4]{Cl_2 \text{ in}} ReCl_4(PPh_3)_2$$

The Re≡N bond in ReNCl$_2$(PEt$_2$Ph)$_3$ exhibits electron-donor properties, like those of a nitrile, in forming adducts with the acceptors BX$_3$ (X = F, Cl, or Br) and PtCl$_2$(PEt$_3$)$_2$. ReNBr$_2$(PPh$_3$)$_2$ reacts with KCN to give K$_2$[ReN(CN)$_4$]·H$_2$O.

There is considerably variation in the Re≡N bond lengths, ranging from 1.53 Å in [Re^{V}N(CN)$_4$]$^{2-}$ to 1.79 Å in Re^{V}NCl$_2$(PEt$_2$Ph)$_3$; in [Re^{VI}N(NCS)$_5$]$^{2-}$ the bond length is 1.657 Å.[33]

The Re=NR bonds in, e.g., Re^V(NR)Cl$_3$(PEt$_2$Ph)$_2$, are 1.685 to 1.710 Å.

[33]   M. A. A. F. de C. T. Carrondo *et al.*, *J.C.S. Dalton*, **1978**, 844.

Rhenium also forms a series of N_2 complexes of the type $Re(N_2)X(PR_3)_4$ (X = Cl or Br; PR_3 is either a monophosphine or half of a chelating diphosphine) from which a paramagnetic rhenium(II) complex $[ReCl(N_2)(PR_3)_4]Cl$ can be obtained on oxidation with chlorine.[34]

The dinitrogen ligand can act also as a donor to give bridged species, e.g., $(PhMe_2P)_4ClReNNCrCl_3(THF)_2$ and $[(PhMe_2P)_4ClReNN]_2MoCl_4$, the latter having a linear Re—N—N—Mo group.[35]

A number of complexes with R_3P or R_3As ligands are known, examples being $ReCl_2$ diars and the nitric oxide compounds $ReCl_2(NO)(PPh_3)_2$.[36] Interaction of $TcCl_6^{2-}$ and $PPh(OEt)_2$ gives $trans\text{-}TcCl_2[PPh(OEt)_2]_4$.[37]

Trivalent complexes include the octahedral monomeric species $ReCl_3(RCN)\text{-}(PPh_3)_2$, $ReCl_3py_3$, $ReCl_3Py_2PPh_3$, $ReCl(acac)_2PPh_3$, and the ion $[ReCl_4(MeCN)_2]^-$ mentioned above.

Tetravalent Complexes. Besides the important MX_6^{2-} species mentioned above, there are a few other Re^{IV} and Tc^{IV} species. The addition of halogens to $ReX_3(PR_3)_3$ (X = Cl or Br), as well as reductive interaction of $ReCl_5$ with PPh_3 in acetone, give the $ReX_4(PR_3)_2$ species, which can be obtained also in other ways. $trans\text{-}[ReCl_4(PEt_2Ph)_2]$ has $\mu_{eff} = 3.64$ BM at 20°, in accord with its being an octahedral complex of a d^3 ion. $Re(CO)_5Cl$, on treatment with diars, gives $Re(CO)_3diarsCl$, which can be chlorinated to give $ReCl_4$ diars. The tetrahydrofuranate $ReCl_4(THF)_2$ and some similar sulfide complexes are also obtained from $ReCl_5$ by direct interaction.

For technetium, direct reaction of ligands with $TcCl_4$ gives complexes such as $TcCl_4(PPh_3)_2$, $[TcCl_4dipy]$, and $[TcCl_2dipy_2]Cl_2$.

Sulfur Complexes. The compound $Re(S_2C_2Ph_2)_3$ has a trigonal prismatic structure. Interaction of K_2ReCl_6 with KSCN gives the aggregate $[Re_4S_4(CN)_{12}]^{4-}$, which has a typical M_4S_4 core.[38a] Another unusual species is $Na_4Re_6^{III}S_{10}(S_2)$,[38b] which has a regular octahedron of Re atoms inside a cube of S atoms with $\mu\text{-}S$ and $\mu\text{-}S_2$ bridges between the octahedra.

22-D-10. Hydrido Complexes[39]

Especially for rhenium, there is an extensive chemistry of complexes with M—H bonds of the following types.

1. Polyhydrides with three to eight hydrogen atoms such as $ReH_5(PPh_3)_3$ and $ReH_7(PPh_3)_2$.

2. Carbonyl hydrides that may be monomeric like $HRe(CO)_5$, or clusters like $Re_3H_3(CO)_{12}$ or $[Re_3H_2(CO)_{12}]^-$.

34   J. Chatt et al., J.C.S. Dalton, **1973**, 612.
35   P. D. Cradwick, J.C.S. Dalton, **1976**, 1934.
36   R. W. Adams et al., J.C.S. Dalton, **1974**, 1075.
37   U. Mazzi et al., Inorg. Chem., **1977**, 16, 1042.
38a M. Laing et al., J.C.S. Chem. Comm., **1977**, 221.
38b S. Chin and W. R. Robinson, J.C.S. Chem. Comm., **1978**, 879.
39   D. Giusto, Inorg. Chim. Acta Rev., **1972**, 91.

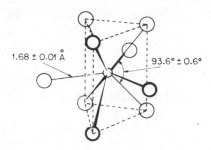

Fig. 22-D-9. The structure of ReH_9^{2-}; the Re—H distance of 1.68 Å is consistent with the sum of single bond radii (Re = 1.28 Å).

3. The dimeric hydrides $Re_2H_8(PR_3)_4$, which have four bridge hydrogen atoms[40] can be prepared best by reaction of BH_4^- with Re_2Cl_8 in presence of PR_3.[16]

4. The hydride $(\eta\text{-}C_5H_5)_2ReH$, which is a base, giving $(\eta\text{-}C_5H_5)_2ReH_2^+$.

5. The unique anions MH_9^{2-}. The ReH_9^{2-} ion is made by reduction of ReO_4^- with an excess of sodium in ethanol. Although the structure is that shown in Fig. 22-D-9, the ^1H nmr spectrum has but a single line because of stereochemical nonrigidity.

The ReH_9^{2-} ion is a source of other hydrido species,[41] e.g.,

$$(Et_4N)_2ReH_9 \xrightarrow[\text{N}_2]{\text{diphos}} ReH(N_2)diphos_2 \underset{C_2H_4}{\overset{CO}{<}} \begin{array}{l} ReH(CO)diphos_2 \\ ReH(C_2H_4)diphos_2 \end{array}$$

22-D-11. Carbonyls and Organometallic Compounds[42]

In addition to the binary carbonyl $M_2(CO)_{10}$ and cluster carbonyls, there is a range of M^I compounds such as, $MX(CO)_{5-n}L_n$, where X can be F,[43]Cl, Br, I, Me, Ph, etc., and L can be py, R_3P, etc.

The cleavage of the rhenium(II) dimers $Re_2X_4(PR_3)_4$ by CO leads to paramagnetic rhenium(II) carbonyls, $trans\text{-}ReX_2(CO)_2(PR_3)_2$.[44]

There is also a variety of nitric oxide compounds[45] in different oxidation states, one example being the ion $[Re^{III}Cl_4(NO)py]^-$, which is octahedral with trans-pyridine and NO groups.

Rhenium forms a number of σ-bonded alkyls and aryls, in all oxidation states except V.[46] Some of the chemistry is shown in Fig. 22-D-10. The structures of several of these species have been determined, including trirhenium(III) cluster

[40] R. Bau, et al., J. Am. Chem. Soc., 1977, **99**, 3872.

[41] M. E. Tulley and A. P. Ginsberg, J. Am. Chem. Soc., 1973, **95**, 2042.

[42] H. C. Lewis, Jr., and B. W. Storhoff, J. Organomet. Chem., 1972, **43**, 1 (review).

[43] D. M. Bruce et al., J.C.S. Dalton, **1976**, 2230.

[44] R. A. Walton et al., Inorg. Chem., 1978, **17**, 2383.

[45] G. Ciani et al., J.C.S. Dalton, **1978**, 798, and references therein.

[46] G. Wilkinson et al., J.C.S. Dalton, **1979**, 361; **1978**, 446; **1977**, 1063; Nouv. J. Chim., **1977**, 389.

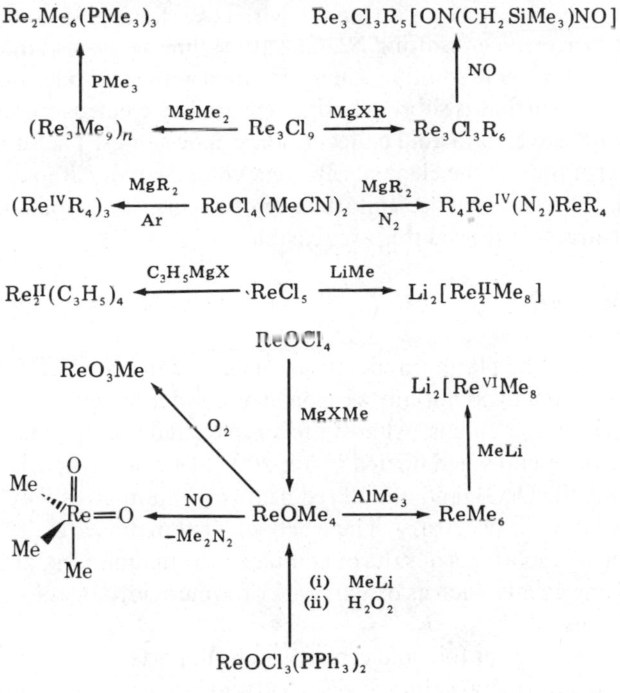

$Re_2Me_6(PMe_3)_3$ $\qquad\qquad$ $Re_3Cl_3R_5[ON(CH_2SiMe_3)NO]$

$\uparrow PMe_3$ $\qquad\qquad\qquad\qquad$ $\uparrow NO$

$(Re_3Me_9)_n \xleftarrow{\ MgMe_2\ } Re_3Cl_9 \xrightarrow{\ MgXR\ } Re_3Cl_3R_6$

$(Re^{IV}R_4)_3 \xleftarrow[Ar]{\ MgR_2\ } ReCl_4(MeCN)_2 \xrightarrow[N_2]{\ MgR_2\ } R_4Re^{IV}(N_2)ReR_4$

$Re_2^{II}(C_3H_5)_4 \xleftarrow{\ C_3H_5MgX\ } \cdot ReCl_5 \xrightarrow{\ LiMe\ } Li_2[Re_2^{II}Me_8]$

$ReOCl_4$

ReO_3Me $\qquad\qquad\qquad\qquad$ $Li_2[Re^{VI}Me_8]$

$\nwarrow O_2 \qquad\qquad \downarrow MgXMe \qquad\qquad\qquad \uparrow MeLi$

$Me \underset{Me}{\overset{O}{\underset{|}{\overset{\|}{\underset{Me}{Re}}}}}{=}O \xleftarrow[-Me_2N_2]{\ NO\ } ReOMe_4 \xrightarrow{\ AlMe_3\ } ReMe_6$

$\uparrow$ (i) MeLi / (ii) H_2O_2

$ReOCl_3(PPh_3)_2$

Fig. 22-D-10. Synthesis of some rhenium alkyls (R = CH₂SiMe₃).

alkyls (see Fig. 27-2) and the tetraallyl.[47] The phenyl compound $(C_6H_5)_3Re$-$(PPhEt_2)_2$ is *thp* with axial phosphines.[48]

Finally there are cyclopentadienyl, arene, and alkene complexes of various types.

22-E. THE PLATINUM METALS

22-E-1. General Remarks

Ruthenium, osmium, rhodium, iridium, palladium, and platinum are the six heaviest members of Group VIII. They are rare elements; platinum is the most common with an abundance of *ca.* $10^{-6}\%$, whereas the others have abundances *ca.* $10^{-7}\%$ of the earth's crust. They occur as metals, often as alloys such as "osmiridium," and in sulfide, arsenide, and other ores. They are commonly associated with copper, silver, and gold. The main sources of the metals are South Africa, the USSR, and Canada; in all ores the "values" of the platinum metals are in grams per ton, and concentration by gravitation and flotation is required. Extraction methods depend

[47] F. A. Cotton, and M. W. Extine, *J. Am. Chem. Soc.*, 1978, **100**, 3788.
[48] W. E. Carroll and R. Bau, *J.C.S. Chem. Comm.*, **1978**, 825.

on the ore, but the concentrate is smelted with coke, lime, and sand and bessem-erized in a converter. The resulting Ni–Cu sulfide "matte" is cast into anodes. On electrolysis in sulfuric acid solution, Cu is deposited at the cathode, and Ni remains in solution, from which it is subsequently recovered by electrodeposition, and the platinum metals, silver, and gold collect in the anode slimes. The subsequent pro-cedures for separation of the elements are very complicated. Although most of the separations involve classical precipitations or crystallizations, some ion-exchange and solvent-extraction procedures are feasible.

22-E-2.   The Metals

Some properties of the platinum metals are collected in Table 22-E-1.

The metals are obtained initially as sponge or powder by ignition of ammonium salts of the hexachloro anions. Almost all complex and binary compounds of the elements give the metal when heated above 200° in air or oxygen; osmium is oxi-dized to the volatile OsO_4, and at dull red heat ruthenium gives RuO_2, so that re-duction in hydrogen is necessary. The finely divided metals are also obtained by reduction of acidic solutions of salts or complexes by magnesium, zinc, hydrogen, or other reducing agents, such as oxalic acid or formic acid, or by electrolysis under proper conditions.

The metals, as gauze or foil, and especially on supports such as charcoal or alu-mina on to which the metal salts are absorbed and reduced *in situ* under specified conditions, are widely used as catalysts for an extremely large range of reactions in the gas phase or in solution. One of the biggest uses of platinum is for the re-forming of hydrocarbons.[1] Commercial uses in homogeneous reactions are fewer, but palladium is used in the Smidt process and rhodium in hydroformylation and in acetic acid synthesis (Chapter 30). Industrially as well as in the laboratory, catalytic reductions are especially important.

Platinum or its alloys are used for electrical contacts, for printed circuitry, and for plating.

TABLE 22-E-1
Some Properties of the Platinum Metals

Element	M.p. (°C)	Form	Best solvent
Ru	2334	Gray-white, brittle, fairly hard	Alkaline oxidizing fusion
Os	~3050	Gray-white, brittle, fairly hard	Alkaline oxidizing fusion
Rh	1960	Silver-white, soft, ductile	Hot conc. H_2SO_4; conc. HCl + $NaClO_3$ at 125–150°
Ir	2443	Silver-white, hard, brittle	Conc. HCl + $NaClO_3$ at 125–150°
Pd	1552	Gray-white, lustrous, malleable, ductile	Conc. HNO_3, HCl + Cl_2
Pt	1769	Gray-white, lustrous, malleable, ductile	Aqua regia

[1]   *Hydrocarbon Processing, Refining Handbook,* September 1978.

Ru and Os are unaffected by mineral acids below ~100° and are best dissolved by alkaline oxidizing fusion (e.g., NaOH + Na_2O_2, $KClO_3$). Rh and Ir are extremely resistant to attack by acids, neither metal dissolving even in aqua regia when in the massive state. Finely divided rhodium can be dissolved in aqua regia or hot concentrated H_2SO_4. Both metals also dissolve in concentrated HCl under pressure of oxygen or in presence of sodium chlorate in a sealed tube at 125 to 150°. At red heat Cl_2 leads to the trichlorides.

Pd and Pt are rather more reactive than the other metals. Pd is dissolved by nitric acid, giving $Pd^{IV}(NO_3)_2(OH)_2$; in the massive state the attack is slow, but it is accelerated by oxygen and oxides of nitrogen. As sponge, Pd also dissolves slowly in HCl in presence of chlorine or oxygen. Platinum is considerably more resistant to acids and is not attacked by any single mineral acid, although it readily dissolves in aqua regia and even slowly in HCl in presence of air since

$$PtCl_4^{2-} + 2e = Pt + 4Cl^- \qquad E° = -0.75 \text{ V}$$

$$PtCl_6^{2-} + 2e = PtCl_4^{2-} + 2Cl^- \qquad E° = 0.77 \text{ V}$$

Platinum is not the inert material that it is often considered to be. There are at least 70 oxidation-reduction and decomposition reactions that are catalyzed by metallic platinum. Examples are the Ce^{IV}–Br^- reaction and the decomposition of N_2H_4 to N_2 and NH_3. It is possible to predict whether catalysis can occur from a knowledge of the electrochemical properties of the reacting couples.

Both Pd and Pt are rapidly attacked by fused alkali oxides, and especially by their peroxides, and by F_2 and Cl_2 at red heat. It is of importance in the use of platinum for laboratory equipment that on heating it combines with, e.g., elemental P, Si, Pb, As, Sb, S, and Se, so that the metal is attacked when compounds of these elements are heated in contact with platinum under reducing conditions.

Both Pd and Pt are capable of absorbing large volumes of molecular hydrogen, and Pd is used for the purification of H_2 by diffusion (see p. 251).

22-E-3. General Remarks on the Chemistry of the Platinum Metals

The chemistries of these elements have some common features, but there are nevertheless wide variations depending on differing stabilities of oxidation states, stereochemistries, etc. The principal areas of general similarity are as follows.

1. *Binary Compounds.* There are a large number of oxides, sulfides, phosphides, etc., but the most important are the halides.

2. *Aqueous Chemistry.* This chemistry is almost exclusively that of complex compounds. Aqua ions of Ru^{II}, Ru^{III}, Rh^{III}, and Pd^{II} exist, but complex ions are formed in presence of anions other than ClO_4^-, BF_4^-, or p-toluene-sulfonate, etc. The precise nature of many supposedly simple solutions (e.g., of rhodium sulfate) is complicated and often unknown.

A vast array of complex ions, predominantly with halide or nitrogen donor ligands, are water soluble. Exchange and kinetic studies have been made with many of these because of interest in (a) trans effects, especially with square Pt^{II}, (b) differences in substitution mechanisms between the ions of the three transition metal series, and (c) the unusually rapid electron transfer process with heavy metal complex ions.

Although the species involved may often not be fully identified, much potential information has been collected from polarographic and other studies.[2]

3. *Compounds with π-Acid Ligands.*

(a) Binary carbonyls are formed by all but Pd and Pt, the majority of them polynuclear. Substituted polynuclear carbonyls are known for Pd and Pt, and all six elements give carbonyl halides and a wide variety of carbonyl complexes containing other ligands and carbonyl anions.

(b) Nitric oxide complexes are a feature of the chemistry of Ru.

(c) An especially widely studied area is the formation of complexes with trialkyl- and triarylphosphines and related phosphites, and to a lesser extent with R_3As and R_2S. The most important are those with triphenylphosphine and methyl-substituted phosphines (e.g., $PPhMe_2$). The latter are more soluble in organic solvents than PPh_3 complexes, and have proved particularly useful for the determination of configuration by nmr.

Mixed complexes of PR_3 with CO, alkenes, halides, and hydride ligands in at least one oxidation state are common for all of the elements.

(d) All these elements have a strong tendency to form bonds to carbon, especially with alkenes and alkynes; Pt^{II}, Pt^{IV}, and to a lesser extent Pd^{II} have a strong tendency to form σ bonds, and Pd^{II} very readily forms π-allyl species.

(e) A highly characteristic feature is the formation of hydrido complexes, and M—H bonds may be formed when the metal halides in higher oxidation states are reduced, especially in presence of tertiary phosphines or other ligands. Hydrogen abstraction from reaction media such as alcohols or dimethylformamide is common.

(f) For the d^8 ions Rh^I, Ir^I, Pd^{II}, and Pt^{II}, the normal coordination is square (though five-coordinate species are fairly common) and oxidative-addition reactions are of great importance.

Platinum metal chemistry is an exceedingly active area of research, and even omitting patents, which are very numerous, the research papers number in the many hundreds per year.

BINARY COMPOUNDS

22-E-4.   Oxides, Sulfides, Phosphides, etc.

Oxides.   The best known anhydrous oxides are listed in Table 22-E-2; the tetraoxides of Ru and Os are discussed later (p. 914). The oxides are generally rather inert to aqueous acids, are reduced to the metal by hydrogen and dissociate

[2]   R. N. Goldberg and L. G. Hepler, *Chem. Rev.,* 1968, **68,** 229 (an authoritative collection of thermodynamic data on compounds of the Pt metals and their oxidation-reduction potentials; also contains much descriptive chemistry).

TABLE 22-E-2
Anhydrous Oxides of Platinum Metals[a]

Oxide	Color/form	Structure	Comment
RuO_2	Blue-black	Rutile	From O_2 on Ru at 1250° or $RuCl_3$ at 500–700°; usually O-defective
RuO_4	Orange-yellow crystals, m.p. 25°, b.p. 100°	Tetrahedral molecules	See Section 22-F-2
OsO_2	Coppery	Rutile	Heat Os in NO or OsO_4 or dry $OsO_2 \cdot nH_2O$
OsO_4	Colorless crystals, m.p. 40°, b.p. 101°	Tetrahedral molecules	Normal product of heating Os in air; see Section 22-F-2
Rh_2O_3	Brown	Corundum	Heat Rh^{III} nitrate or $Rh_2O_3(aq)$
RhO_2	Black	Rutile	Heat $Rh_2O_3(aq)$ at 700–800° in high-pressure O_2
Ir_2O_3	Brown		Impure by heating $K_2IrCl_6 + Na_2CO_3$
IrO_2	Black	Rutile	Normal product of Ir + O_2; dissociates > 1100°
PdO	Black		From Pd + O_2; dissociates 875°; insoluble in all acids
PtO_2	Brown		Dehydrate $PtO_2(aq)$; dec. 650°

[a] In oxygen at 800–1500°, gaseous oxides exist: RuO_3, OsO_3, RhO_2, IrO_3, PtO_2. A number of other solids of uncertain nature exist: RuO_3, OsO_3, Ru_2O_5, Rh_2O_5, Os_2O_3, Ru_2O_3. Two forms of PtO_2, and Pt_3O_4 are also known.

on heating. A number of mixed metal oxides (e.g., $BaRuO_3$, $CaIrO_3$, $Tl_2Pt_2O_7$) are known[3a]; platinum and palladium "bronzes" of formula $M_x^I Pt_3O_4$ ($x = 0$–1) are also known.[3b]

Hydrous oxides are commonly precipitated when NaOH is added to aqueous metal solutions, but they are difficult to free from alkali ions and sometimes readily become colloidal. When freshly precipitated, they may be soluble in acids, but only with great difficulty or not at all after aging.

The black precipitate, probably $Ru_2O_3 \cdot nH_2O$, from Ru^{III} chloride solutions is readily oxidized by air, probably to the black $RuO_{2+x} \cdot yH_2O$, which is formed on reduction of RuO_4 or RuO_4^{2-} solutions by alcohol, hydrogen, etc. Reduction of OsO_4 or addition of OH^- to $OsCl_6^{2-}$ solutions gives $OsO_2 \cdot nH_2O$.

$Rh_2O_3 \cdot nH_2O$ is formed as a yellow precipitate from Rh^{III} solutions. In base solution, powerful oxidants convert it into $RhO_2 \cdot nH_2O$; the latter loses oxygen on dehydration. $Ir_2O_3 \cdot nH_2O$ can be obtained only in moist atmospheres; it is at least partially oxidized by air to $IrO_2 \cdot nH_2O$, which is formed either by action of mild oxidants on $Ir_2O_3 \cdot nH_2O$ or by addition of OH^- to $IrCl_6^{2-}$ in presence of H_2O_2. The precipitation of the oxides of Rh and Ir from buffered $NaHCO_3$ solutions by the action of ClO_2^- or BrO_3^- provides a rather selective separation of these elements.

$PdO \cdot nH_2O$ is a yellow gelatinous precipitate that dries in air to a brown, less hydrated form and at 100° loses more water, eventually becoming black; it cannot be dehydrated completely without loss of oxygen.

[3a] See, e.g., A. W. Sleight, *Mater. Res. Bull.,* 1974, **9,** 1177; W. D. Komar and D. J. Machin, *J. Less Comm. Met.,* 1978, **61,** 91.
[3b] D. Cahen *et al., Inorg. Chem.,* 1974, **13,** 110, 1377.

When $PtCl_6^{2-}$ is boiled with Na_2CO_3, red-brown $PtO_2\cdot nH_2O$ is obtained. It dissolves in acids and also in strong alkalis to give what can be regarded as solutions of hexahydroxoplatinate, $[Pt(OH)_6]^{2-}$. The hydrous oxide becomes insoluble on heating to *ca.* 200°. The brown material formed by fusion of $NaNO_3$ and chloroplatinic acid at *ca.* 550° followed by extraction of soluble salts with water is known as Adams's catalyst and is widely used in organic chemistry for catalytic reductions.

A very unstable Pt^{II} hydrous oxide is obtained by addition of OH^- to $PtCl_4^{2-}$; after drying in CO_2 at 120 to 150° it approximates to $Pt(OH)_2$, but at higher temperatures gives PtO_2 and Pt.

Sulfides, Phosphides, and Similar Compounds. Direct interaction of the metal and other elements such as S, Se, Te, P, As, Bi, Sn, or Pb under selected conditions produces dark, often semimetallic solids that are resistant to acids other than nitric. These products may be stoichiometric compounds and/or nonstoichiometric phases depending on the conditions of preparation.

The chalcogenides[4] and phosphides[5] are generally rather similar to those of other transition metals; indeed many of the phosphides, for example, are isostructural with those of the iron group, viz., Ru_2P with Co_2P; RuP with FeP and CoP; RhP_3, PdP_3 with CoP_3 and NiP_3.

Sulfides can also be obtained by passing H_2S into platinum metal salt solutions. Thus from $PtCl_4^{2-}$ and $PtCl_6^{2-}$ are obtained PtS and PtS_2, respectively; from Pd^{II} solutions PdS, which when heated with S gives PdS_2; the Rh^{III} and Ir^{III} sulfides are assumed to be $M_2S_3\cdot nH_2O$, but exact compositions are uncertain.

22-E-5. Halides of the Platinum Metals[6a]

We discuss here primarily the binary halides. All the platinum metals form halogeno complexes in one or more oxidation states and these, as well as the hydrated halides, which are closely related to them, are discussed under the respective elements.

Fluorides. These are listed in Table 22-E-3. The most interesting are the *hexafluorides,* of which only that of Pd is yet unknown.

They are prepared by fluorination under pressure (4–6 atm) of the metal at elevated temperatures and are purified by vacuum distillation. Platinum wire ignited in fluorine by an electron current continues to react exothermally to give red vapors of PtF_6.

The hexafluorides decrease in stability in the order W > Re > Os > Ir > Pt, and Ru > Rh, dissociating into fluorine and lower fluorides. PtF_6 is one of the most powerful oxidizing agents known; it reacts with oxygen and xenon to give $O_2^+PtF_6^-$ and $Xe(PtF_6)_n$ (pp. 494, 583). The volatility of the compounds also decreases with increasing mass.

---

[4] A. Wold, in *Platinum Group Metals and Compounds,* ACS Advances in Chemistry Series, No. 98, 1971.

[5] See, e.g., D. J. Braun and W. Jeitschko, *Z. Anorg. Allg. Chem.,* 1978, **445,** 157.

[6a] J. H. Canterford and R. Colton, *Halides of the Second and Third Row Transition Series,* Wiley-Interscience, 1968 (an exhaustive reference on halides and halide complexes); G. Thiele and K. Broderson, *Fortschr. Chem. Forsch.,* 1968, **10,** 631 (structural chemistry of Pt halides).

TABLE 22-E-3
Fluorides of the Platinum Metals

II	III	IV	V	VI
—	RuF_3 Brown	RuF_4 Sandy yellow	RuF_5 Dark green[a] m.p. 86.5°; b.p. 227°	RuF_6 Dark brown, m.p. 54°
—	—	OsF_4 Yellow- brown	OsF_5 Blue, m.p. 70°; b.p. 225.9°	OsF_6[b] Pale yellow,[a] m.p. 33.2°; b.p. 47°
—	RhF_3 Red	RhF_4 Purple-red	RhF_5 Dark red	RhF_6 Black
—	IrF_3 Black	IrF_4 Red-brown	IrF_5 Yellow-green, m.p. 104°	IrF_6 Yellow, m.p. 44.8°; b.p. 53.6°
PdF_2 Violet	—[c]	PdF_4 Brick red	—	—
—	—	PtF_4 Yellow- brown	PtF_5 Deep red, m.p. 80°	PtF_6 Dark red, m.p. 61.3°; b.p. 69.1°

[a] Colorless vapor.

[b] OsF_7, made under drastic conditions (500°, 400 atm, F_2), dissociates above $-100°$ and is stable only under high-pressure F_2 (O. Glemser et al., Chem. Ber., 1966, **99**, 2652); there is some evidence for OsF_8.

[c] PdF_3 is $Pd^{2+}PdF_6^{2-}$.

All the hexafluorides are extraordinarily reactive and corrosive substances and normally must be handled in Ni or Monel apparatus, although quartz can be used if necessary. Only PtF_6 and RhF_6 actually react with glass (even when rigorously dry) at room temperature. In addition to thermal dissociation, ultraviolet radiation causes decomposition to lower fluorides, even OsF_6 giving OsF_5. The vapors hydrolyze with water vapor, and liquid water reacts violently, e.g., IrF_6 gives HF, O_2, O_3 and $IrO_2(aq)$; OsF_6 gives OsO_4, HF, and OsF_6^-. The hexafluorides are octahedral, and their magnetic and spectral properties have been studied in detail.

The *pentafluorides* can be obtained by controlled fluorination of the metal. Thus action of F_2 (7 atm) at 370° on Ru gives RuF_5, and fluorination of Rh at 400° gives RhF_5, Ir at ~360° gives IrF_5, and $PtCl_2$ at 350° gives PtF_5.

However OsF_5 and IrF_5 are best made by reduction of the hexafluoride with Si or H_2 in liquid HF,[6b] or for OsF_5, by $W(CO)_6$.[7] The pentafluorides are very reactive, hydrolyzable substances. In the condensed phases these fluorides may be tetramers or chain polymers. The structure[8] of $(RhF_5)_4$ is shown in Fig. 22-E-1; the Rh—F—Rh bridges are bent, unlike the linear ones in, e.g., $(NbF_5)_4$ (p. 834). The polymers appear to have *cis*-F_2 bridges.[9] In the gas phase, oligomers are also present, but depolymerization occurs[7] to give *tbp* monomers that may be accompanied by color changes; thus blue $(OsF_5)_n$ gives a colorless vapor.

[6b] R. T. Paine and L. B. Asprey, Inorg. Chem., 1974, **13**, 1111.

[7] W. E. Falconer et al., J. Fluorine Chem., 1974, **4**, 213.

[8] N. Bartlett et al., Inorg. Chem., 1973, **12**, 2640.

[9] T. Cyr, Can. J. Spectrosc., 1974, **19**, 136.

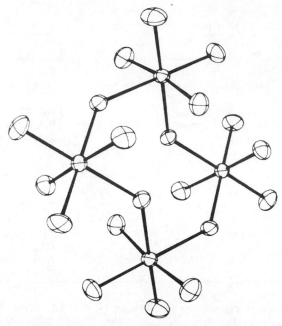

Fig. 22-E-1. Structure of rhodium pentafluoride. [Reproduced by permission from Ref. 8.]

The *tetrafluorides*[6a],[10] may be obtained by reactions such as

$$10RuF_5 + I_2 = 10RuF_4 + 2IF_5$$
$$2IrF_5 + H_2 = 2IrF_4 + 2HF$$

$$RhCl_3 \xrightarrow{BrF_3(l)} RhF_4 \cdot 2BrF_3 \xrightarrow{heat} RhF_4$$

$$Pt \xrightarrow{BrF_3(l)} PtF_4 \cdot 2BrF_3 \xrightarrow{Heat} PtF_4 \xleftarrow{F_2} PtBr_4$$

$$PdBr_2 \xrightarrow{BrF_3(l)} Pd^{II}Pd^{IV}F_6 \xrightarrow{F_2 \cdot 100 \text{ psi, } 150°} PdF_4$$

The formation of BrF_3 adducts as above is a fairly common feature in the preparation of heavy metal fluorides; such adducts may be ionic, i.e., $[BrF_2^+]_2 \cdot MF_6^{2-}$, but it is more likely that they are fluoride-bridged species of the type (22-E-I) or singly bridged polymers.

(22-E-I)

[10] N. Bartlett *et al., Inorg. Chem.,* 1978, **17,** 748; *J. Inorg. Nucl. Chem., Suppl.* 1976; *Compt. Rend. C,* 1974, **278,** 1501.

Like the other fluorides, the tetrafluorides are violently hydrolyzed by water. The structure of diamagnetic PdF_4 indicates approximately octahedral coordination of the metal with two $cis\text{-}Pd$—F and four bridging Pd—F—Pd groups. The other tetrafluorides are probably similar; IrF_4 and RhF_4 are paramagnetic.

Trifluorides. RuF_3 is best obtained by reduction:

$$5RuF_5 + 2Ir \xrightarrow{250°} 5RuF_3 + 2IrF_5$$

Fluorination of $RhCl_3$ at 500° gives RhF_3. The solid is unaffected by water or bases. IrF_3, which can be obtained only indirectly by reduction of IrF_6 (e.g., by Ir at 50°), is also relatively inert to water. The Rh and Ir trifluorides have a slightly distorted ReO_3 structure.

Difluorides. Palladous fluoride can be obtained by the reaction

$$Pd^{II}Pd^{IV}F_6 + SeF_4 \xrightarrow{reflux} 2PdF_2 + SeF_6$$

It is the only simple compound of Pd^{II} that is paramagnetic, and the moment is consistent with the observed octahedral coordination. The Pd^{2+} ion also occurs in $Pd^{II}Pd^{IV}F_6$, $Pd^{II}Sn^{IV}F_6$, and $Pd^{II}Ge^{IV}F_6$, which can be obtained by addition of BrF_3 to mixtures of $PdBr_2$ and, e.g., $SnBr_4$.

Chlorides, Bromides, and Iodides. The anhydrous halides (other than fluorides—see Table 22-E-3) are listed in Table 22-E-4; they are normally obtained by direct interaction under selected conditions. The higher halides form the lower halides on heating.

Except for those of Pd and Pt, the halides are generally insoluble in water, rather inert, and of little utility for the preparation of complex compounds. Hydrated halides, discussed later, are normally used for this purpose. We discuss only some of the more important chlorides.

Ruthenium trichloride has two forms. Interaction of the metal with $Cl_2 + CO$ at 370° gives the β form, which is converted into black leaflets of the α form at 450° in Cl_2. The latter has a layer lattice and is antiferromagnetic. The so-called *iodide*, which is precipitated from aqueous ruthenium chloride solutions by KI, invariably contains strong OH bands in its infrared spectrum and probably has OH bridges in the lattice.

Osmium tetrachloride is formed when an excess of Cl_2 is used at temperatures above 650°; otherwise a mixture with the *trichloride* is formed. The latter is obtained when $OsCl_4$ is decomposed at 470° in a flow system with a low pressure of chlorine. $OsCl_4$ in its orthorhombic high-temperature form has infinite chains of octahedra sharing opposite edges with no structural indication of metal-metal bonding.[11]

Rhodium trichloride has a layer lattice isostructural with $AlCl_3$ and is exceedingly inert. However when $RhCl_3 \cdot 3H_2O$ (see below) is dehydrated in dry HCl at 180°, the red product is much more reactive and dissolves in water or tetrahydrofuran; this property is lost on heating at 300°.

Palladium dichloride and platinum dichloride both exist in two forms, which may be obtained in the following ways:

TABLE 22-E-4

Anhydrous Chlorides, Bromides, and Iodides of Platinum Metals[a]

Oxidation state	Ru	Os	Rh	Ir	Pd	Pt
II	—	—	—	—	$PdCl_2$[b] Red	$PtCl_2$[b] Black-red
	—	—	—	—	$PdBr_2$ Red-black	$PtBr_2$ Brown
	—	—[c]	—	—	PdI_2 Black	PtI_2 Black
III	$RuCl_3$[d]	$OsCl_3$ Dark gray	$RhCl_3$ Red	$IrCl_3$ Brown-red	—	$PtCl_3$; Green-black
	$RuBr_3$ Dark brown	$OsBr_3$ Black	$RhBr_3$ Dark red	$IrBr_3$ Yellow	—	$PtBr_3$ Green-black
	RuI_3 Black	OsI_3 Black	RhI_3(?) Black	IrI_3 Black	—	PtI_3(?) Black
IV	—[e]	$OsCl_4$ Black	—	$IrCl_4$(?)	—	$PtCl_4$ Red-brown
	—	$OsBr_4$ Black	—	$IrBr_4$(?)	—	$PtBr_4$ Dark red
	—	—	—	IrI_4(?)	—	PtI_4 Brown-black

[a] There is some evidence for gray metallic OsI, and lower halides of Rh and Ir.
[b] Two or more polymorphs; $PtCl_2$ yellowish-brown when powdered.
[c] Alleged lower iodides are mixtures of OsI_3 and oxides (H. Schäfer *et al.*, *Z. Anorg. Allg. Chem.*, 1971, **383**, 49).
[d] Two forms: α-$RuCl_3$ (black) or β-$RuCl_3$ (brown).
[e] Some evidence for existence of $RuCl_4$ in vapor.

$$Pd \xrightarrow{Cl_2} \begin{cases} \xrightarrow{>550°} \alpha\text{-}PdCl_2 \\ \xrightarrow{<550°} \beta\text{-}PdCl_2 \end{cases}$$

α-$PdCl_2$ $\xrightarrow{\text{Slow}}$ β-$PdCl_2$

$$H_2PtCl_6\cdot 6H_2O \xrightarrow{Cl_2 \sim 500°} PtCl_4 \xrightarrow{>350°} \beta\text{-}PtCl_2$$

$$\beta\text{-}PtCl_2 \xrightarrow[\text{1-2 days}]{500°} \alpha\text{-}PtCl_2$$

$$Pt \xrightarrow[\text{gradient}]{Cl_2\, 650° \rightarrow 500°} \alpha\text{-}PtCl_2$$

Unlike the nickel halides or PdF_2, which are ionic and paramagnetic, these chlorides are molecular or polymeric and diamagnetic, and Born-Haber calculations indicate that ionic lattices would be endothermic.

The β-forms are isomorphous and have the molecular structure 22-E-II with

[11] F. A. Cotton and C. E. Rice, *Inorg. Chem.*, 1977, **16**, 1865.

M_6Cl_{12} units; the structures appear to be stabilized mainly by the halogen bridges rather than by metal-metal bonds. The molecular behavior is shown by the facts that Pt_6Cl_{12} is soluble in benzene and can be sublimed.[12] The adducts[13] of Pt_6Cl_{12} with benzene, CS_2, $CHCl_3$, etc., are clathrates. On heating with $AlCl_3$, Pt_6Cl_{12} gives a purple vapor of an aluminum chloride complex similar to those noted earlier (p. 322).[14]

The structure of β-$PdCl_2$ is that of a flat chain (22-E-III), but that of β-$PtCl_2$ is uncertain.

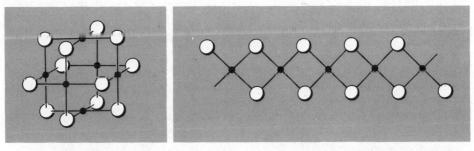

22-E-II   22-E-III

Palladous chloride is soluble in hydrochloric acid, forming the ion $[PdCl_4]^{2-}$; it also reacts with many ligands L, such as amines, benzonitrile, and phosphines, to give complexes of the types L_2PdCl_2 and $[LPdCl_2]_2$. $PtCl_2$ is similar.

The greenish-black *trichloride* $PtCl_3$ contains both Pt^{II} and Pt^{IV} with units of $[Pt_6Cl_{12}]$ and an infinite chain, $(1/\infty)[PtCl_2^a Cl_{4/2}^b]$ containing distorted $PtCl_6$ octahedra linked by common edges, similar to the chain of PtI_4. The chloride and the similar $PtBr_3$ and PtI_3 are made by thermal gradient reactions of Pt and halogen.

Platinic chloride is commonly made by heating chloroplatinic acid to 300° in Cl_2, but is best made by interaction of Pt and SO_2Cl_2.[15] The reddish-brown crystals are readily soluble in water, presumably giving ions such as $[PtCl_4(OH)_2]^{2-}$, and in polar solvents. The structure is not yet known but PtI_4 has PtI_6 octahedra linked by iodide bridges with two *cis*-non-bridging I atoms.

Oxohalides and Halogeno Complexes.   A number of fluoro and other oxohalides of the platinum metals have been characterized: $RuOF_4$, OsO_3F_2, $OsOF_5$, $PtO_xF_{3-x}(?)$, $OsOCl_4$, and Os_2OCl_6. These, and fluoro complex anions, are of small importance or utility, but some of the chloro complex anions are discussed in later sections.

[12]   A. E. Schweizer and G. T. Kerr, *Inorg. Chem.,* 1978, **17**, 2326.
[13]   M. F. Pilbrow, *J.C.S. Dalton,* **1975**, 2432.
[14]   G. N. Papatheodoro, *Inorg. Chem.,* 1973, **12**, 1899.
[15]   M. Degner *et al., Trans. Met. Chem.,* 1975, **1**, 41.

22-F. RUTHENIUM AND OSMIUM[1]

22-F-1. General Remarks: Stereochemistry

The chemistry of ruthenium and osmium bears little resemblance to that of iron except in compounds such as sulfides or phosphides, and in complexes with ligands such as CO, PR_3, η-C_5H_5. The higher oxidation states VI and VIII are much more readily obtained than for iron, and there is an extensive and important chemistry of the tetraoxides (MO_4), oxohalides, and oxo anions. There are analogies between the chemistries of Ru, Os, and Re, especially in oxo, nitrogen, and nitrido compounds.

For ruthenium, the principal lower oxidation states are 0, II, and III, whereas for osmium they are 0, II, III, and IV. For neither element is the I oxidation state very important.

The oxidation states and stereochemistries are summarized in Table 22-F-1.

The 0 State, d^8. The chemistry in this state is primarily one of the metal carbonyls; mononuclear and polynuclear carbonyls are known for both elements. Both types undergo substitution reactions, and in the polynuclear species the clusters are often retained. They also undergo protonation reactions, and a variety of hydrido species are known. We do not deal explicitly with these compounds, but some of the chemistry is noted elsewhere (Chapters 25–27, 29, and 30), and certain aspects are described later in this section.

The II State, d^6. An enormous number of Ru and Os complexes with CO, PR_3, and similar π-acid ligands are known. For other ligands, the main chemistry is that of chloro, ammonia, and other amine ligands, and again large numbers of complexes exist. The aqua ion $[Ru(H_2O)_6]^{2+}$ has been prepared, but it is readily oxidized to $[Ru(H_2O)_6]^{3+}$. For osmium the best characterized complexes are those with aromatic amines.

All Ru^{II} and Os^{II} complexes are octahedral and diamagnetic as expected for the t_{2g}^6 configuration. Although these compounds are fairly labile, the reactions often proceed with retention of configuration, suggesting an associative mechanism.

The III State, d^5. There is an extensive chemistry with both π-acid and σ-donor ligands. Ruthenium(III) species are more common than those of Os^{III}. All the complexes are of low-spin type with one unpaired electron and are octahedral.

The IV State, d^4. In this state most complexes are neutral or anionic, although a few cationic species such as $[Os(diars)_2X_2]^{2+}$ are known; but compared to the II and III states relatively few complexes have been prepared. However it is an important state for Os, where $[OsX_6]^{2-}$ ions are very stable.

Ru^{IV} and Os^{IV} complexes all have octahedral or distorted octahedral structures, thus should have t_{2g}^4 electron configurations. This configuration is especially subject to anomalous magnetic behavior when the spin-orbit coupling constant of the metal

[1]   W. P. Griffith, *The Chemistry of the Rarer Platinum Metals; Os, Ru, Ir and Rh*, Wiley-Interscience, 1967; S. A. Cotton and F. A. Hart, *The Heavy Transition Elements*, Wiley-Halstead, 1975.

TABLE 22-F-1
Oxidation States and Stereochemistry of Ruthenium and Osmium

Oxidation state	Coordination number	Geometry	Examples
Ru^{-II}	4	Tetrahedral(?)	$Ru(CO)_4^{2-}$(?), $[Ru(diphos)_2]^{2-}$
Ru^0, Os^0	5	Tbp	$Ru(CO)_5$, $Os(CO)_5$, $Ru(CO)_3(PPh_3)_2$
Ru^I, d^7	6^a		$[\eta^5\text{-}C_5H_3Ru(CO)_2]_2$, $[Os(CO)_4X]_2$
Ru^{II}, Os^{II}, d^6	5	See text	$RuCl(PPh_3)_3$
	5	Tbp	$RuHCl(PPh_3)_3$
	6^b	Octahedral	$[RuNOCl_5]^{2-}$, $[Ru(bipy)_3]^{2+}$, $[Ru(NH_3)_6]^{2+}$, $[Os(CN)_6]^{4-}$, $RuCl_2CO(PEtPh_2)_3$, $OsHCl(diphos)_2$
Ru^{III}, Os^{III}, d^5	$6^{b,c}$	Octahedral	$[Ru(NH_3)_5Cl]^{2+}$, $[RuCl_5H_2O]^{2-}$, $[Os(dipy)_3]^{3+}$, K_3RuF_6, $[OsCl_6]^3$
Ru^{IV}, Os^{IV}, d^4	$6^{b,c}$	Octahedral	K_2OsCl_6, K_2RuCl_6, $[Os(diars)_2X_2]^{2+}$, $RuO_2{}^d$
	7	?	$OsH_4(PMe_2Ph)_3$
Ru^V, Os^V, d^3	5 in vapor(?)		RuF_5
	6	Octahedral	$KRuF_6$, $NaOsF_6$, $(RuF_5)_4$
Ru^{VI}, Os^{VI}, d^2	4	Tetrahedral	RuO_4^{2-}
	5	?	$OsOCl_4$
	6^c	Octahedral	RuF_6, OsF_6, $[OsO_2Cl_4]^{2-}$, $[OsO_2(OH)_4]^{2-}$, $[OsNCl_5]^{2-}$
Ru^{VII}, Os^{VII}, d^1	4	Tetrahedral	RuO_4^-
	6	Octahedral	$OsOF_5$
Ru^{VIII}, Os^{VIII}, d^0	4	Tetrahedral	RuO_4, OsO_4, $[OsO_3N]^-$
	5	?	OsO_3F_2
		Tbp	$OsO_4NC_7H_{13}$ (see text)
	6	Octahedral	$[OsO_3F_3]^-$, $[OsO_4(OH)_2]^{2-}$

^{a} If η^5-C_5H_5 assumed to occupy three coordination sites.

^{b} Most common states for Ru.

^{c} Most common states for Os.

^{d} Metal-metal bond present.

ion becomes high as it is in Os^{IV} (p. 650). The chief effect in this case is that the effective magnetic moment is brought far below the spin-only value (2.8 BM), typical values for Os^{IV} complexes at room temperature being in the range 1.2–1.7 BM. As the temperature is lowered, μ_{eff} decreases as the square root of the absolute temperature. Ru^{IV} complexes have almost normal moments at room temperature (2.7–2.9 BM), but these also decrease with $T^{1/2}$ as the temperature is lowered.

Ruthenium and *osmium* are readily recovered and separated by utilzing the high volatility of their tetraoxides, which can be distilled from aqueous solutions. Nitric acid is sufficient to oxidize osmium compounds, but for ruthenium more powerful oxidants are required. Since OsO_4 is the commercial source, and since the usual starting material for the preparation of ruthenium compounds is "$RuCl_3$·$3H_2O$," obtained by reducing RuO_4 with concentrated HCl, we discuss the high oxidation states first.

22-F-2. Oxo Compounds of Ruthenium and Osmium (VI), (VII), and (VIII)

Ruthenium and osmium tetraoxides and oxo anions provide some of the more unusual and useful features of the chemistry. The major compounds or ions are shown in Table 22-F-2.

TABLE 22-F-2
Some Oxo Compounds and Ions of Ru and Os

VIII	VII	VI
RuO_4	RuO_4^-	RuO_4^{2-}
		$RuO_2Cl_4^{2-}$
OsO_4		
OsO_3N^-		$[OsO_2X_4]^{2-\ a}$, $[OsO_2en_2]^{2+}$
$OsO_4X_2^{2-\ a}$	$OsOF_5$	$[OsO_2(OH)_2X_2]^{2-\ b}$

^{a} X = F or OH.
^{b} X = Cl, CN, or $\frac{1}{2}$Ox, etc.

The Tetraoxides RuO_4 and OsO_4. These volatile, crystalline solids (Table 22-E-2) are both toxic substances with characteristic, penetrating, ozonelike odors. OsO_4 is a particular hazard to the eyes because of its ready reduction by organic matter to a black oxide, a property utilized in its employment in dilute aqueous solution as a biological fixative.

Ruthenium tetraoxide is obtained when acid ruthenium solutions are treated with oxidizing agents such as MnO_4^-, $AuCl_4^-$, BrO_3^-, or Cl_2; the oxide can be distilled from the solutions or swept out by a gas stream. It may also be obtained by distillation from concentrated perchloric acid solutions or by passing Cl_2 into a melt of ruthenate in NaOH.

Osmium tetraoxide can be obtained by burning osmium or by oxidation of osmium solutions with nitric acid, peroxodisulfate in sulfuric acid, or similar agents. OsO_4 forms very stable five-coordinate *tbp* adducts with quinuclidine, hexamethylene tetramine and, other N donors.[2a]

Both compounds have a tetrahedral structure. They are extremely soluble in CCl_4 and can be extracted from aqueous solutions by it. RuO_4 is quite soluble in dilute sulfuric acid giving golden yellow solutions; OsO_4 is sparingly soluble. The tetraoxides are powerful oxidizing agents. Above $\sim180°$, RuO_4 can explode, giving RuO_2 and O_2, and it is decomposed slowly by light; OsO_4 is more stable in both respects.

Osmium tetraoxide reacts[2b] with olefins to form a number of useful 1,2-diolato complexes. These include the mononuclear ones such as 22-F-Ia and 22-F-Ib, as well as binuclear ones, that exist in syn and anti forms,[2c] 22-F-IIa, b; all of these can be reduced by Na_2SO_3 to give *cis*-diols. The oxide can also be used catalytically for the same conversion in presence of H_2O_2 or ClO_3^-.

[2a] W. P. Griffith *et al.*, *Inorg. Chim. Acta*, 1978, **31**, L413.
[2b] R. J. Collins *et al.*, *J.C.S. Dalton*, **1974**, 1094.
[2c] L. G. Marzilli *et al.*, *Inorg. Chem.*, 1976, **15**, 1661.

(22-F-Ia)   (22-F-Ib)

(22-F-IIa)   (22-F-IIb)

Ruthenium tetraoxide reacts more vigorously with organic substances but also has some uses as an oxidant; a convenient catalytic method uses $RuCl_3$ in sodium hypochlorite solutions.

Both RuO_4 and OsO_4 are soluble in alkali hydroxide solutions, but the behaviors are quite different. RuO_4 is reduced by hydroxide first to perruthenate(VII), which in turn is further reduced to ruthenate(VI):

$$4RuO_4 + 4OH^- \rightarrow 4RuO_4^- + 2H_2O + O_2$$
$$4RuO_4^- + 4OH^- \rightarrow 4RuO_4^{2-} + 2H_2O + O_2$$

On the other hand, OsO_4 gives the ion $[OsO_4(OH)_2]^{2-}$, discussed below. This difference between Ru and Os appears to be due to the ability of the $5d$ metal oxo anion to increase the coordination shell. Similar behavior occurs for ReO_4^- which in concentrated alkali gives yellow *meso*-perrhenate:

$$ReO_4^- + 2OH^- = [ReO_4(OH)_2]^{3-} = ReO_5^{3-} + H_2O$$

Ruthenates(VI) and -(VII). There is a close similarity between Ru and Mn in the oxo anions, both MO_4^- and MO_4^{2-} being known.

The fusion of Ru or its compounds with alkali in presence of an oxidizing agents gives a green melt containing the *perruthenate* ion RuO_4^-. Because of the high alkali concentration, on dissolution in water a deep orange solution of the stable ruthenate(VI) ion RuO_4^{2-} is obtained. However if RuO_4 is collected in ice-cold 1 *M* KOH, black crystals of $KRuO_4$ can be obtained, which are stable when dry. Perruthenate solutions, which are a yellowish green, are reduced by hydroxide ion, and kinetic studies suggest that unstable intermediates with coordinated OH^- are involved—this contrasts with the case of $3d$ metal oxo anions, where there is no evidence for addition of OH^-. Since H_2O_2 is also formed in the reduction and RuO_4^- is incompletely reduced to RuO_4^{2-} by H_2O_2, a step such as

$$[RuO_4(OH)_2]^{2-} \rightarrow RuO_4^{2-} + H_2O_2$$

is plausible.

The tetrahedral RuO_4^{2-} ion is moderately stable in alkaline solution. It is paramagnetic with two unpaired electrons, in contrast to osmate(VI). It may be noted that most ruthenium species in alkaline solution are specifically oxidized to RuO_4^{2-} by $KMnO_4$; hypochlorite gives a mixture of the RuO_4^- and RuO_4^{2-} ions, and Br_2 gives RuO_4^-. The RuO_4^- ion can be conveniently reduced to RuO_4^{2-} by iodide ion, although further reduction can occur with an excess of I^-. RuO_4^{2-} has oxidizing properties that find use in organic chemistry.[3]

Osmates. OsO_4 is moderately soluble in water and its absorption spectrum in the solution is the same as in hexane, indicating that it is still tetrahedral. However in strong alkaline solution coordination of OH^- ion occurs and a deep red solution is formed:

$$OsO_4 + 2OH^- \rightarrow [OsO_4(OH)_2]^{2-}$$

from which red salts such as $K_2[OsO_4(OH)_2]$ can be isolated. These "perosmates" or "osmenates" have *trans*-hydroxo groups. The reduction of such perosmate solutions by alcohol or other agents gives the osmate(VI) ion, which is pink in aqueous solutions but blue in methanol; its salts are also obtained in the alkaline oxidative fusion of the metal; in solution and in salts the ion is octahedral $[OsO_2(OH)_4]^{2-}$. Unlike the corresponding RuO_4^{2-} ion, it is diamagnetic. The diamagnetism of the ion, its substituted derivatives such as $[OsO_2Cl_4]^{2-}$ and $[RuO_2Cl_4]^{2-}$, all of which have *trans*-dioxo groups, can be explained in terms of ligand field theory. If the z axis passes through the two oxide ligands and the x and y axes through OH, there will be a tetragonal splitting of the e_g level into two singlets $d_{x^2-y^2}$ and d_{z^2}, whereas the t_{2g} level gives a singlet d_{xy} and a doublet $d_{xz} d_{yz}$. The oxide ligands will form $Os{=}O$ bonds by π overlap mainly with d_{xz} and d_{yz}, thus destabilizing those orbitals, leaving a low-lying d_{xy} orbital that will be occupied by the two electrons, leading to diamagnetism.

When OsO_4 is reduced by EtOH in presence of pyridine, the binuclear complex 22-F-III is obtained.[4] This compound reacts with nucleosides via the 2′ and 3′ hydroxyl groups of the sugar ring to give complexes[5] like 22-F-IV.

(22-F-III)   (22-F-IV)

With tRNA a similar reaction occurs and has provided a way to introduce heavy atoms for structure determination.

Other Oxo Species. When RuO_4 is treated with gaseous HCl and Cl_2, hygroscopic crystals of $(H_3O)_2[RuO_2Cl_4]$ are produced from which Rb and Cs salts can be obtained. The ion is hydrolyzed by water:

$$2Cs_2RuO_2Cl_4 + 2H_2O \rightarrow RuO_4 + RuO_2 + 4CsCl + 4HCl$$

[3]   M. Schroder and W. P. Griffith, *J.C.S. Chem. Comm.,* **1979,** 58.

[4]   W. P. Griffith and R. Rossetti, *J.C.S. Dalton,* **1972,** 1449; see also E. J. Behrman *et al., Inorg. Chem.,* 1979, **18,** 1364.

[5]   J. F. Conn *et al., J. Am. Chem. Soc.,* 1974, **96,** 7152; F. B. Daniel and E. J. Behrman, *J. Am. Chem. Soc.,* 1975, **97,** 7352.

but there is evidence for other Ru^{VI} species in solution. A green ion, possibly $[RuO_2(SO_4)_2]^{2-}$, can be obtained by reducing RuO_4 in dilute H_2SO_4 with Na_2SO_3, Fe^{2+}, or Ru^{IV}; it decomposes within a few hours to Ru^{IV}.

The osmate ion $[OsO_2(OH)_4]^{2-}$ can undergo substitution reactions with various ions such as Cl^-, Br^-, CN^-, $C_2O_4^{2-}$, and NO_2^-, to give orange or red crystalline salts, sometimes referred to as osmyl derivatives. They can also be obtained directly from OsO_4 with which, for example, aqueous KCN gives the salt $K_2[OsO_2(CN)_4]$. This particular ion is unaffected by hydrochloric or sulfuric acid, but the other oxo anions are not very stable in aqueous solutions, although they are considerably more stable than the ruthenyl salts mentioned above.

The $[OsO_2(OH)_4]^{2-}$ ion reacts with ethylenediamine to give[6] *trans*-$[OsO_2en_2]^{2+}$. From this ion other interesting complexes may be made. Reduction with Zn/Hg affords $[OsH_2en_2]^{2+}$ from which aqueous HX produces[7] the $[Os en_2X_2]^{2+}$ ions.

22-F-3. Ruthenium and Osmium Halo Complexes

In view of the importance of halo complexes in aqueous solution, we discuss those in different oxidation states together.

Ruthenium Chloro Complexes. Figure 22-F-1 gives some common reactions of the chloro species.

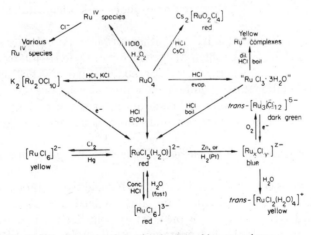

Fig. 22-F-1. Some reactions of ruthenium chloro complexes.

Ruthenium(IV) Chloro Complexes. The reduction of RuO_4 by HCl in presence of KCl gives red crystals of $K_4[Ru_2OCl_{10}]$, which was the first example of the ions $M_2OX_{10}^{4-}$ (X = Cl or Br; M = Ru or Os). This complex has the structure shown in Fig. 22-F-2. The diamagnetism of these ions can be understood by using a simple MO treatment for the M—O—M group. If we assume that these atoms lie along the z axis of a coordinate system, and that the ligand field around each M^{IV} ion

6   J. M. Malin *et al., Inorg. Chem.*, 1971, **10**, 2403; 1977, **16**, 615.
7   A. L. Coelho and J. M. Malin, *Inorg. Chim. Acta,* 1975, **14**, L41.

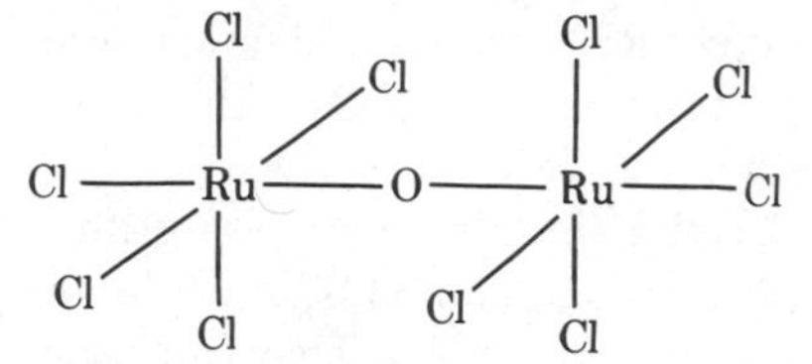

Ru—O, 1.8 Å ; Ru—Cl, 2.34 Å
∠RuORu, 180°; ∠ClRuO, 90°

Fig. 22-F-2. The structure of the $[Ru_2Cl_{10}O]^{4-}$ ion.

is essentially octahedral, the ions will then each have a $d^2_{xy}d_{xz}d_{yz}$ configuration before interaction with the oxygen. By interaction of the d_{xz} orbitals on each M^{IV} and the p_x orbital of oxygen, three three-center MOs, one bonding, one approximately nonbonding, and one antibonding, will be formed. The four electrons (one from each M^{IV} and two from oxygen) will occupy the lower two of these MOs. The same kind of interaction will occur in the Md_{yz}—Op_y—Md_{yz} set of orbitals; thus all electron spins become paired. The Os complexes are less hydrolytically stable than the Ru ones but are essentially the same structurally and electronically.[8]

The reduction of $[Ru_2OCl_{10}]^{2-}$ usually proceeds to Ru^{III} species, so that the simple *hexachlororuthenate*(IV) ion or its salts are best prepared by the oxidation of Ru^{III} chloro species by chlorine. Although the ion is yellow in solution, the salts are dark brown or purple; they are isomorphous with other octahedral Os, Ir, Pd, and Pt salts.

In aqueous solution $[RuCl_6]^{2-}$ is rather unstable. Perchlorate solutions containing Ru^{IV}, which may be obtained by reducing RuO_4 in $HClO_4$ by H_2O_2, contain a variety of poorly characterized species, probably in part polynuclear and possibly with mixed oxidation states; addition of Cl^- to such solutions produces a series of color changes whose sequence and persistence times depend on pH and the Cl^- concentration. Polynuclear species, and chloro, aqua, and hydroxo species, are doubtless involved.

Ruthenium(III) Chloro Complexes.   These are among the best characterized complexes.

When RuO_4 is collected in concentrated HCl and the solution is evaporated, a dark red, deliquescent, crystalline material is obtained. This commercial product, usually called "$RuCl_3·3H_2O$," is the starting point for the preparation of most ruthenium compounds. Though containing Ru^{III} species such as $RuCl_3·(H_2O)_3$, it appears also to contain some polynuclear Ru^{IV} conplexes. It is readily soluble in water, ethanol, acetone, and simple solvents, so that reactions with organic-soluble compounds such as phosphines and alkenes can be readily induced. Although the deep red aqueous solutions initially do not precipitate AgCl on treatment with Ag^+, they darken rapidly owing to hydrolysis. When dilute HCl solutions are heated or are shaken with Hg (as a reductant for Ru^{IV}), the resulting yellow solutions contain not only the aqua ion $[Ru(H_2O)_6]^{3+}$ but chloro series such as cis-$RuCl_3(H_2O)_3$,

[8]   J. San Filippo et al., Inorg. Chem., 1976, **15**, 269; 1977, **16**, 1016.

mer-$RuCl_3(H_2O)_3$, $[RuCl(H_2O)_5]^{2+}$, and *cis*- and *trans*-$[RuCl_2(H_2O)_4]^+$. These can be separated by ion exchange and identified by their electronic spectra. Only with very high concentrations of Cl^- is $[RuCl_6]^{3-}$ formed. The rate of replacement of Cl^- by H_2O increases with the number of Cl^- ions present: the aquation of $[RuCl_6]^{3-}$ to $[RuCl_5H_2O]^{2-}$ is of the order of seconds in water, but the half-reaction time for conversion of $[RuCl(H_2O)_5]^{2+}$ into $[Ru(H_2O)_6]^{3+}$ is about a year. The yellow *trans*-$[RuCl_2(H_2O)_4]^+$ and green *trans*-$[RuCl_4(H_2O)_2]^-$ can be obtained by oxidation of the deep blue complexes to be discussed just below.

Ruthenium(III) chloro species catalyze the reduction of Fe^{3+} by H_2O_2 and also the hydration of acetylenes. With CO or formic acid, Ru^{III} chloro species are formed, as discussed below.

Ruthenium(II) Chloro Complexes. It has long been known that deep inky-blue solutions are obtained when solutions of Ru^{III} chloro complexes in HCl solution are reduced electrolytically or chemically, e.g., by Ti^{3+}, or by H_2 (2 atm) in presence of platinum black. Blue solutions are also obtained on treating the acetate $Ru_2(CO_2CH_3)_4Cl$ (oxidation state of Ru, + 2.5) and the hexammine $[Ru^{II}(NH_3)_6]Cl_2$ with HCl. The constitution of the blue species is not settled, and there may be more than one. It seems clear that they (or it) are polynuclear with ruthenium in a mean oxidation state between +2 and +3. A blue ammine salt appears to be $[Ru_2(NH_3)_6Cl_4(H_2O)]Cl$.

The chloride solutions are air sensitive and form *trans*-$[RuCl_4(H_2O)_2]^-$; in absence of air, oxidation by water gives hydrogen and yellow *trans*-$[RuCl_2(H_2O)_4]^+$. The blue solutions prepared by reduction of $RuCl_3 \cdot 3H_2O$ in methanol provide a useful source for the preparation of other Ru^{II} and Ru^{III} complexes. Electrolytic reduction of solutions of $RuCl_3 \cdot 3H_2O$, from which much Cl^- has been removed by addition of $AgBF_4$, followed by ion-exchange separation gives solutions containing pink $[Ru(H_2O)_6]^{2+}$; this is oxidized by air or ClO_4^-:

$$[Ru(H_2O)_6]^{3+} + e = [Ru(H_2O)_6]^{2+} \qquad E^0 = 0.23 \text{ V}$$

Other Halo Complexes. We have emphasized the chloro complexes of ruthenium because they are most common and important. There are bromo analogues of many of them, and osmium has generally similar though less extensive chemistry. A few specific facts are worth mentioning here.

$[OsX_6]^{n-}$ *complexes* are numerous. Most important is the $[OsCl_6]^{2-}$ ion obtained by reducing OsO_4 in HCl using alcohol or Fe^{2+}. Its salts are orange to brown and it can be reduced to $[OsCl_6]^{3-}$, but unlike the Ru analogue there is no evidence for further reduction to $[OsCl_6]^{4-}$. The $[OsCl_6]^{3-}$ ion is rather unstable and easily hydrolyzed to a hydrous oxide. By exploiting the trans effect, stereospecific reactions have been devised to yield particular isomers[9] of the mixed halo ions $[OsCl_x\text{-}Br_yI_z]^{2-}$ $(x + y + z = 6)$. The four $[OsX_6]^{4-}$ ions with X = F, Cl, Br, and I, have all been well characterized electronically; they are paramagnetic with t_{2g}^4 ground configurations,[10] as expected.

9   W. Preetz *et al.*, *Z. Anorg. Allg. Chem.*, 1974, **407**, 1, and earlier references therein.
10   K.-I. Ikeda and S. Maeda, *Inorg. Chem.*, 1978, **17**, 2698; P. N. Schatz *et al.*, *Inorg. Chem.*, 1978, **17**, 2689.

The $[RuF_6]^{1-,2-,3-}$ ions are all known and have been thoroughly studied spectroscopically.[11]

22-F-4. Complexes with Nitrogen Donor Atoms

These complexes are largely formed by the metals in the II and III oxidation states, and by and large the ruthenium complexes are far more thoroughly investigated. It is most convenient to divide the material according to ligand types (i.e., NH_3 vs. aromatic amines) and to treat the nitrosyl complexes separately in the next section.

Ammonia Complexes.[12a]   The orange hexammine of Ru^{II} is obtained as $[Ru(NH_3)_6]Cl_2$ by Zn dust reduction of a strongly ammoniacal solution of "$RuCl_3\cdot3H_2O$" containing excess NH_4Cl. As obtained by this method it is often contaminated by the nitrogen complex $[Ru(NH_3)_5N_2]Cl_2$ (see below). The hexammine is a reductant,

$$[Ru(NH_3)_6]^{3+} + e = [Ru(NH_3)_6]^{2+} \qquad E^0 = 0.24 \text{ V}$$

but it is sufficiently substitution inert for the electron transfer to proceed by an outer-sphere mechanism. The aquation of $[Ru(NH_3)_6]^{2+}$ is dependent on the H^+ ion concentration, unlike aquation reactions of, e.g., $[Cr(NH_3)_6]^{3+}$. The aqua pentammine ion $[Ru(NH_3)_5(H_2O)]^{2+}$ is a starting material for much other Ru^{II} chemistry. It reacts readily with N_2O to give the only known complex containing N_2O as a ligand. Though conclusive proof is lacking, the mode of bonding appears to involve a linear Ru—N—N—O chain.[12b] It also reacts with a variety of compounds containing sulfur[13] to give the $[Ru(NH_3)_5SSRu(NH_3)_5]^{4+}$ ion, in which there is a trans, planar Ru—S—S—Ru bridge.[14] It has been suggested that this cation is best regarded as a Ru^{II}/Ru^{III} complex bridged by S_2^- with the odd electron delocalized.

It is also possible to prepare many more highly substituted complexes from the aqua pentammine ion. For example, a series of *trans*-$[Ru(NH_3)_4(H_2O)L]^{2+}$ ions has been used to establish a trans-labilizing order for ligands L in octahedral complexes.[15] The specific rates of second-order reactions to replace water by the ligand isonicotinamide (isn), giving *trans*-$[Ru(NH_3)_4L(isn)]^{2+}$, are in the order N_2, CO < isn < py < imidazole < NH_3 < OH^- < $P(OEt)_3$ < CN^- < SO_3^{2-}.

The Ru^{II} ion is low spin (t_{2g}^6) in all its ammine complexes and is a remarkably good π donor.[16] This is shown not only by the formation of N_2, CO, and similar complexes, but also by the fact that in nitrile complexes $[Ru(NH_3)_5NCR]^{2+}$, the CN stretching frequency is substantially lower than in the free nitrile, whereas in

[11]   G. C. Allen, G. A. El-Sharkawy, and K. D. Warren, *Inorg. Chem.,* 1973, **12**, 2231.

[12a]   H. Taube, *Survey of Progress in Chemistry,* Vol. 6, Academic Press, 1974.

[12b]   F. Bottomley and W. V. F. Brooks, *Inorg. Chem.,* 1978, **17**, 501.

[13]   H. Taube, *et al., J. Am. Chem. Soc.,* 1973, **95**, 4758; *Inorg. Chem.,* 1979, **18**, 2212.

[14]   R. C. Elder and M. Trkula, *Inorg. Chem.,* 1977, **16**, 1048.

[15]   D. W. Franco and H. Taube, *Inorg. Chem.,* 1978, **17**, 571.

[16]   H. Taube, *et al., Inorg. Chem.,* 1979, **18**, 2216,

other metal nitrile complexes, even in $[Ru^{III}(NH_3)_5NCR]^{3+}$, the frequency is invariably higher.

Doubtless the most celebrated ammine complex of Ru^{II} is the nitrogen complex $[Ru(NH_3)_5N_2]^{2+}$, which was the first N_2 complex recognized to exist and triggered the rapid development of the chemistry of such complexes (cf. p. 86). This particular complex, which doubtless owes its great stability to the strong π-donor property of the Ru^{II} ion just mentioned, can be prepared in many ways. With the $[Ru(NH_3)_5H_2O]^{2+}$ ion there are the following two reactions:

$$Ru(NH_3)_5(H_2O)^{2+} + N_2 \rightleftharpoons Ru(NH_3)_5N_2^{2+} \qquad K \sim 3 \times 10^4$$
$$Ru(NH_3)_5N_2^{2+} + Ru(NH_3)_5H_2O^{2+} \rightleftharpoons [(NH_3)_5Ru-N-N-Ru(NH_3)_5]^{4+} \quad K \sim 7 \times 10^3$$

The second reaction, forming the bridged ion, exemplifies the donor properties of coordinated N_2. The Ru—N—N—Ru group is nearly linear and the N—N distance (1.124 Å) is only slightly longer than in N_2 itself (1.0976 Å). The bonding can be described by MO theory similar to that for Ru—O—Ru discussed above. The N_2O complex is very rapidly reduced by Cr^{2+} to form the N_2 complex, the overall reaction being

$$[Ru(NH_3)_5N_2O]^{2+} + 2Cr^{2+} + 2H^+ \rightarrow [Ru(NH_3)_5N_2]^{2+} + 2Cr^{3+} + H_2O$$

It is evident that coordination of N_2O lowers the N—O bond strength, but the precise course of the reduction is still uncertain. Azido Ru^{III} ammines are unstable:

$$[Ru(NH_3)_5N_3]^{2+} \rightarrow [Ru(NH_3)_5N_2]^{2+} + \tfrac{1}{2}N_2$$

but in presence of acid, bridged dimers are also formed, and a nitrene complex Ru=NH may be involved.

Still another reaction leading to $[Ru(NH_3)_5N_2]^{2+}$ is the following[17]:

$$[Ru(NH_3)_6]^{3+} + NO + OH^- \rightarrow [Ru(NH_3)_5N_2]^{2+} + 2H_2O$$

It is interesting that in acid solution one gets only the simple replacement of NH_3 by NO:

$$[Ru(NH_3)_6]^{3+} + NO + H^+ \rightarrow [Ru(NH_3)_5NO]^{3+} + NH_4^+$$

There are also N_2 complexes of osmium, $[Os(NH_3)_5N_2]^{2+}$ and cis-$[Os(NH_3)_4(N_2)_2]^{2+}$, which are generally similar to their ruthenium analogues.[18]

Ruthenium(III), with a t_{2g}^5 configuration, is, in contrast to Ru^{II}, a very good π acceptor. This is dramatically demonstrated by the rates of base hydrolysis of free and coordinated nitriles:

$$[Ru(NH_3)_5NCR]^{3+} + H_2O \rightarrow [Ru(NH_3)_5NH_2C(O)R]^{3+}$$
$$NCR + H_2O \rightarrow H_2N\underset{\underset{O}{\|}}{C}R$$

The reaction of the coordinated RCN is 10^8 to 10^9 times faster because the π-ac-

17   S. D. Pell and J. N. Armor, *J. Am. Chem. Soc.*, 1973, **95**, 7625.

18   T. Matsubara, M. Bergkamp, and P. C. Ford, *Inorg. Chem.*, 1978, **17**, 1604.

ceptor character of the $Ru(NH_3)_5^{3+}$ moiety stabilizes the transition state or intermediate resulting from attack of OH^- on the carbon atom by taking electron density from the $C\equiv N$ bond.[19]

The reduction of ammines of the type $[Ru(NH_3)_5L]^{3+}$ with Cr^{2+} and other reducing agents has been studied in detail. The reactions are similar to those of Co^{III} complexes except that the electron enters the t_{2g} rather than the e_g level and the resulting Ru^{II} complexes are diamagnetic. Bridged species are probably intermediates.

In some reactions the rate-determining step appears to be attack on the d-electron density of the metal atom. Thus the hexaammine $[Ru(NH_3)_6]^{3+}$ undergoes aquation only very slowly at room temperature but reacts rapidly with NO, as noted above.

The redox behavior of a range of $[Ru(NH_3)_5L]^{2+,3+}$ complexes has been studied in detail and found to respond in the expected way to the inductive effects of L.[20a]

"Ruthenium Red." A characteristic of ruthenium complex ammine chemistry is the formation of highly colored red or brown species usually referred to as ruthenium reds. Thus if commercial ruthenium chloride is treated with ammonia in air for several days, a red solution is obtained. Alternatively, if Ru^{III} chloro complexes are reduced by refluxing ethanol and the resulting solution is treated with ammonia and exposed to air at 90° with addition of more ammonia at intervals, again a red solution is obtained. Crystallization of the solutions gives ruthenium red, $[Ru_3O_2(NH_3)_{14}]Cl_6\cdot4H_2O$. This contains an essentially linear trinuclear ion with oxygen bridges between the metal atoms, with $Ru-O = 1.85$ Å and $Ru-N = 2.13$ Å.[20b]

$$[(NH_3)_5Ru^{III}-O-Ru^{IV}(NH_3)_4-O-Ru^{III}(NH_3)_5]^{6+}$$

The ethylenediamine derivative, $[(NH_3)_5RuORu(en)_2ORu(NH_3)_5]Cl_6$, is similar.

Since the average oxidation state of Ru is $3\frac{1}{3}$, the metal atoms must be in different formal oxidation states. The diamagnetism can be ascribed to $Ru-O-Ru$ π-bonding as in the $[M_2OX_{10}]^{4-}$ ions. The ion above can be oxidized in acid solution by air, Fe^{3+}, or Ce^{4+} to a brown paramagnetic ion of the same constitution but with charge +7. It is likely that there are corresponding trinuclear chloro complexes, such as $[Ru_3O_2Cl_6(H_2O)_6]$, in the violet aqueous solutions of $RuCl_3\cdot3H_2O$.

Some of the above-mentioned and other important reactions of Ru^{II} and Ru^{III} ammine complexes are summarized in Fig. 22-F-3.

Aromatic Amine Complexes. Complexes with aromatic amine ligands, which have π systems, have distinctly different features of interest from the ammine complexes. The most important species are those containing 2,2'-bipyridyl (dipy) or 1,10-phenanthroline (phen), which form tris chelates.

[19]   A. W. Zanella and P. C. Ford, *Inorg. Chem.,* 1975, **14**, 42.
[20a] T. Matsubara and P. C. Ford, *Inorg. Chem.* 1976, **15**, 1107.
[20b] A. C. Skapski, personal communication.

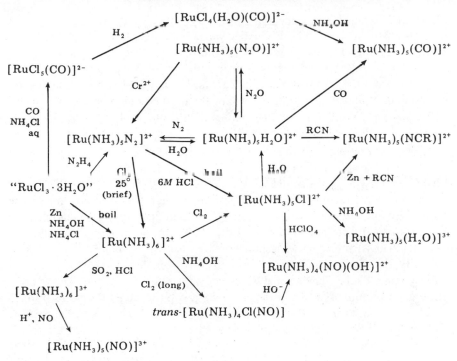

Fig. 22-F-3. Some reactions of ruthenium ammines in aqueous solutions.

$[Ru(bipy)_3]^{2+}$ is extensively used as a sensitizer in photodriven chemical and physical processes.[21] The electronic structure of the excited ion $*[Ru(bipy)_3]^{2+}$ appears to result from transfer of an electron from a metal t_{2g} orbital to a ligand $\pi*$ orbital.[22] This excited state can be described as $Ru^{III}\text{-}L^-$, and it reverts to the ground state by photoemission without chemical reaction unless suitable reactants called quenchers are present, as discussed shortly. This is in distinct contrast to the photochemical behavior of the simple ammines, which respond to photoexcitation[23] by promptly aquating or oxidizing:

$$[Ru(NH_3)_5L]^{2+} + H_2O \xrightarrow{h\nu} [Ru(NH_3)_5H_2O]^{2+} \text{ or } [Ru(NH_3)_4L(H_2O)]^{2+}$$

$$[Ru(NH_3)_5L]^{2+} + H^+ \xrightarrow{h\nu} [Ru(NH_3)_5L]^{3+} + \tfrac{1}{2} H_2$$

The quenching of the emission from the $*[Ru(bipy)_3]^{2+}$, $*[Ru(phen)_3]^{2+}$, and related ions has been extensively studied. There are three important quenching processes as summarized by the following general equation:

$$*[RuL_3]^{2+} + \begin{cases} Q_1 \to [RuL_3]^+ + Q_1^+ \\ Q_2 \to [RuL_3]^{3+} + Q_2^- \\ Q_3 \to [RuL_3]^{2+} + Q_3^* \end{cases}$$

[21] S. Anderson and K. R. Seddon, *J. Chem. Res. (S)*, **1979**, 74; J.-M. Lehn et al., *Nouv. J. Chim.*, 1979, **3**, 423, 449; D. G. Whitten et al., *J. Am. Chem. Soc.*, 1979, **101**, 4007.

[22] J. Van Houten and R. J. Watts, *Inorg. Chem.*, 1978, **17**, 3381.

[23] T. Matsubara and P. C. Ford, *Inorg. Chem.*, 1978, **17**, 1747.

Q_1, Q_2, and Q_3 represent quenching agents that work by reduction, oxidation, or energy transfer, respectively. Reductive quenchers include Eu^{2+},[24] a series of pentacyanoferrate(II) complexes,[25] and others.[26] Oxidative quenching is given by $[Co(NH_3)_5py]^{3+}$ [27] and energy transfer[28] by $[Ni(CN)_4]^{2-}$. The $*[RuL_3]^{2+}$ complexes are much stronger oxidizing agents (by *ca.* 2 V) than the ground state species, and reductive quenching is perhaps the most common process.

$[Os(bipy)_3]^{2+}$ is similar to its Ru analogue photochemically but has been less extensively studied.[29]

A few other complexes with aromatic amine ligands are notable. Replacement of bipy by phosphine ligands, as in $[Ru(bipy)_2(PR_3)X]^+$ or $[Ru(bipy)_2-(Ph_2PCH_2PPh_2)]^{2+}$, gives complexes that are even more resistant to oxidation by further stabilizing the metal $\pi("t_{2g}")$ electrons through Ru—P π bonding.[30] The oxo-bridged dinuclear complexes such as $[(bipy)_2RuORu(bipy)_2]^{4+}$ exhibit strong antiferromagnetic coupling ($2J = -119$ to -173 cm^{-1}) through the essentially linear Ru—O—Ru bridges with short (*ca.* 1.88 Å) Ru—O bonds.[31a]

Finally, the simple ion $[Rupy_6]^{2+}$ has recently been described.[31b]

The Creutz-Taube Complex and Related Complexes. There is a series of complexes whose common feature is the presence of two (or more) Ru atoms bridged by bidentate ligands through which some, potentially adjustable, degree of electron transfer can take place. Such complexes have been extensively studied[32] in recent years because of the information they may give about the processes of electron exchange in bimolecular redox reactions (cf. Chapter 28), and about the general character of electron transmission through chemical systems, including the spectroscopic phenomena associated with such processes.

A point of focus for much of this work has been the complex 22-F-V, often called the Creutz-Taube complex, after its discoverers.[33] As depicted in 22-F-V it is formally a mixed-valence (Ru^{II}-Ru^{III}) species; the +4 (Ru^{II}-Ru^{II}) and +6 (Ru^{III}-Ru^{III}) ions are also known and can be isolated as salts of various large anions such as tosylate, perchlorate, or dithionate.[34]

The question at issue in the Creutz-Taube ion, and similar ones, is whether the Ru valences of II and III are localized (trapped), one at each end, thus making the ends nonequivalent, or whether there is delocalization through the bridging ligand

24   C. Creutz, *Inorg. Chem.,* 1978, **17,** 1046.
25   H. E. Toma and C. Creutz, *Inorg. Chem.,* 1977, **16,** 545.
26   C. Creutz and N. Sutin, *Inorg. Chem.,* 1976, **15,** 496.
27   K. R. Leopold and A. Haim, *Inorg. Chem.,* 1978, **17,** 1753.
28   J. N. Demas and A. W. Adamson, *J. Am. Chem. Soc.,* 1973, **95,** 5159.
29   J. N. Demas *et al., J. Am. Chem. Soc.,* 1975, **97,** 3838.
30   B. P. Sullivan, D. J. Salmon, and T. J. Meyer, *Inorg. Chem.,* 1978, **17,** 3334.
31a T. R. Weaver *et al., J. Am. Chem. Soc.,* 1975, **97,** 3039; D. W. Phelps, E. M. Kahn, and D. J. Hodgson, *Inorg. Chem.,* 1975, **14,** 2486.
31b J. L. Templeton, *J. Am. Chem. Soc.,* 1979, **101,** 4906.
32   T. J. Meyer, ACS Advances in Chemistry Series No. 150, 1976, Chapter 7; *Ann. N.Y. Acad. Sci.,* 1978, **313,** 481; H. Taube, *Ann. N.Y. Acad. Sci.,* 1978, **313,** 496.
33   C. Creutz and H. Taube, *J. Am. Chem. Soc.,* 1973, **95,** 1086.
34   T. Emilsson and V. S. Srinivasan, *Inorg. Chem.,* 1978, **17,** 491.

$$\left[(NH_3)_5RuN\bigcirc NRu(NH_3)_5\right]^{5+}.$$

(22-F-V)

to make the whole ion symmetrical. For the Creutz-Taube ion itself it appears on the basis of spectroscopic data[35] that the valence states are either not trapped or only barely so. An absorption band in the near infrared has been assigned as an internal electron transfer transition and from its width and frequency the rate of the process was calculated to be ca. 10^9 sec^{-1}, but this is subject to some uncertainty. Crystal structure data on one of its salts are most consistent with the assumption that the ion is symmetrical.[36] However even with trapped valences only small differences in bond lengths for Ru^{II}—N and Ru^{III}—N bonds would be expected, and disordering of a slightly unsymmetrical ion cannot be ruled out conclusively.

Though the nature of the Creutz-Taube ion itself has been controversial, the related ion 22-F-VI is clearly a trapped-valence case.[37]

$$\left[(bipy)_2ClRuN\bigcirc NRuCl(bipy)_2\right]^{3+}$$

(22-F-VI)

In the necessarily unsymmetrical ion 22-F-VII, the Ru^{III} is definitely localized at the pentammine end. This ion provides a good model for study of the intervalence transfer absorption bands characteristic of trapped-valence species.[38]

$$\left[(NH_3)_5RuN\bigcirc NRuCl(bipy)_2\right]^{4+}$$

(22-F-VII)

The intervalence transfer absorption band corresponds to the energy required for the process

$$[Ru^{II}-Ru^{III}]^{3+} \xrightarrow{h\nu} {}^*[Ru^{III}-Ru^{II}]^{3+}.$$

This process absorbs energy even when both metal atoms have the same ligand set because the product contains its Ru^{II} in a ligand arrangement (bond lengths and angles) characteristic of Ru^{III} and its Ru^{III} in a environment suited to Ru^{II}. According to the Franck-Condon principle, the electron "jumps" instantaneously, and only later can vibrational relaxation allow the environmental relaxation processes to catch up. Actually, the differences in environments, mainly Ru—N distances, are probably quite small, and that is why only the low energy (ca. 100 kJ

[35]   J. K. Beattie, N. S. Hush, and P. R. Taylor, *Inorg. Chem.*, 1976, **15**, 992; T. C. Strekas and T. G. Spiro, *Inorg. Chem.*, 1976, **15**, 974.
[36]   J. K. Beattie *et al.*, *J.C.S. Dalton*, **1977**, 1121.
[37]   T. J. Meyer *et al.*, *J. Am. Chem. Soc.*, 1977, **99**, 1064.
[38]   T. J. Meyer *et al.*, *Inorg. Chem.*, 1976, **15**, 1457.

mol^{-1}) of infrared radiation suffices to cause the jump.

Some ions similar to 22-F-VI but with different bridging ligands, such as 22-F-VIII through 22-F-X, have also been shown to have trapped valences.[39]

(22-F-VIII)   (22-F-IX)   (22-F-X)

A series of extended species of this type (22-F-XI) has been reported, with $x = 1$ to 4. As these are oxidized, the terminal Ru atoms are oxidized to Ru^{III} first and, except for the trinuclear species, the site of oxidation is well localized, with no detectable end-to-end electron transfer.[40]

(22-F-XI)

Bridging by μ-Cl atoms does not appear to be effective compared to that by pyrazine or related aromatic diamines. Thus when [(bipy)$_2$Ru(μ-Cl)$_2$Ru(bipy)$_2$]$^{2+}$ is oxidized it gives a transient 3+ ion that cleaves spontaneously to give [Ru(bipy)$_2$Cl$_2$]$^+$ and [Ru(bipy)$_2$L$_2$]$^{2+}$, where L is a solvent molecule. Evidently there is only weak interaction between discrete Ru^{II} and Ru^{III} sites in the 3+ dimer.[41]

22-F-5. Nitric Oxide Complexes of Ruthenium[42] and Osmium

The formation of nitric oxide complexes is a marked feature of ruthenium chemistry; those of Os have been less well studied, but where known they are even more stable than the Ru analogues.

The group RuNO can occur in both anionic and cationic octahedral complexes in which it is remarkably stable, being able to presist through a variety of substitution and oxidation-reduction reactions. Ruthenium solutions or compounds that have at any time been treated with nitric acid can be suspected of containing nitric oxide bound to the metal. The presence of NO may be detected by infrared absorption $ca.$ 1930 to 1845 cm^{-1}.

Almost any ligand can be present along with the RuNO group; phosphines are considered in the next section, but conventional complexes are [Ru(NO)Cl$_5$]$^{2-}$, [Ru(NO)(NH$_3$)$_4$Cl]$^{2+}$, and Ru(NO)[S$_2$CNMe$_2$]$_3$. The complexes can be obtained in a variety of ways, and the source of NO can be HNO$_3$, NO, NO$_2$, or NO$_2^-$. A

[39]   M. J. Powers and T. J. Meyer, *Inorg. Chem.*, 1978, **17**, 2955.
[40]   A. von Kameke, G. M. Tom, and H. Taube, *Inorg. Chem.*, 1978, **17**, 1790.
[41]   T. J. Meyer *et al.*, *Inorg. Chem.*, 1978, **17**, 2211.
[42]   F. Bottomley, *Coord. Chem. Rev.*, 1978, **26**, 7.

few examples will illustrate the preparative methods. If RuO_4 in ~$8M$ HCl is evaporated with HNO_3, a purple solution is obtained from which the addition of ammonium chloride precipitates the salt $(NH_4)_2[RuNOCl_5]$. If this salt is boiled with ammonia, it is converted into the golden yellow salt $[RuNO(NH_3)_4Cl]Cl_2$. When commercial "ruthenium chloride" in HCl solution is heated with NO and NO_2, a plum-colored solution is obtained from which brick red $RuNOCl_3 \cdot 5H_2O$ can be isolated. The addition of base to the solution gives a dark brown, gelatinous precipitate of $RuNO(OH)_3 \cdot H_2O$. When this hydroxide is boiled with $8M$ HNO_3 and the solution evaporated, red solutions are obtained from which ion-exchange separation has allowed identification of species such as $[Ru(NO)(NO_3)_4H_2O]^-$, $[Ru(NO)(NO_3)_2(H_2O)_3]^+$, $[Ru(NO)(NO_3)(H_2O)_4]^{2+}$, and $[Ru(NO)(H_2O)_5]^{3+}$. Other complex anions are present as well as neutral species; the main one, $[Ru(NO)(NO_3)_3(H_2O)_2]$, can be extracted into tributyl phosphate.

Some cases are known where the RuNO group is attacked[43a]; one is similar to the well-known attack of OH^- on FeNO to give $FeNO_2$ (Section 29-14), namely,

$$[RuX(bipy)_2(NO)]^{2+} + 2OH^- \rightleftharpoons RuX(bipy)_2NO_2 + H_2O$$

and is reversed by acid. Other reactions with nucleophiles also feature attack on the N atom as in the following reaction,[43b] which is yet another route to the N_2 ammine complex.

$$[Ru(NH_3)_5NO]^{3+} + 2RNH_2 \rightarrow [Ru(NH_3)_5N_2]^{2+} + RNH_3^+ + ROH$$

The vast majority of RuNO complexes are of the general type $Ru(NO)L_5$, in which the metal atom is *formally* in the divalent state, if we postulate electron transfer from NO to the metal as Ru^{III} followed by donation from NO^+. These may be designated $\{RuNO\}^6$ complexes, the superscript 6 indicating a (formal) total of six d electrons. There is a significant number of $\{RuNO\}^8$ and $\{RuNO\}^{10}$ complexes, but these all contain π-acid ligands such as phosphines. In all $\{RuNO\}^6$ complexes the Ru—N—O chains are essentially linear.

For iron there are only a few octahedral $\{FeNO\}^6$ complexes except with cyanide as an associated ligand, and the different behavior of the two elements could be attributed in part to the relatively low stabilization energy of the $t_{2g}^3 e_g^2$ ion for ruthenium and the consequent readiness of Ru^{III} to accept an electron from NO, giving Ru^{II} (t_{2g}^6); the larger size of Ru^{3+} (~0.72) than of Fe^{3+} (~0.64) would also favor better $d\pi$–$p\pi$ overlap for NO π bonding.

22-F-6. Tertiary Phosphine and Related Complexes of Ru and Os

In common with other platinum metals, an intensively studied area is the chemistry involving trialkyl- and triarylphosphines, the corresponding phosphites and, to a lesser extent, the arsines. An extremely wide range of complexes is known, mainly of the II state, although compounds in the 0, III, and less commonly, IV state are

[43a] J. A. McCleverty, *Chem. Rev.,* 1979, **79**, 53.
[43b] C. P. Guengerich and K. Schug, *Inorg. Chem.,* 1978, **17**, 1378.

known; other ligands commonly associated with the PR_3 group are halogens, alkyl and aryl groups, CO, NO, and alkenes.

The main preparative routes are as follows.

1. Interaction of "$RuCl_3 \cdot 3H_2O$," K_2OsCl_6, or other halide species with PR_3 in an alcohol or other solvent. In many of these reactions, either hydride or CO may be abstracted from the solvent molecule, leading to hydro or carbonyl species. Sodium borohydride is also used as reducing agent.

2. Complexes in the 0 oxidation state may be obtained either by reduction of halides such as $RuCl_2(PPh_3)_3$ with Na or Zn in presence of CO or other ligands such as RNC, or by reaction of metal carbonyls with phosphines[44]. Reactions of polynuclear carbonyls such as $Ru_3(CO)_{12}$ with phosphines tend to preserve the cluster structure.

3. For the M^0 and M^{II} species, oxidative-addition reactions with halogens or other molecules give oxidized species.

4. Carbonyl-containing complex ions such as $M(CO)Cl_5^{3-}$, cis-$M(CO)_2Cl_4^{2-}$, and $M(CO)_3Cl_3^-$ are formed by action of CO or formic acid on Ru and Os chloro complexes, and addition of phosphines to such solutions gives replacement products.

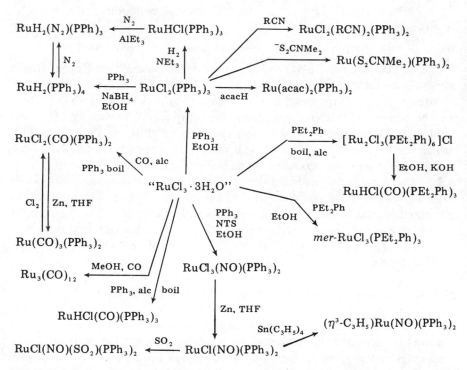

Fig. 22-F-4. Some reactions of tertiary phosphine complexes of ruthenium (alc = 2-methoxy-ethanol).

[44] D. J. Cole-Hamilton and G. Wilkinson, *J. C. S. Dalton*, **1979**, 1283; F. G. A. Stone et al., *J. C. S. Dalton*, **1979**, 1003.

Some typical reactions for ruthenium, which are generally representative also for osmium, are given in Fig. 22-F-4. It is to be noted that phosphines with differing steric and basic properties often give different types of product. The reaction conditions are also critical; for example, for trialkylphosphines and $RuCl_3$ in ethanol, short reaction times give $RuCl_3(PR_3)_3$, whereas prolonged reactions give $[Ru_2Cl_3(PR_3)_6]Cl$.

We can deal here only with a few selected compounds of recent or current interest. The red-brown $RuCl_2(PPh_3)_3$ has an sp structure in which the sixth octahedral position is blocked by an ortho hydrogen atom of one phenyl ring. On treatment with H_2 in base it is converted to $RuHCl(PPh_3)_3$, which has a tbp structure and is one of the most active catalysts known for homogeneous hydrogenation of alkenes, with remarkable specificity towards 1-alkenes.

Both Ru^{II} and Os^{II} form a series of complexes containing PR_3 and CO as ligands along with nitrate ions[45a] or trifluoroacetate ions.[45b] Some of the latter, of general formula $Ru(CO)(O_2CCF_3)_2(PR_3)_2$, function as dehydrogenation catalysts,[45] converting alcohols to aldehydes or ketones; the catalytic cycle includes the intermediate $RuH(CO)(O_2CCF_3)(PR_3)_2$, from which CF_3CO_2H regenerates the dicarboxylato complex with evolution of hydrogen. The hydrido intermediate can be isolated as a yellow, air-sensitive material.

Osmium(III) forms a variety of OsX_3L_3 type complexes with X = Cl, Br, I, and L = PR_3 and AsR_3. These undergo ligand replacement reactions, for example, with isocyanides, to give $OsCl_3(AsR_3)(CNR')_2$.[46] $OsCl_3(PEt_3)_3$ reacts with Zn in tetrahydrofuran under N_2 to give $OsCl_2(N_2)(PEt_3)_2$. There is no exact Ru analogue to this N_2 complex, but by using chelating phosphines nitrogen complexes for Fe, Ru, and Os of the type $trans$-$M(H)(N_2)(diphos)_2^+$ can be isolated.

Osmium(II) forms a large number of octahedral complexes[47] with diars, such as $trans$-$[Os(diars)_2Br(NO)]^{2+}$ and $[Os(diars)_2Cl(N_2)]^+$, and there is the following interesting reversible reaction of an Os^{II} complex with SO_2[48]:

$$
\begin{array}{c}
SO_2 \\
\downarrow \quad CO \\
R_3P\!-\!Os\!-\!PR_3 . \underset{+SO_2}{\overset{-SO_2}{\rightleftarrows}} OsHCl(CO)(PR_3)_2 \\
Cl \quad | \\
H
\end{array}
$$

β-Diketonate complexes of Ru^{II} and Os^{II} are obtained[49a] by reaction of H(dike) with hydrido phosphine or carbonyl phosphine complexes, and are of the types $M(dike)_2(PR_3)_2$ or $M(dike)(PR_3)_2(CO)X$ (X = Cl, H).

The dihydrides $RuH_2(PPh_3)_n$ (n = 3 or 4) may be obtained by borohydride reduction. They undergo reversible additions of H_2, CO, N_2, etc., and will act as hydrogenation catalysts for olefins,[49b] although orthometallated species (see

45a P. B. Critchlow and S. D. Robinson, *Inorg. Chem.*, 1978, **17**, 1896, 1902.
45b S. D. Robinson *et al.*, *Inorg. Chem.*, 1979, **18**, 2055.
46   A. Araneo *et al.*, *Inorg. Chem.*, 1977, **16**, 1197.
47   F. Bottomley and E. M. R. Kiremire, *J.C.S. Dalton*, **1977**, 1125.
48   E. R. Ryan and G. J. Kubas, *Inorg. Chem.*, 1978, **17**, 637.
49a M. A. M. Queiros and S. D. Robinson, *Inorg. Chem.*, 1978, **17**, 310.
49b S. Komiya and A. Yamamoto, *J. Mol. Catal.*, 1979, **5**, 279.

Sections 29-5,29-6) such as $HRu(C_2H_4)(C_6H_4PPh_2)(PPh_3)_2$ produced by interaction of the hydrides with alkenes are doubtless involved.[49c] Some of the alk-
ylarylphosphines and their phosphite analogues, e.g., $H_2Ru[P(OMe)_3]_4$, provide
examples of the unusual nonrigid octahedral molecules (see Chapter 28).

A number of the complexes, especially of Ru^{III}, have halide-bridged structures,
some with three bridges, as in $[Ru_2Cl_3(PEt_2Ph)_6]^+$. The Ru^{III} and Os^{III} complexes
have octahedrally coordinated metal atoms.

The compounds in the IV state are mainly those of osmium, one interesting
species being $OsH_4(PMe_2Ph)_3$ which is one of a series:

$$WH_6(PR_3)_3, ReH_5(PR_3)_3, OsH_4(PR_3)_3, IrH_3(PR_3)_3$$

where the coordination number decreases from 9 to 6.

Finally there are nitrosyl complexes in several oxidation states (cf. Fig. 22-F-4),
examples being $RuCl(CO)(NO)(PPh_3)_2$ and $OsCl(NO)_2(PPh_3)_2^+$. The latter
and its analogue $RuCl(NO)_2(PPh_3)_2^+$, which is made by action of $NO^+PF_6^-$ on
$RuCl(NO)(PPh_3)_2$ [a compound similar to $IrCl(CO)(PPh_3)_2$ in its oxidative-
addition chemistry] are of particular interest because they have both a linear and
a bent Ru—N—O group.

22-F-7. Other Ruthenium(II) and (III) Complexes

In addition to the complexes discussed above, there are a variety with oxygen ligands
of which the oxalates, e.g., $[Ru(ox)_3]^{3-}$, acetylacetonate, $Ru(acac)_3$, $Ru-Cl_2(Me_2SO)_4$, and the $[Ru(Me_2SO)_6]^{2+}$ ion[50] are well characterized. There are
also the $Ru(S_2CNR_2)_3$ compounds that have distorted octahedral (D_3) structures
that are intermediate between trigonal prismatic and trigonal antiprismatic. These
molecules are fluxional, undergoing trigonal twists that interconvert the enantiomers.[51] Oxidation of these tris-dithiocarbamates by iodine gives seven-coordinate
Ru^{IV} complexes; the methyl one has been shown to have pentagonal bipyramidal
coordination with the I atom axial.[52] On the other hand oxidation of $Ru(S_2CNEt_2)_3$
by O_2 in presence of BF_3 gives the structurally complicated dinuclear cation[53]
22-F-XII, which is thought to contain a Ru—Ru bond (2.74 Å); see also Section
4-33.

(22-F-XII)

[49c] G. Wilkinson et al., Nouv. J. Chim., **1977**, 141; J.C.S. Dalton, **1978**, 1739.
[50] A. R. Davis et al., Inorg. Chem., 1978, **17**, 1965.
[51] L. Pignolet, Inorg. Chem., 1974, **13**, 2051.
[52] B. M. Mattson and L. H. Pignolet, Inorg. Chem., 1977, **16**, 488.
[53] B. M. Mattson, J. R. Heiman, and L. H. Pignolet, Inorg. Chem., 1976, **15**, 565.

Ruthenium(III) forms a basic acetate with a $[Ru_3O(O_2CCH_3)_6(H_2O)_3]^+$ ion, as do many other M^{3+} ions (p. 154), but this one has more complex redox properties than most others. It can assume charges from -2 to $+3$ and it appears that even though there is no significant direct Ru—Ru bonding, strong interaction of the metal atoms through the central oxygen atom gives rise to clusterlike redox properties.[54] The $+1$ trinuclear cation also reacts in DMF with up to 3 moles of hydrogen, the reaction proceeding through several postulated intermediates such as $[HRu_3O(O_2CCH_3)_5(DMF)_3](CH_3CO_2)$ to $Ru_2(O_2CCH_3)_2(CO)_2(PR_3)_2$ and $[HRu(CO)_3]_n$. The CO is presumed to derive from the DMF.[55] Formulas and structures for this chemistry are mostly speculative.

Diruthenium Tetracarboxylates. The reaction of "$RuCl_3 \cdot 3H_2O$" with a mixture of RCO_2H and $(RCO)_2O$ in presence of alkali chloride gives rise to compounds of the general formula $Ru_2(O_2CR)_4Cl$. From these a number of other compounds may be derived, and in all the key feature is a $[Ru_2(O_2CR)_4]^+$ ion, loosely coordinated (or linked to other such ions) by Cl^-, H_2O, or other ligands.[56a] The structure of the $[Ru_2(O_2CR)_4]^+$ units (22-F-XIII) is essentially similar to that of $[Mo_2(O_2CR)_4]$ and other such species having strong metal-to-metal bonds. The

(22-F-XIII)

$[Ru_2(O_2CR)_4]^+$ ions all have three unpaired electrons, and MO calculations[56b] indicate that there is effectively an Ru—Ru bond of order $2\frac{1}{2}$, which is consistent with the Ru—Ru distances of ca. 2.27 Å, and Ru—Ru stretching frequencies[57] of ca. 330 cm^{-1}. The puzzling thing about these ions is why the intermediate oxidation state ($+2\frac{1}{2}$) is so stable; they have not been successfully converted to isolable oxidized or reduced products. It should be stressed that the unpaired electrons are fully delocalized, as shown by structural and epr data.[58]

54 T. J. Meyer *et al., J. Am. Chem. Soc.,* 1979, **101**, 2916.
55 S. A. Fouda, and G. L. Rempel, *Inorg. Chem.,* 1979, **18**, 1.
56a A. Bino, F. A. Cotton and T. R. Felthouse, *Inorg. Chem.,* 1979, **18**, 2599.
56b J. G. Norman, Jr., and H. J. Kolari, *J. Am. Chem. Soc.,* 1978, **100**, 791.
57 R. J. Clark and M. L. Franks, *J.C.S. Dalton,* 1976, 1825.
58 F. A. Cotton and E. Pedersen, *Inorg. Chem.,* 1975, **14**, 388.

22-F-8.   Complexes of Ruthenium(V) and Osmium(V), d^3

The V oxidation state is unfavorable and there is no simple compound, save the fluorides and a few complexes. The octahedral hexafluoro complexes can be prepared by various nonaqueous reactions, of which the following are representative:

$$RuCl_3 + M^ICl + F_2 \xrightarrow{300°} M^I[RuF_6]$$
$$Ru + M^{II}Cl_2 + BrF_3 \rightarrow M^{II}[RuF_6]_2$$
$$OsCl_4 + M^ICl + BrF_3 \rightarrow M^I[OsF_6]$$

The colors vary with the mode of preparation, probably owing to presence of traces of impurities. For example, $KRuF_6$ samples prepared by high-temperature fluorination are pale blue, whereas those from bromine trifluoride solution may be pale pink or cream.

The fluororuthenates(V) dissolve in water with evolution of oxygen, undergoing reductions to $[RuF_6]^{2-}$ and also producing traces of RuO_4. The osmium salts dissolve in water without reaction, but when base is added oxygen is evolved and $[OsF_6]^{2-}$ is formed.

The $[MF_6]^-$ ions have t_{2g}^3 configurations with three unpaired electrons. Their magnetic moments are independent of temperature, averaging ~3.7 BM for the $[RuF_6]^-$ salts and ~3.2 BM for the $[OsF_6]^-$ salts. The differences from the spin-only moment (3.87 BM) may be due in part to certain second-order spin-orbit coupling effects, but since observed moments are perhaps lower than can be explained by this process alone, probably antiferromagnetic interactions also contribute.

The osmium(V) complex $OsOCl_3(PPh_3)_2$, which was purported to be formed on reacting osmium tetroxide with triphenylphosphine and hydrochloric acid in refluxing ethanol, has been shown to be a mixture of $OsO_2Cl_2(PPh_3)_2$ and trans-$OsCl_4(PPh_3)_2$.[59]

22-F-9.   Nitrido Complexes of Ru and Os

The osmiamate ion $[OsO_3N]^-$ was the first example of a complex ion in which nitrogen is bound to a transition metal by a multiple bond. When OsO_4 in KOH solutions is treated with strong ammonia, the yellowish-brown color of $[OsO_4(OH)_2]^{2-}$ changes to yellow, and from the solution orange-yellow crystals of $K[OsO_3N]$ can be obtained. This ion has C_{3v} symmetry, and the infrared spectrum shows three main bands, at 1023, 858, and 890 cm^{-1}, the first of these being displaced on isotopic substitution with ^{15}N, which confirms the assignment as the Os—N stretching frequency; the high value suggests considerable Os—N multiple-bond character, and we can formally write this Os≡N.

[59]   D. J. Salmon and R. A. Walton, *Inorg. Chem.*, 1978, **17**, 2379.

Although the osmiamate ion is stable in alkaline solution, it is readily reduced by HCl or HBr, and from the resulting red solutions red crystals of salts of the $[OsNX_4]^-$, $[OsNX_4(H_2O)]^-$, and $[OsNX_5]^{2-}$ ions can be obtained. These diamagnetic ions all have C_{4v} symmetry and very short $Os\equiv N$ bonds.[60] The electronic spectra support the assignment of triple bonds and the two d electrons are assumed to occupy the d_{xy} orbital, all four of the other d orbitals being strongly engaged in forming σ or π bonds to the Cl or N atoms.[61] Some ruthenium analogues are known but have been less studied. Further reduction with acidified stannous chloride gives the anion $[Os^{III}(NH_3)Cl_5]^{2-}$. Also nucleophilic attack by tertiary phosphines on the N atom in $MNCl_3(AsPh_3)_2$ gives the imidato complexes having an M—N—PR_3 group.[62a]

Nitrido-bridged complexes of both Os and Ru are also known, examples being $K_3[(H_2O)Cl_4RuNRuCl_4(H_2O)]$ obtained by reduction of $K_2[RuCl_5(NO)]$ with HCl and $SnCl_2$ and $[Cl(NH_3)_4RuNRu(NH_3)_4Cl]^{3+}$ obtained by treating the chloro anion with NH_3. Structure and bonding in these species are analogous to those in the $[Ru_2OCl_{10}]^{4-}$ ion. There are also some larger complexes containing two nitrido bridges, such as 22-F-XIV, which is a mixed-valence (V–IV–V) species.[62b]

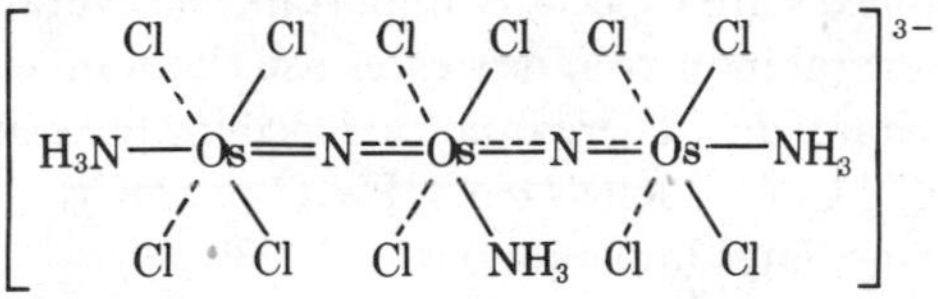

(22-F-XIV)

Finally $OsCl_3 \cdot x\,H_2O$ reacts with excess $Na(S_2CNR_2)$ to give good yields of nitrido-bridged species $Os_2N(S_2CNR_2)_5$; the structure of the R = Me compound is shown as 22-F-XV. The source of the nitrogen is uncertain. If, as suggested, it is the excess dithiocarbamate ion, the reaction is highly unusual.[63]

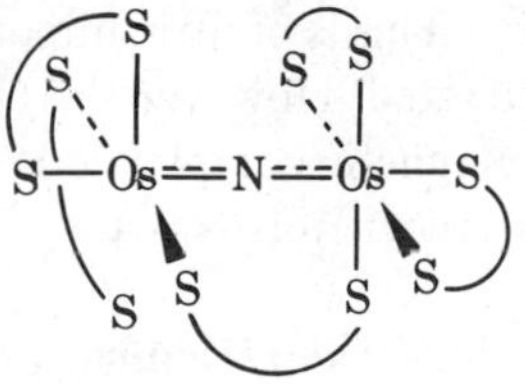

(22-F-XV)

[60]   S. R. Fletcher et al., Inorg. Nucl. Chem. Lett., 1973, **9**, 1117.
[61]   H. B. Gray et al., Inorg. Chem., 1976, **15**, 1747.
[62a] D. Pawson and W. P. Griffith, J.C.S. Dalton, 1975, 417.
[62b] W. P. Griffith and D. Pawson, J.C.S. Dalton, 1973, 1315.
[63]   L. H. Pignolet et al., Inorg. Chem., 1979, **18**, 1261.

22.G. RHODIUM AND IRIDIUM[1]

22-G-1. General Remarks: Stereochemistry

Rhodium and iridium differ from ruthenium and osmium in not forming oxo anions or volatile oxides. Their chemistry centers mainly around the oxidation states —I, 0, II, and III for rhodium and I, III, and IV for iridium. Oxidation states exceeding IV are limited to hexafluorides and to salts of the IrF_6^- ion.

The —I and 0 Oxidation States. These are mainly concerned with the carbonylate anions, polynuclear carbonyls, and substituted carbonyls with ligands such as PPh_3 (Chapters 25 and 26). The compound Rh^0dipy_2, obtained by electrochemical reduction of Rh^{III} dipyridyl complexes, is paramagnetic and presumably monomeric.[2]

The I Oxidation State, d^8. Both square and five-coordinate diamagnetic species exist, mainly with CO, tertiary phosphines, and alkenes as ligands and commonly with halide or H^- ions. Oxidative-addition reactions leading to Rh^{III} and Ir^{III} species are an important feature of the chemistry (Section 29-3).

Rhodium(I) complexes are especially important in several catalytic reactions (Chapter 30), and several industrial processes use rhodium catalysts.

The II Oxidation State, d^7. There is no evidence for the existence of complexes comparable to those of Co^{2+}, such as $Co(NH_3)_6^{2+}$ or $CoCl_4^{2-}$, although such Rh^{II} species may be intermediates in reductions.

The best defined species are certain phosphine-stabilized ones with metal-to-carbon bonds, the bridged carboxylates $Rh_2(OOCR)_4$, carbonate, $[Rh_2(CO_3)_4]^{4-}$, and the ion Rh_2^{4+} (aq).

The III Oxidation State, d^6. Both elements form a wide range of "normal," octahedral, and diamagnetic complexes with nitrogen and oxo ligands. In addition, extensive series of complexes with CO, PR_3, and similar ligands are known, many of which are obtained by oxidative addition from the square M^I species.

The IV Oxidation State, d^5. This is of little importance for Rh, and only a few complexes have been characterized. However the IV state for Ir is well defined, and a number of stable paramagnetic complex ions exist.

The oxidation states and stereochemistries are summarized in Table 22-G-1.

22-G-2. Complexes of Rhodium(I) and Iridium(I), d^8

There is a very extensive chemistry for both Rh and Ir in the I state, but it is exclusively one involving π-acid ligands such as CO, PR_3, and alkenes. Some of this chemistry is discussed in Chapters 25 to 30.

Square, tetrahedral, and five-coordinate species are formed. The latter are

[1] W. P. Griffith, *The Chemistry of the Rarer Platinum Metals,* Wiley-Interscience, 1967; B. R. James, *Coord. Chem. Rev.,* 1966, **1,** 505 (reactions and catalytic properties of Rh^I and Rh^{III} in solution).

[2] H. Caldararu, *et al., J. Am. Chem. Soc.,* 1976, **98,** 4455.

TABLE 22-G-1
Oxidation States and Stereochemistry of Rhodium and Iridium

Oxidation state	Coordination number	Geometry	Examples
Rh^{-1}, Ir^{-1}, d^{10}	4	Tetrahedral	$Rh(CO)_4^-$, $Ir(CO)_3PPh_3^-$
Rh^0, Ir^0, d^9	4	Tetrahedral	$IrNO(PPh_3)_3$
		?	$Rhdipy_2$
Rh^I, Ir^I, d^8	3	Planar?	$RhCl(PCy_3)_2$
		T-shape	$Rh(PPh_3)_3^+ClO_4^-$ (see text)
	$4^{a,b}$	Planar	$RhCl(PMe_3)_3$, $[RhCl(CO)_2]_2$, $IrCl(CO)(PR_3)_2$
		Tetrahedral	$[Rh(PMe_3)_4]^+$
	5	Tbp	$HRh(diphos)_2$, $HIrCO(PPh_3)_3$, $HRh(PF_3)_4$, $(C_8H_{10})_2RhSnCl_3$
Rh^{II}, d^7	?	?	$[Rh_2I_2(CNPh)_8]^{2+}$
	4	Square	$[Rh\{S_2C_2(CN)_2\}_2]^{2-}$, $RhCl_2[P(o\text{-}MeC_6H_4)_3]_2$
	5	?	$[Rh(bipy)_2Cl]^+$
	5	Cu^{II} acetate struct.	$[Rh(OCOR)_2]_2$
	6	Cu^{II} acetate struct.	$[Ph_3PRh(OCOCH_3)_2]_2$
Rh^{III}, Ir^{III}, d^6	5	Tbp	$IrH_3(PR)_3]_2$
	5	Sp	$RhI_2(CH_3)(PPh_3)_2$
	$6^{a,b}$	Octahedral	$[Rh(H_2O)_6]^{3+}$, $RhCl_6^{3-}$, $IrH_3(PPh_3)_3$, $RhCl_3(PEt_3)_3$, $IrCl_6^{3-}$, $[Rh(diars)_2Cl_2]^+$, RhF_3, $IrF_3(ReO_3$ type)
Rh^{IV}, Ir^{IV}, d^5	6^b	Octahedral	K_2RhF_6, $[Ir(C_2O_4)_3]^{2-}$, $IrCl_6^{2-}$, IrO_2 (rutile type)
Ir^V, d^4	6	Octahedral	$CsIrF_6$
	7	?	$IrH_5(PPhEt_2)_2$
Rh^{VI}, Ir^{VI}, d^3	6	Octahedral	RhF_6, IrF_6

[a] Most common states for Rh.
[b] Most common states for Ir.
[c] R. A. Jones et al., J.C.S. Chem. Comm., **1979**, 489.

usually produced by addition of neutral ligands to the former, e.g.,

$$trans\text{-}IrCl(CO)(PPh_3)_2 + CO \rightleftharpoons IrCl(CO)_2(PPh_3)_2$$

The criteria for relative stability of five- and four-coordinate species are by no means fully established. Substitution reactions of square species, which are often rapid, proceed by an associative pathway involving five-coordinate intermediates, e.g.,

$$RhCl(C_8H_{12})SbR_3 + amine \rightleftharpoons RhCl(C_8H_{12})amine + SbR_3$$

The Rh^I and Ir^I complexes are invariably prepared by some form of reduction, either of similar M^{III} complexes or of halide complexes such as $RhCl_3 \cdot 3H_2O$ or K_2IrCl_6 in presence of the complexing ligand. As noted under ruthenium, alcohols, aldehydes, or formic acid may furnish CO and/or H under certain circumstances; the ligand itself may also act as a reducing agent.

Most of the square Rh^I and Ir^I complexes undergo oxidative addition reactions

(Section 29-3), and this constitutes a way of making M^{III} complexes with π-bonding ligands.

Rhodium. Some preparations and reactions of Rh^I complexes are shown in Figs. 22-G-1 and 22-G-6 (p. 945). We can discuss only a few of the more important compounds.

Tetracarbonyldichlorodirhodium $[Rh(CO)_2Cl]_2$. This is most easily obtained by passing CO saturated with ethanol over $RhCl_3 \cdot 3H_2O$ at *ca.* 100°, when it sublimes as red needles. It has the structure shown in Fig. 22-G-2, where the coordination around each Rh atom is planar, and there are bridging chlorides with a marked dihedral angle, along the Cl-Cl line.

Electronic factors within the molecule result in the pronounced bending in the dimer; there seems to be only a very weak metal-metal interaction[3] [3.12 Å, cf. 2.73 Å in $Rh_4(CO)_{12}$] and even weaker interaction between adjacent molecules in the lattice (3.31 Å).[4] The carbonyl chloride is an excellent source of other rhodium(I) species, and the halogen bridges are readily cleaved[5] by a wide variety of donor ligands to give *cis*-dicarbonyl complexes, e.g.,

$$[Rh(CO)_2Cl]_2 + 2L \rightarrow 2RhCl(CO)_2L$$
$$[Rh(CO)_2Cl]_2 + 2Cl^- \rightarrow 2[Rh(CO)_2Cl_2]^-$$
$$[Rh(CO)_2Cl]_2 + acac \rightarrow 2Rh(CO)_2(acac) + 2Cl^-$$

Some of the complexes produced thus may be made directly from rhodium trichloride however.

trans-Chlorocarbonylbis(triphenylphosphine)rhodium, trans-RhCl(CO)- $(PPh_3)_2$. Although not as widely studied as its iridium analogue, this is an important compound and is an intermediate in the preparation of the more important complex $RhH(CO)(PPh_3)_3$, discussed below.

The yellow crystalline complex is best obtained by the reduction of $RhCl_3 \cdot 3H_2O$ in ethanol, with formaldehyde as both a source of CO and as reductant. It is also formed by PPh_3 bridge cleavage from $[Rh(CO)_2Cl]_2$, and by action of CO on $RhCl(PPh_3)_3$ (see below).

Although it is readily oxidized by Cl_2 to $RhCl_2(CO)(PPh_3)_2$, the oxidative adducts are generally less stable than those of *trans-*$IrCl(CO)(PPh_3)_2$, and the equilibria such as

$$trans\text{-}RhCl(CO)(PPh_3)_2 + HCl \rightleftharpoons Rh^{III}HCl_2(CO)(PPh_3)_2$$

generally lie well to the left-hand side.

Other Rhodium Carbonyl Species. In addition to $Rh(CO)_2Cl_2^-$ noted above, other carbonyl anions are known and are best made by the action of CO or formic acid on $RhCl_3$ solutions. In the CO reduction there is an intermediate Rh^{III} complex

[3] Weak Rh—Rh bond formation also accounts for the oligomerization of $[Rh(RNC)_4]^+$ species. See page 941. Other Rh^I species also have stacks with Rh—Rh interactions; see P. W. de Haven and V. L. Goedken, *Inorg. Chem.*, 1979, **18**, 827.

[4] J. G. Norman, Jr., and D. J. Gmur, *J. Am. Chem. Soc.*, 1977, **99**, 1446.

[5] A. J. Pribula and R. S. Drago, *J. Am. Chem. Soc.*, 1976, **98**, 2784, 5129; P. Uguagliati *et al.*, *Inorg. Chim. Acta*, 74, **9**, 20 (extensive review).

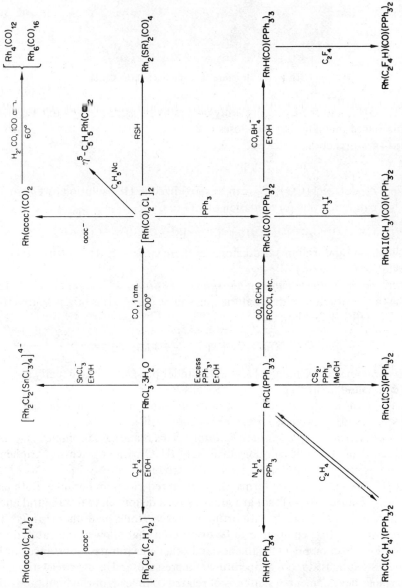

Fig. 22-G-1. Some preparations and reactions of rhodium(I) compounds.

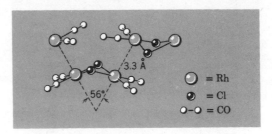

Fig. 22-G-2. The structure of crystalline $[Rh(CO)_2Cl]_2$.

$[RhCl_5CO]^{2-}$, and reduction probably occurs by transfer of H_2O to CO, giving a formato intermediate that then loses CO_2.

The overall reaction is hence

$$Rh^{III} + H_2O + 3CO \rightarrow Rh^{I}(CO)_2 + CO_2 + 2H^+$$

The final product, $Rh(CO)_2Cl_2^-$, can be reoxidized in HCl solution by O_2, so that there is a catalytic cycle for oxidation of CO to CO_2:

$$[Rh(CO)_2Cl_2]^- + O_2 + 2H^+ + 3Cl^- \rightarrow [Rh(CO)Cl_5]^{2-} + CO_2 + H_2O$$

In rhodium (and iridium) β-diketonates there is distinct Rh---Rh interaction in the lattice (see below).

Hydridocarbonyltris(triphenylphosphine)rhodium. This yellow crystalline solid has a *tbp* structure with equatorial phosphine groups. It is best prepared from $RhCl(CO)(PPh_3)_2$ by the reaction

$$trans\text{-}RhCl(CO)(PPh_3)_2 + PPh_3 \xrightarrow[\text{EtOH}]{\text{NaBH}_4} RhH(CO)(PPh_3)_3$$

but it is also formed by action of $CO + H_2$ under pressure with virtually any rhodium compound in presence of an excess of PPh_3.

Although the complex undergoes a range of reactions its main importance is as a hydroformylation catalyst for alkenes (Chapter 30).

Chlorotris(triphenylphosphine)rhodium. This remarkable complex is formed by reduction of ethanolic solutions of $RhCl_3\cdot3H_2O$ with an excess of triphenylphosphine; triphenylphosphine oxide is a by-product.

$RhCl(PPh_3)_3$ exists in two forms, the normal red-violet and orange. Both have structures (Figure 22-G-3) that are square with a distortion to tetrahedral and in both there are close contacts with ortho hydrogen atoms on a phenyl ring.[6] The complex was the first compound to be discovered that allowed the catalytic hydrogenation (Section 30-1) of alkenes and other unsaturated substances in homogeneous solutions at room temperature and pressure, and its discovery stimulated an enormous development in synthesis of related complexes of rhodium (and other metals) with tertiary phosphine ligands.[7] Not only monophosphine but chelate

---

[6]   M. J. Bennett and P. B. Donaldson, *Inorg. Chem.,* 1977, **16**, 655.
[7]   See C. A. McAuliffe, Ed., *Transition Metal Complexes of Phosphorus, Arsenic and Antimony Ligands,* Macmillan, 1973.

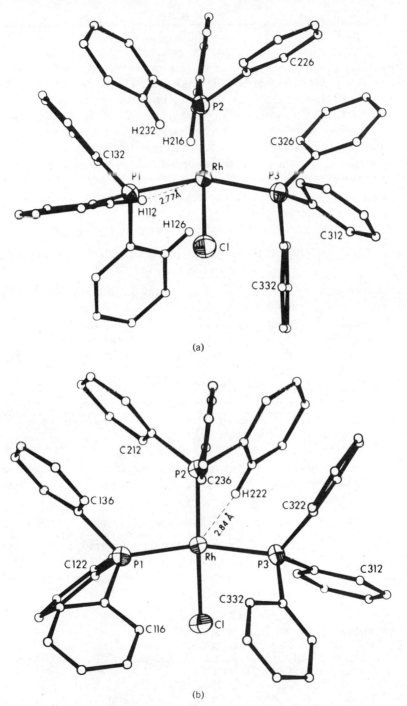

(a)

(b)

Fig. 22-G-3. Structures of (a) red and (b) orange forms of RhCl(PPh$_3$)$_3$. [Reproduced by permission from M. J. Bennett and P. B. Donaldson, *Inorg. Chem.,* 1977, **16,** 655.]

phosphine complexes[8] such as $RhCl[PhP(CH_2CH_2CH_2PPh_2)_2]$ behave similarly.

Although the catalytic hydrogenation requires a three-coordinate species, $RhCl(PPh_3)_2$, such species are rare. Examples are $HRh(PBu_3^t)_2$[9a] and $RhCl(PCy_3)_2$.[9b] Treatment of $RhCl(PPh_3)_3$ with $TlClO_4$ in acetone gives the solvated salt $[Rh(acetone)(PPh_3)_3]^+ClO_4$, which can be recrystallized to give $[Rh(PPh_3)_3]^+ClO_4^-$. The latter has an unusual structure[10] with a T-shaped planar geometry in which there are two normal phosphines; the other PPh_3 group is distorted and there is some interaction between Rh and the C atom of a phenyl bound to P. It is believed that the distortion towards a T shape is of electronic origin, to allow diamagnetism.

Tris(triphenylphosphine)chlororhodium undergoes a wide variety of oxidative-addition and other reactions (Fig. 22-G-1).

It also readily abstracts CO from other molecules stoichiometrically, e.g.,

$$\eta\text{-}C_5H_5Fe(CO)_2(COR) + RhCl(PPh_3)_3 \rightarrow \eta\text{-}C_5H_5Fe(CO)(PPh_3)(R) + RhCl(CO)(PPh_3)_2 + CO$$
$$(R = \text{adamantyl})$$

In all such cases $trans\text{-}RhCl(CO)(PPh_3)_2$ is formed. Catalytic decarbonylation of aldehydes and acyl halides where $RhCl(PPh_3)_3$ is used as starting material doubtless involve the carbonyl complex first formed. (Section 30-11).

Both $RhCl(PPh_3)_3$ and $Rh_2Cl_2(PPh_3)_4$ react with molecular oxygen and can act as oxidation catalysts for cyclohexene and other molecules, probably via a free radical reaction (Section 30-15).

The oxygen compounds $RhCl(O_2)(PPh_3)_2$ and $RhCl(O_2)(PPh_3)_2 \cdot 2CH_2Cl_2$, which contains H-bonded dichloromethane, have been structurally characterized.[11] The oxidized products (Ph_3PO is also produced) can be reconverted to $RhCl(PPh_3)_3$ by refluxing in C_2H_5OH with excess PPh_3.

Iridium. The most important of Ir^I complexes are $trans\text{-}IrCl(CO)(PPh_3)_2$ and its analogues with other phosphines. These compounds have been much studied because they provide some of the clearest examples of oxidative-addition reactions, since the equilibria

$$trans\text{-}IrX(CO)(PR_3)_2 + AB \rightleftharpoons Ir^{III}XAB(CO)(PR_3)_2$$

lie well to the oxidized side and the oxidized compounds are usually stable octahedral species, unlike many of their rhodium analogues.

It is of interest to note that the complex from o-tolylphosphine, unlike the PPh_3 and m- or p-tolylphosphine complexes, is inert to H_2, O_2, SO_2, and

---

[8]   D. W. Meek et al., Inorg. Chem., 1976, **15**, 1365.

[9a]   T. Yoshida et al., J.C.S. Chem. Comm., **1978**, 855.

[9b]   H. L. M. Van Gaal and F. L. A. Van den Beterom, J. Organomet. Chem., 1977, **134**, 237.

[10]   C. A. Reed et al., J. Am. Chem. Soc., 1977, **99**, 7076.

[11]   M. J. Bennett and P. B. Donaldson, Inorg. Chem., 1977, **16**, 1581; G. L. Geoffroy and M. E. Keeney, Inorg. Chem., 1977, **16**, 205; see also H. Mimoun et al., J. Am. Chem. Soc., 1978, **100**, 5437; J.C.S. Chem. Comm., **1978**, 559.

$(CN)_2C=C(CN)_2$. This is attributed to the steric effect of the o-methyl group that blocks the apical sites, thus preventing reaction.[12]

The complexes are usually made by refluxing sodium chloroiridate and phosphine in 2-methoxyethanol or diethylene glycol under an atmosphere of CO.

The carbonyl is readily converted into the five-coordinate hydride:

$$trans\text{-}IrCl(CO)(PPh_3)_2 \overset{CO}{\rightleftarrows} IrCl(CO)_2(PPh_3)_2 \xrightarrow[\text{EtOH}]{NaBH_4} IrH(CO)_2(PPh_3)_2$$

and this is of interest in that it is much more stable than its rhodium analogue, hence allows many prototypes for intermediates in the hydroformylation sequence to be isolated.

Again like rhodium, iridium forms alkene complexes. Examples are the cyclooctene or 1,5-cyclooctadiene (COD) compounds, e.g., $[IrCl(COD)_2]_2$, formed by boiling $(NH_4)_2IrCl_6$ with the olefin in alcohols; this product can be converted into $IrCH_3(COD)(PMe_2Ph)_2$, which shows unusual fluxional behavior. Ethylene forms the unusual five-coordinate $IrCl(C_2H_4)_4$. Finally $IrCl(PPh_3)_3$ can be obtained by action of PPh_3 on $[IrCl(COD)]_2$. This product differs from $RhCl(PPh_3)_3$ in reacting irreversibly with H_2 to give $IrClH_2(PPh_3)_3$. Since this octahedral species does not dissociate in solution, it does not act as a hydrogenation catalyst for olefins at 25°, although it will do so under ultraviolet irradiation. By contrast, the bis species $IrCl(PPh_3)_2$, which is made *in situ* by action of PPh_3 on the cyclooctene complex $[IrCl(C_8H_{14})_2]_2$, is an active catalyst. These observations clearly show the necessity for having a vacant site for coordination of olefin on the hydrido complex.

Both rhodium and iridium form complexes with *isocyanides*, e.g.,

$$RhCl(CO)(PPh_3)_2 \xrightarrow{PhNC} Rh(CNPh)_4Cl + CO + 2PPh_3$$

This rhodium complex is yellow in dilute solutions in acetonitrile but blue when concentrated. This has been interpreted[13] as involving dissociation and cluster formation:

$$2Rh(CNPh)_4^+ \rightleftarrows [Rh(CNPh)_4]_2^{2+} \rightleftarrows [Rh(CNPh)_4]_3^{3+} \dots$$

Cationic and Anionic Complexes.   For both rhodium and iridium, a variety of related cationic and some anionic species (which may be either four or five-coordinate) are known, e.g.,

$$trans\text{-}IrCl(CO)(PR_3)_2 + CO \begin{cases} \xrightarrow{Na/Hg} Na[Ir(CO)_3(PR_3)] \\ \xrightarrow{NaClO_4} [Ir(CO)_3(PR_3)_2]ClO_4 \end{cases}$$

$$trans\text{-}IrCl(CO)(PR_3)_2 \xrightarrow{diphos} [Ir(diphos)_2]Cl$$

$$Rh(diene)acac \xrightarrow{HClO_4,\ PPh_3} [Rh(diene)(PPh_3)_2](ClO_4) + acacH$$

12   R. Brady *et al.*, *Inorg. Chem.*, 1975, **14**, 2669.
13   H. B. Gray *et al.*, *J. Am. Chem. Soc.*, 1978, **100**, 485; K. R. Mann *et al.*, *J. Am. Chem. Soc.*, 1975, **97**, 3553.

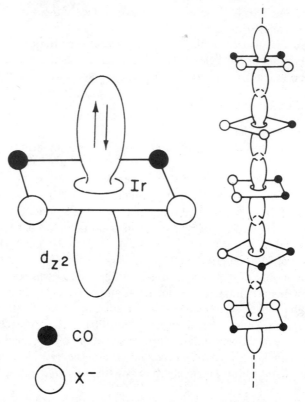

Fig. 22-G-4. Linear chain structure of $Ir(CO)_2X_2^{0.6-}$ showing overlap of d_{z^2} orbitals. [Reproduced by permission from A. P. Ginsberg *et al., Inorg. Chem.,* 1976, **15**, 514.]

The cationic species, e.g., $[Rh(diphos)_2]^+$, also undergo oxidative-addition reactions just like the neutral species and may also have catalytic activity.[14] For iridium there is an extensive chemistry with NO, and the cationic nitrosyl $[IrCl(NO)(PPh_3)_2]^+$, which is isoelectronic with $IrCl(CO)(PPh_3)_2$, undergoes typical oxad chemistry.

Some iridium carbonyl anionic complexes have been studied because like other square compounds, they are stacked so that linear chains of metal-metal bonded atoms are formed (Chapter 26), and such compounds are potential one-dimensional electrical conductors (Fig. 22-G-4). Holes in the *d* bands that allow conduction, arising from metal-metal interaction, may be formed by partial oxidation.

Typical complexes such as $K_{0.98}[Ir(CO)_2Cl_{2.42}]\cdot 0.2MeCOMe$ and $K_{0.6}[Ir\text{-}(CO)_2Cl_2]\cdot 0.5H_2O$, which form conducting bronzelike needles, can be obtained by carbonylation of K_2IrCl_6.[15]

[14]   See, e.g., D. H. Doughty and L. H. Pignolet, *J. Am. Chem. Soc.,* 1978, **100**, 7083; J. T. Mague and E. J. Davis, *Inorg. Chem.,* 1977, **16**, 131; H. C. Clark and K. J. Reimer, *Inorg. Chem.,* 1975, **14**, 2133; J. S. Miller and K. G. Caulton, *Inorg. Chem.,* 1975, **14**, 2296; R. D. Gillard *et al., J.C.S. Dalton,* **1975**, 133.

[15]   A. P. Ginsberg *et al., Inorg. Chem.,* 1976, **15**, 514, 2540.

22-G-3. Complexes of Rhodium(III) and Iridium(III), d^6

Both Rh and Ir form a large number of octahedral complexes, cationic, neutral, and anionic; in contrast to Co^{III} complexes, reduction of Rh^{III} or Ir^{III} does not give rise to divalent complexes (except in a few special cases). Thus depending on the nature of the ligands and on the conditions, reduction may lead to the metal— usually with halogens, water, or amine ligands present—or to hydridic species of M^{III} or to M^I when π-bonding ligands are involved.

Though similar to Co^{III} in giving complex anions with CN^- and NO_2^-, Rh and Ir differ in readily giving octahedral complexes with halides, e.g., $[RhCl_5H_2O]^{2-}$ and $[IrCl_6]^{3-}$, and with oxygen ligands such as oxalate and EDTA.

The cationic and neutral complexes of all three elements are generally kinetically inert, but the anionic complexes of Rh^{III} are usually labile. By contrast, anionic Ir^{III} complexes are inert, and the preparation of such complexes is significantly harder than for the corresponding Rh species.

Rhodium complex cations have proved particularly suitable for studying *trans* effects in octahedral complexes.

In their magnetic and spectral properties the Rh^{III} complexes are fairly simple. All the complexes, and indeed all compounds of rhodium(III), are diamagnetic. This includes even the $[RhF_6]^{3-}$ ion, of which the cobalt analogue constitutes the only example of a high-spin Co^{III}, Rh^{III}, or Ir^{III} ion in octahedral coordination. Thus the inherent tendency of the octahedral d^6 configuration to adopt the low-spin t_{2g}^6 arrangement, together with the relatively high ligand field strengths prevailing in these complexes of tripositive higher transition series ions, as well as the fact that all $4d^n$ and $5d^n$ configurations are more prone to spin pairing than their $3d^n$ analogues, provide a combination of factors that evidently leaves no possibility of there being any high-spin octahedral complex of Rh^{III} or Ir^{III}.

The visible spectra of Rh^{III} complexes have the same explanation as do those of Co^{III} complexes. As illustrated in Fig. 22-G-5 for the $[Rh(NH_3)_6]^{3+}$ and $[RhCl_6]^{3-}$ ions, there are in general two bands toward the blue end of the visible

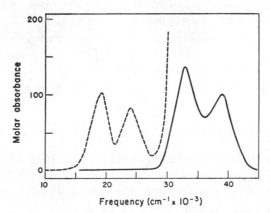

Fig. 22-G-5. The visible spectra of the $[RhCl_6]^{3-}$ (dashed curve) and the $[Rh(NH_3)_6]^{3+}$ (solid curve) ions.

region, which together with any additional absorption in the blue due to charge-transfer transitions, are responsible for the characteristic orange, red, yellow, or brown colors of rhodium(III) compounds. These bands are assigned as transitions from the $^1A_{1g}$ ground state to the $^1T_{1g}$ and $^1T_{2g}$ upper states just as in the energy level diagram for Fe^{II} and Co^{III}. The spectra of Ir^{III} complexes have a similar interpretation.

The Rhodium and Iridium Aqua Ions.   Unlike cobalt, rhodium gives a stable yellow aqua ion $[Rh(H_2O)_6]^{3+}$. It is obtained by dissolution of $Rh_2O_3(aq)$ in cold mineral acids, or, as the perchlorate by repeated evaporation of $HClO_4$ solutions of $RhCl_3(aq)$. Exchange studies with $H_2^{18}O$ confirm the hydration number as 5.9 $\pm$ 0.4. The ion is acidic, $pK_a \sim 3.3$, giving $[Rh(H_2O)_5OH]^{2+}$ in solutions less than about 0.1 M in acid. The crystalline deliquescent perchlorate is isomorphous with other salts containing octahedral cations, e.g., $[Co(NH_3)_6](ClO_4)_3$. The aqua ion also occurs in alums $M^IRh(SO_4)_2{\cdot}12H_2O$, and in the yellow sulfate $Rh_2(SO_4)_3{\cdot}$ $12H_2O$, obtained by vacuum evaporation at $0°$ of solutions of $Rh_2O_3(aq)$ in H_2SO_4. A red sulfate $Rh_2(SO_4)_3{\cdot}6H_2O$, obtained by evaporation of the yellow solutions at $100°$, gives no precipitate with Ba^{2+} ion and is presumably a sulfato complex.

The air-sensitive aqua iridium(III) ion[16a] can be made with some difficulty by dissolution of a hydrous oxide in $HClO_4$ and the salt $[Ir(H_2O)_6](ClO_4)_3$ has been isolated. A sulfite $Ir_2(SO_3)_3{\cdot}6H_2O$ crystallizes from solutions of $Ir_2O_3(aq)$ in water saturated with SO_2, and a sulfate may be isolated from sulfuric acid solutions of the hydrous oxide with exclusion of air, but the structures of these compounds are unknown.

The Rhodium(III)-Chloride System.   One of the most important of Rh^{III} compounds and the usual starting material for the preparation of rhodium complexes (see Figs. 22-G-1 and 22-G-6) is the dark red, crystalline, deliquescent trichloride, $RhCl_3{\cdot}nH_2O$; n is usually 3 or 4. This is obtained by dissolving hydrous Rh_2O_3 in aqueous hydrochloric acid and evaporating the hot solutions. It is very soluble in water and alcohols, giving red-brown solutions.

On boiling the aqueous solutions $[Rh(H_2O)_6]^{3+}$ is formed, and on heating with excess HCl, the rose-pink hexachlororhodate ion $[RhCl_6]^{3-}$. The complexes intermediate between these two extremes, including the various isomers, have been isolated and their interconversions studied.[16b] The steric course of the reactions is governed by the trans effect of chloride. For example, on aquation $[RhCl_6]^{3-}$ produces only cis compounds ending with the very stable neutral fac-$RhCl_3(H_2O)_3$:

$$[RhCl_6]^{3-} \underset{\xrightarrow{}}{\overset{H_2O}{\rightleftharpoons}} [RhCl_5(H_2O)]^{2-} \underset{\xrightarrow{}}{\overset{H_2O}{\rightleftharpoons}} cis\text{-}[RhCl_4(H_2O)_2]^- \underset{\xrightarrow{}}{\overset{H_2O}{\rightleftharpoons}} fac\text{-}RhCl_3(H_2O)_3$$

Hexahalogenorhodates may be obtained also by heating Rh metal and alkali metal halide in Cl_2 (plus a little carbon), extracting the melt, and crystallizing. With very large cations halogen-bridged species like $Rh_2Cl_9^{3-}$ can be obtained.[17]

16a   H. Gamsjäger and P. Beutler, *J. C. S. Dalton,* **1979**, 1415.

16b   K. E. Hyde, H. Kelm, and D. A. Palmer, *Inorg. Chem.,* 1978, **17**, 1647.

17   F. A. Cotton and D. A. Ucko, *Inorg. Chim. Acta,* 1972, **6**, 161.

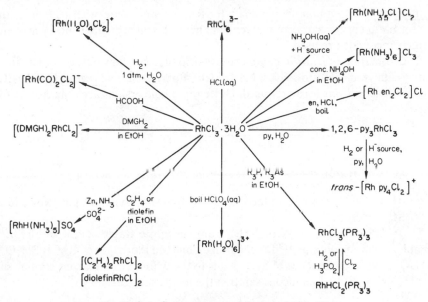

Fig. 22-G-6. Some reactions of rhodium trichloride.

Iridium(III)-Halide Complexes. Several dark green, hydrated Ir[III] halides are obtained by dissolution of $Ir_2O_3(aq)$ in the appropriate acid. Several species formed by aquation of $[IrCl_6]^{3-}$, e.g., $[Ir(H_2O)Cl_5]^{2-}$, $[Ir(H_2O)_2Cl_4]^-$, and $[Ir(H_2O)_3Cl_3]$ have been studied in great detail.

Cationic Complexes. Both Rh and Ir give cobaltlike ammines of the types $[ML_6]^{3+}$, $[ML_5X]^{2+}$, and $[ML_4X_2]^+$, of which $[Rh(NH_3)_5Cl]Cl_2$ is a typical example. The salts are made in various ways, but usually by the interaction of aqueous solutions of $RhCl_3(aq)$ with the ligand.

The formation of complex ions from $RhCl_3(aq)$, $[Rh(H_2O)Cl_5]^{2-}$, or $[RhCl_6]^{3-}$ is often catalyzed by the addition of reducing agents that can furnish hydride ions; ligands such as ethylenediamine may also themselves act in this way. The effect of ethanol was discovered by Delépine long before the general nature of such catalysis was recognized. It now appears that many rhodium complexes have been made only because ethanol was used as a solvent. One example of the catalysis is the action of pyridine, which with $RhCl_3(aq)$ gives mainly $Rhpy_3Cl_3$ and with aqueous $[Rh(H_2O)Cl_5]^{2-}$ gives $[Rhpy_2Cl_4]^-$. On addition of alcohol, hydrazine, BH_4^-, or other reducing substances—even molecular hydrogen at 25° and ≤1 atm—conversion into *trans*-$[Rhpy_4Cl_2]^+$ rapidly occurs. Kinetic studies of this reaction suggest that Rh[I] complexes, rather than hydridic ones, are involved in the catalysis, since $[Rh(CO)_2Cl]_2$ and $Rh(acac)(CO)_2$ are more effective than hydride-producing substances. Further evidence comes from the reaction of $[Rh(H_2O)_6]^{3+}$ in ethanolic ClO_4^- solution with 2,2'-bipyridine, when an air-sensitive brown complex $[Rh^Ibipy_2]ClO_4\cdot3H_2O$ can be isolated.

The reduction of $[Rhen_2Cl_2]^+$ at a mercury electrode does not give a Rh[II] species as first thought, but a mercury complex[18] $\{[en_2Rh]_2Hg\}^+$; the formation of bonds

18   J. Gulens and F. C. Anson, *Inorg. Chem.*, 1973, **12**, 2568.

to mercury is a common phenomenon for Rh, Ru, and some other elements, and the formation of mercury complexes can always be suspected whenever mercury or its compounds are used.

The formation of *iridium* complexes is normally very slow but, as for rhodium, can be catalyzed. Thus to convert $Na_2Ir^{IV}Cl_6$ into py_3IrCl_3 and *trans*-$[Irpy_4Cl_2]Cl$, a bomb reaction was formerly used. Quite rapid conversions are obtained as follows:

$$Na_2IrCl_6 \xrightarrow[\text{boil 30 min}]{NaH_2PO_2(aq) + py} cis\text{-}py_3IrCl_3 \xrightarrow{6\ hr} [Irpy_4Cl_2]Cl$$

$$H_2IrCl_6 \xrightarrow[\text{2-methoxyethanol, boil 10 min}]{HCl\ in} IrCl_6^{3-} \xrightarrow[\text{1 hr}]{+\ py,\ boil} mer\text{-}py_3IrCl_3$$

trans-$[Iren_2Cl_2]^+$ can be similarly obtained by using hypophosphorous acid as catalyst.

Hydrido Complexes. With NH_3 or amines, quite stable octahedral hydrido complexes can be obtained[19] for rhodium. Thus the reduction of $RhCl_3 \cdot 3H_2O$ in NH_4OH by Zn in presence of SO_4^{2-} leads to the white, air-stable, crystalline salt $[RhH(NH_3)_5]SO_4$. In aqueous solution the ion dissociates:

$$[RhH(NH_3)_5]^{2+} + H_2O = [RhH(NH_3)_4H_2O]^{2+} + NH_3 \qquad K \sim 2 \times 10^{-4}$$

Various substitution reactions with other amines can be carried out, and with alkenes remarkably stable alkyl derivatives, e.g., $[RhC_2H_5(NH_3)_5]SO_4$, can be obtained. The structure[20] of $[RhH(NH_3)_5](ClO_4)_2$ shows a distinct hydridic trans weakening effect (0.165 Å) on the trans Rh—N bond. The cyanide $[RhH(CN)_5]^{3-}$ is also known.

For complexes with tertiary phosphines, carbon monoxide, etc., ligands, innumerable hydrido complexes of Rh and Ir are known, not only in the I state but in the III state; those for Ir^{III} are particularly stable. Most are obtained by oxidative addition of HX or H_2 to the corresponding M^I species. One example is *mer*-$IrH_3(PPh_3)_3$, which though thermally stable, loses hydrogen on irradiation[21] to give the orthometallated (Section-29-6) product:

$$IrH_3(PPh_3)_3 \xrightarrow{h\nu} 2H_2 + Ir^I(o\text{-}C_6H_4PPh_2)(PPh_3)_2$$

Neutral Complexes. Interaction of acetylacetone and hydrous Rh_2O_3 gives the trisacetylacetonate, which has been resolved into enantiomeric forms. It undergoes a variety of electrophilic substitution reactions of the coordinated ligand, such as chlorination. The stereochemistry and racemization of the *cis* and *trans* isomers of the unsymmetrical trifluoroacetylacetonate have been studied by nmr spectroscopy; the compound is extremely stable to isomerization.

Neutral complexes with CO, PR_3, pyridines, etc., as ligands may be made directly from $RhCl_3 \cdot 3H_2O$ or Na_3IrCl_6, but as noted above, they are also commonly and often readily obtained by oxidative addition of the M^I complexes. Typical formulas

19　J. F. Endicott *et al.*, *Inorg. Chem.*, 1979, **18**, 450.
20　B. A. Coyle and J. A. Ibers, *Inorg. Chem.*, 1972, **11**, 1105.
21　G. L. Geoffroy and R. Pierantozzi, *J. Am. Chem. Soc.*, 1976, **98**, 8054.

are MCl_3L_3, $MHCl_2I_3$, and $MCl_3(CO)L_2$. The complex $RhCl_3py_2(DMSO)$ has remarkable activity against leukemia in mice.[22] The dimethylsulfoxide complex of iridium $IrCl_3(OSMe_2)_3$ catalyzes, *via* a hydrido intermediate, the reduction of cyclohexanones to axial alcohols and the hydrogenation of α,β-unsaturated ketones.

22-G-4. Complexes of Rhodium(II) and Iridium(II), d^7

Complexes of rhodium(II) are restricted to the following:

1. Short-lived species formed by pulse radiolysis or flash photolysis of $[Rh(NH_3)_5Cl]^{2+}$ [23a] and a species formed upon flash photolysis of $Rh(CF_3CO-CH_2COCH_3)_3$.[23b]

2. Paramagnetic square complexes of the bulky phosphines such as that of tricyclohexylphosphine $[RhCl_2(PCy_3)_2]$.[24] Traces of paramagnetic materials, presumably Rh^{II} complexes, are present in preparations of $RhCl(PPh_3)_3$.

3. Diamagnetic species that have metal-metal bonds, which we discuss in more detail.

The most accessible complex is the acetate $Rh_2(CO_2Me)_4$, obtained on heating sodium acetate with $RhCl_3 \cdot 3H_2O$ in methanol. It has the common tetra-bridged acetate structure with a short Rh—Rh bond (2.386 Å) that despite its shortness, is a single bond.[25] This and similar carboxylates are green. The end positions can be readily occupied by donor ligands[26]; with oxygen the donor adducts are green or blue, but with π acids such as PPh_3 they are red. The carboxylates and their adducts appear to be antitumor[27] agents. The acetate can be oxidized to a purple mixed-valence (II, III) ion $[Rh_2(CO_2Me)_4(H_2O)_2]^+$, electrolytically or by ceric ion[28]; this ion also has a short Rh—Rh bond (2.316 Å). The carboxylates also catalyze the cyclopropanation of alkenes with alkyl diazoacetates[29a] and the hydrogenation of olefins.[29b]

The Rh_2^{4+} (aq) ion can be obtained by Cr^{2+} reduction of $[Rh(H_2O)_5Cl]^{2+}$ and by action of HBF_4 on the acetate in methanol, which removes acetate as methyl acetate. From this ion can be obtained $Rh_2(SO_4)_4^{4-}$ and $Rh_2(CO_3)_4^{4-}$ which have SO_4 or CO_3 bridges.[30] The dimethylglyoxime complex $Rh_2(dmgH)_4(PPh_3)_2 \cdot$

22   P. Colamarino and P. Orioli, *J.C.S. Dalton*, **1976**, 845.

23a J. Lilie *et al.*, *Inorg. Chem.*, 1975, **14**, 2129.

23b G. Ferraudi *et al.*, *J.C.S. Chem. Comm.*, **1979**, 15.

24   H. L. M. Van Gaal *et al.*, *Inorg. Chim. Acta*, 1977, **23**, 43; F. G. Moers *et al.*, *J. Inorg. Nucl. Chem.*, 1973, **35**, 1915.

25   D. S. Martin *et al.*, *Inorg. Chem.*, 1979, **18**, 475, G. G. Christoph and Y.-B. Koh; *J. Am. Chem. Soc.*, 1979, **101**, 1422.

26   J. L. Bear *et al.*, *Inorg. Chim. Acta*, 1979, **32**, 132; Y.-B. Koh and G. G. Christoph, *Inorg. Chem.*, 1979, **18**, 1122.

27   G. Pneumatikakis and N. Hadjiliadis, *J.C.S. Dalton*, 1979, 596.

28   J. J. Ziolkowski *et al.*, *J.C.S. Chem. Comm.*, **1977**, 760; R. D. Cannon *et al.*, *J.C.S. Chem. Comm.*, **1976**, 31; C. R. Wilson and H. Taube, *Inorg. Chem.*, 1975, **14**, 2276.

29a A. J. Hubert and A. Noels, *Synthesis*, 1976, **9**, 600.

29b B. C. Y. Hui *et al.*, *Inorg. Chem.*, 1973, **12**, 757.

30   C. R. Wilson and H. Taube, *Inorg. Chem.*, 1975, **14**, 405.

$H_2O\cdot C_3H_7OH$ also has a Rh—Rh bond but *no* bridge groups. The diamagnetic metal-metal bonded *isocyanides* may be obtained by the reaction

$$Rh^I(CNR)_4^+ + Rh^{III}X_2(CNR)_4 \rightleftharpoons Rh^{II}X_2(CNR)_8^{2+}$$

With diphosphines that can bridge two metal atoms, rhodium(II) complexes have been obtained by oxidation of the corresponding rhodium(I) complexes.[31]

There are also Rh_2^{II} dimeric species of N_4 macrocycles[32a] and porphyrins, e.g., $[OepRh]_2$.[32b]

Iridium(II) complexes are few, one example probably being the paramagnetic species[33] $Ir(CO_2R)_2(AsPh_3)_2(CNC_6H_4Me)$; a formally Ir^{II} species[34] is

22-G-5. Complexes of Rhodium(IV) and Iridium(IV), d^5

Rhodium. Oxidation of Rh^{III} sulfate solutions with O_3 or with sodium bismuthate gives red solutions that may contain Rh^{IV}. Higher states, even V and VI, have been postulated in reactions of Rh^{III} with hypobromite, although this seems unlikely.

The only well-defined species are the halides RhF_6^{2-} and $RhCl_6^{2-}$. The yellow, readily hydrolyzed salts of the former are obtained when $RhCl_3$ and an alkali chloride are treated with F_2 or BrF_3. The magnetic moments of *ca.* 1.8 BM are consistent with a t_{2g}^5 configuration.

The dark green compound Cs_2RhCl_6 is made by oxidation of ice-cold solutions of $RhCl_6^{3-}$ by Cl_2 in presence of CsCl. It is isomorphous with Cs_2PtCl_6. The salt decomposes in water.

Iridium. By contrast, the IV state is comparatively stable for iridium. Crystalline ions IrX_6^{2-} (X = F, Cl, and Br), as well as a variety of aquated complex ions such

31   A. L. Balch, *J. Am. Chem. Soc.*, 1976, **98**, 8049.
32a   V. L. Goedken *et al., J. Am. Chem. Soc.*, 1978, **100**, 1003.
32b   H. Ogoshi *et al., J. Am. Chem. Soc.*, 1977, **99**, 3869.
33   A. Aràneo *et al., J.C.S. Dalton*, **1975**, 2039.
34   A. Thorez *et al., J.C.S. Chem. Comm.*, **1977**, 518.

as $[IrCl_3(H_2O_3)]^+$, $[IrCl_5(H_2O)]^-$, and $[IrCl_4(H_2O)_2]$ have been characterized.
The N- and O-centered triangular species contain iridium in a mean oxidation state
$3\frac{2}{3}$, or III, III, IV. One example, known for a long time is the green ion $[Ir_3N-$
$(SO_4)_6(H_2O)_3]^{4-}$, made by boiling Na_3IrCl_6 and $(NH_4)_2SO_4$ in concentrated
H_2SO_4; another is the acetate[35] $[Ir_3O(CO_2Me)_6py_3]^+$.

Hexachloroiridates(IV) can be made by chlorinating a mixture of iridium
powder and an alkali metal chloride, or, in solution, by adding the alkali metal
chloride to a suspension of hydrous IrO_2 in aqueous HCl. The black crystalline
sodium salt Na_2IrCl_6, which is very soluble in water, is the usual starting material
for the preparation of other Ir^{IV} complexes.

The so-called chloroiridic acid is made by treating the ammonium salt with aqua
regia; it is soluble in ether and hydroxylated solvents and is probably
$(H_3O)_2IrCl_6 \cdot 4H_2O$.

In basic solution the dark red-brown $IrCl_6^{2-}$ is rather unstable, undergoing
spontaneous reduction within minutes to pale yellow-green $IrCl_6^{3-}$:

$$2IrCl_6^{2-} + 2OH^- \rightleftharpoons 2IrCl_6^{3-} + \frac{1}{2}O_2 + H_2O$$

From known potentials the *acid* reaction can be written:

$$2IrCl_6^{2-} + H_2O = 2IrCl_6^{3-} + \frac{1}{2}O_2 + 2H^+ \qquad K = 7 \times 10^{-8} \text{ atm}^{1/2} \text{ mol}^2 \text{ l}^{-2} (25°)$$

Thus in strong acid, say $12M$ HCl, $IrCl_6^{3-}$ is partially oxidized to $IrCl_6^{2-}$ in the cold
and completely on heating, whereas in strong base (pH > 11) $IrCl_6^{2-}$ is rapidly and
quantitatively reduced to $IrCl_6^{3-}$.

$IrCl_6^{2-}$ is readily and quantitatively reduced to $IrCl_6^{3-}$ by KI or sodium oxalate.
In neutral solutions slow reduction of $IrCl_6^{2-}$ occurs spontaneously. A variety of
organic compounds can be oxidized by $IrCl_6^{2-}$.

Octahedral Ir^{IV}, t_{2g}^5, has one unpaired electron. For pure $IrCl_6^{2-}$ salts the μ_{eff}
values are low (1.6–1.7 BM) owing to antiferromagnetic interactions; on dilution
with isomorphous $PtCl_6^{2-}$ salts, normal values are found.

22-G-6 Complexes of Rhodium(V) and Iridium(V), d^4

Only the hexafluoro ions MF_6^- of Rh^V and Ir^V are known. The salts are made by
reactions such as

$$RhF_5 + CsF \xrightarrow{IF_5} CsRhF_6$$

$$IrBr_3 + CsCl \xrightarrow{BrF_3} CsIrF_6$$

The red-brown Rh salt is isomorphous with $CsPtF_6$. The iridium salts are pink,
with magnetic moments *ca.* 1.25 BM at 273°; they are temperature dependent,
suggesting strong spin-orbit coupling and possibly antiferromagnetic interaction.
They dissolve in water, evolving O_2 and being reduced to IrF_6^{2-}. The only other
known complexes are the multihydrides $IrH_5(PR_3)_2$.

35   S. Uemura *et al., J.C.S. Dalton,* **1973,** 2565.

22-H. PALLADIUM AND PLATINUM[1]

22-H-1. General Remarks; Stereochemistry

The principal oxidation states of Pd and Pt are II and IV, but there is important chemistry in the I state, where M—M bonds are involved, and in the 0 state, where tertiary phosphine, CO, or other π-acid ligands always occur and Pt_3 and Pt_4 clusters are also found. Formal negative oxidation states occur in certain remarkable cluster-type carbonyl anions of general formula $[Pt_3(CO)_3(\mu\text{-}CO)_3]_n^{2-}$. The higher states V and VI occur only in a few fluoro compounds (p. 906). By contrast with Ni, the III state is virtually unknown, the only clear example being the ion [bis-(diphenylglyoximato)Pt]ClO$_4$.[2a]

The II State, d^8. The Pd^{2+} ion occurs in PdF_2 (p. 909) and is paramagnetic. In aqueous solution, however, the $[Pd(H_2O)_4]^{2+}$ ion is diamagnetic and is presumably square. In general, however, Pd^{II} and Pt^{II} complexes are square or five-coordinate and are diamagnetic. They can be of all possible types, e.g., ML_4^{2+}, ML_3X^+, *cis*- and *trans*-ML_2X_2, MLX_3^-, and MX_4^-, where X is uninegative and L is a neutral ligand. Similar types with chelate acido or other chelate ligands are common.

As a rule, Pd^{II} and Pt^{II} show a preference for nitrogen (in aliphatic amines and in NO_2), halogens, cyanide, and heavy donor atoms (e.g., P, As, S, Se), and relatively small affinity for oxygen and fluorine. The strong binding of the heavy atom donors is due in great measure to the formation of metal-ligand π bonds by overlap of filled $d\pi$ orbitals (d_{xz}, d_{xy}, and d_{yz}) on the metal with empty $d\pi$ orbitals in the valence shells of the heavy atoms.

Cyanide ions, nitro groups, and carbon monoxide are also bound in a manner involving π bonding, which results in these cases from overlap of filled metal $d\pi$ orbitals with empty $p\pi$ antibonding molecular orbitals of these ligands. In such complexes, there is usually considerable similarity of Ni to Pd and Pt.

The formation of cationic species even with non-π-bonding ligands and anionic species with halide ions contrasts with the chemistry of the isoelectronic Rh^I and Ir^I where most of the complexes involve π bonding. The difference is presumably a reflection of the higher charge. Furthermore, although Pd^{II} and Pt^{II} species add neutral molecules to give five- and six-coordinate species, they do so with much less ease; also the oxidative-addition reactions characteristic of square d^8 complexes tend to be reversible except with strong oxidants, presumably owing to the greater promotional energy for M^{II}—M^{IV} than for M^I—M^{III}.

[1]   F. R. Hartley, *The Chemistry of Platinum and Palladium.* Applied Science Publishers, 1973; D. M. Roundhill, *Adv. Organomet. Chem.,* 1975, **13**, 274; P. M. Henry, *Adv. Organomet. Chem.,* 1975, **13**, 363; P. M. Maitlis, *The Organic Chemistry of Palladium,* Vols. 1, 2, Academic Press, 1971; U. Belluco *et al., Organometallic and Coordination Chemistry of Platinum,* Academic Press, 1974; *Platinum Group Metals and Compounds,* ACS Advances in Chemistry Series No. 98, 1971.

[2a] H. Endres *et al., Z. Naturforsch.,* 1978, **33B**, 843.

TABLE 22-H-I

Oxidation States and Stereochemistry of Palladium and Platinum

Oxidation state	Coordination number	Geometry	Examples
Pd^0, Pt^0, d^{10}	3	Planar	$Pd(PPh_3)_3$
	4	Distorted	$Pt(CO)(PPh_3)_3$
	4	Tetrahedral	$Pt(Ph_2PCH_2CH_2PPh_2)_2$, $Pd(PF_3)_4$
Pd^{II}, Pt^{II}, d^8	$4^{a,b}$	Planar	$[PdCl_2]_n$, $[Pd(NH_3)_4]Cl_2$, PdO, PtO, $PtCl_4^{2-}$, $PtHBr(PEt_3)_2$, $[Pd(CN)_4]^{2-}$, PtS, $[Pd\ py_2Cl]_n$, PdS, $Pt(PEt_3)_2(C_6F_5)_2$
	5	*Tbp*	$[Pd(diars)_2Cl]^+$, $[Pt(SnCl_3)_5]^{3-}$
	6	Octahedral	PdF_2(rutile type), $[Pt(NO)Cl_5]^{2-}$, $Pd(diars)_2I_2$, $Pd(DMGH)_2{}^c$
Pd^{IV}, Pt^{IV}, d^6	6^b	Octahedral	$[Pt(en)_2Cl_2]^{2+}$, $PdCl_6^{2-}$, $[Pt(NH_3)_6]^{4+}$, $[Me_3PtCl]_4$
Pt^V, d^5	6	Octahedral	$[PtF_5]_4$
		Octahedral	PtF_6^-
Pt^{VI}, d^4	6	Octahedral	PtF_6

[a] Most common states for Pd.

[b] Most common states for Pt.

[c] Has planar set of N atoms with weak Pd—Pd bonds completing a distorted octahedron.

Palladium(II) complexes are somewhat less stable in both the thermodynamic and the kinetic sense than their Pt^{II} analogues, but otherwise the two series of complexes are usually similar. The kinetic inertness of the Pt^{II} (and also Pt^{IV}) complexes has allowed them to play a very important role in the development of coordination chemistry. Many studies of geometrical isomerism and reaction mechanisms have had a profound influence on our understanding of complexes. Both elements readily give allylic species, whereas platinum more commonly forms σ-bonded and alkene and alkyne complexes

The IV State, d^6. Although Pd^{IV} compounds exist, they are generally less stable than those of Pt^{IV}. The coordination number is invariably 6. The substitution reactions of platinum(IV) complexes are greatly accelerated by presence of Pt^{II} species. Solutions also readily undergo photochemical reactions in light.

The oxidation states and stereochemistries are summarized in Table 22-H-1.

22-H-2. Complexes of Palladium(II) and Platinum(II), d^8

Halogeno Anions. The ions MCl_4^{2-} are among the most important species because their salts are commonly used as source materials for the preparation of other complexes in the II and the 0 oxidation states. The yellowish $PdCl_4^{2-}$ ion is formed when $PdCl_2$ is dissolved in aqueous HCl or when $PdCl_6^{2-}$ is reduced with Pd sponge. The red $PtCl_4^{2-}$ ion is normally made by reduction of $PtCl_6^{2-}$ with a stoichiometric amount of hydrazine hydrochloride, oxalic acid, or other reducing agent. The sodium salt cannot be obtained pure. In water solvolysis of $PtCl_4^{2-}$ is extensive but the rate is slow:

$$PtCl_4^{2-} + H_2O = PtCl_3(H_2O)^- + Cl^- \qquad K = 1.34 \times 10^{-2} \, M \, (25°, \mu \, 0.5)$$
$$PtCl_3(H_2O) \; + H_2O = PtCl_2(H_2O)_2 + Cl^- \qquad K = 1.1 \times 10^{-3} \, M$$

so that a $10^{-3} \, M$ solution of K_2PtCl_4 at equilibrium contains only 5% of $PtCl_4^{2-}$ with 53% of mono- and 42% of bisaqua species.

For both metals, bromo and iodo complex anions occur, and if large cations such as Et_4N^+ are used, salts of halogeno-bridged ions $M_2X_6^{2-}$ may be obtained. Both MX_4^{2-} and $M_2X_6^{2-}$ are square, but in crystals the ions in K_2MCl_4 are stacked one above the other. However unlike other stacks containing MCl_4^{2-} ions discussed below, the M—M distances (Pd, 4.10 Å; Pt, 4.13 Å) are too large for any chemical bonding; similarly in the dimeric ions there is no evidence for metal-metal inter-action.

The $[M(CN)_4]^{2-}$ ions are extremely strong complexes thermodynamically, with $\log \beta_4$ values of ca. 63 and >65 for Pd and Pt, respectively.[2b]

Neutral Complexes. There is an enormous number of Pd and Pt complexes of the general formula $MXYL_1L_2$, where X and Y are anionic groups and L are neutral donor ligands such as NR_3, PR_3, SR_2, CO, and alkenes. In addition to ions such as Cl^- and SCN^-, X or Y may also be H or an alkyl or aryl group. A common palladium compound often used as a source material is bis(benzonitrile)dichloro-palladium $[PdCl_2(C_6H_5CN)_2]$, made by dissolving $PdCl_2$ in the ligand and crys-tallizing.

Besides the mononuclear species, there is a considerable number of *bridged bi-nuclear complexes* of the type 22-H-I, of which 22-H-II is a specific example. For the triphenylarsinepalladium analogue of 22-H-II, linkage isomers are known and the mode of bonding is solvent dependent.

(22-H-I)   (22-H-II)   (22-H-III)

For Pt^{II}, the bridging tendencies of the anions are in the order $SnCl_3^- < RSO_2^- < Cl^- < Br^- < I^- < R_2PO^- < SR^- < PR_2^-$. The strong tendency of SR and PR_2 groups to form the four-membered rings (22-H-III) may be accounted for by de-localized bonding arising from overlap of filled metal d_{xz} and d_{yz} orbitals with empty d orbitals on sulfur or phosphorus.

Bridged halide complexes are the commonest encountered, and bridged species are quite generally subject to cleavage by donor ligands to give mononuclear species, e.g.,

2b R. D. Hancock and A. Evers, *Inorg. Chem.*, 1976, **15**, 995.

When the bridges are Cl⁻ or Br⁻, the equilibria generally lie toward the mononuclear complexes. It might be supposed that such bridge-splitting reactions should give trans products, and these are probably the initial products of most cleavage reactions. However the relative stabilities of the cis and trans isomers of L_2PdX_2 and L_2PtX_2 complexes vary greatly, depending on the identities of L and X.[2c] A major difference between square complexes of the two metals is that for $Pt(PR_3)_2X_2$ complexes cis-trans isomerization normally proceeds extremely slowly unless catalyzed by excess PR_3, whereas the analogous isomerization of Pd[II] complexes proceeds rapidly to give equilibrium mixtures. There is evidence that these isomerization reactions generally proceed through five-coordinate intermediates where either excess PR_3 or solvent molecules add to generate the intermediate.[3]

Hydrido Complexes. There are a number of stable square Pt[II] complexes containing only one hydride ligand [e.g., $PtHClPR_3)_2$], but complexes with two hydride ligands trans-$[PtH_2(PR_3)_2]$ are stable only if PR_3 is a very bulky ligand.[4a] All stable hydrido complexes have trans structures. Most of the studies have been on platinum compounds, since the comparable palladium (and also nickel) hydrido complexes are usually less stable thermally; some compounds such as trans-$PdClH(PR_3)_2$ (R = cyclohexyl or Ph) have been made.

The phosphine and arsine hydrides are obtained from the corresponding halides (the cis isomer is usually most reactive) by the action of a variety of hydrogen-transfer agents such as KOH in ethanol, H_2 at 50 atm/95°, $LiAlH_4$ in THF or, most conveniently, 90% aqueous hydrazine, e.g.,

$$cis\text{-}PtBr_2(Et_3P)_2 \rightarrow trans\text{-}PtHBr(Et_3P)_2$$

The KOH-alcohol reduction is a general one for the preparation of hydrido or hydridocarbonyl complexes, but detailed mechanistic studies are lacking (see Section 29-13).

The hydrido compounds of Pt[II] are usually air-stable, colorless, crystalline solids, soluble in organic solvents and sublimable. Their chemical reactions resemble those of other hydrido species. They will also add HCl or CH_3I to give octahedral Pt[IV] complexes, but these usually readily lose the added molecules.

Cationic hydrido species can also be obtained by reactions of the type:

$$trans\text{-}PtHCl(AsPh_3)_2 + NaClO_4 + CO \rightarrow [PtH(CO)(AsPh_3)_2]ClO_4 + NaCl$$

There are also binuclear species, $[Pt_2H_3(PR_3)_4]^+$, in which the terminal and bridge hydride ligands are in rapid exchange.[4b]

The nmr and ir spectra of hydrido species have been of interest because of the information that can be obtained concerning trans effects. There is a strong dependence of Pt—H stretching frequencies, proton chemical shifts, and ^{195}Pt—H and ^{31}P—H coupling constants, depending on the ligand trans to hydride. For ex-

2c A. W. Verstuyft, L. W. Cary, and J. H. Nelson, *Inorg. Chem.,* 1976, **15**, 3161.
3   D. A. Redfield, L. W. Cary, and J. H. Nelson, *Inorg. Chem.,* 1975, **14**, 50.
4a B. L. Shaw and M. F. Uttley, *Chem. Comm.,* **1974**, 918.
4b L. M. Venanzi *et al., Angew. Chem., Int. Ed.,* 1979, **18**, 155; S. Otsuka *et al., Inorg. Chem.,* 1979, **18**, 2239.

ample, the Pt—H stretching frequencies and chemical shifts increase for the *trans* ligand in the order $Cl < Br < I < NCS < SnCl_3 < CN$.

Other Pd^{II} and Pt^{II} Complexes.   There are a number of cationic species, of which the ammine $[M(NH_3)_4]^{2+}$ and $[Men_2]^{2+}$ are particularly well known.

Although Pd^{II} and Pt^{II} are generally viewed as having low affinity for F^- and oxygen donors, in the latter case there are some notable exceptions. There are various μ-OH dimers and trimers that have outstanding stability.[5,6] There are also sulfoxide complexes in which, depending on the particular sulfoxide used, there may be *S*-bonded, *O*-bonded, or a mixture of *S*- and *O*-bonded ligands.[7]

Palladium, but not platinum, forms an aqua ion, and brown, deliquescent salts such as $[Pd(H_2O)_4](ClO_4)_2$ are obtained from solutions of PdO in dilute noncomplexing acids. In $3.94M$ $HClO_4$ at $25°$ the Pd/Pd^{2+} potential is 0.98 V. A sublimable anhydrous nitrate is also known.

The conventional β-diketone complexes, with two *O,O*-bonded ligands are known, but β-dike ligands also form a number of Pd and Pt complexes in which the metals are attached to the central carbon atom[8] as in the following reaction.

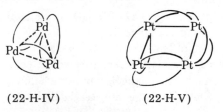

$$L = PR_3, \text{ py, } Et_2HN$$

Both metals form *anhydrous acetates*. Palladous acetate has been thoroughly studied as a catalyst for vinyl acetate synthesis (Section 30-14). It is a brown crystalline substance that acts to some extent like Hg^{II} and Pb^{IV} acetates in attacking benzene and other aromatic hydrocarbons in acid media. Thus in acetic acid, it specifically attacks the side chain of toluene.

Both these acetates are polynuclear, but they have quite different structures. $Pd_3(O_2CCH_3)_6$ has the triangular structure 22-H-IV with Pd—Pd distances (*ca.* 3.15 Å) indicative of no direct bonding; $Pt_4(O_2CCH_3)_8$ has structure 22-H-V, with Pt—Pt distances (2.50 Å) consistent with the occurrence of single bonds between the metal atoms.[9] Heating $Pd_3(O_2CCH_3)_6$ in benzene causes dissociation to a

(22-H-IV)   (22-H-V)

5   B. Lippert *et al., Inorg. Chem.,* 1978, **17,** 2971.
6   G. W. Bushnell *et al., Can. J. Chem.,* 1972, **50,** 3694.
7   B. B. Wayland *et al., Inorg. Chem.,* 1972, **11,** 1280.
8   S. Kawaguchi *et al., Inorg. Chem.,* 1977, **16,** 1730; 1978, **17,** 910.
9   M. Corrondo and A. C. Skapski, *J.C.S. Chem. Comm.,* **1976,** 410.

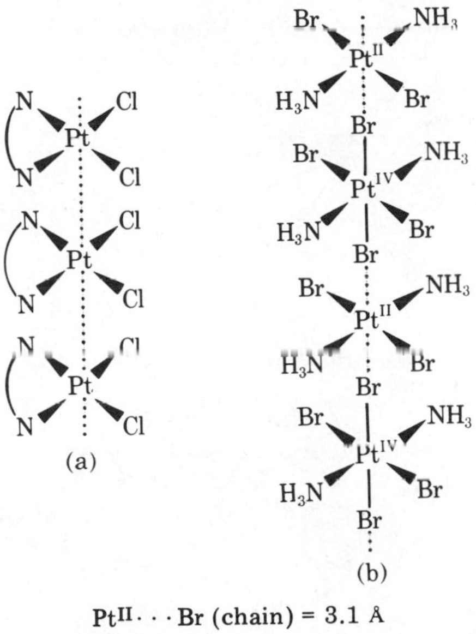

(a)

(b)

$Pt^{II} \cdots Br$ (chain) = 3.1 Å

Fig. 22-H-1. (a) Linear stacks of planar $PtenCl_2$ molecules. (b) Chains of alternating Pt^{II} and Pt^{IV} atoms with bridging bromide ions in $Pt(NH_3)_2Br_3$.

monomer, and similar cleavage of the trimer can be achieved by the action of donor ligands on the acetate and other carboxylates to give yellow trans complexes $Pd(OOCR)_2L_2$. $Pt_4(O_2CCII_3)_8$ is not cleaved by ligands under comparable conditions.

In many cases square complexes of Pd^{II} or Pt^{II} are packed in crystals to form infinite chains of metal atoms (Fig. 22-H-I) close enough to interact electronically with one another, giving rise to marked spectral dichroism and electrical conductivity. These substances are discussed in more detail in Chapter 26.

There is a related class of compound with chainlike structures that contain both M^{II} and M^{IV} but differ from the above in that the metal units are linked by halide bridges. There is similar behavior in that there is high electrical conductivity along the direction of the $—Cl—M^{II}—Cl—M^{IV}—$ chains, e.g., in $[Pd^{II}(NH_3)_2Cl_2]$-$[Pd^{IV}(NH_3)_2Cl_4]$. Thus Wolfram's red salt has octahedral $[Pt(EtNH_2)_4Cl_2]^+$ and planar $[Pt(EtNH_2)_4]^{2+}$ ions linked in chains, the other four Cl^- ions being within the lattice. The structure in Fig. 22-H-I is typical. These compounds are characterized by intense absorption bands, polarized parallel to the chain axis, and assigned to intervalence electron transfer transitions. Several such compounds show exceptionally strong resonance Raman spectra.[10]

Metal-Ligand Interactions. There is also good evidence that there can be interaction between a bound ligand and the metal atom either inter- or intramolecularly.

In the complex *trans*-$PdI_2(PMe_2Ph)_2$ the α-H atoms of the phenyl group of the

[10]   R. J. H. Clark and M. L. Franks, *J.C.S. Dalton*, **1977**, 198.

coordinated phosphine occupy an axial position (cf. pp. 929) and the *trans*-axial position is occupied by an iodine atom of an adjacent molecule, so that a quasi-seven-coordinate complex (22-H-VI) results.

(22-H-VI)

In ammine complexes such as $PtCl_2(NH_3)_2$, infrared evidence suggests that anomalous N—H stretching frequencies can be attributed to a type of hydrogen-bonding interaction between H and the filled d_{xy} or d_{xz} orbital of the metal. Palladium complexes do not show this effect, probably owing to the smaller spatial extension of the $4d$ orbitals.

Finally, another case is the complex ion $[Pd(Et_4dien)Cl]^+$, where the ethyl groups block off the axial positions so that kinetically, in substitution reactions, the ion behaves as an octahedral rather than a square complex.

Five-Coordinate Complexes of Pd^{II} and Pt^{II}. It is generally agreed that many substitution and isomerization reactions of square complexes normally proceed by an associative path involving distorted five-coordinate intermediates, and there is good evidence for solvation, for example, of $PtCl_2(Bu_3^nP)_2$ by CH_3CN. The cis-trans isomerization of $PtCl_2(Bu_3^nP)_2$ and similar Pd complexes, where the isomerization is immeasurably slow in absence of an excess of phosphine, is very fast when phosphine is present. The isomerization doubtless proceeds by pseudorotation of the five-coordinate state. In this case an ionic mechanism is unlikely, since polar solvents actually slow the reaction. Photochemical isomerizations, on the other hand, appear to proceed through tetrahedral intermediates, whereas thermal isomerizations involve an ionic mechanism.

There are a number of quite stable, isolable five-coordinate complexes of Pd^{II} and Pt^{II}. First, multifunctional ligands of a tripod type (p. 76) have afforded a number of discrete five-coordinate species[11] such as that shown in 22-H-VII.

The $[PdCl_4]^{2-}$ and $[Pdphen_2]^{2+}$ ions associate with Cl$^-$ in solution, although the Pt analogues appear not to do so. An isolable five-coordinate Pt^{II} complex with

[11] C. Senoff, *Inorg. Chem.*, 1978, **17**, 2320.

(22-H-VII)

monodentate ligands is $[Pt(SnCl_3)_5]^{3-}$, obtainable as R_4N^+ or R_4P^+ salts from the red solutions obtained by adding an excess of $SnCl_3^-$ to $PtCl_4^{2-}$ in 3 M HCl or $SnCl_2$ to ethanolic solutions of Na_2PtCl_4. The $[Pt(SnCl_3)_5]^{3-}$ ion may exist only in the solid state. The nature of the $SnCl_3^- - PtCl_4^{2-}$ solutions is exceedingly complex and depends on the concentrations, acidity, temperature, and time; solutions doubtless contain mixtures of the various species $[PtCl_n (SnCl_3)_{5-n}]^{3-}$ and others.

The Pt–Sn complexes catalyze the hydrogenation of ethylene and some other olefinic compounds; this action is doubtless connected with the ready dissociation of the complexes in solution, promoted by the *trans* effect of $SnCl_3^-$, which leaves vacant sites for coordination of olefin and of hydrogen.

Finally a number of dithio complexes such as dithiocarbamates and dithiophosphates form adducts with phosphine complexes, e.g., $Pt(S_2CNEt_2)$-$(PMePh_2)$. Some of these show nmr behavior in solution suggesting equilibrium between planar and five-coordinate species.

Octahedral Complexes of Pd^{II} and Pt^{II}. These are very few, and although they may be octahedral in the solid state, e.g., *trans*-$MI_2(diars)_2$, dissociation probably occurs in solution.

Phosphine Complexes. The complexes of tertiary phosphines with Pd^{II} and Pt^{II} have been extensively studied; many complexes are known, and these have often been studied in detail, especially from the nmr viewpoint. Thus ideas of π-bonding and *trans* effects can be obtained through $^{31}P - {}^{195}Pt$ coupling constants. These coupling constants are a sensitive function of the trans ligand and relatively insensitive to the cis ligand; thus they are useful in making structural assignments. For example, *trans*-$Pt(C_6H_5)Cl(PEt_3)_2$ has only a single resonance with J_{Pt-P} 2800 Hz, whereas the cis isomer has two ^{31}P resonances with J_{Pt-P} values of 1580 and 4140 Hz. Orders of trans effects of ligands obtained in this way are similar to those obtained by other methods.

Alkyls and Aryls. These substances are usually white, air-stable, crystalline solids, obtained by action of lithium alkyls or Grignard reagents on the halides. A typical example is *trans*-$PtBrCH_3(PEt_3)_2$. The alkyls can also be made by addition of alkenes at Pt—H bonds or by insertion reactions into other Pt—C bonds. Many closely related compounds with Pt or Pd bound to Si, Ge, or other metallic elements have been made, some typical ones being those with Pt—SiR_3 groups. They can be made by the action of sodium salts on the platinum halides or by addition of halides to Pd^0 or Pt^0 compounds (see below).

22-H-3. Complexes of Palladium(IV) and Platinum(IV), d^6

Palladium. The Pd^{IV} complexes are more stable than simple Pd^{IV} compounds, but only a few are known, and apart from Pd$(NO_3)_2(OH)_2$ formed on dissolution of Pd in concentrated nitric acid, they are mainly the octahedral halide anions. The fluoro complexes of Ni, Pd, and Pt are all very similar and are rapidly hydrolyzed by water. The chloro and bromo ions are stable to hydrolysis but are decomposed by hot water to give the Pd^{II} complex and halogen. The red PdCl$_6^{2-}$ ion is formed when Pd is dissolved in aqua regia or when PdCl$_4^{2-}$ solutions are treated with chlorine.

Oxidation of K$_2$PdCl$_4$ by persulfate in presence of KCN gives the yellow salt K$_2$[Pd(CN)$_6$].

The diamine complexes such as Pdpy$_2$Cl$_4$, which is obtained as a deep orange crystalline powder when Pdpy$_2$Cl$_2$ suspended in chloroform is treated with chlorine, are of marginal stability. They lose chlorine or bromine rapidly in moist air. Other complexes such as Pd(NH$_3$)$_2$(NO$_2$)$_2$Cl$_2$ are stable.

Platinum. In marked contrast to Pd^{IV}, platinum(IV) forms many thermally stable and kinetically inert complexes. So far as is known, Pt^{IV} complexes are invariably octahedral and, in fact, Pt^{IV} has such a pronounced tendency to be six-coordinated that in some of its compounds quite unusual structures are adopted. An apparent exception to the rule is η^5-C$_5$H$_5$Pt(CH$_3$)$_3$, but as with other η^5-C$_5$H$_5$ complexes, the ring can be considered as occupying three positions of an octahedron. In several interesting examples of this tendency of Pt^{IV} to be six-coordinate, novel bonding is required for such coordination to be achieved (see below).

The most extensive and typical series of Pt^{IV} complexes are those that span the entire range from the hexammines [PtAm$_6$]X$_4$, including all intermediates such as [PtAm$_4$X$_2$]X$_2$ and M^I[PtAmX$_5$], to M$_2^I$[PtX$_6$]. Some of these are particularly notable as examples of the classical evidence that led Werner to assign the coordination number 6 to Pt^{IV}. The amines that occur in these complexes include ammonia, hydrazine, hydroxylamine, and ethylenediamine, and the acido groups include the halogen and the thiocyanate, hydroxide, and nitro groups. Although not all these groups are known to occur in all possible combinations in all types of compound, it can be said that with a few exceptions, they are generally interchangeable.

The most important Pt^{IV} compounds are salts of the red *hexachloroplatinate* ion PtCl$_6^{2-}$. The "acid," commonly referred to as chloroplatinic acid, has the composition H$_2$[PtCl$_6$]·nH$_2$O (n = 2, 4, 6) and contains H$_3$O$^+$, H$_5$O$_2^+$, and H$_7$O$_2^+$ ions.[12a] On thermal decomposition[12b] it eventually gives mainly the metal, but volatile Pt$_6$Cl$_{12}$ (β-PtCl$_2$) is one of the intermediates. The acid or its Na or K salts are the normal starting materials for the preparation of many Pt compounds. The ion is formed on dissolving Pt in aqua regia or HCl saturated with Cl$_2$.

Platinum(IV)-to-Carbon Bonds. A characteristic feature of Pt^{IV} is the stability of compounds with a (CH$_3$)$_3$Pt group. In all these Pt^{IV} is octahedrally coordinated. Thus the halides are tetrameric, e.g., [Me$_3$PtCl]$_4$, with three-way halogen bridges.

[12a] U. Greher and A. Schmidt, *Z. Anorg. Allg. Chem.*, 1978, **444**, 97.
[12b] A. E. Schweizer and G. T. Kerr, *Inorg. Chem.*, 1978, **17**, 2326.

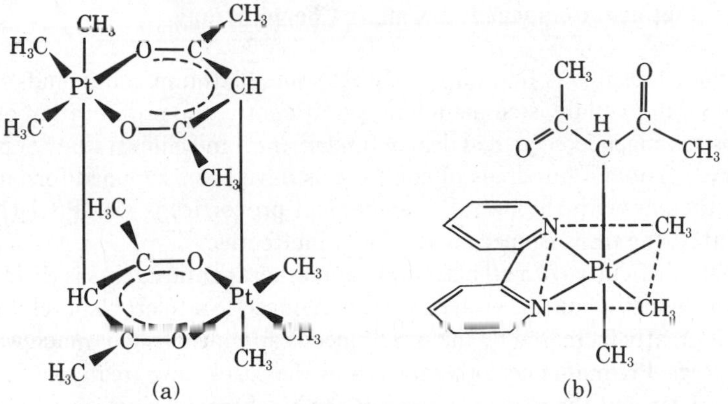

Fig. 22-H-2. (a) The molecular structure of trimethylplatinum acetylacetonate dimer, showing how the Pt^{IV} attains octahedral coordination. (b) Schematic representation of the molecular structure of the bipyridine adduct of trimethylplatinum(IV) acetylacetonate.

In aqueous solutions the very stable octahedral ion $[Me_3Pt(H_2O)_3]^+$ is formed with noncoordinating anions such as BF_4^- or ClO_4^-. Tetramethylplatinum does not exist, and what was thought to be this compound is actually $[Me_3Pt(OH)]_4$.

Trimethylplatinum acetylacetonate, $[(CH_3)_3Pt(O_2C_5H_7)]_2$, long known to be a dimer in noncoordinating solvents, has the structure shown in Fig. 22-H-2a. The acetylacetone functions as a tridentate ligand, the third donor atom being the middle carbon atom of each ring.

In the monomeric compound $(CH_3)_3Pt(dipy)(O_2C_5H_7)$, six- rather than seven-coordination is achieved by the formation of only one bond to the acetylacetonate ion (Fig. 22-H-2b) as in the Pd^{II} complex noted earlier. As evidence of the great strength of the Pt—C bond, we point out that in the preparation of this compound from $[(CH_3)_3Pt(O_2C_5H_7)]_2$ by the action of bipyridine, it is the Pt—O rather than Pt—C bonds that are broken.

The reaction of $[Pt(NH_3)_6]Cl_4$ with β-diketones in basic solution gives N—N-bonded β-diiminato complexes[13] such as 22-H-VIII and 22-H-IX.

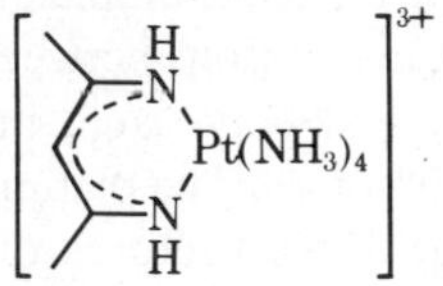

(22-H-VIII) 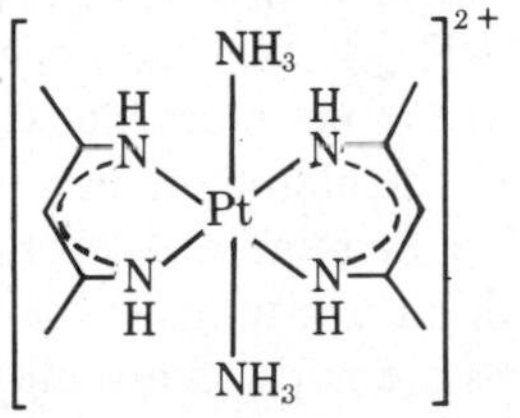   (22-H-IX)

A series of mono- and dihydride complexes of six-coordinate Pt^{IV} such as cis,cis-$[PtH_2Cl_2(PR_3)_2]$ are obtained[14] by oxidative addition of HX to four-coordinate $PtHCl(PR_3)_2$ and in other ways.

[13]   A. M. Sargeson, G. W. Everett, Jr., et al., J. Am. Chem. Soc., 1976, **98**, 8041; Inorg. Chem., 1978, **17**, 1304.
[14]   D. W. W. Anderson, E. A. V. Ebsworth, and D. W. H. Rankin, J.C.S. Dalton, **1973**, 854.

22-H-4.   Platinum Compounds in Cancer Chemotherapy

During the 1960s it was first observed that some platinum compounds have the ability to inhibit cell division including, significantly, the proliferation of cancer cells. This has engendered a great deal of fundamental and clinical study of platinum chemistry.[15] Though hundreds of compounds have been screened for antitumor activity, the one with the overall best clinical properties is $cis\text{-}[PtCl_2(NH_3)_2]$. Remarkably, the trans isomer is essentially ineffective.

A serious difficulty with all platinum chemotherapeutic agents is their toxicity toward the kidneys, but this problem can be reduced to a tolerable level if the flow of urine is greatly increased by the simultaneous administering of much water and diuretic drugs. Platinum chemotherapy is now the best known treatment for certain types of cancer, but the mode of action at the biochemical level is still poorly understood. One view is that the platinum compounds directly inhibit cell division by binding to DNA and/or RNA. Another is that the cancer cells are rendered more susceptible to destruction by the body's own immune system.

Efforts continue to find new and better platinum complexes for chemotherapeutic use. The principal requirements are (*a*) lower toxicity and equal or better activity as compared to $cis\text{-}[PtCl_2(NH_3)_2]$, and (*b*) the maintaining of suitable solubility properties. For example, 22-H-X meets the first two criteria but has insufficient solubility in tissue fluids, but 22-H-XI is a promising rival to $cis\text{-}[PtCl_2(NH_3)_2]$.

(22-H-X)   (22-H-XI)

Platinum Blues.[16]   Hydrolysis of the acetonitrile complex of divalent platinum, $cis\text{-}PtCl_2(NCCH_3)_2$, in presence of silver salts produces a blue product first designated as "Platinblau." This material, once thought to have the composition $Pt^{II}(CH_3CONH)_2 \cdot H_2O$, is now considered to be an acetamido complex of Pt^{IV}.

A second type of platinum blue is formed in the reactions between the aquated products of *cis*-dichlorodiammineplatinum(II) and polyuracil, uracil, 1-methyluracil, uridine, thymine, and other, related pyrimidines. The platinum pyrimidine blues also exhibit antitumor activity. Although the structures of the classical platinum blues are not well established directly, a strong clue has been afforded by the isolation and characterization of a reaction product of $cis\text{-}[Pt(NH_3)_2(OH)_2]^{2+}$ with α-pyridone (22-H-XII), a ligand that models the features

[15]   T. A. Connors and J. J. Roberts, Eds., *Platinum Coordination Complexes in Chemotherapy*, Springer, 1974; B. Rosenberg *et al.*, *Cancer Chemother. Rep.*, 1975, **59**, 287, 589; M. J. Cleare, *Coord. Chem. Rev.*, 1974, **12**, 349; A. P. Zipp and G. S. Zipp, *J. Chem. Educ.*, 1977, **54**, 739.

[16]   S. J. Lippard *et al.*, *J. Am. Chem. Soc.*, 1977, **99**, 2827; 1978, **100**, 3785; 1979, **101**, 1434.

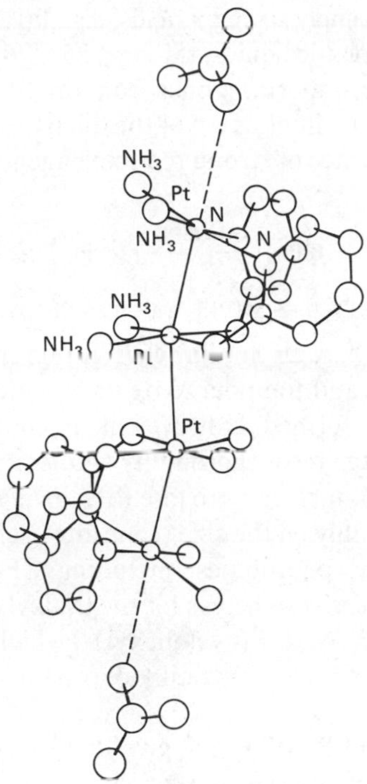

Fig. 22-H-3. Structure of the α-pyridone complex that resembles the platinum blues.

of the above-mentioned pyrimidines. The product of this reaction, $[Pt_4(NH_3)_8(C_5NH_4O)_4](NO_3)_5$, has the structure shown in Fig. 22-H-3.

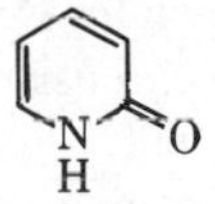

(22-H-XII)

22-H-5. Complexes of Palladium(0) and Platinum(0), d^{10}

There is a reasonably extensive chemistry of the 0 state, which is of importance especially for triphenylphosphine complexes; the latter undergo a wide variety of oxidative-addition reactions.

In general, the compounds are similar to those of Ni^0 except that no analogue of $Ni(CO)_4$ is known. This is believed to be due to a poorer tendency to π bonding, possibly associated with the greater ionization enthalpies of Pd and Pt. It is significant that in presence of triphenylphosphine, which is a better σ donor but poorer π acceptor, the carbonyl compounds are relatively stable, e.g., $Pd(CO)(PPh_3)_3$ and $Pt(CO)_2(PPh_3)_2$. Some Pt^0 compounds are metal atom cluster compounds and are discussed in Section 22-H-6.

The Pd^0 and Pt^0 compounds all contain tertiary phosphines as ligands. Some involve ligands that are extremely strong π acids, e.g., $Pt(PF_3)_4$ and $Pt[P(CF_3)_2F]_4$. These two compounds are volatile liquids, stable at $25°$; $Pd(PF_3)_4$ decomposes above $-20°$. However the most important ones contain triaryl- and even trialkyl-phosphines. They are made by interaction of the dihalides, or $[MX_4]^{2-}$ complexes, with the phosphines in presence of strong reducing agents,[17a] as illustrated by the following reactions:

$$K_2MCl_4 + 4PPh_3 \xrightarrow{N_2H_4,\ EtOH} M(PPh_3)_4$$

$$MCl_2 + 4PEt_3 \xrightarrow{K} M(PEt_3)_4 \xrightarrow{\Delta} M(PEt_3)_2$$

All the compounds have broadly similar chemical properties, centering around their tendency to lose PR_3 and to undergo oxidative addition reactions whereby Pd^{II} and Pt^{II} complexes are formed. However the Pt compounds, especially those of PPh_3, have been by far the most thoroughly studied.[17b]

The tendency of $Pt(PR_3)_4$ molecules to lose PR_3, giving $Pt(PR_3)_3$ and $Pt(PR_3)_2$, appears to depend considerably on the size (cone angle, cf. p. 89) of the phosphine. For triaryl and alkyldiaryl phosphines, including CF_3Ph_2P,[18] dissociation of $Pt(PR_3)_4$ to $Pt(PR_3)_2$ is extensive, whereas for the trialkyl or dialkylaryl phosphines, dissociation does not occur. With $P(cyclohexyl)_3$, which has an extremely large cone angle, and a few other extremely bulky phosphines,[19] the $Pt(PR_3)_2$ species can be isolated.

Substantial dissociation of $Pt(PPh_3)_4$ to give $Pt(Ph_3)_3$ occurs in solution at $25°$, but the extent of dissociation of $Pt(PPh_3)_3$ or $Pt(PPh_3)_2(C_2H_4)$ to give $Pt(PPh_3)_2$ is too slight to allow detection of the latter by nmr.[20] However $Pt(PPh_3)_2$ has been shown kinetically to play a significant role in some reactions.[21]

The oxidative-addition reactions of the $Pt(PR_3)_n$ molecules, in which some PR_3 ligands are lost are extremely numerous and varied. A few that are illustrative of the scope of this chemistry[22] are shown for $Pt(PPh_3)_3$ in Fig. 22-H-4. The dioxygen complex $Pt(PPh_3)_2O_2$ itself has an extensive and important chemistry. It acts as an oxidant toward CO, SO_2, PPh_3, etc.

It should be noted that the description of all the reactions in Fig. 22-H-4 as oxidative-addition reactions is not without difficulties, since the degree of net charge transfer to the addend may not always be sufficient to justify such a description. Clearly when HCl, RI, etc. are added with bond breaking, the description is apt. For $Pt(PR_3)_2O_2$ and $Pt(PR_3)_2(C_2H_4)$, calculations[23] suggest that the former is properly considered to contain Pt^{II} but that the latter is essentially at π complex of Pt^0.

[17a] R. A. Schunn, *Inorg. Chem.*, 1976, **15**, 208.

[17b] V. I. Sokolov and O. A. Reutov, *Coord. Chem. Rev.*, 1978, **27**, 89 (reactions of Pd and Pt zerovalent compounds with non-transition metal organometallics).

[18] T. G. Attig, M. A. A. Beg, and H. C. Clark, *Inorg. Chem.*, 1975, **14**, 2986.

[19] R. B. King and P. N. Kapoor, *Inorg. Chem.*, 1972, **11**, 1524; S. Otsuka et al., *J. Am. Chem. Soc.*, 1974, **96**, 3324.

[20] C. A. Tolman, W. C. Seidel, and D. H. Gerlach, *J. Am. Chem. Soc.*, 1972, **94**, 2669.

[21] J. Halpern and T. A. Weil, *J.C.S. Chem. Comm.*, **1973**, 631.

[22] C. G. Pierpont and H. H. Downs, *Inorg. Chem.*, 1975, **14**, 343; A. L. Balch et al., *Inorg. Chem.*, 1975, **14**, 2327.

[23] J. G. Norman, Jr., *Inorg. Chem.*, 1977, **16**, 1328.

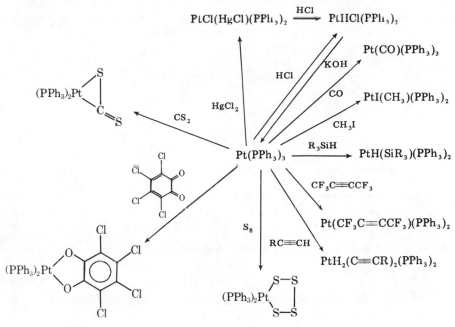

Fig. 22-H-4. Oxidative-addition and related reactions of tris(triphenylphosphine) platinum. Reactions with Pt(PPh₃)₄ and the corresponding palladium complexes are similar.

22-H-6. Compounds with Metal-Metal Bonds

Both Pd and Pt form many compounds with M—M bonds, some dinuclear[24] and others cluster compounds. Most of the dinuclear ones have the metal atoms in formal oxidation state I, and a few representative compounds, together with reactions used to prepare them and the M—M distances, are the following:

Pt(1,5-cyclooctadiene)₂ + (CF₃)₂CO ⟶

(CF₃)₂C——O

Pt—Pt

2.58 Å

(Ref. 25)

[PtCl₄]²⁻ + CO ⟶

$$\left[\begin{array}{c} Cl \quad\quad Cl \\ | \quad\quad \\ Cl-Pt-Pt-Cl \\ | \quad\quad | \\ C \quad C \\ O \quad O \end{array} \right]^{2-}$$

2.58 Å

(Ref. 26)

[24] J. R. Boehm and A. L. Balch, *Inorg. Chem.,* 1977, **16,** 778, and many references therein; J. Kuyper and K. Vrieze, *Trans. Met. Chem.,* 1976, **1,** 199, 208.

[25] F. G. A. Stone *et al., J.C.S. Dalton,* **1977,** 278 (this paper gives extensive references to other Pt—Pt bonds).

[26] A. Modinos and P. Woodward, *J.C.S. Dalton,* **1975,** 1516.

$$\text{"Pd(CO)Br"} \xrightarrow{\text{Ph}_2\text{PCH}_2\text{PPh}_2} \begin{array}{c} \text{H}_2 \\ \text{Ph}_2\text{Pd} \overset{\text{C}}{\diagdown} \text{PPh}_2 \\ | \quad\quad | \\ \text{Cl} - \text{Pd} - \text{Pd} - \text{Cl} \\ | \quad\quad | \\ \text{Ph}_2\text{P} \underset{\text{C}}{\diagup} \text{PPh}_2 \\ \text{H}_2 \end{array}$$ (Ref. 27)

2.70 Å

$$\text{Na}_2\text{PdCl}_4 \xrightarrow{\text{excess MeNC, PF}_6^-} \left[\begin{array}{c} \text{Me} \quad\quad \text{Me} \\ \text{N} \quad\quad\quad \text{N} \\ \text{C} \quad\quad\quad \text{C} \\ | \quad\quad\quad\quad \\ \text{MeNC} - \text{Pd} - \text{Pd} - \text{CNMe} \\ | \quad\quad | \\ \text{C} \quad \text{C} \\ \text{N} \quad \text{N} \\ \text{Me} \quad \text{Me} \end{array} \right] (\text{PF}_6)_2$$ (Ref. 24)

2.53 Å

Examples of dinuclear, M—M bonded complexes in oxidation state II are 22-H-XIII[28] and 22-H-XIV.[29]

$S{\frown}S = S_2CCH_2Ph$

(22-H-XIII)

$S{\frown}S = S_2CC_6H_4C_3H_7$

(22-H-XIV)

Small Clusters.   Mainly by thermolysis of $Pt(PPh_3)_3$ or $Pt(PPh_3)_2(C_2H_4)$, a number of small Pt^0 clusters have been obtained.[30–32] The structures of many are still unknown, and some of the formulas may be inaccurate. Examples of those for which the structures are uncertain, although tentative proposals have been made, are: $Pt_3(PPh_3)_9$, $[Pt(PPh_2)(Ph_2PC_6H_4)]_3$, $Pt_4(PPh_3)_4$, and $Pt_3(PPh_3)_5$. For two compounds that can be crystallized from the deep red solutions formed on refluxing $Pt(PPh_3)_4$ in benzene for several days, the formulas 22-H-XV and 22-H-XVI have been determined X-ray crystallographically,[32] and these are presumably representative. The Pt—Pt bond lengths are 2.60 Å in the dinuclear one and 2.79 Å in the trinuclear one; the former may indicate a bond order >1.

27   R. G. Holloway et al., J.C.S. Chem. Comm., 1976, 485.
28   M. Bonamico, G. Dessy, and V. Fares, J.C.S. Dalton, 1977, 2315.
29   J. P. Fackler, Jr., J. Am. Chem. Soc., 1972, 94, 1009.
30   R. Ugo, et al., Inorg. Chem. Acta, 1976, 18, 113.
31   D. M. Blake and L. M. Leung, Inorg. Chem., 1972, 11, 2879.
32   N. J. Taylor, P. C. Chieh, and A. J. Carty, J.C.S. Chem. Comm., 1975, 448.

$$\begin{array}{c} Ph_2 \\ P \end{array}$$

Ph$_3$P—Pt————Pt—PPh$_3$

P

Ph$_2$

(22-H-XV)

Ph$_3$P

Ph$_2$
P

PPh$_3$

Pt   Pt

Ph$_2$P—Pt——PPh$_2$

Ph

(22-H-XVI)

Carbonyl Anion Compounds.[33] By reduction of $Na_2PtCl_6 \cdot 6H_2O$ or $PtCl_2(CO)_2$ under CO (1 atm, 25°C) in the presence of alkali, a series of anions with the general formula $[Pt_3(CO)_6]_n^{2-}$ is formed, where the value of n depends on the ratio of alkali to platinum. Anions with $n = 1, 2, 3, 4, 5, 6$, and 10 and $[Pt_{19}(CO)_{22}]^{4-}$ have been identified, and a number of them have been isolated and structurally characterized by crystallizing them as salts of bulky cations [e.g., $(C_4H_9)_4N^+$, Ph_4P^+]. The structures of three of these, with $n = 2, 3$, and 5, are shown in Fig. 22-H-5. These structures consist of $[Pt_3(CO)_3(\mu\text{-}CO_3)]$ units arranged to form skewed and twisted stacks.

All these $[Pt_3(CO)_6]_n^{2-}$ species are quite reactive, the specific character of their reactivity depending on n, since the greater the value of n, the less "reduced" is the metal. The larger the value of n, the greater is the reactivity toward nucleophilic and reducing agents; with small n it is greatest toward electrophiles and oxidizing agents. Thus the lower members ($n = 1, 2, 3$) react very avidly with oxygen according to the equation

$$(n + 1)[Pt_3(CO)_6]_n^{2-} + \tfrac{1}{2}O_2 + H_2O \rightarrow n[Pt_3(CO)_6]_{n+1}^{2-} + 2OH^-$$

The base so formed favors reductive carbonylation, tending to reverse this change from n to $n + 1$ if CO is present. Moreover, the various species tend to react with each other, viz.,

$$[Pt_3(CO)_6]_2^{2-} + [Pt_3(CO)_6]_4^{2-} = 2[Pt_3(CO)_6]_3^{2-}$$

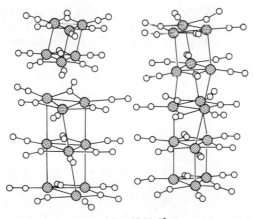

Fig. 22-H-5. The structures of $[Pt_3(CO)_6]_n^{2-}$ ions with $n = 2, 3$, and 5.

[33]   P. Chini, et al., *J. Am. Chem. Soc.*, 1976, **98**, 7225; 1979, **101**, 6110.

The chemistry of these systems is therefore extremely complicated, and the isolation of pure compounds requires great skill.

Curiously, it has proved impossible to isolate any Pd analogues, all attempts having produced metallic palladium only.

22-I. SILVER AND GOLD

Like copper, silver and gold have a single s electron outside a completed d shell, but in spite of the similarity in electronic structures and ionization potentials there are few resemblances between Ag, Au, and Cu, and there are no simple explanations for many of the differences.

Apart from obviously similar stoichiometries of compounds in the same oxidation state (which do not always have the same structure), there are some similarities within the group—or at least between two of the three elements:

1. The metals all crystallize with the same face-centered cubic (ccp) lattice.

2. Cu_2O and Ag_2O have the same body-centered cubic structure where the metal atom has two close oxygen neighbors and every oxygen is tetrahedrally surrounded by four metal atoms.

3. Although the stability constant sequence for halide complexes of many metals is $F > Cl > Br > I$, Cu^I and Ag^I belong to the group of ions of the more noble metals, for which it is the reverse.

4. Cu^I and Ag^I (and to a lesser extent Au^I) form very much the same types of ion and compound, such as $[MCl_2]^-$, $[Et_3AsMI]_4$, and K_2MCl_3.

5. Certain complexes of Cu^{II} and Ag^{II} are isomorphous, and Ag^{III}, Au^{III} and Cu^{III} also give similar complexes.

The only stable cationic aqua ion is Ag^+, and by contrast the Au^+ ion is exceedingly unstable with respect to the disproportionation

$$3Au^+(aq) = Au^{3+}(aq) + 2Au(s) \qquad K \approx 10^{10}$$

Gold(III) is invariably complexed in all solutions, usually as anionic species such as $[AuCl_3OH]^-$. The other oxidation states, Ag^{II}, Ag^{III}, and Au^I, are either unstable to water or exist only in insoluble compounds or complexed species. Intercomparisons of the standard potentials are of limited utility, particularly since these strongly depend on the nature of the anion; some useful ones are:

$$Ag^{2+} \xrightarrow{1.98} Ag^+ \xrightarrow{0.799} Ag$$

$$Ag(CN)_2^- \xrightarrow{-0.31} Ag + 2CN^-$$

$$AuCl_4^- \xrightarrow{1.00} Au + 4Cl^-$$

$$Au(CN)_2^- \xrightarrow{-0.6} Au + 2CN^-$$

Gold(II) occurs *formally* in dithiolene compounds (p. 185) and in the dicarbollyl $[Au(B_9C_2H_{11})_2]^{2-}$ but otherwise possibly exists only as a transient intermediate in reactions.

TABLE 22-I-1
Oxidation States and Stereochemistry of Silver and Gold[a]

Oxidation state	Coordination number	Geometry	Examples
Ag^I, d^{10}	2[b]	Linear	$[Ag(CN)_2]^-$, $[Ag(NH_3)_2]^+$, AgSCN
	3	Trigonal	$(Me_2NC_6H_4PEt_2)_2AgI$
	4[b]	Tetrahedral	$[Ag(SCN)_4]^{3-}$, $[AgIPR_3]_4$, $[AgSCNPPr_3]$, $[Ag(PPh_3)_4]$ ClO_4
	5	Dist. pentagonal plane	$[Ag(L)]^{+2}$ [c]
	3	Pentagonal pyramidal	$[Ag(L)]_2^{3+}$ [c]
	6	Octahedral	AgF, AgCl, AgBr(NaCl struct.)
Ag^{II}, d^9	4	Planar	$[Agpy_4]^{2+}$
	6	Dist. octahedral	$Ag(2,6\text{-pyridinedicarboxylate})_2 \cdot H_2O$
Ag^{III}, d^8	4	Planar	AgF_4^-, half Ag atoms in AgO, $[Ag(ebg)_2]^{3+}$
	6	Octahedral	$[Ag(IO_6)_2]^{7-}$, $CsKAgF_6$
Au^I, d^{10}	2[b]	Linear	$[Au(CN)_2]^-$, $Et_3P \cdot AuC \equiv C \cdot C_6H_5$; $(AuI)_n$
	3	Trigonal	$AuCl(PPh_3)_2$
	4	Tetrahedral	$[Audiars_2]^+$
Au^{II}, d^9	4	Square	$[Au(mnt)_2]^{2-}$ [d]
Au^{III}, d^8	4[b]	Planar	$AuBr_4^-$, Au_2Cl_6, $[(C_2H_5)_2AuBr]_2$, R_3PAuX_3
	5	*Tbp*	$[Au(diars)_2I]^{2+}$
		Sp	$AuCl_3(2,9\text{-Me}_2\text{-}1,10\text{-phen})$,[e] $AuCl(TPP)$[f]
	6	Octahedral	$AuBr_6^{3-}$, *trans*-$[Au(diars)_2I_2]^+$
Au^V, d^6	6	Octahedral	$(Xe_2F_{11})^+(AuF_6)^-$

[a] T. J. Bergendahl, *J. Chem. Educ.*, 1975, **52**, 73; oxidation states of gold; for Au (−1) see text.
[b] Most common states
[c] S. M. Nelson *et al.*, *J. C. S. Chem. Comm.*, **1977**, 167, 370; L is an N_5 macrocycle:

$$L = $$ (with structure: pyridine ring bearing two acetyl groups, N atoms linked by $(CH_2)_3$ chains to NH and HN groups bridged by $(CH_2)_2$)

[d] T. J. Bergendahl and J. W. Waters, *Inorg. Chem.*, 1975, **14**, 2556.
[e] W. T. Robinson, and E. Sinn, *J. C. S. Dalton*, **1976**, 726.
[f] TPP = tetraphenylporphyrin; R. Timkovich and A. Tulinsky, *Inorg. Chem.*, 1977, **16**, 962.

The oxidation states and stereochemistry are summarized in Table 22-I-1.

22-I-1. The Elements

Silver and gold are widely distributed in Nature. They occur as metals and also in numerous sulfide ores, usually accompanied by sulfides of Fe, Cu, Ni, etc. The main sources are South Africa and the USSR. Silver also occurs as *horn silver* (AgCl).

After flotation or other concentration processes, the crucial chemical steps are cyanide leaching and zinc precipitation,[1a] e.g.,

$$4Au + 8KCN + O_2 + 2H_2O = 4KAu(CN)_2 + 4KOH$$
$$2KAu(CN)_2 + Zn \rightarrow K_2Zn(CN)_4 + 2Au$$

Silver and gold[1b] are normally purified by electrolysis.

Silver is a white, lustrous, soft, and malleable metal (m.p. 961°) with the highest known electrical and thermal conductivity. It is chemically less reactive than copper, except toward sulfur and hydrogen sulfide, which rapidly blacken silver surfaces. The metal dissolves in oxidizing acids and in cyanide solutions in presence of oxygen or peroxide.

Gold[2] is a soft, yellow metal (m.p. 1063°) with the highest ductility and malleability of any element. It is chemically unreactive and is not attacked by oxygen or sulfur, but reacts readily with halogens or with solutions containing or generating chlorine such as aqua regia; and it dissolves in cyanide solutions in presence of air or hydrogen peroxide to form $[Au(CN)_2]^-$. The reduction of solutions of $AuCl_4^-$ by various reducing agents may, under suitable conditions, give highly colored solutions containing colloidal gold. The "purple of Cassius" obtained using Sn^{II} is used for coloring ceramics.[3a] Silver atoms that do not aggregate but are trapped in a macrocyclic poly-nitrogen ligand are obtained by photochemical or electrochemical reductions of the Ag^+ complex; the Ag^0 complex can be reoxidized.[3b]

22-I-2. Silver Compounds.[4]

Silver(I), d^{10}. Silver(I) is the normal oxidation state. The Ag^+ ion in water is probably $[Ag(H_2O)_2]^+$, but the water is very labile[5] and no hydrated salts are known.

—Ag—C—N—Ag—C—N—

(22-I-I)   (22-I-II)

$AgNO_3$, $AgClO_3$, and $AgClO_4$ are water soluble, but Ag_2SO_4 and $AgOOCCH_3$ are sparingly so. The salts of oxo anions are primarily ionic, but although the water-insoluble halides AgCl and AgBr have the NaCl structure, there appears

[1a] M. I. Brittan, *Am. Sci.,* 1974, **62,** 402; E. M. Wise, *Gold Recovery, Properties and Applications,* Van Nostrand, 1964.
[1b] E. Schalch and M. T. Nicol, *Gold Bull.,* 1978, **11,** 117.
[2] W. S. Rapson and T. Groenewald, *Gold Usage,* Academic Press, 1978.
[3a] L. B. Hunt, *Gold Bull.,* 1976, **9,** 134.
[3b] E. Pelizzetti *et al., Angew. Chem. Int. Ed.,* 1979, **18,** 630.
[4] L. M. Gedensky and L. G. Hepler, *Engelhard Tech. Bull,.* 1969, **9,** 117 (extensive collection of thermodynamic data).
[5] J. W. Akitt, *J. C. S. Dalton,* **1974,** 175.

to be appreciable covalent character in the Ag···X interactions, whereas in compounds such as AgCN and AgSCN, which have chain structures 22-I-I and 22-I-II, the bonds are considered to be predominantly covalent.

Ag^I and Au^I, along with Cu^I and Hg^{II}, show a pronounced tendency to exhibit linear, 2-fold coordination. This may be due to a relatively small energy difference between the filled d orbitals and the unfilled s orbital ($4d$, $5s$ for Ag^I), which permits extensive hybridization of the d_{z^2} and s orbitals, as shown in Fig. 22-I-1. The

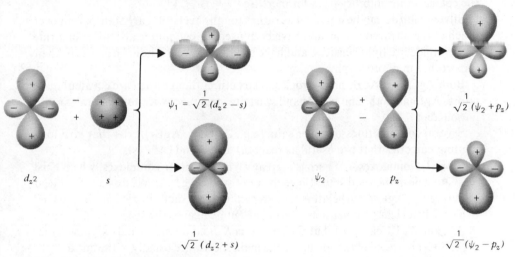

Fig. 22-I-1. The hybrid orbitals formed from a d_{z^2} and an s orbital ψ_1 and ψ_2, and the hybrids that can be formed from ψ_2 and a p_z orbital. In each sketch the z axis is vertical and the actual orbital is the figure generated by rotating the sketch about the z axis.

electron pair initially in the d_{z^2} orbital occupies ψ_1, giving a circular region of relatively high electron density from which ligands are somewhat repelled, and regions above and below this ring in which the electron density is relatively low. Ligands are attracted to the latter regions. By further mixing of ψ_2 with the p_z orbital, two hybrid orbitals suitable for forming a pair of linear covalent bonds can be formed.

Binary Compounds.

Silver(I) Oxide. The addition of alkali hydroxide to Ag$^+$ solutions produces a dark brown precipitate that is difficult to free from alkali ions. It is strongly basic, and its aqueous suspensions are alkaline:

$$\tfrac{1}{2}AgO(s) + \tfrac{1}{2}H_2O = Ag^+ + OH^- \quad \log K = -7.42 \ (25°, 3M \ NaClO_4)$$
$$\tfrac{1}{2}Ag_2O(s) + \tfrac{1}{2}H_2O = AgOH \quad \log K = -5.75$$

they absorb carbon dioxide from the air to give Ag$_2$CO$_3$. The oxide decomposes above ~160° and is readily reduced to the metal by hydrogen. Silver oxide is more soluble in strongly alkaline solution than in water, and AgOH and Ag(OH)$_2^-$ are formed. The treatment of water-soluble halides with a suspension of silver oxide is a useful way of preparing hydroxides, since the silver halides are insoluble. Analogously to copper (p. 802) and gold, alkali metal silver oxides contain Ag$_4$O$_4^{4-}$ units.

Silver(I) Sulfide.   The action of hydrogen sulfide on argentous solutions gives black Ag_2S, which is the least soluble in water of all silver compounds (log $K_{sp} \approx$ 50). The black coating often found on silver articles is the sulfide; this can be readily reduced by contact with aluminum in dilute sodium carbonate solution.

Silver(I) Halides.   The *fluoride* is unique in forming hydrates such as $AgF \cdot 4H_2O$, which are obtained by crystallizing solutions of Ag_2O in aqueous HF. The other well-known halides are precipitated by the addition of X^- to Ag^+ solutions; the color and insolubility in water increase Cl < Br < I.[6]

Silver chloride can be obtained as rather tough sheets that are transparent over much of the infrared region and have been used for cell materials. Silver chloride and bromide are light sensitive and have been intensively studied because of their importance in photography.

Both AgCl and AgBr have a rock salt structure, though they are covalent insulators. AgI has both zinc blende and wurtzite structures with tetrahedral coordination about Ag.[7a]

Several combinations of silver salts (e.g., $AgNO_3$–AgBr) exist that give low-melting eutectics that are useful as reaction media and catalysts.[7b]

Silver(I) Complexes.   There is a great variety of silver complexes, which exist in solution or in the solid state. Since the most stable Ag^+ complexes have the linear structure L—Ag—L$^+$, chelating ligands cannot form such simple ions, hence tend to give polynuclear complex ions. For monodentate ligands the species AgL^+, AgL_2^+, AgL_3^+, and AgL_4^+ can exist, but the constants K_1 and K_2 are usually high, whereas K_3 and K_4 are relatively small; hence the main species are usually of the linear AgL_2 type. The coordination number, however, is sensitive to the nature of the ligand and anions, and a variety of types can occur because of the possibilities of sp^2 and sp^3 bonding of Ag^+, in addition to the linear hybridization discussed earlier.[7c]

Complexes are not well known for oxygen ligands. Ligands with the donor atoms N, P, As, S, Se, etc., give many complex compounds, but it is doubtful whether π bonding of the type important in the earlier transition groups makes any substantial contribution to the bonding where P, S, etc., are the donor atoms. Thus from studies on the complex formation by donor atoms in acids such as $RC_6H_4As(CH_2COOH)_2$, which gives only a 1:1 complex, and similar P, S, and Se compounds, the greater stability of these complexes than of their N analogues has been attributed to the polarizing power of Ag^+ and the high polarizability of the "soft" donor atoms.

The nitrile complex $[Ag(MeCN)_4]^+$ is especially stable and MeCN cannot be displaced, e.g., by alkenes. Like copper (p. 806) silver forms polyhedral thiolates,[8] e.g., $[Ag(SC_6H_{11})]_{12}$ and $[Ag_5(SPh)_7]^{2-}$.

Halogeno Complexes.   The insoluble halides such as AgCl are appreciably soluble in concentrated HNO_3 and HCl. They also dissolve in solutions of high

6   R. Ramette, *J. Chem. Educ.,* 1960, **37,** 348 (solubility of AgCl).

7a   M. R. V. Sahyun, *J. Chem. Educ.,* 1977, **54,** 143.

7b   L. A. Paquette *et al., Tetrahedron Lett.,* **1978,** 1963.

7c   See, e.g., phosphine complexes, E. L. Muetterties and C. W. Allegranti, *J. Am. Chem. Soc.,* 1970, **92,** 4114.

8   I. G. Dance, *Austr. J. Chem.,* 1978, **31,** 2195.

halide ion concentration[9] and in solutions of $AgNO_3$ or $AgClO_4$ in CH_3CN, $(CH_3)_2SO$, acetone, and other solvents. Complex anions or cations are thus formed:

$$AgX + nX^- \rightleftharpoons AgX_{n+1}^{n-}$$
$$AgX + nAg^+ \rightleftharpoons Ag_{n+1}X^{n+}.$$

The stability of the ions is generally $I > Br > Cl$. In Cs_2AgI_3 there are chains of AgI_4 sharing corners, whereas in $[Me_4N]Ag_2I_3$ the AgI_4 tetrahedra share edges. Salts of $Ag_4I_5^-$ and $Ag_{13}I_{15}^{2-}$ have channel structures and behave as solid electrolytes.[10] The linear ion $AgBr_2^-$ is also known.[11]

The halides also dissolve, giving complexes, in solutions of NH_3, CN^-, and $S_2O_3^{2-}$, although AgI is but sparingly soluble in ammonia. Silver cyanide, which is also insoluble in water, dissolves in CN^- and also in liquid HF, where an HCN donor complex $AgNCH^+$ is formed by protonation.

Phosphine Complexes. Complexes of tertiary phosphines or phosphites and arsine analogues have been much studied. Like the corresponding copper(I) complexes, they may have either a limiting cubane or chair structure (see p. 805) or distortions thereof depending on the steric factors involved. The complex (AgIPPh₃)₄ occurs in both isomeric forms.[12] The chair form is the most stable.

A halogen-bridged dimer $(AgBrPPh_3)_2$ is also known, but for the smaller copper atom, triphenylphosphine dimers occur only when the bridge group is N_3^- or NCS^-, where these linear anions reduce the steric crowding.[13] The nitrate $AgNO_3PPh_3$ differs in forming chains where each oxygen of NO_3 is bound to Ag, two oxygens acting as a chelate to one Ag, the other oxygen bound to the next.[14]

Organo Compounds of Silver(I)

Unsaturated Hydrocarbon Complexes.[15] Virtually all alkenes and many aromatic[16] compounds form complexes when the hydrocarbon is shaken with aqueous solutions of soluble silver salts. Equilibrium constants for the formation of the complexes with alkenes and other compounds with double bonds have been measured.[15,17] Similar complexes are formed between $AgBF_4$ and aliphatic or aromatic ketones in CH_2Cl_2.[18]

In all these cases, the order of stability is generally $AgBF_4 > AgClO_4 > AgNO_3$.

Di- or polyalkenes commonly give crystalline complexes that may have Ag^+ bound to one, two, or three double bonds. The structures of many such compounds

9   J. I. Kim and H. Duschner, *J. Inorg. Nucl. Chem.,* 1977, **39**, 471.
10 S. Geller, *Acc. Chem. Res.,* 1978, **11**, 87.
11 J. A. Cras *et al., J. Crystallogr. Mol. Struct.,* 1971, **1**, 155.
12 B.-K. Teo and J. C. Calabrese, *Inorg. Chem.,* 1976, **15**, 2467, 2474; M. R. Churchill *et al., Inorg. Chem.,* 1976, **15**, 2752.
13 B.-K. Teo and J. C. Calabrese, *J. C. S. Chem. Comm.,* **1976**, 185.
14 R. A. Stein and C. Knobler, *Inorg. Chem.,* 1977, **16**, 242.
15 F. M. Hartley, *Chem. Rev.,* **1973**, 163.
16 R. Gut and J. Rueede, *J. Organomet. Chem.,* 1977, **128**, 89.
17 J. V. Crookes and A. A. Wolf, *J. C. S. Dalton,* **1973**, 1241; F. B. Hartley *et al., J. C. S. Dalton,* **1977**, 469.
18 De L. D. Crist *et al., J. Am. Chem. Soc.,* 1974, **96**, 4923.

are known. Because of differing stabilities, the formation of crystalline complexes can be useful for the purification of particular alkenes, or for separation of mixtures (e.g., 1,3-,1,4-, and 1,5-cyclooctadienes) or of the optical isomers of α- and β-pinene. Silver nitrate–ethane-1,2-diol columns have also been used for gas-liquid chromatographic separations of alkenes.

Organosilver(I) Compounds.   The silver(I) alkyls and aryls are less stable than those of copper (p. 808), but comparatively stable fluoroalkyls can be obtained by reactions such as

$$AgF + CF_3CF{=}CF_3 \xrightarrow{\text{MeCN}} Ag\{CF(CF_3)_2\}(MeCN)$$

These compounds are polymeric when nonsolvated.

Like copper, silver also forms compounds with acetylenes that have σ bonds to the metal. Thus action of acetylene on Ag^+ solutions gives a yellow precipitate:

$$C_2H_2 + Ag^+ = AgC{\equiv}CAg + 2H^+$$

Substituted acetylenes give white solids that are insoluble and polymeric, probably owing to π bonding between Ag and the triple bond, as in the copper compounds (p. 810).

Silver(II) Compounds,[19] d^9.   The Ag^{2+} ion and numerous Ag^{II} complexes are known, but only few binary compounds. The oxide of stoichiometry AgO is actually $Ag^IAg^{III}O_2$ (see below).

Silver(II) fluoride is obtained as a dark brown solid by fluorination of AgF or other Ag compounds at elevated temperatures. It is evidently an Ag^{II} compound, although it is antiferromagnetic, with a magnetic moment well below that expected for one unpaired electron. It is a useful fluorinating agent. It is hydrolyzed rapidly by water.

Silver(II) fluorosulfate[20] is made by heating Ag with $S_2O_6F_2$ at 70° and is stable to 210°.

The Silver(II) Ion and Silver(II) Complexes.   Although the disproportionations

$$2Cu^+ = Cu + Cu^{2+} \quad \text{and} \quad 3Au^+ = 2Au + Au^{3+}$$

have been long known, only recently has a ligand-induced disproportionation of Ag^+ been discovered. The low heat of hydration of Ag^{2+} makes the equilibrium unfavorable in water:

$$2Ag^+ = Ag + Ag^{2+} \quad K \approx 10^{-20}$$

However using a tetraazamacrocyclic ligand (L) this has been achieved in both H_2O and molar CN^- solutions, and a paramagnetic complex AgL^{2+}, has been isolated.[21] Normally, however, Ag^{II} species can only be obtained by oxidation.

The aqua ion $[Ag(H_2O)_4]^{2+}$, which is paramagnetic with an unpaired electron,

[19]   H. N. Po, *Coord. Chem. Rev.,* 1976, **20,** 171; J. A. McMillan, *Chem. Rev,* 1962, **62,** 65.
[20]   P. C. Leung, and F. Aubke, *Inorg. Nucl. Chem. Lett.,* **1977,** 263.
[21]   M. O. Kestner and A. L. Allred, *J. Am. Chem. Soc.,* 1972, **94,** 7189; E. K. Barefield and M. T. Mocella, *Inorg. Chem.,* 1973, **12,** 2829.

is obtained in $HClO_4$ or HNO_3 solution by oxidation of Ag^+ with ozone or by dissolution of AgO in acid.

The potentials for the Ag^{2+}/Ag^+ couple, $+2.00$ V in 4M $HClO_4$ and $+1.93$ V in $4M$ HNO_3, show that Ag^{2+} is a powerful oxidizing agent. There is evidence for complexing by NO_3^-, SO_4^{2-}, and ClO_4^- in solution, and the electronic spectra in $HClO_4$ solutions are dependent on acid concentration, for example. The ion is reduced by water, even in strongly acid solution, but the mechanism is complicated.

Many oxidations (e.g., of oxalate) by the peroxodisulfate ion are catalyzed by Ag^+ ion, and the kinetics are best interpreted by assuming initial oxidation to Ag^{2+}, which is then reduced by the substrate. Decarboxylation of carboxylic acids is also promoted by Ag^{II} complexes such as 22-I-III.

(22-I-III)

Numerous complexes of Ag^{II} are known, and they are normally prepared by peroxosulfate oxidation of Ag^+ solutions containing the complexing ligand.

With neutral ligands cationic species such as $[Ag\ py_4]^{2+}$, $[Ag(dipy)_2]^{2+}$, and $[Ag(phen)_2]^{2+}$ form crystalline salts, whereas with uninegative chelating ligands such as 2-pyridinecarboxylate, neutral species such as 22-I-III are obtained.

The Ag^{II} complexes have $\mu_{eff} = 1.75-2.2$ BM, consistent with the d^9 configuration, and their electronic spectra accord with square coordination. The salt $[Ag\ py_4]S_2O_8$ and the bispicolinate 22-I-III are isomorphous with the planar copper(II) analogues. Exceptions to the rule are the unusual 2,3- and 2,6-pyridinedicarboxylates, $Ag(C_7H_4NO_4)_2\cdot H_2O$, which have a very distorted octahedral structure.[22] The Ca^{2+} or Ba^{2+} salts of the square ion AgF_4^{2-} are paramagnetic as expected.[23]

Silver(III), d^8 Compounds.[19] A black oxide that is not readily purified but is probably Ag_2O_3 is obtained by anodic oxidation of Ag^I in alkaline solution. The black so-called silver(II) oxide noted above is better characterized; it is made by oxidation of Ag_2O in NaOH solution with $S_2O_8^{2-}$ or, as single crystals, by controlled electrolysis of 2M $AgNO_3$ solutions.

The oxide is a semiconductor, is stable to $\sim100°$, and dissolves in acids evolving oxygen but giving some Ag^{2+} in solution. It is a powerful oxidizing agent. Since AgO is diamagnetic, it cannot in fact be Ag^{II} oxide. Neutron diffraction shows that it is $Ag^I Ag^{III}O_2$ with two types of silver atom in the lattice, one with linear coordination to two oxygen atoms (Ag^I) and the other square with respect to oxygen (Ag^{III}). When AgO is dissolved in acid Ag^{II} is formed, probably according to the equation

22   D. P. Murther and R. A. Walton, *Inorg. Chem.*, **1973**, *12*, 1279.
23   G. C. Allen and R. F. McMeeking, *J. C. S. Dalton*, **1976**, 1063.

$$AgO^+ + Ag^+ + 2H^+ \rightarrow 2Ag^{2+} + H_2O$$

but, in presence of complexing agents in alkaline solution, Ag^{III} complexes are obtained (see below). The separation of Ag^I and Ag^{III} can be made by the reaction

$$4AgO + 6KOH + 4KIO_4 \rightarrow 2K_5H_2[Ag^{III}(IO_6)_2] + Ag_2O + H_2O$$

Unusual salts of stoichiometry $Ag_7O_8^+ HF_2^-$, which are obtained as black needles by electrolysis of aqueous solutions of AgF, contain a polyhedral in $Ag_6O_8^+$ that acts as a clathrate to enclose Ag^+ and HF_2^-. The salts thus contain Ag^I and Ag^{III} with an average oxidation state of $2^3/_7$.

Complexes of Silver(III). The anodic oxidation of silver metal in strong KOH solution gives the yellow tetrahydroxoargentate(III) ion $[Ag(OH)_4]^-$, which has a half-life of *ca.* 100 min in $1.2M$ NaOH, (<30 min in $0.1M$ NaOH), decomposing to AgO and O_2.[24] The anion reacts rapidly with tellurate or iodate to give very stable complexes,[25] which are usually obtained by direct oxidation of Ag^+ in alkaline solutions by $K_2S_2O_8$ in presence of periodate or tellurate ions. Representative are the ions $\{Ag[TeO_4(OH)_2]_2\}^{5-}$ and $\{Ag[IO_4(OH)]_2\}^{5-}$; copper(III) complexes (p. 819) are analogous, and all the ions appear to have Te—OH or I—OH groups as in 22-I-IV.

In acid solution,[25] the AgO^+ ion is extensively hydrolyzed even in 1.5 to $6M$ acid. An Ag^{III} complex of remarkable stability is the one with ethylenedibiguanide (22-I-V), which is obtained as the red sulfate when Ag_2SO_4 is treated with aqueous potassium peroxodisulfate in the presence of ethylenedibiguanidinium sulfate. The hydroxide, nitrate, and perchlorate have been prepared metathetically. These salts are diamagnetic and oxidize two equivalents of iodide ion per gram-atom of silver. Silver(III) porphyrin complexes are also stable.[26]

(22-I-IV)   (22-I-V)

Less accessible are the yellow fluoro complexes, (e.g., $KAgF_4$ or Cs_2KAgF_6), obtained by action of F_2 at 300° on a stoichiometric mixture of the alkali chlorides and silver nitrate. The AgF_6^{3-} ion appears to be octahedral according to its magnetism and electronic absorption spectrum.

[24] L. J. Kirschenbaum and L. Mrozowski, *Inorg. Chem.*, 1978, **17**, 3718.
[25] A. Balikungeri *et al.*, *Inorg. Chim. Acta*, 1977, **22**, 7.
[26] D. Karweik *et al.*, *J. Am. Chem. Soc.*, 1974, **96**, 591.

Silver Clusters. No compounds with Ag–Ag bonds have been firmly established, and the extent of metal-metal interaction in the thiolate or aryl aggregates noted above is uncertain. It has been claimed[27] that partial reduction of Ag^+ in zeolites can give Ag_6^+ clusters. There are, however, a number of compounds that have silver bound to another transition metal.

22-I-3.   Gold Compounds.[28]

Binary Compounds. The interaction of gold with sodium in a cryptate (p. 163), with lithium vapor followed by dissolution in liquid ammonia, or of AuCl and K in NH_3, gives compounds M^+Au^-. These have been shown, e.g., by electron spectroscopy for chemical analysis (esca) measurements on CsAu, to have gold in the -1 oxidation state.[29]

In the liquid state they conduct ionically.

Oxides. Only Au_2O_3 is known. Although addition of base to $AuCl_4^-$ solutions gives $Au_2O_3 \cdot nH_2O$ as an amorphous brown precipitate, it decomposes on heating to Au, O_2, and H_2O. However by hydrothermal methods, polycrystalline and single crystal brown to ruby red Au_2O_3 can be made.[30] The hydrous oxide is weakly acidic and dissolves in strong base, giving probably $Au(OH)_4^-$. Alkali gold oxides (e.g., $Cs_4Au_4O_4$) are similar to Cu and Ag analogues and contain $Au_4O_4^{4-}$.

Halides. *Gold*(III) *fluoride* is best made by fluorination of Au_2Cl_6 at 300° and forms orange crystals, which decompose to the metal at 500°. It has a unique structure with square AuF_4 units linked into a chain by cis fluoride bridges; the F atoms of adjacent chains interact weakly with the axial sites.

The *chloride* and *bromide*, which form red crystals, are made by direct interaction, but Au_2Cl_6 is best made by the reaction

$$2H_3O^+AuCl_4^- + 2SOCl_2 = 2SO_2 + 6HCl + Au_2Cl_6$$

Both are dimers in the solid and in the vapor. They dissolve in water, undergoing hydrolysis, but in HX the ions AuX_4^- are formed. Gold(III)chloride (or $AuCl_4^-$) is a powerful oxidant,[31] being reduced to Au; it will, for example, oxidize ruthenium complexes in aqueous solution to RuO_4.

Gold(I)*chloride* formed from Au_2Cl_6 at 160° is a yellow powder that dissociates at higher temperatures. It is decomposed by water. In the crystal there are chains with linear Cl—Au—Cl bonds.

A black, air-sensitive chloride Au_4Cl_8, which is made by the reaction in liquid $SOCl_2$

27   Y. Kin and K. Seff, *J. Am. Chem. Soc.,* 1978, **100**, 175.
28   R. J. Puddephat, *The Chemistry of Gold,* Elsevier, 1978; L. M. Gedansky and L. G. Hepler, *Engelhard Tech. Bull.,* 1969, **10**, 5 (thermochemistry and potential data for compounds and aqueous solutions); H. Schmidbauer, *Angew. Chem., Int. Ed.,* **1976,** 728 (an extensive review of gold compounds); for reviews, patents, etc., see *Gold Bulletin,* Chamber of Mines of South Africa.
29   J. Knecht *et al., J. C. S., Chem. Comm.,* **1978,** 905; W. J. Peer and J. J. Lagowski, *J. Am. Chem. Soc.,* 1978, **100**, 6280; A. J. Bond *et al., J. Am. Chem. Soc.,* 1978, **100**, 7768; B. Busse and K. G. Weil, *Angew. Chem. Int. Ed.,* 1979, **18**, 629.
30   E. Schwarzmann *et al., Z. Naturforsch.,* 1976, **31b**, 135.
31   K. G. Moodley and M. J. Nicol, *J. C. S. Dalton,* **1977,** 993.

$$Au_2Cl_6 + 2Au(CO)Cl \rightarrow Au_4Cl_8 + 2CO$$

has structure 22-I-VI with both square Au^{III} and linear Au^I atoms.[32]

(22-I-VI)

Gold(I) Complexes. Although the Au^+ ion cannot exist in high concentrations, changes in the potential on diluting $[Au(MeCN)_2]^+$ with dilute $HClO_4$ allow estimation of the Au/Au^+ potential ($E^0 = +1.695$ V). This indicates that Au^+ is a strong reducing agent; it can survive for up to 5 min.[33]

The most important gold(I) complex in aqueous solution is the ion $Au(CN)_2^-$ noted earlier. Salts such as $KAu(CN)_2$ are known, and the free acid $HAu(CN)_2$ has a hydrogen-bonded lattice with Au—CN—H—NC—Au links. A similar halide ion $AuBr_2^-$ is known.[11] The thiosulfate, $Na_3[Au(SSO_3)_2]\cdot2H_2O$ has an almost linear S—Au—S group.[34]

Numerous complexes with P, As, and especially S donor and less commonly N donors can be obtained.[35] Compounds are of the types $AuXL$, $AuXL_2$, $AuXL_3$, and AuX_2^+ and these can be distinguished, for phosphorus ligands, by ^{31}P nmr, or by ^{197}Au Mössbauer spectra. The bonding appears to involve only $6s$ and $6p$ orbitals, and even with π-acid ligands σ effects dominate and there is negligible Au—L π bonding. Polymeric complexes with four-coordinate metal like those of Cu and Ag are not formed.

The carbonyl[36] is made in $SOCl_2$ by the reaction

$$AuCl_3 + 2CO = AuCl(CO) + COCl_2$$

Its reaction with Au_2Cl_6 has been noted above.

Gold Cluster Compounds.[37] The gold compounds used for decorating china and glass articles and known as "liquid golds" are made by interaction of gold(III) chloro complexes with sulfurized terpenes or resins and are very soluble in organic

32   D. B. Dell'Amico et al., J. C. S. Chem. Comm., **1977**, 31.
33   P. R. Johnson et al., J. C. S. Chem. Comm., **1978**, 606.
34   H. Ruben et al., Inorg. Chem., 1974, **13**, 1836; R. F. Baggio and S. Baggio, J. Inorg. Nucl. Chem., 1973, **35**, 3191.
35   J. J. Guy et al., J. C. S. Dalton, **1977**, 8; C. A. McAuliffe et al., J. C. S. Dalton, **1977**, 1426; P. G. Jones, J. C. S. Dalton, **1977**, 1430, 1434, 1440; C. B. Colburn et al., J. C. S. Chem. Comm., **1979**, 218.
36   D. B. Dell'Amico et al., J. C. S. Dalton, **1976**, 1829.
37   F. A. Vollenbroek et al., Inorg. Chem., 1978, **17**, 1345; J. C. S. Chem. Comm., **1978**, 907; **1979**, 387; D. M. P. Mingos, J. C. S. Dalton, **1976**, 1163, and references therein; M. Manassero et al., J. C. S. Chem. Comm., **1979**, 385.

solvents. Doubtless they were the first known, though unrecognized, thiolate gold aggregates.

The structures of several polynuclear gold cluster compounds have been established. The earliest were those obtained by reduction of $AuXPR_3$ by $NaBH_4$. Two complexes have an Au_{11} cluster (Fig. 22-I-2) consisting, formally at any rate, of

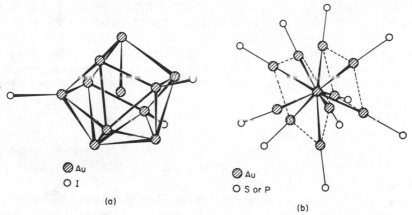

Fig. 22-I-2. The structures of Au_{11} cluster compounds: (a) $Au_{11}I_3[P(p\text{-}ClC_6H_4)_3]_7$ and (b) $Au_{11}(SCN)_3(PPh_3)_7$. [Reproduced by permission from P. Bellon, M. Manassero, and M. Sansone, *J. C. S. Dalton*, **1972**, 1481.]

three Au^I and nine Au^0 atoms. The structure can be roughly described as an icosahedral one with a central gold atom; (each of the nine apices is occupied by a gold atom, and the remaining three are occupied by a single gold atom.

In addition, ionic compounds with Au_6, Au_8, and Au_9 clusters are known, e.g., the yellow $\{Au_6[(p\text{-}C_6H_4Me)_3P]_6\}^{2+}$, which has an octahedron of Au atoms each carrying one phosphine, and green $\{Au_9[P(p\text{-}C_6H_4Me)_3]_8\}^{3+}$, which has a central Au atom with the others in two orthogonal squares, and $[Au_9(PPh_3)_8]^{3+}$.

The "liquid gold" and other sulfur aggregates are probably clusters, as are complexes of thiomalate and *N*-acetyl-L-cysteine[38] that have stoichiometry $[Au_4(SR)_7]^{3-}$, but structural evidence is lacking.

Finally there are numerous compounds with Au bonds to other metals and dimers with Au—Au bonds. Examples are $(CO)_4(NO)FeAuPPh_3$ and $(CO)_5\text{-}Re(AuPPh_3)$, which are made by interaction of the sodium salts of carbonylate anions with complexes such as $AuXPR_3$, e.g.,

$$NaCo(CO)_4 + AuClPPh_3 \xrightarrow{\text{THF}} (CO)_4CoAuPPh_3.$$

The dithiocarbamates $[Au^I(S_2CNR_2)]_2$ are dimeric with both an Au—Au bond and a bridge-dithiocarbamate group; the related dithiophosphinate $\{Au[S_2P(O\text{-}i\text{-}C_3H_7)]\}_2$ is similar, but the dimers are weakly linked into chains by long Au—Au bonds (3 Å).[39a] Compound 22-I-VII has a ten-membered ring.[39b]

[38] A. A. Isab and P. J. Sadler, *J. C. S. Chem. Comm.*, **1976**, 1051.
[39a] S. L. Lawton *et al.*, *Inorg. Chem.*, **1972**, 11, 2227.
[39b] W. S. Crane and H. Beall, *Inorg. Chim Acta*, 1978, **31**, L469.

(22-I-VII)

(22-I-VIII)

Gold(II) Complexes. Although there are several compounds like Au_4Cl_8 referred to earlier that could be formulated as gold(II) species, most are mixed-valence species with Au^I and Au^{III}, e.g., $Au^IAu^{III}(SO_4)_2$. Few authentic Au^{II} complexes are known.[40] One is 22-I-VIII[39b]; others are the maleonitriledithiolato (see p 185) complex $Au(mnt)_2^{2-}$, but the electron is largely in ligand-based orbitals, a phthalocyanine, and the compounds with Au—C bonds such as $[R_3PAuCH_2PR_3]^+$ that we discuss below.

Gold(III) Complexes. Gold(III) is isoelectronic with Pt(II), and its complexes therefore show many structural similarities to those of Pt(II).

Halogeno Complexes. Salts of the tetrafluoroaurate ion are obtained by the action of BrF_3 on a mixture of gold and an alkali chloride; $KAuF_4$ is isomorphous with $KBrF_4$ and has a square AuF_4^- ion.

When gold is dissolved in aqua regia or Au_2Cl_6 is dissolved in HCl and the solution of $AuCl_4^-$ is evaporated, *chloroauric acid* can be obtained as yellow crystals $[H_3O]^+[AuCl_4]^-\cdot3H_2O$. Other water-soluble salts such as $KAuCl_4$ and $NaAuCl_4\cdot2H_2O$ are readily obtained. In water, hydrolysis to $AuCl_3OH^-$ occurs. From the dilute hydrochloric acid solutions the gold can be solvent extracted with a very high partition coefficient into ethyl acetate or diethyl ether; the species in the organic solvent appears to be $[AuCl_3OH]^-$, which is presumably associated in an ion pair with an oxonium ion. Gold is readily recovered from such solutions (e.g., by precipitation with SO_2).

Other Anionic Complexes. There are several other anions such as $[Au(SO_3F)_4]^-$,[41a] $[Au(CN)_4]^-$, $[Au(NO_3)_4]^-$ in which NO_3 is unidentate, and sulfato complexes. Gold(III) differs from Pt^{II} in its greater affinity for oxygen donor ligands.

Cationic Complexes. There are numerous four-coordinate complexes such as $[AuCl_2py_2]Cl$ and $[AuphenCl_2]Cl$. Chloroauric acid reacts with diethylenetriamine to give the ammonium tetrachloroaurate, $[Au\ dienCl]Cl_2$ or $[Au(dienH)Cl]Cl$, depending on the concentration and pH. The kinetics of substitution of various anions in $[AudienCl]^{2+}$ have been compared with those for planar Pt^{II}; there is evidence that axial interactions occur here also in solution, e.g.,

40 T. J. Bergendahl and J. H. Waters, *Inorg. Chem.*, 1975, **14**, 2556; R. L. Schlupp and A. H. Maki, *Inorg. Chem.*, 1974, **13**, 44.
41a K. C. Lee and F. Aubke, *Inorg. Chem.*, 1979, **18**, 389.

Electronic spectra have been interpreted on an MO basis.

Gold(III) complexes have been obtained by using a chelating diarsine ligand from the reaction with sodium tetrachloroaurate(III) in presence of sodium iodide. The iodide $[Au(diars)_2I_2]I$ and other cations $[Au(diars)_2I]^{2+}$ and $[Au(diars)_2]^{3+}$ can be obtained. It is held that these are species with six,- five,- and four-coordination for Au^{III} with octahedral, trigonal-bipyramidal and planar structures, respectively. Chelating phosphine complexes also exist.

Gold(V), d^6, Compounds. The interaction of fluorine and oxygen on gold at 8 atm pressure and 350° gives the salt $O_2^+AuF_6^-$, which on heating gives the pentafluoride AuF_5, a dark red solid. This diamagnetic compound is polymeric with fluoride bridges.[41b] The octahedral AuF_6^- ion, t_{2g}^6, can also be obtained in other salts, e.g.,

$$CsAuF_4 + F_2 = CsAuF_6$$

Organogold Compounds.[28,42] Alkyl derivatives of gold were among the first organometallic compounds of transition metals to be prepared. Both gold(I) and gold(III) compounds with σ bonds to carbon, as well as olefin complexes, are known.

Gold(I) complexes are mainly of the type RAuL, where L is a stabilizing ligand such as R_2S, R_3P, or RNC. They are made from the corresponding halides by action of LiR or RMgX. Acetylides such as $(R_3PAuC\equiv CR')_n$ are also known.

Gold compounds are readily oxidized and undergo oxidative addition reactions (see below). Although $Au(C_6F_5)_3$ is stable in ether, gold(III) compounds are usually stable only when additional ligands are present. Thus for $(CH_3)_3AuPPh_3$ the strength of the Au—P bond inhibits dissociation, which is prerequisite for reductive elimination (see Chapter 27), viz.,

$$Me_3AuPPh_3 \underset{-PPh_3}{\rightleftharpoons} [Me_3Au] \longrightarrow CH_3{-}CH_3 + \text{``MeAu''} \xrightarrow{PPh_3} MeAuPPh_3$$

The main Au^{III} chemistry is of the dialkyl species R_2AuX, especially of the dimethyls, which form ions like $[Me_2Au(OH)_2]^-$ or $[Me_2AuBr_2]^-$. The halides themselves are dimers. From the alkyls $[R_2AuL_2]^+$, elimination of alkane R—R on heating is fastest when L is bulky,[43] e.g., $PPh_3 > PMePh_2 > PMe_2Ph > PMe_3$ and $R_3Sb > R_3As > R_3P$.

Although the formation of Au^{III} from Au^I compounds could be anticipated, oxidative-addition reactions such as

$$MeAu^IPR_3 + CH_3I = (Me)_2Au^{III}IPR_3$$

are slow and are complicated mechanistically. Additions to anionic Au^I species proceed more readily,[44] e.g.,

$$RAuPEt_3 + R'Li = Li[AuRR'PEt_3] \xrightarrow{R''X} RR'R''AuPEt_3 + LiX$$

[41b] M. J. Vasile *et al.*, *J. C. S. Dalton*, **1976**, 351; N. Bartlett *et al.*, *Rev. Chim. Miner.*, 1976, **13**, 82; *Inorg. Chem.*, 1974, **13**, 775.

[42] R. J. Puddephatt, *Gold Bull.*, 1977, **10**, 108.

[43] P. L. Kuch and R. S. Tobias, *J. Organomet. Chem.*, 1976, **122**, 429.

[44] A. Tamaki and J. K. Kochi, *J. C. S. Dalton*, **1973**, 2620.

Interaction of the triphenylphosphine complexes of gold(I) and gold(III), MeAuPPh$_3$, and Me$_3$AuPPh$_3$, respectively, with methyllithium in ether gives the methylate ions [AuMe$_2$]$^-$ and [AuMe$_4$]$^-$, which can be isolated as lithium amine salts[45] and are the most stable organogold(I) compounds.

A number of carbene complexes and a sulfur ylid, Me$_2$AuSMe$_2$,[46] are known, but more important are the heterocyclic alkyls formed by phosphorus ylids (see Chapter 27) in reactions of the type:

$$2AuXL + 2Me_3P{=}CH_2 = Me_2P\underset{CH_2-Au-CH_2}{\overset{CH_2-Au-CH_2}{\diagup\diagdown}}PMe_2 + 2L + 2Me_4PX$$

The heterocycles can also undergo oxidative addition of halogens or CH$_3$I to give formally Au^{II} complexes[47]:

$$Me_2P\underset{CH_2-Au-CH_2}{\overset{CH_2-Au-CH_2}{\diagup\diagdown}}PMe_2 + Cl_2 \longrightarrow$$

Biochemistry of Gold.[48] Gold is not a necessity for any living organism, though some plants concentrate the element. Various gold(I) thiols are biologically active, and the sodium gold thiomalate and thioglucose derivatives that are water soluble have been used as anti-inflammatory drugs in rheumatoid arthritis treatment.

Gold has also commonly been used for labeling enzymes and proteins for X-ray diffraction study, since it can be readily located. The usual labeling agents are Au(CN)$_2^-$, AuCl$_4^-$, and AuI$_4^-$.

[45] G. W. Eice and R. S. Tobias, *Inorg. Chem.*, 1976, **15**, 489; S. Komiya et al., *J. Am. Chem. Soc.*, 1977, **99**, 8440.
[46] J. P. Fackler and C. Paparizos, *J. Am. Chem. Soc.*, 1977, **99**, 2364.
[47] H. Schmidbauer et al., *Z. Naturforsch.*, 1978, **33b**, 1325; *Chem. Ber.*, 1977, **110**, 2751, 2758, 2236.
[48] N. A. Malik et al., *J. C. S. Chem. Comm.*, **1978**, 711; D. H. Brown et al., *J. C. S. Dalton*, **1978**, 199; P. J. Sadler, *Gold Bull.* 1976, **9**, 110; *Struc. Bonding*, 1976, **29**, 171.

CHAPTER TWENTY-THREE

The Lanthanides; also Scandium and Yttrium

GENERAL

23-1.   Lanthanides and the Lanthanide Contraction

The lanthanides—or lanthanoids, as they are sometimes called—are, strictly, the fourteen elements that follow lanthanum in the Periodic Table and in which the fourteen $4f$ electrons are successively added to the lanthanum configuration.

Since these $4f$ electrons are *relatively* uninvolved in bonding, the main result is that all these highly electropositive elements have, as their prime oxidation state, the M^{3+} ion. The radius of this ion decreases with increasing Z from La, thus constituting the "lanthanide contraction" (see below).

Since the term "lanthanide" is used to indicate that these elements form a closely allied group, for the chemistry of which lanthanum is the prototype, the term is usually taken as including lanthanum itself. Table 23-1 gives some properties of the atoms and ions. The electronic configurations are not all known with complete certainty owing to the great complexity of the electronic spectra of the atoms and ions and the attendant difficulty of analysis.

Since *yttrium,* which lies above La in Transition Group III and has a similar +3 ion with a noble gas core, has both atomic and ionic radii lying close to the corresponding values for terbium and dysprosium (because of the lanthanide contraction), this element is also considered here. It is generally found in Nature along with the lanthanides and resembles Tb^{III} and Dy^{III} in its compounds. The lighter element in Group III, *scandium,* has a smaller ionic radius, so that its chemical behavior is intermediate between that of aluminum and of the lanthanides, but it is also considered in this chapter.

The term "lanthanide contraction" was used in discussing the elements of the third transition series, since it has certain important effects on their properties. It

TABLE 23-1
Some Properties of Lanthanide Atoms and Ions[a]

| Atomic number | Name | Symbol | Electronic configuration[b] | | $E^0(V)$[c] | Radius[d] |
			Atom	M^{3+}		M^{3+} (Å)
57	Lanthanum	La	$5d6s^2$	[Xe]	−2.37	1.061
58	Cerium	Ce	$4f^15d^16s^2$	$4f^1$	−2.34	1.034
59	Praseodymium	Pr	$4f^36s^2$	$4f^2$	−2.35	1.013
60	Neodymium	Nd	$4f^46s^2$	$4f^3$	−2.32	0.995
61	Promethium	Pm	$4f^56s^2$	$4f^4$	−2.29	0.979
62	Samarium	Sm	$4f^66s^2$	$4f^5$	−2.30	0.964
63	Europium	Eu	$4f^76s^2$	$4f^6$	−1.99	0.950
64	Gadolinium	Gd	$4f^75d6s^2$	$4f^7$	−2.29	0.938
65	Terbium	Tb	$4f^96s^2$	$4f^8$	−2.30	0.923
66	Dysprosium	Dy	$4f^{10}6s^2$	$4f^9$	−2.29	0.908
67	Holmium	Ho	$4f^{11}6s^2$	$4f^{10}$	−2.33	0.894
68	Erbium	Er	$4f^{12}6s^2$	$4f^{11}$	−2.31	0.881
69	Thulium	Tm	$4f^{13}6s^2$	$4f^{12}$	−2.31	0.869
70	Ytterbium	Yb	$4f^{14}6s^2$	$4f^{13}$	−2.22	0.858
71	Lutetium	Lu	$4f^{14}5d6s^2$	$4f^{14}$	−2.30	0.848

[a] For detailed thermochemical properties of Y, La, and lanthanide elements and ions, see L. R. Morss, *Chem. Rev.,* 1976, **76,** 827. Ionization enthalpies are given by W. C. Martin *et al., J. Phys. Chem., Ref. Data,* 1974, **3,** 771.

[b] Only the valence shell electrons, that is, those outside the [Xe] shell, are given.

[c] For $M^{3+} + 3e = M$, D. A. Johnson, *J.C.S. Dalton,* **1974,** 1671; L. J. Nugent, *J. Inorg. Nucl. Chem.,* 1975, **37,** 1767. Other values: Sc = −1.88 V, Y = −2.37 V.

[d] Values for six-coordination. Sc^{+3} = 0.68 Å, Y^{+3} = 0.88 Å. Radii for different coordination numbers for La are 1.10(7), 1.18(8), 1.20(9), 1.28(10), 1.32(12); for other elements, see S. P. Sinha, *Struct. Bonding,* 1976, **25,** 69.

consists of a significant and steady decrease in the size of the atoms and ions with increasing atomic number; that is, La has the greatest, and Lu the smallest radius (Table 23-1; see also Fig. 24-2). Note that the radius of La^{3+} is about 0.18 Å larger than that of Y^{3+}. Thus if the fourteen lanthanide elements did not intervene, we might have expected Hf^{4+} to have a radius $\sim$0.2 Å greater than that of Zr^{4+}; instead, the shrinkage, amounting to 0.21 Å, almost exactly wipes out this expected increase and results in almost identical radii for Hf^{4+} and Zr^{4+}, as noted previously.

The cause of this contraction is the same as the cause of the less spectacular ones that occur in the *d*-block transition series, namely, the imperfect shielding of one electron by another in the same subshell. As we proceed from La to Lu, the nuclear charge and the number of 4*f* electrons increase by one at each step. The shielding of one 4*f* electron by another is very imperfect (much more than with *d* electrons) owing to the shapes of the orbitals, so that at each increase the effective nuclear charge experienced by each 4*f* electron increases, thus causing a reduction in size of the entire $4f^n$ shell. The accumulation of these successive contractions is the total lanthanide contraction.

It should be noted also that the decrease, though steady, is not quite regular, the biggest decreases occurring with the first f electrons added; there also appears to be a larger decrease after f^7, that is, between Tb and Gd. Certain chemical properties of lanthanide compounds show corresponding divergences from regularity as a consequence of the ionic size (see below).

23-2. Oxidation States

The sum of the first three ionization enthalpies is comparatively low, so that the elements are highly electropositive. They readily form +3 ions in solids like oxides and in aqua ions, $[M(H_2O)_n]^{3+}$, and complexes. However cerium can give Ce^{4+} and Sm, Eu, and Yb, the M^{2+} ions, in aqueous solutions and in solids. Other elements can give the +4 state in solids, whereas the reduction of MX_3 with M can give not only MX_2 but in certain cases other more complicated reduced species that may have metal-metal bonds.

Formerly the existence of +2 and +4 oxidation states was simply ascribed to the extra stability associated with the formation of stable empty $4f^0$ (Ce^{4+}), half-filled $4f^7$ (Eu^{2+}, Tb^{4+}), and filled $4f^{14}$ (Yb^{2+}) subshells. This argument is unconvincing when we note that Sm and Tm give M^{2+} species having f^6 and f^{13} configurations but no M^+ ion, whereas Pr and Nd give M^{4+} ions with configurations f^1 and f^2 but no penta- or hexavalent species. The idea that stability is already favored by the mere *approach* to an f^0, f^7, or f^{14} configuration, even though such a configuration is not actually attained, is of dubious validity. The existence of the various oxidation states is best interpreted by consideration of the ionization enthalpies, enthalpies of sublimation of the metals, lattice energies, etc., in Born-Haber cycles.[1] For the +2 state, as we shall see later, there is a good correlation with the enthalpies of sublimation of the metals.

23-3. Magnetic and Spectral Properties

Several aspects of the magnetic and spectral behavior of the lanthanides differ fundamentally from that of the d-block transition elements. The basic reason for the differences is that the electrons responsible for the properties of lanthanide ions are $4f$ electrons, and the $4f$ orbitals are very effectively shielded from the influence of external forces by the overlying $5s^2$ and $5p^6$ shells. Hence the states arising from the various $4f^n$ configurations are only slightly affected by the surroundings of the ions and remain practically invariant for a given ion in all its compounds.

The states of the $4f^n$ configurations are all given, to a useful approximation, by the Russell-Saunders coupling scheme. In addition, the spin-orbit coupling constants are quite large (order of 1000 cm^{-1}). The result of all this is that with only a few exceptions, the lanthanide ions have ground states with a single well-defined value of the total angular momentum J, with the next lowest J state at energies many

See L. R. Morss, *Chem. Rev.,* 1976, **76**, 827; D. A. Johnson, *Adv. Inorg. Chem. Radiochem.,* 1977, **20**, 1, and *Thermodynamic Aspects of Inorganic Chemistry,* Cambridge University Press, 1968, for thermodynamic data.

times kT (at ordinary temperatures equal to ~200 cm⁻¹) above, hence virtually unpopulated.

Thus the susceptibilities and magnetic moments should be given straightforwardly by formulas considering only this one well-defined J state, and indeed such calculations give results that are, with only two exceptions, in excellent agreement

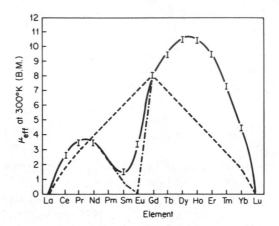

Fig. 23-1. Measured and calculated effective magnetic moments of lanthanide M^{3+} ions at 300° K: I's are ranges of experimental values; solid curve gives values calculated for appropriate J ground states with allowance for the Sm and Eu anomalies; dash-dot curve gives values calculated without allowance for the Sm and Eu anomalies; dashed curve gives calculated spin-only values.

with experimental values (Fig. 23-1). For Sm^{3+} and Eu^{3+}, it turns out that the first excited J state is sufficiently close to the ground state for this state (and in the case of Eu^{3+} even the second and third excited states) to be appreciably populated at ordinary temperatures. Since these excited states have higher J values than the ground state, the actual magnetic moments are higher than those calculated by considering the ground states only. Calculations taking into account the population of excited states afford results in excellent agreement with experiment (Fig. 23-1).

It should be emphasized that magnetic behavior depending on J values is qualitatively different from that depending on S values—that is, the "spin-only" behavior—which gives a fair approximation for many of the d-block transition elements. Only for the f^0, f^7, and f^{14} cases, where there is no orbital angular momentum ($J = S$), do the two treatments give the same answer. For the lanthanides the external fields do not either appreciably split the free-ion terms or quench the orbital angular momentum.

Because the f orbitals are so well shielded from the surroundings of the ions, the various states arising from the f^n configurations are split by external fields only to the extent of ~100 cm⁻¹. Thus when electronic transitions, called f–f transitions, occur from one J state of an f^n configuration to another J state of this configuration, the absorption bands are *extremely sharp*. They are similar to those for free atoms and are quite unlike the broad bands observed for the d–d transitions. Virtually all the absorption bands found in the visible and near-ultraviolet spectra of the lanthanide $+3$ ions have this linelike character. Although the absorption bands normally show little dependence on the nature of the ligand, certain of the lanthanide M^{3+} ions—those with three or more f electrons or equivalent holes—do have one or more transitions that show an increase in intensity when H_2O is replaced by other ligands. There are also bands due to ligand-metal charge transfer with reducing ligands.

The colors and electronic ground states of the M^{3+} ions are given in Table 23-2;

TABLE 23-2
Colors and Electronic Ground States of the M^{3+} Ions

Ion	Ground state	Color	Ion	Ground state
La	$^1S_0$	Colorless	Lu	$^1S_0$
Ce	$^2F_{5/2}$	Colorless	Yb	$^2F_{7/2}$
Pr	$^3H_4$	Green	Tm	$^3H_6$
Nd	$^4I_{9/2}$	Lilac	Er	$^4I_{15/2}$
Pm	$^5I_4$	Pink; yellow	Ho	$^5I_8$
Sm	$^6H_{5/2}$	Yellow	Dy	$^6H_{15/2}$
Eu	$^7F_0$	Pale pink	Tb	$^7F_6$
Gd	$^8S_{7/2}$	Colorless	Gd	$^8S_{7/2}$

the color sequence in the La-Gd series is accidentally repeated in the series Lu-Gd. As implied by the earlier discussion, insofar as the colors are due to f–f transitions, they are virtually independent of the environment of the ions.

An important feature in the spectroscopic behavior is that of fluorescence or luminescence of certain lanthanide ions, notably Y and Eu, when used as activators in lanthanide oxide, silicate, or transition metal oxide lattices. Oxide phosphors are used in color television tubes. Certain of the $+2$ ions trapped in CaF_2 lattices and neodymium glasses or organic cation salts of complex anions such as [Eu β-dike$_4$]$^-$, show laser activity.[2]

The europium isotope ^{151}Eu gives Mössbauer spectra.[3] Because of this and other spectroscopic features such as the long relaxation time of Gd^{3+} and Eu^{2+} in electron spin resonance, and the fluorescence properties of the lanthanide ions, they can be used as probes in biological systems.

23-4. Coordination Numbers and Stereochemistry[4]

The coordination numbers and stereochemistry of the ions are given in Table 23-3. In both ionic crystals and in complexes, coordination numbers exceeding 6 are the general rule rather than the exception. Indeed the number of lanthanide compounds in which the coordination number of 6 has been unequivocally established is small; many complexes that could have been so formulated are known to have solvent molecules bound to the metal, leading to the common coordination numbers 7, 8,

2   R. Reisfeld and C. K. Jørgensen, *Lasers and Excited States of Rare Earths*, Springer, 1977 (720 references).
3   P. Glentworth *et al.*, *J.C.S. Dalton*, **1973**, 546.
4   S. P. Sinha, *Struct. Bonding*, 1976, **25**, 69 (review of high coordination numbers).

TABLE 23-3
Oxidation States, Coordination Numbers, and Stereochemistry of Lanthanide Ions

Oxidation state	Coordination number	Geometry	Examples
+2	6	NaCl type	EuTe, SmO, YbSe
	6	CdI_2 type	YbI_2
	8	CaF_2 type	SmF_2
+3	3	Pyramidal	$M[N(SiMe_3)_2]_3$ [a]
	4	Dist. tetrahedral	$La[N(SiMe_3)_2]_3OPPh_3$ [a]
		Tetrahedral	$[Lu(mesityl)_4]^-$, $[Y(CH_2SiMe_3)_4]^-$
	6	Octahedral	$[Er(NCS)_6]^{3-}$, $[Sc(NCS)_2bipy_2]^+$, MX_6^{3-}
	6	$AlCl_3$ type	MCl_3(Tb-Lu)
	6	Dist. trigonal prism	$Pr[S_2P(C_6H_{11})_2]_3$ [b]
	7	Monocapped trigonal prism	Gd_2S_3, $Y(acac)_3 \cdot H_2O$
	7	Capped dist. octahedron	$Y(PhCOC_6H_4COMe)_3 \cdot H_2O$
	7	ZrO_2 type	ScOF
	8	Dist. square antiprism	$Y(acac)_3 \cdot 3H_2O$; $La(acac)_3(H_2O)_2$
	8	Dodecahedral	$Cs[Y(CF_3COCHCOCF_3)_4]$
	8	Dist. dodecahedral	$Na[Lu(S_2CNEt_2)_4]$ [b]
	8	Cubic	$[La(bipyO_2)_4]^+$ [c]
	8	Bicapped trigonal prism	Gd_2S_3; MX_3(PuBr$_3$ type)
	9	Tricapped dist. trigonal prism	$[Nd(H_2O)_9]^{3+}$, $Y(OH)_3$, $K[La\ edta] \cdot 8H_2O$, $La_2(SO_4)_3 \cdot 9H_2O$
	9	Complex	LaF_3, MCl_3(La-Gd), $[Sc(NO_3)_5]^{2-}$
	10	Complex	$La_2(CO_3)_3 \cdot 8H_2O$
	10	Bicapped dodecahedron	$Ce(NO_3)_5^{2-}$; $La(NO_3)_2(DMSO)_4$
	11	Complex	$La(NO_3)_3(H_2O)_5 \cdot H_2O$ [d]
	12	Dist. icosahedron	$Ce(NO_3)_6^{3-}$, $[Pr(1,8-naphthyridine]^{3+}$ [e]
+4	6	Octahedral	Cs_2CeCl_6
	8	Archim. antiprism	$Ce(acac)_4$
	8	Dist. square antiprism (chains)	$(NH_4)_2CeF_6$
	8	CaF_2 type	CeO_2
	10	Complex	$Ce(NO_3)_4(OPPh_3)_2$
	12	Dist. icosahedron	$(NH_4)_2[Ce(NO_3)_6]$

[a] R. A. Anderson et al., Inorg. Chem., 1978, **17**, 2317; D. C. Bradley et al., J.C.S. Dalton, **1977**, 1166.

[b] Y. Meseri et al., J.C.S. Dalton, **1977**, 725, cf. also M. Ciampolini and N. Nardi, J.C.S. Dalton, **1977**, 2121.

[c] J. S. Wood et al., Inorg. Chem., 1978, **17**, 3702.

[d] B. Eriksson et al., J.C.S. Chem. Comm., **1978**, 616.

[e] A. Clearfield et al., Inorg. Chem., 1977, **16**, 911.

and 9. Compared to the d-block transition elements the lanthanide ions form comparatively few complexes with ligands other than oxygen, and even those with nitrogen ligands are often readily hydrolyzed. Although there is some chemistry of organometallic compounds, it is not as extensive as with d-block elements, and compounds with CO or NO are usually unstable (see, however, page 995) since π bonding can play little role. These differences exist partly because of the unavailability of the f orbitals for formation of hybrid orbitals, which might lead to covalent bond strength, and partly because the lanthanide ions are rather large (radii 0.85–1.06 Å) compared with those of the transition elements (e.g., Cr^{3+} and Fe^{3+}, radii 0.60—0.65 Å), which lowers electrostatic forces of attraction.

23-5. Variations Through the Lanthanide Series

Despite the general similarities to be expected from the electronic structures and sizes of the elements and their ions, there are, nevertheless, some important variations in going from La to Lu.

1. *The +3 Ions.* As a result of the lanthanide contraction, changes in coordination through the series La–Lu are to be expected.

In the *solid state* there may be a pronounced change in structural type at one or more points in the series; at such changeover points a compound may have two or more crystal forms. Examples are the following:

(a) The anhydrous halides MCl_3 of La–Gd are all nine-coordinate with a UCl_3-type lattice; $TbCl_3$ and one form of $DyCl_3$ are of the $PuBr_3$ type, and the Tb–Lu chlorides have the octahedral $AlCl_3$-type structure.

(b) In the M_2S_3 series there are three main structure types: La–Dy (orthorhombic, eight- or seven-coordinate); Dy–Tm (monoclinic, seven- or six-coordinate; Ho_2S_3 actually has half its atoms six- and half seven-coordinate); Yb and Lu (corundum type, six-coordinate).

(c) The enthalpies of sublimation of certain volatile chelates show irregular decreases from Pr to Lu, but with a plateau at Gd.

In solutions the coordination numbers of ions are not always easy to determine. X-Ray[5] and nmr data show that in aqueous solutions of La, Nd, and Gd, even in $10M$ HCl, the average species is eight-coordinate, e.g., $[La(H_2O)_8]^{3+}$ or $[NdCl(H_2O)_7]^{3+}$. Although other types of solution study suggest that at the Lu^{3+} end the coordination number may rise to 9, it seems that there is no very firm evidence for changes in coordination number in solution.

2. *The Metals.* The chemical properties of the metals are noted below, but here we point out that some physical properties, most importantly the enthalpy of sublimation (Fig. 23-2) show considerable differences, especially for the elements that have the greatest tendency to exist in the divalent state.

Unlike the other lanthanide metals, where the $6s$ and $5d$ electrons populate the conduction bonds, and the $4f$ electrons remain bound to the M^{3+} ions, for Eu and Yb only two electrons enter the conduction bands, thus leaving larger cores and affording lower binding forces. Note that the metallic radius of Eu (twelve-coordination) of 2.041 Å can be compared with those of Ca (1.976 Å), Ba (2.236 Å, and the next lanthanide Gd (1.802 Å).[6] It may be noted also that corresponding to the variation in enthalpies of sublimation from 176 to 431 kJ mole^{-1} there is a 10^9 range in the actual vapor pressures of the metals at a given temperature.

5   D. L. Wertz et al., J. Am. Chem. Soc., 1976, **98**, 5125; Inorg. Chem., 1977, **16**, 1225.
6 ˙ W. H. Zachariasen, J. Inorg. Nucl. Chem., 1973, **35**, 3487.

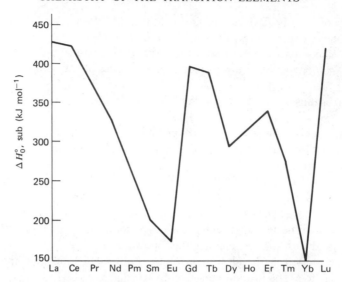

Fig. 23-2. Enthalpies of sublimation of the lanthanide metals. [Data from Ref. 7.]

3.   *Compounds.*   The variation in sublimation energies of the metals is paralleled by dissociation energies of certain refractory compounds such as the monoxides, monosulfides, and dicarbides.[7] These periodic variations are a result of the differing relative energies required to promote a $4f$ electron to the $5d$ level in successive lanthanide atoms.

By contrast, heats of reaction of solid compounds where lanthanide atoms are in combined states in *both* reactants and products show fluctuations of only *ca.* 40 kJ mol^{-1}.

23-6.   Occurrence and Isolation

The lanthanide elements were originally known as the rare earths from their occurrence in oxide (or, in old usage, earth) mixtures. They are not rare elements, and the absolute abundances in the lithosphere are relatively high.

Even the scarcest, thulium, is as common as bismuth ($\sim 2 \times 10^{-5}$ wt %) and more common than As, Cd, Hg, or Se, which are not usually considered rare. Substantial deposits are located in Scandinavia, India, the Soviet Union, and the United States, with a wide occurrence in smaller deposits in many other places. Many minerals make up these deposits, one of the most important being *monazite,* which usually occurs as a heavy dark sand of variable composition. Monazite is essentially a lanthanide orthophosphate, but significant amounts of thorium (up to 30%) occur in most monazite sands. The distribution of the individual lanthanides in minerals is usually such that La, Ce, Pr, and Nd make up about 90% with Y and the heavier elements together constituting the remainder. Monazite and other minerals carrying

7   E. D. Cater, *J. Chem. Educ.,* 1978, **55,** 697.

lanthanides in the +3 oxidation state are usually poor in Eu which, because of its relatively strong tendency to give the +2 state, is often more concentrated in the calcium group minerals.

Promethium[8] occurs in Nature only in traces in uranium ores (4×10^{-15} g kg^{-1}) as a spontaneous fission fragment of ^{238}U. Milligram quantities of pink $^{147}Pm^{3+}$ salts can be isolated by ion-exchange methods from fission products from nuclear reactors where $^{147}Pm(\beta^-$, 2.64 years) is formed.

The lanthanides are separated from most other elements by precipitation of oxalates or fluorides from nitric acid solution. The elements are separated from each other by ion exchange, which is carried out commercially on a large scale. Cerium and europium are normally first removed, the former by oxidation to Ce^{IV} and removal by precipitation of the iodate, which is insoluble in $6M$ HNO_3, or by solvent extraction, and the latter by reduction to Eu^{2+} and removal by precipitation as insoluble $EuSO_4$.

The ion-exchange behavior depends primarily on the hydrated ionic radius, and La should be most tightly bound and Lu least; hence the elution order is Lu $\rightarrow$ La. This trend is accentuated by use of appropriate complexing agents at an appropriate pH; the ion of smallest radius also forms the strongest complexes; hence the preference for the aqueous phase is enhanced. Once of the best ligands is α-hydroxy-isobutyric acid $(CH_3)_2CH(OH)COOH$, at 25° and pH 3.35, but $EDTAH_4$ and other hydroxo or amino carboxylic acids can also be used. The eluates are treated with oxalate ion, and the insoluble oxalates are ignited to the oxides.

Although Ce^{IV} (also Zr^{IV}, Th^{IV}, and Pu^{IV}) is readily extracted from nitric acid solutions by tributyl phosphate dissolved in kerosene or other inert solvent, thus can be readily separated from the +3 lanthanide ions, the trivalent lanthanide nitrates can also be extracted under suitable conditions with various phosphate esters or acids. Extractability under given conditions increases with increasing atomic number; it is higher in strong acid or high nitrate concentrations.

23-7.   The Metals

The lighter metals (La-Gd) are obtained by reduction of the trichlorides with Ca at 1000° or more, whereas for others (Tb, Dy, Ho, Er, Tm, and also Y) the trifluorides are used because the chlorides are too volatile. Pm is made by reduction of PmF_3 with Li. Trichlorides of Eu, Sm, and Yb are reduced only to the dihalides by Ca, but the metals can be prepared by reduction of the oxides M_2O_3 with La at high temperatures. One of the main uses of the lighter metals, mostly cerium (with 20% Fe) is in lighter flints.[9]

The metals are silvery-white and very reactive. They all react directly with water, slowly in the cold, rapidly on heating, to liberate hydrogen. Their high potentials (Table 23-1) show their electropositive character. They tarnish readily in air and all burn easily to give the sesquioxides, except cerium, which gives CeO_2. Yttrium is remarkably resistant to air even up to 1000° owing to formation of a protective

8   F. Weigel, *Chem. Zeit.* 1978, **102**, 339.
9   F. C. Hentze, Jr., and G. G. Long, *J. Chem. Educ.,* 1976, **53**, 651.

oxide coating. The metals react exothermically with hydrogen, though heating to 300 to 400° is often required to initiate the reaction. The resulting phases MH_2 and MH_3, which are usually in a defect state, have remarkable thermal stability, in some cases up to 900° (see p. 250). The metals also react readily with C, N_2, Si, P, S, halogens, and other nonmetals at elevated temperatures.

Metallic Eu and Yb dissolve in liquid ammonia at −78° to give blue solutions, golden when concentrated. The spectra of the blue solutions, which decolorize slowly, are those expected for M^{2+} and solvated electrons.

THE TRIVALENT STATE

23-8.　Binary Compounds

The trivalent state is the characteristic one for all the lanthanides. They form *oxides* M_2O_3, which resemble the Ca-Ba group oxides and absorb carbon dioxide and water from the air to form carbonates and hydroxides, respectively. The *hydroxides* $M(OH)_3$ are definite compounds, not merely hydrous oxides, and may be obtained crystalline by aging M_2O_3 in strong alkali at high temperature and pressure. They have hexagonal structures with nine-coordinate tricapped trigonal prismatic coordination.[10] The basicities of the hydroxides decrease with increasing atomic number, as would be expected from the decrease in ionic radius. The hydroxides are precipitated from aqueous solutions by ammonia or dilute alkalis as gelatinous precipitates. They are not amphoteric.

Among the *halides*,[11] the fluorides are of particular importance because of their insolubility. Addition of hydrofluoric acid or fluoride ions precipitates the fluorides from M^{3+} solutions even $3M$ in nitric acid and is a characteristic test for lanthanide ions. The fluorides, particularly of the heavier lanthanides, are slightly soluble in an excess of HF owing to complex formation. They may be redissolved in $3M$ nitric acid saturated with boric acid, which removes F^- as BF_4^-. The chlorides are soluble in water, from which they crystallize as hydrates, the La-Nd group often with $7H_2O$, and the Nd-Lu group (including Y) with $6H_2O$; other hydrates may also be obtained.

The anhydrous *chlorides* cannot easily be obtained from the hydrates because when heated these lose hydrochloric acid (to give the oxochlorides, MOCl) more readily than they lose water. (However scandium and cerium give Sc_2O_3 and CeO_2, respectively). The chlorides are made by heating oxides with ammonium chloride:

$$M_2O_3 + 6NH_4Cl \xrightarrow{\sim 300°} 2MCl_3 + 3H_2O + 6NH_3$$

or as methanolates $MCl_3 \cdot 4CH_3OH$, by treating the hydrates with 2,2-dimethox-

10　　G. W. Beall *et al., J. Inorg. Nucl. Chem.*, 1977, **39**, 65.
11　　J. C. Taylor, *Coord. Chem. Rev.*, 1976, **20** 197 (structures of halides and oxohalides).

ypropane (cf. p. 550). The halides can also be prepared from the metals by action of $HCl(g)$, Br_2, or I_2. At high temperatures they react with glass:

$$2MX_3 + SiO_2 = 2MOX + SiX_4$$

The bromides and iodides are rather similar to the chlorides.

Numerous other binary compounds are obtained by direct interaction at elevated temperatures; examples are the semiconducting sulfides M_2S_3, which can also be made by reaction of MCl_2 with H_2S at 1100°, Group V compounds MX (X = N, P, As, Sb, or Bi) that have the NaCl structure, borides MB_4 and MB_6, and carbides MC_2 and M_2C_3 (pp. 294, 362). The "hydrides" have been noted in Chapter 6. A representative system[12] is that of Gd, which ranges from an electrically conducting fluorite structure $GdH_{1.8-2.3}$ to a semiconductor $GdH_{2.85-3.0}$

23-9. Oxo Salts

Hydrated salts of common acids, which contain the ions $[M(H_2O)_n]^{3+}$, are readily obtained by dissolving the oxide in acid and crystallizing.

Double salts are very common, the most important being the double nitrates and double sulfates, such as $2M(NO_3)_3 \cdot 3Mg(NO_3)_2 \cdot 24H_2O$, $M(NO_3)_3 \cdot 2NH_4NO_3 \cdot 4H_2O$, and $M_2(SO_4)_3 \cdot 3Na_2SO_4 \cdot 12H_2O$. The solubilities of double sulfates of this type fall roughly into two classes: the cerium group La-Eu, and the yttrium group Gd-Lu and Y. Those of the Ce group are only sparingly soluble in sodium sulfate, whereas those of the Y group are appreciably soluble. Thus a fairly rapid separation of the entire group of lanthanides into two subgroups is possible. Various of the double nitrates were used in the past for further separations by fractional crystallization procedures.

The precipitation of the *oxalates* from dilute nitric acid solution is a quantitative and fairly specific separation procedure for the lanthanides, which can be determined gravimetrically in this way, with subsequent ignition to the oxides. The actual nature of the oxalate precipitate depends on conditions. In nitric acid solutions, where the main ion is Hox^-, ammonium ion gives double salts $NH_4M ox_2 \cdot yH_2O$ (y = 1 or 3). In neutral solution, ammonium oxalate gives the normal oxalate with lighter, but mixtures with heavier, lanthanides. Washing the double salts with $0.01M$ HNO_3 gives, with some ions, the normal oxalates. The phosphates are sparingly soluble in dilute acid solution. Although carbonates exist, many are basic; the normal carbonates are best made by hydrolysis of the chloroacetates:

$$2M(C_2Cl_3O_2)_3 + (x + 3)H_2O \rightarrow M_2(CO_3)_3 \cdot xH_2O + 3CO_2 + 6CHCl_3$$

$La_2(CO_3)_3 \cdot 8H_2O$ has a complex structure with ten-coordinate La, and both uni- and bidentate carbonate ions.

23-10. Complexes

The *aqua ions* M^{3+} are hydrolyzed in water:

12 S. J. Lyle and P. T. Walsh, *J.C.S. Dalton*, **1978**, 601.

$$[M(H_2O)_n]^{3+} + H_2O \rightleftharpoons [M(OH)(H_2O)_{n-1}]^{2+} + H_3O^+$$

and the tendency to hydrolysis increases with increasing atomic number, as would be expected from the contraction in radii. Yttrium also gives predominantly MOH^{2+} but also $M_2(OH)_2^{4+}$ ions; for Ce^{3+} however, only about 1% of the metal ion is hydrolyzed without forming a precipitate, and in this case the main equilibrium appears to be

$$3Ce^{3+} + 5H_2O \rightarrow [Ce_3(OH)_5]^{4+} + 5H^+$$

Halogeno Complexes.   In aqueous solutions rather weak complexes MF^{2+}(aq) are formed with fluoride ion, but there is little evidence for complex anion formation; this is a distinction as a group from the actinide elements, which do form complexes in strong HCl solutions. However by the use of nonaqueous media such as ethanol or acetonitrile, salts of the weak complexes MX_6^{3-} can be prepared. The iodo complexes are exceedingly weak, dissociating in nonaqueous solvents even in presence of an excess of I^-, and they are attacked by moisture and oxygen.

In presence of $AlCl_3$ the anhydrous chlorides are surprisingly volatile, and this has been attributed to the formation of complexes, e.g.,

$$NdCl_3(s) + 2Al_2Cl_6(g) \rightarrow NdAl_4Cl_{15}(g)$$

Oxygen Ligands.   By far the most stable and common of lanthanide complexes are those with chelating oxygen ligands. The use of EDTA-type anions and hydroxo acids such as tartaric or citric, for the formation of water-soluble complexes is of great importance in ion-exchange separations, as noted above. All of these can be assumed to have coordination numbers exceeding 6, as in $[La(OH_2)_4EDTAH]\cdot 3H_2O$.

The *β-diketonates* have been extensively studied, particularly since some of the fluorinated derivatives give complexes that are volatile and suitable for gas-chromatographic separation.

The preparation of β-diketonates by conventional methods *invariably* gives hydrated or solvated species such as $[M(acac)_3]\cdot C_2H_5OH\cdot 3H_2O$. Mono adducts $M(β\text{-dike})_3L$ are seven-coordinate.[13] Anhydrous species obtained by vacuum dehydration appear to be polymeric, not octahedral.[14] The β-diketonates can complex further to give anionic species such as the eight-coordinate[15] thenoyltrifluoroacetate $[Nd(tta)_4]^-$. The alkali metal salts of $[M(β\text{-dike})_4]^-$ are sometimes appreciably volatile and can be sublimed.

An important use of β-diketonate complexes that are soluble in organic solvents, such as those derived from 1,1,2,2,3,3-heptafluoro-7,7-dimethyl-4,6-octanedione, especially of Eu and Pr, is for *shift reagents* in nmr spectrometry[16]. The paramagnetic complex deshields the protons of complicated molecules, and vastly improved separation of the resonance lines may be obtained. The reagents can also be used for the determination of conformation of nonrigid molecules.[17] The effect

13   D. L. Kepert, *J.C.S. Dalton,* **1974,** 617.
14   I. B. Liss and W. G. Bos, *J. Inorg. Nucl. Chem.,* 1977, **39,** 443.
15   J. G. Leipoldt *et al., J. Inorg. Nucl. Chem.,* 1977, **39,** 301.
16   A. F. Cockerell *et al., Chem. Rev.,* 1973, **73,** 553; R. E. Sievers, Ed., *N.m.r. Shift Reagents,* Academic Press, 1973; G. R. Sullivan, in *Topics in Stereochemistry,* Vol. 10, Wiley, 1978, p. 287.
17   G. R. Sullivan, *J. Am. Chem. Soc.,* 1976, **98,** 7162.

depends on the interaction between the shift reagent and substrate, but few complexes as such have been isolated. One is a complex between Co acac$_3$ and a β-diketonate of europium.[18]

Complexes of monodentate oxygen ligands are less stable than those of chelates and tend to dissociate in aqueous solution, but many crystalline compounds or salts have been obtained from the lanthanide salts in ethanolic solutions with hexamethylphosphoramide, which gives six-coordinate species, e.g., Pr(hmpa)$_3$Cl$_3$[19] and [M(hmpa)$_6$](ClO$_4$)$_3$, with triphenylphosphine oxide or triphenyl arsine oxide, and pyridine-*N*-oxides, e.g., M(NO$_3$)$_3$(OAsPh$_3$)$_4$, and [M(PyO)$_8$](ClO$_4$)$_3$, and with DMSO, e.g., (DMSO)$_n$M(NO$_3$)$_3$.

Crown ethers[20] also give complexes of the types M(NO$_3$)$_3$(benzo-15-crown-5) for La–Sm, and M(NO$_3$)$_3$(benzo-15-crown-5)·3H$_2$O·acetone for Sm–Lu; dibenzo-18-crown-6 gives M(NO$_3$)$_3$·L only for La–Nd. Cyclohexyl-18-crown-6 gives La(NO$_3$)$_3$L, which is twelve-coordinate with three bidentate nitrates and six oxygen atoms of the crown bound to La^{3+}.

Many of the complexes above have coordinated anions, and in aqueous solution species such as M(NO$_3$)$^{2+}$ and M(SO$_4$)$^+$ are known. Lanthanide nitrato complexes [M(NO$_3$)$_5$]$^{2-}$ and [M(NO$_3$)$_6$]$^{3-}$ can be isolated,[21] as well as a variety of carbonato complexes, such as the ten coordinate Ho(H$_2$O)$_4$(HCO$_3$)$_3$·2H$_2$O.[22]

The *alcoxide* Nd$_6$(OCHMe$_2$)$_{17}$Cl has been found to have a remarkable structure (Fig. 23-3) with six Nd atoms around a central Cl (cf. the copper complex discussed on page 817) with both edge- and face-bridging OR groups.[23]

Nitrogen Ligands.[24] The bulky dialkylamido derivatives M[N(SiMe$_3$)$_2$]$_3$ of Sc, Nd, Eu, and Gd are pyramidal[25a] by contrast with other trivalent compounds of N(SiMe$_3$)$_2^-$, e.g., Ti[N(SiMe$_3$)$_2$]$_3$, which are planar.[25b] As three-coordinate monomers, they readily form adducts with donor ligands, and these are again unusual in being four or five-coordinate.

More conventional complexes of *amines* such as en, dien, and dipy are known, together with *N*-bonded *thiocyanato* complexes. With few exceptions they have high coordination numbers. Some examples of such complexes are: Men$_3$Cl$_3$, [Mdien$_4$(NO$_3$)](NO$_3$)$_2$, [M(terpy)$_3$](ClO$_4$)$_3$, [M(terpy)(H$_2$O)$_5$Cl]Cl·3H$_2$O, and M(NCS)$_3$(OPPh$_3$)$_4$. In aqueous solution thiocyanato complexes have appreciable formation constants, and SCN$^-$ can be used as elutant for ion-exchange separations. The [M(NCS)$_6$]$^{3-}$ ions are octahedral.

[18]   L. F. Lindoy *et al.*, *J. Am. Chem. Soc.*, 1977, **99**, 5863; *J.C.S. Chem. Comm.*, **1977**, 778.

[19]   L. J. Randonovich and M. D. Glick, *J. Inorg. Nucl. Chem.*, 1973, **35**, 2745.

[20]   G. A. Catton *et al.*, *J.C.S. Dalton*, **1978**, 181; R. M. Izatt *et al.*, *J. Am. Chem. Soc.*, 1974, **96**, 3118; R. B. King, and P. R. Heckley, *J. Am. Chem. Soc.*, 1974, **96**, 3118.

[21]   F. A. Hart *et al.*, *J.C.S. Chem. Comm.*, **1978**, 549; I. M. Walter and D. H. Weedon, *Inorg. Chem.*, 1973, **12**, 773.

[22]   W. J. Rohrbaugh and R. A. Jacobsen, *Inorg. Chem.*, 1974, **13**, 253.

[23]   R. A. Andersen *et al.*, *Inorg. Chem.*, 1978, **17**, 1962.

[24]   J. H. Forsberg, *Coord. Chem. Rev.*, 1973, **10**, 226.

[25a] R. A. Andersen *et al.*, *Inorg. Chem.*, 1978, **17**, 2317; D. C. Bradley *et al.*, *J.C.S. Dalton*, **1977**, 1166.

[25b] R. Altman *et al.*, *J. Organomet. Chem.*, 1978, **162**, 283.

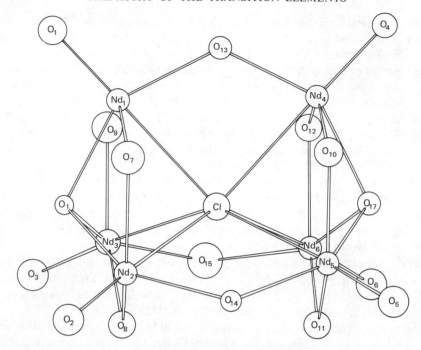

Fig. 23-3. The structure of the Nd, Cl, and O framework in $Nd_6(OCHMe_2)_{17}Cl$. [Reproduced by permission from Ref. 23.]

Porphyrin[26a] *and macrocyclic*[26b] *complexes* can be made by interaction of the hydrated acetylacetonate with H_2porphyrin in 1,2,4-trichlorobenzene. An example is $M(porph)(\beta\text{-dike})$(23-I). These complexes are very resistant to hydrolysis, and this allows the introduction of lanthanide complexes of naturally occurring porphyrins into heme proteins, via reconstitution reactions. The lanthanide ions with their characteristic magnetic and spectral properties are then potential probes for study of natural systems.[26a] Some sulfonated porphyrin complexes are water-soluble shift reagents.

Sulfur Ligands. There are complexes of dithiocarbamates and dithiophosphinates[27] of the types, e.g., $M(dtc)_3$ and $[M(dtc)_4]^-$; they are obtained from MCl_3 and the sodium salt of the anion in ethanol. In the dithiocarbamate anions the lanthanide ion is coordinated by eight sulfur atoms in a distorted dodecahedron, and this structure persists in MeCN solution. The ion $[Pr(S_2PMe_2)_4]^-$ has a distorted tetragonal antiprism; $Pr[S_2P(C_6H_{11})_2]_3$ has a distorted trigonal prism.

[26a] W. de W. Horrocks *et al., J. Am. Chem. Soc.,* 1979, **101**, 334, *Biochem. Biophys. Res. Commun.,* 1975, **64**, 317; *Proc. Nat. Acad. Sci. (U.S.),* 1975, **72**, 4764, E. Nieboer, *Struct. Bonding,* 1975, **22**, 1.

[26b] F. A. Hart, *et al., J. C. S. Chem. Commun.,* **1979**, 114.

[27]   M. Ciampolini *et al., J.C.S. Dalton,* **1977**, 379; A. A. Pinkerton *et al., J.C.S. Dalton,* **1977**, 85, 725; **1976**, 2464, 2466.

(23-I)

Organometallic Compounds.[28]   A variety of organo compounds of lanthanides are known, the main types being cyclopentadienides and cyclooctatetraenides. Some compounds with M—C σ bonds are known.

The cyclopentadienides such as $Pr(C_5H_5)_3$, $Sm(C_5H_5)Cl_2 \cdot 3THF$, and $[(C_5H_5)_2MCl]_2$, in which the bonding is essentially ionic, are thermally stable but air and water sensitive. The structures of these compounds in the solid state are complicated,[29] and there appears to be the maximum number of cation-anion contacts compatible with the sizes and shapes of the ions. Thus in $Nd(MeC_5H_4)_3$ each Nd atom has three η^5-rings and these units are linked into a tetramer by η^1 bonding to one of the rings. In $[(MeC_5H_4)_2YbCl]_2$ there are symmetrical double chlorine bridges between the metal atoms.[30] The tendency to maximize the coordination as above also leads to Lewis acid behavior of $(C_5H_5)_3M$, and solvates are readily formed, as well as adducts with isocyanides and even with metal carbonyl or nitrosyl compounds, where the CO appear to be bound to the metal by oxygen.[31]

The anionic cyclooctatetraenide compounds such as $[Kdiglyme][Ce(C_8H_8)_2]$[32] have typical sandwich structures (see p. 100).

The interaction of the halides $[(C_5H_5)_2MCl]_2$ with sodium borohydride in tetrahydrofuran gives *borohydrides* $(C_5H_5)_2MBH_4 \cdot THF$, where on going from Sm (1.09 Å) to Yb (0.98 Å) there is a change in the bonding of BH_4^- from tridentate to bidentate (cf. p. 108); removal of the solvent gives a polymer.[33]

[28]   T. J. Marks. *Prog. Inorg. Chem.*, 1978, **24**, 51; H. Gysling and M. Tsutsui, *Adv. Organomet. Chem.*, 1970, **9**, 361; M. Tsutsui, N. Ely, and R. Dubois, *Acc. Chem. Res.*, 1976, **9** 217; E. C. Baker, G. W. Halstead, and K. N. Raymond, *Struct. Bonding*, 1976, **25**, 23 (structures and bonding); S. A. Cotton, in *Organometallic Chemistry Research*, Vol. 3, Elsevier, 1977.

[29]   E. C. Baker and K. N. Raymond, *Inorg. Chem.*, 1977, **16**, 2710; J. H. Burns *et al.*, *Inorg. Chem.*, 1974, **13**, 1961.

[30]   E. C. Baker *et al.*, *Inorg. Chem.*, 1975, **14**, 1377.

[31]   A. E. Crease and P. Legzdins, *J.C.S. Dalton*, **1973**, 1501.

[32]   C. W. De Kock *et al.*, *Inorg. Chem.*, 1978, **17**, 625.

[33]   T. J. Marks and G. W. Grynkewich, *Inorg. Chem.*, 1976, **15**, 1302. See also T. J. Marks and J. R. Kolb, *Chem. Rev.*, 1977, **77**, 263, for review on borohydrides.

σ-Bonded Compounds. Thermally stable, though air- and water-sensitive compounds, of the type $(C_5H_5)_2MR$ can also be obtained from the halide and lithium alkyls, aryls, or acetylides.[34] Interaction with $LiAlMe_4$ gives 23-II, which with pyridine gives 23-III.[35]

$$\text{(23-II)} \qquad\qquad\qquad \text{(23-III)}$$

With aryls and with bulky alkyl groups such as *t*-butyl or CH_2SiMe_3, compounds such as $[Li(THF)_4][M(C_4H_9)_4]$, $M(CH_2SiMe_3)_3(THF)_3$, and $Yb(C_6F_5)_2(THF)_4$ have been isolated[36]; with Yb and Lu methylate anions $[Li(tmed)]_3[MMe_6]$, have also been made[37] and are thermally stable to 140°. All these compounds are air and water sensitive. Pure carbonyls are isolable only in matrices at low temperatures.[38]

SCANDIUM[39]

23-11. The Element

Scandium, $[Ar]3d4s^2$, is the congener of Al in Group III and the first member of the Sc, Y, La, Ac group. Its ionic radius ($\sim$0.7 Å) is considerably smaller than the radii of Y and the lanthanides, hence its chemistry more closely resembles that of its congener aluminum. To illustrate this similarity, we note that most scandium compounds are six-coordinate. Higher coordination numbers are very rare, though eight-coordination occurs[40a] in solid complexes such as $Na_5[Sc(CO_3)_4]\cdot6H_2O$ and in the tropolonate ion $[ScT_4]^-$. There is some uncertainty in the solvation of the aqua ion, but in trimethylphosphate[40b] there is the ion $[Sc(tmp)_6]^{3+}$.

34 N. M. Ely and M. Tsutsui, *Inorg. Chem.,* 1975, **14**, 2680.
35 D. G. H. Ballard *et al., J.C.S. Chem. Comm.,* **1978**, 994; G. R. Scollary, *Aus. J. Chem.,* 1978, **31**, 411; J. Holton *et al., J.C.S. Dalton,* **1979** 45, 54.
36 A. L. Wayda and W. J. Evans, *J. Am. Chem. Soc.,* 1978, **100**, 7119; J. L. Atwood *et al., J.C.S. Chem. Comm.,* **1978**, 140; G. B. Deacon *et al., J. Organomet. Chem.,* 1977, **135**, 103; H. Schumann and J. Müller, *J. Organomet. Chem.,* 1978, **146**, C5.
37 H. Schumann and J. Müller, *Angew Chem., Int. Ed.,* 1978, **17**, 276.
38 J. L. Slater *et al., Inorg. Chem.,* 1973, **12**, 1918.
39 C. T. Horowitz, Ed., *Scandium,* Academic Press, 1975 (occurrence, chemistry, physics, metallurgy, etc.); G. A. Melson and R. W. Stotz, *Coord Chem. Rev.,* 1971, **7**, 133 (coordination chemistry).
40a A. R. Davis and F. W. B. Einstein, *Inorg. Chem.,* 1974, **13**, 1880; T. J. Anderson *et al., Inorg. Chem.,* 1974, **13**, 1884.
40b S. F. Lincoln *et al., J.C.S. Chem. Comm.,* **1978**, 1047.

Scandium is quite common, its abundance being comparable to that of As and twice that of B. It is not readily available, owing partly to a lack of rich sources and partly to the difficulty of separation. The separation from the lanthanides can be achieved by using a cation-exchange method and oxalic acid as eluant.

Oxides. The oxide Sc_2O_3 is less basic than those of the lanthanides and is similar to Al_2O_3. Addition of base to Sc^{3+} solutions gives the hydrous oxide $Sc_2O_3 \cdot nH_2O$, but a hydroxide $ScO(OH)$, with the same structure as $AlO(OH)$, is known. The hydrous oxide is amphoteric, dissolving in concentrated $NaOH$. The salt $Na_3[Sc(OH)_6] \cdot 2H_2O$ can be crystallized, but the anion is stable only at hydroxide ion concentrations $>8M$. "Scandates" such as $LiScO_2$ can also be made by fusing Sc_2O_3 and alkali oxides.

The Sc^{3+} ion is more readily hydrolyzed than the lanthanide ions; in perchlorate solutions the main species are $ScOH^{2+}$, $Sc_2(OH)_2^{4+}$, $Sc_3(OH)_4^{5+}$, and $Sc_3(OH)_5^{4+}$, but in chloride media species such as $ScCl^{2+}$ also occur.

Halides. The fluoride is insoluble in water but dissolves readily in an excess of HF or in NH_4F to give fluoro complexes such as ScF_6^{3-}, and the similarity to Al is confirmed by the existence of a cryolite phase Na_3ScF_6, as well as $NaScF_4$ in the $NaF-ScF_3$ system. The chloride $ScCl_3$, and $ScBr_3$, can be obtained by P_2O_5 dehydration of hydrated halides;[41] the anhydrous chloride sublimes at a much lower temperature than the lanthanide halides, and this can be associated with the different structure of the solid, which is isomorphous with $FeCl_3$. Unlike $AlCl_3$ it does not act as a Friedel-Crafts catalyst.

Oxo Salts and Complexes. Simple hydrated oxo salts are known, as well as some double salts such as $K_2SO_4 \cdot Sc_2(SO_4)_3 \cdot nH_2O$, which is very insoluble in K_2SO_4 solution. Ammonium double salts such as the tartrate, phosphate, and oxalate are also insoluble in water.

The β-diketonates resemble those of Al rather than of the lanthanides. Thus the acetylacetonate is normally anhydrous and may be sublimed around 200°; it has a distorted octahedral structure.[42]

The TTA complex can be extracted from aqueous solutions at pH 1.5 to 2 by benzene, and the 8-quinolinolate (cf. Al) can be quantitatively extracted by $CHCl_3$; Sc^{3+} can also be extracted from aqueous sulfate solutions by a quaternary ammonium salt.

Organometallic compounds of Sc and Y show some differences from those of the lanthanides, being more like Al. They include $Sc(C_5H_5)_3$, $(C_5H_5)(C_8H_8)Sc$, $KSc(C_8H_8)_2$, $Sc(C_6H_5)_3$, and $Sc(C\equiv CPh)_3$.[43]

Lower Oxidation States. The prolonged interaction of Sc and $ScCl_3$ at high temperatures gives various reduced species some of which have been characterized.[44] A black laminar phase $ScCl$ is isostructural with $ZrBr$ and has a sheet structure in which there is Sc—Sc bonding; Sc_7Cl_{10} has two parallel chains of Sc_6,

41 R. W. Stotz and G. W. Melson, *Inorg. Chem.*, 1972, **11**, 1720.
42 T. J. Anderson et al., *Inorg. Chem.*, 1974, **13**, 158.
43 A. Westerhof and H. J. De Liefde Meijer, *J. Organomet. Chem.*, 1976 **116**, 319.
44 K. R. Poeppelmeier and J. D. Corbett, *J. Am. Chem. Soc.*, 1978, **100**, 5039.

octahedra sharing a common edge. In $Sc^{3+}Sc_6Cl_{12}^{3-}$ the anion is similar to the M_6Cl_{12} clusters of Nb and Ta (Fig 22-B-5), and again there are Sc—Sc bonds of roughly the same length as in the metal. In Sc_5Cl_8, which approximates to $(ScCl_2^+)_\infty(Sc_4Cl_6^-)_\infty$, there are separate chains of metal octahedra sharing trans edges as shown in Fig. 23-4.

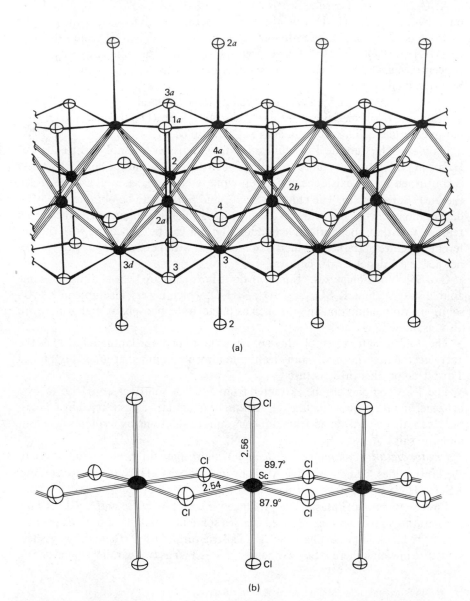

(a)

(b)

Fig. 23-4. (a) The anionic polymetal chain in Sc_5Cl_8 with the b axis horizontal. Solid ellipsoids represent Sc; open, Cl. (b) The chloride-bridged chain about the isolated Sc^{III} atom. [Reproduced by permission from K. R. Poeppelmeier and J. D. Corbett, *J. Am. Chem. Soc.*, 1978, **100**, 5039.]

THE IV OXIDATION STATE

23-12. Cerium(IV)

Cerium (IV) is the only tetrapositive lanthanide species sufficiently stable to exist in aqueous solution as well as in solid compounds. The terms "ceric" and "cerous" are commonly used to designate the IV and III valence states, respectively.

The only binary solid compounds of Ce^{IV} are the oxide CeO_2, the hydrous oxide $CeO_2 \cdot nH_2O$, and the fluoride CeF_4. The dioxide CeO_2, white when pure, is obtained by heating cerium metal $Ce(OH)_3$, or any of several Ce^{III} salts of oxo acids such as the oxalate, carbonate, or nitrate, in air or oxygen; it is a rather inert substance, not attacked by either strong acids or alkalis; it can, however, be dissolved by acids in the presence of reducing agents (H_2O_2, Sn^{II}, etc.), giving Ce^{III} solutions. Hydrous ceric oxide $CeO_2 \cdot nH_2O$ is a yellow, gelatinous precipitate obtained on treating Ce^{IV} solutions with bases; it redissolves fairly easily in acids. CeF_4 is prepared by treating anhydrous $CeCl_3$ or CeF_2 with fluorine at room temperature; it is relatively inert to cold water and is reduced to CeF_3 by hydrogen at 200 to 300°.

Ce^{IV} in solution is obtained by treatment of Ce^{III} solutions with very powerful oxidizing agents, for example, peroxodisulfate or bismuthate in nitric acid. The aqueous chemistry of Ce^{IV} is similar to that of Zr, Hf, and, particularly, tetravalent actinides. Thus Ce^{IV} gives phosphates insoluble in $4M$ HNO_3 and iodates insoluble in $6M$ HNO_3, as well as an insoluble oxalate. The phosphate and iodate precipitations can be used to separate Ce^{IV} from the trivalent lanthanides. Ce^{IV} is also much more readily extracted into organic solvents by tributyl phosphate and similar extractants than are the M^{III} lanthanide ions.

The phosphates have been studied because they act as ion exchangers, but they are complicated, with stoichiometries of the type $Ce(OH)_x(PO_4)_x$-$(HPO_4)_{2-2x} \cdot yH_2O$.[45]

The hydrated ion $[Ce(H_2O)_n]^{4+}$ is a fairly strong acid, and except at very low pH, hydrolysis and polymerization occur. It is probable that the $[Ce(H_2O)_n]^{4+}$ ion exists only in concentrated perchloric acid solution. In other acid media there is coordination of anions, which accounts for the dependence of the potential of the Ce^{IV}/Ce^{III} couple on the nature of the acid medium:

$$Ce^{IV} + e = Ce^{III} \qquad E^0 = +1.44V \ (1 \ MH_2SO_4), +1.61V \ (1 \ MHNO_3), +1.70V \ (1 \ MHClO_4)$$

Comparison of the potential in sulfuric acid, where at high SO_4^{2-} concentrations the major species is $[Ce(SO_4)_3]^{2-}$, with that for the oxidation of water:

$$O_2 + 4H^+ + 4e = 2H_2O \qquad E^0 = +1.229V$$

shows that the acid Ce^{IV} solutions commonly used in analysis are metastable. The oxidation of water is kinetically controlled but can be temporarily catalyzed by fresh glass surfaces.

45 R. G. Herman and A. Clearfield, *J. Inorg. Nucl. Chem.*, 1975, **37**, 1697.

Cerium(IV) is used as an oxidant, not only in analysis, but also in organic chemistry, where it is commonly used in acetic acid.[46] The solid acetate, which is bright yellow, can be made by ozone oxidation of Ce^{III} acetate. The solutions oxidize aldehydes and ketones at the α-carbon atom, and benzaldehyde, for example, gives benzoin, while ammonium hexanitratocerate will oxidize toluenes to aldehydes. These oxidations appear to proceed by the initial formation of 1:1 complexes; the complexes of alcohols are red.

Complex anions are formed quite readily and some of the salts previously considered to be double must be reformulated, notably the analytical standard "ceric ammonium nitrate," which can be crystallized from HNO_3. This is actually $(NH_4)_2[Ce(NO_3)_6]$ with bidentate NO_3 groups both in the crystal and in solution. In NH_4F solutions of CeF_4 the solid phase in equilibrium is $(NH_4)_4[CeF_8]$, although $(NH_4)_3CeF_7 \cdot H_2O$ can be grown from 28% NH_4F solutions; when heated, the octafluorocerate gives $(NH_4)_2CeF_6$.

In aqueous solution Ce^{IV} oxidizes concentrated HCl to Cl_2, but the reaction of CeO_2 with HCl in dioxane gives orange needles of the oxonium salt of the $[CeCl_6]^{2-}$ ion; the corresponding pyridinium salt is thermally stable to 120° and is used to prepare ceric alcoxides:

$$(C_5H_5NH)_2CeCl_6 + 4ROH + 6NH_3 \rightarrow Ce(OR)_4 + 2C_5H_5N + 6NH_4Cl$$

The isopropoxide is crystalline, subliming in a vacuum at 170°, but other alcoxides, prepared from the isopropyl compound by alcohol exchange, are nonvolatile and presumably polymerized by Ce—O(R)—Ce bridges.

23-13. Other Tetravalent Compounds

Praseodymium(IV). Only a few solid compounds are known, the commonest being the black nonstoichiometric oxide formed on heating Pr^{III} salts or Pr_2O_3 in air. The oxide system which is often formulated as Pr_6O_{11} is actually very complicated.[47]

When alkali fluorides mixed in the correct stoichiometric ratio with Pr salts are heated in F_2 at 300 to 500°, compounds such as $NaPrF_5$ or Na_2PrF_6 are obtained. The action of dry HF on the latter gives PrF_4, although this cannot be obtained by direct fluorination of PrF_3.

The tetravalence of Pr in these compounds has been established by magnetic, spectral, and X-ray data.

Pr^{IV} is a very powerful oxidizing agent, the Pr^{IV}/Pr^{III} couple being estimated as +2.9 V. This potential is such that Pr^{IV} would oxidize water itself, so that its nonexistence in solution is not surprising. Pr_6O_{11} dissolves in acids to give aqueous Pr^{III} and liberate oxygen, chloride, etc., depending on the acid used.

There is some evidence that $Pr(NO_3)_4$ is partially formed by action of N_2O_5 and O_3 on PrO_2.

[46]   T. L. Ho, *Synthesis,* 1973, **6**, 347.
[47]   R. L. Martin, *J.C.S. Dalton,* **1974**, 1335.

Terbium(IV). The chemistry resembles that of Pr^{IV}. The Tb—O system is complex and nonstoichiometric, and when oxo salts are ignited under ordinary conditions, an oxide of approximately the composition Tb_4O_7 is obtained. This formula $TbO_{1.75}$ is the nearest approach, using small whole numbers, to the true formula of the stable phase obtained, which varies from $TbO_{1.71}$ to $TbO_{1.81}$, depending on the preparative details. For the average formula Tb_4O_7, Tb^{III}, and Tb^{IV} are present in equal amounts. TbO_2, with a fluorite structure, can be obtained by oxidation of Tb_2O_3 with atomic oxygen at 450°. Colorless TbF_4, isostructural with CeF_4 and ThF_4, is obtained by treating TbF_3 with gaseous fluorine at 300 to 400°, and compounds of the type M_nTbF_{n+4} (M = K, Rb, or Cs; $n \geqslant 2$) are known.

No numerical estimate has been given for the Tb^{IV}/Tb^{III} potential, but it must certainly be more positive than +1.23 V, since dissolution of any oxide containing Tb^{IV} gives only Tb^{III} in solution and oxygen is evolved. TbF_4 is even less reactive than CeF_4 and does not react rapidly even with hot water.

Neodymium(IV) and Dysprosium(IV). Claims of the preparation of higher oxides of these elements, supposedly containing Nd^{IV} and Dy^{IV}, are erroneous. Even treatment of Nd_2O_3 with atomic oxygen gives no Nd^{IV}-containing product. Only in the products of fluorination of mixtures of RbCl and CsCl with $NdCl_3$ and $DyCl_3$ is there fair evidence for the existence of Nd^{IV} and Dy^{IV}. Apparently such compounds as Cs_3NdF_7 and Cs_3DyF_7 can be formed, at least partially, in this way.

Europium(IV). The action of O_2 or Cl_2 on Eu^{2+} in a zeolite lattice is said to give Eu^{IV} [48]

LOWER OXIDATION STATES

The existence both in aqueous solutions and in solid compounds of the divalent ions of Sm, Eu, and Yb has long been known. However, other +2 ions can be stabilized in CaF_2 lattices by reduction of MF_3 in CaF_2 by Ca. Also, reduction of the trihalides by the metal can give (a) saltlike dihalides MX_2, (b) for Sc, as we have seen above, and for some other elements, complicated halides with metal-metal bonds, (c) metallic compounds, and (d) other nonstoichiometric phases that exist at high temperatures.

We can consider only the more important features of this complicated chemistry.

23-14. The Divalent State

The variations in the stability of the M^{2+} states can be largely correlated with the trends in the third ionization enthalpies and the enthalpies of sublimation of the metals (see Fig. 23-2) and the stabilities peak at Eu^{2+} and Yb^{2+} [49] Values of E^0

48   R. L. Firor and K. Seff, *J. Am. Chem. Soc.,* 1978, **100**, 976, 978.
49   L. J. Nugent *et al., J. Inorg. Nucl. Chem.,* 1975, **37**, 1767; *J. Phys. Chem.,* 1973, **77**, 1528; L. R. Morss and M. C. McCue, *Inorg. Chem.,* 1975, **14**, 1624; D. A. Johnson, *J.C.S. Dalton,* **1974**, 1671.

TABLE 23-4

Properties of the Lanthanide M^{2+} Ions

Ion	Color	$E^0(V)$	Crystal radius (Å)
Sm^{2+}	Blood red	−1.40	1.11
Eu^{2+}	Colorless	−0.34	1.10
Yb^{2+}	Yellow	−1.04	0.93

for M^{3+}/M^{2+} for the lanthanide elements have been estimated in various ways,[49] and it is clear that in aqueous solutions only Sm^{2+}, Eu^{2+}, and Yb^{2+} can exist. The properties of these ions are given in Table 23-4.

The europium(II) ion can be readily made by reducing aqueous Eu^{3+} solutions with Zn or Mg. The other ions require the use of sodium amalgam, but all three can be prepared by electrolytic reduction in aqueous solution. The dissolution of reactor-irradiated holmium in dilute acid also appears to give Ho^{2+} ion in solution, but the half-life is only 55.5 min under N_2.[50]

The ions Sm^{2+} and Yb^{2+} are quite rapidly oxidized by water as well as by air, but Eu^{2+} solutions are readily handled. Solutions of Eu^{2+} have been much studied to compare electron transfer mechanisms with those of other one-electron reducing agents. The ion is rapidly oxidized by O_2 but the rates of reaction with other ions often differ from expectations based on the standard potential. Thus Eu^{2+} reduces V^{3+} more slowly than does Cr^{2+} ($E^0 = -0.41$ V), but Eu^{2+} and Cr^{2+} will not reduce ClO_4^-, although the weaker reductant V^{2+} ($E^0 = -0.25$ V) will do so. The reasons for such differences are probably connected with the electronic configuration and the ability to form transition states with bridging groups.

The lanthanide M^{2+} ions resemble the Group II ions, especially Ba^{2+}. Thus the sulfates are insoluble, whereas the hydroxides are soluble, and the Eu^{2+} ion can be readily separated from the other lanthanides by Zn reduction followed by the precipitation of other hydroxides by carbonate-free ammonia.

Solid compounds can be made. Europium gives an oxide, mixed oxides,[51] chalcogenides, halides, carbonate, phosphate, etc., which may be obtained by reduction of the corresponding Eu^{3+} compounds or, metathetically, from $EuCl_2$. The metal reacts with liquid ammonia at 50° to give the orange amide $Eu(NH_2)_2$, which gives EuN when heated. The compounds are usually isostructural with the Sr^{2+} or Ba^{2+} analogues. Both anhydrous and hydrated samarium(II) bromide are known.[52]

23-15. Other Reduced Compounds.[53]

Depending on the metal and the halide, interaction of $MX_3 + M$ gives a variety

[50]   D. J. Apers et al., J. Inorg. Nucl. Chem., 1974, 36, 1441.
[51]   K. Sato et al., Inorg. Chem., 1977, 16, 328; J. E. Greedan et al., Inorg. Chem., 1977, 16, 332.
[52]   J. M. Haschke, Inorg. Chem., 1976, 15, 298.
[53]   J. D. Corbett, Rev. Chim. Miner. 1973, 10, 239; J. D. Corbett et al., Inorg. Chem., 1977, 16, 2134; 1975, 14, 426; 1973, 12, 556.

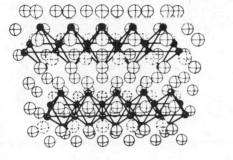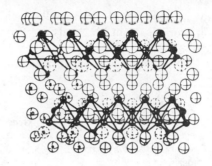

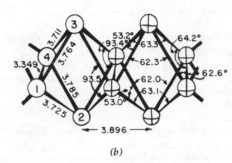

(b)

Fig. 23-5. (a) Stereoscopic view of the structure of Gd_2Cl_2 with gadolinium atoms represented by small solid circles and chlorine atoms by larger open circles. The b axis is horizontal in the plane of the paper to the right; a is approximately toward the viewer. "Bridging" atoms Cl(5) and Cl(10) from a neighboring sheet are shown dashed. (b) Portion of the metal chain, with distances in angstroms and angles in degrees. The pairs of gadolinium atoms 1–4 and 2–3 lie on mirror planes perpendicular to the chain. [Reproduced by permission from D. A. Lokken and J. D. Corbett, *Inorg. Chem.*, 1973, **12**, 556.]

of reduced phases and compounds among which are, of course, the dihalides, but 'here are also compounds such as $PrCl_{2.31}$, $NdI_{1.95}$, and $HoCl_{2.20}$.

Sesquichlorides of unique type are formed by Gd and Tb. In Gd_2Cl_3 or $[Gd_4^{6+}(Cl^-)_6]_n$ there are infinite chains of metal atoms in octahedra sharing opposite edges (Fig. 23-5); chlorine atoms are located over triangles formed by three Gd atoms. These halides can be further reduced at 800° to GdCl and TbCl. The latter materials are graphitelike platelets. They have a layer structure[54] like ZrCl and ScCl where close-packed double layers of metal atoms alternate with double layers of halide atoms (XMMX⋯XMMX) and are related to the CdI_2 type (XMX⋯XMX) (Fig. 1–10).

The reduction of triodides gives products that may be saltlike as with the chlorides, or metallic. Thus La, Ce, Gd, and Pr give metallic iodides that are best formulated $M^{3+}(I^-)_2e$.

Other Compounds. The sulfides of stoichiometry MS (e.g., golden yellow LaS) have the NaCl-type structure and are metallic conductors. They are best formulated

54   A. Simon *et al.*, *Angew Chem., Int. Ed.*, 1976, **15**, 624.

as $M^{3+}(S^{2-})e$ (cf. the metallic iodides above). The sulfides of Sm, Eu, and Yb appear to contain M^{2+} and S^{2-} The carbides MC_2 or $M^{3+}(C_2^{2-})e$ give acetylene on hydrolysis.

The hydrides MH_2 have been discussed (p. 250).

General References

Structure and Bonding, Vols. 13, 22, 25, and 30 contain specialized articles on spectroscopic and other topics.

Asprey, L. B., and B. B. Cunningham, *Prog. Inorg. Chem.,* 1960, **2**, 267. Unusual oxidation states of some actinide and lanthanide elements.

Bagnall, K. W., Ed., *Lanthanides and Actinides,* Butterworth, 1972.

Brown, D., *Halides of the Lanthanides and Actinides,* Wiley Interscience, 1968. An authoritative monograph.

Callow, R. J., *The Industrial Chemistry of the Lanthanons, Yttrium, Thorium and Uranium,* Pergamon Press, 1967.

Callow, R. J., *The Rare Earth Industry,* Pergamon Press, 1966. Sources, recovery, and uses.

Fields, F. R., and T. Moeller, Eds., *Lanthanide-Actinide Chemistry,* ACS Advances in Chemistry Series No. 71, 1967. Conference reports.

Gschneider, K. A. and L. Eyring, Eds., *Handbook on the Physics and Chemistry of Rare Earths,* North Holland, Vol 1 (Metals), 1978.

Henrie, D. E., R. L. Fellows, G. R. Choppin, *Coord. Chem. Rev.,* 1976, **18**, 199. Electronic transitions in $4f$ complexes.

Hirschhorn, I. S., *Chem. Tech.,* **1971**, 314. Uses of lanthanide metals.

Hüfner, S., *Optical Spectra of Transparant Rare Earth Compounds,* Academic Press, 1978.

McCarthy, G. J., and J. J. Rhyne, Eds., *Rare Earths in Modern Science and Technology.* Plenum Press, 1977. Conferences reports.

Misumi, S., S. Kida, and M. Aihari, *Coord Chem. Rev.,* 1968, **3**, 189. Spectra and solution properties.

Moeller, T., *The Chemistry of the Lanthanides,* Reinhold, 1963. An introductory treatment, but authorative and thorough at its level.

Progress in the Science and Technology of the Rare Earths, Vol. 1, 1964. Series with reviews on extraction, solution chemistry, magnetic properties, analysis, halides, oxides, etc.

Sinha, S. P., *Complexes of the Rare Earths,* Pergamon Press, 1966. Emphasizes spectra.

Sinha, S. P., *Europium,* Springer Verlag, 1968.

Spedding F. H., and A. M. Daane, Eds., *The Rare Earths,* Wiley, 1951. Contains detailed discussions of occurrence, extraction procedures, preparation, and properties of metals and alloys; also describes applications.

Taylor, K. N. R., and M. I. Darby, *Physics of Rare Earth Solids,* Chapman & Hall, 1972.

Topp, N. E., *The Chemistry of the Rare Earth Elements,* Elsevier, 1965.

Vickery, R. C., *Analytical Chemistry of the Rare Earths,* Pergamon Press, 1961.

Vickery, R. C., *Chemistry of Yttrium and Scandium,* Pergamon Press, 1961.

Wybourne, B. G., *Spectroscopic Properties of Rare Earths,* Wiley, 1965. Comprehensive discussion of atomic spectra and especially spectra of salts.

Yost, D. M., H., Russell, and C. S. Garner, *The Rare Earth Elements and Their Compounds,* Wiley, 1947. A classical book containing early references; still of value.

CHAPTER TWENTY-FOUR

The Actinide Elements

GENERAL REMARKS

24-1. Occurrence

The actinide or actinoid elements, all of whose isotopes are radioactive, are listed in Table 24-1. Some of their properties are given in Table 24-2. The principal isotopes obtained in macroscopic amounts are given in Table 24-3.

The terrestrial occurrence of Ac, Pa, U, and Th is due to the half-lives of the isotopes ^{235}U, ^{238}U, and ^{232}Th, which are sufficiently long to have enabled the species to persist since genesis. They are the sources of actinium and protactinium formed in the decay series and found in uranium and thorium ores. The half-lives of the most stable isotopes of the trans-uranium elements are such that any pri-

TABLE 24-1
The Actinide Elements and Some of Their Properties

Z	Symbol	Name	Electronic structure	Z	Symbol	Name	Electronic structure
89	Ac	Actinium	$6d7s^2$	100	Fm	Fermium	$5f^{12}7s^2$
90	Th	Thorium	$6d^27s^2$	101	Md	Mendelevium	$5f^{13}7s^2$
91	Pa	Protactinium	$5f^26d7s^2$	102	No	Nobelium	$5f^{14}7s^2$
			or $5f^16d^27s^2$	103	Lr	Lawrencium	$5f^{14}6d7s^2$
92	U	Uranium	$5f^36d7s^2$	104	Rf	Rutherfordium (eka-hafnium)	
93	Np	Neptunium	$5f^57s^2$	105	Ha	Hahnium (eka-tantalum)	
94	Pu	Plutonium	$5f^67s^2$	106		eka-tungsten	
95	Am	Americium	$5f^77s^2$	107		eka-rhenium	
96	Cm	Curium	$5f^76d7s^2$	108		eka-osmium	
97	Bk	Berkelium	$5f^86d7s^2$	109		eka-iridium	
			or $5f^97s^2$	110		eka-platinum	
98	Cf	Californium	$5f^{10}7s^2$	111		eka-gold	
99	Es	Einsteinium	$5f^{11}7s^2$	112		eka-mercury	

TABLE 24-2
Some Properties of the Actinide Elements[a]

Element	Metal		$-E°$ (V)		Ionic radius[b] (Å)	
	m.p. (°C)	Radius (Å)	0—III	0—II	M^{3+}	M^{4+}
Ac	1100	1.898	2.13	—	1.077	—
Th	~1750	1.798	1.17	—	(1.042)	0.99
Pa	1572	1.642	1.49	—	(1.029)	0.94
U	1132	1.542	1.66	—	1.025	0.93
Np	639	1.503	1.79	—	1.012	0.91
Pu	640	1.523	2.00	—	1.00	0.90
Am	1173	1.730	2.07	2.0	0.975	0.89
Cm	1350	1.743	2.06	0.9	0.960	0.88
Bk	986	1.704	1.97	1.6	0.955	—
Cf	900	1.694	2.01	2.2	0.942	—
Es	—	(1.69)	1.98	2.3	0.928	—
Fm	—	(1.94)	1.95	2.5	(0.922)	—
Md	—	(1.94)	1.66		(0.912)	—
No	—	(1.94)	1.78	1.6[c]	(0.900)	
Lr	—	(1.71)	2.06	—	(0.893)	—

[a] For thermodynamic properties see F. David et al. (J. Inorg. Nucl. Chem., 1978, 40, 69), from which radii and $E°$ values are taken. Estimated values in parentheses. Radii of M^{3+} are for coordination number 6.

[b] In MF_6 the M—F distances are U—F = 1.992 Å, Np—F = 1.981 Å, Pu—F = 1.969 Å.

[c] R. E. Meyer et al., J. Inorg. Nucl. Chem., 1976, 38, 1171.

mordial amounts of these elements appear to have disappeared long ago. However neptunium and plutonium have been isolated in traces from uranium minerals in which they are formed continuously by neutron reactions such as

$$^{238}U \xrightarrow{n\gamma} {}^{239}U \xrightarrow{\beta^-} {}^{239}Np \xrightarrow{\beta^-} {}^{239}Pu$$

The neutrons arise from spontaneous fission of ^{235}U or from α,n reactions of light elements present in uranium minerals.

Traces of ^{244}Pu have recently been found in a very old cerium-containing mineral. This plutonium is believed to be primordial in origin, but its formation by cosmic ray interactions is not yet rigorously excluded.

The first trans-uranium elements, neptunium and plutonium, were obtained in tracer amounts from bombardments of uranium by McMillan and Abelson and by Seaborg, McMillan, Kennedy, and Wahl, respectively, in 1940. Both ^{237}Np and ^{239}Pu are obtained in substantial quantities from the uranium fuel elements of nuclear reactors.

Neptunium is converted to ^{238}Pu, a useful heat source. Since ^{239}Pu, like ^{235}U, undergoes fission, it is used as a nuclear fuel; its nuclear properties apparently preclude its use in hydrogen bombs. Certain isotopes of the heavier elements are made by successive neutron capture in ^{239}Pu in high-flux nuclear reactors (>10^{15} neutrons cm^{-2} sec^{-1}). Others are made by the action of accelerated heavy ions of B, C, N, O, or Ne on Pu, Am, or Cm.

The elements above fermium ($Z = 100$) exist only as short-lived species, the most

TABLE 24-3
Principal Actinide Isotopes Available in Macroscopic Amounts[a]

Isotope	Half-life	Source
^{227}Ac	21.7 years	Natural ^{226}Ra$((n\gamma)^{227}$Ra $\xrightarrow[41.2 \text{ min}]{\beta^-} {}^{227}$Ac
^{232}Th	1.39×10^{10} years	Natural; 100% abundance
^{231}Pa	3.28×10^5 years	Natural; 0.34 ppm of U in uranium ores
^{235}U	7.13×10^8 years	Natural 0.7204% abundance
^{238}U	4.50×10^9 years	Natural 99.2739% abundance
^{237}Np	2.20×10^6 years	^{235}U$(n\gamma)^{236}$U$(n\gamma)^{237}$U $\xrightarrow[6.75 \text{ d}]{\beta^-} {}^{237}$Np (and ^{238}U$(n, 2n)^{237}$U)
^{238}Pu	86.4 years	^{237}Np$(n\gamma)^{238}$Np $\xrightarrow{\beta^-} {}^{238}$Pu
^{239}Pu	24,360 years	^{238}U$(n\gamma)^{239}$U $\xrightarrow[23.5 \text{ min}]{\beta} {}^{239}$Np $\xrightarrow[2.35 \text{ days}]{\beta^-} {}^{239}$Pu
^{242}Pu	3.79×10^5 years	Successive $n\gamma$ in ^{239}Pu
^{244}Pu	8.28×10^7 years	Successive $n\gamma$ in ^{239}Pu
^{241}Am	433 years	^{239}Pu$(n\gamma)^{240}$Pu$(n\gamma)^{241}$Pu $\xrightarrow[13.2 \text{ years}]{\beta^-} {}^{241}$Am
^{243}Am	7650 years	Successive $n\gamma$ on ^{230}Pu
^{242}Cm	162.5 days	^{241}Am$(n\gamma)^{242m}$Am $\xrightarrow[16.0 \text{ hr}]{\beta^-} {}^{242}$Cm
^{244}Cm	18.12 years	^{239}Pu$(4n\gamma)^{243}$Pu $\xrightarrow[5.0 \text{ hr}]{\beta^-} {}^{243}Am(n\gamma)^{244}$Am $\xrightarrow[26 \text{ min}]{\beta^-} {}^{244}$Cm
^{249}Bk	314 days	Successive $n\gamma$ on ^{239}Pu
^{252}Cf	2.57 years	Successive $n\gamma$ on ^{242}Pu

[a] ^{237}Np and ^{239}Pu are available in multikilogram quantities; ^{238}Pu, ^{242}Pu, ^{241}Am, ^{243}Am, and ^{244}Cm in amounts of 100 g or above; ^{244}Pu, ^{252}Cf, ^{249}Bk, ^{248}Cm (4.7×10^5 years), ^{253}Es (20 days), ^{254}Es (1.52 years for α) in milligram and ^{257}Fm (94 days) in microgram quantities. Other long-lived isotopes are known but can be obtained in traces by use of accelerators [e.g., ^{247}Bk ($\sim 10^4$ years), ^{252}Es, 471.7 days].

stable being ^{258}Md (53 days), ^{255}No (185 sec), ^{256}Lr (45 sec), and ^{261}Rf (ca. 70 sec). The nuclear properties of isotopes, such as their decay modes and half-lives, can be predicted with some accuracy from nuclear systematics. For elements beyond Pa ($Z = 91$), spontaneous nuclear fission becomes increasingly important and is indeed the limiting factor for the life of a given isotope. The element $Z = 104$ appears to be hafniumlike. Since the $6d$ shell is filled, the subsequent elements up to $Z = 114$ should have chemical properties analogous to those of elements Hf up to Pb.

24-2. Electronic Structures; Comparison with Lanthanides

The atomic spectra of the heavy elements are very complex, and it is difficult to identify levels in terms of quantum numbers and configurations.[1] For chemical behavior the lowest configuration is of greatest importance and the competition between $5f^n 7s^2$ and $5f^{n-1}6d7s^2$ is of interest. Figure 24-1 shows the approximate

---

[1]   For discussion and data, see M. Fred, in *Lanthanide and Actinide Chemistry*, ACS Advances in Chemistry Monograph No. 71, p. 188, 1967.

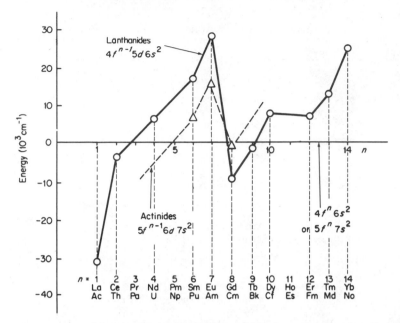

Fig. 24-1. Approximate relative positions of $f^n s^2$ and $f^{n-1} d s^2$ configurations for neutral lanthanide and actinide atoms. [Reproduced by permission from Ref. 1 with additional data kindly provided by Dr. M. Fred.]

relative positions for both the lanthanides and the actinides. For the elements in the first half of the f shell it appears that less energy is required for the promotion of $5f \rightarrow 6d$ than for the $4f \rightarrow 5d$ promotion in the lanthanides; there is thus a greater tendency to supply more bonding electrons with the corollary of higher valences in the actinides. The second half of the actinides resembles the lanthanides more closely.

Another difference is that the $5f$ orbitals have a greater spatial extension relative to the $7s$ and $7p$ orbitals than the $4f$ orbitals have relative to the $6s$ and $6p$ orbitals. The greater spatial extension of the $5f$ orbitals has been shown experimentally; the esr spectrum of UF_3 in a CaF_2 lattice shows structure attributable to the interaction of fluorine nuclei and the electron spin of the U^{3+} ion. This implies a small overlap of $5f$ orbitals with fluorine and constitutes an f covalent contribution to the ionic bonding. With the neodymium ion a similar effect is *not* observed. Because they occupy inner orbitals, the $4f$ electrons in the lanthanides are not accessible for bonding purposes; and virtually no compound in which $4f$ orbitals are used can be said to exist.

In the actinide series, therefore, the energies of the $5f$, $6d$, $7s$, and $7p$ orbitals are about comparable over a range of atomic numbers (especially U—Am), and since the orbitals also overlap spatially, bonding can involve any or all of them. In the chemistries this situation is indicated by the fact that the actinides are much more prone to complex formation than are the lanthanides, where the bonding is almost exclusively ionic. Indeed the actinides can even form complexes with certain

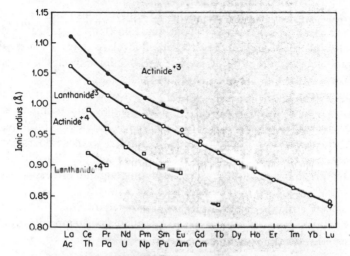

Fig. 24-2. Radii of actinide and lanthanide ions. [Reproduced by permission from D. Brown, Halides of the Lanthanides and Actinides, Wiley-Interscience, 1968.]

π-bonding ligands as well as forming complexes with halide, sulfate, and other ions. The difference from lanthanide chemistry is usually attributed to the contribution of covalent hybrid bonding involving $5f$ electrons.

A further point is that since the energies of the $5f$, $6d$, $7s$, and $7p$ levels are comparable, the energies involved in an electron shifting from one to another, say $5f$ to $6d$, may lie *within* the range of chemical binding energies. Thus the electronic structure of the element in a given oxidation state may vary between compounds and in solution be dependent on the nature of the ligands. It is accordingly also often impossible to say *which* orbitals are being utilized in bonding or to decide meaningfully whether the bonding is covalent or ionic.

24-3. Ionic Radii

The ionic radii for the commonest oxidation states (Table 24-2) are compared with those of the lanthanides in Fig. 24-2. There is clearly an "actinide" contraction, and the similarities in radii of both series correspond to similarities in their chemical behavior for properties that depend on the ionic radius, such as thermodynamic results for hydrolysis of halides. It is also generally the case that similar compounds in the same oxidation state have similar crystal structures that differ only in the parameters.

24-4. Absorption Spectra and Magnetic Properties

The electronic absorption spectra of the actinide ions, like those of the lanthanides, are due to transitions within the $5f^n$ levels and consist of narrow bands, relatively uninfluenced by ligand fields; the intensities are generally about 10 times those of the lanthanide bands.

Spectra involving only one f electron are simple, consisting of only a single transition $^2F_{5/2} - {}^2F_{7/2}$. For the f^7 configuration (Cm^{3+}; cf. Gd^{3+}) the lowest excited state lies about 4 eV above the ground level, so that these ions show only charge-transfer absorption in the ultraviolet. Most actinide species have complicated spectra.[2]

The magnetic properties of the actinide ions are considerably harder to interpret than those of the lanthanide ions. The experimental magnetic moments are usually lower than the values calculated by using Russell-Saunders coupling, and this appears to be due both to ligand field effects similar to those operating in the d transition series and to inadequacy of this coupling scheme. Since 5f orbitals can participate to some extent in covalent bonding, ligand effects are to be expected.

For the ions Pu^{3+} and Am^{3+}, the phenomenon noted for Sm^{3+} and Eu^{3+} is found; since the multiplet levels are comparable to kT, anomalous temperature dependence of the susceptibilities is found.

24-5.   Oxidation States; Stereochemistry

The known oxidation states of the actinides are given in Table 24-4. With the exception of Th and Pa, the common oxidation state, and for trans-americium elements the dominant, oxidation state, is +3, and the behavior is similar to the +3 lanthanides. Thorium and the other elements in the +4 state show resemblances of Hf or Ce^{IV}, whereas Pa and the elements in the +5 state show some resemblances to Ta^V. Exceptions to the latter statement are the +5 state of U, Np, Pu, and Am in the *dioxo ions* MO_2^+; these ions and the MO_2^{2+} ions of the +6 states show unusual and exceptional behavior, as discussed below.

Although there is certainly some doubt concerning the extent of covalent bonding in actinide compounds, the angular distributions and relative strengths of various orbital combinations using f orbitals have been worked out theoretically in a manner similar to that for the schemes for light elements. Examples are: sf, linear; sf^3, tetrahedral; sf^2d, square; and d^2sf^3, octahedral. These hybridizations *could* be considered to hold in PuO_2^{2+}, $NpCl_4$, and UCl_6, for example. However in view of

TABLE 24-4
Oxidation States of the Actinide Elements[a]

Ac	Th	Pa	U	Np	Pu	Am	Cm	Bk	Cf	Es	Fm	Md	No
						2			2	2	2	2	2
3			3	3	3	3	3	3	3	3	3	3	3
	4	4	4	4	4	4	4	4	4				
		5	5	5	5	5							
			6	6	6	6							
				7	7								

[a] Bold number signifies most stable state.

[2]   See, e.g., L. P. Varga *et al.*, *J. Inorg. Nucl. Chem.*, 1973, **35**, 2775, 2787.

TABLE 24-5
Stereochemistry of Actinides

Oxidation state	Coordination number	Geometry[a]	Examples
+3	5	*Tbp*	AcF_3, $BaUF_6$(LaF$_3$-type)
	6	Octahedral	Macac$_3$, $[M(H_2O)_6]^{3+}$
	8	Bicapped trigonal prism	$PuBr_3$, $[AmCl_2(H_2O)_6]^+$
	9	Tricapped trigonal prism	UCl_3, $AmCl_3$ (also La-GdCl$_3$)
+4	4	Distorted	$U(NPh_2)_4$
	5	Dist. *tbp*	$U_2(NEt_2)_8$
	6	Octahedral	UCl_6^{2-}, $UCl_4(hmpa)_2$
	8	Cubic(O_h)	$(Et_4N)_4[U(NCS)_8]$
		Dodecahedral	$[Th ox_4]^{4-}$, $Th(S_2CNEt_2)_4$
		Fluorite struct.	ThO_2, UO_2
		Square antiprism	$ThI_4(s)$, $Uacac_4$, $(NH_4)_2UF_6$, $Cs_4U(NCS)_8$
	9	Capped square antiprism	$Th(trop)_4 \cdot H_2O$
	10	Bicapped square antiprism	$K_4Th ox_4 \cdot 4H_2O$
	10	?	$M(trop)_5^-$ (M = Th or U)
	12	Irreg. icosahedral	$[Th(NO_3)_6]^{2-}$
	14	Complex	$U(BH_4)_4$
+5	6	Octahedral	UF_6^-, α-UF$_5$ (infinite chain)
	7		β-UF$_5$
	8	Cubic(O_h)	Na_3MF_8(M = Pa, U, Np)
	9	Complex	PaF_7^{2-} in K_2PaF_7
+6	6	Octahedral	UF_6, Li_4UO_5 (dist.), UCl_6
	6–8	See text	MO_2^{2+} complexes
	8	?	$M_2^I UF_8$

[a] For detailed discussion of crystal structures, many of which are most complicated, see A. F. Wells, *Structural Inorganic Chemistry,* 4th ed., Oxford University Press, 1975.

the closeness in the energy levels of electrons in the valence shells, and the mutual overlap of orbitals of comparable size in these heavy atoms, several equally valid descriptions can be chosen in a particular case. In such circumstances, the orbitals actually used must be some mixture of all the possible limiting sets, and it is not justified to treat the bonding in terms of any single set, except as a convenient first approximation.

Examples of the stereochemistry of actinide compounds and complexes are given in Table 24-5. For the +3 oxidation state, where the resemblance to the lanthanides is distinct, octahedral coordination is often found, but higher coordination numbers (e.g., 9 in UCl_3) are also common. Eight-coordination is especially a characteristic of the +4 oxidation state. An example here is Th acac$_4$, which is isomorphous with the uranium analogue and has a structure based on a slightly distorted square antiprism. This structure is that predicted on purely electrostatic grounds, and the volatility of the compound is no criterion of covalent bonding but only a reflection of the almost spherical nature of the molecules and valence saturation of the outer atoms. Well-defined ten-coordinate anions $M(tropolonate)_5^-$ have been prepared for Th^{IV} and U^{IV}.

GENERAL CHEMISTRY OF THE ACTINIDES

Given the close similarities in preparations and properties of actinide compounds in a given oxidation state, it is convenient to discuss some general features and to follow this by additional descriptions for the separate elements. Methods of chemical separations of the elements are also discussed.

24-6. The Metals

The metals may all be prepared by a method applicable on either a 10^{-6} g scale, as in the first preparation of Cm, or on a multikilogram scale. This is the reduction of one of the anhydrous fluorides MF_3 or MF_4 with the vapor of Li, Mg, Ca, or Ba at 1100 to 1400°; chlorides or oxides can be reduced similarly. On large scales (e.g., for U) Mg or Ca is normally used.

There are other procedures: thus very pure Th is made from ThI_4 by thermal decomposition (de Boer process). Electrolytic methods are not commonly used, but Th can be obtained from a melt of ThF_4, KCN, and NaCl. Americium has been obtained by a method depending on the volatility, which is greater than that of the other actinides:

$$2La + Am_2O_3 \xrightarrow{1200°} 2Am \uparrow + La_2O_3$$

Curium has also been made on a gram scale by extraction from a melt of $MgCl_2$, MgF_2, and CmO_2 with molten Zn–Mg alloy, the excess of which is then distilled off; uranium can also be obtained as an amalgam, from which it can be recovered, by action of Na/Hg on uranyl acetate.

The melting points of the metals are given in Table 24-2.

Actinium. This silvery-white metal glows in the dark owing to its radioactivity, which also contributes to its disintegration and high reactivity.

Thorium. The metal is white but tarnishes in air. It can be readily machined and forged. It is highly electropositive, resembling the lanthanide metals, and is pyrophoric when finely divided. It is attacked by boiling water, by oxygen at 250°, and by N_2 at 800°. Dilute HF, HNO_3, and H_2SO_4, and concentrated HCl or H_3PO_4, attack thorium only slowly, and concentrated nitric acid makes it passive. The attack of hot $12M$ hydrochloric acid on thorium gives a black residue that appears to be a complex hydride approximating to $ThH(O)(Cl)(H_2O)$ that is very pyrophoric and appears to give ThO on heating in vacuum.[3]

Protactinium. This relatively unreactive, silvery, and malleable metal does react with HCl, Br_2, etc., and with H_2 on heating.[4]

Uranium. For its use in nuclear reactors, uranium must be exceedingly pure and free from elements such as B or Cd, which have high absorption capacities for thermal neutrons.

Uranium is one of the densest metals (19.07 g cm^{-3} at 25°) and has three crys-

3   R. J. Ackerman and E. G. Rauh, *J. Inorg. Nucl. Chem.,* 1973, **35**, 3787.
4   D. Brown *et al., J.C.S. Dalton,* **1977,** 2291.

talline modifications. It forms a wide range of intermetallic compounds (U_6Mn, U_6Ni, USn_3, etc.), but owing to its unique crystal structures, it cannot form extensive ranges of solid solutions. Uranium is chemically reactive and combines directly with most elements. In air the surface is rapidly converted into a yellow and subsequently a black nonprotective film. Powdered uranium is frequently pyrophoric. The reaction with water is complex; boiling water forms UO_2 and hydrogen, the latter reacting with the metal to form a hydride, which causes disintegration. Uranium dissolves rapidly in hydrochloric acid (a black residue often remains; cf. Th) and in nitric acid, but slowly in sulfuric, phosphoric, or hydrofluoric acid. It is unaffected by alkalis. An important reaction of uranium is that with hydrogen, forming the hydride (see p. 251), which is a useful starting material for the synthesis of uranium compounds.

Neptunium.   This silvery metal resembles U but is denser (20.45 g cm^{-3}) and has three modifications.

Plutonium.   The metal is again similar to U chemically; it is pyrophoric and must be handled with extreme care owing to the health hazard. Also, above a certain critical size the pure metal can initiate a nuclear explosion. The metal is unique in having at least six allotropic forms below its melting point, each with a different density, coefficient of expansion, and resistivity; curiously, if the phase expands on heating, the resistance decreases. Plutonium forms numerous alloys.

Americium.   The first actinide metal to resemble a lanthanide, Am melts higher and has a much lower density (13.7 g cm^{-3}) than its predecessors. It is more electropositive than Pu, being comparable to a light lanthanide.

Curium.   The metal resembles its analogue Gd in having a relatively high melting point and in its magnetic properties; it glows red in the dark.

Californium.   The metal radius, like that of Eu is high, suggesting that divalency is involved.

24-7.   The +2 Oxidation State

The +2 oxidation state is an unusual one for the actinides and is at present confined to Am (the $5f$ analogue of Eu) and the Cf-No group. This can be associated with the greater energy of promotion $5f \rightarrow 6d$ than of promotion $4f \rightarrow 5d$ in the lanthanides; thus the +2 state is more stable at the end of the series. Solid compounds of Am and Cf are known, and the solution potentials are given in Table 24-6.

The Md^{2+} ion is especially stable, more so than even Eu^{2+} or Yb^{2+}. Md^{3+} can

TABLE 24-6
Actinide Potentials, $M^{3+} + e = M^{2+}$

Element	E^0 (V)[a]	Element	E^0 (V)[a]
Am	−2.3	Es	−1.3
Cm	−4.4	Fm	1.0
Bk	−2.8	Md	−0.15
Cf	−1.6	No	1.45

[a] L. J. Nugent, *J. Inorg. Nucl. Chem.*, 1975, **37**, 1767.

be reduced by Zn/Hg and can be coprecipitated with sulfates. For nobelium, No^{2+} is the normal valence $(5f^{14})$, and although ^{255}No has a half-life on only 223 sec, ion-exchange study allows an estimate of 1.0 Å for the ionic radius.[5] As expected, the properties of the +2 ions, where known, are similar to those of the +2 lanthanides or Ba^{2+}.

24-8.   The +3 Oxidation State

The common state for all the actinides except Th and Pa, is 3+, and it is the normal state for Ac, Am, and trans-Am elements.

The general chemistry closely resembles that of the lanthanides. The halides MX_3 may be readily prepared and are easily hydrolyzed to MOX. The oxides M_2O_3 are known only for Ac, Pu, and heavier elements. In aqueous solution there are M^{3+} ions, and insoluble hydrated fluorides and oxalates can be precipitated. Isomorphism of crystalline solids is common.

Of the +3 ions, U^{3+} is the most readily oxidized, even by air or more slowly by water.

Since the ionic sizes are comparable for both series, the formation of complex ions and their stability constants are similar. Thus it is difficult to separate actinide from lanthanide elements, though it can be done, as described below (p. 1043), by ion-exchange or solvent-extraction procedures.

24-9.   The +4 Oxidation State

The principal oxidation state for Th is +4; Pa^{IV}, U^{IV}, and Pu^{IV} are reasonably stable in solution; Am^{IV} and Cm^{IV} are much more easily reduced and exist only as complex ions in concentrated fluoride solution of low acidity or, for Am, also in phosphate solutions. The elements after Cf cannot be oxidized. The general chemistry is lanthanidelike, with sparingly soluble hydroxides and hydrated fluorides and phosphates. Other points of importance are: (*a*) the dioxides MO_2 from Th to Cf all have the fluorite lattice, (*b*) the tetrafluorides MF_4 are isostructural with lanthanide tetrafluorides, (*c*) the chlorides and bromides are known only for Th, Pa, U, and Np, presumably owing to the inability of the halogen to oxidize the heavier metals, and for iodides only those of Th, Pa, and U exist, (*d*) oxohalides MOX_2 can be made for Th–Np, e.g., by the reaction:

$$3MX_4 + Sb_2O_3 \xrightarrow{\text{heat}} 3MOX_2 + 2SbX_3 \uparrow 450°$$

(*e*) hydrolysis, complexation, and disproportionation are important in aqueous solution, as discussed below.

24-10.   The +5 Oxidation State

The +5 state is the normal oxidation state for Pa, and there is quite a close resemblance to the chemistry of Nb and Ta.

⁵   R. J. Silva *et al.*, *Inorg. Chem.*, 1974, **13**, 2233.

For the other elements comparatively few solid compounds have been made. The halides are known only for Pa and U. Salts of fluoro anions such as MF_6^-, MF_7^{2-}, and MF_8^{3-} are known for Pa-Pu, although the Np and Pu compounds can be made only by solid state reactions. Oxochlorides $MOCl_3$ are known for Pa, U, and possibly Np.

An important difference from the Nb and Ta group is the formation of the dioxo ion, MO_2^+, which is of great importance for U, Np, Pu, and Am chemistry. These ions are discussed further below; we note here that their stability in aqueous solution is determined by the ease of disproportionation, e.g.:

$$2UO_2^+ + 4H^+ = U^{4+} + UO_2^{2+} + 2H_2O \qquad K = 1.7 \times 10^6$$

24-11. The +6 Oxidation State

In simple compounds the +6 state occurs only in the hexafluorides (MF_6) of U, Np, and Pu, and UCl_6. There are oxohalides (e.g., UOF_4) and various compounds such as $U(OR)_6$. Uranium also forms UOF_5^- and $UOCl_5^-$ ions.[6]

The principal chemistry of the +6 state in solids and in solutions is that of the *dioxo ions* MO_2^{2+} of U, Np, Pu, and Am. These unique ions are discussed below.

24-12. The +7 Oxidation State

The +7 state is known only for Np and Pu, but Am^{VII} has been claimed. The action of ozone on LiOH solutions of neptunates (cf. uranates, p. 1029) gives oxidized solutions from which addition of $[Co(NH_3)_6]^{3+}$ yields the complex Li-$[Co(NH_3)_6][Np_2O_8(OH)_2]\cdot2H_2O$.[7] The anion has a chain structure with NpO_6 octahedra sharing corners. In solution the ion is probably $[NpO_4(OH)_2]^{3-}$. The presence of Np^{VII} has been demonstrated by Mössbauer studies.[8]

The interaction of Li_2O and PuO_2 in oxygen at 430° gives Li_5PuO_6, which dissolves in water to a green unstable solution.

24-13. The Dioxo Ions, MO_2^+ and MO_2^{2+}

The dioxo ions are remarkably stable with respect to the strength of the M—O bond. Unlike some other oxo ions they can persist through a variety of chemical changes, and they behave like cations whose properties are intermediate between those of M^+ or M^{2+} ions of similar size but greater charge. The MO_2 group even appears more or less as an "yl" group in certain oxide and oxo ion structures; furthermore, whereas MoO_2F_2 or WO_2F_2 are molecular halides, in UO_2F_2 there is a linear O—U—O group with F bridges. The stability of UO_2^{2+} and PuO_2^{2+} ions in aqueous solution is shown by the very long half-life for exchange with $H_2^{18}O$ ($>10^4$ hr);

6   K. W. Bagnall *et al.*, *J.C.S. Dalton*, **1973**, 1975.
7   J. H. Burns *et al.*, *Inorg. Chem.*, 1973, **12**, 466.
8   K. Frohlich *et al.*, *J.C.S. Dalton*, **1972**, 991.

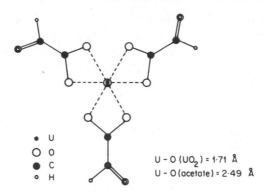

● U
○ O
● C
○ H

U – O (UO_2) = 1·71 Å
U – O (acetate) = 2·49 Å

Fig. 24-3. Structure of the anion in $Na[UO_2(OCOCH_3)_3.]$

the exchange can be catalyzed by the presence of reduced states or, for PuO_2^{2+}, by self-reduction due to radiation effects.

In both crystalline compounds and in solutions the O=M=O group is linear. The ions form complexes with negative ions and neutral molecules. Crystallographic data show that four, five, or six ligand atoms can lie in the equatorial plane of the O—M—O group; the ligand atoms may or may not be entirely coplanar depending on the circumstances. Planar 5- and 6-coordination in the equatorial plane is commonest and appears to give geometry more stable than the puckered hexagonal configurations. Planar 5-coordination best allows rationalization of a number of hydroxide and other structures, as well as the behavior of polynuclear uranyl ions in hydrolyzed solutions. An example is the structure of the anion in the complex salt sodium uranyl acetate shown in Figure 24-3; the carboxylate groups are bidentate and equivalent. Similar structures have been found in other species such as $UO_2(NO_3)_2(H_2O)_2$, $UO_2(NO_3)_2[OP(OEt_3)]_2$, $UO_2(CO_2Me)_2·2H_2O$, $[R_4N]_4[UO_2(CO_3)_3]$, and $[UO_2(OH_2)_5](ClO_4)_2·2H_2O$.[9] The 18-crown-6 ether complex $UO_2(NO_3)_2(H_2O)_2·(2H_2O)(18\text{-crown-6})$ is unusual in that the UO_2^{2+} is *not* within the ether, but there are neutral $UO_2(NO_3)_2(H_2O)_2$ units and separate crown ether molecules linked by H bonding through intermediary water molecules.[10]

For such heavy atoms there are difficulties in accurately locating oxygen atoms and assessing M—O distances, but it is certain that the M—O distances are *not constant*. Thus for UO_2^{2+} the range appears to be *ca.* 1.6–2.0 Å. Accordingly there has been extensive use of infrared data (ν, cm^{-1}) for correlating bond lengths (R, Å), and an equation

$$R_{U-O} = 81.2\nu^{-2/3} + 0.85$$

has been developed using data for salts and uranates.[11]

9   N. W. Alcock and S. Esperås, *J.C.S. Dalton*, **1977**, 893; D. C. Dewan *et al. J.C.S. Dalton*, **1975**, 2171; J. Howatson *et al.*, *J. Inorg. Nucl. Chem.*, 1975, **37**, 1933.
10 P. G. Eller and R. A. Penneman, *Inorg. Chem.*, 1976, **15**, 2439; M. E. Harmon *et al.*, *J.C.S. Chem. Comm.*, **1976**, 396; G. Bombieri *et al.*, *Inorg. Chem. Acta*, 1976, **16**, L23.
11 B. W. Veall *et al.*, *Phys. Rev. B*, 1975, **12**, 5651; see also S. Siegel, *J. Inorg. Nucl. Chem.*, 1978, **40**, 275.

For MO_2^{2+} the bond strengths, as well as chemical stabilities toward reduction, decrease in the order $U > Np > Pu > Am$. It also appears that the force constants for U—O bonds are high, indicating a multiplicity greater than 2. Appropriate d and f orbitals can be combined into molecular orbitals to give one σ plus two π bonds. The MOs are filled at UO_2^{2+}, and succeeding electrons are fed into non-bonding orbitals. The MO scheme allows detailed interpretation of spectroscopic and magnetic data for the oxo ions. It also provides an explanation of the U-Am stability sequence and the nonexistence of PaO_2^{2+}. The latter is connected with the fact that for Pa the $6d$ is higher than the $5f$ level, whereas for $U(5f^3 6d^1 7s^2)$ it is the reverse, so that for Pa, the $5f\sigma$-$2p\sigma$ metal-oxygen overlap is poor. The instability of UO_2^+ is probably also connected with the sensitivity of the energy of the $5f$ electrons to total charge, thus critically affecting the U—O overlap.

24-14. Actinide Ions in Aqueous Solution[12]

The formal reduction potentials of actinide ions in aqueous solution are given in Tables 24-2 and 24-7, from which it is clear that the electropositive character of the metals increases with increasing Z and that the stability of the higher oxidation states decreases. A comparison of various actinide ions is given in Table 24-8. It must be noted also that for comparatively short-lived isotopes decaying by α-emission or spontaneous fission, heating and chemical effects due to the high level of radioactivity occur in both solids and aqueous solutions (e.g., for ^{238}Pu, ^{241}Am, and ^{242}Cm the heat output is calculated as 0.5, 0.1, and 122 W g^{-1}, respectively). Radiation induced decomposition of water leads to H and OH radicals, H_2O_2 production, etc., and in solution higher oxidation states such as Pu^V, Pu^{VI}, and Am^{IV-VI} are reduced. Chemical reactions observable with a short-lived isotope, e.g., ^{242}Cm (163 days), may differ when a longer-lived isotope is used; thus Cm^{IV} can be observed only when ^{244}Cm (17.6 years) is employed.

The possibility of several cationic species introduces complexity into the aqueous chemistries, particularly of U, Np, Pu, and Am. Thus all four oxidation states of Pu can coexist in appreciable concentrations in a solution. The solution chemistries and the oxidation-reduction potentials are further complicated by the formation in the presence of ions other than perchlorate, of cationic, neutral or anionic species. Furthermore, even in solutions of low pH, hydrolysis and the formation of polymeric ions occurs. Third, there is the additional complication of disproportionation of certain ions, which is particularly dependent on the pH.

Since extrapolation to infinite dilution is impossible for most of the actinide ions, owing to hydrolysis—for example, Pu^{4+} cannot exist in solution below $0.05M$ in acid—sometimes only approximate oxidation potentials can be given. The potentials are sensitive to the anions and other conditions.

The actinides have a far greater tendency to complex formation than the lanthanides. Thus there are extensive series of halogeno complexes, and complex ions

[12] A. D. Jones and G. R. Choppin, *Actinide Chem. Rev.,* 1969, **1,** 311 (an extensive review of complex ions); T. W. Newton, *The Kinetics of Oxidation—Reduction Reactions of U, Np, Pu and Am in Aqueous Solution,* ERDA Technical Information Center, Oak Ridge, Tenn., 1975.

TABLE 24-7
(See also Table 24-2)
Formal Reduction Potentials (V) of the Actinides for 1 M Perchloric Acid Solutions at 25°
(brackets indicate estimate)

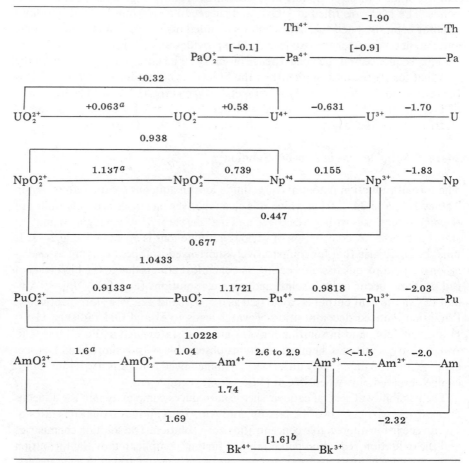

Notes:

1. $PaO_2^+ + 4H^+ = Pa + 2H_2O$, $E = [-1.0]$.

2. Couples involving oxygen transfer, for example, $UO_2^{2+} + 4H^+ + 2e = U^{4+} + 2H_2O$ are *irreversible* and are of course hydrogen ion dependent. Couples such as PuO_2^{2+}/PuO_2^+ *are* reversible.

[a] The true E^0 values for U, Np, Pu, and Am are 0.163, 1.236, 1.013, and 1.7 V, respectively (J. R. Brand and J. W. Cobble, *Inorg. Chem.,* 1970, **9**, 912).

[b] By direct measurement in 0.1 M H_2SO_4, $E = 1.43$ V (R. C. Propst and M. L. Hyder, *J. Inorg. Nucl. Chem.,* 1970, **32**, 2205).

are given with most oxo anions such as NO_3^-, SO_4^{2-}, ox^{2-}, CO_3^{2-}, and phosphate. A vast amount of data exists on complex ion formation in solution, since this has been of primary importance in connection with solvent extraction, ion-exchange behavior, and precipitation reactions involved in the technology of actinide elements. The general tendency to complex ion formation decreases in the direction controlled

TABLE 24-8
The Principal Actinide Ions[a] in Aqueous Solution

Ion	Color[b]	Preparation	Stability
U^{3+}	Red-brown	Na or Zn/Hg on UO_2^{2+}	Slowly oxidized by H_2O, rapidly by air to U^{4+}
Np^{3+}	Purplish	$H_2(Pt)$ or electrolytic	Stable in water; oxidized by air to Np^{4+}
Pu^{3+}	Blue-violet	SO_2, NH_2OH on higher states	Stable to water and air; easily oxidized to Pu^{4+}
Am^{3+}	Pink	I^-, SO_2, etc., on higher states	Stable; difficult to oxidize
Cm^{3+}	Pale yellow		Stable; not oxidized chemically
U^{4+}	Green	Air or O_2 on U^{3+}	Stable; slowly oxidized by air to UO_2^{2+}
Np^{4+}	Yellow-green	SO_2 on NpO_2^+ in H_2SO_4	Stable; slowly oxidized by air to NpO_2^{2+}
Pu^{4+}	Tan	SO_2 or NO_2^- on PuO_2^{2+}	Stable in $6M$ acid; disproportionates in low acid $\rightarrow Pu^3 + PuO_2^{2+}$
Am^{4+} [c]	Pink-red	$Am(OH)_4$ in $15M$ NH_4F	Stable in $15M$ NH_4F; reduced by I^-
Cm^{4+} [c]	Pale yellow	CmF_4 in $15M$ CsF	Stable only 1 hr at $25°$
UO_2^+	?	Transient species	Stability greatest pH 2–4; disproportionates to U^{4+} and UO_2
NpO_2^+	Green	Np^{4+} and hot HNO_3	Stable; disproportionates only in strong acid
PuO_2^+	?	Hydroxylamine on PuO_2^{2+}	Always disproportionates; most stable at low acidity
AmO_2^+	Pale yellow	Am^{3+} with OCl^-, cold $S_2O_8^{2-}$	Disproportionates in strong acid; reduced (2% per hour) by products of own α-radiation
UO_2^{2+}	Yellow	Oxidize U^{4+} with HNO_3, etc.	Very stable; difficult to reduce
NpO_2^{2+}	Pink	⎫	Stable; easily reduced
PuO_2^{2+}	Yellow-pink	⎬ Oxidize lower states with Ce^{4+}, MnO_4^-. O_3 BrO_3^-, etc.	Stable; fairly easy to reduce
AmO_2^{2+}	Rum	⎭	Reduced (4% per hour) by products from own α-radiation

[a] Ac^{3+}, Th^{4+}, Cm^{3+}, and ions of Pa are colorless.

by factors such as ionic size and charge, so that the order is generally $M^{4+} > MO_2^{2+} > M^{3+} > MO_2^+$. For anions the order of complexing ability is generally: uninegative ions, $F^- > NO_3^- > Cl^- > ClO_4^-$; binegative ions, $CO_3^{2-} > ox^{2-} > SO_4^{2-}$.

24-15. Complexes[12,13]

Much of the complex chemistry of the actinides is in aqueous solution, as noted above. However numerous neutral and ionic complexes can be isolated, and there is an extensive chemistry of organometallic and related compounds as discussed below.

[13] U. Casellato, et al., Coord. Chem. Rev., 1978, 26, 85, carboxylic acid complexes); Inorg. Chim. Acta, 1976, 18, 77 (chelate complexes); S. Ramamoorthy and M. Santappa, J. Sci. In. Res., 1972, 31, 69 (coordination chemistry of U^{IV} and U^{VI}); A. E. Comyns, Chem. Rev., 1960, 60, 115 (coordination chemistry of actinides); T. J. Marks and J. R. Kolb, Chem. Rev., 1977, 77, 263 (tetrahydroborate complexes).

Most of the complexes are of the following types:

1.   Adducts of halides with mainly oxygen donor ligands. These may be simple, e.g., MCl_4L_2, or more complex, as $[UCl_2(DMSO)_6]^{2+}$.

2.   Salts of halogeno or oxohalogeno anions.

3.   Chelate complexes of oxo ligands such as oxalate, β-diketonato, tropolonato, or Schiff base ligands. Examples are $Th(ox)_4DMSO$, $UO_2(sal_2en)py$, $Np(trop)_4$.

24-16.   Organometallic Chemistry[14]

Like that of the lanthanides, the organometallic chemistry of the actinides is largely that of compounds containing η^5-C_5H_5, η^8-C_8H_8, and σ-bonded alkyl or aryl groups, although an unstable allyl $U(allyl)_4$ is known. Because of f-orbital participation, the bonding of these groups is much more covalent than is the case for lanthanides. For example, Cp_3UCl does not react with $FeCl_2$ to give ferrocene, whereas the lanthanide cyclopentadienides do so and may be considered as ionic compounds.

Cyclopentadienyls.   The first compound to be made was Cp_3UCl. Many similar derivatives have now been made with other groups replacing Cl, e.g., by interactions with Grignard or lithium reagents:

$$Cp_3UCl + RLi = Cp_3UR$$

where R can be CH_3, C_3H_7, allyl, C_6H_5, etc., as well as BH_4^-; there are also bridged hydrides, e.g., $(Me_5C_5)_2HTh(\mu\text{-}H)_2ThH(Me_5C_5)_2$. All these compounds are air sensitive, though usually thermally stable.[15a] Some compounds show unusual nmr spectra; the fluoride Cp_3UF is monomeric in the vapor but dimeric in benzene due to some H-bonding to the fluorine atom.[15b]

The deep red compound Cp_4U does not, however, have one σ-bonded ring as might have been expected, but has four identical η^5 rings[16] (Fig. 24-4a).

Cyclooctatetraenyls.   Action of $Li_2C_8H_8$ on MCl_4 in tetrahydrofuran gives the neutral sandwich molecules $M(COT)_2$ (Fig. 24-4b). These compounds are thermally stable and sublimable. Compounds can also be obtained by direct interaction of finely divided metal (e.g., from UH_3) and cyclooctatetraene. Reduction with potassium amalgam gives M^{III} species, e.g., $K[Pu(COT)_2]\cdot THF$.[17]

Alkyls and Aryls. σ-Alkyls $(C_5H_5)_3UR$ have been noted above, but so far the only compounds with only σ bonds to the metal are the alkylate anions of U^{IV} and U^V of stoichiometry $Li_2UR_6\cdot8Et_2O$ and $Li_2UR_8\cdot3$ dioxan (R = CH_2SiMe_3, CH_3,

[14]   S. A. Cotton, in *Organomet. Chem. Rev.,* Vol. 3, Elsevier, 1977; T. J. Marks, *Acc. Chem. Res.,* 1976, **9**, 223; M. Tsutsui *et al., Acc. Chem. Res.,* 1976, **9**, 217; K. N. Raymond *et al., Struct. Bonding,* 1976, **25**, 23; T. J. Marks and R. D. Fischer, Eds., *Organometallics of the f-Elements,* Reidel, 1979; E. Cernia and A. Mazzei, *Inorg. Chim. Acta,* 1974, **10**, 239 (chemistry of U^{IV}).

[15a]   See, e.g., T. J. Marks *et al., J. Am. Chem. Soc.,* 1978, **100**, 3939; 1979, **101**, 2656, 5075; R. D. Fischer *et al., Z. Naturforsch.,* 1978, **33B**, 1393.

[15b]   R. R. Ryan *et al., J. Am. Chem. Soc.,* 1977, **99**, 4258.

[16]   J. H. Burns, *J. Am. Chem. Soc.,* 1973, **95**, 3815.

[17]   D. G. Karraker and J. A. Stone, *J. Am. Chem. Soc.,* 1974, **96**, 6885.

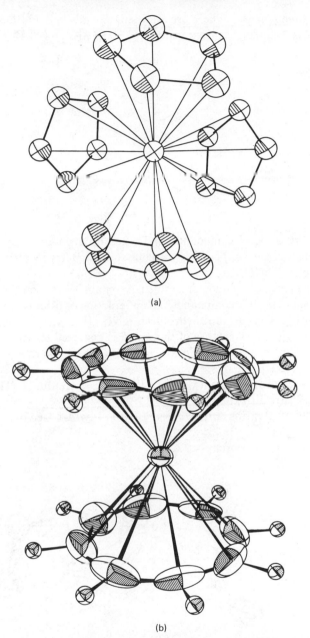

(a)

(b)

Fig. 24-4. Structures of (a) Cp$_4$U viewed down the S_4 molecular axis and (b) M(C$_8$H$_8$)$_2$ (M = Th, U). [Reproduced by permission from K. Raymond *et al., Struct. Bonding,* 1976, **25,** 23.]

etc.).[18] For thorium, however, which behaves much more like Zr or Hf, the tetrahedral benzyl Th(CH$_2$Ph)$_4$ is known.

[18]   E. R. Sigurdson and G. Wilkinson, *J.C.S. Dalton,* **1976,** 812.

Trivalent Compounds.   There are a few compounds in oxidation states lower than IV. One is the unique benzene complex $C_6H_6U(AlCl_4)_3$ (24-I), which contains U^{III}.[19]

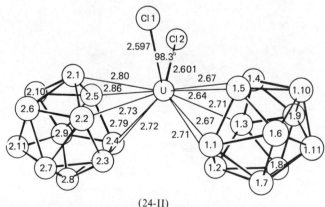

(24-I)

For thorium, a violet compound Cp_3Th has been obtained by reduction of Cp_3ThCl, and a green compound of the same stoichiometry by photolytic β-elimination from $Cp_3ThCHMe_2$.[20]

Related Compounds.   Although they do not contain M—C bonds it is convenient to note here some similar compounds. The interaction of lithium 1,2-dicarbollide (p. 320) with UCl_4 in tetrahydrofuran gives $[UCl_2(B_9C_2H_{11})_2]^{2-}$ as its $[Li(THF)_4]^+$ salt. The structure of this anion is distorted tetrahedral,[21] similar to cyclopentadienyls of the type Cp_2MCl_2 (24-II). The anion reacts with cyclooctatetraene to give $U(C_8H_8)_2$.

Polypyrazolylborate complexes (p. 141) of uranium, $U(HBpz_3)_4$ and $UCl_2(HBpz_3)_2$, are also known.[22]

Finally, there are various *alcoxides* and *dialkylamides* (see p. 1033)

(24-II)

[19]   M. Cesari *et al., Inorg. Chim. Acta*, 1971, **5**, 439.
[20]   T. J. Marks *et al., J. Am. Chem. Soc.,* 1977, **99**, 3877.
[21]   K. N. Raymond *et al., J. Am. Chem. Soc.,* 1977, **99**, 1769.
[22]   K. W. Bagnall *et al., J.C.S. Dalton,* **1975**, 140.

ACTINIUM[23]

24-17.   The Element and Its Compounds

Actinium was originally isolated from uranium minerals in which it occurs in traces, but it is now made on a milligram scale by neutron capture in radium (Table 24-3). The actinium +3 ion is separated from the excess of radium and isotopes of Th, Po, Bi, and Pb formed simultaneously by ion-exchange elution or by solvent extraction with thenoyltrifluoroacetone.

The general chemistry of Ac^{3+} in both solid compounds and solution, where known, is very similar to that of lanthanum, as would be expected from the similarity in position in the Periodic Table and in radii (Ac^{3+}, 1.10; La^{3+}, 1.06 Å) together with the noble gas structure of the ion. Thus actinium is a true member of Group III, the only difference from lanthanum being in the expected increased basicity. The increased basic character is shown by the stronger absorption of the hydrated ion on cation-exchange resins, the poorer extraction of the ion from concentrated nitric acid solutions by tributyl phosphate, and the hydrolysis of the trihalides with water vapor at $\sim 1000°$ to the oxohalides AcOX; the lanthanum halides are hydrolyzed to oxide by water vapor at $1000°$.

The crystal structures of actinium compounds, where they have been studied, for example, in AcH_3, AcF_3, Ac_2S_3, and AcOCl, are the same as those of the analogous lanthanum compounds.

The study of even milligram amounts of actinium is difficult owing to the intense γ-radiation of its decay products that rapidly build up in the initially pure material.

THORIUM[24]

Thorium is widely distributed in Nature, and there are large deposits of the principal mineral *monazite,* a complex phosphate containing uranium, cerium, and other lanthanides. The extraction of thorium from monazite is complicated, the main problems being the destruction of the resistant sand and the separation of thorium from cerium and phosphate. One method involves a digestion with sodium hydroxide; the insoluble hydroxides are removed and dissolved in hydrochloric acid. When the pH of the solution is adjusted to 5.8, all the thorium and uranium, together with about 3% of the lanthanides, are precipitated as hydroxides. The thorium is recovered by tributyl phosphate extraction from $>6M$ hydrochloric acid solution or by extraction with isobutyl methyl or other ketone from nitric acid so-

[23]   C. Keller, *Chem. Ztg.,* 1977, **101**, 500.
[24]   J. F. Smith *et al., Thorium, Preparation and Properties,* Iowa State University Press, 1975 (largely metallurgy).

lutions in presence of an excess of a salt such as aluminum nitrate as "salting-out" agent.

24-18. Binary Compounds of Thorium

Some typical thorium compounds are listed in Table 24-9.

Oxide and Hydroxide. The only oxide, ThO_2, is obtained by ignition of oxo acid salts or of the hydroxide. The latter is insoluble in an excess of alkali hydroxides, although it is readily peptized by heating it with Th^{4+} or Fe^{3+} ions or dilute acids; the colloid exists as fibers that are coiled into spheres in concentrated sols but uncoil on dilution. Addition of hydrogen peroxide to Th^{4+} salts gives a highly insoluble white precipitate of variable composition, which contains an excess of anions in addition to peroxide; the composition is approximately $Th(O_2)_{3.2}X_{0.5}^-O_{0.15}$, but it is usually referred to as thorium peroxide.

Halides. The anhydrous halides may be prepared by dry reactions, e.g.,

$$ThO_2 + 4HF(g) \xrightarrow{600°} ThF_4 + 2H_2O$$

$$ThO_2 + CCl_4 \xrightarrow{600°} ThCl_4 + CO_2$$

All are white crystalline solids, and except for ThF_4, all can be sublimed in a vacuum at 500 to 600°. The hydrated tetrafluoride is precipitated by aqueous hydrofluoric acid from Th^{4+} solutions; it can be dehydrated by heat in an atmosphere of hydrogen fluoride. The other halides are soluble in acid and are partially hydrolyzed by water. They behave as Lewis acids and form complexes with ammonia, amines, ketones, and donor molecules generally.

The *oxohalides* $ThOX_2$ can be obtained by interactions of ThO_2 and ThX_4 at 600°; they appear to have —Th—O—Th—O chains.

TABLE 24-9
Some Thorium Compounds

Compound	Form	M. p. (°C)	Properties
ThO_2	White, crystalline; fluorite structure	3220	Stable, refractory, soluble in HF + HNO_3
ThN	Refractory solid	2500	Slowly hydrolyzed by water
ThS_2	Purple solid	1905	Metal like; soluble in acids
$ThCl_4$	Tetragonal white crystals	770	Soluble in and hydrolyzed by H_2O, Lewis acid
$Th(NO_3)_4 \cdot 5H_2O$	White crystals, orthorhombic		Very soluble in H_2O, alcohols, ketones, ethers
$Th(IO_3)_4$	White crystals		Precipitated from 50% HNO_3: very insoluble
$Th(C_5H_7O_2)_4$	White crystals	171	Sublimes in a vacuum 160°
$Th(BH_4)_4$	White crystals	204	Sublimes in a vacuum about 40°
$Th(C_2O_4)_2 \cdot 6H_2O$	White crystals		Precipitated from up to $2M$ HNO_3

Other Binary Compounds. Various borides, sulfides, carbides, nitrides, etc., have been obtained by direct interaction of the elements at elevated temperatures. Like other actinide and lanthanide metals, thorium also reacts at elevated temperatures with hydrogen. Products with a range of compositions can be obtained, but two definite phases, ThH_2 and Th_4H_{15}, have been characterized.

24-19. Oxo Salts, Aqueous Solutions, and Complexes of Thorium

Thorium salts of strong mineral acids usually have varying amounts of water of crystallization. The most common salt and the usual starting material for preparation of other thorium compounds is the nitrate, $Th(NO_3)_4\cdot 5H_2O$.[25] This salt is very soluble in water as well as in alcohols, ketones, ethers, and esters. Various reagents give insoluble precipitates with thorium solutions, the most important being hydroxide, peroxide, fluoride, iodate, oxalate, and phosphate; the last four give precipitates even from strongly acid ($6M$) solutions and provide useful separations of thorium from elements other than those having +3 or +4 cations with similar properties.

The thorium ion Th^{4+} is more resistant to hydrolysis than other 4+ ions but undergoes extensive hydrolysis in aqueous solution at pH higher than ~3; the species formed are complex and dependent on the conditions of pH, nature of anions, concentration, etc. Binuclear species with hydroxo bridges have been characterized, and there are more complicated polymers.[26] In crystals of $Th(OH)_4$ or $Th(OH)_2CrO_4\cdot H_2O$ there are chainlike structures with $Th(OH)_2^{2+}$ as the repeating unit.

The high charge on Th^{4+} makes it susceptible to complex formation, and in solutions with anions other than perchlorate, complexed species, which may additionally be partially hydrolyzed and polymeric, are formed. Equilibrium constants for reactions such as the following have been measured:

$$Th^{4+} + nCl^- = ThCl_n^{4-n}$$
$$Th^{4+} + NO_3^- = Th(NO_3)^{3+}$$
$$Th^{4+} + 2HSO_4^- = Th(HSO_4)(SO_4)^+ + H^+$$

A number of salts of *complex anions* have been isolated; some of the more important are $K_4Th\ ox_4\cdot 4H_2O$ and $M^{II}[Th(NO_3)_6]\cdot 8H_2O$, where the NO_3 groups are bidentate,[27] $(NH_4)_4ThF_8$, and complexes of EDTA and related acids.

Neutral complexes are formed by 8-quinolinol and β-diketones; an example of the latter type is the tetrakis-(1,3-diphenylacetylacetonate), which has a square-antiprismatic structure.

The dithiocarbamates, thiocarbamates, and carbamates can be obtained by action of CS_2, COS, and CO_2, respectively, on $Th(NR_2)_4$.[28]

There are also *adducts* such as $ThCl_4phen_2$ and $Th(NO_3)_4\cdot 2OPPh_3$.

[25] B. Eriksson *et al., J.C.S. Chem. Comm.,* **1978**, 616.
[26] M. Magini *et al., Acta Chem. Scand.,* 1976, **30A**, 437.
[27] See, e.g., N. W. Alcock, *J.C.S. Dalton,* **1978**, 638.
[28] K. W. Bagnall and E. Yanir, *J. Inorg. Nucl. Chem.,* 1974, **36**, 777.

24-20. Lower Oxidation States

There is no evidence for the existence of any low oxidation state in solution, and little, if any, for its existence in the solid state.

A diamagnetic, gold-colored solid that is air sensitive and has low electrical resistance is obtained on heating ThI_4 and Th at 800°. Despite the stoichiometry ThI_2, this compound has a complex layer structure corresponding to the formulation $Th^{4+}(I^-)_2(e^-)_2$; thus it is similar to the lanthanide "lower" iodides (p. 1003). Sulfides of stoichiometry ThS and Th_2S_3 are doubtless similar, with Th^{4+} and S^{2-} ions and electrons in conduction bands.

PROTACTINIUM

Protactinium as ^{231}Pa occurs in pitchblende, but even the richest ores contain only about 1 part of Pa in 10^7. The isolation of protactinium from residues in the extraction of uranium from its minerals is difficult, as indeed is the study of protactinium chemistry generally, owing to the extreme tendency of the compounds to hydrolyze. In aqueous solution polymeric ionic species and colloidal particles are formed, and these are carried on precipitates and adsorbed on vessels; in solutions other than those containing appreciable amounts of mineral acids or complexing agents or ions such as F^-, the difficulties are almost insuperable.

Protactinium can be recovered from solutions 2 to $8M$ in nitric or hydrochloric acids by extraction with tributyl phosphate, isobutyl methyl ketone, or other organic solvents. The protactinium can be stripped from the solvent by aqueous acid fluoride solutions; the addition to these solutions of Al^{3+} ion or boric acid, which form stronger complexes with fluoride ion than protactinium, then allows reextraction and further purification of protactinium. Anion-exchange procedures involving mixtures of hydrofluoric and hydrochloric acid as elutants can also be used, since in these solutions protactinium fluoro or chloro anions are formed. About 125 g of protactinium was isolated in a twelve-stage process from 60 tons of accumulated sludges of uranium extraction from Congo ore by the United Kingdom Atomic Energy Authority; previously only about 1 g had ever been isolated. The method involved leaching of Pa from the residues with $4M$ HNO_3—$0.5M$ HF, followed by extraction of Pa^V from these solutions by 20% tributyl phosphate in kerosene. After collection of the Pa on a hydrous alumina precipitate, it was purified further by extraction from HCl—HF solution with dibutyl ketone, anion-exchange separation from HCl solution, and finally precipitation from dilute H_2SO_4 by H_2O_2.

In quantity, protactinium can be precipitated from $2M$ HNO_3 solutions in which the hydroxide has limited solubility.[29]

24-21. Protactinium(V) Compounds[30]

The *pentoxide* Pa_2O_5, obtained by ignition of other compounds in air, has a cubic

[29] D. Brown and B. Whittaker, *J. Less Common. Met.,* 1978, **61,** 161.

[30] D. Brown, *Adv. Inorg. Chem. Radiochem.,* 1969, **12,** 1 (general review of Pa); R. Muxart, R. Guillaumont, and G. Boussières, *Actinides Rev.,* 1969, **1,** 233 (solid Pa^V and Pa^{IV} compounds).

lattice; heating it *in vacuo* affords a black suboxide phase $PaO_{2.3}$, and finally PaO_2, but the real situation is more complex.

The hydrous pentoxide is similar to that of niobium, and spectra suggest that M—O—M bonds are present.

The *pentafluoride* is obtained as a white hygroscopic solid by fluorination of PaF_4; it is less volatile than the pentafluorides of V, Nb, and Ta but does sublime *in vacuo* above 500°. It is very soluble in $1M$ or stronger HF, but evaporation of aqueous solutions gives only mixtures. Action of HF gas on hydrous Pa_2O_5 at 60° gives $PaF_5 \cdot H_2O$; on heating this gives oxofluorides.

The *pentahalides* can be obtained[31] by action of $SOCl_2$, Br_2, or I_2 respectively on a "carbide" made by reduction of Pa_2O_5 with carbon at 1700°. $PaCl_5$ sublimes at 160° *in vacuo*; it is readily hydrolyzed to oxo halides and is soluble in THF. It has a structure quite unlike that of UCl_5 or $TaCl_5$ (p. 835) with infinite chains of irregular pentagonal-bipyramidal $PaCl_7$ groups sharing edges.

The oxochloroanion $[PaOCl_5]^{2-}$ has a Pa=O bond.[32]

Aqueous Chemistry.[33] The chemistry of Pa in solution is somewhat like that of Nb and Ta, but Pa is even less tractable because of hydrolysis and the formation of colloidal hydroxo species.

In perchloric acid, cationic species, probably $PaO(OH)_2^+$ and $PaO(OH)^{2+}$, exist; but when Cl^-, NO_3^- or other complexing anions are present, a whole range of species from cationic to anionic may exist, depending on the conditions. A number of anionic complexes are well established. The fluoro complexes, which resist hydrolysis, have been well studied, and salts of the ions PaF_6^-, PaF_7^{2-}, and PaF_8^{3-} have been isolated. In K_2PaF_7 there are PaF_9 groups linked by double, unsymmetrical bridges into infinite chains, and Na_3PaF_8, which is isostructural with the U and Np analogues, has the Pa at the center of a slightly distorted cube of F atoms; in $RbPaF_6$ the Pa atom is also eight-coordinate.

Salts of chloro and bromo anions of Pa^V can be made by interaction of PaX_5 with MX in solvents such as $SOCl_2$ or CH_3CN.

Other stable complex anions are those formed by $C_2O_4^{2-}$, SO_4^{2-}, citrate, and tartrate. Neutral complexes with β-diketones such as TTA and with alkyl phosphates may be extracted from aqueous solutions by benzene or CCl_4.

24-22. Protactinium(IV) Compounds

The fluoride PaF_4 is high melting and insoluble in HNO_3—HF solution; and this salt and $PaCl_4$ are isomorphous with corresponding Th and U halides. PaO_2 is isomorphous with the dioxides of Th—Am inclusive.

The lower oxidation state can also be obtained in aqueous solution by reduction of Pa^V solutions with Cr^{2+} or Zn amalgam, but the solutions are rapidly oxidized by air. The solutions of $PaCl_4$ in HCl, H_2SO_4, and $HClO_4$ have very similar absorption spectra, being similar to $Ce^{III}(4f^1)$. Furthermore, the absorption and esr

[31] D. Brown *et al.*, *J.C.S. Dalton*, **1976**, 1336.

[32] D. Brown *et al.*, *J.C.S. Dalton*, **1972**, 857.

[33] R. Guillaumont, G. Bouissières, and R. Muxart, *Actinides Rev.*, 1968, **1**, 135 (extensive review of Pa^V and Pa^{IV} in solutions).

spectra of Pa^{IV} incorporated in Cs_2ZrCl_6 are more compatible with a $5f^1$ config-uration for Pa^{IV} than with a $6d^1$ configuration, and similar studies where U^{IV} is incorporated again indicate $5f^2$ configuration for the ion. Chloro, bromo, and other complex ions of Pa^{IV} have been characterized.

URANIUM[34a]

Uranium was discovered by Klaproth in 1789. Until the discovery of uranium fission by Hahn and Strassman in 1939, uranium had little commercial importance: its ores were sources of radium, and small quantities were used for coloring glass and ceramics, but the bulk of the uranium was discarded. Uranium is important as a nuclear fuel; its chemical importance lies in its being the prototype for the suc-ceeding three elements.

Although it is more abundant than Ag, Hg, Cd, or Bi, uranium is widely dis-seminated with relatively few economically workable deposits. The most important ores are the oxide, *uraninite* (one form is *pitchblende*), which has variable com-position approximating to UO_2, and uranium vanadates.

The methods of extraction of uranium are numerous and complex, but the final stages usually employ the extraction of uranyl nitrate from aqueous solutions by ether or some other organic solvent.

24-23. Uranium Compounds

The chemistry of uranium compounds has been studied in great detail, and only the more important aspects can be described here. Generally the stoichiometries, structures, and properties of Np, Pu, and Am compounds are similar to those of uranium; in the III and the IV states the properties are similar to those of lanthanide compounds.

Uranium Oxides. The U—O system is one of the most complex oxide systems known, owing in part to the multiplicity of oxidation states of comparable stability; deviations from stoichiometry are the rule rather than the exception, and stoi-chiometric formulas must be considered as ideal compositions. In the dioxide UO_2, for example, about 10% excess oxygen atoms can be added before any notable structural change is observable, and the UO_2 phase extends from UO_2 to $\sim UO_{2.25}$. The main oxides are: UO_2, brown-black; U_3O_8, greenish black; and UO_3, or-ange-yellow.[34b]

The *trioxide* UO_3 is obtained by decomposition at 350° of uranyl nitrate or, better, of "ammonium diuranate" (see below). One polymorph has a structure that can be considered as uranyl ion linked by U—O—U bonds through the equatorial

[34a] M. H. Rand and O. Kubaschewski, *The Thermodynamic Properties of Uranium Compounds,* Wiley, New York, 1963; E. H. P. Cordfunke, *The Chemistry of Uranium,* Elsevier, 1969 (a monograph, including nuclear applications); *The Recovery of Uranium,* International Atomic Energy Agency, Proceedings Series, STI-PUB-262, 1971.

[34b] G. C. Allen et al., *J.C.S. Dalton,* **1974,** 1296.

oxygens to give layers. The same type of structure occurs also in UO_2F_2 (F bridges) and certain uranates. The other oxides can be obtained by the reactions

$$3UO_3 \xrightarrow{700°} U_3O_8 + \tfrac{1}{2}O_2$$

$$UO_3 + CO \xrightarrow{350°} UO_2 + CO_2$$

The black U_3O_8 begins to lose oxygen at *ca.* 600° in absence of oxygen.[35]

All the oxides readily dissolve in nitric acid to give UO_2^{2+} salts. The addition of hydrogen peroxide to uranyl solution at pH 2.5–3.5 gives a pale yellow precipitate, of formula approximately $UO_4 \cdot 2H_2O$. The U^{VI} peroxo system is exceedingly complex; this particular peroxide is best formulated as $UO_2^{2+}(O_2^{2-}) \cdot 2H_2O$; on treatment with NaOH and H_2O_2 it gives the very stable salt, $Na_4[UO_2(O_2)_3] \cdot 9H_2O$, whose anion consists of linear UO_2 with three peroxo groups in the equatorial plane.

Uranates. The fusion of uranium oxides with alkali or alkaline earth carbonates, or thermal decomposition of salts of the uranyl acetate anion, gives orange or yellow materials generally referred to as uranates,[36] e.g.,

$$2UO_3 + Li_2CO_3 \rightarrow Li_2U_2O_7 + CO_2$$
$$Li_2U_2O_7 + Li_2CO_3 \rightarrow 2Li_2UO_4 + CO_2$$
$$Li_2UO_4 + Li_2CO_3 \rightarrow Li_4UO_5 + CO_2$$

Other metal oxides can also be incorporated and such ternary substances are best regarded as mixed oxides. The uranates are generally of stoichiometry $M_2^I U_x O_{3x+1}$, but $M_4^I UO_5$; $M_3^{II} UO_6$, etc., are known. In contrast to Mo or W, there appear to be no iso- or heteropoly anions for U in solution. A useful material obtained by addition of aqueous NH_3 to $UO_2(NO_3)_2$ solutions is the so-called ammonium diuranate. This is mainly a hydrated uranyl hydroxide containing NH_4^+. Below 580° it gives UO_3 and above U_3O_8.[37]

Uranates do not contain discrete ions such as UO_4^{2-}. They have octahedral U^{VI} with unsymmetrical oxygen coordination such that two U–O bonds are short (*ca.* 1.92 Å), constituting a sort of uranyl group,[38] with other longer U—O bonds in the plane normal to this UO_2 axis linked into chains or layers. However Na_4UO_5 has strings of UO_6 octahedra sharing opposite corners, to give infinite —U—O—U—O chains with a planar UO_4 group normal to the chain; the U—O bonds in the chain are longer than those in the UO_2 group.

Other Binary Compounds. Direct reaction of uranium with B, C, Si, N, P, As, Sb, Se, S, Te, etc., leads to semimetallic compounds that are often nonstoichiometric, resembling the oxides. Some of them (e.g., the silicides) are chemically inert, and the sulfides, notably US, can be used as refractories.

[35] R. J. Ackermann *et al., J. Inorg. Nucl. Chem.,* 1977, **39,** 75.

[36] See, e.g., G. C. Allen and A. J. Griffith, *J.C.S. Dalton,* **1979,** 315.

[37] W. I. Stuart, *J. Inorg. Nucl. Chem.,* 1976, **38,** 1378; A. H. Le Page and A. G. Fare, *J. Inorg. Nucl. Chem.,* 1974, **36,** 87; V. Urbánek, *et al., J. Inorg. Nucl. Chem.,* 1979; **41,** 537.

[38] See E. Gebert *et al., J. Inorg. Nucl. Chem.,* 1978, **40,** 65 (for Li_2UO_4).

TABLE 24-10
Uranium Halides[a]

+3	+4	+5	+6
UF_3, green	UF_4, green	UF_5, white-blue	UF_6, colorless
UCl_3, red	UCl_4, green	U_2Cl_{10}, red-brown	UCl_6, green
UBr_3, red	UBr_4, brown	UBr_5, dark red[b]	—
UI_3, black	UI_4, black	—	—

[a] Other fluorides in addition to UF_4 and UF_6 are known; U_2F_9, U_4F_{14}, and U_5F_{22} are black; Pa also forms a fluoride of uncertain stoichiometry, and Pu gives a red solid Pu_4F_{17}.

[b] A. Blair and H. Ihle, *J. Inorg. Nucl. Chem.*, 1973, **35**, 3795.

Uranium Halides.[39] The principal halides are listed in Table 24-10; they have been studied in great detail, and chemical, structural, and thermodynamic properties are well known.

Fluorides. The *trifluoride* UF_3, a high-melting, nonvolatile, crystalline solid resembling the lanthanum fluorides, is insoluble in water or dilute acids; the best preparation is by the reaction of UF_4 and UN at 950°.[40]

The hydrated *tetrafluoride* can be obtained by precipitation from U^{4+} solution, and the anhydrous fluoride by reactions such as:

$$UO_2 \xrightarrow[500-600°]{C_2Cl_4F_2} UF_4$$

The anhydrous fluoride can also be obtained by thermal decomposition of NH_4UF_5 or similar salts formed by electrolytic reduction of UO_2^+ in fluoride solution.[41]

The nonvolatile solid *tetrafluoride* is insoluble in water but is readily soluble in solutions of oxidizing agents.

Uranium hexafluoride is the most important fluoride and is made on a large scale, since it is the compound used in gas diffusion plants for the separation of uranium isotopes. Accordingly its properties have been intensively studied. It is made by the reactions:

$$UO_2 + 4HF = UF_4 + 2H_2O$$
$$3UF_4 + 2ClF_3 = 3UF_6 + Cl_2$$
$$Cl_2 + 3F_2 = 2ClF_3$$

The hexafluoride forms colorless crystals (m.p. 64.1°) with a vapor pressure of 115 mm at 25°. It is octahedral in the gas and nearly so in the crystal.[42]

UF_6 is rapidly hydrolyzed by water. It is a powerful fluorinating agent, converting many compounds into fluoro derivatives [e.g., CS_2 into SF_4, $(CF_3)_2S_3$ etc.], and in chlorofluorocarbons it can be used as a selective oxidant, e.g., for the cleavage of ethers.[43]

[39] J. C. Taylor, *Coord. Chem. Rev.*, 1976, **20**, 197 (structural systematics of halides and oxohalides).

[40] H. Tagawa, *J. Inorg. Nucl. Chem.*, 1975, **37**, 731.

[41] E. R. Russell and M. L. Hyder, *Inorg. Nucl. Chem. Lett.*, 1977, **13**, 175.

[42] J. H. Levy *et al.*, *J.C.S. Dalton*, **1976**, 219.

[43] G. A. Olah and J. Welch, *J. Am. Chem. Soc.*, 1978, **100**, 5396.

$$PhCH_2OMe \xrightarrow{UF_6} PhCHO + MeF \xrightarrow{-UF_4-HF + H_2O} PhCHO + MeOH$$

It also oxidizes Cu, Cd, and Tl in MeCN to give solutions of the corresponding $[UF_6]^-$ salts.[44] The intermediate fluorides UF_5, U_2F_9, and U_4F_{14} are made by interaction of UF_6 and UF_4; they disproportionate quite readily, e.g.,

$$3UF_4 \rightleftharpoons U_2F_9(s) + UF_6(g)$$

UF_5 is best made by photochemical reduction of UF_6 with CO; it has a polymeric chain structure;[45] U_2F_9 has crystallographically identical U atoms, each nine-coordinate; the black color evidently results from charge-transfer transitions giving formally +4 and +5 atoms in the excited state.

Chlorides. The *trichloride* can be made only under anhydrous conditions, usually by the action of HCl on UH_3 (p. 251); U^{3+} solutions are readily oxidized (see below).

UCl_3 gives $UCl_3 \cdot MeCN$ and $UCl_3 \cdot THF$, and there are a few U^{III} complexes of amine ligands.[46a] There are also 18-crown-6-ether complexes of UCl_3 [and $U(BH_4)_3$].[46b]

The *tetrachloride* is readily obtained by liquid-phase chlorination of UO_3 by refluxing with hexachloropropene. The primary product is believed to be UCl_6, which decomposes thermally. UCl_4 is soluble in polar organic solvents and in water. It is the usual starting material for synthesis of other U^{IV} compounds. UCl_4 forms various adducts with oxygen and nitrogen donors[47] with two to seven ligand molecules, but only a few of the structures are known. $UCl_4 \cdot (DMSO)_3$ is $[UCl_2(DMSO)_6]UCl_6$ and $UCl_4 \cdot 6Me_3PO$ is $[UCl(OPMe_3)_6]Cl_3$, but *cis*-$UCl_4(OPPh_3)_2$ and *trans*-$UCl_4(hmpa)_2$ are octahedral.

Uranium *hexachloride* is made by chlorination of a mixture of U_3O_8 and carbon at 380°.[48a] The green crystals sublime in vacuum. In CH_2Cl_2 solution it slowly decomposes to the red-brown volatile crystalline *pentachloride* U_2Cl_{10}, which can also be made by chlorination of UCl_4. In the crystal U_2Cl_{10} has a structure similar to Ta_2Cl_{10} or Mo_2Cl_{10}. It forms a volatile complex with Al_2Cl_6 of stoichiometry $UAlCl_8$.[48b] Both halides are rapidly hydrolyzed by water.

Halogeno Complexes. All the halides can form halogeno complexes, those with F^- and Cl^- being the best known. They can be obtained by interaction of the halide and alkali halides in melts or in solvents such as $SOCl_2$, or in the case of fluorides sometimes in aqueous solution.

Fluoro complexes of U^{IV} can be made by dissolution of UF_4 in RbF. Although

[44]   J. A. Berry, et al., *J.C.S. Dalton*, **1976**, 272.
[45]   G. W. Halstead et al., *Inorg. Chem.*, 1978, **17**, 2967; P. G. Eller, et al., *Inorg. Chim. Acta*, 1979, **37**, 129.
[46a] D. C. Moody and J. D. Odun, *J. Inorg. Nucl. Chem.*, 1979, **41**, 533.
[46b] D. C. Moody et al., *Inorg. Chem.*, 1979, **18**, 208.
[47]   G. Bombieri et al., *J.C.S. Dalton*, **1976**, 735; **1975**, 1873; *J.C.S. Chem. Comm.*, **1975**, 188: M. C. Caiva and L. R. Nassimbeni, *J. Inorg. Nucl. Chem.*, 1977, **39**, 455; J. G. H. du Preeze et al., *J.C.S. Dalton*, **1977**, 1062.
[48a] W. Kolitsch and U. Müller, *Z. Anorg. Allg. Chem.*, 1974, **410**, 21.
[48b] G. N. Papatheodorou and D. A. Buttry, *Inorg. Nucl. Chem. Lett.*, 1979, **15**, 51.

the UF_7^{3-} and UF_8^{4-} ions are more common, the octahedral UF_6^{2-} ion is known.[49]

Although U^V is usually unstable (see below) in aqueous solution, UF_5 dissolves in 48% HF to give blue solutions that are only slowly oxidized in air and from which can be crystallized $(H_3O)UF_6 \cdot 1.5H_2O$ and by addition of Rb or Cs fluorides, the blue salts MUF_6. The latter are best made by action of ClF_5 on MCl in liquid HF.

For U^{VI} the ions UF_7^- and UF_8^{2-} can be made by the reaction

$$UF_6 + NaF = NaUF_7$$
$$UF_6 + 2NOF = (NO)_2UF_8$$

The alkali metal heptafluoro uranates(VI) decompose on heating,[50] e.g.,

$$2NaUF_7 \xrightleftharpoons{>100°} UF_6 + Na_2UF_8$$

$$Na_2UF_8 \xrightarrow{>300} 2NaF + UF_6$$

The alkali salts are also soluble in MeCN and propylene carbonate. The geometry of the ions appears to be quite easily distorted by either solvation or crystalline field forces, presumably because of the possiblity of different conformations having similar energies.

Chloro complex anions for U^{III}, U^{IV}, and U^V are known in salts such as $KUCl_4 \cdot 5H_2O$, K_2UCl_6, $(Me_4N)UCl_6$, and $(Me_4N)_3UCl_8$. U_2Cl_{10} reacts with PCl_5 to give $[PCl_4]^+[UCl_6]^-$.

When U_3O_8 is boiled with hexachloropropene a dark red complex of trichloroacryloyl chloride (L) of stoichiometry UCl_5L is formed. Other adducts can be made from this by displacement reactions with, e.g., $SOCl_2$.

Uranium and other actinide *pseudohalide* complexes can be made, e.g., when UCl_4 is treated with KSCN in MeCN giving the $[U(NCS)_8]^{4-}$ ion.[51]

Oxohalides.   The stable uranyl compounds UO_2X_2 are soluble in water. They are made by reactions[52] such as

$$UCl_4 + O_2 \xrightarrow{350°} UO_2Cl_2 + Cl_2$$

$$UO_3 + 2HF \xrightarrow{400°} UO_2F_2 + H_2O$$

$$U_3O_8 + 3Cl_2 = 3UO_2Cl_2 + O_2$$

$$UO_2Cl_2 \xrightarrow{450°} (UO_2)_2Cl_3 \xrightarrow{500°} UO_2Cl$$

UO_2Cl_2 forms an adduct with two moles of $OPPh_3$.[53] There is also a yellow-orange *oxo fluoride* UOF_4, formed by interaction of UF_6 with SiO_2 in liquid HF. It has

[49]   J. J. Ryan, *et al., Inorg. Chem.,* 1974, **13**, 214; W. Wagner *et al., Inorg. Chem.,* 1977, **16**, 1021.

[50]   R. Bougon *et al., Inorg. Chem.,* 1976, **15**, 2532.

[51]   G. Bombieri *et al., J.C.S. Dalton,* **1975,** 1520.

[52]   E. H. P. Cordfunke *et al., J. Inorg. Nucl. Chem.,* 1977, **39**, 2189; 1974, **36**, 1291.

[53]   G. Bombieri *et al., J.C.S. Dalton,* **1978,** 677.

a polymeric structure in which units with pentagonal bipyramidally coordinated U atoms are linked by F bridges.[54]

Other Compounds. The *tetraalcoxides* $U(OR)_4$ (R = Me, Et, and Bu^t) are thermally stable but easily hydrolyzed compounds obtained from UCl_4. The pentaalcoxides can be made from UF_5[55a] or by oxidation of $U(Me)_4$ and $U(OEt)_4$ with Br_2 in presence of NaOR. Higher alcoxides can be obtained by alcohol exchange with $U(OEt)_5$. The pentamethoxide is a trimer, but other pentaalcoxides are dimers $U_2(OR)_{10}$, with two alcoxo bridges. They form complex anions, e.g., $Na[U(OEt)_6]$.

The uranium(IV) alcoxides[55a] can be of the type $UO_2(OR)_2$, $U(OMe)F_5$, $U(OR)_6$, and there are the related compounds $UF_x(OTeF_5)_{6-x}(x = 0-5)$. The isopropoxide $U(OPr^i)_6$ gives adducts with lithium, magnesium, and aluminum alkyls, in which the oxygen atoms of the alcoxide groups coordinate to Li, Mg, or Al.[18] The *dialkylamides,*[55b] $U(NR_2)_4$ are thermally stable, and like $Th(NR_2)_4$ insert CS_2, COS, and CO_2 to give, e.g., the eight-coordinate dithiocarbamates and carbamates. The green compound $U(NEt_2)_4$ (m.p. 35°) has a unique dialkylamido bridged structure, with five-coordinate U^{IV} in the crystal 24-III, although it is

(24-III)

monomeric in benzene. Interaction with *N,N'*-dimethylethylenediamine gives an even stranger molecule, $U_3(MeNCH_2CH_2NMe)_6$, which has linear U—U—U with three bridges between the atoms. The compound $U(NPh_2)_4$, which is monomeric and four-coordinate in the crystal, an oxobridged complex, and compounds $RU[N(SiMe_3)_2]_3$, R = H, Cl, Me, and BH_4, are also known.[55c] Finally we note the polymeric *carboxylates* of stoichiometry $U(CO_2R)_4$[56] and the dark green borohydride $U(BH_4)_4$ which is also polymeric, with a complex helical structure containing fourteen-coordinate U.[57a]

24-24. Aqueous Chemistry of Uranium

Uranium ions in aqueous solution can give very complex species because in addition to the four oxidation states, complexing reactions with all ions other than ClO_4^-

54 R. J. Paine *et al., Inorg. Chem.,* 1975, **14,** 1113.

55a P. J. Vergamini, *J.C.S. Chem Comm.,* **1979,** 54; T. J. Marks *et al., J. Am. Chem. Soc.,* 1979, **101,** 1036.

55b F. Calderazzo *et al., Inorg. Chem.,* 1978, **17,** 471; J. G. Reynolds *et al., Inorg. Chem.,* 1977, **16,** 599, 1090, 1858, 2822.

55c R. A. Andersen *et al., Inorg. Chem.,* 1979, **18,** 1221, 1507; *J. Am. Chem. Soc.,* 1979, **101,** 2782.

56 K. W. Bagnall and O. V. Lopez, *J.C.S. Dalton,* **1976,** 1109.

57a R. R. Rietz *et al., Inorg. Chem.,* 1978, **17,** 653, 658, 661; R. H. Banks *et al., J. Am. Chem. Soc.,* 1978, **100,** 1957, (Pa, Np, Pu).

as well as hydrolytic reactions leading to polymeric ions occur under appropriate conditions. The formal potentials for $1M$ $HClO_4$ have been given in Table 24-7. In presence of other anions the values differ: thus for the U^{4+}/U^{3+} couple in $1M$ $HClO_4$ the potential is -0.631 V, but in $1M$ HCl it is -0.640 V. The simple ions and their properties are also listed in Table 24-8. Because of hydrolysis, aqueous solutions of uranium salts have an acid reaction that increases in the order $U^{3+} <$ $UO_2^{2+} < U^{4+}$. The uranyl and U^{4+} solutions have been particularly well studied. The main hydrolyzed species[57b] of UO_2^{2+} at 25° are UO_2OH^+ $(UO_2)_2(OH)_2^{2+}$, and $(UO_2)_3(OH)_5^+$, but the system is a complex one and the species present depend on the medium; at higher temperatures the monomer is most stable but the rate of hydrolysis to UO_3 of course increases. The solubility of large amounts of UO_3 in UO_2^{2+} solutions is also attributable to formation of UO_2OH^+ and polymerized hydroxo-bridged species.

The U^{3+} ion, which is readily obtained in perchlorate or chloride solutions by reduction of UO_2^{2+} electrolytically or with zinc amalgam, is a powerful reducing agent.[58] Its oxidation by I_2 or Br_2 occurs by an outer-sphere mechanism.[59] The solutions of U^{3+} in $1M$ HCl are stable for days, but in more acid solution, spontaneous oxidation occurs more rapidly. From fluoride solutions $UF_3 \cdot H_2O$ can be obtained, and $U_2(SO_4)_3 \cdot 5H_2O$ from sulfate solutions.[60]

The U^{4+} ion is only slightly hydrolyzed in molar acid solutions:

$$U^{4+} + H_2O \rightleftharpoons U(OH)^{3+} + H^+ \qquad K_{25°} = 0.027 \ (1M \ HClO_4, \ NaClO_4)$$

but it can also give polynuclear species in less acid solutions.

The U^{4+} ion gives insoluble precipitates with F^-, PO_4^{3-}, and IO_3^- from acid solutions (cf. Th^{4+}).

The uranium(V) oxo ion UO_2^+ is extraordinarily unstable toward disproportionation:

$$2UO_2^+ + 4H^+ = U^{4+} + UO_2^{2+} + 2H_2O$$

Its reactions can be studied by stopped-flow techniques, since the reduction of UO_2^{2+} by Eu^{2+} is fast, whereas further reduction of UO_2^+ is slow.[61] The disproportionation is retarded by excess U^{VI}, probably because a U^V—U^{VI} complex forms. The ion is most stable in the pH range 2.0–4.0, where the disproportionation reaction to give U^{4+} and UO_2^{2+} is negligibly slow. By contrast, reduction of UO_2^{2+} in dimethyl sulfoxide gives UO_2^+ in concentrations sufficiently high to allow the spectrum to be obtained, and disproportionation occurs with a half-life of about an hour. As noted above, U^V can be stabilized in HF solutions as UF_6^-, as well as in concentrated Cl^- and CO_3^{2-} solutions.

Spectroscopic and other studies have shown that in aqueous solutions of UO_2^{2+}

[57b]   R. N. Sylva and M. R. Davidson, *J.C.S. Dalton*, **1979**, 465.
[58]   See e.g., M. K. Loar *et al.*, *Inorg. Chem.*, 1978, **17**, 330; A. Ekstrom *et al.*, *Inorg. Chem.*, 1975, **14**, 1035.
[59]   A. Adegite *et al.*, *J.C.S. Dalton*, **1977**, 833.
[60]   R. Barnard *et al.*, *J.C.S. Dalton*, **1973**, 604; **1972**, 1932.
[61]   A. Ekstrom, *Inorg. Chem.*, 1974, **13**, 2237.

and U^{4+}, complex ions are often readily formed, for example,

$$U^{4+} + Cl^- \rightleftharpoons UCl^{3+} \qquad K = 1.21\ (\mu = 2.0;\ 25°)$$
$$U^{4+} + 2HSO_4^- \rightleftharpoons U(SO_4)_2 + 2H^+ \qquad K = 7.4 \times 10^3\ (\mu = 2.0;\ 25°)$$
$$UO_2^{2+} + Cl^- \rightleftharpoons UO_2Cl^+ \qquad K = 0.88\ (\mu = 2.0;\ 25°)$$
$$UO_2^{2+} + 2SO_4^{2-} \rightleftharpoons UO_2(SO_4)_2^{2-} \qquad K = 7.1 \times 10^2\ (\mu = 2.0;\ 25°)$$

Nitrate complexes also exist, and solutions of U^{IV} contain $[UNO_3(H_2O)_4]^{3+}$ and similar species; From concentrated HNO_3, white R_4N^+ salts of $[U(NO_3)_6]^{2-}$ can be precipitated.

The nature of the reduction of UO_2^{2+}, especially by Cr^{2+}, has been studied and it appears that there is a bright green intermediate complex ion, probably $[(H_2O)_5Cr^{III}{-}O{-}U^VO(H_2O)_n]^{4+}$, which reacts further to give Cr^{III} and U^{IV}. A similar intermediate occurs in the reduction of PuO_2^{2+}, and for Np an intermediate has been suggested by ion exchange. It is pertinent to note that the reverse process, oxidation of U^{4+} by various agents, has been studied in detail; this is possible only because of the slow exchange of UO_2^{2+} with water. Using ^{18}O tracer it was found that PbO_2, H_2O_2, or MnO_2 gave UO_2^{2+}, where virtually all the O came from the solid oxidant, whereas for O_2 and O_3 only one oxygen atom is transferred from the oxidant to U^{IV}.

Complex ions are also formed with citrate and anions of other organic acids, thiocyanate, and phosphate. The phosphates are important in view of the occurrence of uranium in phosphate minerals, and species such as $UO_2H_2PO_4^+$ and $UO_2H_3PO_4^{2+}$, and at high concentrations anionic complexes are known.

24-25. Uranyl Salts

The only common uranium salts are the uranyls, and the most important one is the nitrate, which crystallizes with six, three, or two molecules of water depending on whether it is obtained from dilute, concentrated, or fuming nitric acid. The most unusual and significant property of the nitrate is its solubility in numerous ethers, alcohols, ketones, and esters—it distributes itself between the organic and an aqueous phase. The nitrate is also readily extracted from aqueous solutions, and this operation has become classical for the separation and purification of uranium, since, except for the other actinide MO_2^{2+} ions, few other metal nitrates have any extractability. A great deal of information is available, and phase diagrams for the $UO_2(NO_3)_2$—H_2O—solvent systems have been determined. The effect of added salts, for example, $Ca(NO_3)_2$ or NH_4NO_3, as "salting-out" agents is to increase substantially the extraction ratio to technically usable values. Studies of the organic phase have shown that $UO_2(NO_3)_2$ is accompanied into the solvents by $4H_2O$ molecules, but there is little or no ionization, and the nitrate is undoubtedly coordinated in the equatorial plane of the UO_2 system. An important extractant for uranyl nitrate that does not require a salting-out agent for useful ratios is tributyl phosphate. Anhydrous uranyl nitrate is obtained by the reactions

$$U + N_2O_4(l) \xrightarrow{\text{MeCN}} UO_2(NO_3)_2 \cdot N_2O_4 \cdot 2MeCN \xrightarrow{163°} UO_2(NO_3)_2$$

The structure of $UO_2(NO_3)_2(THF)_2$ is typical for uranyl nitrate compounds with the chelate NO_3 and the oxygen atoms of THF (trans) in the plane of the hexagonal bipyramid.[62]

Other uranyl salts are given by organic acids, sulfate, halides, etc.; the water-soluble acetate in presence of an excess of sodium acetate in dilute acetic acid gives a crystalline precipitate of $NaUO_2(OCOCH_3)_3$.

One of the characteristic features of UO_2^{2+} compounds is fluorescence, and uranyl oxalate is used as an actinometer.[63]

NEPTUNIUM,[64] PLUTONIUM,[65] AND AMERICIUM

24-26.   Isolation of the Elements

Although several isotopes of the elements Np, Pu, and Am are known, the ones that can be obtained in macroscopic amounts are given in Table 24-3. Both ^{237}Np and ^{239}Pu are found in the uranium fuel elements of nuclear reactors, from which plutonium is isolated on a kilogram scale. Americium is produced from intense neutron irradiations of pure plutonium. The problems involved in the extraction of these elements include the recovery of the expensive starting material and the removal of hazardous fission products that are formed simultaneously in amounts comparable to the amounts of the synthetic elements themselves. Not only are the chemical problems themselves quite formidable, but the handling of highly radioactive solutions or solids (in the case of plutonium, a potent carcinogen, the exceedingly high toxicity is an additional hazard, since microgram amounts are potentially lethal) has necessitated the development of remote control operations. For the large-scale extractions from fuel elements, detailed studies of the effects of radiation on structural and process materials have also been required. There are numerous procedures for the separation of Np, Pu, and Am, variously involving precipitation, solvent extraction, differential volatility of compounds, and so on, and we can give only the briefest outline. The most important separation methods are based on the following chemistry.

1.   *Stabilities of Oxidation States.*   The stabilities of the major ions involved are: $UO_2^{2+} > NpO_2^{2+} > PuO_2^{2+} > AmO_2^{2+}$; $Am^{3+} > Pu^{3+} \gg Np^{3+}$, U^{4+}. It is thus possible (see also Table 24-8) by choice of suitable oxidizing or reducing agents to obtain a solution containing the elements in different oxidation states; they can then be separated by precipitation or solvent extraction. For example, Pu can be oxidized to PuO_2^{2+} while Am remains as Am^{3+}—the former could be removed by solvent extraction or the latter by precipitation of AmF_3.

[62]   J. G. Reynolds *et al., Inorg. Chem.,* 1977, **16,** 3357.
[63]   H. D. Burrows and T. J. Kemp, *Chem. Soc. Rev.,* 1974, **3,** 139; E. Rabinowitch and R. Linn Belford, *Spectroscopy and Photochemistry of Uranyl Compounds,* MacMillan, 1964.
[64]   S. K. Patel *et al., Coord. Chem. Rev.,* 1978, **25,** 133 (aqueous chemistry).
[65]   M. Taube, *Plutonium: A General Survey,* Elsevier, 1974; J. M. Cleveland, *The Chemistry of Plutonium,* Gordon and Breach, 1970; F. Oetting, *Chem. Revs.,* 1967, **67,** 261 (thermodynamics of plutonium compounds).

   2. *Extractability into Organic Solvents.* As noted previously, the MO_2^{2+} ions can be extracted from nitrate solutions into organic solvents. The M^{4+} ions can be extracted into tributyl phosphate in kerosene from $6M$ nitric acid solutions; the M^{3+} ions can be similarly extracted from 10 to $16M$ nitric acid; and neighboring actinides can be separated by a choice of conditions.

   3. *Precipitation Reactions.* Only M^{3+} and M^{4+} give insoluble fluorides or phosphates from acid solutions; the higher oxidation states either give no precipitate or can be prevented from precipitation by complex formation with sulfate or other ions.

   4. *Ion-Exchange Procedures.* Although ion-exchange procedures, both cationic and anionic, can be used to separate the actinide ions, they are best suited for small amounts of material. Since they have found most use in the separation of the trans-americium elements, these procedures are discussed below.

   The following examples are for the separation of plutonium from uranium; similar procedures involving the same basic principles have been devised to separate Np and Am. The initial starting material in plutonium extraction is a solution of the uranium fuel element (plus its aluminum or other protective jacket) in nitric acid. The combination of oxidation-reduction cycles coupled with solvent extraction and/or precipitation methods removes the bulk of fission products (FPs); however certain elements—notably ruthenium, which forms cationic, neutral, and anionic nitrosyl complexes—may require special elimination steps. The initial uranyl nitrate solution contains Pu^{4+} since nitric acid cannot oxidize this to Pu^V or Pu^{VI}.

   1. *Isobutyl Methyl Ketone (Hexone) Method.* This is shown in Scheme 24-1.

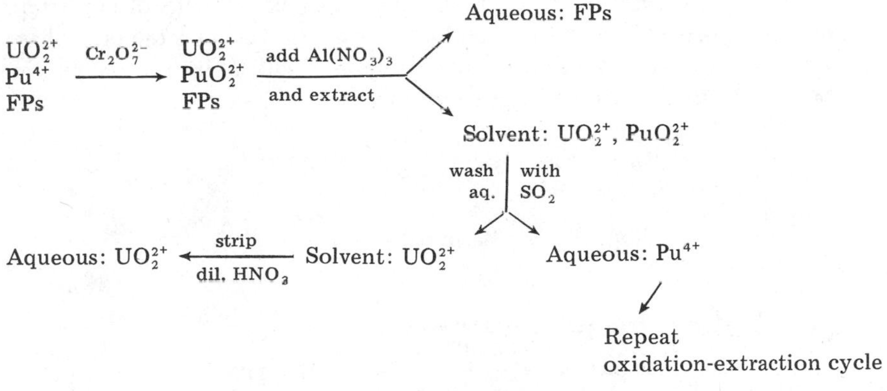

Scheme 24-1

   2. *Tributyl Phosphate Method.* The extraction coefficients from $6M$ nitric acid solutions into 30% tributylphosphate (TBP) in kerosine are $Pu^{4+} > PuO_2^{2+}$; $Np^{4+} \sim NpO_2^+ \gg Pu^{3+}$; $UO_2^{2+} > NpO_2^+ > PuO_2^{2+}$; the M^{3+} ions have very low extraction coefficients in $6M$ acid, but from $12M$ hydrochloric acid or $16M$ nitric acid the extraction increases and the order is $Np < Pu < Am < Cm < Bk$.

Thus in the U—Pu separation, after addition of NO_2^- to adjust all the plutonium to Pu^{4+}, we have Scheme 24-2.

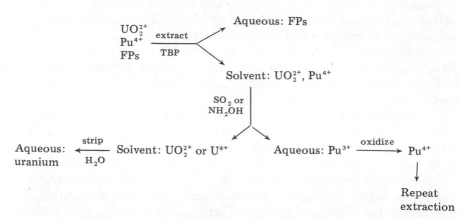

Scheme 24-2

The extraction of ^{237}Np involves similar principles of adjustment of oxidation state and solvent extraction; Pu is reduced by ferrous sulfamate plus hydrazine to unextractable Pu^{III}, while Np^{IV} remains in the solvent from which it is differentially stripped by water to separate it from U.

3. *Lanthanum Fluoride Cycle.* This classical procedure was first developed by McMillan and Abelson for the isolation of neptunium, but it is applicable elsewhere and is of great utility. For the U—Pu separation we have Scheme 24-3. The cycle shown is repeated with progressively smaller amounts of lanthanum carrier and smaller volume of solution until plutonium becomes the bulk phase. This fluoride cycle has also been used in combination with an initial precipitation step for Pu^{4+} with bismuth phosphate as a carrier.

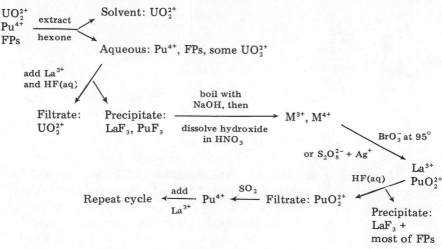

Scheme 24-3

24-27.   Binary Compounds

Oxides.   All three oxide systems have various solid solutions and other non-stoichiometric complications. The monoxides are interstitial compounds. The important oxides of Np, Pu, and Am are the *dioxides,* which are obtained on heating the nitrates or hydroxides of any oxidation state in air; they are isostructural with UO_2. Ordinarily, PuO_2 is nonstoichiometric and may have different colors, but ignition at 1200° gives the stoichiometric oxide. The higher oxide of Np appears to be Np_2O_5, and Np_3O_8 does not exist.[66]

The action of ozone on suspensions of the M^{IV} hydroxides gives rise to the hydrated *trioxides,* brown $NpO_3 \cdot 2H_2O$ and $NpO_3 \cdot H_2O$, and red-gold $PuO_3 \cdot 0.8H_2O$, but in contrast to U, which also gives $UO_3 \cdot 0.8H_2O$ no anhydrous trioxide is known. Above 300° black Np_2O_5 is obtained. $NpO_3 \cdot 2H_2O$ and Np_2O_5 can also be made by oxidation in $LiClO_4$ melts. Reduction of black AmO_2 with hydrogen at 600° gives the pink dimorphic Am_2O_3, which is the first lanthanidelike sesquioxide in the actinide series.

Halides.   The halides are listed in Table 24-11.

The Np, Pu, and Am halides, which are isostructural with and chemically similar to those of uranium, clearly show the decrease in stability of compounds in the higher oxidation states, and this trend continues in the succeeding elements. The preparative methods used are also similar to those for uranium, for example.

$$NpO_2 + \tfrac{1}{2}H_2 + 3HF(g) \xrightarrow{500°} NpF_3 + 2H_2O$$

$$PuF_4 + F_2 \xrightarrow{500°} PuF_6$$

$$AmO_2 + 2CCl_4 \xrightarrow{800°} AmCl_3 + 2COCl_2 + \tfrac{1}{2}Cl_2$$

TABLE 24-11
Halides of Np, Pu, and Am[a]

$+2$[b]	$+3$	$+4$	$+6$
	NpF_3, purple-black	NpF_4, green	NpF_6, orange, m.p. 55.1°
	PuF_3, purple	PuF_4, brown	PuF_6, red-brown, m.p. 51.6° [c]
	AmF_3, pink	AmF_4, tan	—
	$NpCl_3$, white	$NpCl_4$, red-brown	
	$PuCl_3$, emerald	—	
$AmCl_2$, black	$AmCl_3$, pink	—	
	$NpBr_3$, green	$NpBr_4$, red-brown	
	$PuBr_3$, green	—	
$AmBr_2$, black	$AmBr_3$, white	—	
	NpI_3, brown	—	
	PuI_3, green	—	
AmI_2, black	AmI_3, yellow	—	

[a] Certain oxohalides $M^{III}OX$, M^VOF_3, and $M^{VI}O_2F_2$ are known.
[b] R. D. Baybarz *et al., J. Inorg. Nuc. Chem.,* 1973, **35,** 483; 1972, **34,** 3427.
[c] Unlike the situation for U, intermediate fluorides are not formed in conversion of MF_4 into MF_6.

[66]   J. A. Fahey *et al., J. Inorg. Nucl. Chem.,* 1976, **38,** 495.

The fluorides MF_3 and MF_4 can be precipitated from aqueous solutions in hydrated form. The hexafluorides have been much studied because they are volatile; the melting points and stabilities decrease in the order $U > Np > Pu$. PuF_6 is so very much less stable than UF_6 that at equilibrium, the partial pressure of PuF_6 is only 0.004% of the fluorine pressure. Hence PuF_6 formed by fluorination of PuF_4 at 750° must be quenched immediately by a liquid nitrogen probe. The compound also undergoes self-destruction by α-radiation damage, especially in the solid; it must also be handled with extreme care owing to the toxicity of Pu. PuF_6 contains two nonbonding $5f$ electrons and should be paramagnetic; however like UF_6, where all the valence electrons are involved in bonding, it shows only a small temperature-independent paramagnetism. This observation has been explained by ligand field splitting of the f levels to give a lower-lying orbital that is doubly occupied.

NpF_6 has a $5f^1$ configuration according to esr and absorption spectra, the octahedral field splitting the sevenfold orbital degeneracy of the $5f$ electron and leaving a ground state that has only spin degeneracy. This quenching of the orbital angular momentum is similar to that in the first-row d-transition group. It provides further evidence for the closeness of the energy levels of the $5f$ and the valence electrons in actinides, in contrast to the much lower energies of the $4f$ electrons in lanthanides. NpF_6 is slightly distorted in the solid, and its magnetic behavior depends on its environment when diluted with UF_6.

Other Compounds.   A substantial number of compounds, particularly of plutonium, are known, and most of them closely resemble their uranium analogues. The hydride systems of Np, Pu, and Am are more like that of thorium than that of uranium and are complex. Thus nonstoichiometry up to $MH_{2.7}$ is found in addition to stoichiometric hydrides such as PuH_2 and AmH_2.

As with uranium, many complex salts are known [e.g., Cs_2PuCl_6, $NaPuF_5$, $KPuO_2F_3$, $NaPu(SO_4)_2 \cdot 7H_2O$, and $Cs_2Np(NO_3)_6$]. A simple solid hydrated nitrate, $Pu(NO_3)_4$, is obtained by evaporation of Pu^{IV} nitrate solution; at 150 to 180° in air this gives $PuO_2(NO_3)_2$. Pu^V also occurs in salts of the PuF_7^{2-} and PuF_6^- ions.

24-28.   Aqueous Chemistry of Neptunium, Plutonium,[67] and Americium

The formal reduction potentials of Np, Pu, and Am were given in Table 24-7 and the general stabilities of the ions in Table 24-8.

Aqueous solutions of Pu^{VI}, Am^V, Am^{VI}, and especially Am^{IV} undergo rapid self-reduction because of their α-radiation.

For Np the potentials of the four oxidation states are separated, like those of uranium, but in this case the NpO_2^+ state is comparatively stable. Earlier evidence that NpO_2^{2+} was reduced by Cl^- has been shown to be due to catalysis by platinum, and the rate is very slow. With Pu, however, the potentials are not well separated, and in $1M$ $HClO_4$ all four species can coexist in appreciable concentrations; PuO_2^+ becomes increasingly stable with decreasing acidity, since the couples are strongly

67   J. M. Cleveland, *Coord. Chem. Rev.*, 1970, **5**, 101 (an extensive review of aqueous complexes of plutonium).

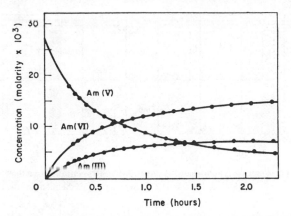

Fig. 24-5. Disproportionation of AmO_2^+ in $6M$ perchloric acid at $25°$. Net reaction $3AmO_2^+ + 4H^+$ $= 2AmO_2^{2+} + Am^{3+} + 2H_2O$. [Reproduced by permission from J. S. Coleman, Inorg. Chem., 1963, 2; 53.]

hydrogen ion dependent. The Am ions stable enough to exist in finite concentrations are Am^{3+}, AmO_2^+, and AmO_2^{2+}; the Am^{3+} ion is the usual state, since powerful oxidation is required to achieve the higher oxidation states. Alkaline solutions are more favorable for the stabilization of Am^{IV}, and for $1M$ basic solution the $Am(OH)_4$—$Am(OH)_3$ couple has a value of $+0.5$ V, nearly 2 V less than for the Am^{4+}/Am^{3+} couple in acid solution. Thus pink $Am(OH)_3$ can be readily converted into black $Am(OH)_4$ or hydrous AmO_2 by the action of hypochlorite. This black hydroxide is also soluble in $13M$ ammonium fluoride solutions to give stable solutions from which $(NH_4)_4AmF_8$ can be precipitated; the anion in this salt probably has the square-antiprism structure, as does AmF_4.

As with uranium, the solution chemistry is complicated owing to hydrolysis and polynuclear ion formation, complex formation with anions other than perchlorate, and disproportionation reactions of some oxidation states. The tendency of ions to displace a proton from water increases with increasing charge and decreasing ion radius, so that the tendency to hydrolysis increases in the same order for each oxidation state, that is, Am > Pu > Np > U and $M^{4+} > MO_2^{2+} > M^{3+} > MO_2^+$; simple ions such as NpO_2OH^+ or $PuOH^{3+}$ are known, in addition to polymeric species that in the case of plutonium can have molecular weights up to 10^{10}.

The complexing tendencies decrease, on the whole, in the same order as the hydrolytic tendencies. The formation of complexes shifts the oxidation potentials, sometimes influencing the relative stabilities of oxidation states; thus the formation of sulfate complexes of Np^{4+} and NpO_2^{2+} is strong enough to cause disproportionation of NpO_2^+. The disproportionation reactions have been studied in some detail; Figs. 24-5 and 24-6 illustrate some of the complexities involved.

Typical of these disproportionations are the following at low acidity:

$$3Pu^{4+} + 2H_2O \rightleftharpoons PuO_2^{2+} + 2Pu^{3+} + 4H^+$$
$$2Pu^{4+} + 2H_2O \rightleftharpoons PuO_2^+ + Pu^{3+} + 4H^+$$
$$PuO_2^+ + Pu^{4+} \rightleftharpoons PuO_2^{2+} + Pu^{3+}$$

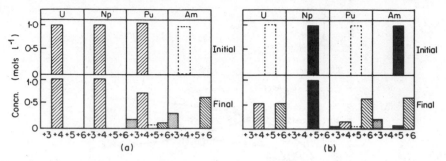

Fig. 24-6. Disproportionation reactions of (a) tetrapositive (b) pentapositive ions in $1M$ acid at 25°. [Reproduced by permission from J. J. Katz and G. T. Seaborg, The Chemistry of the Actinide Elements, Methuen, 1957.]

At 25° and $1M$ $HClO_4$, we have

$$K = \frac{[Pu^{VI}][Pu^{III}]}{[Pu^{V}][Pu^{IV}]} = 10.7$$

which indicates that measurable amounts of all four states can be present.

An example of complex formation is provided by carbonate; for Am this provides a useful separation from Cm, since $Cm(OH)_3$ is insoluble in $NaHCO_3$ and cannot be oxidized to soluble complexes. However treatment of Am^{3+} in $2M$ Na_2CO_3 with O_3 at 25° gives a red-brown Am^{VI} carbonate complex anion of uncertain composition; yet at 90°, reduction to the ion $[Am^{V}O_2CO_3]^-$ occurs unless $S_2O_8^{2-}$ is present. The Am^{VI}/Am^{V} couple in $0.1M$ $NaHCO_3$ is estimated to be *ca*. 1 V. Carbonate complexes of Np^{V}, Pu^{V}, and Am^{V} can also be obtained by oxidation of dilute HNO_3 solutions with O_3, reduction of MO_2^{2+} with KI, and addition of $KHCO_3$. The $KM^{V}O_2CO_3$ salts are isostructural, with layers held together by K^+ ions.

Hexacoordinate AmF_6^{2-} salts also exist. The precipitation reactions of Np, Pu, and Am are generally similar to those of uranium in the corresponding oxidation states, for example, of $NaM^{VI}O_2(OCOCH_3)_3$ or MF_3. A tetraaza (N_4) ligand that has an extremely high complexity constant ($>10^{52}$) for Pu^{IV} has been found, and it may have some use in scavenging Pu from body tissues, etc.[68a]

THE TRANS–AMERICIUM ELEMENTS

The isotope ^{242}Cm was first isolated among the products of α-bombardment of ^{239}Pu, and its discovery actually preceded that of americium. Isotopes of the other elements were first identified in products from the first hydrogen bomb explosion (1952) or in cyclotron bombardments. Although Cm, Bk, and Cf have been obtained in macro amounts (Table 24-3), much of the chemical information has been obtained on the tracer scale. The remaining elements have been characterized only

[68a] F. W. Weitl, *J. Am. Chem. Soc.*, 1978, **100**, 1170.

by their chemical behavior on the tracer scale in conjunction with their specific nuclear decay characteristics.

For these elements the correspondence of the actinide and the lanthanide series becomes most clearly revealed. The position of curium corresponds to that of gadolinium where the f shell is half-filled. For curium the +3 oxidation state is the normal state in solution, although unlike gadolinium, a solid tetrafluoride CmF_4 has been obtained. Berkelium has +3 and +4 oxidation states, as would be expected from its position relative to terbium, but the +4 state of terbium does not exist in solution, whereas for Bk it does. Although CfF_4 and CfO_2 have been made, the remaining elements have only the +3 and +2 states. The great similarity between the +3 ions of Am and the trans-americium elements has meant that the more conventional chemical operations successful for the separation of the previous actinide elements are inadequate and most of the separations require the highly selective procedures of ion-exchange discussed below; solvent extraction of the M^{3+} ions from 10 to $16M$ nitric acid by tributyl phosphate also gives reasonable separations.

24-29. Ion-Exchange Separations

Ion-exchange has been indispensable in the characterization of the trans-americium elements and is also important for some of the preceding elements, particularly for tracer quantities of material. We have seen in the case of the lanthanides (Chapter 23) that the +3 ions can be eluted from a cation-exchange column by various complexing agents, such as buffered citrate, lactate, or 2-hydroxybutyrate solutions, and that the elution order follows the order of the hydrated ionic radii so that the lutetium is eluted first and lanthanum last.

By detailed comparison with the elution of lanthanide ions and by extrapolating data for the lighter actinides such as Np^{3+} or Pu^{3+}, the order of elution of the heavier actinides can be accurately forecast. Even a few *atoms* of the element can be identified because of the characteristic nuclear radiation.

The main problems in the separations are (*a*) separation of the actinides as a group from the lanthanide ions (which are formed as fission fragments in the bombardments which produce the actinides) and (*b*) separation of the actinide elements from one another.

The former problem can be solved by the use of concentrated hydrochloric acid as eluting agent; since the actinide ions form chloride complexes more easily, they are desorbed first from a cation-exchange resin, thus affecting a *group* separation; conversely, the actinides are more strongly adsorbed on anion-exchange resins. Although some of the actinide ions are themselves separated in the concentrated hydrochloric acid elutions on cationic columns, the resolution is not too satisfactory, particularly for Cf and Es. A more effective group separation employs $10M$ LiCl as eluant for a moderately cross-linked, strongly basic anion-exchange column operating at elevated temperatures (up to $\sim90°$). In addition to affording a lanthanide-actinide separation, fractionation of the actinide elements into groups Pu, Am—Cm, Bk, and Cf—Es can be obtained. Except for unexplained reversals ob-

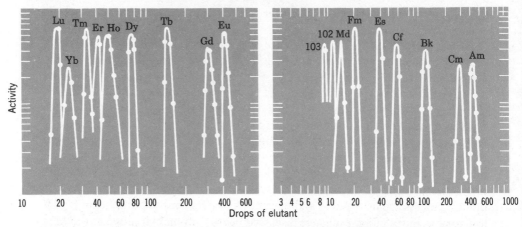

Fig. 24-7. Elution of lanthanide +3 ions (left) and actinide +3 ions (right) from Dowex 50 cation-exchange resin. Buffered ammonium 2-hydroxybutyrate was the eluant. The predicted positions of elements 102 and 103 (unobserved here) are also shown. [Reproduced by permission from J. J. Katz and G. T. Seaborg, The Chemistry of the Actinide Elements, Methuen, 1957.]

served in the elution order of Gd and Ho, and of Cm and Es, the elution sequences proceed in the order of increasing Z, with La the least strongly absorbed. Trivalent actinides (Am, Cm) can also be separated from lanthanides by solvent extraction.[68b]

The actinide ions are effectively separated from each other by elution with citrate or similar eluant; some typical elution curves in which the relative positions of the corresponding lanthanides are also given are displayed in Fig. 24-7. It will be noted that a very striking similarity occurs in the spacings of corresponding elements in the two series. There is a distinct break between Gd and Tb and between Cm and Bk, which can be attributed to the small change in ionic radius occasioned by the half-filling of the $4f$ and $5f$ shells, respectively. The elution order is not always as regular as that in Fig. 24-7. With some complexing agents (e.g., thiocyanate) more complicated elution orders are found, but nevertheless useful purifications (e.g., of Am) can be developed.

After separation by ion exchange, the actinides may be precipitated by fluoride or oxalate in macroscopic amounts or collected on an insoluble fluoride precipitate for trace quantities.

24-30. Compounds of the Elements

Curium. Solid curium compounds are known, for example, CmF_3, CmF_4, $CmCl_3$, $CmBr_3$, white Cm_2O_3 (m.p. 2265°) and black CmO_2. Where X-ray structural studies have been made—and these are difficult, since amounts of the order of 0.5×10^{-6} g must be used to avoid fogging of the film by radioactivity and because of destruction of the lattice by emitted particles—the compounds are isomorphous with other actinide compounds.

[68b] G. B. Kasting et al., J. Inorg. Nucl. Chem., 1979, **41**, 745.

In view of the position of Cm in the actinide series, numerous experiments have been made to ascertain whether Cm has only the +3 state in solution; no evidence for a lower state has been found. Concerning the +4 state, the potential of the Cm^{4+}/Cm^{3+} couple must be greater than that of Am^{4+}/Am^{3+}, which is 2.6 to 2.9 V, so that solutions of Cm^{4+} must be unstable. When CmF_4, prepared by dry fluorination of CmF_3, is treated with $15M$ CsF at 0°, a pale yellow solution is obtained that appears to contain Cm^{4+} as a fluoro complex. The solution exists for only an hour or so at 10° owing to reduction by the effects of α-radiation; its spectrum resembles that of the isoelectronic Am^{3+} ion.

The solution reactions of Cm^{3+} closely resemble those of the lanthanide and actinide +3 ions, and the fluoride, oxalate, phosphate, iodate, and hydroxide are insoluble. There is some evidence for complexing in solution, although the complexes appear to be weaker than those of preceding elements.

Magnetic measurements on CmF_3 diluted in LaF_3 and also the close resemblance of the absorption spectra of CmF_3 and GdF_3 support the hypothesis that the ion has the $5f^7$ configuration.

Berkelium. As the analogue of Tb, berkelium could be expected to show the +4 state and does so, not only in solid compounds but also in solution. Thus Bk^{3+} solutions can be oxidized by BrO_3^-, and the Bk^{4+} ion can be coprecipitated with Ce^{4+} or Zr^{4+} as phosphate or iodate. The ion can also be extracted (cf. Ce^{4+}) by hexane solutions of bis(2-ethylhexyl) hydrogen phosphate or similar complexing agents. Bk^{IV} is a somewhat weaker oxidant than Ce^{IV}.

Comparatively few solid compounds are known, viz., BkF_3 (LaF_3 type), $BkCl_3$ (UCl_3 type), BkOCl, BkF_4 (UF_4 type), Bk_2O_3, and BkO_2 (fluorite type), and the hydrate $[BkCl_2(H_2O)_6]Cl$. The resemblance to Ce^{IV} is also shown by the isolation of orange Cs_2BkCl_6; this is obtained by action of CsCl and Cl_2 on a concentrated HCl solution of the green hydrous oxide, which is in turn obtained by oxidizing Bk^{III} in $2M$ H_2SO_4 with BrO_3^- at 90° and adding ammonia.

Californium. Both CfF_4[69] and CfO_2[70] are known, and in the +3 state there are halides, oxohalides, oxide, and sulfide, and in the +2 state $CfBr_2$ and CfI_2.[71]

Einsteinium. The only isotope available in multimicrogram quantities is ^{259}Es, and its short half-life (20.45 days) creates experimental difficulties. $EsCl_3$, $EsBr_3$, EsOCl, and Es_2O_3[72] have been characterized.

Elements 100 to 103 and the Trans-Actinides.[73] The elements 100 to 103 appear to have the +3 state, but the +2 state seems to be more stable than the +2 state at the end of the lanthanide series. Thus Md^{3+} is readily reduced to Md^{2+}, and for No the +2 state is the more stable. The complexing of No^{2+} with carboxylate[74]

69 R. G. Haire and L. B. Asprey, *Inorg. Nucl. Chem. Lett.*, 1973, **9**, 869.
70 R. D. Baybarz et al., *J. Inorg. Nucl. Chem.*, 1972, **34**, 557.
71 J. R. Peterson and R. D. Baybarz, *Inorg. Nucl. Chem. Lett.*, 1972, **8**, 423; J. F. Wild, UCRL-77709, 1975.
72 R. G. Haire and R. D. Baybarz, *J. Inorg. Nucl. Chem.*, 1973, **35**, 489.
73 C. Keller, *Chem. Ztg.*, 1978, **102**, 437; G. T. Seaborg, *Chem. Eng. News*, April 16, 1979, p. 46; M.A.K. Lodhi, *Superheavy Elements*, Pergamon Press, 1978.
74 W. J. McDowell et al., *J. Inorg. Nucl. Chem.*, 1976, **38**, 1207.

indicates that it resembles Sr^{2+}. The half-lives of isotopes of elements 104, 105, and 106 are exceedingly short, the longest lived being $^{261}104$ (*ca.* 1 min), $^{262}105$ (*ca.* 40 sec), and $^{263}106$ (*ca.* 1 sec).

The elements from 104 onward should show characteristic group behavior, and indeed ion-exchange studies with ^{261}Rf indicate hafniumlike behavior.

There is still dispute regarding the first discovery of an isotope of element 104, and there is no confirmed report of heavier elements.

General References

Actinides, Vol. 5, *Comprehensive Inorganic Chemistry,* Pergamon Press, 1973.

Asprey, L. B., and R. A. Penneman, *Chem. Eng. News,* July 31, 1967, p. 75. A good general review.

Aten, A. H. W., Jr., Ed. *Actinide Reviews.* Elsevier, Vol. 1, 1968. Various topics including nuclear properties, separations, chemistry.

Bagnall, K. W., *Coord. Chem. Rev.,* 1967, **2,** 145. Coordination chemistry of actinide halides.

Bagnall, K. W., *The Actinide Elements,* Elsevier, 1972.

Bagnall, K. W., Ed., *Lanthanides and Actinides,* Butterworths, 1972.

Bulman, R. A., *Struct. Bonding,* 1978, **34,** 39. Chemistry of trans-uranium elements in biosphere.

Brown, D., *Halides of Lanthanides and Actinides,* Wiley-Interscience, 1968. An exhaustive monograph.

Comyns, A. E., *Chem. Rev.,* 1960, **60,** 115. A comprehensive review of the coordination chemistry of the actinides.

Friedman, A. M., Ed., *Actinides in the Environment,* ACS Symposium Series No. 35, 19XX.

Gmelins Handbuch der Anorganischen Chimie: Trans-uranium elements; New Supplements Series

    Part A   *Elements,* Vols. 7a, 7b, 8, 1973–1974
    Part B   *Metals and Alloys,* Vols. 31, 38, 39, 1970–1977
    Part C   *Compounds,* Vol. 4, 1972
    Part D   *Chemistry in Solution,* Vols. 20, 21, 1975

Hyde, E. K., I. Perlman, and G. T. Seaborg, *The Nuclear Properties of the Heavy Elements,* Vols. I–III, Prentice-Hall, 1964. Comprehensive reference treatise on nuclear structure, radioactive properties, and fission.

Katz, J. J., and G. T. Seaborg, *The Chemistry of the Actinide Elements,* Methuen, 1957. A lucidly written reference book.

Keller, C., *Angew. Chem., Int. Ed.,* 1965, **4,** 903. Synthesis of trans-curium elements by heavy ion bombardments.

Keller, C., *The Chemistry of the Transuranium Elements,* Verlag Chemie, 1971. Comprehensive monograph.

Kuroda, P. K., *Acc. Chem. Res.,* 1979, **12,** 73. ^{244}Pu in Nature.

Lanthanide/Actinide Chemistry, ACS Advances in Chemistry Series No. 71, 1967. Symposium reports.

Makarov, E. S., *Crystal Chemistry of Simple Compounds of U, Th, Pu, and Np* (transl.) Consultants Bureau, 1959.

Martin, F. S., and G. L. Miles, *Chemical Processing of Nuclear Fuels,* Butterworths, 1958. Procedures for isolating actinides from reactor fuels.

Roberts, L. E. J., *Q. Rev.,* 1961, **15,** 442. The actinide oxides.

Seaborg, G. T., *Man-Made Transuranium Elements,* Prentice-Hall, 1963. Lucid and well-illustrated introduction enriched with historical development.

Seaborg, G. T., *Transuranium Elements, Products of Modern Alchemy,* Academic Press, 1978.

Yaffé, L., Ed., *Nuclear Chemistry,* 2 vols., Academic Press, 1968.

4

SPECIAL TOPICS

CHAPTER TWENTY-FIVE

Metal Carbonyls and Other Complexes with π-Acceptor Ligands

A characteristic feature of the d-group transition metals is their ability to form complexes with a variety of neutral molecules such as carbon monoxide, isocyanides, substituted phosphines, arsines, stibines or sulfides, nitric oxide, various molecules with delocalized π orbitals such as pyridine, 2,2′-bipyridine, and 1,10-phenanthroline, and with certain ligands containing 1,2-dithioketone or 1,2-dithiolene groups, such as the dithiomaleonitrile anion. Very diverse types of complex exist, ranging from binary molecular compounds such as $Cr(CO)_6$ or $Ni(PF_3)_4$ through mixed species such as $Co(CO)_3NO$ and $(C_6H_5)_3PFe(CO)_4$, to complex ions such as $[Fe(CN)_5CO]^{3-}$, $[Mo(CO)_5I]^-$, $[Mn(CNR)_6]^+$, $[Vphen_3]^-$, and $\{Ni[S_2C_2(CN)_2]_2\}^{2-}$.

In many of these complexes the metal atoms are in low-positive, zero, or negative formal oxidation states. It is a characteristic of the ligands now under discussion that they can stabilize low oxidation states; this property is associated with the fact that these ligands possess vacant π orbitals in addition to lone pairs. These vacant orbitals accept electron density from filled metal orbitals to form a type of π bonding that supplements the σ bonding arising from lone-pair donation; high electron density on the metal atom—of necessity in low oxidation states—thus can be delocalized onto the ligands. The ability of ligands to accept electron density into low-lying empty π orbitals can be called π-*acidity,* the word "acidity" being used in the Lewis sense.

The basic bonding and structural considerations pertinent to metal carbonyls and nitrosyls, and the related N_2, RNC, and R_3P complexes were presented in Sections 3-7 through 3-13. This chapter builds upon that foundation to describe in detail the important types of compounds and their chemical and physical properties.

We can note at this point that the stoichiometries of many, though not all, of the complexes can be predicted by use of the *noble gas formalism* or *18-electron rule*. This requires that the number of valence electrons possessed by the metal atom plus the number of pairs of σ electrons contributed by the ligands be equal to the number of electrons in the succeeding noble gas atom, namely, eighteen. This is simply a phenomenological way of formulating the tendency of the metal atom to use its valence orbitals nd, $(n + 1)s$, and $(n + 1)p$, as fully as possible, in forming bonds to ligands. It is of considerable utility in the design of new compounds, particularly of metal carbonyls, nitrosyls, and isocyanides, and their substitution products, but it is by no means infallible. It fails altogether for the bipyridine and dithioolefin type of ligand, and there are numerous exceptions even among carbonyls such as $V(CO)_6$ and the stable $[M(CO)_2(diphos)_2]^+$ (M = Mo or W) ions.

In general these compounds have to be prepared by indirect methods from other compounds, although it is sometimes possible to combine metal and ligand directly. Ni is most reactive, combining directly with CO, CH_3PCl_2, and 1,2-bis(diethylphosphino)benzene. Co and Pd also combine with the last of these, and the metals Fe, Co, Mo, W, Rh, and Ru also combine with CO, but except for Ni and Fe, the reactions are too sluggish to be of practical value.

CARBON MONOXIDE COMPLEXES

The most important π-acceptor ligand is carbon monoxide. Many of its complexes are of considerable structural interest as well as having importance industrially and in catalytic and other reactions. Carbonyl derivatives of at least one type are known for all the transition metals.

25-1.　Mono- and Polynuclear Metal Carbonyls

We deal in this section with the compounds containing only metal atoms and CO ligands.

Mononuclear Carbonyls.　These compounds (listed in Table 25-1) are all hydrophobic liquids or volatile solids, soluble to varying degrees in nonpolar solvents. The M—C—O chains are essentially linear, and the bonding is of the type discussed in Section 3-7. All these molecules obey the eighteen-electron rule except $V(CO)_6$, which has a seventeen-electron configuration.

$V(CO)_6$.　Because this mononuclear species does not achieve an eighteen-electron configuration, it was for a while thought to dimerize under at least some conditions. It appears that this is not so, presumably because of steric resistance to increasing the coordination number of the metal atom. For an octahedral structure the d^5 configuration gives a $^2T_{2g}$ ground state that should be subject to Jahn-Teller distortion. In the vapor, electron diffraction results are consistent with O_h symmetry,[1] albeit with large vibrational amplitudes. The crystal structure at 244°K is built of octahedral molecules showing only marginal distortion from O_h

---

[1]　D. G. Schmidling, *J. Mol. Struct.*, 1975, **24**, 1.

TABLE 25-1
Mononuclear Binary Metal Carbonyls

Compound	Color and form	Structure[a]	Comments
$V(CO)_6$	Black solid; d. 70°; sublimes in vacuum	Octahedral,[b] $V—C = 2.008\ (3)$	Paramagnetic $(1e^-)$; yellow-orange in solution
$Cr(CO)_6$ $Mo(CO)_6$ $W(CO)_6$	Colorless crystals; all sublime readily	Octahedral: $Cr—C = 1.913(2)^c$ $Mo—C = 2.06(2)$ $W—C = 2.06(4)$	
$Fe(CO)_5$	Yellow liquid; m.p. $-20°$, b.p. $103°$	Tbp; $Fe—C^d =$ $1.810(2)$ ax, $1.833(2)$ eq	
$Ru(CO)_5$	Colorless liquid; m.p. $-22°$	Tbp (by ir)	Very volatile and difficult to obtain pure
$Os(CO)_5$	Colorless liquid; m.p. $-15°$	Tbp (by ir)	Very volatile and difficult to obtain pure
$Ni(CO)_4$	Colorless liquid; m.p. $-25°$	Tetrahedral $Ni—C = 1.838(2)^e$	Very toxic; musty smell; flammable; decomposes readily to metal

[a] M—C bond lengths are in angstroms. Figures in parentheses are error estimates, occurring in the least significant digit.

[b] See text for discussion of possible distortions.

[c] B. Rees and A. Mitschler, *J. Am. Chem. Soc.,* 1976, **98**, 7918.

[d] B. Beagley and D. G. Schmidling, *J. Mol. Struct.,* 1974, **22**, 466.

[e] L. Hedberg, T. Iijima, K. Hedberg, *J. Chem. Phys.,* 1979, **70**, 3224.

symmetry[2]; the magnetic moment corresponds to one unpaired electron. However at 66°K there is a phase transition and antiferromagnetic exchange is observed, until at 4.2°K all paramagnetism is lost; dimer formation is not believe to occur, however.[3] A dimer is obtained along with the monomer by cocondensing V atoms and CO at about 10°K, and a $(OC)_5V(\mu\text{-}CO)_2V(CO)_5$ structure has been proposed[4] following that earlier attributed to the mixed CO/CN anion[5] $[(OC)_4(NC)V(\mu\text{-}CN)_2V(CO)_4(CN)]^{4-}$.

$Cr(CO)_6$, $Mo(CO)_6$, $W(CO)_6$. These constitute the only complete family of carbonyls that are all stable and in common use. No polynuclear group VI carbonyls are known.

$Fe(CO)_5$, $Ru(CO)_5$, $Os(CO)_5$. These form the only other complete set, but the Ru and Os compounds are difficult to prepare, unstable, and seldom encountered.

$Ni(CO)_4$. This is the only carbonyl formed by the Ni, Pd, Pt group that is stable under normal conditions, although there is evidence for $Pd(CO)_4$ and $Pt_3(CO)_4$ in noble gas matrices at *ca.* 20°K.[6]

2   S. Bellard, K. A. Rubinson and G. M. Sheldrick, *Acta Crystallogr., B,* 1979, **35**, 271.

3   J. C. Bernier and O. Kahn, *Chem. Phys. Lett.,* 1973, **19**, 414.

4   T. A. Ford *et al., Inorg. Chem.,* 1976, **15**, 1666; T. C. Devore and H. F. Franzen, *Inorg. Chem.,* 1976, **15**, 1318.

5   D. Rehder, *J. Organomet. Chem.,* 1972, **37**, 303.

6   S. C. Tripathi *et al., Inorg. Chim. Acta,* 1976, **17**, 257 (review of Rh, Ir, Pd, Pt carbonyls).

TABLE 25-2
Binuclear Binary Metal Carbonyls

Compound	Color and form	Structure	Comments
$Mn_2(CO)_{10}$	Yellow; m.p. 154°	$(OC)_5MM(CO)_5$,	Sublime readily;
$Tc_2(CO)_{10}$	White; m.p. 160°	with D_{5d}	weak, reactive
$Re_2(CO)_{10}$	White, m.p. 177°	symmetry	M—M bonds
$Fe_2(CO)_9$	Shiny golden plates	Confacial bioctahedron	Insoluble in organic media
$Ru_2(CO)_9(?)^a$	Dark orange; dec. at RT	Unknown	
$Os_2(CO)_9{}^a$	Orange-yellow; m.p. 64–67°		Decomposes in few days at −20°

^{a} See Ref. 9.

Binuclear Carbonyls. These are listed in Table 25-2. $Mn_2(CO)_{10}$ (as well as its Tc and Re congeners) and $Co_2(CO)_8$ are the simplest molecules satisfying the eighteen-electron rule that these metals can form because they have odd atomic numbers. Thus $Mn(CO)_5$ and $Co(CO)_4$ are radicals with seventeen-electron

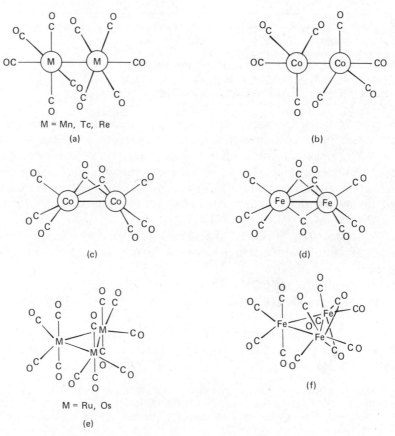

Fig. 25-1. The structures of some di- and polynuclear metal carbonyls.

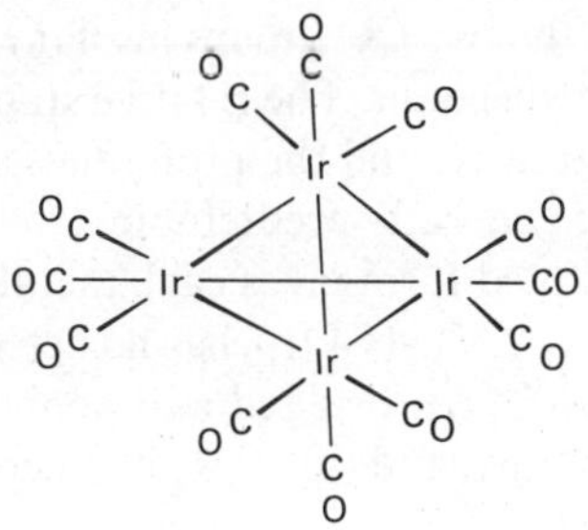

(n)

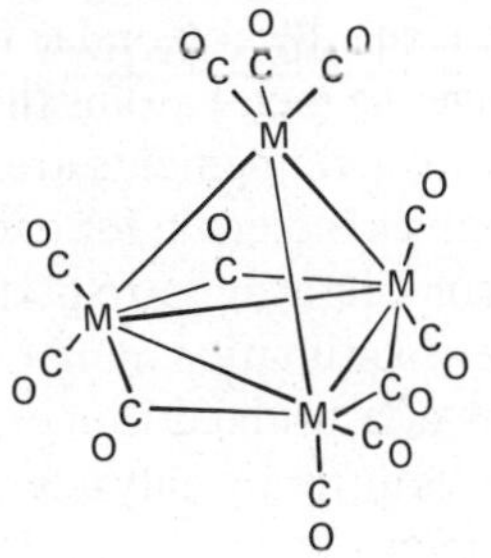

(h)

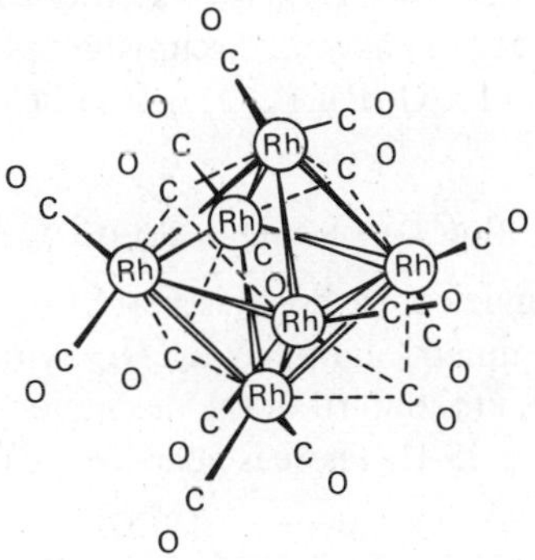

(i)

Fig. 25-1. Con't.

configurations, but by dimerizing through the formation of M—M single bonds, they complete their valence shells. In the case of the $M_2(CO)_{10}$ molecules, the formation of this bond completes the structure, which is that shown in Fig. 25-1a. For $Co_2(CO)_8$ the situation is more interesting, and the $(OC)_4Co-Co(CO)_4$ structure is only one of three that exist in equilibrium in solution.[7] Two of these are represented in Fig. 25-1b, and Fig. 25-1c; the structure of the third isomer is unknown. The crystalline compound contains molecules of structure c only, and this structure predominates in solution at low temperature.

$Co_2(CO)_8$ provides the first example encountered in this discussion of the occurrence of bridging carbonyl groups, and more important, of their occurrence in pairs. Moreover the occurrence of two forms of $Co_2(CO)_8$ illustrates the following important principle of carbonyl structure; the two structural units 25-Ia and 25-Ib

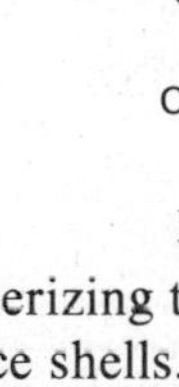

(25–Ia) 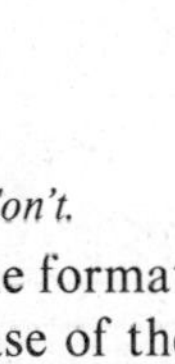   (25–Ib)

7 G. Bor and K. Noack, *J. Organometal. Chem.*, 1974, **64**, 367; S. Onaka and D. F. Shriver, *Inorg. Chem.*, 1976, **15**, 915; G. Bor, U. K. Dietler, and K. Noack, *J.C.S. Chem. Comm.*, **1976**, 914.

represent equally acceptable ways of distributing the two CO groups insofar as they each give the metal atoms the same formal electron count. The relative stabilities of the two arrangements are unlikely to differ greatly, and the preference shown in any given molecule for one or the other is not generally predictable. Doubtless steric considerations are pertinent, since the bridged form gives each metal atom a higher coordination number; this may explain why $Mn_2(CO)_{10}$ has no detectable amount of a bridged isomer. In the case of $Co_2(CO)_8$, bridged and nonbridged structures differ by only a few kJ mol^{-1} in free energy, and this close balance is not uncommon.

The Tc and Re analogues of $Mn_2(CO)_{10}$, as well as the mixed species MnRe-$(CO)_{10}$, are quite stable, and $Re_2(CO)_{10}$ is relatively common, whereas no Rh or Ir analogues of $Co_2(CO)_8$ exist in normal circumstances, though it has been shown that under high pressures of CO, $Rh_4(CO)_{12}$ (see below) dissociates reversibly according to the equation

$$Rh_4(CO)_{12} + 4CO = 2Rh_2(CO)_8$$

The other well-known binuclear carbonyl is $Fe_2(CO)_9$, whose structure[8] is shown in Fig. 25-1d. This very symmetrical molecule (D_{3d} symmetry) has three bridging CO groups. An alternative, in the spirit of the discussion of 25-Ia and 25-Ib, would be a singly bridged structure 25-II. There is no evidence for this in $Fe_2(CO)_9$ under

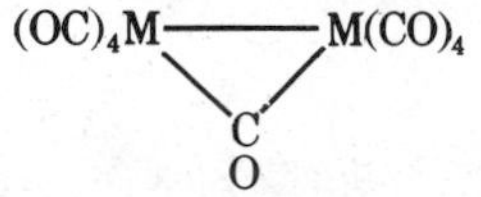

$$(OC)_4M\overset{\displaystyle\diagdown}{}\underset{\displaystyle\underset{O}{C}}{}M(CO)_4$$

(25–II)

any circumstances, but it appears that $Os_2(CO)_9$ and perhaps $Ru_2(CO)_9$ may have that structure.[9] It is clear that $Os_2(CO)_9$ does not have the D_{3d} structure, and the ir spectrum is consistent with one of type 25-II. It has been obtained by photolysis of $Os(CO)_5$ in heptane at $-40°$ and is stable for only a few days even at $-20°$.

Polynuclear Carbonyls.   There are relatively few of these compounds, which contain only metal atoms and CO groups, but the number of polynuclear species containing organic ligands as well is enormous. Other large classes of polynuclear compounds, some of which are considered in Section 25-7, are the metal carbonyl hydrides [e.g., $H_4Ru_4(CO)_{12}$] and carbonylate anions [e.g., $Pt_{18}(CO)_{36}^{2-}$], some of which, like that just mentioned, are of enormous size. We deal here only with the true binary carbonyls.

$M_3(CO)_{12}$ **Compounds.**   These are formed by Fe, Ru, and Os and include not only the homonuclear ones but several (though not yet all possible) mixed ones, e.g., $FeRu_2(CO)_{12}$ and $Fe_2Os(CO)_{12}$. All have the structure shown in Fig. 25-le except $Fe_3(CO)_{12}$, which has a structure approximately as shown in Fig. 25-lf. The former contains only terminal CO groups and has D_{3h} symmetry (in the homonuclear cases), whereas the $Fe_3(CO)_{12}$ structure is a highly unsymmetrical one. It can be thought of as derived from the D_{3h} structure by the rotation of one CO

8   F. A. Cotton and J. M. Troup, *J.C.S. Dalton*, **1974**, 800.
9   J. R. Moss and W. A. G. Graham, *J.C.S. Dalton*, **1977**, 95.

group on each of two metal atoms into bridging positions across the bond between those two atoms, with accompanying rotations of the residual $Fe(CO)_3$ moieties containing those metal atoms. It is, however, even more complex than this, since the CO bridges are not symmetrical. We return to this point in Section 25-4.

$M_4(CO)_{12}$ *Compounds.* These are formed by Co, Rh, and Ir and are the most stable binary carbonyls formed by the two latter elements. There are two structures for these molecules, each consisting of a tetrahedron of metal atoms. For $Ir_4(CO)_{12}$ the deployment of CO groups (Fig. 25-1g) is such as to conserve the full symmetry of a tetrahedron: there are three terminal CO groups on each metal atom with each Ir—C bond approximately trans to an Ir—Ir bond. For the Co and Rh compounds the structure has lower symmetry (Fig. 25-1h). One metal atom has three terminal CO groups, as in $Ir_4(CO)_{12}$, but the remaining nine CO groups occupy both symmetrical bridging positions and terminal positions around the triangle formed by the other three metal atoms; the molecule has C_{3v} symmetry. Five mixed species $M_nM'_{4-n}(CO)_{12}$ are also known,[10] and infrared spectra (see Section 25-9) show that they all have structures of the type shown in Fig. 25-1h.

Larger Clusters. The $M_6(CO)_{16}$ type is formed by Co, Rh, and Ir, and the Rh compound has the structure shown in Fig. 24-1i. There is an octahedron of Rh atoms with two terminal CO groups on each one, while the remaining four CO groups occupy triply bridging positions over four of the triangular faces of the octahedron. The whole arrangement is such as to preserve the elements of T_d symmetry.

The only other well-defined set of polynuclear binary metal carbonyls consists of those formed by osmium: $Os_5(CO)_{16}$, $Os_6(CO)_{18}$, $Os_7(CO)_{21}$ and $Os_8(CO)_{23}$. All these are discussed in some detail in Section 26-3.

25-2. Metal Nitrosyls

As noted in Sec. 3-11, the NO ligand in a linear M—N—O moiety can be regarded as a three-electron donor in the same formal sense as CO is a two-electron donor. On this basis a number of homoleptic nitrosyls [i.e., $M_x(NO)_y$] and mixed carbonyl nitrosyls can be foreseen from the formulas of the metal carbonyls just discussed.

Few homoleptic nitrosyls are known. Only $Cr(CO)_4$ is well documented.[11] Photolysis of $Cr(CO)_6$ in pentane in presence of excess NO gives a red solution from which the red-black solid is isolated. Spectral data indicate a tetrahedral structure, which would be expected by analogy with the isoelectronic $Ni(CO)_4$ [and $Mn(CO)_3CO$, to be described presently]. There is also a report[12] of $Co(NO)_3$, but this has never been confirmed. In another report the reaction of NO with $Co_2(CO)_8$ in hexane has been shown to give $Co(NO)_2NO_2$,[13] whose identity was proved by X-ray crystallography. Clarification of the status of $Co(NO)_3$ is needed. No other

10 P. Chini et al., *J. Organomet. Chem.*, 1973, **59**, 379.
11 M. Herberhold and A. Razavi, *Angew. Chem. Int. Ed.*, 1972, **11**, 1092; B. I. Swanson and S. K. Satija, *J.C.S. Chem. Comm.*, **1973**, 40.
12 I. H. Sabherwal and A. B. Burg, *J.C.S. Chem. Comm.*, **1970**, 1001.
13 C. E. Strouse and B. I. Swanson, *J.C.S. Chem. Comm.*, **1971**, 55.

homoleptic nitrosyls satisfying the eighteen-electron rule have been claimed, although there are old claims, lacking verification, for $Fe(NO)_4$ and $Ru(NO)_4$.

Mixed nitrosyl-carbonyls are more numerous. $Co(CO)_3NO$ and $Fe(CO)_2(NO)_2$, which are isoelectronic (and quasi-isostructural) with $Ni(CO)_4$, have long been known. For Mn there are two possibilities, consistent with the eighteen-electron rule, viz; $Mn(CO)_4NO$ and $Mn(NO)_3CO$, and both are known. The infrared spectra support the expected *tbp* and quasi-tetrahedral (C_{3v}) structures and show also that in the former the NO group is an equatorial position giving C_{2v} symmetry.[14] Although the $Cr(CO)_3(NO)_2$ is an obvious possibility, the only evidence for it was obtained[15] in an argon matrix at $10°K$, where the ir spectrum suggests that the NO groups occupy axial positions (D_{3h} symmetry).

As noted earlier (Section 3-11) doubly and even triply bridging NO groups are known. There are indeed many of the former in organometallic compounds, but we shall not discuss these further here.

23-3.   Homoleptic Compounds with Other π-Acceptor Ligands[16]

Isocyanides. The ability of RNC ligands to play the same role as CO has already been noted (Section 3-12) and a few compounds mentioned. Mononuclear complexes such as $M(CNR)_6$ [M = Cr, Mo, W] and $Ni(CNR)_4$ are well known. It appears likely that a far broader range of RNC analogous of metal carbonyls may be available. Thus $Fe_2(CNEt)_9$ has recently been prepared,[17] and shown to resemble $Fe_2(CO)_9$ closely; it has the same triply bridged structure, and nmr spectra show that like carbonyls in general, (Section 28-15) it readily undergoes intramolecular interchange of bridging and terminal ligands.

Phosphines, Arsines, Sulfides, etc. Complexes containing these ligands attached to low- or zerovalent metals in combination with CO are, of course, known by the thousands. In almost any CO-containing complex one or more CO groups can be replaced by an R_3P, R_3As, R_2S, etc. ligand. Basic aspects of this chemistry have been discussed in Sections 3-10 and 4-20. We note here only the homoleptic compounds with such ligands, which are naturally less numerous; however a considerable number of them can be isolated. The greater the electronegativity of the R group (or the average electronegativity for a mixed set of R groups), the more the R_3P type ligand can mimic the π-acceptor behavior of CO.

PF_3 is most nearly like CO, and many $M_x(PF_3)_y$ compounds are known. The first was $Ni(PF_3)_4$, which can be prepared by direct reaction of PF_3 with nickel. It is significant that $Pd(PF_3)_4$ and $Pt(PF_3)_4$ have also been made even though the analogous carbonyls do not appear to be stable at ordinary temperatures. Similarly, the entire sets of $M(PF_3)_5$ (M = Fe, Ru, Os) and $M(PF_3)_6$ (M = Cr, Mo, W) are also known. With Co, Rh, and Ir, the PF_3 ligand again gives compounds not known

[14]   A. J. Rest and D. J. Taylor, *J.C.S. Chem. Comm.*, **1977**, 717.
[15]   S. K. Satija *et al.*, *Inorg. Chem.*, 1978, **17**, 1737.
[16]   L. Malatesta and S. Cenini, *Zerovalent Compounds of Metals*, Academic Press, 1974.
[17]   F. G. A. Stone *et al.*, *J.C.S. Chem. Comm.*, **1978**, 1000.

with CO, namely, the entire set of $M_2(PF_3)_8$ molecules,[18] all of which presumably have the nonbridged structure analogous to that in Fig. 25-1b for $Co_2(CO)_8$.

Numerous other phosphine-type ligands and a few arsine and sulfide ones can also form stable homoleptic zerovalent metal complexes. These include chelating ones such as $(F_2P)_2NCH_3$, which gives the complete set of $M(LL)_3$ (M = Cr, Mo, W) compounds,[19] and several $R_2P(o\text{-}C_6H_4)PR_2$ ligands, as well as various phosphites that give[20], for example, $Fe[P(OMe)_3]_5$ and $Ni[P(OMe)_3]_4$.

25-4. Additional Structural Aspects of Metal Carbonyls

The structures of most metal carbonyl compounds can be formulated using only the following structural elements:

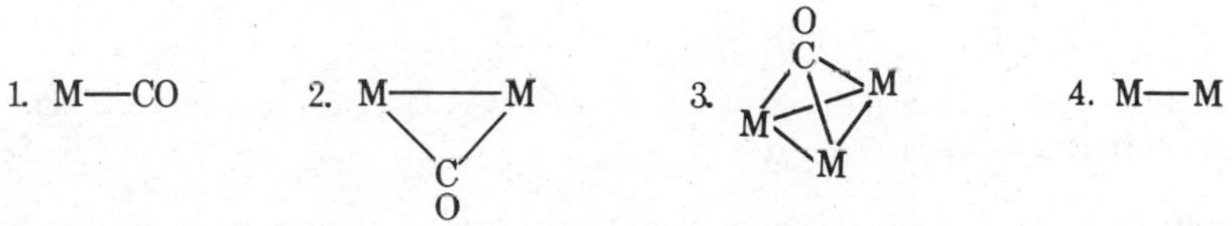

However as the number of metal carbonyl structures determined increases (and there is presently an exponential rate of growth), it becomes clear that there are also some structural elements besides the four shown above that have fairly general importance. One of these is the unsymmetrical bridge or *semibridging CO group*.

The Semibridging CO Group. This is characterized structurally by two features, as shown in 25-III: the two M C distances differ, and the M—C O angles are

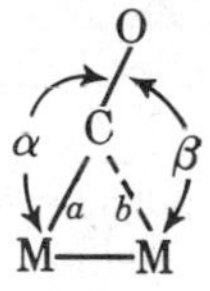

(25– III)

unequal. There appear to be at least three circumstances under which these occur.

1. *Cyclic Sets of Semibridging COs.* It has already been noted that each of the nonbridged and symmetrically double bridged arrangements, 25-Ia and 25-Ib, respectively, gives the metal atoms the same formal electron count and that the arrangements appear, in general, to differ little in inherent stability. In many cases it appears that not only are these two limiting forms of similar stability, but all the intermediate, symmetrical stages by which one can be converted to the other are also of similar energy. This idea is represented graphically in Fig. 25-2, which shows three possible ways in which the energy might vary as a function of the degree of bridge formation. In A the unbridged structure constitutes the preferred minimum

[18] M. A. Bennett and D. J. Patmore, *Inorg. Chem.*, 1971, **10**, 2387.
[19] R. B. King and J. Gimeno, *Inorg. Chem.*, 1978, **17**, 2390.
[20] C. A. Tolman, L. W. Yarbrough, and J. G. Verkade, *Inorg. Chem.*, 1977, **16**, 479.

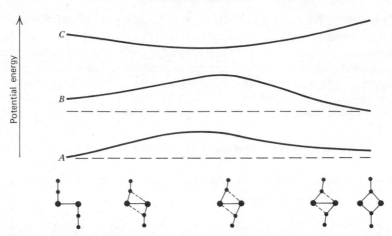

Fig. 25-2. Some ways in which the potential energy of an $M_2(CO)_2$ group may vary as a function of the degree of bridge formation.

and the fully bridged structure a less stable minimum; all symmetric intermediate stages of bridge formation are less stable. Situation B is similar to A except that the relative stabilities of the fully bridged and nonbridged structures are reversed. In case C the most stable conformations are those in which the bridges are incompletely formed. All these intermediate stages maintain a balanced charge distribution over the M—M unit because the two CO groups have at all points achieved equal but opposite degrees of bridge formation.

Case C is probably the least likely of the three shown in Fig. 25-2, but it is not logically impossible, since it occurs in the case of $Fe_3(CO)_{12}$ where the two bridges across one Fe—Fe bond are formed incompletely but in the balanced way indicated by the sketches in Fig. 25-2.

The partial formation of bridges in such a way as to keep the metal atoms equivalent is "a game any number can play," provided the system is cyclic. Thus for an equilateral triangle of metal atoms an arrangement of the sort shown in 25-IV

b---a
a $\;$ b
α $\;$ a
b---a
α

(25–IV)

is "balanced" and just such an arrangement, with mean values of $a \approx 1.82$ Å, $b \approx 2.28$ Å and $\alpha \approx 155°$ is found in the carbonylate anion $[Fe_4(CO)_{13}]^{2-}$. (see Section 25-6). In $Rh_4(CO)_{12}$, on the other hand, $a = b$ and we have a set of fully formed bridges.

2. *Semibridging COs in Inherently Unsymmetrical Environments.* In a number of cases only one CO group is semibridging, and in a few there are two, but they are both closest to the same metal atom. Obviously such cases do not represent the sort of situation just discussed: equivalence of the metal atoms is not

possible. In the majority of these cases an explanation for the semibridging posture of one or more CO ligands can be given in terms of an inequality of metal atom charges caused by the distribution of the other ligands in the molecules.[21] A typical example is shown in 25-V, where the circled CO group is distinctly bent (Fe_1—

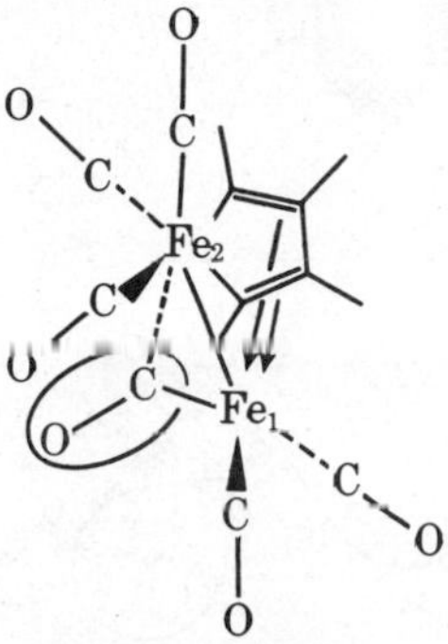

(25-V)

$C—O \approx 165°$) and the two Fe—C distances are ~1.80 and ~2.35 Å; the former distance is approximately normal for a terminal CO group, but the latter is too short to be merely a nonbonded contact. This may be explained in the following way. Ignoring for the moment the presence of the Fe—Fe bond, we can count electrons around each iron atom in the following way: Fe_1 has eight electrons of its own, acquires six more from the three CO groups and four more from the double bonds; this gives it eighteen. Fe_2 has eight electrons of its own, plus six from the three CO groups, and only two more from the two Fe—C bonds. This gives it only sixteen. If each metal atom is to have an eighteen-electron valence shell, the Fe—Fe bond must be considered as a dative bond and written as $\overset{+}{F}_1 \rightarrow \bar{F}e_2$ with resultant separation of charges. The semibridging CO group then serves to mitigate this charge separation by taking charge from Fe_2 in the way shown in Fig. 25-3.

Another example of this type of semibridging CO group is found in 25-VI. The case can be analyzed as follows. We begin with a structure of the type 25-II and

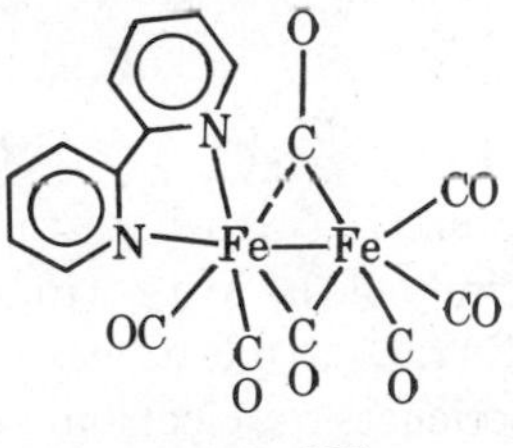

(25-VI)

assume that the dipyridyl ligand replaces two CO groups on one of the iron atoms. This causes the substituted iron atom to bear considerably more negative charge than the other one because dipy is a much better donor and a much less effective π acceptor than two CO groups. One of the CO groups on the other iron atom then

[21]   F. A. Cotton, *Prog. Inorg. Chem.*, 1976, **21**, 1.

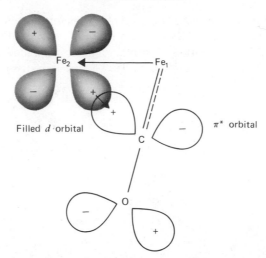

Fig. 25-3. Partial transference of electron density from a filled d orbital of one metal atom into one of the π^* orbitals of a CO group that is principally bonded to an adjacent metal atom.

moves into a semibridging position to help remove some of this excessive electron density.

3. *Semibridging COs Caused by Steric Crowding.* There are a few cases in which a lone semibridging CO group occurs without any apparent electronic cause or function. It is presumed that the CO group is forced into this position by steric crowding in the molecule. Perhaps the most unambiguous examples are provided by molecules of formula $(\eta^5\text{-}C_5H_5)_2Mo_2(CO)_4(RC{\equiv}CR)$, which might have been expected to have the symmetric structure 25-VIIa, whereas they actually have the

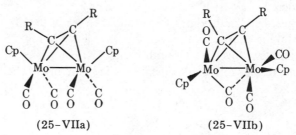

(25–VIIa) (25–VIIb)

unsymmetrical, twisted structure 25-VIIb.[22] In this structure one CO group is apparently forced into the vicinity of the Mo—Mo bond, opposite to the bridging $RC{\equiv}CR$ group, to allow the Cp rings to move away from the R groups. Once there, however, this CO group experiences an electronic interaction such that its CO stretching frequency is lowered (to *ca.* 1840 cm^{-1}) just as in cases 1 and 2.

Side-on Bonding of CO and CNR. This has never been observed for CO in the simple form represented by 25-VIIIa, but it is known in the situation depicted in 25-VIIIb, that is, in combination with terminal bonding, in the binuclear molecule[23]

[22] W. I. Bailey, Jr., *et al., J. Am. Chem. Soc.,* 1978, **100,** 5764.
[23] (a) R. Colton and C. J. Commons, *Aust. J. Chem.,* 1975, **28,** 1673; (b) C. J. Commons and B. F. Hoskins, *Aust. J. Chem.,* 1975, **28,** 1663.

$$M \leftarrow \overset{\displaystyle C}{\underset{\displaystyle O}{\|}}$$

(25–VIIIa)

(25–VIIIb)

$(OC)_2Mn(PH_2PCH_2PPh_2)_2(\mu\text{-}CO)Mn(CO)_2$. Such a CO is a four-electron donor, providing two σ electrons to one metal atom and two π electrons to the other. Recently an analogue of this molecule having a $\mu\text{-}CNC_6H_4CH_3$ ligand of the same sort has been reported.[24]

25-5. Preparation of Metal Carbonyls

Only for nickel and iron is direct reaction of metal with carbon monoxide feasible. Finely divided nickel will react at room temperature; an appreciable rate of reaction with iron requires elevated temperatures and pressures. In all other cases the carbonyls are prepared from metal compounds under reductive conditions. Common reducing agents are sodium, aluminum alkyls, or CO itself, sometimes mixed with hydrogen. The detailed procedures, however, are so enormously varied[25] that we can give here only a few illustrative examples.

Dicobalt octacarbonyl is prepared by the reaction at 250 to 300 atm pressure and 120 to 200°:

$$2CoCO_3 + 2H_2 + 8CO \rightarrow Co_2(CO)_8 + 2CO_2 + 2H_2O$$

Other binary carbonyls are made from the metal halides. The general method is to treat them, usually in suspension in an organic solvent such as tetrahydrofuran, with carbon monoxide at 200 to 300 atm pressure and temperatures up to 300° in presence of a reducing agent. A variety of reducing agents have been employed— electropositive metals like Na, Al, or Mg, trialkylaluminums, copper, or the sodium ketyl of benzophenone (Ph_2CONa). The detailed course of the reactions is not well known, but when organometallic reducing agents are employed it is likely that unstable organo derivatives of the transition metal are formed as intermediates. Vanadium carbonyl is most easily obtained by the reaction

$$VCl_3 + CO + Na(\text{excess}) \xrightarrow[\text{120° 5000 psi}]{\text{diglyme}} [Na\,\text{diglyme}_2][V(CO)_6] \xrightarrow[\text{then sublime 50°}]{H_3PO_4} V(CO)_6$$

Some carbonyls, e.g., $Os(CO)_5$, $Tc_2(CO)_{10}$ and $Re_2(CO)_{10}$ may be prepared from the oxides $(OsO_4, Tc_2O_7, Re_2O_7)$ by action of CO at high temperatures ($\sim$300°) and pressures ($\sim$300 atm).

Metal acetylacetonates in organic solvents often form suitable starting materials. For example,

$$Ru(C_5H_7O_2)_3 \xrightarrow[\text{150°}]{CO + H_2 \text{ at } 200 \text{ atm}} Ru_3(CO)_{12}$$

[24]   L. S. Benner, M. M. Olmstead, and A. L. Balch, *J. Organomet. Chem.*, 1978, **159**, 289.
[25]   See E. W. Abel and F. G. A. Stone, *Q. Rev.*, 1970, **24**, 498, for a more extensive survey.

The binuclear carbonyl $Fe_2(CO)_9$ is obtained as orange mica-like plates by photolysis of $Fe(CO)_5$ in hydrocarbon solvents. The green $Fe_3(CO)_{12}$ is best made by acidification of a polynuclear carbonylate anion (see below), which in turn is obtained from $Fe(CO)_5$ by the action of organic amines such as triethylamine.

The rational and systematic synthesis of mixed metal di- and polynuclear carbonyls poses special problems. Procedures in which potential constituents are mixed and caused to react randomly to give a more or less statistical mixture of all possible products are generally unattractive because separation is usually difficult and inefficient, even by chromatography. For lack of anything better, however, many mixed metal carbonyl compounds are currently obtainable only by such reactions.

The most widely applicable basis for rational synthesis depends on the reaction of carbonylate anions with metal carbonyl halides causing elimination of halide ion. The general reaction may be represented by:

$$[M_n(CO)_x]^- + XM'(CO)_y \xrightarrow{\pm zCO} M_nM'(CO)_{x+y\pm z} + X^-$$

We describe both of these classes of compound in detail presently, but the following reactions may be cited to illustrate this synthetic approach, which is so designed that like metal atoms cannot react with each other:

$$[Mn(CO)_5]^- + BrRe(CO)_5 \longrightarrow (OC)_5MnRe(CO)_5 + Br^-$$

$$2[Co(CO)_4]^- + \textit{trans-}PtCl_2py_2 \longrightarrow (OC)_4Co-\underset{\underset{py}{|}}{\overset{\overset{py}{|}}{Pt}}-Co(CO)_4 + 2Cl^-$$

A variation on this procedure, leading to mixed metal carbonyl hydrides, is illustrated by the following:

$$Os_3(CO)_{12} + Re(CO)_5^- \xrightarrow{then\ H^+} H_3ReOs_3(CO)_{12}$$

$$Ru_3(CO)_{12} + Co(CO)_4^- \xrightarrow{then\ H^+} HCoRu_3(CO)_{13}$$

25-6.   Carbonylate Anions and Carbonyl Hydrides[26]

It was first shown by W. Hieber that if iron pentacarbonyl is treated with aqueous alkali, it dissolves to give an initially yellow solution containing the ion $HFe(CO)_4^-$, which on acidification gives a thermally unstable gas $H_2Fe(CO)_4$. Carbonylate anions have been obtained from most of the carbonyls. Some do not give hydrides on acidification; however the carbonylate ions of Mn, Re, Fe, and Co certainly do.

The carbonylate anions can be obtained in a number of ways—by treating car-

[26]   J. E. Ellis, *J. Organomet. Chem.*, 1975, **86**, 1; R. Bau, Ed., *Transition Metal Hybrides,* ACS Advances in Chemistry Series No. 167, 1978.

bonyls with aqueous or alcoholic alkali hydroxide or with amines, sulfoxides or other
Lewis bases, by cleaving metal-metal bonds with sodium, or in special cases by
refluxing carbonyls with salts in an ether medium. Illustrative examples are

$$Fe(CO)_5 + 3NaOH(aq) \rightarrow Na[HFe(CO)_4](aq) + Na_2CO_3(aq) + H_2O$$

$$Co_2(CO)_8 + 2Na/Hg \xrightarrow{\text{THF}} 2Na[Co(CO)_4]$$

$$Mn_2(CO)_{10} + 2Li \xrightarrow{\text{THF}} 2Li[Mn(CO)_5]$$

$$Mo(CO)_6 + KI \xrightarrow{\text{diglyme}} [K(diglyme)_3]^+[Mo(CO)_3I]^- + CO$$

$$2Co^{2+}(aq) + 11CO + 12OH^- \xrightarrow{\text{KCN (aq)}} 2[Co(CO)_4]^- + 3CO_3^{2-} + 6H_2O$$

$$TaCl_5 + 6CO + Na(excess) \xrightarrow[\text{3000–5000 psi}]{\text{diglyme, } \sim 100°} [Na(diglyme)_2]^+[Ta(CO)_6]^-$$

The simpler carbonylate ions, like the binary carbonyls, obey the noble gas formalism, and their stoichiometries can thus be predicted. The ions are usually fairly
readily oxidized by air; the alkali metal salts are soluble in water, from which they
can be precipitated by large cations such as $[Co(NH_3)_6]^{3+}$ or $[Ph_4As]^+$.

In addition to mononuclear carbonylate anions, a variety of polynuclear species
have been obtained. The iron carbonylate ions, which have been much studied, are
obtained by the action of aqueous alkali or Lewis bases on binary carbonyls or in
other ways, for example,

$$Fe_2(CO)_9 + 4OH^- \rightarrow [Fe_2(CO)_8]^{2-} + CO_3^{2-} + 2H_2O$$

$$2Fe_3(CO)_{12} + 7OH^- \rightarrow [Fe_3(CO)_{11}]^{2-} + [HFe_3(CO)_{11}]^- + 2CO_3^{2-} + 3H_2O$$

$$Fe(CO)_5 + Et_3N \xrightarrow{\text{H}_2\text{O, 80°}} [Et_3NH][HFe_3(CO)_{11}]$$

$$[HF_3(CO)_{11}]^- + Fe(CO)_5 \rightarrow [HFc_4(CO)_{13}]^- + 3CO$$

The $Fe_2(CO)_8^{2-}$ ion has a direct Fe—Fe bond with no bridges; the structure is essentially that shown in Fig. 25-1b.[27] The $HFe_3(CO)_{11}^-$ ion has the structure shown
in Fig. 25-4a, which can be considered analogous to that of $Fe_3(CO)_{12}$ with one
bridging CO group replaced by a bridging H^-. The structure of the $Fe_4(CO)_{13}^{2-}$
ion (Fig. 25-4b) provided one of the earliest examples of a cyclic set of semibridging
carbonyl groups, as already discussed in Section 25-4.

The simplest of the iron carbonylate ions $[Fe(CO)_4]^{2-}$ has been extensively
studied. In the salt $[Na(crypt-222)]_2Fe(CO)_4$ it has an essentially undistorted
tetrahedral structure,[28] as expected by analogy to the isoelectronic $Cr(NO)_4$ and
$Ni(CO)_4$. Figure 25-5 shows some of its reactions, which may be taken as typical
for metal carbonylate anions. It will be seen that for the most part, these reactions
are consequences of the nucleophilicity of the carbonylate anion. The compound
$Na_2Fe(CO)_4 \cdot 1.5$ dioxane can be used in a pseudo-Grignard type of chemistry to

[27]   R. Bau et al., J. Am. Chem. Soc., 1974, 96, 5285.
[28]   R. Bau et al., J. Am. Chem. Soc., 1977, 99, 1104.

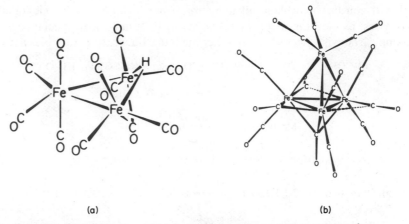

(a)   (b)

Fig. 25-4. The structures of (a) the $HFe_3(CO)_{11}^-$ ion and (b) the $Fe_4(CO)_{13}^{2-}$ ion.

synthesize aldehydes, ketones, acids, and esters from alkyl halides[29] as in the following reactions:

$$Na_2Fe(CO)_4 + RX \longrightarrow [RFe(CO)_4]^-$$

$$[RCFe(CO)_4]^-$$

with branches to R'X / CO, and below:

RCHO   RCO_2H   RCOOR'

(arrows labeled H$^+$, O_2/H_2O, I_2, R'OH)

There is now a large number of polynuclear carbonylate anions formed particularly by Co, Rh, and Pt. Since these are of interest more because they are metal atom cluster species than because of their chemistry as carbonylate anions, the subject is discussed in Chapter 26.

Carbonyl Hydrides. In many cases hydrides corresponding to the carbonylate anions can be isolated. A few of the simpler ones are listed in Table 25-3 along with their main properties.

Many of the carbonyl hydrides are rather unstable substances. They can be obtained by acidification of the appropriate alkali carbonylates or in other ways. Examples of the preparations are

$$NaCo(CO)_4 + H^+(aq) \rightarrow HCo(CO)_4 + Na^+(aq)$$

$$Fe(CO)_4I_2 \xrightarrow{\text{NaBH}_4 \text{ in THF}} H_2Fe(CO)_4$$

$$Mn_2(CO)_{10} + H_2 \xrightarrow[200°]{200 \text{ atm}} 2HMn(CO)_5$$

$$Co + 4CO + \tfrac{1}{2}H_2 \xrightarrow[150°]{50 \text{ atm}} HCo(CO)_4$$

[29] J. P. Collman, *Acc. Chem. Res.,* 1975, **8**, 342.

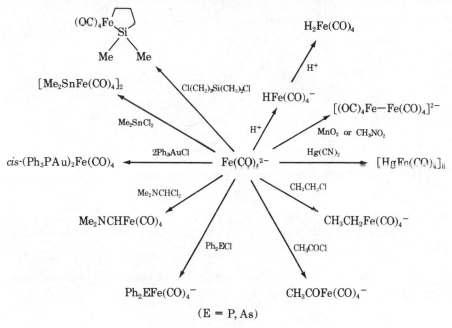

Fig. 25-5. Some reactions of [Fe(CO)₄]²⁻.

The iron and cobalt carbonyl hydrides form pale yellow solids or liquids at low temperatures and in the liquid state begin to decompose above about −10° and −20°, respectively; they are relatively more stable in the gas phase, however, particularly when diluted with carbon monoxide. They both have revolting odors and are readily oxidized by air. $HMn(CO)_5$ is appreciably more stable.

TABLE 25-3
Some Carbonyl Hydrides and Their Properties

Compound	Form	M.p. (°C)	M—H stretching frequency (cm⁻¹)	τ Value[a]	Comment
$HMn(CO)_5$	Colorless liquid	−25	1783	17.5	Stable liquid at 25°; weakly acidic
$H_2Fe(CO)_4$	Yellow liquid, colorless gas	−70	?	21.1	Decomposes at −10° giving H_2 + red $H_2Fe_2(CO)_8$
$H_2Fe_3(CO)_{11}$	Dark red liquid		?	25	
$HCo(CO)_4$	Yellow liquid, colorless gas	−26	~1934	20	Decomposes above m.p. giving H_2 + $Co_2(CO)_8$
$HW(CO)_3(\eta\text{-}C_5H_5)$	Yellow crystals	69	1854	17.5	Stable short time in air
$[HFe(CO)_3(PPh_3)_2]^+$	Yellow	—	?	17.6	Formed by protonation of $Fe(CO)_3(PPh_3)_2$ in H_2SO_4

[a] τ value is position of high-resolution proton magnetic resonance line in parts per million referred to tetramethylsilane reference as 10.00.

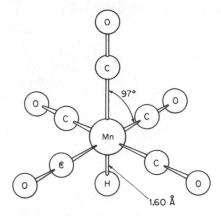

Fig. 25-6. The structure of the $HMn(CO)_5$ molecule, showing both the stereochemical activity of the hydrogen atom and the metal-to-hydrogen distance, which approximates to the sum of normal covalent radii.

The hydrides are not very soluble in water, but in water they behave as acids, ionizing to give the carbonylate ions:

$$\begin{array}{ll} HMn(CO)_5 = H^+ + [Mn(CO)_5]^- & pK \sim 7 \\ H_2Fe(CO)_4 = H^+ + [HFe(CO)_4]^- & pK_1 \sim 4 \\ [HFe(CO)_4]^- = H^+ + [Fe(CO)_4]^{2-} & pK_2 \sim 13 \\ HCo(CO)_4 = H^+ + [Co(CO)_4]^- & \text{strong acid} \end{array}$$

Like hydride complexes in general, the carbonyl hydrides generally exhibit sharp M—H stretching bands in the infrared and proton nuclear resonance absorptions at very high τ values as shown in Table 25-3.

The structural role of hydrogen in the mononuclear carbonyl hydrides and in other cases where it is nonbridging is straightforward. It occupies a distinct coordination site, and the M—H distance is approximately the sum of covalent radii (i.e., 1.6—1.7 Å). A good example is provided by $HMn(CO)_5$ (Fig. 25-6). These terminal hydrogen ligand atoms are stereochemically the same in carbonyl hydrides as they are in hydride complexes generally, though chemically they differ from most of the latter in being characteristically acidic because of the ability of carbonylate anions to delocalize the negative charge.

Bridging hydrogen atoms in bi- and polynuclear metal carbonyl hydrides play a much more subtle and not as yet fully understood structural role.[30a] It is only within the last few years that structural information necessary to the understanding of these questions has begun to become available, mainly as a result of neutron diffraction studies where the positions of hydrogen atoms can be determined directly and accurately.[30b]

Even bridged dinuclear species display surprising features. When an X-ray study of $[Et_4N][Cr_2(CO)_{10}H]$ showed that the $Cr_2(CO)_{10}$ skeleton has an eclipsed structure with D_{4h} symmetry and a Cr···Cr distance of 3.41 Å, it seemed reasonable

[30a] R. Hoffmann et al., J. Am. Chem. Soc., 1978, 100, 6088.
[30b] R. Bau et al., Acc. Chem. Res., 1979, 12, 176.

to assume that the hydrogen atom would be located along the Cr···Cr, line probably at the midpoint. A later neutron diffraction investigation has shown, however, that the bridging H atom, though equidistant from the two metal atoms, lies about 0.3 Å off the Cr···Cr line. The analogous $[W_2(CO)_{10}H]^-$ ion has also been studied by neutron diffraction. In $[Et_4N][W_2(CO)_{10}H]$ the anion is bent (i.e., the $W_2(CO)_{10}$ skeleton does not have a 4-fold symmetry axis), and the two W—H distances are very different (viz. 1.72 and 2.07 Å). It is almost as though a $[W(CO)_5H]^-$ ion is attached through the H atom to a square-pyramidal $W(CO)_5$ group. However in $[Ph_4P][W_2(CO)_{10}H]$, the W—H distances are equal (1.86 Å). The one feature that all these structures seem to share is that the H atom does not lie at the point of intersection of the local 4-fold axes of the M(CO)$_5$ groups, but outside of it (25-IX). It seems, therefore, that the bridge bonding system involves some direct M—M overlap as well as M—H bonding.

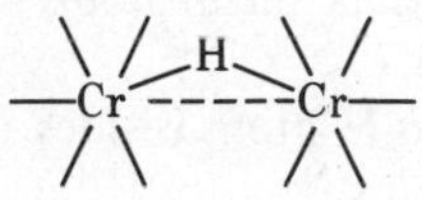

(25–IXa) 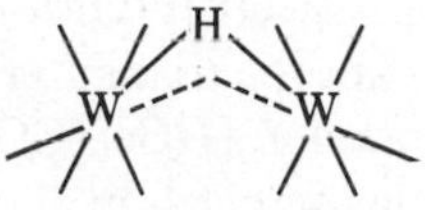 (25–IXb)

Another type of binuclear carbonyl hydride is represented by the anion $[H_2W_2(CO)_8]^{2-}$ and the isoelectronic neutral molecule $H_2Re_2(CO)_8$, which have the structure 25-X.

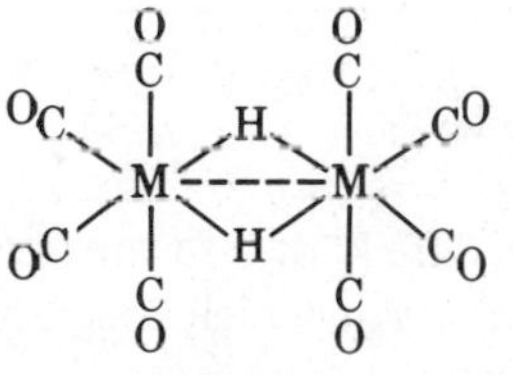

(25–X)

The structures of a number of tri- and polynuclear metal carbonyl hydrides have now been established,[26,30,31] and some are shown schematically as 25-XI to 25-XVI. Inspection of these six structures, which are presumably typical, shows that hydrogen atoms prefer bridging to terminal positions; only in 25-XII is there a terminal

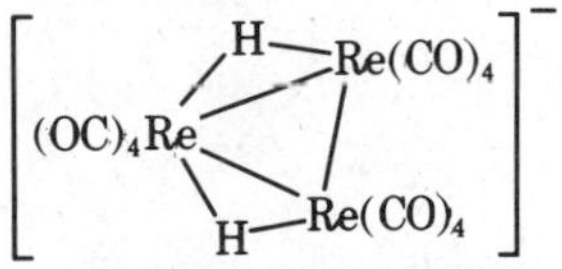

(25–XI) 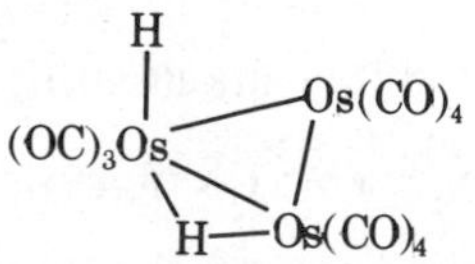 (25–XII) 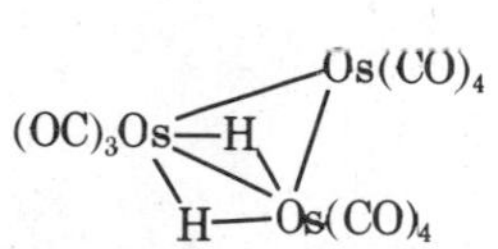 (25–XIII)

H atom. The structures 25-XV and 25-XVI present an interesting contrast. In 25-XV we have the symmetrical disposition of four H atoms on the four tetrahedron faces, thus preserving full tetrahedral symmetry. In 25-XVI, on the other hand, the H atoms are disposed in the least symmetrical way, on four edges; note that

[31] R. D. Wilson and R. Bau, *J. Am. Chem. Soc.*, 1976, **98**, 4687; J. R. Shapley *et al., J. Am. Chem. Soc.*, 1977, **99**, 7384; A. J. Schultz *et al., Inorg. Chem.*, 1979, **18**, 319.

$$Re(CO)_3$$

H

$$(OC)_3Re$$ $$Re(CO)_3$$

$$Re(CO)_3$$

$$Ru(CO)_3$$

H

$$(OC)_3Ru$$ $$Ru(CO)_3$$

H

$$Ru(CO)_3$$

$$(OC)_3Os\overset{Os(CO)_4}{\underset{H-Os(CO)_4}{\overset{|}{\underset{|}{-CH_2}}}}_H$$

(25-XIV)   (25-XV)   (25-XVI)

if two opposite rather than two adjacent edges had been left unbridged, the symmetry would have been D_{2d} and the four metal atoms would have been equivalent. In addition to 25-XV, there are other examples of triply bridging H atoms, one being provided by $HFeCo_3(CO)_9[P(OMe)_3]_3$, where the metal atoms form a tetrahedron and the H atom caps the triangular Co_3 face.

In the hydrido anion $[HCo_6(CO)_{15}]^-$ the H atom resides in the center of a somewhat irregular octahedron of cobalt atoms.[26]

In conclusion it should be emphasized that although we have dealt here specifically with the metal carbonyl hydrides, these are only one type among many transition metal hydrido complexes. Other types have been mentioned under the chemistry of individual elements and are discussed further in Chapters 27 and 29.

25-7. Metal Carbonyl Halides

Carbonyl halides $M_x(CO)_yX_z$ are known for most of the elements forming binary carbonyls but also for Pd, Pt, and Au, which do not form binary carbonyls; Cu^I and Ag^I carbonyl halides also exist.

The carbonyl halides are obtained either by the direct interaction of metal halides and carbon monoxide, usually at high pressure, or in a few cases by the cleavage by halogens of polynuclear carbonyls, for example,

$$Mn_2(CO)_{10} + Br_2(1) \xrightarrow{40°} 2Mn(CO)_5Br \underset{CO\ 150\ atm}{\overset{in\ petroleum\ 120°}{\rightleftharpoons}} [Mn(CO)_4Br]_2 + 2CO$$

$$RuI_3 + 2CO \xrightarrow{220°} [Ru(CO)_2I_2]_n + \tfrac{1}{2}I_2$$

$$2PtCl_2 + 2CO \rightarrow [Pt(CO)Cl_2]_2$$

A few examples of the halides and some of their properties are listed in Table 25-4. Carbonyl halide anions are also known; they are often derived by reaction of ionic halides with metal carbonyls or substituted carbonyls:

$$M(CO)_6 + R_4N^+X^- \xrightarrow{diglyme} R_4N^+[M(CO)_5X]^- + CO \qquad M = Cr, Mo, W$$

$$Mn_2(CO)_{10} + 2R_4N^+X^- \rightarrow (R_4N^+)_2[Mn_2(CO)_8X_2]^{2-} + 2CO$$
$$(R_4N^+)_2[Mn_2(CO)_8X_2]^{2-} + 2R_4N^+Y^- \rightarrow 2(R_4N^+)_2[Mn(CO)_4XY]^{2-}$$
$$M(CO)_4(bipy) + 2KCN \rightarrow K_2[M(CO)_4(CN)_2] + bipy \qquad M = Cr, Mo, W$$

TABLE 25-4
Some Examples of Carbonyl Halide Complexes

Compound	Form	M.p. (°C)	Comment
$Mn(CO)_5Cl$	Pale yellow crystals	Sublimes	Loses CO at 120° in organic solvents; can be substituted by pyridine, etc.
$[Re(CO)_4Cl]_2$	White crystals	Dec. >250	Halogen bridges cleavable by donor ligands or by CO (pressure)
$[Ru(CO)_2I_2]_n$	Orange powder	Stable >200	Halide bridges cleavable by ligands
$[Pt(CO)Cl_2]_2$	Yellow crystals	195; sublimes	Hydrolyzed H_2O; PCl_3 replaces CO

The structures of the carbonyl halides present little problem; when they are dimeric or polymeric they are invariably bridged through the halogen atoms and *not* by carbonyl bridges, for example, in 25-XVII and 25-XVIII. The halogen bridges can

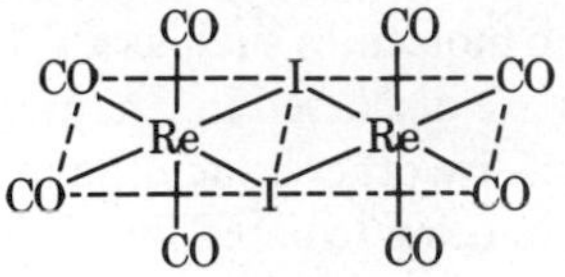

(25–XVII) 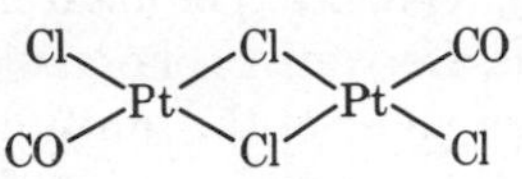   (25–XVIII)

be broken by numerous donor ligands such as pyridine, substituted phosphines, and isocyanides. The breaking of halogen bridges by other donor ligands is not of course confined to the carbonyl halides, and other bridged halides such as those given by olefins can be cleaved. As an example we may cite the reaction

$$[Mn(CO)_4I]_2 + 4py \rightarrow 2Mn(CO)_3Ipy_2 + 2CO$$

The initial product of the cleavage in this reaction is 25-XIX, but the reaction can proceed further, and the product 25-XX is isolated. This occurs because in 25-XIX two of the CO groups are *trans* to each other, thus will be competing across

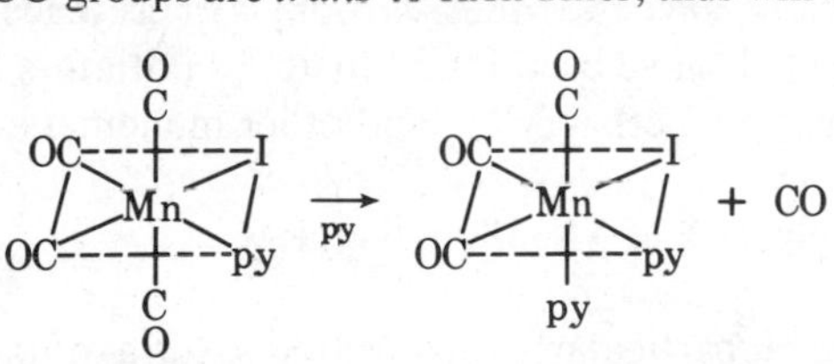

(25–XIX)   (25–XX)

the metal atoms for the same metal π-bonding orbitals. Hence in the presence of any ligand like a nitrogen, phosphorus, or arsenic donor, of *lower* π-bonding requirement or capacity compared to CO, one of the *trans* CO groups will be displaced. It follows that the two pyridine (or other) ligands inserted must appear in the *cis* position to each other. This type of labilization of groups in certain stereochemical situations will be discussed under the *trans effect* (Section 28-7). We can also note that in 25-XX, which is resistant to further displacement of CO, three of the octahedral positions are occupied by essentially non-π-bonding ligands, so that the remaining three CO groups must be responsible for the delocalization of the negative charge on the metal atom; they will, however, now have the exclusive use of the electrons in the d_{xy}, d_{yz}, and d_{xz} metal orbitals for π bonding, hence the metal-carbon bonding is about at a maximum in 25-XX and similar derivatives.

25-8. Reactions of Metal Carbonyls

The number of carbonyls and the variety of their reactions is so enormous that only a few types of reaction can be mentioned. For $Mo(CO)_6$ and $Fe(CO)_5$, Fig. 25-7 gives a suggestion of the extensive chemistry that any individual carbonyl typically has. Other examples are encountered in succeeding chapters. A discussion of reaction mechanisms can be found in Section 28-15.

The most important general reactions of carbonyls are those in which CO groups are displaced by other ligands. These may be individual donor molecules, with varying degrees of back-acceptor ability themselves [e.g., PX_3, PR_3, $P(OR)_3$, SR_2, NR_3, OR_2, RNC, etc.] or unsaturated organic molecules such as C_6H_6 or cycloheptatriene. Derivatives of organic molecules are discussed in Chapter 27.

Although many of the substitution reactions by other π-acid ligands proceed thermally (temperatures up to 200° in some cases being required for the less reactive carbonyls), it is sometimes more convenient to obtain a particular product by photochemical methods; in some cases, substitution proceeds readily *only* under irradiation. For example, the thermal reactions of $Fe(CO)_5$ and triphenylphosphine or triphenylarsine (L) give mixtures, whereas photochemically, $Fe(CO)_4L$ and $Fe(CO)_3L_2$ can be obtained quite simply. Manganese carbonyl and η^5-$C_5H_5Mn(CO)_3$ are usually quite resistent to substitution reactions, but the former under irradiation gives $[Mn(CO)_4PR_3]_2$. In the very rapid photochemical production of acetylene and olefin complexes from $Mo(CO)_6$ and $W(CO)_6$ it is believed that $M(CO)_5$ radicals are the initiating species; even in absence of other ligands, bright yellow solutions are produced when Cr, Mo, and W hexacarbonyls are irradiated in various solvents. Metal carbonyls in presence of organic halogen compounds such as CCl_4 have been found to act as initiators for the free-radical polymerization of methyl methacrylate and other monomers.

25-9. Vibrational Spectra of Metal Carbonyls[32]

The vibrational spectra, particularly the infrared spectra, of metal carbonyls have proved to be a rich and convenient source of information concerning both structure and bonding. Carbon-13 nmr spectra also have some utility in this respect, but they are much less easily measured and often the structure is revealed only by difficult low-temperature measurements because of fluxional behavior (Section 28-15). We now survey and illustrate the principal uses of ir spectra.

[32] Cf. P. S. Braterman, *Struct. Bonding,* 1976, **26**, 1, and references to other comprehensive reviews cited therein, and Ref. 33.

Structural Diagnosis. Perhaps the most important day-to-day use of infrared spectra in the inorganic research laboratory is in deducing the structures of molecules containing carbonyl groups. This may be done in a number of ways.

1. *Detecting Bridging CO Groups.* For neutral molecules, bridging CO groups

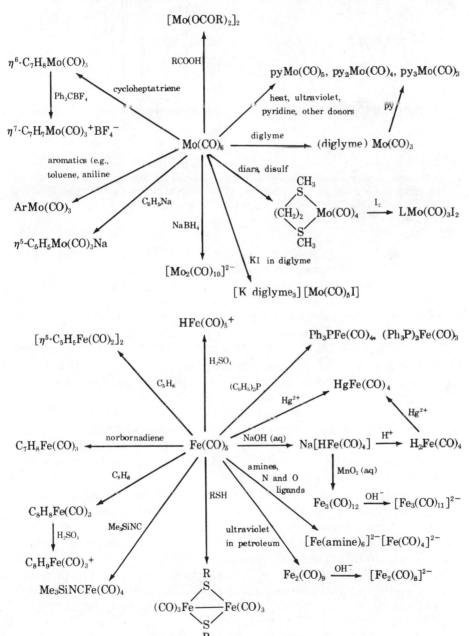

Fig. 25-7. Some reactions of molybdenum and iron carbonyls.

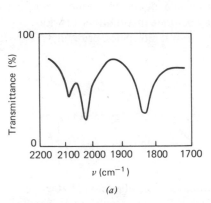 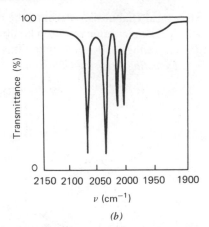

Fig. 25-8. The ir spectra of the CO stretching region of (a) solid $Fe_2(CO)_9$ and (b) $Os_3(CO)_{12}$ in solution. Note the greater sharpness of the solution spectra. The most desirable spectra are those obtained in nonpolar solvents or in the gas phase. [Data from F. A. Cotton, in *Modern Coordination Chemistry*, J. Lewis and R. G. Wilkins, Eds., Wiley-Interscience, 1960, and D. K. Huggins, N. Flitcroft, and H. D. Kaesz, *Inorg. Chem.*, 1965, **4**, 166.]

absorb in the range 1700–1860 cm^{-1} and terminal ones generally absorb at higher frequencies (1850–2125 cm^{-1}). Figure 25-8 illustrates how these properties may be used to infer structures. It is evident that $Fe_2(CO)_9$ has strong absorption bands in both the terminal and the bridging regions. From this alone it could be inferred that the structure must contain both types of CO groups; X-ray study shows that this is so. For $Os_3(CO)_{12}$ several structures consistent with the general rules of valence can be envisioned; some of these would have bridging CO groups, while one (that shown in Fig. 25-1, which is the actual one) does not. The infrared spectrum alone, Fig. 25-8b), shows that no structure with bridging CO groups is acceptable, since there is no absorption band below 2000 cm^{-1}.

As another example, consider the molecule $Co_2Ir_2(CO)_{12}$. Since $Ir_4(CO)_{12}$ has a structure without bridges, whereas $Co_4(CO)_{12}$ has one with three CO bridges (structures g and h, respectively, in Fig. 25-1), the structure of the mixed metal compound might be either one of these. The ir spectrum in Fig. 25-9 clearly indicates that it must be the one with bridges.

In using the positions of CO stretching bands to infer the presence of bridging CO groups, it is necessary to keep certain conditions in mind. The frequencies of terminal CO stretches can be quite low if (*a*) there are a number of ligands present that are good donors but poor π acceptors, or (*b*) there is a net negative charge on the molecule. In either case, back-donation to the CO groups becomes very extensive, thus increasing the M—C bond orders, decreasing the C—O bond orders, and driving the CO stretching frequencies down. Thus in $Mo(dien)(CO)_3$ one of the CO stretching bands is as low as 1760 cm^{-1} and in the $Fe(CO)_4^{2-}$ ion there is a band at 1790 cm^{-1}.

2.   *Molecular Symmetry from the Number of Bands.*   It is often possible to infer the symmetry of the arrangement of the CO groups from the number of CO

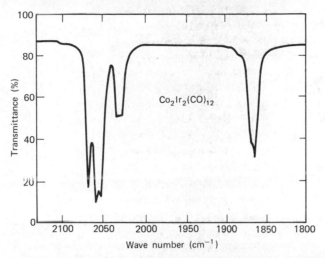

Fig. 25-9. The ir spectrum of $Co_2Ir_2(CO)_{12}$ in hexane solution, as reported in Ref. 10.

stretching bands that are found in the infrared spectrum, though a certain amount of judgment and experience is necessary to avoid errors consistently. The procedure consists in first determining from the mathematical and physical requirements of symmetry how many CO stretching bands ought to appear in the ir spectrum for each of several possible structures.[33] The experimental observations are then compared with the predictions, and structures for which the predictions disagree with observation are considered to be eliminated. In favorable cases there will be only one possible structure remaining. In carrying out this procedure, due regard must be given to the possibilities of bands being weak or superposed and, of course, the correct model must be among those considered. The reliability of the procedure can usually be increased if the behavior of approximate force constants and the relative intensities of the bands are also considered.

To illustrate the procedure, consider the *cis* and *trans* isomers of an $ML_2(CO)_4$ molecule. Figure 25-10 shows the approximate forms of the CO stretching vibrations and also indicates those that are expected to absorb infrared radiation, when only the symmetry of the $M(CO)_4$ portion of the molecule is considered. When L $= (C_2H_5)_3P$, the two isomeric compounds can be isolated. One has four infrared bands (2016, 1915, 1900, 1890 cm^{-1}) and is thus the *cis* isomer; the other shows only one strong band (1890 cm^{-1}) and is thus the *trans* isomer.

It may also be noted that since no major interaction is to be expected between the CO stretching motions in two $M(CO)_4$ groups if they are connected only through the two heavy metal atoms, $Os_3(CO)_{12}$ should have the four-band spectrum of a *cis*-$ML_2(CO)_4$ molecule, and as seen in Fig. 25-8 it does.

3. *Bond Angles from Relative Intensities.* To a good approximation the rel-

---

[33] See P. S. Braterman, *Metal Carbonyl Spectra,* Academic Press, 1975, and S. F. A. Kettle, *Topics in Current Chem.*, 1977, **71**, 111, for more technical discussions of all points relating to carbonyl spectra.

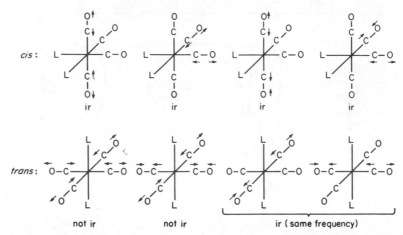

Fig. 25-10. Schematic indication of the forms of the CO stretching vibrations of *cis*- and *trans*-$ML_2(CO)_4$ molecules. For the *cis* isomer all four are distinct and can absorb ir radiation. For the *trans* isomer, two are equivalent and have the same frequency, forming a degenerate vibration; only this form can absorb ir radiation.

ative intensities of different CO stretching modes can be calculated by using a very simple model. In this model, each CO oscillator is treated as a dipole vector, and the total dipole vector for the entire vibrational mode is taken to be the vector sum of these individual vectors. Since the intensities are proportional to the squares of the dipole vectors, there is a relationship of the intensities to the angles between C—O bond directions.

For the case of two CO groups (Fig. 25-11) the ratio of the intensities of the symmetric and antisymmetric bands is given by

$$R_{sym}/R_{asym} = \left\{\frac{2r\cos\theta}{2r\sin\theta}\right\}^2 = \cotan^2\theta$$

Figure 25-12 shows the ir spectra of the two isomeric ions of Fig. 25-13. Both, of course, have two CO stretching bands, so that no decision between them can be made on this simple basis. However the intensity ratios are quite different in the two cases. As will be discussed below, the band of higher frequency can be assigned in both cases to the symmetric mode. From careful measurement of the areas, the intensity ratios given on the spectra are obtained. From each of these the angle 2θ can be obtained, *viz.*,

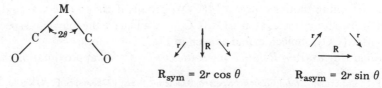

$$R_{sym} = 2r\cos\theta \qquad R_{asym} = 2r\sin\theta$$

Fig. 25-11. Diagrams showing how the dipole vectors, r, of individual CO groups combine to give the dipole vectors R for the symmetric and the antisymmetric mode of vibration of an $M(CO)_2$ moiety.

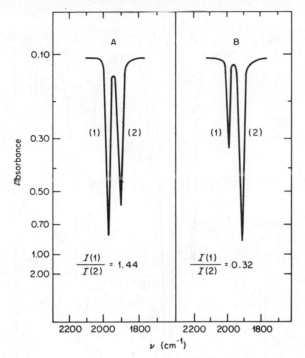

Fig. 25-12. The CO stretching bands and their relative intensities for the two cations of Fig. 25-13. [Data from F. A. Cotton and C. M. Lukehart, *J. Am. Chem. Soc.,* 1971, **93**, 2672, and unpublished work.]

$$2\theta = 2\sqrt{\operatorname{arccotan} 1.44} = 79°$$

$$2\theta = 2\sqrt{\operatorname{arccotan} 0.32} = 121°$$

It is obvious that the angles 79° and 121° can be associated only with the *cis* and *trans* isomers, respectively.

Another illustration of this type of argument is provided by the $[RhCl(CO)_2]_2$ molecule (Section 22-G-2), which has the bent structure shown schematically in 25-XXI in the crystal and illustrated in Fig. 22-G-2. Since the reason for bending is not obvious, it might be tempting to attribute it to intermolecular forces in the crystal. The infrared spectrum of the molecule in solution (Fig. 25-14), however,

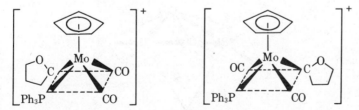

Fig. 25-13. The two isomers of $[\eta^5\text{-}C_5H_5Mo(CO)_2PPh_3C_4H_6O]^+$. The C_4H_6O ligand is a "coordinated carbine," discussed in Section XX.

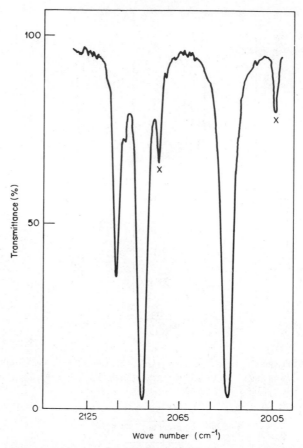

Fig. 25-14. The ir spectrum of $Rh_2Cl_2(CO)_4$. The small peaks marked X are ^{13}CO satellites. The large peaks are the three inactive fundamentals.

shows qualitatively and quantitatively that the structure is essentially identical with that observed in the crystal. The appearance of three bands shows that the molecule is bent. Moreover, from the ratios of the intensities of these bands, the dihedral angle is calculated to be $58 \pm 2°$, which is not significantly different from the value of $56°$ found for ϕ in the crystal.

(25–XXI)

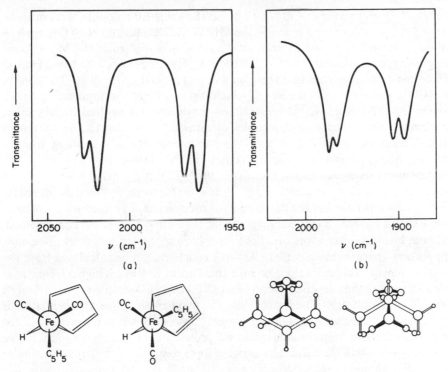

Fig. 25-15. Infrared spectra that demonstrate the presence of conformational isomers. (a) For the molecule $(\eta^5\text{-}C_5H_5)Fe(CO)_2(\eta^1\text{-}C_5H_5)$. [Data from F. A. Cotton and T. J. Marks, *J. Am. Chem. Soc.*, 1969, **91**, 7523.] (b) For the molecule $(\eta^5\text{-}C_5H_5)Mo(CO)_2C_3H_5$. [Data from R. B. King, *Inorg. Chem.*, 1966, **5**, 2242, and A. Davison and W. C. Rode, *Inorg. Chem.*, 1967, **6**, 2124.]

4.   *Detecting Conformers.*   The number of bands in the carbonyl stretching region for certain molecules can often be used to show that two conformational isomers are present. Thus in the molecule $(\eta^5\text{-}C_5H_5)Fe(CO)_2(\eta^1\text{-}C_5H_5)$ one would tend to expect only two bands because of the symmetric and the antisymmetric modes. In fact, there are four bands (Fig. 25-15a). This shows that there must be nearly equal populations of two conformations, presumably those shown below the spectrum. Similarly, for $(\eta^5\text{-}C_5H_5)Mo(CO)_2C_3H_5$, the appearance of two sets of bands indicates the presence of the two conformers shown below the spectrum (Fig. 25-15b).

Force Constants and Bonding.   To make the fullest use of vibrational spectra, Raman as well as infrared data should be obtained and mathematical analysis should be conducted to obtain values of the force constants. The main use for force constants is in seeking a quantitative understanding of bonding relationships.

A rigorous vibrational analysis of even a simple mononuclear metal carbonyl molecule is complex. It is for the Group VI carbonyls $M(CO)_6$ [M = Cr, Mo, or W] and $Ni(CO)_4$ that the most sophisticated treatments have been reported. The C—O force constants (mdyne Å^{-1}) found in these four cases (for the molecules in the gas phase) are: Cr, 17.2; Mo, 17.3; W, 17.2; and Ni, 17.9. The force constant for the CO triple bond in CO^+ is 19.8 and the value for a double bond must be 12–13. Thus the CO bond orders in the $M(CO)_6$ molecules and $Ni(CO)_4$ must be about 2.65 and 2.75, respectively. In the same four molecules the M—C bond stretching force constants are: Cr, 2.08; Mo, 1.96; W, 2.36; Ni, 2.02. Since the CO force constants suggested that there may be less π bonding for $Ni(CO)_4$ than for $Cr(CO)_6$, the similarity of the M—C force constants might appear to indicate more σ bonding to Ni than to Cr. However unless the differences involved are very large, arguments of this kind must be regarded circumspectly, because CO bond strength is influenced not only by the π back-bonding but also by the σ bonding, since as already noted, the σ-donor orbital is somewhat CO antibonding.

The effort required to obtain results of the kind just discussed is so great that there have been many attempts to devise some simple method of calculating useful force constants from limited data and by simple means. The most widely used of these simple methods is that commonly called the Cotton-Kraihanzel (CK) method, although very similar approximations have been suggested by others. The most important approximation is that CO force constants can be calculated from the CO stretching frequencies alone because these are at very much higher frequencies (>1850 cm^{-1}) than all other vibrations (<700 cm^{-1}) in simple metal carbonyls as well as in most substituted carbonyls. The main effect of this approximation is that the force constants so obtained are not "absolute"; but in a series of related molecules the shift from absolute values will probably be essentially constant. Hence relative values, thus the differences between force constants for different but similar molecules, should be rather reliably given.

Beyond this, there are other assumptions in the CK method that are of doubtful validity in the light of the results of the rigorous vibrational analyses. To simplify the equations and to make the application simple, two further approximations are used: (1) the actual fundamental frequencies are used, neglecting the fact that these vibrations are not truly harmonic; (2) the interaction constants between *cis*- and *trans*-CO groups are assumed to be related in a simple way. This assumed relationship is predicated on the idea that the coupling between CO groups is due entirely to the electronic effects of interaction with the metal d orbitals. The first assumption is certainly not true; the effects of anharmonicity of the vibrations amount to 20 to 30 cm^{-1}, which is comparable to the separations between bands due to different modes of vibration, typically $\leqslant 100$ cm^{-1}. As for the cases of the couplings, they are probably due only in part to electronic effects and undoubtedly arise in part from direct electrostatic interaction between the oscillating dipoles of the different CO groups.

One of the ideas inherent in the CK method, empirically valid in a qualitative sense and extremely useful in interpreting the spectra of $M(CO)_n$ moieties, is that the coupling constant for stretching of two CO groups on the same metal atom is positive. Physically, this means that modes in which CO groups stretch in phase have higher frequencies than those in which they stretch out of phase. Another way to state this is that the stretching of one CO group makes it harder to stretch another at the same time. This is to be expected from either the CK approximation, where only the influence of back-bonding is considered, or from a consideration of dipole interactions; therefore even if both factors contribute, this is an expected and understandable rule. It was on this basis that in assigning the spectra in Fig. 25-12 we could safely assume that the upper bands were due to the symmetric modes (those in which the two CO groups stretch or contract simultaneously).

_____

General References

(See also Chapters 26, 27, 29, 30)

Hoffman, R. *et al., J. Amer. Chem. Soc.,* 1976, **98,** 7240; 1979, **101,** 3456, MO study of M_2L_n, M_3L_n includes carbonyls.

Inorganic Reaction Mechanisms and *Organometallic Chemistry,* Specialist-Reports, Chemical Society. Annual reports include carbonyls.

Wender, I. and P. Pino, Eds. *Organic Syntheses via Metal Carbonyls,* Wiley, Vol. 1, 1968 has much data on binary carbonyls.

CHAPTER TWENTY–SIX

Metal-to-Metal Bonds and Metal Atom Clusters

26-1. Introduction

As is well known, one of the most important concepts in chemistry, and perhaps the greatest one in inorganic chemistry, is the concept of the coordination complex, which was developed by Alfred Werner in the decades around 1900. This concept enabled Werner to make sense of an enormous amount of experimental data that had no explanation in the structural and bonding theory that had been worked out previously for compounds of carbon and other main group elements.

The essential idea in the coordination theory of Werner is that a metal ion surrounds itself with ligands and that the nature of the ligands, the character of the metal-ligand bonds, and the geometrical arrangement of the ligands around the metal atom determine the physical and chemical properties of the compound. There is no place in the Wernerian scheme for direct bonding between metal atoms.

Werner recognized the existence of polynuclear complexes, of course, and devoted a great many papers to the elucidation of their properties. However these were viewed simply as a conjunction of two or more mononuclear complexes having some shared ligand atoms. The properties of these complexes were still attributed to the metal-ligand interactions, and direct metal-to-metal (M—M) interactions were not considered. This approach was entirely justified because in the compounds being considered there were no direct M—M interactions of chemical significance.

With the passage of time, however, substances were discovered whose behavior did not appear to be understandable in Wernerian terms. Thus as early as 1907 the compound "$TaCl_2 \cdot 2H_2O$" was reported and by 1913 was shown to be $Ta_6Cl_{14} \cdot 7H_2O$. During the 1920s polynuclear halide compounds of molybdenum were discovered; it was recognized that their chemistry is unlike that of "Werner complexes," but it was not satisfactorily explained.

It was, of course, only with the advent of X-ray crystallography and its development to the stage of applicability to rather large and complex structures that the proper recognition of non-Wernerian compounds—those with one or more direct M—M bonds—became possible. The earliest experimental results were provided by C. Brosset,[1] who showed in 1935 that $[W_2Cl_9]^{3-}$, a chloro complex of tungsten, has its metal atoms separated by only about 2.5 Å, and again in 1945 and 1946 when he showed that several molybdenum(II) chloride compounds contained octahedral groups of Mo atoms with M—M distances of only about 2.6 Å.

Brosset's results did not trigger any noticeable response. In 1950 an unconventional type of X-ray diffraction experiment carried out on aqueous solutions demonstrated that $Ta_6Cl_{14} \cdot 7H_2O$, the corresponding bromide, and the niobium analogues also contained octahedral groups of metal atoms close enough together to warrant the conclusion that "metal-metal bonds are present in all of these compounds."[2] Even these results, added to the previous ones, did not cause anyone to look more closely or more generally at compounds with M—M bonds.

It was not until the early 1960s that some sense of the possible generality of such chemistry began to develop.[3] A key development was the discovery of the $[Re_3Cl_{12}]^{3-}$ ion,[4,5] since this led to the first general discussion[5] of the existence of the entire class of "metal atom cluster" compounds.

A metal atom cluster may be defined as a group of two or more metal atoms in which there are substantial and direct bonds between the metal atoms. Thus a metal atom cluster complex is more than just an ordinary polynuclear complex, and it is unfortunate that occasionally some authors have used the term "cluster" indiscriminately to designate polynuclear complexes of the classical (Werner) type in which there is no significant amount of direct M—M bonding. There are, of course, some borderline cases, that is, compounds in which the metal-to-metal distances and other properties are such that one cannot say unequivocally whether M—M bonding is "chemically significant." That does not negate the value of recognizing the existence of an enormous number of true, unambiguous metal atom cluster compounds in which the M—M bonds are as important as, or more important than, any other bonds in determining the chemistry and properties of the substance. The term "cluster" should be reserved for such compounds.

The history briefly sketched would be incomplete without mentioning the emergence of another class of cluster compounds, namely, those of the metal carbonyl type. It was with the structure determination of $Fe_2(CO)_9$ in 1938 that the close (ca. 2.5 Å) approach of metal atoms in a polynuclear carbonyl-type compound was first observed. Perhaps the most crucial observation in establishing the reality and importance of M—M bonds in this area was the determination of the $Mn_2(CO)_{10}$ structure,[6] where for the first time in the carbonyl field a direct M—M

[1] C. Brosset, *Arki. Kemi, Mineral. Geol.*, 1946, **A22**, No. 11, and earlier references therein.

[2] P. A. Vaughan, J. H. Sturdivant, and L. Pauling, *J. Am. Chem. Soc.*, 1950, **72**, 5477.

[3] B. R. Penfold, in *Perspectives in Structural Chemistry*, Vol. II, J. D. Dunitz and J. A. Ibers, Eds., Wiley, 1968 (covers work in this period in detail).

[4] W. T. Robinson, J. E. Fergusson, and B. R. Penfold, *Proc. Chem. Soc.*, **1963**, 116.

[5] J. A. Bertrand, F. A. Cotton and W. A. Dollase, *J. Am. Chem. Soc.*, 1963, **85**, 1349; *Inorg. Chem.*, 1963, **2**, 1166.

[6] L. F. Dahl, E. Ishishi, and R. E. Rundle, *J. Chem. Phys.*, 1957, **26**, 1750.

bond unsupported by any bridges was seen. There are now scores of such cluster-type compounds comprising neutral carbonyls, carbonylate anions, carbonyl hydrides, isocyanides, phosphites, and many mixed organocarbonyl-type molecules.

The two general types of metal cluster compound, which may be loosely called the "lower halide" type and the "carbonyl" type, differ from each other in many ways, and there is very little in the way of chemical reactions to interrelate them. This chapter discusses both types in some detail, but separately. The lack of any established relationship between them means that this can be done quite conveniently.

The discussion of the noncarbonyl-type clusters is concerned mainly with understanding their electronic structures. This is so closely related to the problem of M—M multiple bonds that we then move directly to that topic, and finally to a discussion of the close relationships between the clusters and the M—M multiple bonds.

We also include in this chapter a discussion of the "one-dimensional" solids in which the stacking pattern leads to the formation of infinite chains of metal atoms closely enough bonded to engender metallic properties.

At least a few examples of each of the types of compound discussed in this chapter have been encountered earlier in the presentation of the chemistry of individual elements. The objectives here, however, are to give a unified view, to introduce a number of additional examples, and to deal with questions of bonding that cut across the chemistry of a number of elements.

There are two very broad generalizations pertinent to this whole area of M—M bonding. One is that M—M bond formation is most likely to occur when the metal atoms are in their lower oxidation states. Thus the most extensive groups of compounds are the carbonyl clusters in which metal oxidation numbers are zero or even negative (in carbonylate anions), and the lower halide type of clusters, where oxidation numbers are usually +2 to +3. However it must be pointed out that important cases of M—M bond formation are found with formal oxidation numbers as high as +4, as in the very stable and numerous trinuclear molybdenum(IV) and tungsten(IV) compounds. It can be said without qualification that M—M bonds have never been found for metal atoms in formal oxidation states of +5 or higher. Presumably high charge causes contraction of the valence orbitals to the extent that overlap with another set of similarly contracted metal atom orbitals is too small to allow effective bond formation.

The second generalization is that for any given group in the Periodic Table, the tendency to form M—M bonds is usually greater for the heavier elements. There are, however, enough exceptions to each of these "rules" that they should be invoked very cautiously.

METAL CARBONYL CLUSTERS

Since metal carbonyl compounds contain metal atoms in very low oxidation states (even formally negative ones in the case of carbonyl hydrides and carbonylate anions), it is not surprising that we find among them many species containing M—M bonds and metal atom clusters. Indeed, the largest and structurally most complex cluster compounds are of these types. The simpler ones, namely, the triangular trinuclear ones such as $Ru_3(CO)_{12}$ and the tetrahedral tetranuclear ones such as $Co_4(CO)_{12}$, were introduced in Chapter 25. From a bonding point of view, these simpler clusters generally offer no problem. In each of the two cases just cited, for example, the electronic structures are easily understandable in terms of an electron-pair bond between each adjacent pair of metal atoms, with each metal atom attaining an eighteen-electron configuration. The sections that follow examine several classes of metal carbonyl type cluster compounds that are more complex structurally and less easily understood electronically.

26-2. Octahedral Cobalt, Rhodium, and Ruthenium Clusters; Wade's Rule

We turn now to metal carbonyl cluster species that are larger and more complex than the M_2, M_3, and M_4 species. A good example is the $Rh_6(CO)_{16}$ compound, briefly noted in Section 25-1, which has the structure in Fig. 25-1i. In this molecule, and others discussed here and in the next section, the bonding in the metal skeleton cannot be described in terms of localized two-center M—M bonds, with the attainment of eighteen-electron configurations by the metal atoms. In $Rh_6(CO)_{16}$, for example, if we say that each Rh atom acquires four electrons from the two terminal CO groups and four more from the four bonds it forms to other Rh atoms, these eight electrons plus the nine that belong to the Rh atom to begin with make seventeen. Therefore at this point, before we have taken account of the contribution of the four triply bridging CO groups, which may be considered to contribute two electrons each, we may say that since each Rh atom is short one electron, the Rh_6 cluster as a whole is short six electrons. However the four triply bridging CO groups contribute a total of eight electrons. The molecule thus contains two electrons too many from the point of view of achieving eighteen-electron configurations by means of localized bond formation. Essentially the same problem is posed by the carbonyl anion $[Co_6(CO)_{14}]^{4-}$, by $H_2Ru_6(CO)_{18}$, and by the carbonyl carbide $Ru_6(CO)_{17}C$, in which the C atom resides approximately at the center of the Ru_6 octahedron (Fig. 26-1a).

In an effort to resolve these and similar "electron counting" problems in clusters,

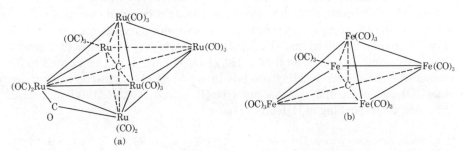

Fig. 26-1. The structures of two metal carbonyl carbide cluster compounds. (a) $Ru_6(CO)_{17}C$. (b) $Fe_5(CO)_{15}C$.

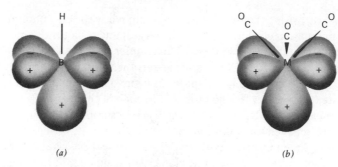

Fig. 26-2. (a) The three orbitals p_x, p_y, and an s, p_z, or sp_z hybrid available for cluster bonding in a $[B_nH_n]^2$ ion. (b) Analogous orbitals for an $M(CO)_3$ group. The shapes shown for two of these represent an arbitrary choice among several possibilities.

a qualitative picture derived from an earlier treatment of the $[B_nH_n]^{2-}$ ions has been suggested by Wade.[7] This approach lacks a rigorous basis and is known to fail in some cases, but it possesses enough utility and has been invoked often enough in the research literature to justify a brief account of it here. The application to boron cage compounds was briefly noted in Section 9-6.

The point of departure for $[B_nH_n]^{2-}$ systems was the recognition that in each B—H unit two electrons and one boron orbital are used to form a B—H bond, thus leaving three orbitals and two electrons to be used in forming the B_n cage. This is indicated in Fig. 26-2a. One of the orbitals is an s, p_z, or sp_z hybrid that will point toward the center of the polyhedron; the other two, p_x and p_y, will lie more or less "on the surface" of the polyhedron. For larger B_nH_n polyhedra the $2n$ "surface" orbitals will overlap to generate n bonding orbitals and n antibonding orbitals. The members of the set of n centrepetally directed orbitals strongly overlap at the center of the polyhedron to generate one strongly bonding orbital and $n - 1$ weakly bonding, nonbonding, or antibonding orbitals. Thus in the B_nH_n cage there are $n + 1$ bonding orbitals. Since there are only n electron pairs brought in by the n B—H units, an additional electron pair is called for, and this is why the $[B_nH_n]^{2-}$ ions (or $B_{n-2}C_2H_n$ molecules) are stable rather than B_nH_n itself.

It may then be argued that an $M(CO)_3$ unit can be treated analogously to the B—H unit. Let us say that the atom M is Ru, and note that although boron has but four valence shell orbitals, a transition metal has nine. Three of these are used to form M—C σ bonds; three more as well as three electron pairs are used to form π bonds to the CO groups. This leaves the Ru atom with three valence shell orbitals (Fig. 26-2b) and two electrons. The situation is clearly very analogous to that with the B—H group. Again, a total of $n + 1$ electron pairs would be needed to fill completely the cluster bonding orbitals in an $M_n(CO)_{3n}$ structure. Thus for $Ru_6(CO)_{18}$ we expect the stable species to be $[Ru_6(CO)_{18}]^{2-}$, and this is, of course, the anion corresponding to $H_2Ru_6(CO)_{18}$.

[7] K. Wade, *Electron Deficient Compounds*, Nelson, 1971; *Adv. Inorg. Chem. Radiochem.*, 1976, **18**, 1.

The $[Ru_6(CO)_{18}]^{2-}$ or $H_2Ru_6(CO)_{18}$ case can also be viewed in the following way. The total number of electrons in the molecule totals 86, when we count $6 \times 8 = 48$ from the Ru atoms, $18 \times 2 = 36$ from the CO groups, and two more from the charge on the H atoms. If we simply think in terms 86 electrons as the "right" number for an octahedral cluster of this sort, then $[Co_6(CO)_{14}]^{4-}$, $(6 \times 9) + (14 \times 2) + 4$, $Rh_6(CO)_{16}$, $(6 \times 9) + (16 \times 2)$, and $Ru_6(CO)_{17}C$, $(6 \times 8) + (17 \times 2) + 4$, are clearly "entitled" to have octahedral structures. In the latter case (Fig. 26-1a) we treat the C atom as having contributed all four of its valence electrons to the total of 86. It may be noted that for $[Co_6(CO)_{14}]^{4-}$ an extended Hückel MO calculation provides support for this picture.[8]

The overall generalization that may be drawn from this is that for clusters with five or more metal atoms, *closo* structures, such as the *tbp* or octahedron, should be stable when there are $n + 1$ electron pairs available for cluster bonding (or, in the special case of the octahedral species, a total of 86 electrons). As with the boranes, if there are more than $n + 1$ electron pairs, the *closo* structure should open up to a *nido* or an *arachno* structure (see Section 9-6).

An example of this opening up is, perhaps, provided by the structure of $Fe_5(CO)_{15}C$ (Fig. 26-1b), which has a square-pyramidal skeleton of iron atoms. Treating this species as $[Fe_5(CO)_{15}]^{4-}$ plus C^{4+}, we figure the total number of electrons as $(5 \times 8) + (15 \times 2) + 4 = 74$. Allowing, as in the $[Ru_6(CO)_{18}]^{2-}$ case, six electron pairs for Fe—CO σ and π bonding in each $Fe(CO)_3$ unit, there remain fourteen electrons, or seven pairs, for cluster bonding. This is one pair in excess of the $n + 1 = 5 + 1 = 6$ appropriate for a *closo* (*tbp*) Fe_5 structure. A *nido* structure therefore results.

The next section gives a few more cases in which Wade's rule provides at least a partial understanding of some structural properties.

26-3. Higher Osmium Carbonyl Clusters

When $Os_3(CO)_{12}$ is heated in a sealed tube CO is evolved and cluster compounds containing five to eight osmium atoms are formed. These in turn can be converted to carbonylate anions and carbonyl hydrides, as summarized in Fig. 26-3. The reactions of the carbonyls with base to produce carbonylate anions and the protonation

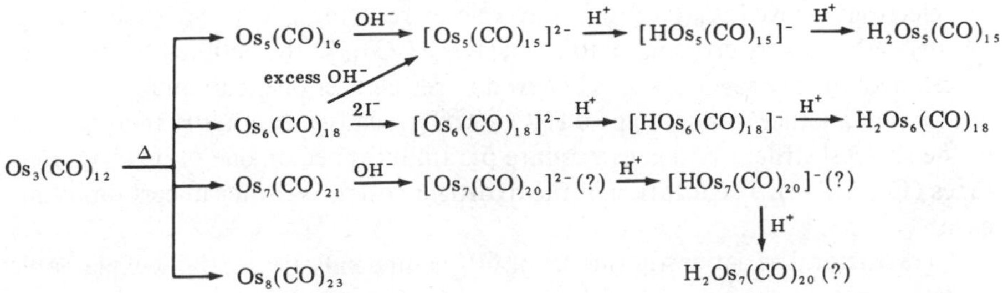

Fig. 26-3. Processes leading from $Os_3(CO)_{12}$ to higher cluster species of osmium.

[8] D. M. P. Mingos, *J.C.S. Dalton,* **1974,** 133.

of these to give the carbonyl hydrides are reactions typical of metal carbonyl chemistry (see Section 25-6), but some of the structural changes that occur are surprising and of unusual interest.

In addition to the species shown in Fig. 26-3, two others that are pertinent are $H_2Os_5(CO)_{16}$, obtained in small yield when $Os_3(CO)_{12}$ is pyrolyzed in presence of water, and $[Os_5(CO)_{15}C]$, which presumably has the same structure as $Fe_5(CO)_{15}C$.

$Os_5(CO)_{16}$ has the structure[9a] shown in Fig. 26-4a. The Os_5 skeleton is an irregular *tbp* in which the bonds to the Os atom bearing four CO groups are all about 0.12 Å longer than the other five Os—Os bonds. One cannot satisfactorily rationalize this structure on the basis of the eighteen-electron rule and two-electron Os—Os bonds. It is easily seen that although the two equatorial $Os(CO)_3$ metal atoms could be assigned eighteen-electron configurations, the axial ones would have only seventeen and the $Os(CO)_4$ metal atom would have 20. The species $[Os_5(CO)_{15}]^{2-}$, $[HOs_5(CO)_{15}]^-$, and $H_2Os_5(CO)_{15}$, which are isoelectronic with $Os_5(CO)_{16}$, all appear to have more regular *tbp* structures in which each metal atom has three CO groups and H atoms when present probably lie over equatorial edges.[9b]

The compound $H_2Os_5(CO)_{16}$ (Fig. 26-4b) has quite a different structure from the Os_5 species just discussed. The hydrogen atoms are believed to bridge the two edges of the tetrahedron that face the $Os(CO)_4$ group.

Wade's rule can be invoked to rationalize at least qualitatively, these structures. In the same sense that the $[Fe_5(CO)_{15}]^{4-}$ anion was said to have one electron pair too many to be stable in a *closo* structure, the $[Os_5(CO)_{15}]^{2-}$ ion has the right number, six, and so, of course, do the "isoelectronic" $Os_5(CO)_{16}$, $[HOs_5(CO)_{15}]^-$, and $H_2Os_5(CO)_{15}$ species. However $H_2Os_5(CO)_{16}$, like $[Fe_5(CO)_{15}]^{4-}$, has seven skeletal electron pairs and the *closo* structure opens up along one of the M—M bonds. In this case, however, it opens along one of the six slant bonds instead of one of the three equatorial bonds. There is no obvious reason for this difference except, perhaps, the different steric requirements of the C atom in the iron compound as against 2H and CO in the $H_2Os_5(CO)_{16}$ molecule.

The structure of $Os_6(CO)_{18}$ is shown in Fig. 26-4c. The Os_6 skeleton has been described as a bicapped tetrahedron or as a monocapped *tbp*. The salient point is that the metal atoms do not form an octahedron. However, upon the addition of two electrons to give $[Os_6(CO)_{18}]^{2-}$ the skeleton rearranges to a octahedron[10] (Fig. 26-4d). When singly protonated to give $[HOs_6(CO)_{18}]^-$ the octahedral structure is retained; the hydrogen atom is believed to reside over one triangular face. The addition of another proton, to give $H_2Os_6(CO)_{18}$, causes another rearrangement of the skeletal structure to give a square pyramid capped on one of its triangular faces (Fig. 26-4e). The location of the hydrogen atoms is rather uncertain in this case.

The structural variations in this set of Os_6 compounds are partially explainable

[9a] B. E. Reichert and G. M. Sheldrick, *Acta Crystallogr. B.*, 1977, **33**, 173.
[9b] J. Lewis *et al.*, *J.C.S. Chem. Comm.*, **1976**, 807.
[10] J. Lewis *et al.*, *J.C.S. Chem. Comm.*, **1976**, 883.

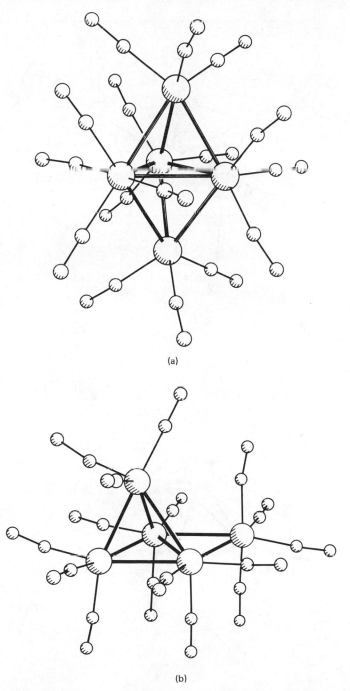

(a)

(b)

Fig. 26-4. Structures of some osmium carbonyl cluster species with five, six, and seven metal atoms: (a) $Os_5(CO)_{16}$, (b) $H_2Os_5(CO)_{16}$, (c) $Os_6(CO)_{18}$, (d) $[Os_6(CO)_{18}]^{2-}$, (e) $H_2Os_6(CO)_{18}$, and (f) $Os_7(CO)_{21}$.

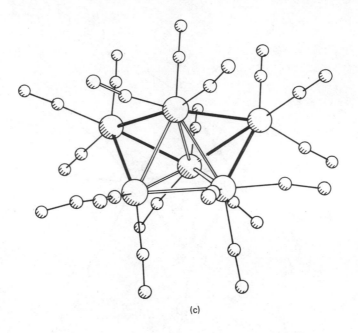

(c)

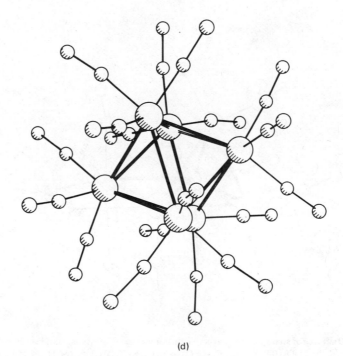

(d)

Fig. 26-4. *Con't.*

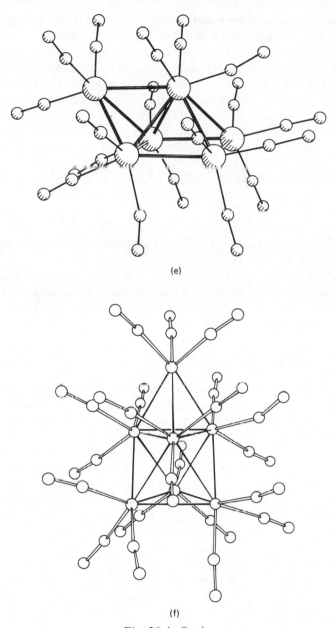

(e)

(f)

Fig. 26-4. *Con't.*

by Wade's rule. $Os_6(CO)_{18}$ itself does not have the $n + 1$ ($= 7$) electron pairs required to stabilize an octahedral skeleton; it has only six pairs. In this way one can perhaps understand why $Os_6(CO)_{18}$ *does not* have an octahedral structure. An "explanation" for why it *does* have the capped *tbp* structure has been suggested.[7] The six skeletal electron pairs are appropriate for a *tbp*, which is formed by five of the metal atoms. The extra metal atom then caps one triangular face of the *tbp* "where the three vacant orbitals that it can formally furnish for cluster bonding enable it to bond to the three metal atoms defining the face without modifying the number of bonding MO's for the rest of the cluster." With the addition of two electrons to $Os_6(CO)_{18}$ to give $[Os_6(CO)_{18}]^{2-}$ the necessary seven pairs of skeletal electrons are now available to stabilize an octahedral structure. Protonation to $[HOs_6(CO)_{18}]^-$ should not cause any change, and none is observed.

However it is not at all clear why a second proton cannot be added to a triangular face opposite the one where the first proton resides, so that an octahedral structure is retained. Indeed, with $H_2Ru_7(CO)_{18}$ this is just what happens. Clearly there are important factors involved here, such as the change from the second to the third transition series, that neither Wade's rule nor any other correlation yet suggested takes account of.

Another curious and unexplained structural problem among these compounds is the presence of the H atom in $[HRu_6(CO)_{18}]^-$ inside the Ru_6 octahedron[11] rather than on a face as in $[HOs_6(CO)_{18}]^-$.

$Os_7(CO)_{21}$ has a capped octahedral skeleton[12] (Fig. 26-4f), which can be explained by Wade's rule in the same way as the capped *tbp* structure of $Os_6(CO)_{18}$ was "explained." The structures of the remaining Os_7 cluster compounds as well as that of $Os_8(CO)_{23}$ are not yet known.

26-4. Other High-Nuclearity Carbonyl Clusters[13]

It is by no means desirable or possible to cover the field of large metal carbonyl cluster compounds comprehensively in this book. However, in addition to the high nuclearity (≥ 5) clusters already mentioned, a few others of unusual interest are briefly reviewed here.

Platinum Carbonyl Anions. We referred to these in Section 22-H-6. We draw attention here to several points. First there are no known palladium analogues of any sort to these platinum compounds. Second, though there are nickel carbonyl anions of high nuclearity, to be discussed presently, they do not possess the extraordinary distorted prismatic structures (Fig. 22-H-5) shown by these platinum compounds. It is not clear why these platinum systems prefer the distorted prismatic type of structure rather than an antiprismatic one, since the latter would straightforwardly minimize the nonbonded repulsive forces. It should also be noted that there is a marked difference between the Pt—Pt distances within the triangular $Pt_3(CO)_3(\mu\text{-}CO)_3$ units (*ca.* 2.65 Å) and those connecting these units with one another (*ca.* 3.05 Å).

Nickel Carbonylate Anions. There is abundant evidence that reduction of $Ni(CO)_4$ generates a large number of polynuclear carbonate anions, but most of them have not yet been isolated and characterized.[14] Reduction with alkali metals

11 J. Lewis *et al.*, *J.C.S. Chem. Comm.*, **1976**, 945.
12 J. Lewis, R. Mason, *et al.*, *J.C.S. Chem. Comm.*, **1977**, 385.
13 P. Chini, G. Longoni, and V. G. Albano, *Adv. Organomet. Chem.*, **1976**, **14**, 285.
14 P. Chini *et al.*, *Inorg. Chem.*, **1976**, **15**, 3025, 3029.

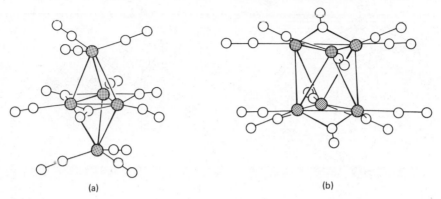

Fig. 26-5. The structures of the nickel carbonylate ions (a) $[Ni_5(CO)_{12}]^{2-}$ and (b) $[Ni_6(CO)_{12}]^{2-}$

in THF or with alkali hydroxides in methanol gives a mixture of $[Ni_5(CO)_{12}]^{2-}$ and $[Ni_6(CO)_{12}]^{2-}$, and these are related by the easily reversible equilibrium:

$$[Ni_5(CO)_{12}]^{2-} + Ni(CO)_4 = [Ni_6(CO)_{12}]^{2-} + 4CO$$

Yellow   Red

The structures of these two anions are shown in Fig. 26-5. It may be noted that the $[Ni_5(CO)_{12}]^{2-}$ ion, which has 76 electrons, is not isoelectronic with $[Os_5(CO)_{15}]^{2-}$, which has 72, but nonetheless it has a *tbp closo* structure; $[Ni_6(CO)_{12}]^{2-}$ is, however, an 86-electron species, like others with octahedral structures discussed earlier. Higher species are believed to include $[Ni_8(CO)_{12}]^{2-}$, $[Ni_8(CO)_{14}H_2]^{2-}$, $[Ni_9(CO)_{18}]^{2-}$, and $[Ni_{11}(CO)_{20}H_2]^{2-}$, or at least, these are the formulas currently assigned.

Other Rhodium Clusters. Besides $Rh_6(CO)_{16}$ and several other related Rh_6 clusters, rhodium has recently yielded a number of others that are remarkable for their size and complexity.

One fascinating type contains clusters of Rh atoms in which there are internal metal atoms,[15] bound only to other metal atoms. Figure 26-6 shows three of these. The Rh_{13} cluster has been structurally characterized in the form of the hydrido anions $[Rh_{13}(CO)_{12}(\mu_2\text{-CO})_{12}H_{2,3}]^{3,2-}$. The arrangement of the metal atoms is that of the smallest possible representative unit of a hexagonally close-packed metal; this ball of metal atoms is then "coated" with CO groups. There is one terminal CO group on each outer Rh atom and half of the 24 external Rh—Rh edges have μ_2-CO groups. The average Rh—Rh distance is 2.81 Å, and there are no significant differences between the internal and surface Rh—Rh bonds. The positions of the hydrogen atoms are still not known.

Figure 26-6b shows the Rh_{14} core of a $[Rh_{14}(CO)_{25}]^{4-}$ ion. This has a 4-fold symmetry axis and can be considered as a fragment of another important type of metal structure (Section 1-3), the body-centered cubic lattice. It has nine terminal and sixteen edge-bridging CO groups. The Rh—Rh distances are shown in the figure.

[15] S. Martinengo *et al., J. Am. Chem. Soc.,* 1978, **100**, 7096.

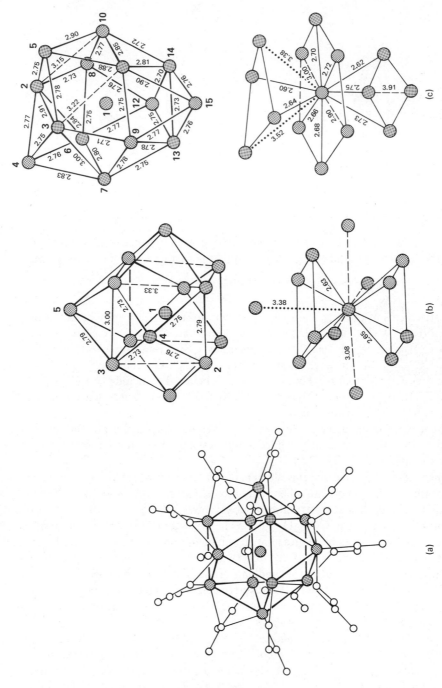

Fig. 26-6. The configurations of the metal atom cores in three high-nuclearity rhodium clusters that have internal metal atoms: (a) $Rh_{13}(CO)_{24}H_{2,3}]^{3,2-}$, (b) $[Rh_{14}(CO)_{25}]^{4-}$, and (c) $[Rh_{15}(CO)_{27}]^{3-}$.

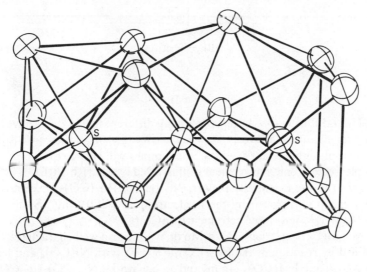

Fig. 26-7. Diagram of $[Rh_{17}(S)_2(CO)_{32}]^{3-}$, omitting the carbon monoxide ligands.

A third cluster of this type is found in the $[Rh_{15}(CO)_{27}]^{3-}$ ion whose Rh_{15} cluster appears in Fig. 26-6c. The packing cannot be described simply as one specific metal lattice fragment. In terms of the arrangement of neighbors around the central rhodium atom, the bottom part is like the *hcp* arrangement, with three and then six nearest neighbors, but the upper part deviates considerably from this. There are thirteen terminal and fourteen edge-bridging CO groups.

Rhodium has also afforded some large clusters in which heteroatoms are encapsulated. There are two carbido clusters $[Rh_{15}(CO)_{28}(C)_2]^-$ and $Rh_{12}(CO)_{25}C_2$; in the former there are two C atoms well separated from each other in the centers of roughly octahedral sets of Rh atoms, and in the latter the two carbon atoms are bonded together (1.47 Å) in the center of a complicated, unsymmetrical *nido* cluster.[13] More recently the $[Rh_{17}(CO)_{32}(S)_2]^{3-}$ anion has been made and structurally characterized.[16] As shown in Fig. 26-7, the Rh_{17} cluster has D_{4d} symmetry with one internal Rh atom on the 4-fold symmetry axis. The sulfur atoms are also on this axis, each one occupying the approximate center of a capped square antiprism.

The coupling of metal carbonyl clusters through one M—M bond has not been widely observed, but a clear example[13] is provided by the anion $[Rh_{12}(CO)_{30}]^{2-}$, which consists of two $Rh_6(CO)_{10}(\mu_3\text{-}CO)_4$ halves that are essentially the same as an $Rh_6(CO)_{16}$ molecule with two terminal CO groups removed from one Rh atom. The two halves are then joined through the unit 26-I. The Rh—Rh bond, the bridging carbonyls, and the two negative charges provide a total electron count at these rhodium atoms equal to the four electrons that two terminal CO groups would contribute.

[16] J. L. Vidal *et al., Inorg. Chem.,* 1978, **17**, 2574.

$$\begin{array}{c} O \\ \| \\ C \\ \diagup \quad \diagdown \\ Rh \!\!-\!\!-\!\!-\!\! Rh \\ \diagdown \quad \diagup \\ C \\ \| \\ O \end{array}$$

(26-I)

Mixed Metal Clusters. Clusters containing different metal atoms have already been mentioned among clusters of low nuclearity (p. 1062); they are also known among those of high nuclearity, as a few examples will suffice to show.

A simple isoelectronic replacement is illustrated by the $[Rh_5Pt(CO)_{15}]^-$ anion. This has a structure essentially the same as that of $Rh_6(CO)_{16}$ except that one $Rh(CO)_2$ group is replaced by the isoelectronic $Pt(CO)^-$ unit. Similarly, the $[Co_4Ni_2(CO)_{14}]^{2-}$ ion is isoelectronic with $[Co_6(CO)_{15}]^{2-}$.

The $[M_2Ni_3(CO)_{16}]^{2-}$ ions, (M = Mo or W) are also well characterized. They are obtained by reaction of the $[M_2(CO)_{10}]^{2-}$ ions with $Ni(CO)_4$ and have *tbp* structures with axial $M(CO)_5$ groups and an equatorial $Ni_3(CO)_3(\mu_2\text{-}CO)_3$ unit 26-II.

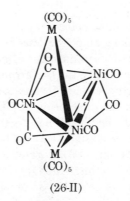

(26-II)

M—M BONDS IN NONCARBONYL COMPOUNDS

We have already noted that during the course of the first 70 years of this century compounds that did not appear to fit the pattern of Wernerian coordination chemistry were discovered and were eventually recognized to contain M—M bonds. The chemistry and structures of most of these have been summarized under the chemistry of the individual elements in Chapters 21 and 22. The next few sections explain the chemical bonding in these compounds and emphasize the electronic and structural features common to them all.

We concentrate on only a few types of compound, first because these are among the most important ones, but second because our understanding of them has reached a point that allows us to say a good deal with certainty about their electronic structures. The species to be discussed are:

M_6X_8 clusters, e.g., $[Mo_6Cl_8]^{4+}$, $[W_6Cl_8]^{4+}$

M_6X_{12} clusters, e.g., $[Nb_6Cl_{12}]^{2+}$, $[Ta_6Br_{12}]^{4+}$

Re_3X_9 clusters, e.g., Re_3Cl_9, $[Re_3Cl_{12}]^{3-}$

$X_3M{\equiv}MX_3$ triple bonds, e.g., $Mo_2(NMe)_6$, $W_2(CH_2SiMe_3)_6$

$X_4M{\equiv}MX_4$ quadruple bonds, e.g., $[Re_2X_8]^{2-}$, $[Mo_2(SO_4)_4]^{4-}$, $Cr_2(RNCR'O)_4$

In all these compounds the metal atoms have formal oxidation numbers of $+2$ to $+3$, whereas in the carbonyl-type compounds the formal oxidation states are mostly zero or negative. This difference is quite important in understanding the bonding, and it is the basis for a considerable difference between the chemical properties of the two broad classes.

As a transition metal atom is ionized, the difference in energy between its valence shell d orbitals and the outer s and p orbitals increases considerably. Therefore in the carbonyl-type clusters it is pertinent to think about the bonding in terms of nine valence shell orbitals for each metal atom, namely, the five nd orbitals plus the $(n + 1)s$ and the three $(n + 1)p$ orbitals. However for the compounds involving formally $+2$ and $+3$ metal atoms the $(n + 1)s$ and $(n + 1)p$ orbitals play only a minor role, and the bonding may be analyzed almost entirely in terms of the nd-orbital contributions.

26-5. The Octahedral and Re_3 Clusters

Figure 26-8 shows the structures we are concerned with in attempting to understand the bonding in these compounds. There have been a number of attempts to formulate the bonding in these species, but we shall consider only three. The first question to ask is whether localized electron-pair bonds can be assigned to all M—M pairs within bonding distance. In the case of the $[Mo_6Cl_8]^{4+}$ unit we have twelve such pairs (the twelve edges of the octahedron) and each metal atom Mo^{II} provides four electrons; with 24 electrons it is then possible to assign an electron-pair bond to each Mo—Mo pair. For the $[Nb_6Cl_{12}]^{2+}$ system we again have twelve M—M pairs but only eight electron pairs, since the six Nb atoms with a total of 30 electrons have lost fourteen. It is therefore necessary to assign only fractional (two-thirds) bond orders to each Nb—Nb pair. Moreover, the $[Nb_6Cl_{12}]^{2+}$ unit is known to undergo oxidation by one and two electrons to give species in which bond orders of 15/24 and 14/24 would have to be assigned. With Re_3Cl_9 or $[Re_3Cl_{12}]^{3-}$ we find that each Re^{III} has four electrons to contribute, making a total of six electron pairs for the three Re—Re bonds. This requires us to regard these as double bonds.

The bond orders of 1, 2/3, or 2 for $[Mo_6Cl_8]^{4+}$, $[Nb_6Cl_{12}]^{2+}$, and Re_3Cl_9, respectively, are plausible when compared to M—M distances of, respectively, $\sim$2.61, $\sim$2.82, and 2.45 Å in typical compounds. However the need to assign a fractional bond order to the niobium cluster (which is even variable depending on the extent of further oxidation) suggests that an analysis of the bonding in terms of molecular orbitals delocalized over the entire cluster might be advantageous. Such an approach

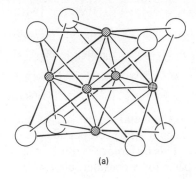

(a)

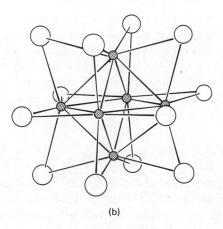

(b)

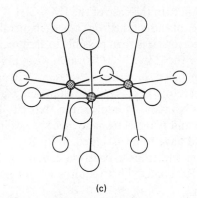

(c)

Fig. 26-8. Structures of the (a) M_6X_8, (b) M_6X_{12}, and (c) $Re_3X_9L_3$ type clusters.

would be attractive in the other cases as well for the eventual purpose of trying to assign their electronic spectra.

As a first attempt to devise MOs for clusters such as these, one may consider only the metal d orbitals, and consider how their overlaps could give rise to cluster

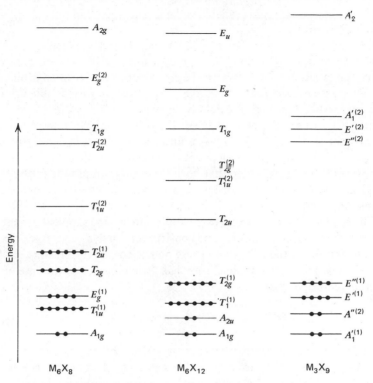

Fig. 26-9. Approximate energies of MOs derived from d-orbital overlap calculations for three metal atom cluster species. The electrons shown are those present for the best known stable species (see text).

MOs.[17] This simple approach has been applied to all three of the cluster species under discussion. One d orbital on each metal atom must be dedicated primarily to M—Cl bonding, leaving four on each one for M—M bond formation. The energies of the MOs that are formed are assumed to be proportional to the magnitudes of the overlap integrals, which is the type of approximation invoked in the extended Hückel, Wolfsberg-Helmholz, and angular overlap methods for conventional complexes. The results for the three clusters under discussion are shown in Fig. 26-9. It can be seen that in each case there is a clean separation for the MOs into bonding and antibonding sets. Moreover, in each case the bonding sets have just the right capacity for the number of electrons available.

For the M_6X_8 case there are four clearly antibonding MOs, one nonbonding MO, and five bonding MOs. The latter set is comprised of the triply degenerate $T_{1u}^{(1)}$, $T_{2u}^{(1)}$, and T_{2g} orbitals, the doubly degenerate $E_g^{(1)}$ orbital and the singly degenerate A_{1g} orbital; these can hold twelve electron pairs, precisely the number available in the $[Mo_6Cl_8]^{4+}$ and $[W_6Cl_8]^{4+}$ clusters and their derivatives.

For the M_6X_{12} case the four bonding orbitals A_{1g}, A_{2u}, $T_{1u}^{(1)}$, and $T_{2g}^{(1)}$ hold

[17] F. A. Cotton and T. E. Haas, *Inorg. Chem.*, 1964, **3**, 10.

precisely the sixteen electrons available in the $[Nb_6Cl_{12}]^{2+}$ and $[Ta_6Cl_{12}]^{2+}$ clusters and their derivatives.

In the M_3X_9 case the four bonding MOs hold the twelve electrons available in Re_3Cl_9 or $[Re_3Cl_{12}]^{3-}$. In this case, we may conveniently carry the analysis even further. The labels of the MOs have the significance that primed orbitals are of σ character with respect to the M_3 plane, whereas those with double primes have π character. Also, E orbitals are doubly degenerate and A orbitals are singly degenerate. We can see therefore that the filled bonding MOs correspond to the presence a set of three Re—Re σ bonds and a set of three Re—Re π bonds.

Only in recent years has it been possible to make more rigorous calculations in which all MOs, including those that have their main amplitude in the region of metal-ligand bonds and ligand lone pairs, are included. This has been done mostly by employing the self-consistent field, $X\alpha$, scattered wave (SCF-$X\alpha$-SW) method.[18] The results of these elaborate calculations support the simple d-orbital overlap pictures in their essential features and confirm that such d–d interactions are the main contributors to the stability of these noncarbonyl-type clusters. As Section 26-6 indicates, a very similar conclusion has been obtained in the case of M—M triple and quadruple bonds.

26-6. Multiple Bonds[19]

Section 26-5 mentioned the existence of M—M double bonds in the Re_3X_9-type cluster compounds. There are a few other cases in which double bonds are also believed to exist, although in all these others there are bridging ligands, and this makes the M—M bond order not entirely unambiguous, since some type or degree of interaction *through* the ligands cannot be ruled out with certainty. In the case of $W_2S_2(S_2CNEt_2)_4$ (p. 881) there is reasonably direct evidence that the bond is truly of order 2. Our concern here, however, is with triple and quadruple bonds between metal atoms. These occur only with transition metals and result from the overlap of d orbitals.

We show in Fig. 26-10 the five possible overlaps between the sets of d orbitals on two metal atoms: these overlaps make it possible for two "naked" metal atoms to form a σ bond, two π bonds, and two δ bonds. The relative values of the overlaps are such that the σ interaction should be very strong, the π interactions of intermediate strength, and the δ interactions rather weak. The set of bonding and antibonding orbitals is shown in the column marked M_2 in Fig. 26-11, arranged in accord with these expected relative degrees of overlap.

When ligand atoms are attached to the metal atoms, this ordering of the MOs is altered according to the ligand arrangement. It is convenient to consider the quadruple bonds first.

Quadruple Bonds. If four ligands are brought up to each metal atom along its local x, $-x$, y, and $-y$ directions, the eight lobes of the two $d_{x^2-y^2}$ orbitals will

[18] F. A. Cotton and G. G. Stanley, *Chem. Phys. Lett.*, 1978, **58**, 450.

[19] F. A. Cotton, *Chem. Soc. Rev.*, 1975, **4**, 27; *Acc. Chem. Res.*, 1978, **11**, 225; M. H. Chisholm and F. A. Cotton, *Acc. Chem. Res.*, 1978, **11**, 356.

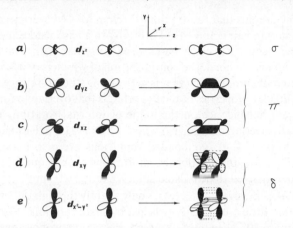

Fig. 26-10. The σ, π, and δ overlaps between two sets of d orbitals on adjacent metal atoms with the internuclear axis as the z axis.

become engaged in the formation of metal-to-ligand σ bonds. Therefore one member of the δ set of M_2 drops to lower energy and becomes an $ML\sigma$ orbital; at the same time one member of the δ^* pair rises in energy and becomes an $ML\sigma^*$ orbital. The exact extent to which these $ML\sigma$ and $ML\sigma^*$ orbitals move relative to the other M—M bonding and antibonding orbitals will vary from case to case, and the arrangement shown in the M_2L_8 column of Fig. 26-11 is only one of many possibilities. In any case, however, the $ML\sigma$ orbit will be filled by electrons that contribute to M—L bonding and will play no further role in M—M bonding.

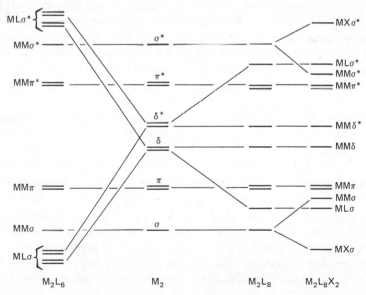

Fig. 26-11. Energy level diagrams showing schematically how d-orbital overlaps between two metal atoms (M_2) can be modified by bonding of ligands to give triple bonds in M_2L_6, strong quadruple bonds in M_2L_8, and weaker quadruple bonds in $M_2L_8X_2$.

The remaining orbitals that result primarily from M—M overlaps are the MMσ, MMπ (a degenerate pair), and MMδ orbitals. In a case where the metal atoms each have four d electrons to contribute, we can fill these four orbitals and obtain a metal-metal quadruple bond, the configuration of which is $\sigma^2\pi^4\delta^2$. Such a bond has two characteristic properties: (1) it is very strong, therefore very short, and (2) because of the angular properties of the d_{xy} orbitals that overlap to form the δ bond, it has an inherent dependence on the angle of internal rotation. The δ bond is strongest (d_{xy}–d_{xy} overlap maximizes) when the two ML$_4$ halves have an eclipsed relationship. However L$\cdots$L nonbonded repulsions are also maximized in this conformation. Therefore the rotational conformation about a quadruple bond might in some cases be expected to be twisted somewhat away from the exactly eclipsed one. Indeed, the d_{xy}–d_{xy} overlap decreases only slightly through the first few degrees of rotation, so that little δ-bond energy is lost by small rotations. Several examples of rotations of up to 20° have been observed; the majority of quadruple bonds are essentially eclipsed, however.

Since the existence of a quadruple bond was first recognized and explained in 1964 in the case of the [Re$_2$Cl$_8$]$^{2-}$ ion, whose structure was depicted in Fig. 22-D-5, hundreds of compounds containing such bonds have been prepared. They are formed by the elements Cr, Mo, W, Tc, and Re, and of these molybdenum has given the greatest number of compounds.

In recent years extensive theoretical and spectroscopic studies have provided quantitative support for the schematic picture of the quadruple bond just discussed. As in the noncarbonyl clusters described earlier, the only broadly applicable, first principles method of calculation for these species has been the SCF-Xα-SW method. Figure 26-12 shows the results obtained by this method for the Mo$_2$(O$_2$CH)$_4$ molecule, which has the type structure shown in 22-C-XII. The quantitative results show that the strength of the Mo—Mo bond components all increase in the Mo$_2$(O$_2$CH)$_4$ molecule over what they are in the Mo$_2^{4+}$ ion. This is mainly because the electron density donated by the HCO$_2^-$ ligands reduces the charge on the Mo$_2$ unit, thus allowing the metal d orbitals to expand and overlap better.

It can be seen in Fig. 26-12 that the calculation predicts that the highest filled level is the Mo—Mo δ-bonding orbital. This is followed by the Mo—Mo π-bonding level; then there is a plethora of ML and L levels with the Mo—Mo σ-bonding level in among them. This arrangement receives clear support from experimental photoelectron spectra[20a], which can be measured for the volatile Mo$_2$(O$_2$CR)$_4$ compounds in the vapor phase. These are shown in Fig. 26-13 for the molecules with R = CH$_3$, H, and CF$_3$. It is evident that each spectrum has two distinct bands that correspond to the predicted δ and π ionizations and then, at higher energies, many close or overlapping bands, just as the calculation would predict. Moreover, as the electron-donating ability (inductive effect) of the R groups increases from CF$_3$ to H to CH$_3$, the ease of detachment of the electrons increases, as would also be expected.

[20a] J. C. Green et al., J.C.S. Dalton, **1979**, 1057.

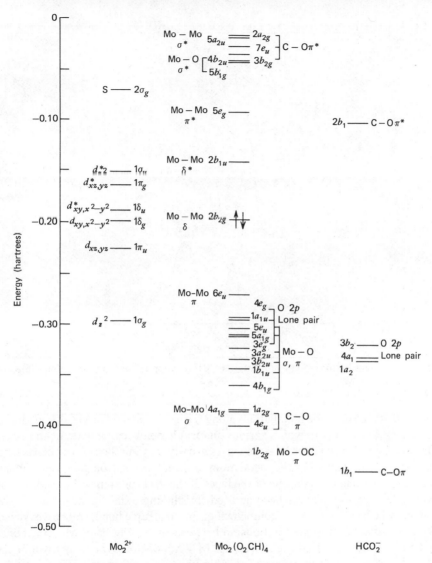

Fig. 26-12. A molecular orbital diagram based on SCF-Xα-SW calculations for $Mo_2(O_2CH)_4$. [Reproduced by permission from J. G. Norman *et al.*, *Inorg. Chem.*, 1977, **16**, 987.] One hartree equals 27.2 eV.

To show that this more detailed picture of the quadruple bond is in harmony with the simple orbital overlap scheme of Fig. 26-10, let us turn to contour diagrams of the principal Mo—Mo π-bonding and Mo—Mo σ-bonding orbitals (Fig. 26-14), as obtained in the SCF-Xα-SW calculation on $Mo_2(O_2CH)_4$ that we have been discussing. It is clear from these diagrams that the M—M bonds are, indeed, the result of overlap of metal d orbitals, essentially as indicated in Fig. 26-10.

Finally it is necessary to comment on the role played by axial ligands, that is,

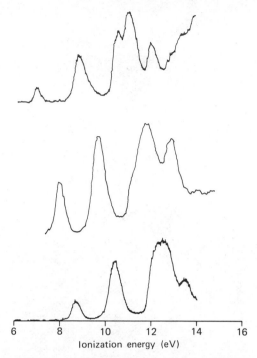

Fig. 26-13. Photoelectron spectra for $Mo_2(O_2CCH_3)_4$ (upper), $Mo_2(O_2CH)_4$ (middle), and $Mo_2(O_2CCF_3)_4$ (lower).

donors that approach the metal atoms along the extensions of the M—M quadruple bond axis. The extent to which various quadruple bonds participate in and are in turn affected by axial coordination varies greatly, but in all cases the behavior is qualitatively the same. For the metal atom to accept an electron pair from a donor approaching along the axis, one of the lobes of the d_{z^2} orbital must be used. To the extent that the d_{z^2} orbitals become used for binding axial ligands, the M—M σ-bond, which is the strongest component of the quadruple bond, will be weakened. We thus expect an inverse relationship between the strengths of M—X_{ax} bonds and M—M bonds, so that the shorter the M—X_{ax} distance the longer will be the M—M distance. A relationship of this kind has been observed in every case for which relevant data are available.

The formulation of this relationship is expressed in terms of molecular orbitals at the extreme right of Fig. 26-11. The binding of the axial ligands X causes the M—Mσ orbital to mix with the σ orbital on X that contains the donor pair of electrons. This, in effect splits the M—Mσ orbital into a more stable one that has mainly MXσ character and a less stable one that has mainly MMσ character. Roughly speaking, the energy gained by M—X bond formation is offset by weakening of the M—Mσ bond.

In most cases—that is, for the many Mo—Mo and Re—Re quadruple bonds—the tendency to bind axial ligands is small; they are in many cases not

Fig. 26-14. Electron density contour diagrams for the MOs of $Mo_2(O_2CH)_4$ that are principally responsible for M—M σ bonding (left) and M—M π bonding (right). [Reproduced by permission from J. G. Norman, Jr., *Inorg. Chem.*, 1977, **16**, 987.]

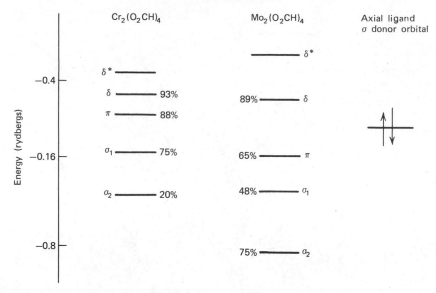

Fig. 26-15. Energies and metal *d*-orbital percentages for MOs involved in Cr—Cr and Mo—Mo bonding in $M_2(O_2CH)_4$ molecules.

present at all, and when they are, they are bound only weakly with little effect on the length of the MM bond. For example, the Mo—Mo bond length in $Mo_2(O_2CCF_3)_4$ is 2.090 ± 0.004 Å; in the dipyridine adduct[20b] $Mo_2(O_2CCF_3)_4 \cdot 2py$ it increases slightly to 2.129 ± 0.002 Å. The Mo—N distance is quite long (2.55 Å), showing that this bond is very weak. This is in accord with the fact that the pyridine can be easily pumped off.

As we have noted in the case of Cr—Cr quadruple bonds (p. 725) there is an enormous variation in their bond lengths. Recent work[21] suggests that this is largely controlled by the axial ligands. In all $Cr_2(O_2CR)_4$ compounds there are axial ligands, either separate ones as in, for example, $Cr_2(O_2CCH_3)_4(H_2O)_2$, or as a result of intermolecular bridging in the crystals; all these compounds have long Cr—Cr bonds (viz., ~2.3 to ~2.5 Å). On the other hand, the compounds with very strong and short (<2.0 Å) Cr—Cr quadruple bonds all lack axial ligands because the steric properties of the ligands block the axial positions. The great sensitivity of the Cr—Cr quadruple bond to axial ligands has been elucidated by SCF-Xα-SW calculations.[22] The key results are shown in Fig. 26-15.

Although all the Mo—Mo bonding interactions are somewhat stronger than the corresponding Cr—Cr interactions, the big difference between the two systems is in their M—Mσ bonding. In the $M_2(O_2CH)_4$ molecules there are, of course, a number of MOs with the same symmetry as the simple $d_{z^2}^{(1)} + d_{z^2}^{(2)}$ combination. Therefore this interaction, which is principally responsible for M—M bonding,

[20b] F. A. Cotton and J. G. Norman, *J. Am. Chem. Soc.*, 1972, **94**, 5697.
[21] A. Bino, F. A. Cotton, and W. Kaim, *J. Am. Chem. Soc.*, 1979, **101**, 2506.
[22] F. A. Cotton and G. G. Stanley, *Inorg. Chem.*, 1977, **16**, 2668.

may be spread through several MOs. The calculations show that it is in each case largely concentrated in two such MOs, and these are both included in Fig. 26-15, labeled σ_1 and σ_2. The percentage listed beside each of the filled MOs give the percentage of metal d character, and it can be seen that the Cr_2 and Mo_2 cases differ markedly in this respect for the σ_1 and σ_2 orbitals. For the Mo_2 compound, the lower σ_2 orbital has the greater amount of metal d character and is the primary contributor to Mo—Mo σ bonding. This orbital is so low in energy that it does not interact appreciably with the filled σ orbitals of potentially axial ligands, which are much higher. In the Cr_2 case, however, the upper σ_1 orbital carries most of the M—Mσ bonding, and its energy is such that strong interaction with the axial ligand σ orbitals is to be expected, as long as there is no steric reason for these axial ligands not to approach.

Triple Bonds. A surprisingly large number of types of these exist. One obvious way to obtain a triple bond is to carry out a two-electron oxidation of a quadruple bond. Attempts to do this with simple $[M_2X_8]^{n-}$ species such as $[Re_2Cl_8]^{2-}$ or $[Mo_2Cl_8]^{4-}$ lead to structural rearrangements, from which species such as $[Cl_3Re(\mu\text{-}Cl)_3ReCl_3]^-$ and $[Cl_3Mo(\mu\text{-}H)(\mu\text{-}Cl)_2MoCl_3]^{3-}$ are obtained. However with certain bridged complexes structural integrity is preserved during the removal of one or two electrons. Thus the $[Mo_2(SO_4)_4]^{4-}$ ion (Mo—Mo = 2.11 Å) is oxidized by air to the $[Mo_2(SO_4)_4]^{3-}$ ion (Mo—Mo = 2.16 Å), which has a $\sigma^2\pi^4\delta$ configuration.[23] Further oxidation of this species has not been reported, but the oxidation of $[Mo_2Cl_8]^{4-}$ by O_2 in H_3PO_4 solution leads directly[24] to the very stable $[Mo_2(HPO_4)_4]^{2-}$ ion (Mo—Mo = 2.22 Å).

The allyl compound $Re_2(C_3H_5)_4$ (26-III) provides a unique but clear example of a $\sigma^2\pi^4$ triple bond between rhenium atoms.[25]

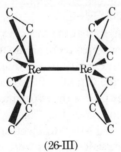

(26-III)

A second way of generating triple bonds from quadruple ones is by addition of electrons to the δ^* orbital. As already noted (p. 892) the $[Tc_2Cl_8]^{3-}$ ion is readily isolated; this has a $\sigma^2\pi^4\delta^2\delta^*$ configuration. Although no compound containing the analogous rhenium species has been isolated, two-electron reduction of $[Re_2Cl_8]^{2-}$ takes place readily when this is treated with alkyl phosphines, giving the compounds $[Re_2Cl_4(PR_3)_4]$. On exposure to air or other mild oxidizing agents these will lose one electron to give the $[Re_2Cl_4(PR_3)_4]^+$ ions. It is noteworthy that $[Re_2\text{-}$

[23] A. Bino and F. A. Cotton, *Inorg. Chem.,* 1979, **18,** 1159.
[24] A. Bino and F. A. Cotton, *Angew. Chem., Int. Ed.,* 1979, **18,** 332.
[25] F. A. Cotton and M. W. Extine, *J. Am. Chem. Soc.,* 1978, **100,** 3788.

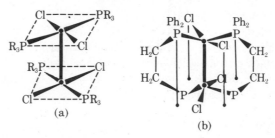

Fig. 26-16. The structures of (a) $Re_2Cl_4(PR_3)_4$ and (b) $Re_2Cl_4(Ph_2PCH_2CH_2PPh_2)$.

$Cl_4(PEt_3)_4$] has an eclipsed conformation even though, with a $\sigma^2\pi^4\delta^2\delta^{*2}$ config-uration, there is no net δ bond. This is because, as shown in Fig. 26-16a, the ste-reochemistry is dominated by the requirement of staggering the bulky PEt_3 groups with respect to one another. When these four groups are staggered, the conformation as a whole is eclipsed. When the four PR_3 groups are replaced by two $Ph_2PCH_2CH_2PPh_2$ ligands, which give six-membered rings that inherently tend to have chair conformations, the rotational conformation about the $Re{\equiv}Re$ bond does, in fact, become essentially staggered[26] (Fig. 26-16b).

By far the largest and most important class of molecules containing triple bonds are those of the type $X_3M{\equiv}MX_3$, in which M may be Mo or W and X is a univalent group. A representative list of such compounds was given in Table 22-C-III, and the structures of two of them appear in Fig. 26-17. Qualitative formulation of the bonding in M_2X_6 compounds is shown in Fig. 26-11. Starting again with the M_2 unit with its ten MOs formed by d-orbital overlaps, we find that the six ligands form bonds to the metal atoms using to some extent all the d orbitals, but primarily the δ type. As indicated schematically in Fig. 26-11, this removes δ- and δ^*-type orbitals from the center of the diagram, so that the $M—M\pi$ and $M—M\sigma$ bonding orbitals are the highest filled orbitals (provided there are no filled nonbonding ligand orbitals that have comparable energy). Quantitative calculations have confirmed all the essentials of this picture, and the validity of these calculations is, in turn, verified by the excellent prediction they give of the experimentally measured photoelectron spectrum of an $Mo_2(OR)_6$ compound (Fig. 26-18). The first two peaks are due to the ionization of the $M—M\pi$ and $M—M\sigma$ electrons. The discrepancy in the 9-10 eV region is expected since the calculations were made for $R = H$ and the mea-surements for $R = CH_2CMe_3$.

The chemistry of these $X_3M{\equiv}MX_3$ compounds has been discussed (p. 876) and can be understood fully using the theoretical discussion of bonding just given. It is notable that triple bonds of this kind are as readily formed and as stable with tungsten as with molybenum, in sharp contrast to the situation with quadruple bonds. It is also notable that no $Cr{\equiv}Cr$ bonds of this type have been observed.

One other type of triply bonded compound deserving mention consists of the cyclopentadienyl compounds $CpM(CO)_2{\equiv}M(CO)_2Cp$. These are obtained by pyrolysis of the $CpM(CO)_3—M(CO)_3Cp$ compounds with loss of 2CO. They are

26 F. A. Cotton, G. G. Stanley, and R. A. Walton, *Inorg. Chem.*, 1978, **17**, 2099.

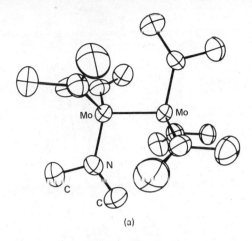

(a)

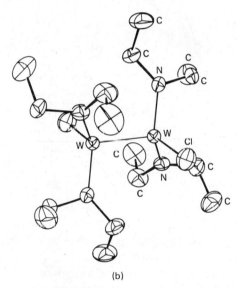

(b)

Fig. 26-17. The structures of two representative compounds containing $\sigma^2\pi^4$-type triple metal—metal bonds: (a) $Mo_2(NMe_2)_6$ and (b) $W_2(NEt_2)_4Cl_2$. All hydrogen atoms are omitted.

quite reactive toward a variety of electron donors such as CO (to regenerate the hexacarbonyls) and acetylenes.[27]

Other M—M Bonds. The electronic structures of certain binuclear, carboxylato-bridged compounds of ruthenium and rhodium can profitably be considered here. As indicated in the introduction of the quadruple bond, it takes eight electrons to fill the set of bonding orbitals σ, π, and δ, formed by overlapping metal d orbitals in an M_2X_8 or $M_2(O_2CR)_4$ type of molecule or ion. If there are more than eight

[27] W. I. Bailey *et al., J. Am. Chem. Soc.,* 1978, **100**, 5764.

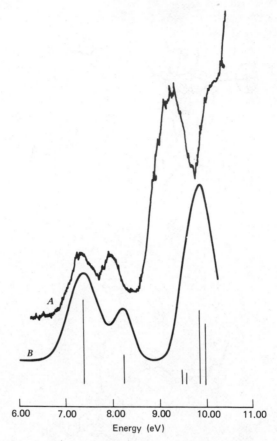

Fig. 26-18. The experimental (*A*) and calculated (*B*) photoelectron spectra for $Mo_2(OCH_2CMe_3)_6$.

electrons, it will be necessary to put the additional ones in antibonding orbitals. We have already mentioned the addition of one or two electrons to the $\delta*$ orbital to give M—M bond orders of 3.5 and 3.0.

In the case of the carboxylato-bridged diruthenium compounds (p. 931) the usual composition is $Ru_2(O_2CR)_4X$ and the number of electrons to be allocated to the M—M bonding and antibonding orbitals is eleven. Therefore with eight bonding electrons and three antibonding electrons, the net bond order should be $5/2 = 2.5$. This is compatible with the Ru—Ru distances of *ca.* 2.28 Å for such molecules. However according to the level order indicated in Fig. 26-11 (which is confirmed by quantitative calculations for compounds of Mo, Tc, and Re), one would expect a $\sigma^2\pi^4\delta^2\delta*^2\pi*$ configuration with one unpaired electron, whereas experiment shows that there are three unpaired electrons. An SCF-Xα-SW calculation for the $Ru_2(O_2CH)_4Cl$ molecule[28] shows that the $\pi*$ levels are in this case located slightly below the $\delta*$ level, so that the predicted configuration is $\sigma^2\pi^4\delta^2\pi*^2\delta*$, in agreement with experiment.

[28] J. G. Norman, Jr., and H. G. Kolari, *J. Am. Chem. Soc.,* 1978, **100**, 791.

For the $Rh_2(O_2CR)_4L_2$ molecules there are fourteen electrons to occupy M—M bonding and antibonding orbitals. Thus there should be four bonding pairs and three antibonding pairs giving a net bond order of 1.0. The detailed calculation for this case[28] gives the configuration $\delta^2\pi^4\delta^2\pi^*4\delta^{*2}$.

26-7. Relation of Clusters to Multiple Bonds

Although it has been convenient to discuss cluster and multiply bonded binuclear species separately, it is also necessary to point out the relationship between them. There is, in a formal way, continuous gradation from high-nuclearity clusters as one extreme to high M—M bond orders as the other.

To appreciate this we note first that the local ligand environment of a metal atom is much the same in the M_6X_8, M_6X_{12}, M_3X_9 clusters and in the X_4MMX_4 quadruply bonded dimers. In each case a metal atom is coordinated by a square set of ligands; the MX_4 group is generally slightly pyramidal, with the ligands forming the base. In addition, there is a tendency in all cases for a fifth ligand to occupy a position along the 4-fold axis of this pyramid, just below the base. We have in each case, the structural element 26-IV.

(26-IV)

The four structural types just mentioned are obtained by combining such elements in different ways, the differences being in the extent and type of bridging by the X groups. In both the octahedral clusters all X ligands are bridges, triple ones in M_6X_8, and double ones in M_6X_{12}. In M_3X_9 only two of the four X ligands are bridges, and in X_4MMX_4 none are bridges.

These structural differences go hand in hand with differences in the way in which M—M bonds are formed. In the MX_4 or MX_4X' unit the metal atom has available four d orbitals with which to form M—M bonds, as shown earlier. Let us suppose we are also dealing with a specific group, such as MoX_4^{2-} or ReX_4^-, in which the metal atom also has four electrons available. What are the possible ways in which these four orbitals and four electrons may be used? At one extreme we have the possibility of forming four single bonds to four other metal atoms. This is, in fact, exactly what happens in the Mo_6X_8-type cluster. At the other extreme the four orbitals and four electrons may all be used to form one quadruple bond to one other metal atom, as in $[Mo_2X_8]^{4-}$ and $[Re_2X_8]^{2-}$. There is then the intermediate case in which two double bonds are formed; this is what occurs in the Re_3X_9 systems. This set of three species is indicated by 26-Va to 26-Vc.

M_6X_8	M_3X_9	M_2X_8
Four M—M single bonds per M	Two M—M double bonds per M	One M—M quadruple bond per M
(26-Va)	(26-Vb)	(26-Vc)

The question of what determines the relative stabilities of these possible arrangements has not been answered in detail. It is likely that avoidance of excessively high total charge plays a part. This could explain why $[Mo_6Cl_{14}]^{2-}$ and $[Mo_2Cl_8]^{4-}$ are stable, but not the molybdenum analogue of $[Re_3Cl_{12}]^{3-}$, since that would be $[Mo_3Cl_{12}]^{6-}$, with perhaps too much negative charge for stability. With rhenium the known chloro compounds are $[Re_2Cl_8]^{2-}$ and $[Re_3Cl_{12}]^{3-}$, but the rhenium analogue of $[Mo_6Cl_{14}]^{2-}$, which would be $[Re_6Cl_{14}]^{4+}$, has not been seen. Interestingly, the compound $Na_4Re_6S_{10}(S_2)$ has recently been discovered and shown to contain an $[Re_6S_8]^{2+}$ unit[29] to which four S^{2-} ligands and two S_2^{2-} ligands, each of which bridges to an adjacent cluster, are coordinated in the six outer positions. By substituting S^{2-} for Cl^-, the unacceptably high charge on a central $[Re_6Cl_8]^{10+}$ has been avoided. Thus for Re^{III} the entire set of structures 26-V is known.

The possibility that the X_3M units of $X_3M{\equiv}MX_3$ might also be able to form tetrahedral $(X_3M)_4$ molecules is an obvious, analogous idea, but to date there has been no report of such a compound.

26-8. One-Dimensional Solids[30]

The substances of interest in this section are mainly those in which planar complexes of platinum and iridium are arranged in infinite stacks in the crystals, as illustrated in Fig. 26-19 for $[Pt(CN)_4]^{n+}$ ions in several of the important compounds. In this way chains of metal atoms are created and there is direct bonding between the metal atoms, so that a kind of one-dimensional metal can be created if the metal atoms approach one another closely enough.

We discuss first the compounds that have been most intensively studied, namely, $K_2Pt(CN)_4 \cdot 3H_2O$ and its partly oxidized derivatives. These are listed in Table 26-1.[31] As indicated, $K_2Pt(CN)_4 \cdot 3H_2O$ itself does *not* have interesting properties. It is white and a nonconductor of electricity. The platinum valence is integral (+2.0) and the Pt···Pt distances are so long that no significant M—M bonding would be expected. As long ago as 1842 it was observed that under oxidizing conditions one

TABLE 26-1
Some Tetracyanoplatinate Compounds with Stacked Anions[a]

Complex	Pt valence	Pt—Pt (Å)	Color	Conductivity (Ω^{-1} cm^{-1})
Pt metal	0	2.775	Metallic	$\sim 9.4 \times 10^4$
$K_2[Pt(CN)_4] \cdot 3H_2O$	+2.0	3.48	White	5×10^{-7}
$K_2[Pt(CN)_4]Br_{0.3} \cdot 3H_2O$	+2.3	2.88	Bronze	4–830
$K_2[Pt(CN)_4]Cl_{0.3} \cdot 3H_2O$	+2.3	2.87	Bronze	~ 200
$K_{1.75}[Pt(CN)_4] \cdot 1.5H_2O$	+2.25	2.96	Bronze	~ 70–100
$Cs_2[Pt(CN)_4](FHF)_{0.39}$	+2.39	2.83	Gold	Unknown

[a] From Ref. 31.

[29] S. Chen and W. R. Robinson, *J.C.S. Chem. Comm.*, **1978**, 879.

[30] J. S. Miller and A. J. Epstein, *Prog. Inorg. Chem.*, 1976, **20**, 1; J. S. Miller, ACS Advances in Chemistry Series No. 150, 1976.

[31] G. D. Stucky, A. J. Schultz, and J. M. Williams, *Ann. Rev. Mater. Sci.*, 1977, **7**, 301.

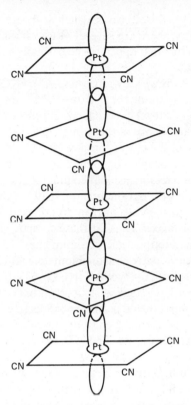

Fig. 26-19. Diagram of the stacking of $[Pt(CN)_4]^{n-}$ ions showing how metal d_{z^2} orbitals can overlap. Note alternating 45° rotation.

could obtain what appeared to be bronze-colored forms of this compound, and the bronze specimens were later found to be electrical conductors. It has since been shown that these bronze, conducting materials contain about 0.3 Br^- or Cl^- ions per Pt and that the oxidation number of Pt in them is therefore about 2.3. As Table 26-1 shows, this fractional oxidation is accompanied by an enormous decrease in the stacking distance, so that Pt—Pt separations approach to within 0.11 Å of those in metallic platinum. Detailed calculations[32] show that a kind of one-dimensional band structure (Section 1-3) is set up and that electrical conductance and other metallic properties result from this. The calculations suggest that the observed stoichiometries give optimum M—M bonding. The electrical conductivity arises from partial filling of a band, just as in the normal three-dimensional case.

The substances containing *ca.* 0.3 Cl or 0.3 Br in the crystals are sometimes called Krogmann salts. The partial oxidation of $[Pt(CN)_4]^{2-}$ may also be done in such a way as to obtain a cation-deficient product like $K_{1.75}[Pt(CN)_4] \cdot 1.5 H_2O$, as also indicated in Table 26-1.

[32] M.-H. Whangbo and R. Hoffmann, *J. Am. Chem. Soc.,* 1978, **101**, 6093, and earlier references therein.

Besides the $[Pt(CN)_4]^{n-}$ ($n < 2$) cases, which, it should be emphasized, we have covered only superficially, there are several other established examples of good one-dimensional conductors that appear to be basically similar but have been less thoroughly investigated.

Partial oxidation of $[Pt(ox)_2]^{2-}$ by nitric acid, Cl_2, H_2O_2, or other agents yields copper-colored needles containing the $[Pt(ox)_2]^{\sim 1.64-}$ ion, many cation-deficient salts of which, e.g., $Mg_{0.82}[Pt(ox)_2] \cdot x H_2O$, have been isolated and studied. Pt—Pt distances in these compounds are in the range 2.80–2.85 Å, and their properties are comparable to those of the cyano compounds.

Halo carbonyliridate compounds containing, in effect, partially oxidized Ir^I have also been shown to form one-dimensional conductors (see Fig. 22-G-4). Ir^I is, of course, isoelectronic with Pt^{II}, but the greater spatial extension of the d orbitals of Ir^I might be expected to allow stronger interactions and better band formation at similar internuclear distances. Thus far these compounds are very incompletely characterized as compared to the cyanoplatinum ones. They include "$Ir(CO)_3X$" (X = Cl, Br), which are evidently nonstoichiometric and have iridium in a mean oxidation state > 1, and salts of the type $M_x^I Ir(CO)_2Cl_2$ ($0.5 < x < 1.0$). There are indications that the latter have stacked structures, but details are still fuzzy.

Certain nonstoichiometric mixed platinum oxides (e.g., $Na_x Pt_3O_4$, $0 < x < 1$) have complex structures with chains of Pt atoms running in all three crystal directions and Pt—Pt distances of ca 2.80 Å. This is broadly reminiscent of the structure of $Hg_{2.86}AsF_6$ (Fig. 19-1).

In conclusion it should be noted that structures in which planar complexes are arranged in stacks to give infinite chains of metal atoms are not uncommon in the coordination chemistry of d^8 metals; a few have already been mentioned (p. 955). In these cases the optical properties of the crystals are usually quite anisotropic, with strong absorption bands polarized in the direction of the metal atom chains, but the M—M distances are long (3.2–3.5 Å), and there is no intrinsic electrical conductivity. We have already mentioned the completely stoichiometric salts of $[Pt(CN)_4]^{2-}$. Other examples of such substances include Magnus's green salt $[Pt(NH_3)_4][PtCl_4]$ and a number of analogous compounds such as $[Pt(NH_2Me)_4][PtBr_4]$, as well as $Rh(CO)_2(acac)$ and bis(dimethylglyoximato)-nickel.

General References

Bassett, J. M., and R. Ugo, *Aspects of Homogeneous Catalysis,* Vol. 3, p. 138. D. Reidel, 1977. Relation between clusters and small particles.

Chini, P., and B. T. Heaton, *Topics Current Chem.* 1977, **71**, 3. M_4 carbonyls (253 refs).

Gillespie, R. J. and J. Passmore, *Adv. Inorg. Chem. Radiochem.,* 1975, **17**, 49. Homopolyatomic cations of the elements.

Johnson, B. F. G., ed., *Transition Metal Clusters,* Wiley, 1979.

Muetterties, E. L. et al., *Chem. Rev.,* 1979, **79**, 91. Clusters and surfaces (295 refs).

Annual Reviews of *Organometallic Chemistry and Organometallic Chemistry.* Chemical Society Specialist Reports. Annual reviews deal with clusters with CO and organic ligands.

Templeton, J. L., *Progr. Inorg. Chem.,* 1979, **26**, 211. M-M quadruple bonds.

Vahrenkamp, H., *Struct. Bond.,* 1977, **32**, 1. Clusters with organic ligands.

CHAPTER TWENTY–SEVEN

Transition Metal Compounds with Bonds to Hydrogen and Carbon

This chapter considers complexes of transition metals that have bonds to hydrogen (M—H) and to carbon (M—CR_3, M=CR_2, and M≡CR). We also discuss complexes that contain coordinated alkenes, alkynes, arenes, etc., or organic groups that are derived from such unsaturated substances.

Many of the compounds discussed here contain CO or other π-acid ligands of the types discussed in Chapter 25 and some are cluster compounds as discussed in Chapter 26. Also many complexes of these types are involved in the reactions discussed in Chapters 28 to 30. This chapter, therefore, ties together the neighboring ones so that the entire set, from 25 to 30, form a more or less coherent unit. The reader will note a certain amount of overlap and interlocking. However we emphasize that throughout these chapters there are numerous examples of attacks on coordinated ligands of all types that may be intra- or intermolecular. Instead of dealing with ligands separately, it would have been possible to classify such attacks on coordinated ligands generally under the conventional headings used by organic chemists (viz., nucleophilic, electrophilic, and radical).

COMPOUNDS WITH TRANSITION METAL TO HYDROGEN BONDS[1]

27-1. General Comments: Methods of Synthesis

Complexes of transition metals with M—H bonds are of critical importance in many catalytic reactions, and the formation of such bonds on metal surfaces is

---

[1] *Transition Metal Hydrides,* R. Bau, Ed., ACS Advances in Chemistry Series, No. 167, 1978 (collection of papers including metal hydrides); E. L. Muetterties, Ed., *Transition Metal Hydrides,* Dekker, 1971; H. D. Kaesz and R. B. Saillant, *Chem. Rev.,* 1972, **72,** 231; J. P. McCue, *Coord. Chem. Rev.,* 1973, **10,** 265; D. M. Roundhill, *Adv. Organomet. Chem.,* 1975, **13,** 273 (Ni, Pd, Pt); D. Giusto, *Inorg. Chim. Acta Rev.,* **1972,** 91, (Re); G. L. Geoffroy and J. R. Lehman, *Adv. Inorg. Chem. Radiochem.,* 1977, **20,** 190, (Ru, Rh, Ir); R. Bau and T. F. Koetzle, *Pure Appl. Chem.,* 1978, **50,** 55 (neutron diffraction).

clearly essential for the high catalytic activity of metals such as Fe, Co, Ni, Pd, and Pt for hydrogenation, C—H bond cleavage, and other reactions.

The first known complexes with M—H bonds were the hydrido carbonyls $H_2Fe(CO)_4$ and $HCo(CO)_4$ made in the 1930s by W. Hieber. Their structures and the nature of the M—H bond were not known until relatively recently.[2]

Compounds with M—H bonds can be synthesized in a variety of ways such as the following.

1. *Action of Hydride Sources on Metal Complexes.* The hydride sources can be organic compounds such as alcohols or aldehydes that have readily transferable hydrogen atoms, water, borohydrides, etc.

Borohydrides do not always give M—H compounds, since borohydride complexes may be formed instead; these can be recognized by an ir band *ca.* 2500 cm^{-1}. Some representative syntheses are

$$\text{\textit{trans}-PtCl}_2(\text{PEt}_3)_2 \xrightarrow{\text{KOH-EtOH}} \text{\textit{trans}-PtClH}(\text{PEt}_3)_2$$

$$[\text{Rh en}_2\text{Cl}_2]^+ \xrightarrow{\text{BH}_4^-} [\text{Rh en}_2\text{ClH}]^+$$

$$\text{Rh}_{aq}^{3+} + \text{SO}_4^{2-} + \text{NH}_3 + \text{Zn} \rightarrow [\text{Rh(NH}_3)_5\text{H}]\text{SO}_4$$

2. *Reactions Involving Molecular Hydrogen.* Hydrogen can add to coordinately unsaturated compounds oxidatively (Section 29-2) or, in presence of bases, heterolytic cleavage may occur. Examples are:

$$\text{RhCl(PPh}_3)_3 + \text{H}_2 \rightleftharpoons \text{RhCl(H)}_2(\text{PPh}_3)_3$$
$$\text{RuCl}_2(\text{PPh}_3)_3 + \text{H}_2 + \text{Et}_3\text{N} \rightarrow \text{RuClH(PPh}_3)_3 + \text{Et}_3\text{NHCl}$$

Transition metal alkyls can also undergo hydrogenolysis, e.g.,

$$(\eta\text{-C}_5\text{H}_5)_2\text{ZrMe}_2 \xrightarrow[80°]{\text{H}_2/60 \text{ atm}} (\eta\text{-C}_5\text{H}_5)_2\text{ZrH}_2 + 2\text{CH}_4$$

3. *Additions of Hydrogen Compounds HX.*[3] These can add also oxidatively, e.g.,

$$\text{\textit{trans}-IrCl(CO)(PPh}_3)_2 + \text{HSiCl}_3 = \text{IrCl(H)(SiCl}_3)(\text{CO})(\text{PPh}_3)_2$$
$$\text{\textit{trans}-PtCl}_2(\text{PEt}_3)_2 + \text{HCl} = \text{PtCl}_3\text{H(PEt}_3)_2$$

Complexes that have lone pairs or putative lone pairs can be protonated by strong acids with noncoordinating anions, e.g.,

$$(\eta\text{-C}_5\text{H}_5)_2\text{MoH}_2 + \text{H}^+\text{BF}_4^- = (\eta\text{-C}_5\text{H}_5)_2\text{MoH}_3^+\text{BF}_4^-$$
$$\text{Fe(CO)}_3(\text{PPh}_3)_2 + \text{H}_2\text{SO}_4 \rightarrow \text{HFe(CO)}_3(\text{PPh}_3)_2^+\text{HSO}_4^-$$
$$\text{Co(CO)}_4^- + \text{H}^+ \rightleftharpoons \text{HCo(CO)}_4$$

$$(\eta\text{-C}_5\text{H}_5)(\text{CO})_3\text{W-W(CO)}_3(\eta\text{-C}_5\text{H}_5) + \text{H}^+\text{BF}_4 = [(\eta\text{-C}_5\text{H}_5)(\text{CO})_3\overset{\overset{\displaystyle H}{\diagup\diagdown}}{W}\text{—W(CO)}_3\eta\text{-C}_5\text{H}_5)]^+\text{BF}_4^-$$

4. *By Intramolecular Hydrogen Transfer.* Reactions such as the cyclometallation reaction (Section 27-2) can form M—H and M—C bonds by an intra-

---

[2]   See E. A. McNeill and F. R. Scholer, *J. Am. Chem. Soc.,* 1977, **99**, 6243.

[3]   For Pt, see D. M. Roundhill, ACS Advances in Chemistry Series, No. 167, 1978, p. 160.

molecular oxidative addition, e.g.,

$$IrH(CO)(PPh_3)_3 \xrightarrow{\text{heat}} Ir(H)_2(CO)(o\text{-}C_6H_4PPh_2)(PPh_3)_2$$

Intramolecular hydride transfers can also occur where water, alkenes, alkynes, etc., can act as H sources, e.g.,

$$Mo(N_2)(diphos)_2 + CH_3CH{=}CH_2 \rightarrow MoH(\eta^3\text{-}CH_2CHCH_2)\,(CH_3CH{=}CH_2)(diphos)_2$$
$$Ru(styrene)_2(PPh_3)_2 + H_2O \rightarrow RuH(OH)(PPh_3)_2 + 2\ styrene$$

Metal-hydrogen bonds can be detected by infrared and ^{1}H nmr spectra. In the latter, a resonance in the region $\delta = 5$ to 50 (Me$_4$Si $= 0.0$) is typical of M—H. Nmr study has allowed detection of hydrido species not only in aqueous and organic solvent solutions but even in media like concentrated H_2SO_4 and "super" acids.

27-2. Types of Hydrido Species

Transition metal hydrido complexes can be classified into the following types.

1. *Pure Hydrido Species.* The only well-characterized example is the ion ReH$_9^{2-}$ (p. 900).

2. *Mononuclear Species.* These are known for many types of ligand, both π-bonding and non-π-bonding. Some examples are [Rh(NH$_3$)$_5$H]$^{2+}$, RuH$_2$(PPh$_3$)$_4$, [Co(CN)$_5$H]$^{3-}$, (η-C$_5$H$_5$)$_2$WH$_2$, PtHCl(PEt$_3$)$_2$, and RhH(CO)-(PPh$_3$)$_3$. Those with phosphine ligands are probably the most intensively studied, and most of the *polyhydride* complexes are phosphine species, e.g., MoH$_4$(PPhEt$_2$)$_4$, ReH$_7$(PPhEt$_2$)$_2$. Some poorly defined polyhydrides may be formed by action of Grignard reagents on metal halides, an example being FeH$_6$Mg$_4$X$_4$(THF)$_8$.[4]

3. *Binuclear and Cluster Hydrides.* This class contains mainly carbon monoxide compounds; both neutral and anionic species exist, many of great complexity. Examples are HFe$_3$(CO)$_9$SPr^i, [H$_6$Re$_4$(CO)$_{12}$]$^{2-}$, and H(CH$_3$)Os(CO)$_8$. There are a few compounds that do not contain CO, e.g., H$_3$Rh$_3$[P(OMe)$_3$]$_6$,[5] and heterometallic clusters such as H$_2$FeRuOs$_2$(CO)$_{13}$ are known.[6]

Terminal Metal-Hydride Bonds. These call for little comment. In all cases where X-ray or neutron diffraction studies have been made, the hydrogen atom occupies a bond position, the M—H bond distances are those expected by addition of single bond radii[7] and lie in the range 1.5–1.7 Å, and the M—H stretching frequencies are in the region 2100–1500 cm^{-1}.

There is no apparent correlation between physical parameters such as bond distances, the δ value of the high field line in the nmr, or the ir stretching frequency, and chemical properties of the hydrido complex.[8] The features that determine the chemical behavior are similar to those that determine the properties of all hydrido

---

4 S. G. Gibbins, *Inorg. Chem.,* 1977, **16,** 2571.
5 V. W. Day *et al., J. Am. Chem. Soc.,* 1977, **99,** 8091.
6 G. L. Geoffroy and W. L. Gladfelter, *J. Am. Chem. Soc.,* 1977, **99,** 7565; J. R. Shapley *et al., J. Am. Chem. Soc.,* 1977, **99,** 8064.
7 See, e.g., M. Cowie and M. J. Bennett, *Inorg. Chem.,* 1977, **16,** 2321, 2325.
8 See, e.g., T. Mikamoko, *J. Organomet. Chem.,* 1977, **134,** 33 (Pt—H).

compounds H–X and depend on the nature of X⁻. Thus although $HCo(CO)_4$ is a strong acid in water, $HCo(CO)_3(PEt_3)$ is weaker by orders of magnitude. Compounds with lone pairs often act as bases [e.g., $(\eta\text{-}C_5H_5)_2WH_2$ or $(\eta\text{-}C_5H_5)_2ReH$] and add protons to form additional M—H bonds. Compounds with M—H bonds usually react with halogens, and halocarbons, to form M—X bonds and, e.g., $CHCl_3$ from CCl_4. The reactions with organic halides proceed via radical chains.[9a] The reduction of ketones has been used in an attempt to determine the hydridic nature (i.e., M^+—H^-) of metal hydrides;[9b] this appears to be greatest for the Ti–Hf group metals.

The most important reaction of M—H bonds is that with unsaturated substances containing C=C or C≡C bonds, which is discussed in detail later (Sections 27-4, 29-3).

Hydrogen-Bridge Bonding[10a]. This may take a number of forms, of which the following are important:

It is uncertain whether many strictly linear symmetrical bridges actually occur. Neutron diffraction studies[10b] to date have always found at least small deviations from linearity.

Bent bridge bonds are well established[11] in complexes such as $HCr_2(CO)_{10}^-$, $HW_2(CO)_8(NO)P(OMe)_3$, $H_2Os_3(CO)_{10}$, and $Mo_2Cl_8H^{3-}$. In these compounds the M—H—M bond angles range from ca. 78° to 108° and the M—H stretching frequencies have been correlated with the interbond angle.[12] The asymmetric stretches range ca. 1200–1700 cm⁻¹.

As an example, in a tungsten compound the bent bridge 27-I has a W—H distance greater than that of a terminal W—H (1.77 Å) bond, and a useful description of the bonding is that of a "protonated metal-metal bond" with a closed $3c$-$2e$ bond symbolized as 27-II. The ion $HW_2(CO)_{10}^-$ appears to exist in both linear and bent

$$\text{(27-I)} \qquad\qquad \text{(27-II)}$$

9a R. G. Bergmann et al., J. Am. Chem. Soc., 1978, **100**, 635.

9b J. A. Labinger and K. H. Komadina, J. Organomet. Chem., 1978, **155**, C25.

10a R. Hoffmann et al., J. Am. Chem. Soc., 1979, **101**, 3141.

10b R. Bau, Acc. Chem. Res., 1979, **12**, 176; R. W. Broach and J. M. Williams, J. Am. Chem. Soc., 1979, **101**, 314.

11 See, e.g., V. Katović and R. E. McCarley, Inorg. Chem., 1978, **17**, 1268; M. P. Brown et al., J.C.S. Dalton, **1978**, 516; J. L. Petersen, Inorg. Chem., 1978, **17**, 3460, 1308; M. R. Churchill et al., Inorg. Chem., 1979, **18**, 156, 171, 843, 848, 1926; A. G. Orpen et al., J.C.S. Chem. Comm., **1978**, 723; M. Y. Darensbourg et al., J. Am. Chem. Soc., 1979, **101**, 2631.

12 M. W. Howard et al., J.C.S. Chem. Comm., **1979**, 18.

forms,[13] suggesting that the W—H—W link is easily deformable. The W—W distance stays fairly constant, and the axial ligand and the W vectors point to the center of the WHW triangle, *not* at the H atom. There appears to be a common overlap in the center of the triangle where all three orbitals overlap, in both the bent and linear bonds (27-III, 27-IV). In first-row metal compounds such as

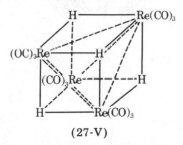

(27-III)   (27-IV)

$HCr_2(CO)_{10}^-$, the bond is only slightly bent, with the H atom only 0.6 Å from the Cr—Cr axis. Although in most bent systems there is a metal-metal bond, in $HRu_3(CO)_9(C{\equiv}CBu^t)$ there appears to be very little, if any, bonding.[14]

Species with more than one bridging hydrogen atom can be regarded naively as having protonated metal-metal multiple bonds,[15a] but MO descriptions have been given.[10a,15b]

Some examples of $(\mu_2\text{-}H)_n$ complexes are the following:

$$(PhEt_2P)_2(H)_2Re(\mu\text{-}H)_4Re(H)_2(PEt_2Ph)_2$$
$$[(CO)_3Re(\mu\text{-}H)_3Re(CO)_3]^-$$
$$[(Me_5C_5)Ir(\mu\text{-}H)_3Ir(C_5Me_5)]^+$$
$$[(CO)_4W(\mu\text{-}H)_2W(CO)_4]^{2-}$$

Cluster and Encapsulated Hydrides.[16]   There are many remarkable cluster polyhydrido compounds, some neutral like $H_4Ru_4(CO)_{12}$, others anionic like $[H_6Re_4(CO)_{12}]^{2-}$. Hydrogen atoms can here occupy μ_3-bridge positions.[17] Thus $H_4Re_4(CO)_{12}$ (27-V), which is one of the few clusters not obeying the eighteen

(27-V)

electron rule, has μ_3 bridges, whereas in $H_4Ru_4(CO)_{12}$, which does obey the rule, the H atoms are *edge* bridging, $Ru(\mu_2\text{-}H)Ru$. Face bridging is also found in

13   R. Bau *et al., J. Organomet. Chem.,* 1975, **91,** C49.
14   M. Catti *et al., J.C.S. Dalton,* **1977,** 2260.
15a R. Bau *et al., J. Am. Chem. Soc.,* 1977, **99,** 3872.
15b B. L. Barnett *et al., Chem. Ber.,* 1977, **110,** 3900.
16   P. Chini and B. Heaton, *Topics Curr. Chem.,* 1977, **71,** 1; J. Evans, *Adv. Organomet. Chem.,* 1977, **17,** 319; A. P. Humphries and H. D. Kaesz, *Prog. Inorg. Chem.,* 1979, **25,** 145.
17   R. D. Wilson and R. Bau, *J. Am. Chem. Soc.,* 1976, **98,** 4687; R. H. Crabtree *et al., J. Organomet. Chem.,* 1978, **161,** C67.

$H_4W_4(CO)_{12}(OH)_4$. Nmr studies show that the hydrogen atoms in clusters may be mobile and scramble by edge ⇌ terminal hydride exchange.[18]

The binding of hydrogen in certain cluster compounds presents an unusual problem. There is evidence that in addition to normal and bridging bonds, hydrogen atoms can be located *inside* (i.e., *encapsulated*) in a sufficiently large metal polyhedron of 10 to 20 atoms. Thus in the rhodium cluster anion[19a] $[Rh_{13}(CO)_{24}H_{5-n}]^{n-}$ ($n = 2, 3$; Fig. 27-1a), the ${}^1H$ nmr spectrum (Fig. 27-1b) shows that a single H atom is coupled to all thirteen rhodium atoms (${}^{103}Rh$, spin $1/2$). The hydrogen atoms are thus rapidly migrating *inside* the hexagonal close-packed metal framework. It appears that the H atoms must be protonic, since the atomic radius of a hydrogen atom (0.37 Å) is much greater than the size of the tetrahedral and octahedral holes of the rhodium faces inside the cluster.

Proof of the interstitial bonding comes from an X-ray structure of $[(Ph_3P)_2N]^+[HCo_6(CO)_{15}]^-$ (Fig. 27-1c), where the central H is six-coordinate,[19b] and neutron diffraction study of the ruthenium compound in Fig. 27-1d.

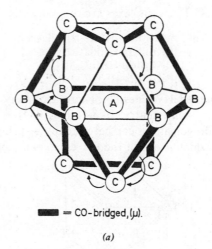

■■■ = CO-bridged, (μ).

(a)

Fig. 27-1. (a) Schematic structure of $[Rh_{13}(CO)_{24}H_{5-n}]^{n-}$ ($n = 2, 3$); arrows show one of the three pathways that may be used for edge-terminal CO exchange. (b) The measured and computed high field ${}^1H$ nmr spectrum of $[Rh_{13}(CO)_{24}H_2]^{3-}$ ion. [Reproduced by permission from P. Chini *et al., J.C.S. Chem. Comm.*, **1977**, 39.] (c) (p. 1120) The structure of the $[HCo_6(CO)_{15}]^-$ anion showing the six-coordinate H atom at its center. The cluster has approximate C_{2v} symmetry, with ten terminal, one symmetrically bridging, and four asymmetrically bridging carbonyl groups. Average H—Co and Co—Co distances are 1.82(1) and 2.58(1) Å, respectively. [Reproduced by permission from P. Chini *et al., Angew. Chem., Int. Ed.*, 1979, **18**, 80]. (d) (p. 1121) The structure of $[HRu_6(CO)_{18}]^-$; H position is not shown but is the hydrogen atom has been detected inside using neutron diffraction (J. Lewis, personal communication). [Reproduced by permission from J. Lewis *et al., J.C.S. Chem. Comm.*, **1976**, 945.]

[18] J. R. Shapley *et al., J. Am. Chem. Soc.*, 1977, **99**, 7385.
[19a] P. Chini *et al., J. Am. Chem. Soc.*, 1978, **100**, 7096: *J.C.S. Chem. Comm.*, **1977**, 39, *Chim. Ind. (Milan)*, 1978, **60**, 989.
[19b] P. Chini *et al., Angew Chem., Int. Ed.*, 1979, **18**, 80.

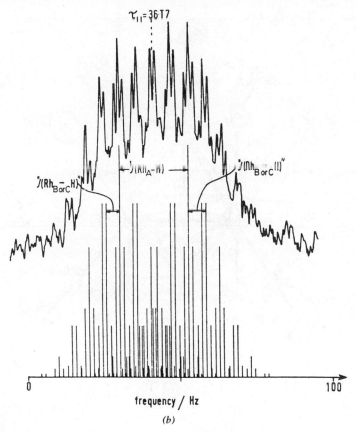

Fig. 27-1. *Con't.*

COMPOUNDS WITH TRANSITION METAL TO CARBON σ BONDS[20]

27-3. General Remarks

Early efforts to prepare simple alkyls or aryls of transition metals, such as diethyl-iron or -nickel, showed that such compounds were generally unstable under ordinary conditions, although they might be present in solutions at low temperatures. Despite the synthesis of $[Me_3PtI]_4$ by Pope and Peachy in 1907, it was assumed until quite recently that metal-carbon bonds are weak.[20c] When it was found that the presence of ligands such as η-C_5H_5, CO or PR_3 allowed the synthesis of thermally stable

[20] (a) R. R. Schrock and G. W. Parshall, *Chem. Rev.,* 1976, **76,** 243; (b) P. J. Davidson, M. F. Lappert, and R. Pearce, *Chem. Rev.,* 1976, **76,** 219; (c) G. W. Parshall and J. J. Mrowca, *Adv. Organomet. Chem.,* 1968, **7,** 157; (d) M. R. Churchill, in *Perspectives in Structural Chemistry,* Vol. III, J. D. Dunitz and J. A. Ibers, Eds., Wiley, 1970, p. 91; (e) R. Taube, H. Drevs, and D. Steinborn, *Z. Chem.,* 1978, **18,** 425; P. S. Braterman and R. J. Cross, *Chem. Soc. Rev.,* 1973, **2,** 271.

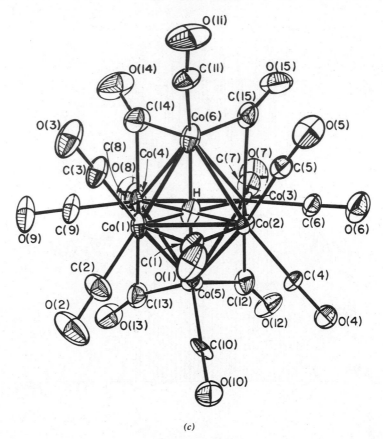

(c)

Fig. 27-1. *Con't.*

compounds, it was considered that the presence of such ligands promoted better overlap between sp^3 hybridized carbon and metal orbitals implying increased bond strength. As in examples from other areas of chemistry, there was a distinct failure to distinguish clearly between thermodynamic and kinetic stability.

Although data are still not extensive,[21] transition metal to carbon single-bond energies appear to be quite reasonable, ranging *ca.* 160–350 kJ mol^{-1}. The instability of many simple binary alkyl and aryl compounds MR_n is of kinetic origin. For alkyls such as C_2H_5 and C_3H_7 a common and well-established reaction is the β-hydride transfer–alkene elimination reaction:

$$M + \tfrac{1}{2}H_2 + RCH{=}CH_2$$

(27-VI)

[21] J. A. Connor, *Topics Curr. Chem.,* 1977, **71**, 71.

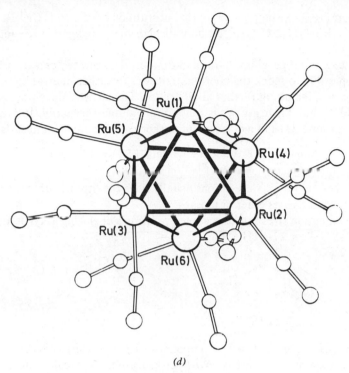

(d)

Fig. 27-1. Con't.

This reaction, which can be reversible under certain circumstances, is important in the synthesis of M—C bonds from M—H bonds and unsaturated compounds. It is discussed in detail in Section 29-3. However we note here that in any compound $L_nMCH_2CH_2R$ the alkyl group occupies one coordination site, whereas in the hydrido-alkene intermediate 27-VI, *two* sites are involved. Hence if all the ligands on the metal are firmly bound and cannot dissociate, i.e., the compound is coordinately saturated (Section 27-1), there is no low-energy pathway for decomposition. A well-studied example[22] of a decomposition that involves PPh_3 dissociation is

$$(\eta\text{-}C_5H_5)(CO)(PPh_3)\text{Fe-alkyl} \rightarrow (\eta\text{-}C_5H_5)(CO)(PPh_3)\text{FeH} + \text{alkene}$$

The difference in modes of decomposition of two similar alkyls, only one of which has a readily transferable β-hydrogen atom, is illustrated by comparison of $Bu_3^nPCuCH_2CH_2CH_2CH_3$ and $Bu_3^nPCuCH_2CMePh_2$. The former decomposes to give Cu, H_2, but-1-ene, and butane; the latter gives as products $PhCMe_3$, $PhCH_2Me_2$, $PhCH_2C(Me)=CH_2$, and $(PhMe_2CCH_2)_2$, which can come only from homolytic fission of the Cu—C bond and subsequent reactions of a free radical.

The stabilization of a metal alkyl by ligands blocking the coordination sites is illustrated as follows.

22   D. L. Reger and E. C. Culbertson, *Inorg. Chem.*, 1977, **16**, 3104; *J. Am. Chem. Soc.*, 1976, **98**, 2789.

1.   There are stable compounds of the substitution inert metals Cr^{III}, Co^{III}, and Rh^{III}, e.g., $[Rh(NH_3)_5C_2H_5]^{2+}$, where displacement of ligands cannot readily occur.

2.   Since a metal peralkyl that is coordinately unsaturated can act as a Lewis acid, it is possible to block the coordination sites by either neutral ligands or by anionic ligands, including further alkyl groups. Thus $TiMe_4$ decomposes readily above $-80°C$, but $TiMe_4(dipy)$ is much more stable thermally. Interaction of lithium alkyls or aryls can give often quite thermally stable lithium salts, e.g.,

$$MnCl_2 + 3LiCH_2SiMe_3 = Li[Mn(CH_2SiMe_3)_3] + 2LiCl$$
$$ReMe_6 + 2LiMe = Li_2[ReMe_8]$$

Note that the lithium alkylate anions usually have the lithium atoms solvated by donor molecules such as ether or amines.

3.   Coordination sites can be occupied by using a chelating alkyl or aryl group such as 27-VII[23a] or 27-VIII.[23b]

(27-VII)   (27-VIII)

A way of obtaining stable alkyl compounds without the use of blocking ligands is to have an alkyl group that has no β-hydrogen atoms. Typical of these so-called *elimination stabilized alkyls* are those with groups such as trimethylsilylmethyl, $-CH_2SiMe_3$, neopentyl, $-CH_2CMe_3$, benzyl, $-CH_2Ph$, and bridgehead 1-norbornyl (27-IX), and the chelates 27-X to 27-XIV.[24] It should be noted, however, that these alkyl groups are also very bulky and that steric factors are probably

(27-IX)   (27-X)   (27-XI)

(27-XII)   (27-XIII)   (27-XIV)

[23a] L. E. Manzer, *J. Am. Chem. Soc.*, 1978, **100**, 8068; *Inorg. Chem.*, 1978, **17**, 1552. B. L. Shaw et al., *J.C.S. Chem. Comm.*, **1979**, 498.

[23b] P. D. Brotherton et al., *J.C.S. Dalton*, **1976**, 1193; K. P. Wainwright and S. B. Wild, *J.C.S. Dalton*, **1976**, 262.

[24] H. Schmidbauer et al., *Angew. Chem., Int. Ed.*, 1978, **17**, 126; *Pure Appl. Chem.*, 1978, **50**, 19; *Acc. Chem. Res.*, 1975, **8**, 62.

important for the existence of compounds like $Co(1\text{-norbornyl})_4$ and in the thermal stability of such alkyls.

The *methyl* group is, in principle, in this class also, but although some per-methyls[25] are known, notably $ReMe_6$, WMe_6, and $TaMe_5$, they are not very stable and indeed can decompose explosively. In the cases studied, methane is formed on decomposition, e.g.,

$$WMe_6 = 3CH_4 + \text{"}W(CH_2)_3\text{"}$$

The nature of the residual black solids is unknown, as is the mechanism. This may well be free radical in nature, a methyl radical formed by homolytic fission, then abstracting hydrogen from the remaining $M—CH_3$ groups.

Aryls cannot undergo the β-hydride shift but even so, these are not very stable. A common decomposition mode is the formation of biaryls, probably by reductive elimination (p. 1138).

Aryl compounds that have orthosubstituents (27-XV, 27-XVI) are more stable than others, probably because the ortho group blocks a coordination site; for the same reason, probably, perchlorinated aryls [e.g., $(PhMe_2P)_2Ni(C_6Cl_5)_2$[26a]] are very stable. An unusual type of orthosubstituted aryl gives compounds of the type 27-XVII.[26b] Although perfluorinated alkyls might have been expected to be stable, relatively few are known; silver perfluoroalkyls and copper(I) perfluoroaryls are probably polymeric. There are numerous perfluoroalkyls having CO or other ligands present.[26c]

(27-XV) (27-XVI) (27-XVII)

As we discuss later in this chapter, there are also compounds with multiple metal-carbon bonds, and Table 27-1 summarizes all the main metal-carbon bond types based on the hybridization at the carbon atom but excluding the π-complex organometallic compounds.

27-4. Metal σ Alkyl and Aryl Types

1. *Per- or Homoleptic Alkyls.* These are compounds that have only transition metal to carbon σ bonds. They may be
   (a) *Mononuclear,* such as $Ti(CH_2Ph)_4$, $Cr(CH_2SiMe_3)_4$, or $Mn(1\text{-nor-}$ bornyl$)_4$.

[25] G. Wilkinson et al., J.C.S. Dalton, **1976**, 2235, 1488; R. R. Schrock, J. Organomet. Chem., 1976, **122**, 209.

[26a] M. Wada et al., Inorg. Chem., 1977, **16**, 446.

[26b] B. L. Shaw et al. J.C.S. Dalton, **1976**, 2053.

[26c] See, e.g., D. H. Hensley and R. P. Stewart, Jr., Inorg. Chem., 1978, **17**, 905.

TABLE 27-1
Summary of Major Types of Transition Metal to Carbon Bonds Excluding π Complexes

Carbon Hybridization	Ligand		
	Terminal		Bridging
sp^3	M—R$_3$	Alkyl	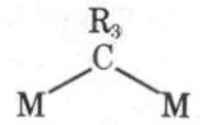 Three-center μ-alkyl
			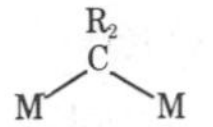 μ-Alkylidene
			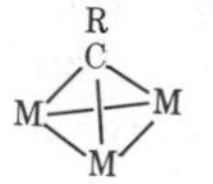 μ-Alkylidyne
sp^2	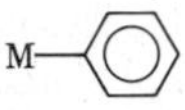 M—⬡	 Aryl	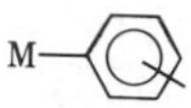 M—⬡
	M=CR$_2$	Carbene or alkylidene	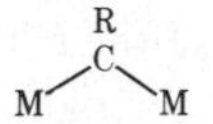 μ-Alkylidyne
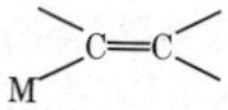	Vinyl	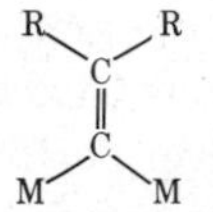 Vinyl	μ-Vinylidene
	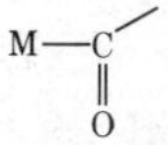 M—C(O)	Acyl	
sp	M≡CR	Carbyne or alkylidyne 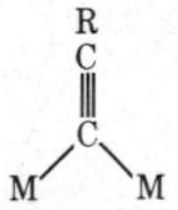 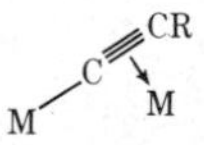	
	M—C≡CR	Acetylide	
	M=C=CR	Vinylidine	

(b) *Binuclear with metal-metal bonds,* such as Mo$_2$(CH$_2$SiMe$_3$)$_6$ (27-XVIII).

(c) *Polymeric with bridging alkyl groups,* as [Mn(CH$_2$SiMe$_3$)$_2$]$_n$ (27-XIX), which has the same type of structure as BeX$_2$ with tetrahedral Mn^{II} atoms.[27]

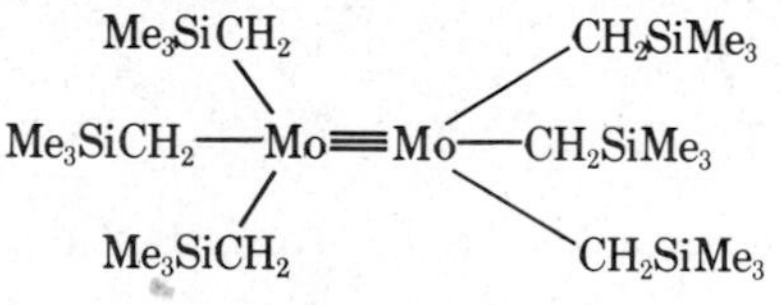

(27-XVIII)

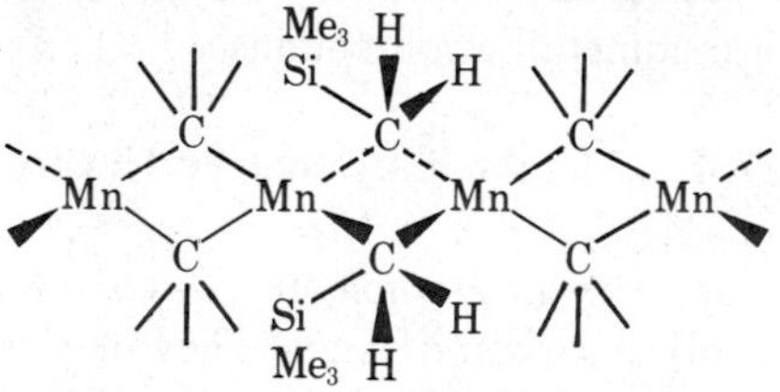

(27-XIX)

Alkyl bridges are found in other compounds such as $(\eta$-C$_5$H$_5)_2$Y-$(\mu$-Me$)_2$Y$(\eta$-C$_5$H$_5)_2$, $(\eta$-C$_5$H$_5)_2$Y$(\mu$-Me$)_2$AlMe$_2$,[28] and Re$_3(\mu$-Me$)_3$Me$_6$-(PEt$_2$Ph)$_2$[29] (Fig. 27-2a).

[27] R. A. Andersen et al., J.C.S. Dalton, **1976,** 2204.
[28] J. Holton et al., J.C.S. Dalton, **1979,** 45, 54.
[29] K. Mertis et al., J.C.S. Dalton, **1980,** 334.

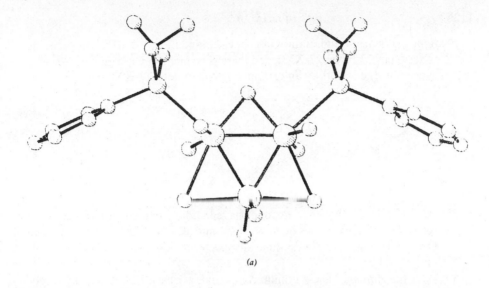

(a)

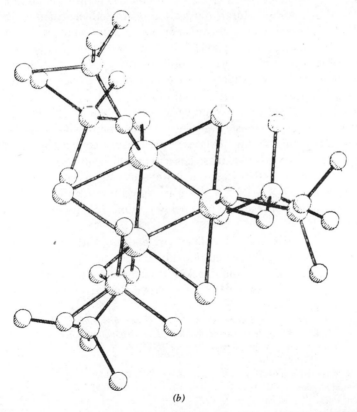

(b)

Fig. 27-2. The structure of the cluster alkyls (a) $Re_3Me_9(PEt_2Ph)_2$ and (b) $Re_3Cl_3(CH_2SiMe_3)_6$. [Reproduced by permission from M. B. Hursthouse *et al.*, *J.C.S. Dalton*, **1978**, 1334, **1980**, 334.]

Bridging aryls are found mainly in copper compounds.[30] The bridges may be of the types 27-XXa, 27-XXb, and there is no metal-metal bonding. Another type of bridge is that by an orthophenylene moiety 27-XXI.[30b]

(27-XXa)   (27-XXb)

(27-XXI)

2. *Mixed Alkyls.* These compounds have halide or oxide ligands in addition to alkyl groups, examples being the mononuclear $TaCl_2Me_3$, WCl_5Me, and $O{=}V(CH_2SiMe_3)_3$. An example of a cluster halide is $Re_3Cl_3(CH_2SiMe_3)_6$ (Fig. 27-2b).

3. *Alkylate Anions.* These lithium compounds are usually, but not exclusively, methyl compounds; a few arylate anions are known. They may be mononuclear or dinuclear with metal-metal bonds. The lithium atom is usually solvated by tetrahydrofuran, dioxan, tetramethylethylenediamine, or similar solvents. There is also evidence for strong interaction of the Li atoms with the hydrogen atoms of the alkyl groups bound to the metal as in lithium alkyls themselves (Section 7-8), and accordingly the C—H stretching frequencies may be usually low. Representative compounds are amine "salts" of the anions $MnMe_4^{2-}$,[27] $ReMe_8^{2-}$,[25] and $PtMe_6^{2-}$.[31] There are also species with metal-metal bonds,[32] an example of which is shown in Fig. 27-3. Some halogenated aryl anions[33] are $[Co(C_6F_5)_4]^{2-}$ and $[Ni(C_6Cl_5)_4]^{2-}$.

The most important lithium alkylates[34a] are those of copper (Section 21-H-2), such as $LiCuMe_2$, which find extensive use in organic synthesis.[34b]

4. *Chelate Alkyls.* Apart from the phosphorus ylid type of chelate alkyl noted above and shown in 27-X to 27-XV, there are only a few metallocycles that do not have other ligands present (see below). One example is $[Cr_2(CH_2-CH_2CH_2CH_2)_4]^{4-}$.

5. *Ligand-Stabilized Alkyls.* There are enormous numbers of compounds with M—C bonds that contain additional ligands, either π-bonding type or not, that occupy coordination sites, thus blocking decomposition pathways. Cobaloximes (Section 21-F-9) such as 27-XXI,[35] which provide models for vitamin B_{12} coen-

[30] (a) J. G. Noltes *et al., J.C.S. Dalton,* **1978,** 1800; *J. Organomet. Chem.,* 1979, **171,** C39; *J. Org. Chem.,* 1977, **42,** 2047; G. J. M. van der Kerk *et al., J. Organomet. Chem.,* 1978, **144,** 255. See also D. Thoennes and E. Weiss, *Chem. Ber.,* 1978, **111,** 3726; (b) M. D. Rausch *et al., J. Am. Chem. Soc.,* 1977, **99,** 7870.

[31] G. W. Rice and R. S. Tobias, *J. Am. Chem. Soc.,* 1977, **99,** 2141.

[32] A. P. Sattelberger and J. P. Fackler, *J. Am. Chem. Soc.,* 1977, **99,** 1258.

[33] R. Uson *et al. J.C.S. Chem. Comm.,* **1977,** 789.

[34a] E. C. Ashby and J. J. Watkins, *J. Am. Chem. Soc.,* 1977, **99,** 5312.

[34b] H. O. House, *Acc. Chem. Res.,* 1976, **9,** 59.

[35] A. Bigotto *et al., J.C.S. Dalton,* **1976,** 96.

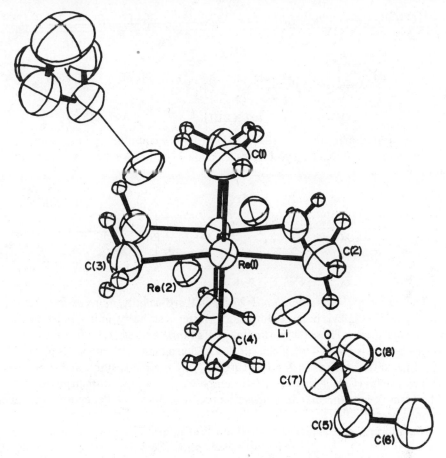

Fig. 27-3. The structure of $[Li(OC_4H_{10})]_2[Re_2(CH_3)_8]$; note the position of the Li ions close to the hydrogen atoms of the methyl groups. Li ions are similarly positioned in analogous $M_2(CH_3)_8^{4-}$ (M = Cr, Mo, W) compounds. [From F. A. Cotton *et al.*, *J. Am. Chem. Soc.*, 1976, **98**, 6922.]

zyme, contain non-π-bonding ligands. Others are alkyls of the substitution-inert ions Cr^{III}, Co^{III}, and Rh^{III}, such as $[RhC_2H_5(NH_3)_5]^{2+}$ noted earlier.

(27-XXI)

Metallocyclic complexes usually have other ligands present and are of types 27-XXII to 27-XXIV.

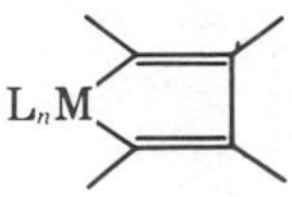

(27-XXII) 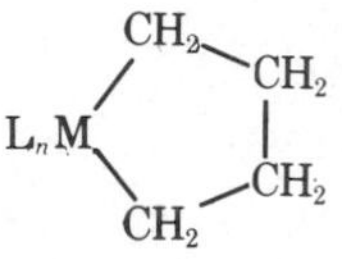　　　　　　(27-XXIII) 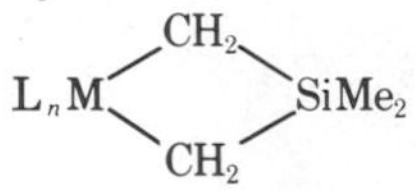　　　　　　(27-XXIVʃ

Unsaturated metallocycles (27-XXII) are quite common and are usually obtained in reactions involving acetylenes (Section 27-10).[36] A representative reaction is:

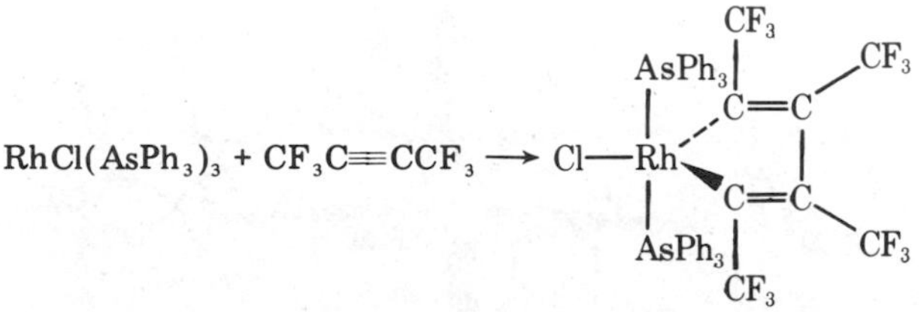

$$RhCl(AsPh_3)_3 + CF_3C\equiv CCF_3 \longrightarrow$$

The *saturated ring compounds* 27-XXIII are usually more stable than the corresponding dialkyls because β-hydride transfer is sterically inhibited by the more rigid ring alkyl. Those of platinum, palladium, and nickel are of the types $(R_3P)_2M(CH_2)_n$ (n = 4–6).[37] Substituted cyclopentadienyl compounds of Ta, Mo, W, Rh, and Ir,[38] such as $(\eta\text{-}Me_5C_5)Rh(CH_2)_5$ are thermally stable at 25° or above. Others such as $(C_5H_5)_2Ti(CH_2)_4$[39a] decompose at low temperature to give ethylene. The nature of the thermal decomposition products, however, depends on the other ligands present.[39b]

Metallocycles are believed to be important intermediates in certain catalytic reactions, notably the alkene metathesis reaction (Section 30-4). A nickel metallocycle $(R_3P)Ni(CH_2)_4$ catalyzes the reaction:

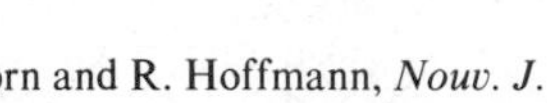

Although the metallocycles are usually made by alkylation using $Mg(CH_2)_n$ or $Li(CH_2)_nLi$, metallocycles of type 27-XXIV such as $(Me_3P)_4RuCH_2SiMe_2CH_2$, which are formed presumably by hydride transfer from the γ-position of an intermediate alkyl $CH_2Si(CH_3)_3$, are also thermally very stable.[40]

6. *Compounds with CR_2 and CR bridges.* Compounds with CR_2 bridges are

36　For references, see D. L. Thorn and R. Hoffmann, *Nouv. J. Chim.,* 1979, **3**, 39.
37　See e.g., J. Ramaram and J. A. Ibers, *J. Am. Chem. Soc.,* 1978, **100**, 829; G. B. Young, and G. M. Whitesides, *J. Am. Chem. Soc.,* 1978, **100**, 5808; P. Diversi *et al., J.C.S. Chem. Comm.,* **1978**, 735; R. H. Grubbs *et al., J. Am. Chem. Soc.,* 1978, **100**, 1300, 2418, 7416, 7418; R. J. Al Essa *et al., J. Am. Chem. Soc.,* 1979, **101**, 364.
38　See, e.g., P. Diversi *et al., J. Organomet. Chem.,* 1979, **165**, 253; M. L. H. Green, *Pure Appl. Chem.,* 1978, **50**, 27; S. J. McLain and R. R. Schrock, *J. Am. Chem., Soc.,* 1978, **100**, 1315.
39a　G. M. Whitesides *et al., J. Am. Chem. Soc.,* 1976, **98**, 6529.
39b　P. S. Braterman, *J.C.S. Chem. Comm.,* **1979**, 70.
40　R. A. Andersen *et al., J.C.S. Dalton,* **1978**, 446.

known to have π-bonding ligands present. Examples are 27-XXV, 27-XXVI, and 27-XXVII.

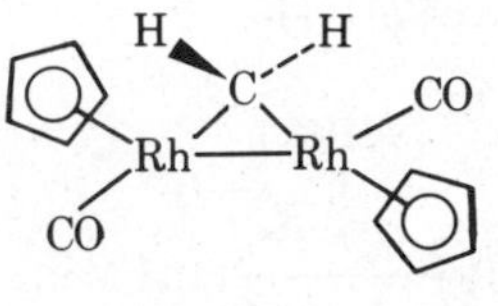

(27-XXV) 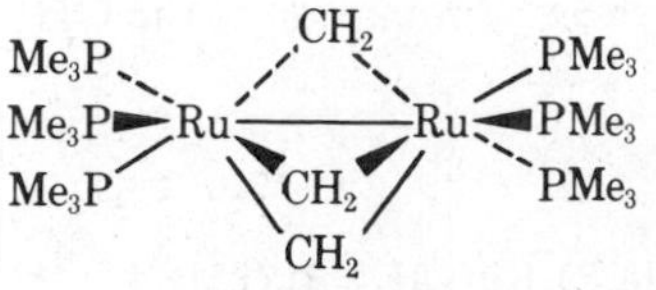 (27-XXVI)

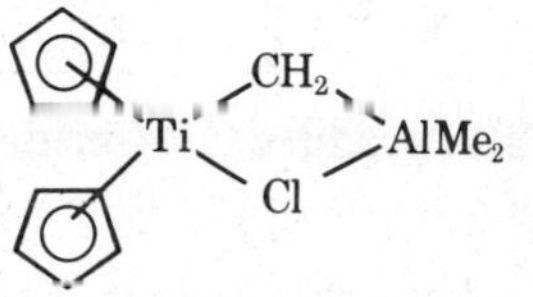

(27-XXVII)

Single-bridged compounds can be obtained by addition of CH_2N_2 or other diazo compounds to complexes having metal-metal bonds, or by reactions involving N-alkyl-N-nitrosoureas,[41a] e.g.,

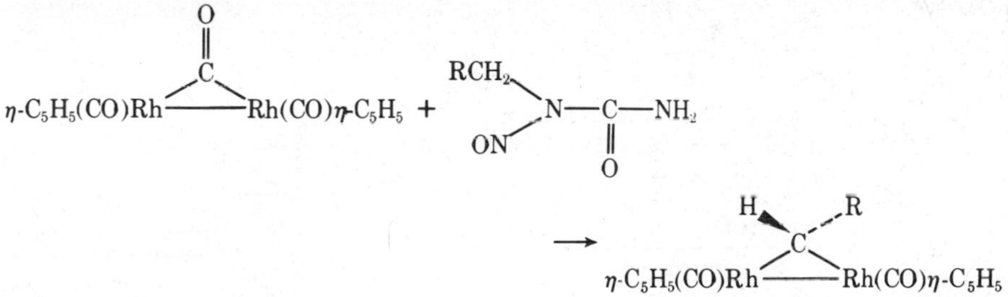

They can also be obtained by reaction of low-valent metal carbene complexes with other metal compounds,[41b] e.g.,

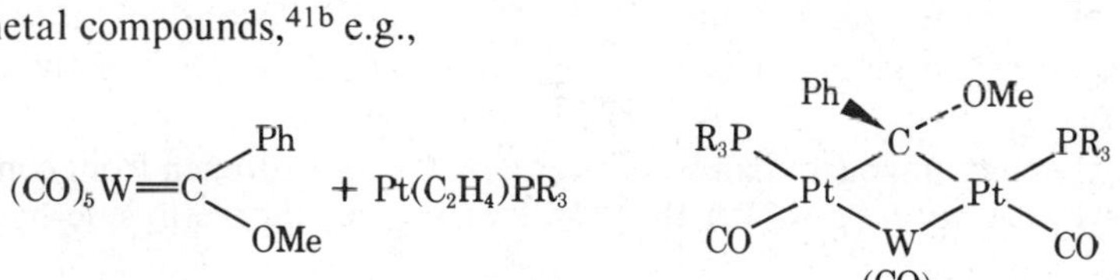

The compound 27-XXVI is obtained[42] by interaction of a ruthenium acetate with methyllithium, and presumably α-hydrogen transfer occurs, viz.,

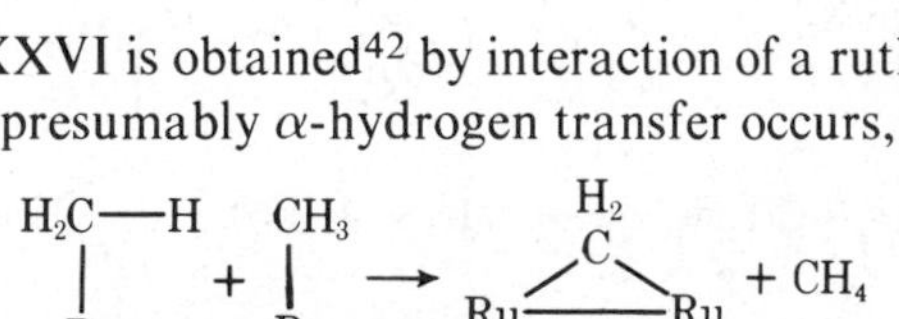

[41a] W. A. Herrmann et al., Chem. Ber., 1978, **111**, 1077, Angew Chem., Int. Ed., 1978, **17**, 800; 1977, **16**, 334; R. J. Puddephat et al., J.C.S. Chem. Comm., **1978**, 749; P. Hofmann, Angew. Chem. Int. Ed., **1979, 18,** 554.

[41b] F. G. A. Stone et al., J.C.S. Chem. Comm., **1979,** 45; E. O. Fischer et al., Z. Naturforsch., 1977, **32b,** 648.

[42] R. A. Jones et al., J. Am. Chem. Soc., 1979, **101,** 4128.

The compound is important in showing the relation between bridging methyl and bridging methylene compounds (cf. the Os_3 compounds, 1140), since we have the reaction where protonation of the CH_2 bridge gives $\mu\text{-}CH_3$.

$$(Me_3P)_3Ru(CH_2)_3Ru(PMe_3)_3 \underset{-H^+}{\overset{+H^+}{\rightleftharpoons}} \left[(Me_3P)_3Ru \overset{\overset{\textstyle H_3}{\textstyle C}}{\underset{\underset{\textstyle H_2}{\textstyle C}\quad CH_2}{\diamond}} Ru(PMe_3)_3 \right]^+$$

Aluminum trimethyl can also produce CH_2-bridged species in certain cases, as in 27-XXVII[43] by the reaction

$$(\eta\text{-}C_5H_5)_2TiCl_2 + 2AlMe_3 \rightarrow (\eta\text{-}C_5H_5)_2Ti(\mu\text{-}Cl)(\mu\text{-}CH_2)AlMe_2 + CH_4 + AlClMe_2$$

Probably the best known μ-alkylidyne compounds with CR bridges are those of the type 27-XXVIII. The compound $Co_3(CO)_9CH$ can be made by interaction of $Co_2(CO)_8$ with $CHCl_3$. Cobalt compounds have an extensive chemistry[44,45] including redox reactions. There appears to be significant delocalization of electron

(27-XXVIIIa)   (27-XXVIIIb)

density in the region of the apical carbon atom that allows transmission of electronic effects between the cobalt atoms and the R groups.[45] Accordingly it seems best not to regard the bridging carbon as sp^3 hybridized, although formally it is convenient to do so.

μ-Alkylidyne compounds can be obtained by loss of hydrogen from compounds with alkyl groups (cf. p. 1140) that have weakly acidic methylene hydrogens, e.g.,

$$NbCl_5 + Me_3SiCH_2MgX \rightarrow (Me_3SiCH_2)_2Nb \overset{\overset{\textstyle SiMe_3}{\textstyle C}}{\underset{\underset{\textstyle SiMe_3}{\textstyle C}}{\diamond}} Nb(CH_2SiMe_3)_2$$

43 F. N. Tebbe et al., J. Am. Chem. Soc., 1978, 100, 3611.
44 A. M. Bond et al., Inorg. Chem., 1979, 18, 1413; D. Seyferth et al., J. Organometal. Chem. 1979, 178, 227; R. A. Epstein et al., Inorg. Chem., 1979, 18, 942.
45 D. C. Miller and T. B. Brill, J. Am. Chem. Soc., 1978, 100, 240, Inorg. Chem., 1978, 17, 240.

Compounds of this type are known for Nb, Ta, Mo, W, and Re.[46a] Just as interaction of $M=CR_2$ complexes can lead to μ-CR_2 compounds, as noted above, interaction of low-valent carbyne compounds[46b] can give μ-CR species, e.g.,

$$(\eta\text{-}C_5H_5)(CO)_2W\equiv CR + Pt(C_2H_4)_2(PMe_3)_2 \longrightarrow$$

A number of carbonyl cluster compounds also have CR or CR_2 bridges (e.g., 27 XXVIIIb).[46c] An important triosmium compound with a CH_2 bridge is discussed below (p. 1140).

7. *Acyl Compounds.*[47] These compounds are especially important intermediates in many catalytic reactions involving carbon monoxide (Section 30-3), since they may be produced by "insertion" reactions (Section 29-8) such as

$$CH_3Mn(CO)_5 + CO \longrightarrow CH_3\underset{\underset{O}{\|}}{C}\text{—}Mn(CO)_5$$

Normal acyls are σ-bonded and the charge distribution is usually

$$\overset{\delta+}{M}\text{—}\underset{\underset{O}{\|}}{\overset{\delta-}{C}}\text{—}R$$

although in $[CpCo(PMe_3)_2COMe]^+$ it is apparently the opposite.[48]

The early transition metals Ti, Zr, and Mo can give η^2-acyls[49a] 27-XXIX. An unusual bridged acyl is also known (27-XXX).

(27-XXIX)   (27-XXX)

The early transition metal and also uranium alkyls give other nonclassical types of CO insertion products,[49b] namely, the following:

[46a] G. Wilkinson et al., J.C.S. Dalton, **1980**, in press.

[46b] F. G. A. Stone et al., J.C.S. Chem. Comm., **1979**, 42.

[46c] C. R. Eady et al., J.C.S. Chem. Comm., **1978**, 421; J.C.S. Dalton, **1977**, 477.

[47] F. Calderazzo, Angew. Chem., Int. Ed., **1977**, **16**, 299.

[48] H. Werner and W. Hofmann, Angew. Chem., Int. Ed., **1978**, **17**, 464.

[49a] G. Fachinette et. al., J.C.S. Dalton, **1977**, 1946, 2297; E. Carmona-Guzman et al., J.C.S. Chem. Comm., **1978**, 465.

[49b] T. J. Marks et al., J. Am. Chem. Soc., **1978**, **100**, 7112.

From $(Me_5C_5)_2UMe_2$

From $(Me_5C_5)_2Zr(CH_2SiMe_3)_2$

From $(Me_5C_5)_2ZrClCH_2SiMe_3$

Related to acyls are the η^2-*iminoacyls* 27-XXXI, where the keto oxygen is replaced by $=NR'$[50a] and η^2-*thioacyls*[50b] (27-XXXII) and η^2-ketones.[50c]

Comparison of bond distances in $M—CH_3$ and $MCOCH_3$ suggests that there is some $d\pi-p\pi$ character in the $M—C$ bond of the acyl.

(27-XXXI)   (27-XXXII)

A special case of acyls is represented by *formyls,* which are discussed separately (Section 29-4).

8.   *Other Compounds with Metal-Carbon Single Bonds.*   Vast numbers of compounds have one or more single bonds to carbon, often in conjunction with π bonding through olefinic groups, and some extraordinary types of compound have been characterized. Some relatively simple types are the following.

Aminomethyl compounds[51a] with an iminium group acting as a three-electron donor (27-XXXIII) and the similar sulfur[51b] (27-XXXIV) and phosphorus[52] (27-XXXV) groups.

50a   R. D. Adams and D. F. Chodosh, *Inorg. Chem.,* 1978, **17,** 41.
50b   G. R. Clark, *et al., J. Organomet. Chem.,* 1978, **157,** C23.
50c   W. A. Herrmann *et al., J. Am. Chem. Soc.,* 1979, **101,** 3133.
51a   H. D. Kaesz *et al., J. Am. Chem. Soc.,* 1977, **16,** 3193, 3201; C. K. Poon *et al., J.C.S. Dalton,* **1978,** 1180, **1977,** 1247; H. Brunner *et al., Angew Chem., Int. Ed.,* 1978, **17,** 453; C. W. Fong and G. Wilkinson, *J.C.S. Dalton,* **1975,** 1110; E. W. Abel and R. J. Rowley, *J.C.S. Dalton,* **1975,** 1096.
51b   G. Yoshida *et al., Chem. Lett.,* **1977,** 1386.
52   H.-H. Karsch *et al., Chem. Ber.,* 1977, **110,** 2200, 2213, 2222; E. L. Muetterties *et al., J. Am. Chem. Soc.,* 1978, **100,** 6966.

$$\text{M} \underset{\text{NR}_2}{\overset{\text{CH}_2}{\diagdown}} \qquad \text{M} \underset{\text{S}\diagdown_{\text{Me}}}{\overset{\text{CH}_2}{\diagdown}} \qquad \text{M} \underset{\text{PMe}_2}{\overset{\text{CH}_2}{\diagdown}}$$

(27-XXXIII) (27-XXXIV) (27-XXXV)

Vinyls have M—CH=CH$_2$, and *σ-allyls* have M—CH$_2$—CH=CH$_2$, groups. The latter are discussed later in this chapter because they may also have a η^3-delocalized type of bonding and allyls are important intermediates in many reactions involving diene complexes.

Acetylides have M—C≡CR groups as in anions such as [Ni(C≡CR)$_4$]$^{2-}$ [53a] or η-C$_5$H$_5$(CO)$_3$CrC≡CPh.[53b]

Finally we note that at least in Pt^{II} compounds there seems to be little difference in the trans effect of carbon bound sp^3 in —CH$_2$SiMe$_3$, sp^2 in —CH=CH$_2$, and sp in —C≡CPh. The actual values are high and comparable to those of H and SiMe$_3$.[54a] The M—C bonds for perfluoro alkyl compounds are shorter than those for corresponding alkyls [e.g., in the similar complexes *trans*-PtCl(R)-(PMePh$_2$)$_2$].[54b] However the extent of the $d\pi$–σ back-bonding postulated to explain this is controversial.

27-5. Synthesis of Compounds with Metal-Carbon σ Bonds

Compounds with M—C σ bonds can be obtained in innumerable ways, some unique. The following are the major routes.

1. *Alkylation by Other Metal Alkyls.* The interaction of Grignard reagents, alkyls of Li, Mg, Zn, Al, Sn, Hg, etc., with metal halides, alcoxides, or acetates, or complexes thereof may give alkyl or aryl compounds.

Mercury, tin, and zinc alkyls commonly give only partial alkylation,[55] e.g.,

$$\text{TaCl}_5 + \text{ZnMe}_2 \rightarrow \text{Me}_3\text{TaCl}_2$$
$$\text{WCl}_6 + \text{HgMe}_2 \rightarrow \text{MeWCl}_5$$

For aluminum alkyls, only *one* alkyl group is normally transferred, e.g.,

$$\text{WCl}_6 + 6\text{AlMe}_3 = \text{WMe}_6 + 6\text{AlCl}_2\text{Me}$$

The reactions with metal halides often lead to reduction unless elimination-stabilized alkyls are employed. Even then the reactions can be complicated. Solvent effects are important, and traces of oxygen can have a substantial effect.

The use of lithium alkyls in excess can lead to the formation of alkylate anions, e.g.,

$$\text{CrCl}_3(\text{THF})_3 \xrightarrow{\text{LiCH}_2\text{SiMe}_3} \text{Li}[\text{Cr}(\text{CH}_2\text{SiMe}_3)_4] \xrightarrow{-e} \text{Cr}(\text{CH}_2\text{SiMe}_3)_4$$

[53a] See, e.g., R. Nast and H. P. Müller, *Chem. Ber.*, 1978, **111**, 1627.
[53b] A. N. Nesmeyanov *et al.*, *J. Organomet. Chem.*, 1979, **166**, 217.
[54a] C. J. Cardin *et al.*, *J.C.S. Dalton*, **1978**, 46.
[54b] M. A. Bennett *et al.*, *Inorg. Chem.*, 1979, **18**, 1061, 1071.
[55] See, e.g., C. Santini-Scampucci and J. G. Riess, *J.C.S. Dalton*, **1976**, 195.

The nature and mechanism of many of the alkylation reactions is poorly known.

2. *Interaction of Sodium Salts of Anions with Alkyl or Aryl Halides.* This is a common synthetic method for use when sodium salts of carbonylate or other anions can be made. Examples are:

$$\eta\text{-}C_5H_5W(CO)_3^-Na^+ + CH_3I = \eta\text{-}C_5H_5W(CO)_3Me + NaI$$
$$Mn(CO)_5^-Na^+ + CH_3I = CH_3Mn(CO)_5 + NaI$$

This method can also be used to synthesize acyls and compounds with bonds to Si, Ge, Sn, etc., e.g.,

$$\eta\text{-}C_5H_5W(CO)_3^-Na^+ + ClSiMe_3 = \eta\text{-}C_5H_5W(CO)_3SiMe_3 + NaCl$$

3. *Oxidative-Addition Reactions.* For metal complexes that can undergo the oxidative-addition reaction (Section 29-2), this is a useful method for synthesis of alkyls, aryls, acyls, and compounds with bonds to other elements. A representative reaction of organic halides is

$$IrCl(CO)(PPh_3)_2 + CH_3I = Ir(CH_3)(Cl)(I)(CO)(PPh_3)_2$$

A modification is the addition of *oxonium salts,* e.g.,

$$IrCl(CO)(PPh_3)_2 + Me_3O^+BF_4^- = [Ir(Me)(Cl)(CO)(PPh_3)_2]^+BF_4^- + Me_2O$$

The oxidative addition to coordinately unsaturated metal complexes of compounds with C=C, C≡C, and other multiple bonds like C=O, C=S, and C=N, can lead to M—C σ bonds, e.g.,

4. *Metal Atom Reactions.*[56] The cocondensation of metal vapors with organic molecules such as alkyl or aryl halides can be regarded as a type of oxidative-addition reaction. Examples are:

$$Pd + CF_3I = (PdICF_3)_n$$
$$Pd + RX + 2PR_3 = PdR(X)(PR_3)_2$$
$$Ni + C_6F_5Br + C_6H_5Me \rightarrow NiBr_2 + (C_6F_5)_2Ni(\eta\text{-}C_6H_5Me)$$

5. *Insertion Reactions.* These are discussed in detail in Section 29-3, but the following are examples of M—C syntheses:

(*a*) Insertion of unsaturated compounds into M—H bonds:

$$[Rh(NH_3)_5H]^{2+} + CH_2=CH_2 = [Rh(NH_3)_5C_2H_5]^{2+}$$
$$HRh(CO)(PPh_3)_3 + CF_2=CF_2 = Rh(CF_2CF_2H)(CO)(PPh_3)_2 + PPh_3$$

$$(\eta\text{-}C_5H_5)_2ZrH_2 + 2HC\equiv CPh = (\eta\, C_5H_5)_2Zr\left(\!\begin{array}{c} H \\ C=C \\ Ph \end{array}\!\right)_2$$

$$(\eta\text{-}C_5H_5)_2W(CO)_3H + CH_2N_2 = (\eta\text{-}C_5H_5)_2W(CO)_3CH_3 + N_2$$

(*b*) Insertion of unsaturated compounds into M—R bonds:

$$L_nMR + {>}C{=}C{<} \rightarrow L_nM-C$$

$$L_nMR + -C\equiv C- \rightarrow L_nM{>}C=C{<}\,^R$$

$$L_nMR + CO \rightarrow L_nM-\underset{\underset{O}{\|}}{C}-R$$

6. *Elimination Reactions.* These reactions are the reverse of insertion reactions. They involve the elimination from appropriate compounds such as acyls, carboxylates, sulfinates, azo compounds, or hydrides of CO, CO_2, SO_2, N_2, H_2, etc., by action of heat or light with the concomitant formation of a metal-to-carbon σ bond. Examples are:

$$Me\text{-}C_6H_4\text{-}\underset{\underset{O}{\|}}{C}\text{-}Mn(CO)_5 \rightarrow Me\text{-}C_6H_4\text{-}Mn(CO)_5 + CO$$

$$Ni(bipy)(CO_2Ph)_2 \rightarrow Ni(bipy)Ph_2 + 2CO_2$$
$$(\eta\text{-}C_5H_5)Fe(CO)_2SO_2R \rightarrow (\eta\text{-}C_5H_5)Fe(CO)_2R + SO_2$$

$$PtHCl(PEt_3)_2 + PhN_2^+ \xrightarrow{KOH} (Et_3P)_2PtCl(N=NPh) \rightarrow (Et_3P)_2PtClPh + N_2$$

[56] P. L. Timms and T. W. Turney, *Adv. Organomet. Chem.*, 1977, **15**, 53; K. Klabunde *et al.*, *J. Am. Chem. Soc.*, 1978, **100**, 1313; *Acc. Chem. Res.*, 1975, **8**, 393; P. S. Skell and M. J. McGlinchey, *Angew. Chem., Int. Ed.*, 1975, **14**, 195. R. Lagow *et al.*, *J. Am. Chem. Soc.*, 1979, **101**, 3229.

In a special case of elimination hydrogen transfer occurs between a ligand and a group such as H or CH_3 on the metal. This type of reaction, which may involve oxidative addition, is known as *cyclometallation*. It is especially common for complexes of aryl phosphines and phosphites and is discussed in detail in Section 29-2. Examples are:

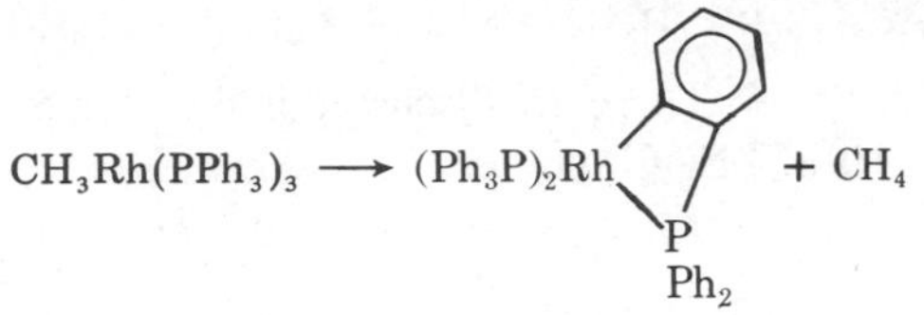

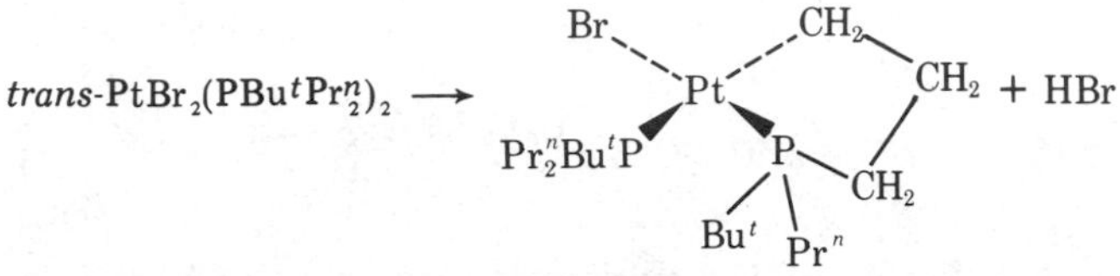

$$CH_3Rh(PPh_3)_3 \longrightarrow (Ph_3P)_2Rh \underset{\underset{Ph_2}{P}}{\diagup} + CH_4$$

$$trans\text{-}PtBr_2(PBu^tPr_2^n)_2 \longrightarrow \underset{Pr_2^nBu^tP}{\overset{Br}{\diagdown}} Pt \underset{\underset{Bu^t \quad Pr^n}{P}}{\overset{CH_2}{\diagup}} \overset{CH_2}{\diagdown}{CH_2} + HBr$$

7. *π-Complex and σ-M—C Transitions.*[57]
    (a) Nucleophilic attacks on the π complexes, discussed later in this chapter, commonly involve π–σ transitions. There are many cases of such reaction, examples being:

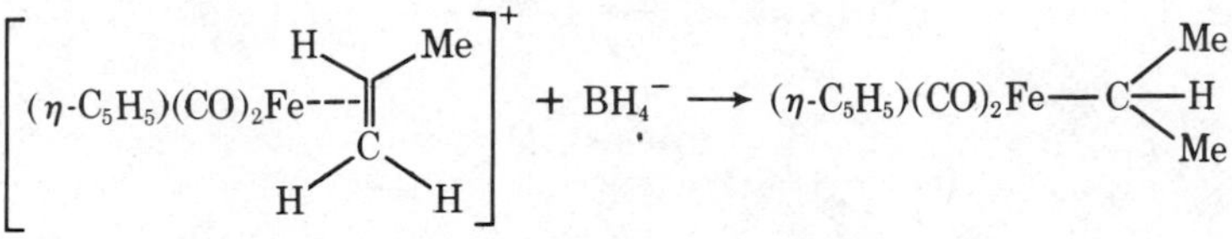

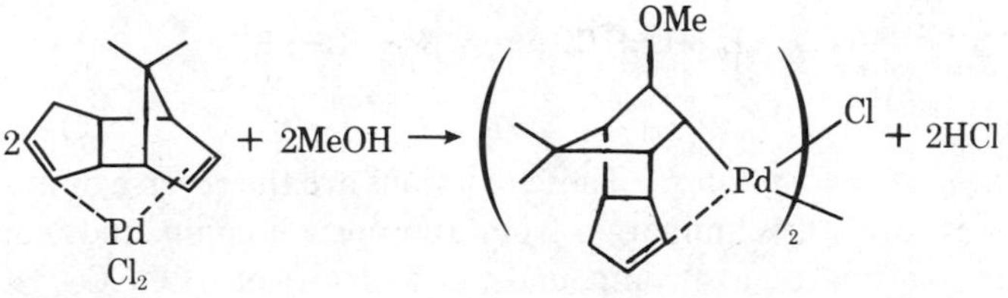

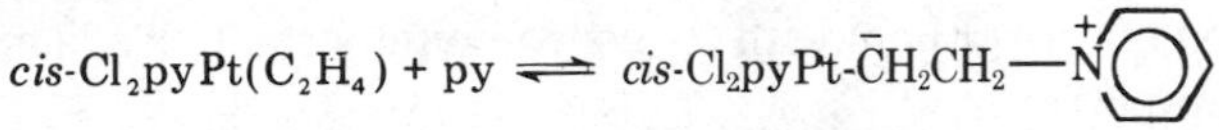

$$\left[(\eta\text{-}C_5H_5)(CO)_2Fe\text{---}\underset{\underset{H}{\overset{}{|}}}{\overset{H \diagdown \quad \diagup Me}{\underset{C}{||}}} \right]^{+} + BH_4^{-} \longrightarrow (\eta\text{-}C_5H_5)(CO)_2Fe\text{---}C\overset{Me}{\underset{Me}{\diagdown}}H$$

$$2 \bigg(\underset{\underset{Cl_2}{Pd}}{} \bigg) + 2MeOH \longrightarrow \bigg(\underset{Pd}{\overset{OMe}{}} \bigg)_2 \overset{Cl}{} + 2HCl$$

$$cis\text{-}Cl_2pyPt(C_2H_4) + py \;\rightleftharpoons\; cis\text{-}Cl_2pyPt\text{-}\bar{C}H_2CH_2\text{---}\overset{+}{N}\bigcirc$$

(b) Allyl groups can undergo π–σ transformations that are usually promoted by Lewis bases such as PR_3 or MeNC,[58] viz.,

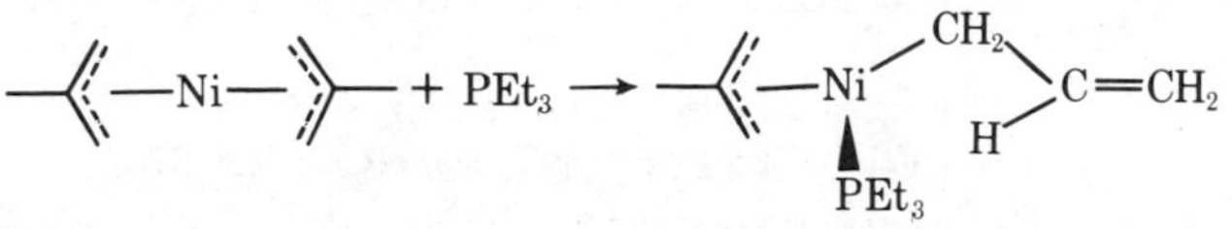

$$\text{---}\bigg(\text{---}Ni\text{---}\bigg) + PEt_3 \longrightarrow \text{---}\bigg(\text{---}Ni\overset{CH_2}{\underset{\underset{PEt_3}{\overset{\blacktriangle}{}}}{\diagup}}\underset{H}{\overset{}{\diagdown}}C{=}CH_2$$

8. *Ylid Syntheses.* The synthesis of ylid compounds is as follows.
    (a) From trimethylphosphanemethylene and metal salts:

[57] M. Tsutsui and A. Courtnay, *Adv. Organomet. Chem.,* 1977, **16**, 241; M. D. Johnson, *Acc. Chem. Res.,* 1978, **11**, 57.
[58] See, e.g., G. Carturan *et al., Inorg. Chim. Acta.,* 1978, **26**, 1.

$$Ag^+ + Me_3P—CH_2 \longrightarrow Me_2P \underset{CH_2—Ag—CH_2}{\overset{CH_2—Ag—CH_2}{<\qquad>}} PMe_2$$

(*b*) From phosphonium salts and lithium alkylate anions:

$$4Me_4PCl + Li_4[Mo_2Me_8] = Mo_2[(CH_2)_2PMe_2]_4 + 4CH_4 + 4LiCl$$

(*c*) From lithium phosphonium methylide and metal salts:

$$Me_4PCl + 2MeLi = Me_2P(CH_2)_2Li + LiCl + 2CH_4$$

$$3Me_2P(CH_2)_2Li + CrCl_3(THF)_3 = Cr \left(\overset{CH_2}{\underset{CH_2}{<\qquad>}} PMe_2 \right)_3 + 3LiCl$$

(*d*) From phosphanes:

$$[(\eta\text{-}C_5H_5)_2W(C_2H_4)]^+ \xrightarrow[PF_6^-]{MeI} [(\eta\text{-}C_5H_5)_2W(C_2H_4)Me]^+ \xrightarrow[-C_2H_4]{+PMe_2Ph}$$

$$\left[(\eta\text{-}C_5H_5)_2W \underset{H}{\overset{CH_2PMe_2Ph}{<\qquad}} \right]^+$$

(*e*) From ylids and carbonyl compounds:

$$MnBr(CO)_5 + Ph_3PCCPPh_3 \rightarrow (CO)_4BrMn—C≡C\overset{+}{P}Ph_3 + PPh_3$$
$$Ph_3PCr(CO)_5 + Ph_3P=CH_2 \rightarrow (CO)_5\overset{-}{C}r\text{-}\overset{+}{C}H_2PPh_3 + PPh_3$$

9.   *Paramagnetic Metal Ion–Free-Radical Reactions.* These involve interaction of paramagnetic complex ions,[59] examples being:

$$[Cr(H_2O)_6]^{2+} + Bu^tOOH = [CrBu^t(H_2O)_5]^{2+}$$
$$[Cr(H_2O)_6]^{2+} + CHCl_3 = [Cr(H_2O)_5Cl]^{2+} + [Cr(H_2O)_5CHCl_2]^+$$
$$Co(dmgH)_2py + CCl_4 = Cl_3CCo(dmgH)_2py + ClCo(dmgH)_2py$$

$$[Co(CN)_5]^3 + HC≡CH = \left[(CN)_5Co \underset{H}{\overset{H}{>}}C=C\overset{H}{\underset{Co(CN)_5}{<}} \right]^{6-}$$

10.   *Oxidative Deamination of Organic Hydrazines.* Examples are:

$$2AuCl_4^- + 2PhNHNH_2HCl = Ph_2AuCl_2^- + AuCl_2^- + 2N_2 + 6HCl$$

[59]   See, e.g., J. H. Espenson *et al., Inorg. Chem.,* 1979, **18**, 1246.

11. *Condensations or Reactions of Compounds with Active Methylene Groups.*
Examples[60] are:

$$trans\text{-}(Ph_3P)_2(CH_3)PtOH + CH_3NO_2 = trans\text{-}(Ph_3P)_2(CH_3)PtCH_2NO_2 + H_2O$$

27-6. Decomposition of Compounds with Metal-Carbon σ Bonds[20a,20b,61]

Two modes of thermal decomposition of metal alkyl compounds have been mentioned, namely, β-hydrogen transfer and homolytic fission, which is accompanied by radical chain reactions. Photochemical decompositions are usually radical.[62]

Many reactions depend on the nature of ligands and solvents and may be very sensitive to traces of water, oxygen, and other impurities. However the following additional decomposition pathways are established.

1. *Intramolecular Reductive Eliminations.* These are the reverse of oxidative-addition reactions (Section 29-2). They may be reversible in some cases. An example of an irreversible reductive elimination is the reaction[63] in Me₂SO:

If the same reaction is carried out in benzene or decalin it proceeds by an *inter-molecular* pathway.

Reductive coupling of alkyl groups can also be induced by oxidation,[64] and in such cases paramagnetic intermediates are clearly involved:

$$Et_2Nibipy \xrightarrow{-e} Et_2Nibipy^+ \rightarrow n\text{-}C_4H_{10}$$

Under normal circumstances no radicals or ions appear to be involved, and the concerted reaction is probably initiated by vibrational modes bringing the carbon

60   T. Yoshida *et al., J.C.S. Dalton,* **1976,** 993; D. Cummins *et al., J.C.S. Dalton,* **1976,** 130.
61   R. D. W. Kemmitt and M. A. R. Smith, *Inorg. React. Mech., Spec. Rep. Chem. Soc.;* M. C. Baird, *J. Organomet. Chem.,* 1974, **64,** 289; P. S. Braterman, and R. J. Cross, *Chem. Soc. Rev.,* 1973, **2,** 271; J. R. Norton, *Acc. Chem. Res.,* 1979, **12,** 139. For electrophilic cleavage of M—C bonds, see T. G. Attig *et al., J. Am. Chem. Soc.,* 1979, **101,** 619, and M. D. Johnson, *Acc. Chem. Res.,* 1978, **11,** 57.
62   See, e.g., P. W. N. M. Van Leeuwen *et al., J. Organomet. Chem.,* 1977, **142,** 233.
63   S. Komiya *et al., J. Am. Chem. Soc.,* 1976, **98,** 7255.
64   T. T. Tsou and J. K. Kochi, *J. Am. Chem. Soc.,* 1978, **100,** 1634; J. K. Kochi, ACS Symposium Series, Vol. 55, 1977, p. 167; G. W. Daub, *Prog. Inorg. Chem.,* 1977, **22,** 409.

atoms of the alkyl groups into close proximity.[65] Aryls similarly decompose[66] possibly via a transition state such as 27-**XXXVI**.

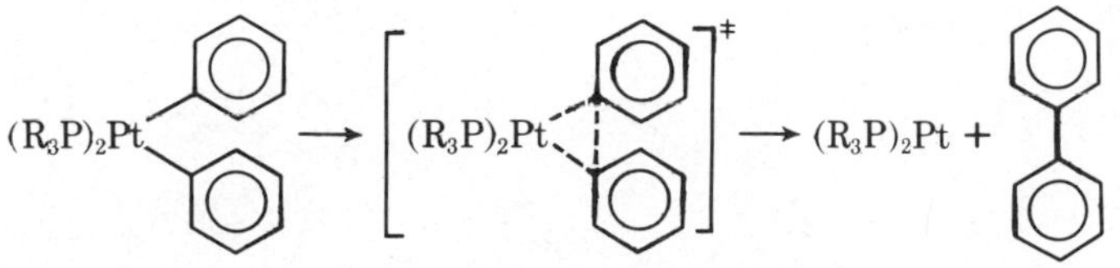

(27-XXXVI)

The decomposition of alkyl hydrides is of particular interest because of the relation to processes involved in catalytic hydrogenations (Section 30-1). For Os-(CO)$_4$H(CH$_3$), *inter*molecular elimination occurs, but for the complexes *cis*-PtH(R)(PPh$_3$)$_2$, kinetic studies[67] show that alkane elimination is *intra*molecular, with the rate-determining step being, e.g.,

$$cis\text{-PtH(CH}_3\text{)(PPh}_3\text{)}_2 \rightarrow \text{Pt(PPh}_3\text{)}_2 + \text{CH}_4$$

The reactivity decreases for the R groups in the order

$$\text{C}_6\text{H}_5 > \text{C}_2\text{H}_5 > \text{CH}_3 > \text{CH}_2\text{CH=CH}_2.$$

In some other cases more complicated behavior is found, and in the decomposition of (η-Me$_5$C$_5$)$_2$ZrH(Bu^i), the elimination involves the cyclopentadienyl ring.[68]

2. *α-Hydrogen Transfer.* We mentioned earlier that permethyls such as WMe$_6$ give methane when they decompose. Since the metal atom is in its maximum oxidation state, an internal oxidative addition:

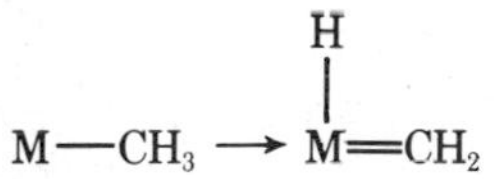

$$\text{M—CH}_3 \rightarrow \overset{\overset{\displaystyle H}{\displaystyle |}}{\text{M=CH}_2}$$

is unlikely, and the decomposition either must be free radical or it must occur by way of a four-center transition state:

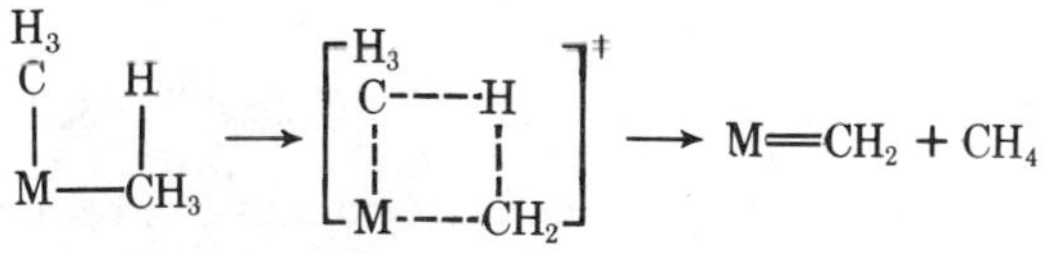

There are well-established cases[69] of the removal of α-hydrogen atoms, and this process is important in the synthesis of the alkylidene complexes discussed below. The decomposition of the benzyl Ta(CH$_2$Ph)$_5$ for example, proceeds by intramolecular α-hydrogen transfer.

A particularly interesting case of transfer from a methyl group that can be studied

[65] M. P. Brown *et al., J.C.S. Dalton,* **1974,** 2457, 1613.
[66] P. S. Braterman *et al., J.C.S. Dalton,* **1977,** 1892.
[67] J. Halpern *et al., J. Am. Chem. Soc.,* 1978, **100,** 2915.
[68] J. E. Bercaw *et al., J. Am. Chem. Soc.,* 1978, **100,** 5966.
[69] R. R. Schrock *et al., J. Am. Chem. Soc.,* 1978, **100,** 3359; *J. Organomet. Chem.,* 1978, **152,** C53; F. N. Tebbe *et al., J. Am. Chem. Soc.,* 1978, **100,** 3611; N. J. Cooper and M. L. H. Green, *J.C.S. Dalton,* **1979,** 1120.

spectroscopically is the following involving $HOs_3(CO)_{10}(CH_3)$. There is good evidence that the methyl group is unsymmetrical and that there is definite Os . . . H—C interaction:[70]

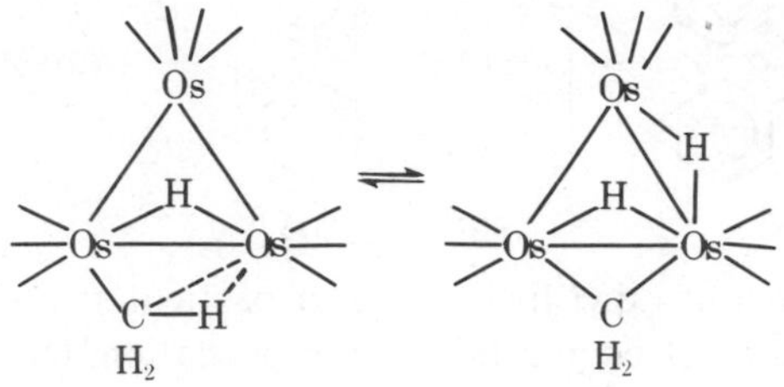

Other cases involving strong interaction between a C—H group and a metal are discussed in Section 29-6.

27-7.   Compounds with Transition Metal–Carbon Multiple Bonds

There are now well-established compounds that have metal-to-carbon double and triple bonds. We distinguish two classes: compounds in high oxidation states and those in low oxidation states that differ noticeably in their reactivity.

We deal first with the compounds obtained by loss of α-hydrogen atoms as discussed in the last section.

Alkylidene Complexes.   The first examples of compounds made via loss of α-hydrogen atoms had bridge groups[71] and could be formulated as having M=C bonds:

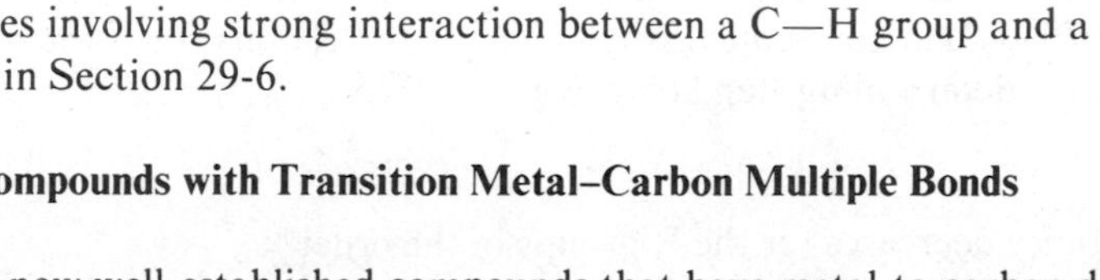

$$NbCl_5 + Me_3SiCH_2Li \longrightarrow$$

Since they have RC bridges, however, such compounds can be formulated as alkylidyne complexes (see p. 1142).

However a variety of mononuclear complexes, mostly again of niobium and tantalum[72,73] can be made by reactions such as:

[70] J. R. Shapley et al., Inorg. Chem., 1979, **18**, 319; R. B. Calvert and J. R. Shapley, J. Am. Chem. Soc., 1978, **100**, 7727.

[71] W. Mowat and G. Wilkinson, J.C.S. Dalton, **1973**, 1120; M. H. Chisholm et al., Inorg. Chem., 1978, **17**, 656.

[72] R. R. Schrock et al., Acc. Chem. Res., 1979, **12**, 98; J. Organomet. Chem., 1979, **171**, 43; J. Am. Chem. Soc., 1979, **101**, 1593, 3210; 1978, **100**, 648, 2309, 2386, 3359, 3793, 6774.

[73] M. R. Churchill et al., Inorg. Chem., 1979, **18**, 1930; 1978, **17**, 1957; J.C.S. Chem. Comm., **1978**, 1048.

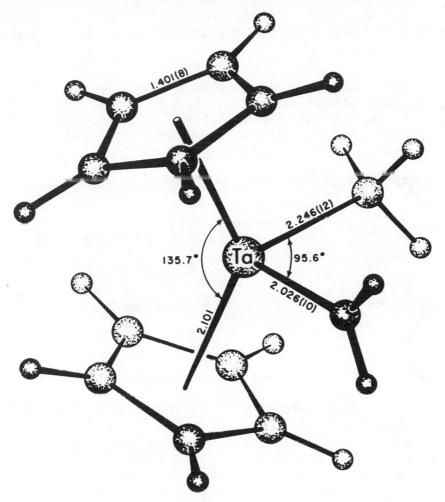

Fig. 27-4. The structure of $(\eta\text{-}C_5H_5)_2Ta(CH_3)(CH_2)$. [Reproduced by permission from L. H. Guggenberger and R. R. Schrock, *J. Am. Chem. Soc.*, 1975, **97**, 6578.]

$$(Me_3CCH_2)_3TaCl_2 \xrightarrow{Me_3CCH_2Li} (Me_3CCH_2)_3Ta{=}C\overset{H}{\underset{CMe_3}{\diagdown}}$$

$$(\eta\text{-}C_5H_5)_2TaMe_3 \xrightarrow{Ph_3C^+BF_4^-} (\eta\text{-}C_5H_5)_2TaMe_2^+ \xrightarrow{NaOMe} (\eta\text{-}C_5H_5)_2Ta\overset{CH_3}{\underset{CH_2}{\diagup}}$$

The structure of the methylene complex $(\eta\text{-}C_5H_5)_2Ta(CH_3)(CH_2)$ is shown in Fig. 27-4; note that there is a direct comparison of Ta—CH$_3$ (2.246 Å) vs. Ta=CH$_2$ (2.026 Å) bond lengths.

The reactions of the alkylidene complexes[72] suggest that the carbon atom bound to the metal is *nucleophilic* rather than electrophilic as in the carbene complexes to be described below. The high-valence metal compounds can be compared to phosphanemethylenes:

$$L_n Ta = CH_2 : R_3 P = CH_2$$

The M=C π bond to electropositive metals in high oxidation states is best regarded as being formed by donation of π-electron density from a filled sp^2 orbital on carbon to an empty orbital on the metal.

Alkylidyne Complexes.[74]   These have metal-carbon triple bonds. One example is the anion obtained by deprotonation of an alkylidene:

$$(Me_3CCH_2)_3Ta = C \begin{matrix} H \\ \diagdown \\ CMe_3 \end{matrix} \xrightarrow{\ Bu^n Li\ } [Lidmp][(Me_3CCH_2)_3Ta \equiv C - CMe_3]$$

where dmp = N,N'-dimethylpiperazine.

Other complexes containing M=C and/or M≡C bonds and in formal high oxidation states are the following.

$$(Me_3CCH_2)_3W \equiv C - CMe_3$$
$$(\eta\text{-}MeC_5H_4)(PMe_3)_2ClTa \equiv C - Ph$$

(27-XXXVII)

Compound 27-**XXXVII** has an approximately *tbp* structure and formally contains W^{VI}.

Low-Valent Metal "Carbene" and "Carbyne" Complexes.[75]   Compounds of a different type with the metal in a low oxidation state were first synthesized by E. O. Fischer and his co-workers. They are known generally as carbene complexes. Although they could be formulated as having a M=C double bond 27-**XXXVIIIa**, other canonical forms (27-**XXXVIIIb**–27-**XXXVIId**) evidently are involved. Many of the earlier known complexes have —OR or NR$_2$ groups attached to carbon, and electron flow from lone pairs on O or N leads to a contribution of the type 27-**XXXVIIIc,d** through O—C or N—C $p\pi \rightarrow p\pi$ bonding. However several complexes without such groups are known.

(27-**XXXVIIIa**)   (27-**XXXVIIIb**)   (27-**XXXVIIIc**)   (27-**XXXVIIId**)

[74]   R. R. Schrock *et al., J. Am. Chem. Soc.,* 1978, **100**, 5962, 5964, 6774; 1975, **97**, 2935; M. R. Churchill and W. J. Youngs, *Inorg. Chem.,* 1979, **18**, 171; *J.C.S. Chem. Comm.,* **1979**, 321; **1978**, 1048.

[75]   J. A. Connor, *Organometallic Chemistry, Specialist Rep., Chemical Society;* E. O. Fischer, *Adv. Organomet. Chem.,* 1976, **14**, 1; D. J. Cardin *et al., Chem. Soc. Rev.,* 1973, **2**, 99, *Chem. Rev.,* 1972, **72**, 545; F. A. Cotton and C. M. Lukehart, *Prog. Inorg. Chem.,* 1972, **16**, 487; E. O. Fischer *et al., Chem. Ber.,* 1979, **112**, 1320.

The important difference from the alkylidene complexes is that the carbon atom bound to the metal is *electrophilic* and is readily attacked by nucleophilic reagents; see, e.g., the exchange reaction below and the reaction[76]

$$(CO)_5Cr{=}C\begin{array}{l}\text{OMe}\\ \text{Me}\end{array} + \text{:CNR} \longrightarrow (CO)_5Cr\text{---}\overset{\overset{\displaystyle MeO\diagdown C \diagup Me}{\|}}{\underset{\underset{\displaystyle R}{\diagup}N}{\overset{C}{\|}}}$$

Ketenimine complex

The structures of many carbene complexes have been determined, representative examples being 27-XXXIX. The M—CRR′ grouping is *always planar,* and the

$$cis\text{-}(OC)_4(Ph_3P)Cr\overset{2.00(2)}{\text{—}}C\overset{\overset{\displaystyle 1.32(2)\,OCH_3}{\diagup}}{\underset{\underset{\displaystyle CH_3}{\diagdown}}{\text{—}1.53(3)}}$$

(27-XXXIXa)

$$(OC)_5Cr\overset{2.16(1)}{\text{—}}C\overset{\overset{\displaystyle 1.31(1)\,N(CH_3)_2}{\diagup}}{\underset{\underset{\displaystyle CH_3}{\diagdown}}{\text{—}1.50(1)}}$$

(27-XXXIXb)

M—C distances indicate considerable shortening consistent with multiple-bond character. The C—O and C—N distances are also shortened, as expected with π bonding.

Carbene complexes can be synthesized by the following methods:

1. *From Metal Carbonyls.* Nucleophilic attack by lithium alkyls on coordinated carbon monoxide (cf. Section 29-4) gives an anion that can be attacked by an electrophile, e.g.,

$$Cr(CO)_6 + LiR \longrightarrow \left[(OC)_5Cr\text{—}C\begin{array}{l}\diagup O\\ \diagdown R\end{array}\right]^- \overset{R_3'O^+}{\longrightarrow}$$

$$(OC)_5Cr{=}C\begin{array}{l}\diagup OR'\\ \diagdown R\end{array} \overset{LiR''}{\longrightarrow} \left[(CO)_5Cr\text{—}\overset{\overset{\displaystyle OR'}{|}}{\underset{\underset{\displaystyle R}{|}}{C}}\text{—}R''\right]^-$$

From these alcoxocarbenes, others may be obtained by exchange reactions, e.g.,

$$(OC)_5Cr{=}C\begin{array}{l}\diagup OMe\\ \diagdown Me\end{array} + NH_2Et = (OC)_5Cr{=}C\begin{array}{l}\diagup NHEt\\ \diagdown Me\end{array} + MeOH$$

Nucleophilic attacks can also be made on thiocarbonyls:

$$(CO)_5WCS + HNR_2 \longrightarrow (OC)_5W{=}C\begin{array}{l}\diagup SH\\ \diagdown NR_2\end{array}$$

76   C. G. Kreiter and R. Amman, *Chem. Ber.*, 1978, **111**, 1223.

2. *Nucleophilic Attacks on Coordinated Isocyanides.* These reactions are also discussed in Section 29-4, but an example[77] is:

$$cis\text{-}Cl_2(PEt_3)Pt\text{-}CNPh + EtOH \longrightarrow cis\text{-}Cl_2(PEt_3)Pt{=}C\overset{\displaystyle OEt}{\underset{\displaystyle NHPh}{\diagup}}$$

Unidentate, chelate, bis,[78a] and tetrakis[78b] carbene complexes can be so obtained. An example is the reaction that gives a square, formally platinum(II) complex:

$$[Pt(CNMe)_4]^{2+} \xrightarrow{\text{MeNH}_2} \left[Pt{=}C\overset{\displaystyle NHMe}{\underset{\displaystyle NHMe}{\diagdown}}\right]_4^{2+}$$

3. *Cleavage of Electron-Rich Alkenes.*[79] These alkenes are of the type 27-XL and react as follows:

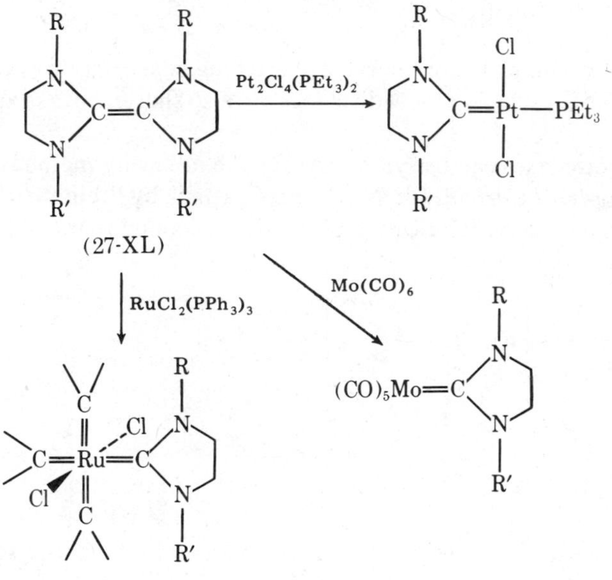

(27-XL)

4. *Intramolecular Cyclization Reactions.* An example of these is the reaction:

[77] E. M. Bradley *et al., J.C.S. Dalton,* **1976,** 1930; L. Calligaro *et al., J. Organomet. Chem.,* 1977, **142,** 105.

[78a] P. R. Branson *et al., J.C.S. Dalton,* **1976,** 12.

[78b] S. Z. Goldberg *et al., Inorg. Chem.,* 1977, **16,** 1502.

[79] M. F. Lappert *et al., J.C.S. Dalton,* **1978,** 826, 837.

Not isolable

5. *From Polyhalogenated Hydrocarbons.* Some unusual halogenated carbene compounds[80a] can be obtained by reduction of Fe^{III} porphyrins in presence of CCl_4 or other chlorinated hydrocarbons, including DDT (this may explain the toxicity of DDT), e.g., the tetraphenylporphyrin complex reacts as follows:

$$(TPP)FeCl + CCl_4 \xrightarrow[\text{MeOH}]{\text{Zn, etc.}} (TPP) Fe{=}CCl_2$$

A difluoro carbene complex[80b] has been made by the reaction

$$(\eta\text{-}C_5H_5)(CO)_3MoCF_3 + SbF_5 \xrightarrow{SO_2(l)} [\eta\text{-}C_5H_5(CO)_3Mo{=}CF_2]^+ + SbF_6^-$$

but the ion could not be isolated.

6. *From $Me_2NCHCl^+Cl^-$ and $Me_2NCH_2^+Cl^-$ Salts.*[81] These salts can form secondary carbene complexes by reactions of the following types:

7. *By Trapping of a Free Carbene.* When the compound $(PhS)_3CH$ is treated with butyllithium, an equilibrium is set up:

$$LiC(SPh)_3 \rightleftharpoons LiSPh + :C(SPh)_2$$

and the dithiocarbene can then react:[82]

[80a] D. Mansuy *et al., J. Am. Chem. Soc.,* 1978, **100**, 3213; *J.C.S. Chem. Comm.,* **1977**, 648.

[80b] D. L. Reger and M. D. Jukes, *J. Organomet. Chem.,* 1978, **153**, 67.

[81] A. J. Hartshorne *et al., J.C.S. Dalton,* **1978**, 348.

[82] H. G. Raubenheimer and H. E. Swanepoel, *J. Organomet. Chem.,* 1977, **141**, C21.

In addition to the carbenes above, there are many that do not have a heteroatom acting as a stabilizing donor. One example[83] is the following:

$$(\eta\text{-}C_5H_5)(CO)_2Fe\text{---}C\overset{H}{\underset{Ph}{\overset{}{\diagdown}}}OMe \xrightarrow{H^+} \left[(\eta\text{-}C_5H_5)(CO)_2Fe\text{=}C\overset{H}{\underset{Ph}{\diagdown}}\right]^+ + MeOH$$

The ion can be considered to be a carbene rather than an alkylidene and can be compared with $[(\eta\text{-}C_5H_5)Fe(CO)_3]^+$.

Metallocarbenes have often been postulated as intermediates in catalytic reactions of olefins, and the reactions of olefins with carbenes have been studied as a model for the metathesis reaction (Section 30-5), using compounds such as $(CO)_5WCPh_2$ or the unisolable $(CO)_5WCHPh$.[84]

However in some types of metathesis reaction, which proceed by formation of four-membered rings, it seems likely that alkylidene rather than carbene-type intermediates are involved. Indeed, primary alkylidenes such as

$$(\eta\text{-}C_5H_5)Cl_2Ta\text{=}C\overset{H}{\underset{CMe_3}{\diagdown}}$$

react with ethylene to give 4,4-dimethylpentene,[85] probably by a reaction of the type:

$$L_nTa^V\text{=}C\overset{H}{\underset{CMe_3}{\diagdown}} \xrightarrow{C_2H_4} \underset{L_nTa^V\text{---}C\overset{}{\underset{CMe_3}{\diagdown}}}{\overset{CH_2\text{---}CH_2}{\overset{|\quad\quad|}{}}}H \longrightarrow CH_2\text{=}CHCH_2CMe_3$$

These catalytic reactions are discussed further in Section 30-5.

Carbyne Complexes.[86] These bear the same relation to carbene complexes as alkylidynes do to alkylidenes. They can be made from carbene complexes, e.g., by the reaction

$$\textit{cis-}(OC)_4(Me_3P)Cr\text{=}C\overset{OMe}{\underset{Me}{\diagdown}} \xrightarrow[\text{pentane}]{BX_3} (OC)_3(Me_3P)XCr\text{≡}CMe + CO + (BX_2OMe)$$

Structural studies on carbyne complexes show that in some cases the W—C—C group is not linear [e.g., $\textit{trans-}I(CO)_4W\text{≡}CPh$ has an angle of 162°], whereas in other compounds [e.g., $\textit{trans-}I(CO)_4Cr\text{≡}CCH_3$] the Cr—C—C group is linear. The force constants for M≡C bonds are comparable to those for M≡N compounds.

[83]   M. Brookhart and G. O. Nelson, *J. Am. Chem. Soc.*, 1977, **99**, 6099.
[84]   C. P. Casey *et al., Inorg. Chem.*, 1977, **16**, 3059; *J. Am. Chem. Soc.*, 1977, **99**, 6097; 1976, **98**, 608.
[85]   R. R. Schrock *et al., J. Am. Chem. Soc.*, 1977, **99**, 3519.
[86]   E. O. Fischer, *J. Organomet. Chem.*, 1978, **149**, C57; **153**, C41; *Chem. Ber.*, 1979, **112**, 1320; **111**, 3525, 3740; *Angew. Chem., Int. Ed.*, 1978, **17**, 50; *Adv. Organomet. Chem.*, 1976, **14**, 1; *J. Organomet. Chem.*, 1975, **100**, 59.

Vinylidene and Related Complexes. In addition to alkylidenes there are some metallocumulenes, namely, *vinylidenes*,[87,88] (e.g., 27-XLI) and *allenylidenes*[89] (e.g., 27-XLII).

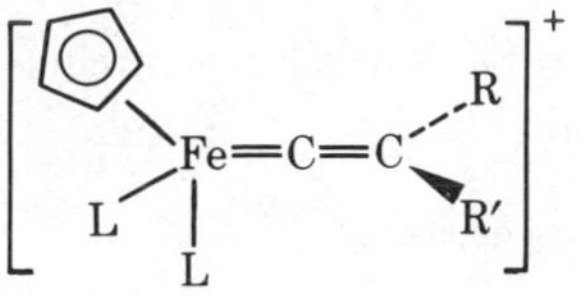

$$(27\text{-}XLI) \qquad\qquad (27\text{-}XLII)$$

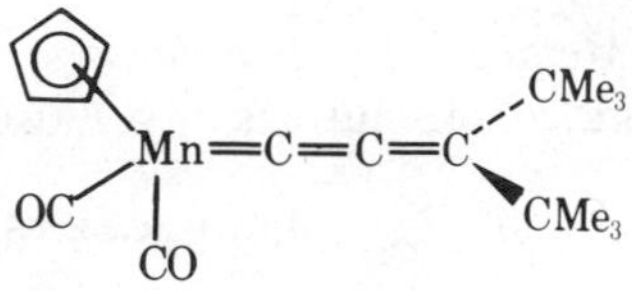

The vinylidene 27-XLI is especially interesting because the following reaction scheme (where $M = C_5H_5FeL_2$) illustrates the interrelationship between species with metal-to-carbon single, double, and triple bonds[87]:

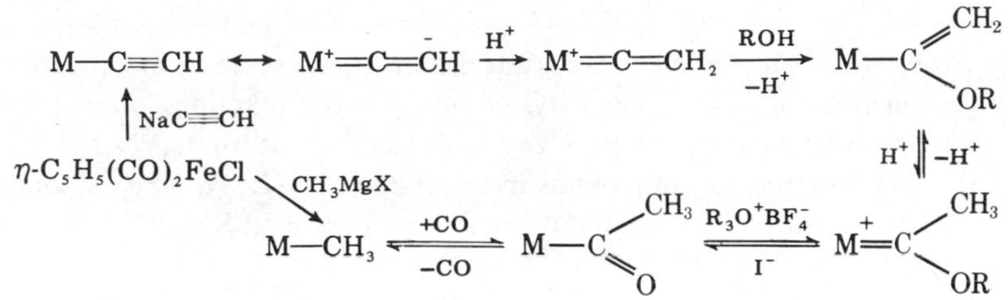

π COMPLEXES

There is a truly vast chemistry of π complexes of transition metals involving all types of unsaturated organic molecules.

The general features of π complexes and the nature of the bonding in them were covered in Chapter 3. Here we discuss mainly methods of synthesis and reactions of the principal classes of compound.

27-8. Alkene Complexes[90]

Alkene complexes are usually made by direct interaction of the alkene with transition metal ions in solution, with metal carbonyl, hydrido, or halide complexes,

[87] R. P. Hughes et al., J. Organomet. Chem., 1979, **172**, C29; A. Davison et al., J. Am. Chem. Soc., 1978, **100**, 7763; J. Organomet. Chem., 1979, **166**, C13.

[88] N. E. Kolobova et al., J. Organomet. Chem., 1977, **137**, 69.

[89] H. Berke, Angew. Chem., Int. Ed., 1976, **15**, 624.

[90] (a) F. R. Hartley, Chem. Rev., 1973, **73**, 163 (thermodynamic data for alkene and alkyne complexes); (b) A. J. Birch and I. D. Jenkins, in Transition Metal Organometallics in Organic Synthesis, Vol. 1, Academic Press, 1976, p. 2 (conjugated di- and triolefin complexes); (c) M. Herberhold, Metal π-Complexes, Vol. II, Parts 1 and 2 (complexes with monoalkene ligands), Elsevier, 1972, 1974; (d) F. L. Bowden and R. Giles, Coord. Chem. Rev., 1976, **20**, 81 (allenes); (e) R. Jones, Chem. Rev., 1968, **68**, 75 (π complexes of substituted alkenes); (f) M. R. Churchill and S. A. Julis, Inorg. Chem., 1978, **17**, 145 (η^4-cis-diene complexes); (g) G. I. Fray and R. G. Saxton, The Chemistry of Cyclooctatetraene and Its Derivatives, Cambridge University Press, 1978; (h) R. Hoffmann, et al., J. Am. Chem. Soc., 1979, **101**, 3801, 3812.

and so forth. Sometimes reducing agents such as Zn or Al alkyls may be used. Substitution reactions (e.g., of metal carbonyl complexes) may occur thermally or in some cases only under irradiation with light. Some representative syntheses are the following.

1. *From Metal Salts of Halide Complexes.* The silver ion reacts with alkenes in aqueous or alcohol solutions[91] (see also p. 971)

$$Ag_{aq}^+ + alkene \rightleftharpoons [Ag\ alkene]_{aq}^+$$

Weakly bound ligands can also be displaced by alkene, especially chelating ones like cycloocta-1,5-diene (1,5-COD)

$$(RCN)_2PdCl_2 + \text{[diene]} \longrightarrow \text{[complex]} PdCl_2 + 2RCN$$

2. *From Metal Carbonyls.* Direct interaction of alkenes with metal carbonyls is a common route to alkene carbonyl complexes: the reactions often proceed thermally, but irradiation to induce CO dissociation can be used also. There is an especially large number of compounds in which the $Fe(CO)_3$ moiety is bound to a diene, commonly a 1,3-conjugated diene system (see, e.g., Ref. 90b). Some examples are:

$$Cr(CO)_6 + \text{[diene]} \longrightarrow \text{[complex]} \underset{(CO)_3}{Cr} + 3CO$$

$$Fe(CO)_5 + \text{[diene]} \longrightarrow \text{[complex]} \underset{(CO)_3}{Fe} + 2CO$$

$$\eta\text{-}C_5H_5Mn(CO)_3 + C_2F_4 \xrightarrow{h\nu} \eta\text{-}C_5H_5Mn(CO)_2C_2F_4 + CO$$

3. *From Metal Hydrides.* Polyhydrides can react in part to hydrogenate the alkene, which may then complex[92]:

$$IrH_5(PPh_3)_2 + C_2H_4 \rightarrow Ir(C_2H_4)_2(o\text{-}C_6H_4PPh_2)(PPh_3)$$

[91] R. J. Laub and H. Purnell, *J.C.S. Perkin, II*, **1978**, 895.
[92] (a) M. C. Clerici et al., *J. Organomet. Chem.*, 1975, **84**, 379; (b) D. J. Cole-Hamilton and G. Wilkinson, *Nouv. J. Chim.*, **1977**, 141; (c) M. Tachikawa et al., *J. Am. Chem. Soc.*, 1976, **98**, 4651.

$$RuH_4(PPh_3)_3 \xrightarrow{C_4H_6} \diagdown\diagdown\diagup\diagup\!-Ru(PPh_3)_2$$

$$\diagdown C_6H_5CH{=}CH_2$$

$$(C_6H_5CH{=}CH_2)_2Ru(PPh_3)_2$$

$$Os_3(CO)_{10}H_2 \xrightarrow{C_4H_6} Os_3(CO)_{10}(C_4H_6)$$

4. *By Using Reducing Agents.* A variety of reducing agents can be used to form alkene complexes from metal halides or halide complexes. Aluminum alkyls have been commonly used,[92] but zinc,[94] Li$_2$C$_8$H$_8$[96f] and other reducing agents can be employed. Examples are:

$$RuCl_3aq + 2 \,\bigcirc \xrightarrow{Zn, \,EtOH} \bigcirc\!\!-Ru\!\!-\bigcirc$$

$$Ni(acac)_2 + \xrightarrow{AlBu_3^i} Ni$$

$$(R_3P)_2PtO_2 + \rangle\!\!=\!\!\langle \xrightarrow{NaBH_4} (R_3P)_2P\!\!-\!\!\langle$$

$$(1,5\text{-}COD)PtCl_2 + Li_2C_8H_9 \xrightarrow{COD} (1,5\text{-}COD)_2Pt$$

5. *From Metals.* Cocondensation reactions of metal vapors with organic substances have been mentioned (p. 1135). Alkene complexes can be obtained also, e.g.,

$$Fe + 1,5\text{-}COD \rightarrow Fe(1,5\text{-}COD)_2$$

The technique of matrix isolation whereby metal atoms or alkenes (or other substrates such as N$_2$) are condensed in an argon matrix at low temperatures allows study of unstable alkene complexes such as Ni(C$_2$H$_4$) or Ni(C$_2$H$_4$)$_2$.[96]

6. *From Other Alkene Complexes.* Exchange reactions can sometimes be used effectively,[95] e.g.,

$$(1,5\text{-}COD)_2Pt + 3C_2H_4 = (C_2H_4)_3Pt + 2(1,5\text{-}COD)$$

However interaction with conjugated dienes may take a different course.[97]

[93] See, e.g., M. L. H. Green and J. Knight, *J.C.S. Dalton,* **1976,** 213.

[94] P. Pertici *et al., J.C.S. Chem. Comm.,* **1975,** 846.

[95] M. Green *et al., J.C.S. Dalton,* **1977,** 271, 278, 1006, 1010.

[96] See, e.g., G. A. Ozin *et al., J.C.S. Dalton,* **1976,** 6508.

[97] See, e.g., (a) P. W. Jolly and G. Wilke, *The Organic Chemistry of Nickel,* Vols. I, II, Academic Press, 1974, 1975; (b) G. K. Barker *et al., J. Am. Chem. Soc.,* 1976, **98,** 3373.

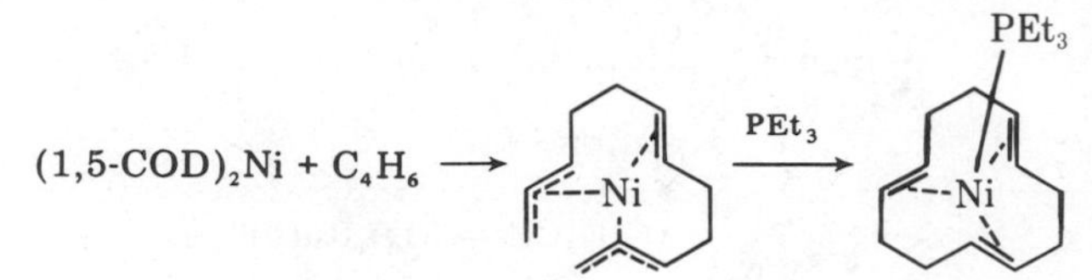

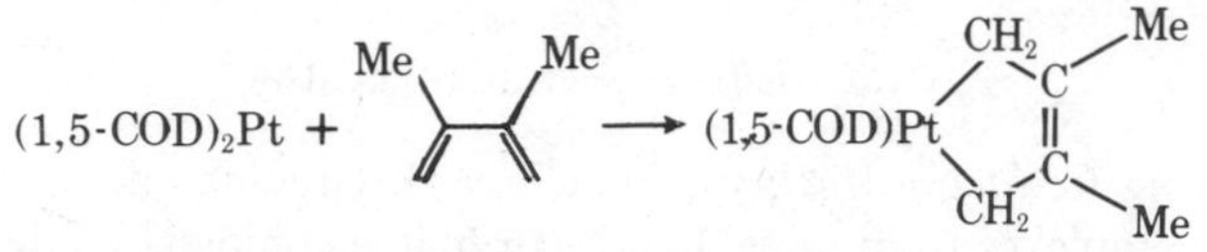

$$(1,5\text{-COD})_2\text{Ni} + C_4H_6 \longrightarrow \xrightarrow{\text{PEt}_3}$$

$$(1,5\text{-COD})_2\text{Pt} + \quad\text{Me}\diagdown\diagup\text{Me} \longrightarrow (1,5\text{-COD})\text{Pt}$$

Also certain complexes of cyclooctatetraene or cyclooctatrienes may undergo dehydrogenation to give complexes of the unstable hydrocarbon pentalene.[98]

7. *In Acetylene Reactions.* Acetylene reactions are discussed below, but an example that gives an alkene complex is:

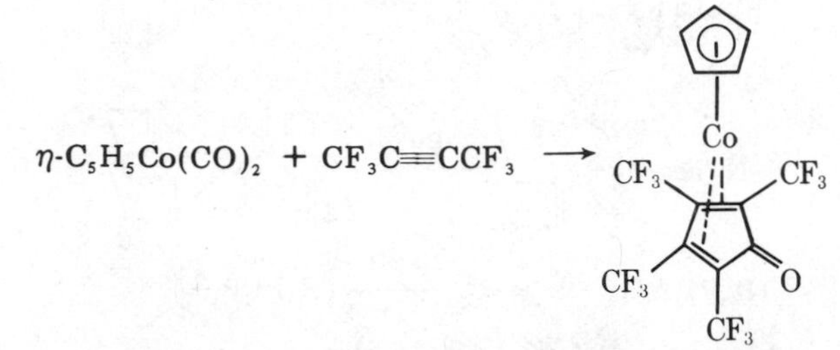

$$\eta\text{-}C_5H_5\text{Co(CO)}_2 + CF_3C\equiv CCF_3 \longrightarrow$$

8. *Hydride Abstraction from Alkyl Compounds.* In certain cases hydride ions can be removed from an alkyl chain to leave a coordinated alkene. This reaction may be reversed by hydride transfer from borohydride:

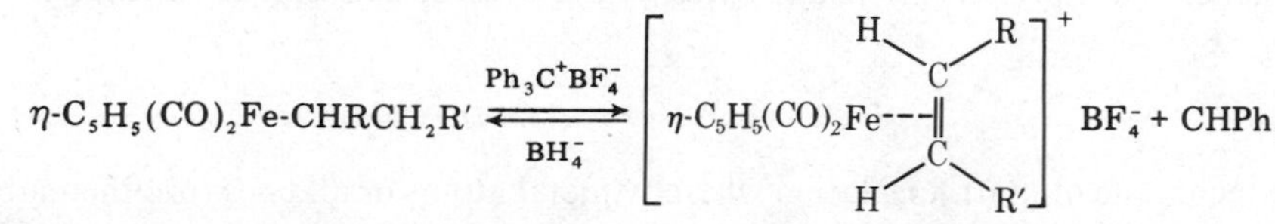

$$\eta\text{-}C_5H_5(\text{CO})_2\text{Fe-CHRCH}_2\text{R}' \underset{\text{BH}_4^-}{\overset{\text{Ph}_3\text{C}^+\text{BF}_4^-}{\rightleftharpoons}} \left[\eta\text{-}C_5H_5(\text{CO})_2\text{Fe--}\| \begin{matrix}\text{H}\diagup\text{C}\diagdown\text{R}\\\text{C}\\\text{H}\diagup\ \diagdown\text{R}'\end{matrix} \right]^+ \text{BF}_4^- + \text{CHPh}_3$$

There are related reactions involving the addition to, or abstraction of *protons* from, the organic group. Thus β-oxoalkyl compounds give ions in which the enol form is stabilized by bonding to the metal atoms[99]:

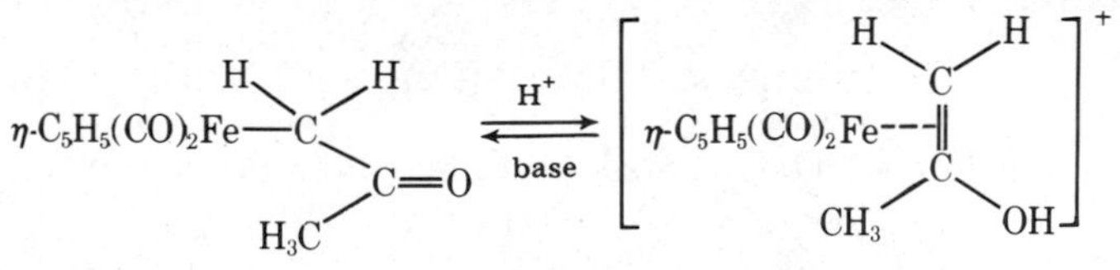

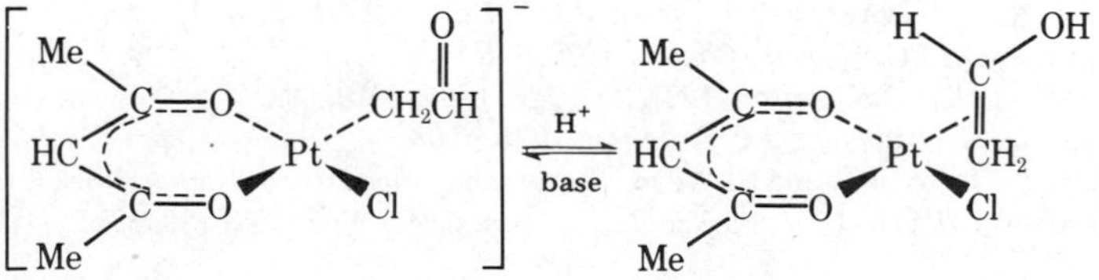

98   F. G. A. Stone *et al.*, *J.C.S. Dalton*, 1976, 1813.
99   F. A. Cotton *et al.*, *J. Am. Chem. Soc.*, 1973, **95**, 2483.

A similar case occurs with cyanoalkyls, which may be reversibly protonated:

$$\eta\text{-}C_5H_5(CO)_2Fe\text{---}\underset{\underset{CH_3}{|}}{\overset{\overset{C\equiv N}{|}}{C}}\text{---}H \underset{\text{base}}{\overset{H^+(D^+)}{\rightleftharpoons}} \left[\eta\text{-}C_5H_5(CO)_2Fe\text{---}\right]^+$$

and, finally, σ-allylic complexes may be protonated:

$$\eta\text{-}C_5H_5(CO)_3Mo\text{---}CH_2\text{---}CH\text{=}CH_2 \overset{H^+}{\longrightarrow} \left[\eta\text{-}C_5H_5(CO)_2Mo\text{---}\right]^+$$

9. Nucleophilic Attacks on Carbocyclic Complexes.[100] Cyclopentadiene compounds may be attacked by nucleophilic reagents to give complexes of conjugated alkene. These reactions may be reversible on addition of H^+, as one of the following examples shows.

The structure of the manganese cyclopentadiene complex shows that the methyl group is in the exo position.[101a] This confirms previous deductions based on spectroscopic evidence that nucleophilic attack occurs directly on the ring in the exo position, not initially on the metal atom. Had the latter occurred, the added group should have been in the endo position.[101b]

There are many other special syntheses of alkene complexes, but of greater importance are reactions involving π-allyl complexes, considered in the next section.

---

[100] For discussion and references, see R. Hoffmann and P. Hofmann, *J..Am. Chem. Soc.*, 1976, **98**, 598; S. G. Davies *et al.*, *Tetrahedron Rep.*, 1979, No. 57.
[101a] G. Evrard *et al.*, *Inorg. Chem.*, 1976, **15**, 52.
[101b] For some rules concerning nucleophilic attacks on ring systems and allyl and alkene compounds, see M. L. H. Green *et al. Nouv. J. Chim.*, **1977**, 445.

27-9. η-Allyl and Other Enyl Complexes[102]

π-, η^3-, or $\eta(1$-$3)$-Allyl Complexes.   These complexes can be made in several ways as follows:

1.   *From Allyl Grignard Reagents or Allyl Halides.*   Examples are the following:

$$NiCl_2 + 2C_3H_5MgBr \xrightarrow[-10°]{ether} Ni + 2MgX_2$$

σ-Allyl π-Allyl

Note that the second example shows the initial formation of a σ- or η^1-allyl compound that is subsequently converted into a η^3-allyl with loss of CO. There are numerous examples of $\eta^1 \rightleftharpoons \eta^3$ allylic conversions involving displacement of CO, PR_3, Cl^-, or other ligands.[103]

2.   *By Hydrogen Transfer from Alkene Complexes and Loss of HCl or H_2.* In some cases direct interaction of an alkene gives an allyl through loss of allylic hydrogen atoms:

$$RuH_4(PPh_3)_2 + CH_3CH{=}CH_2 =$$

The latter reaction is believed to proceed by way of the transfer of hydrogen from propylene (see also Section 29-2) in an oxidative-addition reaction, since such reactions have been observed spectroscopically in certain cases,[104] e.g., the reversible reaction

[102]   R. Baker, *Chem. Rev.,* 1973, **73**, 487; G. P. Chiusoli and L. Cassar, in *Organic Synthesis via Metal Carbonyls,* Vol. 2, I. Wender and P. Pino, Eds., Wiley, 1977. See also Ref 90b.

[103]   See, e.g., G. Carturan et al., *Inorg. Chim. Acta,* 1978, **26**, 1.

[104]   J. A. Osborn et al., *J. Am. Chem. Soc.,* 1975, **97**, 3871; E. O. Sherman, Jr., and P. R. Schreiner, *J.C.S. Chem. Comm.,* **1978**, 223.

Another example is the electrophilic attack of palladium acetate on hexene, which is accompanied by loss of H^+ from the allylic carbon,[105] viz.,

$$Pd_3(CO_2Me)_6 + hexene \rightarrow Pd_3(\eta_3\text{-hexenyl})(CO_2Me)_4$$

3. *Protonation of 1,3-Diene Complexes.*[90b] Protonation may occur in a variety of ways: (*a*) by addition of an acid such as HCl to a butadiene complex (reaction 27-1), (*b*) by reversible addition of a proton only (reaction 27-2), or (*c*) by reaction of a diene with a metal complex containing a transition metal–hydrogen bond (reaction 27-3).

Evidence suggests that the proton attacks from the endo (metal) side of the alkene, but this has not been definitely proved (as it has been in the case of proton attack on various noncoordinated double bonds).

There can also be differences in behavior depending on the metal,[106] e.g., for the cobalt and rhodium cycloocta-1,5-diene complexes, protonation occurs on the ring, viz.,

whereas for the iridium analogue an Ir—H bond is formed and the 1,5-diene isomerizes to a coordinated 1,3-diene, viz.,

Allyl complexes are intermediates in a great variety of reactions, some of which are mentioned in subsequent discussions. There is an extensive reaction chemistry

[105]   R. G. Brown *et al.*, *J.C.S. Dalton*, **1977**, 176, 183.
[106]   J. Evans *et al.*, *J.C.S. Dalton*, **1977**, 510.

of the complexes in organic synthesis, especially where conjugated alkenes interact with metal hydrido species:

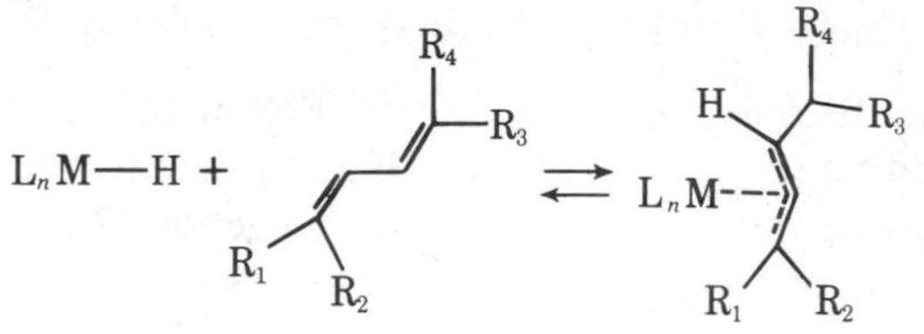

Al. 'ls are involved in cyclization reactions such as

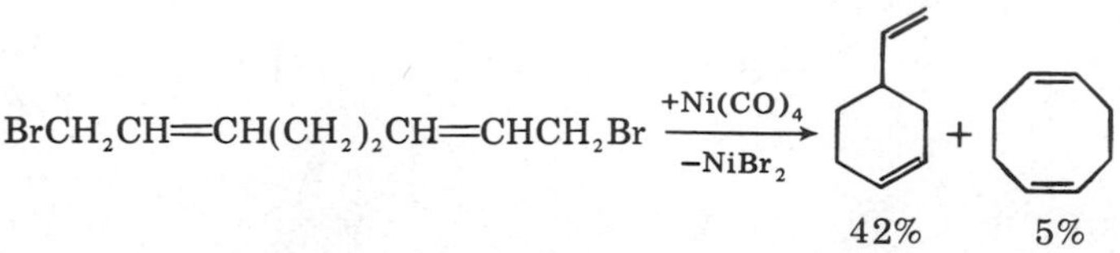

$$BrCH_2CH=CH(CH_2)_2CH=CHCH_2Br \xrightarrow[-NiBr_2]{+Ni(CO)_4}$$

42%   5%

The participation of an η-allylnickel species in this reaction is strongly supported by the isolation of an intermediate and its further reaction with CO as in 27-4:

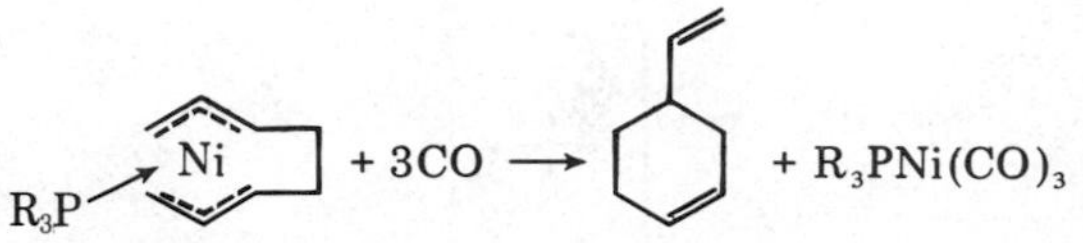

$$+ 3CO \longrightarrow \quad + R_3PNi(CO)_3 \qquad (27\text{-}4)$$

η^3-Allymetal groups are also found in a wide variety of compounds containing more complex ligands, as building blocks in more elaborate structures. The following examples are illustrative.

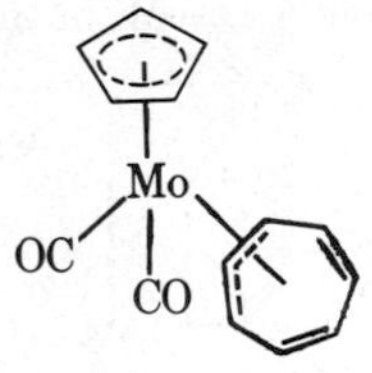

(27-XLIII)

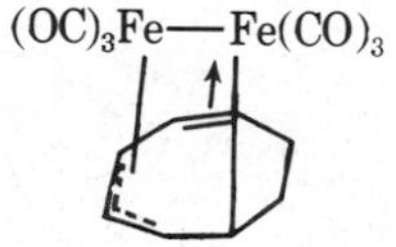

(27-XLIV)

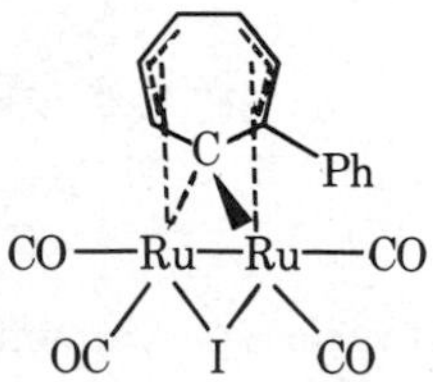

(27-XLV)

(27-XLVI)

In 27-XLV one carbon atom bridges two Ru atoms, and each Ru is bound to a η^3-allylic portion of the seven-membered ring.[107]

Finally, note that η^3-benzyl compounds of the type 27-XLVII are known.[108] Such species are probably intermediates in the reactions of benzyl compounds with metal complexes.

(27-XLVII)

$\eta(1-5)$ **Dienyl Complexes.** Cyclic alkenes with six, seven, eight, or more carbon atoms can give rise to $\eta(1-5)$-dienyl systems.

The most important are $\eta(1-5)$-cyclohexadienyls. These can be generated by addition or deletion of H^- from suitable systems as indicated by reactions 27-5 and 27-6. A considerable number of other complexes of the C_6H_7 group and substituted derivatives thereof are known.

The structure of the manganese complex formed in reaction 27-6 has been confirmed crystallographically.

A special case of 1,5-hexadienyl complexes consists of those with the $>CH_2$ groups replaced by $C=O$ (Section 4-25), $C-NR$, $C=CR_2$,[109a] or $SiMe_2$.[109b] In all of these compounds the five carbon atoms over which there is electron delocalization are essentially planar; the $>CH_2$, $>CNR$ or $>CO$ groups are bent out of this plane away from the metal, largely because of electronic factors.[100] The same bending away also occurs in the C_5H_6 complexes noted earlier (p. 1151).

107 J. A. Howard and P. Woodward, *J.C.S. Dalton,* **1977,** 366.
108 Y. Becker and J. K. Stille, *J. Am. Chem. Soc.,* 1978, **100,** 845.
109a J. F. Helling and W. A. Hendrickson, *J. Organomet. Chem.,* 1977, **141,** 99; D. W. Clack and L. A. P. Kane-Maguire, *J. Organomet. Chem.,* 1978, **145,** 201.
109b W. Fink, *Helv. Chim. Acta,* 1976, **59,** 276.

27-10.   Alkyne and Alkyne-Derived Complexes[110]

Acetylenes can be bound to metals using one or both sets of π-bonds. Complexes of the type 27-XLVIII represent the two extremes of η^2 bonding as discussed in Chapter 3, page 98. Examples of two such complexes are shown in Fig. 27-5.

(27-XLVIIIa)   (27-XLVIIIb)

The C—C bond lengths in η^2 acetylene compounds of type 27-XLVIIIb range from 1.285 Å in $(\eta\text{-}C_5H_5)_2Ti(CO)(Ph_2C_2)$ to 1.32 Å in $(Ph_3P)_2Pt(Ph_2C_2)$.[111] The C—C stretching frequencies also reflect the reduction in the bond order. Note that in all cases the R groups bend away from the metal. More commonly, however, the acetylene acts as a bridge between two metals joined by a metal-metal bond (27-XLIXa). Although the bonding could be conceived of as ethylenelike using the two sets of π bonds at right angles, there is structural evidence that rehybridization occurs to form relatively strong σ-bonds, as shown in 27-XLIXb.[112] Another form of bridge is 27-L.[112c]

(27-XLIXa)   (27-XLIXb)   (27-L)

Complexes of acetylenes are usually made by direct interaction with metal carbonyl or other complexes, e.g.,

[110] (a) S. Otsuka and A. Nakamura, *Adv. Organomet. Chem.,* 1976, **14**, 245 (acetylene and allene complexes: implications in homogeneous catalysis); (b) K. P. C. Vollhardt, *Acc. Chem. Res.,* 1977, **10**, 1 (metal-catalyzed acetylene cyclizations); (c) D. L. Thorn and R.Hoffmann, *Nouv. J. Chim.,* 1979, **3**, 39 (on MO theory of metallocycles, but contains many references to acetylene chemistry); (d) D. Seyferth *et al.,* in *Transition Metal Organometallics in Organic Synthesis,* vol. 2, H. Alper, Ed., Academic Press, 1978.

[111] For references, see G. Fachinetti *et al., J.C.S. Dalton,* **1978**, 1398.

[112] (a) See E. L. Muetterties *et al., J. Am. Chem. Soc.,* 1978, **100**, 2090 for structural data and references; (b) W. I. Bailey, Jr., *et al., J. Am. Chem. Soc.,* 1978, **100**, 5764; (c) R. S. Dickson *et al., J. Organomet. Chem.,* 1979, **166**, 385, *J.C.S. Chem. Comm.,* **1979**, 278.

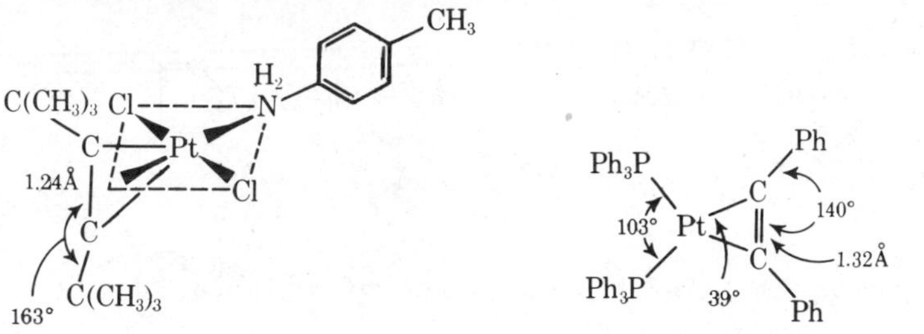

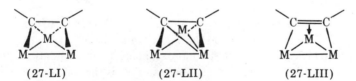

(a) (b)

Fig. 27-5. (a) The structure of a platinum complex in which the alkyne fills essentially the same role as an alkene. (b) The structure of a complex in which diphenylacetylene can be regarded as a divalent bidentate ligand.

$$(\eta\text{-}C_5H_5)_2Rh_2(CO)_3 + CF_3C{\equiv}CCF_3 \rightarrow (\eta\text{-}C_5H_5)_2Rh_2(CO)_2(CF_3C_2CF_3)$$
$$Pt(PPh_3)_3 + RC{\equiv}CR \rightarrow (Ph_3P)_2Pt(RC{\equiv}CR) + PPh_3$$
$$Co_2(CO)_8 + RC{\equiv}CR \rightarrow Co_2(CO)_6(\mu\text{-}RC{\equiv}CR) + 2CO$$
$$Ni(1,5\text{-}COD)_2 + RC{\equiv}CR \rightarrow (1,5\text{-}COD)_2Ni_2(\mu RC{\equiv}CR)$$

and acetylenes will also add across certain metal triple bonds, as in the reaction

$$(\eta\text{-}C_5H_5)_2Mo_2(CO)_4 + RC{\equiv}CR' \rightarrow (\eta\text{-}C_5H_5)_2Mo_2(CO)_4(\mu\text{-}RC{\equiv}CR')$$

Some complexes can be obtained by reduction of halides in presence of acetylenes,[113] e.g.,

$$(\eta\text{-}C_5H_5)_2MoCl_2 + C_2H_2 \xrightarrow{\text{Na/Hg}} (\eta\text{-}C_5H_5)_2Mo(C_2H_2)$$

Other types of compound in which alkynes can act as bridges are clusters[114] of the types 27-LI to 27-LIII.

(27-LI) (27-LII) (27-LIII)

Alkyne-Derived Compounds. Especially on interaction with metal carbonyls, but also with other complexes [e.g., $PdCl_2(PhCN)_2$], alkynes commonly react, and in doing so lose their individuality, forming carbocyclic or metallocyclic systems that may also incorporate CO, if CO is present. Some examples have been noted earlier in this chapter, but others are known.[110c,115] Some examples are the following:

[113] J. L. Thomas, *Inorg. Chem.*, 1978, **17**, 1507.
[114] See, e.g., A. J. Deeming, *J. Organomet. Chem.*, 1978, **150**, 123.
[115] See, e.g., M. Green et al., *J.C.S. Dalton*, **1978**, 1067; J. L. Atwood et al., *J. Am. Chem. Soc.*, 1976, **98**, 2454; P. M. Maitlis et al., *J.C.S. Dalton*, **1979**, 167, 178; E. L. Muetterties et al., *J. Am. Chem. Soc.*, 1978, **100**, 6966; S. Aime et al., *J.C.S. Dalton*, **1979**, 1155.

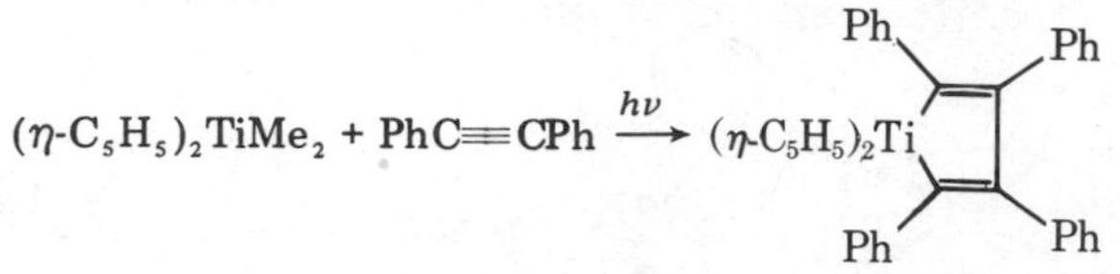

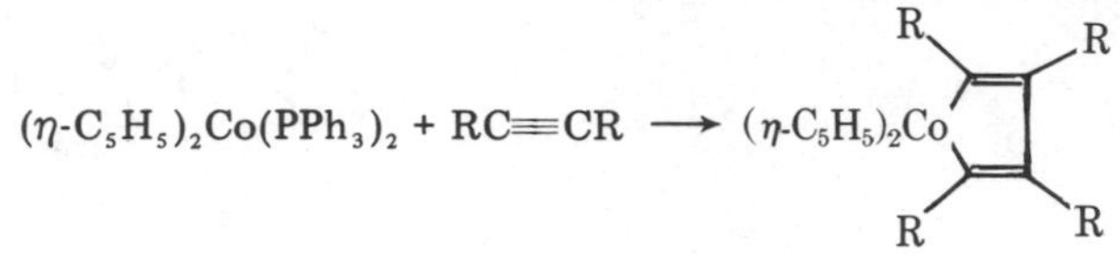

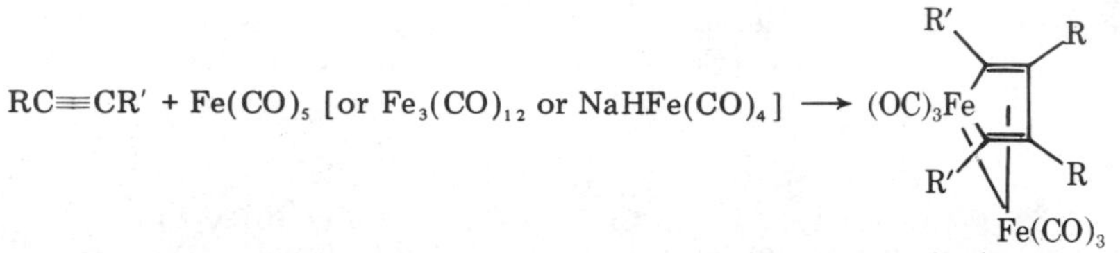

$$(\eta\text{-}C_5H_5)_2TiMe_2 + PhC\equiv CPh \xrightarrow{h\nu} (\eta\text{-}C_5H_5)_2Ti\begin{array}{c}Ph \quad Ph \\ \\ Ph \quad Ph\end{array}$$

$$(\eta\text{-}C_5H_5)_2Co(PPh_3)_2 + RC\equiv CR \longrightarrow (\eta\text{-}C_5H_5)_2Co\begin{array}{c}R \quad R \\ \\ R \quad R\end{array}$$

$$RC\equiv CR' + Fe(CO)_5 \text{ [or } Fe_3(CO)_{12} \text{ or } NaHFe(CO)_4] \longrightarrow (OC)_3Fe\begin{array}{c}R' \quad R \\ \\ R' \quad R \\ Fe(CO)_3\end{array}$$

R, R' = H, CH_3, Ph, or OH

$$Fe(CO)_5 + 2C_2H_2 \longrightarrow \underset{Fe(CO)_3}{\bigcirc}=O \quad + \quad \text{[Fe complex]}$$

$$Fe(CO)_5 + 2C_2(CH_3)_2 \xrightarrow{h\nu} O=\underset{\substack{H_3C \quad CH_3 \\ Fe(CO)_3}}{\bigcirc}=O$$

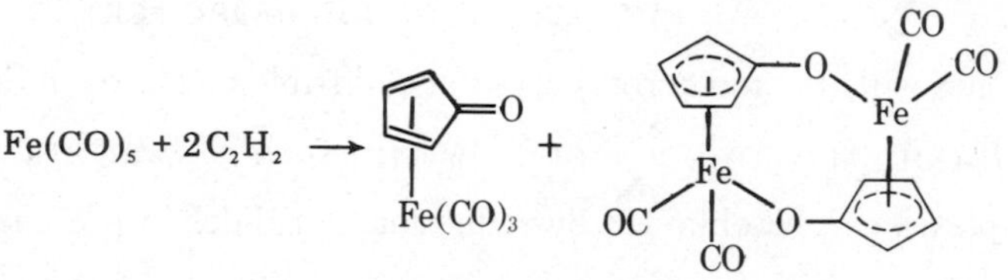

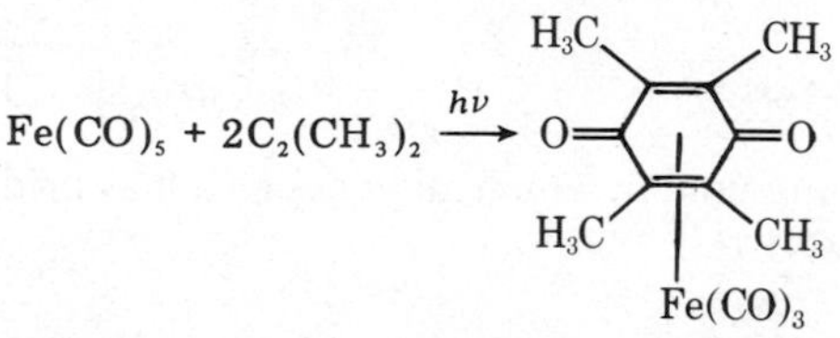

Many of the cyclopentadienone and quinone complexes can, of course be made directly from a metal carbonyl and the cyclic ketone itself.

There are many cases in which still more complex organic ligands are elaborated from alkyne starting materials. A striking example is the following:

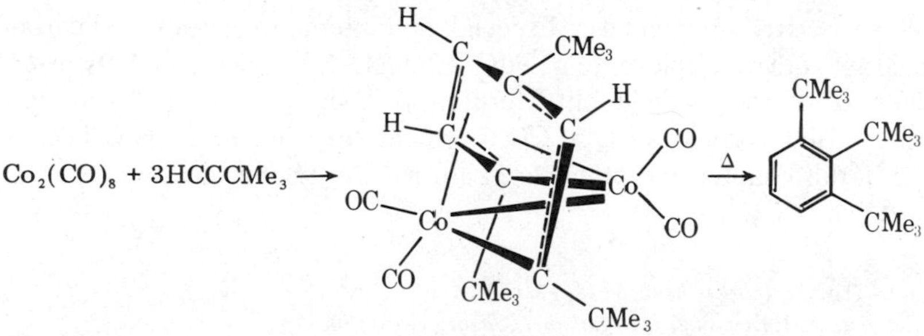

$$Co_2(CO)_8 + 3HCCCMe_3 \longrightarrow \text{[cluster complex]} \xrightarrow{\Delta} \text{[arene with CMe}_3\text{ substituents]}$$

(27-LIV)

In certain cases, the linking of acetylene units to give complexes such as 27-LIV may be confirmed by isolation of the intermediates. Thus in the addition of acet-

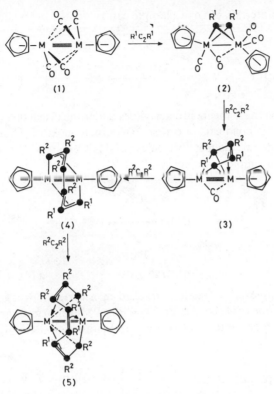

Fig. 27-6. Acetylene linking on dichromium and dimolybdenum complexes. [Reproduced by permission from F. G. A. Stone *et al., J.C.S. Chem. Comm.,* **1978**, 221.]

ylenes to triply bonded species noted above, the complexes with two and four acetylene units have been structurally characterized.[116] The reaction sequence is then that shown in Fig. 27-6.

The formation of four-membered carbocyclic compounds from acetylenes is noted later in this chapter. Finally, just as electron-rich alkenes can be cleaved to give carbenes (p. 1144), so an unusual cleavage of acetylenes occurs[117]:

$$Fe(CO)_5 + Et_2NC{\equiv}CNEt_2 \longrightarrow$$

In addition to forming metal compounds, acetylenes undergo an extremely wide range of reactions catalyzed by metal complexes, both with themselves, and with other multiple bonds, to give oligomers and polymers.[118] Some of these reactions

[116] F. G. A. Stone *et al., J.C.S. Chem. Comm.,* **1978**, 221.
[117] G. G. Cash *et al., J.C.S. Chem. Comm.,* **1977**, 30.
[118] See S. Otsuka and A. Nakamura, *Adv. Organomet. Chem.,* 1976, **14**, 245.

are noted in Chapter 30, but one example can be quoted here, namely, a pyridine synthesis[119]:

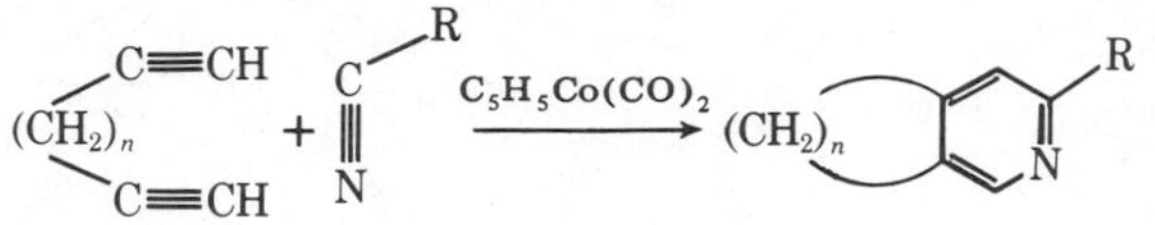

Acetylides. We have mentioned acetylides under the discussion of metal-carbon σ bonds. The triple bond can, of course, participate in further bonding as a two- or four-electron donor. Copper acetylides (21-H-XVII, p. 811) are polymeric for this reason, and a simple example[120] is 27-LV.

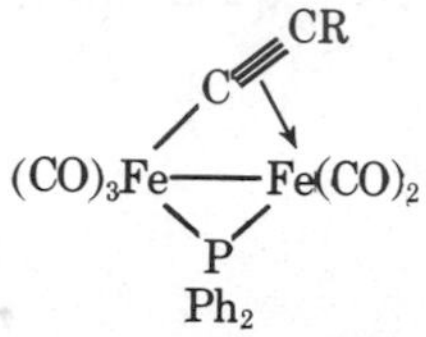

(27-LV)

Acetylide complexes are readily attacked by a variety of reagents.[121] For example, attack on the triple bond in $Pt(C{\equiv}CR)_2(PMe_2Ph)_2$ gives an alcoxycarbene probably by way of an intermediate vinyl carbonium ion, $Pt\overset{+}{C}{=}CHR$, viz.,

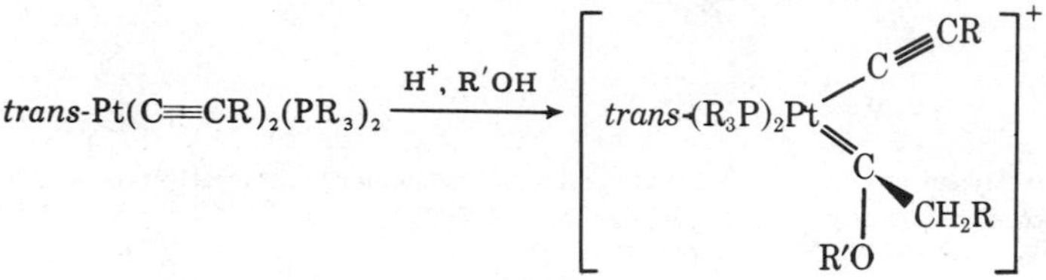

CARBOCYCLIC π COMPLEXES[122]

The general features and bonding in these compounds were discussed in Chapter 3 (p. 99).

The first carbocyclic compound to be recognized, the now-celebrated molecule dicyclopentadienyl iron(II) or ferrocene, was shown to have a sandwich structure with η^5- or π-bonded rings. The only true sandwich compounds with parallel rings are those of the types $(\eta\text{-}C_5H_5)_2M$ and $(\eta\text{-}C_5H_5)_2M^+$; others like $(\eta\text{-}C_5H_5)_2TiCl_2$ have rings at an angle. True sandwich compounds are also formed by arenes [e.g., $(C_6H_6)_2Cr$], by the $C_7H_7^-$ and $C_8H_8^{2-}$ ions in $(\eta\text{-}C_5H_5)(\eta\text{-}C_7H_7)V$ or $(\eta\text{-}C_8H_8)_2U$ respectively, and by some heterocyclic compounds.

[119] A. Naiman and K. P. C. Vollhardt, *Angew. Chem., Int. Ed.,* 1977, **16**, 708.

[120] W. F. Smith *et al., Inorg. Chem.,* 1977, **16**, 1593; A. J. Carty *et al., J. Am. Chem. Soc.,* 1978, **100**, 3051.

[121] M. L. Chisholm *et al., Inorg. Chem.,* 1977, **16**, 677, 698, 2177.

[122] P. L. Pauson, *Coordination Chemistry,* Vol. 17, IUPAC, Pergamon Press, 1977 (the first 25 years).

All carbocyclic or heterocyclic compounds can form compounds with only one π ring, the remainder of the coordination sites being occupied by CO, NO, PR_3, Cl, etc.

27-11. Three- and Four-Membered Rings[123a]

3-Rings. Compounds of nickel and cobalt[123b] are known for C_3Ph_3 and are made by the reactions

$$C_3Ph_3BF_4 + Co_2(CO)_8 \rightarrow (\eta C_3Ph_3)Co(CO)_3$$

$$Ni(CO)_4 + C_3Ph_3X \cdot (\eta\text{-}C_3Ph_3)Ni(CO)_2X \xrightarrow{py} (\eta\text{-}C_3Ph_3)NiXpy_2\text{-}py$$

$$(\eta\text{-}C_3Ph_3)NiXpy_2\text{-}py \xrightarrow{C_5H_5Tl} (\eta\text{-}C_3Ph_3)(\eta\text{-}C_5H_5)Ni$$

X-Ray crystallographic studies have confirmed the symmetrical *trihapto* character of the $(C_3Ph_3)Ni$ bonding.

4-Rings. It is well known that cyclobutadiene (C_4H_4) is antiaromatic and unstable in the free state, but it can be stabilized by bonding to a metal atom with suitable electron configuration. The first such compound prepared was the methyl-substituted nickel compound obtained as in Fig. 27-7.

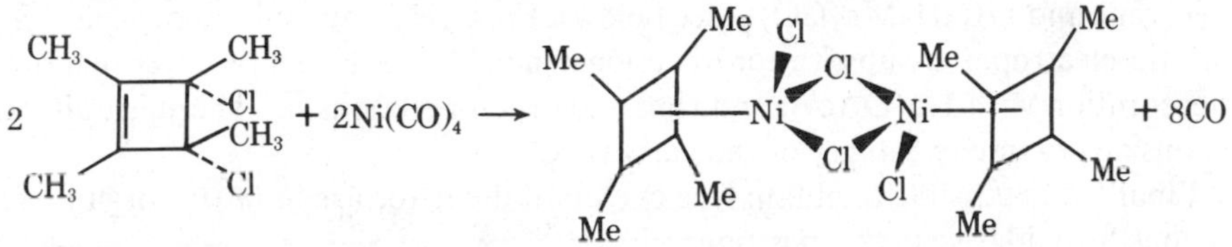

Fig. 27-7. The preparation and structure of $[(CH_3C)_4NiCl_2]_2$.

This chloride can then be reduced in presence of bipyridine to give the nickel(0) complex (Me_4C_4)Nibipy, but only recently has the first true sandwich compound been obtained[124] as a blue solid resistant to air, water, and CO:

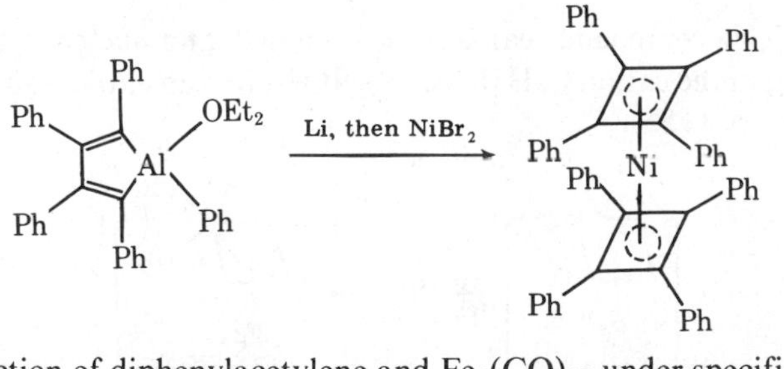

The interaction of diphenylacetylene and $Fe_3(CO)_{12}$ under specified conditions gives $Ph_4C_4Fe(CO)_3$; diacetylenes also react[125]:

123a A. Efraty, *Chem. Rev.,* 1977, **77**, 691; R. Pettit, *J. Organomet. Chem.,* 1975, **100**, 205.
123b T. Chiang *et al., Inorg. Chem.,* 1979, **18**, 1687.
124 H. Hoberg *et al., Angew. Chem., Int. Ed.,* 1978, **17**, 950, 123.
125 M. D. Rausch *et al., J.C.S. Chem. Comm.,* **1978**, 187.

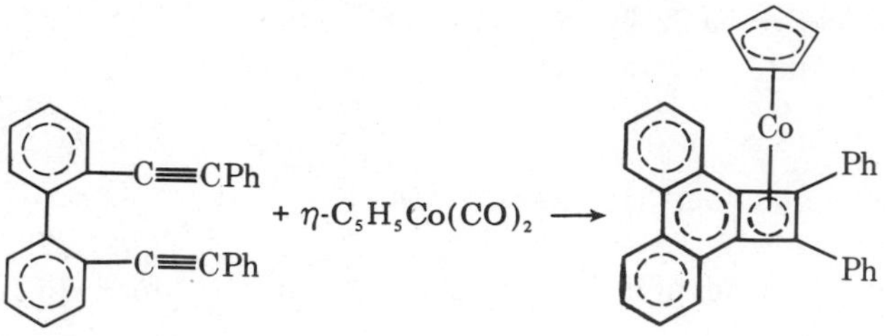

Unsubstituted cyclobutadiene compounds can be made by the reactions[126]:

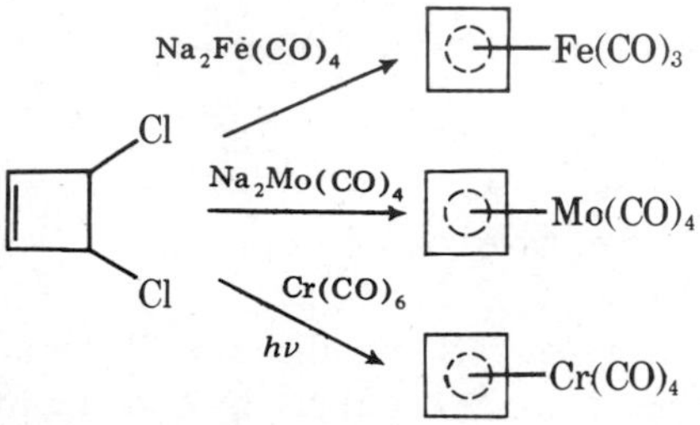

Structural studies on $C_4H_4Fe(CO)_3$ and $C_4H_4Co(\eta\text{-}C_5H_5)$[127] show a square ring; the iron compound has a C—C bond distance of 1.46 Å. Like the C_5H_5 ring in ferrocene and $\eta\text{-}C_5H_5Mn(CO)_3$ (see below), the C_4H_4 ring undergoes typical aromatic electrophilic substitution reactions such as the Friedel-Crafts reaction, mercuration with $Hg(OAc)_2$, and ring expansion reactions.[128] There is quite an extensive organic chemistry of the derivatives.

Finally it is possible to obtain free cyclobutadiene for use in *in situ* organic reactions by oxidative decomposition:

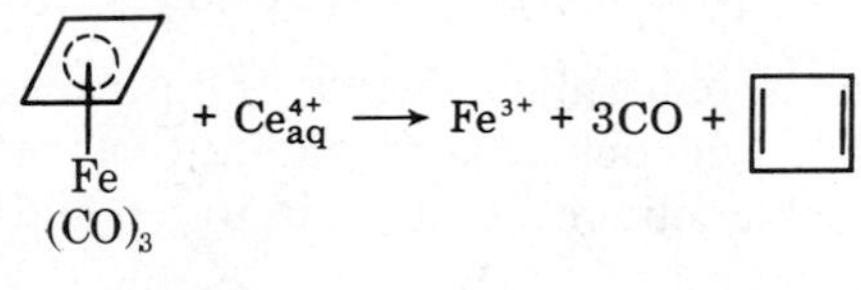

Cationic C_4H_4 compounds can also be obtained, and nucleophilic addition to the C_4H_4 ring in the cation $C_4H_4Fe(CO)_2NO^+$ gives an *exo-η^3-allyl* (cf. attacks on other rings, p. 1155).[129]

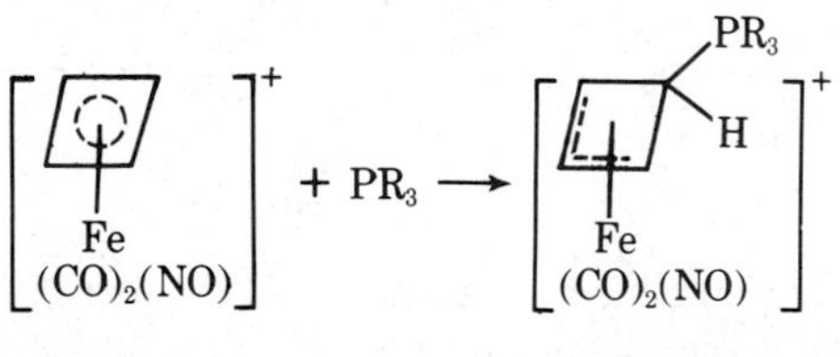

[126]   R. H. Grubbs and T. A. Pancoast, *Synth. React. Inorg. Metal. Org. Chem.*, 1978, **8**, 1.
[127]   P. E. Riley and R. E. Davis, *J. Organomet. Chem.*, 1976, **113**, 157.
[128]   A. Bond *et al.*, *J.C.S. Dalton*, **1977**, 2372.
[129]   D. A. Sweigart *et al.*, *J. Organomet. Chem.*, 1978, **152**, 187; A. Efraty *et al.*, *Inorg. Chem.*, 1977, **16**, 2522; C. E. Chidsey *et al.*; *J. Organomet. Chem.*, 1979, **169**, C12.

27-12.   Five-Membered Rings[130]

Cyclopentadienyl compounds are by far the most important of all carbocyclic π complexes, and the $C_5H_5^-$ ligand or substituted derivatives such as $MeC_5H_4^-$ or $Me_5C_5^-$[131] are widely used.

The formation and synthetic methods for cyclopentadienyl compounds depends on the fact that cyclopentadiene is a weak acid ($pK_a \sim 20$) and with strong bases gives salts of the symmetrical cyclopentadienide ion $C_5H_5^-$. Synthetic methods are as follows.

1.   The most general method is the formation of the sodium salt by action of Na or NaH on C_5H_6 in tetrahydrofuran and the subsequent reaction of this solution with metal halides, carbonyls, etc.

$$2C_5H_6 + Na = C_5H_5^- + Na^+ + H_2 \text{ (main reaction)}$$
$$3C_5H_6 + 2Na = 2C_5H_5^- + 2Na^+ + C_5H_8$$
$$FeCl_2 + 2C_5H_5Na = (\eta\text{-}C_5H_5)_2Fe + 2NaCl$$
$$W(CO)_6 + C_5H_5Na = Na^+[\eta\text{-}C_5H_5W(CO)_3]^- + 3CO$$

An alternative is to use the thallium salt, which is readily prepared and stored:

$$TlOH + C_5H_6 = TlC_5H_5 + H_2O$$

2.   A strong organic base, preferably diethylamine in excess, can be used as an acceptor for HCl, e.g.,

$$2C_5H_6 + CoCl_2 + 2Et_2NH \xrightarrow{THF} (\eta\text{-}C_5H_5)_2Co + 2Et_2NH_2Cl$$

3.   In some cases direct interaction of cyclopentadiene or even dicyclopentadiene with metals or metal carbonyls gives the complex, e.g.,

$$2C_5H_6(g) + Mg(s) \xrightarrow{\Delta} (C_5H_5)_2Mg + H_2$$

$$Fe(CO)_5 + C_{10}H_{12} \rightarrow [(\eta\text{-}C_5H_5)Fe(CO)_2]_2$$

Since the $C_5H_5^-$ anion functions as a uninegative ligand, the compounds are of the type $[(\eta\text{-}C_5H_5)_2M]X_{n-2}$, where the oxidation state of the metal M is n and X is a uninegative ion. Hence in the II oxidation state we obtain neutral, sublimable, and organic solvent-soluble molecules like $(\eta\text{-}C_5H_5)_2Fe$ and $(\eta\text{-}C_5H_5)_2Cr$, and in III, IV, and V oxidation states species such as $(\eta\text{-}C_5H_5)_2Co^+$, $(\eta\text{-}C_5H_5)_2TiCl_2$, and $(\eta\text{-}C_5H_5)_2NbBr_3$, respectively. Some representative compounds are given in Table 27-2.

[130]   P. C. Bharara et al., J. Organomet. Chem., Library, No. 5, 1977, 259 (C_5H_5 compounds with simple ligands); J. W. Lauther and R. Hoffmann, J. Am. Chem. Soc., 1976, 98, 1729 (bonding in Cp$_2$M compounds); E. Maslowsky, Jr., J. Chem. Educ., 1978, 55, 276 (structures); J. D. L. Holloway and W. E. Geiger, Jr., J. Am. Chem. Soc., 1979, 101, 2038 (electron transfer reactions); J. T. Sheats, Organometallic Chemistry Reviews, Vol. 7, Elsevier, 1979 (cobalt compounds).

[131]   D. Feitler and G. M. Whitesides, Inorg. Chem., 1976, 15, 466; R. S. Threlkel and J. E. Bercaw, J. Organomet. Chem., 1977, 136, 1; M. L. H. Green and R. D. Pardy, J.C.S. Dalton, 1979, 354; P. M. Maitlis et al., J.C.S. Dalton, 1979, 371–387.

TABLE 27-2
Some Di-η^5-cyclopentadienylmetal Compounds

Compound	Appearance; m.p. (°C)	Unpaired electrons	Other properties[a]
$(\eta^5\text{-}C_5H_5)_2Fe$	Orange crystals; 174	0	Oxidized by $Ag^+(aq)$, dil. HNO_3; $Cp_2Fe^+ = Cp_2Fe$, $E^0 = -0.3$ V (vs. SCE); stable thermally to >500°
$(\eta^5\text{-}C_5H_5)_2Cr$	Scarlet crystals; 173	2	Very air sensitive; soluble in HCl; giving C_5H_6 and blue cation, probably $[\eta^5\text{-}C_5H_5CrCl(H_2O)_n]^+$
$(\eta^5\text{-}C_5H_5)_2Ni$	Bright green; decomposes at 173°	2	Fairly air stable as solid; oxidized to Cp_2Ni^+; NO gives CpNiNO; Na/Hg in C_2H_5OH gives $CpNiC_5H_7$
$(\eta^5\text{-}C_5H_5)_2Co^+$	Yellow ion in aqueous solution	0	Forms numerous salts and a stable, strong base (absorbs CO_2 from air); thermally stable to ~400°
$(\eta^5\text{-}C_5H_5)_2TiCl_2$	Bright red crystals; 230	0	Slightly soluble in H_2O giving Cp_2TiOH^+; C_6H_5Li gives $Cp_2Ti(C_6H_5)_2$; reducible to Cp_2TiCl; Al alkyls give polymerization catalyst
$(\eta^5\text{-}C_5H_5)_2WH_2$	Yellow crystals; 163	0	Moderately stable in air, soluble benzene, etc.; soluble in acids giving $Cp_2WH_3^+$ ion

[a] $Cp = \eta\text{-}C_5H_5$.

Neutral compounds are known for V, Cr, Mn, Fe, Co and Ni. The manganese (p. 749) and titanium (p. 706) compounds are anomalous. Although Cp_2Mn does have the sandwich structure, it behaves as an ionic cyclopentadienide and is, for example, miscible in all proportions with Cp_2Mg, whereas Cp_2Fe and the others are not.

Although Zr and Hf form compounds much like Ti, the other second- and third-row transition metals tend to behave differently from the first row and, except for Ru and Os, do not form ferrocene analogues. However like "Cp_2W" they may exist as intermediates in reactions.[132a] Such reactive intermediates have "carbene"-like behavior.[132b] Although Cp_2V is stable, it illustrates this behavior by undergoing oxidative-addition reactions with halogens, CS_2, acetylenes, etc., as well as adding CO, e.g.,

[132a] M. L. H. Green et al., J.C.S. Chem. Comm., 1978, 99.

[132b] See, e.g., G. Fachinetti et al., J.C.S. Dalton, 1977, 2297; 1976, 203, 1046; Inorg. Chem., 1979, 18, 2282.

The cationic species, several of which can exist in acidic aqueous solutions, behave like other large unipositive ions (e.g., Cs^+) and can be precipitated by silicotungstate, $PtCl_6^{2-}$, BPh_4^-, and other large anions. The $(\eta-C_5H_5)_2Co^+$ ion is remarkably stable and is unaffected by concentrated sulfuric and nitric acids, even on heating.

Chiral complexes[133] that have titanium as the sole center of chirality or possess chiral centers by virtue of asymmetric C_5H_5 ring or metal substituents are known. Some examples are

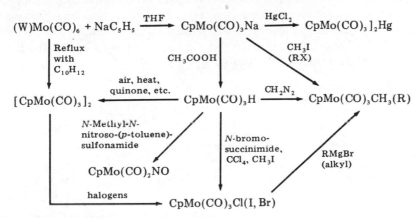

We cannot discuss the extensive chemistry of *monocyclopentadienyl compounds*, but some representative reactions and syntheses are shown in Fig. 27-8.[134]

Fig. 27-8. Some reactions of monocyclopentadienyl compounds of molybdenum and tungsten.

A particularly widely used compound is the iron carbonyl made by the reaction:

$$Fe(CO)_5 + C_{10}H_{12} \xrightarrow{200°} [CpFe(CO)_2]_2$$

This carbonyl-bridged dimer can be cleaved by sodium to give the anion $CpFe(CO)_2^-$, which is one of the strongest of organometallic nucleophiles. Reactions of this ion have allowed the synthesis of chiral compounds[135] such as 27-LVI.

133   J. Tirouflet *et al., J. Organomet. Chem.*, 1975, **101**, 71; *J. Am. Chem. Soc.*, 1975, **97**, 6272.

134   For references and examples, see M. L. H. Green *et al., J.C.S. Dalton*, **1977**, 2189; B. E. R. Schilling *et al., J. Am. Chem. Soc.*, 1979, **101**, 585, 592; A. R. Pinhas and R. Hoffmann, *Inorg. Chem.*, 1979, **18**, 654.

135   G. M. Reisner *et al., Inorg. Chem.*, 1978, **17**, 783; T. C. Flood *et al., J. Am. Chem. Soc.*, 1978, **100**, 7271.

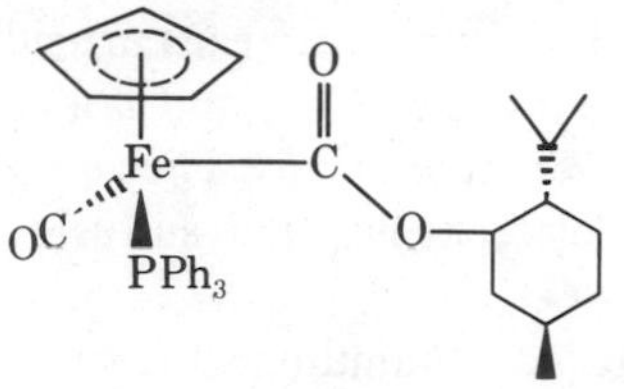

(27-LVI)

Reactions of the C_5H_5 Ring. The C—C bond orders in η-C_5H_5 compounds resemble those of benzene. Most cyclopentadienyl compounds do not however survive the reaction conditions involved in aromatic substitution and similar reactions.

For *ferrocene*, which to a large extent does survive, there is a vast organic chemistry.[136] It undergoes Friedel-Crafts acylation, sulfonation, and metallation[137] by butyllithium, etc. The Friedel-Crafts acylation occurs 3×10^6 times faster than with benzene. The cyclopentadienyl carbonyl η-$C_5H_5Mn(CO)_3$, which has been given the trivial name *cymantrene*, also has an extensive derivative chemistry.

In certain reactions of the molecules the metal atom is directly involved. One example is the intramolecular bonding of ferrocene alcohols, another is protonation by very strong acids to give a cation with bent rings and an Fe—H bond, Cp_2FeH^+.[138] The protonation of Cp_2Ni is discussed later (p. 1172). It has been proposed that electrophilic attacks on Cp_2Fe and $CpMn(CO)_3$ proceed via initial attack on the metal:

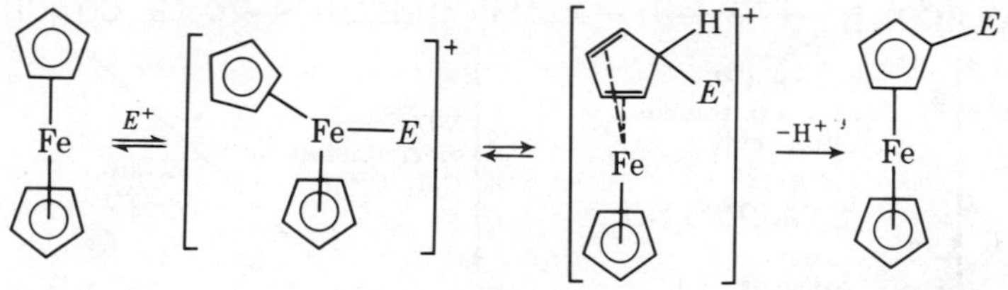

The intermediate cation with the electrophile E bound to the metal rearranges to a cyclopentadiene complex with E in the endo (i.e., metal side) position, which then loses a proton to give the substituted ferrocene.

However since ferrocene is such a weak base to H^+, it is also possible that except in strong acids, attack is on the ring as in benzene and other aromatic systems.

For the monocyclopentadienyl $CpCo(PMe_3)_2$ the metal atom is extremely nucleophilic and gives the hydride $[CpCoH(PMe_3)_2]^+$ with acids as weak as H_2O or MeOH![139] It also reacts with acyl chlorides to give $[CpCo(COR)(PMe_3)_2]^+$.

Ionic Cyclopentadienides. The sodium salts noted above and compounds of other

136   Gmelin, *Handbuch der anorganische Chemie,* Vol. 49, Part A. See also, *Annual Reviews of Organometallic Chemistry.*

137   W. Walczak *et al., J. Am. Chem. Soc.,* 1978, **100,** 6382; A. G. Osborne and R. H. Whiteley, *J. Organomet. Chem.,* 1978, **162,** 79.

138   G. Cerichelli *et al., J. Organomet. Chem.,* 1977, **127,** 357; T. E. Bitterwolf and A. C. Ling, *J. Organomet. Chem.,* 1977, **141,** 355.

139   H. Werner and W. Hoffmann, *Angew. Chem., Int. Ed.,* 1978, **17,** 464.

electropositive metals such as Mg, Ca, Sr, Y, and the lanthanides are typically very reactive toward air, water, and other substances with active hydrogen atoms.

The divalent compounds such as $(C_5H_5)_2Mg$ have a ferrocenelike structure that is the one most favored on electrostatic grounds. Structure is *not* a criterion of bond type in cyclopentadienyl compounds. There has been much discussion of the structure of $(C_5H_5)_2Be$; but it appears to have one σ ring and one π ring in a sort of "slipped" sandwich fashion.[140]

Other Types of Cyclopentadienyl. There are numerous σ-cyclopentadienyl compounds in which the metal is bound to only one carbon of the ring.[141a] Such compounds show unusual nmr spectra and are nonrigid ("ring whizzers") as discussed in Chapter 28. A good example is Cp_4Zr or $(\eta^5\text{-}Cp)_3Zr(\eta^1\text{-}Cp)$.[141b]

In some complexes, the C_5H_5 group can act as a bridge between two metal-metal bonded atoms,[142] e.g.,

Pt + Pt(PPr$^i_3$)$_2$ → (Pr$^i_3$P)Pt—Pt(PPr$^i_3$)

Me   Me

Finally, in $(C_5H_5)_2W(CO)_2$ one ring may be regarded as η^3—if the eighteen-electron rule is to be obeyed.[143a]

Cyclopentadienyl compounds have, so far, found few uses. $(CH_3C_5H_4)Mn(CO)_3$, a yellow liquid, is the only practical alternative to tetraethyl lead as an antiknock agent in gasoline.[143b] Dicyclopentadienylchromium has been used to make Cr on alumina catalysts for ethylene polymerization. Dicyclopentadienylzirconium hydrides are used in the hydrozirconation reaction (Section 29-10). Ferrocene has been used as a combustion catalyst, and the ferrocene unit is built into polymers of various types.

27-13. Six-Membered Rings[144]

The first $(C_6H_6)M$-containing compounds were prepared as long ago as 1919, but their true identities were not recognized until 35 years later. A series of chromium arene compounds was first obtained by Hein by the reaction of $CrCl_3$ with C_6H_5MgBr; they were formulated by Hein as "polyphenylchromium" compounds,

140   N.-S. Chiu and L. Schäfer, *J. Am. Chem. Soc.*, 1978, **100**, 2604.
141a See, e.g., A. J. Campbell *et al., Inorg. Chem.*, 1976, **15**, 1326.
141b R. D. Rogers *et al., J. Am. Chem. Soc.*, 1978, **100**, 5238.
142   H. Werner and H.-J. Kraus, *J.C.S. Chem. Comm.*, **1979**, 814.
143a G. Hüttner *et al., J. Organomet. Chem.*, 1978, **145**, 329.
143b *Chem. Tech.*, **1978**, 484.
144   W. E. Silverthorne, *Adv. Organomet. Chem.*, 1975, **13**, 48; V. Graves and J. J. Lagowski, *Inorg. Chem.*, 1976, **15**, 577 (Cr); M. A. Bennett, in *Rodd's Chemistry of Carbon Compounds*, 2nd ed., Part B, 1974; R. G. Gastinger and K. J. Klabunde, *Trans. Met. Chem.*, 1979, **41**, 1 (Group VIII).

viz., $(Ph)_n Cr^{0,1+}$ ($n = 2$, 3, or 4). They actually contain "sandwich"-bonded $C_6 H_6$ and $C_6 H_5 - C_6 H_5$ groups, as, for example, in 27-LVII.

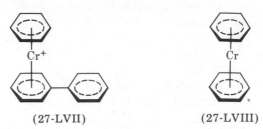

(27-LVII)   (27-LVIII)

The prototype neutral compound *dibenzenechromium* [$(C_6 H_6)_2 Cr$: 27-LVIII] has also been obtained from the Grignard reaction of $CrCl_3$, but a more effective method of wider applicability to other metals is the direct interaction of an aromatic hydrocarbon and a transition metal halide in presence of Al powder as a reducing agent and halogen acceptor and $AlCl_3$ as a Friedel-Crafts-type activator. Although the neutral species are formed directly in the case of chromium, the usual procedure is to hydrolyze the reaction mixture with dilute acid, which gives the cations $(C_6 H_6)_2 Cr^+$, $(mesitylene)_2 Ru^{2+}$, etc. In several cases these cations can be reduced to the neutral molecules by reducing agents such as hypophosphorous acid.

Dibenzenechromium, which forms dark brown crystals, is much more sensitive to air than is ferrocene, with which it is isoelectronic; it does not survive the reaction conditions of aromatic substitution. X-Ray and electron diffraction measurements on dibenzenechromium and other arene complexes show that the C—C bond lengths are equivalent.

A method by which some benzenoid complexes may be obtained involves cyclization of doubly substituted acetylenes, *viz.*,

$$(C_6 H_5)_2 Mn + 6 CH_3 C \equiv CCH_3 \rightarrow [C_6 (CH_3)_6]_2 Mn$$

Bisarene compounds can also be obtained by metal atom cocondensation reactions.[145a] These are useful for compounds that are difficult or impossible to make other ways, e.g., $(C_6 H_6)_2 Ti$, $(p\text{-}C_6 H_4 F_2)_2 V$, $(C_6 H_6)_2 Nb$, $(C_6 H_6)_2 W$, and $(C_6 H_6)_2 Ru$. Just as $(C_5 H_5)_2 MX_n$ compounds have bent sandwich structures, the compounds $(C_6 H_6)_2 Hf(PMe_3)$[145b] and $(C_6 H_5 F)_2 WH^+$ have the rings at an angle.

Although dibenzenechromium will undergo metallation with LiBuTMED, as well as base-catalyzed H–D exchange, other benzenelike reactions usually lead to decomposition of neutral bisarene compounds, since the electron-rich metal atom is usually attacked preferentially. However fluorinated compounds can undergo nucleophilic substitutions with loss of fluoride ion, as does $C_6 F_6$ itself,[146] e.g.,

[145a] P. L. Timms and R. B. King, *J.C.S. Chem. Comm.*, **1978**, 898; K. J. Klabunde et al., *Trans. Metal Chem.*, 1979, **4**, 1; M. L. H. Green et al., *J.C.S. Dalton*, **1975**, 1419; *J.C.S. Chem. Comm.*, **1978**, 72, 431; H. F. Efner et al., *J. Organomet. Chem.*, 1978, **146**, 45.

[145b] F. G. N. Cloke and M. L. H. Green, *J.C.S. Chem. Comm.*, **1979**, 127.

[146] M. J. McGlinchey et al., *J. Organometallic Chem.*, 1977, **141**, 85; *J. Am. Chem. Soc.*, 1976, **98**, 2271.

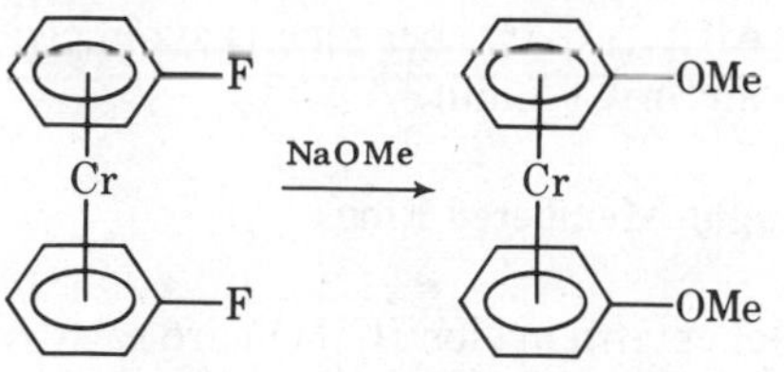

$(C_6H_6)_2W$ is reversibly protonated by dilute acids to give $(C_6H_6)_2WH^+$. Cations like $(\eta\text{-}C_6H_6)_2Fe^{2+}$ or $\eta\text{-}C_6H_6Mn(CO)_3^+$ can undergo additions of C and H nucleophiles; thus the latter with H^- gives η-1-5-cyclohexadienylmanganesetricarbonyl (see p. 1155).

There is a variety of *monoarene* compounds. These have been made by reactions such as

$$C_6H_6 + Mn(CO)_5Cl + AlCl_3 \rightarrow C_6H_6Mn(CO)_3^+AlCl_4^- + 2CO$$
$$C_6H_6 + Cr(CO)_6 \rightarrow C_6H_6Cr(CO)_3 + 3CO$$
$$Co + C_6F_5Br + C_6H_5Me \xrightarrow{\text{cocondense}} (\eta\text{-MeC}_6H_5)Co(C_6F_5)_2{}^{147}$$

Finally certain phenyl compounds, notably $PPh_3{}^{148a}$ and the BPh_4^- ion,[148b] can be π-bonded in compounds such as $[RuH(\eta\text{-}C_6H_5PPh_2)(PPh_3)_2]^+$ and (C_7H_8)-$Rh(\eta\text{-}C_6H_5BPh_3)$.

η^4-**Arenes.** Arenes need not always be symmetrically bound, and where η^6 bonding would lead to the eighteen-electron rule being broken, bent, η^4, or diene-type complexes such as 27-LIX can be formed.[149a]

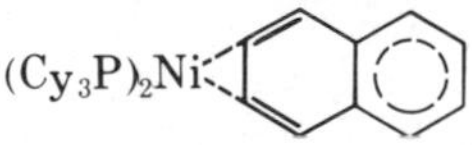

(27-LIXa)

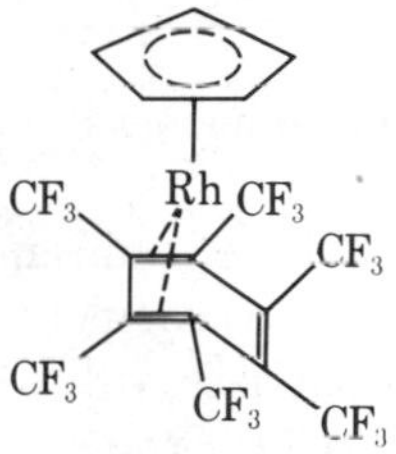

(27-LIXb)

The compound $(\eta^6\text{-}C_6Me_6)Ru(\eta^4\text{-}C_6Me_6)$ is fluxional with interchanging rings. Such species are believed to be intermediates in the homogeneous catalytic hydrogenation of arenes.[149b]

Weakly bound complexes are also formed between arenes and Ag^+, Cu^I, Sn^{II}, etc., in which the metal ion is mostly centered over one double bond of the arene.

[147] B. B. Anderson *et al., J. Am. Chem. Soc.,* 1976, **98,** 5350.

[148a] D. J. Cole-Hamilton *et al., J.C.S. Dalton,* **1976,** 1995.

[148b] L. A. Oro *et al., J. Organomet. Chem.,* 1978, **148,** 81; M. B. Hossain and D. Van der Helm, *Inorg. Chem.,* 1978, **17,** 2893.

[149a] D. J. Brauer and C. Krüger, *Inorg. Chem.,* 1977, **16,** 882; J. O. Albright *et al., J. Am. Chem. Soc.,* 1979, **101,** 611.

[149b] E. L. Muetterties *et al., Inorg. Chem.,* 1979, **18,** 881; *J. Am. Chem. Soc.,* 1978, **100,** 7425.

However in $C_6H_6Pb(AlCl_6)_2 \cdot C_6H_6$, benzene is symmetrically bound; a similar tin(II) compound has an $Sn_2Cl_2^{2+}$ unit.[150]

27-14. Seven- and Eight-Membered Rings

Complexes of the cycloheptatrienyl ion $(C_7H_7^+)$ are relatively scarce.[151] They can be obtained by reactions such as:

$$C_7H_8Mo(CO)_3 + Ph_3C^+BF_4 \rightarrow [\eta\text{-}C_7H_7Mo(CO)_3]^+BF_4^- + Ph_3CH$$

$$\eta\text{-}C_5H_5V(CO)_4 + C_7H_8 \xrightarrow{\text{reflux}} (\eta\text{-}C_5H_5)V(\eta\text{-}C_7H_7)$$

Metal atom syntheses using Ti, V, Cr, Fe, etc., and cycloheptatriene give compounds of the type 27-LX with an η-1–5-dienyl as well as an η-C_7H_7 ring.[152]

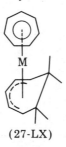

(27-LX)

The interaction of dilithium cyclooctatetraenide and halides of uranium(IV), cerium(IV), and other f-group halides gives compounds like *uranocene* [(η-$C_8H_8)_2U$], which has a true sandwich structure[153] (p. 1021).

27-15. Metal Compounds of Heterocycles

Heterocycles that have significant aromatic character can also form sandwich-type bonds to metals. Thus compounds containing thiophene,[154] e.g., $C_4H_4SCr(CO)_3$ and the anion of pyrrole, viz., $C_4H_4NMn(CO)_3$, are analogous to $C_6H_6Cr(CO)_3$ and η-$C_5H_5Mn(CO)_3$, respectively. *Azaferrocene* (η-cyclopentadienyl-η-pyrro-lyliron) has also been obtained as red crystals (m.p. 114°) of lower thermal stability than ferrocene:

$$\eta\text{-}C_5H_5Fe(CO)_2I + C_4H_4NK$$

$$\xrightarrow{\text{benzene}}$$

$$\eta\text{-}C_5H_5Fe\text{-}\eta\text{-}C_4H_4N$$

$$\xrightarrow{\text{THF}}$$

$$FeCl_2 + C_5H_5Na + C_4H_4NNa$$

[150] E. L. Amma, *et al., Inorg. Chem.,* 1979, **18,** 751.

[151] E. F. Ashworth *et al., J.C.S. Dalton,* **1977,** 1693; M. Green *et al., J.C.S. Dalton,* **1977,** 1755.

[152] P. L. Timms and T. W. Turney, *J.C.S. Dalton,* **1976,** 2021.

[153] J. A. Butcher *et al., J. Am. Chem. Soc.,* 1978, **100,** 1012; C. A. Harmon *et al., Inorg. Chem.,* 1977, **16,** 2143; K. D. Warren, *Struct. Bonding,* 1977, **33,** 97 (theory of f-orbital sandwiches).

[154] M. J. H. Russell *et al., J.C.S. Dalton,* **1978,** 857.

Azaferrocene appears to be isomorphous with ferrocene, so that in the crystal the N atoms must be randomly placed in any of the ten positions occupied by carbon in the ferrocene molecule.

Phosphaferrocenes[155a] have been made by reactions involving cleavage of a P—Ph bond, e.g.,

Arsaferrocenes are also known.[155b] There are also planar η^6 compounds[156] such as 27-LXI. Other η^6 heterocyclic compounds can be made from *pyridines* with

(27-LXI)

orthosubstituents to inhibit coordination through nitrogen, such as $Cr(\eta\text{-}Me_2C_5H_5N)_2$.[157]

Boron-containing compounds are those from *boroles*[158] such as 27-LXII and 27-LXIII and the compounds from Δ^4,1,3,2-diazaborolene (27-LXIV)[159a] and 1,3-diborolene (27-LXV).[159b]

(27-LXII)   **(27-LXIII)**

155a G. De Lanzon *et al., J. Organomet. Chem.,* 1978, **156,** C33.
155b G. Thiollet *et al., Inorg. Chim. Acta,* 1979, **32,** L67.
156   L. Lückoff and K. Dimoth, *Angew. Chem., Int. Ed.,* 1976, **15,** 503; T. Debaerdemacker, *Angew. Chem., Int. Ed.,* 1976, **15,** 504.
157   L. H. Simons *et al., J. Am. Chem. Soc.,* 1976, **98,** 1044.
158   R. Goetze and H. Nöth, *J. Organomet. Chem.,* 1978, **145,** 151; G. E. Herberich *et al., Angew. Chem., Int. Ed.,* 1977, **16,** 42; C. W. Allen and D. E. Palmer, *J. Chem. Educ.,* 1978, **55,** 497.
159a G. W. Schmidt and J. Schulze, *Angew. Chem., Int. Ed.,* 1977, **16,** 249.
159b W. Siebert *et al., Chem. Ber.,* 1978, **111,** 356; *Z. Naturforsch.,* 1978, **33b,** 1410.

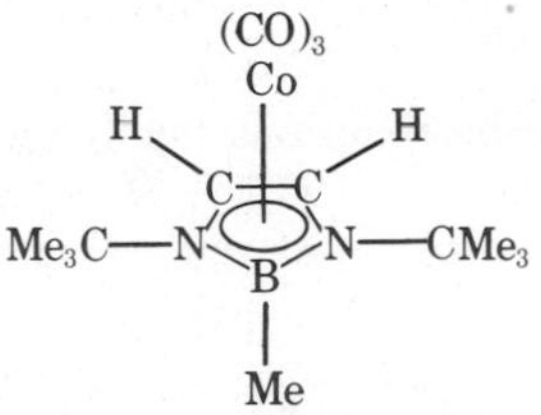

(27-LXIV) 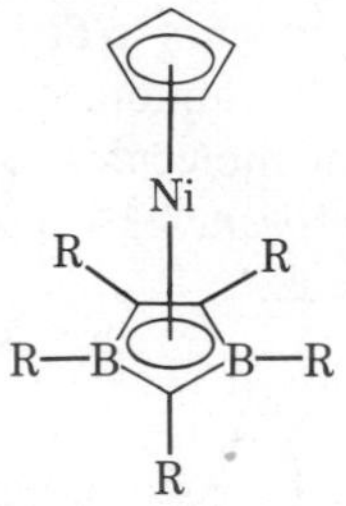   (27-LXV)

The *borazines* or *borazoles* (Section 8-12) can be regarded as analogues of arenes and such compounds as $(B_3N_3R_6)Cr(CO)_3$ are indeed formed. However the ligand rings are nonplanar and they are poorer π acceptors than arenes.[160] Similar compounds are formed by other boron ring compounds, e.g., triphenylborthiine gives the complex $(Ph_3B_3S_3)Mo(CO)_3$.[161]

27-16.   Multidecker Sandwich Compounds[162]

Various Lewis acids such as BF_3 react with $(C_5H_5)_2Ni$ to give a stable cationic species that proved to be a triple-decker sandwich (27-LXVI). The mechanism of

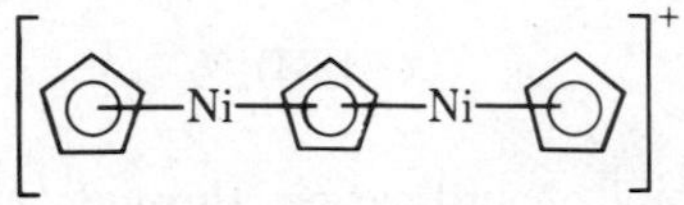

(27-LXVI)

formation, in which the π-C_5H_5 rings are labilized, appears to be similar to that involved in electrophilic attacks on ferrocene discussed earlier, except that direct attack on the ring is proposed rather than initial attack on the metal atom:

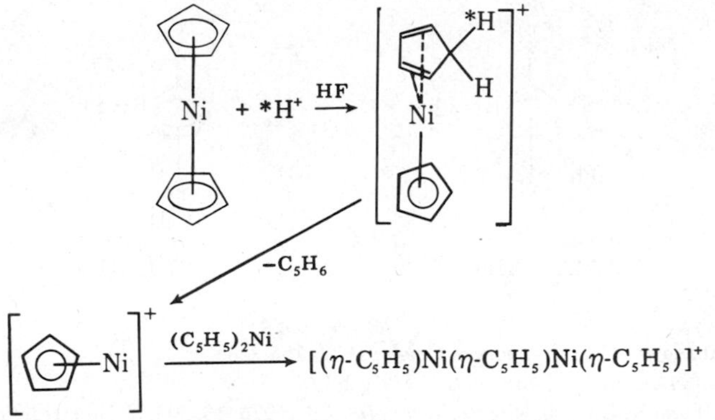

160   J. J. Lagowski, *Coord. Chem. Rev.*, 1977, **22**, 185; M. Scott *et al.*, *Inorg. Chim. Acta*, 1977, **25**, 261.

161   H. Nöth and U. Schuschardt, *J. Organomet. Chem.*, 1977, **134**, 297.

162   H. Werner *et al.*, *Angew. Chem. Int. Ed.*, 1977, **16**, 1; *J. Organomet. Chem.*, 1977, **141**, 339; M. Bochmann *et al.*, *Angew. Chem., Int. Ed.*, 1978, **17**, 863.

The intermediate $C_5H_5NiC_5H_6^+$ ion is stable in liquid HF solution for only a few minutes, but nmr study shows that the added proton is exo rather than endo, as it would be if initial attack on Ni had occurred to give an Ni—H bond. The cation $[C_5H_5Ni]^+$ can be isolated as its BF_4^- or SbF_6^- salt, and these salts react quantitatively with $(C_5H_5)_2Ni$ to give the triple-decker cation.

Just as cyclopentadienyl (Cp_2M) or arene (Ar_2M) compounds tend to follow the eighteen-electron rule, it has been suggested[163] that polynuclear sandwich compounds will follow a 30 or 34 total electron rule; 27-LVI for example has 34 electrons.

Other types of "triple deckers" are now known, for example, 27-LXVII[164a] and 27-LXVIII,[164b] both of which contain $C_8H_8^{2-}$ rings.

(27-LXVII)   (27-LXVIII)

So far, there is one example of a "quadruple decker,"[165] 27-LXIX, although this contains end carbonyls that presumably in principle could be replaced by a η^6 arene.

(27-LXIX)

Other types of complex that are not, strictly speaking triple deckers, are compounds obtained by reactions such as:

$$(\eta\text{-}C_5H_5)_2Co \xrightarrow[90-130°]{HP(O)(OMe)_2} (\eta\text{-}C_5H_5)_2Co_3[OP(OMe)_2]_6$$

These have structures such as 27-LXX and 27-LXXI[166] (R groups omitted for clarity), where the metals are bridged by M—P—O—M bonds.

163   J. W. Lauher et al., J. Am. Chem. Soc., 1977, 99, 3219.
164a   P. S. Skell et al., J. Am. Chem. Soc., 1978, 100, 999.
164b   J. Moraczewski and W. E. Geiger, Jr., J. Am. Chem. Soc., 1978, 100, 7429.
165   W. Siebert et al., Angew. Chem., Int. Ed., 1978, 17, 527.
166   W. Kläui and K. Dehnicke, Chem. Ber., 1978, 111, 451; H. Werner and T. N. Khac, Inorg. Chem. Acta, 1978, 30, L347; W. Kläui, J.C.S. Chem. Comm., 1979, 700.

(27-LXX)

(27-LXXI)

METALLOCARBABORANE AND METALLOBORANE COMPLEXES[167a]

In addition to the π complexes of various boron heterocycles noted above, there is a vast and rapidly expanding chemistry of transition metal complexes that contain ligands derived from carbaboranes and boranes. Many of these compounds also contain η-C_5H_5, CO, or other ligands, and they are quite closely related to the π complexes discussed earlier in this chapter.

27-17. Metallocarbaborane Complexes[167b]

Their is a very close structural and electronic resemblance between certain carbaborane anions (p. 319) and the cyclopentadienide ion. Thus the $B_9C_2H_{11}^{2-}$ anion (27-LXXII; there is an other isomer) can be compared with C_5H_5 (27-LXXIII) as shown:

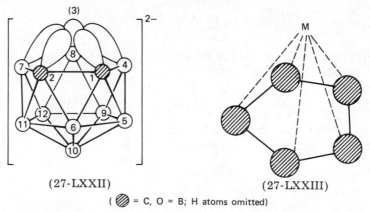

(27-LXXII)   (27-LXXIII)

(⬤ = C, ○ = B; H atoms omitted)

[167a] R. N. Grimes, *Coord. Chem. Rev.*, 1979, **28**, 47.
[167b] K. P. Callahan and M. F. Hawthorne, *Adv. Organomet. Chem.*, 1976, **14**, 145; R. N. Grimes, *Organometallic Reactions and Synthesis*, Vol. 6, Plenum Press 1977, Chapter 2.

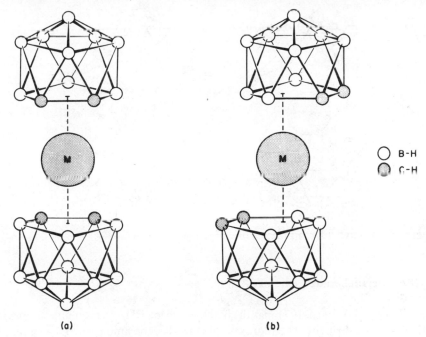

Fig. 27-9. The structures of some bis(dicarbollide) metal complexes: (a) the symmetrical structure and (b) the "slipped" structure.

The isomeric $B_9C_2H_{11}^{2-}$ ions combine readily with transition metal ions such as Fe^{2+} and Co^{3+} that have d^6 configurations. Compounds* are formed that are isoelectronic with $(C_5H_5)_2Fe$ and $(C_5H_5)_2Co^+$, that is, $(B_9C_2H_{11})_2Fe^{2-}$ and $(B_9C_2H_{11})_2Co^-$, respectively. The iron complex undergoes reversible oxidation analogous to that of $(C_5H_5)_2Fe$. The structures of these $(B_9C_2H_{11})_2M^{n-}$ systems are mostly as shown in Fig. 27-9a and have essentially D_{5d} symmetry, if the difference between B and C atoms is ignored. In the cases of the $(B_9C_2H_{11})_2Cu^{2-}$ and $(B_9C_2H_{11})_2Cu^-$ species, and probably some others, there is a "slippage" of the ligands (Fig. 27-9b), caused by the presence of too many electrons to be accommodated in the bonding MO's of the symmetrical structure.

In addition to true "sandwich" compounds, there are also compounds with CO or other ligands and only one ring; some examples are given in Fig. 27-10 (p. 1177), where the resemblance to $\eta\text{-}C_5H_5$ compounds such as $(\eta\text{-}C_5H_5)Mn(CO)_3$ will be recognized.

Other series of carbaborane complexes are known. For example, the anion $B_7C_2H_{11}^{2-}$ (which has isomers) gives the complex $[Co(B_7C_2H_9)_2]^-$, two of whose isomers are:

* Since systematic nomenclature for compounds of $B_9C_2H_{11}^{2-}$ is unwieldy, the trivial name "dicarbollide" from the Spanish *olla* meaning "pot," is sometimes used.

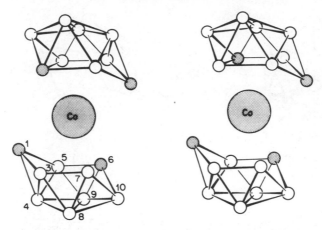

There are also hydrido complexes like $H_2Fe(Me_2C_2B_4H_4)$.[168]

27-18. Metalloborane Complexes[169]

Chapter 4 pointed out (p. 107) that the borohydride ion BH_4^- forms many complexes with M—H—B bridges. The more complicated borane anions (p. 317) similarly form metal complexes, many of which have direct M—B bonding as well as M—H—B bonding.

A simple example is the compound $B_2H_6Fe_2(CO)_6$ (27-LXXIV), which is isoelectronic with the acetylene complex (cf. p. 1156) $C_2H_2Co_2(CO)_6$,[170]

(27-LXXIV)

(27-LXXV)

and like this complex has a B—B unit at 90° to the Fe—Fe bond.

Another relatively simple compound (27-LXXV) has Fe–B bonds[171]; it is obtained by the reaction:

$$K_2Fe(CO)_4 + 3BH_3THF = K[H_2B(\mu\text{-}H)(\mu\text{-}Fe(CO)_4)BH_2] + KBH_4 + 3THF$$

For the higher boranes or their anions, the synthetic routes include interactions

168   R. N. Grimes *et al., Inorg. Chem.,* 1977, **16,** 3094.
169   R. N. Grimes, *Acc. Chem. Res.,* 1978, **11,** 420.
170   E. L. Anderson and T. P. Fehlner, *J. Am. Chem. Soc.,* 1978, **100,** 4606.
171   G. Medford and S. G. Shore, *J. Am. Chem. Soc.,* 1978, **100,** 3952.

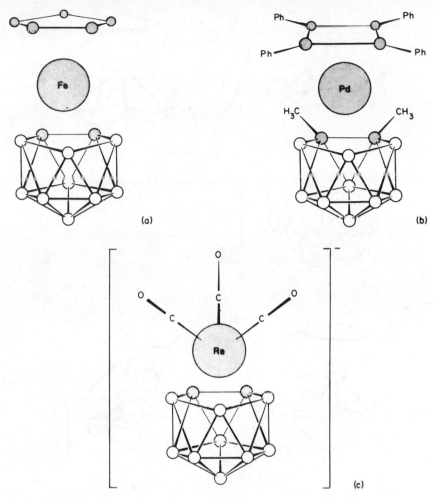

Fig. 27-10. Some complexes containing one dicarbollide ion and ligands of other types: (a) $C_5H_5^-$, (b) C_4Ph_4, and (c) three CO groups.

with metal hydrides or alkyls in which H_2 or alkane is lost, interactions with metal halide complexes, etc.

Some representative reactions are

$$(\eta\text{-}C_5H_5)_2Ni + nido\text{-}B_{10}H_{13}^- \xrightarrow{\text{Na/Hg}} nido\text{-}[C_5H_5NiB_{10}H_{12}]^-$$

$$2B_{10}H_{14} + CdEt_2 \xrightarrow{\text{Et}_2\text{O}} [(Et_2O)_2CdB_{10}H_{12}]_2 + 2C_2H_6$$

$$2B_6H_{10} + [PtCl_3(C_2H_4)]^- = (B_6H_{10})_2PtCl_2 + C_2H_4 + Cl^-$$

$$KB_5H_6 + CuCl(PPh_3)_3 = (B_5H_8)Cu(PPh_3)_2 + KCl + PPh_3$$

$$2BrB_5H_8 + IrBr(CO)(PMe_3)_2 = (B_5H_6)IrBr_2(CO)(PMe_3)_2$$

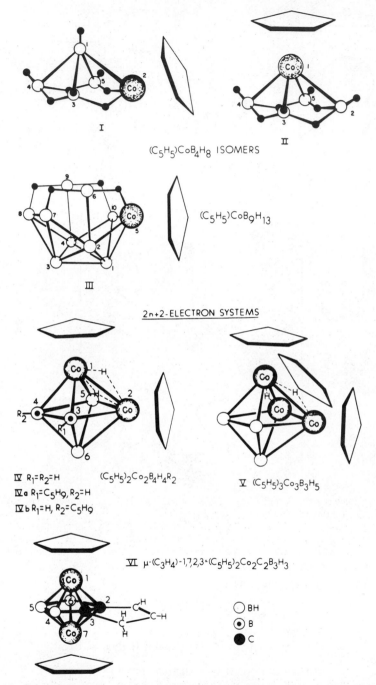

Fig. 27-11. Cobaltoborane clusters of the $2n + 2$ and $2n + 4$ systems. Structure V is established by crystallographic study; others are based on ^{11}B and ^{1}H nmr data. [Reproduced by permission from V. R. Miller, R. Weiss, and R. N. Grimes, *J. Am. Chem. Soc.*, 1977, **99**, 5646.]

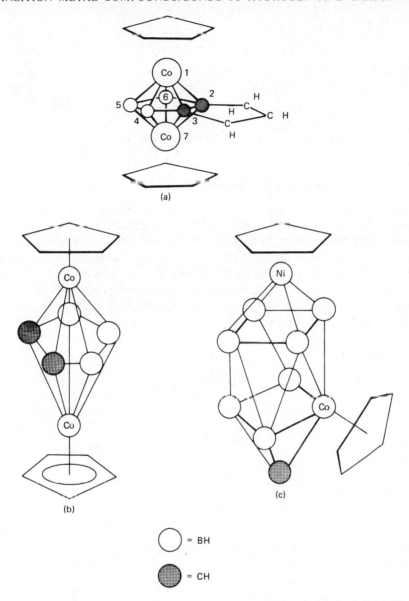

Fig. 27-12. The structures of metallocarbaborane cluster compounds (a) (μ-C_3H_4)-1,7,2,3-(η-C_5H_5)$_2$Co$_2$C$_2$B$_3$H$_3$, (b) (C$_5$H$_5$)$_2$Co$_2$B$_2$B$_3$H$_5$, and (c) (C$_5$H$_5$)$_2$–NiCoB$_7$H$_8$.

The structures of two such compounds (B$_6$H$_{10}$)$_2$PtCl$_2$ (27-LXXVI) and (B$_5$H$_8$)Cu(PPh$_3$)$_2$ (27-LXXVII) show again the M—B bonding.[172] In the platinum compound the Pt atom bridges the unique B^4, B^5 atoms in B$_6$H$_{10}$ (see Fig. 9-6) to give a 3c–2e type bond, and there is a similar bond to B$_5$H$_8$ in the copper case.

172 N. N. Greenwood *et al.*, *J.C.S. Dalton*, **1977**, 37; **1979**, 117.

Cl
B····|····B⁵
   >Pt< / B
B◄ ▼B⁴ / B
Cl   \\ B
        \\
        B

(27-LXXVI)

          B¹
       B³◢ ◣B⁴
PPh₃···|···|
      \ ·· ▲B²—B⁵
     Cu
     |
     PPh₃

(27-LXXVII)

B—H—B bridges:
B_3—B_4
B_2—B_5
B_4—B_5

27-19. Metalloborane and -carbaborane Clusters

There is a wide range of compounds that contain more than one transition metal atom and $M_x B_y$ or $M_x B_y C_z$ skeletons. These heteroclusters[169,173] bear a close relation to the homocluster compounds discussed in Chapter 26. The structures can be classed into those having metalloboron skeletons with $2n$, $2n + 2$, or $2n + 4$ valence electrons. Some examples of the structures are shown in Fig. 27-11, and examples of clusters that have carbon atoms in addition appear in Fig. 27-12.

We cannot go into details of synthetic methods but the compounds are commonly made by interaction of mixtures, e.g., in a simple case,

$$NaC_5H_5 + NaB_5H_8 + CoCl_2 \xrightarrow{\text{THF}} (\eta\text{-}C_5H_5)CoB_5H_8$$

Many of these cluster compounds undergo quite complicated reactions. Thus on heating, intramolecular rearrangements can occur whereby the metal atoms migrate to different positions and isomers are formed. They can also undergo cage expansion reactions on treatment with other metal complexes such as $Fe(CO)_5$, and also redox reactions.[174]

---

[173] See K. Wade, *Chem. Br.*, 1975, **11**, 177; *Adv. Inorg. Chem. Radiochem.*, 1976, **18**, 1.
[174] D. E. Brennan and W. E. Geiger, Jr., *J. Am. Chem. Soc.*, 1979, **101**, 3399.

General References

See also references for Chapters 25, 26, 28, 29, and 30.

Annual Reviews

Organometallic Chemistry, Specialist Report, Chemical Society.
Journal of Organometallic Chemistry. The annual reviews appear as sections of the journal. In *J. Organomet. Chem.*, 1978, **151**, 313, M. I. Bruce provides a survey of structures (2475) with references to the period 1968–1976; 1977 structures are given in 1979, **167**, 361.
Advances in Organometallic Chemistry. Note that in volumes beginning in 1972, M. I. Bruce gives guides to the literature for the period 1950–1970.
Molecular Structures by Diffraction Methods, Specialist Report, Chemical Society.
Organometallic Reactions and Synthesis, E. I. Becker and M. Tsutsui, Eds., Plenum Press.

Texts, Monographs, etc.

Alper, H., Ed., *Transition Metal Organometallics in Organic Synthesis,* Vol. 1, Academic Press; 1976. Alkenes, carbenes; Vol. 2, 1978, acetylenes and arenes.

Belluco, U., *Organometallic and Coordination Chemistry of Platinum,* 1974.

Blackborow, J. R., and D. Young, *Metal Vapour Synthesis in Organometallic Chemistry,* Springer, 1979.

Black, D. St. C., W. R. Jackson, and J. M. Swan, in *Comprehensive Organic Chemistry,* Part 15-6, Vol. 3, Pergamon Press, 1979.

Bowder, F. L., and R. Giles, *Coord. Chem. Rev.,* 1976, **20,** 81.

Brewster, J. H., *Aspects of Mechanism and Organometallic Chemistry,* Plenum, 1979.

Brunner, H., *Topics Cur. Chem.,* 1975, **56,** 67. Chiral organotransition metal compounds.

Carraher, C. E., Jr., J. E. Sheats, and C. V. Pittman, Jr., *Organometallic Polymers,* Academic Press, 1978.

Connor, J. A. *Topics Curr. Chem.,* 1977, **71,** 71. Thermochemistry of metal alkyls and carbonyls.

Greenwood, N. N., *et al., Index of Vibrational Spectra of Inorganic and Organometallic Compounds,* 2 Vols., Butterworths, 1975.

Haaland, A., *Topics Curr. Chem.,* 1974, **53,** 1. Electron diffraction studies.

Hartley, F. R., *The Chemistry of Palladium and Platinum.* Applied Science Publishers, 1973. Contains some organochemistry.

Heck, R. F., *Organotransition Metal Chemistry: A Mechanistic Approach,* Academic Press, 1974. A good general text.

Henry, P. M., *Adv. Organomet. Chem.,* 1975, **13,** 363. Pd.

Houghton, R. P., *Metal Complexes in Organic Chemistry,* Cambridge University Press, 1979.

Hoffmann, R., *et al., Pure Appl. Chem.,* 1978, **50,** 1. Conformational phenomena in alkene complexes.

Houben-Weyl, *Methoden der Organische Chemie,* 4th ed., Vol. V 2a, Thieme, 1977. Metal chemistry of alkynes, allenes, and cumulenes.

Ishii, Y., and M. Tsutsui, Eds., *Organotransition Metal Chemistry,* Plenum Press, 1975. Conference reports.

Johnson, M. D., *Acc. Chem. Res.,* 1978, **11,** 57. Reactions of electrophiles with M—C σ bonds.

Jolly, P. W., and G. Wilke, *The Organic Chemistry of Nickel,* Vol. 1, 1974, Vol. II (*Organic Synthesis*), 1975, Academic Press.

Jones, D. N., *Comprehensive Organic Chemistry,* Vol. 3, Pergamon Press. Includes transition metals.

Jukes, A. E., *Adv. Organomet. Chem.,* 1974, **12,** 215. Cu.

Kochi, J. K., *Organometallic Mechanisms and Catalysis,* Academic Press, 1978.

Koerner von Gustorf, E. A., *et al., The Organic Chemistry of Iron,* Vol. I, Academic Press, 1978.

Kozikowski, A. P., and H. F. Wetter, *Synthesis,* **1976,** 561. Transition metals in organic synthesis.

Maitlis, P. M., *The Organic Chemistry of Palladium,* Vols. I, II, Academic Press, 1971.

Maslowsky, E., *Vibrational Spectra of Organometallic Compounds,* Wiley, 1977.

Mingos, D. P., *Adv. Organomet. Chem.,* 1977, **15,** 1. Theory.

Nease, E. W., and H. Rosenberg, *Metallocene Polymers,* Dekker, 1970.

New Synthetic Methods, Vol. 1, Verlag Chemie, 1975. These volumes contain some transition metal chemistry, especially Vol. 3.

Omae, I., *Chem. Rev.,* 1979, *79,* 287. (Organometallic intramolecular coordination compounds containing nitrogen ligands).

Rylander, P. N., *Organic Syntheses with Noble Metal Catalysts,* Academic Press, 1973.

Seyferth, D., Ed., *New Applications of Organometallic Reagents in Organic Synthesis,* Elsvier, 1976. Symposium reports containing use of alkyl and arene metal complexe.

Sneedon, R. P. A., *Organochromium Compounds,* Academic Press, 1975.

Sutherland, R. G., *Organometallic Chemistry Review,* Vol. 3, Elsevier. π-Arene–π-cyclopentadienyl iron cations and relative systems.

Thorn, D. L. and R. Hoffmann, *Inorg. Chem.,* 1978, **71,** 126. MO study of $LM_2(CO)_6$ (L = dienes, cyclobutadiene, pentalene, and other π systems).

Timms, P. L., and T. W. Turney, *Adv. Organomet. Chem.,* 1977, **15,** 53. Metal atom syntheses.

Trost, B. M., *Tetrahedron Report* No. 32, 1977, p. 2615. Organopalladium intermediates.

Wailes, P. C., R. S. P. Coutts, and H. Weigold, *Organometallic Chemistry of Titanium, Zirconium and Hafnium,* Academic Press, 1974.

Wrighton, M. S., Ed., *Inorganic and Organometallic Photochemistry.* ACS Advances in Chemistry Series, No. 168, 1978.

CHAPTER TWENTY–EIGHT

Reaction Mechanisms and Molecular Rearrangements in Complexes

28-1. Introduction

Processes in which complexes undergo ligand exchange, oxidation, reduction, or rearrangement of the coordination sphere are the bases for the chemical properties of all the metallic elements. A delineation of the mechanisms by which these processes occur is therefore vital to a full understanding and interpretation of the chemistry of complexes. However the design and interpretation of experiments capable of providing mechanistic information is a very difficult and subtle task, and not all interesting reactions make good subjects for kinetic and mechanistic study. One of the important criteria in choosing reactions for kinetic study is that they have rates convenient for measurement by the available techniques.

Whether a reaction has a rate convenient for measurement depends on what type of technique can be applied. The techniques can be classified into three broad categories; reaction rates for which they are generally used are indicated by the half-times:

1. Static methods ($t_{1/2} \geqslant 1$ min).
2. Flow or rapid-mixing techniques (1 min $\geqslant t_{1/2} \geqslant 10^{-3}$ sec).
3. Relaxation methods ($t_{1/2} \leqslant 10^{-1}$ sec).

The static methods are the classical ones in which reactants are mixed simply by pouring them both into one vessel, and the progress of the reaction is followed by observation of the time variation of some physical or chemical observable (e.g., light absorption, gas evolution, pH, isotopic exchange). Flow and rapid mixing techniques differ mainly in achieving rapid mixing (in $\sim 10^{-3}$ sec) of the reactants but use many of the same observational techniques as in static measurements. Relaxation methods depend on either (a) creating a single disturbance in a state of equilibrium in a very short period of time (usually by a temperature or pressure jump) and following the process of relaxation to an equilibrium state by a combination of spectrophotometric and fast electronic recording devices, or (b) continuous disturbances by ultrasonic waves or radiofrequency signals in presence of a magnetic field (i.e., nmr). The latter methods are capable of following the very fastest reactions, and in many cases rate constants up to the limit ($\sim 10^{10}$ sec^{-1}) set by diffusion processes have been measured by ultrasonic methods.

It should be noted that nmr has the unique capability of following certain processes in which there is no net chemical or physical change, provided they involve the passage of observable nuclei among two or more sites in each of which their chemical shift is different. In this way certain rapid intramolecular rearrangements can be studied that would otherwise not even be seen.

The *direct* result of a kinetic study can at best be a *rate law,* that is, an equation showing how the velocity v of a reaction at a given temperature and in a given medium varies as a function of the concentration of the reactants. Certain constants, k_i, called *rate constants,* will appear in the rate law. For example, a rate law for the reaction

$$A + B = C + D$$

might be

$$v = k_a[A] + k_{ab}[A][B] + k_{ab}[A][B][H^+]^{-1}$$

This would mean that the reaction occurs by three detectable paths, one influenced only by [A], a second influenced by [A] and [B], and a third that depends also on pH. The third term shows that not only A and B, but also [OH$^-$] (since this is related inversely to [H$^+$]), participates in the activated complex when this path is followed.

The ultimate purpose of a rate and mechanism study is usually to *interpret* the rate law correctly so as to determine the correct *mechanism* for the reaction. By "mechanism," we mean a specification of what species actually combine to produce activated complexes, and what steps occur before and/or after the formation of the activated complex.*

One final preliminary point to be stressed is the difference between the kinetic and thermodynamic properties (especially "stability" in the loose sense) of a complex. The ability of a particular complex to engage in reactions that result in replacing one or more ligands in its coordination sphere by others is called its *lability.* Those complexes for which reactions of this type are very rapid are called *labile,* whereas those for which such reactions proceed only slowly or not at all are called *inert.* It is important to emphasize that these two terms refer to rates of reactions and should not be confused with the terms "stable" and "unstable," which refer to the thermodynamic tendency of species to exist under equilibrium conditions. A simple example of this distinction is provided by the $[Co(NH_3)_6]^{3+}$ ion, which will persist for days in an acid medium because of its kinetic inertness or lack

---

* We assume that the reader is familiar with the basic concepts of chemical kinetics as taught, for example, in undergraduate physical chemistry courses.

of lability even though it is thermodynamically unstable, as the following equilibrium constant shows:

$$[Co(NH_3)_6]^{3+} + 6H_3O^+ = [Co(H_2O)_6]^{3+} + 6NH_4^+ \qquad K \sim 10^{25}$$

In contrast, the stability of $[Ni(CN)_4]^{2-}$ is extremely high,

$$[Ni(CN)_4]^{2-} = Ni^{2+} + 4CN^- \qquad K \sim 10^{-22}$$

but the rate of exchange of CN^- ions with isotopically labeled CN^- added to the solution is immeasurably fast by ordinary techniques. Of course this lack of any necessary relation between thermodynamic stability and kinetic lability is to be found generally in chemistry, but its appreciation here is especially important.

This chapter emphasizes mainly the processes occurring in the more classical types of complexes as they are formed and react in aqueous solution or in similarly polar solvents. However we also devote some attention to some mechanistic problems peculiar to the metal carbonyls, and we discuss in some detail the types of intermolecular rearrangement involved in the phenomena known variously as stereochemical nonrigidity, fluxionality, polytopal rearrangement, and ligand scrambling.

LIGAND REPLACEMENT REACTIONS

28-2. Possible Mechanisms for Ligand Replacement Reactions

There are two extreme possibilities for ligand replacement reactions. In one the bond to the leaving ligand would be entirely severed to generate a five-coordinate intermediate. The rate of this slow step would determine the rate of the entire reaction, since capture of the new ligand by the five-coordinate intermediate would be very rapid. This mechanism has often been called the S_N1 mechanism, using a term originally coined by organic chemists, where S_N stands for "substitution nucleophilic" and the 1 indicates that the rate-determining step is unimolecular. It can be summarized in eq. 28-1.

$$[ML_6]^{n+} \xrightarrow{\text{slow}} L + [ML_5]^{n+} \xrightarrow[\text{fast}]{+L'} [ML_5L']^{n+} \qquad (28\text{-}1)$$

Five-coordinate intermediate

The total loss of energy of one M—L bond before there is any compensatory gain due to the formation of the new M—L' bond is the key feature here.

At the other extreme is the so-called S_N2 mechanism, in which the new ligand L' becomes fully bound, thus generating a seven-coordinate intermediate, before one of the original ligands is lost. This is summarized in eq. 28-2. The symbol S_N2 indicates that the rate-determining step is bimolecular.

$$[ML_6]^{n+} + L' \xrightarrow{\text{slow}} [ML_6L']^{n+} \xrightarrow{\text{fast}} [MLL']^{n+} + L \qquad (28\text{-}2)$$

Although these extreme mechanisms, and their symbols S_N1 and S_N2, played a role in earlier literature, they are rarely used today because (a) they are too in-flexible to deal with the subtle variations in real mechanisms, and (b) they tend to suggest relationships between observed rate laws and mechanisms that are often erroneous. For example, there are ways in which a reaction might have a rate law $R = k[ML_6][L']$ and still have a mechanism in which M—L bond breaking is the process that plays the largest role in determining the rate.

Contemporary discussions of the mechanisms of ligand replacement reactions usually employ the following classification.

D Mechanism. This is comparable to the S_N1 limit but does not carry the same implications as to observed rate law. The transient intermediate of reduced coor-dination number is assumed to live long enough to be able to discriminate between potential ligands in its vicinity, including the one just lost L, the new one L', and also solvent molecules S. The rate need *not* be independent of [L'] even though M—L dissociation is complete before M—L' bond formation, since at very low [L'] there will be competition by L and L' for the intermediate and only at high [L'] will the rate be the same regardless of the identity or concentration of L'.

To see the basis for the last statement, we write out the steps explicitly:

$$ML_6 \underset{k_{-1}}{\overset{k_1}{\rightleftarrows}} ML_5 + L \tag{28-3a}$$

$$ML_5 + L' \overset{k_2}{\rightarrow} ML_5L' \tag{28-3b}$$

By employing the well-known steady state approximation for $[ML_5]$, we obtain for the rate law:

$$R = \frac{k_1 k_2 [ML_6][L']}{k_{-1} + k_2[L']} \tag{28-4a}$$

The rate clearly depends in general on [L'] as well as on $[ML_6]$, but when [L'] becomes very large k_{-1}, becomes negligible compared to $k_2[L']$, and $k_2[L']$ may be canceled from numerator and denominator, leaving

$$R_{[L']\rightarrow\infty} = k_1[ML_6] \tag{28-4b}$$

I_d Mechanism. This is the dissociative interchange mechanism. The transition state involves considerable extension of an M—L bond (but not its complete rup-ture), together with some incipient interaction with the incoming ligand L'. This incipient interaction can be described as a kind of weak complex ML_6,L', which positions the new ligand L' to enter the (first) coordination sphere immediately on the departure of L. We may represent this by the scheme in eq. 28-5.

$$ML_6 + L' \overset{K}{\longrightarrow} ML_6,L' \overset{k}{\longrightarrow} [L_5M \cdots L,L']^{\ddagger} \longrightarrow ML_5L' + L \tag{28-5}$$

The species ML_6,L' is called an *outer-sphere complex* or, if ML_6 is a cation and L' an anion, an *ion pair*.

I_a Mechanism. This is the associative interchange mechanism. Again, there is interchange of ligands between the inner and next-nearest coordination spheres, but here the interaction between M and L' is much more advanced in the transition state; M···L' bonding is important in defining the activated complex.

A Mechanism. In this extreme there is a fully formed intermediate complex ML_6L', which then dissociates.

In the real world there are processes whose mechanisms define a continuous gradation from D through increasingly associated interchanges to A. In any given real case it is difficult and often futile to try to give an exact description; the experimental criteria are seldom if ever definitive. The experimental data that do provide the best clues to the mechanistic type for a reaction are the following.

Rate Law. A first-order rate law is sometimes indicative of a unimolecular (dissociative) reaction and a second-order rate law indicative of a bimolecular (associative) reaction, but the use of such evidence requires great caution. We have already seen (eq. 28-3, 28-4) that the rate of a purely dissociative process may show a dependence on [L']. For I_d processes such a dependence is always expected. The rate law for the process represented in eq. 28-5 must be eq. 28-6 because of the rapid pre-equilibrium formation of the outer-sphere complex ML_6,L'.

$$R = Kk[ML_6][L'] \qquad (28\text{-}6)$$

A second and very common situation is solvent participation. That is, there may be a "hidden" step in which solvent, especially an excellent nucleophile like water, present at a concentration of *ca.* $55M$, participates:

$$ML_6 + H_2O \rightarrow ML_5(H_2O) + L \qquad \text{Slow} \qquad (28\text{-}7a)$$
$$ML_5(H_2O) + L' \rightarrow ML_5L' + H_2O \qquad \text{Fast} \qquad (28\text{-}7b)$$

The rate law might show *no* dependence on L', but the process by which the first step proceeds could still be a highly *associative* one.

We may also mention here the process of conjugate base formation. Whenever a rate law containing the factor [OH^-] is found, there is a question whether OH^- actually attacks the metal atom in a genuinely associative-type mechanism or whether it appears in the rate law because it rapidly deprotonates a ligand forming a conjugate base, CB, which then reacts as in the sequence of eq. 28-8.

$$[Co(NH_3)_5Cl]^{2+} + OH^- = [Co(NH_3)_4(NH_2)Cl]^+ + H_2O \qquad \text{Fast} \qquad (28\text{-}8a)$$

$$[Co(NH_3)_4(NH_2)Cl]^+ \xrightarrow[\text{then } H^+]{+Y^-} [Co(NH_3)_5Y]^{2+} + Cl^- \qquad \text{Slow} \qquad (28\text{-}8b)$$

In summary, rate laws must be interpreted with extreme caution.

Activation Parameters. By studying a reaction over a temperature range, one can evaluate the enthalpy and entropy changes required to form the activated complex, $\Delta H^{\ddagger}$ and $\Delta S^{\ddagger}$. A more difficult but increasingly common experiment is to measure the rate as a function of pressure and from the equation $(\partial \ln k/\partial P)_T = \Delta V^{\ddagger}/RT$ determine the volume of activation $\Delta V^{\ddagger}$. These three quantities $\Delta H^{\ddagger}$, $\Delta S^{\ddagger}$, and $\Delta V^{\ddagger}$ can be valuable guides to visualizing the transition state, but they too must be used judiciously. For example, a reaction with large positive values of $\Delta H^{\ddagger}$ and $\Delta S^{\ddagger}$ and a positive $\Delta V^{\ddagger}$ can scarcely be other than dissociative, but things are sometimes not so one-sided, as will be seen later.

Other Criteria. The many other data that are sometimes helpful include the dependence of rate on ionic strength and on solvent composition. The latter approach, however, often creates more problems than it solves. Careful measurement of trends in rates and activation parameters during variance of the incoming ligand L′ or observation of the properties (e.g., steric hindrance) of the nonreplaceable ligands can sometimes be informative.

28-3. Water-Exchange Rates

Our knowledge of water-exchange rates depends largely on the results of relaxation measurements, since nearly all the reactions are very fast. As an important special case, we first consider the rates at which aqua ions exchange water molecules with solvent water. Except for $Cr(H_2O)_6^{3+}$, with a half-time $\sim 3.5 \times 10^5$ sec and an activation energy of 112 kJ mol^{-1}, and $Rh(H_2O)_6^{3+}$, which is still slower ($E_{act} \approx 137$ kJ mol^{-1}), these reactions are all fast as shown in Fig. 28-1. It will be seen that though they are all "fast" a range of some 10 orders of magnitude is spanned. These data therefore provide a good base for tackling the question of what factors influence the rates of reaction of similar complexes of different metal ions.

First, by considering the alkali and alkaline earth ions, the influence of size and charge may be seen. Within each group the rate of exchange increases with size, and for M^+ and M^{2+} ions of similar size, the one of lower charge exchanges most rapidly. Since $M—OH_2$ bond strength should increase with charge and decrease with size of the metal ion, these correlations suggest that the transition state for the exchange reaction is attained by breaking an existing $M—OH_2$ bond to a much

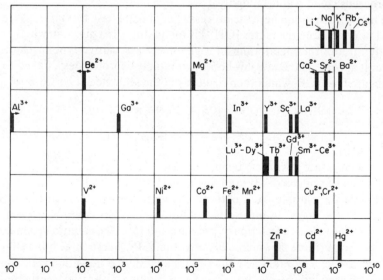

Fig. 28-1. Characteristic rate constants (sec^{-1}) for substitutions of innersphere H_2O of various aqua ions. [Adapted from M. Eigen, *Pure Appl. Chem.*, 1963, **6**, 105, with revised data kindly provided by M. Eigen. See also H. D. Bennett and B. F. Caldin, *J. Chem. Soc., A*, **1971**, 2918.]

greater extent than a new one is formed; that is, that the mechanism is essentially dissociative.

Referring again to Fig. 28-1 it will be seen that other series, that is, (Al^{3+}, Ga^{3+}, In^{3+}), (Sc^{3+}, Y^{3+}), and (Zn^{2+}, Cd^{2+}, Hg^{2+}), also obey the radius rule. There are, however, several cases in which two ions of about the same size disobey the charge rule (i.e., the more highly charged ion exchanges faster). Such exceptions are thought to be due to differences in coordination numbers, but this is not quite certain.

It will also be noted that the divalent transition metal ions do not follow the charge and radius rules very well. There are at least two additional factors involved here. With Cu^{2+} and Cr^{2+} the coordination polyhedron is not a regular octahedron, rather, two bonds are very much longer and weaker than the other four. By way of these, the rate of exchange is increased. Second, for most transition metal ions the rates of ligand-exchange reactions are influenced by the changes in d-electron energies as the coordination changes from that in the reactant to that in the transition state. This will always increase the activation energy, hence decrease the rate, but the magnitude of the effect is not monotonically related to the atomic number; rather, it varies irregularly from ion to ion.

Figure 28-2a shows the variation in rates of water exchange for the divalent ions of the first transition series plotted in a different way. The important trends in this behavior may be fitted quite well using a simple argument employing molecular orbital stabilization energies (MOSEs: see Section 20-19). If we assume an essentially dissociative mechanism, we need to calculate the differences between the MOSEs for octahedral ML_6 and square pyramidal ML_5 complexes for each of the M^{2+} ions. The method of calculating the former has been explained (p. 684) and their values listed (Table 20-8). MOSEs for sp $[ML_5]^{2+}$ can be calculated similarly, and the differences are found to vary as shown in Fig. 28-2c. If we add to this plot the rapidly increasing s and p orbital bonding contributions, as shown by the parallel straight lines in Fig. 20-47, we obtain the graph presented in Fig. 28-2b. This should give the variation in activation energies for essentially dissociative water-exchange processes for the M^{2+} ions, and it is seen to match the rate variations reasonably well.[1]

In summary, it appears clear that the exchange of water molecules between the inner (first) coordination sphere and the solvent occurs typically by a dissociative or dissociative interchange process.

28-4. Formation of Complexes from Aqua Ions

Extensive studies of the rates at which an aqua ion combines with a ligand to form a complex have revealed the following remarkable general rules:

1. The rates for a given ion show little or no dependence (less than a factor of 10) on the identity of the ligand.

[1]   J. K. Burdett, *Adv. Inorg. Chem. Radiochem.*, 1978, **21**, 113.

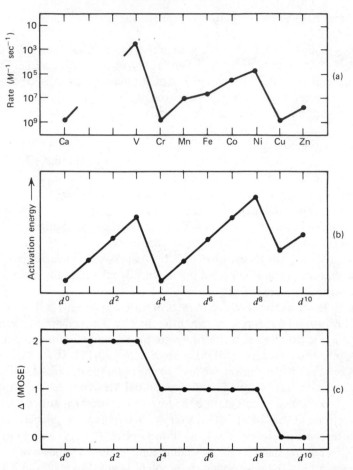

Fig. 28-2. (a) Experimental values for the rates of water exchange by the M^{2+}(aq) ions of the first transition series. (b) Estimated activation energies. (c) Molecular orbital stabilization energy differences.

2. The rates for each ion are practically the same as the rate of water exchange for that ion, usually ≤10 times slower.

It is believed that the only reasonable explanation for these observations is that the formation reactions proceed in two steps, the first being formation of the aqua ion–ligand outer-sphere complex, followed by elimination of H_2O from the aqua ion in the same manner as in the water-exchange process. This is the mechanism described by eq. 28-5, with L = H_2O and L′ representing the incoming ligand.

The rate law for this process would be that of eq. 28-9a. Under the usual experimental conditions [L′] is small and K is usually small, and the experimental rate law will appear to be that of eq. 28-9b with an apparent second-order rate constant k_{app}, which is really the product of K and the true dissociative interchange constant k.

$$R = \frac{Kk[M(H_2O)_6][L']}{1 + K[L']} \qquad (28\text{-}9a)$$

$$R = k_{app}[M(H_2O)_6][L'] = Kk[M(H_2O)_6][L'] \qquad (28\text{-}9b)$$

When the outer-sphere association constant K has been measured or estimated, k can be determined as k_{app}/K. The resulting values of k are usually very similar to the water-exchange rate constant for the $M(H_2O)_6^{n+}$ ion; the activation enthalpies and entropies are also essentially the same. Thus once the outer-sphere complex $M(H_2O)_6,L'$ has been formed, a dissociative loss of one H_2O is the rate-controlling step, just as with the water-exchange process. As an illustration, the value of k_{app} for the formation of $Ni(CH_3PO_3)^+$ is $2.9 \times 10^5 \ M^{-1} \sec^{-1}$, and the outer-sphere association constant K has an estimated value of $40 \ M^{-1}$; from these one obtains $k = 7 \times 10^3 \sec^{-1}$. A direct measurement by ultrasonic relaxation gave 15×10^3 $\sec^{-1}$. These values of k may be compared to the rate constant for water exchange k_{ex}, which is $30 \times 10^3 \sec^{-1}$. It is not unreasonable that k_{ex} should be 2 to 4 times greater than k for the complex, since there is a statistical factor favoring the presence of an outer-sphere water molecule in the right place to move in as an inner-sphere water molecule leaves.

Anation Reactions. It appears that ligand displacement reactions (eq. 28-10) rarely proceed directly. Instead, especially in aqueous solution, X is first replaced by H_2O; then Y attacks the aqua complex (eq. 28-11) in a reaction called an anation reaction. Although these reactions resemble the processes of forming complexes from aqua ions in the sense that a ligand replaces H_2O in the first coordination sphere, they cannot necessarily be assumed to have similar mechanisms. Indeed, the varying character of the other n ligands L may well cause great variations in mechanism. Anation reactions have proved difficult to elucidate for a number of reasons—for example, the formation of ion pairs (outer-sphere complexes).

$$[L_nMX] + Y \rightarrow [L_nMY] + X \qquad (28\text{-}10)$$

$$[L_nM(H_2O)] + Y \rightarrow [L_nMY] + H_2O \qquad (28\text{-}11)$$

To avoid the ion-pairing problem, an anionic complex such as $[Co(CN)_5(H_2O)]^{2-}$ may be used. Studies of anations of this ion have shown that the mechanisms all involve a rate-determining I_d step that is essentially at the pure D limit (eq. 28-12a), giving an intermediate $Co(CN)_5^{2-}$ ion having a long enough lifetime to discriminate between various ligands present in the solution.

$$Co(CN)_5H_2O^{2-} = Co(CN)_5^{2-} + H_2O \qquad (28\text{-}12a)$$
$$Co(CN)_5^{2-} + X^- = Co(CN)_5X^{3-} \qquad (28\text{-}12b)$$

Volumes of activation have provided evidence in favor of I_d or I_a mechanisms in other cases,[2] despite possible complications from ion pairing that might occur in a preliminary step. The following results were reported for several reactions:

[2]   R. van Eldik, D. A. Palmer, and H. Kelm, *Inorg. Chem.*, 1979, **18**, 1520.

$$[Co(NH_3)_5(H_2O)]^{3+} + Cl^- \rightarrow [Co(NH_3)_5Cl]^{2+} + H_2O \qquad \Delta V^{\ddagger} = 1.4 \pm 0.8 \text{ cm}^3 \text{ mol}^{-1}$$
$$[Co(NH_3)_5(H_2O)]^{3+} + SO_4^{2-} \rightarrow [Co(NH_3)_5SO_4]^+ + H_2O \qquad \Delta V^{\ddagger} = 2.3 \pm 1.8 \text{ cm}^3 \text{ mol}^{-1}$$
$$[Rh(NH_3)_5(H_2O)]^{3+} + Cl^- \rightarrow [Rh(NH_3)_5Cl]^{2+} + H_2O \qquad \Delta V^{\ddagger} = 3.0 \pm 0.7 \text{ cm}^3 \text{ mol}^{-1}$$
$$[Cr(NH_3)_5(H_2O)]^{3+} + NCS^- \rightarrow [Cr(NH_3)_5NCS]^{2+} + H_2O \qquad \Delta V^{\ddagger} = -4.9 \pm 0.6 \text{ cm}^3 \text{ mol}^{-1}$$

The first three positive $\Delta V^{\ddagger}$ values indicate an I_d mechanism, and may be compared with the value of 1.2 ± 0.2 cm^3 mol^{-1} for the water-exchange reaction of $[Co(NH_3)_5(H_2O)]^{3+}$, where an I_d mechanism is indicated by much other evidence. The negative $\Delta V^{\ddagger}$ for the chromium complex suggests an I_a mechanism in that case.

28-5. Aquation and Base Hydrolysis

The replacement of a ligand by H_2O is called *aquation*, as illustrated in eq. 28-13 for a pentammine cobalt(III) complex.

$$[Co(NH_3)_5X]^{2+} + H_2O \rightarrow [Co(NH_3)_5(H_2O)]^{3+} + X^- \qquad (28\text{-}13)$$

The rates of such reactions are pH dependent and generally follow the rate law

$$v = k_A[L_5CoX] + k_B[L_5CoX][OH^-] \qquad (28\text{-}14)$$

In general, k_B (for *base hydrolysis*) is some 10^4 times k_A (for *acid hydrolysis*). The interpretation of this rate law has occasioned an enormous amount of experimental study and discussion, but there is not yet a complete and generally accepted interpretation. Here, we can but touch on a few main aspects of the problem.

Acid Hydrolysis. We turn first to the term $k_A[L_5CoX]$. Since the entering ligand is H_2O, which is present in high (~ 55.5 M) and effectively constant concentration, the rate law tells us *nothing* about the order in H_2O; the means for deciding whether this is an associative or a dissociative process must be sought elsewhere.

Among the most thoroughly studied systems are those involving $[Co(NH_3)_5X]$. There are various kinds of data, some of which favor an essentially dissociative mechanism, but the question can perhaps most safely be described as unresolved. To illustrate the work that has been done, the following points may be mentioned:

1. The variation of rates with the identity of X correlates well with the variation in thermodynamic stability of the complexes. This indicates that breaking the Co—X bond is important in reaching the transition state.

2. In a series of complexes where X is a carboxylate ion, there is not only the correlation of higher rates with lower basicity of the RCOO$^-$ group, but an *absence* of any slowing down due to increased size of R, after due allowance for the basicity effect. For an I_d mechanism, increased size of R should decrease the rate, at least if the attack were on the same side as X, although an attack on the *opposite* side is not excluded by these data.

3. In the case where X is H_2O, that is, for the water-exchange reaction, the pressure dependence of the rate has been measured and the volume of activation found to be $+1.2$ ml mol^{-1}. This result is most consistent with a transition state

in which the initial $Co—OH_2$ bond is stretched quite far while formation of a new $Co—*OH_2$ bond is only beginning to occur, that is, the I_d mechanism.

4. For an extreme D mechanism the five-coordinate intermediate $Co(NH_3)_5^{3+}$ would be generated, and its behavior would be independent of its source. When the ions Hg^{2+}, Ag^+, and Tl^{3+} were used to assist in removal of Cl^-, Br^-, and I^- because of their high affinity for these halide ions, the ratio $H_2^{18}O/H_2^{16}O$ in the product was studied. For a genuine $Co(NH_3)_5^{3+}$ intermediate this ratio should be >1 and constant regardless of the identity of X. When the assisting cation was Hg^{2+} the ratio 1.012 was observed for all three $[Co(NH_3)_5X]^{3+}$ ions, indicating the existence of $Co(NH_3)_5^{3+}$ as an intermediate. However with Ag^+ the ratio varied (1.009, 1.007, 1.010) indicating that $Co(NH_3)_5^{3+}$ does not have a completely independent existence in this case. For Tl^{3+} the ratios were 0.996, 0.993, and 1.003, showing considerable deviation from a pure D mechanism. In these experiments there is, however, the question of whether the entering water molecule in the assisted aquations comes from the bulk of the solvent or from the coordination sphere of the assisting metal ion. Thus like many another mechanistic study, this one is tricky to interpret.

5. A means of generating $Co(NH_3)_5^{3+}$ has been found by using the reaction 28-15, where azide is the sixth ligand:

$$[Co(NH_3)_5N_3]^{2+} + HNO_2 = [Co(NH_3)_5N_3NO]^{3+} = Co(NH_3)_5^{3+} + N_2 + N_2O \quad (28\text{-}15)$$

The relative rates of reaction of this with various anions (e.g., Cl^-, Br^-, SCN^-, F^-, HSO_4^-, $H_2PO_4^-$) and with H_2O were studied. The agreement between these results and those in the reaction 28-16:

$$Co(NH_3)_5(H_2O)^{3+} + X^- = Co(NH_3)_5X^{2+} + H_2O \quad (28\text{-}16)$$

was close, thus indicating that 28-16 also involves the intermediate $Co(NH_3)_5^{3+}$ or something of similar reactivity. By the principle of microscopic reversibility, this intermediate must also participate in the reverse of 28-16, that is, in the hydrolysis reaction itself. However other experiments are considered to show that the usual aquation reactions (e.g., that of $[Co(NH_3)_5NO_3]^{2+}$) *cannot* proceed through the same intermediate as that generated by oxidation of $[Co(NH_3)_5N_3]^{2+}$.

For the reaction 28-17, where L–L represents a bidentate amine, it has been found that the rate is increased by increasing bulk of the ligands, a result not in agreement with an I_a mechanism but consistent with an I_d mechanism.

$$[Co(L\text{-}L)_2Cl_2]^+ + H_2O = [Co(L\text{-}L)_2Cl(H_2O)]^{2+} + Cl^- \quad (28\text{-}17)$$

Another class of hydrolyses that have been extensively studied are those of *trans*-Co^{III} en_2AX species, in which the leaving group is X^-. The variation in rates and stereochemistry (*cis* or *trans*) of products as functions of the nature of A have been examined, and certain informative correlations established. When A is NH_3 or NO_2^- the data indicate that the mechanism is I_a, whereas for A = OH^-, Cl^-, N_3^-, and NCS^- an I_d mechanism is postulated. The assignments of mechanism in these cases depend heavily on detailed consideration of the stereochemical possibilities for the intermediates or activated complexes; thus they are indirect though apparently reliable.

Base Hydrolysis. The interpretation of a term of the type $k_B[ML_5X][OH^-]$ in a rate law for base hydrolysis has long been disputed. It could, of course, be interpreted as representing a genuine A process, OH^- making a nucleophilic attack on Co^{III}. However the possibility of a D-CB mechanism must also be considered. There are arguments on both sides, and of course the mechanism may vary for different complexes. Studies of base hydrolysis in octahedral complexes have so far dealt mainly with those of Co^{III}, and it is now reasonably sure that for these the predominant mechanism is, indeed, D-CB, with an intermediate that is probably of *tbp* geometry. The following discussion largely centers around Co^{III} complexes, but it should be pointed out that when the metal is changed from Co^{III} the reactivity patterns and, presumably, the mechanistic aspects are considerably changed. Detailed investigations of these other systems are still to be made.

It may be noted first that base hydrolysis of Co^{III} complexes is generally very much faster than acid hydrolysis, i.e., $k_B \gg k_A$ in eq. 23-13. This, in itself, provides evidence against a simple A mechanism, and therefore in favor of the D-CB mechanism because there is no reason to expect OH^- to be uniquely capable of nucleophilic attack on the metal. In the reactions of square complexes (see below) it turns out to be a distinctly inferior nucleophile toward Pt^{II}.

The validity of an D-CB mechanism can be examined in terms of three aspects of the overall process: (1) the acid-base behavior of the reacting complex, (2) the structure of the five-coordinate intermediate, (3) the ability of the amido or hydroxo group, which results from deprotonation of an amino or H_2O ligand, to stabilize such an intermediate.

The D-CB mechanism, of course, requires that the reacting complex have at least one protonic hydrogen atom on a nonleaving ligand and that the rate of reaction of this hydrogen be fast compared to the rate of ligand displacement. It has been found that the rates of proton exchange in many complexes subject to rapid base hydrolysis are in fact some 10^5 times faster than the hydrolysis itself [e.g., in $Co(NH_3)_5Cl^{2+}$ and $Co\ en_2NH_3Cl^{2+}$]. Such observations are in keeping with the D-CB mechanism but afford no positive proof of it. In the case of the complex 28-I the rates of proton exchange and base hydrolysis were found to be similar. It was further found that there was more exchange in the product than in the reactant and that this additional exchange did not occur subsequent to the act of base hydrolysis. Therefore it was concluded that the proton exchange formed an integral part of the base hydrolysis reaction.

(28-I)

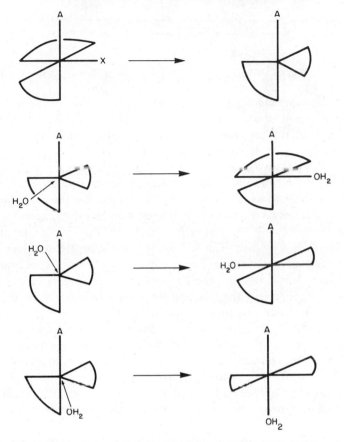

Fig. 28-3. Possible means by which a *tbp* intermediate in base hydrolysis by the *D*-CB mechanism could yield three isomeric products.

There have been a number of experiments supporting the idea that the conjugate base does react dissociatively to produce a reactive, five-coordinate intermediate with a lifetime sufficient for its reactivity to be independent of its origin, and for its characteristic pattern of discrimination among available entering ligands to be manifested. It is a common feature of base hydrolysis of the simpler Co^{III} amine systems (e.g., Co en_2AX) that there is considerable stereochemical change. For example,

$$\Lambda\text{-}cis\text{-}[Co\ en_2Cl_2]^+ + OH^- \rightarrow [Co\ en_2Cl(OH)]^+ + Cl^-$$

63% *trans*

21% Λ-*cis*

16% Δ-*cis*

This sort of result is best accommodated by postulating a *tbp* intermediate, on which the attack of H_2O can occur in several ways, each leading to one of these isomers. One possible intermediate, and the lines of attack on it, are shown in Fig. 28-3. If loss of X were to leave a square-pyramidal intermediate, it seems more likely that

Fig. 28-4. How an amide group could promote the dissociation of another ligand X.

there would be complete retention of stereochemistry. In the case of 28-I, where the rigidity of the macrocyclic ligand would favor the formation of a square-pyramidal intermediate, the *trans* configuration is, in fact, fully retained in the product.

Finally, there is the question of why the conjugate base so readily dissociates to release the ligand X. In view of the very low acidity of coordinated amines, the concentration of the conjugate base is a very small fraction of the total concentration of the complex. Thus its reactivity is enormously greater, by a factor far in excess of the mere ratio of k_B/k_A. It can be estimated that the ratio of the rates of aquation of $[Co(NH_3)_4NH_2Cl]^+$ and $[Co(NH_3)_5Cl]^{2+}$ must be greater than 10^6. Two features of the conjugate base have been considered in efforts to account for this reactivity. First, there is the obvious charge effect. The conjugate base has a charge that is one unit less positive than the complex from which it is derived. Though it is difficult to construct a rigorous argument, it seems entirely unlikely that the charge effect, in itself, can account for the enormous rate difference involved. It has been proposed that the amide ligand could labilize the leaving group X by a combination of electron repulsion in the ground state and a π-bonding contribution to the stability of the five-coordinate intermediate, as suggested in Fig. 28-4. However there are observations that some workers consider to contradict this explanation also; thus the question of why the conjugate base is hyperreactive remains unsettled.

It should also be noted that some results obtained in nonaqueous solvents also support the *D*-CB mechanism. Thus in dimethylsulfoxide reactions of the type

$$Co\ en_2NO_2Cl^+ + Y^- = Co\ en_2NO_2Y^+ + Cl^-$$
$$(Y = N_3^-, NO_2^-, or\ SCN^-)$$

are slow, with half-times in hours, but when traces of OH^- or piperidine are added the half-times are reduced to minutes. Since it was also shown that reaction of Co $en_2NO_2OH^+$ with Y^- is slow, an *A* mechanism with this as an intermediate is ruled out, and a genuine conjugate-base mechanism must prevail here.

Further interesting evidence for the conjugate base (though it does not bear on whether the actual aquation is *D* or *A*) comes from a study of the activity of OOH^- in base hydrolysis. Since OOH^- is a weaker base but a better nucleophile toward metal ions than OH^-, base hydrolysis by OOH^- compared to OH^- should proceed more slowly if its function is to form the conjugate base by removing a proton, but faster if it attacks the metal in a genuine *A* process. Experimental data are in agreement with the former.

Finally, in complexes having no protonic hydrogen, acceleration by base should not be observed according to the *D*-CB mechanism. This is in general true (as for 2,2'-bipyridine complexes, e.g.), but in a few cases a reaction of the first-order in

OH^- is observed nonetheless. One of these is the hydrolysis of $Co(EDTA)^-$ by OH^-. The formation of the seven-coordinate intermediate 28-II has been proposed, since it is also found that the complex racemizes with a first-order dependence on OH^-, but with a rate faster than that of hydrolysis. The seven-coordinate species could revert to $Co(EDTA)^-$ again without hydrolysis occurring, but with con- comitant racemization.

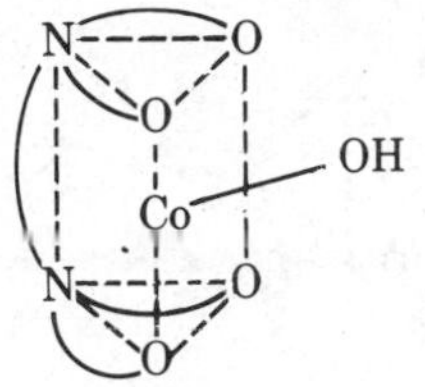

(28-II)

28-6. Attack on Ligands

In some known reactions ligand exchange does not involve the breaking of metal- ligand bonds, but instead bonds within the ligands themselves are broken and re- formed. One well-known case is the aquation of carbonato complexes. When iso- topically labeled water ($H_2{*}O$) is used, it is found that no ${*}O$ gets into the coordi- nation sphere of the ion during aquation,

$$[Co(NH_3)_5OCO_2]^+ + 2H_3{*}O^+ \rightarrow [Co(NH_3)_5(H_2O)]^{3+} + 2H_2{*}O + CO_2$$

The most likely path for this reaction involves proton attack on the oxygen atom bonded to Co followed by expulsion of CO_2 and then protonation of the hydroxo complex (eq. 28-18). Similarly, in the reaction of NO_2^- with pentaammineaquo-

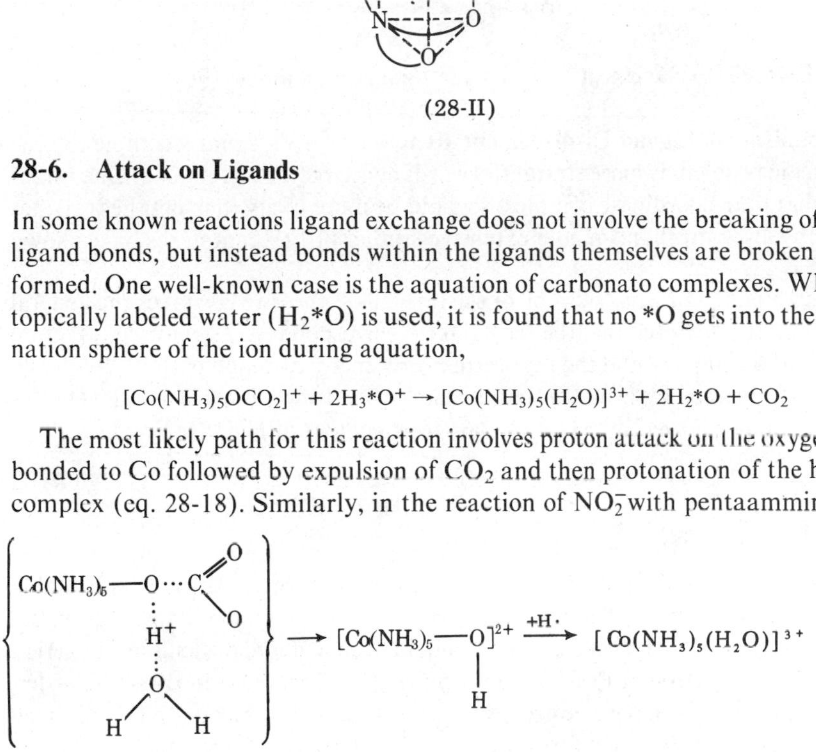

Transition state

cobalt(III) ion, isotopic labeling studies show that the oxygen originally in the bound H_2O turns up in the bound NO_2^-. This remarkable result is explained by the reaction sequence 28-19:

$$2NO_2^- + 2H^+ = N_2O_3 + H_2O \qquad (29\text{-}19a)$$

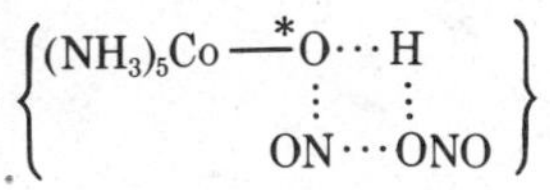

Transition state

$$\xrightarrow{\text{fast}} HNO_2 + [Co(NH_3)_5{*}ONO]^{2+} \xrightarrow{\text{slow}} [Co(NH_3)_5(NO{*}O)^{2+} \qquad (29\text{-}19b)$$

Another classic example of attack on ligands, in this case nucleophilic, is found in the initial step involved in the preparation of metal carbene complexes from metal carbonyls (see also Section 29-13), as shown in the sequence 28-20:

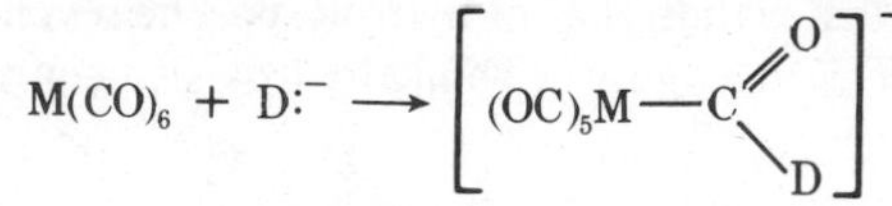

$$M(CO)_6 + D:^- \longrightarrow \left[(OC)_5M-C\underset{D}{\overset{O}{\diagup}} \right]^- \qquad (28\text{-}20a)$$

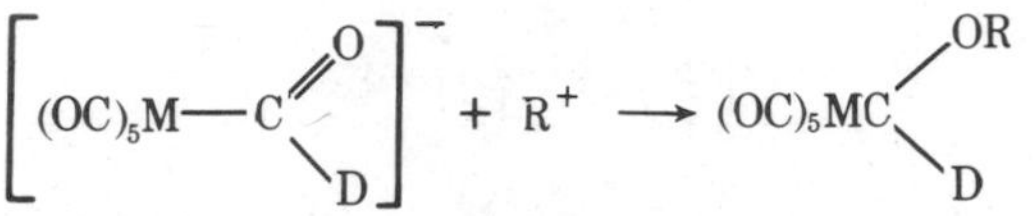

$$\left[(OC)_5M-C\underset{D}{\overset{O}{\diagup}} \right]^- + R^+ \longrightarrow (OC)_5MC\underset{D}{\overset{OR}{\diagup}} \qquad (28\text{-}20b)$$

28-7. Ligand-Displacement Reactions in Square Complexes

Mechanism of Ligand-Displacement Reactions. For square complexes the mechanistic problem is more straightforward, hence better understood. One might expect that four-coordinate complexes would be more likely than octahedral ones to react by an A mechanism, and extensive studies of Pt^{II} complexes have shown that this is so.

For reactions in aqueous solution of the type 28-21 the rate law takes the general form 28-22. It is believed that the second term corresponds to a genuine A reaction of Y with the complex, and the first term represents a two-step path in which one Cl^- is first replaced by H_2O (probably also by an A mechanism) as the rate-determining step followed by relatively fast replacement of H_2O by Y.

$$PtL_nCl_{4-n} + Y = PtL_nCl_{3-n}Y + Cl^- \qquad (28\text{-}21)$$
$$v = k[PtL_nCl_{4-n}] + k'[PtL_nCl_{4-n}][Y] \qquad (28\text{-}22)$$

It has been found that the rates of reaction 28-21 for the series of four complexes in which $L = NH_3$ and $Y = H_2O$ vary by only a factor of 2. This is a remarkably small variation, since the charge on the complex changes from -2 to $+1$ as n goes from 0 to 3. Since Pt—Cl bond breaking should become more difficult in this series, whereas the attraction of Pt for a nucleophile should increase in the same order, the virtual constancy in the rate argues for an A process in which both Pt—Cl bond breaking and Pt$\cdots$OH$_2$ bond formation are of comparable importance.

A general representation of the stereochemical course of displacement reactions of square complexes is given in Fig. 28-5. This process is entirely stereospecific: *cis* and *trans* starting materials lead, respectively, to *cis* and *trans* products.

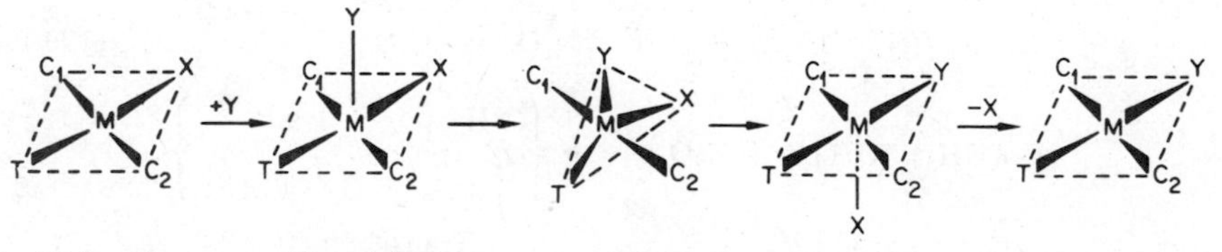

Fig. 28-5. The course of ligand displacement at a planar complex and the *tbp* five-coordinate structure.

Whether any of the three intermediate configurations possesses enough stability to be regarded as an actual intermediate rather than merely a phase of the activated complex remains uncertain. Since the starting complex possesses an empty valence shell orbital with which a fifth Pt—ligand bond could be formed (see Section 22-H-1 for a discussion of isolable five- and six-coordinated Pt^{II} complexes), the first alternative requires consideration.

It is interesting that the rates of reaction of the series of complexes [MCl-(o-tolyl)(PtEt$_3$)$_2$] with pyridine vary enormously with change in the metal M. The relative rates for Ni, Pd, and Pt are $5 \times 10^6:10^5:1$, which seems to be in accord with the relative ease with which these metal ions increase their coordination numbers from 4 to 6, as this is inferred from their general chemical behavior.

Although the evidence is less than complete, it appears likely that the A mechanism is valid for the reactions of square complexes other than those of Pt^{II}, such as those of Ni^{II}, Pd^{II}, Rh^{I}, Ir^{I}, and Au^{III}.

The order of nucleophilic strength of entering ligands (i.e., the order of the rate constants k' in eq. 28-22) for substitution reactions on Pt^{II} is

$$F^- \sim H_2O \sim OH^- < Cl^- < Br^- \sim NH_3 \sim \text{olefins} < C_6H_5NH_2 < C_5H_5N$$
$$< NO_2^- < N_3^- < I^- \sim SCN^- \sim R_3P$$

This order of nucleophilicity toward Pt^{II} does not correlate with the order of the ligands in terms of basicity, redox potentials, or other forms of reactivity, but it is remarkably consistent for a variety of substrates. It can be expressed in the form of a linear free energy relationship, similar to those employed for many types of organic reactions. One first defines the quantity n^0:

$$n^0 = \log(k'/k)$$

where k' and k are as defined in eq. 28-22 when the complex in eq. 28-21 is *trans*-[Pt py$_2$Cl$_2$] in methanol at 30°. We then write

$$\log K' = sn^0 + \log k$$

Log k, the "intrinsic reactivity," varies from one reacting complex to another, as does s, the "discrimination factor," but n^0 values are practically invarient with substrate.

It was observed long ago that *cis-trans* isomerization in planar complexes is catalyzed by traces of free ligands. Since a single displacement reaction is, as mentioned above, stereospecific and conserves stereochemistry, this is best explained in terms of the two-stage mechanism shown in Fig. 28-6.

The *trans* Effect. This particular feature of ligand replacement reactions in square complexes is of less importance in reactions of octahedral complexes except in some special cases where CO (or NO) is present as a ligand, or M=O or M≡N bonds are present. Most work has been done with Pt^{II} complexes, which are numerous and varied and have fairly convenient rates of reaction. Consider the general reaction 28-23:

$$[PtLX_3]^- + Y^- \rightarrow [PtLX_2Y]^- + X^- \tag{28-23}$$

Sterically, there are two possible reaction products, with *cis* and *trans* orientation

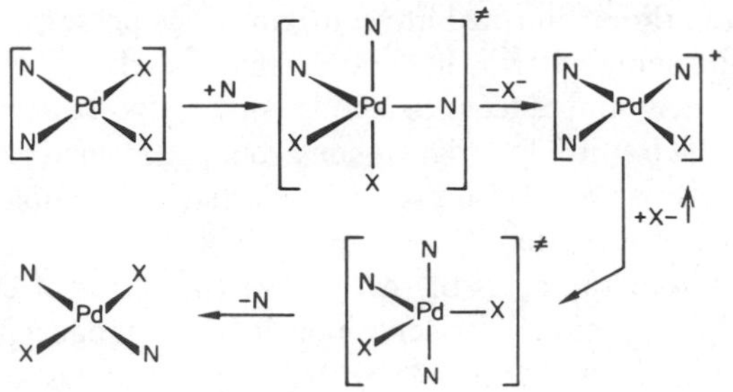

Fig. 28-6. Two-stage mechanism for the catalytic isomerization of *cis*-[Pd(amine)$_2$X$_2$] complexes to the corresponding trans isomers.

of Y with respect to L. It has been observed that the relative proportions of the *cis* and *trans* products vary appreciably with the ligand L. Moreover, in reactions of the type 28-24 either or both of the indicated isomers may be produced. It is found

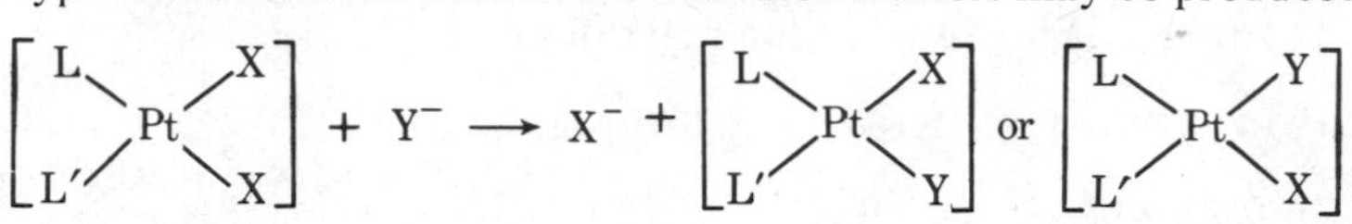

$$\left[\begin{array}{c} L \\ L' \end{array} Pt \begin{array}{c} X \\ X \end{array} \right] + Y^- \longrightarrow X^- + \left[\begin{array}{c} L \\ L' \end{array} Pt \begin{array}{c} X \\ Y \end{array} \right] \text{ or } \left[\begin{array}{c} L \\ L' \end{array} Pt \begin{array}{c} Y \\ X \end{array} \right] \qquad (28\text{-}24)$$

that both in these types of reaction and in others, a fairly extensive series of ligands may be arranged in the same order with respect to their ability to facilitate substitution in the position *trans* to themselves. This phenomenon is known as the *trans effect*. The approximate order of increasing *trans* influence is:

$$H_2O, OH^-, NH_3, py < Cl^-, Br^- < -SCN^-, I^-, NO_2^-,$$
$$C_6H_5^- < SC(NH_2)_2, CH_3^- < H^-, PR_3 < C_2H_4, CN^-, CO$$

It is to be emphasized that the *trans* effect is here defined solely as a kinetic phenomenon. It is the effect of a coordinated group on the rate of substitution at the position *trans* to itself in a square or octahedral complex.

The *trans* effect series has proved very useful in rationalizing known synthetic procedures and in devising new ones. As an example, consider the synthesis of the *cis* and *trans* isomers of [Pt(NH$_3$)$_2$Cl$_2$]. The synthesis of the *cis* isomer is accomplished by treatment of the [PtCl$_4$]$^{2-}$ ion with ammonia (reaction 28-25). Since Cl$^-$ has a greater *trans*-directing influence than does NH$_3$, substitution of NH$_3$

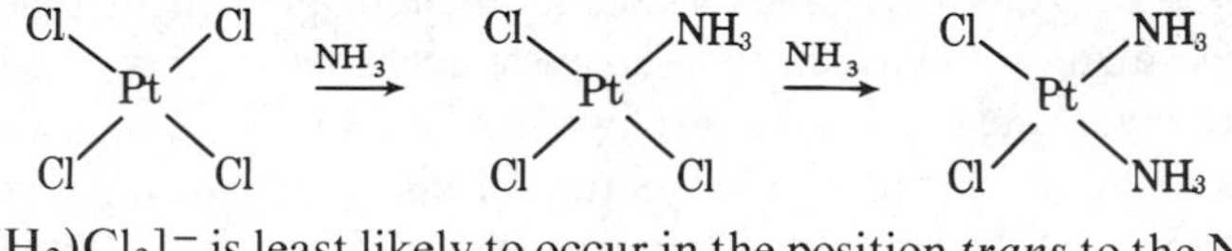

$$\begin{array}{c} Cl \\ Cl \end{array} Pt \begin{array}{c} Cl \\ Cl \end{array} \xrightarrow{NH_3} \begin{array}{c} Cl \\ Cl \end{array} Pt \begin{array}{c} NH_3 \\ Cl \end{array} \xrightarrow{NH_3} \begin{array}{c} Cl \\ Cl \end{array} Pt \begin{array}{c} NH_3 \\ NH_3 \end{array} \qquad (28\text{-}25)$$

into [Pt(NH$_3$)Cl$_3$]$^-$ is least likely to occur in the position *trans* to the NH$_3$ already present; thus the *cis* isomer is favored. The *trans* isomer is made by treating [Pt(NH$_3$)$_4$]$^{2+}$ with Cl$^-$ (reaction 28-26). Here the superior *trans*-directing influence of Cl$^-$ causes the second Cl$^-$ to enter *trans* to the first one, producing *trans*-[Pt(NH$_3$)$_2$Cl$_2$].

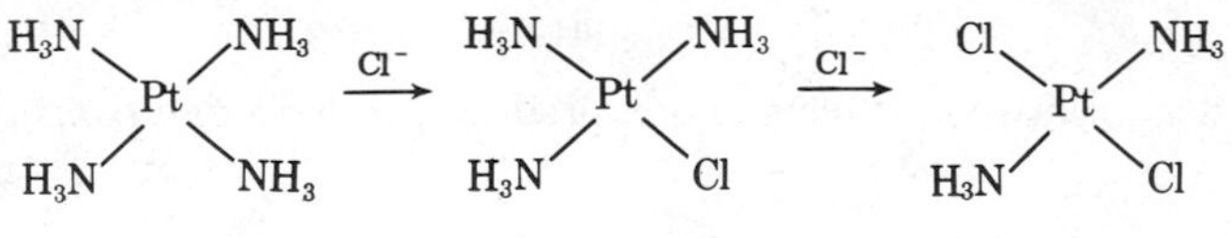

$$\begin{array}{c} H_3N \\ H_3N \end{array} Pt \begin{array}{c} NH_3 \\ NH_3 \end{array} \xrightarrow{Cl^-} \begin{array}{c} H_3N \\ H_3N \end{array} Pt \begin{array}{c} NH_3 \\ Cl \end{array} \xrightarrow{Cl^-} \begin{array}{c} Cl \\ H_3N \end{array} Pt \begin{array}{c} NH_3 \\ Cl \end{array} \qquad (28\text{-}26)$$

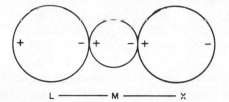

Fig. 28-7. Distribution of dipoles according to the polarization theory of the trans effect.

All theorizing about the *trans* effect must recognize that since it is a kinetic phenomenon, depending on activation energies, the stabilities of both the ground state and the activated complex are relevant. It is in principle possible for the activation energy to be affected by changes in one or the other of these energies or by changes in both.

The earliest attempt to explain the *trans* effect, the so-called polarization theory, is primarily concerned with effects in the ground state. It postulates a charge distribution as shown in Fig. 28-7. The primary charge on the metal ion induces a dipole in the ligand L, which in turn induces a dipole in the metal. The orientation of this dipole on the metal is such as to repel negative charge in the *trans* ligand X. Hence X is less attracted by the metal atom because of the presence of L. This theory would lead to the expectation that the magnitude of the *trans* effect of L and its polarizability should be monotonically related, and for some ligands in the *trans* effect series (e.g., H^-, $I^- > Cl^-$) such a correlation is observed. In effect this theory says that the *trans* effect is attributable to a ground state weakening of the bond to the ligand that is to be displaced.

An alternative theory of the *trans* effect was developed with special reference to the activity of ligands such as phosphines, CO, and olefins, which are known to be strong π acids. This model attributes their effectiveness primarily to their ability to stabilize a five-coordinate transition state or intermediate. Of course this model is relevant only if the reactions are bimolecular; there is good evidence that this is so in the vast majority of cases, if not in all. Figure 28-8 shows how the ability of a ligand to withdraw metal $d\pi$-electron density into its own empty π or π^* orbitals could enhance the stability of a species in which both the incoming ligand Y and the outgoing ligand X are simultaneously bound to the metal atom.

There is evidence to show that even when stabilization of a five-coordinate activated complex may be important, there is still a ground state effect—a weakening and polarization of the *trans* bond. In the anion $C_2H_4PtCl_3^-$ the Pt—Cl bond *trans*

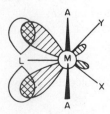

Fig. 28-8. Postulated *tbp* five-coordinate activated complex for reaction of Y with *trans*-MA₂LX, to displace X.

to ethylene is slightly longer than the *cis* ones, the Pt—*trans*-Cl stretching frequency is lower than the average of the two Pt—*cis*-Cl frequencies, and the nuclear quadrupole resonance (nqr) spectrum indicates that the *trans*-Cl atom is more ionically bonded.

The present consensus of opinion among workers in the field appears to be that in each case over the entire series of ligands whose *trans* effect has been studied, both the ground state bond weakening and the activated state stabilizing roles may be involved to some extent. For a hydride ion or a methyl group it is probable that we have the extreme of pure, ground state bond weakening. With the olefins the ground state effect may play a secondary role compared to activated state stabilization, although the relative importance of the two effects in such cases remains a subject for speculation, and further studies are needed.

28-8. Some Reactions of Metal Carbonyls

Chapter 25 was devoted to the metal carbonyls and related compounds, and a number of their reactions were mentioned. Perhaps the most general type of reaction for all metal carbonyls is that in which a CO ligand is replaced by another ligand.[3] A number of other reactions that involve, or lead to, organometallic compounds are discussed mechanistically in the next two chapters, especially Sections 29-8, 29-13, and 30-7 through 30-11. The fluxional character of metal carbonyls is covered later in this chapter (Section 28-15).

This section discusses some of the mechanistic features of ligand-exchange reactions. Our discussion is restricted to octahedral systems, since these provide excellent opportunities to examine stereoselectivity and they have been the most thoroughly studied. It may be mentioned that in some ways mechanistic study of metal carbonyl reactions is easier than for the complexes discussed earlier in this chapter. For instance, the compounds are generally soluble in nonpolar, nonligating solvents, and ion-pair formation is seldom a problem.

Reactions of the hexacarbonyls of Cr, Mo, and W with a variety of ligands (28-27a) have been extensively studied,[4] and in general they follow a two-term rate law (28-27b). The first term corresponds to an essentially limiting D process in which an $M(CO)_5$ intermediate is formed. Coordinately unsaturated intermediates are, in general, far more common and important in the reactions of metal carbonyls than in those of the conventional ionic complexes. The thermodynamic parameters of activation for the first-order path with $L = PhCH_2NH_2$ are $\Delta H_1^{\ddagger} = 150 \pm 12$ kJ mol^{-1} and $\Delta S^{\ddagger} = +80 \pm 37$ J deg^{-1} mol^{-1}, both of which are in good accord with the idea of a dissociative mechanism.

$$M(CO)_6 + L \rightarrow M(CO)_5L + CO \tag{28-27a}$$

$$-\frac{d[M(CO)_6]}{dt} = k_1[M(CO)_6] + k_2[M(CO)_6][L] \tag{28-27b}$$

The transition state for the bimolecular pathway is more difficult to define un-

3   G. R. Dobson, *Acc. Chem. Res.*, 1976, **9**, 300.
4   J. E. Pardue and G. R. Dobson, *Inorg. Chim. Acta*, 1976, **20**, 207.

equivocally. For the same ligand $PhCH_2NH_2$, the activation parameters are $\Delta H_2^{\ddagger}$ = 113 ± 8 kJ mol^{-1} and $\Delta S_2^{\ddagger} = -16 \pm 24$ J deg^{-1} mol^{-1}. These values and other evidence suggest that the second reaction pathway is a dissociative interchange I_d, in which M—CO bond breaking is well advanced in the activated complex but at the same time the ligand L is closely associated with the $M(CO)_6$ molecule.

Substitution reactions of $LM(CO)_5$ molecules often show remarkable stereo-specificity.[5] Each CO group cis to L is about 10 times more labile than the trans one; this is believed to be mainly because a ligand L that is less a π acid than CO stabilizes the pyramidal intermediate 28-III more than 28-IV, rather than because of the relative weakening of the cis M—CO bonds in the ground state, although that factor may contribute. The formation and stability of 28-III rather 28-IV is

(28-III)   (28-IV)

clearly shown by reactions of the type 28-28, where the principle of microscopic

$$LM(CO)_5 + {}^*CO \rightarrow cis\text{-}LM(CO)_4({}^*CO) + CO \qquad (28\text{-}28)$$

reversibility requires that if incoming CO ends up cis to L, the outgoing CO must have come from a cis position. With other entering ligands L', the products are mainly cis as in 28-29, although the reaction

$$LM(CO)_5 + L' \rightarrow cis\text{-}M(CO)_4LL' + CO \qquad (28\text{-}29)$$

conditions employed in actual preparations often lead to considerable isomerization to the trans isomers, which are generally more thermodynamically stable.

The stereoselectivity of reactions 28-28 or 28-29 tells nothing about whether the intermediate 28-III is rigid in the sense that the remaining four CO groups do not interchange among themselves. It is likely that they do so fairly rapidly. In the absence of a ligand L that is different from CO, scrambling definitely occurs as shown by the fact[6] that $cis\text{-}Mo(CO)_4({}^{13}CO)(C_5H_{10}NH)$ reacts with ^{13}CO to afford a statistical mixture of cis- and $trans\text{-}Mo(CO)_4({}^{13}CO)_2$.

Preferential labilization of a cis ligand coupled with retention of the configuration 28-III in the five-coordinate intermediate has some interesting consequences and applications. For example, it provides a convenient synthesis of $cis\text{-}Mo(CO)_4L_2$ species under conditions so mild that racemization is negligible.[7] The compound $cis\text{-}Mo(CO)_4(C_5H_{10}NH)_2$ is easily prepared free of trans isomer. It then reacts smoothly ($40°$; 10–15 min) with L to give $cis\text{-}Mo(CO)_4L(C_5H_{10}NH)$ and then $cis\text{-}Mo(CO)_4L_2$. In each step the weakest bond, Mo—N, is preferentially broken and the remaining noncarbonyl ligand controls the stereochemistry.

5   J. D. Atwood and T. L. Brown, *J. Am. Chem. Soc.,* 1976, **98**, 3160.
6   D. J. Darensbourg, M. Y. Darensbourg, and R. J. Dennenberg, *J. Am. Chem. Soc.,* 1971, **93**, 2807.
7   D. J. Darensbourg and R. L. Kump, *Inorg. Chem.,* 1978, **17**, 2680.

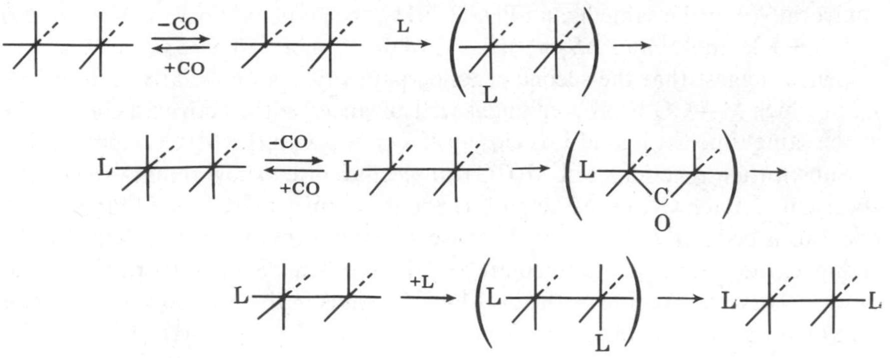

Fig. 28-9. A scheme to account for the course of reactions (28–30) and (28–31).

Another interesting example of the cis labilization principle, with some added complications, is in the stereospecific reactions 28-30 and 28-31, where L = PPh$_3$.

$$Mn_2(CO)_{10} + L \rightarrow ax\text{-}L(OC)_4MnMn(CO)_5 + CO \qquad (28\text{-}30a)$$

$$\text{rate} = k_1[Mn_2(CO)_{10}] + k_2[L][Mn_2(CO)_{10}] \qquad (28\text{-}30b)$$

$$ax\text{-}L(OC)_4MnMn(CO)_5 + L = ax,ax\text{-}L(OC)_4MnMn(CO)_4L + CO \qquad (23\text{-}31a)$$

$$R = k_3[Mn_2(CO)_9L] \qquad (28\text{-}31b)$$

It is found that $k_3 = 50k_1$. These results can be explained by the scheme in Fig. 28-9. One Mn(CO)$_5$ group has a cis labilizing effect on the other, but the first entering ligand L is too bulky to remain in the cis position, and there is prompt rearrangement to the trans (axial) position. The cis labilizing effect of L causes Mn$_2$(CO)$_9$L to lose CO at least 50 times faster than Mn$_2$(CO)$_{10}$, but for steric reasons the second L cannot enter the open site. CO transfer via a bridged intermediate occurs rapidly, and the second L therefore substitutes on the second Mn atom.

Similarly, Ru$_3$(CO)$_{12}$ reacts with PPh$_3$ to give only 1,2,3-Ru$_3$(CO)$_9$L$_3$ with no detectable mono- or disubstituted intermediates, and the process is zero order in PPh$_3$. However with L = P(OPh)$_3$, which is a poorer cis labilizer, the intermediates can be obtained.

Some metal carbonyl hydrides, e.g., HMn(CO)$_5$, HCo(CO)$_4$, are exceptionally reactive toward ligands that replace a CO group. This has been attributed to a special mechanism involving hydride migration; e.g.,

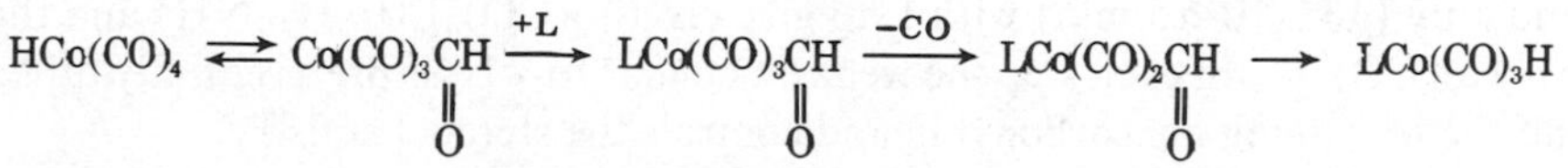

$$HCo(CO)_4 \rightleftharpoons Co(CO)_3\underset{\underset{O}{\|}}{C}H \xrightarrow{+L} LCo(CO)_3\underset{\underset{O}{\|}}{C}H \xrightarrow{-CO} LCo(CO)_2\underset{\underset{O}{\|}}{C}H \rightarrow LCo(CO)_3H$$

The intermediates here, which contain a formyl group, are similar to the reaction products in the case of the alkyl migration reaction (cf. Section 29-8).

Although HMn(CO)$_5$ reacts via this path, HRe(CO)$_5$ does not, nor does it

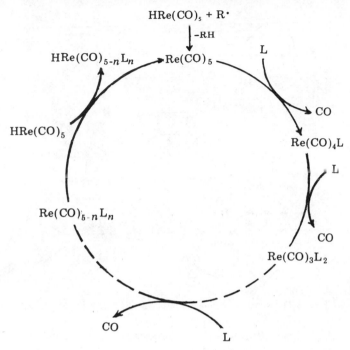

Fig. 28-10. Radical chain pathway, initiated by R', for substitution of CO's by L's in HRe(CO)$_5$.

readily react by initial loss of CO. It does, however, react readily by a radical chain pathway,[8] as shown in Fig. 28-10.

Other ways to obtain stereospecifically labeled metal carbonyl derivatives involve the use of certain olefin-substituted starting compounds. A good illustration[9] is provided in Fig. 28-11 (p. 1206). The key intermediate **2** provides access to a variety of *cis*-Mo(CO)$_4$L$_2$ products including those specifically labeled with ^{13}CO in one of the positions cis to both substituents.

ELECTRON TRANSFER PROCESSES[10]

The electron transfer processes can be divided into two main classes: (1) those in which the electron transfer effects no net chemical change, and (2) those in which there is a chemical change. The former, called *electron-exchange* processes, can be followed only indirectly, as by isotopic labeling or by nmr. The latter are the usual oxidation-reduction reactions and can be followed by many standard chemical and

8 B. H. Byers and T. L. Brown, *J. Am. Chem. Soc.,* 1975, **97**, 947.
9 D. J. Darensbourg, L. J. Todd, and J. P. Hickey, *J. Organomet. Chem.,* 1977, **137**, C1.
10 H. Taube, *Electron Transfer Reactions of Complex Ions in Solution,* Academic Press, 1970; A. Haim, *Acc. Chem. Res.,* 1975, **8**, 264; R. G. Linck, in *Survey of Progress in Chemistry,* Vol. 7, A. F. Scott, Ed., Academic Press, 1976, p. 89.

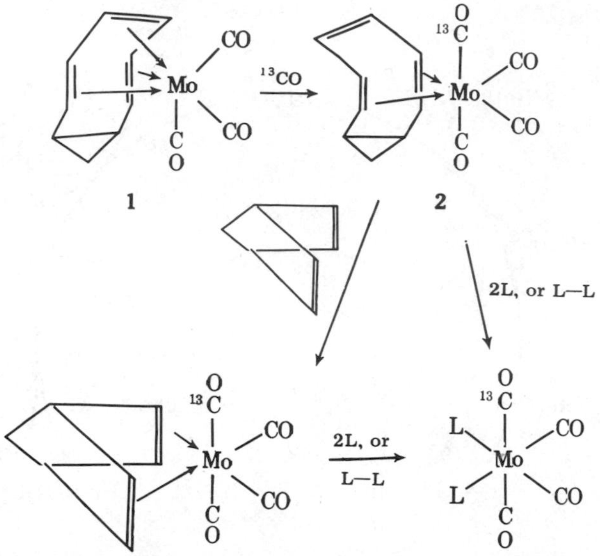

Fig. 28-11. Reactions giving stereospecifically labelled $L_2Mo(CO)_4$.

physical methods. The electron-exchange processes are of interest because of their particular suitability for theoretical study.

There are two well-established general mechanisms for electron transfer processes. In the first, called the *outer-sphere mechanism,* each complex retains its own full coordination shell, and the electron must pass through both. This, of course, is a purely formal statement in that we do not imply that the "same" electron leaves one metal atom and arrives at the other. In the second case, the *inner-sphere mechanism,* the two complexes form an intermediate in which at least one ligand is shared, that is, belongs simultaneously to both coordination shells.

28-9. Outer-Sphere Reactions

The outer-sphere mechanism is certain to be the correct one when *both* species participating in the reaction undergo ligand-exchange reactions more slowly than they participate in the electron transfer process. An example is the reaction

$$[Fe^{II}(CN)_6]^{4-} + [Ir^{IV}Cl_6]^{2-} \rightarrow [Fe^{III}(CN)_6]^{3-} + [Ir^{III}Cl_6]^{3-}$$

where both reactants are classified as inert ($t_{1/2}$ for aquation in 0.1 M solution >1 msec), but the redox reaction has a rate constant of $\sim 10^5$ l mol^{-1} sec^{-1} at 25°.

For reactions of the electron-exchange type a plot of the energy vs. reaction coordinate takes the symmetrical form shown in Fig. 28-12. The energy of activation E_{act} is made up of three parts: the electrostatic energy (repulsive for species of like charge), the energy required to distort the coordination shells of both species, and the energy required to modify the solvent structure about each species. Of these, the energy required to distort may not be immediately obvious, but it is crucial.

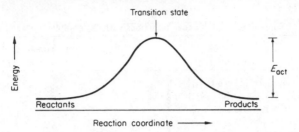

Fig. 28-12. Graph of energy versus reaction coordinate for an electron-exchange reaction in which reactants and products are identical.

The transition state for electron exchange will be one in which each species has the same dimensions. This is so because a transition state for a process in which there is no adjustment of bond lengths prior to the electron jump would necessarily have a much higher energy. Suppose for an M^{II}–M^{III} exchange the electron jumped while both ions were in their normal configurations. This would produce an M^{II} complex with bonds compressed *all the way* to the length appropriate to an M^{III} complex and an M^{III} complex with the bonds lengthened *all the way* to the length of those in the M^{II} complex. This would be the zenith of energy, and as the bonds readjusted, the energy of the exchanging pair would drop to the initial energy of the systems. However this zenith of energy is obviously higher than it would be if the reacting ions first adjusted their configurations so that each met the other one *only halfway* and then exchanged the electron. The more the two reacting species differ initially in their sizes, the higher will be the activation energy.

Table 28-1 lists some electron-exchange reactions believed to proceed by the outer-sphere mechanism, though for the Co^{II}—Co^{III} reactions this might not be correct, since one of the reactants (i.e., the Co^{II} partner) undergoes ligand substitution rapidly. The range covered by the rate constants is very large, extending from $\sim 10^{-4}$ up to perhaps nearly the limit of diffusion control ($\sim 10^{9}$). It is possible to account qualitatively for the observed variation in rates in terms of the second contribution mentioned to the activation energy.

In the first seven cases of Table 28-1, the two species differ by only one electron in an orbital that is approximately nonbonding with respect to the metal-ligand interaction (see Chapter 20). Therefore the lengths of the metal-ligand bonds should be practically the same in the two participating species, and the contribution to the activation energy of stretching and contracting bonds should be small. For the MnO_4^-–MnO_4^{2-} case, the electron concerned is not in a strictly nonbonding orbital. In the three cases of slow electron exchange there is a considerable difference in the metal-ligand bond lengths. However there is also a change in the extent of electron spin pairing among the nonexchanging electrons on each metal ion. Since it is possible that this could affect the rate of the process either through the activation energy or by influencing the frequency factor (the transmission coefficient, in terms of the absolute theory of rate processes), the significance of the Co^{II}–Co^{III} results is not entirely clear.

The importance of the energy required to change metal-ligand bond distances

TABLE 28-1

Rates of Some Electron-Exchange Reactions with Outer-Sphere Mechanisms

Reactants	Rate constants $(1 \text{ mole}^{-1} \text{ sec}^{-1})$
$[\text{Fe(bipy)}_3]^{2+}, [\text{Fe(bipy)}_3]^{3+}$ $[\text{Mn(CN)}_6]^{3-}, [\text{Mn(CN)}_6]^{4-}$ $[\text{Mo(CN)}_8]^{3-}, [\text{Mo(CN)}_8]^{4-}$ $[\text{W(CN)}_8]^{3-}, [\text{W(CN)}_8]^{4-}$ $[\text{IrCl}_6]^{2-}, [\text{IrCl}_6]^{3-}$ $[\text{Os(bipy)}_3]^{2+}, [\text{Os(bipy)}_3]^{3+}$	$>10^6$ at 25°
$[\text{Fe(CN)}_6]^{3-}, [\text{Fe(CN)}_6]^{4-}$	Second order, $\sim10^5$ at 25°
$[\text{MnO}_4]^-, [\text{MnO}_4]^{2-}$	Second order, $\sim10^3$ at 0°
$[\text{Co en}_3]^{2+}, [\text{Co en}_3]^{3+}$ $[\text{Co(NH}_3)_6]^{2+}, [\text{Co(NH}_3)_6]^{3+}$ $[\text{Co(C}_2\text{O}_4)_3]^{3-}, [\text{Co(C}_2\text{O}_4)_3]^{4-}$	Second order, $\sim10^{-4}$ at 25°

is suggested by the fact that V^{2+} and Cr^{2+} both appear to react with substitution-inert $[\text{Co(NH}_3)_6]^{3+}$ by an outer-sphere mechanism. However V^{2+} reacts the faster, even though the redox potential is more favorable with Cr^{2+}. This can be understood because the $Cr^{II}(t_{2g}^3 e_g)$ to $Cr^{III}(t_{2g}^3)$ oxidation presumably requires more reorganization of bond lengths than does the $V^{II}(t_{2g}^3)$ to $V^{III}(t_{2g}^2)$ oxidation. It is also possible, however, that the rate difference is due to the different degrees of orbital overlap by e_g and t_{2g} orbitals with the oxidant.

The rate of electron transfer between $[\text{Ru(phen)}_3]^{3+,2+}$ and $[\text{Ru(bipy)}_3]^{2+,3+}$ is essentially energy neutral even though the partners are chemically distinguishable.[11] It has a rate constant around 10^9 mol^{-1} sec^{-1}, as would be expected, since the Ru—N distances scarcely change in passing from t_{2g}^5 to t_{2g}^6 configurations.

In electron transfer reactions between two dissimilar ions, in which there is a net decrease in free energy, the rates are generally higher than in comparable electron-exchange processes. In other words, one factor favoring rapid electron transfer is the thermodynamic favorability of the overall reaction. This generalization seems to apply not only to the outer-sphere processes now under discussion but also to the inner-sphere mechanism to be discussed shortly.

In several cases the rate constants for reactions in Table 28-1 have been found to depend on the identity and concentration of cations present in the solution. The general effect is an increase in rate with an increase in concentration of the cations, but certain cations are particularly effective. The general effect can be attributed to the formation of ion pairs, which then decrease the electrostatic contribution to the activation energy. Certain specific effects, found for example in the $\text{MnO}_4^- - \text{MnO}_4^{2-}$ and $[\text{Fe(CN)}_6]^{4-} - [\text{Fe(CN)}_6]^{3-}$ systems, are less easily interpreted with certainty. The effect of $[\text{Co(NH}_3)_6]^{3+}$ on the former is thought to be due to ion pairing, greatly enhanced by the high charge. There is no evidence that the cations participate in the actual electron transfer, though this may be so in some cases.

The Marcus Theory. Much effort has been devoted to attempts to provide a

[11] R. C. Young, F. R. Keene, and T. J. Meyer, *J. Am. Chem. Soc.,* 1977, **99**, 2468.

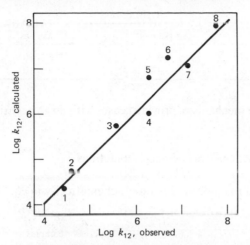

Fig. 28-13. Rate constants calculated from the Marcus cross-relation versus measured constants: 1, $Mo(CN)_8^{3-}-Fe(CN)_6^{4-}$; 2, $Fe(CN)_6^{3-}-W(CN)_8^{4-}$; 3, $IrCl_6^{2-}-Fe(CN)_6^{4-}$; 4, $IrCl_6^{2-}-Mo(CN)_8^{4-}$; 5, $Ce(IV)-Fe(CN)_6^{4-}$; 6, $Mo(CN)_8^{3-}-W(CN)_8^{4-}$; 7, $Ce(IV)-Mo(CN)_8^{4-}$; 8, $IrCl_6^{2-}-W(CN)_8^{4-}$.

quantitative theory that embodies the qualitative principles just discussed. The most successful treatment is that of R. A. Marcus,[12] who treated the problem in essentially classical statistical mechanical terms. The part of Marcus' theory having the broadest practical value is the *Marcus cross-relation,* which can be stated as follows:

$$k_{12} = (k_{11}k_{22}K_{12}f)^{1/2} \qquad (28\text{-}32)$$

where k_{12} and K_{12} are the rate and equilibrium constants for a given electron transfer reaction, and k_{11} and k_{22} are the rate constants for the "component" electron-*exchange* reactions (for which, of course, $K_{11} = K_{22} = 1$). The factor f is defined by eq. 28-33, where Z is the collision frequency

$$\log f = \frac{(\log K_{12})^2}{4\log(k_{11}k_{22}/Z^2)} \qquad (28\text{-}33)$$

for hypothetically uncharged reactant ions.

Figure 28-13 plots some observed values of k_{12} against those calculated by eq. 28-32. Clearly the agreement is remarkably good, considering that four orders of magnitude are covered. However there are some unresolved problems with the theory.[13] It is known, for example, that it gives poorer and poorer agreement as the value of K_{12} increases. This is illustrated by the data in Table 28-2, for two reactions with quite negative values of ΔF.[14] It should be emphasized, however, that the Marcus cross-relation is often justifiably invoked to supply corroboration[15]

12   R. A. Marcus, *Ann. Rev. Phys. Chem.,* 1964, **15**, 155.
13   J. F. Endicott and D. P. Rillema, *J. Am. Chem. Soc.,* 1972, **94**, 394; *Inorg. Chem.,* 1972, **11**, 2361; W. Schmickler, *Ber. Bunsenges. Phys. Chem.,* 1973, **77**, 991; S. G. Christov, *Ber. Bunsenges. Phys. Chem.,* 1975, **79**, 357.
14   A. Ekstrom, A. B. McLaren, and L. E. Smythe, *Inorg. Chem.,* 1976, **15**, 2853.
15   D. H. Huchital and J. Lepore, *Inorg. Chem.,* 1978, **17**, 1134.

TABLE 28-2

Reaction	ΔF (kJ mol^{-1})	k_{12} obs	k_{12} calc
$Fe^{III} + V^{II} \rightarrow Fe^{II} + V^{III}$	-96	3×10^4	1×10^6
$Fe^{III} + U^{III} \rightarrow Fe^{II} + U^{IV}$	-135	7×10^5	1×10^9

for an outer-sphere mechanism, provided cases with large negative ΔF values are avoided.

28-10.   Ligand-Bridged (Inner-Sphere) Reactions

The ligand-bridged reaction is illustrated schematically in eq. 28-34.

$$MX^{m+} + N^{n+} \rightarrow [M^{m+}-X-N^{n+}] \rightarrow [M^{(m-1)+} \underset{a}{-} X \underset{b}{-} N^{(n+1)+}] \rightarrow \text{products} \quad (28\text{-}34)$$

$$\text{Precursor complex} \qquad\qquad \text{Successor complex}$$

The essential features of the mechanism are:

1. A bridged binuclear complex, the *precursor complex,* is formed first by sharing some ligand X, which may use different atoms, even quite remote ones, to form the bonds to M and N. The formation of this complex requires N^{n+} to lose (at least) one of its initial ligands (not shown) to make room for the new bond to X.

2. A net electron transfer from N^{n+} to M^{m+} takes place giving the *successor complex.*

3. The *successor complex* collapses to give the products of the reaction.

In the recognition of this mechanism and pioneering studies of it by H. Taube and his students, reactions of the type 28-35 played a key role.

$$[Co(NH_3)_5X]^{2+} + Cr^{2+}(aq) + 5H^+ = [Cr(H_2O)_5X]^{2+} + Co^{2+}(aq) + 5NH_4^+ \quad (28\text{-}35)$$

$$(X = F^-, Cl^-, Br^-, I^-, SO_4^{2-}, NCS^-, N_3^-, PO_4^{3-}, P_2O_7^{4-}, CH_3COO^-,$$

$$C_3H_7COO^-, \text{crotonate, succinate, oxalate, maleate})$$

The significance and success of these experiments rest on the following facts. The Co^{III} complex is not labile though the Cr^{II} aqua ion is, whereas in the products the $[Cr(H_2O)_5X]^{2+}$ ion is not labile but the Co^{II} aqua ion is. It is found that the transfer of X from $[Co(NH_3)_5X]^{2+}$ to $[Cr(H_2O)_5X]^{2+}$ is quantitative. The most reasonable explanation for these facts is a mechanism such as that illustrated in 28-36.

$$Cr^{II}(H_2O)_6^{2+} + Co^{III}(NH_3)_5Cl^{2+} \rightarrow [(H_2O)_5Cr^{II}ClCo^{III}(NH_3)_5]^{4+}$$

$$\Big\updownarrow \begin{array}{l}\text{electron} \\ \text{transfer}\end{array} \qquad (28\text{-}36)$$

$$Cr(H_2O)_5Cl^{2+} + Co(NH_3)_5(H_2O)^{2+} \rightarrow [H_2O)_5Cr^{III}ClCo^{II}(NH_3)_5]^{4+}$$

$$\Big\downarrow H^+, H_2O$$

$$Co(H_2O)_6^{2+} + 5NH_4^+$$

Since all Cr^{III} species, including $Cr(H_2O)_6^{3+}$ and $Cr(H_2O)_5Cl^{2+}$, are substitution inert, the quantitative production of $Cr(H_2O)_5Cl^{2+}$ must imply that electron transfer, $Cr^{II} \rightarrow Co^{III}$, and Cl^- transfer from Co to Cr are mutually interdependent acts, neither possible without the other. Postulation of the binuclear, chloro-bridged intermediate appears to be the only chemically credible way to explain this.

In reactions between Cr^{2+} and CrX^{2+} and between Cr^{2+} and $Co(NH_3)_5X^{2+}$, which are inner sphere, the rates decrease as X is varied in the order $I^- > Br^- > Cl^- > F^-$. This seems reasonable if ability to "conduct" the transferred electron is associated with polarizability of the bridging group, and it was once thought that this order might be diagnostic of the mechanism. However the opposite order is found for $Fe^{2+}/Co(NH_3)_5X^{2+}$ and $Eu^{2+}/Co(NH_3)_5X^{2+}$ reactions; the $Eu^{2+}/Cr(H_2O)_5X^{2+}$ reactions give the order first mentioned, thus showing that the order is not simply a function of the reducing ion used.

In these classic examples of the inner-sphere mechanism it is the transfer of the bridging ligand X from cobalt to chromium that provides the definitive evidence for its operation. Although such a transfer may usually be expected, and usually occurs, it should be carefully noted that it is not an essential feature; it was not mentioned in the list of the three essential steps given earlier. Whether in the successor complex it is bond a or bond b (or both) that breaks depends on their relative strengths and labilities. Of course, if bond b breaks, there is no ligand transfer, hence no easy, direct proof that the precursor complex did indeed exist. An example in which bond b breaks is afforded by the reduction of $Ru(NH_3)_5[NC_5H_4C(NH_2)-O]^{3+}$, where the large ligand is nicotinamide coordinated to Ru^{III} by its pyridine nitrogen atom, by Cr^{2+} (aq), which attaches to the amido oxygen atom to form the precursor complex. After electron transfer we have the successor complex 28-V.

$$\left[(NH_3)_5Ru^{II}N\bigcirc\!-\!C \begin{array}{c} O \longrightarrow Cr^{III}(H_2O)_5 \\ \backslash \\ NH_2 \end{array} \right]^{5+}$$

$$(28\text{-}V)$$

The rate of Ru^{II}—N bond cleavage is even slower than that of Cr^{III}—O bond cleavage; consequently the reaction products are $Ru(NH_3)_5[NC_5H_4C(NH_2)O]^{2+}$ and Cr^{3+} (aq) and there is no transfer of the bridging ligand.

Actually, in the example just cited, the really unusual and interesting feature is that the successor complex is so stable that it is an observable intermediate. This is a rare situation, but there are other examples. For the reaction of $[Fe(CN)_6]^{3-}$ with $[Co(CN)_5]^{3-}$ the bridged anion $[(NC)_5Fe\text{-}CN\text{-}Co(CN)_5]^{6-}$ can be precipitated as its barium salt.

There are also reactions in which the precursor complex is stable enough to constitute an observable intermediate. Among these are 28-VI, which forms in the reduction of $Co(NH_3)_5(NTA)$ by Fe^{2+} (aq) and has a half-life of ca. 5 sec,[16] and 28-VII formed in the reduction of $[Co(NH_3)_5(4,4'\text{-bipyridyl})]^{3+}$ by the $[Fe(CN)_5(H_2O)]^{3-}$ ion.[17] An intermediate with half-life <0.1 sec, believed to be

[16] R. D. Cannon and J. Gardiner, *Inorg. Chem.*, 1974, **13**, 390.
[17] D. Gaswick and A. Haim, *J. Am. Chem. Soc.*, 1974, **96**, 7845.

the precursor complex, has also been observed in the reaction of $V^{IV}O(EDTA)$ with $V^{II}(EDTA)$.[18]

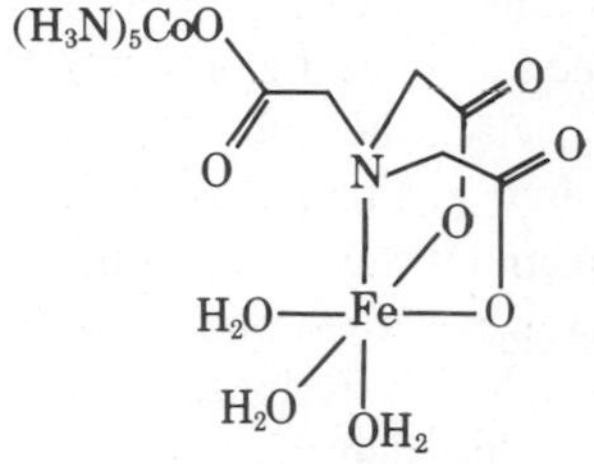

(28-VI)

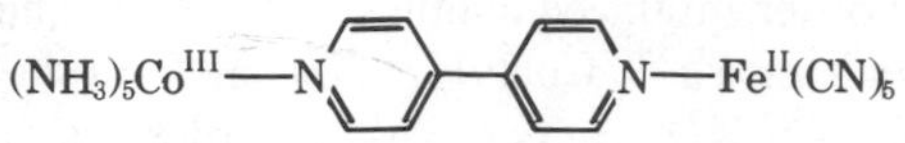

(28-VII)

A series of complexes in which the rates of electron transfer as such (i.e., the conversion of the preformed precursor complex to the successor complex) can be studied is typified by 28-VIII, which contains Co^{III} and Ru^{II}. The +3 cation containing Co^{III} and Ru^{III} can be prepared and isolated. When redissolved and treated with Eu^{II} or $[Ru(NH_3)_6]^{2+}$, extremely rapid oxidation by an outer-sphere mechanism at Ru occurs to give the +2 ion 28-VIII, whose conversion to the successor Co^{II}, Ru^{III} complex can then be observed.[19]

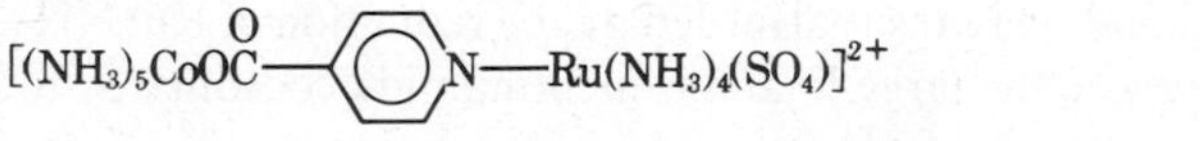

(28-VIII)

There is also the subtle question of the "intimate" mechanism of electron transfer by the inner-sphere path, that is, a detailed idea of how electron density is shifted from the reductant to the oxidant, once the bridged binuclear intermediate has been formed. Basically, two types of "intimate" mechanism have been considered:

1. A "chemical" mechanism, in which an electron is transferred to the bridging group, thus reducing it to a radical anion, whereupon an electron-hopping process eventually carries the electron to the oxidant metal ion.

2. A tunneling mechanism, whereby the electron simply passes from reductant to oxidant by quantum-mechanical tunneling through the barrier constituted by the bridging ligand.

In using organic bridging groups to investigate this question, the problem early arises of distinguishing between adjacent and remote attack by the reductant on the potential bridging group. In the case of benzoate ion as bridging group, attack must be on the coordinated carboxyl group, and there is evidence to show that it actually occurs on the carbonyl oxygen atom, as shown below.

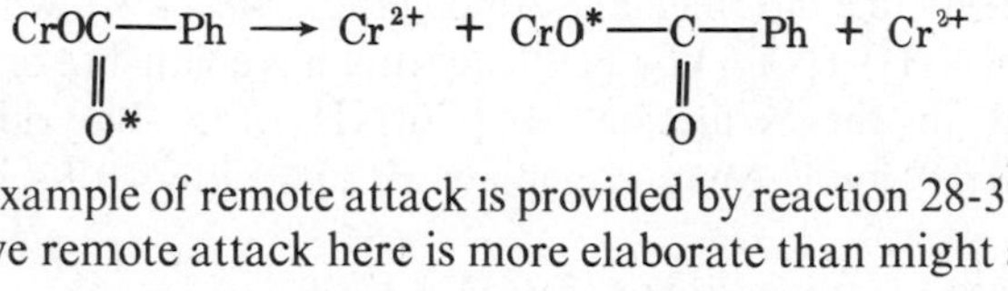

A definitive example of remote attack is provided by reaction 28-37. The evidence required to prove remote attack here is more elaborate than might at first sight be

[18] F. J. Kristine, D. R. Gard, and R. E. Shepherd, *J.C.S. Chem. Comm.,* **1976**, 994.
[19] S. S. Isied and H. Taube, *J. Am. Chem. Soc.,* 1973, **95,** 8198.

$$\left[(NH_3)_5Co-N\bigcirc-CONH_2\right]^{3+} + Cr^{2+} \xrightarrow{+H^+}$$

$$\left[(H_2O)_5CrO=C-\bigcirc NH\atop NH_2\right]^{4+} + Co^{2+} \quad (28\text{-}37)$$

suspected. The mere fact that the Cr^{III} product contains the amide-bound ligand does not assure that remote attack occurred as the rate-determining step. It is necessary to exclude the possibility that 28-IX might initially be formed and then isomerized by unreacted Cr^{2+}, as illustrated. In fact, the equilibrium represented

$$\left[Cr-N\bigcirc C=O\atop NH_2\right]^{3+} + Cr^{2+} = Cr^{2+} + \left[N\bigcirc-C=OCr\atop NH_2\right]^{3+}$$

$$(28\text{-}IX)$$

is established only very slowly, and it lies well to the left (pyridine being a much better ligand than an amide). In addition, reaction 28-38 proceeds much more slowly than reaction 28-37, and exclusively by an outer-sphere mechanism; all chromium appears as $Cr(H_2O)_6^{3+}$. Thus direct remote attack seems certain in the case of the p-amido ligand.

$$\left[(NH_3)_5Co-N\bigcirc\right]^{3+} + Cr^{2+} \xrightarrow{H^+} Cr(H_2O)_6^{3+} + Co^{2+} + \bigcirc NH^+ \quad (28\text{-}38)$$

An indication that the chemical intimate mechanism can be operative in remote attack is afforded by the following rate data. The first two pairs of reactions are inner sphere but presumably involve tunneling as the intimate mechanism. They indicate that tunneling to Co^{3+} is characteristically 10^6 times faster than to Cr^{3+}. The small rate ratio in the last pair of reactions then strongly implies that the rates are primarily set by the rate of reduction of the bridging ligand, this rate being only a second-order function of what metal ion is attached to the far end.

Reactants	Rate ratio, Co/Cr
$Co(NH_3)_5F^{2+}/Cr^{2+}$ $Cr(NH_3)_5F^{2+}/Cr^{2+}$	$\sim 10^6$
$Co(NH_3)_5OH^{2+}/Cr^{2+}$ $Cr(NH_3)_5OH^{2+}/Cr^{2+}$	$\sim 10^6$
$Co(NH_3)_5(N\bigcirc CONH_2)^{3+}/Cr^{2+}$ $Cr(NH_3)_5(N\bigcirc CONH_2)^{3+}/Cr^{2+}$	~ 10

Doubly Bridged Inner-Sphere Transfer. Only a few electron transfer reactions are known to proceed through doubly bridged intermediates, and the factors conducive to such a pathway are not well understood. It is, of course, unlikely that two trans ligands could simultaneous bridge, so that investigations have focused on cis complexes. An early example was reduction of cis-Co en$_2(N_3)_2^+$ or its tetraammine analogue by Cr^{2+}. Approximately 1.2 and 1.4 azide ligands per Cr^{III} are found in the respective products, which would seem to require the frequent occurrence of double-bridge formation with subsequent transfer of the bridges, though this is not mandatory.

A more recent example[20] merits discussion because it illustrates both the means of demonstrating the double-bridge pathway and the lack of an understanding of why it often does not occur even when circumstances seem favorable. The reduction of $Co(acac)_2en^+$ by Cr^{2+} gives the following product distribution for the Cr^{3+} species: $Cr(H_2O)_6^{3+}$, 31%; $Cr(H_2O)_4acac^{2+}$, 38%; and $Cr(H_2O)_2(acac)_2^+$, 31%. It can be concluded that the reduction goes about equally by each of three paths: (1) outer sphere to produce $Cr(H_2O)_6^{3+}$, (2) singly bridged inner sphere to transfer one acac, (3) doubly bridged inner sphere, to give $Cr(H_2O)_2(acac)_2^+$ with the transition state looking roughly as in 28-X. Curiously, the reduction of $Co(acac)_3$ by Cr^{2+} takes place 66% by the outer-sphere path, 34% by a singly bridged path, and not to any detectable extent by the doubly bridged path.

(28-X)

28-11. Mechanistic Criteria

It is not always easy to determine whether a particular electron transfer reaction proceeds by an inner-sphere or an outer-sphere path. As noted at the beginning of Section 28-9, if the electron transfer is between two species for both of which the rates of ligand exchange are distinctly slower than the rate of electron transfer, an outer-sphere mechanism is certain. On the other hand, when quantitative ligand transfer accompanies electron transfer, no explanation other than an inner-sphere mechanism seems tenable. However when such powerful indicators of mechanism do not apply, resort must be had to other criteria, with less conclusive results.

There is, of course, no reason for both mechanisms not to play a role in some cases, and the reduction of $[IrCl_6]^{2-}$ by Cr^{2+} (aq) is an example of this.[21]

[20] R. J. Balahura and N. A. Lewis, *J.C.S. Chem. Comm.,* **1976,** 268.
[21] A. G. Sykes and R. N. F. Thorneley, *J. Chem. Soc., A,* **1970,** 232.

As already noted, good agreement between an observed rate and that calculated by the Marcus cross-relation is often used as evidence in favor of the outer-sphere mechanism, but this, of course, is negative evidence.

Volumes of activation can be indicative of mechanism, since those for the outer-sphere mechanism can be estimated with reasonable confidence.[22] For the reaction of $[Fe(H_2O)_6]^{2+}$ with $[Fe(H_2O)_6]^{3+}$ the estimated value of $\Delta V^{\ddagger}$ is -14.4 cm^3 mol^{-1} and that measured is -12 ± 2 cm^3 mol^{-1}. On the other hand, for the reaction of $[Fe(H_2O)_6]^{2+}$ with $[Fe(H_2O)_5(OH)]^{2+}$ the calculated $\Delta V^{\ddagger}$ is -11.4 cm^3 mol^{-1}, but the measured value is $+0.8 \pm 0.9$ cm^3 mol^{-1} and this, together with the expected ability of the OH^- group to serve as a good bridging ligand, suggests that an inner-sphere mechanism operates in this case. Similarly, for reaction of $[Fe(H_2O)_6]^{2+}$ with the series of $[Co(NH_3)_5X]^{2+}$ cations ($X = F, Cl, Br, N_3$), calculated $\Delta V^{\ddagger}$ values for outer-sphere reactions lie between -10.6 and -12.8 cm^3 mol^{-1}, whereas the measured values are $+11, +8, +8$, and $+14$ cm^3 mol^{-1}, respectively, which strongly suggests an inner-sphere mechanism in all cases.

For aqua ions, or complexes containing H_2O as a ligand, a marked increase in rate with increase in pH is indicative that an inner-sphere mechanism involving bridging OH is at least a contributing mechanism, since outer-sphere reactions show little sensitivity to pH. The H_2O ligand has been found to be a poor bridging group, but deprotonation with increasing pH increases the concentration of M—OH groups that can react rapidly by an inner-sphere mechanism.

28-12. Two-Electron Transfers and Other Redox Reactions

Two-Electron Transfers. There are some elements that have stable oxidation states differing by two electrons, without a stable state in between. It has been shown that in the majority of these cases, if not in all, two-electron transfers occur. The Pt^{II}–Pt^{IV} system (discussed briefly below) and the Tl^I Tl^{III} system[23] have been studied in some detail. For the latter in aqueous perchlorate solution the rate law is

$$v = k_1[Tl^+][Tl^{3+}] + k_2[Tl^+][TlOH^{2+}]$$

In presence of other anions more complicated rate laws are found, indicating that two-electron transfers occur through various Tl^{3+} complexes.

A number of other redox reactions also appear to proceed by two-electron transfers, examples being:

$$Sn^{II} + Tl^{III} \rightarrow Sn^{IV} + Tl^I$$
$$Sn^{II} + Hg^{II} \rightarrow Sn^{IV} + Hg^0$$
$$V^{II} + Tl^{III} \rightarrow V^{IV} + Tl^I$$

All these reactions are *complementary,* meaning that in the overall stoichiometry the oxidant gains and the reductant loses two electrons.

Noncomplementary Reactions. Reactions of a noncomplementary type, that is, the number of electrons gained by one ion of one species is not equal to the

22 D. R. Stranks, *Pure Appl. Chem., 15th Int. Conf. Coord. Chem.,* 1974, **38**, 303.
23 H. A. Schwarz et al., *J. Phys. Chem.,* 1974, **78**, 488.

number lost by one ion of the other, must have multistep mechanisms, since ternary activated complexes are not likely and are not supported by any experimental evidence. This, in turn, means that some relatively unstable intermediate must be generated. For example, the overall reaction of Fe^{2+} with Tl^{3+} (eq. 28-39) might have as its initial step either a one-electron transfer to generate Fe^{3+} and the unstable Tl^{2+} or a two-electron transfer to generate Tl^+ and the unstable Fe^{4+}. It has been reported that the addition of Fe^{3+} reduces the rate of reaction, but the addition of Tl^+ is without effect. These and other observations indicate the mechanism to be that of 28-40a and 28-40b.

$$Tl^{3+}(aq) + 2Fe^{2+}(aq) \rightarrow Tl^+(aq) + 2Fe^{3+}(aq) \tag{28-39}$$

$$Fe^{2+}(aq) + Tl^{3+}(aq) \rightarrow Fe^{3+}(aq) + Tl^{2+}(aq) \tag{28-40a}$$

$$Fe^{2+}(aq) + Tl^{2+}(aq) \rightarrow Fe^{3+}(aq) + Tl^+(aq) \tag{28-40b}$$

Another noncomplementary reaction that has been studied is 28-41, where the

$$Sn^{II} + 2Fe^{III} \rightarrow Sn^{IV} + 2Fe^{II} \tag{28-41}$$

evidence shows successive one-electron transfers with Sn^{III} as the transient intermediate. Although this reaction is very slow when ClO_4^- is the only available anion, in presence of halide ions it is still first order in Fe^{3+} but goes much faster, perhaps because of intermediate oxidation of halide ions to species such as X_2^-. The cluster ion $Ta_6Cl_{12}^{3+}$ is an effective catalyst, presumably because it has three successive oxidation levels $(+2, +3, +4)$ conveniently accessible. Thus the following mechanism has been proposed:

$$Fe^{3+} + Ta_6Cl_{12}^{3+} \rightarrow Fe^{2+} + Ta_6Cl_{12}^{4+}$$
$$Ta_6Cl_{12}^{4+} + Sn^{2+} \rightarrow Sn^{4+} + Ta_6Cl_{12}^{2+}$$
$$Fe^{3+} + Ta_6Cl_{12}^{2+} \rightarrow Fe^{2+} + Ta_6Cl_{12}^{3+}$$

Ligand Exchange via Electron Exchange. When a metal atom forms cations in two oxidation states, one giving labile complexes and the other inert complexes, substitution reactions of the latter can be accelerated by the presence of trace quantities of the former. For example, the reactions of type 28-42 that are catalyzed by a trace of Cr^{2+} must occur as shown in eq. 28-43, in view of the complete retention of X by Cr^{III} while the NH_3's are completely lost.

$$[Cr(NH_3)_5X]^{2+} + 5H^+ = [Cr(H_2O)_5X]^{2+} + 5NH_4^+$$
$$(X = F^-, Cl^-, Br^-, I^-) \tag{28-42}$$

$$[Cr(NH_3)_5X]^{2+} + {}^*Cr^{2+}\,(aq) \longrightarrow$$
$$\{[(H_3N)_5Cr{-}X{-}{}^*Cr(H_2O)_5]^{4+}\} \longrightarrow [{}^*Cr(H_2O)_5X]^{2+} + \{Cr(NH_3)_5^{2+}\}$$

$$\text{Transition state} \qquad\qquad\qquad\qquad\qquad \text{rapidly} \downarrow +5H^+$$

$$5NH_4^+ + Cr^{2+}\,(aq)$$

$$\tag{28-43}$$

There are various other cases, especially the $[Cr(H_2O)_5X]^{2+}$–$[Cr(H_2O)_6]^{2+}$ exchanges, in which retention of the X groups shows that they must be bridges in the activated complex. Also, when Fe^{3+} is reduced by Cr^{2+} in presence of halide ions, the chromium(III) is produced as $[Cr(H_2O)_5X]^{2+}$. Similar phenomena are found also in the Co^{II}–Co^{III} aqua system and fairly generally in Pt^{II}–Pt^{IV} systems.

Pt^{II} catalyzes the exchange of chloride ion with $[Pt(NH_3)_4Cl_2]^{2+}$ in accordance with the following rate law:

$$v = k[Pt^{II}][Pt^{IV}][Cl^-]$$

The mechanism proposed to explain this is:

$$Pt(NH_3)_4^{2+} + {}^*Cl^- = Pt(NH_3)_4{}^*Cl^+ \text{ (fast preequilibrium)}$$

$$Pt(NH_3)_4Cl_2^{2+} + Pt(NH_3)_4{}^*Cl^+ \longrightarrow$$

$$
\begin{bmatrix}
& H_3N & NH_3 & H_3N & NH_3 & \\
& \diagdown\;\diagup & & \diagdown\;\diagup & \\
Cl\!-\!\!-\!\!-\!\!-Pt\!-\!-\!-\!-Cl\!-\!-\!-\!-Pt\!-\!\!-\!\!-\!\!-{}^*Cl \\
& \diagup\;\diagdown & & \diagup\;\diagdown & \\
& H_3N & NH_3 & H_3N & NH_3 &
\end{bmatrix}^{3+}
$$

$$\longrightarrow Pt(NH_3)_4Cl^+ + Pt(NH_3)_4{}^*ClCl^{2+}$$

The structure of the proposed activated complex, or intermediate, is very plausible, being quite comparable to the structures found in crystals of several compounds containing equal molar quantities of Pt^{II} and Pt^{IV} (see Fig. 22-H-1). There is also considerable kinetic evidence corroborating this mechanism. It is likely that traces of Pt^{II} generated by adventitious reducing agents (traces of other metal ions, organic matter, etc.) or photochemically play a role in many reactions of Pt^{IV} complexes. It is also known that traces of other metals, notably Ir, which have several oxidation states, can catalyze Pt^{IV} reactions.

Reductions by Hydrated Electrons.[24] The hydrated electron is a powerful reducing agent with a redox potential estimated to be -2.7 V. It is an ephemeral species with a half-life of $<10^{-3}$ sec, but it may be generated by pulse radiolysis at roughly millimolar concentrations. It reacts very rapidly, generally with second-order rate constants of 10^8 to 5×10^{10}, which indicate that the reactions are at or near the diffusion-controlled limit. Because of its great reducing power and great speed, the hydrated electron is highly useful for generating in aqueous media unstable low-valent cations whose decomposition or reaction kinetics can then be studied.

STEREOCHEMICAL NONRIGIDITY

Most molecules have a single, well-defined nuclear configuration. The atoms execute approximately harmonic vibrations about their equilibrium positions, but in other respects the structures may be considered rigid. There are, however, a significant number of cases in which molecular vibrations or intramolecular rear-

---

[24] See J. C. Sullivan et al., J. Phys. Chem., 1976, **80**, 1684 for examples and references.

rangements carry a molecule from one nuclear configuration into another. When such processes occur at a rate permitting detection by at least some physical or chemical method, the molecules are designated as *stereochemically nonrigid*. In some cases the two or more configurations are not chemically equivalent and the process of interconversion is called *isomerization* or *tautomerization*. In other cases the two or more configurations are chemically equivalent, and this type of stereochemically nonrigid molecule is called *fluxional*.

The rearrangement processes involved in stereochemically nonrigid molecules are of particular interest when they take place rapidly, although there is a continuous gradation of rates and no uniquely defined line of demarcation can be said to exist between "fast" and "slow" processes. The question of the speed of rearrangement most often derives its significance when considered in relation to the *time scale* of the various physical methods of studying molecular structure. In some of these methods, such as electronic and vibrational spectroscopy and gas phase electron diffraction, the act of observation of a given molecule is completed in such an extremely short time ($<10^{-11}$ sec) that processes of rearrangement may seldom if ever be fast enough to influence the results. Thus for a fluxional molecule, where all configurations are equivalent, there will be nothing in the observations to indicate the fluxional character. For interconverting tautomers the two (or more) tautomers will each be registered independently, and there will be nothing in the observations to show that they are interconverting.

It is the technique of nmr spectroscopy that most commonly reveals the occurrence of stereochemical nonrigidity, since its time scale is typically in the range 10^{-2}–10^{-5} sec. The rearrangements involved in stereochemically nonrigid behavior are rate processes with activation energies. When these activation energies are in the range 25–100 kJ mol^{-1} the rates of the rearrangements can be brought into the range of 10^2–10^5 sec^{-1} at temperatures between $+150°$ and $-150°$. Thus by proper choice of temperature, many such rearrangements can be controlled so that they are slow enough at lower temperatures to allow detection of individual molecules, or environments within the molecules, and rapid enough at higher temperatures for the signals from the different molecules or environments to be averaged into a single line at the mean position. Thus by studying nmr spectra over a suitable temperature range, the rearrangement processes can be examined in much detail.[25]

28-13. Stereochemically Nonrigid Coordination Compounds[26]

Coordination polyhedra are usually thought of in essentially static terms, that is, as if there are no intramolecular interchanges of ligands. In many cases, especially for octahedral complexes, this is valid, but there is a growing body of evidence that nonrigidity, particularly fluxionality, is not uncommon. In fact, for five-coordinate

[25]   L. M. Jackman and F. A. Cotton, Eds., *Dynamic Nuclear Magnetic Resonance Spectroscopy,* Academic Press, 1975.
[26]   J. P. Jesson and E. L. Muetterties, in Ref. 25, p. 253.

Fig. 28-14. The inversion of a pyramidal molecule WXYZ. Note that if X, Y, and Z are all different, the *invertomers* are enantiomorphous.

complexes and most of those with coordination numbers of 7 or higher, nonrigidity is the rule rather than the exception.

A common type of fluxional behavior is the inversion of pyramidal molecules (Fig. 28-14). In the cases of NH_3 and other simple noncyclic amines the activation energies, which are equal to the difference between the energies of the pyramidal ground configurations and the planar transition states, are quite low (24–30 kJ mol^{-1}) and the rates of inversion extremely high (e.g., 2.4×10^{10} sec^{-1} for NH_3). Actually in the case of NH_3 the inversion occurs mainly by quantum mechanical tunneling through the barrier rather than by passage over it. In most cases, however, passage over a barrier (i.e., a normal activated rate process) is operative. With phosphines, arsines, R_3S^+, and R_2SO species the barriers are much higher (>100 kJ mol^{-1}), and inversions are slow enough to allow separation of enantiomers in cases such as $RR'R''P$ and $RR'SO$.

Among four-coordinate transition metal complexes fluxional behavior based on planar tetrahedral interconversions is of considerable importance. This is especially true of nickel(II) complexes, where planar complexes of the type $Ni(R_3P)_2X_2$ have been shown to undergo planar $\rightleftharpoons$ tetrahedral rearrangements with activation energies of about 45 kJ mol^{-1} and rates of $\sim10^5$ sec^{-1} at about room temperature.

Trigonal-Bipyramidal Molecules. A class of fluxional molecules of great importance are those with a *tbp* configuration. Because of their importance and the great amount of information on them, they are discussed here at some length. When all five appended groups are identical single atoms, as in AB_5, the symmetry of the molecule is D_{3h}. The two apical atoms B_1 and B_2 (Fig. 28-15) are equivalent but distinct from the three equatorial atoms B_3, B_4, B_5, which are equivalent among themselves. In general, experiments such as measuring nmr spectra of B nuclei, which can sense directly the kind of environmental difference represented by B_1, B_2 versus B_3, B_4, B_5, should indicate the presence of two sorts of B nuclei in *tbp*

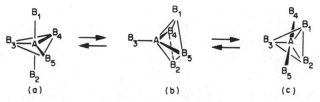

(a)   (b)   (c)

Fig. 28-15. The *tbp-sp-tbp* interconversion, the so-called Berry mechanism or pseudorotation for five-coordinate molecules.

molecules. In many cases, e.g., the ^{13}C spectrum of $Fe(CO)_5$, and the ^{19}F spectrum of PF_5 (to name the two cases where such observations were first made), all five B nuclei appear to be equivalent in the nmr spectrum, even though other experimental data with a shorter time scale, such as diffraction experiments and vibrational spectroscopy, confirm the *tbp* structure.

All the ligands in the nmr spectrum in the cases above appear to be equivalent because they pass rapidly between the axial and equatorial sites. Theory shows that if two nuclei occupying sites whose resonance frequencies ν_1 and ν_2 differ by $\Delta\nu$ sec^{-1} change places at a frequency greater than $\Delta\nu$ sec^{-1}, only one resonance at $(\nu_1 + \nu_2)/2$ will be observed. Obviously a ligand can move from an axial to an equatorial site only if there is a simultaneous shift of a ligand from an equatorial site to an axial one. With this in mind, it is then clear that there are only two types of intramolecular* exchange processes possible: (1) those in which each step involves only one axial and one equatorial ligand, and (2) those in which both axial ligands simultaneously exchange with two equatorial ones. In cases for which direct evidence has been obtained, the second type of process (2-for-2 exchange) is indicated.

It is important to realize that the nmr experiment can never do more than distinguish between two algebraically different permutations (i.e., 1-for-1 or 2-for-2, as above); it can never reveal the detailed pathways of the atoms. Two plausible, idealized pathways have been suggested for the 2-for-2 rearrangement of a *tbp* molecule. One of them, first suggested by R. S. Berry in 1960, is shown in Fig. 28-15. Not only do the *tbp* and *sp* configurations of an AB_5 molecule tend to differ little in energy, but, as Berry pointed out, they can also be interconverted by relatively small and simple angle deformation motions and in this way axial and equatorial vertices of the *tbp* may be interchanged. As shown in Fig. 28-15, the *sp* intermediate (b) is reached by simultaneous closing of the B_1AB_2 angle from 180° and opening of the B_4AB_5 angle from 120° so that both attain the same intermediate value, thus giving a square set of atoms, B_1, B_2, B_4, B_5, all equivalent to each other. This *sp* configuration may then return to a *tbp* configuration in either of two ways, one of which simply recovers the original while the other, as shown, places the erstwhile axial atoms B_1, B_2 in equatorial positions and the erstwhile equatorial atoms B_4, B_5 in the axial positions. Note that B_3 remains an equatorial atom and also that the molecule after the interchange is, effectively, rotated by 90° about the A—B_3 axis. Because of this apparent, but not real rotation, the Berry mechanism is often called a pseudorotation and the atom B_3 is called the pivot atom. Of course, the process can be repeated with B_4 or B_5 as the pivot atom, so that B_3 too will change to an axial position.

A second process that also results in a 2-for-2 exchange, called the "turnstile rotation" for obvious reasons, is shown in Fig. 28-16. As already noted, no choice between these is possible on the basis of the nmr spectra themselves for an AB_5-

---

* In both PF_5 and $Fe(CO)_5$ the persistence of $^{31}P—^{19}F$ and $^{57}Fe—^{13}C$ coupling rules out dissociative or bimolecular processes, and there is no reason to doubt that the overwhelming majority if not all fluxional *tbp* molecules rearrange intramolecularly.

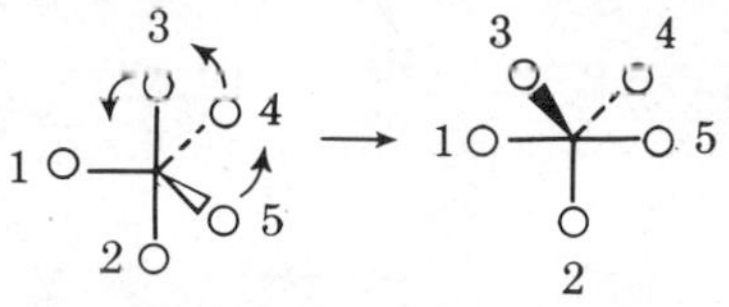

Fig. 28-16. The turnstile rotation.

molecule, nor has any other experiment been devised to do so. Theory,[27,28] however, suggests that the Berry process will usually have a significantly lower activation energy and will thus predominate. From another point of view, the turnstile process would appear to be a vibrationally excited modification of the Berry process.[28]

Systems with Coordination Number 6 or More. The octahedron is usually rather rigid, and fluxional or rapid tautomeric rearrangements generally do not occur in octahedral complexes unless metal-ligand bond breaking is involved. Among the few exceptions are certain iron and ruthenium complexes of the type $M(PR_3)_4H_2$. The *cis* and *trans* isomers of $Fe[PPh(OEt)_2]_4H_2$, for example, have separate, well-resolved signals at $-50°$ that broaden and collapse as the temperature is raised until at $+60°$ there is a single sharp multiplet indicative of rapid interconversion of the two isomeric structures. The preservation of the $^{31}P-^1H$ couplings affords proof that the rearrangement process is nondissociative. The distortion modes postulated to account for the interconversions are shown in Fig. 28-17. The rearrangement of "octahedral" bis and tris chelate complexes is considered in the next section.

Stereochemical nonrigidity, especially if it is fluxional, seems likely to be consistently characteristic of complexes with coordination numbers of 7 or greater. All seven-coordinate complexes so far investigated by nmr techniques have shown ligand-atom equivalence even though there is no plausible structure for a seven-coordinate complex that would give static or instantaneous equivalence. Thus ReF_7 and IF_7, for example, are presumed to be fluxional.

Eight-coordinate structures are characteristically nonrigid.[29] Those with do-

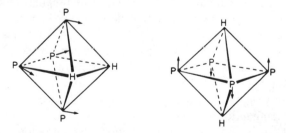

Fig. 28-17. The types of distortion postulated to lead to interconversion of cis and trans isomers of $Fe[PPh(OEt)_2]_4H_2$.

[27] A. Strich, *Inorg. Chem.*, 1978, **17**, 942.
[28] J. A. Altmann, K. Yates, and I. G. Csizmadia, *J. Am. Chem. Soc.*, 1976, **98**, 1450.
[29] E. L. Muetterties, *Inorg. Chem.*, 1973, **12**, 1963.

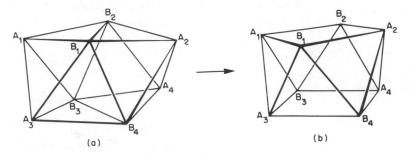

Fig. 28-18. How a dodecahedral set (a) of eight ligands may easily rearrange to a square-antiprismatic configuration (b).

decahedral structures (Fig. 28-18), in which there are two nonequivalent sets of ligands, can interchange ligands rapidly among these sets. A very ready means of doing this can be envisioned, although there is no proof that this mechanism is correct. In a dodecahedron, the A and B vertices differ in that each A vertex has only four next-nearest neighbors, whereas a B vertex has five. This is because a given A vertex is adjacent to three B vertices and one other A vertex, but a given B vertex is adjacent to three A vertices and two other B vertices. Note that if the B_1—B_2 and B_3—B_4 edges of the dodecahedron are lengthened so that $A_1B_2A_2B_1$ and $A_3B_3A_4B_4$ become square sets, the dodecahedron is transformed into a square antiprism (Fig. 28-18b). It would be possible to transform it into a differently labeled square antiprism by lengthening the B_1—B_4 and B_2—B_3 edges appropriately. The square antiprism can then return to the dodecahedron whence it came or, by having the A_1—A_2 and A_3—A_4 pairs approach each other, it may become a dodecahedron in which the former B ligands are A ligands and *vice versa*. Thus, just as the *sp* is a transition state or short-lived intermediate in the interconversion of differently permuted sets of *tbp* ligands, so the square antiprism may serve as a connecting link between differently permuted dodecahedral sets.

The key feature of the process we have just discussed is the opening of a bond shared by two adjacent triangular faces to create a rectangular face, followed by closing of a new bond across the other pair of vertices to generate a new pair of triangles. In the case just discussed two such triangle-diamond-triangle rearrangements must occur together, as shown, to regenerate the original type of polygon. A little reflection will show that the Berry process is also of this type, except that only one triangle-diamond-triangle transformation at a time is required. It is considered likely that this elementary process forms the basis for fluxional behavior in most triangulated polyhedra.

In the case of nine-coordinate species, where the ligands adopt the D_{3h} capped trigonal prism arrangement shown in Fig. 28-19, there is also an easy pathway for interchanging ligands of the two sets, and for species such as ReH_9^{2-}, $ReH_8PR_3^-$, and $ReH_7(PR_3)_2$ attempts to detect by nmr the presence of hydrogen atoms in two different environments have failed. Figure 28-19 shows the probable form of the rearrangement that causes the rapid exchange.

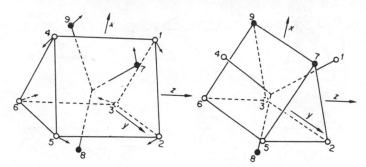

Fig. 28-19. The postulated pathway by which the ligands may pass from one type of vertex to the other in the D_{3h} tricapped trigonal prism.

28-14. Isomerization and Racemization of Trischelate Complexes[30]

Trischelate complexes exist in enantiomeric configurations Λ and Δ about the metal atom (Fig. 3-3), and when the chelating ligand is unsymmetrical (i.e., when it has two different ends), there are also geometrical isomers, *cis* and *trans*. Each geometrical isomer exists in enantiomeric forms; thus there are four different molecules.

In the case of tris complexes with symmetrical ligands, the process of inversion (interconversion of enantiomers) is of considerable interest. When the metal ions are of the inert type, it is often possible to resolve the complex; then the process of racemization can be followed by measurement of optical rotation as a function of time. Possible pathways for racemization fall into two broad classes: those without bond rupture and those with bond rupture.

There are two pathways without bond rupture that have been widely discussed. One is the trigonal, or Bailar, twist and the other is the rhombic, or Ray-Dutt, twist, in Fig. 28-20a and b, respectively.

The simplest dissociative pathways, in which one end of one ligand becomes detached from the metal atom appear in Fig. 28-20c to f. The intermediate may be five-coordinate, with either *tbp* or *sp* geometry, and the dangling ligand may occupy either an axial or an equatorial (or basal) position. In the case of the *sp* intermediates, it is possible that a solvent molecule might temporarily occupy a position in the coordination shell.

It should also be noted that in addition to the pathways shown in Fig. 28-20, it is conceivable that an associative process in which a solvent molecule S becomes coordinated might occur in some cases.[31] A seven-coordinate $M(LL)_3S$ species would probably undergo rapid intramolecular rearrangement, but considerable activation energy might be required to form it in the first place. There is no evidence for such a process as yet.

It has proved extremely difficult to determine unequivocally which of the various

[30] R. H. Holm, in Ref. 25, p. 317; N. Serpone and D. G. Bickley, *Prog. Inorg. Chem.*, 1972, **17**, 391.
[31] D. L. Kepert, *Inorg. Chem.*, 1974, **13**, 2758.

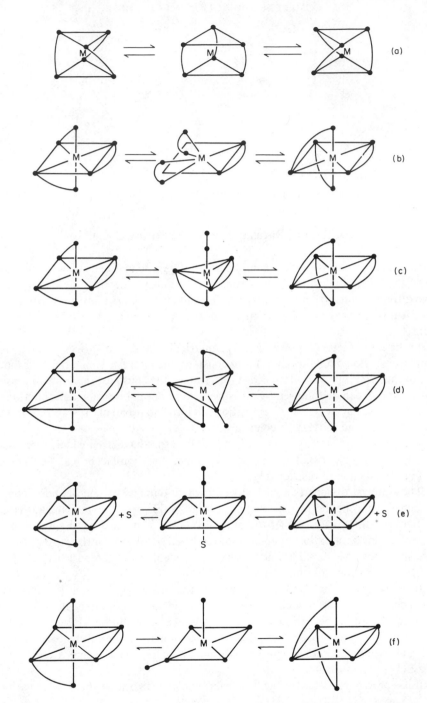

Fig. 28-20. Five possible modes of intramolecular racemization of a trischelate complex: (a) the trigonal shift, (b) the rhombic shift, (c, d) pathways with *tbp* intermediates, (e, f) ring-opening pathways with *sp* intermediates.

pathways shown is the principal one in a particular case. One of the earliest studies dealt with $[Cr(C_2O_4)_3]^{3-}$. For racemization of this complex it is likely that the mechanism is of a ring-opening type, since it has been shown that *all* oxalate oxygen atoms exchange with solvent water at a rate faster than that for oxalate exchange but almost equal to that of racemization.

A considerable amount of effort has been devoted to M(dike)$_3$ complexes because by using unsymmetrical diketonate ligands, the processes of isomerization and racemization can be studied simultaneously. Since isomerization can occur *only* by a dissociative pathway, it is often possible to exploit well-designed experiments to yield information on the pathways for both isomerization and racemization.

To illustrate the approach, let us consider some of the data and deductions for the system Co[CH$_3$COCHCOCH(CH$_3$)$_2$]$_3$, measured in C$_6$H$_5$Cl. It was found that both the isomerization and the racemization are intramolecular processes, which occur at approximately the same rate and with activation energies that are identical within experimental error. It thus appears likely that the two processes have the same transition state. This excludes a twist mechanism as the principal pathway for racemization. Moreover, it was found that isomerization occurs mainly with inversion of configuration. This imposes a considerable restriction on the acceptable pathways. Detailed consideration of the stereochemical consequences of the various dissociative pathways, and combinations thereof, leads to the conclusion that for this system the major pathway is through a *tbp* intermediate with the dangling ligand in an axial position.

In the case of the much more labile trisdiketonate complexes of aluminum and gallium, the techniques of study are more complex, since it is impossible to isolate even partially resolved samples. The most probable mechanisms for these labile complexes appear to be certain twist processes along with bond rupture to give *sp* transition states.

Evidence for the trigonal twist mechanism has been obtained in a few complexes in which the "bite" of the ligand is small, thus causing the ground state configuration to have a small twist angle ϕ, as defined in Fig. 2-10. The complex 28-XI is an example. A twist mechanism is preferred in these cases because the structure is

(28-XI)

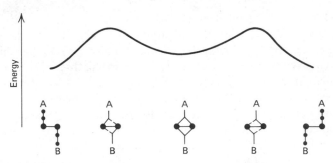

Fig. 28-21. Potential energy curve for the concerted exchange of two CO ligands groups via a bridging intermediate.

considerably distorted away from an octahedral configuration of sulfur atoms toward a trigonal-prismatic configuration. Thus the transition state, which is trigonal prismatic, is probably more energetically accessible than it would be if the complex had an essentially regular octahedral ground configuration.

28-15. Metal Carbonyl Compounds

It has now been conclusively shown that di- and polynuclear metal carbonyl compounds have a marked general tendency to engage in a type of fluxional behavior called *carbonyl scrambling*.[32] This type of behavior arises because of some of the inherent properties of metal to CO bonding. As shown in Section 25-4, especially in Fig. 25-2, the energy of a binuclear system consisting of two metal atoms and two CO groups does not in general vary a great deal (<30 kJ mol^{-1}) over the entire range of configurations from that in which there is one terminal CO ligand on each metal atom to that in which both CO groups are forming symmetrical bridges. The important thing that we now add to the picture presented in Fig. 25-2 is an emphasis on the symmetricality of the system, so that we can take case A from Fig. 25-2 and obtain Fig. 28-21. In this case the terminal arrangement is more stable than the bridging one, so that the overall process allows $CO(A)$ to pass from the left metal atom to the one on the right while $CO(B)$ is simultaneously making exactly the opposite journey. This sort of process may also occur around a three-membered ring (or even a larger one). The converse case, where a bridged arrangement is more stable than the terminal one, as in curve B of Fig. 25-2, may also be recognized. Concerted processes of this type account for most of the known cases in which CO groups are scrambled over a skeleton of two or more metal atoms. Let us examine a few illustrations.

The $Cp_2Fe_2(CO)_4$ molecule ($Cp = \eta\text{-}C_5H_5$) exists in solution as a mixture of cis and trans isomers with bridging CO groups, as shown in eq. 28-44. The ^{1}H nmr

32   R. D. Adams and F. A. Cotton, Ref. 25, page 489; S. Aime and L. Milone, *Prog. NMR Spectrosc.,* 1977, **11**, 183; J. Evans, *Adv. Organomet. Chem.,* 1976, **16**, 319; J. Lewis and B. F. G. Johnson, *Pure Appl. Chem.,* 1975, **44**, 43.

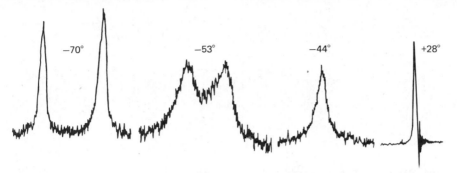

Fig. 28-22. The ^{1}H nmr spectra of the system *cis*- and *trans*-$[\eta^5\text{-}C_5H_5)Fe(CO)_2]_2$ at several temperatures.

(28-44)

cis   trans

resonances for the rings should appear at different positions and, as shown in Fig. 28-22, this is observed at $-70°$. However at $+28°$ only a single sharp signal at the intermediate position is seen. Clearly, between $-70°$ and room temperature some process by which the cis and trans isomers are interconverted becomes very rapid. This process cannot be a simple rotation because of the central rigid ring system. The nmr spectrum of the ^{13}C atoms of the CO groups shows that the cis-trans interconversion is accompanied by interchange of the CO groups between bridging and terminal positions. The explanation for both these processes is that the CO bridges open in a concerted way to give a nonbridged $Cp(OC)_2Fe\text{—}Fe(CO)_2Cp$ intermediate in which rotation about the Fe—Fe bond takes place. This rotation will be followed by a reclosing of bridges, but that may happen to produce either a cis or a trans isomer, regardless of which one was present before bridge opening. In addition, the CO groups that swing into bridge positions need not be the same ones that were there at the outset, so that bridge/terminal interchange will also result.

Another example[33] of CO scrambling is provided by the molecule 23-XII, which has four different types of CO group, *a–d,* as indicated: In a ^{13}C nmr spectrum, only at $-139°$ or lower are all the different resonances seen. When the temperature

33   F. A. Cotton *et al., J Am. Chem. Soc.,* 1977, **99**, 3293.

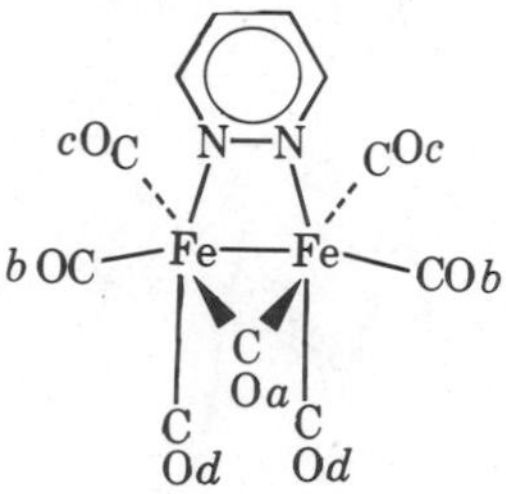

(28-XII)

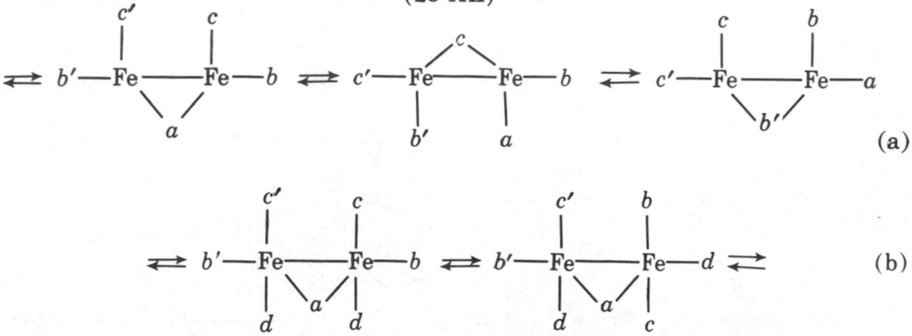

(a)

(b)

Fig. 28-23. (a) Cyclic mode of rearrangement of the five coplanar CO groups. (b) Mode of rearrangement of three terminal CO groups on the same iron atom.

is raised to about $-60°$ only two resonances, in an intensity ratio of 5:2, are seen. This is because the five approximately coplanar CO groups (types a, b, and c) are rapidly cycling around over the five available sites, as in Fig. 28-23a. Between $-60°$ and room temperature, this two-line spectrum collapses to a one-line spectrum as the five in-plane CO groups come into rapid exchange with the other two, probably by a process in which each set of three a, b, and c carbonyl groups on each iron atom rapidly rotates as in Fig. 28-23b. The combined effect of both rapid processes is to move all CO groups over all seven sites rapidly, thus making them all appear equivalent in the nmr spectrum, even though there are four distinct types at any instant.

This example introduces another rather common type of CO scrambling process, namely, localized rotation of the CO groups in an $M(CO)_3$ unit that is bound to some other portion of a large molecule. In virtually every known case, this rotation will occur rapidly below the decomposition temperature of the compound, regardless of whether there is any internuclear scrambling of the CO groups. For example, in 28-XIII the three CO groups are all nonequivalent, and below room temperature three distinct signals are observed, but by 60°C they have all collapsed. In this case, however, the substance decomposes before the combined signal can become sharp enough to be observed. An activation energy for rotation of about 66 kJ mol^{-1} can be estimated for this rotation. On the other hand there are cases of activation energy as low as 25 kJ mol^{-1}, as, for example, in 28-XIV. The polynuclear carbonyl

$Os_6(CO)_{18}$ provides an interesting example of three such processes in the same molecule (see Fig. 26-4c for the structure). For each of the three distinct types of $Os(CO)_3$ unit, the ^{13}CO signals collapse to a singlet at a different temperature, which ranges from $-100°$ to $-10°$ to $+40°$. It has not been possible to determine which type of $Os(CO)_3$ group is associated with which signal, however.

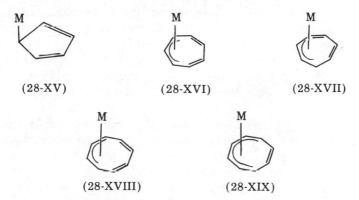

(28-XIII)   (28-XIV)

28-16.   Fluxional Organometallic Compounds[34]

Fluxionality is characteristic of certain classes of organometallic compound and is found sporadically in others. The phenomenon is seen characteristically in compounds containing conjugated cyclic polyolefins such as cyclopentadienyl, cycloheptadienyl, or cyclooctatetraene, to name the three most common ones, attached to a metal atom through at least one but less than all of their carbon atoms, as illustrated by the partial structures 28-XV to 28-XIX.

M   M   M

(28-XV)   (28-XVI)   (28-XVII)

M   M

(28-XVIII)   (28-XIX)

In each of these structures there are several structurally different ring carbon atoms and hydrogen atoms; thus complex nmr spectra would be expected. For example, the $(\eta^1\text{-}C_5H_5)M$ system 28-XV should have a complex downfield multiplet of the $AA'BB'$ type and relative intensity 4 for the four olefinic H atoms and an upfield multiplet of relative intensity 1 for the H atom attached to the same carbon

[34]   F. A. Cotton in Ref. 25, p. 377; K. Vrieze, in Ref. 25, p. 441; J. W. Faller, *Adv. Organomet. Chem.*, 1978, **16**, 211.

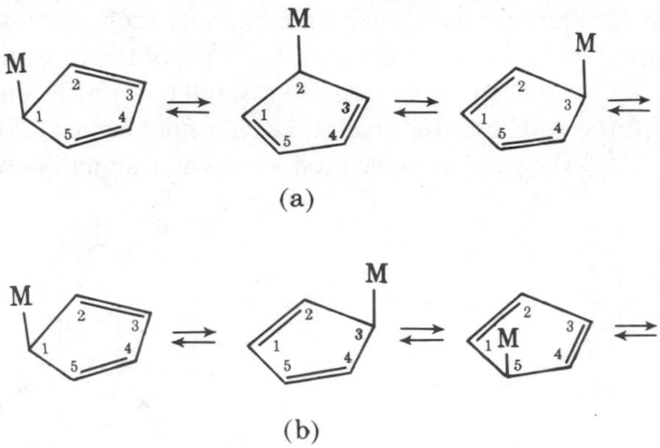

(a)

(b)

Fig. 28-24. Two rearrangement pathways for a $(\eta'\text{-}C_5H_5)M$ moiety. (a) 1,2 shifts. (b) 1,3 shifts.

atom as the metal atom. In fact, at room temperature, nearly all compounds containing moieties of these kinds exhibit only one sharp singlet in the ^1H or ^{13}C spectrum for the entire organic ligand. The explanation for this is that at higher temperatures the place of attachment of the metal atom to the ring is shifting rapidly over some or all members of the set of such equivalent points, thus conferring time-average equivalence on the C and H atoms. Two such hopping processes for the $(\eta^1\text{-}C_5H_5)M$ case are illustrated in Fig. 28-24. Because of the nature of the motion of the ring relative to the metal atom, these systems are commonly called "ring whizzers."

In each of the ring whizzer systems full characterization requires experimental information on three points: (1) proof of structure for the type of configuration in which the molecule rests between hops; (2) estimates of the rate of hopping at various temperatures, from which activation parameters can be obtained; (3) a knowledge of the pathway used in the hopping process. As the $(\eta^1\text{-}C_5H_5)M$ case shows, there is usually more than one set of jumps that can be considered a priori plausible.

On all these points information is provided by nmr spectra recorded at lower temperatures. When a temperature is reached such that the spectrum remains the same at still lower temperature, it is generally safe to assume that this spectrum (the slow-exchange spectrum) can be used to infer the structure of the molecule between jumps. In dozens of cases structural moieties of the types 28-XV to 28-XIX were expected, and their presence was confirmed in this way. Of course X-ray study of the crystalline compounds can usually also be carried out, and in every case the X-ray and low-temperature nmr results have been in agreement. If nmr spectra are then recorded at 10 to 20° temperature intervals between the slow-exchange limit and the higher temperatures and each one is matched to a computer-calculated spectrum for a given rate of rearrangement, the activation parameters can be evaluated by plotting the rates and temperatures in the usual ways.

In quite a few cases it can be shown that the appearance of the intermediate spectra is different for different hopping patterns (such as the 1,2-shifts and 1,3-shifts in Fig. 28-24), and a decision can therefore be made between them. In all cases of $(\eta^1\text{-}C_5H_5)M$ systems, observations of this kind coupled with other data have favored the 1,2-shift pathway. For systems of types 28-XVI to 28-XVIII the rearrangements also occur most easily by 1,2-shifts; it must be noted that additional pathways with higher activation energies may well come into play at higher temperatures, but they will not be detectable.

Direct proof that 1,2-shifts do not always afford the easiest pathway has recently been provided by a somewhat different type of nmr experiment called the spin saturation method. In this way it has been shown that for at least one system of type 28-XIX, namely, $(\eta^6\text{-}C_8H_8)Cr(CO)_3$, ring whizzing proceeds by both 1,3- and 1,2-shifts,[35] with the former having the lower activation energy.

In the case of 28-XX the usual nmr line shape experiment fails to detect the 1,3-shift interconversion shown, indicating either that it does not occur at all or that it has an activation energy greater than about 80 kJ mol^{-1}. A spin saturation experiment has indicated that this process does in fact take place with a free energy of activation of about 92 kJ mol^{-1}, but it is not clear whether it goes by a direct 1,3-shift.[36]

$$\text{Fe(CO)}_3 \qquad \text{Fe(CO)}_3$$

H$_2$ H$_2$

(28-XXa)   (28-XXb)

Although we have called the unitary step in the fluxional pathway for $(\eta^1\text{-}C_5H_5)M$ systems a 1,2-shift, it might equally well have been called a 1,5-shift. Physically there would be no point in doing so, but there are arguments based on the general theory of sigmatropic shifts for thinking of it in that sense. If we go to a system with a larger ring, for example, to a $(\eta^1\text{-}C_7H_7)M$ system, the 1,5-shift is now physically distinguishable from the 1,2-shift. For one such system, viz., $Ph_3Sn(\eta^1\text{-}C_7H_7)$, it has been shown that the fluxional pathway is, indeed, a sequence of 1,5-shifts.[37]

Allyl complexes are characteristically fluxional, the principal pathway being the $\eta^3\text{-}\eta^1\text{-}\eta^3$ process shown in eq. 28-45.

There are also compounds in which ligands exchange their modes of bonding. This is illustrated, for example, by $Ti(\eta^1\text{-}C_5H_5)_2(\eta^5\text{-}C_5H_5)_2$ as shown in eq. 28-46.

[35] J. A. Gibson and B. E. Mann, *J.C.S. Dalton,* **1979,** 1021.
[36] B. E. Mann, *J. Organomet. Chem.,* 1977, **141,** C33.
[37] B. E. Mann et al., *J. Organomet. Chem.,* 1978, **162,** 137.

$$(28\text{-}45)$$

$$(28\text{-}46)$$

General References

Bamford, C. H., and C. F. H. Tipper, Eds., *Comprehensive Chemical Kinetics*, Vol. 6, Elsevier, 1972. Reactions of nonmetallic inorganic compounds.

Basolo, F., and R. G. Pearson, *Mechanisms of Inorganic Reactions*, 2nd ed., Wiley, 1967.

Benson, D., *Mechanisms of Inorganic Reactions in Solution*, McGraw-Hill, 1968.

Burgess, J., *Metal Ions in Solution*, Wiley, 1978. Includes several good survey chapters on kinetics.

Cannon, R. D., *Adv. Inorg. Chem. Radiochem.*, 1978, **21**, 179. Reorganization energies in electron transfer processes.

Cooke, D. W., *Inorganic Reaction Mechanisms*, Chemical Society Monograph for Teachers, No. 33.

Edwards, J. O., Ed., *Inorganic Reaction Mechanisms*, Vols. 13, 17, Wiley-Interscience, 1970, 1972. Part I contains authoritative reviews on cobalt binuclear complexes, fast reactions, peroxide reactions, redox processes, electron transfer, and substitution in square d^8 complexes. Part 2 covers excited states, cation solvation, and isomerization of six-coordinate chelates.

Edwards, J. O., *Inorganic Reaction Mechanisms*, Benjamin, 1974.

Eigen, M., and R. G. Wilkins, in *Mechanisms of Inorganic Reactions*, ACS Advances in Chemistry Series, No. 49, p. 55. A survey of complex formation studies, with extensive references to literature on these and other fast reactions of complexes.

Evans, J. W., *Adv. Organomet. Chem.*, 1979, **16**, 319. Molecular rearrangements in polynuclear complexes including carbonyls.

Faller, J. W., *Adv. Organomet. Chem.*, 1978, **16**, 211. Fluxional and nonrigid π complexes.

Inorganic Reaction Mechanisms, Specialist Chemical Reports, Chemical Society. Authoritative annual surveys.

Pearson, R. G., and P. C. Ellgen, "Mechanisms of Inorganic Reactions in Solution," in *Physical Chemistry, An Advanced Treatise,* Vol. VII, H. Eyring, Ed., Academic Press, 1975.

Sykes, A. G., *Kinetics of Inorganic Reactions,* Pergamon Press, 1966. A good short introduction.

Tobe, M. L., *Inorganic Reaction Mechanisms,* Nelson, 1972.

Wilkins, R. G., *The Study of the Kinetics and Mechanism of Reactions of Transition Metal Complexes,* Allyn & Bacon, 1974.

CHAPTER TWENTY–NINE

Transition Metal to Carbon Bonds in Synthesis

This chapter discusses a number of stoichiometric reactions, some of which play an important role in catalytic reactions involving alkenes, CO, and H_2, which are covered in Chapter 30.

We first introduce the idea of coordinative unsaturation.[1]

If two substances A and B are to react at a metal atom contained in a complex in solution, clearly there must be vacant sites for their coordination. In heterogeneous reactions the surface atoms of metals, metal oxides, halides, etc., are necessarily *coordinately unsaturated;* but when even intrinsically coordinately unsaturated complexes such as square d^8 species are in solution, solvent molecules will occupy the remaining sites and these will have to be displaced by reacting molecules, thus:

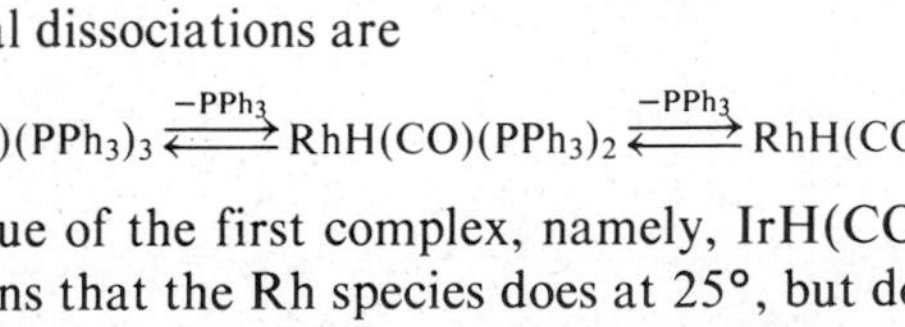

In five- or six-coordinated metal complexes, coordination sites may be made available by dissociation of one or more ligands either thermally or photochemically. Examples of thermal dissociations are

$$RhH(CO)(PPh_3)_3 \underset{-PPh_3}{\rightleftarrows} RhH(CO)(PPh_3)_2 \underset{-PPh_3}{\rightleftarrows} RhH(CO)(PPh_3)$$

The iridium analogue of the first complex, namely, $IrH(CO)(PPh_3)_3$, does not catalyze the reactions that the Rh species does at 25°, but does this when dissociation is induced either by heat or by ultraviolet irradiation.

Dissociation may also be promoted by a change in oxidation state as in oxidative-addition reactions discussed below. Thus although the complex $RhCl(PPh_3)_3$

---

[1]   C. A. Tolman, *Chem. Soc. Rev.,* 1972, **1**, 337.

(Section 22-G-2) is not appreciably dissociated in solution, under hydrogen it gives *cis*-RhH$_2$Cl(PPh$_3$)$_3$, which can dissociate one PPh$_3$ to provide a vacant site:

$$Rh^{I}Cl(PPh_3)_3 \underset{-H_2}{\overset{+H_2}{\rightleftharpoons}} Rh^{III}Cl(H)_2(PPh_3)_3 \underset{+PPh_3}{\overset{-PPh_3}{\rightleftharpoons}} RhCl(H)_2(PPh_3)_2$$

Sites may be made more readily available by the use of ligands with high trans effects. For example, the reaction

$$PtCl_4^{2-} + C_2H_4 \rightarrow [PtCl_3C_2H_4]^- + Cl^-$$

is quite slow, but it is accelerated by addition of tin(II) chloride, which forms PtCl$_3$(SnCl$_3$)$^{2-}$ where the Cl *trans* to SnCl$_3$ is labilized.

For phosphorus ligands (see pp. 87 and 147) in particular, steric effects are especially important,[2a] and compounds with bulky ligands have a greater tendency to dissociate.[2b] This is illustrated by the equilibria (see also p. 90):

$$NiL_4 \rightleftharpoons NiL_3 + L \quad K_d \ 25°, \ benzene$$

where for

$$P\left(O-\!\!\left\langle\!\!\bigcirc\!\!\right\rangle\!\!-CH_3\right)_3,$$

whose cone angle (see p. 89) is 128°, $K_d = 6 \times 10^{-10}$ M, whereas for PMePh$_2$ (136°) $K_d = 5 \times 10^{-2}$ M and for PPh$_3$ (145°) dissociation appears to be complete and no Ni(PPh$_3$)$_4$ could be detected in solution.

Steric effects of bulky ligands are by no means confined to dissociative equilibria but often have a determining influence on reactions such as oxidative addition and in catalytic reactions.[2] It is, however, not always easy to distinguish between steric and electronic effects in substituted phosphine ligands.

It was noted in Chapter 3 that many compounds of π-bonding ligands such as CO or PPh$_3$ obey the noble gas or eighteen electron rule. It has been proposed[1] that in many catalytic reactions sixteen- and eighteen-electron species are involved. However this view is an oversimplification, since in many catalytic cycles discussed in Chapter 30, coordinately unsaturated fourteen-electron species [e.g., RhCl(PPh$_3$)$_2$] are involved. More important is that postulation of the involvement of only even-electron, diamagnetic species completely over looks the fact that many catalytic and stoichiometric reactions that were originally thought to proceed by two-electron steps are, in fact, *radical reactions* involving one-electron changes. This is especially true for oxidative-addition reactions discussed later, but some insertion reactions (p. 1248) and carbonyl substitution[3] reactions are radical.

29-1. The Acid-Base Behavior of Metal Atoms in Complexes

Protonation and Lewis Base Behavior.[4]   In electron-rich complexes the metal atom may have substantial nonbonding electron density located on it and conse-

[2a] C. A. Tolman, *Chem. Rev.*, 1977, **77**, 313.

[2b] See, e.g., J. D. Druliner *et al.*, *J. Am. Chem. Soc.*, 1976, **98**, 2156.

[3]   T. M. Brown *et al.*, *J. Am. Chem. Soc.*, 1977, **99**, 2527.

[4]   J. C. Kotz and D. G. Pedrotty, *Organomet. Chem. Rev., A*, 1969, **4**, 479; D. F. Shriver, *Acc. Chem. Res.*, 1970, **3**, 231.

quently may be attacked by the proton or by other electrophilic reagents.

The first known example of a compound that could be protonated at the metal was $(\eta\text{-}C_5H_5)_2ReH$, an eighteen-electron molecule, which is a base comparable in strength to ammonia, cf.

$$H_3N + H^+ \rightleftharpoons H_4N^+$$
$$(\eta\text{-}C_5H_5)_2HRe + H^+ \rightleftharpoons (\eta\text{-}C_5H_5)_2H_2Re^+$$

The other similar angular compounds with lone pairs, $(\eta\text{-}C_5H_5)_2MoH_2$ and $(\eta\text{-}C_5H_5)_2WH_2$, can be readily protonated, but ferrocene, which has its electrons essentially in bonding orbitals, is an exceedingly weak base and can be protonated only by very strong acids (see p. 1166).

Arene compounds of Cr, Mo, and W such as $(\eta\text{-}C_6H_6)Mo(CO)_2PPhMe_2$ are protonated by trifluoroacetic acid and the cis-trans isomers 29-I are fluxional molecules.[5a]

(29-Ia)　　　　(29-Ib)

Many metal carbonyls, phosphine, and phosphite complexes,[5b] including cluster compounds, can be protonated, and often salts may be isolated. Examples are:

$$Fe(CO)_5 + H^+ \rightleftharpoons FeH(CO)_5^+$$
$$Ni[P(OEt)_3]_4 + H^+ \rightleftharpoons NiH[P(OEt)_3]_4^+$$
$$Ru(CO)_3(PPh_3)_2 + H^+ \rightleftharpoons RuH(CO)_3(PPh_3)_2^+$$
$$Os_3(CO)_{12} + H^+ \rightleftharpoons HOs_3(CO)_{12}^+$$

The acidification of carbonylate anions (p. 1062) can be regarded similarly, e.g.,

$$Mn(CO)_5^- + H^+ \rightleftharpoons HMn(CO)_5$$

Both eighteen- and sixteen-electron compounds that have lone pairs can also form adducts with Lewis acids such as BF_3, e.g.,

$$(\eta\text{-}C_5H_5)_2ReH + BF_3 \rightleftharpoons (\eta\text{-}C_5H_5)_2ReHBF_3$$
$$IrCl(CO)(PPh_3)_2 + BF_3 \rightleftharpoons IrCl(CO)(PPh_3)_2BF_3$$

29-2. Acceptor Properties or Lewis Acidity of Complexes

Coordinately unsaturated compounds, whether transition metal or not, can generally add neutral or anionic nucleophiles, e.g.,

$$PF_5 + F^- \rightleftharpoons PF_6^-$$
$$TiCl_4 + 2OPCl_3 \rightarrow TiCl_4(OPCl_3)_2$$

but it is important to note that even when they are electron rich, coordinately un-

[5a] T. C. Flood et al., J. Am. Chem. Soc., 1977, 99, 4334.
[5b] H. Werner et al., Helv. Chim. Acta, 1977, 60, 326.

saturated species may show Lewis acid *as well as* Lewis base behavior, e.g.,

$$trans\text{-}IrCl(CO)(PPh_3)_2 + CO \rightleftharpoons IrCl(CO)_2(PPh_3)_2$$
$$PdCl_4^{2-} + Cl^- \rightleftharpoons PdCl_5^{3-}$$

That is, sixteen-electron species can add a ligand to become eighteen electron species—the reverse of dissociation as discussed above.

THE OXIDATIVE–ADDITION AND REDUCTIVE–ELIMINATION REACTIONS[6]

29-3.   General Remarks

When a complex behaves *simultaneously* as Lewis acid and as Lewis base we have the so-called oxidative-addition reaction, which can be written generally as

$$L_yM^n + XY \rightarrow L_yM^{n+2}(X)(Y)$$

The reverse reaction can be termed *reductive elimination.* These terms merely describe the reaction and *have no mechanistic implication.* The mechanisms, discussed below, can be extremely complicated and can vary with the nature of the metal-ligand systems and the reactant that is oxidatively added.

For the addition reactions to proceed, we must have (*a*) nonbonding electron density on the metal M, (*b*) two vacant coordination sites on the complex L_yM to allow formation of two new bonds to X and Y, and (*c*) a metal M with its oxidation states separated by two units.

Many reactions of compounds even of nonmetals, not usually thought of as oxidative additions, may be so designated, e.g.,

$$(CH_3)_2S + I_2 \rightleftharpoons (CH_3)_2SI_2$$
$$PF_3 + F_2 \rightleftharpoons PF_5$$
$$SnCl_2 + Cl_2 \rightleftharpoons SnCl_4$$

The most intensively studied reactions for transition metals are those of complexes of metals with the d^8 and d^{10} electron configuration, notably, Fe^0, Ru^0, Os^0; Rh^I, Ir^I; Ni^0, Pd^0, Pt^0, and Pd^{II}, and Pt^{II}. An especially well-studied complex is the square *trans*-IrCl(CO)(PPh_3)_2 (Section 22-G-2), which undergoes reactions such as

$$trans\text{-}Ir^ICl(CO)(PPh_3)_2 + HCl \rightleftharpoons Ir^{III}HCl_2(CO)(PPh_3)_2$$

It will be noted that:

1.   Oxidative addition to sixteen-electron systems such as the coordinately unsaturated square complexes of Rh^I, Ir^I, or Pt^{II} produces eighteen-electron systems. The reverse of such reactions (i.e., reductive elimination) would reduce an eighteen-electron system to a sixteen-electron one.

2.   In additions of molecules such as H_2, HCl, or Cl_2, two new bonds to the metal

[6]   R. G. Pearson, *Symmetry Rules for Chemical Reactions,* Wiley, 1976, Chapter 5; J. K. Stille and K. S. Y. Lau, *Acc. Chem. Res.,* 1978, **11**, 434.

are made and the H—H, H—Cl, or Cl—Cl bond is broken. However molecules that contain *multiple bonds* may be added oxidatively *without* cleavage to form new complexes that have three membered rings. The addition can be looked on as the breaking of one component of the multiple bond, while leaving the others intact, i.e.,

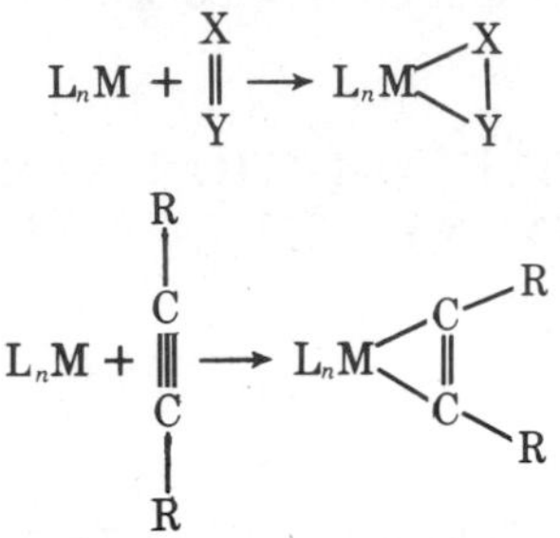

$$L_nM + \underset{Y}{\overset{X}{\|}} \longrightarrow L_nM\begin{matrix} X \\ | \\ Y \end{matrix}$$

$$L_nM + \underset{R}{\overset{R}{\underset{|}{\overset{|}{\underset{C}{\overset{C}{\|}}}}}} \longrightarrow L_nM\begin{matrix} C{-}R \\ \| \\ C{-}R \end{matrix}$$

Specific examples are:

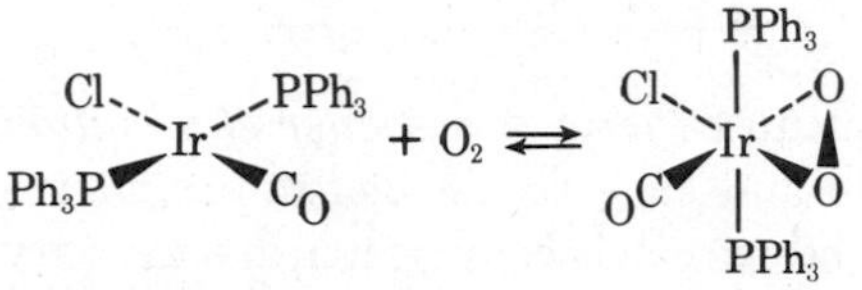

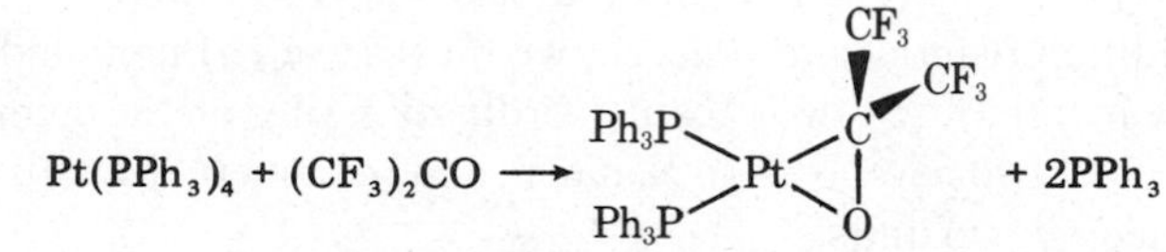

3. Compounds with eighteen electrons cannot undergo oxidative-addition reactions without expulsion of a ligand. This merely means that otherwise the most stable coordination number in the oxidized state would be exceeded. To supplement the last example above are the following cases:

$$Ru^0(CO)_3(PPh_3)_2 + I_2 = Ru^{II}I_2(CO)_2(PPh_3)_2 + CO$$
$$Mo^0(CO)_4bipy + HgCl_2 = Mo^{II}Cl(HgCl)(CO)_3bipy + CO$$

In some cases the oxidative-addition can be photoassisted, the reaction probably involving initial photodissociation, then oxidative addition to a coordinately unsaturated species.[7] Table 29-1 lists molecules that can be oxidatively added either with bond breaking for X—Y molecules or without for those involving multiple bonds.

For addition to sixteen-electron complexes where no ligand loss is involved, there will be an equilibrium reaction

$$L_yM^n + XY \rightleftharpoons L_yM^{n+2}XY$$

Whether the equilibrium lies on the reduced or the oxidized side depends very critically on (*a*) the nature of the metal and its ligands, (*b*) the nature of the added

7   See, e.g., R. A. Faltynek and M. S. Wrighton, *J. Am. Chem. Soc.,* 1978, **100,** 2701.

TABLE 29-1
Substances That Can Be Added Oxidatively[a]

Atoms separate	Atoms remain attached
$X-X$	O_2
H_2,	SO_2
Cl_2, Br_2, I_2, $(SCN)_2$, $RSSR^b$	
	$CF_2{=}CF_2$, $(CN)_2C{=}C(CN)_2$
$C-C$	
	$RC{\equiv}CR'$
Ph_3C-CPh_3, $(CN)_2$, C_6H_5CN,	$RNCS$
$MeC(CN)_3$	
	$RNCO$
$H-X$	
	$RN{=}C{=}NR'$
HCl, HBr, HI, $HClO_4$, C_6F_5OH, C_6H_5SH, H_2S, H_2O^c,	
CH_3OH	
$C_6F_5NH_2$, ⬠NH, $HC{\equiv}CR$, C_5H_6, CH_3CN, HCN,	
HCO_2R, C_6H_6, C_6F_5H, $HSiR_3$, $HSiCl_3$	
$H\text{-}B_{10}C_2HPMe_2$, $H\text{-}B_5H_8$	
	$RCON_3$
$C-X$	
	$R_2C{=}C{=}O$
CH_3I, C_6H_5I; CH_2Cl_2, CCl_4	
CH_3COCl, $C_6H_5CH_2COCl$	
C_6H_5COCl, CF_3COCl	
	CS_2
$M-X$	
	$(CF_3)_2CO$, $(CF_3)_2CS$, CF_3CN
Ph_3PAuCl, $HgCl_2$, $MeHgCl$, R_3SnCl, R_3SiCl, $RGeCl_3$	
H_8B_5Br, Ph_2BX	
Ionic	
$PhN_2^+\ BF_4^-$, $Ph_3C^+\ BF_4^-$	

[a] Oxidative addition can occur with neutral, anionic, and cationic complexes. Additions to more than one metal center are also possible. See R. Poilblanc, *Nouv. J. Chim.*, 1978, **2**, 145.

[b] A. W. Gal *et al.*, *Inorg. Chim. Acta*, 1979, **32**, 235.

[c] T. Yoshida *et al.*, *J. Am. Chem. Soc.*, 1979, **101**, 2027; B. N. Chaudret *et al.*, *J.C.S. Dalton*, **1977**, 1546.

molecule XY and of the M—X and M—Y bonds so formed, and (*c*) the medium in which the reaction is conducted. Qualitatively, for the addition to proceed, E_{MX} + E_{MY} must exceed E_{XY} + P, where E_{MX} and E_{MY} are the free energies of the new bonds to the metal, E_{XY} is the free energy for bond dissociation of XY, and P is the promotional energy for oxidation of the metal. Usually only E_{XY} is known.

The higher oxidation states are usually more stable for the heavier than for the lighter metals, so that, for example, Ir^{III} species are generally more stable than Rh^{III} species. For the ligands, *factors that tend to increase the electron density on the metal make for an increase in oxidizability*. For example, in the reactions of square iridium(I) complexes with carboxylic acids, e.g.,

$$trans\text{-}Ir^IX(CO)L_2 + RCOOH \rightleftharpoons Ir^{III}HX(O_2CR)(CO)L_2$$

the equilibria lie further to the Ir^{III} side in the order X = Cl < Br < I and in the orders L = PPh_3 < $PMePh_2$ < PMe_2Ph < PMe_3 and L = $P(p\text{-}FC_6H_4)_3$ < $P(p\text{-}MeC_6H_4)_3$. There is no direct correlation to be expected between the pK_a of various acids HX *in water* and the propensity to add to metals. For example, HCOOH (pK_a 3.75) and CH_3COSH (pK_a 3.33) have similar acidities, but on addition to $Pt(PPh_3)_4$ the formic acid adduct is very unstable, whereas the thioacetic acid adduct is quite stable. This difference is evidently due to the greater affinity toward platinum of sulfur than of oxygen. Hence both the acidity and the polarizability of the conjugate base are important.

Steric properties of ligands are also important. For example, very bulky ligands such as $PEtBu_2^t$ tend to decrease the ease of oxidative addition, whereas the substitution of an *o*-methoxy group on a phenylphosphine increases the nucleophilicity of the metal by donation[8] (29-II):

(29-II)

Stereochemistry and Mechanism of Addition. When the XY molecule adds without severance of X from Y, the two new bonds to the metal are necessarily in *cis* positions, but when X and Y are separated, the product may be one or more of several isomers with either *cis*- or *trans*-MX and -MY groups, e.g.,

The final product will be the isomer or isomer mixture that is the most stable thermodynamically under the pertaining conditions. The ligands, solvent, temperature, pressure, etc., will have a decisive influence on this. The nature of the final product does not necessarily give a guide to the initial product of the reaction, since isomerization of the initial product may also occur.

29-4. Addition of Specific Molecules

Hydrogen Addition. The addition of hydrogen is a special case in that despite the high H—H bond energy (450 kJ mol⁻¹), oxidative addition often occurs with great facility. The reactions are commonly reversible at 25° and 1 atm pressure,

---

8   E. M. Miller and B. L. Shaw, *J.C.S. Dalton,* **1974**, 480.

the hydrogen being readily removed by sweeping with N_2 or Ar or by evacuation, e.g.,

$$IrCl(CO)(PPh_3)_2 + H_2 \rightleftharpoons Ir(H)_2Cl(CO)(PPh_3)_2$$

The attack on the H_2 molecule is probably initiated by electron density on the metal atom flowing into the antibonding σ^* orbital of the H_2 molecule, thus leading to bond weakening and formation of two cis-M—H bonds via a concerted, three-center addition mechanism:

$$L_nM\!\!\!\diamond + \begin{array}{c} H \\ | \\ H \end{array} \rightleftharpoons \left[L_nM\!\!\!\underset{\cdot\cdot}{\overset{H}{\diamond}}\!\!\!\begin{array}{c} H \\ \vdots \\ H \end{array} \right]^{\ddagger} \rightleftharpoons L_nM\!\!\!\underset{H}{\overset{H}{\diagup}}$$

The kinetics of the reaction with $trans$-$IrCl(CO)(PR_3)_2$ suggest a concerted reaction via a nonpolar transition state, the smallness of the deuterium isotope effect indicating that the Ir—H bond formation is more important than H—H bond breaking.[9a] The reductive elimination of H_2 from dihydrido species formed in the forward direction as above or in other ways, e.g.,[9b]

$$CoH[P(OMe)_3]_4 \xrightarrow{H^+} Co(H)_2[P(OMe)_3]_4^+ \underset{+H_2}{\overset{-H_2}{\rightleftharpoons}} Co[P(OMe)_3]_4^+$$

or the closely related *irreversible* elimination of alkanes from compounds with $MH(R)$ groups [or $M(R)_2$ groups] doubtless proceeds by concerted cis intramolecular transfers as above. However there is evidence in some cases for binuclear elimination reactions [e.g., in $Os(CO)_4HCH_3$, and Os—C homolysis in $Os(CO)_4Me_2$].[10]

HX Additions.   In the addition of HCl or HBr to *trans* $IrCl(CO)(PPh_3)_2$ in nonpolar solvents such as benzene, cis addition is also found, suggesting a concerted mechanism. If wet solvents or polar solvents such as dimethylformamide or benzene-methanol are used, cis-trans mixtures are obtained. In polar media initial protonation of a square complex will produce first a cationic five-coordinate complex, which may then isomerize by an intramolecular mechanism (p. 1219). Coordination of halide ion would finally give the oxidized product:

$$MXL_3 + H^+(solv) \rightarrow MHXL_3^+$$
$$MHXL_3^+ + Cl^-(solv) \rightarrow MHClXL_3$$

Isomers could also result from rapid exchange of halide ions through five-coordinate intermediates. As noted above, there is no correlation between the acidity toward water and the propensity to add oxidatively.

It is possible that in the addition of HX the *halide* ion adds *first*, rather than the proton.[11] This has been confirmed for the reaction of HCl with $IrCl(COD)(PEtPh_2)$ in methanol:

$$IrCl(COD)P_2 \xrightarrow{Cl^-} IrCl_2(COD)P_2^- \xrightarrow{H^+} IrHCl_2(COD)P_2$$

[9a] E. M. Hyde and B. L. Shaw, *J.C.S. Dalton*, **1975**, 765.

[9b] E. L. Muetterties and P. L. Watson, *J. Am. Chem. Soc.*, 1976, **98**, 4665, 6978.

[10] J. R. Norton *et al.*, *J. Am. Chem. Soc.*, 1977, **99**, 295, 5835.

[11] T. V. Ashworth *et al.*, *J.C.S. Dalton*, **1978**, 340; R. H. Crabtree *et al.*, *J. Organomet. Chem.*, 1978, **157**, C13.

Under similar conditions there is *no* reaction using $HClO_4$.

Organic Halides.[12]   There has been intensive study of the mechanisms of metal-carbon bond formation by oxidative addition of organic halides. There is considerable complexity and still some uncertainty and confusion.

In a few instances the addition of alkyl halides appears to be consistent with an S_N2-type reaction,[13] as is usually the case for nucleophilic attacks on alkyl halides in organic chemistry, with the metal acting as the nucleophile:

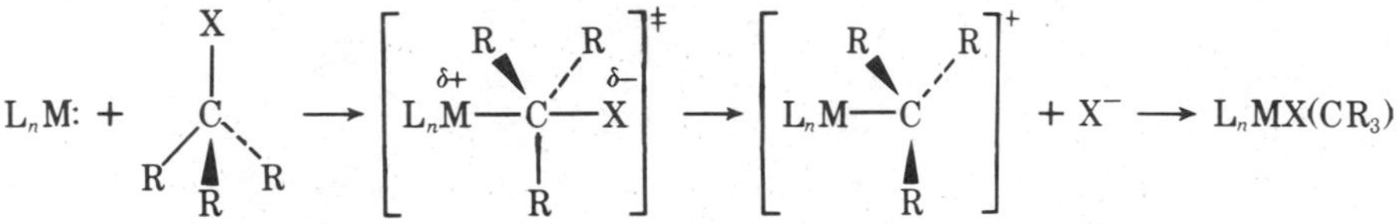

A polar transition state appears to be involved, since there is promotion by polar solvents. Such reactions ought to proceed with inversion of configuration when an optically active halide is used, and this has been observed in a few cases. There are also substantial steric and electronic effects of the phosphines.

The additions[14] to the anion $Fe(CO)_4^{2-}$ as solvent-separated ion pairs in tetrahydrofuran or *N*-methylpyrollidone *appear* to be S_N2, viz.

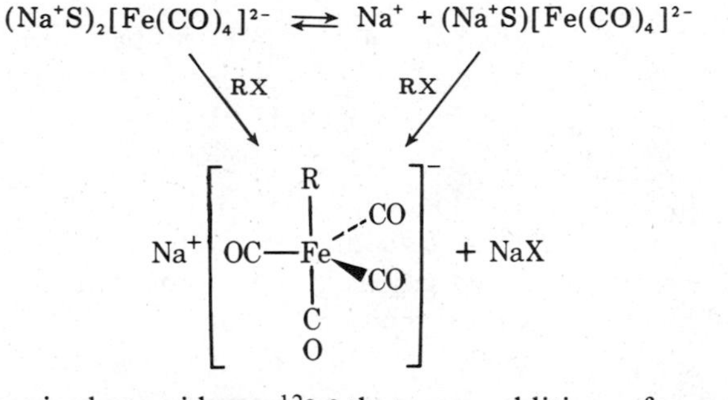

$$(Na^+S)_2[Fe(CO)_4]^{2-} \rightleftharpoons Na^+ + (Na^+S)[Fe(CO)_4]^{2-}$$

However there is clear evidence,[12c–e] that some additions of organic halides involve radicals. The majority of simple bromides and iodides react with Ir^I, Pd^0, and Pt^0 by one-electron paths, although CH_3I, $C_6H_5CH_2Cl$, and $CH_2{=}CH{-}CH_2Cl$ seem generally nonradical. C_6H_5X with Pt^0 (but *not* Ir^I) and $CH_2{=}CHX$ with Pt^0, Pd^0 (but not Ir^I) again are nonradical. Radical paths have not been observed for Rh^I.

In many cases initiation may be due to traces of oxygen, paramagnetic impurities, or light. The steps are thus one-electron rather than two-electron processes.

$$L_nM + R{\cdot} = L_nMR$$
$$L_nMR + RX = L_nMXR + R{\cdot}$$

12   (a) Y. Becker and J. K. Stille, *J. Am. Chem. Soc.,* 1978, **100**, 838; (b) J. K. Stille and K. S. Y. Lau, *Acc. Chem. Res.,* 1977, **10**, 434; (c) J. K. Kochi, *Acc. Chem. Res.,* 1974, **7**, 1351; (d) J. A. Osborn, in *Organotransition Metal Chemistry,* Y. Ishii and M. Tsutsui, Eds., Plenum Press, New York, 1976; (e) M. F. Lappert and P. W. Lednor, *Adv. Organomet. Chem.,* 1966, **14**, 345; M. J. S. Gynane, *et al., J.C.S. Chem. Comm.,* **1978**, 192 (f) R. J. Mureinik *et al., Inorg. Chem.,* 1979, **18**, 915.

13   W. H Thompson and C. T. Sears, Jr., *Inorg. Chem.,* 1977, **16**, 769; J. K. Stille *et al., J. Am. Chem. Soc.,* 1976, **98**, 5832; E. M. Hyde and B. L. Shaw, *J.C.S. Dalton,* **1975**, 765.

14   J. P. Collman *et al., J. Am. Chem. Soc.,* 1977, **99**, 2515.

Many other oxidative additions may be radical. For example, the reaction

$$Mo(CO)_2(diphos)_2 + 2RX = R-R + [Mo(CO)_2(diphos)_2X]X$$

proceeds by a radical reaction involving also a paramagnetic Mo^I species:[15]

$$L_nMo + RX \rightarrow L_nMo^+ + RX^{\pm}$$
$$RX^{\pm} \rightarrow R^{\cdot} + X^-$$

There are no simple generalizations about the mechanisms of oxidative additions. A particular reaction of a metal complex and a substrate may proceed by more than one pathway depending on the conditions of the reaction, and a particular substrate may react with different metal complexes in different ways. However the following seem to be the main types of reaction:

1. Purely ionic mechanisms for ionizable molecules in polar solvents involving five-coordinate intermediates.

2. Concerted three-center additions for hydrogen, certainly, and possibly other substrates of little bond polarity, that lead to new bonds in cis positions.

3. S_N2-type reactions where the metal center acts as the nucleophile.

4. Radical chain reactions involving one-electron processes.

5. Radical pair processes involving one-electron steps.

6. Template reactions.

In reactions with halogens and probably other molecules, the initial interaction is doubtless the formation of a change-transfer complex.

Although we have not specifically discussed the pathways for reductive elimination, we might expect a similar variety of pathways to be available, according to the principle of microscopic reversibility. Relatively few studies have been made on eliminations.

Section 27-6 discussed the decomposition of metal alkyls and aryls by coupling reactions, viz.,

$$L_nM\underset{R}{\overset{R}{\diagdown\diagup}} \longrightarrow L_nM + R-R$$

Other important cis elimination reactions are those for H—H and H—R (R = alkyl, aryl, acyl, etc.). The latter are the final steps in catalytic reactions discussed in Chapter 30.

There are relatively few stable compounds with *cis*-H and R groups, although there are many trans compounds. This instability is illustrated by the fact that the elimination of CH_4 is *three orders of magnitude* greater than the elimination of H_2 in the reactions:

$$cis\text{-}H(CH_3)Co[P(OR)_3]_4^+ \rightarrow CH_4 + Co[P(OR)_3]_4^+$$
$$cis\text{-}H(H)Co[P(OR)_3]_4^+ \rightarrow H_2 + Co[P(OR)_3]_4^+$$

For hydrogen elimination it was also shown that the rate decreased in the order $R = OPr^i > OMe > OEt$.[9b]

The evident favorability of elimination of alkanes clearly accords with the scarcity of C—H oxidative additions of aliphatic hydrocarbons, discussed next.

[15]   J. A. Connor and P. I. Riley, *J.C.S. Chem. Comm.*, **1976**, 634.

29-5. Cleavage of C—H Bonds[16]

The cleavage of C—H bonds is such an important area of oxidative additions that we treat it separately here and in the next section consider a specific form of C—H addition.

There is an increasing number of reactions of organic compounds with transition metal complexes in which C—H bond breaking and M—C bond formation occurs. A particularly common reaction is the so-called cyclometallation reaction discussed below. The reaction of compounds to give C—H bond cleavage is sometimes referred to as the "activation" of C—H bonds.

Oxidative addition can occur with the more acidic compounds such as acetylenes, $RC \equiv CH$, cyclopentadiene, and compounds like $HC(CN)_3$, CH_3NO_2, or CH_3CN[17]:

$$Ir(diphos)_2Cl + CH_3CN \rightleftharpoons IrClH(CH_2CN)(diphos)_2$$

As discussed below, in many cases C_6H_5 or CH_3 groups, as *part of a ligand,* undergo intramolecular oxidative addition, but there are few established cases of *free* hydrocarbons adding.

Aromatic hydrocarbon (e.g., benzene as H and C_6H_5) can be added thermally in a few cases, for example,

$$(\eta\text{-}C_5H_5)_2TaH + C_6H_6 \rightarrow (\eta\text{-}C_5H_5)_2Ta(H)_2C_6H_5$$

The most effective complexes that allow reaction with the C—H bonds of arenes and activated sp^3 C—H bonds in compounds such as CH_3CN and CH_3COCH_3 are some Fe, Ru, and Os complexes.[18] These react not so much as an oxidative addition, although this may be involved, but as a reaction in which there is H^+ attack on the M—C bond, viz.,

$$2HR + \left(\ldots \right) \longrightarrow 2(diphos)_2RuH(R)$$

$$(diphos)_2HFe \text{—} \bigcirc + ArH \rightleftharpoons (diphos)_2HFeAr + C_{10}H_8$$

Arenes and also a C—H bond of $Si(CH_3)_4$ can be added *photochemically* with loss of H_2 to $(\eta\text{-}C_5H_5)_2WH_2$ to give the complexes $(\eta\text{-}C_5H_5)_2W.HR$.[19]

Note also the important oxidative addition or hydride transfer from the CH_3 group of propylene to give a hydrido allyl (pp. 1152, 1255).

There is no established case for addition of alkanes, although exchange reactions

---

[16] G. W. Parshall, *Catalysis,* Vol. 1. Specialist Reports, Chemical Society, 1977.

[17] A. D. English and T. Herskovitz, *J. Am. Chem. Soc.,* 1977, **99,** 1648; M. A. Bennett and T. Yoshida, *J. Am. Chem. Soc.,* 1978, **100,** 1750.

[18] S. D. Ittel *et al., J. Am. Chem. Soc.,* 1979, **101,** 1742; 1978, **100,** 7577; 1976, **98,** 6073.

[19] M. L. H. Green *et al., Nouv. J. Chim.,* 1977, **1,** 187.

between alkanes and D_2 are catalyzed by some metal systems (although it is *not* certain that they are always homogeneous), and oxidative-addition mechanisms have been invoked.[20]

29-6.   Cyclometallation Reactions[16,21]

These are *intra*molecular C—H oxidative additions in which C—H bond breaking and M—C bond formation occurs for alkyl or aryl groups attached to *a coordinated nitrogen or phosphorus ligand*. The result is the formation of a heterocyclic ring. Such reactions may be called *cyclometallation* reactions and the products cyclo-metallated compounds. The hydrogen from C—H may in some cases remain bound to the metal, but more commonly an elimination reaction occurs in which H combines with H, Cl, or CH_3, etc., bound to the metal to generate H_2, HCl, CH_4, etc.

Probably the commonest intramolecular C—H oxidative addition is the so-called orthometallation reaction of aryl groups. It is of great importance in complexes of *triphenylphosphine*[22] and *triphenylphosphite*,[23] but many other compounds such as azobenzene, aromatic ketones and thioketones, and π-cyclopentadienyl compounds can react similarly. Some examples are the following:

$$CH_3Rh(PPh_3)_3 \xrightarrow{heat} CH_4 + Rh(C_6H_4PPh_2)(PPh_3)_2$$

[20]   For references see C. Masters *et al., J.C.S. Dalton,* **1975**, 849, 853, 858; A. E. Shilov, *Pure Appl. Chem.,* 1978, **50,** 725.

[21]   B. Klei and J. M. Teulzen, *J.C.S. Chem. Comm.,* **1978,** 659; F. G. A. Stone *et al., J.C.S. Dalton,* **1978,** 697; M. I. Bruce, *Angew. Chem., Int. Ed.,* 1977, **16,** 73; S. S. Crawford and H. D. Kaesz, *Inorg. Chem.,* 1977, **16,** 3193; H.-P. Abicht and K. Issleib, *Z. Chem.,* 1977, **17,** 1; J. Dehand and M. Pfeffer, *Coord. Chem. Rev.,* 1976, **18,** 327; M. Orchin and D. M. Bollinger, *Struct. Bonding,* 1975, **23,** 182; G. W. Parshall, *Acc. Chem. Res.,* 1975, **8,** 113.

[22]   R. J. McKinney *et al., J. Am. Chem. Soc.,* 1977, **99,** 2988; R. A. Holton and R. G. Davis, *J. Am. Chem. Soc.,* 1977, **99,** 4175; D. J. Cole-Hamilton and G. Wilkinson, *J.C.S. Dalton,* **1977,** 797; F. G. A. Stone *et al., J.C.S. Dalton,* **1976,** 81; L. R. Smith and D. M. Blake, *J. Am. Chem.Soc.,* 1976, **98,** 3302; G. L. Geoffroy and P. Pierantozzi, *J. Am. Chem. Soc.,* 1976, **98,** 8054.

[23]   C. A. Tolman *et al., Inorg. Chem.,* 1978, **17,** 2374; R. P. Stewart *et al., J. Am. Chem. Soc.,* 1976, **98,** 3215; L. W. Gosser, *Inorg. Chem.,* 1974, **14,** 1435.

In a few cases the reactions may be reversible. The hydrido complex formed may be characterized by nmr spectra; an example of such a case is:

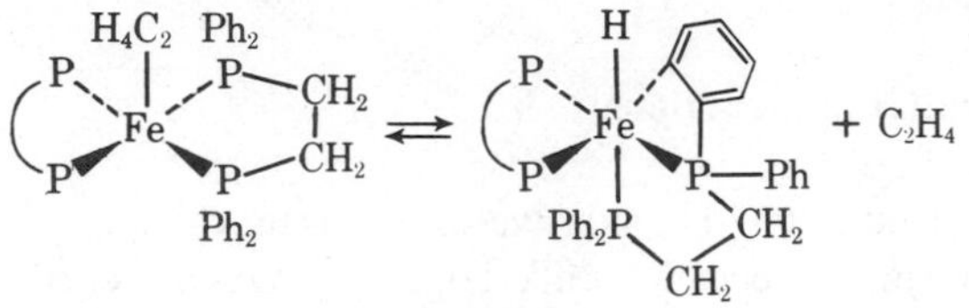

For triphenylphosphine a convenient designation is the following:

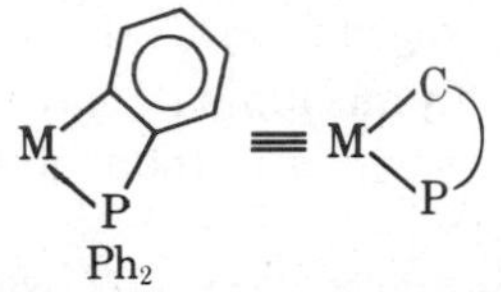

The complex $RuHCl(PPh_3)_2$, which is a sixteen-electron system, can catalyze the exchange of D_2 for the ortho H atom of PPh_3, whereas for the eighteen-electron $RuHCl[P(OPh)_3]_4$ the exchange is slow and is prevented by excess of $P(OPh)_3$. The exchange probably occurs via the reactions

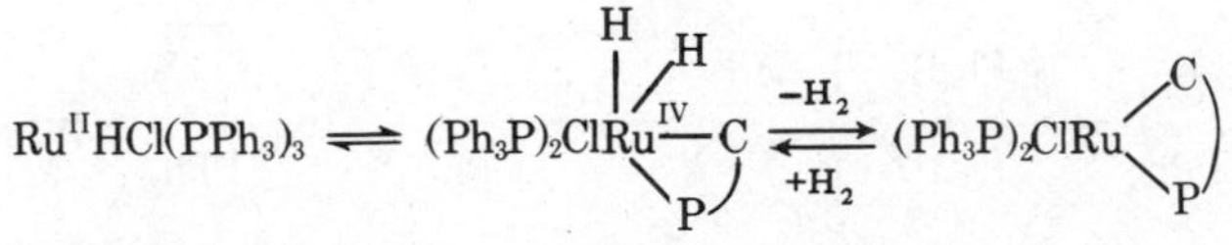

Note that is is not always certain that the reaction involves oxidative addition as shown. Conceivably orthometallation might occur by a concerted process that does not involve a change in oxidation state of the metal, i.e.,

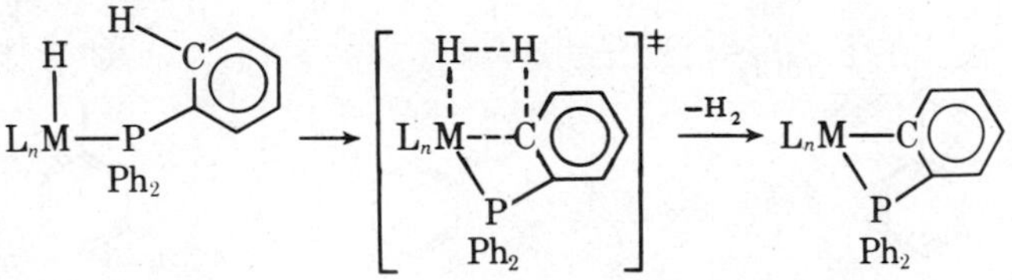

Such a transfer process may be involved[24] in reactions such as:

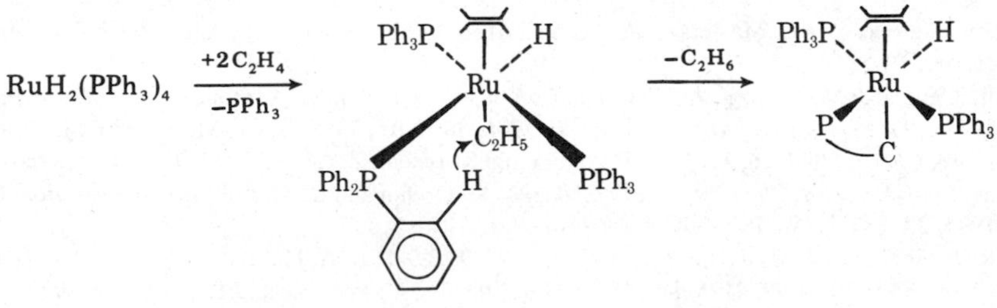

[24]   D. J. Cole–Hamilton and G. Wilkinson, *Nouv. J. Chim.*, 1977, **1**, 141.

Cyclometallation can also occur from alkyl groups,[25] one example being from trimethylphosphine which gives 29-IIIa while a similar example is 29-IIIb from a propylphosphine.

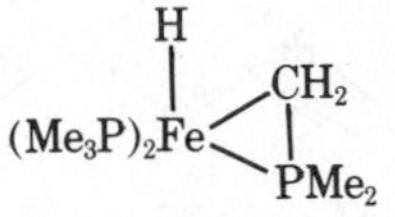

(29-IIIa)

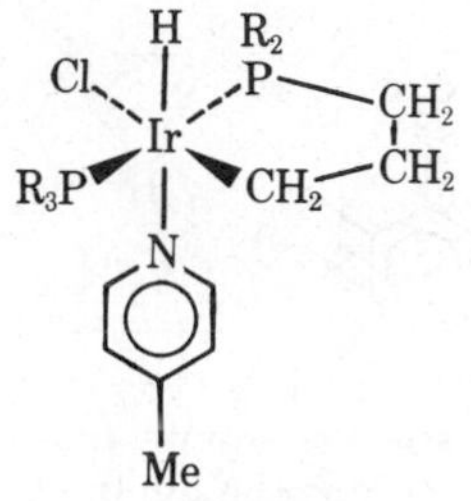

(29-IIIb)

Finally, with regard to triphenylphosphine again, it appears that oxidative addition involving P—C cleavage[26] can occur (see also Section 4–21):

$$L_nM + Ph_2P\text{—}\bigcirc \longrightarrow L_nM\text{—}PPh_2$$

This reaction evidently accounts for the ready formation of phosphido-bridged species by reactions such as:

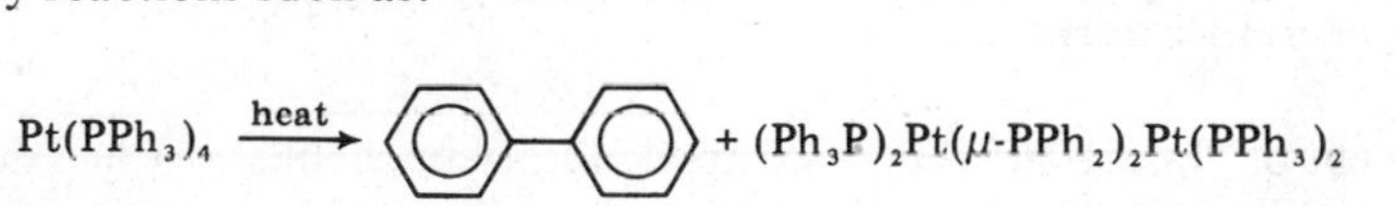

$$Pt(PPh_3)_4 \xrightarrow{\text{heat}} \bigcirc\text{—}\bigcirc + (Ph_3P)_2Pt(\mu\text{-}PPh_2)_2Pt(PPh_3)_2$$

$$HIr(CO)(PPh_3)_3 \longrightarrow C_6H_6 + (CO)(PPh_3)Ir(\mu\text{-}PPh_2)_2Ir(CO)(PPh_3)$$

In reactions of hydrido complexes, benzene has been identified. Further P—C cleavage to give metal clusters with bridging PhP and P groups can occur.

In all these reactions it can be asked, What is the initial interaction between the hydrogen of the C—H bond and the metal?

In a number of cases it appears that there is substantial interaction. An excellent example is the osmium cluster discussed on page 1140. In quite a number of compounds X-ray studies have established that one hydrogen atom of a ligand is in close proximity (2.6–2.9 Å), though not necessarily bonding, to the metal.[27a] An example is the palladium compound[27b] 29-IVa, which *could* be an intermediate on the way to the orthometallated species 29-IVb.

[25]   L. Dahlenburg et al., Chem. Ber., 1978, **111**, 3367; E. L. Muetterties et al., J. Am. Chem. Soc., 1978, **100**, 6966; K. Vrieze et al., J. Organomet. Chem., 1977, **139**, 189.

[26]   K. R. Dixon and A. D. Rattray, Inorg. Chem., 1978, **17**, 1100; B. N. Chaudret et al., J.C.S. Dalton, 1977, 1546; D. R. Fahey and J. E. Mahan, J. Am. Chem. Soc., 1976, **98**, 4499.

[27a] For references see B. L. Shaw, J.C.S. Chem. Comm., **1979**, 104.

[27b] G. P. Khare et al., Inorg. Chem., 1975, **14**, 2475.

(29-IVa)   (29-IVb)

Another example is the ion $\{\eta^3\text{-}C_8H_{13}Fe[P(OMe)_3]_3\}^+$, which has been studied by neutron diffraction.[28] In this complex one of the H atoms of the cyclic hydrocarbon has a very strong interaction (1.879 Å) with the iron atom; this H atom has also a very high field ^1H nmr line at $\delta = -6.54$ ppm. Since covalent Fe–H bonds are estimated to be ca. 1.77 Å, there is clearly some C–H···Fe interaction that could be considered to be a $3c–2e$ bond.

MIGRATION OF ATOMS OR GROUPS FROM METAL TO LIGAND, INSERTION AND EXTRUSION REACTIONS[29]

29-7. General Remarks

The concept of "insertion" is of wide applicability in chemistry when defined as a reaction wherein any atom or group of atoms is inserted between two atoms initially bound together:

$$L_nM\text{—}X + YZ \rightarrow L_nM\text{—}(YZ)\text{—}X$$

Some representative examples are

$$R_3SnNR_2 + CO_2 \rightarrow R_3SnOC(O)NR_2$$
$$Ti(NR_2)_4 + 4CS_2 \rightarrow Ti(S_2CNR_2)_4$$
$$R_3PbR' + SO_2 \rightarrow R_3PbOS(O)R'$$
$$[(NH_3)_5RhH]^{2+} + O_2 \rightarrow [(NH_3)_5RhOOH]^{2+}$$
$$(CO)_5MnCH_3 + CO \rightarrow (CO)_5MnCOCH_3$$

An early example (Berthelot, 1869) is

$$SbCl_5 + 2HC\equiv CH \rightarrow Cl_3Sb(CH=CHCl)_2$$

Representative insertion reactions for transition metals are given in Table 29-2. Although not all insertion reactions are reversible, some are. The reverse reaction is termed *extrusion*.

28   J. M. Williams *et al., J. Am. Chem. Soc.,* 1978, **100,** 7407.
29   H. Berke and R. Hoffmann, *J. Am. Chem. Soc.,* 1978, **100,** 7224 (an extensive theoretical review with many references, especially to CO insertion).

TABLE 29-2

Some Representative Insertion or Group Transfer Reactions

"Inserted" molecule	Bond	Product
CO	M—CR$_3$	MCOCR$_3$
	M—OH	MCOOH
	M—NR$_2$	MCONR$_2$
CO$_2$	M—H	MCOOH
	M—C	MC(O)OR
	M—NR$_2$	MOC(O)NR$_2$ or MO$_2$CR
	M—OH	MOCO$_2$H
CS$_2$	M—M	MSC(S)M
	M—H	MS$_2$CH and MSC(S)H
C$_2$H$_4$	M—H	MC$_2$H$_5$
C$_2$F$_4$	M—H	MCF$_2$CF$_2$H
RC≡CR'	M—H	MC(R)=CH(R')
CH$_2$=C=CH$_2$	M—R	M(η^3-allyl)
SnCl$_2$	M—M	MSn(Cl$_2$)M
RNC	M—H	MCH=NR [a]
	M—R'	MCR'=NR
	M—η^3-C$_3$H$_5$	MC(=NR)(CH$_2$CH=CH$_2$)
RNCS	M—H	M(RNCHS) [b]
RN=C=NR	M—H	M(RNCHNR) [b]
SO$_2$	M—C	MS(R)O$_2$ or MOS(O)R
	M—M [c]	MOS(O)M
	M—(η^3-C$_3$H$_5$)	MSO$_2$CH$_2$CH=CH$_2$
SO$_3$	M—CH$_3$	M—OSO$_2$R
O$_2$	M—CR$_3$	M—OOH
	M CR$_3$	M—OOCR$_3$

[a] The N-aryl or alkyl formimidoyl ligand can also act as a bridge between three Os atoms; see R. D. Adams and N. M. Golembeski, *J. Am. Chem. Soc.*, 78, **100**, 4622.

[b] S. D. Robinson *et al.*, *Inorg. Chem.*, 1977, **16**, 2722, 2728.

[c] Pd, L. S. Benner *et al.*, *J. Organomet. Chem.*, 1978, **153**, C31.

Note that oxidative addition as discussed in the last section *could* be defined as insertion, e.g.,

$$(Ph_3P_3)ClRh + CH_3I \longrightarrow (Ph_3P)ClRh\diagdown\begin{smallmatrix}CH_3\\I\end{smallmatrix}$$

where the Rh atom could be said to be inserted between CH$_3$ and I. A number of oxidative-addition reactions of metals[30] have indeed been termed "insertions," e.g.,

$$Ni + C_6F_5Br + 2Et_3P = C_6F_5—\underset{\underset{PEt_3}{|}}{\overset{\overset{PEt_3}{|}}{Ni}}—Br$$

[30] R. D. Rieke *et al.*, *J. Am. Chem. Soc.*, 1977, **99**, 4159; J. Allison and D. P. Ridge, *J. Am. Chem. Soc.*, 1976, **98**, 7445.

However we define insertion reactions as those exhibiting *no change in the formal oxidation state of the metal.*

29-8. Insertion of Carbon Monoxide[31]

The most intensively studied insertion reactions are those of CO into metal-to-carbon bonds to form acyls; such phenomena have great catalytic and industrial importance (Chapter 30).

Mechanistic studies have been made using mainly $CH_3Mn(CO)_5$ or related compounds. With ^{14}CO as tracer, it has been shown that (*a*) the CO molecule that becomes the acyl-carbonyl is not derived from external CO but is one already coordinated to the metal atom, (*b*) the incoming CO is added *cis* to the acyl group, i.e.,

(*c*) the conversion of alkyl into acyl can be effected by addition of ligands other than CO, e.g.,

The first step involves an equilibrium between the octahedral alkyl and a five-coordinate acyl:

$$CH_3Mn(CO)_5 \rightleftharpoons CH_3\overset{\parallel}{\underset{O}{C}}\!\!-\!\!Mn(CO)_4$$

The coordinately unsaturated acyl then adds the incoming ligand L (CO, PPh_3, etc.) to re-form an octahedral complex

$$CH_3COMn(CO)_4 + L \rightleftharpoons CH_3COMn(CO)_4L$$

Polar solvents can participate in preequilibria and give up to 10^3 or 10^4 increases in rate:

$$CH_3Mn(CO)_5 \underset{-S}{\overset{+S}{\rightleftharpoons}} CH_3COMn(CO)_4S \overset{+L}{\rightarrow} CH_3COMn(CO)_4L$$

[31] J. P. Collman *et al., J. Am. Chem. Soc.,* 1978, **100,** 4766; F. Calderazzo, *Angew Chem., Int. Ed.,* 1977, **16,** 299; A. Wojcicki, *Adv. Organomet. Chem.,* 1973, **11,** 87; G. P. Chiusoli, *Acc. Chem. Res.,* 1973, **6,** 422; G. Henrici-Olivé and S. Olivé, *Trans. Met. Chem.,* 1976, **1,** 77; *Coordination and Catalysis,* Verlag Chemie, 1977.

Such equilibria are well known for complexes of other metals (e.g., for Rh, and Ir species), and equilibrium constants and rates have been measured.[32]

It may be noted that since five-coordinate intermediates can undergo molecular rearrangements, more than one isomer of final products, e.g., $CH_3COMn(CO)_4L$, may be formed.

The CO insertion reaction is best considered as an 1,2-migration of an alkyl group to a coordinated CO ligand in a cis position. The migration probably proceeds through a three-center transition state:

$$
\begin{array}{c}
R_3 \\
C \\
| \\
M\!-\!CO
\end{array}
\longrightarrow
\left[
\begin{array}{c}
R_3 \\
C \\
\diagup \diagdown \\
M\text{-----}CO
\end{array}
\right]^{\ddagger}
\longrightarrow
M\!-\!C\!\!\diagup\!\!\begin{array}{c} CR_3 \\ \\ O \end{array}
$$

There have been several studies of complexes in which chiral groups are present, and in all cases it was found that alkyl transfer proceeds with *retention of configuration*.[33] One example is the insertion of CO into the Fe—C bond of η-C_5H_5Fe-$(CO)_2R$ (R = $CHDCHDCMe_3$) promoted by PPh_3.

Although hydride is usually considered to be a better migrating group than even CH_3, it is in some ways surprising that there is no known example of the insertion to give a formyl group:

$$
L_nMH + CO \rightleftharpoons L_nM\!\!\diagup\!\!\begin{array}{c} H \\ \\ CO \end{array} \rightleftharpoons L_nM\!-\!C\!\!\diagup\!\!\begin{array}{c} H \\ \\ O \end{array}
$$

Such a reaction has been postulated in the Fischer-Tropsch and other CO–H_2 reactions discussed in Chapter 30 and in substitution reactions of $HMn(CO)_5$ and $HCo(CO)_4$ (p. 1204). The equilibrium doubtless lies far to the left, being favored by the formation of both M—H and M—CO bonds. Although a few formyl complexes are known (see later) these are usually unstable, readily forming a hydrido carbonyl.

29-9. Insertion of Sulfur Dioxide[35]

Unlike CO insertion, where only an acyl can be obtained, the products from SO_2 insertion can vary; although the usual product is the *S*-sulfinate 29-IVa, the *O*-alkyl-*S*-sulfoxylate 29-IVb, the *O*-sulfinate (29-IVc, or the *O,O'*-sulfinate 29-IVd may be formed.

[32] M. C. Baird *et al., J. Organomet. Chem.*, 1978, **146**, 71; L. Sacconi *et al., Inorg. Chem.*, 1978, **17**, 718; R. L. Pruett *et al., J. Organomet. Chem.*, 1979, **172**, 405.

[33] See, e.g., N. A. Dunham and M. C. Baird, *J.C.S. Dalton*, 1975, 774; G. M. Whitesides *et al., J. Am. Chem. Soc.*, 1974, **96**, 2814.

[34] (a) T. C. Flood *et al., J. Am. Chem. Soc.*, 1978, **100**, 7278; M. C. Baird *et al., J. Organomet. Chem.*, 1978, **153**, 219; A. Wojcicki, *Adv. Organomet. Chem.*, 1974, **12**, 31; (b) A. E. Crease and M. D. Johnson, *J. Am. Chem. Soc.*, 1978, **100**, 8013.

(29-IVa)   (29-IVb)   (29-IVc)   (29-IVd)

A further type of reaction is

$$(\eta\text{-}C_5H_5)_2Fe(CO)_2CH_2C\equiv CMe \xrightarrow{SO_2} (\eta\text{-}C_5H_5)Fe(CO)_2 \cdots$$

The insertion of SO_2 appears to proceed by several different mechanisms, one of which involves exo attack, as follows:

The insertion of SO_2 into the R—Co groups of cobaloximes (p. 782) is a radical chain reaction[35b] of the type

$$R\text{—}Co = R^{\cdot} + Co^{\cdot}$$
$$Co^{\cdot} + SO_2 \rightarrow CoSO_2^{\cdot}$$
$$CoSO_2^{\cdot} + RCo \rightarrow CoSO_2R + Co^{\cdot}$$

29-10.   Insertion of Alkenes and C—C Unsaturated Compounds[36]

The reaction of transition metal hydrido complexes with alkenes and other unsaturated organic substances is of prime importance in catalytic reactions such as

35   See also references for homogeneous hydrogenation, Chapter 30. D. L. Thorn and R. Hoffmann, *J. Am. Chem. Soc.*, 1978, **100**, 2079 (MO study contains many references); G. Henrici-Olivé and S. Olivé, *Coordination and Catalysis*, Verlag-Chemie, 1977, *Topics Curr. Chem.*, 1976, **67**, 107; G. P. Chiusoli, *Acc. Chem. Res.*, 1973, **6**, 422; C. A. Tolman, in *Transition Metal Hydrides*, E. L. Muetterties, Ed., Dekker, 1971.

36   (a) B. N. Chaudret *et al.*, *Acta. Chem. Scand.*, 1978, **32A**, 763; (b) J. A. Osborn *et al.*, *J. Am. Chem. Soc.*, 1975, **97**, 3861; see also, M. J. D. Aniello, Jr., and E. K. Barefield, *J. Am. Chem. Soc.*, 1978, **100**, 1474; (e) H. Werner and R. Feser, *Angew. Chem., Int. Ed.*, 1979, **18**, 157.

hydrogenation and hydroformylation (Chapter 30). It is one of the major methods for synthesizing metal-to-carbon bonds. We have already discussed the importance of the reverse of the reaction, the β-hydride transfer–alkene elimination reaction in the decomposition of metal alkyls (Sect. 27-3).

The reaction involves transfer of H from the metal to the carbon atom that becomes the β carbon of the new alkyl; it may also be designated as a 1,2-hydride shift. A four-center transition state is commonly postulated:

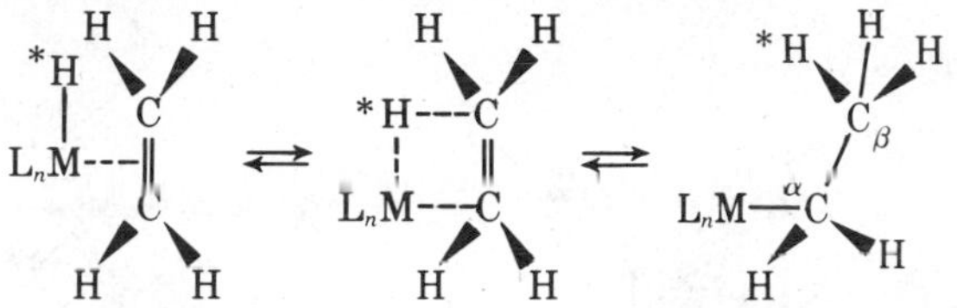

For hydrocarbons such reactions are normally readily reversible. Fluorinated olefins such as $CF_2{=}CF_2$ can give very stable products, and nmr studies confirm the transfer to the β-carbon atom of the alkyl chain as in the diagram above, e.g.,

$$RhH(CO)(PPh_3)_3 + C_2F_4 = Rh(CF_2CF_2H)CO(PPh_3)_2 + PPh_3$$

The first step and probably the rate-determining one,[36a] in such transfer reactions is coordination of alkene:

$$L_nMH + RCH{=}CH_2 \rightleftharpoons L_nMH(RCH{=}CH_2)$$

So far, it has not been possible to measure equilibrium constants for these reactions, although they will probably be rather similar to those for nonhydridic complexes (p. 1147), which can often be measured. The equilibrium constants will depend on the steric and electronic nature of the olefin as well as on the nature of the metal complex. 1-Alkenes appear to have constants $ca.$ 50 times those for 2-alkenes. This complexing reaction is then followed by the rapid and reversible transfer reaction

$$L_nMH(RCH{=}CH_2) \rightleftharpoons L_nM{-}CH_2CH_2R$$

Note that cis transfers are again involved. Note also that although they are not common, there are some stable hydrido alkene complexes that may have H either cis or trans to alkene. The transfer between coordinated alkene and M—H can sometimes be observed by nmr spectra.[36] Thus the complex $Mo(diphos)_2(C_2H_4)_2$ may be protonated by CF_3COOH and the hydrido species so produced in CH_2Cl_2 solution exhibits[36b] the equilibrium:

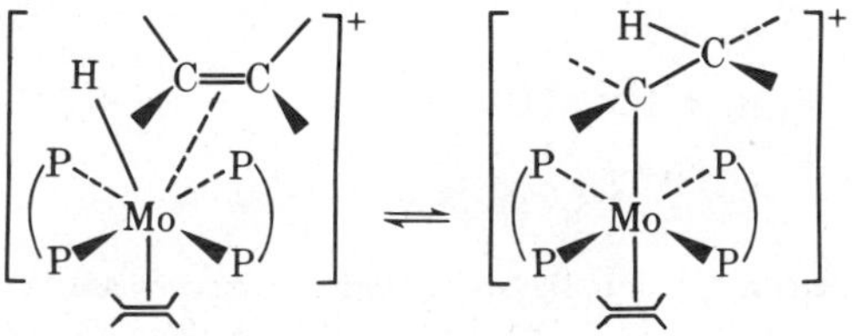

where the H proximate to one coordinated ethylene molecule undergoes a cis site, 1,2-shift to form an ethyl group.

With alkenes other than ethylene, addition of M—H to the double bond in either the Markownikoff or the anti-Markownikoff direction, may occur, viz.,

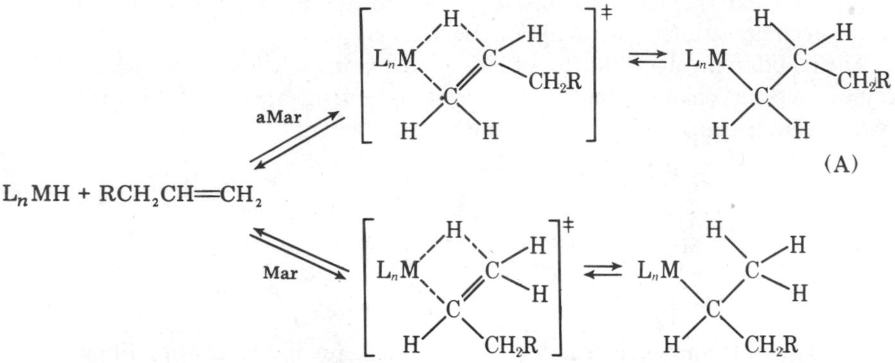

$$L_nMH + RCH_2CH{=}CH_2$$

(A)

(B)

aMar = anti-Markownikoff
Mar = Markownikoff

When the anti-Markownikoff reaction is reversed and alkene eliminated, the original alk-1-ene must be formed, but because of rotation about the C—C bond, the *same* hydrogen need not necessarily be removed from the β carbon, and the possibility of H atom exchange arises. Using a deuterium complex L_nMD, it can be shown that L_nMH is formed by exchange, since the appearance of the high field line of H bound to metal can be observed, e.g.,

$$RhD(CO)(PPh_3)_2 + RCH{=}CH_2 \rightleftarrows RhH(CO)(PPh_3)_2 + RCD{=}CH_2$$

For the secondary or branched chain alkyl (B), formed by Markownikoff addition, there are *two* possibilities for elimination. If the H atom transferred from the β-CH_3 group, again the original alk-1-ene is formed, but if it is transferred from the β-CH_2 of the CH_2R group, then alk-2-ene is formed.

Thus Markownikoff addition provides a mechanism whereby alk-1-enes can be *isomerized* to *cis-trans*-alk-2-enes. Many metal hydrido species [e.g., $HCo(CO)_4$, $HRh(CO)(PPh_3)_3$, or metal halide complexes plus a hydride source] will isomerize alk-1-ene to alk-2-ene catalytically.[37]

The effects of steric factors on the stability of alkyl intermediates in isomerization and other reactions are discussed in Chapter 30.

Finally, note that whereas the hydride-alkene interactions normally appear to proceed via nonradical pathways, in some cases a free-radical mechanism is certainly involved. This is so for the reaction of α-methylstyrene:

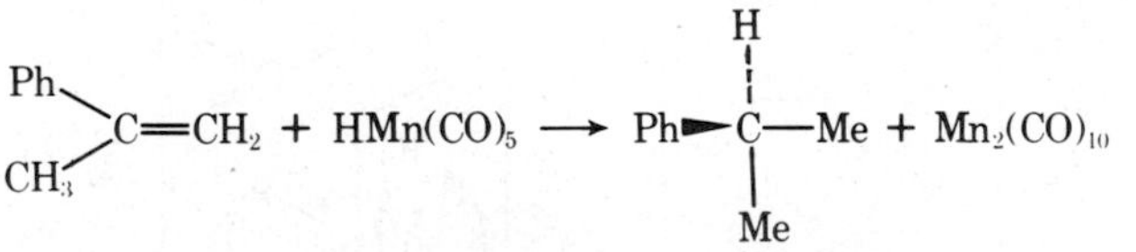

---

[37]   For isomerization references, see J. L. Davidson, *Inorganic Reaction Mechanisms,* Vol. 5, Specialist Report, Chemical Society, 1977; M. J. D'Aniello, Jr. and E. K. Barefield, *J. Am. Chem. Soc.,* 1978, **100**, 1474.

The radical $\cdot Mn(CO)_5$ is involved, and chemically induced dynamic nuclear polarization effects were observed when the reaction was followed by nmr at 70°C.[38a] The reaction between $PtHCl(PR_3)_2$ and some acetylenes also appears to be radical initiated.[38b]

A second important type of hydride shift that it is convenient to mention here is the *1,3-hydride shift*, which can occur from the methyl group of propene to form a hydrido π-allyl complex, doubtless via an intermediate σ allyl:

$$
H_3C^3{-}C^2{\overset{H}{\diagdown}}\ L_nM + {}^1CH_2 \;\rightleftharpoons\; L_nM{-}^1CH_2\ {\overset{{}^3CH_2}{\underset{{}^2CH}{\|}}}\overset{H}{\diagdown} \;\rightleftharpoons\; L_nM{-}\overset{H}{\underset{CH_2}{\diagup}}CH
$$

The reason for the name 1,3-shift should be clear from the numbering given. As also noted in Section 27-9, this reaction has been observed by nmr study[36b] in the reaction of $Mo(diphos)_2(N_2)_2$ with propene:

$$
\text{(structure)} \;\rightleftharpoons\; \text{(structure)}
$$

Certain isomerizations of unsaturated compounds proceed by 1,3- rather than 1,2-shifts,[39] and reactions of propene or other alkenes sometimes produce η^3-allyl species; e.g., $RuH_2(PPh_3)_4$ with propene gives $(\eta^3\text{-}C_3H_5)_2Ru(PPh_3)_2$.[24]

There are also a few cases in which hydrogen is transferred from γ-carbon atoms as in the reaction[40]:

$$
\text{(structure)} \xrightarrow{-Me_4Si} \text{(structure)}
$$

Finally, it is convenient to note here a stoichiometric hydride transfer reaction that has potential utility in organic synthesis. This is the *hydrozirconation reaction*,[41] which uses the hydride $(\eta\text{-}C_5H_5)_2ZrHCl$ or a mixture of $(\eta\text{-}C_5H_5)_2ZrCl_2$ with $AlMe_3$, to give an organo compound by reaction with alkenes, alkynes, 1,3-

[38a] R. L. Sweany and J. Halpern, *J. Am. Chem. Soc.*, 1977, **99**, 8335.

[38b] H. C. Clark and C. S. Wong, *J. Am. Chem. Soc.*, 1977, **99**, 7073.

[39] See, e.g., M. Green and R. P. Hughes, *J.C.S. Dalton*, **1976**, 1907.

[40] R. A. Andersen et al., *J.C.S. Dalton*, **1978**, 446.

[41] J. A. Labinger, *J. Organomet. Chem.*, annual surveys, 1977, 138, 185; E. Negishi et al., *J. Am. Chem. Soc.*, 1978, **100**, 2252, 2254; J. Schwartz et al., *J. Am. Chem. Soc.*, 1977, **99**, 8045; *Angew. Chem., Int. Ed.*, 1976, **15**, 333.

dienes, epoxides, etc. The resulting compound is then made to undergo cleavage of the Zr—C bond by the electrophilic reagents or to insert CO to give acyls, which are then decomposed, e.g.,

$$
RBr \xleftarrow{\;Br_2\;} (C_5H_5)_2Zr \overset{Cl}{\underset{R}{\diagdown}} \xrightarrow{\;CO\;} (C_5H_5)_2Zr \overset{Cl}{\underset{\underset{R}{\overset{|}{C}}=O}{\diagdown}}
$$

with products:
$$\xrightarrow{H^+} RCHO$$
$$\xrightarrow{Br_2/MeOH} RCOMe \;\; (C{=}O)$$
$$\xrightarrow{H_2O_2} RCOH \;\; (C{=}O)$$

29-11. Insertions of Other Molecules

The mechanism of insertion of other molecules is not well known. The insertions of *carbon dioxide* into dialkylamides of Ti, Zr, V, Nb, and Ta,[42] the overall stoichiometries of which are of the type,

$$Ta(NMe_2)_5 + 5CO_2 = Ta(O_2CNMe_2)_5$$

$$W(NMe_2)_6 + 3CO_2 = W(NMe_2)_3(O_2CNMe_2)_3$$

appear to be catalyzed by traces of free amine, so that we have the sequence:

$$Me_2NH + CO_2 \rightleftharpoons Me_2NCO_2H$$
$$L_nMNMe_2 + Me_2NCO_2H \rightleftharpoons L_nMO_2CNMe_2 + HNMe_2$$

Insertions of *nitric oxide*[43] into metal-methyl bonds of metal alkyls depends on whether the alkyl is diamagnetic or paramagnetic. If it is diamagnetic, the reaction involves a radical intermediate

$$Me_5W\text{—}CH_3 \xrightarrow{\;\overset{\bullet}{N}O\;} Me_5W\text{—}O\overset{\overset{\bullet}{N}\text{—}CH_3}{\diagup}$$

which reacts rapidly with the radical $\overset{\bullet}{N}O$ as follows:

$$Me_5W\text{—}O\underset{\underset{\bullet}{N}\text{—}CH_3}{\diagdown} + \overset{\bullet}{N}O \longrightarrow Me_5W\overset{O\text{—}N\diagup^{CH_3}}{\underset{O=N}{\diagdown}}$$

However a paramagnetic alkyl such as $ReO(CH_3)_4$ or $(\eta\text{-}C_5H_5)_2 Nb(CH_3)_2$ cannot so react because the initial species must be diamagnetic, e.g.,

$$\overset{O}{\underset{}{\overset{\|}{Me_3ReCH_3}}} \xrightarrow{\;\overset{\bullet}{N}O\;} \overset{O}{\underset{}{\overset{\|}{Me_3Re\text{—}ON}}}\diagup^{CH_3}$$

The intermediate decomposes to the nitrene MeN, which in turn dimerizes to

[42] M. H. Chisholm and M. W. Extine, *J. Am. Chem. Soc.,* 1977, **99**, 782, 792.
[43] K. Mertis and G. Wilkinson, *J.C.S. Dalton,* **1976**, 1488.

MeN=NMe. Nitrene transfers to alkenes from similar intermediates have been noted in Section 4-18.

$$2Me_3ReON{\overset{Me}{\diagup}} \longrightarrow 2Me_3Re{\overset{O}{\underset{O}{\diagdown N-Me}}} \xrightarrow{-MeN=NMe} 2 \; {\overset{Me}{\underset{Me}{\diagdown}}Re{=}O \atop |} \; Me$$

Group IV metal dihalides $SnCl_2$ and $GeCl_2$ can be inserted into M—C, M—X, and M—M bonds; for insertion into $(\eta-C_5H_5)(CO)_2FeCH_3$ to give $(\eta-C_5H_5)$-$(CO)_2FeSnCl_2CH_3$, a radical chain process is proposed,[44]

REACTIONS OF COORDINATED LIGANDS

29-12. General Remarks

In the last section we considered intramolecular transfer to coordinated ligands. Here we discuss attacks on coordinated ligands by external reagents.[45] Some reactions of coordinated ligands have been mentioned in Chapter 3, 4, and 28, and in Chapter 27 we discussed nucleophilic and electrophilic attacks on π complexes of various sorts.

It is not always easy to prove that the ligand reacts while coordinated or that prior coordination of the reagent is not involved, as in the last section. However in the following cases it seems that intramolecular transfers are not involved.

29-13. Carbon Monoxide.

Metal carbonyl complexes may be attacked by all manner of nucleophiles such as OH^-, py, NR_2^-, H^-, and Me^-.[46]

Hydride. The interaction of H_2 and CO to give organic products (Chapter 30) is of very great importance, and catalytic reactions probably involve attack of H^- on coordinated CO.[47] The attack of hydride ion from $NaBH_4$ in THF on carbonyls can give formyls, hydroxymethyls, or finally methyls[48]:

$$M{-}CO \rightarrow M{-}CHO \rightarrow M{-}CH_2OH \rightarrow M{-}CH_3$$

Thus $(\eta-C_5H_5)(PPh_3)W(CO)_3^+$ is reduced to $(\eta-C_5H_5)(PPh_3)(CO)_2WCH_3$,

[44] J. D. Cotton and G. A. Morris, *J. Organomet. Chem.*, 1978, **145**, 245.
[45] For references see *Inorganic Reaction Mechanisms*, Specialist Reports, Chemical Society.
[46] See, e.g., T. L. Brown and P. A. Bellus, *Inorg. Chem.*, 1978, **17**, 3726.
[47] For references, see J. A. Labinger *et al.*, *J. Am. Chem. Soc.*, 1978, **100**, 3254.
[48] J. A. Gladysz *et al.*, *J. Am. Chem. Soc.*, 1979, **101**, 1589; *Tetrahedron Lett.*, **1978**, 319; C. P. Casey *et al.*, *J. Am. Chem. Soc.*, 1979, **101**, 741, 3371; R. S. Winter *et al.*, *J. Organomet. Chem.*, 1977, **133**, 339; A. R. Cutler, *J. Am. Chem. Soc.*, 1979, **101**, 604; J. R. Sweet and W. A. Graham, *J. Organomet. Chem.*, 1979, **173**, C9.

whereas the reduction of $(\eta\text{-}C_5H_5)(NO)Re(CO)_2^+$ by $NaBH_4$ in THF/water gave the unstable $(\eta\text{-}C_5H_5)(NO)(CO)ReCH_2OH$.

As noted earlier, intramolecular H transfer to CO does not occur to any detectable extent. Nevertheless formyl groups in carbonylate anions are reasonably stable and are formed by attacks on neutral carbonyls by $Na[BH(OMe)_3]$ or $Li[BHEt_3]$ in tetrahydrofuran, e.g.,

$$Fe(CO)_5 + H^- \longrightarrow \left[(CO)_4Fe-C\begin{smallmatrix}\diagup O \\ \diagdown H\end{smallmatrix} \right]^-$$

$$Cr(CO)_6 + H^- \longrightarrow \left[(CO)_5Cr-C\begin{smallmatrix}\diagup O \\ \diagdown H\end{smallmatrix} \right]^-$$

A neutral but unstable formyl complex is formed in the decomposition of a ruthenium methoxide complex[49a]:

| Methoxide | Formaldehyde complex | Formyl complex | Carbonyl complex |

Examples of neutral complexes are $Os(CHO)Cl(CO)_2(PPh_3)_2$, $Os(\eta^2\text{-}CH_2O)(CO)_2(PPh_3)_2$,[49b] and $\eta\text{-}C_5H_5Re(CO)_3CHO$.[49c]

Hydroxide Ion and Water.   The attack of OH^- or of OH_2 on metal carbonyl complexes gives hydrido species by the following type of reaction[50]:

$$Fe(CO)_5 + OH^- \longrightarrow \left[(CO)_4Fe-C\begin{smallmatrix}\diagup O \\ \diagdown H-O\end{smallmatrix} \right]^- \longrightarrow (CO)_4FeH^- + CO_2$$

The hydroxy carbonyl species are normally unstable and undergo β-hydride transfer to give M—H, but the compound $IrCl_2(CO_2H)(CO)(PMe_2Ph)_2$ is stable and has been shown to decompose giving CO_2 and the hydride.[50a]

The oxygen atom of coordinated CO can also undergo exchange with $H_2{}^{18}O$, presumably through attack of H_2O or OH^- on the carbon atoms, e.g.,

[49a] B. N. Chaudret *et al., J.C.S. Dalton,* **1977,** 1546.

[49b] K. L. Brown, *et al., J. Am. Chem. Soc.,* 1979, **101,** 503.

[49c] N. E. Kolobova *et al., Bull. Acad. Sci., USSR. Div. Chem. Sci.,* 1978, **27,** 639; J. A. Gladysz *et al., J. Am. Chem. Soc.,* 1979, **101,** 1589; *J.C.S. Chem. Comm.,* **1979,** 530.

[49d] R. A. Fiato *et al., J. Organomet. Chem.,* 1978, **172,** C4.

[50] (a) M. Y. Darensbourg *et al., Inorg. Chem.,* 1979, **18,** 1401; (b) C. R. Eady *et al., J.C.S. Dalton,* **1977,** 838; (c) R. Pettit *et al., J. Am. Chem. Soc.,* 1979, **101,** 1627.

$$L(CO)_4ReCO^+ + H_2{}^*O \rightleftharpoons L(CO)_4Re-C\overset{*OH}{\underset{O}{\diagup}} + H^+$$

The conversion of CO to CO_2 and the reaction of NO and CO to give N_2O and CO_2 using rhodium complexes also involve water.[51]

Alcoxide Ions.   These attack coordinated CO to form alcoxycarbonyl complexes,[52] and this reaction has been observed for complexes of Mn, Re, Fe, Ru, Os, Co, Rh, Ir, Pd, Pt and Hg, e.g.,

$$[Ir(CO)_3(PPh_3)_2]^+ \xrightarrow[H^+]{CH_3O^-} Ir\left(-C\overset{O}{\underset{OCH_3}{\diagup}}\right)(CO)_2(PPh_3)_2$$

The reaction is important in the synthesis of carboxylic acids and esters from olefins, CO and H_2O, or alcohols (Chapter 30).

Ammonia, Amines.   Attack on cationic carbonyls[52] leads to carbamoyl or carboxamido complexes:

$$L_n\overset{+}{M}-CO + 2NHRR' \rightleftharpoons L_nM-C\overset{O}{\underset{NRR'}{\diagup}} + NH_2RR'^+$$

Thiocarbonyl complexes may be similarly attacked, e.g.,

$$\eta\text{-}C_5H_5(CO)_2FeCS^+ \xrightarrow{MeNH_2} \eta\text{-}C_5H_5(CO)_2Fe-C\overset{S}{\underset{NHMe}{\diagup}}$$

Other nitrogen nucleophiles such as NR_2^- (lithium dialkylamides), hydrazines, and azide ion also attack CO; in the latter case $MC(O)N_3$ readily loses N_2 and rearranges to give MNCO.

Lithium Alkyls.   The interaction of metal carbonyls with lithium alkyls and the conversion of the anions so produced to carbene compounds has been discussed earlier (p. 1143). The acyl tetracarbonylferrates obtained from $Fe(CO)_5$

$$Fe(CO)_5 + LiR = Li^+ \left[(CO)_4Fe-C\overset{O}{\underset{R}{\diagup}}\right]^-$$

are particularly useful as reagents in organic synthesis.[53]

Oxygen Atoms.   The interaction of tertiary amine oxides (R_3NO), with carbonyls results in conversion of CO to CO_2.[54] This is a useful means of providing a labile site on metal carbonyls, since the amine can be readily displaced:

---

[51]   D. E. Hendriksen and R. Eisenberg, *J. Am. Chem. Soc.*, 1976, **98**, 4662.

[52]   R. J. Angelici, *Acc. Chem. Res.*, 1972, **5**, 335.

[53]   C. P. Casey and C. A. Bunnell, *J. Am. Chem. Soc.*, 1976, **98**, 436; see also J. P. Collman *et al.*, *J. Am. Chem. Soc.*, 1978, **100**, 4766.

[54]   J. R. Shapley *et al.*, *J. Am. Chem. Soc.*, 1978, **100**, 2596; J. Elzinga and H. Hogeveen, *J.C.S. Chem. Comm.*, **1977**, 705; F. A. Cotton and B. E. Hanson, *Inorg. Chem.*, 1977, **16**, 2820.

$$(CO)_4Fe=C=O \longrightarrow (CO)_4\overset{-}{Fe}-C\overset{\displaystyle O}{\underset{\displaystyle \overset{O}{\underset{N^+}{|}}}{\diagup}} \xrightarrow{-CO_2} (CO)_4FeNMe_3$$
$$Me_3N^+-O^-$$
$$Me_3$$

Electrophilic Attacks. The bridging CO in metal carbonyls and substituted carbonyls can also be attacked by electrophiles[55] and strong Lewis acids, for example, BX_3 or AlX_3, may give adducts that have $C-O-BX_3$ groups.

Attacks by the powerful electrophiles such as CH_3^+ on the bridging CO of the polynuclear anions $[Fe(CO)_{11}]^{2-}$ or $[HFe_3(CO)_{11}]^-$ gives derivatives of the type $[Fe_3(CO)_{10}C-O-R]^-$ and $HFe_3(CO)_{10}COR$. Similarly, protonation of $[HFe_3(CO)_{11}]^-$ gives $H_2Fe_3(CO)_{11}$ that has a $C-OH$ group.

29-14. Nitric Oxide

The reaction of nitroprusside ion with hydroxide to give a nitro compound has long been known:

$$[Fe(CN)_5NO]^{2-} + OH^- \underset{\longleftarrow}{\overset{slow}{\longrightarrow}}$$

$$\left[Fe(CN)_5N\overset{\displaystyle O}{\underset{\displaystyle \underset{H}{\overset{O}{\diagup}}}{\diagdown}} \right]^{3-} \underset{\longleftarrow}{\overset{OH^-}{\longrightarrow}} [Fe(CN)_5NO_2]^{4-} + H_2O \qquad K = 1.5 \times 10^6 \; L^2 \, mol^{-2}$$

There are now many examples of nucleophilic attacks[56a] on the N of coordinated NO by OH^-, OR^-, amine, and other ligands, e.g., the condensation reaction:

$$[bipy_2ClRuNO]^+ + H_2NPh = [bipy_2ClRu-N=NPh]^+ + H_2O$$

The reactivity of NO depends on the mode of bonding, linear nitrosyls with v-NO in the range 1886–1945 cm^{-1}, reacting as above, while bent nitrosyls (p. 92) undergo electrophilic attacks[56b] such as:

$$OsCl(NO)(CO)L_2 + HCl \rightleftharpoons OsCl_2(NHO)(CO)L_2$$
$$Co(NO)sal_2en + Et_3N + \tfrac{1}{2}O_2 \rightarrow Co(NO_2)(sal_2en)NEt_3$$

[55] H. A. Hodoli and D. F. Shriver, *Inorg. Chem.*, 1979, **18**, 1236, and references therein.

[56a] J. H. Enemark and R. D. Feltham, *Coord. Chem. Rev.*, 1974, **13**, 339; C. Bremard *et al.*, *J.C.S. Dalton*, **1977**, 2307; F. Bottomley, *Acc. Chem. Res.*, 1978, **11**, 158; *Coord. Chem. Rev.*, 1978, **26**, 7; F. Bottomley and E. M. R. Kiremire, *J.C.S. Dalton*, **1977**, 1125; T. J. Meyer *et al.*, *J. Am. Chem. Soc.*, 1977, **99**, 4340; R. Eisenberg and C. D. Meyer, *Acc. Chem. Resv*, 1975, **8**, 26; K. G. Caulton, *Coord. Chem. Rev.*, 1975, **14**, 317; J. A, McCleverty, *Chem. Rev.*, 1979, **79**, 53.

[56b] R. D. Wilson and J. A. Ibers, *Inorg. Chem.* 1479, **18**, 336; F. Bottomley, *Acc. Chem. Res.*, 1978, **11**, 158.

29-15.   Isocyanides

The attack on isocyanides by nucleophiles to give carbene complexes has been noted earlier (p. 1144). Some other examples[57a] are

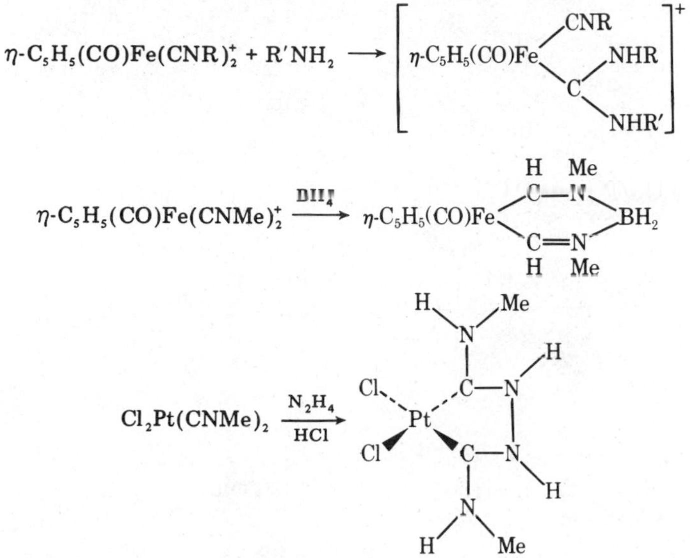

$$\eta\text{-}C_5H_5(CO)Fe(CNR)_2^+ + R'NH_2 \longrightarrow \left[\eta\text{-}C_5H_5(CO)Fe \underset{\underset{\text{NHR}'}{C}}{\overset{CNR}{\underset{NHR}{\diagup}}} \right]^+$$

$$\eta\text{-}C_5H_5(CO)Fe(CNMe)_2^+ \xrightarrow{\ BH_4^-\ } \eta\text{-}C_5H_5(CO)Fe \underset{\underset{H\quad Me}{C=N}}{\overset{\overset{H\quad Me}{C-N}}{\diagdown}} BH_2$$

$$Cl_2Pt(CNMe)_2 \xrightarrow[\text{HCl}]{\ N_2H_4\ } \text{(Pt complex)}$$

Attack on the isocyanide group by protons can also occur[57b]:

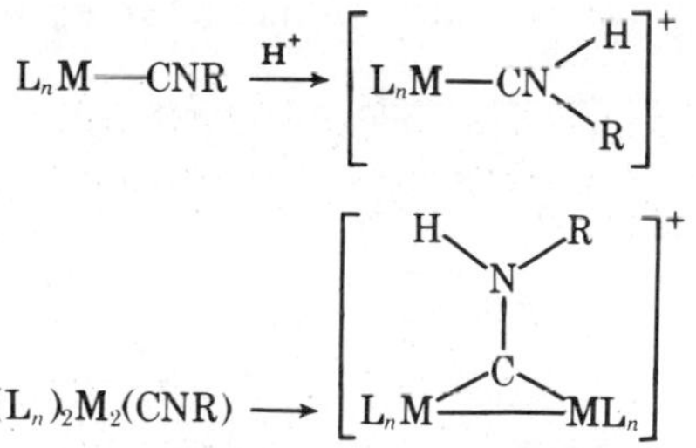

$$L_nM\text{---}CNR \xrightarrow{\ H^+\ } \left[L_nM\text{---}CN\underset{R}{\overset{H}{\diagup}} \right]^+$$

$$(L_n)_2M_2(CNR) \longrightarrow \left[L_nM \underset{\diagdown}{\overset{\diagup}{\underset{\underset{H\diagdown N \diagup R}{C}}{}}} ML_n \right]^+$$

29-16.   Oxygen

Coordinated dioxygen (Chapter 4, p. 155) is kinetically more reactive than is the free molecule. In stoichiometric reactions with attacking species the oxidized entity usually remains bound to the metal. Some examples[58] of simple reactions are the following:

[57] (a) F. Bonati and G. Minghetti, *Inorg. Chim., Acta,* 1974, **9**, 95; B. V. Johnson *et al., J. Organomet. Chem.,* 1978, **154**, 89; I. I. Creaser *et al., J.C.S. Chem. Comm.,* **1978**, 239; (b) F. S. Stephens *et al., J.C.S. Dalton,* **1979**, 23, and references quoted.
[58] D. M. Roundhill *et al., J. Am. Chem. Soc.,* 1978, **100**, 1147; S. Bhaduri *et al., J.C.S. Chem. Comm.,* **1978**, 991.

Detailed studies of the interaction of $Pt(O_2)(PPh_3)_2$ with PPh_3 in ethanol shows that oxidation to $OPPh_3$ occurs by way of intermediates with EtO^-, OH^-, and O_2H^- groups and and that even in nonpolar media hydroperoxide may be generated by trace impurities.[59]

It is clear that the oxidation of organic compounds catalyzed by metals (Chapter 30) proceeds in almost every case by way of peroxo or hydroperoxo intermediates and radical mechanisms.[60]

29-17.   Dinitrogen[61]

Although the attack on coordinated N_2 in biological nitrogen "fixation" is very important, the mechanism is not well understood. Coordinated N_2 is usually inert despite the weakening of the N—N bond on coordination.

A few Mo and W complexes have been attacked stoichiometrically,[62] probably electrophilically, by H^+ or RCO^+, to give hydrazido or aroyldiazenido complexes (p. 128), e.g.,

$$(PhMe_2P)_4Mo(N_2)_2 + 2HX \rightarrow [(PhMe_2P)_4XMoNNH_2]X + N_2$$

$$(diphos)_2W(N_2)_2 + PhCOCl \rightarrow (diphos)_2ClW(NNCOPh) + N_2$$

59   A. Sen and J. Halpern, *J. Am. Chem. Soc.*, 1977, **99**, 8337.
60   See, e.g., A. Sakamoto *et al.*, *J. Catal.*, 1977, **48**, 427.
61   J. Chatt *et al.*, *Chem. Rev.*, 1978, **78**, 589; F. Bottomley and R. C. Burns, Eds., *Treatise on Dinitrogen Fixation*, Wiley, 1979.
62   M. Hidai *et al.*, *J. Am. Chem. Soc.*, 1979, **101**, 3406; G. Butler *et al.*, *Inorg. Chim. Acta*, 1978, **30**, L287; M. Sato *et al.*, *J. Organomet. Chem.*, 1978, **152**, 239.

Nucleophilic attack[63] on coordinated dinitrogen (e.g., by MeLi, PhLi) is also possible. The addition product from MeLi can be treated with trimethyloxonium tetrafluoroborate to give an azomethane complex:

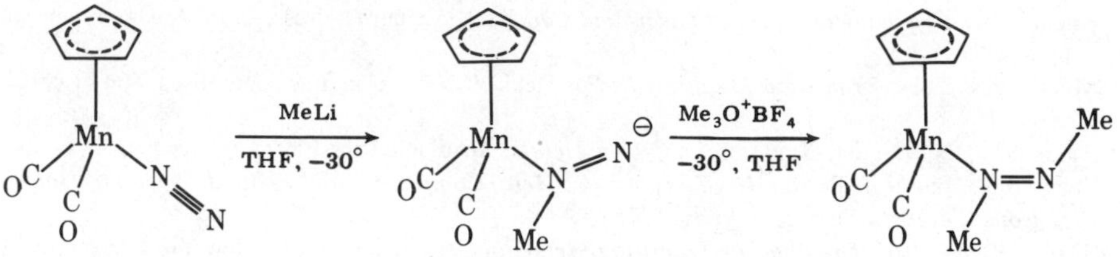

The attack on N_2 in $Mo(N_2)_2(diphos)_2$ by alkyl halides appears to proceed by a *radical* pathway.[64] Alkyl halide is coordinated and instead of undergoing oxidative addition, undergoes C—X bond cleavage. The radical R· then attacks coordinated dinitrogen:

$$(diphos)_2Mo(N_2)_2 \xrightarrow[-N_2]{RX} (diphos)_2Mo(RX)(N_2)$$

$$(diphos)_2Mo(RX)(N_2) \rightarrow (diphos)_2MoX(N_2) + R·$$
$$(diphos)_2MoX(N_2) + ·R \rightarrow (diphos)_2MoX(NNR)$$

A number of other dinitrogen compounds especially of titanium and zirconium give ammonia or hydrazine on acid hydrolysis, but the systems are labile and complicated, and the proposed mechanisms are far from well established.

Finally, we note that a number of metal complexes, e.g., molybdenum thiol complexes in presence of reducing agents or mixed V–Mg hydroxides,[65] will convert N_2 to NH_3 or N_2H_4, but again, the pathways are uncertain.

63 D. Sellmann and W. Weiss, *Angew. Chem., Int. Ed.*, 1978, **17** 269; 1977, **16**, 880; *J. Organomet. Chem.*, 1978, **160**, 165.
64 J. Chatt *et al., J.C.S. Dalton*, **1978**, 1638.
65 P. Sobota and B. Jezowska-Trzebiatowska, *Coord. Chem. Rev.*, 1978, **26**, 71; S. I. Jones *et al., J. Am. Chem. Soc.*, 1978, **100**, 2133; N. T. Denisov *et al., Nouv. J. Chim.*, 1979, **3**, 403.

General References

See also references for Chapter 27 and 30.

Journals and Series

Specialist Reports, Chemical Society. Reports on *Organometallic Chemistry, Inorganic Reaction Mechanisms, Inorganic Chemistry of Transition Elements* (ceased at Volume 6) and *Catalysis*.
Synthesis.   Contains occasional reviews on organic syntheses via metal complexes.

Books and Reviews

Alper, H. *Transition Metal Organometallics in Organic Synthesis,* Vol. 1, Academic Press, 1976.
Augustine, E. L., Ed., *Oxidation—Techniques and in Organic Synthesis,* Dekker, 1969. Use of transition metal and other compounds.
Bird, C. W., *Transition Metal Intermediates in Organic Synthesis,* Logos Press (London), Academic Press (New York), 1967. Comprehensive review with extensive references.

Candlin, J. P., K. A. Taylor, and D. T. Thompson, *Reactions of Transition Metal Complexes,* Elsevier, 1968. General review with extensive references.

Heck, R. F., *Organotransition Metal Chemistry. A Mechanistic Approach,* Academic Press, 1974. An excellent text.

Kochi, J. K., *Organometallic Mechanisms and Catalysis,* Academic Press, 1978. A comprehensive text.

Muetterties, E. L., *Transition Metal Hydrides,* Dekker, 1971. Contains some discussion of catalysis.

Rylander, P. N., *Organic Synthesis via Noble Metals,* Academic Press, 1973.

Shaw, B. L., and N. I. Tucker, *Organotransition Metal Compounds and Related Aspects of Homogeneous Catalysis,* Pergamon Press, 1975.

Slocum, D. W., Ed., *The Place of Transition Metals in Organic Synthesis,* New York Academy of Science, 1977.

Tsuji, J., *Organic Synthesis by means of Transition Metal Complexes,* Springer Verlag, 1975. Short useful text.

Wender, I, and P. Pino, *Organic Syntheses via Metal Carbonyls,* Vol. 1, 1968, Vol. 2, 1976, Wiley.

CHAPTER THIRTY

Transition Metal to Carbon Bonds in Catalysis

Most organic chemicals produced in bulk quantities, the majority of which are oxygenated compounds, are derived from natural gas or petroleum, usually by conversion of these hydrocarbons into olefins such as ethylene, propylene, or butadiene. The use of transition metal complexes for the conversion of unsaturated substances into polymers, alcohols, ketones, carboxylic acids, etc., has been intensively studied and has generated a vast patent literature as well as a great many papers in scientific journals. In particular, the discovery of the low pressure poly merization of ethylene and propylene by Ziegler and Natta led not only to technical syntheses of polyalkenes and rubbers, but also to the wide use of aluminum alkyls as alkylating agents and reductants for metal complexes. Similarly, the discovery by Smidt of palladium-catalyzed oxidation of alkenes stimulated an enormous growth in the use of palladium complexes for a variety of catalytic and stoichiometric reactions of organic compounds.

The term "catalyst" is ambiguous and requires careful use. In heterogeneous reactions—where, for example, a gas mixture is passed over a solid that evidently undergoes no change—the term may have some point insofar as it means a substance added to accelerate a reaction. However homogeneous catalytic reactions in solution are commonly very complex and proceed by way of linked chemical reactions in a closed cycle involving different metal species. The concept of one particular species being "the catalyst," even if it is the one added to initiate or accelerate the reaction, has no real validity. It is convenient, if inaccurate, to call a substance added to accelerate a reaction in solution a "catalyst," provided it is recognized that a catalytic cycle of linked reactions is actually involved.

This chapter deals mainly with homogeneous reactions, since many of these can be studied mechanistically and spectroscopically, and in certain favorable cases intermediates in the reaction cycles, isolated or otherwise, are characterized. The principles on which our understanding of these reactions are based are mainly those discussed in Chapter 29—namely coordinative unsaturation, oxidative-addition and insertion reactions, and attacks on coordinated ligands. Although these principles doubtless have applicability to heterogeneous catalysis, we discuss only the cases of Ziegler-Natta polymerization and the Fischer-Tropsch reaction, which can be understood on these principles. There are currently attempts to correlate and interrelate the mechanisms of homogeneous and heterogeneous catalysis.[1]

From the industrial point of view, heterogeneous systems have great practical advantages over homogeneous ones, notably because of the ease of separation of products from any excess of reactants and from the catalyst. However a number of industrial homogeneous processes have been developed because they may give much higher selectivity in reactions, because they operate under milder conditions of temperature and pressure, or because they have other technical advantages. The major homogeneous processes operated in the United States (and elsewhere) are listed in Table 30-1, and some of these are discussed in this chapter.

We begin, however, with the simpler and better understood reactions and proceed to those of more complexity.

CATALYTIC ADDITION OF MOLECULES TO C—C MULTIPLE BONDS

30-1. Homogeneous Hydrogenation of Unsaturated Compounds[2]

It has been known for more than 40 years that molecular hydrogen can be "activated" homogeneously by solutions of ions such as Ag^+ or MnO_4^- or by complexes such as Cu^{2+} in quinoline or Co^{2+} in aqueous cyanide. Certain of these species can be used as catalysts for the slow reduction of unsaturated substances. Although the intermediacy of species with M—H bonds was often postulated, proof was not obtained, and indeed in the Co^{2+}–CN^- system free radicals appear to be involved.

The first rapid and practical system for the homogeneous reduction of alkenes, alkynes, and other unsaturated substances at 25° and 1 atm pressure used the complex $RhCl(PPh_3)_3$ (Sect 22-G-2) sometimes known as Wilkinson's catalyst.[3a]

Subsequently many other tertiary phosphine complexes of similar types have been studied as catalysts for hydrogenation of C=C and other multiple bonds. They

[1]   H. F. Schaefer, *Acc. Chem. Res.,* 1977, **10,** 287; E. L. Muetterties, *Angew. Chem. Int. Ed.,* 1978, **17,** 545.
[2]   B. R. James, *Homogeneous Hydrogenation,* Wiley, 1973; *Adv. Organomet. Chem.,* 1979, **17,** 319; P. N. Rylander, *Organic Syntheses with Noble Metal Catalysts,* Academic Press, 1973; G. Dolcetti and N. W. Hoffman, *Inorg. Chim. Acta,* 1974, **9,** 209; R. E. Harman et al., *Chem. Rev.,* 1973, **73,** 21; J. F. McQuillan, *Homogeneous Hydrogenation in Organic Chemistry,* Reidel, 1976; A. J. Birch and D. H. Williamson, in *Organic Reactions,* Vol. 24, Wiley, 1976; G. Webb, *Catalysis,* Vol. 2, 1978, Specialist Report, Chemical Society.
[3a] F. H. Jardine *et al., J. Chem. Soc., A.,* **1966,** 1711.

TABLE 30-1

Major Applications of Homogeneous Catalysis in the U.S. Chemical Industry[a]

Application	Approximate 1975 capacity (c) or production (p) (thousands of metric tons)
Carbonylations	
$CH_3CH=CH_2 + CO + H_2 \rightarrow C_3H_7CHO$ (includes other oxo products)	650 p
$RCH=CH_2 + CO + 2H_2 \rightarrow RCH_2CH_2CH_2OH (R > C_9H_{19})$	170 c
$CH_3OH + CO \rightarrow CH_3COOH$	190 c
Monoolefin reactions	
$CH_2=CH_2 + O_2 \rightarrow CH_3CHO$	410 p
$CH_3CH=CH_2 + ROOH \rightarrow CH_3CH-CH_2 + ROH$ (epoxide)	250 c
$CH_2=CH_2 \rightarrow$ Polyethylene (excludes oxide-supported catalysts)	150 c
$CH_2=CH_2 + CH_3CH=CH_2 +$ diene $\rightarrow$ EPDM rubber	85 p
Diene reactions	
$3CH_2=CHCH=CH_2 \rightarrow$ cyclododecatriene	10 p
$C_4H_6 + CH_2=CH_2 \rightarrow$ 1,4-hexadiene	2 p
$C_4H_6 + 2HCN \rightarrow NC(CH_2)_4CN$	70 c
$C_4H_6 \rightarrow$ cis-1,4-polybutadiene (all catalysts)	290 p
Oxidations	
$c\text{-}C_6H_{12} \xrightarrow{O_2} c\text{-}C_6H_{11}OH + c\text{-}C_6H_{10}=O \xrightarrow[HNO_3]{O_2 \text{ or}}$ Adipic acid	610 p
$c\text{-}C_{12}H_{24} \xrightarrow{O_2} c\text{-}C_{12}H_{23}OH + c\text{-}C_{12}H_{22}=O \xrightarrow{HNO_3}$ Dodecanedioic acid	10 c
$CH_3 \langle \rangle CH_3 \xrightarrow{O_2}$ Terephthalic acid and esters	2100 p
$n\text{-}C_4H_{10} \xrightarrow{O_2} CH_3COOH$	470 p
$CH_3CHO \xrightarrow{O_2} CH_3COOH$	335 p
Other reactions	
$CH_2=CHCHClCH_2Cl \rightleftharpoons ClCH_2CH=CHCH_2Cl$	270 p
$ClCH_2CH=CHCH_2Cl + 2NaCN \rightarrow NCCH_2CH=CHCH_2CN$	125 c
$ROOC \langle \rangle COOR + HOCH_2CH_2OH \rightarrow$ Polyester	1900 p

[a] From G. W. Parshall, *J. Mol. Catal.*, 1978, **4**, 243. See also J. M. Tedder, A. Nechvatal, and A. J. Jubb, *Basic Organic Chemistry*, Part 5, *Industrial Products*, Wiley, 1975; A. L. Waddams, *Chemicals from Petroleum*, 3rd ed., Murray, 1973; P. Wiseman, *An Introduction to Industrial Oganic Chemistry*, Applied Science Publishers, 1972.

may be either neutral, e.g., $RuHCl(PPh_3)_3$, or cationic[3b] such as $[Rh(diene)-(PPh_3)_2]^+$ or $[Rh(diphos)]^+$.

There are two main types of compounds.

[3b] R. R. Schrock and J. A. Osborn, *J. Am. Chem. Soc.*, 1976, **98**, 2134, 2143, 4450; J. Halpern *et al.*, *J. Am. Chem. Soc.*, 1977, **99**, 8055.

1. Those without a metal-hydride bond such as $RhCl(PPh_3)_3$ and $[RhS_2(PR_3)_2]^+$ (S = solvent), which react, usually reversibly, with molecular hydrogen.

2. Those with an M—H bond such as $RhH(CO)(PPh_3)_3$ or $RuHCl(PPh_3)_3$, which do not usually react with H_2. Some compounds such as $RuCl_2(PPh_3)_3$ or $RuCl_2(CO)_2(PPh_3)_2$ may form M—H bonds under hydrogenation conditions by heterolytic cleavage of H_2.

Reversible *cis*-Dihydrido Catalysts.   In benzene the distorted square d^8 complex $RhCl(PPh_3)_3$ is red. Despite the steric bulk of the phosphine, it dissociates to only a small extent at 25°:

$$RhCl(PPh_3)_3 \rightleftharpoons RhCl(PPh_3)_2 + PPh_3 \qquad K = 1.4 \times 10^{-4} M$$

but dimerization to an orange, halogen-bridged species can occur, especially at higher temperatures

$$2RhCl(PPh_3)_3 \rightleftharpoons (Ph_3P)_2Rh\underset{Cl}{\overset{Cl}{\diagdown\diagup}}Rh(PPh_3)_2 + 2PPh_3 \qquad K = 3 \times 10^{-4} M$$

Under hydrogen it very rapidly becomes yellow forming a *cis*-dihydridorhodium(III) complex, whose configuration can be established by ^1H and ^{31}P nmr.[3c]

$$RhCl(PPh_3)_3 + H_2 \underset{K = 18 \text{ atm}}{\rightleftharpoons}$$

$$(30\text{-}I) \qquad\qquad (30\text{-}II)$$

It also reacts with olefins to a small extent, the complexity constant for ethylene being the largest[4]:

$$RhCl(PPh_3)_3 + C_2H_4 \rightleftharpoons RhCl(C_2H_4)(PPh_3)_2 + PPh_3 \qquad K = 0.4$$

The octahedral complex 30-I has a PPh_3 group trans to hydrogen; thus it is not surprising that the strong trans effect of H leads to dissociation to give a five-coordinate species, 30-II. This is doubtless fluxional and labile as well as being coordinately unsaturated and able to coordinate alkenes.

There has been some argument whether the hydrogen attacks an alkene complex similar to the ethylene one noted above, or whether an alkene attacks a dihydrido species. The latter seems to be the case, as originally proposed.[3a] In the complex $Rh(H)_2Cl(alkene)(PPh_3)_2$, two successive hydrogen transfer reactions now occur. The first β-hydrogen transfer generates a rhodium alkyl. The second hydrogen is then transferred to the carbon atom of the alkyl chain via a three-centered transition state. The saturated product then merely leaves the sphere of the metal. The entire

[3c]   C. A. Tolman *et al., J. Am. Chem. Soc.,* 1974, **96**, 2762.

[4]   M. H. J. M. de Croon *et al., J. Mol. Catal.,* 1978, **4**, 325; C. Rousseau *et al., J. Mol. Catal.,* 1978, **3**, 309; Y. Otahni *et al., Bull. Chem. Soc. Jap.,* 1977, **50**, 1432; J. Halpern *et al., J. Mol. Catal.,* 1976, **2**, 65.

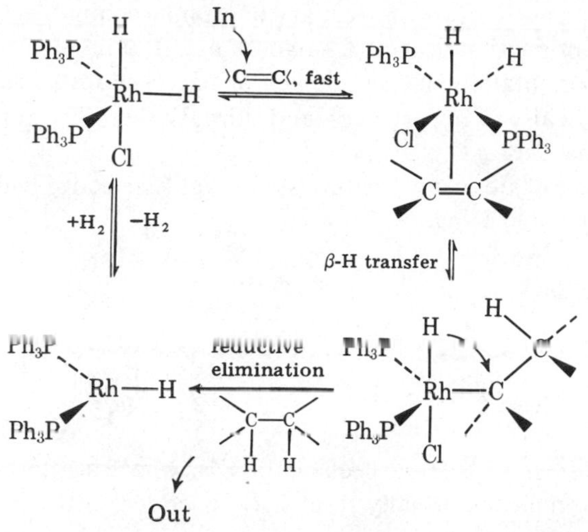

Fig. 30-1. Simplified catalytic cycle for the hydrogenation of C-C multiple bonds by *bis*(triphenyl-phosphine)chlororhodium species as derived from Wilkinson's Catalyst, $RhCl(PPh_3)_3$, or from [(ole-fin)$_2$RhCl]$_2$ + PR$_3$. Possible coordination of solvent molecules is disregarded. Similar cycles operate for cationic species where the *cis*-dihydrido species are of the type $[Rh(H)_2(PR_3)_2S_2]^+$, S = solvent.

second step is, of course, simply the reductive elimination of R H from R—Rh—H. It was shown by generating species with different Rh/PPh$_3$ ratios by reactions such as:.

$$[(cyclooctene)_2RhCl]_2 + 4PPh_3 = 2RhCl(PPh_3)_2 + 4 \text{ cyclooctene}$$

that the optimum catalytic activity requires a 1:2 ratio. Also it was found that more basic and less sterically hindered phosphines (e.g., PMe$_2$Ph) are less effective than PPh$_3$.

Hence the principal catalytic cycle is best described as involving bisphosphine species (Fig. 30-1). For RhCl(PPh$_3$)$_3$, or where excess phosphine is present, other equilibria will be involved, as well as, potentially, the dimeric species [RhCl(PPh$_3$)$_2$]$_2$; the system is quite complicated. Detailed studies have allowed determination of equilibrium and rate constants,[4] and in dilute solutions containing only millimolar concentrations of RhCl(PPh$_3$)$_3$ and large concentrations of alkenes, the cycle in Fig. 30-1 is the main one. Note that the species RhCl(PPh$_3$)$_2$ is a fourteen-electron one.

Several features in the cycle, that not only are typical of catalytic systems but are *requirements* for activity can be noted:

1. The high lability of species that can exist in several forms because of disso-ciative equilibria,

2. The changes in coordination numbers from 3 to 6 (or from fourteen- to eighteen-electron species).

3. The changes in oxidation state from I to III in the oxidative-addition and reductive-elimination steps.

In all catalytic cycles involving tertiary phosphine complexes the steric bulk of the ligand is of prime significance (Chapter 3, p. 89). It influences the dissociative equilibria, the orientation and complexing of the unsaturated substrates (whose steric factors are also of importance), and third, the stability of the intermediate alkyls (see below).

A consequence of steric interactions is that highly selective hydrogenations are possible, one example being

Finally, in contrast to heterogeneous catalysis, where scattering of deuterium throughout the molecule usually results, D_2 is added to the bond and remains there.

Although iridium(I) complexes such as Vaska's compound [*trans*-IrCl(CO)-(PPh$_3$)$_2$: Section 22-G-2] take up hydrogen reversibly, the octahedral species [e.g., Ir(H)$_2$Cl(CO)(PPh$_3$)$_2$] so formed do not eliminate a neutral ligand. Hence they have no vacant site for coordination of unsaturated substances and cannot act as hydrogenation catalysts at 25° and atmospheric pressure, although they may do so when phosphine dissociation is promoted thermally or photochemically.

Monohydrido Complexes. Compounds such as RuHCl(PPh$_3$)$_3$, RhH(CO)-(PPh$_3$)$_3$, RhH(dph)$_2$ (dph = dibenzophosphole[5]), and IrH(CO)(PPh$_3$)$_3$, which have a bound hydrogen, also act as hydrogenation catalysts. The hydride RhH(PPh$_3$)$_3$ is a much more active catalyst than RhCl(PPh$_3$)$_3$.[6] Compounds such as RuCl$_2$(PPh$_3$)$_3$ or RuCl$_2$(CO)(PPh$_3$)$_2$[7] undergo hydrogenolysis to hydrido species in presence of hydrogen halide acceptors, e.g.,

$$RuCl_2(PPh_3)_3 + H_2 + Et_3N = RuHCl(PPh_3)_3 + Et_3NHCl$$

In these cases the catalytic cycle differs from that discussed previously. The cycle for RhH(CO)(PPh$_3$)$_3$ is shown in Fig. 30-2. The complex dissociates in benzene, and both isomerization and hydrogenation of alk-1-enes are suppressed by addition of excess PPh$_3$. By contrast, CoH(CO)(PPh$_3$)$_3$ catalyzes hydrogenation only at *ca.* 150° and 150 atm hydrogen, and IrH(CO)(PPh$_3$)$_3$ will hydrogenate alkenes at *ca.* 50°.[8]

The hydrogenation of alkenes by RhH(CO)(PPh$_3$)$_3$ is unusual in that there is a high selectivity for the hydrogenation of alk-1-ene compared to alk-2-ene. We noted in Section 29-10 (p. 1252) that alk-1-ene can react with a metal-hydrogen bond in either a Markownikoff or an anti-Markownikoff fashion and that a secondary branched or a primary straight-chain alkyl, respectively, is produced. An

[5]   D. E. Budd *et al., Can. J. Chem*, 1974, **32**, 775.
[6]   D. F. Shriver *et al., Inorg. Chem.*, 1978, **17**. 3064, 3069.
[7]   D. R. Fahey, in *Catalysis in Organic Synthesis,* Academic Press, 1976.
[8]   M. G. Burnett *et al., J.C.S. Dalton,* **1974,** 1663.

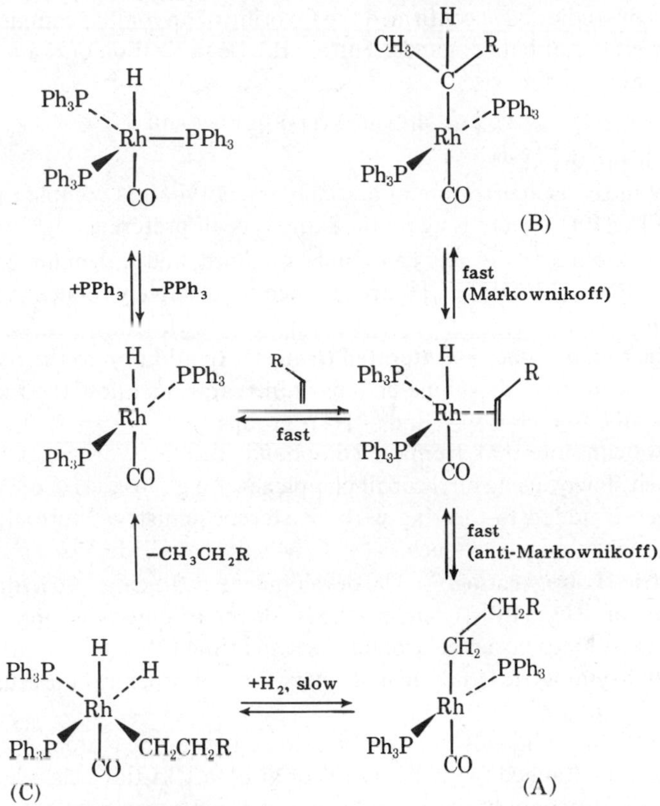

Ph_3P

H

Rh—PPh_3

Ph_3P

CO

H

CH_3 R

C

PPh_3

Rh

Ph_3P

CO

(B)

$+PPh_3$ $-PPh_3$

fast
(Markownikoff)

H

PPh_3

Rh

Ph_3P

CO

fast

R

Ph_3P

H R

Rh

Ph_3P

CO

fast
(anti-Markownikoff)

$-CH_3CH_2R$

CH_2R

CH_2

PPh_3

Rh

Ph_3P

CO

(A)

Ph_3P

H H

Rh

Ph_3P CH_2CH_2R

CO

(C)

$+H_2$, slow

Fig. 30-2. Catalytic cycle of hydrogenation and isomerization of 1-alkenes by $RhH(CO)(PPh_3)_3$ at 25° and 1 atm pressure.

alk-2-ene can, of course, give only a branched-chain alkyl. We can now explain the selectivity as follows.[9] In the square species A and B of Fig. 30-2, the bulky triphenylphosphine groups are in trans positions, and the result is that a primary alkyl (i.e., Rh—CH_2—CH_2—R) will experience much less steric interaction than will a more bulky alkyl (e.g., Rh—$CH(CH_3)CH_2R$). The lowered stability of a secondary alkyl complex means that such a species would undergo the reverse β-hydrogen transfer reaction to give olefin more easily; hence the alkyl would have a shorter lifetime in solution—so short, in fact, that it would not live long enough to undergo the slow rate-determining oxidative addition of hydrogen to give species C. That the secondary alkyl complex is formed simultaneously with the primary alkyl complex from alk-1-ene is shown by comparability of the rates of isomerization (alk-1-ene to alk-2-ene) and hydrogenation.

The steric factors introduced by bulky PPh_3 groups are stressed here because in the hydroformylation reaction discussed later, which is also catalyzed by the same rhodium complex, the high selectivity for the formation of a straight-chain product is of great significance.

9   C. O'Connor and G. Wilkinson, *J. Chem. Soc.*, **1968**, 2665.

Other recent studies have confirmed the favoring of an n-alkyl complex induced by the steric effects of bulky ligands. Thus in the isomerization of branched-chain acyls to give alkyl carbonyls:

$$(Ph_3P)_2Cl_2IrCOR \rightleftharpoons (Ph_3P)_2Cl_2Ir(CO)R$$

the n-alkyl is preferred.[10a]

Similarly in the isomerization of a secondary butyl alkyl complex $(\eta\text{-}C_5H_5)\text{-}Fe(CO)(PPh_3)(Bu^s)$, there is again the same type of preference.[10b]

Cationic monohydrido species can also be obtained, and ruthenium complexes[11] e.g., $[RuH(PPh_3)_n(MeOH)_{5-n}]^+$, are claimed to selectively reduce cyclic dienes to monoenes.

Hydrogenation of Other Unsaturated Groups.   In addition to the reduction of $C=C$ or $C\equiv C$ bonds, phosphine or bipyridine catalysts allow the reduction of $C=O$, $-N=N$, $-CH=N-$, and $-NO_2$ groups.[12]

Finally we can note that aromatic compounds can be hydrogenated homogeneously albeit slowly, using allyl cobalt complexes,[13] e.g., $\eta_3\text{-}C_3H_5Co[P(OMe)_3]_3$; the hydrogen is added to the ring with cis stereochemistry. More effective are ruthenium arene compounds[14] such as $(\eta^6\text{-}C_6Me_6)Ru(\eta^4\text{-}C_6Me_6)$ (see p. 1169).

Asymmetric Hydrogenation.[15]   The development of the catalytic hydrogenation system based on $RhCl(PPh_3)_3$ and methods for the resolution of optical isomers of tertiary phosphines occurred around the same time (1965), and this led to the possibility of asymmetric catalytic hydrogenations of prochiral unsaturated substances.

Tertiary phosphine ligands can have chirality either at the P atom,[16] at a carbon atom on a group attached to the P atom,[17] or at both.[18] A third class has an axial element of chirality only.[19]

[10a]   M. A. Bennett *et al.*, *J. Am. Chem. Soc.*, 1978, **100**, 2737.

[10b]   D. L. Reger and E. C. Culbertson *Inorg. Chem.*, 1977, **16**, 3104.

[11]   R. J. Young and G. Wilkinson, *J.C.S. Dalton*, **1976**, 719; A. Spencer, *J. Organomet. Chem.*, 1976, **93**, 389.

[12]   G. Mestroni *et al.*, *J. Organomet. Chem.*, 1978, **157**, 345; M. Gargano *et al.*, *J. Organomet. Chem.*, 1977, **129**, 239; G. Zassinovitch *et al.*, *J. Organomet. Chem.* 1977, **140**, 63; R. H. Crabtree *et al.*, *J. Organomet. Chem.*, 1977, **141**, 113; J. Solodar, *Chem. Tech.*, **1975**, 421.

[13]   E. L. Muetterties *et al.*, *Accts. Chem. Res.*, 1979, **12**, 324.

[14]   M. A. Bennett *et al.*, *J.C.S. Chem. Commun.*, **1978**, 582, **1979**, 312.

[15]   R. Pearce, *Catalysis*, Vol. 2, 1978, Specialist Report, Chemical Society. J. D. Morrison *et al.*, in *Advances in Catalysis*, Vol. 25, Academic Press, 1976, p. 81; P. N. Rylander and H. Greenfield, Eds., *Catalysis in Organic Synthesis*, Academic Press, 1976; J. W. Scott and D. Valentine, Jr., *Science*, 1974, **184**, 943 (general asymmetric synthesis); L. Marko and B. Heil, *Catal. Rev.*, 1973, **8**, 269; J. D. Morrison, and H. S. Moser, *Asymmetric Organic Reactions*, Prentice-Hall 1971; H. B. Kagan, *Pure Appl. Chem.*, 1975 **43**, 401, H. B. Kagan and J. C. Fiaud. in *Topics in Stereochemistry*, Vol. 10, Wiley, 1978, p. 175.

[16]   W. S. Knowles *et al.*, *J. Am. Chem. Soc.*, 1978, **100**, 7561; 1975, **97**, 2567; K. Tani *et al.*, *J. Am. Chem. Soc.*, 1977, **99**, 7876.

[17]   H. B. Kagan *et al.*, *J. Am. Chem. Soc.*, 1978, **100**, 7556; *J. Organomet. Chem.*, 1975, **91**, 105; R. G. Ball and N. C. Payne, *Inorg. Chem.*, 1977, **16**, 1187; W. R. Cullen and Y. Sugi, *Tetrahedron Lett.*, **1978**, 1635 (sugars).

[18]   C. Fisher and H. S. Mosher, *Tetrahedron Lett.*, **1977**, 2487.

[19]   K. Tamao *et al.*, *Tetrahedron Lett.*, **1977**, 1389; R. H. Grubbs and A. A. De Vries, *Tetrahedron Lett.*, **1977**, 1879.

Some representative ligands are *o*-anisyl cyclohexyl methyl phosphine, (−
or +)-2,3-*O*-isopropylidcnc-2,3-dihydroxy-1,4-bis(diphenylphosphino)-butane
(30-III) usually abbreviated to (+ or −) DIOP, which is readily obtained from the
appropriate tartaric acid, neomenthyldiphenylphosphine (31-IV), and compound
30-V, abbreviated to (+)-camphos. The ligand *trans*-1,2-bis(diphenylphosphi-
noxy)cyclopentane (30-VI) is particularly useful for substrates without functional
groups.[20] The methyl ligand 30-VII is an example similar to 31-IV but chiral at
both P and C. Compound 30-VIII is an example of axial chirality.

(30-III)   (30-IV)   (30-V)

(30-VI)   (30-VII)   (30-VIII)

The asymmetric reducing systems may be of either type discussed above but most
commonly are *cis*-dihydrido rhodium species prepared *in situ* by addition of the
chiral ligand to rhodium complexes such as $[(C_8H_{10})RhCl]_2$ or the cationic rhodium
species noted earlier.[21]

The most important application of asymmetric hydrogenation is in the synthesis
of the drug L-DOPA (dihydroxyphenylalanine) used in treatment of Parkinson's
disease. In general, prochiral olefins of the type $R'CH{=}C(NHR_2)COOH$ can be
reduced to chiral amino acids, and the optical purity achieved may be as high as
95%.

The precise role of the chiral ligand is not well understood, and the basis of selection of the most appropriate ligand and hydrogenation conditions such as solvent,
temperature, and hydrogen pressure for a particular reduction, is empirical. It seems
certain that there are interactions between the coordinated prochiral ligand and
either the metal or the aryl rings of the chiral phosphines, and these interactions
lead to high asymmetric induction.[22]

Interaction involving a carbonyl group has been proved for the DIOP complex
of (Z)-α-benzamidocinnamic acid, where ^{31}P nmr spectra showed that only one
of the two possible diastereoisomers was present[23] (30-IX).

[20] T. Hayashi *et al.*, *Tetrahedron Lett.*, **1977**, 295.
[21] See D. A. Slack and M. C. Baird, *J. Organomet. Chem.*, 1977, **142**, C69.
[22] See B. Bosnich and M. D. Fryzuk, *J. Am. Chem. Soc.*, 1978, **100**, 5491; B. D. Vineyard *et al.*,
      J. Am. Chem. Soc., 1977, **99**, 5946.
[23] J. Halpern *et al.*, *Inorg. Chim. Acta*, 1979, **37**, L477.

$$
\begin{array}{c}
R^2 \\
| \\
O=C \\
\diagup \qquad \diagdown \\
(+-DIOP)Rh \qquad\qquad NH \\
\qquad\qquad Ph\text{-}\text{-} \qquad \blacktriangleleft CO_2R^1 \\
\qquad\qquad | \\
\qquad\qquad H
\end{array}
$$

(30-IX)

The major part of the asymmetric induction arises from the stereoselectivity in the binding step rather than from the relative rates of formation or hydrogenation of the diastereomeric complexes. Oxidative addition of hydrogen is to the olefin complex and not vice versa, as in Fig. 30-1.

Chiral phosphine ligands can be used for asymmetric syntheses in other catalytic systems where tertiary phosphines are employed, e.g., for hydrosilation and hydroformylation of alkenes (see later).

Transfer Hydrogenations.[24] The reductions discussed above use molecular hydrogen. It is also possible, but usually less useful, to use other molecules such as alcohols, aldehydes, dioxan, and tetrahydrofuran that are fairly readily dehydrogenated as hydrogen sources. The hydrogen donor must be able to coordinate to the metal, and β-hydrogen transfer or 1,2- shifts similar to those described earlier are doubtless involved, e.g.,

Typical transfer reactions use rhodium–tertiary phosphine complexes such as $RhCl(PPh_3)_3$ with dioxane.[25] The overall transfer reaction is merely tantamount to breaking C—H bonds in one molecule and forming them in another, viz.,

Ketones and other unsaturated substances can also be reduced by transfer hydrogenation; e.g., the complex $Mo(N_2)_2(diphos)_2$ with hydrogen donors gives $MoH_4(diphos)_2$, which will reduce ketones.[26]

It is convenient to note here that dehydrogenation of alcohols to aldehydes where the hydrogen is evolved as *molecular hydrogen* is catalyzed by the ruthenium complex $Ru(CO_2CF_3)_2(CO)(PPh_3)_2$ in presence of CF_3COOH.[27] The reaction

[24]   G. Brieger and T. J. Nestrick, *Chem. Rev.*, 1974, **74**, 567.
[25]   T. Nishiguchi and K. Fukuzumi, *J. Am. Chem. Soc.*, 1974, **96**, 1892; C. A. Masters *et al.*, *J. Am. Chem. Soc.*, 1976, **98**, 1357.
[26]   T. Tatsumi *et al.*, *Chem. Lett.*, **1977**, 191.
[27]   S. D. Robinson *et al.*, *Inorg. Chem.*, 1978, **17**, 1896; 1977, **16**, 137, 1321.

proceeds via formation of an alcoxide, which then undergoes β-hydrogen transfer (cf. p. 1258).

$$L_nM - OCH_2R \rightarrow L_nMH + RCHO$$

30-2.   Hydrosilation of Unsaturated Compounds[28]

The hydrosilation (cf. hydroboration, p. 321) or hydrosilylation reaction is similar to hydrogenation except that H and SiR_3 from a silane $HSiR_3$, are added across a double bond. Olefins, usually, give the terminal product:

$$RCH=CH_2 + HSiR_3 \equiv RCH_2CH_2SiR_3$$

The reaction was first discovered using chloroplatinic acid as catalyst, and H_2PtCl_6 in ethanol is commonly referred to as Speier's catalyst. It is effective in extremely low concentration (10^{-5} to 10^{-8} mole of Pt per mole of reactant). The reaction is used in silicone technology in various ways; e.g., hydrosilation of acrylonitrile allows incorporation of C_2H_4CN groups into rubbers for self-sealing fuel tanks and for crosslinking reactions in the curing of gums to elastomers.

The reaction may proceed by the initial formation of platinum(II) olefin complexes—indeed such complexes can be used directly—which then undergo addition as shown in Fig. 30-3. However since traces of oxygen are known to have a profound promoter action, it is quite possible that hydrosilation is a radical chain reaction. The stoichiometric oxidative addition of silanes to compounds such as $RhCl(PPh_3)_3$ or trans-$IrCl(CO)(PPh_3)_2$ provides evidence for the first step, e.g.,

$$IrCl(CO)(PPh_3)_2 + R_3SiH = RhCl(H)(SiR_3)(CO)(PPh_3)_2$$

The addition of chiral silanes shows that the addition occurs with retention of configuration.

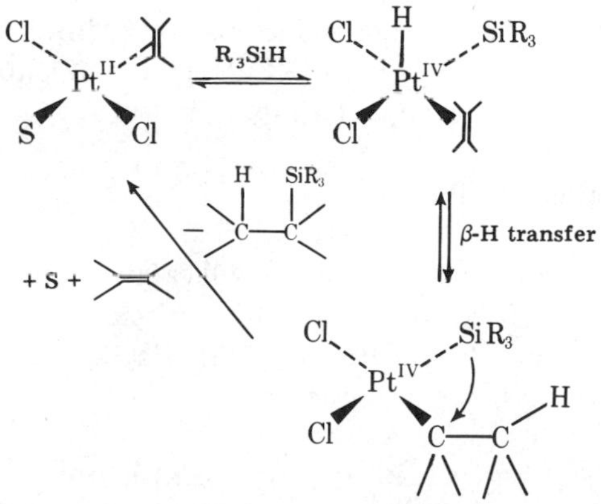

Fig. 30-3. Simplified cycle for hydrosilation of alkenes by platinum complexes. S = solvent.

[28]   Z. V. Belyakova et al., J. Organomet. Chem. Libr., Vol. 5, 1977, J. F. Harrod and A. J. Chalk, in Organic Syntheses via Metal Carbonyls, Vol. 2, I. Wender and P. Pino, Eds., Wiley, 1976; C. S. Cundy et al., Adv. Organomet. Chem., 1973, **11**, 253; J. L. Speier, Adv. Organomet. Chem., 1979, **17**, 407.

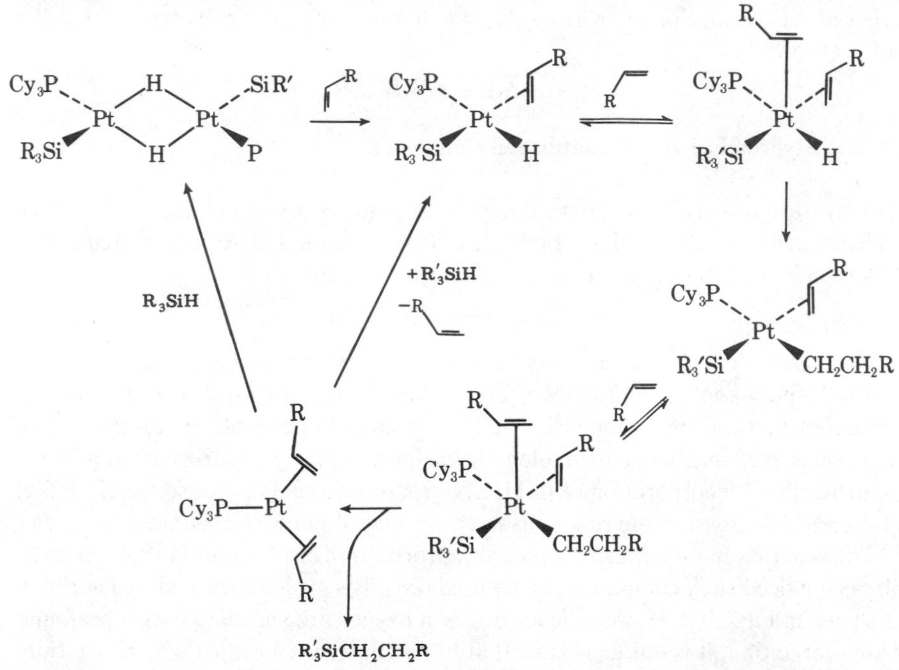

Fig. 30-4. Catalytic cycle for hydrosilation of olefins by platinum tricyclohexylphosphine (Cy₃P) complexes. A similar cycle can be written for acetylenes, the main difference being that H transfer to coordinated olefin gives a Pt-alkyl whereas transfer to coordinated acetylene gives a Pt-vinyl group.

The same types of phosphine complex also catalyze the reaction. One of the earlier studies used $RhCl(PPh_3)_3$, and an effect of oxygen was also observed. A recently proposed catalytic cycle (Fig. 30-4) uses platinum complexes.[29] In addition, $C \equiv C$ bonds,[30] aldehydes, ketones, and imines may by hydrosilated. By use of chiral phosphine complexes (e.g., of Rh^I with DIOP), prochiral ketones such as acetophenone can be hydrosilated asymmetrically.

30-3. Hydrocyanation of Alkenes[31]

Various complexes, especially of Ni and Pd, will catalyze the addition of HCN to olefins to form nitriles. Although HCN is a weak acid, it oxidatively adds quite readily to compounds such as $RhCl(PPh_3)_3$, $Pt(PPh_3)_3$, and $Ni[P(OEt)_3]_4$.[32] In nickel-catalyzed reactions[33] we have the steps:

$$NiL_4 + HCN \underset{+L}{\overset{-L}{\rightleftharpoons}} NiH(CN)L_3 \underset{+L}{\overset{-L}{\rightleftharpoons}} NiH(CN)L_2$$

29 M. Green et al., J.C.S. Dalton, **1977**, 1519, 1529.
30 H. M. Dickers et al., J. Organomet. Chem., 1978, **161**, 91.
31 E. S. Brown, in Organic Syntheses via Metal Carbonyls, Vol. 2, I. Wender and P. Pino, Eds., Wiley, 1976, and in Aspects of Homogeneous Catalysis, Vol. 2, R. Ugo, Ed., Reidel, 1974.
32 J. D. Druliner et al., J. Am. Chem. Soc., 1976, **98**, 2156.
33 B. W. Taylor and H. E. Swift, J. Catal., 1972, **26**, 254.

followed by

$$NiH(CN)L_2 \xrightarrow{RCH=CH_2} (RCH=CH_2)NiH(CN)L_2 \rightarrow RCH_2CH_2Ni(CN)L_2$$

and, completing the cycle:

$$RCH_2CH_2Ni(CN)L_2 \rightarrow RCH_2CH_2CN + NiL_2$$
$$NiL_2 + HCN \rightarrow NiH(CN)L_2$$

As in other additions of HX to alkenes, Markownikoff and anti-Markownikoff products result. The addition of Lewis acids such as $ZnCl_2$ enhances the rate of reaction, probably by increasing the acidity of HCN; also affected is the ratio of straight- to branched-chain isomers.

The most important application is the synthesis from butadiene of adiponitrile, which is then used to make hexamethylenediamine for Nylon 66. Patents (Du Pont) suggest that a nickel phosphite catalyst together with a weak Lewis acid cocatalyst such as $ZnCl_2$ is used. Although details are not available, it could be that 3-pentenenitrile forms first, is isomerized by the Lewis acid to 4-pentenenitrile, which then adds more HCN to give adiponitrile.

$$CH_2=CH-CH=CH_2 \xrightarrow{HCN} CH_2=CH-CH(CN)CH_3 + CH_3CH=CH-CH_2CN$$

$$NC(CH_2)_4CN \xleftarrow{HCN} CH_2=CHCH_2CH_2CN$$

Addition of HCN to butadiene, accompanied by dehydrogenation to give *trans*-$CNCH_2CH=CHCH_2CN$, catalyzed by Cu^{II} halides in EtCN, has also been claimed:[34] the Cu^{II} is slowly reduced to Cu^I, which is then reoxidized by air.

METATHESIS, POLYMERIZATION, AND RELATED REACTIONS OF ALKENES AND ALKYNES

30-4.   Alkene Metathesis[35]

Although the metathesis of alkenes is technically nowhere nearly as important as the polymerization and oligomerization reactions of these compounds, we discuss it here because it is the first reaction whose mechanism has been clearly shown to involve carbene intermediates. It now seems likely that some major polymerization reactions also involve carbene intermediates.

[34]   German Patent 2,642,449.
[35]   T. J. Katz, *Adv. Organomet. Chem.*, 1977, **16**, 283; J. J. Rooney and A. Stewart, *Catalysis*, Vol. 1, Specialist Report, Chemical Society, 1977, p. 277; R. J. Haines and G. J. Leigh, *Chem. Soc. Rev.*, **1975**, 155; W. B. Hughes, *Chem. Tech.*, **1975**, 486; J. C. Mol and J. A. Mouliyn, *Adv. Catal.*, 1975, **24**, 00; G. C. Bailey, *Catal. Rev.*, 1970, **3**, 37. N. Calderon *et al.*, *Angew. Chem., Int. Ed.*, 1976, **15**, 401; *Adv. Organomet. Chem.* 1979, **17**, 449; F. D. Mango, *J. Am. Chem. Soc.*, 1977, **99**, 6117, *Rec. Trav. Chim.*, 1977, **96**, M-5; R. H. Grubbs, *Prog. Inorg. Chem.*, 1978, **24**, 1.

The metathesis or dismutation reaction is reversible and can be extremely rapid:

$$
\begin{array}{c}
CH_2{=}CHR \\
+ \\
CH_2{=}CHR
\end{array}
\rightleftharpoons
\begin{array}{c}
CH_2 \\
\parallel \\
CH_2
\end{array}
+
\begin{array}{c}
CHR \\
\parallel \\
CHR
\end{array}
$$

It was first discovered by Banks and Bailey in heterogeneous systems where olefins were passed at 150 to 500° over $Mo(CO)_6$ or $W(CO)_6$ deposited on Al_2O_3, silica-supported metals, etc. Later a wide variety of homogeneous catalysts mostly of Mo, W, or Re have been used, typical ones being $WCl_6 + EtAlCl_2$ in ethanol, $Mopy_2(NO)_2Cl_2 + AlR_3$, and $MoCl_6 + AlCl_3$ in chlorobenzene.

There has been much speculation on the mechanism, but early ideas were abandoned in favor of a carbene mechanism, first suggested by the French chemists Herrisson and Chauvin. The involvement of metal carbenes (p. 1140) has been demonstrated in several ways. These included (a) direct reaction of olefins with carbene complexes,[36] (b) complicated kinetic and alkene crossing experiments, mainly by Katz and Grubbs, and (c) trapping of metal carbene intermediates by ethyl acrylate.[37]

The direct stoichiometric interaction of pentacarbonyl diphenylcarbenetungsten(0) with olefin is considered to proceed via a tungsten(II) intermediate with two W—C σ bonds and a puckered four-membered ring as shown:

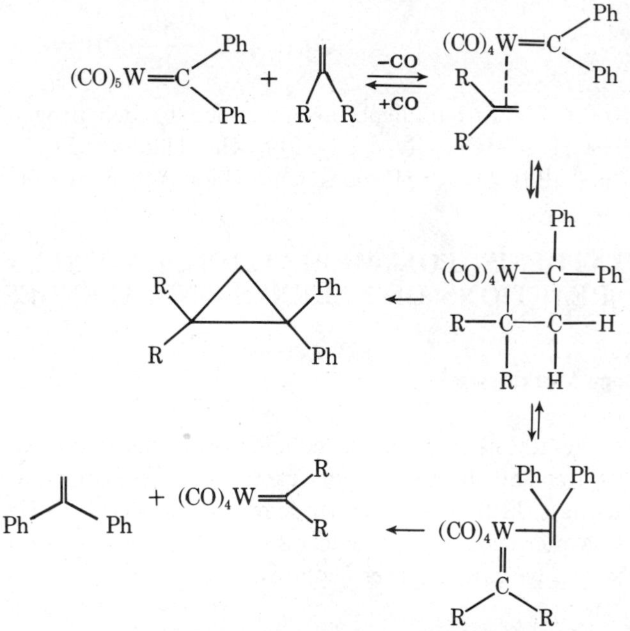

36　C. P. Casey et al., J. Am. Chem. Soc., 1977, **99**, 2533, 6097.

37　C. P. Casey and H. E. Tuinstra, J. Am. Chem. Soc., 1978, **100**, 2270; T. J. Katz and J. McGinnis, J. Am. Chem. Soc., 1977, **99**, 1903; R. H. Grubbs et al., J. Am. Chem. Soc., 1979, **101**, 1499; 1976, **98**, 3478; P. G. Gassman and T. H. Johnson, J. Am. Chem. Soc., 1976, **98**, 6055, et seq., 1977, **99**, 622; J. L. Bilhou et al., J. Am. Chem. Soc., 1977, **99**, 4083; T. J. Katz and W. H. Hersh, Tetrahedron Lett., 1977, 585.

The carbene intermediate formed in the reaction of *cis*-2-butene with $C_6H_5WCl_3$ + $AlCl_3$ in C_6H_5Cl reacts with ethyl acrylate faster than with a simple alkene; thus the isolation of the cyclopropane derivative confirms the intermediacy of the carbenes:

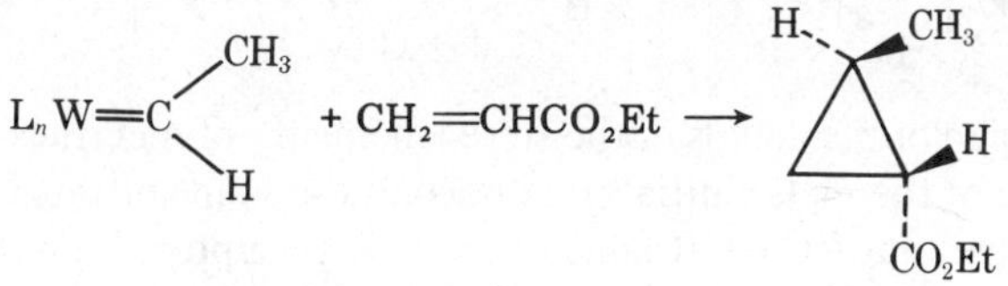

$$L_n W = C\begin{array}{c} CH_3 \\ \\ H \end{array} + CH_2{=}CHCO_2Et \longrightarrow$$

The catalytic systems are thus chain reactions of the type proceeding by a one-carbene complex and a four-center transition state.

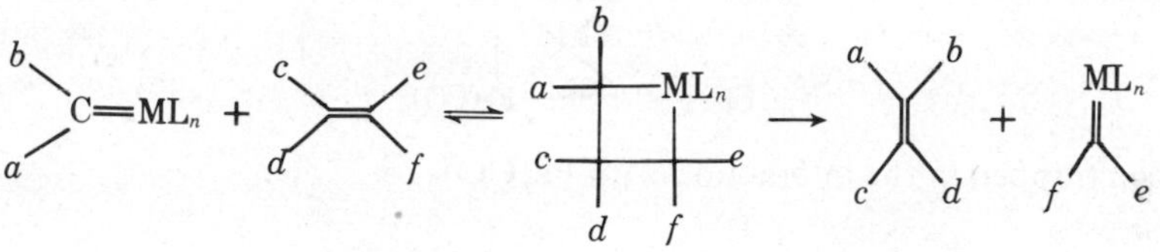

Although generation of the initiating carbene species is most likely by an α-elimination reaction (p. 1139), there are several possible ways, depending on the catalyst-solvent system used,[38] e.g.,

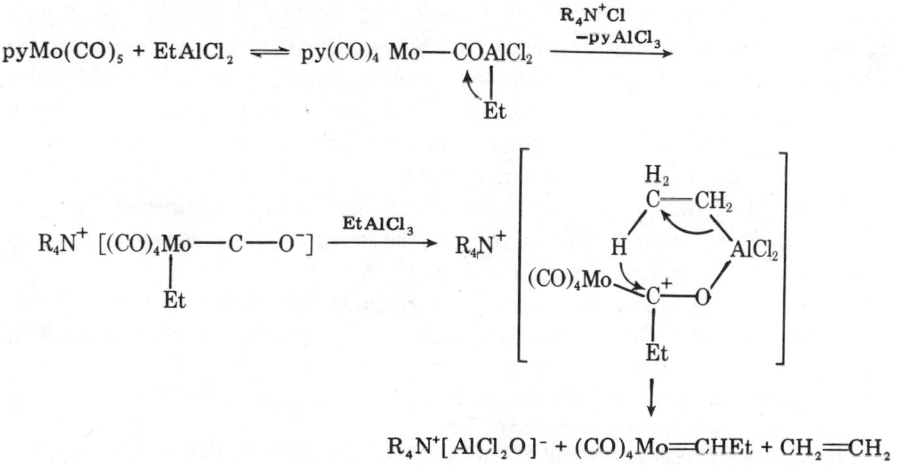

$$pyMo(CO)_5 + EtAlCl_2 \rightleftharpoons py(CO)_4 Mo{-}COAlCl_2 \xrightarrow[-py\,AlCl_3]{R_4N^+Cl}$$

$$R_4N^+ [(CO)_4Mo{-}C{-}O^-] \xrightarrow{EtAlCl_3} R_4N^+$$

$$R_4N^+[AlCl_2O]^- + (CO)_4Mo{=}CHEt + CH_2{=}CH_2$$

Whatever the route by which it is formed, the carbene presumably has a high barrier to rotation, allowing the empty p_z orbital on the carbene to interact with the coordinated olefin. The stereochemistry of the reaction would then be determined by the geometry of the approach:

[38]   V. W. Motz and M. Farone, *Inorg. Chem.*, 1977, **16**, 2544; R. H. Grubbs and C. E. Hoppin, *J.C.S. Chem. Commun.*, **1977**, 634.

Further discussion of the details of the stereochemistry of metathesis, which varies with the nature of the metal initiator, is beyond the scope of this book.

Finally we note that $RC{\equiv}CR$ bonds can undergo rupture[39] on interaction with metal complexes, initially probably to give RC, carbyne units. These units usually polymerize (e.g. to give arenes), but by use of the acetylene $Et_2NC{\equiv}CNEt_2$, an intermediate

has been trapped in the interaction with $Fe(CO)_5$.

30-5. Ziegler-Natta Polymerization of Ethylene and Propylene[40]

The discovery by Ziegler that hydrocarbon solutions of $TiCl_4$ in presence of triethylaluminum give heterogeneous suspensions that polymerize ethylene at 1 atm pressure has led to an extremely diverse chemistry in which aluminum alkyls are used to generate transition metal–alkyl species. At about the same time it was found by the Phillips Petroleum Company that when suspended in inert hydrocarbon solvents, specially activated chromium oxides on an alumina support also polymerize ethylene. A more effective process for C_2H_4 operated in the gas phase (Union Carbide) uses alumina treated with $(\eta\text{-}C_5H_5)_2Cr$. These processes produce a linear high-density polyethylene that differs from the low-density, branched polyethylene produced by free-radical polymerization at very high pressure (Imperial Chemical Industries). These processes will not produce the stereoregular polypropylene, which was Natta's development of the Ziegler method.

The Ziegler-Natta system is also heterogeneous, and the active metal species is a fibrous form of $TiCl_3$ formed *in situ* from $TiCl_4$ and $AlEt_3$. Recent unspecified

[39] R. B. King *et al., The Place of Transition Metals in Organic Synthesis,* New York *Academy of Science* 1977; *J.C.S. Chem. Comm.,* **1977,** 30.

[40] J. A. Moore, Ed., *Macromolecular Synthesis,* Vol. 1, Wiley, 1978; A. D. Gaunt, *Catalysis,* Vol. 1, Specialist Report, Chemical Society, 1977; W. M. Saltman, Ed., *The Stereorubbers,* Wiley, 1977; K. Weissermel *et al., J. Polym. Sci. Polym. Symp.,* No. 51, **1975,** 187, T. Keii, *The Kinetics of Ziegler-Natta Polymerisation,* Chapman & Hall, 1972; J. C. W. Chien, Ed., *Coordination Polymerisation,* Academic Press, 1974; A. Yamamoto and T. Yamamoto, *J. Polym. Sci. Macromol. Rev.,* 1978, **13,** 161 (Polymerisation by transition metal alkyls and hydrides, 229 refs). A. D. Ketley, Ed., *The Stereochemistry of Macromolecules,* Vols. I–III, Arnold (London), Dekker (New York), 1967; L. S. Reich and A. Schindler, *Polymerisation by Organometallic Compounds,* Wiley-Interscience, 1966; G. Natta, Ed., *Stereoregular Polymers,* Vols. I, II, Pergamon Press.

improvements[41] in the catalyst made by the Montedison Company allow the production of *ca.* 2.5×10^5 g of polypropylene per gram of Ti compared to the usual 3×10^3 g g^{-1}.

The interaction of zirconium tetrabenzyl and similar peralkyls with silica and other supports that have reactive –OH groups also gives active catalysts.[42]

For many years the accepted mechanism of polymerization has been that due to Cossee. Basically this is one involving insertion of C_2H_4 or C_3H_8 into a titanium-alkyl group, initially formed from aluminum alkyl, on the surface.

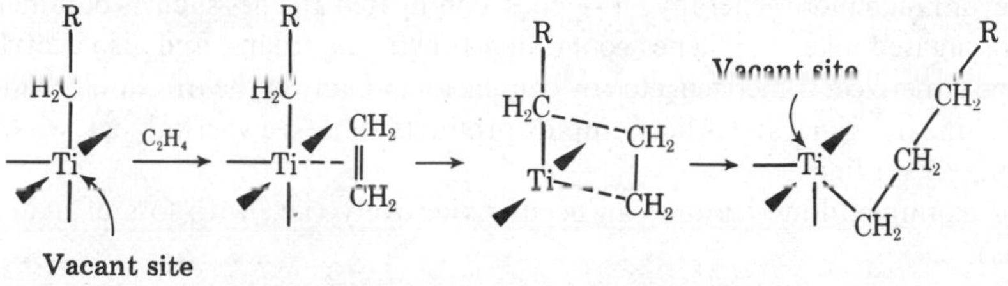

The stereoregularity of polypropylene was said to arise because of the nature of sterically hindered sites on the TiCl$_3$ lattice.

Although the pathway involving insertion of ethylene or propylene into the metal-alkyl bond has always been assumed, only very recently has such insertion been proved for a cobalt trimethylphosphine methyl complex.[43a] An alternative mechanism for not only alkene polymerization, but dimerizations and oligomerizations of alkenes catalyzed by other metals, invokes carbene species generated by α-hydrogen transfers (p. 1139). A scheme for propylene polymerization,[43b] where Ⓟ represents the polymer chain (for initiation Ⓟ could have either CH_3 or H), and [Ti] the titanium atom on the surface is the following:

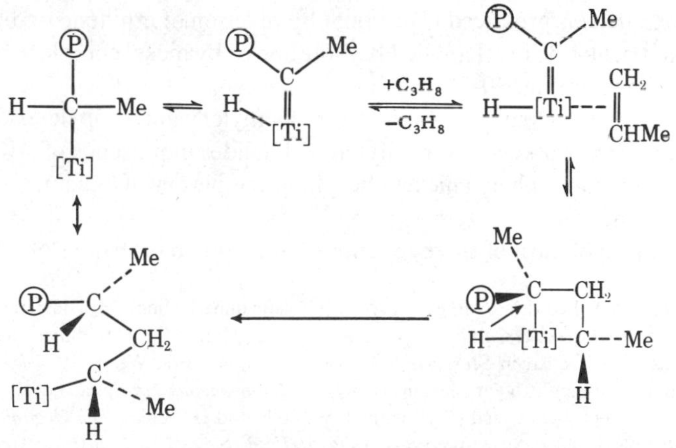

[41] *Chem. Eng. News,* May 2, 1977, p. 28.
[42] D. G. H. Ballard, *Adv. Catal.,* 1973, **23**, 263.
[43a] E. R. Evitt and R. Bergmann, *J. Am. Chem. Soc.,* 1979, **101**, 3973.
[43b] M. L. H. Green, *Pure Appl. Chem.,* 1978, **50**, 27; K. J. Irvin *et al., J.C.S. Chem. Commun.,* **1978**, 604; M. L. H. Green and A. Mahtab, *J.C.S. Dalton,* **1979**, 262.

The stereospecific polymerization of propylene depends on the relative orientations of the substituents on the metallocycle ring. Thus if the CH_3 groups are on the same side of the ring as shown, syndiotactic polymer results—on the opposite side, isotactic.

30-6.   Oligomerization and Related Reactions[44]

The polymerization of alkenes usually occurs only with Ti, Zr, V, and Cr catalysts. However much more generally alk-1-enes, conjugated alkenes such as butadiene,[45] nonconjugated alkenes, alkenes containing functional groups, and also acetylenes may be dimerized, trimerized, etc., by complexes of many of the transition elements, to give linear, cyclic, and other complex products. This is a vast subject, which we can merely outline.

For example, dimerization can occur oxidatively (i.e., with loss of hydrogen atoms):

$$2RCH{=}CHR' \xrightarrow{-2H} RCH{=}\underset{R}{C}{-}\underset{R}{C}{=}CHR'$$

or reductively (i.e., with addition of hydrogen atoms):

$$2RCH{=}CHR' \xrightarrow{+2H} RCH_2CH{-}\underset{R}{CH}{-}\underset{R'}{CH}{-}CH_2R$$

as well as by simple addition of one monomer to another.

There are also related reactions in which two different olefins are codimerized.[46a] Thus 1,4-hexadiene is produced (Du Pont) by reaction of ethylene and butadiene with rhodium trichloride in ethanolic HCl solution or by nickel complexes according to the cycle shown in Fig. 30-5.

The most active catalysts for dimerization of ethylene and propylene, discovered by Wilke and his co-workers, are π-allyl nickel halides in presence of $AlCl_3$. Other metal complexes (e.g., a phenyl nickel phosphine complex) will oligomerize ethylene to linear alk-1-enes.[46b]

The mechanism of most of these reactions has assumed insertions of alkenes into

[44]   L. J. Kricka and A. Ledwith, *Synthesis,* **1974,** 539 (olefin dimerization); M. Hidai and A. Misrow, in *Aspects of Homogeneous Catalysis,* Vol. 2, R. Ugo, Ed., Manfredi, 1974, (dimerisation of acrylic compounds); W. Cooper, in *Stereorubbers,* W. M. Satman, Ed., Wiley, 1977 (polydienes by coordination catalysis); P. Heimbach, in *Aspects of Homogeneous Catalysis,* Vol. 2, R. Ugo, Ed., Reidel, 1974, (nickel-catalyzed reactions); P. W. Jolly and G. Wilke, *The Organic Chemistry of Nickel,* Vols. 1 and 2, Academic Press, 1974, 1975; R. Noyori, in *Transition Metal Organometallics in Organic Synthesis,* Vol. 1, H. Alper, Ed., Academic Press, 1976 (coupling reactions); J. Tsuji, *Acc. Chem. Res.,* 1973, **6,** 8 (butadiene addition reactions); R. P. A. Sneedon, *Organochromium Compounds,* Academic Press, 1975.

[45]   D. H. Richards, *Chem. Soc. Rev.,* 1977, **6,** 235.

[46a] A. C. L. Su, *Adv. Organomet. Chem.,* 1979, **17,** 269.

[46b] W. Keim et al., *Angew. Chem., Int. Ed.,* 1978, **17,** 446.

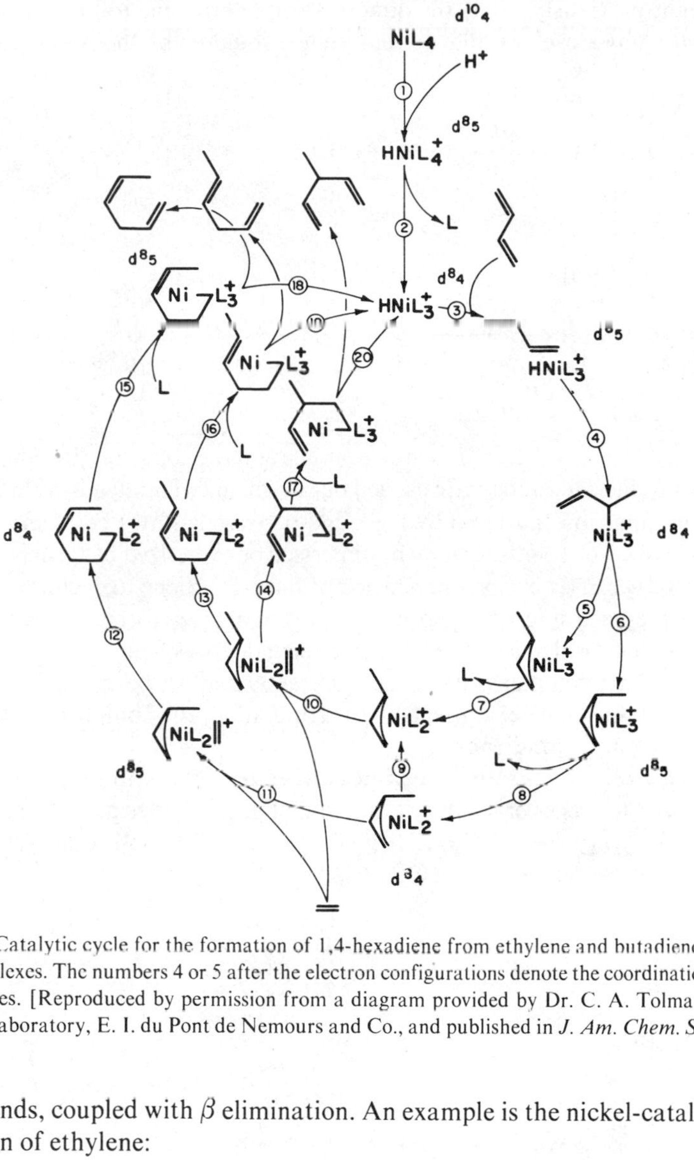

Fig. 30-5. Catalytic cycle for the formation of 1,4-hexadiene from ethylene and butadiene by use of nickel complexes. The numbers 4 or 5 after the electron configurations denote the coordination number of the species. [Reproduced by permission from a diagram provided by Dr. C. A. Tolman, Central Research Laboratory, E. I. du Pont de Nemours and Co., and published in *J. Am. Chem. Soc.,* 1970, **92,** 6777.]

M—C bonds, coupled with β elimination. An example is the nickel-catalyzed dimerization of ethylene:

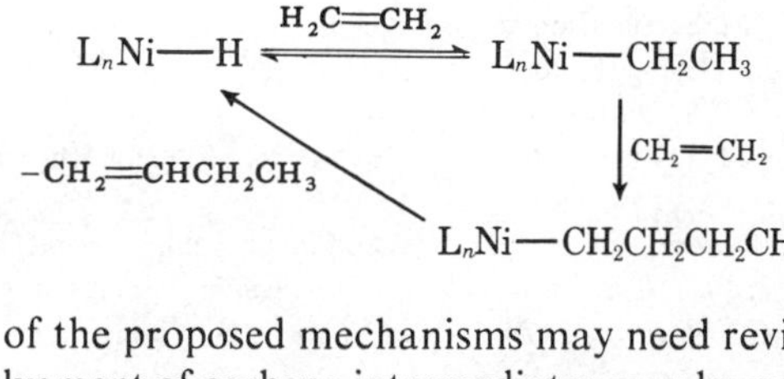

However some of the proposed mechanisms may need revision, since it is now accepted that involvement of carbene intermediates may be more widespread than was thought previously. Thus the dimerization of ethylene to but-1-ene, instead of proceeding by a cycle similar to that above, could go by the route*:

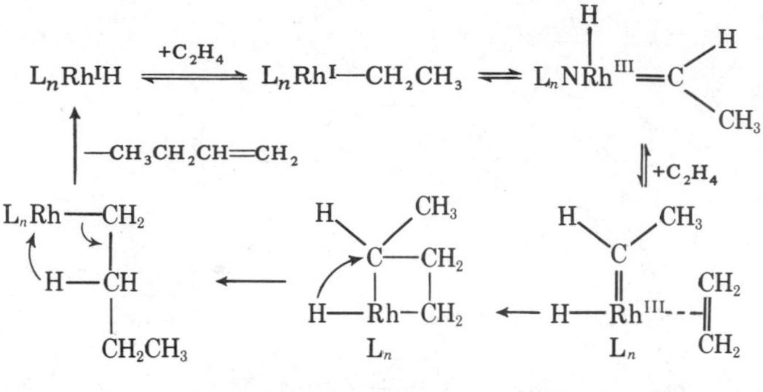

The conversion of 1,3-dienes such as isoprene and the copolymerization of styrene and buta-1,3-diene to rubbers are carried out by anionic (initiated by lithium alkyls) or aqueous emulsion (initiated by Fe^{2+} and peroxodisulfate) polymerization, respectively. However 1,3-dienes may be dimerized or trimerized by nickel complexes. Thus $Ni(cod)_2$ causes cyclotrimerization of buta-1,3-diene to cyclododecatriene, whereas in presence of tertiary phosphines, only dimers are formed. The mechanism of these reactions, which involve allylic intermediates is well understood (Fig. 30-6), and intermediate complexes have been characterized for both nickel[47a] and platinum[47b] reactions, one example being the metallocyclic complex 30-X, derived from $Pt(cod)_2$ and butadiene.

Acetylenic compounds are sometimes dimerized,[48a] but more commonly they are trimerized to arenes or polymerized to high linear or other polymer products.[48b] For example, acetylenes ($RC\equiv CH$) give polymers with molecular weights up to *ca.* 12,000 on reaction with $Mo(CO)_3C_6H_6$.

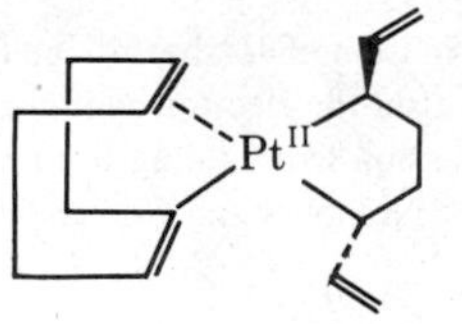

(30-X)

* For alternative metallocycle pathways see R. R. Schrock, et al., *J. Am. Chem. Soc.*, 1979, **101**, 4558, 5099.

47a C. R. Graham and L. M. Stephenson, *J. Am. Chem. Soc.*, 1977, **99**, 7098; H. Schenkühl *et al.*, *Angew. Chem., Int. Ed.*, 1979, **18**, 400.
47b F. G. A. Stone *et al.*, *J.C.S. Dalton*, **1978**, 1839.
48a See, e.g., L. Carlton and G. Read, *J.C.S. Perkin I*, **1978**, 1631, for Rh-catalyzed dimerizations of RC≡CH.
48b S. Sartori *et al.*, *Isr. J. Chem.*, 1977, **15**, 230 (Ni and Pd phosphine complexes); K. P. C. Vollhardt, *Acc. Chem. Res.*, 1977, **10**, 1; L. D. Brown *et al.*, *J. Am. Chem. Soc.*, 1978, **100**, 8232; E. A. Kelley and P. M. Maitlis, *J.C.S. Dalton*, **1979**, 167; L. P. Yuv'eva, *Russ. Chem. Rev.*, 1974, **43**, 48.

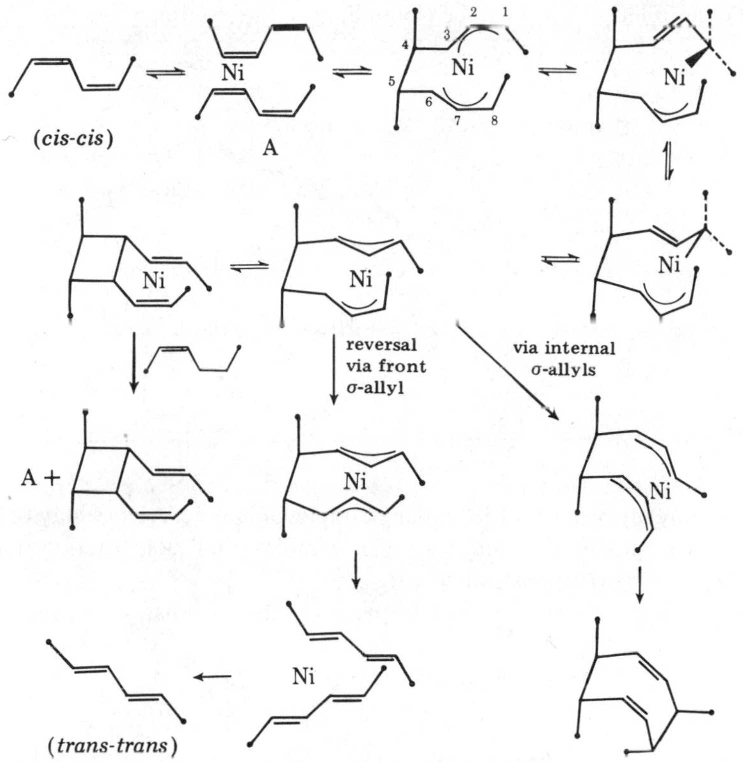

Fig. 30-6. Dimerization of 1,4-disubstituted diene by Ni(cod)$_2$ in presence of R$_3$P (not shown), adapted from reference 47 (a).

Acetylene reactions are usually complicated, and they depend very much on the nature of the metal system used, the nature of the acetylene, and the reaction conditions. For RC≡CH an oxidative addition is commonly the first step, but for disubstituted acetylenes we may have initial reactions of the types[49a]

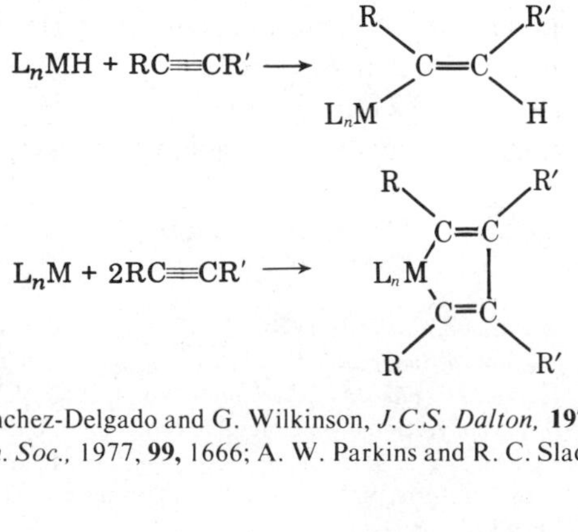

[49a] See, e.g., R. A. Sanchez-Delgado and G. Wilkinson, *J.C.S. Dalton,* **1977,** 804; R. G. Bergman *et al., J. Am. Chem. Soc.,* 1977, **99,** 1666; A. W. Parkins and R. C. Slade, *J.C.S. Dalton,* **1975,** 1352.

In some cases intermediates can be isolated, as in the reaction:

$$IrH(CO)(PPh_3)_3 \xrightarrow[-PPh_3]{R^1C \equiv CR^2}$$

We have already seen (p. 1158) other examples of how acetylenes may be linked together while remaining bound to a metal complex, and such reactions indicate how further polymerizations can occur.

Finally we note a useful synthesis of pyridines that is cobalt catalyzed[49b].

$$R^1CN + 2R^2C \equiv CH \longrightarrow$$

CARBON MONOXIDE REACTIONS[50]

In 1922 Fischer and Tropsch first described the heterogeneously catalyzed conversion of CO and H_2 to hydrocarbons and to oxygenated products. There are now innumerable reactions, both heterogeneous and homogeneous, of carbon monoxide with organic compounds. Much of the early study of "carbonylation" reactions was due to W. Reppe of I. G. Farbenindustrie, and most of the technical processes originated in Germany.

The most important commercial reaction is probably the hydroformylation of olefins, discovered by O. Roelen (Ruhrchemie) in 1938 during studies on Fischer-Tropsch reactions. More than 3 million tons of alcohols is produced each year by this process. Carbonylation of methanol to acetic acid is also an important process.

The only feasible sources of CO and CO–H_2 mixtures (synthesis gas) are natural

49b H. Bönnemann, *Angew. Chem., Int. Ed.,* 1978, **17**, 505, see also Belgian Patent, 859, 768.

50 (a) J. Falbe, *Carbon Monoxide in Organic Synthesis,* Springer-Verlag 1970; (b) I. Wender and P. Pino, Eds., *Organic Synthesis via Metal Carbonyls,* Vol. I, 1968, Vol. 2, 1976, Wiley (various topics including metal carbonyls and organometallic compounds and uses in synthesis); (c) G. Henrici-Olivé and S. Olivé, *Trans. Met. Chem.,* 1976, **1**, 77; (d) P. J. Davidson, R. R. Hignett, and D. T. Thompson, in *Catalysis,* Vol. 1, 1977; P. J. Denney and D. A. Whan, Vol. 2, 1978, Specialist Reports, Chemical Society.

gas and petroleum or coal.[51a] *Synthesis gas* may be obtained by (*a*) controlled combustion of petroleum in oxygen, (*b*) catalytic steam re-forming of methane or light hydrocarbons ("naphtha"),

$$CH_4 + H_2O \rightleftharpoons CO + 3H_2 \qquad \Delta H = +205 \text{ kJ mol}^{-1}$$

or (*c*) gasification of coal with steam and oxygen at $\sim 1500°$. CO_2 is removed by monoethanolamine scrubbing.

Carbon monoxide is also formed in blast furnaces, coke ovens, etc., and as producer gas by passing air over hot coke. There is potential for separating CO from N_2 in such diluted gas streams by complexing with $CuAlCl_4$ in an aromatic solvent.[51b] One critical prerequisite for use of CO or synthesis gas in reactions catalyzed by transition metals is that sulfur compounds such as H_2S or COS formed from sulfur compounds in oil or coal first be removed.

30-7. Reduction of Carbon Monoxide by Hydrogen

We first discuss the reaction of hydrogen and carbon monoxide. This has been much studied, and some thermodynamic data (Table 30-2) show that only the reduction to methane is favorable. Nevertheless, reduction of CO by H_2 to give alcohols is feasible using appropriate catalysts and is the subject of intense study.

The Fischer-Tropsch Process.[52] This is a heterogeneous reaction of H_2 and CO catalyzed by various metals such as Fe, Co, Ni, or Ru on alumina or other support and by cobalt-chromium and other oxides. Using different catalysts and varying the conditions either CH_4 (when the reaction is usually called *methanation*) liquid

TABLE 30-2
Thermodynamic Data for CO–H_2 Reactions (at 500°K)[a]

Reaction	ΔG kJ mol^{-1}	Log K_p
$CO + 3H_2 = CH_4 + H_2O$	-96.22	10.065
$CO + H_2 = HCHO$	50.62	-5.293
$CO + 2H_2 = CH_3OH$	21.23	-2.222
$2CO + 3H_2 = \begin{matrix} CH_2OH \\ \| \\ CH_2OH \end{matrix}$	65.92	-6.891

[a] From D. R. Stull, E. F. Westrum, Jr., and G. C. Sinke, *The Thermodynamics of Organic Compounds,* Wiley, 1969.

[51a] A. L. Waddams, *Chemicals from Petroleum,* 3rd. ed., Murray, 1973; J. M. Tedder, A. Nechvatal, and A. J. Jubb, *Basic Organic Chemistry,* Part 5, Industrial Products, Wiley, 1975; I. Howard-Smith and G. J. Werner, *Coal Conversion Technology,* Noyes Data Corp., (includes gasification, Fischer-Tropsch, methanol synthesis, methanation reactions, etc.).

[51b] *Chem. Tech.,* **1977,** 633.

[52] (a) V. Ponec, *Catal. Rev. Sci. Eng.,* 1978, **18,** 151; (b) H. Koelbel and M. Ralek, *Chemierohst. Kohle,* **1977,** 219, 413 (306 references); (c) H. Schulz *et al., Fuel Process Technol.,* 1977, **1,** 31, 45; *Erdöl Kohle,* 1977, **30,** 123; (d) G. Henrici-Olivé and S. Olivé, *J. Mol. Catal.,* 1978, **3,** 443; *Angew. Chem., Int. Ed.,* 1976, **15,** 136. (e) A. Deluzarche *et al., J. Chem. Res. (S),* **1979,** 136; (f) C. Masters, *Adv. Organomet. Chem.,* 1979, **17,** 61.

hydrocarbons, high molecular weight waxy hydrocarbons, methanol, higher alcohols, olefins etc., can be obtained. Nickel is best for methanation, cobalt for higher alkanes.

Since CO and H_2 have to be obtained from coal or oil, the use of Fischer-Tropsch catalysis for manufacture of synthetic fuels is uneconomic and only in South Africa, where coal is cheap, is the process used.

However the methanation reaction, which is now used to remove small amounts of CO from hydrogen over a nickel catalyst, may sometime have to be used for synthesis of substitute natural gas from coal.

The mechanism of the Fischer-Tropsch type of reaction has been the subject of speculation. The first step has been usually assumed to be

$$\text{MCO} \xrightarrow{H_2} \overset{\overset{\displaystyle H}{|}}{\text{M}}\!\!-\!\!\text{CHO}$$

although there is no evidence to suggest that this reaction can occur. These initial reactions are supposed to be followed by hydrogen transfer, more CO insertions, olefin eliminations, etc.

However for both the methanation reaction on Ni and Fischer-Tropsch reactions on other metals, it is much more likely that a carbene mechanism is involved. It is known[53] that the Group VIII metals especially active for these reactions, cause dissociation of CO at temperatures as low as 100°. The formation of a monolayer of active carbon on the metal surface is not only rapid but evidently essential, although eventually poisoning by a thick carbon layer occurs. The carbon is very active to hydrogen and could well form $M=CH_2$. Two other pieces of evidence support a carbene mechanism.

1. Carbene complexes can be cleaved[54] by hydrogen:

$$(CO)_5W\!=\!C\!\!\begin{smallmatrix}\nearrow Ph\\ \\ \searrow Ph\end{smallmatrix} \xrightarrow[\substack{100° \\ 69 \text{ atm}}]{H_2} H_2C\!\!\begin{smallmatrix}\nearrow Ph\\ \\ \searrow Ph\end{smallmatrix} + W(CO)_6 + \text{other products.}$$

(41%)

2. A carbene complex has been carbonylated and the η^2-ketene complex hydrogenated to give aldehyde[55]:

$$\eta\text{-}C_5H_5(CO)_2MnCPh_2 \xrightarrow{CO} \eta\text{-}C_5H_5(CO)_2Mn\!\!\begin{smallmatrix} \overset{\displaystyle C}{\diagdown} \!\!=\!\! O \\ | \\ C \\ Ph_2 \end{smallmatrix} \xrightarrow{H_2} Ph_2CHCHO$$

53   D. J. Dwyer and G. A. Somorjai, *J. Catal.*, 1978, **52**, 291; P. R. Wentrcek *et al., Am. Chem. Soc. Div. Fuel. Chem., Prep.*, 1976, **21**, 52. P. Biloen and W. M. H. Sachtler, *J. Catal.*, 1979, **58**, 95.
54   C. P. Casey and S. M. Newmann, *J. Am. Chem. Soc.*, 1977, **99**, 1651.
55   W. A. Herrmann and J. Plank, *Angew. Chem., Int. Ed.*, 1978, **17**, 525.

Attempts to carry out Fischer-Tropsch reactions homogeneously have been notably unsuccessful. Trivial amounts of methane have been obtained after prolonged reaction times in interactions of CO with H_2 in presence of metal carbonyls,[56a] although $Ir_4(CO)_{12}$ in an $AlCl_3$–NaCl melt at 180° and 1.5 atm is said to produce CH_4, C_2H_6, and C_3H_8. However the conversion of benzene, CO, and H_2 to monoalkylbenzenes [$C_6H_5(CH_2)_nH$, $n = 1$–5] by $W(CO)_6$ + $AlCl_3$ under pressure at 200° is said to be homogeneous.[56b] If the synthesis of ethyl benzene were effective, it could provide a new route to styrene (by dehydrogenation).

Alcohol Synthesis. The reduction of CO to give *methanol:*

$$CO + 2H_2 = CH_3OH \qquad \Delta H = -92 \text{ kJ mol}^{-1}$$

over copper-promoted ZnO catalyst at *ca.* 250° and 50 atm is used commercially.[57] The reaction provides a means for the utilization (via synthesis gas) of natural gas from oil wells often now wasted by flaring.

Since methanol can be converted to hydrocarbons[58] by use of framework silicates with specially constructed cavity and channel dimensions (Mobil Oil, ZSM-5):

$$nCH_3OH = (CH_2)_n + nH_2O$$

the inherent costs and difficulties of synthesis of liquid hydrocarbons by Fischer-Tropsch synthesis may be avoided where natural gas, which is essentially CH_4, is still available cheaply.

Although it is not an operating process, the reduction of CO by H_2 homogeneously by rhodium carbonyl cluster anions such as $[Rh_{12}(CO)_{34}]^{2-}$ (p. 1091) is described in patents[59]; but the reaction occurs only at *ca.* 250° and at extreme pressures, up to 3000 atm. It gives CH_3OH and other alcohols but, most important, high yields of ethyleneglycol.

Homogeneous hydrogenation of CO to CH_3OH and higher alcohols and their formate esters is also slowly catalyzed by cobalt carbonyls at 200° and 300 atm.[60a] It appears that the mechanism involves $HCo(CO)_4$, *not* metal atom clusters, and probably proceeds by a radical pathway.

$$HCo(CO)_4 + CO \xrightarrow{\text{slow}} H\overset{\cdot}{C}O + {}^{\cdot}Co(CO)_4$$
$$H\overset{\cdot}{C}O + HCo(CO)_4 \xrightarrow{\text{fast}} \text{products}$$

Methanol and methyl formate are also formed using ruthenium carbonyl and

[56a] E. L. Muetterties *et al., J. Am. Chem. Soc.,* 1977, **99**, 2796; 1976, **98**, 1296.

[56b] G. Henrici-Olivé and S. Olivé, *Angew. Chem., Int. Ed.,* 1979, **18**, 77.

[57] E. Supp, *Chem. Tech.,* **1973,** 430.

[58] S. L. Meisel *et al., Chem. Tech.,* **1976,** 86; U. S. Patent (Mobil), 4,025,572, 1977. See also E. M. Flanigen *et al., Nature,* 1978, **271,** 512; *Chem Eng. News,* January 30, 1978, p. 26; E. G. Derouane *et al., J. Mol. Catal.,* **1978,** 453; *J. Catal.,* 1978, **53,** 40; *Chem. Tech.,* **1977,** 588.

[59] R. L. Pruett *et al.,* Union Carbide, Belgian Patents 793086, 815841; U.S. Patents 3833634, 3878214, 3878290, 3878292, 3833634, French Patent 2317261; see also R. L. Pruett, *Ann. N.Y. Acad. Sci.,* 1977, **295,** 239.

[60a] J. W. Rathke and H. M. Feder, *J. Am. Chem. Soc.,* 1978, **100,** 3623.

carbonyl phosphines at 268° and 1300 atm, and mononuclear rather than poly-nuclear species are involved.[60b] It is also known that cobalt carbonyl catalyzes homologation reactions of alcohols, e.g.,

$$CH_3OH + CO + 2H_2 \rightarrow CH_3CH_2OH + H_2O$$

Tertiary phosphines and other metals such as ruthenium may be added, as well as iodide that functions, presumably, in the same way as in the acetic acid synthesis[60c] (p. 1294).

There are some stoichiometric reductions of CO by H_2 using π-cyclopentadi-enylzirconium compounds,[61] (see also p. 1131).

Water Gas Shift Reaction. The reaction

$$H_2O + CO \rightleftharpoons H_2 + CO_2 \qquad \Delta H = -42 \text{ kJ mol}^{-1}$$

is an important heterogeneous reaction ($ZnO + Cr_2O_3$), widely used to generate H_2.[62a] The reaction can proceed homogeneously using various metal carbonyls in basic solution in methanol, ethoxyethanol, tetrahydrofuran, etc.[62b,c]

The mechanism is believed to be of the type:

$$HFe(CO)_4^- + CO = Fe(CO)_5 + H^-$$

$$H^- + H_2O = OH^- + H_2$$

$$OH^- + Fe(CO)_5 = (CO)_4Fe-C\overset{O^-}{\underset{OH}{\diagdown}}$$

$$(CO)_4Fe-C\overset{O^-}{\underset{OH}{\diagdown}} = HFe(CO)_4^- + CO_2$$

Iron and ruthenium carbonyls in acid solution will also slowly catalyze the shift reaction, as will $Pt(PEt_3)_3$, rhodium iodo carbonyl, and $PtCl_2$–$SnCl_2$ complexes.[63]

Finally note that the $CO + H_2O$ systems can be used to replace hydrogen in a number of reductions,[64] e.g.,

60b J. S. Bradley, *J. Am. Chem. Soc.,* 1979, **101**, 7419.

60c See for example U. S. Patent 4,133,966 (Gulf Research).

61 J. E. Bercaw *et al., J. Am. Chem. Soc.,* 1979, **101**, 218. 1978, **100**, 2716, 1976, **98**, 6733; L. I. Shoer and J. Schwarz, *J. Am. Chem. Soc.,* 1977, **99**, 5831; G. G. Fachinetti *et al., J.C.S. Chem. Comm.,* **1978**, 269.

62 (a) J. J. Verdonck *et al., J.C.S. Chem. Comm.,* **1979**, 181; (b) R. B. King *et al., J. Am. Chem. Soc.,* 1978, **100**, 2925; (c) R. M. Laine, *J. Am. Chem. Soc.,* 1978, **100**, 6451, 6527.

63 S. Otsuka *et al., J. Am. Chem. Soc.,* 1978, **100**, 3941; P. C. Ford *et al., J. Am. Chem. Soc.,* 1978, **100**, 4595; C.-H. Cheng, and R. Eisenberg, *J. Am. Chem. Soc.,* 1978, **100**, 5969.

64 K. Cann *et al., J. Am. Chem. Soc.,* 1978, **100**, 3969; H. Kang *et al., J. Am. Chem. Soc.,* 1977, **99**, 8323; K. Murata *et al., J.C.S. Chem. Comm.,* **1979**, 785.

$$RNO_2 + 3CO + H_2O \xrightarrow{Fe(CO)_5} RNH_2 + 3CO_2$$

and in the hydroformylation reaction of olefins, now to be discussed. However the procedure is not especially useful or economic, since expensive CO is lost as CO_2.

30-8. Hydroformylation of Unsaturated Compounds[50,65]

The hydroformylation of unsaturated compounds is an important reaction that is homogeneous, and its mechanism, which is now known with some certainty, provides an excellent example of the principles outlined in Chapter 29.

The name "hydroformylation" stems from the nature of the reaction, which formally involves adding H and the formyl group, CHO, derived from H_2 and CO to an olefin:

$$\diagdown C=C \diagup + H_2 + CO \longrightarrow \diagdown \underset{H}{\overset{}{C}}-\underset{CHO}{\overset{}{C}} \diagup$$

The reaction is also called the *oxo* reaction or process. The main technical use is in the conversion of ethylene, propylene, and higher alk-1-enes into aldehyde and alcohols, but other compounds with C=C and C≡C bonds can be hydroformylated with varying degrees of effectiveness.

The original Roelen reaction uses cobalt as the catalyst, and the commercial process requires a high temperature (150°–180°) and pressure (3000–6000 psi). It also produces a mixture, roughly 3:1, of both linear and branched-chain aldehydes or alcohols. These products are formally the products from anti-Markownikoff or Markownikoff addition, respectively, viz.,

The cobalt process is technically difficult to operate because the cobalt hydrocarbonyl $HCo(CO)_4$ participating in the catalyst cycle is volatile and must be separated from the alcohol products, and the cobalt must be recovered as cobalt sulfate and

[65] P. Pino et al., in Ref. 50b, Vol. 2, p. 43; F. E. Paulik, Catal. Rev., 1972, 6, 49; M. Orchin and W. Rupilius, Cat. Rev., 1972, 6, 85; R. Cornils et al., Hydrocarbon Processing, 1975, 83; R. Fowler et al., Chem. Tech., 1976, 772; R. L. Pruett, Adv. Organomet. Chem., 1979, 17, 1.

recycled. Another disadvantage is that some 15% of the alkene is lost by hydrogenation; alcohol condensation also occurs, and there are ketone by-products.

For ethylene and propylene a more effective process using rhodium triphenylphosphine complexes is now used (Union Carbide). Although the reaction will proceed at 25° and atmospheric pressure, the large-scale process operates at *ca.* 100° and only a few atmospheres pressure. It has additional advantages over the cobalt process in that there is no loss of olefin by hydrogenation, only aldehyde is produced, and most important for propylene, it gives almost entirely *n*-butyraldehyde. Also since the products are removed in the gas stream, no recycling of the catalyst solution is necessary. The C_2H_5CHO is converted either to propanol or to propionic acid and C_3H_7CHO to either *n*-butanol or 2-ethylhexanol.

The main steps in the mechanism, first elucidated by D. S. Breslow and R. F. Heck (Hercules Inc.) for the cobalt system, are as follows:

$$HCo(CO)_4 \rightleftharpoons HCo(CO)_3 + CO \qquad (30\text{-}1)$$
$$RCH{=}CH_2 + HCo(CO)_3 \rightleftharpoons RCH_2CH_2Co(CO)_3 \qquad (30\text{-}2)$$
$$RCH_2CH_2Co(CO)_3 + CO \rightleftharpoons RCH_2CH_2COCo(CO)_3 \qquad (30\text{-}3)$$
$$RCH_2CH_2COCo(CO)_3 + H_2 \rightarrow RCH_2CH_2CHO + HCo(CO)_3 \qquad (30\text{-}4)$$

where eq. 30-2 involves the β-hydrogen transfer to coordinated olefin, eq. 30-3 the insertion of CO to form acyl, and eq. 30-4 the acyl cleavage by hydrogen. Branched-chain aldehyde results from Markownikoff addition of Co—H to the alkene. Noncatalytic reactions that confirm the type of intermediate have also been studied (e.g., using the less labile system of $CH_3Mn(CO)_5$ with CO and H_2).[66]

The cobalt system is difficult to study, but the rhodium system based on $RhH(CO)(PPh_3)_3$ has allowed more detailed investigation. Most rhodium compounds [e.g., $Rh(CO)_2acac$ or $Rh_2(CO_2Me)_4$], when put into the system with PPh_3, CO, and H_2, give the cycle shown in Fig. 30-7.

The initial step is attack of the alkene on the sixteen-electron species $RhH(CO)(PPh_3)_2$ (A in Fig. 30-7), which leads to the alkyl complex B. The latter then undergoes CO addition and insertion to form the acyl derivative C, which subsequently undergoes oxidative addition of molecular hydrogen to give the dihydridoacyl complex D.

The last step, which is the only one in the cycle that involves a change in oxidation state of the metal, is probably rate determining. The final steps are another H transfer to the carbon atom of the acyl group in D, followed by loss of aldehyde and regeneration of the four-coordinate species A. An excess of CO over H_2 inhibits the hydroformylation reaction, probably through the formation of five-coordinate dicarbonyl acyls E, which cannot react with hydrogen.

To obtain very high selectivity for the formation of *n*-aldehyde, very high concentrations of triphenylphosphine are required, evidently to maintain bisphosphine species in the cycle and to suppress dissociation to give monophosphine species. The high selectivity evidently results from the formation of a straight-chain alkyl in species B, and this is promoted, as discussed earlier (p. 1271) in connection with hydrogenations using $RhH(CO)(PPh_3)_3$, by the steric factors due to the bulky

66   R. B. King *et al., J. Am. Chem. Soc.,* 1978, **100,** 1687.

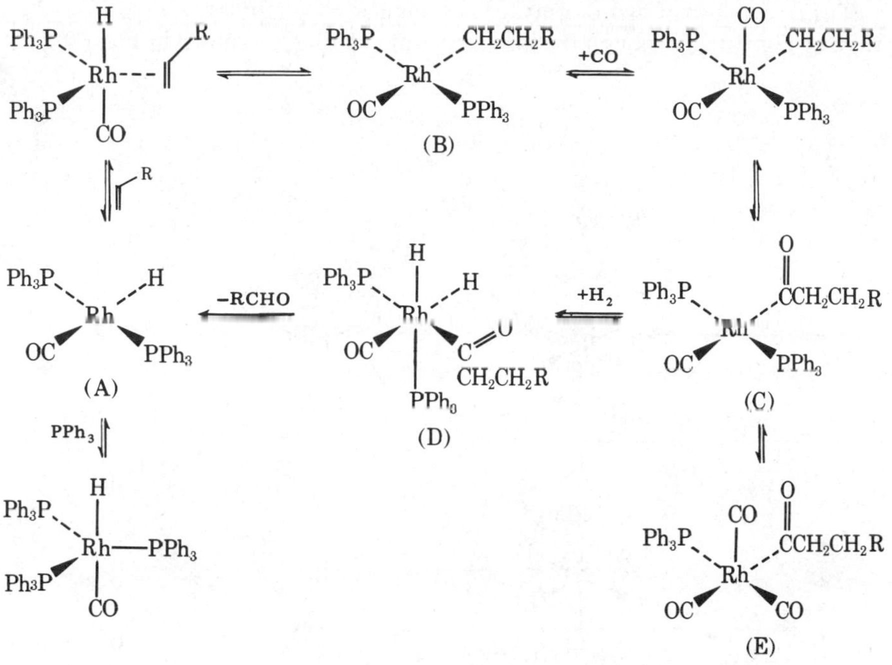

Fig. 30-7. Simplified catalytic cycle for the hydroformylation of alkenes involving triphenylphosphine rhodium complex species. Note that the configurations of the complexes are not known with certainty and that 5-coordinate species are fluxional.

trans-PPh₃ ligands. In absence of excess PPh₃, straight to branched chain aldehydes with a 3:1 ratio are produced as in the cobalt system.

Although the intermediates shown in Fig. 30-7 are too unstable for isolation, it has been possible by using alkenes such as C_2F_4, which gives stable species such as the square alkyl $Rh(C_2F_4H)(CO)(PPh_3)_2$, and by using the similar but much more stable analogous iridium complexes, to characterize analogues for all the species except for the final dihydridoacyl complex; a comparable complex, $IrHCl(COCH_2R)(CO)(PPh_3)_2$, may be isolated, however.

A wide variety of unsaturated substances can be hydroformylated, but conjugated alkenes (e.g., butadiene) may give a number of products including hydrogenated monoaldehydes. The mechanism is then quite different from that for monoalkenes, since addition of M—H to conjugated dienes leads to allylic species, which may be present as σ-bonded intermediates or, commonly as η^3-allyls (cf. p. 1152), e.g.,

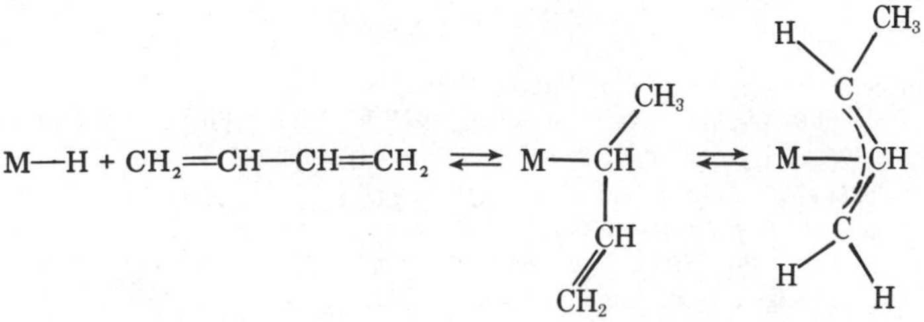

$$M\text{—}H + CH_2\text{=}CH\text{—}CH\text{=}CH_2 \rightleftharpoons M\text{—}CH \rightleftharpoons M\text{—}CH$$

The mechanism of hydroformylation using $Ru(CO)_3(PPh_3)_2$[67] also differs from than of Fig. 30-7, although the basic steps are similar, as shown in Fig. 30-8.

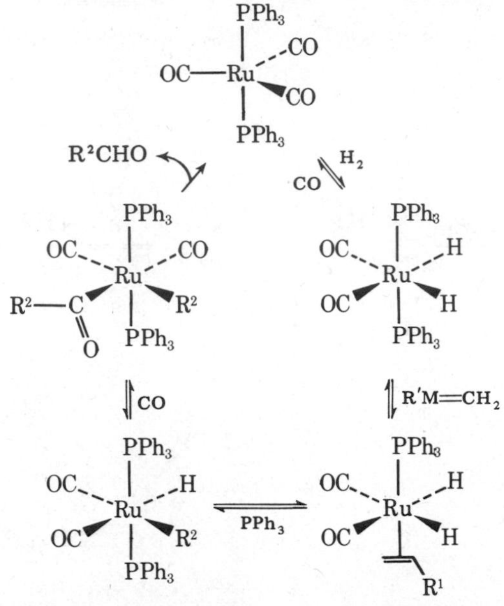

Fig. 30-8. Hydroformylation by $Ru(CO)_3(PPh_3)_2$.

Platinum catalysts,[68a] which are hydrido species $PtH(CO)SnCl_3(PPh_3)_2$, appear to operate by a cycle similar to that for $RhH(CO)(PPh_3)_3$, and the electronic and/or steric requirements for the formation of largely normal aldehyde are provided by $SnCl_3^-$ and PPh_3. Chelating phosphines that can form seven-membered rings [e.g., $Ph_2P(CH_2)_4PPh_2$] give more active catalysts.[68b]

Stereochemical studies on cobalt and rhodium systems indicate that cis addition of H and CHO occurs probably generally.[69] The use of chiral phosphines also allows asymmetric hydroformylation of prochiral compounds.[70]

Hydroformylation can also be carried out using supported homogeneous catalysts (p. 1305) and also some cluster compounds (p. 1304).

30-9. Acetic Acid Synthesis

In 1941 Reppe discovered that the reaction

$$CH_3OH + CO = CH_3COOH$$

67   R. Sanchez-Delgado et al., J.C.S. Dalton, 1976, 399.
68a C.-Y. Hsu and M. Orchin, J. Am. Chem. Soc., 1975, 97, 3553; P. Pino et al., Chim. Ind. (Milan), 1978, 60, 396. U.S. Patent 4,101,564, July, 18 1978 (Celanese).
68b Y. Kawabata et al., J.C.S. Chem. Comm., 1979, 462.
69   A. Stefani et al., J. Am. Chem. Soc., 1977, 98, 1058.
70   C. Consiglio and P. Pino, Helv. Chim. Acta, 1976, 59, 642; C. Botteghi et al., Chim. Ind. (Milan), 1978, 60, 16; I. Ogata et al., Chem. Lett., 1978, 361.

could be catalyzed by carbonyls of Fe, Co, and Ni, in presence of iodides. A catalytic process operating at 210° and 7500 psi was developed (Badische ASF). As in the case of hydroformylation, rhodium catalysts allow much lower pressures and temperatures (*ca.* 100°) to be used, and the reaction will even proceed at 1 atm.

The key reaction, which shows the necessity for iodide cocatalyst, is that generating methyl iodide:

$$CH_3OH + HI \rightleftharpoons CH_3I + H_2O$$

The CH_3I then oxidatively adds to a Rh^I carbonyl iodide anion as in Fig. 30-9; the

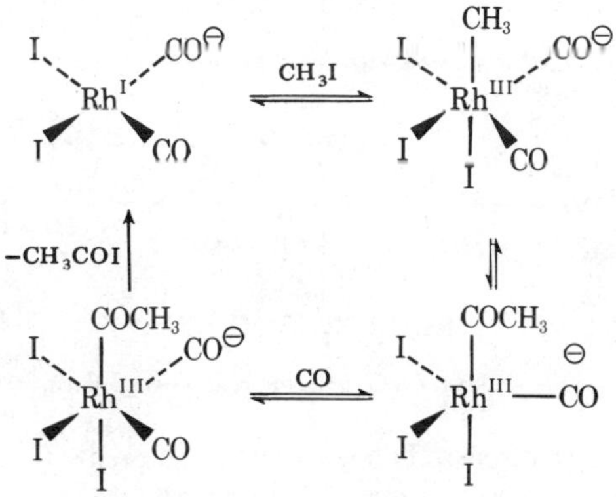

$$CH_3COI + H_2O = CH_3COOH + HI$$

$$CH_3OH + HI \rightleftharpoons CH_3I + H_2O$$

Fig. 30-9. Carbonylation of methanol by rhodium iodo carbonyl anion; oxidative addition of CH_3I is believed to be rate determining. An alternative, less likely, pathway for formation of acetic acid is attack of water on $[RhI_3(CO)_2COCH_3]^-$ to give CH_3COOH and a hydrido species $[RhI_3H(CO)_2]^-$ from which HI is eliminated to re-form $[RhI_2(CO)_2]^-$.

water finally hydrolyzes acetyl iodide, produced in reductive elimination, to complete the cycle.[71] However the reaction is by no means as simple as this, since the rhodium acyl iodide can react with methanol to give CH_3COOCH_3 and HI, and there are also the catalyzed reactions

$$2CH_3OH \rightleftharpoons CH_3OCH_3 + H_2O$$
$$CH_3COOH + CH_3OH \rightleftharpoons CH_3COOCH_3 + H_2O$$

The process (Monsanto) is therefore operated under equilibrium conditions in tri-*n*-octylphosphine oxide as solvent, so as to maximize acetic and minimize dimethylether and ester formation. The carbonylation can also be carried out heterogeneously using rhodium on zeolites.[72]

[71] D. Forster, *J. Am. Chem. Soc.*, 1976, **98**, 846; *Adv. Organometal. Chem.*, 1979, **17**, 255. J. Hjortkjaer *et al., J. Mol. Catal.*, 1978, **4**, 199; T. Matsumoto *et al., Bull. Chem. Soc. Jap.*, 1977, **50**, 2337.
[72] B. Cristensen and M. S. Scurrele, *J.C.S. Faraday*, **1977**, 2036.

30-10.   Other Carbonylation Reactions

There are many other reactions of CO with olefins, acetylenes, and other organic compounds, not all of which are catalytic however, that may give a multiplicity of products. Some examples are given in Fig. 30-10. Many of these reactions proceed

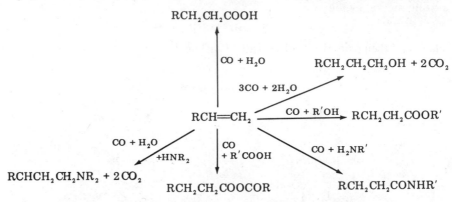

Fig. 30-10.   Carbonylation reactions of olefins.

by steps similar to those discussed above, the characteristic reaction being formation of an acyl, which then is attacked by a molecule having an active hydrogen.

     Unsaturated compounds with functional groups can also be carbonylated,[73] e.g.,

$$CH_2\!\!=\!\!CHCH_2Cl + CO \rightarrow CH_2\!\!=\!\!CHCH_2COCl$$
$$RCH\!\!=\!\!CH_2 + CCl_4 + CO \rightarrow Cl_3CCH_2CH(R)COCl$$

$$C_6H_5CH_2Cl + CO \rightarrow C_6H_5CH_2COCl \xrightarrow{H_2O} C_6H_5CH_2COOH$$

Phenylacetic acid can be conveniently made from benzylchloride by the latter reactions, catalyzed by palladium complexes, and carried out under phase transfer conditions using 40% NaOH, diphenylether as organic phase, $NaCo(CO)_4$, and $PhCH_2NMe_3Cl$ under CO.[74]

     Some other carbonylation reactions are:

$$\bigcirc\!\!\text{NH} + CO \longrightarrow \bigcirc\!\!\text{NCHO}$$

$$HC\!\!\equiv\!\!CH + CO + H_2O \longrightarrow H_2C\!\!=\!\!CHCOOH$$

$$2HC\!\!\equiv\!\!CH + 3CO + H_2O \longrightarrow HO\!\!-\!\!\bigcirc\!\!-\!\!OH + CO_2$$

[73]   T. A. Weil *et al., Catalysis via Metal Carbonyls,* Vol. 2, Wiley, 1976 (C—Cl bonds); G. P. Chiusoli *et al., Acc. Chem. Res.,* **1973,** 422; *Synthesis,* **1973,** 509; J. A. Scheban and I. L. Mador, in *Catalysis in Organic Synthesis,* Academic Press, 1976 (C–Cl bonds); R. F. Heck, *Adv. Catal.,* 1977, **26,** 323 (organic halides with CO, olefins, and acetylenes).

[74]   Montedison, Netherlands Appl., 7,800, 488.

Carbon monoxide can be used for the deoxygenation of nitro arenes,[75a,75b] where Fe, Ru, or Rh complexes are catalysts, e.g.,

$$2PhNO_2 + 4CO = PhN{=}NPh + 4CO_2$$
$$PhNO_2 + 3CO = PhNCO + 2CO_2$$

The first step in these reactions is probably the formation of a nitrene complex[75b] (p. 125):

$$RNO_2 + M(CO)_n \xrightarrow{2CO} RNM(CO)_n + 2CO_2$$

The RN moiety then undergoes further reaction such as hydrogenation, CO insertion, and dimerization.

In presence of trapping agents such as aldehydes or acetylenes, this deoxygenation reaction provides a means for synthesis of nitrogen compounds, e.g.,

The conversion of aromatic nitro compounds to anilines can also be carried out using phase-transfer catalysis by $Fe_3(CO)_{12}$ in aqueous NaOH-benzene and $PhCH_2NEt_3^+Cl^-$.[76] Another interesting deoxygenation reaction of NO that does not proceed even at 450° heterogeneously is

$$2NO + CO = N_2O + CO_2$$

It is catalyzed by $[RhCl(CO)_2]^-$ and $PdCl_2 - CuCl_2$ in HCl.[77]

30-11. Decarbonylation Reactions[78]

The insertion of CO into metal-to-carbon bonds to give an acyl is reversible, as noted

[75a] A. F. M. Iqbal, Chem. Tech., **1974**, 566; J. F. Knifton, in Catalysis in Organic Synthesis, Academic Press 1976.

[75b] See, e.g., S. Aime et al., J.C.S. Dalton, **1978**, 534.

[76] H. Alper et al., Nouv. J. Chim., 1978, **2**, 245; J. Am. Chem. Soc., 1977, **99**, 98; Angew. Chem., Int. Ed., 1977, **16**, 41; R. C. Ryan et al., J. Mol. Catal., 1979, **5**, 319.

[77] M. Kubota et al., J. Am. Chem. Soc., 1978, **100**, 342; R. Eisenberg et al., Inorg. Chem., 1977, **16**, 970.

[78] T. Tsuji, in Organic Synthesis via Metal Carbonyls, Vol. 2, I. Wender and P. Pino, Eds., Wiley, 1977.

earlier (p. 1250). However CO can be irreversibly removed from organic molecules, and these reactions commonly proceed via acyl intermediates. Thus aldehydes, acyl and aroyl halides, etc., can be decarbonylated either stoichiometrically or catalytically by complexes such as $RhCl(PPh_3)_3$ or $RhCl(CO)(PPh_3)_2$.

Examples of stoichiometric decarbonylations are:

$$RhCl(PPh_3)_3 + RCOCl \rightarrow RhCl_2(COR)(PPh_3)_2 \rightleftharpoons RhRCl_2(CO)(PPh_3)_2$$
$$\rightarrow RCl + RhCl(CO)(PPh_3)_2$$

$$Cp(CO)_2Fe-\overset{\overset{\displaystyle O}{\|}}{C}R + RhCl(PPh_3)_3 = Cp(CO)(PPh_3)FeR + RhCl(CO)(PPh_3)_2 + CO$$

The decarbonylations have been well studied both kinetically and from the stereochemical point of view.[79] The principal steps in decarbonylation of acyl chlorides or of aldehydes involve oxidative addition of RCOCl or RCHO to give an acyl, followed by migration of R to the metal and subsequent reductive elimination of RCl or RH.

With triphenylphosphine the rhodium system is catalytic only above 200°, but using a diphosphine complex $[Rh(P-P)_2]^+$, much more efficient catalytic decarbonylation can be achieved[80].

We have already noted the dehydrogenation of alcohols (p. 1274), but a common reaction of alcohols particularly with transition metal halides in presence of phosphines, or with phosphine complexes, is decarbonylation with the concomitant formation of metal carbonyls and/or metal-hydrogen bonds. These reactions proceed via alcoxide intermediates, which then undergo β-hydride transfer and elimination reactions. The interaction of $RuHCl(PPh_3)_3$ with $NaOCH_3$[81] is one example, see Chapter 29, page 1258.

REACTIONS INVOLVING C—C AND C—H BOND BREAKING

Although heterogeneous catalysts such as the platinum metals readily cause cleavage of C—H and C—C bonds, there are relatively few homogeneous catalytic reactions that involve such bond breaking.

30-12. Valence Isomerizations of Strained Hydrocarbons[82]

A number of metal complexes, notably of rhodium(I), palladium(0), and nickel(0)

[79] J. . Suggs, *J. Am. Chem. Soc.,* 1978, **100**, 640; J. K. Stille *et al., J. Am. Chem. Soc.,* 1977, **99**, 5664; D. L. Egglestone *et al., J.C.S. Dalton,* **1977**, 1576; Yu. S. Varshavsky *et al., J. Organomet. Chem.,* 1979, **170**, 81; E. H. Kuhlmann and J. J. Alexander; *Inorg. Chim. Acta;* 1979, **34**, 197.

[80] D. H. Doughty and L. H. Pignolet, *J. Am. Chem. Soc.,* 1978, **100**, 7083.

[81] B. N. Chaudret *et al., J.C.S. Dalton,* **1977**, 1546.

[82] K. C. Bishop, III, *Chem. Rev.,* 1976, **76**, 461; L. A. Paquette *et al., Synthesis,* **1975**, 347; *J. Am. Chem. Soc.,* 1975, **97**, 1084–1124; J. Halpern, in *Organic Synthesis via Metal Carbonyls,* Vol. 2, I. Wender and P. Pino, Eds., Wiley, 1976; F. D. Mango, *Coord. Chem. Rev.,* 1975, **15**, 109.

will induce valence isomerizations in strained hydrocarbons.[83] These reactions are believed to proceed by oxidative addition of a C—C bond. For cyclopropane the reaction with $[Rh(CO)_2Cl]_2$ gives an isolable intermediate:

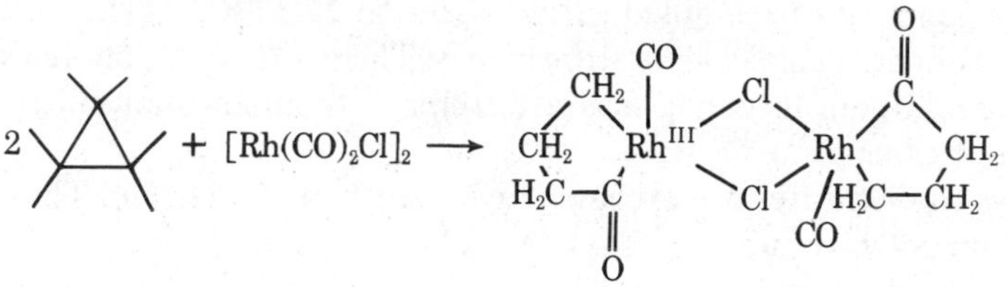

A few other intermediates have been isolated.[84] With noncarbonyl complexes [e.g., $RhCl(PPh_3)_3$] the reaction is held to proceed as follows:

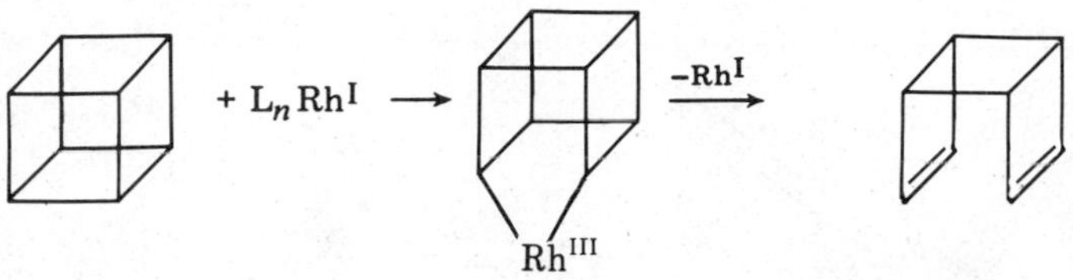

| Cubane | L_n | syn-Tricyclooctadiene |

The Ag^+ ion also catalyzes such isomerizations,[85] but these reactions proceed by electrophilic cleavage of C—C bonds followed by carbonium ion rearrangements:

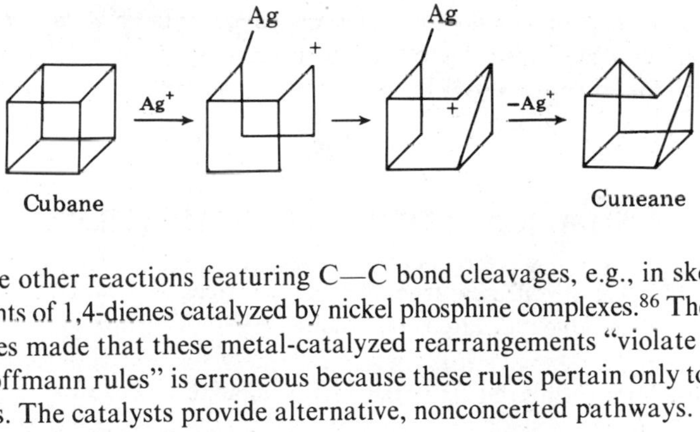

| Cubane | | Cuneane |

There are other reactions featuring C—C bond cleavages, e.g., in skeletal rearrangements of 1,4-dienes catalyzed by nickel phosphine complexes.[86] The statement sometimes made that these metal-catalyzed rearrangements "violate the Woodward-Hoffmann rules" is erroneous because these rules pertain only to concerted processes. The catalysts provide alternative, nonconcerted pathways.

[83]   See, e.g., H. Hogeveen and B. J. Nusse, *J. Am. Chem. Soc.*, 1978, **100**, 3110; R. Noyori *et al.*, *J. Am. Chem. Soc.*, 1976, **98**, 1471; D. L. Beach *et al.*, *J. Organomet. Chem.*, 1977, **142**, 211, 225; L. E. Paquette and M. R. Detty, *Tetrahedron Lett.*, **1978**, 713; P. E. Eaton *et al.*, *J. Am. Chem. Soc.*, 1978, **100**, 2573, 3634; J. Halpern *et al.*, *J. Am. Chem. Soc.*, 1979, **101**, 2694.

[84]   J. Blum and C. Zlotogorski, *Tetrahedron Lett.*, **1978**, 3501; D. B. Brown and V. A. Viens, *J. Organomet. Chem.*, 1977, **142**, 117.

[85]   For references, see L. A. Paquette *et al.*, *Tetrahedron Lett.*, **1978**, 1963.

[86]   R. G. Miller *et al.*, *J. Am. Chem. Soc.*, 1974, **96**, 4211–4235.

30-13. C—H Bond-Breaking Reactions[87]

The intramolecular oxidative addition of both aromatic and aliphatic C—H bonds on bound ligands has been noted earlier (Section 29-5, p. 1244), and there are numerous examples of stoichiometric reactions. There are, so far, but few examples of oxidative additions involving separate aromatic or alkane molecules, and even fewer catalytic ones.

The best studied cases are exchange reactions with deuterium. Thus niobium and tantalum polyhydrides [e.g., $TaH_5(dmpe)_2$] catalyze the benzene-D_2 exchange reaction, probably involving the cycle shown in Fig. 30-11. Both oxidative addition of D_2 and C—H are invoked.

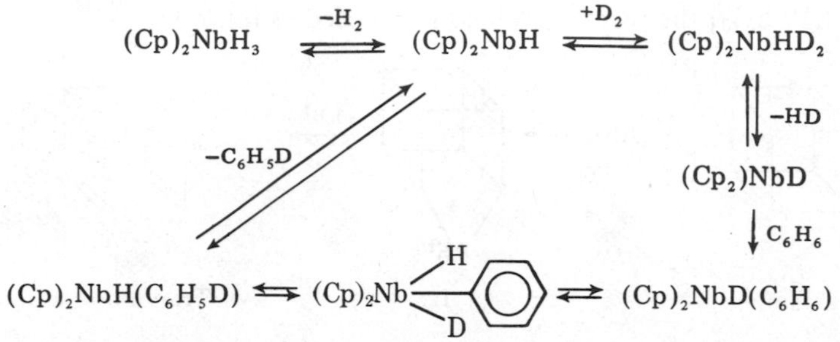

Fig. 30-11. Proposed mechanism for exchange between C_6H_6 and D_2 using (η-C_5H_5)$_2$NbH$_3$. The mechanism for exchange with TaH$_5$(dmpe)$_2$ and other polyhydrides is similar.

It has also been claimed that exchange with saturated hydrocarbons is catalyzed by Rh^{III}, Ir^{III}, and Pt^{II} salts in acetic acid containing D_2O, but it is by no means certain that these reactions are homogeneous. Oxidative addition of *activated* C—H bonds in acetylene, acetonitrile, etc., has been noted (p. 1244). The addition of tetramethylsilane to (η-C_5H_5)$_2$WH$_2$ occurs photochemically[88] to give a species with a W—CH$_2$SiMe$_3$ group.

OXIDATION REACTIONS OF ALKENES

Reactions of a type quite different from the ones described above are those in which alkenes are oxidized to oxygenated compounds.

30-14. Palladium-Catalyzed Oxidations: Oxypalladation Reactions[89]

It has been known since 1894 that ethylene complexes of palladium, e.g.,

[87] D. E. Webster, *Adv. Organomet. Chem.*, 1977, **15**, 147; G. W. Parshall, *Acc. Chem. Res.*, 1975, **8**, 113, and, in *Catalysis*, Vol. 1, Specialist Reports, Chemical Society, 1977; A. E. Shilov and A. A. Shteinman, *Mech. Hydrocarb. React. Symp.*, **1975**, 479; *Coord. Chem. Rev.*, 1977, **24**, 97; M. L. H. Green, *Pure Appl. Chem.*, 1978, **50**, 27.

[88] M. L. H. Green *et al.*, *Nouv. J. Chim.*, 1977, **1**, 187.

[89] P. M. Maitlis, *Organic Chemistry of Palladium*, Vol. 2, Academic Press, 1971; J. M. Davidson, *Catalysis*, Vol. 2, Specialist Report Chemical Society, 1978.

$[(C_2H_4)PdCl_2]_2$, are rapidly decomposed by water to form acetaldehyde and palladium metal. The conversion of this stoichiometric reaction into a cyclic one was achieved by Smidt and his co-workers (at Wacker Chemie). Their main contribution was the linking together of the known individual reactions:

$$C_2H_4 + PdCl_2 + H_2O \rightarrow CH_3CHO + Pd + 2HCl$$
$$Pd + 2CuCl_2 \rightarrow PdCl_2 + 2CuCl$$
$$\underline{2CuCl + 2HCl + \frac{1}{2}O_2 \rightarrow 2CuCl_2 + H_2O}$$
$$C_2H_4 + \frac{1}{2}O_2 \rightarrow CH_3CHO$$

The oxidation of ethylene by palladium(II)-copper(II) chloride solution is essentially quantitative and only low Pd concentrations are required; the process can proceed either in one stage or in two stages—in the latter the reoxidation by O_2 is done separately.

Oxidation of alkenes of the type $RCH{=}CHR'$ or $RCH{=}CH_2$ gives ketones; an important example is the formation of acetone from propene.

When media other than water are used, different but related processes operate. Thus in acetic acid ethylene gives vinyl acetate, whereas vinyl ethers may be formed in alcohols; there are usually competing reactions giving unwanted side products. There has been much industrial study of such processes, and both homogeneous and heterogeneous syntheses of vinyl acetate have been commercialized. The latter process (Hoechst) involves direct oxidation over a palladium catalyst containing alkali acetate on a support:

$$C_2H_4 + CH_3COOH + \frac{1}{2}O_2 = CH_2{=}CHCOOCH_3 + H_2O$$

There has been extensive study of the mechanism of the reaction in aqueous solution containing chloride ion.

The initial equilibria appear to be:

$$[PdCl_4]^{2-} + C_2H_4 \rightleftharpoons [PdCl_3(C_2H_4)]^- + Cl^-$$
$$[PdCl_3(C_2H_4)]^- + H_2O \rightleftharpoons PdCl_2(H_2O)(C_2H_4) + Cl^-$$

It was originally thought that a palladium hydroxo species underwent a cis addition to give a σ-bonded hydroxoalkyl:

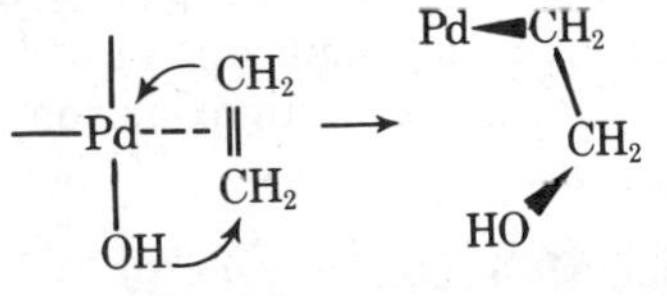

However recent elegant work[90] conclusively shows that there is nucleophilic attack *by external water,* not in the coordination sphere, on the neutral complex $PdCl_2(H_2O)(C_2H_4)$:

$$Cl_2(H_2O)Pd(C_2H_4) + H_2O \rightleftharpoons [Cl_2(H_2O)Pd{-}CH_2CH_2OH]^- + H^+$$

Deuteration studies[90,91] show that the stereochemistry of the addition is trans, i.e.,

90 J. E. Bäckvall, B. Åkermark, and S. O. Ljunggren, *J. Am. Chem. Soc.,* 1979, **101**, 2411.
91 J. K. Stille and R. Divakaruni, *J. Organomet. Chem.,* 1979, **169**, 239.

This reaction is then followed by the rate-determining step:

$$Cl_2(H_2O)Pd–CH_2CH_2OH + H_2O \rightleftharpoons Cl(H_2O)_2PdCH_2CH_2OH + Cl^-$$

This hydroxoalkyl then isomerises by β-hydride transfers finally eliminating acetaldehyde as follows:

Similar mechanisms can be written involving CH_3CO_2H and alcohols as nucleophiles in these media.

The mechanism for the oxidation of Pd metal by Cu^{II} chloro complexes is not well understood, but electron transfer *via* halide bridges (cf. page 1210) is probably involved. The extremely rapid air-oxidation of Cu^I chloro complexes is better known and probably proceeds through an initial oxygen complex:

$$CuCl_2^- + O_2 \rightleftharpoons ClCuO_2 + Cl^-$$

followed by formation of radicals such as O_2^-, OH, or HO_2:

$$ClCuO_2 + H_3O^+ \rightarrow CuCl^+ + HO_2 + H_2O$$

As with Ziegler and Natta's discovery, recognition of the reactivity of palladium complexes has led to a deluge of patents and papers involving all manner of organic substances and types of reactions other than oxidations including additions, carbonylations and polymerizations.[92]

30-15.   Reactions Involving Molecular Oxygen[93]

A wide variety of oxidations of organic compounds are catalyzed by transition metals using molecular oxygen, and we can deal only briefly with this vast subject. In addition to reactions of the Wacker type first discussed, there are two main classes: (*a*) those involving radical autoxidation, and (*b*) those involving reactions

[92]   J. Tsuji, *Adv. Organomet. Chem.,* 1979, **17,** 141.
[93]   J. M. Davidson, *Catalysis,* Vol. 2, Specialist Report, Chemical Society, 1978; J. E. Lyons, in *Aspects of Homogeneous Catalysis,* Vol. 3, R. Ugo, Ed., Reidel, 1977 (review with 524 references).

with hydroperoxides. Only the former involves direct interaction of O_2 with an organic molecule; the latter, like palladium oxidations require oxygen to reoxidize the catalyst after each cycle. Some commercial applications are given in Table 30-1.

Many of the tertiary phosphine complexes noted earlier such as $RuHCl(PPh_3)_3$, $RhCl(PPh_3)_3$, and $Pd(PPh_3)_3$ will react with O_2 to give oxygen complexes (p. 155), and these may react stoichiometrically with SO_2, ketones, etc., to give peroxo complexes. The oxidations of alkenes and other organic compounds catalyzed by such phosphine complexes are not very selective, and the products are the same as those formed in free-radical oxidations.[94] The phosphine ligands are usually also oxidized (e.g., to Ph_3PO), so that the reactions are not generally useful. Although hydroperoxo or other peroxo radicals are probably involved, it has been claimed that the conversion of oct-1-ene and PPh_3 to octan-2-one and Ph_3PO by $RhCl(PPh_3)_3$ and O_2 does not involve radicals but proceeds by direct oxygen transfer.[95]

The complexes $IrCl(CO)(PPh_3)_2$, $Rh_6(CO)_{16}$, and $Pt(PPh_3)_3$ will oxidize ketones, cleaving C—C bonds. Thus cyclohexanone is converted to adipic acid. $Rh_6(CO)_{16}$ also catalyzes the oxidation of CO to CO_2 at $100°$.[96]

Many organic compounds, including both unsaturated and saturated hydrocarbons, are oxidized by molecular oxygen in presence of Cu, Co, and Mn compounds, notably the carboxylates; but these reactions mostly do not involve intermediates with M—C bonds,[97] and they proceed via peroxo species and free-radical reactions.

A recently developed oxidation system uses combinations of $RhCl_3$ and $Cu(ClO_4)_2$ or $Cu(NO_3)_2$ in alcohols,[98] in presence of O_2 to oxidize alkenes to methyl ketones. It is believed to operate by way of five-membered peroxo metallocycles such as 30-X, which decompose to give a methyl ketone and an oxo species (although such species are not otherwise known for Rh), which is then protonated.

(30-X)

94   S. Muto and Y. Kamiya, *J. Catal.,* 1976, **41**, 148; V. C. Choy and C. J. O'Connor, *Coord. Chem. Rev.,* 1972–1973, **9**, 145;

95   R. Tang *et al., J.C.S. Chem. Commun.,* **1979**, 274.

96   D. M. Roundhill *et al., J. Am. Chem. Soc.,* 1977, **99**, 6551.

97   D. J. Huckwell, *Selective Oxidation of Hydrocarbons,* Academic Press, 1974; K. Tanaka, *Chem. Tech.,* **1974**, 555.

98   H. Mimoun *et al., J. Am. Chem. Soc.,* 1978, **100**, 5437; *J.C.S. Chem. Comm.,* **1978**, 559.

Ruthenium trichloride catalyzes the oxidation of primary amines to nitriles and secondary alcohols to ketones at 100° under 2 to 3 atm of O_2.[99a]

Another direct oxidation is that of HCN to oxamide catalyzed by $Cu(NO_3)_2$ in acetic acid at 50 to 80° and 1 atm[99b]:

$$2HCN + \tfrac{1}{2}O_2 + H_2O \longrightarrow \quad \begin{array}{c} CONH_2 \\ | \\ CONH_2 \end{array}$$

which provides a means of converting waste HCN to fertilizer. If the reaction is run in CH_3CN instead of CH_3CO_2H, cyanogen is formed, presumably because the Cu^I acetonitrile complex is less able to catalyze the hydration step.

The direct oxidation of hydrocarbons has, of course, been intensively studied, but in some cases peroxo species can be used. Studies have been made on stoichiometric reactions[100a] of peroxo species such as $O=Mo(O_2)_2(HMPA)$ and alkyl peroxo species of the type 30-X are also probably involved here. A commercial oxidation catalyzed by molybdates is that of propylene[100b] to propylene oxide (Oxirane). This uses t-butylhydroperoxide (made by oxidation of isobutane) to epoxidize propylene:

$$2Me_3CH + 1.5\ O_2 = Me_3COOH + Me_3COH$$

$$Me_3COOH + CH_3CH=CH_2 \longrightarrow \quad \begin{array}{c} CH_3CH-CH_2 \\ \diagdown \diagup \\ O \end{array} + Me_3COH$$

The disadvantage is that almost three times as much t-butanol as propylene oxide is produced, but this can be potentially used to synthesize methacrylic acid[100c] (for methylmethacrylate polymers).

$$Me_3C=CH_2 + Me_3COH + H_2O \xrightarrow{O_2} CH_2=C\begin{array}{c} \diagup CHO \\ \diagdown Me \end{array} \xrightarrow{O_2} CH_2=C\begin{array}{c} \diagup CO_2H \\ \diagdown Me \end{array}.$$

$$\qquad\qquad\qquad\qquad\qquad\qquad\text{Methacrolein}\qquad\qquad\text{Methacrylic acid}$$

OTHER TOPICS

30-16.   Cluster Compounds in Catalysis[101]

All the reaction cycles discussed in the preceding sections proceed via mononuclear

99   (a) F. Mares et al., J.C.S. Chem. Comm., **1978**, 562; (b) W. Riemschneider, Chem. Tech., **1976**, 658.

100 (a) K. B. Sharpless et al., J. Am. Chem. Soc., 1977, **99**, 1990; J. Org. Chem., 1977, **42**, 1587; F. Mares et al., Inorg. Chem., 1978, **17**, 3055; S. Yamada et al., J. Am. Chem. Soc., 1977, **99**, 1988; H. B. Kagen et al., Angew. Chem. Int. Ed., 1979, **18**, 485; cf. also A. F. Noels et al., J. Organomet. Chem., 1979, **166**, 79; (b) K. H. Simmrack, Hydrocarbon Processing, 1978, **57**, 105, (c) Chem. Eng., July 3, 1978, p. 25.

101 A. K. Smith and J. M. Bassett, J. Mol. Catal., 1977, **2**, 229; E. L. Muetterties et al., Science, 1977, **196**, 839; Bull. Soc. Chim. Belg., 1975, **84**, 959; Chem. Rev., 1979, **79**, 91.

complexes. Heterogeneous catalysts such as metals on supports like Al_2O_3 appear to have crystallites containing probably only a small number of atoms. These will readily carry out reactions such as C—H or C—C bond cleavages in saturated hydrocarbons, and CO and N_2 hydrogenations that are not readily, or not at all, achieved by mononuclear metal complexes.[102]

The question arises whether cluster compounds as such actually act as catalysts, since the cluster may be dissociating under the reaction conditions. However there appear to be a few cases involving true cluster catalysis. We have already noted ethyleneglycol synthesis by anionic rhodium carbonyl clusters (p. 1289) and CO oxidation on $Rh_6(CO)_{12}$ (p. 1303), and the conversion of nitrobenzene to aniline (p. 1297) is catalyzed by $Ru_3(CO)_{12}$.

Some other, fairly well-established[103] reactions are the hydrogenation of C—C multiple bonds by hydrido carbonyl clusters and the hydrogenation of acetylenes, isocyanides, and nitriles using $Ni_4(CNCMe_3)_7$, and hydroformylation using $PhCCo_3(CO)_9$ and $(PhP)_2Co_4(CO)_{10}$.

The phosphite clusters $\{HRh[P(OR)_3]_2\}_x$ ($x = 2, 3$) are said to be particularly active for hydrogenation of alkenes[104]; however since they react exothermically with alkenes, it seems most likely that the catalysis is due to a fourteen-electron mononuclear species $HRh[P(OR)_3]_2$.

As far as can be ascertained at present, the mechanisms of reactions on clusters resemble those on mononuclear species. It seems likely that the groups on clusters should be mobile or nonrigid (cf. p. 1226) to provide vacant sites for substrate coordination and to permit catalytic activity. The opening of clusters to provide vacant sites is also likely,[104b] e.g.,

30-17. Heterogenized Homogeneous Catalysts[105]

A major difficulty associated with all homogeneous catalytic systems is the sepa-

102   See, e.g., J. H. Sinfelt, *Acc. Chem. Res.*, 1977, **10**, 15.
103   P. M. Lausarot *et al.*, *Inorg. Chim. Acta*, 1977, **25**, L107; D. La Broue and R. Poilblanc, *J. Mol. Catal.*, 1977, **2**, 329; E. Band *et al.*, *J. Am. Chem. Soc.*, 1977, **99**, 7380; C. U. Pittman, Jr., *et al.*, *J. Am. Chem. Soc.*, 1977, **99**, 1987.
104   (a) V. W. Day *et al.*, *J. Am. Chem. Soc.*, 1977, **99**, 8091; (b) G. Hüttner *et al.*, *Angew Chem. Int. Ed.*, 1979, **18**, 76.
105   M. S. Scurrell, in *Catalysis*, Vol. 2., Specialist Report; Chemical Society., 1978. R. H. Grubbs *et al.*, *J. Mol. Catal.*, 1978, **3**, 259; *Chem. Tech.*, **1977**, 512; D. C. Neckers, *Chem. Tech.*, **1978**, 108; various authors, in *Catalysis in Organic Synthesis*, G. V. Smith, Ed., Academic Press, 1977; L. L. Murell, in *Advances Materials in Catalysis*, J. J. Burton and R. L. Garten, Eds., Academic Press, 1977; F. R. Hartley and P. N. Vezey, *Adv. Organomet. Chem.*, 1977, **15**, 189; J. Manassen and Y. Dror, *J. Mol. Catal.*, 1978, **3**, 227.

ration of products from reactants and catalyst. Since homogeneous systems often allow of much greater selectivity in reactions, attempts have been made to achieve similar benefits by "heterogenizing" known homogeneous catalysts. This has been done by incorporating tertiary phosphine, pyridine, thiol, or other ligands into styrenedivinylbenzene or other polymeric materials, as well as by supporting complexes on carbon, silica, alumina, ion-exchange resins, or molecular sieves.

The anchoring of phosphines to polymers is usually done by first chloromethylating phenyl rings of the polymer by chloro methyl ethers, then reacting the chloro polymers with LiPPh$_2$. The functionalized polymer is exchanged with a typical homogeneous catalyst as follows:

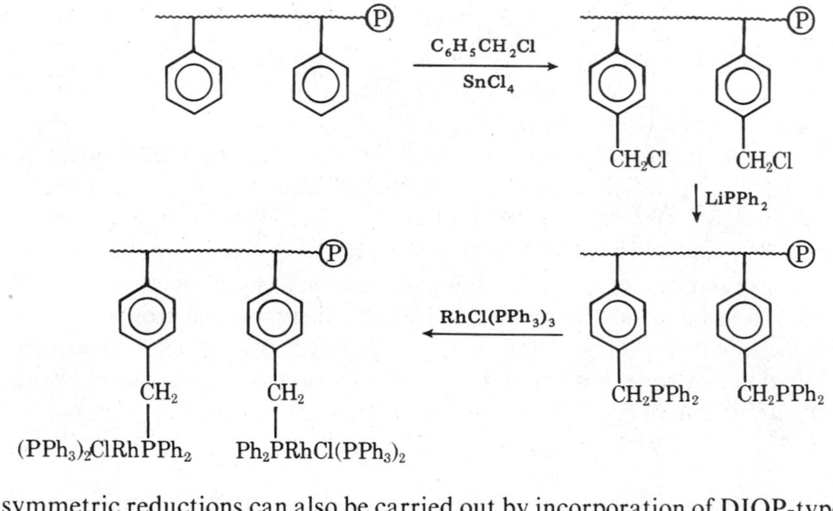

Asymmetric reductions can also be carried out by incorporation of DIOP-type ligands.[106]

Although many reactions (hydrogenation, and hydroformylation[107] of olefins, methanol carbonylation, etc.), can be carried out, often with high selectivity, supported catalysts have disadvantages also. First, if reactions are carried out in solution there may well be an equilibrium:

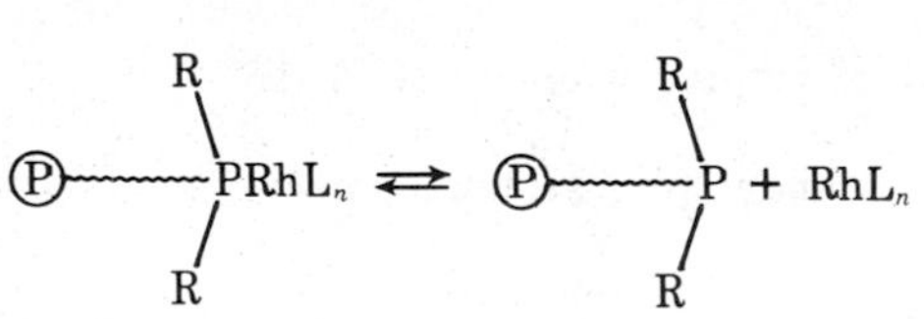

so that leaching of metal complex from the support can occur. With purely gaseous reactants this is not likely. The other disadvantage is that neither the environment nor indeed the nature of the metal on the surface is known with certainty or easy to control.

A study[108] of RhCl(PPh$_3$)$_3$ on a divinylbenzenestyrene polymer by the extended

[106]   J. K. Stille et al., J. Am. Chem. Soc., 1978, **100**, 264, 268.
[107]   C. U. Pittman and A. Hirao, J. Org. Chem., 1978, **43**, 640; H. Arai, J. Catal., 1978, **51**, 135.
[108]   J. Read et al., J. Am. Chem. Soc., 1978, **100**, 2375.

X-ray absorption fine structure (EXAFS) technique suggests that although the bond distances are similar to those in RhCl(PPh$_3$)$_3$), aggregation to form binuclear species has occurred. This could account for the lowered activity in selective hydrogenation.

A further development is to use organic-solvent-soluble functionalized polymers with PPh$_2$ groups.[109]

Semipermeable Membranes and Two-Phase Systems. Another approach to solving the difficulty of separation of products from homogeneous systems is to use reverse osmosis with selectively permeable membranes.[110] Although this method is not yet used on a large scale, certain polyamides or polyimides can allow small molecules to permeate under reverse osmosis conditions while retaining metal complexes. Thus the equilibrium

$$RuH_2(PPh_3)_4 + N_2 \rightleftharpoons RuH_2(N_2)(PPh_3)_3 + PPh_3$$

can be shifted to the right by removal of PPh$_3$. Similarly, adiponitrile obtained by hydrogenation of 1,4-dicyanobutene with RhCl(PPh$_3$)$_3$ can be separated from the complex.

Another alternative is the use of two-phase systems in which the catalyst is in one phase and the substrate and products are in another; a phase transfer catalyst such as long-chain amines may be required. Two-phase catalytic hydrogenation and hydroformylation reactions have been achieved using water-soluble compounds derived from functionalized phosphines of various types[111]; simple examples are:

$$P(m-C_6H_4SO_3H)_3 \quad and \quad P(C_5H_4N)_3$$

There are an increasing number of examples of phase transfer catalysis using transition metal complexes, and one was mentioned earlier in connection with the synthesis of phenylacetic acid (p. 1296).

[109]   E. Bayer and V. Schurig, *Angew. Chem., Int. Ed.*, 1975, **14**, 493; *Chem. Tech.*, **1976**, 212.
[110]   G. W. Parshall *et al., J. Mol. Catal.*, 1977, **2**, 253.
[111]   A. Borowski *et al., Nouv. J. Chim.*, 1978, **2**, 137; G. M. Whitesides *et al., J. Am. Chem. Soc.*, 1979, **101**, 3683; 1978, **100**, 2269, 306; M. T. Beck *et al., Inorg. Chim. Acta*, 1977, **25**, L61.

General References

See also references for Chapters 25, 26, 27, 28, and 29.

Journals and Periodicals

Specialist Reports, Chemical Society. Catalysis. The reports on *Organometallic Chemistry, Inorganic Chemistry of the Transition Elements Chemistry* and *Inorganic Reaction Mechanisms* also have sections on catalysis.
Journal of Molecular Catalysis. Mostly homogeneous catalysis.
Journal of Catalysis. Mostly heterogeneous catalysis.
Advances in Catalysis
Catalyst Reviews, Science and Engineering,
Synthesis. Contains occasional reviews on organic syntheses via metal complexes.
Catalysis in Organic Synthesis, Academic Press.

MONOGRAPHS AND REVIEWS

Advances in Chemistry Series, *Homogeneous Catalysis,* Vol. 1, 1968, Vol. 2, 1974, American Chemical Society.

Alper, H., *Transition Metal Organometallics in Organic Synthesis,* Vol. 2, Academic Press, 1978.

Augustine, E. L., Ed., *Oxidation—Techniques and Applications in Organic Synthesis,* Dekker, 1969. Use of transition metal and other compounds.

Bird, C. W., *Transition Metal Intermediates in Organic Synthesis,* Logos Press (London), Academic Press (New York), 1967. (Comprehensive review with extensive references).

Candlin, J. P., K. A. Taylor, and D. T. Thompson, *Reactions of Transition Metal Complexes,* Elsevier, 1968. General review with extensive references.

Gorewit, B., and M. Tsutsui, *Adv. Catal.,* 1978, **12,** 277. σ-π transitions.

Heck, R. F., *Acc. Chem. Res.,* 1979, **12,** 146. Pd-catalyzed reactions of olefins and organic halides.

Heck, R. F., *Organotransition Metal Chemistry. A Mechanistic Approach,* Academic Press, 1974. An excellent text.

Henrici-Olivé, G., and S. Olivé, *Coordination and Catalysis,* Verlag-Chemie, 1977.

Delmon, B., and G. Jannes, Eds., *Catalysis, Heterogeneous and Homogeneous,* Elsevier, 1975.

Khan, M. M. T., and A. E. Martell, *Homogeneous Catalysis by Metal Complexes,* Academic Press, 1974.

King, R. B., *Inorganic Compounds with Unusual Properties,* II, Adv. Chem. Series, No. 173, Am. Chem. Soc., 1979

Kochi, J. K., *Organometallic Mechanisms and Catalysis,* Academic Press, 1978.

Mango, F. D., *Coord. Chem. Rev.,* 1975, **15,** 109. Oligomerization, cyclo additions, etc.

McCue, J. P., *Coord. Chem. Rev.,* 1973, **10,** 265. Metal hydrides and use in catalysis.

Muetterties, E. L., *Transition Metal Hydrides,* Dekker, 1971. Contains some discussion of catalysis.

Muetterties, E. L., *et al., Chem. Rev.,* 1979, **79,** 91. Clusters and Surfaces in catalysis (295 refs).

Rylander, P. N., *Organic Syntheses via Noble Metals,* Academic Press, 1973.

Rylander, P. N., and H. Greenfield, Eds., *Catalysis in Organic Synthesis,* Academic Press, 1976.

Schrautzer, G. N., Ed., *Transition Metals in Homogeneous Catalysis,* Dekker, 1977.

Shaw, B. L., and N. I. Tucker, *Organotransition Metal Compounds and Related Aspects of Homogeneous Catalysis,* Pergamon Press, 1975.

Tedder, J. M. A. Nechvatal, and A. H. Jubb, *Basic Organic Chemistry,* Part 5, *Industrial Products,* Wiley, 1975.

Tsuji, J., *Organic Syntheses by means of Transition Metal Complexes,* Springer-Verlag, 1975. Short useful text.

Ugo, R., Ed., *Aspects of Homogeneous Catalysis,* Vol. 1, 1970, Vol. 2, 1976. Wiley.

Waddams, A. L., *Chemicals from Petroleum,* 3rd ed., Murray, 1973.

Weissermel, K., and H.-J. Arpe, *Industrielle Organische Chemie,* Verlag-Chemie, 1976.

Wender, I., and P. Pino, Eds., *Organic Syntheses via Metal Carbonyls,* Vol. 1, 1968, Vol. 2, 1976, Wiley.

Wiseman, P., *An Introduction to Industrial Organic Chemistry,* Applied Science Publishers, 1972.

SPECIFIC METALS

Cobalt

Chalk, A. J., and J. F. Harrod, *Adv. Organomet. Chem.,* 1968, **6,** 119. Carbonyls.

Heck, R. F., *Adv. Organomet. Chem.,* 1966, **4,** 243. Carbonyls.

Chromium, Molybdenum, and Tungsten

Farona, M. F., in *Organometallic Reactions and Synthesis,* Vol. 6, Plenum Press 1977, p. 233. Arene tricarbonyls.

Nickel

Jolly, P. W., and G. Wilke, *The Organic Chemistry of Nickel,* Academic Press. Vol. 2 is mainly catalysis.

Palladium

Heck, R. F., *Acc. Chem. Res.,* 1979, **12,** 146.

Maitlis, P. M., *The Organic Chemistry of Palladium,* Vols. 1 and 2, Academic Press, 1971.

Tsuji, J., *Organic Synthesis with Palladium Compounds,* Springer, 1979.

Rhodium

Proc. Symp. Rhodium Homogen. Catal., 1978. Veszprem. Vegyip, Egy., Kozp., Konyutara, Veszprem., Hungary.

Ruthenium

Fahey, D. F., in *Catalysis in Organic Synthesis,* 1976, Academic Press. Use of $RuCl_2(CO)(PPh_3)_2$.

James, B. R., *Inorg. Chim. Acta Revs.,* 1970, **4,** 73.

OTHER REFERENCES

Tsutsui, M., Y. Ishii, and R. Ugo, Eds., *Fundamental Research in Homogeneous Catalysis,* Vol., 1, 1977, Vol. 2, 1978, Plenum Press. Conference reports, various topics.

Hanzlik, *Inorganic Aspects of Biology and Organic Chemistry,* Academic Press, 1976. Chapters on organometallic compounds, metal ions, O_2, and N_2 in catalysis.

Sheldon, R. A., and J. K. Kochi, *Oxid. Combust. Rev.,* 1973, **5,** 135; *Adv. Catal.,* 1976, **25,** 274. Liquid phase oxidations.

Tertiary Phosphines as Catalysts, Suppl. 1., M and T Chemicals Inc.

Tsutsui, M., and A. Courtney, *Adv. Organometallic Chem.,* 1977, **16,** 241. σ-π rearrangements in catalysis.

Witcoff, H., *Chem. Tech.,* **1978,** 252; **1977,** 752. Industrial catalysis.

CHAPTER THIRTY–ONE

Bioinorganic Chemistry

31-1. Introduction

Although biochemistry involves extremely elaborate organic molecules (and also many simple ones) and the array of organic reactions that occur is dazzling in its scope and subtlety, it is important to recognize that biochemistry is not merely an elaboration of organic chemistry. Without the participation of the metallic elements and some others that are not normally involved in organic chemistry, life—at least in its present form—would not exist. The inorganic chemist, especially the coordination chemist, has a contribution to make to the understanding of life that is just as essential, though smaller in scope, as that of the organic chemist. This chapter gives an overview of the principal ways in which metals and some other elements participate in biochemical processes.

Of the 30 elements now recognized as essential to life (Fig. 31-1), seventeen are metals and four more (B, Si, P, Se) are metalloids that might not necessarily be considered as having central roles in organic chemistry. Of the seventeen metallic elements, at least seven appear to be essential to *every* known form of life. The most abundant elements in the biosphere are first iron, then zinc, followed by magnesium, calcium, copper, and potassium.

One of the major roles played by metallic elements in biochemistry is in *metalloenzymes*. This term is applied to enzymes that not only require the participation of a metal ion at the active site to function but bind that metal ion (or ions) strongly even in the resting state. Known metalloenzymes now number several hundred. They may be considered as a subclass (the largest) of the *metalloproteins,* that is, proteins that incorporate one or more metal atoms as a normal part of their structures. This includes, then, not only enzymes but respiratory proteins like hemoglobin and myoglobin, electron transport proteins such as cytochromes and ferridoxins, and metal storage proteins.

In many but not all cases it is possible to remove the metal atoms and then restore them (or replace them by others) without collapse of the overall protein structure. The protein from which the metal ions have been removed is called the *apoprotein,*

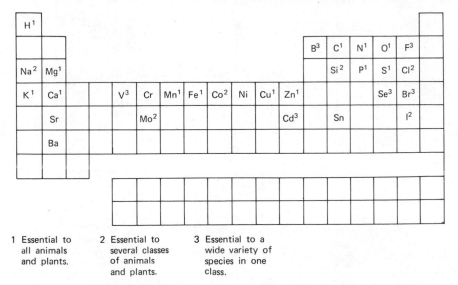

1 Essential to all animals and plants.

2 Essential to several classes of animals and plants.

3 Essential to a wide variety of species in one class.

Fig. 31-1. A periodic table for the biosphere. Superscripts: 1, essential to all animals and plants; 2, essential to several classes of animals and plants; 3, essential to a wide variety of species in one class.

the use of this term usually (though perhaps not invariably) implying that the active metalloprotein can be recovered on restoration of the metal ions.

We shall not attempt to cover even superficially all the elements known to be involved in bioinorganic chemistry. Instead, we discuss only those that are presently recognized to be most important or about which most is known. Based on one or both of these criteria, the element iron comes most readily to mind, and we treat it first, then proceed to briefer discussions of several other elements.

THE BIOCHEMISTRY OF IRON

Iron is truly ubiquitous in living systems. Its versatility is unique. It is at the active center of molecules responsible for oxygen transport and electron transport and it is found in, or with, such diverse metalloenzymes as nitrogenase, various oxidases, hydrogenases, reductases, dehydrogenases, deoxygenases, and dehydrases. In many cases little if anything is known beyond the fact that it is there (hence a vast field of research lies waiting), whereas in others, such as hemoglobin, ferredoxins, and cytochromes, we know the molecular structures and electronic properties of the active sites in considerable detail.

Not only is iron involved in an enormous range of functions, it also is found in the whole gamut of life forms from bacteria to man. Iron is extremely abundant in the earth's crust and it has two readily interconverted oxidation states; doubtless these properties have led to its evolutionary selection for use in many life processes. Table 31-1 shows the principal forms in which iron is found in the human body.

TABLE 31-1
Distribution of Iron-Containing Proteins in a Normal Adult

Protein	Molecular weight of protein	Amount of protein (g)	Amount of iron (g)	% of total body iron	Nature of iron heme (H) or non-heme (N)	Number of iron atoms bound per molecule	Valence state	Function
Hemoglobin	64,500	750	2.60	65	H	4	Fe^{2+}	Oxygen transport in plasma
Myoglobin	17,000	40	0.13	6	H	1	Fe^{2+}	Oxygen storage in muscle
Transferrin	76,000	20	0.007	0.2	N	2	Fe^{3+}	Iron transport via plasma
Ferritin	444,000	2.4	0.52	13	N	0–4,300	Fe^{3+}	Iron storage in cells
Hemosiderin	Not known	1.6	0.48	12	N	5000	Fe^{3+}	Iron storage in cells
Catalase	280,000	5.0	0.004	0.1	H		Fe^{2+}	Metabolism of H_2O_2
Cytochrome *c*	12,500	0.8	0.004	0.1	H	1	Fe^{2+}/Fe^{3+}	Terminal oxidation
Peroxidase	44,100				H			Metabolism of H_2O_2
Cytochromes and oxidase			0.02	<0.5	H		Fe^{2+}/Fe^{3+}	Terminal oxidation
Flavoprotein dehydrogenases, oxidases, and oxygenases					N		Fe^{2+}	Oxidation reactions, incorporation of molecular oxygen

The next few sections treat the biochemistry of iron under two broad headings: storage and transport, and function. The latter is broken into a number of topics because of the great diversity of function and the widely differing chemistries of the iron-containing species.

31-2.   Iron Storage and Transport

As Table 31-1 indicates, there are known to be two principal iron-storage compounds in the human body: ferritin and hemosiderin. Hemosiderin appears to be rather variable in composition and properties and is poorly understood compared to ferritin. It is not discussed further here.

Ferritin.   Ferritin serves as a depot in which surplus iron can be stored within cells in nontoxic form and from which it can be released in usable form as required, either within that cell itself or in other cells of the organism. It is widely distributed throughout the various organs of all mammals, especially high concentrations being found in liver, spleen, and bone marrow.

Ferritin consists of a shell of protein surrounding a core of ferric hydroxyphosphate. The iron-containing core has a diameter of 70 to 75 Å, and the entire structure has a diameter of about 120 Å. Analysis shows that the composition of the core can be represented as predominantly $FeO \cdot OH$ together with 1.0 to 1.5% phosphate. The iron percentage is 57%, and "formulas" such as $(FeO \cdot OH)_8(FeO \cdot H_2PO_4)$ have been suggested. The gross weight of the ferritin unit ("molecule") approaches 900,000; about 418,000 of this is due to the core, which contains approximately 4300 iron atoms. Various suggestions have been made concerning the detailed structure of the core based on comparison with the known structures of hydrated ferric oxides. All involve close-packed layers of oxide ions with ferric ions randomly distributed over the octahedral and tetrahedral interstices. A recent study by EXAFS favors a sheet structure in which all iron atoms are six-coordinate.[1]

The magnetic behavior of ferritin has been extensively studied, and data from Mössbauer spectroscopy are also available. The ferric ions are present in high-spin form and are subject to strong antiferromagnetic coupling, as would be expected for the type of structure discussed above. The protein sheath is made up of 24 identical subunits each consisting of about 163 amino acid residues and having a molecular weight of about 18,500. X-Ray crystallography of apoferritin from horse spleen[2] has shown that the subunit is roughly cylindrical, being about 27 Å in diameter and about twice as long. Each one is made up mainly of four nearly parallel α helices. The arrangement of the 24 subunits to form the nearly spherical apoferritin sheath is highly symmetrical, having full cubic rotation symmetry with 4-fold and 3-fold axes. Whether this full symmetry is strictly retained in ferritin itself is not known.

Transferrin.   This molecule has the function of carrying ingested iron from the

[1]   E. A. Stern et al., J. Am. Chem. Soc., 1979, **101,** 67.
[2]   S. H. Banyard, D. K. Stammers, and P. M. Harrison, Nature, 1978, **271,** 282.

stomach and introducing it into the iron metabolic processes of the body. As iron passes from the stomach (which is acidic) into the blood (pH $\sim$7.4) it is oxidized to Fe^{III} in a process catalyzed by the copper metalloenzyme ceruloplasmin, then picked up by transferrin molecules. These are proteins with a molecular weight of about 80,000, and they contain two metal-binding sites that are similar but not identical in their binding affinity. These sites bind iron only in presence of certain anions, with CO_3^{2-} or HCO_3^- being those apparently preferred *in vivo*. The details of the nature of these binding sites are still unknown. The binding constant is about 10^{26}; therefore transferrin is an extremely efficient scavenger of iron, removing it from such stable complexes as those with phosphate or citrate. Transferrin then becomes bound to the cell wall of an immature red cell, which requires iron, and the transfer is made under control of a complex mechanism for transporting the iron through the cell membrane.

Transferrin also carries iron to ferritin *in vitro* and the process of transfer is a complex one requiring ATP and ascorbic acid and proceeding via Fe^{II}. The nature of the binding sites of transferrin has not been established, but tyrosine, histidine, and the carbonate ion have been suggested as ligands, rather than two hydroxamate ligands that are well established in the bacterial iron transport ligands, discussed next.

Bacterial Iron Transport.[3]   The agents responsible for iron transport into and within bacteria have been intensively studied, and many structural and chemical details have been firmly established. The problem is that iron is not spontaneously available to aerobic organisms in an aqueous environment because of the very low solubility of ferric hydroxide ($pK_{sp} \approx 38$). Thus Fe^{3+} ions at a pH of about 7 have a molar concentration of only about 10^{-18}, and simple diffusion into cells could never suffice to supply their needs. Indeed, simple inward diffusion would not occur, since iron is already more concentrated than this in the living cell. Therefore special chelating agents called *siderophores* are produced by bacteria and ejected into their environment to gather iron and transfer it through the cell wall into the cell. In some cases it appears that the iron is released by the chelator at the wall and it passes through alone, whereas in others the entire complex enters the cell.

The siderophores are rather diverse chemically but have in common the use of chelating, oxygen-donor-type ligands; only in one case is nitrogen known to be a donor atom. A very large fraction of the siderophores that have been characterized employ hydroxamate moieties (31-I) as the ligating units. For example, a type of siderophore called a *ferrichrome* consists of a cyclic hexapeptide in which three

(31-I)   (31-II)

successive amino acid residues have side chains ending in hydroxamate groups.

3   K. N. Raymond and C. J. Carrano, *Acc. Chem. Res.*, 1979, **12**, 183.

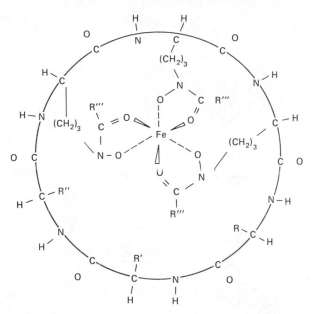

Fig. 31-2. A schematic diagram of coordination of iron in ferrichromes.

These three hydroxamate groups bind iron by forming a trischelate octahedral complex (Fig. 31-2).

Another type of siderophore, especially common in prokaryotes such as enteric bacteria, is called an *enterobactin*; the ligating units are catecholate anions (31-II) that also chelate very effectively. Figures 31-3a and b give a specific example of an enterobactin and a diagram of a model system containing Cr^{3+}, respectively.

The siderophores form Fe^{3+} complexes that are very stable thermodynamically (i.e., they have very high formation constants). It has been estimated that the enterobactin in Fig. 31-3 complexes Fe^{3+} with a formation constant of $\geq 10^{45}$. However these complexes are kinetically labile, and of course such lability is essential so that iron can be easily taken up before transport and released afterward.

The Cr^{3+} analogues with comparable thermodynamic stability accompanied by kinetic inertness can be prepared, and these are useful for studying structural and spectroscopic properties of the siderophore complexes. Even simple model compounds such as the $[Cr(O_2C_6H_4)_3]^{3-}$ ion can be conveniently studied, and they provide useful insight.

Other siderophores that have been characterized structurally include the *ferrimycobactins*,[4] which form very distorted octahedral complexes in which one of the donor atoms is nitrogen rather than oxygen, and a hexapeptide called *ferrichrysin*,[5] which utilizes hydroxamate chelating groups.

[4] E. Hough and D. Rogers, *Biochem. Biophys. Res. Comm.*, 1974, **57**, 73.
[5] R. Norrestarn, B. Stensland, and C.-I. Branden, *J. Mol. Biol.*, 1975, **99**, 501.

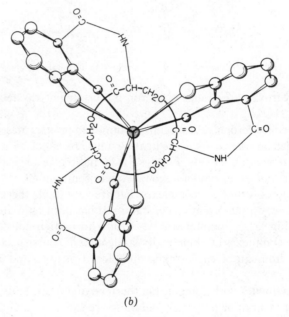

Fig. 31-3. (a) The structure of an enterobactin. (b) The conformation of a Δ-*cis*-trianion complex of Cr^{3+} formed by the same enterobactin. [Reproduced by permission from S. S. Isied, G. Kuo, and K. N. Raymond, *J. Am. Chem. Soc.,* 1976, **98,** 1763.]

31-3. Hemoglobin and Myoglobin

Myoglobin (Mb) consists of one polypeptide chain (globin) with one heme group (the iron porphyrin complex shown as 31-III) embedded therein. The peptide chain contains 150 to 160 amino acid residues, the precise number depending on the species in which it is found. Hemoglobin (Hb), with a molecular weight of *ca.*

$$CO_2^-\quad CO_2^-$$
$$(CH_2)_2\quad (CH_2)_2$$

(31-III)

64,500, consists of four myoglobinlike subunits; these four are similar but not all identical, two being α units and the other β units. Neither the α nor the β units of hemoglobin have amino acid sequences that match the sequence in myoglobin, but nevertheless the ways in which the chains are coiled to give three-dimensional structures (tertiary structure) are quite similar. In each subunit of hemoglobin and in myoglobin, the iron atom is also bonded to the nitrogen atom from the imidazole side chain of a histidine residue. Figure 31-4 gives a schematic representation of the β-subunit of hemoglobin; its essential features are typical of α-subunits and myoglobin as well.

According to the description above, the iron atoms in Hb and Mb when no oxygen is present (the deoxy forms) would be five-coordinate. In fact, there is probably a water molecule loosely bonded in the sixth position (i.e., trans to the histidine nitrogen atom) to complete a distorted octahedron. The iron atom appears to be out of the porphyrin plane toward the histidine. In both deoxy-Mb and deoxy-Hb the iron atoms are high-spin Fe^{II}, with four unpaired electrons.

The function of both Hb and Mb is to bind oxygen, but their physiological roles are very different. Hb picks up oxygen in the lungs and carries it to tissues via the circulatory system. Cellular oxygen is bound by myoglobin molecules that store it until it is required for metabolic action, whereupon they release it to other acceptors. Hb has an additional function, however, and that is to carry CO_2 back to the lungs; this is done by certain amino acid side chains, and the heme groups are not directly involved. Because the circumstances under which Hb and Mb are required to bind and release O_2 are very different, the two substances have quite different binding constants as a function of O_2 partial pressure (Fig. 31-5).

Hemoglobin is not simply a passive carrier of oxygen but an intricate molecular machine. This may be appreciated by comparing its affinity for O_2 to that of myoglobin. For myoglobin (Mb) we have the following simple equilibrium:

$$Mb + O_2 = MbO_2 \qquad K = \frac{[MbO_2]}{[Mb][O_2]}$$

If f represents the fraction of myoglobin molecules bearing oxygen and P represents

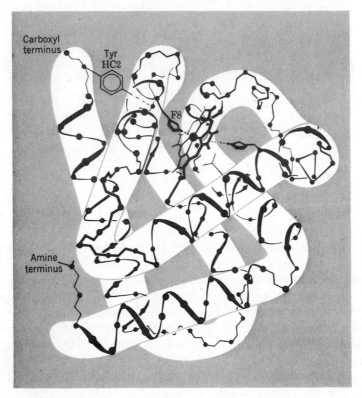

Fig. 31-4. A schematic representation of one of the four subunits of hemoglobin. The continuous black band represents the peptide chain, and the various sections of helix are evident. Dots on the chain represent the α-carbon atoms. The heme group can be seen at upper right center with the iron as a large dot. The coordinated histidine side chain is labeled F8 (meaning the eighth residue of the F helix). [This figure was adapted from one kindly supplied by M. Perutz.]

the equilibrium partial pressure of oxygen, it follows that

$$K = \frac{f}{(1-f)P} \quad \text{and} \quad f = \frac{KP}{1 + KP}$$

This is the equation for the hyperbolic curve labeled Mb in Fig. 31-5. Hemoglobin with its four subunits has more complex behavior; it approximately follows the equation

$$f = \frac{KP^n}{1 + KP^n} \qquad n \approx 2.8$$

where the exact value of n depends on pH. Thus for hemoglobin the oxygen binding curves are sigmoidal as shown in Fig. 31-5. The fact that n exceeds unity can be ascribed physically to the increase caused by attachment of O_2 to one heme group in the binding constant for the next O_2, which in turn increases the constant for the next one and so on. This is called cooperativity. Just as a value of $n = 1$ would represent the lower limit of no cooperativity, a value of $n = 4$ would represent the

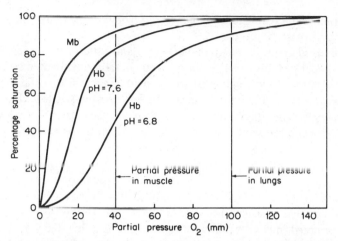

Fig. 31-5. The oxygen-binding curves for myoglobin (Mb) and hemoglobin (Hb), showing also the pH dependence (Bohr effect) for the latter.

maximum of cooperativity, since it would imply that only Hb and $Hb(O_2)_4$ would be participants in the equilibrium.

It will be seen that although Hb is about as good an O_2 binder as Mb at high O_2 pressure, it is much poorer at the lower pressures prevailing in muscle, hence passes its oxygen on to the Mb as required. Moreover, the need for O_2 is greatest in tissues that have already consumed oxygen and simultaneously produced CO_2. The CO_2 lowers the pH, thus causing the Hb to release even more oxygen to the Mb. The pH sensitivity (called the Bohr effect) and the progressive increase of the O_2 binding constants in Hb are due to interactions between the subunits; Mb behaves more simply because it consists of only one unit. It is clear that each of the two is essential in the complete oxygen transport process. Carbon monoxide, PF_3, and a few other substances are toxic because they become bound to the iron atoms of Hb instead of O_2; their effect is one of competitive binding.

Nature of the Heme-Dioxygen Binding. Because of the size and complexity of both myoglobin and hemoglobin it has not been easy to establish either the structural or the electronic character of the bonding of O_2 to heme. As noted in Section 4-24, there are several ways for the O_2 unit to be bound to a metal atom: the possible geometries are side-on (to form a triangle), end-on linear, and end-on bent. The electronic possibilities include a simple electron-pair donation, with retention of essentially neutral O_2 character, one-electron acceptance by O_2 so that there is effectively a coordinated superoxide ion (and Fe^{III}), and two-electron acceptance so that there is effectively a coordinated peroxide ion. Since the last would imply the presence of Fe^{IV}, it need not be considered further.

As far as geometry is concerned, the end-on-bent arrangement is strongly suggested by infrared evidence and by X-ray studies of model compounds.[6] Elec-

---

[6] J. P. Collman, *et al.*, *Proc. Nat. Acad. Sci.* (*U.S.*), 1976, **73**, 3333; J. P. Collman, J. A. Ibers, *et al.*, *J. Am. Chem. Soc.*, 1978, **100**, 6769.

tronically the $Fe^{III}-O_2^-$ formulation has been favored recently by many[6,7] workers but it is not universally accepted.[8]

It was observed many years ago that $Hb(O_2)_4$ and MbO_2 have diamagnetic heme units, and model systems are also diamagnetic in their oxygenated state. Coordination of singlet O_2 to low-spin Fe^{II} would readily explain this. To reconcile the diamagnetism with the $Fe^{III}-O_2^-$ formulation, it is necessary to postulate that there is strong magnetic coupling between the one unpaired electron that a low-spin octahedral Fe^{III} complex would have and the odd electron expected for an O_2^- ion. The credibility of such a coupling is supported by the fact that the $Cr(TPP)(py)O_2$ molecule (TPP = tetraphenylporphyrin) has a magnetic moment of 2.7 BM, indicative of only two unpaired electrons. It is proposed[9] that one of the three d electrons of a Cr^{III} is paired (or very strongly coupled) with the odd electron of O_2^-, and this is regarded as a model for the heme-O_2 situation.

It may be noted in connection with the end-on-bent geometry of the heme-O_2 bonding that in natural hemoproteins the arrangement of amino acid side chains over the heme plane where O_2 is bound is such as to naturally accommodate such a bent Fe—X—X chain but not a linear one. Thus there is a steric discrimination in favor of O_2 as against CO for which a bending of the Fe—C—O chain is destabilizing.

Model Systems. These have been mentioned several times, and it is appropriate to say a few words about how they are designed. They are intended to mimic the natural oxygen-carrying heme proteins while avoiding the difficulties introduced in the study of the natural systems by the great size and other inconvenient properties of the large proteins. The models contain an iron-porphyrin complex but attempt to simulate the role of the protein in simpler ways. It has been fully appreciated for a long time that if only a simple iron-porphyrin complex is used oxygen reacts in an irreversible manner with two such hemelike units, as represented schematically in 31-1.

$$L\!-\!Fe^{II}\!-\!O_2 \;+\; Fe^{II}\!-\!L \;\longrightarrow\; L\!-\!Fe^{III}\!-\!O\!-\!O\!-\!Fe^{III}\!-\!L \;\longrightarrow$$

$$L\!-\!Fe^{III}\!-\!O\!-\!Fe^{III}\!-\!L \qquad (31\text{-}I)$$

In nature this is prevented because individual heme units are attached to the bulky globin (protein) molecules and the close approach required for oxygen bridging is impossible. In the model systems (see also Section 4-24) the bimolecular reaction is usually precluded by some kind of steric hindrance,[10] deliberately introduced in one of several ways.

7 F. Basolo, B. M. Hoffman, and J. A. Ibers, *Acc. Chem. Res.,* 1975, **8,** 384.

8 B. H. Huynk, D. A. Case, and M. Karplus, *J. Am. Chem. Soc.,* 1977, **99,** 6103.

9 C. A. Reed and S. K. Cheung, *Proc. Nat. Acad. Sci. (U.S.),* 1977, **74,** 1780.

10 J. P. Collman, *Acc. Chem. Res.,* 1977, **10,** 265; J. E. Baldwin *et al., J. Am. Chem. Soc.,* 1975, **97,** 227; *J.C.S. Chem. Commun.,* **1976,** 881; A. R. Battersby *et al., J.C.S. Chem. Comm.,* **1976,** 879.

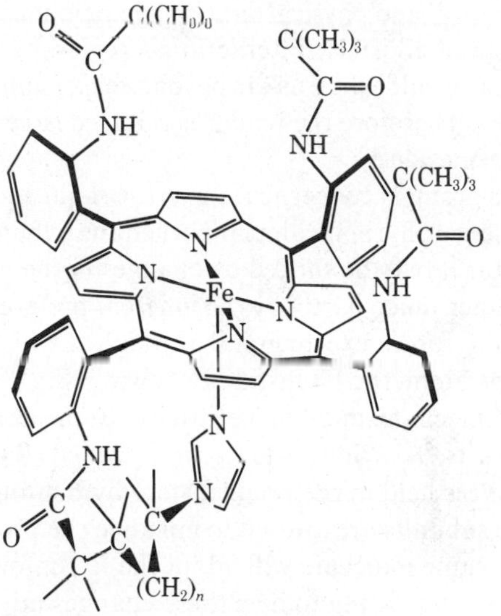

Fig. 31-6. One of the "picket fence" models for a heme-protein oxygen carrier.

1. The "picket fence" approach, in which three or four large groups stand around the iron atom on the side where oxygen is to be bound (31-IV), has been widely exploited. A real example of a "picket fence" model is shown in Fig. 31-6.

2. "Strapped" models (31-V) in which one chain extends over the iron atom, but leaves room for the O_2.

3. "Roofed" models (31-VI) in which there is more complete enclosure of the binding site.

(31-IV)   (31-V)   (31-VI)

As an alternative to steric hindrance, the rate of oxidation can also be markedly reduced by use of low temperatures (*ca.* −50°) and dilute solutions (*ca.* 10^{-4} *M*),[11] but it is usually so advantageous to be able to work at ordinary temperatures that the steric hindrance approach is preferable.

Cooperativity in Hemoglobin. The critical importance of cooperativity in the process of oxygenation of hemoglobin has been emphasized. In some way as each subunit binds an O_2 ligand, one or more of the other subunits acquires an increased affinity for O_2, until $Hb(O_2)_4$ is reached. Much effort has been dedicated to elucidating the mechanism by which this "communication" between the subunits is accomplished, but the picture is still far from clear. There are, in fact, widely differing views on many critical points, and we present here only a brief discussion of the problem with emphasis entirely on the role that may be played by structural changes in the heme group on oxygenation. The problem as a whole encompasses

[11]   C. K. Chang and T. G. Traylor, *J. Am. Chem. Soc.,* 1973, **95**, 5810, 8475, 8477.

a large amount of chemical and physical data on hemoglobin and has to be analyzed in terms of the concept of allosteric interaction; a review of these data and an explanation of these ideas would carry us far beyond reasonable limits for a textbook of inorganic chemistry. Therefore the reader is referred to several excellent review articles for more information.[12]

As far as the heme group is concerned, the critical question is how far the iron atom may lie from the porphyrin mean plane when the subunit is in the deoxy and oxy forms, and how far it may be shifted by changes in the protein conformation. These distances are determined partly by the inherent preferences of the heme unit itself and partly by the forces exerted on it through the histidine residue that is coordinated to the iron atom, the F8 histidine shown in Fig. 31-4. According to the allosteric interaction mechanism of cooperativity, each subunit may be in either a tense (T) state where its O_2-affinity is low or in a relaxed (R) state where it is high. When any one subunit is held in the relaxed state by binding an O_2 molecule, its interactions with the subunits are altered to enhance the probability that still another subunit of that same molecule will adopt the R conformation.

The earliest suggestion relating to how these changes might be communicated laid great stress on a large movement (ca. 0.75 Å) of the iron atom from an initial position well out of the porphyrin plane toward an in-plane location. In this large movement the iron atom was believed to drag the F8 histidine residue with it, and this motion was in turn believed to set off a sequence of conformational changes that resulted in a stabilization of the R state in the other subunits. This, in essence, was Perutz's "trigger mechanism." However subsequent experimental and theoretical studies have suggested that movements of the iron atom are not nearly so large as had been assumed and that the tension in the T state is by no means localized in the heme unit.[13] Thus renewed efforts are now underway to explain, in atomic terms, the cooperativity exhibited by hemoglobin. There is no reason to assume that any simple picture will be able to account fully for the phenomenon,[14] but whatever the ultimate detailed explanation, it is certain that motions of the iron atoms in the heme units will play an important part. Thus continued study of model systems in which the displacement of the iron atom can be controlled and accurately known[6] is certainly worthwhile.

31-4. Cytochromes

Cytochromes are heme proteins that act as electron carriers, linking the oxidation of substrates to the reduction of O_2, as shown schematically in Fig. 31-7. Cytochromes are found in all aerobic forms of life. They are small (MW $\approx$ 12,000) molecules, and they operate by shuttling of the iron atom between Fe^{II} and Fe^{III}. The classification of cytochromes is complex because they differ from one organism

[12]   M. F. Perutz Br. Med. Bull., 1976, **32**, 195; Sci. Am., 1978, **239**, 92.
[13]   Extensive references are given by D. K. White, J. B. Cannon, and T. G. Traylor, J. Am. Chem. Soc., 1979, **101**, 2443.
[14]   See, for example, A. Warshel, Proc. Nat. Acad. Sci. (U.S.), 1977, **74**, 1789.

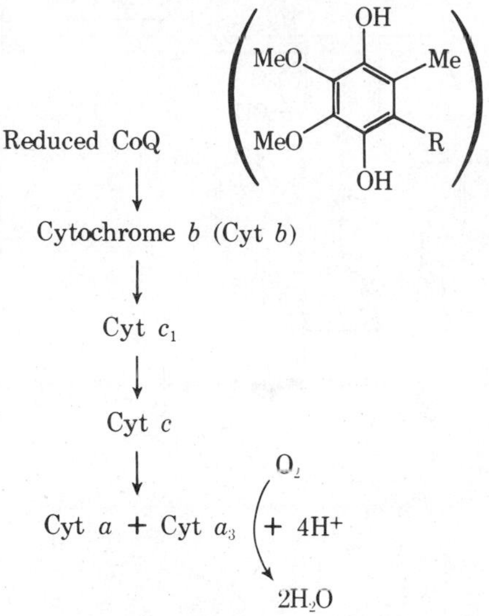

$$\text{Reduced CoQ} \quad \left(\begin{array}{c} \text{OH} \\ \text{MeO} \quad \text{Me} \\ \text{MeO} \quad \text{R} \\ \text{OH} \end{array} \right)$$

Cytochrome b (Cyt b)

$\downarrow$

Cyt c_1

$\downarrow$

Cyt c

$\downarrow$

$$\text{Cyt } a + \text{Cyt } a_3 \left(\begin{array}{c} O_2 \\ + 4H^+ \end{array} \right.$$

$$2H_2O$$

Fig. 31-7. A scheme showing the sequence of cytochromes (Cyt) that intervene between coenzyme Q (CoQ) and the reduction of dioxygen to water.

to another and even from one type of cell to another in a given species. A broadly applicable classification scheme is included in Fig. 31-7. The cytochromes are one-electron carriers; thus two Cyt b molecules are required to react with reduced CoQ to begin the process. At the end, the conversion of O_2 to $2H_2O$ will require four electrons, and it is not known how they are all delivered as needed.

The redox potentials of the intervening cytochromes gradually increase to cover the gap stepwise between that for the oxidation of reduced CoQ and that for reduction of dioxygen. To do this the cytochromes have distinctive structures and properties. In Cyt b, Cyt c_1, and Cyt c the heme group is the same as that in hemoglobin and myoglobin. In Cyt b the heme is not covalently attached to the protein, but is held only by ligation to the iron atom. In Cyt c_1 and Cyt c the heme is covalently connected by thioether linkages (Fig. 31-8a), and the iron atom is coordinated by protein side chains (Fig. 31-8b).

Cyt a and Cyt a_3 exist together as a complex, sometimes called *cytochrome oxidase,* and they contain a heme with several different substituents on the periphery of the porphyrin ring. Cyt a_3 also contains copper, which goes from Cu^{II} to Cu^{I} and thus transfers an electron from the heme of Cyt a_3 to the dioxygen molecule.

Of all the cytochromes, only Cyt c has been a convenient subject for detailed study and much is now known about it, since it is water soluble and easily isolated intact from the mitochondrial membrane. The three-dimensional structures of both the Fe^{II} and Fe^{III} forms have been determined at high resolution. The protein is a single peptide chain of 104 amino acid residues, and it forms a roughly spherical mass about 34 Å in diameter surrounding the heme, so that only about 20% of the edge of the porphyrin ring is exposed. Electron transfer may occur here, or by

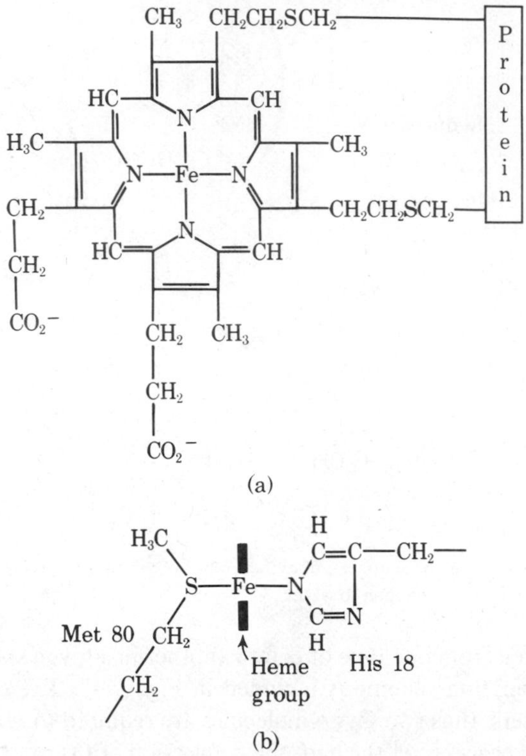

(a)

(b)

Fig. 31-8. (a) Mode of attachment of porphyrin ring to protein; (b) Coordination of the iron atom in cytochrome c.

tunnelling through the protein, by an outer-sphere mechanism. The folding of the protein chain is such that aside from the methionine and histidine residues that coordinate to the iron atom, the side chains that lie inside facing the heme group are the hydrophobic ones. The polar or charged side chains lie on the outside and are arranged in a unique pattern that seems to be designed to allow appropriate matching with both Cyt c_1 from which an electron must be accepted and cytochrome oxidase to which an electron must be transferred.

Cytochrome c seems to be one of the most ancient of biomolecules, having evolved in essentially its present form more than 1.5 billion years ago, even though it is present in all animals and plants, including those that have appeared more recently. It has been found that the cytochrome c of any eucaryotic species (one having cells with nuclei) will react with the cytochrome oxidase of any other species, thus confirming that this electron transfer chain has resisted evolutionary change for a very long time.

Cytochrome P_{450} Enzymes. These heme proteins, found in cell membranes, catalyze the hydroxylation of C—H bonds; the name given to them is thus misleading, since they do not serve the type of electron transfer function just discussed for the "regular" cytochromes, but are actually enzymes. They consist of one heme group in a single-chain protein and have the iron atom coordinated by a cysteine thiolate sulfur atom. They require a source of electrons (an iron-sulfur protein or a simple cytochrome), and the overall reaction they catalyze can be written as follows:

$$RH + O_2 + 2H^+ + 2e \rightarrow ROH + H_2O$$

The inorganic chemistry has been explored to some extent using model compounds, and this provides some of the best evidence for the proposed coordination by cysteine sulfur.[15]

31-5. Iron-Sulfur Proteins[16]

The iron-sulfur proteins are compounds of relatively low molecular weight containing redox centers composed of atoms of both elements with structures of the types shown in Fig. 31-9. The one-iron types are called *rubredoxins,* and the two-iron and the four-iron types are *ferredoxins,* although certain four-iron types are also called high-potential iron-sulfur proteins, or *Hipip's* for short. The reason for this distinction is explained shortly.

Rubredoxins contain one iron atom coordinated tetrahedrally by four mercapto sulfur atoms from the side chains of cysteine amino acid residues of the protein chain. Figure 31-10 shows in more detail how the conformation of the protein chain of one rubredoxin positions the four sulfur atoms to coordinate the iron atom. Rubredoxins are small proteins, consisting of 50 to 60 amino acid residues and having molecular weights of *ca.* 7000.

A rubredoxin is a one-electron donor or acceptor. Its two forms are designated rubredoxin$_{red}$, which contains high-spin Fe^{II} and is essentially colorless, and rubredoxin$_{ox}$, which contains high-spin Fe^{III} and is red. The tetrahedral coordination of Fe^{III} has been directly observed by X-ray crystallography; it is believed that this coordination geometry persists after reduction.

There has been some uncertainty about the existence of severe distortion of the FeS$_4$ tetrahedron in oxidized rubredoxin, but EXAFS results[17] have shown that there is very little distortion, all Fe—S distances being within the range 2.30 ± 0.04 Å. Moreover, model compounds 31-VII have been prepared and studied, and these have essentially regular structures.[18]

(31-VII)

Ferredoxins. The larger four-iron ferredoxins are more thoroughly known than

15 R. H. Holm, J. A. Ibers, *et al., J. Am. Chem. Soc.,* 1976, **98,** 2414.
16 W. Lovenberg, Ed., *Iron-Sulfur Proteins,* Academic Press, Vol. 1, 1973; Vol. 2, 1974; Vol. 3, 1976.
17 P. Eisenberger and B. M. Kinkaid, *Science,* 1978, **200,** 1441 (a review).
18 J. A. Ibers, R. H. Holm, *et al., J. Am. Chem. Soc.,* 1977, **99,** 84.

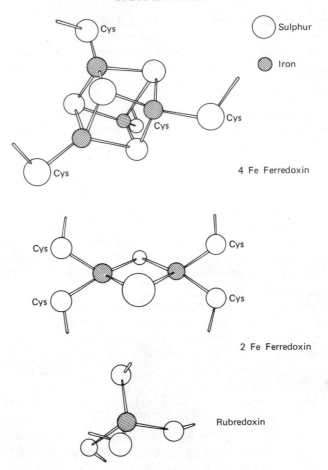

Fig. 31-9. Structures of the 4Fe–4S centers and 2Fe–2S centers in ferredoxins and the 1Fe center in rubredoxin.

the two-iron ones. This is due in considerable measure to the laboratory synthesis of iron-sulfide cluster compounds (see p. 763) that are identical to those found in Nature. In fact, the clusters in the ferredoxins can be removed intact to leave the apoproteins, and then the natural or the equivalent synthetic cores can be reintroduced into the apoproteins.[19] In this way the identity of the clusters in the ferredoxins can be unambiguously determined. Most ferredoxins evidently contain either four-iron or two-iron clusters, but there is now at least one that appears to contain both.[20]

The redox behavior of the two-iron and four-iron clusters is more complex than that of the rubredoxins, as would be expected . The four-iron clusters, both in the ferredoxins and in the isolated state without the surrounding protein, have been

[19]   D. M. Kurtz, G. B. Wong, and R. H. Holm, *J. Am. Chem. Soc.,* 1978, **100,** 6777.
[20]   C. D. Stout, *Nature,* 1979, **279,** 83.

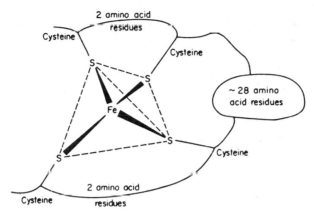

Fig. 31-10. The tetrahedral iron complex found in rubredoxin.

extensively investigated. Solvent and medium effects are important in determining the absolute values of the potentials,[21] but the relative stages of oxidation follow a common pattern in all cases. The essential features of this pattern are summarized in Fig. 31-11: as indicated, the behavior of the high-potential iron proteins (HP) is also understandable in terms of the same three-stage oxidation pattern.

A ferredoxin in its normal oxidized state Fd_{ox} has the same electronic population as HP in its normal reduced state HP_{red}. The ferredoxin corresponding to HP_{ox} is in a superoxidized condition Fd_{S-ox} and has no normal biological role as far as is known. Similarly, HP_{S-red}, superreduced HP, which does not occur in biological systems, is analogous in electronic configuration to normal reduced ferredoxin Fd_{red}. There are considerable esr, structural[22] and nmr data[23] to support this redox scheme.

The potential for the $Fd_{red} \rightleftharpoons Fd_{ox}$ step is typically about -400 mV; that for

$$
\begin{array}{lll}
[Fe_4S_4(SR)_4]^{1-} & HP_{ox} & (Fd_{S-ox}) \\
S = 1/2 & & \\
-e^- \Big\Updownarrow +e^- & & \\
[Fe_4S_4(SR)_4]^{2-} & HP_{red} & Fd_{ox} \\
-e^- \Big\Updownarrow +e^- & & \\
[Fe_4S_4(SR)_4]^{3-} & (HP_{S-red}) & Fd_{red} \\
S = 1/2 & &
\end{array}
$$

Fig. 31-11. The three principal stages of oxidation of the 4-iron clusters in ferredoxin (Fd) and high-potential iron proteins (HP).

[21]   R. H. Holm et al., J. Am. Chem. Soc., 1977, 99, 2549; H. L. Carrell, J. P. Glusker, R. Job, and T. C. Bruice, J. Am. Chem. Soc., 1977, 99, 3683.
[22]   R. H. Holm, J. A. Ibers, et al., J. Am. Chem. Soc., 1978, 100, 5322.
[23]   J. G. Reynolds, E. J. Laskowski, and R. H. Holm, J. Am. Chem. Soc., 1978, 100, 5315.

$HP_{red} \rightleftharpoons HP_{ox}$ is typically about +350 mV. Since different levels of oxidation of the cluster are involved in the two cases, the difference in these potentials and the direction of the difference are understandable without difficulty. What is not at all obvious, however, is why the same type of $[Fe_4S_4(SR)_4]$ species prefers the 1−/2− pair of states in one protein (HP) but the 2−/3− pair of states in another (Fd). It is expected that the explanation of this difference must lie in considerable differences in the polar surroundings of the cluster, especially in the availability of hydrogen bonds, in the two types of protein. Presumably more protonic groups are available in the ferridoxins, thus stabilizing the more intrinsically negative clusters.

31-6. Hemerythrins[24]

In a great variety of marine worms the oxygen-carrying molecules are iron-bearing proteins, but they do not contain porphyrins. They are all presumably similar in chemical nature and are called hemerythrins. The one that has been most widely studied is derived from the saltwater worm (sipunculid), *Goldfingia gouldii*. It has a molecular weight of 108,000 but consists of eight identical subunits, each containing two iron atoms. The oxygen affinity of this particular hemerythrin is not pH sensitive, though others are. Typically hemerythrins bind oxygen 5 to 10 times more strongly then hemoglobin or myoglobin. Each subunit can bind one oxygen molecule; thus the ratio of Fe to O_2 is 2:1 rather than 1:1 as in hemoglobin. The oxygenated substance is violet-pink and the deoxygenated material is colorless. The subunits form a roughly square-antiprismatic array in the octamer, and each consists of 113 amino acid residues arranged in four nearly parallel helical segments 30 to 40 Å long. The iron atoms are held within these four segments.

Medium-resolution X-ray diffraction data are available for several forms of hemerythrin and have given preliminary indications of the environment of the iron atoms. The two iron atoms are close together. In the case of aquamethemerythrin (which contains two Fe^{III}), they are less than 3.4 Å apart and appear to occupy the centers of two octahedra that share a face.[25] It has been suggested tentatively that the three vertices of the shared face consist of an oxygen atom from an aspartate carboxyl group, an amide nitrogen atom of a glutamine residue, and a water molecule. It has been further suggested that the site of the water molecule is the binding site for O_2. This picture is not particularly attractive to the coordination chemist, since H_2O is rarely found as a bridging ligand (though OH^- is), and a site that accommodates H_2O would not seem likely to accommodate O_2 without considerable rearrangement.

In the other case the structural results are based on even fewer data for an azido complex.[26] Here the iron atoms appear to be about 3.45 Å apart and perhaps to

24   I. M. Klotz, G. L. Klippenstein, and W. A. Hendrickson, *Science*, 1976, **192**, 335; D. M. Kurtz, Jr., D. F. Shriver, and I. M. Klotz, *Coord. Chem. Rev.*, 1977, **24**, 145.
25   R. E. Stenkamp, L. C. Sieker, and L. H. Jensen, *Proc. Nat. Acad. Sci. (U.S.)*, 1976, **73**, 349.
26   W. A. Hendrickson, G. L. Klippenstein, and K. B. Ward, *Proc. Nat. Acad. Sci. (U.S.)*, 1975, **72**, 2160.

be bridged by an oxygen atom. Clearly the present structural data are inadequate but improvement may be expected in due course.

Even without detailed structural knowledge of the binding site, a good deal has been learned about the mode of binding of O_2 in hemerythrins by spectroscopic and magnetic studies. The iron atoms in deoxyhemerythrin are definitely Fe^{II}, each having four unpaired electrons. However there is strong antiferromagnetic coupling between them, presumably through a bridging O^{2-} ion. In oxyhemerythrin the reversibly bound dioxygen appears to be present as peroxide (O_2^{2-}), as compared to the coordinated superoxide (O_2^-) situation for hemoglobin. A strong indicator of this difference is the difference between the O—O and Fe—O stretching frequencies in the two cases. For oxyhemoglobin we have $\nu_{O-O} - 1107$ cm^{-1} and $\nu_{Fe-O} = 567$ cm^{-1}, whereas for oxyhemerythrin the corresponding frequencies are 844 cm^{-1} and 504 cm^{-1}. This indicates a much higher degree of electron transfer to the antibonding π^* orbitals of dioxygen in the latter case, which is possible because there are two cooperating iron atoms to do the job. However, much remains to be learned about the hemerythrin system. There is, for example, no clear precedent for the suggested reversible reductive binding of O_2 as peroxide; indeed in many cases this type of binding has been found to be entirely irreversible (i.e., to constitute oxidation rather than oxygenation).

THE BIOCHEMISTRY OF OTHER METALS

31-7. Zinc

Zinc is now recognized to be essential to all forms of life, and a large number of diseases and congenital disorders have been traced to zinc deficiency. In the adult human body there are 2 to 3 g of zinc, as compared to 4 to 6 g of iron and only about 250 mg of copper. Biochemists were somewhat slow to appreciate the presence and importance of zinc because it is colorless, nonmagnetic, and generally, not as easily noticed as iron and copper.

In 1940 carbonic anhydrase was shown to be a zinc enzyme, and in 1955 carboxypeptidase became the second recognized zinc enzyme. Since then more than 80 other zinc enzymes have been reported, and functionally they are of many kinds, including alcohol dehydrogenases, aldolases, peptidases, carboxypeptidases, proteases, phosphatases, transphosphorylases, a transcarbamylase, and DNA- and RNA-polymerases. We discuss here only two of the zinc enzymes, namely, carboxypeptidase and carbonic anhydrase, since our structural and kinetic knowledge for these is extensive and throws considerable light on the role played by the zinc ion in these chemical reactions.

Carboxypeptidase A. This type of enzyme, with a molecular weight of about 34,600, catalyzes the hydrolysis of the terminal peptide bond at the carboxyl end of proteins and other peptide chains. It is also effective toward correspondingly placed ester linkages. There is a marked preference for peptide bonds in which the

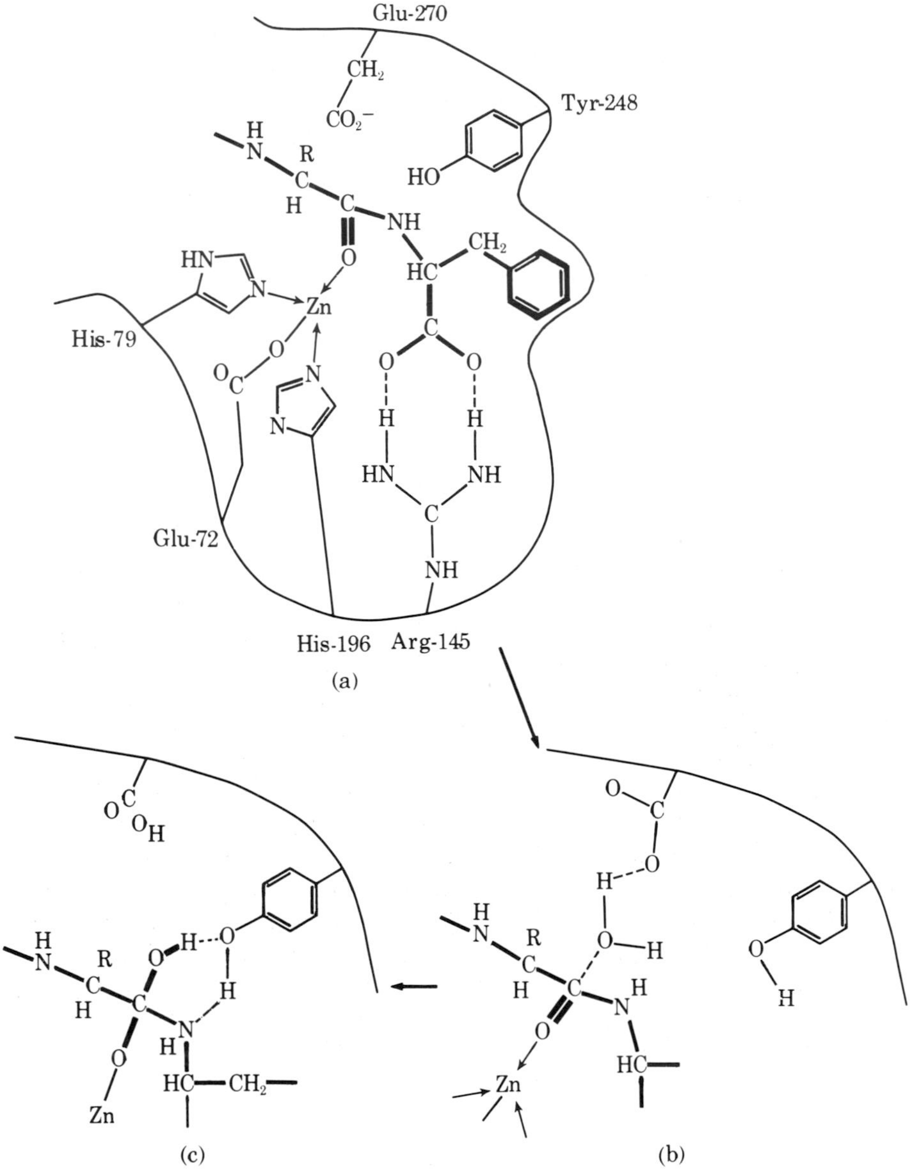

Fig. 31-12. Currently preferred mechanism by which carboxypeptidase-A catalyzes peptide hydrolysis. (Based on Ref. 27.)

side chain of the terminal residue is aromatic or a branched aliphatic chain. Carboxypeptidase A's (CPA's) have been found in the pancreas of many mammals including man, but most studies have been carried out on the bovine enzyme.

The overall structure of the bovine CPA has been determined by X-ray crystallography. The zinc ion is coordinated by two histidine nitrogen atoms and an

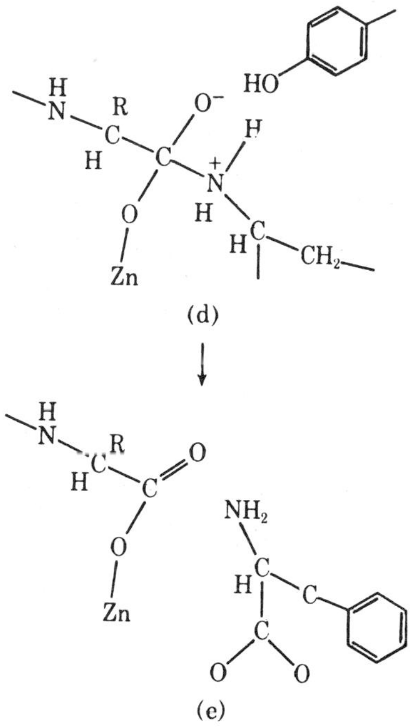

(d)

↓

(e)

Fig. 31-12. *Con't.*

oxygen atom from the carboxyl side chain of a glutamate residue, with a water molecule completing a roughly tetrahedral arrangement. The conformation of the molecule is such as to create a groove in which the zinc ion is found and nearby a pocket that contains no binding groups but is of the right size to accommodate the large nonpolar side chains mentioned above.

By combining the structural information with indications from a variety of chemical studies, it has been possible to devise several hypotheses about the mechanism of catalysis. Up to a certain point there is general agreement. In Fig. 31-12a we see the generally accepted mode of binding of the substrate, which is common to all mechanisms. The key points are as follows:

1. The carbonyl oxygen atom of the peptide link that is to undergo scission is coordinated to the zinc ion in place of the water molecule present in the resting enzyme. This polarizes the CO group, to make the carbon atom more positive, hence more susceptible to nucleophilic attack.

2. The terminal carboxyl group is bound to the guanidyl group of arginine-145 through two hydrogen bonds. This positions the entire terminal residue of the substrate to direct the large nonpolar side chain that is present in the preferred substrates into the hydrophobic pocket. It also places the amide nitrogen atom of the peptide link in proximity to the OH group of tyrosine-248.

There have been two main schools of thought regarding what happens next. According to one, the carboxyl group of glutamate-270 attacks the peptide carbon atom, displacing the peptide nitrogen atom and forming an acid anhydride inter-mediate, which is subsequently hydrolyzed. There is now considerable evidence against this,[27] and the other main type of mechanism (Fig. 31-12b–e) has come to be regarded as essentially correct. The carboxyl anion of glutamate-270 acts to position and deliver a water molecule to the carbonyl carbon atom, as in b. This leads to the situation represented in c, where the Glu-270 carboxyl group has been protonated, and the OH group now attached to the peptide carbonyl carbon atom is hydrogen bonded to the OH group of Tyr-248, which in turn is hydrogen-bonded to the peptide nitrogen atom. Two protons then shift to give the arrangement shown in d, after which there is prompt C—N bond scission to give products as shown in e. These are then released from the enzyme, and a new cycle can begin.

Carbonic Anhydrase. As discussed earlier (p. 366) the following reaction is

$$CO_2(aq) + H_2O = H_2CO_3$$

slow; yet many physiological processes require rapid equilibration of CO_2 with HCO_3^- and H_2CO_3 at physiological pH, which is about 7.0. Therefore enzymes called carbonic anhydrases are found in nearly all phyla, and all appear to contain zinc.

Human carbonic anhydrase has a molecular weight of about 30,000 and occurs in two similar but not identical forms. The molecular structures of both are known from X-ray crystallography. A schematic indication of one of them (the other is very similar) appears in Fig. 31-13a. The zinc ion lies in a deep pocket, and is coordinated by three histidine imidazole nitrogen atoms, as shown in Fig. 31-13b, which also gives an indication of the mode of CO_2 binding. The enzyme enhances the rate of the hydration reaction in either direction by a factor of 10^6 or more.

The enzyme-catalyzed reaction rates are pH dependent in a way that indicates the existence of a group in the enzyme with a pK_a of about 7 that must be depro-tonated to give that form of the enzyme E, which is required for hydration of CO_2. Conversely, an acid form EH^+ is required for the reverse reaction. It is necessary to assume that the substrate for dehydration is HCO_3^-, since the pH dependence of dehydration is the inverse of that for hydration. A reaction involving only neutral species, $CO_2 + H_2O$ to give H_2CO_3, would not give such a relationship. A four-step mechanism that appears to satisfy all available evidence is shown in eq. 31-2.

$$E + CO_2 \rightleftharpoons E \cdots CO_2 \underset{H_2O}{\overset{H_2O}{\rightleftharpoons}} EH^+ \cdots HCO_3^- \rightleftharpoons$$

$$EH^+ + HCO_3^- \rightleftharpoons E + H^+ + HCO_3^- \quad (31\text{-}2)$$

All available data now show that the activity-linked group in the enzyme with a pK_a of 7 is an H_2O molecule coordinated to the zinc ion. In short, the initial complex of CO_2 with the active site is a $ZnOH\cdots CO_2$ complex, where the CO_2 is doubtless positioned by several other interactions so that the oxygen atom of ZnOH approaches the carbon atom of CO_2 as indicated in Fig. 31-13b. The nature of the pocket in which the zinc ion is found lowers the otherwise much higher pK_a of the

27   R. Breslow and D. Wernick, *J. Am. Chem. Soc.*, 1976, **98**, 259.

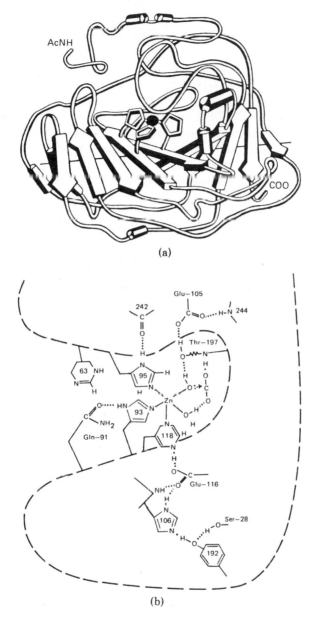

(a)

(b)

Fig. 31-13. Structural aspects of carbonic anhydrase. (a) Schematic indication of the overall molecular conformation. (b) Structure at the active site of the working enzyme (postulated).

$ZnOH_2$ group to about 7, while at the same time leaving the coordinated OH^- sufficiently nucleophilic to attack the carbon atom of the CO_2. The immediate product of this attack is the HCO_3^- ion, and this, in turn, must be the substrate in the reverse reaction.

In effect, the enzyme makes the fast reaction $CO_2 + OH^- = HCO_3^-$ the major pathway of hydration at a pH of 7, whereas normally the slow reaction of CO_2 with H_2O would predominate at a pH of 7 and the reaction with OH^- would not become dominant until pH 10 or greater.

It will be noted in Fig. 31-13 that a fairly complex hydrogen-bonded structure is invoked around the active site and, in particular, a five-coordinate Zn^{2+} ion is believed to be involved. There is a part of the mechanism that is effectively an associative interchange at the Zn^{2+} ion, with the incoming ligand taking a position adjacent to the departing one and temporarily increasing the coordination number to 5. This is presumably a more facile process than a simple dissociative transfer of OH^- to CO_2 with later entry of H_2O.

Metallothioneins. This relatively new class of zinc proteins certainly deserves mention in an inorganic text because of the enormous metal contents involved. These substances may contain zinc, cadmium, mercury, or a mixture (as well as certain other metals such as Fe and Cu), and they have been found in equine and human renal cortex and livers. Similar proteins have also been found in lower vertebrates. These proteins contain 6 to 11% metal by weight. The source of their tremendous capacity to bind metal ions is the large number of cysteines they contain. Renal equine metallothionein has been shown[28] to have a single peptide chain of 61 amino acid residues, of which 20 are cysteine and seven are lysine (also a good metal ion binder). The cysteine residues are clustered into seven groups, separated by stretches of three or more other residues. The total number of metal atoms, typically a mixture of Zn and Cd, is also seven. Thus it appears that the protein has been designed to do what it does, but what purpose this may serve physiologically is not firmly established. Control of metal metabolism or detoxification have been suggested.

31-8. Copper

Copper is the third most abundant metallic element in the body, following iron and zinc. All other animals, as well as man, possesses homeostatic mechanisms for absorption, transport, utilization and excretion of copper, and in man at least two lethal, hereditary disorders of copper metabolism, Wilson's disease and Menkes' kinky hair syndrome, are known.

A considerable number of metalloproteins containing copper have been identified. We have already mentioned the presence of copper in cytochrome oxidases. This section briefly discusses several more of the better-known ones, beginning with some oxidases, which are the type of enzyme in which copper is most often found.

Oxidases. There are a number of these, but in no case has a 3-dimensional structure been determined, probably because they are all rather large molecules and not easily crystallized. There are four, namely, *cytochrome oxidase, laccase, ascorbate oxidase* and *ceruloplasmin,* which have the common feature of promoting the conversion of O_2 to H_2O in such a way that *both* oxygen atoms of O_2 are reduced to H_2O. They all contain four or more copper atoms, presumably because the O_2 to $2H_2O$ reduction is a four-electron process.

---

[28]   J. H. R. Kägi *et al., Proc. Nat. Acad. Sci. (U.S.),* 1976, **73,** 3413.

Laccase has a molecular weight which varies with the source. That from *Polyporous versicolor* (a fungus) has been extensively studied; it has a molecular weight of about 60,000 and contains four copper atoms. Laccases catalyze the oxidation of a number of diamines and diphenols, including catechol, hydroquinone, and paraphenylenediamine. During the reaction, the copper atoms are reduced to Cu^I and reoxidized to Cu^{II} by O_2. The four copper atoms in the laccase of *P. versicolor* are not all equivalent: in the fully oxidized enzyme two are apparently isolated and are detectable by esr, but the other two appear to be strongly antiferromagnetically coupled.

Ceruloplasmin is found in the blood plasma of mammals, birds, reptiles, and amphibians. In all cases an intense blue color is present because of the copper, but compositional details vary. Human ceruloplasmin has a molecular weight of about 150,000 and appears to consist of four peptide chains, two with molecular weights of about 59,000 and two with weights of about 16,000. The number of copper atoms present is 7 ± 1. In many of its characteristics and in its kinetic behavior, ceruloplasmin is similar to laccase. Ceruloplasmin also contains nine or ten heterosaccharide chains per molecule.

Ascorbate oxidase is widely distributed in plants and in some microorganisms. Molecular weights are typically around 140,000, and there are eight copper atoms. It catalyzes the oxidation of ascorbic acid (vitamin C) to dehydroascorbic acid.

Cytochrome oxidase is the only one of the oxidases to contain iron (as heme) as well as copper, and the two occur in a $1:1$ atom ratio; the total number of metal atoms is not established. As noted earlier this enzyme catalyzes the final step in the cytochrome redox chain, where O_2 is reduced to $2H_2O$.

Galactose Oxidase. This enzyme, which catalyzes reaction 31-3, is a relatively simple one (MW = 68,000) consisting of a single polypeptide chain and one copper

$$\text{(31-3)}$$

atom. There is no direct evidence concerning the structure of the active site, but chemical and kinetic studies have led to the suggestion[29] that Cu^{III} plays a key role in the mechanism. Specifically, it has been suggested that a key cycle in the catalytic process is that shown in eq. 31-4, where E—Cu represents the enzyme and its copper atom.

$$\text{(31-4)}$$

The most notable aspect of this proposed mechanism is the occurrence of Cu^{III}. Although this is not a common oxidation state for copper, a number of examples

[29] G. A. Hamilton, *et al.*, *J. Am. Chem. Soc.*, 1978, **100**, 1899.

have been found in its ordinary coordination chemistry (p. 818), and the mechanism is not an unreasonable one. The participation of Cu^{III} in other enzymic processes is a possibility meriting investigation.

The One-Copper Blue Proteins. These are found in plants, and their function seems to be as one-electron transfer agents. They have strongly attracted the interest of inorganic chemists because of their striking spectral characteristics. Since there is only one copper atom, the problem of explaining these characteristics in terms of the composition and geometry of the coordination sphere was thought to be a tractable one based on classical knowledge and model systems. This, however, did not prove to be the case: all "predictions" of the copper(II) environment, such as five-coordination or the numbers and types of S and N ligands, appear to be incorrect, and no entirely satisfactory model system has yet been provided. The only truly reliable insight so far has come from the actual crystallographic structure determination for one of the plastocyanins, as discussed below.

Three of the best known substances of the type under discussion and their molecular weights are the following:

> Azurin from *Pseudomonas aeruginosa* (16,600)
> Plastocyanin from *Populus nigra* (10,500)
> Stellacyanin from *Rhus vernicifera* (16,800)

These and others are all small molecules, as proteins go, and it is likely that a number of crystal structure will eventually become available. That for the particular plastocyanin mentioned already is known (at medium resolution), and work on the azurin structure is in progress.

The electronic spectra of the proteins above, as well as other one-copper blue proteins, are characterized by three electronic absorption bands at approximately 450, 600, and 750 nm; for the 600 nm bands the extinction coefficients are much higher than those for the ligand field bands that ordinarily make up the entire visible spectra of Cu^{II} complexes. The proteins also have very unusual esr spectra, unusually low $A_\parallel$ values, and it is the totality of these spectral characteristics that no model system has yet mimicked, though some have come close.[30] There are strong indications from model systems and general considerations that the intense absorption band at 600 nm is essentially a charge-transfer band, arising in a sulfur-copper σ bond. The ligand concerned must, presumably, be the $—CH_2S^-$ group of a cysteine residue.[31] The presence of at least one cysteine and at least one histidine residue in the coordination sphere of the copper ion is also supported by Raman[32] and nmr[33] evidence.

In the case of the plastocyanin from poplar leaves, a crystallographic structure determination[34] at 2.7 Å resolution has shown that the copper(II) is coordinated by two imidazole nitrogen atoms from histidine residues, the thiol group of a cys-

30   J. S. Thomson, T. J. Marks, and J. A. Ibers, *Proc. Nat. Acad. Sci. (U.S.)*, 1977, **74**, 3114.
31   Y. Suguira and Y. Hirayama, *J. Am. Chem. Soc.*, 1977, **99**, 1581.
32   W. H. Woodruff *et al.*, *J. Am. Chem. Soc.*, 1978, **100**, 5939.
33   E. L. Ulrich and J. L. Markley, *Coord. Chem. Rev.*, 1978, **27**, 109.
34   H. C. Freeman *et al.*, *Nature*, 1978, **272**, 319.

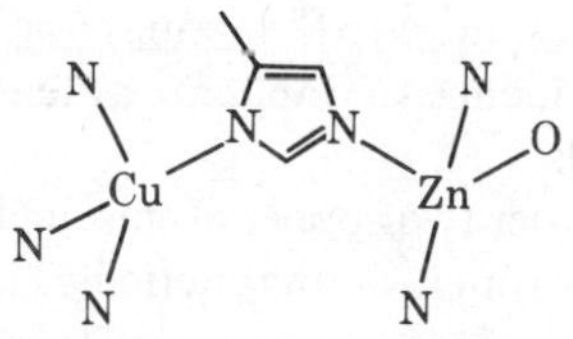

Fig. 31-14. The bimetallic active site of superoxide dismutase.

teine residue, and the sulfur atom of a methionine. The structure is not nearly precise enough to provide a detailed geometry of the coordination sphere, but it appears that the arrangement of the four ligand atoms is a very distorted one, definitely not planar, but also far from tetrahedral. The finding that a methionine sulfur atom is one of the ligands leaves some uncertainty about the ligand set in the stellacyanins, since the peptide chains in these do not contain any methionine residue.

Superoxide Dismutase, A Zn-Cu Metalloenzyme. This enzyme, also sometimes called erythrocuprein because it was first found in erythrocytes (red blood cells), is not fully understood, but one of its functions is to catalyze the prompt disproportionation of superoxide ion O_2^- (eq. 31-5), which, because of its great reactivity, could do enormous damage if allowed to remain in cells. The catalysis is so efficient that the reaction occurs at a rate that is practically diffusion controlled[35a].

$$2O_2^- + 2H^+ \rightarrow H_2O_2 + O_2 \qquad\qquad (31\text{-}5)$$

The three-dimensional structure of the enzyme, which has recently been determined by X-ray crystallography,[35b] shows that the Zn^{2+} and Cu^{2+} ions are close together and bridged by the imidazole ring of a histidine residue (Fig. 31-14): note that the weakly acidic proton of the pyrrole nitrogen atom has been displaced. It is not yet known how this unusual bimetallic active site functions chemically. It seems likely that the O_2^- radical would reduce the Cu^{2+} ion, but what function is served by the zinc ion is quite obscure. It has been suggested that once the copper has been reduced, the imidazolate group dissociates from the Cu^I, is immediately protonated, and transfers this proton to the coordinated peroxide ion, but this is pure speculation.

Hemocyanins.[36]   These are the respiratory (oxygen-carrying) proteins in the blood of certain (but not all) animals in the phyla Mollusca and Arthropoda (e.g., squids, cuttlefish, some snails, including the edible ones, lobsters, and crabs). They are large molecules with molecular weights in the range of 4×10^5 to 9×10^6, and they consist of subunits.

Although there is a good deal of information available on some of their properties, such as oxygen-binding equilibria and spectral characteristics, there is little direct information on the binding sites of the copper atoms. It is known that the binding

[35a] A. M. Michelson, J. M. McCord, and I. Fridovich, Eds., *Superoxide and Superoxide Dismutases*, Academic Press, 1977.

[35b] J. S. Richardson, D. C. Richardson, *et al.*, *Proc. Nat. Acad. Sci. (U.S.)*, 1975, **72**, 1349.

[36]   F. Ghiretti, Ed., *Physiology and Biochemistry of Hemocyanins*, Academic Press, 1968; N. M. Senozan, *J. Chem. Educ.*, 1976, **53**, 684.

is direct (not via heme groups, despite the name) and strong ($k_{diss} \approx 10^{-20}$), and there is chemical and epr evidence to implicate at least two and perhaps four imidazole groups as the ligands.[37,38]

One O_2 molecule is bound per two copper atoms, which are in the oxidation state II. Nevertheless, oxyhemocyanin is diamagnetic because of strong antiferromagnetic coupling (thought to be 625 cm^{-1} or more[39]) between pairs of Cu^{II} ions.

A resonance Raman study of oxyhemocyanins from a mollusc and from an arthropod has shown[40] that despite some differences in details in the two binding sites, the mode of attachment of O_2 to the two metal atoms is essentially similar. In each case there are ν_{O-O} frequencies (744, 749 cm^{-1}) in the range typical for peroxo groups, and there are bands that can be assigned to Cu–N stretching, in keeping with the view that imidazole groups of histidine residues are among those binding the copper atoms to the protein.

The strong electronic absorption band at 345 nm ($\epsilon \approx 10^5$) has been assigned to an excitation involving simultaneous $d \rightarrow d$ transitions on both Cu^{II} ions in the strongly coupled pair.[41]

31-9.   Cobalt

Cobalt has only one important biochemical role, as far as we know at present, namely, in vitamin B$_{12}$. This vitamin is a cofactor for a number of enzymes[42a], virtually all of which catalyze a reaction of the type 31-6.

$$
\begin{array}{ccc}
\text{R} \;\; \text{H} & & \text{H} \;\; \text{R} \\
| \;\;\; | & & | \;\;\; | \\
-\text{C}-\text{C}- & \longrightarrow & -\text{C}-\text{C}- \\
| \;\;\; | & & | \;\;\; | \\
\text{H} \;\; \text{H} & & \text{H} \;\; \text{H}
\end{array}
\tag{31-6}
$$

The adult human body contains 2 to 5 mg of vitamin B$_{12}$ and its derivatives, mainly in the liver. A feature of great significance is that in nearly every case the foregoing type of reaction occurs so that there is no exchange between the migrating hydrogen atom and those in the solvent water.

The structure of vitamin B$_{12}$, which is known conclusively from X-ray crystallography as well as chemical studies, is shown in Fig. 31-15. The bond to the 5′ carbon atom of the deoxyribose moiety is labile, and the ligand attached to cobalt at that coordination site is variable. The vitamin is quite commonly isolated in a form called cyanocobalamin, in which CN occupies this position. The similarity to the porphyrin ring systems of the tetradenate ring, called a *corrin* ring, should be obvious, but there is a key difference, namely, that one of the bridges between pyrrole rings is a direct ring-to-ring bond instead of an HC$\lessapprox$ or CH$_3$C unit.

37   B. Salvato, A. Ghiretti-Magoldi, and F. Ghiretti, *Biochemistry*, 1974, **13**, 4778.
38   A. J. M. Schoot-Uiterkamp *et al*, *Biochem. Biophys. Acta*, 1974, **372**, 407.
39   E. I. Solomon *et al.*, *J. Am. Chem. Soc.*, 1976, **98**, 1029.
40   T. B. Freedman, J. S. Loehr, and T. M. Loehr, *J. Am. Chem. Soc.*, 1976, **98**, 2809.
41   J. A. Larrabee *et al.*, *J. Am. Chem. Soc.*, 1977, **99**, 1979.
42a   B. M. Babior, Ed., *Cobalamin*, Wiley, 1975.

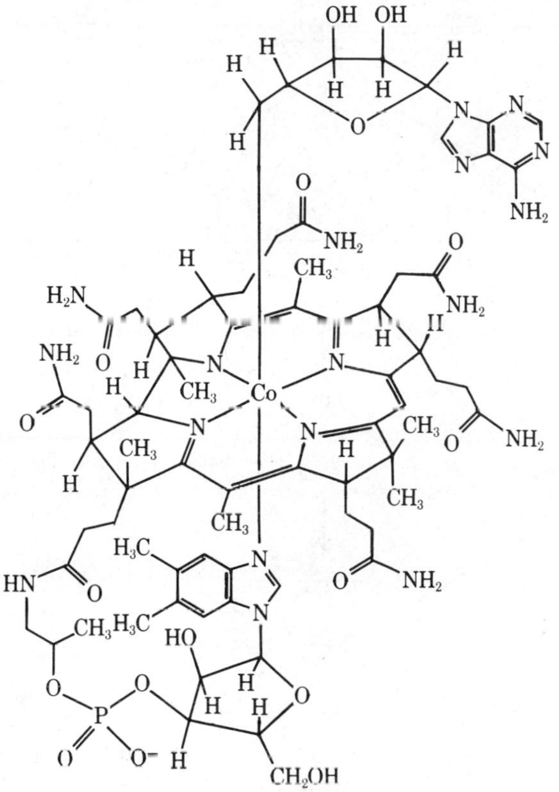

Fig. 31-15. The structure of Vitamin B_{12} coenzyme. Note the Co—C bond to the 5' carbon atom of the adenosyl moiety.

The mechanistic role of vitamin B_{12} is now understood in some of its broad features, but many aspects are still obscure. As a first approximation to the mechanism, the following sequence of steps may be considered, where SH and PH stand for substrate and product, respectively, and —CH_2—R represents the adenosyl group.

$$\text{SH} + \underset{\underset{\text{Co}}{|}}{CH_2-R} \longrightarrow \underset{\underset{\text{Co}}{|}}{S} + CH_3-R \longrightarrow \underset{\underset{\text{Co}}{|}}{P} + CH_3-R \longrightarrow \text{PH} + \underset{\underset{\text{Co}}{|}}{CH_2-R}$$

The substrate transfers H to CH_2—R and forms a Co—C bond while the adenosyl CH_2—R group becomes CH_3—R, but remains in the vicinity. The enzyme then acts on the bound S to cause a 1,2-shift of its R group (31-6) so that we now have P—Co. The first step is then reversed, liberating PH and regenerating the catalyst. According to this mechanism, the H atom of SH has only one chance in 3 of returning to PH, and detailed isotopic labeling studies confirm this, as well as demonstrating the existence of kinetic isotope effects for D and T transfer that are in accord with this mechanism.

That the foregoing mechanism is incomplete is demonstrated by the results of esr studies that show that radicals play an essential role in the mechanism. To take

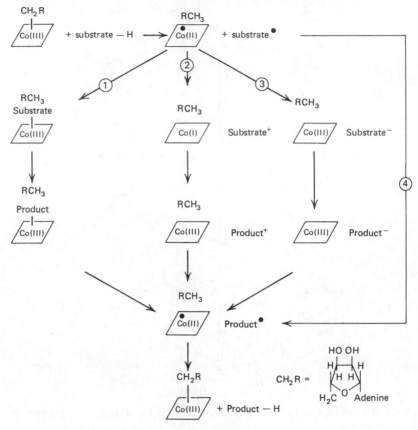

Fig. 31-16. Possible pathways for vitamin B_{12}–coenzyme mechanisms.

account of this, the following modified sequence has been proposed, where curly brackets indicate complexes bound together by the enzyme:

$$
\begin{array}{ccc}
\text{SH} + \underset{\text{Co(III)}}{\text{CH}_2\text{–R}} & \longrightarrow & \underbrace{\text{SH} \quad {}^\bullet\text{CH}_2\text{–R}}_{{}^\bullet\text{Co(II)}} & \longrightarrow & \underbrace{\text{S}^\bullet \quad \text{CH}_3\text{–R}}_{{}^\bullet\text{Co(II)}} \\
\end{array}
$$

$$
\longrightarrow \underbrace{\text{P}^\bullet \quad \text{CH}_3\text{—R}}_{{}^\bullet\text{Co(II)}} \longrightarrow \underbrace{\text{PH} \quad {}^\bullet\text{CH}_2\text{—R}}_{{}^\bullet\text{Co(III)}} \longrightarrow \underset{\text{Co(III)}}{\text{PH} + \text{CH}_2\text{—R}}
$$

As indicated, this scheme implies the formal conversion of Co^{III} to Co^{II} in the various enzyme-bound intermediates. Even this amended scheme omits to show how the substrate radical $S^\bullet$ is converted to the product radical $R^\bullet$. This, of course, is done by the enzyme, and it is possible that different enzymes may use different paths. Figure 31-16 shows several possible paths; current research focuses on seeking evidence to discriminate among these.

One other aspect of cobalt "biochemistry" that should be mentioned is that Co^{2+} can be substituted for Zn^{2+} in a number of enzymes without gross change in activity; carboxypeptidase and carbonic anhydrase are good examples. So far as is known this does not happen in Nature, but it serves as a useful tool for the enzymologist, since the Co^{2+} ion with its spectroscopic and magnetic properties is a useful probe of the enzyme active site.

31-10. Molybdenum[42b]

Molybdenum is the only transition metal below the first series that is known to be essential in living systems. It is an essential trace element for many plants, probably entering them from the soil as molybdate ion. It also has a role in higher animals and humans.

Molybdenum is best known for its essential role in various aspects of biological nitrogen chemistry. Table 31-2 lists several of the enzymes that contain molybdenum. They are all very large proteins, and no structures are yet known. The molybdenum centers in many if not all of these may be quite similar but the nitrogenases probably differ. However the only one for which there is definite information is the nitrogenase from the anaerobic bacteria *Clostridium pasteurianum*.

Nitrogenases. This type of enzyme is found in blue-green algae, in free-living bacteria, and in bacteria that have a symbiotic relationship with leguminous plants, converting N_2 to NH_3 in the soil. The general composition of the *C. pasteurianum* nitrogenase appears to be typical. It consists of two distinct oxygen-sensitive proteins, both of which are essential for activity. The smaller part (MW *ca.* 56,000) contains four Fe and four sulfide-type sulfur atoms, probably in the form of an Fe_4S_4 ferredoxin-type cluster. The larger protein, which consists of four subunits and has an aggregate molecular weight of *ca.* 220,000, contains two Mo atoms, 24 Fe atoms and 24 S^{2-}. There is evidence that the large protein is responsible for binding the N_2 substrate, and it is interesting that acetylene is also efficiently bound and reduced. It is also known that the enzyme will reduce hydrazine to ammonia, thus suggesting that H_2NNH_2 may be an intermediate in N_2 reduction.

Considerable progress has recently been made in defining the chemical state of the molybdenum in nitrogenase. It has been possible to isolate from the Fe–Mo (large) protein a much smaller molecule containing molybdenum, iron, and sulfur in a $1:8:6$ atomic ratio; this substance is called the FeMo cofactor (FeMo-co).[43] Studies of the EXAFS of both the Fe–Mo protein and FeMo-co have shown[44] that the molybdenum atoms occur in Mo—Fe—S clusters, with sulfur atoms as the nearest neighbors to Mo, iron atoms forming a set of next nearest neighbors, and more sulfur atoms in a third shell. Based on these results, cubic $MoFe_3S_4$ structures, similar to the $Fe_4S_4(SR)_4$ structures of the four-iron ferridoxins, become highly likely, and in fact several workers have prepared model systems of this kind and

[42b] E. I. Stiefel, *Prog. Inorg. Chem.*, 1977, **22**, 1; *Proc. Nat. Acad. Sci.* (*U.S.*), 1973, **70**, 988.
[43] V. K. Shah and W. J. Brill, *Proc. Nat. Acad. Sci.* (*U.S.*), 1977, **74**, 3249.
[44] K. O. Hodgson *et al.*, *J. Am. Chem. Soc.*, 1978, **100**, 3398, 3814.

TABLE 31-2
Some Enzymes That Contain Molybdenum

Enzyme	Source	Molecular Weight	Number of Mo atoms	Iron and sulfur content	Reaction catalyzed
Nitrogenase	Bacteria; blue-green algae	$\left\{\begin{array}{l}220,000 \\ 56,000\end{array}\right.$	2 0	$\left.\begin{array}{l}24Fe, 24S \\ 4Fe, 4S\end{array}\right\}$	$N_2 + 6H^+ + 6e^- \rightarrow 2NH_3$
Nitrate reductase	Fungi	228,000	1 or 2	Cytochrome b	$NO_3^- + 2H^+ + 2e^- \rightarrow NO_2^- + H_2O$
Xanthine oxidase	Cow's milk	275,000	2	8Fe, 8S	Xanthine + $H_2O \rightarrow$ uric acid + $2H^+ + 2e$
Xanthine dehydrogenase	Chicken liver	300,000	2	8Fe, 8S	Same as above
Aldehyde oxidase	Rabbit liver	270,000	2	8Fe, 8S	$RCHO + H_2O \rightarrow RCOOH + 2H^+ + 2e-$
Sulfite oxidase	Cow liver	110,000	2	2 Heme	$SO_3^{2-} + H_2O \rightarrow SO_4^{2-} + 2H^+ + 2e^-$

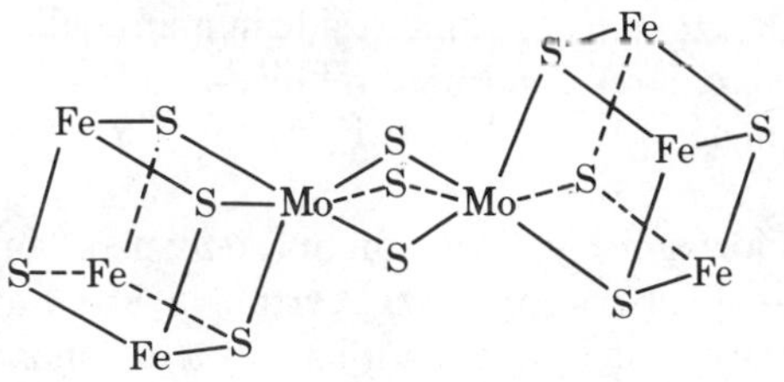

Fig. 31-17. A schematic drawing of the type of Mo-Fe-S clusters in models for the FeMo-Co of nitrogenase.

established their structures by X-ray crystallography.[45,46] The essential features of these structures are shown in Fig. 31-17. There is an RS^- ligand attached to each of the six iron atoms, as in the $Fe_4S_4(SR)_4$ systems. Perhaps the most remarkable and important observation[45] concerning these model systems is that one of them has an Mo EXAFS pattern that is essentially superimposable on that for the FeMo-co.

31-11. Miscellaneous Other Metals

The Alkaline Earths. Only magnesium and calcium have biological roles; the other alkaline earths are more or less toxic.

Magnesium has several important biochemical functions. Its presence in chlorophyll, and the structure and photosynthetic activity of chlorophyll, have been mentioned (p. 284). Animal organisms also require magnesium. For example, an adult human body normally contains about 20 g, of which about half is found in the bones and the other half within cells. The major role of intracellular magnesium is to act as a cofactor for various enzymes that catalyze the hydrolysis or cleavage of polyphosphates. Among these are alkaline phosphatase (which is a zinc metalloenzyme as mentioned earlier), ATPase, hexokinase, and one or more of the deoxyribonucleases. The Mg^{2+} ion functions as a Lewis acid, polarizing the phosphate groups, thereby enhancing the possibility of nucleophilic attack on a terminal phosphorus atom.

Calcium is found almost entirely (*ca.* 98%) in the bones, but it has several other roles as well. It is intimately involved in the process of muscle contraction, in neuron activity, and in at least one part of the visual process. Calcium is also believed to be an integral part of biological membranes. Several enzymes (e.g., α-amylase and thermolysin) use Ca^{2+} ions in support of their structures, and in at least one case, micrococcal nuclease, the Ca^{2+} ion has been shown to participate directly in the active site, where it is coordinated by an octahedral set of oxygen atoms.[47]

There are several proteins whose function appears to be calcium storage and transport, especially in conjunction with its role in muscle contraction. The best

[45]   R. H. Holm *et al., J. Am. Chem. Soc.,* 1978, **100,** 4630.
[46]   C. D. Garner *et al., J.C.S. Chem. Comm.,* **1978,** 740.
[47]   F. A. Cotton and E. E. Hazen, Jr., *Proc. Nat. Acad. Sci. (U.S.),* 1979, **76,** 2551.

known of these are the *troponins,* which reside in mammalian muscle tissue, and the *parvalbumins,* which are found in amphibia and fish. The structure of carp parvalbumin has been determined[48] and shows how Ca^{2+} ions are bound by carboxyl anions.

Vanadium. The biochemistry of vanadium is only poorly understood, although it has been known for at least 50 years that certain invertebrates (mostly ascidian worms) accumulate the element, concentrating it by a factor of over 10^6 with respect to the seawater in which they live. The accumulated vanadium is not free in the blood but is confined to vanadophores, which are distinct regions of vanadium-carrying cells called vanadocytes. These cells are highly unusual in that they contain enough sulfuric acid to have a pH of about 0. The function of the vanadium, if indeed it has a function, is obscure.

Chromium. Only recently has chromium been recognized to be an essential element in higher animals. No requirement is known in lower animals or plants. It forms an essential part of the glucose tolerance factor (GTF) that together with insulin, controls the removal of glucose from the blood. Study of its biochemical action is difficult because of the very low levels of chromium that are present. Some data appear to indicate that the chromium in GTF from brewer's yeast is Cr^{III}, but the nature of the complex is not known. Several other biochemical roles for chromium have also been tentatively indicated.

Manganese. This element appears to be essential in trace amounts to virtually all forms of life. Deficiency diseases of both plants and animals are numerous. The adult human body contains *ca.* 15 mg. However very little is known about the chemical processes in which manganese participates. It is known to be an essential component of the photosynthetic mechanism in green plants. It appears to function as a redox catalyst in the part of the mechanism, photosystem-II, in which water is oxidized to dioxygen. However virtually nothing is known about the chemical details of its participation.[49]

Nickel. Until quite recently nickel stood out among the first-row transition elements as, apparently, lacking any inherent biochemical role. It has been found that a protein containing nickel occurs in the blood serum of man and other mammals. The protein in rabbit serum appears to have one nickel atom per molecule of weight in excess of 5×10^5 daltons. More exciting, perhaps, is the discovery that jack bean urease, which in 1920 became the first enzyme ever to be crystallized, contains 2.0 ± 0.3 atoms of nickel per 105,000 daltons of enzyme[50]. The metal content of urease escaped detection for so many years because of its low concentration combined with the extremely weak visible spectrum of octahedrally coordinated Ni^{II} (p. 787). Nothing is yet known about the role of the nickel ions in urease.

[48]   R. H. Kretzinger *et al., Proc. Nat. Acad. Sci. (U.S.),* 1972, **69,** 581.
[49]   See, for example, R. L. Heath, *Int. Rev. Cytol.,* 1973, **34,** 49.
[50]   N. E. Dixon *et al., J. Am. Chem. Soc.,* 1975, **97,** 4131.

General References

Addison, A. W., *et al.,* Eds., *Biological Aspects of Inorganic Chemistry,* Wiley, 1977.

Dessy, R., J. Dillard, and L. Taylor, *Bioinorganic Chemistry,* ACS Advances in Chemistry Series, Vol. 100, 1971.

Dolphin, D., Ed., *The Porphyrins,* 7 Vols. Academic Press, 1978.

Dunitz, J. D., *et al.,* Eds., *Structure and Bonding,* Vol. 17, Springer. Articles on erythrocuprein, ferritin, iron protein conformations, and calcium-binding proteins.

Eichhorn, G. L., Ed., *Inorganic Biochemistry,* Elsevier, 1973. A comprehensive textbook.

Hughes, M. N., *The Inorganic Chemistry of Biological Processes,* Wiley, 1972.

McAuliffe, C. A., *Techniques and Topics in Bioinorganic Chemistry,* Wiley, 1975.

Mennear, J. H., ed., *Cadmium Toxicity,* Dekker, 1979.

Phipps, D. A., *Metals and Metabolism,* Clarendon Press, 1976.

Raymond, K. N., *Bioinorganic Chemistry—II,* ACS Advances in Chemistry Series No. 162, 1977.

Sigel, H., Ed., *Metal Ions in Biological Systems,* Dekker, Vol. 1, 1974. Series of volumes on all aspects.

Vallee, B. L., and W. E. C. Wacker, *Metalloproteins,* Vol. 5, in the series *The Proteins,* H. Neurath, Ed., Academic Press 1970.

Williams, R. J. P., and J. R. R. F. Da Silva, *New Threads in Bio-Inorganic Chemistry,* Academic Press, 1978.

Appendix 1
Units, Fundamental Constants, and Conversion Factors

The SI Units.   In recent years there has been a concerted effort by many, but not all, scientists to have a set of units, called collectively the Système Internationale d'Unités (SI for short), adopted universally and *in toto*. To some extent this effort has succeeded, and many of the SI units are now commonly seen in the research literature. Tables A-1 and A-2 present brief lists of the ones most pertinent to the subject matter of this book.

It should be noted, however, that very sound objections can be made to many aspects of the SI units,* and the adoption of SI *in toto* and uncritically has not been shown to be either necessary or desirable. In this book we use some (e.g., joules instead of calories) but not others (e.g., we retain angstroms, atmospheres, degrees Celsius, and the familiar magnetic units).

Physical Constants (in SI Units).   Table A-3 presents some that are useful in chemistry.

Conversion Factors.   Some that are frequently used, especially in relating SI units to others, are listed below.

1 atm = 760 mm Hg (= torr) = 1.01325×10^6 dyne cm^{-2} = 101.325 kN m^{-2}

1 bar = 10^6 dyne cm^{-2} = 0.987 atm = 10^5 kN m^{-2}

1 Bohr magneton = 9.273×10^{-24} Am^2 molecule^{-1} = 9.273×10^{-21} erg gauss^{-1}

1 calorie (thermochemical) = 4.184 J

1 coulomb = 0.10000 emu = 2.9979×10^9 esu = 1 amp sec

1 dyne = 10^{-5} N

1 erg = 2.3901×10^{-8} cal = 10^{-7} J

1 eV = 8066 cm^{-1} = 23.06 kcal mol^{-1} = 96.48 kJ mol^{-1} = 1.602×10^{-12} erg = 1.602×10^{-19} J

1 gauss = 10^{-4} T

1 kJ mol^{-1} = 83.54 cm^{-1}

1 mass unit (mu) = 931.5 MeV = 1.660×10^{-24} erg

Molar Magnetic Susceptibility (SI) = Molar Susceptibility (cgs) $\times 4\pi \times 10^{-6}$

$RT(T = 300 \text{ K}) = 2.495$ kJ mol^{-1} = 208.4 cm^{-1}

---

* Cf. A. W. Adamson, *J. Chem. Ed.*, 1978, **55**, 634.

TABLE A-1
Basic SI Units

Physical quantity	Name of unit	Symbol for unit
Length	meter	m
Mass	kilogram	kg
Time	second	s[a]
Electric current	ampere	A
Thermodynamic temperature	kelvin	K
Luminous intensity	candela	cd
Amount of substance	mole	mol

[a] In this book the more familiar "sec" and "hr" are used.

TABLE A-2
Derived SI Units

Physical quantity	Name of unit	Symbol for unit	Definition of unit
Energy	joule	J	$kg\ m^2s^{-2}$
Force	newton	N	$kg\ m\ s^{-2} = J\ m^{-1}$
Power	watt	W	$kg\ m^2s^{-3} = J\ s^{-1}$
Pressure	pascal	Pa	$kg\ m^{-1}s^{-2} = N\ m^{-2}$
Electric charge	coulomb	C	$A\ s$
Electric potential difference	volt	V	$kg\ m^2s^{-3}A^{-1} = J\ A^{-1}s^{-1}$
Electric resistance	ohm	Ω	$kg\ m^2s^{-3}A^{-2} = V\ A^{-1}$
Electric capacitance	farad	F	$A^2s^4kg^{-1}m^{-2} = A\ s\ V^{-1}$
Magnetic flux	weber	Wb	$kg\ m^2s^{-2}A^{-1} = V\ s$
Inductance	henry	H	$kg\ m^2s^{-2}A^{-2} = V\ s\ A^{-1}$
Magnetic flux density	tesla	T	$kg\ s^{-2}A^{-1} = V\ s\ m^{-2}$
	lumen	lm	cd sr
Frequency	herz	Hz	$Hz = sec^{-1}$
Customary temperature, t	degree Celsius	°C	$t[°C] = T[K] - 273.15$

TABLE A-3
Physical Constants[a]

Quantity	Symbol	Value
Avogadro's number	N_A (or L)	$602.2169 \times 10^{21}\ mol^{-1}$
Bohr magneton	μ_B	$9.274096 \times 10^{-24}\ J\ T^{-1}$
Bohr radius	a_0	$52.917715 \times 10^{-12}\ m$
Boltzmann constant	k	$13.80622 \times 10^{-24}\ J\ K^{-1}$
Electric permittivity of free space	ϵ_0	$8.854185276 \times 10^{-12}\ C^2\ N^{-1}\ m^{-2}$
Electron charge	e	$1.6021917 \times 10^{-19}\ C$
Electron magnetic moment, in Bohr magnetons	μ_e/μ_B	1.0011596389
Electron rest mass	m_e	$9.109558 \times 10^{-31}\ kg$
Faraday constant	F	$96.4867 \times 10^3\ C\ mol^{-1}$
Gas constant	R	$8.31434\ J\ K^{-1}\ mol^{-1}$
Magnetic permeability of free space	μ_0	$1.256637062 \times 10^{-6}\ Wb\ A^{-1}\ m^{-1}$
Neutron rest mass	m_n	$1.674920 \times 10^{-27}\ kg$
		$1.0086652\ u$
Planck's constant	h	$662.6196 \times 10^{-36}\ J\ s$
Proton magnetic moment	μ_p	$14.106203 \times 10^{-27}\ J\ T^{-1}$

TABLE A-3
Physical Constants[a]

Quantity	Symbol	Value
Proton magnetic moment, in Bohr magnetons	μ_p/μ_B	$1.52103264 \times 10^{-3}$
Proton rest mass	m_p	1.672614×10^{-27} kg
		1.00727661 u
Rydberg constant	R_a	10.9737312×10^6 m^{-1}
Velocity of light	c	2.997925×10^8 m s^{-1}

[a] From B. N. Taylor, W. H. Parker, and D. N. Langenberg, *Rev. Mod. Phys.*, 1969 **41**, 375.

Appendix 2
Ionization Enthalpies of the Atoms

Definition: The ionization enthalpies of an atom X are the enthalpies of the processes

$$X(g) = X^+(g) + e^-(g) \qquad \Delta H^\circ_{ion}(1)$$
$$X^+(g) = X^{2+}(g) + e^-(g) \qquad \Delta H^\circ_{ion}(2), \text{ etc.}$$

Older chemical literature commonly uses the term "ionization potential," which is $-\Delta H^\circ_{ion}$, usually expressed in electron-volts. Table A-4 gives $-\Delta H^\circ_{ion}(n)$ values.

TABLE A-4
Values of $\Delta H^\circ_{ion}(n)$ (kJ mol^{-1})

	First	Second	Third	Fourth
1 H	1311			
2 He	2372	5249		
3 Li	520.0	7297	11,810	
4 Be	899.1	1758	14,850	21,000
5 B	800.5	2428	2394	25,020
6 C	1086	2353	4618	6512
7 N	1403	2855	4577	7473
8 O	1410	3388	5297	7450
9 F	1681	3375	6045	8409
10 Ne	2080	3963	6130	9363
11 Na	495.8	4561	6913	9543
12 Mg	737.5	1450	7731	10,540
13 Al	577.5	1817	2745	11,580
14 Si	786.3	1577	3228	4355
15 P	1012	1903	2910	4955
16 S	999.3	2260	3380	4562
17 Cl	1255	2297	3850	5160
18 Ar	1520	2665	3950	5771
19 K	418.7	3069	4400	5876
20 Ca	589.6	1146	4942	6500
21 Sc	631	1235	2389	7130
22 Ti	656	1309	2650	4173
23 V	650	1414	2828	4600
24 Cr	652.5	1592	3056	4900

TABLE A-4

Values of ΔH°_{ion} (n) (kJ mol^{-1})

	First	Second	Third	Fourth
25 Mn	717.1	1509	3251	
26 Fe	762	1561	2956	
27 Co	758	1644	3231	
28 Ni	736.5	1752	3489	
29 Cu	745.2	1958	3545	
30 Zn	906.1	1734	3831	
31 Ga	579	1979	2962	6190
32 Ge	760	1537	3301	4410
33 As	947	1798	2735	4830
34 Se	941	2070	3090	4140
35 Br	1142	2080	3460	4560
36 Kr	1351	2370	3560	
37 Rb	402.9	2650	3900	
38 Sr	549.3	1064		5500
39 Y	616	1180	1979	
40 Zr	674.1	1268	2217	3313
41 Nb	664	1381	2416	3700
42 Mo	685	1558	2618	4480
43 Tc	703	1472	2850	
44 Ru	710.6	1617	2746	
45 Rh	720	1744	2996	
46 Pd	804	1874	3177	
47 Ag	730.8	2072	3360	
48 Cd	876.4	1630	3615	
49 In	558.1	1820	2705	5250
50 Sn	708.2	1411	2942	3928
51 Sb	833.5	1590	2440	4250
52 Te	869	1800	3000	3600
53 I	1191	1842		
54 Xe	1169	2050	3100	
55 Cs	375.5	2420		
56 Ba	502.5	964		
57 La	541	1103	1849	
72 Hf	760	1440	2250	3210
73 Ta	760	1560		
74 W	770	1710		
75 Re	759	1600		
76 Os	840	1640		
77 Ir	900			
78 Pt	870	1791		
79 Au	889	1980		
80 Hg	1007	1809	3300	
81 Tl	588.9	1970	2880	4890
82 Pb	715.3	1450	3080	4082
83 Bi	702.9	1609	2465	4370
84 Po	813			
86 Rn	1037			
88 Ra	509.1	978.6		
89 Ac	670	1170		

Appendix 3
Enthalpies of Electron
Attachment for Atoms

Definition: The enthalpy of electron attachment pertains to the process:

$$X(g) + e^- = X^-(g) \qquad \Delta H^{\circ}_{EA}$$

Older chemical literature commonly uses the term "electron affinity," defined as $-\Delta H^{\circ}_{EA}$. Table A-5 give $-\Delta H^{\circ}_{EA}$ values.

TABLE A-5
Values of ΔH°_{EA} (kJ mol^{-1})[a]

Z	Atom	ΔH°_{EA}
1	H	72.77
3	Li	59.8(6)
7	N	96(3)
8	O	141.1(3)
9	F	328.0(3)
11	Na	52.7(5)
15	P	71(1)
16	S	200.42(5)
17	Cl	348.8(4)
19	K	48.36(5)
33	As	77(5)
34	Se	194.96(3)
35	Br	324.6(4)
37	Rb	46.89(5)
51	Sb	101(5)
52	Te	190.15(3)
53	I	295.3(4)
55	Cs	45.5(2)
79	Au	222.8(1)
83	Bi	97(2)
84	Po	183(30)
85	At	270(20)

[a] All taken from H. Hotop and W. C. Lineberger, *J. Phys. Chem. Reference Data,* 1975, **4,** 539, which should be consulted for sources and background.

Appendix 4
Atomic Orbitals

Atomic orbitals may be described in the conventional polar coordinates r, θ, and ϕ. The complete wave function for each one may be factored into a radial part that depends on r and also on the principal quantum number, and an angular part that depends on θ and ϕ but is invariant with principal quantum number. Table A-6 lists angular wave functions, normalized to unity, for s, p, d, and f orbitals.

TABLE A-6
Angular Wave Functions for Atomic Orbitals

Orbital			$A(\theta,\phi)$	
Letter type	Full polynomial	Simplified polynomial	Normalizing factor	Angular function
s			$\dfrac{1/\sqrt{\pi}}{2}$	
p	z		$\dfrac{\sqrt{3/\pi}}{2}$	$\cos\theta$
	x		$\dfrac{\sqrt{3/\pi}}{2}$	$\sin\theta\cos\phi$
	y		$\dfrac{\sqrt{3/\pi}}{2}$	$\sin\theta\sin\phi$
d	$2z^2 - x^2 - y^2$	z^2	$\dfrac{\sqrt{5/\pi}}{4}$	$(3\cos^2\theta - 1)$
	xz		$\dfrac{\sqrt{15/\pi}}{2}$	$\sin\theta\cos\theta\cos\phi$
	yz		$\dfrac{\sqrt{15/\pi}}{2}$	$\sin\theta\cos\theta\sin\phi$
	$x^2 - y^2$		$\dfrac{\sqrt{15/\pi}}{4}$	$\sin^2\theta\cos 2\phi$
	xy		$\dfrac{\sqrt{15/\pi}}{4}$	$\sin^2\theta\sin 2\phi$
f^a	xyz		$\dfrac{\sqrt{105/\pi}}{4}$	$\sin^2\theta\cos\theta\sin 2\phi$
	$x(z^2 - y^2)$		$\dfrac{\sqrt{105/\pi}}{4}$	$\sin\theta\cos\phi(\cos^2\theta - \sin^2\theta\sin^2\phi)$
	$y(z^2 - x^2)$		$\dfrac{\sqrt{105/\pi}}{4}$	$\sin\theta\sin\phi(\cos^2\theta - \sin^2\theta\cos^2\phi)$
	$z(x^2 - y^2)$		$\dfrac{\sqrt{105/\pi}}{4}$	$\sin^2\theta\cos\theta\cos 2\phi$
	$x(5x^2 - 3r^2)$	x^3	$\dfrac{\sqrt{7/\pi}}{4}$	$\sin\theta\cos\phi(5\sin^2\theta\cos^2\phi - 3)$
	$y(5y^2 - 3r^2)$	y^3	$\dfrac{\sqrt{7/\pi}}{4}$	$\sin\theta\sin\phi(5\sin^2\theta\sin^2\phi - 3)$
	$z(5z^2 - 3r^2)$	z^3	$\dfrac{\sqrt{7/\pi}}{4}$	$5\cos^3\theta - 3\cos\theta$

$^a\, r^2 = x^2 + y^2 + z^2.$

Figure A-1 shows the shapes of the $1s$, $2p$, and $3d$ sets of orbitals.
The radial wave functions for the $1s$, $2s$, and $3s$ orbitals are shown in Fig. A-2 both as ψ_r itself and as $4\pi r^2\psi_r^2$, the electron density per spherical shell. Abscissa scale is in Bohr radii (*ca.* 0.529 Å).

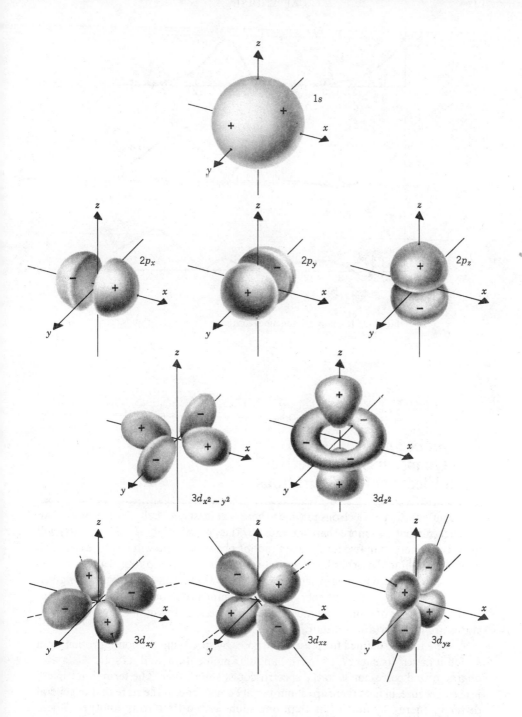

Fig. A-1. The 1s, 2p, and 3d sets of orbitals.

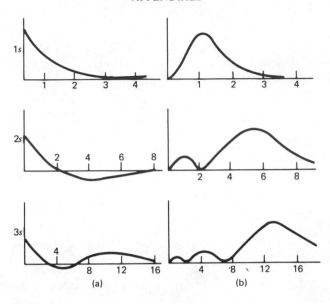

Fig. A-2. Radial wave functions for the $1s$, $2s$, and $3s$ orbitals.

Appendix 5.
The Quantum States Derived
from Electron Configurations

Any configuration of electrons in which there is at least one shell that has more than one electron but also more than one vacancy (i.e., p^2, p^3, p^4, or d^2, d^3, ..., d^8) will give rise to more than two total wave functions because there will be more than two ways to combine the orbital and spin angular momenta into total values for the entire set of electrons. These several *electronic states* that arise from the given configuration have different energies, and in many problems concerning the electronic structures of atoms and molecules it is necessary to know what the possible states are for a certain configuration.

A *state* may be defined for present purposes by specifying the configuration from which it arises, its energy, its orbital angular momentum, and its spin. A state so characterized corresponds to the spectroscopist's *multiplet*. The term "multiplet" is used because, in fact, there are a number of components of the state that in general differ in energy by much less than one such state differs from another. These components of the state or multiplet have different values of the total angular momentum, which is a result of the combination of the orbital and spin angular momenta. Now the three characteristic quantities of a state of the system have magnitudes that are determined by the manner in which the three corresponding quantities for each of the individual electrons combine to produce the resultant quantities for the entire group of electrons. Even for the simplest cases, this is a complicated matter that cannot be dealt with entirely rigorously. However we are fortunate in finding, from experiment, that for the lighter atoms (up to, approximately, the lanthanides) Nature follows a scheme that can be understood to a fairly accurate level of approximation by using a set of relatively simple rules. This set of rules may be designated the *Russell-Saunders* or LS coupling scheme, and it forms the subject of this appendix.

Each electron in an atom has a set of quantum numbers n, l, m_l, m_s; it is the last three that are of concern here. Just as l is used, as $\sqrt{l(l+1)}$, to indicate the orbital angular momentum of a single electron, there is a quantum number L such that $\sqrt{L(L+1)}$ gives the total orbital angular momentum of the atom. The symbol M_l is used to represent a component of L in a reference direction and is analogous to m_l for a single electron. Similarly, we use a quantum number S to represent the total electron spin angular momentum, given by $\sqrt{S(S+1)}$, in analogy to the quantum number s for a single electron. There is the difference here that s is limited to the value $\frac{1}{2}$, whereas S may take any integral or half-integral value beginning with zero. Components of S in a reference direction are designated by M_S, analogous to m_s.

Symbols for the states of atoms are analogous to the symbols for the orbitals of single electrons. Thus the capital letters $S, P, D, F, G, H, \ldots$ are used to designate states with $L = 0, 1, 2, 3, 4, 5, \ldots$. The use of S for both a state and a quantum number is unfortunate but in practice seldom causes any difficulty. The complete symbol for a state also indicates the total spin, but not directly in terms of the value of S. Rather, the number of different M_S values, which is called the *spin multiplicity*, is used. Thus for a state with $S = 1$, the spin multiplicity is 3 because there are three M_S values, $1, 0, -1$. In general the spin multiplicity is $2S + 1$ and is indicated as a left superscript to the symbol for L. The following examples should clarify the usage:

For $M_L = 4, S = \frac{1}{2}$, the symbol is ^2G
For $M_L = 2, S = \frac{3}{2}$, the symbol is ^4D
For $M_L = 0, S = 1$, the symbol is ^3S

In speaking or writing of states with spin multiplicities of $1, 2, 3, 4, 5, 6, \ldots$, we call them respectively, *singlets, doublets, triplets, quartets, quintets, sextets, ...* Thus the three states shown above would be called doublet G, quartet D, and triplet S, respectively.

As in the case of a single electron, we may sometimes be interested in the total angular momentum, that is, the vector sum of L and S. For the entire atom this is designated J. When required, J values are appended to the symbol as right subscripts. For example, a ^4D state may have any of the following J values, the appropriate symbols being as annexed.

L	M_s	J	Symbol
2	$3/2$	$7/2$	$^4D_{7/2}$
$\cdot 2$	$1/2$	$5/2$	$^4D_{5/2}$
2	$-1/2$	$3/2$	$^4D_{3/2}$
2	$-3/2$	$1/2$	$^4D_{1/2}$

To determine what states may actually occur for a given atom or ion, we begin with the following definitions, which represent the essence of the approximation we are using:

$$M_L = m_l^{(1)} + m_l^{(2)} + m_l^{(3)} + \cdots + m_l^{(n)}$$
$$M_S = m_s^{(1)} + m_s^{(2)} + m_s^{(3)} + \cdots + m_s^{(n)}$$

in which $m_l^{(i)}$ and $m_s^{(i)}$ stand for the m_l and m_s values of the ith electron in an atom having a total of n electrons.

In general it is not necessary to pay specific attention to all the electrons in an atom when calculating M_L and M_S, since those groups of electrons that completely fill any one set of orbitals (s, p, d, etc.) collectively contribute zero to M_L and to M_S. For instance, a complete set of p electrons includes two with $m_l = 0$, two with $m_l = 1$, and two with $m_l = -1$; the sum $0 + 0 + 1 + 1 - 1 - 1$ is zero. At the same time, half the electrons have $m_s = 1/2$ and the other half have $m_s = -1/2$, making M_S equal to zero. The generalization to any filled shell should be obvious. Therefore we need only concern ourselves with partly filled shells.

For a partly filled shell there is always more than one way of assigning m_l and m_s values to the various electrons. All ways must be considered except those that either are prohibited by the exclusion principle or are physically redundant, as explained below. For convenience we use symbols in which $+$ and $-$ superscripts represent $m_s = +1/2$ and $m_s = -1/2$, respectively. Thus when the first electron has $m_l = 1$, $m_s = +1/2$, the second electron has $m_l = 2$, $m_s = -1/2$, the third electron has $m_l = 0$, $m_s = +1/2$, etc., we write $(1^+, 2^-, 0^+, \ldots)$. Such a specification of m_l and m_s values of all electrons is called a microstate.

Let us now consider the two configurations $2p3p$ and $2p^2$. In the first case our freedom to assign quantum numbers m_l and m_s to the two electrons is unrestricted by the exclusion principle, since the electrons already differ in their principal quantum numbers. Thus microstates such as $(1^+, 1^+)$ and $(0^-, 0^-)$ are permitted. They are not permitted for the $2p^2$ configuration, however. Second, since the two electrons of the $2p3p$ configuration can be distinguished by their n quantum numbers, two microstates such as $(1^+, 0^-)$ and $(0^-, 1^+)$ are physically different. However for the $2p^2$ configuration such a pair are actually identical, since there is no *physical* distinction between "the first electron" and "the second electron." For the $2p3p$ configuration there are thus $6 \times 6 = 36$ different microstates, and for the $2p^2$ configuration six of these are nullified by the exclusion principle and the remaining 30 consist of pairs that are physically redundant. Hence there are only fifteen microstates for the $2p^2$ configuration.

In Table A-7a the microstates for the $2p^2$ configuration are arranged according to their M_L and M_S values. It is now our problem to deduce from this array the

TABLE A-7
Tabulation of Microstates for a p^2 Configuration

(a)

M_L	M_S		
	1	0	-1
2		$(1^+, 1^-)$	
1	$(1^+, 0^+)$	$(1^+, 0^-)(1^-, 0^+)$	$(1^-, 0^-)$
0	$(1^+, -1^+)$	$(1^+, -1^-)(0^+, 0^-)(1^-, -1^+)$	$(1^-, -1^-)$
-1	$(-1^+, 0^+)$	$(-1^+, 0^-)(-1^-, 0^+)$	$(-1^-, 0^-)$
-2		$(-1^+, -1^-)$	

(b)

M_L	M_S		
	1	0	-1
2			
1	$(1^+, 0^+)$	$(1^-, 0^+)$	$(1^-, 0^-)$
0	$(1^+, -1^+)$	$(1^-, -1^+)(0^+, 0^-)$	$(1^-, -1^-)$
-1	$(-1^+, 0^+)$	$(-1^-, 0^+)$	$(-1^-, 0^-)$
-2			

possible values for L and S. We first note that the maximum and minimum values of M_L are 2 and -2, each of which is associated with $M_S = 0$. These must be the two extreme M_L values derived from a state with $L = 2$ and $S = 0$, namely, a ^1D state. Also belonging to this ^1D state must be microstates with $M_S = 0$ and $M_L = 1, 0,$ and -1. If we now delete a set of five microstates appropriate to the ^1D state, we are left with those shown in Table A-7b. Note that it is not important which of the two or three microstates we have removed from a box that originally contained several, since the microstates occupying the same box actually mix among themselves to give new ones. However the *number* of microstates per box is fixed, whatever their exact descriptions may be. In Table A-7b we see that there are microstates with $M_L = 1, 0, -1$ for each of the M_S values $1, 0, -1$. Nine such microstates constitute the component of a ^3P state. When they are removed, there remains only a single microstate with $M_L = 0$ and $M_S = 0$. This must be associated with a ^1S state of the configuration. Thus the permitted states of the $2p^2$ configuration—or any np^2 configuration—are ^1D, ^3P, and ^1S. It is to be noted that the sum of the degeneracies of these states must be equal to the number of microstates. The ^1S state has neither spin nor orbital degeneracy; its degeneracy number is therefore 1. The ^1D state has no spin degeneracy but is orbitally $2L + 1 = 5$-fold degenerate. The ^3P state has 3-fold spin degeneracy and 3-fold orbital degeneracy, giving it a total degeneracy number of $3 \times 3 = 9$. The sum of these degeneracy numbers is indeed 15.

For the $2p3p$ configuration the allowed states are again of the types S, P, and D, but now there is a singlet and a triplet of each kind. This can be demonstrated by making a table of microstates and proceeding as before. It can be seen perhaps more easily by noting that for *every* combination of $m_l^{(1)}$ and $m_l^{(2)}$ there are four microstates, with spin assignments $++$, $+-$, $-+$, and $--$. One of these, either $+-$ or $-+$, can be taken as belonging to a singlet state and the other three belong then to a triplet state. It will be noted that the sum of the degeneracy numbers for the six states ^3D, ^1D, ^3P, ^1P, ^3S, ^1S is 36, the number of microstates.

For practice in using the LS coupling scheme, the reader may verify, by the method used for np^2, that an nd^2 configuration gives rise to the states ^3F, ^3P, ^1G, ^1D, and ^1S, and that an np^3 configuration gives the states ^4S, ^2P, ^2D.

The method shown for determining the states of an electron configuration will obviously become very cumbersome as the number of electrons increases beyond perhaps 5, but there is, fortunately, a relationship that makes many of the problems with still larger configurations tractable. This relationship is called the *hole formalism,* and with it a partially filled shell of n electrons can be treated either as n electrons or as $N - n$ positrons, where N represents the total capacity of the shell. As far as electrostatic interactions of electrons among themselves are concerned, it makes no difference whether they are all positively charged or all negatively charged, since the energies of interaction are all proportional to the product of two charges. It is actually rather easy to see that the hole formalism must be true, for whenever we select a microstate for n electrons in a shell of capacity N, there remains a set of m_l and m_s values that could be used by $N - n$ electrons.

The several states derived from a particular configuration have different energies. However purely theoretical evaluation of these energy differences is neither easy nor accurate, since they are expressed as certain integrals representing electron-electron repulsions, which cannot be precisely evaluated by computation. However when there are many terms arising from a configuration, it is usually possible to express all the energy differences in terms of only a few integrals. Thus when the energies of just a few of the states have been measured, the others may be estimated with fair accuracy, though not exactly, because the coupling scheme itself is only an approximation. The magnitudes of the energy differences are generally comparable to energies of chemical bonds and chemical reactions. For example, the energies of the ^1D and ^1S excited states of the carbon atom in the configuration $1s^2 2s^2 2p^2$ are ~ 105 and ~ 135 kJ mol^{-1}, respectively, above the ^3P ground state.

As mentioned above, each state of the type ^{2S+1}L actually consists of a group of substates with different values of the quantum number J. The energy differences between these substates are generally an order of magnitude less than the energy differences between the various states themselves, and usually they can be ignored in ordinary problems of chemical interest. However in certain cases—for example, in understanding the magnetic properties of the lanthanides (Chapter 23) and in nearly all problems with the very heavy elements (those of the third transition series and the actinides)—these energy differences are of great significance and cannot be ignored. Indeed, for the very heavy elements they become comparable in magnitude to the energy differences between the ^{2S+1}L states. When this happens, the LS coupling scheme this is assumed to be negligible. Thus, as stated, we assume complex treatments must be used.

The cause of the separation between substates with different J values is direct coupling between the spin and the orbital angular momenta of the electrons. In the *LS* coupling scheme this is assumed to be negligible. Thus, as stated, we assume that the $m_l^{(i)}$ add only to one another and the $m_s^{(i)}$ add only to one another. When coupling between $m_l^{(i)}$ and $m_s^{(i)}$ for each electron becomes *very* strong, it is possible to utilize another relatively simple method to determine states of the electron configuration. In this method, known as the *jj* coupling scheme, one assumes that the states arise from the various combinations of *j* values for each electron. In the *jj* coupling scheme, the quantum numbers M_L and M_S are no longer meaningful and the states are characterized by other quantum numbers. However since *jj* coupling does not find any direct application in this book, we shall not discuss it further here.

Appendix 6
Magnetic Properties of
Chemical Compounds[1]

All the magnetic properties of substances in bulk are ultimately determined by the electrical properties of the subatomic particles, electrons, and nucleons. Because the magnetic effects due to nucleons and nuclei are some 10^{-3} times those due to electrons, they ordinarily have no detectable effect on magnetic phenomena of direct chemical significance. Thus we concentrate entirely on the properties of the electron and on the magnetic properties of matter that result therefrom. There are direct and often sensitive relationships between the magnetic properties of matter in bulk and the number and distribution of unpaired electrons in its various constituent atoms or ions.

There are several kinds of magnetism qualitatively speaking; the salient features of each are summarized in Table A-8. A *paramagnetic* substance is attracted into a magnetic field with a force proportional to the field strength times the field gradient. Paramagnetism is generally caused by the presence in the substance of ions, atoms or molecules having unpaired electrons. Each of these has a definite paramagnetic moment that exists in the absence of any external magnetic field. A

TABLE A-8
Main Types of Magnetic Behavior

Type	Sign of χ_M	Magnitude[a] of χ_M (cgs units)	Dependence of χ_M on H	Origin
Diamagnetism	−	1–500×10^{-6}	Independent	Electron charge
Paramagnetism	+	0–10^{-2}	Independent	Spin and orbital motion of electrons on individual atoms
Ferromagnetism	+	10^{-2}–10^{6}	Dependent	Cooperative interaction between
Antiferromag- netism	+	0–10^{-2}	May be dependent	magnetic moments of individual atoms

[a] Assuming molecular or ionic weights in the range of about 50–1000; χ_M is the susceptibility per mole of substance, as explained on p 626.

[1] Landolt-Börnstein, *Magnetic Properties of Coordination and Organometallic Transition Metal Compounds,* Springer Verlag. E. König, Vol. II/2, 1966: E. König and G. König, Vol. II/8, 1976.

diamagnetic substance is repelled by a magnetic field. All matter has this property to some extent. Diamagnetic behavior is due to small magnetic moments that are induced by the magnetic field but do not exist in the absence of the field. Moments so induced are in opposition to the inducing field, thus causing repulsion. Finally, there are the more complex forms of magnetic behavior known as ferromagnetism and antiferromagnetism, and still others that are not discussed here.

The paramagnetism that originates strictly from the uncoupled spins of electrons, and from orbital angular momentum that may be associated with them, is temperature dependent, as explained shortly. It is noteworthy, however, that in many systems that contain unpaired electrons, as well as in a few (e.g., CrO_4^{2-}) that do not, weak paramagnetism that is *independent* of temperature can arise by a coupling of the ground state of the system with excited states of high energy under the influence of the magnetic field. This *temperature-independent paramagnetism* (TIP) thus resembles diamagnetism in that it is not due to any magnetic dipole existing in the molecule but is induced when the substance is placed in the magnetic field. It also resembles diamagnetism in its temperature independence and in its order of magnitude, viz., 0–500×10^{-6} cgs units per mole. It is often ignored in interpreting the paramagnetic behavior of ions with unpaired electrons, but in work that pretends to accuracy it should not be. Certainly when measured susceptibilities are corrected for diamagnetism (see below), it is illogical not to correct them also for TIP if this is known to occur in the system concerned.

Diamagnetism is a property of all forms of matter. All substances contain at least some, if not all, electrons in closed shells. In closed shells the electron-spin moments and orbital moments of individual electrons balance one another out so that there is no net magnetic moment. However when an atom or molecule is placed in a magnetic field, a small magnetic moment directly proportional to the strength of the field is induced. The electron spins have nothing to do with this induced moment; they remain tightly coupled together in antiparallel pairs. However the planes of the orbitals are tipped slightly so that a small net orbital moment is set up in opposition to the applied field. It is because of this opposition that diamagnetic substances are repelled from magnetic fields.

Even an atom with a permanent magnetic moment will have diamagnetic behavior working in opposition to the paramagnetism when placed in a magnetic field, provided only that the atom has one or more closed shells of electrons. Thus the net paramagnetism measured is slightly less than the true paramagnetism because some of the latter is "canceled out" by the diamagnetism.

Since diamagnetism is usually several orders of magnitude weaker than paramagnetism, substances with unpaired electrons almost always have a net paramagnetism. Of course a very dilute solution of a paramagnetic ion in a diamagnetic solvent such as water may be diamagnetic because of the large ratio of diamagnetic to paramagnetic species in it. Another important feature of diamagnetism is that its magnitude does not vary with temperature. This is because the moment induced depends only on the sizes and shapes of the orbitals in the closed shells, and these are not temperature-dependent.

Magnetic Susceptibility. Chemically useful information is obtained by proper interpretation of measured values of magnetic moments. However magnetic moments are not measured directly. Instead, one measures the magnetic susceptibility of a material, from which it is possible to calculate the magnetic moment of the paramagnetic ion or atom therein.

Magnetic susceptibility is defined in the following way. If a substance is placed in a magnetic field of magnitude H, the flux B, within the substance, is given by

$$B = H + 4\pi I$$

where I is the intensity of magnetization. The ratio B/H, called the magnetic permeability of the material, is given by

$$B/H = 1 + 4\pi(I/H) = 1 + 4\pi\kappa$$

where κ is the magnetic susceptibility per unit volume, or simply the volume susceptibility. The physical significance of this equation is easily seen. The permeability B/H is just the ratio of the density of lines of force within the substance to the density of such lines in the same region in the absence of the specimen. Thus the volume susceptibility of a vacuum is by definition zero, since in a vacuum B/H must be 1. The susceptibility of a diamagnetic substance is negative because lines of force from induced dipoles cancel out some lines of force due to the applied field. For paramagnetic substances the flux is greater within the substance than it would be in a vacuum; thus paramagnetic substances have positive susceptibilities.

There are numerous methods for measuring magnetic susceptibilities, and all depend on measuring the force exerted on a body when it is placed in an inhomogeneous magnetic field. The more paramagnetic the body, the more strongly will it be drawn toward the more intense part of the field.

Magnetic Moments from Magnetic Susceptibilities. It is generally more convenient to discuss magnetic susceptibility on a weight basis than on a volume basis; thus the following relations are used:

$$\kappa/d = \chi$$
$$M\chi = \chi_M$$

In these equations d is the density in g cm^{-3} and M is the molecular weight; χ is called the gram susceptibility and χ_M is called the molar susceptibility. When a value of χ_M is obtained from the measured volume susceptibility κ, it can be corrected for the diamagnetic contribution and for the TIP to give a "corrected" molar susceptibility χ_M^{corr}, which is the most useful quantity in drawing conclusions about electronic structure.

In his classic studies Pierre Curie showed that paramagnetic susceptibilities depend inversely on temperature and often follow or closely approximate the behavior required by the simple equation

$$\chi_M^{corr} = C/T$$

Here T represents the absolute temperature, and C is a constant that is characteristic of the substance and known as its Curie constant. This equation expresses what is known as *Curie's law.*

---

* Actually Curie's law was originally based on χ; that is, the effects of diamagnetism and TIP were neglected. However, its significance and utility are enhanced when these are taken into account.

Now, on theoretical grounds, just such an equation is to be expected. The magnetic field in which the sample is placed tends to align the moments of the paramagnetic atoms or ions; at the same time, thermal agitation tends to randomize the orientations of these individual moments. The situation is entirely analogous to that encountered in the electric polarization of matter containing electric dipoles, with which the student is probably already familiar from a standard physical chemistry course. Applying a straightforward statistical treatment, one obtains the following equation showing how the molar susceptibility of a substance containing independent atoms, ions or molecules, each of magnetic moment μ (in BM), will vary with temperature:

$$\chi_M^{corr} = \frac{N\mu^2/3\,k}{T}$$

where N is Avogadro's number, and k is the Boltzmann constant. Obviously, by comparison with *Curie's law* we have

$$C = N\mu^2/3\,k$$

and at any given temperature

$$\mu = \sqrt{3k/N}\cdot\sqrt{\chi_M^{corr}T}$$

which, on evaluating $\sqrt{3k/N}$ numerically, becomes

$$\mu = 2.84\sqrt{\chi_M^{corr}T}$$

Thus to recapitulate, one first makes a direct measurement of the volume susceptibility of a substance from which χ_M is calculated; in accurate work this result is corrected for diamagnetism and TIP. From this corrected molar susceptibility and the temperature of the measurement, one may calculate the magnetic moment of the ion, atom, or molecule responsible for the paramagnetism.

For a substance that obeys *Curie's law,* a plot of χ_M^{corr} versus $1/T$ should give a straight line of slope C that intersects the origin.

Although there are many substances that within the limits of experimental error do show this behavior, there are also many others for which the line does not go through the origin, but instead looks somewhat like one of those shown in Fig. A-3

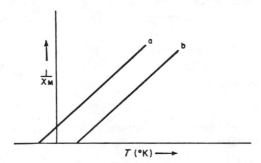

Fig. A-3. Some deviations from the Curie law that may
be fitted to the Curie-Weiss law.

cutting the T axis at a temperature below $0°$ K as in line a or above $0°$ K as in line b. Obviously, such a line can be represented by a slight modification of the Curie equation,

$$\Delta\chi_M^{corr} = \frac{C}{T - \theta}$$

where θ is the temperature at which the line cuts the T axis. This equation expresses what is known as the *Curie-Weiss law,* and θ is known as the *Weiss constant.* Actually, just such an equation can be derived if one assumes, not that the dipoles in the various ions, atoms, or molecules of a solid are completely independent, but instead that the orientation of each one is influenced by the orientations of its neighbors as well as by the field to which it is subjected. Thus the Weiss constant can be thought of as taking account of the interionic or intermolecular interactions, thereby enabling us to eliminate this extraneous effect by computing the magnetic moment from the equation

$$\mu = 2.84 \sqrt{\chi_M^{corr}(T - \theta)}$$

Unfortunately, there are also cases in which magnetic behavior appears to follow the Curie-Weiss equation without the Weiss constant having this simple interpretation. When the Curie law does not accurately fit the data and the applicability of the Curie-Weiss law is in doubt (even though it *may* fit the data), the best practice is to compute a magnetic moment at a given temperature using the Curie law and call this an *effective magnetic moment* μ_{eff}, at the specified temperature.

Ferromagnetism and Antiferromagnetism. In addition to the simple paramagnetism we have discussed, where the Curie or Curie Weiss law is followed and the susceptibility shows no dependence on field strength, there are other forms of paramagnetism in which the dependence on both temperature and field strength is complicated. Two of the most important of these are ferromagnetism and antiferromagnetism. No attempt is made to explain either of these in detail, either phenomenologically or theoretically, but it is important for the student to recognize their salient features. Figure A-4 compares the qualitative temperature-dependence of the susceptibility for simple paramagnetism, ferromagnetism, and antiferromagnetism. Of course Fig. A-4a is just a rough graph of Curie's law. In Fig. A-4b it should be noted that there is a discontinuity at some temperature T_C, called the

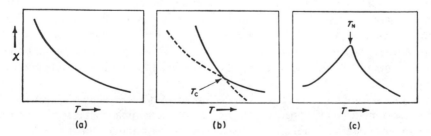

Fig. A-4. Diagrams indicating the qualitative temperature dependence of magnetic susceptibility for (a) simple paramagnetism, (b) ferromagnetism, and (c) antiferromagnetism.

Curie temperature. Above the Curie temperature the substance follows the Curie or the Curie-Weiss law; that is, it is a simple paramagnetic. Below the Curie temperature, however, it varies in a different way with temperature and is also field-strength dependent. For antiferromagnetism there is again a characteristic temperature T_N, called the Néel temperature. Above T_N the substance has the behavior of a simple paramagnetic, but below the Néel temperature the susceptibility *drops* with decreasing temperature.

These peculiarities in the behavior of ferromagnetic and antiferromagnetic substances below their Curie or Néel points are due to interionic interactions, which have magnitudes comparable to the thermal energies at the Curie or Néel temperature and thus become progressively greater than thermal energies as the temperature is further lowered. In the case of antiferromagnetism, the moments of the ions in the lattice tend to align themselves so as to cancel one another out. Above the Néel temperature thermal agitation prevents very effective alignment, and the interactions are manifested only in the form of a Weiss constant. However below the Néel temperature this antiparallel aligning becomes effective and the susceptibility is diminished. In ferromagnetic substances the moments of the separate ions tend to align themselves parallel, thus to reinforce one another. Above the Curie temperature thermal energies are more or less able to randomize the orientations; below T_C, however, the tendency to alignment becomes controlling, and the susceptibility increases much more rapidly with decreasing temperature than it would if the ion moments behaved independently of one another.

Presumably even in the substances we ordinarily regard as simple paramagnetics there are some interionic interactions, however weak; therefore there must be some temperature, however low, below which they will show ferromagnetic or antiferromagnetic behavior, depending on the sign of the interaction. We still do not know precisely why such interactions are so large in some substances that they have Curie or Néel temperature near and even above room temperature. Suffice it to say that in many cases it is certain that the magnetic interactions cannot be direct dipole-dipole interactions, but instead the dipoles are coupled through the electrons of intervening atoms in oxides, sulfides, halides, and similar compounds.

In general ferromagnetic and antiferromagnetic interactions are decreased when the magnetic species are separated from one another physically. Thus when the magnetic behavior of a solid shows the effects of interionic coupling, solutions of the same substance will be free from such interactions. This includes solid solutions; for example, when K_2OsCl_6, which has μ_{eff} per Os atom at 300 °K of 1.44 BM, is contained at a level of $\leqslant 10$ mol % in diamagnetic and isomorphous K_2PtCl_6, its μ_{eff} value at the same temperature rises to 1.94 BM owing to the elimination of antiferromagnetic coupling between the Os^{IV} ions through intervening chlorine atoms.

Localized Coupling. There are many instances of binuclear complexes in which there is magnetic coupling between the two ions in the complex but essentially none, even in the crystalline compound, between one complex and neighboring ones. The magnetic properties of such a complex are usually treated in the following way. The two paramagnetic ions A and B have spins denoted S_A and S_B, and their interaction means that the orientation of each one influences the orientation of the other. The energy of the system is thereby affected, and the Hamiltonian operator must then include a term to take account of this. Regardless of the physical mechanism by which this mutual interaction occurs, its effect on the energy of the system can usually be described to a good approximation by adding the term

$$H' = -2JS_AS_B$$

By convention negative values of J imply antiferromagnetic interaction, that is, one in which the state of lowest energy has the spins S_A and S_B opposed, so as to give the lowest possible magnetic moment.

Index

Absolute configurations, 79
Acceptor properties of complexes, 1236
Acetic acid synthesis, 1294
Acetylenes, 99
  compound dimerizations, 1284
Acetylide complexes, 1160
Acetylides, 362
Acid hydrolysis, 1192
π-Acidity, 1049
π-Acid ligands, 62, 81
  complexes with, 1049
Acids, binary, 233
  Hammett acidity function for, 238
  oxo, 235
    strengths in water, 235
  pure, 238
  strong, properties of, 241-42
  super, 239
Actinides, 1005
  absorption spectra, 1009
  chemistry of, 1012
  comparison with lanthanides, 1007
  complexes, 1019
  dioxo ions, 1015
  electronic structures, 1007
  elements, 1005
  elements-100-103, 1045
  formal reduction potentials for 1M per-
      chloric acid solutions, 1018
  ionic radii, 1009
  ions in aqueous solutions, 1017, 1019
  isotopes available in macroscopic amounts,
      1007
  organometallic compounds, 1020
    alkyls and aryls, 1020
    cyclooctatetraenyls, 1020
    cyclopentadienyls, 1020
    trivalent compounds, 1022
  oxidation states, 1010
    +2, 1013
    +3, 1014
    +4, 1014
    +5, 1014
    +6, 1015

    +7, 1015
    properties, 1005-6
    separation methods, 1036-38
    stereochemistry, 1010-11
Actinium, 1005, 1023
  compounds, 1023
  elemental, 1023
  metal, 1012
  properties, 1005-6
  see also Actinides
Actinoids, 1005. See also Actinides
Activated complex, 1184
Activation parameters, 1187
Acyl compounds, 1131
Acyl tetracarbonylferrates, 1259
Alcohol dehydrogenases, 1329
Alcohols, as ligands, 158
Alcoxides, as ligands, 158, 160-61
Aldehyde oxidase, 1342
Aldolases, 1329
Alfin catalysts, 269
Alkaline earths, biochemistry of, 1343
  halides, 204-5
Alkaline phosphatase, 1343
Alkanesulfonic esters, 533
Alkene complexes, 95, 98, 1147
  in insertion reactions, 1252
  representative syntheses of, 1148-51
Alkene metathesis, 1277
  carbene mechanism, 1278
Alkylation by metal alkyls, 1133
Alkyl groups, bridging, 111
Alkylamidino ligands, 132
Alkylidenephosphoranes, 467
Alkylidyne compounds, 1130
Alkylimido complexes, 125
Alkyl titanates, 700
Alkyne complexes, 98, 1156
Allenylidenes, 1147
Allyl complexes, 1152-55
Alumina, 329
Alluminates, 334
Aluminum, 326
  alkyls, 342

aqua ions, 333
carbide, 332
β-diketonates, 337
halides, 330
hydride, 338
isolation, 328
lower valent compounds, 347
nitrides, 332
nitrogen compounds, 344
occurrence, 328
organometallic compounds, 341
oxide, 329
properties, 326, 328
see also Group III elements
Alums, 334
Amanita muscaria, 709
Ambidentate ligands, 64
Americium, 1005, 1036
aqueous chemistry, 1040
binary compounds, 1039
elemental, 1013
halides, 1039
isolation, 1036
oxides, 1039
see also Actinides
Amidine, 122
Amines, complexes, 118
Aminoiminophosphanes, 464
Aminomethyl compounds, 1132
Ammonia, 414
complexes, 118
reactions, 416
solutions of metals in, 258
Ammonium salts, 256, 417
Amphiboles, 388, 389
α-Amylase, 1343
Anatase, 695
Anation reactions, 1191
Angular overlap model, 652-56
Anorthite, 390
Antabuse, 373
Antifluorite structure, 16
Anti-Markownikoff reaction, 1254
Antimonates, 481
Antimonides, 444
Antimony, 438
allotropes, 444
aqueous cationic chemistry, 470
hydride, 446
lone pairs in, 441
occurrence, 442
oxides, 456
oxo acids and anions, 480
oxo chloride, 459

pentachloride, 450
pentafluoride, 239, 448
planar compounds, 442
properties, 438
sulfides, 458-59
trichloride, 448
trihalides, 447
complexes, 451
trivalent, tartarate complexes, 471
see also Group V elements
Apoferritin, 1313
Apoprotein, 1310
Aqua ions, 65
Aquation, 1192
Arachno structure, 1085
Argon, 577. *See also* Noble gases
Aromatic ring systems, 99
Arsenic, 438
catenation, 444
hydride, 446
lone pairs in, 441
occurrence, 442
oxides, 455
oxo acids and anions, 480
pentachloride, 450
planar compounds, 442
properties, 438
sulfides, 458
trihalides, 447
see also Group V elements
Arsenic acid, 481
Arsenides, 444
Arsine, 446
Arylamidino ligands, 132
Asbestos minerals, 389
Ascidians, 708
Ascorbate oxidase, 1334
Astatine, 542
compounds, 547
properties, 543, 547
see also Group VII elements
Asymmetric hydrogenation, 1272
ATPase, 1343
Azaferrocene, 1170
Azide, 421
as ligand, 132
Azurin, 1336

Back-bonding, 83
Bacterial iron transport, 1314
Baddeleyite, 824
Bailar twist, 1223
Band theory, 7
Barium, 271, 281

binary compounds, 282-83
  occurrence, 281
  titanite, 695
  see also Group II elements
Barytes, 281
Base hydrolysis, 1192, 1194
Bauxite, 328
Benitoite, 388
Bent molecules, 49
Benzoyl peroxide, 499
Berkelium, 1005, 1042
  compounds, 1045
  see also Actinides
Berry pseudorotation mechanism, 1220
Beryl, 276, 388
Beryllium, 271, 274
  alkyls, 279-80
  binary compounds, 276
  chloride, 274
  complex chemistry, 277
  compounds, 274
  covalency and stereochemistry, 274
  elemental, 276
  fluoride, 277
  hydride, 249
  hydroxide, 275
  metallic, 276
  nitride, 277
  nitrogen complexes, 279
  organoberyllium compounds, 279
  oxide, 276
  oxygen ligands, 277
  phthalocyanine, 276
  sulfide, 275
Bicapped square antiprism, 55
Biguanide complexes, 146
Biimidazole complexes, 122
Bioinorganic chemistry, 1310
Biosphere, 1310
  periodic table for, 1311
2,2'-Bipyridine, complexes, 119
Bismuth, 438
  allotropes, 444
  aqueous cationic chemistry, 470
  halo complexes, 451
  lone pairs in, 441
  occurrence, 442
  oxides, 456
  oxo acids and anions, 480
  oxo chloride, 459
  pentafluoride, 449
  properties, 438
  sulfides, 459
  trichloride, 448

trihalides, 447
  see also Group V elements
Bismuthates, 481
Bismuthides, 444
Bispentafluorosulfurperoxide, 500
Bisperfluoromethylperoxide, 500
Boehmite, 329
Bohr effect, 1319
Bohr magnetons, 626
Bonding, chemical, for d-block elements,
    628
  π-complexes, 632
  secondary, 504-5
π-Bonding ligands, 62, 81-95
Boranes, 303
  adducts, 313
  anions, 315
  chemistry of, 313
  complexes, 108
  nomenclature, 309-12
  polyhedral, 314
  properties of, 305
  semitopological scheme, 306, 308
  structure-bonding correlation, 312
  structures of, 307
  structural study by nmr, 308
  with carbon, see Carbaboranes
  with hetero atoms other than carbon, 320
Borane anions, 315
  polyhedral, 317
Borates, 296
  esters, 298
  hydrated, 296
  ions in solution, 296
  tetraalkyl, 322
  tetraaryl, 322
Borax, 292
Borazine, 323
Borazine compounds, 1172
Borazole compounds, 1172
Boric acid, 296
  reactions of, 297
  structure, 297-98
Borides, 293
  types, 293
Born-Haber cycle, 22, 551
Boroles, 1171
Boron, 289
  acceptor behavior, 290
  carbide, 363
  chloride, 300, 303
  compounds, 293
  crystalline, 292
  electronic structure and bonding, 289

1370   INDEX

elemental, 290, 292
fluoride, 300
halides, 300, 302
hydrides, *see* Boranes
isolation, 292
nitride, 295
nitrogen compounds, 323
occurrence, 292
organoboron compounds, 320
oxides, 295
oxygen compounds, 295
properties, 292
sulfide, 295
tetrachloroborates, 302
trichloride, 301
trifluoride, 299
trihalides, 301
triiodide, 302
Boronic acid, 321
Boronous acid, 321
Boroxine ring, 296
Bridging groups, 64
Bromic acid, 559
Bromine, 542
occurrence, 546
oxides, 555
properties, 543
see also Group VII elements
Brookite, 695
1,3-Butadiene, complexes, 98

Cadmium, 6, 273, 589
aqua ions, 599
biological role of, 602
chalconides, 598
complexes, 600
halides, 598-99
formation constants of, 601
hydroxides, 597
organometallic compounds, 608
oxides, 597
oxo salts, 599
properties, 589, 591
selenides, 598
stereochemistry, 590-91
sulfides, 598
tellurides, 598
univalent state, 593
see also Group IIb elements
Cage and cluster structures, 56
four vertices, 56
five vertices, 57
six vertices, 57
seven vertices, 58

eight vertices, 59
nine vertices, 59
ten vertices, 59
eleven vertices, 60
twelve vertices, 60
Calcium, 271, 281
binary compounds, 282-83
biochemistry of, 1343
cyanamide, 369
occurrence, 281
see also Group II elements
Calcium group metals, 8
Californium, 1005, 1042
compounds, 1045
elemental, 1013
see also Actinides
Cancer chemotherapy, 960
Capped dodecahedral geometry, 55
Capped octahedron, 53
Capped square antiprism, 55
Capped trigonal prism, 53
Carbaboranes, 318
anions, 319
preparation, 318
Carbamate complexes, 172-73
Carbamic acid, 367
Carbides, 361
covalent, 362
Group II elements, 283
interstitial, 362
saltlike, 361
Carbocations, *see* Carbonium ions
Carbocyclic π complexes, 1160
η^4-arenes, 1169
eight-membered rings, 1170
five-membered rings, 1163
four-membered rings, 1161
heterocyclic, 1170
multidecker sandwiches, 1172
seven-membered rings, 1170
six-membered rings, 1167
three-membered rings, 1161
Carbodiimides, 477
Carbon, 352
allotropy of, 356
as a ligand atom, 109, 1113
catenation, 354
cyanide complexes, 113
bonding, 114
types, 113
cyanides and related compounds, 367
dioxide, 365
complexes of, 115

insertions of, 1256
disulfide, 371
  complexes of, 114
divalent compounds, 354
electronic configuration, 352
electronic structure, 352
halides, 363
hydrogen bond cleavage, 1244, 1300
monoxide, 365
  as ligand, 82
  in insertion reactions, 1250
  reactions, 1285
  reactions as a coordinated ligand, 1257
nonmolecular compounds, 361
oxides, 364
phase diagram, 357
simple molecular compounds, 363
stability of catenated compounds, 355
suboxide, 365
Carbonato complexes, 115, 170
Carbonic acid, 236, 241, 366
Carbonic anhydrase, 241, 1332
Carbonium ions, 240, 352
Carbonylate anion complexes, 1062
Carbonylation reactions, 1286
  of benzylchloride, 1296
  of methanol, 1295
Carbonyl halides, 364
Carbonyl hydride complexes, 1064
  properties, 1065
Carbonyl, metal, see Metal carbonyls
Carbonyl process, 784
Carbonyl scrambling, 1226
Carboxylate complexes, 170-72
Carboxypeptidase A, 1329
Carnallite, 257, 281
Carnotite, 708
Cassiterite, 692
Catalysis, 1265
  cluster compounds in, 1304
  of organic reactions, acetic acid synthesis,
    1294
  addition to C–C multiple bonds, 1266
  alkene metathesis, 1277
  carbonylation reactions, 1297
  decarbonylation reactions, 1297
  hydrocyanation of alkenes, 1276
  hydroformylation of olefins, 1286,
    1291
  hydrosilation, 1275
  homogeneous hydrogenation, 1266
  in alcohol synthesis, 1289
  oligomerization reactions, 1282
  oxidations of alkenes, 1300

transfer hydrogenations, 1274
  valence isomerizations, 1298
  Ziegler-Natta, 831, 1280
Catalyst, heterogeneous, 1265
  heterogenized homogeneous, 1305
  homogeneous, 1265
    major applications in industry, 1267
  reversible cis-dihydride, 1268
Catalytic cracking, 392
Catenasulfur, 507-8
Cation-exchange resin, 263
Center-of-gravity rule, 679
Cerium, 982
  tetravalent, 999
    as an oxidant, 1000
    complex anions, 1000
    isopropoxide, 1000
  trivalent, 999
  see also Lanthanides
Ceruloplasmin, 1334
Cesium, 253
  preparation, properties, uses, 257
  suboxides, 254, 261
  see also Group I elements
Chabazite, 391
Chalconide ions, 502
Chalconides, metals, 512
Chaoite, 359
Charge-transfer compounds, of halogens,
    547
  L → M transitions in octahedral complexes,
    667
  L → M transitions in tetrahedral com-
    plexes, 668
  M → L transitions, 668
  spectra, 666
  transitions, 666
Chelate, 63
Chelate effect, 71
Chelate rings, conformations of, 78
Chiral, 669
Chirality, 47
Chloric acid, 559
Chlorine, 542
  dioxide, 554
  occurrence, 545
  oxides, 554
  preparation of, 545
  properties, 543
  see also Group VII elements
Chlorites, 559
Chloroauric acid, 978
Chlorophylls, 284
Chloroplatinic acid, 958

Chlorosulfuric acid, 540
Chlorous acid, 559
Chromate, 733
Chromite, 720, 723
Chromium, 719
  binary compounds, 721, 723
  biochemistry of, 1344
  carbonyl, 1051
  compounds, 721
  divalent, 724
    aqua ion, 724
    binuclear compounds, 725
    carboxylates, 725-26
    chemistry of, 724
    mononuclear compounds, 724
    quadruply bonded, 725-27
  elemental, 720
  halides, 721-22
  hexavalent, 733
    chemistry of, 733
    oxo halides, 734
    peroxo complexes, 735
  oxidation states, 720
  oxides, 722
  pentavalent, 732
    chemistry of, 732
    peroxo complexes, 735
  stereochemistry, 720
  tetravalent, 731
    chemistry of, 731
    peroxo complexes, 735
  trivalent, 727
    ammine complexes, 727
    ammonia complexes, 727
    anionic complexes, 728
    complexes, 727
    electronic structure of complexes, 730
    hexaqua ion, 727
    organo complexes, 731
Chromyl fluoride, 735
Cinnabar, 592
Circular dichroism, 671
Citrate complexes, 172
Classical complexes, 62
Clathrates, 226
Clathrochelate ligands, 143-44
Close packing, of anions, 16
  body-centered cubic, 6
  cubic, 5, 6
  hexagonal, 5
  of spheres, 4
Closo structure, 1085
Clostridium pasteurianum, 1341
Cluster structures, see Cage and

cluster structures
C-Nitroso compounds, 143-46
Cobaloximes, 782
Cobalt, 766
  biochemistry of, 1311, 1338
  binary compounds, 766
  clusters, 1083
  complexes with π-acceptor ligands, 780
  compounds, 766
    of isocyanides, 782
    of monomethylcyclopentadiene, 782
    reactions with PR_3 ligands, 781
  divalent, 768
    complexes, 768
    electronic structures of compounds, 770
    five-coordinate complexes, 772
    high-spin octahedral and tetrahedral
      complexes, 770
    ion, 66
    low-spin octahedral complexes, 772
    oxidation by molecular oxygen, 776
    square complexes, 772
    tetrahedral complexes, 768
  elemental, 766
  halides, 767
  oxidation states, 767
  oxides, 766
  pentavalent, 778
  salts, 766, 768
  stereochemistry, 767
  sulfides, 768
  tetravalent, 778
  trivalent, 773
    acetate, 774
    complexes, 773
    electronic structures of complexes, 775
    trifluoromethylsulfonate, 775
  univalent complexes, 779
Cobaltoborane clusters, 1178
Cobaltite, 766
Coesite, 387
Columbite-tantalite series, 831
Complexes, 48, 61
  formation constants of, 685
  stability of, 73
    enthalpy effects, 73
    entropy effects, 73
  stepwise formation of, 67
  with no π bonding, 630
  with π bonding, 632
π-Complexes, 65, 81
  of unsaturated organic molecules,
    95
Complex ion, 48

Condensation reactions for metal-carbon
    bond formation, 1138
Coordinated ligands, reactions of, 1257
Coordination compounds, 47
Coordination geometry, 48
Coordination number, 13, 48
  two, 49
  three, 49
  four, 49
  five, 50
  six, 51
  seven, 53
  eight, 53
  nine, 55
  higher, 55
Coordinative unsaturation, 1234
Copper, 798
  biochemistry of, 1334
  compounds, 800
    catalytic properties of, 820
    divalent, 811
      aqueous chemistry, 815
      binary compounds, 814
      carboxylates, 817-18
      chloro complexes, 813, 815
      complexes, 813, 815
      halides, 814-15
      magnetic properties, 814
      planar complexes, 813
      polynuclear complexes, 819
      salts of oxo acids, 815
      spectral properties, 814
      stereochemistry, 812
      trigonal-bipyramidal coordination, 814
    elemental, 800
    hydrides, 252
    oxidation states, 799
    stereochemistry, 799
    tetravalent complexes, 818
      compounds, 818, 820
      oxides, 818
    trivalent complexes, 818
      compounds, 818
      oxides, 818
      peptide complexes, 819
    univalent, 800
      acetylene compounds, 810
      aggregate compounds, 807
      alcoxides, 802-3
      alkene complexes, 809
      alkyls, 808
      aryls, 808
      binary compounds, 802
      binuclear species, 804
      carbonyls, 810
      carboxylates, 802
      cationic complexes, 809
      chain structures, 805
      complexes, 804
      complex hydrides, 809
      decanuclear complexes, 807
      equilibrium with Cu(II), 801
      hexanuclear complexes, 807
      mononuclear complexes, 804
      octanuclear complexes, 807
      organo compounds, 808
      oxide, 802
      pentanuclear complexes, 807
      sulfate, 802-3
      sulfide, 802
      tetrameric structures, 805
      trifluoromethylsulfonate, 803
Correlation diagrams, 202
Corrin ring, 1338
Corrins, 137
Corundum, 16, 329
Cotton effect, 672
Cotton-Kraihanzel method, 1078
Covalent solids, 23
  compounds, 24
  elements, 23
Creutz-Taube complex, 924-26
Cristobalite, 386, 392
Croconate, 170, 367
Crown ethers, 76, 163, 265, 285, 993
Cryolite, 328, 544
  structure of, 335
Cryptates, 76, 163, 254, 265, 286
Crysoberyl, 330
Crystal field splittings, 646
Crystal field stabilization energy, 682
  of d^n configurations in octahedral com-
    plexes, 683
Crystal field theory, 639
Cubane reactions, 1299
Cube, 45
Cumyl hydroperoxide, 499
Cuneane, 1299
Curium, 1005, 1042
  compounds, 1044
  elemental, 1013
  see also Actinides
Cyanamide, 370
Cyanate, 371
Cyanides, 369
Cyanoborohydride, 316
Cyanogen, 367
  chloride, 370

halides, 370
    reactions of, 369
Cyanuric chloride, 370
Cyclobutadiene complexes, 1161-62
Cyclohexasulfur, 506
Cyclometallation reactions, 130, 1245
Cyclooctasulfur, 506
Cyclooctatetraene, complexes, 105
Cyclopentadienyl compounds, 1163
    ionic cyclopentadienides, 1166
    metallocenes, properties of, 1164
    reactions of the C_5H_5 ring, 1166
    σ-bonded, 1167
    synthetic methods, 1163
Cyclopolyarsines, 468
Cyclopolyphosphines, 468
Cyclosulfurs, 506-7
    oxides, 529
Cyclothiazenium ions, 518
    anions, 518-19
    cations, 518
Cyclotriazine, 420
Cymantrene, 749, 1166
Cytochrome oxidase, 1323, 1334-35
    electron transfer mechanisms, 1323-24
Cytochrome P_{450} enzymes, 1324-25
Cytochromes, 1322

Deacon process, 545
Decaborane, 315
Decarbonylation reactions, 1297
Decavanadate ion, 712
Defect structures, 25
    nonstoichiometric, 25
    stoichiometric, 25
Dehydrogenation catalysts, 929
Denticity, 63
Deoxygenation of nitro arenes, 1297
Deoxyribonucleases, 1343
Deuterium, 215
    oxide, 215-16
Dialkylamido complexes, 123, 125
Diamond, 356
Diaspore, 329
Diastereomeric molecules, 80
Diazene ligands, 128, 130-31
Diazenido ligands, 128-30
Diazine, 420
Dibenzenechromium, 1168
Diborane, 304, 313
    reactions of, 314
Diboron oxide, 303
    tetrahalides, 302
Dicarbollide ion complexes, 1177

Dichlorine heptooxide, 555
Dichlorine hexaoxide, 555
Dichlorine monoxide, 554
Dichlorine tetraoxide, 555
Dichromate, 733
Dienyl complexes, 1155
Difluorodiazene, 434
1,2-Dihydroxoarenes, 161
Diimine complexes, 118
Dimethylsulfoxide, as ligand, 177
Dinitrogen, as ligand, 86, 95
    difluoride, 434
    pentoxide, 427
    reactions of coordinated, 1262
    tetroxide, 426
    trioxide, 425
Diopside, 388
Dioxovanadium ion, 712
Dioxygen, as ligand, 95, 155-56
    chemical properties, 489
    difluoride, 493
    reaction of coordinated, 1261
    reactions with, 1302
Dioxygenyl cation, 494
Diphosphine, 446
Diphosphine ligands, 148
Directed valence theory, 200, 208
Dirhenium compounds, 892
Diruthenium tetracarboxylates, 931
Dissymmetric, 47, 669
Disulfide ion, complex geometries, 179
Disulfides, 514
Disulfites, 532
Disulfur dinitride, 517
Disulfur nitride ion, 517
Dithiocarbamates, 372
    complex geometries, 183
    as ligands, 182-84
Dithiocarboxylates, 182
    complexes, 184
Dithio esters, complexes of, 188
1,3-Dithioketonates, as ligands, 189
1,2-Dithiolenes, as ligands, 185
    complexes, 186
    electronic structures, 187
    syntheses of complexes, 185-86
Dithionic acid, 536
Dithionous acid, 535-36
DNA- and RNA-polymerases, 1329
Dodecahedron, 53
Dolomite, 281
Döppler effect, 750
d orbitals, partly filled, structural and ther-
    modynamic consequences, 676

splitting of, 641, 644
  thermodynamic effects of, 682
dπ-pπ bonds, 208-9
  in less symmetrical molecules, 210-11
  in tetrahedral molecules, 209-10
Dysprosium, 982
  tetravalent, 1001
  see also Lanthanides

Eighteen electron rule, 63, 1050
Einsteinium, 1005, 1042
  compounds, 1045
  see also Actinides
Electrode potentials, 230
Electron deficiency, 304
Electron donor, 63
  two, 63
  three, 63
  four, 63
Electron-exchange processes, 1205
  with ligand exchange, 1216
Electronic absorption spectra, 657
  intensities and line widths, 663
Electronic configurations, high-spin, low-
    spin configurations, 645
Electron-pairing energy, 644-47
Electron paramagnetic resonance, 626
Electron transfer processes, 1205
  mechanistic criteria, 1214
Elimination reactions, in transition metal
    complex synthesis, 1135
Elimination stabilized alkyls, 1122
Enantiomorphs, 79
Enclosure compounds, 226
Energy level diagram for d-orbital splittings,
    641, 643, 662
  for d^2-d^8 ions, 658
  for spy and tbp complexes, 642
Enneatungsto compounds, 859
Enstatite, 388
Enthalpy, of atomization, 8
  of electron attachment, 18
Enterobactin, 1315
  structure of, 1316
Entropy, 72
Enyl complexes, 106
Erbium, 982. *See also* Lanthanides
Erionite, 391
Esters, as ligands, 162
Ethers, as ligands, 162
Ethylene, complexes, 97
  polymerization, 1280
Ethylenediaminetetraacetic acid, 172
Europium, 982

tetravalent, 1001
  see also Lanthanides
Extended Hückel method, 637
Extended X-ray absorption fine structure
    technique (EXAFS), applications,
    1306-7, 1313, 1325

Faraday effect, 674
F centers, 25
Feldspars, 389
Fenske-Hall method, 637
Fermium, 1005, 1045. *See also* Actinides
Ferredoxins, 1325
Ferrichrome, 1314
Ferrichrysin, 1315
Ferrimycobactins, 1315
Ferritin, 1313
Ferrocene, 99, 1160, 1166
  bonding, 100-102
  MO diagram, 639
  properties, 1164
Ferrocyanide, 755
Ferrovanadium, 709
Fisher-Tropsch process, 1287
  carbene mechanism in, 1288
Fluorapatite, 422, 544
Fluorides, Group IV elements, 385
  Group V elements, 448
  Group VII elements, 563
  molecular, 552
  oxygen, 493
Fluorine, 542
  electronic configuration, 543
  occurrence, 544
  organic compounds, 572
    preparative methods, 572
    properties of, 573-75
  oxides, 493, 553-54
  properties, 544
  reactivity, 545
  see also Group VII elements
Fluorite structure, 16
Fluoroborate anions, 301
  as ligands, 193
Fluoroselenic acid, 530
Fluorosulfuric acid, 246
Fluorosulfurous acid, 540
Fluoroxosulfate ion, 571
Fluorspar, 544
Fluxional molecules, 1218
Formamidino ligands, 132
Formation constants, overall, 69
  stepwise, 69
Francium, 254

Free-radical reactions in metal-carbon bond
    synthesis, 1137
Fremy's salt, 431
Frenkel defect, 25
Friedel-Crafts acylation, 1166
Friedel-Crafts catalysts, 335

Gadolinium, 982. *See also* Lanthanides
Galactose oxidase, 820, 1335
Galena, 591
Gallates, 334
Gallium, 326
    alkyls, 345
    aqua ions, 333
    halides, 330
    hydride, 339
    lower valent compounds, 347
    nitrato complexes, 338
    nitrides, 332
    occurrence, isolation, and properties,
        328
    organometallic compounds, 341
    oxide, 330
    see also Group III elements
Garnierite, 784
Gas hydrates, 227
Germanates, 393
    meta, 393
    ortho, 393
Germanes, 384
Germanium, 374
    allotropic forms, 383
    as ligand atom, 116
    dihalides, 396
    divalent state compounds, 380, 396, 400
    hydrides, 384
    hydroxides, 392
    occurrence, isolation, and properties, 382
    oxides, 392
    oxo anions, 393
    tetrachloride, 386
    see also Group IV elements
Gibbsite, 329
Glucose tolerance factor, 1344
Gluthathione peroxidase, 505
Gmelinite, 391
Gold, 966
    binary compounds, 975
    biochemistry of, 980
    cluster compounds, 976
    compounds, 975
    divalent complexes, 978
    elemental, 967
    halides, 975

organo compounds, 979
oxidation states, 967
oxides, 975
pentavalent compounds, 979
stereochemistry, 967
trivalent complexes, 978
univalent complexes, 976
Goldfingia gouldii, 1328
Graphite, 356, 357-61
    electrically conducting intercalation com-
        pounds, 359
    fluoride, 359
    intercalation compounds of, 359
    oxide, 359
Grignard reagents, 286-87, 405
Group, 36
    properties of, 36
    very high symmetry, 44
Group I elements (Li, Na, K, Rb, Cs), 253
    binary compounds, 260
    complexes, 264
    compounds, 260
    hydroxides, 261
    ionic salts, 261
        characteristics, 262
        structures and stabilities, 262
    M^+ ions in solution, 263
    organolithium compounds, 266
    organopotassium compounds, 268
    organosodium compounds, 268
    oxides, 260
    ozonides, 498
    peroxides, 497
    preparation, properties, uses, 257
    properties of, 253
    solutions in liquid ammonia and other sol-
        vents, 258
    superoxides, 498
Group II elements (Be, Mg, Ca, Sr, Ba, Ra),
        271
    binary compounds, 276, 282
    carbides, 283
    complexes, 277, 283
    group relationships, 271
    halides, 277, 282
    hydrides, 247
    ions, 283
    organometallic compounds, 279, 286
    oxides, 276, 282
    oxo salts, 283
    peroxides, 497
    physical parameters, 271
Group IIb elements (Zn, Cd, Hg), 589
    aqua ions, 599

group trends, 589
halides, 598, 603
hydroxides, 597
isolation, 591
occurrence, 591
organometallic compounds, 608
oxides, 597, 602
oxo salts, 599, 604
position in periodic table, 589
properties of, 589, 591
selenides, 598
stereochemistry, 590
sulfides, 598, 602
tellurides, 598
univalent state, 593
Group III elements (Al, Ga, In, Tl), 111,
    326
  adducts, 334
    donor, 341
    neutral, 336
  alcoxides, 337
  aqua ions, 333
  aqueous chemistry, 333
  binary compounds, 329
  cationic complexes, 336
  chelate complexes, 337
  chemistry of the trivalent state, 329
  complex compounds, 333
  complex halides, 334
  complex hydrides, 338
  dithiocarbamates, 337
  electronic structures and valences, 326
  halides, 330
  hydride anions, 339
  isolation, 328
  lower valent compounds, 346
  nitrato compounds, 338
  occurrence, 328
  organometallic compounds, 341
  oxidation states, 328
  oxo salts, 333
  oxygen compounds, 329
  properties, 326, 328
  stereochemistries, 328
  transition metal complexes, 346
Group IV elements (Si, Ge, Sn, Pb), 374
  alcoxides, 395
  alkyl- and aryl-silicon halides, 403
  bond strengths, 374
  carboxylates, 395
  catenation, 374
  chloride oxides, 386
  complexes, 393
  compounds, 383

divalent, oxidation state, 376, 380, 395,
    400
electronegativities, 375
fluorides, 385
group trends, 374
halides, 384-85
multiple bonding, 377
occurrence, isolation, and properties, 382
organo compounds, 400
oxo salts, 395
oxygen compounds, 392
stereochemistry, 381
sulfides, 395
tetravalent, oxidation state, 380, 401
valence, 381
Group V elements (P, As, Sb, Bi), 438
  aromatic heterocycles, 470
  aqueous cationic chemistry, 470
  binary compounds, 444
  chalcogenides, 458
  chlorides, 449
  compounds, 459
    pentavalent, 469
  five-coordinate compounds, 441
  fluorides, 448
  group trends, 438
  halides, 446
  halo complexes, 451
  hydrides, 445
  lower halides, 450
  organic compounds, 465
  oxides, 453
  oxo acids and anions, 472
  oxo halides, 459
  pentahalides, 448
  properties, 438-9
  stereochemistry, 438
  sulfides, 456
Group, VI elements (S, Se, Te, Po), 502
  binary compounds, 511
  d orbital participation, 504
  electronic structure, 502
  group trends, 503
  halides, 519
  halo complexes, 525
  halooxo acids, 540
  hydrides, 511
  metal chalconides, 512
  occurrence, 505
  oxides, 526
  oxo acids, 530
  oxohalides, 537
  properties, 502
  reactions of, 509-11

secondary bonding, 504
stereochemistries, 502-3
structures of, 509
valences, 502
Group VII elements (F, Cl, Br, I, At), 542
  charge-transfer compounds, 547
  diatomic compounds, 562
  electronic structures and valences, 542
  fluorides, 563
  halides, 549
  interhalogen compounds, 562
  interhalogen ions, 562, 566
  oxides, 553
  oxo acids and salts, 556
  oxo halogeno fluorides, 569
  properties, 543
  reaction with H$_2$O and OH$^-$, 556
Guanidines, 147

Haber process, 414
Hafnium, 824
  elemental, 825
  lower oxidation states, 829
  organometallic compounds, 830
  oxidation numbers less than three, 830
  oxidation states, 824
  stereochemistry, 824
  tetravalent, 825
    aqueous chemistry, 827
    borohydrides, 829
    chalcogenides, 827
    compounds, 825
    dithiolene complex anions, 828
    nitrato complexes, 829
    oxide and mixed oxide, 826
    stereochemistry, 826
  trivalent, 829
Hägg compounds, 11
Hahnium, 1005, 1045. See also Actinides
Half-Dawson structures, 859
Halic acids, 559
Halide bridges types, 191
Halide complex ions, as ligands, 193
Halide ions, affinities for metal ions, 189
  complexes of, 189
  stabilities, 190
Halides, alkyl-silicon, 403
  anhydrous, preparation of, 549
  aryl-silicon, 403
  binary ionic, 550
  carbon, 363
  Group II elements, 277, 282
  Group III elements, 330
    complexes, 334

Group IV elements, 384-85
Group V elements, 446
Group VI elements, 519
Group VII elements, 549
  hydrated, dehydration of, 550
  hydrogen, 241-42
  molecular, 552
    reactivity, 553
Halite ions, 559
Halocomplexes, of heavier elements, 206
Halogen fluorides, 563
  physical properties of, 564
  substituted, 566
Halogen nitrates, 436
Halogenosulfuric acids, 540
Halogens, see Group VII elements
  cationic complexes, 568
  oxides, 554
  oxo acids of, 556
Halous acids, 559
Hausmannite, 741
Helium, 577
  phase diagram, 579
  special properties of, 579
  see also Noble gases
Hematite, 751
Heme-dioxygen binding, 1319
  model systems, 1320
Hemerythrins, 1328
Hemimorphite, 388
Hemocyanins, 1337
Hemoglobin, 1316
  cooperativity, 1321
Hemosiderin, 1313
Heterocyclic arene rings, complexes, 100
Heteropoly blues, 861
Heteropolyniobate ions, 746
Heteropolyvanadate ions, 746
Hexachlorocyclotriphosphazene, 461
Hexachloroplatinate, 958
Hexachlororhenate(IV), 890
Hexachlorotechnate, 890
Hexafluorophosphate ion, 478
Hexokinase, 1343
Hole formalism, 658
Holmium, 982. See also Lanthanides
Horn silver, 967
Hybridization theory, 200, 208
Hydrated electrons, reductions by, 1217
Hydrates, 225
  bromine, 227
  crystalline, 225
  gas, 226, 227
  salt, 227

Hydrazine, 418
  complexes, 125, 127-29
Hydrazoic acid, 421
Hydrides, 107, 247
  boron, 303. *See also* Boranes
  classification of, 247
  d-block transition metals, 251, 1115
  Group I elements, 247-49
  Group II elements, 247-49
  Group IV elements, 383
  Group V elements, 445
  Group VI elements, 511
  lanthanide, 250
  molecular, boiling points, 220
  organotin, 404
  reaction with coordinated CO, 1257
  saline, 247-49
Hydride shift (1,3), 1255
β-Hydride transfer-alkene elimination
      reaction, 1120
Hydroboration, 321
Hydroformylation reactions, 1291
  catalytic cycle for, 1293
Hydrogen, 215
  addition to metal complexes, 1240
  as ligand atom, 107, 1113
  bond, 218, 219
    bifurcated, 223
    parameters of, 221
    strengths, 221
    strong, 221-23
    theory of, 223
  bonding of, 217
  bridge bonds, 218
  bridges, 108
  bromide, 242
  chloride, 242
  cyanide, 95, 368
  electrode, 229
  fluoride, 241
    crystalline, 241
  industrial use, 216
  iodide, 242
  ion, 228
  peroxide, 495
    structure of, 496
  properties, 216
  reactions, 217
  sulfide, 512
    as ligand, 177-78
  sulfites, 532
Hydrogenation catalysts, 929
  with unsaturated compounds, 1266
Hydroisomerization, 392

Hydroperoxides, 499
  in organic oxidations, 1303
Hydrosilation reaction, 385
Hydroxamates, ligands, 177
Hydroxide ion, 485
  as ligand, 152
Hydroxonium ion, 228
Hydroxylamine-N,N-disulfonate, 431
Hydrozirconation reaction, 1255
Hypohalous acids, 557
Hypomanganate cyclic esters, 746
Hypomanganates, 746
Hyponitrous acid, 428
Hypophosphorous acid, 472, 473

Ice, 224
  structure of, 224
Icosahedral geometry, 56
Icosahedron, 46
Identity, 32
Ilmenite, 695
  structure, 17
Imidazole complexes, 122
Improper rotation, 30
Indium, 326
  aqua ions, 333
  carboxylates, 337
  halides, 330
  lower valent compounds, 348
  nitrides, 332
  occurrence, isolation, and properties, 328
  organometallic compounds, 341
  see also Group III elements
Inert complexes, 1184
Inert pair effect, 327
Inner-sphere mechanism, 1206
Inner-sphere reactions, 1210
  doubly bridged, 1214
  intimate mechanism, 1212
Insertion reaction, 117
  definition, 1250
  general remarks, 1248
  in synthesis of metal-carbon σ-bonds,
      1135
  representative groups in transfers, 1249
Insulator, 8
Interstitial compounds, 9
  characteristics, 12
  phases, 12
Inversion, 30
Iodic acid, 559
Iodine, 542
  occurrence, 546
  oxides, 555

oxo compounds, 572
properties, 543
trichloride, 563
trivalent, compounds of, 571
see also Group VII elements
Ionic crystals, energetics, 18
Ionic crystal structures, 12, 13, 15
Ionic cyclopentadienides, 1166
Ionic oxide, 4
Ionic halides, 4
Ionic radii, 12
table of, 14
Iridium, 934
anionic complexes, 941
cationic complexes, 941
divalent complexes, 948
general remarks, 934
pentavalent complexes, 949
stereochemistry, 934-35
tetravalent complexes, 948
trivalent, aqua ions, 944
cationic complexes, 945
complexes of, 943
halide complexes, 945
hydrido complexes, 946
neutral complexes, 946
univalent complexes, 940
see also Platinum metals
Iron, 749
binary compounds, 753
biochemistry of, 1311
carbonyl, 1051
reactions, 1071
compounds, 752
with mixed oxidation states, 762
divalent, 754
aqueous chemistry, 754
complexes, 755
coordination chemistry, 754
electronic structures of complexes, 756
five-coordinate complexes, 756
tetrahedral complexes, 756
elemental, 750
halides, 753
hexavalent, 765
higher oxidation states, 765
oxidation states, 751
oxides, 752
proteins, 1312
stereochemistry, 751
storage and transport, 1313
tetravalent, 765
trivalent, 757
aqueous chemistry, 757

complexes, 758
coordination chemistry, 757
electronic structures of compounds,
761
five-coordinate complexes, 762
oxo-bridged complexes, 760
Iron-sulfur clusters, 763
reactions of cluster anions, 764
Iron-sulfur proteins, 1325
Isocyanides, 93
reactions of coordinated, 1261
Isomerization process, definition, 1218
of trischelate complexes, 1223
Isopolymolybdates, 852
Isopolytungstates, 856

Jahn-Teller effects, 678, 680

Keggin structure, 857
β-Ketoenolato complexes, 166
C-bonded complexes, 168-69
Ketones, as ligands, 162, 165
Krogmann salts, 1111
Krypton, 577
difluoride, 587
see also Noble gases

Lability of complexes, 1184
Laccase, 1334
Lamellar compounds, 359
Lanthanide contraction, 822, 981
Lanthanum fluoride cycle, 1038
Lanthanides, 981
aqua ions, 991
binary compounds, 990
borohydrides, 995
chlorides, 990
comparison with actinides, 1007
complexes, 991
amines, 993
crown ethers, 993
halogeno, 992
macrocyclic, 994
nitrogen ligand, 993
oxygen ligand, 992
porphyrins, 994
sulfur ligand, 994
thiocyanates, 993
coordination numbers, 985
diketonates, 992
as shift reagents, 992
divalent, 1001
halides, 990
isolation, 988

magnetic properties, 983
metals, 989
occurrence, 988
organometallic compounds, 995
  σ-bonded compounds, 996
oxalates, 991
oxidation states, 983
oxo salts, 991
properties, 982
reduced compounds, 1002
spectral properties, 983
stereochemistry, 985
sulfides, 1003
tetravalent, 999
trivalent state, 990
variations in the series, 987
Lanthanoids, 981. *See also* Lanthanides
Lanthanum, 982. *See also* Lanthanides
Lapis lazuli, 390, 513
Lattice energy, 19
  calculations, 21
Lawrencium, 1005, 1045. *See also*
    Actinides
Layer structure, 16
Lead, 374
  allotropic forms, 383
  divalent compounds, 380, 399
  hydride, 385
  hydroxides, 399
  occurrence, isolation, and properties, 382
  organo compounds, 400, 405
    $Pb(CH_3)_4$, 405
    $Pb(C_2H_5)_4$, 405
  oxides, 395
  oxo anions, 393
  tetrachlorides, 386
  see also Group IV elements
Ligand displacement reactions, 1198
  mechanism of, 1198
  in square complexes, 1198
Ligand exchange via electron exchange, 1216
Ligand field theory, 639
Ligand replacement reactions, mechanisms
    for, 1185
  classifications, 1186
Ligands, 48, 61
  bidentate, 74
  classes of, 62
  classification by donor atoms, 107
  encapsulating, 77, 143
  higher-dentate, 76
  macrocyclic, 76
  pentadentate, 76
  quadridentate, 75

tridentate, 75
tripod, 76
types and classification of, 74
of unusual reach, 78
Limonite, 751
Linear molecules, 43, 49
Linkage isomerism, 64
Litharge, 399
Lithium, 253
  acetylide, 255
  alkyls, 113, 266
    structures, 267
  aluminum hydride, 339
    reactions of, 340
  aryls, 266
  compounds, 256, 268
  hydride, 248
  nmr, 266
  ore, 388
  preparation, properties, uses, 257
  tetraalkylmanganates, 748
  see also Group I elements
Lutetium, 982. *See also* Lanthanides

Macrocyclic complexes, characteristics, 133
  of alkali metal ions, 265
  containing phosphorus, 149
  containing sulfur, 149
Macrocyclic effect, 73
Macrocyclic ligands, with conjugated π sys-
    tems, 133
  without conjugated π systems, 139
Macrocyclic nitrogen ligands, 132
Macrocyclic polyethers, *see* Crown ethers
Madelung constant, 19
Magnesium, 271, 281
  binary compounds, 282-83
  biochemistry of, 1343
  hydride, 249
  occurrence, 281
  organomagnesium compounds, 286
  see also Group II elements
Magnetic circular dichroism, 674
Magnetic moment, 626
  spin-only, 627
  for transition-metal ions, 628
Magnetic properties, of d-block elements,
    644, 823
  of heavier elements, 650
Magnetism, in transition metal chemistry,
    625
Magnetite, 751
Magnus's green salt,
    1112

Main group compounds, stereochemistry
  and bonding, 195
Manganate ion, 746
Manganese, 736
  biological role, 742, 1344
  carbonyl complex, 1052-53
  compounds, 738
  divalent, 738
    acetylacetonate, 739
    binary compounds, 738
    chemistry of, 738
    complexes, 739
    electronic spectra, 663, 740
    oxide, 738
    phthalocyanines, 740
    porphyrins, 740
    salts, 738
    sulfide, 738
  elemental, 736
  heptavalent, 746
    chemistry of, 747
    oxide, 748
    oxo halides, 748
  hexavalent, chemistry of, 746
  low oxidation states, 749
  organometallic compounds, 748
    alkyls, 748
    cyclopentadienyl compounds, 749
  oxidation states, 737
  pentavalent, 746
    chemistry of, 746
    oxohalide, 746
  stereochemistry, 737
  tetravalent, 744
    binary compounds, 744
    chemistry of, 744
    complexes, 745
    dioxide, 744-45
    sorbitolate complex, 746
    tetradentate Schiff base complex, 745
    tetrafluoride, 745
  trivalent, 741
    acetylacetonate, 742
    binary compounds, 741
    chemistry of, 741
    complexes, 741
    electronic structure of compounds, 743
    fluoride, 741
    gluconate ion-complex, 743
    halogeno complexes, 742
    ion, 741
    Schiff base complexes, 743
Manganic acetate, 742
Manganite, 741

Manganocene, 104
Marcasite structure, 514
Marcus cross-relation, 1209
Marcus theory, 1208
Markownikoff additions, of alkenes, 1254
Massicott, 399
Melamine, 370
Mendelevium, 1005, 1045. See also
  Actinides
Mercuration, 613
Mercuric carboxylates, 604
Mercuric chloride, 603
Mercuric complexes, 604
Mercuric compounds with nitrogen, 606
Mercuric fluoride, 603
Mercuric halide, 603
  and pseudohalide complexes, 605
Mercuric oxide, 602
Mercuric oxo complexes, 605
Mercuric oxo salts, 604
Mercuric sulfide, 602
Mercurinium ions, 614
Mercurous compounds, 593, 596
Mercurous halides, 596
Mercurous ion, 593
Mercurous-mercuric equilibria, 593
Mercurous nitrate, 594, 596
Mercurous perchlorate, 596
Mercury, 273, 589
  compounds with metal-to-mercury bonds,
    607
  divalent compounds, 602
  in oxidation state 0.33, 596
  organometallic compounds, 610
    in environment, 611
  properties, 589, 591
  stereochemistry, 590-91
  trivalent, 608
  univalent, 593, 596
  see also Group IIb elements
Metaborates, 299
Metal-arene complexes, bonding in, 639
Metal arsines, 1056
Metal atom clusters, 729, 881, 1080
  carbonyl compounds, 1082
    Co, Rh, Ru, 1083
    Os, 1085-90
  in catalysis, 1304-5
  high nuclearity, 1090
    Ni, 1090-91
    Pt, 1090
    Rh, 1091-93
  mixed metal, 1094
  noncarbonyl compounds, 1094

octahedral and Re₃ clusters, 1095
relation to multiple bonds, 1109
Metal atom reactions, 1135
Metal complex cap, 144
Metal carbene complexes, 1143
  synthesis of, 1143-46
Metal carbonyl halides, 1068-70
Metal carbonyls, 1049
  additional structural aspects, 1057
    semibridging CO groups, 1057
  binuclear, 1052
  bonding in, 638
  clusters, 1082
    high-nuclearity, 1090
    mixed-metal, 1094
  fluxionality in, 1226
  mononuclear, 1050-51
  polynuclear, 1054
    large clusters, 1055
    $M_4(CO)_{12}$ compounds, 1055
    $M_3(CO)_{12}$ compounds, 1054
  preparation of, 1061
  reactions of, 1070, 1202, 1257-60
  side-on bonding of CO, 1060
  vibrational spectra of, 1070
    force constants and bonding, 1077
    structural diagnosis, 1071
Metal-carbyne complexes, 1146
  preparation of, 1146-47
Metal isocyanides, 1056
  side-on bonding, 1060
Metallic conductance, 8
Metalloborane complexes, 1174, 1176
  clusters, 1180
Metallocarbaborane complexes, 1174
  clusters, 1180
  structure of, 1179
Metallocumulenes, 1147
Metallocycle, 98
Metallocyclic complexes, 1128
Metalloenzymes, 1310
Metallooxaziridines, 145
Metalloproteins, 1310
Metallothioneins, 1334
Metal-metal bonding, 823
Metal-metal bonds, 1080
  carbonyl clusters, 1082
  multiple, 1098
    bonds of orders less than 3, 1107-9
    quadruple, 1098-1105
    triple, 1105
  noncarbonyl compounds, 1094
Metal nitrosyls, 1055
  reactions of, 1260

Metal-olefin bonding, 97
Metals, 4, 5
  bonding, 6
  characteristic properties, 6
  cohesive energies, 8
  structures, 6
Metaphosphates, 476
Metal phosphines, 1056
Metal sulfides, 1056
Metathesis reactions, 1277
Methacrolein, 1304
Methacrylic acid synthesis, 1304
Methanation, 1207
Methyleneamido ligands, 127
Michaelis-Arbusov reaction, 473
Migration of atoms, metal to ligand, 1248
Millerite, 784
Millon's base, 606
Mixed oxide structures, 16
Molecular dissymmetry, 47
Molecular orbital stabilization energy, 684
Molecular rearrangements, 1183
Molecular sieves, 391
Molecular symmetry, 28, 47
Molybdates, 850
Molybdenite, 846
Molybdenum, 844
  aqua ions, 867
  blue oxides, 848
  biochemistry of, 1341
  carbonyl complexes, 877, 1051
    reactions, 1071
  compounds with triple and quadruple
    bonds, 873
    reactions, 875
  cyano complexes, 878
  dinitrogen compounds, 879
  elemental, 846
  halide and halo complexes, 861
  halogeno complexes, 864
  heteropolyacids and their salts, 852, 857
  heteropoly blues, 861
  hexavalent, oxide halides, 866
    oxo complexes, 868, 872
  isopolyacids and their salts, 852
  metal atom cluster compounds, 881
  mixed oxides, 848
  non-oxo complexes, 877
  octahedral clusters, 1095
  organoheteropoly anions, 859
  oxidation states, 845
  oxide halides, 866
  oxides, 847
  oxo acids, 847

pentavalent, oxide halides, 867
  oxo species, 870
phosphine complexes, 877
stereochemistry, 845
sulfides, 847, 850
Molybdic acid, 851
Monazite, 988
Monocyclopentadienyl compounds, 1165
  some reactions of, 1165
Monohydrido complexes, 1270
Mössbauer effect, 749
Mulliken symbols, 659
Multicenter bonding, 304
Myoglobin, 1316

Naphthenic acids, 171
  complexes, 171
Neodymium, 982
  tetravalent, 1001
  see also Lanthanides
Neon, 577. See also Noble gases
Neptunium, 1005, 1036
  aqueous chemistry, 1040
  binary compounds, 1039
  elemental, 1013
  halides, 1039
  isolation, 1036
  oxides, 1039
  see also Actinides
Nickel, 783
  biochemistry of, 1344
  carbonyl, 1051
  compounds, 785
  divalent, 785
    addition of ligands to square complexes,
      791
    anomalous properties, 791
    binary compounds, 785-86
    binuclear compounds with Ni-Ni bonds,
      790
    chemistry of, 785
    conformational changes, 791
    cyanide, 785
    electronic structure, 786
    five-coordinate complexes, 788
    halides, 785
    hydroxide, 785
    isomerism, 793
    monomer-polymer equilibria, 793
    octahedral complexes, 786
    oxide, 785
    planar complexes, 790
    salts of oxo acids, 786
    square-tetrahedral equilibria, 793

stereochemistry, 786
tetrahedral complexes, 788
thermochromism in, 794
elemental, 784
higher oxidation states, 795
  oxides, hydroxides, halides, 795
lower oxidation states, 797
oxidation states, 784
stereochemistry, 784
tetravalent complexes, 795-96
trivalent complexes, 796
  five-coordinate, 797
  with cyclam, 797
  with deprotonated polypeptide amides,
    797
Nickel arsenide structure, 514
Nido structure, 1085
Niobates, 833
Niobium, 831
  cluster compounds, 841
  compounds, 831
  elemental, 831
  lower oxidation states, 843
  octahedral cluster, 1095
  oxidation states, 832
  pentavalent, 831
    alcoxides, 836
    complexes, 836
    compounds, 838
    dialkylamides, 836
    fluorides and fluoride complexes, 834
    halides, 835
    hydrides, 838
    isopolyanions, 833
    organometallic chemistry, 838
    oxide halides, 835
    oxygen compounds, 831
  stereochemistry, 832
  tetravalent, 839
    alcoxides, 840
    complexes, 840
    diketonates, 840
    halide complexes, 840
    halide, 839
    oxides, 839
    sulfides, 839
  trivalent compounds, 841
Nitramide, 429
Nitrate reductase, 1342
Nitrates, 430
  complex geometries, 173
  ir spectrum, 173
Nitrene complexes, 123, 125-26
Nitric acid, 243, 430

gaseous, 244
Nitric oxide, 423
  as ligand, 90
  bent, terminal, 92
  bridging, 92
  in insertion reactions, 1256
  linear-terminal, 91
Nitrides, 414
Nitrido complexes, 123-24
Nitriles, as ligands, 142
  complexes, 119
Nitrite ion, complex geometries, 173
  ir spectrum, 174
Nitroalkane, 174
Nitrogen, 407
  active, 413
  as ligand atom, 118
  binary halides, 432
  catenation, 411
  comparison with phosphorus, 439
  complexes, 118
  compounds, 409, 414
    multiple bonding, 409
  covalence, 408
  dioxide, 426
  electronic configuration, 407
  elemental, 412
  fixation, 413
  four-covalent, 411
  halogen compounds, 432, 434, 436
  hydrides, 414, 420
  hydrogen bonding, 412
  inversion, 409
  multiple bonding, 409
  occurrence and properties, 412
  oxidation numbers, 408
  oxides, 422, 432
  oxo acids, 428
  oxo compounds, 410
  oxo halides, 434
  stereochemistry, 408
  three-covalent, 409
  three-valent, donor properties, 411
  trifluoride, 432
  trihalides, 434
Nitrogenase, model compound, 185, 1341, 1343
Nitrogenases, 845, 1341-43
Nitrogen heterocycles, as ligands, 121
Nitronium ion, 244, 426, 430
  salts, 431
Nitroso disulfonate ions, 431
Nitrosonium ion, 424
Nitrosylazide, 428

Nitrosyl halides, 435
  hypofluorite, 435
Nitrous acid, 429
Nitrous oxide, 422
Nitroxide radicals, 408
Nitrylazide, 428
Nitryl halides, 435
Nobelium, 1005, 1045. *See also* Actinides
Nobel gases, 577
  application, 578
  chemistry of, 580
  group trends, 577
  isolation, 579
  occurrence, 578
  properties, 577
  solid, 4
Noble gas formalism, 63, 1050
Nonclassical ligands, 62
Noncomplementary reactions, 1215
Nonmolecular solids, 3
Nucleotide complexes, 121
Nylon 66, 1277

Octahedron, 45
Oleums, 245
Oligomerization reactions, 1282
One-copper blue proteins, 1336
One-dimensional solids, 1110
Optical activity, 47, 669
  applications, 672
Optical rotatory dispersion, 671
Orbital moment, 626
Orbital splitting, 644
Organic halides in addition reactions, 1242
Organoborane compounds, 320
Organogold compounds, 979
Organometallic compounds, fluxional, 1229
  general survey, 109
  ionic, 110
  nonclassical, 111
  π complexes, 1141
    alkenes, 1147
    alkynes, 1156
    allyl and enyl, 1152
    carbocyclic, 1160
    multidecker sandwiches, 1172
  with metal-carbon σ-bonds, 110, 1119
    alkyl and aryl, 1123
    in catalysis, 1265
    decomposition of, 1138
    in synthesis, 1234
    synthesis of, 1133
  with multiple metal-carbon bonds, 1140
    alkylidenes, 1140

alkylidynes, 1142
  carbenes and carbynes, 1142
Organosilver compounds, 972
Organotin compounds, 404
  catenated linear and cyclic, 404
  hydrides, 404
Organotitanium compounds, 704
Organovanadium compounds, 718
Orthoclase, 390
Orthometallation reaction of aryl groups, 1245
Orthosilicates, 387
Orthophosphoric acid, 472, 474
Osmates, 916
Osmenates, 916
Osmiamate ion, 933
Osmiridium, 901
Osmium, 912
  carbonyl, 1051
  cluster anions, 116
  general remarks, 912
  halo complexes, 917, 919
  nitric oxide complexes, 926
  nitrido complexes, 932
  nitrogen donor ligand complexes, 921
  oxidation states, 913
  oxo compounds, 914
  pentavalent complexes, 932
  stereochemistry, 912-13
  tertiary phosphine and related complexes, 927
  tetrachloride, 909
  tetraoxide, 914
  trichloride, 909
  see also Platinum metals
Outer-sphere complex, 1186
Outer-sphere mechanism, 1206
Outer-sphere reactions, 1206
  Marcus theory, 1208
Oxalato complexes, 170
Oxidases, 1334
Oxidation potential, 230
Oxidative-addition reactions, general remarks, 1237
  mechanisms of, 1240, 1243
  specific molecules, 1240
    hydrogen, 1240
    organic halides, 1242
  stereochemistry of, 1240
  substances which can be added, 1239
  in synthesis of metal-carbon σ bonds, 1134
Oxidative deamination, 1137
Oxide ions, as ligands, 152

Oxides, acidic, 484
  amphoteric, 485
  basic, 484
  Group I elements, 260
  Group II elements, 276, 282
  Group III elements, 329
  Group V elements, 453
  Group VI elements, 526
  Group VII elements, 553
  mixed metal, 16
  types of, 483
Oxime compounds, 143
Oxo anions, as ligands, 170
Oxocarbenium ion, 336
Oxo-centered complexes, 154
Oxo-centered triangles, 154
Oxo compounds, 153
  linear, 154
  pyramidal, 154
  singly bridged, 153
Oxohyponitrite ion, 429
Oxomercuration, 613
Oxonium ions, 231, 486
Oxonium salts, 1134
Oxo process, 1291
Oxovanadium ion, 715
Oxygen, 483
  allotropes, 487
  as ligand atom, 152
  carriers, 776
  catenation, 487
  chemical properties, 489
  complexes, see Dioxygen
  compounds, 493
  covalent compounds, 486
  difluoride, 493
  electronic structure, 483
  elemental, 487
  fluorides, 493
  four-coordinate, 487
  occurrence and properties, 487
  photochemical oxidations, 491
  singlet, 491
  stereochemistry, 486
  three-coordinate, 486
  two-coordinate, 486
  unicoordinate, multiply bonded, 487
Oxypalladation reactions, 1300
Ozone, 488
  chemical properties, 492
Ozonides, 497, 498

Palladium, 950
  alkyl and aryl complexes, 957

anhydrous acetates, 954
catalyzed oxidations, 1300
complexes of, 951, 954
compounds with Pd-Pd bonds, 963
dichloride, 909
five-coordinate complexes, 956
general remarks, 950
halogeno anions, 951
hydrido complexes, 251, 953
neutral complexes, 952
octahedral complexes, 957
oxidation states, 951
phosphine complexes, 937
stereochemistry, 950-51
tetravalent complexes, 958
zerovalent complexes, 961
see also Platinum metals
Parkinson's disease, 1273
Parvalbumins, 1344
Patronite, 708
Pentaborane, 315
Pentafluorosulfur hypofluorite, 523
Pentagonal bipyramid, 53
Pentagonal dodecahedron, 46
Pentagonal planar coordination, 51
Peptidases, 1329
Perbromates, 560
Perbromic acid, 560
Perbromyl fluoride, 570
Perchlorate ion, as ligand, 174-75
ir spectrum, 174-75
Perchlorates, 560
Perchloric acid, 244
Perchloryl fluoride, 570
Perfluoroalkyl halides, 574
Perhalate ions, 559
Perhalogenoalkylmercury compounds,
611
Periodates, 561
Periodic acid, 561
Perlithio compounds, 268
Permanganate ion, 747
chemical properties, 747
preparation, 747
Perosmates, 916
Perovskite, 695
structure, 18
Peroxides, 490, 497
fluorinated, 500
ionic, 497
organic, 499
Peroxoacetic acid, 499
Peroxoborates, 299
Peroxodisulfonyldifluoride, 500

Peroxodisulfuric acid, 534
Peroxodisulfuryl difluoride, 539
Peroxo ligands, 155
geometries of, 156
Peroxomonosulfuric acid, 534
Peroxonitrous acid, 430
Perrhenyl fluorides, 895
Perruthenate ion, 915
Pertechnetyl fluorides, 895
Perutz's trigger mechanism, 1322
Perxenate ions, 585
1,10-Phenanthroline, complexes, 120
Phenoxides, as ligands, 158, 161
Phosphacumulene ylids, 467
Phosphaferrocenes, 1171
Phospham, 465
Phosphatases, 1329
Phosphate esters, 475
in biology, 478
Phosphates, 474
condensed, 475-77
complexes of, 474-5
halogeno, 478
Phosphazenes, 460
cyclic, structures of, 462
Phosphides, 444
Phosphido-bridged ligands, 150-51
Phosphine, 445
ligands, macrocyclic, 149
tertiary, 148
oxides, as ligands, 177
dissymmetric, 466
trialkyl and triaryl, 466-67
Phosphite triesters, 473
Phosphonate ions, 473
Phosphonium compounds, 460
Phosphorous acid, 473
Phosphorus, 438
allotropic forms, 442-43
catenation, 442
comparison with nitrogen, 460
compounds as ligands, 147, 150, 780
heterocyclic, 149
trivalent, 87
nitrogen compounds, 460
nitrogen ylids, 464
occurrence, 442
oxides, 453
as ligands, 150
oxo acids, and anions, 472
as ligands, 175
pentachloride, 449
pentafluoride, 448
pentoxide, 453

properties, 438
radicals, 442
sulfides, 456
    structures of, 457
trichloride, 447
    reactions of, 448
trifluoride, 446
see also Group V elements
Phosphine imidate complexes, 124
Phosphine sulfides, as ligands, 189
Phthalocyanines, 133
Platinic chloride, 911
Platinum, 950
    alkyls and aryls, 957
    anhydrous acetates, 954
    blues, 960
    carbonyl anion compounds, 965
    complexes of, 951, 954
    compounds, in cancer chemotherapy, 960
        with Pt-Pt bonds, 963
    dichloride, 909
    five-coordinate complexes, 956
    general remarks, 950
    halogeno anions, 951
    hydrido complexes, 953
    neutral complexes, 952
    octahedral complexes, 957
    one-dimensional compounds, 1110
    oxidation states, 951
    phosphine complexes, 957
    small clusters, 964
    stereochemistry, 950-51
    tetravalent complexes, 958
        with carbon bonds, 958
    zerovalent complexes, 961
    see also Platinum metals
Platinum metals, 901
    binary compounds, 904
    bromides, 909
    chemistry of, 903
    chlorides, 909
    elements, 902
    fluorides, 906-9
    general remarks, 901
    halides, 906
    halogeno complexes, 911
    iodides, 909
    oxides, 904
        anhydrous, 905
        hydrous, 905
    oxohalides, 911
    phosphides, 904, 906
    properties of, 902
    sulfides, 904, 906

Pitchblende, 1028
Planar geometry, 49
Plastocyanin, 1336
Plumbane, 385
Plutonium, 1005, 1036
    aqueous chemistry, 1040
    binary compounds, 1039
    elemental, 1013
    halides, 1039
    isolation, 1036
    oxides, 1039
    see also Actinides
Point group, 37
Polonium, 502
    dioxide, 528
    halides, 525
    occurrence, 505
    oxides, 526-7
    properties, 502
    reactions, 509
    structure, 509
    see also Group VI elements
Polyhalide ions, principal types, 567
Polyhedral expansion, 315
Polylithio compounds, 268
Polyphenylchromium compounds, 1167
Polyphosphazenes, linear, 462
Polyphosphates, 476
    cyclic, 476
    linear, 476
Polyphosphoric acid, 474, 476
Polyporous versicolor, 1335
Polypyrazolylborate ligands, 141
Polysulfide ions, as ligands, 180
Polythiazyl, 517-18
Polythiocyanogen, 371
Polythionates, 537
Polyvinyl chloride (PVC), 545
Populus nigra, 1336
Porphyrins, 134
    crowned, 136
    picket-fence, 136, 159
    synthetic, 135
Potassium, 253
    alkyls, 268
    aryls, 269
    nitrosodisulfonate, 431
    preparation, properties, uses, 257
    see also Group I elements
Praseodymium, 982
    tetravalent, 1000
    see also Lanthanides
Precursor complex, 1210
Promethium, 982

occurrence, 989
see also Lanthanides
Proper rotation, 29
Propylene polymerization, 1280
Protactinium, 1005, 1026
  aqueous chemistry, 1027
  elemental, 1012
  pentavalent compounds, 1026
  tetravalent compounds, 1027
  see also Actinides
Proteases, 1329
Protonic acids, 228
  strengths of, 223
Prussian blue, 750, 762-63
Pseudomonas aeruginosa, 1336
Purine complexes, 121
Purple of Cassius, 968
Pyramidal geometry, 49
Pyridine N-oxide, as ligand, 177
Pyrite structure, 514
Pyrochlore, 831
Pyrolusite, 744
Pyrosilicate ion, 388
Pyroxenes, 388-89
Pyrrhotite, 784

Quadruple bonds, 725, 874
Quartz, 386, 392
Quinol, 226
Quinone complexes, 162

Racemization of trischelate complexes,
    1223
Radical reactions, 1235
Radium, 271, 273, 281
  binary compounds, 282-83
  occurrence, 281
  see also Group II elements
Radius-ratio rule, 11
Radon, 577. *See also* Noble gases
Rate constants, 1184
Rate law, 1184, 1187
Ratios of successive constants, general
    theory of, 236
Ray-Dutt twist, 1223
Reaction mechanisms, 1183
Red-Al, 340
Redox reactions, 1205, 1215
Reduction potential, 230
Reductive-elimination reactions, 1237
Reflection, 30
Renal equine, 1334
Rhenium, 883
  alkyls, synthesis of, 901

carbonyl compounds, 900, 1054
complex halides, 890
compounds, 887
dioxo complexes, 897
elemental, 886
halides, 888
heptaoxides, 887
hydrido complexes, 899
nitrogen-ligand complexes, 898
organometallic compounds, 900
oxidation states, 844-45
oxides, 887
oxo anions, 894
oxo compounds, 894
oxo halides, 895
oxohalogeno anions, 895
pentachloride, 889
stereochemistry, 844-45
sulfides, 887
sulfur complexes, 899
tetrabromide, 890
tetrachloride, 889
tetraiodide, 890
tetravalent complexes, 899
Rhodium, 934
  anionic complexes, 941
  carbonyl complex, 1054
  cationic complexes, 941, 945
  clusters, 1083
  divalent complexes, 947-48
  general remarks, 934
  pentafluoride, structure of, 908
  stereochemistry, 934-35
  tetravalent complexes, 948
  trichloride, 909
    reactions of, 945
  trivalent, aqua ions, 944
    cationic complexes, 945
    chloride system, 944
    complexes of, 943
    hydrido complexes, 946
    neutral complexes, 946
  univalent complexes, 934
    preparations and reactions, 936-40
  see also Platinum metals
Rhombic distortion, 51
Rhombic twist, 1223
Rhus vernicifera, 1336
Ring whizzers, 1167, 1230
Rochow process, 403
Roehlen reaction, 1291
Roussin ester, 764
Roussin salts, 764
Rubidium, 253

suboxides, 254, 261
see also Group I elements
Rubredoxins, 1325
Ruby, natural or synthetic, 730
Russell-Saunders states, 659
Ruthenates, 915
Ruthenium, 912
    acetate, 931
    ammines, reactions of, 923
    ammonia complexes, 920
    aromatic amine complexes, 922
    carbonyl, 1051
    chloro complexes, 917
    clusters, 1083
    divalent, chloro complexes, 918, 922
    carbonyl, 1051
    chloro complexes, 917
    clusters, 1083
    general remarks, 912
    halo complexes, 917, 919
    nitric oxide complexes, 926
    nitrido complexes, 932
    nitrogen donor ligand complexes, 920
    oxidation states, 913
    oxo compounds, 914, 916
    pentavalent complexes, 932
    red, 922
    stereochemistry, 912-13
    tertiary phosphine and related complexes,
        927
        reactions, 928
    tetraoxide, 914
    tetravalent, chloro complexes, 917
    trichloride, 909
    triiodide, 909
    trivalent, chloro complexes, 918
    see also Platinum metals
Rutherfordium, 1005, 1045. *See also*
        Actinides
Rutile, 695
    structure, 16

Saline hydrides, 247
    properties, 248
Salt hydrates, 227
Samarium, 982. *See also* Lanthanides
Saturated ring compounds, 1128
Scandium, 981, 996
    complexes, 997
    elemental, 996
    halides, 997
    lower oxidation states, 997-98
    organometallic compounds, 997
    oxides, 997

oxo salts, 997
see also Lanthanides
Scheelite, 846
Schiff base ligands, 139
Schottky defect, 25
Sea squirts, 709
Selenic acid, 533-34
Selenium, 502
    dioxide, 528
    halide complexes, 525
    halides, 524
    occurrence, 505
    oxides, 526-7
    oxo acids, 533-34
    oxo halides, 537
    properties, 502
    reactions, 509
    structure, 509
    trioxide, 530
    see also Group VI elements
Selenous acid, 533
Selenyl halides, 537-38
Semibridging CO group, 1057
Semiconductor, n-type, 26
    p-type, 26
Shear structure, 847
Shift reagents, 992
Siderite, 751
Siderophores, 1314
Silanes, 383
Silica, 386
Silicates, 387
    cyclic anions, 388
    framework minerals, 389
    infinite chain anions, 388
    infinite sheet anions, 389
    noncyclic anions, 388
Silicon, 374
    allotropic forms, 383
    as ligand atom, 116
    chlorides, 385
    divalent state, 380
    hydrides, 383
    nitrogen compounds, 395
    occurrence, isolation, and properties, 382
    organo compounds, 400
    oxygen compounds, 386
    radicals, 402
    see also Group IV elements
Silicones, 403
Siloxanes, 403
Silver, 966
    clusters, 975
    compounds, 968

divalent, complexes, 972-73
  compounds, 972
  fluoride, 972
  fluorosulfate, 972
  ion, 972
elemental, 967
oxidation states, 967
stereochemistry, 967
trivalent, complexes, 974
  compounds, 973
univalent, 968
  binary compounds, 969
  complexes, 970
  halides, 970
  halogeno complexes, 970-71
  organo compounds, 971
  oxide, 969
  phosphine complexes, 971
  sulfide, 970
zeolite, 420
Site preference problem, 686
Smaltite, 766
Sodium, 253
  acetylide, 255
  alkyls, 269
  amalgam, 258
  aryls, 269
  bicarbonate, 220
  cryptate complexes, 254, 265
  hydride, 248
  preparation, properties, uses, 257
  tetraphenylborate, 322
  see also Group I elements
Speier's catalyst, 1275
Speier reaction, 385
Speisses, 766
Sphalerite, 591
Spin crossovers, 648-50
Spinal, 330
Spinal structure, 17
  disordered, 17
  inverse, 17
Spin moment, 626
Spodumene, 388
Squarate, 170, 367
Square antiprism, 53
Square complexes, 50
Square pyramidal geometry, 50
Stability constants, see Formation constants
Stannane, 384
Stellacyanin, 1336
Stereochemical nonrigidity, 50, 1217
  in coordination compounds, 1218
    six coordinate and higher, 1221

    trigonal-bipyramidal molecules, 1219
  in metal carbonyl compounds, 1226
  in organometallic compounds, 1229
Steric factors, 89
Stibine, 446
Stishovite, 387, 392
Stolzite, 846
Strati-bisporphyrin, 136
Strontium, 271, 281
  binary compounds, 282-83
  occurrence, 281
  see also Group II elements
Successor complex, 1210
Sulfaminic chloride, 519
Sulfanedisulfonic acid, 537
Sulfanes, 512
Sulfate, as ligand, 175
  ir spectrum, 175-76
Sulfenates, as ligands, 176
Sulfides, ionic, 512
  ions, 512
    as ligands, 177-79
  nonmetallic binary, 515
  transition metal, 514-515
Sulfinates, as ligands, 176, 188
Sulfite oxidase, 1342
Sulfites, 532
  dialkyl, 533
  as ligands, 176, 188
  reactions of, 533
Sulfur, 502
  as ligand donor atom, 177
  catenation, 503
  chlorides, 523
  decafluoride, 522
  dioxide, 528
    as ligand, 181-82
  in insertion reactions, 1251
  fluorides, 519-22
    substituted, 522-23
  hexafluoride, 521
  liquid, 508
  monoclinic, 506
  nitride, 515-17
    as ligand, 187-88
  nitrogen compounds, 515
    structures, 516
  occurrence, 505
  orthorhombic, 506
  oxides, 526
    as ligands, 181
  oxo acids, 530
    as ligands, 175
  oxo halides, 537

peroxo acids, 534
properties, 502
reactions, 509
solid, 506
structural relationships, 506
tetrafluoride, 520
trioxide, 529
vapor, 508
see also Group VI elements
Sulfurdiimines, 132
Sulfuric acid, 245
Sulfurous acid, 532
Sulfuryl halides, 538
Superoxide dismutase, 742, 1337
Superoxide radical ion, 490
Superoxides, 497, 498
Superoxo ligands, 155
bent, 158
bridged, 157
end-on, 158
unidentate, 158
Symmetry element, definition, 31
Symmetry groups, 28, 35
Symmetry operations, general rules, 39
multiplication, 35
systematic listing of, 39
Symmetry and structure, 28
Symmetry, operations and elements, 29
Synergic bonding, 83
Synthesis gas, 365, 1287
Synthetic blood, 574

Tantalates, 833
Tantalum, 831
cluster compounds, 841
compounds, 831
elemental, 831
low oxidation states, 843
oxidation states, 832
pentavalent, 831
alcoxides, 836
chloride, reactions of, 837
complexes, 836
compounds, 838
dialkylamides, 836
fluorides and fluoride complexes, 834
halides, 835
hydrides, 838
isopolyanions, 833
organometallic chemistry, 838
oxide halides, 835
oxygen compounds, 831
stereochemistry, 832
tetravalent, 839

alcoxides, 840
complexes, 840
diketonates, 840
halide complexes, 840
halides, 839
oxides, 839
sulfides, 839
trivalent compounds, 841
Tartar emetic, 172, 471
Tartrato complexes, 172
Technetium, 883
complex halides, 890
compounds, 887
elemental, 886
halides, 888
heptaoxides, 887
nitrogen-ligand complexes, 899
oxidation states, 884-85
oxides, 887
oxo anions, 894
oxo compounds, 894
oxohalides, 895
oxohalogeno anions, 895
stereochemistry, 884-85
sulfides, 887
tetravalent complexes, 899
Teflon, 552
Telluric acid, 533-34
Tellurium, 502
dioxide, 528
halide complexes, 525
halides, 524
occurrence, 505
oxides, 526-527
oxo acids, 533-534
properties, 502
reactions, 509
structure, 510
subhalides, 525
trioxide, 530
see also Group VI elements
Tellurous acid, 533
Template syntheses, 137-39
Terbium, 982
tetravalent, 1001
see also Lanthanides
Tertiary amine oxides, interaction with
metal carbonyls, 1259
Tetraatomic molecules, 204
Tetracyanoethylene, complexes, 97
Tetracyanoplatinate compounds, 1110
Tetrafluoroammonium ion, 433
Tetrafluoroberyllates, 278
Tetrafluoroethylene, 574

Tetrafluorohydrazine, 433
Tetragonal distortion, 51
Tetrahedral geometry, 50
Tetrahedron, 44
Tetrahydridoborate, 107, 315
Tetrametaphosphate, 478
Tetrasulfurtetranitride, 516
  reactions of, 517
  structure of, 516
Thallium, 326
  aqua ions, 333
  carboxylates, 337
  halides, 332
  lower valent compounds, 348
  mixed oxides, 330
  nitrato complexes, 338
  occurrence, isolation, and properties, 328
  organometallic compounds, 341
  oxide, 330
  univalent, 256
  see also Group III elements
Thermochromism, 794
Thermodynamic cycle, 22
Thermodynamic stability, 67
Thermolysin, 1343
Thiaborane anion, 320
Thiocarbamates, 184
Thiocarbonates, 372
Thiocarbonyl complexes, 86
Thiocarboxamido ligands, 188
Thiocyanate, 371
  as ligand, 188
Thiocyanogen, 371
Thioethers, ligands, 180-81
Thiols, as ligands, 177
Thiomolybdate, 851
Thionyl dichloride, 537
  halides, 537-38
Thiosemicarbazides, as ligands, 189
Thiosemicarbazones, as ligands, 189
Thiosulfuric acid, 535
Thiotungstate, 851
Thiourea, bonding of, 188
Thiuram disulfides, 372
Thorium, 1005, 1023
  aqueous solutions, 1025
  binary compounds, 1024
  complexes, 1025
  elemental, 1012
  halides, 1024
  hydroxide, 1024
  lower oxidation states, 1026
  oxide, 1024
  oxohalides, 1024

oxo salts, 1025
  see also Actinides
Thortveitite, 388
Three-center bond model, 201, 208
Thulium, 982. See also Lanthanides
Tin, 374
  allotropic forms, 383
  as a ligand atom, 116
  catenated linear and cyclic organo com-
      pounds, 404
  divalent compounds, 117, 380, 397, 400
  hydrides, 384
  hydroxides, 398
  occurrence, isolation, and properties, 382
  organo compounds, 392, 403
  oxides, 392
  oxo anions, 393
  stannates, 393
  stannous compounds, 397
  tetrachlorides, 386
  tetravalent compounds, 116
  see also Group IV elements
Titanates, 695
Titanium, 692
  compounds, 694
  divalent, 703
    chemistry of, 703
    halides, 703
    oxide, 703
  electronic configuration, 692
  elemental, 693
  organometallic compounds, 704
    chiral tetrahedral compounds, 705
    cyclopentadienyl, 704
    pentamethylcyclopentadienyl, 707
  oxidation states, 693
  stereochemistry, 693
  tetrachloride, 692
  tetravalent, 692
    adducts of TiX_4, 697
    alcoxides, 699
    anionic complexes, 697
    aqueous chemistry, 696
    binary compounds, 694
    chemistry of, 694
    complexes, 696
    compounds, 700
    diketonates, 698
    halides, 694
    nitrogen compounds, 700
    oxide, 695
    oxo salts, 696
    peroxo complexes, 699
    Schiff base complexes, 698

sulfide, 695
trivalent, 701
  binary compounds, 701
  chemistry of, 701
  complexes, 701
  electronic structure, 703
  oxide, 701
  Schiff base complexes, 702
  solution chemistry, 701
  trichloride, 701
Titanocene, 706
p-Toluenesulfonate ion, as ligand, 176
Tourmaline, 292
Trans-actinides, 1045
Trans-americium elements, 1042
  ion-exchange separations, 1043
Transcarbamylase, 1329
Trans effect, 116, 1199-1202
Transferrin, 1313
Transition element complexes
  acid-base behavior, 1235
  d^1 and d^9 systems, 657
  d^2 - d^8 ions, 658
  electronic absorption spectra, 657
  energy level diagrams, 658
    for d^2 configuration, 660
    for a d^8 ion, 661
    for various d^n configurations, 662
  five-coordinate, 648
  formation constants, 685
  formation from aqua ions, 1189
  high-spin, low-spin crossovers, 648
  inner orbital splittings, 676
    effect on ionic radii, 677
    Jahn-Teller effects, 678
  molecular orbital theory, 629
  with no π bonding, 630
  octahedral, 644
    and tetrahedral, relative stability, 686
    electronic absorption spectra, 657
    L→M transitions, 667
    square and tetragonally-distorted, 647
  optical activity, 669
  with π bonding, 632
  reaction mechanisms, 1183
  reactions involving H^+, H^-, and H_2,
      1113
  spectrochemical series, 663
  splitting of Russell-Saunders states,
      tetrahedral, 659
    and octahedral, relative stability, 686
    electronic absorption spectra, 657
    L→M transitions, 668
  trigonal-bipyramidal, 642, 648

Transition element ions, crystal field
  ionic radii, 677
  mean electron pairing energies, 646
  splittings, 646
Transition elements, 619
  characteristics, 619
  definition, 619
  electronic configurations, 624
  first series, 619, 689
    comparison with second and third, 822
    general remarks, 689
    oxidation states, 691
    standard potentials, 691
    trends, 690
  general properties, 619
  metal-metal bonding, 823
  position in Periodic Table, 621
  second series, 619
    comparison with first, 822-23
  third series, 619
    comparison with first, 822-23
Transition metal compounds, π complexes,
      1141
  alkene complexes, 1147
  alkyne complexes, 1156
  allyl and enyl complexes, 1152
  carbocyclic, 1160
  multidecker sandwich compounds, 1172
  with metal to hydrogen bonds, 1113
    methods of synthesis, 1113
    types, 1115
  with metal to carbon bonds, 1119
    alkyl and aryl, 1123
    decomposition of, 1138
    in catalysis, 1265
    in synthesis, 1234
    synthesis of, 1133
  with multiple metal to carbon bonds,
      1140
    alkylidene complexes, 1140
    alkylidyne complexes, 1142
    carbenes and carbynes, 1142
Transition metal hydrido complexes, 1115
  cluster and encapsulated hydrides, 1117
  hydrogen bridge bonding, 1116
  metal-hydride bonds, 1115
Transphosphorylases, 1329
Tremolite, 389
Triatomic molecules, 202-4
1,3-Triazenido complexes, 128, 131-32
Triazine, 420
Tricapped trigonal prism, 55
Tricyclooctadine (syn), 1299
Tridymite, 387

Trifluoroamine oxide, 436
Trifluoromethylsulfonic acid, 246
Trigonal bipyramidal geometry, 50
Trigonal distortion, 51
Trigonal twist, 1223
Trimethylcarbonium ions, 240
Trioxodinitrate(II) ion, 429
Triphenylphosphine complexes, in cyclo-
	metallation reactions, 1245
Triphenylphosphite complexes, in cyclo-
	metallation reactions, 1245
Trirhenium clusters, 1095
Trirhenium nonachloride, 890
	reactions of, 892
	structure, 891
Tritium, 215
Tropolonato complexes, 166, 169-70
Troponins, 1344
Tungstates, 850
Tungsten, 844
	aqua ions, 867
	blue oxides, 848
	bronzes, 849
	carbonyl complexes, 877, 1051
	compounds with triple and quadruple
		bonds, 873
	cyano compounds, 878
	dinitrogen compounds, 879
	elemental, 846
	halide and halo complexes, 861
	halogeno complexes, 864
	heteropoly acids and their salts, 852, 857
	heteropoly blues, 861
	hexavalent, oxide halides, 866
		oxo complexes, 868, 872
	isopolyacids and their salts, 852
	metal atom cluster compounds, 881
	mixed oxides, 848
	non-oxo complexes, 877
	organoheteropoly anions, 859
	oxidation states, 845
	oxide halides, 866
	oxides, 847
	pentavalent, oxide halides, 867
		oxo species, 870
	phosphine complexes, 877
	stereochemistry, 845
	sulfides, 847, 850
Tungstic acid, 851
Tunicates, 709
Turnbull's blue, 750, 762-63
Two-electron transfers, 1215

Ultramarines, 389, 390, 513

Ultraphosphates, 476
Uranates, 1029
Uraninite, 1028
Uranium, 1005, 1028
	aqueous chemistry, 1033
	chlorides, 1031
	complexes, 1033
	compounds, 1028
	elemental, 1012
	fluorides, 1030
	halides, 1030
	halogeno complexes, 1031
	nitrate complexes, 1035
	oxides, 1028
	oxohalides, 1032
	see also Actinides
Uranocene, 1170
Uranyl salts, 1035
Urease, 1344

Valence shell electron pair repulsion model,
	196, 206-8
	rules, 196-99
Valinomycin, 265
Vanadates, 711
Vanadinite, 708
Vanadium, 708
	biochemistry of, 1344
	carbonyl, 1050-51
	compounds, 710
		with carbonyl, 718
	divalent, 717
		aqueous solutions, 718
		chemistry of, 717
		complexes, 718
		salts, 718
	elemental, 708
	halides, 710
	oxidation states, 709
	pentafluoride, 711
	pentavalent, 711
		chemistry of, 711
		complexes, 712
		oxide, 711
		oxo halides, 712
	stereochemistry, 709
	tetrachloride, 710
	tetravalent, 714
		nonoxo compounds, 716
		oxide, 714
		oxo ion, 715
		oxo ion complexes, 715
		oxo-porphyrins, 708
	trivalent, 716

aqua ion, 717
  chemistry of, 716
  complexes, 717
  oxide, 716
Vanadocytes, 1344
Vanadophores, 1344
Van-Arkel-de Boer method, 694, 709
Vaska's compound [trans-IrCl(CO)-
    (PPh$_3$)$_2$], 1270
Vinylidenes, 1147
Vitamin B$_{12}$, 137, 782, 1338

Wackenroder's liquid, 537
Wacker process, 820
Wade's rules, 312, 1083-85
Walsh diagram, 207
Water, 224
  as ligand, 152
  structure of, 225
Water-exchange rates, 1188
Water gas shift reaction, 216, 1290
Water splitting, 216
Wilkinson's catalyst [RhCl(PPh$_3$)$_3$], 1266
  catalytic hydrogenation cycle for, 1269
Wittig reaction, 467
Wolframite, 846
Wolfsberg-Helmholz method, 637
Woodward-Hoffmann rules, 1299
Wulfenite, 846
Wurtzite structure, 16

Xα method, 637
  applied to metal atoms clusters, 1098
  applied to quadruply bonded molecules,
    1100-1105
Xanthine dehydrogenase, 1342
Xanthine oxidase, 1342
Xenon, 577
  chemistry of, 580
  compounds, 580
  difluoride, 581
  fluorides, 581
    compounds, 584
  fluorocations, 583
  hexafluoride, 581
  nitrogen complexes, 586-87
  oxofluorides, 582
  oxygen compounds, 585
  tetrafluoride, 581
  tetroxide, 586
  trioxide, 585
  see also Noble gases

Ylid syntheses, 1136

Ytterbium, 982. See also Lanthanides
Yttrium, 981. See also Lanthanides

Zeeman splitting, 675
Zeise's salt, 95
Zeolites, 389, 390
  in catalysis, 392
  in ion exchange, 391
  in molecular sieves, 391
Ziegler-Natta catalysis, 831
  polymerization, 1280
  reaction, 719
Zinc, 6, 273, 589
  aqua ions, 599
  biological role of, 602, 1329
  chalconides, 598
  complexes, 600
  halides, 598-99
    formation constants of, 601
  hydroxides, 597
  organometallic compounds, 608
  oxides, 597
  oxo salts, 599
  properties, 589, 591
  selenides, 598
  stereochemistry, 590
  sulfides, 598
  tellurides, 598
  univalent state, 593
  see also Group IIb elements
Zinc blende structure, 16
Zintl compounds, 261
Zircon, 824
Zirconates, 826
Zirconium, 824
  elemental, 824
  lower oxidation states, 829
  organometallic compounds, 830
  oxidation numbers less than three,
    830
  oxidation states, 824
  stereochemistry, 824
  tetravalent, 825
    aqueous chemistry, 827
    borohydrides, 829
    chalcogenides, 827
    compounds, 825
    dithiolene complex anions,
      828
    nitrato complexes, 829
    phosphate, 827
    stereochemistry, 826
  trivalent, 829
Zirconyl salts, 828

Periodic Table of the Elements

Period	Group Ia	Group IIa	Group IIIa	Group IVa	Group Va	Group VIa	Group VIIa	Group VIII			Group Ib	Group IIb	Group IIIb	Group IVb	Group Vb	Group VIb	Group VIIb	Group 0
1 1s	1 **H**																1 **H**	2 **He**
2 2s2p	3 **Li**	4 **Be**											5 **B**	6 **C**	7 **N**	8 **O**	9 **F**	10 **Ne**
3 3s3p	11 **Na**	12 **Mg**											13 **Al**	14 **Si**	15 **P**	16 **S**	17 **Cl**	18 **Ar**
4 4s3d 4p	19 **K**	20 **Ca**	21 **Sc**	22 **Ti**	23 **V**	24 **Cr**	25 **Mn**	26 **Fe**	27 **Co**	28 **Ni**	29 **Cu**	30 **Zn**	31 **Ga**	32 **Ge**	33 **As**	34 **Se**	35 **Br**	36 **Kr**
5 5s4d 5p	37 **Rb**	38 **Sr**	39 **Y**	40 **Zr**	41 **Nb**	42 **Mo**	43 **Tc**	44 **Ru**	45 **Rh**	46 **Pd**	47 **Ag**	48 **Cd**	49 **In**	50 **Sn**	51 **Sb**	52 **Te**	53 **I**	54 **Xe**
6 6s (4f) 5d 6p	55 **Cs**	56 **Ba**	57* **La**	72 **Hf**	73 **Ta**	74 **W**	75 **Re**	76 **Os**	77 **Ir**	78 **Pt**	79 **Au**	80 **Hg**	81 **Tl**	82 **Pb**	83 **Bi**	84 **Po**	85 **At**	86 **Rn**
7 7s (5f) 6d	87 **Fr**	88 **Ra**	89** **Ac**															

*Lanthanide series 4f	58 **Ce**	59 **Pr**	60 **Nd**	61 **Pm**	62 **Sm**	63 **Eu**	64 **Gd**	65 **Tb**	66 **Dy**	67 **Ho**	68 **Er**	69 **Tm**	70 **Yb**	71 **Lu**
**Actinide series 5f	90 **Th**	91 **Pa**	92 **U**	93 **Np**	94 **Pu**	95 **Am**	96 **Cm**	97 **Bk**	98 **Cf**	99 **Es**	100 **Fm**	101 **Md**	102 **No**	103 **Lr**